$144.95 per copy (in United States)
Price is subject to change without prior notice.

0040

2010

RSMeans
A division of Reed Construction Data
Construction Publishers & Consultants
63 Smiths Lane
Kingston, MA 02364-3008
(781) 422-5000

Copyright©2009 by RSMeans
A division of Reed Construction Data
All rights reserved.

Printed in the United States of America
ISSN 0271-5945
ISBN 978-0-87629-827-5

The authors, editors, and engineers of RSMeans apply diligence and judgment in locating and using reliable sources for the information published. However, RSMeans makes no express or implied warranty or guarantee in connection with the content of the information contained herein, including the accuracy, correctness, value, sufficiency, or completeness of the data, methods, and other information contained herein. RSMeans makes no express or implied warranty of merchantability or fitness for a particular purpose. RSMeans shall have no liability to any customer or third party for any loss, expense, or damage, including consequential, incidental, special, or punitive damage, including lost profits or lost revenue, caused directly or indirectly by any error or omission, or arising out of, or in connection with, the information contained herein.

No part of this publication may be reproduced, stored in a retrieval system, or transmitted in any form or by any means without prior written permission of RSMeans. The cost data contained in this publication is valuable and proprietary information of RSMeans and others and there is no transfer to you of any ownership rights in the cost data or any license granted to you to create derivative works based on or which utilize the cost data.

RSMeans Repair & Remodeling Cost Data

31ST ANNUAL EDITION

Senior Editor
Bob Mewis, CCC

Contributing Editors
Christopher Babbitt
Ted Baker
Barbara Balboni
Robert A. Bastoni
John H. Chiang, PE
Gary W. Christensen
David G. Drain, PE
Cheryl Elsmore
Robert J. Kuchta
Robert C. McNichols
Melville J. Mossman, PE
Jeannene D. Murphy
Stephen C. Plotner
Eugene R. Spencer
Marshall J. Stetson
Phillip R. Waier, PE

Book Design
Norman R. Forgit

President & CEO
Iain Melville

**Vice President,
Product & Development**
Richard Remington

**Vice President of
Operations**
David C. Walsh

Vice President of Sales
Thomas Kesler

Director, Cost Products
Daimon Bridge

Marketing Director
John M. Shea

Engineering Manager
Bob Mewis, CCC

Cover Design
Paul Robertson
Wayne Engebretson

Production Manager
Michael Kokernak

Production Coordinator
Jill Goodman

Production
Paula Reale-Camelio
Sheryl A. Rose
Jonathan Forgit
Debbie Panarelli
Mary Lou Geary

Technical Support
A.K. Dash
Tom Dion
Roger Hancock
Gary L. Hoitt
Genevieve Medeiros
Kathryn S. Rodriguez
Sharon Proulx

This book is printed on recycled paper,
using soy-based printing ink. This book is recyclable.

Foreword

Our Mission

Since 1942, RSMeans has been actively engaged in construction cost publishing and consulting throughout North America.

Today, more than 65 years after RSMeans began, our primary objective remains the same: to provide you, the construction and facilities professional, with the most current and comprehensive construction cost data possible.

Whether you are a contractor, owner, architect, engineer, facilities manager, or anyone else who needs a reliable construction cost estimate, you'll find this publication to be a highly useful and necessary tool.

With the constant flow of new construction methods and materials today, it's difficult to find the time to look at and evaluate all the different construction cost possibilities. In addition, because labor and material costs keep changing, last year's cost information is not a reliable basis for today's estimate or budget.

That's why so many construction professionals turn to RSMeans. We keep track of the costs for you, along with a wide range of other key information, from city cost indexes . . . to productivity rates . . . to crew composition . . . to contractor's overhead and profit rates.

RSMeans performs these functions by collecting data from all facets of the industry and organizing it in a format that is instantly accessible to you. From the preliminary budget to the detailed unit price estimate, you'll find the data in this book useful for all phases of construction cost determination.

The Staff, the Organization, and Our Services

When you purchase one of RSMeans' publications, you are, in effect, hiring the services of a full-time staff of construction and engineering professionals.

Our thoroughly experienced and highly qualified staff works daily at collecting, analyzing, and disseminating comprehensive cost information for your needs. These staff members have years of practical construction experience and engineering training prior to joining the firm. As a result, you can count on them not only for accurate cost figures, but also for additional background reference information that will help you create a realistic estimate.

The RSMeans organization is always prepared to help you solve construction problems through its variety of data solutions, including online, CD, and print book formats, as well as cost estimating expertise available via our business solutions, training, and seminars.

Besides a full array of construction cost estimating books, RSMeans also publishes a number of other reference works for the construction industry. Subjects include construction estimating and project and business management, special topics such as green building and job order contracting, and a library of facility management references.

In addition, you can access all of our construction cost data electronically using *Means CostWorks*® CD or on the Web using **MeansCostWorks.com.**

What's more, you can increase your knowledge and improve your construction estimating and management performance with an RSMeans construction seminar or in-house training program. These two-day seminar programs offer unparalleled opportunities for everyone in your organization to become updated on a wide variety of construction-related issues.

RSMeans is also a worldwide provider of construction cost management and analysis services for commercial and government owners.

In short, RSMeans can provide you with the tools and expertise for constructing accurate and dependable construction estimates and budgets in a variety of ways.

Robert Snow Means Established a Tradition of Quality That Continues Today

Robert Snow Means spent years building RSMeans, making certain he always delivered a quality product.

Today, at RSMeans, we do more than talk about the quality of our data and the usefulness of our books. We stand behind all of our data, from historical cost indexes to construction materials and techniques to current costs.

If you have any questions about our products or services, please call us toll-free at 1-800-334-3509. Our customer service representatives will be happy to assist you. You can also visit our Web site at **www.rsmeans.com.**

Table of Contents

Foreword	ii
How the Book is Built: An Overview	v
Absolute Essentials for a Quick Start	vi
Estimating with RSMeans Unit Price Cost Data	vii
How To Use the Book: The Details	ix
Unit Price Section	1
Assemblies Section	489
Reference Section	628
Construction Equipment Rental Costs	631
Crew Listings	643
Historical Cost Indexes	673
City Cost Indexes	674
Location Factors	694
Reference Tables	700
Change Orders	736
Square Foot Costs	741
Abbreviations	751
Index	755
Other RSMeans Products and Services	788
Labor Trade Rates including Overhead & Profit	Inside Back Cover

Related RSMeans Products and Services

The engineers at RSMeans suggest the following products and services as companion information resources to *RSMeans Repair & Remodeling Cost Data*:

Construction Cost Data Books
Building Construction Cost Data 2010
Facilities Construction Cost Data 2010
Facilities Maintenance & Repair Cost Data 2010

Reference Books
ADA Compliance Pricing Guide
Building Security: Strategies & Costs
Designing & Building with the IBC
Estimating Building Costs
Estimating Handbook
Green Building: Project Planning & Estimating
How to Estimate with Means Data and CostWorks
Plan Reading & Material Takeoff
Project Scheduling & Management for Construction
Repair & Remodeling Estimating Methods

Seminars and In-House Training
Repair & Remodeling Estimating
Means CostWorks® Training
Means Data for Job Order Contracting (JOC)
Plan Reading & Material Takeoff
Scheduling & Project Management

RSMeans on the Internet
Visit RSMeans at **www.rsmeans.com** for a one-stop portal for the most reliable and current resources available on the market. Here you can request or download **FREE** estimating software demos, visit our bookstore for convenient ordering, and learn more about new publications and companion products.

Also visit **www.reedconstructiondata.com** for RSMeans business solutions, seminars, training, and more.

RSMeans Electronic Data
Get the information found in RSMeans cost books electronically on *Means CostWorks®* CD or on the Web at **MeansCostWorks.com**.

RSMeans Business Solutions
Cost engineers and analysts conduct facility life cycle and benchmark research studies and predictive cost modeling, as well as offer consultation services for conceptual estimating, real property management, and job order contracting. Research studies are designed with the application of proprietary cost and project data from Reed Construction Data and RSMeans' extensive North American databases. Analysts offer building product manufacturers qualitative and quantitative market research, as well as time/motion studies for new products, and Web-based dashboards for market opportunity analysis. Clients are from both the public and private sectors, including federal, state, and municipal agencies, corporations, institutions, construction management firms, hospitals, and associations.

RSMeans for Job Order Contracting (JOC)
Best practice job order contracting (JOC) services support the owner in the development of their program. RSMeans develops all required contracting documents, works with stakeholders, and develops qualified and experienced contractor lists. Means JOCWorks™ software is the best of class in the functional tools that meet JOC specific requirements. RSMeans data is organized to meet JOC cost estimating requirements.

Construction Costs for Software Applications
Over 25 unit price and assemblies cost databases are available through a number of leading estimating and facilities management software providers (listed below). For more information see the "Other RSMeans Products" pages at the back of this publication.
MeansData™ is also available to federal, state, and local government agencies as multi-year, multi-seat licenses.

- 4Clicks-Solutions, LLC
- ArenaSoft Estimating
- Beck Technology
- BSD – Building Systems Design, Inc.
- CMS – Construction Management Software
- Corecon Technologies, Inc.
- CorVet Systems
- Estimating Systems, Inc.
- Hard Dollar Corporation
- MC^2
- Parsons Corporation
- Sage Timberline Office
- UDA Technologies, Inc.
- US Cost, Inc.
- VFA – Vanderweil Facility Advisers
- WinEstimator, Inc.

How the Book Is Built: An Overview

The Construction Specifications Institute (CSI) and Construction Specifications Canada (CSC) have produced the 2004 edition of MasterFormat, a new and updated system of titles and numbers used extensively to organize construction information.

All unit price data in the RSMeans cost data books is now arranged in the 50-division MasterFormat 2004 system.

A Powerful Construction Tool

You have in your hands one of the most powerful construction tools available today. A successful project is built on the foundation of an accurate and dependable estimate. This book will enable you to construct just such an estimate.

For the casual user the book is designed to be:

- quickly and easily understood so you can get right to your estimate.
- filled with valuable information so you can understand the necessary factors that go into the cost estimate.

For the regular user, the book is designed to be:

- a handy desk reference that can be quickly referred to for key costs.
- a comprehensive, fully reliable source of current construction costs and productivity rates so you'll be prepared to estimate any project.
- a source book for preliminary project cost, product selections, and alternate materials and methods.

To meet all of these requirements we have organized the book into the following clearly defined sections.

New: Quick Start

See our new "Quick Start" instructions on the following page to get started right away.

Estimating with RSMeans Unit Price Cost Data

Please see these steps for guidance on completing an estimate using RSMeans unit price cost data.

How To Use the Book: The Details

This section contains an in-depth explanation of how the book is arranged . . . and how you can use it to determine a reliable construction cost estimate. It includes information about how we develop our cost figures and how to completely prepare your estimate.

Unit Price Section

All unit price cost data has been divided into the 50 divisions according to the MasterFormat system of classification and numbering. For a listing of these divisions and an outline of their subdivisions, see the Unit Price Section Table of Contents.

Estimating tips are included at the beginning of each division.

Assemblies Section

The cost data in this section has been organized in an "Assemblies" format. These assemblies are the functional elements of a building and are arranged according to the 7 divisions of the UNIFORMAT II classification system. For a complete explanation of a typical "Assemblies" page, see "How To Use the Assemblies Cost Tables."

Reference Section

This section includes information on Equipment Rental Costs, Crew Listings, Historical Cost Indexes, City Cost Indexes, Location Factors, Reference Tables, Change Orders, Square Foot Costs, and a listing of Abbreviations.

Equipment Rental Costs: This section contains the average costs to rent and operate hundreds of pieces of construction equipment.

Crew Listings: This section lists all the crews referenced in the book. For the purposes of this book, a crew is composed of more than one trade classification and/or the addition of power equipment to any trade classification. Power equipment is included in the cost of the crew. Costs are shown both with the bare labor rates and with the installing contractor's overhead and profit added. For each, the total crew cost per eight-hour day and the composite cost per labor-hour are listed.

Historical Cost Indexes: These indexes provide you with data to adjust construction costs over time.

City Cost Indexes: All costs in this book are U.S. national averages. Costs vary because of the regional economy. You can adjust costs by CSI Division to over 316 locations throughout the U.S. and Canada by using the data in this section.

Location Factors: You can adjust total project costs to over 900 locations throughout the U.S. and Canada by using the data in this section.

Reference Tables: At the beginning of selected major classifications in the Unit Price and Assemblies sections are "reference numbers" shown in a shaded box. These numbers refer you to related information in the Reference Section. In this section, you'll find reference tables, explanations, estimating information that support how we develop the unit price data, technical data, and estimating procedures.

Change Orders: This section includes information on the factors that influence the pricing of change orders.

Square Foot Costs: This section contains costs for 59 different building types that allow you to make a rough estimate for the overall cost of a project or its major components.

Abbreviations: A listing of abbreviations used throughout this book, along with the terms they represent, is included in this section.

Index

A comprehensive listing of all terms and subjects in this book will help you quickly find what you need when you are not sure where it falls in MasterFormat.

The Scope of This Book

This book is designed to be as comprehensive and as easy to use as possible. To that end we have made certain assumptions and limited its scope in two key ways:

1. We have established material prices based on a national average.
2. We have computed labor costs based on a 30-city national average of union wage rates.

For a more detailed explanation of how the cost data is developed, see "How To Use the Book: The Details."

Project Size

The material prices in Means data cost books are "contractor's prices." They are the prices that contractors can expect to pay at the lumberyards, suppliers/distributers warehouses, etc. Small orders of speciality items would be higher than the costs shown, while very large orders, such as truckload lots, would be less. The variation would depend on the size, timing, and negotiating power of the contractor. The labor costs are primarily for new construction or major renovations rather than repairs or minor alterations. **With reasonable exercise of judgment, the figures can be used for any building work.**

Absolute Essentials for a Quick Start

If you feel you are ready to use this book and don't think you will need the detailed instructions that begin on the following page, this Absolute Essentials for a Quick Start page is for you. These steps will allow you to get started estimating in a matter of minutes.

1 Scope

Think through the project that you will be estimating, and identify the many individual work tasks that will need to be covered in your estimate.

2 Quantify

Determine the number of units that will be required for each work task that you identified.

3 Pricing

Locate individual Unit Price line items that match the work tasks you identified. The Unit Price Section Table of Contents that begins on page 1 and the Index in the back of the book will help you find these line items.

4 Multiply

Multiply the Total Incl O&P cost for a Unit Price line item in the book by your quantity for that item. The price you calculate will be an estimate for a completed item of work performed by a subcontractor. Keep adding line items in this manner to build your estimate.

5 Project Overhead

Include project overhead items in your estimate. These items are needed to make the job run and are typically, but not always, provided by the General Contractor. They can be found in Division 1. An alternate method of estimating project overhead costs is to apply a percentage of the total project cost.

Include rented tools not included in crews, waste, rubbish handling, and cleanup.

6 Estimate Summary

Include General Contractor's markup on subcontractors, General Contractor's office overhead and profit, and sales tax on materials and equipment.

Adjust your estimate to the project's location by using the City Cost Indexes or Location Factors found in the Reference Section.

Editors' Note: We urge you to spend time reading and understanding the supporting material in the front of this book. An accurate estimate requires experience, knowledge, and careful calculation. The more you know about how we at RSMeans developed the data, the more accurate your estimate will be. In addition, it is important to take into consideration the reference material in the back of the book such as Equipment Listings, Crew Listings, City Cost Indexes, Location Factors, and Reference Numbers.

Estimating with RSMeans Unit Price Cost Data

Following these steps will allow you to complete an accurate estimate using RSMeans Unit Price cost data.

1 Scope out the project

- Identify the individual work tasks that will need to be covered in your estimate.
- The Unit Price data inside this book has been divided into 44 Divisions according to the CSI MasterFormat 2004—their titles are listed on the back cover of your book.
- Think through the project that you will be estimating, and identify those CSI Divisions that you will need to use in your estimate.
- The Unit Price Section Table of Contents that begins on page 1 may also be helpful when scoping out your project.
- Experienced estimators find it helpful to begin an estimate with Division 2, estimating Division 1 after the full scope of the project is known.

2 Quantify

- Determine the number of units that will be required for each work task that you previously identified.
- Experienced estimators will include an allowance for waste in their quantities. (Waste is not included in RSMeans Unit Price line items unless so stated.)

3 Price the quantities

- Use the Unit Price Table of Contents, and the Index, to locate the first individual Unit Price line item for your estimate.
- Reference Numbers indicated within a Unit Price section refer to additional information that you may find useful.
- The crew indicates who is performing the work for that task. Crew codes are expanded in the Crew Listings in the Reference Section to include all trades and equipment that comprise the crew.
- The Daily Output is the amount of work the crew is expected to do in one day.
- The Labor-Hours value is the amount of time it will take for the average crew member to install one unit of measure.
- The abbreviated Unit designation indicates the unit of measure upon which the crew, productivity, and prices are based.
- Bare Costs are shown for materials, labor, and equipment needed to complete the Unit Price line item. Bare costs do not include waste, project overhead, payroll insurance, payroll taxes, main office overhead, or profit.
- The Total Incl O&P cost is the billing rate or invoice amount of the installing contractor or subcontractor who performs the work for the Unit Price line item.

4 Multiply

- Multiply the total number of units needed for your project by the Total Incl O&P cost for the Unit Price line item.
- Be careful that the final unit of measure for your quantity of units matches the unit of measure in the Unit column in the book.
- The price you calculate will be an estimate for a completed item of work.
- Keep scoping individual tasks, determining the number of units required for those tasks, matching them up with individual Unit Price line items in the book, and multiplying quantities by Total Incl O&P costs. In this manner, keep building your estimate.
- An estimate completed to this point, in this manner, will be priced as if a subcontractor, or set of subcontractors, will perform the work. The estimate does not yet include Project Overhead or Estimate Summary components such as General Contractor markups on subcontracted and self-performed work, General Contractor office overhead and profit, contingency, and location factor.

5 Project Overhead

- Include project overhead items from Division 1 – General Requirements.
- These are items that will be needed to make the job run. These are typically, but not always, provided by the General Contractor. They include, but are not limited to, such items as field personnel, insurance, performance bond, permits, testing, temporary utilities, field office and storage facilities, temporary scaffolding and platforms, equipment mobilization and demobilization, temporary roads and sidewalks, winter protection, temporary barricades and fencing, temporary security, temporary signs, field engineering and layout, final cleaning and commissioning.
- These items should be scoped, quantified, matched to individual Unit Price line items in Division 1, priced and added to your estimate.
- An alternate method of estimating project overhead costs is to apply a percentage of the total project cost, usually 5% to 15% with an average of 10% (see General Conditions, page ix).
- Include other project related expenses in your estimate such as:
  - Rented equipment not itemized in the Crew Listings
  - Rubbish handling throughout the project (see 02 41 19.23)

6 Estimate Summary

- Includes sales tax on materials and equipment.
  Note: Sales tax must be added for materials in subcontracted work.
- Include the General Contractor's markup on self-performed work, usually 5% to 15% with an average of 10%.
- Include the General Contractor's markup on subcontracted work, usually 5% to 15% with an average of 10%.
- Include General Contractor's main office overhead and profit.
- RSMeans gives general guidelines on the General Contractor's main office overhead (see section 01 31 13.50 and Reference Number R013113-50).
- RSMeans gives no guidance on the General Contractor's profit.
- Markups will depend on the size of the General Contractor's operations, his projected annual revenue, the level of risk he is taking on, and on the level of competitiveness in the local area and for this project in particular.
- Include a contingency, usually 3% to 5%.
- Adjust your estimate to the project's location by using the City Cost Indexes or the Location Factors in the Reference Section.
- Look at the rules on the pages for How to Use the City Cost Indexes to see how to apply the Indexes for your location.
- When the proper Index or Factor has been identified for the project's location, convert it to a multiplier by dividing it by 100, then multiply that multiplier by your estimate total cost. Your original estimate total cost will now be adjusted up or down from the national average to a total that is appropriate for your location.

Editors' Notes:

1) *We urge you to spend time reading and understanding the supporting material in the front of this book. An accurate estimate requires experience, knowledge, and careful calculation. The more you know about how we at RSMeans developed the data, the more accurate your estimate will be. In addition, it is important to take into consideration the reference material in the back of the book such as Equipment Listings, Crew Listings, City Cost Indexes, Location Factors, and Reference Numbers.*

2) *Contractors who are bidding or are involved in JOC, DOC, SABER, or IDIQ type contracts are cautioned that workers' compensation insurance, federal and state payroll taxes, waste, project supervision, project overhead, main office overhead, and profit are not included in bare costs. Your coefficient or multiplier must cover these costs.*

How to Use the Book: The Details

What's Behind the Numbers? The Development of Cost Data

The staff at RSMeans continuously monitors developments in the construction industry in order to ensure reliable, thorough, and up-to-date cost information.

While *overall* construction costs may vary relative to general economic conditions, price fluctuations within the industry are dependent upon many factors. Individual price variations may, in fact, be opposite to overall economic trends. Therefore, costs are continually monitored and complete updates are published yearly. Also, new items are frequently added in response to changes in materials and methods.

Costs—$ (U.S.)

All costs represent U.S. national averages and are given in U.S. dollars. The RSMeans City Cost Indexes can be used to adjust costs to a particular location. The City Cost Indexes for Canada can be used to adjust U.S. national averages to local costs in Canadian dollars. No exchange rate conversion is necessary.

[G] The processes or products identified by the green symbol in this publication have been determined to be environmentally responsible and/or resource-efficient solely by the RSMeans engineering staff. The inclusion of the green symbol does not represent compliance with any specific industry association or standard.

Material Costs

The RSMeans staff contacts manufacturers, dealers, distributors, and contractors all across the U.S. and Canada to determine national average material costs. If you have access to current material costs for your specific location, you may wish to make adjustments to reflect differences from the national average. Included within material costs are fasteners for a normal installation. RSMeans engineers use manufacturers' recommendations, written specifications, and/or standard construction practice for size and spacing of fasteners. Adjustments to material costs may be required for your specific application or location. Material costs do not include sales tax.

Labor Costs

Labor costs are based on the average of wage rates from 30 major U.S. cities. Rates are determined from labor union agreements or prevailing wages for construction trades for the current year. Rates, along with overhead and profit markups, are listed on the inside back cover of this book.

- If wage rates in your area vary from those used in this book, or if rate increases are expected within a given year, labor costs should be adjusted accordingly.

Labor costs reflect productivity based on actual working conditions. In addition to actual installation, these figures include time spent during a normal weekday on tasks such as, material receiving and handling, mobilization at site, site movement, breaks, and cleanup.

Productivity data is developed over an extended period so as not to be influenced by abnormal variations and reflects a typical average.

Equipment Costs

Equipment costs include not only rental, but also operating costs for equipment under normal use. The operating costs include parts and labor for routine servicing such as repair and replacement of pumps, filters, and worn lines. Normal operating expendables, such as fuel, lubricants, tires, and electricity (where applicable), are also included. Extraordinary operating expendables with highly variable wear patterns, such as diamond bits and blades, are excluded. These costs are included under materials. Equipment rental rates are obtained from industry sources throughout North America—contractors, suppliers, dealers, manufacturers, and distributors.

Equipment costs do not include operators' wages; nor do they include the cost to move equipment to a job site (mobilization) or from a job site (demobilization).

Equipment Cost/Day—The cost of power equipment required for each crew is included in the Crew Listings in the Reference Section (small tools that are considered as essential everyday tools are not listed out separately). The Crew Listings itemize specialized tools and heavy equipment along with labor trades. The daily cost of itemized equipment included in a crew is based on dividing the weekly bare rental rate by 5 (number of working days per week) and then adding the hourly operating cost times 8 (the number of hours per day). This Equipment Cost/Day is shown in the last column of the Equipment Rental Cost pages in the Reference Section.

Mobilization/Demobilization—The cost to move construction equipment from an equipment yard or rental company to the job site and back again is not included in equipment costs. Mobilization (to the site) and demobilization (from the site) costs can be found in the Unit Price Section. If a piece of equipment is already at the job site, it is not appropriate to utilize mob./demob. costs again in an estimate.

General Conditions

Cost data in this book is presented in two ways: Bare Costs and Total Cost including O&P (Overhead and Profit). General Conditions, when applicable, should also be added to the Total Cost including O&P. The costs for General Conditions are listed in Division 1 of the Unit Price Section and the Reference Section of this book. General Conditions for the *Installing Contractor* may range from 0% to 10% of the Total Cost including O&P. For the *General* or *Prime Contractor*, costs for General Conditions may range from 5% to 15% of the Total Cost including O&P, with a figure of 10% as the most typical allowance.

Overhead and Profit

Total Cost including O&P for the *Installing Contractor* is shown in the last column on both the Unit Price and the Assemblies pages of this book. This figure is the sum of the bare material cost plus 10% for profit, the bare labor cost plus total overhead and profit, and the bare equipment cost plus 10% for profit. Details for the calculation of Overhead and Profit on labor are shown on the inside back cover and in the

ix

Reference Section of this book. (See the "How to Use the Unit Price Pages" for an example of this calculation.)

Factors Affecting Costs

Costs can vary depending upon a number of variables. Here's how we have handled the main factors affecting costs.

Quality—The prices for materials and the workmanship upon which productivity is based represent sound construction work. They are also in line with U.S. government specifications.

Overtime—We have made no allowance for overtime. If you anticipate premium time or work beyond normal working hours, be sure to make an appropriate adjustment to your labor costs.

Productivity—The productivity, daily output, and labor-hour figures for each line item are based on working an eight-hour day in daylight hours in moderate temperatures. For work that extends beyond normal work hours or is performed under adverse conditions, productivity may decrease. (See the section in "How To Use the Unit Price Pages" for more on productivity.)

Size of Project—The size, scope of work, and type of construction project will have a significant impact on cost. Economies of scale can reduce costs for large projects. Unit costs can often run higher for small projects. Costs in this book are intended for the size and type of project as previously described in "How the Book Is Built: An Overview." Costs for projects of a significantly different size or type should be adjusted accordingly.

Location—Material prices in this book are for metropolitan areas. However, in dense urban areas, traffic and site storage limitations may increase costs. Beyond a 20-mile radius of large cities, extra trucking or transportation charges may also increase the material costs slightly. On the other hand, lower wage rates may be in effect. Be sure to consider both of these factors when preparing an estimate, particularly if the job site is located in a central city or remote rural location.

In addition, highly specialized subcontract items may require travel and per-diem expenses for mechanics.

Other Factors—
- season of year
- contractor management
- weather conditions
- local union restrictions
- building code requirements
- availability of:
  - adequate energy
  - skilled labor
  - building materials
- owner's special requirements/restrictions
- safety requirements
- environmental considerations

Unpredictable Factors—General business conditions influence "in-place" costs of all items. Substitute materials and construction methods may have to be employed. These may affect the installed cost and/or life cycle costs. Such factors may be difficult to evaluate and cannot necessarily be predicted on the basis of the job's location in a particular section of the country. Thus, where these factors apply, you may find significant but unavoidable cost variations for which you will have to apply a measure of judgment to your estimate.

Rounding of Costs

In general, all unit prices in excess of $5.00 have been rounded to make them easier to use and still maintain adequate precision of the results. The rounding rules we have chosen are in the following table.

Prices from . . .	Rounded to the nearest . . .
$.01 to $5.00	$.01
$5.01 to $20.00	$.05
$20.01 to $100.00	$.50
$100.01 to $300.00	$1.00
$300.01 to $1,000.00	$5.00
$1,000.01 to $10,000.00	$25.00
$10,000.01 to $50,000.00	$100.00
$50,000.01 and above	$500.00

Final Checklist

Estimating can be a straightforward process provided you remember the basics. Here's a checklist of some of the steps you should remember to complete before finalizing your estimate.

Did you remember to . . .

- factor in the City Cost Index for your locale?
- take into consideration which items have been marked up and by how much?
- mark up the entire estimate sufficiently for your purposes?
- read the background information on techniques and technical matters that could impact your project time span and cost?
- include all components of your project in the final estimate?
- make use of Minimum Labor/Equipment Charges for Small Quantities (see the following page for more details)?
- double check your figures for accuracy?
- call RSMeans if you have any questions about your estimate or the data you've found in our publications?

Remember, RSMeans stands behind its publications. If you have any questions about your estimate . . . about the costs you've used from our books . . . or even about the technical aspects of the job that may affect your estimate, feel free to call the RSMeans editors at 1-800-334-3509.

Using Minimum Labor/Equipment Charges for Small Quantities

Estimating small construction or repair tasks often creates situations in which the quantity of work to be performed is very small. When this occurs, the labor and/or equipment costs to perform the work may be too low to allow for the crew to get to the job, receive instructions, find materials, get set up, perform the work, clean up, and get to the next job. In these situations, the estimator should compare the developed labor and/or equipment costs for performing the work (e.g., quantity x labor and/or equipment costs) with the *"minimum labor/equipment charge"* within that Unit Price section of the book.(These "minimum labor/equipment charge" line items appear only in the Facilities Construction Cost Data and the Repair & Remodeling Cost Data titles).

If the labor and/or equipment costs developed by the estimator are LOWER THAN the *"minimum labor/equipment charge"* listed at the bottom of specific sections of Unit Price costs, the estimator should adjust the developed costs upward to the *"minimum labor/equipment charge."* The proper use of a *"minimum labor/equipment charge"* results in having enough money in the estimate to cover the contractor's higher cost of performing a very small amount of work during a partial workday.

A *"minimum labor/equipment charge"* should be used only when the task being estimated is the only task the crew will perform at the job site that day. If, however, the crew will be able to perform other tasks at the job site that day, the use of a *"minimum labor/equipment charge"* is not appropriate.

08 52 Wood Windows
08 52 10 – Plain Wood Windows

08 52 10.40 Casement Window		Crew	Daily Output	Labor-Hours	Unit	Material	2010 Bare Costs Labor	Equipment	Total	Total Incl O&P
0010	CASEMENT WINDOW, including frame, screen and grilles									
0100	Avg. quality, bldrs. model, 2'-0" x 3'-0" H, dbl. insulated glass	G	1 Carp	10	.800	Ea.	276	33		309	360
0150	Low E glass	G		10	.800		225	33		258	305
0200	2'-0" x 4'-6" high, double insulated glass	G		9	.889		345	37		382	440
0250	Low E glass	G		9	.889		263	37		300	350
0300	2'-4" x 6'-0" high, double insulated glass	G		8	1		400	41.50		441.50	510
0350	Low E glass	G		8	1		465	41.50		506.50	585
0522	Vinyl clad, premium, double insulated glass, 2'-0" x 3'-0"	G		10	.800		266	33		299	345
0524	2'-0" x 4'-0"	G		9	.889		310	37		347	400
0525	2'-0" x 5'-0"	G		8	1		355	41.50		396.50	460
0528	2'-0" x 6'-0"	G		8	1		355	41.50		396.50	460
8100	Metal clad, deluxe, dbl. insul. glass, 2'-0" x 3'-0" high	G	1 Carp	10	.800		272	33		305	355
9000	Minimum labor/equipment charge		1 Carp	3	2.667	Job		111		111	182

Example:

Establish the bid price to install two casement windows. Assume installation of 2' x 4' metal clad windows with insulating glass [Unit Price line number 08 52 10.40 0200], and that this is the only task this crew will perform at the job site that day.

Solution:

Step One — Develop the Bare Labor Cost for this task:

Bare Labor Cost = 2 windows @ $37.00/each = $74.00

Step Two — Evaluate the *"minimum labor/equipment charge"* for this Unit Price section against the developed Bare Labor Cost for this task:

"Minimum labor/equipment charge" = $111.00 (compare with $74.00)

Step Three — Choose to adjust the developed labor cost upward to the *"minimum labor/equipment charge."*

Step Four — Develop the bid price for this task (including O&P):

Add together the marked-up Bare Material Cost for this task and the marked-up *"minimum labor/equipment charge"* for this Unit Price section.

$\quad$ 2 x ($345.00 + 10%) + ($111.00 + 64.2%)

$\quad$ = 2 x ($345.00 + $34.50) + ($111.00 + $71.26)

$\quad$ = 2 x ($379.50) + $182.26

$\quad$ = $759.00 + $182.26

$\quad$ = $941.26

ANSWER: $941.26 is the correct bid price to use. This sum takes into consideration the Material Cost (with 10% for profit) for these two windows, plus the *"minimum labor/equipment charge"* (with O&P included) for this section of the Unit Price book.

Unit Price Section

Table of Contents

Div. No.		Page
	General Requirements	
01 11	Summary of Work	8
01 21	Allowances	8
01 31	Project Management and Coordination	9
01 41	Regulatory Requirements	11
01 45	Quality Control	11
01 51	Temporary Utilities	12
01 52	Construction Facilities	12
01 54	Construction Aids	12
01 55	Vehicular Access and Parking	16
01 56	Temporary Barriers and Enclosures	16
01 58	Project Identification	18
01 71	Examination and Preparation	18
01 74	Cleaning and Waste Management	18
01 91	Commissioning	18
01 93	Facility Maintenance	19
	Existing Conditions	
02 21	Surveys	22
02 32	Geotechnical Investigations	22
02 41	Demolition	22
02 42	Removal and Salvage of Construction Material	28
02 43	Structure Moving	31
02 58	Snow Control	31
02 65	Undergrnd. Storage Tank Remvl.	31
02 81	Transportation and Disposal of Hazardous Materials	32
02 82	Asbestos Remediation	33
02 83	Lead Remediation	36
02 85	Mold Remediation	38
	Concrete	
03 01	Maintenance of Concrete	42
03 05	Common Work Results for Concrete	42
03 11	Concrete Forming	43
03 15	Concrete Accessories	48
03 21	Reinforcing Steel	50
03 22	Welded Wire Fabric Reinforcing	51
03 24	Fibrous Reinforcing	52
03 30	Cast-In-Place Concrete	52
03 31	Structural Concrete	55
03 35	Concrete Finishing	57
03 37	Specialty Placed Concrete	58
03 39	Concrete Curing	58
03 41	Precast Structural Concrete	59
03 45	Precast Architectural Concrete	60
03 48	Precast Concrete Specialties	60
03 51	Cast Roof Decks	60
03 52	Lightweight Concrete Roof Insulation	61
03 54	Cast Underlayment	61
03 63	Epoxy Grouting	61
03 64	Injection Grouting	62
03 81	Concrete Cutting	62
03 82	Concrete Boring	62
	Masonry	
04 01	Maintenance of Masonry	66
04 05	Common Work Results for Masonry	67
04 21	Clay Unit Masonry	72
04 22	Concrete Unit Masonry	75
04 23	Glass Unit Masonry	78
04 27	Multiple-Wythe Unit Masonry	79
04 41	Dry-Placed Stone	79
04 43	Stone Masonry	80
04 54	Refractory Brick Masonry	82
04 57	Masonry Fireplaces	82
04 71	Manufactured Brick Masonry	82
04 72	Cast Stone Masonry	82

Div. No.		Page
	Metals	
05 05	Common Work Results for Metals	86
05 12	Structural Steel Framing	89
05 15	Wire Rope Assemblies	92
05 21	Steel Joist Framing	92
05 31	Steel Decking	94
05 35	Raceway Decking Assemblies	94
05 41	Structural Metal Stud Framing	95
05 42	Cold-Formed Metal Joist Framing	98
05 44	Cold-Formed Metal Trusses	104
05 51	Metal Stairs	105
05 52	Metal Railings	107
05 56	Metal Castings	108
05 58	Formed Metal Fabrications	108
05 71	Decorative Metal Stairs	109
05 73	Decorative Metal Railings	109
	Wood, Plastics & Composites	
06 05	Common Work Results for Wood, Plastics, and Composites	112
06 11	Wood Framing	118
06 12	Structural Panels	127
06 13	Heavy Timber	127
06 15	Wood Decking	128
06 16	Sheathing	128
06 17	Shop-Fabricated Structural Wood	130
06 18	Glued-Laminated Construction	131
06 22	Millwork	132
06 25	Prefinished Paneling	134
06 26	Board Paneling	136
06 43	Wood Stairs and Railings	136
06 44	Ornamental Woodwork	138
06 48	Wood Frames	140
06 49	Wood Screens and Exterior Wood Shutters	141
06 51	Structural Plastic Shapes and Plates	141
06 63	Plastic Railings	142
06 65	Plastic Simulated Wood Trim	142
	Thermal & Moisture Protection	
07 01	Operation and Maint. of Thermal and Moisture Protection	146
07 05	Common Work Results for Thermal and Moisture Protection	146
07 11	Dampproofing	147
07 12	Built-up Bituminous Waterproofing	148
07 13	Sheet Waterproofing	148
07 16	Cementitious and Reactive Waterproofing	149
07 19	Water Repellents	149
07 21	Thermal Insulation	149
07 22	Roof and Deck Insulation	153
07 24	Exterior Insulation and Finish Systems	154
07 25	Weather Barriers	155
07 26	Vapor Retarders	155
07 31	Shingles and Shakes	155
07 32	Roof Tiles	158
07 33	Natural Roof Coverings	159
07 41	Roof Panels	159
07 42	Wall Panels	161
07 46	Siding	162
07 51	Built-Up Bituminous Roofing	164
07 52	Modified Bituminous Membrane Roofing	166
07 53	Elastomeric Membrane Roofing	167
07 54	Thermoplastic Membrane Roofing	167
07 56	Fluid-Applied Roofing	168
07 57	Coated Foamed Roofing	168
07 58	Roll Roofing	168
07 61	Sheet Metal Roofing	169
07 62	Sheet Metal Flashing and Trim	170

Div. No.		Page
07 65	Flexible Flashing	170
07 71	Roof Specialties	172
07 72	Roof Accessories	176
07 81	Applied Fireproofing	178
07 84	Firestopping	178
07 91	Pre-formed Joint Seals	180
07 92	Joint Sealants	181
07 95	Expansion Control	182
	Openings	
08 01	Operation and Maintenance of Openings	184
08 05	Common Work Results for Openings	185
08 11	Metal Doors and Frames	186
08 12	Metal Frames	187
08 13	Metal Doors	188
08 14	Wood Doors	191
08 16	Composite Doors	197
08 17	Integrated Door Opening Assemblies	198
08 31	Access Doors and Panels	199
08 32	Sliding Glass Doors	200
08 33	Coiling Doors and Grilles	201
08 34	Special Function Doors	202
08 36	Panel Doors	203
08 38	Traffic Doors	205
08 41	Entrances and Storefronts	205
08 42	Entrances	207
08 43	Storefronts	207
08 45	Translucent Wall and Roof Assemblies	208
08 51	Metal Windows	209
08 52	Wood Windows	210
08 53	Plastic Windows	217
08 54	Composite Windows	219
08 56	Special Function Windows	220
08 61	Roof Windows	220
08 62	Unit Skylights	220
08 71	Door Hardware	221
08 74	Access Control Hardware	229
08 75	Window Hardware	230
08 79	Hardware Accessories	230
08 81	Glass Glazing	230
08 83	Mirrors	232
08 84	Plastic Glazing	233
08 85	Glazing Accessories	233
08 87	Glazing Surface Films	234
08 88	Special Function Glazing	234
08 91	Louvers	234
08 95	Vents	235
	Finishes	
09 01	Maintenance of Finishes	238
09 05	Common Work Results for Finishes	238
09 21	Plaster and Gypsum Board Assemblies	240
09 22	Supports for Plaster and Gypsum Board	241
09 23	Gypsum Plastering	245
09 24	Portland Cement Plastering	245
09 26	Veneer Plastering	246
09 28	Backing Boards and Underlayments	246
09 29	Gypsum Board	247
09 30	Tiling	249
09 51	Acoustical Ceilings	251
09 53	Acoustical Ceiling Suspension Assemblies	252
09 54	Specialty Ceilings	253
09 62	Specialty Flooring	253
09 63	Masonry Flooring	254

Table of Contents (cont.)

Div. No.		Page
09 64	Wood Flooring	255
09 65	Resilient Flooring	256
09 66	Terrazzo Flooring	258
09 67	Fluid-Applied Flooring	259
09 68	Carpeting	259
09 69	Access Flooring	260
09 72	Wall Coverings	261
09 77	Special Wall Surfacing	262
09 81	Acoustic Insulation	263
09 84	Acoustic Room Components	263
09 91	Painting	263
09 93	Staining and Transparent Finishing	277
09 96	High-Performance Coatings	277
Specialties		
10 11	Visual Display Surfaces	280
10 13	Directories	280
10 14	Signage	281
10 17	Telephone Specialties	282
10 21	Compartments and Cubicles	282
10 22	Partitions	286
10 26	Wall and Door Protection	288
10 28	Toilet, Bath, and Laundry Accessories	288
10 31	Manufactured Fireplaces	290
10 32	Fireplace Specialties	291
10 35	Stoves	291
10 44	Fire Protection Specialties	291
10 51	Lockers	292
10 55	Postal Specialties	293
10 56	Storage Assemblies	293
10 57	Wardrobe and Closet Specialties	293
10 73	Protective Covers	294
10 74	Manufactured Exterior Specialties	295
Equipment		
11 11	Vehicle Service Equipment	298
11 12	Parking Control Equipment	298
11 13	Loading Dock Equipment	299
11 16	Vault Equipment	299
11 17	Teller and Service Equipment	299
11 19	Detention Equipment	300
11 21	Mercantile and Service Equipment	300
11 23	Commercial Laundry and Dry Cleaning Equipment	300
11 24	Maintenance Equipment	301
11 26	Unit Kitchens	301
11 27	Photographic Processing Equipment	302
11 31	Residential Appliances	302
11 33	Retractable Stairs	304
11 41	Food Storage Equipment	304
11 42	Food Preparation Equipment	306
11 43	Food Delivery Carts and Conveyors	306
11 44	Food Cooking Equipment	307
11 46	Food Dispensing Equipment	307
11 47	Ice Machines	308
11 48	Cleaning and Disposal Equipment	308
11 52	Audio-Visual Equipment	309
11 53	Laboratory Equipment	309
11 57	Vocational Shop Equipment	310
11 61	Theater and Stage Equipment	310
11 66	Athletic Equipment	311
11 82	Solid Waste Handling Equipment	312
11 91	Religious Equipment	312
Furnishings		
12 21	Window Blinds	314
12 23	Interior Shutters	314
12 24	Window Shades	314
12 32	Manufactured Wood Casework	314
12 34	Manufactured Plastic Casework	317
12 35	Specialty Casework	317
12 36	Countertops	318

Div. No.		Page
12 48	Rugs and Mats	320
12 51	Office Furniture	320
12 54	Hospitality Furniture	320
12 55	Detention Furniture	321
12 56	Institutional Furniture	321
12 63	Stadium and Arena Seating	322
12 67	Pews and Benches	322
12 93	Site Furnishings	322
Special Construction		
13 05	Common Work Results for Special Construction	326
13 11	Swimming Pools	329
13 12	Fountains	329
13 17	Tubs and Pools	330
13 21	Controlled Environment Rooms	330
13 24	Special Activity Rooms	330
13 28	Athletic and Recreational Special Construction	331
13 34	Fabricated Engineered Structures	331
13 42	Building Modules	332
13 48	Sound, Vibration, and Seismic Control	332
13 49	Radiation Protection	332
Conveying Equipment		
14 11	Manual Dumbwaiters	336
14 12	Electric Dumbwaiters	336
14 21	Electric Traction Elevators	336
14 24	Hydraulic Elevators	337
14 27	Custom Elevator Cabs	338
14 28	Elevator Equipment and Controls	339
14 42	Wheelchair Lifts	340
14 45	Vehicle Lifts	341
14 51	Correspondence and Parcel Lifts	341
14 91	Facility Chutes	341
14 92	Pneumatic Tube Systems	342
Fire Suppression		
21 05	Common Work Results for Fire Suppression	344
21 11	Facility Fire-Suppression Water-Service Piping	344
21 12	Fire-Suppression Standpipes	344
21 13	Fire-Suppression Sprinkler Systems	345
Plumbing		
22 01	Operation and Maintenance of Plumbing	348
22 05	Common Work Results for Plumbing	348
22 07	Plumbing Insulation	350
22 11	Facility Water Distribution	352
22 13	Facility Sanitary Sewerage	364
22 14	Facility Storm Drainage	367
22 15	General Service Compressed-Air Systems	368
22 31	Domestic Water Softeners	368
22 33	Electric Domestic Water Heaters	368
22 34	Fuel-Fired Domestic Water Heaters	369
22 41	Residential Plumbing Fixtures	370
22 42	Commercial Plumbing Fixtures	373
22 43	Healthcare Plumbing Fixtures	375
22 45	Emergency Plumbing Fixtures	375
22 47	Drinking Fountains and Water Coolers	376
22 66	Chemical-Waste Systems for Lab. and Healthcare Facilities	377
Heating Ventilation Air Conditioning		
23 05	Common Work Results for HVAC	380
23 07	HVAC Insulation	381
23 09	Instrumentation and Control for HVAC	381
23 12	Facility Fuel Pumps	382
23 21	Hydronic Piping and Pumps	382

Div. No.		Page
23 31	HVAC Ducts and Casings	383
23 33	Air Duct Accessories	384
23 34	HVAC Fans	386
23 36	Air Terminal Units	388
23 37	Air Outlets and Inlets	388
23 38	Ventilation Hoods	390
23 41	Particulate Air Filtration	390
23 42	Gas-Phase Air Filtration	390
23 43	Electronic Air Cleaners	391
23 51	Breechings, Chimneys, and Stacks	391
23 52	Heating Boilers	392
23 54	Furnaces	394
23 55	Fuel-Fired Heaters	396
23 56	Solar Energy Heating Equipment	396
23 61	Refrigerant Compressors	398
23 62	Packaged Compressor and Condenser Units	398
23 64	Packaged Water Chillers	398
23 65	Cooling Towers	399
23 72	Air-to-Air Energy Recovery Equipment	399
23 73	Indoor Central-Station Air-Handling Units	400
23 74	Packaged Outdoor HVAC Equipment	400
23 81	Decentralized Unitary HVAC Equipment	400
23 82	Convection Heating and Cooling Units	402
23 83	Radiant Heating Units	404
23 84	Humidity Control Equipment	404
23 91	Prefabricated Equipment Supports	404
Electrical		
26 05	Common Work Results for Electrical	406
26 22	Low-Voltage Transformers	421
26 24	Switchboards and Panelboards	421
26 27	Low-Voltage Distribution Equipment	423
26 28	Low-Voltage Circuit Protective Devices	425
26 29	Low-Voltage Controllers	425
26 32	Packaged Generator Assemblies	426
26 51	Interior Lighting	427
26 52	Emergency Lighting	428
26 53	Exit Signs	429
26 55	Special Purpose Lighting	429
26 56	Exterior Lighting	429
26 61	Lighting Systems and Accessories	430
Communications		
27 15	Communications Horizontal Cabling	432
27 41	Audio-Video Systems	432
27 51	Distributed Audio-Video Communications Systems	432
Electronic Safety & Security		
28 13	Access Control	436
28 16	Intrusion Detection	436
28 23	Video Surveillance	437
28 31	Fire Detection and Alarm	437
Earthwork		
31 05	Common Work Results for Earthwork	440
31 06	Schedules for Earthwork	440
31 23	Excavation and Fill	441
31 25	Erosion and Sedimentation Controls	454
31 31	Soil Treatment	455
31 41	Shoring	455
31 43	Concrete Raising	455
31 46	Needle Beams	456

Table of Contents (cont.)

Div. No.		Page
31 48	Underpinning	456
31 62	Driven Piles	456
31 63	Bored Piles	457
Exterior Improvements		
32 01	Operation and Maintenance of Exterior Improvements	460
32 06	Schedules for Exterior Improvements	461
32 11	Base Courses	462
32 12	Flexible Paving	463
32 13	Rigid Paving	464
32 14	Unit Paving	464
32 16	Curbs and Gutters	465
32 17	Paving Specialties	466
32 31	Fences and Gates	467
32 32	Retaining Walls	470

Div. No.		Page
32 84	Planting Irrigation	470
32 91	Planting Preparation	471
32 92	Turf and Grasses	472
32 93	Plants	473
32 94	Planting Accessories	475
32 96	Transplanting	475
Utilities		
33 01	Operation and Maintenance of Utilities	478
33 05	Common Work Results for Utilities	478
33 11	Water Utility Distribution Piping	479
33 12	Water Utility Distribution Equipment	479
33 21	Water Supply Wells	480
33 31	Sanitary Utility Sewerage Piping	480

Div. No.		Page
33 36	Utility Septic Tanks	480
33 41	Storm Utility Drainage Piping	481
33 44	Storm Utility Water Drains	482
33 49	Storm Drainage Structures	482
33 51	Natural-Gas Distribution	482
33 52	Liquid Fuel Distribution	483
33 71	Electrical Utility Transmission and Distribution	484
Transportation		
34 71	Roadway Construction	486
Material Processing & Handling Equipment		
41 22	Cranes and Hoists	488

How to Use the Unit Price Pages

The following is a detailed explanation of a sample entry in the Unit Price Section. Next to each bold number below is the described item with the appropriate component of the sample entry following in parentheses. Some prices are listed as bare costs; others as costs that include overhead and profit of the installing contractor. In most cases, if the work is to be subcontracted, the general contractor will need to add an additional markup (RSMeans suggests using 10%) to the figures in the column "Total Incl. O&P."

1. Division Number/Title
(03 30/Cast-In-Place Concrete)

Use the Unit Price Section Table of Contents to locate specific items. The sections are classified according to the CSI MasterFormat 2004 system.

2. Line Numbers
(03 30 53.40 3920)

Each unit price line item has been assigned a unique 12-digit code based on the CSI MasterFormat classification.

- MasterFormat Division (03)
- MasterFormat Level 2 (03 30 00)
- MasterFormat Level 3

03 30 53.40 3920

- MasterFormat Level 4
- RSMeans 12-Digit Line Number

3. Description
(Concrete In Place, etc.)

Each line item is described in detail. Sub-items and additional sizes are indented beneath the appropriate line items. The first line or two after the main item (in boldface) may contain descriptive information that pertains to all line items beneath this boldface listing.

4. Reference Number Information

R033053-50 You'll see reference numbers shown in shaded boxes at the beginning of some sections. These refer to related items in the Reference Section, visually identified by a vertical gray bar on the page edges.

The relation may be an estimating procedure that should be read before estimating, or technical information.

The "R" designates the Reference Section. The numbers refer to the MasterFormat 2004 classification system.

It is strongly recommended that you review all reference numbers that appear within the section in which you are working.

Note: Not all reference numbers appear in all RSMeans publications.

03 30 Cast-In-Place Concrete
03 30 53 – Miscellaneous Cast-In-Place Concrete

03 30 53.40 Concrete In Place		Crew	Daily Output	Labor-Hours	Unit	Material	2010 Bare Costs Labor	Equipment	Total	Total Incl O&P
3570	6' x 6' x 8" thick		14	3.429		146	141	2.16	289.16	390
3580	8' x 8' x 10" thick		8	6		310	246	3.78	559.78	750
3590	10' x 10' x 12" thick		5	9.600		530	395	6.05	931.05	1,225
3800	Footings (3000 psi), spread under 1 C.Y.	C-14C	28	4	C.Y.	150	160	1.08	311.08	430
3825	1 C.Y. to 5 C.Y.		43	2.605		173	104	.70	277.70	360
3850	Over 5 C.Y.		75	1.493		158	59.50	.40	217.90	272
3900	Footings, strip (3000 psi), 18" x 9", unreinforced	C-14L	40	2.400		112	93	.77	205.77	276
3920	18" x 9", reinforced	C-14C	35	3.200		131	128	.86	259.86	355
3925	20" x 10", unreinforced	C-14L	45	2.133		110	82.50	.68	193.18	257
3930	20" x 10", reinforced	C-14C	40	2.800		126	112	.76	238.76	325
3935	24" x 12", unreinforced	C-14L	55	1.745		110	67.50	.56	178.06	233
3940	24" x 12", reinforced	C-14C	48	2.333		126	93	.63	219.63	292
3945	36" x 12", unreinforced	C-14L	70	1.371		107	53	.44	160.44	205
3950	36" x 12", reinforced	C-14C	60	1.867		122	74.50	.50	197	257

Crew (C-14C)

The "Crew" column designates the typical trade or crew used to install the item. If an installation can be accomplished by one trade and requires no power equipment, that trade and the number of workers are listed (for example, "2 Carpenters"). If an installation requires a composite crew, a crew code designation is listed (for example, "C-14C"). You'll find full details on all composite crews in the Crew Listings.
* For a complete list of all trades utilized in this book and their abbreviations, see the inside back cover.

Crews

Crew No.	Bare Costs				Incl. Subs O&P		Cost Per Labor-Hour	
Crew C-14C	Hr.	Daily		Hr.	Daily		Bare Costs	Incl. O&P
1 Carpenter Foreman (out)	$43.55	$348.40		$71.50	$572.00		$39.90	$65.55
6 Carpenters	41.55	1994.40		68.25	3276.00			
2 Rodmen (reinf.)	46.80	748.80		78.65	1258.40			
4 Laborers	33.10	1059.20		54.35	1739.20			
1 Cement Finisher	39.70	317.60		61.95	495.60			
1 Gas Engine Vibrator		30.40			33.44		.27	.30
112 L.H., Daily Totals		$4498.80			$7374.64		$40.17	$65.84

Productivity: Daily Output (35.0)/ Labor-Hours (3.20)

The "Daily Output" represents the typical number of units the designated crew will install in a normal 8-hour day. To find out the number of days the given crew would require to complete the installation, divide your quantity by the daily output. For example:

Quantity	÷	Daily Output	=	Duration
100 C.Y.	÷	35.0/ Crew Day	=	2.86 Crew Days

The "Labor-Hours" figure represents the number of labor-hours required to install one unit of work. To find out the number of labor-hours required for your particular task, multiply the quantity of the item times the number of labor-hours shown. For example:

Quantity	x	Productivity Rate	=	Duration
100 C.Y.	x	3.20 Labor-Hours/ C.Y.	=	320 Labor-Hours

Unit (C.Y.)

The abbreviated designation indicates the unit of measure upon which the price, production, and crew are based (C.Y. = Cubic Yard). For a complete listing of abbreviations, refer to the Abbreviations Listing in the Reference Section of this book.

Bare Costs:
Mat. (Bare Material Cost) (131)
The unit material cost is the "bare" material cost with no overhead or profit included. Costs shown reflect national average material prices for January of the current year and include delivery to the job site. No sales taxes are included.

Labor (128)
The unit labor cost is derived by multiplying bare labor-hour costs for Crew C-14C by labor-hour units. The bare labor-hour cost is found in the Crew Section under C-14C. (If a trade is listed, the hourly labor cost—the wage rate—is found on the inside back cover.)

Labor-Hour Cost Crew C-14C	x	Labor-Hour Units	=	Labor
$39.90	x	3.20	=	$128.00

Equip. (Equipment) (.86)
Costs for equipment included in each crew are listed in the Crew Listings in the Reference Section. Costs for small tools that are considered as essential everyday tools are covered in the "Overhead" column on the "Installing Contractor's Overhead and Profit" table that lists labor trades, base rates and markups. The unit equipment cost displayed here is derivced by multiplying the bare equipment hourly cost in the Crew Listing by the unit Labor-Hours value.

Equipment Cost Crew C-14C	x	Labor-Hour Units	=	Equip.
$.27	x	3.20	=	$.86

Total (259.86)
The total of the bare costs is the arithmetic total of the three previous columns: mat., labor, and equip.

Material	+	Labor	+	Equip.	=	Total
$131	+	$128	+	$.86	=	$259.86

Total Costs Including O&P

This figure is the sum of the bare material cost plus 10% for profit; the bare labor cost plus total overhead and profit (per the inside back cover or, if a crew is listed, from the crew listings); and the bare equipment cost plus 10% for profit.

Material is Bare Material Cost + 10% = 131 + 13.10	=	$144.10
Labor for Crew C-14C = Labor-Hour Cost (65.55) x Labor-Hour Units (3.20)	=	$209.76
Equip. is Bare Equip. Cost + 10% = .86 + .09	=	$ .95
Total (Rounded)	=	$ 355

Estimating Tips

01 20 00 Price and Payment Procedures

- Allowances that should be added to estimates to cover contingencies and job conditions that are not included in the national average material and labor costs are shown in section 01 21.
- When estimating historic preservation projects (depending on the condition of the existing structure and the owner's requirements), a 15%–20% contingency or allowance is recommended, regardless of the stage of the drawings.

01 30 00 Administrative Requirements

- Before determining a final cost estimate, it is a good practice to review all the items listed in Subdivisions 01 31 and 01 32 to make final adjustments for items that may need customizing to specific job conditions.
- Requirements for initial and periodic submittals can represent a significant cost to the General Requirements of a job. Thoroughly check the submittal specifications when estimating a project to determine any costs that should be included.

01 40 00 Quality Requirements

- All projects will require some degree of Quality Control. This cost is not included in the unit cost of construction listed in each division. Depending upon the terms of the contract, the various costs of inspection and testing can be the responsibility of either the owner or the contractor. Be sure to include the required costs in your estimate.

01 50 00 Temporary Facilities and Controls

- Barricades, access roads, safety nets, scaffolding, security, and many more requirements for the execution of a safe project are elements of direct cost. These costs can easily be overlooked when preparing an estimate. When looking through the major classifications of this subdivision, determine which items apply to each division in your estimate.
- Construction Equipment Rental Costs can be found in the Reference Section in section 01 54 33. Operators' wages are not included in equipment rental costs.
- Equipment mobilization and demobilization costs are not included in equipment rental costs and must be considered separately in section 01 54 36.50.
- The cost of small tools provided by the installing contractor for his workers is covered in the "Overhead" column on the "Installing Contractor's Overhead and Profit" table that lists labor trades, base rates and markups and, therefore, is included in the "Total Incl. O&P" cost of any Unit Price line item. For those users who are constrained by contract terms to use only bare costs, there are two line items in section 01 54 39.70 for small tools as a percentage of bare labor cost. If some of those users are further constrained by contract terms to refrain from using line items from Division 1, you are advised to cover the cost of small tools within your coefficient or multiplier.

01 70 00 Execution and Closeout Requirements

- When preparing an estimate, thoroughly read the specifications to determine the requirements for Contract Closeout. Final cleaning, record documentation, operation and maintenance data, warranties and bonds, and spare parts and maintenance materials can all be elements of cost for the completion of a contract. Do not overlook these in your estimate.

Reference Numbers

Reference numbers are shown in shaded boxes at the beginning of some major classifications. These numbers refer to related items in the Reference Section. The reference information may be an estimating procedure, an alternate pricing method, or technical information.

Note: Not all subdivisions listed here necessarily appear in this publication.

No part of this publication may be reproduced, stored in a retrieval system, or transmitted in any form or by any means without prior written permission of R.S. Means Company, Inc.

01 11 Summary of Work

01 11 31 – Professional Consultants

01 11 31.10 Architectural Fees		Crew	Daily Output	Labor-Hours	Unit	Material	2010 Bare Costs Labor	Equipment	Total	Total Incl O&P
0010	**ARCHITECTURAL FEES** R011110-10									
0020	For new construction									
0060	Minimum				Project					4.90%
0090	Maximum									16%
0100	For alteration work, to $500,000, add to new construction fee									50%
0150	Over $500,000, add to new construction fee									25%

01 11 31.20 Construction Management Fees

0010	**CONSTRUCTION MANAGEMENT FEES**									
0060	For work to $100,000				Project					10%
0070	To $250,000									9%
0090	To $1,000,000									6%
0100	To $5,000,000									5%
0110	To $10,000,000									4%

01 11 31.30 Engineering Fees

0010	**ENGINEERING FEES** R011110-30									
0020	Educational planning consultant, minimum				Project					.50%
0100	Maximum				"					2.50%
0400	Elevator & conveying systems, minimum				Contrct					2.50%
0500	Maximum									5%
1000	Mechanical (plumbing & HVAC), minimum									4.10%
1100	Maximum									10.10%
1200	Structural, minimum				Project					1%
1300	Maximum				"					2.50%

01 11 31.75 Renderings

0010	**RENDERINGS** Color, matted, 20" x 30", eye level,									
0020	1 building, minimum				Ea.	1,950			1,950	2,150
0050	Average					2,775			2,775	3,075
0100	Maximum					4,450			4,450	4,900
1000	5 buildings, minimum					3,900			3,900	4,300
1100	Maximum					7,800			7,800	8,575

01 21 Allowances

01 21 16 – Contingency Allowances

01 21 16.50 Contingencies

0010	**CONTINGENCIES**, Add to estimate									
0020	Conceptual stage				Project					25%
0050	Schematic stage									20%
0100	Preliminary working drawing stage (Design Dev.)									15%
0150	Final working drawing stage									8%

01 21 53 – Factors Allowance

01 21 53.50 Factors

0010	**FACTORS** Cost adjustments R012153-10									
0100	Add to construction costs for particular job requirements									
0500	Cut & patch to match existing construction, add, minimum				Costs	2%	3%			
0550	Maximum					5%	9%			
0800	Dust protection, add, minimum					1%	2%			
0850	Maximum					4%	11%			
1100	Equipment usage curtailment, add, minimum					1%	1%			
1150	Maximum					3%	10%			
1400	Material handling & storage limitation, add, minimum					1%	1%			

01 21 Allowances

01 21 53 – Factors Allowance

01 21 53.50 Factors

		Crew	Daily Output	Labor-Hours	Unit	Material	2010 Bare Costs Labor	Equipment	Total	Total Incl O&P
1450	Maximum				Costs	6%	7%			
1700	Protection of existing work, add, minimum					2%	2%			
1750	Maximum					5%	7%			
2000	Shift work requirements, add, minimum						5%			
2050	Maximum						30%			
2300	Temporary shoring and bracing, add, minimum					2%	5%			
2350	Maximum					5%	12%			
2400	Work inside prisons and high security areas, add, minimum						30%			
2450	Maximum						50%			

01 21 55 – Job Conditions Allowance

01 21 55.50 Job Conditions

		Crew	Daily Output	Labor-Hours	Unit	Material	2010 Bare Costs Labor	Equipment	Total	Total Incl O&P
0010	**JOB CONDITIONS** Modifications to applicable									
0020	cost summaries									
0100	Economic conditions, favorable, deduct				Project				2%	2%
0200	Unfavorable, add								5%	5%
0300	Hoisting conditions, favorable, deduct								2%	2%
0400	Unfavorable, add								5%	5%
0500	General Contractor management, experienced, deduct								2%	2%
0600	Inexperienced, add								10%	10%
0700	Labor availability, surplus, deduct								1%	1%
0800	Shortage, add								10%	10%
0900	Material storage area, available, deduct								1%	1%
1000	Not available, add								2%	2%
1100	Subcontractor availability, surplus, deduct								5%	5%
1200	Shortage, add								12%	12%
1300	Work space, available, deduct								2%	2%
1400	Not available, add								5%	5%

01 21 63 – Taxes

01 21 63.10 Taxes

		Crew	Daily Output	Labor-Hours	Unit	Material	2010 Bare Costs Labor	Equipment	Total	Total Incl O&P
0010	**TAXES**		R012909-80							
0020	Sales tax, State, average				%	5.01%				
0050	Maximum		R012909-85			8.25%				
0200	Social Security, on first $106,800 of wages						7.65%			
0300	Unemployment, combined Federal and State, minimum						.80%			
0350	Average						6.20%			
0400	Maximum						10.96%			

01 31 Project Management and Coordination

01 31 13 – Project Coordination

01 31 13.20 Field Personnel

		Crew	Daily Output	Labor-Hours	Unit	Material	2010 Bare Costs Labor	Equipment	Total	Total Incl O&P
0010	**FIELD PERSONNEL**									
0020	Clerk, average				Week		395		395	605
0180	Project manager, minimum						1,725		1,725	2,650
0200	Average						2,000		2,000	3,075
0220	Maximum						2,275		2,275	3,500
0240	Superintendent, minimum						1,675		1,675	2,575
0260	Average						1,850		1,850	2,850
0280	Maximum						2,100		2,100	3,225

01 31 Project Management and Coordination

01 31 13 – Project Coordination

01 31 13.30 Insurance		Crew	Daily Output	Labor-Hours	Unit	Material	2010 Bare Costs Labor	Equipment	Total	Total Incl O&P
0010	**INSURANCE** R013113-40									
0020	Builders risk, standard, minimum				Job					.24%
0050	Maximum R013113-60									.64%
0200	All-risk type, minimum									.25%
0250	Maximum				↓					.62%
0400	Contractor's equipment floater, minimum				Value					.50%
0450	Maximum				"					1.50%
0600	Public liability, average				Job					2.02%
0810	Workers' compensation & employer's liability									
2000	Range of 35 trades in 50 states, excl. wrecking, min.				Payroll		2.39%			
2100	Average						14.40%			
2200	Maximum				↓		170.80%			

01 31 13.40 Main Office Expense										
0010	**MAIN OFFICE EXPENSE** Average for General Contractors									
0020	As a percentage of their annual volume									
0030	Annual volume to $300,000, minimum				% Vol.					20%
0040	Maximum									30%
0060	To $500,000, minimum									17%
0070	Maximum									22%
0080	To $1,000,000, minimum									16%
0090	Maximum									19%
0110	To $3,000,000, minimum									14%
0120	Maximum									16%
0130	To $5,000,000, minimum									8%
0140	Maximum				↓					10%

01 31 13.50 General Contractor's Mark-Up										
0010	**GENERAL CONTRACTOR'S MARK-UP** on Change Orders									
0200	Extra work, by subcontractors, add				%					10%
0250	By General Contractor, add									15%
0400	Omitted work, by subcontractors, deduct all but									5%
0450	By General Contractor, deduct all but									7.50%
0600	Overtime work, by subcontractors, add									15%
0650	By General Contractor, add				↓					10%

01 31 13.60 Installing Contractor's Main Office Overhead										
0010	**INSTALLING CONTRACTOR'S MAIN OFFICE OVERHEAD** R013113-50									
0020	As percent of direct costs, minimum				%					5%
0050	Average									16%
0100	Maximum				↓					30%

01 31 13.80 Overhead and Profit										
0010	**OVERHEAD & PROFIT** Allowance to add to items in this R013113-50									
0020	book that do not include Subs O&P, average				%					30%
0100	Allowance to add to items in this book that									
0110	do include Subs O&P, minimum				%					5%
0150	Average									10%
0200	Maximum									15%
0290	Typical, by size of project, under $50,000									40%
0310	$50,000 to $100,000									35%
0320	$100,000 to $500,000									25%
0330	$500,000 to $1,000,000				↓					20%

01 41 Regulatory Requirements

01 41 26 – Permits

01 41 26.50 Permits		Crew	Daily Output	Labor-Hours	Unit	Material	2010 Bare Costs Labor	Equipment	Total	Total Incl O&P
0010	**PERMITS**									
0020	Rule of thumb, most cities, minimum				Job					.50%
0100	Maximum				"					2%

01 45 Quality Control

01 45 23 – Testing and Inspecting Services

01 45 23.50 Testing		Crew	Daily Output	Labor-Hours	Unit	Material	2010 Bare Costs Labor	Equipment	Total	Total Incl O&P
0010	**TESTING** and Inspecting Services									
1800	Compressive test, cylinder, delivered to lab, ASTM C 39				Ea.				12	13
1900	Picked up by lab, minimum								14	15
1950	Average								18	20
2000	Maximum								27	30
2200	Compressive strength, cores (not incl. drilling), ASTM C 42								36	40
2300	Patching core holes								22	24
4730	Soil testing									
4735	Soil density, nuclear method, ASTM D2922				Ea.				35	38.67
4740	Sand cone method ASTM D1556								27	30.17
4750	Moisture content, ASTM D 2216								9	10
4780	Permeability test, double ring infiltrometer								500	550
4800	Permeability, var. or constant head, undist., ASTM D 2434								227	250
4850	Recompacted								250	275
4900	Proctor compaction, 4" standard mold, ASTM D 698								123	135
4950	6" modified mold								68	75
5100	Shear tests, triaxial, minimum								409	450
5150	Maximum								545	600
5550	Technician for inspection, per day, earthwork								210	231
5650	Bolting								268	295
5750	Roofing								244	256
5790	Welding								257	283
5820	Non-destructive metal testing, dye penetrant				Day				310	341
5840	Magnetic particle								310	341
5860	Radiography								450	495
5880	Ultrasonic								309	340
6000	Welding certification, minimum				Ea.				91	100
6100	Maximum				"				250	275
7000	Underground storage tank									
7500	Volumetric tightness test ,<=12,000 gal				Ea.				435	478
7510	<=30,000 gal				"				613	675
7600	Vadose zone (soil gas) sampling, 10-40 samples, min.				Day				1,364	1,500
7610	Maximum				"				2,273	2,500
7700	Ground water monitoring incl. drilling 3 wells, min.				Total				4,545	5,000
7710	Maximum				"				6,364	7,000
8000	X-ray concrete slabs				Ea.				182	200

01 51 Temporary Utilities

01 51 13 – Temporary Electricity

01 51 13.80 Temporary Utilities	Crew	Daily Output	Labor-Hours	Unit	Material	2010 Bare Costs Labor	Equipment	Total	Total Incl O&P
0010 **TEMPORARY UTILITIES**									
0100 Heat, incl. fuel and operation, per week, 12 hrs. per day	1 Skwk	100	.080	CSF Flr	27	3.41		30.41	35
0200 24 hrs. per day	"	60	.133		52	5.70		57.70	66.50
0350 Lighting, incl. service lamps, wiring & outlets, minimum	1 Elec	34	.235		2.63	11.55		14.18	20.50
0360 Maximum	"	17	.471		5.70	23		28.70	42
0400 Power for temp lighting only, per month, min/month 6.6 KWH								.92	1.01
0430 Average/month 11.8 KWH								1.65	1.82
0450 Maximum/month 23.6 KWH								3.30	3.63
0600 Power for job duration incl. elevator, etc., minimum								47	51.70
0650 Maximum								110	121

01 52 Construction Facilities

01 52 13 – Field Offices and Sheds

01 52 13.20 Office and Storage Space

		Crew	Daily Output	Labor-Hours	Unit	Material	Labor	Equipment	Total	Total Incl O&P
0010	**OFFICE AND STORAGE SPACE**									
0020	Trailer, furnished, no hookups, 20' x 8', buy	2 Skwk	1	16	Ea.	8,550	680		9,230	10,500
0250	Rent per month					146			146	160
0300	32' x 8', buy	2 Skwk	.70	22.857		12,300	975		13,275	15,100
0350	Rent per month					193			193	213
0400	50' x 10', buy	2 Skwk	.60	26.667		23,300	1,125		24,425	27,500
0450	Rent per month					286			286	315
0500	50' x 12', buy	2 Skwk	.50	32		26,000	1,375		27,375	30,800
0550	Rent per month					360			360	395
0700	For air conditioning, rent per month, add					41			41	45
0800	For delivery, add per mile				Mile	4.50			4.50	4.95
1200	Storage boxes, 20' x 8', buy	2 Skwk	1.80	8.889	Ea.	3,225	380		3,605	4,150
1250	Rent per month					69.50			69.50	76.50
1300	40' x 8', buy	2 Skwk	1.40	11.429		4,225	485		4,710	5,400
1350	Rent per month					92.50			92.50	102

01 54 Construction Aids

01 54 16 – Temporary Hoists

01 54 16.50 Weekly Forklift Crew

		Crew	Daily Output	Labor-Hours	Unit	Material	Labor	Equipment	Total	Total Incl O&P
0010	**WEEKLY FORKLIFT CREW**									
0100	All-terrain forklift, 45' lift, 35' reach, 9000 lb. capacity	A-3P	.20	40	Week		1,650	2,400	4,050	5,225

01 54 19 – Temporary Cranes

01 54 19.50 Daily Crane Crews

		Crew	Daily Output	Labor-Hours	Unit	Material	Labor	Equipment	Total	Total Incl O&P
0010	**DAILY CRANE CREWS** for small jobs, portal to portal									
0100	12-ton truck-mounted hydraulic crane	A-3H	1	8	Day		355	870	1,225	1,525
0200	25-ton	A-3I	1	8			355	1,025	1,380	1,675
0300	40-ton	A-3J	1	8			355	1,250	1,605	1,925
0400	55-ton	A-3K	1	16			660	1,775	2,435	2,975
0500	80-ton	A-3L	1	16			660	2,225	2,885	3,450
0600	100-ton	A-3M	1	16			660	2,425	3,085	3,700

01 54 19.60 Monthly Tower Crane Crew

		Crew	Daily Output	Labor-Hours	Unit	Material	Labor	Equipment	Total	Total Incl O&P
0010	**MONTHLY TOWER CRANE CREW**, excludes concrete footing									
0100	Static tower crane, 130' high, 106' jib, 6200 lb. capacity	A-3N	.05	176	Month		7,825	23,100	30,925	37,600

01 54 Construction Aids

01 54 23 – Temporary Scaffolding and Platforms

01 54 23.60 Pump Staging		Crew	Daily Output	Labor-Hours	Unit	Material	2010 Bare Costs Labor	Equipment	Total	Total Incl O&P
0010	**PUMP STAGING**, Aluminum R015423-20									
0200	24' long pole section, buy				Ea.	415			415	460
0300	18' long pole section, buy					325			325	355
0400	12' long pole section, buy					218			218	240
0500	6' long pole section, buy					115			115	126
0600	6' long splice joint section, buy					85.50			85.50	94
0700	Pump jack, buy					139			139	153
0900	Foldable brace, buy					56.50			56.50	62.50
1000	Workbench/back safety rail support, buy					73.50			73.50	81
1100	Scaffolding planks/workbench, 14" wide x 24' long, buy					680			680	750
1200	Plank end safety rail, buy					370			370	405
1250	Safety net, 22' long, buy				↓	335			335	365
1300	System in place, 50' working height, per use based on 50 uses	2 Carp	84.80	.189	C.S.F.	6.40	7.85		14.25	19.95
1400	100 uses		84.80	.189		3.21	7.85		11.06	16.45
1500	150 uses	↓	84.80	.189	↓	2.15	7.85		10	15.25

01 54 23.70 Scaffolding		Crew	Daily Output	Labor-Hours	Unit	Material	Labor	Equipment	Total	Total Incl O&P
0010	**SCAFFOLDING** R015423-10									
0015	Steel tube, regular, no plank, labor only to erect & dismantle									
0090	Building exterior, wall face, 1 to 5 stories, 6'-4" x 5' frames	3 Carp	8	3	C.S.F.		125		125	205
0200	6 to 12 stories	4 Carp	8	4			166		166	273
0301	13 to 20 stories	5 Clab	8	5			166		166	272
0460	Building interior, wall face area, up to 16' high	3 Carp	12	2			83		83	137
0560	16' to 40' high	↓	10	2.400	↓		99.50		99.50	164
0800	Building interior floor area, up to 30' high		150	.160	C.C.F.		6.65		6.65	10.90
0900	Over 30' high	4 Carp	160	.200	"		8.30		8.30	13.65
0906	Complete system for face of walls, no plank, material only rent/mo				C.S.F.	36			36	39.50
0908	Interior spaces, no plank, material only rent/mo				C.C.F.	3.42			3.42	3.76
0910	Steel tubular, heavy duty shoring, buy									
0920	Frames 5' high 2' wide				Ea.	93			93	102
0925	5' high 4' wide					105			105	116
0930	6' high 2' wide					106			106	117
0935	6' high 4' wide				↓	124			124	136
0940	Accessories									
0945	Cross braces				Ea.	18			18	19.80
0950	U-head, 8" x 8"					21.50			21.50	24
0955	J-head, 4" x 8"					15.80			15.80	17.40
0960	Base plate, 8" x 8"					17.60			17.60	19.35
0965	Leveling jack				↓	38			38	41.50
1000	Steel tubular, regular, buy									
1100	Frames 3' high 5' wide				Ea.	71			71	78
1150	5' high 5' wide					83			83	91.50
1200	6'-4" high 5' wide					104			104	114
1350	7'-6" high 6' wide					179			179	197
1500	Accessories cross braces					20.50			20.50	22.50
1550	Guardrail post					18.60			18.60	20.50
1600	Guardrail 7' section					9.90			9.90	10.90
1650	Screw jacks & plates					27			27	29.50
1700	Sidearm brackets					43			43	47.50
1750	8" casters					35.50			35.50	39
1800	Plank 2" x 10" x 16'-0"					51			51	56
1900	Stairway section					325			325	360
1910	Stairway starter bar				↓	36.50			36.50	40

01 54 Construction Aids

01 54 23 – Temporary Scaffolding and Platforms

01 54 23.70 Scaffolding

		Crew	Daily Output	Labor-Hours	Unit	Material	2010 Bare Costs Labor	Equipment	Total	Total Incl O&P
1920	Stairway inside handrail				Ea.	66			66	72.50
1930	Stairway outside handrail					98			98	108
1940	Walk-thru frame guardrail				▼	47			47	51.50
2000	Steel tubular, regular, rent/mo.									
2100	Frames 3' high 5' wide				Ea.	5			5	5.50
2150	5' high 5' wide					5			5	5.50
2200	6'-4" high 5' wide					5.15			5.15	5.65
2250	7'-6" high 6' wide					7			7	7.70
2500	Accessories, cross braces					1			1	1.10
2550	Guardrail post					1			1	1.10
2600	Guardrail 7' section					1			1	1.10
2650	Screw jacks & plates					2			2	2.20
2700	Sidearm brackets					2			2	2.20
2750	8" casters					8			8	8.80
2800	Outrigger for rolling tower					3			3	3.30
2850	Plank 2" x 10" x 16'-0"					6			6	6.60
2900	Stairway section					40			40	44
2940	Walk-thru frame guardrail				▼	2.50			2.50	2.75
3000	Steel tubular, heavy duty shoring, rent/mo.									
3250	5' high 2' & 4' wide				Ea.	5			5	5.50
3300	6' high 2' & 4' wide					5			5	5.50
3500	Accessories, cross braces					1			1	1.10
3600	U - head, 8" x 8"					1			1	1.10
3650	J - head, 4" x 8"					1			1	1.10
3700	Base plate, 8" x 8"					1			1	1.10
3750	Leveling jack					2			2	2.20
5700	Planks, 2" x 10" x 16'-0", labor only to erect & remove to 50' H	3 Carp	72	.333			13.85		13.85	23
5800	Over 50' high	4 Carp	80	.400	▼		16.60		16.60	27.50

01 54 23.75 Scaffolding Specialties

		Crew	Daily Output	Labor-Hours	Unit	Material	Labor	Equipment	Total	Total Incl O&P
0010	**SCAFFOLDING SPECIALTIES**									
1200	Sidewalk bridge, heavy duty steel posts & beams, including									
1210	parapet protection & waterproofing									
1220	8' to 10' wide, 2 posts	3 Carp	15	1.600	L.F.	31.50	66.50		98	144
1230	3 posts	"	10	2.400	"	48.50	99.50		148	218
1500	Sidewalk bridge using tubular steel									
1510	scaffold frames, including planking	3 Carp	45	.533	L.F.	5.60	22		27.60	42.50
1600	For 2 uses per month, deduct from all above					50%				
1700	For 1 use every 2 months, add to all above					100%				
1900	Catwalks, 20" wide, no guardrails, 7' span, buy				Ea.	145			145	160
2000	10' span, buy					203			203	223
3720	Putlog, standard, 8' span, with hangers, buy					75			75	82.50
3730	Rent per month					10			10	11
3750	12' span, buy					113			113	124
3755	Rent per month					15			15	16.50
3760	Trussed type, 16' span, buy					260			260	286
3770	Rent per month					20			20	22
3790	22' span, buy					310			310	345
3795	Rent per month					30			30	33
4000	7 step					725			725	800
4100	Rolling towers, buy, 5' wide, 7' long, 10' high					1,350			1,350	1,500
4200	For 5' high added sections, to buy, add				▼	207			207	228
4300	Complete incl. wheels, railings, outriggers,									

01 54 Construction Aids

01 54 23 – Temporary Scaffolding and Platforms

01 54 23.75 Scaffolding Specialties

		Crew	Daily Output	Labor-Hours	Unit	Material	2010 Bare Costs Labor	2010 Bare Costs Equipment	Total	Total Incl O&P
4350	21' high, to buy				Ea.	2,250			2,250	2,475
4400	Rent/month = 5% of purchase cost				"	113			113	124

01 54 23.80 Staging Aids

		Crew	Daily Output	Labor-Hours	Unit	Material	Labor	Equipment	Total	Total Incl O&P
0010	**STAGING AIDS** and fall protection equipment									
0100	Sidewall staging bracket, tubular, buy				Ea.	46			46	50.50
0110	Cost each per day, based on 250 days use				Day	.18			.18	.20
0200	Guard post, buy				Ea.	22.50			22.50	25
0210	Cost each per day, based on 250 days use				Day	.09			.09	.10
0300	End guard chains, buy per pair				Pair	31			31	34
0310	Cost per set per day, based on 250 days use				Day	.18			.18	.20
1010	Cost each per day, based on 250 days use				"	.04			.04	.04
1100	Wood bracket, buy				Ea.	16.60			16.60	18.25
1110	Cost each per day, based on 250 days use				Day	.07			.07	.07
2010	Cost per pair per day, based on 250 days use				"	.43			.43	.48
2100	Steel siderail jack, buy per pair				Pair	122			122	134
2110	Cost per pair per day, based on 250 days use				Day	.49			.49	.54
3010	Cost each per day, based on 250 days use				"	.20			.20	.22
3100	Aluminum scaffolding plank, 20" wide x 24' long, buy				Ea.	890			890	975
3110	Cost each per day, based on 250 days use				Day	3.55			3.55	3.91
4000	Nylon full body harness, lanyard and rope grab				Ea.	215			215	236
4010	Cost each per day, based on 250 days use				Day	.86			.86	.95
4100	Rope for safety line, 5/8" x 100' nylon, buy				Ea.	70			70	77
4110	Cost each per day, based on 250 days use				Day	.28			.28	.31
4200	Permanent U-Bolt roof anchor, buy				Ea.	35			35	38.50
4300	Temporary (one use) roof ridge anchor, buy				"	28.50			28.50	31
5000	Installation (setup and removal) of staging aids									
5010	Sidewall staging bracket	2 Carp	64	.250	Ea.		10.40		10.40	17.05
5020	Guard post with 2 wood rails	"	64	.250			10.40		10.40	17.05
5030	End guard chains, set	1 Carp	64	.125			5.20		5.20	8.55
5100	Roof shingling bracket		96	.083			3.46		3.46	5.70
5200	Ladder jack		64	.125			5.20		5.20	8.55
5300	Wood plank, 2" x 10" x 16'	2 Carp	80	.200			8.30		8.30	13.65
5310	Aluminum scaffold plank, 20" x 24'	"	40	.400			16.60		16.60	27.50
5410	Safety rope	1 Carp	40	.200			8.30		8.30	13.65
5420	Permanent U-Bolt roof anchor (install only)	2 Carp	40	.400			16.60		16.60	27.50
5430	Temporary roof ridge anchor (install only)	1 Carp	64	.125			5.20		5.20	8.55

01 54 26 – Temporary Swing Staging

01 54 26.50 Swing Staging

		Crew	Daily Output	Labor-Hours	Unit	Material	Labor	Equipment	Total	Total Incl O&P
0010	**SWING STAGING**, 500 lb cap., 2' wide to 24' long, hand operated									
0020	steel cable type, with 60' cables, buy				Ea.	4,050			4,050	4,450
0030	Rent per month				"	405			405	445
0600	Lightweight (not for masons) 24' long for 150' height,									
0610	manual type, buy				Ea.	4,150			4,150	4,575
0620	Rent per month					415			415	455
0700	Powered, electric or air, to 150' high, buy					13,300			13,300	14,700
0710	Rent per month					935			935	1,025
0780	To 300' high, buy					13,500			13,500	14,800
0800	Rent per month					945			945	1,025
1000	Bosun's chair or work basket 3' x 3.5', to 300' high, electric, buy					8,625			8,625	9,500
1010	Rent per month					605			605	665
2200	Move swing staging (setup and remove)	E-4	2	16	Move		760	73	833	1,450

01 54 Construction Aids

01 54 36 – Equipment Mobilization

01 54 36.50 Mobilization

		Crew	Daily Output	Labor-Hours	Unit	Material	2010 Bare Costs Labor	2010 Bare Costs Equipment	Total	Total Incl O&P
0010	**MOBILIZATION** (Use line item again for demobilization) R015433-10									
0015	Up to 25 mi haul dist (50 mi RT for mob/demob crew)									
0020	Dozer, loader, backhoe, excav., grader, paver, roller, 70 to 150 H.P.	B-34N	4	2	Ea.		66.50	136	202.50	258
0100	Above 150 HP	B-34K	3	2.667			88.50	305	393.50	485
0300	Scraper, towed type (incl. tractor), 6 C.Y. capacity		3	2.667			88.50	305	393.50	485
0400	10 C.Y.		2.50	3.200			106	370	476	575
0600	Self-propelled scraper, 15 C.Y.		2.50	3.200			106	370	476	575
0700	24 C.Y.		2	4			133	460	593	720
0900	Shovel or dragline, 3/4 C.Y.		3.60	2.222			73.50	256	329.50	400
1000	1-1/2 C.Y.		3	2.667			88.50	305	393.50	485
1100	Small equipment, placed in rear of, or towed by pickup truck	A-3A	8	1			32.50	18.90	51.40	73.50
1150	Equip up to 70 HP, on flatbed trailer behind pickup truck	A-3D	4	2			64.50	64	128.50	176
2000	Crane, truck-mounted, up to 75 ton, (driver only, one-way)	1 Eqhv	7.20	1.111			49.50		49.50	77.50
2100	Crane, truck-mounted, over 75 ton	A-3E	2.50	6.400			248	60.50	308.50	460
2200	Crawler-mounted, up to 75 ton	A-3F	2	8			310	475	785	1,025
2300	Over 75 ton	A-3G	1.50	10.667			415	715	1,130	1,450
2500	For each additional 5 miles haul distance, add						10%	10%		
3000	For large pieces of equipment, allow for assembly/knockdown									
3100	For mob/demob of micro-tunneling equip, see Div. 33 05 23.19									
3200	For mob/demob of pile driving equip, see Div. 31 62 19.10									

01 54 39 – Construction Equipment

01 54 39.70 Small Tools

		Crew	Daily Output	Labor-Hours	Unit	Material	Labor	Equipment	Total	Total Incl O&P
0010	**SMALL TOOLS** R013113-50									
0020	As % of contractor's bare labor cost for project, minimum				Total		.50%			
0100	Maximum				"		2%			

01 55 Vehicular Access and Parking

01 55 23 – Temporary Roads

01 55 23.50 Roads and Sidewalks

		Crew	Daily Output	Labor-Hours	Unit	Material	Labor	Equipment	Total	Total Incl O&P
0010	**ROADS AND SIDEWALKS** Temporary									
0050	Roads, gravel fill, no surfacing, 4" gravel depth	B-14	715	.067	S.Y.	4	2.34	.47	6.81	8.70
0100	8" gravel depth	"	615	.078	"	8	2.72	.55	11.27	13.80

01 56 Temporary Barriers and Enclosures

01 56 13 – Temporary Air Barriers

01 56 13.60 Tarpaulins

		Crew	Daily Output	Labor-Hours	Unit	Material	Labor	Equipment	Total	Total Incl O&P
0010	**TARPAULINS**									
0020	Cotton duck, 10 oz. to 13.13 oz. per S.Y., minimum				S.F.	.59			.59	.65
0050	Maximum					.53			.53	.58
0100	Polyvinyl coated nylon, 14 oz. to 18 oz., minimum					.48			.48	.53
0150	Maximum					.68			.68	.75
0200	Reinforced polyethylene 3 mils thick, white					.15			.15	.17
0300	4 mils thick, white, clear or black					.20			.20	.22
0400	5.5 mils thick, clear					.14			.14	.15
0500	White, fire retardant					.35			.35	.39
0600	7.5 mils, oil resistant, fire retardant					.40			.40	.44
0700	8.5 mils, black					.53			.53	.58
0720	Steel reinforced polyethylene, 4 mils thick					.53			.53	.58

01 56 Temporary Barriers and Enclosures

01 56 13 – Temporary Air Barriers

01 56 13.60 Tarpaulins

		Crew	Daily Output	Labor-Hours	Unit	Material	2010 Bare Costs Labor	Equipment	Total	Total Incl O&P
0730	Polyester reinforced w/integral fastening system 11 mils thick				S.F.	1.07			1.07	1.18
0740	Mylar polyester, non-reinforced, 7 mils thick					1.17			1.17	1.29

01 56 13.90 Winter Protection

0010	**WINTER PROTECTION**									
0100	Framing to close openings	2 Clab	750	.021	S.F.	.39	.71		1.10	1.59
0200	Tarpaulins hung over scaffolding, 8 uses, not incl. scaffolding		1500	.011		.25	.35		.60	.86
0250	Tarpaulin polyester reinf. w/integral fastening system 11 mils thick		1600	.010		.80	.33		1.13	1.42
0300	Prefab fiberglass panels, steel frame, 8 uses		1200	.013		.85	.44		1.29	1.66

01 56 23 – Temporary Barricades

01 56 23.10 Barricades

0010	**BARRICADES**									
0020	5' high, 3 rail @ 2" x 8", fixed	2 Carp	20	.800	L.F.	4.41	33		37.41	59.50
0150	Movable	"	30	.533	"	3.54	22		25.54	40.50
0300	Stock units, 6' high, 8' wide, plain, buy				Ea.	435			435	480
0350	With reflective tape, buy				"	525			525	580
0400	Break-a-way 3" PVC pipe barricade									
0410	with 3 ea. 1' x 4' reflectorized panels, buy				Ea.	305			305	335
0500	Plywood with steel legs, 32" wide					72			72	79
0600	Telescoping Christmas tree, 9' high, 5 flags, buy					122			122	134
0800	Traffic cones, PVC, 18" high					9.70			9.70	10.65
0850	28" high					16.40			16.40	18.05
0900	Barrels, 55 gal., with flasher	1 Clab	96	.083		67	2.76		69.76	78.50
1000	Guardrail, wooden, 3' high, 1" x 6", on 2" x 4" posts	2 Carp	200	.080	L.F.	.97	3.32		4.29	6.50
1100	2" x 6", on 4" x 4" posts	"	165	.097		2	4.03		6.03	8.80
1200	Portable metal with base pads, buy					19.30			19.30	21
1250	Typical installation, assume 10 reuses	2 Carp	600	.027		2.50	1.11		3.61	4.57
1300	Barricade tape, polyethylene, 7 mil, 3" wide x 500' long roll				Ea.	25			25	27.50

01 56 26 – Temporary Fencing

01 56 26.50 Temporary Fencing

0010	**TEMPORARY FENCING**									
0020	Chain link, 11 ga, 5' high	2 Clab	400	.040	L.F.	2.75	1.32		4.07	5.20
0100	6' high		300	.053		3.25	1.77		5.02	6.50
0200	Rented chain link, 6' high, to 1000' (up to 12 mo.)		400	.040		4.98	1.32		6.30	7.65
0250	Over 1000' (up to 12 mo.)		300	.053		3.98	1.77		5.75	7.30
0350	Plywood, painted, 2" x 4" frame, 4' high	A-4	135	.178		4.89	7.10		11.99	16.90
0400	4" x 4" frame, 8' high	"	110	.218		9.45	8.70		18.15	24.50
0500	Wire mesh on 4" x 4" posts, 4' high	2 Carp	100	.160		9.65	6.65		16.30	21.50
0550	8' high	"	80	.200		14.75	8.30		23.05	30

01 56 29 – Temporary Protective Walkways

01 56 29.50 Protection

0010	**PROTECTION**									
0020	Stair tread, 2" x 12" planks, 1 use	1 Carp	75	.107	Tread	3.18	4.43		7.61	10.80
0100	Exterior plywood, 1/2" thick, 1 use		65	.123		1.26	5.10		6.36	9.80
0200	3/4" thick, 1 use		60	.133		1.75	5.55		7.30	11
2200	Sidewalks, 2" x 12" planks, 2 uses		350	.023	S.F.	.53	.95		1.48	2.14
2300	Exterior plywood, 2 uses, 1/2" thick		750	.011		.21	.44		.65	.96
2400	5/8" thick		650	.012		.25	.51		.76	1.12
2500	3/4" thick		600	.013		.29	.55		.84	1.23

01 56 Temporary Barriers and Enclosures

01 56 32 – Temporary Security

01 56 32.50 Watchman		Crew	Daily Output	Labor-Hours	Unit	Material	2010 Bare Costs Labor	Equipment	Total	Total Incl O&P
0010	**WATCHMAN**									
0020	Service, monthly basis, uniformed person, minimum				Hr.				25	27.50
0100	Maximum								45.45	50
0200	Person and command dog, minimum								31	34
0300	Maximum				▼				54.55	60
0500	Sentry dog, leased, with job patrol (yard dog), 1 dog				Week				290	319
0600	2 dogs				"				390	429
0800	Purchase, trained sentry dog, minimum				Ea.				1,364	1,500
0900	Maximum				"				2,727	3,000

01 58 Project Identification

01 58 13 – Temporary Project Signage

01 58 13.50 Signs

		Crew	Daily Output	Labor-Hours	Unit	Material	Labor	Equipment	Total	Total Incl O&P
0010	**SIGNS**									
0020	High intensity reflectorized, no posts, buy				S.F.	26.50			26.50	29.50

01 71 Examination and Preparation

01 71 23 – Field Engineering

01 71 23.13 Construction Layout

		Crew	Daily Output	Labor-Hours	Unit	Material	Labor	Equipment	Total	Total Incl O&P
0010	**CONSTRUCTION LAYOUT**									
1100	Crew for layout of building, trenching or pipe laying, 2 person crew	A-6	1	16	Day		660	70	730	1,150
1200	3 person crew	A-7	1	24			1,075	70	1,145	1,800
1400	Crew for roadway layout, 4 person crew	A-8	1	32	▼		1,400	70	1,470	2,325

01 74 Cleaning and Waste Management

01 74 13 – Progress Cleaning

01 74 13.20 Cleaning Up

		Crew	Daily Output	Labor-Hours	Unit	Material	Labor	Equipment	Total	Total Incl O&P
0010	**CLEANING UP**									
0020	After job completion, allow, minimum				Job					.30%
0040	Maximum				"					1%
0052	Cleanup of floor area, continuous, per day, during const.	A-5	16	1.125	M.S.F.	1.70	37	3.75	42.45	67
0100	Final by GC at end of job	"	11.50	1.565	"	2.71	51.50	5.20	59.41	93

01 91 Commissioning

01 91 13 – General Commissioning Requirements

01 91 13.50 Commissioning

		Crew	Daily Output	Labor-Hours	Unit	Material	Labor	Equipment	Total	Total Incl O&P
0010	**COMMISSIONING** Including documentation of design intent									
0100	performance verification, O&M, training, minimum				Project					1%
0150	Maximum				"					1.25%

01 93 Facility Maintenance
01 93 09 – Facility Equipment

01 93 09.50 Moving Equipment		Crew	Daily Output	Labor-Hours	Unit	Material	2010 Bare Costs Labor	2010 Bare Costs Equipment	Total	Total Incl O&P
0010	**MOVING EQUIPMENT**, Remove and reset, 100' distance,									
0020	No obstructions, no assembly or leveling unless noted									
0100	Annealing furnace, 24' overall	B-67	4	4	Ea.		169	77.50	246.50	350
0200	Annealing oven, small		14	1.143			48	22	70	100
0240	Very large		1	16			675	310	985	1,400
0400	Band saw, small		12	1.333			56	26	82	117
0440	Large		8	2			84.50	38.50	123	175
0500	Blue print copy machine		7	2.286			96.50	44	140.50	200
0600	Bonding mill, 6"		7	2.286			96.50	44	140.50	200
0620	12"		6	2.667			112	51.50	163.50	233
0640	18"		4	4			169	77.50	246.50	350
0660	24"		2	8			335	155	490	695
0700	Boring machine (jig)	B-68	7	3.429			145	44	189	276
0800	Bridgeport mill, standard	B-67	14	1.143			48	22	70	100
1000	Calibrator, 6 unit	"	14	1.143			48	22	70	100
1100	Comparitor, bench top	2 Clab	14	1.143			38		38	62
1140	Floor mounted	B-67	7	2.286			96.50	44	140.50	200
1200	Computer, desk top	2 Clab	25	.640			21		21	35
1300	Copy machine	"	25	.640			21		21	35
1500	Deflasher	B-67	14	1.143			48	22	70	100
1600	Degreaser, small		14	1.143			48	22	70	100
1640	Large 24' overall		1	16			675	310	985	1,400
1700	Desk with chair	2 Clab	25	.640			21		21	35
1800	Dial press	B-67	7	2.286			96.50	44	140.50	200
1900	Drafting table	2 Clab	14	1.143			38		38	62
2000	Drill press, bench top	"	14	1.143			38		38	62
2040	Floor mounted	B-67	14	1.143			48	22	70	100
2080	Industrial radial	"	7	2.286			96.50	44	140.50	200
2100	Dust collector, portable	2 Clab	25	.640			21		21	35
2140	Stationary, small	B-67	7	2.286			96.50	44	140.50	200
2180	Stationary, large	"	2	8			335	155	490	695
2300	Electric discharge machine	B-68	7	3.429			145	44	189	276
2400	Environmental chamber walls, including assembly	4 Clab	18	1.778	L.F.		59		59	96.50
2600	File cabinet	2 Clab	25	.640	Ea.		21		21	35
2800	Grinder/sander, pedestal mount	B-67	14	1.143			48	22	70	100
3000	Hack saw, power	2 Clab	24	.667			22		22	36
3100	Hydraulic press	B-67	14	1.143			48	22	70	100
3500	Laminar flow tables	"	14	1.143			48	22	70	100
3600	Lathe, bench	2 Clab	14	1.143			38		38	62
3640	6"	B-67	14	1.143			48	22	70	100
3680	10"		13	1.231			52	24	76	107
3720	12"		12	1.333			56	26	82	117
4000	Milling machine		8	2			84.50	38.50	123	175
4100	Molding press, 25 ton		5	3.200			135	62	197	279
4140	60 ton		4	4			169	77.50	246.50	350
4180	100 ton		2	8			335	155	490	695
4220	150 ton		1.50	10.667			450	206	656	925
4260	200 ton		1	16			675	310	985	1,400
4300	300 ton		.75	21.333			900	415	1,315	1,850
4700	Oil pot stand		14	1.143			48	22	70	100
5000	Press, 10 ton		14	1.143			48	22	70	100
5040	15 ton		12	1.333			56	26	82	117
5080	20 ton		10	1.600			67.50	31	98.50	139

01 93 Facility Maintenance

01 93 09 – Facility Equipment

01 93 09.50 Moving Equipment		Crew	Daily Output	Labor-Hours	Unit	Material	2010 Bare Costs Labor	2010 Bare Costs Equipment	Total	Total Incl O&P
5120	30 ton	B-67	8	2	Ea.		84.50	38.50	123	175
5160	45 ton		6	2.667			112	51.50	163.50	233
5200	60 ton		4	4			169	77.50	246.50	350
5240	75 ton		2.50	6.400			270	124	394	555
5280	100 ton		2	8			335	155	490	695
5500	Raised floor, including assembly	2 Carp	250	.064	S.F.		2.66		2.66	4.37
5600	Rolling mill, 6"	B-67	7	2.286	Ea.		96.50	44	140.50	200
5640	9"		6	2.667			112	51.50	163.50	233
5680	12"		4	4			169	77.50	246.50	350
5720	13"		3.50	4.571			193	88.50	281.50	400
5760	18"		2	8			335	155	490	695
5800	25"		1	16			675	310	985	1,400
6000	Sander, floor stand		14	1.143			48	22	70	100
6100	Screw machine		7	2.286			96.50	44	140.50	200
6200	Shaper, 16"		14	1.143			48	22	70	100
6300	Shear, power assist		4	4			169	77.50	246.50	350
6400	Slitter, 6"		14	1.143			48	22	70	100
6440	8"		13	1.231			52	24	76	107
6480	10"		12	1.333			56	26	82	117
6520	12"		11	1.455			61.50	28	89.50	127
6560	16"		10	1.600			67.50	31	98.50	139
6600	20"		8	2			84.50	38.50	123	175
6640	24"		6	2.667			112	51.50	163.50	233
6800	Snag and tap machine		7	2.286			96.50	44	140.50	200
6900	Solder machine (auto)		7	2.286			96.50	44	140.50	200
7000	Storage cabinet metal, small	2 Clab	36	.444			14.70		14.70	24
7040	Large		25	.640			21		21	35
7100	Storage rack open, small		14	1.143			38		38	62
7140	Large		7	2.286			75.50		75.50	124
7200	Surface bench, small	B-67	14	1.143			48	22	70	100
7240	Large	"	5	3.200			135	62	197	279
7300	Surface grinder, large wet	B-68	5	4.800			204	62	266	390
7500	Time check machine	2 Clab	14	1.143			38		38	62
8000	Welder, 30 KVA (bench)		14	1.143			38		38	62
8100	Work bench with chair		25	.640			21		21	35

Estimating Tips

02 30 00 Subsurface Investigation

In preparing estimates on structures involving earthwork or foundations, all information concerning soil characteristics should be obtained. Look particularly for hazardous waste, evidence of prior dumping of debris, and previous stream beds.

02 41 00 Demolition and Structure Moving

The costs shown for selective demolition do not include rubbish handling or disposal. These items should be estimated separately using RSMeans data or other sources.

- Historic preservation often requires that the contractor remove materials from the existing structure, rehab them, and replace them. The estimator must be aware of any related measures and precautions that must be taken when doing selective demolition and cutting and patching. Requirements may include special handling and storage, as well as security.
- In addition to Subdivision 02 41 00, you can find selective demolition items in each division. Example: Roofing demolition is in Division 7.

02 42 10 Building Deconstruction

This is a new section on the careful demantling and recycling of most of low-rise building materials.

02 56 13 Containment of Hazardous Waste

This is a new section for disposal of hazardous waste on site.

02 80 00 Hazardous Material Disposal/Remediation

This subdivision includes information on hazardous waste handling, asbestos remediation, lead remediation, and mold remediation. See reference R028213-20 and R028319-60 for further guidance in using these unit price lines.

02 91 10 Monitoring Chemical Sampling, Testing Analysis

This is a new section for on-site sampling and testing hazardous waste.

Reference Numbers

Reference numbers are shown in shaded boxes at the beginning of some major classifications. These numbers refer to related items in the Reference Section. The reference information may be an estimating procedure, an alternate pricing method, or technical information.

Note: Not all subdivisions listed here necessarily appear in this publication.

No part of this publication may be reproduced, stored in a retrieval system, or transmitted in any form or by any means without prior written permission of R.S. Means Company, Inc.

02 21 Surveys

02 21 13 – Site Surveys

02 21 13.09 Topographical Surveys		Crew	Daily Output	Labor-Hours	Unit	Material	2010 Bare Costs Labor	Equipment	Total	Total Incl O&P
0010	**TOPOGRAPHICAL SURVEYS**									
0020	Topographical surveying, conventional, minimum	A-7	3.30	7.273	Acre	18	325	21	364	570
0100	Maximum	A-8	.60	53.333	"	55	2,325	116	2,496	3,950

02 21 13.13 Boundary and Survey Markers

0010	**BOUNDARY AND SURVEY MARKERS**									
0300	Lot location and lines, large quantities, minimum	A-7	2	12	Acre	32	540	35	607	940
0320	Average	"	1.25	19.200		51	865	56	972	1,500
0400	Small quantities, maximum	A-8	1	32		68	1,400	70	1,538	2,400

02 32 Geotechnical Investigations

02 32 13 – Subsurface Drilling and Sampling

02 32 13.10 Boring and Exploratory Drilling

		Crew	Daily Output	Labor-Hours	Unit	Material	Labor	Equipment	Total	Total Incl O&P
0010	**BORING AND EXPLORATORY DRILLING**									
0020	Borings, initial field stake out & determination of elevations	A-6	1	16	Day		660	70	730	1,150
0100	Drawings showing boring details				Total		250		250	310
0200	Report and recommendations from P.E.						600		600	750
0300	Mobilization and demobilization, minimum	B-55	4	6			197	261	458	610
0350	For over 100 miles, per added mile		450	.053	Mile		1.75	2.32	4.07	5.40
0600	Auger holes in earth, no samples, 2-1/2" diameter		78.60	.305	L.F.		10	13.30	23.30	31
0650	4" diameter		67.50	.356			11.65	15.50	27.15	36
0800	Cased borings in earth, with samples, 2-1/2" diameter		55.50	.432		20.50	14.20	18.85	53.55	66
0850	4" diameter		32.60	.736		32.50	24	32	88.50	111
1000	Drilling in rock, "BX" core, no sampling	B-56	34.90	.458			17.05	42	59.05	73.50
1050	With casing & sampling		31.70	.505		20.50	18.80	46	85.30	104
1200	"NX" core, no sampling		25.92	.617			23	56.50	79.50	99
1250	With casing and sampling		25	.640		25	24	58.50	107.50	130
1400	Borings, earth, drill rig and crew with truck mounted auger	B-55	1	24	Day		790	1,050	1,840	2,450
1450	Rock using crawler type drill	B-56	1	16	"		595	1,475	2,070	2,550
1500	For inner city borings add, minimum									10%
1510	Maximum									20%

02 41 Demolition

02 41 13 – Selective Site Demolition

02 41 13.15 Hydrodemolition

		Crew	Daily Output	Labor-Hours	Unit	Material	Labor	Equipment	Total	Total Incl O&P
0010	**HYDRODEMOLITION**									
0015	Hydrodemolition, concrete pavement, 4000 PSI, 2" depth	B-5	500	.112	S.F.		4.05	2.60	6.65	9.40
0120	4" depth		450	.124			4.50	2.89	7.39	10.50
0130	6" depth		400	.140			5.05	3.26	8.31	11.80
0410	6000 PSI, 2" depth		410	.137			4.94	3.18	8.12	11.50
0420	4" depth		350	.160			5.80	3.72	9.52	13.45
0430	6" depth		300	.187			6.75	4.34	11.09	15.65
0510	8000 PSI, 2" depth		330	.170			6.15	3.95	10.10	14.30
0520	4" depth		280	.200			7.25	4.65	11.90	16.80
0530	6" depth		240	.233			8.45	5.40	13.85	19.60

02 41 13.17 Demolish, Remove Pavement and Curb

			Crew	Daily Output	Labor-Hours	Unit	Material	Labor	Equipment	Total	Total Incl O&P
0010	**DEMOLISH, REMOVE PAVEMENT AND CURB**	R024119-10									
5010	Pavement removal, bituminous roads, 3" thick		B-38	690	.058	S.Y.		2.15	1.71	3.86	5.35
5050	4" to 6" thick			420	.095			3.53	2.80	6.33	8.80
5100	Bituminous driveways			640	.063			2.32	1.84	4.16	5.75

02 41 Demolition

02 41 13 – Selective Site Demolition

02 41 13.17 Demolish, Remove Pavement and Curb

		Crew	Daily Output	Labor-Hours	Unit	Material	2010 Bare Costs Labor	2010 Bare Costs Equipment	Total	Total Incl O&P
5200	Concrete to 6" thick, hydraulic hammer, mesh reinforced	B-38	255	.157	S.Y.		5.80	4.62	10.42	14.45
5300	Rod reinforced		200	.200			7.40	5.90	13.30	18.45
5400	Concrete, 7" to 24" thick, plain		33	1.212	C.Y.		45	35.50	80.50	112
5500	Reinforced		24	1.667	"		62	49	111	154
5590	Minimum labor/equipment charge		6	6.667	Job		247	196	443	615
5600	With hand held air equipment, bituminous, to 6" thick	B-39	1900	.025	S.F.		.88	.12	1	1.56
5700	Concrete to 6" thick, no reinforcing		1600	.030			1.04	.14	1.18	1.85
5800	Mesh reinforced		1400	.034			1.19	.16	1.35	2.12
5900	Rod reinforced		765	.063			2.18	.29	2.47	3.87
5990	Minimum labor/equipment charge	B-38	6	6.667	Job		247	196	443	615
6000	Curbs, concrete, plain	B-6	360	.067	L.F.		2.39	.94	3.33	4.88
6100	Reinforced		275	.087			3.13	1.23	4.36	6.40
6200	Granite		360	.067			2.39	.94	3.33	4.88
6300	Bituminous		528	.045			1.63	.64	2.27	3.33
6390	Minimum labor/equipment charge		6	4	Job		143	56.50	199.50	293

02 41 13.33 Minor Site Demolition

		Crew	Daily Output	Labor-Hours	Unit	Material	2010 Bare Costs Labor	2010 Bare Costs Equipment	Total	Total Incl O&P
0010	**MINOR SITE DEMOLITION** R024119-10									
0015	No hauling, abandon catch basin or manhole	B-6	7	3.429	Ea.		123	48.50	171.50	251
0020	Remove existing catch basin or manhole, masonry		4	6			215	84.50	299.50	440
0030	Catch basin or manhole frames and covers, stored		13	1.846			66	26	92	136
0040	Remove and reset		7	3.429			123	48.50	171.50	251
0100	Roadside delineators, remove only	B-80	175	.183			6.50	4.26	10.76	15.15
0110	Remove and reset	"	100	.320			11.35	7.45	18.80	26.50
0400	Minimum labor/equipment charge	B-6	4	6	Job		215	84.50	299.50	440
0800	Guiderail, corrugated steel, remove only	B-80A	100	.240	L.F.		7.95	2.96	10.91	16.30
0850	Remove and reset	"	40	.600	"		19.85	7.40	27.25	40.50
0860	Guide posts, remove only	B-80B	120	.267	Ea.		9.35	2.02	11.37	17.40
0870	Remove and reset	B-55	50	.480	"		15.75	21	36.75	49
0890	Minimum labor/equipment charge	2 Clab	4	4	Job		132		132	217
0900	Hydrants, fire, remove only	B-21A	5	8	Ea.		330	131	461	670
0950	Remove and reset	"	2	20	"		825	330	1,155	1,650
0990	Minimum labor/equipment charge	2 Plum	2	8	Job		415		415	645
1000	Masonry walls, block, solid	B-5	1800	.031	C.F.		1.13	.72	1.85	2.62
1200	Brick, solid		900	.062			2.25	1.45	3.70	5.25
1400	Stone, with mortar		900	.062			2.25	1.45	3.70	5.25
1500	Dry set		1500	.037			1.35	.87	2.22	3.13
1600	Median barrier, precast concrete, remove and store	B-3	430	.112	L.F.		3.92	5.60	9.52	12.50
1610	Remove and reset	"	390	.123	"		4.32	6.20	10.52	13.80
1650	Minimum labor/equipment charge	A-1	4	2	Job		66	18.45	84.45	130
2900	Pipe removal, sewer/water, no excavation, 12" diameter	B-6	175	.137	L.F.		4.91	1.93	6.84	10.05
2930	15"-18" diameter		150	.160			5.75	2.25	8	11.75
2960	21"-24" diameter		120	.200			7.15	2.82	9.97	14.65
3000	27"-36" diameter		90	.267			9.55	3.75	13.30	19.55
3200	Steel, welded connections, 4" diameter		160	.150			5.35	2.11	7.46	10.95
3300	10" diameter		80	.300			10.75	4.22	14.97	22
3390	Minimum labor/equipment charge		3	8	Job		287	113	400	585
3500	Railroad track removal, ties and track	B-13	330	.170	L.F.		6.05	2.27	8.32	12.30
3600	Ballast	B-14	500	.096	C.Y.		3.34	.68	4.02	6.20
3700	Remove and re-install, ties & track using new bolts & spikes		50	.960	L.F.		33.50	6.75	40.25	62
3800	Turnouts using new bolts and spikes		1	48	Ea.		1,675	340	2,015	3,100
3890	Minimum labor/equipment charge		5	9.600	Job		335	67.50	402.50	620
4000	Sidewalk removal, bituminous, 2-1/2" thick	B-6	325	.074	S.Y.		2.65	1.04	3.69	5.40

02 41 Demolition

02 41 13 – Selective Site Demolition

02 41 13.33 Minor Site Demolition

		Crew	Daily Output	Labor-Hours	Unit	Material	2010 Bare Costs Labor	Equipment	Total	Total Incl O&P
4050	Brick, set in mortar	B-6	185	.130	S.Y.		4.65	1.83	6.48	9.50
4100	Concrete, plain, 4"		160	.150			5.35	2.11	7.46	10.95
4200	Mesh reinforced	▼	150	.160	▼		5.75	2.25	8	11.75
4290	Minimum labor/equipment charge	B-39	12	4	Job		139	18.70	157.70	248

02 41 13.60 Selective Demolition Fencing

			Crew	Daily Output	Labor-Hours	Unit	Material	Labor	Equipment	Total	Total Incl O&P
0010	**SELECTIVE DEMOLITION FENCING**	R024119-10									
1600	Fencing, barbed wire, 3 strand		2 Clab	430	.037	L.F.		1.23		1.23	2.02
1650	5 strand		"	280	.057			1.89		1.89	3.11
1700	Chain link, posts & fabric, 8' to 10' high, remove only		B-6	445	.054	▼		1.93	.76	2.69	3.95

02 41 16 – Structure Demolition

02 41 16.13 Building Demolition

			Crew	Daily Output	Labor-Hours	Unit	Material	Labor	Equipment	Total	Total Incl O&P
0010	**BUILDING DEMOLITION** Large urban projects, incl. 20 mi. haul	R024119-10									
0011	No foundation or dump fees, C.F. is vol. of building standing										
0020	Steel		B-8	21500	.003	C.F.		.11	.15	.26	.33
0050	Concrete			15300	.004			.15	.21	.36	.47
0080	Masonry			20100	.003			.12	.16	.28	.36
0100	Mixture of types, average		▼	20100	.003			.12	.16	.28	.36
0500	Small bldgs, or single bldgs, no salvage included, steel		B-3	14800	.003			.11	.16	.27	.36
0600	Concrete			11300	.004			.15	.21	.36	.48
0650	Masonry			14800	.003			.11	.16	.27	.36
0700	Wood		▼	14800	.003			.11	.16	.27	.36
0750	For buildings with no interior walls, deduct					▼				50%	
1000	Single family, one story house, wood, minimum					Ea.				4,100	4,800
1020	Maximum									8,200	9,600
1200	Two family, two story house, wood, minimum									6,100	7,200
1220	Maximum									10,900	12,800
1300	Three family, three story house, wood, minimum									8,200	9,600
1320	Maximum									13,600	16,000
5000	For buildings with no interior walls, deduct					▼				50%	

02 41 16.17 Building Demolition Footings and Foundations

			Crew	Daily Output	Labor-Hours	Unit	Material	Labor	Equipment	Total	Total Incl O&P
0010	**BUILDING DEMOLITION FOOTINGS AND FOUNDATIONS**	R024119-10									
0200	Floors, concrete slab on grade,										
0240	4" thick, plain concrete		B-9	500	.080	S.F.		2.68	.45	3.13	4.89
0280	Reinforced, wire mesh			470	.085			2.85	.48	3.33	5.20
0300	Rods			400	.100			3.35	.56	3.91	6.10
0400	6" thick, plain concrete			375	.107			3.57	.60	4.17	6.50
0420	Reinforced, wire mesh			340	.118			3.94	.66	4.60	7.20
0440	Rods			300	.133	▼		4.47	.75	5.22	8.15
1000	Footings, concrete, 1' thick, 2' wide		B-5	300	.187	L.F.		6.75	4.34	11.09	15.65
1080	1'-6" thick, 2' wide			250	.224			8.10	5.20	13.30	18.85
1120	3' wide			200	.280			10.15	6.50	16.65	23.50
1140	2' thick, 3' wide		▼	175	.320			11.60	7.45	19.05	27
1200	Average reinforcing, add									10%	10%
1220	Heavy reinforcing, add					▼				20%	20%
2000	Walls, block, 4" thick		1 Clab	180	.044	S.F.		1.47		1.47	2.42
2040	6" thick			170	.047			1.56		1.56	2.56
2080	8" thick			150	.053			1.77		1.77	2.90
2100	12" thick		▼	150	.053			1.77		1.77	2.90
2200	For horizontal reinforcing, add									10%	10%
2220	For vertical reinforcing, add									20%	20%
2400	Concrete, plain concrete, 6" thick		B-9	160	.250			8.40	1.41	9.81	15.30
2420	8" thick		▼	140	.286			9.55	1.61	11.16	17.45

02 41 Demolition

02 41 16 – Structure Demolition

02 41 16.17 Building Demolition Footings and Foundations

		Crew	Daily Output	Labor-Hours	Unit	Material	2010 Bare Costs Labor	Equipment	Total	Total Incl O&P
2440	10" thick	B-9	120	.333	S.F.		11.15	1.87	13.02	20.50
2500	12" thick	↓	100	.400			13.40	2.25	15.65	24.50
2600	For average reinforcing, add								10%	10%
2620	For heavy reinforcing, add								20%	20%
9000	Minimum labor/equipment charge	A-1	2	4	Job		132	37	169	258

02 41 16.19 Minor Building Deconstruction

			Crew	Daily Output	Labor-Hours	Unit	Material	Labor	Equipment	Total	Total Incl O&P
0010	**MINOR BUILDING DECONSTRUCTION**	R024119-10									
0011	For salvage, avg house										
0225	2 story, pre 1970 house, 1400 SF, labor for all salv mat, min	G	6 Clab	25	1.920	SF Flr.		63.50		63.50	104
0230	Maximum	G	"	15	3.200	"		106		106	174
0235	Salvage of carpet, tackless	G	2 Clab	2400	.007	S.F.		.22		.22	.36
0240	Wood floors, incl denailing and packaging	G	3 Clab	270	.089	"		2.94		2.94	4.83
0260	Wood doors and trim, standard	G	1 Clab	16	.500	Ea.		16.55		16.55	27
0280	Base or cove mouldings, incl denailing and packaging	G		500	.016	L.F.		.53		.53	.87
0320	Closet shelving and trim, incl denailing and packaging	G	↓	18	.444	Set		14.70		14.70	24
0340	Kitchen cabinets, uppers and lowers, prefab type	G	2 Clab	24	.667	L.F.		22		22	36
0360	Kitchen cabinets, uppers and lowers, built-in	G		12	1.333	"		44		44	72.50
0370	Bath fixt, incl toilet, tub, vanity, and med cabinet	G	↓	3	5.333	Set		177		177	290

02 41 19 – Selective Structure Demolition

02 41 19.13 Selective Building Demolition

0010	**SELECTIVE BUILDING DEMOLITION**
0020	Costs related to selective demolition of specific building components
0025	are included under Common Work Results (XX 05 00)
0030	in the component's appropriate division.

02 41 19.16 Selective Demolition, Cutout

			Crew	Daily Output	Labor-Hours	Unit	Material	Labor	Equipment	Total	Total Incl O&P
0010	**SELECTIVE DEMOLITION, CUTOUT**	R024119-10									
0020	Concrete, elev. slab, light reinforcement, under 6 C.F.		B-9	65	.615	C.F.		20.50	3.46	23.96	38
0050	Light reinforcing, over 6 C.F.			75	.533	"		17.85	3	20.85	33
0200	Slab on grade to 6" thick, not reinforced, under 8 S.F.			85	.471	S.F.		15.75	2.64	18.39	29
0250	8 – 16 S.F.		↓	175	.229	"		7.65	1.28	8.93	13.95
0255	For over 16 S.F. see Div. 02 41 16.17 0400										
0600	Walls, not reinforced, under 6 C.F.		B-9	60	.667	C.F.		22.50	3.75	26.25	40.50
0650	6 – 12 C.F.		"	80	.500	"		16.75	2.81	19.56	30.50
0655	For over 12 C.F. see Div. 02 41 16.17 2500										
1000	Concrete, elevated slab, bar reinforced, under 6 C.F.		B-9	45	.889	C.F.		30	5	35	54.50
1050	Bar reinforced, over 6 C.F.			50	.800	"		27	4.50	31.50	49
1200	Slab on grade to 6" thick, bar reinforced, under 8 S.F.			75	.533	S.F.		17.85	3	20.85	33
1250	8 – 16 S.F.		↓	150	.267	"		8.95	1.50	10.45	16.30
1255	For over 16 S.F. see Div. 02 41 16.17 0440										
1400	Walls, bar reinforced, under 6 C.F.		B-9	50	.800	C.F.		27	4.50	31.50	49
1450	6 – 12 C.F.		"	70	.571	"		19.15	3.21	22.36	35
1455	For over 12 C.F. see Div. 02 41 16.17 2500 and 2600										
2000	Brick, to 4 S.F. opening, not including toothing										
2040	4" thick		B-9	30	1.333	Ea.		44.50	7.50	52	82
2060	8" thick			18	2.222			74.50	12.50	87	136
2080	12" thick			10	4			134	22.50	156.50	245
2400	Concrete block, to 4 S.F. opening, 2" thick			35	1.143			38.50	6.40	44.90	70
2420	4" thick			30	1.333			44.50	7.50	52	82
2440	8" thick			27	1.481			49.50	8.35	57.85	90.50
2460	12" thick			24	1.667			56	9.35	65.35	102
2600	Gypsum block, to 4 S.F. opening, 2" thick			80	.500			16.75	2.81	19.56	30.50
2620	4" thick		↓	70	.571	↓		19.15	3.21	22.36	35

02 41 Demolition

02 41 19 – Selective Structure Demolition

02 41 19.16 Selective Demolition, Cutout

		Crew	Daily Output	Labor-Hours	Unit	Material	2010 Bare Costs Labor	2010 Bare Costs Equipment	Total	Total Incl O&P
2640	8" thick	B-9	55	.727	Ea.		24.50	4.09	28.59	44.50
2800	Terra cotta, to 4 S.F. opening, 4" thick		70	.571			19.15	3.21	22.36	35
2840	8" thick		65	.615			20.50	3.46	23.96	38
2880	12" thick		50	.800			27	4.50	31.50	49
4000	For toothing masonry, see Div. 04 01 20.50									
6000	Walls, interior, not including re-framing,									
6010	openings to 5 S.F.									
6100	Drywall to 5/8" thick	1 Clab	24	.333	Ea.		11.05		11.05	18.10
6200	Paneling to 3/4" thick		20	.400			13.25		13.25	21.50
6300	Plaster, on gypsum lath		20	.400			13.25		13.25	21.50
6340	On wire lath		14	.571			18.90		18.90	31
7000	Wood frame, not including re-framing, openings to 5 S.F.									
7200	Floors, sheathing and flooring to 2" thick	1 Clab	5	1.600	Ea.		53		53	87
7310	Roofs, sheathing to 1" thick, not including roofing		6	1.333			44		44	72.50
7410	Walls, sheathing to 1" thick, not including siding		7	1.143			38		38	62
8500	Minimum labor/equipment charge		4	2	Job		66		66	109

02 41 19.18 Selective Demolition, Disposal Only

			Crew	Daily Output	Labor-Hours	Unit	Material	Labor	Equipment	Total	Total Incl O&P
0010	**SELECTIVE DEMOLITION, DISPOSAL ONLY**	R024119-10									
0015	Urban bldg w/salvage value allowed										
0020	Including loading and 5 mile haul to dump										
0200	Steel frame		B-3	430	.112	C.Y.		3.92	5.60	9.52	12.50
0300	Concrete frame			365	.132			4.61	6.60	11.21	14.75
0400	Masonry construction			445	.108			3.79	5.40	9.19	12.10
0500	Wood frame			247	.194			6.80	9.75	16.55	22

02 41 19.19 Selective Demolition, Dump Charges

			Crew	Daily Output	Labor-Hours	Unit	Material	Labor	Equipment	Total	Total Incl O&P
0010	**SELECTIVE DEMOLITION, DUMP CHARGES**	R024119-10									
0020	Dump charges, typical urban city, tipping fees only										
0100	Building construction materials					Ton	90			90	99
0200	Trees, brush, lumber						75			75	83
0300	Rubbish only						80			80	88
0500	Reclamation station, usual charge						90			90	99

02 41 19.21 Selective Demolition, Gutting

			Crew	Daily Output	Labor-Hours	Unit	Material	Labor	Equipment	Total	Total Incl O&P
0010	**SELECTIVE DEMOLITION, GUTTING**	R024119-10									
0020	Building interior, including disposal, dumpster fees not included										
0500	Residential building										
0560	Minimum		B-16	400	.080	SF Flr.		2.69	1.67	4.36	6.25
0580	Maximum		"	360	.089	"		2.99	1.86	4.85	6.95
0900	Commercial building										
1000	Minimum		B-16	350	.091	SF Flr.		3.07	1.91	4.98	7.15
1020	Maximum			250	.128	"		4.30	2.67	6.97	10
3000	Minimum labor/equipment charge			4	8	Job		269	167	436	625

02 41 19.23 Selective Demolition, Rubbish Handling

			Crew	Daily Output	Labor-Hours	Unit	Material	Labor	Equipment	Total	Total Incl O&P
0010	**SELECTIVE DEMOLITION, RUBBISH HANDLING**	R024119-10									
0020	The following are to be added to the demolition prices										
0400	Chute, circular, prefabricated steel, 18" diameter		B-1	40	.600	L.F.	48.50	20.50		69	87
0440	30" diameter		"	30	.800	"	50	27		77	99.50
0600	Dumpster, weekly rental, 1 dump/week, 6 C.Y. capacity (2 Tons)					Week	450			450	495
0700	10 C.Y. capacity (4 Tons)						525			525	578
0725	20 C.Y. capacity (8 Tons)	R024119-20					700			700	770
0800	30 C.Y. capacity (10 Tons)						850			850	935
0840	40 C.Y. capacity (13 Tons)						1,000			1,000	1,100
0900	Alternate pricing for dumpsters										

02 41 Demolition

02 41 19 – Selective Structure Demolition

02 41 19.23 Selective Demolition, Rubbish Handling

		Crew	Daily Output	Labor-Hours	Unit	Material	2010 Bare Costs Labor	2010 Bare Costs Equipment	Total	Total Incl O&P
0910	Delivery, average for all sizes				Ea.	75			75	83
0920	Haul, average for all sizes					200			200	220
0930	Rent per day, average for all sizes					10			10	11
0940	Rent per month, average for all sizes					75			75	83
0950	Disposal fee per ton, average for all sizes				Ton	75			75	83
1000	Dust partition, 6 mil polyethylene, 1" x 3" frame	2 Carp	2000	.008	S.F.	.45	.33		.78	1.04
1080	2" x 4" frame	"	2000	.008	"	.24	.33		.57	.81
2000	Load, haul, dump and return, 50' haul, hand carried	2 Clab	24	.667	C.Y.		22		22	36
2005	Wheeled		37	.432			14.30		14.30	23.50
2040	100' haul, hand carried		16.50	.970			32		32	52.50
2045	Wheeled		25	.640			21		21	35
2080	Over 100' haul, add per 100 L.F., hand carried		35.50	.451			14.90		14.90	24.50
2085	Wheeled		54	.296			9.80		9.80	16.10
2120	In elevators, per 10 floors, add		140	.114			3.78		3.78	6.20
2130	Load, haul, dump and return, 50' haul, incl. 5 riser stair, hand carried		23	.696			23		23	38
2135	Wheeled		35	.457			15.15		15.15	25
2140	6 – 10 riser stairs, hand carried		22	.727			24		24	39.50
2145	Wheeled		34	.471			15.60		15.60	25.50
2150	11 – 20 riser stairs, hand carried		20	.800			26.50		26.50	43.50
2155	Wheeled		31	.516			17.10		17.10	28
2160	21 – 40 riser stairs, hand carried		16	1			33		33	54.50
2165	Wheeled		24	.667			22		22	36
2170	100' haul, incl. 5 riser stair, hand carried		15	1.067			35.50		35.50	58
2175	Wheeled		23	.696			23		23	38
2180	6 – 10 riser stair, hand carried		14	1.143			38		38	62
2185	Wheeled		21	.762			25		25	41.50
2190	11 – 20 riser stair, hand carried		12	1.333			44		44	72.50
2195	Wheeled		18	.889			29.50		29.50	48.50
2200	21 – 40 riser stair, hand carried		8	2			66		66	109
2205	Wheeled		12	1.333			44		44	72.50
2210	Over 100' haul, add per 100 L.F., hand carried		35.50	.451			14.90		14.90	24.50
2215	Wheeled		54	.296			9.80		9.80	16.10
2220	For each additional flight of stairs, 5 risers		550	.029	Flight		.96		.96	1.58
2225	6 – 10 risers		275	.058			1.93		1.93	3.16
2230	11-20 risers		138	.116			3.84		3.84	6.30
2235	21 – 40 risers		69	.232			7.70		7.70	12.60
3000	Loading & trucking, including 2 mile haul, chute loaded	B-16	45	.711	C.Y.		24	14.85	38.85	55.50
3040	Hand loading truck, 50' haul	"	48	.667			22.50	13.90	36.40	52
3080	Machine loading truck	B-17	120	.267			9.40	6.15	15.55	22
3120	Wheeled 50' and ramp dump loaded	2 Clab	24	.667			22		22	36
5000	Haul, per mile, up to 8 C.Y. truck	B-34B	1165	.007			.23	.57	.80	1
5100	Over 8 C.Y. truck	"	1550	.005			.17	.43	.60	.75

02 41 19.25 Selective Demolition, Saw Cutting

			Crew	Daily Output	Labor-Hours	Unit	Material	Labor	Equipment	Total	Total Incl O&P
0010	**SELECTIVE DEMOLITION, SAW CUTTING**	R024119-10									
0015	Asphalt, up to 3" deep		B-89	1050	.015	L.F.	.41	.56	.45	1.42	1.84
0020	Each additional inch of depth		"	1800	.009		.09	.33	.26	.68	.91
1200	Masonry walls, hydraulic saw, brick, per inch of depth		B-89B	300	.053		.43	1.96	2.81	5.20	6.70
1220	Block walls, solid, per inch of depth		"	250	.064		.43	2.35	3.37	6.15	7.95
2000	Brick or masonry w/hand held saw, per inch of depth		A-1	125	.064		.34	2.12	.59	3.05	4.51
5000	Wood sheathing to 1" thick, on walls		1 Carp	200	.040			1.66		1.66	2.73
5020	On roof		"	250	.032			1.33		1.33	2.18
9000	Minimum labor/equipment charge		A-1	2	4	Job		132	37	169	258

02 41 Demolition

02 41 19 – Selective Structure Demolition

02 41 19.27 Selective Demolition, Torch Cutting		Crew	Daily Output	Labor-Hours	Unit	Material	2010 Bare Costs Labor	2010 Bare Costs Equipment	Total	Total Incl O&P
0010	**SELECTIVE DEMOLITION, TORCH CUTTING** R024119-10									
0020	Steel, 1" thick plate	1 Clab	360	.022	L.F.	.23	.74		.97	1.46
0040	1" diameter bar	"	210	.038	Ea.		1.26		1.26	2.07
1000	Oxygen lance cutting, reinforced concrete walls									
1040	12" to 16" thick walls	1 Clab	10	.800	L.F.		26.50		26.50	43.50
1080	24" thick walls	"	6	1.333	"		44		44	72.50
1090	Minimum labor/equipment charge	E-25	2	4	Job		196	49	245	415
1100	See Div. 05 05 21.10									

02 42 Removal and Salvage of Construction Material

02 42 10 – Building Deconstruction

02 42 10.10 Estimated Salvage Value or Savings

			Crew	Daily Output	Labor-Hours	Unit	Material	Labor	Equipment	Total	Total Incl O&P
0010	**ESTIMATED SALVAGE VALUE OR SAVINGS**										
0015	Excludes material handling, packaging, container costs and										
0020	transportation for salvage or disposal										
0050	All Items in section 02 42 10.10 are credit deducts and not costs										
0100	Copper Wire Salvage Value	G				Lb.				1.30	1.30
0110	Disposal Savings	G								.03	.03
0200	Copper Pipe Salvage Value	G								1.34	1.34
0210	Disposal Savings	G								.03	.03
0300	Steel Pipe Salvage Value	G								.06	.06
0310	Disposal Savings	G								.03	.03
0400	Cast Iron Pipe Salvage Value	G								.06	.06
0410	Disposal Savings	G								.03	.03
0500	Steel Doors or Windows Salvage Value	G								.06	.06
0510	Aluminum	G								.66	.66
0520	Disposal Savings	G								.03	.03
0600	Aluminum Siding Salvage Value	G								.49	.49
0630	Disposal Savings	G								.03	.03
0640	Wood Siding (no lead or asbestos)	G				C.Y.				10.92	10.92
0800	Clean Concrete Disposal Savings	G				Ton				60	60
0850	Asphalt Shingles Disposal Savings	G				"				60	60
1000	Wood wall framing clean salvage value	G				M.B.F.				50	50
1010	Painted	G								40	40
1020	Floor framing	G								50	50
1030	Painted	G								40	40
1050	Roof framing	G								50	50
1060	Painted	G								40	40
1100	Wood beams salvage value	G								50	50
1200	Wood framing and beams disposal savings	G				Ton				60	60
1220	Wood sheating and sub-base flooring	G								60	60
1230	Wood wall paneling (1/4 inch thick)	G								60	60
1300	Wood panel 3/4-1 inch thick low salvage value	G				S.F.				.50	.50
1350	high salvage value	G				"				2	2
1400	Disposal savings	G				Ton				60	60
1500	Flooring tongue and groove 25/32 inch thick low salvage value	G				S.F.				.50	.50
1530	High salvage value	G				"				1	1
1560	Disposal savings	G				Ton				60	60
1600	Drywall or sheet rock salvage value	G								20	20
1650	Disposal savings	G								60	60

02 42 Removal and Salvage of Construction Material

02 42 10 – Building Deconstruction

02 42 10.20 Deconstruction of Building Components

		Crew	Daily Output	Labor-Hours	Unit	Material	2010 Bare Costs Labor	Equipment	Total	Total Incl O&P
0010	**DECONSTRUCTION OF BUILDING COMPONENTS**									
0012	Buildings one or two stories only									
0015	Excludes material handling, packaging, container costs and									
0020	transportation for salvage or disposal									
0050	Deconstruction of Plumbing Fixtures									
0100	Wall hung or countertop lavatory	G	2 Clab	16	1	Ea.		33		33	54.50
0110	Single or double compartment kitchen sink	G		14	1.143			38		38	62
0120	Wall hung urinal	G		14	1.143			38		38	62
0130	Floor mounted	G		8	2			66		66	109
0140	Floor mounted water closet	G		16	1			33		33	54.50
0150	Wall hung	G		14	1.143			38		38	62
0160	Water fountain, free standing	G		16	1			33		33	54.50
0170	Wall hung or deck mounted	G		12	1.333			44		44	72.50
0180	Bathtub, steel or fiberglass	G		10	1.600			53		53	87
0190	Cast iron	G		8	2			66		66	109
0200	Shower, single	G		6	2.667			88.50		88.50	145
0210	Group	G	▼	7	2.286	▼		75.50		75.50	124
0300	Deconstruction of Electrical Fixtures									
0310	Surface mount incandescent fixtures	G	2 Clab	48	.333	Ea.		11.05		11.05	18.10
0320	Fluorescent, 2 lamp	G		32	.500			16.55		16.55	27
0330	4 lamp	G		24	.667			22		22	36
0340	Strip Fluorescent, 1 lamp	G		40	.400			13.25		13.25	21.50
0350	2 lamp	G		32	.500			16.55		16.55	27
0400	Recessed drop-in fluorescent fixture, 2 lamp	G		27	.593			19.60		19.60	32
0410	4 lamp	G	▼	18	.889	▼		29.50		29.50	48.50
0500	Deconstruction of appliances									
0510	Cooking stoves	G	2 Clab	26	.615	Ea.		20.50		20.50	33.50
0520	Dishwashers	G	"	26	.615	"		20.50		20.50	33.50
0600	Deconstruction of millwork and trim									
0610	Cabinets, wood	G	2 Carp	40	.400	L.F.		16.60		16.60	27.50
0620	Countertops	G		100	.160	"		6.65		6.65	10.90
0630	Wall paneling, 1 inch thick	G		500	.032	S.F.		1.33		1.33	2.18
0640	Ceiling trim	G		500	.032	L.F.		1.33		1.33	2.18
0650	Wainscoting	G		500	.032	S.F.		1.33		1.33	2.18
0660	Base, 3/4" to 1" thick	G	▼	600	.027	L.F.		1.11		1.11	1.82
0700	Deconstruction of doors and windows									
0710	Doors, wrap, interior, wood, single, no closers	G	2 Carp	21	.762	Ea.	3.50	31.50		35	56
0720	Double	G		13	1.231		7	51		58	91.50
0730	Solid core, single, exterior or interior	G		10	1.600		3.50	66.50		70	113
0740	Double	G		8	2		7	83		90	145
0810	Windows, wrap, wood, single	G	▼	21	.762	▼	3.50	31.50		35	56
0812	with no casement or cladding										
0820	with casement and/or cladding	G	2 Carp	18	.889	Ea.	3.50	37		40.50	64.50
0900	Deconstruction of interior finishes									
0910	Drywall for recycling	G	2 Clab	1775	.009	S.F.		.30		.30	.49
0920	Plaster wall, first floor	G		1775	.009			.30		.30	.49
0930	Second floor	G	▼	1330	.012	▼		.40		.40	.65
1000	Deconstruction of roofing and accessories									
1010	Built-up roofs	G	2 Clab	570	.028	S.F.		.93		.93	1.53
1020	Gutters, facia and rakes	G	"	1140	.014	L.F.		.46		.46	.76
2000	Deconstruction of wood components									
2010	Roof sheeting	G	2 Clab	570	.028	S.F.		.93		.93	1.53

02 42 Removal and Salvage of Construction Material

02 42 10 – Building Deconstruction

02 42 10.20 Deconstruction of Building Components

			Crew	Daily Output	Labor-Hours	Unit	Material	2010 Bare Costs Labor	Equipment	Total	Total Incl O&P
2020	Main roof framing	G	2 Clab	760	.021	L.F.		.70		.70	1.14
2030	Porch roof framing	G	↓	445	.036			1.19		1.19	1.95
2040	Beams 4" x 8"	G	B-1	375	.064			2.16		2.16	3.55
2050	4" x 10"	G		300	.080			2.70		2.70	4.44
2055	4" x 12"	G		250	.096			3.24		3.24	5.30
2060	6" x 8"	G		250	.096			3.24		3.24	5.30
2065	6" x 10"	G		200	.120			4.05		4.05	6.65
2070	6" x 12"	G		170	.141			4.77		4.77	7.85
2075	8" x 12"	G		126	.190			6.45		6.45	10.55
2080	10" x 12"	G	↓	100	.240			8.10		8.10	13.30
2100	Ceiling joists	G	2 Clab	800	.020			.66		.66	1.09
2150	Wall framing, interior	G		1230	.013	↓		.43		.43	.71
2160	Sub-floor	G		2000	.008	S.F.		.26		.26	.43
2170	Floor joists	G		2000	.008	L.F.		.26		.26	.43
2200	Wood siding (no lead or asbestos)	G		1300	.012	S.F.		.41		.41	.67
2300	Wall framing, exterior	G		1600	.010	L.F.		.33		.33	.54
2400	Stair risers	G		53	.302	Ea.		10		10	16.40
2500	Posts	G	↓	800	.020	L.F.		.66		.66	1.09
3000	Deconstruction of exterior brick walls										
3010	Exterior brick walls, first floor	G	2 Clab	200	.080	S.F.		2.65		2.65	4.35
3020	Second floor	G		64	.250	"		8.30		8.30	13.60
3030	Brick chiminey	G	↓	100	.160	C.F.		5.30		5.30	8.70
4000	Deconstruction of concrete										
4010	Slab on grade, 4" thick, plain concrete	G	B-9	500	.080	S.F.		2.68	.45	3.13	4.89
4020	Wire mesh reinforced	G		470	.085			2.85	.48	3.33	5.20
4030	Rod reinforced	G		400	.100			3.35	.56	3.91	6.10
4110	Foundation wall, 6" thick, plain concrete	G		160	.250			8.40	1.41	9.81	15.30
4120	8" thick	G		140	.286			9.55	1.61	11.16	17.45
4130	10" thick	G	↓	120	.333	↓		11.15	1.87	13.02	20.50
9000	Deconstruction process, support equipment as needed										
9010	Daily use, portal to portal, 12-ton truck-mounted hydraulic crane crew	G	A-3H	1	8	Day		355	870	1,225	1,525
9020	Daily use, skid steer and operator	G	A-3C	1	8			330	291	621	840
9030	Daily use, backhoe 48 H. P., operator and labor	G	"	1	8	↓		330	291	621	840

02 42 10.30 Deconstruction Material Handling

			Crew	Daily Output	Labor-Hours	Unit	Material	Labor	Equipment	Total	Total Incl O&P
0010	**DECONSTRUCTION MATERIAL HANDLING**										
0012	Buildings one or two stories only										
0100	Clean and stack brick on pallet	G	2 Clab	1200	.013	Ea.		.44		.44	.72
0200	Haul 50' and load rough lumber up to 2" x 8" size	G		2000	.008	"		.26		.26	.43
0210	Lumber larger than 2" x 8"	G		3200	.005	B.F.		.17		.17	.27
0300	Finish wood for recycling stack and wrap per pallet	G	↓	8	2	Ea.	28	66		94	140
1000	See section 02 41 19.23 for bulk material handling										

02 43 Structure Moving

02 43 13 – Structure Relocation

02 43 13.13 Building Relocation

		Crew	Daily Output	Labor-Hours	Unit	Material	2010 Bare Costs Labor	Equipment	Total	Total Incl O&P
0010	**BUILDING RELOCATION**									
0011	One day move, up to 24' wide									
0020	Reset on new foundation, patch & hook-up, average move				Total					11,220
0040	Wood or steel frame bldg., based on ground floor area	G	B-4	185	.259	S.F.		8.70	2.69	11.39	17.15
0060	Masonry bldg., based on ground floor area	G	"	137	.350			11.70	3.63	15.33	23
0200	For 24' to 42' wide, add										15%
0220	For each additional day on road, add	G	B-4	1	48	Day		1,600	495	2,095	3,175
0240	Construct new basement, move building, 1 day										
0300	move, patch & hook-up, based on ground floor area	G	B-3	155	.310	S.F.	9.10	10.85	15.55	35.50	44.50

02 58 Snow Control

02 58 13 – Snow Fencing

02 58 13.10 Snow Fencing System

		Crew	Daily Output	Labor-Hours	Unit	Material	Labor	Equipment	Total	Total Incl O&P
0010	**SNOW FENCING SYSTEM**									
7001	Snow fence on steel posts 10' O.C., 4' high		B-1	500	.048	L.F.	2.62	1.62		4.24	5.55

02 65 Underground Storage Tank Removal

02 65 10 – Underground Tank and Contaminated Soil Removal

02 65 10.30 Removal of Underground Storage Tanks

			Crew	Daily Output	Labor-Hours	Unit	Material	Labor	Equipment	Total	Total Incl O&P
0010	**REMOVAL OF UNDERGROUND STORAGE TANKS**	R026510-20									
0011	Petroleum storage tanks, non-leaking										
0100	Excavate & load onto trailer										
0110	3000 gal. to 5000 gal. tank	G	B-14	4	12	Ea.		420	84.50	504.50	775
0120	6000 gal. to 8000 gal. tank	G	B-3A	3	13.333			470	340	810	1,125
0130	9000 gal. to 12000 gal. tank	G	"	2	20			700	510	1,210	1,700
0190	Known leaking tank, add					%				100%	100%
0200	Remove sludge, water and remaining product from tank bottom										
0201	of tank with vacuum truck										
0300	3000 gal. to 5000 gal. tank	G	A-13	5	1.600	Ea.		66	155	221	275
0310	6000 gal. to 8000 gal. tank	G		4	2			82.50	194	276.50	340
0320	9000 gal. to 12000 gal. tank	G		3	2.667			110	259	369	455
0390	Dispose of sludge off-site, average					Gal.				6	6.60
0400	Insert inert solid CO$_2$ "dry ice" into tank										
0401	For cleaning/transporting tanks (1.5 lbs./100 gal. cap)	G	1 Clab	500	.016	Lb.	1.72	.53		2.25	2.76
0403	Insert solid carbon dioxide, 1.5 lbs./100 gal.	G	"	400	.020	"	1.72	.66		2.38	2.98
0503	Disconnect and remove piping	G	1 Plum	160	.050	L.F.		2.60		2.60	4.03
0603	Transfer liquids, 10% of volume	G	"	1600	.005	Gal.		.26		.26	.40
0703	Cut accessway into underground storage tank	G	1 Clab	5.33	1.501	Ea.		49.50		49.50	81.50
0813	Remove sludge, wash and wipe tank, 500 gal.	G	1 Plum	8	1			52		52	80.50
0823	3,000 gal.	G		6.67	1.199			62.50		62.50	96.50
0833	5,000 gal.	G		6.15	1.301			67.50		67.50	105
0843	8,000 gal.	G		5.33	1.501			78		78	121
0853	10,000 gal.	G		4.57	1.751			91		91	141
0863	12,000 gal.	G		4.21	1.900			99		99	153
1020	Haul tank to certified salvage dump, 100 miles round trip										
1023	3000 gal. to 5000 gal. tank					Ea.				700	770
1026	6000 gal. to 8000 gal. tank									800	880
1029	9,000 gal. to 12,000 gal. tank									1,000	1,100
1100	Disposal of contaminated soil to landfill										

02 65 Underground Storage Tank Removal

02 65 10 – Underground Tank and Contaminated Soil Removal

02 65 10.30 Removal of Underground Storage Tanks		Crew	Daily Output	Labor-Hours	Unit	Material	2010 Bare Costs Labor	Equipment	Total	Total Incl O&P
1110	Minimum				C.Y.				120	132
1111	Maximum				"				380	418
1120	Disposal of contaminated soil to									
1121	bituminous concrete batch plant									
1130	Minimum				C.Y.				75	83
1131	Maximum				"				120	132
1203	Excavate, pull, & load tank, backfill hole, 8,000 gal. +	G	B-12C	.50	32	Ea.		1,250	2,650	3,900	4,875
1213	Haul tank to certified dump, 100 miles rt, 8,000 gal. +	G	B-34K	1	8			265	920	1,185	1,450
1223	Excavate, pull, & load tank, backfill hole, 500 gal.	G	B-11C	1	16			610	340	950	1,350
1233	Excavate, pull, & load tank, backfill hole, 3,000 – 5,000 gal.	G	B-11M	.50	32			1,225	775	2,000	2,800
1243	Haul tank to certified dump, 100 miles rt, 500 gal.	G	B-34L	1	8			330	240	570	785
2010	Decontamination of soil on site incl poly tarp on top/bottom										
2011	Soil containment berm, and chemical treatment										
2020	Minimum	G	B-11C	100	.160	C.Y.	7.05	6.10	3.38	16.53	21
2021	Maximum	G	"	100	.160		9.15	6.10	3.38	18.63	23.50
2050	Disposal of decontaminated soil, minimum									75	83
2055	Maximum									150	165

02 81 Transportation and Disposal of Hazardous Materials

02 81 20 – Hazardous Waste Handling

02 81 20.10 Hazardous Waste Cleanup/Pickup/Disposal

		Crew	Daily Output	Labor-Hours	Unit	Material	Labor	Equipment	Total	Total Incl O&P
0010	**HAZARDOUS WASTE CLEANUP/PICKUP/DISPOSAL**									
0100	For contractor rental equipment, i.e., Dozer,									
0110	Front end loader, Dump truck, etc., see 01 54 33 Reference Section									
1000	Solid pickup									
1100	55 gal. drums				Ea.				230	253
1120	Bulk material, minimum				Ton				175	193
1130	Maximum				"				580	638
1200	Transportation to disposal site									
1220	Truckload = 80 drums or 25 C.Y. or 18 tons									
1260	Minimum				Mile				3.60	3.96
1270	Maximum				"				6.50	7.15
3000	Liquid pickup, vacuum truck, stainless steel tank									
3100	Minimum charge, 4 hours									
3110	1 compartment, 2200 gallon				Hr.				125	138
3120	2 compartment, 5000 gallon				"				175	193
3400	Transportation in 6900 gallon bulk truck				Mile				6.50	7.15
3410	In teflon lined truck				"				8	8.80
5000	Heavy sludge or dry vacuumable material				Hr.				125	138
6000	Dumpsite disposal charge, minimum				Ton				125	138
6020	Maximum				"				450	495

02 82 Asbestos Remediation

02 82 13 – Asbestos Abatement

02 82 13.41 Asbestos Abatement Equip.

		Crew	Daily Output	Labor-Hours	Unit	Material	2010 Bare Costs Labor	Equipment	Total	Total Incl O&P
0010	**ASBESTOS ABATEMENT EQUIP.** R028213-20									
0011	Equipment and supplies, buy									
0200	Air filtration device, 2000 C.F.M.				Ea.	795			795	875
0250	Large volume air sampling pump, minimum					355			355	390
0260	Maximum					490			490	540
0300	Airless sprayer unit, 2 gun					4,800			4,800	5,300
0350	Light stand, 500 watt					52			52	57.50
0400	Personal respirators									
0410	Negative pressure, 1/2 face, dual operation, min.				Ea.	25.50			25.50	28
0420	Maximum					27.50			27.50	30.50
0450	P.A.P.R., full face, minimum					147			147	162
0460	Maximum					147			147	162
0470	Supplied air, full face, incl. air line, minimum					163			163	179
0480	Maximum					288			288	315
0500	Personnel sampling pump, minimum					221			221	244
1500	Power panel, 20 unit, incl. G.F.I.					570			570	625
1600	Shower unit, including pump and filters					1,175			1,175	1,275
1700	Supplied air system (type C)					9,800			9,800	10,800
1750	Vacuum cleaner, HEPA, 16 gal., stainless steel, wet/dry					1,050			1,050	1,150
1760	55 gallon					1,700			1,700	1,875
1800	Vacuum loader, 9 – 18 ton/hr					90,000			90,000	99,000
1900	Water atomizer unit, including 55 gal. drum					230			230	253
2000	Worker protection, whole body, foot, head cover & gloves, plastic					8.70			8.70	9.60
2500	Respirator, single use					28.50			28.50	31.50
2550	Cartridge for respirator					6.90			6.90	7.60
2570	Glove bag, 7 mil, 50" x 64"					17.25			17.25	19
2580	10 mil, 44" x 60"					74.50			74.50	82
3000	HEPA vacuum for work area, minimum					1,325			1,325	1,475
3050	Maximum					1,400			1,400	1,525
6000	Disposable polyethylene bags, 6 mil, 3 C.F.					2.26			2.26	2.49
6300	Disposable fiber drums, 3 C.F.					16.30			16.30	17.95
6400	Pressure sensitive caution labels, 3" x 5"					3.11			3.11	3.42
6450	11" x 17"					7.05			7.05	7.75
6500	Negative air machine, 1800 C.F.M.					810			810	890

02 82 13.42 Preparation of Asbestos Containment Area

		Crew	Daily Output	Labor-Hours	Unit	Material	Labor	Equipment	Total	Total Incl O&P
0010	**PREPARATION OF ASBESTOS CONTAINMENT AREA**									
0100	Pre-cleaning, HEPA vacuum and wet wipe, flat surfaces	A-9	12000	.005	S.F.	.02	.24		.26	.41
0200	Protect carpeted area, 2 layers 6 mil poly on 3/4" plywood	"	1000	.064		2	2.92		4.92	6.90
0300	Separation barrier, 2" x 4" @ 16", 1/2" plywood ea. side, 8' high	2 Carp	400	.040		2	1.66		3.66	4.93
0310	12' high		320	.050		2	2.08		4.08	5.60
0320	16' high		200	.080		2	3.32		5.32	7.65
0400	Personnel decontam. chamber, 2" x 4" @ 16", 3/4" ply ea. side		280	.057		3	2.37		5.37	7.20
0450	Waste decontam. chamber, 2" x 4" studs @ 16", 3/4" ply ea. side		360	.044		3	1.85		4.85	6.35
0500	Cover surfaces with polyethylene sheeting									
0501	Including glue and tape									
0550	Floors, each layer, 6 mil	A-9	8000	.008	S.F.	.05	.36		.41	.65
0551	4 mil		9000	.007		.04	.32		.36	.56
0560	Walls, each layer, 6 mil		6000	.011		.05	.49		.54	.84
0561	4 mil		7000	.009		.04	.42		.46	.71
0570	For heights above 12', add						20%			
0575	For heights above 20', add						30%			
0580	For fire retardant poly, add					100%				

02 82 Asbestos Remediation

02 82 13 – Asbestos Abatement

02 82 13.42 Preparation of Asbestos Containment Area

		Crew	Daily Output	Labor-Hours	Unit	Material	2010 Bare Costs Labor	Equipment	Total	Total Incl O&P
0590	For large open areas, deduct					10%	20%			
0600	Seal floor penetrations with foam firestop to 36 Sq. In.	2 Carp	200	.080	Ea.	7.60	3.32		10.92	13.85
0610	36 Sq. In. to 72 Sq. In.		125	.128		15.25	5.30		20.55	25.50
0615	72 Sq. In. to 144 Sq. In.		80	.200		30.50	8.30		38.80	47
0620	Wall penetrations, to 36 square inches		180	.089		7.60	3.69		11.29	14.45
0630	36 Sq. In. to 72 Sq. In.		100	.160		15.25	6.65		21.90	27.50
0640	72 Sq. In. to 144 Sq. In.		60	.267		30.50	11.10		41.60	51.50
0800	Caulk seams with latex	1 Carp	230	.035	L.F.	.15	1.45		1.60	2.54
0900	Set up neg. air machine, 1-2k C.F.M. /25 M.C.F. volume	1 Asbe	4.30	1.860	Ea.		84.50		84.50	136

02 82 13.43 Bulk Asbestos Removal

		Crew	Daily Output	Labor-Hours	Unit	Material	Labor	Equipment	Total	Total Incl O&P
0010	**BULK ASBESTOS REMOVAL**									
0020	Includes disposable tools and 2 suits and 1 respirator filter/day/worker									
0100	Beams, W 10 x 19	A-9	235	.272	L.F.	.83	12.40		13.23	21
0110	W 12 x 22		210	.305		.93	13.90		14.83	23.50
0120	W 14 x 26		180	.356		1.08	16.20		17.28	27
0130	W 16 x 31		160	.400		1.22	18.25		19.47	31
0140	W 18 x 40		140	.457		1.39	21		22.39	35
0150	W 24 x 55		110	.582		1.77	26.50		28.27	44.50
0160	W 30 x 108		85	.753		2.29	34.50		36.79	57.50
0170	W 36 x 150		72	.889		2.70	40.50		43.20	68
0200	Boiler insulation		480	.133	S.F.	.48	6.10		6.58	10.25
0210	With metal lath, add				%				50%	
0300	Boiler breeching or flue insulation	A-9	520	.123	S.F.	.37	5.60		5.97	9.40
0310	For active boiler, add				%				100%	
0400	Duct or AHU insulation	A-10B	440	.073	S.F.	.22	3.32		3.54	5.60
0500	Duct vibration isolation joints, up to 24 Sq. In. duct	A-9	56	1.143	Ea.	3.48	52		55.48	87.50
0520	25 Sq. In. to 48 Sq. In. duct		48	1.333		4.06	61		65.06	102
0530	49 Sq. In. to 76 Sq. In. duct		40	1.600		4.87	73		77.87	122
0600	Pipe insulation, air cell type, up to 4" diameter pipe		900	.071	L.F.	.22	3.24		3.46	5.45
0610	4" to 8" diameter pipe		800	.080		.24	3.65		3.89	6.10
0620	10" to 12" diameter pipe		700	.091		.28	4.17		4.45	7
0630	14" to 16" diameter pipe		550	.116		.35	5.30		5.65	8.90
0650	Over 16" diameter pipe		650	.098	S.F.	.30	4.49		4.79	7.55
0700	With glove bag up to 3" diameter pipe		200	.320	L.F.	6.05	14.60		20.65	30
1000	Pipe fitting insulation up to 4" diameter pipe		320	.200	Ea.	.61	9.10		9.71	15.30
1100	6" to 8" diameter pipe		304	.211		.64	9.60		10.24	16.10
1110	10" to 12" diameter pipe		192	.333		1.01	15.20		16.21	25.50
1120	14" to 16" diameter pipe		128	.500		1.52	23		24.52	38
1130	Over 16" diameter pipe		176	.364	S.F.	1.11	16.60		17.71	27.50
1200	With glove bag, up to 8" diameter pipe		75	.853	L.F.	6.55	39		45.55	69.50
2000	Scrape foam fireproofing from flat surface		2400	.027	S.F.	.08	1.22		1.30	2.04
2100	Irregular surfaces		1200	.053		.16	2.43		2.59	4.08
3000	Remove cementitious material from flat surface		1800	.036		.11	1.62		1.73	2.72
3100	Irregular surface		1000	.064		.14	2.92		3.06	4.83
4000	Scrape acoustical coating/fireproofing, from ceiling		3200	.020		.06	.91		.97	1.53
5000	Remove VAT from floor by hand		2400	.027		.08	1.22		1.30	2.04
5100	By machine	A-11	4800	.013		.04	.61	.01	.66	1.03
5150	For 2 layers, add				%				50%	
6000	Remove contaminated soil from crawl space by hand	A-9	400	.160	C.F.	.49	7.30		7.79	12.25
6100	With large production vacuum loader	A-12	700	.091	"	.28	4.17	1.11	5.56	8.25
7000	Radiator backing, not including radiator removal	A-9	1200	.053	S.F.	.16	2.43		2.59	4.08
8000	Cement-asbestos transite board	2 Asbe	1000	.016		.14	.73		.87	1.32

02 82 Asbestos Remediation

02 82 13 – Asbestos Abatement

02 82 13.43 Bulk Asbestos Removal		Crew	Daily Output	Labor-Hours	Unit	Material	2010 Bare Costs Labor	Equipment	Total	Total Incl O&P
8100	Transite shingle siding	A-10B	750	.043	S.F.	.21	1.95		2.16	3.36
8200	Shingle roofing	"	2000	.016		.08	.73		.81	1.26
8250	Built-up, no gravel, non-friable	B-2	1400	.029		.08	.96		1.04	1.66
8300	Asbestos millboard	2 Asbe	1000	.016		.08	.73		.81	1.26
9000	For type B (supplied air) respirator equipment, add				%					10%

02 82 13.44 Demolition In Asbestos Contaminated Area										
0010	**DEMOLITION IN ASBESTOS CONTAMINATED AREA**									
0200	Ceiling, including suspension system, plaster and lath	A-9	2100	.030	S.F.	.09	1.39		1.48	2.33
0210	Finished plaster, leaving wire lath		585	.109		.33	4.99		5.32	8.35
0220	Suspended acoustical tile		3500	.018		.06	.83		.89	1.40
0230	Concealed tile grid system		3000	.021		.06	.97		1.03	1.63
0240	Metal pan grid system		1500	.043		.13	1.95		2.08	3.26
0250	Gypsum board		2500	.026		.08	1.17		1.25	1.96
0260	Lighting fixtures up to 2' x 4'		72	.889	Ea.	2.70	40.50		43.20	68
0400	Partitions, non load bearing									
0410	Plaster, lath, and studs	A-9	690	.093	S.F.	.85	4.23		5.08	7.75
0450	Gypsum board and studs	"	1390	.046	"	.14	2.10		2.24	3.52
9000	For type B (supplied air) respirator equipment, add				%					10%

02 82 13.45 OSHA Testing										
0010	**OSHA TESTING**									
0100	Certified technician, minimum				Day				240	264
0110	Maximum				"				320	352
0200	Personal sampling, PCM analysis, NIOSH 7400, minimum	1 Asbe	8	1	Ea.	2.75	45.50		48.25	76
0210	Maximum	"	4	2	"	3	91		94	149
0300	Industrial hygienist, minimum				Day				320	352
0310	Maximum				"				480	528
1000	Cleaned area samples	1 Asbe	8	1	Ea.	2.63	45.50		48.13	76
1100	PCM air sample analysis, NIOSH 7400, minimum		8	1		30.50	45.50		76	107
1110	Maximum		4	2		3.20	91		94.20	150
1200	TEM air sample analysis, NIOSH 7402, minimum								100	125
1210	Maximum								400	500

02 82 13.46 Decontamination of Asbestos Containment Area										
0010	**DECONTAMINATION OF ASBESTOS CONTAINMENT AREA**									
0100	Spray exposed substrate with surfactant (bridging)									
0200	Flat surfaces	A-9	6000	.011	S.F.	.36	.49		.85	1.18
0250	Irregular surfaces		4000	.016	"	.31	.73		1.04	1.51
0300	Pipes, beams, and columns		2000	.032	L.F.	.56	1.46		2.02	2.96
1000	Spray encapsulate polyethylene sheeting		8000	.008	S.F.	.31	.36		.67	.93
1100	Roll down polyethylene sheeting		8000	.008	"		.36		.36	.59
1500	Bag polyethylene sheeting		400	.160	Ea.	.77	7.30		8.07	12.55
2000	Fine clean exposed substrate, with nylon brush		2400	.027	S.F.		1.22		1.22	1.95
2500	Wet wipe substrate		4800	.013			.61		.61	.98
2600	Vacuum surfaces, fine brush		6400	.010			.46		.46	.73
3000	Structural demolition									
3100	Wood stud walls	A-9	2800	.023	S.F.		1.04		1.04	1.67
3500	Window manifolds, not incl. window replacement		4200	.015			.70		.70	1.11
3600	Plywood carpet protection		2000	.032			1.46		1.46	2.34
4000	Remove custom decontamination facility	A-10A	8	3	Ea.	15	137		152	237
4100	Remove portable decontamination facility	3 Asbe	12	2	"	12.75	91		103.75	160
5000	HEPA vacuum, shampoo carpeting	A-9	4800	.013	S.F.	.05	.61		.66	1.04
9000	Final cleaning of protected surfaces	A-10A	8000	.003	"		.14		.14	.22

02 82 Asbestos Remediation

02 82 13 – Asbestos Abatement

02 82 13.47 Asbestos Waste Pkg., Handling, and Disp.

		Crew	Daily Output	Labor-Hours	Unit	Material	2010 Bare Costs Labor	Equipment	Total	Total Incl O&P
0010	**ASBESTOS WASTE PACKAGING, HANDLING, AND DISPOSAL**									
0100	Collect and bag bulk material, 3 C.F. bags, by hand	A-9	400	.160	Ea.	2.26	7.30		9.56	14.20
0200	Large production vacuum loader	A-12	880	.073		.80	3.32	.88	5	7.15
1000	Double bag and decontaminate	A-9	960	.067		2.26	3.04		5.30	7.35
2000	Containerize bagged material in drums, per 3 C.F. drum	"	800	.080		16.30	3.65		19.95	24
3000	Cart bags 50' to dumpster	2 Asbe	400	.040			1.82		1.82	2.92
5000	Disposal charges, not including haul, minimum				C.Y.				50	55
5020	Maximum				"				170	187
5100	Remove refrigerant from system	1 Stpi	40	.200	Lb.		10.40		10.40	16.10
9000	For type B (supplied air) respirator equipment, add				%					10%

02 82 13.48 Asbestos Encapsulation With Sealants

		Crew	Daily Output	Labor-Hours	Unit	Material	Labor	Equipment	Total	Total Incl O&P
0010	**ASBESTOS ENCAPSULATION WITH SEALANTS**									
0100	Ceilings and walls, minimum	A-9	21000	.003	S.F.	.27	.14		.41	.52
0110	Maximum		10600	.006		.42	.28		.70	.90
0200	Columns and beams, minimum		13300	.005		.27	.22		.49	.65
0210	Maximum		5325	.012		.47	.55		1.02	1.40
0300	Pipes to 12" diameter including minor repairs, minimum		800	.080	L.F.	.37	3.65		4.02	6.25
0310	Maximum		400	.160	"	1.04	7.30		8.34	12.85

02 83 Lead Remediation

02 83 19 – Lead-Based Paint Remediation

02 83 19.23 Encapsulation of Lead-Based Paint

		Crew	Daily Output	Labor-Hours	Unit	Material	Labor	Equipment	Total	Total Incl O&P
0010	**ENCAPSULATION OF LEAD-BASED PAINT**									
0020	Interior, brushwork, trim, under 6"	1 Pord	240	.033	L.F.	2.46	1.21		3.67	4.63
0030	6" to 12" wide		180	.044		3.28	1.62		4.90	6.15
0040	Balustrades		300	.027		1.98	.97		2.95	3.72
0050	Pipe to 4" diameter		500	.016		1.19	.58		1.77	2.23
0060	To 8" diameter		375	.021		1.57	.78		2.35	2.96
0070	To 12" diameter		250	.032		2.36	1.16		3.52	4.45
0080	To 16" diameter		170	.047		3.47	1.71		5.18	6.55
0090	Cabinets, ornate design		200	.040	S.F.	2.97	1.45		4.42	5.60
0100	Simple design		250	.032	"	2.36	1.16		3.52	4.45
0110	Doors, 3' x 7', both sides, incl. frame & trim									
0120	Flush	1 Pord	6	1.333	Ea.	30.50	48.50		79	111
0130	French, 10 – 15 lite		3	2.667		6.05	97		103.05	161
0140	Panel		4	2		36.50	72.50		109	155
0150	Louvered		2.75	2.909		33.50	106		139.50	205
0160	Windows, per interior side, per 15 S.F.									
0170	1 to 6 lite	1 Pord	14	.571	Ea.	21	21		42	56
0180	7 to 10 lite		7.50	1.067		23	39		62	87
0190	12 lite		5.75	1.391		31	50.50		81.50	115
0200	Radiators		8	1		73.50	36.50		110	139
0210	Grilles, vents		275	.029	S.F.	2.15	1.06		3.21	4.05
0220	Walls, roller, drywall or plaster		1000	.008		.59	.29		.88	1.11
0230	With spunbonded reinforcing fabric		720	.011		.68	.40		1.08	1.39
0240	Wood		800	.010		.74	.36		1.10	1.39
0250	Ceilings, roller, drywall or plaster		900	.009		.68	.32		1	1.26
0260	Wood		700	.011		.84	.42		1.26	1.58
0270	Exterior, brushwork, gutters and downspouts		300	.027	L.F.	1.98	.97		2.95	3.72
0280	Columns		400	.020	S.F.	1.47	.73		2.20	2.77

02 83 Lead Remediation

02 83 19 – Lead-Based Paint Remediation

02 83 19.23 Encapsulation of Lead-Based Paint

		Crew	Daily Output	Labor-Hours	Unit	Material	2010 Bare Costs Labor	Equipment	Total	Total Incl O&P
0290	Spray, siding	1 Pord	600	.013	S.F.	.99	.48		1.47	1.86
0300	Miscellaneous									
0310	Electrical conduit, brushwork, to 2" diameter	1 Pord	500	.016	L.F.	1.19	.58		1.77	2.23
0320	Brick, block or concrete, spray		500	.016	S.F.	1.19	.58		1.77	2.23
0330	Steel, flat surfaces and tanks to 12"		500	.016		1.19	.58		1.77	2.23
0340	Beams, brushwork		400	.020		1.47	.73		2.20	2.77
0350	Trusses		400	.020		1.47	.73		2.20	2.77

02 83 19.26 Removal of Lead-Based Paint

		Crew	Daily Output	Labor-Hours	Unit	Material	Labor	Equipment	Total	Total Incl O&P
0010	**REMOVAL OF LEAD-BASED PAINT** R028319-60									
0011	By chemicals, per application									
0050	Baseboard, to 6" wide	1 Pord	64	.125	L.F.	.66	4.54		5.20	7.95
0070	To 12" wide		32	.250	"	1.32	9.10		10.42	15.90
0200	Balustrades, one side		28	.286	S.F.	1.32	10.40		11.72	17.95
1400	Cabinets, simple design		32	.250		1.32	9.10		10.42	15.90
1420	Ornate design		25	.320		1.32	11.65		12.97	19.90
1600	Cornice, simple design		60	.133		1.32	4.85		6.17	9.15
1620	Ornate design		20	.400		5.05	14.55		19.60	28.50
2800	Doors, one side, flush		84	.095		1	3.46		4.46	6.60
2820	Two panel		80	.100		1.32	3.64		4.96	7.20
2840	Four panel		45	.178		1.32	6.45		7.77	11.70
2880	For trim, one side, add		64	.125	L.F.	.66	4.54		5.20	7.95
3000	Fence, picket, one side		30	.267	S.F.	1.32	9.70		11.02	16.85
3200	Grilles, one side, simple design		30	.267		1.32	9.70		11.02	16.85
3220	Ornate design		25	.320		1.32	11.65		12.97	19.90
4400	Pipes, to 4" diameter		90	.089	L.F.	1	3.23		4.23	6.25
4420	To 8" diameter		50	.160		2	5.80		7.80	11.45
4440	To 12" diameter		36	.222		3	8.10		11.10	16.10
4460	To 16" diameter		20	.400		4	14.55		18.55	27.50
4500	For hangers, add		40	.200	Ea.	2.54	7.25		9.79	14.35
4800	Siding		90	.089	S.F.	1.17	3.23		4.40	6.45
5000	Trusses, open		55	.145	SF Face	1.85	5.30		7.15	10.45
6200	Windows, one side only, double hung, 1/1 light, 24" x 48" high		4	2	Ea.	25.50	72.50		98	143
6220	30" x 60" high		3	2.667		34	97		131	192
6240	36" x 72" high		2.50	3.200		41	116		157	230
6280	40" x 80" high		2	4		51	145		196	287
6400	Colonial window, 6/6 light, 24" x 48" high		2	4		51	145		196	288
6420	30" x 60" high		1.50	5.333		68	194		262	385
6440	36" x 72" high		1	8		102	291		393	570
6480	40" x 80" high		1	8		102	291		393	570
6600	8/8 light, 24" x 48" high		2	4		51	145		196	288
6620	40" x 80" high		1	8		102	291		393	570
6800	12/12 light, 24" x 48" high		1	8		102	291		393	570
6820	40" x 80" high		.75	10.667		136	390		526	765
6840	Window frame & trim items, included in pricing above									
9000	Minimum labor/equipment charge	1 Pord	3	2.667	Job		97		97	154

02 85 Mold Remediation

02 85 16 – Mold Remediation Preparation and Containment

02 85 16.40 Mold Remediation Plans and Methods

		Crew	Daily Output	Labor-Hours	Unit	Material	2010 Bare Costs Labor	Equipment	Total	Total Incl O&P
0010	**MOLD REMEDIATION PLANS AND METHODS**									
0020	Initial inspection, average 3 bedroom home				Total				250	275
0030	Average 5 bedroom home								340	374
0040	Testing, air sample each								250	275
0050	Swab sample								150	165
0060	Tape sample								150	165
0070	Post remediation air test								250	275
0080	Mold abatement plan, average 3 bedroom home								1,300	1,430
0090	Average 5 bedroom home								1,580	1,740
0100	Packup & removal of contents, average 3 bedroom home, excl storage								10,500	11,550
0110	Average 5 bedroom home, excl storage								26,400	29,000
0120	For demolition in mold contaminated areas, see Div. 02 85 33.50									
0130	For personal protection equipment, see Div. 02 82 13.41									

02 85 16.50 Preparation of Mold Containment Area

		Crew	Daily Output	Labor-Hours	Unit	Material	Labor	Equipment	Total	Total Incl O&P
0010	**PREPARATION OF MOLD CONTAINMENT AREA**									
0100	Pre-cleaning, HEPA vacuum and wet wipe, flat surfaces	A-9	12000	.005	S.F.	.02	.24		.26	.41
0300	Separation barrier, 2" x 4" @ 16", 1/2" plywood ea. side, 8' high	2 Carp	400	.040		2	1.66		3.66	4.93
0310	12' high		320	.050		2	2.08		4.08	5.60
0320	16' high		200	.080		2	3.32		5.32	7.65
0400	Personnel decontam. chamber, 2" x 4" @ 16", 3/4" ply ea. side		280	.057		3	2.37		5.37	7.20
0450	Waste decontam. chamber, 2" x 4" studs @ 16", 3/4" ply each side		360	.044		3	1.85		4.85	6.35
0500	Cover surfaces with polyethylene sheeting									
0501	Including glue and tape									
0550	Floors, each layer, 6 mil	A-9	8000	.008	S.F.	.06	.36		.42	.66
0551	4 mil		9000	.007		.04	.32		.36	.57
0560	Walls, each layer, 6 mil		6000	.011		.09	.49		.58	.88
0561	4 mil		7000	.009		.07	.42		.49	.75
0570	For heights above 12', add						20%			
0575	For heights above 20', add						30%			
0580	For fire retardant poly, add					100%				
0590	For large open areas, deduct					10%	20%			
0600	Seal floor penetrations with foam firestop to 36 sq. in.	2 Carp	200	.080	Ea.	7.60	3.32		10.92	13.85
0610	36 sq. in. to 72 sq. in.		125	.128		15.25	5.30		20.55	25.50
0615	72 sq. in. to 144 sq. in.		80	.200		30.50	8.30		38.80	47
0620	Wall penetrations, to 36 square inches		180	.089		7.60	3.69		11.29	14.45
0630	36 sq. in. to 72 sq. in.		100	.160		15.25	6.65		21.90	27.50
0640	72 Sq. in. to 144 sq. in.		60	.267		30.50	11.10		41.60	51.50
0800	Caulk seams with latex caulk	1 Carp	230	.035	L.F.	.15	1.45		1.60	2.54
0900	Set up neg. air machine, 1-2k C.F.M. /25 M.C.F. volume	1 Asbe	4.30	1.860	Ea.		84.50		84.50	136

02 85 33 – Removal and Disposal of Materials with Mold

02 85 33.50 Demolition in Mold Contaminated Area

		Crew	Daily Output	Labor-Hours	Unit	Material	Labor	Equipment	Total	Total Incl O&P
0010	**DEMOLITION IN MOLD CONTAMINATED AREA**									
0200	Ceiling, including suspension system, plaster and lath	A-9	2100	.030	S.F.	.09	1.39		1.48	2.33
0210	Finished plaster, leaving wire lath		585	.109		.33	4.99		5.32	8.35
0220	Suspended acoustical tile		3500	.018		.06	.83		.89	1.40
0230	Concealed tile grid system		3000	.021		.06	.97		1.03	1.63
0240	Metal pan grid system		1500	.043		.13	1.95		2.08	3.26
0250	Gypsum board		2500	.026		.08	1.17		1.25	1.96
0255	Plywood		2500	.026		.08	1.17		1.25	1.96
0260	Lighting fixtures up to 2' x 4'		72	.889	Ea.	2.70	40.50		43.20	68
0400	Partitions, non load bearing									
0410	Plaster, lath, and studs	A-9	690	.093	S.F.	.85	4.23		5.08	7.75

02 85 Mold Remediation

02 85 33 – Removal and Disposal of Materials with Mold

02 85 33.50 Demolition in Mold Contaminated Area		Crew	Daily Output	Labor-Hours	Unit	Material	2010 Bare Costs Labor	2010 Bare Costs Equipment	Total	Total Incl O&P
0450	Gypsum board and studs	A-9	1390	.046	S.F.	.14	2.10		2.24	3.52
0465	Carpet & pad		1390	.046	↓	.14	2.10		2.24	3.52
0600	Pipe insulation, air cell type, up to 4" diameter pipe		900	.071	L.F.	.22	3.24		3.46	5.45
0610	4" to 8" diameter pipe		800	.080		.24	3.65		3.89	6.10
0620	10" to 12" diameter pipe		700	.091		.28	4.17		4.45	7
0630	14" to 16" diameter pipe		550	.116	↓	.35	5.30		5.65	8.90
0650	Over 16" diameter pipe	↓	650	.098	S.F.	.30	4.49		4.79	7.55
9000	For type B (supplied air) respirator equipment, add				%					10%

Division Notes

		CREW	DAILY OUTPUT	LABOR-HOURS	UNIT	2010 BARE COSTS				TOTAL INCL O&P
						MAT.	LABOR	EQUIP.	TOTAL	

Estimating Tips

General

- Carefully check all the plans and specifications. Concrete often appears on drawings other than structural drawings, including mechanical and electrical drawings for equipment pads. The cost of cutting and patching is often difficult to estimate. See Subdivision 03 81 for Concrete Cutting, Subdivision 02 41 19.16 for Cutout Demolition, Subdivision 03 05 05.10 for Concrete Demolition, and Subdivision 02 41 19.23 for Rubbish Handling (handling, loading and hauling of debris).
- Always obtain concrete prices from suppliers near the job site. A volume discount can often be negotiated, depending upon competition in the area. Remember to add for waste, particularly for slabs and footings on grade.

03 10 00 Concrete Forming and Accessories

- A primary cost for concrete construction is forming. Most jobs today are constructed with prefabricated forms. The selection of the forms best suited for the job and the total square feet of forms required for efficient concrete forming and placing are key elements in estimating concrete construction. Enough forms must be available for erection to make efficient use of the concrete placing equipment and crew.
- Concrete accessories for forming and placing depend upon the systems used. Study the plans and specifications to ensure that all special accessory requirements have been included in the cost estimate, such as anchor bolts, inserts, and hangers.
- Included within costs for forms-in-place are all necessary bracing and shoring.

03 20 00 Concrete Reinforcing

- Ascertain that the reinforcing steel supplier has included all accessories, cutting, bending, and an allowance for lapping, splicing, and waste. A good rule of thumb is 10% for lapping, splicing, and waste. Also, 10% waste should be allowed for welded wire fabric.
- The unit price items in the subdivision for Reinforcing In Place include the labor to install accessories such as beam and slab bolsters, high chairs, and bar ties and tie wire. The material cost for these accessories is not included; they may be obtained from the Accessories Division.

03 30 00 Cast-In-Place Concrete

- When estimating structural concrete, pay particular attention to requirements for concrete additives, curing methods, and surface treatments. Special consideration for climate, hot or cold, must be included in your estimate. Be sure to include requirements for concrete placing equipment, and concrete finishing.
- For accurate concrete estimating, the estimator must consider each of the following major components individually: forms, reinforcing steel, ready-mix concrete, placement of the concrete, and finishing of the top surface. For faster estimating, Subdivision 03 30 53.40 for Concrete-In-Place can be used; here, various items of concrete work are presented that include the costs of all five major components (unless specifically stated otherwise).

03 40 00 Precast Concrete
03 50 00 Cast Decks and Underlayment

- The cost of hauling precast concrete structural members is often an important factor. For this reason, it is important to get a quote from the nearest supplier. It may become economically feasible to set up precasting beds on the site if the hauling costs are prohibitive.

Reference Numbers

Reference numbers are shown in shaded boxes at the beginning of some major classifications. These numbers refer to related items in the Reference Section. The reference information may be an estimating procedure, an alternate pricing method, or technical information.

Note: Not all subdivisions listed here necessarily appear in this publication.

No part of this publication may be reproduced, stored in a retrieval system, or transmitted in any form or by any means without prior written permission of R.S. Means Company, Inc.

03 01 Maintenance of Concrete

03 01 30 – Maintenance of Cast-In-Place Concrete

03 01 30.62 Concrete Patching		Crew	Daily Output	Labor-Hours	Unit	Material	2010 Bare Costs Labor	Equipment	Total	Total Incl O&P
0010	**CONCRETE PATCHING**									
0100	Floors, 1/4" thick, small areas, regular grout	1 Cefi	170	.047	S.F.	1.18	1.87		3.05	4.22
0150	Epoxy grout	"	100	.080	"	7.20	3.18		10.38	12.85
2000	Walls, including chipping, cleaning and epoxy grout									
2100	1/4" deep	1 Cefi	65	.123	S.F.	7.50	4.89		12.39	15.85
2150	1/2" deep		50	.160		15.05	6.35		21.40	26.50
2200	3/4" deep		40	.200		22.50	7.95		30.45	37.50
9000	Minimum labor/equipment charge		4.50	1.778	Job		70.50		70.50	110

03 05 Common Work Results for Concrete

03 05 05 – Selective Concrete Demolition

03 05 05.10 Selective Demolition, Concrete

			Crew	Daily Output	Labor-Hours	Unit	Material	Labor	Equipment	Total	Total Incl O&P
0010	**SELECTIVE DEMOLITION, CONCRETE**	R024119-10									
0012	Excludes saw cutting, torch cutting, loading or hauling										
0050	Break up into small pieces, minimum reinforcing		B-9	24	1.667	C.Y.		56	9.35	65.35	102
0060	Average reinforcing			16	2.500			84	14.05	98.05	153
0070	Maximum reinforcing			8	5			168	28	196	305
0150	Remove whole pieces, up to 2 tons per piece		E-18	36	1.111	Ea.		51.50	28.50	80	124
0160	2 – 5 tons per piece			30	1.333			62	34.50	96.50	148
0170	5 – 10 tons per piece			24	1.667			77.50	43	120.50	185
0180	10 – 15 tons per piece			18	2.222			103	57	160	247
0250	Precast unit embedded in masonry, up to 1 CF		D-1	16	1			37.50		37.50	60.50
0260	1 – 2 CF			12	1.333			50.50		50.50	81
0270	2 – 5 CF			10	1.600			60.50		60.50	97
0280	5 – 10 CF			8	2			75.50		75.50	121
0990	For hydrodemolition see 02 41 13.15										
1910	Minimum labor/equipment charge		B-9	2	20	Job		670	112	782	1,225

03 05 13 – Basic Concrete Materials

03 05 13.20 Concrete Admixtures and Surface Treatments

		Crew	Daily Output	Labor-Hours	Unit	Material	Labor	Equipment	Total	Total Incl O&P
0010	**CONCRETE ADMIXTURES AND SURFACE TREATMENTS**									
0040	Abrasives, aluminum oxide, over 20 tons				Lb.	1.75			1.75	1.93
0050	1 to 20 tons					1.60			1.60	1.76
0070	Under 1 ton					1.67			1.67	1.84
0100	Silicon carbide, black, over 20 tons					2.44			2.44	2.68
0110	1 to 20 tons					2.57			2.57	2.83
0120	Under 1 ton					2.68			2.68	2.95
0200	Air entraining agent, .7 to 1.5 oz. per bag, 55 gallon drum				Gal.	9.95			9.95	10.95
0220	5 gallon pail					13.75			13.75	15.15
0300	Bonding agent, acrylic latex, 250 S.F. per gallon, 5 gallon pail					19.60			19.60	21.50
0320	Epoxy resin, 80 S.F. per gallon, 3.5 gallon unit					47.50			47.50	52.50
0400	Calcium chloride, 50 lb. bags, TL lots				Ton	580			580	635
0420	Less than truckload lots				Bag	17.95			17.95	19.75
0500	Carbon black, liquid, 2 to 8 lbs. per bag of cement				Lb.	6.45			6.45	7.10
0600	Colors, integral, 2 to 10 lb. per bag of cement, minimum					2.52			2.52	2.77
0610	Average					3.50			3.50	3.85
0620	Maximum					4.88			4.88	5.35
0920	Dustproofing compound, 250 SF/gal, 5 gallon pail				Gal.	7.40			7.40	8.10
1010	Epoxy based, 125 SF/Gal, 5 gallon pail				"	50			50	55
1100	Hardeners, metallic, 55 lb. bags, natural (grey)				Lb.	.85			.85	.93
1200	Colors					1.51			1.51	1.66

03 05 Common Work Results for Concrete

03 05 13 – Basic Concrete Materials

03 05 13.20 Concrete Admixtures and Surface Treatments		Crew	Daily Output	Labor-Hours	Unit	Material	2010 Bare Costs Labor	Equipment	Total	Total Incl O&P
1300	Non-metallic, 55 lb. bags, natural grey				Lb.	.42			.42	.47
1320	Colors					.71			.71	.78
1600	Sealer, hardener and dustproofer, epoxy-based, 125 SF/gal, 5 gallon, min				Gal.	50			50	55
1620	5 gallon pail, maximum					55			55	60.50
1630	Sealer, solvent-based, 250 SF/gal, 55 gallon drum					17.30			17.30	19.05
1640	5 gallon pail					24.50			24.50	27
1650	Sealer, water based, 350 S.F., 55 gallon drum					20.50			20.50	22.50
1660	5 gallon pail					23.50			23.50	26
1900	Set retarder, 100 SF/gallon, 5 gallon pail					19.70			19.70	21.50
2000	Waterproofing, integral 1 lb. per bag of cement				Lb.	1.52			1.52	1.67
2100	Powdered metallic, 40 lbs. per 100 S.F., minimum					1.79			1.79	1.97
2120	Maximum					2.51			2.51	2.76
3000	For integral colors, 2500 psi (5 bag mix)									
3100	Red, yellow or brown, 1.8 lb. per bag, add				C.Y.	22.50			22.50	25
3200	9.4 lb. per bag, add					118			118	130
3400	Black, 1.8 lb. per bag, add					31.50			31.50	34.50
3500	7.5 lb. per bag, add					131			131	144
3700	Green, 1.8 lb. per bag, add					44			44	48.50
3800	7.5 lb. per bag, add					183			183	201

03 05 13.85 Winter Protection		Crew	Daily Output	Labor-Hours	Unit	Material	Labor	Equipment	Total	Total Incl O&P
0010	**WINTER PROTECTION**									
0012	For heated ready mix, add, minimum				C.Y.	4.25			4.25	4.68
0050	Maximum				"	5.30			5.30	5.85
0100	Temporary heat to protect concrete, 24 hours, minimum	2 Clab	50	.320	M.S.F.	520	10.60		530.60	590
0150	Maximum	"	25	.640	"	680	21		701	780
0200	Temporary shelter for slab on grade, wood frame/polyethylene sheeting									
0201	Build or remove, minimum	2 Carp	10	1.600	M.S.F.	215	66.50		281.50	345
0210	Maximum	"	3	5.333	"	261	222		483	650
0710	Electrically, heated pads, 15 watts/S.F., 20 uses, minimum				S.F.	.24			.24	.26
0800	Maximum				"	.39			.39	.43

03 11 Concrete Forming

03 11 13 – Structural Cast-In-Place Concrete Forming

03 11 13.20 Forms In Place, Beams and Girders

			Crew	Daily Output	Labor-Hours	Unit	Material	Labor	Equipment	Total	Total Incl O&P
0010	**FORMS IN PLACE, BEAMS AND GIRDERS**	R031113-40									
0500	Exterior spandrel, job-built plywood, 12" wide, 1 use	R031113-60	C-2	225	.213	SFCA	3.43	8.65		12.08	18
0650	4 use			310	.155		1.12	6.25		7.37	11.55
1000	18" wide, 1 use			250	.192		2.77	7.75		10.52	15.80
1150	4 use			315	.152		.91	6.15		7.06	11.15
1500	24" wide, 1 use			265	.181		2.54	7.35		9.89	14.85
1650	4 use			325	.148		.83	6		6.83	10.70
2000	Interior beam, job-built plywood, 12" wide, 1 use			300	.160		4.03	6.50		10.53	15.10
2150	4 use			377	.127		1.31	5.15		6.46	9.90
2500	24" wide, 1 use			320	.150		2.57	6.05		8.62	12.80
2650	4 use			395	.122		.83	4.92		5.75	9
3000	Encasing steel beam, hung, job-built plywood, 1 use			325	.148		2.87	6		8.87	12.95
3150	4 use			430	.112		.93	4.52		5.45	8.45
9000	Minimum labor/equipment charge		2 Carp	2	8	Job		330		330	545

03 11 Concrete Forming

03 11 13 – Structural Cast-In-Place Concrete Forming

03 11 13.25 Forms In Place, Columns

			Crew	Daily Output	Labor-Hours	Unit	Material	2010 Bare Costs Labor	Equipment	Total	Total Incl O&P
0010	**FORMS IN PLACE, COLUMNS**	R031113-40									
0500	Round fiberglass, 4 use per mo., rent, 12" diameter		C-1	160	.200	L.F.	8.15	7.90		16.05	22
0550	16" diameter	R031113-60		150	.213		9.70	8.40		18.10	24.50
0600	18" diameter			140	.229		10.85	9		19.85	27
0650	24" diameter			135	.237		13.50	9.35		22.85	30
0700	28" diameter			130	.246		15.10	9.70		24.80	32.50
0800	30" diameter			125	.256		15.75	10.10		25.85	34
0850	36" diameter			120	.267		21	10.50		31.50	40.50
1500	Round fiber tube, recycled paper, 1 use, 8" diameter	G		155	.206		1.44	8.15		9.59	14.95
1550	10" diameter	G		155	.206		1.99	8.15		10.14	15.55
1600	12" diameter	G		150	.213		2.33	8.40		10.73	16.35
1650	14" diameter	G		145	.221		3.31	8.70		12.01	17.95
1700	16" diameter	G		140	.229		3.99	9		12.99	19.20
1720	18" diameter	G		140	.229		4.67	9		13.67	19.95
1750	20" diameter	G		135	.237		6.45	9.35		15.80	22.50
1800	24" diameter	G		130	.246		8.15	9.70		17.85	25
1850	30" diameter	G		125	.256		11.85	10.10		21.95	29.50
1900	36" diameter	G		115	.278		14.90	10.95		25.85	34.50
1950	42" diameter	G		100	.320		37	12.60		49.60	61
2000	48" diameter	G		85	.376		41.50	14.85		56.35	70.50
2200	For seamless type, add						15%				
3000	Round, steel, 4 use per mo., rent, regular duty, 14" diameter	G	C-1	145	.221	L.F.	15.15	8.70		23.85	31
3050	16" diameter	G		125	.256		15.50	10.10		25.60	33.50
3100	Heavy duty, 20" diameter	G		105	.305		17	12		29	38.50
3150	24" diameter	G		85	.376		18.65	14.85		33.50	45
3200	30" diameter	G		70	.457		21.50	18.05		39.55	53
3250	36" diameter	G		60	.533		23	21		44	60
3300	48" diameter	G		50	.640		68	25		93	117
3350	60" diameter	G		45	.711		42	28		70	92
4500	For second and succeeding months, deduct						50%				
5000	Job-built plywood, 8" x 8" columns, 1 use		C-1	165	.194	SFCA	2.27	7.65		9.92	15.05
5050	2 use			195	.164		1.30	6.45		7.75	12.05
5100	3 use			210	.152		.91	6		6.91	10.85
5150	4 use			215	.149		.75	5.85		6.60	10.45
5500	12" x 12" columns, 1 use			180	.178		2.11	7		9.11	13.80
5550	2 use			210	.152		1.16	6		7.16	11.15
5600	3 use			220	.145		.84	5.75		6.59	10.35
5650	4 use			225	.142		.69	5.60		6.29	9.95
6000	16" x 16" columns, 1 use			185	.173		2.06	6.80		8.86	13.45
6050	2 use			215	.149		1.11	5.85		6.96	10.85
6100	3 use			230	.139		.83	5.50		6.33	9.90
6150	4 use			235	.136		.67	5.35		6.02	9.55
6500	24" x 24" columns, 1 use			190	.168		2.28	6.65		8.93	13.40
6550	2 use			216	.148		1.25	5.85		7.10	11
6600	3 use			230	.139		.91	5.50		6.41	10
6650	4 use			238	.134		.74	5.30		6.04	9.50
7000	36" x 36" columns, 1 use			200	.160		1.65	6.30		7.95	12.15
7050	2 use			230	.139		.94	5.50		6.44	10.05
7100	3 use			245	.131		.66	5.15		5.81	9.20
7150	4 use			250	.128		.54	5.05		5.59	8.90
7400	Steel framed plywood, based on 50 uses of purchased										
7420	forms, and 4 uses of bracing lumber										

03 11 Concrete Forming

03 11 13 – Structural Cast-In-Place Concrete Forming

03 11 13.25 Forms In Place, Columns

		Crew	Daily Output	Labor-Hours	Unit	Material	2010 Bare Costs Labor	Equipment	Total	Total Incl O&P
7500	8" x 8" column	C-1	340	.094	SFCA	4.51	3.71		8.22	11.05
7550	10" x 10"		350	.091		3.97	3.61		7.58	10.25
7600	12" x 12"		370	.086		3.36	3.41		6.77	9.30
7650	16" x 16"		400	.080		2.62	3.16		5.78	8.10
7700	20" x 20"		420	.076		2.34	3		5.34	7.50
7750	24" x 24"		440	.073		1.66	2.87		4.53	6.55
7755	30" x 30"		440	.073		2.13	2.87		5	7.05
7760	36" x 36"		460	.070		1.90	2.74		4.64	6.60
9000	Minimum labor/equipment charge	2 Carp	2	8	Job		330		330	545

03 11 13.35 Forms In Place, Elevated Slabs

			Crew	Daily Output	Labor-Hours	Unit	Material	Labor	Equipment	Total	Total Incl O&P
0010	**FORMS IN PLACE, ELEVATED SLABS**	R031113-40									
1000	Flat plate, job-built plywood, to 15' high, 1 use		C-2	470	.102	S.F.	4.06	4.13		8.19	11.25
1150	4 use	R031113-60		560	.086		1.32	3.47		4.79	7.15
1500	15' to 20' high ceilings, 4 use			495	.097		1.53	3.93		5.46	8.15
2000	Flat slab, drop panels, job-built plywood, to 15' high, 1 use			449	.107		4.43	4.33		8.76	11.95
2150	4 use			544	.088		1.44	3.57		5.01	7.45
2250	15' to 20' high ceilings, 4 use			480	.100		5	4.05		9.05	12.20
3500	Floor slab, with 1-way joist pans, 1 use			415	.116		6.20	4.68		10.88	14.50
3650	4 use			500	.096		3.19	3.89		7.08	9.90
4500	With 2-way waffle domes, 1 use			405	.119		6.30	4.80		11.10	14.85
4550	4 use			470	.102		3.32	4.13		7.45	10.45
5000	Box out for slab openings, over 16" deep, 1 use			190	.253	SFCA	3.60	10.25		13.85	21
5050	2 use			240	.200	"	1.98	8.10		10.08	15.45
5500	Shallow slab box outs, to 10 S.F.			42	1.143	Ea.	7.20	46.50		53.70	84
5550	Over 10 S.F. (use perimeter)			600	.080	L.F.	.96	3.24		4.20	6.35
6000	Bulkhead forms for slab, with keyway, 1 use, 2 piece			500	.096		1.67	3.89		5.56	8.25
6100	3 piece (see also edge forms)			460	.104		1.79	4.22		6.01	8.90
6200	Slab bulkhead form, 4-1/2" high, exp metal, w/keyway & stakes	G	C-1	1200	.027		1.94	1.05		2.99	3.86
6210	5-1/2" high	G		1100	.029		2.27	1.15		3.42	4.38
6215	7-1/2" high	G		960	.033		2.99	1.31		4.30	5.45
6220	9-1/2" high	G		840	.038		3.39	1.50		4.89	6.20
6500	Curb forms, wood, 6" to 12" high, on elevated slabs, 1 use			180	.178	SFCA	1.03	7		8.03	12.65
6550	2 use			205	.156		.56	6.15		6.71	10.70
6600	3 use			220	.145		.41	5.75		6.16	9.85
6650	4 use			225	.142		.33	5.60		5.93	9.55
7000	Edge forms to 6" high, on elevated slab, 4 use			500	.064	L.F.	.12	2.52		2.64	4.28
7500	Depressed area forms to 12" high, 4 use			300	.107		.80	4.21		5.01	7.80
7550	12" to 24" high, 4 use			175	.183		1.09	7.20		8.29	13.05
8000	Perimeter deck and rail for elevated slabs, straight			90	.356		12.40	14		26.40	36.50
8050	Curved			65	.492		17	19.40		36.40	50.50
8500	Void forms, round fiber, 3" diameter	G		450	.071		.97	2.80		3.77	5.70
8650	8" diameter	G		375	.085		3.85	3.37		7.22	9.80
9000	Minimum labor/equipment charge		2 Carp	2	8	Job		330		330	545

03 11 13.40 Forms In Place, Equipment Foundations

			Crew	Daily Output	Labor-Hours	Unit	Material	Labor	Equipment	Total	Total Incl O&P
0010	**FORMS IN PLACE, EQUIPMENT FOUNDATIONS**	R031113-40									
0020	1 use		C-2	160	.300	SFCA	2.58	12.15		14.73	23
0050	2 use	R031113-60		190	.253		1.42	10.25		11.67	18.35
0100	3 use			200	.240		1.03	9.70		10.73	17.10
0150	4 use			205	.234		.84	9.50		10.34	16.50
9000	Minimum labor/equipment charge		1 Carp	3	2.667	Job		111		111	182

03 11 Concrete Forming

03 11 13 – Structural Cast-In-Place Concrete Forming

03 11 13.45 Forms In Place, Footings

			Crew	Daily Output	Labor-Hours	Unit	Material	2010 Bare Costs Labor	Equipment	Total	Total Incl O&P
0010	**FORMS IN PLACE, FOOTINGS**	R031113-40									
0020	Continuous wall, plywood, 1 use		C-1	375	.085	SFCA	5.25	3.37		8.62	11.30
0150	4 use	R031113-60	"	485	.066	"	1.70	2.60		4.30	6.15
1500	Keyway, 4 use, tapered wood, 2" x 4"		1 Carp	530	.015	L.F.	.13	.63		.76	1.18
1550	2" x 6"		"	500	.016	"	.19	.66		.85	1.30
3000	Pile cap, square or rectangular, job-built plywood, 1 use		C-1	290	.110	SFCA	2.15	4.35		6.50	9.50
3150	4 use			383	.084		.70	3.30		4	6.15
5000	Spread footings, job-built lumber, 1 use			305	.105		1.54	4.14		5.68	8.50
5150	4 use			414	.077		.50	3.05		3.55	5.55
9000	Minimum labor/equipment charge		1 Carp	3	2.667	Job		111		111	182

03 11 13.47 Forms In Place, Gas Station Forms

			Crew	Daily Output	Labor-Hours	Unit	Material	Labor	Equipment	Total	Total Incl O&P
0010	**FORMS IN PLACE, GAS STATION FORMS**										
0050	Curb fascia, with template, 12 ga. steel, left in place, 9" high	G	1 Carp	50	.160	L.F.	11.55	6.65		18.20	23.50
1000	Sign or light bases, 18" diameter, 9" high	G		9	.889	Ea.	72.50	37		109.50	141
1050	30" diameter, 13" high	G		8	1	"	115	41.50		156.50	196
1990	Minimum labor/equipment charge			2	4	Job		166		166	273
2000	Island forms, 10' long, 9" high, 3'-6" wide	G	C-1	10	3.200	Ea.	320	126		446	560
2050	4' wide	G		9	3.556		335	140		475	595
2500	20' long, 9" high, 4' wide	G		6	5.333		535	210		745	935
2550	5' wide	G		5	6.400		555	252		807	1,025
9000	Minimum labor/equipment charge		1 Carp	3	2.667	Job		111		111	182

03 11 13.50 Forms In Place, Grade Beam

			Crew	Daily Output	Labor-Hours	Unit	Material	Labor	Equipment	Total	Total Incl O&P
0010	**FORMS IN PLACE, GRADE BEAM**	R031113-40									
0020	Job-built plywood, 1 use		C-2	530	.091	SFCA	2.30	3.67		5.97	8.55
0150	4 use	R031113-60	"	605	.079	"	.75	3.21		3.96	6.05
9000	Minimum labor/equipment charge		2 Carp	2	8	Job		330		330	545

03 11 13.65 Forms In Place, Slab On Grade

			Crew	Daily Output	Labor-Hours	Unit	Material	Labor	Equipment	Total	Total Incl O&P
0010	**FORMS IN PLACE, SLAB ON GRADE**	R031113-40									
1000	Bulkhead forms w/keyway, wood, 6" high, 1 use	R031113-60	C-1	510	.063	L.F.	.64	2.47		3.11	4.77
1400	Bulkhead form for slab, 4-1/2" high, exp metal, incl keyway & stakes	G		1200	.027		1.94	1.05		2.99	3.86
1410	5-1/2" high	G		1100	.029		2.27	1.15		3.42	4.38
1420	7-1/2" high	G		960	.033		2.99	1.31		4.30	5.45
1430	9-1/2" high	G		840	.038		3.39	1.50		4.89	6.20
2000	Curb forms, wood, 6" to 12" high, on grade, 1 use			215	.149	SFCA	2.21	5.85		8.06	12.10
2150	4 use			275	.116	"	.72	4.59		5.31	8.35
3000	Edge forms, wood, 4 use, on grade, to 6" high			600	.053	L.F.	.27	2.10		2.37	3.75
3050	7" to 12" high			435	.074	SFCA	.59	2.90		3.49	5.40
3500	For depressed slabs, 4 use, to 12" high			300	.107	L.F.	.46	4.21		4.67	7.40
3550	To 24" high			175	.183		.60	7.20		7.80	12.50
4000	For slab blockouts, to 12" high, 1 use			200	.160		.49	6.30		6.79	10.90
4050	To 24" high, 1 use			120	.267		.62	10.50		11.12	17.95
4100	Plastic (extruded), to 6" high, multiple use, on grade			800	.040		7.90	1.58		9.48	11.30
5000	Screed, 24 ga. metal key joint, see Div. 03 15 05.25										
5020	Wood, incl. wood stakes, 1" x 3"		C-1	900	.036	L.F.	.68	1.40		2.08	3.05
5050	2" x 4"			900	.036	"	.59	1.40		1.99	2.95
6000	Trench forms in floor, wood, 1 use			160	.200	SFCA	1.13	7.90		9.03	14.20
6150	4 use			185	.173	"	.37	6.80		7.17	11.60
8760	Void form, corrugated fiberboard, 6" x 12", 10' long	G		240	.133	S.F.	.95	5.25		6.20	9.70
9000	Minimum labor/equipment charge		1 Carp	2	4	Job		166		166	273

03 11 Concrete Forming

03 11 13 – Structural Cast-In-Place Concrete Forming

03 11 13.85 Forms In Place, Walls

			Crew	Daily Output	Labor-Hours	Unit	Material	2010 Bare Costs Labor	2010 Bare Costs Equipment	Total	Total Incl O&P
0010	**FORMS IN PLACE, WALLS**	R031113-10									
0100	Box out for wall openings, to 16" thick, to 10 S.F.		C-2	24	2	Ea.	16.85	81		97.85	152
0150	Over 10 S.F. (use perimeter)	R031113-40	"	280	.171	L.F.	1.44	6.95		8.39	13
0250	Brick shelf, 4" w, add to wall forms, use wall area abv shelf										
0260	1 use	R031113-60	C-2	240	.200	SFCA	1.53	8.10		9.63	15
0350	4 use			300	.160	"	.61	6.50		7.11	11.30
0500	Bulkhead, wood with keyway, 1 use, 2 piece			265	.181	L.F.	1.31	7.35		8.66	13.50
0600	Bulkhead forms with keyway, 1 piece expanded metal, 8" wall	G	C-1	1000	.032		2.99	1.26		4.25	5.35
0610	10" wall	G		800	.040		3.39	1.58		4.97	6.30
0620	12" wall	G		525	.061		4.07	2.40		6.47	8.40
0700	Buttress, to 8' high, 1 use		C-2	350	.137	SFCA	5.25	5.55		10.80	14.90
0850	4 use			480	.100	"	1.74	4.05		5.79	8.55
1000	Corbel or haunch, to 12" wide, add to wall forms, 1 use			150	.320	L.F.	1.74	12.95		14.69	23.50
1150	4 use			180	.267	"	.57	10.80		11.37	18.35
2000	Wall, job-built plywood, to 8' high, 1 use			370	.130	SFCA	2.24	5.25		7.49	11.05
2050	2 use			435	.110		1.38	4.47		5.85	8.85
2100	3 use			495	.097		1	3.93		4.93	7.55
2150	4 use			505	.095		.81	3.85		4.66	7.20
2400	Over 8' to 16' high, 1 use			280	.171		2.51	6.95		9.46	14.15
2450	2 use			345	.139		1	5.65		6.65	10.35
2500	3 use			375	.128		.72	5.20		5.92	9.30
2550	4 use			395	.122		.59	4.92		5.51	8.75
2700	Over 16' high, 1 use			235	.204		2.05	8.25		10.30	15.85
2750	2 use			290	.166		1.13	6.70		7.83	12.25
2800	3 use			315	.152		.82	6.15		6.97	11.05
2850	4 use			330	.145		.67	5.90		6.57	10.40
3000	For architectural finish, add			1820	.026		5.45	1.07		6.52	7.75
4000	Radial, smooth curved, job-built plywood, 1 use			245	.196		1.84	7.95		9.79	15
4150	4 use			335	.143		.60	5.80		6.40	10.20
4200	Below grade, job-built plywood, 1 use			225	.213		1.89	8.65		10.54	16.30
4210	2 use			225	.213		1.04	8.65		9.69	15.35
4220	3 use			225	.213		.85	8.65		9.50	15.15
4230	4 use			225	.213		.62	8.65		9.27	14.90
4600	Retaining wall, battered, job-built plyw'd, to 8' high, 1 use			300	.160		1.49	6.50		7.99	12.30
4750	4 use			390	.123		.49	4.98		5.47	8.75
4900	Over 8' to 16' high, 1 use			240	.200		1.61	8.10		9.71	15.05
5050	4 use			320	.150		.52	6.05		6.57	10.50
5750	Liners for forms (add to wall forms), A.B.S. plastic										
5800	Aged wood, 4" wide, 1 use		1 Carp	256	.031	SFCA	3.79	1.30		5.09	6.30
5820	2 use			256	.031		2.08	1.30		3.38	4.42
5830	3 use			256	.031		1.52	1.30		2.82	3.80
5840	4 use			256	.031		1.23	1.30		2.53	3.48
5900	Fractured rope rib, 1 use			192	.042		5.45	1.73		7.18	8.85
5925	2 use			192	.042		3.01	1.73		4.74	6.15
5950	3 use			192	.042		2.19	1.73		3.92	5.25
6000	4 use			192	.042		1.78	1.73		3.51	4.80
6100	Ribbed, 3/4" deep x 1-1/2" O.C., 1 use			224	.036		5.45	1.48		6.93	8.45
6125	2 use			224	.036		3.01	1.48		4.49	5.75
6150	3 use			224	.036		2.19	1.48		3.67	4.85
6200	4 use			224	.036		1.78	1.48		3.26	4.40
6300	Rustic brick pattern, 1 use			224	.036		3.79	1.48		5.27	6.60
6325	2 use			224	.036		2.08	1.48		3.56	4.73

03 11 Concrete Forming

03 11 13 – Structural Cast-In-Place Concrete Forming

03 11 13.85 Forms In Place, Walls		Crew	Daily Output	Labor-Hours	Unit	Material	2010 Bare Costs Labor	Equipment	Total	Total Incl O&P
6350	3 use	1 Carp	224	.036	SFCA	1.52	1.48		3	4.11
6400	4 use		224	.036		1.23	1.48		2.71	3.79
6500	3/8" striated, random, 1 use		224	.036		3.79	1.48		5.27	6.60
6525	2 use		224	.036		2.08	1.48		3.56	4.73
6550	3 use		224	.036		1.52	1.48		3	4.11
6600	4 use		224	.036		1.23	1.48		2.71	3.79
7500	Lintel or sill forms, 1 use		30	.267		2.29	11.10		13.39	20.50
7560	4 use		37	.216		.74	9		9.74	15.55
7800	Modular prefabricated plywood, based on 20 uses of purchased									
7820	forms, and 4 uses of bracing lumber									
7860	To 8' high	C-2	800	.060	SFCA	1.44	2.43		3.87	5.60
8060	Over 8' to 16' high		600	.080		1.47	3.24		4.71	6.90
8600	Pilasters, 1 use		270	.178		2.60	7.20		9.80	14.65
8660	4 use		385	.125		.85	5.05		5.90	9.25
9475	For elevated walls, add						10%			
9480	For battered walls, 1 side battered, add					10%	10%			
9485	For battered walls, 2 sides battered, add					15%	15%			
9900	Minimum labor/equipment charge	2 Carp	2	8	Job		330		330	545

03 15 Concrete Accessories

03 15 05 – Concrete Forming Accessories

03 15 05.25 Expansion Joints

			Crew	Daily Output	Labor-Hours	Unit	Material	Labor	Equipment	Total	Total Incl O&P
0010	**EXPANSION JOINTS**										
0020	Keyed, cold, 24 ga., incl. stakes, 3-1/2" high	G	1 Carp	200	.040	L.F.	1.71	1.66		3.37	4.61
0050	4-1/2" high	G		200	.040		1.94	1.66		3.60	4.86
0100	5-1/2" high	G		195	.041		2.27	1.70		3.97	5.30
0150	7-1/2" high	G		190	.042		2.99	1.75		4.74	6.15
0160	9-1/2" high	G		185	.043		3.39	1.80		5.19	6.70
0300	Poured asphalt, plain, 1/2" x 1"		1 Clab	450	.018		.66	.59		1.25	1.70
0350	1" x 2"			400	.020		2.64	.66		3.30	3.99
0500	Neoprene, liquid, cold applied, 1/2" x 1"			450	.018		2.22	.59		2.81	3.41
0550	1" x 2"			400	.020		8.85	.66		9.51	10.85
0700	Polyurethane, poured, 2 part, 1/2" x 1"			400	.020		1.30	.66		1.96	2.52
0750	1" x 2"			350	.023		5.20	.76		5.96	6.95
0900	Rubberized asphalt, hot or cold applied, 1/2" x 1"			450	.018		.36	.59		.95	1.37
0950	1" x 2"			400	.020		1.43	.66		2.09	2.66
1100	Hot applied, fuel resistant, 1/2" x 1"			450	.018		.54	.59		1.13	1.56
1150	1" x 2"			400	.020		2.15	.66		2.81	3.45
2000	Premolded, bituminous fiber, 1/2" x 6"		1 Carp	375	.021		.41	.89		1.30	1.91
2050	1" x 12"			300	.027		1.91	1.11		3.02	3.92
2140	Concrete expansion joint, recycled paper and fiber, 1/2" x 6"	G		390	.021		.36	.85		1.21	1.80
2150	1/2" x 12"	G		360	.022		.72	.92		1.64	2.32
2250	Cork with resin binder, 1/2" x 6"			375	.021		1.82	.89		2.71	3.46
2300	1" x 12"			300	.027		7.55	1.11		8.66	10.10
2500	Neoprene sponge, closed cell, 1/2" x 6"			375	.021		1.93	.89		2.82	3.58
2550	1" x 12"			300	.027		7.10	1.11		8.21	9.60
2750	Polyethylene foam, 1/2" x 6"			375	.021		.66	.89		1.55	2.19
2800	1" x 12"			300	.027		2.15	1.11		3.26	4.19
3000	Polyethylene backer rod, 3/8" diameter			460	.017		.05	.72		.77	1.25
3050	3/4" diameter			460	.017		.10	.72		.82	1.30
3100	1" diameter			460	.017		.19	.72		.91	1.40

03 15 Concrete Accessories

03 15 05 – Concrete Forming Accessories

03 15 05.25 Expansion Joints

		Crew	Daily Output	Labor-Hours	Unit	Material	2010 Bare Costs Labor	Equipment	Total	Total Incl O&P
3500	Polyurethane foam, with polybutylene, 1/2" x 1/2"	1 Carp	475	.017	L.F.	.95	.70		1.65	2.20
3550	1" x 1"		450	.018		2.40	.74		3.14	3.85
3750	Polyurethane foam, regular, closed cell, 1/2" x 6"		375	.021		.70	.89		1.59	2.23
3800	1" x 12"		300	.027		2.50	1.11		3.61	4.57
4000	Polyvinyl chloride foam, closed cell, 1/2" x 6"		375	.021		1.90	.89		2.79	3.55
4050	1" x 12"		300	.027		6.55	1.11		7.66	9
4250	Rubber, gray sponge, 1/2" x 6"		375	.021		1.60	.89		2.49	3.22
4300	1" x 12"	▼	300	.027	▼	5.75	1.11		6.86	8.15
4500	Lead wool for joints				Lb.	10			10	11
5000	For installation in walls, add						75%			
5250	For installation in boxouts, add						25%			

03 15 05.70 Shores

			Crew	Daily Output	Labor-Hours	Unit	Material	Labor	Equipment	Total	Total Incl O&P
0010	**SHORES**										
0020	Erect and strip, by hand, horizontal members										
0500	Aluminum joists and stringers	G	2 Carp	60	.267	Ea.		11.10		11.10	18.20
0600	Steel, adjustable beams	G		45	.356			14.75		14.75	24.50
0700	Wood joists			50	.320			13.30		13.30	22
0800	Wood stringers			30	.533			22		22	36.50
1000	Vertical members to 10' high	G		55	.291			12.10		12.10	19.85
1050	To 13' high	G		50	.320			13.30		13.30	22
1100	To 16' high	G		45	.356	▼		14.75		14.75	24.50
1500	Reshoring	G	▼	1400	.011	S.F.	.50	.47		.97	1.33
1600	Flying truss system	G	C-17D	9600	.009	SFCA		.38	.08	.46	.70
1760	Horizontal, aluminum joists, 6-1/4" high x 5' to 21' span, buy	G				L.F.	36			36	39.50
1770	Beams, 7-1/4" high x 4' to 30' span	G				"	47.50			47.50	52.50
1810	Horizontal, steel beam, adjustable, 4' to 7' span, buy	G				Ea.	330			330	360
1830	6' to 10' span	G					330			330	360
1920	9' to 15' span	G					545			545	600
1940	12' to 20' span	G				▼	650			650	715
1970	Steel stringer, W8x10, 4' to 16' span, buy	G				L.F.	25.50			25.50	28
3000	Rent for job duration, aluminum joist @ 2' O.C., per mo	G				SF Flr.	.90			.90	.99
3050	Steel W8x10	G					.64			.64	.70
3060	Steel adjustable	G				▼	.82			.82	.90
3500	#1 post shore, steel, 5'-7" to 9'-6" high, 10000# cap., buy	G				Ea.	410			410	450
3550	#2 post shore, 7'-3" to 12'-10" high, 7800# capacity	G					465			465	515
3600	#3 post shore, 8'-10" to 16'-1" high, 3800# capacity	G					540			540	595
5010	Frame shoring systems, steel, 12000#/leg, buy										
5040	Frame, 2' wide x 6' high	G				Ea.	106			106	117
5250	X-brace	G					18			18	19.80
5550	Base plate	G					17.60			17.60	19.35
5600	Screw jack	G					38			38	41.50
5650	U-head, 8" x 8"	G				▼	21.50			21.50	24

03 15 05.85 Stair Tread Inserts

			Crew	Daily Output	Labor-Hours	Unit	Material	Labor	Equipment	Total	Total Incl O&P
0010	**STAIR TREAD INSERTS**										
0015	Cast iron, abrasive, 3" wide	G	1 Carp	90	.089	L.F.	10.10	3.69		13.79	17.15
0020	4" wide	G		80	.100		13.45	4.16		17.61	21.50
0040	6" wide	G		75	.107		20	4.43		24.43	29.50
0050	9" wide	G		70	.114		30.50	4.75		35.25	41.50
0100	12" wide	G	▼	65	.123		40.50	5.10		45.60	53
0300	Cast aluminum, compared to cast iron, deduct						10%				
0500	Extruded aluminum safety tread, 3" wide	G	1 Carp	75	.107		11.50	4.43		15.93	19.95
0550	4" wide	G	▼	75	.107	▼	15.35	4.43		19.78	24

03 15 Concrete Accessories

03 15 05 – Concrete Forming Accessories

03 15 05.85 Stair Tread Inserts

		Crew	Daily Output	Labor-Hours	Unit	Material	2010 Bare Costs Labor	Equipment	Total	Total Incl O&P
0600	6" wide	G 1 Carp	75	.107	L.F.	23	4.43		27.43	33
0650	9" wide to resurface stairs	G ↓	70	.114	↓	34.50	4.75		39.25	46
1700	Cement fill for pan-type metal treads, plain	1 Cefi	115	.070	S.F.	1.80	2.76		4.56	6.30
1750	Non-slip	"	100	.080	"	1.98	3.18		5.16	7.15

03 15 13 – Waterstops

03 15 13.50 Waterstops

		Crew	Daily Output	Labor-Hours	Unit	Material	Labor	Equipment	Total	Total Incl O&P
0010	**WATERSTOPS**, PVC and Rubber									
0020	PVC, ribbed 3/16" thick, 4" wide	1 Carp	155	.052	L.F.	1.08	2.14		3.22	4.71
0050	6" wide		145	.055		1.98	2.29		4.27	5.95
0500	With center bulb, 6" wide, 3/16" thick		135	.059		1.75	2.46		4.21	5.95
0550	3/8" thick		130	.062		3.16	2.56		5.72	7.70
0600	9" wide x 3/8" thick	↓	125	.064	↓	4.84	2.66		7.50	9.65

03 21 Reinforcing Steel

03 21 10 – Uncoated Reinforcing Steel

03 21 10.60 Reinforcing In Place

			Crew	Daily Output	Labor-Hours	Unit	Material	Labor	Equipment	Total	Total Incl O&P
0015	**REINFORCING IN PLACE**, 50-60 ton lots, A615 Grade 60	R032110-10									
0020	Includes labor, but not material cost, to install accessories										
0030	Made from recycled materials	G									
0102	Beams & Girders, #3 to #7	G	4 Rodm	3200	.010	Lb.	.42	.47		.89	1.25
0152	#8 to #18	G		5400	.006		.42	.28		.70	.93
0202	Columns, #3 to #7	G		3000	.011		.42	.50		.92	1.30
0252	#8 to #18	G		4600	.007		.42	.33		.75	1.01
0402	Elevated slabs, #4 to #7	G		5800	.006		.44	.26		.70	.91
0502	Footings, #4 to #7	G		4200	.008		.42	.36		.78	1.06
0552	#8 to #18	G		7200	.004		.42	.21		.63	.81
0602	Slab on grade, #3 to #7	G		4200	.008		.40	.36		.76	1.04
0702	Walls, #3 to #7	G		6000	.005		.42	.25		.67	.88
0752	#8 to #18	G	↓	8000	.004	↓	.42	.19		.61	.77
0900	For other than 50-60 ton lots										
1000	Under 10 ton job, #3 to #7, add					Ton	25%	10%			
1010	#8 to #18, add						20%	10%			
1050	10 – 50 ton job, #3 to #7, add						10%				
1060	#8 to #18, add						5%				
1100	60 – 100 ton job, #3 - #7, deduct						5%				
1110	#8 to #18, deduct						10%				
1150	Over 100 ton job, #3 - #7, deduct						10%				
1160	#8 - #18, deduct						15%				
2000	Unloading & sorting, add to above		C-5	100	.560			25.50	7.50	33	50.50
2200	Crane cost for handling, add to above, minimum			135	.415			18.90	5.55	24.45	37
2210	Average			92	.609			27.50	8.15	35.65	55
2220	Maximum		↓	35	1.600	↓		73	21.50	94.50	144
2400	Dowels, 2 feet long, deformed, #3	G	2 Rodm	520	.031	Ea.	.33	1.44		1.77	2.78
2410	#4	G		480	.033		.59	1.56		2.15	3.27
2420	#5	G		435	.037		.92	1.72		2.64	3.90
2430	#6	G		360	.044		1.32	2.08		3.40	4.95
2450	Longer and heavier dowels, add	G		725	.022	Lb.	.44	1.03		1.47	2.22
2500	Smooth dowels, 12" long, 1/4" or 3/8" diameter	G		140	.114	Ea.	.74	5.35		6.09	9.80
2520	5/8" diameter	G		125	.128		1.29	6		7.29	11.45
2530	3/4" diameter	G	↓	110	.145	↓	1.60	6.80		8.40	13.20

03 21 Reinforcing Steel

03 21 10 – Uncoated Reinforcing Steel

03 21 10.60 Reinforcing In Place

		Crew	Daily Output	Labor-Hours	Unit	Material	2010 Bare Costs Labor	Equipment	Total	Total Incl O&P
2600	Dowel sleeves for CIP concrete, 2-part system									
2610	Sleeve base, plastic, for 5/8" smooth dowel sleeve, fasten to edge form	1 Rodm	200	.040	Ea.	.49	1.87		2.36	3.69
2615	Sleeve, plastic, 12" long, for 5/8" smooth dowel sleeve, snap onto base		400	.020		1.06	.94		2	2.74
2620	Sleeve base, for 3/4" smooth dowel sleeve		175	.046		.52	2.14		2.66	4.17
2625	Sleeve, 12" long, for 3/4" smooth dowel		350	.023		1.59	1.07		2.66	3.55
2630	Sleeve base, for 1" smooth dowel sleeve		150	.053		.63	2.50		3.13	4.88
2635	Sleeve, 12" long, for 1" smooth dowel		300	.027		2.26	1.25		3.51	4.59
2700	Dowel caps, visual warning only, plastic, #3 to #8	2 Rodm	800	.020		.42	.94		1.36	2.03
2720	#8 to #18		750	.021		1.05	1		2.05	2.84
2750	Impalement protective, plastic, #4 to #9		800	.020		1.62	.94		2.56	3.35
9000	Minimum labor/equipment charge	1 Rodm	4	2	Job		93.50		93.50	157

03 21 10.70 Glass Fiber Reinforced Polymer Bars

		Crew	Daily Output	Labor-Hours	Unit	Material	Labor	Equipment	Total	Total Incl O&P
0010	**GLASS FIBER REINFORCED POLYMER BARS**									
0050	#2 bar, .043 lbs/ft.	4 Rodm	9500	.003	L.F.	.35	.16		.51	.66
0100	#3 bar, .092 lbs/ft.		9300	.003		.48	.16		.64	.80
0150	#4 bar, .160 lbs/ft.		9100	.004		.70	.16		.86	1.05
0200	#5 bar, .258 lbs/ft.		8700	.004		.98	.17		1.15	1.37
0250	#6 bar, .372 lbs/ft.		8300	.004		1.35	.18		1.53	1.79
0300	#7 bar, .497 lbs/ft.		7900	.004		1.70	.19		1.89	2.19
0350	#8 bar, .620 lbs/ft.		7400	.004		2.25	.20		2.45	2.82
0400	#9 bar, .800 lbs/ft.		6800	.005		2.90	.22		3.12	3.56
0450	#10 bar, 1.08 lbs/ft.		5800	.006		3.45	.26		3.71	4.23
0500	For Bends, add per bend				Ea.	1			1	1.10

03 22 Welded Wire Fabric Reinforcing

03 22 05 – Uncoated Welded Wire Fabric

03 22 05.50 Welded Wire Fabric

			Crew	Daily Output	Labor-Hours	Unit	Material	Labor	Equipment	Total	Total Incl O&P
0010	**WELDED WIRE FABRIC** ASTM A185	R032205-30									
0030	Made from recycled materials	G									
0050	Sheets										
0100	6 x 6 - W1.4 x W1.4 (10 x 10) 21 lb. per C.S.F.	G	2 Rodm	35	.457	C.S.F.	12	21.50		33.50	49
0200	6 x 6 - W2.1 x W2.1 (8 x 8) 30 lb. per C.S.F.	G		31	.516		19.25	24		43.25	61.50
0300	6 x 6 - W2.9 x W2.9 (6 x 6) 42 lb. per C.S.F.	G		29	.552		24	26		50	70
0400	6 x 6 - W4 x W4 (4 x 4) 58 lb. per C.S.F.	G		27	.593		32	27.50		59.50	82
0500	4 x 4 - W1.4 x W1.4 (10 x 10) 31 lb. per C.S.F.	G		31	.516		17.75	24		41.75	60
0600	4 x 4 - W2.1 x W2.1 (8 x 8) 44 lb. per C.S.F.	G		29	.552		25	26		51	71
0650	4 x 4 - W2.9 x W2.9 (6 x 6) 61 lb. per C.S.F.	G		27	.593		37	27.50		64.50	87.50
0700	4 x 4 - W4 x W4 (4 x 4) 85 lb. per C.S.F.	G		25	.640		47	30		77	102
0750	Rolls										
0800	2 x 2 - #14 galv., 21 lb/C.S.F., beam & column wrap	G	2 Rodm	6.50	2.462	C.S.F.	40	115		155	238
0900	2 x 2 - #12 galv. for gunite reinforcing	G	"	6.50	2.462	"	35	115		150	233
9000	Minimum labor/equipment charge		1 Rodm	4	2	Job		93.50		93.50	157

03 24 Fibrous Reinforcing

03 24 05 – Reinforcing Fibers

03 24 05.30 Synthetic Fibers		Crew	Daily Output	Labor-Hours	Unit	Material	2010 Bare Costs Labor	Equipment	Total	Total Incl O&P
0010	**SYNTHETIC FIBERS**									
0100	Synthetic fibers, add to concrete				Lb.	4.72			4.72	5.20
0110	1-1/2 lb. per C.Y.				C.Y.	7.30			7.30	8

03 24 05.70 Steel Fibers

0010	**STEEL FIBERS**									
0150	Steel fibers, add to concrete	G			Lb.	.77			.77	.85
0155	25 lb. per C.Y.	G			C.Y.	19.25			19.25	21
0160	50 lb. per C.Y.	G				38.50			38.50	42.50
0170	75 lb. per C.Y.	G				59.50			59.50	65
0180	100 lb. per C.Y.	G				77			77	84.50

03 30 Cast-In-Place Concrete

03 30 53 – Miscellaneous Cast-In-Place Concrete

03 30 53.40 Concrete In Place

		Crew	Daily Output	Labor-Hours	Unit	Material	Labor	Equipment	Total	Total Incl O&P
0010	**CONCRETE IN PLACE**									
0020	Including forms (4 uses), reinforcing steel, concrete, placement,									
0050	and finishing unless otherwise indicated									
0300	Beams (3500 psi), 5 kip per L.F., 10' span	C-14A	15.62	12.804	C.Y.	282	535	49.50	866.50	1,250
0350	25' span	"	18.55	10.782		290	450	41.50	781.50	1,100
0500	Chimney foundations (5000 psi), industrial, minimum	C-14C	32.22	3.476		140	139	.94	279.94	385
0510	Maximum	"	23.71	4.724		162	188	1.28	351.28	490
0700	Columns, square (4000 psi), 12" x 12", minimum reinforcing	C-14A	11.96	16.722		305	700	64.50	1,069.50	1,550
0720	Average reinforcing		10.13	19.743		475	825	76	1,376	1,950
0740	Maximum reinforcing		9.03	22.148		700	925	85.50	1,710.50	2,400
0800	16" x 16", minimum reinforcing		16.22	12.330		246	515	47.50	808.50	1,175
0820	Average reinforcing		12.57	15.911		405	665	61.50	1,131.50	1,625
0840	Maximum reinforcing		10.25	19.512		615	815	75	1,505	2,125
0900	24" x 24", minimum reinforcing		23.66	8.453		211	355	32.50	598.50	850
0920	Average reinforcing		17.71	11.293		360	470	43.50	873.50	1,225
0940	Maximum reinforcing		14.15	14.134		565	590	54.50	1,209.50	1,650
1000	36" x 36", minimum reinforcing		33.69	5.936		188	248	23	459	635
1020	Average reinforcing		23.32	8.576		320	360	33	713	980
1040	Maximum reinforcing		17.82	11.223		530	470	43	1,043	1,400
1200	16" diameter, minimum reinforcing		31.49	6.351		251	265	24.50	540.50	740
1220	Average reinforcing		19.12	10.460		420	435	40.50	895.50	1,225
1240	Maximum reinforcing		13.77	14.524		620	605	56	1,281	1,725
1300	20" diameter, minimum reinforcing		41.04	4.873		254	204	18.75	476.75	635
1320	Average reinforcing		24.05	8.316		405	345	32	782	1,050
1340	Maximum reinforcing		17.01	11.758		620	490	45.50	1,155.50	1,550
1400	24" diameter, minimum reinforcing		51.85	3.857		238	161	14.85	413.85	545
1420	Average reinforcing		27.06	7.391		405	310	28.50	743.50	980
1440	Maximum reinforcing		18.29	10.935		610	455	42	1,107	1,475
1500	36" diameter, minimum reinforcing		75.04	2.665		237	111	10.25	358.25	455
1520	Average reinforcing		37.49	5.335		385	223	20.50	628.50	815
1540	Maximum reinforcing		22.84	8.757		590	365	33.50	988.50	1,275
1900	Elevated slab (4000 psi), flat slab with drops, 125 psf Sup. Load, 20' span	C-14B	38.45	5.410		244	226	20	490	660
1950	30' span		50.99	4.079		252	170	15.10	437.10	575
2100	Flat plate, 125 psf Sup. Load, 15' span		30.24	6.878		226	287	25.50	538.50	745
2150	25' span		49.60	4.194		231	175	15.50	421.50	560
2300	Waffle const., 30" domes, 125 psf Sup. Load, 20' span		37.07	5.611		227	234	21	482	655
2350	30' span		44.07	4.720		212	197	17.45	426.45	575

03 30 Cast-In-Place Concrete

03 30 53 – Miscellaneous Cast-In-Place Concrete

03 30 53.40 Concrete In Place

		Crew	Daily Output	Labor-Hours	Unit	Material	2010 Bare Costs Labor	Equipment	Total	Total Incl O&P
2500	One way joists, 30" pans, 125 psf Sup. Load, 15' span	C-14B	27.38	7.597	C.Y.	274	315	28	617	850
2550	25' span		31.15	6.677		258	278	24.50	560.50	765
2700	One way beam & slab, 125 psf Sup. Load, 15' span		20.59	10.102		248	420	37.50	705.50	1,000
2750	25' span		28.36	7.334		229	305	27	561	780
2900	Two way beam & slab, 125 psf Sup. Load, 15' span		24.04	8.652		236	360	32	628	885
2950	25' span		35.87	5.799		202	242	21.50	465.50	640
3100	Elevated slabs, flat plate, including finish, not									
3110	including forms or reinforcing									
3150	Regular concrete (4000 psi), 4" slab	C-8	2613	.021	S.F.	1.33	.79	.28	2.40	3.03
3200	6" slab		2585	.022		1.96	.79	.29	3.04	3.74
3250	2-1/2" thick floor fill		2685	.021		.85	.77	.28	1.90	2.47
3300	Lightweight, 110# per C.F., 2-1/2" thick floor fill		2585	.022		1.15	.79	.29	2.23	2.84
3400	Cellular concrete, 1-5/8" fill, under 5000 S.F.		2000	.028		.77	1.03	.37	2.17	2.90
3450	Over 10,000 S.F.		2200	.025		.73	.93	.34	2	2.68
3500	Add per floor for 3 to 6 stories high		31800	.002			.06	.02	.08	.13
3520	For 7 to 20 stories high		21200	.003			.10	.03	.13	.20
3540	Equipment pad (3000 psi), 3' x 3' x 6" thick	C-14H	45	1.067	Ea.	42.50	44	.67	87.17	119
3550	4' x 4' x 6" thick		30	1.600		62.50	65.50	1.01	129.01	177
3560	5' x 5' x 8" thick		18	2.667		108	109	1.68	218.68	300
3570	6' x 6' x 8" thick		14	3.429		146	141	2.16	289.16	390
3580	8' x 8' x 10" thick		8	6		310	246	3.78	559.78	750
3590	10' x 10' x 12" thick		5	9.600		530	395	6.05	931.05	1,225
3800	Footings (3000 psi), spread under 1 C.Y.	C-14C	28	4	C.Y.	150	160	1.08	311.08	430
3825	1 C.Y to 5 C.Y.		43	2.605		173	104	.70	277.70	360
3850	Over 5 C.Y.		75	1.493		158	59.50	.40	217.90	272
3900	Footings, strip (3000 psi), 18" x 9", unreinforced	C-14L	40	2.400		112	93	.77	205.77	276
3920	18" x 9", reinforced	C-14C	35	3.200		131	128	.86	259.86	355
3925	20" x 10", unreinforced	C-14L	45	2.133		110	82.50	.68	193.18	257
3930	20" x 10", reinforced	C-14C	40	2.800		126	112	.76	238.76	325
3935	24" x 12", unreinforced	C-14L	55	1.745		110	67.50	.56	178.06	233
3940	24" x 12", reinforced	C-14C	48	2.333		126	93	.63	219.63	292
3945	36" x 12", unreinforced	C-14L	70	1.371		107	53	.44	160.44	205
3950	36" x 12", reinforced	C-14C	60	1.867		122	74.50	.50	197	257
4000	Foundation mat (3000 psi), under 10 C.Y.		38.67	2.896		180	116	.78	296.78	390
4050	Over 20 C.Y.		56.40	1.986		159	79	.54	238.54	305
4200	Wall, free-standing (3000 psi), 8" thick, 8' high	C-14D	45.83	4.364		175	180	16.80	371.80	505
4250	14' high		27.26	7.337		204	305	28.50	537.50	755
4260	12" thick, 8' high		64.32	3.109		155	129	11.95	295.95	395
4270	14' high		40.01	4.999		163	207	19.25	389.25	540
4300	15" thick, 8' high		80.02	2.499		145	103	9.60	257.60	340
4350	12' high		51.26	3.902		145	161	15	321	440
4500	18' high		48.85	4.094		160	169	15.75	344.75	470
4520	Handicap access ramp (4000 psi), railing both sides, 3' wide	C-14H	14.58	3.292	L.F.	273	135	2.07	410.07	525
4525	5' wide		12.22	3.928		283	161	2.47	446.47	575
4530	With 6" curb and rails both sides, 3' wide		8.55	5.614		282	230	3.54	515.54	690
4535	5' wide		7.31	6.566		287	269	4.14	560.14	760
4650	Slab on grade (3500 psi), not including finish, 4" thick	C-14E	60.75	1.449	C.Y.	117	59.50	.51	177.01	227
4700	6" thick	"	92	.957	"	113	39.50	.33	152.83	189
4701	Thickened slab edge (3500 psi), for slab on grade poured									
4702	monolithically with slab; depth is in addition to slab thickness;									
4703	formed vertical outside edge, earthen bottom and inside slope									
4705	8" deep x 8" wide bottom, unreinforced	C-14L	2190	.044	L.F.	3.07	1.70	.01	4.78	6.20
4710	8" x 8", reinforced	C-14C	1670	.067		4.95	2.68	.02	7.65	9.85

03 30 Cast-In-Place Concrete

03 30 53 – Miscellaneous Cast-In-Place Concrete

03 30 53.40 Concrete In Place		Crew	Daily Output	Labor-Hours	Unit	Material	2010 Bare Costs Labor	Equipment	Total	Total Incl O&P
4715	12" deep x 12" wide bottom, unreinforced	C-14L	1800	.053	L.F.	6.40	2.07	.02	8.49	10.45
4720	12" x 12", reinforced	C-14C	1310	.086		9.80	3.41	.02	13.23	16.45
4725	16" deep x 16" wide bottom, unreinforced	C-14L	1440	.067		10.90	2.58	.02	13.50	16.25
4730	16" x 16", reinforced	C-14C	1120	.100		15	3.99	.03	19.02	23
4735	20" deep x 20" wide bottom, unreinforced	C-14L	1150	.083		16.65	3.23	.03	19.91	23.50
4740	20" x 20", reinforced	C-14C	920	.122		22	4.86	.03	26.89	32
4745	24" deep x 24" wide bottom, unreinforced	C-14L	930	.103		23.50	4	.03	27.53	32.50
4750	24" x 24", reinforced	C-14C	740	.151		30.50	6.05	.04	36.59	43.50
4751	Slab on grade (3500 psi), incl. troweled finish, not incl. forms									
4760	or reinforcing, over 10,000 S.F., 4" thick	C-14F	3425	.021	S.F.	1.29	.79	.01	2.09	2.69
4820	6" thick		3350	.021		1.89	.81	.01	2.71	3.37
4840	8" thick		3184	.023		2.59	.85	.01	3.45	4.21
4900	12" thick		2734	.026		3.88	.99	.01	4.88	5.85
4950	15" thick		2505	.029		4.88	1.08	.01	5.97	7.10
5000	Slab on grade (3000 psi), incl. textured finish, not incl. forms									
5001	or reinforcing, 4" thick	C-14G	2873	.019	S.F.	1.26	.72	.01	1.99	2.55
5010	6" thick		2590	.022		1.97	.80	.01	2.78	3.46
5020	8" thick		2320	.024		2.57	.90	.01	3.48	4.27
5200	Lift slab in place above the foundation, incl. forms,									
5210	reinforcing, concrete (4000 psi) and columns, minimum	C-14B	2113	.098	S.F.	6.10	4.10	.36	10.56	13.85
5250	Average		1650	.126		6.60	5.25	.47	12.32	16.40
5300	Maximum		1500	.139		7.15	5.80	.51	13.46	17.95
5500	Lightweight, ready mix, including screed finish only,									
5510	not including forms or reinforcing									
5550	1:4 (2500 psi) for structural roof decks	C-14B	260	.800	C.Y.	135	33.50	2.96	171.46	207
5600	1:6 (3000 psi) for ground slab with radiant heat	C-14F	92	.783		133	29.50	.33	162.83	194
5650	1:3:2 (2000 psi) with sand aggregate, roof deck	C-14B	260	.800		132	33.50	2.96	168.46	203
5700	Ground slab (2000 psi)	C-14F	107	.673		132	25.50	.28	157.78	185
5900	Pile caps (3000 psi), incl. forms and reinf., sq. or rect., under 10 C.Y.	C-14C	54.14	2.069		147	82.50	.56	230.06	299
5950	Over 10 C.Y.		75	1.493		140	59.50	.40	199.90	252
6000	Triangular or hexagonal, under 10 C.Y.		53	2.113		115	84.50	.57	200.07	266
6050	Over 10 C.Y.		85	1.318		127	52.50	.36	179.86	227
6200	Retaining walls (3000 psi), gravity, 4' high see Div. 32 32	C-14D	66.20	3.021		133	125	11.65	269.65	365
6250	10' high		125	1.600		126	66	6.15	198.15	254
6300	Cantilever, level backfill loading, 8' high		70	2.857		153	118	11	282	375
6350	16' high		91	2.198		140	91	8.45	239.45	310
6800	Stairs (3500 psi), not including safety treads, free standing, 3'-6" wide	C-14H	83	.578	LF Nose	5.10	23.50	.36	28.96	45
6850	Cast on ground		125	.384	"	4.03	15.75	.24	20.02	30.50
7000	Stair landings, free standing		200	.240	S.F.	4.20	9.85	.15	14.20	21
7050	Cast on ground		475	.101	"	3.13	4.15	.06	7.34	10.30
9000	Minimum labor/equipment charge	2 Carp	1	16	Job		665		665	1,100

03 31 Structural Concrete

03 31 05 – Normal Weight Structural Concrete

03 31 05.35 Normal Weight Concrete, Ready Mix

		Crew	Daily Output	Labor-Hours	Unit	Material	Labor	Equipment	Total	Total Incl O&P
0010	**NORMAL WEIGHT CONCRETE, READY MIX**, delivered									
0012	Includes local aggregate, sand, Portland cement, and water									
0015	Excludes all additives and treatments R033105-20									
0020	2000 psi				C.Y.	91.50			91.50	101
0100	2500 psi					94			94	103
0150	3000 psi					97			97	107
0200	3500 psi					99.50			99.50	110
0300	4000 psi					103			103	113
0350	4500 psi					106			106	116
0400	5000 psi					109			109	120
0411	6000 psi					124			124	137
0412	8000 psi					203			203	223
0413	10,000 psi					288			288	315
0414	12,000 psi					350			350	380
1000	For high early strength cement, add					10%				
1300	For winter concrete (hot water), add					4.25			4.25	4.68
1400	For hot weather concrete (ice), add					9.10			9.10	10
1410	For mid-range water reducer, add					4.13			4.13	4.54
1420	For high-range water reducer/superplasticizer, add					6.35			6.35	6.95
1430	For retarder, add					2.71			2.71	2.98
1440	For non-Chloride accelerator, add					4.83			4.83	5.30
1450	For Chloride accelerator, per 1%, add					3.28			3.28	3.61
1460	For fiber reinforcing, synthetic (1 Lb./C.Y.), add					6.65			6.65	7.30
1500	For Saturday delivery, add					8.85			8.85	9.70
1510	For truck holding/waiting time past 1st hour per load, add				Hr.	87.50			87.50	96.50
1520	For short load (less than 4 C.Y.), add per load				Ea.	112			112	124
2000	For all lightweight aggregate, add				C.Y.	45%				

03 31 05.70 Placing Concrete

		Crew	Daily Output	Labor-Hours	Unit	Material	Labor	Equipment	Total	Total Incl O&P
0010	**PLACING CONCRETE** R033105-70									
0020	Includes labor and equipment to place, strike off and consolidate									
0050	Beams, elevated, small beams, pumped	C-20	60	1.067	C.Y.		38	13.30	51.30	75.50
0100	With crane and bucket	C-7	45	1.600			57	28.50	85.50	124
0200	Large beams, pumped	C-20	90	.711			25	8.90	33.90	51
0250	With crane and bucket	C-7	65	1.108			39.50	19.85	59.35	86
0400	Columns, square or round, 12" thick, pumped	C-20	60	1.067			38	13.30	51.30	75.50
0450	With crane and bucket	C-7	40	1.800			64.50	32.50	97	140
0600	18" thick, pumped	C-20	90	.711			25	8.90	33.90	51
0650	With crane and bucket	C-7	55	1.309			47	23.50	70.50	102
0800	24" thick, pumped	C-20	92	.696			24.50	8.70	33.20	49.50
0850	With crane and bucket	C-7	70	1.029			37	18.45	55.45	80
1000	36" thick, pumped	C-20	140	.457			16.20	5.70	21.90	32.50
1050	With crane and bucket	C-7	100	.720			25.50	12.90	38.40	55.50
1400	Elevated slabs, less than 6" thick, pumped	C-20	140	.457			16.20	5.70	21.90	32.50
1450	With crane and bucket	C-7	95	.758			27	13.60	40.60	58.50
1500	6" to 10" thick, pumped	C-20	160	.400			14.15	5	19.15	28.50
1550	With crane and bucket	C-7	110	.655			23.50	11.75	35.25	50.50
1600	Slabs over 10" thick, pumped	C-20	180	.356			12.60	4.44	17.04	25.50
1650	With crane and bucket	C-7	130	.554			19.80	9.95	29.75	43
1900	Footings, continuous, shallow, direct chute	C-6	120	.400			13.80	.51	14.31	23
1950	Pumped	C-20	150	.427			15.10	5.35	20.45	30.50
2000	With crane and bucket	C-7	90	.800			28.50	14.35	42.85	62
2100	Footings, continuous, deep, direct chute	C-6	140	.343			11.85	.44	12.29	19.75

03 31 Structural Concrete

03 31 05 – Normal Weight Structural Concrete

03 31 05.70 Placing Concrete

		Crew	Daily Output	Labor-Hours	Unit	Material	2010 Bare Costs Labor	2010 Bare Costs Equipment	Total	Total Incl O&P
2150	Pumped	C-20	160	.400	C.Y.		14.15	5	19.15	28.50
2200	With crane and bucket	C-7	110	.655			23.50	11.75	35.25	50.50
2400	Footings, spread, under 1 C.Y., direct chute	C-6	55	.873			30	1.11	31.11	50
2450	Pumped	C-20	65	.985			35	12.30	47.30	70
2500	With crane and bucket	C-7	45	1.600			57	28.50	85.50	124
2600	Over 5 C.Y., direct chute	C-6	120	.400			13.80	.51	14.31	23
2650	Pumped	C-20	150	.427			15.10	5.35	20.45	30.50
2700	With crane and bucket	C-7	100	.720			25.50	12.90	38.40	55.50
2900	Foundation mats, over 20 C.Y., direct chute	C-6	350	.137			4.74	.17	4.91	7.90
2950	Pumped	C-20	400	.160			5.65	2	7.65	11.35
3000	With crane and bucket	C-7	300	.240			8.60	4.30	12.90	18.60
3200	Grade beams, direct chute	C-6	150	.320			11.05	.41	11.46	18.40
3250	Pumped	C-20	180	.356			12.60	4.44	17.04	25.50
3300	With crane and bucket	C-7	120	.600			21.50	10.75	32.25	46.50
3500	High rise, for more than 5 stories, pumped, add per story	C-20	2100	.030			1.08	.38	1.46	2.17
3510	With crane and bucket, add per story	C-7	2100	.034			1.23	.61	1.84	2.66
3700	Pile caps, under 5 C.Y., direct chute	C-6	90	.533			18.40	.68	19.08	30.50
3750	Pumped	C-20	110	.582			20.50	7.25	27.75	41.50
3800	With crane and bucket	C-7	80	.900			32	16.15	48.15	70
3850	Pile cap, 5 C.Y. to 10 C.Y., direct chute	C-6	175	.274			9.45	.35	9.80	15.80
3900	Pumped	C-20	200	.320			11.35	4	15.35	23
3950	With crane and bucket	C-7	150	.480			17.15	8.60	25.75	37
4000	Over 10 C.Y., direct chute	C-6	215	.223			7.70	.28	7.98	12.85
4050	Pumped	C-20	240	.267			9.45	3.33	12.78	18.95
4100	With crane and bucket	C-7	185	.389			13.90	6.95	20.85	30
4300	Slab on grade, up to 6" thick, direct chute	C-6	110	.436			15.05	.55	15.60	25
4350	Pumped	C-20	130	.492			17.45	6.15	23.60	35
4400	With crane and bucket	C-7	110	.655			23.50	11.75	35.25	50.50
4600	Over 6" thick, direct chute	C-6	165	.291			10.05	.37	10.42	16.75
4650	Pumped	C-20	185	.346			12.25	4.32	16.57	24.50
4700	With crane and bucket	C-7	145	.497			17.75	8.90	26.65	38.50
4900	Walls, 8" thick, direct chute	C-6	90	.533			18.40	.68	19.08	30.50
4950	Pumped	C-20	100	.640			22.50	8	30.50	45.50
5000	With crane and bucket	C-7	80	.900			32	16.15	48.15	70
5050	12" thick, direct chute	C-6	100	.480			16.55	.61	17.16	27.50
5100	Pumped	C-20	110	.582			20.50	7.25	27.75	41.50
5200	With crane and bucket	C-7	90	.800			28.50	14.35	42.85	62
5300	15" thick, direct chute	C-6	105	.457			15.80	.58	16.38	26
5350	Pumped	C-20	120	.533			18.90	6.65	25.55	38
5400	With crane and bucket	C-7	95	.758			27	13.60	40.60	58.50
5600	Wheeled concrete dumping, add to placing costs above									
5610	Walking cart, 50' haul, add	C-18	32	.281	C.Y.		9.35	1.83	11.18	17.40
5620	150' haul, add		24	.375			12.50	2.43	14.93	23
5700	250' haul, add		18	.500			16.65	3.25	19.90	31
5800	Riding cart, 50' haul, add	C-19	80	.113			3.75	1.17	4.92	7.45
5810	150' haul, add		60	.150			5	1.57	6.57	9.90
5900	250' haul, add		45	.200			6.65	2.09	8.74	13.25
9000	Minimum labor/equipment charge	C-6	2	24	Job		830	30.50	860.50	1,375

03 35 Concrete Finishing

03 35 29 – Tooled Concrete Finishing

03 35 29.30 Finishing Floors

		Crew	Daily Output	Labor-Hours	Unit	Material	2010 Bare Costs Labor	2010 Bare Costs Equipment	Total	Total Incl O&P
0010	**FINISHING FLOORS**									
0012	Finishing requires that concrete first be placed, struck off & consolidated									
0015	Basic finishing for various unspecified flatwork									
0100	Bull float only	C-10	4000	.006	S.F.		.23		.23	.36
0125	Bull float & manual float		2000	.012			.45		.45	.71
0150	Bull float, manual float, & broom finish, w/edging & joints		1850	.013			.49		.49	.77
0200	Bull float, manual float & manual steel trowel		1265	.019			.71		.71	1.13
0210	For specified Random Access Floors in ACI Classes 1, 2, 3 and 4 to achieve									
0215	Composite Overall Floor Flatness and Levelness values up to F35/F25									
0250	Bull float, machine float & machine trowel (walk-behind)	C-10C	1715	.014	S.F.		.52	.03	.55	.86
0300	Power screed, bull float, machine float & trowel (walk-behind)	C-10D	2400	.010			.38	.05	.43	.64
0350	Power screed, bull float, machine float & trowel (ride-on)	C-10E	4000	.006			.23	.06	.29	.43
0352	For specified Random Access Floors in ACI Classes 5, 6, 7 and 8 to achieve									
0354	Composite Overall Floor Flatness and Levelness values up to F50/F50									
0356	Add for two-dimensional restraightening after power float	C-10	6000	.004	S.F.		.15		.15	.24
0358	For specified Random or Defined Access Floors in ACI Class 9 to achieve									
0360	Composite Overall Floor Flatness and Levelness values up to F100/F100									
0362	Add for two-dimensional restraightening after bull float & power float	C-10	3000	.008	S.F.		.30		.30	.48
0364	For specified Superflat Defined Access Floors in ACI Class 9 to achieve									
0366	Minimum Floor Flatness and Levelness values of F100/F100									
0368	Add for 2-dim'l restraightening after bull float, power float, power trowel	C-10	2000	.012	S.F.		.45		.45	.71
0370	Minimum labor/equipment charge	"	2	12	Job		450		450	715
0400	Integral topping and finish, using 1:1:2 mix, 3/16" thick	C-10B	1000	.040	S.F.	.10	1.43	.27	1.80	2.70
0450	1/2" thick		950	.042		.27	1.51	.28	2.06	3.03
0500	3/4" thick		850	.047		.41	1.68	.31	2.40	3.49
0600	1" thick		750	.053		.54	1.91	.36	2.81	4.05
0800	Granolithic topping, laid after, 1:1:1-1/2 mix, 1/2" thick		590	.068		.30	2.42	.45	3.17	4.72
0820	3/4" thick		580	.069		.45	2.47	.46	3.38	4.97
0850	1" thick		575	.070		.61	2.49	.46	3.56	5.15
0950	2" thick		500	.080		1.21	2.86	.53	4.60	6.50
9100	Minimum labor/equipment charge	C-10	2	12	Job		450		450	715

03 35 29.35 Control Joints, Saw Cut

		Crew	Daily Output	Labor-Hours	Unit	Material	Labor	Equipment	Total	Total Incl O&P
0010	**CONTROL JOINTS, SAW CUT**									
0100	Sawcut in green concrete									
0120	1" depth	C-27	2000	.008	L.F.	.06	.32	.08	.46	.66
0140	1-1/2" depth		1800	.009		.09	.35	.09	.53	.75
0160	2" depth		1600	.010		.12	.40	.10	.62	.86
0200	Blow construction debris out of control joint	C-28	6000	.001			.05		.05	.08
0300	Joint sealant									
0320	Backer rod, polyethylene, 1/4" diameter	1 Cefi	460	.017	L.F.	.03	.69		.72	1.11
0340	Sealant, polyurethane									
0360	1/4" x 1/4" (308 LF/Gal)	1 Cefi	270	.030	L.F.	.16	1.18		1.34	2.02
0380	1/4" x 1/2" (154 LF/Gal)	"	255	.031	"	.32	1.25		1.57	2.30

03 35 29.60 Finishing Walls

		Crew	Daily Output	Labor-Hours	Unit	Material	Labor	Equipment	Total	Total Incl O&P
0010	**FINISHING WALLS**									
0020	Break ties and patch voids	1 Cefi	540	.015	S.F.	.03	.59		.62	.96
0050	Burlap rub with grout	"	450	.018		.03	.71		.74	1.14
0300	Bush hammer, green concrete	B-39	1000	.048			1.67	.22	1.89	2.97
0350	Cured concrete	"	650	.074			2.57	.35	2.92	4.56
0700	Sandblast, light penetration	E-11	1100	.029		.76	1.09	.21	2.06	2.96
0750	Heavy penetration	"	375	.085		1.53	3.18	.61	5.32	7.90
0850	Grind form fins flush	1 Clab	700	.011	L.F.		.38		.38	.62

03 35 Concrete Finishing

03 35 29 – Tooled Concrete Finishing

03 35 29.60 Finishing Walls		Crew	Daily Output	Labor-Hours	Unit	Material	2010 Bare Costs Labor	Equipment	Total	Total Incl O&P
9000	Minimum labor/equipment charge	C-10	2	12	Job		450		450	715

03 35 33 – Stamped Concrete Finishing

03 35 33.50 Slab Texture Stamping

		Crew	Daily Output	Labor-Hours	Unit	Material	Labor	Equipment	Total	Total Incl O&P
0010	**SLAB TEXTURE STAMPING**									
0050	Stamping requires that concrete first be placed, struck off, consolidated,									
0060	bull floated and free of bleed water. Decorative stamping tasks include:									
0100	Step 1 - first application of dry shake colored hardener	1 Cefi	6400	.001	S.F.	.38	.05		.43	.50
0110	Step 2 - bull float		6400	.001			.05		.05	.08
0130	Step 3 - second application of dry shake colored hardener		6400	.001		.19	.05		.24	.29
0140	Step 4 - bull float, manual float & steel trowel	3 Cefi	1280	.019			.74		.74	1.16
0150	Step 5 - application of dry shake colored release agent	1 Cefi	6400	.001		.09	.05		.14	.17
0160	Step 6 - place, tamp & remove mats	3 Cefi	2400	.010		1.41	.40		1.81	2.17
0170	Step 7 - touch up edges, mat joints & simulated grout lines	1 Cefi	1280	.006			.25		.25	.39
0300	Alternate stamping estimating method includes all tasks above	4 Cefi	800	.040		2.07	1.59		3.66	4.76
0400	Step 8 - pressure wash @ 3000 psi after 24 hours	1 Cefi	1600	.005			.20		.20	.31
0500	Step 9 - roll 2 coats cure/seal compound when dry	"	800	.010		.49	.40		.89	1.16

03 37 Specialty Placed Concrete

03 37 13 – Shotcrete

03 37 13.30 Gunite (Dry-Mix)

		Crew	Daily Output	Labor-Hours	Unit	Material	Labor	Equipment	Total	Total Incl O&P
0010	**GUNITE (DRY-MIX)**									
0020	Applied in 1" layers, no mesh included	C-8	2000	.028	S.F.	.33	1.03	.37	1.73	2.42
0300	Typical in place, including mesh, 2" thick, minimum	C-16	1000	.072		1.01	2.80	.74	4.55	6.45
0350	Maximum		500	.144		1.01	5.60	1.48	8.09	11.85
0500	4" thick, minimum		750	.096		1.66	3.74	.99	6.39	8.95
0550	Maximum		350	.206		1.66	8	2.11	11.77	17.15
0900	Prepare old walls, no scaffolding, minimum	C-10	1000	.024			.90		.90	1.43
0950	Maximum	"	275	.087			3.27		3.27	5.20
1100	For high finish requirement or close tolerance, add, minimum						50%			
1150	Maximum						110%			
9000	Minimum labor/equipment charge	C-10	1	24	Job		900		900	1,425

03 39 Concrete Curing

03 39 13 – Water Concrete Curing

03 39 13.50 Water Curing

		Crew	Daily Output	Labor-Hours	Unit	Material	Labor	Equipment	Total	Total Incl O&P
0010	**WATER CURING**									
0015	With burlap, 4 uses assumed, 7.5 oz.	2 Clab	55	.291	C.S.F.	8.65	9.65		18.30	25.50
0100	10 oz.	"	55	.291	"	15.55	9.65		25.20	33
9000	Minimum labor/equipment charge	1 Clab	5	1.600	Job		53		53	87

03 39 23 – Membrane Concrete Curing

03 39 23.13 Chemical Compound Membrane Concrete Curing

		Crew	Daily Output	Labor-Hours	Unit	Material	Labor	Equipment	Total	Total Incl O&P
0010	**CHEMICAL COMPOUND MEMBRANE CONCRETE CURING**									
0300	Sprayed membrane curing compound	2 Clab	95	.168	C.S.F.	6.10	5.55		11.65	15.90
0700	Curing compound, solvent based, 400 S.F./gal, 55 gallon lots				Gal.	17.30			17.30	19.05
0720	5 gallon lots					24.50			24.50	27
0800	Curing compound, water based, 250 S.F./gal, 55 gallon lots					20.50			20.50	22.50
0820	5 gallon lots					23.50			23.50	26

03 39 Concrete Curing

03 39 23 – Membrane Concrete Curing

03 39 23.23 Sheet Membrane Concrete Curing		Crew	Daily Output	Labor-Hours	Unit	Material	2010 Bare Costs Labor	2010 Bare Costs Equipment	Total	Total Incl O&P
0010	**SHEET MEMBRANE CONCRETE CURING**									
0200	Curing blanket, burlap/poly, 2-ply	2 Clab	70	.229	C.S.F.	16	7.55		23.55	30

03 41 Precast Structural Concrete

03 41 05 – Precast Concrete Members

03 41 05.10 Precast Beams

			Crew	Daily Output	Labor-Hours	Unit	Material	Labor	Equipment	Total	Total Incl O&P
0010	**PRECAST BEAMS**	R034105-30									
0011	L-shaped, 20' span, 12" x 20"		C-11	32	2.250	Ea.	2,000	103	62.50	2,165.50	2,450
1000	Inverted tee beams, add to above, small beams						15%				
1050	Large beams						20%				
1200	Rectangular, 20' span, 12" x 20"		C-11	32	2.250		1,500	103	62.50	1,665.50	1,900
1250	18" x 36"			24	3		2,725	138	83.50	2,946.50	3,325
1300	24" x 44"			22	3.273		3,775	150	91	4,016	4,550
1400	30' span, 12" x 36"			24	3		3,425	138	83.50	3,646.50	4,100
1450	18" x 44"			20	3.600		4,925	165	100	5,190	5,825
1500	24" x 52"			16	4.500		6,575	207	125	6,907	7,725
1600	40' span, 12" x 52"			20	3.600		6,375	165	100	6,640	7,400
1650	18" x 52"			16	4.500		7,575	207	125	7,907	8,825
1700	24" x 52"			12	6		8,775	275	167	9,217	10,300
2000	"T" shaped, 20' span, 12" x 20"			32	2.250		2,600	103	62.50	2,765.50	3,100
2050	18" x 36"			24	3		4,675	138	83.50	4,896.50	5,475
2100	24" x 44"			22	3.273		6,475	150	91	6,716	7,500
2200	30' span, 12" x 36"			24	3		5,825	138	83.50	6,046.50	6,725
2250	18" x 44"			20	3.600		8,375	165	100	8,640	9,600
2300	24" x 52"			16	4.500		11,200	207	125	11,532	12,800
2500	40' span, 12" x 52"			20	3.600		10,800	165	100	11,065	12,200
2550	18" x 52"			16	4.500		12,900	207	125	13,232	14,700
2600	24" x 52"			12	6		14,900	275	167	15,342	17,100

03 41 05.15 Precast Columns

0010	**PRECAST COLUMNS**	R034105-30									
0020	Rectangular to 12' high, small columns		C-11	120	.600	L.F.	106	27.50	16.70	150.20	183
0050	Large columns		"	96	.750	"	184	34.50	21	239.50	286

03 41 13 – Precast Concrete Hollow Core Planks

03 41 13.50 Precast Slab Planks

0010	**PRECAST SLAB PLANKS**	R034105-30									
0020	Prestressed roof/floor members, grouted, solid, 4" thick		C-11	2400	.030	S.F.	5.55	1.38	.83	7.76	9.45
0050	6" thick			2800	.026		6.50	1.18	.72	8.40	10.05
0100	Hollow, 8" thick			3200	.023		7.20	1.03	.63	8.86	10.40
0150	10" thick			3600	.020		7.50	.92	.56	8.98	10.50
0200	12" thick			4000	.018		8	.83	.50	9.33	10.80

03 41 23 – Precast Concrete Stairs

03 41 23.50 Precast Stairs

0010	**PRECAST STAIRS**										
0020	Precast concrete treads on steel stringers, 3' wide		C-12	75	.640	Riser	105	26	8.75	139.75	168
0300	Front entrance, 5' wide with 48" platform, 2 risers			16	3	Flight	460	123	41	624	750
0350	5 risers			12	4		715	164	54.50	933.50	1,100
0500	6' wide, 2 risers			15	3.200		510	131	43.50	684.50	820
0550	5 risers			11	4.364		790	179	59.50	1,028.50	1,225
1200	Basement entrance stairs, steel bulkhead doors, minimum		B-51	22	2.182		1,350	72.50	10.90	1,433.40	1,600

59

03 41 Precast Structural Concrete

03 41 23 – Precast Concrete Stairs

03 41 23.50 Precast Stairs		Crew	Daily Output	Labor-Hours	Unit	Material	2010 Bare Costs Labor	Equipment	Total	Total Incl O&P
1250	Maximum	B-51	11	4.364	Flight	2,225	145	22	2,392	2,700

03 45 Precast Architectural Concrete

03 45 13 – Faced Architectural Precast Concrete

03 45 13.50 Precast Wall Panels

		Crew	Daily Output	Labor-Hours	Unit	Material	Labor	Equipment	Total	Total Incl O&P
0010	**PRECAST WALL PANELS**									
0750	20' x 10', 6" thick, smooth gray	C-11	1400	.051	S.F.	31	2.36	1.43	34.79	40
2200	Fiberglass reinforced cement with urethane core									
2210	R20, 8' x 8', minimum	E-2	750	.075	S.F.	21	3.41	2.14	26.55	31.50
2220	Maximum	"	600	.093	"	31.50	4.26	2.68	38.44	45

03 48 Precast Concrete Specialties

03 48 43 – Precast Concrete Trim

03 48 43.40 Precast Lintels

		Crew	Daily Output	Labor-Hours	Unit	Material	Labor	Equipment	Total	Total Incl O&P
0010	**PRECAST LINTELS**, smooth gray, prestressed, stock units only									
0800	4" wide, 8" high, x 4' long	D-10	28	1.143	Ea.	21	46.50	23.50	91	124
0850	8' long		24	1.333		56.50	54.50	27.50	138.50	179
1000	6" wide, 8" high, x 4' long		26	1.231		31	50.50	25	106.50	143
1050	10' long		22	1.455		83	59.50	30	172.50	220
1200	8" wide, 8" high, x 4' long		24	1.333		37.50	54.50	27.50	119.50	158
1250	12' long		20	1.600		130	65.50	33	228.50	283
1275	For custom sizes, types, colors, or finishes of precast lintels, add					150%				

03 48 43.90 Precast Window Sills

		Crew	Daily Output	Labor-Hours	Unit	Material	Labor	Equipment	Total	Total Incl O&P
0010	**PRECAST WINDOW SILLS**									
0600	Precast concrete, 4" tapers to 3", 9" wide	D-1	70	.229	L.F.	12.25	8.60		20.85	27.50
0650	11" wide		60	.267		16	10.05		26.05	34
0700	13" wide, 3 1/2" tapers to 2 1/2", 12" wall		50	.320		16	12.05		28.05	37

03 51 Cast Roof Decks

03 51 13 – Cementitious Wood Fiber Decks

03 51 13.50 Cementitious/Wood Fiber Planks

		Crew	Daily Output	Labor-Hours	Unit	Material	Labor	Equipment	Total	Total Incl O&P
0010	**CEMENTITIOUS/WOOD FIBER PLANKS**									
0050	Plank, beveled edge, 1" thick	2 Carp	1000	.016	S.F.	2.32	.66		2.98	3.64
0100	1-1/2" thick		975	.016		2.94	.68		3.62	4.35
0150	T & G, 2" thick		950	.017		3.54	.70		4.24	5.05
0200	2-1/2" thick		925	.017		4.14	.72		4.86	5.75
1000	Bulb tee, sub-purlin and grout, 6' span, add	E-1	5000	.005		2.07	.22	.03	2.32	2.68
1100	8' span	"	4200	.006		2.31	.26	.03	2.60	3.04

03 52 Lightweight Concrete Roof Insulation

03 52 16 – Lightweight Insulating Concrete

03 52 16.13 Lightweight Cellular Insulating Concrete		Crew	Daily Output	Labor-Hours	Unit	Material	2010 Bare Costs Labor	Equipment	Total	Total Incl O&P
0010	**LIGHTWEIGHT CELLULAR INSULATING CONCRETE** R035216-10									
0020	Portland cement and foaming agent G	C-8	50	1.120	C.Y.	144	41	14.75	199.75	240

03 52 16.16 Lightweight Aggregate Insulating Concrete

0010	**LIGHTWEIGHT AGGREGATE INSULATING CONCRETE** R035216-10									
0100	Poured vermiculite or perlite, field mix,									
0110	1:6 field mix G	C-8	50	1.120	C.Y.	219	41	14.75	274.75	325
0200	Ready mix, 1:6 mix, roof fill, 2" thick G		10000	.006	S.F.	1.22	.21	.07	1.50	1.75
0250	3" thick G	↓	7700	.007	"	1.83	.27	.10	2.20	2.55

03 54 Cast Underlayment

03 54 16 – Hydraulic Cement Underlayment

03 54 16.50 Cement Underlayment

		Crew	Daily Output	Labor-Hours	Unit	Material	Labor	Equipment	Total	Total Incl O&P
0010	**CEMENT UNDERLAYMENT**									
2510	Underlayment, P.C based self-leveling, 4100 psi, pumped, 1/4"	C-8	20000	.003	S.F.	1.52	.10	.04	1.66	1.87
2520	1/2"		19000	.003		3.04	.11	.04	3.19	3.56
2530	3/4"		18000	.003		4.56	.11	.04	4.71	5.25
2540	1"		17000	.003		6.10	.12	.04	6.26	6.95
2550	1-1/2"	↓	15000	.004		9.10	.14	.05	9.29	10.30
2560	Hand mix, 1/2"	C-18	4000	.002		3.04	.08	.01	3.13	3.49
2610	Topping, P.C. based self-level/dry 6100 psi, pumped, 1/4"	C-8	20000	.003		2.34	.10	.04	2.48	2.77
2620	1/2"		19000	.003		4.68	.11	.04	4.83	5.35
2630	3/4"		18000	.003		7	.11	.04	7.15	7.95
2660	1"		17000	.003		9.35	.12	.04	9.51	10.55
2670	1-1/2"	↓	15000	.004		14.05	.14	.05	14.24	15.70
2680	Hand mix, 1/2"	C-18	4000	.002	↓	4.68	.08	.01	4.77	5.30

03 63 Epoxy Grouting

03 63 05 – Grouting of Dowels and Fasteners

03 63 05.10 Epoxy Only

		Crew	Daily Output	Labor-Hours	Unit	Material	Labor	Equipment	Total	Total Incl O&P
0010	**EPOXY ONLY**									
1500	Chemical anchoring, epoxy cartridge, excludes layout, drilling, fastener									
1530	For fastener 3/4" diam. x 6" embedment	2 Skwk	72	.222	Ea.	4.57	9.45		14.02	20.50
1535	1" diam. x 8" embedment		66	.242		6.85	10.35		17.20	24.50
1540	1-1/4" diam. x 10" embedment		60	.267		13.70	11.35		25.05	33.50
1545	1-3/4" diam. x 12" embedment		54	.296		23	12.60		35.60	45.50
1550	14" embedment		48	.333		27.50	14.20		41.70	53
1555	2" diam. x 12" embedment		42	.381		36.50	16.25		52.75	66.50
1560	18" embedment	↓	32	.500	↓	45.50	21.50		67	85

03 64 Injection Grouting

03 64 23 – Epoxy Injection Grouting

03 64 23.10 Crack Repair

		Crew	Daily Output	Labor-Hours	Unit	Material	2010 Bare Costs Labor	Equipment	Total	Total Incl O&P
0010	**CRACK REPAIR**									
0100	Epoxy injection, 1/8" wide, 12" deep	B-9	80	.500	L.F.	6.95	16.75	2.81	26.51	38
0110	1/4" wide, 12" deep		60	.667		13.95	22.50	3.75	40.20	56
0200	Latex injection, 1/8" wide, 12" deep		100	.400		2.42	13.40	2.25	18.07	27
0210	1/4" wide, 12" deep		75	.533		4.85	17.85	3	25.70	38

03 81 Concrete Cutting

03 81 13 – Flat Concrete Sawing

03 81 13.50 Concrete Floor/Slab Cutting

		Crew	Daily Output	Labor-Hours	Unit	Material	Labor	Equipment	Total	Total Incl O&P
0010	**CONCRETE FLOOR/SLAB CUTTING**									
0400	Concrete slabs, mesh reinforcing, up to 3" deep	B-89	980	.016	L.F.	.50	.60	.49	1.59	2.04
0420	Each additional inch of depth	"	1600	.010	"	.17	.37	.30	.84	1.10

03 81 16 – Track Mounted Concrete Wall Sawing

03 81 16.50 Concrete Wall Cutting

		Crew	Daily Output	Labor-Hours	Unit	Material	Labor	Equipment	Total	Total Incl O&P
0010	**CONCRETE WALL CUTTING**									
0800	Concrete walls, hydraulic saw, plain, per inch of depth	B-89B	250	.064	L.F.	.45	2.35	3.37	6.17	7.95
0820	Rod reinforcing, per inch of depth	"	150	.107	"	.64	3.92	5.60	10.16	13.15

03 82 Concrete Boring

03 82 13 – Concrete Core Drilling

03 82 13.10 Core Drilling

		Crew	Daily Output	Labor-Hours	Unit	Material	Labor	Equipment	Total	Total Incl O&P
0010	**CORE DRILLING**									
0020	Reinf. conc slab, up to 6" thick, incl. bit, layout & set up									
0100	1" diameter core	B-89A	17	.941	Ea.	3.65	35.50	6.70	45.85	69.50
0150	Each added inch thick in same hole, add		1440	.011		.68	.42	.08	1.18	1.52
0300	3" diameter core		16	1		12.20	38	7.10	57.30	82.50
0350	Each added inch thick in same hole, add		720	.022		1.40	.84	.16	2.40	3.08
0500	4" diameter core		15	1.067		15.45	40.50	7.60	63.55	91
0550	Each added inch thick in same hole, add		480	.033		1.72	1.26	.24	3.22	4.20
0700	6" diameter core		14	1.143		20	43.50	8.15	71.65	101
0750	Each added inch thick in same hole, add		360	.044		2.73	1.68	.32	4.73	6.10
0900	8" diameter core		13	1.231		33	46.50	8.75	88.25	122
0950	Each added inch thick in same hole, add		288	.056		3.75	2.10	.40	6.25	8
1100	10" diameter core		12	1.333		26.50	50.50	9.50	86.50	121
1150	Each added inch thick in same hole, add		240	.067		3.75	2.52	.47	6.74	8.75
1300	12" diameter core		11	1.455		31	55	10.35	96.35	135
1350	Each added inch thick in same hole, add		206	.078		4.55	2.94	.55	8.04	10.40
1500	14" diameter core		10	1.600		37	60.50	11.40	108.90	152
1550	Each added inch thick in same hole, add		180	.089		6.20	3.36	.63	10.19	13
1700	18" diameter core		9	1.778		45.50	67.50	12.65	125.65	174
1750	Each added inch thick in same hole, add		144	.111		8.80	4.21	.79	13.80	17.40
1760	For horizontal holes, add to above						20%	20%		
1770	Prestressed hollow core plank, 8" thick									
1780	1" diameter core	B-89A	17.50	.914	Ea.	2.37	34.50	6.50	43.37	66.50
1790	Each added inch thick in same hole, add		3840	.004		.47	.16	.03	.66	.81
1800	3" diameter core		17	.941		4.95	35.50	6.70	47.15	71
1810	Each added inch thick in same hole, add		1920	.008		.85	.32	.06	1.23	1.52
1820	4" diameter core		16.50	.970		6.80	36.50	6.90	50.20	75
1830	Each added inch thick in same hole, add		1280	.013		1.25	.47	.09	1.81	2.25

03 82 Concrete Boring

03 82 13 – Concrete Core Drilling

03 82 13.10 Core Drilling

		Crew	Daily Output	Labor-Hours	Unit	Material	2010 Bare Costs Labor	Equipment	Total	Total Incl O&P
1840	6" diameter core	B-89A	15.50	1.032	Ea.	8.35	39	7.35	54.70	81
1850	Each added inch thick in same hole, add		960	.017		1.42	.63	.12	2.17	2.72
1860	8" diameter core		15	1.067		11.85	40.50	7.60	59.95	87
1870	Each added inch thick in same hole, add		768	.021		1.99	.79	.15	2.93	3.63
1880	10" diameter core		14	1.143		15.70	43.50	8.15	67.35	96.50
1890	Each added inch thick in same hole, add		640	.025		2.07	.95	.18	3.20	4.02
1900	12" diameter core		13.50	1.185		18.35	45	8.45	71.80	102
1910	Each added inch thick in same hole, add		548	.029		3.32	1.11	.21	4.64	5.70
1999	Drilling, core, minimum labor/equipment charge		4	4	Job		151	28.50	179.50	279
3010	Bits for core drill, diamond, premium, 1" diameter				Ea.	108			108	119
3020	3" diameter					234			234	257
3040	4" diameter					260			260	286
3050	6" diameter					415			415	455
3080	8" diameter					585			585	645
3120	12" diameter					885			885	975
3180	18" diameter					1,725			1,725	1,875
3240	24" diameter					2,425			2,425	2,675

03 82 16 – Concrete Drilling

03 82 16.10 Concrete Impact Drilling

		Crew	Daily Output	Labor-Hours	Unit	Material	Labor	Equipment	Total	Total Incl O&P
0010	**CONCRETE IMPACT DRILLING**									
0050	Up to 4" deep in conc/brick floor/wall, incl. bit & layout, no anchor									
0100	Holes, 1/4" diameter	1 Carp	75	.107	Ea.	.07	4.43		4.50	7.35
0150	For each additional inch of depth, add		430	.019		.02	.77		.79	1.29
0200	3/8" diameter		63	.127		.06	5.30		5.36	8.70
0250	For each additional inch of depth, add		340	.024		.02	.98		1	1.63
0300	1/2" diameter		50	.160		.06	6.65		6.71	10.95
0350	For each additional inch of depth, add		250	.032		.02	1.33		1.35	2.20
0400	5/8" diameter		48	.167		.09	6.95		7.04	11.50
0450	For each additional inch of depth, add		240	.033		.02	1.38		1.40	2.30
0500	3/4" diameter		45	.178		.11	7.40		7.51	12.25
0550	For each additional inch of depth, add		220	.036		.03	1.51		1.54	2.51
0600	7/8" diameter		43	.186		.13	7.75		7.88	12.85
0650	For each additional inch of depth, add		210	.038		.03	1.58		1.61	2.64
0700	1" diameter		40	.200		.15	8.30		8.45	13.80
0750	For each additional inch of depth, add		190	.042		.04	1.75		1.79	2.91
0800	1-1/4" diameter		38	.211		.22	8.75		8.97	14.60
0850	For each additional inch of depth, add		180	.044		.06	1.85		1.91	3.09
0900	1-1/2" diameter		35	.229		.32	9.50		9.82	15.95
0950	For each additional inch of depth, add		165	.048		.08	2.01		2.09	3.40
1000	For ceiling installations, add						40%			

Division Notes

		CREW	DAILY OUTPUT	LABOR-HOURS	UNIT	2010 BARE COSTS				TOTAL INCL O&P
						MAT.	LABOR	EQUIP.	TOTAL	

Estimating Tips

04 05 00 Common Work Results for Masonry

- The terms *mortar* and *grout* are often used interchangeably, and incorrectly. Mortar is used to bed masonry units, seal the entry of air and moisture, provide architectural appearance, and allow for size variations in the units. Grout is used primarily in reinforced masonry construction and is used to bond the masonry to the reinforcing steel. Common mortar types are M(2500 psi), S(1800 psi), N(750 psi), and O(350 psi), and conform to ASTM C270. Grout is either fine or coarse and conforms to ASTM C476, and in-place strengths generally exceed 2500 psi. Mortar and grout are different components of masonry construction and are placed by entirely different methods. An estimator should be aware of their unique uses and costs.

- Waste, specifically the loss/droppings of mortar and the breakage of brick and block, is included in all masonry assemblies in this division. A factor of 25% is added for mortar and 3% for brick and concrete masonry units.

- Scaffolding or staging is not included in any of the Division 4 costs. Refer to Subdivision 01 54 23 for scaffolding and staging costs.

04 20 00 Unit Masonry

- The most common types of unit masonry are brick and concrete masonry. The major classifications of brick are building brick (ASTM C62), facing brick (ASTM C216), glazed brick, fire brick, and pavers. Many varieties of texture and appearance can exist within these classifications, and the estimator would be wise to check local custom and availability within the project area. For repair and remodeling jobs, matching the existing brick may be the most important criteria.

- Brick and concrete block are priced by the piece and then converted into a price per square foot of wall. Openings less than two square feet are generally ignored by the estimator because any savings in units used is offset by the cutting and trimming required.

- It is often difficult and expensive to find and purchase small lots of historic brick. Costs can vary widely. Many design issues affect costs, selection of mortar mix, and repairs or replacement of masonry materials. Cleaning techniques must be reflected in the estimate.

- All masonry walls, whether interior or exterior, require bracing. The cost of bracing walls during construction should be included by the estimator, and this bracing must remain in place until permanent bracing is complete. Permanent bracing of masonry walls is accomplished by masonry itself, in the form of pilasters or abutting wall corners, or by anchoring the walls to the structural frame. Accessories in the form of anchors, anchor slots, and ties are used, but their supply and installation can be by different trades. For instance, anchor slots on spandrel beams and columns are supplied and welded in place by the steel fabricator, but the ties from the slots into the masonry are installed by the bricklayer. Regardless of the installation method, the estimator must be certain that these accessories are accounted for in pricing.

Reference Numbers

Reference numbers are shown in shaded boxes at the beginning of some major classifications. These numbers refer to related items in the Reference Section. The reference information may be an estimating procedure, an alternate pricing method, or technical information.

Note: Not all subdivisions listed here necessarily appear in this publication.

04 01 Maintenance of Masonry

04 01 20 – Maintenance of Unit Masonry

04 01 20.20 Pointing Masonry

		Crew	Daily Output	Labor-Hours	Unit	Material	2010 Bare Costs Labor	2010 Bare Costs Equipment	Total	Total Incl O&P
0010	**POINTING MASONRY**									
0300	Cut and repoint brick, hard mortar, running bond	1 Bric	80	.100	S.F.	.53	4.18		4.71	7.30
0320	Common bond		77	.104		.53	4.34		4.87	7.60
0360	Flemish bond		70	.114		.56	4.77		5.33	8.25
0400	English bond		65	.123		.56	5.15		5.71	8.85
0600	Soft old mortar, running bond		100	.080		.53	3.34		3.87	5.95
0620	Common bond		96	.083		.53	3.48		4.01	6.20
0640	Flemish bond		90	.089		.56	3.71		4.27	6.55
0680	English bond		82	.098		.56	4.07		4.63	7.15
0700	Stonework, hard mortar		140	.057	L.F.	.70	2.39		3.09	4.61
0720	Soft old mortar		160	.050	"	.70	2.09		2.79	4.13
1000	Repoint, mask and grout method, running bond		95	.084	S.F.	.70	3.52		4.22	6.40
1020	Common bond		90	.089		.70	3.71		4.41	6.70
1040	Flemish bond		86	.093		.74	3.88		4.62	7.05
1060	English bond		77	.104		.74	4.34		5.08	7.80
2000	Scrub coat, sand grout on walls, minimum		120	.067		2.91	2.78		5.69	7.70
2020	Maximum		98	.082		4.04	3.41		7.45	9.95
9000	Minimum labor/equipment charge		3	2.667	Job		111		111	179

04 01 20.30 Pointing CMU

		Crew	Daily Output	Labor-Hours	Unit	Material	Labor	Equipment	Total	Total Incl O&P
0010	**POINTING CMU**									
0300	Cut and repoint block, hard mortar, running bond	1 Bric	190	.042	S.F.	.22	1.76		1.98	3.07
0310	Stacked bond		200	.040		.22	1.67		1.89	2.93
0600	Soft old mortar, running bond		230	.035		.22	1.45		1.67	2.58
0610	Stacked bond		245	.033		.22	1.36		1.58	2.43

04 01 20.50 Toothing Masonry

		Crew	Daily Output	Labor-Hours	Unit	Material	Labor	Equipment	Total	Total Incl O&P
0010	**TOOTHING MASONRY**									
0500	Brickwork, soft old mortar	1 Clab	40	.200	V.L.F.		6.60		6.60	10.85
0520	Hard mortar		30	.267			8.85		8.85	14.50
0700	Blockwork, soft old mortar		70	.114			3.78		3.78	6.20
0720	Hard mortar		50	.160			5.30		5.30	8.70
9000	Minimum labor/equipment charge		4	2	Job		66		66	109

04 01 30 – Unit Masonry Cleaning

04 01 30.20 Cleaning Masonry

		Crew	Daily Output	Labor-Hours	Unit	Material	Labor	Equipment	Total	Total Incl O&P
0010	**CLEANING MASONRY**									
0200	Chemical cleaning, new construction, brush and wash, minimum	D-1	1000	.016	S.F.	.05	.60		.65	1.02
0220	Average		800	.020		.07	.75		.82	1.29
0240	Maximum		600	.027		.09	1.01		1.10	1.72
0260	Light restoration, minimum		800	.020		.08	.75		.83	1.29
0270	Average		400	.040		.11	1.51		1.62	2.56
0280	Maximum		330	.048		.15	1.83		1.98	3.11
0300	Heavy restoration, minimum		600	.027		.09	1.01		1.10	1.72
0310	Average		400	.040		.13	1.51		1.64	2.58
0320	Maximum		250	.064		.18	2.41		2.59	4.07
0400	High pressure water only, minimum	B-9	2000	.020			.67	.11	.78	1.22
0420	Average		1500	.027			.89	.15	1.04	1.63
0440	Maximum		1000	.040			1.34	.22	1.56	2.45
0800	High pressure water and chemical, minimum		1800	.022		.09	.74	.12	.95	1.46
0820	Average		1200	.033		.14	1.12	.19	1.45	2.19
0840	Maximum		800	.050		.18	1.68	.28	2.14	3.26
1200	Sandblast, wet system, minimum		1750	.023		.51	.77	.13	1.41	1.96
1220	Average		1100	.036		.76	1.22	.20	2.18	3.06
1240	Maximum		700	.057		1.02	1.91	.32	3.25	4.61

04 01 Maintenance of Masonry

04 01 30 – Unit Masonry Cleaning

04 01 30.20 Cleaning Masonry

		Crew	Daily Output	Labor-Hours	Unit	Material	2010 Bare Costs Labor	Equipment	Total	Total Incl O&P
1400	Dry system, minimum	B-9	2500	.016	S.F.	.51	.54	.09	1.14	1.54
1420	Average		1750	.023		.76	.77	.13	1.66	2.24
1440	Maximum		1000	.040		1.02	1.34	.22	2.58	3.57
1800	For walnut shells, add					.50			.50	.55
1820	For corn chips, add					.50			.50	.55
2000	Steam cleaning, minimum	B-9	3000	.013			.45	.07	.52	.81
2020	Average		2500	.016			.54	.09	.63	.98
2040	Maximum		1500	.027			.89	.15	1.04	1.63
4000	Add for masking doors and windows	1 Clab	800	.010		.06	.33		.39	.61
4200	Add for pedestrian protection				Job					10%
9000	Minimum labor/equipment charge	D-4	2	16	"		600	66	666	1,025

04 05 Common Work Results for Masonry

04 05 05 – Selective Masonry Demolition

04 05 05.10 Selective Demolition

		Crew	Daily Output	Labor-Hours	Unit	Material	2010 Bare Costs Labor	Equipment	Total	Total Incl O&P
0010	**SELECTIVE DEMOLITION** R024119-10									
0200	Bond beams, 8" block with #4 bar	2 Clab	32	.500	L.F.		16.55		16.55	27
0300	Concrete block walls, unreinforced, 2" thick		1200	.013	S.F.		.44		.44	.72
0310	4" thick		1150	.014			.46		.46	.76
0320	6" thick		1100	.015			.48		.48	.79
0330	8" thick		1050	.015			.50		.50	.83
0340	10" thick		1000	.016			.53		.53	.87
0360	12" thick		950	.017			.56		.56	.92
0380	Reinforced alternate courses, 2" thick		1130	.014			.47		.47	.77
0390	4" thick		1080	.015			.49		.49	.80
0400	6" thick		1035	.015			.51		.51	.84
0410	8" thick		990	.016			.53		.53	.88
0420	10" thick		940	.017			.56		.56	.93
0430	12" thick		890	.018			.60		.60	.98
0440	Reinforced alternate courses & vertically 48" OC, 4" thick		900	.018			.59		.59	.97
0450	6" thick		850	.019			.62		.62	1.02
0460	8" thick		800	.020			.66		.66	1.09
0480	10" thick		750	.021			.71		.71	1.16
0490	12" thick		700	.023			.76		.76	1.24
1000	Chimney, 16" x 16", soft old mortar	1 Clab	55	.145	C.F.		4.81		4.81	7.90
1020	Hard mortar		40	.200			6.60		6.60	10.85
1030	16" x 20", soft old mortar		55	.145			4.81		4.81	7.90
1040	Hard mortar		40	.200			6.60		6.60	10.85
1050	16" x 24", soft old mortar		55	.145			4.81		4.81	7.90
1060	Hard mortar		40	.200			6.60		6.60	10.85
1080	20" x 20", soft old mortar		55	.145			4.81		4.81	7.90
1100	Hard mortar		40	.200			6.60		6.60	10.85
1110	20" x 24", soft old mortar		55	.145			4.81		4.81	7.90
1120	Hard mortar		40	.200			6.60		6.60	10.85
1140	20" x 32", soft old mortar		55	.145			4.81		4.81	7.90
1160	Hard mortar		40	.200			6.60		6.60	10.85
1200	48" x 48", soft old mortar		55	.145			4.81		4.81	7.90
1220	Hard mortar		40	.200			6.60		6.60	10.85
1250	Metal, high temp steel jacket, 24" diameter	E-2	130	.431	V.L.F.		19.65	12.35	32	48
1260	60" diameter	"	60	.933			42.50	27	69.50	105
1280	Flue lining, up to 12" x 12"	1 Clab	200	.040			1.32		1.32	2.17

04 05 Common Work Results for Masonry

04 05 05 – Selective Masonry Demolition

04 05 05.10 Selective Demolition

		Crew	Daily Output	Labor-Hours	Unit	Material	2010 Bare Costs Labor	Equipment	Total	Total Incl O&P
1282	Up to 24" x 24"	1 Clab	150	.053	V.L.F.		1.77		1.77	2.90
2000	Columns, 8" x 8", soft old mortar		48	.167			5.50		5.50	9.05
2020	Hard mortar		40	.200			6.60		6.60	10.85
2060	16" x 16", soft old mortar		16	.500			16.55		16.55	27
2100	Hard mortar		14	.571			18.90		18.90	31
2140	24" x 24", soft old mortar		8	1			33		33	54.50
2160	Hard mortar		6	1.333			44		44	72.50
2200	36" x 36", soft old mortar		4	2			66		66	109
2220	Hard mortar		3	2.667			88.50		88.50	145
2230	Alternate pricing method, soft old mortar		30	.267	C.F.		8.85		8.85	14.50
2240	Hard mortar		23	.348	"		11.50		11.50	18.90
3000	Copings, precast or masonry, to 8" wide									
3020	Soft old mortar	1 Clab	180	.044	L.F.		1.47		1.47	2.42
3040	Hard mortar	"	160	.050	"		1.66		1.66	2.72
3100	To 12" wide									
3120	Soft old mortar	1 Clab	160	.050	L.F.		1.66		1.66	2.72
3140	Hard mortar	"	140	.057	"		1.89		1.89	3.11
4000	Fireplace, brick, 30" x 24" opening									
4020	Soft old mortar	1 Clab	2	4	Ea.		132		132	217
4040	Hard mortar		1.25	6.400			212		212	350
4100	Stone, soft old mortar		1.50	5.333			177		177	290
4120	Hard mortar		1	8			265		265	435
5000	Veneers, brick, soft old mortar		140	.057	S.F.		1.89		1.89	3.11
5020	Hard mortar		125	.064			2.12		2.12	3.48
5100	Granite and marble, 2" thick		180	.044			1.47		1.47	2.42
5120	4" thick		170	.047			1.56		1.56	2.56
5140	Stone, 4" thick		180	.044			1.47		1.47	2.42
5160	8" thick		175	.046			1.51		1.51	2.48
5400	Alternate pricing method, stone, 4" thick		60	.133	C.F.		4.41		4.41	7.25
5420	8" thick		85	.094	"		3.12		3.12	5.10
9000	Minimum labor/equipment charge		2	4	Job		132		132	217

04 05 13 – Masonry Mortaring

04 05 13.10 Cement

0010	**CEMENT**									
0100	Masonry, 70 lb. bag, T.L. lots				Bag	8.80			8.80	9.70
0150	L.T.L. lots					9.35			9.35	10.30
0200	White, 70 lb. bag, T.L. lots					14.85			14.85	16.35
0250	L.T.L. lots					16.70			16.70	18.35

04 05 13.20 Lime

0010	**LIME**									
0020	Masons, hydrated, 50 lb. bag, T.L. lots				Bag	8.80			8.80	9.70
0050	L.T.L. lots					9.30			9.30	10.25
0200	Finish, double hydrated, 50 lb. bag, T.L. lots					9.90			9.90	10.90
0250	L.T.L. lots					11			11	12.10

04 05 13.30 Mortar

0010	**MORTAR**									
0100	Type M, 1:1:6 mix	1 Brhe	143	.056	C.F.	4.83	1.88		6.71	8.35
0200	Type N, 1:3 mix		143	.056		4.22	1.88		6.10	7.70
0300	Type O, 1:3 mix		143	.056		6.45	1.88		8.33	10.15
0400	Type PM, 1:1:6 mix, 2500 psi		143	.056		6.85	1.88		8.73	10.60
0500	Type S, 1/2:1:4 mix		143	.056		7.25	1.88		9.13	11.05
2000	With portland cement and lime									

04 05 Common Work Results for Masonry

04 05 13 – Masonry Mortaring

04 05 13.30 Mortar

		Crew	Daily Output	Labor-Hours	Unit	Material	2010 Bare Costs Labor	Equipment	Total	Total Incl O&P
2100	Type M, 1:1/4:3 mix	1 Brhe	143	.056	C.F.	8.80	1.88		10.68	12.75
2200	Type N, 1:1:6 mix, 750 psi		143	.056		6.80	1.88		8.68	10.50
2300	Type O, 1:2:9 mix (Pointing Mortar)		143	.056		7.50	1.88		9.38	11.30
2400	Type PL, 1:1/2:4 mix, 2500 psi		143	.056		7.45	1.88		9.33	11.25
2500	Type K, 1:3:12 mix, 75 psi		143	.056		6.45	1.88		8.33	10.15
2600	Type S, 1:1/2:4 mix, 1800 psi		143	.056		9.15	1.88		11.03	13.10
2650	Pre-mixed, type S or N					4.94			4.94	5.45
2700	Mortar for glass block	1 Brhe	143	.056		10	1.88		11.88	14.05
2900	Mortar for Fire Brick, 80 lb. bag, T.L. Lots				Bag	25.50			25.50	28

04 05 16 – Masonry Grouting

04 05 16.30 Grouting

		Crew	Daily Output	Labor-Hours	Unit	Material	Labor	Equipment	Total	Total Incl O&P
0010	**GROUTING**									
0011	Bond beams & lintels, 8" deep, 6" thick, 0.15 C.F. per L.F.	D-4	1480	.022	L.F.	.64	.81	.09	1.54	2.10
0020	8" thick, 0.2 C.F. per L.F.		1400	.023		.99	.86	.09	1.94	2.56
0050	10" thick, 0.25 C.F. per L.F.		1200	.027		1.06	1	.11	2.17	2.89
0060	12" thick, 0.3 C.F. per L.F.		1040	.031		1.28	1.16	.13	2.57	3.39
0200	Concrete block cores, solid, 4" thk., by hand, 0.067 C.F./S.F. of wall	D-8	1100	.036	S.F.	.28	1.40		1.68	2.56
0210	6" thick, pumped, 0.175 C.F. per S.F.	D-4	720	.044		.74	1.67	.18	2.59	3.69
0250	8" thick, pumped, 0.258 C.F. per S.F.		680	.047		1.10	1.77	.19	3.06	4.24
0300	10" thick, pumped, 0.340 C.F. per S.F.		660	.048		1.45	1.82	.20	3.47	4.72
0350	12" thick, pumped, 0.422 C.F. per S.F.		640	.050		1.79	1.88	.21	3.88	5.20
0500	Cavity walls, 2" space, pumped, 0.167 C.F./S.F. of wall		1700	.019		.71	.71	.08	1.50	2
0550	3" space, 0.250 C.F./S.F.		1200	.027		1.06	1	.11	2.17	2.89
0600	4" space, 0.333 C.F. per S.F.		1150	.028		1.42	1.05	.12	2.59	3.36
0700	6" space, 0.500 C.F. per S.F.		800	.040		2.13	1.50	.17	3.80	4.92
0800	Door frames, 3' x 7' opening, 2.5 C.F. per opening		60	.533	Opng.	10.65	20	2.21	32.86	46
0850	6' x 7' opening, 3.5 C.F. per opening		45	.711	"	14.90	26.50	2.94	44.34	62
2000	Grout, C476, for bond beams, lintels and CMU cores		350	.091	C.F.	4.25	3.44	.38	8.07	10.60
9000	Minimum labor/equipment charge	1 Bric	2	4	Job		167		167	269

04 05 19 – Masonry Anchorage and Reinforcing

04 05 19.05 Anchor Bolts

		Crew	Daily Output	Labor-Hours	Unit	Material	Labor	Equipment	Total	Total Incl O&P
0010	**ANCHOR BOLTS**									
0020	Hooked, with nut and washer, 1/2" diam., 8" long	1 Bric	200	.040	Ea.	.80	1.67		2.47	3.57
0030	12" long		190	.042		1.53	1.76		3.29	4.51
0040	5/8" diameter, 8" long		180	.044		1.13	1.86		2.99	4.22
0050	12" long		170	.047		1.24	1.96		3.20	4.53
0060	3/4" diameter, 8" long		160	.050		3.20	2.09		5.29	6.90
0070	12" long		150	.053		4	2.23		6.23	8

04 05 19.16 Masonry Anchors

		Crew	Daily Output	Labor-Hours	Unit	Material	Labor	Equipment	Total	Total Incl O&P
0010	**MASONRY ANCHORS**									
0020	For brick veneer, galv., corrugated, 7/8" x 7", 22 Ga.	1 Bric	10.50	.762	C	9.60	32		41.60	61.50
0100	24 Ga.		10.50	.762		9	32		41	61
0150	16 Ga.		10.50	.762		23.50	32		55.50	77
0200	Buck anchors, galv., corrugated, 16 gauge, 2" bend, 8" x 2"		10.50	.762		148	32		180	214
0250	8" x 3"		10.50	.762		138	32		170	203
0300	Adjustable, rectangular, 4-1/8" wide									
0350	Anchor and tie, 3/16" wire, mill galv.									
0400	2-3/4" eye, 3-1/4" tie	1 Bric	1.05	7.619	M	370	320		690	920
0500	4-3/4" tie		1.05	7.619		415	320		735	965
0520	5-1/2" tie		1.05	7.619		440	320		760	995
0550	4-3/4" eye, 3-1/4" tie		1.05	7.619		410	320		730	960

04 05 Common Work Results for Masonry

04 05 19 – Masonry Anchorage and Reinforcing

04 05 19.16 Masonry Anchors		Crew	Daily Output	Labor-Hours	Unit	Material	2010 Bare Costs Labor	Equipment	Total	Total Incl O&P
0570	4-3/4" tie	1 Bric	1.05	7.619	M	465	320		785	1,025
0580	5-1/2" tie		1.05	7.619		480	320		800	1,025
0670	3/16" diameter		10.50	.762	C	20	32		52	73.50
0680	1/4" diameter		10.50	.762		38	32		70	93
0850	8" long, 3/16" diameter		10.50	.762		23	32		55	76.50
0855	1/4" diameter		10.50	.762		45	32		77	101
1000	Rectangular type, galvanized, 1/4" diameter, 2" x 6"		10.50	.762		65.50	32		97.50	123
1050	4" x 6"		10.50	.762		81.50	32		113.50	141
1100	3/16" diameter, 2" x 6"		10.50	.762		35.50	32		67.50	90.50
1150	4" x 6"		10.50	.762		39	32		71	94
1200	Mesh wall tie, 1/2" mesh, hot dip galvanized									
1400	16 gauge, 12" long, 3" wide	1 Bric	9	.889	C	87.50	37		124.50	156
1420	6" wide		9	.889		130	37		167	203
1440	12" wide		8.50	.941		208	39.50		247.50	292
1500	Rigid partition anchors, plain, 8" long, 1" x 1/8"		10.50	.762		91	32		123	151
1550	1" x 1/4"		10.50	.762		150	32		182	216
1580	1-1/2" x 1/8"		10.50	.762		121	32		153	184
1600	1-1/2" x 1/4"		10.50	.762		246	32		278	320
1650	2" x 1/8"		10.50	.762		157	32		189	223
1700	2" x 1/4"		10.50	.762		288	32		320	365
2000	Column flange ties, wire, galvanized									
2300	3/16" diameter, up to 3" wide	1 Bric	10.50	.762	C	77	32		109	136
2350	To 5" wide		10.50	.762		84	32		116	144
2400	To 7" wide		10.50	.762		90	32		122	150
2600	To 9" wide		10.50	.762		97	32		129	158
2650	1/4" diameter, up to 3" wide		10.50	.762		97.50	32		129.50	158
2700	To 5" wide		10.50	.762		108	32		140	170
2800	To 7" wide		10.50	.762		119	32		151	182
2850	To 9" wide		10.50	.762		128	32		160	192
2900	For hot dip galvanized, add					35%				
4000	Channel slots, 1-3/8" x 1/2" x 8"									
4100	12 gauge, plain	1 Bric	10.50	.762	C	213	32		245	286
4150	16 gauge, galvanized	"	10.50	.762	"	130	32		162	194
4200	Channel slot anchors									
4300	16 gauge, galvanized, 1-1/4" x 3-1/2"				C	49.50			49.50	54.50
4350	1-1/4" x 5-1/2"					62.50			62.50	68.50
4400	1-1/4" x 7-1/2"					71.50			71.50	78.50
4500	1/8" plain, 1-1/4" x 3-1/2"					90			90	99
4550	1-1/4" x 5-1/2"					96			96	106
4600	1-1/4" x 7-1/2"					105			105	115
4700	For corrugation, add					48.50			48.50	53.50
4750	For hot dip galvanized, add					35%				
5000	Dowels									
5100	Plain, 1/4" diameter, 3" long				C	35.50			35.50	39
5150	4" long					39.50			39.50	43.50
5200	6" long					48			48	53
5300	3/8" diameter, 3" long					47			47	51.50
5350	4" long					58.50			58.50	64
5400	6" long					67.50			67.50	74
5500	1/2" diameter, 3" long					68.50			68.50	75.50
5550	4" long					81.50			81.50	89.50
5600	6" long					108			108	119
5700	5/8" diameter, 3" long					94			94	103

04 05 Common Work Results for Masonry

04 05 19 – Masonry Anchorage and Reinforcing

04 05 19.16 Masonry Anchors		Crew	Daily Output	Labor-Hours	Unit	Material	2010 Bare Costs Labor	Equipment	Total	Total Incl O&P
5750	4" long				C	116			116	127
5800	6" long					159			159	175
6000	3/4" diameter, 3" long					117			117	128
6100	4" long					150			150	165
6150	6" long					212			212	233
6300	For hot dip galvanized, add					35%				

04 05 19.26 Masonry Reinforcing Bars

		Crew	Daily Output	Labor-Hours	Unit	Material	Labor	Equipment	Total	Total Incl O&P
0010	**MASONRY REINFORCING BARS**									
0015	Steel bars A615, placed horiz., #3 & #4 bars	1 Bric	450	.018	Lb.	.40	.74		1.14	1.63
0020	#5 & #6 bars		800	.010		.40	.42		.82	1.11
0050	Placed vertical, #3 & #4 bars		350	.023		.40	.95		1.35	1.98
0060	#5 & #6 bars		650	.012		.40	.51		.91	1.27
0500	Joint reinforcing, ladder type, mill std galvanized									
0600	9 ga. sides, 9 ga. ties, 4" wall	1 Bric	30	.267	C.L.F.	9.95	11.15		21.10	29
0650	6" wall		30	.267		11.80	11.15		22.95	31
0700	8" wall		25	.320		12.40	13.35		25.75	35
0750	10" wall		20	.400		13.45	16.70		30.15	42
0800	12" wall		20	.400		14.15	16.70		30.85	42.50
1000	Truss type									
1100	9 ga. sides, 9 ga. ties, 4" wall	1 Bric	30	.267	C.L.F.	15.75	11.15		26.90	35.50
1150	6" wall		30	.267		16.65	11.15		27.80	36.50
1200	8" wall		25	.320		17.40	13.35		30.75	40.50
1250	10" wall		20	.400		18.20	16.70		34.90	47
1300	12" wall		20	.400		19.20	16.70		35.90	48
1500	3/16" sides, 9 ga. ties, 4" wall		30	.267		22.50	11.15		33.65	42.50
1550	6" wall		30	.267		19.65	11.15		30.80	39.50
1600	8" wall		25	.320		20	13.35		33.35	43.50
1650	10" wall		20	.400		21	16.70		37.70	50
1700	12" wall		20	.400		24.50	16.70		41.20	54
2000	3/16" sides, 3/16" ties, 4" wall		30	.267		22.50	11.15		33.65	43
2050	6" wall		30	.267		23	11.15		34.15	43
2100	8" wall		25	.320		23.50	13.35		36.85	47
2150	10" wall		20	.400		25	16.70		41.70	54
2200	12" wall		20	.400		30	16.70		46.70	60
2500	Cavity truss type, galvanized									
2600	9 ga. sides, 9 ga. ties, 4" wall	1 Bric	25	.320	C.L.F.	29.50	13.35		42.85	54
2650	6" wall		25	.320		30	13.35		43.35	54
2700	8" wall		20	.400		30.50	16.70		47.20	60.50
2750	10" wall		15	.533		33	22.50		55.50	72.50
2800	12" wall		15	.533		33	22.50		55.50	72.50
3000	3/16" sides, 9 ga. ties, 4" wall		25	.320		38	13.35		51.35	63.50
3050	6" wall		25	.320		38.50	13.35		51.85	64
3100	8" wall		20	.400		39.50	16.70		56.20	70.50
3150	10" wall		15	.533		40	22.50		62.50	80
3200	12" wall		15	.533		42	22.50		64.50	82
3500	For hot dip galvanizing, add					80%				

04 05 23 – Masonry Accessories

04 05 23.13 Control Joint

		Crew	Daily Output	Labor-Hours	Unit	Material	Labor	Equipment	Total	Total Incl O&P
0010	**CONTROL JOINT**									
0020	Rubber, for double wythe 8" minimum wall (Brick/CMU)	1 Bric	400	.020	L.F.	1.49	.84		2.33	2.98
0025	"T" shaped		320	.025		1.07	1.04		2.11	2.86
0030	Cross-shaped for CMU units		280	.029		1.44	1.19		2.63	3.50

04 05 Common Work Results for Masonry

04 05 23 – Masonry Accessories

04 05 23.13 Control Joint		Crew	Daily Output	Labor-Hours	Unit	Material	2010 Bare Costs Labor	2010 Bare Costs Equipment	Total	Total Incl O&P
0050	PVC, for double wythe 8" minimum wall (Brick/CMU)	1 Bric	400	.020	L.F.	1.09	.84		1.93	2.54
0120	"T" shaped		320	.025		.64	1.04		1.68	2.38
0160	Cross-shaped for CMU units		280	.029		.88	1.19		2.07	2.89

04 05 23.19 Vent Box

		Crew	Daily Output	Labor-Hours	Unit	Material	Labor	Equipment	Total	Total Incl O&P
0010	**VENT BOX**									
0020	Extruded aluminum, 4" deep, 2-3/8" x 8-1/8"	1 Bric	30	.267	Ea.	30.50	11.15		41.65	51.50
0050	5" x 8-1/8"		25	.320		40	13.35		53.35	65.50
0100	2-1/4" x 25"		25	.320		69.50	13.35		82.85	98
0150	5" x 16-1/2"		22	.364		58.50	15.20		73.70	88.50
0200	6" x 16-1/2"		22	.364		81	15.20		96.20	114
0250	7-3/4" x 16-1/2"		20	.400		69	16.70		85.70	103
0400	For baked enamel finish, add					35%				
0500	For cast aluminum, painted, add					60%				
1000	Stainless steel ventilators, 6" x 6"	1 Bric	25	.320		126	13.35		139.35	160
1050	8" x 8"		24	.333		132	13.90		145.90	169
1100	12" x 12"		23	.348		153	14.50		167.50	192
1150	12" x 6"		24	.333		134	13.90		147.90	170
1200	Foundation block vent, galv., 1-1/4" thk, 8" high, 16" long, no damper		30	.267		24.50	11.15		35.65	45
1250	For damper, add					8.20			8.20	9

04 21 Clay Unit Masonry

04 21 13 – Brick Masonry

04 21 13.13 Brick Veneer Masonry

			Crew	Daily Output	Labor-Hours	Unit	Material	Labor	Equipment	Total	Total Incl O&P
0010	**BRICK VENEER MASONRY**, T.L. lots, excl. scaff., grout & reinforcing	R042110-20									
0015	Material costs incl. 3% brick and 25% mortar waste										
2000	Standard, sel. common, 4" x 2-2/3" x 8", (6.75/S.F.)	R042110-50	D-8	230	.174	S.F.	3.61	6.70		10.31	14.70
2020	Standard, red, 4" x 2-2/3" x 8", running bond (6.75/SF)			220	.182		4.67	7		11.67	16.40
2050	Full header every 6th course (7.88/S.F.)			185	.216		5.45	8.35		13.80	19.40
2100	English, full header every 2nd course (10.13/S.F.)			140	.286		6.95	11		17.95	25.50
2150	Flemish, alternate header every course (9.00/S.F.)			150	.267		6.20	10.25		16.45	23.50
2200	Flemish, alt. header every 6th course (7.13/S.F.)			205	.195		4.93	7.50		12.43	17.50
2250	Full headers throughout (13.50/S.F.)			105	.381		9.25	14.65		23.90	33.50
2300	Rowlock course (13.50/S.F.)			100	.400		9.25	15.40		24.65	35
2350	Rowlock stretcher (4.50/S.F.)			310	.129		3.13	4.97		8.10	11.45
2400	Soldier course (6.75/S.F.)			200	.200		4.67	7.70		12.37	17.55
2450	Sailor course (4.50/S.F.)			290	.138		3.13	5.30		8.43	12
2600	Buff or gray face, running bond, (6.75/S.F.)			220	.182		4.94	7		11.94	16.70
2700	Glazed face brick, running bond			210	.190		11.95	7.35		19.30	25
2750	Full header every 6th course (7.88/S.F.)			170	.235		13.95	9.05		23	30
3000	Jumbo, 6" x 4" x 12" running bond (3.00/S.F.)			435	.092		4.87	3.54		8.41	11.05
3050	Norman, 4" x 2-2/3" x 12" running bond, (4.5/S.F.)			320	.125		5.70	4.81		10.51	14
3100	Norwegian, 4" x 3-1/5" x 12" (3.75/S.F.)			375	.107		4.15	4.11		8.26	11.15
3150	Economy, 4" x 4" x 8" (4.50/S.F.)			310	.129		4.19	4.97		9.16	12.60
3200	Engineer, 4" x 3-1/5" x 8" (5.63/S.F.)			260	.154		3.44	5.90		9.34	13.35
3250	Roman, 4" x 2" x 12" (6.00/S.F.)			250	.160		5.75	6.15		11.90	16.25
3300	SCR, 6" x 2-2/3" x 12" (4.50/S.F.)			310	.129		5.25	4.97		10.22	13.75
3350	Utility, 4" x 4" x 12" (3.00/S.F.)			450	.089		4.50	3.42		7.92	10.45
3400	For cavity wall construction, add							15%			
3450	For stacked bond, add							10%			
3500	For interior veneer construction, add							15%			
3550	For curved walls, add							30%			

04 21 Clay Unit Masonry

04 21 13 – Brick Masonry

04 21 13.13 Brick Veneer Masonry	Crew	Daily Output	Labor-Hours	Unit	Material	2010 Bare Costs Labor	Equipment	Total	Total Incl O&P
9000 Minimum labor/equipment charge	D-1	2	8	Job		300		300	485

04 21 13.14 Thin Brick Veneer

0010 **THIN BRICK VENEER**									
0015 Material costs incl. 3% brick and 25% mortar waste									
0020 On & incl. metal panel support sys, modular, 2-2/3" x 5/8" x 8", red	D-7	92	.174	S.F.	8.75	6.15		14.90	19.25
0100 Closure, 4" x 5/8" x 8"		110	.145		8.75	5.15		13.90	17.65
0110 Norman, 2-2/3" x 5/8" x 12"		110	.145		8.95	5.15		14.10	17.85
0120 Utility, 4" x 5/8" x 12"		125	.128		8.95	4.52		13.47	16.90
0130 Emperor, 4" x 3/4" x 16"		175	.091		10.15	3.23		13.38	16.25
0140 Super emperor, 8" x 3/4" x 16"		195	.082		10.15	2.89		13.04	15.70
0150 For L shaped corners with 4" return, add				L.F.	9.25			9.25	10.20
0200 On masonry/plaster back-up, modular, 2-2/3" x 5/8" x 8", red	D-7	137	.117	S.F.	3.95	4.12		8.07	10.80
0210 Closure, 4" x 5/8" x 8"		165	.097		3.95	3.42		7.37	9.70
0220 Norman, 2-2/3" x 5/8" x 12"		165	.097		4.15	3.42		7.57	9.90
0230 Utility, 4" x 5/8" x 12"		185	.086		4.15	3.05		7.20	9.35
0240 Emperor, 4" x 3/4" x 16"		260	.062		5.35	2.17		7.52	9.30
0250 Super emperor, 8" x 3/4" x 16"		285	.056		5.35	1.98		7.33	9
0260 For L shaped corners with 4" return, add				L.F.	9.25			9.25	10.20
0270 For embedment into pre-cast concrete panels, add				S.F.	14.40			14.40	15.85

04 21 13.15 Chimney

0010 **CHIMNEY**, excludes foundation, scaffolding, grout and reinforcing									
0100 Brick, 16" x 16", 8" flue	D-1	18.20	.879	V.L.F.	21	33		54	77
0150 16" x 20" with one 8" x 12" flue		16	1		33.50	37.50		71	97.50
0200 16" x 24" with two 8" x 8" flues		14	1.143		49	43		92	124
0250 20" x 20" with one 12" x 12" flue		13.70	1.168		38.50	44		82.50	114
0300 20" x 24" with two 8" x 12" flues		12	1.333		55.50	50.50		106	142
0350 20" x 32" with two 12" x 12" flues		10	1.600		68	60.50		128.50	172

04 21 13.18 Columns

0010 **COLUMNS**, solid, excludes scaffolding, grout and reinforcing									
0050 Brick, 8" x 8", 9 brick per VLF	D-1	56	.286	V.L.F.	6.05	10.75		16.80	24
0100 12" x 8", 13.5 brick per VLF		37	.432		9.10	16.30		25.40	36
0200 12" x 12", 20 brick per VLF		25	.640		13.50	24		37.50	54
0300 16" x 12", 27 brick per VLF		19	.842		18.20	32		50.20	71
0400 16" x 16", 36 brick per VLF		14	1.143		24.50	43		67.50	96
0500 20" x 16", 45 brick per VLF		11	1.455		30.50	55		85.50	122
0600 20" x 20", 56 brick per VLF		9	1.778		38	67		105	150
0700 24" x 20", 68 brick per VLF		7	2.286		46	86		132	190
0800 24" x 24", 81 brick per VLF		6	2.667		54.50	101		155.50	222
1000 36" x 36", 182 brick per VLF		3	5.333		123	201		324	460
9000 Minimum labor/equipment charge		2	8	Job		300		300	485

04 21 13.30 Oversized Brick

0010 **OVERSIZED BRICK**, excludes scaffolding, grout and reinforcing									
0100 Veneer, 4" x 2.25" x 16"	D-8	387	.103	S.F.	6.40	3.98		10.38	13.45
0105 4" x 2.75" x 16"		412	.097		5.80	3.74		9.54	12.35
0110 4" x 4" x 16"		460	.087		5.30	3.35		8.65	11.25
0120 4" x 8" x 16"		533	.075		6.10	2.89		8.99	11.35
0125 Loadbearing, 6" x 4" x 16", grouted and reinforced		387	.103		8.95	3.98		12.93	16.25
0130 8" x 4" x 16", grouted and reinforced		327	.122		9.45	4.71		14.16	18
0135 6" x 8" x 16", grouted and reinforced		440	.091		8.70	3.50		12.20	15.20
0140 8" x 8" x 16", grouted and reinforced		400	.100		9.55	3.85		13.40	16.75
0145 Curtainwall/reinforced veneer, 6" x 4" x 16"		387	.103		13.25	3.98		17.23	21
0150 8" x 4" x 16"		327	.122		15.75	4.71		20.46	25

04 21 Clay Unit Masonry

04 21 13 – Brick Masonry

04 21 13.30 Oversized Brick		Crew	Daily Output	Labor-Hours	Unit	Material	2010 Bare Costs Labor	Equipment	Total	Total Incl O&P
0155	6" x 8" x 16"	D-8	440	.091	S.F.	13.35	3.50		16.85	20.50
0160	8" x 8" x 16"		400	.100		15.85	3.85		19.70	23.50
0200	For 1 to 3 slots in face, add					15%				
0210	For 4 to 7 slots in face, add					25%				
0220	For bond beams, add					20%				
0230	For bullnose shapes, add					20%				
0240	For open end knockout, add					10%				
0250	For white or gray color group, add					10%				
0260	For 135 degree corner, add					250%				

04 21 13.35 Common Building Brick

0010	**COMMON BUILDING BRICK**, C62, TL lots, material only	R042110-20								
0020	Standard, minimum					M	400			400	440
0050	Average (select)					"	430			430	475

04 21 13.40 Structural Brick

0010	**STRUCTURAL BRICK** C652, Grade SW, incl. mortar, scaffolding not incl.									
0100	Standard unit, 4-5/8" x 2-3/4" x 9-5/8"	D-8	245	.163	S.F.	5.15	6.30		11.45	15.80
0120	Bond beam		225	.178		9.65	6.85		16.50	21.50
0140	V cut bond beam		225	.178		10.45	6.85		17.30	22.50
0160	Stretcher quoin, 5-5/8" x 2-3/4" x 9-5/8"		245	.163		10.10	6.30		16.40	21
0180	Corner quoin		245	.163		11.30	6.30		17.60	22.50
0200	Corner, 45 deg, 4-5/8" x 2-3/4" x 10-7/16"		235	.170		11.70	6.55		18.25	23.50

04 21 13.45 Face Brick

0010	**FACE BRICK** Material Only, C216, TL lots	R042110-20								
0300	Standard modular, 4" x 2-2/3" x 8", minimum					M	585			585	640
0350	Maximum	R042110-50					700			700	770
0450	Economy, 4" x 4" x 8", minimum						815			815	895
0500	Maximum						1,050			1,050	1,150
0510	Economy, 4" x 4" x 12", minimum						1,200			1,200	1,325
0520	Maximum						1,550			1,550	1,725
0550	Jumbo, 6" x 4" x 12", minimum						1,475			1,475	1,625
0600	Maximum						1,875			1,875	2,075
0610	Jumbo, 8" x 4" x 12", minimum						1,475			1,475	1,625
0620	Maximum						1,875			1,875	2,075
0650	Norwegian, 4" x 3-1/5" x 12", minimum						980			980	1,075
0700	Maximum						1,275			1,275	1,400
0710	Norwegian, 6" x 3-1/5" x 12", minimum						1,425			1,425	1,575
0720	Maximum						1,875			1,875	2,050
0850	Standard glazed, plain colors, 4" x 2-2/3" x 8", minimum						1,625			1,625	1,800
0900	Maximum						2,125			2,125	2,325
1000	Deep trim shades, 4" x 2-2/3" x 8", minimum						1,900			1,900	2,100
1050	Maximum						2,175			2,175	2,400
1080	Jumbo utility, 4" x 4" x 12"						1,350			1,350	1,500
1120	4" x 8" x 8"						1,675			1,675	1,850
1140	4" x 8" x 16"						4,100			4,100	4,500
1260	Engineer, 4" x 3-1/5" x 8", minimum						505			505	555
1270	Maximum						730			730	805
1350	King, 4" x 2-3/4" x 10", minimum						460			460	505
1360	Maximum						560			560	620
1400	Norman, 4" x 2-3/4" x 12"						460			460	505
1450	Roman, 4" x 2" x 12"						460			460	505
1500	SCR, 6" x 2-2/3" x 12"						460			460	505
1550	Double, 4" x 5-1/3" x 8"						460			460	505

04 21 Clay Unit Masonry

04 21 13 – Brick Masonry

04 21 13.45 Face Brick

		Crew	Daily Output	Labor-Hours	Unit	Material	2010 Bare Costs Labor	Equipment	Total	Total Incl O&P
1600	Triple, 4" x 5-1/3" x 12"				M	460			460	505
1770	Standard modular, double glazed, 4" x 2-2/3" x 8"					2,275			2,275	2,500
1850	Jumbo, colored glazed ceramic, 6" x 4" x 12"					2,400			2,400	2,650
2050	Jumbo utility, glazed, 4" x 4" x 12"					4,550			4,550	5,000
2100	4" x 8" x 8"					5,325			5,325	5,875
2150	4" x 16" x 8"					6,200			6,200	6,825
2170	For less than truck load lots, add					15			15	16.50
2180	For buff or gray brick, add					16			16	17.60
3050	Used brick, minimum					395			395	430
3100	Maximum					375			375	415
3150	Add for brick to match existing work, minimum					5%				
3200	Maximum					50%				

04 21 26 – Glazed Structural Clay Tile Masonry

04 21 26.10 Structural Facing Tile

		Crew	Daily Output	Labor-Hours	Unit	Material	Labor	Equipment	Total	Total Incl O&P
0010	**STRUCTURAL FACING TILE**, std. colors, excl. scaffolding, grout, reinforcing									
0020	6T series, 5-1/3" x 12", 2.3 pieces per S.F., glazed 1 side, 2" thick	D-8	225	.178	S.F.	7.95	6.85		14.80	19.75
0100	4" thick		220	.182		10.75	7		17.75	23
0150	Glazed 2 sides		195	.205		13.85	7.90		21.75	28
0250	6" thick		210	.190		15.75	7.35		23.10	29
0300	Glazed 2 sides		185	.216		18.70	8.35		27.05	34
0400	8" thick		180	.222		21	8.55		29.55	37
9000	Minimum labor/equipment charge	D-1	2	8	Job		300		300	485

04 21 29 – Terra Cotta Masonry

04 21 29.10 Terra Cotta Masonry Components

		Crew	Daily Output	Labor-Hours	Unit	Material	Labor	Equipment	Total	Total Incl O&P
0010	**TERRA COTTA MASONRY COMPONENTS**									
0020	Coping, split type, not glazed, 9" wide	D-1	90	.178	L.F.	10.40	6.70		17.10	22.50
0100	13" wide		80	.200		15.75	7.55		23.30	29.50
0200	Coping, split type, glazed, 9" wide		90	.178		17.40	6.70		24.10	30
0250	13" wide		80	.200		22.50	7.55		30.05	37
0500	Partition or back-up blocks, scored, in C.L. lots									
0700	Non-load bearing 12" x 12", 3" thick, special order	D-8	550	.073	S.F.	19.30	2.80		22.10	25.50
0850	8" thick		400	.100		11.05	3.85		14.90	18.40
1000	Load bearing, 12" x 12", 4" thick, in walls		500	.080		6.95	3.08		10.03	12.60
1400	8" thick, in walls		400	.100		11.05	3.85		14.90	18.40
9000	Minimum labor/equipment charge	D-1	2	8	Job		300		300	485

04 22 Concrete Unit Masonry

04 22 10 – Concrete Masonry Units

04 22 10.11 Autoclave Aerated Concrete Block

			Crew	Daily Output	Labor-Hours	Unit	Material	Labor	Equipment	Total	Total Incl O&P
0010	**AUTOCLAVE AERATED CONCRETE BLOCK**, excl. scaffolding, grout & reinforcing										
0050	Solid, 4" x 12" x 24", incl mortar	G	D-8	600	.067	S.F.	1.36	2.57		3.93	5.60
0060	6" x 12" x 24"	G		600	.067		2.03	2.57		4.60	6.35
0070	8" x 8" x 24"	G		575	.070		2.72	2.68		5.40	7.30
0080	10" x 12" x 24"	G		575	.070		3.32	2.68		6	7.95
0090	12" x 12" x 24"	G		550	.073		4.07	2.80		6.87	9

04 22 10.14 Concrete Block, Back-Up

		Crew	Daily Output	Labor-Hours	Unit	Material	Labor	Equipment	Total	Total Incl O&P
0010	**CONCRETE BLOCK, BACK-UP**, C90, 2000 psi									
0020	Normal weight, 8" x 16" units, tooled joint 1 side									
0050	Not-reinforced, 2000 psi, 2" thick	D-8	475	.084	S.F.	1.39	3.24		4.63	6.75
0200	4" thick		460	.087		1.56	3.35		4.91	7.10

04 22 Concrete Unit Masonry

04 22 10 – Concrete Masonry Units

04 22 10.14 Concrete Block, Back-Up

		Crew	Daily Output	Labor-Hours	Unit	Material	2010 Bare Costs Labor	Equipment	Total	Total Incl O&P
0300	6" thick	D-8	440	.091	S.F.	2.17	3.50		5.67	8.05
0350	8" thick		400	.100		2.45	3.85		6.30	8.90
0400	10" thick		330	.121		3.09	4.67		7.76	10.90
0450	12" thick	D-9	310	.155		3.51	5.85		9.36	13.25
1000	Reinforced, alternate courses, 4" thick	D-8	450	.089		1.67	3.42		5.09	7.35
1100	6" thick		430	.093		2.30	3.58		5.88	8.30
1150	8" thick		395	.101		2.58	3.90		6.48	9.10
1200	10" thick		320	.125		3.23	4.81		8.04	11.30
1250	12" thick	D-9	300	.160		3.66	6.05		9.71	13.70
9000	Minimum labor/equipment charge	D-1	2	8	Job		300		300	485

04 22 10.16 Concrete Block, Bond Beam

		Crew	Daily Output	Labor-Hours	Unit	Material	Labor	Equipment	Total	Total Incl O&P
0010	**CONCRETE BLOCK, BOND BEAM**, C90, 2000 psi									
0020	Not including grout or reinforcing									
0125	Regular block, 6" thick	D-8	584	.068	L.F.	2.32	2.64		4.96	6.80
0130	8" high, 8" thick	"	565	.071		2.50	2.73		5.23	7.15
0150	12" thick	D-9	510	.094		3.50	3.55		7.05	9.55
0525	Lightweight, 6" thick	D-8	592	.068		2.55	2.60		5.15	7
2000	Including grout and 2 #5 bars									
2100	Regular block, 8" high, 8" thick	D-8	300	.133	L.F.	4.35	5.15		9.50	13.05
2150	12" thick	D-9	250	.192	"	5.90	7.25		13.15	18.15
9000	Minimum labor/equipment charge	D-1	2	8	Job		300		300	485

04 22 10.19 Concrete Block, Insulation Inserts

		Crew	Daily Output	Labor-Hours	Unit	Material	Labor	Equipment	Total	Total Incl O&P
0010	**CONCRETE BLOCK, INSULATION INSERTS**									
0100	Styrofoam, plant installed, add to block prices									
0200	8" x 16" units, 6" thick				S.F.	1.81			1.81	1.99
0250	8" thick					1.81			1.81	1.99
0300	10" thick					2.13			2.13	2.34
0350	12" thick					2.24			2.24	2.46
0500	8" x 8" units, 8" thick					1.49			1.49	1.64
0550	12" thick					1.81			1.81	1.99

04 22 10.23 Concrete Block, Decorative

		Crew	Daily Output	Labor-Hours	Unit	Material	Labor	Equipment	Total	Total Incl O&P
0010	**CONCRETE BLOCK, DECORATIVE**, C90, 2000 psi									
1000	Fluted high strength									
1100	8" x 16" x 4" thick, flutes 1 side,	D-8	345	.116	S.F.	4.17	4.46		8.63	11.80
1150	Flutes 2 sides		335	.119		5.05	4.60		9.65	12.95
1200	8" thick		300	.133		6.60	5.15		11.75	15.50
1250	For special colors, add					.41			.41	.45
1400	Deep grooved, smooth face									
1450	8" x 16" x 4" thick	D-8	345	.116	S.F.	2.72	4.46		7.18	10.20
1500	8" thick	"	300	.133	"	4.69	5.15		9.84	13.40
1600										
4000	Slump block									
4100	4" face height x 16" x 4" thick	D-1	165	.097	S.F.	3.71	3.66		7.37	10
4150	6" thick		160	.100		5	3.77		8.77	11.55
4200	8" thick		155	.103		5.65	3.89		9.54	12.45
4250	10" thick		140	.114		9.95	4.31		14.26	17.90
4300	12" thick		130	.123		10.45	4.64		15.09	18.95
5000	Split rib profile units, 1" deep ribs, 8 ribs									
5100	8" x 16" x 4" thick	D-8	345	.116	S.F.	3.16	4.46		7.62	10.70
5150	6" thick		325	.123		3.64	4.74		8.38	11.60
5200	8" thick		300	.133		4.17	5.15		9.32	12.85
5250	12" thick	D-9	275	.175		4.94	6.60		11.54	16.05

04 22 Concrete Unit Masonry

04 22 10 – Concrete Masonry Units

04 22 10.23 Concrete Block, Decorative		Crew	Daily Output	Labor-Hours	Unit	Material	2010 Bare Costs Labor	Equipment	Total	Total Incl O&P
9000	Minimum labor/equipment charge	D-1	2	8	Job		300		300	485

04 22 10.24 Concrete Block, Exterior

		Crew	Daily Output	Labor-Hours	Unit	Material	Labor	Equipment	Total	Total Incl O&P
0010	**CONCRETE BLOCK, EXTERIOR**, C90, 2000 psi									
0020	Reinforced alt courses, tooled joints 2 sides									
0100	Normal weight, 8" x 16" x 6" thick	D-8	395	.101	S.F.	2.55	3.90		6.45	9.05
0200	8" thick		360	.111		3.67	4.28		7.95	10.95
0250	10" thick		290	.138		4.29	5.30		9.59	13.25
0300	12" thick	D-9	250	.192		4.53	7.25		11.78	16.65
9000	Minimum labor/equipment charge	D-1	2	8	Job		300		300	485

04 22 10.26 Concrete Block Foundation Wall

		Crew	Daily Output	Labor-Hours	Unit	Material	Labor	Equipment	Total	Total Incl O&P
0010	**CONCRETE BLOCK FOUNDATION WALL**, C90/C145									
0050	Normal-weight, cut joints, horiz joint reinf, no vert reinf									
0200	Hollow, 8" x 16" x 6" thick	D-8	455	.088	S.F.	2.61	3.39		6	8.30
0250	8" thick		425	.094		2.90	3.62		6.52	9.05
0300	10" thick		350	.114		3.55	4.40		7.95	11
0350	12" thick	D-9	300	.160		3.98	6.05		10.03	14.10
1000	Reinforced, #4 vert @ 48"									
1100	Hollow, 8" x 16" block, 4" thick	D-8	455	.088	S.F.	2.49	3.39		5.88	8.20
1125	6" thick		445	.090		3.43	3.46		6.89	9.35
1150	8" thick		415	.096		4.08	3.71		7.79	10.45
1200	10" thick		340	.118		5.05	4.53		9.58	12.90
1250	12" thick	D-9	290	.166		5.85	6.25		12.10	16.50
9000	Minimum labor/equipment charge	D-1	2	8	Job		300		300	485

04 22 10.30 Concrete Block, Interlocking

		Crew	Daily Output	Labor-Hours	Unit	Material	Labor	Equipment	Total	Total Incl O&P
0010	**CONCRETE BLOCK, INTERLOCKING**									
0100	Not including grout or reinforcing									
0200	8" x 16" units, 2,000 psi, 8" thick	D-1	245	.065	S.F.	2.72	2.46		5.18	6.95
0300	12" thick	"	220	.073		4.03	2.74		6.77	8.85
0400	Including grout & reinforcing, 8" thick	D-4	245	.131		7.30	4.91	.54	12.75	16.50
0450	12" thick	"	220	.145		8.80	5.45	.60	14.85	19.05
9000	Minimum labor/equipment charge	D-1	2	8	Job		300		300	485

04 22 10.32 Concrete Block, Lintels

		Crew	Daily Output	Labor-Hours	Unit	Material	Labor	Equipment	Total	Total Incl O&P
0010	**CONCRETE BLOCK, LINTELS**, C90, normal weight									
0100	Including grout and horizontal reinforcing									
0200	8" x 8" x 8", 1 #4 bar	D-4	300	.107	L.F.	4.83	4.01	.44	9.28	12.20
0250	2 #4 bars		295	.108		5	4.08	.45	9.53	12.50
1000	12" x 8" x 8", 1 #4 bar		275	.116		6.80	4.37	.48	11.65	15
1150	2 #5 bars		270	.119		7.15	4.46	.49	12.10	15.55
9000	Minimum labor/equipment charge	D-1	2	8	Job		300		300	485

04 22 10.34 Concrete Block, Partitions

		Crew	Daily Output	Labor-Hours	Unit	Material	Labor	Equipment	Total	Total Incl O&P
0010	**CONCRETE BLOCK, PARTITIONS**, excludes scaffolding									
1000	Lightweight block, tooled joints, 2 sides, hollow									
1100	Not reinforced, 8" x 16" x 4" thick	D-8	440	.091	S.F.	1.65	3.50		5.15	7.45
1150	6" thick		410	.098		2.37	3.76		6.13	8.65
1200	8" thick		385	.104		2.88	4		6.88	9.60
1250	10" thick		370	.108		3.49	4.16		7.65	10.55
1300	12" thick	D-9	350	.137		4.34	5.15		9.49	13.05
1500	Reinforced alternate courses, 4" thick	D-8	435	.092		1.72	3.54		5.26	7.60
1600	6" thick		405	.099		2.46	3.80		6.26	8.80
1650	8" thick		380	.105		2.98	4.05		7.03	9.75
1700	10" thick		365	.110		3.59	4.22		7.81	10.75

04 22 Concrete Unit Masonry

04 22 10 – Concrete Masonry Units

04 22 10.34 Concrete Block, Partitions

		Crew	Daily Output	Labor-Hours	Unit	Material	2010 Bare Costs Labor	Equipment	Total	Total Incl O&P
1750	12" thick	D-9	345	.139	S.F.	4.44	5.25		9.69	13.35
4000	Regular block, tooled joints, 2 sides, hollow									
4100	Not reinforced, 8" x 16" x 4" thick	D-8	430	.093	S.F.	1.47	3.58		5.05	7.35
4150	6" thick		400	.100		2.08	3.85		5.93	8.50
4200	8" thick		375	.107		2.36	4.11		6.47	9.20
4250	10" thick		360	.111		3	4.28		7.28	10.20
4300	12" thick	D-9	340	.141		3.42	5.30		8.72	12.30
4500	Reinforced alternate courses, 8" x 16" x 4" thick	D-8	425	.094		1.58	3.62		5.20	7.60
4550	6" thick		395	.101		2.21	3.90		6.11	8.70
4600	8" thick		370	.108		2.49	4.16		6.65	9.45
4650	10" thick		355	.113		3.43	4.34		7.77	10.75
4700	12" thick	D-9	335	.143		3.57	5.40		8.97	12.60
9000	Minimum labor/equipment charge	D-1	2	8	Job		300		300	485

04 22 10.38 Concrete Brick

		Crew	Daily Output	Labor-Hours	Unit	Material	Labor	Equipment	Total	Total Incl O&P
0010	**CONCRETE BRICK**, C55, grade N, type 1									
0100	Regular, 4 x 2-1/4 x 8	D-8	660	.061	Ea.	.45	2.33		2.78	4.25
0125	Rusticated, 4 x 2-1/4 x 8		660	.061		.50	2.33		2.83	4.30
0150	Frog, 4 x 2-1/4 x 8		660	.061		.49	2.33		2.82	4.28
0200	Double, 4 x 4-7/8 x 8		535	.075		.79	2.88		3.67	5.50

04 22 10.44 Glazed Concrete Block

		Crew	Daily Output	Labor-Hours	Unit	Material	Labor	Equipment	Total	Total Incl O&P
0010	**GLAZED CONCRETE BLOCK** C744									
0100	Single face, 8" x 16" units, 2" thick	D-8	360	.111	S.F.	8.25	4.28		12.53	16
0200	4" thick		345	.116		8.50	4.46		12.96	16.55
0250	6" thick		330	.121		9.15	4.67		13.82	17.55
0300	8" thick		310	.129		9.25	4.97		14.22	18.15
0350	10" thick		295	.136		11.10	5.20		16.30	20.50
0400	12" thick	D-9	280	.171		11.80	6.45		18.25	23.50
0700	Double face, 8" x 16" units, 4" thick	D-8	340	.118		12.20	4.53		16.73	20.50
0750	6" thick		320	.125		14.50	4.81		19.31	23.50
0800	8" thick		300	.133		15.20	5.15		20.35	25
1500	Cove base, 8" x 16", 2" thick		315	.127	L.F.	11	4.89		15.89	19.95
1550	4" thick		285	.140		7.15	5.40		12.55	16.55
1600	6" thick		265	.151		7.70	5.80		13.50	17.80
1650	8" thick		245	.163		8.10	6.30		14.40	19.05
9000	Minimum labor/equipment charge	D-1	2	8	Job		300		300	485

04 23 Glass Unit Masonry

04 23 13 – Vertical Glass Unit Masonry

04 23 13.10 Glass Block

		Crew	Daily Output	Labor-Hours	Unit	Material	Labor	Equipment	Total	Total Incl O&P
0010	**GLASS BLOCK**									
0100	Plain, 4" thick, under 1,000 S.F., 6" x 6"	D-8	115	.348	S.F.	25.50	13.40		38.90	49.50
0150	8" x 8"		160	.250		15.55	9.65		25.20	32.50
0160	end block		160	.250		47	9.65		56.65	67
0170	90 deg corner		160	.250		46	9.65		55.65	66
0180	45 deg corner		160	.250		33.50	9.65		43.15	52.50
0200	12" x 12"		175	.229		20	8.80		28.80	36
0210	4" x 8"		160	.250		15.20	9.65		24.85	32
0220	6" x 8"		160	.250		14.60	9.65		24.25	31.50
0300	1,000 to 5,000 S.F., 6" x 6"		135	.296		25	11.40		36.40	46
0350	8" x 8"		190	.211		15.25	8.10		23.35	30

04 23 Glass Unit Masonry

04 23 13 – Vertical Glass Unit Masonry

04 23 13.10 Glass Block		Crew	Daily Output	Labor-Hours	Unit	Material	2010 Bare Costs Labor	Equipment	Total	Total Incl O&P
0400	12" x 12"	D-8	215	.186	S.F.	19.70	7.15		26.85	33
0410	4" x 8"		215	.186		6	7.15		13.15	18.10
0420	6" x 8"		215	.186		5.70	7.15		12.85	17.75
0700	For solar reflective blocks, add					100%				
1000	Thinline, plain, 3-1/8" thick, under 1,000 S.F., 6" x 6"	D-8	115	.348	S.F.	17.35	13.40		30.75	40.50
1050	8" x 8"		160	.250		9.70	9.65		19.35	26
1400	For cleaning block after installation (both sides), add		1000	.040		.11	1.54		1.65	2.60
9000	Minimum labor/equipment charge	D-1	2	8	Job		300		300	485

04 27 Multiple-Wythe Unit Masonry

04 27 10 – Multiple-Wythe Masonry

04 27 10.10 Cornices

		Crew	Daily Output	Labor-Hours	Unit	Material	Labor	Equipment	Total	Total Incl O&P
0010	**CORNICES**									
0110	Face bricks, 12 brick/S.F., minimum	D-1	30	.533	SF Face	5.95	20		25.95	39
0150	15 brick/S.F., maximum		23	.696	"	7.15	26		33.15	50
9000	Minimum labor/equipment charge		1.50	10.667	Job		400		400	645

04 27 10.30 Brick Walls

			Crew	Daily Output	Labor-Hours	Unit	Material	Labor	Equipment	Total	Total Incl O&P
0010	**BRICK WALLS**, including mortar, excludes scaffolding										
0800	Face brick, 4" thick wall, 6.75 brick/S.F.		D-8	215	.186	S.F.	4.61	7.15		11.76	16.55
0850	Common brick, 4" thick wall, 6.75 brick/S.F.	R042110-50		240	.167		3.33	6.40		9.73	13.95
0900	8" thick, 13.50 bricks per S.F.			135	.296		6.90	11.40		18.30	26
1000	12" thick, 20.25 bricks per S.F.			95	.421		10.40	16.20		26.60	37.50
1050	16" thick, 27.00 bricks per S.F.			75	.533		14.05	20.50		34.55	48.50
1200	Reinforced, face brick, 4" thick wall, 6.75 brick/S.F.			210	.190		4.74	7.35		12.09	17
1220	Common brick, 4" thick wall, 6.75 brick/S.F.			235	.170		3.46	6.55		10.01	14.35
1250	8" thick, 13.50 bricks per S.F.			130	.308		7.15	11.85		19	27
1300	12" thick, 20.25 bricks per S.F.			90	.444		10.80	17.10		27.90	39.50
1350	16" thick, 27.00 bricks per S.F.			70	.571		14.60	22		36.60	51.50
9000	Minimum labor/equipment charge		D-1	2	8	Job		300		300	485

04 41 Dry-Placed Stone

04 41 10 – Dry Placed Stone

04 41 10.10 Rough Stone Wall

			Crew	Daily Output	Labor-Hours	Unit	Material	Labor	Equipment	Total	Total Incl O&P
0011	**ROUGH STONE WALL**, Dry										
0012	Dry laid (no mortar), under 18" thick		D-1	60	.267	C.F.	8.80	10.05		18.85	26
0100	Random fieldstone, under 18" thick	G	D-12	60	.533		8.80	20.50		29.30	42.50
0150	Over 18" thick	G	"	63	.508		10.55	19.40		29.95	42.50
0500	Field stone veneer	G	D-8	120	.333	S.F.	9.95	12.85		22.80	31.50
0510	Valley stone veneer	G		120	.333		9.95	12.85		22.80	31.50
0520	River stone veneer	G		120	.333		9.95	12.85		22.80	31.50
0600	Rubble stone walls, in mortar bed, up to 18" thick	G	D-11	75	.320	C.F.	10.65	12.70		23.35	32
9000	Minimum labor/equipment charge		D-1	2	8	Job		300		300	485

04 43 Stone Masonry

04 43 10 – Masonry with Natural and Processed Stone

04 43 10.10 Bluestone

		Crew	Daily Output	Labor-Hours	Unit	Material	2010 Bare Costs Labor	Equipment	Total	Total Incl O&P
0010	**BLUESTONE**, cut to size									
0500	Sills, natural cleft, 10" wide to 6' long, 1-1/2" thick	D-11	70	.343	L.F.	13.35	13.60		26.95	36.50
0550	2" thick		63	.381		14.60	15.15		29.75	40.50
0600	Smooth finish, 1-1/2" thick		70	.343		25	13.60		38.60	49.50
0650	2" thick		63	.381		30	15.15		45.15	57.50
0800	Thermal finish, 1-1/2" thick		70	.343		30	13.60		43.60	55
0850	2" thick		63	.381		35	15.15		50.15	63
1000	Stair treads, natural cleft, 12" wide, 6' long, 1-1/2" thick	D-10	115	.278		30	11.40	5.70	47.10	57.50
1050	2" thick		105	.305		30	12.45	6.25	48.70	60
1100	Smooth finish, 1-1/2" thick		115	.278		15	11.40	5.70	32.10	41
1150	2" thick		105	.305		15	12.45	6.25	33.70	43.50
1300	Thermal finish, 1-1/2" thick		115	.278		15	11.40	5.70	32.10	41
1350	2" thick		105	.305		15	12.45	6.25	33.70	43.50
9000	Minimum labor/equipment charge	D-1	2.50	6.400	Job		241		241	390

04 43 10.45 Granite

		Crew	Daily Output	Labor-Hours	Unit	Material	Labor	Equipment	Total	Total Incl O&P
0010	**GRANITE**, cut to size									
0050	Veneer, polished face, 3/4" to 1-1/2" thick									
0150	Low price, gray, light gray, etc.	D-10	130	.246	S.F.	26	10.05	5.05	41.10	50
0180	Medium price, pink, brown, etc.		130	.246		28.50	10.05	5.05	43.60	53
0220	High price, red, black, etc.		130	.246		41	10.05	5.05	56.10	67
2500	Steps, copings, etc., finished on more than one surface									
2550	Minimum	D-10	50	.640	C.F.	91	26	13.10	130.10	156
2600	Maximum	"	50	.640	"	146	26	13.10	185.10	216
9000	Minimum labor/equipment charge	D-1	2	8	Job		300		300	485

04 43 10.55 Limestone

		Crew	Daily Output	Labor-Hours	Unit	Material	Labor	Equipment	Total	Total Incl O&P
0010	**LIMESTONE**, cut to size									
0020	Veneer facing panels									
0750	5" thick, 5' x 14' panels	D-10	275	.116	S.F.	52.50	4.76	2.38	59.64	67.50
1000	Sugarcube finish, 2" Thick, 3' x 5' panels		275	.116		35	4.76	2.38	42.14	48.50
1050	3" Thick, 4' x 9' panels		275	.116		38	4.76	2.38	45.14	52
1200	4" Thick, 5' x 11' panels		275	.116		41	4.76	2.38	48.14	55
1400	Sugarcube, textured finish, 4-1/2" thick, 5' x 12'		275	.116		45	4.76	2.38	52.14	59.50
1450	5" thick, 5' x 14' panels		275	.116		53.50	4.76	2.38	60.64	68.50
2000	Coping, sugarcube finish, top & 2 sides		30	1.067	C.F.	68	43.50	22	133.50	169
2100	Sills, lintels, jambs, trim, stops, sugarcube finish, average		20	1.600		68	65.50	33	166.50	215
2150	Detailed		20	1.600		68	65.50	33	166.50	215
2300	Steps, extra hard, 14" wide, 6" rise		50	.640	L.F.	41.50	26	13.10	80.60	102
3000	Quoins, plain finish, 6" x 12" x 12"	D-12	25	1.280	Ea.	38	49		87	121
3050	6" x 16" x 24"	"	25	1.280	"	50.50	49		99.50	134
9000	Minimum labor/equipment charge	D-1	2	8	Job		300		300	485

04 43 10.60 Marble

		Crew	Daily Output	Labor-Hours	Unit	Material	Labor	Equipment	Total	Total Incl O&P
0011	**MARBLE**, ashlar, split face, 4" + or - thick, random									
0040	Lengths 1' to 4' & heights 2" to 7-1/2", average	D-8	175	.229	S.F.	17.75	8.80		26.55	33.50
0100	Base, polished, 3/4" or 7/8" thick, polished, 6" high	D-10	65	.492	L.F.	16.15	20	10.10	46.25	61
0300	Carvings or bas relief, from templates, average		80	.400	S.F.	144	16.35	8.20	168.55	193
0350	Maximum		80	.400	"	335	16.35	8.20	359.55	405
1000	Facing, polished finish, cut to size, 3/4" to 7/8" thick									
1050	Average	D-10	130	.246	S.F.	23.50	10.05	5.05	38.60	47.50
1100	Maximum	"	130	.246	"	55	10.05	5.05	70.10	82
2500	Flooring, polished tiles, 12" x 12" x 3/8" thick									
2510	Thin set, average	D-11	90	.267	S.F.	11.90	10.60		22.50	30
2600	Maximum		90	.267		183	10.60		193.60	218

04 43 Stone Masonry

04 43 10 – Masonry with Natural and Processed Stone

04 43 10.60 Marble		Crew	Daily Output	Labor-Hours	Unit	Material	2010 Bare Costs Labor	Equipment	Total	Total Incl O&P
2700	Mortar bed, average	D-11	65	.369	S.F.	10.65	14.65		25.30	35.50
2740	Maximum		65	.369		183	14.65		197.65	225
3500	Thresholds, 3' long, 7/8" thick, 4" to 5" wide, plain	D-12	24	1.333	Ea.	17.15	51		68.15	101
3550	Beveled		24	1.333	"	19.90	51		70.90	104
3700	Window stools, polished, 7/8" thick, 5" wide		85	.376	L.F.	15.95	14.40		30.35	40.50
9000	Minimum labor/equipment charge	D-1	2	8	Job		300		300	485

04 43 10.75 Sandstone or Brownstone

		Crew	Daily Output	Labor-Hours	Unit	Material	Labor	Equipment	Total	Total Incl O&P
0011	**SANDSTONE OR BROWNSTONE**									
0100	Sawed face veneer, 2-1/2" thick, to 2' x 4' panels	D-10	130	.246	S.F.	20	10.05	5.05	35.10	44
0150	4" thick, to 3'-6" x 8' panels		100	.320		20	13.10	6.55	39.65	50.50
0300	Split face, random sizes		100	.320		13.55	13.10	6.55	33.20	43
9000	Minimum labor/equipment charge	D-1	2.50	6.400	Job		241		241	390

04 43 10.80 Slate

		Crew	Daily Output	Labor-Hours	Unit	Material	Labor	Equipment	Total	Total Incl O&P
0010	**SLATE**									
0040	Pennsylvania - blue gray to black									
0050	Vermont - unfading green, mottled green & purple, gray & purple									
0100	Virginia - blue black									
3100	Stair landings, 1" thick, black, clear	D-1	65	.246	S.F.	20	9.30		29.30	37
3200	Ribbon	"	65	.246	"	22	9.30		31.30	39
3500	Stair treads, sand finish, 1" thick x 12" wide									
3550	Under 3 L.F.	D-10	85	.376	L.F.	22	15.40	7.70	45.10	57
3600	3 L.F. to 6 L.F.	"	120	.267	"	24	10.90	5.45	40.35	49.50
3700	Ribbon, sand finish, 1" thick x 12" wide									
3750	To 6 L.F.	D-10	120	.267	L.F.	20	10.90	5.45	36.35	45.50
4000	Stools or sills, sand finish, 1" thick, 6" wide	D-12	160	.200		11.50	7.65		19.15	25
4200	10" wide		90	.356		17.75	13.60		31.35	41.50
4400	2" thick, 6" wide		140	.229		18.50	8.75		27.25	34.50
4600	10" wide		90	.356		29	13.60		42.60	54
4800	For lengths over 3', add					25%				
9000	Minimum labor/equipment charge	D-1	2.50	6.400	Job		241		241	390

04 43 10.85 Window Sill

		Crew	Daily Output	Labor-Hours	Unit	Material	Labor	Equipment	Total	Total Incl O&P
0010	**WINDOW SILL**									
0020	Bluestone, thermal top, 10" wide, 1-1/2" thick	D-1	85	.188	S.F.	16.05	7.10		23.15	29
0050	2" thick		75	.213	"	18.75	8.05		26.80	33.50
0100	Cut stone, 5" x 8" plain		48	.333	L.F.	11.90	12.55		24.45	33
0200	Face brick on edge, brick, 8" wide		80	.200		2.53	7.55		10.08	14.95
0400	Marble, 9" wide, 1" thick		85	.188		8.50	7.10		15.60	21
0900	Slate, colored, unfading, honed, 12" wide, 1" thick		85	.188		17.10	7.10		24.20	30
0950	2" thick		70	.229		24	8.60		32.60	40
9000	Minimum labor/equipment charge	1 Bric	2	4	Job		167		167	269

04 54 Refractory Brick Masonry

04 54 10 – Refractory Brick Work

04 54 10.20 Fire Clay

		Crew	Daily Output	Labor-Hours	Unit	Material	2010 Bare Costs Labor	2010 Bare Costs Equipment	Total	Total Incl O&P
0010	**FIRE CLAY**									
0020	Gray, high duty, 100 lb. bag				Bag	59			59	65
0050	100 lb. drum, premixed (400 brick per drum)				Drum	73			73	80.50

04 57 Masonry Fireplaces

04 57 10 – Brick or Stone Fireplaces

04 57 10.10 Fireplace

		Crew	Daily Output	Labor-Hours	Unit	Material	Labor	Equipment	Total	Total Incl O&P
0010	**FIREPLACE**									
0100	Brick fireplace, not incl. foundations or chimneys									
0110	30" x 29" opening, incl. chamber, plain brickwork	D-1	.40	40	Ea.	535	1,500		2,035	3,025
0200	Fireplace box only (110 brick)	"	2	8	"	176	300		476	680
0300	For elaborate brickwork and details, add					35%	35%			
0400	For hearth, brick & stone, add	D-1	2	8	Ea.	197	300		497	700
0410	For steel angle, damper, cleanouts, add		4	4		138	151		289	395
0600	Plain brickwork, incl. metal circulator		.50	32		1,025	1,200		2,225	3,075
0800	Face brick only, standard size, 8" x 2-2/3" x 4"		.30	53.333	M	535	2,000		2,535	3,825
0900	Stone fireplace, fieldstone, add				SF Face	14.05			14.05	15.50
1000	Cut stone, add				"	15.50			15.50	17.05
9000	Minimum labor/equipment charge	D-1	2	8	Job		300		300	485

04 71 Manufactured Brick Masonry

04 71 10 – Simulated or Manufactured Brick

04 71 10.10 Simulated Brick

		Crew	Daily Output	Labor-Hours	Unit	Material	Labor	Equipment	Total	Total Incl O&P
0010	**SIMULATED BRICK**									
0020	Aluminum, baked on colors	1 Carp	200	.040	S.F.	3.53	1.66		5.19	6.60
0050	Fiberglass panels		200	.040		3.87	1.66		5.53	7
0100	Urethane pieces cemented in mastic		150	.053		6.70	2.22		8.92	11
0150	Vinyl siding panels		200	.040		2.67	1.66		4.33	5.65

04 72 Cast Stone Masonry

04 72 10 – Cast Stone Masonry Features

04 72 10.10 Coping

		Crew	Daily Output	Labor-Hours	Unit	Material	Labor	Equipment	Total	Total Incl O&P
0010	**COPING**, stock units									
0050	Precast concrete, 10" wide, 4" tapers to 3-1/2", 8" wall	D-1	75	.213	L.F.	17.55	8.05		25.60	32.50
0100	12" wide, 3-1/2" tapers to 3", 10" wall		70	.229		17.55	8.60		26.15	33
0150	16" wide, 4" tapers to 3-1/2", 14" wall		60	.267		17.10	10.05		27.15	35
0300	Limestone for 12" wall, 4" thick		90	.178		15.35	6.70		22.05	27.50
0350	6" thick		80	.200		23	7.55		30.55	37.50
0500	Marble, to 4" thick, no wash, 9" wide		90	.178		23	6.70		29.70	36.50
0550	12" wide		80	.200		35	7.55		42.55	50.50
0700	Terra cotta, 9" wide		90	.178		5.60	6.70		12.30	16.95
0750	12" wide		80	.200		9.20	7.55		16.75	22.50
0800	Aluminum, for 12" wall		80	.200		13.30	7.55		20.85	27
9000	Minimum labor/equipment charge		2	8	Job		300		300	485

04 72 Cast Stone Masonry

04 72 20 – Cultured Stone Veneer

04 72 20.10 Cultured Stone Veneer Components		Crew	Daily Output	Labor-Hours	Unit	Material	2010 Bare Costs Labor	Equipment	Total	Total Incl O&P
0010	**CULTURED STONE VENEER COMPONENTS**									
0110	On wood frame and sheathing substrate, random sized cobbles, corner stones	D-8	70	.571	V.L.F.	12.15	22		34.15	49
0120	Field stones		140	.286	S.F.	8.80	11		19.80	27.50
0130	Random sized flats, corner stones		70	.571	V.L.F.	11.95	22		33.95	48.50
0140	Field stones		140	.286	S.F.	10.10	11		21.10	29
0150	Horizontal lined ledgestones, corner stones		75	.533	V.L.F.	12.15	20.50		32.65	46.50
0160	Field stones		150	.267	S.F.	8.80	10.25		19.05	26
0170	Random shaped flats, corner stones		65	.615	V.L.F.	12.15	23.50		35.65	51.50
0180	Field stones		150	.267	S.F.	8.80	10.25		19.05	26
0190	Random shaped/textured face, corner stones		65	.615	V.L.F.	12.15	23.50		35.65	51.50
0200	Field stones		130	.308	S.F.	8.80	11.85		20.65	29
0210	Random shaped river rock, corner stones		65	.615	V.L.F.	12.15	23.50		35.65	51.50
0220	Field stones		130	.308	S.F.	8.80	11.85		20.65	29
0240	On concrete or CMU substrate, random sized cobbles, corner stones		70	.571	V.L.F.	11.50	22		33.50	48
0250	Field stones		140	.286	S.F.	8.50	11		19.50	27
0260	Random sized flats, corner stones		70	.571	V.L.F.	11.30	22		33.30	48
0270	Field stones		140	.286	S.F.	9.75	11		20.75	28.50
0280	Horizontal lined ledgestones, corner stones		75	.533	V.L.F.	11.50	20.50		32	45.50
0290	Field stones		150	.267	S.F.	8.50	10.25		18.75	26
0300	Random shaped flats, corner stones		70	.571	V.L.F.	11.50	22		33.50	48
0310	Field stones		140	.286	S.F.	8.50	11		19.50	27
0320	Random shaped/textured face, corner stones		65	.615	V.L.F.	11.50	23.50		35	50.50
0330	Field stones		130	.308	S.F.	8.50	11.85		20.35	28.50
0340	Random shaped river rock, corner stones		65	.615	V.L.F.	11.50	23.50		35	50.50
0350	Field stones		130	.308	S.F.	8.50	11.85		20.35	28.50
0360	Cultured stone veneer, #15 felt weather resistant barrier	1 Clab	3700	.002	Sq.	5.90	.07		5.97	6.55
0370	Expanded metal lath, diamond, 2.5 lb/sy, galvanized	1 Lath	85	.094	S.Y.	2.94	3.48		6.42	8.70
0390	Water table or window sill, 18" long	1 Bric	80	.100	Ea.	11.85	4.18		16.03	19.75

Division Notes

	CREW	DAILY OUTPUT	LABOR-HOURS	UNIT	2010 BARE COSTS				TOTAL INCL O&P
					MAT.	LABOR	EQUIP.	TOTAL	

Estimating Tips

05 05 00 Common Work Results for Metals

- Nuts, bolts, washers, connection angles, and plates can add a significant amount to both the tonnage of a structural steel job and the estimated cost. As a rule of thumb, add 10% to the total weight to account for these accessories.
- Type 2 steel construction, commonly referred to as "simple construction," consists generally of field-bolted connections with lateral bracing supplied by other elements of the building, such as masonry walls or x-bracing. The estimator should be aware, however, that shop connections may be accomplished by welding or bolting. The method may be particular to the fabrication shop and may have an impact on the estimated cost.

05 12 23 Structural Steel

- Steel items can be obtained from two sources: a fabrication shop or a metals service center. Fabrication shops can fabricate items under more controlled conditions than can crews in the field. They are also more efficient and can produce items more economically. Metal service centers serve as a source of long mill shapes to both fabrication shops and contractors.
- Most line items in this structural steel subdivision, and most items in 05 50 00 Metal Fabrications, are indicated as being shop fabricated. The bare material cost for these shop fabricated items is the "Invoice Cost" from the shop and includes the mill base price of steel plus mill extras, transportation to the shop, shop drawings and detailing where warranted, shop fabrication and handling, sandblasting and a shop coat of primer paint, all necessary structural bolts, and delivery to the job site. The bare labor cost and bare equipment cost for these shop fabricated items is for field installation or erection.
- Line items in Subdivision 05 12 23.40 Lightweight Framing, and other items scattered in Division 5, are indicated as being field fabricated. The bare material cost for these field fabricated items is the "Invoice Cost" from the metals service center and includes the mill base price of steel plus mill extras, transportation to the metals service center, material handling, and delivery of long lengths of mill shapes to the job site. Material costs for structural bolts and welding rods should be added to the estimate. The bare labor cost and bare equipment cost for these items is for both field fabrication and field installation or erection, and include time for cutting, welding and drilling in the fabricated metal items. Drilling into concrete and fasteners to fasten field fabricated items to other work are not included and should be added to the estimate.

05 20 00 Steel Joist Framing

- In any given project the total weight of open web steel joists is determined by the loads to be supported and the design. However, economies can be realized in minimizing the amount of labor used to place the joists. This is done by maximizing the joist spacing, and therefore minimizing the number of joists required to be installed on the job. Certain spacings and locations may be required by the design, but in other cases maximizing the spacing and keeping it as uniform as possible will keep the costs down.

05 30 00 Steel Decking

- The takeoff and estimating of metal deck involves more than simply the area of the floor or roof and the type of deck specified or shown on the drawings. Many different sizes and types of openings may exist. Small openings for individual pipes or conduits may be drilled after the floor/roof is installed, but larger openings may require special deck lengths as well as reinforcing or structural support. The estimator should determine who will be supplying this reinforcing. Additionally, some deck terminations are part of the deck package, such as screed angles and pour stops, and others will be part of the steel contract, such as angles attached to structural members and cast-in-place angles and plates. The estimator must ensure that all pieces are accounted for in the complete estimate.

05 50 00 Metal Fabrications

- The most economical steel stairs are those that use common materials, standard details, and most importantly, a uniform and relatively simple method of field assembly. Commonly available A36 channels and plates are very good choices for the main stringers of the stairs, as are angles and tees for the carrier members. Risers and treads are usually made by specialty shops, and it is most economical to use a typical detail in as many places as possible. The stairs should be pre-assembled and shipped directly to the site. The field connections should be simple and straightforward to be accomplished efficiently, and with minimum equipment and labor.

Reference Numbers

Reference numbers are shown in shaded boxes at the beginning of some major classifications. These numbers refer to related items in the Reference Section. The reference information may be an estimating procedure, an alternate pricing method, or technical information. *Note:* Not all subdivisions listed here necessarily appear in this publication.

05 05 Common Work Results for Metals

05 05 05 – Selective Metals Demolition

05 05 05.10 Selective Demolition, Metals

		Crew	Daily Output	Labor-Hours	Unit	Material	2010 Bare Costs Labor	2010 Bare Costs Equipment	Total	Total Incl O&P
0010	**SELECTIVE DEMOLITION, METALS** R024119-10									
0015	Excludes shores, bracing, cutting, loading, hauling, dumping									
0020	Remove nuts only up to 3/4" diameter	1 Sswk	480	.017	Ea.		.78		.78	1.43
0030	7/8" to 1-1/4" diameter		240	.033			1.56		1.56	2.86
0040	1-3/8" to 2" diameter		160	.050			2.35		2.35	4.29
0060	Unbolt and remove structural bolts up to 3/4" diameter		240	.033			1.56		1.56	2.86
0070	7/8" to 2" diameter		160	.050			2.35		2.35	4.29
0140	Light weight framing members, remove whole or cut up, up to 20 lb		240	.033			1.56		1.56	2.86
0150	21 – 40 lb	2 Sswk	210	.076			3.57		3.57	6.55
0160	41 – 80 lb	3 Sswk	180	.133			6.25		6.25	11.45
0170	81 – 120 lb	4 Sswk	150	.213			10		10	18.30
0230	Structural members, remove whole or cut up, up to 500 lb	E-19	48	.500			23	21.50	44.50	63.50
0240	1/4 – 2 tons	E-18	36	1.111			51.50	28.50	80	124
0250	2 – 5 tons	E-24	30	1.067			49	25	74	114
0260	5 – 10 tons	E-20	24	2.667			122	52.50	174.50	274
0270	10 – 15 tons	E-2	18	3.111			142	89.50	231.50	350
0340	Fabricated item, remove whole or cut up, up to 20 lb	1 Sswk	96	.083			3.91		3.91	7.15
0350	21 – 40 lb	2 Sswk	84	.190			8.95		8.95	16.35
0360	41 – 80 lb	3 Sswk	72	.333			15.65		15.65	28.50
0370	81 – 120 lb	4 Sswk	60	.533			25		25	45.50
0380	121 – 500 lb	E-19	48	.500			23	21.50	44.50	63.50
0390	501 – 1000 lb	"	36	.667			30.50	28.50	59	85
2950	Minimum labor/equipment charge	B-13	2	28	Job		1,000	375	1,375	2,025

05 05 21 – Fastening Methods for Metal

05 05 21.10 Cutting Steel

		Crew	Daily Output	Labor-Hours	Unit	Material	Labor	Equipment	Total	Total Incl O&P
0010	**CUTTING STEEL**									
0020	Hand burning, incl. preparation, torch cutting & grinding, no staging									
0100	Steel to 1/2" thick	E-25	320	.025	L.F.		1.22	.31	1.53	2.58
0150	3/4" thick		260	.031			1.50	.38	1.88	3.17
0200	1" thick		200	.040			1.96	.49	2.45	4.12
9000	Minimum labor/equipment charge		2	4	Job		196	49	245	415

05 05 21.15 Drilling Steel

		Crew	Daily Output	Labor-Hours	Unit	Material	Labor	Equipment	Total	Total Incl O&P
0010	**DRILLING STEEL**									
1910	Drilling & layout for steel, up to 1/4" deep, no anchor									
1920	Holes, 1/4" diameter	1 Sswk	112	.071	Ea.	.09	3.35		3.44	6.25
1925	For each additional 1/4" depth, add		336	.024		.09	1.12		1.21	2.14
1930	3/8" diameter		104	.077		.10	3.61		3.71	6.70
1935	For each additional 1/4" depth, add		312	.026		.10	1.20		1.30	2.31
1940	1/2" diameter		96	.083		.11	3.91		4.02	7.25
1945	For each additional 1/4" depth, add		288	.028		.11	1.30		1.41	2.50
1950	5/8" diameter		88	.091		.15	4.26		4.41	7.95
1955	For each additional 1/4" depth, add		264	.030		.15	1.42		1.57	2.76
1960	3/4" diameter		80	.100		.17	4.69		4.86	8.80
1965	For each additional 1/4" depth, add		240	.033		.17	1.56		1.73	3.04
1970	7/8" diameter		72	.111		.20	5.20		5.40	9.75
1975	For each additional 1/4" depth, add		216	.037		.20	1.74		1.94	3.39
1980	1" diameter		64	.125		.22	5.85		6.07	10.95
1985	For each additional 1/4" depth, add		192	.042		.22	1.95		2.17	3.82
1990	For drilling up, add					40%				
2000	Minimum labor/equipment charge	1 Carp	4	2	Job		83		83	137

05 05 Common Work Results for Metals

05 05 21 – Fastening Methods for Metal

05 05 21.90 Welding Steel

		Crew	Daily Output	Labor-Hours	Unit	Material	2010 Bare Costs Labor	Equipment	Total	Total Incl O&P
0010	**WELDING STEEL**, Structural R050521-20									
0020	Field welding, 1/8" E6011, cost per welder, no oper. engr	E-14	8	1	Hr.	4.73	49	18.25	71.98	115
0200	With 1/2 operating engineer	E-13	8	1.500		4.73	69.50	18.25	92.48	147
0300	With 1 operating engineer	E-12	8	2		4.73	90	18.20	112.93	179
0500	With no operating engineer, 2# weld rod per ton	E-14	8	1	Ton	4.73	49	18.25	71.98	115
0600	8# E6011 per ton	"	2	4		18.90	196	73	287.90	460
0800	With one operating engineer per welder, 2# E6011 per ton	E-12	8	2		4.73	90	18.20	112.93	179
0900	8# E6011 per ton	"	2	8		18.90	360	73	451.90	715
1200	Continuous fillet, stick welding, incl. equipment									
1300	Single pass, 1/8" thick, 0.1#/L.F.	E-14	150	.053	L.F.	.24	2.61	.97	3.82	6.10
1400	3/16" thick, 0.2#/L.F.		75	.107		.47	5.20	1.94	7.61	12.20
1500	1/4" thick, 0.3#/L.F.		50	.160		.71	7.80	2.92	11.43	18.30
1610	5/16" thick, 0.4#/L.F.		38	.211		.95	10.30	3.84	15.09	24
1800	3 passes, 3/8" thick, 0.5#/L.F.		30	.267		1.18	13.05	4.86	19.09	30.50
2010	4 passes, 1/2" thick, 0.7#/L.F.		22	.364		1.66	17.80	6.65	26.11	41.50
2200	5 to 6 passes, 3/4" thick, 1.3#/L.F.		12	.667		3.07	32.50	12.15	47.72	76
2400	8 to 11 passes, 1" thick, 2.4#/L.F.		6	1.333		5.65	65	24.50	95.15	152
2600	For all position welding, add, minimum						20%			
2700	Maximum						300%			
2900	For semi-automatic welding, deduct, minimum						5%			
3000	Maximum						15%			
4000	Cleaning and welding plates, bars, or rods									
4010	to existing beams, columns, or trusses	E-14	12	.667	L.F.	1.18	32.50	12.15	45.83	74
9000	Minimum labor/equipment charge	"	4	2	Job		98	36.50	134.50	219

05 05 23 – Metal Fastenings

05 05 23.15 Chemical Anchors

		Crew	Daily Output	Labor-Hours	Unit	Material	2010 Bare Costs Labor	Equipment	Total	Total Incl O&P
0010	**CHEMICAL ANCHORS**									
0020	Includes layout & drilling									
1430	Chemical anchor, w/rod & epoxy cartridge, 3/4" diam. x 9-1/2" long	B-89A	27	.593	Ea.	9.45	22.50	4.21	36.16	51.50
1435	1" diameter x 11-3/4" long		24	.667		16.55	25	4.74	46.29	64.50
1440	1-1/4" diameter x 14" long		21	.762		29	29	5.40	63.40	84.50
1445	1-3/4" diameter x 15" long		20	.800		50.50	30.50	5.70	86.70	111
1450	18" long		17	.941		60.50	35.50	6.70	102.70	132
1455	2" diameter x 18" long		16	1		82.50	38	7.10	127.60	160
1460	24" long		15	1.067		107	40.50	7.60	155.10	191

05 05 23.20 Expansion Anchors

			Crew	Daily Output	Labor-Hours	Unit	Material	2010 Bare Costs Labor	Equipment	Total	Total Incl O&P
0010	**EXPANSION ANCHORS**										
0100	Anchors for concrete, brick or stone, no layout and drilling										
0200	Expansion shields, zinc, 1/4" diameter, 1-5/16" long, single	G	1 Carp	90	.089	Ea.	.34	3.69		4.03	6.40
0300	1-3/8" long, double	G		85	.094		.47	3.91		4.38	6.90
0400	3/8" diameter, 1-1/2" long, single	G		85	.094		.63	3.91		4.54	7.10
0500	2" long, double	G		80	.100		1.17	4.16		5.33	8.15
0600	1/2" diameter, 2-1/16" long, single	G		80	.100		1.16	4.16		5.32	8.15
0700	2-1/2" long, double	G		75	.107		1.63	4.43		6.06	9.10
0800	5/8" diameter, 2-5/8" long, single	G		75	.107		2.26	4.43		6.69	9.80
0900	2-3/4" long, double	G		70	.114		2.33	4.75		7.08	10.35
1000	3/4" diameter, 2-3/4" long, single	G		70	.114		2.07	4.75		6.82	10.10
1100	3-15/16" long, double	G		65	.123		4.51	5.10		9.61	13.35
2100	Hollow wall anchors for gypsum wall board, plaster or tile										
2300	1/8" diameter, short	G	1 Carp	160	.050	Ea.	.18	2.08		2.26	3.61
2400	Long	G		150	.053		.21	2.22		2.43	3.87

05 05 Common Work Results for Metals

05 05 23 – Metal Fastenings

05 05 23.20 Expansion Anchors		Crew	Daily Output	Labor-Hours	Unit	Material	2010 Bare Costs Labor	Equipment	Total	Total Incl O&P
2500	3/16" diameter, short	G	1 Carp	150	.053	Ea.	.32	2.22		2.54	3.99
2600	Long	G		140	.057		.46	2.37		2.83	4.41
2700	1/4" diameter, short	G		140	.057		.54	2.37		2.91	4.49
2800	Long	G		130	.062		.48	2.56		3.04	4.73
3000	Toggle bolts, bright steel, 1/8" diameter, 2" long	G		85	.094		.18	3.91		4.09	6.60
3100	4" long	G		80	.100		.23	4.16		4.39	7.10
3200	3/16" diameter, 3" long	G		80	.100		.22	4.16		4.38	7.10
3300	6" long	G		75	.107		.31	4.43		4.74	7.65
3400	1/4" diameter, 3" long	G		75	.107		.29	4.43		4.72	7.60
3500	6" long	G		70	.114		.46	4.75		5.21	8.30
3600	3/8" diameter, 3" long	G		70	.114		.70	4.75		5.45	8.55
3700	6" long	G		60	.133		1.20	5.55		6.75	10.40
3800	1/2" diameter, 4" long	G		60	.133		1.69	5.55		7.24	10.95
3900	6" long	G		50	.160		2.02	6.65		8.67	13.10
4000	Nailing anchors										
4100	Nylon nailing anchor, 1/4" diameter, 1" long		1 Carp	3.20	2.500	C	11.30	104		115.30	183
4200	1-1/2" long			2.80	2.857		12.45	119		131.45	209
4300	2" long			2.40	3.333		17.50	139		156.50	247
4400	Metal nailing anchor, 1/4" diameter, 1" long	G		3.20	2.500		16.65	104		120.65	189
4500	1-1/2" long	G		2.80	2.857		22.50	119		141.50	220
4600	2" long	G		2.40	3.333		26	139		165	257
8000	Wedge anchors, not including layout or drilling										
8050	Carbon steel, 1/4" diameter, 1-3/4" long	G	1 Carp	150	.053	Ea.	.29	2.22		2.51	3.96
8100	3-1/4" long	G		140	.057		.38	2.37		2.75	4.32
8150	3/8" diameter, 2-1/4" long	G		145	.055		.34	2.29		2.63	4.15
8200	5" long	G		140	.057		.60	2.37		2.97	4.56
8250	1/2" diameter, 2-3/4" long	G		140	.057		.73	2.37		3.10	4.70
8300	7" long	G		125	.064		1.24	2.66		3.90	5.75
8350	5/8" diameter, 3-1/2" long	G		130	.062		1.15	2.56		3.71	5.45
8400	8-1/2" long	G		115	.070		2.45	2.89		5.34	7.45
8450	3/4" diameter, 4-1/4" long	G		115	.070		2.02	2.89		4.91	6.95
8500	10" long	G		95	.084		4.60	3.50		8.10	10.80
8550	1" diameter, 6" long	G		100	.080		7.70	3.32		11.02	13.95
8575	9" long	G		85	.094		10	3.91		13.91	17.40
8600	12" long	G		75	.107		10.80	4.43		15.23	19.20
8650	1-1/4" diameter, 9" long	G		70	.114		23	4.75		27.75	33.50
8700	12" long	G		60	.133		29.50	5.55		35.05	41.50
8750	For type 303 stainless steel, add						350%				
8800	For type 316 stainless steel, add						450%				
8950	Self-drilling concrete screw, hex washer head, 3/16" diam. x 1-3/4" long	G	1 Carp	300	.027	Ea.	.19	1.11		1.30	2.03
8960	2-1/4" long	G		250	.032		.20	1.33		1.53	2.40
8970	Phillips flat head, 3/16" diam. x 1-3/4" long	G		300	.027		.17	1.11		1.28	2.01
8980	2-1/4" long	G		250	.032		.20	1.33		1.53	2.40
9000	Minimum labor/equipment charge			4	2	Job		83		83	137

05 05 23.30 Lag Screws

0010	**LAG SCREWS**										
0020	Steel, 1/4" diameter, 2" long	G	1 Carp	200	.040	Ea.	.09	1.66		1.75	2.83
0200	1/2" diameter, 3" long	G		130	.062		.48	2.56		3.04	4.73
0300	5/8" diameter, 3" long	G		120	.067		1.11	2.77		3.88	5.75

05 05 23.40 Machinery Anchors

0010	**MACHINERY ANCHORS**, heavy duty, incl. sleeve, floating base nut,
0020	lower stud & coupling nut, fiber plug, connecting stud, washer & nut.

05 05 Common Work Results for Metals

05 05 23 – Metal Fastenings

05 05 23.40 Machinery Anchors		Crew	Daily Output	Labor-Hours	Unit	Material	2010 Bare Costs Labor	Equipment	Total	Total Incl O&P
0030	For flush mounted embedment in poured concrete heavy equip. pads.									
0200	Stud & bolt, 1/2" diameter	G	E-16	40	.400	Ea.	68.50	19.15	3.64	91.29	115
0300	5/8" diameter	G		35	.457		76	22	4.16	102.16	128
0500	3/4" diameter	G		30	.533		87.50	25.50	4.86	117.86	148
0600	7/8" diameter	G		25	.640		95.50	30.50	5.85	131.85	167
0800	1" diameter	G		20	.800		100	38.50	7.30	145.80	188
0900	1-1/4" diameter	G		15	1.067		133	51	9.70	193.70	251

05 05 23.50 Powder Actuated Tools and Fasteners		Crew	Daily Output	Labor-Hours	Unit	Material	Labor	Equipment	Total	Total Incl O&P
0010	**POWDER ACTUATED TOOLS & FASTENERS**										
0020	Stud driver, .22 caliber, buy, minimum					Ea.	219			219	241
0100	Maximum					"	370			370	405
0300	Powder charges for above, low velocity					C	4.99			4.99	5.50
0400	Standard velocity						5.95			5.95	6.55
0600	Drive pins & studs, 1/4" & 3/8" diam., to 3" long, minimum	G	1 Carp	4.80	1.667		4.38	69.50		73.88	119
0700	Maximum	G	"	4	2		9.40	83		92.40	147

05 05 23.70 Structural Blind Bolts		Crew	Daily Output	Labor-Hours	Unit	Material	Labor	Equipment	Total	Total Incl O&P
0010	**STRUCTURAL BLIND BOLTS**										
0100	1/4" diameter x 1/4" grip	G	1 Sswk	240	.033	Ea.	1.24	1.56		2.80	4.22
0150	1/2" grip	G		216	.037		.95	1.74		2.69	4.23
0200	3/8" diameter x 1/2" grip	G		232	.034		1.75	1.62		3.37	4.89
0250	3/4" grip	G		208	.038		1.83	1.80		3.63	5.30
0300	1/2" diameter x 1/2" grip	G		224	.036		5.15	1.67		6.82	8.70
0350	3/4" grip	G		200	.040		5.60	1.88		7.48	9.60
0400	5/8" diameter x 3/4" grip	G		216	.037		8.25	1.74		9.99	12.30
0450	1" grip	G		192	.042		9.05	1.95		11	13.50

05 05 23.87 Weld Studs		Crew	Daily Output	Labor-Hours	Unit	Material	Labor	Equipment	Total	Total Incl O&P
0010	**WELD STUDS**										
0020	1/4" diameter, 2-11/16" long	G	E-10	1120	.014	Ea.	.33	.68	.39	1.40	2.04
0100	4-1/8" long	G		1080	.015		.31	.71	.41	1.43	2.09
0200	3/8" diameter, 4-1/8" long	G		1080	.015		.36	.71	.41	1.48	2.15
0300	6-1/8" long	G		1040	.015		.47	.74	.42	1.63	2.33
9000	Minimum labor/equipment charge		1 Sswk	2	4	Job		188		188	345

05 12 Structural Steel Framing

05 12 23 – Structural Steel for Buildings

05 12 23.05 Canopy Framing		Crew	Daily Output	Labor-Hours	Unit	Material	Labor	Equipment	Total	Total Incl O&P
0010	**CANOPY FRAMING**										
0020	6" and 8" members, shop fabricated	G	E-4	3000	.011	Lb.	1.32	.51	.05	1.88	2.42
9000	Minimum labor/equipment charge		1 Sswk	1	8	Job		375		375	685

05 12 23.10 Ceiling Supports		Crew	Daily Output	Labor-Hours	Unit	Material	Labor	Equipment	Total	Total Incl O&P
0010	**CEILING SUPPORTS**										
1000	Entrance door/folding partition supports, shop fabricated	G	E-4	60	.533	L.F.	22	25.50	2.43	49.93	72.50
1100	Linear accelerator door supports	G		14	2.286		100	108	10.40	218.40	320
1200	Lintels or shelf angles, hung, exterior hot dipped galv.	G		267	.120		15	5.70	.55	21.25	27.50
1250	Two coats primer paint instead of galv.	G		267	.120		13	5.70	.55	19.25	25.50
1400	Monitor support, ceiling hung, expansion bolted	G		4	8	Ea.	350	380	36.50	766.50	1,125
1450	Hung from pre-set inserts	G		6	5.333		375	253	24.50	652.50	900
1600	Motor supports for overhead doors	G		4	8		177	380	36.50	593.50	930
1700	Partition support for heavy folding partitions, without pocket	G		24	1.333	L.F.	50	63	6.10	119.10	178
1750	Supports at pocket only	G		12	2.667		100	126	12.15	238.15	355

05 12 Structural Steel Framing

05 12 23 – Structural Steel for Buildings

05 12 23.10 Ceiling Supports		Crew	Daily Output	Labor-Hours	Unit	Material	2010 Bare Costs Labor	Equipment	Total	Total Incl O&P
2000	Rolling grilles & fire door supports	G E-4	34	.941	L.F.	43	44.50	4.29	91.79	133
2100	Spider-leg light supports, expansion bolted to ceiling slab	G	8	4	Ea.	143	190	18.25	351.25	520
2150	Hung from pre-set inserts	G	12	2.667	"	154	126	12.15	292.15	415
2400	Toilet partition support	G	36	.889	L.F.	50	42	4.05	96.05	136
2500	X-ray travel gantry support	G	12	2.667	"	172	126	12.15	310.15	435

05 12 23.17 Columns, Structural

0010	**COLUMNS, STRUCTURAL** R051223-10									
0015	Made from recycled materials	G								
0020	Shop fab'd for 100-ton, 1-2 story project, bolted connections									
0800	Steel, concrete filled, extra strong pipe, 3-1/2" diameter	E-2	660	.085	L.F.	36.50	3.87	2.44	42.81	49.50
0830	4" diameter		780	.072		40.50	3.27	2.06	45.83	52.50
0890	5" diameter		1020	.055		48	2.50	1.58	52.08	59
0930	6" diameter		1200	.047		64	2.13	1.34	67.47	75
0940	8" diameter		1100	.051		64	2.32	1.46	67.78	75.50
1500	Steel pipe, extra strong, no concrete, 3" to 5" diameter	G	16000	.004	Lb.	1.10	.16	.10	1.36	1.60
3300	Structural tubing, square, A500GrB, 4" to 6" square, light section	G	11270	.005	"	1.10	.23	.14	1.47	1.77
8090	For projects 75 to 99 tons, add				All	10%				
8092	50 to 74 tons, add					20%				
8094	25 to 49 tons, add					30%	10%			
8096	10 to 24 tons, add					50%	25%			
8098	2 to 9 tons, add					75%	50%			
8099	Less than 2 tons, add					100%	100%			
9000	Minimum labor/equipment charge	1 Sswk	1	8	Job		375		375	685

05 12 23.20 Curb Edging

0010	**CURB EDGING**									
0020	Steel angle w/anchors, shop fabricated, on forms, 1" x 1", 0.8#/L.F.	G E-4	350	.091	L.F.	1.44	4.33	.42	6.19	9.95
0300	4" x 4" angles, 8.2#/L.F.	G	275	.116	"	11.45	5.50	.53	17.48	23.50
9000	Minimum labor/equipment charge		4	8	Job		380	36.50	416.50	735

05 12 23.40 Lightweight Framing

0010	**LIGHTWEIGHT FRAMING**									
0015	Made from recycled materials	G								
0200	For load-bearing steel studs see Div. 05 41 13.30									
0400	Angle framing, field fabricated, 4" and larger	G E-3	440	.055	Lb.	.64	2.59	.33	3.56	5.80
0450	Less than 4" angles	G	265	.091		.66	4.31	.55	5.52	9.25
0600	Channel framing, field fabricated, 8" and larger	G	500	.048		.66	2.28	.29	3.23	5.20
0650	Less than 8" channels	G	335	.072		.66	3.41	.44	4.51	7.45
1000	Continuous slotted channel framing system, shop fab, minimum	G	2 Sswk 2400	.007		3.41	.31		3.72	4.32
1200	Maximum	"	1600	.010		3.85	.47		4.32	5.10
1250	Plate & bar stock for reinforcing beams and trusses	G				1.21			1.21	1.33
1300	Cross bracing, rods, shop fabricated, 3/4" diameter	G E-3	700	.034		1.32	1.63	.21	3.16	4.66
1310	7/8" diameter		850	.028		1.32	1.34	.17	2.83	4.10
1320	1" diameter		1000	.024		1.32	1.14	.15	2.61	3.70
1330	Angle, 5" x 5" x 3/8"		2800	.009		1.32	.41	.05	1.78	2.26
1350	Hanging lintels, shop fabricated, average	G	850	.028		1.32	1.34	.17	2.83	4.10
1380	Roof frames, shop fabricated, 3'-0" square, 5' span	G E-2	4200	.013		1.32	.61	.38	2.31	2.94
1400	Tie rod, not upset, 1-1/2" to 4" diameter, with turnbuckle	G 2 Sswk	800	.020		1.43	.94		2.37	3.29
1420	No turnbuckle	G	700	.023		1.38	1.07		2.45	3.47
1500	Upset, 1-3/4" to 4" diameter, with turnbuckle	G	800	.020		1.43	.94		2.37	3.29
1520	No turnbuckle	G	700	.023		1.38	1.07		2.45	3.47
9000	Minimum labor/equipment charge		2	8	Job		375		375	685

05 12 Structural Steel Framing

05 12 23 – Structural Steel for Buildings

05 12 23.45 Lintels		Crew	Daily Output	Labor-Hours	Unit	Material	2010 Bare Costs Labor	Equipment	Total	Total Incl O&P
0010	**LINTELS**									
0015	Made from recycled materials	G								
0020	Plain steel angles, shop fabricated, under 500 lb.	G	1 Bric	550	.015	Lb.	.85	.61		1.46	1.91
0100	500 to 1000 lb.	G		640	.013		.83	.52		1.35	1.75
0200	1,000 to 2,000 lb.	G		640	.013		.80	.52		1.32	1.72
0300	2,000 to 4,000 lb.	G		640	.013		.78	.52		1.30	1.70
0500	For built-up angles and plates, add to above	G					.28			.28	.30
0700	For engineering, add to above						.11			.11	.12
0900	For galvanizing, add to above, under 500 lb.						.27			.27	.29
0950	500 to 2,000 lb.						.24			.24	.27
1000	Over 2,000 lb.						.21			.21	.23
2000	Steel angles, 3-1/2" x 3", 1/4" thick, 2'-6" long	G	1 Bric	47	.170	Ea.	11.90	7.10		19	24.50
2100	4'-6" long	G		26	.308		21.50	12.85		34.35	44
2500	3-1/2" x 3-1/2" x 5/16", 5'-0" long	G		18	.444		31.50	18.55		50.05	65
2600	4" x 3-1/2", 1/4" thick, 5'-0" long	G		21	.381		27.50	15.90		43.40	55.50
2700	9'-0" long	G		12	.667		49	28		77	99
2800	4" x 3-1/2" x 5/16", 7'-0" long	G		12	.667		47.50	28		75.50	97
2900	5" x 3-1/2" x 5/16", 10'-0" long	G		8	1		76.50	42		118.50	151
9000	Minimum labor/equipment charge			4	2	Job		83.50		83.50	134

05 12 23.60 Pipe Support Framing		Crew	Daily Output	Labor-Hours	Unit	Material	Labor	Equipment	Total	Total Incl O&P
0010	**PIPE SUPPORT FRAMING**									
0020	Under 10#/L.F., shop fabricated	G	E-4	3900	.008	Lb.	1.47	.39	.04	1.90	2.37
0200	10.1 to 15#/L.F.	G		4300	.007		1.45	.35	.03	1.83	2.28
0400	15.1 to 20#/L.F.	G		4800	.007		1.43	.32	.03	1.78	2.18
0600	Over 20#/L.F.	G		5400	.006		1.41	.28	.03	1.72	2.09

05 12 23.65 Plates		Crew	Daily Output	Labor-Hours	Unit	Material	Labor	Equipment	Total	Total Incl O&P
0010	**PLATES**									
0015	Made from recycled materials	G								
0020	For connections & stiffener plates, shop fabricated									
0050	1/8" thick (5.1 Lb./S.F.)	G				S.F.	5.60			5.60	6.15
0100	1/4" thick (10.2 Lb./S.F.)	G					11.20			11.20	12.35
0300	3/8" thick (15.3 Lb./S.F.)	G					16.85			16.85	18.50
0400	1/2" thick (20.4 Lb./S.F.)	G					22.50			22.50	24.50
0450	3/4" thick (30.6 Lb./S.F.)	G					33.50			33.50	37
0500	1" thick (40.8 Lb.S.F.)	G					45			45	49.50
2000	Steel plate, warehouse prices, no shop fabrication										
2100	1/4" thick (10.2 Lb./S.F.)	G				S.F.	8.40			8.40	9.25

05 12 23.75 Structural Steel Members		Crew	Daily Output	Labor-Hours	Unit	Material	Labor	Equipment	Total	Total Incl O&P
0010	**STRUCTURAL STEEL MEMBERS** R051223-10									
0015	Made from recycled materials	G								
0020	Shop fab'd for 100-ton, 1-2 story project, bolted connections									
0102	W 6 x 9 R051223-15	G	E-2	600	.093	L.F.	10.90	4.26	2.68	17.84	22.50
0302	W 8 x 10	G		600	.093		12.10	4.26	2.68	19.04	24
0502	x 31	G		550	.102		37.50	4.64	2.92	45.06	53
0702	W 10 x 22	G		600	.093		26.50	4.26	2.68	33.44	40
0902	x 49	G		550	.102		59.50	4.64	2.92	67.06	76.50
1102	W 12 x 16	G		880	.064		19.35	2.90	1.83	24.08	28.50
1302	x 22	G		880	.064		26.50	2.90	1.83	31.23	36.50
1502	x 26	G		880	.064		31.50	2.90	1.83	36.23	41.50
1902	W 14 x 26	G		990	.057		31.50	2.58	1.62	35.70	41
2102	x 30	G		900	.062		36.50	2.84	1.79	41.13	47

05 12 Structural Steel Framing

05 12 23 - Structural Steel for Buildings

05 12 23.75 Structural Steel Members

		Crew	Daily Output	Labor-Hours	Unit	Material	2010 Bare Costs Labor	Equipment	Total	Total Incl O&P
2702	W 16 x 26	E-2	1000	.056	L.F.	31.50	2.55	1.61	35.66	41
2902	x 31		900	.062		37.50	2.84	1.79	42.13	48.50
8490	For projects 75 to 99 tons, add					10%				
8492	50 to 74 tons, add					20%				
8494	25 to 49 tons, add					30%	10%			
8496	10 to 24 tons, add					50%	25%			
8498	2 to 9 tons, add					75%	50%			
8499	Less than 2 tons, add					100%	100%			
9000	Minimum labor/equipment charge	E-2	2	28	Job		1,275	805	2,080	3,125

05 12 23.77 Structural Steel Projects

		Crew	Daily Output	Labor-Hours	Unit	Material	Labor	Equipment	Total	Total Incl O&P
0010	**STRUCTURAL STEEL PROJECTS**									
0015	Made from recycled materials									
0020	Shop fab'd for 100-ton, 1-2 story project, bolted connections									
0700	Offices, hospitals, etc., steel bearing, 1 to 2 stories R050521-20	E-5	10.30	7.767	Ton	2,200	360	170	2,730	3,250
1300	Industrial bldgs., 1 story, beams & girders, steel bearing		12.90	6.202		2,200	286	136	2,622	3,075
1400	Masonry bearing R051223-10		10	8		2,200	370	175	2,745	3,275
1600	1 story with roof trusses, steel bearing		10.60	7.547		2,600	350	165	3,115	3,650
1700	Masonry bearing R051223-15		8.30	9.639		2,600	445	211	3,256	3,875
5390	For projects 75 to 99 tons, add					10%				
5392	50 to 74 tons, add					20%				
5394	25 to 49 tons, add					30%	10%			
5396	10 to 24 tons, add					50%	25%			
5398	2 to 9 tons, add					75%	50%			
5399	Less than 2 tons, add					100%	100%			

05 15 Wire Rope Assemblies

05 15 16 - Steel Wire Rope Assemblies

05 15 16.70 Temporary Cable Safety Railing

		Crew	Daily Output	Labor-Hours	Unit	Material	Labor	Equipment	Total	Total Incl O&P
0010	**TEMPORARY CABLE SAFETY RAILING**, Each 100' strand incl.									
0020	2 eyebolts, 1 turnbuckle, 100' cable, 2 thimbles, 6 clips									
0025	Made from recycled materials									
0100	One strand using 1/4" cable & accessories	2 Sswk	4	4	C.L.F.	189	188		377	555
0200	1/2" cable & accessories	"	2	8	"	400	375		775	1,125

05 21 Steel Joist Framing

05 21 13 - Deep Longspan Steel Joist Framing

05 21 13.50 Deep Longspan Joists

		Crew	Daily Output	Labor-Hours	Unit	Material	Labor	Equipment	Total	Total Incl O&P
0010	**DEEP LONGSPAN JOISTS**									
3010	DLH series, 40-ton job lots, bolted cross bridging, shop primer									
3015	Made from recycled materials									
3020	Spans to 144' (shipped in 2 pieces), minimum	E-7	16	5	Ton	1,100	231	119	1,450	1,775
3040	Average		13	6.154		1,200	284	146	1,630	1,975
3100	Maximum		11	7.273		1,425	335	173	1,933	2,375
3500	For less than 40-ton job lots									
3502	For 30 to 39 tons, add					10%				
3504	20 to 29 tons, add					20%				
3506	10 to 19 tons, add					30%				
3507	5 to 9 tons, add					50%	25%			
3508	1 to 4 tons, add					75%	50%			

05 21 Steel Joist Framing

05 21 13 – Deep Longspan Steel Joist Framing

05 21 13.50 Deep Longspan Joists		Crew	Daily Output	Labor-Hours	Unit	Material	2010 Bare Costs Labor	Equipment	Total	Total Incl O&P
3509	Less than 1 ton, add					100%	100%			
4010	SLH series, 40-ton job lots, bolted cross bridging, shop primer									
4020	Spans to 200' (shipped in 3 pieces), minimum	G	E-7	16	5	Ton	1,075	231	119	1,425	1,750
4040	Average	G		13	6.154		1,225	284	146	1,655	2,025
4060	Maximum	G		11	7.273		1,450	335	173	1,958	2,400
6100	For less than 40-ton job lots										
6102	For 30 to 39 tons, add						10%				
6104	20 to 29 tons, add						20%				
6106	10 to 19 tons, add						30%				
6107	5 to 9 tons, add						50%	25%			
6108	1 to 4 tons, add						75%	50%			
6109	Less than 1 ton, add						100%	100%			

05 21 16 – Longspan Steel Joist Framing

05 21 16.50 Longspan Joists

		Crew	Daily Output	Labor-Hours	Unit	Material	Labor	Equipment	Total	Total Incl O&P
0010	**LONGSPAN JOISTS**									
2000	LH series, 40-ton job lots, bolted cross bridging, shop primer									
2015	Made from recycled materials	G								
2020	Spans to 96', minimum	G	E-7	16	5	Ton	1,025	231	119	1,375	1,675
2040	Average	G		13	6.154		1,125	284	146	1,555	1,925
2080	Maximum	G		11	7.273		1,325	335	173	1,833	2,275
2600	For less than 40-ton job lots										
2602	For 30 to 39 tons, add						10%				
2604	20 to 29 tons, add						20%				
2606	10 to 19 tons, add						30%				
2607	5 to 9 tons, add						50%	25%			
2608	1 to 4 tons, add						75%	50%			
2609	Less than 1 ton, add						100%	100%			

05 21 19 – Open Web Steel Joist Framing

05 21 19.10 Open Web Joists

		Crew	Daily Output	Labor-Hours	Unit	Material	Labor	Equipment	Total	Total Incl O&P
0010	**OPEN WEB JOISTS**									
0015	Made from recycled materials	G								
0020	K series, 40-ton lots, horiz. bridging, spans to 30', shop primer, minimum	G	E-7	15	5.333	Ton	910	246	127	1,283	1,575
0050	Average	G		12	6.667		1,025	310	158	1,493	1,850
0080	Maximum	G		9	8.889		1,225	410	211	1,846	2,300
0410	Span 30' to 50', minimum	G		17	4.706		890	217	112	1,219	1,500
0440	Average	G		17	4.706		1,000	217	112	1,329	1,600
0460	Maximum	G		10	8		1,050	370	190	1,610	2,050
0800	For less than 40-ton job lots										
0802	For 30 to 39 tons, add						10%				
0804	20 to 29 tons, add						20%				
0806	10 to 19 tons, add						30%				
0807	5 to 9 tons, add						50%	25%			
0808	1 to 4 tons, add						75%	50%			
0809	Less than 1 ton, add						100%	100%			
1010	CS series, 40-ton job lots, horizontal bridging, shop primer										
1020	Spans to 30', minimum	G	E-7	15	5.333	Ton	940	246	127	1,313	1,600
1040	Average	G		12	6.667		1,050	310	158	1,518	1,875
1060	Maximum	G		9	8.889		1,225	410	211	1,846	2,300
1500	For less than 40-ton job lots										
1502	For 30 to 39 tons, add						10%				
1504	20 to 29 tons, add						20%				
1506	10 to 19 tons, add						30%				

05 21 Steel Joist Framing

05 21 19 – Open Web Steel Joist Framing

05 21 19.10 Open Web Joists		Crew	Daily Output	Labor-Hours	Unit	Material	2010 Bare Costs Labor	Equipment	Total	Total Incl O&P
1507	5 to 9 tons, add					50%	25%			
1508	1 to 4 tons, add					75%	50%			
1509	Less than 1 ton, add					100%	100%			
6400	Individual steel bearing plate, 6" x 6" x 1/4" with J-hook	G	1 Bric	160	.050	Ea.	6.60	2.09		8.69	10.60
9000	Minimum labor and equipment, bar joists		F-6	1	40	Job		1,550	655	2,205	3,250

05 31 Steel Decking

05 31 13 – Steel Floor Decking

05 31 13.50 Floor Decking

			Crew	Daily Output	Labor-Hours	Unit	Material	Labor	Equipment	Total	Total Incl O&P
0010	**FLOOR DECKING**	R053100-10									
0015	Made from recycled materials	G									
5100	Non-cellular composite decking, galvanized, 1-1/2" deep, 16 gauge	G	E-4	3500	.009	S.F.	2.18	.43	.04	2.65	3.24
5120	18 gauge	G		3650	.009		1.73	.42	.04	2.19	2.70
5140	20 gauge	G		3800	.008		1.41	.40	.04	1.85	2.33
5200	2" deep, 22 gauge	G		3860	.008		1.22	.39	.04	1.65	2.10
5300	20 gauge	G		3600	.009		1.36	.42	.04	1.82	2.31
5400	18 gauge	G		3380	.009		1.69	.45	.04	2.18	2.73
5500	16 gauge	G		3200	.010		2.16	.47	.05	2.68	3.29
5700	3" deep, 22 gauge	G		3200	.010		1.33	.47	.05	1.85	2.38
5800	20 gauge	G		3000	.011		1.50	.51	.05	2.06	2.62
5900	18 gauge	G		2850	.011		1.80	.53	.05	2.38	3.01
6000	16 gauge	G		2700	.012		2.45	.56	.05	3.06	3.79
9000	Minimum labor/equipment charge		1 Sswk	1	8	Job		375		375	685

05 31 23 – Steel Roof Decking

05 31 23.50 Roof Decking

0010	**ROOF DECKING**										
0015	Made from recycled materials	G									
2100	Open type B wide rib, galvanized, 1-1/2" deep, 22 gauge, under 50 sq	G	E-4	4500	.007	S.F.	1.29	.34	.03	1.66	2.08
2400	Over 500 squares	G	"	5100	.006	"	.93	.30	.03	1.26	1.59

05 31 33 – Steel Form Decking

05 31 33.50 Form Decking

0010	**FORM DECKING**										
0015	Made from recycled materials	G									
6100	Slab form, steel, 28 gauge, 9/16" deep, uncoated	G	E-4	4000	.008	S.F.	.98	.38	.04	1.40	1.81
6200	Galvanized	G	"	4000	.008	"	.87	.38	.04	1.29	1.69

05 35 Raceway Decking Assemblies

05 35 13 – Steel Cellular Decking

05 35 13.50 Cellular Decking

0010	**CELLULAR DECKING**										
0015	Made from recycled materials	G									
0200	Cellular units, galv, 2" deep, 20-20 gauge, over 15 squares	G	E-4	1460	.022	S.F.	5.10	1.04	.10	6.24	7.60
0400	3" deep, galvanized, 20-20 gauge	G		1375	.023		5.60	1.10	.11	6.81	8.35
1000	4-1/2" deep, galvanized, 20-18 gauge	G		1100	.029		7.85	1.38	.13	9.36	11.30
1500	For acoustical deck, add						15%				
1900	For multi-story or congested site, add							50%			

05 41 Structural Metal Stud Framing

05 41 13 – Load-Bearing Metal Stud Framing

05 41 13.05 Bracing

		Crew	Daily Output	Labor-Hours	Unit	Material	2010 Bare Costs Labor	Equipment	Total	Total Incl O&P
0010	**BRACING**, shear wall X-bracing, per 10' x 10' bay, one face									
0015	Made of recycled materials	G								
0120	Metal strap, 20 ga x 4" wide	G	2 Carp	18	.889	Ea.	19.35	37		56.35	82
0130	6" wide	G		18	.889		30	37		67	93.50
0160	18 ga x 4" wide	G		16	1		28	41.50		69.50	99
0170	6" wide	G	↓	16	1	↓	41	41.50		82.50	114
0410	Continuous strap bracing, per horizontal row on both faces									
0420	Metal strap, 20 ga x 2" wide, studs 12" O.C.	G	1 Carp	7	1.143	C.L.F.	48.50	47.50		96	131
0430	16" O.C.	G		8	1		48.50	41.50		90	122
0440	24" O.C.	G		10	.800		48.50	33		81.50	108
0450	18 ga x 2" wide, studs 12" O.C.	G		6	1.333		68	55.50		123.50	166
0460	16" O.C.	G		7	1.143		68	47.50		115.50	153
0470	24" O.C.	G	↓	8	1		68	41.50		109.50	144

05 41 13.10 Bridging

		Crew	Daily Output	Labor-Hours	Unit	Material	Labor	Equipment	Total	Total Incl O&P
0010	**BRIDGING**, solid between studs w/1-1/4" leg track, per stud bay									
0015	Made from recycled materials	G								
0200	Studs 12" O.C., 18 ga x 2-1/2" wide	G	1 Carp	125	.064	Ea.	.82	2.66		3.48	5.30
0210	3-5/8" wide	G		120	.067		1	2.77		3.77	5.65
0220	4" wide	G		120	.067		1.07	2.77		3.84	5.75
0230	6" wide	G		115	.070		1.38	2.89		4.27	6.25
0240	8" wide	G		110	.073		1.75	3.02		4.77	6.90
0300	16 ga x 2-1/2" wide	G		115	.070		1.03	2.89		3.92	5.90
0310	3-5/8" wide	G		110	.073		1.26	3.02		4.28	6.35
0320	4" wide	G		110	.073		1.34	3.02		4.36	6.45
0330	6" wide	G		105	.076		1.72	3.17		4.89	7.10
0340	8" wide	G		100	.080		2.20	3.32		5.52	7.85
1200	Studs 16" O.C., 18 ga x 2-1/2" wide	G		125	.064		1.06	2.66		3.72	5.55
1210	3-5/8" wide	G		120	.067		1.29	2.77		4.06	5.95
1220	4" wide	G		120	.067		1.37	2.77		4.14	6.05
1230	6" wide	G		115	.070		1.77	2.89		4.66	6.70
1240	8" wide	G		110	.073		2.24	3.02		5.26	7.45
1300	16 ga x 2-1/2" wide	G		115	.070		1.32	2.89		4.21	6.20
1310	3-5/8" wide	G		110	.073		1.62	3.02		4.64	6.75
1320	4" wide	G		110	.073		1.72	3.02		4.74	6.85
1330	6" wide	G		105	.076		2.21	3.17		5.38	7.65
1340	8" wide	G		100	.080		2.82	3.32		6.14	8.55
2200	Studs 24" O.C., 18 ga x 2-1/2" wide	G		125	.064		1.53	2.66		4.19	6.05
2210	3-5/8" wide	G		120	.067		1.86	2.77		4.63	6.60
2220	4" wide	G		120	.067		1.98	2.77		4.75	6.75
2230	6" wide	G		115	.070		2.55	2.89		5.44	7.55
2240	8" wide	G		110	.073		3.25	3.02		6.27	8.55
2300	16 ga x 2-1/2" wide	G		115	.070		1.91	2.89		4.80	6.85
2310	3-5/8" wide	G		110	.073		2.34	3.02		5.36	7.55
2320	4" wide	G		110	.073		2.48	3.02		5.50	7.70
2330	6" wide	G		105	.076		3.20	3.17		6.37	8.70
2340	8" wide	G	↓	100	.080	↓	4.08	3.32		7.40	9.95
3000	Continuous bridging, per row									
3100	16 ga x 1-1/2" channel thru studs 12" O.C.	G	1 Carp	6	1.333	C.L.F.	45.50	55.50		101	141
3110	16" O.C.	G		7	1.143		45.50	47.50		93	128
3120	24" O.C.	G		8.80	.909		45.50	38		83.50	112
4100	2" x 2" angle x 18 ga, studs 12" O.C.	G		7	1.143		68	47.50		115.50	153
4110	16" O.C.	G	↓	9	.889		68	37		105	136

05 41 Structural Metal Stud Framing

05 41 13 – Load-Bearing Metal Stud Framing

05 41 13.10 Bridging

		Crew	Daily Output	Labor-Hours	Unit	Material	2010 Bare Costs Labor	Equipment	Total	Total Incl O&P
4120	24" O.C.	1 Carp G	12	.667	C.L.F.	68	27.50		95.50	121
4200	16 ga, studs 12" O.C.	G	5	1.600		86	66.50		152.50	204
4210	16" O.C.	G	7	1.143		86	47.50		133.50	173
4220	24" O.C.	G	10	.800		86	33		119	149

05 41 13.25 Framing, Boxed Headers/Beams

		Crew	Daily Output	Labor-Hours	Unit	Material	Labor	Equipment	Total	Total Incl O&P
0010	**FRAMING, BOXED HEADERS/BEAMS**									
0015	Made from recycled materials	G								
0200	Double, 18 ga x 6" deep	2 Carp G	220	.073	L.F.	4.81	3.02		7.83	10.25
0210	8" deep	G	210	.076		5.35	3.17		8.52	11.10
0220	10" deep	G	200	.080		6.45	3.32		9.77	12.55
0230	12" deep	G	190	.084		7.10	3.50		10.60	13.55
0300	16 ga x 8" deep	G	180	.089		6.15	3.69		9.84	12.80
0310	10" deep	G	170	.094		7.35	3.91		11.26	14.45
0320	12" deep	G	160	.100		8	4.16		12.16	15.65
0400	14 ga x 10" deep	G	140	.114		8.55	4.75		13.30	17.20
0410	12" deep	G	130	.123		9.40	5.10		14.50	18.70
1210	Triple, 18 ga x 8" deep	G	170	.094		7.70	3.91		11.61	14.90
1220	10" deep	G	165	.097		9.25	4.03		13.28	16.80
1230	12" deep	G	160	.100		10.20	4.16		14.36	18.10
1300	16 ga x 8" deep	G	145	.110		8.90	4.58		13.48	17.35
1310	10" deep	G	140	.114		10.60	4.75		15.35	19.45
1320	12" deep	G	135	.119		11.60	4.92		16.52	21
1400	14 ga x 10" deep	G	115	.139		11.65	5.80		17.45	22.50
1410	12" deep	G	110	.145		12.90	6.05		18.95	24

05 41 13.30 Framing, Stud Walls

		Crew	Daily Output	Labor-Hours	Unit	Material	Labor	Equipment	Total	Total Incl O&P
0010	**FRAMING, STUD WALLS** w/top & bottom track, no openings,									
0020	Headers, beams, bridging or bracing									
0025	Made from recycled materials	G								
4100	8' high walls, 18 ga x 2-1/2" wide, studs 12" O.C.	2 Carp G	54	.296	L.F.	8.05	12.30		20.35	29
4110	16" O.C.	G	77	.208		6.45	8.65		15.10	21.50
4120	24" O.C.	G	107	.150		4.81	6.20		11.01	15.50
4130	3-5/8" wide, studs 12" O.C.	G	53	.302		9.60	12.55		22.15	31
4140	16" O.C.	G	76	.211		7.65	8.75		16.40	23
4150	24" O.C.	G	105	.152		5.75	6.35		12.10	16.75
4160	4" wide, studs 12" O.C.	G	52	.308		10.05	12.80		22.85	32
4170	16" O.C.	G	74	.216		8.05	9		17.05	23.50
4180	24" O.C.	G	103	.155		6.05	6.45		12.50	17.25
4190	6" wide, studs 12" O.C.	G	51	.314		12.70	13.05		25.75	35.50
4200	16" O.C.	G	73	.219		10.20	9.10		19.30	26
4210	24" O.C.	G	101	.158		7.65	6.60		14.25	19.25
4220	8" wide, studs 12" O.C.	G	50	.320		15.55	13.30		28.85	39
4230	16" O.C.	G	72	.222		12.50	9.25		21.75	29
4240	24" O.C.	G	100	.160		9.45	6.65		16.10	21.50
4300	16 ga x 2-1/2" wide, studs 12" O.C.	G	47	.340		9.50	14.15		23.65	33.50
4310	16" O.C.	G	68	.235		7.50	9.80		17.30	24.50
4320	24" O.C.	G	94	.170		5.55	7.05		12.60	17.70
4330	3-5/8" wide, studs 12" O.C.	G	46	.348		11.35	14.45		25.80	36
4340	16" O.C.	G	66	.242		9	10.05		19.05	26.50
4350	24" O.C.	G	92	.174		6.65	7.25		13.90	19.15
4360	4" wide, studs 12" O.C.	G	45	.356		11.95	14.75		26.70	37.50
4370	16" O.C.	G	65	.246		9.45	10.25		19.70	27
4380	24" O.C.	G	90	.178		7	7.40		14.40	19.85

05 41 Structural Metal Stud Framing

05 41 13 – Load-Bearing Metal Stud Framing

05 41 13.30 Framing, Stud Walls		Crew	Daily Output	Labor-Hours	Unit	Material	2010 Bare Costs Labor	Equipment	Total	Total Incl O&P
4390	6" wide, studs 12" O.C.	G 2 Carp	44	.364	L.F.	15	15.10		30.10	41.50
4400	16" O.C.	G	64	.250		11.90	10.40		22.30	30
4410	24" O.C.	G	88	.182		8.80	7.55		16.35	22
4420	8" wide, studs 12" O.C.	G	43	.372		18.55	15.45		34	46
4430	16" O.C.	G	63	.254		14.75	10.55		25.30	33.50
4440	24" O.C.	G	86	.186		10.95	7.75		18.70	25
5100	10' high walls, 18 ga x 2-1/2" wide, studs 12" O.C.	G	54	.296		9.65	12.30		21.95	30.50
5110	16" O.C.	G	77	.208		7.65	8.65		16.30	22.50
5120	24" O.C.	G	107	.150		5.60	6.20		11.80	16.40
5130	3-5/8" wide, studs 12" O.C.	G	53	.302		11.50	12.55		24.05	33
5140	16" O.C.	G	76	.211		9.10	8.75		17.85	24.50
5150	24" O.C.	G	105	.152		6.70	6.35		13.05	17.80
5160	4" wide, studs 12" O.C.	G	52	.308		12.05	12.80		24.85	34.50
5170	16" O.C.	G	74	.216		9.55	9		18.55	25.50
5180	24" O.C.	G	103	.155		7.05	6.45		13.50	18.35
5190	6" wide, studs 12" O.C.	G	51	.314		15.20	13.05		28.25	38.50
5200	16" O.C.	G	73	.219		12.05	9.10		21.15	28.50
5210	24" O.C.	G	101	.158		8.90	6.60		15.50	20.50
5220	8" wide, studs 12" O.C.	G	50	.320		18.65	13.30		31.95	42.50
5230	16" O.C.	G	72	.222		14.80	9.25		24.05	31.50
5240	24" O.C.	G	100	.160		11	6.65		17.65	23
5300	16 ga x 2-1/2" wide, studs 12" O.C.	G	47	.340		11.45	14.15		25.60	35.50
5310	16" O.C.	G	68	.235		9	9.80		18.80	26
5320	24" O.C.	G	94	.170		6.50	7.05		13.55	18.75
5330	3-5/8" wide, studs 12" O.C.	G	46	.348		13.70	14.45		28.15	38.50
5340	16" O.C.	G	66	.242		10.75	10.05		20.80	28.50
5350	24" O.C.	G	92	.174		7.80	7.25		15.05	20.50
5360	4" wide, studs 12" O.C.	G	45	.356		14.45	14.75		29.20	40.50
5370	16" O.C.	G	65	.246		11.35	10.25		21.60	29.50
5380	24" O.C.	G	90	.178		8.25	7.40		15.65	21
5390	6" wide, studs 12" O.C.	G	44	.364		18.10	15.10		33.20	45
5400	16" O.C.	G	64	.250		14.25	10.40		24.65	32.50
5410	24" O.C.	G	88	.182		10.35	7.55		17.90	24
5420	8" wide, studs 12" O.C.	G	43	.372		22.50	15.45		37.95	50
5430	16" O.C.	G	63	.254		17.60	10.55		28.15	36.50
5440	24" O.C.	G	86	.186		12.85	7.75		20.60	27
6190	12' high walls, 18 ga x 6" wide, studs 12" O.C.	G	41	.390		17.75	16.20		33.95	46
6200	16" O.C.	G	58	.276		13.95	11.45		25.40	34
6210	24" O.C.	G	81	.198		10.20	8.20		18.40	24.50
6220	8" wide, studs 12" O.C.	G	40	.400		21.50	16.60		38.10	51.50
6230	16" O.C.	G	57	.281		17.10	11.65		28.75	38
6240	24" O.C.	G	80	.200		12.50	8.30		20.80	27.50
6390	16 ga x 6" wide, studs 12" O.C.	G	35	.457		21	19		40	54.50
6400	16" O.C.	G	51	.314		16.55	13.05		29.60	40
6410	24" O.C.	G	70	.229		11.90	9.50		21.40	28.50
6420	8" wide, studs 12" O.C.	G	34	.471		26	19.55		45.55	61
6430	16" O.C.	G	50	.320		20.50	13.30		33.80	44.50
6440	24" O.C.	G	69	.232		14.75	9.65		24.40	32
6530	14 ga x 3-5/8" wide, studs 12" O.C.	G	34	.471		20	19.55		39.55	54
6540	16" O.C.	G	48	.333		15.70	13.85		29.55	40.50
6550	24" O.C.	G	65	.246		11.30	10.25		21.55	29
6560	4" wide, studs 12" O.C.	G	33	.485		21.50	20		41.50	56.50
6570	16" O.C.	G	47	.340		16.60	14.15		30.75	41.50

05 41 Structural Metal Stud Framing

05 41 13 – Load-Bearing Metal Stud Framing

05 41 13.30 Framing, Stud Walls

			Crew	Daily Output	Labor-Hours	Unit	Material	2010 Bare Costs Labor	Equipment	Total	Total Incl O&P
6580	24" O.C.	G	2 Carp	64	.250	L.F.	11.90	10.40		22.30	30
6730	12 ga x 3-5/8" wide, studs 12" O.C.	G		31	.516		28	21.50		49.50	65.50
6740	16" O.C.	G		43	.372		21.50	15.45		36.95	49
6750	24" O.C.	G		59	.271		15.10	11.25		26.35	35
6760	4" wide, studs 12" O.C.	G		30	.533		30	22		52	69.50
6770	16" O.C.	G		42	.381		23	15.85		38.85	51.50
6780	24" O.C.	G		58	.276		16.25	11.45		27.70	36.50
7390	16' high walls, 16 ga x 6" wide, studs 12" O.C.	G		33	.485		27.50	20		47.50	63
7400	16" O.C.	G		48	.333		21	13.85		34.85	46.50
7410	24" O.C.	G		67	.239		15	9.90		24.90	33
7420	8" wide, studs 12" O.C.	G		32	.500		33.50	21		54.50	71
7430	16" O.C.	G		47	.340		26	14.15		40.15	52
7440	24" O.C.	G		66	.242		18.55	10.05		28.60	37
7560	14 ga x 4" wide, studs 12" O.C.	G		31	.516		27.50	21.50		49	65.50
7570	16" O.C.	G		45	.356		21.50	14.75		36.25	48
7580	24" O.C.	G		61	.262		15.05	10.90		25.95	34.50
7590	6" wide, studs 12" O.C.	G		30	.533		34.50	22		56.50	74.50
7600	16" O.C.	G		44	.364		27	15.10		42.10	54.50
7610	24" O.C.	G		60	.267		18.95	11.10		30.05	39
7760	12 ga x 4" wide, studs 12" O.C.	G		29	.552		39	23		62	80.50
7770	16" O.C.	G		40	.400		30	16.60		46.60	60.50
7780	24" O.C.	G		55	.291		21	12.10		33.10	43
7790	6" wide, studs 12" O.C.	G		28	.571		49	23.50		72.50	93
7800	16" O.C.	G		39	.410		37.50	17.05		54.55	69.50
7810	24" O.C.	G		54	.296		26	12.30		38.30	49
8590	20' high walls, 14 ga x 6" wide, studs 12" O.C.	G		29	.552		42.50	23		65.50	84
8600	16" O.C.	G		42	.381		32.50	15.85		48.35	62
8610	24" O.C.	G		57	.281		23	11.65		34.65	44
8620	8" wide, studs 12" O.C.	G		28	.571		52	23.50		75.50	96
8630	16" O.C.	G		41	.390		40	16.20		56.20	70.50
8640	24" O.C.	G		56	.286		28	11.85		39.85	50.50
8790	12 ga x 6" wide, studs 12" O.C.	G		27	.593		60.50	24.50		85	108
8800	16" O.C.	G		37	.432		46.50	17.95		64.45	80.50
8810	24" O.C.	G		51	.314		32	13.05		45.05	56.50
8820	8" wide, studs 12" O.C.	G		26	.615		74	25.50		99.50	124
8830	16" O.C.	G		36	.444		56.50	18.45		74.95	92.50
8840	24" O.C.	G		50	.320		39	13.30		52.30	65
9000	Minimum labor/equipment charge			4	4	Job		166		166	273

05 42 Cold-Formed Metal Joist Framing

05 42 13 – Cold-Formed Metal Floor Joist Framing

05 42 13.05 Bracing

			Crew	Daily Output	Labor-Hours	Unit	Material	Labor	Equipment	Total	Total Incl O&P
0010	**BRACING**, continuous, per row, top & bottom										
0015	Made from recycled materials	G									
0120	Flat strap, 20 ga x 2" wide, joists at 12" O.C.	G	1 Carp	4.67	1.713	C.L.F.	50.50	71		121.50	173
0130	16" O.C.	G		5.33	1.501		49	62.50		111.50	156
0140	24" O.C.	G		6.66	1.201		47	50		97	134
0150	18 ga x 2" wide, joists at 12" O.C.	G		4	2		67.50	83		150.50	212
0160	16" O.C.	G		4.67	1.713		66.50	71		137.50	190
0170	24" O.C.	G		5.33	1.501		65	62.50		127.50	174

05 42 Cold-Formed Metal Joist Framing

05 42 13 – Cold-Formed Metal Floor Joist Framing

05 42 13.10 Bridging

		Crew	Daily Output	Labor-Hours	Unit	Material	2010 Bare Costs Labor	2010 Bare Costs Equipment	Total	Total Incl O&P
0010	**BRIDGING**, solid between joists w/1-1/4" leg track, per joist bay									
0015	Made from recycled materials G									
0230	Joists 12" O.C., 18 ga track x 6" wide G	1 Carp	80	.100	Ea.	1.38	4.16		5.54	8.35
0240	8" wide G		75	.107		1.75	4.43		6.18	9.25
0250	10" wide G		70	.114		2.16	4.75		6.91	10.20
0260	12" wide G		65	.123		2.50	5.10		7.60	11.15
0330	16 ga track x 6" wide G		70	.114		1.72	4.75		6.47	9.70
0340	8" wide G		65	.123		2.20	5.10		7.30	10.80
0350	10" wide G		60	.133		2.70	5.55		8.25	12.05
0360	12" wide G		55	.145		3.11	6.05		9.16	13.40
0440	14 ga track x 8" wide G		60	.133		2.78	5.55		8.33	12.15
0450	10" wide G		55	.145		3.41	6.05		9.46	13.70
0460	12" wide G		50	.160		3.94	6.65		10.59	15.25
0550	12 ga track x 10" wide G		45	.178		5	7.40		12.40	17.65
0560	12" wide G		40	.200		5.65	8.30		13.95	19.90
1230	16" O.C., 18 ga track x 6" wide G		80	.100		1.77	4.16		5.93	8.80
1240	8" wide G		75	.107		2.24	4.43		6.67	9.75
1250	10" wide G		70	.114		2.77	4.75		7.52	10.85
1260	12" wide G		65	.123		3.20	5.10		8.30	11.90
1330	16 ga track x 6" wide G		70	.114		2.21	4.75		6.96	10.25
1340	8" wide G		65	.123		2.82	5.10		7.92	11.50
1350	10" wide G		60	.133		3.47	5.55		9.02	12.90
1360	12" wide G		55	.145		3.99	6.05		10.04	14.35
1440	14 ga track x 8" wide G		60	.133		3.56	5.55		9.11	13
1450	10" wide G		55	.145		4.37	6.05		10.42	14.75
1460	12" wide G		50	.160		5.05	6.65		11.70	16.45
1550	12 ga track x 10" wide G		45	.178		6.40	7.40		13.80	19.20
1560	12" wide G		40	.200		7.25	8.30		15.55	21.50
2230	24" O.C., 18 ga track x 6" wide G		80	.100		2.55	4.16		6.71	9.65
2240	8" wide G		75	.107		3.25	4.43		7.68	10.85
2250	10" wide G		70	.114		4.01	4.75		8.76	12.20
2260	12" wide G		65	.123		4.63	5.10		9.73	13.50
2330	16 ga track x 6" wide G		70	.114		3.20	4.75		7.95	11.30
2340	8" wide G		65	.123		4.08	5.10		9.18	12.90
2350	10" wide G		60	.133		5	5.55		10.55	14.60
2360	12" wide G		55	.145		5.80	6.05		11.85	16.30
2440	14 ga track x 8" wide G		60	.133		5.15	5.55		10.70	14.75
2450	10" wide G		55	.145		6.35	6.05		12.40	16.90
2460	12" wide G		50	.160		7.30	6.65		13.95	18.95
2550	12 ga track x 10" wide G		45	.178		9.30	7.40		16.70	22.50
2560	12" wide G		40	.200		10.50	8.30		18.80	25

05 42 13.25 Framing, Band Joist

		Crew	Daily Output	Labor-Hours	Unit	Material	Labor	Equipment	Total	Total Incl O&P
0010	**FRAMING, BAND JOIST** (track) fastened to bearing wall									
0015	Made from recycled materials G									
0220	18 ga track x 6" deep G	2 Carp	1000	.016	L.F.	1.12	.66		1.78	2.33
0230	8" deep G		920	.017		1.43	.72		2.15	2.76
0240	10" deep G		860	.019		1.76	.77		2.53	3.21
0320	16 ga track x 6" deep G		900	.018		1.41	.74		2.15	2.76
0330	8" deep G		840	.019		1.80	.79		2.59	3.28
0340	10" deep G		780	.021		2.21	.85		3.06	3.83
0350	12" deep G		740	.022		2.54	.90		3.44	4.28
0430	14 ga track x 8" deep G		750	.021		2.27	.89		3.16	3.95

05 42 Cold-Formed Metal Joist Framing

05 42 13 – Cold-Formed Metal Floor Joist Framing

05 42 13.25 Framing, Band Joist

			Crew	Daily Output	Labor-Hours	Unit	Material	Labor	Equipment	Total	Total Incl O&P
0440	10" deep	G	2 Carp	720	.022	L.F.	2.78	.92		3.70	4.58
0450	12" deep	G		700	.023		3.21	.95		4.16	5.10
0540	12 ga track x 10" deep	G		670	.024		4.08	.99		5.07	6.10
0550	12" deep	G		650	.025		4.62	1.02		5.64	6.80

05 42 13.30 Framing, Boxed Headers/Beams

			Crew	Daily Output	Labor-Hours	Unit	Material	Labor	Equipment	Total	Total Incl O&P
0010	**FRAMING, BOXED HEADERS/BEAMS**										
0015	Made from recycled materials	G									
0200	Double, 18 ga x 6" deep	G	2 Carp	220	.073	L.F.	4.81	3.02		7.83	10.25
0210	8" deep	G		210	.076		5.35	3.17		8.52	11.10
0220	10" deep	G		200	.080		6.45	3.32		9.77	12.55
0230	12" deep	G		190	.084		7.10	3.50		10.60	13.55
0300	16 ga x 8" deep	G		180	.089		6.15	3.69		9.84	12.80
0310	10" deep	G		170	.094		7.35	3.91		11.26	14.45
0320	12" deep	G		160	.100		8	4.16		12.16	15.65
0400	14 ga x 10" deep	G		140	.114		8.55	4.75		13.30	17.20
0410	12" deep	G		130	.123		9.40	5.10		14.50	18.70
0500	12 ga x 10" deep	G		110	.145		11.30	6.05		17.35	22.50
0510	12" deep	G		100	.160		12.50	6.65		19.15	24.50
1210	Triple, 18 ga x 8" deep	G		170	.094		7.70	3.91		11.61	14.90
1220	10" deep	G		165	.097		9.25	4.03		13.28	16.80
1230	12" deep	G		160	.100		10.20	4.16		14.36	18.10
1300	16 ga x 8" deep	G		145	.110		8.90	4.58		13.48	17.35
1310	10" deep	G		140	.114		10.60	4.75		15.35	19.45
1320	12" deep	G		135	.119		11.60	4.92		16.52	21
1400	14 ga x 10" deep	G		115	.139		12.40	5.80		18.20	23
1410	12" deep	G		110	.145		13.65	6.05		19.70	25
1500	12 ga x 10" deep	G		90	.178		16.55	7.40		23.95	30.50
1510	12" deep	G		85	.188		18.40	7.80		26.20	33

05 42 13.40 Framing, Joists

			Crew	Daily Output	Labor-Hours	Unit	Material	Labor	Equipment	Total	Total Incl O&P
0010	**FRAMING, JOISTS**, no band joists (track), web stiffeners, headers,										
0020	Beams, bridging or bracing										
0025	Made from recycled materials	G									
0030	Joists (2" flange) and fasteners, materials only										
0220	18 ga x 6" deep	G				L.F.	1.48			1.48	1.63
0230	8" deep	G					1.76			1.76	1.94
0240	10" deep	G					2.07			2.07	2.28
0320	16 ga x 6" deep	G					1.82			1.82	2
0330	8" deep	G					2.18			2.18	2.40
0340	10" deep	G					2.54			2.54	2.80
0350	12" deep	G					2.89			2.89	3.18
0430	14 ga x 8" deep	G					2.75			2.75	3.03
0440	10" deep	G					3.17			3.17	3.49
0450	12" deep	G					3.61			3.61	3.97
0540	12 ga x 10" deep	G					4.62			4.62	5.10
0550	12" deep	G					5.25			5.25	5.80
1010	Installation of joists to band joists, beams & headers, labor only										
1220	18 ga x 6" deep		2 Carp	110	.145	Ea.		6.05		6.05	9.95
1230	8" deep			90	.178			7.40		7.40	12.15
1240	10" deep			80	.200			8.30		8.30	13.65
1320	16 ga x 6" deep			95	.168			7		7	11.50
1330	8" deep			70	.229			9.50		9.50	15.60
1340	10" deep			60	.267			11.10		11.10	18.20

05 42 Cold-Formed Metal Joist Framing

05 42 13 – Cold-Formed Metal Floor Joist Framing

05 42 13.40 Framing, Joists		Crew	Daily Output	Labor-Hours	Unit	Material	2010 Bare Costs Labor	Equipment	Total	Total Incl O&P
1350	12" deep	2 Carp	55	.291	Ea.		12.10		12.10	19.85
1430	14 ga x 8" deep		65	.246			10.25		10.25	16.80
1440	10" deep		45	.356			14.75		14.75	24.50
1450	12" deep		35	.457			19		19	31
1540	12 ga x 10" deep		40	.400			16.60		16.60	27.50
1550	12" deep		30	.533			22		22	36.50
9000	Minimum labor/equipment charge		4	4	Job		166		166	273

05 42 13.45 Framing, Web Stiffeners

			Crew	Daily Output	Labor-Hours	Unit	Material	Labor	Equipment	Total	Total Incl O&P
0010	**FRAMING, WEB STIFFENERS** at joist bearing, fabricated from										
0020	Stud piece (1-5/8" flange) to stiffen joist (2" flange)										
0025	Made from recycled materials	G									
2120	For 6" deep joist, with 18 ga x 2-1/2" stud	G	1 Carp	120	.067	Ea.	1.78	2.77		4.55	6.50
2130	3-5/8" stud	G		110	.073		1.97	3.02		4.99	7.10
2140	4" stud	G		105	.076		1.90	3.17		5.07	7.30
2150	6" stud	G		100	.080		2.08	3.32		5.40	7.75
2160	8" stud	G		95	.084		2.14	3.50		5.64	8.10
2220	8" deep joist, with 2-1/2" stud	G		120	.067		1.95	2.77		4.72	6.70
2230	3-5/8" stud	G		110	.073		2.12	3.02		5.14	7.30
2240	4" stud	G		105	.076		2.08	3.17		5.25	7.50
2250	6" stud	G		100	.080		2.28	3.32		5.60	7.95
2260	8" stud	G		95	.084		2.46	3.50		5.96	8.45
2320	10" deep joist, with 2-1/2" stud	G		110	.073		2.76	3.02		5.78	8
2330	3-5/8" stud	G		100	.080		3.03	3.32		6.35	8.80
2340	4" stud	G		95	.084		2.99	3.50		6.49	9.05
2350	6" stud	G		90	.089		3.24	3.69		6.93	9.60
2360	8" stud	G		85	.094		3.30	3.91		7.21	10.05
2420	12" deep joist, with 2-1/2" stud	G		110	.073		2.92	3.02		5.94	8.15
2430	3-5/8" stud	G		100	.080		3.17	3.32		6.49	8.95
2440	4" stud	G		95	.084		3.10	3.50		6.60	9.15
2450	6" stud	G		90	.089		3.40	3.69		7.09	9.80
2460	8" stud	G		85	.094		3.67	3.91		7.58	10.45
3130	For 6" deep joist, with 16 ga x 3-5/8" stud	G		100	.080		2.07	3.32		5.39	7.70
3140	4" stud	G		95	.084		2.05	3.50		5.55	8
3150	6" stud	G		90	.089		2.25	3.69		5.94	8.50
3160	8" stud	G		85	.094		2.38	3.91		6.29	9
3230	8" deep joist, with 3-5/8" stud	G		100	.080		2.29	3.32		5.61	7.95
3240	4" stud	G		95	.084		2.24	3.50		5.74	8.20
3250	6" stud	G		90	.089		2.49	3.69		6.18	8.80
3260	8" stud	G		85	.094		2.67	3.91		6.58	9.35
3330	10" deep joist, with 3-5/8" stud	G		85	.094		3.13	3.91		7.04	9.85
3340	4" stud	G		80	.100		3.19	4.16		7.35	10.35
3350	6" stud	G		75	.107		3.47	4.43		7.90	11.10
3360	8" stud	G		70	.114		3.63	4.75		8.38	11.80
3430	12" deep joist, with 3-5/8" stud	G		85	.094		3.42	3.91		7.33	10.15
3440	4" stud	G		80	.100		3.35	4.16		7.51	10.55
3450	6" stud	G		75	.107		3.72	4.43		8.15	11.40
3460	8" stud	G		70	.114		3.99	4.75		8.74	12.20
4230	For 8" deep joist, with 14 ga x 3-5/8" stud	G		90	.089		2.97	3.69		6.66	9.30
4240	4" stud	G		85	.094		3.03	3.91		6.94	9.75
4250	6" stud	G		80	.100		3.28	4.16		7.44	10.45
4260	8" stud	G		75	.107		3.52	4.43		7.95	11.20
4330	10" deep joist, with 3-5/8" stud	G		75	.107		4.18	4.43		8.61	11.90

05 42 Cold-Formed Metal Joist Framing

05 42 13 – Cold-Formed Metal Floor Joist Framing

05 42 13.45 Framing, Web Stiffeners

			Crew	Daily Output	Labor-Hours	Unit	Material	2010 Bare Costs Labor	Equipment	Total	Total Incl O&P
4340	4" stud	G	1 Carp	70	.114	Ea.	4.14	4.75		8.89	12.35
4350	6" stud	G		65	.123		4.56	5.10		9.66	13.40
4360	8" stud	G		60	.133		4.76	5.55		10.31	14.35
4430	12" deep joist, with 3-5/8" stud	G		75	.107		4.44	4.43		8.87	12.20
4440	4" stud	G		70	.114		4.52	4.75		9.27	12.80
4450	6" stud	G		65	.123		4.90	5.10		10	13.80
4460	8" stud	G		60	.133		5.25	5.55		10.80	14.90
5330	For 10" deep joist, with 12 ga x 3-5/8" stud	G		65	.123		4.40	5.10		9.50	13.25
5340	4" stud	G		60	.133		4.54	5.55		10.09	14.10
5350	6" stud	G		55	.145		5	6.05		11.05	15.45
5360	8" stud	G		50	.160		5.50	6.65		12.15	16.95
5430	12" deep joist, with 3-5/8" stud	G		65	.123		4.88	5.10		9.98	13.75
5440	4" stud	G		60	.133		4.79	5.55		10.34	14.35
5450	6" stud	G		55	.145		5.45	6.05		11.50	15.95
5460	8" stud	G		50	.160		6.30	6.65		12.95	17.80

05 42 23 – Cold-Formed Metal Roof Joist Framing

05 42 23.05 Framing, Bracing

0010	**FRAMING, BRACING**										
0015	Made from recycled materials	G									
0020	Continuous bracing, per row										
0100	16 ga x 1-1/2" channel thru rafters/trusses @ 16" O.C.	G	1 Carp	4.50	1.778	C.L.F.	45.50	74		119.50	171
0120	24" O.C.	G		6	1.333		45.50	55.50		101	141
0300	2" x 2" angle x 18 ga, rafters/trusses @ 16" O.C.	G		6	1.333		68	55.50		123.50	166
0320	24" O.C.	G		8	1		68	41.50		109.50	144
0400	16 ga, rafters/trusses @ 16" O.C.	G		4.50	1.778		86	74		160	216
0420	24" O.C.	G		6.50	1.231		86	51		137	179

05 42 23.10 Framing, Bridging

0010	**FRAMING, BRIDGING**										
0015	Made from recycled materials	G									
0020	Solid, between rafters w/1-1/4" leg track, per rafter bay										
1200	Rafters 16" O.C., 18 ga x 4" deep	G	1 Carp	60	.133	Ea.	1.37	5.55		6.92	10.60
1210	6" deep	G		57	.140		1.77	5.85		7.62	11.55
1220	8" deep	G		55	.145		2.24	6.05		8.29	12.40
1230	10" deep	G		52	.154		2.77	6.40		9.17	13.55
1240	12" deep	G		50	.160		3.20	6.65		9.85	14.40
2200	24" O.C., 18 ga x 4" deep	G		60	.133		1.98	5.55		7.53	11.30
2210	6" deep	G		57	.140		2.55	5.85		8.40	12.40
2220	8" deep	G		55	.145		3.25	6.05		9.30	13.50
2230	10" deep	G		52	.154		4.01	6.40		10.41	14.90
2240	12" deep	G		50	.160		4.63	6.65		11.28	16

05 42 23.50 Framing, Parapets

0010	**FRAMING, PARAPETS**										
0015	Made from recycled materials	G									
0100	3' high installed on 1st story, 18 ga x 4" wide studs, 12" O.C.	G	2 Carp	100	.160	L.F.	5.05	6.65		11.70	16.45
0110	16" O.C.	G		150	.107		4.28	4.43		8.71	12
0120	24" O.C.	G		200	.080		3.53	3.32		6.85	9.35
0200	6" wide studs, 12" O.C.	G		100	.160		6.40	6.65		13.05	17.95
0210	16" O.C.	G		150	.107		5.45	4.43		9.88	13.30
0220	24" O.C.	G		200	.080		4.51	3.32		7.83	10.40
1100	Installed on 2nd story, 18 ga x 4" wide studs, 12" O.C.	G		95	.168		5.05	7		12.05	17.05
1110	16" O.C.	G		145	.110		4.28	4.58		8.86	12.25
1120	24" O.C.	G		190	.084		3.53	3.50		7.03	9.65

05 42 Cold-Formed Metal Joist Framing

05 42 23 – Cold-Formed Metal Roof Joist Framing

05 42 23.50 Framing, Parapets

		Crew	Daily Output	Labor-Hours	Unit	Material	2010 Bare Costs Labor	Equipment	Total	Total Incl O&P
1200	6" wide studs, 12" O.C.	G 2 Carp	95	.168	L.F.	6.40	7		13.40	18.55
1210	16" O.C.	G	145	.110		5.45	4.58		10.03	13.55
1220	24" O.C.	G	190	.084		4.51	3.50		8.01	10.70
2100	Installed on gable, 18 ga x 4" wide studs, 12" O.C.	G	85	.188		5.05	7.80		12.85	18.40
2110	16" O.C.	G	130	.123		4.28	5.10		9.38	13.10
2120	24" O.C.	G	170	.094		3.53	3.91		7.44	10.30
2200	6" wide studs, 12" O.C.	G	85	.188		6.40	7.80		14.20	19.90
2210	16" O.C.	G	130	.123		5.45	5.10		10.55	14.40
2220	24" O.C.	G	170	.094		4.51	3.91		8.42	11.35

05 42 23.60 Framing, Roof Rafters

		Crew	Daily Output	Labor-Hours	Unit	Material	Labor	Equipment	Total	Total Incl O&P
0010	**FRAMING, ROOF RAFTERS**									
0015	Made from recycled materials	G								
0100	Boxed ridge beam, double, 18 ga x 6" deep	G 2 Carp	160	.100	L.F.	4.81	4.16		8.97	12.15
0110	8" deep	G	150	.107		5.35	4.43		9.78	13.20
0120	10" deep	G	140	.114		6.45	4.75		11.20	14.90
0130	12" deep	G	130	.123		7.10	5.10		12.20	16.20
0200	16 ga x 6" deep	G	150	.107		5.45	4.43		9.88	13.30
0210	8" deep	G	140	.114		6.15	4.75		10.90	14.55
0220	10" deep	G	130	.123		7.35	5.10		12.45	16.45
0230	12" deep	G	120	.133		8	5.55		13.55	17.90
1100	Rafters, 2" flange, material only, 18 ga x 6" deep	G				1.48			1.48	1.63
1110	8" deep	G				1.76			1.76	1.94
1120	10" deep	G				2.07			2.07	2.28
1130	12" deep	G				2.40			2.40	2.64
1200	16 ga x 6" deep	G				1.82			1.82	2
1210	8" deep	G				2.18			2.18	2.40
1220	10" deep	G				2.54			2.54	2.80
1230	12" deep	G				2.89			2.89	3.18
2100	Installation only, ordinary rafter to 4:12 pitch, 18 ga x 6" deep	2 Carp	35	.457	Ea.		19		19	31
2110	8" deep		30	.533			22		22	36.50
2120	10" deep		25	.640			26.50		26.50	43.50
2130	12" deep		20	.800			33		33	54.50
2200	16 ga x 6" deep		30	.533			22		22	36.50
2210	8" deep		25	.640			26.50		26.50	43.50
2220	10" deep		20	.800			33		33	54.50
2230	12" deep		15	1.067			44.50		44.50	73
8100	Add to labor, ordinary rafters on steep roofs						25%			
8110	Dormers & complex roofs						50%			
8200	Hip & valley rafters to 4:12 pitch						25%			
8210	Steep roofs						50%			
8220	Dormers & complex roofs						75%			
8300	Hip & valley jack rafters to 4:12 pitch						50%			
8310	Steep roofs						75%			
8320	Dormers & complex roofs						100%			
9000	Minimum labor/equipment charge	2 Carp	4	4	Job		166		166	273

05 42 23.70 Framing, Soffits and Canopies

		Crew	Daily Output	Labor-Hours	Unit	Material	Labor	Equipment	Total	Total Incl O&P
0010	**FRAMING, SOFFITS & CANOPIES**									
0015	Made from recycled materials	G								
0130	Continuous ledger track @ wall, studs @ 16" O.C., 18 ga x 4" wide	G 2 Carp	535	.030	L.F.	.91	1.24		2.15	3.04
0140	6" wide	G	500	.032		1.18	1.33		2.51	3.47
0150	8" wide	G	465	.034		1.50	1.43		2.93	4
0160	10" wide	G	430	.037		1.85	1.55		3.40	4.57

05 42 Cold-Formed Metal Joist Framing

05 42 23 – Cold-Formed Metal Roof Joist Framing

05 42 23.70 Framing, Soffits and Canopies

			Crew	Daily Output	Labor-Hours	Unit	Material	2010 Bare Costs Labor	Equipment	Total	Total Incl O&P
0230	Studs @ 24" O.C., 18 ga x 4" wide	G	2 Carp	800	.020	L.F.	.87	.83		1.70	2.33
0240	6" wide	G		750	.021		1.12	.89		2.01	2.70
0250	8" wide	G		700	.023		1.43	.95		2.38	3.13
0260	10" wide	G		650	.025		1.76	1.02		2.78	3.62
1000	Horizontal soffit and canopy members, material only										
1030	1-5/8" flange studs, 18 ga x 4" deep	G				L.F.	1.20			1.20	1.32
1040	6" deep	G					1.51			1.51	1.66
1050	8" deep	G					1.84			1.84	2.02
1140	2" flange joists, 18 ga x 6" deep	G					1.69			1.69	1.86
1150	8" deep	G					2.02			2.02	2.22
1160	10" deep	G					2.36			2.36	2.60
4030	Installation only, 18 ga, 1-5/8" flange x 4" deep		2 Carp	130	.123	Ea.		5.10		5.10	8.40
4040	6" deep			110	.145			6.05		6.05	9.95
4050	8" deep			90	.178			7.40		7.40	12.15
4140	2" flange, 18 ga x 6" deep			110	.145			6.05		6.05	9.95
4150	8" deep			90	.178			7.40		7.40	12.15
4160	10" deep			80	.200			8.30		8.30	13.65
6010	Clips to attach facia to rafter tails, 2" x 2" x 18 ga angle	G	1 Carp	120	.067		.81	2.77		3.58	5.45
6020	16 ga angle	G	"	100	.080		1.01	3.32		4.33	6.55
9000	Minimum labor/equipment charge		2 Carp	4	4	Job		166		166	273

05 44 Cold-Formed Metal Trusses

05 44 13 – Cold-Formed Metal Roof Trusses

05 44 13.60 Framing, Roof Trusses

			Crew	Daily Output	Labor-Hours	Unit	Material	Labor	Equipment	Total	Total Incl O&P
0010	**FRAMING, ROOF TRUSSES**										
0015	Made from recycled materials	G									
0020	Fabrication of trusses on ground, Fink (W) or King Post, to 4:12 pitch										
0120	18 ga x 4" chords, 16' span	G	2 Carp	12	1.333	Ea.	56	55.50		111.50	153
0130	20' span	G		11	1.455		70	60.50		130.50	177
0140	24' span	G		11	1.455		84	60.50		144.50	192
0150	28' span	G		10	1.600		98	66.50		164.50	217
0160	32' span	G		10	1.600		112	66.50		178.50	232
0250	6" chords, 28' span	G		9	1.778		123	74		197	257
0260	32' span	G		9	1.778		141	74		215	276
0270	36' span	G		8	2		159	83		242	310
0280	40' span	G		8	2		176	83		259	330
1120	5:12 to 8:12 pitch, 18 ga x 4" chords, 16' span	G		10	1.600		64	66.50		130.50	180
1130	20' span	G		9	1.778		80	74		154	209
1140	24' span	G		9	1.778		96	74		170	227
1150	28' span	G		8	2		112	83		195	260
1160	32' span	G		8	2		128	83		211	278
1250	6" chords, 28' span	G		7	2.286		141	95		236	310
1260	32' span	G		7	2.286		161	95		256	335
1270	36' span	G		6	2.667		181	111		292	380
1280	40' span	G		6	2.667		202	111		313	405
2120	9:12 to 12:12 pitch, 18 ga x 4" chords, 16' span	G		8	2		80	83		163	225
2130	20' span	G		7	2.286		100	95		195	266
2140	24' span	G		7	2.286		120	95		215	288
2150	28' span	G		6	2.667		140	111		251	335
2160	32' span	G		6	2.667		160	111		271	360
2250	6" chords, 28' span	G		5	3.200		176	133		309	410

05 44 Cold-Formed Metal Trusses

05 44 13 – Cold-Formed Metal Roof Trusses

05 44 13.60 Framing, Roof Trusses

		Crew	Daily Output	Labor-Hours	Unit	Material	2010 Bare Costs Labor	Equipment	Total	Total Incl O&P
2260	32' span [G]	2 Carp	5	3.200	Ea.	202	133		335	440
2270	36' span [G]		4	4		227	166		393	520
2280	40' span [G]		4	4		252	166		418	550
4900	Minimum labor/equipment charge		4	4	Job		166		166	273
5120	Erection only of roof trusses, to 4:12 pitch, 16' span	F-6	48	.833	Ea.		32.50	13.65	46.15	67.50
5130	20' span		46	.870			33.50	14.25	47.75	70
5140	24' span		44	.909			35	14.90	49.90	73.50
5150	28' span		42	.952			37	15.60	52.60	77
5160	32' span		40	1			38.50	16.40	54.90	81
5170	36' span		38	1.053			41	17.25	58.25	85.50
5180	40' span		36	1.111			43	18.20	61.20	90
5220	5:12 to 8:12 pitch, 16' span		42	.952			37	15.60	52.60	77
5230	20' span		40	1			38.50	16.40	54.90	81
5240	24' span		38	1.053			41	17.25	58.25	85.50
5250	28' span		36	1.111			43	18.20	61.20	90
5260	32' span		34	1.176			45.50	19.30	64.80	95
5270	36' span		32	1.250			48.50	20.50	69	101
5280	40' span		30	1.333			51.50	22	73.50	108
5320	9:12 to 12:12 pitch, 16' span		36	1.111			43	18.20	61.20	90
5330	20' span		34	1.176			45.50	19.30	64.80	95
5340	24' span		32	1.250			48.50	20.50	69	101
5350	28' span		30	1.333			51.50	22	73.50	108
5360	32' span		28	1.429			55.50	23.50	79	116
5370	36' span		26	1.538			59.50	25	84.50	125
5380	40' span		24	1.667			64.50	27.50	92	135
9000	Minimum labor/equipment charge		2	20	Job		775	330	1,105	1,600

05 51 Metal Stairs

05 51 13 – Metal Pan Stairs

05 51 13.50 Pan Stairs

		Crew	Daily Output	Labor-Hours	Unit	Material	2010 Bare Costs Labor	Equipment	Total	Total Incl O&P
0010	**PAN STAIRS**, shop fabricated, steel stringers									
0015	Made from recycled materials [G]									
0200	Cement fill metal pan, picket rail, 3'-6" wide [G]	E-4	35	.914	Riser	415	43.50	4.17	462.67	540
0300	4'-0" wide [G]		30	1.067		470	50.50	4.86	525.36	615
0350	Wall rail, both sides, 3'-6" wide [G]		53	.604		315	28.50	2.75	346.25	405
1500	Landing, steel pan, conventional [G]		160	.200	S.F.	55	9.50	.91	65.41	79

05 51 16 – Metal Floor Plate Stairs

05 51 16.50 Floor Plate Stairs

		Crew	Daily Output	Labor-Hours	Unit	Material	2010 Bare Costs Labor	Equipment	Total	Total Incl O&P
0010	**FLOOR PLATE STAIRS**, shop fabricated, steel stringers									
0015	Made from recycled materials [G]									
0400	Cast iron tread and pipe rail, 3'-6" wide [G]	E-4	35	.914	Riser	440	43.50	4.17	487.67	570
0450	4'-0" wide [G]		30	1.067		470	50.50	4.86	525.36	615
0475	5'-0" wide [G]		28	1.143		510	54	5.20	569.20	665
0500	Checkered plate tread, industrial, 3'-6" wide [G]		28	1.143		275	54	5.20	334.20	410
0550	Circular, for tanks, 3'-0" wide [G]		33	.970		305	46	4.42	355.42	425
0600	For isolated stairs, add						100%			
0800	Custom steel stairs, 3'-6" wide, minimum [G]	E-4	35	.914		415	43.50	4.17	462.67	540
0810	Average [G]		30	1.067		550	50.50	4.86	605.36	705
0900	Maximum [G]		20	1.600		690	76	7.30	773.30	900
1100	For 4' wide stairs, add					5%	5%			

05 51 Metal Stairs

05 51 16 – Metal Floor Plate Stairs

05 51 16.50 Floor Plate Stairs

		Crew	Daily Output	Labor-Hours	Unit	Material	2010 Bare Costs Labor	Equipment	Total	Total Incl O&P
1300	For 5' wide stairs, add				Riser	10%	10%			

05 51 19 – Metal Grating Stairs

05 51 19.50 Grating Stairs

			Crew	Daily Output	Labor-Hours	Unit	Material	Labor	Equipment	Total	Total Incl O&P
0010	**GRATING STAIRS**, shop fabricated, steel stringers, safety nosing on treads										
0015	Made from recycled materials	G									
0020	Grating tread and pipe railing, 3'-6" wide	G	E-4	35	.914	Riser	275	43.50	4.17	322.67	390
0100	4'-0" wide	G		30	1.067	"	360	50.50	4.86	415.36	495
9000	Minimum labor/equipment charge		↓	2	16	Job		760	73	833	1,450

05 51 23 – Metal Fire Escapes

05 51 23.25 Fire Escapes

			Crew	Daily Output	Labor-Hours	Unit	Material	Labor	Equipment	Total	Total Incl O&P
0010	**FIRE ESCAPES**, shop fabricated										
0200	2' wide balcony, 1" x 1/4" bars 1-1/2" O.C.	G	1 Sswk	5	1.600	L.F.	48	75		123	190
0400	1st story cantilevered stair, standard	G	"	.09	88.889	Ea.	2,000	4,175		6,175	9,825
0500	Cable counterweight	G				"	1,875			1,875	2,050
0700	Platform & fixed stair, 36" x 40"	G	1 Sswk	.17	47.059	Flight	890	2,200		3,090	5,000
0900	For 3'-6" wide escapes, add to above						100%	150%			

05 51 23.50 Fire Escape Stairs

			Crew	Daily Output	Labor-Hours	Unit	Material	Labor	Equipment	Total	Total Incl O&P
0010	**FIRE ESCAPE STAIRS**, shop fabricated										
0020	One story, disappearing, stainless steel	G	2 Sswk	20	.800	V.L.F.	212	37.50		249.50	300
0100	Portable ladder					Ea.	99			99	109
1100	Fire escape, galvanized steel, 8'-0" to 10'-4" ceiling	G	2 Carp	1	16		1,525	665		2,190	2,775
1110	10'-6" to 13'-6" ceiling	G	"	1	16	↓	2,000	665		2,665	3,300

05 51 33 – Metal Ladders

05 51 33.13 Vertical Metal Ladders

			Crew	Daily Output	Labor-Hours	Unit	Material	Labor	Equipment	Total	Total Incl O&P
0010	**VERTICAL METAL LADDERS**, shop fabricated										
0015	Made from recycled materials	G									
0020	Steel, 20" wide, bolted to concrete, with cage	G	E-4	50	.640	V.L.F.	74	30.50	2.92	107.42	140
0100	Without cage	G		85	.376		38	17.85	1.72	57.57	76.50
0300	Aluminum, bolted to concrete, with cage	G		50	.640		120	30.50	2.92	153.42	191
0400	Without cage	G		85	.376	↓	55.50	17.85	1.72	75.07	95.50
9000	Minimum labor/equipment charge		↓	2	16	Job		760	73	833	1,450

05 51 33.16 Inclined Metal Ladders

			Crew	Daily Output	Labor-Hours	Unit	Material	Labor	Equipment	Total	Total Incl O&P
0010	**INCLINED METAL LADDERS**, shop fabricated										
3900	Industrial ships ladder, 3' W, grating treads, 2 line pipe rail	G	E-4	30	1.067	Riser	149	50.50	4.86	204.36	262
4000	Aluminum	G	"	30	1.067	"	229	50.50	4.86	284.36	350

05 51 33.23 Alternating Tread Ladders

			Crew	Daily Output	Labor-Hours	Unit	Material	Labor	Equipment	Total	Total Incl O&P
0010	**ALTERNATING TREAD LADDERS**, shop fabricated										
0015	Made from recycled materials	G									
1350	Alternating tread stair, 56/68°, steel, standard paint color	G	2 Sswk	50	.320	V.L.F.	202	15		217	250
1360	Non-standard paint color	G		50	.320		228	15		243	279
1370	Galvanized steel	G		50	.320		235	15		250	287
1380	Stainless steel	G		50	.320		340	15		355	400
1390	68°, aluminum	G	↓	50	.320	↓	248	15		263	300

05 52 Metal Railings

05 52 13 – Pipe and Tube Railings

05 52 13.50 Railings, Pipe

		Crew	Daily Output	Labor-Hours	Unit	Material	2010 Bare Costs Labor	Equipment	Total	Total Incl O&P
0010	**RAILINGS, PIPE**, shop fab'd, 3'-6" high, posts @ 5' O.C.									
0015	Made from recycled materials	G								
0020	Aluminum, 2 rail, satin finish, 1-1/4" diameter	G	E-4	160	.200	L.F.	31.50	9.50	.91	41.91	53
0030	Clear anodized	G		160	.200		39	9.50	.91	49.41	61
0040	Dark anodized	G		160	.200		44	9.50	.91	54.41	66.50
0080	1-1/2" diameter, satin finish	G		160	.200		37.50	9.50	.91	47.91	60
0090	Clear anodized	G		160	.200		42	9.50	.91	52.41	64.50
0100	Dark anodized	G		160	.200		46.50	9.50	.91	56.91	69.50
0140	Aluminum, 3 rail, 1-1/4" diam., satin finish	G		137	.234		48	11.05	1.07	60.12	74
0150	Clear anodized	G		137	.234		60	11.05	1.07	72.12	87
0160	Dark anodized	G		137	.234		66.50	11.05	1.07	78.62	94
0200	1-1/2" diameter, satin finish	G		137	.234		57.50	11.05	1.07	69.62	84.50
0210	Clear anodized	G		137	.234		65.50	11.05	1.07	77.62	93
0220	Dark anodized	G		137	.234		71.50	11.05	1.07	83.62	100
0500	Steel, 2 rail, on stairs, primed, 1-1/4" diameter	G		160	.200		20.50	9.50	.91	30.91	41
0520	1-1/2" diameter	G		160	.200		22.50	9.50	.91	32.91	43.50
0540	Galvanized, 1-1/4" diameter	G		160	.200		28.50	9.50	.91	38.91	50
0560	1-1/2" diameter	G		160	.200		32	9.50	.91	42.41	53.50
0580	Steel, 3 rail, primed, 1-1/4" diameter	G		137	.234		30.50	11.05	1.07	42.62	55
0600	1-1/2" diameter	G		137	.234		32.50	11.05	1.07	44.62	57
0620	Galvanized, 1-1/4" diameter	G		137	.234		43	11.05	1.07	55.12	68.50
0640	1-1/2" diameter	G		137	.234		50.50	11.05	1.07	62.62	77
0700	Stainless steel, 2 rail, 1-1/4" diam. #4 finish	G		137	.234		90.50	11.05	1.07	102.62	121
0720	High polish	G		137	.234		146	11.05	1.07	158.12	182
0740	Mirror polish	G		137	.234		183	11.05	1.07	195.12	222
0760	Stainless steel, 3 rail, 1-1/2" diam., #4 finish	G		120	.267		137	12.65	1.22	150.87	175
0770	High polish	G		120	.267		226	12.65	1.22	239.87	273
0780	Mirror finish	G		120	.267		275	12.65	1.22	288.87	330
0900	Wall rail, alum. pipe, 1-1/4" diam., satin finish	G		213	.150		17.95	7.10	.69	25.74	33.50
0905	Clear anodized	G		213	.150		22	7.10	.69	29.79	38
0910	Dark anodized	G		213	.150		26.50	7.10	.69	34.29	43
0915	1-1/2" diameter, satin finish	G		213	.150		19.85	7.10	.69	27.64	36
0920	Clear anodized	G		213	.150		25	7.10	.69	32.79	41.50
0925	Dark anodized	G		213	.150		31	7.10	.69	38.79	48
0930	Steel pipe, 1-1/4" diameter, primed	G		213	.150		12.45	7.10	.69	20.24	27.50
0935	Galvanized	G		213	.150		18.05	7.10	.69	25.84	33.50
0940	1-1/2" diameter	G		176	.182		12.85	8.60	.83	22.28	31
0945	Galvanized	G		213	.150		18.15	7.10	.69	25.94	33.50
0955	Stainless steel pipe, 1-1/2" diam., #4 finish	G		107	.299		72.50	14.20	1.36	88.06	107
0960	High polish	G		107	.299		148	14.20	1.36	163.56	190
0965	Mirror polish	G		107	.299		174	14.20	1.36	189.56	219
2000	Aluminum pipe & picket railing, double top rail, pickets @ 4-1/2" OC										
2010	36" high, straight & level	G	2 Sswk	80	.200	L.F.	81	9.40		90.40	106
2020	Curved & level	G		60	.267		111	12.50		123.50	145
2030	Straight & sloped	G		40	.400		93	18.75		111.75	137
9000	Minimum labor/equipment charge		1 Sswk	2	4	Job		188		188	345

05 52 16 – Industrial Railings

05 52 16.50 Railings, Industrial

		Crew	Daily Output	Labor-Hours	Unit	Material	Labor	Equipment	Total	Total Incl O&P
0010	**RAILINGS, INDUSTRIAL**, shop fab'd, 3'-6" high, posts @ 5' O.C.									
0020	2 rail, 3'-6" high, 1-1/2" pipe	G	E-4	255	.125	L.F.	22.50	5.95	.57	29.02	36.50
0200	For 4" high kick plate, 10 gauge, add	G					4.72			4.72	5.20
0500	For curved rails, add						30%	30%			

05 52 Metal Railings

05 52 16 – Industrial Railings

05 52 16.50 Railings, Industrial

		Crew	Daily Output	Labor-Hours	Unit	Material	2010 Bare Costs Labor	Equipment	Total	Total Incl O&P
9000	Minimum labor/equipment charge	1 Sswk	2	4	Job		188		188	345

05 56 Metal Castings

05 56 13 – Metal Construction Castings

05 56 13.50 Construction Castings

			Crew	Daily Output	Labor-Hours	Unit	Material	Labor	Equipment	Total	Total Incl O&P
0010	**CONSTRUCTION CASTINGS**										
0020	Manhole covers and frames, see Div. 33 44 13.13										
0100	Column bases, cast iron, 16" x 16", approx. 65 lb.	G	E-4	46	.696	Ea.	135	33	3.17	171.17	212
0200	32" x 32", approx. 256 lb.	G		23	1.391	"	505	66	6.35	577.35	685
0600	Miscellaneous C.I. castings, light sections, less than 150 lb.	G		3200	.010	Lb.	2.09	.47	.05	2.61	3.22
1300	Special low volume items	G		3200	.010	"	3.63	.47	.05	4.15	4.91

05 58 Formed Metal Fabrications

05 58 23 – Formed Columns

05 58 23.10 Aluminum Columns

			Crew	Daily Output	Labor-Hours	Unit	Material	Labor	Equipment	Total	Total Incl O&P
0010	**ALUMINUM COLUMNS**										
0015	Made from recycled materials	G									
0020	Aluminum, extruded, stock units, no cap or base, 6" diameter	G	E-4	240	.133	L.F.	11.65	6.30	.61	18.56	25
0100	8" diameter	G	"	170	.188		15	8.90	.86	24.76	33.50
0500	For square columns, add to column prices above						50%				

05 58 25 – Formed Lamp Posts

05 58 25.40 Lamp Posts

			Crew	Daily Output	Labor-Hours	Unit	Material	Labor	Equipment	Total	Total Incl O&P
0010	**LAMP POSTS**										
0020	Aluminum, 7' high, stock units, post only	G	1 Carp	16	.500	Ea.	31.50	21		52.50	68.50
0100	Mild steel, plain	G		16	.500	"	28	21		49	65
9000	Minimum labor/equipment charge			4	2	Job		83		83	137

05 58 27 – Formed Guards

05 58 27.90 Window Guards

			Crew	Daily Output	Labor-Hours	Unit	Material	Labor	Equipment	Total	Total Incl O&P
0010	**WINDOW GUARDS**, shop fabricated										
0015	Expanded metal, steel angle frame, permanent	G	E-4	350	.091	S.F.	19.10	4.33	.42	23.85	29.50
0025	Steel bars, 1/2" x 1/2", spaced 5" O.C.	G	"	290	.110	"	13.25	5.25	.50	19	24.50
0030	Hinge mounted, add	G				Opng.	38.50			38.50	42
0040	Removable type, add	G				"	24.50			24.50	26.50
0050	For galvanized guards, add					S.F.	35%				
0070	For pivoted or projected type, add						105%	40%			
0100	Mild steel, stock units, economy	G	E-4	405	.079		5.20	3.75	.36	9.31	12.95
0200	Deluxe	G		405	.079		10.65	3.75	.36	14.76	19
0400	Woven wire, stock units, 3/8" channel frame, 3' x 5' opening	G		40	.800	Opng.	140	38	3.65	181.65	228
0500	4' x 6' opening	G		38	.842		224	40	3.84	267.84	325
0800	Basket guards for above, add	G					192			192	211
1000	Swinging guards for above, add	G					66			66	72.50
9000	Minimum labor/equipment charge		1 Sswk	2	4	Job		188		188	345

05 71 Decorative Metal Stairs

05 71 13 – Fabricated Metal Spiral Stairs

05 71 13.50 Spiral Stairs

		Crew	Daily Output	Labor-Hours	Unit	Material	2010 Bare Costs Labor	Equipment	Total	Total Incl O&P
0010	**SPIRAL STAIRS**, shop fabricated									
1810	Spiral aluminum, 5'-0" diameter, stock units	E-4 [G]	45	.711	Riser	500	33.50	3.24	536.74	615
1820	Custom units	[G]	45	.711		945	33.50	3.24	981.74	1,125
1900	Spiral, cast iron, 4'-0" diameter, ornamental, minimum	[G]	45	.711		445	33.50	3.24	481.74	555
1920	Maximum	[G]	25	1.280		605	60.50	5.85	671.35	785
2000	Spiral, steel, industrial checkered plate, 4' diameter	[G]	45	.711		445	33.50	3.24	481.74	555
2200	Stock units, 6'-0" diameter	[G]	40	.800		540	38	3.65	581.65	670
3110	Spiral steel, stock units, primed, flat metal tread, 3'-6" dia	2 Carp [G]	1.60	10	Flight	1,250	415		1,665	2,025
3120	4'-0" dia	[G]	1.45	11.034		1,425	460		1,885	2,325
3130	4'-6" dia	[G]	1.35	11.852		1,575	490		2,065	2,525
3140	5'-0" dia	[G]	1.25	12.800		1,700	530		2,230	2,750
3210	Galvanized, 3'-6" dia	[G]	1.60	10		1,575	415		1,990	2,425
3220	4'-0" dia	[G]	1.45	11.034		1,800	460		2,260	2,750
3230	4'-6" dia	[G]	1.35	11.852		2,000	490		2,490	3,000
3240	5'-0" dia	[G]	1.25	12.800		2,150	530		2,680	3,250
3310	Checkered plate tread, 3'-6" dia	[G]	1.45	11.034		1,450	460		1,910	2,350
3320	4'-0" dia	[G]	1.35	11.852		1,650	490		2,140	2,600
3330	4'-6" dia	[G]	1.25	12.800		1,800	530		2,330	2,850
3340	5'-0" dia	[G]	1.15	13.913		1,950	580		2,530	3,100
3410	Galvanized, 3'-6" dia	[G]	1.45	11.034		1,850	460		2,310	2,800
3420	4'-0" dia	[G]	1.35	11.852		2,075	490		2,565	3,075
3430	4'-6" dia	[G]	1.25	12.800		2,250	530		2,780	3,375
3440	5'-0" dia	[G]	1.15	13.913		2,425	580		3,005	3,625
3510	Red oak tread on flat metal, 3'-6" dia		1.35	11.852		2,550	490		3,040	3,625
3520	4'-0" dia		1.25	12.800		2,825	530		3,355	3,975
3530	4'-6" dia		1.15	13.913		3,025	580		3,605	4,300
3540	5'-0" dia		1.05	15.238		3,325	635		3,960	4,700

05 73 Decorative Metal Railings

05 73 16 – Wire Rope Decorative Metal Railings

05 73 16.10 Cable Railings

		Crew	Daily Output	Labor-Hours	Unit	Material	Labor	Equipment	Total	Total Incl O&P
0010	**CABLE RAILINGS**, with 316 stainless steel 1 x 19 cable, 3/16" diameter									
0015	Made from recycled materials	[G]								
0100	1-3/4" diameter stainless steel posts x 42" high, cables 4" OC	[G] 2 Sswk	25	.640	L.F.	56.50	30		86.50	117

05 73 23 – Ornamental Railings

05 73 23.50 Railings, Ornamental

		Crew	Daily Output	Labor-Hours	Unit	Material	Labor	Equipment	Total	Total Incl O&P
0010	**RAILINGS, ORNAMENTAL**, shop fab'd, 3'-6" high, posts @ 5' O.C.									
0020	Bronze or stainless, minimum	1 Sswk [G]	24	.333	L.F.	35	15.65		50.65	67
0100	Maximum	[G]	9	.889		345	41.50		386.50	455
0200	Aluminum ornamental rail, minimum	[G]	15	.533		35	25		60	84
0300	Maximum	[G]	8	1		105	47		152	201
0400	Hand-forged wrought iron, minimum	[G]	12	.667		109	31.50		140.50	177
0500	Maximum	[G]	8	1		325	47		372	445
0600	Composite metal/wood/glass, minimum		6	1.333		202	62.50		264.50	335
0700	Maximum		5	1.600		405	75		480	580
9000	Minimum labor/equipment charge		2	4	Job		188		188	345

Division Notes

	CREW	DAILY OUTPUT	LABOR-HOURS	UNIT	2010 BARE COSTS MAT.	LABOR	EQUIP.	TOTAL	TOTAL INCL O&P

Estimating Tips

06 05 00 Common Work Results for Wood, Plastics, and Composites

- Common to any wood-framed structure are the accessory connector items such as screws, nails, adhesives, hangers, connector plates, straps, angles, and hold-downs. For typical wood-framed buildings, such as residential projects, the aggregate total for these items can be significant, especially in areas where seismic loading is a concern. For floor and wall framing, the material cost is based on 10 to 25 lbs. per MBF. Hold-downs, hangers, and other connectors should be taken off by the piece.

06 10 00 Carpentry

- Lumber is a traded commodity and therefore sensitive to supply and demand in the marketplace. Even in "budgetary" estimating of wood-framed projects, it is advisable to call local suppliers for the latest market pricing.
- Common quantity units for wood-framed projects are "thousand board feet" (MBF). A board foot is a volume of wood, 1" x 1' x 1', or 144 cubic inches. Board-foot quantities are generally calculated using nominal material dimensions—dressed sizes are ignored. Board foot per lineal foot of any stick of lumber can be calculated by dividing the nominal cross-sectional area by 12. As an example, 2,000 lineal feet of 2 x 12 equates to 4 MBF by dividing the nominal area, 2 x 12, by 12, which equals 2, and multiplying by 2,000 to give 4,000 board feet. This simple rule applies to all nominal dimensioned lumber.
- Waste is an issue of concern at the quantity takeoff for any area of construction. Framing lumber is sold in even foot lengths, i.e., 10', 12', 14', 16', and depending on spans, wall heights, and the grade of lumber, waste is inevitable. A rule of thumb for lumber waste is 5%–10% depending on material quality and the complexity of the framing.
- Wood in various forms and shapes is used in many projects, even where the main structural framing is steel, concrete, or masonry. Plywood as a back-up partition material and 2x boards used as blocking and cant strips around roof edges are two common examples. The estimator should ensure that the costs of all wood materials are included in the final estimate.

06 20 00 Finish Carpentry

- It is necessary to consider the grade of workmanship when estimating labor costs for erecting millwork and interior finish. In practice, there are three grades: premium, custom, and economy. The RSMeans daily output for base and case moldings is in the range of 200 to 250 L.F. per carpenter per day. This is appropriate for most average custom-grade projects. For premium projects, an adjustment to productivity of 25%–50% should be made, depending on the complexity of the job.

Reference Numbers

Reference numbers are shown in shaded boxes at the beginning of some major classifications. These numbers refer to related items in the Reference Section. The reference information may be an estimating procedure, an alternate pricing method, or technical information.

Note: Not all subdivisions listed here necessarily appear in this publication.

No part of this publication may be reproduced, stored in a retrieval system, or transmitted in any form or by any means without prior written permission of R.S. Means Company, Inc.

06 05 Common Work Results for Wood, Plastics, and Composites

06 05 05 – Selective Wood and Plastics Demolition

06 05 05.10 Selective Demolition Wood Framing

		Crew	Daily Output	Labor-Hours	Unit	Material	2010 Bare Costs Labor	2010 Bare Costs Equipment	Total	Total Incl O&P
0010	**SELECTIVE DEMOLITION WOOD FRAMING** R024119-10									
0100	Timber connector, nailed, small	1 Clab	96	.083	Ea.		2.76		2.76	4.53
0110	Medium		60	.133			4.41		4.41	7.25
0120	Large		48	.167			5.50		5.50	9.05
0130	Bolted, small		48	.167			5.50		5.50	9.05
0140	Medium		32	.250			8.30		8.30	13.60
0150	Large		24	.333			11.05		11.05	18.10
2958	Beams, 2" x 6"	2 Clab	1100	.015	L.F.		.48		.48	.79
2960	2" x 8"		825	.019			.64		.64	1.05
2965	2" x 10"		665	.024			.80		.80	1.31
2970	2" x 12"		550	.029			.96		.96	1.58
2972	2" x 14"		470	.034			1.13		1.13	1.85
2975	4" x 8"	B-1	413	.058			1.96		1.96	3.22
2980	4" x 10"		330	.073			2.46		2.46	4.03
2985	4" x 12"		275	.087			2.95		2.95	4.84
3000	6" x 8"		275	.087			2.95		2.95	4.84
3040	6" x 10"		220	.109			3.68		3.68	6.05
3080	6" x 12"		185	.130			4.38		4.38	7.20
3120	8" x 12"		140	.171			5.80		5.80	9.50
3160	10" x 12"		110	.218			7.35		7.35	12.10
3162	Alternate pricing method		1.10	21.818	M.B.F.		735		735	1,200
3170	Blocking, in 16" OC wall framing, 2" x 4"	1 Clab	600	.013	L.F.		.44		.44	.72
3172	2" x 6"		400	.020			.66		.66	1.09
3174	In 24" OC wall framing, 2" x 4"		600	.013			.44		.44	.72
3176	2" x 6"		400	.020			.66		.66	1.09
3178	Alt method, wood blocking removal from wood framing		.40	20	M.B.F.		660		660	1,075
3179	Wood blocking removal from steel framing		.36	22.222	"		735		735	1,200
3180	Bracing, let in, 1" x 3", studs 16" OC		1050	.008	L.F.		.25		.25	.41
3181	Studs 24" OC		1080	.007			.25		.25	.40
3182	1" x 4", studs 16" OC		1050	.008			.25		.25	.41
3183	Studs 24" OC		1080	.007			.25		.25	.40
3184	1" x 6", studs 16" OC		1050	.008			.25		.25	.41
3185	Studs 24" OC		1080	.007			.25		.25	.40
3186	2" x 3", studs 16" OC		800	.010			.33		.33	.54
3187	Studs 24" OC		830	.010			.32		.32	.52
3188	2" x 4", studs 16" OC		800	.010			.33		.33	.54
3189	Studs 24" OC		830	.010			.32		.32	.52
3190	2" x 6", studs 16" OC		800	.010			.33		.33	.54
3191	Studs 24" OC		830	.010			.32		.32	.52
3192	2" x 8", studs 16" OC		800	.010			.33		.33	.54
3193	Studs 24" OC		830	.010			.32		.32	.52
3194	"T" shaped metal bracing, studs at 16" OC		1060	.008			.25		.25	.41
3195	Studs at 24" OC		1200	.007			.22		.22	.36
3196	Metal straps, studs at 16" OC		1200	.007			.22		.22	.36
3197	Studs at 24" OC		1240	.006			.21		.21	.35
3200	Columns, round, 8' to 14' tall		40	.200	Ea.		6.60		6.60	10.85
3202	Dimensional lumber sizes	2 Clab	1.10	14.545	M.B.F.		480		480	790
3250	Blocking, between joists	1 Clab	320	.025	Ea.		.83		.83	1.36
3252	Bridging, metal strap, between joists		320	.025	Pr.		.83		.83	1.36
3254	Wood, between joists		320	.025	"		.83		.83	1.36
3260	Door buck, studs, header & access., 8' high 2" x 4" wall, 3' wide		32	.250	Ea.		8.30		8.30	13.60
3261	4' wide		32	.250			8.30		8.30	13.60
3262	5' wide		32	.250			8.30		8.30	13.60

06 05 Common Work Results for Wood, Plastics, and Composites

06 05 05 – Selective Wood and Plastics Demolition

06 05 05.10 Selective Demolition Wood Framing		Crew	Daily Output	Labor-Hours	Unit	Material	2010 Bare Costs Labor	Equipment	Total	Total Incl O&P
3263	6' wide	1 Clab	32	.250	Ea.		8.30		8.30	13.60
3264	8' wide		30	.267			8.85		8.85	14.50
3265	10' wide		30	.267			8.85		8.85	14.50
3266	12' wide		30	.267			8.85		8.85	14.50
3267	2" x 6" wall, 3' wide		32	.250			8.30		8.30	13.60
3268	4' wide		32	.250			8.30		8.30	13.60
3269	5' wide		32	.250			8.30		8.30	13.60
3270	6' wide		32	.250			8.30		8.30	13.60
3271	8' wide		30	.267			8.85		8.85	14.50
3272	10' wide		30	.267			8.85		8.85	14.50
3273	12' wide		30	.267			8.85		8.85	14.50
3274	Window buck, studs, header & access, 8' high 2" x 4" wall, 2' wide		24	.333			11.05		11.05	18.10
3275	3' wide		24	.333			11.05		11.05	18.10
3276	4' wide		24	.333			11.05		11.05	18.10
3277	5' wide		24	.333			11.05		11.05	18.10
3278	6' wide		24	.333			11.05		11.05	18.10
3279	7' wide		24	.333			11.05		11.05	18.10
3280	8' wide		22	.364			12.05		12.05	19.75
3281	10' wide		22	.364			12.05		12.05	19.75
3282	12' wide		22	.364			12.05		12.05	19.75
3283	2" x 6" wall, 2' wide		24	.333			11.05		11.05	18.10
3284	3' wide		24	.333			11.05		11.05	18.10
3285	4' wide		24	.333			11.05		11.05	18.10
3286	5' wide		24	.333			11.05		11.05	18.10
3287	6' wide		24	.333			11.05		11.05	18.10
3288	7' wide		24	.333			11.05		11.05	18.10
3289	8' wide		22	.364			12.05		12.05	19.75
3290	10' wide		22	.364			12.05		12.05	19.75
3291	12' wide		22	.364			12.05		12.05	19.75
3400	Fascia boards, 1" x 6"		500	.016	L.F.		.53		.53	.87
3440	1" x 8"		450	.018			.59		.59	.97
3480	1" x 10"		400	.020			.66		.66	1.09
3490	2" x 6"		450	.018			.59		.59	.97
3500	2" x 8"		400	.020			.66		.66	1.09
3510	2" x 10"		350	.023			.76		.76	1.24
3610	Furring, on wood walls or ceiling		4000	.002	S.F.		.07		.07	.11
3620	On masonry or concrete walls or ceiling		1200	.007	"		.22		.22	.36
3800	Headers over openings, 2 @ 2" x 6"		110	.073	L.F.		2.41		2.41	3.95
3840	2 @ 2" x 8"		100	.080			2.65		2.65	4.35
3880	2 @ 2" x 10"		90	.089			2.94		2.94	4.83
3885	Alternate pricing method		.26	30.651	M.B.F.		1,025		1,025	1,675
3920	Joists, 1" x 4"		1250	.006	L.F.		.21		.21	.35
3930	1" x 6"		1135	.007			.23		.23	.38
3940	1" x 8"		1000	.008			.26		.26	.43
3950	1" x 10"		895	.009			.30		.30	.49
3960	1" x 12"		765	.010			.35		.35	.57
4200	2" x 4"	2 Clab	1000	.016			.53		.53	.87
4230	2" x 6"		970	.016			.55		.55	.90
4240	2" x 8"		940	.017			.56		.56	.93
4250	2" x 10"		910	.018			.58		.58	.96
4280	2" x 12"		880	.018			.60		.60	.99
4281	2" x 14"		850	.019			.62		.62	1.02
4282	Composite joists, 9-1/2"		960	.017			.55		.55	.91

06 05 Common Work Results for Wood, Plastics, and Composites
06 05 05 – Selective Wood and Plastics Demolition

06 05 05.10 Selective Demolition Wood Framing		Crew	Daily Output	Labor-Hours	Unit	Material	2010 Bare Costs Labor	Equipment	Total	Total Incl O&P
4283	11-7/8"	2 Clab	930	.017	L.F.		.57		.57	.93
4284	14"		897	.018			.59		.59	.97
4285	16"		865	.019			.61		.61	1.01
4290	Wood joists, alternate pricing method		1.50	10.667	M.B.F.		355		355	580
4500	Open web joist, 12" deep		500	.032	L.F.		1.06		1.06	1.74
4505	14" deep		475	.034			1.11		1.11	1.83
4510	16" deep		450	.036			1.18		1.18	1.93
4520	18" deep		425	.038			1.25		1.25	2.05
4530	24" deep		400	.040			1.32		1.32	2.17
4550	Ledger strips, 1" x 2"	1 Clab	1200	.007			.22		.22	.36
4560	1" x 3"		1200	.007			.22		.22	.36
4570	1" x 4"		1200	.007			.22		.22	.36
4580	2" x 2"		1100	.007			.24		.24	.40
4590	2" x 4"		1000	.008			.26		.26	.43
4600	2" x 6"		1000	.008			.26		.26	.43
4601	2" x 8 or 2" x 10"		800	.010			.33		.33	.54
4602	4" x 6"		600	.013			.44		.44	.72
4604	4" x 8"		450	.018			.59		.59	.97
5400	Posts, 4" x 4"	2 Clab	800	.020			.66		.66	1.09
5405	4" x 6"		550	.029			.96		.96	1.58
5410	4" x 8"		440	.036			1.20		1.20	1.98
5425	4" x 10"		390	.041			1.36		1.36	2.23
5430	4" x 12"		350	.046			1.51		1.51	2.48
5440	6" x 6"		400	.040			1.32		1.32	2.17
5445	6" x 8"		350	.046			1.51		1.51	2.48
5450	6" x 10"		320	.050			1.66		1.66	2.72
5455	6" x 12"		290	.055			1.83		1.83	3
5480	8" x 8"		300	.053			1.77		1.77	2.90
5500	10" x 10"		240	.067			2.21		2.21	3.62
5660	Tongue and groove floor planks		2	8	M.B.F.		265		265	435
5750	Rafters, ordinary, 16" OC, 2" x 4"		880	.018	S.F.		.60		.60	.99
5755	2" x 6"		840	.019			.63		.63	1.04
5760	2" x 8"		820	.020			.65		.65	1.06
5770	2" x 10"		820	.020			.65		.65	1.06
5780	2" x 12"		810	.020			.65		.65	1.07
5785	24" OC, 2" x 4"		1170	.014			.45		.45	.74
5786	2" x 6"		1117	.014			.47		.47	.78
5787	2" x 8"		1091	.015			.49		.49	.80
5788	2" x 10"		1091	.015			.49		.49	.80
5789	2" x 12"		1077	.015			.49		.49	.81
5795	Rafters, ordinary, 2" x 4" (alternate method)		862	.019	L.F.		.61		.61	1.01
5800	2" x 6" (alternate method)		850	.019			.62		.62	1.02
5840	2" x 8" (alternate method)		837	.019			.63		.63	1.04
5855	2" x 10" (alternate method)		825	.019			.64		.64	1.05
5865	2" x 12" (alternate method)		812	.020			.65		.65	1.07
5870	Sill plate, 2" x 4"	1 Clab	1170	.007			.23		.23	.37
5871	2" x 6"		780	.010			.34		.34	.56
5872	2" x 8"		586	.014			.45		.45	.74
5873	Alternate pricing method		.78	10.256	M.B.F.		340		340	555
5885	Ridge board, 1" x 4"	2 Clab	900	.018	L.F.		.59		.59	.97
5886	1" x 6"		875	.018			.61		.61	.99
5887	1" x 8"		850	.019			.62		.62	1.02
5888	1" x 10"		825	.019			.64		.64	1.05

06 05 Common Work Results for Wood, Plastics, and Composites

06 05 05 – Selective Wood and Plastics Demolition

06 05 05.10 Selective Demolition Wood Framing		Crew	Daily Output	Labor-Hours	Unit	Material	2010 Bare Costs Labor	Equipment	Total	Total Incl O&P
5889	1" x 12"	2 Clab	800	.020	L.F.		.66		.66	1.09
5890	2" x 4"		900	.018			.59		.59	.97
5892	2" x 6"		875	.018			.61		.61	.99
5894	2" x 8"		850	.019			.62		.62	1.02
5896	2" x 10"		825	.019			.64		.64	1.05
5898	2" x 12"		800	.020			.66		.66	1.09
5900	Hip & valley rafters, 2" x 6"		500	.032			1.06		1.06	1.74
5940	2" x 8"		420	.038			1.26		1.26	2.07
6050	Rafter tie, 1" x 4"		1250	.013			.42		.42	.70
6052	1" x 6"		1135	.014			.47		.47	.77
6054	2" x 4"		1000	.016			.53		.53	.87
6056	2" x 6"		970	.016			.55		.55	.90
6070	Sleepers, on concrete, 1" x 2"	1 Clab	4700	.002			.06		.06	.09
6075	1" x 3"		4000	.002			.07		.07	.11
6080	2" x 4"		3000	.003			.09		.09	.15
6085	2" x 6"		2600	.003			.10		.10	.17
6086	Sheathing from roof, 5/16"	2 Clab	1600	.010	S.F.		.33		.33	.54
6088	3/8"		1525	.010			.35		.35	.57
6090	1/2"		1400	.011			.38		.38	.62
6092	5/8"		1300	.012			.41		.41	.67
6094	3/4"		1200	.013			.44		.44	.72
6096	Board sheathing from roof		1400	.011			.38		.38	.62
6100	Sheathing, from walls, 1/4"		1200	.013			.44		.44	.72
6110	5/16"		1175	.014			.45		.45	.74
6120	3/8"		1150	.014			.46		.46	.76
6130	1/2"		1125	.014			.47		.47	.77
6140	5/8"		1100	.015			.48		.48	.79
6150	3/4"		1075	.015			.49		.49	.81
6152	Board sheathing from walls		1500	.011			.35		.35	.58
6158	Subfloor, with boards		1050	.015			.50		.50	.83
6160	Plywood, 1/2" thick		768	.021			.69		.69	1.13
6162	5/8" thick		760	.021			.70		.70	1.14
6164	3/4" thick		750	.021			.71		.71	1.16
6165	1-1/8" thick		720	.022			.74		.74	1.21
6166	Underlayment, particle board, 3/8" thick	1 Clab	780	.010			.34		.34	.56
6168	1/2" thick		768	.010			.34		.34	.57
6170	5/8" thick		760	.011			.35		.35	.57
6172	3/4" thick		750	.011			.35		.35	.58
6200	Stairs and stringers, minimum	2 Clab	40	.400	Riser		13.25		13.25	21.50
6240	Maximum	"	26	.615	"		20.50		20.50	33.50
6300	Components, tread	1 Clab	110	.073	Ea.		2.41		2.41	3.95
6320	Riser		80	.100	"		3.31		3.31	5.45
6390	Stringer, 2" x 10"		260	.031	L.F.		1.02		1.02	1.67
6400	2" x 12"		260	.031			1.02		1.02	1.67
6410	3" x 10"		250	.032			1.06		1.06	1.74
6420	3" x 12"		250	.032			1.06		1.06	1.74
6590	Wood studs, 2" x 3"	2 Clab	3076	.005			.17		.17	.28
6600	2" x 4"		2000	.008			.26		.26	.43
6640	2" x 6"		1600	.010			.33		.33	.54
6720	Wall framing, including studs, plates and blocking, 2" x 4"	1 Clab	600	.013	S.F.		.44		.44	.72
6740	2" x 6"		480	.017	"		.55		.55	.91
6750	Headers, 2" x 4"		1125	.007	L.F.		.24		.24	.39
6755	2" x 6"		1125	.007			.24		.24	.39

06 05 Common Work Results for Wood, Plastics, and Composites

06 05 05 – Selective Wood and Plastics Demolition

06 05 05.10 Selective Demolition Wood Framing		Crew	Daily Output	Labor-Hours	Unit	Material	2010 Bare Costs Labor	Equipment	Total	Total Incl O&P
6760	2" x 8"	1 Clab	1050	.008	L.F.		.25		.25	.41
6765	2" x 10"		1050	.008			.25		.25	.41
6770	2" x 12"		1000	.008			.26		.26	.43
6780	4" x 10"		525	.015			.50		.50	.83
6785	4" x 12"		500	.016			.53		.53	.87
6790	6" x 8"		560	.014			.47		.47	.78
6795	6" x 10"		525	.015			.50		.50	.83
6797	6" x 12"		500	.016	▼		.53		.53	.87
8000	Soffit, T & G wood		520	.015	S.F.		.51		.51	.84
8010	Hardboard, vinyl or aluminum	▼	640	.013			.41		.41	.68
8030	Plywood	2 Carp	315	.051	▼		2.11		2.11	3.47
9000	Minimum labor/equipment charge	1 Clab	4	2	Job		66		66	109

06 05 05.20 Selective Demolition Millwork and Trim

0010	**SELECTIVE DEMOLITION MILLWORK AND TRIM** R024119-10									
1000	Cabinets, wood, base cabinets, per L.F.	2 Clab	80	.200	L.F.		6.60		6.60	10.85
1020	Wall cabinets, per L.F.	"	80	.200	"		6.60		6.60	10.85
1060	Remove and reset, base cabinets	2 Carp	18	.889	Ea.		37		37	60.50
1070	Wall cabinets	"	20	.800	"		33		33	54.50
1100	Steel, painted, base cabinets	2 Clab	60	.267	L.F.		8.85		8.85	14.50
1120	Wall cabinets		60	.267	"		8.85		8.85	14.50
1200	Casework, large area		320	.050	S.F.		1.66		1.66	2.72
1220	Selective		200	.080	"		2.65		2.65	4.35
1500	Counter top, minimum		200	.080	L.F.		2.65		2.65	4.35
1510	Maximum	▼	120	.133			4.41		4.41	7.25
1550	Remove and reset, minimum	2 Carp	50	.320			13.30		13.30	22
1560	Maximum	"	40	.400	▼		16.60		16.60	27.50
2000	Paneling, 4' x 8' sheets	2 Clab	2000	.008	S.F.		.26		.26	.43
2100	Boards, 1" x 4"		700	.023			.76		.76	1.24
2120	1" x 6"		750	.021			.71		.71	1.16
2140	1" x 8"		800	.020			.66		.66	1.09
3000	Trim, baseboard, to 6" wide		1200	.013	L.F.		.44		.44	.72
3040	Greater than 6" and up to 12" wide		1000	.016			.53		.53	.87
3080	Remove and reset, minimum	2 Carp	400	.040			1.66		1.66	2.73
3090	Maximum	"	300	.053			2.22		2.22	3.64
3100	Ceiling trim	2 Clab	1000	.016			.53		.53	.87
3120	Chair rail		1200	.013			.44		.44	.72
3140	Railings with balusters		240	.067	▼		2.21		2.21	3.62
3160	Wainscoting	▼	700	.023	S.F.		.76		.76	1.24
9000	Minimum labor/equipment charge	1 Clab	4	2	Job		66		66	109

06 05 23 – Wood, Plastic, and Composite Fastenings

06 05 23.10 Nails

0010	**NAILS**, material only, based upon 50# box purchase									
0020	Copper nails, plain				Lb.	10			10	11
0400	Stainless steel, plain					7.55			7.55	8.30
0500	Box, 3d to 20d, bright					1.26			1.26	1.39
0520	Galvanized					1.77			1.77	1.95
0600	Common, 3d to 60d, plain					1.30			1.30	1.43
0700	Galvanized					1.56			1.56	1.72
0800	Aluminum					4.55			4.55	5
1000	Annular or spiral thread, 4d to 60d, plain					1.80			1.80	1.98
1200	Galvanized					3			3	3.30
1400	Drywall nails, plain				▼	.85			.85	.94

06 05 Common Work Results for Wood, Plastics, and Composites

06 05 23 – Wood, Plastic, and Composite Fastenings

06 05 23.10 Nails

		Crew	Daily Output	Labor-Hours	Unit	Material	2010 Bare Costs Labor	Equipment	Total	Total Incl O&P
1600	Galvanized				Lb.	1.75			1.75	1.93
1800	Finish nails, 4d to 10d, plain					1.40			1.40	1.54
2000	Galvanized					1.79			1.79	1.97
2100	Aluminum					4.10			4.10	4.51
2300	Flooring nails, hardened steel, 2d to 10d, plain					2.55			2.55	2.81
2400	Galvanized					3.88			3.88	4.27
2500	Gypsum lath nails, 1-1/8", 13 ga. flathead, blued					2.68			2.68	2.95
2600	Masonry nails, hardened steel, 3/4" to 3" long, plain					2.45			2.45	2.70
2700	Galvanized					3.72			3.72	4.09
2900	Roofing nails, threaded, galvanized					1.63			1.63	1.79
3100	Aluminum					4.80			4.80	5.30
3300	Compressed lead head, threaded, galvanized					2.47			2.47	2.72
3600	Siding nails, plain shank, galvanized					2			2	2.20
3800	Aluminum					4.11			4.11	4.52
5000	Add to prices above for cement coating					.11			.11	.12
5200	Zinc or tin plating					.14			.14	.15
5500	Vinyl coated sinkers, 8d to 16d					.62			.62	.68

06 05 23.50 Wood Screws

		Crew	Daily Output	Labor-Hours	Unit	Material	2010 Bare Costs Labor	Equipment	Total	Total Incl O&P
0010	**WOOD SCREWS**									
0020	#8 x 1" long, steel				C	5.10			5.10	5.60
0100	Brass					10.30			10.30	11.35
0600	#10, 2" long, steel					6.15			6.15	6.80
0700	Brass					24			24	26.50
1500	#12, 3" long, steel					12.10			12.10	13.30

06 05 23.60 Timber Connectors

		Crew	Daily Output	Labor-Hours	Unit	Material	2010 Bare Costs Labor	Equipment	Total	Total Incl O&P
0010	**TIMBER CONNECTORS**									
0100	Connector plates, steel, with bolts, straight	2 Carp	75	.213	Ea.	25.50	8.85		34.35	42.50
0110	Tee, 7 gauge		50	.320		29.50	13.30		42.80	54
0120	T-Strap, 14 gauge, 12" x 8" x 2"		50	.320		29.50	13.30		42.80	54
0150	Anchor plates, 7 ga, 9" x 7"		75	.213		25.50	8.85		34.35	42.50
0200	Bolts, machine, sq. hd. with nut & washer, 1/2" diameter, 4" long	1 Carp	140	.057		.86	2.37		3.23	4.85
0300	7-1/2" long		130	.062		1.50	2.56		4.06	5.85
0500	3/4" diameter, 7-1/2" long		130	.062		4.39	2.56		6.95	9.05
0610	Machine bolts, w/nut, washer, 3/4" diam., 15" L, HD's & beam hangers		95	.084		8.10	3.50		11.60	14.70
0800	Drilling bolt holes in timber, 1/2" diameter		450	.018	Inch		.74		.74	1.21
0900	1" diameter		350	.023	"		.95		.95	1.56
1100	Framing anchor, angle, 3" x 3" x 1-1/2", 12 ga		175	.046	Ea.	2.20	1.90		4.10	5.55
1150	Framing anchors, 18 gauge, 4-1/2" x 2-3/4"		175	.046		2.20	1.90		4.10	5.55
1160	Framing anchors, 18 gauge, 4-1/2" x 3"		175	.046		2.20	1.90		4.10	5.55
1170	Clip anchors plates, 18 gauge, 12" x 1-1/8"		175	.046		2.20	1.90		4.10	5.55
1250	Holdowns, 3 gauge base, 10 gauge body		8	1		21	41.50		62.50	91.50
1260	Holdowns, 7 gauge 11-1/16" x 3-1/4"		8	1		21	41.50		62.50	91.50
1270	Holdowns, 7 gauge 14-3/8" x 3-1/8"		8	1		21	41.50		62.50	91.50
1275	Holdowns, 12 gauge 8" x 2-1/2"		8	1		21	41.50		62.50	91.50
1300	Joist and beam hangers, 18 ga. galv., for 2" x 4" joist		175	.046		.61	1.90		2.51	3.79
1400	2" x 6" to 2" x 10" joist		165	.048		1.17	2.01		3.18	4.60
1600	16 ga. galv., 3" x 6" to 3" x 10" joist		160	.050		2.49	2.08		4.57	6.15
1700	3" x 10" to 3" x 14" joist		160	.050		4.14	2.08		6.22	7.95
1800	4" x 6" to 4" x 10" joist		155	.052		2.58	2.14		4.72	6.35
1900	4" x 10" to 4" x 14" joist		155	.052		4.23	2.14		6.37	8.15
2000	Two-2" x 6" to two-2" x 10" joists		150	.053		3.52	2.22		5.74	7.50
2100	Two-2" x 10" to two-2" x 14" joists		150	.053		3.94	2.22		6.16	7.95

06 05 Common Work Results for Wood, Plastics, and Composites

06 05 23 – Wood, Plastic, and Composite Fastenings

06 05 23.60 Timber Connectors		Crew	Daily Output	Labor-Hours	Unit	Material	2010 Bare Costs Labor	Equipment	Total	Total Incl O&P
2300	3/16" thick, 6" x 8" joist	1 Carp	145	.055	Ea.	54.50	2.29		56.79	64
2400	6" x 10" joist		140	.057		57	2.37		59.37	66.50
2500	6" x 12" joist		135	.059		59.50	2.46		61.96	69
2700	1/4" thick, 6" x 14" joist		130	.062		61.50	2.56		64.06	72
2900	Plywood clips, extruded aluminum H clip, for 3/4" panels					.20			.20	.22
3000	Galvanized 18 ga. back-up clip					.16			.16	.18
3200	Post framing, 16 ga. galv. for 4" x 4" base, 2 piece	1 Carp	130	.062		13.80	2.56		16.36	19.35
3300	Cap		130	.062		19.15	2.56		21.71	25
3500	Rafter anchors, 18 ga. galv., 1-1/2" wide, 5-1/4" long		145	.055		.41	2.29		2.70	4.22
3600	10-3/4" long		145	.055		1.23	2.29		3.52	5.10
3800	Shear plates, 2-5/8" diameter		120	.067		2	2.77		4.77	6.75
3900	4" diameter		115	.070		4.75	2.89		7.64	10
4000	Sill anchors, embedded in concrete or block, 25-1/2" long		115	.070		10.70	2.89		13.59	16.50
4100	Spike grids, 3" x 6"		120	.067		.81	2.77		3.58	5.45
4400	Split rings, 2-1/2" diameter		120	.067		1.65	2.77		4.42	6.35
4500	4" diameter		110	.073		2.50	3.02		5.52	7.70
4550	Tie plate, 20 gauge, 7" x 3 1/8"		110	.073		2.50	3.02		5.52	7.70
4560	Tie plate, 20 gauge, 5" x 4 1/8"		110	.073		2.50	3.02		5.52	7.70
4575	Twist straps, 18 gauge, 12" x 1 1/4"		110	.073		2.50	3.02		5.52	7.70
4580	Twist straps, 18 gauge, 16" x 1 1/4"		110	.073		2.50	3.02		5.52	7.70
4600	Strap ties, 20 ga., 2-1/16" wide, 12 13/16" long		180	.044		.80	1.85		2.65	3.91
4700	Strap ties, 16 ga., 1-3/8" wide, 12" long		180	.044		.80	1.85		2.65	3.91
4800	21-5/8" x 1-1/4"		160	.050		2.49	2.08		4.57	6.15
5000	Toothed rings, 2-5/8" or 4" diameter		90	.089		1.49	3.69		5.18	7.70
5200	Truss plates, nailed, 20 gauge, up to 32' span		17	.471	Truss	10.85	19.55		30.40	44
5400	Washers, 2" x 2" x 1/8"				Ea.	.34			.34	.37
5500	3" x 3" x 3/16"				"	.89			.89	.98
9000	Minimum labor/equipment charge	1 Carp	4	2	Job		83		83	137

06 05 23.70 Rough Hardware

0010	**ROUGH HARDWARE**, average percent of carpentry material									
0020	Minimum				Job	.50%				
0200	Maximum				"	1.50%				

06 05 23.80 Metal Bracing

0010	**METAL BRACING**									
0302	Let-in, "T" shaped, 22 ga. galv. steel, studs at 16" O.C.	1 Carp	580	.014	L.F.	.75	.57		1.32	1.77
0402	Studs at 24" O.C.		600	.013		.75	.55		1.30	1.74
0502	16 ga. galv. steel straps, studs at 16" O.C.		600	.013		.97	.55		1.52	1.98
0602	Studs at 24" O.C.		620	.013		.97	.54		1.51	1.95

06 11 Wood Framing

06 11 10 – Framing with Dimensional, Engineered or Composite Lumber

06 11 10.02 Blocking

		Crew	Daily Output	Labor-Hours	Unit	Material	Labor	Equipment	Total	Total Incl O&P
0010	**BLOCKING**									
2600	Miscellaneous, to wood construction									
2620	2" x 4"	1 Carp	.17	47.059	M.B.F.	375	1,950		2,325	3,600
2625	Pneumatic nailed		.21	38.095		375	1,575		1,950	3,000
2660	2" x 8"		.27	29.630		420	1,225		1,645	2,475
2665	Pneumatic nailed		.33	24.242		420	1,000		1,420	2,100
2720	To steel construction									
2740	2" x 4"	1 Carp	.14	57.143	M.B.F.	375	2,375		2,750	4,300

06 11 Wood Framing

06 11 10 – Framing with Dimensional, Engineered or Composite Lumber

06 11 10.02 Blocking		Crew	Daily Output	Labor-Hours	Unit	Material	2010 Bare Costs Labor	Equipment	Total	Total Incl O&P
2780	2" x 8"	1 Carp	.21	38.095	M.B.F.	420	1,575		1,995	3,050
9000	Minimum labor/equipment charge	"	4	2	Job		83		83	137
06 11 10.04 Bracing										
0012	**BRACING** Let-in, with 1" x 6" boards, studs @ 16" O.C.	1 Carp	150	.053	C.L.F.	.64	2.22		2.86	4.34
0202	Studs @ 24" O.C.	"	230	.035	"	.64	1.45		2.09	3.07
06 11 10.06 Bridging										
0012	**BRIDGING** Wood, for joists 16" O.C., 1" x 3"	1 Carp	130	.062	C. Pr.	.51	2.56		3.07	4.76
0017	Pneumatic nailed		170	.047		.51	1.96		2.47	3.77
0102	2" x 3" bridging		130	.062		.40	2.56		2.96	4.64
0107	Pneumatic nailed		170	.047		.40	1.96		2.36	3.65
0302	Steel, galvanized, 18 ga., for 2" x 10" joists at 12" O.C.		130	.062		1.58	2.56		4.14	5.95
0402	24" O.C.		140	.057		1.69	2.37		4.06	5.75
0902	Compression type, 16" O.C., 2" x 8" joists		200	.040		1.64	1.66		3.30	4.53
1002	2" x 12" joists		200	.040		1.64	1.66		3.30	4.53
06 11 10.10 Beam and Girder Framing										
0010	**BEAM AND GIRDER FRAMING** R061110-30									
1000	Single, 2" x 6"	2 Carp	700	.023	L.F.	.38	.95		1.33	1.97
1005	Pneumatic nailed		812	.020		.38	.82		1.20	1.75
1020	2" x 8"		650	.025		.56	1.02		1.58	2.29
1025	Pneumatic nailed		754	.021		.56	.88		1.44	2.06
1040	2" x 10"		600	.027		.72	1.11		1.83	2.62
1045	Pneumatic nailed		696	.023		.72	.96		1.68	2.37
1060	2" x 12"		550	.029		1.06	1.21		2.27	3.16
1065	Pneumatic nailed		638	.025		1.06	1.04		2.10	2.88
1080	2" x 14"		500	.032		2	1.33		3.33	4.38
1085	Pneumatic nailed		580	.028		2	1.15		3.15	4.08
1100	3" x 8"		550	.029		2.08	1.21		3.29	4.28
1120	3" x 10"		500	.032		2.77	1.33		4.10	5.25
1140	3" x 12"		450	.036		3.54	1.48		5.02	6.30
1160	3" x 14"		400	.040		4.48	1.66		6.14	7.65
1170	4" x 6"	F-3	1100	.036		2.39	1.53	.60	4.52	5.80
1180	4" x 8"		1000	.040		2.88	1.68	.66	5.22	6.65
1200	4" x 10"		950	.042		3.64	1.77	.69	6.10	7.65
1220	4" x 12"		900	.044		6.10	1.87	.73	8.70	10.55
1240	4" x 14"		850	.047		7.15	1.98	.77	9.90	11.90
2000	Double, 2" x 6"	2 Carp	625	.026		.75	1.06		1.81	2.58
2005	Pneumatic nailed		725	.022		.75	.92		1.67	2.34
2020	2" x 8"		575	.028		1.12	1.16		2.28	3.13
2025	Pneumatic nailed		667	.024		1.12	1		2.12	2.87
2040	2" x 10"		550	.029		1.45	1.21		2.66	3.58
2045	Pneumatic nailed		638	.025		1.45	1.04		2.49	3.30
2060	2" x 12"		525	.030		2.12	1.27		3.39	4.41
2065	Pneumatic nailed		610	.026		2.12	1.09		3.21	4.12
2080	2" x 14"		475	.034		4	1.40		5.40	6.70
2085	Pneumatic nailed		551	.029		4	1.21		5.21	6.40
3000	Triple, 2" x 6"		550	.029		1.13	1.21		2.34	3.23
3005	Pneumatic nailed		638	.025		1.13	1.04		2.17	2.95
3020	2" x 8"		525	.030		1.67	1.27		2.94	3.92
3025	Pneumatic nailed		609	.026		1.67	1.09		2.76	3.63
3040	2" x 10"		500	.032		2.17	1.33		3.50	4.57
3045	Pneumatic nailed		580	.028		2.17	1.15		3.32	4.27
3060	2" x 12"		475	.034		3.18	1.40		4.58	5.80

06 11 Wood Framing

06 11 10 – Framing with Dimensional, Engineered or Composite Lumber

06 11 10.10 Beam and Girder Framing

		Crew	Daily Output	Labor-Hours	Unit	Material	2010 Bare Costs Labor	Equipment	Total	Total Incl O&P
3065	Pneumatic nailed	2 Carp	551	.029	L.F.	3.18	1.21		4.39	5.50
3080	2" x 14"		450	.036		6	1.48		7.48	9.05
3085	Pneumatic nailed		522	.031		6	1.27		7.27	8.70
9000	Minimum labor/equipment charge	1 Carp	2	4	Job		166		166	273

06 11 10.12 Ceiling Framing

		Crew	Daily Output	Labor-Hours	Unit	Material	Labor	Equipment	Total	Total Incl O&P
0010	**CEILING FRAMING**									
6000	Suspended, 2" x 3"	2 Carp	1000	.016	L.F.	.27	.66		.93	1.38
6050	2" x 4"		900	.018		.25	.74		.99	1.48
6100	2" x 6"		800	.020		.38	.83		1.21	1.78
6150	2" x 8"		650	.025		.56	1.02		1.58	2.29
9000	Minimum labor/equipment charge	1 Carp	4	2	Job		83		83	137

06 11 10.14 Posts and Columns

		Crew	Daily Output	Labor-Hours	Unit	Material	Labor	Equipment	Total	Total Incl O&P
0010	**POSTS AND COLUMNS**									
0100	4" x 4"	2 Carp	390	.041	L.F.	1.55	1.70		3.25	4.50
0150	4" x 6"		275	.058		2.39	2.42		4.81	6.60
0200	4" x 8"		220	.073		2.88	3.02		5.90	8.15
0250	6" x 6"		215	.074		3.19	3.09		6.28	8.60
0300	6" x 8"		175	.091		4.28	3.80		8.08	10.95
0350	6" x 10"		150	.107		6.60	4.43		11.03	14.60
9000	Minimum labor/equipment charge	1 Carp	2	4	Job		166		166	273

06 11 10.18 Joist Framing

		Crew	Daily Output	Labor-Hours	Unit	Material	Labor	Equipment	Total	Total Incl O&P
0010	**JOIST FRAMING** R061110-30									
2002	Joists, 2" x 4"	2 Carp	1250	.013	L.F.	.25	.53		.78	1.14
2007	Pneumatic nailed		1438	.011		.25	.46		.71	1.03
2100	2" x 6"		1250	.013		.38	.53		.91	1.28
2105	Pneumatic nailed		1438	.011		.38	.46		.84	1.17
2152	2" x 8"		1100	.015		.56	.60		1.16	1.60
2157	Pneumatic nailed		1265	.013		.56	.53		1.09	1.47
2202	2" x 10"		900	.018		.72	.74		1.46	2.01
2207	Pneumatic nailed		1035	.015		.72	.64		1.36	1.86
2252	2" x 12"		875	.018		1.06	.76		1.82	2.42
2257	Pneumatic nailed		1006	.016		1.06	.66		1.72	2.26
2302	2" x 14"		770	.021		2	.86		2.86	3.62
2307	Pneumatic nailed		886	.018		2	.75		2.75	3.43
2352	3" x 6"		925	.017		1.39	.72		2.11	2.71
2402	3" x 10"		780	.021		2.77	.85		3.62	4.45
2452	3" x 12"		600	.027		3.54	1.11		4.65	5.70
2502	4" x 6"		800	.020		2.39	.83		3.22	4
2552	4" x 10"		600	.027		3.64	1.11		4.75	5.85
2602	4" x 12"		450	.036		6.10	1.48		7.58	9.15
2607	Sister joist, 2" x 6"		800	.020		.38	.83		1.21	1.78
2608	Pneumatic nailed		960	.017		.38	.69		1.07	1.55
2612	2" x 8"		640	.025		.56	1.04		1.60	2.32
2613	Pneumatic nailed		768	.021		.56	.87		1.43	2.03
2617	2" x 10"		535	.030		.72	1.24		1.96	2.84
2618	Pneumatic nailed		642	.025		.72	1.04		1.76	2.50
2622	2" x 12"		455	.035		1.06	1.46		2.52	3.57
2627	Pneumatic nailed		546	.029		1.06	1.22		2.28	3.17
3000	Composite wood joist 9-1/2" deep		.90	17.778	M.L.F.	1,575	740		2,315	2,950
3010	11-1/2" deep		.88	18.182		1,700	755		2,455	3,100
3020	14" deep		.82	19.512		2,325	810		3,135	3,900
3030	16" deep		.78	20.513		3,075	850		3,925	4,775

06 11 Wood Framing

06 11 10 – Framing with Dimensional, Engineered or Composite Lumber

06 11 10.18 Joist Framing

		Crew	Daily Output	Labor-Hours	Unit	Material	2010 Bare Costs Labor	2010 Bare Costs Equipment	Total	Total Incl O&P
4000	Open web joist 12" deep	2 Carp	.88	18.182	M.L.F.	2,775	755		3,530	4,300
4010	14" deep		.82	19.512		2,975	810		3,785	4,600
4020	16" deep		.78	20.513		2,950	850		3,800	4,650
4030	18" deep		.74	21.622		3,175	900		4,075	4,975
6000	Composite rim joist, 1-1/4" x 9-1/2"		.90	17.778		1,700	740		2,440	3,100
6010	1-1/4" x 11-1/2"		.88	18.182		2,125	755		2,880	3,575
6020	1-1/4" x 14-1/2"		.82	19.512		2,425	810		3,235	4,000
6030	1-1/4" x 16-1/2"		.78	20.513		3,100	850		3,950	4,825
9000	Minimum labor/equipment charge	1 Carp	4	2	Job		83		83	137

06 11 10.24 Miscellaneous Framing

		Crew	Daily Output	Labor-Hours	Unit	Material	2010 Bare Costs Labor	2010 Bare Costs Equipment	Total	Total Incl O&P
0010	**MISCELLANEOUS FRAMING**									
2002	Firestops, 2" x 4"	2 Carp	780	.021	L.F.	.25	.85		1.10	1.67
2007	Pneumatic nailed		952	.017		.25	.70		.95	1.42
2102	2" x 6"		600	.027		.38	1.11		1.49	2.23
2107	Pneumatic nailed		732	.022		.38	.91		1.29	1.90
5002	Nailers, treated, wood construction, 2" x 4"		800	.020		.35	.83		1.18	1.75
5007	Pneumatic nailed		960	.017		.35	.69		1.04	1.52
5102	2" x 6"		750	.021		.56	.89		1.45	2.07
5107	Pneumatic nailed		900	.018		.56	.74		1.30	1.82
5122	2" x 8"		700	.023		1.03	.95		1.98	2.70
5127	Pneumatic nailed		840	.019		1.03	.79		1.82	2.44
5202	Steel construction, 2" x 4"		750	.021		.35	.89		1.24	1.84
5222	2" x 6"		700	.023		.56	.95		1.51	2.17
5242	2" x 8"		650	.025		1.03	1.02		2.05	2.82
7002	Rough bucks, treated, for doors or windows, 2" x 6"		400	.040		.56	1.66		2.22	3.34
7007	Pneumatic nailed		480	.033		.56	1.38		1.94	2.88
7102	2" x 8"		380	.042		1.03	1.75		2.78	4.01
7107	Pneumatic nailed		456	.035		1.03	1.46		2.49	3.53
8000	Stair stringers, 2" x 10"		130	.123		.72	5.10		5.82	9.20
8100	2" x 12"		130	.123		1.06	5.10		6.16	9.55
8150	3" x 10"		125	.128		2.77	5.30		8.07	11.80
8200	3" x 12"		125	.128		3.54	5.30		8.84	12.65
8870	Laminated structural lumber, 1-1/4" x 11-1/2"		130	.123		2.11	5.10		7.21	10.70
8880	1-1/4" x 14-1/2"		130	.123		2.43	5.10		7.53	11.10
9000	Minimum labor/equipment charge	1 Carp	4	2	Job		83		83	137

06 11 10.26 Partitions

		Crew	Daily Output	Labor-Hours	Unit	Material	2010 Bare Costs Labor	2010 Bare Costs Equipment	Total	Total Incl O&P
0010	**PARTITIONS**									
0020	Single bottom and double top plate, no waste, std. & better lumber									
0182	2" x 4" studs, 8' high, studs 12" O.C.	2 Carp	80	.200	L.F.	3.01	8.30		11.31	16.95
0187	12" O.C., pneumatic nailed		96	.167		3.01	6.95		9.96	14.70
0202	16" O.C.		100	.160		2.46	6.65		9.11	13.60
0207	16" O.C., pneumatic nailed		120	.133		2.46	5.55		8.01	11.80
0302	24" O.C.		125	.128		1.91	5.30		7.21	10.85
0307	24" O.C., pneumatic nailed		150	.107		1.91	4.43		6.34	9.40
0382	10' high, studs 12" O.C.		80	.200		3.56	8.30		11.86	17.55
0387	12" O.C., pneumatic nailed		96	.167		3.56	6.95		10.51	15.30
0402	16" O.C.		100	.160		2.87	6.65		9.52	14.05
0407	16" O.C., pneumatic nailed		120	.133		2.87	5.55		8.42	12.25
0502	24" O.C.		125	.128		2.19	5.30		7.49	11.15
0507	24" O.C., pneumatic nailed		150	.107		2.19	4.43		6.62	9.70
0582	12' high, studs 12" O.C.		65	.246		4.10	10.25		14.35	21.50
0587	12" O.C., pneumatic nailed		78	.205		4.10	8.50		12.60	18.50

06 11 Wood Framing

06 11 10 – Framing with Dimensional, Engineered or Composite Lumber

06 11 10.26 Partitions

		Crew	Daily Output	Labor-Hours	Unit	Material	2010 Bare Costs Labor	Equipment	Total	Total Incl O&P
0602	16" O.C.	2 Carp	80	.200	L.F.	3.28	8.30		11.58	17.25
0607	16" O.C., pneumatic nailed		96	.167		3.28	6.95		10.23	15
0701	24" O.C.		100	.160		2.46	6.65		9.11	13.60
0702										
0706	24" O.C., pneumatic nailed	2 Carp	120	.133	L.F.	2.46	5.55		8.01	11.80
0782	2" x 6" studs, 8' high, studs 12" O.C.		70	.229		4.56	9.50		14.06	20.50
0787	12" O.C., pneumatic nailed		84	.190		4.56	7.90		12.46	18
0802	16" O.C.		90	.178		3.73	7.40		11.13	16.25
0807	16" O.C., pneumatic nailed		108	.148		3.73	6.15		9.88	14.20
0902	24" O.C.		115	.139		2.90	5.80		8.70	12.70
0907	24" O.C., pneumatic nailed		138	.116		2.90	4.82		7.72	11.10
0982	10' high, studs 12" O.C.		70	.229		5.40	9.50		14.90	21.50
0987	12" O.C., pneumatic nailed		84	.190		5.40	7.90		13.30	18.95
1002	16" O.C.		90	.178		4.35	7.40		11.75	16.95
1007	16" O.C., pneumatic nailed		108	.148		4.35	6.15		10.50	14.90
1102	24" O.C.		115	.139		3.32	5.80		9.12	13.15
1107	24" O.C., pneumatic nailed		138	.116		3.32	4.82		8.14	11.55
1182	12' high, studs 12" O.C.		55	.291		6.20	12.10		18.30	26.50
1187	12" O.C., pneumatic nailed		66	.242		6.20	10.05		16.25	23.50
1202	16" O.C.		70	.229		4.98	9.50		14.48	21
1207	16" O.C., pneumatic nailed		84	.190		4.98	7.90		12.88	18.45
1302	24" O.C.		90	.178		3.73	7.40		11.13	16.25
1307	24" O.C., pneumatic nailed		108	.148		3.73	6.15		9.88	14.20
1402	For horizontal blocking, 2" x 4", add		600	.027		.27	1.11		1.38	2.12
1502	2" x 6", add		600	.027		.41	1.11		1.52	2.28
1600	For openings, add		250	.064			2.66		2.66	4.37
1702	Headers for above openings, material only, add				B.F.	.46			.46	.51
9000	Minimum labor/equipment charge	1 Carp	4	2	Job		83		83	137

06 11 10.28 Porch or Deck Framing

		Crew	Daily Output	Labor-Hours	Unit	Material	Labor	Equipment	Total	Total Incl O&P
0010	**PORCH OR DECK FRAMING**									
0100	Treated lumber, posts or columns, 4" x 4"	2 Carp	390	.041	L.F.	1.56	1.70		3.26	4.51
0110	4" x 6"		275	.058		2.30	2.42		4.72	6.50
0120	4" x 8"		220	.073		4.74	3.02		7.76	10.15
0130	Girder, single, 4" x 4"		675	.024		1.56	.98		2.54	3.33
0140	4" x 6"		600	.027		2.30	1.11		3.41	4.35
0150	4" x 8"		525	.030		4.74	1.27		6.01	7.30
0160	Double, 2" x 4"		625	.026		.72	1.06		1.78	2.55
0170	2" x 6"		600	.027		1.16	1.11		2.27	3.10
0180	2" x 8"		575	.028		2.13	1.16		3.29	4.24
0190	2" x 10"		550	.029		2.78	1.21		3.99	5.05
0200	2" x 12"		525	.030		3.69	1.27		4.96	6.15
0210	Triple, 2" x 4"		575	.028		1.09	1.16		2.25	3.09
0220	2" x 6"		550	.029		1.74	1.21		2.95	3.91
0230	2" x 8"		525	.030		3.19	1.27		4.46	5.60
0240	2" x 10"		500	.032		4.17	1.33		5.50	6.75
0250	2" x 12"		475	.034		5.55	1.40		6.95	8.40
0260	Ledger, bolted 4' O.C., 2" x 4"		400	.040		.47	1.66		2.13	3.24
0270	2" x 6"		395	.041		.68	1.68		2.36	3.50
0280	2" x 8"		390	.041		1.15	1.70		2.85	4.07
0300	2" x 12"		380	.042		1.92	1.75		3.67	4.98
0310	Joists, 2" x 4"		1250	.013		.36	.53		.89	1.27
0320	2" x 6"		1250	.013		.58	.53		1.11	1.51

06 11 Wood Framing

06 11 10 – Framing with Dimensional, Engineered or Composite Lumber

06 11 10.28 Porch or Deck Framing

		Crew	Daily Output	Labor-Hours	Unit	Material	2010 Bare Costs Labor	Equipment	Total	Total Incl O&P
0330	2" x 8"	2 Carp	1100	.015	L.F.	1.07	.60		1.67	2.17
0340	2" x 10"		900	.018		1.40	.74		2.14	2.75
0350	2" x 12"		875	.018		1.68	.76		2.44	3.09
0440	Balusters, square, 2" x 2"		660	.024		.33	1.01		1.34	2.01
0450	Turned, 2" x 2"		420	.038		.44	1.58		2.02	3.08
0460	Stair stringer, 2" x 10"		130	.123		1.40	5.10		6.50	9.95
0470	2" x 12"		130	.123		1.68	5.10		6.78	10.25
0480	Stair treads, 1" x 4"		140	.114		1.56	4.75		6.31	9.50
0490	2" x 4"		140	.114		.36	4.75		5.11	8.20
0500	2" x 6"		160	.100		.55	4.16		4.71	7.45
0510	5/4" x 6"		160	.100		1.10	4.16		5.26	8.05
0520	Turned handrail post, 4" x 4"		64	.250	Ea.	32	10.40		42.40	52
0530	Lattice panel, 4' x 8', 1/2"		1600	.010	S.F.	.68	.42		1.10	1.43
0535	3/4"		1600	.010	"	1.02	.42		1.44	1.80
0540	Cedar, posts or columns, 4" x 4"		390	.041	L.F.	3.01	1.70		4.71	6.10
0550	4" x 6"		275	.058		5.75	2.42		8.17	10.25
0560	4" x 8"		220	.073		9.70	3.02		12.72	15.65
0800	Decking, 1" x 4"		550	.029		1.65	1.21		2.86	3.80
0810	2" x 4"		600	.027		3.40	1.11		4.51	5.55
0820	2" x 6"		640	.025		5.55	1.04		6.59	7.80
0830	5/4" x 6"		640	.025		3.73	1.04		4.77	5.80
0840	Railings and trim, 1" x 4"		600	.027		1.65	1.11		2.76	3.63
0860	2" x 4"		600	.027		3.40	1.11		4.51	5.55
0870	2" x 6"		600	.027		5.55	1.11		6.66	7.90
0920	Stair treads, 1" x 4"		140	.114		1.65	4.75		6.40	9.60
0930	2" x 4"		140	.114		3.40	4.75		8.15	11.55
0940	2" x 6"		160	.100		5.55	4.16		9.71	12.95
0950	5/4" x 6"		160	.100		3.73	4.16		7.89	10.95
0980	Redwood, posts or columns, 4" x 4"		390	.041		6.40	1.70		8.10	9.85
0990	4" x 6"		275	.058		11.20	2.42		13.62	16.25
1000	4" x 8"		220	.073		20.50	3.02		23.52	27.50
1280	Railings and trim, 1" x 4"		600	.027		1.17	1.11		2.28	3.10
1310	2" x 6"		600	.027		7.50	1.11		8.61	10.05
1420	Alternative decking, wood/plastic composite, 5/4" x 6" [G]		640	.025		3.01	1.04		4.05	5.05
1430	Vinyl, 1-1/2" x 5-1/2"		640	.025		4.15	1.04		5.19	6.30
1440	1" x 4" square edge fir		550	.029		1.57	1.21		2.78	3.71
1450	1" x 4" tongue and groove fir		450	.036		1.41	1.48		2.89	3.98
1460	1" x 4" mahogany		550	.029		2.51	1.21		3.72	4.75
1462	5/4" x 6" PVC		550	.029		2.72	1.21		3.93	4.98
1470	Accessories, joist hangers, 2" x 4"	1 Carp	160	.050	Ea.	.61	2.08		2.69	4.08
1480	2" x 6" through 2" x 12"	"	150	.053		1.17	2.22		3.39	4.93
1530	Post footing, incl excav, backfill, tube form & concrete, 4' deep, 8" dia	F-7	12	2.667		10.60	99.50		110.10	175
1540	10" diameter		11	2.909		15.70	109		124.70	195
1550	12" diameter		10	3.200		21	119		140	219

06 11 10.30 Roof Framing

		Crew	Daily Output	Labor-Hours	Unit	Material	Labor	Equipment	Total	Total Incl O&P
0010	**ROOF FRAMING** R061110-30									
2000	Fascia boards, 2" x 8"	2 Carp	225	.071	L.F.	.56	2.95		3.51	5.45
2100	2" x 10"		180	.089		.72	3.69		4.41	6.85
5000	Rafters, to 4 in 12 pitch, 2" x 6", ordinary		1000	.016		.38	.66		1.04	1.50
5060	2" x 8", ordinary		950	.017		.56	.70		1.26	1.76
5250	Composite rafter, 9-1/2" deep		575	.028		1.56	1.16		2.72	3.62
5260	11-1/2" deep		575	.028		1.69	1.16		2.85	3.76

06 11 Wood Framing

06 11 10 – Framing with Dimensional, Engineered or Composite Lumber

06 11 10.30 Roof Framing

		Crew	Daily Output	Labor-Hours	Unit	Material	2010 Bare Costs Labor	2010 Bare Costs Equipment	Total	Total Incl O&P
5300	Hip and valley rafters, 2" x 6", ordinary	2 Carp	760	.021	L.F.	.38	.87		1.25	1.85
5360	2" x 8", ordinary		720	.022		.56	.92		1.48	2.13
5540	Hip and valley jacks, 2" x 6", ordinary		600	.027		.38	1.11		1.49	2.23
5600	2" x 8", ordinary		490	.033		.56	1.36		1.92	2.84
5761	For slopes steeper than 4 in 12, add						30%			
5770	For dormers or complex roofs, add						50%			
5780	Rafter tie, 1" x 4", #3	2 Carp	800	.020	L.F.	.44	.83		1.27	1.85
5800	Ridge board, #2 or better, 1" x 6"		600	.027		.64	1.11		1.75	2.52
5820	1" x 8"		550	.029		.84	1.21		2.05	2.91
5840	1" x 10"		500	.032		1.17	1.33		2.50	3.46
5860	2" x 6"		500	.032		.38	1.33		1.71	2.59
5880	2" x 8"		450	.036		.56	1.48		2.04	3.04
5900	2" x 10"		400	.040		.72	1.66		2.38	3.53
5920	Roof cants, split, 4" x 4"		650	.025		1.55	1.02		2.57	3.38
5940	6" x 6"		600	.027		3.19	1.11		4.30	5.35
5960	Roof curbs, untreated, 2" x 6"		520	.031		.38	1.28		1.66	2.51
5980	2" x 12"		400	.040		1.06	1.66		2.72	3.90
6000	Sister rafters, 2" x 6"		800	.020		.38	.83		1.21	1.78
6020	2" x 8"		640	.025		.56	1.04		1.60	2.32
6040	2" x 10"		535	.030		.72	1.24		1.96	2.84
6060	2" x 12"		455	.035		1.06	1.46		2.52	3.57
9000	Minimum labor/equipment charge	1 Carp	4	2	Job		83		83	137

06 11 10.32 Sill and Ledger Framing

		Crew	Daily Output	Labor-Hours	Unit	Material	Labor	Equipment	Total	Total Incl O&P
0010	**SILL AND LEDGER FRAMING**									
2002	Ledgers, nailed, 2" x 4"	2 Carp	755	.021	L.F.	.25	.88		1.13	1.72
2052	2" x 6"		600	.027		.38	1.11		1.49	2.23
2102	Bolted, not including bolts, 3" x 6"		325	.049		1.39	2.05		3.44	4.89
2152	3" x 12"		233	.069		3.54	2.85		6.39	8.60
2602	Mud sills, redwood, construction grade, 2" x 4"		895	.018		2.26	.74		3	3.71
2622	2" x 6"		780	.021		3.40	.85		4.25	5.15
4002	Sills, 2" x 4"		600	.027		.25	1.11		1.36	2.09
4052	2" x 6"		550	.029		.38	1.21		1.59	2.40
4082	2" x 8"		500	.032		.56	1.33		1.89	2.79
4202	Treated, 2" x 4"		550	.029		.35	1.21		1.56	2.37
4222	2" x 6"		500	.032		.56	1.33		1.89	2.79
4242	2" x 8"		450	.036		1.03	1.48		2.51	3.57
4402	4" x 4"		450	.036		1.52	1.48		3	4.11
4422	4" x 6"		350	.046		2.25	1.90		4.15	5.60
4462	4" x 8"		300	.053		4.68	2.22		6.90	8.80
4481	4" x 10"		260	.062		5.85	2.56		8.41	10.65
9000	Minimum labor/equipment charge	1 Carp	4	2	Job		83		83	137

06 11 10.34 Sleepers

		Crew	Daily Output	Labor-Hours	Unit	Material	Labor	Equipment	Total	Total Incl O&P
0010	**SLEEPERS**									
0100	On concrete, treated, 1" x 2"	2 Carp	2350	.007	L.F.	.33	.28		.61	.82
0150	1" x 3"		2000	.008		.53	.33		.86	1.13
0200	2" x 4"		1500	.011		.35	.44		.79	1.11
0250	2" x 6"		1300	.012		.56	.51		1.07	1.45
9000	Minimum labor/equipment charge	1 Carp	4	2	Job		83		83	137

06 11 10.36 Soffit and Canopy Framing

		Crew	Daily Output	Labor-Hours	Unit	Material	Labor	Equipment	Total	Total Incl O&P
0010	**SOFFIT AND CANOPY FRAMING**									
1002	Canopy or soffit framing, 1" x 4"	2 Carp	900	.018	L.F.	.44	.74		1.18	1.69
1042	1" x 8"		750	.021		.84	.89		1.73	2.38

06 11 Wood Framing

06 11 10 – Framing with Dimensional, Engineered or Composite Lumber

06 11 10.36 Soffit and Canopy Framing

		Crew	Daily Output	Labor-Hours	Unit	Material	2010 Bare Costs Labor	2010 Bare Costs Equipment	Total	Total Incl O&P
1102	2" x 4"	2 Carp	620	.026	L.F.	.25	1.07		1.32	2.03
1142	2" x 8"		500	.032		.56	1.33		1.89	2.79
1202	3" x 4"		500	.032		.83	1.33		2.16	3.09
1242	3" x 10"		300	.053		2.77	2.22		4.99	6.70
9000	Minimum labor/equipment charge	1 Carp	4	2	Job		83		83	137

06 11 10.38 Treated Lumber Framing Material

					Unit	Material			Total	Total Incl O&P
0010	**TREATED LUMBER FRAMING MATERIAL**									
0100	2" x 4"				M.B.F.	520			520	570
0110	2" x 6"					555			555	615
0120	2" x 8"					775			775	855
0130	2" x 10"					810			810	890
0140	2" x 12"					900			900	990
0200	4" x 4"					1,150			1,150	1,250
0210	4" x 6"					1,125			1,125	1,250
0220	4" x 8"					1,750			1,750	1,925

06 11 10.40 Wall Framing

			Crew	Daily Output	Labor-Hours	Unit	Material	Labor	Equipment	Total	Total Incl O&P
0010	**WALL FRAMING**	R061110-30									
0100	Door buck, studs, header, access, 8' H, 2" x 4" wall, 3' W		1 Carp	32	.250	Ea.	10.40	10.40		20.80	28.50
0110	4' wide			32	.250		11.15	10.40		21.55	29.50
0120	5' wide			32	.250		13.80	10.40		24.20	32.50
0130	6' wide			32	.250		14.95	10.40		25.35	33.50
0140	8' wide			30	.267		19.90	11.10		31	40
0150	10' wide			30	.267		29.50	11.10		40.60	50.50
0160	12' wide			30	.267		57	11.10		68.10	80.50
0170	2" x 6" wall, 3' wide			32	.250		14.50	10.40		24.90	33
0180	4' wide			32	.250		15.25	10.40		25.65	34
0190	5' wide			32	.250		17.90	10.40		28.30	37
0200	6' wide			32	.250		19.05	10.40		29.45	38
0210	8' wide			30	.267		24	11.10		35.10	44.50
0220	10' wide			30	.267		34	11.10		45.10	55
0230	12' wide			30	.267		61	11.10		72.10	85
0240	Window buck, studs, header & access, 8' high 2" x 4" wall, 2' wide			24	.333		11.05	13.85		24.90	35
0250	3' wide			24	.333		12.90	13.85		26.75	37
0260	4' wide			24	.333		14.15	13.85		28	38.50
0270	5' wide			24	.333		16.80	13.85		30.65	41.50
0280	6' wide			24	.333		18.65	13.85		32.50	43.50
0300	8' wide			22	.364		25	15.10		40.10	52.50
0310	10' wide			22	.364		35.50	15.10		50.60	64
0320	12' wide			22	.364		64.50	15.10		79.60	96
0330	2" x 6" wall, 2' wide			24	.333		16.20	13.85		30.05	41
0340	3' wide			24	.333		18.30	13.85		32.15	43
0350	4' wide			24	.333		19.60	13.85		33.45	44.50
0360	5' wide			24	.333		22.50	13.85		36.35	47.50
0370	6' wide			24	.333		24.50	13.85		38.35	50
0380	7' wide			24	.333		29.50	13.85		43.35	55.50
0390	8' wide			22	.364		32	15.10		47.10	60
0400	10' wide			22	.364		43	15.10		58.10	72
0410	12' wide			22	.364		72.50	15.10		87.60	105
2002	Headers over openings, 2" x 6"		2 Carp	360	.044	L.F.	.38	1.85		2.23	3.44
2007	2" x 6", pneumatic nailed			432	.037		.38	1.54		1.92	2.94
2052	2" x 8"			340	.047		.56	1.96		2.52	3.82
2057	2" x 8", pneumatic nailed			408	.039		.56	1.63		2.19	3.29

06 11 Wood Framing

06 11 10 – Framing with Dimensional, Engineered or Composite Lumber

06 11 10.40 Wall Framing		Crew	Daily Output	Labor-Hours	Unit	Material	2010 Bare Costs Labor	2010 Bare Costs Equipment	Total	Total Incl O&P
2100	2" x 10"	2 Carp	320	.050	L.F.	.72	2.08		2.80	4.21
2105	2" x 10", pneumatic nailed		384	.042		.72	1.73		2.45	3.64
2152	2" x 12"		300	.053		1.06	2.22		3.28	4.81
2157	2" x 12", pneumatic nailed		360	.044		1.06	1.85		2.91	4.20
2202	4" x 12"		190	.084		6.10	3.50		9.60	12.45
2207	4" x 12", pneumatic nailed		228	.070		6.10	2.92		9.02	11.50
2252	6" x 12"		140	.114		10.15	4.75		14.90	19
5002	Plates, untreated, 2" x 3"		850	.019		.27	.78		1.05	1.57
5007	2" x 3", pneumatic nailed		1020	.016		.27	.65		.92	1.36
5022	2" x 4"		800	.020		.25	.83		1.08	1.64
5027	2" x 4", pneumatic nailed		960	.017		.25	.69		.94	1.41
5041	2" x 6"		750	.021		.38	.89		1.27	1.87
5046	2" x 6", pneumatic nailed		900	.018		.38	.74		1.12	1.62
5122	Studs, 8' high wall, 2" x 3"		1200	.013		.27	.55		.82	1.20
5127	2" x 3", pneumatic nailed		1440	.011		.27	.46		.73	1.05
5142	2" x 4"		1100	.015		.25	.60		.85	1.26
5147	2" x 4", pneumatic nailed		1320	.012		.25	.50		.75	1.10
5162	2" x 6"		1000	.016		.38	.66		1.04	1.50
5167	2" x 6", pneumatic nailed		1200	.013		.38	.55		.93	1.32
5182	3" x 4"		800	.020		.83	.83		1.66	2.28
5187	3" x 4", pneumatic nailed		960	.017		.83	.69		1.52	2.05
8200	For 12' high walls, deduct						5%			
8220	For stub wall, 6' high, add						20%			
8240	3' high, add						40%			
8250	For second story & above, add						5%			
8300	For dormer & gable, add						15%			
9000	Minimum labor/equipment charge	1 Carp	4	2	Job		83		83	137

06 11 10.42 Furring										
0010	**FURRING**									
0012	Wood strips, 1" x 2", on walls, on wood	1 Carp	550	.015	L.F.	.25	.60		.85	1.26
0017	Pneumatic nailed		710	.011		.25	.47		.72	1.04
0300	On masonry		495	.016		.25	.67		.92	1.37
0400	On concrete		260	.031		.25	1.28		1.53	2.37
0600	1" x 3", on walls, on wood		550	.015		.34	.60		.94	1.37
0607	Pneumatic nailed		710	.011		.34	.47		.81	1.15
0700	On masonry		495	.016		.34	.67		1.01	1.48
0800	On concrete		260	.031		.34	1.28		1.62	2.48
0850	On ceilings, on wood		350	.023		.34	.95		1.29	1.94
0857	Pneumatic nailed		450	.018		.34	.74		1.08	1.59
0900	On masonry		320	.025		.34	1.04		1.38	2.09
0950	On concrete		210	.038		.34	1.58		1.92	2.98
9000	Minimum labor/equipment charge		4	2	Job		83		83	137

06 11 10.44 Grounds										
0010	**GROUNDS**									
0020	For casework, 1" x 2" wood strips, on wood	1 Carp	330	.024	L.F.	.25	1.01		1.26	1.92
0100	On masonry		285	.028		.25	1.17		1.42	2.19
0200	On concrete		250	.032		.25	1.33		1.58	2.45
0400	For plaster, 3/4" deep, on wood		450	.018		.25	.74		.99	1.48
0500	On masonry		225	.036		.25	1.48		1.73	2.70
0600	On concrete		175	.046		.25	1.90		2.15	3.39
0700	On metal lath		200	.040		.25	1.66		1.91	3
9000	Minimum labor/equipment charge		4	2	Job		83		83	137

06 12 Structural Panels

06 12 10 – Structural Insulated Panels

06 12 10.10 OSB Faced Panels

			Daily	Labor-			2010 Bare Costs			Total
		Crew	Output	Hours	Unit	Material	Labor	Equipment	Total	Incl O&P
0010	**OSB FACED PANELS**									
0100	Structural insul. panels, 7/16" OSB both faces, EPS insul, 3-5/8" T G	F-3	2075	.019	S.F.	2.91	.81	.32	4.04	4.87
0110	5-5/8" thick G		1725	.023		3.33	.98	.38	4.69	5.65
0120	7-3/8" thick G		1425	.028		3.69	1.18	.46	5.33	6.50
0130	9-3/8" thick G		1125	.036		4.13	1.50	.58	6.21	7.60
0140	7/16" OSB one face, EPS insul, 3-5/8" thick G		2175	.018		2.77	.77	.30	3.84	4.64
0150	5-5/8" thick G		1825	.022		3.05	.92	.36	4.33	5.25
0160	7-3/8" thick G		1525	.026		3.33	1.10	.43	4.86	5.95
0170	9-3/8" thick G		1225	.033		3.64	1.38	.54	5.56	6.85
0190	7/16" OSB - 1/2" GWB faces, EPS insul, 3-5/8" T G		2075	.019		2.87	.81	.32	4	4.83
0200	5-5/8" thick G		1725	.023		3.29	.98	.38	4.65	5.65
0210	7-3/8" thick G		1425	.028		3.45	1.18	.46	5.09	6.25
0220	9-3/8" thick G		1125	.036		3.87	1.50	.58	5.95	7.35
0240	7/16" OSB - 1/2" MRGWB faces, EPS insul, 3-5/8" T G		2075	.019		2.87	.81	.32	4	4.83
0250	5-5/8" thick G		1725	.023		3.29	.98	.38	4.65	5.65
0260	7-3/8" thick G		1425	.028		3.45	1.18	.46	5.09	6.25
0270	9-3/8" thick G		1125	.036		3.87	1.50	.58	5.95	7.35
0300	For 1/2" GWB added to OSB skin, add G					.60			.60	.66
0310	For 1/2" MRGWB added to OSB skin, add G					.75			.75	.83
0320	For one T1-11 skin, add to OSB-OSB G					.74			.74	.81
0330	For one 19/32" CDX skin, add to OSB-OSB G					.69			.69	.76
0500	Structural insulated panel, 7/16" OSB both sides, straw core									
0510	4-3/8" T, walls (w/sill, splines, plates) G	F-6	2400	.017	S.F.	7.60	.65	.27	8.52	9.70
0520	Floors (w/splines) G		2400	.017		7.60	.65	.27	8.52	9.70
0530	Roof (w/splines) G		2400	.017		7.60	.65	.27	8.52	9.70
0550	7-7/8" T, walls (w/sill, splines, plates) G		2400	.017		10.95	.65	.27	11.87	13.40
0560	Floors (w/splines) G		2400	.017		10.95	.65	.27	11.87	13.40
0570	Roof (w/splines) G		2400	.017		10.95	.65	.27	11.87	13.40

06 13 Heavy Timber

06 13 23 – Heavy Timber Construction

06 13 23.10 Heavy Framing

		Crew	Daily Output	Labor-Hours	Unit	Material	Labor	Equipment	Total	Total Incl O&P
0010	**HEAVY FRAMING**									
0020	Beams, single 6" x 10"	2 Carp	1.10	14.545	M.B.F.	1,600	605		2,205	2,750
0100	Single 8" x 16"		1.20	13.333	"	1,975	555		2,530	3,075
0202	Built from 2" lumber, multiple 2" x 14"		900	.018	B.F.	.86	.74		1.60	2.15
0212	Built from 3" lumber, multiple 3" x 6"		700	.023		.93	.95		1.88	2.58
0222	Multiple 3" x 8"		800	.020		1.04	.83		1.87	2.52
0232	Multiple 3" x 10"		900	.018		1.11	.74		1.85	2.43
0242	Multiple 3" x 12"		1000	.016		1.18	.66		1.84	2.39
0252	Built from 4" lumber, multiple 4" x 6"		800	.020		1.19	.83		2.02	2.68
0262	Multiple 4" x 8"		900	.018		1.08	.74		1.82	2.40
0272	Multiple 4" x 10"		1000	.016		1.09	.66		1.75	2.29
0282	Multiple 4" x 12"		1100	.015		1.53	.60		2.13	2.67
0292	Columns, structural grade, 1500f, 4" x 4"		450	.036	L.F.	1.70	1.48		3.18	4.30
0302	6" x 6"		225	.071		3.95	2.95		6.90	9.20
0402	8" x 8"		240	.067		7.20	2.77		9.97	12.45
0502	10" x 10"		90	.178		12.10	7.40		19.50	25.50
0602	12" x 12"		70	.229		18.35	9.50		27.85	35.50
0802	Floor planks, 2" thick, T & G, 2" x 6"		1050	.015	B.F.	1.05	.63		1.68	2.20
0902	2" x 10"		1100	.015		1.05	.60		1.65	2.15

06 13 Heavy Timber

06 13 23 – Heavy Timber Construction

06 13 23.10 Heavy Framing		Crew	Daily Output	Labor-Hours	Unit	Material	2010 Bare Costs Labor	2010 Bare Costs Equipment	Total	Total Incl O&P
1102	3" thick, 3" x 6"	2 Carp	1050	.015	B.F.	1.25	.63		1.88	2.42
1202	3" x 10"		1100	.015		1.25	.60		1.85	2.37
1402	Girders, structural grade, 12" x 12"		800	.020		1.53	.83		2.36	3.05
1502	10" x 16"		1000	.016		2.43	.66		3.09	3.76
2302	Roof purlins, 4" thick, structural grade		1050	.015		1.33	.63		1.96	2.50
2502	Roof trusses, add timber connectors, Division 06 05 23.60		450	.036		1.22	1.48		2.70	3.77
9000	Minimum labor/equipment charge	1 Carp	2	4	Job		166		166	273

06 15 Wood Decking

06 15 16 – Wood Roof Decking

06 15 16.10 Solid Wood Roof Decking		Crew	Daily Output	Labor-Hours	Unit	Material	2010 Bare Costs Labor	2010 Bare Costs Equipment	Total	Total Incl O&P
0010	**SOLID WOOD ROOF DECKING**									
0400	Cedar planks, 3" thick	2 Carp	320	.050	S.F.	7.70	2.08		9.78	11.90
0500	4" thick		250	.064		10.30	2.66		12.96	15.70
0702	Douglas fir, 3" thick		320	.050		3.28	2.08		5.36	7
0802	4" thick		250	.064		4.36	2.66		7.02	9.15
1002	Hemlock, 3" thick		320	.050		3.28	2.08		5.36	7
1102	4" thick		250	.064		4.36	2.66		7.02	9.15
1302	Western white spruce, 3" thick		320	.050		3.28	2.08		5.36	7
1402	4" thick		250	.064		4.36	2.66		7.02	9.15
9000	Minimum labor/equipment charge	1 Carp	2	4	Job		166		166	273

06 15 23 – Laminated Wood Decking

06 15 23.10 Laminated Roof Deck		Crew	Daily Output	Labor-Hours	Unit	Material	2010 Bare Costs Labor	2010 Bare Costs Equipment	Total	Total Incl O&P
0010	**LAMINATED ROOF DECK**									
0020	Pine or hemlock, 3" thick	2 Carp	425	.038	S.F.	3.36	1.56		4.92	6.25
0100	4" thick		325	.049		4.47	2.05		6.52	8.30
0300	Cedar, 3" thick		425	.038		4.11	1.56		5.67	7.10
0400	4" thick		325	.049		5.25	2.05		7.30	9.10
0600	Fir, 3" thick		425	.038		3.43	1.56		4.99	6.35
0700	4" thick		325	.049		4.29	2.05		6.34	8.10
9000	Minimum labor/equipment charge	1 Carp	3	2.667	Job		111		111	182

06 16 Sheathing

06 16 23 – Subflooring

06 16 23.10 Subfloor		Crew	Daily Output	Labor-Hours	Unit	Material	2010 Bare Costs Labor	2010 Bare Costs Equipment	Total	Total Incl O&P
0010	**SUBFLOOR** R061636-20									
0011	Plywood, CDX, 1/2" thick	2 Carp	1500	.011	SF Flr.	.42	.44		.86	1.19
0015	Pneumatic nailed		1860	.009		.42	.36		.78	1.05
0102	5/8" thick		1350	.012		.51	.49		1	1.37
0107	Pneumatic nailed		1674	.010		.51	.40		.91	1.21
0202	3/4" thick		1250	.013		.58	.53		1.11	1.51
0207	Pneumatic nailed		1550	.010		.58	.43		1.01	1.34
0302	1-1/8" thick, 2-4-1 including underlayment		1050	.015		1.50	.63		2.13	2.69
0440	With boards, 1" x 6", S4S, laid regular		900	.018		1.40	.74		2.14	2.74
0452	1" x 8", laid regular		1000	.016		1.34	.66		2	2.56
0462	Laid diagonal		850	.019		1.34	.78		2.12	2.75
0502	1" x 10", laid regular		1100	.015		1.47	.60		2.07	2.61
0602	Laid diagonal		900	.018		1.47	.74		2.21	2.83
8990	Subfloor adhesive, 3/8" bead	1 Carp	2300	.003	L.F.	.09	.14		.23	.34

06 16 Sheathing

06 16 23 – Subflooring

06 16 23.10 Subfloor		Crew	Daily Output	Labor-Hours	Unit	Material	2010 Bare Costs Labor	Equipment	Total	Total Incl O&P
9000	Minimum labor/equipment charge	1 Carp	4	2	Job		83		83	137

06 16 26 – Underlayment

06 16 26.10 Wood Product Underlayment

			Crew	Daily Output	Labor-Hours	Unit	Material	Labor	Equipment	Total	Total Incl O&P
0010	**WOOD PRODUCT UNDERLAYMENT**	R061636-20									
0030	Plywood, underlayment grade, 3/8" thick		2 Carp	1500	.011	SF Flr.	.77	.44		1.21	1.58
0080	Pneumatic nailed			1860	.009		.77	.36		1.13	1.44
0102	1/2" thick			1450	.011		.80	.46		1.26	1.63
0107	Pneumatic nailed			1798	.009		.80	.37		1.17	1.49
0202	5/8" thick			1400	.011		1.17	.47		1.64	2.07
0207	Pneumatic nailed			1736	.009		1.17	.38		1.55	1.92
0302	3/4" thick			1300	.012		1.23	.51		1.74	2.19
0306	Pneumatic nailed			1612	.010		1.23	.41		1.64	2.03
0502	Particle board, 3/8" thick			1500	.011		.36	.44		.80	1.13
0507	Pneumatic nailed			1860	.009		.36	.36		.72	.99
0602	1/2" thick			1450	.011		.39	.46		.85	1.18
0607	Pneumatic nailed			1798	.009		.39	.37		.76	1.04
0802	5/8" thick			1400	.011		.50	.47		.97	1.33
0807	Pneumatic nailed			1736	.009		.50	.38		.88	1.18
0902	3/4" thick			1300	.012		.65	.51		1.16	1.56
0907	Pneumatic nailed			1612	.010		.65	.41		1.06	1.40
0955	Particleboard, 100% recycled straw/wheat, 4' x 8' x 1/4"	G		1450	.011	S.F.	.26	.46		.72	1.04
0960	4' x 8' x 3/8"	G		1450	.011		.39	.46		.85	1.18
0965	4' x 8' x 1/2"	G		1350	.012		.53	.49		1.02	1.39
0970	4' x 8' x 5/8"	G		1300	.012		.63	.51		1.14	1.53
0975	4' x 8' x 3/4"	G		1250	.013		.73	.53		1.26	1.67
0980	4' x 8' x 1"	G		1150	.014		.98	.58		1.56	2.03
0985	4' x 8' x 1-1/4"	G		1100	.015		1.15	.60		1.75	2.26
1102	Hardboard, underlayment grade, 4' x 4', .215" thick			1500	.011	SF Flr.	.57	.44		1.01	1.36
9000	Minimum labor/equipment charge		1 Carp	4	2	Job		83		83	137

06 16 36 – Wood Panel Product Sheathing

06 16 36.10 Sheathing

			Crew	Daily Output	Labor-Hours	Unit	Material	Labor	Equipment	Total	Total Incl O&P
0010	**SHEATHING**	R061110-30									
0012	Plywood on roofs, CDX										
0032	5/16" thick		2 Carp	1600	.010	S.F.	.53	.42		.95	1.26
0037	Pneumatic nailed	R061636-20		1952	.008		.53	.34		.87	1.14
0052	3/8" thick			1525	.010		.36	.44		.80	1.12
0057	Pneumatic nailed			1860	.009		.36	.36		.72	.99
0102	1/2" thick			1400	.011		.42	.47		.89	1.24
0103	Pneumatic nailed			1708	.009		.42	.39		.81	1.10
0202	5/8" thick			1300	.012		.51	.51		1.02	1.40
0207	Pneumatic nailed			1586	.010		.51	.42		.93	1.25
0302	3/4" thick			1200	.013		.58	.55		1.13	1.55
0307	Pneumatic nailed			1464	.011		.58	.45		1.03	1.39
0502	Plywood on walls with exterior CDX, 3/8" thick			1200	.013		.36	.55		.91	1.31
0507	Pneumatic nailed			1488	.011		.36	.45		.81	1.13
0602	1/2" thick			1125	.014		.42	.59		1.01	1.43
0607	Pneumatic nailed			1395	.011		.42	.48		.90	1.24
0702	5/8" thick			1050	.015		.51	.63		1.14	1.60
0707	Pneumatic nailed			1302	.012		.51	.51		1.02	1.40
0802	3/4" thick			975	.016		.58	.68		1.26	1.76
0807	Pneumatic nailed			1209	.013		.58	.55		1.13	1.54
0840	Oriented strand board, 7/16" thick	G		1400	.011		.22	.47		.69	1.02

06 16 Sheathing

06 16 36 – Wood Panel Product Sheathing

06 16 36.10 Sheathing

		Crew	Daily Output	Labor-Hours	Unit	Material	2010 Bare Costs Labor	Equipment	Total	Total Incl O&P
0845	Pneumatic nailed	G 2 Carp	1736	.009	S.F.	.22	.38		.60	.87
0846	1/2" thick	G	1325	.012		.22	.50		.72	1.06
0847	Pneumatic nailed	G	1643	.010		.22	.40		.62	.90
0852	5/8" thick	G	1250	.013		.34	.53		.87	1.24
0857	Pneumatic nailed	G	1550	.010		.34	.43		.77	1.07
1000	For shear wall construction, add						20%			
1200	For structural 1 exterior plywood, add				S.F.	10%				
1402	With boards, on roof 1" x 6" boards, laid horizontal	2 Carp	725	.022		1.40	.92		2.32	3.04
1502	Laid diagonal		650	.025		1.40	1.02		2.42	3.21
1702	1" x 8" boards, laid horizontal		875	.018		1.34	.76		2.10	2.72
1802	Laid diagonal		725	.022		1.34	.92		2.26	2.98
2000	For steep roofs, add						40%			
2200	For dormers, hips and valleys, add					5%	50%			
2402	Boards on walls, 1" x 6" boards, laid regular	2 Carp	650	.025		1.40	1.02		2.42	3.21
2502	Laid diagonal		585	.027		1.40	1.14		2.54	3.40
2702	1" x 8" boards, laid regular		765	.021		1.34	.87		2.21	2.90
2802	Laid diagonal		650	.025		1.34	1.02		2.36	3.15
2852	Gypsum, weatherproof, 1/2" thick		1050	.015		.38	.63		1.01	1.46
2900	With embedded glass mats		1100	.015		.49	.60		1.09	1.53
3000	Wood fiber, regular, no vapor barrier, 1/2" thick		1200	.013		.56	.55		1.11	1.53
3100	5/8" thick		1200	.013		.71	.55		1.26	1.69
3300	No vapor barrier, in colors, 1/2" thick		1200	.013		.75	.55		1.30	1.74
3400	5/8" thick		1200	.013		.80	.55		1.35	1.79
3600	With vapor barrier one side, white, 1/2" thick		1200	.013		.55	.55		1.10	1.52
3700	Vapor barrier 2 sides, 1/2" thick		1200	.013		.77	.55		1.32	1.76
3800	Asphalt impregnated, 25/32" thick		1200	.013		.28	.55		.83	1.22
3850	Intermediate, 1/2" thick		1200	.013		.20	.55		.75	1.13
9000	Minimum labor/equipment charge	1 Carp	2	4	Job		166		166	273

06 17 Shop-Fabricated Structural Wood

06 17 33 – Wood I-Joists

06 17 33.10 Wood and Composite I-Joists

		Crew	Daily Output	Labor-Hours	Unit	Material	Labor	Equipment	Total	Total Incl O&P
0010	**WOOD AND COMPOSITE I-JOISTS**									
0100	Plywood webs, incl. bridging & blocking, panels 24" O.C.									
1200	15' to 24' span, 50 psf live load	F-5	2400	.013	SF Flr.	1.78	.56		2.34	2.88
1300	55 psf live load		2250	.014		1.93	.60		2.53	3.10
1400	24' to 30' span, 45 psf live load		2600	.012		2.66	.52		3.18	3.77
1500	55 psf live load		2400	.013		3.50	.56		4.06	4.77
1600	Tubular steel open webs, 45 psf, 24" O.C., 40' span	F-3	6250	.006		2.24	.27	.10	2.61	3.02
1700	55' span		7750	.005		2.17	.22	.08	2.47	2.83
1800	70' span		9250	.004		2.82	.18	.07	3.07	3.48
1900	85 psf live load, 26' span		2300	.017		2.63	.73	.29	3.65	4.39

06 17 53 – Shop-Fabricated Wood Trusses

06 17 53.10 Roof Trusses

		Crew	Daily Output	Labor-Hours	Unit	Material	Labor	Equipment	Total	Total Incl O&P
0010	**ROOF TRUSSES**									
0100	Fink (W) or King post type, 2'-0" O.C.									
0200	Metal plate connected, 4 in 12 slope									
0210	24' to 29' span	F-3	3000	.013	SF Flr.	2.10	.56	.22	2.88	3.46
0300	30' to 43' span		3000	.013		2.28	.56	.22	3.06	3.66
0400	44' to 60' span		3000	.013		2.52	.56	.22	3.30	3.92

06 17 Shop-Fabricated Structural Wood

06 17 53 – Shop-Fabricated Wood Trusses

06 17 53.10 Roof Trusses	Crew	Daily Output	Labor-Hours	Unit	Material	2010 Bare Costs Labor	Equipment	Total	Total Incl O&P
0700 Glued and nailed, add				SF Flr.	50%				

06 18 Glued-Laminated Construction

06 18 13 – Glued-Laminated Beams

06 18 13.20 Laminated Framing

		Crew	Daily Output	Labor-Hours	Unit	Material	Labor	Equipment	Total	Total Incl O&P
0010	**LAMINATED FRAMING**									
0020	30 lb., short term live load, 15 lb. dead load									
0200	Straight roof beams, 20' clear span, beams 8' O.C.	F-3	2560	.016	SF Flr.	1.76	.66	.26	2.68	3.29
0300	Beams 16' O.C.		3200	.013		1.27	.53	.20	2	2.49
0500	40' clear span, beams 8' O.C.		3200	.013		3.36	.53	.20	4.09	4.79
0600	Beams 16' O.C.		3840	.010		2.75	.44	.17	3.36	3.93
0800	60' clear span, beams 8' O.C.	F-4	2880	.017		5.80	.69	.42	6.91	7.95
0900	Beams 16' O.C.	"	3840	.013		4.31	.52	.31	5.14	5.95
1100	Tudor arches, 30' to 40' clear span, frames 8' O.C.	F-3	1680	.024		7.55	1	.39	8.94	10.35
1200	Frames 16' O.C.	"	2240	.018		5.90	.75	.29	6.94	8.05
1400	50' to 60' clear span, frames 8' O.C.	F-4	2200	.022		8.15	.91	.55	9.61	11
1500	Frames 16' O.C.		2640	.018		6.90	.75	.46	8.11	9.30
1700	Radial arches, 60' clear span, frames 8' O.C.		1920	.025		7.60	1.04	.63	9.27	10.70
1800	Frames 16' O.C.		2880	.017		5.85	.69	.42	6.96	8.05
2000	100' clear span, frames 8' O.C.		1600	.030		7.90	1.24	.76	9.90	11.50
2100	Frames 16' O.C.		2400	.020		6.90	.83	.50	8.23	9.50
2300	120' clear span, frames 8' O.C.		1440	.033		10.45	1.38	.84	12.67	14.65
2400	Frames 16' O.C.		1920	.025		9.55	1.04	.63	11.22	12.85
2600	Bowstring trusses, 20' O.C., 40' clear span	F-3	2400	.017		4.72	.70	.27	5.69	6.65
2700	60' clear span	F-4	3600	.013		4.24	.55	.34	5.13	5.90
2800	100' clear span		4000	.012		6	.50	.30	6.80	7.75
2900	120' clear span		3600	.013		6.45	.55	.34	7.34	8.30
3100	For premium appearance, add to S.F. prices					5%				
3300	For industrial type, deduct					15%				
3500	For stain and varnish, add					5%				
3900	For 3/4" laminations, add to straight					25%				
4100	Add to curved					15%				
4300	Alternate pricing method: (use nominal footage of									
4310	components). Straight beams, camber less than 6"	F-3	3.50	11.429	M.B.F.	2,600	480	187	3,267	3,875
4400	Columns, including hardware		2	20		2,800	840	330	3,970	4,800
4600	Curved members, radius over 32'		2.50	16		2,875	675	262	3,812	4,550
4700	Radius 10' to 32'		3	13.333		2,850	560	219	3,629	4,275
4900	For complicated shapes, add maximum					100%				
5100	For pressure treating, add to straight					35%				
5200	Add to curved					45%				
6000	Laminated veneer members, southern pine or western species									
6050	1-3/4" wide x 5-1/2" deep	2 Carp	480	.033	L.F.	2.39	1.38		3.77	4.90
6100	9-1/2" deep		480	.033		3.64	1.38		5.02	6.25
6150	14" deep		450	.036		5.60	1.48		7.08	8.60
6200	18" deep		450	.036		7.25	1.48		8.73	10.45
6300	Parallel strand members, southern pine or western species									
6350	1-3/4" wide x 9-1/4" deep	2 Carp	480	.033	L.F.	4.50	1.38		5.88	7.20
6400	11-1/4" deep		450	.036		5.60	1.48		7.08	8.65
6450	14" deep		400	.040		6.65	1.66		8.31	10.05
6500	3-1/2" wide x 9-1/4" deep		480	.033		15.15	1.38		16.53	18.95
6550	11-1/4" deep		450	.036		18.10	1.48		19.58	22.50

06 18 Glued-Laminated Construction

06 18 13 – Glued-Laminated Beams

06 18 13.20 Laminated Framing		Crew	Daily Output	Labor-Hours	Unit	Material	2010 Bare Costs Labor	Equipment	Total	Total Incl O&P
6600	14" deep	2 Carp	400	.040	L.F.	22.50	1.66		24.16	27
6650	7" wide x 9-1/4" deep		450	.036		30.50	1.48		31.98	36
6700	11-1/4" deep		420	.038		38	1.58		39.58	44
6750	14" deep		400	.040		44.50	1.66		46.16	51.50
9000	Minimum labor/equipment charge	F-3	2.50	16	Job		675	262	937	1,400

06 22 Millwork

06 22 13 – Standard Pattern Wood Trim

06 22 13.15 Moldings, Base

		Crew	Daily Output	Labor-Hours	Unit	Material	Labor	Equipment	Total	Total Incl O&P
0010	**MOLDINGS, BASE**									
0500	Base, stock pine, 9/16" x 3-1/2"	1 Carp	240	.033	L.F.	2.22	1.38		3.60	4.71
0550	9/16" x 4-1/2"		200	.040		2.46	1.66		4.12	5.45
0561	Base shoe, oak, 3/4" x 1"		240	.033		1.13	1.38		2.51	3.51
9000	Minimum labor/equipment charge		4	2	Job		83		83	137

06 22 13.30 Moldings, Casings

		Crew	Daily Output	Labor-Hours	Unit	Material	Labor	Equipment	Total	Total Incl O&P
0010	**MOLDINGS, CASINGS**									
0090	Apron, stock pine, 5/8" x 2"	1 Carp	250	.032	L.F.	1.22	1.33		2.55	3.52
0110	5/8" x 3-1/2"		220	.036		1.62	1.51		3.13	4.26
0300	Band, stock pine, 11/16" x 1-1/8"		270	.030		.82	1.23		2.05	2.92
0350	11/16" x 1-3/4"		250	.032		1.06	1.33		2.39	3.35
0700	Casing, stock pine, 11/16" x 2-1/2"		240	.033		1.33	1.38		2.71	3.73
0750	11/16" x 3-1/2"		215	.037		1.62	1.55		3.17	4.32
0760	Door & window casing, exterior, 1-1/4" x 2"		200	.040		2.14	1.66		3.80	5.10
0770	Finger jointed, 1-1/4" x 2"		200	.040		.99	1.66		2.65	3.82
9000	Minimum labor/equipment charge		4	2	Job		83		83	137

06 22 13.35 Moldings, Ceilings

		Crew	Daily Output	Labor-Hours	Unit	Material	Labor	Equipment	Total	Total Incl O&P
0010	**MOLDINGS, CEILINGS**									
0600	Bed, stock pine, 9/16" x 1-3/4"	1 Carp	270	.030	L.F.	1.02	1.23		2.25	3.14
0650	9/16" x 2"		240	.033		1.22	1.38		2.60	3.61
1200	Cornice molding, stock pine, 9/16" x 1-3/4"		330	.024		1.02	1.01		2.03	2.77
1300	9/16" x 2-1/4"		300	.027		1.24	1.11		2.35	3.18
2400	Cove scotia, stock pine, 9/16" x 1-3/4"		270	.030		.67	1.23		1.90	2.76
2500	11/16" x 2-3/4"		255	.031		1.86	1.30		3.16	4.19
2600	Crown, stock pine, 9/16" x 3-5/8"		250	.032		2.30	1.33		3.63	4.71
2700	11/16" x 4-5/8"		220	.036		3.87	1.51		5.38	6.75
9000	Minimum labor/equipment charge		4	2	Job		83		83	137

06 22 13.40 Moldings, Exterior

		Crew	Daily Output	Labor-Hours	Unit	Material	Labor	Equipment	Total	Total Incl O&P
0010	**MOLDINGS, EXTERIOR**									
1500	Cornice, boards, pine, 1" x 2"	1 Carp	330	.024	L.F.	.25	1.01		1.26	1.92
1700	1" x 6"		250	.032		.64	1.33		1.97	2.88
2000	1" x 12"		180	.044		1.42	1.85		3.27	4.59
2200	Three piece, built-up, pine, minimum		80	.100		1.33	4.16		5.49	8.30
2300	Maximum		65	.123		3.42	5.10		8.52	12.15
3000	Corner board, sterling pine, 1" x 4"		200	.040		.79	1.66		2.45	3.60
3100	1" x 6"		200	.040		1.05	1.66		2.71	3.89
3350	Fascia, sterling pine, 1" x 6"		250	.032		1.05	1.33		2.38	3.34
3370	1" x 8"		225	.036		1.43	1.48		2.91	4
3400	Trim, back band, 11/16" x 1-1/16"		250	.032		.83	1.33		2.16	3.09
3500	Casing, 11/16" x 4-1/4"		250	.032		2.74	1.33		4.07	5.20
3600	Crown, 11/16" x 4-1/4"		250	.032		3.85	1.33		5.18	6.40

06 22 Millwork

06 22 13 – Standard Pattern Wood Trim

06 22 13.40 Moldings, Exterior

		Crew	Daily Output	Labor-Hours	Unit	Material	2010 Bare Costs Labor	2010 Bare Costs Equipment	Total	Total Incl O&P
3700	1" x 4" and 1" x 6" railing w/balusters 4" O.C.	1 Carp	22	.364	L.F.	22.50	15.10		37.60	49.50
3800	Insect screen framing stock, 1-1/16" x 1-3/4"		395	.020		2.03	.84		2.87	3.61
4100	Verge board, sterling pine, 1" x 4"		200	.040		.79	1.66		2.45	3.60
4200	1" x 6"		200	.040		1.05	1.66		2.71	3.89
4300	2" x 6"		165	.048		2.10	2.01		4.11	5.60
4400	2" x 8"		165	.048		2.86	2.01		4.87	6.45
4700	For redwood trim, add					200%				
9000	Minimum labor/equipment charge	1 Carp	4	2	Job		83		83	137

06 22 13.45 Moldings, Trim

		Crew	Daily Output	Labor-Hours	Unit	Material	Labor	Equipment	Total	Total Incl O&P
0010	**MOLDINGS, TRIM**									
0200	Astragal, stock pine, 11/16" x 1-3/4"	1 Carp	255	.031	L.F.	1.22	1.30		2.52	3.48
0250	1-5/16" x 2-3/16"		240	.033		2.57	1.38		3.95	5.10
0800	Chair rail, stock pine, 5/8" x 2-1/2"		270	.030		1.43	1.23		2.66	3.59
0900	5/8" x 3-1/2"		240	.033		2.26	1.38		3.64	4.76
1000	Closet pole, stock pine, 1-1/8" diameter		200	.040		1.05	1.66		2.71	3.89
1100	Fir, 1-5/8" diameter		200	.040		1.93	1.66		3.59	4.85
3300	Half round, stock pine, 1/4" x 1/2"		270	.030		.24	1.23		1.47	2.28
3350	1/2" x 1"		255	.031		.62	1.30		1.92	2.82
3400	Handrail, fir, single piece, stock, hardware not included									
3450	1-1/2" x 1-3/4"	1 Carp	80	.100	L.F.	2.03	4.16		6.19	9.10
3470	Pine, 1-1/2" x 1-3/4"		80	.100		2.03	4.16		6.19	9.10
3500	1-1/2" x 2-1/2"		76	.105		3.04	4.37		7.41	10.55
3600	Lattice, stock pine, 1/4" x 1-1/8"		270	.030		.42	1.23		1.65	2.48
3700	1/4" x 1-3/4"		250	.032		.65	1.33		1.98	2.90
3800	Miscellaneous, custom, pine, 1" x 1"		270	.030		.35	1.23		1.58	2.40
3900	1" x 3"		240	.033		1.05	1.38		2.43	3.42
4100	Birch or oak, nominal 1" x 1"		240	.033		.45	1.38		1.83	2.76
4200	Nominal 1" x 3"		215	.037		1.35	1.55		2.90	4.02
4400	Walnut, nominal 1" x 1"		215	.037		.58	1.55		2.13	3.18
4500	Nominal 1" x 3"		200	.040		1.74	1.66		3.40	4.64
4700	Teak, nominal 1" x 1"		215	.037		1.60	1.55		3.15	4.30
4800	Nominal 1" x 3"		200	.040		4.79	1.66		6.45	8
4900	Quarter round, stock pine, 1/4" x 1/4"		275	.029		.22	1.21		1.43	2.23
4950	3/4" x 3/4"		255	.031		.60	1.30		1.90	2.80
5600	Wainscot moldings, 1-1/8" x 9/16", 2' high, minimum		76	.105	S.F.	11.15	4.37		15.52	19.50
5700	Maximum		65	.123	"	20	5.10		25.10	31
9000	Minimum labor/equipment charge		4	2	Job		83		83	137

06 22 13.50 Moldings, Window and Door

		Crew	Daily Output	Labor-Hours	Unit	Material	Labor	Equipment	Total	Total Incl O&P
0010	**MOLDINGS, WINDOW AND DOOR**									
2800	Door moldings, stock, decorative, 1-1/8" wide, plain	1 Carp	17	.471	Set	45	19.55		64.55	81.50
2900	Detailed		17	.471	"	87.50	19.55		107.05	129
2960	Clear pine door jamb, no stops, 11/16" x 4-9/16"		240	.033	L.F.	5.15	1.38		6.53	7.90
3150	Door trim set, 1 head and 2 sides, pine, 2-1/2 wide		12	.667	Opng.	22.50	27.50		50	70.50
3170	3-1/2" wide		11	.727	"	27.50	30		57.50	80
3250	Glass beads, stock pine, 3/8" x 1/2"		275	.029	L.F.	.31	1.21		1.52	2.33
3270	3/8" x 7/8"		270	.030		.41	1.23		1.64	2.47
4850	Parting bead, stock pine, 3/8" x 3/4"		275	.029		.36	1.21		1.57	2.39
4870	1/2" x 3/4"		255	.031		.45	1.30		1.75	2.64
5000	Stool caps, stock pine, 11/16" x 3-1/2"		200	.040		2.27	1.66		3.93	5.25
5100	1-1/16" x 3-1/4"		150	.053		3.56	2.22		5.78	7.55
5300	Threshold, oak, 3' long, inside, 5/8" x 3-5/8"		32	.250	Ea.	9	10.40		19.40	27
5400	Outside, 1-1/2" x 7-5/8"		16	.500	"	38	21		59	76

06 22 Millwork

06 22 13 – Standard Pattern Wood Trim

06 22 13.50 Moldings, Window and Door		Crew	Daily Output	Labor-Hours	Unit	Material	2010 Bare Costs Labor	Equipment	Total	Total Incl O&P
5900	Window trim sets, including casings, header, stops,									
5910	stool and apron, 2-1/2" wide, minimum	1 Carp	13	.615	Opng.	32.50	25.50		58	77.50
5950	Average		10	.800		38	33		71	96.50
6000	Maximum		6	1.333		62.50	55.50		118	160
9000	Minimum labor/equipment charge		4	2	Job		83		83	137

06 22 13.60 Moldings, Soffits

	06 22 13.60 Moldings, Soffits		Crew	Daily Output	Labor-Hours	Unit	Material	Labor	Equipment	Total	Incl O&P
0010	**MOLDINGS, SOFFITS**	R061110-30									
0200	Soffits, pine, 1" x 4"		2 Carp	420	.038	L.F.	.44	1.58		2.02	3.08
0210	1" x 6"			420	.038		.64	1.58		2.22	3.30
0220	1" x 8"			420	.038		.84	1.58		2.42	3.52
0230	1" x 10"			400	.040		1.17	1.66		2.83	4.01
0240	1" x 12"			400	.040		1.42	1.66		3.08	4.29
0250	STK cedar, 1" x 4"			420	.038		.52	1.58		2.10	3.17
0260	1" x 6"			420	.038		.84	1.58		2.42	3.52
0270	1" x 8"			420	.038		1.31	1.58		2.89	4.04
0280	1" x 10"			400	.040		1.89	1.66		3.55	4.81
0290	1" x 12"			400	.040		2.63	1.66		4.29	5.60
1000	Exterior AC plywood, 1/4" thick			420	.038	S.F.	.68	1.58		2.26	3.35
1050	3/8" thick			420	.038		.77	1.58		2.35	3.45
1100	1/2" thick			420	.038		.80	1.58		2.38	3.48
1150	Polyvinyl chloride, white, solid		1 Carp	230	.035		1.09	1.45		2.54	3.57
1160	Perforated		"	230	.035		1.09	1.45		2.54	3.57
9000	Minimum labor/equipment charge		2 Carp	5	3.200	Job		133		133	218

06 25 Prefinished Paneling

06 25 13 – Prefinished Hardboard Paneling

06 25 13.10 Paneling, Hardboard

	06 25 13.10 Paneling, Hardboard		Crew	Daily Output	Labor-Hours	Unit	Material	Labor	Equipment	Total	Incl O&P
0010	**PANELING, HARDBOARD**										
0050	Not incl. furring or trim, hardboard, tempered, 1/8" thick	G	2 Carp	500	.032	S.F.	.39	1.33		1.72	2.61
0100	1/4" thick	G		500	.032		.53	1.33		1.86	2.76
0300	Tempered pegboard, 1/8" thick	G		500	.032		.44	1.33		1.77	2.66
0400	1/4" thick	G		500	.032		.67	1.33		2	2.92
0600	Untempered hardboard, natural finish, 1/8" thick	G		500	.032		.38	1.33		1.71	2.60
0700	1/4" thick	G		500	.032		.46	1.33		1.79	2.69
0900	Untempered pegboard, 1/8" thick	G		500	.032		.38	1.33		1.71	2.60
1000	1/4" thick	G		500	.032		.42	1.33		1.75	2.64
1200	Plastic faced hardboard, 1/8" thick	G		500	.032		.67	1.33		2	2.92
1300	1/4" thick	G		500	.032		.85	1.33		2.18	3.12
1500	Plastic faced pegboard, 1/8" thick	G		500	.032		.61	1.33		1.94	2.85
1600	1/4" thick	G		500	.032		.75	1.33		2.08	3.01
1800	Wood grained, plain or grooved, 1/4" thick, minimum	G		500	.032		.57	1.33		1.90	2.81
1900	Maximum	G		425	.038		1.19	1.56		2.75	3.88
2100	Moldings for hardboard, wood or aluminum, minimum			500	.032	L.F.	.38	1.33		1.71	2.60
2200	Maximum			425	.038	"	1.08	1.56		2.64	3.76
9000	Minimum labor/equipment charge		1 Carp	2	4	Job		166		166	273

06 25 16 – Prefinished Plywood Paneling

06 25 16.10 Paneling, Plywood

	06 25 16.10 Paneling, Plywood		Crew	Daily Output	Labor-Hours	Unit	Material	Labor	Equipment	Total	Incl O&P
0010	**PANELING, PLYWOOD**	R061636-20									
2400	Plywood, prefinished, 1/4" thick, 4' x 8' sheets										
2410	with vertical grooves. Birch faced, minimum		2 Carp	500	.032	S.F.	.89	1.33		2.22	3.16

06 25 Prefinished Paneling

06 25 16 – Prefinished Plywood Paneling

06 25 16.10 Paneling, Plywood

		Crew	Daily Output	Labor-Hours	Unit	Material	2010 Bare Costs Labor	2010 Bare Costs Equipment	Total	Total Incl O&P
2420	Average	2 Carp	420	.038	S.F.	1.35	1.58		2.93	4.09
2430	Maximum		350	.046		1.97	1.90		3.87	5.30
2600	Mahogany, African		400	.040		2.51	1.66		4.17	5.50
2700	Philippine (Lauan)		500	.032		1.08	1.33		2.41	3.37
2900	Oak or Cherry, minimum		500	.032		2.10	1.33		3.43	4.49
3000	Maximum		400	.040		3.22	1.66		4.88	6.25
3200	Rosewood		320	.050		4.58	2.08		6.66	8.45
3400	Teak		400	.040		3.22	1.66		4.88	6.25
3600	Chestnut		375	.043		4.77	1.77		6.54	8.15
3800	Pecan		400	.040		2.06	1.66		3.72	5
3900	Walnut, minimum		500	.032		2.75	1.33		4.08	5.20
3950	Maximum		400	.040		5.20	1.66		6.86	8.50
4000	Plywood, prefinished, 3/4" thick, stock grades, minimum		320	.050		1.24	2.08		3.32	4.77
4100	Maximum		224	.071		5.40	2.97		8.37	10.80
4300	Architectural grade, minimum		224	.071		3.98	2.97		6.95	9.25
4400	Maximum		160	.100		6.10	4.16		10.26	13.55
4600	Plywood, "A" face, birch, V.C., 1/2" thick, natural		450	.036		1.89	1.48		3.37	4.51
4700	Select		450	.036		2.06	1.48		3.54	4.70
4900	Veneer core, 3/4" thick, natural		320	.050		1.99	2.08		4.07	5.60
5000	Select		320	.050		2.23	2.08		4.31	5.85
5200	Lumber core, 3/4" thick, natural		320	.050		2.99	2.08		5.07	6.70
5500	Plywood, knotty pine, 1/4" thick, A2 grade		450	.036		1.63	1.48		3.11	4.22
5600	A3 grade		450	.036		2.06	1.48		3.54	4.70
5800	3/4" thick, veneer core, A2 grade		320	.050		2.11	2.08		4.19	5.75
5900	A3 grade		320	.050		2.38	2.08		4.46	6.05
6100	Aromatic cedar, 1/4" thick, plywood		400	.040		2.08	1.66		3.74	5
6200	1/4" thick, particle board		400	.040		1.01	1.66		2.67	3.84
9000	Minimum labor/equipment charge	1 Carp	2	4	Job		166		166	273

06 25 26 – Panel System

06 25 26.10 Panel Systems

		Crew	Daily Output	Labor-Hours	Unit	Material	2010 Bare Costs Labor	2010 Bare Costs Equipment	Total	Total Incl O&P
0010	**PANEL SYSTEMS**									
0100	Raised panel, eng. wood core w/wood veneer, std., paint grade	2 Carp	300	.053	S.F.	11.90	2.22		14.12	16.75
0110	Oak veneer		300	.053		19.70	2.22		21.92	25
0120	Maple veneer		300	.053		25	2.22		27.22	31
0130	Cherry veneer		300	.053		32	2.22		34.22	38.50
0300	Class I fire rated, paint grade		300	.053		14.25	2.22		16.47	19.30
0310	Oak veneer		300	.053		23.50	2.22		25.72	29.50
0320	Maple veneer		300	.053		30	2.22		32.22	36.50
0330	Cherry veneer		300	.053		38.50	2.22		40.72	45.50
0510	Beadboard, 5/8" MDF, standard, primed		300	.053		8.40	2.22		10.62	12.90
0520	Oak veneer, unfinished		300	.053		14.60	2.22		16.82	19.70
0530	Maple veneer, unfinished		300	.053		17.25	2.22		19.47	22.50
0610	Rustic paneling, 5/8" MDF, standard, maple veneer, unfinished		300	.053		22	2.22		24.22	28

06 26 Board Paneling

06 26 13 – Profile Board Paneling

06 26 13.10 Paneling, Boards		Crew	Daily Output	Labor-Hours	Unit	Material	2010 Bare Costs Labor	Equipment	Total	Total Incl O&P
0010	**PANELING, BOARDS**									
6400	Wood board paneling, 3/4" thick, knotty pine	2 Carp	300	.053	S.F.	1.44	2.22		3.66	5.20
6500	Rough sawn cedar		300	.053		1.85	2.22		4.07	5.70
6700	Redwood, clear, 1" x 4" boards		300	.053		4.96	2.22		7.18	9.10
6900	Aromatic cedar, closet lining, boards		275	.058		3.32	2.42		5.74	7.60
9000	Minimum labor/equipment charge	1 Carp	2	4	Job		166		166	273

06 43 Wood Stairs and Railings

06 43 13 – Wood Stairs

06 43 13.20 Prefabricated Wood Stairs

		Crew	Daily Output	Labor-Hours	Unit	Material	Labor	Equipment	Total	Total Incl O&P
0010	**PREFABRICATED WOOD STAIRS**									
0100	Box stairs, prefabricated, 3'-0" wide									
0110	Oak treads, up to 14 risers	2 Carp	39	.410	Riser	85	17.05		102.05	122
0600	With pine treads for carpet, up to 14 risers	"	39	.410	"	55	17.05		72.05	88.50
1100	For 4' wide stairs, add				Flight	25%				
1550	Stairs, prefabricated stair handrail with balusters	1 Carp	30	.267	L.F.	75	11.10		86.10	100
1700	Basement stairs, prefabricated, pine treads									
1710	Pine risers, 3' wide, up to 14 risers	2 Carp	52	.308	Riser	55	12.80		67.80	81.50
4000	Residential, wood, oak treads, prefabricated		1.50	10.667	Flight	1,100	445		1,545	1,925
4200	Built in place		.44	36.364	"	1,750	1,500		3,250	4,400
4400	Spiral, oak, 4'-6" diameter, unfinished, prefabricated,									
4500	incl. railing, 9' high	2 Carp	1.50	10.667	Flight	3,625	445		4,070	4,700
9000	Minimum labor/equipment charge	"	3	5.333	Job		222		222	365

06 43 13.40 Wood Stair Parts

		Crew	Daily Output	Labor-Hours	Unit	Material	Labor	Equipment	Total	Total Incl O&P
0010	**WOOD STAIR PARTS**									
0020	Pin top balusters, 1-1/4", oak, 34"	1 Carp	96	.083	Ea.	7.80	3.46		11.26	14.30
0030	38"		96	.083		9.85	3.46		13.31	16.55
0040	42"		96	.083		12.10	3.46		15.56	19
0050	Poplar, 34"		96	.083		3.05	3.46		6.51	9.05
0060	38"		96	.083		3.62	3.46		7.08	9.70
0070	42"		96	.083		8.55	3.46		12.01	15.15
0080	Maple, 34"		96	.083		4.90	3.46		8.36	11.10
0090	38"		96	.083		6.20	3.46		9.66	12.50
0100	42"		96	.083		7.60	3.46		11.06	14.05
0130	Primed, 34"		96	.083		2.85	3.46		6.31	8.85
0140	38"		96	.083		3.62	3.46		7.08	9.70
0150	42"		96	.083		4.42	3.46		7.88	10.55
0180	Box top balusters, 1-1/4", oak, 34"		60	.133		10.85	5.55		16.40	21
0190	38"		60	.133		12.50	5.55		18.05	23
0200	42"		60	.133		12.95	5.55		18.50	23.50
0210	Poplar, 34"		60	.133		8.65	5.55		14.20	18.60
0220	38"		60	.133		9.50	5.55		15.05	19.55
0230	42"		60	.133		10	5.55		15.55	20
0240	Maple, 34"		60	.133		11.90	5.55		17.45	22
0250	38"		60	.133		13.80	5.55		19.35	24.50
0260	42"		60	.133		14.20	5.55		19.75	24.50
0290	Primed, 34"		60	.133		6.95	5.55		12.50	16.75
0300	38"		60	.133		8	5.55		13.55	17.90
0310	42"		60	.133		8.35	5.55		13.90	18.30
0340	Square balusters, cut from lineal stock, pine, 1-1/16" x 1-1/16"		180	.044	L.F.	1.32	1.85		3.17	4.48
0350	1-5/16" x 1-5/16"		180	.044		1.70	1.85		3.55	4.90

06 43 Wood Stairs and Railings

06 43 13 – Wood Stairs

06 43 13.40 Wood Stair Parts

		Crew	Daily Output	Labor-Hours	Unit	Material	2010 Bare Costs Labor	Equipment	Total	Total Incl O&P
0360	1-5/8" x 1-5/8"	1 Carp	180	.044	L.F.	2.82	1.85		4.67	6.15
0370	Turned newel, oak, 3-1/2" square, 48" high		8	1	Ea.	94	41.50		135.50	172
0380	62" high		8	1		138	41.50		179.50	221
0390	Poplar, 3-1/2" square, 48" high		8	1		52	41.50		93.50	126
0400	62" high		8	1		75	41.50		116.50	151
0410	Maple, 3-1/2" square, 48" high		8	1		66	41.50		107.50	141
0420	62" high		8	1		93	41.50		134.50	171
0430	Square newel, oak, 3-1/2" square, 48" high		8	1		74	41.50		115.50	150
0440	58" high		8	1		78	41.50		119.50	155
0450	Poplar, 3-1/2" square, 48" high		8	1		49	41.50		90.50	123
0460	58" high		8	1		54	41.50		95.50	128
0470	Maple, 3" square, 48" high		8	1		66	41.50		107.50	141
0480	58" high		8	1		70	41.50		111.50	146
0490	Railings, oak, minimum		96	.083	L.F.	11	3.46		14.46	17.80
0500	Average		96	.083		14	3.46		17.46	21
0510	Maximum		96	.083		14.50	3.46		17.96	21.50
0520	Maple, minimum		96	.083		15	3.46		18.46	22
0530	Average		96	.083		15.25	3.46		18.71	22.50
0540	Maximum		96	.083		15.50	3.46		18.96	23
0550	Oak, for bending rail, minimum		48	.167		21	6.95		27.95	34.50
0560	Average		48	.167		22	6.95		28.95	35.50
0570	Maximum		48	.167		23	6.95		29.95	37
0580	Maple, for bending rail, minimum		48	.167		23	6.95		29.95	36.50
0590	Average		48	.167		25	6.95		31.95	38.50
0600	Maximum		48	.167		27	6.95		33.95	41
0610	Risers, oak, 3/4" x 8", 36" long		80	.100	Ea.	18	4.16		22.16	26.50
0620	42" long		80	.100		21	4.16		25.16	30
0630	48" long		80	.100		24	4.16		28.16	33.50
0640	54" long		80	.100		27	4.16		31.16	36.50
0650	60" long		80	.100		30	4.16		34.16	40
0660	72" long		80	.100		36	4.16		40.16	46.50
0670	Poplar, 3/4" x 8", 36" long		80	.100		11.10	4.16		15.26	19.05
0680	42" long		80	.100		12.95	4.16		17.11	21
0690	48" long		80	.100		14.80	4.16		18.96	23
0700	54" long		80	.100		16.65	4.16		20.81	25
0710	60" long		80	.100		18.50	4.16		22.66	27.50
0720	72" long		80	.100		22	4.16		26.16	31.50
0730	Pine, 1" x 8", 36" long		80	.100		2.51	4.16		6.67	9.60
0740	42" long		80	.100		2.93	4.16		7.09	10.05
0750	48" long		80	.100		3.34	4.16		7.50	10.55
0760	54" long		80	.100		3.76	4.16		7.92	11
0770	60" long		80	.100		4.18	4.16		8.34	11.45
0780	72" long		80	.100		5	4.16		9.16	12.35
0790	Treads, oak, no returns, 1-1/32" x 11-1/2" x 36" long		32	.250		32	10.40		42.40	52
0800	42" long		32	.250		37.50	10.40		47.90	58
0810	48" long		32	.250		42.50	10.40		52.90	64
0820	54" long		32	.250		48	10.40		58.40	70
0830	60" long		32	.250		53.50	10.40		63.90	75.50
0840	72" long		32	.250		64	10.40		74.40	87.50
0850	Mitred return one end, 1-1/32" x 11-1/2" x 36" long		24	.333		45	13.85		58.85	72.50
0860	42" long		24	.333		52.50	13.85		66.35	81
0870	48" long		24	.333		60	13.85		73.85	89
0880	54" long		24	.333		67.50	13.85		81.35	97.50

06 43 Wood Stairs and Railings

06 43 13 – Wood Stairs

06 43 13.40 Wood Stair Parts

		Crew	Daily Output	Labor-Hours	Unit	Material	Labor	Equipment	Total	Total Incl O&P
0890	60" long	1 Carp	24	.333	Ea.	75	13.85		88.85	106
0900	72" long		24	.333		90	13.85		103.85	122
0910	Mitred return two ends, 1-1/32" x 11-1/2" x 36" long		12	.667		49	27.50		76.50	99.50
0920	42" long		12	.667		57	27.50		84.50	109
0930	48" long		12	.667		65.50	27.50		93	118
0940	54" long		12	.667		73.50	27.50		101	127
0950	60" long		12	.667		81.50	27.50		109	136
0960	72" long		12	.667		98	27.50		125.50	154
0970	Starting step, oak, 48", bullnose		8	1		176	41.50		217.50	263
0980	Double end bullnose		8	1		275	41.50		316.50	375
1030	Skirt board, pine, 1" x 10"		55	.145	L.F.	1.17	6.05		7.22	11.25
1040	1" x 12"		52	.154	"	1.42	6.40		7.82	12.05
1050	Oak landing tread, 1-1/16" thick		54	.148	S.F.	8.25	6.15		14.40	19.20
1060	Oak cove molding		96	.083	L.F.	1.45	3.46		4.91	7.30
1070	Oak stringer molding		96	.083	"	2.75	3.46		6.21	8.75
1090	Rail bolt, 5/16" x 3-1/2"		48	.167	Ea.	2.75	6.95		9.70	14.45
1100	5/16" x 4-1/2"		48	.167		2.75	6.95		9.70	14.45
1120	Newel post anchor		16	.500		13.50	21		34.50	49
1130	Tapered plug, 1/2"		240	.033		.30	1.38		1.68	2.60
1140	1"		240	.033		.40	1.38		1.78	2.71
9000	Minimum labor/equipment charge		3	2.667	Job		111		111	182

06 43 16 – Wood Railings

06 43 16.10 Wood Handrails and Railings

		Crew	Daily Output	Labor-Hours	Unit	Material	Labor	Equipment	Total	Total Incl O&P
0010	**WOOD HANDRAILS AND RAILINGS**									
0020	Custom design, architectural grade, hardwood, minimum	1 Carp	38	.211	L.F.	9.50	8.75		18.25	25
0100	Maximum		30	.267		50	11.10		61.10	73
0300	Stock interior railing with spindles 4" O.C., 4' long		40	.200		37	8.30		45.30	54
0400	8' long		48	.167		37	6.95		43.95	52
9000	Minimum labor/equipment charge		3	2.667	Job		111		111	182

06 44 Ornamental Woodwork

06 44 19 – Wood Grilles

06 44 19.10 Grilles

		Crew	Daily Output	Labor-Hours	Unit	Material	Labor	Equipment	Total	Total Incl O&P
0010	**GRILLES** and panels, hardwood, sanded									
0020	2' x 4' to 4' x 8', custom designs, unfinished, minimum	1 Carp	38	.211	S.F.	54	8.75		62.75	74
0050	Average		30	.267		62.50	11.10		73.60	87
0100	Maximum		19	.421		71	17.50		88.50	107
9000	Minimum labor/equipment charge		2	4	Job		166		166	273

06 44 33 – Wood Mantels

06 44 33.10 Fireplace Mantels

		Crew	Daily Output	Labor-Hours	Unit	Material	Labor	Equipment	Total	Total Incl O&P
0010	**FIREPLACE MANTELS**									
0015	6" molding, 6' x 3'-6" opening, minimum	1 Carp	5	1.600	Opng.	228	66.50		294.50	360
0100	Maximum		5	1.600		410	66.50		476.50	565
0300	Prefabricated pine, colonial type, stock, deluxe		2	4		1,250	166		1,416	1,675
0400	Economy		3	2.667		530	111		641	765
9000	Minimum labor/equipment charge		3	2.667	Job		111		111	182

06 44 33.20 Fireplace Mantel Beam

		Crew	Daily Output	Labor-Hours	Unit	Material	Labor	Equipment	Total	Total Incl O&P
0010	**FIREPLACE MANTEL BEAM**									
0020	Rough texture wood, 4" x 8"	1 Carp	36	.222	L.F.	6.90	9.25		16.15	23
0100	4" x 10"		35	.229	"	12.80	9.50		22.30	29.50

06 44 Ornamental Woodwork

06 44 33 – Wood Mantels

06 44 33.20 Fireplace Mantel Beam

		Crew	Daily Output	Labor-Hours	Unit	Material	2010 Bare Costs Labor	Equipment	Total	Total Incl O&P
0300	Laminated hardwood, 2-1/4" x 10-1/2" wide, 6' long	1 Carp	5	1.600	Ea.	110	66.50		176.50	230
0400	8' long		5	1.600	"	150	66.50		216.50	274
0600	Brackets for above, rough sawn		12	.667	Pr.	10	27.50		37.50	56.50
0700	Laminated		12	.667	"	15	27.50		42.50	62
9000	Minimum labor/equipment charge		4	2	Job		83		83	137

06 44 39 – Wood Posts and Columns

06 44 39.10 Decorative Beams

		Crew	Daily Output	Labor-Hours	Unit	Material	Labor	Equipment	Total	Total Incl O&P
0010	**DECORATIVE BEAMS**									
0020	Rough sawn cedar, non-load bearing, 4" x 4"	2 Carp	180	.089	L.F.	1	3.69		4.69	7.15
0100	4" x 6"		170	.094		1.42	3.91		5.33	7.95
0200	4" x 8"		160	.100		1.89	4.16		6.05	8.95
0300	4" x 10"		150	.107		2.37	4.43		6.80	9.90
0400	4" x 12"		140	.114		2.84	4.75		7.59	10.90
0500	8" x 8"		130	.123		4.96	5.10		10.06	13.85
9000	Minimum labor/equipment charge	1 Carp	3	2.667	Job		111		111	182

06 44 39.20 Columns

		Crew	Daily Output	Labor-Hours	Unit	Material	Labor	Equipment	Total	Total Incl O&P
0010	**COLUMNS**									
0050	Aluminum, round colonial, 6" diameter	2 Carp	80	.200	V.L.F.	16	8.30		24.30	31.50
0100	8" diameter		62.25	.257		19.30	10.70		30	38.50
0200	10" diameter		55	.291		23.50	12.10		35.60	46
0250	Fir, stock units, hollow round, 6" diameter		80	.200		26.50	8.30		34.80	42.50
0300	8" diameter		80	.200		31	8.30		39.30	48
0350	10" diameter		70	.229		39.50	9.50		49	59
0400	Solid turned, to 8' high, 3-1/2" diameter		80	.200		9	8.30		17.30	23.50
0500	4-1/2" diameter		75	.213		14	8.85		22.85	30
0600	5-1/2" diameter		70	.229		18.40	9.50		27.90	35.50
0800	Square columns, built-up, 5" x 5"		65	.246		15.50	10.25		25.75	34
0900	Solid, 3-1/2" x 3-1/2"		130	.123		9	5.10		14.10	18.30
1600	Hemlock, tapered, T & G, 12" diam, 10' high		100	.160		53	6.65		59.65	69
1700	16' high		65	.246		79.50	10.25		89.75	104
1900	10' high, 14" diameter		100	.160		121	6.65		127.65	144
2000	18' high		65	.246		107	10.25		117.25	135
2200	18" diameter, 12' high		65	.246		153	10.25		163.25	186
2300	20' high		50	.320		148	13.30		161.30	185
2500	20" diameter, 14' high		40	.400		159	16.60		175.60	203
2600	20' high		35	.457		149	19		168	194
2800	For flat pilasters, deduct					33%				
3000	For splitting into halves, add				Ea.	116			116	128
4000	Rough sawn cedar posts, 4" x 4"	2 Carp	250	.064	V.L.F.	6.30	2.66		8.96	11.30
4100	4" x 6"		235	.068		15	2.83		17.83	21
4200	6" x 6"		220	.073		26.50	3.02		29.52	34
4300	8" x 8"		200	.080		60	3.32		63.32	71.50
9000	Minimum labor/equipment charge	1 Carp	3	2.667	Job		111		111	182

06 48 Wood Frames

06 48 13 – Exterior Wood Door Frames

06 48 13.10 Exterior Wood Door Frames and Accessories

		Crew	Daily Output	Labor-Hours	Unit	Material	2010 Bare Costs Labor	2010 Bare Costs Equipment	Total	Total Incl O&P
0010	**EXTERIOR WOOD DOOR FRAMES AND ACCESSORIES**									
0400	Exterior frame, incl. ext. trim, pine, 5/4 x 4-9/16" deep	2 Carp	375	.043	L.F.	5.55	1.77		7.32	9
0420	5-3/16" deep		375	.043		11.35	1.77		13.12	15.40
0440	6-9/16" deep		375	.043		8.50	1.77		10.27	12.25
0600	Oak, 5/4 x 4-9/16" deep		350	.046		10.70	1.90		12.60	14.90
0620	5-3/16" deep		350	.046		11.80	1.90		13.70	16.05
0640	6-9/16" deep		350	.046		14.50	1.90		16.40	19.05
1000	Sills, 8/4 x 8" deep, oak, no horns		100	.160		17	6.65		23.65	29.50
1020	2" horns		100	.160		17.10	6.65		23.75	29.50
1040	3" horns		100	.160		17.10	6.65		23.75	29.50
1100	8/4 x 10" deep, oak, no horns		90	.178		20.50	7.40		27.90	34.50
1120	2" horns		90	.178		21	7.40		28.40	35.50
1140	3" horns		90	.178		21	7.40		28.40	35.50
2000	Exterior, colonial, frame & trim, 3' opng., in-swing, minimum		22	.727	Ea.	315	30		345	400
2010	Average		21	.762		470	31.50		501.50	565
2020	Maximum		20	.800		1,075	33		1,108	1,225
2100	5'-4" opening, in-swing, minimum		17	.941		360	39		399	460
2120	Maximum		15	1.067		1,075	44.50		1,119.50	1,250
2140	Out-swing, minimum		17	.941		370	39		409	470
2160	Maximum		15	1.067		1,100	44.50		1,144.50	1,300
2400	6'-0" opening, in-swing, minimum		16	1		345	41.50		386.50	450
2420	Maximum		10	1.600		1,100	66.50		1,166.50	1,325
2460	Out-swing, minimum		16	1		375	41.50		416.50	485
2480	Maximum		10	1.600		1,300	66.50		1,366.50	1,525
2600	For two sidelights, add, minimum		30	.533	Opng.	355	22		377	425
2620	Maximum		20	.800	"	1,125	33		1,158	1,300
2700	Custom birch frame, 3'-0" opening		16	1	Ea.	205	41.50		246.50	295
2750	6'-0" opening		16	1		305	41.50		346.50	405
2900	Exterior, modern, plain trim, 3' opng., in-swing, minimum		26	.615		35.50	25.50		61	81
2920	Average		24	.667		42	27.50		69.50	91.50
2940	Maximum		22	.727		50.50	30		80.50	105

06 48 16 – Interior Wood Door Frames

06 48 16.10 Interior Wood Door Jamb and Frames

		Crew	Daily Output	Labor-Hours	Unit	Material	2010 Bare Costs Labor	2010 Bare Costs Equipment	Total	Total Incl O&P
0010	**INTERIOR WOOD DOOR JAMB AND FRAMES**									
3000	Interior frame, pine, 11/16" x 3-5/8" deep	2 Carp	375	.043	L.F.	5.55	1.77		7.32	9
3020	4-9/16" deep		375	.043		6.50	1.77		8.27	10.05
3200	Oak, 11/16" x 3-5/8" deep		350	.046		4.35	1.90		6.25	7.90
3220	4-9/16" deep		350	.046		4.50	1.90		6.40	8.05
3240	5-3/16" deep		350	.046		4.58	1.90		6.48	8.15
3400	Walnut, 11/16" x 3-5/8" deep		350	.046		7.30	1.90		9.20	11.15
3420	4-9/16" deep		350	.046		7.50	1.90		9.40	11.35
3440	5-3/16" deep		350	.046		7.90	1.90		9.80	11.75
3600	Pocket door frame		16	1	Ea.	74.50	41.50		116	150
3800	Threshold, oak, 5/8" x 3-5/8" deep		200	.080	L.F.	3.83	3.32		7.15	9.65
3820	4-5/8" deep		190	.084		4.58	3.50		8.08	10.80
3840	5-5/8" deep		180	.089		7.15	3.69		10.84	13.90
9000	Minimum labor/equipment charge	1 Carp	4	2	Job		83		83	137

06 49 Wood Screens and Exterior Wood Shutters

06 49 19 – Exterior Wood Shutters

06 49 19.10 Shutters, Exterior		Crew	Daily Output	Labor-Hours	Unit	Material	2010 Bare Costs Labor	Equipment	Total	Total Incl O&P
0010	**SHUTTERS, EXTERIOR**									
0012	Aluminum, louvered, 1'-4" wide, 3'-0" long	1 Carp	10	.800	Pr.	60.50	33		93.50	121
0400	6'-8" long		9	.889		122	37		159	195
1000	Pine, louvered, primed, each 1'-2" wide, 3'-3" long		10	.800		105	33		138	171
1100	4'-7" long		10	.800		142	33		175	211
1250	Each 1'-4" wide, 3'-0" long		10	.800		107	33		140	173
1350	5'-3" long		10	.800		161	33		194	232
1500	Each 1'-6" wide, 3'-3" long		10	.800		112	33		145	178
1600	4'-7" long		10	.800		157	33		190	228
1610	Door blinds, 6'-9" long 1'-3" wide		9	.889		196	37		233	276
1615	1' - 6" wide		9	.889		216	37		253	298
1620	Hemlock, louvered, 1'-2" wide, 5'-7" long		10	.800		172	33		205	244
1630	Each 1'-4" wide, 2'-2" long		10	.800		107	33		140	173
1640	3'-0" long		10	.800		107	33		140	173
1650	3'-3" long		10	.800		116	33		149	182
1660	3'-11" long		10	.800		132	33		165	200
1670	4'-3" long		10	.800		128	33		161	196
1680	5'-3" long		10	.800		160	33		193	231
1690	5'-11" long		10	.800		181	33		214	254
1700	Door blinds, 6'-9" long, each 1'-3" wide		9	.889		182	37		219	261
1710	1'-6" wide		9	.889		196	37		233	276
1720	Hemlock, solid raised panel, each 1'-4" wide, 3'-3" long		10	.800		174	33		207	246
1730	3'-11" long		10	.800		203	33		236	278
1740	4'-3" long		10	.800		220	33		253	297
1750	4'-7" long		10	.800		237	33		270	315
1760	4'-11" long		10	.800		258	33		291	340
1770	5'-11" long		10	.800		294	33		327	380
1800	Door blinds, 6'-9" long, each 1'-3" wide		9	.889		330	37		367	425
1900	1'-6" wide		9	.889		360	37		397	455
2500	Polystyrene, solid raised panel, each 1'-4" wide, 3'-3" long		10	.800		40.50	33		73.50	99
2700	4'-7" long		10	.800		51	33		84	111
4500	Polystyrene, louvered, each 1'-2" wide, 3'-3" long		10	.800		27.50	33		60.50	85
4600	4'-7" long		10	.800		34	33		67	92
4750	5'-3" long		10	.800		39	33		72	97.50
4850	6'-8" long		9	.889		46.50	37		83.50	112
6000	Vinyl, louvered, each 1'-2" x 4'-7" long		10	.800		34	33		67	92
6200	Each 1'-4" x 6'-8" long		9	.889		47.50	37		84.50	113
9000	Minimum labor/equipment charge		4	2	Job		83		83	137

06 51 Structural Plastic Shapes and Plates

06 51 13 – Plastic Lumber

06 51 13.10 Recycled Plastic Lumber

			Crew	Daily Output	Labor-Hours	Unit	Material	Labor	Equipment	Total	Total Incl O&P
0010	**RECYCLED PLASTIC LUMBER**										
4000	Sheeting, recycled plastic, black or white, 4' x 8' x 1/8"	G	2 Carp	1100	.015	S.F.	1.56	.60		2.16	2.71
4010	4' x 8' x 3/16"	G		1100	.015		2.33	.60		2.93	3.55
4020	4' x 8' x 1/4"	G		950	.017		3.11	.70		3.81	4.57
4030	4' x 8' x 3/8"	G		950	.017		4.67	.70		5.37	6.30
4040	4' x 8' x 1/2"	G		900	.018		6.20	.74		6.94	8.05
4050	4' x 8' x 5/8"	G		900	.018		7.70	.74		8.44	9.65
4060	4' x 8' x 3/4"	G		850	.019		9.15	.78		9.93	11.40
4070	Add for colors	G				Ea.	5%			5%	6%

06 51 Structural Plastic Shapes and Plates

06 51 13 – Plastic Lumber

06 51 13.10 Recycled Plastic Lumber		Crew	Daily Output	Labor-Hours	Unit	Material	2010 Bare Costs Labor	Equipment	Total	Total Incl O&P
8500	100% recycled plastic, var colors, NLB, 2" x 2"	G			L.F.	1.41			1.41	1.55
8510	2" x 4"	G				2.95			2.95	3.25
8520	2" x 6"	G				4.63			4.63	5.10
8530	2" x 8"	G				6.35			6.35	6.95
8540	2" x 10"	G				9.25			9.25	10.15
8550	5/4" x 4"	G				3.02			3.02	3.32
8560	5/4" x 6"	G				3			3	3.30
8570	1" x 6"	G				2.28			2.28	2.51
8580	1/2" x 8"	G				2.46			2.46	2.71
8590	2" x 10" T & G	G				8.60			8.60	9.45
8600	3" x 10" T & G	G				14.25			14.25	15.70
8610	Add for premium colors	G				20%			20%	22%

06 63 Plastic Railings

06 63 10 – Plastic (PVC) Railings

06 63 10.10 Plastic Railings

		Crew	Daily Output	Labor-Hours	Unit	Material	Labor	Equipment	Total	Total Incl O&P
0010	**PLASTIC RAILINGS**									
0100	Horizontal PVC handrail with balusters, 3-1/2" wide, 36" high	1 Carp	96	.083	L.F.	24	3.46		27.46	32
0150	42" high		96	.083		27	3.46		30.46	35
0200	Angled PVC handrail with balusters, 3-1/2" wide, 36" high		72	.111		28	4.62		32.62	38
0250	42" high		72	.111		30.50	4.62		35.12	41
0300	Post sleeve for 4 x 4 post		96	.083		12	3.46		15.46	18.90
0400	Post cap for 4 x 4 post, flat profile		48	.167	Ea.	11.40	6.95		18.35	24
0450	Newel post style profile		48	.167		21	6.95		27.95	34.50
0500	Raised corbeled profile		48	.167		23	6.95		29.95	36.50
0550	Post base trim for 4 x 4 post		96	.083		12.80	3.46		16.26	19.80

06 65 Plastic Simulated Wood Trim

06 65 10 – PVC Trim

06 65 10.10 PVC Trim, Exterior

		Crew	Daily Output	Labor-Hours	Unit	Material	Labor	Equipment	Total	Total Incl O&P
0010	**PVC TRIM, EXTERIOR**									
0100	Cornerboards, 5/4" x 6" x 6"	1 Carp	240	.033	L.F.	5.70	1.38		7.08	8.55
0110	Door / window casing, 1" x 4"		200	.040		1.30	1.66		2.96	4.16
0120	1" x 6"		200	.040		2.05	1.66		3.71	4.99
0130	1" x 8"		195	.041		2.70	1.70		4.40	5.75
0140	1" x 10"		195	.041		3.42	1.70		5.12	6.55
0150	1" x 12"		190	.042		4.19	1.75		5.94	7.50
0160	5/4" x 4"		195	.041		1.73	1.70		3.43	4.70
0170	5/4" x 6"		195	.041		2.70	1.70		4.40	5.75
0180	5/4" x 8"		190	.042		3.55	1.75		5.30	6.80
0190	5/4" x 10"		190	.042		4.53	1.75		6.28	7.85
0200	5/4" x 12"		185	.043		5.50	1.80		7.30	9
0210	Fascia, 1" x 4"		250	.032		1.30	1.33		2.63	3.61
0220	1" x 6"		250	.032		2.05	1.33		3.38	4.44
0230	1" x 8"		225	.036		2.70	1.48		4.18	5.40
0240	1" x 10"		225	.036		3.42	1.48		4.90	6.20
0250	1" x 12"		200	.040		4.19	1.66		5.85	7.35
0260	5/4" x 4"		240	.033		1.73	1.38		3.11	4.17
0270	5/4" x 6"		240	.033		2.70	1.38		4.08	5.25

06 65 Plastic Simulated Wood Trim

06 65 10 – PVC Trim

06 65 10.10 PVC Trim, Exterior		Crew	Daily Output	Labor-Hours	Unit	Material	2010 Bare Costs Labor	Equipment	Total	Total Incl O&P
0280	5/4" x 8"	1 Carp	215	.037	L.F.	3.55	1.55		5.10	6.45
0290	5/4" x 10"		215	.037		4.53	1.55		6.08	7.50
0300	5/4" x 12"		190	.042		5.50	1.75		7.25	8.90
0310	Frieze, 1" x 4"		250	.032		1.30	1.33		2.63	3.61
0320	1" x 6"		250	.032		2.05	1.33		3.38	4.44
0330	1" x 8"		225	.036		2.70	1.48		4.18	5.40
0340	1" x 10"		225	.036		3.42	1.48		4.90	6.20
0350	1" x 12"		200	.040		4.19	1.66		5.85	7.35
0360	5/4" x 4"		240	.033		1.73	1.38		3.11	4.17
0370	5/4" x 6"		240	.033		2.70	1.38		4.08	5.25
0380	5/4" x 8"		215	.037		3.55	1.55		5.10	6.45
0390	5/4" x 10"		215	.037		4.53	1.55		6.08	7.50
0400	5/4" x 12"		190	.042		5.50	1.75		7.25	8.90
0410	Rake, 1" x 4"		200	.040		1.30	1.66		2.96	4.16
0420	1" x 6"		200	.040		2.05	1.66		3.71	4.99
0430	1" x 8"		190	.042		2.70	1.75		4.45	5.85
0440	1" x 10"		190	.042		3.42	1.75		5.17	6.65
0450	1" x 12"		180	.044		4.19	1.85		6.04	7.65
0460	5/4" x 4"		195	.041		1.73	1.70		3.43	4.70
0470	5/4" x 6"		195	.041		2.70	1.70		4.40	5.75
0480	5/4" x 8"		185	.043		3.55	1.80		5.35	6.85
0490	5/4" x 10"		185	.043		4.53	1.80		6.33	7.95
0500	5/4" x 12"		175	.046		5.50	1.90		7.40	9.15
0510	Rake trim, 1" x 4"		225	.036		1.30	1.48		2.78	3.86
0520	1" x 6"		225	.036		2.05	1.48		3.53	4.69
0560	5/4" x 4"		220	.036		1.73	1.51		3.24	4.38
0570	5/4" x 6"		220	.036		2.70	1.51		4.21	5.45
0610	Soffit, 1" x 4"	2 Carp	420	.038		1.30	1.58		2.88	4.03
0620	1" x 6"		420	.038		2.05	1.58		3.63	4.86
0630	1" x 8"		420	.038		2.70	1.58		4.28	5.55
0640	1" x 10"		400	.040		3.42	1.66		5.08	6.50
0650	1" x 12"		400	.040		4.19	1.66		5.85	7.35
0660	5/4" x 4"		410	.039		1.73	1.62		3.35	4.56
0670	5/4" x 6"		410	.039		2.70	1.62		4.32	5.65
0680	5/4" x 8"		410	.039		3.55	1.62		5.17	6.55
0690	5/4" x 10"		390	.041		4.53	1.70		6.23	7.80
0700	5/4" x 12"		390	.041		5.50	1.70		7.20	8.85

Division Notes

		CREW	DAILY OUTPUT	LABOR-HOURS	UNIT	2010 BARE COSTS				TOTAL INCL O&P
						MAT.	LABOR	EQUIP.	TOTAL	

Estimating Tips

07 10 00 Dampproofing and Waterproofing

- Be sure of the job specifications before pricing this subdivision. The difference in cost between waterproofing and dampproofing can be great. Waterproofing will hold back standing water. Dampproofing prevents the transmission of water vapor. Also included in this section are vapor retarding membranes.

07 20 00 Thermal Protection

- Insulation and fireproofing products are measured by area, thickness, volume or R-value. Specifications may give only what the specific R-value should be in a certain situation. The estimator may need to choose the type of insulation to meet that R-value.

07 30 00 Steep Slope Roofing
07 40 00 Roofing and Siding Panels

- Many roofing and siding products are bought and sold by the square. One square is equal to an area that measures 100 square feet. This simple change in unit of measure could create a large error if the estimator is not observant. Accessories necessary for a complete installation must be figured into any calculations for both material and labor.

07 50 00 Membrane Roofing
07 60 00 Flashing and Sheet Metal
07 70 00 Roofing and Wall Specialties and Accessories

- The items in these subdivisions compose a roofing system. No one component completes the installation, and all must be estimated. Built-up or single-ply membrane roofing systems are made up of many products and installation trades. Wood blocking at roof perimeters or penetrations, parapet coverings, reglets, roof drains, gutters, downspouts, sheet metal flashing, skylights, smoke vents, and roof hatches all need to be considered along with the roofing material. Several different installation trades will need to work together on the roofing system. Inherent difficulties in the scheduling and coordination of various trades must be accounted for when estimating labor costs.

07 90 00 Joint Protection

- To complete the weather-tight shell, the sealants and caulkings must be estimated. Where different materials meet—at expansion joints, at flashing penetrations, and at hundreds of other locations throughout a construction project—they provide another line of defense against water penetration. Often, an entire system is based on the proper location and placement of caulking or sealants. The detailed drawings that are included as part of a set of architectural plans show typical locations for these materials. When caulking or sealants are shown at typical locations, this means the estimator must include them for all the locations where this detail is applicable. Be careful to keep different types of sealants separate, and remember to consider backer rods and primers if necessary.

Reference Numbers

Reference numbers are shown in shaded boxes at the beginning of some major classifications. These numbers refer to related items in the Reference Section. The reference information may be an estimating procedure, an alternate pricing method, or technical information.

Note: Not all subdivisions listed here necessarily appear in this publication.

07 01 Operation and Maint. of Thermal and Moisture Protection

07 01 50 – Maintenance of Membrane Roofing

07 01 50.10 Roof Coatings

		Crew	Daily Output	Labor-Hours	Unit	Material	2010 Bare Costs Labor	Equipment	Total	Total Incl O&P
0010	**ROOF COATINGS**									
0012	Asphalt, brush grade, material only				Gal.	8.05			8.05	8.85
0200	Asphalt base, fibered aluminum coating	G				9.75			9.75	10.75
0300	Asphalt primer, 5 gallon				↓	7.80			7.80	8.60
0600	Coal tar pitch, 200 lb. barrels				Ton	1,025			1,025	1,125
0700	Tar roof cement, 5 gal. lots				Gal.	13.20			13.20	14.50
0800	Glass fibered roof & patching cement, 5 gallon				"	8			8	8.80
0900	Reinforcing glass membrane, 450 S.F./roll				Ea.	51			51	56
1000	Neoprene roof coating, 5 gal, 2 gal/sq				Gal.	25.50			25.50	28
1100	Roof patch & flashing cement, 5 gallon					9.70			9.70	10.65
1200	Roof resaturant, glass fibered, 3 gal/sq				↓	7.90			7.90	8.70

07 05 Common Work Results for Thermal and Moisture Protection

07 05 05 – Selective Demolition

07 05 05.10 Selective Demo., Thermal and Moist. Protection

			Crew	Daily Output	Labor-Hours	Unit	Material	2010 Bare Costs Labor	Equipment	Total	Total Incl O&P
0010	**SELECTIVE DEMOLITION, THERMAL AND MOISTURE PROTECTION**										
0020	Caulking/sealant, to 1" x 1" joint	R024119-10	1 Clab	600	.013	L.F.		.44		.44	.72
0120	Downspouts, including hangers			350	.023	"		.76		.76	1.24
0220	Flashing, sheet metal			290	.028	S.F.		.91		.91	1.50
0420	Gutters, aluminum or wood, edge hung			240	.033	L.F.		1.10		1.10	1.81
0520	Built-in			100	.080	"		2.65		2.65	4.35
0620	Insulation, air/vapor barrier			3500	.002	S.F.		.08		.08	.12
0670	Batts or blankets			1400	.006	C.F.		.19		.19	.31
0720	Foamed or sprayed in place		2 Clab	1000	.016	B.F.		.53		.53	.87
0770	Loose fitting		1 Clab	3000	.003	C.F.		.09		.09	.15
0870	Rigid board			3450	.002	B.F.		.08		.08	.13
1120	Roll roofing, cold adhesive			12	.667	Sq.		22		22	36
1170	Roof accessories, adjustable metal chimney flashing			9	.889	Ea.		29.50		29.50	48.50
1325	Plumbing vent flashing			32	.250	"		8.30		8.30	13.60
1375	Ridge vent strip, aluminum			310	.026	L.F.		.85		.85	1.40
1620	Skylight to 10 S.F.			8	1	Ea.		33		33	54.50
2120	Roof edge, aluminum soffit and fascia			570	.014	L.F.		.46		.46	.76
2170	Concrete coping, up to 12" wide		2 Clab	160	.100			3.31		3.31	5.45
2220	Drip edge		1 Clab	1650	.005			.16		.16	.26
2270	Gravel stop			1650	.005			.16		.16	.26
2370	Sheet metal coping, up to 12" wide		↓	240	.033	↓		1.10		1.10	1.81
2470	Roof insulation board, over 2" thick		B-2	7800	.005	B.F.		.17		.17	.28
2520	Up to 2" thick		"	3900	.010	S.F.		.34		.34	.56
2620	Roof ventilation, louvered gable vent		1 Clab	16	.500	Ea.		16.55		16.55	27
2670	Remove, roof hatch		G-3	15	2.133			87.50		87.50	140
2675	Rafter vents		1 Clab	960	.008	↓		.28		.28	.45
2720	Soffit vent and/or fascia vent			575	.014	L.F.		.46		.46	.76
2775	Soffit vent strip, aluminum, 3" to 4" wide			160	.050			1.66		1.66	2.72
2820	Roofing accessories, shingle moulding, to 1" x 4"		↓	1600	.005			.17		.17	.27
2870	Cant strip		B-2	2000	.020			.67		.67	1.10
2920	Concrete block walkway		1 Clab	230	.035	↓		1.15		1.15	1.89
3070	Roofing, felt paper, 15#			70	.114	Sq.		3.78		3.78	6.20
3125	#30 felt		↓	30	.267	"		8.85		8.85	14.50
3170	Asphalt shingles, 1 layer		B-2	3500	.011	S.F.		.38		.38	.63
3180	2 layers			1750	.023	"		.77		.77	1.26
3370	Modified bitumen			26	1.538	Sq.		51.50		51.50	84.50

07 05 Common Work Results for Thermal and Moisture Protection

07 05 05 – Selective Demolition

07 05 05.10 Selective Demo., Thermal and Moist. Protection

		Crew	Daily Output	Labor-Hours	Unit	Material	2010 Bare Costs Labor	Equipment	Total	Total Incl O&P
3420	Built-up, no gravel, 3 ply	B-2	25	1.600	Sq.		53.50		53.50	88
3470	4 ply		21	1.905	▼		64		64	105
3620	5 ply		1600	.025	S.F.		.84		.84	1.38
3720	5 ply, with gravel		890	.045			1.51		1.51	2.47
3725	Gravel removal, minimum		5000	.008			.27		.27	.44
3730	Maximum		2000	.020			.67		.67	1.10
3870	Fiberglass sheet		1200	.033			1.12		1.12	1.83
4120	Slate shingles		1900	.021	▼		.71		.71	1.16
4170	Ridge shingles, clay or slate		2000	.020	L.F.		.67		.67	1.10
4320	Single ply membrane, attached at seams		52	.769	Sq.		26		26	42.50
4370	Ballasted		75	.533			17.85		17.85	29.50
4420	Fully adhered	▼	39	1.026	▼		34.50		34.50	56.50
4550	Roof hatch, 2'-6" x 3'-0"	1 Clab	10	.800	Ea.		26.50		26.50	43.50
4670	Wood shingles	B-2	2200	.018	S.F.		.61		.61	1
4820	Sheet metal roofing	"	2150	.019			.62		.62	1.02
4970	Siding, horizontal wood clapboards	1 Clab	380	.021			.70		.70	1.14
5025	Exterior insulation finish system	"	120	.067			2.21		2.21	3.62
5070	Tempered hardboard, remove and reset	1 Carp	380	.021			.87		.87	1.44
5120	Tempered hardboard sheet siding	"	375	.021	▼		.89		.89	1.46
5170	Metal, corner strips	1 Clab	850	.009	L.F.		.31		.31	.51
5225	Horizontal strips		444	.018	S.F.		.60		.60	.98
5320	Vertical strips		400	.020			.66		.66	1.09
5520	Wood shingles		350	.023			.76		.76	1.24
5620	Stucco siding		360	.022			.74		.74	1.21
5670	Textured plywood		725	.011			.37		.37	.60
5720	Vinyl siding		510	.016	▼		.52		.52	.85
5770	Corner strips		900	.009	L.F.		.29		.29	.48
5870	Wood, boards, vertical	▼	400	.020	S.F.		.66		.66	1.09
5920	Waterproofing, protection / drain board	2 Clab	3900	.004	B.F.		.14		.14	.22
5970	Over 1/2" thick		1750	.009	S.F.		.30		.30	.50
6020	To 1/2" thick	▼	2000	.008	"		.26		.26	.43
9000	Minimum labor/equipment charge	1 Clab	2	4	Job		132		132	217

07 11 Dampproofing

07 11 13 – Bituminous Dampproofing

07 11 13.10 Bituminous Asphalt Coating

		Crew	Daily Output	Labor-Hours	Unit	Material	Labor	Equipment	Total	Total Incl O&P
0010	**BITUMINOUS ASPHALT COATING**									
0030	Brushed on, below grade, 1 coat	1 Rofc	665	.012	S.F.	.20	.43		.63	.97
0100	2 coat		500	.016		.40	.57		.97	1.44
0300	Sprayed on, below grade, 1 coat, 25.6 S.F./gal.		830	.010		.20	.34		.54	.82
0400	2 coat, 20.5 S.F./gal.		500	.016		.39	.57		.96	1.43
0600	Troweled on, asphalt with fibers, 1/16" thick		500	.016		.42	.57		.99	1.47
0700	1/8" thick		400	.020		.75	.71		1.46	2.08
1000	1/2" thick		350	.023	▼	2.44	.81		3.25	4.11
9000	Minimum labor/equipment charge	▼	3	2.667	Job		94.50		94.50	166

07 12 Built-up Bituminous Waterproofing

07 12 13 – Built-Up Asphalt Waterproofing

07 12 13.20 Membrane Waterproofing

		Crew	Daily Output	Labor-Hours	Unit	Material	2010 Bare Costs Labor	Equipment	Total	Total Incl O&P
0010	**MEMBRANE WATERPROOFING**									
0012	On slabs, 1 ply, felt, mopped	G-1	3000	.019	S.F.	.38	.62	.16	1.16	1.68
0015	On walls, 1 ply, felt, mopped		3000	.019		.38	.62	.16	1.16	1.68
0100	On slabs, 1 ply, glass fiber fabric, mopped		2100	.027		.39	.88	.23	1.50	2.23
0105	On walls, 1 ply, glass fiber fabric, mopped		2100	.027		.39	.88	.23	1.50	2.23
0300	On slabs, 2 ply, felt, mopped		2500	.022		.77	.74	.19	1.70	2.35
0305	On walls, 2 ply, felt, mopped		2500	.022		.77	.74	.19	1.70	2.35
0400	On slabs, 2 ply, glass fiber fabric, mopped		1650	.034		.88	1.12	.29	2.29	3.27
0405	On walls, 2 ply, glass fiber fabric, mopped		1650	.034		.88	1.12	.29	2.29	3.27
0600	On slabs, 3 ply, felt, mopped		2100	.027		1.15	.88	.23	2.26	3.07
0605	On walls, 3 ply, felt, mopped		2100	.027		1.15	.88	.23	2.26	3.07
0700	On slabs, 3 ply, glass fiber fabric, mopped		1550	.036		1.17	1.19	.31	2.67	3.73
0705	On walls, 3 ply, glass fiber fabric, mopped		1550	.036		1.17	1.19	.31	2.67	3.73
0710	Asphaltic hardboard protection board, 1/8" thick	2 Rofc	500	.032		.40	1.13		1.53	2.44
0715	1/4" thick		450	.036		.68	1.26		1.94	2.97
1000	1/4" EPS membrane protection board		3500	.005		.29	.16		.45	.59
1050	3/8" thick		3500	.005		.33	.16		.49	.64
1060	1/2" thick		3500	.005		.38	.16		.54	.69
1070	Fiberglass fabric, black, 20/10 mesh		116	.138	Sq.	20	4.88		24.88	30.50
1080	White, 20/10 mesh		116	.138	"	20	4.88		24.88	30.50
9000	Minimum labor/equipment charge		2	8	Job		283		283	500

07 13 Sheet Waterproofing

07 13 53 – Elastomeric Sheet Waterproofing

07 13 53.10 Elastomeric Sheet Waterproofing and Access.

		Crew	Daily Output	Labor-Hours	Unit	Material	2010 Bare Costs Labor	Equipment	Total	Total Incl O&P
0010	**ELASTOMERIC SHEET WATERPROOFING AND ACCESS.**									
0090	EPDM, plain, 45 mils thick	2 Rofc	580	.028	S.F.	1.31	.98		2.29	3.16
0100	60 mils thick		570	.028		1.36	.99		2.35	3.25
0300	Nylon reinforced sheets, 45 mils thick		580	.028		1.27	.98		2.25	3.12
0400	60 mils thick		570	.028		1.58	.99		2.57	3.49
0600	Vulcanizing splicing tape for above, 2" wide				C.L.F.	48			48	53
0700	4" wide				"	110			110	121
0900	Adhesive, bonding, 60 SF per gal				Gal.	21			21	23
1000	Splicing, 75 SF per gal				"	31			31	34
1200	Neoprene sheets, plain, 45 mils thick	2 Rofc	580	.028	S.F.	1.68	.98		2.66	3.57
1300	60 mils thick		570	.028		2.87	.99		3.86	4.91
1500	Nylon reinforced, 45 mils thick		580	.028		1.75	.98		2.73	3.65
1600	60 mils thick		570	.028		2.25	.99		3.24	4.23
1800	120 mils thick		500	.032		4	1.13		5.13	6.40
1900	Adhesive, splicing, 150 S.F. per gal. per coat				Gal.	20			20	22
2100	Fiberglass reinforced, fluid applied, 1/8" thick	2 Rofc	500	.032	S.F.	1.90	1.13		3.03	4.09
2200	Polyethylene and rubberized asphalt sheets, 1/8" thick		550	.029		.76	1.03		1.79	2.65
2400	Polyvinyl chloride sheets, plain, 10 mils thick		580	.028		.17	.98		1.15	1.91
2500	20 mils thick		570	.028		.28	.99		1.27	2.06
2700	30 mils thick		560	.029		.39	1.01		1.40	2.21
3000	Adhesives, trowel grade, 40-100 SF per gal				Gal.	22			22	24
3100	Brush grade, 100-250 SF per gal.				"	22			22	24
3300	Bitumen modified polyurethane, fluid applied, 55 mils thick	2 Rofc	665	.024	S.F.	.78	.85		1.63	2.36
3600	Vinyl plastic, sprayed on, 25 to 40 mils thick		475	.034	"	1.21	1.19		2.40	3.43
9000	Minimum labor/equipment charge		2	8	Job		283		283	500

07 16 Cementitious and Reactive Waterproofing

07 16 16 – Crystalline Waterproofing

07 16 16.20 Cementitious Waterproofing

		Crew	Daily Output	Labor-Hours	Unit	Material	2010 Bare Costs Labor	Equipment	Total	Total Incl O&P
0010	**CEMENTITIOUS WATERPROOFING**									
0020	1/8" application, sprayed on	G-2A	1000	.024	S.F.	1.64	.76	.69	3.09	3.86
0030	2 coat, cementitious/metallic slurry, troweled, 1/4" thick	1 Cefi	2.48	3.226	C.S.F.	24	128		152	227
0040	3 coat, 3/8" thick		1.84	4.348		42.50	173		215.50	315
0050	4 coat, 1/2" thick	↓	1.20	6.667	↓	57	265		322	480

07 19 Water Repellents

07 19 19 – Silicone Water Repellents

07 19 19.10 Silicone Based Water Repellents

		Crew	Daily Output	Labor-Hours	Unit	Material	Labor	Equipment	Total	Total Incl O&P
0010	**SILICONE BASED WATER REPELLENTS**									
0020	Water base liquid, roller applied	2 Rofc	7000	.002	S.F.	.65	.08		.73	.86
0200	Silicone or stearate, sprayed on CMU, 1 coat	1 Rofc	4000	.002		.36	.07		.43	.52
0300	2 coats		3000	.003	↓	.73	.09		.82	.97
9000	Minimum labor/equipment charge	↓	3	2.667	Job		94.50		94.50	166

07 21 Thermal Insulation

07 21 13 – Board Insulation

07 21 13.10 Rigid Insulation

			Crew	Daily Output	Labor-Hours	Unit	Material	Labor	Equipment	Total	Total Incl O&P
0010	**RIGID INSULATION**, for walls										
0040	Fiberglass, 1.5#/CF, unfaced, 1" thick, R4.1	G	1 Carp	1000	.008	S.F.	.44	.33		.77	1.03
0060	1-1/2" thick, R6.2	G		1000	.008		.64	.33		.97	1.25
0080	2" thick, R8.3	G		1000	.008		.69	.33		1.02	1.31
0120	3" thick, R12.4	G		800	.010		.80	.42		1.22	1.56
0370	3#/CF, unfaced, 1" thick, R4.3	G		1000	.008		.49	.33		.82	1.09
0390	1-1/2" thick, R6.5	G		1000	.008		.95	.33		1.28	1.60
0400	2" thick, R8.7	G		890	.009		1.15	.37		1.52	1.88
0420	2-1/2" thick, R10.9	G		800	.010		1.01	.42		1.43	1.79
0440	3" thick, R13	G		800	.010		1.01	.42		1.43	1.79
0520	Foil faced, 1" thick, R4.3	G		1000	.008		.92	.33		1.25	1.56
0540	1-1/2" thick, R6.5	G		1000	.008		1.36	.33		1.69	2.05
0560	2" thick, R8.7	G		890	.009		1.71	.37		2.08	2.49
0580	2-1/2" thick, R10.9	G		800	.010		2.02	.42		2.44	2.90
0600	3" thick, R13	G		800	.010		2.19	.42		2.61	3.09
1500	Foamglass, 1-1/2" thick, R4.5	G		800	.010		1.40	.42		1.82	2.22
1550	3" thick, R9	G	↓	730	.011	↓	3.45	.46		3.91	4.55
1600	Isocyanurate, 4' x 8' sheet, foil faced, both sides										
1610	1/2" thick	G	1 Carp	800	.010	S.F.	.31	.42		.73	1.02
1620	5/8" thick	G		800	.010		.34	.42		.76	1.05
1630	3/4" thick	G		800	.010		.36	.42		.78	1.08
1640	1" thick	G		800	.010		.55	.42		.97	1.29
1650	1-1/2" thick	G		730	.011		.67	.46		1.13	1.49
1660	2" thick	G		730	.011		.84	.46		1.30	1.67
1670	3" thick	G		730	.011		1.90	.46		2.36	2.84
1680	4" thick	G		730	.011		2.14	.46		2.60	3.10
1700	Perlite, 1" thick, R2.77	G		800	.010		.30	.42		.72	1.01
1750	2" thick, R5.55	G		730	.011		.60	.46		1.06	1.41
1900	Extruded polystyrene, 25 PSI compressive strength, 1" thick, R5	G		800	.010		.51	.42		.93	1.24
1940	2" thick R10	G		730	.011		1.04	.46		1.50	1.89
1960	3" thick, R15	G	↓	730	.011	↓	1.40	.46		1.86	2.29

07 21 Thermal Insulation

07 21 13 – Board Insulation

07 21 13.10 Rigid Insulation

		Crew	Daily Output	Labor-Hours	Unit	Material	2010 Bare Costs Labor	Equipment	Total	Total Incl O&P
2100	Expanded polystyrene, 1" thick, R3.85 [G]	1 Carp	800	.010	S.F.	.36	.42		.78	1.08
2120	2" thick, R7.69 [G]		730	.011		.72	.46		1.18	1.54
2140	3" thick, R11.49 [G]		730	.011		1.08	.46		1.54	1.94
9000	Minimum labor/equipment charge		4	2	Job		83		83	137

07 21 13.13 Foam Board Insulation

		Crew	Daily Output	Labor-Hours	Unit	Material	Labor	Equipment	Total	Total Incl O&P
0010	**FOAM BOARD INSULATION**									
0600	Polystyrene, expanded, 1" thick, R4 [G]	1 Carp	680	.012	S.F.	.36	.49		.85	1.20
0700	2" thick, R8 [G]		675	.012	"	.72	.49		1.21	1.60
9000	Minimum labor/equipment charge		4	2	Job		83		83	137

07 21 16 – Blanket Insulation

07 21 16.10 Blanket Insulation for Floors/Ceilings

		Crew	Daily Output	Labor-Hours	Unit	Material	Labor	Equipment	Total	Total Incl O&P
0010	**BLANKET INSULATION FOR FLOORS/CEILINGS**									
0020	Including spring type wire fasteners									
2000	Fiberglass, blankets or batts, paper or foil backing									
2100	3-1/2" thick, R13 [G]	1 Carp	700	.011	S.F.	.38	.47		.85	1.20
2150	6-1/4" thick, R19 [G]		600	.013		.48	.55		1.03	1.44
2210	9-1/2" thick, R 30 [G]		500	.016		.69	.66		1.35	1.85
2220	12" thick, R 38 [G]		475	.017		1	.70		1.70	2.25
3000	Unfaced, 3-1/2" thick, R13 [G]		600	.013		.30	.55		.85	1.24
3010	6-1/4" thick, R19 [G]		500	.016		.45	.66		1.11	1.59
3020	9-1/2" thick, R30 [G]		450	.018		.69	.74		1.43	1.97
3030	12" thick, R38 [G]		425	.019		.81	.78		1.59	2.17
9000	Minimum labor/equipment charge		4	2	Job		83		83	137

07 21 16.20 Blanket Insulation for Walls

		Crew	Daily Output	Labor-Hours	Unit	Material	Labor	Equipment	Total	Total Incl O&P
0010	**BLANKET INSULATION FOR WALLS**									
0020	Kraft faced fiberglass, 3-1/2" thick, R11, 15" wide [G]	1 Carp	1350	.006	S.F.	.28	.25		.53	.71
0030	23" wide [G]		1600	.005		.28	.21		.49	.65
0060	R13, 11" wide [G]		1150	.007		.33	.29		.62	.84
0080	15" wide [G]		1350	.006		.33	.25		.58	.76
0100	23" wide [G]		1600	.005		.33	.21		.54	.70
0110	R-15, 11" wide [G]		1150	.007		.45	.29		.74	.98
0120	15" wide [G]		1350	.006		.45	.25		.70	.90
0130	23" wide [G]		1600	.005		.45	.21		.66	.84
0140	6" thick, R19, 11" wide [G]		1150	.007		.43	.29		.72	.95
0160	15" wide [G]		1350	.006		.43	.25		.68	.87
0180	23" wide [G]		1600	.005		.43	.21		.64	.81
0182	R21, 11" wide [G]		1150	.007		.65	.29		.94	1.20
0184	15" wide [G]		1350	.006		.65	.25		.90	1.12
0186	23" wide [G]		1600	.005		.65	.21		.86	1.06
0188	9" thick, R-30, 11" wide [G]		985	.008		.70	.34		1.04	1.32
0200	15" wide [G]		1150	.007		.70	.29		.99	1.25
0220	23" wide [G]		1350	.006		.70	.25		.95	1.17
0230	12" thick, R38, 11" wide [G]		985	.008		1	.34		1.34	1.65
0240	15" wide [G]		1150	.007		1	.29		1.29	1.58
0260	23" wide [G]		1350	.006		1	.25		1.25	1.50
0410	Foil faced fiberglass, 3-1/2" thick, R13, 11" wide [G]		1150	.007		.54	.29		.83	1.07
0420	15" wide [G]		1350	.006		.54	.25		.79	.99
0440	23" wide [G]		1600	.005		.54	.21		.75	.93
0442	R15, 11" wide [G]		1150	.007		.56	.29		.85	1.10
0444	15" wide [G]		1350	.006		.56	.25		.81	1.02
0446	23" wide [G]		1600	.005		.56	.21		.77	.96
0448	6" thick, R19, 11" wide [G]		1150	.007		.77	.29		1.06	1.33

07 21 Thermal Insulation

07 21 16 – Blanket Insulation

07 21 16.20 Blanket Insulation for Walls

		Crew	Daily Output	Labor-Hours	Unit	Material	2010 Bare Costs Labor	Equipment	Total	Total Incl O&P
0460	15" wide [G]	1 Carp	1350	.006	S.F.	.77	.25		1.02	1.25
0480	23" wide [G]		1600	.005		.77	.21		.98	1.19
0482	R-21, 11" wide [G]		1150	.007		.82	.29		1.11	1.38
0484	15" wide [G]		1350	.006		.82	.25		1.07	1.30
0486	23" wide [G]		1600	.005		.82	.21		1.03	1.24
0488	9" thick, R-30, 11" wide [G]		985	.008		.89	.34		1.23	1.53
0500	9" thick, R30, 15" wide [G]		1150	.007		.89	.29		1.18	1.46
0550	23" wide [G]		1350	.006		.89	.25		1.14	1.38
0560	12" thick, R-38, 11" wide [G]		985	.008		.90	.34		1.24	1.54
0570	15" wide [G]		1150	.007		.90	.29		1.19	1.47
0580	23" wide [G]		1350	.006		.90	.25		1.15	1.39
0620	Unfaced fiberglass, 3-1/2" thick, R-13, 11" wide [G]		1150	.007		.30	.29		.59	.81
0820	15" wide [G]		1350	.006		.30	.25		.55	.73
0830	23" wide [G]		1600	.005		.30	.21		.51	.67
0832	R15, 11" wide [G]		1150	.007		.44	.29		.73	.96
0836	23" wide [G]		1600	.005		.44	.21		.65	.82
0838	6" thick, R19, 11" wide [G]		1150	.007		.45	.29		.74	.98
0860	15" wide [G]		1150	.007		.45	.29		.74	.98
0880	23" wide [G]		1350	.006		.45	.25		.70	.90
0882	R-21, 11" wide [G]		1150	.007		.46	.29		.75	.99
0886	15" wide [G]		1350	.006		.46	.25		.71	.91
0888	23" wide [G]		1600	.005		.46	.21		.67	.85
0890	9" thick, R30, 11" wide [G]		985	.008		.64	.34		.98	1.25
0900	15" wide [G]		1150	.007		.64	.29		.93	1.18
0920	23" wide [G]		1350	.006		.64	.25		.89	1.10
0930	12" thick, R38, 11" wide [G]		985	.008		.81	.34		1.15	1.44
0940	15" wide [G]		1000	.008		.81	.33		1.14	1.44
0960	23" wide [G]		1150	.007		.81	.29		1.10	1.37
1300	Mineral fiber batts, kraft faced									
1320	3-1/2" thick, R12 [G]	1 Carp	1600	.005	S.F.	.38	.21		.59	.76
1340	6" thick, R19 [G]		1600	.005		.48	.21		.69	.87
1380	10" thick, R30 [G]		1350	.006		.73	.25		.98	1.20
1850	Friction fit wire insulation supports, 16" O.C.		960	.008	Ea.	.05	.35		.40	.63
9000	Minimum labor/equipment charge		4	2	Job		83		83	137

07 21 23 – Loose-Fill Insulation

07 21 23.10 Poured Loose-Fill Insulation

		Crew	Daily Output	Labor-Hours	Unit	Material	Labor	Equipment	Total	Total Incl O&P
0010	**POURED LOOSE-FILL INSULATION**									
0020	Cellulose fiber, R3.8 per inch [G]	1 Carp	200	.040	C.F.	.60	1.66		2.26	3.39
0040	Ceramic type (perlite), R3.2 per inch [G]		200	.040		1.75	1.66		3.41	4.66
0080	Fiberglass wool, R4 per inch [G]		200	.040		.46	1.66		2.12	3.24
0100	Mineral wool, R3 per inch [G]		200	.040		.39	1.66		2.05	3.16
0300	Polystyrene, R4 per inch [G]		200	.040		3.09	1.66		4.75	6.15
0400	Vermiculite or perlite, R2.7 per inch [G]		200	.040		1.75	1.66		3.41	4.66
9000	Minimum labor/equipment charge		4	2	Job		83		83	137

07 21 23.20 Masonry Loose-Fill Insulation

		Crew	Daily Output	Labor-Hours	Unit	Material	Labor	Equipment	Total	Total Incl O&P
0010	**MASONRY LOOSE-FILL INSULATION**, vermiculite or perlite									
0100	In cores of concrete block, 4" thick wall, .115 CF/SF [G]	D-1	4800	.003	S.F.	.20	.13		.33	.42
0200	6" thick wall, .175 CF/SF [G]		3000	.005		.31	.20		.51	.66
0300	8" thick wall, .258 CF/SF [G]		2400	.007		.45	.25		.70	.90
0400	10" thick wall, .340 CF/SF [G]		1850	.009		.60	.33		.93	1.17
0500	12" thick wall, .422 CF/SF [G]		1200	.013		.74	.50		1.24	1.62
0550	For sand fill, deduct from above [G]					70%				

07 21 Thermal Insulation

07 21 23 – Loose-Fill Insulation

07 21 23.20 Masonry Loose-Fill Insulation		Crew	Daily Output	Labor-Hours	Unit	Material	2010 Bare Costs Labor	Equipment	Total	Total Incl O&P
0600	Poured cavity wall, vermiculite or perlite, water repellant	G	D-1	250	.064	C.F.	1.75	2.41		4.16	5.80
0700	Foamed in place, urethane in 2-5/8" cavity	G	G-2A	1035	.023	S.F.	.42	.73	.67	1.82	2.46
0800	For each 1" added thickness, add	G	"	2372	.010	"	.12	.32	.29	.73	1

07 21 26 – Blown Insulation

07 21 26.10 Blown Insulation

0010	**BLOWN INSULATION** Ceilings, with open access										
0020	Cellulose, 3-1/2" thick, R13	G	G-4	5000	.005	S.F.	.19	.16	.08	.43	.57
0030	5-3/16" thick, R19	G		3800	.006		.28	.21	.10	.59	.78
0050	6-1/2" thick, R22	G		3000	.008		.36	.27	.13	.76	.99
0100	8-11/16" thick, R30	G		2600	.009		.48	.31	.15	.94	1.21
1000	Fiberglass, 5.5" thick, R11	G		3800	.006		.19	.21	.10	.50	.68
1050	6" thick, R12	G		3000	.008		.23	.27	.13	.63	.84
1100	8.8" thick, R19	G		2200	.011		.33	.37	.18	.88	1.17
1200	10" thick, R22	G		1800	.013		.38	.45	.22	1.05	1.40
1300	11.5" thick, R26	G		1500	.016		.46	.54	.27	1.27	1.69
1350	13" thick, R30	G		1400	.017		.50	.58	.28	1.36	1.81
1450	16" thick, R38	G		1145	.021		.61	.71	.35	1.67	2.21
1500	20" thick, R49	G		920	.026		.77	.88	.43	2.08	2.77
2000	Mineral wool, 4" thick, R12	G		3500	.007		.23	.23	.11	.57	.76
2050	6" thick, R17	G		2500	.010		.26	.32	.16	.74	1
2100	9" thick, R23	G		1750	.014		.34	.46	.23	1.03	1.38
9000	Minimum labor/equipment charge			4	6	Job		203	99.50	302.50	445

07 21 27 – Reflective Insulation

07 21 27.10 Reflective Insulation Options

0010	**REFLECTIVE INSULATION OPTIONS**										
0020	Aluminum foil on reinforced scrim	G	1 Carp	19	.421	C.S.F.	14.20	17.50		31.70	44
0100	Reinforced with woven polyolefin	G		19	.421		17.20	17.50		34.70	47.50
0500	With single bubble air space, R8.8	G		15	.533		25.50	22		47.50	64.50
0600	With double bubble air space, R9.8	G		15	.533		26.50	22		48.50	65.50
9000	Minimum labor/equipment charge			4	2	Job		83		83	137

07 21 29 – Sprayed Insulation

07 21 29.10 Sprayed-On Insulation

0010	**SPRAYED-ON INSULATION**										
0020	Fibrous/cementitious, finished wall, 1" thick, R3.7	G	G-2	2050	.012	S.F.	.26	.41	.06	.73	1.01
0100	Attic, 5.2" thick, R19	G		1550	.015	"	.42	.54	.09	1.05	1.42
0200	Fiberglass, R4 per inch, vertical	G		2050	.012	B.F.	.15	.41	.06	.62	.89
0210	Horizontal	G		1550	.015	"	.15	.54	.09	.78	1.13
0300	Closed cell, spray polyurethane foam, 2 pounds per cubic foot density										
0310	1" thick	G	G-2A	6000	.004	S.F.	.41	.13	.12	.66	.80
0320	2" thick	G		3000	.008		.82	.25	.23	1.30	1.58
0330	3" thick	G		2000	.012		1.23	.38	.35	1.96	2.39
0335	3-1/2" thick	G		1715	.014		1.44	.44	.40	2.28	2.78
0340	4" thick	G		1500	.016		1.64	.51	.46	2.61	3.19
0350	5" thick	G		1200	.020		2.05	.63	.58	3.26	3.98
0355	5-1/2" thick	G		1090	.022		2.26	.70	.63	3.59	4.38
0360	6" thick	G		1000	.024		2.46	.76	.69	3.91	4.77
9000	Minimum labor/equipment charge		G-2	2	12	Job		415	66.50	481.50	745

07 22 Roof and Deck Insulation

07 22 16 – Roof Board Insulation

07 22 16.10 Roof Deck Insulation		Crew	Daily Output	Labor-Hours	Unit	Material	2010 Bare Costs Labor	Equipment	Total	Total Incl O&P
0010	**ROOF DECK INSULATION**									
0020	Fiberboard low density, 1/2" thick R1.39 G	1 Rofc	1000	.008	S.F.	.24	.28		.52	.76
0030	1" thick R2.78 G		800	.010		.42	.35		.77	1.08
0080	1 1/2" thick R4.17 G		800	.010		.64	.35		.99	1.32
0100	2" thick R5.56 G		800	.010		.85	.35		1.20	1.56
0110	Fiberboard high density, 1/2" thick R1.3 G		1000	.008		.22	.28		.50	.74
0120	1" thick R2.5 G		800	.010		.44	.35		.79	1.10
0130	1-1/2" thick R3.8 G		800	.010		.66	.35		1.01	1.35
0200	Fiberglass, 3/4" thick R2.78 G		1000	.008		.56	.28		.84	1.12
0400	15/16" thick R3.70 G		1000	.008		.74	.28		1.02	1.31
0460	1-1/16" thick R4.17 G		1000	.008		.93	.28		1.21	1.52
0600	1-5/16" thick R5.26 G		1000	.008		1.26	.28		1.54	1.89
0650	2-1/16" thick R8.33 G		800	.010		1.35	.35		1.70	2.11
0700	2-7/16" thick R10 G		800	.010		1.55	.35		1.90	2.33
1500	Foamglass, 1-1/2" thick R4.5 G		800	.010		1.57	.35		1.92	2.35
1530	3" thick R9 G		700	.011	▼	2.89	.40		3.29	3.89
1600	Tapered for drainage G		600	.013	B.F.	1.16	.47		1.63	2.11
1650	Perlite, 1/2" thick R1.32 G		1050	.008	S.F.	.34	.27		.61	.85
1655	3/4" thick R2.08 G		800	.010		.37	.35		.72	1.03
1660	1" thick R2.78 G		800	.010		.46	.35		.81	1.13
1670	1-1/2" thick R4.17 G		800	.010		.48	.35		.83	1.15
1680	2" thick R5.56 G		700	.011		.80	.40		1.20	1.59
1685	2-1/2" thick R6.67 G		700	.011	▼	.95	.40		1.35	1.76
1690	Tapered for drainage G		800	.010	B.F.	.73	.35		1.08	1.42
1700	Polyisocyanurate, 2#/CF density, 3/4" thick G		1500	.005	S.F.	.44	.19		.63	.81
1705	1" thick G		1400	.006		.51	.20		.71	.92
1715	1-1/2" thick G		1250	.006		.62	.23		.85	1.08
1725	2" thick G		1100	.007		.86	.26		1.12	1.40
1735	2-1/2" thick G		1050	.008		1.01	.27		1.28	1.59
1745	3" thick G		1000	.008		1.24	.28		1.52	1.86
1755	3-1/2" thick G		1000	.008	▼	1.95	.28		2.23	2.65
1765	Tapered for drainage G		1400	.006	B.F.	1.95	.20		2.15	2.51
1900	Extruded Polystyrene									
1910	15 PSI compressive strength, 1" thick, R5 G	1 Rofc	1500	.005	S.F.	.49	.19		.68	.87
1920	2" thick, R10 G		1250	.006		.62	.23		.85	1.08
1930	3" thick R15 G		1000	.008		1.27	.28		1.55	1.90
1932	4" thick R20 G		1000	.008	▼	1.64	.28		1.92	2.30
1934	Tapered for drainage G		1500	.005	B.F.	.53	.19		.72	.91
1940	25 PSI compressive strength, 1" thick R5 G		1500	.005	S.F.	.67	.19		.86	1.07
1942	2" thick R10 G		1250	.006		1.27	.23		1.50	1.80
1944	3" thick R15 G		1000	.008		1.93	.28		2.21	2.62
1946	4" thick R20 G		1000	.008	▼	2.72	.28		3	3.49
1948	Tapered for drainage G		1500	.005	B.F.	.56	.19		.75	.95
1950	40 psi compressive strength, 1" thick R5 G		1500	.005	S.F.	.51	.19		.70	.89
1952	2" thick R10 G		1250	.006		.96	.23		1.19	1.46
1954	3" thick R15 G		1000	.008		1.40	.28		1.68	2.04
1956	4" thick R20 G		1000	.008	▼	1.88	.28		2.16	2.57
1958	Tapered for drainage G		1400	.006	B.F.	.71	.20		.91	1.14
1960	60 PSI compressive strength, 1" thick R5 G		1450	.006	S.F.	.59	.20		.79	.99
1962	2" thick R10 G		1200	.007		1.05	.24		1.29	1.58
1964	3" thick R15 G		975	.008		1.55	.29		1.84	2.22
1966	4" thick R20 G		950	.008	▼	2.17	.30		2.47	2.92
1968	Tapered for drainage G		1400	.006	B.F.	.85	.20		1.05	1.30

07 22 Roof and Deck Insulation

07 22 16 – Roof Board Insulation

07 22 16.10 Roof Deck Insulation

		Crew	Daily Output	Labor-Hours	Unit	Material	2010 Bare Costs Labor	Equipment	Total	Total Incl O&P
2010	Expanded polystyrene, 1#/CF density, 3/4" thick R2.89 G	1 Rofc	1500	.005	S.F.	.27	.19		.46	.63
2020	1" thick R3.85 G		1500	.005		.36	.19		.55	.73
2100	2" thick R7.69 G		1250	.006		.72	.23		.95	1.19
2110	3" thick R11.49 G		1250	.006		1.08	.23		1.31	1.59
2120	4" thick R15.38 G		1200	.007		1.44	.24		1.68	2
2130	5" thick R19.23 G		1150	.007		1.80	.25		2.05	2.41
2140	6" thick R23.26 G		1150	.007		2.16	.25		2.41	2.81
2150	Tapered for drainage G		1500	.005	B.F.	.54	.19		.73	.92
2400	Composites with 2" EPS									
2410	1" fiberboard G	1 Rofc	950	.008	S.F.	1.24	.30		1.54	1.89
2420	7/16" oriented strand board G		800	.010		.94	.35		1.29	1.65
2430	1/2" plywood G		800	.010		1.14	.35		1.49	1.87
2440	1" perlite G		800	.010		1.18	.35		1.53	1.92
2450	Composites with 1-1/2" polyisocyanurate									
2460	1" fiberboard G	1 Rofc	800	.010	S.F.	1.14	.35		1.49	1.87
2470	1" perlite G		850	.009		1.08	.33		1.41	1.78
2480	7/16" oriented strand board G		800	.010		.84	.35		1.19	1.54
9000	Minimum labor/equipment charge		3.25	2.462	Job		87		87	153

07 24 Exterior Insulation and Finish Systems

07 24 13 – Polymer Based Exterior Insulation and Finish Systems

07 24 13.10 Exterior Insulation and Finish Systems

		Crew	Daily Output	Labor-Hours	Unit	Material	2010 Bare Costs Labor	Equipment	Total	Total Incl O&P
0010	**EXTERIOR INSULATION AND FINISH SYSTEMS**									
0095	Field applied, 1" EPS insulation G	J-1	390	.103	S.F.	2.30	3.69	.35	6.34	8.75
0100	With 1/2" cement board sheathing G		268	.149		2.95	5.35	.51	8.81	12.35
0105	2" EPS insulation G		390	.103		2.66	3.69	.35	6.70	9.15
0110	With 1/2" cement board sheathing G		268	.149		3.31	5.35	.51	9.17	12.75
0115	3" EPS insulation G		390	.103		3.02	3.69	.35	7.06	9.55
0120	With 1/2" cement board sheathing G		268	.149		3.67	5.35	.51	9.53	13.15
0125	4" EPS insulation G		390	.103		3.38	3.69	.35	7.42	9.95
0130	With 1/2" cement board sheathing G		268	.149		4.67	5.35	.51	10.53	14.25
0140	Premium finish add		1265	.032		.31	1.14	.11	1.56	2.27
0150	Heavy duty reinforcement add		914	.044		1.08	1.57	.15	2.80	3.85
0160	2.5#/S.Y. metal lath substrate add	1 Lath	75	.107	S.Y.	2.37	3.94		6.31	8.80
0170	3.4#/S.Y. metal lath substrate add	"	75	.107	"	2.58	3.94		6.52	9.05
0180	Color or texture change,	J-1	1265	.032	S.F.	.83	1.14	.11	2.08	2.84
0190	With substrate leveling base coat	1 Plas	530	.015		.83	.56		1.39	1.81
0210	With substrate sealing base coat	1 Pord	1224	.007		.08	.24		.32	.47
0370	V groove shape in panel face				L.F.	.60			.60	.66
0380	U groove shape in panel face				"	.79			.79	.87
0433	Crack repair, acrylic rubber, fluid applied, 20 mils thick	1 Plas	350	.023	S.F.	1.43	.85		2.28	2.93
0437	50 mils thick, reinforced	"	200	.040	"	2.62	1.49		4.11	5.25
0440	For higher than one story, add						25%			

07 25 Weather Barriers

07 25 10 – Weather Barriers or Wraps

07 25 10.10 Weather Barriers

		Crew	Daily Output	Labor-Hours	Unit	Material	2010 Bare Costs Labor	Equipment	Total	Total Incl O&P
0010	**WEATHER BARRIERS**									
0400	Asphalt felt paper, 15#	1 Carp	37	.216	Sq.	5.90	9		14.90	21
0401	Per square foot	"	3700	.002	S.F.	.06	.09		.15	.21
0450	Housewrap, exterior, spun bonded polypropylene									
0470	Small roll	1 Carp	3800	.002	S.F.	.23	.09		.32	.39
0480	Large roll	"	4000	.002	"	.13	.08		.21	.28
2100	Asphalt felt roof deck vapor barrier, class 1 metal decks	1 Rofc	37	.216	Sq.	21.50	7.65		29.15	37.50
2200	For all other decks	"	37	.216		15.85	7.65		23.50	31
2800	Asphalt felt, 50% recycled content, 15 lb, 4 sq per roll	1 Carp	36	.222		4.77	9.25		14.02	20.50
2810	30 lb, 2 sq per roll	"	36	.222		9.55	9.25		18.80	25.50
3000	Building wrap, spunbonded polyethylene	2 Carp	8000	.002	S.F.	.13	.08		.21	.28
9960	Minimum labor/equipment charge	1 Carp	2	4	Job		166		166	273

07 26 Vapor Retarders

07 26 10 – Above-Grade Vapor Retarders

07 26 10.10 Vapor Retarders

			Crew	Daily Output	Labor-Hours	Unit	Material	2010 Bare Costs Labor	Equipment	Total	Total Incl O&P
0010	**VAPOR RETARDERS**										
0020	Aluminum and kraft laminated, foil 1 side	G	1 Carp	37	.216	Sq.	8.80	9		17.80	24.50
0100	Foil 2 sides	G		37	.216		9.50	9		18.50	25
0600	Polyethylene vapor barrier, standard, .002" thick	G		37	.216		1.09	9		10.09	15.95
0700	.004" thick	G		37	.216		3.39	9		12.39	18.50
0900	.006" thick	G		37	.216		5.20	9		14.20	20.50
1200	.010" thick	G		37	.216		6.70	9		15.70	22
1300	Clear reinforced, fire retardant, .008" thick	G		37	.216		10.15	9		19.15	26
1350	Cross laminated type, .003" thick	G		37	.216		7.10	9		16.10	22.50
1400	.004" thick	G		37	.216		7.80	9		16.80	23.50
1800	Reinf. waterproof, .002" polyethylene backing, 1 side			37	.216		5.65	9		14.65	21
1900	2 sides			37	.216		7.45	9		16.45	23
2400	Waterproofed kraft with sisal or fiberglass fibers, minimum			37	.216		6.10	9		15.10	21.50
2500	Maximum			37	.216		15.10	9		24.10	31.50
9950	Minimum labor/equipment charge			4	2	Job		83		83	137

07 31 Shingles and Shakes

07 31 13 – Asphalt Shingles

07 31 13.10 Asphalt Roof Shingles

		Crew	Daily Output	Labor-Hours	Unit	Material	2010 Bare Costs Labor	Equipment	Total	Total Incl O&P
0010	**ASPHALT ROOF SHINGLES**									
0100	Standard strip shingles									
0150	Inorganic, class A, 210-235 lb/sq	1 Rofc	5.50	1.455	Sq.	73.50	51.50		125	171
0155	Pneumatic nailed		7	1.143		73.50	40.50		114	152
0200	Organic, class C, 235-240 lb/sq		5	1.600		64	56.50		120.50	170
0205	Pneumatic nailed		6.25	1.280		64	45.50		109.50	150
0250	Standard, laminated multi-layered shingles									
0300	Class A, 240-260 lb/sq	1 Rofc	4.50	1.778	Sq.	86.50	63		149.50	207
0305	Pneumatic nailed		5.63	1.422		86.50	50.50		137	184
0350	Class C, 260-300 lb/square, 4 bundles/square		4	2		84	71		155	218
0355	Pneumatic nailed		5	1.600		84	56.50		140.50	193
0400	Premium, laminated multi-layered shingles									
0450	Class A, 260-300 lb, 4 bundles/sq	1 Rofc	3.50	2.286	Sq.	100	81		181	253
0455	Pneumatic nailed		4.37	1.831		100	65		165	224

07 31 Shingles and Shakes

07 31 13 – Asphalt Shingles

07 31 13.10 Asphalt Roof Shingles		Crew	Daily Output	Labor-Hours	Unit	Material	2010 Bare Costs Labor	Equipment	Total	Total Incl O&P
0500	Class C, 300-385 lb/square, 5 bundles/square	1 Rofc	3	2.667	Sq.	132	94.50		226.50	310
0505	Pneumatic nailed		3.75	2.133		132	75.50		207.50	278
0800	#15 felt underlayment		64	.125		5.90	4.43		10.33	14.25
0825	#30 felt underlayment		58	.138		11.95	4.88		16.83	22
0850	Self adhering polyethylene and rubberized asphalt underlayment		22	.364		60	12.85		72.85	88.50
0900	Ridge shingles		330	.024	L.F.	1.90	.86		2.76	3.60
0905	Pneumatic nailed		412.50	.019	"	1.90	.69		2.59	3.30
1000	For steep roofs (7 to 12 pitch or greater), add						50%			
9000	Minimum labor/equipment charge	1 Rofc	3	2.667	Job		94.50		94.50	166

07 31 16 – Metal Shingles

07 31 16.10 Aluminum Shingles

		Crew	Daily Output	Labor-Hours	Unit	Material	Labor	Equipment	Total	Total Incl O&P
0010	**ALUMINUM SHINGLES**									
0020	Mill finish, .019 thick	1 Carp	5	1.600	Sq.	195	66.50		261.50	325
0100	.020" thick	"	5	1.600		188	66.50		254.50	315
0300	For colors, add					16			16	17.60
0600	Ridge cap, .024" thick	1 Carp	170	.047	L.F.	2.10	1.96		4.06	5.50
0700	End wall flashing, .024" thick		170	.047		1.71	1.96		3.67	5.10
0900	Valley section, .024" thick		170	.047		2.81	1.96		4.77	6.30
1000	Starter strip, .024" thick		400	.020		1.36	.83		2.19	2.87
1200	Side wall flashing, .024" thick		170	.047		1.71	1.96		3.67	5.10
9000	Minimum labor/equipment charge		3	2.667	Job		111		111	182

07 31 16.20 Steel Shingles

		Crew	Daily Output	Labor-Hours	Unit	Material	Labor	Equipment	Total	Total Incl O&P
0010	**STEEL SHINGLES**									
0012	Galvanized, 26 gauge	1 Rots	2.20	3.636	Sq.	232	129		361	485
0200	24 gauge	"	2.20	3.636		244	129		373	495
0300	For colored galvanized shingles, add					55			55	60.50
0500	For 1" factory applied polystyrene insulation, add					40			40	44
9000	Minimum labor/equipment charge	1 Rots	3	2.667	Job		95		95	167

07 31 19 – Mineral-Fiber Cement Shingles

07 31 19.10 Fiber Cement Shingles

		Crew	Daily Output	Labor-Hours	Unit	Material	Labor	Equipment	Total	Total Incl O&P
0010	**FIBER CEMENT SHINGLES**									
0012	Field shingles, 16" x 9.35", 500 lb per square	1 Rofc	2.20	3.636	Sq.	335	129		464	595
0200	Shakes, 16" x 9.35", 550 lb per square		2.20	3.636	"	305	129		434	560
0300	Hip & ridge, 4.75 x 14"		1	8	C.L.F.	830	283		1,113	1,425
0400	Hexagonal, 16" x 16"		3	2.667	Sq.	230	94.50		324.50	420
0500	Square, 16" x 16"		3	2.667		205	94.50		299.50	390
2000	For steep roofs (7/12 pitch or greater), add						50%			
9000	Minimum labor/equipment charge	1 Rofc	3	2.667	Job		94.50		94.50	166

07 31 26 – Slate Shingles

07 31 26.10 Slate Roof Shingles

			Crew	Daily Output	Labor-Hours	Unit	Material	Labor	Equipment	Total	Total Incl O&P
0010	**SLATE ROOF SHINGLES**										
0100	Buckingham Virginia black, 3/16" - 1/4" thick	G	1 Rots	1.75	4.571	Sq.	475	163		638	810
0200	1/4" thick	G		1.75	4.571		475	163		638	810
0900	Pennsylvania black, Bangor, #1 clear	G		1.75	4.571		490	163		653	825
1200	Vermont, unfading, green, mottled green	G		1.75	4.571		480	163		643	810
1300	Semi-weathering green & gray	G		1.75	4.571		350	163		513	665
1400	Purple	G		1.75	4.571		425	163		588	750
1500	Black or gray	G		1.75	4.571		460	163		623	790
2500	Slate roof repair, extensive replacement			1	8		590	285		875	1,150
2600	Repair individual pieces, scattered			19	.421	Ea.	5.90	15		20.90	33
9000	Minimum labor/equipment charge			3	2.667	Job		95		95	167

07 31 Shingles and Shakes

07 31 29 – Wood Shingles and Shakes

07 31 29.13 Wood Shingles

		Crew	Daily Output	Labor-Hours	Unit	Material	2010 Bare Costs Labor	2010 Bare Costs Equipment	Total	Total Incl O&P
0010	**WOOD SHINGLES** R061110-30									
0012	16" No. 1 red cedar shingles, 5" exposure, on roof	1 Carp	2.50	3.200	Sq.	310	133		443	560
0015	Pneumatic nailed		3.25	2.462		310	102		412	510
0200	7-1/2" exposure, on walls		2.05	3.902		207	162		369	495
0205	Pneumatic nailed		2.67	2.996		207	124		331	430
0300	18" No. 1 red cedar perfections, 5-1/2" exposure, on roof		2.75	2.909		265	121		386	490
0305	Pneumatic nailed		3.57	2.241		265	93		358	445
0600	Resquared, and rebutted, 5-1/2" exposure, on roof		3	2.667		236	111		347	440
0605	Pneumatic nailed		3.90	2.051		236	85		321	400
0900	7-1/2" exposure, on walls		2.45	3.265		173	136		309	415
0905	Pneumatic nailed		3.18	2.516		173	105		278	360
1000	Add to above for fire retardant shingles					55			55	60.50
1050	18" long					55			55	60.50
1060	Preformed ridge shingles	1 Carp	400	.020	L.F.	3.60	.83		4.43	5.35
2000	White cedar shingles, 16" long, extras, 5" exposure, on roof		2.40	3.333	Sq.	174	139		313	420
2005	Pneumatic nailed		3.12	2.564		174	107		281	365
2050	5" exposure on walls		2	4		174	166		340	465
2055	Pneumatic nailed		2.60	3.077		174	128		302	400
2100	7-1/2" exposure, on walls		2	4		124	166		290	410
2105	Pneumatic nailed		2.60	3.077		124	128		252	345
2150	"B" grade, 5" exposure on walls		2	4		150	166		316	440
2155	Pneumatic nailed		2.60	3.077		150	128		278	375
2300	For 15# organic felt underlayment on roof, 1 layer, add		64	.125		5.90	5.20		11.10	15
2400	2 layers, add		32	.250		11.75	10.40		22.15	30
2600	For steep roofs (7/12 pitch or greater), add to above						50%			
2700	Panelized systems, No.1 cedar shingles on 5/16" CDX plywood									
2800	On walls, 8' strips, 7" or 14" exposure	2 Carp	700	.023	S.F.	5.55	.95		6.50	7.65
3500	On roofs, 8' strips, 7" or 14" exposure	1 Carp	3	2.667	Sq.	555	111		666	790
3505	Pneumatic nailed		4	2	"	555	83		638	745
9000	Minimum labor/equipment charge		3	2.667	Job		111		111	182

07 31 29.16 Wood Shakes

		Crew	Daily Output	Labor-Hours	Unit	Material	2010 Bare Costs Labor	2010 Bare Costs Equipment	Total	Total Incl O&P
0010	**WOOD SHAKES**									
1100	Hand-split red cedar shakes, 1/2" thick x 24" long, 10" exp. on roof	1 Carp	2.50	3.200	Sq.	165	133		298	400
1105	Pneumatic nailed		3.25	2.462		165	102		267	350
1110	3/4" thick x 24" long, 10" exp. on roof		2.25	3.556		165	148		313	425
1115	Pneumatic nailed		2.92	2.740		165	114		279	370
1200	1/2" thick, 18" long, 8-1/2" exp. on roof		2	4		146	166		312	435
1205	Pneumatic nailed		2.60	3.077		146	128		274	370
1210	3/4" thick x 18" long, 8 1/2" exp. on roof		1.80	4.444		146	185		331	465
1215	Pneumatic nailed		2.34	3.419		146	142		288	395
1255	10" exp. on walls		2	4		141	166		307	430
1260	10" exposure on walls, pneumatic nailed		2.60	3.077		141	128		269	365
1700	Add to above for fire retardant shakes, 24" long					55			55	60.50
1800	18" long					55			55	60.50
1810	Ridge shakes	1 Carp	350	.023	L.F.	3.60	.95		4.55	5.50

07 32 Roof Tiles

07 32 13 – Clay Roof Tiles

07 32 13.10 Clay Tiles

		Crew	Daily Output	Labor-Hours	Unit	Material	2010 Bare Costs Labor	2010 Bare Costs Equipment	Total	Total Incl O&P
0010	**CLAY TILES**									
0200	Lanai tile or Classic tile, 158 pc per sq G	1 Rots	1.65	4.848	Sq.	500	173		673	855
0300	Americana, 158 pc per sq, most colors G		1.65	4.848		660	173		833	1,025
0350	Green, gray or brown G		1.65	4.848		625	173		798	990
0400	Blue G		1.65	4.848		625	173		798	990
0600	Spanish tile, 171 pc per sq, red G		1.80	4.444		325	158		483	635
0800	Buff, green, gray, brown G		1.80	4.444		600	158		758	940
0900	Glazed white		1.80	4.444		665	158		823	1,000
1100	Mission tile, 192 pc per sq, machine scored finish, red G		1.15	6.957		740	248		988	1,250
1700	French tile, 133 pc per sq, smooth finish, red G		1.35	5.926		675	211		886	1,100
1750	Blue or green G		1.35	5.926		870	211		1,081	1,325
1800	Norman black 317 pc per sq G		1	8		1,050	285		1,335	1,650
2200	Williamsburg tile, 158 pc per sq, aged cedar G		1.35	5.926		730	211		941	1,175
2250	Gray or green		1.35	5.926		595	211		806	1,025
3000	For steep roofs (7/12 pitch or greater), add to above						50%			
3010	Clay tile, #15 felt underlayment	1 Rofc	64	.125		5.90	4.43		10.33	14.25
3020	Clay tile, #30 felt underlayment		58	.138		11.95	4.88		16.83	22
3040	Clay tile, polyethylene and rubberized asph. underlayment		22	.364		60	12.85		72.85	88.50
9000	Minimum labor/equipment charge	1 Rots	3	2.667	Job		95		95	167

07 32 16 – Concrete Roof Tiles

07 32 16.10 Concrete Tiles

		Crew	Daily Output	Labor-Hours	Unit	Material	Labor	Equipment	Total	Total Incl O&P
0010	**CONCRETE TILES**									
0020	Corrugated, 13" x 16-1/2", 90 per sq, 950 lb per sq									
0050	Earthtone colors, nailed to wood deck	1 Rots	1.35	5.926	Sq.	103	211		314	485
0150	Blues		1.35	5.926		104	211		315	485
0200	Greens		1.35	5.926		104	211		315	485
0250	Premium colors		1.35	5.926		229	211		440	620
0500	Shakes, 13" x 16-1/2", 90 per sq, 950 lb per sq									
0600	All colors, nailed to wood deck	1 Rots	1.50	5.333	Sq.	270	190		460	630
1500	Accessory pieces, ridge & hip, 10" x 16-1/2", 8 lbs. each	"	120	.067	Ea.	3.64	2.37		6.01	8.20
1700	Rake, 6-1/2" x 16-3/4", 9 lbs. each					3.64			3.64	4
1800	Mansard hip, 10" x 16-1/2", 9.2 lbs. each					3.64			3.64	4
1900	Hip starter, 10" x 16-1/2", 10.5 lbs. each					11.45			11.45	12.60
2000	3 or 4 way apex, 10" each side, 11.5 lbs. each					13.25			13.25	14.55
9000	Minimum labor/equipment charge	1 Rots	3	2.667	Job		95		95	167

07 32 19 – Metal Roof Tiles

07 32 19.10 Metal Roof Tiles

		Crew	Daily Output	Labor-Hours	Unit	Material	Labor	Equipment	Total	Total Incl O&P
0010	**METAL ROOF TILES**									
0020	Accessories included, .032" thick aluminum, mission tile	1 Carp	2.50	3.200	Sq.	815	133		948	1,125
0200	Spanish tiles		3	2.667	"	560	111		671	795
9000	Minimum labor/equipment charge		3	2.667	Job		111		111	182

07 33 Natural Roof Coverings

07 33 63 – Vegetated Roofing

07 33 63.10 Green Roof Systems

			Crew	Daily Output	Labor-Hours	Unit	Material	2010 Bare Costs Labor	Equipment	Total	Total Incl O&P
0010	**GREEN ROOF SYSTEMS**										
0020	Soil mixture for green roof 30% sand, 55% gravel, 15% soil	G									
0100	Hoist and spread soil mixture 4 inch depth up to five stories tall roof	G	B-13B	4000	.014	S.F.	.22	.50	.30	1.02	1.38
0150	6 inch depth	G		2667	.021		.33	.75	.45	1.53	2.07
0200	8 inch depth	G		2000	.028		.44	1	.60	2.04	2.77
0250	10 inch depth	G		1600	.035		.55	1.25	.76	2.56	3.46
0300	12 inch depth	G		1335	.042		.66	1.50	.91	3.07	4.15
0350	Mobilization 55 ton crane to site	G	1 Eqhv	3.60	2.222	Ea.		98.50		98.50	155
0355	Hoisting cost to five stories per day (Avg. 28 picks per day)	G	B-13B	1	56	Day		2,000	1,200	3,200	4,550
0360	Mobilization or demobilization, 100 ton crane to site driver & escort	G	A-3E	2.50	6.400	Ea.		248	60.50	308.50	460
0365	Hoisting cost six to ten stories per day (Avg. 21 picks per day)	G	B-13C	1	56	Day		2,000	1,725	3,725	5,125
0370	Hoist and spread soil mixture 4 inch depth six to ten stories tall roof	G		4000	.014	S.F.	.22	.50	.43	1.15	1.53
0375	6 inch depth	G		2667	.021		.33	.75	.65	1.73	2.28
0380	8 inch depth	G		2000	.028		.44	1	.87	2.31	3.05
0385	10 inch depth	G		1600	.035		.55	1.25	1.08	2.88	3.82
0390	12 inch depth	G		1335	.042		.66	1.50	1.30	3.46	4.58
0400	Green roof edging treated lumber 4" x 4" no hoisting included	G	2 Carp	400	.040	L.F.	1.56	1.66		3.22	4.44
0410	4" x 6"	G		400	.040		2.30	1.66		3.96	5.25
0420	4" x 8"	G		360	.044		4.74	1.85		6.59	8.25
0430	4" x 6" double stacked	G		300	.053		4.60	2.22		6.82	8.70
0500	Green roof edging redwood lumber 4" x 4" no hoisting included	G		400	.040		6.40	1.66		8.06	9.80
0510	4" x 6"	G		400	.040		11.20	1.66		12.86	15.05
0520	4" x 8"	G		360	.044		20.50	1.85		22.35	25.50
0530	4" x 6" double stacked	G		300	.053		11.20	2.22		13.42	16
0600	Planting sedum, light soil, potted, 2-1/4" diameter, two per SF	G	1 Clab	420	.019	S.F.	4.98	.63		5.61	6.55
0610	one per SF	G	"	840	.010		2.49	.32		2.81	3.26
0630	Planting sedum mat per SF including shipping (4000 SF Minimum)	G	4 Clab	4000	.008		6	.26		6.26	7
0640	Installation sedum mat system (no soil required) per SF (4000 SF minimum)	G	"	4000	.008		8.30	.26		8.56	9.60
0645	Note: pricing of sedum mats shipped in full truck loads (4000-5000 SF)	G									

07 41 Roof Panels

07 41 13 – Metal Roof Panels

07 41 13.10 Aluminum Roof Panels

		Crew	Daily Output	Labor-Hours	Unit	Material	2010 Bare Costs Labor	Equipment	Total	Total Incl O&P
0010	**ALUMINUM ROOF PANELS**									
0020	Corrugated or ribbed, .0155" thick, natural	G-3	1200	.027	S.F.	.96	1.10		2.06	2.81
0300	Painted		1200	.027		1.40	1.10		2.50	3.29
0400	Corrugated, .018" thick, on steel frame, natural finish		1200	.027		1.27	1.10		2.37	3.15
0600	Painted		1200	.027		1.55	1.10		2.65	3.46
0700	Corrugated, on steel frame, natural, .024" thick		1200	.027		1.80	1.10		2.90	3.73
0800	Painted		1200	.027		2.19	1.10		3.29	4.16
0900	.032" thick, natural		1200	.027		2.32	1.10		3.42	4.30
1200	Painted		1200	.027		2.97	1.10		4.07	5
9000	Minimum labor/equipment charge	1 Rofc	3	2.667	Job		94.50		94.50	166

07 41 13.20 Steel Roofing Panels

		Crew	Daily Output	Labor-Hours	Unit	Material	2010 Bare Costs Labor	Equipment	Total	Total Incl O&P
0010	**STEEL ROOFING PANELS**									
0012	Corrugated or ribbed, on steel framing, 30 ga galv	G-3	1100	.029	S.F.	1.68	1.20		2.88	3.76
0100	28 ga		1050	.030		1.79	1.25		3.04	3.97
0300	26 ga		1000	.032		1.85	1.32		3.17	4.14
0400	24 ga		950	.034		2.91	1.38		4.29	5.40
0600	Colored, 28 ga		1050	.030		1.68	1.25		2.93	3.85
0700	26 ga		1000	.032		1.99	1.32		3.31	4.29

07 41 Roof Panels

07 41 13 – Metal Roof Panels

07 41 13.20 Steel Roofing Panels		Crew	Daily Output	Labor-Hours	Unit	Material	2010 Bare Costs Labor	Equipment	Total	Total Incl O&P
0710	Flat profile, 1-3/4" standing seams, 10" wide, standard finish, 26 ga	G-3	1000	.032	S.F.	3.89	1.32		5.21	6.40
0715	24 ga		950	.034		4.51	1.38		5.89	7.15
0720	22 ga		900	.036		5.55	1.46		7.01	8.50
0725	Zinc aluminum alloy finish, 26 ga		1000	.032		3.05	1.32		4.37	5.45
0730	24 ga		950	.034		3.64	1.38		5.02	6.20
0735	22 ga		900	.036		4.17	1.46		5.63	6.90
0740	12" wide, standard finish, 26 ga		1000	.032		3.88	1.32		5.20	6.35
0745	24 ga		950	.034		5.10	1.38		6.48	7.80
0750	Zinc aluminum alloy finish, 26 ga		1000	.032		4.41	1.32		5.73	6.95
0755	24 ga		950	.034		3.63	1.38		5.01	6.20
0840	Flat profile, 1" x 3/8" batten, 12" wide, standard finish, 26 ga		1000	.032		3.42	1.32		4.74	5.85
0845	24 ga		950	.034		4.01	1.38		5.39	6.60
0850	22 ga		900	.036		4.80	1.46		6.26	7.65
0855	Zinc aluminum alloy finish, 26 ga		1000	.032		3.28	1.32		4.60	5.70
0860	24 ga		950	.034		3.66	1.38		5.04	6.25
0865	22 ga		900	.036		4.23	1.46		5.69	7
0870	16-1/2" wide, standard finish, 24 ga		950	.034		3.95	1.38		5.33	6.55
0875	22 ga		900	.036		4.42	1.46		5.88	7.20
0880	Zinc aluminum alloy finish, 24 ga		950	.034		3.45	1.38		4.83	6
0885	22 ga		900	.036		3.85	1.46		5.31	6.55
0890	Flat profile, 2" x 2" batten, 12" wide, standard finish, 26 ga		1000	.032		3.93	1.32		5.25	6.40
0895	24 ga		950	.034		4.69	1.38		6.07	7.35
0900	22 ga		900	.036		5.75	1.46		7.21	8.70
0905	Zinc aluminum alloy finish, 26 ga		1000	.032		3.66	1.32		4.98	6.15
0910	24 ga		950	.034		4.18	1.38		5.56	6.80
0915	22 ga		900	.036		4.87	1.46		6.33	7.70
0920	16-1/2" wide, standard finish, 24 ga		950	.034		4.32	1.38		5.70	6.95
0925	22 ga		900	.036		5.05	1.46		6.51	7.90
0930	Zinc aluminum alloy finish, 24 ga		950	.034		3.91	1.38		5.29	6.50
0935	22 ga		900	.036		4.45	1.46		5.91	7.25
9000	Minimum labor/equipment charge	1 Rofc	2	4	Job		142		142	249

07 41 33 – Plastic Roof Panels

07 41 33.10 Fiberglass Panels

		Crew	Daily Output	Labor-Hours	Unit	Material	Labor	Equipment	Total	Total Incl O&P
0010	**FIBERGLASS PANELS**									
0012	Corrugated panels, roofing, 8 oz per S.F.	G-3	1000	.032	S.F.	1.58	1.32		2.90	3.84
0100	12 oz per S.F.		1000	.032		3.43	1.32		4.75	5.85
0300	Corrugated siding, 6 oz per S.F.		880	.036		1.58	1.49		3.07	4.12
0400	8 oz per S.F.		880	.036		1.58	1.49		3.07	4.12
0500	Fire retardant		880	.036		3.21	1.49		4.70	5.90
0600	12 oz. siding, textured		880	.036		3.32	1.49		4.81	6.05
0700	Fire retardant		880	.036		4.39	1.49		5.88	7.20
0900	Flat panels, 6 oz per S.F., clear or colors		880	.036		1.83	1.49		3.32	4.39
1100	Fire retardant, class A		880	.036		3.18	1.49		4.67	5.90
1300	8 oz per S.F., clear or colors		880	.036		2.38	1.49		3.87	5
9000	Minimum labor/equipment charge	1 Rofc	2	4	Job		142		142	249

07 42 Wall Panels

07 42 13 – Metal Wall Panels

07 42 13.10 Mansard Panels

		Crew	Daily Output	Labor-Hours	Unit	Material	2010 Bare Costs Labor	Equipment	Total	Total Incl O&P
0010	**MANSARD PANELS**									
0600	Aluminum, stock units, straight surfaces	1 Shee	115	.070	S.F.	3.35	3.42		6.77	9.05
0700	Concave or convex surfaces		75	.107	"	3.67	5.25		8.92	12.25
0800	For framing, to 5' high, add		115	.070	L.F.	3.67	3.42		7.09	9.40
0900	Soffits, to 1' wide		125	.064	S.F.	1.82	3.14		4.96	6.90
9000	Minimum labor/equipment charge		2.50	3.200	Job		157		157	246

07 42 13.20 Aluminum Siding Panels

		Crew	Daily Output	Labor-Hours	Unit	Material	Labor	Equipment	Total	Total Incl O&P
0010	**ALUMINUM SIDING PANELS**									
0012	Corrugated, on steel framing, .019 thick, natural finish	G-3	775	.041	S.F.	1.47	1.70		3.17	4.33
0100	Painted		775	.041		1.58	1.70		3.28	4.45
0400	Farm type, .021" thick on steel frame, natural		775	.041		1.48	1.70		3.18	4.34
0600	Painted		775	.041		1.58	1.70		3.28	4.45
0700	Industrial type, corrugated, on steel, .024" thick, mill		775	.041		2.06	1.70		3.76	4.98
0900	Painted		775	.041		2.22	1.70		3.92	5.15
1000	.032" thick, mill		775	.041		2.37	1.70		4.07	5.30
1200	Painted		775	.041		2.89	1.70		4.59	5.90
1300	V-Beam, on steel frame, .032" thick, mill		775	.041		2.67	1.70		4.37	5.65
1500	Painted		775	.041		3.06	1.70		4.76	6.10
1600	.040" thick, mill		775	.041		3.24	1.70		4.94	6.25
1800	Painted		775	.041		3.80	1.70		5.50	6.90
3800	Horizontal, colored clapboard, 8" wide, plain	2 Carp	515	.031		1.73	1.29		3.02	4.02
3810	Insulated		515	.031		1.86	1.29		3.15	4.17
3830	8" embossed, painted		515	.031		1.73	1.29		3.02	4.02
3840	Insulated		515	.031		2	1.29		3.29	4.32
3860	12" painted, smooth		600	.027		1.73	1.11		2.84	3.72
3870	Insulated		600	.027		1.96	1.11		3.07	3.98
3890	12" embossed, painted		600	.027		2.05	1.11		3.16	4.08
3900	Insulated		515	.031		1.86	1.29		3.15	4.17
4000	Vertical board & batten, colored, non-insulated		515	.031		1.53	1.29		2.82	3.80
4200	For simulated wood design, add					.10			.10	.11
4300	Corners for above, outside	2 Carp	515	.031	V.L.F.	3.05	1.29		4.34	5.50
4500	Inside corners	"	515	.031	"	1.39	1.29		2.68	3.65
9000	Minimum labor/equipment charge	1 Carp	3	2.667	Job		111		111	182

07 42 13.30 Steel Siding

		Crew	Daily Output	Labor-Hours	Unit	Material	Labor	Equipment	Total	Total Incl O&P
0010	**STEEL SIDING**									
0020	Beveled, vinyl coated, 8" wide	1 Carp	265	.030	S.F.	1.97	1.25		3.22	4.23
0050	10" wide	"	275	.029		2.11	1.21		3.32	4.31
0080	Galv, corrugated or ribbed, on steel frame, 30 gauge	G-3	800	.040		1.32	1.64		2.96	4.07
0100	28 gauge		795	.040		1.38	1.65		3.03	4.16
0300	26 gauge		790	.041		1.94	1.67		3.61	4.79
0400	24 gauge		785	.041		1.95	1.68		3.63	4.82
0600	22 gauge		770	.042		2.25	1.71		3.96	5.20
0700	Colored, corrugated/ribbed, on steel frame, 10 yr fnsh, 28 ga.		800	.040		2.05	1.64		3.69	4.88
0900	26 gauge		795	.040		2.14	1.65		3.79	4.99
1000	24 gauge		790	.041		2.49	1.67		4.16	5.40
1020	20 gauge		785	.041		3.16	1.68		4.84	6.15
9000	Minimum labor/equipment charge	1 Carp	3	2.667	Job		111		111	182

07 46 Siding

07 46 23 – Wood Siding

07 46 23.10 Wood Board Siding

		Crew	Daily Output	Labor-Hours	Unit	Material	2010 Bare Costs Labor	Equipment	Total	Total Incl O&P
0010	**WOOD BOARD SIDING**									
3200	Wood, cedar bevel, A grade, 1/2" x 6"	1 Carp	295	.027	S.F.	4.73	1.13		5.86	7.05
3300	1/2" x 8"		330	.024		5.55	1.01		6.56	7.75
3500	3/4" x 10", clear grade		375	.021		6.10	.89		6.99	8.15
3600	"B" grade		375	.021		5.20	.89		6.09	7.15
3800	Cedar, rough sawn, 1" x 4", A grade, natural		220	.036		3.77	1.51		5.28	6.65
3900	Stained		220	.036		4.17	1.51		5.68	7.05
4100	1" x 12", board & batten, #3 & Btr., natural		420	.019		3.94	.79		4.73	5.65
4200	Stained		420	.019		4.20	.79		4.99	5.90
4400	1" x 8" channel siding, #3 & Btr., natural		330	.024		2.34	1.01		3.35	4.22
4500	Stained		330	.024		2.50	1.01		3.51	4.40
4700	Redwood, clear, beveled, vertical grain, 1/2" x 4"		220	.036		3.10	1.51		4.61	5.90
4750	1/2" x 6"		295	.027		3.10	1.13		4.23	5.25
4800	1/2" x 8"		330	.024		3.24	1.01		4.25	5.20
5000	3/4" x 10"		375	.021		3.32	.89		4.21	5.10
5200	Channel siding, 1" x 10", B grade		375	.021		3.32	.89		4.21	5.10
5250	Redwood, T&G boards, B grade, 1" x 4"		220	.036		3.10	1.51		4.61	5.90
5270	1" x 8"		330	.024		3.24	1.01		4.25	5.20
5400	White pine, rough sawn, 1" x 8", natural		330	.024		2.20	1.01		3.21	4.07
5500	Stained		330	.024		2.60	1.01		3.61	4.51
9000	Minimum labor/equipment charge		2	4	Job		166		166	273

07 46 29 – Plywood Siding

07 46 29.10 Plywood Siding Options

		Crew	Daily Output	Labor-Hours	Unit	Material	Labor	Equipment	Total	Total Incl O&P
0010	**PLYWOOD SIDING OPTIONS**									
0900	Plywood, medium density overlaid, 3/8" thick	2 Carp	750	.021	S.F.	1.14	.89		2.03	2.71
1000	1/2" thick		700	.023		1.22	.95		2.17	2.90
1100	3/4" thick		650	.025		1.55	1.02		2.57	3.39
1600	Texture 1-11, cedar, 5/8" thick, natural		675	.024		2.43	.98		3.41	4.29
1700	Factory stained		675	.024		1.95	.98		2.93	3.77
1900	Texture 1-11, fir, 5/8" thick, natural		675	.024		1.37	.98		2.35	3.13
2000	Factory stained		675	.024		1.73	.98		2.71	3.52
2050	Texture 1-11, S.Y.P., 5/8" thick, natural		675	.024		1.19	.98		2.17	2.93
2100	Factory stained		675	.024		1.24	.98		2.22	2.98
2200	Rough sawn cedar, 3/8" thick, natural		675	.024		1.20	.98		2.18	2.94
2300	Factory stained		675	.024		.96	.98		1.94	2.68
2500	Rough sawn fir, 3/8" thick, natural		675	.024		.76	.98		1.74	2.46
2600	Factory stained		675	.024		.96	.98		1.94	2.68
2800	Redwood, textured siding, 5/8" thick		675	.024		1.92	.98		2.90	3.73
3000	Polyvinyl chloride coated, 3/8" thick		750	.021		1.11	.89		2	2.68
9000	Minimum labor/equipment charge	1 Carp	2	4	Job		166		166	273

07 46 33 – Plastic Siding

07 46 33.10 Vinyl Siding

		Crew	Daily Output	Labor-Hours	Unit	Material	Labor	Equipment	Total	Total Incl O&P
0010	**VINYL SIDING**									
3995	Clapboard profile, woodgrain texture, .048 thick, double 4	2 Carp	495	.032	S.F.	.92	1.34		2.26	3.22
4000	Double 5		550	.029		.92	1.21		2.13	3
4005	Single 8		495	.032		.96	1.34		2.30	3.27
4010	Single 10		550	.029		1.15	1.21		2.36	3.26
4015	.044 thick, double 4		495	.032		.84	1.34		2.18	3.13
4020	Double 5		550	.029		.84	1.21		2.05	2.91
4025	.042 thick, double 4		495	.032		.84	1.34		2.18	3.13
4030	Double 5		550	.029		.84	1.21		2.05	2.91
4035	Cross sawn texture, .040 thick, double 4		495	.032		.66	1.34		2	2.94

07 46 Siding

07 46 33 – Plastic Siding

07 46 33.10 Vinyl Siding

		Crew	Daily Output	Labor-Hours	Unit	Material	2010 Bare Costs Labor	Equipment	Total	Total Incl O&P
4040	Double 5	2 Carp	550	.029	S.F.	.66	1.21		1.87	2.72
4045	Smooth texture, .042 thick, double 4		495	.032		.75	1.34		2.09	3.04
4050	Double 5		550	.029		.75	1.21		1.96	2.82
4055	Single 8		495	.032		.96	1.34		2.30	3.27
4060	Cedar texture, .044 thick, double 4		495	.032		.92	1.34		2.26	3.22
4065	Double 6		600	.027		.92	1.11		2.03	2.83
4070	Dutch lap profile, woodgrain texture, .048 thick, double 5		550	.029		.92	1.21		2.13	3
4075	.044 thick, double 4.5		525	.030		.92	1.27		2.19	3.09
4080	.042 thick, double 4.5		525	.030		.75	1.27		2.02	2.91
4085	.040 thick, double 4.5		525	.030		.66	1.27		1.93	2.81
4090	Shingle profile, random grooves, double 7		400	.040		2.57	1.66		4.23	5.55
4095	Triple 5		400	.040		2.57	1.66		4.23	5.55
4100	Shake profile, 10" wide		400	.040		2.19	1.66		3.85	5.15
4105	Vertical pattern, .046 thick, double 5		550	.029		1.22	1.21		2.43	3.33
4110	.044 thick, triple 3		550	.029		1.37	1.21		2.58	3.50
4115	.040 thick, triple 4		550	.029		1.22	1.21		2.43	3.33
4120	.040 thick, triple 2.66		550	.029		1.47	1.21		2.68	3.61
4125	Insulation, fan folded extruded polystyrene, 1/4"		2000	.008		.25	.33		.58	.83
4130	3/8"		2000	.008		.27	.33		.60	.85
4135	Accessories, J channel, 5/8" pocket		700	.023	L.F.	.43	.95		1.38	2.03
4140	3/4" pocket		695	.023		.48	.96		1.44	2.09
4145	1-1/4" pocket		680	.024		.61	.98		1.59	2.28
4150	Flexible, 3/4" pocket		600	.027		1.81	1.11		2.92	3.81
4155	Under sill finish trim		500	.032		.45	1.33		1.78	2.67
4160	Vinyl starter strip		700	.023		.36	.95		1.31	1.95
4165	Aluminum starter strip		700	.023		.27	.95		1.22	1.85
4170	Window casing, 2-1/2" wide, 3/4" pocket		510	.031		1.14	1.30		2.44	3.39
4175	Outside corner, woodgrain finish, 4" face, 3/4" pocket		700	.023		1.75	.95		2.70	3.49
4180	5/8" pocket		700	.023		1.67	.95		2.62	3.40
4185	Smooth finish, 4" face, 3/4" pocket		700	.023		1.84	.95		2.79	3.58
4190	7/8" pocket		690	.023		1.78	.96		2.74	3.54
4195	1-1/4" pocket		700	.023		1.23	.95		2.18	2.91
4200	Soffit and fascia, 1' overhang, solid		120	.133		3.37	5.55		8.92	12.80
4205	Vented		120	.133		3.37	5.55		8.92	12.80
4207	18" overhang, solid		110	.145		3.98	6.05		10.03	14.30
4208	Vented		110	.145		3.98	6.05		10.03	14.30
4210	2' overhang, solid		100	.160		4.58	6.65		11.23	15.95
4215	Vented		100	.160		4.58	6.65		11.23	15.95
4217	3' overhang, solid		100	.160		5.80	6.65		12.45	17.25
4218	Vented		100	.160		5.80	6.65		12.45	17.25
4220	Colors for siding and soffits, add				S.F.	.16			.16	.18
4225	Colors for accessories and trim, add				L.F.	.32			.32	.35
9000	Minimum labor/equipment charge	1 Carp	3	2.667	Job		111		111	182

07 46 46 – Mineral-Fiber Cement Siding

07 46 46.10 Fiber Cement Siding

		Crew	Daily Output	Labor-Hours	Unit	Material	Labor	Equipment	Total	Total Incl O&P
0010	**FIBER CEMENT SIDING**									
0020	Lap siding, 5/16" thick, 6" wide, 4-3/4" exposure, smooth texture	2 Carp	415	.039	S.F.	1.29	1.60		2.89	4.04
0025	Woodgrain texture		415	.039		1.29	1.60		2.89	4.04
0030	7-1/2" wide, 6-1/4" exposure, smooth texture		425	.038		1.36	1.56		2.92	4.07
0035	Woodgrain texture		425	.038		1.36	1.56		2.92	4.07
0040	8" wide, 6-3/4" exposure, smooth texture		425	.038		1.51	1.56		3.07	4.23
0045	Roughsawn texture		425	.038		1.51	1.56		3.07	4.23

07 46 Siding

07 46 46 – Mineral-Fiber Cement Siding

07 46 46.10 Fiber Cement Siding

		Crew	Daily Output	Labor-Hours	Unit	Material	2010 Bare Costs Labor	Equipment	Total	Total Incl O&P
0050	9-1/2" wide, 8-1/4" exposure, smooth texture	2 Carp	440	.036	S.F.	1.45	1.51		2.96	4.08
0055	Woodgrain texture		440	.036		1.45	1.51		2.96	4.08
0060	12" wide, 10-3/8" exposure, smooth texture		455	.035		1.34	1.46		2.80	3.87
0065	Woodgrain texture		455	.035		1.34	1.46		2.80	3.87
0070	Panel siding, 5/16" thick, smooth texture		750	.021		1.12	.89		2.01	2.70
0075	Stucco texture		750	.021		1.12	.89		2.01	2.70
0080	Grooved woodgrain texture		750	.021		1.12	.89		2.01	2.70
0085	V - grooved woodgrain texture		750	.021		1.12	.89		2.01	2.70
0090	Wood starter strip		400	.040	L.F.	.53	1.66		2.19	3.31

07 46 73 – Soffit

07 46 73.10 Soffit Options

		Crew	Daily Output	Labor-Hours	Unit	Material	Labor	Equipment	Total	Total Incl O&P
0010	**SOFFIT OPTIONS**									
0012	Aluminum, residential, .020" thick	1 Carp	210	.038	S.F.	1.62	1.58		3.20	4.38
0100	Baked enamel on steel, 16 or 18 gauge		105	.076		5.35	3.17		8.52	11.10
0300	Polyvinyl chloride, white, solid		230	.035		1.09	1.45		2.54	3.57
0400	Perforated		230	.035		1.09	1.45		2.54	3.57
0500	For colors, add					.13			.13	.14
9000	Minimum labor/equipment charge	1 Carp	3	2.667	Job		111		111	182

07 51 Built-Up Bituminous Roofing

07 51 13 – Built-Up Asphalt Roofing

07 51 13.10 Built-Up Roofing Components

		Crew	Daily Output	Labor-Hours	Unit	Material	Labor	Equipment	Total	Total Incl O&P
0010	**BUILT-UP ROOFING COMPONENTS**									
0012	Asphalt saturated felt, #30, 2 square per roll	1 Rofc	58	.138	Sq.	11.95	4.88		16.83	22
0200	#15, 4 sq per roll, plain or perforated, not mopped		58	.138		5.90	4.88		10.78	15.05
0300	Roll roofing, smooth, #65		15	.533		9.65	18.90		28.55	44
0500	#90		15	.533		34.50	18.90		53.40	71.50
0520	Mineralized		15	.533		34.50	18.90		53.40	71.50
0540	D.C. (Double coverage), 19" selvage edge		10	.800		44	28.50		72.50	98
0580	Adhesive (lap cement)				Gal.	4.79			4.79	5.25
0600	Steep, flat or dead level asphalt, 10 ton lots, bulk				Ton	490			490	540
0800	Packaged				"	830			830	915
9000	Minimum labor/equipment charge	1 Rofc	4	2	Job		71		71	125

07 51 13.13 Cold-Applied Built-Up Asphalt Roofing

		Crew	Daily Output	Labor-Hours	Unit	Material	Labor	Equipment	Total	Total Incl O&P
0010	**COLD-APPLIED BUILT-UP ASPHALT ROOFING**									
0020	3 ply system, installation only (components listed below)	G-5	50	.800	Sq.		25.50	3.77	29.27	49.50
0100	Spunbond poly. fabric, 1.35 oz/SY, 36"W, 10.8 Sq/roll				Ea.	175			175	193
0200	49" wide, 14.6 Sq./roll					244			244	268
0300	2.10 oz./S.Y., 36" wide, 10.8 Sq./roll					266			266	293
0400	49" wide, 14.6 Sq./roll					360			360	395
0500	Base & finish coat, 3 gal./Sq., 5 gal./can				Gal.	5			5	5.50
0600	Coating, ceramic granules, 1/2 Sq./bag				Ea.	16.75			16.75	18.45
0700	Aluminum, 2 gal./Sq.				Gal.	13.50			13.50	14.85
0800	Emulsion, fibered or non-fibered, 4 gal./Sq.				"	4.70			4.70	5.15

07 51 13.20 Built-Up Roofing Systems

		Crew	Daily Output	Labor-Hours	Unit	Material	Labor	Equipment	Total	Total Incl O&P
0010	**BUILT-UP ROOFING SYSTEMS**									
0120	Asphalt flood coat with gravel/slag surfacing, not including									
0140	Insulation, flashing or wood nailers									
0200	Asphalt base sheet, 3 plies #15 asphalt felt, mopped	G-1	22	2.545	Sq.	94	84	21.50	199.50	275
0350	On nailable decks		21	2.667		98	88	22.50	208.50	288

07 51 Built-Up Bituminous Roofing

07 51 13 – Built-Up Asphalt Roofing

07 51 13.20 Built-Up Roofing Systems

		Crew	Daily Output	Labor-Hours	Unit	Material	2010 Bare Costs Labor	Equipment	Total	Total Incl O&P
0500	4 plies #15 asphalt felt, mopped	G-1	20	2.800	Sq.	130	92.50	24	246.50	330
0550	On nailable decks		19	2.947		115	97.50	25	237.50	325
0700	Coated glass base sheet, 2 plies glass (type IV), mopped		22	2.545		91.50	84	21.50	197	273
0850	3 plies glass, mopped		20	2.800		110	92.50	24	226.50	310
0950	On nailable decks		19	2.947		103	97.50	25	225.50	315
1100	4 plies glass fiber felt (type IV), mopped		20	2.800		135	92.50	24	251.50	335
1150	On nailable decks		19	2.947		121	97.50	25	243.50	335
1200	Coated & saturated base sheet, 3 plies #15 asph. felt, mopped		20	2.800		104	92.50	24	220.50	305
1250	On nailable decks		19	2.947		96.50	97.50	25	219	305
1300	4 plies #15 asphalt felt, mopped		22	2.545		121	84	21.50	226.50	305
2000	Asphalt flood coat, smooth surface									
2200	Asphalt base sheet & 3 plies #15 asphalt felt, mopped	G-1	24	2.333	Sq.	100	77	19.85	196.85	268
2400	On nailable decks		23	2.435		92.50	80.50	20.50	193.50	267
2600	4 plies #15 asphalt felt, mopped		24	2.333		117	77	19.85	213.85	287
2700	On nailable decks		23	2.435		110	80.50	20.50	211	286
2900	Coated glass fiber base sheet, mopped, and 2 plies of									
2910	glass fiber felt (type IV)	G-1	25	2.240	Sq.	86.50	74	19.05	179.55	246
3100	On nailable decks		24	2.333		81	77	19.85	177.85	248
3200	3 plies, mopped		23	2.435		104	80.50	20.50	205	280
3300	On nailable decks		22	2.545		97.50	84	21.50	203	279
3800	4 plies glass fiber felt (type IV), mopped		23	2.435		123	80.50	20.50	224	300
3900	On nailable decks		22	2.545		116	84	21.50	221.50	299
4000	Coated & saturated base sheet, 3 plies #15 asph. felt, mopped		24	2.333		98.50	77	19.85	195.35	266
4200	On nailable decks		23	2.435		91.50	80.50	20.50	192.50	265
4300	4 plies #15 organic felt, mopped		22	2.545		116	84	21.50	221.50	299
4500	Coal tar pitch with gravel/slag surfacing									
4600	4 plies #15 tarred felt, mopped	G-1	21	2.667	Sq.	183	88	22.50	293.50	380
4800	3 plies glass fiber felt (type IV), mopped	"	19	2.947	"	150	97.50	25	272.50	365
5000	Coated glass fiber base sheet, and 2 plies of									
5010	glass fiber felt, (type IV) mopped	G-1	19	2.947	Sq.	148	97.50	25	270.50	365
5300	On nailable decks		18	3.111		130	103	26.50	259.50	355
5600	4 plies glass fiber felt (type IV), mopped		21	2.667		206	88	22.50	316.50	405
5800	On nailable decks		20	2.800		188	92.50	24	304.50	395

07 51 13.30 Cants

		Crew	Daily Output	Labor-Hours	Unit	Material	Labor	Equipment	Total	Total Incl O&P
0010	**CANTS**									
0012	Lumber, treated, 4" x 4" cut diagonally	1 Rofc	325	.025	L.F.	1.52	.87		2.39	3.21
0100	Foamglass		325	.025		2.23	.87		3.10	3.99
0300	Mineral or fiber, trapezoidal, 1" x 4" x 48"		325	.025		.19	.87		1.06	1.75
0400	1-1/2" x 5-5/8" x 48"		325	.025		.31	.87		1.18	1.88
9000	Minimum labor/equipment charge		4	2	Job		71		71	125

07 51 13.40 Felts

		Crew	Daily Output	Labor-Hours	Unit	Material	Labor	Equipment	Total	Total Incl O&P
0010	**FELTS**									
0012	Glass fibered roofing felt, #15, not mopped	1 Rofc	58	.138	Sq.	6.55	4.88		11.43	15.80
0300	Base sheet, #45, channel vented		58	.138		26	4.88		30.88	37
0400	#50, coated		58	.138		14	4.88		18.88	24
0500	Cap, mineral surfaced		58	.138		34.50	4.88		39.38	46.50
0600	Flashing membrane, #65		16	.500		9.65	17.70		27.35	41.50
0800	Coal tar fibered, #15, no mopping		58	.138		8	4.88		12.88	17.40
0900	Asphalt felt, #15, 4 sq per roll, no mopping		58	.138		5.90	4.88		10.78	15.05
1100	#30, 2 sq per roll		58	.138		11.95	4.88		16.83	22
1200	Double coated, #33		58	.138		9.90	4.88		14.78	19.50
1400	#40, base sheet		58	.138		11.30	4.88		16.18	21

07 51 Built-Up Bituminous Roofing

07 51 13 – Built-Up Asphalt Roofing

07 51 13.40 Felts

		Crew	Daily Output	Labor-Hours	Unit	Material	2010 Bare Costs Labor	Equipment	Total	Total Incl O&P
1450	Coated and saturated	1 Rofc	58	.138	Sq.	9.90	4.88		14.78	19.45
1500	Tarred felt, organic, #15, 4 sq rolls		58	.138		15.60	4.88		20.48	26
1550	#30, 2 sq roll		58	.138		22.50	4.88		27.38	33.50
1700	Add for mopping above felts, per ply, asphalt, 24 lb per sq	G-1	192	.292		9.95	9.65	2.48	22.08	30.50
1800	Coal tar mopping, 30 lb per sq		186	.301		15.30	9.95	2.56	27.81	37
1900	Flood coat, with asphalt, 60 lb per sq		60	.933		25	31	7.95	63.95	91
2000	With coal tar, 75 lb per sq		56	1		38	33	8.50	79.50	109
9000	Minimum labor/equipment charge	1 Rofc	4	2	Job		71		71	125

07 51 13.50 Walkways for Built-Up Roofs

		Crew	Daily Output	Labor-Hours	Unit	Material	Labor	Equipment	Total	Incl O&P
0010	**WALKWAYS FOR BUILT-UP ROOFS**									
0020	Asphalt impregnated, 3' x 6' x 1/2" thick	1 Rofc	400	.020	S.F.	1.70	.71		2.41	3.12
0100	3' x 3' x 3/4" thick		400	.020	"	3.13	.71		3.84	4.69
0600	100% recycled rubber, 3' x 4' x 3/8" [G]		400	.020	L.F.	6.30	.71		7.01	8.20
0610	3' x 4' x 1/2" [G]		400	.020		6.75	.71		7.46	8.70
0620	3' x 4' x 3/4" [G]		400	.020		8.40	.71		9.11	10.50
9000	Minimum labor/equipment charge		2.75	2.909	Job		103		103	181

07 52 Modified Bituminous Membrane Roofing

07 52 13 – Atactic-Polypropylene-Modified Bituminous Membrane Roofing

07 52 13.10 APP Modified Bituminous Membrane

			Crew	Daily Output	Labor-Hours	Unit	Material	Labor	Equipment	Total	Incl O&P
0010	**APP MODIFIED BITUMINOUS MEMBRANE**	R075213-30									
0020	Base sheet, #15 glass fiber felt, nailed to deck		1 Rofc	58	.138	Sq.	7.45	4.88		12.33	16.80
0030	Spot mopped to deck		G-1	295	.190		11.55	6.30	1.61	19.46	25.50
0040	Fully mopped to deck		"	192	.292		16.50	9.65	2.48	28.63	38
0050	#15 organic felt, nailed to deck		1 Rofc	58	.138		6.80	4.88		11.68	16.10
0060	Spot mopped to deck		G-1	295	.190		10.85	6.30	1.61	18.76	25
0070	Fully mopped to deck		"	192	.292		15.85	9.65	2.48	27.98	37
2100	APP mod., smooth surf. cap sheet, poly. reinf., torched, 160 mils		G-5	2100	.019	S.F.	.48	.61	.09	1.18	1.71
2150	170 mils			2100	.019		.55	.61	.09	1.25	1.79
2200	Granule surface cap sheet, poly. reinf., torched, 180 mils			2000	.020		.59	.64	.09	1.32	1.88
2250	Smooth surface flashing, torched, 160 mils			1260	.032		.48	1.02	.15	1.65	2.49
2300	170 mils			1260	.032		.55	1.02	.15	1.72	2.57
2350	Granule surface flashing, torched, 180 mils			1260	.032		.59	1.02	.15	1.76	2.61
2400	Fibrated aluminum coating		1 Rofc	3800	.002		.10	.07		.17	.24

07 52 16 – Styrene-Butadiene-Styrene Modified Bituminous Membrane Roofing

07 52 16.10 SBS Modified Bituminous Membrane

		Crew	Daily Output	Labor-Hours	Unit	Material	Labor	Equipment	Total	Incl O&P
0010	**SBS MODIFIED BITUMINOUS MEMBRANE**									
0080	SBS modified, granule surf cap sheet, polyester rein., mopped									
1500	Glass fiber reinforced, mopped, 160 mils	G-1	2000	.028	S.F.	.49	.93	.24	1.66	2.43
1600	Smooth surface cap sheet, mopped, 145 mils		2100	.027		.49	.88	.23	1.60	2.34
1700	Smooth surface flashing, 145 mils		1260	.044		.49	1.47	.38	2.34	3.55
1800	150 mils		1260	.044		.48	1.47	.38	2.33	3.54
1900	Granular surface flashing, 150 mils		1260	.044		.58	1.47	.38	2.43	3.65
2000	160 mils		1260	.044		.77	1.47	.38	2.62	3.86

07 53 Elastomeric Membrane Roofing

07 53 16 – Chlorosulfonate Polyethylene Roofing

07 53 16.10 Chlorosulfonated Polyethylene Roofing	Crew	Daily Output	Labor-Hours	Unit	Material	2010 Bare Costs Labor	Equipment	Total	Total Incl O&P
0010 **CHLOROSULFONATED POLYETHYLENE ROOFING**									
0800 Chlorosulfonated polyethylene-hypalon (CSPE), 45 mils,									
0900 0.29 P.S.F., fully adhered	G-5	26	1.538	Sq.	185	49.50	7.25	241.75	299
1200 Mechanically attached	"	35	1.143	"	186	36.50	5.40	227.90	275

07 53 23 – Ethylene-Propylene-Diene-Monomer Roofing

07 53 23.20 Ethylene-Propylene-Diene-Monomer Roofing

	Crew	Daily Output	Labor-Hours	Unit	Material	Labor	Equipment	Total	Total Incl O&P
0010 **ETHYLENE-PROPYLENE-DIENE-MONOMER ROOFING (E.P.D.M.)**									
3500 Ethylene-propylene-diene-monomer (EPDM), 45 mils, 0.28 P.S.F.									
3600 Loose-laid & ballasted with stone (10 P.S.F.)	G-5	51	.784	Sq.	76	25	3.69	104.69	132
3700 Mechanically attached		35	1.143		68.50	36.50	5.40	110.40	145
3800 Fully adhered with adhesive	↓	26	1.538	↓	104	49.50	7.25	160.75	209
4500 60 mils, 0.40 P.S.F.									
4600 Loose-laid & ballasted with stone (10 P.S.F.)	G-5	51	.784	Sq.	90	25	3.69	118.69	148
4700 Mechanically attached		35	1.143		81.50	36.50	5.40	123.40	160
4800 Fully adhered with adhesive	↓	26	1.538		117	49.50	7.25	173.75	223
4810 45 mil, .28 PSF, membrane only					43			43	47.50
4820 60 mil, .40 PSF, membrane only					55			55	60.50
4850 Seam tape for membrane, 3" x 100' roll				Ea.	48			48	53
4900 Batten strips, 10' sections					2.80			2.80	3.08
4910 Cover tape for batten strips, 6" x 100' roll				↓	110			110	121
4930 Plate anchors				M	105			105	116
4970 Adhesive for fully adhered systems, 60 S.F./gal.				Gal.	21			21	23

07 53 29 – Polyisobutylene Roofing

07 53 29.10 Polyisobutylene Roofing

	Crew	Daily Output	Labor-Hours	Unit	Material	Labor	Equipment	Total	Total Incl O&P
0010 **POLYISOBUTYLENE ROOFING**									
7500 Polyisobutylene (PIB), 100 mils, 0.57 P.S.F.									
7600 Loose-laid & ballasted with stone/gravel (10 P.S.F.)	G-5	51	.784	Sq.	181	25	3.69	209.69	248
7700 Partially adhered with adhesive		35	1.143		227	36.50	5.40	268.90	320
7800 Hot asphalt attachment		35	1.143		217	36.50	5.40	258.90	310
7900 Fully adhered with contact cement	↓	26	1.538	↓	234	49.50	7.25	290.75	355

07 54 Thermoplastic Membrane Roofing

07 54 19 – Polyvinyl-Chloride Roofing

07 54 19.10 Polyvinyl-Chloride Roofing (P.V.C.)

	Crew	Daily Output	Labor-Hours	Unit	Material	Labor	Equipment	Total	Total Incl O&P
0010 **POLYVINYL-CHLORIDE ROOFING (P.V.C.)**									
8200 Heat welded seams									
8700 Reinforced, 48 mils, 0.33 P.S.F.									
8750 Loose-laid & ballasted with stone/gravel (12 P.S.F.)	G-5	51	.784	Sq.	97	25	3.69	125.69	156
8800 Mechanically attached		35	1.143		90	36.50	5.40	131.90	169
8850 Fully adhered with adhesive	↓	26	1.538	↓	118	49.50	7.25	174.75	225
8860 Reinforced, 60 mils, .40 P.S.F.									
8870 Loose-laid & ballasted with stone/gravel (12 P.S.F.)	G-5	51	.784	Sq.	96.50	25	3.69	125.19	155
8880 Mechanically attached		35	1.143		89.50	36.50	5.40	131.40	169
8890 Fully adhered with adhesive	↓	26	1.538	↓	118	49.50	7.25	174.75	224

07 54 23 – Thermoplastic Polyolefin Roofing

07 54 23.10 Thermoplastic Polyolefin Roofing (T.P.O)

	Crew	Daily Output	Labor-Hours	Unit	Material	Labor	Equipment	Total	Total Incl O&P
0010 **THERMOPLASTIC POLYOLEFIN ROOFING (T.P.O.)**									
0100 45 mils, loose laid & ballasted with stone(1/2 ton/sq.)	G-5	51	.784	Sq.	90.50	25	3.69	119.19	149
0120 Fully adhered	↓	25	1.600	↓	80.50	51.50	7.55	139.55	187

07 54 Thermoplastic Membrane Roofing

07 54 23 – Thermoplastic Polyolefin Roofing

07 54 23.10 Thermoplastic Polyolefin Roofing (T.P.O)	Crew	Daily Output	Labor-Hours	Unit	Material	2010 Bare Costs Labor	Equipment	Total	Total Incl O&P
0140 Mechanically attached	G-5	34	1.176	Sq.	80	38	5.55	123.55	161
0160 Self adhered		35	1.143		88	36.50	5.40	129.90	167
0180 60 mil membrane, heat welded seams, ballasted		50	.800		107	25.50	3.77	136.27	168
0200 Fully adhered		25	1.600		96	51.50	7.55	155.05	205
0220 Mechanically attached		34	1.176		99	38	5.55	142.55	182
0240 Self adhered		35	1.143		109	36.50	5.40	150.90	190

07 56 Fluid-Applied Roofing

07 56 10 – Fluid-Applied Roofing Elastomers

07 56 10.10 Elastomeric Roofing

		Crew	Daily Output	Labor-Hours	Unit	Material	Labor	Equipment	Total	Total Incl O&P
0010	ELASTOMERIC ROOFING									
0110	Acrylic rubber, fluid applied, 20 mils thick	G-5	2000	.020	S.F.	2.04	.64	.09	2.77	3.47
0120	50 mils, reinforced		1200	.033		3.16	1.07	.16	4.39	5.55
0130	For walking surface, add		900	.044		.96	1.43	.21	2.60	3.80
0300	Hypalon neoprene, fluid applied, 20 mil thick, not-reinforced	G-1	1135	.049		2.57	1.63	.42	4.62	6.15
0600	Non-woven polyester, reinforced		960	.058		2.61	1.93	.50	5.04	6.80
0700	5 coat neoprene deck, 60 mil thick, under 10,000 SF		325	.172		5.75	5.70	1.46	12.91	17.95
0900	Over 10,000 SF		625	.090		5.35	2.96	.76	9.07	11.95
1300	Vinyl plastic traffic deck, sprayed, 2 to 4 mils thick		625	.090		1.68	2.96	.76	5.40	7.90
1500	Vinyl and neoprene membrane traffic deck		1550	.036		1.77	1.19	.31	3.27	4.39
9000	Minimum labor/equipment charge	1 Rofc	2	4	Job		142		142	249

07 57 Coated Foamed Roofing

07 57 13 – Sprayed Polyurethane Foam Roofing

07 57 13.10 Sprayed Polyurethane Foam Roofing (S.P.F.)

		Crew	Daily Output	Labor-Hours	Unit	Material	Labor	Equipment	Total	Total Incl O&P
0010	SPRAYED POLYURETHANE FOAM ROOFING (S.P.F.)									
0100	Primer for metal substrate (when required)	G-2A	3000	.008	S.F.	.42	.25	.23	.90	1.14
0200	Primer for non-metal substrate (when required)		3000	.008		.16	.25	.23	.64	.86
0300	Closed cell spray, polyurethane foam, 3 lbs per CF density, 1", R6.7		15000	.002		.55	.05	.05	.65	.75
0400	2", R13.4		13125	.002		1.10	.06	.05	1.21	1.37
0500	3", R18.6		11485	.002		1.66	.07	.06	1.79	2
0550	4", R24.8		10080	.002		2.21	.08	.07	2.36	2.64
0700	Spray-on silicone coating		2500	.010		.92	.30	.28	1.50	1.83
0800	Warranty 5-20 year manufacturer's									.15
0900	Warranty 20 year, no dollar limit									.20

07 58 Roll Roofing

07 58 10 – Asphalt Roll Roofing

07 58 10.10 Roll Roofing

		Crew	Daily Output	Labor-Hours	Unit	Material	Labor	Equipment	Total	Total Incl O&P
0010	ROLL ROOFING									
0100	Asphalt, mineral surface									
0200	1 ply #15 organic felt, 1 ply mineral surfaced									
0300	Selvage roofing, lap 19", nailed & mopped	G-1	27	2.074	Sq.	64	68.50	17.65	150.15	210
0400	3 plies glass fiber felt (type IV), 1 ply mineral surfaced									
0500	Selvage roofing, lapped 19", mopped	G-1	25	2.240	Sq.	103	74	19.05	196.05	265
0600	Coated glass fiber base sheet, 2 plies of glass fiber									
0700	Felt (type IV), 1 ply mineral surfaced selvage									

07 58 Roll Roofing

07 58 10 – Asphalt Roll Roofing

07 58 10.10 Roll Roofing		Crew	Daily Output	Labor-Hours	Unit	Material	2010 Bare Costs Labor	Equipment	Total	Total Incl O&P
0800	Roofing, lapped 19", mopped	G-1	25	2.240	Sq.	111	74	19.05	204.05	273
0900	On nailable decks	"	24	2.333	"	101	77	19.85	197.85	269
1000	3 plies glass fiber felt (type III), 1 ply mineral surfaced									
1100	Selvage roofing, lapped 19", mopped	G-1	25	2.240	Sq.	103	74	19.05	196.05	265

07 61 Sheet Metal Roofing

07 61 13 – Standing Seam Sheet Metal Roofing

07 61 13.10 Standing Seam Sheet Metal Roofing, Field Fab.

		Crew	Daily Output	Labor-Hours	Unit	Material	Labor	Equipment	Total	Total Incl O&P
0010	**STANDING SEAM SHEET METAL ROOFING, FIELD FABRICATED**									
0400	Copper standing seam roofing, over 10 squares, 16 oz, 125 lb per sq	1 Shee	1.30	6.154	Sq.	1,075	300		1,375	1,675
0600	18 oz, 140 lb per sq		1.20	6.667		1,200	325		1,525	1,800
0700	20 oz, 150 lb per sq		1.10	7.273		1,300	355		1,655	1,975
1200	For abnormal conditions or small areas, add					25%	100%			
1300	For lead-coated copper, add					25%				
1400	Lead standing seam roofing, 5 lb. per S.F.	1 Shee	1.30	6.154		600	300		900	1,125
1500	Zinc standing seam roofing, .020" thick		1.20	6.667		1,075	325		1,400	1,700
1510	.027" thick		1.15	6.957		1,550	340		1,890	2,225
1520	.032" thick		1.10	7.273		1,825	355		2,180	2,550
1530	.040" thick		1.05	7.619		2,200	375		2,575	2,975
9000	Minimum labor/equipment charge		2	4	Job		196		196	305

07 61 16 – Batten Seam Sheet Metal Roofing

07 61 16.10 Batten Seam Sheet Metal Roofing, Field Fabricated

		Crew	Daily Output	Labor-Hours	Unit	Material	Labor	Equipment	Total	Total Incl O&P
0009	**BATTEN SEAM SHEET METAL ROOFING, FIELD FABRICATED**									
0012	Copper batten seam roofing, over 10 sq, 16 oz, 130 lb per sq	1 Shee	1.10	7.273	Sq.	1,075	355		1,430	1,750
0020	Lead batten seam roofing, 5 lb. per SF		1.20	6.667		600	325		925	1,175
0100	Zinc/copper alloy batten seam roofing, .020 thick		1.20	6.667		1,075	325		1,400	1,700
0200	Copper roofing, batten seam, over 10 sq, 18 oz, 145 lb per sq		1	8		1,200	395		1,595	1,925
0300	20 oz, 160 lb per sq		1	8		1,300	395		1,695	2,050
0500	Stainless steel batten seam roofing, type 304, 28 gauge		1.20	6.667		690	325		1,015	1,275
0600	26 gauge		1.15	6.957		695	340		1,035	1,300
0800	Zinc, copper alloy roofing, batten seam, .027" thick		1.15	6.957		1,550	340		1,890	2,225
0900	.032" thick		1.10	7.273		1,825	355		2,180	2,550
1000	.040" thick		1.05	7.619		2,200	375		2,575	2,975
9000	Minimum labor/equipment charge		2	4	Job		196		196	305

07 61 19 – Flat Seam Sheet Metal Roofing

07 61 19.10 Flat Seam Sheet Metal Roofing, Field Fabricated

		Crew	Daily Output	Labor-Hours	Unit	Material	Labor	Equipment	Total	Total Incl O&P
0010	**FLAT SEAM SHEET METAL ROOFING, FIELD FABRICATED**									
0900	Copper flat seam roofing, over 10 squares, 16 oz, 115 lb per sq	1 Shee	1.20	6.667	Sq.	1,075	325		1,400	1,700
0950	18 oz, 130 lb/sq		1.15	6.957		1,200	340		1,540	1,825
1000	20 oz, 145 lb per sq		1.10	7.273		1,300	355		1,655	1,975
1008	Zinc flat seam roofing, .020" thick		1.20	6.667		1,075	325		1,400	1,700
1010	.027" thick		1.15	6.957		1,550	340		1,890	2,225
1020	.032" thick		1.12	7.143		1,825	350		2,175	2,550
1100	Lead flat seam roofing, 5 lb per SF		1.30	6.154		600	300		900	1,125
9000	Minimum labor/equipment charge		2.75	2.909	Job		143		143	223

07 62 Sheet Metal Flashing and Trim

07 62 10 – Sheet Metal Trim

07 62 10.10 Sheet Metal Cladding

		Crew	Daily Output	Labor-Hours	Unit	Material	2010 Bare Costs Labor	Equipment	Total	Total Incl O&P
0010	**SHEET METAL CLADDING**									
0100	Aluminum, up to 6 bends, .032" thick, window casing	1 Carp	180	.044	S.F.	.87	1.85		2.72	3.99
0200	Window sill		72	.111	L.F.	.87	4.62		5.49	8.55
0300	Door casing		180	.044	S.F.	.87	1.85		2.72	3.99
0400	Fascia		250	.032		.87	1.33		2.20	3.14
0500	Rake trim		225	.036		.87	1.48		2.35	3.39
0700	.024" thick, window casing		180	.044		1.55	1.85		3.40	4.74
0800	Window sill		72	.111	L.F.	1.55	4.62		6.17	9.30
0900	Door casing		180	.044	S.F.	1.55	1.85		3.40	4.74
1000	Fascia		250	.032		1.55	1.33		2.88	3.89
1100	Rake trim		225	.036		1.55	1.48		3.03	4.14
1200	Vinyl coated aluminum, up to 6 bends, window casing		180	.044		1.03	1.85		2.88	4.16
1300	Window sill		72	.111	L.F.	1.03	4.62		5.65	8.75
1400	Door casing		180	.044	S.F.	1.03	1.85		2.88	4.16
1500	Fascia		250	.032		1.03	1.33		2.36	3.31
1600	Rake trim		225	.036		1.03	1.48		2.51	3.56

07 65 Flexible Flashing

07 65 10 – Sheet Metal Flashing

07 65 10.10 Sheet Metal Flashing and Counter Flashing

		Crew	Daily Output	Labor-Hours	Unit	Material	Labor	Equipment	Total	Total Incl O&P
0010	**SHEET METAL FLASHING AND COUNTER FLASHING**									
0011	Including up to 4 bends									
0020	Aluminum, mill finish, .013" thick	1 Rofc	145	.055	S.F.	.67	1.95		2.62	4.18
0030	.016" thick		145	.055		.79	1.95		2.74	4.31
0060	.019" thick		145	.055		1.02	1.95		2.97	4.56
0100	.032" thick		145	.055		1.48	1.95		3.43	5.05
0200	.040" thick		145	.055		2.21	1.95		4.16	5.85
0300	.050" thick		145	.055		2.52	1.95		4.47	6.20
0325	Mill finish 5" x 7" step flashing, .016" thick		1920	.004	Ea.	.16	.15		.31	.44
0350	Mill finish 12" x 12" step flashing, .016" thick		1600	.005	"	.61	.18		.79	.98
0400	Painted finish, add				S.F.	.37			.37	.41
1000	Mastic-coated 2 sides, .005" thick	1 Rofc	330	.024		1.67	.86		2.53	3.35
1100	.016" thick		330	.024		1.84	.86		2.70	3.53
1600	Copper, 16 oz, sheets, under 1000 lbs.		115	.070		7.55	2.46		10.01	12.65
1700	Over 4000 lbs.		155	.052		6.75	1.83		8.58	10.65
1900	20 oz sheets, under 1000 lbs.		110	.073		9.05	2.57		11.62	14.50
2000	Over 4000 lbs.		145	.055		8.15	1.95		10.10	12.40
2200	24 oz sheets, under 1000 lbs.		105	.076		11.30	2.70		14	17.20
2300	Over 4000 lbs.		135	.059		10.15	2.10		12.25	14.90
2500	32 oz sheets, under 1000 lbs.		100	.080		14.10	2.83		16.93	20.50
2600	Over 4000 lbs.		130	.062		12.70	2.18		14.88	17.80
2700	W shape for valleys, 16 oz, 24" wide		100	.080	L.F.	13.50	2.83		16.33	19.85
5800	Lead, 2.5 lb. per SF, up to 12" wide		135	.059	S.F.	4.55	2.10		6.65	8.70
5900	Over 12" wide		135	.059		4.55	2.10		6.65	8.70
8650	Copper, 16 oz		100	.080		6	2.83		8.83	11.60
8900	Stainless steel sheets, 32 ga, .010" thick		155	.052		3.79	1.83		5.62	7.40
9000	28 ga, .015" thick		155	.052		4.70	1.83		6.53	8.35
9100	26 ga, .018" thick		155	.052		5.70	1.83		7.53	9.45
9200	24 ga, .025" thick		155	.052		7.40	1.83		9.23	11.35
9290	For mechanically keyed flashing, add					40%				
9320	Steel sheets, galvanized, 20 gauge	1 Rofc	130	.062	S.F.	1.19	2.18		3.37	5.15

07 65 Flexible Flashing

07 65 10 – Sheet Metal Flashing

07 65 10.10 Sheet Metal Flashing and Counter Flashing		Crew	Daily Output	Labor-Hours	Unit	Material	2010 Bare Costs Labor	Equipment	Total	Total Incl O&P
9340	30 gauge	1 Rofc	160	.050	S.F.	.50	1.77		2.27	3.67
9400	Terne coated stainless steel, .015" thick, 28 ga		155	.052		7.35	1.83		9.18	11.30
9500	.018" thick, 26 ga		155	.052		8.30	1.83		10.13	12.30
9600	Zinc and copper alloy (brass), .020" thick		155	.052		5.15	1.83		6.98	8.85
9700	.027" thick		155	.052		6.90	1.83		8.73	10.80
9800	.032" thick		155	.052		8.05	1.83		9.88	12.05
9900	.040" thick		155	.052		9.85	1.83		11.68	14
9950	Minimum labor/equipment charge		3	2.667	Job		94.50		94.50	166

07 65 12 – Fabric and Mastic Flashings

07 65 12.10 Fabric and Mastic Flashing and Counter Flashing

		Crew	Daily Output	Labor-Hours	Unit	Material	Labor	Equipment	Total	Total Incl O&P
0010	**FABRIC AND MASTIC FLASHING AND COUNTER FLASHING**									
1300	Asphalt flashing cement, 5 gallon				Gal.	10.80			10.80	11.85
4900	Fabric, asphalt-saturated cotton, specification grade	1 Rofc	35	.229	S.Y.	2.37	8.10		10.47	16.85
5000	Utility grade		35	.229		1.50	8.10		9.60	15.90
5200	Open-mesh fabric, saturated, 40 oz per S.Y.		35	.229		1.67	8.10		9.77	16.10
5300	Close-mesh fabric, saturated, 17 oz per S.Y.		35	.229		1.75	8.10		9.85	16.20
5500	Fiberglass, resin-coated		35	.229		1.43	8.10		9.53	15.80
5600	Asphalt-coated, 40 oz per S.Y.		35	.229		9.75	8.10		17.85	25
8500	Shower pan, bituminous membrane, 7 oz		155	.052	S.F.	2	1.83		3.83	5.40

07 65 13 – Laminated Sheet Flashing

07 65 13.10 Laminated Sheet Flashing

		Crew	Daily Output	Labor-Hours	Unit	Material	Labor	Equipment	Total	Total Incl O&P
0010	**LAMINATED SHEET FLASHING**, Including up to 4 bends									
0500	Fabric-backed 2 sides, .004" thick	1 Rofc	330	.024	S.F.	1.41	.86		2.27	3.06
0700	.005" thick		330	.024		1.67	.86		2.53	3.35
0750	Mastic-backed, self adhesive		460	.017		3.46	.62		4.08	4.89
0800	Mastic-coated 2 sides, .004" thick		330	.024		1.41	.86		2.27	3.06
2800	Copper, paperbacked 1 side, 2 oz		330	.024		1.42	.86		2.28	3.07
2900	3 oz		330	.024		1.85	.86		2.71	3.55
3100	Paperbacked 2 sides, 2 oz		330	.024		1.43	.86		2.29	3.08
3150	3 oz		330	.024		1.84	.86		2.70	3.53
3200	5 oz		330	.024		2.76	.86		3.62	4.55
3250	7 oz		330	.024		4.50	.86		5.36	6.45
3400	Mastic-backed 2 sides, copper, 2 oz		330	.024		1.69	.86		2.55	3.37
3500	3 oz		330	.024		2.08	.86		2.94	3.80
3700	5 oz		330	.024		3.05	.86		3.91	4.87
3800	Fabric-backed 2 sides, copper, 2 oz		330	.024		1.79	.86		2.65	3.48
4000	3 oz		330	.024		2.31	.86		3.17	4.05
4100	5 oz		330	.024		3.14	.86		4	4.96
4300	Copper-clad stainless steel, .015" thick, under 500 lbs.		115	.070		5.25	2.46		7.71	10.10
4400	Over 2000 lbs.		155	.052		5.05	1.83		6.88	8.75
4600	.018" thick, under 500 lbs.		100	.080		6.90	2.83		9.73	12.60
4700	Over 2000 lbs.		145	.055		5.05	1.95		7	9
6100	Lead-coated copper, fabric-backed, 2 oz		330	.024		2.42	.86		3.28	4.17
6200	5 oz		330	.024		2.78	.86		3.64	4.57
6400	Mastic-backed 2 sides, 2 oz		330	.024		1.89	.86		2.75	3.59
6500	5 oz		330	.024		2.34	.86		3.20	4.08
6700	Paperbacked 1 side, 2 oz		330	.024		1.63	.86		2.49	3.30
6800	3 oz		330	.024		1.92	.86		2.78	3.62
7000	Paperbacked 2 sides, 2 oz		330	.024		1.69	.86		2.55	3.37
7100	5 oz		330	.024		2.75	.86		3.61	4.54
8550	3 ply copper and fabric, 3 oz		155	.052		2.31	1.83		4.14	5.75
8600	7 oz		155	.052		3.25	1.83		5.08	6.80

07 65 Flexible Flashing

07 65 13 – Laminated Sheet Flashing

07 65 13.10 Laminated Sheet Flashing		Crew	Daily Output	Labor-Hours	Unit	Material	2010 Bare Costs Labor	Equipment	Total	Total Incl O&P
8700	Lead on copper and fabric, 5 oz	1 Rofc	155	.052	S.F.	2.78	1.83		4.61	6.30
8800	7 oz		155	.052		5.25	1.83		7.08	9
9300	Stainless steel, paperbacked 2 sides, .005" thick	↓	330	.024	↓	3.34	.86		4.20	5.20

07 65 19 – Plastic Sheet Flashing

07 65 19.10 Plastic Sheet Flashing and Counter Flashing

		Crew	Daily Output	Labor-Hours	Unit	Material	Labor	Equipment	Total	Total Incl O&P
0010	**PLASTIC SHEET FLASHING AND COUNTER FLASHING**									
7300	Polyvinyl chloride, black, .010" thick	1 Rofc	285	.028	S.F.	.23	.99		1.22	2
7400	.020" thick		285	.028		.33	.99		1.32	2.11
7600	.030" thick		285	.028		.42	.99		1.41	2.21
7700	.056" thick		285	.028		1.03	.99		2.02	2.88
7900	Black or white for exposed roofs, .060" thick	↓	285	.028	↓	2.25	.99		3.24	4.23
8060	PVC tape, 5" x 45 mils, for joint covers, 100 L.F./roll				Ea.	120			120	132

07 65 23 – Rubber Sheet Flashing

07 65 23.10 Rubber Sheet Flashing and Counterflashing

		Crew	Daily Output	Labor-Hours	Unit	Material	Labor	Equipment	Total	Total Incl O&P
0010	**RUBBER SHEET FLASHING AND COUNTERFLASHING**									
8100	Rubber, butyl, 1/32" thick	1 Rofc	285	.028	S.F.	1.07	.99		2.06	2.93
8200	1/16" thick		285	.028		1.57	.99		2.56	3.48
8300	Neoprene, cured, 1/16" thick		285	.028		2.03	.99		3.02	3.98
8400	1/8" thick	↓	285	.028	↓	4.09	.99		5.08	6.25

07 71 Roof Specialties

07 71 16 – Manufactured Counterflashing Systems

07 71 16.10 Roof Drain Boot

		Crew	Daily Output	Labor-Hours	Unit	Material	Labor	Equipment	Total	Total Incl O&P
0010	**ROOF DRAIN BOOT**									
0100	Cast iron, 2" x 3"	1 Shee	125	.064	L.F.	114	3.14		117.14	130
0200	3" x 4"		125	.064		227	3.14		230.14	255
0300	4" x 5"		125	.064		238	3.14		241.14	267
0400	4" x 6"	↓	125	.064		286	3.14		289.14	320

07 71 19 – Manufactured Gravel Stops and Fascias

07 71 19.10 Gravel Stop

		Crew	Daily Output	Labor-Hours	Unit	Material	Labor	Equipment	Total	Total Incl O&P
0010	**GRAVEL STOP**									
0020	Aluminum, .050" thick, 4" face height, mill finish	1 Shee	145	.055	L.F.	6.10	2.71		8.81	11
0080	Duranodic finish		145	.055		5.90	2.71		8.61	10.75
0100	Painted		145	.055		6.85	2.71		9.56	11.75
0300	6" face height		135	.059		6.55	2.91		9.46	11.80
0350	Duranodic finish		135	.059		6.95	2.91		9.86	12.20
0400	Painted		135	.059		8.10	2.91		11.01	13.45
0600	8" face height		125	.064		8.20	3.14		11.34	13.95
0650	Duranodic finish		125	.064		8	3.14		11.14	13.70
0700	Painted		125	.064		8.15	3.14		11.29	13.85
0900	12" face height, .080 thick, 2 piece		100	.080		10.95	3.93		14.88	18.20
0950	Duranodic finish		100	.080		10.05	3.93		13.98	17.20
1000	Painted		100	.080		11.90	3.93		15.83	19.25
1500	Polyvinyl chloride, 6" face height		135	.059		4.45	2.91		7.36	9.45
1600	9" face height		125	.064		5.25	3.14		8.39	10.70
1800	Stainless steel, 24 ga., 6" face height		135	.059		11.45	2.91		14.36	17.15
1900	12" face height		100	.080		24	3.93		27.93	32.50
2100	20 ga., 6" face height		135	.059		13	2.91		15.91	18.85
2200	12" face height	↓	100	.080	↓	27.50	3.93		31.43	36

07 71 Roof Specialties

07 71 19 – Manufactured Gravel Stops and Fascias

07 71 19.10 Gravel Stop		Crew	Daily Output	Labor-Hours	Unit	Material	2010 Bare Costs Labor	Equipment	Total	Total Incl O&P
9000	Minimum labor/equipment charge	1 Shee	3.50	2.286	Job		112		112	176

07 71 19.30 Fascia

0010	**FASCIA**									
0100	Aluminum, reverse board and batten, .032" thick, colored, no furring incl	1 Shee	145	.055	S.F.	5.95	2.71		8.66	10.80
0200	Residential type, aluminum	1 Carp	200	.040	L.F.	1.21	1.66		2.87	4.06
0220	Vinyl	"	200	.040	"	1.30	1.66		2.96	4.16
0300	Steel, galv and enameled, stock, no furring, long panels	1 Shee	145	.055	S.F.	4	2.71		6.71	8.65
0600	Short panels		115	.070	"	5.45	3.42		8.87	11.35
9000	Minimum labor/equipment charge		4	2	Job		98		98	154

07 71 23 – Manufactured Gutters and Downspouts

07 71 23.10 Downspouts

0010	**DOWNSPOUTS**									
0020	Aluminum 2" x 3", .020" thick, embossed	1 Shee	190	.042	L.F.	1.04	2.07		3.11	4.37
0100	Enameled		190	.042		1.59	2.07		3.66	4.98
0300	Enameled, .024" thick, 2" x 3"		180	.044		1.84	2.18		4.02	5.45
0400	3" x 4"		140	.057		2.16	2.81		4.97	6.75
0600	Round, corrugated aluminum, 3" diameter, .020" thick		190	.042		1.42	2.07		3.49	4.79
0700	4" diameter, .025" thick		140	.057		2.39	2.81		5.20	7
0900	Wire strainer, round, 2" diameter		155	.052	Ea.	3.32	2.53		5.85	7.60
1000	4" diameter		155	.052		6.10	2.53		8.63	10.70
1200	Rectangular, perforated, 2" x 3"		145	.055		2.30	2.71		5.01	6.75
1300	3" x 4"		145	.055		3.32	2.71		6.03	7.90
1500	Copper, round, 16 oz., stock, 2" diameter		190	.042	L.F.	6.65	2.07		8.72	10.60
1600	3" diameter		190	.042		6.25	2.07		8.32	10.10
1800	4" diameter		145	.055		9	2.71		11.71	14.15
1900	5" diameter		130	.062		14.20	3.02		17.22	20.50
2100	Rectangular, corrugated copper, stock, 2" x 3"		190	.042		7.35	2.07		9.42	11.35
2200	3" x 4"		145	.055		8.60	2.71		11.31	13.70
2400	Rectangular, plain copper, stock, 2" x 3"		190	.042		7	2.07		9.07	10.95
2500	3" x 4"		145	.055		9.10	2.71		11.81	14.25
2700	Wire strainers, rectangular, 2" x 3"		145	.055	Ea.	5.50	2.71		8.21	10.30
2800	3" x 4"		145	.055		6.60	2.71		9.31	11.50
3000	Round, 2" diameter		145	.055		6.90	2.71		9.61	11.85
3100	3" diameter		145	.055		6.90	2.71		9.61	11.85
3300	4" diameter		145	.055		12.70	2.71		15.41	18.25
3400	5" diameter		115	.070		16.90	3.42		20.32	24
3600	Lead-coated copper, round, stock, 2" diameter		190	.042	L.F.	22.50	2.07		24.57	27.50
3700	3" diameter		190	.042		22	2.07		24.07	27
3900	4" diameter		145	.055		23.50	2.71		26.21	29.50
4000	5" diameter, corrugated		130	.062		24.50	3.02		27.52	31.50
4200	6" diameter, corrugated		105	.076		30	3.74		33.74	39
4300	Rectangular, corrugated, stock, 2" x 3"		190	.042		12.70	2.07		14.77	17.20
4500	Plain, stock, 2" x 3"		190	.042		11	2.07		13.07	15.35
4600	3" x 4"		145	.055		12	2.71		14.71	17.45
4800	Steel, galvanized, round, corrugated, 2" or 3" diameter, 28 gauge		190	.042		2.24	2.07		4.31	5.70
4900	4" diameter, 28 gauge		145	.055		2.73	2.71		5.44	7.25
5100	5" diameter, 28 gauge		130	.062		3.53	3.02		6.55	8.60
5200	26 gauge		130	.062		3.75	3.02		6.77	8.85
5400	6" diameter, 28 gauge		105	.076		4	3.74		7.74	10.25
5500	26 gauge		105	.076		5.45	3.74		9.19	11.85
5700	Rectangular, corrugated, 28 gauge, 2" x 3"		190	.042		1.28	2.07		3.35	4.64
5800	3" x 4"		145	.055		2.16	2.71		4.87	6.60

07 71 Roof Specialties

07 71 23 – Manufactured Gutters and Downspouts

07 71 23.10 Downspouts

		Crew	Daily Output	Labor-Hours	Unit	Material	2010 Bare Costs Labor	Equipment	Total	Total Incl O&P
6000	Rectangular, plain, 28 gauge, galvanized, 2" x 3"	1 Shee	190	.042	L.F.	1.28	2.07		3.35	4.64
6100	3" x 4"		145	.055		2.64	2.71		5.35	7.15
6300	Epoxy painted, 24 gauge, corrugated, 2" x 3"		190	.042		2	2.07		4.07	5.45
6400	3" x 4"		145	.055	↓	2.50	2.71		5.21	7
6600	Wire strainers, rectangular, 2" x 3"		145	.055	Ea.	12	2.71		14.71	17.45
6700	3" x 4"		145	.055		14	2.71		16.71	19.65
6900	Round strainers, 2" or 3" diameter		145	.055		3	2.71		5.71	7.55
7000	4" diameter		145	.055	↓	5.60	2.71		8.31	10.40
9000	Minimum labor/equipment charge	↓	4	2	Job		98		98	154

07 71 23.20 Downspout Elbows

		Crew	Daily Output	Labor-Hours	Unit	Material	Labor	Equipment	Total	Total Incl O&P
0010	**DOWNSPOUT ELBOWS**									
0020	Aluminum, 2" x 3", embossed	1 Shee	100	.080	Ea.	1.31	3.93		5.24	7.60
0100	Enameled		100	.080		1.31	3.93		5.24	7.60
0200	3" x 4", .025" thick, embossed		100	.080		3.99	3.93		7.92	10.55
0300	Enameled		100	.080		5.15	3.93		9.08	11.80
0400	Round corrugated, 3", embossed, .020" thick		100	.080		2.27	3.93		6.20	8.65
0500	4", .025" thick		100	.080		6.50	3.93		10.43	13.30
0600	Copper, 16 oz. round, 2" diameter		100	.080		13.40	3.93		17.33	21
0700	3" diameter		100	.080		11.60	3.93		15.53	18.90
0800	4" diameter		100	.080		16.95	3.93		20.88	25
1000	2" x 3" corrugated		100	.080		6.50	3.93		10.43	13.30
1100	3" x 4" corrugated		100	.080	↓	13.15	3.93		17.08	20.50
9000	Minimum labor/equipment charge	↓	4	2	Job		98		98	154

07 71 23.30 Gutters

		Crew	Daily Output	Labor-Hours	Unit	Material	Labor	Equipment	Total	Total Incl O&P
0010	**GUTTERS**									
0012	Aluminum, stock units, 5" K type, .027" thick, plain	1 Shee	120	.067	L.F.	2.75	3.27		6.02	8.15
0100	Enameled		120	.067		2.44	3.27		5.71	7.80
0300	5" K type type, .032" thick, plain		120	.067		2.74	3.27		6.01	8.10
0400	Enameled		120	.067		2.90	3.27		6.17	8.30
0700	Copper, half round, 16 oz, stock units, 4" wide		120	.067		6.95	3.27		10.22	12.75
0900	5" wide		120	.067		7.35	3.27		10.62	13.15
1000	6" wide		115	.070		9.95	3.42		13.37	16.30
1200	K type, 16 oz, stock, 4" wide		120	.067		6.80	3.27		10.07	12.60
1300	5" wide		120	.067		8.20	3.27		11.47	14.10
1500	Lead coated copper, half round, stock, 4" wide		120	.067		9.45	3.27		12.72	15.50
1600	6" wide		115	.070		15.60	3.42		19.02	22.50
1800	K type, stock, 4" wide		120	.067		11.35	3.27		14.62	17.60
1900	5" wide		120	.067		16.35	3.27		19.62	23
2100	Stainless steel, half round or box, stock, 4" wide		120	.067		5.90	3.27		9.17	11.55
2200	5" wide		120	.067		6.35	3.27		9.62	12.05
2400	Steel, galv, half round or box, 28 ga, 5" wide, plain		120	.067		1.46	3.27		4.73	6.70
2500	Enameled		120	.067		1.51	3.27		4.78	6.75
2700	26 ga, stock, 5" wide		120	.067		1.59	3.27		4.86	6.85
2800	6" wide	↓	120	.067		2.99	3.27		6.26	8.40
3000	Vinyl, O.G., 4" wide	1 Carp	110	.073		1.08	3.02		4.10	6.15
3100	5" wide		110	.073		1.38	3.02		4.40	6.50
3200	4" half round, stock units	↓	110	.073	↓	1.08	3.02		4.10	6.15
3250	Joint connectors				Ea.	2.88			2.88	3.17
3300	Wood, clear treated cedar, fir or hemlock, 3" x 4"	1 Carp	100	.080	L.F.	7.95	3.32		11.27	14.20
3400	4" x 5"	"	100	.080	"	9.45	3.32		12.77	15.85
9000	Minimum labor/equipment charge	1 Shee	3.75	2.133	Job		105		105	164

07 71 Roof Specialties

07 71 23 – Manufactured Gutters and Downspouts

07 71 23.35 Gutter Guard

		Crew	Daily Output	Labor-Hours	Unit	Material	2010 Bare Costs Labor	2010 Bare Costs Equipment	Total	Total Incl O&P
0010	**GUTTER GUARD**									
0020	6" wide strip, aluminum mesh	1 Carp	500	.016	L.F.	2.18	.66		2.84	3.49
0100	Vinyl mesh		500	.016	"	2.58	.66		3.24	3.93
9000	Minimum labor/equipment charge		4	2	Job		83		83	137

07 71 26 – Reglets

07 71 26.10 Reglets and Accessories

		Crew	Daily Output	Labor-Hours	Unit	Material	2010 Bare Costs Labor	2010 Bare Costs Equipment	Total	Total Incl O&P
0010	**REGLETS AND ACCESSORIES**									
0020	Aluminum, .025" thick, in concrete parapet	1 Carp	225	.036	L.F.	1.39	1.48		2.87	3.96
0100	Copper, 10 oz.		225	.036		2.62	1.48		4.10	5.30
0300	16 oz.		225	.036		5.10	1.48		6.58	8.10
0400	Galvanized steel, 24 gauge		225	.036		.94	1.48		2.42	3.46
0600	Stainless steel, .020" thick		225	.036		3.15	1.48		4.63	5.90
0700	Zinc and copper alloy, 20 oz.		225	.036		2.62	1.48		4.10	5.30
0900	Counter flashing for above, 12" wide, .032 aluminum	1 Shee	150	.053		1.63	2.62		4.25	5.90
1000	Copper, 10 oz.		150	.053		4.95	2.62		7.57	9.55
1200	16 oz.		150	.053		4.80	2.62		7.42	9.40
1300	Galvanized steel, .020" thick		150	.053		1.17	2.62		3.79	5.40
1500	Stainless steel, .020" thick		150	.053		5.10	2.62		7.72	9.70
1600	Zinc and copper alloy, 20 oz.		150	.053		4.78	2.62		7.40	9.35
9000	Minimum labor/equipment charge	1 Carp	3	2.667	Job		111		111	182

07 71 29 – Manufactured Roof Expansion Joints

07 71 29.10 Expansion Joints

		Crew	Daily Output	Labor-Hours	Unit	Material	2010 Bare Costs Labor	2010 Bare Costs Equipment	Total	Total Incl O&P
0010	**EXPANSION JOINTS**									
0300	Butyl or neoprene center with foam insulation, metal flanges									
0400	Aluminum, .032" thick for openings to 2-1/2"	1 Rofc	165	.048	L.F.	11.65	1.72		13.37	15.80
0600	For joint openings to 3-1/2"		165	.048		13.60	1.72		15.32	17.95
0610	For joint openings to 5"		165	.048		15.50	1.72		17.22	20
0620	For joint openings to 8"		165	.048		26	1.72		27.72	31.50
0700	Copper, 16 oz. for openings to 2-1/2"		165	.048		18.95	1.72		20.67	24
0900	For joint openings to 3-1/2"		165	.048		22	1.72		23.72	27
0910	For joint openings to 5"		165	.048		25.50	1.72		27.22	31
0920	For joint openings to 8"		165	.048		38	1.72		39.72	45
1000	Galvanized steel, 26 ga. for openings to 2-1/2"		165	.048		9.55	1.72		11.27	13.55
1200	For joint openings to 3-1/2"		165	.048		11.30	1.72		13.02	15.45
1210	For joint openings to 5"		165	.048		14.15	1.72		15.87	18.60
1220	For joint openings to 8"		165	.048		24	1.72		25.72	29.50
1300	Lead-coated copper, 16 oz. for openings to 2-1/2"		165	.048		33.50	1.72		35.22	40
1500	For joint openings to 3-1/2"		165	.048		40	1.72		41.72	47
1600	Stainless steel, .018", for openings to 2-1/2"		165	.048		14.55	1.72		16.27	19.05
1800	For joint openings to 3-1/2"		165	.048		16.60	1.72		18.32	21.50
1810	For joint openings to 5"		165	.048		20.50	1.72		22.22	25.50
1820	For joint openings to 8"		165	.048		31	1.72		32.72	37.50
1900	Neoprene, double-seal type with thick center, 4-1/2" wide		125	.064		13.60	2.27		15.87	19
1950	Polyethylene bellows, with galv steel flat flanges		100	.080		5.25	2.83		8.08	10.80
1960	With galvanized angle flanges		100	.080		5.80	2.83		8.63	11.40
2000	Roof joint with extruded aluminum cover, 2"	1 Shee	115	.070		36.50	3.42		39.92	45.50
2100	Roof joint, plastic curbs, foam center, standard	1 Rofc	100	.080		13.30	2.83		16.13	19.65
2200	Large		100	.080		17.75	2.83		20.58	24.50
2300	Transitions, regular, minimum		10	.800	Ea.	115	28.50		143.50	176
2350	Maximum		4	2		146	71		217	286
2400	Large, minimum		9	.889		168	31.50		199.50	241

07 71 Roof Specialties

07 71 29 – Manufactured Roof Expansion Joints

07 71 29.10 Expansion Joints		Crew	Daily Output	Labor-Hours	Unit	Material	2010 Bare Costs Labor	Equipment	Total	Total Incl O&P
2450	Maximum	1 Rofc	3	2.667	Ea.	176	94.50		270.50	360
2500	Roof to wall joint with extruded aluminum cover	1 Shee	115	.070	L.F.	31	3.42		34.42	39.50
2650										
2700	Wall joint, closed cell foam on PVC cover, 9" wide	1 Rofc	125	.064	L.F.	4.74	2.27		7.01	9.20
2800	12" wide	"	115	.070	"	5.35	2.46		7.81	10.25
9000	Minimum labor/equipment charge	1 Shee	3	2.667	Job		131		131	205

07 71 43 – Drip Edge

07 71 43.10 Drip Edge, Rake Edge, Ice Belts

		Crew	Daily Output	Labor-Hours	Unit	Material	Labor	Equipment	Total	Total Incl O&P
0010	**DRIP EDGE, RAKE EDGE, ICE BELTS**									
0020	Aluminum, .016" thick, 5" wide, mill finish	1 Carp	400	.020	L.F.	.38	.83		1.21	1.79
0100	White finish		400	.020		.38	.83		1.21	1.79
0200	8" wide, mill finish		400	.020		.46	.83		1.29	1.88
0300	Ice belt, 28" wide, mill finish		100	.080		3.85	3.32		7.17	9.70
0310	Vented, mill finish		400	.020		2.76	.83		3.59	4.41
0320	Painted finish		400	.020		2.86	.83		3.69	4.52
0400	Galvanized, 5" wide		400	.020		.48	.83		1.31	1.90
0500	8" wide, mill finish		400	.020		.69	.83		1.52	2.13
0510	Rake edge, aluminum, 1-1/2" x 1-1/2"		400	.020		.26	.83		1.09	1.66
0520	3-1/2" x 1-1/2"		400	.020		.43	.83		1.26	1.84
9000	Minimum labor/equipment charge		4	2	Job		83		83	137

07 72 Roof Accessories

07 72 23 – Relief Vents

07 72 23.10 Roof Vents

		Crew	Daily Output	Labor-Hours	Unit	Material	Labor	Equipment	Total	Total Incl O&P
0010	**ROOF VENTS**									
0020	Mushroom shape, for built-up roofs, aluminum	1 Rofc	30	.267	Ea.	68.50	9.45		77.95	92
0100	PVC, 6" high		30	.267	"	28.50	9.45		37.95	48
9000	Minimum labor/equipment charge		2.75	2.909	Job		103		103	181

07 72 23.20 Vents

		Crew	Daily Output	Labor-Hours	Unit	Material	Labor	Equipment	Total	Total Incl O&P
0010	**VENTS**									
0100	Soffit or eave, aluminum, mill finish, strips, 2-1/2" wide	1 Carp	200	.040	L.F.	.40	1.66		2.06	3.17
0200	3" wide		200	.040		.44	1.66		2.10	3.21
0300	Enamel finish, 3" wide		200	.040		.52	1.66		2.18	3.30
0400	Mill finish, rectangular, 4" x 16"		72	.111	Ea.	1.43	4.62		6.05	9.15
0500	8" x 16"		72	.111		2.15	4.62		6.77	9.95
2420	Roof ventilator	Q-9	16	1		50.50	44		94.50	125

07 72 26 – Ridge Vents

07 72 26.10 Ridge Vents and Accessories

		Crew	Daily Output	Labor-Hours	Unit	Material	Labor	Equipment	Total	Total Incl O&P
0010	**RIDGE VENTS AND ACCESSORIES**									
0100	Aluminum strips, mill finish	1 Rofc	160	.050	L.F.	2.62	1.77		4.39	6
0150	Painted finish		160	.050	"	3.74	1.77		5.51	7.25
0200	Connectors		48	.167	Ea.	3.68	5.90		9.58	14.45
0300	End caps		48	.167	"	1.37	5.90		7.27	11.90
0400	Galvanized strips		160	.050	L.F.	2.26	1.77		4.03	5.60
0430	Molded polyethylene, shingles not included		160	.050	"	3.05	1.77		4.82	6.50
0440	End plugs		48	.167	Ea.	1.37	5.90		7.27	11.90
0450	Flexible roll, shingles not included		160	.050	L.F.	2.25	1.77		4.02	5.60
2300	Ridge vent strip, mill finish	1 Shee	155	.052	"	2.73	2.53		5.26	6.95

07 72 Roof Accessories

07 72 33 – Roof Hatches

07 72 33.10 Roof Hatch Options

	07 72 33.10 Roof Hatch Options	Crew	Daily Output	Labor-Hours	Unit	Material	2010 Bare Costs Labor	Equipment	Total	Total Incl O&P
0010	**ROOF HATCH OPTIONS**									
0500	2'-6" x 3', aluminum curb and cover	G-3	10	3.200	Ea.	985	132		1,117	1,275
0520	Galvanized steel curb and aluminum cover		10	3.200		595	132		727	865
0540	Galvanized steel curb and cover		10	3.200		500	132		632	760
0600	2'-6" x 4'-6", aluminum curb and cover		9	3.556		850	146		996	1,175
0800	Galvanized steel curb and aluminum cover		9	3.556		600	146		746	895
0900	Galvanized steel curb and cover		9	3.556		825	146		971	1,150
1100	4' x 4' aluminum curb and cover		8	4		1,675	164		1,839	2,075
1120	Galvanized steel curb and aluminum cover		8	4		1,675	164		1,839	2,075
1140	Galvanized steel curb and cover		8	4		1,600	164		1,764	2,025
1200	2'-6" x 8'-0", aluminum curb and cover		6.60	4.848		1,675	199		1,874	2,150
1400	Galvanized steel curb and aluminum cover		6.60	4.848		1,600	199		1,799	2,075
1500	Galvanized steel curb and cover	↓	6.60	4.848		1,200	199		1,399	1,650
1800	For plexiglass panels, 2'-6" x 3'-0", add to above				↓	400			400	440
9000	Minimum labor/equipment charge	2 Carp	2	8	Job		330		330	545

07 72 36 – Smoke Vents

07 72 36.10 Smoke Hatches

	07 72 36.10 Smoke Hatches	Crew	Daily Output	Labor-Hours	Unit	Material	Labor	Equipment	Total	Total Incl O&P
0010	**SMOKE HATCHES**									
0200	For 3'-0" long, add to roof hatches from Division 07 72 33.10				Ea.	25%	5%			
0250	For 4'-0" long, add to roof hatches from Division 07 72 33.10					20%	5%			
0300	For 8'-0" long, add to roof hatches from Division 07 72 33.10				↓	10%	5%			

07 72 36.20 Smoke Vent Options

	07 72 36.20 Smoke Vent Options	Crew	Daily Output	Labor-Hours	Unit	Material	Labor	Equipment	Total	Total Incl O&P
0010	**SMOKE VENT OPTIONS**									
0100	4' x 4' aluminum cover and frame	G-3	13	2.462	Ea.	1,900	101		2,001	2,250
0200	Galvanized steel cover and frame		13	2.462		1,675	101		1,776	1,975
0300	4' x 8' aluminum cover and frame		8	4		2,600	164		2,764	3,100
0400	Galvanized steel cover and frame	↓	8	4	↓	2,200	164		2,364	2,675
9000	Minimum labor/equipment charge	2 Carp	2	8	Job		330		330	545

07 72 53 – Snow Guards

07 72 53.10 Snow Guard Options

	07 72 53.10 Snow Guard Options	Crew	Daily Output	Labor-Hours	Unit	Material	Labor	Equipment	Total	Total Incl O&P
0010	**SNOW GUARD OPTIONS**									
0100	Slate & asphalt shingle roofs, fastened with nails	1 Rofc	160	.050	Ea.	9.40	1.77		11.17	13.45
0200	Standing seam metal roofs, fastened with set screws		48	.167		14.60	5.90		20.50	26.50
0300	Surface mount for metal roofs, fastened with solder		48	.167	↓	7.70	5.90		13.60	18.85
0400	Double rail pipe type, including pipe	↓	130	.062	L.F.	22.50	2.18		24.68	29

07 72 73 – Pitch Pockets

07 72 73.10 Pitch Pockets, Variable Sizes

	07 72 73.10 Pitch Pockets, Variable Sizes	Crew	Daily Output	Labor-Hours	Unit	Material	Labor	Equipment	Total	Total Incl O&P
0010	**PITCH POCKETS, VARIABLE SIZES**									
0100	Adjustable, 4" to 7", welded corners, 4" deep	1 Rofc	48	.167	Ea.	12.90	5.90		18.80	24.50
0200	Side extenders, 6"	"	240	.033	"	1.97	1.18		3.15	4.25

07 72 80 – Vents

07 72 80.30 Vent Options

	07 72 80.30 Vent Options	Crew	Daily Output	Labor-Hours	Unit	Material	Labor	Equipment	Total	Total Incl O&P
0010	**VENT OPTIONS**									
0020	Plastic, for insulated decks, 1 per M.S.F., minimum	1 Rofc	40	.200	Ea.	19	7.10		26.10	33.50
0100	Maximum		20	.400		38	14.15		52.15	67
0300	Aluminum	↓	30	.267		19	9.45		28.45	37.50
0800	Polystyrene baffles, 12" wide for 16" O.C. rafter spacing	1 Carp	90	.089		.40	3.69		4.09	6.50
0900	For 24" O.C. rafter spacing		110	.073	↓	.75	3.02		3.77	5.80
9000	Minimum labor/equipment charge	↓	3	2.667	Job		111		111	182

07 81 Applied Fireproofing

07 81 16 – Cementitious Fireproofing

07 81 16.10 Sprayed Cementitious Fireproofing		Crew	Daily Output	Labor-Hours	Unit	Material	2010 Bare Costs Labor	Equipment	Total	Total Incl O&P
0010	**SPRAYED CEMENTITIOUS FIREPROOFING**									
0050	Not incl tamping or canvas protection									
0100	1" thick, on flat plate steel	G-2	3000	.008	S.F.	.53	.28	.04	.85	1.08
0200	Flat decking		2400	.010		.53	.35	.06	.94	1.20
0400	Beams		1500	.016		.53	.56	.09	1.18	1.57
0500	Corrugated or fluted decks		1250	.019		.79	.67	.11	1.57	2.06
0700	Columns, 1-1/8" thick		1100	.022		.59	.76	.12	1.47	2
0800	2-3/16" thick		700	.034		1.13	1.19	.19	2.51	3.37
0850	For tamping, add						10%			
0900	For canvas protection, add	G-2	5000	.005	S.F.	.07	.17	.03	.27	.38
9000	Minimum labor/equipment charge	"	3	8	Job		278	44	322	495

07 84 Firestopping

07 84 13 – Penetration Firestopping

07 84 13.10 Firestopping			Crew	Daily Output	Labor-Hours	Unit	Material	2010 Bare Costs Labor	Equipment	Total	Total Incl O&P
0010	**FIRESTOPPING**	R078413-30									
0100	Metallic piping, non insulated										
0110	Through walls, 2" diameter		1 Carp	16	.500	Ea.	12.40	21		33.40	47.50
0120	4" diameter			14	.571		18.90	23.50		42.40	60
0130	6" diameter			12	.667		25.50	27.50		53	73.50
0140	12" diameter			10	.800		45	33		78	104
0150	Through floors, 2" diameter			32	.250		7.50	10.40		17.90	25.50
0160	4" diameter			28	.286		10.80	11.85		22.65	31.50
0170	6" diameter			24	.333		14.20	13.85		28.05	38.50
0180	12" diameter			20	.400		24	16.60		40.60	54
0190	Metallic piping, insulated										
0200	Through walls, 2" diameter		1 Carp	16	.500	Ea.	17.55	21		38.55	53.50
0210	4" diameter			14	.571		24	23.50		47.50	65.50
0220	6" diameter			12	.667		30.50	27.50		58	79
0230	12" diameter			10	.800		50	33		83	110
0240	Through floors, 2" diameter			32	.250		12.70	10.40		23.10	31
0250	4" diameter			28	.286		15.95	11.85		27.80	37
0260	6" diameter			24	.333		19.35	13.85		33.20	44.50
0270	12" diameter			20	.400		24	16.60		40.60	54
0280	Non metallic piping, non insulated										
0290	Through walls, 2" diameter		1 Carp	12	.667	Ea.	51	27.50		78.50	102
0300	4" diameter			10	.800		64	33		97	125
0310	6" diameter			8	1		89	41.50		130.50	167
0330	Through floors, 2" diameter			16	.500		40	21		61	78
0340	4" diameter			6	1.333		49.50	55.50		105	146
0350	6" diameter			6	1.333		59.50	55.50		115	157
0370	Ductwork, insulated & non insulated, round										
0380	Through walls, 6" diameter		1 Carp	12	.667	Ea.	26	27.50		53.50	74
0390	12" diameter			10	.800		51.50	33		84.50	112
0400	18" diameter			8	1		84	41.50		125.50	161
0410	Through floors, 6" diameter			16	.500		14.20	21		35.20	49.50
0420	12" diameter			14	.571		26	23.50		49.50	67.50
0430	18" diameter			12	.667		45	27.50		72.50	95
0440	Ductwork, insulated & non insulated, rectangular										
0450	With stiffener/closure angle, through walls, 6" x 12"		1 Carp	8	1	Ea.	21.50	41.50		63	92
0460	12" x 24"			6	1.333		28.50	55.50		84	123

07 84 Firestopping

07 84 13 – Penetration Firestopping

07 84 13.10 Firestopping

		Crew	Daily Output	Labor-Hours	Unit	Material	2010 Bare Costs Labor	Equipment	Total	Total Incl O&P
0470	24" x 48"	1 Carp	4	2	Ea.	81.50	83		164.50	227
0480	With stiffener/closure angle, through floors, 6" x 12"		10	.800		11.65	33		44.65	67.50
0490	12" x 24"		8	1		21	41.50		62.50	91.50
0500	24" x 48"		6	1.333		41	55.50		96.50	136
0510	Multi trade openings									
0520	Through walls, 6" x 12"	1 Carp	2	4	Ea.	45	166		211	325
0530	12" x 24"	"	1	8		182	330		512	745
0540	24" x 48"	2 Carp	1	16		730	665		1,395	1,900
0550	48" x 96"	"	.75	21.333		2,925	885		3,810	4,675
0560	Through floors, 6" x 12"	1 Carp	2	4		45	166		211	325
0570	12" x 24"	"	1	8		182	330		512	745
0580	24" x 48"	2 Carp	.75	21.333		730	885		1,615	2,250
0590	48" x 96"	"	.50	32		2,925	1,325		4,250	5,400
0600	Structural penetrations, through walls									
0610	Steel beams, W8 x 10	1 Carp	8	1	Ea.	28.50	41.50		70	100
0620	W12 x 14		6	1.333		45	55.50		100.50	141
0630	W21 x 44		5	1.600		90.50	66.50		157	209
0640	W36 x 135		3	2.667		220	111		331	425
0650	Bar joists, 18" deep		6	1.333		41.50	55.50		97	137
0660	24" deep		6	1.333		51.50	55.50		107	148
0670	36" deep		5	1.600		77.50	66.50		144	195
0680	48" deep		4	2		90.50	83		173.50	237
0690	Construction joints, floor slab at exterior wall									
0700	Precast, brick, block or drywall exterior									
0710	2" wide joint	1 Carp	125	.064	L.F.	6.45	2.66		9.11	11.45
0720	4" wide joint	"	75	.107	"	12.90	4.43		17.33	21.50
0730	Metal panel, glass or curtain wall exterior									
0740	2" wide joint	1 Carp	40	.200	L.F.	15.30	8.30		23.60	30.50
0750	4" wide joint	"	25	.320	"	21	13.30		34.30	45
0760	Floor slab to drywall partition									
0770	Flat joint	1 Carp	100	.080	L.F.	6.35	3.32		9.67	12.40
0780	Fluted joint		50	.160		12.90	6.65		19.55	25
0790	Etched fluted joint		75	.107		8.40	4.43		12.83	16.55
0800	Floor slab to concrete/masonry partition									
0810	Flat joint	1 Carp	75	.107	L.F.	14.20	4.43		18.63	23
0820	Fluted joint	"	50	.160	"	16.80	6.65		23.45	29.50
0830	Concrete/CMU wall joints									
0840	1" wide	1 Carp	100	.080	L.F.	7.75	3.32		11.07	14
0850	2" wide		75	.107		14.20	4.43		18.63	23
0860	4" wide		50	.160		27	6.65		33.65	41
0870	Concrete/CMU floor joints									
0880	1" wide	1 Carp	200	.040	L.F.	3.88	1.66		5.54	7
0890	2" wide		150	.053		7.10	2.22		9.32	11.45
0900	4" wide		100	.080		13.55	3.32		16.87	20.50

07 91 Pre-formed Joint Seals

07 91 13 – Compression Seals

07 91 13.10 Compression Seals		Crew	Daily Output	Labor-Hours	Unit	Material	2010 Bare Costs Labor	Equipment	Total	Total Incl O&P
0010	**COMPRESSION SEALS**									
4900	Compression seals, O-ring type cord, 1/4"	1 Bric	472	.017	L.F.	.47	.71		1.18	1.66
4910	1/2"		440	.018		1.78	.76		2.54	3.18
4920	3/4"		424	.019		3.61	.79		4.40	5.25
4930	1"		408	.020		6.95	.82		7.77	8.95
4940	1-1/4"		384	.021		13.60	.87		14.47	16.35
4950	1-1/2"		368	.022		17.20	.91		18.11	20.50
4960	1-3/4"		352	.023		36	.95		36.95	41
4970	2"		344	.023		63.50	.97		64.47	71.50

07 91 16 – Joint Gaskets

07 91 16.10 Joint Gaskets		Crew	Daily Output	Labor-Hours	Unit	Material	Labor	Equipment	Total	Total Incl O&P
0010	**JOINT GASKETS**									
4400	Joint gaskets, neoprene, closed cell w/adh, 1/8" x 3/8"	1 Bric	240	.033	L.F.	.26	1.39		1.65	2.53
4500	1/4" x 3/4"		215	.037		.58	1.55		2.13	3.14
4700	1/2" x 1"		200	.040		1.30	1.67		2.97	4.12
4800	3/4" x 1-1/2"		165	.048		1.44	2.02		3.46	4.84

07 91 23 – Backer Rods

07 91 23.10 Backer Rods		Crew	Daily Output	Labor-Hours	Unit	Material	Labor	Equipment	Total	Total Incl O&P
0010	**BACKER RODS**									
0032	Backer rod, polyethylene, 1/4" diameter	1 Bric	460	.017	L.F.	.03	.73		.76	1.20
0052	1/2" diameter		460	.017		.07	.73		.80	1.24
0072	3/4" diameter		460	.017		.10	.73		.83	1.28
0092	1" diameter		460	.017		.19	.73		.92	1.38

07 91 26 – Joint Fillers

07 91 26.10 Joint Fillers		Crew	Daily Output	Labor-Hours	Unit	Material	Labor	Equipment	Total	Total Incl O&P
0010	**JOINT FILLERS**									
4360	Butyl rubber filler, 1/4" x 1/4"	1 Bric	290	.028	L.F.	.14	1.15		1.29	2.01
4365	1/2" x 1/2"		250	.032		.57	1.34		1.91	2.77
4370	1/2" x 3/4"		210	.038		.85	1.59		2.44	3.50
4375	3/4" x 3/4"		230	.035		1.28	1.45		2.73	3.75
4380	1" x 1"		180	.044		1.71	1.86		3.57	4.86
4390	For coloring, add					12%				
4980	Polyethylene joint backing, 1/4" x 2"	1 Bric	2.08	3.846	C.L.F.	12	161		173	271
4990	1/4" x 6"		1.28	6.250	"	36	261		297	460
5600	Silicone, room temp vulcanizing foam seal, 1/4" x 1/2"		1312	.006	L.F.	.29	.25		.54	.73
5610	1/2" x 1/2"		656	.012		.57	.51		1.08	1.45
5620	1/2" x 3/4"		442	.018		.86	.76		1.62	2.17
5630	3/4" x 3/4"		328	.024		1.29	1.02		2.31	3.06
5640	1/8" x 1"		1312	.006		.29	.25		.54	.73
5650	1/8" x 3"		442	.018		.86	.76		1.62	2.17
5670	1/4" x 3"		295	.027		1.72	1.13		2.85	3.71
5680	1/4" x 6"		148	.054		3.44	2.26		5.70	7.40
5690	1/2" x 6"		82	.098		6.90	4.07		10.97	14.10
5700	1/2" x 9"		52.50	.152		10.30	6.35		16.65	21.50
5710	1/2" x 12"		33	.242		13.75	10.10		23.85	31.50

07 92 Joint Sealants

07 92 13 – Elastomeric Joint Sealants

07 92 13.10 Masonry Joint Sealants

		Crew	Daily Output	Labor-Hours	Unit	Material	2010 Bare Costs Labor	Equipment	Total	Total Incl O&P
0010	**MASONRY JOINT SEALANTS**, 1/2" x 1/2" joint									
0050	Re-caulk only, oil base	1 Bric	225	.036	L.F.	.35	1.48		1.83	2.77
0100	Acrylic latex		205	.039		.16	1.63		1.79	2.80
0200	Polyurethane		200	.040		4.17	1.67		5.84	7.25
0300	Silicone		195	.041		3.41	1.71		5.12	6.50
1000	Cut out and re-caulk, oil base		145	.055		.35	2.30		2.65	4.08
1050	Acrylic latex		130	.062		.16	2.57		2.73	4.31
1100	Polyurethane		125	.064		4.17	2.67		6.84	8.90
1150	Silicone		120	.067		3.41	2.78		6.19	8.25
9000	Minimum labor/equipment charge		4	2	Job		83.50		83.50	134

07 92 13.20 Caulking and Sealant Options

		Crew	Daily Output	Labor-Hours	Unit	Material	Labor	Equipment	Total	Total Incl O&P
0010	**CAULKING AND SEALANT OPTIONS**									
0050	Latex acrylic based, bulk				Gal.	27			27	29.50
0055	Bulk in place 1/4" x 1/4" bead	1 Bric	300	.027	L.F.	.09	1.11		1.20	1.88
0060	1/4" x 3/8"		294	.027		.14	1.14		1.28	1.99
0065	1/4" x 1/2"		288	.028		.19	1.16		1.35	2.08
0075	3/8" x 3/8"		284	.028		.21	1.18		1.39	2.12
0080	3/8" x 1/2"		280	.029		.28	1.19		1.47	2.23
0085	3/8" x 5/8"		276	.029		.35	1.21		1.56	2.34
0095	3/8" x 3/4"		272	.029		.42	1.23		1.65	2.44
0100	1/2" x 1/2"		275	.029		.38	1.21		1.59	2.37
0105	1/2" x 5/8"		269	.030		.47	1.24		1.71	2.52
0110	1/2" x 3/4"		263	.030		.57	1.27		1.84	2.67
0115	1/2" x 7/8"		256	.031		.66	1.30		1.96	2.83
0120	1/2" x 1"		250	.032		.76	1.34		2.10	2.98
0125	3/4" x 3/4"		244	.033		.85	1.37		2.22	3.14
0130	3/4" x 1"		225	.036		1.14	1.48		2.62	3.64
0135	1" x 1"		200	.040		1.52	1.67		3.19	4.36
0190	Cartridges				Gal.	22.50			22.50	24.50
0200	11 fl. oz cartridge				Ea.	1.93			1.93	2.12
0500	1/4" x 1/2"	1 Bric	288	.028	L.F.	.16	1.16		1.32	2.04
0600	1/2" x 1/2"		275	.029		.32	1.21		1.53	2.30
0800	3/4" x 3/4"		244	.033		.71	1.37		2.08	2.98
0900	3/4" x 1"		225	.036		.95	1.48		2.43	3.43
1000	1" x 1"		200	.040		1.18	1.67		2.85	3.99
1400	Butyl based, bulk				Gal.	26			26	28.50
1500	Cartridges				"	32			32	35
1700	1/4" x 1/2", 154 L.F./gal.	1 Bric	288	.028	L.F.	.17	1.16		1.33	2.06
1800	1/2" x 1/2", 77 L.F./gal.	"	275	.029	"	.34	1.21		1.55	2.32
2300	Polysulfide compounds, 1 component, bulk				Gal.	51			51	56
2400	Cartridges				"	54.50			54.50	60
2600	1 or 2 component, in place, 1/4" x 1/4", 308 L.F./gal.	1 Bric	300	.027	L.F.	.17	1.11		1.28	1.97
2700	1/2" x 1/4", 154 L.F./gal.		288	.028		.33	1.16		1.49	2.24
2900	3/4" x 3/8", 68 L.F./gal.		272	.029		.75	1.23		1.98	2.80
3000	1" x 1/2", 38 L.F./gal.		250	.032		1.35	1.34		2.69	3.63
3200	Polyurethane, 1 or 2 component				Gal.	50			50	55
3300	Cartridges				"	64			64	70.50
3500	Bulk, in place, 1/4" x 1/4"	1 Bric	300	.027	L.F.	.16	1.11		1.27	1.97
3655	1/2" x 1/4"		288	.028		.32	1.16		1.48	2.23
3800	3/4" x 3/8", 68 L.F./gal.		272	.029		.74	1.23		1.97	2.78
3900	1" x 1/2"		250	.032		1.30	1.34		2.64	3.58
4100	Silicone rubber, bulk				Gal.	41			41	45

07 92 Joint Sealants

07 92 13 – Elastomeric Joint Sealants

07 92 13.20 Caulking and Sealant Options	Crew	Daily Output	Labor-Hours	Unit	Material	2010 Bare Costs Labor	Equipment	Total	Total Incl O&P
4200 Cartridges				Gal.	41			41	45
4300 Bulk in place, 1/4" x 1/2", 154 L.F./gal.	1 Bric	235	.034	L.F.	.27	1.42		1.69	2.58

07 92 16 – Rigid Joint Sealants

07 92 16.10 Rigid Joint Sealants

0010 **RIGID JOINT SEALANTS**									
5802 Tapes, sealant, P.V.C. foam adhesive, 1/16" x 1/4"				L.F.	.06			.06	.06
5902 1/16" x 1/2"					.08			.08	.09
5952 1/16" x 1"					.14			.14	.15
6002 1/8" x 1/2"					.09			.09	.10

07 92 19 – Acoustical Joint Sealants

07 92 19.10 Acoustical Sealant

0010 **ACOUSTICAL SEALANT**									
0020 Acoustical sealant, elastomeric, cartridges				Ea.	2.25			2.25	2.48
0025 In place, 1/4" x 1/4"	1 Bric	300	.027	L.F.	.09	1.11		1.20	1.89
0030 1/4" x 1/2"		288	.028		.18	1.16		1.34	2.07
0035 1/2" x 1/2"		275	.029		.37	1.21		1.58	2.35
0040 1/2" x 3/4"		263	.030		.55	1.27		1.82	2.65
0045 3/4" x 3/4"		244	.033		.83	1.37		2.20	3.11
0050 1" x 1"		200	.040		1.47	1.67		3.14	4.31

07 95 Expansion Control

07 95 13 – Expansion Joint Cover Assemblies

07 95 13.50 Expansion Joint Assemblies

	Crew	Daily Output	Labor-Hours	Unit	Material	Labor	Equipment	Total	Total Incl O&P
0010 **EXPANSION JOINT ASSEMBLIES**									
0200 Floor cover assemblies, 1" space, aluminum	1 Sswk	38	.211	L.F.	20.50	9.85		30.35	40.50
0300 Bronze		38	.211		41.50	9.85		51.35	63.50
0500 2" space, aluminum		38	.211		24.50	9.85		34.35	45
0600 Bronze		38	.211		44.50	9.85		54.35	67
0800 Wall and ceiling assemblies, 1" space, aluminum		38	.211		12.15	9.85		22	31.50
0900 Bronze		38	.211		38.50	9.85		48.35	60.50
1100 2" space, aluminum		38	.211		20.50	9.85		30.35	40.50
1200 Bronze		38	.211		47.50	9.85		57.35	70.50
1400 Floor to wall assemblies, 1" space, aluminum		38	.211		18.55	9.85		28.40	38.50
1500 Bronze or stainless		38	.211		47	9.85		56.85	69.50
1700 Gym floor angle covers, aluminum, 3" x 3" angle		46	.174		16.05	8.15		24.20	32.50
1800 3" x 4" angle		46	.174		18.95	8.15		27.10	36
2000 Roof closures, aluminum, flat roof, low profile, 1" space		57	.140		37	6.60		43.60	53
2100 High profile		57	.140		45.50	6.60		52.10	62
2300 Roof to wall, low profile, 1" space		57	.140		20.50	6.60		27.10	34.50
2400 High profile		57	.140		26.50	6.60		33.10	41
9000 Minimum labor/equipment charge		2	4	Job		188		188	345

Estimating Tips

08 10 00 Doors and Frames

All exterior doors should be addressed for their energy conservation.

- Most metal doors and frames look alike, but there may be significant differences among them. When estimating these items, be sure to choose the line item that most closely compares to the specification or door schedule requirements regarding:
  — type of metal
  — metal gauge
  — door core material
  — fire rating
  — finish

- Wood and plastic doors vary considerably in price. The primary determinant is the veneer material. Lauan, birch, and oak are the most common veneers. Other variables include the following:
  — hollow or solid core
  — fire rating
  — flush or raised panel
  — finish

08 30 00 Specialty Doors and Frames

- There are many varieties of special doors, and they are usually priced per each. Add frames, hardware, or operators required for a complete installation.

08 40 00 Entrances, Storefronts, and Curtain Walls

- Glazed curtain walls consist of the metal tube framing and the glazing material. The cost data in this subdivision is presented for the metal tube framing alone or the composite wall. If your estimate requires a detailed takeoff of the framing, be sure to add the glazing cost.

08 50 00 Windows

- Most metal windows are delivered preglazed. However, some metal windows are priced without glass. Refer to 08 80 00 Glazing for glass pricing. The grade C indicates commercial grade windows, usually ASTM C-35.
- All wood windows and vinyl are priced preglazed. The glazing is insulating glass. Some wood windows may have single pane float glass. Add the cost of screens and grills if required, and not already included.

08 70 00 Hardware

- Hardware costs add considerably to the cost of a door. The most efficient method to determine the hardware requirements for a project is to review the door schedule.
- Door hinges are priced by the pair, with most doors requiring 1-1/2 pairs per door. The hinge prices do not include installation labor because it is included in door installation. Hinges are classified according to the frequency of use.

08 80 00 Glazing

- Different openings require different types of glass. The three most common types are:
  — float
  — tempered
  — insulating

- Most exterior windows are glazed with insulating glass. Entrance doors and window walls, where the glass is less than 18" from the floor, are generally glazed with tempered glass. Interior windows and some residential windows are glazed with float glass.

- Coastal communities are starting to require the use of impact-resistant glass.

- The insulation or 'u' value is a strong consideration, along with solar heat gain.

Reference Numbers

Reference numbers are shown in shaded boxes at the beginning of some major classifications. These numbers refer to related items in the Reference Section. The reference information may be an estimating procedure, an alternate pricing method, or technical information.

Note: Not all subdivisions listed here necessarily appear in this publication.

No part of this publication may be reproduced, stored in a retrieval system, or transmitted in any form or by any means without prior written permission of R.S. Means Company, Inc.

08 01 Operation and Maintenance of Openings

08 01 53 – Operation and Maintenance of Plastic Windows

08 01 53.81 Solid Vinyl Replacement Windows		Crew	Daily Output	Labor-Hours	Unit	Material	2010 Bare Costs Labor	Equipment	Total	Total Incl O&P
0010	**SOLID VINYL REPLACEMENT WINDOWS** R085313-20									
0020	Double hung, insulated glass, up to 83 united inches	G	2 Carp	8	2	Ea.	330	83		413	495
0040	84 to 93	G		8	2		360	83		443	530
0060	94 to 101	G		6	2.667		360	111		471	575
0080	102 to 111	G		6	2.667		380	111		491	595
0100	112 to 120	G		6	2.667		410	111		521	635
0120	For each united inch over 120, add	G		800	.020	Inch	5.20	.83		6.03	7.05
0140	Casement windows, one operating sash, 42 to 60 united inches	G		8	2	Ea.	212	83		295	370
0160	61 to 70	G		8	2		241	83		324	400
0180	71 to 80	G		8	2		261	83		344	425
0200	81 to 96	G		8	2		277	83		360	440
0220	Two operating sash, 58 to 78 united inches	G		8	2		425	83		508	600
0240	79 to 88	G		8	2		450	83		533	630
0260	89 to 98	G		8	2		490	83		573	675
0280	99 to 108	G		6	2.667		515	111		626	745
0300	109 to 121	G		6	2.667		555	111		666	790
0320	Two operating.one fixed sash, 73 to 108 united inches	G		8	2		670	83		753	870
0340	109 to 118	G		8	2		705	83		788	910
0360	119 to 128	G		6	2.667		725	111		836	975
0380	129 to 138	G		6	2.667		775	111		886	1,025
0400	139 to 156	G		6	2.667		815	111		926	1,075
0420	Four operating sash, 98 to 118 united inches	G		8	2		960	83		1,043	1,175
0440	119 to 128	G		8	2		1,025	83		1,108	1,250
0460	129 to 138	G		6	2.667		1,100	111		1,211	1,375
0480	139 to 148	G		6	2.667		1,150	111		1,261	1,425
0500	149 to 168	G		6	2.667		1,225	111		1,336	1,525
0520	169 to 178	G		6	2.667		1,325	111		1,436	1,625
0560	Fixed picture window, up to 63 united inches	G		8	2		144	83		227	295
0580	64 to 83	G		8	2		170	83		253	325
0600	84 to 101	G		8	2		220	83		303	380
0620	For each united inch over 101, add	G		900	.018	Inch	2.50	.74		3.24	3.96
0640	Picture window opt., low E glazing, up to 101 united inches	G				Ea.	21			21	23
0660	102 to 124	G					26			26	28.50
0680	124 and over	G					38			38	42
0700	Options, low E glazing, up to w/gas 101 united inches	G					32.50			32.50	36
0720	102 to 124	G					32.50			32.50	36
0740	124 and over	G					32.50			32.50	36
0760	Muntins, between glazing, square, per lite						4			4	4.40
0780	Diamond shape, per full or partial diamond						5			5	5.50
0800	Cellulose fiber insulation, poured into sash balance cavity	G	1 Carp	36	.222	C.F.	.60	9.25		9.85	15.80
0820	Silicone caulking at perimeter	G	"	800	.010	L.F.	.13	.42		.55	.83

08 05 Common Work Results for Openings

08 05 05 – Selective Windows and Doors Demolition

08 05 05.10 Selective Demolition Doors

		Crew	Daily Output	Labor-Hours	Unit	Material	2010 Bare Costs Labor	2010 Bare Costs Equipment	Total	Total Incl O&P
0010	**SELECTIVE DEMOLITION DOORS** R024119-10									
0200	Doors, exterior, 1-3/4" thick, single, 3' x 7' high	1 Clab	16	.500	Ea.		16.55		16.55	27
0210	Door demo, doors, exterior, 1-3/4" thick, single, 3'-0" x 8'-0"		10	.800			26.50		26.50	43.50
0215	Door demo, doors, exterior, double 1-3/4" thick, 3'-0" x 8'-0"		6	1.333			44		44	72.50
0220	Double, 6' x 7' high		12	.667			22		22	36
0500	Interior, 1-3/8" thick, single, 3' x 7' high		20	.400			13.25		13.25	21.50
0520	Double, 6' x 7' high		16	.500			16.55		16.55	27
0700	Bi-folding, 3' x 6'-8" high		20	.400			13.25		13.25	21.50
0720	6' x 6'-8" high		18	.444			14.70		14.70	24
0900	Bi-passing, 3' x 6'-8" high		16	.500			16.55		16.55	27
0940	6' x 6'-8" high		14	.571			18.90		18.90	31
1500	Remove and reset, minimum	1 Carp	8	1			41.50		41.50	68.50
1520	Maximum		6	1.333			55.50		55.50	91
2000	Frames, including trim, metal		8	1			41.50		41.50	68.50
2200	Wood	2 Carp	32	.500			21		21	34
2201	Alternate pricing method	1 Carp	200	.040	L.F.		1.66		1.66	2.73
3000	Special doors, counter doors	2 Carp	6	2.667	Ea.		111		111	182
3100	Double acting		10	1.600			66.50		66.50	109
3200	Floor door (trap type), or access type		8	2			83		83	137
3300	Glass, sliding, including frames		12	1.333			55.50		55.50	91
3400	Overhead, commercial, 12' x 12' high		4	4			166		166	273
3440	up to 20' x 16' high		3	5.333			222		222	365
3445	up to 35' x 30' high		1	16			665		665	1,100
3500	Residential, 9' x 7' high		8	2			83		83	137
3540	16' x 7' high		7	2.286			95		95	156
3600	Remove and reset, minimum		4	4			166		166	273
3620	Maximum		2.50	6.400			266		266	435
3700	Roll-up grille		5	3.200			133		133	218
3800	Revolving door		2	8			330		330	545
3900	Storefront swing door		3	5.333			222		222	365
3902	Cafe/bar swing door	2 Clab	8	2			66		66	109
4000	Residential lockset, exterior	1 Carp	28	.286			11.85		11.85	19.50
4224	Pocket door	"	8	1			41.50		41.50	68.50
5590	Remove mail slot	1 Clab	45	.178			5.90		5.90	9.65
7100	Remove double swing pneumatic doors, openers and sensors	2 Skwk	.50	32	Opng.		1,375		1,375	2,200
7570	Remove shock absorbing door	2 Sswk	1.90	8.421	"		395		395	720
9000	Minimum labor/equipment charge	1 Carp	4	2	Job		83		83	137

08 05 05.20 Selective Demolition of Windows

		Crew	Daily Output	Labor-Hours	Unit	Material	Labor	Equipment	Total	Total Incl O&P
0010	**SELECTIVE DEMOLITION OF WINDOWS** R024119-10									
0200	Aluminum, including trim, to 12 S.F.	1 Clab	16	.500	Ea.		16.55		16.55	27
0240	To 25 S.F.		11	.727			24		24	39.50
0280	To 50 S.F.		5	1.600			53		53	87
0320	Storm windows/screens, to 12 S.F.		27	.296			9.80		9.80	16.10
0360	To 25 S.F.		21	.381			12.60		12.60	20.50
0400	To 50 S.F.		16	.500			16.55		16.55	27
0600	Glass, minimum		200	.040	S.F.		1.32		1.32	2.17
0620	Maximum		150	.053	"		1.77		1.77	2.90
1000	Steel, including trim, to 12 S.F.		13	.615	Ea.		20.50		20.50	33.50
1020	To 25 S.F.		9	.889			29.50		29.50	48.50
1040	To 50 S.F.		4	2			66		66	109
2000	Wood, including trim, to 12 S.F.		22	.364			12.05		12.05	19.75
2020	To 25 S.F.		18	.444			14.70		14.70	24

08 05 Common Work Results for Openings

08 05 05 – Selective Windows and Doors Demolition

08 05 05.20 Selective Demolition of Windows		Crew	Daily Output	Labor-Hours	Unit	Material	2010 Bare Costs Labor	Equipment	Total	Total Incl O&P
2060	To 50 S.F.	1 Clab	13	.615	Ea.		20.50		20.50	33.50
2065	To 180 S.F.	↓	8	1	↓		33		33	54.50
4410	Remove skylight, plstc domes, flush/curb mtd	G-3	395	.081	S.F.		3.33		3.33	5.30
5020	Remove and reset window, minimum	1 Carp	6	1.333	Ea.		55.50		55.50	91
5040	Average		4	2			83		83	137
5080	Maximum	↓	2	4	↓		166		166	273
6000	Screening only	1 Clab	4000	.002	S.F.		.07		.07	.11
9000	Minimum labor/equipment charge	"	4	2	Job		66		66	109

08 11 Metal Doors and Frames

08 11 16 – Aluminum Doors and Frames

08 11 16.10 Entrance Doors and Frames

		Crew	Daily Output	Labor-Hours	Unit	Material	Labor	Equipment	Total	Total Incl O&P
0010	**ENTRANCE DOORS AND FRAMES** Aluminum, narrow stile									
0011	Including standard hardware, clear finish, no glass									
0020	3'-0" x 7'-0" opening	2 Sswk	2	8	Ea.	745	375		1,120	1,500
0030	3'-6" x 7'-0" opening		2	8		695	375		1,070	1,450
0100	3'-0" x 10'-0" opening, 3' high transom		1.80	8.889		1,175	415		1,590	2,050
0200	3'-6" x 10'-0" opening, 3' high transom		1.80	8.889		1,225	415		1,640	2,100
0280	5'-0" x 7'-0" opening		2	8		1,175	375		1,550	1,950
0300	6'-0" x 7'-0" opening		1.30	12.308	↓	1,075	575		1,650	2,225
0400	6'-0" x 10'-0" opening, 3' high transom		1.10	14.545	Pr.	1,525	680		2,205	2,925
0420	7'-0" x 7'-0" opening		1	16	"	1,250	750		2,000	2,750
0520	3'-0" x 7'-0" opening, wide stile		2	8	Ea.	890	375		1,265	1,675
0540	3'-6" x 7'-0" opening		2	8		1,000	375		1,375	1,800
0560	5'-0" x 7'-0" opening		2	8	↓	1,425	375		1,800	2,225
0580	6'-0" x 7'-0" opening		1.30	12.308	Pr.	1,450	575		2,025	2,625
0600	7'-0" x 7'-0" opening	↓	1	16	"	1,600	750		2,350	3,125
1100	For full vision doors, with 1/2" glass, add				Leaf	55%				
1200	For non-standard size, add					67%				
1300	Light bronze finish, add					36%				
1400	Dark bronze finish, add					18%				
1500	For black finish, add					36%				
1600	Concealed panic device, add				↓	910			910	1,000
1700	Electric striker release, add				Opng.	252			252	277
1800	Floor check, add				Leaf	630			630	695
1900	Concealed closer, add				"	505			505	555
2000	Flush 3' x 7' Insulated, 12"x 12" lite, clear finish	2 Sswk	2	8	Ea.	1,275	375		1,650	2,075
9000	Minimum labor/equipment charge	2 Carp	4	4	Job		166		166	273

08 11 63 – Metal Screen and Storm Doors and Frames

08 11 63.23 Aluminum Screen and Storm Doors and Frames

		Crew	Daily Output	Labor-Hours	Unit	Material	Labor	Equipment	Total	Total Incl O&P
0010	**ALUMINUM SCREEN AND STORM DOORS AND FRAMES**									
0020	Combination storm and screen									
0400	Clear anodic coating, 6'-8" x 2'-6" wide	2 Carp	15	1.067	Ea.	172	44.50		216.50	262
0420	2'-8" wide		14	1.143		156	47.50		203.50	249
0440	3'-0" wide	↓	14	1.143	↓	156	47.50		203.50	249
0500	For 7' door height, add					5%				
1000	Mill finish, 6'-8" x 2'-6" wide	2 Carp	15	1.067	Ea.	225	44.50		269.50	320
1020	2'-8" wide		14	1.143		225	47.50		272.50	325
1040	3'-0" wide	↓	14	1.143		246	47.50		293.50	350
1100	For 7'-0" door, add				↓	5%				

08 11 Metal Doors and Frames

08 11 63 – Metal Screen and Storm Doors and Frames

08 11 63.23 Aluminum Screen and Storm Doors and Frames		Crew	Daily Output	Labor-Hours	Unit	Material	2010 Bare Costs Labor	2010 Bare Costs Equipment	Total	Total Incl O&P
1500	White painted, 6'-8" x 2'-6" wide	2 Carp	15	1.067	Ea.	282	44.50		326.50	385
1520	2'-8" wide		14	1.143		278	47.50		325.50	385
1540	3'-0" wide	↓	14	1.143		292	47.50		339.50	400
1600	For 7'-0" door, add				↓	5%				
2000	Wood door & screen, see Div. 08 14 33.20									
9000	Minimum labor/equipment charge	1 Carp	4	2	Job		83		83	137

08 12 Metal Frames

08 12 13 – Hollow Metal Frames

08 12 13.13 Standard Hollow Metal Frames

		Crew	Daily Output	Labor-Hours	Unit	Material	Labor	Equipment	Total	Total Incl O&P
0010	**STANDARD HOLLOW METAL FRAMES**									
0020	16 ga., up to 5-3/4" jamb depth									
0025	6'-8" high, 3'-0" wide, single	2 Carp	16	1	Ea.	136	41.50		177.50	218
0028	3'-6" wide, single		16	1		142	41.50		183.50	226
0030	4'-0" wide, single		16	1		142	41.50		183.50	226
0040	6'-0" wide, double		14	1.143		195	47.50		242.50	292
0045	8'-0" wide, double		14	1.143		203	47.50		250.50	300
0100	7'-0" high, 3'-0" wide, single		16	1		140	41.50		181.50	223
0110	3'-6" wide, single		16	1		147	41.50		188.50	231
0112	4'-0" wide, single		16	1		147	41.50		188.50	231
0140	6'-0" wide, double		14	1.143		185	47.50		232.50	282
0145	8'-0" wide, double		14	1.143		219	47.50		266.50	320
0160	6'-0" wide, double		14	1.143		185	47.50		232.50	282
1000	16 ga., up to 4-7/8" deep, 7'-0" H, 3'-0" W, single		16	1		154	41.50		195.50	239
1140	6'-0" wide, double		14	1.143		171	47.50		218.50	266
1200	16 ga., 8-3/4" deep, 7'-0" H, 3'-0" W, single		16	1		163	41.50		204.50	248
2800	14 ga., up to 3-7/8" deep, 7'-0" high, 3'-0" wide, single		16	1		149	41.50		190.50	232
2840	6'-0" wide, double		14	1.143		174	47.50		221.50	269
3000	14 ga., up to 5-3/4" deep, 6'-8" high, 3'-0" wide, single		16	1		140	41.50		181.50	223
3002	3'-6" wide, single		16	1		183	41.50		224.50	270
3005	4'-0" wide, single		16	1		183	41.50		224.50	270
3600	up to 5-3/4" jamb depth, 7'-0" high, 4'-0" wide, single		15	1.067		140	44.50		184.50	227
3620	6'-0" wide, double		12	1.333		182	55.50		237.50	291
3640	8'-0" wide, double		12	1.333		179	55.50		234.50	288
3700	8'-0" high, 4'-0" wide, single		15	1.067		169	44.50		213.50	259
3740	8'-0" wide, double		12	1.333		260	55.50		315.50	375
4000	6-3/4" deep, 7'-0" high, 4'-0" wide, single		15	1.067		198	44.50		242.50	291
4020	6'-0" wide, double		12	1.333		243	55.50		298.50	360
4040	8'-0", wide double		12	1.333		251	55.50		306.50	365
4100	8'-0" high, 4'-0" wide, single		15	1.067		208	44.50		252.50	300
4140	8'-0" wide, double		12	1.333		208	55.50		263.50	320
4400	8-3/4" deep, 7'-0" high, 4'-0" wide, single		15	1.067		176	44.50		220.50	266
4440	8'-0" wide, double		12	1.333		237	55.50		292.50	350
4500	8'-0" high, 4'-0" wide, single		15	1.067		177	44.50		221.50	267
4540	8'-0" wide, double	↓	12	1.333		208	55.50		263.50	320
4900	For welded frames, add					60			60	66
5400	14 ga., "B" label, up to 5-3/4" deep, 7'-0" high, 4'-0" wide, single	2 Carp	15	1.067		194	44.50		238.50	286
5440	8'-0" wide, double		12	1.333		252	55.50		307.50	370
5800	6-3/4" deep, 7'-0" high, 4'-0" wide, single		15	1.067		208	44.50		252.50	300
5840	8'-0" wide, double		12	1.333		360	55.50		415.50	485
6200	8-3/4" deep, 7'-0" high, 4'-0" wide, single	↓	15	1.067		289	44.50		333.50	395

08 12 Metal Frames

08 12 13 – Hollow Metal Frames

08 12 13.13 Standard Hollow Metal Frames

		Crew	Daily Output	Labor-Hours	Unit	Material	2010 Bare Costs Labor	Equipment	Total	Total Incl O&P
6240	8'-0" wide, double	2 Carp	12	1.333	Ea.	335	55.50		390.50	460
6300	For "A" label use same price as "B" label									
6400	For baked enamel finish, add					30%	15%			
6500	For galvanizing, add					15%				
6600	For hospital stop, add				Ea.	282			282	310
7900	Transom lite frames, fixed, add	2 Carp	155	.103	S.F.	52	4.29		56.29	64
8000	Movable, add	"	130	.123	"	65	5.10		70.10	80
9000	Minimum labor/equipment charge	1 Carp	4	2	Job		83		83	137

08 12 13.25 Channel Metal Frames

		Crew	Daily Output	Labor-Hours	Unit	Material	Labor	Equipment	Total	Total Incl O&P
0010	**CHANNEL METAL FRAMES**									
0020	Steel channels with anchors and bar stops									
0100	6" channel @ 8.2#/L.F., 3' x 7' door, weighs 150#	E-4	13	2.462	Ea.	198	117	11.20	326.20	445
0200	8" channel @ 11.5#/L.F., 6' x 8' door, weighs 275#		9	3.556		365	169	16.20	550.20	730
0300	8' x 12' door, weighs 400#		6.50	4.923		530	233	22.50	785.50	1,025
0800	For frames without bar stops, light sections, deduct					15%				
0900	Heavy sections, deduct					10%				
9000	Minimum labor/equipment charge	E-4	4	8	Job		380	36.50	416.50	735

08 13 Metal Doors

08 13 13 – Hollow Metal Doors

08 13 13.13 Standard Hollow Metal Doors

			Crew	Daily Output	Labor-Hours	Unit	Material	Labor	Equipment	Total	Total Incl O&P
0010	**STANDARD HOLLOW METAL DOORS**	R081313-20									
0015	Flush, full panel, hollow core										
0020	1-3/8" thick, 20 ga., 2'-0"x 6'-8"	G	2 Carp	20	.800	Ea.	320	33		353	405
0040	2'-8" x 6'-8"	G		18	.889		330	37		367	420
0060	3'-0" x 6'-8"	G		17	.941		330	39		369	430
0100	3'-0" x 7'-0"			17	.941		345	39		384	440
0120	For vision lite, add						85			85	93.50
0140	For narrow lite, add						92			92	101
0320	Half glass, 20 ga., 2'-0" x 6'-8"	G	2 Carp	20	.800		470	33		503	570
0340	2'-8" x 6'-8"	G		18	.889		480	37		517	585
0360	3'-0" x 6'-8"	G		17	.941		480	39		519	595
0400	3'-0" x 7'-0"	G		17	.941		560	39		599	680
0410	1-3/8" thick, 18 ga., 2'-0"x 6'-8"	G		20	.800		380	33		413	475
0420	3'-0" x 6'-8"	G		17	.941		385	39		424	490
0425	3'-0" x 7'-0"	G		17	.941		400	39		439	505
0450	For vision lite, add						85			85	93.50
0452	For narrow lite, add						92			92	101
0460	Half glass, 18 ga., 2'-0" x 6'-8"	G	2 Carp	20	.800		530	33		563	640
0465	2'-8" x 6'-8"	G		18	.889		550	37		587	665
0470	3'-0" x 6'-8"	G		17	.941		535	39		574	655
0475	3'-0" x 7'-0"	G		17	.941		550	39		589	670
0500	Hollow core, 1-3/4" thick, full panel, 20 ga., 2'-8" x 6'-8"	G		18	.889		375	37		412	470
0520	3'-0" x 6'-8"	G		17	.941		335	39		374	435
0640	3'-0" x 7'-0"	G		17	.941		390	39		429	495
0680	4'-0" x 7'-0"	G		15	1.067		535	44.50		579.50	665
0700	4'-0" x 8'-0"	G		13	1.231		610	51		661	755
1000	18 ga., 2'-8" x 6'-8"	G		17	.941		400	39		439	505
1020	3'-0" x 6'-8"	G		16	1		385	41.50		426.50	495
1120	3'-0" x 7'-0"	G		17	.941		445	39		484	555
1180	4'-0" x 7'-0"	G		14	1.143		535	47.50		582.50	670

08 13 Metal Doors

08 13 13 – Hollow Metal Doors

08 13 13.13 Standard Hollow Metal Doors

		Crew	Daily Output	Labor-Hours	Unit	Material	2010 Bare Costs Labor	Equipment	Total	Total Incl O&P
1200	4'-0" x 8'-0" G	2 Carp	17	.941	Ea.	540	39		579	660
1212	For vision lite, add					85			85	93.50
1214	For narrow lite, add					92			92	101
1230	Half glass, 20 ga., 2'-8" x 6'-8" G	2 Carp	20	.800		430	33		463	530
1240	3'-0" x 6'-8" G		18	.889		420	37		457	525
1260	3'-0" x 7'-0" G		18	.889		425	37		462	525
1320	18 ga., 2'-8" x 6'-8" G		18	.889		460	37		497	565
1340	3'-0" x 6'-8" G		17	.941		455	39		494	565
1360	3'-0" x 7'-0" G		17	.941		465	39		504	575
1380	4'-0" x 7'-0" G		15	1.067		580	44.50		624.50	710
1400	4'-0" x 8'-0" G		14	1.143		660	47.50		707.50	810
1720	Insulated, 1-3/4" thick, full panel, 18 ga., 3'-0" x 6'-8" G		15	1.067		400	44.50		444.50	520
1740	2'-8" x 7'-0" G		16	1		420	41.50		461.50	530
1760	3'-0" x 7'-0" G		15	1.067		415	44.50		459.50	530
1800	4'-0" x 8'-0" G		13	1.231		605	51		656	750
1805	For vision lite, add					85			85	93.50
1810	For narrow lite, add					92			92	101
1820	Half glass, 18 ga., 3'-0" x 6'-8" G	2 Carp	16	1		535	41.50		576.50	660
1840	2'-8" x 7'-0" G		17	.941		545	39		584	665
1860	3'-0" x 7'-0" G		16	1		585	41.50		626.50	715
1900	4'-0" x 8'-0" G		14	1.143		655	47.50		702.50	800
2000	For bottom louver, add					245			245	269
2020	For baked enamel finish, add					30%	15%			
2040	For galvanizing, add					15%				
2540	For soundproofing STC 40, add				Ea.	1,725			1,725	1,900
2560	For 3 hour door, add				"	370			370	405
9000	Minimum labor/equipment charge	1 Carp	4	2	Job		83		83	137

08 13 13.15 Metal Fire Doors

			Crew	Daily Output	Labor-Hours	Unit	Material	Labor	Equipment	Total	Total Incl O&P
0010	**METAL FIRE DOORS**	R081313-20									
0015	Steel, flush, "B" label, 90 minute										
0020	Full panel, 20 ga., 2'-0" x 6'-8"		2 Carp	20	.800	Ea.	340	33		373	430
0040	2'-8" x 6'-8"			18	.889		345	37		382	440
0060	3'-0" x 6'-8"			17	.941		345	39		384	445
0080	3'-0" x 7'-0"			17	.941		360	39		399	460
0140	18 ga., 3'-0" x 6'-8"			16	1		395	41.50		436.50	505
0160	2'-8" x 7'-0"			17	.941		415	39		454	520
0180	3'-0" x 7'-0"			16	1		395	41.50		436.50	505
0200	4'-0" x 7'-0"			15	1.067		520	44.50		564.50	645
0220	For "A" label, 3 hour, 18 ga., use same price as "B" label										
0240	For vision lite, add					Ea.	130			130	143
0520	Flush, "B" label 90 min., composite, 20 ga., 2'-0" x 6'-8"		2 Carp	18	.889		430	37		467	535
0540	2'-8" x 6'-8"			17	.941		435	39		474	545
0560	3'-0" x 6'-8"			16	1		440	41.50		481.50	550
0580	3'-0" x 7'-0"			16	1		450	41.50		491.50	565
0640	Flush, "A" label 3 hour, composite, 18 ga., 3'-0" x 6'-8"			15	1.067		495	44.50		539.50	620
0660	2'-8" x 7'-0"			16	1		515	41.50		556.50	640
0680	3'-0" x 7'-0"			15	1.067		505	44.50		549.50	630
0700	4'-0" x 7'-0"			14	1.143		630	47.50		677.50	775
9000	Minimum labor/equipment charge		1 Carp	4	2	Job		83		83	137

08 13 13.20 Residential Steel Doors

0010	**RESIDENTIAL STEEL DOORS**
0020	Prehung, insulated, exterior

08 13 Metal Doors

08 13 13 – Hollow Metal Doors

08 13 13.20 Residential Steel Doors

		Crew	Daily Output	Labor-Hours	Unit	Material	2010 Bare Costs Labor	Equipment	Total	Total Incl O&P
0030	Embossed, full panel, 2'-8" x 6'-8" G	2 Carp	17	.941	Ea.	293	39		332	385
0040	3'-0" x 6'-8" G		15	1.067		272	44.50		316.50	370
0060	3'-0" x 7'-0" G		15	1.067		320	44.50		364.50	425
0070	5'-4" x 6'-8", double G		8	2		575	83		658	770
0220	Half glass, 2'-8" x 6'-8"		17	.941		290	39		329	385
0240	3'-0" x 6'-8"		16	1		290	41.50		331.50	390
0260	3'-0" x 7'-0"		16	1		340	41.50		381.50	445
0270	5'-4" x 6'-8", double		8	2		585	83		668	775
1320	Flush face, full panel, 2'-8" x 6'-8"		16	1		230	41.50		271.50	320
1340	3'-0" x 6'-8"		15	1.067		230	44.50		274.50	325
1360	3'-0" x 7'-0"		15	1.067		290	44.50		334.50	395
1380	5'-4" x 6'-8", double		8	2		470	83		553	650
1420	Half glass, 2'-8" x 6'-8"		17	.941		289	39		328	385
1440	3'-0" x 6'-8"		16	1		289	41.50		330.50	390
1460	3'-0" x 7'-0"		16	1		335	41.50		376.50	440
1480	5'-4" x 6'-8", double		8	2		555	83		638	745
1500	Sidelight, full lite, 1'-0" x 6'-8" with grille					236			236	260
1510	1'-0" x 6'-8", low e					257			257	282
1520	1'-0" x 6'-8", half lite					254			254	280
1530	1'-0" x 6'-8", half lite, low e					262			262	289
2300	Interior, residential, closet, bi-fold, 6'-8" x 2'-0" wide	2 Carp	16	1		158	41.50		199.50	243
2330	3'-0" wide		16	1		197	41.50		238.50	286
2360	4'-0" wide		15	1.067		255	44.50		299.50	355
2400	5'-0" wide		14	1.143		310	47.50		357.50	420
2420	6'-0" wide		13	1.231		330	51		381	450
9000	Minimum labor/equipment charge	1 Carp	4	2	Job		83		83	137

08 13 16 – Aluminum Doors

08 13 16.10 Commercial Aluminum Doors

		Crew	Daily Output	Labor-Hours	Unit	Material	Labor	Equipment	Total	Total Incl O&P
0010	**COMMERCIAL ALUMINUM DOORS**, no glazing									
0020	Incl. hinges, push/pull, deadlock, cyl., threshold									
1000	Narrow stile, no glazing, standard hardware, 3'-0" x 7'-0", single	2 Carp	3	5.333	Ea.	590	222		812	1,000
1200	Pair of 3'-0" x 7'-0"		1.70	9.412	Pr.	1,175	390		1,565	1,925
1500	3'-6" x 7'-0", single		3	5.333	Ea.	685	222		907	1,125
2100	Medium stile, 3'-0" x 7'-0", single		3	5.333	"	720	222		942	1,150
2200	Pair of 3'-0" x 7'-0"		1.70	9.412	Pr.	1,450	390		1,840	2,225
2300	3'-6" x 7'-0", single		3	5.333	Ea.	930	222		1,152	1,400
5000	Flush panel doors, pair of 2'-6" x 7'-0"	2 Sswk	2	8	Pr.	1,325	375		1,700	2,125
5050	3'-0" x 7'-0", single		2.50	6.400	Ea.	755	300		1,055	1,375
5100	Pair of 3'-0" x 7'-0"		2	8	Pr.	1,425	375		1,800	2,225
5150	3'-6" x 7'-0", single		2.50	6.400	Ea.	900	300		1,200	1,550

08 14 Wood Doors

08 14 13 – Carved Wood Doors

08 14 13.10 Types of Wood Doors, Carved		Crew	Daily Output	Labor-Hours	Unit	Material	2010 Bare Costs Labor	Equipment	Total	Total Incl O&P
0010	**TYPES OF WOOD DOORS, CARVED**									
3000	Solid wood, 1-3/4" thick stile and rail									
3020	Mahogany, 3'-0" x 7'-0", minimum	2 Carp	14	1.143	Ea.	945	47.50		992.50	1,125
3030	Maximum		10	1.600		1,500	66.50		1,566.50	1,750
3040	3'-6" x 8'-0", minimum		10	1.600		1,250	66.50		1,316.50	1,475
3050	Maximum		8	2		1,850	83		1,933	2,175
3100	Pine, 3'-0" x 7'-0", minimum		14	1.143		450	47.50		497.50	575
3110	Maximum		10	1.600		750	66.50		816.50	935
3120	3'-6" x 8'-0", minimum		10	1.600		885	66.50		951.50	1,075
3130	Maximum		8	2		1,800	83		1,883	2,100
3200	Red oak, 3'-0" x 7'-0", minimum		14	1.143		1,700	47.50		1,747.50	1,950
3210	Maximum		10	1.600		2,025	66.50		2,091.50	2,325
3220	3'-6" x 8'-0", minimum		10	1.600		2,200	66.50		2,266.50	2,525
3230	Maximum		8	2		3,200	83		3,283	3,650
4000	Hand carved door, mahogany									
4020	3'-0" x 7'-0", minimum	2 Carp	14	1.143	Ea.	1,600	47.50		1,647.50	1,825
4030	Maximum		11	1.455		3,600	60.50		3,660.50	4,050
4040	3'-6" x 8'-0", minimum		10	1.600		2,500	66.50		2,566.50	2,850
4050	Maximum		8	2		3,500	83		3,583	3,975
4200	Rose wood, 3'-0" x 7'-0", minimum		14	1.143		5,000	47.50		5,047.50	5,575
4210	Maximum		11	1.455		13,600	60.50		13,660.50	15,100
4220	3'-6" x 8'-0", minimum		10	1.600		5,700	66.50		5,766.50	6,375
4280	For 6'-8" high door, deduct from 7'-0" door					34			34	37.50
4400	For custom finish, add					400			400	440
4600	Side light, mahogany, 7'-0" x 1'-6" wide, minimum	2 Carp	18	.889		900	37		937	1,050
4610	Maximum		14	1.143		2,625	47.50		2,672.50	2,975
4620	8'-0" x 1'-6" wide, minimum		14	1.143		1,700	47.50		1,747.50	1,950
4630	Maximum		10	1.600		1,950	66.50		2,016.50	2,250
4640	Side light, oak, 7'-0" x 1'-6" wide, minimum		18	.889		1,100	37		1,137	1,250
4650	Maximum		14	1.143		1,900	47.50		1,947.50	2,175
4660	8'-0" x 1'-6" wide, minimum		14	1.143		1,000	47.50		1,047.50	1,175
4670	Maximum		10	1.600		1,900	66.50		1,966.50	2,200

08 14 16 – Flush Wood Doors

08 14 16.09 Smooth Wood Doors

		Crew	Daily Output	Labor-Hours	Unit	Material	Labor	Equipment	Total	Total Incl O&P
0010	**SMOOTH WOOD DOORS**									
0015	Flush, interior, 1-3/8", 7 ply, hollow core									
0020	Lauan face, 2'-0" x 6'-8"	2 Carp	17	.941	Ea.	32	39		71	99.50
0080	3'-0" x 6'-8"		17	.941		52	39		91	122
0100	4'-0" x 6'-8"		16	1		126	41.50		167.50	208
0108	1-3/4", 3'-0" x 7'-0"		16	1		103	41.50		144.50	182
0112	Pair of 3'-0" x 7'-0"		9	1.778	Pr.	182	74		256	320
0120	Birch face, 2'-0" x 6'-8"		17	.941	Ea.	45	39		84	114
0140	2'-6" x 6'-8"		17	.941		82.50	39		121.50	155
0180	3'-0" x 6'-8"		17	.941		94	39		133	167
0200	4'-0" x 6'-8"		16	1		158	41.50		199.50	243
0210	1-3/4", 3'-0" x 7'-0"		16	1		103	41.50		144.50	182
0214	Pair of 3'-0" x 7'-0"		9	1.778	Pr.	197	74		271	335
0220	Oak face, 2'-0" x 6'-8"		17	.941	Ea.	86.50	39		125.50	159
0240	2'-6" x 6'-8"		17	.941		94	39		133	167
0280	3'-0" x 6'-8"		17	.941		101	39		140	175
0300	4'-0" x 6'-8"		16	1		125	41.50		166.50	206
0320	Walnut face, 2'-0" x 6'-8"		17	.941		171	39		210	252

08 14 Wood Doors

08 14 16 – Flush Wood Doors

08 14 16.09 Smooth Wood Doors		Crew	Daily Output	Labor-Hours	Unit	Material	2010 Bare Costs Labor	Equipment	Total	Total Incl O&P
0340	2'-6" x 6'-8"	2 Carp	17	.941	Ea.	175	39		214	256
0380	3'-0" x 6'-8"		17	.941		182	39		221	265
0400	4'-0" x 6'-8"	↓	16	1		205	41.50		246.50	295
0430	For 7'-0" high, add					16			16	17.60
0440	For 8'-0" high, add					30			30	33
0480	For prefinishing, clear, add					35			35	38.50
0500	For prefinishing, stain, add					45			45	49.50
1320	M.D. overlay on hardboard, 2'-0" x 6'-8"	2 Carp	17	.941		94	39		133	167
1340	2'-6" x 6'-8"		17	.941		94	39		133	167
1380	3'-0" x 6'-8"		17	.941		112	39		151	187
1400	4'-0" x 6'-8"	↓	16	1		160	41.50		201.50	245
1420	For 7'-0" high, add					10			10	11
1440	For 8'-0" high, add					25			25	27.50
1720	H.P. plastic laminate, 2'-0" x 6'-8"	2 Carp	16	1		224	41.50		265.50	315
1740	2'-6" x 6'-8"		16	1		224	41.50		265.50	315
1780	3'-0" x 6'-8"		15	1.067		278	44.50		322.50	380
1800	4'-0" x 6'-8"	↓	14	1.143		355	47.50		402.50	475
1820	For 7'-0" high, add					10			10	11
1840	For 8'-0" high, add					25			25	27.50
2020	5 ply particle core, lauan face, 2'-6" x 6'-8"	2 Carp	15	1.067		82	44.50		126.50	163
2040	3'-0" x 6'-8"		14	1.143		85	47.50		132.50	172
2080	3'-0" x 7'-0"		13	1.231		93	51		144	186
2100	4'-0" x 7'-0"		12	1.333		110	55.50		165.50	212
2120	Birch face, 2'-6" x 6'-8"		15	1.067		94	44.50		138.50	176
2140	3'-0" x 6'-8"		14	1.143		103	47.50		150.50	191
2180	3'-0" x 7'-0"		13	1.231		104	51		155	198
2200	4'-0" x 7'-0"		12	1.333		127	55.50		182.50	231
2220	Oak face, 2'-6" x 6'-8"		15	1.067		103	44.50		147.50	186
2240	3'-0" x 6'-8"		14	1.143		114	47.50		161.50	203
2280	3'-0" x 7'-0"		13	1.231		117	51		168	213
2300	4'-0" x 7'-0"		12	1.333		143	55.50		198.50	248
2320	Walnut face, 2'-0" x 6'-8"		15	1.067		114	44.50		158.50	198
2340	2'-6" x 6'-8"		14	1.143		130	47.50		177.50	221
2380	3'-0" x 6'-8"		13	1.231		147	51		198	246
2400	4'-0" x 6'-8"	↓	12	1.333		192	55.50		247.50	300
2440	For 8'-0" high, add					30			30	33
2460	For 8'-0" high walnut, add					16			16	17.60
2480	For solid wood core, add					35			35	38.50
2720	For prefinishing, clear, add					23			23	25.50
2740	For prefinishing, stain, add					50			50	55
3320	M.D. overlay on hardboard, 2'-6" x 6'-8"	2 Carp	14	1.143		104	47.50		151.50	192
3340	3'-0" x 6'-8"		13	1.231		110	51		161	205
3380	3'-0" x 7'-0"		12	1.333		112	55.50		167.50	214
3400	4'-0" x 7'-0"	↓	10	1.600		145	66.50		211.50	269
3440	For 8'-0" height, add					32			32	35
3460	For solid wood core, add					40			40	44
3720	H.P. plastic laminate, 2'-6" x 6'-8"	2 Carp	13	1.231		152	51		203	251
3740	3'-0" x 6'-8"		12	1.333		175	55.50		230.50	284
3780	3'-0" x 7'-0"		11	1.455		180	60.50		240.50	298
3800	4'-0" x 7'-0"	↓	8	2		218	83		301	375
3840	For 8'-0" height, add					32			32	35
3860	For solid wood core, add					38			38	42
4000	Exterior, flush, solid wood stave core, birch, 1-3/4" x 7'-0" x 2'-6"	2 Carp	15	1.067		162	44.50		206.50	251

08 14 Wood Doors

08 14 16 – Flush Wood Doors

08 14 16.09 Smooth Wood Doors

		Crew	Daily Output	Labor-Hours	Unit	Material	2010 Bare Costs Labor	2010 Bare Costs Equipment	Total	Total Incl O&P
4020	2'-8" wide	2 Carp	15	1.067	Ea.	170	44.50		214.50	260
4040	3'-0" wide		14	1.143		185	47.50		232.50	282
4100	Oak faced 1-3/4" x 7'-0" x 2'-6" wide		15	1.067		180	44.50		224.50	271
4120	2'-8" wide		15	1.067		195	44.50		239.50	288
4140	3'-0" wide		14	1.143		206	47.50		253.50	305
4160	Walnut faced, 1-3/4" x 6'-8" x 3'-0" wide	1 Carp	17	.471		276	19.55		295.55	335
4180	3'-6" wide	"	17	.471		375	19.55		394.55	445
4200	Walnut faced, 1-3/4" x 7'-0" x 2'-6" wide	2 Carp	15	1.067		266	44.50		310.50	365
4220	2'-8" wide		15	1.067		280	44.50		324.50	385
4240	3'-0" wide		14	1.143		290	47.50		337.50	400
4300	For 6'-8" high door, deduct from 7'-0" door					16			16	17.60
5000	Wood doors, for vision lite, add					85			85	93.50
5010	Wood doors, for narrow lite, add					92			92	101
5015	Wood doors, for bottom (or top) louver, add					245			245	269
9000	Minimum labor/equipment charge	1 Carp	4	2	Job		83		83	137

08 14 16.10 Wood Doors Decorator

		Crew	Daily Output	Labor-Hours	Unit	Material	Labor	Equipment	Total	Total Incl O&P
0010	**WOOD DOORS DECORATOR**									
0040	7 ply hollow core lauan face, 2'-6" x 6'-8"	2 Carp	17	.941	Ea.	37.50	39		76.50	105

08 14 16.20 Wood Fire Doors

		Crew	Daily Output	Labor-Hours	Unit	Material	Labor	Equipment	Total	Total Incl O&P
0010	**WOOD FIRE DOORS**									
0020	Particle core, 7 face plys, "B" label,									
0040	1 hour, birch face, 1-3/4" x 2'-6" x 6'-8"	2 Carp	14	1.143	Ea.	380	47.50		427.50	500
0080	3'-0" x 6'-8"		13	1.231		390	51		441	515
0090	3'-0" x 7'-0"		12	1.333		410	55.50		465.50	540
0100	4'-0" x 7'-0"		12	1.333		510	55.50		565.50	650
0140	Oak face, 2'-6" x 6'-8"		14	1.143		335	47.50		382.50	450
0180	3'-0" x 6'-8"		13	1.231		350	51		401	470
0190	3'-0" x 7'-0"		12	1.333		365	55.50		420.50	490
0200	4'-0" x 7'-0"		12	1.333		480	55.50		535.50	620
0240	Walnut face, 2'-6" x 6'-8"		14	1.143		450	47.50		497.50	575
0280	3'-0" x 6'-8"		13	1.231		470	51		521	600
0290	3'-0" x 7'-0"		12	1.333		490	55.50		545.50	630
0300	4'-0" x 7'-0"		12	1.333		610	55.50		665.50	760
0440	M.D. overlay on hardboard, 2'-6" x 6'-8"		15	1.067		300	44.50		344.50	405
0480	3'-0" x 6'-8"		14	1.143		315	47.50		362.50	425
0490	3'-0" x 7'-0"		13	1.231		330	51		381	450
0500	4'-0" x 7'-0"		12	1.333		400	55.50		455.50	530
0740	90 minutes, birch face, 1-3/4" x 2'-6" x 6'-8"		14	1.143		320	47.50		367.50	430
0780	3'-0" x 6'-8"		13	1.231		305	51		356	420
0790	3'-0" x 7'-0"		12	1.333		375	55.50		430.50	505
0800	4'-0" x 7'-0"		12	1.333		470	55.50		525.50	610
0840	Oak face, 2'-6" x 6'-8"		14	1.143		425	47.50		472.50	545
0880	3'-0" x 6'-8"		13	1.231		435	51		486	560
0890	3'-0" x 7'-0"		12	1.333		450	55.50		505.50	585
0900	4'-0" x 7'-0"		12	1.333		580	55.50		635.50	730
0940	Walnut face, 2'-6" x 6'-8"		14	1.143		390	47.50		437.50	510
0980	3'-0" x 6'-8"		13	1.231		400	51		451	525
0990	3'-0" x 7'-0"		12	1.333		460	55.50		515.50	595
1000	4'-0" x 7'-0"		12	1.333		610	55.50		665.50	760
1140	M.D. overlay on hardboard, 2'-6" x 6'-8"		15	1.067		350	44.50		394.50	460
1180	3'-0" x 6'-8"		14	1.143		370	47.50		417.50	490
1190	3'-0" x 7'-0"		13	1.231		400	51		451	525

08 14 Wood Doors

08 14 16 – Flush Wood Doors

08 14 16.20 Wood Fire Doors		Crew	Daily Output	Labor-Hours	Unit	Material	2010 Bare Costs Labor	Equipment	Total	Total Incl O&P
1200	4'-0" x 7'-0"	2 Carp	12	1.333	Ea.	450	55.50		505.50	585
1240	For 8'-0" height, add					65			65	71.50
1260	For 8'-0" height walnut, add					80			80	88
2200	Custom architectural "B" label, flush, 1-3/4" thick, birch,									
2210	Solid core									
2220	2'-6" x 7'-0"	2 Carp	15	1.067	Ea.	296	44.50		340.50	400
2260	3'-0" x 7'-0"		14	1.143		305	47.50		352.50	415
2300	4'-0" x 7'-0"		13	1.231		400	51		451	520
2420	4'-0" x 8'-0"		11	1.455		430	60.50		490.50	570
2480	For oak veneer, add					50%				
2500	For walnut veneer, add					75%				
9000	Minimum labor/equipment charge	1 Carp	4	2	Job		83		83	137

08 14 23 – Clad Wood Doors

08 14 23.13 Metal-Faced Wood Doors

		Crew	Daily Output	Labor-Hours	Unit	Material	Labor	Equipment	Total	Total Incl O&P
0010	**METAL-FACED WOOD DOORS**									
0020	Interior, flush type, 3' x 7'	2 Carp	4.30	3.721	Opng.	240	155		395	520
9000	Minimum labor/equipment charge	1 Carp	2	4	Job		166		166	273

08 14 23.20 Tin Clad Wood Doors

0010	**TIN CLAD WOOD DOORS**									
0020	3 ply, 6' x 7', double sliding, doors only	2 Carp	1	16	Opng.	1,900	665		2,565	3,200
1000	For electric operator, add	1 Elec	2	4	"	3,250	196		3,446	3,875
9000	Minimum labor/equipment charge	2 Carp	1	16	Job		665		665	1,100

08 14 33 – Stile and Rail Wood Doors

08 14 33.10 Wood Doors Paneled

		Crew	Daily Output	Labor-Hours	Unit	Material	Labor	Equipment	Total	Total Incl O&P
0010	**WOOD DOORS PANELED**									
0020	Interior, six panel, hollow core, 1-3/8" thick									
0040	Molded hardboard, 2'-0" x 6'-8"	2 Carp	17	.941	Ea.	56	39		95	126
0060	2'-6" x 6'-8"		17	.941		60	39		99	130
0070	2'-8" x 6'-8"		17	.941		63	39		102	134
0080	3'-0" x 6'-8"		17	.941		66	39		105	137
0140	Embossed print, molded hardboard, 2'-0" x 6'-8"		17	.941		60	39		99	130
0160	2'-6" x 6'-8"		17	.941		60	39		99	130
0180	3'-0" x 6'-8"		17	.941		66	39		105	137
0540	Six panel, solid, 1-3/8" thick, pine, 2'-0" x 6'-8"		15	1.067		145	44.50		189.50	233
0560	2'-6" x 6'-8"		14	1.143		162	47.50		209.50	256
0580	3'-0" x 6'-8"		13	1.231		136	51		187	234
1020	Two panel, bored rail, solid, 1-3/8" thick, pine, 1'-6" x 6'-8"		16	1		263	41.50		304.50	360
1040	2'-0" x 6'-8"		15	1.067		345	44.50		389.50	455
1060	2'-6" x 6'-8"		14	1.143		395	47.50		442.50	515
1340	Two panel, solid, 1-3/8" thick, fir, 2'-0" x 6'-8"		15	1.067		150	44.50		194.50	238
1360	2'-6" x 6'-8"		14	1.143		200	47.50		247.50	298
1380	3'-0" x 6'-8"		13	1.231		405	51		456	530
1740	Five panel, solid, 1-3/8" thick, fir, 2'-0" x 6'-8"		15	1.067		270	44.50		314.50	370
1760	2'-6" x 6'-8"		14	1.143		415	47.50		462.50	535
1780	3'-0" x 6'-8"		13	1.231		415	51		466	540
9000	Minimum labor/equipment charge	1 Carp	4	2	Job		83		83	137

08 14 33.20 Wood Doors Residential

		Crew	Daily Output	Labor-Hours	Unit	Material	Labor	Equipment	Total	Total Incl O&P
0010	**WOOD DOORS RESIDENTIAL**									
0200	Exterior, combination storm & screen, pine									
0260	2'-8" wide	2 Carp	10	1.600	Ea.	286	66.50		352.50	425
0280	3'-0" wide		9	1.778		295	74		369	445

08 14 Wood Doors

08 14 33 – Stile and Rail Wood Doors

08 14 33.20 Wood Doors Residential		Crew	Daily Output	Labor-Hours	Unit	Material	2010 Bare Costs Labor	Equipment	Total	Total Incl O&P
0300	7'-1" x 3'-0" wide	2 Carp	9	1.778	Ea.	335	74		409	490
0400	Full lite, 6'-9" x 2'-6" wide		11	1.455		296	60.50		356.50	425
0420	2'-8" wide		10	1.600		296	66.50		362.50	435
0440	3'-0" wide		9	1.778		300	74		374	450
0500	7'-1" x 3'-0" wide		9	1.778		325	74		399	475
0700	Dutch door, pine, 1-3/4" x 6'-8" x 2'-8" wide, minimum		12	1.333		675	55.50		730.50	835
0720	Maximum		10	1.600		825	66.50		891.50	1,025
0800	3'-0" wide, minimum		12	1.333		690	55.50		745.50	845
0820	Maximum		10	1.600		900	66.50		966.50	1,100
1000	Entrance door, colonial, 1-3/4" x 6'-8" x 2'-8" wide		16	1		460	41.50		501.50	575
1020	6 panel pine, 3'-0" wide		15	1.067		430	44.50		474.50	550
1100	8 panel pine, 2'-8" wide		16	1		570	41.50		611.50	700
1120	3'-0" wide		15	1.067		555	44.50		599.50	685
1200	For tempered safety glass lites, (min of 2) add					74.50			74.50	82
1300	Flush, birch, solid core, 1-3/4" x 6'-8" x 2'-8" wide	2 Carp	16	1		104	41.50		145.50	183
1320	3'-0" wide		15	1.067		114	44.50		158.50	198
1350	7'-0" x 2'-8" wide		16	1		113	41.50		154.50	193
1360	3'-0" wide		15	1.067		121	44.50		165.50	206
1380	For tempered safety glass lites, add					102			102	112
1720	Carved mahogany 3'-0" x 6'-8"	2 Carp	15	1.067		670	44.50		714.50	810
1800	Lauan, solid core, 1-3/4" x 7'-0" x 2'-4" wide		16	1		106	41.50		147.50	185
1810	2'-6" wide		15	1.067		109	44.50		153.50	192
1820	2'-8" wide		9	1.778		113	74		187	245
1830	3'-0" wide		16	1		124	41.50		165.50	205
1840	3'-4" wide		16	1		207	41.50		248.50	296
1850	Pair of 3'-0" wide		15	1.067	Pr.	247	44.50		291.50	345
2700	Interior, closet, bi-fold, w/hardware, no frame or trim incl.									
2720	Flush, birch, 6'-6" or 6'-8" x 2'-6" wide	2 Carp	13	1.231	Ea.	51	51		102	140
2740	3'-0" wide		13	1.231		55	51		106	144
2760	4'-0" wide		12	1.333		97.50	55.50		153	198
2780	5'-0" wide		11	1.455		100	60.50		160.50	210
2800	6'-0" wide		10	1.600		126	66.50		192.50	248
2820	Flush, hardboard, primed, 6'-8" x 2'-6" wide		13	1.231		46.50	51		97.50	135
2840	3'-0" wide		13	1.231		52.50	51		103.50	142
2860	4'-0" wide		12	1.333		83.50	55.50		139	183
2880	5'-0" wide		11	1.455		92.50	60.50		153	201
2900	6'-0" wide		10	1.600		135	66.50		201.50	257
3000	Raised panel pine, 6'-6" or 6'-8" x 2'-6" wide		13	1.231		178	51		229	280
3020	3'-0" wide		13	1.231		273	51		324	385
3040	4'-0" wide		12	1.333		305	55.50		360.50	425
3060	5'-0" wide		11	1.455		365	60.50		425.50	500
3080	6'-0" wide		10	1.600		400	66.50		466.50	550
3200	Louvered, pine 6'-6" or 6'-8" x 2'-6" wide		13	1.231		137	51		188	235
3220	3'-0" wide		13	1.231		197	51		248	300
3240	4'-0" wide		12	1.333		222	55.50		277.50	335
3260	5'-0" wide		11	1.455		256	60.50		316.50	380
3280	6'-0" wide		10	1.600		280	66.50		346.50	415
4400	Bi-passing closet, incl. hardware and frame, no trim incl.									
4420	Flush, lauan, 6'-8" x 4'-0" wide	2 Carp	12	1.333	Opng.	169	55.50		224.50	276
4440	5'-0" wide		11	1.455		184	60.50		244.50	300
4460	6'-0" wide		10	1.600		198	66.50		264.50	325
4600	Flush, birch, 6'-8" x 4'-0" wide		12	1.333		217	55.50		272.50	330
4620	5'-0" wide		11	1.455		209	60.50		269.50	330

08 14 Wood Doors

08 14 33 – Stile and Rail Wood Doors

08 14 33.20 Wood Doors Residential

		Crew	Daily Output	Labor-Hours	Unit	Material	2010 Bare Costs Labor	Equipment	Total	Total Incl O&P
4640	6'-0" wide	2 Carp	10	1.600	Opng.	254	66.50		320.50	390
4800	Louvered, pine, 6'-8" x 4'-0" wide		12	1.333		400	55.50		455.50	530
4820	5'-0" wide		11	1.455		390	60.50		450.50	525
4840	6'-0" wide		10	1.600		515	66.50		581.50	680
5000	Paneled, pine, 6'-8" x 4'-0" wide		12	1.333		500	55.50		555.50	640
5020	5'-0" wide		11	1.455		405	60.50		465.50	545
5040	6'-0" wide		10	1.600		595	66.50		661.50	765
6100	Folding accordion, closet, including track and frame									
6120	Vinyl, 2 layer, stock see Div. 10 22 26.13	2 Carp	10	1.600	Ea.	61	66.50		127.50	176
6200	Rigid PVC	"	10	1.600	"	33.50	66.50		100	146
6220	For custom partition, add					25%	10%			
7310	Passage doors, flush, no frame included									
7320	Hardboard, hollow core, 1-3/8" x 6'-8" x 1'-6" wide	2 Carp	18	.889	Ea.	42.50	37		79.50	107
7330	2'-0" wide		18	.889		42	37		79	107
7340	2'-6" wide		18	.889		46.50	37		83.50	112
7350	2'-8" wide		18	.889		48	37		85	114
7360	3'-0" wide		17	.941		51	39		90	120
7420	Lauan, hollow core, 1-3/8" x 6'-8" x 1'-6" wide		18	.889		31	37		68	94.50
7440	2'-0" wide		18	.889		31	37		68	94.50
7450	2'-4" wide		18	.889		34.50	37		71.50	98.50
7460	2'-6" wide		18	.889		34.50	37		71.50	98.50
7480	2'-8" wide		18	.889		36	37		73	100
7500	3'-0" wide		17	.941		37	39		76	105
7700	Birch, hollow core, 1-3/8" x 6'-8" x 1'-6" wide		18	.889		38	37		75	103
7720	2'-0" wide		18	.889		41	37		78	106
7740	2'-6" wide		18	.889		49	37		86	114
7760	2'-8" wide		18	.889		49.50	37		86.50	115
7780	3'-0" wide		17	.941		51	39		90	120
8000	Pine louvered, 1-3/8" x 6'-8" x 1'-6" wide		19	.842		101	35		136	169
8020	2'-0" wide		18	.889		128	37		165	201
8040	2'-6" wide		18	.889		139	37		176	214
8060	2'-8" wide		18	.889		146	37		183	222
8080	3'-0" wide		17	.941		167	39		206	247
8300	Pine paneled, 1-3/8" x 6'-8" x 1'-6" wide		19	.842		111	35		146	180
8320	2'-0" wide		18	.889		130	37		167	204
8330	2'-4" wide		18	.889		150	37		187	226
8340	2'-6" wide		18	.889		158	37		195	234
8360	2'-8" wide		18	.889		161	37		198	238
8380	3'-0" wide		17	.941		167	39		206	247
8490	3'-0" wide		17	.941		252	39		291	340
8804	Pocket door, 6 panel pine, 2'-6" x 6'-8"		10.50	1.524		221	63.50		284.50	345
8814	2'-8" x 6'-8"		10.50	1.524		233	63.50		296.50	360
8824	3'-0" x 6'-8"		10.50	1.524		240	63.50		303.50	370
9000	Birch, solid core, 1-3/8" x 7'-0" x 2'-4" wide		16	1		109	41.50		150.50	189
9020	2'-8" wide		16	1		116	41.50		157.50	197
9040	3'-0" wide		16	1		125	41.50		166.50	207
9060	3'-4" wide		15	1.067		144	44.50		188.50	232
9080	Pair of 3'-0" wide		9	1.778	Pr.	248	74		322	395
9100	Lauan, solid core, 1-3/8" x 7'-0" x 2'-4" wide		16	1	Ea.	101	41.50		142.50	180
9120	2'-8" wide		16	1		107	41.50		148.50	187
9140	3'-0" wide		16	1		114	41.50		155.50	194
9160	3'-4" wide		15	1.067		122	44.50		166.50	207
9180	Pair of 3'-0" wide		9	1.778	Pr.	207	74		281	350

08 14 Wood Doors

08 14 33 – Stile and Rail Wood Doors

08 14 33.20 Wood Doors Residential		Crew	Daily Output	Labor-Hours	Unit	Material	2010 Bare Costs Labor	2010 Bare Costs Equipment	Total	Total Incl O&P
9200	Hardboard, solid core, 1-3/8" x 7'-0" x 2'-4" wide	2 Carp	16	1	Ea.	124	41.50		165.50	206
9220	2'-8" wide		16	1		128	41.50		169.50	210
9240	3'-0" wide		16	1		133	41.50		174.50	215
9260	3'-4" wide		15	1.067		148	44.50		192.50	236
9900	Minimum labor/equipment charge	1 Carp	4	2	Job		83		83	137

08 14 40 – Interior Cafe Doors

08 14 40.10 Cafe Style Doors

		Crew	Daily Output	Labor-Hours	Unit	Material	Labor	Equipment	Total	Total Incl O&P
0010	**CAFE STYLE DOORS**									
6520	Interior cafe doors, 2'-6" opening, stock, panel pine	2 Carp	16	1	Ea.	188	41.50		229.50	275
6540	3'-0" opening	"	16	1	"	195	41.50		236.50	284
6550	Louvered pine									
6560	2'-6" opening	2 Carp	16	1	Ea.	164	41.50		205.50	249
8000	3'-0" opening		16	1		175	41.50		216.50	261
8010	2'-6" opening, hardwood		16	1		238	41.50		279.50	330
8020	3'-0" opening		16	1		264	41.50		305.50	360
9000	Minimum labor/equipment charge	1 Carp	4	2	Job		83		83	137

08 16 Composite Doors

08 16 13 – Fiberglass Doors

08 16 13.10 Entrance Doors, Fiberous Glass

			Crew	Daily Output	Labor-Hours	Unit	Material	Labor	Equipment	Total	Total Incl O&P
0010	**ENTRANCE DOORS, FIBEROUS GLASS**										
0020	Exterior, fiberglass, door, 2'-8" wide x 6'-8" high	G	2 Carp	15	1.067	Ea.	252	44.50		296.50	350
0040	3'-0" wide x 6'-8" high	G		15	1.067		252	44.50		296.50	350
0060	3'-0" wide x 7'-0" high	G		15	1.067		450	44.50		494.50	570
0080	3'-0" wide x 6'-8" high, with two lites	G		15	1.067		294	44.50		338.50	400
0100	3'-0" wide x 8'-0" high, with two lites	G		15	1.067		490	44.50		534.50	610
0110	Half glass, 3'-0" wide x 6'-8" high	G		15	1.067		294	44.50		338.50	400
0120	3'-0" wide x 6'-8" high, low e	G		15	1.067		325	44.50		369.50	430
0130	3'-0" wide x 8'-0" high	G		15	1.067		580	44.50		624.50	710
0140	3'-0" wide x 8'-0" high, low e	G		15	1.067		640	44.50		684.50	775
0150	Side lights, 1'-0" wide x 6'-8" high,	G					247			247	271
0160	1'-0" wide x 6'-8" high, low e	G					256			256	281
0180	1'-0" wide x 6'-8" high, full glass	G					275			275	305
0190	1'-0" wide x 6'-8" high, low e	G					284			284	315

08 16 14 – French Doors

08 16 14.10 Exterior Doors With Glass Lites

		Crew	Daily Output	Labor-Hours	Unit	Material	Labor	Equipment	Total	Total Incl O&P
0010	**EXTERIOR DOORS WITH GLASS LITES**									
0020	French, Fir, 1-3/4", 3'-0"wide x 6'-8" high	2 Carp	12	1.333	Ea.	460	55.50		515.50	595
0025	double		12	1.333		925	55.50		980.50	1,125
0030	Maple, 1-3/4", 3'-0"wide x 6'-8" high		12	1.333		545	55.50		600.50	690
0035	double		12	1.333		1,100	55.50		1,155.50	1,300
0040	Cherry, 1-3/4", 3'-0"wide x 6'-8" high		12	1.333		625	55.50		680.50	775
0045	double		12	1.333		1,250	55.50		1,305.50	1,475
0100	Mahogany, 1-3/4", 3'-0"wide x 8'-0" high		10	1.600		510	66.50		576.50	670
0105	double		10	1.600		1,025	66.50		1,091.50	1,225
0110	Fir, 1-3/4", 3'-0"wide x 8'-0" high		10	1.600		870	66.50		936.50	1,075
0115	double		10	1.600		1,750	66.50		1,816.50	2,025
0120	Oak, 1-3/4", 3'-0"wide x 8'-0" high		10	1.600		1,800	66.50		1,866.50	2,075
0125	double		10	1.600		3,600	66.50		3,666.50	4,050

08 17 Integrated Door Opening Assemblies

08 17 23 – Integrated Wood Door Opening Assemblies

08 17 23.10 Pre-Hung Doors

		Crew	Daily Output	Labor-Hours	Unit	Material	2010 Bare Costs Labor	Equipment	Total	Total Incl O&P
0010	**PRE-HUNG DOORS**									
0300	Exterior, wood, comb. storm & screen, 6'-9" x 2'-6" wide	2 Carp	15	1.067	Ea.	289	44.50		333.50	395
0320	2'-8" wide		15	1.067		289	44.50		333.50	395
0340	3'-0" wide		15	1.067		296	44.50		340.50	400
0360	For 7'-0" high door, add					24			24	26.50
1600	Entrance door, flush, birch, solid core									
1620	4-5/8" solid jamb, 1-3/4" x 6'-8" x 2'-8" wide	2 Carp	16	1	Ea.	285	41.50		326.50	385
1640	3'-0" wide		16	1		350	41.50		391.50	455
1642	5-5/8" jamb		16	1		325	41.50		366.50	425
1680	For 7'-0" high door, add					21			21	23
2000	Entrance door, colonial, 6 panel pine									
2020	4-5/8" solid jamb, 1-3/4" x 6'-8" x 2'-8" wide	2 Carp	16	1	Ea.	530	41.50		571.50	650
2040	3'-0" wide	"	16	1		530	41.50		571.50	650
2060	For 7'-0" high door, add					52.50			52.50	58
2200	For 5-5/8" solid jamb, add					40.50			40.50	44.50
2270	Pine french door, Pair, 6'-8" x 5'-0" wide, including frame	2 Carp	7	2.286	Pr.	1,250	95		1,345	1,525
2280	French door, Two 6'-8" x 3'-0" wide, 15 lites	"	14	1.143	Ea.	1,200	47.50		1,247.50	1,400
4000	Interior, passage door, 4-5/8" solid jamb									
4400	Lauan, flush, solid core, 1-3/8" x 6'-8" x 2'-6" wide	2 Carp	17	.941	Ea.	182	39		221	264
4420	2'-8" wide		17	.941		182	39		221	264
4440	3'-0" wide		16	1		195	41.50		236.50	284
4600	Hollow core, 1-3/8" x 6'-8" x 2'-6" wide		17	.941		120	39		159	196
4620	2'-8" wide		17	.941		120	39		159	196
4640	3'-0" wide		16	1		135	41.50		176.50	218
4700	For 7'-0" high door, add					25			25	27.50
5000	Birch, flush, solid core, 1-3/8" x 6'-8" x 2'-6" wide	2 Carp	17	.941		187	39		226	270
5020	2'-8" wide		17	.941		195	39		234	279
5040	3'-0" wide		16	1		217	41.50		258.50	310
5200	Hollow core, 1-3/8" x 6'-8" x 2'-6" wide		17	.941		147	39		186	226
5220	2'-8" wide		17	.941		153	39		192	232
5240	3'-0" wide		16	1		152	41.50		193.50	237
5280	For 7'-0" high door, add					22			22	24
5500	Hardboard paneled, 1-3/8" x 6'-8" x 2'-6" wide	2 Carp	17	.941		142	39		181	220
5520	2'-8" wide		17	.941		150	39		189	229
5540	3'-0" wide		16	1		147	41.50		188.50	231
6000	Pine paneled, 1-3/8" x 6'-8" x 2'-6" wide		17	.941		245	39		284	335
6020	2'-8" wide		17	.941		265	39		304	355
6040	3'-0" wide		16	1		270	41.50		311.50	365
6500	For 5-5/8" solid jamb, add					19			19	21
6520	For split jamb, deduct					21			21	23
8500	Pocket door frame with lauan, flush, hollow core, 1-3/8" x 3'-0" x 6'-8"	1 Carp	17	.471		245	19.55		264.55	300
9000	Minimum labor/equipment charge	"	4	2	Job		83		83	137

08 31 Access Doors and Panels

08 31 13 – Access Doors and Frames

08 31 13.10 Types of Framed Access Doors	Crew	Daily Output	Labor-Hours	Unit	Material	2010 Bare Costs Labor	Equipment	Total	Total Incl O&P
0010 **TYPES OF FRAMED ACCESS DOORS**									
1000 Fire rated door with lock									
1100 Metal, 12" x 12"	1 Carp	10	.800	Ea.	158	33		191	229
1150 18" x 18"		9	.889		205	37		242	287
1200 24" x 24"		9	.889		325	37		362	420
1250 24" x 36"		8	1		335	41.50		376.50	435
1300 24" x 48"		8	1		410	41.50		451.50	525
1350 36" x 36"		7.50	1.067		495	44.50		539.50	620
1400 48" x 48"		7.50	1.067		635	44.50		679.50	770
1600 Stainless steel, 12" x 12"		10	.800		282	33		315	365
1650 18" x 18"		9	.889		410	37		447	510
1700 24" x 24"		9	.889		500	37		537	610
1750 24" x 36"		8	1		640	41.50		681.50	775
2000 Flush door for finishing									
2100 Metal 8" x 8"	1 Carp	10	.800	Ea.	45	33		78	104
2150 12" x 12"	"	10	.800	"	50	33		83	110
3000 Recessed door for acoustic tile									
3100 Metal, 12" x 12"	1 Carp	4.50	1.778	Ea.	69	74		143	197
3150 12" x 24"		4.50	1.778		90	74		164	220
3200 24" x 24"		4	2		121	83		204	270
3250 24" x 36"		4	2		155	83		238	310
4000 Recessed door for drywall									
4100 Metal 12" x 12"	1 Carp	6	1.333	Ea.	77	55.50		132.50	176
4150 12" x 24"		5.50	1.455		114	60.50		174.50	225
4200 24" x 36"		5	1.600		182	66.50		248.50	310
6000 Standard door									
6100 Metal, 8" x 8"	1 Carp	10	.800	Ea.	40	33		73	98.50
6150 12" x 12"		10	.800		45	33		78	104
6200 18" x 18"		9	.889		62	37		99	129
6250 24" x 24"		9	.889		82	37		119	151
6300 24" x 36"		8	1		120	41.50		161.50	201
6350 36" x 36"		8	1		146	41.50		187.50	230
6500 Stainless steel, 8" x 8"		10	.800		80	33		113	143
6550 12" x 12"		10	.800		107	33		140	173
6600 18" x 18"		9	.889		198	37		235	279
6650 24" x 24"		9	.889		260	37		297	345
7010 Aluminum cover, Ceiling hatches, 2'-6" x 2'-6", single leaf, st fr	G-3	11	2.909		540	120		660	785
9000 Minimum labor/equipment charge	1 Carp	4	2	Job		83		83	137

08 31 13.20 Bulkhead/Cellar Doors

	Crew	Daily Output	Labor-Hours	Unit	Material	Labor	Equipment	Total	Total Incl O&P
0010 **BULKHEAD/CELLAR DOORS**									
0020 Steel, not incl. sides, 44" x 62"	1 Carp	5.50	1.455	Ea.	490	60.50		550.50	640
0100 52" x 73"		5.10	1.569		660	65		725	830
0500 With sides and foundation plates, 57" x 45" x 24"		4.70	1.702		700	70.50		770.50	885
0600 42" x 49" x 51"		4.30	1.860		790	77.50		867.50	995
9000 Minimum labor/equipment charge		2	4	Job		166		166	273

08 31 13.30 Commercial Floor Doors

	Crew	Daily Output	Labor-Hours	Unit	Material	Labor	Equipment	Total	Total Incl O&P
0010 **COMMERCIAL FLOOR DOORS**									
0020 Aluminum tile, steel frame, one leaf, 2' x 2' opng.	2 Sswk	3.50	4.571	Opng.	775	214		989	1,250
0050 3'-6" x 3'-6" opening		3.50	4.571		1,475	214		1,689	2,025
0500 Double leaf, 4' x 4' opening		3	5.333		1,575	250		1,825	2,175
0550 5' x 5' opening		3	5.333		2,800	250		3,050	3,525
9000 Minimum labor/equipment charge		2	8	Job		375		375	685

08 31 Access Doors and Panels

08 31 13 – Access Doors and Frames

08 31 13.35 Industrial Floor Doors

		Crew	Daily Output	Labor-Hours	Unit	Material	2010 Bare Costs Labor	Equipment	Total	Total Incl O&P
0010	**INDUSTRIAL FLOOR DOORS**									
0020	Steel 300 psf L.L., single leaf, 2' x 2', 175#	2 Sswk	6	2.667	Opng.	735	125		860	1,050
0050	3' x 3' opening, 300#		5.50	2.909		1,000	136		1,136	1,350
0300	Double leaf, 4' x 4' opening, 455#		5	3.200		1,625	150		1,775	2,075
0350	5' x 5' opening, 645#		4.50	3.556		2,450	167		2,617	3,000
1000	Aluminum, 300 psf L.L., single leaf, 2' x 2', 60#		6	2.667		740	125		865	1,050
1050	3' x 3' opening, 100#		5.50	2.909		1,100	136		1,236	1,450
1500	Double leaf, 4' x 4' opening, 160#		5	3.200		1,950	150		2,100	2,425
1550	5' x 5' opening, 235#		4.50	3.556		2,675	167		2,842	3,250
9000	Minimum labor/equipment charge		2	8	Job		375		375	685

08 32 Sliding Glass Doors

08 32 13 – Sliding Aluminum-Framed Glass Doors

08 32 13.10 Sliding Aluminum Doors

		Crew	Daily Output	Labor-Hours	Unit	Material	Labor	Equipment	Total	Total Incl O&P
0010	**SLIDING ALUMINUM DOORS**									
0350	Aluminum, 5/8" tempered insulated glass, 6' wide									
0400	Premium	2 Carp	4	4	Ea.	1,550	166		1,716	2,000
0450	Economy		4	4		870	166		1,036	1,225
0500	8' wide, premium		3	5.333		1,625	222		1,847	2,150
0550	Economy		3	5.333		1,400	222		1,622	1,900
0600	12' wide, premium		2.50	6.400		2,950	266		3,216	3,675
0650	Economy		2.50	6.400		1,500	266		1,766	2,075
0700	Non-insulated, 6' wide		4	4		1,075	166		1,241	1,450
4000	Aluminum, baked on enamel, temp glass, 6'-8" x 10'-0" wide		4	4		970	166		1,136	1,350
4020	Insulating glass, 6'-8" x 6'-0" wide		4	4		860	166		1,026	1,225
4040	8'-0" wide		3	5.333		1,000	222		1,222	1,475
4060	10'-0" wide		2	8		1,275	330		1,605	1,950
4080	Anodized, temp glass, 6'-8" x 6'-0" wide		4	4		440	166		606	760
4100	8'-0" wide		3	5.333		560	222		782	980
4120	10'-0" wide		2	8		630	330		960	1,250
4200	Vinyl clad, anodized, temp glass, 6'-8" x 6'-0" wide		4	4		410	166		576	725
4240	8'-0" wide		3	5.333		570	222		792	990
4260	10'-0" wide		2	8		620	330		950	1,225
4280	Insulating glass, 6'-8" x 6'-0' wide		4	4		995	166		1,161	1,375
4300	8'-0" wide		3	5.333		920	222		1,142	1,375
4320	10'-0" wide		4	4		1,500	166		1,666	1,925

08 32 19 – Sliding Wood-Framed Glass Doors

08 32 19.10 Sliding Wood Doors

		Crew	Daily Output	Labor-Hours	Unit	Material	Labor	Equipment	Total	Total Incl O&P
0010	**SLIDING WOOD DOORS**									
0020	Wood, 5/8" tempered insul. glass, 6' wide, premium	2 Carp	4	4	Ea.	1,400	166		1,566	1,800
0100	Economy		4	4		1,100	166		1,266	1,500
0150	8' wide, wood, premium		3	5.333		1,775	222		1,997	2,325
0200	Economy		3	5.333		1,475	222		1,697	2,000
0250	12' wide, wood, vinyl clad		2.50	6.400		2,925	266		3,191	3,650
0300	Economy		2.50	6.400		2,400	266		2,666	3,075
9000	Minimum labor/equipment charge	1 Carp	2	4	Job		166		166	273

08 32 19.15 Sliding Glass Vinyl-Clad Wood Doors

			Crew	Daily Output	Labor-Hours	Unit	Material	Labor	Equipment	Total	Total Incl O&P
0010	**SLIDING GLASS VINYL-CLAD WOOD DOORS**										
0012	Vinyl clad, 1" insul. glass, 6'-0" x 6'-10" high	G	2 Carp	4	4	Opng.	1,575	166		1,741	2,025
0020	Glass, sliding vinyl clad, insul. glass, 6'-0" x 6'-8",	G		4	4	Ea.	1,525	166		1,691	1,950

08 32 Sliding Glass Doors

08 32 19 – Sliding Wood-Framed Glass Doors

08 32 19.15 Sliding Glass Vinyl-Clad Wood Doors		Crew	Daily Output	Labor-Hours	Unit	Material	2010 Bare Costs Labor	Equipment	Total	Total Incl O&P
0030	6'-0" x 8'-0" high	G 2 Carp	4	4	Ea.	1,950	166		2,116	2,425
0050	5'-0" x 6'-8" high	G	4	4		1,475	166		1,641	1,900
0100	8'-0" x 6'-10" high	G	4	4	Opng.	2,000	166		2,166	2,475
0150	8'-0" x 8'-0" high	G	4	4		2,175	166		2,341	2,675
0500	4 leaf, 9'-0" x 6'-10" high	G	3	5.333		3,175	222		3,397	3,875
0550	9'-0" x 8'-0" high	G	3	5.333	Ea.	3,675	222		3,897	4,425
0600	12'-0" x 6'-10" high	G	3	5.333	Opng.	3,800	222		4,022	4,550
9000	Minimum labor/equipment charge	1 Carp	4	2	Job		83		83	137

08 33 Coiling Doors and Grilles

08 33 13 – Coiling Counter Doors

08 33 13.10 Counter Doors, Coiling Type

		Crew	Daily Output	Labor-Hours	Unit	Material	Labor	Equipment	Total	Total Incl O&P
0010	**COUNTER DOORS, COILING TYPE**									
0020	Manual, incl. frm and hdwe, galv. stl., 4' roll-up, 6' long	2 Carp	2	8	Opng.	1,150	330		1,480	1,800
0300	Galvanized steel, UL label		1.80	8.889		1,200	370		1,570	1,925
0600	Stainless steel, 4' high roll-up, 6' long		2	8		2,100	330		2,430	2,850
0700	10' long		1.80	8.889		2,450	370		2,820	3,300
2000	Aluminum, 4' high, 4' long		2.20	7.273		1,275	300		1,575	1,900
2020	6' long		2	8		1,425	330		1,755	2,100
2040	8' long		1.90	8.421		1,650	350		2,000	2,375
2060	10' long		1.80	8.889		1,725	370		2,095	2,500
2080	14' long		1.40	11.429		2,500	475		2,975	3,525
2100	6' high, 4' long		2	8		1,425	330		1,755	2,125
2120	6' long		1.60	10		1,450	415		1,865	2,275
2140	10' long		1.40	11.429		1,975	475		2,450	2,950
9000	Minimum labor/equipment charge	1 Carp	2	4	Job		166		166	273

08 33 16 – Coiling Counter Grilles

08 33 16.10 Coiling Grilles

		Crew	Daily Output	Labor-Hours	Unit	Material	Labor	Equipment	Total	Total Incl O&P
0010	**COILING GRILLES**									
2020	Aluminum, manual operated, mill finish	2 Sswk	82	.195	S.F.	27.50	9.15		36.65	47
2040	Bronze anodized		82	.195	"	43.50	9.15		52.65	65
2060	Steel, manual operated, 10' x 10' high		1	16	Opng.	2,450	750		3,200	4,075
2080	15' x 8' high		.80	20	"	2,850	940		3,790	4,850
3000	For safety edge bottom bar, electric, add				L.F.	49			49	54
8000	For motor operation, add	2 Sswk	5	3.200	Opng.	1,250	150		1,400	1,650
9000	Minimum labor/equipment charge	"	1	16	Job		750		750	1,375

08 33 23 – Overhead Coiling Doors

08 33 23.10 Rolling Service Doors

		Crew	Daily Output	Labor-Hours	Unit	Material	Labor	Equipment	Total	Total Incl O&P
0010	**ROLLING SERVICE DOORS** Steel, manual, 20 ga., incl. hardware									
0050	8' x 8' high	2 Sswk	1.60	10	Ea.	1,050	470		1,520	2,025
0130	12' x 12' high, standard		1.20	13.333		1,750	625		2,375	3,075
0160	10' x 20' high, standard		.50	32		2,300	1,500		3,800	5,275
2000	Class A fire doors, manual, 20 ga., 8' x 8' high		1.40	11.429		1,400	535		1,935	2,525
2100	10' x 10' high		1.10	14.545		1,900	680		2,580	3,350
2200	20' x 10' high		.80	20		3,900	940		4,840	6,025
2300	12' x 12' high		1	16		3,000	750		3,750	4,675
2400	20' x 12' high		.80	20		4,225	940		5,165	6,375
2500	14' x 14' high		.60	26.667		3,300	1,250		4,550	5,900
2600	20' x 16' high		.50	32		5,300	1,500		6,800	8,575
2700	10' x 20' high		.40	40		4,100	1,875		5,975	7,925

08 33 Coiling Doors and Grilles

08 33 23 – Overhead Coiling Doors

08 33 23.10 Rolling Service Doors

		Crew	Daily Output	Labor-Hours	Unit	Material	2010 Bare Costs Labor	Equipment	Total	Total Incl O&P
3000	For 18 ga. doors, add				S.F.	1.35			1.35	1.49
3300	For enamel finish, add				"	1.60			1.60	1.76
3600	For safety edge bottom bar, pneumatic, add				L.F.	19.90			19.90	22
4000	For weatherstripping, extruded rubber, jambs, add					12.90			12.90	14.20
4100	Hood, add					7.80			7.80	8.55
4200	Sill, add					4.46			4.46	4.91
4500	Motor operators, to 14' x 14' opening	2 Sswk	5	3.200	Ea.	1,025	150		1,175	1,400
4700	For fire door, additional fusible link, add				"	24.50			24.50	27
9000	Minimum labor/equipment charge	2 Sswk	1	16	Job		750		750	1,375

08 34 Special Function Doors

08 34 13 – Cold Storage Doors

08 34 13.10 Doors for Cold Area Storage

		Crew	Daily Output	Labor-Hours	Unit	Material	Labor	Equipment	Total	Total Incl O&P
0010	**DOORS FOR COLD AREA STORAGE**									
0020	Single, 20 ga. galvanized steel									
0300	Horizontal sliding, 5' x 7', manual operation, 3.5" thick	2 Carp	2	8	Ea.	3,150	330		3,480	4,000
0400	4" thick		2	8		3,825	330		4,155	4,775
0500	6" thick		2	8		3,200	330		3,530	4,075
0800	5' x 7', power operation, 2" thick		1.90	8.421		5,225	350		5,575	6,325
0900	4" thick		1.90	8.421		5,350	350		5,700	6,450
1000	6" thick		1.90	8.421		6,100	350		6,450	7,275
1300	9' x 10', manual operation, 2" insulation		1.70	9.412		4,250	390		4,640	5,325
1400	4" insulation		1.70	9.412		4,350	390		4,740	5,425
1500	6" insulation		1.70	9.412		5,225	390		5,615	6,400
1800	Power operation, 2" insulation		1.60	10		7,200	415		7,615	8,600
1900	4" insulation		1.60	10		7,425	415		7,840	8,825
2000	6" insulation		1.70	9.412		8,400	390		8,790	9,900
2300	For stainless steel face, add					22%				
3000	Hinged, lightweight, 3' x 7'-0", galvanized 1 face, 2" thick	2 Carp	2	8	Ea.	1,350	330		1,680	2,025
3050	4" thick		1.90	8.421		1,700	350		2,050	2,425
3300	Aluminum doors, 3' x 7'-0", 4" thick		1.90	8.421		1,250	350		1,600	1,950
3350	6" thick		1.40	11.429		2,200	475		2,675	3,200
3600	Stainless steel, 3' x 7'-0", 4" thick		1.90	8.421		1,625	350		1,975	2,350
3650	6" thick		1.40	11.429		2,775	475		3,250	3,825
3900	Painted, 3' x 7'-0", 4" thick		1.90	8.421		1,200	350		1,550	1,900
3950	6" thick		1.40	11.429		2,200	475		2,675	3,200
5000	Bi-parting, electric operated									
5010	6' x 8' opening, galv. faces, 4" thick for cooler	2 Carp	.80	20	Opng.	6,800	830		7,630	8,850
5050	For freezer, 4" thick		.80	20		7,500	830		8,330	9,625
5300	For door buck framing and door protection, add		2.50	6.400		600	266		866	1,100
6000	Galvanized batten door, galvanized hinges, 4' x 7'		2	8		1,700	330		2,030	2,425
6050	6' x 8'		1.80	8.889		2,325	370		2,695	3,150
6500	Fire door, 3 hr., 6' x 8', single slide		.80	20		7,800	830		8,630	9,950
6550	Double, bi-parting		.70	22.857		12,200	950		13,150	15,000
9000	Minimum labor/equipment charge	1 Carp	2	4	Job		166		166	273

08 34 36 – Darkroom Doors

08 34 36.10 Various Types of Darkroom Doors

		Crew	Daily Output	Labor-Hours	Unit	Material	Labor	Equipment	Total	Total Incl O&P
0010	**VARIOUS TYPES OF DARKROOM DOORS**									
0015	Revolving, standard, 2 way, 36" diameter	2 Carp	3.10	5.161	Opng.	2,250	214		2,464	2,825
0020	41" diameter		3.10	5.161		2,425	214		2,639	3,025

08 34 Special Function Doors

08 34 36 – Darkroom Doors

08 34 36.10 Various Types of Darkroom Doors

		Crew	Daily Output	Labor-Hours	Unit	Material	2010 Bare Costs Labor	Equipment	Total	Total Incl O&P
0050	3 way, 51" diameter	2 Carp	1.40	11.429	Opng.	3,075	475		3,550	4,150
1000	4 way, 49" diameter		1.40	11.429		3,625	475		4,100	4,775
2000	Hinged safety, 2 way, 41" diameter		2.30	6.957		2,825	289		3,114	3,600
2500	3 way, 51" diameter		1.40	11.429		3,625	475		4,100	4,775
3000	Pop out safety, 2 way, 41" diameter		3.10	5.161		3,600	214		3,814	4,300
4000	3 way, 51" diameter		1.40	11.429		3,550	475		4,025	4,675
5000	Wheelchair-type, pop out, 51" diameter		1.40	11.429		3,525	475		4,000	4,650
5020	72" diameter		.90	17.778		7,225	740		7,965	9,175

08 34 73 – Sound Control Door Assemblies

08 34 73.10 Acoustical Doors

		Crew	Daily Output	Labor-Hours	Unit	Material	Labor	Equipment	Total	Total Incl O&P
0010	**ACOUSTICAL DOORS**									
0020	Including framed seals, 3' x 7', wood, 27 STC rating	2 Carp	1.50	10.667	Ea.	800	445		1,245	1,600
0100	Steel, 40 STC rating		1.50	10.667		2,800	445		3,245	3,800
0200	45 STC rating		1.50	10.667		3,300	445		3,745	4,350
0300	48 STC rating		1.50	10.667		3,800	445		4,245	4,900
0400	52 STC rating		1.50	10.667		4,300	445		4,745	5,450
9000	Minimum labor/equipment charge	1 Carp	4	2	Job		83		83	137

08 36 Panel Doors

08 36 13 – Sectional Doors

08 36 13.10 Overhead Commercial Doors

		Crew	Daily Output	Labor-Hours	Unit	Material	Labor	Equipment	Total	Total Incl O&P
0010	**OVERHEAD COMMERCIAL DOORS**									
1000	Stock, sectional, heavy duty, wood, 1-3/4" thick, 8' x 8' high	2 Carp	2	8	Ea.	775	330		1,105	1,400
1200	12' x 12' high		1.50	10.667		1,675	445		2,120	2,550
1300	Chain hoist, 14' x 14' high		1.30	12.308		2,550	510		3,060	3,650
1600	20' x 16' high		.65	24.615		5,975	1,025		7,000	8,250
2100	For medium duty custom door, deduct					5%	5%			
2150	For medium duty stock doors, deduct					10%	5%			
2300	Fiberglass and aluminum, heavy duty, sectional, 12' x 12' high	2 Carp	1.50	10.667	Ea.	2,575	445		3,020	3,575
2450	Chain hoist, 20' x 20' high	"	.50	32		6,500	1,325		7,825	9,325
2900	For electric trolley operator, 1/3 H.P., to 12' x 12', add	1 Carp	2	4		900	166		1,066	1,250
2950	Over 12' x 12', 1/2 H.P., add	"	1	8		1,025	330		1,355	1,675
9000	Minimum labor/equipment charge	2 Carp	1.50	10.667	Job		445		445	730

08 36 13.20 Residential Garage Doors

		Crew	Daily Output	Labor-Hours	Unit	Material	Labor	Equipment	Total	Total Incl O&P
0010	**RESIDENTIAL GARAGE DOORS**									
0050	Hinged, wood, custom, double door, 9' x 7'	2 Carp	4	4	Ea.	585	166		751	920
0070	16' x 7'		3	5.333		965	222		1,187	1,425
0200	Overhead, sectional, incl. hardware, fiberglass, 9' x 7', standard		5.28	3.030		735	126		861	1,000
0220	Deluxe		5.28	3.030		935	126		1,061	1,225
0300	16' x 7', standard		6	2.667		1,550	111		1,661	1,875
0320	Deluxe		6	2.667		1,975	111		2,086	2,350
0500	Hardboard, 9' x 7', standard		8	2		550	83		633	740
0520	Deluxe		8	2		775	83		858	990
0600	16' x 7', standard		6	2.667		1,125	111		1,236	1,425
0620	Deluxe		6	2.667		1,350	111		1,461	1,650
0700	Metal, 9' x 7', standard		5.28	3.030		580	126		706	845
0720	Deluxe		8	2		855	83		938	1,075
0800	16' x 7', standard		3	5.333		730	222		952	1,175
0820	Deluxe		6	2.667		1,375	111		1,486	1,675
0900	Wood, 9' x 7', standard		8	2		720	83		803	925

08 36 Panel Doors

08 36 13 – Sectional Doors

08 36 13.20 Residential Garage Doors

		Crew	Daily Output	Labor-Hours	Unit	Material	2010 Bare Costs Labor	Equipment	Total	Total Incl O&P
0920	Deluxe	2 Carp	8	2	Ea.	2,050	83		2,133	2,375
1000	16' x 7', standard		6	2.667		1,425	111		1,536	1,750
1020	Deluxe		6	2.667		2,975	111		3,086	3,450
1800	Door hardware, sectional	1 Carp	4	2		320	83		403	485
1810	Door tracks only		4	2		138	83		221	289
1820	One side only		7	1.143		110	47.50		157.50	199
3000	Swing-up, including hardware, fiberglass, 9' x 7', standard	2 Carp	8	2		980	83		1,063	1,200
3020	Deluxe		8	2		790	83		873	1,000
3100	16' x 7', standard		6	2.667		1,225	111		1,336	1,525
3120	Deluxe		6	2.667		980	111		1,091	1,250
3200	Hardboard, 9' x 7', standard		8	2		345	83		428	515
3220	Deluxe		8	2		455	83		538	640
3300	16' x 7', standard		6	2.667		480	111		591	710
3320	Deluxe		6	2.667		715	111		826	965
3400	Metal, 9' x 7', standard		8	2		380	83		463	555
3420	Deluxe		8	2		835	83		918	1,050
3500	16' x 7', standard		6	2.667		600	111		711	840
3520	Deluxe		6	2.667		960	111		1,071	1,225
3600	Wood, 9' x 7', standard		8	2		450	83		533	630
3620	Deluxe		8	2		935	83		1,018	1,150
3700	16' x 7', standard		6	2.667		725	111		836	980
3720	Deluxe		6	2.667		1,025	111		1,136	1,300
3900	Door hardware only, swing up	1 Carp	4	2		145	83		228	297
3920	One side only		7	1.143		70	47.50		117.50	155
4000	For electric operator, economy, add		8	1		410	41.50		451.50	520
4100	Deluxe, including remote control		8	1		600	41.50		641.50	730
4500	For transmitter/receiver control, add to operator				Total	105			105	116
4600	Transmitters, additional				"	50			50	55
6000	Replace section, on sectional door, fiberglass, 9' x 7'	1 Carp	4	2	Ea.	445	83		528	620
6020	16' x 7'		3.50	2.286		525	95		620	735
6200	Hardboard, 9' x 7'		4	2		113	83		196	261
6220	16' x 7'		3.50	2.286		214	95		309	390
6300	Metal, 9' x 7'		4	2		181	83		264	335
6320	16' x 7'		3.50	2.286		295	95		390	480
6500	Wood, 9' x 7'		4	2		114	83		197	262
6520	16' x 7'		3.50	2.286		221	95		316	400
9000	Minimum labor/equipment charge		2.50	3.200	Job		133		133	218

08 36 19 – Multi-Leaf Vertical Lift Doors

08 36 19.10 Sectional Vertical Lift Doors

		Crew	Daily Output	Labor-Hours	Unit	Material	Labor	Equipment	Total	Total Incl O&P
0010	**SECTIONAL VERTICAL LIFT DOORS**									
0020	Motorized, 14 ga. stl, incl., frm and ctrl pnl									
0050	16' x 16' high	L-10	.50	48	Ea.	21,300	2,250	1,300	24,850	28,800
0100	10' x 20' high		1.30	18.462		31,000	865	505	32,370	36,200
0120	15' x 20' high		1.30	18.462		38,300	865	505	39,670	44,300
0140	20' x 20' high		1	24		44,700	1,125	655	46,480	52,000
0160	25' x 20' high		1	24		51,000	1,125	655	52,780	58,500
0170	32' x 24' high		.75	32		46,000	1,500	875	48,375	54,000
0180	20' x 25' high		1	24		52,500	1,125	655	54,280	60,000
0200	25' x 25' high		.70	34.286		61,000	1,600	935	63,535	71,000
0220	25' x 30' high		.70	34.286		66,000	1,600	935	68,535	76,500
0240	30' x 30' high		.70	34.286		77,000	1,600	935	79,535	88,500
0260	35' x 30' high		.70	34.286		87,000	1,600	935	89,535	99,500

08 38 Traffic Doors

08 38 19 – Rigid Traffic Doors

08 38 19.20 Double Acting Swing Doors		Crew	Daily Output	Labor-Hours	Unit	Material	2010 Bare Costs Labor	Equipment	Total	Total Incl O&P
0010	**DOUBLE ACTING SWING DOORS**									
0020	Including frame, closer, hardware and vision panel									
1000	.063" aluminum, 7'-0" high, 4'-0" wide	2 Carp	4.20	3.810	Pr.	1,900	158		2,058	2,350
1025	6'-0" wide		4	4		2,100	166		2,266	2,575
1050	6'-8" wide		4	4		2,300	166		2,466	2,800
2000	Solid core wood, 3/4" thick, metal frame, stainless steel									
2010	base plate, 7' high opening, 4' wide	2 Carp	4	4	Pr.	2,200	166		2,366	2,700
2050	7' wide		3.80	4.211	"	2,500	175		2,675	3,025
9000	Minimum labor/equipment charge		2	8	Job		330		330	545

08 38 19.30 Shock Absorbing Doors										
0010	**SHOCK ABSORBING DOORS**									
0020	Rigid, no frame, 1-1/2" thick, 5' x 7'	2 Sswk	1.90	8.421	Opng.	1,400	395		1,795	2,275
0100	8' x 8'		1.80	8.889		2,000	415		2,415	2,950
0500	Flexible, no frame, insulated, .16" thick, economy, 5' x 7'		2	8		1,750	375		2,125	2,600
0600	Deluxe		1.90	8.421		2,625	395		3,020	3,600
1000	8' x 8' opening, economy		2	8		2,750	375		3,125	3,700
1100	Deluxe		1.90	8.421		3,500	395		3,895	4,575
9000	Minimum labor/equipment charge		2	8	Job		375		375	685

08 41 Entrances and Storefronts

08 41 13 – Aluminum-Framed Entrances and Storefronts

08 41 13.20 Tube Framing		Crew	Daily Output	Labor-Hours	Unit	Material	2010 Bare Costs Labor	Equipment	Total	Total Incl O&P
0010	**TUBE FRAMING**, For window walls and store fronts, aluminum stock									
0050	Plain tube frame, mill finish, 1-3/4" x 1-3/4"	2 Glaz	103	.155	L.F.	8.70	6.25		14.95	19.65
0150	1-3/4" x 4"		98	.163		11.90	6.55		18.45	23.50
0200	1-3/4" x 4-1/2"		95	.168		13.80	6.75		20.55	26
0250	2" x 6"		89	.180		20.50	7.25		27.75	34
0350	4" x 4"		87	.184		23	7.40		30.40	37.50
0400	4-1/2" x 4-1/2"		85	.188		25	7.55		32.55	39.50
0450	Glass bead		240	.067		2.60	2.68		5.28	7.15
1000	Flush tube frame, mill finish, 1/4" glass, 1-3/4" x 4", open header		80	.200		11.45	8.05		19.50	25.50
1050	Open sill		82	.195		9.30	7.85		17.15	23
1100	Closed back header		83	.193		16.35	7.75		24.10	30.50
1150	Closed back sill		85	.188		15.60	7.55		23.15	29.50
1160	Tube fmg, spandrel cover both sides, alum 1" wide	1 Sswk	85	.094	S.F.	86	4.41		90.41	103
1170	Tube fmg, spandrel cover both sides, alum 2" wide	"	85	.094	"	33	4.41		37.41	44.50
1200	Vertical mullion, one piece	2 Glaz	75	.213	L.F.	17.15	8.60		25.75	32.50
1250	Two piece		73	.219		18.20	8.80		27	34
1300	90° or 180° vertical corner post		75	.213		27.50	8.60		36.10	44.50
1400	1-3/4" x 4-1/2", open header		80	.200		13.90	8.05		21.95	28
1450	Open sill		82	.195		11.60	7.85		19.45	25.50
1500	Closed back header		83	.193		16.90	7.75		24.65	31
1550	Closed back sill		85	.188		16.35	7.55		23.90	30
1600	Vertical mullion, one piece		75	.213		18.30	8.60		26.90	34
1650	Two piece		73	.219		19.30	8.80		28.10	35
1700	90° or 180° vertical corner post		75	.213		19.70	8.60		28.30	35.50
2000	Flush tube frame, mil fin., ins. glass w/thml brk, 2" x 4-1/2", open header		75	.213		14.50	8.60		23.10	29.50
2050	Open sill		77	.208		12.25	8.35		20.60	27
2100	Closed back header		78	.205		13.75	8.25		22	28.50
2150	Closed back sill		80	.200		14.70	8.05		22.75	29
2200	Vertical mullion, one piece		70	.229		15.25	9.20		24.45	31.50

08 41 Entrances and Storefronts

08 41 13 – Aluminum-Framed Entrances and Storefronts

08 41 13.20 Tube Framing

		Crew	Daily Output	Labor-Hours	Unit	Material	2010 Bare Costs Labor	Equipment	Total	Total Incl O&P
2250	Two piece	2 Glaz	68	.235	L.F.	16.45	9.45		25.90	33.50
2300	90° or 180° vertical corner post		70	.229		15.70	9.20		24.90	32
5000	Flush tube frame, mill fin., thermal brk., 2-1/4"x 4-1/2", open header		74	.216		15.15	8.70		23.85	30.50
5050	Open sill		75	.213		13.45	8.60		22.05	28.50
5100	Vertical mullion, one piece		69	.232		16.50	9.30		25.80	33
5150	Two piece		67	.239		19.15	9.60		28.75	36.50
5200	90° or 180° vertical corner post		69	.232		16.85	9.30		26.15	33.50
6980	Door stop (snap in)	↓	380	.042	↓	3	1.69		4.69	6
7000	For joints, 90°, clip type, add				Ea.	23			23	25
7050	Screw spline joint, add					20.50			20.50	22.50
7100	For joint other than 90°, add				↓	43			43	47
8000	For bronze anodized aluminum, add					15%				
8020	For black finish, add					27%				
8050	For stainless steel materials, add					350%				
8100	For monumental grade, add					52%				
8150	For steel stiffener, add	2 Glaz	200	.080	L.F.	10.40	3.22		13.62	16.60
8200	For 2 to 5 stories, add per story				Story		6%			
9000	Minimum labor/equipment charge	2 Glaz	2	8	Job		320		320	515

08 41 19 – Stainless-Steel-Framed Entrances and Storefronts

08 41 19.10 Stainless-Steel and Glass Entrance Unit

		Crew	Daily Output	Labor-Hours	Unit	Material	Labor	Equipment	Total	Total Incl O&P
0010	**STAINLESS-STEEL AND GLASS ENTRANCE UNIT**, narrow stiles									
0020	3' x 7' opening, including hardware, minimum	2 Sswk	1.60	10	Opng.	6,100	470		6,570	7,550
0050	Average		1.40	11.429		6,500	535		7,035	8,125
0100	Maximum	↓	1.20	13.333		7,000	625		7,625	8,850
1000	For solid bronze entrance units, statuary finish, add					62%				
1100	Without statuary finish, add				↓	45%				
2000	Balanced doors, 3' x 7', economy	2 Sswk	.90	17.778	Ea.	8,200	835		9,035	10,600
2100	Premium		.70	22.857	"	14,100	1,075		15,175	17,500
9000	Minimum labor/equipment charge	↓	2	8	Job		375		375	685

08 41 26 – All-Glass Entrances and Storefronts

08 41 26.10 Window Walls Aluminum, Stock

		Crew	Daily Output	Labor-Hours	Unit	Material	Labor	Equipment	Total	Total Incl O&P
0010	**WINDOW WALLS ALUMINUM, STOCK**, including glazing									
0020	Minimum	H-2	160	.150	S.F.	40	5.65		45.65	53
0050	Average		140	.171		55	6.50		61.50	71
0100	Maximum	↓	110	.218	↓	155	8.25		163.25	183
0500	For translucent sandwich wall systems, see Div. 07 41 33.10									
0850	Cost of the above walls depends on material,									
0860	finish, repetition, and size of units.									
0870	The larger the opening, the lower the S.F. cost									
1200	Double glazed acoustical window wall for airports,									
1220	including 1" thick glass with 2" x 4-1/2" tube frame	H-2	40	.600	S.F.	112	22.50		134.50	160

08 42 Entrances

08 42 26 – All-Glass Entrances

08 42 26.10 Swinging Glass Doors		Crew	Daily Output	Labor-Hours	Unit	Material	2010 Bare Costs Labor	Equipment	Total	Total Incl O&P
0010	**SWINGING GLASS DOORS**									
0020	Including hardware, 1/2" thick, tempered, 3' x 7' opening	2 Glaz	2	8	Opng.	2,075	320		2,395	2,800
0100	6' x 7' opening		1.40	11.429	"	3,900	460		4,360	5,050
9000	Minimum labor/equipment charge		2	8	Job		320		320	515

08 42 29 – Automatic Entrances

08 42 29.23 Sliding Automatic Entrances

		Crew	Daily Output	Labor-Hours	Unit	Material	Labor	Equipment	Total	Total Incl O&P
0010	**SLIDING AUTOMATIC ENTRANCES** 12' x 7'-6" opng., 5' x 7' door, 2 way traffic									
0020	Mat or electronic activated, panic pushout, incl. operator & hardware,									
0030	not including glass or glazing	2 Glaz	.70	22.857	Opng.	7,800	920		8,720	10,100
9000	Minimum labor/equipment charge	"	.70	22.857	Job		920		920	1,475

08 42 36 – Balanced Door Entrances

08 42 36.10 Balanced Entrance Doors

		Crew	Daily Output	Labor-Hours	Unit	Material	Labor	Equipment	Total	Total Incl O&P
0010	**BALANCED ENTRANCE DOORS**									
0020	Hardware & frame, alum. & glass, 3' x 7', econ.	2 Sswk	.90	17.778	Ea.	5,900	835		6,735	8,000
0150	Premium		.70	22.857	"	7,300	1,075		8,375	9,975
9000	Minimum labor/equipment charge		1	16	Job		750		750	1,375

08 43 Storefronts

08 43 13 – Aluminum-Framed Storefronts

08 43 13.10 Aluminum-Framed Entrance Doors and Frames

		Crew	Daily Output	Labor-Hours	Unit	Material	Labor	Equipment	Total	Total Incl O&P
0010	**ALUMINUM-FRAMED ENTRANCE DOORS AND FRAMES**									
0015	Standard hardware and glass stops but no glazing									
0020	Entrance door, 3' x 7' opening, clear anodized finish	2 Sswk	7	2.286	Opng.	355	107		462	590
0040	Bronze finish		7	2.286		375	107		482	610
0060	Black finish		7	2.286		420	107		527	655
0200	3'-6" x 7'-0", mill finish		7	2.286		405	107		512	640
0220	Bronze finish		7	2.286		505	107		612	750
0240	Black finish		7	2.286		540	107		647	790
0500	6' x 7' opening, clear finish		6	2.667		715	125		840	1,025
0520	Bronze finish		6	2.667		750	125		875	1,050
0600	Door Frame 3'-0" x 7'-0", mill finish		6	2.667		305	125		430	565
0620	Bronze finish		6	2.667		355	125		480	620
0640	Black finish		6	2.667		390	125		515	655
0700	3'-6" x 7'-0", mill finish		6	2.667		272	125		397	530
0720	Bronze finish		6	2.667		272	125		397	530
0740	Black finish		6	2.667		272	125		397	530
0800	6'-0" x 7'-0", mill finish		6	2.667		272	125		397	530
0820	Bronze finish		6	2.667		278	125		403	535
0840	Black finish		6	2.667		298	125		423	560
1000	With 3' high transom above, 3' x 7' opening, clear finish		5.50	2.909		410	136		546	700
1050	Bronze finish		5.50	2.909		435	136		571	725
1100	Black finish		5.50	2.909		465	136		601	760
1300	3'-6" x 7'-0" opening, clear finish		5.50	2.909		320	136		456	605
1320	Bronze finish		5.50	2.909		330	136		466	615
1340	Black finish		5.50	2.909		350	136		486	635
1500	6' x 7' opening, clear finish		5.50	2.909		500	136		636	800
1550	Bronze finish		5.50	2.909		520	136		656	825
1600	Black finish		5.50	2.909		580	136		716	885
9000	Minimum labor/equipment charge		4	4	Job		188		188	345

08 43 Storefronts

08 43 13 – Aluminum-Framed Storefronts

08 43 13.20 Storefront Systems

		Crew	Daily Output	Labor-Hours	Unit	Material	2010 Bare Costs Labor	Equipment	Total	Total Incl O&P
0010	**STOREFRONT SYSTEMS**, aluminum frame clear 3/8" plate glass									
0020	incl. 3' x 7' door with hardware (400 sq. ft. max. wall)									
0500	Wall height to 12' high, commercial grade	2 Glaz	150	.107	S.F.	19.30	4.29		23.59	28
0600	Institutional grade		130	.123		24	4.95		28.95	34.50
0700	Monumental grade		115	.139		36.50	5.60		42.10	49
1000	6' x 7' door with hardware, commercial grade		135	.119		19.50	4.76		24.26	29
1100	Institutional grade		115	.139		27	5.60		32.60	38.50
1200	Monumental grade		100	.160		49.50	6.45		55.95	64.50
1500	For bronze anodized finish, add					15%				
1600	For black anodized finish, add					35%				
1700	For stainless steel framing, add to monumental					76%				
9000	Minimum labor/equipment charge	2 Glaz	1	16	Job		645		645	1,025

08 43 29 – Sliding Storefronts

08 43 29.10 Sliding Panels

		Crew	Daily Output	Labor-Hours	Unit	Material	Labor	Equipment	Total	Total Incl O&P
0010	**SLIDING PANELS**									
0020	Mall fronts, aluminum & glass, 15' x 9' high	2 Glaz	1.30	12.308	Opng.	3,100	495		3,595	4,200
0100	24' x 9' high		.70	22.857		4,400	920		5,320	6,325
0200	48' x 9' high, with fixed panels		.90	17.778		8,200	715		8,915	10,200
0500	For bronze finish, add					17%				
9000	Minimum labor/equipment charge	2 Glaz	1	16	Job		645		645	1,025

08 45 Translucent Wall and Roof Assemblies

08 45 10 – Translucent Roof Assemblies

08 45 10.10 Skyroofs

		Crew	Daily Output	Labor-Hours	Unit	Material	Labor	Equipment	Total	Total Incl O&P
0010	**SKYROOFS**, Translucent panels, 2-3/4" thick									
0020	Under 5000 S.F.	G-3	395	.081	SF Hor.	32	3.33		35.33	41
0100	Over 5000 S.F.		465	.069		30	2.83		32.83	37.50
0300	Continuous vaulted, semi-circular, to 8' wide, double glazed [G]		145	.221		50	9.05		59.05	69.50
0400	Single glazed		160	.200		34	8.20		42.20	50.50
0600	To 20' wide, single glazed		175	.183		37.50	7.50		45	53.50
0700	Over 20' wide, single glazed		200	.160		43.50	6.60		50.10	58
0900	Motorized opening type, single glazed, 1/3 opening		145	.221		44.50	9.05		53.55	63.50
1000	Full opening		130	.246		51	10.10		61.10	72
1200	Pyramid type units, self-supporting, to 30' clear opening,									
1300	square or circular, single glazed, minimum	G-3	200	.160	SF Hor.	26	6.60		32.60	39
1310	Average		165	.194		35.50	7.95		43.45	51.50
1400	Maximum		130	.246		51	10.10		61.10	72
1500	Grid type, 4' to 10' modules, single glass glazed, minimum		200	.160		34	6.60		40.60	48
1550	Maximum		128	.250		53.50	10.30		63.80	75
1600	Preformed acrylic, minimum		300	.107		41	4.38		45.38	52
1650	Maximum		175	.183		56	7.50		63.50	73.50
9000	Minimum labor/equipment charge	2 Carp	8	2	Job		83		83	137

08 51 Metal Windows

08 51 13 – Aluminum Windows

08 51 13.10 Aluminum Sash

		Crew	Daily Output	Labor-Hours	Unit	Material	2010 Bare Costs Labor	2010 Bare Costs Equipment	Total	Total Incl O&P
0010	**ALUMINUM SASH**									
0020	Stock, grade C, glaze & trim not incl., casement	2 Sswk	200	.080	S.F.	36	3.75		39.75	46.50
0050	Double hung		200	.080		36.50	3.75		40.25	47
0100	Fixed casement		200	.080		15.65	3.75		19.40	24
0150	Picture window		200	.080		16.65	3.75		20.40	25
0200	Projected window		200	.080		33	3.75		36.75	43
0250	Single hung		200	.080		15.80	3.75		19.55	24.50
0300	Sliding		200	.080		20.50	3.75		24.25	29.50
1000	Mullions for above, tubular		240	.067	L.F.	5.45	3.13		8.58	11.65
2950	Single glazing for above, add	2 Glaz	200	.080	S.F.	6.85	3.22		10.07	12.70
3000	Double glazing for above, add		200	.080		11.45	3.22		14.67	17.75
3100	Triple glazing for above, add		85	.188		13.55	7.55		21.10	27
9000	Minimum labor/equipment charge	1 Sswk	2	4	Job		188		188	345

08 51 13.20 Aluminum Windows

		Crew	Daily Output	Labor-Hours	Unit	Material	Labor	Equipment	Total	Total Incl O&P
0010	**ALUMINUM WINDOWS**, incl. frame and glazing, commercial grade									
1000	Stock units, casement, 3'-1" x 3'-2" opening	2 Sswk	10	1.600	Ea.	355	75		430	525
1050	Add for storms					115			115	126
1600	Projected, with screen, 3'-1" x 3'-2" opening	2 Sswk	10	1.600		335	75		410	505
1700	Add for storms					112			112	123
2000	4'-5" x 5'-3" opening	2 Sswk	8	2		380	94		474	590
2100	Add for storms					120			120	132
2500	Enamel finish windows, 3'-1" x 3'-2"	2 Sswk	10	1.600		340	75		415	510
2600	4'-5" x 5'-3"		8	2		385	94		479	595
3000	Single hung, 2' x 3' opening, enameled, standard glazed		10	1.600		198	75		273	355
3100	Insulating glass		10	1.600		240	75		315	400
3300	2'-8" x 6'-8" opening, standard glazed		8	2		350	94		444	555
3400	Insulating glass		8	2		450	94		544	665
3700	3'-4" x 5'-0" opening, standard glazed		9	1.778		286	83.50		369.50	465
3800	Insulating glass		9	1.778		315	83.50		398.50	500
4000	Sliding aluminum, 3' x 2' opening, standard glazed		10	1.600		209	75		284	365
4100	Insulating glass		10	1.600		224	75		299	385
4300	5' x 3' opening, standard glazed		9	1.778		320	83.50		403.50	500
4400	Insulating glass		9	1.778		370	83.50		453.50	560
4600	8' x 4' opening, standard glazed		6	2.667		335	125		460	600
4700	Insulating glass		6	2.667		545	125		670	825
5000	9' x 5' opening, standard glazed		4	4		510	188		698	905
5100	Insulating glass		4	4		820	188		1,008	1,250
5500	Sliding, with thermal barrier and screen, 6' x 4', 2 track		8	2		695	94		789	935
5700	4 track		8	2		880	94		974	1,150
6000	For above units with bronze finish, add					12%				
6200	For installation in concrete openings, add					6%				
6400										

08 51 23 – Steel Windows

08 51 23.10 Steel Sash

		Crew	Daily Output	Labor-Hours	Unit	Material	Labor	Equipment	Total	Total Incl O&P
0010	**STEEL SASH** Custom units, glazing and trim not included									
0100	Casement, 100% vented	2 Sswk	200	.080	S.F.	48	3.75		51.75	59.50
0200	50% vented		200	.080		44	3.75		47.75	55.50
0300	Fixed		200	.080		30	3.75		33.75	40
1000	Projected, commercial, 40% vented		200	.080		48.50	3.75		52.25	60
1100	Intermediate, 50% vented		200	.080		54.50	3.75		58.25	67
1500	Industrial, horizontally pivoted		200	.080		50.50	3.75		54.25	62.50
1600	Fixed		200	.080		29	3.75		32.75	39

08 51 Metal Windows

08 51 23 – Steel Windows

08 51 23.10 Steel Sash

		Crew	Daily Output	Labor-Hours	Unit	Material	2010 Bare Costs Labor	Equipment	Total	Total Incl O&P
2000	Industrial security sash, 50% vented	2 Sswk	200	.080	S.F.	54	3.75		57.75	66.50
2100	Fixed		200	.080		44.50	3.75		48.25	56
2500	Picture window		200	.080		28.50	3.75		32.25	38.50
3000	Double hung		200	.080		56	3.75		59.75	68.50
5000	Mullions for above, open interior face		240	.067	L.F.	9.80	3.13		12.93	16.45
5100	With interior cover		240	.067	"	16.35	3.13		19.48	23.50
5200	Single glazing for above, add	2 Glaz	200	.080	S.F.	6.65	3.22		9.87	12.45
6000	Double glazing for above, add		200	.080		12.15	3.22		15.37	18.55
6100	Triple glazing for above, add		85	.188		11.85	7.55		19.40	25
9000	Minimum labor/equipment charge	1 Sswk	2	4	Job		188		188	345

08 51 23.20 Steel Windows

		Crew	Daily Output	Labor-Hours	Unit	Material	Labor	Equipment	Total	Total Incl O&P
0010	**STEEL WINDOWS** Stock, including frame, trim and insul. glass R085123-10									
1000	Custom units, double hung, 2'-8" x 4'-6" opening	2 Sswk	12	1.333	Ea.	680	62.50		742.50	865
1100	2'-4" x 3'-9" opening		12	1.333		565	62.50		627.50	735
1500	Commercial projected, 3'-9" x 5'-5" opening		10	1.600		1,200	75		1,275	1,425
1600	6'-9" x 4'-1" opening		7	2.286		1,575	107		1,682	1,925
2000	Intermediate projected, 2'-9" x 4'-1" opening		12	1.333		670	62.50		732.50	850
2100	4'-1" x 5'-5" opening		10	1.600		1,350	75		1,425	1,600
9000	Minimum labor/equipment charge	1 Sswk	3	2.667	Job		125		125	229

08 51 66 – Metal Window Screens

08 51 66.10 Screens

		Crew	Daily Output	Labor-Hours	Unit	Material	Labor	Equipment	Total	Total Incl O&P
0010	**SCREENS**									
0020	For metal sash, aluminum or bronze mesh, flat screen	2 Sswk	1200	.013	S.F.	4.10	.63		4.73	5.65
0500	Wicket screen, inside window		1000	.016		6.30	.75		7.05	8.30
0800	Security screen, aluminum frame with stainless steel cloth		1200	.013		22.50	.63		23.13	26
0900	Steel grate, painted, on steel frame		1600	.010		12.50	.47		12.97	14.60
1000	For solar louvers, add		160	.100		23.50	4.69		28.19	34
4000	See Div. 05 58 27.90									

08 52 Wood Windows

08 52 10 – Plain Wood Windows

08 52 10.20 Awning Window

		Crew	Daily Output	Labor-Hours	Unit	Material	Labor	Equipment	Total	Total Incl O&P
0010	**AWNING WINDOW**, Including frame, screens and grilles									
0100	Average quality, builders model, 34" x 22", double insulated glass	1 Carp	10	.800	Ea.	253	33		286	335
0200	Low E glass		10	.800		257	33		290	335
0300	40" x 28", double insulated glass		9	.889		310	37		347	400
0400	Low E Glass		9	.889		330	37		367	425
0500	48" x 36", double insulated glass		8	1		450	41.50		491.50	565
0600	Low E glass		8	1		475	41.50		516.50	590
1000	Vinyl clad, 34" x 22"		10	.800		259	33		292	340
1100	40" x 22"		10	.800		283	33		316	365
1200	36" x 28"		9	.889		300	37		337	390
1300	36" x 36"		9	.889		335	37		372	430
1400	48" x 28"		8	1		360	41.50		401.50	470
1500	60" x 36"		8	1		505	41.50		546.50	625
2200	Metal clad, 36" x 25"		9	.889		269	37		306	355
2300	40" x 30"		9	.889		335	37		372	430
2400	48" x 28"		8	1		345	41.50		386.50	450
2500	60" x 36"		8	1		365	41.50		406.50	475
4000	Impact windows, minimum, add					60%				

08 52 Wood Windows

08 52 10 – Plain Wood Windows

08 52 10.20 Awning Window		Crew	Daily Output	Labor-Hours	Unit	Material	2010 Bare Costs Labor	2010 Bare Costs Equipment	Total	Total Incl O&P
4010	Impact windows, maximum, add				Ea.	160%				
9000	Minimum labor/equipment charge	1 Carp	4	2	Job		83		83	137

08 52 10.30 Wood Windows

			Crew	Daily Output	Labor-Hours	Unit	Material	Labor	Equipment	Total	Total Incl O&P
0010	**WOOD WINDOWS**, double hung										
0020	Including frame, double insulated glass, screens and grilles										
0040	Double hung, 2'-2" x 3'-4" high		2 Carp	15	1.067	Ea.	206	44.50		250.50	300
0060	2'-2" x 4'-4"			14	1.143		223	47.50		270.50	325
0080	2'-6" x 3'-4"			13	1.231		214	51		265	320
0100	2'-6" x 4'-0"			12	1.333		226	55.50		281.50	340
0120	2'-6" x 4'-8"			12	1.333		244	55.50		299.50	360
0140	2'-10" x 3'-4"			10	1.600		219	66.50		285.50	350
0160	2'-10" x 4'-0"			10	1.600		242	66.50		308.50	375
0180	3'-7" x 3'-4"			9	1.778		248	74		322	395
0200	3'-7" x 5'-4"			9	1.778		320	74		394	470
0220	3'-10" x 5'-4"			8	2		495	83		578	680
6000	Replacement sash, double hung, double glazing, to 12 S.F.		1 Carp	64	.125	S.F.	25	5.20		30.20	36.50
6100	12 S.F. to 20 S.F.			94	.085	"	19.50	3.54		23.04	27.50
7000	Sash, single lite, 2'-0" x 2'-0" high			20	.400	Ea.	49	16.60		65.60	81
7050	2'-6" x 2'-0" high			19	.421		60	17.50		77.50	94.50
7100	2'-6" x 2'-6" high			18	.444		67.50	18.45		85.95	105
7150	3'-0" x 2'-0" high			17	.471		88	19.55		107.55	129

08 52 10.40 Casement Window

			Crew	Daily Output	Labor-Hours	Unit	Material	Labor	Equipment	Total	Total Incl O&P
0010	**CASEMENT WINDOW**, including frame, screen and grilles										
0100	Avg. quality, bldrs. model, 2'-0" x 3'-0" H, dbl. insulated glass	G	1 Carp	10	.800	Ea.	276	33		309	360
0150	Low E glass	G		10	.800		225	33		258	305
0200	2'-0" x 4'-6" high, double insulated glass	G		9	.889		345	37		382	440
0250	Low E glass	G		9	.889		263	37		300	350
0260	Casement 4'-2" x 4'-2" double insulated glass	G		11	.727		855	30		885	990
0270	4'-0" x 4'-0" Low E glass	G		11	.727		465	30		495	565
0290	6'-4" x 5'-7" Low E glass	G		9	.889		1,100	37		1,137	1,250
0300	2'-4" x 6'-0" high, double insulated glass	G		8	1		400	41.50		441.50	510
0350	Low E glass	G		8	1		465	41.50		506.50	585
0522	Vinyl clad, premium, double insulated glass, 2'-0" x 3'-0"	G		10	.800		266	33		299	345
0524	2'-0" x 4'-0"	G		9	.889		310	37		347	400
0525	2'-0" x 5'-0"	G		8	1		355	41.50		396.50	460
0528	2'-0" x 6'-0"	G		8	1		355	41.50		396.50	460
0600	3'-0" x 5'-0"			8	1		650	41.50		691.50	785
0700	4'-0" x 3'-0"			8	1		715	41.50		756.50	860
0710	4'-0" x 4'-0"	G		8	1		610	41.50		651.50	745
0720	4'-8" x 4'-0"			8	1		675	41.50		716.50	815
0730	4'-8" x 5'-0"			6	1.333		770	55.50		825.50	940
0740	4'-8" x 6'-0"			6	1.333		870	55.50		925.50	1,050
0750	6'-0" x 4'-0"			6	1.333		790	55.50		845.50	955
0800	6'-0" x 5'-0"			6	1.333		885	55.50		940.50	1,050
0900	5'-6" x 5'-6"	G	2 Carp	15	1.067		1,400	44.50		1,444.50	1,600
8100	. Metal clad, deluxe, dbl. insul. glass, 2'-0" x 3'-0" high	G	1 Carp	10	.800		272	33		305	355
8120	2'-0" x 4'-0" high	G		9	.889		305	37		342	395
8140	2'-0" x 5'-0" high	G		8	1		320	41.50		361.50	420
8160	2'-0" x 6'-0" high	G		8	1		355	41.50		396.50	460
8190	For installation, add per leaf							15%			
8200	For multiple leaf units, deduct for stationary sash										
8220	2' high					Ea.	22.50			22.50	24.50

08 52 Wood Windows

08 52 10 – Plain Wood Windows

08 52 10.40 Casement Window		Crew	Daily Output	Labor-Hours	Unit	Material	2010 Bare Costs Labor	Equipment	Total	Total Incl O&P
8240	4'-6" high				Ea.	25.50			25.50	28
8260	6' high					33.50			33.50	37
8300	Impact windows, minimum, add					60%				
8310	Impact windows, maximum, add					160%				
9000	Minimum labor/equipment charge		1 Carp	3	2.667	Job		111		111	182

08 52 10.50 Double Hung

			Crew	Daily Output	Labor-Hours	Unit	Material	Labor	Equipment	Total	Total Incl O&P
0010	**DOUBLE HUNG**, Including frame, screens and grilles										
0100	Avg. quality, bldrs. model, 2'-0" x 3'-0" high, dbl insul. glass	G	1 Carp	10	.800	Ea.	184	33		217	257
0150	Low E glass	G		10	.800		208	33		241	284
0200	3'-0" x 4'-0" high, double insulated glass	G		9	.889		257	37		294	345
0250	Low E glass	G		9	.889		278	37		315	365
0300	4'-0" x 4'-6" high, double insulated glass	G		8	1		315	41.50		356.50	415
0350	Low E glass	G		8	1		340	41.50		381.50	445
1000	Vinyl clad, premium, double insulated glass, 2'-6" x 3'-0"	G		10	.800		225	33		258	305
1005	2'-6" x 4'-0"	G		10	.800		276	33		309	360
1100	3'-0" x 3'-6"	G		10	.800		261	33		294	340
1200	3'-0" x 4'-0"	G		9	.889		370	37		407	465
1300	3'-0" x 4'-6"	G		9	.889		340	37		377	430
1400	3'-0" x 5'-0"	G		8	1		315	41.50		356.50	415
1490	3'-4" x 5'-0"	G		8	1		355	41.50		396.50	460
1500	3'-6" x 6'-0"	G		8	1		370	41.50		411.50	475
1520	4'-0" x 5'-0"	G		7	1.143		550	47.50		597.50	685
1530	4'-0" x 6'-0"	G		7	1.143		685	47.50		732.50	835
2000	Metal clad, deluxe, dbl. insul. glass, 2'-6" x 3'-0" high	G		10	.800		265	33		298	345
2100	3'-0" x 3'-6" high	G		10	.800		305	33		338	390
2200	3'-0" x 4'-0" high	G		9	.889		320	37		357	410
2300	3'-0" x 4'-6" high	G		9	.889		335	37		372	430
2400	3'-0" x 5'-0" high	G		8	1		365	41.50		406.50	470
2500	3'-6" x 6'-0" high	G		8	1		440	41.50		481.50	550
8000	Impact windows, minimum, add						60%				
8010	Impact windows, maximum, add						160%				
9000	Minimum labor/equipment charge		1 Carp	3	2.667	Job		111		111	182

08 52 10.55 Picture Window

		Crew	Daily Output	Labor-Hours	Unit	Material	Labor	Equipment	Total	Total Incl O&P
0010	**PICTURE WINDOW**, Including frame and grilles									
0100	Average quality, bldrs. model, 3'-6" x 4'-0" high, dbl insulated glass	2 Carp	12	1.333	Ea.	345	55.50		400.50	470
0150	Low E glass		12	1.333		410	55.50		465.50	540
0200	4'-0" x 4'-6" high, double insulated glass		11	1.455		490	60.50		550.50	640
0250	Low E glass		11	1.455		520	60.50		580.50	670
0300	5'-0" x 4'-0" high, double insulated glass		11	1.455		570	60.50		630.50	725
0350	Low E glass		11	1.455		590	60.50		650.50	750
0400	6'-0" x 4'-6" high, double insulated glass		10	1.600		610	66.50		676.50	785
0450	Low E glass		10	1.600		620	66.50		686.50	795

08 52 10.65 Wood Sash

		Crew	Daily Output	Labor-Hours	Unit	Material	Labor	Equipment	Total	Total Incl O&P
0010	**WOOD SASH**, Including glazing but not trim									
0050	Custom, 5'-0" x 4'-0", 1" dbl. glazed, 3/16" thick lites	2 Carp	3.20	5	Ea.	167	208		375	525
0100	1/4" thick lites		5	3.200		172	133		305	405
0200	1" thick, triple glazed		5	3.200		395	133		528	650
0300	7'-0" x 4'-6" high, 1" double glazed, 3/16" thick lites		4.30	3.721		400	155		555	695
0400	1/4" thick lites		4.30	3.721		450	155		605	750
0500	1" thick, triple glazed		4.30	3.721		515	155		670	820
0600	8'-6" x 5'-0" high, 1" double glazed, 3/16" thick lites		3.50	4.571		540	190		730	900
0700	1/4" thick lites		3.50	4.571		590	190		780	960

08 52 Wood Windows

08 52 10 – Plain Wood Windows

08 52 10.65 Wood Sash		Crew	Daily Output	Labor-Hours	Unit	Material	2010 Bare Costs Labor	Equipment	Total	Total Incl O&P
0800	1" thick, triple glazed	2 Carp	3.50	4.571	Ea.	595	190		785	965
0900	Window frames only, based on perimeter length				L.F.	3.83			3.83	4.21
1200	Window sill, stock, per lineal foot					8.25			8.25	9.10
1250	Casing, stock					3.15			3.15	3.47

08 52 10.70 Sliding Windows

0010	**SLIDING WINDOWS**									
0100	Average quality, bldrs. model, 3'-0" x 3'-0" high, double insulated	G	1 Carp	10	.800	Ea.	272	33	305	355
0120	Low E glass	G		10	.800		293	33	326	375
0200	4'-0" x 3'-6" high, double insulated	G		9	.889		325	37	362	420
0220	Low E glass	G		9	.889		350	37	387	445
0300	6'-0" x 5'-0" high, double insulated	G		8	1		460	41.50	501.50	580
0320	Low E glass	G		8	1		500	41.50	541.50	620
9000	Minimum labor/equipment charge			3	2.667	Job		111	111	182

08 52 13 – Metal-Clad Wood Windows

08 52 13.10 Awning Windows, Metal-Clad

0010	**AWNING WINDOWS, METAL-CLAD**									
2000	Metal clad, awning deluxe, double insulated glass, 34" x 22"		1 Carp	10	.800	Ea.	242	33	275	320
2100	40" x 22"		"	10	.800	"	285	33	318	370

08 52 13.35 Picture and Sliding Windows Metal-Clad

0010	**PICTURE AND SLIDING WINDOWS METAL-CLAD**									
2000	Metal clad, dlx picture, dbl. insul. glass, 4'-0" x 4'-0" high		2 Carp	12	1.333	Ea.	365	55.50	420.50	495
2100	4'-0" x 6'-0" high			11	1.455		535	60.50	595.50	690
2200	5'-0" x 6'-0" high			10	1.600		595	66.50	661.50	765
2300	6'-0" x 6'-0" high			10	1.600		680	66.50	746.50	860
2400	Metal clad, dlx sliding, double insulated glass, 3'-0" x 3'-0" high	G	1 Carp	10	.800		330	33	363	415
2420	4'-0" x 3'-6" high	G		9	.889		400	37	437	500
2440	5'-0" x 4'-0" high	G		9	.889		480	37	517	585
2460	6'-0" x 5'-0" high	G		8	1		705	41.50	746.50	845
9000	Minimum labor/equipment charge		2 Carp	2.75	5.818	Job		242	242	395

08 52 16 – Plastic-Clad Wood Windows

08 52 16.10 Bow Window

0010	**BOW WINDOW** Including frames, screens, and grilles									
0020	End panels operable									
1000	Bow type, casement, wood, bldrs mdl, 8' x 5' dbl insltd glass, 4 panel		2 Carp	10	1.600	Ea.	1,500	66.50	1,566.50	1,750
1050	Low E glass			10	1.600		1,325	66.50	1,391.50	1,550
1100	10'-0" x 5'-0", double insulated glass, 6 panels			6	2.667		1,350	111	1,461	1,675
1200	Low E glass, 6 panels			6	2.667		1,450	111	1,561	1,775
1300	Vinyl clad, bldrs model, double insulated glass, 6'-0" x 4'-0", 3 panel			10	1.600		1,025	66.50	1,091.50	1,225
1340	9'-0" x 4'-0", 4 panel			8	2		1,350	83	1,433	1,625
1380	10'-0" x 6'-0", 5 panels			7	2.286		2,250	95	2,345	2,625
1420	12'-0" x 6'-0", 6 panels			6	2.667		2,925	111	3,036	3,400
1600	Metal clad, casement, bldrs mdl, 6'-0" x 4'-0", dbl insltd gls, 3 panels			10	1.600		920	66.50	986.50	1,100
1640	9'-0" x 4'-0", 4 panels			8	2		1,250	83	1,333	1,500
1680	10'-0" x 5'-0", 5 panels			7	2.286		1,725	95	1,820	2,050
1720	12'-0" x 6'-0", 6 panels			6	2.667		2,400	111	2,511	2,825
2000	Bay window, builders model, 8' x 5' dbl insul glass,			10	1.600		1,825	66.50	1,891.50	2,125
2050	Low E glass,			10	1.600		2,225	66.50	2,291.50	2,550
2100	12'-0" x 6'-0", double insulated glass, 6 panels			6	2.667		2,300	111	2,411	2,700
2200	Low E glass			6	2.667		2,350	111	2,461	2,775
2300	Vinyl clad, premium, double insulated glass, 8'-0" x 5'-0"			10	1.600		1,750	66.50	1,816.50	2,025
2340	10'-0" x 5'-0"			8	2		2,275	83	2,358	2,625

08 52 Wood Windows

08 52 16 – Plastic-Clad Wood Windows

08 52 16.10 Bow Window

		Crew	Daily Output	Labor-Hours	Unit	Material	2010 Bare Costs Labor	Equipment	Total	Total Incl O&P
2380	10'-0" x 6'-0"	2 Carp	7	2.286	Ea.	2,525	95		2,620	2,925
2420	12'-0" x 6'-0"		6	2.667		3,200	111		3,311	3,700
2600	Metal clad, deluxe, dbl insul. glass, 8'-0" x 5'-0" high, 4 panels		10	1.600		1,625	66.50		1,691.50	1,875
2640	10'-0" x 5'-0" high, 5 panels		8	2		1,750	83		1,833	2,050
2680	10'-0" x 6'-0" high, 5 panels		7	2.286		2,050	95		2,145	2,425
2720	12'-0" x 6'-0" high, 6 panels		6	2.667		2,850	111		2,961	3,325
3000	Double hung, bldrs. model, bay, 8' x 4' high, dbl insulated glass		10	1.600		1,300	66.50		1,366.50	1,525
3050	Low E glass		10	1.600		1,375	66.50		1,441.50	1,625
3100	9'-0" x 5'-0" high, double insulated glass		6	2.667		1,400	111		1,511	1,700
3200	Low E glass		6	2.667		1,475	111		1,586	1,800
3300	Vinyl clad, premium, double insulated glass, 7'-0" x 4'-6"		10	1.600		1,325	66.50		1,391.50	1,575
3340	8'-0" x 4'-6"		8	2		1,375	83		1,458	1,625
3380	8'-0" x 5'-0"		7	2.286		1,425	95		1,520	1,725
3420	9'-0" x 5'-0"		6	2.667		1,475	111		1,586	1,800
3600	Metal clad, deluxe, dbl insul. glass, 7'-0" x 4'-0" high		10	1.600		1,225	66.50		1,291.50	1,450
3640	8'-0" x 4'-0" high		8	2		1,275	83		1,358	1,525
3680	8'-0" x 5'-0" high		7	2.286		1,325	95		1,420	1,600
3720	9'-0" x 5'-0" high		6	2.667		1,400	111		1,511	1,725
9000	Minimum labor/equipment charge		2.50	6.400	Job		266		266	435

08 52 16.20 Half Round, Vinyl Clad

		Crew	Daily Output	Labor-Hours	Unit	Material	Labor	Equipment	Total	Total Incl O&P
0010	**HALF ROUND, VINYL CLAD**, double insulated glass, incl. grille									
0800	14" height x 24" base	2 Carp	9	1.778	Ea.	405	74		479	565
1040	15" height x 25" base		8	2		375	83		458	550
1060	16" height x 28" base		7	2.286		405	95		500	600
1080	17" height x 29" base		7	2.286		420	95		515	615
2000	19" height x 33" base	1 Carp	6	1.333		450	55.50		505.50	585
2100	20" height x 35" base		6	1.333		450	55.50		505.50	585
2200	21" height x 37" base		6	1.333		475	55.50		530.50	615
2250	23" height x 41" base	2 Carp	6	2.667		520	111		631	750
2300	26" height x 48" base		6	2.667		535	111		646	765
2350	30" height x 56" base		6	2.667		615	111		726	855
3000	36" height x 67" base	1 Carp	4	2		1,050	83		1,133	1,275
3040	38" height x 71" base	2 Carp	5	3.200		975	133		1,108	1,300
3050	40" height x 75" base	"	5	3.200		1,275	133		1,408	1,625
5000	Elliptical, 71" x 16"	1 Carp	11	.727		975	30		1,005	1,125
5100	95" x 21"	"	10	.800		1,375	33		1,408	1,575

08 52 16.40 Transom Windows

		Crew	Daily Output	Labor-Hours	Unit	Material	Labor	Equipment	Total	Total Incl O&P
0010	**TRANSOM WINDOWS**									
0050	Vinyl clad, premium, double insulated glass, 32" x 8"	1 Carp	16	.500	Ea.	176	21		197	227
0100	36" x 8"		16	.500		186	21		207	239
0110	36" x 12"		16	.500		199	21		220	253
2000	Custom sizes, up to 350 sq. in.		12	.667		240	27.50		267.50	310
2100	351 to 750 sq. in.		12	.667		305	27.50		332.50	380
2200	751 to 1150 sq. in.		11	.727		375	30		405	465
2300	1151 to 1450 sq. in.		11	.727		430	30		460	525
2400	1451 to 1850 sq. in.	2 Carp	12	1.333		495	55.50		550.50	635
2500	1851 to 2250 sq. in.		12	1.333		565	55.50		620.50	715
2600	2251 to 2650 sq. in.		11	1.455		635	60.50		695.50	800
2700	2651 to 3050 sq. in.		11	1.455		640	60.50		700.50	805
2800	3051 to 3450 sq. in.		11	1.455		705	60.50		765.50	875
2900	3451 to 3850 sq. in.		10	1.600		760	66.50		826.50	945
3000	3851 to 4250 sq. in.		10	1.600		815	66.50		881.50	1,000

08 52 Wood Windows

08 52 16 – Plastic-Clad Wood Windows

08 52 16.40 Transom Windows		Crew	Daily Output	Labor-Hours	Unit	Material	2010 Bare Costs Labor	Equipment	Total	Total Incl O&P
3100	4251 to 4650 sq. in.	2 Carp	10	1.600	Ea.	830	66.50		896.50	1,025
3200	4651 to 5050 sq. in.		10	1.600		880	66.50		946.50	1,075
3300	5051 to 5450 sq. in.		9	1.778		950	74		1,024	1,175
3400	5451 to 5850 sq. in.		9	1.778		920	74		994	1,125
3600	6251 to 6650 sq. in.		8	2		1,100	83		1,183	1,350
3700	6651 to 7050 sq. in.		8	2		1,175	83		1,258	1,400

08 52 16.45 Trapezoid Windows

		Crew	Daily Output	Labor-Hours	Unit	Material	Labor	Equipment	Total	Total Incl O&P
0010	**TRAPEZOID WINDOWS**									
0900	20" base x 44" leg x 53" leg	2 Carp	13	1.231	Ea.	375	51		426	500
1000	24" base x 90" leg x 102" leg		8	2		640	83		723	840
3000	36" base x 40" leg x 22" leg		12	1.333		410	55.50		465.50	540
3010	36" base x 44" leg x 25" leg		13	1.231		430	51		481	560
3050	36" base x 26" leg x 48" leg		9	1.778		440	74		514	605
3100	36" base x 42" legs, 50" peak		9	1.778		520	74		594	690
3200	36" base x 60" leg x 81" leg		11	1.455		680	60.50		740.50	850
4320	44" base x 23" leg x 56" leg		11	1.455		535	60.50		595.50	690
4350	44" base x 59" leg x 92" leg		10	1.600		800	66.50		866.50	990
4500	46" base x 15" leg x 46" leg		8	2		405	83		488	580
4550	46" base x 16" leg x 48" leg		8	2		430	83		513	610
4600	46" base x 50" leg x 80" leg		7	2.286		650	95		745	870
6600	66" base x 12" leg x 42" leg		8	2		550	83		633	740
6650	66" base x 12" legs, 28" peak		9	1.778		440	74		514	605
6700	68" base x 23" legs, 31" peak		8	2		550	83		633	740

08 52 16.70 Vinyl Clad, Premium, Dbl. Insulated Glass

			Crew	Daily Output	Labor-Hours	Unit	Material	Labor	Equipment	Total	Total Incl O&P
0010	**VINYL CLAD, PREMIUM, DBL. INSULATED GLASS**										
1000	Sliding, 3'-0" x 3'-0"	G	1 Carp	10	.800	Ea.	575	33		608	685
1050	4'-0" x 3'-6"	G		9	.889		670	37		707	795
1100	5'-0" x 4'-0"	G		9	.889		860	37		897	1,000
1150	6'-0" x 5'-0"	G		8	1		1,075	41.50		1,116.50	1,275

08 52 50 – Window Accessories

08 52 50.10 Window Grille or Muntin

		Crew	Daily Output	Labor-Hours	Unit	Material	Labor	Equipment	Total	Total Incl O&P
0010	**WINDOW GRILLE OR MUNTIN**, snap in type									
0020	Standard pattern interior grilles									
2000	Wood, awning window, glass size 28" x 16" high	1 Carp	30	.267	Ea.	24	11.10		35.10	44.50
2060	44" x 24" high		32	.250		36	10.40		46.40	56.50
2100	Casement, glass size, 20" x 36" high		30	.267		29	11.10		40.10	49.50
2180	20" x 56" high		32	.250		40	10.40		50.40	61
2200	Double hung, glass size, 16" x 24" high		24	.333	Set	48	13.85		61.85	76
2280	32" x 32" high		34	.235	"	128	9.80		137.80	157
2500	Picture, glass size, 48" x 48" high		30	.267	Ea.	114	11.10		125.10	143
2580	60" x 68" high		28	.286	"	160	11.85		171.85	196
2600	Sliding, glass size, 14" x 36" high		24	.333	Set	26.50	13.85		40.35	52
2680	36" x 36" high		22	.364	"	39	15.10		54.10	68
9000	Minimum labor/equipment charge		5	1.600	Job		66.50		66.50	109

08 52 66 – Wood Window Screens

08 52 66.10 Wood Screens

		Crew	Daily Output	Labor-Hours	Unit	Material	Labor	Equipment	Total	Total Incl O&P
0010	**WOOD SCREENS**									
0020	Over 3 S.F., 3/4" frames	2 Carp	375	.043	S.F.	4.37	1.77		6.14	7.70
0100	1-1/8" frames		375	.043		7.70	1.77		9.47	11.40
0200	Rescreen wood frame		500	.032		1.09	1.33		2.42	3.38
9000	Minimum labor/equipment charge	1 Carp	4	2	Job		83		83	137

08 52 Wood Windows

08 52 69 – Wood Storm Windows

08 52 69.10 Storm Windows		Crew	Daily Output	Labor-Hours	Unit	Material	2010 Bare Costs Labor	Equipment	Total	Total Incl O&P
0010	**STORM WINDOWS**, aluminum residential									
0300	Basement, mill finish, incl. fiberglass screen									
0320	1'-10" x 1'-0" high	2 Carp	30	.533	Ea.	35	22		57	75
0340	2'-9" x 1'-6" high		30	.533		38	22		60	78.50
0360	3'-4" x 2'-0" high	↓	30	.533	↓	44	22		66	85
1600	Double-hung, combination, storm & screen									
1700	Custom, clear anodic coating, 2'-0" x 3'-5" high	2 Carp	30	.533	Ea.	90	22		112	136
1720	2'-6" x 5'-0" high		28	.571		110	23.50		133.50	160
1740	4'-0" x 6'-0" high		25	.640		226	26.50		252.50	293
1800	White painted, 2'-0" x 3'-5" high		30	.533		100	22		122	147
1820	2'-6" x 5'-0" high		28	.571		160	23.50		183.50	215
1840	4'-0" x 6'-0" high		25	.640		273	26.50		299.50	345
2000	Average quality, clear anodic coating, 2'-0" x 3'-5" high		30	.533		85	22		107	130
2020	2'-6" x 5'-0" high		28	.571		107	23.50		130.50	156
2040	4'-0" x 6'-0" high		25	.640		126	26.50		152.50	183
2400	White painted, 2'-0" x 3'-5" high		30	.533		85	22		107	130
2420	2'-6" x 5'-0" high		28	.571		90	23.50		113.50	138
2440	4'-0" x 6'-0" high		25	.640		98	26.50		124.50	152
2600	Mill finish, 2'-0" x 3'-5" high		30	.533		75	22		97	119
2620	2'-6" x 5'-0" high		28	.571		85	23.50		108.50	133
2640	4'-0" x 6'-8" high	↓	25	.640	↓	99	26.50		125.50	153
4000	Picture window, storm, 1 lite, white or bronze finish									
4020	4'-6" x 4'-6" high	2 Carp	25	.640	Ea.	120	26.50		146.50	176
4040	5'-8" x 4'-6" high		20	.800		144	33		177	213
4400	Mill finish, 4'-6" x 4'-6" high		25	.640		126	26.50		152.50	183
4420	5'-8" x 4'-6" high	↓	20	.800	↓	144	33		177	213
4600	3 lite, white or bronze finish									
4620	4'-6" x 4'-6" high	2 Carp	25	.640	Ea.	152	26.50		178.50	211
4640	5'-8" x 4'-6" high		20	.800		166	33		199	238
4800	Mill finish, 4'-6" x 4'-6" high		25	.640		144	26.50		170.50	202
4820	5'-8" x 4'-6" high	↓	20	.800	↓	152	33		185	222
6000	Sliding window, storm, 2 lite, white or bronze finish									
6020	3'-4" x 2'-7" high	2 Carp	28	.571	Ea.	110	23.50		133.50	160
6040	4'-4" x 3'-3" high		25	.640		150	26.50		176.50	209
6060	5'-4" x 6'-0" high	↓	20	.800	↓	240	33		273	320
8000	PVC framed									
8100	Double-hung, combination, storm & screen									
8120	2'-6" x 3'-5"	2 Carp	15	1.067	Ea.	93.50	44.50		138	176
8140	4' x 6'	"	12.50	1.280	"	128	53		181	229
8600	Single lite picture storm									
8620	4'-6" x 4'-6"	2 Carp	12.50	1.280	Ea.	153	53		206	256
8640	5'-8" x 4'-6"	"	10	1.600	"	162	66.50		228.50	287
9000	Magnetic interior storm window									
9100	3/16" plate glass	1 Glaz	107	.075	S.F.	5.15	3.01		8.16	10.50
9410	Minimum labor/equipment charge	1 Carp	4	2	Job		83		83	137

08 53 Plastic Windows

08 53 13 – Vinyl Windows

08 53 13.20 Vinyl Single Hung Windows

			Crew	Daily Output	Labor-Hours	Unit	Material	2010 Bare Costs Labor	Equipment	Total	Total Incl O&P
0010	**VINYL SINGLE HUNG WINDOWS**										
0100	Grids, low E, J fin, ext. jambs, 21" x 53"	G	2 Carp	18	.889	Ea.	162	37		199	239
0110	21" x 57"	G		17	.941		166	39		205	247
0120	21" x 65"	G		16	1		173	41.50		214.50	259
0130	25" x 41"	G		20	.800		157	33		190	228
0140	25" x 49"	G		18	.889		169	37		206	247
0150	25" x 57"	G		17	.941		173	39		212	254
0160	25" x 65"	G		16	1		180	41.50		221.50	267
0170	29" x 41"	G		18	.889		163	37		200	240
0180	29" x 53"	G		18	.889		175	37		212	254
0190	29" x 57"	G		17	.941		179	39		218	261
0200	29" x 65"	G		16	1		185	41.50		226.50	273
0210	33" x 41"	G		20	.800		169	33		202	241
0220	33" x 53"	G		18	.889		182	37		219	261
0230	33" x 57"	G		17	.941		185	39		224	268
0240	33" x 65"	G		16	1		196	41.50		237.50	285
0250	37" x 41"	G		20	.800		178	33		211	251
0260	37" x 53"	G		18	.889		191	37		228	271
0270	37" x 57"	G		17	.941		195	39		234	279
0280	37" x 65"	G		16	1		203	41.50		244.50	292

08 53 13.30 Vinyl Double Hung Windows

			Crew	Daily Output	Labor-Hours	Unit	Material	Labor	Equipment	Total	Total Incl O&P
0010	**VINYL DOUBLE HUNG WINDOWS**										
0100	Grids, low E, J fin, ext. jambs, 21" x 53"	G	2 Carp	18	.889	Ea.	186	37		223	266
0102	21" x 37"	G		18	.889		165	37		202	243
0104	21" x 41"	G		18	.889		170	37		207	248
0106	21" x 49"	G		18	.889		178	37		215	257
0110	21" x 57"	G		17	.941		190	39		229	273
0120	21" x 65"	G		16	1		198	41.50		239.50	287
0128	25" x 37"	G		20	.800		175	33		208	248
0130	25" x 41"	G		20	.800		179	33		212	252
0140	25" x 49"	G		18	.889		184	37		221	263
0145	25" x 53"	G		18	.889		192	37		229	272
0150	25" x 57"	G		17	.941		192	39		231	275
0160	25" x 65"	G		16	1		204	41.50		245.50	293
0162	25" x 69"	G		16	1		212	41.50		253.50	300
0164	25" x 77"	G		16	1		225	41.50		266.50	315
0168	29" x 37"	G		18	.889		180	37		217	259
0170	29" x 41"	G		18	.889		184	37		221	263
0172	29" x 49"	G		18	.889		193	37		230	273
0180	29" x 53"	G		18	.889		197	37		234	278
0190	29" x 57"	G		17	.941		202	39		241	286
0200	29" x 65"	G		16	1		217	41.50		258.50	310
0202	29" x 69"	G		16	1		217	41.50		258.50	310
0205	29" x 77"	G		16	1		230	41.50		271.50	320
0208	33" x 37"	G		20	.800		185	33		218	259
0210	33" x 41"	G		20	.800		189	33		222	263
0215	33" x 49"	G		20	.800		199	33		232	274
0220	33" x 53"	G		18	.889		204	37		241	285
0230	33" x 57"	G		17	.941		208	39		247	293
0240	33" x 65"	G		16	1		214	41.50		255.50	305
0242	33" x 69"	G		16	1		227	41.50		268.50	320
0246	33" x 77"	G		16	1		238	41.50		279.50	330

08 53 Plastic Windows

08 53 13 – Vinyl Windows

08 53 13.30 Vinyl Double Hung Windows

		Crew	Daily Output	Labor-Hours	Unit	Material	Labor	Equipment	Total	Total Incl O&P
0250	37" x 41"	2 Carp [G]	20	.800	Ea.	191	33		224	265
0255	37" x 49"	[G]	20	.800		204	33		237	279
0260	37" x 53"	[G]	18	.889		212	37		249	294
0270	37" x 57"	[G]	17	.941		217	39		256	305
0280	37" x 65"	[G]	16	1		222	41.50		263.50	315
0282	37" x 69"	[G]	16	1		227	41.50		268.50	320
0286	37" x 77"	[G]	16	1		248	41.50		289.50	340
0300	Solid vinyl, average quality, double insulated glass, 2'-0" x 3'-0"	1 Carp [G]	10	.800		279	33		312	360
0310	3'-0" x 4'-0"	[G]	9	.889		195	37		232	276
0320	4'-0" x 4'-6"	[G]	8	1		293	41.50		334.50	390
0330	Premium, double insulated glass, 2'-6" x 3'-0"	[G]	10	.800		214	33		247	291
0340	3'-0" x 3'-6"	[G]	9	.889		245	37		282	330
0350	3'-0" x 4'-0"	[G]	9	.889		256	37		293	345
0360	3'-0" x 4'-6"	[G]	9	.889		248	37		285	335
0370	3'-0" x 5'-0"	[G]	8	1		268	41.50		309.50	365
0380	3'-6" x 6'-0"	[G]	8	1		300	41.50		341.50	400

08 53 13.40 Vinyl Casement Windows

		Crew	Daily Output	Labor-Hours	Unit	Material	Labor	Equipment	Total	Total Incl O&P
0010	**VINYL CASEMENT WINDOWS**									
0100	Grids, low E, J fin, ext. jambs, 1 lt, 21" x 41"	2 Carp [G]	20	.800	Ea.	242	33		275	320
0110	21" x 47"	[G]	20	.800		264	33		297	345
0120	21" x 53"	[G]	20	.800		286	33		319	370
0128	24" x 35"	[G]	19	.842		233	35		268	315
0130	24" x 41"	[G]	19	.842		253	35		288	335
0140	24" x 47"	[G]	19	.842		274	35		309	360
0150	24" x 53"	[G]	19	.842		296	35		331	385
0158	28" x 35"	[G]	19	.842		248	35		283	330
0160	28" x 41"	[G]	19	.842		268	35		303	355
0170	28" x 47"	[G]	19	.842		290	35		325	380
0180	28" x 53"	[G]	19	.842		320	35		355	410
0184	28" x 59"	[G]	19	.842		325	35		360	420
0188	Two lites, 33" x 35"	[G]	18	.889		400	37		437	500
0190	33" x 41"	[G]	18	.889		425	37		462	530
0200	33" x 47"	[G]	18	.889		460	37		497	565
0210	33" x 53"	[G]	18	.889		490	37		527	600
0212	33" x 59"	[G]	18	.889		520	37		557	630
0215	33" x 72"	[G]	18	.889		540	37		577	655
0220	41" x 41"	[G]	18	.889		465	37		502	575
0230	41" x 47"	[G]	18	.889		500	37		537	610
0240	41" x 53"	[G]	17	.941		530	39		569	645
0242	41" x 59"	[G]	17	.941		560	39		599	680
0246	41" x 72"	[G]	17	.941		585	39		624	705
0250	47" x 41"	[G]	17	.941		470	39		509	585
0260	47" x 47"	[G]	17	.941		500	39		539	615
0270	47" x 53"	[G]	17	.941		530	39		569	650
0272	47" x 59"	[G]	17	.941		585	39		624	705
0280	56" x 41"	[G]	15	1.067		505	44.50		549.50	630
0290	56" x 47"	[G]	15	1.067		530	44.50		574.50	660
0300	56" x 53"	[G]	15	1.067		580	44.50		624.50	715
0302	56" x 59"	[G]	15	1.067		605	44.50		649.50	745
0310	56" x 72"	[G]	15	1.067		660	44.50		704.50	800
0340	Solid vinyl, premium, double insulated glass, 2'-0" x 3'-0" high	1 Carp [G]	10	.800		252	33		285	330
0360	2'-0" x 4'-0" high	[G]	9	.889		280	37		317	370

08 53 Plastic Windows

08 53 13 – Vinyl Windows

08 53 13.40 Vinyl Casement Windows		Crew	Daily Output	Labor-Hours	Unit	Material	2010 Bare Costs Labor	Equipment	Total	Total Incl O&P
0380	2'-0" x 5'-0" high	G 1 Carp	8	1	Ea.	291	41.50		332.50	390

08 53 13.50 Vinyl Picture Windows

		Crew	Daily Output	Labor-Hours	Unit	Material	Labor	Equipment	Total	Total Incl O&P
0010	**VINYL PICTURE WINDOWS**									
0100	Grids, low E, J fin, ext. jambs, 33" x 47"	2 Carp	12	1.333	Ea.	238	55.50		293.50	355
0110	35" x 71"		12	1.333		310	55.50		365.50	430
0120	41" x 47"		12	1.333		252	55.50		307.50	370
0130	41" x 71"		12	1.333		320	55.50		375.50	445
0140	47" x 47"		12	1.333		277	55.50		332.50	395
0150	47" x 71"		11	1.455		350	60.50		410.50	485
0160	53" x 47"		11	1.455		300	60.50		360.50	430
0170	53" x 71"		11	1.455		315	60.50		375.50	445
0180	59" x 47"		11	1.455		340	60.50		400.50	475
0190	59" x 71"		11	1.455		365	60.50		425.50	500
0200	71" x 47"		10	1.600		375	66.50		441.50	525
0210	71" x 71"		10	1.600		395	66.50		461.50	545

08 53 13.60 Vinyl Half Round Windows

		Crew	Daily Output	Labor-Hours	Unit	Material	Labor	Equipment	Total	Total Incl O&P
0010	**VINYL HALF ROUND WINDOWS**, Including grille, j fin low E, ext. jambs									
0100	10" height x 20" base	2 Carp	8	2	Ea.	475	83		558	655
0110	15" height x 30" base		8	2		425	83		508	605
0120	17" height x 34" base		7	2.286		287	95		382	470
0130	19" height x 38" base		7	2.286		310	95		405	495
0140	19" height x 33" base		7	2.286		555	95		650	770
0150	24" height x 48" base	1 Carp	6	1.333		320	55.50		375.50	445
0160	25" height x 50" base	"	6	1.333		755	55.50		810.50	920
0170	30" height x 60" base	2 Carp	6	2.667		635	111		746	880

08 54 Composite Windows

08 54 13 – Fiberglass Windows

08 54 13.10 Fiberglass Single Hung Windows

			Crew	Daily Output	Labor-Hours	Unit	Material	Labor	Equipment	Total	Total Incl O&P
0010	**FIBERGLASS SINGLE HUNG WINDOWS**										
0100	Grids, low E, 18" x 24"	G	2 Carp	18	.889	Ea.	335	37		372	430
0110	18" x 40"	G		17	.941		300	39		339	395
0130	24" x 40"	G		20	.800		330	33		363	420
0230	36" x 36"	G		17	.941		370	39		409	470
0250	36" x 48"	G		20	.800		405	33		438	500
0260	36" x 60"	G		18	.889		445	37		482	550
0280	36" x 72"	G		16	1		470	41.50		511.50	585
0290	48" x 40"	G		16	1		470	41.50		511.50	585

08 56 Special Function Windows

08 56 63 – Detention Windows

08 56 63.13 Visitor Cubicle Windows		Crew	Daily Output	Labor-Hours	Unit	Material	2010 Bare Costs Labor	Equipment	Total	Total Incl O&P
0010	**VISITOR CUBICLE WINDOWS**									
4000	Visitor cubicle, vision panel, no intercom	E-4	2	16	Ea.	2,925	760	73	3,758	4,675

08 61 Roof Windows

08 61 13 – Metal Roof Windows

08 61 13.10 Roof Windows

		Crew	Daily Output	Labor-Hours	Unit	Material	Labor	Equipment	Total	Total Incl O&P
0010	**ROOF WINDOWS**, fixed high perf tmpd glazing, metallic framed									
0020	46" x 21-1/2", Flashed for shinge roof	1 Carp	8	1	Ea.	245	41.50		286.50	340
0100	46" x 28"		8	1		272	41.50		313.50	370
0125	57" x 44"		6	1.333		335	55.50		390.50	460
0130	72" x 28"		7	1.143		335	47.50		382.50	450
0150	Fixed, laminated tempered glazing, 46" x 21-1/2"		8	1		415	41.50		456.50	525
0175	46" x 28"		8	1		460	41.50		501.50	575
0200	57" x 44"		6	1.333		430	55.50		485.50	565
0500	Vented Flashing set for shingled roof, 46" x 21-1/2"		7	1.143		415	47.50		462.50	535
0525	46" x 28"		6	1.333		460	55.50		515.50	595
0550	57" x 44"		5	1.600		575	66.50		641.50	745
0560	72" x 28"		5	1.600		575	66.50		641.50	745
0575	Flashing set for low pitched roof, 46" x 21-1/2"		7	1.143		475	47.50		522.50	605
0600	46" x 28"		7	1.143		525	47.50		572.50	655
0625	57" x 44"		5	1.600		650	66.50		716.50	825
0650	Flashing set for curb 46" x 21-1/2"		7	1.143		550	47.50		597.50	685
0675	46" x 28"		7	1.143		600	47.50		647.50	740
0700	57" x 44"		5	1.600		740	66.50		806.50	920

08 62 Unit Skylights

08 62 13 – Domed Unit Skylights

08 62 13.10 Domed Skylights

			Crew	Daily Output	Labor-Hours	Unit	Material	Labor	Equipment	Total	Total Incl O&P
0010	**DOMED SKYLIGHTS**										
0020	Skylight, fixed dome type, 22" x 22"	G	G-3	12	2.667	Ea.	200	110		310	395
0030	22" x 46"	G		10	3.200		229	132		361	460
0040	30" x 30"	G		12	2.667		230	110		340	430
0050	30" x 46"	G		10	3.200		290	132		422	530
0110	Fixed, double glazed, 22" x 27"	G		12	2.667		213	110		323	410
0120	22" x 46"	G		10	3.200		269	132		401	505
0130	44" x 46"	G		10	3.200		400	132		532	650
0210	Operable, double glazed, 22" x 27"	G		12	2.667		310	110		420	515
0220	22" x 46"	G		10	3.200		365	132		497	610
0230	44" x 46"	G		10	3.200		770	132		902	1,050
9000	Minimum labor/equipment charge			2	16	Job		660		660	1,050

08 62 13.20 Skylights

			Crew	Daily Output	Labor-Hours	Unit	Material	Labor	Equipment	Total	Total Incl O&P
0010	**SKYLIGHTS**, Plastic domes, flush or curb mounted ten or										
0100	more units										
0300	Nominal size under 10 S.F., double	G	G-3	130	.246	S.F.	28	10.10		38.10	47
0400	Single			160	.200		24.50	8.20		32.70	40
0600	10 S.F. to 20 S.F., double	G		315	.102		24	4.18		28.18	33
0700	Single			395	.081		24	3.33		27.33	31.50
0900	20 S.F. to 30 S.F., double			395	.081		22	3.33		25.33	30
1000	Single			465	.069		19.70	2.83		22.53	26

08 62 Unit Skylights

08 62 13 – Domed Unit Skylights

08 62 13.20 Skylights

		Crew	Daily Output	Labor-Hours	Unit	Material	2010 Bare Costs Labor	Equipment	Total	Total Incl O&P
1200	30 S.F. to 65 S.F., double	G-3	465	.069	S.F.	22	2.83		24.83	29
1300	Single		610	.052		17	2.16		19.16	22
1500	For insulated 4" curbs, double, add					27%				
1600	Single, add					30%				
1800	For integral insulated 9" curbs, double, add					30%				
1900	Single, add					40%				
2120	Ventilating insulated plexiglass dome with									
2130	curb mounting, 36" x 36"	G-3	12	2.667	Ea.	460	110		570	680
2150	52" x 52"		12	2.667		640	110		750	880
2160	28" x 52"		10	3.200		470	132		602	725
2170	36" x 52"		10	3.200		520	132		652	780
2180	For electric opening system, add					300			300	330
2200	Field fabricated, factory type, aluminum and wire glass	G-3	120	.267	S.F.	16	10.95		26.95	35
2300	Insulated safety glass with aluminum frame		160	.200		90	8.20		98.20	112
2400	Sandwich panels, fiberglass, for walls, 1-9/16" thick, to 250 SF		200	.160		16.20	6.60		22.80	28.50
2500	250 SF and up		265	.121		14.60	4.96		19.56	24
2700	As above, but for roofs, 2-3/4" thick, to 250 SF		295	.108		23.50	4.46		27.96	32.50
2800	250 SF and up		330	.097		19	3.99		22.99	27.50

08 71 Door Hardware

08 71 13 – Automatic Door Operators

08 71 13.10 Automatic Openers Commercial

		Crew	Daily Output	Labor-Hours	Unit	Material	2010 Bare Costs Labor	Equipment	Total	Total Incl O&P
0010	**AUTOMATIC OPENERS COMMERCIAL**									
0020	Pneumatic, incl opener, motion sens, control box, tubing, compressor									
0050	For single swing door, per opening	2 Skwk	.80	20	Ea.	4,200	850		5,050	6,000
0100	Pair, per opening		.50	32	Opng.	6,900	1,375		8,275	9,800
1000	For single sliding door, per opening		.60	26.667		4,600	1,125		5,725	6,875
1300	Bi-parting pair		.50	32		7,000	1,375		8,375	9,900
1420	Electronic door opener incl motion sens, 12V control box, motor									
1450	For single swing door, per opening	2 Skwk	.80	20	Opng.	3,500	850		4,350	5,225
1500	Pair, per opening		.50	32		5,700	1,375		7,075	8,475
1600	For single sliding door, per opening		.60	26.667		3,800	1,125		4,925	6,000
1700	Bi-parting pair		.50	32		5,800	1,375		7,175	8,575
1750	Handicap actuator buttons, 2, including 12V DC wiring, add	1 Carp	1.50	5.333	Pr.	415	222		637	820

08 71 20 – Hardware

08 71 20.15 Hardware

		Crew	Daily Output	Labor-Hours	Unit	Material	2010 Bare Costs Labor	Equipment	Total	Total Incl O&P
0009	**HARDWARE**									
0010	Average percentage for hardware, total job cost									
0050	Maximum									4%
0500	Total hardware for building, average distribution					85%	15%			
1000	Door hardware, apartment, interior				Door	490			490	540
1300	Average, door hardware, motel/hotel interior					635			635	700
1500	Hospital bedroom, minimum					595			595	650
2000	Maximum					710			710	780
2250	School, single exterior, incl. lever, incl. panic device					1,075			1,075	1,200
2500	Single interior, regular use, lever included					600			600	660
2600	Average, door hdwe, school, classroom, ANSI F88, incl. lever					780			780	860
2850	Stairway, single interior					515			515	570
3100	Double exterior, with panic device				Pr.	2,025			2,025	2,250
3600	Toilet, public, single interior				Door	210			210	231

08 71 Door Hardware

08 71 20 – Hardware

08 71 20.15 Hardware		Crew	Daily Output	Labor-Hours	Unit	Material	2010 Bare Costs Labor	Equipment	Total	Total Incl O&P
6020	Add for fire alarm door holder, electro-magnetic	1 Elec	4	2	Ea.	99	98		197	260
08 71 20.30 Door Closers										
0010	**DOOR CLOSERS**									
0020	Adjustable backcheck, 3 way mount, all sizes, regular arm	1 Carp	6	1.333	Ea.	166	55.50		221.50	273
0040	Hold open arm		6	1.333		210	55.50		265.50	320
0100	Fusible link		6.50	1.231		140	51		191	238
0200	Non sized, regular arm		6	1.333		160	55.50		215.50	267
0240	Hold open arm		6	1.333		195	55.50		250.50	305
0400	4 way mount, non sized, regular arm		6	1.333		210	55.50		265.50	320
0440	Hold open arm		6	1.333		220	55.50		275.50	335
2000	Backcheck and adjustable power, hinge face mount									
2010	All sizes, regular arm	1 Carp	6.50	1.231	Ea.	200	51		251	305
2040	Hold open arm		6.50	1.231		240	51		291	350
2400	Top jamb mount, all sizes, regular arm		6	1.333		195	55.50		250.50	305
2440	Hold open arm		6	1.333		205	55.50		260.50	315
2800	Top face mount, all sizes, regular arm		6.50	1.231		190	51		241	293
2840	Hold open arm		6.50	1.231		208	51		259	315
4000	Backcheck, overhead concealed, all sizes, regular arm		5.50	1.455		210	60.50		270.50	330
4040	Concealed arm		5	1.600		225	66.50		291.50	355
4400	Compact overhead, concealed, all sizes, regular arm		5.50	1.455		340	60.50		400.50	475
4440	Concealed arm		5	1.600		380	66.50		446.50	530
4800	Concealed in door, all sizes, regular arm		5.50	1.455		350	60.50		410.50	485
4840	Concealed arm		5	1.600		370	66.50		436.50	515
4900	Floor concealed, all sizes, single acting		2.20	3.636		375	151		526	665
4940	Double acting		2.20	3.636		300	151		451	580
5000	For cast aluminum cylinder, deduct					30			30	33
5040	For delayed action, add					40			40	44
5080	For fusible link arm, add					25			25	27.50
5120	For shock absorbing arm, add					45			45	49.50
5160	For spring power adjustment, add					35			35	38.50
6000	Closer-holder, hinge face mount, all sizes, exposed arm	1 Carp	6.50	1.231		160	51		211	260
7000	Electronic closer-holder, hinge facemount, concealed arm		5	1.600		260	66.50		326.50	395
7400	With built-in detector		5	1.600		660	66.50		726.50	835
9000	Minimum labor/equipment charge		4	2	Job		83		83	137
08 71 20.31 Door Closers										
0010	**DOOR CLOSERS**									
0015	Door closer, rack and pinion	1 Carp	6.50	1.231	Ea.	155	51		206	255
08 71 20.35 Panic Devices										
0010	**PANIC DEVICES**									
0015	For rim locks, single door exit only	1 Carp	6	1.333	Ea.	495	55.50		550.50	635
0020	Outside key and pull		5	1.600		540	66.50		606.50	700
0200	Bar and vertical rod, exit only		5	1.600		715	66.50		781.50	895
0210	Outside key and pull		4	2		790	83		873	1,000
0400	Bar and concealed rod		4	2		590	83		673	785
0600	Touch bar, exit only		6	1.333		500	55.50		555.50	640
0610	Outside key and pull		5	1.600		570	66.50		636.50	740
0700	Touch bar and vertical rod, exit only		5	1.600		645	66.50		711.50	820
0710	Outside key and pull		4	2		745	83		828	955
1000	Mortise, bar, exit only		4	2		530	83		613	715
1600	Touch bar, exit only		4	2		605	83		688	800
2000	Narrow stile, rim mounted, bar, exit only		6	1.333		635	55.50		690.50	790
2010	Outside key and pull		5	1.600		690	66.50		756.50	865

08 71 Door Hardware

08 71 20 – Hardware

08 71 20.35 Panic Devices		Crew	Daily Output	Labor-Hours	Unit	Material	2010 Bare Costs Labor	Equipment	Total	Total Incl O&P
2200	Bar and vertical rod, exit only	1 Carp	5	1.600	Ea.	655	66.50		721.50	830
2210	Outside key and pull		4	2		655	83		738	855
2400	Bar and concealed rod, exit only		3	2.667		760	111		871	1,025
3000	Mortise, bar, exit only		4	2		575	83		658	765
3600	Touch bar, exit only		4	2		830	83		913	1,050
4000	Double doors, exit only		2	4	Pr.	755	166		921	1,100
4500	Exit & entrance		2	4	"	910	166		1,076	1,275
9000	Minimum labor/equipment charge		2.50	3.200	Job		133		133	218

08 71 20.40 Lockset										
0010	**LOCKSET**, Standard duty									
0020	Non-keyed, passage, w/sect.trim	1 Carp	12	.667	Ea.	53.50	27.50		81	105
0100	Privacy		12	.667		62	27.50		89.50	114
0400	Keyed, single cylinder function		10	.800		86	33		119	149
0420	Hotel		8	1		130	41.50		171.50	212
0500	Lever handled, keyed, single cylinder function		10	.800		156	33		189	227
1000	Heavy duty with sectional trim, non-keyed, passages		12	.667		140	27.50		167.50	200
1100	Privacy		12	.667		180	27.50		207.50	244
1400	Keyed, single cylinder function		10	.800		215	33		248	292
1420	Hotel		8	1		320	41.50		361.50	420
1600	Communicating		10	.800		240	33		273	320
1690	For re-core cylinder, add					38			38	42
1700	Residential, interior door, minimum	1 Carp	16	.500		22	21		43	58
1720	Maximum		8	1		55	41.50		96.50	129
1800	Exterior, minimum		14	.571		40	23.50		63.50	83
1810	Average		8	1		72	41.50		113.50	148
1820	Maximum		8	1		160	41.50		201.50	245
9000	Minimum labor/equipment charge		6	1.333	Job		55.50		55.50	91

08 71 20.41 Dead Locks										
0010	**DEAD LOCKS**									
0011	Mortise heavy duty outside key (security item)	1 Carp	9	.889	Ea.	260	37		297	345
0020	Double cylinder		9	.889		265	37		302	355
0100	Medium duty, outside key		10	.800		140	33		173	209
0110	Double cylinder		10	.800		155	33		188	226
1000	Tubular, standard duty, outside key		10	.800		65	33		98	126
1010	Double cylinder		10	.800		82	33		115	145
1200	Night latch, outside key		10	.800		81	33		114	144

08 71 20.42 Mortise Locksets										
0010	**MORTISE LOCKSETS**, Comm., wrought knobs & full escutcheon trim									
0020	Non-keyed, passage, minimum	1 Carp	9	.889	Ea.	182	37		219	261
0030	Maximum		8	1		294	41.50		335.50	395
0040	Privacy, minimum		9	.889		195	37		232	276
0050	Maximum		8	1		320	41.50		361.50	420
0100	Keyed, office/entrance/apartment, minimum		8	1		226	41.50		267.50	320
0110	Maximum		7	1.143		380	47.50		427.50	500
0120	Single cylinder, typical, minimum		8	1		194	41.50		235.50	282
0130	Maximum		7	1.143		350	47.50		397.50	465
0200	Hotel, minimum		7	1.143		225	47.50		272.50	325
0210	Maximum		6	1.333		370	55.50		425.50	495
0300	Communication, double cylinder, minimum		8	1		220	41.50		261.50	310
0310	Maximum		7	1.143		290	47.50		337.50	400
1000	Wrought knobs and sectional trim, non-keyed, passage, minimum		10	.800		120	33		153	187
1010	Maximum		9	.889		235	37		272	320

08 71 Door Hardware

08 71 20 – Hardware

08 71 20.42 Mortise Locksets		Crew	Daily Output	Labor-Hours	Unit	Material	2010 Bare Costs Labor	2010 Bare Costs Equipment	Total	Total Incl O&P
1040	Privacy, minimum	1 Carp	10	.800	Ea.	141	33		174	210
1050	Maximum		9	.889		256	37		293	345
1100	Keyed, entrance, office/apartment, minimum		9	.889		216	37		253	299
1110	Maximum		8	1		300	41.50		341.50	400
1120	Single cylinder, typical, minimum		9	.889		210	37		247	292
1130	Maximum		8	1		291	41.50		332.50	390
2000	Cast knobs and full escutcheon trim									
2010	Non-keyed, passage, minimum	1 Carp	9	.889	Ea.	256	37		293	345
2020	Maximum		8	1		410	41.50		451.50	520
2040	Privacy, minimum		9	.889		310	37		347	400
2050	Maximum		8	1		440	41.50		481.50	555
2120	Keyed, single cylinder, typical, minimum		8	1		315	41.50		356.50	415
2130	Maximum		7	1.143		490	47.50		537.50	620
2200	Hotel, minimum		7	1.143		335	47.50		382.50	450
2210	Maximum		6	1.333		595	55.50		650.50	745
3000	Cast knob and sectional trim, non-keyed, passage, minimum		10	.800		200	33		233	275
3010	Maximum		10	.800		390	33		423	485
3040	Privacy, minimum		10	.800		225	33		258	305
3050	Maximum		10	.800		400	33		433	495
3100	Keyed, office/entrance/apartment, minimum		9	.889		255	37		292	340
3110	Maximum		9	.889		400	37		437	500
3120	Single cylinder, typical, minimum		9	.889		255	37		292	340
3130	Maximum		9	.889		490	37		527	600
3190	For re-core cylinder, add					35			35	38.50
3800	Cipher lockset w/key pad (security item)	1 Carp	13	.615		810	25.50		835.50	935
3900	Cipher lockset, with dial for swinging doors (security item)		13	.615		1,725	25.50		1,750.50	1,950
3920	with dial for swinging doors & drill resistant plate (security item)		12	.667		2,100	27.50		2,127.50	2,350
3950	Cipher lockset with dial for safe/vault door (security item)		12	.667		1,400	27.50		1,427.50	1,600
3980	Keyless, pushbutton type									
4000	Residential/light commercial, deadbolt, standard	1 Carp	9	.889	Ea.	125	37		162	199
4010	Heavy duty		9	.889		142	37		179	217
4020	Industrial, heavy duty, with deadbolt		9	.889		284	37		321	375
4030	Key override		9	.889		320	37		357	410
4040	Lever activated handle		9	.889		335	37		372	430
4050	Key override		9	.889		375	37		412	475
4060	Double sided pushbutton type		8	1		695	41.50		736.50	830
4070	Key override		8	1		690	41.50		731.50	825

08 71 20.45 Peepholes		Crew	Daily Output	Labor-Hours	Unit	Material	Labor	Equipment	Total	Total Incl O&P
0010	**PEEPHOLES**									
2010	Peephole	1 Carp	32	.250	Ea.	15.80	10.40		26.20	34.50

08 71 20.50 Door Stops		Crew	Daily Output	Labor-Hours	Unit	Material	Labor	Equipment	Total	Total Incl O&P
0010	**DOOR STOPS**									
0020	Holder & bumper, floor or wall	1 Carp	32	.250	Ea.	34.50	10.40		44.90	55
1300	Wall bumper, 4" diameter, with rubber pad, aluminum		32	.250		10.95	10.40		21.35	29
1600	Door bumper, floor type, aluminum		32	.250		7	10.40		17.40	25
1900	Plunger type, door mounted		32	.250		28	10.40		38.40	47.50
9000	Minimum labor/equipment charge		6	1.333	Job		55.50		55.50	91

08 71 20.55 Push-Pull Plates		Crew	Daily Output	Labor-Hours	Unit	Material	Labor	Equipment	Total	Total Incl O&P
0010	**PUSH-PULL PLATES**									
0100	Push plate, .050 thick, 4" x 16", aluminum	1 Carp	12	.667	Ea.	9.15	27.50		36.65	55.50
0500	Bronze		12	.667		21	27.50		48.50	69
1500	Pull handle and push bar, aluminum		11	.727		131	30		161	194

08 71 Door Hardware

08 71 20 – Hardware

08 71 20.55 Push-Pull Plates

		Crew	Daily Output	Labor-Hours	Unit	Material	2010 Bare Costs Labor	Equipment	Total	Total Incl O&P
2000	Bronze	1 Carp	10	.800	Ea.	169	33		202	240
3000	Push plate both sides, aluminum		14	.571		18.25	23.50		41.75	59
3500	Bronze		13	.615		43	25.50		68.50	89.50
4000	Door pull, designer style, cast aluminum, minimum		12	.667		71	27.50		98.50	124
5000	Maximum		8	1		370	41.50		411.50	475
6000	Cast bronze, minimum		12	.667		87	27.50		114.50	141
7000	Maximum		8	1		405	41.50		446.50	515
8000	Walnut, minimum		12	.667		67	27.50		94.50	119
9000	Maximum		8	1		370	41.50		411.50	475
9800	Minimum labor/equipment charge		5	1.600	Job		66.50		66.50	109

08 71 20.60 Entrance Locks

		Crew	Daily Output	Labor-Hours	Unit	Material	Labor	Equipment	Total	Total Incl O&P
0010	**ENTRANCE LOCKS**									
0015	Cylinder, grip handle deadlocking latch	1 Carp	9	.889	Ea.	132	37		169	206
0020	Deadbolt		8	1		160	41.50		201.50	245
0100	Push and pull plate, dead bolt		8	1		150	41.50		191.50	234
0900	For handicapped lever, add					170			170	187

08 71 20.65 Thresholds

		Crew	Daily Output	Labor-Hours	Unit	Material	Labor	Equipment	Total	Total Incl O&P
0010	**THRESHOLDS**									
0011	Threshold 3' long saddles aluminum	1 Carp	48	.167	L.F.	4.16	6.95		11.11	16
0100	Aluminum, 8" wide, 1/2" thick		12	.667	Ea.	39	27.50		66.50	88.50
0500	Bronze		60	.133	L.F.	40	5.55		45.55	53
0600	Bronze, panic threshold, 5" wide, 1/2" thick		12	.667	Ea.	66	27.50		93.50	118
0700	Rubber, 1/2" thick, 5-1/2" wide		20	.400		38	16.60		54.60	69.50
0800	2-3/4" wide		20	.400		16	16.60		32.60	45
9000	Minimum labor/equipment charge		4	2	Job		83		83	137

08 71 20.75 Door Hardware Accessories

		Crew	Daily Output	Labor-Hours	Unit	Material	Labor	Equipment	Total	Total Incl O&P
0010	**DOOR HARDWARE ACCESSORIES**									
0050	Door closing coordinator, 36" (for paired openings up to 56")	1 Carp	8	1	Ea.	94	41.50		135.50	172
0060	48" (for paired openings up to 84")		8	1		101	41.50		142.50	180
0070	56" (for paired openings up to 96")		8	1		111	41.50		152.50	191
1000	Knockers, brass, standard		16	.500		43.50	21		64.50	82
1100	Deluxe		10	.800		126	33		159	193
4100	Deluxe		18	.444		44	18.45		62.45	79
4500	Rubber door silencers		540	.015		.33	.62		.95	1.37
9000	Minimum labor/equipment charge		6	1.333	Job		55.50		55.50	91

08 71 20.80 Hasps

		Crew	Daily Output	Labor-Hours	Unit	Material	Labor	Equipment	Total	Total Incl O&P
0010	**HASPS**, steel assembly									
0015	3"	1 Carp	26	.308	Ea.	4.04	12.80		16.84	25.50
0020	4-1/2"		13	.615		5.15	25.50		30.65	47.50
0040	6"		12.50	.640		8.60	26.50		35.10	53

08 71 20.90 Hinges

			Crew	Daily Output	Labor-Hours	Unit	Material	Labor	Equipment	Total	Total Incl O&P
0010	**HINGES**	R087120-10									
0012	Full mortise, avg. freq., steel base, USP, 4-1/2" x 4-1/2"					Pr.	25.50			25.50	28
0100	5" x 5", USP						42.50			42.50	46.50
0200	6" x 6", USP						90			90	99
0400	Brass base, 4-1/2" x 4-1/2", US10						52.50			52.50	57.50
0500	5" x 5", US10						76			76	84
0600	6" x 6", US10						130			130	143
0800	Stainless steel base, 4-1/2" x 4-1/2", US32						77			77	85
0900	For non removable pin, add (security item)					Ea.	4.47			4.47	4.92
0910	For floating pin, driven tips, add						3.15			3.15	3.47

08 71 Door Hardware

08 71 20 – Hardware

08 71 20.90 Hinges

		Crew	Daily Output	Labor-Hours	Unit	Material	2010 Bare Costs Labor	Equipment	Total	Total Incl O&P
0930	For hospital type tip on pin, add				Ea.	13.80			13.80	15.20
0940	For steeple type tip on pin, add				▼	12			12	13.20
0950	Full mortise, high frequency, steel base, 3-1/2" x 3-1/2", US26D				Pr.	27			27	29.50
1000	4-1/2" x 4-1/2", USP					61			61	67.50
1100	5" x 5", USP					57			57	62.50
1200	6" x 6", USP					140			140	154
1400	Brass base, 3-1/2" x 3-1/2", US4					46.50			46.50	51.50
1430	4-1/2" x 4-1/2", US10					82			82	90.50
1500	5" x 5", US10					121			121	133
1600	6" x 6", US10					175			175	192
1800	Stainless steel base, 4-1/2" x 4-1/2", US32					127			127	140
1810	5" x 4-1/2", US32				▼	180			180	198
1930	For hospital type tip on pin, add				Ea.	12.30			12.30	13.50
1950	Full mortise, low frequency, steel base, 3-1/2" x 3-1/2", US26D				Pr.	11.20			11.20	12.35
2000	4-1/2" x 4-1/2", USP					12.55			12.55	13.85
2100	5" x 5", USP					31			31	34
2200	6" x 6", USP					62			62	68
2300	4-1/2" x 4-1/2", US3					18.10			18.10	19.90
2310	5" x 5", US3					43.50			43.50	48
2400	Brass base, 4-1/2" x 4-1/2", US10					43.50			43.50	48
2500	5" x 5", US10					66			66	72.50
2800	Stainless steel base, 4-1/2" x 4-1/2", US32				▼	75			75	82.50

08 71 20.91 Special Hinges

		Crew	Daily Output	Labor-Hours	Unit	Material	2010 Bare Costs Labor	Equipment	Total	Total Incl O&P
0010	**SPECIAL HINGES**									
0015	Paumelle, high frequency									
0020	Steel base, 6" x 4-1/2", US10				Pr.	145			145	160
0100	Bronze base, 5" x 4-1/2", US10					181			181	199
0200	Paumelle, average frequency, steel base, 4-1/2" x 3-1/2", US10					98.50			98.50	108
0400	Olive knuckle, low frequency, brass base, 6" x 4-1/2", US10				▼	165			165	181
1000	Electric hinge with concealed conductor, average frequency									
1010	Steel base, 4-1/2" x 4-1/2", US26D				Pr.	315			315	345
1100	Bronze base, 4-1/2" x 4-1/2", US26D				"	320			320	350
1200	Electric hinge with concealed conductor, high frequency									
1210	Steel base, 4-1/2" x 4-1/2", US26D				Pr.	240			240	264
1600	Double weight, 800 lb., steel base, removable pin, 5" x 6", USP					132			132	145
1700	Steel base-welded pin, 5" x 6", USP					162			162	178
1800	Triple weight, 2000 lb., steel base, welded pin, 5" x 6", USP					151			151	166
2000	Pivot reinf., high frequency, steel base, 7-3/4" door plate, USP					182			182	200
2200	Bronze base, 7-3/4" door plate, US10				▼	229			229	252
3000	Swing clear, full mortise, full or half surface, high frequency,									
3010	Steel base, 5" high, USP				Pr.	159			159	175
3200	Swing clear, full mortise, average frequency									
3210	Steel base, 4-1/2" high, USP				Pr.	127			127	139
4000	Wide throw, average frequency, steel base, 4-1/2" x 6", USP					96.50			96.50	106
4200	High frequency, steel base, 4-1/2" x 6", USP					148			148	163
4440	US26D				▼	88.50			88.50	97.50
4600	Spring hinge, single acting, 6" flange, steel				Ea.	54			54	59.50
4700	Brass					95.50			95.50	105
4900	Double acting, 6" flange, steel					96.50			96.50	106
4950	Brass				▼	157			157	173
5000	T-strap, galvanized, 4"				Pr.	19.15			19.15	21
5010	6"				▼	30.50			30.50	33.50

08 71 Door Hardware

08 71 20 – Hardware

08 71 20.91 Special Hinges

		Crew	Daily Output	Labor-Hours	Unit	Material	2010 Bare Costs Labor	Equipment	Total	Total Incl O&P
5020	8"				Pr.	45.50			45.50	50
9000	Continuous hinge, steel, full mortise, heavy duty	2 Carp	64	.250	L.F.	22.50	10.40		32.90	42

08 71 20.95 Kick Plates

		Crew	Daily Output	Labor-Hours	Unit	Material	Labor	Equipment	Total	Total Incl O&P
0010	**KICK PLATES**									
0020	Stainless steel, .050, 16 ga, 8" x 28", US32	1 Carp	15	.533	Ea.	34	22		56	74
0030	8" x 30"		15	.533		35	22		57	75
0040	8" x 34"		15	.533		39.50	22		61.50	80
0050	10" x 28"		15	.533		66.50	22		88.50	110
0060	10" x 30"		15	.533		71	22		93	115
0070	10" x 34"		15	.533		80.50	22		102.50	125
0080	Mop/Kick, 4" x 28"		15	.533		29.50	22		51.50	68.50
0090	4" x 30"		15	.533		31.50	22		53.50	71
0100	4" x 34"		15	.533		35.50	22		57.50	75.50
0110	6" x 28"		15	.533		38	22		60	78.50
0120	6" x 30"		15	.533		40.50	22		62.50	81.50
0130	6" x 34"		15	.533		46	22		68	87.50
0500	Bronze, .050", 8" x 28"		15	.533		46.50	22		68.50	87.50
0510	8" x 30"		15	.533		62.50	22		84.50	105
0520	8" x 34"		15	.533		71	22		93	115
0530	10" x 28"		15	.533		66.50	22		88.50	110
0540	10" x 30"		15	.533		71	22		93	115
0550	10" x 34"		15	.533		80.50	22		102.50	125
0560	Mop/Kick, 4" x 28"		15	.533		31	22		53	70.50
0570	4" x 30"		15	.533		33	22		55	73
0580	4" x 34"		15	.533		35	22		57	75
0590	6" x 28"		15	.533		43.50	22		65.50	84.50
0600	6" x 30"		15	.533		50	22		72	91.50
0610	6" x 34"		15	.533		50	22		72	91.50
1000	Acrylic, .125", 8" x 26"		15	.533		26	22		48	65
1010	8" x 36"		15	.533		36	22		58	76
1020	8" x 42"		15	.533		42	22		64	82.50
1030	10" x 26"		15	.533		32.50	22		54.50	72.50
1040	10" x 36"		15	.533		45	22		67	86
1050	10" x 42"		15	.533		68.50	22		90.50	112
1060	Mop/Kick, 4" x 26"		15	.533		15.60	22		37.60	53.50
1070	4" x 23"		15	.533		21.50	22		43.50	60
1080	4" x 42"		15	.533		24.50	22		46.50	63.50
1090	6" x 26"		15	.533		23	22		45	62
1100	6" x 36"		15	.533		31.50	22		53.50	71.50
1110	6" x 42"		15	.533		37	22		59	77
2000	Aluminum, .050, with 3 beveled edges, 10" x 28"		15	.533		23	22		45	62
2010	10" x 30"		15	.533		25	22		47	64
2020	10" x 34"		15	.533		26	22		48	65
2040	10" x 38"		15	.533		28	22		50	67
9000	Minimum labor/equipment charge		6	1.333	Job		55.50		55.50	91

08 71 21 – Astragals

08 71 21.10 Exterior Mouldings, Astragals

		Crew	Daily Output	Labor-Hours	Unit	Material	Labor	Equipment	Total	Total Incl O&P
0010	**EXTERIOR MOULDINGS, ASTRAGALS**									
0400	One piece, overlapping cadmium plated steel, flat, 3/16" x 2"	1 Carp	90	.089	L.F.	3.50	3.69		7.19	9.90
0600	Prime coated steel, flat, 1/8" x 3"		90	.089		4.70	3.69		8.39	11.20
0800	Stainless steel, flat, 3/32" x 1-5/8"		90	.089		17.90	3.69		21.59	26
1000	Aluminum, flat, 1/8" x 2"		90	.089		3.55	3.69		7.24	9.95

08 71 Door Hardware

08 71 21 – Astragals

08 71 21.10 Exterior Mouldings, Astragals

	08 71 21.10 Exterior Mouldings, Astragals	Crew	Daily Output	Labor-Hours	Unit	Material	2010 Bare Costs Labor	Equipment	Total	Total Incl O&P
1200	Nail on, "T" extrusion	1 Carp	120	.067	L.F.	.90	2.77		3.67	5.55
1300	Vinyl bulb insert		105	.076		1.30	3.17		4.47	6.65
1600	Screw on, "T" extrusion		90	.089		5.15	3.69		8.84	11.70
1700	Vinyl insert		75	.107		3.45	4.43		7.88	11.10
2000	"L" extrusion, neoprene bulbs		75	.107		2.05	4.43		6.48	9.55
2100	Neoprene sponge insert		75	.107		6	4.43		10.43	13.90
2200	Magnetic		75	.107		9.90	4.43		14.33	18.20
2400	Spring hinged security seal, with cam		75	.107		6.40	4.43		10.83	14.35
2600	Spring loaded locking bolt, vinyl insert		45	.178		8.65	7.40		16.05	21.50
2800	Neoprene sponge strip, "Z" shaped, aluminum		60	.133		4.20	5.55		9.75	13.70
2900	Solid neoprene strip, nail on aluminum strip		90	.089		3.45	3.69		7.14	9.85
3000	One piece stile protection									
3020	Neoprene fabric loop, nail on aluminum strips	1 Carp	60	.133	L.F.	.80	5.55		6.35	10
3110	Flush mounted aluminum extrusion, 1/2" x 1-1/4"		60	.133		3.32	5.55		8.87	12.75
3140	3/4" x 1-3/8"		60	.133		4.22	5.55		9.77	13.75
3160	1-1/8" x 1-3/4"		60	.133		6.65	5.55		12.20	16.40
3300	Mortise, 9/16" x 3/4"		60	.133		3.60	5.55		9.15	13.05
3320	13/16" x 1-3/8"		60	.133		3.85	5.55		9.40	13.35
3600	Spring bronze strip, nail on type		105	.076		3.10	3.17		6.27	8.60
3620	Screw on, with retainer		75	.107		2.50	4.43		6.93	10.05
3800	Flexible stainless steel housing, pile insert, 1/2" door		105	.076		6.80	3.17		9.97	12.70
3820	3/4" door		105	.076		7.65	3.17		10.82	13.60
4000	Extruded aluminum retainer, flush mount, pile insert		105	.076		2.41	3.17		5.58	7.85
4080	Mortise, felt insert		90	.089		4.28	3.69		7.97	10.75
4160	Mortise with spring, pile insert		90	.089		3.35	3.69		7.04	9.75
4400	Rigid vinyl retainer, mortise, pile insert		105	.076		2.40	3.17		5.57	7.85
4600	Wool pile filler strip, aluminum backing		105	.076		2.45	3.17		5.62	7.90
5000	Two piece overlapping astragal, extruded aluminum retainer									
5010	Pile insert	1 Carp	60	.133	L.F.	3.10	5.55		8.65	12.50
5020	Vinyl bulb insert		60	.133		2.05	5.55		7.60	11.35
5040	Vinyl flap insert		60	.133		6.20	5.55		11.75	15.95
5060	Solid neoprene flap insert		60	.133		6.15	5.55		11.70	15.85
5080	Hypalon rubber flap insert		60	.133		6.30	5.55		11.85	16
5090	Snap on cover, pile insert		60	.133		7.15	5.55		12.70	16.95
5400	Magnetic aluminum, surface mounted		60	.133		24.50	5.55		30.05	36
5500	Interlocking aluminum, 5/8" x 1" neoprene bulb insert		45	.178		3.91	7.40		11.31	16.45
5600	Adjustable aluminum, 9/16" x 21/32", pile insert		45	.178		18.20	7.40		25.60	32
5790	For vinyl bulb, deduct					.60			.60	.66
5800	Magnetic, adjustable, 9/16" x 21/32"	1 Carp	45	.178		23.50	7.40		30.90	37.50
6000	Two piece stile protection									
6010	Cloth backed rubber loop, 1" gap, nail on aluminum strips	1 Carp	45	.178	L.F.	4	7.40		11.40	16.55
6040	Screw on aluminum strips		45	.178		6.15	7.40		13.55	18.90
6100	1-1/2" gap, screw on aluminum extrusion		45	.178		5.50	7.40		12.90	18.20
6240	Vinyl fabric loop, slotted aluminum extrusion, 1" gap		45	.178		2	7.40		9.40	14.35
6300	1-1/4" gap		45	.178		5.80	7.40		13.20	18.55

08 71 25 – Weatherstripping

08 71 25.10 Mechanical Seals, Weatherstripping

		Crew	Daily Output	Labor-Hours	Unit	Material	Labor	Equipment	Total	Total Incl O&P
0010	**MECHANICAL SEALS, WEATHERSTRIPPING**									
1000	Doors, wood frame, interlocking, for 3' x 7' door, zinc	1 Carp	3	2.667	Opng.	15.70	111		126.70	199
1100	Bronze		3	2.667		25	111		136	210
1300	6' x 7' opening, zinc		2	4		17	166		183	292
1400	Bronze		2	4		32	166		198	310

08 71 Door Hardware

08 71 25 – Weatherstripping

08 71 25.10 Mechanical Seals, Weatherstripping

		Crew	Daily Output	Labor-Hours	Unit	Material	2010 Bare Costs Labor	Equipment	Total	Total Incl O&P
1700	Wood frame, spring type, bronze									
1800	3' x 7' door	1 Carp	7.60	1.053	Opng.	22	43.50		65.50	96
1900	6' x 7' door	"	7	1.143	"	28	47.50		75.50	109
2200	Metal frame, spring type, bronze									
2300	3' x 7' door	1 Carp	3	2.667	Opng.	36.50	111		147.50	223
2400	6' x 7' door	"	2.50	3.200	"	46.50	133		179.50	269
2500	For stainless steel, spring type, add					133%				
2700	Metal frame, extruded sections, 3' x 7' door, aluminum	1 Carp	3	2.667	Opng.	45	111		156	232
2800	Bronze		3	2.667		114	111		225	305
3100	6' x 7' door, aluminum		1.50	5.333		57.50	222		279.50	430
3200	Bronze		1.50	5.333		136	222		358	515
3500	Threshold weatherstripping									
3650	Door sweep, flush mounted, aluminum	1 Carp	25	.320	Ea.	13.50	13.30		26.80	37
3700	Vinyl		25	.320		15.90	13.30		29.20	39.50
5000	Garage door bottom weatherstrip, 12' aluminum, clear		14	.571		21.50	23.50		45	62.50
5010	Bronze		14	.571		83	23.50		106.50	131
5050	Bottom protection, Rubber		14	.571		22	23.50		45.50	63.50
5100	Threshold		14	.571		90.50	23.50		114	139
9000	Minimum labor/equipment charge		3	2.667	Job		111		111	182

08 74 Access Control Hardware

08 74 13 – Card Key Access Control Hardware

08 74 13.50 Card Key Access

		Crew	Daily Output	Labor-Hours	Unit	Material	2010 Bare Costs Labor	Equipment	Total	Total Incl O&P
0010	**CARD KEY ACCESS**, does not include door hardware									
0020	Card type, 1 time zone, minimum				Ea.	750			750	820
0040	Maximum					1,775			1,775	1,950
0060	3 time zones, minimum					2,200			2,200	2,425
0080	Maximum					2,450			2,450	2,700
0100	System with printer, and control console, 3 zones				Total	9,150			9,150	10,100
0120	6 zones				"	12,000			12,000	13,200
0140	For each door, minimum, add				Ea.	1,400			1,400	1,550
0160	Maximum, add				"	2,000			2,000	2,200

08 74 19 – Biometric Identity Access Control Hardware

08 74 19.50 Biometric Identity Access

		Crew	Daily Output	Labor-Hours	Unit	Material	2010 Bare Costs Labor	Equipment	Total	Total Incl O&P
0010	**BIOMETRIC IDENTITY ACCESS**									
0220	Hand geometry scanner, mem of 512 users, excl striker/powr	1 Elec	3	2.667	Ea.	2,025	131		2,156	2,425
0230	Memory upgrade for, adds 9,700 user profiles		8	1		264	49		313	365
0240	Adds 32,500 user profiles		8	1		580	49		629	715
0250	Prison type, memory of 256 users, excl striker, power		3	2.667		2,600	131		2,731	3,050
0260	Memory upgrade for, adds 3,300 user profiles		8	1		211	49		260	310
0270	Adds 9,700 user profiles		8	1		425	49		474	540
0280	Adds 27,900 user profiles		8	1		580	49		629	715
0290	All weather, mem of 512 users, excl striker/pwr		3	2.667		3,900	131		4,031	4,475
0300	Facial & fingerprint scanner, combination unit, excl striker/power		3	2.667		4,200	131		4,331	4,825
0310	Access for, for initial setup, excl striker/power		3	2.667		1,000	131		1,131	1,300

08 75 Window Hardware

08 75 10 – Window Handles and Latches

08 75 10.10 Handles and Latches		Crew	Daily Output	Labor-Hours	Unit	Material	2010 Bare Costs Labor	2010 Bare Costs Equipment	Total	Total Incl O&P
0010	**HANDLES AND LATCHES**									
1000	Handles, surface mounted, aluminum	1 Carp	24	.333	Ea.	3.98	13.85		17.83	27.50
1020	Brass		24	.333		7.40	13.85		21.25	31
1040	Chrome		24	.333		6.20	13.85		20.05	30
1500	Recessed, aluminum		12	.667		2.19	27.50		29.69	48
1520	Brass		12	.667		2.73	27.50		30.23	48.50
1540	Chrome		12	.667		2.26	27.50		29.76	48
2000	Latches, aluminum		20	.400		2.89	16.60		19.49	30.50
2020	Brass		20	.400		3.97	16.60		20.57	32
2040	Chrome		20	.400		2.47	16.60		19.07	30
9000	Minimum labor/equipment charge		6	1.333	Job		55.50		55.50	91

08 75 30 – Weatherstripping

08 75 30.10 Mechanical Weather Seals		Crew	Daily Output	Labor-Hours	Unit	Material	Labor	Equipment	Total	Total Incl O&P
0010	**MECHANICAL WEATHER SEALS**, Window, double hung, 3' X 5'									
0020	Zinc	1 Carp	7.20	1.111	Opng.	12.80	46		58.80	90
0100	Bronze		7.20	1.111		27	46		73	106
0200	Vinyl V strip		7	1.143		4.95	47.50		52.45	83.50
0500	As above but heavy duty, zinc		4.60	1.739		17.60	72.50		90.10	138
0600	Bronze		4.60	1.739		30	72.50		102.50	152
9000	Minimum labor/equipment charge	1 Clab	4.60	1.739	Job		57.50		57.50	94.50

08 79 Hardware Accessories

08 79 20 – Door Accessories

08 79 20.10 Door Hardware Accessories		Crew	Daily Output	Labor-Hours	Unit	Material	Labor	Equipment	Total	Total Incl O&P
0010	**DOOR HARDWARE ACCESSORIES**									
0140	Door bolt, surface, 4"	1 Carp	32	.250	Ea.	9	10.40		19.40	27
0160	Door latch	"	12	.667	"	8.35	27.50		35.85	54.50
0200	Sliding closet door									
0220	Track and hanger, single	1 Carp	10	.800	Ea.	54	33		87	114
0240	Double		8	1		73	41.50		114.50	149
0260	Door guide, single		48	.167		24	6.95		30.95	38
0280	Double		48	.167		33	6.95		39.95	48
0600	Deadbolt and lock cover plate, brass or stainless steel		30	.267		27	11.10		38.10	47.50
0620	Hole cover plate, brass or chrome		35	.229		7.15	9.50		16.65	23.50
2240	Mortise lockset, passage, lever handle		9	.889		194	37		231	274
4000	Security chain, standard		18	.444		7.90	18.45		26.35	39

08 81 Glass Glazing

08 81 10 – Float Glass

08 81 10.10 Various Types and Thickness of Float Glass		Crew	Daily Output	Labor-Hours	Unit	Material	Labor	Equipment	Total	Total Incl O&P
0010	**VARIOUS TYPES AND THICKNESS OF FLOAT GLASS**									
0020	3/16" Plain	2 Glaz	130	.123	S.F.	4.58	4.95		9.53	13
0200	Tempered, clear		130	.123		6.15	4.95		11.10	14.75
0300	Tinted		130	.123		7.25	4.95		12.20	15.90
0600	1/4" thick, clear, plain		120	.133		5.35	5.35		10.70	14.50
0700	Tinted		120	.133		7.40	5.35		12.75	16.75
0800	Tempered, clear		120	.133		7.60	5.35		12.95	16.95
0900	Tinted		120	.133		9.90	5.35		15.25	19.50
1600	3/8" thick, clear, plain		75	.213		9.35	8.60		17.95	24

08 81 Glass Glazing

08 81 10 – Float Glass

08 81 10.10 Various Types and Thickness of Float Glass

		Crew	Daily Output	Labor-Hours	Unit	Material	2010 Bare Costs Labor	Equipment	Total	Total Incl O&P
1700	Tinted	2 Glaz	75	.213	S.F.	14.80	8.60		23.40	30
1800	Tempered, clear		75	.213		15.40	8.60		24	30.50
1900	Tinted		75	.213		17.15	8.60		25.75	32.50
2200	1/2" thick, clear, plain		55	.291		18.50	11.70		30.20	39.50
2300	Tinted		55	.291		26	11.70		37.70	47.50
2400	Tempered, clear		55	.291		22.50	11.70		34.20	44
2500	Tinted		55	.291		26.50	11.70		38.20	48
2800	5/8" thick, clear, plain		45	.356		26	14.30		40.30	51.50
2900	Tempered, clear		45	.356		29.50	14.30		43.80	55.50
3200	3/4" thick, clear, plain		35	.457		33.50	18.40		51.90	66
3300	Tempered, clear		35	.457		39	18.40		57.40	72.50
3600	1" thick, clear, plain		30	.533		55.50	21.50		77	95.50
8900	For low emissivity coating for 3/16" & 1/4" only, add to above					16%				
9000	Minimum labor/equipment charge	1 Glaz	2	4	Job		161		161	258

08 81 17 – Fire Glass

08 81 17.10 Fire Resistant Glass

		Crew	Daily Output	Labor-Hours	Unit	Material	Labor	Equipment	Total	Total Incl O&P
0010	**FIRE RESISTANT GLASS**									
0020	Fire Glass Minimum	2 Glaz	40	.400	S.F.	25	16.10		41.10	53.50
0030	Mid Range		40	.400		65	16.10		81.10	97.50
0050	High End		40	.400		300	16.10		316.10	355

08 81 20 – Vision Panels

08 81 20.10 Full Vision

		Crew	Daily Output	Labor-Hours	Unit	Material	Labor	Equipment	Total	Total Incl O&P
0010	**FULL VISION**, window system with 3/4" glass mullions									
0020	Up to 10' high	H-2	130	.185	S.F.	59	7		66	76
0100	10' to 20' high, minimum		110	.218		62	8.25		70.25	82
0150	Average		100	.240		67	9.10		76.10	88.50
0200	Maximum		80	.300		75.50	11.35		86.85	101
9000	Minimum labor/equipment charge	1 Glaz	2	4	Job		161		161	258

08 81 25 – Glazing Variables

08 81 25.10 Applications of Glazing

			Crew	Daily Output	Labor-Hours	Unit	Material	Labor	Equipment	Total	Total Incl O&P
0010	**APPLICATIONS OF GLAZING**	R088110-10									
0500	For high rise glazing, exterior, add per S.F. per story					S.F.					.48
0600	For glass replacement, add					"		100%			
0700	For gasket settings, add					L.F.	5.55			5.55	6.10
0900	For sloped glazing, add					S.F.		25%			
2000	Fabrication, polished edges, 1/4" thick					Inch	.48			.48	.53
2100	1/2" thick						1.21			1.21	1.33
2500	Mitered edges, 1/4" thick						1.21			1.21	1.33
2600	1/2" thick						1.96			1.96	2.16

08 81 30 – Insulating Glass

08 81 30.10 Reduce Heat Transfer Glass

			Crew	Daily Output	Labor-Hours	Unit	Material	Labor	Equipment	Total	Total Incl O&P
0010	**REDUCE HEAT TRANSFER GLASS**										
0015	2 lites 1/8" float, /2" thk under 15 S.F.										
0020	Clear	R088110-10	G	2 Glaz	95	.168	S.F.	9.25	6.75		16	21
0100	Tinted		G		95	.168		12.90	6.75		19.65	25
0200	2 lites 3/16" float, for 5/8" thk unit, 15 to 30 S.F., clear		G		90	.178		12.95	7.15		20.10	26
0400	1" thk, dbl. glazed, 1/4" float, 30-70 S.F., clear		G		75	.213		15.50	8.60		24.10	31
0500	Tinted		G		75	.213		22	8.60		30.60	38.50
2000	Both lites, light & heat reflective		G		85	.188		29.50	7.55		37.05	44.50
2500	Heat reflective, film inside, 1" thick unit, clear		G		85	.188		26	7.55		33.55	40.50

08 81 Glass Glazing

08 81 30 – Insulating Glass

08 81 30.10 Reduce Heat Transfer Glass

		Crew	Daily Output	Labor-Hours	Unit	Material	2010 Bare Costs Labor	Equipment	Total	Total Incl O&P
2600	Tinted	2 Glaz G	85	.188	S.F.	28	7.55		35.55	43
3000	Film on weatherside, clear, 1/2" thick unit	G	95	.168		18.50	6.75		25.25	31.50
3100	5/8" thick unit	G	90	.178		18.50	7.15		25.65	32
3200	1" thick unit	G	85	.188		25.50	7.55		33.05	40
3350	Minimum	G 1 Glaz	50	.160		11.90	6.45		18.35	23.50
3360	Maximum	G "	25	.320		12.70	12.85		25.55	34.50
3370	Reflective or tinted, add	G				3.80			3.80	4.18
5000	Spectrally selective film, on ext, blocks solar gain/allows 70% of light	G 2 Glaz	95	.168		9.80	6.75		16.55	21.50
9000	Minimum labor/equipment charge	1 Glaz	2	4	Job		161		161	258

08 81 40 – Plate Glass

08 81 40.10 Plate Glass

0010	**PLATE GLASS** Twin ground, polished,									
0020	3/16" thick	2 Glaz	100	.160	S.F.	4.83	6.45		11.28	15.65
0100	1/4" thick		94	.170		6.60	6.85		13.45	18.30
0200	3/8" thick		60	.267		11.40	10.70		22.10	30
0300	1/2" thick		40	.400		22	16.10		38.10	50.50

08 81 55 – Window Glass

08 81 55.10 Sheet Glass

0010	**SHEET GLASS** (window), clear float, stops, putty bed									
0015	1/8" thick, clear float	2 Glaz	480	.033	S.F.	4.35	1.34		5.69	6.95
0500	3/16" thick, clear		480	.033		5.35	1.34		6.69	8.05
0600	Tinted		480	.033		7.15	1.34		8.49	10
0700	Tempered		480	.033		8.65	1.34		9.99	11.65
2000	Replace broken window lite, 1/8" glass (9 S.F. maximum)	1 Glaz	48	.167		4.65	6.70		11.35	15.85
2100	1/4" plate (16 S.F. maximum)	"	48	.167		5.50	6.70		12.20	16.80
9000	Minimum labor/equipment charge	2 Glaz	5	3.200	Job		129		129	207

08 81 65 – Wire Glass

08 81 65.10 Glass Reinforced With Wire

0010	**GLASS REINFORCED WITH WIRE**									
0012	1/4" thick rough obscure	2 Glaz	135	.119	S.F.	14.10	4.76		18.86	23
1000	Polished wire, 1/4" thick, diamond, clear		135	.119		19.40	4.76		24.16	29
1500	Pinstripe, obscure		135	.119		21	4.76		25.76	30.50

08 83 Mirrors

08 83 13 – Mirrored Glass Glazing

08 83 13.10 Mirrors

0010	**MIRRORS**, No frames, wall type, 1/4" plate glass, polished edge									
0100	Up to 5 S.F.	2 Glaz	125	.128	S.F.	9.10	5.15		14.25	18.25
0200	Over 5 S.F.		160	.100		8.80	4.02		12.82	16.15
0500	Door type, 1/4" plate glass, up to 12 S.F.		160	.100		8	4.02		12.02	15.25
1000	Float glass, up to 10 S.F., 1/8" thick		160	.100		5.25	4.02		9.27	12.20
1100	3/16" thick		150	.107		6.35	4.29		10.64	13.90
1500	12" x 12" wall tiles, square edge, clear		195	.082		1.90	3.30		5.20	7.40
1600	Veined		195	.082		4.86	3.30		8.16	10.65
2000	1/4" thick, stock sizes, one way transparent		125	.128		18	5.15		23.15	28
2010	Bathroom, unframed, laminated		160	.100		15	4.02		19.02	23
2500	Tempered		160	.100		16.65	4.02		20.67	25

08 83 13.15 Reflective Glass

0010	**REFLECTIVE GLASS**									

08 83 Mirrors

08 83 13 – Mirrored Glass Glazing

08 83 13.15 Reflective Glass		Crew	Daily Output	Labor-Hours	Unit	Material	2010 Bare Costs Labor	Equipment	Total	Total Incl O&P
0100	1/4" float with fused metallic oxide fixed	G 2 Glaz	115	.139	S.F.	15.40	5.60		21	26
0500	1/4" float glass with reflective applied coating	G "	115	.139	"	12.85	5.60		18.45	23

08 84 Plastic Glazing

08 84 10 – Plexiglass Glazing

08 84 10.10 Plexiglass Acrylic

		Crew	Daily Output	Labor-Hours	Unit	Material	Labor	Equipment	Total	Total Incl O&P
0010	**PLEXIGLASS ACRYLIC**, clear, masked,									
0020	1/8" thick, cut sheets	2 Glaz	170	.094	S.F.	5	3.78		8.78	11.60
0200	Full sheets		195	.082		2.60	3.30		5.90	8.15
0500	1/4" thick, cut sheets		165	.097		8.70	3.90		12.60	15.80
0600	Full sheets		185	.086		4.75	3.48		8.23	10.85
0900	3/8" thick, cut sheets		155	.103		15.90	4.15		20.05	24
1000	Full sheets		180	.089		8.60	3.57		12.17	15.20
1300	1/2" thick, cut sheets		135	.119		18.30	4.76		23.06	27.50
1400	Full sheets		150	.107		17.80	4.29		22.09	26.50
1700	3/4" thick, cut sheets		115	.139		65	5.60		70.60	80.50
1800	Full sheets		130	.123		37.50	4.95		42.45	49.50
2100	1" thick, cut sheets		105	.152		73.50	6.15		79.65	90.50
2200	Full sheets		125	.128		45.50	5.15		50.65	58.50
3000	Colored, 1/8" thick, cut sheets		170	.094		15.10	3.78		18.88	22.50
3200	Full sheets		195	.082		9.80	3.30		13.10	16.10
3500	1/4" thick, cut sheets		165	.097		17	3.90		20.90	25
3600	Full sheets		185	.086		11.55	3.48		15.03	18.30
4000	Mirrors, untinted, cut sheets, 1/8" thick		185	.086		7.10	3.48		10.58	13.40
4200	1/4" thick		180	.089		11.50	3.57		15.07	18.40

08 84 20 – Polycarbonate

08 84 20.10 Thermoplastic

		Crew	Daily Output	Labor-Hours	Unit	Material	Labor	Equipment	Total	Total Incl O&P
0010	**THERMOPLASTIC**, clear, masked, cut sheets									
0020	1/8" thick	2 Glaz	170	.094	S.F.	9.70	3.78		13.48	16.75
0500	3/16" thick		165	.097		11.10	3.90		15	18.45
1000	1/4" thick		155	.103		12.50	4.15		16.65	20.50
1500	3/8" thick		150	.107		22	4.29		26.29	31.50
9000	Minimum labor/equipment charge	1 Glaz	2	4	Job		161		161	258

08 85 Glazing Accessories

08 85 10 – Miscellaneous Glazing Accessories

08 85 10.10 Glazing Gaskets

		Crew	Daily Output	Labor-Hours	Unit	Material	Labor	Equipment	Total	Total Incl O&P
0010	**GLAZING GASKETS**, Neoprene for glass tongued mullion									
0015	1/4" glass	2 Glaz	200	.080	L.F.	.65	3.22		3.87	5.85
0020	3/8"		200	.080		.83	3.22		4.05	6.05
0040	1/2"		200	.080		.94	3.22		4.16	6.20
0060	3/4"		180	.089		1.36	3.57		4.93	7.25
0080	1"		180	.089		1.11	3.57		4.68	6.95
1000	Glazing compound, wood	1 Glaz	58	.138		.82	5.55		6.37	9.80
1005	Glazing compound, per window, up to 30 L.F.					.82			.82	.90
1006	Glazing compound, per window, up to 30 L.F. E.A.				Ea.	24.50			24.50	27
1020	Metal	1 Glaz	58	.138	L.F.	.82	5.55		6.37	9.80

08 87 Glazing Surface Films

08 87 13 – Solar Control Films

08 87 13.10 Solar Films On Glass		Crew	Daily Output	Labor-Hours	Unit	Material	2010 Bare Costs Labor	Equipment	Total	Total Incl O&P
0010	**SOLAR FILMS ON GLASS** (glass not included)									
2000	Minimum	G	2 Glaz	180	.089	S.F.	6.20	3.57		9.77	12.60
2050	Maximum	G	"	225	.071	"	14.40	2.86		17.26	20.50

08 87 53 – Security Films

08 87 53.10 Security Film

			Crew	Daily Output	Labor-Hours	Unit	Material	Labor	Equipment	Total	Total Incl O&P
0010	**SECURITY FILM**, clear, 32000psi tensile strength, adhered to glass	R088110-10									
0100	.002" thick, daylight installation		H-2	950	.025	S.F.	1.85	.96		2.81	3.58
0150	.004" thick, daylight installation			800	.030		2	1.13		3.13	4.03
0200	.006" thick, daylight installation			700	.034		2.15	1.30		3.45	4.47
0210	Install for anchorage			600	.040		2.39	1.51		3.90	5.10
0400	.007" thick, daylight installation			600	.040		2.25	1.51		3.76	4.93
0410	Install for anchorage			500	.048		2.50	1.82		4.32	5.70
0500	.008" thick, daylight installation			500	.048		2.65	1.82		4.47	5.85
0510	Install for anchorage			500	.048		2.94	1.82		4.76	6.20
0600	.015" thick, daylight installation			400	.060		4.40	2.27		6.67	8.50
0610	Install for anchorage		▼	400	.060	▼	2.94	2.27		5.21	6.90
0900	Security film anchorage, mechanical attachment and cover plate		H-3	370	.043	L.F.	8.70	1.55		10.25	12.10
0950	Security film anchorage, wet glaze structural caulking		1 Glaz	225	.036	"	.83	1.43		2.26	3.21
1000	Adhered security film removal		1 Clab	275	.029	S.F.		.96		.96	1.58

08 88 Special Function Glazing

08 88 56 – Ballistics-Resistant Glazing

08 88 56.10 Laminated Glass

		Crew	Daily Output	Labor-Hours	Unit	Material	Labor	Equipment	Total	Total Incl O&P
0010	**LAMINATED GLASS**									
0020	Clear float .03" vinyl 1/4"	2 Glaz	90	.178	S.F.	11.35	7.15		18.50	24
0100	3/8" thick		78	.205		20	8.25		28.25	36
0200	.06" vinyl, 1/2" thick		65	.246		23.50	9.90		33.40	42
1000	5/8" thick		90	.178		28	7.15		35.15	42
2000	Bullet-resisting, 1-3/16" thick, to 15 S.F.		16	1		77	40		117	150
2100	Over 15 S.F.		16	1		94	40		134	168
2500	2-1/4" thick, to 15 S.F.		12	1.333		91	53.50		144.50	186
2600	Over 15 S.F.		12	1.333		81	53.50		134.50	175
2700	Level 2 (.357 magnum)		12	1.333		63	53.50		116.50	156
2750	Level 3A (.44 magnum)		12	1.333		68	53.50		121.50	161
2800	Level 4 (AK-47)		12	1.333		89	53.50		142.50	184
2850	Level 5 (M-16)		12	1.333		99	53.50		152.50	195
2900	Level 3 (7.62 Armor Piercing)	▼	12	1.333	▼	118	53.50		171.50	216

08 91 Louvers

08 91 19 – Fixed Louvers

08 91 19.10 Aluminum Louvers

		Crew	Daily Output	Labor-Hours	Unit	Material	Labor	Equipment	Total	Total Incl O&P
0010	**ALUMINUM LOUVERS**									
0020	Aluminum with screen, residential, 8" x 8"	1 Carp	38	.211	Ea.	10.95	8.75		19.70	26.50
0100	12" x 12"		38	.211		14	8.75		22.75	30
0200	12" x 18"		35	.229		13.95	9.50		23.45	31
0250	14" x 24"		30	.267		25	11.10		36.10	45.50
0300	18" x 24"		27	.296		27	12.30		39.30	49.50
0500	24" x 30"		24	.333		32	13.85		45.85	58
0700	Triangle, adjustable, small	▼	20	.400	▼	32.50	16.60		49.10	63.50

08 91 Louvers

08 91 19 – Fixed Louvers

08 91 19.10 Aluminum Louvers		Crew	Daily Output	Labor-Hours	Unit	Material	2010 Bare Costs Labor	Equipment	Total	Total Incl O&P
0800	Large	1 Carp	15	.533	Ea.	40.50	22		62.50	81
2100	Midget, aluminum, 3/4" deep, 1" diameter		85	.094		.73	3.91		4.64	7.20
2150	3" diameter		60	.133		1.78	5.55		7.33	11.05
2200	4" diameter		50	.160		3.64	6.65		10.29	14.90
2250	6" diameter		30	.267		3.74	11.10		14.84	22.50

08 95 Vents

08 95 13 – Soffit Vents

08 95 13.10 Wall Louvers

		Crew	Daily Output	Labor-Hours	Unit	Material	Labor	Equipment	Total	Total Incl O&P
0010	**WALL LOUVERS**									
2330	Soffit vent, continuous, 3" wide, aluminum, mill finish	1 Carp	200	.040	L.F.	.48	1.66		2.14	3.26
2340	Baked enamel finish		200	.040	"	.52	1.66		2.18	3.30
2400	Under eaves vent, aluminum, mill finish, 16" x 4"		48	.167	Ea.	1.33	6.95		8.28	12.85
2500	16" x 8"		48	.167	"	1.54	6.95		8.49	13.10

08 95 16 – Wall Vents

08 95 16.10 Louvers

		Crew	Daily Output	Labor-Hours	Unit	Material	Labor	Equipment	Total	Total Incl O&P
0010	**LOUVERS**									
0020	Redwood, 2'-0" diameter, full circle	1 Carp	16	.500	Ea.	169	21		190	220
0100	Half circle		16	.500		189	21		210	242
0200	Octagonal		16	.500		129	21		150	176
0300	Triangular, 5/12 pitch, 5'-0" at base		16	.500		189	21		210	242
7000	Vinyl gable vent, 8" x 8"		38	.211		11.55	8.75		20.30	27
7020	12" x 12"		38	.211		24	8.75		32.75	41
7080	12" x 18"		35	.229		31	9.50		40.50	49.50
7200	18" x 24"		30	.267		37.50	11.10		48.60	59.50
9000	Minimum labor/equipment charge		3.50	2.286	Job		95		95	156

Division Notes

	CREW	DAILY OUTPUT	LABOR-HOURS	UNIT	2010 BARE COSTS				TOTAL INCL O&P
					MAT.	LABOR	EQUIP.	TOTAL	

Estimating Tips

General

- Room Finish Schedule: A complete set of plans should contain a room finish schedule. If one is not available, it would be well worth the time and effort to obtain one.

09 20 00 Plaster and Gypsum Board

- Lath is estimated by the square yard plus a 5% allowance for waste. Furring, channels, and accessories are measured by the linear foot. An extra foot should be allowed for each accessory miter or stop.
- Plaster is also estimated by the square yard. Deductions for openings vary by preference, from zero deduction to 50% of all openings over 2 feet in width. The estimator should allow one extra square foot for each linear foot of horizontal interior or exterior angle located below the ceiling level. Also, double the areas of small radius work.
- Drywall accessories, studs, track, and acoustical caulking are all measured by the linear foot. Drywall taping is figured by the square foot. Gypsum wallboard is estimated by the square foot. No material deductions should be made for door or window openings under 32 S.F.

09 60 00 Flooring

- Tile and terrazzo areas are taken off on a square foot basis. Trim and base materials are measured by the linear foot. Accent tiles are listed per each. Two basic methods of installation are used. Mud set is approximately 30% more expensive than thin set. In terrazzo work, be sure to include the linear footage of embedded decorative strips, grounds, machine rubbing, and power cleanup.
- Wood flooring is available in strip, parquet, or block configuration. The latter two types are set in adhesives with quantities estimated by the square foot. The laying pattern will influence labor costs and material waste. In addition to the material and labor for laying wood floors, the estimator must make allowances for sanding and finishing these areas unless the flooring is prefinished.
- Sheet flooring is measured by the square yard. Roll widths vary, so consideration should be given to use the most economical width, as waste must be figured into the total quantity. Consider also the installation methods available, direct glue down or stretched.

09 70 00 Wall Finishes

- Wall coverings are estimated by the square foot. The area to be covered is measured, length by height of wall above baseboards, to calculate the square footage of each wall. This figure is divided by the number of square feet in the single roll which is being used. Deduct, in full, the areas of openings such as doors and windows. Where a pattern match is required allow 25%–30% waste.

09 80 00 Acoustic Treatment

- Acoustical systems fall into several categories. The takeoff of these materials should be by the square foot of area with a 5% allowance for waste. Do not forget about scaffolding, if applicable, when estimating these systems.

09 90 00 Painting and Coating

- A major portion of the work in painting involves surface preparation. Be sure to include cleaning, sanding, filling, and masking costs in the estimate.
- Protection of adjacent surfaces is not included in painting costs. When considering the method of paint application, an important factor is the amount of protection and masking required. These must be estimated separately and may be the determining factor in choosing the method of application.

Reference Numbers

Reference numbers are shown in shaded boxes at the beginning of some major classifications. These numbers refer to related items in the Reference Section. The reference information may be an estimating procedure, an alternate pricing method, or technical information.

Note: Not all subdivisions listed here necessarily appear in this publication.

No part of this publication may be reproduced, stored in a retrieval system, or transmitted in any form or by any means without prior written permission of R.S. Means Company, Inc.

09 01 Maintenance of Finishes

09 01 70 – Maintenance of Wall Finishes

09 01 70.10 Gypsum Wallboard Repairs

		Crew	Daily Output	Labor-Hours	Unit	Material	2010 Bare Costs Labor	2010 Bare Costs Equipment	Total	Total Incl O&P
0010	**GYPSUM WALLBOARD REPAIRS**									
0100	Fill and sand, pin/nail holes	1 Carp	960	.008	Ea.		.35		.35	.57
0110	Screw head pops		480	.017			.69		.69	1.14
0120	Dents, up to 2" square		48	.167		.01	6.95		6.96	11.40
0130	2" to 4" square		24	.333		.03	13.85		13.88	23
0140	Cut square, patch, sand and finish, holes, up to 2" square		12	.667		.04	27.50		27.54	45.50
0150	2" to 4" square		11	.727		.09	30		30.09	49.50
0160	4" to 8" square		10	.800		.22	33		33.22	54.50
0170	8" to 12" square		8	1		.42	41.50		41.92	69
0180	12" to 32" square		6	1.333		2.62	55.50		58.12	94
0210	16" by 48"		5	1.600		2.01	66.50		68.51	111
0220	32" by 48"		4	2		3.36	83		86.36	141
0230	48" square		3.50	2.286		4.73	95		99.73	161
0240	60" square		3.20	2.500		8.30	104		112.30	180
0500	Skim coat surface with joint compound		1600	.005	S.F.	.03	.21		.24	.37
9000	Minimum labor/equipment charge		2	4	Job		166		166	273

09 05 Common Work Results for Finishes

09 05 05 – Selective Finishes Demolition

09 05 05.10 Selective Demolition, Ceilings

			Crew	Daily Output	Labor-Hours	Unit	Material	Labor	Equipment	Total	Total Incl O&P
0010	**SELECTIVE DEMOLITION, CEILINGS**	R024119-10									
0200	Ceiling, drywall, furred and nailed or screwed		2 Clab	800	.020	S.F.		.66		.66	1.09
0220	On metal frame			760	.021			.70		.70	1.14
0240	On suspension system, including system			720	.022			.74		.74	1.21
1000	Plaster, lime and horse hair, on wood lath, incl. lath			700	.023			.76		.76	1.24
1020	On metal lath			570	.028			.93		.93	1.53
1100	Gypsum, on gypsum lath			720	.022			.74		.74	1.21
1120	On metal lath			500	.032			1.06		1.06	1.74
1200	Suspended ceiling, mineral fiber, 2' x 2' or 2' x 4'			1500	.011			.35		.35	.58
1250	On suspension system, incl. system			1200	.013			.44		.44	.72
1500	Tile, wood fiber, 12" x 12", glued			900	.018			.59		.59	.97
1540	Stapled			1500	.011			.35		.35	.58
1580	On suspension system, incl. system			760	.021			.70		.70	1.14
2000	Wood, tongue and groove, 1" x 4"			1000	.016			.53		.53	.87
2040	1" x 8"			1100	.015			.48		.48	.79
2400	Plywood or wood fiberboard, 4' x 8' sheets			1200	.013			.44		.44	.72
9000	Minimum labor/equipment charge		1 Clab	2	4	Job		132		132	217

09 05 05.20 Selective Demolition, Flooring

			Crew	Daily Output	Labor-Hours	Unit	Material	Labor	Equipment	Total	Total Incl O&P
0010	**SELECTIVE DEMOLITION, FLOORING**	R024119-10									
0200	Brick with mortar		2 Clab	475	.034	S.F.		1.11		1.11	1.83
0400	Carpet, bonded, including surface scraping			2000	.008			.26		.26	.43
0480	Tackless			9000	.002			.06		.06	.10
0550	Carpet tile, releasable adhesive			5000	.003			.11		.11	.17
0560	Permanent adhesive			1850	.009			.29		.29	.47
0600	Composition, acrylic or epoxy			400	.040			1.32		1.32	2.17
0700	Concrete, scarify skin		A-1A	225	.036			1.51	.96	2.47	3.51
0800	Resilient, sheet goods		2 Clab	1400	.011			.38		.38	.62
0820	For gym floors		"	900	.018			.59		.59	.97
0850	Vinyl or rubber cove base		1 Clab	1000	.008	L.F.		.26		.26	.43
0860	Vinyl or rubber cove base, molded corner		"	1000	.008	Ea.		.26		.26	.43
0870	For glued and caulked installation, add to labor							50%			

09 05 Common Work Results for Finishes

09 05 05 – Selective Finishes Demolition

09 05 05.20 Selective Demolition, Flooring

		Crew	Daily Output	Labor-Hours	Unit	Material	2010 Bare Costs Labor	2010 Bare Costs Equipment	Total	Total Incl O&P
0900	Vinyl composition tile, 12" x 12"	2 Clab	1000	.016	S.F.		.53		.53	.87
2000	Tile, ceramic, thin set		675	.024			.78		.78	1.29
2020	Mud set		625	.026			.85		.85	1.39
2200	Marble, slate, thin set		675	.024			.78		.78	1.29
2220	Mud set		625	.026			.85		.85	1.39
2600	Terrazzo, thin set		450	.036			1.18		1.18	1.93
2620	Mud set		425	.038			1.25		1.25	2.05
2640	Terrazzo, cast in place	▼	300	.053			1.77		1.77	2.90
3000	Wood, block, on end	1 Carp	400	.020			.83		.83	1.37
3200	Parquet		450	.018			.74		.74	1.21
3400	Strip flooring, interior, 2-1/4" x 25/32" thick		325	.025			1.02		1.02	1.68
3500	Exterior, porch flooring, 1" x 4"		220	.036			1.51		1.51	2.48
3800	Subfloor, tongue and groove, 1" x 6"		325	.025			1.02		1.02	1.68
3820	1" x 8"		430	.019			.77		.77	1.27
3840	1" x 10"		520	.015			.64		.64	1.05
4000	Plywood, nailed		600	.013			.55		.55	.91
4100	Glued and nailed		400	.020			.83		.83	1.37
4200	Hardboard, 1/4" thick	▼	760	.011			.44		.44	.72
8000	Remove flooring, bead blast, minimum	A-1A	1000	.008			.34	.22	.56	.79
8100	Maximum		400	.020			.85	.54	1.39	1.97
8150	Mastic only	▼	1500	.005	▼		.23	.14	.37	.53
9000	Minimum labor/equipment charge	1 Clab	4	2	Job		66		66	109

09 05 05.30 Selective Demolition, Walls and Partitions

		Crew	Daily Output	Labor-Hours	Unit	Material	2010 Bare Costs Labor	2010 Bare Costs Equipment	Total	Total Incl O&P
0010	**SELECTIVE DEMOLITION, WALLS AND PARTITIONS** R024119-10									
0020	Walls, concrete, reinforced	B-39	120	.400	C.F.		13.90	1.87	15.77	24.50
0025	Plain	"	160	.300			10.45	1.40	11.85	18.55
0100	Brick, 4" to 12" thick	B-9	220	.182	▼		6.10	1.02	7.12	11.10
0200	Concrete block, 4" thick		1000	.040	S.F.		1.34	.22	1.56	2.45
0280	8" thick		810	.049			1.65	.28	1.93	3.03
0300	Exterior stucco 1" thick over mesh	▼	3200	.013			.42	.07	.49	.77
1000	Drywall, nailed or screwed	1 Clab	1000	.008			.26		.26	.43
1010	2 layers		400	.020			.66		.66	1.09
1020	Glued and nailed		900	.009			.29		.29	.48
1500	Fiberboard, nailed		900	.009			.29		.29	.48
1520	Glued and nailed		800	.010			.33		.33	.54
1568	Plenum barrier, sheet lead		300	.027			.88		.88	1.45
2000	Movable walls, metal, 5' high		300	.027			.88		.88	1.45
2020	8' high	▼	400	.020			.66		.66	1.09
2200	Metal or wood studs, finish 2 sides, fiberboard	B-1	520	.046			1.56		1.56	2.56
2250	Lath and plaster		260	.092			3.12		3.12	5.10
2300	Plasterboard (drywall)		520	.046			1.56		1.56	2.56
2350	Plywood	▼	450	.053			1.80		1.80	2.96
2800	Paneling, 4' x 8' sheets	1 Clab	475	.017			.56		.56	.92
3000	Plaster, lime and horsehair, on wood lath		400	.020			.66		.66	1.09
3020	On metal lath		335	.024			.79		.79	1.30
3400	Gypsum or perlite, on gypsum lath		410	.020			.65		.65	1.06
3420	On metal lath		300	.027	▼		.88		.88	1.45
3450	Plaster, interior gypsum, acoustic, or cement		60	.133	S.Y.		4.41		4.41	7.25
3500	Stucco, on masonry		145	.055			1.83		1.83	3
3510	Commercial 3-coat		80	.100			3.31		3.31	5.45
3520	Interior stucco		25	.320	▼		10.60		10.60	17.40
3760	Tile, ceramic, on walls, thin set		300	.027	S.F.		.88		.88	1.45

09 05 Common Work Results for Finishes

09 05 05 – Selective Finishes Demolition

09 05 05.30 Selective Demolition, Walls and Partitions		Crew	Daily Output	Labor-Hours	Unit	Material	2010 Bare Costs Labor	Equipment	Total	Total Incl O&P
3765	Mud set	1 Clab	250	.032	S.F.		1.06		1.06	1.74
3800	Toilet partitions, slate or marble		5	1.600	Ea.		53		53	87
3820	Metal or plastic	↓	8	1	"		33		33	54.50
5000	Wallcovering, vinyl	1 Pape	700	.011	S.F.		.42		.42	.66
5010	With release agent		1500	.005			.19		.19	.31
5025	Wallpaper, 2 layers or less, by hand		250	.032			1.17		1.17	1.86
5035	3 layers or more		165	.048			1.77		1.77	2.81
5040	Designer	↓	480	.017	↓		.61		.61	.97
9000	Minimum labor/equipment charge	1 Clab	4	2	Job		66		66	109

09 21 Plaster and Gypsum Board Assemblies

09 21 16 – Gypsum Board Assemblies

09 21 16.23 Gypsum Board Shaft Wall Assemblies

		Crew	Daily Output	Labor-Hours	Unit	Material	Labor	Equipment	Total	Total Incl O&P
0010	**GYPSUM BOARD SHAFT WALL ASSEMBLIES**									
0020	Cavity type on 25 ga. J-track & C-H studs, 24" O.C.									
0030	1" thick coreboard wall liner on shaft side									
0040	2-hour assembly with double layer									
0060	5/8" fire rated gypsum board on room side	2 Carp	220	.073	S.F.	1.67	3.02		4.69	6.80
0100	3-hour assembly with triple layer									
0300	5/8" fire rated gypsum board on room side	2 Carp	180	.089	S.F.	1.39	3.69		5.08	7.60
0400	4-hour assembly, 1" coreboard, 5/8" fire rated gypsum board									
0600	and 3/4" galv. metal furring channels, 24" O.C., with									
0700	Double layer 5/8" fire rated gypsum board on room side	2 Carp	110	.145	S.F.	1.30	6.05		7.35	11.40
0900	For taping & finishing, add per side	1 Carp	1050	.008	"	.04	.32		.36	.57
1000	For insulation, see Div. 07 21									
5200	For work over 8' high, add	2 Carp	3060	.005	S.F.		.22		.22	.36
5300	For distribution cost over 3 stories high, add per story	"	6100	.003	"		.11		.11	.18

09 21 16.33 Partition Wall

		Crew	Daily Output	Labor-Hours	Unit	Material	Labor	Equipment	Total	Total Incl O&P
0010	**PARTITION WALL** Stud wall, 8' to 12' high									
0050	1/2", interior, gypsum board, std, tape & finish 2 sides									
0500	Installed on and incl., 2" x 4" wood studs, 16" O.C.	2 Carp	310	.052	S.F.	.84	2.14		2.98	4.45
1000	Metal studs, NLB, 25 ga., 16" O.C., 3-5/8" wide		350	.046		.89	1.90		2.79	4.10
1200	6" wide		330	.048		1.05	2.01		3.06	4.47
1400	Water resistant, on 2" x 4" wood studs, 16" O.C.		310	.052		1.06	2.14		3.20	4.69
1600	Metal studs, NLB, 25 ga., 16" O.C., 3-5/8" wide		350	.046		1.11	1.90		3.01	4.34
1800	6" wide		330	.048		1.27	2.01		3.28	4.71
2000	Fire res., 2 layers, 1-1/2 hr., on 2" x 4" wood studs, 16" O.C.		210	.076		1.48	3.17		4.65	6.85
2200	Metal studs, NLB, 25 ga., 16" O.C., 3-5/8" wide		250	.064		1.53	2.66		4.19	6.05
2400	6" wide		230	.070		1.69	2.89		4.58	6.60
2600	Fire & water res., 2 layers, 1-1/2 hr., 2" x 4" studs, 16" O.C.		210	.076		1.48	3.17		4.65	6.85
2800	Metal studs, NLB, 25 ga., 16" O.C., 3-5/8" wide		250	.064		1.53	2.66		4.19	6.05
3000	6" wide	↓	230	.070	↓	1.69	2.89		4.58	6.60
3200	5/8", interior, gypsum board, std, tape & finish 2 sides									
3400	Installed on and including 2" x 4" wood studs, 16" O.C.	2 Carp	300	.053	S.F.	.92	2.22		3.14	4.66
3600	24" O.C.		330	.048		.86	2.01		2.87	4.26
3800	Metal studs, NLB, 25 ga., 16" O.C., 3-5/8" wide		340	.047		.97	1.96		2.93	4.27
4000	6" wide		320	.050		1.13	2.08		3.21	4.66
4200	24" O.C., 3-5/8" wide		360	.044		.88	1.85		2.73	4
4400	6" wide		340	.047		1	1.96		2.96	4.31
4800	Water resistant, on 2" x 4" wood studs, 16" O.C.		300	.053		1.22	2.22		3.44	4.99
5000	24" O.C.	↓	330	.048	↓	1.16	2.01		3.17	4.59

09 21 Plaster and Gypsum Board Assemblies

09 21 16 – Gypsum Board Assemblies

09 21 16.33 Partition Wall		Crew	Daily Output	Labor-Hours	Unit	Material	2010 Bare Costs Labor	Equipment	Total	Total Incl O&P
5200	Metal studs, NLB, 25 ga. 16" O.C., 3-5/8" wide	2 Carp	340	.047	S.F.	1.27	1.96		3.23	4.60
5400	6" wide		320	.050		1.43	2.08		3.51	4.99
5600	24" O.C., 3-5/8" wide		360	.044		1.18	1.85		3.03	4.33
5800	6" wide		340	.047		1.30	1.96		3.26	4.64
6000	Fire resistant, 2 layers, 2 hr., on 2" x 4" wood studs, 16" O.C.		205	.078		1.38	3.24		4.62	6.85
6200	24" O.C.		235	.068		1.38	2.83		4.21	6.15
6400	Metal studs, NLB, 25 ga., 16" O.C., 3-5/8" wide		245	.065		1.49	2.71		4.20	6.10
6600	6" wide		225	.071		1.65	2.95		4.60	6.65
6800	24" O.C., 3-5/8" wide		265	.060		1.40	2.51		3.91	5.65
7000	6" wide		245	.065		1.52	2.71		4.23	6.15
7200	Fire & water resistant, 2 layers, 2 hr., 2" x 4" studs, 16" O.C.		205	.078		1.44	3.24		4.68	6.95
7400	24" O.C.		235	.068		1.38	2.83		4.21	6.15
7600	Metal studs, NLB, 25 ga., 16" O.C., 3-5/8" wide		245	.065		1.49	2.71		4.20	6.10
7800	6" wide		225	.071		1.65	2.95		4.60	6.65
8000	24" O.C., 3-5/8" wide		265	.060		1.40	2.51		3.91	5.65
8200	6" wide		245	.065		1.52	2.71		4.23	6.15
8600	1/2" blueboard, mesh tape both sides									
8620	Installed on and including 2" x 4" wood studs, 16" O.C.	2 Carp	300	.053	S.F.	.92	2.22		3.14	4.66
8640	Metal studs, NLB, 25 ga. 16" O.C., 3-5/8" wide		340	.047		.97	1.96		2.93	4.27
8660	6" wide		320	.050		1.13	2.08		3.21	4.66
9000	Exterior, 1/2" gypsum sheathing, 1/2" gypsum finished, interior,									
9100	including foil faced insulation, metal studs, 20 ga.									
9200	16" O.C., 3-5/8" wide	2 Carp	290	.055	S.F.	1.61	2.29		3.90	5.55
9400	6" wide		270	.059		1.77	2.46		4.23	6
9600	Partitions, for work over 8' high, add		1530	.010			.43		.43	.71

09 22 Supports for Plaster and Gypsum Board

09 22 03 – Fastening Methods for Finishes

09 22 03.20 Drilling Plaster/Drywall

		Crew	Daily Output	Labor-Hours	Unit	Material	Labor	Equipment	Total	Total Incl O&P
0010	**DRILLING PLASTER/DRYWALL**									
1100	Drilling & layout for drywall/plaster walls, up to 1" deep, no anchor									
1200	Holes, 1/4" diameter	1 Carp	150	.053	Ea.	.01	2.22		2.23	3.65
1300	3/8" diameter		140	.057		.01	2.37		2.38	3.91
1400	1/2" diameter		130	.062		.01	2.56		2.57	4.21
1500	3/4" diameter		120	.067		.01	2.77		2.78	4.57
1600	1" diameter		110	.073		.02	3.02		3.04	4.98
1700	1-1/4" diameter		100	.080		.03	3.32		3.35	5.50
1800	1-1/2" diameter		90	.089		.04	3.69		3.73	6.10
1900	For ceiling installations, add						40%			

09 22 13 – Metal Furring

09 22 13.13 Metal Channel Furring

		Crew	Daily Output	Labor-Hours	Unit	Material	Labor	Equipment	Total	Total Incl O&P
0010	**METAL CHANNEL FURRING**									
0030	Beams and columns, 7/8" channels, galvanized, 12" O.C.	1 Lath	155	.052	S.F.	.34	1.91		2.25	3.37
0050	16" O.C.		170	.047		.28	1.74		2.02	3.04
0070	24" O.C.		185	.043		.19	1.60		1.79	2.71
0100	Ceilings, on steel, 7/8" channels, galvanized, 12" O.C.		210	.038		.31	1.41		1.72	2.55
0300	16" O.C.		290	.028		.28	1.02		1.30	1.91
0400	24" O.C.		420	.019		.19	.70		.89	1.31
0600	1-5/8" channels, galvanized, 12" O.C.		190	.042		.41	1.56		1.97	2.90
0700	16" O.C.		260	.031		.37	1.14		1.51	2.20

09 22 Supports for Plaster and Gypsum Board

09 22 13 – Metal Furring

09 22 13.13 Metal Channel Furring

		Crew	Daily Output	Labor-Hours	Unit	Material	2010 Bare Costs Labor	Equipment	Total	Total Incl O&P
0900	24" O.C.	1 Lath	390	.021	S.F.	.25	.76		1.01	1.46
0930	7/8" channels with sound isolation clips, 12" O.C.		120	.067		1.37	2.46		3.83	5.40
0940	16" O.C.		100	.080		1.91	2.96		4.87	6.75
0950	24" O.C.		165	.048		1.25	1.79		3.04	4.18
0960	1-5/8" channels, galvanized, 12" O.C.		110	.073		1.48	2.69		4.17	5.85
0970	16" O.C.		100	.080		2	2.96		4.96	6.85
0980	24" O.C.		155	.052		1.31	1.91		3.22	4.44
1000	Walls, 7/8" channels, galvanized, 12" O.C.		235	.034		.31	1.26		1.57	2.32
1200	16" O.C.		265	.030		.28	1.12		1.40	2.06
1300	24" O.C.		350	.023		.19	.84		1.03	1.53
1500	1-5/8" channels, galvanized, 12" O.C.		210	.038		.41	1.41		1.82	2.67
1600	16" O.C.		240	.033		.37	1.23		1.60	2.34
1800	24" O.C.		305	.026		.25	.97		1.22	1.79
1920	7/8" channels with sound isolation clips, 12" O.C.		125	.064		1.37	2.36		3.73	5.25
1940	16" O.C.		100	.080		1.91	2.96		4.87	6.75
1950	24" O.C.		150	.053		1.25	1.97		3.22	4.47
1960	1-5/8" channels, galvanized, 12" O.C.		115	.070		1.48	2.57		4.05	5.65
1970	16" O.C.		95	.084		2	3.11		5.11	7.10
1980	24" O.C.		140	.057		1.31	2.11		3.42	4.76
9000	Minimum labor/equipment charge		4	2	Job		74		74	116

09 22 16 – Non-Structural Metal Framing

09 22 16.13 Metal Studs and Track

		Crew	Daily Output	Labor-Hours	Unit	Material	Labor	Equipment	Total	Total Incl O&P
0010	**METAL STUDS AND TRACK**									
1600	Non-load bearing, galv, 8' high, 25 ga. 1-5/8" wide, 16" O.C.	1 Carp	619	.013	S.F.	.25	.54		.79	1.15
1610	24" O.C.		950	.008		.18	.35		.53	.77
1620	2-1/2" wide, 16" O.C.		613	.013		.30	.54		.84	1.22
1630	24" O.C.		938	.009		.22	.35		.57	.83
1640	3-5/8" wide, 16" O.C.		600	.013		.34	.55		.89	1.29
1650	24" O.C.		925	.009		.26	.36		.62	.87
1660	4" wide, 16" O.C.		594	.013		.35	.56		.91	1.30
1670	24" O.C.		925	.009		.26	.36		.62	.87
1680	6" wide, 16" O.C.		588	.014		.51	.57		1.08	1.50
1690	24" O.C.		906	.009		.39	.37		.76	1.02
1700	20 ga. studs, 1-5/8" wide, 16" O.C.		494	.016		.34	.67		1.01	1.49
1710	24" O.C.		763	.010		.26	.44		.70	1
1720	2-1/2" wide, 16" O.C.		488	.016		.39	.68		1.07	1.55
1730	24" O.C.		750	.011		.30	.44		.74	1.06
1740	3-5/8" wide, 16" O.C.		481	.017		.43	.69		1.12	1.61
1750	24" O.C.		738	.011		.32	.45		.77	1.09
1760	4" wide, 16" O.C.		475	.017		.50	.70		1.20	1.70
1770	24" O.C.		738	.011		.38	.45		.83	1.16
1780	6" wide, 16" O.C.		469	.017		.60	.71		1.31	1.81
1790	24" O.C.		725	.011		.45	.46		.91	1.24
2000	Non-load bearing, galv, 10' high, 25 ga. 1-5/8" wide, 16" O.C.		495	.016		.23	.67		.90	1.36
2100	24" O.C.		760	.011		.17	.44		.61	.91
2200	2-1/2" wide, 16" O.C.		490	.016		.28	.68		.96	1.42
2250	24" O.C.		750	.011		.21	.44		.65	.96
2300	3-5/8" wide, 16" O.C.		480	.017		.32	.69		1.01	1.50
2350	24" O.C.		740	.011		.24	.45		.69	1
2400	4" wide, 16" O.C.		475	.017		.33	.70		1.03	1.51
2450	24" O.C.		740	.011		.24	.45		.69	1.01
2500	6" wide, 16" O.C.		470	.017		.49	.71		1.20	1.70

09 22 Supports for Plaster and Gypsum Board

09 22 16 – Non-Structural Metal Framing

09 22 16.13 Metal Studs and Track		Crew	Daily Output	Labor-Hours	Unit	Material	2010 Bare Costs Labor	2010 Bare Costs Equipment	Total	Total Incl O&P
2550	24" O.C.	1 Carp	725	.011	S.F.	.36	.46		.82	1.15
2600	20 ga. studs, 1-5/8" wide, 16" O.C.		395	.020		.33	.84		1.17	1.74
2650	24" O.C.		610	.013		.24	.54		.78	1.15
2700	2-1/2" wide, 16" O.C.		390	.021		.37	.85		1.22	1.81
2750	24" O.C.		600	.013		.28	.55		.83	1.21
2800	3-5/8" wide, 16" OC		385	.021		.40	.86		1.26	1.86
2850	24" O.C.		590	.014		.30	.56		.86	1.26
2900	4" wide, 16" O.C.		380	.021		.48	.87		1.35	1.97
2950	24" O.C.		590	.014		.35	.56		.91	1.32
3000	6" wide, 16" O.C.		375	.021		.56	.89		1.45	2.08
3050	24" O.C.		580	.014		.42	.57		.99	1.40
3060	Non-load bearing, galv, 12' high, 25 ga. 1-5/8" wide, 16" O.C.		413	.019		.22	.80		1.02	1.56
3070	24" O.C.		633	.013		.16	.53		.69	1.04
3080	2-1/2" wide, 16" O.C.		408	.020		.27	.81		1.08	1.64
3090	24" O.C.		625	.013		.20	.53		.73	1.09
3100	3-5/8" wide, 16" O.C.		400	.020		.31	.83		1.14	1.71
3110	24" O.C.		617	.013		.23	.54		.77	1.14
3120	4" wide, 16" O.C.		396	.020		.31	.84		1.15	1.73
3130	24" O.C.		617	.013		.23	.54		.77	1.14
3140	6" wide, 16" O.C.		392	.020		.47	.85		1.32	1.90
3150	24" O.C.		604	.013		.34	.55		.89	1.27
3160	20 ga. studs, 1-5/8" wide, 16" O.C.		329	.024		.31	1.01		1.32	2
3170	24" O.C.		508	.016		.23	.65		.88	1.32
3180	2-1/2" wide, 16" O.C.		325	.025		.36	1.02		1.38	2.07
3190	24" O.C.		500	.016		.26	.66		.92	1.38
3200	3-5/8" wide, 16" O.C.		321	.025		.39	1.04		1.43	2.12
3210	24" O.C.		492	.016		.28	.68		.96	1.42
3220	4" wide, 16" O.C.		317	.025		.46	1.05		1.51	2.22
3230	24" O.C.		492	.016		.33	.68		1.01	1.48
3240	6" wide, 16" O.C.		313	.026		.54	1.06		1.60	2.33
3250	24" O.C.		483	.017		.39	.69		1.08	1.56
5000	Load bearing studs, see Div. 05 41 13.30									
9000	Minimum labor/equipment charge	1 Carp	4	2	Job		83		83	137

09 22 26 – Suspension Systems

09 22 26.13 Ceiling Suspension Systems

		Crew	Daily Output	Labor-Hours	Unit	Material	Labor	Equipment	Total	Total Incl O&P
0010	**CEILING SUSPENSION SYSTEMS** For gypsum board or plaster									
8000	Suspended ceilings, including carriers									
8200	1-1/2" carriers, 24" O.C. with:									
8300	7/8" channels, 16" O.C.	1 Lath	275	.029	S.F.	.47	1.07		1.54	2.21
8320	24" O.C.		310	.026		.38	.95		1.33	1.92
8400	1-5/8" channels, 16" O.C.		205	.039		.57	1.44		2.01	2.89
8420	24" O.C.		250	.032		.44	1.18		1.62	2.35
8600	2" carriers, 24" O.C. with:									
8700	7/8" channels, 16" O.C.	1 Lath	250	.032	S.F.	.53	1.18		1.71	2.44
8720	24" O.C.		285	.028		.44	1.04		1.48	2.11
8800	1-5/8" channels, 16" O.C.		190	.042		.62	1.56		2.18	3.12
8820	24" O.C.		225	.036		.50	1.31		1.81	2.61

09 22 36 – Lath

09 22 36.13 Gypsum Lath

			Crew	Daily Output	Labor-Hours	Unit	Material	Labor	Equipment	Total	Total Incl O&P
0010	**GYPSUM LATH**	R092000-50									
0020	Plain or perforated, nailed, 3/8" thick		1 Lath	85	.094	S.Y.	5.05	3.48		8.53	11
0100	1/2" thick			80	.100		4.95	3.70		8.65	11.25

09 22 Supports for Plaster and Gypsum Board

09 22 36 – Lath

09 22 36.13 Gypsum Lath		Crew	Daily Output	Labor-Hours	Unit	Material	2010 Bare Costs Labor	Equipment	Total	Total Incl O&P
0300	Clipped to steel studs, 3/8" thick	1 Lath	75	.107	S.Y.	5.05	3.94		8.99	11.75
0400	1/2" thick		70	.114		4.95	4.22		9.17	12.10
1500	For ceiling installations, add		216	.037			1.37		1.37	2.15
1600	For columns and beams, add		170	.047	↓		1.74		1.74	2.73
9000	Minimum labor/equipment charge	↓	4.25	1.882	Job		69.50		69.50	109

09 22 36.23 Metal Lath										
0010	**METAL LATH**	R092000-50								
0020	Diamond, expanded, 2.5 lb. per S.Y., painted				S.Y.	4.52			4.52	4.97
0100	Galvanized					2.94			2.94	3.23
0300	3.4 lb. per S.Y., painted					3.99			3.99	4.39
0400	Galvanized					4.53			4.53	4.98
0600	For 15# asphalt sheathing paper, add					.53			.53	.58
0900	Flat rib, 1/8" high, 2.75 lb., painted					3.69			3.69	4.06
1000	Foil backed					3.60			3.60	3.96
1200	3.4 lb. per S.Y., painted					4.71			4.71	5.20
1300	Galvanized					4.19			4.19	4.61
1500	For 15# asphalt sheathing paper, add					.53			.53	.58
1800	High rib, 3/8" high, 3.4 lb. per S.Y., painted					5.70			5.70	6.25
1900	Galvanized				↓	5.70			5.70	6.25
2400	3/4" high, painted, .60 lb. per S.F.				S.F.	.59			.59	.65
2500	.75 lb. per S.F.				"	1.27			1.27	1.40
2800	Stucco mesh, painted, 3.6 lb.				S.Y.	3.32			3.32	3.65
3000	K-lath, perforated, absorbent paper, regular					4.43			4.43	4.87
3100	Heavy duty					5.25			5.25	5.75
3300	Waterproof, heavy duty, grade B backing					5.10			5.10	5.65
3400	Fire resistant backing					5.65			5.65	6.25
3600	2.5 lb. diamond painted, on wood framing, on walls	1 Lath	85	.094		4.52	3.48		8	10.40
3700	On ceilings		75	.107		4.52	3.94		8.46	11.15
3900	3.4 lb. diamond painted, on wood framing, on walls		80	.100		4.71	3.70		8.41	11
4000	On ceilings		70	.114		4.71	4.22		8.93	11.85
4200	3.4 lb. diamond painted, wired to steel framing		75	.107		4.71	3.94		8.65	11.40
4300	On ceilings		60	.133		4.71	4.93		9.64	12.95
4600	Cornices, wired to steel		35	.229		4.71	8.45		13.16	18.45
4800	Screwed to steel studs, 2.5 lb.		80	.100		4.52	3.70		8.22	10.75
4900	3.4 lb.		75	.107		3.99	3.94		7.93	10.60
5100	Rib lath, painted, wired to steel, on walls, 2.5 lb.		75	.107		3.69	3.94		7.63	10.25
5200	3.4 lb.		70	.114		5.70	4.22		9.92	12.90
5400	4.0 lb.	↓	65	.123		5.70	4.55		10.25	13.45
5500	For self-furring lath, add					.10			.10	.11
5700	Suspended ceiling system, incl. 3.4 lb. diamond lath, painted	1 Lath	15	.533		4.28	19.70		23.98	35.50
5800	Galvanized	"	15	.533	↓	7.30	19.70		27	39
6000	Hollow metal stud partitions, 3.4 lb. painted lath both sides									
6010	Non-load bearing, 25 ga., w/rib lath 2-1/2" studs, 12" O.C.	1 Lath	20.30	.394	S.Y.	14.60	14.55		29.15	39
6300	16" O.C.		21.10	.379		13.90	14		27.90	37.50
6350	24" O.C.		22.70	.352		13.25	13		26.25	35
6400	3-5/8" studs, 16" O.C.		19.50	.410		14.30	15.15		29.45	39.50
6600	24" O.C.		20.40	.392		13.50	14.50		28	38
6700	4" studs, 16" O.C.		20.40	.392		14.35	14.50		28.85	39
6900	24" O.C.		21.60	.370		13.55	13.70		27.25	36.50
7000	6" studs, 16" O.C.		19.50	.410		15.75	15.15		30.90	41.50
7100	24" O.C.		21.10	.379		14.60	14		28.60	38
7200	L.B. partitions, 16 ga., w/rib lath, 2-1/2" studs, 16" O.C.	↓	20	.400	↓	13.85	14.80		28.65	38.50

09 22 Supports for Plaster and Gypsum Board

09 22 36 – Lath

09 22 36.23 Metal Lath		Crew	Daily Output	Labor-Hours	Unit	Material	2010 Bare Costs Labor	Equipment	Total	Total Incl O&P
7300	3-5/8" studs, 16 ga.	1 Lath	19.70	.406	S.Y.	15.50	15		30.50	40.50
7500	4" studs, 16 ga.		19.50	.410		16	15.15		31.15	41.50
7600	6" studs, 16 ga.		18.70	.428		18.60	15.80		34.40	45.50
9000	Minimum labor/equipment charge		4.25	1.882	Job		69.50		69.50	109

09 23 Gypsum Plastering

09 23 13 – Acoustical Gypsum Plastering

09 23 13.10 Perlite or Vermiculite Plaster

0010	**PERLITE OR VERMICULITE PLASTER** R092000-50									
0020	In 100 lb. bags, under 200 bags				Bag	14.10			14.10	15.50
0300	2 coats, no lath included, on walls	J-1	92	.435	S.Y.	3.78	15.65	1.49	20.92	31
0400	On ceilings		79	.506		3.78	18.20	1.73	23.71	35
0900	3 coats, no lath included, on walls		74	.541		6.30	19.45	1.85	27.60	40
1000	On ceilings		63	.635		6.30	23	2.17	31.47	46
1700	For irregular or curved surfaces, add to above						30%			
1800	For columns and beams, add to above						50%			
1900	For soffits, add to ceiling prices						40%			
9000	Minimum labor/equipment charge	1 Plas	1	8	Job		298		298	475

09 23 20 – Gypsum Plaster

09 23 20.10 Gypsum Plaster On Walls and Ceilings

0010	**GYPSUM PLASTER ON WALLS AND CEILINGS** R092000-50									
0020	80# bag, less than 1 ton				Bag	15			15	16.50
0300	2 coats, no lath included, on walls	J-1	105	.381	S.Y.	3.74	13.70	1.30	18.74	27.50
0400	On ceilings		92	.435		3.74	15.65	1.49	20.88	30.50
0900	3 coats, no lath included, on walls		87	.460		5.35	16.50	1.57	23.42	34
1000	On ceilings		78	.513		5.35	18.45	1.75	25.55	37.50
1600	For irregular or curved surfaces, add						30%			
1800	For columns & beams, add						50%			
9000	Minimum labor/equipment charge	1 Plas	1	8	Job		298		298	475

09 24 Portland Cement Plastering

09 24 23 – Portland Cement Stucco

09 24 23.40 Stucco

0010	**STUCCO** R092000-50									
0015	3 coats 1" thick, float finish, with mesh, on wood frame	J-2	63	.762	S.Y.	5.60	27.50	2.17	35.27	52
0100	On masonry construction, no mesh incl.	J-1	67	.597		2.28	21.50	2.04	25.82	39
0300	For trowel finish, add	1 Plas	170	.047			1.76		1.76	2.80
0400	For 3/4" thick, on masonry, deduct	J-1	880	.045		.61	1.63	.16	2.40	3.44
0600	For coloring and special finish, add, minimum		685	.058		.39	2.10	.20	2.69	3.99
0700	Maximum		200	.200		1.36	7.20	.68	9.24	13.70
0900	For soffits, add	J-2	155	.310		2.11	11.20	.88	14.19	21
1000	Exterior stucco, with bonding agent, 3 coats, on walls, no mesh incl.	J-1	200	.200		3.54	7.20	.68	11.42	16.10
1200	Ceilings		180	.222		3.54	8	.76	12.30	17.45
1300	Beams		80	.500		3.54	17.95	1.71	23.20	34.50
1500	Columns		100	.400		3.54	14.40	1.37	19.31	28.50
1550	Minimum labor/equipment charge	1 Plas	1	8	Job		298		298	475
1600	Mesh, painted, nailed to wood, 1.8 lb.	1 Lath	60	.133	S.Y.	5.90	4.93		10.83	14.20
1800	3.6 lb.		55	.145		3.32	5.35		8.67	12.10
1900	Wired to steel, painted, 1.8 lb.		53	.151		5.90	5.60		11.50	15.20

09 24 Portland Cement Plastering

09 24 23 – Portland Cement Stucco

09 24 23.40 Stucco		Crew	Daily Output	Labor-Hours	Unit	Material	2010 Bare Costs Labor	Equipment	Total	Total Incl O&P
2100	3.6 lb.	1 Lath	50	.160	S.Y.	3.32	5.90		9.22	12.95
9000	Minimum labor/equipment charge	↓	4	2	Job		74		74	116

09 26 Veneer Plastering

09 26 13 – Gypsum Veneer Plastering

09 26 13.20 Blueboard

		Crew	Daily Output	Labor-Hours	Unit	Material	Labor	Equipment	Total	Total Incl O&P
0010	**BLUEBOARD** For use with thin coat									
0100	plaster application (see Div. 09 26 13.80)									
1000	3/8" thick, on walls or ceilings, standard, no finish included	2 Carp	1900	.008	S.F.	.26	.35		.61	.86
1100	With thin coat plaster finish		875	.018		.34	.76		1.10	1.63
1400	On beams, columns, or soffits, standard, no finish included		675	.024		.30	.98		1.28	1.95
1450	With thin coat plaster finish		475	.034		.38	1.40		1.78	2.72
3000	1/2" thick, on walls or ceilings, standard, no finish included		1900	.008		.29	.35		.64	.89
3100	With thin coat plaster finish		875	.018		.37	.76		1.13	1.66
3300	Fire resistant, no finish included		1900	.008		.29	.35		.64	.89
3400	With thin coat plaster finish		875	.018		.37	.76		1.13	1.66
3450	On beams, columns, or soffits, standard, no finish included		675	.024		.33	.98		1.31	1.99
3500	With thin coat plaster finish		475	.034		.42	1.40		1.82	2.76
3700	Fire resistant, no finish included		675	.024		.33	.98		1.31	1.99
3800	With thin coat plaster finish		475	.034		.42	1.40		1.82	2.76
5000	5/8" thick, on walls or ceilings, fire resistant, no finish included		1900	.008		.29	.35		.64	.89
5100	With thin coat plaster finish		875	.018		.37	.76		1.13	1.66
5500	On beams, columns, or soffits, no finish included		675	.024		.33	.98		1.31	1.99
5600	With thin coat plaster finish		475	.034		.42	1.40		1.82	2.76
6000	For high ceilings, over 8' high, add		3060	.005			.22		.22	.36
6500	For over 3 stories high, add per story	↓	6100	.003	↓		.11		.11	.18
9000	Minimum labor/equipment charge	1 Carp	2	4	Job		166		166	273

09 26 13.80 Thin Coat Plaster

			Crew	Daily Output	Labor-Hours	Unit	Material	Labor	Equipment	Total	Total Incl O&P
0010	**THIN COAT PLASTER**	R092000-50									
0012	1 coat veneer, not incl. lath		J-1	3600	.011	S.F.	.08	.40	.04	.52	.77
1000	In 50 lb. bags					Bag	11.30			11.30	12.45

09 28 Backing Boards and Underlayments

09 28 13 – Cementitious Backing Boards

09 28 13.10 Cementitious Backerboard

		Crew	Daily Output	Labor-Hours	Unit	Material	Labor	Equipment	Total	Total Incl O&P
0010	**CEMENTITIOUS BACKERBOARD**									
0070	Cementitious backerboard, on floor, 3' x 4' x 1/2" sheets	2 Carp	525	.030	S.F.	.69	1.27		1.96	2.84
0080	3' x 5' x 1/2" sheets		525	.030		.72	1.27		1.99	2.87
0090	3' x 6' x 1/2" sheets		525	.030		.64	1.27		1.91	2.79
0100	3' x 4' x 5/8" sheets		525	.030		.93	1.27		2.20	3.10
0110	3' x 5' x 5/8" sheets		525	.030		.92	1.27		2.19	3.09
0120	3' x 6' x 5/8" sheets		525	.030		.75	1.27		2.02	2.90
0150	On wall, 3' x 4' x 1/2" sheets		350	.046		.69	1.90		2.59	3.88
0160	3' x 5' x 1/2" sheets		350	.046		.72	1.90		2.62	3.91
0170	3' x 6' x 1/2" sheets		350	.046		.64	1.90		2.54	3.83
0180	3' x 4' x 5/8" sheets		350	.046		.93	1.90		2.83	4.14
0190	3' x 5' x 5/8" sheets		350	.046		.92	1.90		2.82	4.13
0200	3' x 6' x 5/8" sheets		350	.046		.75	1.90		2.65	3.94
0250	On counter, 3' x 4' x 1/2" sheets	↓	180	.089	↓	.69	3.69		4.38	6.80

09 28 Backing Boards and Underlayments

09 28 13 – Cementitious Backing Boards

09 28 13.10 Cementitious Backerboard		Crew	Daily Output	Labor-Hours	Unit	Material	2010 Bare Costs Labor	Equipment	Total	Total Incl O&P
0260	3' x 5' x 1/2" sheets	2 Carp	180	.089	S.F.	.72	3.69		4.41	6.85
0270	3' x 6' x 1/2" sheets		180	.089		.64	3.69		4.33	6.75
0300	3' x 4' x 5/8" sheets		180	.089		.93	3.69		4.62	7.05
0310	3' x 5' x 5/8" sheets		180	.089		.92	3.69		4.61	7.05
0320	3' x 6' x 5/8" sheets		180	.089		.75	3.69		4.44	6.85

09 29 Gypsum Board

09 29 10 – Gypsum Board Panels

09 29 10.30 Gypsum Board			Crew	Daily Output	Labor-Hours	Unit	Material	2010 Bare Costs Labor	Equipment	Total	Total Incl O&P
0010	**GYPSUM BOARD** on walls & ceilings	R092910-10									
0100	Nailed or screwed to studs unless otherwise noted										
0150	3/8" thick, on walls, standard, no finish included		2 Carp	2000	.008	S.F.	.24	.33		.57	.81
0200	On ceilings, standard, no finish included			1800	.009		.24	.37		.61	.87
0250	On beams, columns, or soffits, no finish included			675	.024		.24	.98		1.22	1.88
0300	1/2" thick, on walls, standard, no finish included			2000	.008		.24	.33		.57	.81
0350	Taped and finished (level 4 finish)			965	.017		.28	.69		.97	1.44
0390	With compound skim coat (level 5 finish)			775	.021		.33	.86		1.19	1.77
0400	Fire resistant, no finish included			2000	.008		.28	.33		.61	.86
0450	Taped and finished (level 4 finish)			965	.017		.32	.69		1.01	1.48
0490	With compound skim coat (level 5 finish)			775	.021		.37	.86		1.23	1.81
0500	Water resistant, no finish included			2000	.008		.35	.33		.68	.94
0550	Taped and finished (level 4 finish)			965	.017		.39	.69		1.08	1.56
0590	With compound skim coat (level 5 finish)			775	.021		.44	.86		1.30	1.89
0600	Prefinished, vinyl, clipped to studs			900	.018		.60	.74		1.34	1.87
0700	Mold resistant, no finish included			2000	.008		.42	.33		.75	1.01
0710	Taped and finished (level 4 finish)			965	.017		.46	.69		1.15	1.64
0720	With compound skim coat (level 5 finish)			775	.021		.51	.86		1.37	1.97
1000	On ceilings, standard, no finish included			1800	.009		.24	.37		.61	.87
1050	Taped and finished (level 4 finish)			765	.021		.28	.87		1.15	1.74
1090	With compound skim coat (level 5 finish)			610	.026		.33	1.09		1.42	2.15
1100	Fire resistant, no finish included			1800	.009		.28	.37		.65	.92
1150	Taped and finished (level 4 finish)			765	.021		.32	.87		1.19	1.78
1195	With compound skim coat (level 5 finish)			610	.026		.37	1.09		1.46	2.19
1200	Water resistant, no finish included			1800	.009		.35	.37		.72	1
1250	Taped and finished (level 4 finish)			765	.021		.39	.87		1.26	1.86
1290	With compound skim coat (level 5 finish)			610	.026		.44	1.09		1.53	2.27
1310	Mold resistant, no finish included			1800	.009		.42	.37		.79	1.07
1320	Taped and finished (level 4 finish)			765	.021		.46	.87		1.33	1.94
1330	With compound skim coat (level 5 finish)			610	.026		.51	1.09		1.60	2.35
1500	On beams, columns, or soffits, standard, no finish included			675	.024		.28	.98		1.26	1.92
1550	Taped and finished (level 4 finish)			475	.034		.28	1.40		1.68	2.61
1590	With compound skim coat (level 5 finish)			540	.030		.33	1.23		1.56	2.38
1600	Fire resistant, no finish included			675	.024		.28	.98		1.26	1.93
1650	Taped and finished (level 4 finish)			475	.034		.32	1.40		1.72	2.65
1690	With compound skim coat (level 5 finish)			540	.030		.37	1.23		1.60	2.42
1700	Water resistant, no finish included			675	.024		.40	.98		1.38	2.06
1750	Taped and finished (level 4 finish)			475	.034		.39	1.40		1.79	2.73
1790	With compound skim coat (level 5 finish)			540	.030		.44	1.23		1.67	2.50
1800	Mold resistant, no finish included			675	.024		.48	.98		1.46	2.15
1810	Taped and finished (level 4 finish)			475	.034		.46	1.40		1.86	2.81
1820	With compound skim coat (level 5 finish)			540	.030		.51	1.23		1.74	2.58

09 29 Gypsum Board

09 29 10 – Gypsum Board Panels

09 29 10.30 Gypsum Board		Crew	Daily Output	Labor-Hours	Unit	Material	2010 Bare Costs Labor	Equipment	Total	Total Incl O&P
2000	5/8" thick, on walls, standard, no finish included	2 Carp	2000	.008	S.F.	.28	.33		.61	.86
2050	Taped and finished (level 4 finish)		965	.017		.32	.69		1.01	1.48
2090	With compound skim coat (level 5 finish)		775	.021		.37	.86		1.23	1.81
2100	Fire resistant, no finish included		2000	.008		.27	.33		.60	.85
2150	Taped and finished (level 4 finish)		965	.017		.31	.69		1	1.47
2195	With compound skim coat (level 5 finish)		775	.021		.36	.86		1.22	1.80
2200	Water resistant, no finish included		2000	.008		.43	.33		.76	1.02
2250	Taped and finished (level 4 finish)		965	.017		.47	.69		1.16	1.65
2290	With compound skim coat (level 5 finish)		775	.021		.52	.86		1.38	1.98
2300	Prefinished, vinyl, clipped to studs		900	.018		.71	.74		1.45	1.99
2510	Mold resistant, no finish included		2000	.008		.49	.33		.82	1.09
2520	Taped and finished (level 4 finish)		965	.017		.53	.69		1.22	1.72
2530	With compound skim coat (level 5 finish)		775	.021		.58	.86		1.44	2.04
3000	On ceilings, standard, no finish included		1800	.009		.28	.37		.65	.92
3050	Taped and finished (level 4 finish)		765	.021		.32	.87		1.19	1.78
3090	With compound skim coat (level 5 finish)		615	.026		.37	1.08		1.45	2.18
3100	Fire resistant, no finish included		1800	.009		.27	.37		.64	.91
3150	Taped and finished (level 4 finish)		765	.021		.31	.87		1.18	1.77
3190	With compound skim coat (level 5 finish)		615	.026		.36	1.08		1.44	2.17
3200	Water resistant, no finish included		1800	.009		.43	.37		.80	1.08
3250	Taped and finished (level 4 finish)		765	.021		.47	.87		1.34	1.95
3290	With compound skim coat (level 5 finish)		615	.026		.52	1.08		1.60	2.35
3300	Mold resistant, no finish included		1800	.009		.49	.37		.86	1.15
3310	Taped and finished (level 4 finish)		765	.021		.53	.87		1.40	2.02
3320	With compound skim coat (level 5 finish)		615	.026		.58	1.08		1.66	2.41
3500	On beams, columns, or soffits, no finish included		675	.024		.32	.98		1.30	1.97
3550	Taped and finished (level 4 finish)		475	.034		.37	1.40		1.77	2.71
3590	With compound skim coat (level 5 finish)		380	.042		.42	1.75		2.17	3.33
3600	Fire resistant, no finish included		675	.024		.31	.98		1.29	1.96
3650	Taped and finished (level 4 finish)		475	.034		.36	1.40		1.76	2.69
3690	With compound skim coat (level 5 finish)		380	.042		.36	1.75		2.11	3.26
3700	Water resistant, no finish included		675	.024		.49	.98		1.47	2.16
3750	Taped and finished (level 4 finish)		475	.034		.54	1.40		1.94	2.90
3790	With compound skim coat (level 5 finish)		380	.042		.52	1.75		2.27	3.44
3800	Mold resistant, no finish included		675	.024		.56	.98		1.54	2.24
3810	Taped and finished (level 4 finish)		475	.034		.61	1.40		2.01	2.97
3820	With compound skim coat (level 5 finish)		380	.042		.58	1.75		2.33	3.50
4000	Fireproofing, beams or columns, 2 layers, 1/2" thick, incl finish		330	.048		.60	2.01		2.61	3.97
4010	Mold resistant		330	.048		.88	2.01		2.89	4.28
4050	5/8" thick		300	.053		.62	2.22		2.84	4.33
4060	Mold resistant		300	.053		1.06	2.22		3.28	4.81
4100	3 layers, 1/2" thick		225	.071		.88	2.95		3.83	5.80
4110	Mold resistant		225	.071		1.30	2.95		4.25	6.30
4150	5/8" thick		210	.076		.94	3.17		4.11	6.25
4160	Mold resistant		210	.076		1.60	3.17		4.77	6.95
5050	For 1" thick coreboard on columns		480	.033		.55	1.38		1.93	2.88
5100	For foil-backed board, add					.14			.14	.15
5200	For work over 8' high, add	2 Carp	3060	.005			.22		.22	.36
5270	For textured spray, add	2 Lath	1600	.010		.05	.37		.42	.64
5300	For distribution cost over 3 stories high, add per story	2 Carp	6100	.003			.11		.11	.18
5350	For finishing inner corners, add		950	.017	L.F.	.09	.70		.79	1.25
5355	For finishing outer corners, add		1250	.013		.21	.53		.74	1.10
5500	For acoustical sealant, add per bead	1 Carp	500	.016		.04	.66		.70	1.14

09 29 Gypsum Board

09 29 10 – Gypsum Board Panels

09 29 10.30 Gypsum Board

		Crew	Daily Output	Labor-Hours	Unit	Material	2010 Bare Costs Labor	Equipment	Total	Total Incl O&P
5550	Sealant, 1 quart tube				Ea.	6.85			6.85	7.55
5600	Sound deadening board, 1/4" gypsum	2 Carp	1800	.009	S.F.	.33	.37		.70	.97
5650	1/2" wood fiber	"	1800	.009	"	.82	.37		1.19	1.51
9000	Minimum labor/equipment charge	1 Carp	2	4	Job		166		166	273

09 29 15 – Gypsum Board Accessories

09 29 15.10 Accessories, Gypsum Board

		Crew	Daily Output	Labor-Hours	Unit	Material	Labor	Equipment	Total	Total Incl O&P
0010	**ACCESSORIES, GYPSUM BOARD**									
0020	Casing bead, galvanized steel	1 Carp	2.90	2.759	C.L.F.	23.50	115		138.50	214
0100	Vinyl		3	2.667		22.50	111		133.50	207
0300	Corner bead, galvanized steel, 1" x 1"		4	2		14.85	83		97.85	153
0400	1-1/4" x 1-1/4"		3.50	2.286		16.60	95		111.60	174
0600	Vinyl		4	2		17.80	83		100.80	157
0900	Furring channel, galv. steel, 7/8" deep, standard		2.60	3.077		25	128		153	238
1000	Resilient		2.55	3.137		25.50	130		155.50	242
1100	J trim, galvanized steel, 1/2" wide		3	2.667		25	111		136	210
1120	5/8" wide		2.95	2.712		24	113		137	212
1500	Z stud, galvanized steel, 1-1/2" wide		2.60	3.077		36	128		164	250
9000	Minimum labor/equipment charge		3	2.667	Job		111		111	182

09 30 Tiling

09 30 13 – Ceramic Tiling

09 30 13.10 Ceramic Tile

		Crew	Daily Output	Labor-Hours	Unit	Material	Labor	Equipment	Total	Total Incl O&P
0010	**CERAMIC TILE**									
0600	Cove base, 4-1/4" x 4-1/4" high, mud set	D-7	91	.176	L.F.	3.75	6.20		9.95	13.80
0700	Thin set		128	.125		3.64	4.41		8.05	10.90
0900	6" x 4-1/4" high, mud set		100	.160		3.34	5.65		8.99	12.45
1000	Thin set		137	.117		3.06	4.12		7.18	9.80
1200	Sanitary cove base, 6" x 4-1/4" high, mud set		93	.172		3.84	6.05		9.89	13.70
1300	Thin set		124	.129		3.59	4.55		8.14	11.05
1500	6" x 6" high, mud set		84	.190		4.48	6.70		11.18	15.45
1600	Thin set		117	.137		4.20	4.82		9.02	12.15
2400	Bullnose trim, 4-1/4" x 4-1/4", mud set		82	.195		3.55	6.90		10.45	14.65
2500	Thin set		128	.125		3.11	4.41		7.52	10.30
2700	6" x 4-1/4" bullnose trim, mud set		84	.190		2.70	6.70		9.40	13.45
2800	Thin set		124	.129		2.50	4.55		7.05	9.85
3000	Floors, natural clay, random or uniform, thin set, color group 1		183	.087	S.F.	4.19	3.08		7.27	9.45
3100	Color group 2		183	.087		4.51	3.08		7.59	9.80
3300	Porcelain type, 1 color, color group 2, 1" x 1"		183	.087		4.71	3.08		7.79	10
3310	2" x 2" or 2" x 1", thin set		190	.084		4.98	2.97		7.95	10.15
3350	For random blend, 2 colors, add					.88			.88	.97
3360	4 colors, add					1.24			1.24	1.36
4300	Specialty tile, 4-1/4" x 4-1/4" x 1/2", decorator finish	D-7	183	.087		10	3.08		13.08	15.80
4500	Add for epoxy grout, 1/16" joint, 1" x 1" tile		800	.020		.62	.71		1.33	1.78
4600	2" x 2" tile		820	.020		.59	.69		1.28	1.72
4800	Pregrouted sheets, walls, 4-1/4" x 4-1/4", 6" x 4-1/4"									
4810	and 8-1/2" x 4-1/4", 4 S.F. sheets, silicone grout	D-7	240	.067	S.F.	4.73	2.35		7.08	8.85
5100	Floors, unglazed, 2 S.F. sheets,									
5110	Urethane adhesive	D-7	180	.089	S.F.	4.71	3.14		7.85	10.10
5400	Walls, interior, thin set, 4-1/4" x 4-1/4" tile		190	.084		2.34	2.97		5.31	7.20
5500	6" x 4-1/4" tile		190	.084		2.71	2.97		5.68	7.60

09 30 Tiling

09 30 13 – Ceramic Tiling

09 30 13.10 Ceramic Tile

		Crew	Daily Output	Labor-Hours	Unit	Material	2010 Bare Costs Labor	Equipment	Total	Total Incl O&P
5700	8-1/2" x 4-1/4" tile	D-7	190	.084	S.F.	3.78	2.97		6.75	8.80
5800	6" x 6" tile		200	.080		3.04	2.82		5.86	7.75
5810	8" x 8" tile		225	.071		4.06	2.51		6.57	8.40
5820	12" x 12" tile		300	.053		3.47	1.88		5.35	6.75
5830	16" x 16" tile		500	.032		3.75	1.13		4.88	5.90
6000	Decorated wall tile, 4-1/4" x 4-1/4", minimum		270	.059		3.51	2.09		5.60	7.10
6100	Maximum		180	.089		41.50	3.14		44.64	51
6600	Crystalline glazed, 4-1/4" x 4-1/4", mud set, plain		100	.160		4.28	5.65		9.93	13.50
6700	4-1/4" x 4-1/4", scored tile		100	.160		5.25	5.65		10.90	14.55
6900	6" x 6" plain		93	.172		5.30	6.05		11.35	15.30
7000	For epoxy grout, 1/16" joints, 4-1/4" tile, add		800	.020		.39	.71		1.10	1.53
7200	For tile set in dry mortar, add		1735	.009			.33		.33	.51
7300	For tile set in Portland cement mortar, add		290	.055		.27	1.95		2.22	3.34
9300	Ceramic tiles, recycled glass, standard colors, 2" x 2" thru 6" x 6"	G	190	.084		17	2.97		19.97	23.50
9310	6" x 6"	G	200	.080		17	2.82		19.82	23
9320	8" x 8"	G	225	.071		18.45	2.51		20.96	24.50
9330	12"x12"	G	300	.053		18.45	1.88		20.33	23.50
9340	Earthtones, 2" x 2" to 4" x 8"	G	190	.084		21.50	2.97		24.47	28
9350	6" x 6"	G	200	.080		21.50	2.82		24.32	28
9360	8" x 8"	G	225	.071		23	2.51		25.51	29
9370	12" x 12"	G	300	.053		23	1.88		24.88	28
9380	Deep colors, 2" x 2" to 4" x 8"	G	190	.084		31	2.97		33.97	38.50
9390	6" x 6"	G	200	.080		31	2.82		33.82	38.50
9400	8" x 8"	G	225	.071		32	2.51		34.51	39.50
9410	12" x 12"	G	300	.053		32	1.88		33.88	38.50
9500	Minimum labor/equipment charge		3.25	4.923	Job		174		174	271

09 30 16 – Quarry Tiling

09 30 16.10 Quarry Tile

		Crew	Daily Output	Labor-Hours	Unit	Material	Labor	Equipment	Total	Total Incl O&P
0010	**QUARRY TILE**									
0100	Base, cove or sanitary, mud set, to 5" high, 1/2" thick	D-7	110	.145	L.F.	5.30	5.15		10.45	13.85
0300	Bullnose trim, red, mud set, 6" x 6" x 1/2" thick		120	.133		4.36	4.70		9.06	12.15
0400	4" x 4" x 1/2" thick		110	.145		4.78	5.15		9.93	13.25
0600	4" x 8" x 1/2" thick, using 8" as edge		130	.123		4.24	4.34		8.58	11.45
0700	Floors, mud set, 1,000 S.F. lots, red, 4" x 4" x 1/2" thick		120	.133	S.F.	7.50	4.70		12.20	15.60
0900	6" x 6" x 1/2" thick		140	.114		6.85	4.03		10.88	13.80
1000	4" x 8" x 1/2" thick		130	.123		7.50	4.34		11.84	15.05
1300	For waxed coating, add					.67			.67	.74
1500	For non-standard colors, add					.40			.40	.44
1600	For abrasive surface, add					.47			.47	.52
1800	Brown tile, imported, 6" x 6" x 3/4"	D-7	120	.133		8.40	4.70		13.10	16.60
1900	8" x 8" x 1"		110	.145		9.05	5.15		14.20	17.95
2100	For thin set mortar application, deduct		700	.023			.81		.81	1.26
2700	Stair tread, 6" x 6" x 3/4", plain		50	.320		5.05	11.30		16.35	23
2800	Abrasive		47	.340		5.30	12		17.30	24.50
3000	Wainscot, 6" x 6" x 1/2", thin set, red		105	.152		4.09	5.40		9.49	12.90
3100	Non-standard colors		105	.152		4.54	5.40		9.94	13.40
3300	Window sill, 6" wide, 3/4" thick		90	.178	L.F.	5.10	6.25		11.35	15.40
3400	Corners		80	.200	Ea.	5.70	7.05		12.75	17.25
9000	Minimum labor/equipment charge		3.25	4.923	Job		174		174	271

09 30 Tiling

09 30 29 – Metal Tiling

09 30 29.10 Metal Tile

		Crew	Daily Output	Labor-Hours	Unit	Material	2010 Bare Costs Labor	Equipment	Total	Total Incl O&P
0010	**METAL TILE** 4' x 4' sheet, 24 ga., tile pattern, nailed									
0200	Stainless steel	2 Carp	512	.031	S.F.	24.50	1.30		25.80	29
0400	Aluminized steel	"	512	.031	"	13.15	1.30		14.45	16.60
9000	Minimum labor/equipment charge	1 Carp	4	2	Job		83		83	137

09 51 Acoustical Ceilings

09 51 23 – Acoustical Tile Ceilings

09 51 23.10 Suspended Acoustic Ceiling Tiles

		Crew	Daily Output	Labor-Hours	Unit	Material	Labor	Equipment	Total	Total Incl O&P
0010	**SUSPENDED ACOUSTIC CEILING TILES**, not including									
0100	suspension system									
0300	Fiberglass boards, film faced, 2' x 2' or 2' x 4', 5/8" thick	1 Carp	625	.013	S.F.	.70	.53		1.23	1.64
0400	3/4" thick		600	.013		1.61	.55		2.16	2.68
0500	3" thick, thermal, R11		450	.018		1.53	.74		2.27	2.89
0600	Glass cloth faced fiberglass, 3/4" thick		500	.016		2.87	.66		3.53	4.25
0700	1" thick		485	.016		2.18	.69		2.87	3.53
0820	1-1/2" thick, nubby face		475	.017		2.66	.70		3.36	4.08
1110	Mineral fiber tile, lay-in, 2' x 2' or 2' x 4', 5/8" thick, fine texture		625	.013		.55	.53		1.08	1.48
1115	Rough textured		625	.013		1.23	.53		1.76	2.22
1125	3/4" thick, fine textured		600	.013		1.50	.55		2.05	2.56
1130	Rough textured		600	.013		1.79	.55		2.34	2.88
1135	Fissured		600	.013		2.45	.55		3	3.61
1150	Tegular, 5/8" thick, fine textured		470	.017		1.45	.71		2.16	2.76
1155	Rough textured		470	.017		1.90	.71		2.61	3.25
1165	3/4" thick, fine textured		450	.018		2.08	.74		2.82	3.50
1170	Rough textured		450	.018		2.35	.74		3.09	3.80
1175	Fissured		450	.018		3.63	.74		4.37	5.20
1180	For aluminum face, add					6.05			6.05	6.65
1185	For plastic film face, add					.94			.94	1.03
1190	For fire rating, add					.45			.45	.50
1300	Mirror tile, acrylic, 2' x 2'	1 Carp	500	.016		8.70	.66		9.36	10.65
1900	Eggcrate, acrylic, 1/2" x 1/2" x 1/2" cubes		500	.016		1.82	.66		2.48	3.09
2100	Polystyrene eggcrate, 3/8" x 3/8" x 1/2" cubes		510	.016		1.53	.65		2.18	2.75
2200	1/2" x 1/2" x 1/2" cubes		500	.016		1.75	.66		2.41	3.02
2400	Luminous panels, prismatic, acrylic		400	.020		2.22	.83		3.05	3.81
2500	Polystyrene		400	.020		1.14	.83		1.97	2.62
2700	Flat white acrylic		400	.020		3.86	.83		4.69	5.60
2800	Polystyrene		400	.020		2.65	.83		3.48	4.29
3000	Drop pan, white, acrylic		400	.020		5.65	.83		6.48	7.60
3100	Polystyrene		400	.020		4.73	.83		5.56	6.55
3600	Perforated aluminum sheets, .024" thick, corrugated, painted		490	.016		2.25	.68		2.93	3.59
3700	Plain		500	.016		3.76	.66		4.42	5.25
3720	Mineral fiber, 24" x 24" or 48", reveal edge, painted, 5/8" thick		600	.013		1.38	.55		1.93	2.43
3740	3/4" thick		575	.014		2.24	.58		2.82	3.41
5020	66 – 78% recycled content, 3/4" thick G		600	.013		1.82	.55		2.37	2.91
5040	Mylar, 42% recycled content, 3/4" thick G		600	.013		4.28	.55		4.83	5.60
9000	Minimum labor/equipment charge		4	2	Job		83		83	137

09 51 23.30 Suspended Ceilings, Complete

		Crew	Daily Output	Labor-Hours	Unit	Material	Labor	Equipment	Total	Total Incl O&P
0010	**SUSPENDED CEILINGS, COMPLETE** Including standard									
0100	suspension system but not incl. 1-1/2" carrier channels									
0600	Fiberglass ceiling board, 2' x 4' x 5/8", plain faced	1 Carp	500	.016	S.F.	1.38	.66		2.04	2.61

09 51 Acoustical Ceilings

09 51 23 – Acoustical Tile Ceilings

09 51 23.30 Suspended Ceilings, Complete		Crew	Daily Output	Labor-Hours	Unit	Material	2010 Bare Costs Labor	Equipment	Total	Total Incl O&P
0700	Offices, 2' x 4' x 3/4"	1 Carp	380	.021	S.F.	2.29	.87		3.16	3.96
0800	Mineral fiber, on 15/16" T bar susp. 2' x 2' x 3/4" lay-in board		345	.023		2.35	.96		3.31	4.16
0810	2' x 4' x 5/8" tile		380	.021		1.23	.87		2.10	2.79
0820	Tegular, 2' x 2' x 5/8" tile on 9/16" grid		250	.032		2.46	1.33		3.79	4.88
0830	2' x 4' x 3/4" tile		275	.029		2.92	1.21		4.13	5.20
0900	Luminous panels, prismatic, acrylic		255	.031		2.90	1.30		4.20	5.35
1200	Metal pan with acoustic pad, steel		75	.107		4.21	4.43		8.64	11.95
1300	Painted aluminum		75	.107		2.93	4.43		7.36	10.50
1500	Aluminum, degreased finish		75	.107		4.85	4.43		9.28	12.65
1600	Stainless steel		75	.107		9.10	4.43		13.53	17.30
1800	Tile, Z bar suspension, 5/8" mineral fiber tile		150	.053		2.16	2.22		4.38	6
1900	3/4" mineral fiber tile		150	.053		2.31	2.22		4.53	6.20
2402	For strip lighting, see Div. 26 51 13.50									
2500	For rooms under 500 S.F., add				S.F.		25%			
9000	Minimum labor/equipment charge	1 Carp	2	4	Job		166		166	273

09 51 53 – Direct-Applied Acoustical Ceilings

09 51 53.10 Ceiling Tile

		Crew	Daily Output	Labor-Hours	Unit	Material	Labor	Equipment	Total	Total Incl O&P
0010	**CEILING TILE**, Stapled or cemented									
0100	12" x 12" or 12" x 24", not including furring									
0600	Mineral fiber, vinyl coated, 5/8" thick	1 Carp	300	.027	S.F.	1.79	1.11		2.90	3.79
0700	3/4" thick		300	.027		1.91	1.11		3.02	3.92
0900	Fire rated, 3/4" thick, plain faced		300	.027		1.40	1.11		2.51	3.36
1000	Plastic coated face		300	.027		1.18	1.11		2.29	3.12
1200	Aluminum faced, 5/8" thick, plain		300	.027		1.51	1.11		2.62	3.48
3000	Wood fiber tile, 1/2" thick		400	.020		.92	.83		1.75	2.38
3100	3/4" thick		400	.020		1.17	.83		2	2.66
3300	For flameproofing, add					.10			.10	.11
3400	For sculptured 3 dimensional, add					.29			.29	.32
3900	For ceiling primer, add					.13			.13	.14
4000	For ceiling cement, add					.37			.37	.41
9000	Minimum labor/equipment charge	1 Carp	4	2	Job		83		83	137

09 53 Acoustical Ceiling Suspension Assemblies

09 53 23 – Metal Acoustical Ceiling Suspension Assemblies

09 53 23.30 Ceiling Suspension Systems

			Crew	Daily Output	Labor-Hours	Unit	Material	Labor	Equipment	Total	Total Incl O&P
0010	**CEILING SUSPENSION SYSTEMS** for boards and tile										
0050	Class A suspension system, 15/16" T bar, 2' x 4' grid		1 Carp	800	.010	S.F.	.68	.42		1.10	1.43
0300	2' x 2' grid			650	.012		.85	.51		1.36	1.77
0310	25% recycled steel, 2' x 4' grid	G		800	.010		.79	.42		1.21	1.55
0320	2' x 2' grid	G		650	.012		.99	.51		1.50	1.93
0350	For 9/16" grid, add						.16			.16	.18
0360	For fire rated grid, add						.09			.09	.10
0370	For colored grid, add						.21			.21	.23
0400	Concealed Z bar suspension system, 12" module		1 Carp	520	.015		.69	.64		1.33	1.81
0600	1-1/2" carrier channels, 4' O.C., add		"	470	.017		.11	.71		.82	1.28
0700	Carrier channels for ceilings with										
0900	recessed lighting fixtures, add		1 Carp	460	.017	S.F.	.19	.72		.91	1.40
1200	Seismic bracing, 2-1/2" compression post with four bracing wires			50	.160	Ea.	2.74	6.65		9.39	13.90
9000	Minimum labor/equipment charge			5	1.600	Job		66.50		66.50	109

09 54 Specialty Ceilings

09 54 33 – Decorative Panel Ceilings

09 54 33.20 Metal Panel Ceilings		Crew	Daily Output	Labor-Hours	Unit	Material	2010 Bare Costs Labor	Equipment	Total	Total Incl O&P
0010	**METAL PANEL CEILINGS**									
0020	Lay-in or screwed to furring, not including grid									
0100	Tin ceilings, 2' x 2' or 2' x 4', bare steel finish	2 Carp	300	.053	S.F.	2.37	2.22		4.59	6.25
0120	Painted white finish		300	.053	"	3.59	2.22		5.81	7.60
0140	Copper, chrome or brass finish		300	.053	L.F.	5.95	2.22		8.17	10.20
0200	Cornice molding, 2-1/2" to 3-1/2" wide, 4' long, bare steel finish		200	.080	S.F.	2.25	3.32		5.57	7.90
0220	Painted white finish		200	.080		3.27	3.32		6.59	9.05
0240	Copper, chrome or brass finish		200	.080		4.51	3.32		7.83	10.40
0320	5" to 6-1/2" wide, 4' long, bare steel finish		150	.107		3.12	4.43		7.55	10.75
0340	Painted white finish		150	.107		4.51	4.43		8.94	12.25
0360	Copper, chrome or brass finish		150	.107		7	4.43		11.43	15
0420	Flat molding, 3-1/2" to 5" wide, 4' long, bare steel finish		250	.064		3.44	2.66		6.10	8.15
0440	Painted white finish		250	.064		4.86	2.66		7.52	9.70
0460	Copper, chrome or brass finish		250	.064		10.05	2.66		12.71	15.40

09 62 Specialty Flooring

09 62 19 – Laminate Flooring

09 62 19.10 Floating Floor

		Crew	Daily Output	Labor-Hours	Unit	Material	Labor	Equipment	Total	Total Incl O&P
0010	**FLOATING FLOOR**									
8300	Floating floor, laminate, wood pattern strip, complete	1 Clab	133	.060	S.F.	4.25	1.99		6.24	7.95
8310	Components, T & G wood composite strips					3.66			3.66	4.03
8320	Film					.15			.15	.16
8330	Foam					.30			.30	.33
8340	Adhesive					.17			.17	.18
8350	Installation kit					.17			.17	.19
8360	Trim, 2" wide x 3' long				L.F.	2.79			2.79	3.07
8370	Reducer moulding				"	5.85			5.85	6.40

09 62 23 – Bamboo Flooring

09 62 23.10 Flooring, Bamboo

			Crew	Daily Output	Labor-Hours	Unit	Material	Labor	Equipment	Total	Total Incl O&P
0010	**FLOORING, BAMBOO**										
8600	Flooring, wood, bamboo strips, unfinished, 5/8" x 4" x 3'	G	1 Carp	255	.031	S.F.	4.44	1.30		5.74	7
8610	5/8" x 4" x 4'	G		275	.029		4.61	1.21		5.82	7.05
8620	5/8" x 4" x 6'	G		295	.027		5.05	1.13		6.18	7.40
8630	Finished, 5/8" x 4" x 3'	G		255	.031		4.89	1.30		6.19	7.55
8640	5/8" x 4" x 4'	G		275	.029		5.10	1.21		6.31	7.65
8650	5/8" x 4" x 6'	G		295	.027		4.96	1.13		6.09	7.30
8660	Stair treads, unfinished, 1-1/16" x 11-1/2" x 4'	G		18	.444	Ea.	42.50	18.45		60.95	77.50
8670	Finished, 1-1/16" x 11-1/2" x 4'	G		18	.444		75.50	18.45		93.95	114
8680	Stair risers, unfinished, 5/8" x 7-1/2" x 4'	G		18	.444		15.75	18.45		34.20	48
8690	Finished, 5/8" x 7-1/2" x 4'	G		18	.444		30	18.45		48.45	63.50
8700	Stair nosing, unfinished, 6' long	G		16	.500		35	21		56	72.50
8710	Finished, 6' long	G		16	.500		40	21		61	78

09 63 Masonry Flooring

09 63 13 – Brick Flooring

09 63 13.10 Miscellaneous Brick Flooring

		Crew	Daily Output	Labor-Hours	Unit	Material	2010 Bare Costs Labor	Equipment	Total	Total Incl O&P
0010	**MISCELLANEOUS BRICK FLOORING**									
0020	Acid-proof shales, red, 8" x 3-3/4" x 1-1/4" thick	D-7	.43	37.209	M	860	1,325		2,185	3,000
0050	2-1/4" thick	D-1	.40	40		1,000	1,500		2,500	3,525
0200	Acid-proof clay brick, 8" x 3-3/4" x 2-1/4" thick G		.40	40		895	1,500		2,395	3,400
0250	9" x 4-1/2" x 3" G		95	.168	S.F.	4.12	6.35		10.47	14.75
0260	Cast ceramic, pressed, 4" x 8" x 1/2", unglazed	D-7	100	.160		6.25	5.65		11.90	15.70
0270	Glazed		100	.160		8.35	5.65		14	17.95
0280	Hand molded flooring, 4" x 8" x 3/4", unglazed		95	.168		8.25	5.95		14.20	18.40
0290	Glazed		95	.168		10.35	5.95		16.30	20.50
0300	8" hexagonal, 3/4" thick, unglazed		85	.188		9.05	6.65		15.70	20.50
0310	Glazed		85	.188		16.35	6.65		23	28.50
0400	Heavy duty industrial, cement mortar bed, 2" thick, not incl. brick	D-1	80	.200		.83	7.55		8.38	13.05
0450	Acid-proof joints, 1/4" wide	"	65	.246		1.44	9.30		10.74	16.50
0500	Pavers, 8" x 4", 1" to 1-1/4" thick, red	D-7	95	.168		3.64	5.95		9.59	13.30
0510	Ironspot	"	95	.168		5.15	5.95		11.10	14.95
0540	1-3/8" to 1-3/4" thick, red	D-1	95	.168		3.51	6.35		9.86	14.05
0560	Ironspot		95	.168		5.10	6.35		11.45	15.80
0580	2-1/4" thick, red		90	.178		3.57	6.70		10.27	14.75
0590	Ironspot		90	.178		5.55	6.70		12.25	16.90
0700	Paver, adobe brick, 6" x 12", 1/2" joint G		42	.381		.90	14.35		15.25	24
0710	Mexican red, 12" x 12" G	1 Tilf	48	.167		1.61	6.55		8.16	12
0720	Saltillo, 12" x 12" G	"	48	.167		1.13	6.55		7.68	11.50
0800	For sidewalks and patios with pavers, see Div. 32 14 16.10									
0870	For epoxy joints, add	D-1	600	.027	S.F.	2.72	1.01		3.73	4.61
0880	For Furan underlayment, add	"	600	.027		2.25	1.01		3.26	4.10
0890	For waxed surface, steam cleaned, add	A-1H	1000	.008		.20	.26	.06	.52	.72
9000	Minimum labor/equipment charge	1 Bric	2	4	Job		167		167	269

09 63 40 – Stone Flooring

09 63 40.10 Marble

		Crew	Daily Output	Labor-Hours	Unit	Material	Labor	Equipment	Total	Total Incl O&P
0010	**MARBLE**									
0020	Thin gauge tile, 12" x 6", 3/8", white Carara	D-7	60	.267	S.F.	11.15	9.40		20.55	27
0100	Travertine		60	.267		12.20	9.40		21.60	28
0200	12" x 12" x 3/8", thin set, floors		60	.267		6.05	9.40		15.45	21.50
0300	On walls		52	.308		9.35	10.85		20.20	27.50
1000	Marble threshold, 4" wide x 36" long x 5/8" thick, white		60	.267	Ea.	6.05	9.40		15.45	21.50
9000	Minimum labor/equipment charge		3	5.333	Job		188		188	294

09 63 40.20 Slate Tile

		Crew	Daily Output	Labor-Hours	Unit	Material	Labor	Equipment	Total	Total Incl O&P
0010	**SLATE TILE**									
0020	Vermont, 6" x 6" x 1/4" thick, thin set	D-7	180	.089	S.F.	6.80	3.14		9.94	12.35
9000	Minimum labor/equipment charge	"	3	5.333	Job		188		188	294

09 64 Wood Flooring

09 64 16 – Wood Block Flooring

09 64 16.10 End Grain Block Flooring

		Crew	Daily Output	Labor-Hours	Unit	Material	2010 Bare Costs Labor	Equipment	Total	Total Incl O&P
0010	**END GRAIN BLOCK FLOORING**									
0020	End grain flooring, coated, 2" thick	1 Carp	295	.027	S.F.	3.29	1.13		4.42	5.45
0400	Natural finish, 1" thick, fir		125	.064		3.40	2.66		6.06	8.10
0600	1-1/2" thick, pine		125	.064		3.34	2.66		6	8.05
0700	2" thick, pine		125	.064		4.64	2.66		7.30	9.45
9000	Minimum labor/equipment charge		2	4	Job		166		166	273

09 64 23 – Wood Parquet Flooring

09 64 23.10 Wood Parquet

		Crew	Daily Output	Labor-Hours	Unit	Material	Labor	Equipment	Total	Total Incl O&P
0010	**WOOD PARQUET** flooring									
5200	Parquetry, standard, 5/16" thick, not incl. finish, oak, minimum	1 Carp	160	.050	S.F.	4.41	2.08		6.49	8.25
5300	Maximum		100	.080		4.95	3.32		8.27	10.90
5500	Teak, minimum		160	.050		4.79	2.08		6.87	8.65
5600	Maximum		100	.080		8.35	3.32		11.67	14.65
5650	13/16" thick, select grade oak, minimum		160	.050		9.20	2.08		11.28	13.55
5700	Maximum		100	.080		14	3.32		17.32	21
5800	Custom parquetry, including finish, minimum		100	.080		15.45	3.32		18.77	22.50
5900	Maximum		50	.160		20.50	6.65		27.15	33.50
6700	Parquetry, prefinished white oak, 5/16" thick, minimum		160	.050		3.06	2.08		5.14	6.80
6800	Maximum		100	.080		6.55	3.32		9.87	12.65
7000	Walnut or teak, parquetry, minimum		160	.050		5.35	2.08		7.43	9.30
7100	Maximum		100	.080		9.35	3.32		12.67	15.75
7200	Acrylic wood parquet blocks, 12" x 12" x 5/16",									
7210	Irradiated, set in epoxy	1 Carp	160	.050	S.F.	7.80	2.08		9.88	12

09 64 29 – Wood Strip and Plank Flooring

09 64 29.10 Wood

		Crew	Daily Output	Labor-Hours	Unit	Material	Labor	Equipment	Total	Total Incl O&P
0010	**WOOD** R061110-30									
0020	Fir, vertical grain, 1" x 4", not incl. finish, grade B & better	1 Carp	255	.031	S.F.	2.74	1.30		4.04	5.15
0100	C grade & better		255	.031		2.58	1.30		3.88	4.98
0300	Flat grain, 1" x 4", not incl. finish, B & better		255	.031		3.13	1.30		4.43	5.60
0400	C & better		255	.031		3.01	1.30		4.31	5.45
4000	Maple, strip, 25/32" x 2-1/4", not incl. finish, select		170	.047		5.25	1.96		7.21	9
4100	#2 & better		170	.047		3.32	1.96		5.28	6.85
4300	33/32" x 3-1/4", not incl. finish, #1 grade		170	.047		4.19	1.96		6.15	7.80
4400	#2 & better		170	.047		3.73	1.96		5.69	7.30
4600	Oak, white or red, 25/32" x 2-1/4", not incl. finish									
4700	#1 common	1 Carp	170	.047	S.F.	2.40	1.96		4.36	5.85
4900	Select quartered, 2-1/4" wide		170	.047		2.94	1.96		4.90	6.45
5000	Clear		170	.047		3.83	1.96		5.79	7.40
6100	Prefinished, white oak, prime grade, 2-1/4" wide		170	.047		4.79	1.96		6.75	8.45
6200	3-1/4" wide		185	.043		6.15	1.80		7.95	9.70
6400	Ranch plank		145	.055		6.75	2.29		9.04	11.15
6500	Hardwood blocks, 9" x 9", 25/32" thick		160	.050		5.75	2.08		7.83	9.75
7400	Yellow pine, 3/4" x 3-1/8", T & G, C & better, not incl. finish		200	.040		1.99	1.66		3.65	4.92
7500	Refinish wood floor, sand, 2 cts poly, wax, soft wood, min.	1 Clab	400	.020		.78	.66		1.44	1.95
7600	Hard wood, max		130	.062		1.16	2.04		3.20	4.62
7800	Sanding and finishing, 2 coats polyurethane		295	.027		.78	.90		1.68	2.33
7900	Subfloor and underlayment, see Div. 06 16									
8015	Transition molding, 2 1/4" wide, 5' long	1 Carp	19.20	.417	Ea.	15	17.30		32.30	45
9000	Minimum labor/equipment charge	"	2	4	Job		166		166	273

09 65 Resilient Flooring

09 65 10 – Resilient Tile Underlayment

09 65 10.10 Latex Underlayment

		Crew	Daily Output	Labor-Hours	Unit	Material	2010 Bare Costs Labor	Equipment	Total	Total Incl O&P
0010	**LATEX UNDERLAYMENT**									
3600	Latex underlayment, 1/8" thk., cementitious for resilient flooring	1 Tilf	160	.050	S.F.	.79	1.97		2.76	3.94
4000	Liquid, fortified				Gal.	38.50			38.50	42

09 65 13 – Resilient Base and Accessories

09 65 13.13 Resilient Base

		Crew	Daily Output	Labor-Hours	Unit	Material	2010 Bare Costs Labor	Equipment	Total	Total Incl O&P
0010	**RESILIENT BASE**									
0800	Base, cove, rubber or vinyl									
1100	Standard colors, 0.080" thick, 2-1/2" high	1 Tilf	315	.025	L.F.	.80	1		1.80	2.44
1150	4" high		315	.025		.94	1		1.94	2.59
1200	6" high		315	.025		1.30	1		2.30	2.99
1450	1/8" thick, 2-1/2" high		315	.025		.80	1		1.80	2.44
1500	4" high		315	.025		.73	1		1.73	2.36
1550	6" high		315	.025		1.35	1		2.35	3.05
1600	Corners, 2-1/2" high		315	.025	Ea.	1.50	1		2.50	3.21
1630	4" high		315	.025		2.85	1		3.85	4.70
1660	6" high		315	.025		2.95	1		3.95	4.81

09 65 13.23 Resilient Stair Treads and Risers

		Crew	Daily Output	Labor-Hours	Unit	Material	2010 Bare Costs Labor	Equipment	Total	Total Incl O&P
0010	**RESILIENT STAIR TREADS AND RISERS**									
0300	Rubber, molded tread, 12" wide, 5/16" thick, black	1 Tilf	115	.070	L.F.	12	2.74		14.74	17.50
0400	Colors		115	.070		12.60	2.74		15.34	18.15
0600	1/4" thick, black		115	.070		11.85	2.74		14.59	17.35
0700	Colors		115	.070		12.60	2.74		15.34	18.15
0900	Grip strip safety tread, colors, 5/16" thick		115	.070		16.20	2.74		18.94	22
1000	3/16" thick		120	.067		11.30	2.62		13.92	16.55
1200	Landings, smooth sheet rubber, 1/8" thick		120	.067	S.F.	5.95	2.62		8.57	10.65
1300	3/16" thick		120	.067	"	7.50	2.62		10.12	12.35
1500	Nosings, 3" wide, 3/16" thick, black		140	.057	L.F.	3.20	2.25		5.45	7.05
1600	Colors		140	.057		3.15	2.25		5.40	7
1800	Risers, 7" high, 1/8" thick, flat		250	.032		6.20	1.26		7.46	8.80
1900	Coved		250	.032		5.10	1.26		6.36	7.55
2100	Vinyl, molded tread, 12" wide, colors, 1/8" thick		115	.070		5.80	2.74		8.54	10.70
2200	1/4" thick		115	.070		7.70	2.74		10.44	12.75
2300	Landing material, 1/8" thick		200	.040	S.F.	5.80	1.57		7.37	8.85
2400	Riser, 7" high, 1/8" thick, coved		175	.046	L.F.	3.35	1.80		5.15	6.50
2500	Tread and riser combined, 1/8" thick		80	.100	"	9	3.94		12.94	16.05
9000	Minimum labor/equipment charge		3	2.667	Job		105		105	164

09 65 16 – Resilient Sheet Flooring

09 65 16.10 Rubber and Vinyl Sheet Flooring

			Crew	Daily Output	Labor-Hours	Unit	Material	2010 Bare Costs Labor	Equipment	Total	Total Incl O&P
0010	**RUBBER AND VINYL SHEET FLOORING**										
5500	Linoleum, sheet goods	G	1 Tilf	360	.022	S.F.	4.55	.87		5.42	6.35
5900	Rubber, sheet goods, 36" wide, 1/8" thick			120	.067		6.45	2.62		9.07	11.20
5950	3/16" thick			100	.080		8.75	3.15		11.90	14.55
6000	1/4" thick			90	.089		10.30	3.50		13.80	16.80
8000	Vinyl sheet goods, backed, .065" thick, minimum			250	.032		3.50	1.26		4.76	5.80
8050	Maximum			200	.040		3.61	1.57		5.18	6.45
8100	.080" thick, minimum			230	.035		3.60	1.37		4.97	6.10
8150	Maximum			200	.040		5.10	1.57		6.67	8.05
8200	.125" thick, minimum			230	.035		4.10	1.37		5.47	6.65
8250	Maximum			200	.040		6.10	1.57		7.67	9.15
8700	Adhesive cement, 1 gallon per 200 to 300 S.F.					Gal.	25			25	27.50
8800	Asphalt primer, 1 gallon per 300 S.F.						12			12	13.20

09 65 Resilient Flooring

09 65 16 – Resilient Sheet Flooring

09 65 16.10 Rubber and Vinyl Sheet Flooring		Crew	Daily Output	Labor-Hours	Unit	Material	2010 Bare Costs Labor	Equipment	Total	Total Incl O&P
8900	Emulsion, 1 gallon per 140 S.F.				Gal.	15.50			15.50	17.05

09 65 19 – Resilient Tile Flooring

09 65 19.10 Miscellaneous Resilient Tile Flooring

			Crew	Daily Output	Labor-Hours	Unit	Material	Labor	Equipment	Total	Total Incl O&P
0010	**MISCELLANEOUS RESILIENT TILE FLOORING**										
2200	Cork tile, standard finish, 1/8" thick	G	1 Tilf	315	.025	S.F.	6	1		7	8.15
2250	3/16" thick	G		315	.025		6	1		7	8.15
2300	5/16" thick	G		315	.025		7.45	1		8.45	9.75
2350	1/2" thick	G		315	.025		8.80	1		9.80	11.20
2500	Urethane finish, 1/8" thick	G		315	.025		7.10	1		8.10	9.40
2550	3/16" thick	G		315	.025		7.45	1		8.45	9.75
2600	5/16" thick	G		315	.025		9.55	1		10.55	12.05
2650	1/2" thick	G		315	.025		12.85	1		13.85	15.70
6050	Rubber tile, marbleized colors, 12" x 12", 1/8" thick			400	.020		6.45	.79		7.24	8.30
6100	3/16" thick			400	.020		9.50	.79		10.29	11.70
6300	Special tile, plain colors, 1/8" thick			400	.020		9.50	.79		10.29	11.70
6350	3/16" thick			400	.020		9.65	.79		10.44	11.85
7000	Vinyl composition tile, 12" x 12", 1/16" thick			500	.016		.92	.63		1.55	1.99
7050	Embossed			500	.016		2.17	.63		2.80	3.37
7100	Marbleized			500	.016		2.17	.63		2.80	3.37
7150	Solid			500	.016		2.79	.63		3.42	4.05
7200	3/32" thick, embossed			500	.016		1.38	.63		2.01	2.50
7250	Marbleized			500	.016		2.49	.63		3.12	3.72
7300	Solid			500	.016		2.31	.63		2.94	3.52
7350	1/8" thick, marbleized			500	.016		1.64	.63		2.27	2.78
7400	Solid			500	.016		1.80	.63		2.43	2.96
7450	Conductive			500	.016		5.45	.63		6.08	6.95
7500	Vinyl tile, 12" x 12", .050" thick, minimum			500	.016		3	.63		3.63	4.28
7550	Maximum			500	.016		6	.63		6.63	7.60
7600	1/8" thick, minimum			500	.016		4.35	.63		4.98	5.75
7650	Solid colors			500	.016		4.95	.63		5.58	6.45
7700	Marbleized or Travertine pattern			500	.016		5.65	.63		6.28	7.20
7750	Florentine pattern			500	.016		6	.63		6.63	7.60
7800	Maximum			500	.016		12.70	.63		13.33	14.95
9500	Minimum labor/equipment charge			4	2	Job		78.50		78.50	123

09 65 33 – Conductive Resilient Flooring

09 65 33.10 Conductive Rubber and Vinyl Flooring

		Crew	Daily Output	Labor-Hours	Unit	Material	Labor	Equipment	Total	Total Incl O&P
0010	**CONDUCTIVE RUBBER AND VINYL FLOORING**									
1700	Conductive flooring, rubber tile, 1/8" thick	1 Tilf	315	.025	S.F.	5.10	1		6.10	7.15
1800	Homogeneous vinyl tile, 1/8" thick	"	315	.025	"	7.15	1		8.15	9.40

09 66 Terrazzo Flooring

09 66 13 – Portland Cement Terrazzo Flooring

09 66 13.10 Portland Cement Terrazzo		Crew	Daily Output	Labor-Hours	Unit	Material	2010 Bare Costs Labor	Equipment	Total	Total Incl O&P
0010	**PORTLAND CEMENT TERRAZZO**, cast-in-place									
0020	Cove base, 6" high, 16ga. zinc	1 Mstz	20	.400	L.F.	3.32	15.60		18.92	28
0100	Curb, 6" high and 6" wide		6	1.333		5.25	52		57.25	87
0300	Divider strip for floors, 14 ga., 1-1/4" deep, zinc		375	.021		1.37	.83		2.20	2.81
0400	Brass		375	.021		2.69	.83		3.52	4.26
0600	Heavy top strip 1/4" thick, 1-1/4" deep, zinc		300	.027		1.88	1.04		2.92	3.69
1200	For thin set floors, 16 ga., 1/2" x 1/2", zinc		350	.023		.80	.89		1.69	2.27
1500	Floor, bonded to concrete, 1-3/4" thick, gray cement	J-3	75	.213	S.F.	2.87	7.55	3.56	13.98	18.90
1600	White cement, mud set		75	.213		3.27	7.55	3.56	14.38	19.30
1800	Not bonded, 3" total thickness, gray cement		70	.229		3.59	8.10	3.81	15.50	21
1900	White cement, mud set		70	.229		3.92	8.10	3.81	15.83	21
9000	Minimum labor/equipment charge	1 Mstz	1	8	Job		310		310	485

09 66 13.30 Terrazzo, Precast

		Crew	Daily Output	Labor-Hours	Unit	Material	Labor	Equipment	Total	Total Incl O&P
0010	**TERRAZZO, PRECAST**									
0020	Base, 6" high, straight	1 Mstz	70	.114	L.F.	10.40	4.46		14.86	18.40
0100	Cove		60	.133		12.15	5.20		17.35	21.50
0300	8" high, straight		60	.133		10.85	5.20		16.05	20
0400	Cove		50	.160		15.95	6.25		22.20	27.50
0600	For white cement, add					.43			.43	.47
0700	For 16 ga. zinc toe strip, add					1.63			1.63	1.79
0900	Curbs, 4" x 4" high	1 Mstz	40	.200		30.50	7.80		38.30	45.50
1000	8" x 8" high	"	30	.267		35.50	10.40		45.90	55.50
1200	Floor tiles, non-slip, 1" thick, 12" x 12"	D-1	60	.267	S.F.	18.30	10.05		28.35	36
1300	1-1/4" thick, 12" x 12"		60	.267		19.60	10.05		29.65	37.50
1500	16" x 16"		50	.320		21.50	12.05		33.55	43
1600	1-1/2" thick, 16" x 16"		45	.356		19.45	13.40		32.85	43
1800	For Venetian terrazzo, add					5.85			5.85	6.40
1900	For white cement, add					.55			.55	.61
2400	Stair treads, 1-1/2" thick, non-slip, three line pattern	2 Mstz	70	.229	L.F.	39.50	8.90		48.40	57.50
2500	Nosing and two lines		70	.229		39.50	8.90		48.40	57.50
2700	2" thick treads, straight		60	.267		43	10.40		53.40	63.50
2800	Curved		50	.320		56	12.50		68.50	81
3000	Stair risers, 1" thick, to 6" high, straight sections		60	.267		9.80	10.40		20.20	27
3100	Cove		50	.320		14.35	12.50		26.85	35.50
3300	Curved, 1" thick, to 6" high, vertical		48	.333		19.55	13		32.55	42
3400	Cove		38	.421		37	16.40		53.40	66.50
3600	Stair tread and riser, single piece, straight, minimum		60	.267		50	10.40		60.40	71.50
3700	Maximum		40	.400		65	15.60		80.60	95.50
3900	Curved tread and riser, minimum		40	.400		70	15.60		85.60	102
4000	Maximum		32	.500		88.50	19.50		108	128
4200	Stair stringers, notched, 1" thick		25	.640		28.50	25		53.50	70.50
4300	2" thick		22	.727		34	28.50		62.50	81.50
4500	Stair landings, structural, non-slip, 1-1/2" thick		85	.188	S.F.	31	7.35		38.35	46
4600	3" thick		75	.213		44.50	8.30		52.80	61.50
4800	Wainscot, 12" x 12" x 1" tiles	1 Mstz	12	.667		6.50	26		32.50	47.50
4900	16" x 16" x 1-1/2" tiles	"	8	1		13.70	39		52.70	76
9500	Minimum labor/equipment charge	1 Tilf	2	4	Job		157		157	246

09 66 16 – Terrazzo Floor Tile

09 66 16.10 Tile or Terrazzo Base

		Crew	Daily Output	Labor-Hours	Unit	Material	Labor	Equipment	Total	Total Incl O&P
0010	**TILE OR TERRAZZO BASE**									
0020	Scratch coat only	1 Mstz	150	.053	S.F.	.41	2.08		2.49	3.70
0500	Scratch and brown coat only		75	.107	"	.78	4.16		4.94	7.35

09 66 Terrazzo Flooring

09 66 16 – Terrazzo Floor Tile

09 66 16.10 Tile or Terrazzo Base	Crew	Daily Output	Labor-Hours	Unit	Material	2010 Bare Costs Labor	Equipment	Total	Total Incl O&P
9000 Minimum labor/equipment charge	1 Mstz	1	8	Job		310		310	485

09 67 Fluid-Applied Flooring

09 67 20 – Epoxy-Marble Chip Flooring

09 67 20.13 Elastomeric Liquid Flooring

0010 **ELASTOMERIC LIQUID FLOORING**									
0020 Cementitious acrylic, 1/4" thick	C-6	520	.092	S.F.	1.45	3.19	.12	4.76	6.95
0200 Methyl methachrylate, 1/4" thick	C-8A	3000	.016	"	5.40	.57		5.97	6.85

09 67 20.16 Epoxy Terrazzo

0010 **EPOXY TERRAZZO**									
1800 Epoxy terrazzo, 1/4" thick, chemical resistant, minimum	J-3	200	.080	S.F.	5.25	2.84	1.34	9.43	11.70
1900 Maximum	"	150	.107	"	7.70	3.78	1.78	13.26	16.35

09 67 20.26 Quartz Flooring

0010 **QUARTZ FLOORING**									
0600 Epoxy, with colored quartz chips, broadcast, minimum	C-6	675	.071	S.F.	2.49	2.46	.09	5.04	6.85
0700 Maximum		490	.098		3.02	3.38	.12	6.52	8.95
0900 Trowelled, minimum		560	.086		3.21	2.96	.11	6.28	8.45
1000 Maximum		480	.100		4.69	3.45	.13	8.27	10.90
1200 Heavy duty epoxy topping, 1/4" thick,									
1300 500 to 1,000 S.F.	C-6	420	.114	S.F.	4.67	3.95	.15	8.77	11.70
1500 1,000 to 2,000 S.F.		450	.107		3.87	3.68	.14	7.69	10.40
1600 Over 10,000 S.F.		480	.100		3.62	3.45	.13	7.20	9.70

09 68 Carpeting

09 68 05 – Carpet Accessories

09 68 05.11 Flooring Transition Strip

0010 **FLOORING TRANSITION STRIP**									
0107 Clamp down brass divider, 12' strip, vinyl to carpet	1 Tilf	31.25	.256	Ea.	20.50	10.05		30.55	38.50
0117 Vinyl to hard surface	"	31.25	.256	"	20.50	10.05		30.55	38.50

09 68 10 – Carpet Pad

09 68 10.10 Commercial Grade Carpet Pad

0010 **COMMERCIAL GRADE CARPET PAD**									
9000 Sponge rubber pad, minimum	1 Tilf	150	.053	S.Y.	3.97	2.10		6.07	7.65
9100 Maximum		150	.053		10	2.10		12.10	14.30
9200 Felt pad, minimum		150	.053		4.14	2.10		6.24	7.85
9300 Maximum		150	.053		7.05	2.10		9.15	11.05
9400 Bonded urethane pad, minimum		150	.053		4.57	2.10		6.67	8.35
9500 Maximum		150	.053		7.80	2.10		9.90	11.90
9600 Prime urethane pad, minimum		150	.053		2.62	2.10		4.72	6.15
9700 Maximum		150	.053		4.84	2.10		6.94	8.60

09 68 13 – Tile Carpeting

09 68 13.10 Carpet Tile

0010 **CARPET TILE**									
0100 Tufted nylon, 18" x 18", hard back, 20 oz.	1 Tilf	150	.053	S.Y.	22	2.10		24.10	28
0110 26 oz.		150	.053		38	2.10		40.10	45
0200 Cushion back, 20 oz.		150	.053		28	2.10		30.10	34
0210 26 oz.		150	.053		43.50	2.10		45.60	51.50
1100 Tufted, 24" x 24", hard back, 24 oz. nylon		80	.100		29	3.94		32.94	37.50

09 68 Carpeting

09 68 13 – Tile Carpeting

09 68 13.10 Carpet Tile

		Crew	Daily Output	Labor-Hours	Unit	Material	2010 Bare Costs Labor	Equipment	Total	Total Incl O&P
1180	35 oz.	1 Tilf	80	.100	S.Y.	34	3.94		37.94	43.50
5060	42 oz.	↓	80	.100	↓	45.50	3.94		49.44	56

09 68 16 – Sheet Carpeting

09 68 16.10 Sheet Carpet

		Crew	Daily Output	Labor-Hours	Unit	Material	Labor	Equipment	Total	Total Incl O&P
0010	**SHEET CARPET**									
0700	Nylon, level loop, 26 oz., light to medium traffic	1 Tilf	75	.107	S.Y.	27	4.20		31.20	36
0720	28 oz., light to medium traffic		75	.107		28	4.20		32.20	37
0900	32 oz., medium traffic		75	.107		33	4.20		37.20	43
1100	40 oz., medium to heavy traffic		75	.107		49.50	4.20		53.70	61
2920	Nylon plush, 30 oz., medium traffic		57	.140		25.50	5.50		31	36.50
3000	36 oz., medium traffic		75	.107		34	4.20		38.20	44
3100	42 oz., medium to heavy traffic		70	.114		38	4.50		42.50	49
3200	46 oz., medium to heavy traffic		70	.114		44.50	4.50		49	56
3300	54 oz., heavy traffic		70	.114		49	4.50		53.50	60.50
4500	50 oz., medium to heavy traffic		75	.107		116	4.20		120.20	135
4700	Patterned, 32 oz., medium to heavy traffic		70	.114		115	4.50		119.50	134
4900	48 oz., heavy traffic	↓	70	.114	↓	118	4.50		122.50	137
5000	For less than full roll (approx. 1500 S.F.), add					25%				
5100	For small rooms, less than 12' wide, add						25%			
5200	For large open areas (no cuts), deduct						25%			
5600	For bound carpet baseboard, add	1 Tilf	300	.027	L.F.	2.40	1.05		3.45	4.28
5610	For stairs, not incl. price of carpet, add	"	30	.267	Riser		10.50		10.50	16.40
5620	For borders and patterns, add to labor						18%			
8950	For tackless, stretched installation, add padding to above									
9850	For brand-named specific fiber, add				S.Y.	25%				
9900	Carpet cleaning machine, rent				Day				34.50	37.75
9910	Minimum labor/equipment charge	1 Tilf	3	2.667	Job		105		105	164

09 68 20 – Athletic Carpet

09 68 20.10 Indoor Athletic Carpet

		Crew	Daily Output	Labor-Hours	Unit	Material	Labor	Equipment	Total	Total Incl O&P
0010	**INDOOR ATHLETIC CARPET**									
3700	Polyethylene, in rolls, no base incl., landscape surfaces	1 Tilf	275	.029	S.F.	3.20	1.14		4.34	5.30
3800	Nylon action surface, 1/8" thick		275	.029		3.36	1.14		4.50	5.50
3900	1/4" thick		275	.029		4.84	1.14		5.98	7.10
4000	3/8" thick		275	.029		6.10	1.14		7.24	8.50
5500	Polyvinyl chloride, sheet goods for gyms, 1/4" thick		80	.100		7.75	3.94		11.69	14.70
5600	3/8" thick	↓	60	.133	↓	8.75	5.25		14	17.85

09 69 Access Flooring

09 69 13 – Rigid-Grid Access Flooring

09 69 13.10 Access Floors

		Crew	Daily Output	Labor-Hours	Unit	Material	Labor	Equipment	Total	Total Incl O&P
0010	**ACCESS FLOORS**									
0015	Access floor package including panel, pedestal, stringers & laminate cover									
0100	Computer room, greater than 6,000 S.F.	4 Carp	750	.043	S.F.	8.65	1.77		10.42	12.40
0110	Less than 6,000 S.F.	2 Carp	375	.043		9.95	1.77		11.72	13.85
0120	Office, greater than 6,000 S.F.	4 Carp	1050	.030		4.56	1.27		5.83	7.10
0250	Panels, particle board or steel, 1250# load, no covering, under 6,000 S.F.	2 Carp	600	.027		4.24	1.11		5.35	6.50
0300	Over 6,000 S.F.		640	.025		3.72	1.04		4.76	5.80
0400	Aluminum, 24" panels	↓	500	.032		31.50	1.33		32.83	37
0600	For carpet covering, add					8.55			8.55	9.40
0700	For vinyl floor covering, add					6.70			6.70	7.35

09 69 Access Flooring

09 69 13 – Rigid-Grid Access Flooring

09 69 13.10 Access Floors		Crew	Daily Output	Labor-Hours	Unit	Material	2010 Bare Costs Labor	Equipment	Total	Total Incl O&P
0900	For high pressure laminate covering, add				S.F.	5.45			5.45	6
0910	For snap on stringer system, add	2 Carp	1000	.016		1.44	.66		2.10	2.67
0950	Office applications, steel or concrete panels,									
0960	no covering, over 6,000 S.F.	2 Carp	960	.017	S.F.	10.10	.69		10.79	12.25
1050	Pedestals, 6" to 12"	"	85	.188	Ea.	7.95	7.80		15.75	21.50

09 72 Wall Coverings

09 72 23 – Wallpapering

09 72 23.10 Wallpaper

		Crew	Daily Output	Labor-Hours	Unit	Material	Labor	Equipment	Total	Total Incl O&P
0010	**WALLPAPER** including sizing; add 10-30 percent waste @ takeoff									
0050	Aluminum foil	1 Pape	275	.029	S.F.	1.02	1.06		2.08	2.81
0100	Copper sheets, .025" thick, vinyl backing		240	.033		5.45	1.22		6.67	7.95
0300	Phenolic backing		240	.033		7.05	1.22		8.27	9.75
0600	Cork tiles, light or dark, 12" x 12" x 3/16"		240	.033		4.47	1.22		5.69	6.85
0700	5/16" thick		235	.034		3.80	1.24		5.04	6.15
0900	1/4" basketweave		240	.033		5.85	1.22		7.07	8.40
1000	1/2" natural, non-directional pattern		240	.033		7.85	1.22		9.07	10.60
1100	3/4" natural, non-directional pattern		240	.033		10.80	1.22		12.02	13.80
1200	Granular surface, 12" x 36", 1/2" thick		385	.021		1.26	.76		2.02	2.60
1300	1" thick		370	.022		1.63	.79		2.42	3.05
1500	Polyurethane coated, 12" x 12" x 3/16" thick		240	.033		3.95	1.22		5.17	6.30
1600	5/16" thick		235	.034		5.65	1.24		6.89	8.20
1800	Cork wallpaper, paperbacked, natural		480	.017		1.99	.61		2.60	3.16
1900	Colors		480	.017		2.78	.61		3.39	4.03
2100	Flexible wood veneer, 1/32" thick, plain woods		100	.080		2.36	2.92		5.28	7.25
2200	Exotic woods		95	.084		3.58	3.08		6.66	8.85
2400	Gypsum-based, fabric-backed, fire resistant									
2500	for masonry walls, minimum, 21 oz./S.Y.	1 Pape	800	.010	S.F.	.82	.37		1.19	1.48
2600	Average		720	.011		1.23	.41		1.64	1.99
2700	Maximum (small quantities)		640	.013		1.37	.46		1.83	2.24
2750	Acrylic, modified, semi-rigid PVC, .028" thick	2 Carp	330	.048		1.13	2.01		3.14	4.55
2800	.040" thick	"	320	.050		1.49	2.08		3.57	5.05
3000	Vinyl wall covering, fabric-backed, lightweight, type 1 (12-15 oz./S.Y.)	1 Pape	640	.013		.69	.46		1.15	1.49
3300	Medium weight, type 2 (20-24 oz./S.Y.)		480	.017		.82	.61		1.43	1.87
3400	Heavy weight, type 3 (28 oz./S.Y.)		435	.018		1.64	.67		2.31	2.87
3600	Adhesive, 5 gal. lots (18 S.Y./Gal.)				Gal.	10.50			10.50	11.55
3700	Wallpaper, average workmanship, solid pattern, low cost paper	1 Pape	640	.013	S.F.	.37	.46		.83	1.14
3900	basic patterns (matching required), avg. cost paper		535	.015		.71	.55		1.26	1.65
4000	Paper at $85 per double roll, quality workmanship		435	.018		2.47	.67		3.14	3.79
4200	Grass cloths with lining paper, minimum		400	.020		.78	.73		1.51	2.02
4300	Maximum		350	.023		2.49	.84		3.33	4.07
5990	Wallpaper removal, 1 layer, minimum		800	.010		.06	.37		.43	.65
6000	Wallpaper removal, 3 layer, maximum		400	.020		.14	.73		.87	1.31
9000	Minimum labor/equipment charge		2	4	Job		146		146	232

09 77 Special Wall Surfacing

09 77 33 – Fiberglass Reinforced Panels

09 77 33.10 Fiberglass Reinforced Plastic Panels		Crew	Daily Output	Labor-Hours	Unit	Material	2010 Bare Costs Labor	2010 Bare Costs Equipment	Total	Total Incl O&P
0010	**FIBERGLASS REINFORCED PLASTIC PANELS**, .090" thick									
0020	On walls, adhesive mounted, embossed surface	2 Carp	640	.025	S.F.	1.23	1.04		2.27	3.06
0030	Smooth surface		640	.025		1.51	1.04		2.55	3.37
0040	Fire rated, embossed surface		640	.025		2.19	1.04		3.23	4.12
0050	Nylon rivet mounted, on drywall, embossed surface		480	.033		1.15	1.38		2.53	3.54
0060	Smooth surface		480	.033		1.43	1.38		2.81	3.84
0070	Fire rated, embossed surface		480	.033		2.11	1.38		3.49	4.59
0080	On masonry, embossed surface		320	.050		1.15	2.08		3.23	4.68
0090	Smooth surface		320	.050		1.43	2.08		3.51	4.98
0100	Fire rated, embossed surface		320	.050		2.11	2.08		4.19	5.75
0110	Nylon rivet and adhesive mounted, on drywall, embossed surface		240	.067		1.30	2.77		4.07	6
0120	Smooth surface		240	.067		1.58	2.77		4.35	6.30
0130	Fire rated, embossed surface		240	.067		2.26	2.77		5.03	7.05
0140	On masonry, embossed surface		190	.084		1.30	3.50		4.80	7.20
0150	Smooth surface		190	.084		1.58	3.50		5.08	7.50
0160	Fire rated, embossed surface		190	.084		2.26	3.50		5.76	8.25
0170	For moldings add	1 Carp	250	.032	L.F.	.28	1.33		1.61	2.49
0180	On ceilings, for lay in grid system, embossed surface		400	.020	S.F.	1.23	.83		2.06	2.72
0190	Smooth surface		400	.020		1.51	.83		2.34	3.03
0200	Fire rated, embossed surface		400	.020		2.19	.83		3.02	3.78

09 77 43 – Panel Systems

09 77 43.20 Slatwall Panels and Accessories		Crew	Daily Output	Labor-Hours	Unit	Material	2010 Bare Costs Labor	2010 Bare Costs Equipment	Total	Total Incl O&P
0010	**SLATWALL PANELS AND ACCESSORIES**									
0100	Slatwall panel, 4' x 8' x 3/4" T, MDF, paint grade	1 Carp	500	.016	S.F.	1.46	.66		2.12	2.70
0110	Melamine finish		500	.016		2.09	.66		2.75	3.39
0120	High pressure plastic laminate finish		500	.016		3.41	.66		4.07	4.84
0130	Aluminum channel inserts, add					3.34			3.34	3.67
0200	Accessories, corner forms, 8' L				L.F.	4.49			4.49	4.94
0210	T-connector, 8' L					5.50			5.50	6.05
0220	J-mold, 8' L					1.50			1.50	1.65
0230	Edge cap, 8' L					1.08			1.08	1.19
0240	Finish end cap, 8' L					3.39			3.39	3.73
0300	Display hook, metal, 4" L				Ea.	1			1	1.10
0310	6" L					1.09			1.09	1.20
0320	8" L					1.16			1.16	1.28
0330	10" L					1.31			1.31	1.44
0340	12" L					1.45			1.45	1.60
0350	Acrylic, 4" L					.91			.91	1
0360	6" L					1.05			1.05	1.16
0370	8" L					1.10			1.10	1.21
0380	10" L					1.12			1.12	1.23
0400	Waterfall hanger, metal, 12" - 16"					5.85			5.85	6.40
0410	Acrylic					9.85			9.85	10.85
0500	Shelf bracket, metal, 8"					6.35			6.35	6.95
0510	10"					6.55			6.55	7.20
0520	12"					6.75			6.75	7.45
0530	14"					7.15			7.15	7.85
0540	16"					8.60			8.60	9.45
0550	Acrylic, 8"					3.16			3.16	3.48
0560	10"					3.53			3.53	3.88
0570	12"					3.91			3.91	4.30
0580	14"					4.43			4.43	4.87

09 77 Special Wall Surfacing

09 77 43 – Panel Systems

09 77 43.20 Slatwall Panels and Accessories	Crew	Daily Output	Labor-Hours	Unit	Material	2010 Bare Costs Labor	Equipment	Total	Total Incl O&P
0600 Shelf, acrylic, 12" x 16" x 1/4"				Ea.	26			26	28.50
0610 12" x 24" x 1/4"				↓	43.50			43.50	47.50

09 81 Acoustic Insulation

09 81 16 – Acoustic Blanket Insulation

09 81 16.10 Sound Attenuation Blanket

	Crew	Daily Output	Labor-Hours	Unit	Material	Labor	Equipment	Total	Total Incl O&P
0010 **SOUND ATTENUATION BLANKET**									
0020 Blanket, 1" thick	1 Carp	925	.009	S.F.	.25	.36		.61	.87
0500 1-1/2" thick		920	.009		.25	.36		.61	.87
1000 2" thick		915	.009		.34	.36		.70	.97
1500 3" thick	↓	910	.009		.50	.37		.87	1.15
3400 Urethane plastic foam, open cell, on wall, 2" thick	2 Carp	2050	.008		3.05	.32		3.37	3.89
3500 3" thick		1550	.010		4.05	.43		4.48	5.15
3600 4" thick		1050	.015		5.70	.63		6.33	7.30
3700 On ceiling, 2" thick		1700	.009		3.04	.39		3.43	3.98
3800 3" thick		1300	.012		4.05	.51		4.56	5.30
3900 4" thick	↓	900	.018	↓	5.70	.74		6.44	7.45
4000 Nylon matting 0.4" thick, with carbon black spinerette									
4010 plus polyester fabric, on floor	D-7	4000	.004	S.F.	2.26	.14		2.40	2.71
4200 Fiberglass reinf. backer board underlayment, 7/16" thick, on floor	"	800	.020	"	1.90	.71		2.61	3.19
9000 Minimum labor/equipment charge	1 Carp	5	1.600	Job		66.50		66.50	109

09 84 Acoustic Room Components

09 84 13 – Fixed Sound-Absorptive Panels

09 84 13.10 Fixed Panels

	Crew	Daily Output	Labor-Hours	Unit	Material	Labor	Equipment	Total	Total Incl O&P
0010 **FIXED PANELS** Perforated steel facing, painted with									
0100 Fiberglass or mineral filler, no backs, 2-1/4" thick, modular									
0200 space units, ceiling or wall hung, white or colored	1 Carp	100	.080	S.F.	10	3.32		13.32	16.45
0300 Fiberboard sound deadening panels, 1/2" thick	"	600	.013	"	.32	.55		.87	1.26
0500 Fiberglass panels, 4' x 8' x 1" thick, with									
0600 glass cloth face for walls, cemented	1 Carp	155	.052	S.F.	7	2.14		9.14	11.20
0700 1-1/2" thick, dacron covered, inner aluminum frame,									
0710 wall mounted	1 Carp	300	.027	S.F.	8.75	1.11		9.86	11.40
0900 Mineral fiberboard panels, fabric covered, 30" x 108",									
1000 3/4" thick, concealed spline, wall mounted	1 Carp	150	.053	S.F.	6.25	2.22		8.47	10.55
9000 Minimum labor/equipment charge	"	4	2	Job		83		83	137

09 91 Painting

09 91 03 – Paint Restoration

09 91 03.20 Sanding

	Crew	Daily Output	Labor-Hours	Unit	Material	Labor	Equipment	Total	Total Incl O&P
0010 **SANDING** and puttying interior trim, compared to									
0100 Painting 1 coat, on quality work				L.F.		100%			
0300 Medium work						50%			
0400 Industrial grade				↓		25%			
0500 Surface protection, placement and removal									
0510 Basic drop cloths	1 Pord	6400	.001	S.F.		.05		.05	.07
0520 Masking with paper	↓	800	.010	↓	.06	.36		.42	.65

09 91 Painting

09 91 03 – Paint Restoration

09 91 03.20 Sanding		Crew	Daily Output	Labor-Hours	Unit	Material	2010 Bare Costs Labor	Equipment	Total	Total Incl O&P
0530	Volume cover up (using plastic sheathing, or building paper)	1 Pord	16000	.001	S.F.		.02		.02	.03
09 91 03.30 Exterior Surface Preparation										
0010	**EXTERIOR SURFACE PREPARATION**									
0015	Doors, per side, not incl. frames or trim									
0020	Scrape & sand									
0030	Wood, flush	1 Pord	616	.013	S.F.		.47		.47	.75
0040	Wood, detail		496	.016			.59		.59	.93
0050	Wood, louvered		280	.029			1.04		1.04	1.65
0060	Wood, overhead		616	.013			.47		.47	.75
0070	Wire brush									
0080	Metal, flush	1 Pord	640	.013	S.F.		.45		.45	.72
0090	Metal, detail		520	.015			.56		.56	.89
0100	Metal, louvered		360	.022			.81		.81	1.28
0110	Metal or fibr., overhead		640	.013			.45		.45	.72
0120	Metal, roll up		560	.014			.52		.52	.82
0130	Metal, bulkhead		640	.013			.45		.45	.72
0140	Power wash, based on 2500 lb. operating pressure									
0150	Metal, flush	A-1H	2240	.004	S.F.		.12	.03	.15	.22
0160	Metal, detail		2120	.004			.12	.03	.15	.23
0170	Metal, louvered		2000	.004			.13	.03	.16	.25
0180	Metal or fibr., overhead		2400	.003			.11	.03	.14	.21
0190	Metal, roll up		2400	.003			.11	.03	.14	.21
0200	Metal, bulkhead		2200	.004			.12	.03	.15	.23
0400	Windows, per side, not incl. trim									
0410	Scrape & sand									
0420	Wood, 1-2 lite	1 Pord	320	.025	S.F.		.91		.91	1.44
0430	Wood, 3-6 lite		280	.029			1.04		1.04	1.65
0440	Wood, 7-10 lite		240	.033			1.21		1.21	1.92
0450	Wood, 12 lite		200	.040			1.45		1.45	2.31
0460	Wood, Bay / Bow		320	.025			.91		.91	1.44
0470	Wire brush									
0480	Metal, 1-2 lite	1 Pord	480	.017	S.F.		.61		.61	.96
0490	Metal, 3-6 lite		400	.020			.73		.73	1.15
0500	Metal, Bay/Bow		480	.017			.61		.61	.96
0510	Power wash, based on 2500 lb. operating pressure									
0520	1-2 lite	A-1H	4400	.002	S.F.		.06	.01	.07	.12
0530	3-6 lite		4320	.002			.06	.01	.07	.12
0540	7-10 lite		4240	.002			.06	.01	.07	.12
0550	12 lite		4160	.002			.06	.01	.07	.12
0560	Bay/Bow		4400	.002			.06	.01	.07	.12
0600	Siding, scrape and sand, light=10-30%, med.=30-70%									
0610	Heavy=70-100% of surface to sand									
0650	Texture 1-11, light	1 Pord	480	.017	S.F.		.61		.61	.96
0660	Med.		440	.018			.66		.66	1.05
0670	Heavy		360	.022			.81		.81	1.28
0680	Wood shingles, shakes, light		440	.018			.66		.66	1.05
0690	Med.		360	.022			.81		.81	1.28
0700	Heavy		280	.029			1.04		1.04	1.65
0710	Clapboard, light		520	.015			.56		.56	.89
0720	Med.		480	.017			.61		.61	.96
0730	Heavy		400	.020			.73		.73	1.15
0740	Wire brush									

09 91 Painting

09 91 03 – Paint Restoration

09 91 03.30 Exterior Surface Preparation

		Crew	Daily Output	Labor-Hours	Unit	Material	2010 Bare Costs Labor	2010 Bare Costs Equipment	Total	Total Incl O&P
0750	Aluminum, light	1 Pord	600	.013	S.F.		.48		.48	.77
0760	Med.		520	.015			.56		.56	.89
0770	Heavy	↓	440	.018	↓		.66		.66	1.05
0780	Pressure wash, based on 2500 lb. operating pressure									
0790	Stucco	A-1H	3080	.003	S.F.		.09	.02	.11	.16
0800	Aluminum or vinyl		3200	.003			.08	.02	.10	.16
0810	Siding, masonry, brick & block	↓	2400	.003	↓		.11	.03	.14	.21
1300	Miscellaneous, wire brush									
1310	Metal, pedestrian gate	1 Pord	100	.080	S.F.		2.91		2.91	4.62
8000	For chemical washing, see Div. 04 01 30									
8010	For steam cleaning, see Div. 04 01 30.20									

09 91 03.40 Interior Surface Preparation

		Crew	Daily Output	Labor-Hours	Unit	Material	Labor	Equipment	Total	Total Incl O&P
0010	**INTERIOR SURFACE PREPARATION**									
0020	Doors, per side, not incl. frames or trim									
0030	Scrape & sand									
0040	Wood, flush	1 Pord	616	.013	S.F.		.47		.47	.75
0050	Wood, detail		496	.016			.59		.59	.93
0060	Wood, louvered	↓	280	.029	↓		1.04		1.04	1.65
0070	Wire brush									
0080	Metal, flush	1 Pord	640	.013	S.F.		.45		.45	.72
0090	Metal, detail		520	.015			.56		.56	.89
0100	Metal, louvered	↓	360	.022	↓		.81		.81	1.28
0110	Hand wash									
0120	Wood, flush	1 Pord	2160	.004	S.F.		.13		.13	.21
0130	Wood, detailed		2000	.004			.15		.15	.23
0140	Wood, louvered		1360	.006			.21		.21	.34
0150	Metal, flush		2160	.004			.13		.13	.21
0160	Metal, detail		2000	.004			.15		.15	.23
0170	Metal, louvered	↓	1360	.006	↓		.21		.21	.34
0400	Windows, per side, not incl. trim									
0410	Scrape & sand									
0420	Wood, 1-2 lite	1 Pord	360	.022	S.F.		.81		.81	1.28
0430	Wood, 3-6 lite		320	.025			.91		.91	1.44
0440	Wood, 7-10 lite		280	.029			1.04		1.04	1.65
0450	Wood, 12 lite		240	.033			1.21		1.21	1.92
0460	Wood, Bay/Bow	↓	360	.022	↓		.81		.81	1.28
0470	Wire brush									
0480	Metal, 1-2 lite	1 Pord	520	.015	S.F.		.56		.56	.89
0490	Metal, 3-6 lite		440	.018			.66		.66	1.05
0500	Metal, Bay/Bow	↓	520	.015	↓		.56		.56	.89
0600	Walls, sanding, light=10-30%, medium - 30-70%,									
0610	heavy=70-100% of surface to sand									
0650	Walls, sand									
0660	Gypsum board or plaster, light	1 Pord	3077	.003	S.F.		.09		.09	.15
0670	Gypsum board or plaster, medium		2160	.004			.13		.13	.21
0680	Gypsum board or plaster, heavy		923	.009			.32		.32	.50
0690	Wood, T&G, light		2400	.003			.12		.12	.19
0700	Wood, T&G, med.		1600	.005			.18		.18	.29
0710	Wood, T&G, heavy	↓	800	.010	↓		.36		.36	.58
0720	Walls, wash									
0730	Gypsum board or plaster	1 Pord	3200	.003	S.F.		.09		.09	.14
0740	Wood, T&G	↓	3200	.003	↓		.09		.09	.14

09 91 Painting

09 91 03 – Paint Restoration

09 91 03.40 Interior Surface Preparation

		Crew	Daily Output	Labor-Hours	Unit	Material	2010 Bare Costs Labor	Equipment	Total	Total Incl O&P
0750	Masonry, brick & block, smooth	1 Pord	2800	.003	S.F.		.10		.10	.17
0760	Masonry, brick & block, coarse	↓	2000	.004	↓		.15		.15	.23
8000	For chemical washing, see Div. 04 01 30									
8010	For steam cleaning, see Div. 04 01 30.20									
9010	Minimum labor/equipment charge	1 Pord	3	2.667	Job		97		97	154

09 91 03.41 Scrape After Fire Damage

		Crew	Daily Output	Labor-Hours	Unit	Material	Labor	Equipment	Total	Total Incl O&P
0010	**SCRAPE AFTER FIRE DAMAGE**									
0050	Boards, 1" x 4"	1 Pord	336	.024	L.F.		.87		.87	1.37
0060	1" x 6"		260	.031			1.12		1.12	1.78
0070	1" x 8"		207	.039			1.40		1.40	2.23
0080	1" x 10"		174	.046			1.67		1.67	2.65
0500	Framing, 2" x 4"		265	.030			1.10		1.10	1.74
0510	2" x 6"		221	.036			1.32		1.32	2.09
0520	2" x 8"		190	.042			1.53		1.53	2.43
0530	2" x 10"		165	.048			1.76		1.76	2.80
0540	2" x 12"		144	.056			2.02		2.02	3.21
1000	Heavy framing, 3" x 4"		226	.035			1.29		1.29	2.04
1010	4" x 4"		210	.038			1.38		1.38	2.20
1020	4" x 6"		191	.042			1.52		1.52	2.42
1030	4" x 8"		165	.048			1.76		1.76	2.80
1040	4" x 10"		144	.056			2.02		2.02	3.21
1060	4" x 12"		131	.061	↓		2.22		2.22	3.52
2900	For sealing, minimum		825	.010	S.F.	.14	.35		.49	.71
2920	Maximum	↓	460	.017	"	.28	.63		.91	1.31
3000	For sandblasting, see Div. 04 01 30.20									
3020										
9000	Minimum labor/equipment charge	1 Pord	3	2.667	Job		97		97	154

09 91 13 – Exterior Painting

09 91 13.30 Fences

		Crew	Daily Output	Labor-Hours	Unit	Material	Labor	Equipment	Total	Total Incl O&P
0010	**FENCES**									
0100	Chain link or wire metal, one side, water base									
0110	Roll & brush, first coat	1 Pord	960	.008	S.F.	.05	.30		.35	.54
0120	Second coat		1280	.006		.05	.23		.28	.42
0130	Spray, first coat		2275	.004		.05	.13		.18	.26
0140	Second coat	↓	2600	.003	↓	.05	.11		.16	.24
0150	Picket, water base									
0160	Roll & brush, first coat	1 Pord	865	.009	S.F.	.06	.34		.40	.59
0170	Second coat		1050	.008		.06	.28		.34	.50
0180	Spray, first coat		2275	.004		.06	.13		.19	.26
0190	Second coat	↓	2600	.003	↓	.06	.11		.17	.24
0200	Stockade, water base									
0210	Roll & brush, first coat	1 Pord	1040	.008	S.F.	.06	.28		.34	.50
0220	Second coat		1200	.007		.06	.24		.30	.44
0230	Spray, first coat		2275	.004		.06	.13		.19	.26
0240	Second coat	↓	2600	.003	↓	.06	.11		.17	.24
9000	Minimum labor/equipment charge		2	4	Job		145		145	231

09 91 13.42 Miscellaneous, Exterior

		Crew	Daily Output	Labor-Hours	Unit	Material	Labor	Equipment	Total	Total Incl O&P
0010	**MISCELLANEOUS, EXTERIOR**									
0100	Railing, ext., decorative wood, incl. cap & baluster									
0110	Newels & spindles @ 12" O.C.									
0120	Brushwork, stain, sand, seal & varnish									
0130	First coat	1 Pord	90	.089	L.F.	.69	3.23		3.92	5.90

09 91 Painting

09 91 13 – Exterior Painting

09 91 13.42 Miscellaneous, Exterior

		Crew	Daily Output	Labor-Hours	Unit	Material	2010 Bare Costs Labor	Equipment	Total	Total Incl O&P
0140	Second coat	1 Pord	120	.067	L.F.	.69	2.42		3.11	4.61
0150	Rough sawn wood, 42" high, 2" x 2" verticals, 6" O.C.									
0160	Brushwork, stain, each coat	1 Pord	90	.089	L.F.	.22	3.23		3.45	5.40
0170	Wrought iron, 1" rail, 1/2" sq. verticals									
0180	Brushwork, zinc chromate, 60" high, bars 6" O.C.									
0190	Primer	1 Pord	130	.062	L.F.	.59	2.24		2.83	4.20
0200	Finish coat		130	.062		.18	2.24		2.42	3.75
0210	Additional coat		190	.042		.21	1.53		1.74	2.66
0220	Shutters or blinds, single panel, 2' x 4', paint all sides									
0230	Brushwork, primer	1 Pord	20	.400	Ea.	.60	14.55		15.15	23.50
0240	Finish coat, exterior latex		20	.400		.43	14.55		14.98	23.50
0250	Primer & 1 coat, exterior latex		13	.615		.89	22.50		23.39	36.50
0260	Spray, primer		35	.229		.88	8.30		9.18	14.15
0270	Finish coat, exterior latex		35	.229		.90	8.30		9.20	14.20
0280	Primer & 1 coat, exterior latex		20	.400		.93	14.55		15.48	24
0290	For louvered shutters, add				S.F.	10%				
0300	Stair stringers, exterior, metal									
0310	Roll & brush, zinc chromate, to 14", each coat	1 Pord	320	.025	L.F.	.06	.91		.97	1.51
0320	Rough sawn wood, 4" x 12"									
0330	Roll & brush, exterior latex, each coat	1 Pord	215	.037	L.F.	.06	1.35		1.41	2.22
0340	Trellis/lattice, 2" x 2" @ 3" O.C. with 2" x 8" supports									
0350	Spray, latex, per side, each coat	1 Pord	475	.017	S.F.	.06	.61		.67	1.04
0450	Decking, ext., sealer, alkyd, brushwork, sealer coat		1140	.007		.09	.26		.35	.51
0460	1st coat		1140	.007		.09	.26		.35	.50
0470	2nd coat		1300	.006		.06	.22		.28	.42
0500	Paint, alkyd, brushwork, primer coat		1140	.007		.07	.26		.33	.49
0510	1st coat		1140	.007		.16	.26		.42	.58
0520	2nd coat		1300	.006		.12	.22		.34	.48
0600	Sand paint, alkyd, brushwork, 1 coat		150	.053		.11	1.94		2.05	3.20
9000	Minimum labor/equipment charge		2	4	Job		145		145	231

09 91 13.60 Siding Exterior

		Crew	Daily Output	Labor-Hours	Unit	Material	2010 Bare Costs Labor	Equipment	Total	Total Incl O&P
0010	**SIDING EXTERIOR**, Alkyd (oil base)									
0450	Steel siding, oil base, paint 1 coat, brushwork	2 Pord	2015	.008	S.F.	.13	.29		.42	.60
0500	Spray		4550	.004		.19	.13		.32	.41
0800	Paint 2 coats, brushwork		1300	.012		.25	.45		.70	.99
1000	Spray		2750	.006		.17	.21		.38	.53
1200	Stucco, rough, oil base, paint 2 coats, brushwork		1300	.012		.25	.45		.70	.99
1400	Roller		1625	.010		.27	.36		.63	.86
1600	Spray		2925	.005		.28	.20		.48	.63
1800	Texture 1-11 or clapboard, oil base, primer coat, brushwork		1300	.012		.10	.45		.55	.82
2000	Spray		4550	.004		.10	.13		.23	.31
2400	Paint 2 coats, brushwork		810	.020		.37	.72		1.09	1.54
2600	Spray		2600	.006		.41	.22		.63	.80
3400	Stain 2 coats, brushwork		950	.017		.15	.61		.76	1.13
4000	Spray		3050	.005		.17	.19		.36	.48
4200	Wood shingles, oil base primer coat, brushwork		1300	.012		.09	.45		.54	.81
4400	Spray		3900	.004		.08	.15		.23	.33
5000	Paint 2 coats, brushwork		810	.020		.31	.72		1.03	1.48
5200	Spray		2275	.007		.29	.26		.55	.73
6500	Stain 2 coats, brushwork		950	.017		.15	.61		.76	1.13
7000	Spray		2660	.006		.21	.22		.43	.58
8000	For latex paint, deduct					10%				

09 91 Painting

09 91 13 – Exterior Painting

09 91 13.60 Siding Exterior

		Crew	Daily Output	Labor-Hours	Unit	Material	2010 Bare Costs Labor	Equipment	Total	Total Incl O&P
8100	For work over 12' H, from pipe scaffolding, add				S.F.		15%			
8200	For work over 12' H, from extension ladder, add						25%			
8300	For work over 12' H, from swing staging, add						35%			
9000	Minimum labor/equipment charge	1 Pord	2	4	Job		145		145	231

09 91 13.62 Siding, Misc.

		Crew	Daily Output	Labor-Hours	Unit	Material	Labor	Equipment	Total	Total Incl O&P
0010	**SIDING, MISC.**, latex paint									
0100	Aluminum siding									
0110	Brushwork, primer	2 Pord	2275	.007	S.F.	.05	.26		.31	.47
0120	Finish coat, exterior latex		2275	.007		.04	.26		.30	.46
0130	Primer & 1 coat exterior latex		1300	.012		.11	.45		.56	.83
0140	Primer & 2 coats exterior latex		975	.016		.15	.60		.75	1.12
0150	Mineral fiber shingles									
0160	Brushwork, primer	2 Pord	1495	.011	S.F.	.10	.39		.49	.73
0170	Finish coat, industrial enamel		1495	.011		.15	.39		.54	.78
0180	Primer & 1 coat enamel		810	.020		.25	.72		.97	1.41
0190	Primer & 2 coats enamel		540	.030		.39	1.08		1.47	2.14
0200	Roll, primer		1625	.010		.11	.36		.47	.69
0210	Finish coat, industrial enamel		1625	.010		.16	.36		.52	.75
0220	Primer & 1 coat enamel		975	.016		.27	.60		.87	1.24
0230	Primer & 2 coats enamel		650	.025		.43	.89		1.32	1.89
0240	Spray, primer		3900	.004		.08	.15		.23	.33
0250	Finish coat, industrial enamel		3900	.004		.13	.15		.28	.39
0260	Primer & 1 coat enamel		2275	.007		.22	.26		.48	.65
0270	Primer & 2 coats enamel		1625	.010		.35	.36		.71	.96
0280	Waterproof sealer, first coat		4485	.004		.07	.13		.20	.28
0290	Second coat		5235	.003		.06	.11		.17	.25
0300	Rough wood incl. shingles, shakes or rough sawn siding									
0310	Brushwork, primer	2 Pord	1280	.013	S.F.	.12	.45		.57	.85
0320	Finish coat, exterior latex		1280	.013		.08	.45		.53	.81
0330	Primer & 1 coat exterior latex		960	.017		.20	.61		.81	1.18
0340	Primer & 2 coats exterior latex		700	.023		.28	.83		1.11	1.63
0350	Roll, primer		2925	.005		.16	.20		.36	.50
0360	Finish coat, exterior latex		2925	.005		.09	.20		.29	.42
0370	Primer & 1 coat exterior latex		1790	.009		.26	.33		.59	.80
0380	Primer & 2 coats exterior latex		1300	.012		.35	.45		.80	1.10
0390	Spray, primer		3900	.004		.14	.15		.29	.39
0400	Finish coat, exterior latex		3900	.004		.07	.15		.22	.32
0410	Primer & 1 coat exterior latex		2600	.006		.21	.22		.43	.58
0420	Primer & 2 coats exterior latex		2080	.008		.28	.28		.56	.75
0430	Waterproof sealer, first coat		4485	.004		.12	.13		.25	.34
0440	Second coat		4485	.004		.07	.13		.20	.28
0450	Smooth wood incl. butt, T&G, beveled, drop or B&B siding									
0460	Brushwork, primer	2 Pord	2325	.007	S.F.	.09	.25		.34	.50
0470	Finish coat, exterior latex		1280	.013		.08	.45		.53	.81
0480	Primer & 1 coat exterior latex		800	.020		.17	.73		.90	1.33
0490	Primer & 2 coats exterior latex		630	.025		.24	.92		1.16	1.74
0500	Roll, primer		2275	.007		.10	.26		.36	.52
0510	Finish coat, exterior latex		2275	.007		.08	.26		.34	.50
0520	Primer & 1 coat exterior latex		1300	.012		.18	.45		.63	.91
0530	Primer & 2 coats exterior latex		975	.016		.27	.60		.87	1.24
0540	Spray, primer		4550	.004		.08	.13		.21	.28
0550	Finish coat, exterior latex		4550	.004		.07	.13		.20	.28

09 91 Painting

09 91 13 – Exterior Painting

09 91 13.62 Siding, Misc.

		Crew	Daily Output	Labor-Hours	Unit	Material	2010 Bare Costs Labor	2010 Bare Costs Equipment	Total	Total Incl O&P
0560	Primer & 1 coat exterior latex	2 Pord	2600	.006	S.F.	.15	.22		.37	.51
0570	Primer & 2 coats exterior latex		1950	.008		.22	.30		.52	.71
0580	Waterproof sealer, first coat		5230	.003		.07	.11		.18	.25
0590	Second coat		5980	.003		.07	.10		.17	.22
0600	For oil base paint, add					10%				
9000	Minimum labor/equipment charge	1 Pord	2	4	Job		145		145	231

09 91 13.70 Doors and Windows, Exterior

		Crew	Daily Output	Labor-Hours	Unit	Material	Labor	Equipment	Total	Total Incl O&P
0010	**DOORS AND WINDOWS, EXTERIOR**									
0100	Door frames & trim, only									
0110	Brushwork, primer	1 Pord	512	.016	L.F.	.05	.57		.62	.96
0120	Finish coat, exterior latex		512	.016		.05	.57		.62	.96
0130	Primer & 1 coat, exterior latex		300	.027		.11	.97		1.08	1.66
0135	2 coats, exterior latex, both sides		15	.533	Ea.	4.74	19.40		24.14	36
0140	Primer & 2 coats, exterior latex		265	.030	L.F.	.16	1.10		1.26	1.92
0150	Doors, flush, both sides, incl. frame & trim									
0160	Roll & brush, primer	1 Pord	10	.800	Ea.	4.16	29		33.16	50.50
0170	Finish coat, exterior latex		10	.800		4.04	29		33.04	50.50
0180	Primer & 1 coat, exterior latex		7	1.143		8.20	41.50		49.70	75
0190	Primer & 2 coats, exterior latex		5	1.600		12.25	58		70.25	106
0200	Brushwork, stain, sealer & 2 coats polyurethane		4	2		24	72.50		96.50	142
0210	Doors, French, both sides, 10-15 lite, incl. frame & trim									
0220	Brushwork, primer	1 Pord	6	1.333	Ea.	2.08	48.50		50.58	79.50
0230	Finish coat, exterior latex		6	1.333		2.02	48.50		50.52	79
0240	Primer & 1 coat, exterior latex		3	2.667		4.10	97		101.10	159
0250	Primer & 2 coats, exterior latex		2	4		6	145		151	238
0260	Brushwork, stain, sealer & 2 coats polyurethane		2.50	3.200		8.80	116		124.80	195
0270	Doors, louvered, both sides, incl. frame & trim									
0280	Brushwork, primer	1 Pord	7	1.143	Ea.	4.16	41.50		45.66	70.50
0290	Finish coat, exterior latex		7	1.143		4.04	41.50		45.54	70.50
0300	Primer & 1 coat, exterior latex		4	2		8.20	72.50		80.70	124
0310	Primer & 2 coats, exterior latex		3	2.667		12	97		109	167
0320	Brushwork, stain, sealer & 2 coats polyurethane		4.50	1.778		24	64.50		88.50	130
0330	Doors, panel, both sides, incl. frame & trim									
0340	Roll & brush, primer	1 Pord	6	1.333	Ea.	4.16	48.50		52.66	81.50
0350	Finish coat, exterior latex		6	1.333		4.04	48.50		52.54	81.50
0360	Primer & 1 coat, exterior latex		3	2.667		8.20	97		105.20	163
0370	Primer & 2 coats, exterior latex		2.50	3.200		12	116		128	198
0380	Brushwork, stain, sealer & 2 coats polyurethane		3	2.667		24	97		121	181
0400	Windows, per ext. side, based on 15 SF									
0410	1 to 6 lite									
0420	Brushwork, primer	1 Pord	13	.615	Ea.	.82	22.50		23.32	36.50
0430	Finish coat, exterior latex		13	.615		.80	22.50		23.30	36.50
0440	Primer & 1 coat, exterior latex		8	1		1.62	36.50		38.12	59.50
0450	Primer & 2 coats, exterior latex		6	1.333		2.37	48.50		50.87	79.50
0460	Stain, sealer & 1 coat varnish		7	1.143		3.47	41.50		44.97	70
0470	7 to 10 lite									
0480	Brushwork, primer	1 Pord	11	.727	Ea.	.82	26.50		27.32	43
0490	Finish coat, exterior latex		11	.727		.80	26.50		27.30	43
0500	Primer & 1 coat, exterior latex		7	1.143		1.62	41.50		43.12	68
0510	Primer & 2 coats, exterior latex		5	1.600		2.37	58		60.37	95
0520	Stain, sealer & 1 coat varnish		6	1.333		3.47	48.50		51.97	81
0530	12 lite									

09 91 Painting

09 91 13 – Exterior Painting

09 91 13.70 Doors and Windows, Exterior

		Crew	Daily Output	Labor-Hours	Unit	Material	2010 Bare Costs Labor	Equipment	Total	Total Incl O&P
0540	Brushwork, primer	1 Pord	10	.800	Ea.	.82	29		29.82	47
0550	Finish coat, exterior latex		10	.800		.80	29		29.80	47
0560	Primer & 1 coat, exterior latex		6	1.333		1.62	48.50		50.12	79
0570	Primer & 2 coats, exterior latex		5	1.600		2.37	58		60.37	95
0580	Stain, sealer & 1 coat varnish		6	1.333		3.40	48.50		51.90	80.50
0590	For oil base paint, add					10%				
9000	Minimum labor/equipment charge	1 Pord	2	4	Job		145		145	231

09 91 13.80 Trim, Exterior

		Crew	Daily Output	Labor-Hours	Unit	Material	Labor	Equipment	Total	Total Incl O&P
0010	**TRIM, EXTERIOR**									
0100	Door frames & trim (see Doors, interior or exterior)									
0110	Fascia, latex paint, one coat coverage									
0120	1" x 4", brushwork	1 Pord	640	.013	L.F.	.02	.45		.47	.74
0130	Roll		1280	.006		.02	.23		.25	.38
0140	Spray		2080	.004		.01	.14		.15	.24
0150	1" x 6" to 1" x 10", brushwork		640	.013		.06	.45		.51	.78
0160	Roll		1230	.007		.06	.24		.30	.45
0170	Spray		2100	.004		.05	.14		.19	.27
0180	1" x 12", brushwork		640	.013		.06	.45		.51	.78
0190	Roll		1050	.008		.06	.28		.34	.51
0200	Spray		2200	.004		.05	.13		.18	.26
0210	Gutters & downspouts, metal, zinc chromate paint									
0220	Brushwork, gutters, 5", first coat	1 Pord	640	.013	L.F.	.06	.45		.51	.79
0230	Second coat		960	.008		.06	.30		.36	.55
0240	Third coat		1280	.006		.05	.23		.28	.41
0250	Downspouts, 4", first coat		640	.013		.06	.45		.51	.79
0260	Second coat		960	.008		.06	.30		.36	.55
0270	Third coat		1280	.006		.05	.23		.28	.41
0280	Gutters & downspouts, wood									
0290	Brushwork, gutters, 5", primer	1 Pord	640	.013	L.F.	.05	.45		.50	.78
0300	Finish coat, exterior latex		640	.013		.05	.45		.50	.77
0310	Primer & 1 coat exterior latex		400	.020		.11	.73		.84	1.27
0320	Primer & 2 coats exterior latex		325	.025		.16	.89		1.05	1.60
0330	Downspouts, 4", primer		640	.013		.05	.45		.50	.78
0340	Finish coat, exterior latex		640	.013		.05	.45		.50	.77
0350	Primer & 1 coat exterior latex		400	.020		.11	.73		.84	1.27
0360	Primer & 2 coats exterior latex		325	.025		.08	.89		.97	1.51
0370	Molding, exterior, up to 14" wide									
0380	Brushwork, primer	1 Pord	640	.013	L.F.	.07	.45		.52	.79
0390	Finish coat, exterior latex		640	.013		.06	.45		.51	.78
0400	Primer & 1 coat exterior latex		400	.020		.13	.73		.86	1.29
0410	Primer & 2 coats exterior latex		315	.025		.13	.92		1.05	1.61
0420	Stain & fill		1050	.008		.09	.28		.37	.54
0430	Shellac		1850	.004		.09	.16		.25	.35
0440	Varnish		1275	.006		.10	.23		.33	.47
9000	Minimum labor/equipment charge		2	4	Job		145		145	231

09 91 13.90 Walls, Masonry (CMU), Exterior

		Crew	Daily Output	Labor-Hours	Unit	Material	Labor	Equipment	Total	Total Incl O&P
0350	**WALLS, MASONRY (CMU), EXTERIOR**									
0360	Concrete masonry units (CMU), smooth surface									
0370	Brushwork, latex, first coat	1 Pord	640	.013	S.F.	.06	.45		.51	.78
0380	Second coat		960	.008		.05	.30		.35	.53
0390	Waterproof sealer, first coat		736	.011		.24	.40		.64	.90
0400	Second coat		1104	.007		.24	.26		.50	.69

09 91 Painting

09 91 13 – Exterior Painting

09 91 13.90 Walls, Masonry (CMU), Exterior

		Crew	Daily Output	Labor-Hours	Unit	Material	2010 Bare Costs Labor	Equipment	Total	Total Incl O&P
0410	Roll, latex, paint, first coat	1 Pord	1465	.005	S.F.	.07	.20		.27	.40
0420	Second coat		1790	.004		.05	.16		.21	.32
0430	Waterproof sealer, first coat		1680	.005		.24	.17		.41	.54
0440	Second coat		2060	.004		.24	.14		.38	.49
0450	Spray, latex, paint, first coat		1950	.004		.05	.15		.20	.30
0460	Second coat		2600	.003		.04	.11		.15	.23
0470	Waterproof sealer, first coat		2245	.004		.24	.13		.37	.48
0480	Second coat		2990	.003		.24	.10		.34	.42
0490	Concrete masonry unit (CMU), porous									
0500	Brushwork, latex, first coat	1 Pord	640	.013	S.F.	.12	.45		.57	.85
0510	Second coat		960	.008		.06	.30		.36	.54
0520	Waterproof sealer, first coat		736	.011		.24	.40		.64	.90
0530	Second coat		1104	.007		.24	.26		.50	.69
0540	Roll latex, first coat		1465	.005		.09	.20		.29	.42
0550	Second coat		1790	.004		.06	.16		.22	.32
0560	Waterproof sealer, first coat		1680	.005		.24	.17		.41	.54
0570	Second coat		2060	.004		.24	.14		.38	.49
0580	Spray latex, first coat		1950	.004		.06	.15		.21	.31
0590	Second coat		2600	.003		.04	.11		.15	.23
0600	Waterproof sealer, first coat		2245	.004		.24	.13		.37	.48
0610	Second coat		2990	.003		.24	.10		.34	.42
9000	Minimum labor/equipment charge		2	4	Job		145		145	231

09 91 23 – Interior Painting

09 91 23.20 Cabinets and Casework

		Crew	Daily Output	Labor-Hours	Unit	Material	2010 Bare Costs Labor	Equipment	Total	Total Incl O&P
0010	**CABINETS AND CASEWORK**									
1000	Primer coat, oil base, brushwork	1 Pord	650	.012	S.F.	.07	.45		.52	.79
2000	Paint, oil base, brushwork, 1 coat		650	.012		.10	.45		.55	.83
3000	Stain, brushwork, wipe off		650	.012		.07	.45		.52	.79
4000	Shellac, 1 coat, brushwork		650	.012		.07	.45		.52	.79
4500	Varnish, 3 coats, brushwork, sand after 1st coat		325	.025		.24	.89		1.13	1.68
5000	For latex paint, deduct					10%				
9010	Minimum labor/equipment charge	1 Pord	2	4	Job		145		145	231

09 91 23.33 Doors and Windows, Interior Alkyd (Oil Base)

		Crew	Daily Output	Labor-Hours	Unit	Material	2010 Bare Costs Labor	Equipment	Total	Total Incl O&P
0010	**DOORS AND WINDOWS, INTERIOR ALKYD (OIL BASE)**									
0500	Flush door & frame, 3' x 7', oil, primer, brushwork	1 Pord	10	.800	Ea.	2.90	29		31.90	49
1000	Paint, 1 coat		10	.800		4.03	29		33.03	50.50
1400	Stain, brushwork, wipe off		18	.444		1.57	16.15		17.72	27
1600	Shellac, 1 coat, brushwork		25	.320		1.52	11.65		13.17	20
1800	Varnish, 3 coats, brushwork, sand after 1st coat		9	.889		5	32.50		37.50	57
2000	Panel door & frame, 3' x 7', oil, primer, brushwork		6	1.333		2.72	48.50		51.22	80
2200	Paint, 1 coat		6	1.333		4.03	48.50		52.53	81.50
2600	Stain, brushwork, panel door, 3' x 7', not incl. frame		16	.500		1.57	18.20		19.77	30.50
2800	Shellac, 1 coat, brushwork		22	.364		1.52	13.20		14.72	22.50
3000	Varnish, 3 coats, brushwork, sand after 1st coat		7.50	1.067		5	39		44	67
4400	Windows, including frame and trim, per side									
4600	Colonial type, 6/6 lites, 2' x 3', oil, primer, brushwork	1 Pord	14	.571	Ea.	.43	21		21.43	33.50
5800	Paint, 1 coat		14	.571		.64	21		21.64	33.50
6200	3' x 5' opening, 6/6 lites, primer coat, brushwork		12	.667		1.07	24		25.07	39.50
6400	Paint, 1 coat		12	.667		1.59	24		25.59	40.50
6800	4' x 8' opening, 6/6 lites, primer coat, brushwork		8	1		2.29	36.50		38.79	60
7000	Paint, 1 coat		8	1		3.39	36.50		39.89	61
8000	Single lite type, 2' x 3', oil base, primer coat, brushwork		33	.242		.43	8.80		9.23	14.45

09 91 Painting

09 91 23 – Interior Painting

09 91 23.33 Doors and Windows, Interior Alkyd (Oil Base)

		Crew	Daily Output	Labor-Hours	Unit	Material	2010 Bare Costs Labor	Equipment	Total	Total Incl O&P
8200	Paint, 1 coat	1 Pord	33	.242	Ea.	.64	8.80		9.44	14.70
8600	3' x 5' opening, primer coat, brushwork		20	.400		1.07	14.55		15.62	24
8800	Paint, 1 coat		20	.400		1.59	14.55		16.14	25
9200	4' x 8' opening, primer coat, brushwork		14	.571		2.29	21		23.29	35.50
9400	Paint, 1 coat		14	.571		3.39	21		24.39	36.50
9900	Minimum labor/equipment charge		2	4	Job		145		145	231

09 91 23.35 Doors and Windows, Interior Latex

		Crew	Daily Output	Labor-Hours	Unit	Material	Labor	Equipment	Total	Total Incl O&P
0010	**DOORS & WINDOWS, INTERIOR LATEX**									
0100	Doors, flush, both sides, incl. frame & trim									
0110	Roll & brush, primer	1 Pord	10	.800	Ea.	3.78	29		32.78	50
0120	Finish coat, latex		10	.800		4.09	29		33.09	50.50
0130	Primer & 1 coat latex		7	1.143		7.85	41.50		49.35	74.50
0140	Primer & 2 coats latex		5	1.600		11.75	58		69.75	105
0160	Spray, both sides, primer		20	.400		3.98	14.55		18.53	27.50
0170	Finish coat, latex		20	.400		4.30	14.55		18.85	27.50
0180	Primer & 1 coat latex		11	.727		8.35	26.50		34.85	51
0190	Primer & 2 coats latex		8	1		12.40	36.50		48.90	71
0200	Doors, French, both sides, 10-15 lite, incl. frame & trim									
0210	Roll & brush, primer	1 Pord	6	1.333	Ea.	1.89	48.50		50.39	79
0220	Finish coat, latex		6	1.333		2.05	48.50		50.55	79.50
0230	Primer & 1 coat latex		3	2.667		3.94	97		100.94	158
0240	Primer & 2 coats latex		2	4		5.85	145		150.85	237
0260	Doors, louvered, both sides, incl. frame & trim									
0270	Roll & brush, primer	1 Pord	7	1.143	Ea.	3.78	41.50		45.28	70
0280	Finish coat, latex		7	1.143		4.09	41.50		45.59	70.50
0290	Primer & 1 coat, latex		4	2		7.65	72.50		80.15	123
0300	Primer & 2 coats, latex		3	2.667		11.95	97		108.95	167
0320	Spray, both sides, primer		20	.400		3.98	14.55		18.53	27.50
0330	Finish coat, latex		20	.400		4.30	14.55		18.85	27.50
0340	Primer & 1 coat, latex		11	.727		8.35	26.50		34.85	51
0350	Primer & 2 coats, latex		8	1		12.70	36.50		49.20	71.50
0360	Doors, panel, both sides, incl. frame & trim									
0370	Roll & brush, primer	1 Pord	6	1.333	Ea.	3.98	48.50		52.48	81.50
0380	Finish coat, latex		6	1.333		4.09	48.50		52.59	81.50
0390	Primer & 1 coat, latex		3	2.667		7.85	97		104.85	163
0400	Primer & 2 coats, latex		2.50	3.200		11.95	116		127.95	198
0420	Spray, both sides, primer		10	.800		3.98	29		32.98	50.50
0430	Finish coat, latex		10	.800		4.30	29		33.30	50.50
0440	Primer & 1 coat, latex		5	1.600		8.35	58		66.35	102
0450	Primer & 2 coats, latex		4	2		12.70	72.50		85.20	129
0460	Windows, per interior side, based on 15 SF									
0470	1 to 6 lite									
0480	Brushwork, primer	1 Pord	13	.615	Ea.	.75	22.50		23.25	36.50
0490	Finish coat, enamel		13	.615		.81	22.50		23.31	36.50
0500	Primer & 1 coat enamel		8	1		1.55	36.50		38.05	59
0510	Primer & 2 coats enamel		6	1.333		2.36	48.50		50.86	79.50
0530	7 to 10 lite									
0540	Brushwork, primer	1 Pord	11	.727	Ea.	.75	26.50		27.25	43
0550	Finish coat, enamel		11	.727		.81	26.50		27.31	43
0560	Primer & 1 coat enamel		7	1.143		1.55	41.50		43.05	67.50
0570	Primer & 2 coats enamel		5	1.600		2.36	58		60.36	95
0590	12 lite									

09 91 Painting

09 91 23 – Interior Painting

09 91 23.35 Doors and Windows, Interior Latex

		Crew	Daily Output	Labor-Hours	Unit	Material	2010 Bare Costs Labor	2010 Bare Costs Equipment	Total	Total Incl O&P
0600	Brushwork, primer	1 Pord	10	.800	Ea.	.75	29		29.75	47
0610	Finish coat, enamel		10	.800		.81	29		29.81	47
0620	Primer & 1 coat enamel		6	1.333		1.55	48.50		50.05	78.50
0630	Primer & 2 coats enamel		5	1.600		2.36	58		60.36	95
0650	For oil base paint, add					10%				
9000	Minimum labor/equipment charge	1 Pord	2	4	Job		145		145	231

09 91 23.39 Doors and Windows, Interior Latex, Zero Voc

			Crew	Daily Output	Labor-Hours	Unit	Material	Labor	Equipment	Total	Total Incl O&P
0010	**DOORS & WINDOWS, INTERIOR LATEX, ZERO VOC**										
0100	Doors flush, both sides, incl. frame & trim										
0110	Roll & brush, primer	G	1 Pord	10	.800	Ea.	5.05	29		34.05	51.50
0120	Finish coat, latex	G		10	.800		5.25	29		34.25	52
0130	Primer & 1 coat latex	G		7	1.143		10.30	41.50		51.80	77.50
0140	Primer & 2 coats latex	G		5	1.600		15.25	58		73.25	109
0160	Spray, both sides, primer	G		20	.400		5.35	14.55		19.90	29
0170	Finish coat, latex	G		20	.400		5.50	14.55		20.05	29
0180	Primer & 1 coat latex	G		11	.727		10.90	26.50		37.40	54
0190	Primer & 2 coats latex	G		8	1		16.15	36.50		52.65	75.50
0200	Doors, French, both sides, 10-15 lite, incl. frame & trim										
0210	Roll & brush, primer	G	1 Pord	6	1.333	Ea.	2.53	48.50		51.03	80
0220	Finish coat, latex	G		6	1.333		2.62	48.50		51.12	80
0230	Primer & 1 coat latex	G		3	2.667		5.15	97		102.15	160
0240	Primer & 2 coats latex	G		2	4		7.60	145		152.60	239
0360	Doors, panel, both sides, incl. frame & trim										
0370	Roll & brush, primer	G	1 Pord	6	1.333	Ea.	5.35	48.50		53.85	83
0380	Finish coat, latex	G		6	1.333		5.25	48.50		53.75	83
0390	Primer & 1 coat, latex	G		3	2.667		10.30	97		107.30	165
0400	Primer & 2 coats, latex	G		2.50	3.200		15.55	116		131.55	202
0420	Spray, both sides, primer	G		10	.800		5.35	29		34.35	52
0430	Finish coat, latex	G		10	.800		5.50	29		34.50	52
0440	Primer & 1 coat, latex	G		5	1.600		10.90	58		68.90	105
0450	Primer & 2 coats, latex	G		4	2		16.45	72.50		88.95	133
0460	Windows, per interior side, based on 15 SF										
0470	1 to 6 lite										
0480	Brushwork, primer	G	1 Pord	13	.615	Ea.	1	22.50		23.50	36.50
0490	Finish coat, enamel	G		13	.615		1.03	22.50		23.53	36.50
0500	Primer & 1 coat enamel	G		8	1		2.03	36.50		38.53	59.50
0510	Primer & 2 coats enamel	G		6	1.333		3.07	48.50		51.57	80.50
9000	Minimum labor/equipment charge			2	4	Job		145		145	231

09 91 23.40 Floors, Interior

		Crew	Daily Output	Labor-Hours	Unit	Material	Labor	Equipment	Total	Total Incl O&P
0010	**FLOORS, INTERIOR**									
0100	Concrete paint, latex									
0110	Brushwork									
0120	1st coat	1 Pord	975	.008	S.F.	.15	.30		.45	.64
0130	2nd coat		1150	.007		.10	.25		.35	.51
0140	3rd coat		1300	.006		.08	.22		.30	.44
0150	Roll									
0160	1st coat	1 Pord	2600	.003	S.F.	.20	.11		.31	.40
0170	2nd coat		3250	.002		.12	.09		.21	.27
0180	3rd coat		3900	.002		.09	.07		.16	.22
0190	Spray									
0200	1st coat	1 Pord	2600	.003	S.F.	.17	.11		.28	.37
0210	2nd coat		3250	.002		.09	.09		.18	.24

09 91 Painting

09 91 23 – Interior Painting

09 91 23.40 Floors, Interior		Crew	Daily Output	Labor-Hours	Unit	Material	2010 Bare Costs Labor	Equipment	Total	Total Incl O&P
0220	3rd coat	1 Pord	3900	.002	S.F.	.08	.07		.15	.20
0300	Acid stain and sealer									
0310	Stain, one coat	1 Pord	650	.012	S.F.	.11	.45		.56	.83
0320	Two coats		570	.014		.22	.51		.73	1.05
0330	Acrylic sealer, one coat		2600	.003		.15	.11		.26	.35
0340	Two coats		1400	.006		.30	.21		.51	.66

09 91 23.52 Miscellaneous, Interior

		Crew	Daily Output	Labor-Hours	Unit	Material	Labor	Equipment	Total	Total Incl O&P
0010	**MISCELLANEOUS, INTERIOR**									
2400	Floors, conc./wood, oil base, primer/sealer coat, brushwork	2 Pord	1950	.008	S.F.	.06	.30		.36	.54
2450	Roller		5200	.003		.06	.11		.17	.25
2600	Spray		6000	.003		.06	.10		.16	.22
2650	Paint 1 coat, brushwork		1950	.008		.10	.30		.40	.58
2800	Roller		5200	.003		.10	.11		.21	.29
2850	Spray		6000	.003		.11	.10		.21	.27
3000	Stain, wood floor, brushwork, 1 coat		4550	.004		.07	.13		.20	.28
3200	Roller		5200	.003		.08	.11		.19	.27
3250	Spray		6000	.003		.08	.10		.18	.24
3400	Varnish, wood floor, brushwork		4550	.004		.08	.13		.21	.29
3450	Roller		5200	.003		.08	.11		.19	.27
3600	Spray		6000	.003		.09	.10		.19	.25
3800	Grilles, per side, oil base, primer coat, brushwork	1 Pord	520	.015		.14	.56		.70	1.05
3850	Spray		1140	.007		.15	.26		.41	.57
3920	Paint 2 coats, brushwork		325	.025		.41	.89		1.30	1.87
3940	Spray		650	.012		.47	.45		.92	1.23
5000	Pipe, 1" - 4" diameter, primer or sealer coat, oil base, brushwork	2 Pord	1250	.013	L.F.	.06	.47		.53	.81
5100	Spray		2165	.007		.06	.27		.33	.50
5350	Paint 2 coats, brushwork		775	.021		.15	.75		.90	1.35
5400	Spray		1240	.013		.16	.47		.63	.92
6300	13" - 16" diameter, primer or sealer coat, brushwork		310	.052		.25	1.88		2.13	3.26
6450	Spray		540	.030		.33	1.08		1.41	2.08
6500	Paint 2 coats, brushwork		195	.082		.58	2.98		3.56	5.35
6550	Spray		310	.052		.65	1.88		2.53	3.69
7000	Trim, wood, incl. puttying, under 6" wide									
7200	Primer coat, oil base, brushwork	1 Pord	650	.012	L.F.	.04	.45		.49	.75
7250	Paint, 1 coat, brushwork		650	.012		.05	.45		.50	.77
7450	3 coats		325	.025		.15	.89		1.04	1.59
7500	Over 6" wide, primer coat, brushwork		650	.012		.07	.45		.52	.79
7550	Paint, 1 coat, brushwork		650	.012		.11	.45		.56	.83
7650	3 coats		325	.025		.31	.89		1.20	1.76
8000	Cornice, simple design, primer coat, oil base, brushwork		650	.012	S.F.	.07	.45		.52	.79
8250	Paint, 1 coat		650	.012		.11	.45		.56	.83
8350	Ornate design, primer coat		350	.023		.07	.83		.90	1.40
8400	Paint, 1 coat		350	.023		.11	.83		.94	1.44
8600	Balustrades, primer coat, oil base, brushwork		520	.015		.07	.56		.63	.97
8800										
8900	Trusses and wood frames, primer coat, oil base, brushwork	1 Pord	800	.010	S.F.	.07	.36		.43	.66
8950	Spray		1200	.007		.07	.24		.31	.46
9220	Paint 2 coats, brushwork		500	.016		.21	.58		.79	1.15
9240	Spray		600	.013		.23	.48		.71	1.02
9260	Stain, brushwork, wipe off		600	.013		.07	.48		.55	.85
9280	Varnish, 3 coats, brushwork		275	.029		.24	1.06		1.30	1.94
9350	For latex paint, deduct					10%				

09 91 Painting

09 91 23 – Interior Painting

09 91 23.52 Miscellaneous, Interior		Crew	Daily Output	Labor-Hours	Unit	Material	2010 Bare Costs Labor	Equipment	Total	Total Incl O&P
9900	Minimum labor/equipment charge	1 Pord	2	4	Job		145		145	231

09 91 23.62 Electrostatic Painting

		Crew	Daily Output	Labor-Hours	Unit	Material	Labor	Equipment	Total	Total Incl O&P
0010	**ELECTROSTATIC PAINTING**									
0100	In Shop									
0200	Flat Surfaces (lockers, casework, elev doors. etc)									
0300	One coat	1 Pord	200	.040	S.F.	.46	1.45		1.91	2.82
0400	Two coats	"	120	.067	"	.67	2.42		3.09	4.59
0500	Irregular Surfaces (furniture, door frames, etc)									
0600	One coat	1 Pord	150	.053	S.F.	.46	1.94		2.40	3.59
0700	Two coats	"	100	.080	"	.67	2.91		3.58	5.35
0800	On Site									
0900	Flat Surfaces (lockers, casework, elev doors. etc)									
1000	One coat	1 Pord	150	.053	S.F.	.46	1.94		2.40	3.59
1100	Two coats	"	100	.080	"	.67	2.91		3.58	5.35
1200	Irregular Surfaces (furniture, door frames, etc)									
1300	One coat	1 Pord	115	.070	S.F.	.46	2.53		2.99	4.52
1400	Two coats	"	70	.114	"	.67	4.15		4.82	7.35

09 91 23.72 Walls and Ceilings, Interior

		Crew	Daily Output	Labor-Hours	Unit	Material	Labor	Equipment	Total	Total Incl O&P
0010	**WALLS AND CEILINGS, INTERIOR**									
0100	Concrete, drywall or plaster, oil base, primer or sealer coat									
0200	Smooth finish, brushwork	1 Pord	1150	.007	S.F.	.07	.25		.32	.48
0240	Roller		1350	.006		.07	.22		.29	.41
0280	Spray		2750	.003		.05	.11		.16	.23
0300	Sand finish, brushwork		975	.008		.07	.30		.37	.54
0340	Roller		1150	.007		.08	.25		.33	.48
0380	Spray		2275	.004		.06	.13		.19	.26
0800	Paint 2 coats, smooth finish, brushwork		680	.012		.18	.43		.61	.88
0840	Roller		800	.010		.19	.36		.55	.79
0880	Spray		1625	.005		.16	.18		.34	.46
0900	Sand finish, brushwork		605	.013		.18	.48		.66	.96
0940	Roller		1020	.008		.19	.29		.48	.66
0980	Spray		1700	.005		.16	.17		.33	.45
1600	Glaze coating, 2 coats, spray, clear		1200	.007		.49	.24		.73	.92
1640	Multicolor		1200	.007		.95	.24		1.19	1.43
1660	Painting walls, complete, including surface prep, primer &									
1670	2 coats finish, on drywall or plaster, with roller	1 Pord	325	.025	S.F.	.18	.89		1.07	1.62
1700	For latex paint, deduct					10%				
1800	For ceiling installations, add						25%			
2000	Masonry or concrete block, oil base, primer or sealer coat									
2100	Smooth finish, brushwork	1 Pord	1224	.007	S.F.	.06	.24		.30	.45
2180	Spray		2400	.003		.09	.12		.21	.29
2200	Sand finish, brushwork		1089	.007		.10	.27		.37	.53
2280	Spray		2400	.003		.09	.12		.21	.29
2800	Paint 2 coats, smooth finish, brushwork		756	.011		.20	.38		.58	.83
2880	Spray		1360	.006		.18	.21		.39	.54
2900	Sand finish, brushwork		672	.012		.20	.43		.63	.91
2980	Spray		1360	.006		.18	.21		.39	.54
3600	Glaze coating, 3 coats, spray, clear		900	.009		.70	.32		1.02	1.28
3620	Multicolor		900	.009		1.10	.32		1.42	1.72
4000	Block filler, 1 coat, brushwork		425	.019		.13	.68		.81	1.24
4100	Silicone, water repellent, 2 coats, spray		2000	.004		.26	.15		.41	.51
4120	For latex paint, deduct					10%				

09 91 Painting

09 91 23 – Interior Painting

09 91 23.72 Walls and Ceilings, Interior

		Crew	Daily Output	Labor-Hours	Unit	Material	2010 Bare Costs Labor	Equipment	Total	Total Incl O&P
8200	For work 8' - 15' H, add				S.F.		10%			
8300	For work over 15' H, add						20%			
8400	For light textured surfaces, add						10%			
8410	Heavy textured, add						25%			
9900	Minimum labor/equipment charge	1 Pord	2	4	Job		145		145	231

09 91 23.74 Walls and Ceilings, Interior, Zero VOC Latex

			Crew	Daily Output	Labor-Hours	Unit	Material	Labor	Equipment	Total	Total Incl O&P
0010	**WALLS AND CEILINGS, INTERIOR, ZERO VOC LATEX**										
0100	Concrete, dry wall or plaster, latex, primer or sealer coat										
0200	Smooth finish, brushwork	G	1 Pord	1150	.007	S.F.	.07	.25		.32	.47
0240	Roller	G		1350	.006		.06	.22		.28	.41
0280	Spray	G		2750	.003		.05	.11		.16	.22
0300	Sand finish, brushwork	G		975	.008		.06	.30		.36	.54
0340	Roller	G		1150	.007		.07	.25		.32	.48
0380	Spray	G		2275	.004		.05	.13		.18	.26
0800	Paint 2 coats, smooth finish, brushwork	G		680	.012		.13	.43		.56	.82
0840	Roller	G		800	.010		.13	.36		.49	.72
0880	Spray	G		1625	.005		.11	.18		.29	.40
0900	Sand finish, brushwork	G		605	.013		.12	.48		.60	.89
0940	Roller	G		1020	.008		.13	.29		.42	.59
0980	Spray	G		1700	.005		.11	.17		.28	.39
1800	For ceiling installations, add	G						25%			
8200	For work 8' - 15' H, add							10%			
8300	For work over 15' H, add							20%			
9900	Minimum labor/equipment charge		1 Pord	2	4	Job		145		145	231

09 91 23.75 Dry Fall Painting

		Crew	Daily Output	Labor-Hours	Unit	Material	Labor	Equipment	Total	Total Incl O&P
0010	**DRY FALL PAINTING**									
0100	Sprayed on walls, gypsum board or plaster									
0220	One coat	1 Pord	2600	.003	S.F.	.05	.11		.16	.23
0250	Two coats		1560	.005		.10	.19		.29	.41
0280	Concrete or textured plaster, one coat		1560	.005		.05	.19		.24	.35
0310	Two coats		1300	.006		.10	.22		.32	.46
0340	Concrete block, one coat		1560	.005		.05	.19		.24	.35
0370	Two coats		1300	.006		.10	.22		.32	.46
0400	Wood, one coat		877	.009		.05	.33		.38	.58
0430	Two coats		650	.012		.10	.45		.55	.82
0440	On ceilings, gypsum board or plaster									
0470	One coat	1 Pord	1560	.005	S.F.	.05	.19		.24	.35
0500	Two coats		1300	.006		.10	.22		.32	.46
0530	Concrete or textured plaster, one coat		1560	.005		.05	.19		.24	.35
0560	Two coats		1300	.006		.10	.22		.32	.46
0570	Structural steel, bar joists or metal deck, one coat		1560	.005		.05	.19		.24	.35
0580	Two coats		1040	.008		.10	.28		.38	.55
9900	Minimum labor/equipment charge		2	4	Job		145		145	231

09 93 Staining and Transparent Finishing

09 93 23 – Interior Staining and Finishing

09 93 23.10 Varnish

		Crew	Daily Output	Labor-Hours	Unit	Material	2010 Bare Costs Labor	Equipment	Total	Total Incl O&P
0010	**VARNISH**									
0012	1 coat + sealer, on wood trim, brush, no sanding included	1 Pord	400	.020	S.F.	.07	.73		.80	1.23
0100	Hardwood floors, 2 coats, no sanding included, roller		1890	.004	"	.15	.15		.30	.41
9000	Minimum labor/equipment charge		4	2	Job		72.50		72.50	115

09 96 High-Performance Coatings

09 96 23 – Graffiti-Resistant Coatings

09 96 23.10 Graffiti Resistant Treatments

		Crew	Daily Output	Labor-Hours	Unit	Material	Labor	Equipment	Total	Total Incl O&P
0010	**GRAFFITI RESISTANT TREATMENTS**, sprayed on walls									
0100	Non-sacrificial, permanent non-stick coating, clear, on metals	1 Pord	2000	.004	S.F.	2.06	.15		2.21	2.50
0200	Concrete		2000	.004		2.35	.15		2.50	2.81
0300	Concrete block		2000	.004		3.03	.15		3.18	3.57
0400	Brick		2000	.004		3.44	.15		3.59	4.01
0500	Stone		2000	.004		3.44	.15		3.59	4.01
0600	Unpainted wood		2000	.004		3.97	.15		4.12	4.60
2000	Semi-permanent cross linking polymer primer, on metals		2000	.004		.40	.15		.55	.67
2100	Concrete		2000	.004		.48	.15		.63	.76
2200	Concrete block		2000	.004		.60	.15		.75	.89
2300	Brick		2000	.004		.48	.15		.63	.76
2400	Stone		2000	.004		.48	.15		.63	.76
2500	Unpainted wood		2000	.004		.67	.15		.82	.96
3000	Top coat, on metals		2000	.004		.43	.15		.58	.70
3100	Concrete		2000	.004		.49	.15		.64	.76
3200	Concrete block		2000	.004		.68	.15		.83	.98
3300	Brick		2000	.004		.57	.15		.72	.85
3400	Stone		2000	.004		.57	.15		.72	.85
3500	Unpainted wood		2000	.004		.68	.15		.83	.98
5000	Sacrificial, water based, on metal		2000	.004		.24	.15		.39	.50
5100	Concrete		2000	.004		.24	.15		.39	.50
5200	Concrete block		2000	.004		.24	.15		.39	.50
5300	Brick		2000	.004		.24	.15		.39	.50
5400	Stone		2000	.004		.24	.15		.39	.50
5500	Unpainted wood		2000	.004		.24	.15		.39	.50
8000	Cleaner for use after treatment									
8100	Towels or wipes, per package of 30				Ea.	.63			.63	.70
8200	Aerosol spray, 24 oz. can				"	18			18	19.80

09 96 56 – Epoxy Coatings

09 96 56.20 Wall Coatings

		Crew	Daily Output	Labor-Hours	Unit	Material	Labor	Equipment	Total	Total Incl O&P
0010	**WALL COATINGS**									
0100	Acrylic glazed coatings, minimum	1 Pord	525	.015	S.F.	.28	.55		.83	1.19
0200	Maximum		305	.026		.60	.95		1.55	2.17
0300	Epoxy coatings, minimum		525	.015		.37	.55		.92	1.29
0400	Maximum		170	.047		1.12	1.71		2.83	3.95
2400	Sprayed perlite or vermiculite, 1/16" thick, minimum		2935	.003		.24	.10		.34	.42
2500	Maximum		640	.013		.68	.45		1.13	1.47
2700	Vinyl plastic wall coating, minimum		735	.011		.30	.40		.70	.96
2800	Maximum		240	.033		.75	1.21		1.96	2.75
3000	Urethane on smooth surface, 2 coats, minimum		1135	.007		.24	.26		.50	.67
3100	Maximum		665	.012		.53	.44		.97	1.27

Division Notes

	CREW	DAILY OUTPUT	LABOR-HOURS	UNIT	MAT.	LABOR	EQUIP.	TOTAL	TOTAL INCL O&P

Estimating Tips

General

- The items in this division are usually priced per square foot or each.
- Many items in Division 10 require some type of support system or special anchors that are not usually furnished with the item. The required anchors must be added to the estimate in the appropriate division.
- Some items in Division 10, such as lockers, may require assembly before installation. Verify the amount of assembly required. Assembly can often exceed installation time.

10 20 00 Interior Specialties

- Support angles and blocking are not included in the installation of toilet compartments, shower/dressing compartments, or cubicles. Appropriate line items from Divisions 5 or 6 may need to be added to support the installations.
- Toilet partitions are priced by the stall. A stall consists of a side wall, pilaster, and door with hardware. Toilet tissue holders and grab bars are extra.
- The required acoustical rating of a folding partition can have a significant impact on costs. Verify the sound transmission coefficient rating of the panel priced to the specification requirements.
- Grab bar installation does not include supplemental blocking or backing to support the required load. When grab bars are installed at an existing facility, provisions must be made to attach the grab bars to solid structure.

Reference Numbers

Reference numbers are shown in shaded boxes at the beginning of some major classifications. These numbers refer to related items in the Reference Section. The reference information may be an estimating procedure, an alternate pricing method, or technical information.

Note: Not all subdivisions listed here necessarily appear in this publication.

No part of this publication may be reproduced, stored in a retrieval system, or transmitted in any form or by any means without prior written permission of R.S. Means Company, Inc.

10 11 Visual Display Surfaces

10 11 13 – Chalkboards

10 11 13.13 Fixed Chalkboards

		Crew	Daily Output	Labor-Hours	Unit	Material	2010 Bare Costs Labor	Equipment	Total	Total Incl O&P
0010	**FIXED CHALKBOARDS** Porcelain enamel steel									
3900	Wall hung									
4000	Aluminum frame and chalktrough									
4300	3' x 5'	2 Carp	15	1.067	Ea.	248	44.50		292.50	345
4600	4' x 12'	"	13	1.231	"	515	51		566	650
4700	Wood frame and chalktrough									
4800	3' x 4'	2 Carp	16	1	Ea.	188	41.50		229.50	276
5300	4' x 8'	"	13	1.231	"	350	51		401	470
5400	Liquid chalk, white porcelain enamel, wall hung									
5420	Deluxe units, aluminum trim and chalktrough									
5450	4' x 4'	2 Carp	16	1	Ea.	216	41.50		257.50	305
5550	4' x 12'	"	12	1.333	"	500	55.50		555.50	640
5700	Wood trim and chalktrough									
5900	4' x 4'	2 Carp	16	1	Ea.	550	41.50		591.50	675
6200	4' x 8'	↓	14	1.143	"	760	47.50		807.50	915
9000	Minimum labor/equipment charge		3	5.333	Job		222		222	365

10 11 13.43 Portable Chalkboards

		Crew	Daily Output	Labor-Hours	Unit	Material	Labor	Equipment	Total	Total Incl O&P
0010	**PORTABLE CHALKBOARDS**									
0100	Freestanding, reversible									
0120	Economy, wood frame, 4' x 6'									
0140	Chalkboard both sides				Ea.	525			525	575
0160	Chalkboard one side, cork other side				"	500			500	555
0200	Standard, lightweight satin finished aluminum, 4' x 6'									
0220	Chalkboard both sides				Ea.	565			565	620
0240	Chalkboard one side, cork other side				"	595			595	655
0300	Deluxe, heavy duty extruded aluminum, 4' x 6'									
0320	Chalkboard both sides				Ea.	990			990	1,100
0340	Chalkboard one side, cork other side				"	890			890	980

10 11 23 – Tackboards

10 11 23.10 Fixed Tackboards

		Crew	Daily Output	Labor-Hours	Unit	Material	Labor	Equipment	Total	Total Incl O&P
0010	**FIXED TACKBOARDS**									
2120	Prefabricated, 1/4" cork, 3' x 5' with aluminum frame	2 Carp	16	1	Ea.	113	41.50		154.50	193
2140	Wood frame	"	16	1	"	143	41.50		184.50	226
2300	Glass enclosed cabinets, alum., cork panel, hinged doors									
2600	4' x 7', 3 door	2 Carp	10	1.600	Ea.	1,650	66.50		1,716.50	1,925
9000	Minimum labor/equipment charge	"	4	4	Job		166		166	273

10 13 Directories

10 13 10 – Building Directories

10 13 10.10 Directory Boards

		Crew	Daily Output	Labor-Hours	Unit	Material	Labor	Equipment	Total	Total Incl O&P
0010	**DIRECTORY BOARDS**									
0050	Plastic, glass covered, 30" x 20"	2 Carp	3	5.333	Ea.	243	222		465	630
0100	36" x 48"		2	8		745	330		1,075	1,375
0900	Outdoor, weatherproof, black plastic, 36" x 24"		2	8		720	330		1,050	1,325
1000	36" x 36"	↓	1.50	10.667	↓	830	445		1,275	1,650
9000	Minimum labor/equipment charge	1 Carp	1	8	Job		330		330	545

10 14 Signage

10 14 19 – Dimensional Letter Signage

10 14 19.10 Exterior Signs

		Crew	Daily Output	Labor-Hours	Unit	Material	2010 Bare Costs Labor	2010 Bare Costs Equipment	Total	Total Incl O&P
0010	**EXTERIOR SIGNS**									
3900	Plaques, custom, 20" x 30", for up to 450 letters, cast aluminum	2 Carp	4	4	Ea.	1,050	166		1,216	1,425
4000	Cast bronze		4	4		1,550	166		1,716	2,000
4200	30" x 36", up to 900 letters cast aluminum		3	5.333		1,925	222		2,147	2,500
4300	Cast bronze		3	5.333		3,700	222		3,922	4,425
5100	Exit signs, 24 ga. alum., 14" x 12" surface mounted	1 Carp	30	.267		41.50	11.10		52.60	63.50
5200	10" x 7"	"	20	.400		25	16.60		41.60	55
6400	Replacement sign faces, 6" or 8"	1 Clab	50	.160		48	5.30		53.30	61.50
9000	Minimum labor/equipment charge	1 Carp	4	2	Job		83		83	137

10 14 23 – Panel Signage

10 14 23.13 Engraved Interior Panel Signage

		Crew	Daily Output	Labor-Hours	Unit	Material	Labor	Equipment	Total	Total Incl O&P
0010	**ENGRAVED INTERIOR PANEL SIGNAGE**									
1010	Flexible door sign, adhesive back, w/Braille, 5/8" letters, 4" x 4"	1 Clab	32	.250	Ea.	26.50	8.30		34.80	43
1050	6" x 6"		32	.250		39.50	8.30		47.80	57
1100	8" x 2"		32	.250		27.50	8.30		35.80	43.50
1150	8" x 4"		32	.250		35	8.30		43.30	52
1200	8" x 8"		32	.250		48	8.30		56.30	66.50
1250	12" x 2"		32	.250		35	8.30		43.30	52
1300	12" x 6"		32	.250		50	8.30		58.30	68.50
1350	12" x 12"		32	.250		99	8.30		107.30	123
1500	Graphic symbols, 2" x 2"		32	.250		12	8.30		20.30	27
1550	6" x 6"		32	.250		30	8.30		38.30	46.50
1600	8" x 8"		32	.250		30	8.30		38.30	46.50

10 14 53 – Traffic Signage

10 14 53.20 Traffic Signs

		Crew	Daily Output	Labor-Hours	Unit	Material	Labor	Equipment	Total	Total Incl O&P
0010	**TRAFFIC SIGNS**									
0012	Stock, 24" x 24", no posts, .080" alum. reflectorized	B-80	70	.457	Ea.	71.50	16.20	10.65	98.35	117
0100	High intensity		70	.457		71.50	16.20	10.65	98.35	117
0300	30" x 30", reflectorized		70	.457		146	16.20	10.65	172.85	198
0400	High intensity		70	.457		146	16.20	10.65	172.85	198
0600	Guide and directional signs, 12" x 18", reflectorized		70	.457		41.50	16.20	10.65	68.35	83
0700	High intensity		70	.457		43.50	16.20	10.65	70.35	85.50
0900	18" x 24", stock signs, reflectorized		70	.457		43.50	16.20	10.65	70.35	85.50
1000	High intensity		70	.457		43.50	16.20	10.65	70.35	85.50
1200	24" x 24", stock signs, reflectorized		70	.457		53.50	16.20	10.65	80.35	96
1300	High intensity		70	.457		59.50	16.20	10.65	86.35	103
1500	Add to above for steel posts, galvanized, 10'-0" upright, bolted		200	.160		32.50	5.65	3.73	41.88	49.50
1600	12'-0" upright, bolted		140	.229		51	8.10	5.30	64.40	75
1800	Highway road signs, aluminum, over 20 S.F., reflectorized		350	.091	S.F.	28	3.24	2.13	33.37	38.50
2000	High intensity		350	.091		28	3.24	2.13	33.37	38.50
2200	Highway, suspended over road, 80 S.F. min., reflectorized		165	.194		28	6.85	4.52	39.37	47
2300	High intensity		165	.194		28	6.85	4.52	39.37	47

10 17 Telephone Specialties

10 17 16 – Telephone Enclosures

10 17 16.10 Commercial Telephone Enclosures		Crew	Daily Output	Labor-Hours	Unit	Material	2010 Bare Costs Labor	Equipment	Total	Total Incl O&P
0010	**COMMERCIAL TELEPHONE ENCLOSURES**									
0300	Shelf type, wall hung, minimum	2 Carp	5	3.200	Ea.	1,050	133		1,183	1,375
0400	Maximum		5	3.200		2,350	133		2,483	2,800
0600	Booth type, painted steel, indoor or outdoor, minimum		1.50	10.667		3,475	445		3,920	4,550
0700	Maximum (stainless steel)		1.50	10.667		11,600	445		12,045	13,400
1900	Outdoor, drive-up type, wall mounted		4	4		930	166		1,096	1,300
2000	Post mounted, stainless steel posts		3	5.333		1,425	222		1,647	1,950
9000	Minimum labor/equipment charge		4	4	Job		166		166	273

10 21 Compartments and Cubicles

10 21 13 – Toilet Compartments

10 21 13.13 Metal Toilet Compartments

		Crew	Daily Output	Labor-Hours	Unit	Material	Labor	Equipment	Total	Total Incl O&P
0010	**METAL TOILET COMPARTMENTS**									
0110	Cubicles, ceiling hung									
0200	Powder coated steel	2 Carp	4	4	Ea.	445	166		611	765
0500	Stainless steel	"	4	4		1,400	166		1,566	1,825
0600	For handicap units, incl. 52" grab bars, add					405			405	445
0900	Floor and ceiling anchored									
1000	Powder coated steel	2 Carp	5	3.200	Ea.	535	133		668	810
1300	Stainless steel	"	5	3.200		1,525	133		1,658	1,900
1400	For handicap units, incl. 52" grab bars, add					287			287	315
1610	Floor anchored									
1700	Powder coated steel	2 Carp	7	2.286	Ea.	535	95		630	745
2000	Stainless steel	"	7	2.286		1,475	95		1,570	1,775
2100	For handicap units, incl. 52" grab bars, add					285			285	315
2200	For juvenile units, deduct					41.50			41.50	45.50
2450	Floor anchored, headrail braced									
2500	Powder coated steel	2 Carp	6	2.667	Ea.	470	111		581	695
2804	Stainless steel	"	4.60	3.478		1,175	145		1,320	1,525
2900	For handicap units, incl. 52" grab bars, add					330			330	365
3000	Wall hung partitions, powder coated steel	2 Carp	7	2.286		625	95		720	840
3300	Stainless steel	"	7	2.286		1,650	95		1,745	1,975
3400	For handicap units, incl. 52" grab bars, add					330			330	365
4000	Screens, entrance, floor mounted, 58" high, 48" wide									
4200	Powder coated steel	2 Carp	15	1.067	Ea.	233	44.50		277.50	330
4500	Stainless steel	"	15	1.067	"	895	44.50		939.50	1,050
4650	Urinal screen, 18" wide									
4704	Powder coated steel	2 Carp	6.15	2.602	Ea.	180	108		288	375
5004	Stainless steel	"	6.15	2.602	"	580	108		688	820
5100	Floor mounted, head rail braced									
5300	Powder coated steel	2 Carp	8	2	Ea.	209	83		292	365
5600	Stainless steel	"	8	2	"	575	83		658	765
5750	Pilaster, flush									
5800	Powder coated steel	2 Carp	10	1.600	Ea.	274	66.50		340.50	410
6100	Stainless steel		10	1.600		670	66.50		736.50	845
6300	Post braced, powder coated steel		10	1.600		175	66.50		241.50	300
6600	Stainless steel		10	1.600		450	66.50		516.50	605
6700	Wall hung, bracket supported									
6800	Powder coated steel	2 Carp	10	1.600	Ea.	305	66.50		371.50	445
7100	Stainless steel		10	1.600		269	66.50		335.50	405
7400	Flange supported, powder coated steel		10	1.600		104	66.50		170.50	224

10 21 Compartments and Cubicles

10 21 13 – Toilet Compartments

10 21 13.13 Metal Toilet Compartments		Crew	Daily Output	Labor-Hours	Unit	Material	2010 Bare Costs Labor	Equipment	Total	Total Incl O&P
7700	Stainless steel	2 Carp	10	1.600	Ea.	305	66.50		371.50	445
7800	Wedge type, powder coated steel		10	1.600		134	66.50		200.50	257
8100	Stainless steel	↓	10	1.600	↓	565	66.50		631.50	730
9000	Minimum labor/equipment charge	1 Carp	2.50	3.200	Job		133		133	218

10 21 13.14 Metal Toilet Compartment Components

		Crew	Daily Output	Labor-Hours	Unit	Material	Labor	Equipment	Total	Total Incl O&P
0010	**METAL TOILET COMPARTMENT COMPONENTS**									
0100	Pilasters									
0110	Overhead braced, powder coated steel, 7" wide x 82" high	2 Carp	22.20	.721	Ea.	75	30		105	132
0120	Stainless steel		22.20	.721		189	30		219	257
0130	Floor braced, powder coated steel, 7" wide x 70" high		23.30	.687		116	28.50		144.50	175
0140	Stainless steel		23.30	.687		277	28.50		305.50	350
0150	Ceiling hung, powder coated steel, 7" wide x 83" high		13.30	1.203		123	50		173	217
0160	Stainless steel		13.30	1.203		300	50		350	415
0170	Wall hung, powder coated steel, 3" wide x 58" high		18.90	.847		115	35		150	185
0180	Stainless steel	↓	18.90	.847	↓	166	35		201	240
0200	Panels									
0210	Powder coated steel, 31" wide x 58" high	2 Carp	18.90	.847	Ea.	128	35		163	198
0220	Stainless steel		18.90	.847		400	35		435	500
0230	Powder coated steel, 53" wide x 58" high		18.90	.847		163	35		198	237
0240	Stainless steel		18.90	.847		530	35		565	640
0250	Powder coated steel, 63" wide x 58" high		18.90	.847		195	35		230	273
0260	Stainless steel	↓	18.90	.847	↓	645	35		680	770
0300	Doors									
0310	Powder coated steel, 24" wide x 58" high	2 Carp	14.10	1.135	Ea.	141	47		188	233
0320	Stainless steel		14.10	1.135		385	47		432	500
0330	Powder coated steel, 26" wide x 58" high		14.10	1.135		144	47		191	236
0340	Stainless steel		14.10	1.135		390	47		437	510
0350	Powder coated steel, 28" wide x 58" high		14.10	1.135		152	47		199	246
0360	Stainless steel		14.10	1.135		410	47		457	530
0370	Powder coated steel, 36" wide x 58" high		14.10	1.135		175	47		222	271
0380	Stainless steel	↓	14.10	1.135	↓	470	47		517	600
0400	Headrails									
0410	For powder coated steel, 62" long	2 Carp	65	.246	Ea.	23.50	10.25		33.75	43
0420	Stainless steel		65	.246		23.50	10.25		33.75	43
0430	For powder coated steel, 84" long		50	.320		32	13.30		45.30	57
0440	Stainless steel		50	.320		30.50	13.30		43.80	55.50
0450	For powder coated steel, 120" long		30	.533		44	22		66	84.50
0460	Stainless steel	↓	30	.533	↓	44	22		66	84.50
9000	Minimum labor/equipment charge	1 Carp	4	2	Job		83		83	137

10 21 13.16 Plastic-Laminate-Clad Toilet Compartments

			Crew	Daily Output	Labor-Hours	Unit	Material	Labor	Equipment	Total	Total Incl O&P
0010	**PLASTIC-LAMINATE-CLAD TOILET COMPARTMENTS**										
0110	Cubicles, ceiling hung										
0300	Plastic laminate on particle board	♿	2 Carp	4	4	Ea.	560	166		726	890
0600	For handicap units, incl. 52" grab bars, add					"	405			405	445
0900	Floor and ceiling anchored										
1100	Plastic laminate on particle board	♿	2 Carp	5	3.200	Ea.	765	133		898	1,050
1400	For handicap units, incl. 52" grab bars, add					"	287			287	315
1610	Floor mounted										
1800	Plastic laminate on particle board		2 Carp	7	2.286	Ea.	555	95		650	765
2450	Floor mounted, headrail braced										
2600	Plastic laminate on particle board	♿	2 Carp	6	2.667	Ea.	800	111		911	1,050
3400	For handicap units, incl. 52" grab bars, add					↓	330			330	365

283

10 21 Compartments and Cubicles

10 21 13 – Toilet Compartments

10 21 13.16 Plastic-Laminate-Clad Toilet Compartments

		Crew	Daily Output	Labor-Hours	Unit	Material	2010 Bare Costs Labor	Equipment	Total	Total Incl O&P
4300	Entrance screen, floor mtd., plas. lam., 58" high, 48" wide	2 Carp	15	1.067	Ea.	580	44.50		624.50	715
4800	Urinal screen, 18" wide, ceiling braced, plastic laminate		8	2		174	83		257	330
5400	Floor mounted, headrail braced		8	2		220	83		303	380
5900	Pilaster, flush, plastic laminate		10	1.600		485	66.50		551.50	640
6400	Post braced, plastic laminate		10	1.600		285	66.50		351.50	425
6700	Wall hung, bracket supported									
6900	Plastic laminate on particle board	2 Carp	10	1.600	Ea.	91	66.50		157.50	209
7450	Flange supported									
7500	Plastic laminate on particle board	2 Carp	10	1.600	Ea.	220	66.50		286.50	350
9000	Minimum labor/equipment charge	1 Carp	2.50	3.200	Job		133		133	218

10 21 13.17 Plastic-Laminate Clad Toilet Compartment Components

		Crew	Daily Output	Labor-Hours	Unit	Material	Labor	Equipment	Total	Total Incl O&P
0010	**PLASTIC-LAMINATE CLAD TOILET COMPARTMENT COMPONENTS**									
0100	Pilasters									
0110	Overhead braced, 7" wide x 82" high	2 Carp	22.20	.721	Ea.	100	30		130	159
0130	Floor anchored, 7" wide x 70" high		23.30	.687		98.50	28.50		127	155
0150	Ceiling hung, 7" wide x 83" high		13.30	1.203		97	50		147	189
0180	Wall hung, 3" wide x 58" high		18.90	.847		87.50	35		122.50	155
0200	Panels									
0210	31" wide x 58" high	2 Carp	18.90	.847	Ea.	135	35		170	206
0230	51" wide x 58" high		18.90	.847		188	35		223	265
0250	63" wide x 58" high		18.90	.847		223	35		258	305
0300	Doors									
0310	24" wide x 58" high	2 Carp	14.10	1.135	Ea.	135	47		182	227
0330	26" wide x 58" high		14.10	1.135		140	47		187	231
0350	28" wide x 58" high		14.10	1.135		145	47		192	237
0370	36" wide x 58" high		14.10	1.135		172	47		219	267
0400	Headrails									
0410	62" long	2 Carp	65	.246	Ea.	25.50	10.25		35.75	45
0430	84" long		60	.267		30.50	11.10		41.60	51.50
0450	120" long		30	.533		44	22		66	84.50
9000	Minimum labor/equipment charge	1 Carp	4	2	Job		83		83	137

10 21 13.19 Plastic Toilet Compartments

		Crew	Daily Output	Labor-Hours	Unit	Material	Labor	Equipment	Total	Total Incl O&P
0010	**PLASTIC TOILET COMPARTMENTS**									
0110	Cubicles, ceiling hung									
0250	Phenolic	2 Carp	4	4	Ea.	905	166		1,071	1,275
0600	For handicap units, incl. 52" grab bars, add				"	405			405	445
0900	Floor and ceiling anchored									
1050	Phenolic	2 Carp	5	3.200	Ea.	940	133		1,073	1,250
1400	For handicap units, incl. 52" grab bars, add				"	287			287	315
1610	Floor mounted									
1750	Phenolic	2 Carp	7	2.286	Ea.	1,350	95		1,445	1,625
2100	For handicap units, incl. 52" grab bars, add					285			285	315
2200	For juvenile units, deduct					41.50			41.50	45.50
2450	Floor mounted, headrail braced									
2550	Phenolic	2 Carp	6	2.667	Ea.	895	111		1,006	1,175
9000	Minimum labor/equipment charge	1 Carp	2.50	3.200	Job		133		133	218

10 21 13.20 Plastic Toilet Compartment Components

		Crew	Daily Output	Labor-Hours	Unit	Material	Labor	Equipment	Total	Total Incl O&P
0010	**PLASTIC TOILET COMPARTMENT COMPONENTS**									
0100	Pilasters									
0110	Overhead braced, polymer plastic, 7" wide x 82" high	2 Carp	22.20	.721	Ea.	158	30		188	223
0120	Phenolic		22.20	.721		153	30		183	217
0130	Floor braced, polymer plastic, 7" wide x 70" high		23.30	.687		165	28.50		193.50	228

10 21 Compartments and Cubicles

10 21 13 – Toilet Compartments

10 21 13.20 Plastic Toilet Compartment Components

		Crew	Daily Output	Labor-Hours	Unit	Material	2010 Bare Costs Labor	2010 Bare Costs Equipment	Total	Total Incl O&P
0140	Phenolic	2 Carp	23.30	.687	Ea.	142	28.50		170.50	203
0150	Ceiling hung, polymer plastic, 7" wide x 83" high		13.30	1.203		141	50		191	237
0160	Phenolic		13.30	1.203		161	50		211	259
0180	Wall hung, phenolic, 3" wide x 58" high		18.90	.847		90.50	35		125.50	158
0200	Panels									
0210	Polymer plastic, 31" high x 55" high	2 Carp	18.90	.847	Ea.	330	35		365	425
0220	Phenolic, 31" wide x 58" high		18.90	.847		259	35		294	345
0230	Polymer plastic, 51" wide x 55" high		18.90	.847		505	35		540	615
0240	Phenolic, 51" wide x 58" high		18.90	.847		395	35		430	490
0250	Polymer plastic, 63" wide x 55" high		18.90	.847		610	35		645	730
0260	Phenolic, 63" wide x 58" high		18.90	.847		465	35		500	575
0300	Doors									
0310	Polymer plastic, 24" wide x 55" high	2 Carp	14.10	1.135	Ea.	263	47		310	365
0320	Phenolic, 24" wide x 58" high		14.10	1.135		289	47		336	400
0330	Polymer plastic, 26" high x 55" high		14.10	1.135		279	47		326	385
0340	Phenolic, 26" wide x 58" high		14.10	1.135		305	47		352	415
0350	Polymer plastic, 28" wide x 55" high		14.10	1.135		295	47		342	405
0360	Phenolic, 28" wide x 58" high		14.10	1.135		325	47		372	435
0370	Polymer plastic, 36" wide x 55" high		14.10	1.135		370	47		417	490
0380	Phenolic, 36" wide x 58" high		14.10	1.135		405	47		452	525
0400	Headrails									
0410	For polymer plastic, 62" long	2 Carp	65	.246	Ea.	32	10.25		42.25	52.50
0420	Phenolic		65	.246		23	10.25		33.25	42.50
0430	For polymer plastic, 84" long		50	.320		35.50	13.30		48.80	61
0440	Phenolic		50	.320		28.50	13.30		41.80	53.50
0450	For polymer plastic, 120" long		30	.533		50.50	22		72.50	92
0460	Phenolic		30	.533		38	22		60	78.50
9000	Minimum labor/equipment charge	1 Carp	4	2	Job		83		83	137

10 21 13.40 Stone Toilet Compartments

			Crew	Daily Output	Labor-Hours	Unit	Material	Labor	Equipment	Total	Total Incl O&P
0010	**STONE TOILET COMPARTMENTS**										
0100	Cubicles, ceiling hung, marble		2 Marb	2	8	Ea.	1,450	320		1,770	2,125
0600	For handicap units, incl. 52" grab bars, add	♿					405			405	445
0800	Floor & ceiling anchored, marble		2 Marb	2.50	6.400		1,825	255		2,080	2,400
1400	For handicap units, incl. 52" grab bars, add	♿					287			287	315
1600	Floor mounted, marble		2 Marb	3	5.333		1,075	213		1,288	1,525
2400	Floor mounted, headrail braced, marble		"	3	5.333		1,025	213		1,238	1,475
2900	For handicap units, incl. 52" grab bars, add	♿					330			330	365
4100	Entrance screen, floor mounted marble, 58" high, 48" wide		2 Marb	9	1.778		675	71		746	860
4600	Urinal screen, 18" wide, ceiling braced, marble		D-1	6	2.667		675	101		776	905
5100	Floor mounted, head rail braced										
5200	Marble		D-1	6	2.667	Ea.	575	101		676	790
5700	Pilaster, flush, marble			9	1.778		750	67		817	935
6200	Post braced, marble			9	1.778		745	67		812	930
9000	Minimum labor/equipment charge		1 Carp	2.50	3.200	Job		133		133	218

10 21 16 – Shower and Dressing Compartments

10 21 16.10 Partitions, Shower

		Crew	Daily Output	Labor-Hours	Unit	Material	Labor	Equipment	Total	Total Incl O&P
0010	**PARTITIONS, SHOWER** Floor mounted, no plumbing									
0400	Cabinet, one piece, fiberglass, 32" x 32"	2 Carp	5	3.200	Ea.	475	133		608	740
0420	36" x 36"		5	3.200		530	133		663	800
0440	36" x 48"		5	3.200		1,375	133		1,508	1,750
0460	Acrylic, 32" x 32"		5	3.200		325	133		458	575
0480	36" x 36"		5	3.200		980	133		1,113	1,300

10 21 Compartments and Cubicles

10 21 16 – Shower and Dressing Compartments

10 21 16.10 Partitions, Shower		Crew	Daily Output	Labor-Hours	Unit	Material	2010 Bare Costs Labor	Equipment	Total	Total Incl O&P
0500	36" x 48"	2 Carp	5	3.200	Ea.	2,100	133		2,233	2,550
0520	Shower door for above, clear plastic, 24" wide	1 Carp	8	1		162	41.50		203.50	247
0540	28" wide		8	1		184	41.50		225.50	271
0560	Tempered glass, 24" wide		8	1		175	41.50		216.50	261
0580	28" wide		8	1		197	41.50		238.50	286
4100	Shower doors, economy plastic, 24" wide	1 Shee	9	.889		132	43.50		175.50	214
4200	Tempered glass door, economy		8	1		219	49		268	320
4700	Deluxe, tempered glass, chrome on brass frame, minimum		8	1		235	49		284	335
4800	Maximum		1	8		910	395		1,305	1,625
9990	Minimum labor/equipment charge		2.50	3.200	Job		157		157	246

10 22 Partitions

10 22 16 – Folding Gates

10 22 16.10 Security Gates

		Crew	Daily Output	Labor-Hours	Unit	Material	Labor	Equipment	Total	Total Incl O&P
0010	**SECURITY GATES** For roll up type, see Div. 08 33 13.10									
0300	Scissors type folding gate, ptd. steel, single, 6-1/2' high, 5-1/2' wide	2 Sswk	4	4	Opng.	214	188		402	580
0350	6-1/2' wide		4	4		213	188		401	580
0400	7-1/2' wide		4	4		231	188		419	600
0600	Double gate, 8' high, 8' wide		2.50	6.400		340	300		640	925
0650	10' wide		2.50	6.400		400	300		700	990
0700	12' wide		2	8		530	375		905	1,275
0750	14' wide		2	8		605	375		980	1,350
0900	Door gate, folding steel, 4' wide, 61" high		4	4		115	188		303	470
1000	71" high		4	4		105	188		293	460
1200	81" high		4	4		114	188		302	470
1300	Window gates, 2' to 4' wide, 31" high		4	4		58	188		246	410
1500	55" high		3.75	4.267		90	200		290	465
1600	79" high		3.50	4.571		105	214		319	505

10 22 19 – Demountable Partitions

10 22 19.43 Demountable Composite Partitions

		Crew	Daily Output	Labor-Hours	Unit	Material	Labor	Equipment	Total	Total Incl O&P
0010	**DEMOUNTABLE COMPOSITE PARTITIONS**, add for doors									
0100	Do not deduct door openings from total L.F.									
0900	Demountable gypsum system on 2" to 2-1/2"									
1000	steel studs, 9' high, 3" to 3-3/4" thick									
1200	Vinyl clad gypsum	2 Carp	48	.333	L.F.	58	13.85		71.85	86.50
1300	Fabric clad gypsum		44	.364		144	15.10		159.10	183
1500	Steel clad gypsum		40	.400		156	16.60		172.60	200
1600	1.75 system, aluminum framing, vinyl clad hardboard,									
1800	paper honeycomb core panel, 1-3/4" to 2-1/2" thick									
1900	9' high	2 Carp	48	.333	L.F.	97	13.85		110.85	130
2100	7' high		60	.267		87	11.10		98.10	114
2200	5' high		80	.200		73.50	8.30		81.80	94.50
2250	Unitized gypsum system									
2300	Unitized panel, 9' high, 2" to 2-1/2" thick									
2350	Vinyl clad gypsum	2 Carp	48	.333	L.F.	125	13.85		138.85	160
2400	Fabric clad gypsum	"	44	.364	"	206	15.10		221.10	251
2500	Unitized mineral fiber system									
2510	Unitized panel, 9' high, 2-1/4" thick, aluminum frame									
2550	Vinyl clad mineral fiber	2 Carp	48	.333	L.F.	124	13.85		137.85	160
2600	Fabric clad mineral fiber	"	44	.364	"	185	15.10		200.10	229

10 22 Partitions

10 22 19 – Demountable Partitions

10 22 19.43 Demountable Composite Partitions

		Crew	Daily Output	Labor-Hours	Unit	Material	2010 Bare Costs Labor	Equipment	Total	Total Incl O&P
2800	Movable steel walls, modular system									
2900	Unitized panels, 9' high, 48" wide									
3100	Baked enamel, pre-finished	2 Carp	60	.267	L.F.	141	11.10		152.10	173
3200	Fabric clad steel	"	56	.286	"	203	11.85		214.85	244
5500	For acoustical partitions, add, minimum				S.F.	2.22			2.22	2.44
5550	Maximum				"	10.35			10.35	11.40
5700	For doors, see Div. 08 11 & 08 16									
5800	For door hardware, see Div. 08 71									
6100	In-plant modular office system, w/prehung hollow core door									
6200	3" thick polystyrene core panels									
6250	12' x 12', 2 wall	2 Clab	3.80	4.211	Ea.	3,650	139		3,789	4,250
6300	4 wall		1.90	8.421		4,650	279		4,929	5,550
6350	16' x 16', 2 wall		3.60	4.444		4,850	147		4,997	5,575
6400	4 wall		1.80	8.889		6,525	294		6,819	7,650
9000	Minimum labor/equipment charge	2 Carp	3	5.333	Job		222		222	365

10 22 23 – Portable Partitions, Screens, and Panels

10 22 23.13 Wall Screens

		Crew	Daily Output	Labor-Hours	Unit	Material	Labor	Equipment	Total	Total Incl O&P
0010	**WALL SCREENS**, divider panels, free standing, fiber core									
0020	Fabric face straight									
0100	3'-0" long, 4'-0" high	2 Carp	100	.160	L.F.	131	6.65		137.65	155
0200	5'-0" high		90	.178		115	7.40		122.40	138
0500	6'-0" high		75	.213		117	8.85		125.85	144
0900	5'-0" long, 4'-0" high		175	.091		79	3.80		82.80	93.50
1000	5'-0" high		150	.107		90.50	4.43		94.93	107
1500	6"-0" high		125	.128		93.50	5.30		98.80	112
1600	6'-0" long, 5'-0" high		162	.099		78	4.10		82.10	93
3200	Economical panels, fabric face, 4'-0" long, 5'-0" high		132	.121		50.50	5.05		55.55	64
3250	6'-0" high		112	.143		54	5.95		59.95	69
3300	5'-0" long, 5'-0" high		150	.107		42	4.43		46.43	53.50
3350	6'-0" high		125	.128		46	5.30		51.30	59.50
3450	Acoustical panels, 60 to 90 NRC, 3'-0" long, 5'-0" high		90	.178		89.50	7.40		96.90	111
3550	6'-0" high		75	.213		99.50	8.85		108.35	124
3600	5'-0" long, 5'-0" high		150	.107		70	4.43		74.43	84.50
3650	6'-0" high		125	.128		68	5.30		73.30	83.50
3700	6'-0" long, 5'-0" high		162	.099		61.50	4.10		65.60	74.50
3750	6'-0" high		138	.116		56	4.82		60.82	69.50
3800	Economy acoustical panels, 40 N.R.C., 4'-0" long, 5'-0" high		132	.121		50.50	5.05		55.55	64
3850	6'-0" high		112	.143		54	5.95		59.95	69
3900	5'-0" long, 6'-0" high		125	.128		46	5.30		51.30	59.50
3950	6'-0" long, 5'-0" high		162	.099		36	4.10		40.10	46.50
9000	Minimum labor/equipment charge		3	5.333	Job		222		222	365

10 22 26 – Operable Partitions

10 22 26.13 Accordion Folding Partitions

		Crew	Daily Output	Labor-Hours	Unit	Material	Labor	Equipment	Total	Total Incl O&P
0010	**ACCORDION FOLDING PARTITIONS**									
0100	Vinyl covered, over 150 S.F., frame not included									
0300	Residential, 1.25 lb. per S.F., 8' maximum height	2 Carp	300	.053	S.F.	19.35	2.22		21.57	25
0400	Commercial, 1.75 lb. per S.F., 8' maximum height		225	.071		22	2.95		24.95	29.50
0900	Acoustical, 3 lb. per S.F., 17' maximum height		100	.160		25.50	6.65		32.15	39
1200	5 lb. per S.F., 20' maximum height		95	.168		35.50	7		42.50	50.50
1500	Vinyl clad wood or steel, electric operation, 5.0 psf		160	.100		48.50	4.16		52.66	60
1900	Wood, non-acoustic, birch or mahogany, to 10' high		300	.053		26	2.22		28.22	32
9000	Minimum labor/equipment charge		4	4	Job		166		166	273

10 22 Partitions

10 22 26 – Operable Partitions

10 22 26.33 Folding Panel Partitions		Crew	Daily Output	Labor-Hours	Unit	Material	2010 Bare Costs Labor	Equipment	Total	Total Incl O&P
0010	**FOLDING PANEL PARTITIONS**, acoustic, wood									
0100	Vinyl faced, to 18' high, 6 psf, minimum	2 Carp	60	.267	S.F.	51.50	11.10		62.60	74.50
0150	Average		45	.356		61.50	14.75		76.25	92
0200	Maximum		30	.533		79.50	22		101.50	124
0400	Formica or hardwood finish, minimum		60	.267		53	11.10		64.10	76.50
0500	Maximum		30	.533		56.50	22		78.50	99
0600	Wood, low acoustical type, 4.5 psf, to 14' high		50	.320		38.50	13.30		51.80	64.50
9000	Minimum labor/equipment charge		4	4	Job		166		166	273

10 26 Wall and Door Protection

10 26 13 – Corner Guards

10 26 13.10 Metal Corner Guards

		Crew	Daily Output	Labor-Hours	Unit	Material	Labor	Equipment	Total	Total Incl O&P
0010	**METAL CORNER GUARDS**									
0020	Steel angle w/anchors, 1" x 1" x 1/4", 1.5#/L.F.	2 Carp	160	.100	L.F.	5.05	4.16		9.21	12.40
0100	2" x 2" x 1/4" angles, 3.2#/L.F.	"	150	.107	"	8.95	4.43		13.38	17.10
9000	Minimum labor/equipment charge	1 Carp	2	4	Job		166		166	273

10 28 Toilet, Bath, and Laundry Accessories

10 28 13 – Toilet Accessories

10 28 13.13 Commercial Toilet Accessories

		Crew	Daily Output	Labor-Hours	Unit	Material	Labor	Equipment	Total	Total Incl O&P
0010	**COMMERCIAL TOILET ACCESSORIES**									
0200	Curtain rod, stainless steel, 5' long, 1" diameter	1 Carp	13	.615	Ea.	34.50	25.50		60	80
0300	1-1/4" diameter		13	.615		24	25.50		49.50	68.50
0400	Diaper changing station, horizontal, wall mounted, plastic		10	.800		217	33		250	293
0500	Dispenser units, combined soap & towel dispensers,									
0510	mirror and shelf, flush mounted	1 Carp	10	.800	Ea.	310	33		343	395
0600	Towel dispenser and waste receptacle,									
0610	18 gallon capacity	1 Carp	10	.800	Ea.	263	33		296	345
0800	Grab bar, straight, 1-1/4" diameter, stainless steel, 18" long		24	.333		30	13.85		43.85	56.50
0900	24" long		23	.348		31.50	14.45		45.95	58
1000	30" long		22	.364		32.50	15.10		47.60	60.50
1100	36" long		20	.400		34.50	16.60		51.10	65.50
1105	42" long		20	.400		37.50	16.60		54.10	69
1200	1-1/2" diameter, 24" long		23	.348		28.50	14.45		42.95	55
1300	36" long		20	.400		32.50	16.60		49.10	63.50
1310	42" long		18	.444		43	18.45		61.45	78
1500	Tub bar, 1-1/4" diameter, 24" x 36"		14	.571		75	23.50		98.50	122
1600	Plus vertical arm		12	.667		85.50	27.50		113	140
1900	End tub bar, 1" diameter, 90° angle, 16" x 32"		12	.667		96	27.50		123.50	151
2010	Tub/shower/toilet, 2-wall, 36" x 24"		12	.667		96	27.50		123.50	152
2300	Hand dryer, surface mounted, electric, 115 volt, 20 amp		4	2		390	83		473	560
2400	230 volt, 10 amp		4	2		655	83		738	855
2600	Hat and coat strip, stainless steel, 4 hook, 36" long		24	.333		71	13.85		84.85	102
2700	6 hook, 60" long		20	.400		112	16.60		128.60	151
3000	Mirror, with stainless steel 3/4" square frame, 18" x 24"		20	.400		56.50	16.60		73.10	90
3100	36" x 24"		15	.533		105	22		127	153
3200	48" x 24"		10	.800		152	33		185	223
3300	72" x 24"		6	1.333		210	55.50		265.50	320
3500	With 5" stainless steel shelf, 18" x 24"		20	.400		189	16.60		205.60	236

10 28 Toilet, Bath, and Laundry Accessories

10 28 13 – Toilet Accessories

10 28 13.13 Commercial Toilet Accessories

		Crew	Daily Output	Labor-Hours	Unit	Material	2010 Bare Costs Labor	Equipment	Total	Total Incl O&P
3600	36" x 24"	1 Carp	15	.533	Ea.	202	22		224	259
3700	48" x 24"		10	.800		214	33		247	291
3800	72" x 24"		6	1.333		264	55.50		319.50	380
4100	Mop holder strip, stainless steel, 5 holders, 48" long		20	.400		78	16.60		94.60	114
4200	Napkin/tampon dispenser, recessed		15	.533		540	22		562	630
4250	Napkin receptacle, recessed		6.50	1.231		142	51		193	240
4300	Robe hook, single, regular		36	.222		11.65	9.25		20.90	28
4400	Heavy duty, concealed mounting		36	.222		13.35	9.25		22.60	30
4600	Soap dispenser, chrome, surface mounted, liquid		20	.400		42.50	16.60		59.10	74.50
4700	Powder		20	.400		50.50	16.60		67.10	83
5000	Recessed stainless steel, liquid		10	.800		126	33		159	194
5100	Powder		10	.800		200	33		233	275
5300	Soap tank, stainless steel, 1 gallon		10	.800		200	33		233	275
5400	5 gallon		5	1.600		275	66.50		341.50	410
5600	Shelf, stainless steel, 5" wide, 18 ga., 24" long		24	.333		57	13.85		70.85	86
5700	48" long		16	.500		134	21		155	181
5800	8" wide shelf, 18 ga., 24" long		22	.364		63	15.10		78.10	94.50
5900	48" long		14	.571		116	23.50		139.50	167
6000	Toilet seat cover dispenser, stainless steel, recessed		20	.400		162	16.60		178.60	206
6050	Surface mounted		15	.533		30.50	22		52.50	70
6100	Toilet tissue dispenser, surface mounted, SS, single roll		30	.267		17.40	11.10		28.50	37.50
6200	Double roll		24	.333		24	13.85		37.85	49.50
6290	Toilet seat	1 Plum	40	.200		30	10.40		40.40	49
6400	Towel bar, stainless steel, 18" long	1 Carp	23	.348		39.50	14.45		53.95	67
6500	30" long		21	.381		111	15.85		126.85	149
6700	Towel dispenser, stainless steel, surface mounted		16	.500		35	21		56	72.50
6800	Flush mounted, recessed		10	.800		237	33		270	315
7000	Towel holder, hotel type, 2 guest size		20	.400		43.50	16.60		60.10	75.50
7200	Towel shelf, stainless steel, 24" long, 8" wide		20	.400		56	16.60		72.60	89
7400	Tumbler holder, tumbler only		30	.267		40	11.10		51.10	62
7500	Soap, tumbler & toothbrush		30	.267		18.30	11.10		29.40	38
7700	Wall urn ash receiver, surface mount, 11" long		12	.667		72	27.50		99.50	125
8000	Waste receptacles, stainless steel, with top, 13 gallon		10	.800		295	33		328	380
8100	36 gallon		8	1		325	41.50		366.50	430
9000	Minimum labor/equipment charge		5	1.600	Job		66.50		66.50	109

10 28 16 – Bath Accessories

10 28 16.20 Medicine Cabinets

		Crew	Daily Output	Labor-Hours	Unit	Material	Labor	Equipment	Total	Total Incl O&P
0010	**MEDICINE CABINETS**									
0020	With mirror, st. st. frame, 16" x 22", unlighted	1 Carp	14	.571	Ea.	66.50	23.50		90	112
0100	Wood frame		14	.571		119	23.50		142.50	170
0300	Sliding mirror doors, 20" x 16" x 4-3/4", unlighted		7	1.143		110	47.50		157.50	199
0400	24" x 19" x 8-1/2", lighted		5	1.600		159	66.50		225.50	284
0600	Triple door, 30" x 32", unlighted, plywood body		7	1.143		279	47.50		326.50	385
0700	Steel body		7	1.143		310	47.50		357.50	420
0900	Oak door, wood body, beveled mirror, single door		7	1.143		160	47.50		207.50	254
1000	Double door		6	1.333		375	55.50		430.50	500
1200	Hotel cabinets, stainless, with lower shelf, unlighted		10	.800		202	33		235	277
1300	Lighted		5	1.600		300	66.50		366.50	440
9000	Minimum labor/equipment charge		4	2	Job		83		83	137

10 31 Manufactured Fireplaces

10 31 13 – Manufactured Fireplace Chimneys

10 31 13.10 Fireplace Chimneys		Crew	Daily Output	Labor-Hours	Unit	Material	2010 Bare Costs Labor	Equipment	Total	Total Incl O&P
0010	**FIREPLACE CHIMNEYS**									
0500	Chimney dbl. wall, all stainless, over 8'-6", 7" diam., add to fireplace	1 Carp	33	.242	V.L.F.	71.50	10.05		81.55	95
0600	10" diameter, add to fireplace		32	.250		109	10.40		119.40	137
0700	12" diameter, add to fireplace		31	.258		136	10.70		146.70	167
0800	14" diameter, add to fireplace		30	.267		183	11.10		194.10	219
1000	Simulated brick chimney top, 4' high, 16" x 16"		10	.800	Ea.	258	33		291	340
1100	24" x 24"		7	1.143	"	480	47.50		527.50	610

10 31 13.20 Chimney Accessories

0010	**CHIMNEY ACCESSORIES**									
0020	Chimney screens, galv., 13" x 13" flue	1 Bric	8	1	Ea.	54	42		96	126
0050	24" x 24" flue		5	1.600		124	67		191	244
0200	Stainless steel, 13" x 13" flue		8	1		330	42		372	425
0250	20" x 20" flue		5	1.600		450	67		517	600
2400	Squirrel and bird screens, galvanized, 8" x 8" flue		16	.500		46.50	21		67.50	85
2450	13" x 13" flue		12	.667		50.50	28		78.50	101
9000	Minimum labor/equipment charge		3.50	2.286	Job		95.50		95.50	153

10 31 16 – Manufactured Fireplace Forms

10 31 16.10 Fireplace Forms

0010	**FIREPLACE FORMS**									
1800	Fireplace forms, no accessories, 32" opening	1 Bric	3	2.667	Ea.	650	111		761	895
1900	36" opening		2.50	3.200		830	134		964	1,125
2000	40" opening		2	4		1,100	167		1,267	1,475
2100	78" opening		1.50	5.333		1,600	223		1,823	2,100

10 31 23 – Prefabricated Fireplaces

10 31 23.10 Fireplace, Prefabricated

0010	**FIREPLACE, PREFABRICATED**, free standing or wall hung									
0100	With hood & screen, minimum	1 Carp	1.30	6.154	Ea.	1,200	256		1,456	1,750
0150	Average		1	8		1,700	330		2,030	2,400
0200	Maximum		.90	8.889		3,375	370		3,745	4,300
1500	Simulated logs, gas fired, 40,000 BTU, 2' long, minimum		7	1.143	Set	450	47.50		497.50	575
1600	Maximum		6	1.333		980	55.50		1,035.50	1,175
1700	Electric, 1,500 BTU, 1'-6" long, minimum		7	1.143		175	47.50		222.50	271
1800	11,500 BTU, maximum		6	1.333		395	55.50		450.50	520
2000	Fireplace, built-in, 36" hearth, radiant		1.30	6.154	Ea.	595	256		851	1,075
2100	Recirculating, small fan		1	8		925	330		1,255	1,575
2150	Large fan		.90	8.889		1,925	370		2,295	2,700
2200	42" hearth, radiant		1.20	6.667		835	277		1,112	1,375
2300	Recirculating, small fan		.90	8.889		1,075	370		1,445	1,800
2350	Large fan		.80	10		1,925	415		2,340	2,775
2400	48" hearth, radiant		1.10	7.273		1,800	300		2,100	2,475
2500	Recirculating, small fan		.80	10		2,225	415		2,640	3,125
2550	Large fan		.70	11.429		3,450	475		3,925	4,575
3000	See through, including doors		.80	10		2,925	415		3,340	3,900
3200	Corner (2 wall)		1	8		2,950	330		3,280	3,775

10 32 Fireplace Specialties

10 32 13 – Fireplace Dampers

10 32 13.10 Dampers		Crew	Daily Output	Labor-Hours	Unit	Material	2010 Bare Costs Labor	Equipment	Total	Total Incl O&P
0010	**DAMPERS**									
0800	Damper, rotary control, steel, 30" opening	1 Bric	6	1.333	Ea.	79.50	55.50		135	177
0850	Cast iron, 30" opening		6	1.333		88.50	55.50		144	187
0880	36" opening		6	1.333		96.50	55.50		152	196
0900	48" opening		6	1.333		137	55.50		192.50	241
0920	60" opening		6	1.333		310	55.50		365.50	430
1000	84" opening, special order		5	1.600		795	67		862	975
1050	96" opening, special order		4	2		805	83.50		888.50	1,025
1200	Steel plate, poker control, 60" opening		8	1		281	42		323	375
1250	84" opening, special order		5	1.600		510	67		577	670
1400	"Universal" type, chain operated, 32" x 20" opening		8	1		218	42		260	305
1450	48" x 24" opening		5	1.600		325	67		392	465

10 32 23 – Fireplace Doors

10 32 23.10 Doors		Crew	Daily Output	Labor-Hours	Unit	Material	Labor	Equipment	Total	Total Incl O&P
0010	**DOORS**									
0400	Cleanout doors and frames, cast iron, 8" x 8"	1 Bric	12	.667	Ea.	37.50	28		65.50	86
0450	12" x 12"		10	.800		45.50	33.50		79	104
0500	18" x 24"		8	1		131	42		173	211
0550	Cast iron frame, steel door, 24" x 30"		5	1.600		283	67		350	415
1600	Dutch Oven door and frame, cast iron, 12" x 15" opening		13	.615		115	25.50		140.50	168
1650	Copper plated, 12" x 15" opening		13	.615		224	25.50		249.50	289

10 35 Stoves

10 35 13 – Heating Stoves

10 35 13.10 Woodburning Stoves		Crew	Daily Output	Labor-Hours	Unit	Material	Labor	Equipment	Total	Total Incl O&P
0010	**WOODBURNING STOVES**									
0015	Cast iron, minimum	2 Carp	1.30	12.308	Ea.	1,175	510		1,685	2,150
0020	Average		1	16		1,600	665		2,265	2,850
0030	Maximum		.80	20		2,850	830		3,680	4,525
0050	For gas log lighter, add					42			42	46.50

10 44 Fire Protection Specialties

10 44 13 – Fire Extinguisher Cabinets

10 44 13.53 Fire Equipment Cabinets		Crew	Daily Output	Labor-Hours	Unit	Material	Labor	Equipment	Total	Total Incl O&P
0010	**FIRE EQUIPMENT CABINETS**, not equipped, 20 ga. steel box									
0040	recessed, D.S. glass in door, box size given									
1000	Portable extinguisher, single, 8" x 12" x 27", alum. door & frame	Q-12	8	2	Ea.	124	90.50		214.50	277
1100	Steel door and frame	"	8	2	"	93	90.50		183.50	242
3000	Hose rack assy., 1-1/2" valve & 100' hose, 24" x 40" x 5-1/2"									
3200	Steel door and frame	Q-12	6	2.667	Ea.	291	121		412	505
4000	Hose rack assy., 2-1/2" x 1-1/2" valve, 100' hose, 24" x 40" x 8"									
4200	Steel door and frame	Q-12	6	2.667	Ea.	208	121		329	415
5000	Hose rack assy., 2-1/2" x 1-1/2" valve, 100' hose									
5010	and extinguisher, 30" x 40" x 8"									
5200	Steel door and frame	Q-12	5	3.200	Ea.	229	145		374	475

10 44 Fire Protection Specialties

10 44 16 – Fire Extinguishers

10 44 16.13 Portable Fire Extinguishers	Crew	Daily Output	Labor-Hours	Unit	Material	2010 Bare Costs Labor	Equipment	Total	Total Incl O&P
0010 **PORTABLE FIRE EXTINGUISHERS**									
0140 CO_2, with hose and "H" horn, 10 lb.				Ea.	191			191	210
1000 Dry chemical, pressurized									
1040 Standard type, portable, painted, 2-1/2 lb.				Ea.	31.50			31.50	34.50
1080 10 lb.					69.50			69.50	76.50
1100 20 lb.					113			113	124
1120 30 lb.					375			375	415
2000 ABC all purpose type, portable, 2-1/2 lb.					21			21	23.50
2080 9-1/2 lb.					44			44	48

10 51 Lockers

10 51 13 – Metal Lockers

10 51 13.10 Lockers

		Crew	Daily Output	Labor-Hours	Unit	Material	Labor	Equipment	Total	Total Incl O&P
0011	**LOCKERS** Steel, baked enamel, pre-assembled									
0110	Single tier box locker, 12" x 15" x 72"	1 Shee	20	.400	Ea.	221	19.65		240.65	274
0120	18" x 15" x 72"		20	.400		240	19.65		259.65	294
0130	12" x 18" x 72"		20	.400		255	19.65		274.65	310
0140	18" x 18" x 72"		20	.400		251	19.65		270.65	305
0410	Double tier, 12" x 15" x 36"		30	.267		233	13.10		246.10	277
0420	18" x 15" x 36"		30	.267		269	13.10		282.10	315
0430	12" x 18" x 36"		30	.267		251	13.10		264.10	297
0440	18" x 18" x 36"		30	.267		249	13.10		262.10	295
0500	Two person, 18" x 15" x 72"		20	.400		273	19.65		292.65	330
0510	18" x 18" x 72"		20	.400		285	19.65		304.65	345
0520	Duplex, 15" x 15" x 72"		20	.400		299	19.65		318.65	360
0530	15" x 21" x 72"		20	.400		325	19.65		344.65	385
1100	Wire meshed wardrobe, floor. mtd., open front varsity type		7.50	1.067		234	52.50		286.50	340
2400	16-person locker unit with clothing rack									
2500	72 wide x 15" deep x 72" high	1 Shee	15	.533	Ea.	465	26		491	550
2550	18" deep		15	.533	"	610	26		636	715
3250	Rack w/24 wire mesh baskets		1.50	5.333	Set	345	262		607	790
3260	30 baskets		1.25	6.400		267	315		582	785
3270	36 baskets		.95	8.421		350	415		765	1,025
3280	42 baskets		.80	10		385	490		875	1,200
3600	For hanger rods, add				Ea.	1.86			1.86	2.05
3650	For number plate kit, 100 plates #1 - #100, add	1 Shee	4	2		60	98		158	220
3700	For locker base, closed front panel		90	.089		6.45	4.36		10.81	13.95
3710	End panel, bolted		36	.222		8.20	10.90		19.10	26
3800	For sloping top, 12" wide		24	.333		22.50	16.35		38.85	50.50
3810	15" wide		24	.333		23.50	16.35		39.85	51
3820	18" wide		24	.333		25	16.35		41.35	53
3850	Sloping top end panel, 12" deep		72	.111		8.55	5.45		14	17.95
3860	15" deep		72	.111		9.50	5.45		14.95	19
3870	18" deep		72	.111		10.10	5.45		15.55	19.65
3900	For finish end panels, steel, 60" high, 15" deep		12	.667		50	32.50		82.50	106
3910	72" high, 12" deep		12	.667		50	32.50		82.50	106
3920	18" deep		12	.667		59.50	32.50		92	117
5000	For 'ready to assemble' lockers,									
5010	Add to labor						75%			
5020	Deduct from material					20%				

10 51 Lockers

10 51 13 – Metal Lockers

10 51 13.10 Lockers

		Crew	Daily Output	Labor-Hours	Unit	Material	2010 Bare Costs Labor	Equipment	Total	Total Incl O&P
9000	Minimum labor/equipment charge	1 Shee	2.50	3.200	Job		157		157	246

10 55 Postal Specialties

10 55 23 – Mail Boxes

10 55 23.10 Commercial Mail Boxes

		Crew	Daily Output	Labor-Hours	Unit	Material	Labor	Equipment	Total	Total Incl O&P
0010	**COMMERCIAL MAIL BOXES**									
0020	Horiz., key lock, 5"H x 6"W x 15"D, alum., rear load	1 Carp	34	.235	Ea.	35	9.80		44.80	54.50
0100	Front loading		34	.235		35	9.80		44.80	54.50
0200	Double, 5"H x 12"W x 15"D, rear loading		26	.308		71.50	12.80		84.30	100
0300	Front loading		26	.308		66	12.80		78.80	93.50
0500	Quadruple, 10"H x 12"W x 15"D, rear loading		20	.400		91	16.60		107.60	128
0600	Front loading		20	.400		108	16.60		124.60	147
1600	Vault type, horizontal, for apartments, 4" x 5"		34	.235		41	9.80		50.80	61.50
1700	Alphabetical directories, 120 names		10	.800		127	33		160	195
1830	Lobby collection boxes, aluminum	2 Shee	5	3.200		2,025	157		2,182	2,475
1840	Bronze or stainless	"	4.50	3.556		2,500	175		2,675	3,025
1900	Letter slot, residential	1 Carp	20	.400		88	16.60		104.60	125
2000	Post office type		8	1		105	41.50		146.50	185
9000	Minimum labor/equipment charge		5	1.600	Job		66.50		66.50	109

10 56 Storage Assemblies

10 56 13 – Metal Storage Shelving

10 56 13.10 Shelving

		Crew	Daily Output	Labor-Hours	Unit	Material	Labor	Equipment	Total	Total Incl O&P
0010	**SHELVING**									
0020	Metal, industrial, cross-braced, 3' wide, 12" deep	1 Sswk	175	.046	SF Shlf	6.60	2.14		8.74	11.15
0100	24" deep		330	.024		4.71	1.14		5.85	7.30
2200	Wide span, 1600 lb. capacity per shelf, 6' wide, 24" deep		380	.021		13.35	.99		14.34	16.45
2400	36" deep		440	.018		8.25	.85		9.10	10.60
9000	Minimum labor/equipment charge	1 Carp	4	2	Job		83		83	137

10 57 Wardrobe and Closet Specialties

10 57 23 – Closet and Utility Shelving

10 57 23.19 Wood Closet and Utility Shelving

		Crew	Daily Output	Labor-Hours	Unit	Material	Labor	Equipment	Total	Total Incl O&P
0010	**WOOD CLOSET AND UTILITY SHELVING**									
0020	Pine, clear grade, no edge band, 1" x 8"	1 Carp	115	.070	L.F.	2.79	2.89		5.68	7.80
0100	1" x 10"		110	.073		3.48	3.02		6.50	8.80
0200	1" x 12"		105	.076		4.19	3.17		7.36	9.80
0600	Plywood, 3/4" thick with lumber edge, 12" wide		75	.107		1.58	4.43		6.01	9.05
0700	24" wide		70	.114		2.81	4.75		7.56	10.90
0900	Bookcase, clear grade pine, shelves 12" O.C., 8" deep, per SF shelf		70	.114	S.F.	9.10	4.75		13.85	17.80
1000	12" deep shelves		65	.123	"	13.60	5.10		18.70	23.50
1200	Adjustable closet rod and shelf, 12" wide, 3' long		20	.400	Ea.	7.80	16.60		24.40	36
1300	8' long		15	.533	"	16.40	22		38.40	54.50
1500	Prefinished shelves with supports, stock, 8" wide		75	.107	L.F.	3.14	4.43		7.57	10.75
1600	10" wide		70	.114	"	4.27	4.75		9.02	12.50
9000	Minimum labor/equipment charge		4	2	Job		83		83	137

10 73 Protective Covers

10 73 13 – Awnings

10 73 13.10 Awnings, Fabric		Crew	Daily Output	Labor-Hours	Unit	Material	2010 Bare Costs Labor	Equipment	Total	Total Incl O&P
0010	**AWNINGS, FABRIC**									
0020	Including acrylic canvas and frame, standard design									
0100	Door and window, slope, 3' high, 4' wide	1 Carp	4.50	1.778	Ea.	690	74		764	880
0110	6' wide		3.50	2.286		890	95		985	1,125
0120	8' wide		3	2.667		1,100	111		1,211	1,375
0200	Quarter round convex, 4' wide		3	2.667		1,075	111		1,186	1,350
0210	6' wide		2.25	3.556		1,400	148		1,548	1,775
0220	8' wide		1.80	4.444		1,700	185		1,885	2,175
0300	Dome, 4' wide		7.50	1.067		415	44.50		459.50	530
0310	6' wide		3.50	2.286		935	95		1,030	1,175
0320	8' wide		2	4		1,650	166		1,816	2,100
0350	Elongated dome, 4' wide		1.33	6.015		1,550	250		1,800	2,100
0360	6' wide		1.11	7.207		1,850	299		2,149	2,550
0370	8' wide		1	8		2,175	330		2,505	2,950
1000	Entry or walkway, peak, 12' long, 4' wide	2 Carp	.90	17.778		4,925	740		5,665	6,625
1010	6' wide		.60	26.667		7,575	1,100		8,675	10,200
1020	8' wide		.40	40		10,500	1,650		12,150	14,200
1100	Radius with dome end, 4' wide		1.10	14.545		3,725	605		4,330	5,100
1110	6' wide		.70	22.857		6,000	950		6,950	8,150
1120	8' wide		.50	32		8,525	1,325		9,850	11,600
2000	Retractable lateral arm awning, manual									
2010	To 12' wide, 8' - 6" projection	2 Carp	1.70	9.412	Ea.	1,100	390		1,490	1,875
2020	To 14' wide, 8' - 6" projection		1.10	14.545		1,300	605		1,905	2,425
2030	To 19' wide, 8' - 6" projection		.85	18.824		1,750	780		2,530	3,200
2040	To 24' wide, 8' - 6" projection		.67	23.881		2,225	990		3,215	4,075
2050	Motor for above, add	1 Carp	2.67	3		955	125		1,080	1,250
3000	Patio/deck canopy with frame									
3010	12' wide, 12' projection	2 Carp	2	8	Ea.	1,575	330		1,905	2,275
3020	16' wide, 14' projection	"	1.20	13.333		2,450	555		3,005	3,575
9000	For fire retardant canvas, add					7%				
9010	For lettering or graphics, add					35%				
9020	For painted or coated acrylic canvas, deduct					8%				
9030	For translucent or opaque vinyl canvas, add					10%				
9040	For 6 or more units, deduct					20%	15%			

10 73 16 – Canopies

10 73 16.10 Canopies, Residential

		Crew	Daily Output	Labor-Hours	Unit	Material	Labor	Equipment	Total	Total Incl O&P
0010	**CANOPIES, RESIDENTIAL** Prefabricated									
0500	Carport, free standing, baked enamel, alum., .032", 40 psf									
0520	16' x 8', 4 posts	2 Carp	3	5.333	Ea.	4,275	222		4,497	5,075
0600	20' x 10', 6 posts		2	8		4,450	330		4,780	5,450
0605	30' x 10', 8 posts		2	8		6,700	330		7,030	7,900
1000	Door canopies, extruded alum., .032", 42" projection, 4' wide	1 Carp	8	1		274	41.50		315.50	370
1020	6' wide	"	6	1.333		355	55.50		410.50	485
1040	8' wide	2 Carp	9	1.778		480	74		554	650
1060	10' wide		7	2.286		705	95		800	930
1080	12' wide		5	3.200		810	133		943	1,100
1200	54" projection, 4' wide	1 Carp	8	1		320	41.50		361.50	425
1220	6' wide	"	6	1.333		400	55.50		455.50	530
1240	8' wide	2 Carp	9	1.778		460	74		534	625
1260	10' wide		7	2.286		760	95		855	990
1280	12' wide		5	3.200		810	133		943	1,100
1300	Painted, add					20%				

10 73 Protective Covers

10 73 16 – Canopies

10 73 16.10 Canopies, Residential

		Crew	Daily Output	Labor-Hours	Unit	Material	2010 Bare Costs Labor	Equipment	Total	Total Incl O&P
1310	Bronze anodized, add				Ea.	50%				
3000	Window awnings, aluminum, window 3' high, 4' wide	1 Carp	10	.800		217	33		250	294
3020	6' wide	"	8	1		263	41.50		304.50	360
3040	9' wide	2 Carp	9	1.778		310	74		384	460
3060	12' wide	"	5	3.200		360	133		493	615
3100	Window, 4' high, 4' wide	1 Carp	10	.800		162	33		195	233
3120	6' wide	"	8	1		230	41.50		271.50	320
3140	9' wide	2 Carp	9	1.778		385	74		459	540
3160	12' wide	"	5	3.200		410	133		543	675
3200	Window, 6' high, 4' wide	1 Carp	10	.800		222	33		255	300
3220	6' wide	"	8	1		340	41.50		381.50	445
3240	9' wide	2 Carp	9	1.778		900	74		974	1,100
3260	12' wide	"	5	3.200		1,375	133		1,508	1,750
3400	Roll-up aluminum, 2'-6" wide	1 Carp	14	.571		190	23.50		213.50	248
3420	3' wide		12	.667		195	27.50		222.50	260
3440	4' wide		10	.800		234	33		267	310
3460	6' wide		8	1		261	41.50		302.50	355
3480	9' wide	2 Carp	9	1.778		370	74		444	530
3500	12' wide	"	5	3.200		475	133		608	745
3600	Window awnings, canvas, 24" drop, 3' wide	1 Carp	30	.267	L.F.	48.50	11.10		59.60	71.50
3620	4' wide		40	.200		42	8.30		50.30	59.50
3700	30" drop, 3' wide		30	.267		51.50	11.10		62.60	75
3720	4' wide		40	.200		44.50	8.30		52.80	62.50
3740	5' wide		45	.178		40	7.40		47.40	56
3760	6' wide		48	.167		37	6.95		43.95	52
3780	8' wide		48	.167		33.50	6.95		40.45	48
3800	10' wide		50	.160		41	6.65		47.65	56.50
3900	Repair canvas, minimum		8	1	Ea.		41.50		41.50	68.50
3920	Maximum		4	2	"		83		83	137
9000	Minimum labor/equipment charge		3	2.667	Job		111		111	182

10 73 16.20 Metal Canopies

		Crew	Daily Output	Labor-Hours	Unit	Material	Labor	Equipment	Total	Total Incl O&P
0010	**METAL CANOPIES**									
0020	Wall hung, .032", aluminum, prefinished, 8' x 10'	K-2	1.30	18.462	Ea.	2,150	790	228	3,168	4,000
0300	8' x 20'		1.10	21.818		4,275	930	269	5,474	6,650
1000	12' x 20'		1	24		5,475	1,025	296	6,796	8,175
1050										
2300	Aluminum entrance canopies, flat soffit, .032"									
2500	3'-6" x 4'-0", clear anodized	2 Carp	4	4	Ea.	900	166		1,066	1,275
9000	Minimum labor/equipment charge	"	2	8	Job		330		330	545

10 74 Manufactured Exterior Specialties

10 74 23 – Cupolas

10 74 23.10 Wood Cupolas

		Crew	Daily Output	Labor-Hours	Unit	Material	Labor	Equipment	Total	Total Incl O&P
0010	**WOOD CUPOLAS**									
0020	Stock units, pine, painted, 18" sq., 28" high, alum. roof	1 Carp	4.10	1.951	Ea.	162	81		243	310
0100	Copper roof		3.80	2.105		208	87.50		295.50	375
0300	23" square, 33" high, aluminum roof		3.70	2.162		345	90		435	530
0400	Copper roof		3.30	2.424		465	101		566	675
0600	30" square, 37" high, aluminum roof		3.70	2.162		495	90		585	695
0700	Copper roof		3.30	2.424		590	101		691	815
0900	Hexagonal, 31" wide, 46" high, copper roof		4	2		730	83		813	940

10 74 Manufactured Exterior Specialties

10 74 23 – Cupolas

10 74 23.10 Wood Cupolas		Crew	Daily Output	Labor-Hours	Unit	Material	2010 Bare Costs Labor	Equipment	Total	Total Incl O&P
1000	36" wide, 50" high, copper roof	1 Carp	3.50	2.286	Ea.	1,275	95		1,370	1,550
1200	For deluxe stock units, add to above					25%				
1400	For custom built units, add to above					50%	50%			
9000	Minimum labor/equipment charge	1 Carp	2.75	2.909	Job		121		121	199

10 74 29 – Steeples

10 74 29.10 Prefabricated Steeples

		Crew	Daily Output	Labor-Hours	Unit	Material	Labor	Equipment	Total	Total Incl O&P
0010	**PREFABRICATED STEEPLES**									
4000	Steeples, translucent fiberglass, 30" square, 15' high	F-3	2	20	Ea.	5,450	840	330	6,620	7,725
4150	25' high	"	1.80	22.222		6,350	935	365	7,650	8,925
4600	Aluminum, baked finish, 16" square, 14' high					4,200			4,200	4,625
4640	35' high, 8' base					27,400			27,400	30,200
4680	152' high, custom					495,000			495,000	544,500

10 74 33 – Weathervanes

10 74 33.10 Residential Weathervanes

		Crew	Daily Output	Labor-Hours	Unit	Material	Labor	Equipment	Total	Total Incl O&P
0010	**RESIDENTIAL WEATHERVANES**									
0020	Residential types, minimum	1 Carp	8	1	Ea.	205	41.50		246.50	294
0100	Maximum		2	4	"	1,700	166		1,866	2,150
9000	Minimum labor/equipment charge		4	2	Job		83		83	137

Estimating Tips
General
- The items in this division are usually priced per square foot or each. Many of these items are purchased by the owner for installation by the contractor. Check the specifications for responsibilities and include time for receiving, storage, installation, and mechanical and electrical hookups in the appropriate divisions.
- Many items in Division 11 require some type of support system that is not usually furnished with the item. Examples of these systems include blocking for the attachment of casework and support angles for ceiling-hung projection screens. The required blocking or supports must be added to the estimate in the appropriate division.
- Some items in Division 11 may require assembly or electrical hookups. Verify the amount of assembly required or the need for a hard electrical connection and add the appropriate costs.

Reference Numbers
Reference numbers are shown in shaded boxes at the beginning of some major classifications. These numbers refer to related items in the Reference Section. The reference information may be an estimating procedure, an alternate pricing method, or technical information.

Note: Not all subdivisions listed here necessarily appear in this publication.

11 11 Vehicle Service Equipment

11 11 13 – Compressed-Air Vehicle Service Equipment

11 11 13.10 Compressed Air Equipment		Crew	Daily Output	Labor-Hours	Unit	Material	2010 Bare Costs Labor	Equipment	Total	Total Incl O&P
0010	**COMPRESSED AIR EQUIPMENT**									
0030	Compressors, electric, 1-1/2 H.P., standard controls	L-4	1.50	16	Ea.	400	625		1,025	1,450
0550	Dual controls		1.50	16		715	625		1,340	1,775
0600	5 H.P., 115/230 volt, standard controls		1	24		2,050	935		2,985	3,775
0650	Dual controls		1	24		2,175	935		3,110	3,925

11 11 19 – Vehicle Lubrication Equipment

11 11 19.10 Lubrication Equipment

0010	**LUBRICATION EQUIPMENT**									
3000	Lube equipment, 3 reel type, with pumps, not including piping	L-4	.50	48	Set	7,900	1,875		9,775	11,700

11 11 33 – Vehicle Spray Painting Equipment

11 11 33.10 Spray Painting Equipment

0010	**SPRAY PAINTING EQUIPMENT**									
4000	Spray painting booth, 26' long, complete	L-4	.40	60	Ea.	15,900	2,325		18,225	21,200

11 12 Parking Control Equipment

11 12 13 – Parking Key and Card Control Units

11 12 13.10 Parking Control Units

		Crew	Daily Output	Labor-Hours	Unit	Material	Labor	Equipment	Total	Total Incl O&P
0010	**PARKING CONTROL UNITS**									
5100	Card reader	1 Elec	2	4	Ea.	1,825	196		2,021	2,300
5120	Proximity with customer display	2 Elec	1	16		5,500	785		6,285	7,250
6000	Parking control software, minimum	1 Elec	.50	16		21,900	785		22,685	25,300
6020	Maximum	"	.20	40		96,000	1,950		97,950	108,500

11 12 16 – Parking Ticket Dispensers

11 12 16.10 Ticket Dispensers

0010	**TICKET DISPENSERS**									
5900	Ticket spitter with time/date stamp, standard	2 Elec	2	8	Ea.	5,975	390		6,365	7,150
5920	Mag stripe encoding	"	2	8	"	18,600	390		18,990	21,000

11 12 26 – Parking Fee Collection Equipment

11 12 26.13 Parking Fee Coin Collection Equipment

0010	**PARKING FEE COIN COLLECTION EQUIPMENT**									
5200	Cashier booth, average	B-22	1	30	Ea.	9,650	1,150	207	11,007	12,700
5300	Collector station, pay on foot	2 Elec	.20	80		109,500	3,925		113,425	126,500
5320	Credit card only	"	.50	32		20,200	1,575		21,775	24,700

11 12 26.23 Fee Equipment

0010	**FEE EQUIPMENT**									
5600	Fee computer	1 Elec	1.50	5.333	Ea.	13,400	261		13,661	15,100

11 12 33 – Parking Gates

11 12 33.13 Parking Gate Equipment

0010	**PARKING GATE EQUIPMENT**									
5000	Barrier gate with programmable controller	2 Elec	3	5.333	Ea.	3,350	261		3,611	4,075
5020	Industrial		3	5.333		4,575	261		4,836	5,425
5500	Exit verifier		1	16		17,400	785		18,185	20,400
5700	Full sign, 4" letters	1 Elec	2	4		1,225	196		1,421	1,625
5800	Inductive loop	2 Elec	4	4		169	196		365	485
5950	Vehicle detector, microprocessor based	1 Elec	3	2.667		390	131		521	630

11 13 Loading Dock Equipment

11 13 13 – Loading Dock Bumpers

11 13 13.10 Dock Bumpers		Crew	Daily Output	Labor-Hours	Unit	Material	2010 Bare Costs Labor	2010 Bare Costs Equipment	Total	Total Incl O&P
0010	**DOCK BUMPERS** Bolts not included									
0012	2" x 6" to 4" x 8", average	1 Carp	300	.027	B.F.	.63	1.11		1.74	2.51
0050	Bumpers, rubber blocks 4-1/2" thick, 10" high, 14" long		26	.308	Ea.	49	12.80		61.80	74.50
0200	24" long		22	.364		86.50	15.10		101.60	120
0300	36" long		17	.471		91.50	19.55		111.05	133
0500	12" high, 14" long		25	.320		91	13.30		104.30	122

11 13 16 – Loading Dock Seals and Shelters

11 13 16.10 Dock Seals and Shelters

		Crew	Daily Output	Labor-Hours	Unit	Material	Labor	Equipment	Total	Total Incl O&P
0010	**DOCK SEALS AND SHELTERS**									
6200	Shelters, fabric, for truck or train, scissor arms, minimum	1 Carp	1	8	Ea.	1,500	330		1,830	2,200
6300	Maximum	"	.50	16	"	1,950	665		2,615	3,250

11 13 19 – Stationary Loading Dock Equipment

11 13 19.10 Dock Equipment

		Crew	Daily Output	Labor-Hours	Unit	Material	Labor	Equipment	Total	Total Incl O&P
0010	**DOCK EQUIPMENT**									
2200	Dock boards, heavy duty, 60" x 60", aluminum, 5,000 lb. capacity				Ea.	1,175			1,175	1,275
4200	Platform lifter, 6' x 6', portable, 3,000 lb. capacity				"	8,300			8,300	9,150

11 16 Vault Equipment

11 16 16 – Safes

11 16 16.10 Commercial Safes

		Crew	Daily Output	Labor-Hours	Unit	Material	Labor	Equipment	Total	Total Incl O&P
0010	**COMMERCIAL SAFES**									
0800	Money, "B" label, 9" x 14" x 14"				Ea.	525			525	580
0900	Tool resistive, 24" x 24" x 20"					3,575			3,575	3,925
1050	Tool and torch resistive, 24" x 24" x 20"					9,150			9,150	10,100

11 17 Teller and Service Equipment

11 17 13 – Teller Equipment Systems

11 17 13.10 Bank Equipment

		Crew	Daily Output	Labor-Hours	Unit	Material	Labor	Equipment	Total	Total Incl O&P
0010	**BANK EQUIPMENT**									
0020	Alarm system, police	2 Elec	1.60	10	Ea.	4,675	490		5,165	5,900
0400	Bullet resistant teller window, 44" x 60"	1 Glaz	.60	13.333		3,275	535		3,810	4,450
0500	48" x 60"	"	.60	13.333		4,175	535		4,710	5,425
3000	Counters for banks, frontal only	2 Carp	1	16	Station	1,750	665		2,415	3,025
3100	Complete with steel undercounter		.50	32	"	3,400	1,325		4,725	5,925
5400	Partitions, bullet-resistant, 1-3/16" glass, 8' high		10	1.600	L.F.	189	66.50		255.50	315
5600	Pass thru, bullet-res. window, painted steel, 24" x 36"	2 Sswk	1.60	10	Ea.	2,300	470		2,770	3,375

11 19 Detention Equipment

11 19 30 – Detention Cell Equipment

11 19 30.10 Cell Equipment	Crew	Daily Output	Labor-Hours	Unit	Material	2010 Bare Costs Labor	2010 Bare Costs Equipment	Total	Total Incl O&P
0010 **CELL EQUIPMENT**									
3000 Toilet apparatus including wash basin, average	L-8	1.50	13.333	Ea.	3,350	580		3,930	4,650

11 21 Mercantile and Service Equipment

11 21 13 – Cash Registers and Checking Equipment

11 21 13.10 Checkout Counter

	Crew	Daily Output	Labor-Hours	Unit	Material	Labor	Equipment	Total	Total Incl O&P
0010 **CHECKOUT COUNTER**									
0020 Supermarket conveyor, single belt	2 Clab	10	1.600	Ea.	2,900	53		2,953	3,275
0100 Double belt, power take-away		9	1.778		4,175	59		4,234	4,700
0400 Double belt, power take-away, incl. side scanning		7	2.286		4,900	75.50		4,975.50	5,525
0800 Warehouse or bulk type		6	2.667		5,800	88.50		5,888.50	6,525
1000 Scanning system, 2 lanes, w/registers, scan gun & memory				System	15,200			15,200	16,700
1100 10 lanes, single processor, full scan, with scales				"	144,500			144,500	158,500
2000 Register, restaurant, minimum				Ea.	615			615	675
2100 Maximum					2,700			2,700	2,975
2150 Store, minimum					615			615	675
2200 Maximum					2,700			2,700	2,975

11 21 53 – Barber and Beauty Shop Equipment

11 21 53.10 Barber Equipment

	Crew	Daily Output	Labor-Hours	Unit	Material	Labor	Equipment	Total	Total Incl O&P
0010 **BARBER EQUIPMENT**									
0020 Chair, hydraulic, movable, minimum	1 Carp	24	.333	Ea.	420	13.85		433.85	485
0050 Maximum	"	16	.500		2,975	21		2,996	3,300
0200 Wall hung styling station with mirrors, minimum	L-2	8	2		370	73		443	525
0300 Maximum	"	4	4		1,975	146		2,121	2,425
0500 Sink, hair washing basin, rough plumbing not incl.	1 Plum	8	1		425	52		477	545
1000 Sterilizer, liquid solution for tools					138			138	151
1100 Total equipment, rule of thumb, per chair, minimum	L-8	1	20		1,575	875		2,450	3,150
1150 Maximum	"	1	20		4,400	875		5,275	6,275

11 23 Commercial Laundry and Dry Cleaning Equipment

11 23 13 – Dry Cleaning Equipment

11 23 13.13 Dry Cleaners

	Crew	Daily Output	Labor-Hours	Unit	Material	Labor	Equipment	Total	Total Incl O&P
0010 **DRY CLEANERS** Not incl. rough-in									
2000 Dry cleaners, electric, 20 lb. capacity	L-1	.20	80	Ea.	33,300	4,050		37,350	42,900
2050 25 lb. capacity		.17	94.118		45,100	4,750		49,850	57,000
2100 30 lb. capacity		.15	106		47,600	5,400		53,000	61,000
2150 60 lb. capacity		.09	177		73,000	8,975		81,975	94,000

11 23 16 – Drying and Conditioning Equipment

11 23 16.13 Dryers

	Crew	Daily Output	Labor-Hours	Unit	Material	Labor	Equipment	Total	Total Incl O&P
0010 **DRYERS**, Not including rough-in									
1500 Industrial, 30 lb. capacity	1 Plum	2	4	Ea.	2,975	208		3,183	3,600
1600 50 lb. capacity	"	1.70	4.706		3,200	245		3,445	3,900
4700 Lint collector, ductwork not included, 8,000 to 10,000 C.F.M.	Q-10	.30	80		8,575	3,675		12,250	15,200

11 23 19 – Finishing Equipment

11 23 19.13 Folders and Spreaders

	Crew	Daily Output	Labor-Hours	Unit	Material	Labor	Equipment	Total	Total Incl O&P
0010 **FOLDERS AND SPREADERS**									
3500 Folders, blankets & sheets, minimum	1 Elec	.17	47.059	Ea.	30,600	2,300		32,900	37,300
3700 King size with automatic stacker		.10	80		55,500	3,925		59,425	67,000

11 23 Commercial Laundry and Dry Cleaning Equipment

11 23 19 – Finishing Equipment

11 23 19.13 Folders and Spreaders

		Crew	Daily Output	Labor-Hours	Unit	Material	2010 Bare Costs Labor	2010 Bare Costs Equipment	Total	Total Incl O&P
3800	For conveyor delivery, add	1 Elec	.45	17.778	Ea.	6,000	870		6,870	7,950

11 23 23 – Commercial Ironing Equipment

11 23 23.13 Irons and Pressers

0010	**IRONS AND PRESSERS**									
4500	Ironers, institutional, 110", single roll	1 Elec	.20	40	Ea.	29,700	1,950		31,650	35,700

11 23 26 – Commercial Washers and Extractors

11 23 26.13 Washers and Extractors

0010	**WASHERS AND EXTRACTORS**, not including rough-in									
6000	Combination washer/extractor, 20 lb. capacity	L-6	1.50	8	Ea.	5,350	410		5,760	6,525
6100	30 lb. capacity		.80	15		8,500	765		9,265	10,500
6200	50 lb. capacity		.68	17.647		9,925	900		10,825	12,300
6300	75 lb. capacity		.30	40		18,600	2,050		20,650	23,700
6350	125 lb. capacity		.16	75		24,900	3,825		28,725	33,300

11 23 33 – Coin-Operated Laundry Equipment

11 23 33.13 Coin Operated Washers and Dryers

0010	**COIN OPERATED WASHERS AND DRYERS**									
0990	Dryer, gas fired									
1000	Commercial, 30 lb. capacity, coin operated, single	1 Plum	3	2.667	Ea.	3,100	139		3,239	3,625
1100	Double stacked	"	2	4	"	6,275	208		6,483	7,225
5290	Clothes washer									
5300	Commercial, coin operated, average	1 Plum	3	2.667	Ea.	1,125	139		1,264	1,475

11 24 Maintenance Equipment

11 24 19 – Vacuum Cleaning Systems

11 24 19.10 Vacuum Cleaning

0010	**VACUUM CLEANING**									
0020	Central, 3 inlet, residential	1 Skwk	.90	8.889	Total	1,025	380		1,405	1,750
0200	Commercial		.70	11.429		1,225	485		1,710	2,100
0400	5 inlet system, residential		.50	16		1,450	680		2,130	2,700
0600	7 inlet system, commercial		.40	20		1,625	850		2,475	3,150
4010	Rule of thumb: First 1200 S.F., installed								1,325	1,450
4020	For each additional S.F., add				S.F.					.24

11 26 Unit Kitchens

11 26 13 – Metal Unit Kitchens

11 26 13.10 Commercial Unit Kitchens

0010	**COMMERCIAL UNIT KITCHENS**									
1500	Combination range, refrigerator and sink, 30" wide, minimum	L-1	2	8	Ea.	925	405		1,330	1,650
1550	Maximum	"	1	16	"	3,675	810		4,485	5,300
1640	Combination range, refrigerator, sink, microwave									
1660	Oven and ice maker	L-1	.80	20	Ea.	4,450	1,000		5,450	6,450

11 27 Photographic Processing Equipment

11 27 13 – Darkroom Processing Equipment

11 27 13.10 Darkroom Equipment

		Crew	Daily Output	Labor-Hours	Unit	Material	2010 Bare Costs Labor	Equipment	Total	Total Incl O&P
0010	**DARKROOM EQUIPMENT**									
0020	Developing sink, 5" deep, 24" x 48"	Q-1	2	8	Ea.	5,625	375		6,000	6,750
0050	48" x 52"		1.70	9.412		5,675	440		6,115	6,900
0200	10" deep, 24" x 48"		1.70	9.412		7,775	440		8,215	9,225
0250	24" x 108"		1.50	10.667		10,200	500		10,700	12,100
0500	Dryers, dehumidified filtered air, 36" x 25" x 68" high	L-7	6	4.667		5,075	188		5,263	5,875
0550	48" x 25" x 68" high		5	5.600		10,000	225		10,225	11,400
2000	Processors, automatic, color print, minimum		4	7		15,700	281		15,981	17,800
2050	Maximum		.60	46.667		24,700	1,875		26,575	30,300
2300	Black and white print, minimum		2	14		11,900	565		12,465	14,000
2350	Maximum		.80	35		61,500	1,400		62,900	70,000
2600	Manual processor, 16" x 20" maximum print size		2	14		9,450	565		10,015	11,300
2650	20" x 24" maximum print size		1	28		8,975	1,125		10,100	11,700
3000	Viewing lights, 20" x 24"		6	4.667		380	188		568	720
3100	20" x 24" with color correction		6	4.667		535	188		723	895
3500	Washers, round, minimum sheet 11" x 14"	Q-1	2	8		3,400	375		3,775	4,325
3550	Maximum sheet 20" x 24"		1	16		3,475	750		4,225	4,975
3800	Square, minimum sheet 20" x 24"		1	16		3,125	750		3,875	4,575
3900	Maximum sheet 50" x 56"		.80	20		4,850	935		5,785	6,775
4500	Combination tank sink, tray sink, washers, with									
4510	Dry side tables, average	Q-1	.45	35.556	Ea.	10,600	1,675		12,275	14,300

11 31 Residential Appliances

11 31 13 – Residential Kitchen Appliances

11 31 13.13 Cooking Equipment

		Crew	Daily Output	Labor-Hours	Unit	Material	Labor	Equipment	Total	Total Incl O&P
0010	**COOKING EQUIPMENT**									
0020	Cooking range, 30" free standing, 1 oven, minimum	2 Clab	10	1.600	Ea.	345	53		398	465
0050	Maximum	"	4	4		1,900	132		2,032	2,325
0350	Built-in, 30" wide, 1 oven, minimum	1 Elec	6	1.333		680	65.50		745.50	850
0400	Maximum	2 Carp	2	8		1,500	330		1,830	2,200
0900	Countertop cooktops, 4 burner, standard, minimum	1 Elec	6	1.333		300	65.50		365.50	430
0950	Maximum		3	2.667		600	131		731	860
1250	Microwave oven, minimum		4	2		110	98		208	272
1300	Maximum		2	4		420	196		616	760
5380	Oven, built in, standard		4	2		485	98		583	680
5390	Deluxe		2	4		2,000	196		2,196	2,500

11 31 13.23 Refrigeration Equipment

		Crew	Daily Output	Labor-Hours	Unit	Material	Labor	Equipment	Total	Total Incl O&P
0010	**REFRIGERATION EQUIPMENT**									
5450	Refrigerator, no frost, 6 C.F.	2 Clab	15	1.067	Ea.	230	35.50		265.50	310
5500	Refrigerator, no frost, 10 C.F. to 12 C.F. minimum		10	1.600		445	53		498	575
5600	Maximum		6	2.667		450	88.50		538.50	640
6790	Energy-star qualified, 18 CF, minimum [G]	2 Carp	4	4		495	166		661	820
6795	Maximum [G]		2	8		1,300	330		1,630	1,975
6797	21.7 CF, minimum [G]		4	4		815	166		981	1,175
6799	Maximum [G]		4	4		1,300	166		1,466	1,700

11 31 13.33 Kitchen Cleaning Equipment

		Crew	Daily Output	Labor-Hours	Unit	Material	Labor	Equipment	Total	Total Incl O&P
0010	**KITCHEN CLEANING EQUIPMENT**									
2750	Dishwasher, built-in, 2 cycles, minimum	L-1	4	4	Ea.	212	202		414	545
2800	Maximum		2	8		320	405		725	975
3100	Energy-star qualified, minimum [G]		4	4		340	202		542	685

11 31 Residential Appliances

11 31 13 – Residential Kitchen Appliances

11 31 13.33 Kitchen Cleaning Equipment		Crew	Daily Output	Labor-Hours	Unit	Material	2010 Bare Costs Labor	Equipment	Total	Total Incl O&P
3110	Maximum	G L-1	2	8	Ea.	1,250	405		1,655	2,000

11 31 13.43 Waste Disposal Equipment
0010	**WASTE DISPOSAL EQUIPMENT**									
1750	Compactor, residential size, 4 to 1 compaction, minimum	1 Carp	5	1.600	Ea.	535	66.50		601.50	695
1800	Maximum	"	3	2.667		830	111		941	1,100
3300	Garbage disposal, sink type, minimum	L-1	10	1.600		72.50	81		153.50	205
3350	Maximum	"	10	1.600		191	81		272	335

11 31 13.53 Kitchen Ventilation Equipment
0010	**KITCHEN VENTILATION EQUIPMENT**									
4150	Hood for range, 2 speed, vented, 30" wide, minimum	L-3	5	3.200	Ea.	53.50	145		198.50	290
4200	Maximum		3	5.333		820	242		1,062	1,275
4300	42" wide, minimum		5	3.200		234	145		379	490
4330	Custom		5	3.200		1,500	145		1,645	1,900
4350	Maximum		3	5.333		1,850	242		2,092	2,400
4400	Ventless hood, 2 speed, 30" wide, minimum	1 Elec	8	1		68	49		117	151
4450	Maximum		7	1.143		785	56		841	950
4510	36" wide, minimum		7	1.143		64	56		120	157
4550	Maximum		6	1.333		690	65.50		755.50	860
4580	42" wide, minimum		6	1.333		195	65.50		260.50	315
4600	Maximum		5	1.600		360	78.50		438.50	515
4650	For vented 1 speed, deduct from maximum					45.50			45.50	50
4700										

11 31 23 – Residential Laundry Appliances

11 31 23.13 Washers
0010	**WASHERS**									
5000	Residential, 4 cycle, average	1 Plum	3	2.667	Ea.	800	139		939	1,100
6750	Energy star qualified, front loading, minimum	G	3	2.667		480	139		619	745
6760	Maximum	G	1	8		1,700	415		2,115	2,525
6764	Top loading, minimum	G	3	2.667		575	139		714	850
6766	Maximum	G	3	2.667		1,050	139		1,189	1,375

11 31 23.23 Dryers
0010	**DRYERS**									
0500	Gas fired residential, 16 lb. capacity, average	1 Plum	3	2.667	Ea.	625	139		764	905
6770	Electric, front loading, energy-star qualified, minimum	G L-2	3	5.333		360	195		555	715
6780	Maximum	G "	2	8		1,300	293		1,593	1,900

11 31 33 – Miscellaneous Residential Appliances

11 31 33.13 Sump Pumps
0010	**SUMP PUMPS**									
6400	Cellar drainer, pedestal, 1/3 H.P., molded PVC base	1 Plum	3	2.667	Ea.	119	139		258	345
6450	Solid brass	"	2	4	"	263	208		471	615
6460	Sump pump, see also Div. 22 14 29.16									

11 33 Retractable Stairs

11 33 10 – Disappearing Stairs

11 33 10.10 Disappearing Stairway		Crew	Daily Output	Labor-Hours	Unit	Material	2010 Bare Costs Labor	Equipment	Total	Total Incl O&P
0010	**DISAPPEARING STAIRWAY** No trim included									
0020	One piece, yellow pine, 8'-0" ceiling	2 Carp	4	4	Ea.	240	166		406	535
0030	9'-0" ceiling		4	4		273	166		439	575
0040	10'-0" ceiling		3	5.333		335	222		557	730
0050	11'-0" ceiling		3	5.333		455	222		677	865
0060	12'-0" ceiling		3	5.333		455	222		677	865
0100	Custom grade, pine, 8'-6" ceiling, minimum	1 Carp	4	2		138	83		221	288
0150	Average		3.50	2.286		174	95		269	345
0200	Maximum		3	2.667		257	111		368	465
0250										
0500	Heavy duty, pivoted, from 7'-7" to 12'-10" floor to floor	1 Carp	3	2.667	Ea.	650	111		761	895
0600	16'-0" ceiling		2	4		1,325	166		1,491	1,725
0800	Economy folding, pine, 8'-6" ceiling		4	2		121	83		204	270
0900	9'-6" ceiling		4	2		130	83		213	280
1100	Automatic electric, aluminum, floor to floor height, 8' to 9'	2 Carp	1	16		8,250	665		8,915	10,200
1400	11' to 12'		.90	17.778		8,875	740		9,615	11,000
1700	14' to 15'		.70	22.857		9,475	950		10,425	12,000
9000	Minimum labor/equipment charge	1 Carp	2	4	Job		166		166	273

11 41 Food Storage Equipment

11 41 13 – Refrigerated Food Storage Cases

11 41 13.10 Refrigerated Food Cases

		Crew	Daily Output	Labor-Hours	Unit	Material	Labor	Equipment	Total	Total Incl O&P
0010	**REFRIGERATED FOOD CASES**									
0030	Dairy, multi-deck, 12' long	Q-5	3	5.333	Ea.	10,400	249		10,649	11,900
0100	For rear sliding doors, add					1,500			1,500	1,650
0200	Delicatessen case, service deli, 12' long, single deck	Q-5	3.90	4.103		7,025	192		7,217	8,025
0300	Multi-deck, 18 S.F. shelf display		3	5.333		6,450	249		6,699	7,450
0400	Freezer, self-contained, chest-type, 30 C.F.		3.90	4.103		7,550	192		7,742	8,600
0500	Glass door, upright, 78 C.F.		3.30	4.848		9,750	226		9,976	11,100
0600	Frozen food, chest type, 12' long		3.30	4.848		7,050	226		7,276	8,100
0700	Glass door, reach-in, 5 door		3	5.333		13,500	249		13,749	15,200
0800	Island case, 12' long, single deck		3.30	4.848		7,975	226		8,201	9,125
0900	Multi-deck		3	5.333		16,900	249		17,149	19,000
1000	Meat case, 12' long, single deck		3.30	4.848		5,800	226		6,026	6,725
1050	Multi-deck		3.10	5.161		9,975	241		10,216	11,400
1100	Produce, 12' long, single deck		3.30	4.848		7,675	226		7,901	8,775
1200	Multi-deck		3.10	5.161		8,375	241		8,616	9,575

11 41 13.20 Refrigerated Food Storage Equipment

			Crew	Daily Output	Labor-Hours	Unit	Material	Labor	Equipment	Total	Total Incl O&P
0010	**REFRIGERATED FOOD STORAGE EQUIPMENT**										
2350	Cooler, reach-in, beverage, 6' long		Q-1	6	2.667	Ea.	4,525	125		4,650	5,175
4300	Freezers, reach-in, 44 C.F.			4	4		2,700	187		2,887	3,275
4500	68 C.F.			3	5.333		3,650	250		3,900	4,375
4600	Freezer, pre-fab, 8' x 8' w/refrigeration		2 Carp	.45	35.556		10,900	1,475		12,375	14,400
4620	8' x 12'			.35	45.714		12,900	1,900		14,800	17,300
4640	8' x 16'			.25	64		15,600	2,650		18,250	21,500
4660	8' x 20'			.17	94.118		18,700	3,900		22,600	27,000
4680	Reach-in, 1 compartment		Q-1	4	4		1,650	187		1,837	2,125
4685	Energy star rated	G	R-18	7.80	3.333		2,250	128		2,378	2,650
4700	2 compartment		Q-1	3	5.333		3,425	250		3,675	4,150
4705	Energy star rated	G	R-18	6.20	4.194		3,200	161		3,361	3,775
4710	3 compartment		Q-1	3	5.333		4,850	250		5,100	5,725

11 41 Food Storage Equipment

11 41 13 – Refrigerated Food Storage Cases

11 41 13.20 Refrigerated Food Storage Equipment

		Crew	Daily Output	Labor-Hours	Unit	Material	2010 Bare Costs Labor	Equipment	Total	Total Incl O&P
4715	Energy star rated G	R-18	5.60	4.643	Ea.	4,325	178		4,503	5,025
8320	Refrigerator, reach-in, 1 compartment		7.80	3.333		2,100	128		2,228	2,525
8325	Energy star rated G		7.80	3.333		2,250	128		2,378	2,650
8330	2 compartment		6.20	4.194		3,125	161		3,286	3,675
8335	Energy star rated G		6.20	4.194		3,200	161		3,361	3,775
8340	3 compartment		5.60	4.643		4,175	178		4,353	4,875
8345	Energy star rated G		5.60	4.643		4,325	178		4,503	5,025
8350	Pre-fab, with refrigeration, 8' x 8'	2 Carp	.45	35.556		6,675	1,475		8,150	9,775
8360	8' x 12'		.35	45.714		7,975	1,900		9,875	11,900
8370	8' x 16'		.25	64		11,400	2,650		14,050	16,900
8380	8' x 20'		.17	94.118		14,400	3,900		18,300	22,300
8390	Pass-thru/roll-in, 1 compartment	R-18	7.80	3.333		4,000	128		4,128	4,600
8400	2 compartment		6.24	4.167		5,875	160		6,035	6,725
8410	3 compartment		5.60	4.643		7,525	178		7,703	8,550
8420	Walk-in, alum, door & floor only, no refrig, 6' x 6' x 7'-6"	2 Carp	1.40	11.429		8,675	475		9,150	10,300
8430	10' x 6' x 7'-6"		.55	29.091		12,400	1,200		13,600	15,600
8440	12' x 14' x 7'-6"		.25	64		17,600	2,650		20,250	23,700
8450	12' x 20' x 7'-6"		.17	94.118		21,400	3,900		25,300	30,000
8460	Refrigerated cabinets, mobile					3,575			3,575	3,925
8470	Refrigerator/freezer, reach-in, 1 compartment	R-18	5.60	4.643		4,700	178		4,878	5,450
8480	2 compartment	"	4.80	5.417		6,625	208		6,833	7,625

11 41 13.30 Wine Cellar

		Crew	Daily Output	Labor-Hours	Unit	Material	Labor	Equipment	Total	Total Incl O&P
0010	**WINE CELLAR**, refrigerated, Redwood interior, carpeted, walk-in type									
0020	6'-8" high, including racks									
0200	80" W x 48" D for 900 bottles	2 Carp	1.50	10.667	Ea.	3,350	445		3,795	4,400
0250	80" W x 72" D for 1300 bottles		1.33	12.030		4,425	500		4,925	5,675
0300	80" W x 94" D for 1900 bottles		1.17	13.675		5,725	570		6,295	7,225
0400	80" W x 124" D for 2500 bottles		1	16		6,925	665		7,590	8,725

11 41 33 – Food Storage Shelving

11 41 33.20 Metal Food Storage Shelving

		Crew	Daily Output	Labor-Hours	Unit	Material	Labor	Equipment	Total	Total Incl O&P
0010	**METAL FOOD STORAGE SHELVING**									
8600	Stainless steel shelving, louvered 4-tier, 20" x 3'	1 Clab	6	1.333	Ea.	1,225	44		1,269	1,425
8605	20" x 4'		6	1.333		1,750	44		1,794	2,000
8610	20" x 6'		6	1.333		2,550	44		2,594	2,875
8615	24" x 3'		6	1.333		1,600	44		1,644	1,850
8620	24" x 4'		6	1.333		1,875	44		1,919	2,125
8625	24" x 6'		6	1.333		2,725	44		2,769	3,050
8630	Flat 4-tier, 20" x 3'		6	1.333		1,100	44		1,144	1,275
8635	20" x 4'		6	1.333		1,325	44		1,369	1,525
8640	20" x 5'		6	1.333		1,525	44		1,569	1,750
8645	24" x 3'		6	1.333		1,250	44		1,294	1,450
8650	24" x 4'		6	1.333		2,000	44		2,044	2,275
8655	24" x 6'		6	1.333		2,425	44		2,469	2,725
8700	Galvanized shelving, louvered 4-tier, 20" x 3'		6	1.333		625	44		669	760
8705	20" x 4'		6	1.333		695	44		739	840
8710	20" x 6'		6	1.333		870	44		914	1,025
8715	24" x 3'		6	1.333		555	44		599	685
8720	24" x 4'		6	1.333		745	44		789	895
8725	24" x 6'		6	1.333		1,075	44		1,119	1,250
8730	Flat 4-tier, 20" x 3'		6	1.333		570	44		614	700
8735	20" x 4'		6	1.333		625	44		669	765
8740	20" x 6'		6	1.333		875	44		919	1,050

11 41 Food Storage Equipment

11 41 33 – Food Storage Shelving

11 41 33.20 Metal Food Storage Shelving

		Crew	Daily Output	Labor-Hours	Unit	Material	2010 Bare Costs Labor	Equipment	Total	Total Incl O&P
8745	24" x 3'	1 Clab	6	1.333	Ea.	600	44		644	735
8750	24" x 4'		6	1.333		635	44		679	775
8755	24" x 6'		6	1.333		910	44		954	1,075
8760	Stainless steel dunnage rack, 24" x 3'		8	1		415	33		448	510
8765	24" x 4'		8	1		800	33		833	935
8770	Galvanized dunnage rack, 24" x 3'		8	1		157	33		190	228
8775	24" x 4'	▼	8	1	▼	184	33		217	258

11 42 Food Preparation Equipment

11 42 10 – Commercial Food Preparation Equipment

11 42 10.10 Choppers, Mixers and Misc. Equipment

		Crew	Daily Output	Labor-Hours	Unit	Material	Labor	Equipment	Total	Total Incl O&P
0010	**CHOPPERS, MIXERS AND MISC. EQUIPMENT**									
1700	Choppers, 5 pounds	R-18	7	3.714	Ea.	1,725	142		1,867	2,125
1720	16 pounds		5	5.200		2,125	199		2,324	2,675
1740	35 to 40 pounds	▼	4	6.500		3,150	249		3,399	3,850
1840	Coffee brewer, 5 burners	1 Plum	3	2.667		1,225	139		1,364	1,575
1850	Coffee urn, twin 6 gallon urns		2	4		2,475	208		2,683	3,050
1860	Single, 3 gallon	▼	3	2.667		1,825	139		1,964	2,250
3000	Fast food equipment, total package, minimum	6 Skwk	.08	600		184,500	25,600		210,100	244,000
3100	Maximum	"	.07	685		251,000	29,200		280,200	323,000
3800	Food mixers, 20 quarts	L-7	7	4		2,775	161		2,936	3,300
3850	40 quarts		5.40	5.185		6,950	208		7,158	8,000
3900	60 quarts		5	5.600		8,725	225		8,950	9,975
4040	80 quarts		3.90	7.179		12,400	289		12,689	14,100
4080	130 quarts		2.20	12.727		17,100	510		17,610	19,600
4100	Floor type, 20 quarts		15	1.867		3,175	75		3,250	3,600
4120	60 quarts		14	2		7,925	80.50		8,005.50	8,850
4140	80 quarts		12	2.333		10,600	94		10,694	11,900
4160	140 quarts	▼	8.60	3.256		18,700	131		18,831	20,800
6700	Peelers, small	R-18	8	3.250		1,525	125		1,650	1,875
6720	Large	"	6	4.333		4,625	166		4,791	5,350
6800	Pulper/extractor, close coupled, 5 HP	1 Plum	1.90	4.211		3,525	219		3,744	4,225
8580	Slicer with table	R-18	9	2.889	▼	4,850	111		4,961	5,500

11 43 Food Delivery Carts and Conveyors

11 43 13 – Food Delivery Carts

11 43 13.10 Mobile Carts, Racks and Trays

		Crew	Daily Output	Labor-Hours	Unit	Material	Labor	Equipment	Total	Total Incl O&P
0010	**MOBILE CARTS, RACKS AND TRAYS**									
1650	Cabinet, heated, 1 compartment, reach-in	R-18	5.60	4.643	Ea.	3,475	178		3,653	4,100
1655	Pass-thru roll-in		5.60	4.643		5,125	178		5,303	5,900
1660	2 compartment, reach-in	▼	4.80	5.417		7,450	208		7,658	8,500
1670	Mobile					3,175			3,175	3,500
6850	Mobile rack w/pan slide					1,275			1,275	1,400
9180	Tray and silver dispenser, mobile	1 Clab	16	.500	▼	840	16.55		856.55	950

11 44 Food Cooking Equipment

11 44 13 – Commercial Ranges

11 44 13.10 Cooking Equipment

	Cooking Equipment		Crew	Daily Output	Labor-Hours	Unit	Material	2010 Bare Costs Labor	Equipment	Total	Total Incl O&P
0010	**COOKING EQUIPMENT**										
0020	Bake oven, gas, one section		Q-1	8	2	Ea.	4,650	93.50		4,743.50	5,275
0300	Two sections			7	2.286		10,500	107		10,607	11,700
0600	Three sections			6	2.667		13,500	125		13,625	15,000
0900	Electric convection, single deck		L-7	4	7		6,425	281		6,706	7,500
1300	Broiler, without oven, standard		Q-1	8	2		3,550	93.50		3,643.50	4,050
1550	Infrared		L-7	4	7		9,325	281		9,606	10,800
4750	Fryer, with twin baskets, modular model		Q-1	7	2.286		1,500	107		1,607	1,825
5000	Floor model, on 6" legs		"	5	3.200		2,275	150		2,425	2,725
5100	Extra single basket, large						96			96	105
5170	Energy star rated, 50 lb. capacity	G	R-18	4	6.500		3,775	249		4,024	4,550
5175	85 lb. capacity	G	"	4	6.500		7,675	249		7,924	8,850
5300	Griddle, SS, 24" plate, w/4" legs, elec, 208V, 3 phase, 3' long		Q-1	7	2.286		1,125	107		1,232	1,425
5550	4' long		"	6	2.667		1,525	125		1,650	1,875
6200	Iced tea brewer		1 Plum	3.44	2.326		675	121		796	930
6350	Kettle, w/steam jacket, tilting, w/positive lock, SS, 20 gallons		L-7	7	4		7,125	161		7,286	8,075
6600	60 gallons		"	6	4.667		9,225	188		9,413	10,500
6900	Range, restaurant type, 6 burners and 1 standard oven, 36" wide		Q-1	7	2.286		2,325	107		2,432	2,725
6950	Convection			7	2.286		5,125	107		5,232	5,825
7150	2 standard ovens, 24" griddle, 60" wide			6	2.667		7,700	125		7,825	8,675
7200	1 standard, 1 convection oven			6	2.667		6,800	125		6,925	7,675
7450	Heavy duty, single 34" standard oven, open top			5	3.200		5,575	150		5,725	6,350
7500	Convection oven			5	3.200		4,650	150		4,800	5,325
7700	Griddle top			6	2.667		3,250	125		3,375	3,775
7750	Convection oven			6	2.667		6,800	125		6,925	7,675
8850	Steamer, electric 27 KW		L-7	7	4		7,875	161		8,036	8,925
9100	Electric, 10 KW or gas 100,000 BTU		"	5	5.600		5,725	225		5,950	6,650
9150	Toaster, conveyor type, 16-22 slices per minute						1,200			1,200	1,325
9160	Pop-up, 2 slot						570			570	630
9200	For deluxe models of above equipment, add						75%				
9400	Rule of thumb: Equipment cost based										
9410	on kitchen work area										
9420	Office buildings, minimum		L-7	77	.364	S.F.	82	14.60		96.60	114
9450	Maximum			58	.483		138	19.40		157.40	184
9550	Public eating facilities, minimum			77	.364		107	14.60		121.60	142
9600	Maximum			46	.609		175	24.50		199.50	232
9750	Hospitals, minimum			58	.483		110	19.40		129.40	154
9800	Maximum			39	.718		203	29		232	271

11 46 Food Dispensing Equipment

11 46 16 – Service Line Equipment

11 46 16.10 Commercial Food Dispensing Equipment

		Crew	Daily Output	Labor-Hours	Unit	Material	Labor	Equipment	Total	Total Incl O&P
0010	**COMMERCIAL FOOD DISPENSING EQUIPMENT**									
1050	Butter pat dispenser	1 Clab	13	.615	Ea.	910	20.50		930.50	1,025
1100	Bread dispenser, counter top		13	.615		865	20.50		885.50	990
1900	Cup and glass dispenser, drop in		4	2		975	66		1,041	1,175
1920	Disposable cup, drop in		16	.500		425	16.55		441.55	495
2650	Dish dispenser, drop in, 12"		11	.727		1,975	24		1,999	2,200
2660	Mobile		10	.800		2,375	26.50		2,401.50	2,650
3300	Food warmer, counter, 1.2 KW					670			670	740
3550	1.6 KW					1,950			1,950	2,150

11 46 Food Dispensing Equipment

11 46 16 – Service Line Equipment

11 46 16.10 Commercial Food Dispensing Equipment		Crew	Daily Output	Labor-Hours	Unit	Material	2010 Bare Costs Labor	Equipment	Total	Total Incl O&P
3600	Well, hot food, built-in, rectangular, 12" x 20"	R-30	10	2.600	Ea.	630	102		732	855
3610	Circular, 7 qt		10	2.600		335	102		437	535
3620	Refrigerated, 2 compartments		10	2.600		2,675	102		2,777	3,125
3630	3 compartments		9	2.889		3,325	114		3,439	3,850
3640	4 compartments		8	3.250		3,850	128		3,978	4,450
4720	Frost cold plate		9	2.889		16,700	114		16,814	18,500
5700	Hot chocolate dispenser	1 Plum	4	2		1,075	104		1,179	1,325
6250	Jet spray dispenser	R-18	4.50	5.778		3,100	221		3,321	3,750
6300	Juice dispenser, concentrate	"	4.50	5.778		2,100	221		2,321	2,675
6690	Milk dispenser, bulk, 2 flavor	R-30	8	3.250		1,450	128		1,578	1,800
6695	3 flavor	"	8	3.250		2,000	128		2,128	2,400
8800	Serving counter, straight	1 Carp	40	.200	L.F.	785	8.30		793.30	875
8820	Curved section	"	30	.267	"	980	11.10		991.10	1,100
8825	Solid surface, see Div. 12 36 61.16									
8830	Soft serve ice cream machine, medium	R-18	11	2.364	Ea.	10,300	90.50		10,390.50	11,500
8840	Large	"	9	2.889	"	18,000	111		18,111	20,000

11 47 Ice Machines

11 47 10 – Commercial Ice Machines

11 47 10.10 Commercial Ice Equipment

		Crew	Daily Output	Labor-Hours	Unit	Material	Labor	Equipment	Total	Total Incl O&P
0010	**COMMERCIAL ICE EQUIPMENT**									
5800	Ice cube maker, 50 pounds per day	Q-1	6	2.667	Ea.	1,550	125		1,675	1,900
5900	250 pounds per day		1.20	13.333		2,550	625		3,175	3,775
6050	500 pounds per day		4	4		3,125	187		3,312	3,725
6060	With bin		1.20	13.333		2,850	625		3,475	4,100
6090	1000 pounds per day, with bin		1	16		4,975	750		5,725	6,625
6100	Ice flakers, 300 pounds per day		1.60	10		3,150	470		3,620	4,200
6120	600 pounds per day		.95	16.842		4,000	790		4,790	5,600
6130	1000 pounds per day		.75	21.333		4,400	1,000		5,400	6,375
6140	2000 pounds per day		.65	24.615		19,500	1,150		20,650	23,300
6160	Ice storage bin, 500 pound capacity	Q-5	1	16		1,000	745		1,745	2,250
6180	1000 pound	"	.56	28.571		1,950	1,325		3,275	4,225

11 48 Cleaning and Disposal Equipment

11 48 13 – Commercial Dishwashers

11 48 13.10 Dishwashers

			Crew	Daily Output	Labor-Hours	Unit	Material	Labor	Equipment	Total	Total Incl O&P
0010	**DISHWASHERS**										
2700	Dishwasher, commercial, rack type										
2720	10 to 12 racks per hour		Q-1	3.20	5	Ea.	4,300	234		4,534	5,100
2730	Energy star rated, 35-40 racks/hour	G		1.30	12.308		4,650	575		5,225	6,000
2740	50-60 racks/hour	G		1.30	12.308		8,525	575		9,100	10,300
2800	Automatic, 190 to 230 racks per hour		L-6	.35	34.286		13,100	1,750		14,850	17,100
2820	235 to 275 racks per hour			.25	48		30,300	2,450		32,750	37,100
2840	8,750 to 12,500 dishes per hour			.10	120		49,500	6,125		55,625	64,000
2950	Dishwasher hood, canopy type		L-3A	10	1.200	L.F.	800	54.50		854.50	970
2960	Pant leg type		"	2.50	4.800	Ea.	7,875	218		8,093	9,025
5200	Garbage disposal 1.5 HP, 100 GPH		L-1	4.80	3.333		1,800	168		1,968	2,225
5210	3 HP, 120 GPH			4.60	3.478		2,175	176		2,351	2,675
5220	5 HP, 250 GPH			4.50	3.556		3,550	180		3,730	4,175

11 48 Cleaning and Disposal Equipment

11 48 13 – Commercial Dishwashers

11 48 13.10 Dishwashers		Crew	Daily Output	Labor-Hours	Unit	Material	2010 Bare Costs Labor	Equipment	Total	Total Incl O&P
6750	Pot sink, 3 compartment	1 Plum	7.25	1.103	L.F.	925	57.50		982.50	1,125
6760	Pot washer, small		1.60	5	Ea.	8,850	260		9,110	10,100
6770	Large		1.20	6.667		15,100	345		15,445	17,100
9170	Trash compactor, small, up to 125 lb. compacted weight	L-4	4	6		21,900	234		22,134	24,400
9175	Large, up to 175 lb. compacted weight	"	3	8		26,700	310		27,010	29,900

11 52 Audio-Visual Equipment

11 52 13 – Projection Screens

11 52 13.10 Projection Screens, Wall or Ceiling Hung

		Crew	Daily Output	Labor-Hours	Unit	Material	Labor	Equipment	Total	Total Incl O&P
0010	**PROJECTION SCREENS, WALL OR CEILING HUNG**, matte white									
0100	Manually operated, economy	2 Carp	500	.032	S.F.	5.85	1.33		7.18	8.65
0300	Intermediate		450	.036		6.80	1.48		8.28	9.95
0400	Deluxe		400	.040		9.45	1.66		11.11	13.10
9000	Minimum labor/equipment charge		3	5.333	Job		222		222	365

11 52 16 – Projectors

11 52 16.10 Movie Equipment

		Crew	Daily Output	Labor-Hours	Unit	Material	Labor	Equipment	Total	Total Incl O&P
0011	**MOVIE EQUIPMENT**									
0020	Changeover, minimum				Ea.	465			465	510
3000	Projection screens, rigid, in wall, acrylic, 1/4" thick	2 Glaz	195	.082	S.F.	42	3.30		45.30	51.50
3100	1/2" thick	"	130	.123	"	48.50	4.95		53.45	61.50
3700	Sound systems, incl. amplifier, mono, minimum	1 Elec	.90	8.889	Ea.	3,300	435		3,735	4,300
3800	Dolby/Super Sound, maximum		.40	20		18,100	980		19,080	21,400
4100	Dual system, 2 channel, front surround, minimum		.70	11.429		4,625	560		5,185	5,950
4200	Dolby/Super Sound, 4 channel, maximum		.40	20		16,500	980		17,480	19,700
5700	Seating, painted steel, upholstered, minimum	2 Carp	35	.457		132	19		151	176
5800	Maximum	"	28	.571		420	23.50		443.50	505

11 53 Laboratory Equipment

11 53 33 – Emergency Safety Appliances

11 53 33.13 Emergency Equipment

		Crew	Daily Output	Labor-Hours	Unit	Material	Labor	Equipment	Total	Total Incl O&P
0010	**EMERGENCY EQUIPMENT**									
1400	Safety equipment, eye wash, hand held				Ea.	405			405	445
1450	Deluge shower				"	740			740	815

11 53 43 – Service Fittings and Accessories

11 53 43.13 Fittings

		Crew	Daily Output	Labor-Hours	Unit	Material	Labor	Equipment	Total	Total Incl O&P
0010	**FITTINGS**									
1600	Sink, one piece plastic, flask wash, hose, free standing	1 Plum	1.60	5	Ea.	1,800	260		2,060	2,375
1610	Epoxy resin sink, 25" x 16" x 10"	"	2	4	"	198	208		406	545
1630	Steel table, open underneath				L.F.	430			430	470
1640	Doors underneath				"	635			635	700
1900	Utility cabinet, stainless steel				Ea.	715			715	790
1950	Utility table, acid resistant top with drawers	2 Carp	30	.533	L.F.	147	22		169	198
8000	Alternate pricing method: as percent of lab furniture									
8050	Installation, not incl. plumbing & duct work				% Furn.				20%	22%
8100	Plumbing, final connections, simple system								9.09%	10%
8110	Moderately complex system								13.64%	15%
8120	Complex system								18.18%	20%
8150	Electrical, simple system								9.09%	10%

11 53 Laboratory Equipment
11 53 43 – Service Fittings and Accessories

11 53 43.13 Fittings		Crew	Daily Output	Labor-Hours	Unit	Material	2010 Bare Costs Labor	Equipment	Total	Total Incl O&P
8160	Moderately complex system				% Furn.				18.18%	20%
8170	Complex system				↓				31.90%	35%

11 57 Vocational Shop Equipment
11 57 10 – Shop Equipment
11 57 10.10 Vocational School Shop Equipment

		Crew	Daily Output	Labor-Hours	Unit	Material	Labor	Equipment	Total	Total Incl O&P
0010	**VOCATIONAL SCHOOL SHOP EQUIPMENT**									
0020	Benches, work, wood, average	2 Carp	5	3.200	Ea.	490	133		623	760
0100	Metal, average		5	3.200		450	133		583	715
0400	Combination belt & disc sander, 6"		4	4		1,375	166		1,541	1,800
0700	Drill press, floor mounted, 12", 1/2 H.P.	↓	4	4		405	166		571	720
0800	Dust collector, not incl. ductwork, 6" diameter	1 Shee	1.10	7.273		4,125	355		4,480	5,100
0810	Dust collector bag, 20" diameter	"	5	1.600		310	78.50		388.50	465
1000	Grinders, double wheel, 1/2 H.P.	2 Carp	5	3.200		247	133		380	490
1300	Jointer, 4", 3/4 H.P.		4	4		1,250	166		1,416	1,650
1600	Kilns, 16 C.F., to 2000°		4	4		2,300	166		2,466	2,800
1900	Lathe, woodworking, 10", 1/2 H.P.		4	4		735	166		901	1,075
2200	Planer, 13" x 6"		4	4		970	166		1,136	1,350
2500	Potter's wheel, motorized		4	4		1,175	166		1,341	1,575
2800	Saws, band, 14", 3/4 H.P.		4	4		950	166		1,116	1,325
3100	Metal cutting band saw, 14"		4	4		2,175	166		2,341	2,650
3400	Radial arm saw, 10", 2 H.P.		4	4		1,250	166		1,416	1,650
3700	Scroll saw, 24"		4	4		580	166		746	915
4000	Table saw, 10", 3 H.P.		4	4		2,100	166		2,266	2,575
4300	Welder AC arc, 30 amp capacity	↓	4	4	↓	2,100	166		2,266	2,600

11 61 Theater and Stage Equipment
11 61 23 – Folding and Portable Stages
11 61 23.10 Portable Stages

		Crew	Daily Output	Labor-Hours	Unit	Material	Labor	Equipment	Total	Total Incl O&P
0010	**PORTABLE STAGES**									
5000	Stages, portable with steps, folding legs, stock, 8" high				SF Stg.	27.50			27.50	30.50
5100	16" high					23.50			23.50	26
5200	32" high					38			38	41.50
5300	40" high				↓	59.50			59.50	65

11 61 33 – Rigging Systems and Controls
11 61 33.10 Controls

		Crew	Daily Output	Labor-Hours	Unit	Material	Labor	Equipment	Total	Total Incl O&P
0010	**CONTROLS**									
0050	Control boards with dimmers and breakers, minimum	1 Elec	1	8	Ea.	12,300	390		12,690	14,100
0150	Maximum	"	.20	40	"	126,000	1,950		127,950	141,500
0160										

11 61 43 – Stage Curtains
11 61 43.10 Curtains

		Crew	Daily Output	Labor-Hours	Unit	Material	Labor	Equipment	Total	Total Incl O&P
0010	**CURTAINS**									
0500	Curtain track, straight, light duty	2 Carp	20	.800	L.F.	27	33		60	84
0700	Curved sections		12	1.333	"	173	55.50		228.50	281
1000	Curtains, velour, medium weight	↓	600	.027	S.F.	7.90	1.11		9.01	10.50

11 66 Athletic Equipment

11 66 13 – Exercise Equipment

11 66 13.10 Physical Training Equipment

		Crew	Daily Output	Labor-Hours	Unit	Material	2010 Bare Costs Labor	Equipment	Total	Total Incl O&P
0010	**PHYSICAL TRAINING EQUIPMENT**									
0020	Abdominal rack, 2 board capacity				Ea.	470			470	515
0050	Abdominal board, upholstered					525			525	575
0200	Bicycle trainer, minimum					695			695	765
0300	Deluxe, electric					4,150			4,150	4,575
0400	Barbell set, chrome plated steel, 25 lbs.					244			244	269
0420	100 lbs.					370			370	405
0450	200 lbs.				↓	715			715	785
0500	Weight plates, cast iron, per lb.				Lb.	5			5	5.50
0520	Storage rack, 10 station				Ea.	790			790	870
0600	Circuit training apparatus, 12 machines minimum	2 Clab	1.25	12.800	Set	27,600	425		28,025	31,000
0700	Average		1	16		33,800	530		34,330	38,100
0800	Maximum	↓	.75	21.333		40,100	705		40,805	45,300
0820	Dumbbell set, cast iron, with rack and 5 pair				↓	580			580	640
0900	Squat racks	2 Clab	5	3.200	Ea.	765	106		871	1,025
4150	Exercise equipment, bicycle trainer					595			595	655
4180	Chinning bar, adjustable, wall mounted	1 Carp	5	1.600		425	66.50		491.50	580
4200	Exercise ladder, 16' x 1'-7", suspended	L-2	3	5.333		1,475	195		1,670	1,950
4210	High bar, floor plate attached	1 Carp	4	2		1,100	83		1,183	1,350
4240	Parallel bars, adjustable		4	2		2,100	83		2,183	2,425
4270	Uneven parallel bars, adjustable	↓	4	2	↓	2,675	83		2,758	3,050
4280	Wall mounted, adjustable	L-2	1.50	10.667	Set	770	390		1,160	1,500
4300	Rope, ceiling mounted, 18' long	1 Carp	3.66	2.186	Ea.	257	91		348	430
4330	Side horse, vaulting		5	1.600		1,700	66.50		1,766.50	1,975
4360	Treadmill, motorized, deluxe, training type	↓	5	1.600		3,500	66.50		3,566.50	3,950
4390	Weight lifting multi-station, minimum	2 Clab	1	16		395	530		925	1,300

11 66 23 – Gymnasium Equipment

11 66 23.13 Basketball Equipment

		Crew	Daily Output	Labor-Hours	Unit	Material	Labor	Equipment	Total	Total Incl O&P
0010	**BASKETBALL EQUIPMENT**									
1000	Backstops, wall mtd., 6' extended, fixed, minimum	L-2	1	16	Ea.	1,150	585		1,735	2,200
1100	Maximum		1	16		1,675	585		2,260	2,800
1200	Swing up, minimum		1	16		1,400	585		1,985	2,475
1250	Maximum		1	16		5,350	585		5,935	6,850
1300	Portable, manual, heavy duty, spring operated		1.90	8.421		13,300	310		13,610	15,100
1400	Ceiling suspended, stationary, minimum		.78	20.513		2,850	750		3,600	4,375
1450	Fold up, with accessories, maximum	↓	.40	40		6,100	1,475		7,575	9,125
1600	For electrically operated, add	1 Elec	1	8	↓	2,000	390		2,390	2,800
5800	Wall pads, 1-1/2" thick, standard (not fire rated)	2 Carp	640	.025	S.F.	6.80	1.04		7.84	9.20

11 66 23.19 Boxing Ring

		Crew	Daily Output	Labor-Hours	Unit	Material	Labor	Equipment	Total	Total Incl O&P
0010	**BOXING RING**									
4100	Elevated, 22' x 22'	L-4	.10	240	Ea.	7,225	9,350		16,575	23,200
4110	For cellular plastic foam padding, add		.10	240		850	9,350		10,200	16,100
4120	Floor level, including posts and ropes only, 20' x 20'		.80	30		3,000	1,175		4,175	5,200
4130	Canvas, 30' x 30'	↓	5	4.800		2,575	187		2,762	3,125

11 66 23.47 Gym Mats

		Crew	Daily Output	Labor-Hours	Unit	Material	Labor	Equipment	Total	Total Incl O&P
0010	**GYM MATS**									
5500	2" thick, naugahyde covered				S.F.	3.59			3.59	3.95
5600	Vinyl/nylon covered					6.20			6.20	6.85
6000	Wrestling mats, 1" thick, heavy duty				↓	5.50			5.50	6.05

11 66 Athletic Equipment

11 66 43 – Interior Scoreboards

11 66 43.10 Scoreboards		Crew	Daily Output	Labor-Hours	Unit	Material	2010 Bare Costs Labor	Equipment	Total	Total Incl O&P
0010	**SCOREBOARDS**									
7000	Baseball, minimum	R-3	1.30	15.385	Ea.	4,000	745	106	4,851	5,675
7200	Maximum		.05	400		17,500	19,300	2,775	39,575	52,000
7300	Football, minimum		.86	23.256		5,175	1,125	161	6,461	7,575
7400	Maximum		.20	100		15,400	4,825	690	20,915	25,200
7500	Basketball (one side), minimum		2.07	9.662		2,325	465	67	2,857	3,350
7600	Maximum		.30	66.667		3,400	3,225	460	7,085	9,225
7700	Hockey-basketball (four sides), minimum		.25	80		5,500	3,850	555	9,905	12,600
7800	Maximum		.15	133		5,575	6,425	920	12,920	17,100

11 66 53 – Gymnasium Dividers

11 66 53.10 Divider Curtains

		Crew	Daily Output	Labor-Hours	Unit	Material	Labor	Equipment	Total	Total Incl O&P
0010	**DIVIDER CURTAINS**									
4500	Gym divider curtain, mesh top, vinyl bottom, manual	L-4	500	.048	S.F.	8.15	1.87		10.02	12.05
4700	Electric roll up	L-7	400	.070	"	9.95	2.81		12.76	15.50

11 82 Solid Waste Handling Equipment

11 82 26 – Waste Compactors and Destructors

11 82 26.10 Compactors

		Crew	Daily Output	Labor-Hours	Unit	Material	Labor	Equipment	Total	Total Incl O&P
0010	**COMPACTORS**									
0020	Compactors, 115 volt, 250#/hr., chute fed	L-4	1	24	Ea.	11,200	935		12,135	13,900
0100	Hand fed		2.40	10		8,225	390		8,615	9,650
1000	Heavy duty industrial compactor, 0.5 C.Y. capacity		1	24		7,300	935		8,235	9,575
1050	1.0 C.Y. capacity		1	24		11,100	935		12,035	13,800
1400	For handling hazardous waste materials, 55 gallon drum packer, std.					17,300			17,300	19,000
1410	55 gallon drum packer w/HEPA filter					21,600			21,600	23,700
1420	55 gallon drum packer w/charcoal & HEPA filter					28,800			28,800	31,700
1430	All of the above made explosion proof, add					13,100			13,100	14,500

11 91 Religious Equipment

11 91 13 – Baptistries

11 91 13.10 Baptistry

		Crew	Daily Output	Labor-Hours	Unit	Material	Labor	Equipment	Total	Total Incl O&P
0010	**BAPTISTRY**									
0150	Fiberglass, 3'-6" deep, x 13'-7" long,									
0160	steps at both ends, incl. plumbing, minimum	L-8	1	20	Ea.	3,825	875		4,700	5,625
0200	Maximum	"	.70	28.571		6,250	1,250		7,500	8,900
0250	Add for filter, heater and lights					1,375			1,375	1,500

11 91 23 – Sanctuary Equipment

11 91 23.10 Sanctuary Furnishings

		Crew	Daily Output	Labor-Hours	Unit	Material	Labor	Equipment	Total	Total Incl O&P
0010	**SANCTUARY FURNISHINGS**									
0020	Altar, wood, custom design, plain	1 Carp	1.40	5.714	Ea.	1,925	237		2,162	2,500
0050	Deluxe		.20	40	"	9,225	1,650		10,875	12,900
5000	Wall cross, aluminum, extruded, 2" x 2" section		34	.235	L.F.	132	9.80		141.80	161
5150	4" x 4" section		29	.276		190	11.45		201.45	228
5300	Bronze, extruded, 1" x 2" section		31	.258		260	10.70		270.70	305
5350	2-1/2" x 2-1/2" section		34	.235		395	9.80		404.80	450

Estimating Tips

General

- The items in this division are usually priced per square foot or each. Most of these items are purchased by the owner and placed by the contractor. Do not assume the items in Division 12 will be purchased and installed by the contractor. Check the specifications for responsibilities and include receiving, storage, installation, and mechanical and electrical hookups in the appropriate divisions.

- Some items in this division require some type of support system that is not usually furnished with the item. Examples of these systems include blocking for the attachment of casework and heavy drapery rods. The required blocking must be added to the estimate in the appropriate division.

Reference Numbers

Reference numbers are shown in shaded boxes at the beginning of some major classifications. These numbers refer to related items in the Reference Section. The reference information may be an estimating procedure, an alternate pricing method, or technical information.

Note: Not all subdivisions listed here necessarily appear in this publication.

12 21 Window Blinds

12 21 13 – Horizontal Louver Blinds

12 21 13.13 Metal Horizontal Louver Blinds	Crew	Daily Output	Labor-Hours	Unit	Material	2010 Bare Costs Labor	Equipment	Total	Total Incl O&P
0010 **METAL HORIZONTAL LOUVER BLINDS**									
0020 Horizontal, 1" aluminum slats, solid color, stock	1 Carp	590	.014	S.F.	4.75	.56		5.31	6.20

12 23 Interior Shutters

12 23 10 – Wood Interior Shutters

12 23 10.10 Wood Interior Shutters

	Crew	Daily Output	Labor-Hours	Unit	Material	Labor	Equipment	Total	Total Incl O&P
0010 **WOOD INTERIOR SHUTTERS**, louvered									
0200 Two panel, 27" wide, 36" high	1 Carp	5	1.600	Set	124	66.50		190.50	245
0300 33" wide, 36" high		5	1.600		162	66.50		228.50	288
0500 47" wide, 36" high		5	1.600		216	66.50		282.50	345
1000 Four panel, 27" wide, 36" high		5	1.600		147	66.50		213.50	270
1100 33" wide, 36" high		5	1.600		180	66.50		246.50	305
1300 47" wide, 36" high		5	1.600		255	66.50		321.50	390

12 23 10.13 Wood Panels

	Crew	Daily Output	Labor-Hours	Unit	Material	Labor	Equipment	Total	Total Incl O&P
0010 **WOOD PANELS**									
3000 Wood folding panels with movable louvers, 7" x 20" each	1 Carp	17	.471	Pr.	47	19.55		66.55	83.50
4000 Fixed louver type, stock units, 8" x 20" each		17	.471		84	19.55		103.55	125
4450 18" x 40" each		17	.471		118	19.55		137.55	162

12 24 Window Shades

12 24 13 – Roller Window Shades

12 24 13.10 Shades

	Crew	Daily Output	Labor-Hours	Unit	Material	Labor	Equipment	Total	Total Incl O&P
0010 **SHADES**									
0020 Basswood, roll-up, stain finish, 3/8" slats	1 Carp	300	.027	S.F.	13.95	1.11		15.06	17.15
0030 Double layered, heat reflective		685	.012		8.75	.49		9.24	10.40
0200 7/8" slats		300	.027		13.15	1.11		14.26	16.30
0900 Mylar, single layer, non-heat reflective		685	.012		6.70	.49		7.19	8.15
1000 Double layered, heat reflective		685	.012		9.25	.49		9.74	11
1100 Triple layered, heat reflective		685	.012		10.80	.49		11.29	12.65
5000 Thermal, roll up, R-4		44	.182		11.25	7.55		18.80	25
5030 R-10.7		44	.182		12.95	7.55		20.50	26.50
5050 Magnetic clips, set of 20				Set	24			24	26.50

12 32 Manufactured Wood Casework

12 32 16 – Manufactured Plastic-Laminate-Clad Casework

12 32 16.20 Plastic Laminate Casework Doors

	Crew	Daily Output	Labor-Hours	Unit	Material	Labor	Equipment	Total	Total Incl O&P
0010 **PLASTIC LAMINATE CASEWORK DOORS**									
1000 For casework frames, see Div. 12 32 23.15									
1100 For casework hardware, see Div. 12 32 23.35									
6000 Plastic laminate on particle board									
6100 12" wide, 18" high	1 Carp	25	.320	Ea.	10.50	13.30		23.80	33.50
6140 30" high		23	.348		17.50	14.45		31.95	43
6500 18" wide, 18" high		24	.333		15.75	13.85		29.60	40.50
6600 30" high		22	.364		26.50	15.10		41.60	54

12 32 16.25 Plastic Laminate Drawer Fronts

	Crew	Daily Output	Labor-Hours	Unit	Material	Labor	Equipment	Total	Total Incl O&P
0010 **PLASTIC LAMINATE DRAWER FRONTS**									
2800 Plastic laminate on particle board front									

12 32 Manufactured Wood Casework

12 32 16 – Manufactured Plastic-Laminate-Clad Casework

12 32 16.25 Plastic Laminate Drawer Fronts		Crew	Daily Output	Labor-Hours	Unit	Material	2010 Bare Costs Labor	Equipment	Total	Total Incl O&P
3000	4" high, 12" wide	1 Carp	17	.471	Ea.	3.51	19.55		23.06	36
3200	18" wide	"	16	.500	"	5.25	21		26.25	40

12 32 23 – Hardwood Casework

12 32 23.10 Manufactured Wood Casework, Stock Units

		Crew	Daily Output	Labor-Hours	Unit	Material	Labor	Equipment	Total	Total Incl O&P
0010	**MANUFACTURED WOOD CASEWORK, STOCK UNITS**									
0300	Built-in drawer units, pine, 18" deep, 32" high, unfinished									
0400	Minimum	2 Carp	53	.302	L.F.	145	12.55		157.55	181
0500	Maximum	"	40	.400	"	180	16.60		196.60	226
0700	Kitchen base cabinets, hardwood, not incl. counter tops,									
0710	24" deep, 35" high, prefinished									
0800	One top drawer, one door below, 12" wide	2 Carp	24.80	.645	Ea.	229	27		256	296
0820	15" wide		24	.667		238	27.50		265.50	310
0840	18" wide		23.30	.687		259	28.50		287.50	330
0860	21" wide		22.70	.705		272	29.50		301.50	345
0880	24" wide		22.30	.717		310	30		340	395
1000	Four drawers, 12" wide		24.80	.645		240	27		267	310
1020	15" wide		24	.667		243	27.50		270.50	315
1040	18" wide		23.30	.687		268	28.50		296.50	340
1060	24" wide		22.30	.717		305	30		335	385
1200	Two top drawers, two doors below, 27" wide		22	.727		330	30		360	415
1220	30" wide		21.40	.748		365	31		396	450
1240	33" wide		20.90	.766		385	32		417	475
1260	36" wide		20.30	.788		400	33		433	495
1280	42" wide		19.80	.808		425	33.50		458.50	525
1300	48" wide		18.90	.847		460	35		495	565
1500	Range or sink base, two doors below, 30" wide		21.40	.748		305	31		336	385
1520	33" wide		20.90	.766		330	32		362	415
1540	36" wide		20.30	.788		345	33		378	435
1560	42" wide		19.80	.808		370	33.50		403.50	460
1580	48" wide		18.90	.847		385	35		420	485
1800	For sink front units, deduct					145			145	160
2000	Corner base cabinets, 36" wide, standard	2 Carp	18	.889		540	37		577	655
2100	Lazy Susan with revolving door	"	16.50	.970		725	40.50		765.50	860
4000	Kitchen wall cabinets, hardwood, 12" deep with two doors									
4050	12" high, 30" wide	2 Carp	24.80	.645	Ea.	205	27		232	269
4100	36" wide		24	.667		240	27.50		267.50	310
4400	15" high, 30" wide		24	.667		165	27.50		192.50	227
4420	33" wide		23.30	.687		226	28.50		254.50	296
4440	36" wide		22.70	.705		240	29.50		269.50	310
4450	42" wide		22.70	.705		288	29.50		317.50	365
4700	24" high, 30" wide		23.30	.687		275	28.50		303.50	350
4720	36" wide		22.70	.705		305	29.50		334.50	385
4740	42" wide		22.30	.717		350	30		380	435
5000	30" high, one door, 12" wide		22	.727		185	30		215	254
5020	15" wide		21.40	.748		208	31		239	279
5040	18" wide		20.90	.766		229	32		261	305
5060	24" wide		20.30	.788		267	33		300	345
5300	Two doors, 27" wide		19.80	.808		294	33.50		327.50	380
5320	30" wide		19.30	.829		305	34.50		339.50	390
5340	36" wide		18.80	.851		350	35.50		385.50	445
5360	42" wide		18.50	.865		380	36		416	480
5380	48" wide		18.40	.870		430	36		466	535

12 32 Manufactured Wood Casework

12 32 23 – Hardwood Casework

12 32 23.10 Manufactured Wood Casework, Stock Units		Crew	Daily Output	Labor-Hours	Unit	Material	2010 Bare Costs Labor	Equipment	Total	Total Incl O&P
6000	Corner wall, 30" high, 24" wide	2 Carp	18	.889	Ea.	233	37		270	315
6050	30" wide		17.20	.930		264	38.50		302.50	355
6100	36" wide		16.50	.970		295	40.50		335.50	390
6500	Revolving Lazy Susan		15.20	1.053		415	43.50		458.50	525
7000	Broom cabinet, 84" high, 24" deep, 18" wide		10	1.600		575	66.50		641.50	745
7500	Oven cabinets, 84" high, 24" deep, 27" wide		8	2		860	83		943	1,075
7750	Valance board trim		396	.040	L.F.	12.40	1.68		14.08	16.40
9000	For deluxe models of all cabinets, add					40%				
9500	For custom built in place, add					25%	10%			
9550	Rule of thumb, kitchen cabinets not including									
9560	appliances & counter top, minimum	2 Carp	30	.533	L.F.	146	22		168	198
9600	Maximum	"	25	.640	"	310	26.50		336.50	390
9700	Minimum labor/equipment charge	1 Carp	3	2.667	Job		111		111	182

12 32 23.15 Manufactured Wood Casework Frames

		Crew	Daily Output	Labor-Hours	Unit	Material	Labor	Equipment	Total	Total Incl O&P
0010	**MANUFACTURED WOOD CASEWORK FRAMES**									
0050	Base cabinets, counter storage, 36" high									
0100	One bay, 18" wide	1 Carp	2.70	2.963	Ea.	134	123		257	350
0400	Two bay, 36" wide		2.20	3.636		204	151		355	470
1100	Three bay, 54" wide		1.50	5.333		243	222		465	630
2800	Bookcases, one bay, 7' high, 18" wide		2.40	3.333		157	139		296	400
3500	Two bay, 36" wide		1.60	5		228	208		436	590
4100	Three bay, 54" wide		1.20	6.667		375	277		652	870
5100	Coat racks, one bay, 7' high, 24" wide		4.50	1.778		157	74		231	294
5300	Two bay, 48" wide		2.75	2.909		219	121		340	440
5800	Three bay, 72" wide		2.10	3.810		320	158		478	615
6100	Wall mounted cabinet, one bay, 24" high, 18" wide		3.60	2.222		86.50	92.50		179	247
6800	Two bay, 36" wide		2.20	3.636		126	151		277	385
7400	Three bay, 54" wide		1.70	4.706		157	196		353	495
8400	30" high, one bay, 18" wide		3.60	2.222		94	92.50		186.50	255
9000	Two bay, 36" wide		2.15	3.721		125	155		280	390
9400	Three bay, 54" wide		1.60	5		156	208		364	510
9800	Wardrobe, 7' high, single, 24" wide		2.70	2.963		173	123		296	390
9880	Partition & adjustable shelves, 48" wide		1.70	4.706		220	196		416	560
9950	Partition, adjustable shelves & drawers, 48" wide		1.40	5.714		330	237		567	755
9970	Minimum labor/equipment charge		4	2	Job		83		83	137

12 32 23.20 Manufactured Hardwood Casework Doors

		Crew	Daily Output	Labor-Hours	Unit	Material	Labor	Equipment	Total	Total Incl O&P
0010	**MANUFACTURED HARDWOOD CASEWORK DOORS**									
2000	Glass panel, hardwood frame									
2200	12" wide, 18" high	1 Carp	34	.235	Ea.	18.75	9.80		28.55	36.50
2600	30" high		32	.250		31.50	10.40		41.90	51.50
4450	18" wide, 18" high		32	.250		28	10.40		38.40	48
4550	30" high		29	.276		47	11.45		58.45	70.50
5000	Hardwood, raised panel									
5100	12" wide, 18" high	1 Carp	16	.500	Ea.	24	21		45	60.50
5200	30" high		15	.533		40	22		62	80.50
5500	18" wide, 18" high		15	.533		36	22		58	76
5600	30" high		14	.571		60	23.50		83.50	105
9000	Minimum labor/equipment charge		4	2	Job		83		83	137

12 32 23.25 Manufactured Wood Casework Drawer Fronts

		Crew	Daily Output	Labor-Hours	Unit	Material	Labor	Equipment	Total	Total Incl O&P
0010	**MANUFACTURED WOOD CASEWORK DRAWER FRONTS**									
0100	Solid hardwood front									
1000	4" high, 12" wide	1 Carp	17	.471	Ea.	2.83	19.55		22.38	35

12 32 Manufactured Wood Casework

12 32 23 – Hardwood Casework

12 32 23.25 Manufactured Wood Casework Drawer Fronts		Crew	Daily Output	Labor-Hours	Unit	Material	2010 Bare Costs Labor	2010 Bare Costs Equipment	Total	Total Incl O&P
1200	18" wide	1 Carp	16	.500	Ea.	4.25	21		25.25	38.50
9000	Minimum labor/equipment charge	↓	4	2	Job		83		83	137

12 32 23.30 Manufactured Wood Casework Vanities

0010	**MANUFACTURED WOOD CASEWORK VANITIES**									
8000	Vanity bases, 2 doors, 30" high, 21" deep, 24" wide	2 Carp	20	.800	Ea.	227	33		260	305
8050	30" wide		16	1		262	41.50		303.50	355
8100	36" wide		13.33	1.200		350	50		400	465
8150	48" wide	↓	11.43	1.400		415	58		473	555
9000	For deluxe models of all vanities, add to above					40%				
9500	For custom built in place, add to above				↓	25%	10%			

12 32 23.35 Manufactured Wood Casework Hardware

0010	**MANUFACTURED WOOD CASEWORK HARDWARE**									
1000	Catches, minimum	1 Carp	235	.034	Ea.	1	1.41		2.41	3.42
1040	Maximum	"	80	.100	"	6.25	4.16		10.41	13.70
2000	Door/drawer pulls, handles									
2200	Handles and pulls, projecting, metal, minimum	1 Carp	160	.050	Ea.	4.52	2.08		6.60	8.40
2240	Maximum		68	.118		9.55	4.89		14.44	18.60
2300	Wood, minimum		160	.050		4.76	2.08		6.84	8.65
2340	Maximum		68	.118		8.75	4.89		13.64	17.65
2600	Flush, metal, minimum		160	.050		4.76	2.08		6.84	8.65
2640	Maximum		68	.118	↓	8.75	4.89		13.64	17.65
3000	Drawer tracks/glides, minimum		48	.167	Pr.	8.05	6.95		15	20.50
3040	Maximum		24	.333		23.50	13.85		37.35	49
4000	Cabinet hinges, minimum		160	.050		2.74	2.08		4.82	6.40
4040	Maximum	↓	68	.118	↓	10.35	4.89		15.24	19.45

12 34 Manufactured Plastic Casework

12 34 16 – Manufactured Solid-Plastic Casework

12 34 16.10 Outdoor Casework

0010	**OUTDOOR CASEWORK**									
0020	Cabinet, base, sink/range, 36"	2 Carp	20.30	.788	Ea.	1,625	33		1,658	1,825
0100	Base, 36"		20.30	.788		2,325	33		2,358	2,600
0200	Filler strip, 1" x 30"		158	.101		35	4.21		39.21	45.50
0210	Filler strip, 2" x 30"	↓	158	.101	↓	45	4.21		49.21	56.50

12 35 Specialty Casework

12 35 53 – Laboratory Casework

12 35 53.13 Metal Laboratory Casework

0010	**METAL LABORATORY CASEWORK**									
0020	Cabinets, base, door units, metal	2 Carp	18	.889	L.F.	192	37		229	272
0300	Drawer units		18	.889		430	37		467	530
0700	Tall storage cabinets, open, 7' high		20	.800		410	33		443	505
0900	With glazed doors		20	.800		490	33		523	595
1300	Wall cabinets, metal, 12-1/2" deep, open		20	.800		135	33		168	204
1500	With doors	↓	20	.800	↓	281	33		314	365

12 36 Countertops

12 36 16 – Metal Countertops

12 36 16.10 Stainless Steel Countertops	Crew	Daily Output	Labor-Hours	Unit	Material	2010 Bare Costs Labor	Equipment	Total	Total Incl O&P
0010 **STAINLESS STEEL COUNTERTOPS**									
3200 Stainless steel, custom	1 Carp	24	.333	S.F.	143	13.85		156.85	180

12 36 19 – Wood Countertops

12 36 19.10 Maple Countertops

	Crew	Daily Output	Labor-Hours	Unit	Material	Labor	Equipment	Total	Incl O&P
0010 **MAPLE COUNTERTOPS**									
2900 Solid, laminated, 1-1/2" thick, no splash	1 Carp	28	.286	L.F.	62.50	11.85		74.35	88
3000 With square splash		28	.286	"	74	11.85		85.85	101
3400 Recessed cutting block with trim, 16" x 20" x 1"		8	1	Ea.	72.50	41.50		114	148

12 36 23 – Plastic Countertops

12 36 23.13 Plastic-Laminate-Clad Countertops

	Crew	Daily Output	Labor-Hours	Unit	Material	Labor	Equipment	Total	Incl O&P
0010 **PLASTIC-LAMINATE-CLAD COUNTERTOPS**									
0020 Stock, 24" wide w/backsplash, minimum	1 Carp	30	.267	L.F.	9.10	11.10		20.20	28
0100 Maximum		25	.320		19.50	13.30		32.80	43.50
0300 Custom plastic, 7/8" thick, aluminum molding, no splash		30	.267		21	11.10		32.10	41
0400 Cove splash		30	.267		27.50	11.10		38.60	48
0600 1-1/4" thick, no splash		28	.286		31	11.85		42.85	53.50
0700 Square splash		28	.286		30	11.85		41.85	52.50
0900 Square edge, plastic face, 7/8" thick, no splash		30	.267		26	11.10		37.10	47
1000 With splash		30	.267		33.50	11.10		44.60	54.50
1200 For stainless channel edge, 7/8" thick, add					2.75			2.75	3.03
1300 1-1/4" thick, add					3.22			3.22	3.54
1500 For solid color suede finish, add					2.68			2.68	2.95
1700 For end splash, add				Ea.	17.85			17.85	19.65
1900 For cut outs, standard, add, minimum	1 Carp	32	.250		3.57	10.40		13.97	21
2000 Maximum		8	1		5.95	41.50		47.45	75
2100 Postformed, including backsplash and front edge		30	.267	L.F.	10.70	11.10		21.80	30
2110 Mitred, add		12	.667	Ea.		27.50		27.50	45.50
2200 Built-in place, 25" wide, plastic laminate		25	.320	L.F.	14.30	13.30		27.60	37.50
9000 Minimum labor/equipment charge		3.75	2.133	Job		88.50		88.50	146

12 36 33 – Tile Countertops

12 36 33.10 Ceramic Tile Countertops

	Crew	Daily Output	Labor-Hours	Unit	Material	Labor	Equipment	Total	Incl O&P
0010 **CERAMIC TILE COUNTERTOPS**									
2300 Ceramic tile mosaic	1 Carp	25	.320	L.F.	31	13.30		44.30	56

12 36 40 – Stone Countertops

12 36 40.10 Natural Stone Countertops

	Crew	Daily Output	Labor-Hours	Unit	Material	Labor	Equipment	Total	Incl O&P
0010 **NATURAL STONE COUNTERTOPS**									
2500 Marble, stock, with splash, 1/2" thick, minimum	1 Bric	17	.471	L.F.	38	19.65		57.65	73.50
2700 3/4" thick, maximum	"	13	.615	"	95.50	25.50		121	147

12 36 53 – Laboratory Countertops

12 36 53.10 Laboratory Countertops and Sinks

	Crew	Daily Output	Labor-Hours	Unit	Material	Labor	Equipment	Total	Incl O&P
0010 **LABORATORY COUNTERTOPS AND SINKS**									
0020 Countertops, not incl. base cabinets, acid-proof, minimum	2 Carp	82	.195	S.F.	33	8.10		41.10	49.50
0030 Maximum		70	.229		42.50	9.50		52	62
0040 Stainless steel		82	.195		98	8.10		106.10	121

12 36 61 – Simulated Stone Countertops

12 36 61.16 Solid Surface Countertops

	Crew	Daily Output	Labor-Hours	Unit	Material	Labor	Equipment	Total	Incl O&P
0010 **SOLID SURFACE COUNTERTOPS**, Acrylic polymer									
0020 Pricing for orders of 100 L.F. or greater									
0100 25" wide, solid colors	2 Carp	28	.571	L.F.	44.50	23.50		68	88

12 36 Countertops

12 36 61 – Simulated Stone Countertops

12 36 61.16 Solid Surface Countertops

		Crew	Daily Output	Labor-Hours	Unit	Material	2010 Bare Costs Labor	Equipment	Total	Total Incl O&P
0200	Patterned colors	2 Carp	28	.571	L.F.	56.50	23.50		80	101
0300	Premium patterned colors		28	.571		70.50	23.50		94	117
0400	With silicone attached 4" backsplash, solid colors		27	.593		49	24.50		73.50	94.50
0500	Patterned colors		27	.593		62	24.50		86.50	109
0600	Premium patterned colors		27	.593		77	24.50		101.50	126
0700	With hard seam attached 4" backsplash, solid colors		23	.696		49	29		78	102
0800	Patterned colors		23	.696		62	29		91	116
0900	Premium patterned colors		23	.696		77	29		106	133
1000	Pricing for order of 51 – 99 L.F.									
1100	25" wide, solid colors	2 Carp	24	.667	L.F.	51.50	27.50		79	102
1200	Patterned colors		24	.667		65	27.50		92.50	117
1300	Premium patterned colors		24	.667		81.50	27.50		109	135
1400	With silicone attached 4" backsplash, solid colors		23	.696		56.50	29		85.50	110
1500	Patterned colors		23	.696		71.50	29		100.50	126
1600	Premium patterned colors		23	.696		89	29		118	145
1700	With hard seam attached 4" backsplash, solid colors		20	.800		56.50	33		89.50	117
1800	Patterned colors		20	.800		71.50	33		104.50	133
1900	Premium patterned colors		20	.800		89	33		122	152
2000	Pricing for order of 1 – 50 L.F.									
2100	25" wide, solid colors	2 Carp	20	.800	L.F.	60	33		93	121
2200	Patterned colors		20	.800		76.50	33		109.50	139
2300	Premium patterned colors		20	.800		95.50	33		128.50	160
2400	With silicone attached 4" backsplash, solid colors		19	.842		66	35		101	130
2500	Patterned colors		19	.842		83.50	35		118.50	150
2600	Premium patterned colors		19	.842		104	35		139	173
2700	With hard seam attached 4" backsplash, solid colors		15	1.067		66	44.50		110.50	146
2800	Patterned colors		15	1.067		83.50	44.50		128	165
2900	Premium patterned colors		15	1.067		104	44.50		148.50	188
3000	Sinks, pricing for order of 100 or greater units									
3100	Single bowl, hard seamed, solid colors, 13" x 17"	1 Carp	3	2.667	Ea.	300	111		411	510
3200	10" x 15"		7	1.143		139	47.50		186.50	231
3300	Cutouts for sinks		8	1			41.50		41.50	68.50
3400	Sinks, pricing for order of 51 – 99 units									
3500	Single bowl, hard seamed, solid colors, 13" x 17"	1 Carp	2.55	3.137	Ea.	345	130		475	595
3600	10" x 15"		6	1.333		160	55.50		215.50	267
3700	Cutouts for sinks		7	1.143			47.50		47.50	78
3800	Sinks, pricing for order of 1 – 50 units									
3900	Single bowl, hard seamed, solid colors, 13" x 17"	1 Carp	2	4	Ea.	405	166		571	720
4000	10" x 15"		4.55	1.758		188	73		261	325
4100	Cutouts for sinks		5.25	1.524			63.50		63.50	104
4200	Cooktop cutouts, pricing for 100 or greater units		4	2		22	83		105	162
4300	51 – 99 units		3.40	2.353		25.50	98		123.50	189
4400	1 – 50 units		3	2.667		30	111		141	215

12 36 61.17 Solid Surface Vanity Tops

		Crew	Daily Output	Labor-Hours	Unit	Material	Labor	Equipment	Total	Total Incl O&P
0010	**SOLID SURFACE VANITY TOPS**									
0015	Solid surface, center bowl, 17" x 19"	1 Carp	12	.667	Ea.	189	27.50		216.50	254
0020	19" x 25"		12	.667		229	27.50		256.50	298
0030	19" x 31"		12	.667		278	27.50		305.50	350
0040	19" x 37"		12	.667		325	27.50		352.50	400
0050	22" x 25"		10	.800		202	33		235	277
0060	22" x 31"		10	.800		236	33		269	315
0070	22" x 37"		10	.800		274	33		307	355

12 36 Countertops

12 36 61 – Simulated Stone Countertops

12 36 61.17 Solid Surface Vanity Tops

		Crew	Daily Output	Labor-Hours	Unit	Material	2010 Bare Costs Labor	Equipment	Total	Total Incl O&P
0080	22" x 43"	1 Carp	10	.800	Ea.	315	33		348	400
0090	22" x 49"		10	.800		345	33		378	435
0110	22" x 55"		8	1		395	41.50		436.50	505
0120	22" x 61"		8	1		450	41.50		491.50	565
0220	Double bowl, 22" x 61"		8	1		505	41.50		546.50	630
0230	Double bowl, 22" x 73"		8	1		705	41.50		746.50	845
0240	For aggregate colors, add					35%				
0250	For faucets and fittings, see Div. 22 41 39.10									

12 36 61.19 Engineered Stone Countertops

		Crew	Daily Output	Labor-Hours	Unit	Material	Labor	Equipment	Total	Total Incl O&P
0010	**ENGINEERED STONE COUNTERTOPS**									
0100	25" wide, 4" backsplash, color group A, minimum	2 Carp	15	1.067	L.F.	17	44.50		61.50	91.50
0110	Maximum		15	1.067		42	44.50		86.50	119
0120	Color group B, minimum		15	1.067		22	44.50		66.50	97
0130	Maximum		15	1.067		49	44.50		93.50	127
0140	Color group C, minimum		15	1.067		29.50	44.50		74	106
0150	Maximum		15	1.067		60	44.50		104.50	139
0160	Color group D, minimum		15	1.067		37	44.50		81.50	114
0170	Maximum		15	1.067		70	44.50		114.50	150

12 48 Rugs and Mats

12 48 13 – Entrance Floor Mats and Frames

12 48 13.13 Entrance Floor Mats

			Crew	Daily Output	Labor-Hours	Unit	Material	Labor	Equipment	Total	Total Incl O&P
0010	**ENTRANCE FLOOR MATS**										
2000	Recycled rubber tire tile, 12" x 12" x 3/8" thick	G	1 Clab	125	.064	S.F.	8.50	2.12		10.62	12.85
2510	Natural cocoa fiber, 1/2" thick	G		125	.064		8.10	2.12		10.22	12.45
2520	3/4" thick	G		125	.064		6.25	2.12		8.37	10.35
2530	1" thick	G		125	.064		6.90	2.12		9.02	11.10

12 51 Office Furniture

12 51 16 – Case Goods

12 51 16.13 Metal Case Goods

		Crew	Daily Output	Labor-Hours	Unit	Material	Labor	Equipment	Total	Total Incl O&P
0010	**METAL CASE GOODS**									
0020	Desks, 29" high, double pedestal, 30" x 60", metal, minimum				Ea.	500			500	550
0030	Maximum				"	1,200			1,200	1,325

12 54 Hospitality Furniture

12 54 13 – Hotel and Motel Furniture

12 54 13.10 Hotel Furniture

		Crew	Daily Output	Labor-Hours	Unit	Material	Labor	Equipment	Total	Total Incl O&P
0010	**HOTEL FURNITURE**									
0020	Standard quality set, minimum				Room	2,325			2,325	2,550
0200	Maximum				"	8,300			8,300	9,125

12 54 16 – Restaurant Furniture

12 54 16.20 Furniture, Restaurant

		Crew	Daily Output	Labor-Hours	Unit	Material	Labor	Equipment	Total	Total Incl O&P
0010	**FURNITURE, RESTAURANT**									
0020	Bars, built-in, front bar	1 Carp	5	1.600	L.F.	250	66.50		316.50	385
0200	Back bar		5	1.600	"	182	66.50		248.50	310
0500	Booth unit, molded plastic, stub wall and 2 seats, minimum		2	4	Set	355	166		521	665

12 54 Hospitality Furniture

12 54 16 – Restaurant Furniture

12 54 16.20 Furniture, Restaurant		Crew	Daily Output	Labor-Hours	Unit	Material	2010 Bare Costs Labor	Equipment	Total	Total Incl O&P
0600	Maximum	1 Carp	1.50	5.333	Set	1,225	222		1,447	1,700
0800	Booth seat, upholstered, foursome, single (end) minimum		5	1.600	Ea.	650	66.50		716.50	825
0900	Maximum		4	2		870	83		953	1,100
1000	Foursome, double, minimum		4	2		1,075	83		1,158	1,300
1100	Maximum		3	2.667		1,500	111		1,611	1,825
1300	Circle booth, upholstered, 1/4 circle, minimum		3	2.667		1,175	111		1,286	1,475
1400	Maximum		2	4		1,625	166		1,791	2,050
1500	3/4 circle, minimum		1.50	5.333		4,625	222		4,847	5,450
1600	Maximum		1	8		5,600	330		5,930	6,700

12 55 Detention Furniture

12 55 13 – Detention Bunks

12 55 13.13 Cots

		Crew	Daily Output	Labor-Hours	Unit	Material	Labor	Equipment	Total	Total Incl O&P
0010	**COTS**									
2500	Bolted, single, painted steel	E-4	20	1.600	Ea.	330	76	7.30	413.30	510
2700	Stainless steel	"	20	1.600	"	960	76	7.30	1,043.30	1,200

12 56 Institutional Furniture

12 56 43 – Dormitory Furniture

12 56 43.10 Dormitory Furnishings

		Crew	Daily Output	Labor-Hours	Unit	Material	Labor	Equipment	Total	Total Incl O&P
0010	**DORMITORY FURNISHINGS**									
0300	Bunkable bed, twin, minimum				Ea.	300			300	330
0320	Maximum				"	685			685	750

12 56 51 – Library Furniture

12 56 51.10 Library Furnishings

		Crew	Daily Output	Labor-Hours	Unit	Material	Labor	Equipment	Total	Total Incl O&P
0010	**LIBRARY FURNISHINGS**									
1710	Carrels, hardwood, 36" x 24", minimum	1 Carp	5	1.600	Ea.	460	66.50		526.50	615
1720	Maximum		4	2		930	83		1,013	1,150
1730	Metal, minimum		5	1.600		251	66.50		317.50	385
1740	Maximum		4	2		590	83		673	785
6010	Bookshelf, metal, 90" high, 10" shelf, double face		11.50	.696	L.F.	118	29		147	178
6020	Single face		12	.667	"	113	27.50		140.50	170
6050	For 8" shelving, subtract from above					10%				
6070	For 42" high with countertop, subtract from above					20%				

12 56 70 – Healthcare Furniture

12 56 70.10 Furniture, Hospital

		Crew	Daily Output	Labor-Hours	Unit	Material	Labor	Equipment	Total	Total Incl O&P
0010	**FURNITURE, HOSPITAL**									
0020	Beds, manual, minimum				Ea.	755			755	830
0100	Maximum				"	2,575			2,575	2,825
1100	Patient wall systems, not incl. plumbing, minimum				Room	1,075			1,075	1,200
1200	Maximum				"	2,000			2,000	2,200

12 63 Stadium and Arena Seating

12 63 13 – Stadium and Arena Bench Seating

12 63 13.13 Bleachers		Crew	Daily Output	Labor-Hours	Unit	Material	2010 Bare Costs Labor	Equipment	Total	Total Incl O&P
0010	**BLEACHERS**									
3000	Telescoping, manual to 15 tier, minimum	F-5	65	.492	Seat	78.50	20.50		99	120
3100	Maximum		60	.533		118	22.50		140.50	166
3300	16 to 20 tier, minimum		60	.533		188	22.50		210.50	244
3400	Maximum		55	.582		235	24.50		259.50	299
3600	21 to 30 tier, minimum		50	.640		196	27		223	259
3700	Maximum		40	.800		235	33.50		268.50	315
3900	For integral power operation, add, minimum	2 Elec	300	.053		39	2.61		41.61	47
4000	Maximum	"	250	.064		62.50	3.14		65.64	74
5000	Benches, folding, in wall, 14' table, 2 benches	L-4	2	12	Set	665	465		1,130	1,500

12 67 Pews and Benches

12 67 13 – Pews

12 67 13.13 Sanctuary Pews

		Crew	Daily Output	Labor-Hours	Unit	Material	Labor	Equipment	Total	Total Incl O&P
0010	**SANCTUARY PEWS**									
1500	Bench type, hardwood, minimum	1 Carp	20	.400	L.F.	74	16.60		90.60	109
1550	Maximum	"	15	.533		147	22		169	198
1570	For kneeler, add					18.15			18.15	19.95

12 93 Site Furnishings

12 93 23 – Trash and Litter Receptors

12 93 23.10 Trash Receptacles

			Crew	Daily Output	Labor-Hours	Unit	Material	Labor	Equipment	Total	Total Incl O&P
0010	**TRASH RECEPTACLES**										
0500	Recycled plastic, var colors, round, 32 gal., 28" x 38" H	G	2 Clab	5	3.200	Ea.	291	106		397	495
0510	32 gal., 31" x 32" H	G	"	5	3.200	"	360	106		466	570

12 93 33 – Manufactured Planters

12 93 33.10 Planters

		Crew	Daily Output	Labor-Hours	Unit	Material	Labor	Equipment	Total	Total Incl O&P
0010	**PLANTERS**									
0012	Concrete, sandblasted, precast, 48" diameter, 24" high	2 Clab	15	1.067	Ea.	655	35.50		690.50	780
0100	Fluted, precast, 7' diameter, 36" high		10	1.600		1,100	53		1,153	1,300
0300	Fiberglass, circular, 36" diameter, 24" high		15	1.067		475	35.50		510.50	580
0400	60" diameter, 24" high		10	1.600		800	53		853	965
9000	Minimum labor/equipment charge	1 Clab	2	4	Job		132		132	217

12 93 43 – Site Seating and Tables

12 93 43.13 Site Seating

		Crew	Daily Output	Labor-Hours	Unit	Material	Labor	Equipment	Total	Total Incl O&P
0010	**SITE SEATING**									
0012	Seating, benches, park, precast conc, w/backs, wood rails, 4' long	2 Clab	5	3.200	Ea.	420	106		526	640
0100	8' long		4	4		800	132		932	1,100
0300	Fiberglass, without back, one piece, 4' long		10	1.600		610	53		663	755
0400	8' long		7	2.286		1,225	75.50		1,300.50	1,475
0500	Steel barstock pedestals w/backs, 2" x 3" wood rails, 4' long		10	1.600		1,050	53		1,103	1,225
0510	8' long		7	2.286		1,250	75.50		1,325.50	1,500
0520	3" x 8" wood plank, 4' long		10	1.600		1,050	53		1,103	1,250
0530	8' long		7	2.286		1,225	75.50		1,300.50	1,475
0540	Backless, 4" x 4" wood plank, 4' square		10	1.600		1,025	53		1,078	1,200
0550	8' long		7	2.286		975	75.50		1,050.50	1,200
0600	Aluminum pedestals, with backs, aluminum slats, 8' long		8	2		335	66		401	480
0610	15' long		5	3.200		585	106		691	820

12 93 Site Furnishings

12 93 43 – Site Seating and Tables

12 93 43.13 Site Seating		Crew	Daily Output	Labor-Hours	Unit	Material	2010 Bare Costs Labor	Equipment	Total	Total Incl O&P
0620	Portable, aluminum slats, 8' long	2 Clab	8	2	Ea.	335	66		401	480
0630	15' long		5	3.200		735	106		841	985
0800	Cast iron pedestals, back & arms, wood slats, 4' long		8	2		295	66		361	435
0820	8' long		5	3.200		980	106		1,086	1,250
0840	Backless, wood slats, 4' long		8	2		545	66		611	710
0860	8' long		5	3.200		930	106		1,036	1,200
1700	Steel frame, fir seat, 10' long		10	1.600		240	53		293	350
9000	Minimum labor/equipment charge		2	8	Job		265		265	435

Division Notes

	CREW	DAILY OUTPUT	LABOR-HOURS	UNIT	MAT.	LABOR	EQUIP.	TOTAL	TOTAL INCL O&P

Estimating Tips

General

- The items and systems in this division are usually estimated, purchased, supplied, and installed as a unit by one or more subcontractors. The estimator must ensure that all parties are operating from the same set of specifications and assumptions, and that all necessary items are estimated and will be provided. Many times the complex items and systems are covered, but the more common ones, such as excavation or a crane, are overlooked for the very reason that everyone assumes nobody could miss them. The estimator should be the central focus and be able to ensure that all systems are complete.

- Another area where problems can develop in this division is at the interface between systems. The estimator must ensure, for instance, that anchor bolts, nuts, and washers are estimated and included for the air-supported structures and pre-engineered buildings to be bolted to their foundations. Utility supply is a common area where essential items or pieces of equipment can be missed or overlooked due to the fact that each subcontractor may feel it is another's responsibility. The estimator should also be aware of certain items which may be supplied as part of a package but installed by others, and ensure that the installing contractor's estimate includes the cost of installation. Conversely, the estimator must also ensure that items are not costed by two different subcontractors, resulting in an inflated overall estimate.

13 30 00 Special Structures

- The foundations and floor slab, as well as rough mechanical and electrical, should be estimated, as this work is required for the assembly and erection of the structure. Generally, as noted in the book, the pre-engineered building comes as a shell. Pricing is based on the size and structural design parameters stated in the reference section. Additional features, such as windows and doors, must also be included by the estimator. Here again, the estimator must have a clear understanding of the scope of each portion of the work and all the necessary interfaces.

Reference Numbers

Reference numbers are shown in shaded boxes at the beginning of some major classifications. These numbers refer to related items in the Reference Section. The reference information may be an estimating procedure, an alternate pricing method, or technical information.

Note: Not all subdivisions listed here necessarily appear in this publication.

No part of this publication may be reproduced, stored in a retrieval system, or transmitted in any form or by any means without prior written permission of R.S. Means Company, Inc.

13 05 Common Work Results for Special Construction

13 05 05 – Selective Special Construction Demolition

13 05 05.10 Selective Demolition, Air Supported Structures

		Crew	Daily Output	Labor-Hours	Unit	Material	2010 Bare Costs Labor	Equipment	Total	Total Incl O&P
0010	**SELECTIVE DEMOLITION, AIR SUPPORTED STRUCTURES**									
0020	Tank covers, scrim, dbl layer, vinyl poly w/hdw, blower & controls									
0050	Round and rectangular R024119-10	B-2	9000	.004	S.F.		.15		.15	.24
0100	Warehouse structures									
0120	Poly/vinyl fabric, 28 oz, incl tension cables & inflation system	4 Clab	9000	.004	SF Flr.		.12		.12	.19
0150	Reinforced vinyl, 12 oz., 3000 S.F.	"	5000	.006			.21		.21	.35
0200	12,000 to 24,000 S.F.	8 Clab	20000	.003			.11		.11	.17
0250	Tedlar vinyl fabric, 28 oz. w/liner, to 3000 S.F.	4 Clab	5000	.006			.21		.21	.35
0300	12,000 to 24,000 S.F.	8 Clab	20000	.003			.11		.11	.17
0350	Greenhouse/shelter, woven polyethylene with liner									
0400	3000 S.F.	4 Clab	5000	.006	SF Flr.		.21		.21	.35
0450	12,000 to 24,000 S.F.	8 Clab	20000	.003			.11		.11	.17
0500	Tennis/gymnasium, poly/vinyl fabric, 28 oz., incl thermal liner	4 Clab	9000	.004			.12		.12	.19
0600	Stadium/convention center, teflon coated fiberglass, incl thermal liner	9 Clab	40000	.002			.06		.06	.10
0700	Doors, air lock, 15' long, 10' x 10'	2 Carp	1.50	10.667	Ea.		445		445	730
0720	15' x 15'		.80	20			830		830	1,375
0750	Revolving personnel door, 6' dia. x 6'-6" high		1.50	10.667			445		445	730

13 05 05.20 Selective Demolition, Garden Houses

		Crew	Daily Output	Labor-Hours	Unit	Material	Labor	Equipment	Total	Total Incl O&P
0010	**SELECTIVE DEMOLITION, GARDEN HOUSES**									
0020	Prefab, wood, excl foundation, average	2 Clab	400	.040	SF Flr.		1.32		1.32	2.17

13 05 05.25 Selective Demolition, Geodesic Domes

		Crew	Daily Output	Labor-Hours	Unit	Material	Labor	Equipment	Total	Total Incl O&P
0010	**SELECTIVE DEMOLITION, GEODESIC DOMES**									
0050	Shell only, interlocking plywood panels, 30' diameter	F-5	3.20	10	Ea.		420		420	690
0060	34' diameter		2.30	13.913			585		585	960
0070	39' diameter		2	16			675		675	1,100
0080	45' diameter	F-3	2.20	18.182			765	298	1,063	1,575
0090	55' diameter		2	20			840	330	1,170	1,725
0100	60' diameter		2	20			840	330	1,170	1,725
0110	65' diameter		1.60	25			1,050	410	1,460	2,175

13 05 05.30 Selective Demolition, Greenhouses

		Crew	Daily Output	Labor-Hours	Unit	Material	Labor	Equipment	Total	Total Incl O&P
0010	**SELECTIVE DEMOLITION, GREENHOUSES**									
0020	Resi-type, free standing, excl foundations, 9' long x 8' wide	2 Clab	160	.100	SF Flr.		3.31		3.31	5.45
0030	9' long x 11' wide		170	.094			3.12		3.12	5.10
0040	9' long x 14' wide		220	.073			2.41		2.41	3.95
0050	9' long x 17' wide		320	.050			1.66		1.66	2.72
0060	Lean-to type, 4' wide		64	.250			8.30		8.30	13.60
0070	7' wide		120	.133			4.41		4.41	7.25
0080	Geodesic hemisphere, 1/8" plexiglass glazing, 8' diam.		4	4	Ea.		132		132	217
0090	24' dia		.80	20			660		660	1,075
0100	48' dia		.40	40			1,325		1,325	2,175

13 05 05.35 Selective Demolition, Hangars

		Crew	Daily Output	Labor-Hours	Unit	Material	Labor	Equipment	Total	Total Incl O&P
0010	**SELECTIVE DEMOLITION, HANGARS**									
0020	T type hangars, prefab, steel, galv roof & walls, incl doors, excl fndtn	E-2	2550	.022	SF Flr.		1	.63	1.63	2.45
0030	Circular type, prefab, steel frame, plastic skin, incl foundation, 80' diam	"	.50	112	Total		5,100	3,225	8,325	12,500

13 05 05.45 Selective Demolition, Lightning Protection

		Crew	Daily Output	Labor-Hours	Unit	Material	Labor	Equipment	Total	Total Incl O&P
0010	**SELECTIVE DEMOLITION, LIGHTNING PROTECTION**									
0020	Air terminal & base, copper, 3/8" diam. x 10", to 75' h	1 Clab	16	.500	Ea.		16.55		16.55	27
0030	1/2" diam. x 12", over 75' h		16	.500			16.55		16.55	27
0050	Aluminum, 1/2" diam. x 12", to 75' h		16	.500			16.55		16.55	27
0060	5/8" diam. x 12", over 75' h		16	.500			16.55		16.55	27
0070	Cable, copper, 220 lb per thousand feet, to 75' high		640	.013	L.F.		.41		.41	.68

13 05 Common Work Results for Special Construction

13 05 05 – Selective Special Construction Demolition

13 05 05.45 Selective Demolition, Lightning Protection

		Crew	Daily Output	Labor-Hours	Unit	Material	2010 Bare Costs Labor	Equipment	Total	Total Incl O&P
0080	375 lb per thousand feet, over 75' high	1 Clab	460	.017	L.F.		.58		.58	.95
0090	Aluminum, 101 lb per thousand feet, to 75' high		560	.014			.47		.47	.78
0100	199 lb per thousand feet, over 75' high		480	.017			.55		.55	.91
0110	Arrester, 175 V AC, to ground		16	.500	Ea.		16.55		16.55	27
0120	650 V AC, to ground		13	.615	"		20.50		20.50	33.50

13 05 05.50 Selective Demolition, Pre-Engineered Steel Buildings

		Crew	Daily Output	Labor-Hours	Unit	Material	Labor	Equipment	Total	Total Incl O&P
0010	**SELECTIVE DEMOLITION, PRE-ENGINEERED STEEL BUILDINGS**									
0500	Pre-engd steel bldgs, rigid frame, clear span & multi post, excl salvage									
0550	3,500 to 7,500 S.F	L-10	1000	.024	SF Flr.		1.12	.66	1.78	2.68
0600	7,501 to 12,500 S.F		1500	.016			.75	.44	1.19	1.79
0650	12,500 S.F or greater		1650	.015			.68	.40	1.08	1.63
0700	Pre-engd steel building components									
0710	Entrance canopy, including frame 4' x 4'	E-24	8	4	Ea.		184	93.50	277.50	430
0720	4' x 8'	"	7	4.571			210	107	317	485
0730	H.M doors, self framing, single leaf	2 Skwk	8	2			85		85	138
0740	Double leaf		5	3.200			136		136	220
0760	Gutter, eave type		600	.027	L.F.		1.14		1.14	1.84
0770	Sash, single slide, double slide or fixed		24	.667	Ea.		28.50		28.50	46
0780	Skylight, fiberglass, to 30 S.F.		16	1			42.50		42.50	69
0785	Roof vents, circular, 12" to 24" diameter		12	1.333			57		57	92
0790	Continuous, 10' long		8	2			85		85	138
0900	Shelters, aluminum frame									
0910	Acrylic glazing, 3' x 9' x 8' high	2 Skwk	2	8	Ea.		340		340	550
0920	9' x 12' x 8' high	"	1.50	10.667	"		455		455	735

13 05 05.60 Selective Demolition, Silos

		Crew	Daily Output	Labor-Hours	Unit	Material	Labor	Equipment	Total	Total Incl O&P
0010	**SELECTIVE DEMOLITION, SILOS**									
0020	Conc stave, indstrl, conical/sloping bott, excl fndtn, 12' diam., 35' h	E-24	.18	177	Ea.		8,150	4,150	12,300	19,000
0030	16' diam., 45' h		.12	266			12,200	6,225	18,425	28,500
0040	25' dia, 75' h		.08	400			18,400	9,350	27,750	42,800
0050	Steel, factory fabricated, 30,000 gal cap, painted or epoxy lined	L-5	2	28			1,300	375	1,675	2,750

13 05 05.65 Selective Demolition, Sound Control

		Crew	Daily Output	Labor-Hours	Unit	Material	Labor	Equipment	Total	Total Incl O&P
0010	**SELECTIVE DEMOLITION, SOUND CONTROL**									
0120	Acoustical enclosure, 4" thick walls & ceiling panels, 8 lbs/SF	3 Carp	144	.167	SF Surf		6.95		6.95	11.40
0130	10.5 lbs/SF		128	.188			7.80		7.80	12.80
0140	Reverb chamber, parallel walls, 4" thick		120	.200			8.30		8.30	13.65
0150	Skewed walls, parallel roof, 4" thick		110	.218			9.05		9.05	14.90
0160	Skewed walls/roof, 4" layer/air space		96	.250			10.40		10.40	17.05
0170	Sound-absorbing panels, painted metal, 2'-6" x 8', under 1,000 SF		430	.056			2.32		2.32	3.81
0180	Over 1,000 SF		480	.050			2.08		2.08	3.41
0190	Flexible transparent curtain, clear	3 Shee	430	.056			2.74		2.74	4.29
0192	50% clear, 50% foam		430	.056			2.74		2.74	4.29
0194	25% clear, 75% foam		430	.056			2.74		2.74	4.29
0196	100% foam		430	.056			2.74		2.74	4.29
0200	Audio-masking sys, incl speakers, amplfr, signal gnrtr, clng mntd									
0205	Ceiling mounted, 5,000 SF	2 Elec	4800	.003	S.F.		.16		.16	.25
0210	10,000 SF		5600	.003			.14		.14	.22
0220	Plenum mounted, 5,000 SF		7600	.002			.10		.10	.16
0230	10,000 SF		8800	.002			.09		.09	.14

13 05 05.70 Selective Demolition, Special Purpose Rooms

		Crew	Daily Output	Labor-Hours	Unit	Material	Labor	Equipment	Total	Total Incl O&P
0010	**SELECTIVE DEMOLITION, SPECIAL PURPOSE ROOMS**									
0100	Audiometric rooms, under 500 S.F. surface	4 Carp	200	.160	SF Surf		6.65		6.65	10.90
0110	Over 500 S.F. surface	"	240	.133	"		5.55		5.55	9.10

13 05 Common Work Results for Special Construction

13 05 05 – Selective Special Construction Demolition

13 05 05.70 Selective Demolition, Special Purpose Rooms		Crew	Daily Output	Labor-Hours	Unit	Material	2010 Bare Costs Labor	Equipment	Total	Total Incl O&P
0200	Clean rooms, 12' x 12' soft wall, class 100	1 Carp	.30	26.667	Ea.		1,100		1,100	1,825
0210	Class 1000		.30	26.667			1,100		1,100	1,825
0220	Class 10,000		.35	22.857			950		950	1,550
0230	Class 100,000		.35	22.857			950		950	1,550
0300	Darkrooms, shell complete, 8' high	2 Carp	220	.073	SF Flr.		3.02		3.02	4.96
0310	12' high		110	.145	"		6.05		6.05	9.95
0350	Darkrooms doors, mini-cylindrical, revolving		4	4	Ea.		166		166	273
0400	Music room, practice modular		140	.114	SF Surf		4.75		4.75	7.80
0500	Refrigeration structures and finishes									
0510	Wall finish, 2 coat portland cement plaster, 1/2" thick	1 Clab	200	.040	S.F.		1.32		1.32	2.17
0520	Fiberglass panels, 1/8" thick		400	.020			.66		.66	1.09
0530	Ceiling finish, polystyrene plastic, 1" to 2" thick		500	.016			.53		.53	.87
0540	4" thick		450	.018			.59		.59	.97
0550	Refrigerator, prefab aluminum walk-in, 7'-6" high, 6' x 6' OD	2 Carp	100	.160	SF Flr.		6.65		6.65	10.90
0560	10' x 10' OD		160	.100			4.16		4.16	6.85
0570	Over 150 S.F.		200	.080			3.32		3.32	5.45
0600	Sauna, prefabricated, including heater & controls, 7' high, to 30 S.F.		120	.133			5.55		5.55	9.10
0610	To 40 S.F.		140	.114			4.75		4.75	7.80
0620	To 60 S.F.		175	.091			3.80		3.80	6.25
0630	To 100 S.F.		220	.073			3.02		3.02	4.96
0640	To 130 S.F.		250	.064			2.66		2.66	4.37

13 05 05.75 Selective Demolition, Storage Tanks		Crew	Daily Output	Labor-Hours	Unit	Material	Labor	Equipment	Total	Total Incl O&P
0010	**SELECTIVE DEMOLITION, STORAGE TANKS**									
0500	Steel tank, single wall, above ground, not incl fdn, pumps or piping									
0510	Single wall, 275 gallon	Q-1	3	5.333	Ea.		250		250	385
0520	550 thru 2,000 gallon	B-34P	2	12			510	315	825	1,150
0530	5,000 thru 10,000 gallon	B-34Q	2	12			515	625	1,140	1,500
0540	15,000 thru 30,000 gallon	B-34S	2	16			725	1,800	2,525	3,100
0600	Steel tank, double wall, above ground not incl fdn, pumps & piping									
0620	500 thru 2,000 gallon	B-34P	2	12	Ea.		510	315	825	1,150

13 05 05.85 Selective Demolition, Swimming Pool Equip		Crew	Daily Output	Labor-Hours	Unit	Material	Labor	Equipment	Total	Total Incl O&P
0010	**SELECTIVE DEMOLITION, SWIMMING POOL EQUIP**									
0020	Diving stand, stainless steel, 3 meter	2 Clab	3	5.333	Ea.		177		177	290
0030	1 meter		5	3.200			106		106	174
0040	Diving board, 16' long, aluminum		5.40	2.963			98		98	161
0050	Fiberglass		5.40	2.963			98		98	161
0070	Ladders, heavy duty, stainless steel, 2 tread		14	1.143			38		38	62
0080	4 tread		12	1.333			44		44	72.50
0090	Lifeguard chair, stainless steel, fixed		5	3.200			106		106	174
0100	Slide, tubular, fiberglass, aluminum handrails & ladder, 5', straight		4	4			132		132	217
0110	8', curved		6	2.667			88.50		88.50	145
0120	10', curved		3	5.333			177		177	290
0130	12' straight, with platform		2.50	6.400			212		212	350
0140	Removable access ramp, stainless steel		4	4			132		132	217
0150	Removable stairs, stainless steel, collapsible		4	4			132		132	217

13 05 05.90 Selective Demolition, Tension Structures		Crew	Daily Output	Labor-Hours	Unit	Material	Labor	Equipment	Total	Total Incl O&P
0010	**SELECTIVE DEMOLITION, TENSION STRUCTURES**									
0020	Tension structure, steel/alum frame, fabric shell, 60' clear span, 6,000 SF	B-41	2000	.022	SF Flr.		.75	.15	.90	1.39
0030	12,000 SF		2200	.020			.68	.14	.82	1.27
0040	80' clear span, 20,800 SF		2440	.018			.62	.12	.74	1.14
0050	100' clear span, 10,000 SF	L-5	4350	.013			.60	.17	.77	1.27
0060	26,000 SF	"	4600	.012			.57	.16	.73	1.20

13 05 Common Work Results for Special Construction

13 05 05 – Selective Special Construction Demolition

13 05 05.95 Selective Demo, X-Ray/Radio Freq Protection

		Crew	Daily Output	Labor-Hours	Unit	Material	2010 Bare Costs Labor	Equipment	Total	Total Incl O&P
0010	**SELECTIVE DEMO, X-RAY/RADIO FREQ PROTECTION**									
0020	Shielding lead, lined door frame, excl hdwe, 1/16" thick	1 Clab	4.80	1.667	Ea.		55		55	90.50
0030	Lead sheets, 1/16" thick	2 Clab	270	.059	S.F.		1.96		1.96	3.22
0050	Lead shielding, 1/4" thick		270	.059			1.96		1.96	3.22
0060	1/2" thick		240	.067			2.21		2.21	3.62
0070	Lead glass, 1/4" thick, 2.0 mm LE, 12" x 16"	2 Glaz	16	1	Ea.		40		40	64.50
0080	24" x 36"		8	2			80.50		80.50	129
0090	36" x 60"		4	4			161		161	258
0100	Lead glass window frame, with 1/16" lead & voice passage, 36" x 60"		4	4			161		161	258
0110	Lead glass window frame, 24" x 36"		8	2			80.50		80.50	129
0120	Lead gypsum board, 5/8" thick with 1/16" lead	2 Clab	320	.050	S.F.		1.66		1.66	2.72
0130	1/8" lead		280	.057			1.89		1.89	3.11
0140	1/32" lead		400	.040			1.32		1.32	2.17
0150	Butt joints, 1/8" lead or thicker, 2" x 7' long batten strip		480	.033	Ea.		1.10		1.10	1.81
0160	X-ray protection, average radiography room, up to 300 SF, 1/16" lead, min		.50	32	Total		1,050		1,050	1,750
0170	Maximum		.30	53.333			1,775		1,775	2,900
0180	Deep therapy X-ray room, 250 KV cap, up to 300 SF, 1/4" lead, min		.20	80			2,650		2,650	4,350
0190	Maximum		.12	133			4,425		4,425	7,250
0880	Radio frequency shielding, prefab or screen-type copper or steel, minimum		360	.044	SF Surf		1.47		1.47	2.42
0890	Average		310	.052			1.71		1.71	2.81
0895	Maximum		290	.055			1.83		1.83	3

13 11 Swimming Pools

13 11 13 – Below-Grade Swimming Pools

13 11 13.50 Swimming Pools

		Crew	Daily Output	Labor-Hours	Unit	Material	Labor	Equipment	Total	Total Incl O&P
0011	**SWIMMING POOLS**, Outdoor, incl. equip. & houses, minimum				SF Surf				32.73	36
0300	Maximum								61.82	68
0400	Residential, incl. equipment, permanent type, minimum								10.91	12
0700	Maximum								22.73	25
0900	Municipal, including equipment only, over 5000 S.F., minimum								22.73	25
1000	Maximum								46.36	51
1300	Motel or apt., incl. equipment only, under 5000 S.F., minimum								20	22
1400	Maximum								30.91	34

13 12 Fountains

13 12 13 – Exterior Fountains

13 12 13.10 Yard Fountains

		Crew	Daily Output	Labor-Hours	Unit	Material	Labor	Equipment	Total	Total Incl O&P
0010	**YARD FOUNTAINS**									
0100	Outdoor fountain, 48" high with bowl and figures	2 Clab	2	8	Ea.	253	265		518	715

13 17 Tubs and Pools

13 17 13 – Redwood Hot Tub System

13 17 13.10 Redwood Hot Tub System	Crew	Daily Output	Labor-Hours	Unit	Material	2010 Bare Costs Labor	Equipment	Total	Total Incl O&P
0010 **REDWOOD HOT TUB SYSTEM**									
7050 4' diameter x 4' deep	Q-1	1	16	Ea.	1,900	750		2,650	3,250
7150 6' diameter x 4' deep		.80	20		2,950	935		3,885	4,700
7200 8' diameter x 4' deep		.80	20		4,100	935		5,035	5,950

13 17 33 – Whirlpool Tubs

13 17 33.10 Whirlpool Bath

	Crew	Daily Output	Labor-Hours	Unit	Material	Labor	Equipment	Total	Total Incl O&P
0010 **WHIRLPOOL BATH**									
6000 Whirlpool, bath with vented overflow, molded fiberglass									
6100 66" x 48" x 24"	Q-1	1	16	Ea.	3,225	750		3,975	4,700
6400 72" x 36" x 24"		1	16		3,175	750		3,925	4,625
6500 60" x 30" x 21"		1	16		2,725	750		3,475	4,150
6600 72" x 42" x 22"		1	16		4,250	750		5,000	5,825
6700 83" x 65"		.30	53.333		5,550	2,500		8,050	9,975
6710 For color add					10%				
6711 For designer colors and trim add					25%				

13 21 Controlled Environment Rooms

13 21 26 – Cold Storage Rooms

13 21 26.50 Refrigeration

	Crew	Daily Output	Labor-Hours	Unit	Material	Labor	Equipment	Total	Total Incl O&P
0010 **REFRIGERATION**									
0020 Curbs, 12" high, 4" thick, concrete	2 Carp	58	.276	L.F.	4.37	11.45		15.82	23.50
6300 Rule of thumb for complete units, w/o doors & refrigeration, cooler		146	.110	SF Flr.	127	4.55		131.55	148
6400 Freezer		109.60	.146	"	150	6.05		156.05	175

13 24 Special Activity Rooms

13 24 16 – Saunas

13 24 16.50 Saunas and Heaters

	Crew	Daily Output	Labor-Hours	Unit	Material	Labor	Equipment	Total	Total Incl O&P
0010 **SAUNAS AND HEATERS**									
0020 Prefabricated, incl. heater & controls, 7' high, 6' x 4', C/C	L-7	2.20	12.727	Ea.	4,625	510		5,135	5,925
1700 Door only, cedar, 2' x 6', with tempered insulated glass window	2 Carp	3.40	4.706		610	196		806	990
1800 Prehung, incl. jambs, pulls & hardware	"	12	1.333		615	55.50		670.50	765
2500 Heaters only (incl. above), wall mounted, to 200 C.F.					605			605	665
4480 For additional equipment, see Div. 11 66 13.10									

13 24 26 – Steam Baths

13 24 26.50 Steam Baths and Components

	Crew	Daily Output	Labor-Hours	Unit	Material	Labor	Equipment	Total	Total Incl O&P
0010 **STEAM BATHS AND COMPONENTS**									
0020 Heater, timer & head, single, to 140 C.F.	1 Plum	1.20	6.667	Ea.	1,400	345		1,745	2,100
0500 To 300 C.F.	"	1.10	7.273		1,625	380		2,005	2,350
2000 Multiple, motels, apts., 2 baths, w/blow-down assm., 500 C.F.	Q-1	1.30	12.308		5,900	575		6,475	7,400
2500 4 baths	"	.70	22.857		7,450	1,075		8,525	9,850

13 28 Athletic and Recreational Special Construction

13 28 33 – Athletic and Recreational Court Walls

13 28 33.50 Sport Court		Crew	Daily Output	Labor-Hours	Unit	Material	2010 Bare Costs Labor	Equipment	Total	Total Incl O&P
0010	**SPORT COURT**									
0020	Floors, No. 2 & better maple, 25/32" thick				SF Flr.				6.05	6.65
0300	Squash, regulation court in existing building, minimum				Court	36,800			36,800	40,400
0400	Maximum				"	41,000			41,000	45,000
0450	Rule of thumb for components:									
0470	Walls	3 Carp	.15	160	Court	11,000	6,650		17,650	23,000
0500	Floor	"	.25	96		8,725	4,000		12,725	16,100
0550	Lighting	2 Elec	.60	26.667	↓	2,100	1,300		3,400	4,300

13 34 Fabricated Engineered Structures

13 34 13 – Glazed Structures

13 34 13.13 Greenhouses

		Crew	Daily Output	Labor-Hours	Unit	Material	Labor	Equipment	Total	Total Incl O&P
0010	**GREENHOUSES**, Shell only, stock units, not incl. 2' stub walls,									
0020	foundation, floors, heat or compartments									
0300	Residential type, free standing, 8'-6" long x 7'-6" wide	2 Carp	59	.271	SF Flr.	38	11.25		49.25	60
0400	10'-6" wide		85	.188		27	7.80		34.80	42.50
0600	13'-6" wide		108	.148		23	6.15		29.15	35.50
0700	17'-0" wide		160	.100		27	4.16		31.16	37
0900	Lean-to type, 3'-10" wide		34	.471		33.50	19.55		53.05	69
1000	6'-10" wide		58	.276		38	11.45		49.45	61
1050	8'-0" wide	↓	60	.267	↓	35.50	11.10		46.60	57.50
1060										
1100	Wall mounted to existing window, 3' x 3'	1 Carp	4	2	Ea.	1,500	83		1,583	1,775
1120	4' x 5'	"	3	2.667	"	1,900	111		2,011	2,275
3900	For cooling, add, minimum				SF Flr.	2.81			2.81	3.09
4000	Maximum					6.95			6.95	7.65
4200	For heaters, 13.6 MBH, add					5.35			5.35	5.85
4300	60 MBH, add				↓	2			2	2.20
4500	For benches, 2' x 3'-6", add				SF Hor.	24			24	26.50
4600	3' x 10', add				S.F.	12.95			12.95	14.25
4800	For controls, add, minimum				Total	2,400			2,400	2,650
4900	Maximum				"	14,300			14,300	15,700
5100	For humidification equipment, add				M.C.F.	6.15			6.15	6.75
5200	For vinyl shading, add				S.F.	1.29			1.29	1.42

13 34 13.19 Swimming Pool Enclosures

		Crew	Daily Output	Labor-Hours	Unit	Material	Labor	Equipment	Total	Total Incl O&P
0010	**SWIMMING POOL ENCLOSURES** Translucent, free standing									
0020	not including foundations, heat or light									
0200	Economy, minimum	2 Carp	200	.080	SF Hor.	14.80	3.32		18.12	21.50
0300	Maximum		100	.160		43	6.65		49.65	58.50
0400	Deluxe, minimum		100	.160		50.50	6.65		57.15	66.50
0600	Maximum	↓	70	.229	↓	227	9.50		236.50	266

13 34 23 – Fabricated Structures

13 34 23.16 Fabricated Control Booths

		Crew	Daily Output	Labor-Hours	Unit	Material	Labor	Equipment	Total	Total Incl O&P
0010	**FABRICATED CONTROL BOOTHS**									
0100	Guard House, prefab conc w/bullet resistant doors & windows, roof & wiring									
0110	8' x 8', Level III	L-10	1	24	Ea.	22,000	1,125	655	23,780	26,900
0120	8' x 8', Level IV	"	1	24	"	27,500	1,125	655	29,280	33,000

13 34 23.45 Kiosks

		Crew	Daily Output	Labor-Hours	Unit	Material	Labor	Equipment	Total	Total Incl O&P
0010	**KIOSKS**									
0020	Round, 5' diameter, 8' high, 1/4" fiberglass wall				Ea.	5,800			5,800	6,375

13 34 Fabricated Engineered Structures

13 34 23 – Fabricated Structures

13 34 23.45 Kiosks		Crew	Daily Output	Labor-Hours	Unit	Material	2010 Bare Costs Labor	Equipment	Total	Total Incl O&P
0100	1" insulated double wall, fiberglass				Ea.	6,600			6,600	7,250
0500	Rectangular, 5' x 9', 7'-6" high, 1/4" fiberglass wall					8,450			8,450	9,300
0600	1" insulated double wall, fiberglass				↓	10,000			10,000	11,000

13 34 63 – Natural Fiber Construction

13 34 63.50 Straw Bale Construction

0010	**STRAW BALE CONSTRUCTION**									
2000	Straw bales wall, incl labor & material complete, minimum	G			SF Flr.					130
2010	Maximum	G			"					160
2020	Straw bales in walls w/modified post and beam frame	G	2 Carp	320	.050	S.F.	4.11	2.08		6.19	7.95

13 42 Building Modules

13 42 63 – Detention Cell Modules

13 42 63.16 Steel Detention Cell Modules

0010	**STEEL DETENTION CELL MODULES**									
2000	Cells, prefab., 5' to 6' wide, 7' to 8' high, 7' to 8' deep,									
2010	bar front, cot, not incl. plumbing		E-4	1.50	21.333	Ea.	9,650	1,000	97.50	10,747.50	12,600

13 48 Sound, Vibration, and Seismic Control

13 48 13 – Manufactured Sound and Vibration Control Components

13 48 13.50 Audio Masking

		Crew	Daily Output	Labor-Hours	Unit	Material	Labor	Equipment	Total	Total Incl O&P
0010	**AUDIO MASKING**, acoustical enclosure, 4" thick wall and ceiling									
0020	8# per S.F., up to 12' span	3 Carp	72	.333	SF Surf	32	13.85		45.85	58.50
0300	Better quality panels, 10.5# per S.F.		64	.375		36.50	15.60		52.10	65.50
0400	Reverb-chamber, 4" thick, parallel walls	↓	60	.400	↓	45.50	16.60		62.10	77.50

13 49 Radiation Protection

13 49 13 – Lead Sheet

13 49 13.50 Lead Sheets

0010	**LEAD SHEETS**									
0300	Lead sheets, 1/16" thick	2 Lath	135	.119	S.F.	7.15	4.38		11.53	14.80
0400	1/8" thick		120	.133		14.45	4.93		19.38	23.50
0500	Lead shielding, 1/4" thick		135	.119		34	4.38		38.38	44.50
0550	1/2" thick	↓	120	.133	↓	59.50	4.93		64.43	73.50
1200	X-ray protection, average radiography or fluoroscopy									
1210	room, up to 300 S.F. floor, 1/16" lead, minimum	2 Lath	.25	64	Total	8,075	2,375		10,450	12,600
1500	Maximum, 7'-0" walls	"	.15	106	"	9,700	3,950		13,650	16,900
1600	Deep therapy X-ray room, 250 kV capacity,									
1800	up to 300 S.F. floor, 1/4" lead, minimum	2 Lath	.08	200	Total	22,600	7,400		30,000	36,400
1900	Maximum, 7'-0" walls		.06	266	"	27,800	9,850		37,650	46,100
1999	Minimum labor/equipment charge	↓	4.50	3.556	Job		131		131	206

13 49 21 – Lead Glazing

13 49 21.50 Lead Glazing

0010	**LEAD GLAZING**									
0600	Lead glass, 1/4" thick, 2.0 mm LE, 12" x 16"	2 Glaz	13	1.231	Ea.	265	49.50		314.50	370
0700	24" x 36"	"	8	2	"	980	80.50		1,060.50	1,200
2000	X-ray viewing panels, clear lead plastic									

13 49 Radiation Protection

13 49 21 - Lead Glazing

13 49 21.50 Lead Glazing		Crew	Daily Output	Labor-Hours	Unit	Material	2010 Bare Costs Labor	Equipment	Total	Total Incl O&P
2010	7 mm thick, 0.3 mm LE, 2.3 lbs/S.F.	H-3	139	.115	S.F.	143	4.13		147.13	165
2020	12 mm thick, 0.5 mm LE, 3.9 lbs/S.F.		82	.195		194	7		201	224
2030	18 mm thick, 0.8 mm LE, 5.9 lbs/S.F.		54	.296		210	10.65		220.65	248
2040	22 mm thick, 1.0 mm LE, 7.2 lbs/S.F.		44	.364		215	13.05		228.05	257
2050	35 mm thick, 1.5 mm LE, 11.5 lbs/S.F.		28	.571		241	20.50		261.50	298
2060	46 mm thick, 2.0 mm LE, 15.0 lbs/S.F.	▼	21	.762	▼	315	27.50		342.50	395
2090	For panels 12 S.F. to 48 S.F., add crating charge				Ea.					50

13 49 23 - Modular Shielding Partitions

13 49 23.50 Modular Shielding Partitions

		Crew	Daily Output	Labor-Hours	Unit	Material	Labor	Equipment	Total	Total Incl O&P
0010	**MODULAR SHIELDING PARTITIONS**									
4000	X-ray barriers, modular, panels mounted within framework for									
4002	attaching to floor, wall or ceiling, upper portion is clear lead									
4005	plastic window panels 48"H, lower portion is opaque leaded									
4008	steel panels 36"H, structural supports not incl.									
4010	1-section barrier, 36"W x 84"H overall									
4020	0.5 mm LE panels	H-3	6.40	2.500	Ea.	4,525	90		4,615	5,125
4030	0.8 mm LE panels		6.40	2.500		4,825	90		4,915	5,450
4040	1.0 mm LE panels		5.33	3.002		4,900	108		5,008	5,575
4050	1.5 mm LE panels	▼	5.33	3.002	▼	5,225	108		5,333	5,925
4060	2-section barrier, 72"W x 84"H overall									
4070	0.5 mm LE panels	H-3	4	4	Ea.	9,250	144		9,394	10,400
4080	0.8 mm LE panels		4	4		9,775	144		9,919	11,000
4090	1.0 mm LE panels		3.56	4.494		9,975	161		10,136	11,300
5000	1.5 mm LE panels	▼	3.20	5	▼	11,300	180		11,480	12,800
5010	3-section barrier, 108"W x 84"H overall									
5020	0.5 mm LE panels	H-3	3.20	5	Ea.	13,900	180		14,080	15,600
5030	0.8 mm LE panels		3.20	5		14,700	180		14,880	16,400
5040	1.0 mm LE panels		2.67	5.993		14,900	215		15,115	16,800
5050	1.5 mm LE panels	▼	2.46	6.504	▼	16,200	234		16,434	18,200
7000	X-ray barriers, mobile, mounted within framework w/casters on									
7005	bottom, clear lead plastic window panels on upper portion,									
7010	opaque on lower, 30"W x 75"H overall, incl. framework									
7020	24"H upper w/0.5 mm LE, 48"H lower w/0.8 mm LE	1 Carp	16	.500	Ea.	2,400	21		2,421	2,675
7030	48"W x 75"H overall, incl. framework									
7040	36"H upper w/0.5 mm LE, 36"H lower w/0.8 mm LE	1 Carp	16	.500	Ea.	4,425	21		4,446	4,875
7050	36"H upper w/1.0 mm LE, 36"H lower w/1.5 mm LE	"	16	.500	"	5,250	21		5,271	5,800
7060	72"W x 75"H overall, incl. framework									
7070	36"H upper w/0.5 mm LE, 36"H lower w/0.8 mm LE	1 Carp	16	.500	Ea.	5,225	21		5,246	5,775
7080	36"H upper w/1.0 mm LE, 36"H lower w/1.5 mm LE	"	16	.500	"	6,550	21		6,571	7,225

Division Notes

		CREW	DAILY OUTPUT	LABOR-HOURS	UNIT	2010 BARE COSTS				TOTAL INCL O&P
						MAT.	LABOR	EQUIP.	TOTAL	

Estimating Tips

General

- Many products in Division 14 will require some type of support or blocking for installation not included with the item itself. Examples are supports for conveyors or tube systems, attachment points for lifts, and footings for hoists or cranes. Add these supports in the appropriate division.

14 10 00 Dumbwaiters
14 20 00 Elevators

- Dumbwaiters and elevators are estimated and purchased in a method similar to buying a car. The manufacturer has a base unit with standard features. Added to this base unit price will be whatever options the owner or specifications require. Increased load capacity, additional vertical travel, additional stops, higher speed, and cab finish options are items to be considered. When developing an estimate for dumbwaiters and elevators, remember that some items needed by the installers may have to be included as part of the general contract. Examples are:
  - shaftway
  - rail support brackets
  - machine room
  - electrical supply
  - sill angles
  - electrical connections
  - pits
  - roof penthouses
  - pit ladders

  Check the job specifications and drawings before pricing.

- Installation of elevators and handicapped lifts in historic structures can require significant additional costs. The associated structural requirements may involve cutting into and repairing finishes, mouldings, flooring, etc. The estimator must account for these special conditions.

14 30 00 Escalators and Moving Walks

- Escalators and moving walks are specialty items installed by specialty contractors. There are numerous options associated with these items. For specific options, contact a manufacturer or contractor. In a method similar to estimating dumbwaiters and elevators, you should verify the extent of general contract work and add items as necessary.

14 40 00 Lifts
14 90 00 Other Conveying Equipment

- Products such as correspondence lifts, chutes, and pneumatic tube systems, as well as other items specified in this subdivision, may require trained installers. The general contractor might not have any choice as to who will perform the installation or when it will be performed. Long lead times are often required for these products, making early decisions in scheduling necessary.

Reference Numbers

Reference numbers are shown in shaded boxes at the beginning of some major classifications. These numbers refer to related items in the Reference Section. The reference information may be an estimating procedure, an alternate pricing method, or technical information.

Note: Not all subdivisions listed here necessarily appear in this publication.

14 11 Manual Dumbwaiters

14 11 10 – Hand Operated Dumbwaiters

14 11 10.20 Manual Dumbwaiters		Crew	Daily Output	Labor-Hours	Unit	Material	2010 Bare Costs Labor	Equipment	Total	Total Incl O&P
0010	**MANUAL DUMBWAITERS**									
0020	2 stop, minimum	2 Elev	.75	21.333	Ea.	2,925	1,325		4,250	5,250
0100	Maximum		.50	32	"	6,650	1,975		8,625	10,400
0300	For each additional stop, add		.75	21.333	Stop	1,050	1,325		2,375	3,200

14 12 Electric Dumbwaiters

14 12 10 – Dumbwaiters

14 12 10.10 Electric Dumbwaiters

		Crew	Daily Output	Labor-Hours	Unit	Material	Labor	Equipment	Total	Total Incl O&P
0010	**ELECTRIC DUMBWAITERS**									
0020	2 stop, minimum	2 Elev	.13	123	Ea.	7,375	7,600		14,975	19,800
0100	Maximum		.11	145	"	22,200	8,975		31,175	38,200
0600	For each additional stop, add		.54	29.630	Stop	3,250	1,825		5,075	6,375
0750	Correspondence lift, 1 floor, 2 stop, 45 lb capacity	2 Elec	.20	80	Ea.	8,350	3,925		12,275	15,200

14 21 Electric Traction Elevators

14 21 13 – Electric Traction Freight Elevators

14 21 13.10 Electric Traction Freight Elevators and Options

			Crew	Daily Output	Labor-Hours	Unit	Material	Labor	Equipment	Total	Total Incl O&P
0010	**ELECTRIC TRACTION FREIGHT ELEVATORS AND OPTIONS**	R142000-10									
0425	Electric freight, base unit, 4000 lb, 200 fpm, 4 stop, std. fin.		2 Elev	.05	320	Ea.	98,000	19,700		117,700	138,000
0450	For 5000 lb capacity, add	R142000-40					5,525			5,525	6,100
0500	For 6000 lb capacity, add						17,300			17,300	19,100
0525	For 7000 lb capacity, add						21,300			21,300	23,400
0550	For 8000 lb capacity, add						28,900			28,900	31,800
0575	For 10000 lb capacity, add						32,200			32,200	35,400
0600	For 12000 lb capacity, add						39,000			39,000	42,900
0625	For 16000 lb capacity, add						49,700			49,700	54,500
0650	For 20000 lb capacity, add						59,000			59,000	65,000
0675	For increased speed, 250 fpm, add						22,200			22,200	24,500
0700	300 fpm, geared electric, add						28,200			28,200	31,000
0725	350 fpm, geared electric, add						34,100			34,100	37,500
0750	400 fpm, geared electric, add						39,600			39,600	43,600
0775	500 fpm, gearless electric, add						47,200			47,200	52,000
0800	600 fpm, gearless electric, add						57,000			57,000	63,000
0825	700 fpm, gearless electric, add						68,500			68,500	75,000
0850	800 fpm, gearless electric, add						79,000			79,000	87,000
0875	For class "B" loading, add						6,500			6,500	7,150
0900	For class "C-1" loading, add						9,000			9,000	9,900
0925	For class "C-2" loading, add						10,200			10,200	11,300
0950	For class "C-3" loading, add						12,900			12,900	14,200
0975	For travel over 40 V.L.F., add		2 Elev	7.25	2.207	V.L.F.	540	136		676	805
1000	For number of stops over 4, add		"	.27	59.259	Stop	3,975	3,650		7,625	10,000

14 21 23 – Electric Traction Passenger Elevators

14 21 23.10 Electric Traction Passenger Elevators and Options

		Crew	Daily Output	Labor-Hours	Unit	Material	Labor	Equipment	Total	Total Incl O&P
0010	**ELECTRIC TRACTION PASSENGER ELEVATORS AND OPTIONS**									
1625	Electric pass., base unit, 2000 lb, 200 fpm, 4 stop, std. fin.	2 Elev	.05	320	Ea.	88,000	19,700		107,700	127,500
1650	For 2500 lb capacity, add					3,675			3,675	4,025
1675	For 3000 lb capacity, add					4,375			4,375	4,825
1700	For 3500 lb capacity, add					7,325			7,325	8,050
1725	For 4000 lb capacity, add					6,875			6,875	7,575

14 21 Electric Traction Elevators

14 21 23 – Electric Traction Passenger Elevators

14 21 23.10 Electric Traction Passenger Elevators and Options	Crew	Daily Output	Labor-Hours	Unit	Material	2010 Bare Costs Labor	Equipment	Total	Total Incl O&P
1750 For 4500 lb capacity, add				Ea.	8,700			8,700	9,575
1775 For 5000 lb capacity, add					10,500			10,500	11,500
1800 For increased speed, 250 fpm, geared electric, add					6,950			6,950	7,650
1825 300 fpm, geared electric, add					10,500			10,500	11,600
1850 350 fpm, geared electric, add					11,900			11,900	13,100
1875 400 fpm, geared electric, add					15,200			15,200	16,700
1900 500 fpm, gearless electric, add					31,200			31,200	34,400
1925 600 fpm, gearless electric, add					34,600			34,600	38,100
1950 700 fpm, gearless electric, add					39,700			39,700	43,700
1975 800 fpm, gearless electric, add					45,400			45,400	49,900
2000 For travel over 40 V.L.F., add	2 Elev	7.25	2.207	V.L.F.	650	136		786	925
2025 For number of stops over 4, add		.27	59.259	Stop	7,775	3,650		11,425	14,200
2400 Electric hospital, base unit, 4000 lb, 200 fpm, 4 stop, std fin.		.05	320	Ea.	76,000	19,700		95,700	114,500
2425 For 4500 lb capacity, add					5,450			5,450	5,975
2450 For 5000 lb capacity, add					6,525			6,525	7,175
2475 For increased speed, 250 fpm, geared electric, add					7,025			7,025	7,725
2500 300 fpm, geared electric, add					10,400			10,400	11,500
2525 350 fpm, geared electric, add					11,900			11,900	13,100
2550 400 fpm, geared electric, add					15,000			15,000	16,500
2575 500 fpm, gearless electric, add					30,900			30,900	33,900
2600 600 fpm, gearless electric, add					35,300			35,300	38,800
2625 700 fpm, gearless electric, add					40,000			40,000	44,000
2650 800 fpm, gearless electric, add					46,100			46,100	50,500
2675 For travel over 40 V.L.F., add	2 Elev	7.25	2.207	V.L.F.	665	136		801	945
2700 For number of stops over 4, add	"	.27	59.259	Stop	4,050	3,650		7,700	10,100

14 21 33 – Electric Traction Residential Elevators

14 21 33.20 Residential Elevators

		Crew	Daily Output	Labor-Hours	Unit	Material	Labor	Equipment	Total	Total Incl O&P
0010	**RESIDENTIAL ELEVATORS**									
7000	Residential, cab type, 1 floor, 2 stop, minimum	2 Elev	.20	80	Ea.	11,000	4,925		15,925	19,700
7100	Maximum		.10	160		18,600	9,875		28,475	35,700
7200	2 floor, 3 stop, minimum		.12	133		16,400	8,225		24,625	30,600
7300	Maximum		.06	266		26,700	16,500		43,200	54,500

14 24 Hydraulic Elevators

14 24 13 – Hydraulic Freight Elevators

14 24 13.10 Hydraulic Freight Elevators and Options

		Crew	Daily Output	Labor-Hours	Unit	Material	Labor	Equipment	Total	Total Incl O&P
0010	**HYDRAULIC FREIGHT ELEVATORS AND OPTIONS**									
1025	Hydraulic freight, base unit, 2000 lb, 50 fpm, 2 stop, std. fin.	2 Elev	.10	160	Ea.	78,500	9,875		88,375	101,000
1050	For 2500 lb capacity, add					4,925			4,925	5,425
1075	For 3000 lb capacity, add					8,725			8,725	9,600
1100	For 3500 lb capacity, add					15,800			15,800	17,400
1125	For 4000 lb capacity, add					16,900			16,900	18,600
1150	For 4500 lb capacity, add					19,800			19,800	21,800
1175	For 5000 lb capacity, add					24,900			24,900	27,400
1200	For 6000 lb capacity, add					31,100			31,100	34,200
1225	For 7000 lb capacity, add					37,400			37,400	41,100
1250	For 8000 lb capacity, add					41,100			41,100	45,200
1275	For 10000 lb capacity, add					52,000			52,000	57,000
1300	For 12000 lb capacity, add					68,000			68,000	75,000
1325	For 16000 lb capacity, add					93,000			93,000	102,500

14 24 Hydraulic Elevators

14 24 13 – Hydraulic Freight Elevators

14 24 13.10 Hydraulic Freight Elevators and Options		Crew	Daily Output	Labor-Hours	Unit	2010 Bare Costs			Total	Total Incl O&P
						Material	Labor	Equipment		
1350	For 20000 lb capacity, add				Ea.	113,500			113,500	125,000
1375	For increased speed, 100 fpm, add					2,150			2,150	2,350
1400	125 fpm, add					4,625			4,625	5,075
1425	150 fpm, add					5,500			5,500	6,050
1450	175 fpm, add					9,375			9,375	10,300
1475	For class "B" loading, add					6,475			6,475	7,125
1500	For class "C-1" loading, add					8,975			8,975	9,850
1525	For class "C-2" loading, add					10,200			10,200	11,200
1550	For class "C-3" loading, add					12,900			12,900	14,200
1575	For travel over 20 V.L.F., add	2 Elev	7.25	2.207	V.L.F.	1,675	136		1,811	2,050
1600	For number of stops over 2, add	"	.27	59.259	Stop	4,550	3,650		8,200	10,600

14 24 23 – Hydraulic Passenger Elevators

14 24 23.10 Hydraulic Passenger Elevators and Options

		Crew	Daily Output	Labor-Hours	Unit	Material	Labor	Equipment	Total	Total Incl O&P
0010	**HYDRAULIC PASSENGER ELEVATORS AND OPTIONS**									
2050	Hyd. pass., base unit, 1500 lb, 100 fpm, 2 stop, std. fin.	2 Elev	.10	160	Ea.	36,900	9,875		46,775	56,000
2075	For 2000 lb capacity, add					2,000			2,000	2,200
2100	For 2500 lb. capacity, add					3,125			3,125	3,425
2125	For 3000 lb capacity, add					4,100			4,100	4,525
2150	For 3500 lb capacity, add					5,950			5,950	6,550
2175	For 4000 lb capacity, add					7,300			7,300	8,050
2200	For 4500 lb capacity, add					8,600			8,600	9,475
2225	For 5000 lb capacity, add					11,300			11,300	12,500
2250	For increased speed, 125 fpm, add					2,950			2,950	3,250
2275	150 fpm, add					3,700			3,700	4,075
2300	175 fpm, add					5,525			5,525	6,075
2325	200 fpm, add					7,650			7,650	8,400
2350	For travel over 12 V.L.F., add	2 Elev	7.25	2.207	V.L.F.	890	136		1,026	1,200
2375	For number of stops over 2, add		.27	59.259	Stop	2,550	3,650		6,200	8,450
2725	Hydraulic hospital, base unit, 4000 lb, 100 fpm, 2 stop, std. fin.		.10	160	Ea.	59,500	9,875		69,375	80,500
2775	For 4500 lb capacity, add					6,775			6,775	7,450
2800	For 5000 lb capacity, add					8,925			8,925	9,825
2825	For increased speed, 125 fpm, add					3,425			3,425	3,750
2850	150 fpm, add					4,100			4,100	4,525
2875	175 fpm, add					5,975			5,975	6,550
2900	200 fpm, add					7,175			7,175	7,900
2925	For travel over 12 V.L.F., add	2 Elev	7.25	2.207	V.L.F.	745	136		881	1,025
2950	For number of stops over 2, add	"	.27	59.259	Stop	4,075	3,650		7,725	10,100

14 27 Custom Elevator Cabs

14 27 13 – Custom Elevator Cab Finishes

14 27 13.10 Cab Finishes

		Crew	Daily Output	Labor-Hours	Unit	Material	Labor	Equipment	Total	Total Incl O&P
0010	**CAB FINISHES**									
3325	Passenger elevator cab finishes (based on 3500 lb cab size)									
3350	Acrylic panel ceiling				Ea.	735			735	810
3375	Aluminum eggcrate ceiling					1,200			1,200	1,325
3400	Stainless steel doors					2,925			2,925	3,225
3425	Carpet flooring					605			605	665
3450	Epoxy flooring					460			460	505
3475	Quarry tile flooring					825			825	910
3500	Slate flooring					1,450			1,450	1,600

14 27 Custom Elevator Cabs

14 27 13 – Custom Elevator Cab Finishes

14 27 13.10 Cab Finishes	Crew	Daily Output	Labor-Hours	Unit	Material	2010 Bare Costs Labor	Equipment	Total	Total Incl O&P
3525 Textured rubber flooring				Ea.	200			200	220
3550 Stainless steel walls					5,125			5,125	5,650
3575 Stainless steel returns at door				▼	1,200			1,200	1,300
4450 Hospital elevator cab finishes (based on 3500 lb cab size)									
4475 Aluminum eggcrate ceiling				Ea.	1,200			1,200	1,300
4500 Stainless steel doors					2,675			2,675	2,950
4525 Epoxy flooring					370			370	405
4550 Quarry tile flooring					845			845	930
4575 Textured rubber flooring					660			660	730
4600 Stainless steel walls					5,125			5,125	5,650
4625 Stainless steel returns at door				▼	1,200			1,200	1,300

14 28 Elevator Equipment and Controls

14 28 10 – Elevator Equipment and Control Options

14 28 10.10 Elevator Controls and Doors	Crew	Daily Output	Labor-Hours	Unit	Material	2010 Bare Costs Labor	Equipment	Total	Total Incl O&P
0010 **ELEVATOR CONTROLS AND DOORS**									
2975 Passenger elevator options									
3000 2 car group automatic controls	2 Elev	.66	24.242	Ea.	6,600	1,500		8,100	9,575
3025 3 car group automatic controls		.44	36.364		12,800	2,250		15,050	17,600
3050 4 car group automatic controls		.33	48.485		26,100	3,000		29,100	33,400
3075 5 car group automatic controls		.26	61.538		47,900	3,800		51,700	58,500
3100 6 car group automatic controls		.22	72.727		98,500	4,475		102,975	115,000
3125 Intercom service		3	5.333		1,375	330		1,705	2,000
3150 Duplex car selective collective		.66	24.242		12,100	1,500		13,600	15,600
3175 Center opening 1 speed doors		2	8		1,800	495		2,295	2,725
3200 Center opening 2 speed doors		2	8		2,525	495		3,020	3,525
3225 Rear opening doors (opposite front)		2	8		3,000	495		3,495	4,075
3250 Side opening 2 speed doors		2	8		4,725	495		5,220	5,950
3275 Automatic emergency power switching		.66	24.242		1,150	1,500		2,650	3,575
3300 Manual emergency power switching	▼	8	2		480	123		603	720
3625 Hall finishes, stainless steel doors					2,000			2,000	2,200
3650 Stainless steel frames					1,450			1,450	1,600
3675 12 month maintenance contract								3,600	3,960
3700 Signal devices, hall lanterns	2 Elev	8	2		535	123		658	780
3725 Position indicators, up to 3		9.40	1.702		640	105		745	865
3750 Position indicators, per each over 3	▼	32	.500		620	31		651	730
3775 High speed heavy duty door opener					4,200			4,200	4,600
3800 Variable voltage, O.H. gearless machine, min.	2 Elev	.16	100		69,000	6,175		75,175	85,500
3815 Maximum		.07	228		102,000	14,100		116,100	133,500
3825 Basement installed geared machine	▼	.33	48.485	▼	71,500	3,000		74,500	83,000
3850 Freight elevator options									
3875 Doors, bi-parting	2 Elev	.66	24.242	Ea.	9,825	1,500		11,325	13,100
3900 Power operated door and gate	"	.66	24.242		22,800	1,500		24,300	27,300
3925 Finishes, steel plate floor					2,450			2,450	2,700
3950 14 ga. 1/4" x 4' steel plate walls					1,850			1,850	2,050
3975 12 month maintenance contract								3,600	3,960
4000 Signal devices, hall lanterns	2 Elev	8	2		535	123		658	780
4025 Position indicators, up to 3		9.40	1.702		645	105		750	870
4050 Position indicators, per each over 3		32	.500		620	31		651	730
4075 Variable voltage basement installed geared machine	▼	.66	24.242		18,800	1,500		20,300	23,000
4100 Hospital elevator options									

14 28 Elevator Equipment and Controls

14 28 10 – Elevator Equipment and Control Options

14 28 10.10 Elevator Controls and Doors		Crew	Daily Output	Labor-Hours	Unit	Material	2010 Bare Costs Labor	Equipment	Total	Total Incl O&P
4125	2 car group automatic controls	2 Elev	.66	24.242	Ea.	6,550	1,500		8,050	9,500
4150	3 car group automatic controls		.44	36.364		12,900	2,250		15,150	17,700
4175	4 car group automatic controls		.33	48.485		26,300	3,000		29,300	33,500
4200	5 car group automatic controls		.26	61.538		48,000	3,800		51,800	58,500
4225	6 car group automatic controls		.22	72.727		98,500	4,475		102,975	115,000
4250	Intercom service		3	5.333		1,375	330		1,705	2,000
4275	Duplex car selective collective		.66	24.242		12,100	1,500		13,600	15,600
4300	Center opening 1 speed doors		2	8		1,800	495		2,295	2,725
4325	Center opening 2 speed doors		2	8		2,450	495		2,945	3,450
4350	Rear opening doors (opposite front)		2	8		3,000	495		3,495	4,050
4375	Side opening 2 speed doors		2	8		3,775	495		4,270	4,900
4400	Automatic emergency power switching		.66	24.242		1,125	1,500		2,625	3,525
4425	Manual emergency power switching	▼	8	2		470	123		593	705
4675	Hall finishes, stainless steel doors					2,000			2,000	2,200
4700	Stainless steel frames					1,450			1,450	1,600
4725	12 month maintenance contract								3,600	3,960
4750	Signal devices, hall lanterns	2 Elev	8	2		535	123		658	780
4775	Position indicators, up to 3		9.40	1.702		640	105		745	865
4800	Position indicators, per each over 3	▼	32	.500		620	31		651	730
4825	High speed heavy duty door opener					4,200			4,200	4,600
4850	Variable voltage, O.H. gearless machine, min.	2 Elev	.16	100		69,000	6,175		75,175	85,500
4865	Maximum		.07	228		102,000	14,100		116,100	133,500
4875	Basement installed geared machine	▼	.33	48.485	▼	26,100	3,000		29,100	33,300
5000	Drilling for piston, casing included, 18" diameter	B-48	80	.700	V.L.F.	46	26	37	109	133

14 42 Wheelchair Lifts

14 42 13 – Inclined Wheelchair Lifts

14 42 13.10 Inclined Wheelchair Lifts and Stairclimbers

		Crew	Daily Output	Labor-Hours	Unit	Material	Labor	Equipment	Total	Total Incl O&P
0010	**INCLINED WHEELCHAIR LIFTS AND STAIRCLIMBERS**									
7700	Stair climber (chair lift), single seat, minimum	2 Elev	1	16	Ea.	5,325	985		6,310	7,400
7800	Maximum		.20	80		7,350	4,925		12,275	15,700
8700	Stair lift, minimum		1	16		14,500	985		15,485	17,400
8900	Maximum	▼	.20	80	▼	22,900	4,925		27,825	32,800

14 42 16 – Vertical Wheelchair Lifts

14 42 16.10 Wheelchair Lifts

		Crew	Daily Output	Labor-Hours	Unit	Material	Labor	Equipment	Total	Total Incl O&P
0010	**WHEELCHAIR LIFTS**									
8000	Wheelchair lift, minimum	2 Elev	1	16	Ea.	7,325	985		8,310	9,575
8500	Maximum	"	.50	32	"	17,300	1,975		19,275	22,000

14 45 Vehicle Lifts

14 45 10 – Hydraulic Vehicle Lifts

14 45 10.10 Hydraulic Lifts

		Crew	Daily Output	Labor-Hours	Unit	Material	2010 Bare Costs Labor	Equipment	Total	Total Incl O&P
0010	**HYDRAULIC LIFTS**									
0200	Double post, 9000 lb frame	L-4	1.15	20.870	Ea.	5,425	810		6,235	7,275
0500	Four post, 50,000 lb frame		.18	133		57,500	5,200		62,700	71,500
2200	Hoists, single post, 8000 lb capacity, swivel arms		.40	60		4,975	2,325		7,300	9,275
2400	Two posts, adjustable frames, 11,000 lb capacity		.25	96		6,400	3,725		10,125	13,100
2500	24,000 lb capacity		.15	160		8,550	6,225		14,775	19,500
2700	7500 lb capacity, frame supports		.50	48		7,125	1,875		9,000	10,900
2800	Four post, roll on ramp		.50	48		6,400	1,875		8,275	10,100
2810	Hydraulic lifts, above ground, 2 post, clear floor, 6000 lb cap		2.67	8.989		7,025	350		7,375	8,300
2815	9000 lb capacity		2.29	10.480		16,700	410		17,110	19,000
2820	15,000 lb capacity		2	12		18,800	465		19,265	21,500
2825	30,000 lb capacity		1.60	15		41,600	585		42,185	46,700
2830	4 post, ramp style, 25,000 lb capacity		2	12		16,300	465		16,765	18,700
2835	35,000 lb capacity		1	24		76,000	935		76,935	85,000
2840	50,000 lb capacity		1	24		85,000	935		85,935	95,000
2845	75,000 lb capacity		1	24		98,500	935		99,435	110,000
2850	For drive thru tracks, add, minimum					1,050			1,050	1,150
2855	Maximum					1,800			1,800	1,975
2860	Ramp extensions, 3' (set of 2)					860			860	945
2865	Rolling jack platform					3,000			3,000	3,300
2870	Elec/hyd jacking beam					8,000			8,000	8,800
2880	Scissor lift, portable, 6000 lb capacity					7,850			7,850	8,625

14 51 Correspondence and Parcel Lifts

14 51 10 – Electric Correspondence and Parcel Lifts

14 51 10.10 Correspondence Lifts

		Crew	Daily Output	Labor-Hours	Unit	Material	Labor	Equipment	Total	Total Incl O&P
0010	**CORRESPONDENCE LIFTS**									
0020	1 floor, 2 stop, 25 lb capacity, electric	2 Elev	.20	80	Ea.	5,975	4,925		10,900	14,200
0100	Hand, 5 lb capacity	"	.20	80	"	2,250	4,925		7,175	10,100

14 91 Facility Chutes

14 91 33 – Laundry and Linen Chutes

14 91 33.10 Chutes

		Crew	Daily Output	Labor-Hours	Unit	Material	Labor	Equipment	Total	Total Incl O&P
0011	**CHUTES**, linen, trash or refuse									
0050	Aluminized steel, 16 ga., 18" diameter	2 Shee	3.50	4.571	Floor	1,275	224		1,499	1,775
0100	24" diameter		3.20	5		1,350	246		1,596	1,875
0400	Galvanized steel, 16 ga., 18" diameter		3.50	4.571		1,000	224		1,224	1,450
0500	24" diameter		3.20	5		1,125	246		1,371	1,625
0800	Stainless steel, 18" diameter		3.50	4.571		2,650	224		2,874	3,250
0900	24" diameter		3.20	5		2,875	246		3,121	3,525

14 91 82 – Trash Chutes

14 91 82.10 Trash Chutes and Accessories

		Crew	Daily Output	Labor-Hours	Unit	Material	Labor	Equipment	Total	Total Incl O&P
0010	**TRASH CHUTES AND ACCESSORIES**									
9000	Minimum labor/equipment charge	1 Shee	1	8	Job		395		395	615

14 92 Pneumatic Tube Systems

14 92 10 – Conventional, Automatic and Computer Controlled Pneumatic Tube Systems

14 92 10.10 Pneumatic Tube Systems	Crew	Daily Output	Labor-Hours	Unit	Material	2010 Bare Costs Labor	Equipment	Total	Total Incl O&P
0010 **PNEUMATIC TUBE SYSTEMS**									
0020 100' long, single tube, 2 stations, stock									
0100 3" diameter	2 Stpi	.12	133	Total	6,600	6,925		13,525	18,000
0300 4" diameter	"	.09	177	"	7,450	9,225		16,675	22,500
0400 Twin tube, two stations or more, conventional system									
0700 3" round	2 Stpi	46	.348	L.F.	15.25	18.05		33.30	45
1050 Add for blower		2	8	System	5,200	415		5,615	6,375
1110 Plus for each round station, add		7.50	2.133	Ea.	585	111		696	815
1200 Alternate pricing method: base cost, minimum		.75	21.333	Total	6,375	1,100		7,475	8,750
1300 Maximum		.25	64	"	12,700	3,325		16,025	19,100
1500 Plus total system length, add, minimum		93.40	.171	L.F.	8.20	8.90		17.10	23
1600 Maximum		37.60	.426	"	24.50	22		46.50	60.50

Division 21 - Fire Suppression

Estimating Tips

Pipe for fire protection and all uses is located in Subdivision 22 11 13.

The labor adjustment factors listed in Subdivision 22 01 02.20 also apply to Division 21.

Many, but not all, areas in the U.S. require backflow protection in the fire system. It is advisable to check local building codes for specific requirements.

For your reference, the following is a list of the most applicable Fire Codes and Standards which may be purchased from the NFPA, 1 Batterymarch Park, Quincy, MA 02169-7471.

NFPA 1: Uniform Fire Code
NFPA 10: Portable Fire Extinguishers
NFPA 11: Low-, Medium-, and High-Expansion Foam
NFPA 12: Carbon Dioxide Extinguishing Systems (Also companion 12A)
NFPA 13: Installation of Sprinkler Systems (Also companion 13D, 13E, and 13R)
NFPA 14: Installation of Standpipe and Hose Systems
NFPA 15: Water Spray Fixed Systems for Fire Protection
NFPA 16: Installation of Foam-Water Sprinkler and Foam-Water Spray Systems
NFPA 17: Dry Chemical Extinguishing Systems (Also companion 17A)
NFPA 18: Wetting Agents
NFPA 20: Installation of Stationary Pumps for Fire Protection
NFPA 22: Water Tanks for Private Fire Protection
NFPA 24: Installation of Private Fire Service Mains and their Appurtenances
NFPA 25: Inspection, Testing and Maintenance of Water-Based Fire Protection

Reference Numbers

Reference numbers are shown in shaded boxes at the beginning of some major classifications. These numbers refer to related items in the Reference Section. The reference information may be an estimating procedure, an alternate pricing method, or technical information.

Note: Not all subdivisions listed here necessarily appear in this publication.

Note: **Trade Service,** *in part, has been used as a reference source for some of the material prices used in Division 21.*

21 05 Common Work Results for Fire Suppression

21 05 23 – General-Duty Valves for Water-Based Fire-Suppression Piping

21 05 23.50 General-Duty Valves for Water-Based Fire Supp.	Crew	Daily Output	Labor-Hours	Unit	Material	2010 Bare Costs Labor	Equipment	Total	Total Incl O&P
0010 **GENERAL-DUTY VALVES FOR WATER-BASED FIRE SUPPRESSION PIPING**									
6500 Check, swing, C.I. body, brass fittings, auto. ball drip									
6520 4" size	Q-12	3	5.333	Ea.	255	242		497	655
9990 Minimum labor/equipment charge	1 Spri	3	2.667	Job		134		134	208

21 11 Facility Fire-Suppression Water-Service Piping

21 11 16 – Facility Fire Hydrants

21 11 16.50 Fire Hydrants for Buildings

	Crew	Daily Output	Labor-Hours	Unit	Material	Labor	Equipment	Total	Total Incl O&P
0010 **FIRE HYDRANTS FOR BUILDINGS**									
3750 Hydrants, wall, w/caps, single, flush, polished brass									
3800 2-1/2" x 2-1/2"	Q-12	5	3.200	Ea.	175	145		320	415
4350 Double, projecting, polished brass									
4400 2-1/2" x 2-1/2" x 4"	Q-12	5	3.200	Ea.	209	145		354	455

21 11 19 – Fire-Department Connections

21 11 19.50 Connections for the Fire-Department

	Crew	Daily Output	Labor-Hours	Unit	Material	Labor	Equipment	Total	Total Incl O&P
0010 **CONNECTIONS FOR THE FIRE-DEPARTMENT**									
7140 Standpipe connections, wall, w/plugs & chains									
7280 Double, flush, polished brass									
7300 2-1/2" x 2-1/2" x 4"	Q-12	5	3.200	Ea.	425	145		570	695
9990 Minimum labor/equipment charge	1 Spri	4	2	Job		101		101	156

21 12 Fire-Suppression Standpipes

21 12 13 – Fire-Suppression Hoses and Nozzles

21 12 13.50 Fire Hoses and Nozzles

	Crew	Daily Output	Labor-Hours	Unit	Material	Labor	Equipment	Total	Total Incl O&P
0010 **FIRE HOSES AND NOZZLES**									
2200 Hose, less couplings									
2260 Synthetic jacket, lined, 300 lb. test, 1-1/2" diameter	Q-12	2600	.006	L.F.	2.39	.28		2.67	3.06
2280 2-1/2" diameter	"	2200	.007	"	3.98	.33		4.31	4.89
5600 Nozzles, brass									
5620 Adjustable fog, 3/4" booster line				Ea.	91			91	100
5640 1-1/2" leader line					66			66	72.50
5660 2-1/2" direct connection				↓	172			172	189
9900 Minimum labor/equipment charge	1 Plum	2	4	Job		208		208	325

21 12 19 – Fire-Suppression Hose Racks

21 12 19.50 Fire Hose Racks

	Crew	Daily Output	Labor-Hours	Unit	Material	Labor	Equipment	Total	Total Incl O&P
0010 **FIRE HOSE RACKS**									
2600 Hose rack, swinging, for 1-1/2" diameter hose,									
2620 Enameled steel, 50' & 75' lengths of hose	Q-12	20	.800	Ea.	48.50	36.50		85	109

21 12 23 – Fire-Suppression Hose Valves

21 12 23.70 Fire Hose Valves

	Crew	Daily Output	Labor-Hours	Unit	Material	Labor	Equipment	Total	Total Incl O&P
0010 **FIRE HOSE VALVES**									
3000 Gate, hose, wheel handle, N.R.S., rough brass, 1-1/2"	1 Spri	12	.667	Ea.	111	33.50		144.50	174
3040 2-1/2", 300 lb.	"	7	1.143	"	156	57.50		213.50	260

21 13 Fire-Suppression Sprinkler Systems

21 13 13 – Wet-Pipe Sprinkler Systems

21 13 13.50 Wet-Pipe Sprinkler System Components	Crew	Daily Output	Labor-Hours	Unit	Material	2010 Bare Costs Labor	2010 Bare Costs Equipment	Total	Total Incl O&P
0010 **WET-PIPE SPRINKLER SYSTEM COMPONENTS**									
1220 Water motor, complete with gong	1 Spri	4	2	Ea.	305	101		406	490
1800 Firecycle system, controls, includes panel,									
1820 batteries, solenoid valves and pressure switches	Q-13	1	32	Ea.	10,800	1,525		12,325	14,300
1860 Detector	1 Spri	16	.500	"	525	25		550	615
1900 Flexible sprinkler head connectors									
1910 Braided stainless steel hose with mounting bracket									
1920 1/2" and 1" outlet size									
1930 24" length	1 Spri	32	.250	Ea.	24	12.60		36.60	46
1940 36" length		30	.267		28	13.45		41.45	51.50
1950 48" length		26	.308		31.50	15.50		47	58.50
1960 60" length		22	.364		35.50	18.35		53.85	67.50
1970 72" length	↓	18	.444	↓	39	22.50		61.50	77.50
1982 May replace hard-pipe armovers									
1984 For wet, pre-action, deluge or dry pipe systems									
2600 Sprinkler heads, not including supply piping									
3700 Standard spray, pendent or upright, brass, 135°F to 286°F									
3740 1/2" NPT, 1/2" orifice	1 Spri	16	.500	Ea.	7.35	25		32.35	47
3790 For riser & feeder piping, see Div. 22 11 13.00									

21 13 16 – Dry-Pipe Sprinkler Systems

21 13 16.50 Dry-Pipe Sprinkler System Components

	Crew	Daily Output	Labor-Hours	Unit	Material	Labor	Equipment	Total	Total Incl O&P
0010 **DRY-PIPE SPRINKLER SYSTEM COMPONENTS**									
0800 Air compressor for dry pipe system, automatic, complete									
0820 280 gal. system capacity, 3/4 HP	1 Spri	1.30	6.154	Ea.	770	310		1,080	1,325
2600 Sprinkler heads, not including supply piping									
2640 Dry, pendent, 1/2" orifice, 3/4" or 1" NPT									
2660 1/2" to 6" length	1 Spri	14	.571	Ea.	90.50	29		119.50	144
2700 15-1/4" to 18" length	"	14	.571		106	29		135	162
8200 Dry pipe valve, incl. trim and gauges, 3" size	Q-12	2	8		1,950	365		2,315	2,700
8220 4" size	"	1	16	↓	2,100	725		2,825	3,425

21 13 39 – Foam-Water Systems

21 13 39.50 Foam-Water System Components

	Crew	Daily Output	Labor-Hours	Unit	Material	Labor	Equipment	Total	Total Incl O&P
0010 **FOAM-WATER SYSTEM COMPONENTS**									
2600 Sprinkler heads, not including supply piping									
3600 Foam-water, pendent or upright, 1/2" NPT	1 Spri	12	.667	Ea.	130	33.50		163.50	194

Division Notes

	CREW	DAILY OUTPUT	LABOR-HOURS	UNIT	2010 BARE COSTS				TOTAL INCL O&P
					MAT.	LABOR	EQUIP.	TOTAL	

Estimating Tips

22 10 00 Plumbing Piping and Pumps

This subdivision is primarily basic pipe and related materials. The pipe may be used by any of the mechanical disciplines, i.e., plumbing, fire protection, heating, and air conditioning.

- The labor adjustment factors listed in Subdivision 22 01 02.20 apply throughout Divisions 21, 22, and 23. CAUTION: the correct percentage may vary for the same items. For example, the percentage add for the basic pipe installation should be based on the maximum height that the craftsman must install for that particular section. If the pipe is to be located 14' above the floor but it is suspended on threaded rod from beams, the bottom flange of which is 18' high (4' rods), then the height is actually 18' and the add is 20%. The pipe coverer, however, does not have to go above the 14', and so his or her add should be 10%.
- Most pipe is priced first as straight pipe with a joint (coupling, weld, etc.) every 10' and a hanger usually every 10'. There are exceptions with hanger spacing such as for cast iron pipe (5') and plastic pipe (3 per 10'). Following each type of pipe there are several lines listing sizes and the amount to be subtracted to delete couplings and hangers. This is for pipe that is to be buried or supported together on trapeze hangers. The reason that the couplings are deleted is that these runs are usually long, and frequently longer lengths of pipe are used. By deleting the couplings, the estimator is expected to look up and add back the correct reduced number of couplings.
- When preparing an estimate, it may be necessary to approximate the fittings. Fittings usually run between 25% and 50% of the cost of the pipe. The lower percentage is for simpler runs, and the higher number is for complex areas, such as mechanical rooms.
- For historic restoration projects, the systems must be as invisible as possible, and pathways must be sought for pipes, conduit, and ductwork. While installations in accessible spaces (such as basements and attics) are relatively straightforward to estimate, labor costs may be more difficult to determine when delivery systems must be concealed.

22 40 00 Plumbing Fixtures

- Plumbing fixture costs usually require two lines: the fixture itself and its "rough-in, supply, and waste."
- In the Assemblies Section (Plumbing D2010) for the desired fixture, the System Components Group at the center of the page shows the fixture on the first line. The rest of the list (fittings, pipe, tubing, etc.) will total up to what we refer to in the Unit Price section as "Rough-in, supply, waste, and vent." Note that for most fixtures we allow a nominal 5' of tubing to reach from the fixture to a main or riser.
- Remember that gas- and oil-fired units need venting.

Reference Numbers

Reference numbers are shown in shaded boxes at the beginning of some major classifications. These numbers refer to related items in the Reference Section. The reference information may be an estimating procedure, an alternate pricing method, or technical information.

Note: Not all subdivisions listed here necessarily appear in this publication.

Note: **Trade Service,** *in part, has been used as a reference source for some of the material prices used in Division 22.*

22 01 Operation and Maintenance of Plumbing

22 01 02 – Labor Adjustments

22 01 02.20 Labor Adjustment Factors

		Crew	Daily Output	Labor-Hours	Unit	Material	2010 Bare Costs Labor	Equipment	Total	Total Incl O&P
0010	**LABOR ADJUSTMENT FACTORS**, (For Div. 21,22 and 23)									
1000	Add to labor for elevated installation (Above floor level)									
1080	10' to 14.5' high						10%			
1100	15' to 19.5' high						20%			
1120	20' to 24.5' high						25%			
1140	25' to 29.5' high						35%			
1160	30' to 34.5' high						40%			
1180	35' to 39.5' high						50%			
1200	40' and higher						55%			
2000	Add to labor for crawl space									
2100	3' high						40%			
2140	4' high						30%			
3000	Add to labor for multi-story building									
3100	Add per floor for floors 3 thru 19						2%			
3140	Add per floor for floors 20 and up						4%			
4000	Add to labor for working in existing occupied buildings									
4100	Hospital						35%			
4140	Office building						25%			
4180	School						20%			
4220	Factory or warehouse						15%			
4260	Multi dwelling						15%			
5000	Add to labor, miscellaneous									
5100	Cramped shaft						35%			
5140	Congested area						15%			
5180	Excessive heat or cold						30%			
9000	Labor factors, The above are reasonable suggestions, however									
9010	each project should be evaluated for its own peculiarities.									
9100	Other factors to be considered are:									
9140	Movement of material and equipment through finished areas									
9180	Equipment room									
9220	Attic space									
9260	No service road									
9300	Poor unloading/storage area									
9340	Congested site area/heavy traffic									

22 05 Common Work Results for Plumbing

22 05 05 – Selective Plumbing Demolition

22 05 05.10 Plumbing Demolition

			Crew	Daily Output	Labor-Hours	Unit	Material	Labor	Equipment	Total	Total Incl O&P
0010	**PLUMBING DEMOLITION**	R220105-10									
1020	Fixtures, including 10' piping										
1100	Bathtubs, cast iron	R024119-10	1 Plum	4	2	Ea.		104		104	161
1120	Fiberglass			6	1.333			69.50		69.50	108
1140	Steel			5	1.600			83.50		83.50	129
1200	Lavatory, wall hung			10	.800			41.50		41.50	64.50
1220	Counter top			8	1			52		52	80.50
1300	Sink, single compartment			8	1			52		52	80.50
1320	Double compartment			7	1.143			59.50		59.50	92
1400	Water closet, floor mounted			8	1			52		52	80.50
1420	Wall mounted			7	1.143			59.50		59.50	92
1500	Urinal, floor mounted			4	2			104		104	161
1520	Wall mounted			7	1.143			59.50		59.50	92

22 05 Common Work Results for Plumbing

22 05 05 – Selective Plumbing Demolition

22 05 05.10 Plumbing Demolition		Crew	Daily Output	Labor-Hours	Unit	Material	2010 Bare Costs Labor	Equipment	Total	Total Incl O&P
1600	Water fountains, free standing	1 Plum	8	1	Ea.		52		52	80.50
1620	Wall or deck mounted		6	1.333	↓		69.50		69.50	108
2000	Piping, metal, up thru 1-1/2" diameter		200	.040	L.F.		2.08		2.08	3.23
2050	2" thru 3-1/2" diameter	↓	150	.053			2.78		2.78	4.30
2100	4" thru 6" diameter	2 Plum	100	.160			8.35		8.35	12.90
2150	8" thru 14" diameter	"	60	.267			13.90		13.90	21.50
2153	16" thru 20" diameter	Q-18	70	.343			16.60	.84	17.44	26.50
2155	24" thru 26" diameter		55	.436			21	1.06	22.06	33.50
2156	30" thru 36" diameter	↓	40	.600	↓		29	1.46	30.46	46.50
2250	Water heater, 40 gal.	1 Plum	6	1.333	Ea.		69.50		69.50	108
6000	Remove and reset fixtures, minimum		6	1.333			69.50		69.50	108
6100	Maximum		4	2	↓		104		104	161
9000	Minimum labor/equipment charge	↓	2	4	Job		208		208	325

22 05 23 – General-Duty Valves for Plumbing Piping

22 05 23.20 Valves, Bronze

		Crew	Daily Output	Labor-Hours	Unit	Material	Labor	Equipment	Total	Total Incl O&P
0010	**VALVES, BRONZE**									
1020	Angle, 150 lb., rising stem, threaded									
1070	3/4"	1 Plum	20	.400	Ea.	127	21		148	172
1080	1"		19	.421		182	22		204	235
1100	1-1/2"		13	.615		305	32		337	385
1110	2"	↓	11	.727	↓	490	38		528	600
1380	Ball, 150 psi, threaded									
1460	3/4"	1 Plum	20	.400	Ea.	17.80	21		38.80	52
1750	Check, swing, class 150, regrinding disc, threaded									
1850	1/2"	1 Plum	24	.333	Ea.	45.50	17.35		62.85	77
1860	3/4"		20	.400		65.50	21		86.50	105
1870	1"		19	.421		94.50	22		116.50	138
1890	1-1/2"		13	.615		158	32		190	224
1900	2"	↓	11	.727	↓	233	38		271	315
2850	Gate, N.R.S., soldered, 125 psi									
2920	1/2"	1 Plum	24	.333	Ea.	32	17.35		49.35	62
2940	3/4"		20	.400		36.50	21		57.50	72.50
2950	1"		19	.421		52	22		74	91
2970	1-1/2"		13	.615		88.50	32		120.50	147
2980	2"	↓	11	.727	↓	125	38		163	197
4850	Globe, class 150, rising stem, threaded									
4950	1/2"	1 Plum	24	.333	Ea.	69	17.35		86.35	103
4960	3/4"	"	20	.400	"	92.50	21		113.50	135
5600	Relief, pressure & temperature, self-closing, ASME, threaded									
5640	3/4"	1 Plum	28	.286	Ea.	124	14.85		138.85	160
5650	1"		24	.333		190	17.35		207.35	235
5660	1-1/4"		20	.400		360	21		381	430
5670	1-1/2"		18	.444		735	23		758	845
5680	2"	↓	16	.500	↓	755	26		781	870
6400	Pressure, water, ASME, threaded									
6440	3/4"	1 Plum	28	.286	Ea.	70	14.85		84.85	100
6450	1"		24	.333		149	17.35		166.35	191
6470	1-1/2"	↓	18	.444	↓	325	23		348	395
6900	Reducing, water pressure									
6940	1/2"	1 Plum	24	.333	Ea.	225	17.35		242.35	274
6960	1"	"	19	.421	"	350	22		372	420
8350	Tempering, water, sweat connections									

22 05 Common Work Results for Plumbing

22 05 23 – General-Duty Valves for Plumbing Piping

22 05 23.20 Valves, Bronze		Crew	Daily Output	Labor-Hours	Unit	Material	2010 Bare Costs Labor	Equipment	Total	Total Incl O&P
8400	1/2"	1 Plum	24	.333	Ea.	74.50	17.35		91.85	109
8440	3/4"	"	20	.400	"	91	21		112	133
8650	Threaded connections									
8700	1/2"	1 Plum	24	.333	Ea.	112	17.35		129.35	150
8740	3/4"		20	.400	"	390	21		411	465
9000	Minimum labor/equipment charge		4	2	Job		104		104	161

22 05 23.60 Valves, Plastic

		Crew	Daily Output	Labor-Hours	Unit	Material	Labor	Equipment	Total	Total Incl O&P
0010	**VALVES, PLASTIC**									
0020										
1150	Ball, PVC, socket or threaded, single union									
1280	2"	1 Plum	17	.471	Ea.	70.50	24.50		95	116
1650	CPVC, socket or threaded, single union									
1770	2"	1 Plum	17	.471	Ea.	132	24.50		156.50	183
3150	Ball check, PVC, socket or threaded									
3290	2"	1 Plum	17	.471	Ea.	106	24.50		130.50	155
9000	Minimum labor/equipment charge	"	3.50	2.286	Job		119		119	184

22 05 76 – Facility Drainage Piping Cleanouts

22 05 76.10 Cleanouts

		Crew	Daily Output	Labor-Hours	Unit	Material	Labor	Equipment	Total	Total Incl O&P
0010	**CLEANOUTS**									
0060	Floor type									
0080	Round or square, scoriated nickel bronze top									
0100	2" pipe size	1 Plum	10	.800	Ea.	155	41.50		196.50	235
0140	4" pipe size	"	6	1.333	"	232	69.50		301.50	365
0980	Round top, recessed for terrazzo									
1000	2" pipe size	1 Plum	9	.889	Ea.	155	46.50		201.50	242
1100	4" pipe size	"	4	2		232	104		336	415
1120	5" pipe size	Q-1	6	2.667		295	125		420	520
9000	Minimum labor/equipment charge	1 Plum	3	2.667	Job		139		139	215

22 05 76.20 Cleanout Tees

		Crew	Daily Output	Labor-Hours	Unit	Material	Labor	Equipment	Total	Total Incl O&P
0010	**CLEANOUT TEES**									
0100	Cast iron, B&S, with countersunk plug									
0220	3" pipe size	1 Plum	3.60	2.222	Ea.	230	116		346	430
0240	4" pipe size	"	3.30	2.424	"	286	126		412	510
0500	For round smooth access cover, same price									
4000	Plastic, tees and adapters. Add plugs									
4010	ABS, DWV									
4020	Cleanout tee, 1-1/2" pipe size	1 Plum	15	.533	Ea.	4.62	28		32.62	48
9000	Minimum labor/equipment charge	"	2.75	2.909	Job		151		151	235

22 07 Plumbing Insulation

22 07 16 – Plumbing Equipment Insulation

22 07 16.10 Insulation for Plumbing Equipment

			Crew	Daily Output	Labor-Hours	Unit	Material	Labor	Equipment	Total	Total Incl O&P
0010	**INSULATION FOR PLUMBING EQUIPMENT**										
2900	Domestic water heater wrap kit										
2920	1-1/2" with vinyl jacket, 20-60 gal.	G	1 Plum	8	1	Ea.	16.05	52		68.05	98
2925	50 to 80 gallons	G	"	8	1	"	19.15	52		71.15	102

22 07 19 – Plumbing Piping Insulation

22 07 19.10 Piping Insulation

0010	**PIPING INSULATION**

22 07 Plumbing Insulation

22 07 19 – Plumbing Piping Insulation

22 07 19.10 Piping Insulation

		Crew	Daily Output	Labor-Hours	Unit	Material	2010 Bare Costs Labor	2010 Bare Costs Equipment	Total	Total Incl O&P
0100	Rule of thumb, as a percentage of total mechanical costs				Job				10%	
0110	Insulation req'd is based on the surface size/area to be covered									
2930	Insulated protectors, (ADA)									
2935	For exposed piping under sinks or lavatories									
2940	Vinyl coated foam, velcro tabs									
2945	P Trap, 1-1/4" or 1-1/2"	1 Plum	32	.250	Ea.	18.50	13		31.50	40.50
2960	Valve and supply cover									
2965	1/2", 3/8", and 7/16" pipe size	1 Plum	32	.250	Ea.	17.60	13		30.60	39.50
2970	Extension drain cover									
2975	1-1/4", or 1-1/2" pipe size	1 Plum	32	.250	Ea.	19.30	13		32.30	41.50
2980	Tailpiece offset (wheelchair)									
2985	1-1/4" pipe size	1 Plum	32	.250	Ea.	21.50	13		34.50	43.50
4000	Pipe covering (price copper tube one size less than IPS)									
6120	2" wall, 1/2" iron pipe size	Q-14	145	.110	L.F.	4.28	4.52		8.80	11.95
6210	4" iron pipe size	"	125	.128	"	7.80	5.25		13.05	16.95
6600	Fiberglass, with all service jacket									
6840	1" wall, 1/2" iron pipe size	Q-14	240	.067	L.F.	.95	2.73		3.68	5.45
6860	3/4" iron pipe size		230	.070		1.04	2.85		3.89	5.70
6870	1" iron pipe size		220	.073		1.12	2.98		4.10	6
6880	1-1/4" iron pipe size		210	.076		1.21	3.12		4.33	6.35
6890	1-1/2" iron pipe size		210	.076		1.31	3.12		4.43	6.45
6900	2" iron pipe size		200	.080		1.41	3.28		4.69	6.80
7080	1-1/2" wall, 1/2" iron pipe size		230	.070		1.81	2.85		4.66	6.55
7140	2" iron pipe size		190	.084		2.44	3.45		5.89	8.25
7160	3" iron pipe size		170	.094		2.74	3.86		6.60	9.20
7879	Rubber tubing, flexible closed cell foam									
8100	1/2" wall, 1/4" iron pipe size	1 Asbe	90	.089	L.F.	.50	4.05		4.55	7.05
8120	3/8" iron pipe size		90	.089		.55	4.05		4.60	7.10
8130	1/2" iron pipe size		89	.090		.61	4.09		4.70	7.20
8140	3/4" iron pipe size		89	.090		.68	4.09		4.77	7.30
8150	1" iron pipe size		88	.091		.75	4.14		4.89	7.50
8170	1-1/2" iron pipe size		87	.092		1.06	4.19		5.25	7.85
8180	2" iron pipe size		86	.093		1.38	4.24		5.62	8.30
8200	3" iron pipe size		85	.094		1.95	4.29		6.24	9.05
8220	4" iron pipe size		80	.100		2.89	4.56		7.45	10.50
8300	3/4" wall, 1/4" iron pipe size		90	.089		.77	4.05		4.82	7.35
8330	1/2" iron pipe size		89	.090		1.01	4.09		5.10	7.65
8340	3/4" iron pipe size		89	.090		1.23	4.09		5.32	7.90
8350	1" iron pipe size		88	.091		1.41	4.14		5.55	8.20
8360	1-1/4" iron pipe size		87	.092		1.90	4.19		6.09	8.80
8370	1-1/2" iron pipe size		87	.092		2.15	4.19		6.34	9.05
8380	2" iron pipe size		86	.093		2.52	4.24		6.76	9.55
8444	1" wall, 1/2" iron pipe size		86	.093		1.93	4.24		6.17	8.90
8445	3/4" iron pipe size		84	.095		2.34	4.34		6.68	9.50
8446	1" iron pipe size		84	.095		2.72	4.34		7.06	9.95
8447	1-1/4" iron pipe size		82	.098		3.08	4.44		7.52	10.55
8448	1-1/2" iron pipe size		82	.098		3.58	4.44		8.02	11.10
8449	2" iron pipe size		80	.100		4.78	4.56		9.34	12.55
8450	2-1/2" iron pipe size		80	.100		6.25	4.56		10.81	14.15
8456	Rubber insulation tape, 1/8" x 2" x 30'				Ea.	11.40			11.40	12.50
9600	Minimum labor/equipment charge	1 Plum	4	2	Job		104		104	161

22 11 Facility Water Distribution

22 11 13 – Facility Water Distribution Piping

22 11 13.14 Pipe, Brass

		Crew	Daily Output	Labor-Hours	Unit	Material	2010 Bare Costs Labor	Equipment	Total	Total Incl O&P
0010	**PIPE, BRASS**, Plain end									
0900	Field threaded, coupling & clevis hanger assembly 10' O.C.									
0920	Regular weight									
1120	1/2" diameter	1 Plum	48	.167	L.F.	5.80	8.70		14.50	19.85
1140	3/4" diameter		46	.174		7.85	9.05		16.90	22.50
1160	1" diameter		43	.186		11.45	9.70		21.15	27.50
1180	1-1/4" diameter	Q-1	72	.222		17.50	10.40		27.90	35.50
1200	1-1/2" diameter		65	.246		21	11.55		32.55	41
1220	2" diameter		53	.302		29	14.15		43.15	54
9000	Minimum labor/equipment charge	1 Plum	4	2	Job		104		104	161

22 11 13.16 Pipe Fittings, Brass

		Crew	Daily Output	Labor-Hours	Unit	Material	2010 Bare Costs Labor	Equipment	Total	Total Incl O&P
0010	**PIPE FITTINGS, BRASS**, Rough bronze, threaded.									
0020										
1000	Standard wt., 90° Elbow									
1100	1/2"	1 Plum	12	.667	Ea.	14.05	34.50		48.55	69.50
1120	3/4"		11	.727		18.75	38		56.75	79
1140	1"		10	.800		30.50	41.50		72	98
1160	1-1/4"	Q-1	17	.941		49	44		93	123
1180	1-1/2"	"	16	1		61	47		108	140
1500	45° Elbow, 1/8"	1 Plum	13	.615		17.25	32		49.25	68.50
1580	1/2"		12	.667		17.25	34.50		51.75	73
1600	3/4"		11	.727		25	38		63	86
1620	1"		10	.800		42	41.50		83.50	111
1640	1-1/4"	Q-1	17	.941		66.50	44		110.50	142
1660	1-1/2"	"	16	1		83.50	47		130.50	165
2000	Tee, 1/8"	1 Plum	9	.889		16.45	46.50		62.95	89.50
2080	1/2"		8	1		16.45	52		68.45	98.50
2100	3/4"		7	1.143		23.50	59.50		83	118
2120	1"		6	1.333		42	69.50		111.50	155
2140	1-1/4"	Q-1	10	1.600		72.50	75		147.50	196
2160	1-1/2"	"	9	1.778		82	83.50		165.50	220
2500	Coupling, 1/8"	1 Plum	26	.308		11.75	16		27.75	38
2580	1/2"		15	.533		11.75	28		39.75	56
2600	3/4"		14	.571		16.45	29.50		45.95	64
2620	1"		13	.615		28	32		60	80.50
2640	1-1/4"	Q-1	22	.727		47	34		81	105
2660	1-1/2"		20	.800		61	37.50		98.50	125
2680	2"		18	.889		101	41.50		142.50	176
9000	Minimum labor/equipment charge	1 Plum	4	2	Job		104		104	161

22 11 13.23 Pipe/Tube, Copper

		Crew	Daily Output	Labor-Hours	Unit	Material	2010 Bare Costs Labor	Equipment	Total	Total Incl O&P
0010	**PIPE/TUBE, COPPER**, Solder joints									
1000	Type K tubing, couplings & clevis hanger assemblies 10' O.C.									
1100	1/4" diameter	1 Plum	84	.095	L.F.	2.80	4.96		7.76	10.80
1200	1" diameter		66	.121		7.95	6.30		14.25	18.55
1260	2" diameter		40	.200		20	10.40		30.40	38
2000	Type L tubing, couplings & clevis hanger assemblies 10' O.C.									
2100	1/4" diameter	1 Plum	88	.091	L.F.	1.60	4.73		6.33	9.10
2120	3/8" diameter		84	.095		2.27	4.96		7.23	10.20
2140	1/2" diameter		81	.099		2.49	5.15		7.64	10.70
2160	5/8" diameter		79	.101		3.87	5.25		9.12	12.40
2180	3/4" diameter		76	.105		3.83	5.50		9.33	12.70
2200	1" diameter		68	.118		5.95	6.10		12.05	16.05

22 11 Facility Water Distribution

22 11 13 – Facility Water Distribution Piping

22 11 13.23 Pipe/Tube, Copper

		Crew	Daily Output	Labor-Hours	Unit	Material	Labor	Equipment	Total	Total Incl O&P
2220	1-1/4" diameter	1 Plum	58	.138	L.F.	8.35	7.20		15.55	20.50
2240	1-1/2" diameter		52	.154		10.75	8		18.75	24
2260	2" diameter		42	.190		16.70	9.90		26.60	34
2280	2-1/2" diameter	Q-1	62	.258		26	12.10		38.10	47.50
2300	3" diameter		56	.286		35.50	13.40		48.90	59.50
2320	3-1/2" diameter		43	.372		49	17.45		66.45	81
2340	4" diameter		39	.410		60.50	19.20		79.70	96.50
2360	5" diameter		34	.471		130	22		152	177
2380	6" diameter	Q-2	40	.600		182	29		211	245
2410	For other than full hard temper, add					21%				
2590	For silver solder, add						15%			
3000	Type M tubing, couplings & clevis hanger assemblies 10' O.C.									
3140	1/2" diameter	1 Plum	84	.095	L.F.	1.86	4.96		6.82	9.75
3180	3/4" diameter		78	.103		2.93	5.35		8.28	11.45
3200	1" diameter		70	.114		4.72	5.95		10.67	14.40
4000	Type DWV tubing, couplings & clevis hanger assemblies 10' O.C.									
4100	1-1/4" diameter	1 Plum	60	.133	L.F.	9.80	6.95		16.75	21.50
4120	1-1/2" diameter		54	.148		9.25	7.70		16.95	22
4140	2" diameter		44	.182		12.55	9.45		22	28.50
4160	3" diameter	Q-1	58	.276		23.50	12.90		36.40	46
4180	4" diameter		40	.400		42	18.75		60.75	75
4200	5" diameter		36	.444		117	21		138	162
4220	6" diameter	Q-2	42	.571		169	28		197	229
9000	Minimum labor/equipment charge	1 Plum	4	2	Job		104		104	161

22 11 13.25 Pipe/Tube Fittings, Copper

		Crew	Daily Output	Labor-Hours	Unit	Material	Labor	Equipment	Total	Total Incl O&P
0010	**PIPE/TUBE FITTINGS, COPPER**, Wrought unless otherwise noted									
0040	Solder joints, copper x copper									
0070	90° elbow, 1/4"	1 Plum	22	.364	Ea.	4.59	18.95		23.54	34.50
0100	1/2"		20	.400		1.45	21		22.45	34
0120	3/4"		19	.421		3.27	22		25.27	37.50
0130	1"		16	.500		8	26		34	49.50
0250	45° elbow, 1/4"		22	.364		8.60	18.95		27.55	39
0270	3/8"		22	.364		7	18.95		25.95	37
0280	1/2"		20	.400		2.67	21		23.67	35.50
0290	5/8"		19	.421		13.35	22		35.35	48.50
0300	3/4"		19	.421		4.68	22		26.68	39
0310	1"		16	.500		11.75	26		37.75	53.50
0320	1-1/4"		15	.533		15.90	28		43.90	60.50
0330	1-1/2"		13	.615		19.15	32		51.15	70.50
0340	2"		11	.727		32	38		70	93.50
0350	2-1/2"	Q-1	13	1.231		68	57.50		125.50	165
0360	3"		13	1.231		101	57.50		158.50	201
0370	3-1/2"		10	1.600		170	75		245	305
0380	4"		9	1.778		215	83.50		298.50	365
0390	5"		6	2.667		835	125		960	1,100
0400	6"	Q-2	9	2.667		1,325	130		1,455	1,650
0450	Tee, 1/4"	1 Plum	14	.571		9.65	29.50		39.15	56.50
0470	3/8"		14	.571		7.35	29.50		36.85	54
0480	1/2"		13	.615		2.49	32		34.49	52
0490	5/8"		12	.667		16	34.50		50.50	71.50
0500	3/4"		12	.667		6	34.50		40.50	60.50
0510	1"		10	.800		18.55	41.50		60.05	85

22 11 Facility Water Distribution

22 11 13 − Facility Water Distribution Piping

22 11 13.25 Pipe/Tube Fittings, Copper		Crew	Daily Output	Labor-Hours	Unit	Material	2010 Bare Costs Labor	Equipment	Total	Total Incl O&P
0520	1-1/4"	1 Plum	9	.889	Ea.	25	46.50		71.50	99
0530	1-1/2"		8	1		39	52		91	124
0540	2"	▼	7	1.143		61	59.50		120.50	159
0550	2-1/2"	Q-1	8	2		123	93.50		216.50	280
0560	3"		7	2.286		187	107		294	370
0570	3-1/2"		6	2.667		545	125		670	790
0580	4"		5	3.200		455	150		605	730
0590	5"	▼	4	4		1,475	187		1,662	1,925
0600	6"	Q-2	6	4		2,025	194		2,219	2,525
0612	Tee, reducing on the outlet, 1/4"	1 Plum	15	.533		17.65	28		45.65	62.50
0613	3/8"		15	.533		14.55	28		42.55	59
0614	1/2"		14	.571		15.15	29.50		44.65	62.50
0615	5/8"		13	.615		26	32		58	78
0616	3/4"		12	.667		8.50	34.50		43	63.50
0617	1"		11	.727		23	38		61	83.50
0618	1-1/4"		10	.800		32	41.50		73.50	99.50
0619	1-1/2"		9	.889		29	46.50		75.50	104
0620	2"	▼	8	1		52.50	52		104.50	139
0621	2-1/2"	Q-1	9	1.778		143	83.50		226.50	287
0622	3"		8	2		158	93.50		251.50	320
0623	4"		6	2.667		305	125		430	530
0624	5"	▼	5	3.200		1,475	150		1,625	1,850
0625	6"	Q-2	7	3.429		1,925	167		2,092	2,375
0626	8"	"	6	4		7,825	194		8,019	8,925
0630	Tee, reducing on the run, 1/4"	1 Plum	15	.533		19.45	28		47.45	64.50
0631	3/8"		15	.533		26	28		54	71.50
0632	1/2"		14	.571		17.90	29.50		47.40	65.50
0633	5/8"		13	.615		26	32		58	78
0634	3/4"		12	.667		7.05	34.50		41.55	62
0635	1"		11	.727		22	38		60	83
0636	1-1/4"		10	.800		36	41.50		77.50	104
0637	1-1/2"		9	.889		62	46.50		108.50	140
0638	2"	▼	8	1		86	52		138	175
0639	2-1/2"	Q-1	9	1.778		186	83.50		269.50	335
0640	3"		8	2		287	93.50		380.50	460
0641	4"		6	2.667		605	125		730	860
0642	5"	▼	5	3.200		1,400	150		1,550	1,775
0643	6"	Q-2	7	3.429		2,150	167		2,317	2,600
0644	8"	"	6	4		7,825	194		8,019	8,925
0650	Coupling, 1/4"	1 Plum	24	.333		1.06	17.35		18.41	28
0670	3/8"		24	.333		1.41	17.35		18.76	28.50
0680	1/2"		22	.364		1.10	18.95		20.05	30.50
0690	5/8"		21	.381		3.32	19.85		23.17	34
0700	3/4"		21	.381		2.36	19.85		22.21	33
0710	1"		18	.444		4.41	23		27.41	41
0715	1-1/4"		17	.471		7.85	24.50		32.35	46.50
0716	1-1/2"		15	.533		10.35	28		38.35	54.50
0718	2"	▼	13	.615		17.30	32		49.30	68.50
0721	2-1/2"	Q-1	15	1.067		37	50		87	118
0722	3"		13	1.231		55.50	57.50		113	151
0724	3-1/2"		8	2		106	93.50		199.50	262
0726	4"		7	2.286		116	107		223	293
0728	5"	▼	6	2.667	▼	286	125		411	510

22 11 Facility Water Distribution

22 11 13 – Facility Water Distribution Piping

22 11 13.25 Pipe/Tube Fittings, Copper

		Crew	Daily Output	Labor-Hours	Unit	Material	2010 Bare Costs Labor	Equipment	Total	Total Incl O&P
0731	6"	Q-2	8	3	Ea.	460	146		606	730
2000	DWV, solder joints, copper x copper									
2030	90° Elbow, 1-1/4"	1 Plum	13	.615	Ea.	14.05	32		46.05	65
2050	1-1/2"		12	.667		21	34.50		55.50	77
2070	2"	↓	10	.800		27.50	41.50		69	95
2090	3"	Q-1	10	1.600		68	75		143	191
2100	4"	"	9	1.778		390	83.50		473.50	560
2250	Tee, Sanitary, 1-1/4"	1 Plum	9	.889		28	46.50		74.50	102
2270	1-1/2"		8	1		34.50	52		86.50	119
2290	2"	↓	7	1.143		40.50	59.50		100	137
2310	3"	Q-1	7	2.286		164	107		271	345
2330	4"	"	6	2.667		415	125		540	655
2400	Coupling, 1-1/4"	1 Plum	14	.571		6.60	29.50		36.10	53.50
2420	1-1/2"		13	.615		8.20	32		40.20	58.50
2440	2"	↓	11	.727		11.35	38		49.35	71
2460	3"	Q-1	11	1.455		22	68		90	130
2480	4"	"	10	1.600	↓	70	75		145	193
9000	Minimum labor/equipment charge	1 Plum	4	2	Job		104		104	161

22 11 13.44 Pipe, Steel

		Crew	Daily Output	Labor-Hours	Unit	Material	Labor	Equipment	Total	Total Incl O&P
0010	**PIPE, STEEL** R221113-50									
0020	All pipe sizes are to Spec. A-53 unless noted otherwise									
0050	Schedule 40, threaded, with couplings, and clevis hanger									
0060	assemblies sized for covering, 10' O.C.									
0540	Black, 1/4" diameter	1 Plum	66	.121	L.F.	2.78	6.30		9.08	12.85
0560	1/2" diameter		63	.127		3.61	6.60		10.21	14.20
0570	3/4" diameter		61	.131		4.28	6.85		11.13	15.30
0580	1" diameter	↓	53	.151		6.25	7.85		14.10	19.05
0590	1-1/4" diameter	Q-1	89	.180		7.95	8.40		16.35	22
0600	1-1/2" diameter		80	.200		9.45	9.35		18.80	25
0610	2" diameter		64	.250		12.60	11.70		24.30	32
0620	2-1/2" diameter		50	.320		19.45	15		34.45	44.50
0630	3" diameter		43	.372		25	17.45		42.45	54.50
0640	3-1/2" diameter		40	.400		34	18.75		52.75	66.50
0650	4" diameter	↓	36	.444	↓	36.50	21		57.50	73
1280	All pipe sizes are to Spec. A-53 unless noted otherwise									
1281	Schedule 40, threaded, with couplings and clevis hanger									
1282	assemblies sized for covering, 10' O.C.									
1290	Galvanized, 1/4" diameter	1 Plum	66	.121	L.F.	3.80	6.30		10.10	14
1310	1/2" diameter		63	.127		5.15	6.60		11.75	15.90
1320	3/4" diameter		61	.131		5	6.85		11.85	16.10
1330	1" diameter	↓	53	.151		8.45	7.85		16.30	21.50
1350	1-1/2" diameter	Q-1	80	.200		12.90	9.35		22.25	28.50
1360	2" diameter		64	.250		17.25	11.70		28.95	37
1370	2-1/2" diameter		50	.320		29	15		44	54.50
1380	3" diameter		43	.372		37	17.45		54.45	67.50
1400	4" diameter	↓	36	.444	↓	52	21		73	90
2000	Welded, sch. 40, on yoke & roll hanger assy's, sized for covering, 10' O.C.									
2040	Black, 1" diameter	Q-15	93	.172	L.F.	3.75	8.05	.63	12.43	17.30
2070	2" diameter		61	.262		6.55	12.30	.96	19.81	27.50
2090	3" diameter		43	.372		11.40	17.45	1.36	30.21	41
2110	4" diameter		37	.432		15.95	20.50	1.58	38.03	51
2120	5" diameter	↓	32	.500		21.50	23.50	1.83	46.83	62.50

22 11 Facility Water Distribution

22 11 13 – Facility Water Distribution Piping

22 11 13.44 Pipe, Steel

		Crew	Daily Output	Labor-Hours	Unit	Material	2010 Bare Costs Labor	Equipment	Total	Total Incl O&P
2130	6" diameter	Q-16	36	.667	L.F.	26.50	32.50	1.63	60.63	81
9990	Minimum labor/equipment charge	1 Plum	3	2.667	Job		139		139	215

22 11 13.45 Pipe Fittings, Steel, Threaded

			Crew	Daily Output	Labor-Hours	Unit	Material	Labor	Equipment	Total	Total Incl O&P
0010	**PIPE FITTINGS, STEEL, THREADED**	R221113-50									
0020	Cast Iron										
0040	Standard weight, black										
0060	90° Elbow, straight										
0070	1/4"		1 Plum	16	.500	Ea.	6.95	26		32.95	48
0080	3/8"			16	.500		10.05	26		36.05	51.50
0090	1/2"			15	.533		4.41	28		32.41	48
0100	3/4"			14	.571		4.60	29.50		34.10	51
0110	1"			13	.615		5.45	32		37.45	55.50
0130	1-1/2"		Q-1	20	.800		10.70	37.50		48.20	70
0140	2"			18	.889		16.70	41.50		58.20	83
0150	2-1/2"			14	1.143		40	53.50		93.50	127
0160	3"			10	1.600		65.50	75		140.50	188
0180	4"			6	2.667		122	125		247	330
0500	Tee, straight										
0510	1/4"		1 Plum	10	.800	Ea.	10.90	41.50		52.40	76.50
0520	3/8"			10	.800		10.60	41.50		52.10	76
0540	3/4"			9	.889		8	46.50		54.50	80.50
0550	1"			8	1		7.15	52		59.15	88.50
0570	1-1/2"		Q-1	13	1.231		16.90	57.50		74.40	108
0580	2"			11	1.455		23.50	68		91.50	132
0590	2-1/2"			9	1.778		61	83.50		144.50	197
0600	3"			6	2.667		94	125		219	297
0620	4"			4	4		183	187		370	490
0700	Standard weight, galvanized cast iron										
0720	90° Elbow, straight										
0730	1/4"		1 Plum	16	.500	Ea.	11.65	26		37.65	53.50
0740	3/8"			16	.500		11.65	26		37.65	53.50
0750	1/2"			15	.533		8.85	28		36.85	52.50
0760	3/4"			14	.571		13.05	29.50		42.55	60.50
0770	1"			13	.615		15.05	32		47.05	66
0790	1-1/2"		Q-1	20	.800		32	37.50		69.50	93.50
0800	2"			18	.889		47.50	41.50		89	117
0810	2-1/2"			14	1.143		97.50	53.50		151	190
0820	3"			10	1.600		149	75		224	279
0840	4"			6	2.667		272	125		397	495
1100	Tee, straight										
1110	1/4"		1 Plum	10	.800	Ea.	22	41.50		63.50	88.50
1120	3/8"			10	.800		14.90	41.50		56.40	81
1140	3/4"			9	.889		18.75	46.50		65.25	92
1150	1"			8	1		20.50	52		72.50	103
1170	1-1/2"		Q-1	13	1.231		47	57.50		104.50	142
1180	2"			11	1.455		59	68		127	171
1190	2-1/2"			9	1.778		127	83.50		210.50	268
1200	3"			6	2.667		330	125		455	555
1220	4"			4	4		320	187		507	645
5000	Malleable iron, 150 lb.										
5020	Black										
5040	90° elbow, straight										

22 11 Facility Water Distribution

22 11 13 – Facility Water Distribution Piping

22 11 13.45 Pipe Fittings, Steel, Threaded		Crew	Daily Output	Labor-Hours	Unit	Material	2010 Bare Costs Labor	Equipment	Total	Total Incl O&P
5060	1/4"	1 Plum	16	.500	Ea.	3.42	26		29.42	44.50
5070	3/8"		16	.500		3.42	26		29.42	44.50
5090	3/4"		14	.571		2.87	29.50		32.37	49
5100	1"	↓	13	.615		4.98	32		36.98	55
5120	1-1/2"	Q-1	20	.800		10.80	37.50		48.30	70
5130	2"		18	.889		18.60	41.50		60.10	85
5140	2-1/2"		14	1.143		41.50	53.50		95	129
5150	3"		10	1.600		60.50	75		135.50	183
5170	4"	↓	6	2.667	↓	130	125		255	335
5450	Tee, straight									
5470	1/4"	1 Plum	10	.800	Ea.	4.98	41.50		46.48	70
5480	3/8"		10	.800		4.98	41.50		46.48	70
5500	3/4"		9	.889		4.55	46.50		51.05	76.50
5510	1"	↓	8	1		7.80	52		59.80	89
5520	1-1/4"	Q-1	14	1.143		12.65	53.50		66.15	97
5530	1-1/2"		13	1.231		15.70	57.50		73.20	107
5540	2"		11	1.455		27	68		95	136
5550	2-1/2"		9	1.778		57.50	83.50		141	193
5560	3"		6	2.667		85	125		210	288
5570	3-1/2"		5	3.200		191	150		341	445
5580	4"	↓	4	4	↓	205	187		392	515
5650	Coupling									
5670	1/4"	1 Plum	19	.421	Ea.	4.26	22		26.26	38.50
5680	3/8"		19	.421		4.26	22		26.26	38.50
5690	1/2"		19	.421		3.28	22		25.28	37.50
5700	3/4"		18	.444		3.84	23		26.84	40
5710	1"	↓	15	.533		5.75	28		33.75	49.50
5730	1-1/2"	Q-1	24	.667		10.05	31		41.05	59.50
5740	2"		21	.762		14.90	35.50		50.40	72
5750	2-1/2"		18	.889		41	41.50		82.50	110
5760	3"		14	1.143		56	53.50		109.50	145
5780	4"	↓	10	1.600		112	75		187	239
6000	For galvanized elbows, tees, and couplings add					20%				
9990	Minimum labor/equipment charge	1 Plum	4	2	Job		104		104	161
22 11 13.47 Pipe Fittings, Steel										
0010	**PIPE FITTINGS, STEEL**, Flanged, Welded & Special									
3000	Weld joint, butt, carbon steel, standard weight									
3040	90° elbow, long radius									
3050	1/2" pipe size	Q-15	16	1	Ea.	45	47	3.66	95.66	126
3060	3/4" pipe size		16	1		45	47	3.66	95.66	126
3070	1" pipe size		16	1		21	47	3.66	71.66	99.50
3100	2" pipe size		10	1.600		22.50	75	5.85	103.35	147
3120	3" pipe size		7	2.286		33.50	107	8.35	148.85	212
3130	4" pipe size		5	3.200		48	150	11.70	209.70	298
3136	5" pipe size	↓	4	4		75	187	14.65	276.65	390
3140	6" pipe size	Q-16	5	4.800	↓	106	233	11.70	350.70	490
3350	Tee, straight									
3360	1/2" pipe size	Q-15	10	1.600	Ea.	110	75	5.85	190.85	243
3370	3/4" pipe size		10	1.600		110	75	5.85	190.85	243
3380	1" pipe size		10	1.600		54.50	75	5.85	135.35	182
3410	2" pipe size		6	2.667		54.50	125	9.75	189.25	265
3430	3" pipe size	↓	4	4	↓	84	187	14.65	285.65	400

22 11 Facility Water Distribution

22 11 13 – Facility Water Distribution Piping

22 11 13.47 Pipe Fittings, Steel		Crew	Daily Output	Labor-Hours	Unit	Material	2010 Bare Costs Labor	Equipment	Total	Total Incl O&P
3440	4" pipe size	Q-15	3	5.333	Ea.	118	250	19.50	387.50	535
3446	5" pipe size	↓	2.50	6.400		195	300	23.50	518.50	705
3450	6" pipe size	Q-16	3	8	↓	203	390	19.50	612.50	845
9990	Minimum labor/equipment charge	Q-15	3	5.333	Job		250	19.50	269.50	405

22 11 13.48 Pipe, Fittings and Valves, Steel, Grooved-Joint

		Crew	Daily Output	Labor-Hours	Unit	Material	Labor	Equipment	Total	Total Incl O&P
0010	**PIPE, FITTINGS AND VALVES, STEEL, GROOVED-JOINT**									
0012	Fittings are ductile iron. Steel fittings noted.									
0020	Pipe includes coupling & clevis type hanger assemblies, 10' O.C.									
1000	Schedule 40, black									
1040	3/4" diameter	1 Plum	71	.113	L.F.	5.30	5.85		11.15	14.90
1050	1" diameter		63	.127		3.38	6.60		9.98	13.95
1060	1-1/4" diameter		58	.138		4.43	7.20		11.63	15.95
1070	1-1/2" diameter		51	.157		5.10	8.15		13.25	18.25
1080	2" diameter	↓	40	.200		6.30	10.40		16.70	23
1090	2-1/2" diameter	Q-1	57	.281		9.15	13.15		22.30	30.50
1100	3" diameter		50	.320		11.50	15		26.50	35.50
1110	4" diameter		45	.356		16.35	16.65		33	44
1120	5" diameter	↓	37	.432	↓	22.50	20.50		43	56.50
4000	Elbow, 90° or 45°, painted									
4030	3/4" diameter	1 Plum	50	.160	Ea.	40	8.35		48.35	57
4040	1" diameter		50	.160		21.50	8.35		29.85	36.50
4050	1-1/4" diameter		40	.200		21.50	10.40		31.90	39.50
4060	1-1/2" diameter		33	.242		21.50	12.60		34.10	43
4070	2" diameter	↓	25	.320		21.50	16.65		38.15	49.50
4080	2-1/2" diameter	Q-1	40	.400		21.50	18.75		40.25	52.50
4100	4" diameter		25	.640		41.50	30		71.50	92
4110	5" diameter	↓	20	.800	↓	99	37.50		136.50	167
4250	For galvanized elbows, add					26%				
4690	Tee, painted									
4700	3/4" diameter	1 Plum	38	.211	Ea.	43.50	10.95		54.45	64.50
4740	1" diameter		33	.242		33	12.60		45.60	56
4750	1-1/4" diameter		27	.296		33	15.40		48.40	60.50
4760	1-1/2" diameter		22	.364		33	18.95		51.95	66
4770	2" diameter	↓	17	.471		33	24.50		57.50	74.50
4780	2-1/2" diameter	Q-1	27	.593		33	28		61	79.50
4800	4" diameter		17	.941		70	44		114	146
4810	5" diameter	↓	13	1.231	↓	163	57.50		220.50	270
4900	For galvanized tees, add					24%				
9990	Minimum labor/equipment charge	1 Plum	4	2	Job		104		104	161

22 11 13.74 Pipe, Plastic

		Crew	Daily Output	Labor-Hours	Unit	Material	Labor	Equipment	Total	Total Incl O&P
0010	**PIPE, PLASTIC**									
1800	PVC, couplings 10' O.C., clevis hanger assemblies, 3 per 10'									
1820	Schedule 40									
1860	1/2" diameter	1 Plum	54	.148	L.F.	1.65	7.70		9.35	13.75
1870	3/4" diameter		51	.157		1.89	8.15		10.04	14.75
1880	1" diameter		46	.174		2.18	9.05		11.23	16.45
1890	1-1/4" diameter		42	.190		2.72	9.90		12.62	18.35
1900	1-1/2" diameter	↓	36	.222		2.92	11.55		14.47	21
1910	2" diameter	Q-1	59	.271		3.51	12.70		16.21	23.50
1920	2-1/2" diameter		56	.286		5.40	13.40		18.80	26.50
1930	3" diameter		53	.302		6.95	14.15		21.10	29.50
1940	4" diameter	↓	48	.333	↓	9.60	15.60		25.20	34.50

22 11 Facility Water Distribution

22 11 13 – Facility Water Distribution Piping

22 11 13.74 Pipe, Plastic

		Crew	Daily Output	Labor-Hours	Unit	Material	2010 Bare Costs Labor	Equipment	Total	Total Incl O&P
1950	5" diameter	Q-1	43	.372	L.F.	13.95	17.45		31.40	42.50
1960	6" diameter	↓	39	.410	↓	17.80	19.20		37	49.50
4100	DWV type, schedule 40, couplings 10' O.C., clevis hanger assy's, 3 per 10'									
4210	ABS, schedule 40, foam core type									
4212	Plain end black									
4214	1-1/2" diameter	1 Plum	39	.205	L.F.	1.93	10.70		12.63	18.70
4216	2" diameter	Q-1	62	.258		2.35	12.10		14.45	21.50
4218	3" diameter		56	.286		4.42	13.40		17.82	25.50
4220	4" diameter		51	.314		6.10	14.70		20.80	29.50
4222	6" diameter	↓	42	.381	↓	12.35	17.85		30.20	41
4240	To delete coupling & hangers, subtract									
4244	1-1/2" diam. to 6" diam.					43%	48%			
4400	PVC									
4410	1-1/4" diameter	1 Plum	42	.190	L.F.	1.99	9.90		11.89	17.55
4460	2" diameter	Q-1	59	.271		2.41	12.70		15.11	22.50
4470	3" diameter		53	.302		4.51	14.15		18.66	27
4480	4" diameter		48	.333		6.10	15.60		21.70	31
4490	6" diameter	↓	39	.410	↓	11.10	19.20		30.30	42
5300	CPVC, socket joint, couplings 10' O.C., clevis hanger assemblies, 3 per 10'									
5302	Schedule 40									
5304	1/2" diameter	1 Plum	54	.148	L.F.	2.35	7.70		10.05	14.55
5305	3/4" diameter		51	.157		2.85	8.15		11	15.80
5306	1" diameter		46	.174		3.76	9.05		12.81	18.20
5307	1-1/4" diameter		42	.190		4.88	9.90		14.78	20.50
5308	1-1/2" diameter	↓	36	.222		5.75	11.55		17.30	24
5309	2" diameter	Q-1	59	.271		7.15	12.70		19.85	27.50
5310	2-1/2" diameter		56	.286		12.25	13.40		25.65	34
5311	3" diameter	↓	53	.302	↓	14.90	14.15		29.05	38.50
5360	CPVC, threaded, couplings 10' O.C., clevis hanger assemblies, 3 per 10'									
5380	Schedule 40									
5460	1/2" diameter	1 Plum	54	.148	L.F.	3.28	7.70		10.98	15.55
5470	3/4" diameter		51	.157		3.88	8.15		12.03	16.90
5480	1" diameter		46	.174		4.76	9.05		13.81	19.30
5490	1-1/4" diameter		42	.190		5.85	9.90		15.75	22
5500	1-1/2" diameter	↓	36	.222		6.60	11.55		18.15	25
5510	2" diameter	Q-1	59	.271		8.20	12.70		20.90	28.50
5520	2-1/2" diameter		56	.286		13.35	13.40		26.75	35
5530	3" diameter	↓	53	.302	↓	16.50	14.15		30.65	40
5560										
9900	Minimum labor/equipment charge	1 Plum	4	2	Job		104		104	161

22 11 13.76 Pipe Fittings, Plastic

		Crew	Daily Output	Labor-Hours	Unit	Material	2010 Bare Costs Labor	Equipment	Total	Total Incl O&P
0010	**PIPE FITTINGS, PLASTIC**									
2700	PVC (white), schedule 40, socket joints									
2760	90° elbow, 1/2"	1 Plum	33.30	.240	Ea.	.41	12.50		12.91	19.85
2810	2"	Q-1	36.40	.440	"	2.47	20.50		22.97	34.50
4500	DWV, ABS, non pressure, socket joints									
4540	1/4 Bend, 1-1/4"	1 Plum	20.20	.396	Ea.	2.16	20.50		22.66	34.50
4560	1-1/2"	"	18.20	.440		1.56	23		24.56	37
4570	2"	Q-1	33.10	.483	↓	2.30	22.50		24.80	37.50
4800	Tee, sanitary									
4820	1-1/4"	1 Plum	13.50	.593	Ea.	2.58	31		33.58	51
4830	1-1/2"	"	12.10	.661	↓	2.08	34.50		36.58	56

22 11 Facility Water Distribution

22 11 13 − Facility Water Distribution Piping

22 11 13.76 Pipe Fittings, Plastic

		Crew	Daily Output	Labor-Hours	Unit	Material	2010 Bare Costs Labor	Equipment	Total	Total Incl O&P
4840	2"	Q-1	20	.800	Ea.	3.21	37.50		40.71	61.50
5000	DWV, PVC, schedule 40, socket joints									
5040	1/4 bend, 1-1/4"	1 Plum	20.20	.396	Ea.	3.69	20.50		24.19	36
5060	1-1/2"	"	18.20	.440		1.10	23		24.10	36.50
5070	2"	Q-1	33.10	.483		1.74	22.50		24.24	37
5080	3"		20.80	.769		5.15	36		41.15	61.50
5090	4"		16.50	.970		10.10	45.50		55.60	81.50
5100	6"	↓	10.10	1.584		35.50	74		109.50	154
5105	8"	Q-2	9.30	2.581		72	125		197	273
5110	1/4 bend, long sweep, 1-1/2"	1 Plum	18.20	.440		2.65	23		25.65	38.50
5112	2"	Q-1	33.10	.483		2.86	22.50		25.36	38
5114	3"		20.80	.769		6.75	36		42.75	63.50
5116	4"	↓	16.50	.970		12.90	45.50		58.40	84.50
5215	8"	Q-2	9.30	2.581		54	125		179	254
5250	Tee, sanitary 1-1/4"	1 Plum	13.50	.593		3.19	31		34.19	51.50
5254	1-1/2"	"	12.10	.661		2	34.50		36.50	55.50
5255	2"	Q-1	20	.800		2.94	37.50		40.44	61
5256	3"		13.90	1.151		7.55	54		61.55	92
5257	4"		11	1.455		13.10	68		81.10	120
5259	6"	↓	6.70	2.388		57.50	112		169.50	236
5261	8"	Q-2	6.20	3.871		190	188		378	500
5264	2" x 1-1/2"	Q-1	22	.727		2.42	34		36.42	55.50
5266	3" x 1-1/2"		15.50	1.032		5.45	48.50		53.95	81
5268	4" x 3"		12.10	1.322		16.65	62		78.65	114
5271	6" x 4"	↓	6.90	2.319		61.50	109		170.50	236
5314	Combination Y & 1/8 bend, 1-1/2"	1 Plum	12.10	.661		4.34	34.50		38.84	58.50
5315	2"	Q-1	20	.800		5.75	37.50		43.25	64.50
5317	3"		13.90	1.151		14.30	54		68.30	99
5318	4"		11	1.455	↓	28.50	68		96.50	137
5324	Combination Y & 1/8 bend, reducing									
5325	2" x 2" x 1-1/2"	Q-1	22	.727	Ea.	6.25	34		40.25	60
5327	3" x 3" x 1-1/2"		15.50	1.032		10.80	48.50		59.30	87
5328	3" x 3" x 2"		15.30	1.046		8.50	49		57.50	85.50
5329	4" x 4" x 2"	↓	12.20	1.311		13.80	61.50		75.30	110
5331	Wye, 1-1/4"	1 Plum	13.50	.593		4.06	31		35.06	52.50
5332	1-1/2"	"	12.10	.661		3.06	34.50		37.56	57
5333	2"	Q-1	20	.800		3.58	37.50		41.08	62
5334	3"		13.90	1.151		9.65	54		63.65	94
5335	4"		11	1.455		17.55	68		85.55	125
5336	6"	↓	6.70	2.388		53	112		165	231
5337	8"	Q-2	6.20	3.871		155	188		343	460
5341	2" x 1-1/2"	Q-1	22	.727		4.39	34		38.39	58
5342	3" x 1-1/2"		15.50	1.032		6.30	48.50		54.80	82
5343	4" x 3"		12.10	1.322		14.25	62		76.25	112
5344	6" x 4"	↓	6.90	2.319		39	109		148	211
5345	8" x 6"	Q-2	6.40	3.750		117	182		299	410
5347	Double wye, 1-1/2"	1 Plum	9.10	.879		6.90	46		52.90	78.50
5348	2"	Q-1	16.60	.964		7.45	45		52.45	78
5349	3"		10.40	1.538		19.20	72		91.20	133
5350	4"	↓	8.25	1.939	↓	39	91		130	184
5353	Double wye, reducing									
5354	2" x 2" x 1-1/2" x 1-1/2"	Q-1	16.80	.952	Ea.	6.80	44.50		51.30	76.50
5355	3" x 3" x 2" x 2"	↓	10.60	1.509	↓	14.35	70.50		84.85	126

22 11 Facility Water Distribution
22 11 13 – Facility Water Distribution Piping

22 11 13.76 Pipe Fittings, Plastic		Crew	Daily Output	Labor-Hours	Unit	Material	2010 Bare Costs Labor	Equipment	Total	Total Incl O&P
5356	4" x 4" x 3" x 3"	Q-1	8.45	1.893	Ea.	31	88.50		119.50	171
5357	6" x 6" x 4" x 4"		7.25	2.207		81.50	103		184.50	250
5410	Reducer bushing, 2" x 1-1/4"		36.50	.438		1.95	20.50		22.45	34
5412	3" x 1-1/2"		27.30	.586		4.53	27.50		32.03	47.50
5414	4" x 2"		18.20	.879		8.75	41		49.75	73.50
5416	6" x 4"		11.10	1.441		23.50	67.50		91	131
5418	8" x 6"	Q-2	10.20	2.353		50.50	114		164.50	233
5500	CPVC, Schedule 80, threaded joints									
5540	90° Elbow, 1/4"	1 Plum	32	.250	Ea.	10.40	13		23.40	31.50
5560	1/2"		30.30	.264		6.05	13.75		19.80	28
5570	3/4"		26	.308		9.05	16		25.05	35
5580	1"		22.70	.352		12.65	18.35		31	42.50
5590	1-1/4"		20.20	.396		24.50	20.50		45	59
5600	1-1/2"		18.20	.440		26.50	23		49.50	64.50
5610	2"	Q-1	33.10	.483		35	22.50		57.50	73.50
5620	2-1/2"		24.20	.661		109	31		140	168
5630	3"		20.80	.769		117	36		153	185
5730	Coupling, 1/4"	1 Plum	32	.250		13.25	13		26.25	34.50
5732	1/2"		30.30	.264		13.25	13.75		27	36
5734	3/4"		26	.308		15.90	16		31.90	42.50
5736	1"		22.70	.352		17.60	18.35		35.95	48
5738	1-1/4"		20.20	.396		21	20.50		41.50	55.50
5740	1-1/2"		18.20	.440		23	23		46	60.50
5742	2"	Q-1	33.10	.483		27	22.50		49.50	64.50
5744	2-1/2"		24.20	.661		48	31		79	101
5746	3"		20.80	.769		56	36		92	118
5900	CPVC, Schedule 80, socket joints									
5904	90° Elbow, 1/4"	1 Plum	32	.250	Ea.	9.70	13		22.70	30.50
5906	1/2"		30.30	.264		3.81	13.75		17.56	25.50
5908	3/4"		26	.308		4.85	16		20.85	30.50
5910	1"		22.70	.352		7.70	18.35		26.05	37
5912	1-1/4"		20.20	.396		16.65	20.50		37.15	50.50
5914	1-1/2"		18.20	.440		18.60	23		41.60	56
5916	2"	Q-1	33.10	.483		22.50	22.50		45	59.50
5918	2-1/2"		24.20	.661		51.50	31		82.50	105
5920	3"		20.80	.769		58.50	36		94.50	121
5930	45° Elbow, 1/4"	1 Plum	32	.250		14.40	13		27.40	36
5932	1/2"		30.30	.264		4.65	13.75		18.40	26.50
5934	3/4"		26	.308		6.70	16		22.70	32.50
5936	1"		22.70	.352		10.70	18.35		29.05	40.50
5938	1-1/4"		20.20	.396		21	20.50		41.50	55
5940	1-1/2"		18.20	.440		21.50	23		44.50	59
5942	2"	Q-1	33.10	.483		24	22.50		46.50	61.50
5944	2-1/2"		24.20	.661		49.50	31		80.50	103
5946	3"		20.80	.769		63.50	36		99.50	126
5990	Coupling, 1/4"	1 Plum	32	.250		10.35	13		23.35	31.50
5992	1/2"		30.30	.264		4.02	13.75		17.77	26
5994	3/4"		26	.308		5.60	16		21.60	31
5996	1"		22.70	.352		7.55	18.35		25.90	37
5998	1-1/4"		20.20	.396		11.35	20.50		31.85	44.50
6000	1-1/2"		18.20	.440		14.25	23		37.25	51
6002	2"	Q-1	33.10	.483		16.55	22.50		39.05	53.50
6004	2-1/2"		24.20	.661		37	31		68	88.50

22 11 Facility Water Distribution

22 11 13 – Facility Water Distribution Piping

22 11 13.76 Pipe Fittings, Plastic		Crew	Daily Output	Labor-Hours	Unit	Material	2010 Bare Costs Labor	Equipment	Total	Total Incl O&P
6006	3"	Q-1	20.80	.769	Ea.	40	36		76	100
9900	Minimum labor/equipment charge	1 Plum	4	2	Job		104		104	161

22 11 19 – Domestic Water Piping Specialties

22 11 19.10 Flexible Connectors

		Crew	Daily Output	Labor-Hours	Unit	Material	Labor	Equipment	Total	Total Incl O&P
0010	**FLEXIBLE CONNECTORS**, Corrugated, 7/8" O.D., 1/2" I.D.									
0050	Gas, seamless brass, steel fittings									
0200	12" long	1 Plum	36	.222	Ea.	15.10	11.55		26.65	34.50
0220	18" long		36	.222		18.70	11.55		30.25	38.50
0240	24" long		34	.235		22	12.25		34.25	43.50
0260	30" long		34	.235		24	12.25		36.25	45.50
9000	Minimum labor/equipment charge		4	2	Job		104		104	161

22 11 19.18 Mixing Valve

0010	**MIXING VALVE**, Automatic, water tempering.									
0040	1/2" size	1 Stpi	19	.421	Ea.	470	22		492	555
0050	3/4" size		18	.444	"	470	23		493	555
9000	Minimum labor/equipment charge		5	1.600	Job		83		83	129

22 11 19.38 Water Supply Meters

0010	**WATER SUPPLY METERS**									
2000	Domestic/commercial, bronze									
2020	Threaded									
2060	5/8" diameter, to 20 GPM	1 Plum	16	.500	Ea.	42	26		68	86.50
2080	3/4" diameter, to 30 GPM		14	.571		76.50	29.50		106	130
2100	1" diameter, to 50 GPM		12	.667		116	34.50		150.50	182
2300	Threaded/flanged									
2340	1-1/2" diameter, to 100 GPM	1 Plum	8	1	Ea.	284	52		336	390
9000	Minimum labor/equipment charge	"	3.25	2.462	Job		128		128	199

22 11 19.42 Backflow Preventers

0010	**BACKFLOW PREVENTERS**, Includes valves									
0020	and four test cocks, corrosion resistant, automatic operation									
1000	Double check principle									
1010	Threaded, with ball valves									
1020	3/4" pipe size	1 Plum	16	.500	Ea.	245	26		271	310
1030	1" pipe size		14	.571		282	29.50		311.50	355
1040	1-1/2" pipe size		10	.800		740	41.50		781.50	875
1050	2" pipe size		7	1.143		880	59.50		939.50	1,050
1080	Threaded, with gate valves									
1100	3/4" pipe size	1 Plum	16	.500	Ea.	945	26		971	1,100
1120	1" pipe size		14	.571		955	29.50		984.50	1,100
1140	1-1/2" pipe size		10	.800		1,225	41.50		1,266.50	1,425
1160	2" pipe size		7	1.143		1,500	59.50		1,559.50	1,750
4000	Reduced pressure principle									
4100	Threaded, bronze, valves are ball									
4120	3/4" pipe size	1 Plum	16	.500	Ea.	365	26		391	440
4140	1" pipe size		14	.571		390	29.50		419.50	475
4160	1-1/2" pipe size		10	.800		735	41.50		776.50	875
4180	2" pipe size		7	1.143		825	59.50		884.50	1,000
5000	Flanged, bronze, valves are OS&Y									
5060	2-1/2" pipe size	Q-1	5	3.200	Ea.	3,525	150		3,675	4,100
5080	3" pipe size		4.50	3.556		3,700	167		3,867	4,325
5100	4" pipe size		3	5.333		4,650	250		4,900	5,475
5120	6" pipe size	Q-2	3	8		6,725	390		7,115	8,000

22 11 Facility Water Distribution

22 11 19 – Domestic Water Piping Specialties

22 11 19.42 Backflow Preventers

		Crew	Daily Output	Labor-Hours	Unit	Material	2010 Bare Costs Labor	Equipment	Total	Total Incl O&P
5600	Flanged, iron, valves are OS&Y									
5660	2-1/2" pipe size	Q-1	5	3.200	Ea.	1,925	150		2,075	2,350
5680	3" pipe size		4.50	3.556		2,950	167		3,117	3,500
5700	4" pipe size		3	5.333		3,800	250		4,050	4,550
5720	6" pipe size	Q-2	3	8		5,375	390		5,765	6,500
5800	Rebuild 4" diameter reduced pressure BFP	1 Plum	4	2		290	104		394	480
5810	6" diameter		2.66	3.008		305	157		462	585
5820	8" diameter		2	4		380	208		588	745
5830	10" diameter		1.60	5		445	260		705	895
9010	Minimum labor/equipment charge		2	4	Job		208		208	325

22 11 19.50 Vacuum Breakers

		Crew	Daily Output	Labor-Hours	Unit	Material	Labor	Equipment	Total	Total Incl O&P
0010	**VACUUM BREAKERS**									
0013	See also backflow preventers Div. 22 11 19.42									
1000	Anti-siphon continuous pressure type									
1010	Max. 150 PSI - 210°F									
1020	Bronze body									
1030	1/2" size	1 Stpi	24	.333	Ea.	172	17.30		189.30	216
1040	3/4" size		20	.400		172	21		193	221
1050	1" size		19	.421		178	22		200	229
1060	1-1/4" size		15	.533		355	27.50		382.50	435
1070	1-1/2" size		13	.615		425	32		457	520
1080	2" size		11	.727		440	38		478	545
1200	Max. 125 PSI with atmospheric vent									
1210	Brass, in-line construction									
1220	1/4" size	1 Stpi	24	.333	Ea.	59.50	17.30		76.80	92.50
1230	3/8" size	"	24	.333		59.50	17.30		76.80	92.50
1260	For polished chrome finish, add					13%				
2000	Anti-siphon, non-continuous pressure type									
2010	Hot or cold water 125 PSI - 210°F									
2020	Bronze body									
2030	1/4" size	1 Stpi	24	.333	Ea.	38	17.30		55.30	68.50
2040	3/8" size		24	.333		38	17.30		55.30	68.50
2050	1/2" size		24	.333		43	17.30		60.30	74.50
2060	3/4" size		20	.400		52.50	21		73.50	89.50
2070	1" size		19	.421		79.50	22		101.50	122
2080	1-1/4" size		15	.533		139	27.50		166.50	196
2090	1-1/2" size		13	.615		164	32		196	230
2100	2" size		11	.727		255	38		293	340
2110	2-1/2" size		8	1		730	52		782	885
2120	3" size		6	1.333		975	69		1,044	1,175
2150	For polished chrome finish, add					50%				

22 11 19.54 Water Hammer Arresters/Shock Absorbers

		Crew	Daily Output	Labor-Hours	Unit	Material	Labor	Equipment	Total	Total Incl O&P
0010	**WATER HAMMER ARRESTERS/SHOCK ABSORBERS**									
0490	Copper									
0500	3/4" male I.P.S. For 1 to 11 fixtures	1 Plum	12	.667	Ea.	20.50	34.50		55	76.50
0600	1" male I.P.S., For 12 to 32 fixtures		8	1		41.50	52		93.50	127
0700	1-1/4" male I.P.S. For 33 to 60 fixtures		8	1		44	52		96	129
0800	1-1/2" male I.P.S. For 61 to 113 fixtures		8	1		63	52		115	150
0900	2" male I.P.S. For 114 to 154 fixtures		8	1		92	52		144	182
1000	2-1/2" male I.P.S. For 155 to 330 fixtures		4	2		285	104		389	475
9000	Minimum labor/equipment charge		3.50	2.286	Job		119		119	184

22 11 Facility Water Distribution

22 11 19 – Domestic Water Piping Specialties

22 11 19.64 Hydrants

		Crew	Daily Output	Labor-Hours	Unit	Material	2010 Bare Costs Labor	2010 Bare Costs Equipment	Total	Total Incl O&P
0010	**HYDRANTS**									
0050	Wall type, moderate climate, bronze, encased									
0200	3/4" IPS connection	1 Plum	16	.500	Ea.	600	26		626	700
1000	Non-freeze, bronze, exposed									
1100	3/4" IPS connection, 4" to 9" thick wall	1 Plum	14	.571	Ea.	405	29.50		434.50	490
1120	10" to 14" thick wall	"	12	.667		520	34.50		554.50	625
1280	For anti-siphon type, add					106			106	117
9000	Minimum labor/equipment charge	1 Plum	3	2.667	Job		139		139	215

22 13 Facility Sanitary Sewerage

22 13 16 – Sanitary Waste and Vent Piping

22 13 16.20 Pipe, Cast Iron

			Crew	Daily Output	Labor-Hours	Unit	Material	Labor	Equipment	Total	Total Incl O&P
0010	**PIPE, CAST IRON**, Soil, on clevis hanger assemblies, 5' O.C.	R221113-50									
0020	Single hub, service wt., lead & oakum joints 10' O.C.										
2120	2" diameter		Q-1	63	.254	L.F.	7	11.90		18.90	26
2140	3" diameter			60	.267		9.80	12.50		22.30	30
2160	4" diameter			55	.291		12.70	13.65		26.35	35
2180	5" diameter		Q-2	76	.316		17.65	15.35		33	43.50
2200	6" diameter		"	73	.329		21.50	15.95		37.45	48.50
2220	8" diameter		Q-3	59	.542		33.50	27		60.50	78.50
4000	No hub, couplings 10' O.C.										
4100	1-1/2" diameter		Q-1	71	.225	L.F.	7.05	10.55		17.60	24
4120	2" diameter			67	.239		7.35	11.20		18.55	25.50
4140	3" diameter			64	.250		10.15	11.70		21.85	29.50
4160	4" diameter			58	.276		13.20	12.90		26.10	34.50
4180	5" diameter		Q-2	83	.289		18.50	14.05		32.55	42.50
9000	Minimum labor/equipment charge		1 Plum	4	2	Job		104		104	161

22 13 16.30 Pipe Fittings, Cast Iron

			Crew	Daily Output	Labor-Hours	Unit	Material	Labor	Equipment	Total	Total Incl O&P
0010	**PIPE FITTINGS, CAST IRON**, Soil	R221113-50									
0040	Hub and spigot, service weight, lead & oakum joints										
0080	1/4 bend, 2"		Q-1	16	1	Ea.	14.85	47		61.85	89
0120	3"			14	1.143		19.90	53.50		73.40	105
0140	4"			13	1.231		31	57.50		88.50	124
0160	5"		Q-2	18	1.333		40.50	65		105.50	145
0180	6"		"	17	1.412		54	68.50		122.50	166
0200	8"		Q-3	11	2.909		163	144		307	400
0340	1/8 bend, 2"		Q-1	16	1		10.55	47		57.55	84
0350	3"			14	1.143		16.60	53.50		70.10	101
0360	4"			13	1.231		24	57.50		81.50	116
0380	5"		Q-2	18	1.333		33.50	65		98.50	137
0400	6"		"	17	1.412		41	68.50		109.50	151
0420	8"		Q-3	11	2.909		123	144		267	360
0500	Sanitary tee, 2"		Q-1	10	1.600		21	75		96	139
0540	3"			9	1.778		33.50	83.50		117	166
0620	4"			8	2		41	93.50		134.50	190
0700	5"		Q-2	12	2		82	97		179	241
0800	6"		"	11	2.182		93	106		199	266
0880	8"		Q-3	7	4.571		246	227		473	620
0954	10" x 6"		"	8	4		330	198		528	665
5990	No hub										

22 13 Facility Sanitary Sewerage

22 13 16 – Sanitary Waste and Vent Piping

22 13 16.30 Pipe Fittings, Cast Iron		Crew	Daily Output	Labor-Hours	Unit	Material	2010 Bare Costs Labor	Equipment	Total	Total Incl O&P
6000	Cplg. & labor required at joints not incl. in fitting									
6010	price. Add 1 coupling per joint for installed price									
6020	1/4 Bend, 1-1/2"				Ea.	7.70			7.70	8.45
6060	2"					8.45			8.45	9.30
6080	3"					11.60			11.60	12.80
6120	4"					16.80			16.80	18.50
6184	1/4 Bend, long sweep, 1-1/2"					18.20			18.20	20
6186	2"					18.20			18.20	20
6188	3"					21.50			21.50	23.50
6189	4"					34.50			34.50	38
6190	5"					63.50			63.50	70
6191	6"					77.50			77.50	85
6192	8"					187			187	206
6193	10"					335			335	370
6200	1/8 Bend, 1-1/2"					6.45			6.45	7.10
6210	2"					7.20			7.20	7.95
6212	3"					9.75			9.75	10.70
6214	4"					12.30			12.30	13.55
6380	Sanitary Tee, tapped, 1-1/2"					14.20			14.20	15.65
6382	2" x 1-1/2"					12.80			12.80	14.10
6384	2"					14.35			14.35	15.80
6386	3" x 2"					20			20	22
6388	3"					37			37	40.50
6390	4" x 1-1/2"					17.70			17.70	19.50
6392	4" x 2"					20			20	22
6393	4"					20			20	22
6394	6" x 1-1/2"					44			44	48.50
6396	6" x 2"					45			45	49.50
6459	Sanitary Tee, 1-1/2"					10.70			10.70	11.80
6460	2"					11.60			11.60	12.80
6470	3"					14.20			14.20	15.65
6472	4"					26			26	28.50
8000	Coupling, standard (by CISPI Mfrs.)									
8020	1-1/2"	Q-1	48	.333	Ea.	14.45	15.60		30.05	40
8040	2"		44	.364		14.45	17.05		31.50	42.50
8080	3"		38	.421		17.25	19.75		37	49.50
8120	4"		33	.485		20.50	22.50		43	57.50
9000	Minimum labor/equipment charge	1 Plum	4	2	Job		104		104	161

22 13 16.60 Traps

		Crew	Daily Output	Labor-Hours	Unit	Material	Labor	Equipment	Total	Total Incl O&P
0010	**TRAPS**									
0030	Cast iron, service weight									
0050	Running P trap, without vent									
1100	2"	Q-1	16	1	Ea.	107	47		154	191
1150	4"	"	13	1.231		107	57.50		164.50	208
1160	6"	Q-2	17	1.412		465	68.50		533.50	620
3000	P trap, B&S, 2" pipe size	Q-1	16	1		25.50	47		72.50	101
3040	3" pipe size	"	14	1.143		38	53.50		91.50	125
3900										
4700	Copper, drainage, drum trap									
4840	3" x 6" swivel, 1-1/2" pipe size	1 Plum	16	.500	Ea.	201	26		227	262
5100	P trap, standard pattern									
5200	1-1/4" pipe size	1 Plum	18	.444	Ea.	93.50	23		116.50	139

22 13 Facility Sanitary Sewerage

22 13 16 – Sanitary Waste and Vent Piping

22 13 16.60 Traps

		Crew	Daily Output	Labor-Hours	Unit	Material	2010 Bare Costs Labor	2010 Bare Costs Equipment	Total	Total Incl O&P
5240	1-1/2" pipe size	1 Plum	17	.471	Ea.	90	24.50		114.50	137
5260	2" pipe size		15	.533		194	28		222	256
5280	3" pipe size		11	.727		335	38		373	430
9000	Minimum labor/equipment charge		3	2.667	Job		139		139	215

22 13 16.80 Vent Flashing and Caps

		Crew	Daily Output	Labor-Hours	Unit	Material	Labor	Equipment	Total	Total Incl O&P
0010	**VENT FLASHING AND CAPS**									
0120	Vent caps									
0140	Cast iron									
0180	2-1/2" - 3-5/8" pipe	1 Plum	21	.381	Ea.	40	19.85		59.85	74.50
0190	4" - 4-1/8" pipe	"	19	.421	"	48	22		70	87
0900	Vent flashing									
1000	Aluminum with lead ring									
1050	3" pipe	1 Plum	17	.471	Ea.	12.30	24.50		36.80	51.50
1060	4" pipe	"	16	.500	"	14.85	26		40.85	57
1350	Copper with neoprene ring									
1440	2" pipe	1 Plum	18	.444	Ea.	21	23		44	59
1450	3" pipe		17	.471		24.50	24.50		49	65
1460	4" pipe		16	.500		27	26		53	70.50
9000	Minimum labor/equipment charge		4	2	Job		104		104	161

22 13 19 – Sanitary Waste Piping Specialties

22 13 19.13 Sanitary Drains

		Crew	Daily Output	Labor-Hours	Unit	Material	Labor	Equipment	Total	Total Incl O&P
0010	**SANITARY DRAINS**									
0400	Deck, auto park, C.I., 13" top									
0440	3", 4", 5", and 6" pipe size	Q-1	8	2	Ea.	1,125	93.50		1,218.50	1,400
0480	For galvanized body, add				"	610			610	675
0500										
2000	Floor, medium duty, C.I., deep flange, 7" dia top									
2040	2" and 3" pipe size	Q-1	12	1.333	Ea.	158	62.50		220.50	271
2080	For galvanized body, add					75.50			75.50	83
2120	For polished bronze top, add					89			89	98
2500	Heavy duty, cleanout & trap w/bucket, C.I., 15" top									
2540	2", 3", and 4" pipe size	Q-1	6	2.667	Ea.	5,175	125		5,300	5,875
2560	For galvanized body, add					1,325			1,325	1,450
2580	For polished bronze top, add					570			570	630

22 13 19.14 Floor Receptors

		Crew	Daily Output	Labor-Hours	Unit	Material	Labor	Equipment	Total	Total Incl O&P
0010	**FLOOR RECEPTORS**, For connection to 2", 3" & 4" diameter pipe									
0200	12-1/2" square top, 25 sq in open area	Q-1	10	1.600	Ea.	765	75		840	955
0300	For grate with 4" diam. x 3-3/4" high funnel, add					137			137	150
0400	For grate with 6" diameter x 6" high funnel, add					176			176	193
0700	For acid-resisting bucket, add					211			211	233
0900	For stainless steel mesh bucket liner, add					170			170	187
2000	12-5/8" diameter top, 40 sq. in. open area	Q-1	10	1.600		605	75		680	780
2100	For options, add same prices as square top									
3000	8" x 4" rectangular top, 7.5 sq. in. open area	Q-1	14	1.143	Ea.	585	53.50		638.50	730
3100	For trap primer connections, add				"	69.50			69.50	76
9000	Minimum labor/equipment charge	Q-1	3	5.333	Job		250		250	385

22 13 23 – Sanitary Waste Interceptors

22 13 23.10 Interceptors

		Crew	Daily Output	Labor-Hours	Unit	Material	Labor	Equipment	Total	Total Incl O&P
0010	**INTERCEPTORS**									
0150	Grease, cast iron, 4 GPM, 8 lb. fat capacity	1 Plum	4	2	Ea.	935	104		1,039	1,175
0200	7 GPM, 14 lb. fat capacity		4	2		1,300	104		1,404	1,575

22 13 Facility Sanitary Sewerage

22 13 23 – Sanitary Waste Interceptors

22 13 23.10 Interceptors		Crew	Daily Output	Labor-Hours	Unit	Material	2010 Bare Costs Labor	Equipment	Total	Total Incl O&P
1000	10 GPM, 20 lb. fat capacity	1 Plum	4	2	Ea.	1,525	104		1,629	1,825
1160	100 GPM, 200 lb. fat capacity	Q-1	2	8		11,500	375		11,875	13,300
1240	300 GPM, 600 lb. fat capacity	"	1	16	↓	25,500	750		26,250	29,300
9000	Minimum labor/equipment charge	1 Plum	3	2.667	Job		139		139	215

22 13 29 – Sanitary Sewerage Pumps

22 13 29.14 Sewage Ejector Pumps

		Crew	Daily Output	Labor-Hours	Unit	Material	Labor	Equipment	Total	Total Incl O&P
0010	**SEWAGE EJECTOR PUMPS**, With operating and level controls									
0100	Simplex system incl. tank, cover, pump 15' head									
0500	37 gal PE tank, 12 GPM, 1/2 HP, 2" discharge	Q-1	3.20	5	Ea.	440	234		674	850
0510	3" discharge		3.10	5.161		480	242		722	900
0530	87 GPM, .7 HP, 2" discharge		3.20	5		670	234		904	1,100
0540	3" discharge		3.10	5.161		725	242		967	1,175
0600	45 gal. coated stl tank, 12 GPM, 1/2 HP, 2" discharge		3	5.333		780	250		1,030	1,250
0610	3" discharge		2.90	5.517		810	258		1,068	1,300
0630	87 GPM, .7 HP, 2" discharge		3	5.333		1,000	250		1,250	1,475
0640	3" discharge		2.90	5.517		1,050	258		1,308	1,550
0660	134 GPM, 1 HP, 2" discharge		2.80	5.714		1,075	268		1,343	1,625
0680	3" discharge		2.70	5.926		1,150	278		1,428	1,675
0700	70 gal. PE tank, 12 GPM, 1/2 HP, 2" discharge		2.60	6.154		855	288		1,143	1,375
0710	3" discharge		2.40	6.667		910	310		1,220	1,475
0730	87 GPM, 0.7 HP, 2" discharge		2.50	6.400		1,100	300		1,400	1,675
0740	3" discharge		2.30	6.957		1,150	325		1,475	1,775
0760	134 GPM, 1 HP, 2" discharge		2.20	7.273		1,200	340		1,540	1,825
0770	3" discharge		2	8	↓	1,275	375		1,650	1,975
9000	Minimum labor/equipment charge	↓	2.50	6.400	Job		300		300	465

22 14 Facility Storm Drainage

22 14 26 – Facility Storm Drains

22 14 26.13 Roof Drains

		Crew	Daily Output	Labor-Hours	Unit	Material	Labor	Equipment	Total	Total Incl O&P
0010	**ROOF DRAINS**									
0140	Cornice, C.I., 45° or 90° outlet									
0200	3" and 4" pipe size	Q-1	12	1.333	Ea.	259	62.50		321.50	380
0260	For galvanized body, add					59.50			59.50	65.50
0280	For polished bronze dome, add				↓	46			46	51
3860	Roof, flat metal deck, C.I. body, 12" C.I. dome									
3890	3" pipe size	Q-1	14	1.143	Ea.	310	53.50		363.50	425
3900	4" pipe size	"	13	1.231	"	310	57.50		367.50	430
4620	Main, all aluminum, 12" low profile dome									
4640	2", 3" and 4" pipe size	Q-1	14	1.143	Ea.	345	53.50		398.50	465
9000	Minimum labor/equipment charge	1 Plum	4	2	Job		104		104	161

22 14 29 – Sump Pumps

22 14 29.16 Submersible Sump Pumps

		Crew	Daily Output	Labor-Hours	Unit	Material	Labor	Equipment	Total	Total Incl O&P
0010	**SUBMERSIBLE SUMP PUMPS**									
7000	Sump pump, automatic									
7100	Plastic, 1-1/4" discharge, 1/4 HP	1 Plum	6	1.333	Ea.	215	69.50		284.50	345
7140	1/3 HP		5	1.600		175	83.50		258.50	320
7180	1-1/2" discharge, 1/2 HP		4	2		238	104		342	425
7500	Cast iron, 1-1/4" discharge, 1/4 HP		6	1.333		166	69.50		235.50	290
7540	1/3 HP		6	1.333		195	69.50		264.50	325
7560	1/2 HP	↓	5	1.600	↓	236	83.50		319.50	390

22 14 Facility Storm Drainage

22 14 29 – Sump Pumps

22 14 29.16 Submersible Sump Pumps

		Crew	Daily Output	Labor-Hours	Unit	Material	2010 Bare Costs Labor	Equipment	Total	Total Incl O&P
9000	Minimum labor/equipment charge	1 Plum	4	2	Job		104		104	161

22 15 General Service Compressed-Air Systems

22 15 19 – General Service Packaged Air Compressors and Receivers

22 15 19.10 Air Compressors

		Crew	Daily Output	Labor-Hours	Unit	Material	Labor	Equipment	Total	Total Incl O&P
0010	**AIR COMPRESSORS**									
5250	Air, reciprocating air cooled, splash lubricated, tank mounted									
5300	Single stage, 1 phase, 140 psi									
5303	1/2 HP, 30 gal tank	1 Stpi	3	2.667	Ea.	1,600	138		1,738	2,000
5305	3/4 HP, 30 gal tank		2.60	3.077		1,650	160		1,810	2,075
5307	1 HP, 30 gal tank		2.20	3.636		2,000	189		2,189	2,500

22 31 Domestic Water Softeners

22 31 13 – Residential Domestic Water Softeners

22 31 13.10 Residential Water Softeners

		Crew	Daily Output	Labor-Hours	Unit	Material	Labor	Equipment	Total	Total Incl O&P
0010	**RESIDENTIAL WATER SOFTENERS**									
7350	Water softener, automatic, to 30 grains per gallon	2 Plum	5	3.200	Ea.	380	167		547	680
7400	To 100 grains per gallon	"	4	4	"	750	208		958	1,150

22 33 Electric Domestic Water Heaters

22 33 13 – Instantaneous Electric Domestic Water Heaters

22 33 13.20 Instantaneous Electric Point-Of-use Water Heaters

		Crew	Daily Output	Labor-Hours	Unit	Material	Labor	Equipment	Total	Total Incl O&P
0010	**INSTANTANEOUS ELECTRIC POINT-OF-USE WATER HEATERS**									
8965	Point of use, electric, glass lined									
8969	Energy saver									
8970	2.5 gal. single element	1 Plum	2.80	2.857	Ea.	225	149		374	475
8971	4 gal. single element		2.80	2.857		340	149		489	605
8974	6 gal. single element		2.50	3.200		365	167		532	660
8975	10 gal. single element		2.50	3.200		385	167		552	685
8976	15 gal. single element		2.40	3.333		405	174		579	715
8977	20 gal. single element		2.40	3.333		440	174		614	755
8978	30 gal. single element		2.30	3.478		490	181		671	820
8979	40 gal. single element		2.20	3.636		615	189		804	970
8988	Commercial (ASHRAE energy std. 90)									
8989	6 gallon	1 Plum	2.50	3.200	Ea.	650	167		817	975
8990	10 gallon		2.50	3.200		690	167		857	1,025
8991	15 gallon		2.40	3.333		735	174		909	1,075
8992	20 gallon		2.40	3.333		785	174		959	1,125
8993	30 gallon		2.30	3.478		1,650	181		1,831	2,100
8995	Under the sink, copper, w/bracket									
8996	2.5 gallon	1 Plum	4	2	Ea.	335	104		439	525
9000	Minimum labor/equipment charge	"	1.75	4.571	Job		238		238	370

22 33 30 – Residential, Electric Domestic Water Heaters

22 33 30.13 Residential, Small-Capacity Elec. Water Heaters

		Crew	Daily Output	Labor-Hours	Unit	Material	Labor	Equipment	Total	Total Incl O&P
0010	**RESIDENTIAL, SMALL-CAPACITY ELECTRIC DOMESTIC WATER HEATERS**									
0050										
1000	Residential, electric, glass lined tank, 5 yr, 10 gal., single element	1 Plum	2.30	3.478	Ea.	385	181		566	705

22 33 Electric Domestic Water Heaters

22 33 30 – Residential, Electric Domestic Water Heaters

22 33 30.13 Residential, Small-Capacity Elec. Water Heaters		Crew	Daily Output	Labor-Hours	Unit	Material	2010 Bare Costs Labor	Equipment	Total	Total Incl O&P
1060	30 gallon, double element	1 Plum	2.20	3.636	Ea.	575	189		764	930
1080	40 gallon, double element		2	4		615	208		823	1,000
1100	52 gallon, double element		2	4		690	208		898	1,075
1120	66 gallon, double element		1.80	4.444		930	231		1,161	1,375
1140	80 gallon, double element		1.60	5		1,025	260		1,285	1,550

22 33 33 – Light-Commercial Electric Domestic Water Heaters

22 33 33.10 Commercial Electric Water Heaters

0010	**COMMERCIAL ELECTRIC WATER HEATERS**									
4000	Commercial, 100° rise. NOTE: for each size tank, a range of									
4010	heaters between the ones shown are available									
4020	Electric									
4100	5 gal., 3 kW, 12 GPH, 208V	1 Plum	2	4	Ea.	2,725	208		2,933	3,325
4160	50 gal., 36 kW, 148 GPH, 208V	"	1.80	4.444		6,350	231		6,581	7,350
4480	400 gal., 210 kW, 860 GPH, 480V	Q-1	1	16		38,000	750		38,750	43,000

22 34 Fuel-Fired Domestic Water Heaters

22 34 13 – Instantaneous, Tankless, Gas Domestic Water Heaters

22 34 13.10 Instantaneous, Tankless, Gas Water Heaters

0010	**INSTANTANEOUS, TANKLESS, GAS WATER HEATERS**									
9410	Natural gas/propane, incl vent, 3.2 GPM [G]	1 Plum	2	4	Ea.	535	208		743	915

22 34 30 – Residential Gas Domestic Water Heaters

22 34 30.13 Residential, Atmos, Gas Domestic Wtr Heaters

0010	**RESIDENTIAL, ATMOSPHERIC, GAS DOMESTIC WATER HEATERS**									
2000	Gas fired, foam lined tank, 10 yr, vent not incl.									
2040	30 gallon	1 Plum	2	4	Ea.	895	208		1,103	1,300
2060	40 gallon		1.90	4.211		895	219		1,114	1,325
2100	75 gallon		1.50	5.333		1,225	278		1,503	1,775

22 34 36 – Commercial Gas Domestic Water Heaters

22 34 36.13 Commercial, Atmos., Gas Domestic Water Htrs.

0010	**COMMERCIAL, ATMOSPHERIC, GAS DOMESTIC WATER HEATERS**									
6000	Gas fired, flush jacket, std. controls, vent not incl.									
6040	75 MBH input, 73 GPH	1 Plum	1.40	5.714	Ea.	1,575	297		1,872	2,175
6060	98 MBH input, 95 GPH		1.40	5.714		4,575	297		4,872	5,500
6180	200 MBH input, 192 GPH		.60	13.333		7,650	695		8,345	9,475

22 34 46 – Oil-Fired Domestic Water Heaters

22 34 46.10 Residential Oil-Fired Water Heaters

0010	**RESIDENTIAL OIL-FIRED WATER HEATERS**									
3000	Oil fired, glass lined tank, 5 yr, vent not included, 30 gallon	1 Plum	2	4	Ea.	665	208		873	1,050
3040	50 gallon	"	1.80	4.444	"	1,050	231		1,281	1,500

22 34 46.20 Commercial Oil-Fired Water Heaters

0010	**COMMERCIAL OIL-FIRED WATER HEATERS**									
8000	Oil fired, glass lined, UL listed, std. controls, vent not incl.									
8060	140 gal., 140 MBH input, 134 GPH	Q-1	2.13	7.512	Ea.	14,900	350		15,250	16,900
8080	140 gal., 199 MBH input, 191 GPH		2	8		15,400	375		15,775	17,500
8160	140 gal., 540 MBH input, 519 GPH		.96	16.667		21,000	780		21,780	24,300
8280	201 gal., 1250 MBH input, 1200 GPH	Q-2	1.22	19.672		32,900	955		33,855	37,600

22 41 Residential Plumbing Fixtures

22 41 13 – Residential Water Closets, Urinals, and Bidets

22 41 13.40 Water Closets

		Crew	Daily Output	Labor-Hours	Unit	Material	2010 Bare Costs Labor	Equipment	Total	Total Incl O&P
0010	**WATER CLOSETS** R224000-40									
0032	For automatic flush, see Div. 22 42 39.10 0972									
0150	Tank type, vitreous china, incl. seat, supply pipe w/stop									
0200	Wall hung									
0400	Two piece, close coupled	Q-1	5.30	3.019	Ea.	595	141		736	875
0960	For rough-in, supply, waste, vent and carrier		2.73	5.861		910	275		1,185	1,425
1000	Floor mounted		5.30	3.019		540	141		681	810
1100	Two piece, close coupled		5.30	3.019		258	141		399	505
1110	Two piece, close coupled, dual flush		5.30	3.019		256	141		397	500
1140	Two piece, close coupled, 1.28 gpf, ADA [G]		5.30	3.019		295	141		436	545
1960	For color, add					30%				
1980	For rough-in, supply, waste and vent	Q-1	3.05	5.246	Ea.	283	246		529	690

22 41 16 – Residential Lavatories and Sinks

22 41 16.10 Lavatories

		Crew	Daily Output	Labor-Hours	Unit	Material	Labor	Equipment	Total	Total Incl O&P
0010	**LAVATORIES**, With trim, white unless noted otherwise R224000-40									
0020										
0500	Vanity top, porcelain enamel on cast iron									
0600	20" x 18"	Q-1	6.40	2.500	Ea.	260	117		377	465
0640	33" x 19" oval	"	6.40	2.500	"	610	117		727	855
0860	For color, add					25%				
1000	Cultured marble, 19" x 17", single bowl	Q-1	6.40	2.500	Ea.	190	117		307	390
1120	25" x 22", single bowl		6.40	2.500		238	117		355	445
1900	Stainless steel, self-rimming, 25" x 22", single bowl, ledge		6.40	2.500		345	117		462	560
1960	17" x 22", single bowl		6.40	2.500		340	117		457	550
2600	Steel, enameled, 20" x 17", single bowl		5.80	2.759		184	129		313	405
2900	Vitreous china, 20" x 16", single bowl		5.40	2.963		278	139		417	520
2960	20" x 17", single bowl		5.40	2.963		194	139		333	430
3580	Rough-in, supply, waste and vent for all above lavatories		2.30	6.957		252	325		577	780
4000	Wall hung									
4040	Porcelain enamel on cast iron, 16" x 14", single bowl	Q-1	8	2	Ea.	455	93.50		548.50	645
4180	20" x 18", single bowl	"	8	2	"	355	93.50		448.50	535
4580	For color, add					30%				
6000	Vitreous china, 18" x 15", single bowl with backsplash	Q-1	7	2.286	Ea.	249	107		356	440
6960	Rough-in, supply, waste and vent for above lavatories	"	1.66	9.639	"	385	450		835	1,125
7000	Pedestal type									
7600	Vitreous china, 27" x 21", white	Q-1	6.60	2.424	Ea.	565	114		679	795
7610	27" x 21", colored		6.60	2.424		675	114		789	915
7620	27" x 21", premium color		6.60	2.424		740	114		854	985
7660	26" x 20", white		6.60	2.424		515	114		629	740
7670	26" x 20", colored		6.60	2.424		610	114		724	845
7680	26" x 20", premium color		6.60	2.424		670	114		784	910
7700	24" x 18", white		6.60	2.424		480	114		594	705
7710	24" x 18", colored		6.60	2.424		570	114		684	800
7720	24" x 18", premium color		6.60	2.424		625	114		739	860
7760	21" x 18", white		6.60	2.424		233	114		347	430
7770	21" x 18", colored		6.60	2.424		250	114		364	450
7990	Rough-in, supply, waste and vent for pedestal lavatories		1.66	9.639		385	450		835	1,125
9000	Minimum labor/equipment charge	1 Plum	3	2.667	Job		139		139	215

22 41 16.30 Sinks

		Crew	Daily Output	Labor-Hours	Unit	Material	Labor	Equipment	Total	Total Incl O&P
0010	**SINKS**, With faucets and drain									
2000	Kitchen, counter top style, P.E. on C.I., 24" x 21" single bowl	Q-1	5.60	2.857	Ea.	257	134		391	490
2100	31" x 22" single bowl		5.60	2.857		300	134		434	535

22 41 Residential Plumbing Fixtures

22 41 16 – Residential Lavatories and Sinks

22 41 16.30 Sinks

		Crew	Daily Output	Labor-Hours	Unit	Material	2010 Bare Costs Labor	Equipment	Total	Total Incl O&P
2200	32" x 21" double bowl	Q-1	4.80	3.333	Ea.	385	156		541	665
3000	Stainless steel, self rimming, 19" x 18" single bowl		5.60	2.857		555	134		689	815
3100	25" x 22" single bowl		5.60	2.857		615	134		749	885
3200	33" x 22" double bowl		4.80	3.333		895	156		1,051	1,225
3300	43" x 22" double bowl		4.80	3.333		1,050	156		1,206	1,400
4000	Steel, enameled, with ledge, 24" x 21" single bowl		5.60	2.857		180	134		314	405
4100	32" x 21" double bowl		4.80	3.333		207	156		363	470
4960	For color sinks except stainless steel, add					10%				
4980	For rough-in, supply, waste and vent, counter top sinks	Q-1	2.14	7.477		272	350		622	845
5790	For rough-in, supply, waste & vent, sinks	"	1.85	8.649		272	405		677	930

22 41 19 – Residential Bathtubs

22 41 19.10 Baths

			Crew	Daily Output	Labor-Hours	Unit	Material	Labor	Equipment	Total	Total Incl O&P
0010	**BATHS**	R224000-40									
0100	Tubs, recessed porcelain enamel on cast iron, with trim										
0180	48" x 42"		Q-1	4	4	Ea.	1,650	187		1,837	2,125
0220	72" x 36"		"	3	5.333	"	2,225	250		2,475	2,825
0300	Mat bottom										
0380	5' long		Q-1	4.40	3.636	Ea.	945	170		1,115	1,325
0560	Corner 48" x 44"			4.40	3.636		2,150	170		2,320	2,650
2000	Enameled formed steel, 4'-6" long			5.80	2.759		385	129		514	625
4600	Module tub & showerwall surround, molded fiberglass										
4610	5' long x 34" wide x 76" high		Q-1	4	4	Ea.	755	187		942	1,125
4750	Handicap with 1-1/2" OD grab bar, antiskid bottom										
4760	60" x 32-3/4" x 72" high		Q-1	4	4	Ea.	750	187		937	1,125
4770	60" x 30" x 71" high with molded seat			3.50	4.571		740	214		954	1,150
9600	Rough-in, supply, waste and vent, for all above tubs, add			2.07	7.729		315	360		675	905
9900	Minimum labor/equipment charge			3	5.333	Job		250		250	385

22 41 23 – Residential Shower Receptors and Basins

22 41 23.20 Showers

		Crew	Daily Output	Labor-Hours	Unit	Material	Labor	Equipment	Total	Total Incl O&P
0011	**SHOWERS**, Stall, with drain only									
0020										
1520	32" square	Q-1	5	3.200	Ea.	885	150		1,035	1,200
1530	36" square		4.80	3.333		2,175	156		2,331	2,625
1540	Terrazzo receptor, 32" square		5	3.200		1,300	150		1,450	1,650
1580	36" corner angle		4.80	3.333		1,675	156		1,831	2,075
3000	Fiberglass, one piece, with 3 walls, 32" x 32" square		5.50	2.909		515	136		651	780
3100	36" x 36" square		5.50	2.909		530	136		666	795
4200	Rough-in, supply, waste and vent for above showers		2.05	7.805		390	365		755	995

22 41 23.40 Shower System Components

			Crew	Daily Output	Labor-Hours	Unit	Material	Labor	Equipment	Total	Total Incl O&P
0010	**SHOWER SYSTEM COMPONENTS**										
5500	Head, water economizer	G	1 Plum	24	.333	Ea.	49.50	17.35		66.85	81.50

22 41 36 – Residential Laundry Trays

22 41 36.10 Laundry Sinks

		Crew	Daily Output	Labor-Hours	Unit	Material	Labor	Equipment	Total	Total Incl O&P
0010	**LAUNDRY SINKS**, With trim									
0020	Porcelain enamel on cast iron, black iron frame									
0050	24" x 21", single compartment	Q-1	6	2.667	Ea.	465	125		590	705
0100	26" x 21", single compartment	"	6	2.667	"	450	125		575	690
2000	Molded stone, on wall hanger or legs									
2020	22" x 23", single compartment	Q-1	6	2.667	Ea.	146	125		271	355
2100	45" x 21", double compartment	"	5	3.200	"	259	150		409	515
3000	Plastic, on wall hanger or legs									

22 41 Residential Plumbing Fixtures

22 41 36 – Residential Laundry Trays

22 41 36.10 Laundry Sinks		Crew	Daily Output	Labor-Hours	Unit	Material	2010 Bare Costs Labor	Equipment	Total	Total Incl O&P
3020	18" x 23", single compartment	Q-1	6.50	2.462	Ea.	141	115		256	335
3100	20" x 24", single compartment		6.50	2.462		141	115		256	335
3200	36" x 23", double compartment		5.50	2.909		167	136		303	395
3300	40" x 24", double compartment		5.50	2.909		248	136		384	485
5000	Stainless steel, counter top, 22" x 17" single compartment		6	2.667		61.50	125		186.50	262
5200	33" x 22", double compartment		5	3.200		73.50	150		223.50	315
9600	Rough-in, supply, waste and vent, for all laundry sinks		2.14	7.477		272	350		622	845
9810	Minimum labor/equipment charge	1 Plum	3	2.667	Job		139		139	215

22 41 39 – Residential Faucets, Supplies and Trim

22 41 39.10 Faucets and Fittings

		Crew	Daily Output	Labor-Hours	Unit	Material	Labor	Equipment	Total	Total Incl O&P
0010	**FAUCETS AND FITTINGS**									
0150	Bath, faucets, diverter spout combination, sweat	1 Plum	8	1	Ea.	86.50	52		138.50	176
0200	For integral stops, IPS unions, add					109			109	120
0420	Bath, press-bal mix valve w/diverter, spout, shower hd, arm/flange	1 Plum	8	1		155	52		207	252
0500	Drain, central lift, 1-1/2" IPS male		20	.400		44.50	21		65.50	81
0600	Trip lever, 1-1/2" IPS male		20	.400		45	21		66	82
0810	Bidet									
0812	Fitting, over the rim, swivel spray/pop-up drain	1 Plum	8	1	Ea.	212	52		264	315
1000	Kitchen sink faucets, top mount, cast spout		10	.800		61.50	41.50		103	132
1100	For spray, add		24	.333		16.15	17.35		33.50	45
1200	Wall type, swing tube spout		10	.800		74.50	41.50		116	146
1300	Single control lever handle									
1310	With pull out spray									
1320	Polished chrome	1 Plum	10	.800	Ea.	157	41.50		198.50	237
2000	Laundry faucets, shelf type, IPS or copper unions		12	.667		49.50	34.50		84	109
2100	Lavatory faucet, centerset, without drain		10	.800		44.50	41.50		86	114
2120	With pop-up drain		6.66	1.201		80.50	62.50		143	186
2210	Porcelain cross handles and pop-up drain									
2220	Polished chrome	1 Plum	6.66	1.201	Ea.	126	62.50		188.50	236
2230	Polished brass	"	6.66	1.201	"	189	62.50		251.50	305
2260	Single lever handle and pop-up drain									
2270	Black nickel	1 Plum	6.66	1.201	Ea.	220	62.50		282.50	340
2280	Polished brass		6.66	1.201		220	62.50		282.50	340
2290	Polished chrome		6.66	1.201		172	62.50		234.50	286
2800	Self-closing, center set		10	.800		131	41.50		172.50	209
2810	Automatic sensor and operator, with faucet head G		6.15	1.301		370	67.50		437.50	515
4000	Shower by-pass valve with union		18	.444		85.50	23		108.50	130
4100	Shower arm with flange and head		22	.364		71.50	18.95		90.45	108
4140	Shower, hand held, pin mount, massage action, chrome		22	.364		101	18.95		119.95	141
4142	Polished brass		22	.364		119	18.95		137.95	160
4144	Shower, hand held, wall mtd, adj. spray, 2 wall mounts, chrome		20	.400		147	21		168	195
4146	Polished brass		20	.400		251	21		272	310
4148	Shower, hand held head, bar mounted 24", adj. spray, chrome		20	.400		194	21		215	246
4150	Polished brass		20	.400		385	21		406	455
4200	Shower thermostatic mixing valve, concealed		8	1		365	52		417	485
4204	For inlet strainer, check, and stops, add					45			45	49.50
4220	Shower pressure balancing mixing valve,									
4230	With shower head, arm, flange and diverter tub spout									
4240	Chrome	1 Plum	6.14	1.303	Ea.	222	68		290	350
4250	Polished brass		6.14	1.303		279	68		347	410
4260	Satin		6.14	1.303		279	68		347	410
4270	Polished chrome/brass		6.14	1.303		229	68		297	355

22 41 Residential Plumbing Fixtures

22 41 39 – Residential Faucets, Supplies and Trim

22 41 39.10 Faucets and Fittings		Crew	Daily Output	Labor-Hours	Unit	Material	2010 Bare Costs Labor	Equipment	Total	Total Incl O&P
5000	Sillcock, compact, brass, IPS or copper to hose	1 Plum	24	.333	Ea.	8.05	17.35		25.40	36
6000	Stop and waste valves, bronze									
6100	Angle, solder end 1/2"	1 Plum	24	.333	Ea.	8.85	17.35		26.20	36.50
6110	3/4"		20	.400		10.35	21		31.35	44
6300	Straightway, solder end 3/8"		24	.333		6.15	17.35		23.50	34
6310	1/2"		24	.333		2.81	17.35		20.16	30
6410	Straightway, threaded 1/2"		24	.333		3.82	17.35		21.17	31
6420	3/4"		20	.400		6.35	21		27.35	39.50
6430	1"		19	.421		12.60	22		34.60	48
7800	Water closet, wax gasket		96	.083		1.42	4.34		5.76	8.25
7820	Gasket toilet tank to bowl		32	.250		1.24	13		14.24	21.50
7830	Replacement diaphragm washer assy for ballcock valve		12	.667		1.24	34.50		35.74	55.50
7850	Dual flush valve		12	.667		25	34.50		59.50	81.50
8000	Water supply stops, polished chrome plate									
8200	Angle, 3/8"	1 Plum	24	.333	Ea.	7.95	17.35		25.30	36
8300	1/2"		22	.364		7.95	18.95		26.90	38.50
8400	Straight, 3/8"		26	.308		8.35	16		24.35	34
8500	1/2"		24	.333		8.35	17.35		25.70	36
8600	Water closet, angle, w/flex riser, 3/8"		24	.333		30.50	17.35		47.85	60.50
9000	Minimum labor/equipment charge		4	2	Job		104		104	161

22 42 Commercial Plumbing Fixtures

22 42 13 – Commercial Water Closets, Urinals, and Bidets

22 42 13.30 Urinals

		Crew	Daily Output	Labor-Hours	Unit	Material	Labor	Equipment	Total	Total Incl O&P
0010	**URINALS**									
3000	Wall hung, vitreous china, with hanger & self-closing valve									
3100	Siphon jet type	Q-1	3	5.333	Ea.	293	250		543	710
3120	Blowout type		3	5.333		435	250		685	865
3140	Water saving .5 gpf G		3	5.333		715	250		965	1,175
3300	Rough-in, supply, waste & vent		2.83	5.654		240	265		505	675
5000	Stall type, vitreous china, includes valve		2.50	6.400		665	300		965	1,200
6980	Rough-in, supply, waste and vent		1.99	8.040		435	375		810	1,075
9000	Minimum labor/equipment charge		4	4	Job		187		187	290

22 42 13.40 Water Closets

		Crew	Daily Output	Labor-Hours	Unit	Material	Labor	Equipment	Total	Total Incl O&P
0010	**WATER CLOSETS**									
3000	Bowl only, with flush valve, seat, 1.6 gpf unless noted									
3100	Wall hung	Q-1	5.80	2.759	Ea.	305	129		434	535
3200	For rough-in, supply, waste and vent, single WC		2.56	6.250		950	293		1,243	1,500
3360	With floor outlet, 1.28 gpf G		5.80	2.759		565	129		694	820
3362	With floor outlet, 1.28 gpf, ADA G		5.80	2.759		595	129		724	855
9000	Minimum labor/equipment charge		4	4	Job		187		187	290

22 42 16 – Commercial Lavatories and Sinks

22 42 16.10 Handwasher-Dryer Module

		Crew	Daily Output	Labor-Hours	Unit	Material	Labor	Equipment	Total	Total Incl O&P
0010	**HANDWASHER-DRYER MODULE**									
0030	Wall mounted									
0040	With electric dryer									
0050	Sensor operated	Q-1	8	2	Ea.	3,650	93.50		3,743.50	4,150
0110	Sensor operated (ADA)		8	2		3,650	93.50		3,743.50	4,175
0140	Sensor operated (ADA), surface mounted		8	2		4,400	93.50		4,493.50	5,000
0150	With paper towels									

22 42 Commercial Plumbing Fixtures

22 42 16 – Commercial Lavatories and Sinks

22 42 16.10 Handwasher-Dryer Module

		Crew	Daily Output	Labor-Hours	Unit	Material	2010 Bare Costs Labor	Equipment	Total	Total Incl O&P
0180	Sensor operated	Q-1	8	2	Ea.	3,550	93.50		3,643.50	4,050

22 42 16.14 Lavatories

| 0010 | **LAVATORIES**, With trim, white unless noted otherwise | | | | | | | | | |
| 0020 | Commercial lavatories same as residential. See Div. 22 41 16.10 | | | | | | | | | |

22 42 16.40 Service Sinks

0010	**SERVICE SINKS**									
6650	Service, floor, corner, P.E. on C.I., 28" x 28"	Q-1	4.40	3.636	Ea.	695	170		865	1,025
6790	For rough-in, supply, waste & vent, floor service sinks		1.64	9.756	"	730	455		1,185	1,500
9000	Minimum labor/equipment charge	↓	4	4	Job		187		187	290

22 42 23 – Commercial Shower Receptors and Basins

22 42 23.30 Group Showers

| 0010 | **GROUP SHOWERS** | | | | | | | | | |
| 9000 | Minimum labor/equipment charge | Q-1 | 4 | 4 | Job | | 187 | | 187 | 290 |

22 42 33 – Wash Fountains

22 42 33.20 Commercial Wash Fountains

0010	**COMMERCIAL WASH FOUNTAINS**									
1900	Group, foot control									
2000	Precast terrazzo, circular, 36" diam., 5 or 6 persons	Q-2	3	8	Ea.	4,200	390		4,590	5,200
2100	54" diameter for 8 or 10 persons		2.50	9.600		5,225	465		5,690	6,450
2400	Semi-circular, 36" diam. for 3 persons		3	8		3,850	390		4,240	4,850
2500	54" diam. for 4 or 5 persons	↓	2.50	9.600	↓	4,650	465		5,115	5,850
5610	Group, infrared control, barrier free									
5614	Precast terrazzo									
5620	Semi-circular 36" diam. for 3 persons	Q-2	3	8	Ea.	5,875	390		6,265	7,075
5630	46" diam. for 4 persons		2.80	8.571		6,425	415		6,840	7,700
5640	Circular, 54" diam. for 8 persons, button control	↓	2.50	9.600		7,700	465		8,165	9,200
5700	Rough-in, supply, waste and vent for above wash fountains	Q-1	1.82	8.791	↓	420	410		830	1,100
9000	Minimum labor/equipment charge	Q-2	3	8	Job		390		390	600

22 42 39 – Commercial Faucets, Supplies, and Trim

22 42 39.10 Faucets and Fittings

0010	**FAUCETS AND FITTINGS**									
0840	Flush valves, with vacuum breaker									
0850	Water closet									
0860	Exposed, rear spud	1 Plum	8	1	Ea.	172	52		224	271
0870	Top spud		8	1		161	52		213	258
0880	Concealed, rear spud		8	1		180	52		232	279
0890	Top spud		8	1		149	52		201	245
0900	Wall hung		8	1		159	52		211	256
0910	Dual flush flushometer		12	.667		130	34.50		164.50	197
0912	Flushometer retrofit kit	↓	18	.444	↓	50	23		73	91
0920	Urinal									
0930	Exposed, stall	1 Plum	8	1	Ea.	161	52		213	258
0940	Wall, (washout)		8	1		161	52		213	258
0950	Pedestal, top spud		8	1		161	52		213	258
0960	Concealed, stall		8	1		145	52		197	240
0970	Wall (washout)	↓	8	1	↓	156	52		208	253
0971	Automatic flush sensor and operator for [G]									
0972	urinals or water closets, standard [G]	1 Plum	8	1	Ea.	415	52		467	540
0980	High efficiency water saving									
0984	Water closets, 1.28 gpf [G]	1 Plum	8	1	Ea.	395	52		447	515

22 42 Commercial Plumbing Fixtures

22 42 39 – Commercial Faucets, Supplies, and Trim

22 42 39.10 Faucets and Fittings

		Crew	Daily Output	Labor-Hours	Unit	Material	2010 Bare Costs Labor	Equipment	Total	Total Incl O&P
0988	Urinals, .5 gpf [G]	1 Plum	8	1	Ea.	395	52		447	515
2790	Faucets for lavatories									
2800	Self-closing, center set	1 Plum	10	.800	Ea.	131	41.50		172.50	209
2810	Automatic sensor and operator, with faucet head		6.15	1.301		370	67.50		437.50	515
3000	Service sink faucet, cast spout, pail hook, hose end		14	.571		80	29.50		109.50	134

22 42 39.30 Carriers and Supports

		Crew	Daily Output	Labor-Hours	Unit	Material	Labor	Equipment	Total	Total Incl O&P
0010	**CARRIERS AND SUPPORTS**, For plumbing fixtures									
0020										
3000	Lavatory, concealed arm									
3050	Floor mounted, single									
3100	High back fixture	1 Plum	6	1.333	Ea.	395	69.50		464.50	540
3200	Flat slab fixture		6	1.333		340	69.50		409.50	485
3220	Paraplegic		6	1.333		440	69.50		509.50	595
6980	Water closet, siphon jet									
7000	Horizontal, adjustable, caulk									
7040	Single, 4" pipe size	1 Plum	5.33	1.501	Ea.	755	78		833	950
7060	5" pipe size		5.33	1.501		955	78		1,033	1,175
7100	Double, 4" pipe size		5	1.600		1,100	83.50		1,183.50	1,350
7120	5" pipe size		5	1.600		1,600	83.50		1,683.50	1,875
8200	Water closet, residential									
8220	Vertical centerline, floor mount									
8240	Single, 3" caulk, 2" or 3" vent	1 Plum	6	1.333	Ea.	455	69.50		524.50	610
8260	4" caulk, 2" or 4" vent		6	1.333	"	575	69.50		644.50	745
9990	Minimum labor/equipment charge		3.50	2.286	Job		119		119	184

22 43 Healthcare Plumbing Fixtures

22 43 39 – Healthcare Faucets, Supplies, and Trim

22 43 39.10 Faucets and Fittings

		Crew	Daily Output	Labor-Hours	Unit	Material	Labor	Equipment	Total	Total Incl O&P
0010	**FAUCETS AND FITTINGS**									
2850	Medical, bedpan cleanser, with pedal valve,	1 Plum	12	.667	Ea.	760	34.50		794.50	895
2860	With screwdriver stop valve		12	.667		395	34.50		429.50	490
2870	With self-closing spray valve		12	.667		243	34.50		277.50	320
2900	Faucet, gooseneck spout, wrist handles, grid drain		10	.800		134	41.50		175.50	213
2940	Mixing valve, knee action, screwdriver stops		4	2		425	104		529	630

22 45 Emergency Plumbing Fixtures

22 45 13 – Emergency Showers

22 45 13.10 Emergency Showers

		Crew	Daily Output	Labor-Hours	Unit	Material	Labor	Equipment	Total	Total Incl O&P
0010	**EMERGENCY SHOWERS**, Rough-in not included									
0020										
5000	Shower, single head, drench, ball valve, pull, freestanding	Q-1	4	4	Ea.	340	187		527	665
5200	Horizontal or vertical supply		4	4		490	187		677	830
6000	Multi-nozzle, eye/face wash combination		4	4		615	187		802	965
6400	Multi-nozzle, 12 spray, shower only		4	4		1,600	187		1,787	2,050
6600	For freeze-proof, add		6	2.667		365	125		490	600
9000	Minimum labor/equipment charge		3	5.333	Job		250		250	385

22 45 Emergency Plumbing Fixtures

22 45 16 – Eyewash Equipment

22 45 16.10 Eyewash Safety Equipment	Crew	Daily Output	Labor-Hours	Unit	Material	2010 Bare Costs Labor	Equipment	Total	Total Incl O&P
0010 **EYEWASH SAFETY EQUIPMENT**, Rough-in not included									
1000 Eye wash fountain									
1400 Plastic bowl, pedestal mounted	Q-1	4	4	Ea.	262	187		449	580

22 47 Drinking Fountains and Water Coolers

22 47 13 – Drinking Fountains

22 47 13.10 Drinking Water Fountains

		Crew	Daily Output	Labor-Hours	Unit	Material	Labor	Equipment	Total	Total Incl O&P
0010	**DRINKING WATER FOUNTAINS,** For connection to cold water supply									
1000	Wall mounted, non-recessed									
2700	Stainless steel, single bubbler, no back	1 Plum	4	2	Ea.	920	104		1,024	1,150
2740	With back		4	2		970	104		1,074	1,225
2780	Dual handle & wheelchair projection type		4	2		670	104		774	900
2820	Dual level for handicapped type		3.20	2.500		1,350	130		1,480	1,700
3980	For rough-in, supply and waste, add		2.21	3.620		190	188		378	500
4000	Wall mounted, semi-recessed									
4200	Poly-marble, single bubbler	1 Plum	4	2	Ea.	920	104		1,024	1,175
4600	Stainless steel, satin finish, single bubbler	"	4	2	"	730	104		834	960
6000	Wall mounted, fully recessed									
6400	Poly-marble, single bubbler	1 Plum	4	2	Ea.	1,525	104		1,629	1,825
6800	Stainless steel, single bubbler		4	2		1,250	104		1,354	1,525
7580	For rough-in, supply and waste, add		1.83	4.372		190	228		418	565
7590										
7600	Floor mounted, pedestal type									
8600	Enameled iron, heavy duty service, 2 bubblers	1 Plum	2	4	Ea.	2,400	208		2,608	2,950
8880	For freeze-proof valve system, add		2	4		385	208		593	750
8900	For rough-in, supply and waste, add		1.83	4.372		190	228		418	565
9000	Minimum labor/equipment charge		2	4	Job		208		208	325

22 47 16 – Pressure Water Coolers

22 47 16.10 Electric Water Coolers

		Crew	Daily Output	Labor-Hours	Unit	Material	Labor	Equipment	Total	Total Incl O&P
0010	**ELECTRIC WATER COOLERS**									
0100	Wall mounted, non-recessed									
0140	4 GPH	Q-1	4	4	Ea.	575	187		762	920
0160	8 GPH, barrier free, sensor operated		4	4		830	187		1,017	1,200
1000	Dual height, 8.2 GPH		3.80	4.211		1,975	197		2,172	2,475
1040	14.3 GPH		3.80	4.211		1,075	197		1,272	1,475
3300	Semi-recessed, 8.1 GPH		4	4		600	187		787	950
4600	Floor mounted, flush-to-wall									
4640	4 GPH	1 Plum	3	2.667	Ea.	715	139		854	1,000
4980	For stainless steel cabinet, add					128			128	141
5000	Dual height, 8.2 GPH	1 Plum	2	4		1,000	208		1,208	1,425
9000	Minimum labor/equipment charge	"	2	4	Job		208		208	325

22 66 Chemical-Waste Systems for Lab. and Healthcare Facilities

22 66 53 – Laboratory Chemical-Waste and Vent Piping

22 66 53.30 Glass Pipe

	22 66 53.30 Glass Pipe	Crew	Daily Output	Labor-Hours	Unit	Material	2010 Bare Costs Labor	Equipment	Total	Total Incl O&P
0010	**GLASS PIPE**, Borosilicate, couplings & clevis hanger assemblies, 10' O.C.									
0020	Drainage									
1100	1-1/2" diameter	Q-1	52	.308	L.F.	13.20	14.40		27.60	37
1120	2" diameter		44	.364		17.45	17.05		34.50	45.50
1140	3" diameter		39	.410		23.50	19.20		42.70	56
1160	4" diameter		30	.533		42.50	25		67.50	85.50
1180	6" diameter		26	.615		78	29		107	130
9000	Minimum labor/equipment charge	1 Plum	4	2	Job		104		104	161

22 66 53.40 Pipe Fittings, Glass

	22 66 53.40 Pipe Fittings, Glass	Crew	Daily Output	Labor-Hours	Unit	Material	Labor	Equipment	Total	Total Incl O&P
0010	**PIPE FITTINGS, GLASS**									
0020	Drainage, beaded ends									
0040	Coupling & labor required at joints not incl. in fitting									
0050	price. Add 1 per joint for installed price									
0070	90° Bend or sweep, 1-1/2"				Ea.	29			29	32
0090	2"					36			36	39.50
0100	3"					60.50			60.50	66.50
0110	4"					96.50			96.50	106
0120	6" (sweep only)					294			294	325
0350	Tee, single sanitary, 1-1/2"					47			47	51.50
0370	2"					47			47	51.50
0380	3"					68.50			68.50	75
0390	4"					125			125	138
0400	6"					335			335	370
0500	Coupling, stainless steel, TFE seal ring									
0520	1-1/2"	Q-1	32	.500	Ea.	20.50	23.50		44	59
0530	2"		30	.533		25.50	25		50.50	67
0540	3"		25	.640		35	30		65	85
0550	4"		23	.696		60	32.50		92.50	117
0560	6"		20	.800		134	37.50		171.50	206
9000	Minimum labor/equipment charge	1 Plum	4	2	Job		104		104	161

22 66 53.60 Corrosion Resistant Pipe

	22 66 53.60 Corrosion Resistant Pipe	Crew	Daily Output	Labor-Hours	Unit	Material	Labor	Equipment	Total	Total Incl O&P
0010	**CORROSION RESISTANT PIPE**, No couplings or hangers									
0020	Iron alloy, drain, mechanical joint									
1000	1-1/2" diameter	Q-1	70	.229	L.F.	28.50	10.70		39.20	48
1100	2" diameter		66	.242		29	11.35		40.35	49.50
1120	3" diameter		60	.267		37.50	12.50		50	61
1140	4" diameter		52	.308		58.50	14.40		72.90	87
2980	Plastic, epoxy, fiberglass filament wound, B&S joint									
3000	2" diameter	Q-1	62	.258	L.F.	9.60	12.10		21.70	29.50
3100	3" diameter		51	.314		11.35	14.70		26.05	35.50
3120	4" diameter		45	.356		16.45	16.65		33.10	44
3160	8" diameter	Q-2	38	.632		36.50	30.50		67	87.50
3200	12" diameter	"	28	.857		60	41.50		101.50	131
9800	Minimum labor/equipment charge	1 Plum	4	2	Job		104		104	161

22 66 53.70 Pipe Fittings, Corrosion Resistant

	22 66 53.70 Pipe Fittings, Corrosion Resistant	Crew	Daily Output	Labor-Hours	Unit	Material	Labor	Equipment	Total	Total Incl O&P
0010	**PIPE FITTINGS, CORROSION RESISTANT**									
0030	Iron alloy									
0050	Mechanical joint									
0060	1/4 Bend, 1-1/2"	Q-1	12	1.333	Ea.	38.50	62.50		101	140
0080	2"		10	1.600		63	75		138	186
0090	3"		9	1.778		75.50	83.50		159	212
0100	4"		8	2		87	93.50		180.50	241

22 66 Chemical-Waste Systems for Lab. and Healthcare Facilities

22 66 53 – Laboratory Chemical-Waste and Vent Piping

22 66 53.70 Pipe Fittings, Corrosion Resistant		Crew	Daily Output	Labor-Hours	Unit	Material	2010 Bare Costs Labor	Equipment	Total	Total Incl O&P
0160	Tee and Y, sanitary, straight									
0170	1-1/2"	Q-1	8	2	Ea.	42	93.50		135.50	191
0180	2"		7	2.286		56	107		163	228
0190	3"		6	2.667		87	125		212	290
0200	4"		5	3.200		160	150		310	410
0360	Coupling, 1-1/2"		14	1.143		29.50	53.50		83	116
0380	2"		12	1.333		33.50	62.50		96	134
0390	3"		11	1.455		35	68		103	145
0400	4"		10	1.600		40	75		115	160
3000	Epoxy, filament wound									
3030	Quick-lock joint									
3040	90° Elbow, 2"	Q-1	28	.571	Ea.	81	27		108	131
3060	3"		16	1		93.50	47		140.50	176
3070	4"		13	1.231		128	57.50		185.50	231
3190	Tee, 2"		19	.842		194	39.50		233.50	274
3200	3"		11	1.455		235	68		303	365
3210	4"		9	1.778		281	83.50		364.50	440
9000	Minimum labor/equipment charge	1 Plum	4	2	Job		104		104	161

Estimating Tips

The labor adjustment factors listed in Subdivision 22 01 02.20 also apply to Division 23.

23 10 00 Facility Fuel Systems

- The prices in this subdivision for above- and below-ground storage tanks do not include foundations or hold-down slabs, unless noted. The estimator should refer to Divisions 3 and 31 for foundation system pricing. In addition to the foundations, required tank accessories, such as tank gauges, leak detection devices, and additional manholes and piping, must be added to the tank prices.

23 50 00 Central Heating Equipment

- When estimating the cost of an HVAC system, check to see who is responsible for providing and installing the temperature control system. It is possible to overlook controls, assuming that they would be included in the electrical estimate.
- When looking up a boiler, be careful on specified capacity. Some manufacturers rate their products on output while others use input.
- Include HVAC insulation for pipe, boiler, and duct (wrap and liner).
- Be careful when looking up mechanical items to get the correct pressure rating and connection type (thread, weld, flange).

23 70 00 Central HVAC Equipment

- Combination heating and cooling units are sized by the air conditioning requirements. (See Reference No. R236000-20 for preliminary sizing guide.)
- A ton of air conditioning is nominally 400 CFM.
- Rectangular duct is taken off by the linear foot for each size, but its cost is usually estimated by the pound. Remember that SMACNA standards now base duct on internal pressure.
- Prefabricated duct is estimated and purchased like pipe: straight sections and fittings.
- Note that cranes or other lifting equipment are not included on any lines in Division 23. For example, if a crane is required to lift a heavy piece of pipe into place high above a gym floor, or to put a rooftop unit on the roof of a four-story building, etc., it must be added. Due to the potential for extreme variation—from nothing additional required to a major crane or helicopter—we feel that including a nominal amount for "lifting contingency" would be useless and detract from the accuracy of the estimate. When using equipment rental cost data from RSMeans, do not forget to include the cost of the operator(s).

Reference Numbers

Reference numbers are shown in shaded boxes at the beginning of some major classifications. These numbers refer to related items in the Reference Section. The reference information may be an estimating procedure, an alternate pricing method, or technical information.

Note: Not all subdivisions listed here necessarily appear in this publication.

Note: **Trade Service,** *in part, has been used as a reference source for some of the material prices used in Division 23.*

23 05 Common Work Results for HVAC

23 05 05 – Selective HVAC Demolition

23 05 05.10 HVAC Demolition		Crew	Daily Output	Labor-Hours	Unit	Material	2010 Bare Costs Labor	2010 Bare Costs Equipment	Total	Total Incl O&P
0010	**HVAC DEMOLITION** R220105-10									
0100	Air conditioner, split unit, 3 ton	Q-5	2	8	Ea.		375		375	580
0150	Package unit, 3 ton R024119-10	Q-6	3	8	"		385		385	600
0298	Boilers									
0300	Electric, up thru 148 kW	Q-19	2	12	Ea.		570		570	880
0310	150 thru 518 kW	"	1	24			1,150		1,150	1,750
0320	550 thru 2000 kW	Q-21	.40	80			3,875		3,875	6,000
0330	2070 kW and up	"	.30	106			5,175		5,175	8,000
0340	Gas and/or oil, up thru 150 MBH	Q-7	2.20	14.545			720		720	1,125
0350	160 thru 2000 MBH		.80	40			1,975		1,975	3,050
0360	2100 thru 4500 MBH		.50	64			3,175		3,175	4,900
0370	4600 thru 7000 MBH		.30	106			5,275		5,275	8,175
0380	7100 thru 12,000 MBH		.16	200			9,875		9,875	15,300
0390	12,200 thru 25,000 MBH		.12	266			13,200		13,200	20,400
1000	Ductwork, 4" high, 8" wide	1 Clab	200	.040	L.F.		1.32		1.32	2.17
1020	10" wide		190	.042			1.39		1.39	2.29
1040	14" wide		180	.044			1.47		1.47	2.42
1100	6" high, 8" wide		165	.048			1.60		1.60	2.63
1120	12" wide		150	.053			1.77		1.77	2.90
1140	18" wide		135	.059			1.96		1.96	3.22
1200	10" high, 12" wide		125	.064			2.12		2.12	3.48
1220	18" wide		115	.070			2.30		2.30	3.78
1240	24" wide		110	.073			2.41		2.41	3.95
1300	12"-14" high, 16"-18" wide		85	.094			3.12		3.12	5.10
1320	24" wide		75	.107			3.53		3.53	5.80
1340	48" wide		71	.113			3.73		3.73	6.10
1400	18" high, 24" wide		67	.119			3.95		3.95	6.50
1420	36" wide		63	.127			4.20		4.20	6.90
1440	48" wide		59	.136			4.49		4.49	7.35
1500	30" high, 36" wide		56	.143			4.73		4.73	7.75
1520	48" wide		53	.151			5		5	8.20
1540	72" wide		50	.160			5.30		5.30	8.70
1550	Duct heater, electric strip	1 Elec	8	1	Ea.		49		49	75.50
1850	Minimum labor/equipment charge	1 Clab	3	2.667	Job		88.50		88.50	145
2200	Furnace, electric	Q-20	2	10	Ea.		450		450	705
2300	Gas or oil, under 120 MBH	Q-9	4	4			177		177	277
2340	Over 120 MBH	"	3	5.333			236		236	370
2800	Heat pump, package unit, 3 ton	Q-5	2.40	6.667			310		310	480
2840	Split unit, 3 ton		2	8			375		375	580
3000	Mechanical equipment, light items. Unit is weight, not cooling.		.90	17.778	Ton		830		830	1,275
3600	Heavy items		1.10	14.545	"		680		680	1,050
9000	Minimum labor/equipment charge	Q-6	3	8	Job		385		385	600

23 05 23 – General-Duty Valves for HVAC Piping

23 05 23.30 Valves, Iron Body

		Crew	Daily Output	Labor-Hours	Unit	Material	Labor	Equipment	Total	Total Incl O&P
0010	**VALVES, IRON BODY**									
1020	Butterfly, wafer type, gear actuator, 200 lb.									
1030	2"	1 Plum	14	.571	Ea.	167	29.50		196.50	229
1060	4"	Q-1	5	3.200	"	202	150		352	455
1650	Gate, 125 lb., N.R.S.									
2150	Flanged									
2240	2-1/2"	Q-1	5	3.200	Ea.	540	150		690	825
2260	3"		4.50	3.556		605	167		772	925

23 05 Common Work Results for HVAC

23 05 23 – General-Duty Valves for HVAC Piping

23 05 23.30 Valves, Iron Body		Crew	Daily Output	Labor-Hours	Unit	Material	2010 Bare Costs Labor	Equipment	Total	Total Incl O&P
2280	4"	Q-1	3	5.333	Ea.	865	250		1,115	1,325
3550	OS&Y, 125 lb., flanged									
3680	4"	Q-1	3	5.333	Ea.	525	250		775	960
3690	5"	Q-2	3.40	7.059		865	345		1,210	1,475
3700	6"	"	3	8	↓	865	390		1,255	1,550
9000	Minimum labor/equipment charge	1 Plum	3	2.667	Job		139		139	215

23 07 HVAC Insulation

23 07 13 – Duct Insulation

23 07 13.10 Duct Thermal Insulation

			Crew	Daily Output	Labor-Hours	Unit	Material	Labor	Equipment	Total	Total Incl O&P
0010	**DUCT THERMAL INSULATION**										
0100	Rule of thumb, as a percentage of total mechanical costs					Job				10%	
0110	Insulation req'd is based on the surface size/area to be covered										
3000	Ductwork										
3020	Blanket type, fiberglass, flexible										
3140	FSK vapor barrier wrap, .75 lb. density										
3160	1" thick	G	Q-14	350	.046	S.F.	.20	1.87		2.07	3.23
3170	1-1/2" thick	G		320	.050		.23	2.05		2.28	3.54
3180	2" thick	G		300	.053		.28	2.19		2.47	3.82
3190	3" thick	G		260	.062		.40	2.52		2.92	4.49
3200	4" thick	G		242	.066		.60	2.71		3.31	5
3212	Vinyl Jacket, .75 lb. density, 1-1/2" thick	G	↓	320	.050	↓	.23	2.05		2.28	3.54
9600	Minimum labor/equipment charge		1 Stpi	4	2	Job		104		104	161

23 07 16 – HVAC Equipment Insulation

23 07 16.10 HVAC Equipment Thermal Insulation

0010	**HVAC EQUIPMENT THERMAL INSULATION**										
0100	Rule of thumb, as a percentage of total mechanical costs					Job				10%	
0110	Insulation req'd is based on the surface size/area to be covered										

23 09 Instrumentation and Control for HVAC

23 09 33 – Electric and Electronic Control System for HVAC

23 09 33.10 Electronic Control Systems

0010	**ELECTRONIC CONTROL SYSTEMS**										
9000	Minimum labor/equipment charge		1 Plum	8	1	Job		52		52	80.50

23 09 53 – Pneumatic and Electric Control System for HVAC

23 09 53.10 Control Components

			Crew	Daily Output	Labor-Hours	Unit	Material	Labor	Equipment	Total	Total Incl O&P
0010	**CONTROL COMPONENTS**										
5000	Thermostats										
5030	Manual		1 Shee	8	1	Ea.	44	49		93	125
5040	1 set back, electric, timed	G		8	1		33.50	49		82.50	114
5050	2 set back, electric, timed	G		8	1		143	49		192	234
5200	24 hour, automatic, clock	G	↓	8	1	↓	133	49		182	223
6000	Valves, motorized zone										
6100	Sweat connections, 1/2" C x C		1 Stpi	20	.400	Ea.	156	21		177	204
6110	3/4" C x C			20	.400		157	21		178	205
6120	1" C x C		↓	19	.421	↓	204	22		226	258
9000	Minimum labor/equipment charge		1 Plum	4	2	Job		104		104	161

23 12 Facility Fuel Pumps

23 12 16 – Facility Gasoline Dispensing Pumps

23 12 16.20 Fuel Dispensing Equipment	Crew	Daily Output	Labor-Hours	Unit	Material	2010 Bare Costs Labor	Equipment	Total	Total Incl O&P
0010 **FUEL DISPENSING EQUIPMENT**									
1100 Product dispenser with vapor recovery for 6 nozzles, installed, not									
1110 including piping to storage tanks				Ea.	22,500			22,500	24,800

23 21 Hydronic Piping and Pumps

23 21 20 – Hydronic HVAC Piping Specialties

23 21 20.34 Dielectric Unions

0010 **DIELECTRIC UNIONS**, Standard gaskets for water and air.									
0020 250 psi maximum pressure									
0280 Female IPT to sweat, straight									
0340 3/4" pipe size	1 Plum	20	.400	Ea.	5.05	21		26.05	38
0780 Female IPT to female IPT, straight									
0800 1/2" pipe size	1 Plum	24	.333	Ea.	9.20	17.35		26.55	37
0840 3/4" pipe size		20	.400	"	10.40	21		31.40	44
9000 Minimum labor/equipment charge		4	2	Job		104		104	161

23 21 20.46 Expansion Tanks

0010 **EXPANSION TANKS**									
2000 Steel, liquid expansion, ASME, painted, 15 gallon capacity	Q-5	17	.941	Ea.	520	44		564	645
2040 30 gallon capacity		12	1.333		580	62.50		642.50	735
2080 60 gallon capacity		8	2		815	93.50		908.50	1,050
2120 100 gallon capacity		6	2.667		1,175	125		1,300	1,500
3000 Steel ASME expansion, rubber diaphragm, 19 gal. cap. accept.		12	1.333		2,150	62.50		2,212.50	2,475
3020 31 gallon capacity		8	2		2,375	93.50		2,468.50	2,775
3040 61 gallon capacity		6	2.667		3,350	125		3,475	3,900
3080 119 gallon capacity		4	4		3,625	187		3,812	4,275
9000 Minimum labor/equipment charge		4	4	Job		187		187	289

23 21 20.70 Steam Traps

0010 **STEAM TRAPS**									
0030 Cast iron body, threaded									
0040 Inverted bucket									
0050 1/2" pipe size	1 Stpi	12	.667	Ea.	131	34.50		165.50	198
0100 1" pipe size		9	.889		350	46		396	455
0120 1-1/4" pipe size		8	1		530	52		582	660
1000 Float & thermostatic, 15 psi									
1010 3/4" pipe size	1 Stpi	16	.500	Ea.	105	26		131	156
1020 1" pipe size		15	.533		127	27.50		154.50	182
1040 1-1/2" pipe size		9	.889		223	46		269	320
1060 2" pipe size		6	1.333		410	69		479	555
9000 Minimum labor/equipment charge		4	2	Job		104		104	161

23 21 20.76 Strainers, Y Type, Bronze Body

0010 **STRAINERS, Y TYPE, BRONZE BODY**									
0050 Screwed, 150 lb., 1/4" pipe size	1 Stpi	24	.333	Ea.	17.30	17.30		34.60	46
0100 1/2" pipe size		20	.400		23	21		44	57
0120 3/4" pipe size		19	.421		26	22		48	62.50
0140 1" pipe size		17	.471		30.50	24.50		55	72
0160 1-1/2" pipe size		14	.571		66	29.50		95.50	119
0180 2" pipe size		13	.615		88.50	32		120.50	147
0182 3" pipe size		12	.667		670	34.50		704.50	790
0220 3" pipe size	Q-5	16	1		835	46.50		881.50	995
0240 4" pipe size	"	15	1.067		1,900	50		1,950	2,175

23 21 Hydronic Piping and Pumps

23 21 20 – Hydronic HVAC Piping Specialties

23 21 20.76 Strainers, Y Type, Bronze Body

		Crew	Daily Output	Labor-Hours	Unit	Material	2010 Bare Costs Labor	Equipment	Total	Total Incl O&P
1000	Flanged, 150 lb., 1-1/2" pipe size	1 Stpi	11	.727	Ea.	435	38		473	540
1020	2" pipe size	"	8	1		590	52		642	730
1030	2-1/2" pipe size	Q-5	5	3.200		875	149		1,024	1,200
1040	3" pipe size		4.50	3.556		1,075	166		1,241	1,425
1060	4" pipe size		3	5.333		1,625	249		1,874	2,175
1100	6" pipe size	Q-6	3	8		3,125	385		3,510	4,025
1106	8" pipe size	"	2.60	9.231		3,425	445		3,870	4,475
1500	For 300 lb rating, add					40%				
9000	Minimum labor/equipment charge	1 Stpi	3.75	2.133	Job		111		111	172

23 21 23 – Hydronic Pumps

23 21 23.13 In-Line Centrifugal Hydronic Pumps

		Crew	Daily Output	Labor-Hours	Unit	Material	Labor	Equipment	Total	Total Incl O&P
0010	**IN-LINE CENTRIFUGAL HYDRONIC PUMPS**									
0600	Bronze, sweat connections, 1/40 HP, in line									
0640	3/4" size	Q-1	16	1	Ea.	192	47		239	284
1000	Flange connection, 3/4" to 1-1/2" size									
1040	1/12 HP	Q-1	6	2.667	Ea.	495	125		620	740
1060	1/8 HP		6	2.667		855	125		980	1,125
1100	1/3 HP		6	2.667		955	125		1,080	1,250
1140	2" size, 1/6 HP		5	3.200		1,025	150		1,175	1,350
1180	2-1/2" size, 1/4 HP		5	3.200		1,525	150		1,675	1,925
1220	3" size, 1/4 HP		4	4		1,600	187		1,787	2,075
9000	Minimum labor/equipment charge		3.25	4.923	Job		231		231	355

23 31 HVAC Ducts and Casings

23 31 13 – Metal Ducts

23 31 13.13 Rectangular Metal Ducts

			Crew	Daily Output	Labor-Hours	Unit	Material	Labor	Equipment	Total	Total Incl O&P
0010	**RECTANGULAR METAL DUCTS**	R233100-20									
0020	Fabricated rectangular, includes fittings, joints, supports,										
0100	Aluminum, alloy 3003-H14, under 100 lb.		Q-10	75	.320	Lb.	3.72	14.65		18.37	27
0110	100 to 500 lb.			80	.300		2.48	13.75		16.23	24
0120	500 to 1,000 lb.			95	.253		2.31	11.60		13.91	20.50
0500	Galvanized steel, under 200 lb.			235	.102		.47	4.68		5.15	7.80
0520	200 to 500 lb.			245	.098		.46	4.49		4.95	7.50
0540	500 to 1,000 lb.			255	.094		.45	4.31		4.76	7.25
0560	1,000 to 2,000 lb.			265	.091		.45	4.15		4.60	7
0580	Over 5,000 lb.			285	.084		.43	3.86		4.29	6.50
0590											
1000	Stainless steel, type 304, under 100 lb.		Q-10	165	.145	Lb.	4.35	6.65		11	15.25
1020	100 to 500 lb.			175	.137		4.02	6.30		10.32	14.25
1030	500 to 1,000 lb.			190	.126		3.68	5.80		9.48	13.10

23 31 13.16 Round and Flat-Oval Spiral Ducts

		Crew	Daily Output	Labor-Hours	Unit	Material	Labor	Equipment	Total	Total Incl O&P
0010	**ROUND AND FLAT-OVAL SPIRAL DUCTS**									
9990	Minimum labor/equipment charge	1 Shee	3	2.667	Job		131		131	205

23 31 13.19 Metal Duct Fittings

		Crew	Daily Output	Labor-Hours	Unit	Material	Labor	Equipment	Total	Total Incl O&P
0010	**METAL DUCT FITTINGS**									
0050	Air extractors, 12" x 4"	1 Shee	24	.333	Ea.	16.10	16.35		32.45	43.50
0100	8" x 6"		22	.364		16.10	17.85		33.95	46
0200	20" x 8"		16	.500		36.50	24.50		61	78.50
0240	18" x 10"		14	.571		35.50	28		63.50	83
0280	24" x 12"		10	.800		49.50	39.50		89	116

23 31 HVAC Ducts and Casings

23 31 16 – Nonmetal Ducts

23 31 16.13 Fibrous-Glass Ducts		Crew	Daily Output	Labor-Hours	Unit	Material	2010 Bare Costs Labor	2010 Bare Costs Equipment	Total	Total Incl O&P
0010	**FIBROUS-GLASS DUCTS** R233100-20									
9990	Minimum labor/equipment charge	1 Shee	3	2.667	Job		131		131	205
23 31 16.19 PVC Ducts										
0010	**PVC DUCTS**									
9990	Minimum labor/equipment charge	1 Shee	3	2.667	Job		131		131	205

23 33 Air Duct Accessories

23 33 13 – Dampers

23 33 13.13 Volume-Control Dampers

		Crew	Daily Output	Labor-Hours	Unit	Material	Labor	Equipment	Total	Total Incl O&P
0010	**VOLUME-CONTROL DAMPERS**									
5990	Multi-blade dampers, opposed blade, 8" x 6"	1 Shee	24	.333	Ea.	20	16.35		36.35	47.50
5994	8" x 8"		22	.364		21	17.85		38.85	51
5996	10" x 10"		21	.381		24.50	18.70		43.20	56.50
6000	12" x 12"		21	.381		27.50	18.70		46.20	59.50
6020	12" x 18"		18	.444		36.50	22		58.50	74.50
6030	14" x 10"		20	.400		26.50	19.65		46.15	60
6031	14" x 14"		17	.471		32.50	23		55.50	71.50
6033	16" x 12"		17	.471		32.50	23		55.50	71.50
6035	16" x 16"		16	.500		40.50	24.50		65	83
6037	18" x 16"		15	.533		44.50	26		70.50	90
6038	18" x 18"		15	.533		48	26		74	94
6070	20" x 16"		14	.571		48	28		76	97
6072	20" x 20"		13	.615		57.50	30		87.50	111
6074	22" x 18"		14	.571		57.50	28		85.50	108
6076	24" x 16"		11	.727		57	35.50		92.50	119
6078	24" x 20"		8	1		67	49		116	151
6080	24" x 24"		8	1		78.50	49		127.50	164
6110	26" x 26"		6	1.333		86.50	65.50		152	197
6133	30" x 30"	Q-9	6.60	2.424		124	107		231	305
6135	32" x 32"		6.40	2.500		144	111		255	330
6180	48" x 36"		5.60	2.857		237	126		363	460
7500	Variable volume modulating motorized damper, incl. elect. mtr.									
7504	8" x 6"	1 Shee	15	.533	Ea.	109	26		135	161
7506	10" x 6"		14	.571		109	28		137	164
7510	10" x 10"		13	.615		112	30		142	172
7520	12" x 12"		12	.667		117	32.50		149.50	180
7522	12" x 16"		11	.727		118	35.50		153.50	186
7524	16" x 10"		12	.667		115	32.50		147.50	178
7526	16" x 14"		10	.800		120	39.50		159.50	193
7528	16" x 18"		9	.889		125	43.50		168.50	206
7542	18" x 18"		8	1		127	49		176	217
7544	20" x 14"		8	1		130	49		179	220
7546	20" x 18"		7	1.143		136	56		192	238
7560	24" x 12"		8	1		132	49		181	222
7562	24" x 18"		7	1.143		143	56		199	245
7564	24" x 24"		6	1.333		148	65.50		213.50	264
7568	28" x 10"		7	1.143		132	56		188	233
7590	30" x 14"		5	1.600		140	78.50		218.50	277
7600	30" x 18"		4	2		180	98		278	350
7610	30" x 24"		3.80	2.105		239	103		342	425

23 33 Air Duct Accessories

23 33 13 – Dampers

23 33 13.13 Volume-Control Dampers

		Crew	Daily Output	Labor-Hours	Unit	Material	2010 Bare Costs Labor	Equipment	Total	Total Incl O&P
7700	For thermostat, add	1 Shee	8	1	Ea.	40.50	49		89.50	122
8000	Multi-blade dampers, parallel blade									
8100	8" x 8"	1 Shee	24	.333	Ea.	73	16.35		89.35	106
8140	16" x 10"		20	.400		94	19.65		113.65	134
8160	18" x 12"		18	.444		107	22		129	152
8220	28" x 16"		10	.800		134	39.50		173.50	209

23 33 13.16 Fire Dampers

		Crew	Daily Output	Labor-Hours	Unit	Material	Labor	Equipment	Total	Total Incl O&P
0010	**FIRE DAMPERS**									
3000	Fire damper, curtain type, 1-1/2 hr rated, vertical, 6" x 6"	1 Shee	24	.333	Ea.	21.50	16.35		37.85	49.50
3020	8" x 6"		22	.364		21.50	17.85		39.35	52
3240	16" x 14"		18	.444		29.50	22		51.50	66.50
3400	24" x 20"		8	1		42	49		91	123

23 33 19 – Duct Silencers

23 33 19.10 Duct Silencers

		Crew	Daily Output	Labor-Hours	Unit	Material	Labor	Equipment	Total	Total Incl O&P
0009	**DUCT SILENCERS**									
0010	Silencers, noise control for air flow, duct				MCFM	55.50			55.50	61

23 33 23 – Turning Vanes

23 33 23.13 Air Turning Vanes

		Crew	Daily Output	Labor-Hours	Unit	Material	Labor	Equipment	Total	Total Incl O&P
0010	**AIR TURNING VANES**									
9900	Minimum labor/equipment charge	1 Shee	4	2	Job		98		98	154

23 33 46 – Flexible Ducts

23 33 46.10 Flexible Air Ducts

			Crew	Daily Output	Labor-Hours	Unit	Material	Labor	Equipment	Total	Total Incl O&P
0010	**FLEXIBLE AIR DUCTS**	R233100-20									
1280	Add to labor for elevated installation										
1282	of prefabricated (purchased) ductwork										
1283	10' to 15' high							10%			
1284	15' to 20' high							20%			
1285	20' to 25' high							25%			
1286	25' to 30' high							35%			
1287	30' to 35' high							40%			
1288	35' to 40' high							50%			
1289	Over 40' high							55%			
1300	Flexible, coated fiberglass fabric on corr. resist. metal helix										
1400	pressure to 12" (WG) UL-181										
1500	Non-insulated, 3" diameter		Q-9	400	.040	L.F.	1.02	1.77		2.79	3.89
1540	5" diameter			320	.050		1.25	2.21		3.46	4.84
1560	6" diameter			280	.057		1.53	2.53		4.06	5.65
1580	7" diameter			240	.067		1.82	2.95		4.77	6.60
1600	8" diameter			200	.080		2.10	3.54		5.64	7.85
1640	10" diameter			160	.100		2.61	4.42		7.03	9.75
1660	12" diameter			120	.133		3.17	5.90		9.07	12.70
1900	Insulated, 1" thick, PE jacket, 3" diameter	G		380	.042		2.15	1.86		4.01	5.30
1910	4" diameter	G		340	.047		2.15	2.08		4.23	5.60
1920	5" diameter	G		300	.053		2.15	2.36		4.51	6.05
1940	6" diameter	G		260	.062		2.38	2.72		5.10	6.85
1960	7" diameter	G		220	.073		2.78	3.21		5.99	8.10
1980	8" diameter	G		180	.089		3	3.93		6.93	9.45
2020	10" diameter	G		140	.114		3.51	5.05		8.56	11.75
2040	12" diameter	G		100	.160		4.36	7.05		11.41	15.85
2060	14" diameter	G		80	.200		5.25	8.85		14.10	19.65
9990	Minimum labor/equipment charge		1 Shee	3	2.667	Job		131		131	205

23 33 Air Duct Accessories

23 33 53 – Duct Liners

23 33 53.10 Duct Liner Board

			Crew	Daily Output	Labor-Hours	Unit	Material	2010 Bare Costs Labor	Equipment	Total	Total Incl O&P
0010	**DUCT LINER BOARD**										
3490	Board type, fiberglass liner, 3 lb. density										
3500	Fire resistant, black pigmented, 1 side										
3520	1" thick	G	Q-14	150	.107	S.F.	.77	4.37		5.14	7.85
3540	1-1/2" thick	G	"	130	.123	"	.97	5.05		6.02	9.15
3940	Board type, non-fibrous foam										
3950	Temperature, bacteria and fungi resistant										
3960	1" thick	G	Q-14	150	.107	S.F.	2.25	4.37		6.62	9.50
3970	1-1/2" thick	G		130	.123		3	5.05		8.05	11.40
3980	2" thick	G		120	.133		3.60	5.45		9.05	12.70

23 34 HVAC Fans

23 34 13 – Axial HVAC Fans

23 34 13.10 Axial Flow HVAC Fans

		Crew	Daily Output	Labor-Hours	Unit	Material	Labor	Equipment	Total	Total Incl O&P
0010	**AXIAL FLOW HVAC FANS**									
0020	Air conditioning and process air handling									
0030	Axial flow, compact, low sound, 2.5" S.P.									
0050	3,800 CFM, 5 HP	Q-20	3.40	5.882	Ea.	4,500	266		4,766	5,375
0120	15,600 CFM, 10 HP	"	1.60	12.500	"	7,875	565		8,440	9,550

23 34 14 – Blower HVAC Fans

23 34 14.10 Blower Type HVAC Fans

		Crew	Daily Output	Labor-Hours	Unit	Material	Labor	Equipment	Total	Total Incl O&P
0010	**BLOWER TYPE HVAC FANS**									
2500	Ceiling fan, right angle, extra quiet, 0.10" S.P.									
2540	210 CFM	Q-20	19	1.053	Ea.	283	47.50		330.50	385
2580	885 CFM		16	1.250		710	56.50		766.50	870
2620	2,960 CFM		11	1.818		1,300	82		1,382	1,575
2640	For wall or roof cap, add	1 Shee	16	.500		240	24.50		264.50	305

23 34 16 – Centrifugal HVAC Fans

23 34 16.10 Centrifugal Type HVAC Fans

		Crew	Daily Output	Labor-Hours	Unit	Material	Labor	Equipment	Total	Total Incl O&P
0010	**CENTRIFUGAL TYPE HVAC FANS**									
0200	In-line centrifugal, supply/exhaust booster									
0220	aluminum wheel/hub, disconnect switch, 1/4" S.P.									
0240	500 CFM, 10" diameter connection	Q-20	3	6.667	Ea.	1,150	300		1,450	1,750
0280	1,520 CFM, 16" diameter connection		2	10		1,325	450		1,775	2,150
0320	3,480 CFM, 20" diameter connection		.80	25		1,725	1,125		2,850	3,625
0326	5,080 CFM, 20" diameter connection		.75	26.667		1,875	1,200		3,075	3,925
4500	Corrosive fume resistant, plastic									
4600	roof ventilators, centrifugal, V belt drive, motor									
4620	1/4" S.P., 250 CFM, 1/4 HP	Q-20	6	3.333	Ea.	3,875	151		4,026	4,475
4640	895 CFM, 1/3 HP		5	4		4,200	181		4,381	4,900
4660	1630 CFM, 1/2 HP		4	5		4,975	226		5,201	5,825
4680	2240 CFM, 1 HP		3	6.667		5,175	300		5,475	6,175
5000	Utility set, centrifugal, V belt drive, motor									
5020	1/4" S.P., 1200 CFM, 1/4 HP	Q-20	6	3.333	Ea.	4,275	151		4,426	4,925
5040	1520 CFM, 1/3 HP		5	4		1,650	181		1,831	2,075
5060	1850 CFM, 1/2 HP		4	5		1,700	226		1,926	2,225
5080	2180 CFM, 3/4 HP		3	6.667		1,950	300		2,250	2,625
7100	Direct drive, 320 CFM, 11" sq. damper		7	2.857		600	129		729	860
7120	600 CFM, 11" sq. damper		6	3.333		750	151		901	1,050

23 34 HVAC Fans

23 34 16 – Centrifugal HVAC Fans

23 34 16.10 Centrifugal Type HVAC Fans		Crew	Daily Output	Labor-Hours	Unit	Material	2010 Bare Costs Labor	Equipment	Total	Total Incl O&P
9900	Minimum labor/equipment charge	1 Elec	4	2	Job		98		98	151

23 34 23 – HVAC Power Ventilators

23 34 23.10 HVAC Power Circulators and Ventilators

		Crew	Daily Output	Labor-Hours	Unit	Material	Labor	Equipment	Total	Total Incl O&P
0010	**HVAC POWER CIRCULATORS AND VENTILATORS**									
6650	Residential, bath exhaust, grille, back draft damper									
6660	50 CFM	Q-20	24	.833	Ea.	44.50	37.50		82	108
6670	110 CFM		22	.909		71.50	41		112.50	143
6680	Light combination, squirrel cage, 100 watt, 70 CFM		24	.833		80.50	37.50		118	147
6700	Light/heater combination, ceiling mounted									
6710	70 CFM, 1450 watt	Q-20	24	.833	Ea.	114	37.50		151.50	184
6800	Heater combination, recessed, 70 CFM		24	.833		46.50	37.50		84	110
6820	With 2 infrared bulbs		23	.870		70	39.50		109.50	138
6900	Kitchen exhaust, grille, complete, 160 CFM		22	.909		83.50	41		124.50	156
6910	180 CFM		20	1		70.50	45		115.50	149
6920	270 CFM		18	1.111		128	50		178	219
6940	Residential roof jacks and wall caps									
6944	Wall cap with back draft damper									
6946	3" & 4" dia. round duct	1 Shee	11	.727	Ea.	19.05	35.50		54.55	77
6948	6" dia. round duct	"	11	.727	"	46	35.50		81.50	107
6958	Roof jack with bird screen and back draft damper									
6960	3" & 4" dia. round duct	1 Shee	11	.727	Ea.	18.30	35.50		53.80	76
6962	3-1/4" x 10" rectangular duct	"	10	.800	"	34	39.50		73.50	99
6980	Transition									
6982	3-1/4" x 10" to 6" dia. round	1 Shee	20	.400	Ea.	19.95	19.65		39.60	52.50
8020	Attic, roof type									
8030	Aluminum dome, damper & curb									
8040	6" diameter, 300 CFM	1 Elec	16	.500	Ea.	400	24.50		424.50	480
8050	7" diameter, 450 CFM		15	.533		435	26		461	520
8060	9" diameter, 900 CFM		14	.571		480	28		508	570
8080	12" diameter, 1000 CFM (gravity)		10	.800		495	39		534	605
8090	16" diameter, 1500 CFM (gravity)		9	.889		595	43.50		638.50	720
8100	20" diameter, 2500 CFM (gravity)		8	1		730	49		779	880
8110	26" diameter, 4000 CFM (gravity)		7	1.143		885	56		941	1,050
8120	32" diameter, 6500 CFM (gravity)		6	1.333		1,225	65.50		1,290.50	1,450
8130	38" diameter, 8000 CFM (gravity)		5	1.600		1,800	78.50		1,878.50	2,100
8140	50" diameter, 13,000 CFM (gravity)		4	2		2,625	98		2,723	3,025
8160	Plastic, ABS dome									
8180	1050 CFM	1 Elec	14	.571	Ea.	146	28		174	203
8200	1600 CFM	"	12	.667	"	218	32.50		250.50	290
8240	Attic, wall type, with shutter, one speed									
8250	12" diameter, 1000 CFM	1 Elec	14	.571	Ea.	315	28		343	390
8260	14" diameter, 1500 CFM		12	.667		340	32.50		372.50	425
8270	16" diameter, 2000 CFM		9	.889		385	43.50		428.50	490
8290	Whole house, wall type, with shutter, one speed									
8300	30" diameter, 4800 CFM	1 Elec	7	1.143	Ea.	825	56		881	990
8310	36" diameter, 7000 CFM		6	1.333		900	65.50		965.50	1,100
8320	42" diameter, 10,000 CFM		5	1.600		1,000	78.50		1,078.50	1,225
8330	48" diameter, 16,000 CFM		4	2		1,250	98		1,348	1,525
8340	For two speed, add					75.50			75.50	83
8350	Whole house, lay-down type, with shutter, one speed									
8360	30" diameter, 4500 CFM	1 Elec	8	1	Ea.	880	49		929	1,050
8370	36" diameter, 6500 CFM		7	1.143		945	56		1,001	1,125

23 34 HVAC Fans

23 34 23 – HVAC Power Ventilators

23 34 23.10 HVAC Power Circulators and Ventilators		Crew	Daily Output	Labor-Hours	Unit	Material	2010 Bare Costs Labor	Equipment	Total	Total Incl O&P
8380	42" diameter, 9000 CFM	1 Elec	6	1.333	Ea.	1,050	65.50		1,115.50	1,250
8390	48" diameter, 12,000 CFM	↓	5	1.600	↓	1,175	78.50		1,253.50	1,425
8440	For two speed, add					56.50			56.50	62
8450	For 12 hour timer switch, add	1 Elec	32	.250	↓	56.50	12.25		68.75	81

23 36 Air Terminal Units

23 36 13 – Constant-Air-Volume Units

23 36 13.10 Constant Volume Mixing Boxes

0010	**CONSTANT VOLUME MIXING BOXES**									
5180	Mixing box, includes electric or pneumatic motor									
5200	Constant volume, 150 to 270 CFM	Q-9	12	1.333	Ea.	635	59		694	785
5210	270 to 600 CFM	"	11	1.455	"	650	64.50		714.50	815

23 36 16 – Variable-Air-Volume Units

23 36 16.10 Variable Volume Mixing Boxes

0010	**VARIABLE VOLUME MIXING BOXES**									
5180	Mixing box, includes electric or pneumatic motor									
5500	VAV Cool only, pneumatic, pressure independent 300 to 600 CFM	Q-9	11	1.455	Ea.	410	64.50		474.50	555
5510	500 to 1000 CFM		9	1.778		425	78.50		503.50	590
5520	800 to 1600 CFM		9	1.778		435	78.50		513.50	605
5530	1100 to 2000 CFM		8	2		445	88.50		533.50	630
5540	1500 to 3000 CFM		7	2.286		475	101		576	685
5550	2000 to 4000 CFM	↓	6	2.667	↓	480	118		598	715
5560	For electric, w/thermostat, pressure dependent, add					23			23	25.50

23 37 Air Outlets and Inlets

23 37 13 – Diffusers, Registers, and Grilles

23 37 13.10 Diffusers

0010	**DIFFUSERS**, Aluminum, opposed blade damper unless noted									
0100	Ceiling, linear, also for sidewall									
0120	2" wide	1 Shee	32	.250	L.F.	43	12.30		55.30	66
0160	4" wide		26	.308		57.50	15.10		72.60	87
0180	6" wide		24	.333		71.50	16.35		87.85	104
0200	8" wide		22	.364	↓	83.50	17.85		101.35	120
0500	Perforated, 24" x 24" lay-in panel size, 6" x 6"		16	.500	Ea.	57.50	24.50		82	102
0520	8" x 8"		15	.533		59	26		85	106
0530	9" x 9"		14	.571		63	28		91	114
0540	10" x 10"		14	.571		65	28		93	116
0560	12" x 12"		12	.667		66	32.50		98.50	124
0590	16" x 16"		11	.727		109	35.50		144.50	176
0600	18" x 18"		10	.800		115	39.50		154.50	189
0610	20" x 20"		10	.800		133	39.50		172.50	208
0620	24" x 24"		9	.889		165	43.50		208.50	251
1000	Rectangular, 1 to 4 way blow, 6" x 6"		16	.500		46.50	24.50		71	89.50
1010	8" x 8"		15	.533		55.50	26		81.50	102
1014	9" x 9"		15	.533		61	26		87	109
1016	10" x 10"		15	.533		65	26		91	113
1020	12" x 6"		15	.533		67.50	26		93.50	115
1040	12" x 9"		14	.571		73.50	28		101.50	125
1060	12" x 12"	↓	12	.667		73	32.50		105.50	132

23 37 Air Outlets and Inlets

23 37 13 – Diffusers, Registers, and Grilles

23 37 13.10 Diffusers

		Crew	Daily Output	Labor-Hours	Unit	Material	2010 Bare Costs Labor	Equipment	Total	Total Incl O&P
1070	14" x 6"	1 Shee	13	.615	Ea.	73	30		103	128
1074	14" x 14"		12	.667		103	32.50		135.50	164
1150	18" x 18"		9	.889		128	43.50		171.50	210
1170	24" x 12"		10	.800		136	39.50		175.50	212
2000	T bar mounting, 24" x 24" lay-in frame, 6" x 6"		16	.500		42.50	24.50		67	85.50
2020	8" x 8"		14	.571		44.50	28		72.50	92.50
2040	12" x 12"		12	.667		52	32.50		84.50	109
2060	16" x 16"		11	.727		70	35.50		105.50	133
2080	18" x 18"		10	.800		78	39.50		117.50	148
6000	For steel diffusers instead of aluminum, deduct					10%				
9000	Minimum labor/equipment charge	1 Shee	4	2	Job		98		98	154

23 37 13.30 Grilles

		Crew	Daily Output	Labor-Hours	Unit	Material	Labor	Equipment	Total	Total Incl O&P
0010	**GRILLES**									
0020	Aluminum									
1000	Air return, 6" x 6"	1 Shee	26	.308	Ea.	17.55	15.10		32.65	43
1020	10" x 6"		24	.333		17.55	16.35		33.90	45
1080	16" x 8"		22	.364		22	17.85		39.85	52.50
1100	12" x 12"		22	.364		24	17.85		41.85	54
1120	24" x 12"		18	.444		33	22		55	70
1300	48" x 24"		12	.667		122	32.50		154.50	185
9000	Minimum labor/equipment charge		4	2	Job		98		98	154

23 37 13.60 Registers

		Crew	Daily Output	Labor-Hours	Unit	Material	Labor	Equipment	Total	Total Incl O&P
0010	**REGISTERS**									
0020										
0980	Air supply									
1000	Ceiling/wall, O.B. damper, anodized aluminum									
1010	One or two way deflection, adj. curved face bars									
1140	14" x 8"	1 Shee	17	.471	Ea.	51.50	23		74.50	93
3000	Baseboard, hand adj. damper, enameled steel									
3012	8" x 6"	1 Shee	26	.308	Ea.	13.90	15.10		29	39
3020	10" x 6"		24	.333		15.10	16.35		31.45	42
3040	12" x 5"		23	.348		16.40	17.10		33.50	44.50
3060	12" x 6"		23	.348		16.40	17.10		33.50	44.50
3080	12" x 8"		22	.364		24	17.85		41.85	54
3100	14" x 6"		20	.400		17.80	19.65		37.45	50
9000	Minimum labor/equipment charge		4	2	Job		98		98	154

23 37 23 – HVAC Gravity Ventilators

23 37 23.10 HVAC Gravity Air Ventilators

		Crew	Daily Output	Labor-Hours	Unit	Material	Labor	Equipment	Total	Total Incl O&P
0010	**HVAC GRAVITY AIR VENTILATORS**, Incl. base and damper									
9000	Minimum labor/equipment charge	1 Plum	2	4	Job		208		208	325

23 38 Ventilation Hoods

23 38 13 – Commercial-Kitchen Hoods

23 38 13.10 Hood and Ventilation Equipment		Crew	Daily Output	Labor-Hours	Unit	Material	2010 Bare Costs Labor	Equipment	Total	Total Incl O&P
0010	HOOD AND VENTILATION EQUIPMENT									
2970	Exhaust hood, sst, gutter on all sides, 4' x 4' x 2'	1 Carp	1.80	4.444	Ea.	4,325	185		4,510	5,075
2980	4' x 4' x 7'	"	1.60	5		6,800	208		7,008	7,825
7950	Hood fire protection system, minimum	Q-1	3	5.333		4,300	250		4,550	5,100
8050	Maximum	"	1	16		33,800	750		34,550	38,400

23 41 Particulate Air Filtration

23 41 13 – Panel Air Filters

23 41 13.10 Panel Type Air Filters

0010	PANEL TYPE AIR FILTERS									
0020										
2950	Mechanical media filtration units									
3000	High efficiency type, with frame, non-supported	G				MCFM	45			45	49.50
3100	Supported type	G				"	60			60	66
5500	Throwaway glass or paper media type					Ea.	3.35			3.35	3.69

23 41 16 – Renewable-Media Air Filters

23 41 16.10 Disposable Media Air Filters

0010	DISPOSABLE MEDIA AIR FILTERS									
5000	Renewable disposable roll					MCFM	250			250	275

23 41 19 – Washable Air Filters

23 41 19.10 Permanent Air Filters

0010	PERMANENT AIR FILTERS									
4500	Permanent washable	G				MCFM	20			20	22

23 41 23 – Extended Surface Filters

23 41 23.10 Expanded Surface Filters

0010	EXPANDED SURFACE FILTERS									
4000	Medium efficiency, extended surface	G				MCFM	5.50			5.50	6.05

23 42 Gas-Phase Air Filtration

23 42 13 – Activated-Carbon Air Filtration

23 42 13.10 Charcoal Type Air Filtration

0010	CHARCOAL TYPE AIR FILTRATION								
0050	Activated charcoal type, full flow				MCFM	600			600	660
0060	Full flow, impregnated media 12" deep					225			225	248
0070	HEPA filter & frame for field erection					300			300	330
0080	HEPA filter-diffuser, ceiling install.					275			275	305

23 42 16 – Chemically-Impregnated Adsorption Air Filtration

23 42 16.10 Chemical Adsorption Air Filtration

0010	CHEMICAL ADSORPTION AIR FILTRATION									
9000	Minimum labor/equipment charge		1 Shee	2.25	3.556	Job		175		175	273

23 43 Electronic Air Cleaners

23 43 13 – Washable Electronic Air Cleaners

23 43 13.10 Electronic Air Cleaners	Crew	Daily Output	Labor-Hours	Unit	Material	2010 Bare Costs Labor	Equipment	Total	Total Incl O&P
0010 **ELECTRONIC AIR CLEANERS**									
2000 Electronic air cleaner, duct mounted									
2150 400 – 1000 CFM	1 Shee	2.30	3.478	Ea.	1,675	171		1,846	2,125
2200 1000 – 1400 CFM		2.20	3.636		1,525	179		1,704	1,950
2250 1400 – 2000 CFM	↓	2.10	3.810	↓	1,525	187		1,712	1,975

23 51 Breechings, Chimneys, and Stacks

23 51 13 – Draft Control Devices

23 51 13.16 Vent Dampers

	Crew	Daily Output	Labor-Hours	Unit	Material	Labor	Equipment	Total	Total Incl O&P
0010 **VENT DAMPERS**									
5000 Vent damper, bi-metal, gas, 3" diameter	Q-9	24	.667	Ea.	40	29.50		69.50	90
5010 4" diameter		24	.667		40	29.50		69.50	90
5020 5" diameter		23	.696		40	31		71	92
5030 6" diameter		22	.727		40	32		72	94.50
5040 7" diameter		21	.762		42.50	33.50		76	99.50
5050 8" diameter	↓	20	.800	↓	46	35.50		81.50	107
9000 Minimum labor/equipment charge	1 Shee	4	2	Job		98		98	154

23 51 13.19 Barometric Dampers

	Crew	Daily Output	Labor-Hours	Unit	Material	Labor	Equipment	Total	Total Incl O&P
0010 **BAROMETRIC DAMPERS**									
1000 Barometric, gas fired system only, 6" size for 5" and 6" pipes	1 Shee	20	.400	Ea.	59.50	19.65		79.15	96
1040 8" size, for 7" and 8" pipes	"	18	.444	"	82.50	22		104.50	125
2000 All fuel, oil, oil/gas, coal									
2020 10" for 9" and 10" pipes	1 Shee	15	.533	Ea.	138	26		164	192
3260 For thermal switch for above, add	"	24	.333	"	57	16.35		73.35	88

23 51 23 – Gas Vents

23 51 23.10 Gas Chimney Vents

	Crew	Daily Output	Labor-Hours	Unit	Material	Labor	Equipment	Total	Total Incl O&P
0010 **GAS CHIMNEY VENTS,** Prefab metal, U.L. listed									
0020 Gas, double wall, galvanized steel									
0080 3" diameter	Q-9	72	.222	V.L.F.	5.35	9.80		15.15	21.50
0100 4" diameter		68	.235		6.70	10.40		17.10	23.50
0120 5" diameter		64	.250		7.80	11.05		18.85	26
0140 6" diameter		60	.267		9.10	11.80		20.90	28.50
0160 7" diameter		56	.286		13.35	12.65		26	34.50
0180 8" diameter		52	.308		14.85	13.60		28.45	38
0200 10" diameter		48	.333		34	14.75		48.75	60
0220 12" diameter		44	.364		45	16.05		61.05	74.50
0260 16" diameter	↓	40	.400		103	17.70		120.70	141
0300 20" diameter	Q-10	36	.667		139	30.50		169.50	201
0340 24" diameter	"	32	.750	↓	245	34.50		279.50	325

23 51 26 – All-Fuel Vent Chimneys

23 51 26.30 All-Fuel Vent Chimneys, Double Wall, St. Stl.

	Crew	Daily Output	Labor-Hours	Unit	Material	Labor	Equipment	Total	Total Incl O&P
0010 **ALL-FUEL VENT CHIMNEYS, DOUBLE WALL, STAINLESS STEEL**									
7800 All fuel, double wall, stainless steel, 6" diameter	Q-9	60	.267	V.L.F.	56.50	11.80		68.30	80.50
7802 7" diameter		56	.286		73	12.65		85.65	100
7804 8" diameter		52	.308		86.50	13.60		100.10	117
7806 10" diameter		48	.333		125	14.75		139.75	160
7808 12" diameter		44	.364		167	16.05		183.05	209
7810 14" diameter	↓	42	.381		219	16.85		235.85	268
8000 All fuel, double wall, stainless steel fittings									
8010 Roof support 6" diameter	Q-9	30	.533	Ea.	103	23.50		126.50	151

23 51 Breechings, Chimneys, and Stacks

23 51 26 – All-Fuel Vent Chimneys

23 51 26.30 All-Fuel Vent Chimneys, Double Wall, St. Stl.

		Crew	Daily Output	Labor-Hours	Unit	Material	2010 Bare Costs Labor	Equipment	Total	Total Incl O&P
8020	7" diameter	Q-9	28	.571	Ea.	117	25.50		142.50	168
8030	8" diameter		26	.615		130	27		157	186
8040	10" diameter		24	.667		150	29.50		179.50	211
8050	12" diameter		22	.727		181	32		213	250
8060	14" diameter		21	.762		229	33.50		262.50	305
8100	Elbow 15°, 6" diameter		30	.533		165	23.50		188.50	219
8120	7" diameter		28	.571		186	25.50		211.50	245
8140	8" diameter		26	.615		212	27		239	276
8160	10" diameter		24	.667		278	29.50		307.50	350
8180	12" diameter		22	.727		335	32		367	420
8200	14" diameter		21	.762		405	33.50		438.50	500
8300	Insulated tee with insulated tee cap, 6" diameter		30	.533		156	23.50		179.50	209
8340	7" diameter		28	.571		205	25.50		230.50	265
8360	8" diameter		26	.615		231	27		258	297
8380	10" diameter		24	.667		325	29.50		354.50	405
8400	12" diameter		22	.727		455	32		487	550
8420	14" diameter		21	.762		600	33.50		633.50	715
8500	Joist shield, 6" diameter		30	.533		47.50	23.50		71	89.50
8510	7" diameter		28	.571		51.50	25.50		77	96.50
8520	8" diameter		26	.615		63.50	27		90.50	113
8530	10" diameter		24	.667		86	29.50		115.50	141
8540	12" diameter		22	.727		107	32		139	169
8550	14" diameter		21	.762		133	33.50		166.50	199
8600	Round top, 6" diameter		30	.533		53	23.50		76.50	95
8620	7" diameter		28	.571		72	25.50		97.50	119
8640	8" diameter		26	.615		97	27		124	150
8660	10" diameter		24	.667		180	29.50		209.50	244
8680	12" diameter		22	.727		249	32		281	325
8700	14" diameter		21	.762		330	33.50		363.50	420
8800	Adjustable roof flashing, 6" diameter		30	.533		63	23.50		86.50	106
8820	7" diameter		28	.571		72	25.50		97.50	119
8840	8" diameter		26	.615		78	27		105	129
8860	10" diameter		24	.667		100	29.50		129.50	156
8880	12" diameter		22	.727		130	32		162	193
8900	14" diameter		21	.762		162	33.50		195.50	231

23 51 33 – Insulated Sectional Chimneys

23 51 33.10 Prefabricated Insulated Sectional Chimneys

		Crew	Daily Output	Labor-Hours	Unit	Material	Labor	Equipment	Total	Total Incl O&P
0010	**PREFABRICATED INSULATED SECTIONAL CHIMNEYS**									
9990	Minimum labor/equipment charge	Q-9	3	5.333	Job		236		236	370

23 52 Heating Boilers

23 52 13 – Electric Boilers

23 52 13.10 Electric Boilers, ASME

		Crew	Daily Output	Labor-Hours	Unit	Material	Labor	Equipment	Total	Total Incl O&P
0010	**ELECTRIC BOILERS, ASME**, Standard controls and trim.									
1000	Steam, 6 KW, 20.5 MBH	Q-19	1.20	20	Ea.	3,500	950		4,450	5,325
1160	60 KW, 205 MBH		1	24		6,025	1,150		7,175	8,375
1300	296 KW, 1010 MBH		.45	53.333		13,200	2,525		15,725	18,400
1400	592 KW, 2020 MBH	Q-21	.34	94.118		25,100	4,575		29,675	34,700
1540	1110 KW, 3788 MBH	"	.19	168		35,100	8,175		43,275	51,000
2000	Hot water, 7.5 KW, 25.6 MBH	Q-19	1.30	18.462		3,750	875		4,625	5,475

23 52 Heating Boilers

23 52 13 – Electric Boilers

23 52 13.10 Electric Boilers, ASME

		Crew	Daily Output	Labor-Hours	Unit	Material	2010 Bare Costs Labor	Equipment	Total	Total Incl O&P
2040	30 KW, 102 MBH	Q-19	1.20	20	Ea.	3,975	950		4,925	5,850
2060	45 KW, 164 MBH		1.20	20		4,475	950		5,425	6,400
2070	60 KW, 205 MBH		1.20	20		4,725	950		5,675	6,675
2080	75 KW, 256 MBH		1.10	21.818		5,075	1,025		6,100	7,200
2100	90 KW, 307 MBH	▼	1.10	21.818	▼	5,475	1,025		6,500	7,625
9000	Minimum labor/equipment charge	Q-20	1	20	Job		905		905	1,400

23 52 23 – Cast-Iron Boilers

23 52 23.20 Gas-Fired Boilers

		Crew	Daily Output	Labor-Hours	Unit	Material	Labor	Equipment	Total	Total Incl O&P
0010	**GAS-FIRED BOILERS,** Natural or propane, standard controls, packaged									
1000	Cast iron, with insulated jacket									
2000	Steam, gross output, 81 MBH	Q-7	1.40	22.857	Ea.	1,950	1,125		3,075	3,900
2060	163 MBH		.90	35.556		2,825	1,750		4,575	5,850
2200	440 MBH		.51	62.500		5,525	3,100		8,625	10,900
2320	1,875 MBH		.30	106		18,300	5,275		23,575	28,300
2400	3570 MBH		.18	181		26,500	8,975		35,475	43,100
2540	6,970 MBH	▼	.10	320	▼	86,500	15,800		102,300	119,500
2800										
3000	Hot water, gross output, 80 MBH	Q-7	1.46	21.918	Ea.	1,775	1,075		2,850	3,650
3020	100 MBH		1.35	23.704		2,300	1,175		3,475	4,350
3040	122 MBH		1.10	29.091		2,200	1,450		3,650	4,650
3060	163 MBH		1	32		2,800	1,575		4,375	5,550
3200	440 MBH		.58	54.983		5,350	2,725		8,075	10,100
3320	2,000 MBH		.26	125		17,900	6,175		24,075	29,300
3540	6,970 MBH	▼	.09	359		94,000	17,800		111,800	131,000
7000	For tankless water heater, add					10%				
7050	For additional zone valves up to 312 MBH add				▼	150			150	165
9900	Minimum labor/equipment charge	Q-6	1	24	Job		1,150		1,150	1,800

23 52 23.30 Gas/Oil Fired Boilers

		Crew	Daily Output	Labor-Hours	Unit	Material	Labor	Equipment	Total	Total Incl O&P
0010	**GAS/OIL FIRED BOILERS,** Combination with burners and controls, packaged									
1000	Cast iron with insulated jacket									
2000	Steam, gross output, 720 MBH	Q-7	.43	74.074	Ea.	13,900	3,650		17,550	21,000
2140	2,700 MBH		.19	165		27,500	8,200		35,700	43,000
2380	6,970 MBH	▼	.09	372	▼	95,000	18,400		113,400	133,000
2900	Hot water, gross output									
2910	200 MBH	Q-6	.62	39.024	Ea.	7,775	1,900		9,675	11,500
2920	300 MBH		.49	49.080		7,775	2,375		10,150	12,200
2930	400 MBH		.41	57.971		9,100	2,800		11,900	14,400
2940	500 MBH	▼	.36	67.039		9,800	3,250		13,050	15,800
3000	584 MBH	Q-7	.44	72.072		8,125	3,575		11,700	14,500
3060	1,460 MBH		.28	113		22,900	5,600		28,500	33,900
3300	13,500 MBH, 403.3 BHP	▼	.04	727	▼	140,500	35,900		176,400	210,000

23 52 23.40 Oil-Fired Boilers

		Crew	Daily Output	Labor-Hours	Unit	Material	Labor	Equipment	Total	Total Incl O&P
0010	**OIL-FIRED BOILERS,** Standard controls, flame retention burner, packaged									
1000	Cast iron, with insulated flush jacket									
2000	Steam, gross output, 109 MBH	Q-7	1.20	26.667	Ea.	1,975	1,325		3,300	4,225
2060	207 MBH		.90	35.556		2,700	1,750		4,450	5,700
2180	1,084 MBH		.38	85.106		9,400	4,200		13,600	16,800
2200	1,360 MBH		.33	98.160		10,800	4,850		15,650	19,400
2240	2,175 MBH		.24	133		17,300	6,625		23,925	29,400
2340	4,360 MBH		.15	214		37,300	10,600		47,900	57,500
2460	6,970 MBH	▼	.09	363	▼	88,000	18,000		106,000	125,000
3000	Hot water, same price as steam									

23 52 Heating Boilers

23 52 26 – Steel Boilers

23 52 26.40 Oil-Fired Boilers

		Crew	Daily Output	Labor-Hours	Unit	Material	Labor	Equipment	Total	Total Incl O&P
0010	**OIL-FIRED BOILERS**, Standard controls, flame retention burner									
5000	Steel, with insulated flush jacket									
7000	Hot water, gross output, 103 MBH	Q-6	1.60	15	Ea.	1,725	725		2,450	3,025
7020	122 MBH		1.45	16.506		1,825	800		2,625	3,250
7040	137 MBH		1.36	17.595		1,925	850		2,775	3,450
7060	168 MBH		1.30	18.405		2,025	890		2,915	3,600
7080	225 MBH		1.22	19.704		2,700	955		3,655	4,425
7100	315 MBH		.96	25.105		5,525	1,225		6,750	7,975
7120	420 MBH		.70	34.483		5,975	1,675		7,650	9,150
9000	Minimum labor/equipment charge		1.75	13.714	Job		665		665	1,025

23 52 88 – Burners

23 52 88.10 Replacement Type Burners

		Crew	Daily Output	Labor-Hours	Unit	Material	Labor	Equipment	Total	Total Incl O&P
0010	**REPLACEMENT TYPE BURNERS**									
0990	Residential, conversion, gas fired, LP or natural									
1000	Gun type, atmospheric input 50 to 225 MBH	Q-1	2.50	6.400	Ea.	650	300		950	1,175
1020	100 to 400 MBH		2	8		1,075	375		1,450	1,750
1040	300 to 1000 MBH		1.70	9.412		3,300	440		3,740	4,300
3000	Flame retention oil fired assembly, input									
3040	2.0 to 5.0 GPH	Q-1	2	8	Ea.	345	375		720	960

23 54 Furnaces

23 54 13 – Electric-Resistance Furnaces

23 54 13.10 Electric Furnaces

		Crew	Daily Output	Labor-Hours	Unit	Material	Labor	Equipment	Total	Total Incl O&P
0010	**ELECTRIC FURNACES**, Hot air, blowers, std. controls									
0011	not including gas, oil or flue piping									
1000	Electric, UL listed									
1020	10.2 MBH	Q-20	5	4	Ea.	475	181		656	800
1100	34.1 MBH	"	4.40	4.545	"	615	205		820	995

23 54 16 – Fuel-Fired Furnaces

23 54 16.13 Gas-Fired Furnaces

		Crew	Daily Output	Labor-Hours	Unit	Material	Labor	Equipment	Total	Total Incl O&P
0010	**GAS-FIRED FURNACES**									
3000	Gas, AGA certified, upflow, direct drive models									
3020	45 MBH input	Q-9	4	4	Ea.	575	177		752	910
3040	60 MBH input		3.80	4.211		615	186		801	965
3060	75 MBH input		3.60	4.444		685	196		881	1,050
3100	100 MBH input		3.20	5		695	221		916	1,100
3120	125 MBH input		3	5.333		700	236		936	1,150
3130	150 MBH input		2.80	5.714		705	253		958	1,175
3140	200 MBH input		2.60	6.154		2,425	272		2,697	3,075

23 54 16.16 Oil-Fired Furnaces

		Crew	Daily Output	Labor-Hours	Unit	Material	Labor	Equipment	Total	Total Incl O&P
0010	**OIL-FIRED FURNACES**									
6000	Oil, UL listed, atomizing gun type burner									
6020	56 MBH output	Q-9	3.60	4.444	Ea.	1,850	196		2,046	2,325
6040	95 MBH output		3.40	4.706		1,875	208		2,083	2,400
6060	134 MBH output		3.20	5		1,950	221		2,171	2,500
6080	151 MBH output		3	5.333		2,275	236		2,511	2,875
6100	200 MBH input		2.60	6.154		2,325	272		2,597	2,975
9000	Minimum labor/equipment charge		2.75	5.818	Job		257		257	400

23 54 Furnaces

23 54 24 – Furnace Components for Cooling

23 54 24.10 Furnace Components and Combinations	Crew	Daily Output	Labor-Hours	Unit	Material	2010 Bare Costs Labor	Equipment	Total	Total Incl O&P
0010 **FURNACE COMPONENTS AND COMBINATIONS**									
0080 Coils, A/C evaporator, for gas or oil furnaces									
0090 Add-on, with holding charge									
0100 Upflow									
0120 1-1/2 ton cooling	Q-5	4	4	Ea.	158	187		345	465
0130 2 ton cooling		3.70	4.324		188	202		390	520
0140 3 ton cooling		3.30	4.848		238	226		464	610
0150 4 ton cooling		3	5.333		330	249		579	745
0160 5 ton cooling		2.70	5.926		385	277		662	855
0300 Downflow									
0330 2-1/2 ton cooling	Q-5	3	5.333	Ea.	232	249		481	640
0340 3-1/2 ton cooling		2.60	6.154		276	287		563	750
0350 5 ton cooling		2.20	7.273		385	340		725	950
0600 Horizontal									
0630 2 ton cooling	Q-5	3.90	4.103	Ea.	260	192		452	585
0640 3 ton cooling		3.50	4.571		296	213		509	655
0650 4 ton cooling		3.20	5		375	234		609	770
0660 5 ton cooling		2.90	5.517		375	258		633	810
2000 Cased evaporator coils for air handlers									
2100 1-1/2 ton cooling	Q-5	4.40	3.636	Ea.	325	170		495	620
2110 2 ton cooling		4.10	3.902		325	182		507	640
2120 2-1/2 ton cooling		3.90	4.103		345	192		537	675
2130 3 ton cooling		3.70	4.324		385	202		587	735
2140 3-1/2 ton cooling		3.50	4.571		400	213		613	770
2150 4 ton cooling		3.20	5		480	234		714	890
2160 5 ton cooling		2.90	5.517		555	258		813	1,000
3010 Air handler, modular									
3100 With cased evaporator cooling coil									
3120 1-1/2 ton cooling	Q-5	3.80	4.211	Ea.	1,025	197		1,222	1,425
3130 2 ton cooling		3.50	4.571		1,175	213		1,388	1,600
3140 2-1/2 ton cooling		3.30	4.848		1,175	226		1,401	1,625
3150 3 ton cooling		3.10	5.161		1,275	241		1,516	1,775
3160 3-1/2 ton cooling		2.90	5.517		1,325	258		1,583	1,850
3170 4 ton cooling		2.50	6.400		1,475	299		1,774	2,100
3180 5 ton cooling		2.10	7.619		1,675	355		2,030	2,375
3500 With no cooling coil									
3520 1-1/2 ton coil size	Q-5	12	1.333	Ea.	795	62.50		857.50	970
3530 2 ton coil size		10	1.600		815	74.50		889.50	1,025
3540 2-1/2 ton coil size		10	1.600		890	74.50		964.50	1,100
3554 3 ton coil size		9	1.778		905	83		988	1,125
3560 3-1/2 ton coil size		9	1.778		970	83		1,053	1,200
3570 4 ton coil size		8.50	1.882		1,150	88		1,238	1,375
3580 5 ton coil size		8	2		1,300	93.50		1,393.50	1,600
4000 With heater									
4120 5 kW, 17.1 MBH	Q-5	16	1	Ea.	690	46.50		736.50	835
4130 7.5 kW, 25.6 MBH		15.60	1.026		810	48		858	965
4140 10 kW, 34.2 MBH		15.20	1.053		930	49		979	1,100
4150 12.5 KW, 42.7 MBH		14.80	1.081		1,050	50.50		1,100.50	1,225

23 55 Fuel-Fired Heaters

23 55 13 – Fuel-Fired Duct Heaters

23 55 13.16 Gas-Fired Duct Heaters

		Crew	Daily Output	Labor-Hours	Unit	Material	2010 Bare Costs Labor	Equipment	Total	Total Incl O&P
0010	**GAS-FIRED DUCT HEATERS**, Includes burner, controls, stainless steel									
0020	heat exchanger. Gas fired, electric ignition									
1000	Outdoor installation, with power venter									
1120	225 MBH output	Q-5	2.50	6.400	Ea.	5,225	299		5,524	6,225
1160	375 MBH output		1.60	10		6,425	465		6,890	7,800
1180	450 MBH output	↓	1.40	11.429	↓	7,450	535		7,985	9,025

23 55 33 – Fuel-Fired Unit Heaters

23 55 33.16 Gas-Fired Unit Heaters

		Crew	Daily Output	Labor-Hours	Unit	Material	Labor	Equipment	Total	Total Incl O&P
0010	**GAS-FIRED UNIT HEATERS**, Cabinet, grilles, fan, ctrls., burner, no piping									
0022	thermostat, no piping. For flue see Div. 23 51 23.10									
1000	Gas fired, floor mounted									
1100	60 MBH output	Q-5	10	1.600	Ea.	805	74.50		879.50	1,000
1180	180 MBH output		6	2.667		1,275	125		1,400	1,600
2000	Suspension mounted, propeller fan, 20 MBH output		8.50	1.882		985	88		1,073	1,200
2040	60 MBH output		7	2.286		1,200	107		1,307	1,500
2100	130 MBH output		5	3.200		1,500	149		1,649	1,875
2240	320 MBH output	↓	2	8		2,650	375		3,025	3,475
2500	For powered venter and adapter, add					355			355	390
5000	Wall furnace, 17.5 MBH output	Q-5	6	2.667		715	125		840	980
5020	24 MBH output		5	3.200		740	149		889	1,050
5040	35 MBH output	↓	4	4	↓	895	187		1,082	1,275
9000	Minimum labor/equipment charge	↓	3.50	4.571	Job		213		213	330

23 56 Solar Energy Heating Equipment

23 56 16 – Packaged Solar Heating Equipment

23 56 16.40 Solar Heating Systems

			Crew	Daily Output	Labor-Hours	Unit	Material	Labor	Equipment	Total	Total Incl O&P
0010	**SOLAR HEATING SYSTEMS**										
0020	System/Package prices, not including connecting										
0030	pipe, insulation, or special heating/plumbing fixtures	G									
0500	Hot water, standard package, low temperature	G									
0580	2 collectors, circulator, fittings, 120 gal. tank	G	Q-1	.40	40	Ea.	2,325	1,875		4,200	5,450

23 56 19 – Solar Heating Components

23 56 19.50 Solar Heating Ancillary

			Crew	Daily Output	Labor-Hours	Unit	Material	Labor	Equipment	Total	Total Incl O&P
0010	**SOLAR HEATING ANCILLARY**										
2250	Controller, liquid temperature	G	1 Plum	5	1.600	Ea.	96	83.50		179.50	234
2300	Circulators, air	G									
2310	Blowers										
2580	8" diameter, 150 CFM	G	Q-9	16	1	Ea.	39.50	44		83.50	113
2660	Shutter/damper	G		12	1.333		49	59		108	146
2670	Shutter motor	G	↓	16	1		119	44		163	200
2800	Circulators, liquid, 1/25 HP, 5.3 GPM	G	Q-1	14	1.143		268	53.50		321.50	380
2820	1/20 HP, 17 GPM	G	"	12	1.333	↓	154	62.50		216.50	267
3000	Collector panels, air with aluminum absorber plate										
3010	Wall or roof mount										
3040	Flat black, plastic glazing										
3080	4' x 8'	G	Q-9	6	2.667	Ea.	635	118		753	880
3300	Collector panels, liquid with copper absorber plate										
3320	Black chrome, tempered glass glazing										
3330	Alum. frame, 4' x 8', 5/32" single glazing	G	Q-1	9.50	1.684	Ea.	945	79		1,024	1,175
3450	Flat black, alum. frame, 3.5' x 7.5'	G		9	1.778		690	83.50		773.50	890

23 56 Solar Energy Heating Equipment

23 56 19 – Solar Heating Components

23 56 19.50 Solar Heating Ancillary		Crew	Daily Output	Labor-Hours	Unit	Material	2010 Bare Costs Labor	Equipment	Total	Total Incl O&P
3500	4' x 8'	G	Q-1	5.50	2.909	Ea.	865	136		1,001	1,150
3600	Liquid, full wetted, plastic, alum. frame, 3' x 10'	G		5	3.200		234	150		384	490
3650	Collector panel mounting, flat roof or ground rack	G		7	2.286		217	107		324	405
3670	Roof clamps	G		70	.229	Set	2.35	10.70		13.05	19.20
3700	Roof strap, teflon	G	1 Plum	205	.039	L.F.	19.95	2.03		21.98	25
3900	Differential controller with two sensors										
3930	Thermostat, hard wired	G	1 Plum	8	1	Ea.	79.50	52		131.50	168
4050	Pool valve system	G	"	2.50	3.200	"	264	167		431	550
4150	Sensors										
4220	Freeze prevention	G	1 Plum	32	.250	Ea.	22	13		35	44
4300	Heat exchanger										
4315	includes coil, blower, circulator										
4316	and controller for DHW and space hot air										
4380	70 MBH	G	Q-1	3.50	4.571	Ea.	335	214		549	695
4400	80 MBH	G	"	3	5.333	"	445	250		695	875
4580	Fluid to fluid package includes two circulating pumps										
4590	expansion tank, check valve, relief valve										
4600	controller, high temperature cutoff and sensors	G	Q-1	2.50	6.400	Ea.	695	300		995	1,225
4650	Heat transfer fluid										
4700	Propylene glycol, inhibited anti-freeze	G	1 Plum	28	.286	Gal.	12.85	14.85		27.70	37
4800	Solar storage tanks, knocked down										
5020	7' x 10'-6" = 459 C.F./3000 gallons	G	Q-10	.80	30	Ea.	14,700	1,375		16,075	18,400
5210	7' x 14' = 613 C.F./4000 gallons	G		.60	40		17,700	1,825		19,525	22,400
5230	10'-6" x 14' = 919 C.F./6000 gallons	G		.40	60		21,500	2,750		24,250	27,900
5250	14' x 17'-6" = 1531 C.F./10,000 gallons	G	Q-11	.30	106		28,100	5,000		33,100	38,800
7000	Solar control valves and vents										
7070	Air eliminator, automatic 3/4" size	G	1 Plum	32	.250	Ea.	28.50	13		41.50	51.50
7090	Air vent, automatic, 1/8" fitting	G		32	.250		12.65	13		25.65	34
7100	Manual, 1/8" NPT	G		32	.250		2.78	13		15.78	23
7120	Backflow preventer, 1/2" pipe size	G		16	.500		58.50	26		84.50	105
7130	3/4" pipe size	G		16	.500		106	26		132	158
7150	Balancing valve, 3/4" pipe size	G		20	.400		43	21		64	79.50
7180	Draindown valve, 1/2" copper tube	G		9	.889		202	46.50		248.50	294
7200	Flow control valve, 1/2" pipe size	G		22	.364		107	18.95		125.95	148
7220	Expansion tank, up to 5 gal.	G		32	.250		61.50	13		74.50	87.50
7400	Pressure gauge, 2" dial	G		32	.250		22	13		35	44.50
7450	Relief valve, temp. and pressure 3/4" pipe size	G		30	.267		15.60	13.90		29.50	38.50
7500	Solenoid valve, normally closed										
7750	Vacuum relief valve, 3/4" pipe size	G	1 Plum	32	.250	Ea.	26.50	13		39.50	49
7800	Thermometers										
8250	Water storage tank with heat exchanger and electric element										
8270	66 gal. with 2" x 2 lb. density insulation	G	1 Plum	1.60	5	Ea.	970	260		1,230	1,475
8300	80 gal. with 2" x 2 lb. density insulation	G	"	1.60	5	"	1,100	260		1,360	1,600
8500	Water storage module, plastic										
8600	Tubular, 12" diameter, 4' high	G	1 Carp	48	.167	Ea.	103	6.95		109.95	124
8610	12" diameter, 8' high	G	"	40	.200		159	8.30		167.30	189
8650	Cap, 12" diameter	G					19			19	21
9000	Minimum labor/equipment charge	G	1 Plum	2	4	Job		208		208	325

23 61 Refrigerant Compressors

23 61 16 – Reciprocating Refrigerant Compressors

23 61 16.10 Reciprocating Compressors	Crew	Daily Output	Labor Hours	Unit	Material	2010 Bare Costs Labor	Equipment	Total	Total Incl O&P
0010 **RECIPROCATING COMPRESSORS**									
0990 Refrigeration, recip. hermetic, switches & protective devices									
1000 10 ton	Q-5	1	16	Ea.	8,850	745		9,595	10,900
1100 20 ton	Q-6	.72	33.333		13,400	1,625		15,025	17,200
1400 50 ton	"	.20	120		17,500	5,800		23,300	28,200
1600 130 ton	Q-7	.21	152		22,100	7,525		29,625	36,000

23 62 Packaged Compressor and Condenser Units

23 62 13 – Packaged Air-Cooled Refrigerant Compressor and Condenser Units

23 62 13.10 Packaged Air-Cooled Refrig. Condensing Units

	Crew	Daily Output	Labor Hours	Unit	Material	Labor	Equipment	Total	Total Incl O&P
0010 **PACKAGED AIR-COOLED REFRIGERANT CONDENSING UNITS**									
0020 Condensing unit									
0030 Air cooled, compressor, standard controls									
0050 1.5 ton	Q-5	2.50	6.400	Ea.	1,175	299		1,474	1,775
0200 2.5 ton		1.70	9.412		1,300	440		1,740	2,125
0300 3 ton		1.30	12.308		1,325	575		1,900	2,375
0400 4 ton		.90	17.778		1,775	830		2,605	3,225
0500 5 ton		.60	26.667		2,125	1,250		3,375	4,275

23 64 Packaged Water Chillers

23 64 19 – Reciprocating Water Chillers

23 64 19.10 Reciprocating Type Water Chillers

	Crew	Daily Output	Labor Hours	Unit	Material	Labor	Equipment	Total	Total Incl O&P
0010 **RECIPROCATING TYPE WATER CHILLERS**, With standard controls									
0494 Water chillers, integral air cooled condenser									
0980 Water cooled, multiple compressor, semi-hermetic, tower not incl.									
1100 50 ton cooling	Q-7	.28	113	Ea.	30,500	5,625		36,125	42,200
1160 100 ton cooling		.18	179		58,500	8,875		67,375	78,500
1200 145 ton cooling		.16	202		77,500	10,000		87,500	101,000

23 64 23 – Scroll Water Chillers

23 64 23.10 Scroll Water Chillers

	Crew	Daily Output	Labor Hours	Unit	Material	Labor	Equipment	Total	Total Incl O&P
0010 **SCROLL WATER CHILLERS**, With standard controls									
0490 Packaged w/integral air cooled condenser, 15 ton cool	Q-7	.37	86.486	Ea.	17,400	4,275		21,675	25,700
0500 20 ton cooling		.34	94.118		23,600	4,650		28,250	33,100
0520 40 ton cooling		.30	108		31,700	5,350		37,050	43,100
0680 Scroll water cooled, single compressor, hermetic, tower not incl.									
0760 10 ton cooling	Q-6	.36	67.039	Ea.	6,125	3,250		9,375	11,800

23 64 26 – Rotary-Screw Water Chillers

23 64 26.10 Rotary-Screw Type Water Chillers

	Crew	Daily Output	Labor Hours	Unit	Material	Labor	Equipment	Total	Total Incl O&P
0010 **ROTARY-SCREW TYPE WATER CHILLERS**, With standard controls									
0110 Screw, liquid chiller, air cooled, insulated evaporator									
0120 130 ton	Q-7	.14	228	Ea.	81,000	11,300		92,300	106,500
0124 160 ton	"	.13	246	"	100,000	12,200		112,200	129,000

23 64 33 – Direct Expansion Water Chillers

23 64 33.10 Direct Expansion Type Water Chillers

	Crew	Daily Output	Labor Hours	Unit	Material	Labor	Equipment	Total	Total Incl O&P
0010 **DIRECT EXPANSION TYPE WATER CHILLERS**, With standard controls									
9000 Minimum labor/equipment charge	Q-6	1	24	Job		1,150		1,150	1,800

23 65 Cooling Towers

23 65 13 – Forced-Draft Cooling Towers

23 65 13.10 Forced-Draft Type Cooling Towers

		Crew	Daily Output	Labor-Hours	Unit	Material	2010 Bare Costs Labor	2010 Bare Costs Equipment	Total	Total Incl O&P
0010	**FORCED-DRAFT TYPE COOLING TOWERS**, Packaged units									
0070	Galvanized steel									
0080	Induced draft, crossflow									
0100	Vertical, belt drive, 61 tons	Q-6	90	.267	TonAC	163	12.90		175.90	199
0150	100 ton		100	.240		156	11.60		167.60	189
0200	115 ton		109	.220		145	10.65		155.65	177
0250	131 ton		120	.200		144	9.70		153.70	173
0260	162 ton		132	.182		142	8.80		150.80	170
1000	For higher capacities, use multiples									
1500	Induced air, double flow									
1900	Vertical, gear drive, 167 ton	Q-6	126	.190	TonAC	138	9.20		147.20	166
2000	297 ton		129	.186		86	9		95	109
2100	582 ton		132	.182		69.50	8.80		78.30	90
2200	1016 ton		150	.160		62.50	7.75		70.25	81
3500	For pumps and piping, add		38	.632		81.50	30.50		112	137
4000	For absorption systems, add					75%	75%			
5000	Fiberglass tower on galvanized steel support structure									
5010	Draw thru									
5100	100 ton	Q-6	1.40	17.143	Ea.	10,100	830		10,930	12,400
5120	120 ton		1.20	20		11,700	970		12,670	14,400
5140	140 ton		1	24		12,700	1,150		13,850	15,800
5160	160 ton		.80	30		14,100	1,450		15,550	17,800
5180	180 ton		.65	36.923		16,000	1,800		17,800	20,400
5200	200 ton		.48	50		18,100	2,425		20,525	23,700
5300	For stainless steel support structure, add					30%				
5360	For higher capacities, use multiples of each size									
6000	Stainless steel									
6010	Induced draft, crossflow, horizontal, belt drive									
6100	57 ton	Q-6	1.50	16	Ea.	21,200	775		21,975	24,600
6120	91 ton		.99	24.242		25,700	1,175		26,875	30,000
6140	111 ton		.43	55.814		38,500	2,700		41,200	46,600
6160	126 ton		.22	109		38,500	5,275		43,775	50,500
6170	Induced draft, crossflow, vertical, gear drive									
6172	167 ton	Q-6	.75	32	Ea.	48,900	1,550		50,450	56,500
6174	297 ton		.43	55.814		54,500	2,700		57,200	64,000
6176	582 ton		.23	104		86,000	5,050		91,050	102,500
6180	1016 ton		.15	160		135,000	7,750		142,750	160,500
9000	Minimum labor/equipment charge		1	24	Job		1,150		1,150	1,800

23 72 Air-to-Air Energy Recovery Equipment

23 72 13 – Heat-Wheel Air-to-Air Energy-Recovery Equipment

23 72 13.10 Heat-Wheel Air-to-Air Energy Recovery Equip.

		Crew	Daily Output	Labor-Hours	Unit	Material	2010 Bare Costs Labor	2010 Bare Costs Equipment	Total	Total Incl O&P
0010	**HEAT-WHEEL AIR-TO-AIR ENERGY RECOVERY EQUIPMENT**									
0100	Air to air									
9000	Minimum labor/equipment charge	1 Shee	4	2	Job		98		98	154

23 73 Indoor Central-Station Air-Handling Units

23 73 39 – Indoor, Direct Gas-Fired Heating and Ventilating Units

23 73 39.10 Make-Up Air Unit		Crew	Daily Output	Labor-Hours	Unit	Material	2010 Bare Costs Labor	Equipment	Total	Total Incl O&P
0010	**MAKE-UP AIR UNIT**									
0020	Indoor suspension, natural/LP gas, direct fired,									
0032	standard control. For flue see Div. 23 51 23.10									
0040	70°F temperature rise, MBH is input									
0100	75 MBH input	Q-6	3.60	6.667	Ea.	2,400	325		2,725	3,150
0220	225 MBH input	↓	2.40	10	↓	3,300	485		3,785	4,375
0300	400 MBH input	↓	1.60	15	↓	7,450	725		8,175	9,300

23 74 Packaged Outdoor HVAC Equipment

23 74 23 – Packaged, Outdoor, Heating-Only Makeup-Air Units

23 74 23.16 Packaged, Ind.-Fired, In/Outdoor

		Crew	Daily Output	Labor-Hours	Unit	Material	Labor	Equipment	Total	Total Incl O&P
0010	**PACKAGED, IND.-FIRED, IN/OUTDOOR**, Heat-Only Makeup-Air Units									
9000	Minimum labor/equipment charge	Q-6	2.75	8.727	Job		425		425	655

23 74 33 – Packaged, Outdoor, Heating and Cooling Makeup Air-Conditioners

23 74 33.10 Roof Top Air Conditioners

		Crew	Daily Output	Labor-Hours	Unit	Material	Labor	Equipment	Total	Total Incl O&P
0010	**ROOF TOP AIR CONDITIONERS**, Standard controls, curb, economizer									
0020										
1000	Single zone, electric cool, gas heat									
1100	3 ton cooling, 60 MBH heating	Q-5	.70	22.857	Ea.	3,475	1,075		4,550	5,475
1120	4 ton cooling, 95 MBH heating	"	.61	26.403		4,225	1,225		5,450	6,550
1160	10 ton cooling, 200 MBH heating	Q-6	.67	35.982		10,600	1,750		12,350	14,300
1220	30 ton cooling, 540 MBH heating	Q-7	.47	68.376		25,500	3,375		28,875	33,300
1240	40 ton cooling, 675 MBH heating	"	.35	91.168	↓	33,400	4,500		37,900	43,700
2000	Multizone, electric cool, gas heat, economizer									
2120	20 ton cooling, 360 MBH heating	Q-7	.53	60.038	Ea.	61,000	2,975		63,975	72,000
2180	30 ton cooling, 540 MBH heating		.37	85.562		94,000	4,225		98,225	109,500
2210	50 ton cooling, 540 MBH heating		.23	142		133,500	7,025		140,525	157,500
2220	70 ton cooling, 1500 MBH heating		.16	198		144,500	9,825		154,325	174,000
2240	80 ton cooling, 1500 MBH heating		.14	228		165,000	11,300		176,300	199,000
2280	105 ton cooling, 1500 MBH heating	↓	.11	290		190,500	14,400		204,900	232,000
2400	For hot water heat coil, deduct					5%				
2500	For steam heat coil, deduct					2%				
2600	For electric heat, deduct				↓	3%	5%			
9000	Minimum labor/equipment charge	Q-5	1.50	10.667	Job		500		500	770

23 81 Decentralized Unitary HVAC Equipment

23 81 19 – Self-Contained Air-Conditioners

23 81 19.10 Window Unit Air Conditioners

		Crew	Daily Output	Labor-Hours	Unit	Material	Labor	Equipment	Total	Total Incl O&P
0010	**WINDOW UNIT AIR CONDITIONERS**									
4500	10,000 BTUH	1 Carp	6	1.333	Ea.	515	55.50		570.50	655
4520	12,000 BTUH	L-2	8	2	"	630	73		703	815
9000	Minimum labor/equipment charge	1 Carp	2	4	Job		166		166	273

23 81 19.20 Self-Contained Single Package

		Crew	Daily Output	Labor-Hours	Unit	Material	Labor	Equipment	Total	Total Incl O&P
0010	**SELF-CONTAINED SINGLE PACKAGE**									
0100	Air cooled, for free blow or duct, not incl. remote condenser									
0200	3 ton cooling	Q-5	1	16	Ea.	3,300	745		4,045	4,775
0210	4 ton cooling	"	.80	20		3,575	935		4,510	5,400
0240	10 ton cooling	Q-7	1	32	↓	6,775	1,575		8,350	9,900

23 81 Decentralized Unitary HVAC Equipment

23 81 19 – Self-Contained Air-Conditioners

23 81 19.20 Self-Contained Single Package		Crew	Daily Output	Labor-Hours	Unit	Material	2010 Bare Costs Labor	Equipment	Total	Total Incl O&P
0280	30 ton cooling	Q-7	.80	40	Ea.	21,600	1,975		23,575	26,900
1000	Water cooled for free blow or duct, not including tower									
1010	Constant volume									
1100	3 ton cooling	Q-6	1	24	Ea.	3,250	1,150		4,400	5,375
1120	5 ton cooling	"	1	24		4,275	1,150		5,425	6,500
1140	10 ton cooling	Q-7	.90	35.556		8,300	1,750		10,050	11,900
1180	30 ton cooling	"	.70	45.714		34,200	2,250		36,450	41,100
1240	60 ton cooling	Q-8	.30	106		63,000	5,275	195	68,470	77,500
9000	Minimum labor/equipment charge	Q-5	1	16	Job		745		745	1,150

23 81 23 – Computer-Room Air-Conditioners

23 81 23.10 Computer Room Units

		Crew	Daily Output	Labor-Hours	Unit	Material	Labor	Equipment	Total	Total Incl O&P
0010	**COMPUTER ROOM UNITS**									
1000	Air cooled, includes remote condenser but not									
1020	interconnecting tubing or refrigerant									
1160	6 ton	Q-5	.30	53.333	Ea.	30,400	2,500		32,900	37,300
1240	10 ton	"	.25	64		33,800	3,000		36,800	41,800
1300	20 ton	Q-6	.26	92.308		44,500	4,475		48,975	56,000
1320	22 ton	"	.24	100		45,100	4,850		49,950	57,000

23 81 26 – Split-System Air-Conditioners

23 81 26.10 Split Ductless Systems

		Crew	Daily Output	Labor-Hours	Unit	Material	Labor	Equipment	Total	Total Incl O&P
0010	**SPLIT DUCTLESS SYSTEMS**									
0100	Cooling only, single zone									
0110	Wall mount									
0120	3/4 ton cooling	Q-5	2	8	Ea.	655	375		1,030	1,300
0130	1 ton cooling		1.80	8.889		730	415		1,145	1,450
0140	1-1/2 ton cooling		1.60	10		1,150	465		1,615	1,975
0150	2 ton cooling		1.40	11.429		1,300	535		1,835	2,250
1000	Ceiling mount									
1020	2 ton cooling	Q-5	1.40	11.429	Ea.	2,475	535		3,010	3,550
1030	3 ton cooling	"	1.20	13.333	"	3,200	625		3,825	4,500
3000	Multizone									
3010	Wall mount									
3020	2 @ 3/4 ton cooling	Q-5	1.80	8.889	Ea.	2,100	415		2,515	2,975
5000	Cooling/Heating									
5010	Wall mount									
5110	1 ton cooling	Q-5	1.70	9.412	Ea.	765	440		1,205	1,525
5120	1-1/2 ton cooling	"	1.50	10.667	"	1,175	500		1,675	2,075
7000	Accessories for all split ductless systems									
7010	Add for ambient frost control	Q-5	8	2	Ea.	179	93.50		272.50	340
7020	Add for tube/wiring kit									
7030	15' kit	Q-5	32	.500	Ea.	40	23.50		63.50	80
7036	25' kit		28	.571		95	26.50		121.50	147
7040	35' kit		24	.667		129	31		160	190
7050	50' kit		20	.800		167	37.50		204.50	242

23 81 43 – Air-Source Unitary Heat Pumps

23 81 43.10 Air-Source Heat Pumps

		Crew	Daily Output	Labor-Hours	Unit	Material	Labor	Equipment	Total	Total Incl O&P
0010	**AIR-SOURCE HEAT PUMPS**, Not including interconnecting tubing									
1000	Air to air, split system, not including curbs, pads, fan coil and ductwork									
1012	Outside condensing unit only, for fan coil see Div. 23 82 19.10									
1020	2 ton cooling, 8.5 MBH heat @ 0°F	Q-5	2	8	Ea.	2,125	375		2,500	2,900
1040	3 ton cooling, 13 MBH heat @ 0°F		1.20	13.333		2,650	625		3,275	3,900

23 81 Decentralized Unitary HVAC Equipment

23 81 43 – Air-Source Unitary Heat Pumps

	23 81 43.10 Air-Source Heat Pumps	Crew	Daily Output	Labor-Hours	Unit	Material	2010 Bare Costs Labor	Equipment	Total	Total Incl O&P
1080	7.5 ton cooling, 33 MBH heat @ 0°F	Q-5	.45	35.556	Ea.	5,625	1,650		7,275	8,750
1120	15 ton cooling, 64 MBH heat @ 0°F	Q-6	.50	48		10,400	2,325		12,725	15,100
1130	20 ton cooling, 85 MBH heat @ 0°F	↓	.35	68.571		15,600	3,325		18,925	22,400
1140	25 ton cooling, 119 MBH heat @ 0°F	↓	.25	96	↓	18,800	4,650		23,450	27,900
1500	Single package, not including curbs, pads, or plenums									
1520	2 ton cooling, 6.5 MBH heat @ 0°F	Q-5	1.50	10.667	Ea.	2,750	500		3,250	3,800
1560	3 ton cooling, 10 MBH heat @ 0°F	"	1.20	13.333	"	3,275	625		3,900	4,575

23 81 46 – Water-Source Unitary Heat Pumps

	23 81 46.10 Water Source Heat Pumps	Crew	Daily Output	Labor-Hours	Unit	Material	Labor	Equipment	Total	Total Incl O&P
0010	**WATER SOURCE HEAT PUMPS**, Not incl. connecting tubing or water source									
2000	Water source to air, single package									
2100	1 ton cooling, 13 MBH heat @ 75°F	Q-5	2	8	Ea.	1,475	375		1,850	2,200
2140	2 ton cooling, 19 MBH heat @ 75°F		1.70	9.412		1,600	440		2,040	2,450
2220	5 ton cooling, 29 MBH heat @ 75°F		.90	17.778	↓	2,625	830		3,455	4,150
9000	Minimum labor/equipment charge	↓	1.75	9.143	Job		425		425	660

23 82 Convection Heating and Cooling Units

23 82 16 – Air Coils

23 82 16.10 Flanged Coils

		Crew	Daily Output	Labor-Hours	Unit	Material	Labor	Equipment	Total	Total Incl O&P
0010	**FLANGED COILS**									
1500	Hot water heating, 1 row, 24" x 48"	Q-5	4	4	Ea.	1,450	187		1,637	1,900
9000	Minimum labor/equipment charge	1 Plum	1	8	Job		415		415	645

23 82 16.20 Duct Heaters

		Crew	Daily Output	Labor-Hours	Unit	Material	Labor	Equipment	Total	Total Incl O&P
0010	**DUCT HEATERS**, Electric, 480 V, 3 Ph.									
0020	Finned tubular insert, 500°F									
0100	8" wide x 6" high, 4.0 kW	Q-20	16	1.250	Ea.	705	56.50		761.50	865
0120	12" high, 8.0 kW		15	1.333		1,175	60		1,235	1,375
0140	18" high, 12.0 kW		14	1.429		1,650	64.50		1,714.50	1,900
0160	24" high, 16.0 kW		13	1.538		2,100	69.50		2,169.50	2,425
0180	30" high, 20.0 kW		12	1.667		2,575	75.50		2,650.50	2,950
0300	12" wide x 6" high, 6.7 kW		15	1.333		750	60		810	920
0320	12" high, 13.3 kW		14	1.429		1,200	64.50		1,264.50	1,425
0340	18" high, 20.0 kW	↓	13	1.538	↓	1,700	69.50		1,769.50	1,975
8000	To obtain BTU multiply kW by 3413									
9900	Minimum labor/equipment charge	1 Shee	4	2	Job		98		98	154

23 82 19 – Fan Coil Units

23 82 19.10 Fan Coil Air Conditioning

		Crew	Daily Output	Labor-Hours	Unit	Material	Labor	Equipment	Total	Total Incl O&P
0010	**FAN COIL AIR CONDITIONING** Cabinet mounted, filters, controls									
0020										
0100	Chilled water, 1/2 ton cooling	Q-5	8	2	Ea.	710	93.50		803.50	925
0120	1 ton cooling		6	2.667		815	125		940	1,100
0160	2.5 ton cooling		5	3.200		1,525	149		1,674	1,900
0180	3 ton cooling	↓	4	4	↓	1,725	187		1,912	2,200
0262	For hot water coil, add					40%	10%			
0940	Direct expansion, for use w/air cooled condensing unit, 1.5 ton cooling	Q-5	5	3.200	Ea.	690	149		839	990
1000	5 ton cooling	"	3	5.333		1,325	249		1,574	1,825
1040	10 ton cooling	Q-6	2.60	9.231		2,775	445		3,220	3,750
1060	20 ton cooling	"	.70	34.286		5,375	1,650		7,025	8,475
1512	For condensing unit add see Div. 23 62									
3000	Chilled water, horizontal unit, housing, 2 pipe, fan control, no valves									

23 82 Convection Heating and Cooling Units

23 82 19 – Fan Coil Units

23 82 19.10 Fan Coil Air Conditioning

		Crew	Daily Output	Labor-Hours	Unit	Material	2010 Bare Costs Labor	Equipment	Total	Total Incl O&P
3100	1/2 ton cooling	Q-5	8	2	Ea.	1,075	93.50		1,168.50	1,350
3110	1 ton cooling		6	2.667		1,300	125		1,425	1,650
3120	1.5 ton cooling		5.50	2.909		1,575	136		1,711	1,925
3130	2 ton cooling		5.25	3.048		1,875	142		2,017	2,300
3140	3 ton cooling		4	4		2,175	187		2,362	2,700
3150	3.5 ton cooling		3.80	4.211		2,400	197		2,597	2,925
3160	4 ton cooling		3.80	4.211		2,400	197		2,597	2,925
4000	With electric heat, 2 pipe									
4100	1/2 ton cooling	Q-5	8	2	Ea.	1,325	93.50		1,418.50	1,600
4105	3/4 ton cooling		7	2.286		1,450	107		1,557	1,775
4110	1 ton cooling		6	2.667		1,600	125		1,725	1,975
4120	1.5 ton cooling		5.50	2.909		1,775	136		1,911	2,150
4130	2 ton cooling		5.25	3.048		2,250	142		2,392	2,700
4135	2.5 ton cooling		4.80	3.333		3,100	156		3,256	3,650
4140	3 ton cooling		4	4		4,275	187		4,462	5,000
4150	3.5 ton cooling		3.80	4.211		5,125	197		5,322	5,925
4160	4 ton cooling		3.60	4.444		5,925	208		6,133	6,825
9000	Minimum labor/equipment charge		3.75	4.267	Job		199		199	310

23 82 29 – Radiators

23 82 29.10 Hydronic Heating

		Crew	Daily Output	Labor-Hours	Unit	Material	Labor	Equipment	Total	Total Incl O&P
0010	**HYDRONIC HEATING**, Terminal units, not incl. main supply pipe									
1000	Radiation									
3000	Radiators, cast iron									
3100	Free standing or wall hung, 6 tube, 25" high	Q-5	96	.167	Section	36	7.80		43.80	51.50
9000	Minimum labor/equipment charge	"	3.20	5	Job		234		234	360

23 82 36 – Finned-Tube Radiation Heaters

23 82 36.10 Finned Tube Radiation

		Crew	Daily Output	Labor-Hours	Unit	Material	Labor	Equipment	Total	Total Incl O&P
0010	**FINNED TUBE RADIATION**, Terminal units, not incl. main supply pipe									
1150	Fin tube, wall hung, 14" slope top cover, with damper									
1200	1-1/4" copper tube, 4-1/4" alum. fin	Q-5	38	.421	L.F.	40.50	19.65		60.15	75.50
1250	1-1/4" steel tube, 4-1/4" steel fin		36	.444		38.50	21		59.50	74.50
1255	2" steel tube, 4-1/4" steel fin		32	.500		41.50	23.50		65	81.50
1310	Baseboard, pkgd, 1/2" copper tube, alum. fin, 7" high		60	.267		7.55	12.45		20	27.50
1320	3/4" copper tube, alum. fin, 7" high		58	.276		8	12.90		20.90	29
1340	1" copper tube, alum. fin, 8-7/8" high		56	.286		25	13.35		38.35	48
1360	1-1/4" copper tube, alum. fin, 8-7/8" high		54	.296		26	13.85		39.85	50
1500	Note: fin tube may also require corners, caps, etc.									

23 82 39 – Unit Heaters

23 82 39.16 Propeller Unit Heaters

		Crew	Daily Output	Labor-Hours	Unit	Material	Labor	Equipment	Total	Total Incl O&P
0010	**PROPELLER UNIT HEATERS**									
3950	Unit heaters, propeller, 115 V 2 psi steam, 60°F entering air									
4000	Horizontal, 12 MBH	Q-5	12	1.333	Ea.	335	62.50		397.50	465
4060	43.9 MBH		8	2		505	93.50		598.50	700
4240	286.9 MBH		2	8		1,600	375		1,975	2,325
4250	326.0 MBH		1.90	8.421		1,825	395		2,220	2,600
4260	364 MBH		1.80	8.889		1,900	415		2,315	2,750
4270	404 MBH		1.60	10		2,500	465		2,965	3,475
4300	For vertical diffuser, add					174			174	191
4310	Vertical flow, 40 MBH	Q-5	11	1.455		500	68		568	655
4326	131.0 MBH	"	4	4		780	187		967	1,150
4354	420 MBH, (460 V)	Q-6	1.80	13.333		2,000	645		2,645	3,200

23 82 Convection Heating and Cooling Units

23 82 39 – Unit Heaters

23 82 39.16 Propeller Unit Heaters		Crew	Daily Output	Labor-Hours	Unit	Material	2010 Bare Costs Labor	Equipment	Total	Total Incl O&P
4358	500 MBH, (460 V)	Q-6	1.71	14.035	Ea.	2,650	680		3,330	3,950
4362	570 MBH, (460 V)		1.40	17.143		3,725	830		4,555	5,375
4366	620 MBH, (460 V)		1.30	18.462		4,275	895		5,170	6,075
4370	960 MBH, (460 V)		1.10	21.818		7,200	1,050		8,250	9,525

23 83 Radiant Heating Units

23 83 33 – Electric Radiant Heaters

23 83 33.10 Electric Heating

		Crew	Daily Output	Labor-Hours	Unit	Material	Labor	Equipment	Total	Total Incl O&P
0010	**ELECTRIC HEATING**, not incl. conduit or feed wiring									
1300	Baseboard heaters, 2' long, 375 watt	1 Elec	8	1	Ea.	40	49		89	120
1400	3' long, 500 watt		8	1		46	49		95	126
1600	4' long, 750 watt		6.70	1.194		54	58.50		112.50	150
1800	5' long, 935 watt		5.70	1.404		68	69		137	181
2000	6' long, 1125 watt		5	1.600		74	78.50		152.50	202
2400	8' long, 1500 watt		4	2		92	98		190	252
2600	9' long, 1680 watt		3.60	2.222		104	109		213	281
2800	10' long, 1875 watt		3.30	2.424		193	119		312	395
2950	Wall heaters with fan, 120 to 277 volt									
3190	1500 watt	1 Elec	4	2	Ea.	134	98		232	298
3230	2500 watt		3.50	2.286		335	112		447	540
3250	4000 watt		2.70	2.963		495	145		640	770
3600	Thermostats, integral		16	.500		30	24.50		54.50	70.50
3800	Line voltage, 1 pole		8	1		31	49		80	110
3810	2 pole		8	1		29.50	49		78.50	108
9990	Minimum labor/equipment charge		4	2	Job		98		98	151

23 84 Humidity Control Equipment

23 84 13 – Humidifiers

23 84 13.10 Humidifier Units

		Crew	Daily Output	Labor-Hours	Unit	Material	Labor	Equipment	Total	Total Incl O&P
0010	**HUMIDIFIER UNITS**									
0520	Steam, room or duct, filter, regulators, auto. controls, 220 V									
0560	22 lb. per hour	Q-5	5	3.200	Ea.	2,600	149		2,749	3,075
0580	33 lb. per hour		4	4		2,650	187		2,837	3,225
0620	100 lb. per hour		3	5.333		3,900	249		4,149	4,675
9000	Minimum labor/equipment charge		3.50	4.571	Job		213		213	330

23 91 Prefabricated Equipment Supports

23 91 10 – Prefabricated Curbs, Pads and Stands

23 91 10.10 Prefabricated Pads and Stands

		Crew	Daily Output	Labor-Hours	Unit	Material	Labor	Equipment	Total	Total Incl O&P
0010	**PREFABRICATED PADS AND STANDS**									
6000	Pad, fiberglass reinforced concrete with polystyrene foam core									
6050	Condenser, 2" thick, 20" x 38"	1 Shee	8	1	Ea.	23.50	49		72.50	103
6090	24" x 36"	"	12	.667		30	32.50		62.50	84
6280	36" x 36"	Q-9	8	2		43.50	88.50		132	186
6300	36" x 40"		7	2.286		49.50	101		150.50	213
6320	36" x 48"		7	2.286		58	101		159	222

Estimating Tips

26 05 00 Common Work Results for Electrical

- Conduit should be taken off in three main categories—power distribution, branch power, and branch lighting—so the estimator can concentrate on systems and components, therefore making it easier to ensure all items have been accounted for.
- For cost modifications for elevated conduit installation, add the percentages to labor according to the height of installation, and only to the quantities exceeding the different height levels, not to the total conduit quantities.
- Remember that aluminum wiring of equal ampacity is larger in diameter than copper and may require larger conduit.
- If more than three wires at a time are being pulled, deduct percentages from the labor hours of that grouping of wires.
- When taking off grounding systems, identify separately the type and size of wire, and list each unique type of ground connection.
- The estimator should take the weights of materials into consideration when completing a takeoff. Topics to consider include: How will the materials be supported? What methods of support are available? How high will the support structure have to reach? Will the final support structure be able to withstand the total burden? Is the support material included or separate from the fixture, equipment, and material specified?
- Do not overlook the costs for equipment used in the installation. If scaffolding or highlifts are available in the field, contractors may use them in lieu of the proposed ladders and rolling staging.

26 20 00 Low-Voltage Electrical Transmission

- Supports and concrete pads may be shown on drawings for the larger equipment, or the support system may be only a piece of plywood for the back of a panelboard. In either case, it must be included in the costs.

26 40 00 Electrical and Cathodic Protection

- When taking off cathodic protections systems, identify the type and size of cable, and list each unique type of anode connection.

26 50 00 Lighting

- Fixtures should be taken off room by room, using the fixture schedule, specifications, and the ceiling plan. For large concentrations of lighting fixtures in the same area, deduct the percentages from labor hours.

Reference Numbers

Reference numbers are shown in shaded boxes at the beginning of some major classifications. These numbers refer to related items in the Reference Section. The reference information may be an estimating procedure, an alternate pricing method, or technical information.

Note: Not all subdivisions listed here necessarily appear in this publication.

Note: **Trade Service,** *in part, has been used as a reference source for some of the material prices used in Division 26.*

26 05 Common Work Results for Electrical

26 05 05 – Selective Electrical Demolition

26 05 05.10 Electrical Demolition		Crew	Daily Output	Labor-Hours	Unit	Material	2010 Bare Costs Labor	2010 Bare Costs Equipment	Total	Total Incl O&P
0010	**ELECTRICAL DEMOLITION** R260105-30									
0020	Conduit to 15' high, including fittings & hangers									
0100	Rigid galvanized steel, 1/2" to 1" diameter R024119-10	1 Elec	242	.033	L.F.		1.62		1.62	2.49
0120	1-1/4" to 2"	"	200	.040			1.96		1.96	3.01
0140	2-1/2" to 3-1/2"	2 Elec	302	.053			2.60		2.60	3.99
0200	Electric metallic tubing (EMT), 1/2" to 1"	1 Elec	394	.020			.99		.99	1.53
0220	1-1/4" to 1-1/2"		326	.025			1.20		1.20	1.85
0240	2" to 3"		236	.034			1.66		1.66	2.55
0400	Wiremold raceway, including fittings & hangers									
0420	No. 3000	1 Elec	250	.032	L.F.		1.57		1.57	2.41
0440	No. 4000	"	217	.037	"		1.81		1.81	2.78
0500	Channels, steel, including fittings & hangers									
0520	3/4" x 1-1/2"	1 Elec	308	.026	L.F.		1.27		1.27	1.96
0540	1-1/2" x 1-1/2"	"	269	.030	"		1.46		1.46	2.24
0600	Copper bus duct, indoor, 3 phase									
0610	Including hangers & supports									
0620	225 amp	2 Elec	135	.119	L.F.		5.80		5.80	8.90
0640	400 amp		106	.151			7.40		7.40	11.35
0660	600 amp		86	.186			9.10		9.10	14
0680	1000 amp		60	.267			13.05		13.05	20
0700	1600 amp		40	.400			19.60		19.60	30
0720	3000 amp		10	1.600			78.50		78.50	120
0800	Plug-in switches, 600V 3 ph, incl. disconnecting									
0820	wire, conduit terminations, 30 amp	1 Elec	15.50	.516	Ea.		25.50		25.50	39
0840	60 amp		13.90	.576			28		28	43.50
0850	100 amp		10.40	.769			37.50		37.50	58
0860	200 amp		6.20	1.290			63		63	97
0940	1200 amp	2 Elec	2	8			390		390	600
0960	1600 amp	"	1.70	9.412			460		460	710
1010	Safety switches, 250 or 600V, incl. disconnection									
1050	of wire & conduit terminations									
1100	30 amp	1 Elec	12.30	.650	Ea.		32		32	49
1120	60 amp		8.80	.909			44.50		44.50	68.50
1140	100 amp		7.30	1.096			53.50		53.50	82.50
1160	200 amp		5	1.600			78.50		78.50	120
1210	Panel boards, incl. removal of all breakers,									
1220	conduit terminations & wire connections									
1230	3 wire, 120/240V, 100A, to 20 circuits	1 Elec	2.60	3.077	Ea.		151		151	232
1240	200 amps, to 42 circuits	2 Elec	2.60	6.154			300		300	465
1260	4 wire, 120/208V, 125A, to 20 circuits	1 Elec	2.40	3.333			163		163	251
1270	200 amps, to 42 circuits	2 Elec	2.40	6.667			325		325	500
1300	Transformer, dry type, 1 phase, incl. removal of									
1320	supports, wire & conduit terminations									
1340	1 kVA	1 Elec	7.70	1.039	Ea.		51		51	78
1360	5 kVA	"	4.70	1.702			83.50		83.50	128
1420	75 kVA	2 Elec	2.50	6.400			315		315	480
1440	3 phase to 600V, primary									
1460	3 kVA	1 Elec	3.85	2.078	Ea.		102		102	156
1480	15 kVA	2 Elec	4.20	3.810			187		187	287
1500	30 kVA	"	3.50	4.571			224		224	345
1530	112.5 kVA	R-3	2.90	6.897			335	47.50	382.50	570
1560	500 kVA		1.40	14.286			690	98.50	788.50	1,175
1570	750 kVA		1.10	18.182			880	126	1,006	1,500

26 05 Common Work Results for Electrical

26 05 05 – Selective Electrical Demolition

26 05 05.10 Electrical Demolition		Crew	Daily Output	Labor-Hours	Unit	Material	2010 Bare Costs Labor	2010 Bare Costs Equipment	Total	Total Incl O&P
1600	Pull boxes & cabinets, sheet metal, incl. removal									
1620	of supports and conduit terminations									
1640	6" x 6" x 4"	1 Elec	31.10	.257	Ea.		12.60		12.60	19.35
1660	12" x 12" x 4"		23.30	.343			16.80		16.80	26
1720	Junction boxes, 4" sq. & oct.		80	.100			4.90		4.90	7.55
1740	Handy box		107	.075			3.66		3.66	5.65
1760	Switch box		107	.075			3.66		3.66	5.65
1780	Receptacle & switch plates		257	.031			1.53		1.53	2.34
1800	Wire, THW-THWN-THHN, removed from									
1810	in place conduit, to 15' high									
1830	#14	1 Elec	65	.123	C.L.F.		6.05		6.05	9.25
1840	#12		55	.145			7.15		7.15	10.95
1850	#10		45.50	.176			8.60		8.60	13.25
1880	#4	2 Elec	53	.302			14.80		14.80	22.50
1890	#3		50	.320			15.70		15.70	24
1910	1/0		33.20	.482			23.50		23.50	36.50
1920	2/0		29.20	.548			27		27	41.50
1930	3/0		25	.640			31.50		31.50	48
1980	400 kcmil		17	.941			46		46	71
1990	500 kcmil		16.20	.988			48.50		48.50	74.50
2000	Interior fluorescent fixtures, incl. supports									
2010	& whips, to 15' high									
2100	Recessed drop-in 2' x 2', 2 lamp	2 Elec	35	.457	Ea.		22.50		22.50	34.50
2120	2' x 4', 2 lamp		33	.485			24		24	36.50
2140	2' x 4', 4 lamp		30	.533			26		26	40
2160	4' x 4', 4 lamp		20	.800			39		39	60
2180	Surface mount, acrylic lens & hinged frame									
2200	1' x 4', 2 lamp	2 Elec	44	.364	Ea.		17.80		17.80	27.50
2220	2' x 2', 2 lamp		44	.364			17.80		17.80	27.50
2260	2' x 4', 4 lamp		33	.485			24		24	36.50
2280	4' x 4', 4 lamp		23	.696			34		34	52.50
2300	Strip fixtures, surface mount									
2320	4' long, 1 lamp	2 Elec	53	.302	Ea.		14.80		14.80	22.50
2340	4' long, 2 lamp		50	.320			15.70		15.70	24
2360	8' long, 1 lamp		42	.381			18.65		18.65	28.50
2380	8' long, 2 lamp		40	.400			19.60		19.60	30
2400	Pendant mount, industrial, incl. removal									
2410	of chain or rod hangers, to 15' high									
2420	4' long, 2 lamp	2 Elec	35	.457	Ea.		22.50		22.50	34.50
2440	8' long, 2 lamp	"	27	.593	"		29		29	44.50
2460	Interior incandescent, surface, ceiling									
2470	or wall mount, to 12' high									
2480	Metal cylinder type, 75 Watt	2 Elec	62	.258	Ea.		12.65		12.65	19.45
2500	150 Watt	"	62	.258	"		12.65		12.65	19.45
2520	Metal halide, high bay									
2540	400 Watt	2 Elec	15	1.067	Ea.		52.50		52.50	80.50
2560	1000 Watt		12	1.333			65.50		65.50	100
2580	150 Watt, low bay		20	.800			39		39	60
2600	Exterior fixtures, incandescent, wall mount									
2620	100 Watt	2 Elec	50	.320	Ea.		15.70		15.70	24
2640	Quartz, 500 Watt		33	.485			24		24	36.50
2660	1500 Watt		27	.593			29		29	44.50
2680	Wall pack, mercury vapor									

26 05 Common Work Results for Electrical

26 05 05 – Selective Electrical Demolition

26 05 05.10 Electrical Demolition		Crew	Daily Output	Labor-Hours	Unit	Material	2010 Bare Costs Labor	Equipment	Total	Total Incl O&P
2700	175 Watt	2 Elec	25	.640	Ea.		31.50		31.50	48
2720	250 Watt	"	25	.640	"		31.50		31.50	48
9000	Minimum labor/equipment charge	1 Elec	4	2	Job		98		98	151
9900	Add to labor for higher elevated installation									
9910	15' to 20' high, add						10%			
9920	20' to 25' high, add						20%			
9930	25' to 30' high, add						25%			
9940	30' to 35' high, add						30%			
9950	35' to 40' high, add						35%			
9960	Over 40' high, add						40%			

26 05 19 – Low-Voltage Electrical Power Conductors and Cables

26 05 19.20 Armored Cable

		Crew	Daily Output	Labor-Hours	Unit	Material	Labor	Equipment	Total	Total Incl O&P
0010	**ARMORED CABLE**									
0050	600 volt, copper (BX), #14, 2 conductor, solid	1 Elec	2.40	3.333	C.L.F.	70.50	163		233.50	330
0100	3 conductor, solid		2.20	3.636		111	178		289	395
0150	#12, 2 conductor, solid		2.30	3.478		71.50	170		241.50	340
0200	3 conductor, solid		2	4		114	196		310	425
0250	#10, 2 conductor, solid		2	4		130	196		326	445
0300	3 conductor, solid		1.60	5		179	245		424	570
0350	#8, 3 conductor, solid		1.30	6.154		335	300		635	835
9010	600 volt, copper (MC) steel clad, #14, 2 wire		2.40	3.333		74	163		237	335
9020	3 wire		2.20	3.636		114	178		292	400
9040	#12, 2 wire		2.30	3.478		75	170		245	345
9050	3 wire		2	4		126	196		322	440
9070	#10, 2 wire		2	4		158	196		354	475
9080	3 wire		1.60	5		219	245		464	615
9100	#8, 2 wire, stranded		1.80	4.444		282	218		500	645
9110	3 wire, stranded		1.30	6.154		420	300		720	925
9900	Minimum labor/equipment charge		4	2	Job		98		98	151

26 05 19.50 Mineral Insulated Cable

		Crew	Daily Output	Labor-Hours	Unit	Material	Labor	Equipment	Total	Total Incl O&P
0010	**MINERAL INSULATED CABLE** 600 volt									
0100	1 conductor, #12	1 Elec	1.60	5	C.L.F.	350	245		595	760
1500	2 conductor, #12		1.40	5.714		725	280		1,005	1,225
1600	#10		1.20	6.667		880	325		1,205	1,475
1800	#8		1.10	7.273		1,100	355		1,455	1,750
2000	#6		1.05	7.619		1,400	375		1,775	2,125
2100	#4	2 Elec	2	8		1,850	390		2,240	2,625
2200	3 conductor, #12	1 Elec	1.20	6.667		915	325		1,240	1,500
2400	#10		1.10	7.273		1,050	355		1,405	1,725
2600	#8		1.05	7.619		1,275	375		1,650	1,975
2800	#6		1	8		1,650	390		2,040	2,425
3000	#4	2 Elec	1.80	8.889		2,025	435		2,460	2,925
3100	4 conductor, #12	1 Elec	1.20	6.667		975	325		1,300	1,575
3200	#10		1.10	7.273		1,150	355		1,505	1,825
3400	#8		1	8		1,500	390		1,890	2,250
3600	#6		.90	8.889		1,875	435		2,310	2,750
3620	7 conductor, #12		1.10	7.273		1,225	355		1,580	1,900
3640	#10		1	8		1,550	390		1,940	2,300
3800	Terminations, 600 volt, 1 conductor, #12		8	1	Ea.	15.45	49		64.45	92.50
5500	2 conductor, #12		6.70	1.194		15.45	58.50		73.95	107
5600	#10		6.40	1.250		23.50	61.50		85	120
5800	#8		6.20	1.290		23.50	63		86.50	123

26 05 Common Work Results for Electrical

26 05 19 – Low-Voltage Electrical Power Conductors and Cables

26 05 19.50 Mineral Insulated Cable		Crew	Daily Output	Labor-Hours	Unit	Material	2010 Bare Costs Labor	Equipment	Total	Total Incl O&P
6000	#6	1 Elec	5.70	1.404	Ea.	23.50	69		92.50	132
6200	#4		5.30	1.509		51.50	74		125.50	171
6400	3 conductor, #12		5.70	1.404		23.50	69		92.50	132
6500	#10		5.50	1.455		23.50	71.50		95	136
6600	#8		5.20	1.538		23.50	75.50		99	142
6800	#6		4.80	1.667		23.50	81.50		105	152
7200	#4		4.60	1.739		51.50	85		136.50	188
7400	4 conductor, #12		4.60	1.739		25.50	85		110.50	159
7500	#10		4.40	1.818		25.50	89		114.50	165
7600	#8		4.20	1.905		25.50	93.50		119	171
8400	#6		4	2		54	98		152	211
8500	7 conductor, #12		3.50	2.286		26	112		138	201
8600	#10		3	2.667		56	131		187	263
8800	Crimping tool, plier type					55			55	60.50
9000	Stripping tool					207			207	228
9200	Hand vise					62			62	68
9500	Minimum labor/equipment charge	1 Elec	4	2	Job		98		98	151

26 05 19.55 Non-Metallic Sheathed Cable										
0010	**NON-METALLIC SHEATHED CABLE** 600 volt									
0100	Copper with ground wire, (Romex)									
0150	#14, 2 conductor	1 Elec	2.70	2.963	C.L.F.	23	145		168	248
0200	3 conductor		2.40	3.333		32	163		195	286
0250	#12, 2 conductor		2.50	3.200		34	157		191	279
0300	3 conductor		2.20	3.636		48.50	178		226.50	330
0350	#10, 2 conductor		2.20	3.636		54	178		232	335
0400	3 conductor		1.80	4.444		77.50	218		295.50	420
0450	#8, 3 conductor		1.50	5.333		121	261		382	535
0500	#6, 3 conductor		1.40	5.714		196	280		476	645
0550	SE type SER aluminum cable, 3 RHW and									
0600	1 bare neutral, 3 #8 & 1 #8	1 Elec	1.60	5	C.L.F.	120	245		365	505
0650	3 #6 & 1 #6	"	1.40	5.714		136	280		416	580
0700	3 #4 & 1 #6	2 Elec	2.40	6.667		152	325		477	665
0750	3 #2 & 1 #4		2.20	7.273		224	355		579	795
0800	3 #1/0 & 1 #2		2	8		340	390		730	975
0850	3 #2/0 & 1 #1		1.80	8.889		400	435		835	1,100
0900	3 #4/0 & 1 #2/0		1.60	10		570	490		1,060	1,375
6500	Service entrance cap for copper SEU									
6700	150 amp	1 Elec	10	.800	Ea.	23.50	39		62.50	86
6800	200 amp		8	1	"	36	49		85	115
9000	Minimum labor/equipment charge		4	2	Job		98		98	151

26 05 19.90 Wire										
0010	**WIRE** R260533-22									
0020	600 volt type THW, copper, solid, #14	1 Elec	13	.615	C.L.F.	7.80	30		37.80	55
0030	#12		11	.727		11.65	35.50		47.15	68
0040	#10		10	.800		18.25	39		57.25	80
0140	#8		8	1		33	49		82	112
0160	#6		6.50	1.231		54.50	60.50		115	153
0180	#4	2 Elec	10.60	1.509		86.50	74		160.50	209
0200	#3		10	1.600		107	78.50		185.50	237
0220	#2		9	1.778		134	87		221	282
0240	#1		8	2		174	98		272	340
0260	1/0		6.60	2.424		211	119		330	415

26 05 Common Work Results for Electrical

26 05 19 – Low-Voltage Electrical Power Conductors and Cables

26 05 19.90 Wire		Crew	Daily Output	Labor-Hours	Unit	Material	2010 Bare Costs Labor	Equipment	Total	Total Incl O&P
0280	2/0	2 Elec	5.80	2.759	C.L.F.	265	135		400	500
0300	3/0		5	3.200		335	157		492	605
0350	4/0		4.40	3.636		420	178		598	735
0400	250 kcmil	3 Elec	6	4		495	196		691	845
0420	300 kcmil		5.70	4.211		595	206		801	965
0450	350 kcmil		5.40	4.444		695	218		913	1,100
0480	400 kcmil		5.10	4.706		790	231		1,021	1,225
0490	500 kcmil		4.80	5		960	245		1,205	1,425
0510	750 kcmil		3.30	7.273		2,125	355		2,480	2,900
0540	Aluminum, stranded, #6	1 Elec	8	1		22	49		71	99.50
0560	#4	2 Elec	13	1.231		28	60.50		88.50	124
0580	#2		10.60	1.509		37.50	74		111.50	156
0600	#1		9	1.778		54.50	87		141.50	194
0620	1/0		8	2		65.50	98		163.50	223
0640	2/0		7.20	2.222		77.50	109		186.50	253
0680	3/0		6.60	2.424		96	119		215	289
0700	4/0		6.20	2.581		107	126		233	310
0720	250 kcmil	3 Elec	8.70	2.759		131	135		266	350
0740	300 kcmil		8.10	2.963		180	145		325	420
0760	350 kcmil		7.50	3.200		183	157		340	440
0780	400 kcmil		6.90	3.478		214	170		384	495
0800	500 kcmil		6	4		236	196		432	560
0850	600 kcmil		5.70	4.211		299	206		505	645
0880	700 kcmil		5.10	4.706		345	231		576	735
9000	Minimum labor/equipment charge	1 Elec	4	2	Job		98		98	151

26 05 26 – Grounding and Bonding for Electrical Systems

26 05 26.80 Grounding

		Crew	Daily Output	Labor-Hours	Unit	Material	Labor	Equipment	Total	Total Incl O&P
0010	**GROUNDING**									
0030	Rod, copper clad, 8' long, 1/2" diameter	1 Elec	5.50	1.455	Ea.	16.75	71.50		88.25	128
0040	5/8" diameter		5.50	1.455		18.40	71.50		89.90	130
0050	3/4" diameter		5.30	1.509		30.50	74		104.50	148
0080	10' long, 1/2" diameter		4.80	1.667		19.65	81.50		101.15	148
0090	5/8" diameter		4.60	1.739		21.50	85		106.50	155
0100	3/4" diameter		4.40	1.818		36.50	89		125.50	177
0130	15' long, 3/4" diameter		4	2		92	98		190	252
0260	Wire ground bare armored, #8-1 conductor		2	4	C.L.F.	119	196		315	430
0280	#4-1 conductor		1.60	5		194	245		439	590
0390	Bare copper wire, stranded, #8		11	.727		30.50	35.50		66	88.50
0400	#6		10	.800		53.50	39		92.50	119
0600	#2	2 Elec	10	1.600		134	78.50		212.50	268
0800	3/0		6.60	2.424		305	119		424	520
1000	4/0		5.70	2.807		385	138		523	635
1200	250 kcmil	3 Elec	7.20	3.333		455	163		618	750
1800	Water pipe ground clamps, heavy duty									
2000	Bronze, 1/2" to 1" diameter	1 Elec	8	1	Ea.	26	49		75	104
2100	1-1/4" to 2" diameter		8	1		34	49		83	113
2200	2-1/2" to 3" diameter		6	1.333		57	65.50		122.50	163
2800	Brazed connections, #6 wire		12	.667		13.45	32.50		45.95	65
3000	#2 wire		10	.800		18	39		57	80
3100	3/0 wire		8	1		27	49		76	105
3200	4/0 wire		7	1.143		31	56		87	120
3400	250 kcmil wire		5	1.600		36	78.50		114.50	160

26 05 Common Work Results for Electrical

26 05 26 – Grounding and Bonding for Electrical Systems

26 05 26.80 Grounding

		Crew	Daily Output	Labor-Hours	Unit	Material	2010 Bare Costs Labor	Equipment	Total	Total Incl O&P
3600	500 kcmil wire	1 Elec	4	2	Ea.	44.50	98		142.50	200
9000	Minimum labor/equipment charge		4	2	Job		98		98	151

26 05 33 – Raceway and Boxes for Electrical Systems

26 05 33.05 Conduit

			Crew	Daily Output	Labor-Hours	Unit	Material	Labor	Equipment	Total	Total Incl O&P
0010	**CONDUIT** To 15' high, includes 2 terminations, 2 elbows,	R260533-22									
0020	11 beam clamps, and 11 couplings per 100 L.F.										
1161	Field bends, 45° to 90°, 1/2" diameter		1 Elec	53	.151	Ea.		7.40		7.40	11.35
1162	3/4" diameter			47	.170			8.35		8.35	12.80
1163	1" diameter			44	.182			8.90		8.90	13.70
1164	1-1/4" diameter			23	.348			17.05		17.05	26
1165	1-1/2" diameter			21	.381			18.65		18.65	28.50
1166	2" diameter			16	.500			24.50		24.50	37.50
1991	Field bends, 45° to 90°, 1/2" diameter			44	.182			8.90		8.90	13.70
1992	3/4" diameter			40	.200			9.80		9.80	15.05
1993	1" diameter			36	.222			10.90		10.90	16.75
1994	1-1/4" diameter			19	.421			20.50		20.50	31.50
1995	1-1/2" diameter			18	.444			22		22	33.50
1996	2" diameter			13	.615			30		30	46.50
2500	Steel, intermediate conduit (IMC), 1/2" diameter			100	.080	L.F.	1.90	3.92		5.82	8.10
2530	3/4" diameter			90	.089		2.26	4.36		6.62	9.20
2550	1" diameter			70	.114		3.13	5.60		8.73	12.05
2570	1-1/4" diameter			65	.123		4.13	6.05		10.18	13.80
2600	1-1/2" diameter			60	.133		4.95	6.55		11.50	15.50
2630	2" diameter			50	.160		6.30	7.85		14.15	19
2650	2-1/2" diameter			40	.200		12.35	9.80		22.15	28.50
2670	3" diameter		2 Elec	60	.267		16.70	13.05		29.75	38.50
2700	3-1/2" diameter			54	.296		21.50	14.50		36	46.50
2730	4" diameter			50	.320		24	15.70		39.70	50.50
2731	Field bends, 45° to 90°, 1/2" diameter		1 Elec	44	.182	Ea.		8.90		8.90	13.70
2732	3/4" diameter			40	.200			9.80		9.80	15.05
2733	1" diameter			36	.222			10.90		10.90	16.75
2734	1-1/4" diameter			19	.421			20.50		20.50	31.50
2735	1-1/2" diameter			18	.444			22		22	33.50
2736	2" diameter			13	.615			30		30	46.50
5000	Electric metallic tubing (EMT), 1/2" diameter			170	.047	L.F.	.62	2.31		2.93	4.22
5020	3/4" diameter			130	.062		.96	3.02		3.98	5.70
5040	1" diameter			115	.070		1.70	3.41		5.11	7.10
5060	1-1/4" diameter			100	.080		2.77	3.92		6.69	9.05
5080	1-1/2" diameter			90	.089		3.59	4.36		7.95	10.65
5100	2" diameter			80	.100		4.67	4.90		9.57	12.70
5120	2-1/2" diameter			60	.133		11.15	6.55		17.70	22.50
5140	3" diameter		2 Elec	100	.160		13.25	7.85		21.10	26.50
5160	3-1/2" diameter			90	.178		16.80	8.70		25.50	32
5180	4" diameter			80	.200		18.60	9.80		28.40	35.50
5200	Field bends, 45° to 90°, 1/2" diameter		1 Elec	89	.090	Ea.		4.40		4.40	6.75
5220	3/4" diameter			80	.100			4.90		4.90	7.55
5240	1" diameter			73	.110			5.35		5.35	8.25
5260	1-1/4" diameter			38	.211			10.30		10.30	15.85
5280	1-1/2" diameter			36	.222			10.90		10.90	16.75
5300	2" diameter			26	.308			15.10		15.10	23
5320	Offsets, 1/2" diameter			65	.123			6.05		6.05	9.25
5340	3/4" diameter			62	.129			6.30		6.30	9.70

26 05 Common Work Results for Electrical

26 05 33 – Raceway and Boxes for Electrical Systems

26 05 33.05 Conduit

		Crew	Daily Output	Labor-Hours	Unit	Material	2010 Bare Costs Labor	2010 Bare Costs Equipment	Total	Total Incl O&P
5360	1" diameter	1 Elec	53	.151	Ea.		7.40		7.40	11.35
5380	1-1/4" diameter		30	.267			13.05		13.05	20
5400	1-1/2" diameter		28	.286			14		14	21.50
7600	EMT, "T" fittings with covers, 1/2" diameter, set screw		16	.500		14.90	24.50		39.40	54
9990	Minimum labor/equipment charge		4	2	Job		98		98	151

26 05 33.45 Wireway

		Crew	Daily Output	Labor-Hours	Unit	Material	Labor	Equipment	Total	Total Incl O&P
0010	**WIREWAY** to 15' high									
0100	Screw cover, NEMA 1 w/fittings and supports, 2-1/2" x 2-1/2"	1 Elec	45	.178	L.F.	11.45	8.70		20.15	26
0200	4" x 4"	"	40	.200		12.65	9.80		22.45	29
0400	6" x 6"	2 Elec	60	.267		19.70	13.05		32.75	41.50
0600	8" x 8"	"	40	.400		33	19.60		52.60	66.50
4475	Screw cover, NEMA 3R w/fittings and supports, 4" x 4"	1 Elec	36	.222		27	10.90		37.90	46.50
4480	6" x 6"	2 Elec	55	.291		34	14.25		48.25	59.50
4485	8" x 8"		36	.444		55	22		77	94
4490	12" x 12"		18	.889		77	43.50		120.50	152

26 05 33.50 Outlet Boxes

		Crew	Daily Output	Labor-Hours	Unit	Material	Labor	Equipment	Total	Total Incl O&P
0010	**OUTLET BOXES**									
0020	Pressed steel, octagon, 4"	1 Elec	20	.400	Ea.	3.36	19.60		22.96	33.50
0060	Covers, blank		64	.125		1.40	6.15		7.55	10.95
0100	Extension rings		40	.200		5.60	9.80		15.40	21
0150	Square, 4"		20	.400		2.38	19.60		21.98	32.50
0200	Extension rings		40	.200		5.65	9.80		15.45	21.50
0250	Covers, blank		64	.125		1.53	6.15		7.68	11.10
0300	Plaster rings		64	.125		3.09	6.15		9.24	12.80
0652	Switchbox		24	.333		5.40	16.35		21.75	31
1100	Concrete, floor, 1 gang		5.30	1.509		89.50	74		163.50	213
9000	Minimum labor/equipment charge		4	2	Job		98		98	151

26 05 33.60 Outlet Boxes, Plastic

		Crew	Daily Output	Labor-Hours	Unit	Material	Labor	Equipment	Total	Total Incl O&P
0010	**OUTLET BOXES, PLASTIC**									
0050	4" diameter, round with 2 mounting nails	1 Elec	25	.320	Ea.	3.28	15.70		18.98	27.50
0100	Bar hanger mounted		25	.320		6	15.70		21.70	30.50
0200	4", square with 2 mounting nails		25	.320		4.96	15.70		20.66	29.50
0300	Plaster ring		64	.125		2.08	6.15		8.23	11.70
0400	Switch box with 2 mounting nails, 1 gang		30	.267		2.24	13.05		15.29	22.50
0500	2 gang		25	.320		4.56	15.70		20.26	29
0600	3 gang		20	.400		7.30	19.60		26.90	38
0700	Old work box		30	.267		4.32	13.05		17.37	25
9000	Minimum labor/equipment charge		4	2	Job		98		98	151

26 05 33.65 Pull Boxes

		Crew	Daily Output	Labor-Hours	Unit	Material	Labor	Equipment	Total	Total Incl O&P
0010	**PULL BOXES**									
0100	Sheet metal, pull box, NEMA 1, type SC, 6" W x 6" H x 4" D	1 Elec	8	1	Ea.	13.65	49		62.65	90.50
0200	8" W x 8" H x 4" D		8	1		18.70	49		67.70	96
0300	10" W x 12" H x 6" D		5.30	1.509		33	74		107	151
0400	16" W x 20" H x 8" D		4	2		124	98		222	287
0500	20" W x 24" H x 8" D		3.20	2.500		134	123		257	335
0600	24" W x 36" H x 8" D		2.70	2.963		207	145		352	450
0650	Pull box, hinged, NEMA 1, 6" W x 6" H x 4" D		8	1		14	49		63	91
0800	12" W x 16" H x 6" D		4.70	1.702		46	83.50		129.50	179
1000	20" W x 20" H x 6" D		3.60	2.222		92.50	109		201.50	269
1200	20" W x 20" H x 8" D		3.20	2.500		185	123		308	390
1400	24" W x 36" H x 8" D		2.70	2.963		296	145		441	550
1600	24" W x 42" H x 8" D		2	4		450	196		646	795

26 05 Common Work Results for Electrical

26 05 33 – Raceway and Boxes for Electrical Systems

26 05 33.95 Cutting and Drilling

		Crew	Daily Output	Labor-Hours	Unit	Material	2010 Bare Costs Labor	Equipment	Total	Total Incl O&P
0010	**CUTTING AND DRILLING**									
0100	Hole drilling to 10' high, concrete wall									
0110	8" thick, 1/2" pipe size	R-31	12	.667	Ea.	5	32.50	4.45	41.95	60.50
0120	3/4" pipe size		12	.667		5	32.50	4.45	41.95	60.50
0130	1" pipe size		9.50	.842		9.80	41.50	5.65	56.95	80.50
0140	1-1/4" pipe size		9.50	.842		9.80	41.50	5.65	56.95	80.50
0150	1-1/2" pipe size		9.50	.842		9.80	41.50	5.65	56.95	80.50
0160	2" pipe size		4.40	1.818		15	89	12.15	116.15	167
0170	2-1/2" pipe size		4.40	1.818		15	89	12.15	116.15	167
0180	3" pipe size		4.40	1.818		15	89	12.15	116.15	167
0190	3-1/2" pipe size		3.30	2.424		18.90	119	16.20	154.10	222
0200	4" pipe size		3.30	2.424		18.90	119	16.20	154.10	222
0500	12" thick, 1/2" pipe size		9.40	.851		7.75	41.50	5.70	54.95	79
0520	3/4" pipe size		9.40	.851		7.75	41.50	5.70	54.95	79
0540	1" pipe size		7.30	1.096		14.05	53.50	7.30	74.85	106
0560	1-1/4" pipe size		7.30	1.096		14.05	53.50	7.30	74.85	106
0570	1-1/2" pipe size		7.30	1.096		14.05	53.50	7.30	74.85	106
0580	2" pipe size		3.60	2.222		20.50	109	14.85	144.35	206
0590	2-1/2" pipe size		3.60	2.222		20.50	109	14.85	144.35	206
0600	3" pipe size		3.60	2.222		20.50	109	14.85	144.35	206
0610	3-1/2" pipe size		2.80	2.857		26	140	19.10	185.10	265
0630	4" pipe size		2.50	3.200		26	157	21.50	204.50	293
0650	16" thick, 1/2" pipe size		7.60	1.053		10.45	51.50	7.05	69	99
0670	3/4" pipe size		7	1.143		10.45	56	7.65	74.10	106
0690	1" pipe size		6	1.333		18.35	65.50	8.90	92.75	130
0710	1-1/4" pipe size		5.50	1.455		18.35	71.50	9.70	99.55	141
0730	1-1/2" pipe size		5.50	1.455		18.35	71.50	9.70	99.55	141
0750	2" pipe size		3	2.667		26	131	17.80	174.80	250
0770	2-1/2" pipe size		2.70	2.963		26	145	19.80	190.80	274
0790	3" pipe size		2.50	3.200		26	157	21.50	204.50	294
0810	3-1/2" pipe size		2.30	3.478		32.50	170	23	225.50	325
0830	4" pipe size		2	4		32.50	196	26.50	255	365
0850	20" thick, 1/2" pipe size		6.40	1.250		13.15	61.50	8.35	83	118
0870	3/4" pipe size		6	1.333		13.15	65.50	8.90	87.55	124
0890	1" pipe size		5	1.600		22.50	78.50	10.70	111.70	157
0910	1-1/4" pipe size		4.80	1.667		22.50	81.50	11.15	115.15	163
0930	1-1/2" pipe size		4.60	1.739		22.50	85	11.60	119.10	169
0950	2" pipe size		2.70	2.963		32	145	19.80	196.80	280
0970	2-1/2" pipe size		2.40	3.333		32	163	22.50	217.50	310
0990	3" pipe size		2.20	3.636		32	178	24.50	234.50	335
1010	3-1/2" pipe size		2	4		39.50	196	26.50	262	375
1030	4" pipe size		1.70	4.706		39.50	231	31.50	302	435
1050	24" thick, 1/2" pipe size		5.50	1.455		15.90	71.50	9.70	97.10	138
1070	3/4" pipe size		5.10	1.569		15.90	77	10.50	103.40	147
1090	1" pipe size		4.30	1.860		27	91	12.45	130.45	183
1110	1-1/4" pipe size		4	2		27	98	13.35	138.35	195
1130	1-1/2" pipe size		4	2		27	98	13.35	138.35	195
1150	2" pipe size		2.40	3.333		37.50	163	22.50	223	315
1170	2-1/2" pipe size		2.20	3.636		37.50	178	24.50	240	340
1190	3" pipe size		2	4		37.50	196	26.50	260	370
1210	3-1/2" pipe size		1.80	4.444		46.50	218	29.50	294	420
1230	4" pipe size		1.50	5.333		46.50	261	35.50	343	490

26 05 Common Work Results for Electrical

26 05 33 – Raceway and Boxes for Electrical Systems

26 05 33.95 Cutting and Drilling		Crew	Daily Output	Labor-Hours	Unit	Material	2010 Bare Costs Labor	Equipment	Total	Total Incl O&P
1500	Brick wall, 8" thick, 1/2" pipe size	R-31	18	.444	Ea.	5	22	2.97	29.97	42.50
1520	3/4" pipe size		18	.444		5	22	2.97	29.97	42.50
1540	1" pipe size		13.30	.602		9.80	29.50	4.02	43.32	60.50
1560	1-1/4" pipe size		13.30	.602		9.80	29.50	4.02	43.32	60.50
1580	1-1/2" pipe size		13.30	.602		9.80	29.50	4.02	43.32	60.50
1600	2" pipe size		5.70	1.404		15	69	9.40	93.40	133
1620	2-1/2" pipe size		5.70	1.404		15	69	9.40	93.40	133
1640	3" pipe size		5.70	1.404		15	69	9.40	93.40	133
1660	3-1/2" pipe size		4.40	1.818		18.90	89	12.15	120.05	171
1680	4" pipe size		4	2		18.90	98	13.35	130.25	187
1700	12" thick, 1/2" pipe size		14.50	.552		7.75	27	3.69	38.44	54
1720	3/4" pipe size		14.50	.552		7.75	27	3.69	38.44	54
1740	1" pipe size		11	.727		14.05	35.50	4.86	54.41	76
1760	1-1/4" pipe size		11	.727		14.05	35.50	4.86	54.41	76
1780	1-1/2" pipe size		11	.727		14.05	35.50	4.86	54.41	76
1800	2" pipe size		5	1.600		20.50	78.50	10.70	109.70	154
1820	2-1/2" pipe size		5	1.600		20.50	78.50	10.70	109.70	154
1840	3" pipe size		5	1.600		20.50	78.50	10.70	109.70	154
1860	3-1/2" pipe size		3.80	2.105		26	103	14.05	143.05	203
1880	4" pipe size		3.30	2.424		26	119	16.20	161.20	229
1900	16" thick, 1/2" pipe size		12.30	.650		10.45	32	4.34	46.79	65.50
1920	3/4" pipe size		12.30	.650		10.45	32	4.34	46.79	65.50
1940	1" pipe size		9.30	.860		18.35	42	5.75	66.10	91.50
1960	1-1/4" pipe size		9.30	.860		18.35	42	5.75	66.10	91.50
1980	1-1/2" pipe size		9.30	.860		18.35	42	5.75	66.10	91.50
2000	2" pipe size		4.40	1.818		26	89	12.15	127.15	179
2010	2-1/2" pipe size		4.40	1.818		26	89	12.15	127.15	179
2030	3" pipe size		4.40	1.818		26	89	12.15	127.15	179
2050	3-1/2" pipe size		3.30	2.424		32.50	119	16.20	167.70	237
2070	4" pipe size		3	2.667		32.50	131	17.80	181.30	257
2090	20" thick, 1/2" pipe size		10.70	.748		13.15	36.50	4.99	54.64	76.50
2110	3/4" pipe size		10.70	.748		13.15	36.50	4.99	54.64	76.50
2130	1" pipe size		8	1		22.50	49	6.70	78.20	108
2150	1-1/4" pipe size		8	1		22.50	49	6.70	78.20	108
2170	1-1/2" pipe size		8	1		22.50	49	6.70	78.20	108
2190	2" pipe size		4	2		32	98	13.35	143.35	201
2210	2-1/2" pipe size		4	2		32	98	13.35	143.35	201
2230	3" pipe size		4	2		32	98	13.35	143.35	201
2250	3-1/2" pipe size		3	2.667		39.50	131	17.80	188.30	264
2270	4" pipe size		2.70	2.963		39.50	145	19.80	204.30	289
2290	24" thick, 1/2" pipe size		9.40	.851		15.90	41.50	5.70	63.10	88
2310	3/4" pipe size		9.40	.851		15.90	41.50	5.70	63.10	88
2330	1" pipe size		7.10	1.127		27	55	7.55	89.55	123
2350	1-1/4" pipe size		7.10	1.127		27	55	7.55	89.55	123
2370	1-1/2" pipe size		7.10	1.127		27	55	7.55	89.55	123
2390	2" pipe size		3.60	2.222		37.50	109	14.85	161.35	224
2410	2-1/2" pipe size		3.60	2.222		37.50	109	14.85	161.35	224
2430	3" pipe size		3.60	2.222		37.50	109	14.85	161.35	224
2450	3-1/2" pipe size		2.80	2.857		46.50	140	19.10	205.60	287
2470	4" pipe size		2.50	3.200		46.50	157	21.50	225	315
2480										
3000	Knockouts to 8' high, metal boxes & enclosures									
3020	With hole saw, 1/2" pipe size	1 Elec	53	.151	Ea.		7.40		7.40	11.35

26 05 Common Work Results for Electrical

26 05 33 – Raceway and Boxes for Electrical Systems

26 05 33.95 Cutting and Drilling		Crew	Daily Output	Labor-Hours	Unit	Material	2010 Bare Costs Labor	Equipment	Total	Total Incl O&P
3040	3/4" pipe size	1 Elec	47	.170	Ea.		8.35		8.35	12.80
3050	1" pipe size		40	.200			9.80		9.80	15.05
3060	1-1/4" pipe size		36	.222			10.90		10.90	16.75
3070	1-1/2" pipe size		32	.250			12.25		12.25	18.85
3080	2" pipe size		27	.296			14.50		14.50	22.50
3090	2-1/2" pipe size		20	.400			19.60		19.60	30
4010	3" pipe size		16	.500			24.50		24.50	37.50
4030	3-1/2" pipe size		13	.615			30		30	46.50
4050	4" pipe size		11	.727			35.50		35.50	55
4070	With hand punch set, 1/2" pipe size		40	.200			9.80		9.80	15.05
4090	3/4" pipe size		32	.250			12.25		12.25	18.85
4110	1" pipe size		30	.267			13.05		13.05	20
4130	1-1/4" pipe size		28	.286			14		14	21.50
4150	1-1/2" pipe size		26	.308			15.10		15.10	23
4170	2" pipe size		20	.400			19.60		19.60	30
4190	2-1/2" pipe size		17	.471			23		23	35.50
4200	3" pipe size		15	.533			26		26	40
4220	3-1/2" pipe size		12	.667			32.50		32.50	50
4240	4" pipe size		10	.800			39		39	60
4260	With hydraulic punch, 1/2" pipe size		44	.182			8.90		8.90	13.70
4280	3/4" pipe size		38	.211			10.30		10.30	15.85
4300	1" pipe size		38	.211			10.30		10.30	15.85
4320	1-1/4" pipe size		38	.211			10.30		10.30	15.85
4340	1-1/2" pipe size		38	.211			10.30		10.30	15.85
4360	2" pipe size		32	.250			12.25		12.25	18.85
4380	2-1/2" pipe size		27	.296			14.50		14.50	22.50
4400	3" pipe size		23	.348			17.05		17.05	26
4420	3-1/2" pipe size		20	.400			19.60		19.60	30

26 05 39 – Underfloor Raceways for Electrical Systems

26 05 39.30 Conduit In Concrete Slab

		Crew	Daily Output	Labor-Hours	Unit	Material	Labor	Equipment	Total	Total Incl O&P
0010	**CONDUIT IN CONCRETE SLAB** Including terminations,									
0020	fittings and supports									
3230	PVC, schedule 40, 1/2" diameter	1 Elec	270	.030	L.F.	.67	1.45		2.12	2.97
3250	3/4" diameter		230	.035		.81	1.70		2.51	3.51
3270	1" diameter		200	.040		1.11	1.96		3.07	4.23
3300	1-1/4" diameter		170	.047		1.55	2.31		3.86	5.25
3330	1-1/2" diameter		140	.057		1.85	2.80		4.65	6.35
3350	2" diameter		120	.067		2.38	3.27		5.65	7.60
4350	Rigid galvanized steel, 1/2" diameter		200	.040		2.26	1.96		4.22	5.50
4400	3/4" diameter		170	.047		2.50	2.31		4.81	6.30
4450	1" diameter		130	.062		3.54	3.02		6.56	8.50
4500	1-1/4" diameter		110	.073		4.73	3.56		8.29	10.70
4600	1-1/2" diameter		100	.080		5.65	3.92		9.57	12.20
4800	2" diameter		90	.089		7.30	4.36		11.66	14.70
9000	Minimum labor/equipment charge		4	2	Job		98		98	151

26 05 39.40 Conduit In Trench

		Crew	Daily Output	Labor-Hours	Unit	Material	Labor	Equipment	Total	Total Incl O&P
0010	**CONDUIT IN TRENCH** Includes terminations and fittings									
0020	Does not include excavation or backfill, see Div. 31 23 16.00									
0200	Rigid galvanized steel, 2" diameter	1 Elec	150	.053	L.F.	7.05	2.61		9.66	11.75
0400	2-1/2" diameter	"	100	.080		14.05	3.92		17.97	21.50
0600	3" diameter	2 Elec	160	.100		17.35	4.90		22.25	26.50
0800	3-1/2" diameter		140	.114		22.50	5.60		28.10	33.50

26 05 Common Work Results for Electrical

26 05 39 – Underfloor Raceways for Electrical Systems

26 05 39.40 Conduit In Trench		Crew	Daily Output	Labor-Hours	Unit	Material	2010 Bare Costs Labor	Equipment	Total	Total Incl O&P
1000	4" diameter	2 Elec	100	.160	L.F.	25.50	7.85		33.35	40
1200	5" diameter		80	.200		53.50	9.80		63.30	74
1400	6" diameter		60	.267		78.50	13.05		91.55	107
9000	Minimum labor/equipment charge	1 Elec	4	2	Job		98		98	151

26 05 80 – Wiring Connections

26 05 80.10 Motor Connections

		Crew	Daily Output	Labor-Hours	Unit	Material	Labor	Equipment	Total	Incl O&P
0010	**MOTOR CONNECTIONS**									
0020	Flexible conduit and fittings, 115 volt, 1 phase, up to 1 HP motor	1 Elec	8	1	Ea.	9.55	49		58.55	86
9000	Minimum labor/equipment charge	"	4	2	Job		98		98	151

26 05 90 – Residential Applications

26 05 90.10 Residential Wiring

		Crew	Daily Output	Labor-Hours	Unit	Material	Labor	Equipment	Total	Incl O&P
0010	**RESIDENTIAL WIRING**									
0020	20' avg. runs and #14/2 wiring incl. unless otherwise noted									
1000	Service & panel, includes 24' SE-AL cable, service eye, meter,									
1010	Socket, panel board, main bkr., ground rod, 15 or 20 amp									
1020	1-pole circuit breakers, and misc. hardware									
1100	100 amp, with 10 branch breakers	1 Elec	1.19	6.723	Ea.	565	330		895	1,125
1110	With PVC conduit and wire		.92	8.696		630	425		1,055	1,350
1120	With RGS conduit and wire		.73	10.959		805	535		1,340	1,725
1150	150 amp, with 14 branch breakers		1.03	7.767		880	380		1,260	1,550
1170	With PVC conduit and wire		.82	9.756		1,025	480		1,505	1,850
1180	With RGS conduit and wire		.67	11.940		1,350	585		1,935	2,400
1200	200 amp, with 18 branch breakers	2 Elec	1.80	8.889		1,150	435		1,585	1,950
1220	With PVC conduit and wire		1.46	10.959		1,300	535		1,835	2,250
1230	With RGS conduit and wire		1.24	12.903		1,750	630		2,380	2,900
1800	Lightning surge suppressor for above services, add	1 Elec	32	.250		51.50	12.25		63.75	75.50
2000	Switch devices									
2100	Single pole, 15 amp, Ivory, with a 1-gang box, cover plate,									
2110	Type NM (Romex) cable	1 Elec	17.10	.468	Ea.	10.05	23		33.05	46
2120	Type MC (BX) cable		14.30	.559		27.50	27.50		55	72.50
2130	EMT & wire		5.71	1.401		32.50	68.50		101	142
2150	3-way, #14/3, type NM cable		14.55	.550		14.05	27		41.05	57
2170	Type MC cable		12.31	.650		38	32		70	90.50
2180	EMT & wire		5	1.600		36	78.50		114.50	160
2200	4-way, #14/3, type NM cable		14.55	.550		30.50	27		57.50	75
2220	Type MC cable		12.31	.650		54	32		86	109
2230	EMT & wire		5	1.600		52.50	78.50		131	178
2250	S.P., 20 amp, #12/2, type NM cable		13.33	.600		16.85	29.50		46.35	63.50
2270	Type MC cable		11.43	.700		32	34.50		66.50	88
2280	EMT & wire		4.85	1.649		41	81		122	169
2290	S.P. rotary dimmer, 600W, no wiring		17	.471		20.50	23		43.50	58.50
2300	S.P. rotary dimmer, 600W, type NM cable		14.55	.550		25.50	27		52.50	69.50
2320	Type MC cable		12.31	.650		43	32		75	96
2330	EMT & wire		5	1.600		49	78.50		127.50	174
2350	3-way rotary dimmer, type NM cable		13.33	.600		32.50	29.50		62	81
2370	Type MC cable		11.43	.700		50	34.50		84.50	108
2380	EMT & wire		4.85	1.649		56.50	81		137.50	186
2400	Interval timer wall switch, 20 amp, 1-30 min., #12/2									
2410	Type NM cable	1 Elec	14.55	.550	Ea.	50	27		77	96.50
2420	Type MC cable		12.31	.650		58.50	32		90.50	114
2430	EMT & wire		5	1.600		74	78.50		152.50	202
2500	Decorator style									

26 05 Common Work Results for Electrical

26 05 90 – Residential Applications

26 05 90.10 Residential Wiring

		Crew	Daily Output	Labor-Hours	Unit	Material	2010 Bare Costs Labor	Equipment	Total	Total Incl O&P
2510	S.P., 15 amp, type NM cable	1 Elec	17.10	.468	Ea.	14.90	23		37.90	51.50
2520	Type MC cable		14.30	.559		32.50	27.50		60	77.50
2530	EMT & wire		5.71	1.401		37.50	68.50		106	147
2550	3-way, #14/3, type NM cable		14.55	.550		18.95	27		45.95	62.50
2570	Type MC cable		12.31	.650		43	32		75	96
2580	EMT & wire		5	1.600		41	78.50		119.50	165
2600	4-way, #14/3, type NM cable		14.55	.550		35	27		62	80
2620	Type MC cable		12.31	.650		59	32		91	114
2630	EMT & wire		5	1.600		57.50	78.50		136	183
2650	S.P., 20 amp, #12/2, type NM cable		13.33	.600		21.50	29.50		51	69
2670	Type MC cable		11.43	.700		37	34.50		71.50	93.50
2680	EMT & wire		4.85	1.649		45.50	81		126.50	174
2700	S.P., slide dimmer, type NM cable		17.10	.468		33	23		56	71.50
2720	Type MC cable		14.30	.559		50.50	27.50		78	97.50
2730	EMT & wire		5.71	1.401		57	68.50		125.50	169
2770	Type MC cable		14.30	.559		44.50	27.50		72	91
2780	EMT & wire		5.71	1.401		51	68.50		119.50	162
2800	3-way touch dimmer, type NM cable		13.33	.600		45.50	29.50		75	95
2820	Type MC cable		11.43	.700		63	34.50		97.50	122
2830	EMT & wire		4.85	1.649		69.50	81		150.50	200
3100	S.P. switch/15 amp recpt., Ivory, 1-gang box, plate									
3110	Type NM cable	1 Elec	11.43	.700	Ea.	21.50	34.50		56	76
3120	Type MC cable		10	.800		39	39		78	103
3130	EMT & wire		4.40	1.818		45.50	89		134.50	187
3150	S.P. switch/pilot light, type NM cable		11.43	.700		22	34.50		56.50	77
3170	Type MC cable		10	.800		39.50	39		78.50	104
3180	EMT & wire		4.43	1.806		46	88.50		134.50	187
3190	2-S.P. switches, 2-#14/2, no wiring		14	.571		8.25	28		36.25	52
3200	2-S.P. switches, 2-#14/2, type NM cables		10	.800		25	39		64	87.50
3220	Type MC cable		8.89	.900		52	44		96	126
3230	EMT & wire		4.10	1.951		49	95.50		144.50	201
3250	3-way switch/15 amp recpt., #14/3, type NM cable		10	.800		29	39		68	92
3270	Type MC cable		8.89	.900		53	44		97	127
3280	EMT & wire		4.10	1.951		51	95.50		146.50	204
3300	2-3 way switches, 2-#14/3, type NM cables		8.89	.900		38	44		82	110
3320	Type MC cable		8	1		78	49		127	162
3330	EMT & wire		4	2		58.50	98		156.50	216
3350	S.P. switch/20 amp recpt., #12/2, type NM cable		10	.800		33.50	39		72.50	97
3370	Type MC cable		8.89	.900		42.50	44		86.50	115
3380	EMT & wire		4.10	1.951		57.50	95.50		153	211
3400	Decorator style									
3410	S.P. switch/15 amp recpt., type NM cable	1 Elec	11.43	.700	Ea.	26.50	34.50		61	81.50
3420	Type MC cable		10	.800		44	39		83	108
3430	EMT & wire		4.40	1.818		50	89		139	192
3450	S.P. switch/pilot light, type NM cable		11.43	.700		27	34.50		61.50	82.50
3470	Type MC cable		10	.800		44.50	39		83.50	109
3480	EMT & wire		4.40	1.818		51	89		140	193
3500	2-S.P. switches, 2-#14/2, type NM cables		10	.800		30	39		69	93
3520	Type MC cable		8.89	.900		57	44		101	131
3530	EMT & wire		4.10	1.951		54	95.50		149.50	207
3550	3-way/15 amp recpt., #14/3, type NM cable		10	.800		34	39		73	97.50
3580	EMT & wire		4.10	1.951		56	95.50		151.50	209
3650	2-3 way switches, 2-#14/3, type NM cables		8.89	.900		43	44		87	116

26 05 Common Work Results for Electrical

26 05 90 – Residential Applications

26 05 90.10 Residential Wiring		Crew	Daily Output	Labor-Hours	Unit	Material	2010 Bare Costs Labor	Equipment	Total	Total Incl O&P
3670	Type MC cable	1 Elec	8	1	Ea.	83	49		132	167
3680	EMT & wire		4	2		63.50	98		161.50	221
3700	S.P. switch/20 amp recpt., #12/2, type NM cable		10	.800		38.50	39		77.50	103
3720	Type MC cable		8.89	.900		47	44		91	120
3730	EMT & wire		4.10	1.951		62.50	95.50		158	216
4000	Receptacle devices									
4010	Duplex outlet, 15 amp recpt., Ivory, 1-gang box, plate									
4015	Type NM cable	1 Elec	14.55	.550	Ea.	8.55	27		35.55	51
4020	Type MC cable		12.31	.650		26	32		58	77.50
4030	EMT & wire		5.33	1.501		31	73.50		104.50	147
4050	With #12/2, type NM cable		12.31	.650		10.80	32		42.80	61
4070	Type MC cable		10.67	.750		26	36.50		62.50	85.50
4080	EMT & wire		4.71	1.699		34.50	83		117.50	166
4100	20 amp recpt., #12/2, type NM cable		12.31	.650		19.85	32		51.85	71
4120	Type MC cable		10.67	.750		35	36.50		71.50	95.50
4130	EMT & wire		4.71	1.699		44	83		127	176
4140	For GFI see Div. 26 05 90.10 line 4300 below									
4150	Decorator style, 15 amp recpt., type NM cable	1 Elec	14.55	.550	Ea.	13.40	27		40.40	56.50
4170	Type MC cable		12.31	.650		31	32		63	83
4180	EMT & wire		5.33	1.501		36	73.50		109.50	153
4200	With #12/2, type NM cable		12.31	.650		15.70	32		47.70	66.50
4220	Type MC cable		10.67	.750		31	36.50		67.50	90.50
4230	EMT & wire		4.71	1.699		39.50	83		122.50	172
4250	20 amp recpt. #12/2, type NM cable		12.31	.650		24.50	32		56.50	76
4270	Type MC cable		10.67	.750		40	36.50		76.50	101
4280	EMT & wire		4.71	1.699		48.50	83		131.50	182
4300	GFI, 15 amp recpt., type NM cable		12.31	.650		43	32		75	96.50
4320	Type MC cable		10.67	.750		60.50	36.50		97	123
4330	EMT & wire		4.71	1.699		65.50	83		148.50	200
4350	GFI with #12/2, type NM cable		10.67	.750		45.50	36.50		82	107
4370	Type MC cable		9.20	.870		61	42.50		103.50	133
4380	EMT & wire		4.21	1.900		69.50	93		162.50	219
4400	20 amp recpt., #12/2 type NM cable		10.67	.750		47.50	36.50		84	109
4420	Type MC cable		9.20	.870		63	42.50		105.50	135
4430	EMT & wire		4.21	1.900		71.50	93		164.50	222
4500	Weather-proof cover for above receptacles, add		32	.250		5.25	12.25		17.50	24.50
4550	Air conditioner outlet, 20 amp-240 volt recpt.									
4560	30' of #12/2, 2 pole circuit breaker									
4570	Type NM cable	1 Elec	10	.800	Ea.	60.50	39		99.50	127
4580	Type MC cable		9	.889		79.50	43.50		123	155
4590	EMT & wire		4	2		84.50	98		182.50	244
4600	Decorator style, type NM cable		10	.800		65	39		104	132
4620	Type MC cable		9	.889		84.50	43.50		128	160
4630	EMT & wire		4	2		89	98		187	249
4650	Dryer outlet, 30 amp-240 volt recpt., 20' of #10/3									
4660	2 pole circuit breaker									
4670	Type NM cable	1 Elec	6.41	1.248	Ea.	66	61		127	167
4680	Type MC cable		5.71	1.401		80	68.50		148.50	194
4690	EMT & wire		3.48	2.299		81	113		194	262
4700	Range outlet, 50 amp-240 volt recpt., 30' of #8/3									
4710	Type NM cable	1 Elec	4.21	1.900	Ea.	93	93		186	245
4720	Type MC cable		4	2		163	98		261	330
4730	EMT & wire		2.96	2.703		111	132		243	325

26 05 Common Work Results for Electrical

26 05 90 – Residential Applications

26 05 90.10 Residential Wiring

		Crew	Daily Output	Labor-Hours	Unit	Material	2010 Bare Costs Labor	Equipment	Total	Total Incl O&P
4750	Central vacuum outlet, Type NM cable	1 Elec	6.40	1.250	Ea.	62.50	61.50		124	163
4770	Type MC cable		5.71	1.401		90	68.50		158.50	205
4780	EMT & wire		3.48	2.299		88.50	113		201.50	271
4800	30 amp-110 volt locking recpt., #10/2 circ. bkr.									
4810	Type NM cable	1 Elec	6.20	1.290	Ea.	72.50	63		135.50	177
4820	Type MC cable		5.40	1.481		104	72.50		176.50	227
4830	EMT & wire		3.20	2.500		99	123		222	297
4900	Low voltage outlets									
4910	Telephone recpt., 20' of 4/C phone wire	1 Elec	26	.308	Ea.	10.35	15.10		25.45	34.50
4920	TV recpt., 20' of RG59U coax wire, F type connector	"	16	.500	"	19.30	24.50		43.80	58.50
4950	Door bell chime, transformer, 2 buttons, 60' of bellwire									
4970	Economy model	1 Elec	11.50	.696	Ea.	75	34		109	135
4980	Custom model		11.50	.696		119	34		153	184
4990	Luxury model, 3 buttons		9.50	.842		320	41.50		361.50	420
6000	Lighting outlets									
6050	Wire only (for fixture), type NM cable	1 Elec	32	.250	Ea.	6.45	12.25		18.70	26
6070	Type MC cable		24	.333		17.05	16.35		33.40	44
6080	EMT & wire		10	.800		20.50	39		59.50	82.50
6100	Box (4"), and wire (for fixture), type NM cable		25	.320		16.25	15.70		31.95	42
6120	Type MC cable		20	.400		27	19.60		46.60	59.50
6130	EMT & wire		11	.727		30.50	35.50		66	88.50
6200	Fixtures (use with lines 6050 or 6100 above)									
6210	Canopy style, economy grade	1 Elec	40	.200	Ea.	34	9.80		43.80	52.50
6220	Custom grade		40	.200		52.50	9.80		62.30	73
6250	Dining room chandelier, economy grade		19	.421		85	20.50		105.50	125
6270	Luxury grade		15	.533		565	26		591	660
6310	Kitchen fixture (fluorescent), economy grade		30	.267		63.50	13.05		76.55	90
6320	Custom grade		25	.320		173	15.70		188.70	214
6350	Outdoor, wall mounted, economy grade		30	.267		32	13.05		45.05	55
6360	Custom grade		30	.267		111	13.05		124.05	142
6370	Luxury grade		25	.320		250	15.70		265.70	299
6410	Outdoor PAR floodlights, 1 lamp, 150 watt		20	.400		35	19.60		54.60	68.50
6420	2 lamp, 150 watt each		20	.400		58	19.60		77.60	94
6425	Motion sensing, 2 lamp, 150 watt each		20	.400		70	19.60		89.60	107
6430	For infrared security sensor, add		32	.250		117	12.25		129.25	148
6450	Outdoor, quartz-halogen, 300 watt flood		20	.400		42	19.60		61.60	76.50
6600	Recessed downlight, round, pre-wired, 50 or 75 watt trim		30	.267		43.50	13.05		56.55	67.50
6610	With shower light trim		30	.267		53.50	13.05		66.55	79
6620	With wall washer trim		28	.286		64.50	14		78.50	92.50
6630	With eye-ball trim		28	.286		64.50	14		78.50	92.50
6640	For direct contact with insulation, add					2.20			2.20	2.42
6700	Porcelain lamp holder	1 Elec	40	.200		4.20	9.80		14	19.65
6710	With pull switch		40	.200		5.05	9.80		14.85	20.50
6750	Fluorescent strip, 1-20 watt tube, wrap around diffuser, 24"		24	.333		54	16.35		70.35	84.50
6760	1-34 watt tube, 48"		24	.333		69.50	16.35		85.85	102
6770	2-34 watt tubes, 48"		20	.400		82.50	19.60		102.10	121
6780	With residential ballast		20	.400		92.50	19.60		112.10	132
6800	Bathroom heat lamp, 1-250 watt		28	.286		46	14		60	72
6810	2-250 watt lamps		28	.286		71.50	14		85.50	100
6820	For timer switch, see Div. 26 05 90.10 line 2400									
6900	Outdoor post lamp, incl. post, fixture, 35' of #14/2									
6910	Type NMC cable	1 Elec	3.50	2.286	Ea.	202	112		314	395
6920	Photo-eye, add		27	.296		35	14.50		49.50	61

26 05 Common Work Results for Electrical

26 05 90 – Residential Applications

26 05 90.10 Residential Wiring		Crew	Daily Output	Labor-Hours	Unit	Material	2010 Bare Costs Labor	Equipment	Total	Total Incl O&P
6950	Clock dial time switch, 24 hr., w/enclosure, type NM cable	1 Elec	11.43	.700	Ea.	61	34.50		95.50	120
6970	Type MC cable		11	.727		78.50	35.50		114	142
6980	EMT & wire	↓	4.85	1.649	↓	83.50	81		164.50	216
7000	Alarm systems									
7050	Smoke detectors, box, #14/3, type NM cable	1 Elec	14.55	.550	Ea.	34	27		61	79
7070	Type MC cable		12.31	.650		51.50	32		83.50	106
7080	EMT & wire	↓	5	1.600		49.50	78.50		128	175
7090	For relay output to security system, add				↓	13			13	14.30
8000	Residential equipment									
8050	Disposal hook-up, incl. switch, outlet box, 3' of flex									
8060	20 amp-1 pole circ. bkr., and 25' of #12/2									
8070	Type NM cable	1 Elec	10	.800	Ea.	30.50	39		69.50	93.50
8080	Type MC cable		8	1		48	49		97	128
8090	EMT & wire	↓	5	1.600		56.50	78.50		135	183
8100	Trash compactor or dishwasher hook-up, incl. outlet box,									
8110	3' of flex, 15 amp-1 pole circ. bkr., and 25' of #14/2									
8120	Type NM cable	1 Elec	10	.800	Ea.	22.50	39		61.50	85
8130	Type MC cable		8	1		43	49		92	123
8140	EMT & wire	↓	5	1.600		49	78.50		127.50	174
8150	Hot water sink dispensor hook-up, use line 8100									
8200	Vent/exhaust fan hook-up, type NM cable	1 Elec	32	.250	Ea.	6.45	12.25		18.70	26
8220	Type MC cable		24	.333		17.05	16.35		33.40	44
8230	EMT & wire	↓	10	.800	↓	20.50	39		59.50	82.50
8250	Bathroom vent fan, 50 CFM (use with above hook-up)									
8260	Economy model	1 Elec	15	.533	Ea.	25	26		51	67.50
8270	Low noise model		15	.533		32	26		58	75
8280	Custom model	↓	12	.667	↓	117	32.50		149.50	179
8300	Bathroom or kitchen vent fan, 110 CFM									
8310	Economy model	1 Elec	15	.533	Ea.	69.50	26		95.50	117
8320	Low noise model	"	15	.533	"	86	26		112	135
8350	Paddle fan, variable speed (w/o lights)									
8360	Economy model (AC motor)	1 Elec	10	.800	Ea.	106	39		145	177
8370	Custom model (AC motor)		10	.800		182	39		221	260
8380	Luxury model (DC motor)		8	1		360	49		409	470
8390	Remote speed switch for above, add	↓	12	.667		30	32.50		62.50	83
8500	Whole house exhaust fan, ceiling mount, 36", variable speed									
8510	Remote switch, incl. shutters, 20 amp-1 pole circ. bkr.									
8520	30' of #12/2, type NM cable	1 Elec	4	2	Ea.	1,025	98		1,123	1,275
8530	Type MC cable		3.50	2.286		1,050	112		1,162	1,325
8540	EMT & wire	↓	3	2.667	↓	1,050	131		1,181	1,375
8600	Whirlpool tub hook-up, incl. timer switch, outlet box									
8610	3' of flex, 20 amp-1 pole GFI circ. bkr.									
8620	30' of #12/2, type NM cable	1 Elec	5	1.600	Ea.	124	78.50		202.50	256
8630	Type MC cable		4.20	1.905		136	93.50		229.50	292
8640	EMT & wire	↓	3.40	2.353		144	115		259	335
8650	Hot water heater hook-up, incl. 1-2 pole circ. bkr., box;									
8660	3' of flex, 20' of #10/2, type NM cable	1 Elec	5	1.600	Ea.	32	78.50		110.50	156
8670	Type MC cable		4.20	1.905		55.50	93.50		149	204
8680	EMT & wire	↓	3.40	2.353		52	115		167	235
9000	Heating/air conditioning									
9050	Furnace/boiler hook-up, incl. firestat, local on-off switch									
9060	Emergency switch, and 40' of type NM cable	1 Elec	4	2	Ea.	54.50	98		152.50	211
9070	Type MC cable	↓	3.50	2.286	↓	81.50	112		193.50	262

Reed Construction Data
The leader in construction information and BIM solutions

AIA StrategicPartner

Reed Construction Data, Inc. is a leading provider of construction information and building information modeling (BIM) solutions. The company's portfolio of information products and services is designed specifically to help construction industry professionals advance their business with timely and accurate project, product, and cost data. Reed Construction Data is a division of Reed Business Information, a member of the Reed Elsevier PLC group of companies.

Cost Information
RSMeans, the undisputed market leader in construction costs, provides current cost and estimating information through its innovative MeansCostworks.com® web-based solution. In addition, RSMeans publishes annual cost books, estimating software, and a rich library of reference books. RSMeans also conducts a series of professional seminars and provides construction cost consulting for owners, manufacturers, designers, and contractors to sharpen personal skills and maximize the effective use of cost estimating and management tools.

Project Data
Reed Construction Data assembles one of the largest databases of public and private project data for use by contractors, distributors, and building product manufacturers in the U.S. and Canadian markets. In addition, Reed Construction Data is the North American construction community's premier resource for project leads and bid documents. Reed Bulletin and Reed CONNECT™ provide project leads and project data through all stages of construction for many of the country's largest public sector, commercial, industrial, and multi-family residential projects.

Research and Analytics
Reed Construction Data's forecasting tools cover most aspects of the construction business in the U.S. and Canada. With a vast network of resources, Reed Construction Data is uniquely qualified to give you the information you need to keep your business profitable.

SmartBIM Solutions
Reed Construction Data has emerged as the leader in the field of building information modeling (BIM) with products and services that have helped to advance the evolution of BIM. Through an in-depth SmartBIM Object Creation Program, BPMs can rely on Reed to create high-quality, real world objects embedded with superior cost data (RSMeans). In addition, Reed has made it easy for architects to manage these manufacturer-specific objects as well as generic objects in Revit with the SmartBIM Library.

SmartBuilding Index
The leading industry source for product research, product documentation, BIM objects, design ideas, and source locations. Search, select, and specify available building products with our online directories of manufacturer profiles, MANU-SPEC, SPEC-DATA, guide specs, manufacturer catalogs, building codes, historical project data, and BIM objects.

SmartBuilding Studio
A vibrant online resource library which brings together a comprehensive catalog of commercial interior finishes and products under a single standard for high-definition imagery, product data, and searchable attributes.

Associated Construction Publications (ACP)
Reed Construction Data's regional construction magazines cover the nation through a network of 14 regional magazines. Serving the construction market for more than 100 years, our magazines are a trusted source of news and information in the local and national construction communities.

For more information, please visit our website at www.reedconstructiondata.com

The leader in construction information and BIM solutions.

26 05 Common Work Results for Electrical

26 05 90 – Residential Applications

26 05 90.10 Residential Wiring

		Crew	Daily Output	Labor-Hours	Unit	Material	2010 Bare Costs Labor	Equipment	Total	Total Incl O&P
9080	EMT & wire	1 Elec	1.50	5.333	Ea.	89.50	261		350.50	500
9100	Air conditioner hook-up, incl. local 60 amp disc. switch									
9110	3' sealtite, 40 amp, 2 pole circuit breaker									
9130	40' of #8/2, type NM cable	1 Elec	3.50	2.286	Ea.	216	112		328	410
9140	Type MC cable		3	2.667		315	131		446	545
9150	EMT & wire	↓	1.30	6.154	↓	252	300		552	745
9200	Heat pump hook-up, 1-40 & 1-100 amp 2 pole circ. bkr.									
9210	Local disconnect switch, 3' sealtite									
9220	40' of #8/2 & 30' of #3/2									
9230	Type NM cable	1 Elec	1.30	6.154	Ea.	545	300		845	1,075
9240	Type MC cable		1.08	7.407		745	365		1,110	1,375
9250	EMT & wire	↓	.94	8.511	↓	615	415		1,030	1,325
9500	Thermostat hook-up, using low voltage wire									
9520	Heating only, 25' of #18-3	1 Elec	24	.333	Ea.	9.85	16.35		26.20	36
9530	Heating/cooling, 25' of #18-4	"	20	.400	"	12.30	19.60		31.90	43.50

26 22 Low-Voltage Transformers

26 22 13 – Low-Voltage Distribution Transformers

26 22 13.10 Transformer, Dry-Type

		Crew	Daily Output	Labor-Hours	Unit	Material	Labor	Equipment	Total	Total Incl O&P
0010	**TRANSFORMER, DRY-TYPE**									
0050	Single phase, 240/480 volt primary, 120/240 volt secondary									
0100	1 kVA	1 Elec	2	4	Ea.	272	196		468	600
0300	2 kVA		1.60	5		410	245		655	825
0500	3 kVA		1.40	5.714		505	280		785	985
0700	5 kVA	↓	1.20	6.667		700	325		1,025	1,275
0900	7.5 kVA	2 Elec	2.20	7.273		970	355		1,325	1,625
1100	10 kVA		1.60	10		1,250	490		1,740	2,125
1300	15 kVA	↓	1.20	13.333	↓	1,700	655		2,355	2,850
2300	3 phase, 480 volt primary 120/208 volt secondary									
2310	Ventilated, 3 kVA	1 Elec	1	8	Ea.	885	390		1,275	1,575
2700	6 kVA		.80	10		1,225	490		1,715	2,100
2900	9 kVA	↓	.70	11.429		1,375	560		1,935	2,375
3100	15 kVA	2 Elec	1.10	14.545		1,850	715		2,565	3,125
3300	30 kVA		.90	17.778		2,150	870		3,020	3,725
3500	45 kVA	↓	.80	20	↓	2,600	980		3,580	4,350
9000	Minimum labor/equipment charge	1 Elec	1	8	Job		390		390	600

26 24 Switchboards and Panelboards

26 24 16 – Panelboards

26 24 16.20 Panelboard and Load Center Circuit Breakers

		Crew	Daily Output	Labor-Hours	Unit	Material	Labor	Equipment	Total	Total Incl O&P
0010	**PANELBOARD AND LOAD CENTER CIRCUIT BREAKERS**									
0050	Bolt-on, 10,000 amp IC, 120 volt, 1 pole									
0100	15 to 50 amp	1 Elec	10	.800	Ea.	16	39		55	77.50
0200	60 amp		8	1		16	49		65	93
0300	70 amp	↓	8	1	↓	30.50	49		79.50	109
0350	240 volt, 2 pole									
0400	15 to 50 amp	1 Elec	8	1	Ea.	35	49		84	114
0500	60 amp		7.50	1.067		35	52.50		87.50	119
0600	80 to 100 amp	↓	5	1.600	↓	90.50	78.50		169	220

26 24 Switchboards and Panelboards

26 24 16 – Panelboards

26 24 16.20 Panelboard and Load Center Circuit Breakers		Crew	Daily Output	Labor-Hours	Unit	Material	2010 Bare Costs Labor	Equipment	Total	Total Incl O&P
0700	3 pole, 15 to 60 amp	1 Elec	6.20	1.290	Ea.	111	63		174	219
0800	70 amp		5	1.600		140	78.50		218.50	274
0900	80 to 100 amp		3.60	2.222		159	109		268	340
1000	22,000 amp I.C., 240 volt, 2 pole, 70 – 225 amp		2.70	2.963		685	145		830	975
1100	3 pole, 70 – 225 amp		2.30	3.478		760	170		930	1,100
1200	14,000 amp I.C., 277 volts, 1 pole, 15 – 30 amp		8	1		42.50	49		91.50	122
1300	22,000 amp I.C., 480 volts, 2 pole, 70 – 225 amp		2.70	2.963		685	145		830	975
1400	3 pole, 70 – 225 amp		2.30	3.478		845	170		1,015	1,200
2060	Plug-in tandem, 120/240 V, 2-15 A, 1 pole		11	.727		33	35.50		68.50	91.50
2070	1-15 A & 1-20 A		11	.727		33	35.50		68.50	91.50
2080	2-20 A		11	.727		33	35.50		68.50	91.50
2082	Arc fault circuit interrupter, 120/240 V, 1-15 A & 1-20 A, 1 pole		11	.727		85	35.50		120.50	149
9000	Minimum labor/equipment charge		3	2.667	Job		131		131	201

26 24 16.30 Panelboards Commercial Applications

		Crew	Daily Output	Labor-Hours	Unit	Material	Labor	Equipment	Total	Total Incl O&P
0010	**PANELBOARDS COMMERCIAL APPLICATIONS**									
0050	NQOD, w/20 amp 1 pole bolt-on circuit breakers									
0100	3 wire, 120/240 volts, 100 amp main lugs									
0150	10 circuits	1 Elec	1	8	Ea.	510	390		900	1,150
0200	14 circuits		.88	9.091		595	445		1,040	1,350
0250	18 circuits		.75	10.667		650	525		1,175	1,525
0300	20 circuits		.65	12.308		735	605		1,340	1,725
0600	4 wire, 120/208 volts, 100 amp main lugs, 12 circuits		1	8		575	390		965	1,225
0650	16 circuits		.75	10.667		660	525		1,185	1,525
0700	20 circuits		.65	12.308		770	605		1,375	1,775
0750	24 circuits		.60	13.333		835	655		1,490	1,925
0800	30 circuits		.53	15.094		965	740		1,705	2,175
0850	225 amp main lugs, 32 circuits	2 Elec	.90	17.778		1,075	870		1,945	2,550
0900	34 circuits		.84	19.048		1,125	935		2,060	2,650
0950	36 circuits		.80	20		1,150	980		2,130	2,750
1000	42 circuits		.68	23.529		1,275	1,150		2,425	3,175
1200	NEHB, w/20 amp, 1 pole bolt-on circuit breakers									
1250	4 wire, 277/480 volts, 100 amp main lugs, 12 circuits	1 Elec	.88	9.091	Ea.	1,100	445		1,545	1,875
1300	20 circuits	"	.60	13.333		1,625	655		2,280	2,775
1350	225 amp main lugs, 24 circuits	2 Elec	.90	17.778		1,875	870		2,745	3,400
1400	30 circuits		.80	20		2,250	980		3,230	3,975
1450	36 circuits		.72	22.222		2,625	1,100		3,725	4,550
1600	NQOD panel, w/20 amp, 1 pole, circuit breakers									
2000	4 wire, 120/208 volts with main circuit breaker									
2050	100 amp main, 24 circuits	1 Elec	.47	17.021	Ea.	1,050	835		1,885	2,425
2100	30 circuits	"	.40	20		1,200	980		2,180	2,800
2200	225 amp main, 32 circuits	2 Elec	.72	22.222		2,000	1,100		3,100	3,875
2250	42 circuits		.56	28.571		2,200	1,400		3,600	4,575
2300	400 amp main, 42 circuits		.48	33.333		2,975	1,625		4,600	5,775
2350	600 amp main, 42 circuits		.40	40		4,425	1,950		6,375	7,850
2400	NEHB, with 20 amp, 1 pole circuit breaker									
2450	4 wire, 277/480 volts with main circuit breaker									
2500	100 amp main, 24 circuits	1 Elec	.42	19.048	Ea.	2,175	935		3,110	3,800
2550	30 circuits	"	.38	21.053		2,550	1,025		3,575	4,375
2600	225 amp main, 30 circuits	2 Elec	.72	22.222		3,200	1,100		4,300	5,200
2650	42 circuits	"	.56	28.571		3,950	1,400		5,350	6,500
9000	Minimum labor/equipment charge	1 Elec	1	8	Job		390		390	600

26 27 Low-Voltage Distribution Equipment

26 27 13 – Electricity Metering

26 27 13.10 Meter Centers and Sockets		Crew	Daily Output	Labor-Hours	Unit	Material	2010 Bare Costs Labor	Equipment	Total	Total Incl O&P
0010	**METER CENTERS AND SOCKETS**									
0100	Sockets, single position, 4 terminal, 100 amp	1 Elec	3.20	2.500	Ea.	44	123		167	237
0200	150 amp		2.30	3.478		51	170		221	320
0300	200 amp		1.90	4.211		74	206		280	395
0400	Transformer rated, 20 amp		3.20	2.500		124	123		247	325
0500	Double position, 4 terminal, 100 amp		2.80	2.857		180	140		320	415
0600	150 amp		2.10	3.810		212	187		399	520
0700	200 amp		1.70	4.706		435	231		666	835
9000	Minimum labor/equipment charge		3	2.667	Job		131		131	201

26 27 16 – Electrical Cabinets and Enclosures

26 27 16.10 Cabinets		Crew	Daily Output	Labor-Hours	Unit	Material	Labor	Equipment	Total	Total Incl O&P
0010	**CABINETS**									
7000	Cabinets, current transformer									
7050	Single door, 24" H x 24" W x 10" D	1 Elec	1.60	5	Ea.	206	245		451	600
7100	30" H x 24" W x 10" D		1.30	6.154		210	300		510	695
7150	36" H x 24" W x 10" D		1.10	7.273		215	355		570	785
7200	30" H x 30" W x 10" D		1	8		225	390		615	850
7250	36" H x 30" W x 10" D		.90	8.889		305	435		740	1,000
7300	36" H x 36" W x 10" D		.80	10		325	490		815	1,125
7500	Double door, 48" H x 36" W x 10" D		.60	13.333		670	655		1,325	1,725
7550	24" H x 24" W x 12" D		1	8		274	390		664	900
9990	Minimum labor/equipment charge		2	4	Job		196		196	300

26 27 23 – Indoor Service Poles

26 27 23.40 Surface Raceway		Crew	Daily Output	Labor-Hours	Unit	Material	Labor	Equipment	Total	Total Incl O&P
0010	**SURFACE RACEWAY**									
0090	Metal, straight section									
0100	No. 500	1 Elec	100	.080	L.F.	1.03	3.92		4.95	7.15
0110	No. 700		100	.080		1.16	3.92		5.08	7.30
0400	No. 1500, small pancake		90	.089		2.10	4.36		6.46	9
0600	No. 2000, base & cover, blank		90	.089		2.15	4.36		6.51	9.05
0800	No. 3000, base & cover, blank		75	.107		4.10	5.25		9.35	12.55
1000	No. 4000, base & cover, blank		65	.123		6.70	6.05		12.75	16.60
1200	No. 6000, base & cover, blank		50	.160		11.10	7.85		18.95	24.50
2400	Fittings, elbows, No. 500		40	.200	Ea.	1.85	9.80		11.65	17.10
2800	Elbow cover, No. 2000		40	.200		3.50	9.80		13.30	18.90
2880	Tee, No. 500		42	.190		3.60	9.35		12.95	18.30
2900	No. 2000		27	.296		11.70	14.50		26.20	35.50
3000	Switch box, No. 500		16	.500		12.40	24.50		36.90	51
3400	Telephone outlet, No. 1500		16	.500		14.50	24.50		39	53.50
3600	Junction box, No. 1500		16	.500		9.90	24.50		34.40	48.50
3800	Plugmold wired sections, No. 2000									
4000	1 circuit, 6 outlets, 3 ft. long	1 Elec	8	1	Ea.	36	49		85	115
4100	2 circuits, 8 outlets, 6 ft. long	"	5.30	1.509	"	59.50	74		133.50	180
9300	Non-metallic, straight section									
9310	7/16" x 7/8", base & cover, blank	1 Elec	160	.050	L.F.	1.68	2.45		4.13	5.60
9320	Base & cover w/adhesive		160	.050		1.84	2.45		4.29	5.80
9340	7/16" x 1-5/16", base & cover, blank		145	.055		1.84	2.70		4.54	6.15
9350	Base & cover w/adhesive		145	.055		2.15	2.70		4.85	6.50
9370	11/16" x 2-1/4", base & cover, blank		130	.062		2.60	3.02		5.62	7.50
9380	Base & cover w/adhesive		130	.062		2.98	3.02		6	7.90
9385	1-11/16" x 5-1/4", two compartment base & cover w/screws		80	.100		7.35	4.90		12.25	15.65
9400	Fittings, elbows, 7/16" x 7/8"		50	.160	Ea.	1.84	7.85		9.69	14.05

26 27 Low-Voltage Distribution Equipment

26 27 23 – Indoor Service Poles

26 27 23.40 Surface Raceway		Crew	Daily Output	Labor-Hours	Unit	Material	2010 Bare Costs Labor	Equipment	Total	Total Incl O&P
9410	7/16" x 1-5/16"	1 Elec	45	.178	Ea.	1.91	8.70		10.61	15.50
9420	11/16" x 2-1/4"		40	.200		2.08	9.80		11.88	17.35
9425	1-11/16" x 5-1/4"		28	.286		2.08	14		16.08	24
9430	Tees, 7/16" x 7/8"		35	.229		2.42	11.20		13.62	19.85
9440	7/16" x 1-5/16"		32	.250		2.49	12.25		14.74	21.50
9450	11/16" x 2-1/4"		30	.267		2.52	13.05		15.57	23
9455	1-11/16" x 5-1/4"		24	.333		17.90	16.35		34.25	44.50
9460	Cover clip, 7/16" x 7/8"		80	.100		.48	4.90		5.38	8.10
9470	7/16" x 1-5/16"		72	.111		.43	5.45		5.88	8.80
9480	11/16" x 2-1/4"		64	.125		.74	6.15		6.89	10.20
9484	1-11/16" x 5-1/4"		42	.190		2.63	9.35		11.98	17.25
9486	Wire clip, 1-11/16" x 5-1/4"		68	.118		.45	5.75		6.20	9.35
9490	Blank end, 7/16" x 7/8"		50	.160		.69	7.85		8.54	12.80
9500	7/16" x 1-5/16"		45	.178		.77	8.70		9.47	14.25
9510	11/16" x 2-1/4"		40	.200		1.17	9.80		10.97	16.35
9515	1-11/16" x 5-1/4"		38	.211		1.17	10.30		11.47	17.15
9520	Round fixture box, 5.5" dia x 1"		25	.320		11.10	15.70		26.80	36
9530	Device box, 1 gang		30	.267		4.90	13.05		17.95	25.50
9540	2 gang		25	.320		7.30	15.70		23	32
9990	Minimum labor/equipment charge		5	1.600	Job		78.50		78.50	120

26 27 26 – Wiring Devices

26 27 26.20 Wiring Devices Elements

		Crew	Daily Output	Labor-Hours	Unit	Material	Labor	Equipment	Total	Total Incl O&P
0010	**WIRING DEVICES ELEMENTS**									
0200	Toggle switch, quiet type, single pole, 15 amp	1 Elec	40	.200	Ea.	5.30	9.80		15.10	21
0500	20 amp		27	.296		7.40	14.50		21.90	30.50
0550	Rocker, 15 amp		40	.200		6.20	9.80		16	22
0560	20 amp		27	.296		14.25	14.50		28.75	38
0600	3 way, 15 amp		23	.348		7.65	17.05		24.70	34.50
0850	Rocker, 15 amp		23	.348		8.80	17.05		25.85	35.50
0860	20 amp		18	.444		20.50	22		42.50	56
0900	4 way, 15 amp		15	.533		25	26		51	67.50
1030	Rocker, 15 amp		15	.533		34	26		60	77.50
1040	20 amp		11	.727		50.50	35.50		86	111
1650	Dimmer switch, 120 volt, incandescent, 600 watt, 1 pole	G	16	.500		13.25	24.50		37.75	52
2460	Receptacle, duplex, 120 volt, grounded, 15 amp		40	.200		1.37	9.80		11.17	16.55
2470	20 amp		27	.296		10.40	14.50		24.90	34
2480	Ground fault interrupting, 15 amp		27	.296		36	14.50		50.50	62
2482	20 amp		27	.296		38	14.50		52.50	64.50
2490	Dryer, 30 amp		15	.533		6.25	26		32.25	47
2500	Range, 50 amp		11	.727		14.60	35.50		50.10	71
2600	Wall plates, stainless steel, 1 gang		80	.100		2.50	4.90		7.40	10.30
2800	2 gang		53	.151		4.25	7.40		11.65	16.05
3200	Lampholder, keyless		26	.308		12.80	15.10		27.90	37
3400	Pullchain with receptacle		22	.364		13.60	17.80		31.40	42.50
9000	Minimum labor/equipment charge		4	2	Job		98		98	151

26 27 73 – Door Chimes

26 27 73.10 Doorbell System

		Crew	Daily Output	Labor-Hours	Unit	Material	Labor	Equipment	Total	Total Incl O&P
0010	**DOORBELL SYSTEM**, incl. transformer, button & signal									
1000	Door chimes, 2 notes, minimum	1 Elec	16	.500	Ea.	30	24.50		54.50	70.50
1020	Maximum		12	.667		158	32.50		190.50	224
1100	Tube type, 3 tube system		12	.667		234	32.50		266.50	305
1180	4 tube system		10	.800		375	39		414	470

26 27 Low-Voltage Distribution Equipment

26 27 73 – Door Chimes

26 27 73.10 Doorbell System		Crew	Daily Output	Labor-Hours	Unit	Material	2010 Bare Costs Labor	Equipment	Total	Total Incl O&P
1900	For transformer & button, minimum add	1 Elec	5	1.600	Ea.	17	78.50		95.50	139
1960	Maximum, add		4.50	1.778		51	87		138	190
3000	For push button only, minimum		24	.333		3.44	16.35		19.79	29
3100	Maximum		20	.400		27	19.60		46.60	59.50
9000	Minimum labor/equipment charge		4	2	Job		98		98	151

26 28 Low-Voltage Circuit Protective Devices

26 28 16 – Enclosed Switches and Circuit Breakers

26 28 16.10 Circuit Breakers

		Crew	Daily Output	Labor-Hours	Unit	Material	Labor	Equipment	Total	Incl O&P
0010	**CIRCUIT BREAKERS** (in enclosure)									
0100	Enclosed (NEMA 1), 600 volt, 3 pole, 30 amp	1 Elec	3.20	2.500	Ea.	630	123		753	885
0200	60 amp		2.80	2.857		630	140		770	910
0400	100 amp		2.30	3.478		720	170		890	1,050
0600	225 amp		1.50	5.333		1,675	261		1,936	2,225
0700	400 amp	2 Elec	1.60	10		2,850	490		3,340	3,875
9000	Minimum labor/equipment charge	1 Elec	4	2	Job		98		98	151

26 28 16.20 Safety Switches

		Crew	Daily Output	Labor-Hours	Unit	Material	Labor	Equipment	Total	Incl O&P
0010	**SAFETY SWITCHES**									
0100	General duty 240 volt, 3 pole NEMA 1, fusible, 30 amp	1 Elec	3.20	2.500	Ea.	113	123		236	315
0200	60 amp		2.30	3.478		192	170		362	475
0300	100 amp		1.90	4.211		330	206		536	675
0400	200 amp		1.30	6.154		710	300		1,010	1,250
0500	400 amp	2 Elec	1.80	8.889		1,775	435		2,210	2,650
9990	Minimum labor/equipment charge	1 Elec	3	2.667	Job		131		131	201

26 28 16.40 Time Switches

		Crew	Daily Output	Labor-Hours	Unit	Material	Labor	Equipment	Total	Incl O&P
0010	**TIME SWITCHES**									
0100	Single pole, single throw, 24 hour dial	1 Elec	4	2	Ea.	107	98		205	269
0200	24 hour dial with reserve power		3.60	2.222		460	109		569	670
0300	Astronomic dial		3.60	2.222		184	109		293	370
0400	Astronomic dial with reserve power		3.30	2.424		595	119		714	840
0500	7 day calendar dial		3.30	2.424		146	119		265	345
0600	7 day calendar dial with reserve power		3.20	2.500		655	123		778	910
0700	Photo cell 2000 watt		8	1		19	49		68	96.50
1080	Load management device, 4 loads		2	4		1,100	196		1,296	1,500
1100	8 loads		1	8		1,800	390		2,190	2,575
9000	Minimum labor/equipment charge		3.50	2.286	Job		112		112	172

26 29 Low-Voltage Controllers

26 29 13 – Enclosed Controllers

26 29 13.40 Relays

		Crew	Daily Output	Labor-Hours	Unit	Material	Labor	Equipment	Total	Incl O&P
0010	**RELAYS** Enclosed (NEMA 1)									
0100	2 pole, 12 amp	1 Elec	5	1.600	Ea.	96.50	78.50		175	226
0200	4 pole, 10 amp	"	4.50	1.778	"	128	87		215	275

26 29 23 – Variable-Frequency Motor Controllers

26 29 23.10 Variable Frequency Drives/Adj. Frequency Drives

			Crew	Daily Output	Labor-Hours	Unit	Material	Labor	Equipment	Total	Incl O&P
0010	**VARIABLE FREQUENCY DRIVES/ADJ. FREQUENCY DRIVES**										
0100	Enclosed (NEMA 1), 460 volt, for 3 HP motor size	G	1 Elec	.80	10	Ea.	1,000	490		1,490	1,850
0110	5 HP motor size	G		.80	10		1,150	490		1,640	2,000

26 29 Low-Voltage Controllers

26 29 23 – Variable-Frequency Motor Controllers

26 29 23.10 Variable Frequency Drives/Adj. Frequency Drives		Crew	Daily Output	Labor-Hours	Unit	Material	2010 Bare Costs Labor	Equipment	Total	Total Incl O&P
0120	7.5 HP motor size G	1 Elec	.67	11.940	Ea.	1,375	585		1,960	2,400
0130	10 HP motor size G		.67	11.940		1,625	585		2,210	2,675
0140	15 HP motor size G	2 Elec	.89	17.978		1,925	880		2,805	3,475
0150	20 HP motor size G		.89	17.978		2,300	880		3,180	3,900
0160	25 HP motor size G		.67	23.881		2,850	1,175		4,025	4,925
0170	30 HP motor size G		.67	23.881		3,525	1,175		4,700	5,675
0180	40 HP motor size G		.67	23.881		4,200	1,175		5,375	6,425
0190	50 HP motor size G		.53	30.189		5,250	1,475		6,725	8,050
0200	60 HP motor size G	R-3	.56	35.714		6,275	1,725	247	8,247	9,825
0210	75 HP motor size G		.56	35.714		7,250	1,725	247	9,222	10,900
0220	100 HP motor size G		.50	40		8,450	1,925	276	10,651	12,600
0230	125 HP motor size G		.50	40		9,400	1,925	276	11,601	13,700
0240	150 HP motor size G		.50	40		10,800	1,925	276	13,001	15,200
0250	200 HP motor size G		.42	47.619		14,000	2,300	330	16,630	19,300
1100	Custom-engineered, 460 volt, for 3 HP motor size G	1 Elec	.56	14.286		2,350	700		3,050	3,650
1110	5 HP motor size G		.56	14.286		2,350	700		3,050	3,650
1120	7.5 HP motor size G		.47	17.021		2,450	835		3,285	3,975
1130	10 HP motor size G		.47	17.021		2,575	835		3,410	4,100
1140	15 HP motor size G	2 Elec	.62	25.806		3,225	1,275		4,500	5,500
1150	20 HP motor size G		.62	25.806		3,625	1,275		4,900	5,950
1160	25 HP motor size G		.47	34.043		4,250	1,675		5,925	7,250
1170	30 HP motor size G		.47	34.043		5,300	1,675		6,975	8,400
1180	40 HP motor size G		.47	34.043		6,275	1,675		7,950	9,475
1190	50 HP motor size G		.37	43.243		7,200	2,125		9,325	11,200
1200	60 HP motor size G	R-3	.39	51.282		10,900	2,475	355	13,730	16,100
1210	75 HP motor size G		.39	51.282		11,600	2,475	355	14,430	16,900
1220	100 HP motor size G		.35	57.143		12,600	2,750	395	15,745	18,600
1230	125 HP motor size G		.35	57.143		13,500	2,750	395	16,645	19,600
1240	150 HP motor size G		.35	57.143		15,500	2,750	395	18,645	21,700
1250	200 HP motor size G		.29	68.966		20,200	3,325	475	24,000	27,900
2000	For complex & special design systems to meet specific									
2010	requirements, obtain quote from vendor.									

26 32 Packaged Generator Assemblies

26 32 13 – Engine Generators

26 32 13.16 Gas-Engine-Driven Generator Sets

		Crew	Daily Output	Labor-Hours	Unit	Material	Labor	Equipment	Total	Total Incl O&P
0010	**GAS-ENGINE-DRIVEN GENERATOR SETS**									
0020	Gas or gasoline operated, includes battery,									
0050	charger, muffler & transfer switch									
0200	3 phase 4 wire, 277/480 volt, 7.5 kW	R-3	.83	24.096	Ea.	7,925	1,175	167	9,267	10,700
0300	11.5 kW		.71	28.169		11,200	1,350	195	12,745	14,700
0400	20 kW		.63	31.746		13,200	1,525	219	14,944	17,200
0500	35 kW		.55	36.364		15,800	1,750	251	17,801	20,300

26 51 Interior Lighting

26 51 13 – Interior Lighting Fixtures, Lamps, and Ballasts

26 51 13.50 Interior Lighting Fixtures		Crew	Daily Output	Labor-Hours	Unit	Material	2010 Bare Costs Labor	Equipment	Total	Total Incl O&P
0010	**INTERIOR LIGHTING FIXTURES** Including lamps, mounting									
0030	hardware and connections									
0100	Fluorescent, C.W. lamps, troffer, recess mounted in grid, RS									
0130	Grid ceiling mount									
0200	Acrylic lens, 1'W x 4'L, two 40 watt	1 Elec	5.70	1.404	Ea.	53.50	69		122.50	165
0300	2'W x 2'L, two U40 watt		5.70	1.404		57	69		126	169
0600	2'W x 4'L, four 40 watt		4.70	1.702		65.50	83.50		149	200
0910	Acrylic lens, 1'W x 4'L, two 32 watt T-8 G		5.70	1.404		67	69		136	180
0930	2'W x 2'L, two U32 watt T-8 G		5.70	1.404		80	69		149	194
0940	2'W x 4'L, two 32 watt T-8 G		5.30	1.509		72	74		146	194
0950	2'W x 4'L, three 32 watt T-8 G		5	1.600		76.50	78.50		155	204
0960	2'W x 4'L, four 32 watt T-8 G		4.70	1.702		78.50	83.50		162	215
1000	Surface mounted, RS									
1030	Acrylic lens with hinged & latched door frame									
1100	1'W x 4'L, two 40 watt	1 Elec	7	1.143	Ea.	74	56		130	168
1200	2'W x 2'L, two U40 watt		7	1.143		79.50	56		135.50	173
1500	2'W x 4'L, four 40 watt		5.30	1.509		94.50	74		168.50	218
2100	Strip fixture									
2130	Surface mounted									
2200	4' long, one 40 watt, RS	1 Elec	8.50	.941	Ea.	31.50	46		77.50	106
2300	4' long, two 40 watt, RS	"	8	1		34.50	49		83.50	114
2600	8' long, one 75 watt, SL	2 Elec	13.40	1.194		47	58.50		105.50	142
2700	8' long, two 75 watt, SL	"	12.40	1.290		57	63		120	160
3000	Strip, pendent mounted, industrial, white porcelain enamel									
3100	4' long, two 40 watt, RS	1 Elec	5.70	1.404	Ea.	52	69		121	163
3200	4' long, two 60 watt, HO	"	5	1.600		81	78.50		159.50	209
3300	8' long, two 75 watt, SL	2 Elec	8.80	1.818		96	89		185	243
4220	Metal halide, integral ballast, ceiling, recess mounted									
4230	prismatic glass lens, floating door									
4240	2'W x 2'L, 250 watt	1 Elec	3.20	2.500	Ea.	310	123		433	530
4250	2'W x 2'L, 400 watt	2 Elec	5.80	2.759		360	135		495	605
4260	Surface mounted, 2'W x 2'L, 250 watt	1 Elec	2.70	2.963		335	145		480	595
4270	400 watt	2 Elec	4.80	3.333		395	163		558	685
4280	High bay, aluminum reflector,									
4290	Single unit, 400 watt	2 Elec	4.60	3.478	Ea.	405	170		575	705
4300	Single unit, 1000 watt		4	4		580	196		776	940
4310	Twin unit, 400 watt		3.20	5		805	245		1,050	1,250
4320	Low bay, aluminum reflector, 250W DX lamp	1 Elec	3.20	2.500		355	123		478	580
4340	High pressure sodium integral ballast ceiling, recess mounted									
4350	prismatic glass lens, floating door									
4360	2'W x 2'L, 150 watt lamp	1 Elec	3.20	2.500	Ea.	380	123		503	605
4370	2'W x 2'L, 400 watt lamp	2 Elec	5.80	2.759		450	135		585	705
4380	Surface mounted, 2'W x 2'L, 150 watt lamp	1 Elec	2.70	2.963		460	145		605	735
4390	400 watt lamp	2 Elec	4.80	3.333		515	163		678	820
4400	High bay, aluminum reflector,									
4410	Single unit, 400 watt lamp	2 Elec	4.60	3.478	Ea.	370	170		540	670
4430	Single unit, 1000 watt lamp	"	4	4		535	196		731	890
4440	Low bay, aluminum reflector, 150 watt lamp	1 Elec	3.20	2.500		320	123		443	545
4450	Incandescent, high hat can, round alzak reflector, prewired									
4470	100 watt	1 Elec	8	1	Ea.	70	49		119	153
4480	150 watt		8	1		102	49		151	188
4500	300 watt		6.70	1.194		236	58.50		294.50	350
4600	Square glass lens with metal trim, prewired									

26 51 Interior Lighting

26 51 13 - Interior Lighting Fixtures, Lamps, and Ballasts

26 51 13.50 Interior Lighting Fixtures		Crew	Daily Output	Labor-Hours	Unit	Material	2010 Bare Costs Labor	Equipment	Total	Total Incl O&P
4630	100 watt	1 Elec	6.70	1.194	Ea.	54	58.50		112.50	150
4700	200 watt		6.70	1.194		95	58.50		153.50	195
6010	Vapor tight, incandescent, ceiling mounted, 200 watt		6.20	1.290		70.50	63		133.50	175
6100	Fluorescent, surface mounted, 2 lamps, 4'L, RS, 40 watt		3.20	2.500		113	123		236	315
6850	Vandalproof, surface mounted, fluorescent, two 32 watt T-8 G		3.20	2.500		247	123		370	460
6860	Incandescent, one 150 watt		8	1		73	49		122	156
6900	Mirror light, fluorescent, RS, acrylic enclosure, two 40 watt		8	1		116	49		165	204
6910	One 40 watt		8	1		94	49		143	179
6920	One 20 watt		12	.667		74	32.50		106.50	132
7500	Ballast replacement, by weight of ballast, to 15' high									
7520	Indoor fluorescent, less than 2 lbs.	1 Elec	10	.800	Ea.	25	39		64	87.50
7540	Two 40W, watt reducer, 2 to 5 lbs.		9.40	.851		40	41.50		81.50	108
7560	Two F96 slimline, over 5 lbs.		8	1		75.50	49		124.50	159
7580	Vaportite ballast, less than 2 lbs.		9.40	.851		25	41.50		66.50	91.50
7600	2 lbs. to 5 lbs.		8.90	.899		40	44		84	112
7620	Over 5 lbs.		7.60	1.053		75.50	51.50		127	163
7630	Electronic ballast for two tubes		8	1		36.50	49		85.50	116
7640	Dimmable ballast one lamp G		8	1		108	49		157	195
7650	Dimmable ballast two-lamp G		7.60	1.053		108	51.50		159.50	199
9000	Minimum labor/equipment charge		3	2.667	Job		131		131	201

26 51 13.70 Residential Fixtures		Crew	Daily Output	Labor-Hours	Unit	Material	Labor	Equipment	Total	Total Incl O&P
0010	**RESIDENTIAL FIXTURES**									
0400	Fluorescent, interior, surface, circline, 32 watt & 40 watt	1 Elec	20	.400	Ea.	103	19.60		122.60	143
0500	2' x 2', two U 32 watt T8		8	1		138	49		187	227
0700	Shallow under cabinet, two 20 watt		16	.500		47.50	24.50		72	90
0900	Wall mounted, 4'L, two 32 watt T8, with baffle		10	.800		155	39		194	231
2000	Incandescent, exterior lantern, wall mounted, 60 watt		16	.500		40	24.50		64.50	81.50
2100	Post light, 150W, with 7' post		4	2		117	98		215	280
2500	Lamp holder, weatherproof with 150W PAR		16	.500		30.50	24.50		55	71
2550	With reflector and guard		12	.667		64	32.50		96.50	121
2600	Interior pendent, globe with shade, 150 watt		20	.400		174	19.60		193.60	221
9000	Minimum labor/equipment charge		4	2	Job		98		98	151

26 52 Emergency Lighting

26 52 13 - Emergency Lighting Equipments

26 52 13.10 Emergency Lighting and Battery Units		Crew	Daily Output	Labor-Hours	Unit	Material	Labor	Equipment	Total	Total Incl O&P
0010	**EMERGENCY LIGHTING AND BATTERY UNITS**									
0300	Emergency light units, battery operated									
0350	Twin sealed beam light, 25 watt, 6 volt each									
0500	Lead battery operated	1 Elec	4	2	Ea.	129	98		227	292
0700	Nickel cadmium battery operated		4	2	"	635	98		733	850
9000	Minimum labor/equipment charge		4	2	Job		98		98	151

26 53 Exit Signs

26 53 13 – Exit Lighting

26 53 13.10 Exit Lighting Fixtures	Crew	Daily Output	Labor-Hours	Unit	Material	2010 Bare Costs Labor	Equipment	Total	Total Incl O&P
0010 **EXIT LIGHTING FIXTURES**									
0080 Exit light ceiling or wall mount, incandescent, single face	1 Elec	8	1	Ea.	42.50	49		91.50	123
0100 Double face		6.70	1.194	"	49	58.50		107.50	144
9000 Minimum labor/equipment charge		4	2	Job		98		98	151

26 55 Special Purpose Lighting

26 55 59 – Display Lighting

26 55 59.10 Track Lighting	Crew	Daily Output	Labor-Hours	Unit	Material	Labor	Equipment	Total	Total Incl O&P
0010 **TRACK LIGHTING**									
0080 Track, 1 circuit, 4' section	1 Elec	6.70	1.194	Ea.	49.50	58.50		108	144
0100 8' section	2 Elec	10.60	1.509		81	74		155	204
0300 3 circuits, 4' section	1 Elec	6.70	1.194		68	58.50		126.50	165
0400 8' section	2 Elec	10.60	1.509		106	74		180	231
9000 Minimum labor/equipment charge	1 Elec	3	2.667	Job		131		131	201

26 56 Exterior Lighting

26 56 23 – Area Lighting

26 56 23.10 Exterior Fixtures	Crew	Daily Output	Labor-Hours	Unit	Material	Labor	Equipment	Total	Total Incl O&P
0010 **EXTERIOR FIXTURES** With lamps									
0200 Wall mounted, incandescent, 100 watt	1 Elec	8	1	Ea.	31.50	49		80.50	110
0400 Quartz, 500 watt		5.30	1.509		56	74		130	176
1100 Wall pack, low pressure sodium, 35 watt		4	2		234	98		332	410
1150 55 watt		4	2		278	98		376	455
1160 High pressure sodium, 70 watt		4	2		230	98		328	405
1170 150 watt		4	2		251	98		349	425
1180 Metal Halide, 175 watt		4	2		255	98		353	430
1190 250 watt		4	2		290	98		388	470
1195 400 watt		4	2		345	98		443	525

26 56 33 – Walkway Lighting

26 56 33.10 Walkway Luminaire	Crew	Daily Output	Labor-Hours	Unit	Material	Labor	Equipment	Total	Total Incl O&P
0010 **WALKWAY LUMINAIRE**									
9000 Minimum labor/equipment charge	1 Elec	3.75	2.133	Job		105		105	161

26 56 36 – Flood Lighting

26 56 36.20 Floodlights	Crew	Daily Output	Labor-Hours	Unit	Material	Labor	Equipment	Total	Total Incl O&P
0010 **FLOODLIGHTS** with ballast and lamp,									
1400 pole mounted, pole not included									
2250 Low pressure sodium, 55 watt	1 Elec	2.70	2.963	Ea.	565	145		710	845
2270 90 watt	"	2	4	"	620	196		816	985

26 61 Lighting Systems and Accessories

26 61 23 - Lamps Applications

26 61 23.10 Lamps		Crew	Daily Output	Labor-Hours	Unit	Material	2010 Bare Costs Labor	Equipment	Total	Total Incl O&P
0010	**LAMPS**									
0080	Fluorescent, rapid start, cool white, 2' long, 20 watt	1 Elec	1	8	C	355	390		745	990
0100	4' long, 40 watt		.90	8.889		300	435		735	1,000
0170	4' long, 34 watt energy saver	G	.90	8.889		350	435		785	1,050
0176	2' long, T8, 17 W energy saver	G	1	8		500	390		890	1,150
0178	3' long, T8, 25 W energy saver	G	.90	8.889		460	435		895	1,175
0180	4' long, T8, 32 watt energy saver	G	.90	8.889		320	435		755	1,025
0560	Twin tube compact lamp	G	.90	8.889		585	435		1,020	1,325
0570	Double twin tube compact lamp	G	.80	10		1,300	490		1,790	2,200
0600	Mercury vapor, mogul base, deluxe white, 100 watt		.30	26.667		3,100	1,300		4,400	5,425
0800	400 watt		.30	26.667		3,875	1,300		5,175	6,250
1000	Metal halide, mogul base, 175 watt		.30	26.667		4,125	1,300		5,425	6,550
1100	250 watt		.30	26.667		4,675	1,300		5,975	7,150
1350	High pressure sodium, 70 watt		.30	26.667		4,800	1,300		6,100	7,275
1370	150 watt		.30	26.667		5,150	1,300		6,450	7,675
1500	Low pressure sodium, 35 watt		.30	26.667		10,700	1,300		12,000	13,800
1600	90 watt		.30	26.667		12,500	1,300		13,800	15,800
1800	Incandescent, interior, A21, 100 watt		1.60	5		200	245		445	595
1900	A21, 150 watt		1.60	5		192	245		437	585
2300	R30, 75 watt		1.30	6.154		605	300		905	1,125
2500	Exterior, PAR 38, 75 watt		1.30	6.154		1,475	300		1,775	2,075
2600	PAR 38, 150 watt		1.30	6.154		1,725	300		2,025	2,375
9000	Minimum labor/equipment charge		4	2	Job		98		98	151

Estimating Tips

27 20 00 Data Communications
27 30 00 Voice Communications
27 40 00 Audio-Video Communications

When estimating material costs for special systems, it is always prudent to obtain manufacturers' quotations for equipment prices and special installation requirements which will affect the total costs.

Reference Numbers

Reference numbers are shown in shaded boxes at the beginning of some major classifications. These numbers refer to related items in the Reference Section. The reference information may be an estimating procedure, an alternate pricing method, or technical information.

Note: Not all subdivisions listed here necessarily appear in this publication.

Note: **Trade Service,** *in part, has been used as a reference source for some of the material prices used in Division 27.*

27 15 Communications Horizontal Cabling

27 15 10 – Special Communications Cabling

27 15 10.23 Sound and Video Cables and Fittings		Crew	Daily Output	Labor-Hours	Unit	Material	2010 Bare Costs Labor	Equipment	Total	Total Incl O&P
0010	**SOUND AND VIDEO CABLES & FITTINGS**									
1250	Nonshielded, #22-2 conductor	1 Elec	10	.800	C.L.F.	13.70	39		52.70	75

27 15 13 – Communications Copper Horizontal Cabling

27 15 13.13 Communication Cables

		Crew	Daily Output	Labor-Hours	Unit	Material	Labor	Equipment	Total	Total Incl O&P
0010	**COMMUNICATION CABLES**									
5000	High performance unshielded twisted pair (UTP)									
7000	Category 5, #24, 4 pair solid, PVC jacket	1 Elec	7	1.143	C.L.F.	14.50	56		70.50	102
7100	4 pair solid, plenum		7	1.143		44.50	56		100.50	135
7200	4 pair stranded, PVC jacket		7	1.143		21.50	56		77.50	110
7210	Category 5e, #24, 4 pair solid, PVC jacket		7	1.143		13.90	56		69.90	101
7212	4 pair solid, plenum		7	1.143		41	56		97	131
7214	4 pair stranded, PVC jacket		7	1.143		23	56		79	112
7240	Category 6, #24, 4 pair solid, PVC jacket		7	1.143		24	56		80	113
7242	4 pair solid, plenum		7	1.143		68	56		124	161
7244	4 pair stranded, PVC jacket		7	1.143		30	56		86	119
7300	Connector, RJ-45, category 5		80	.100	Ea.	1.26	4.90		6.16	8.95
7302	Shielded RJ-45, category 5		72	.111		3.60	5.45		9.05	12.30
7310	Jack, UTP RJ-45, category 3		72	.111		3.66	5.45		9.11	12.40
7312	Category 5		65	.123		4.74	6.05		10.79	14.45
7314	Category 5e		65	.123		4.74	6.05		10.79	14.45
7316	Category 6		65	.123		4.74	6.05		10.79	14.45

27 41 Audio-Video Systems

27 41 19 – Portable Audio-Video Equipment

27 41 19.10 T.V. Systems

		Crew	Daily Output	Labor-Hours	Unit	Material	Labor	Equipment	Total	Total Incl O&P
0010	**T.V. SYSTEMS**, not including rough-in wires, cables & conduits									
5000	T.V. Antenna only, minimum	1 Elec	6	1.333	Ea.	45	65.50		110.50	150
5100	Maximum	"	4	2	"	189	98		287	360

27 51 Distributed Audio-Video Communications Systems

27 51 16 – Public Address and Mass Notification Systems

27 51 16.10 Public Address System

		Crew	Daily Output	Labor-Hours	Unit	Material	Labor	Equipment	Total	Total Incl O&P
0010	**PUBLIC ADDRESS SYSTEM**									
0100	Conventional, office	1 Elec	5.33	1.501	Speaker	129	73.50		202.50	254
0200	Industrial		2.70	2.963	"	248	145		393	495
9000	Minimum labor/equipment charge		3.50	2.286	Job		112		112	172

27 51 19 – Sound Masking Systems

27 51 19.10 Sound System

		Crew	Daily Output	Labor-Hours	Unit	Material	Labor	Equipment	Total	Total Incl O&P
0010	**SOUND SYSTEM**, not including rough-in wires, cables & conduits									
2000	Intercom, 25 station capacity, master station	2 Elec	2	8	Ea.	1,950	390		2,340	2,750
2020	11 station capacity	"	4	4		910	196		1,106	1,300
3600	House telephone, talking station	1 Elec	1.60	5		445	245		690	865
3800	Press to talk, release to listen	"	5.30	1.509		104	74		178	228
4000	System-on button					62			62	68
4200	Door release	1 Elec	4	2		111	98		209	273
4400	Combination speaker and microphone		8	1		189	49		238	284
4600	Termination box		3.20	2.500		59.50	123		182.50	254
4800	Amplifier or power supply		5.30	1.509		680	74		754	865

27 51 Distributed Audio-Video Communications Systems

27 51 19 – Sound Masking Systems

27 51 19.10 Sound System		Crew	Daily Output	Labor-Hours	Unit	Material	2010 Bare Costs Labor	Equipment	Total	Total Incl O&P
5000	Vestibule door unit	1 Elec	16	.500	Name	125	24.50		149.50	176
5200	Strip cabinet		27	.296	Ea.	237	14.50		251.50	284
5400	Directory		16	.500	"	112	24.50		136.50	161
9000	Minimum labor/equipment charge	↓	3.50	2.286	Job		112		112	172

Division Notes

	CREW	DAILY OUTPUT	LABOR-HOURS	UNIT	2010 BARE COSTS				TOTAL INCL O&P
					MAT.	LABOR	EQUIP.	TOTAL	

Estimating Tips

- When estimating material costs for electronic safety and security systems, it is always prudent to obtain manufacturers' quotations for equipment prices and special installation requirements that affect the total cost.
- Fire alarm systems consist of control panels, annunciator panels, battery with rack, charger, and fire alarm actuating and indicating devices. Some fire alarm systems include speakers, telephone lines, door closer controls, and other components. Be careful not to overlook the costs related to installation for these items. Also be aware of costs for integrated automation instrumentation and terminal devices, control equipment, control wiring, and programming.
- Security equipment includes items such as CCTV, access control, and other detection and identification systems to perform alert and alarm functions. Be sure to consider the costs related to installation for this security equipment, such as for integrated automation instrumentation and terminal devices, control equipment, control wiring, and programming.

Reference Numbers

Reference numbers are shown in shaded boxes at the beginning of some major classifications. These numbers refer to related items in the Reference Section. The reference information may be an estimating procedure, an alternate pricing method, or technical information.

Note: Not all subdivisions listed here necessarily appear in this publication.

No part of this publication may be reproduced, stored in a retrieval system, or transmitted in any form or by any means without prior written permission of R.S. Means Company, Inc.

28 13 Access Control

28 13 53 − Security Access Detection

28 13 53.13 Security Access Metal Detectors

		Crew	Daily Output	Labor-Hours	Unit	Material	2010 Bare Costs Labor	Equipment	Total	Total Incl O&P
0010	**SECURITY ACCESS METAL DETECTORS**									
0240	Metal detector, hand-held, wand type, unit only				Ea.	81.50			81.50	90
0250	Metal detector, walk through portal type, single zone	1 Elec	2	4		2,750	196		2,946	3,325
0260	Multi-zone	"	2	4		3,500	196		3,696	4,150

28 13 53.16 Security Access X-Ray Equipment

		Crew	Daily Output	Labor-Hours	Unit	Material	Labor	Equipment	Total	Total Incl O&P
0010	**SECURITY ACCESS X-RAY EQUIPMENT**									
0290	X-ray machine, desk top, for mail/small packages/letters	1 Elec	4	2	Ea.	3,000	98		3,098	3,450
0300	Conveyor type, incl monitor, minimum		2	4		14,000	196		14,196	15,700
0310	Maximum		2	4		25,000	196		25,196	27,800
0320	X-ray machine, large unit, for airports, incl monitor, min	2 Elec	1	16		35,000	785		35,785	39,700
0330	Maximum	"	.50	32		60,000	1,575		61,575	68,500

28 13 53.23 Security Access Explosive Detection Equipment

		Crew	Daily Output	Labor-Hours	Unit	Material	Labor	Equipment	Total	Total Incl O&P
0010	**SECURITY ACCESS EXPLOSIVE DETECTION EQUIPMENT**									
0270	Explosives detector, walk through portal type	1 Elec	2	4	Ea.	3,500	196		3,696	4,150
0280	Hand-held, battery operated				"				25,500	28,100

28 16 Intrusion Detection

28 16 16 − Intrusion Detection Systems Infrastructure

28 16 16.50 Intrusion Detection

		Crew	Daily Output	Labor-Hours	Unit	Material	Labor	Equipment	Total	Total Incl O&P
0010	**INTRUSION DETECTION**, not including wires & conduits									
0100	Burglar alarm, battery operated, mechanical trigger	1 Elec	4	2	Ea.	272	98		370	450
0200	Electrical trigger		4	2		325	98		423	505
0400	For outside key control, add		8	1		82	49		131	166
0600	For remote signaling circuitry, add		8	1		122	49		171	211
0800	Card reader, flush type, standard		2.70	2.963		910	145		1,055	1,225
1000	Multi-code		2.70	2.963		1,175	145		1,320	1,525
1200	Door switches, hinge switch		5.30	1.509		57.50	74		131.50	177
1400	Magnetic switch		5.30	1.509		67.50	74		141.50	189
1600	Exit control locks, horn alarm		4	2		340	98		438	525
1800	Flashing light alarm		4	2		380	98		478	570
2000	Indicating panels, 1 channel		2.70	2.963		360	145		505	620
2200	10 channel	2 Elec	3.20	5		1,225	245		1,470	1,725
2400	20 channel		2	8		2,400	390		2,790	3,250
2600	40 channel		1.14	14.035		4,375	690		5,065	5,850
2800	Ultrasonic motion detector, 12 volt	1 Elec	2.30	3.478		225	170		395	510
3000	Infrared photoelectric detector		2.30	3.478		186	170		356	465
3200	Passive infrared detector		2.30	3.478		278	170		448	565
3400	Glass break alarm switch		8	1		91.50	49		140.50	176
3420	Switchmats, 30" x 5'		5.30	1.509		83	74		157	206
3440	30" x 25'		4	2		199	98		297	370
3460	Police connect panel		4	2		239	98		337	415
3480	Telephone dialer		5.30	1.509		375	74		449	530
3500	Alarm bell		4	2		102	98		200	264
3520	Siren		4	2		143	98		241	310
3540	Microwave detector, 10' to 200'		2	4		655	196		851	1,025
3560	10' to 350'		2	4		1,900	196		2,096	2,400

28 23 Video Surveillance

28 23 23 – Video Surveillance Systems Infrastructure

28 23 23.50 Video Surveillance Equipments	Crew	Daily Output	Labor-Hours	Unit	Material	2010 Bare Costs Labor	Equipment	Total	Total Incl O&P
0010 **VIDEO SURVEILLANCE EQUIPMENTS**									
0200 Video cameras, wireless, hidden in exit signs, clocks, etc, incl receiver	1 Elec	3	2.667	Ea.	149	131		280	365
0210 Accessories for VCR, single camera		3	2.667		570	131		701	825
0220 For multiple cameras		3	2.667		1,225	131		1,356	1,550
0230 Video cameras, wireless, for under vehicle searching, complete	↓	2	4	↓	9,975	196		10,171	11,300

28 31 Fire Detection and Alarm

28 31 23 – Fire Detection and Alarm Annunciation Panels and Fire Stations

28 31 23.50 Alarm Panels and Devices

		Crew	Daily Output	Labor-Hours	Unit	Material	Labor	Equipment	Total	Total Incl O&P
0010	**ALARM PANELS AND DEVICES**, not including wires & conduits									
3594	Fire, alarm control panel									
3600	4 zone	2 Elec	2	8	Ea.	730	390		1,120	1,400
3800	8 zone		1	16		760	785		1,545	2,025
4000	12 zone	↓	.67	23.988		2,300	1,175		3,475	4,325
4200	Battery and rack	1 Elec	4	2		395	98		493	585
4400	Automatic charger		8	1		560	49		609	695
4600	Signal bell		8	1		70.50	49		119.50	153
4800	Trouble buzzer or manual station		8	1		80.50	49		129.50	164
5600	Strobe and horn		5.30	1.509		147	74		221	275
5800	Fire alarm horn		6.70	1.194		58.50	58.50		117	155
6000	Door holder, electro-magnetic		4	2		99	98		197	260
6200	Combination holder and closer		3.20	2.500		119	123		242	320
6600	Drill switch		8	1		360	49		409	470
6800	Master box		2.70	2.963		6,175	145		6,320	7,025
7000	Break glass station		8	1		65	49		114	147
7800	Remote annunciator, 8 zone lamp	↓	1.80	4.444		201	218		419	555
8000	12 zone lamp	2 Elec	2.60	6.154		345	300		645	845
8200	16 zone lamp	"	2.20	7.273	↓	345	355		700	930

28 31 43 – Fire Detection Sensors

28 31 43.50 Fire and Heat Detectors

		Crew	Daily Output	Labor-Hours	Unit	Material	Labor	Equipment	Total	Total Incl O&P
0010	**FIRE & HEAT DETECTORS**									
5000	Detector, rate of rise	1 Elec	8	1	Ea.	49	49		98	130
5100	Fixed temperature	"	8	1	"	49	49		98	130

28 31 46 – Smoke Detection Sensors

28 31 46.50 Smoke Detectors

		Crew	Daily Output	Labor-Hours	Unit	Material	Labor	Equipment	Total	Total Incl O&P
0010	**SMOKE DETECTORS**									
5200	Smoke detector, ceiling type	1 Elec	6.20	1.290	Ea.	126	63		189	236
5400	Duct type	"	3.20	2.500	"	315	123		438	535

Division Notes

	CREW	DAILY OUTPUT	LABOR-HOURS	UNIT	2010 BARE COSTS				TOTAL INCL O&P
					MAT.	LABOR	EQUIP.	TOTAL	

Estimating Tips

31 05 00 Common Work Results for Earthwork

- Estimating the actual cost of performing earthwork requires careful consideration of the variables involved. This includes items such as type of soil, whether water will be encountered, dewatering, whether banks need bracing, disposal of excavated earth, and length of haul to fill or spoil sites, etc. If the project has large quantities of cut or fill, consider raising or lowering the site to reduce costs, while paying close attention to the effect on site drainage and utilities.
- If the project has large quantities of fill, creating a borrow pit on the site can significantly lower the costs.
- It is very important to consider what time of year the project is scheduled for completion. Bad weather can create large cost overruns from dewatering, site repair, and lost productivity from cold weather.

Reference Numbers

Reference numbers are shown in shaded boxes at the beginning of some major classifications. These numbers refer to related items in the Reference Section. The reference information may be an estimating procedure, an alternate pricing method, or technical information.

Note: Not all subdivisions listed here necessarily appear in this publication.

31 05 Common Work Results for Earthwork

31 05 13 – Soils for Earthwork

31 05 13.10 Borrow		Crew	Daily Output	Labor-Hours	Unit	Material	2010 Bare Costs Labor	Equipment	Total	Total Incl O&P
0010	**BORROW**									
0020	Spread, 200 H.P. dozer, no compaction, 2 mi. RT haul									
0200	Common borrow	B-15	600	.047	C.Y.	7.40	1.68	4.21	13.29	15.50
0700	Screened loam		600	.047		30	1.68	4.21	35.89	40.50
0800	Topsoil, weed free	↓	600	.047		23	1.68	4.21	28.89	33
0900	For 5 mile haul, add	B-34B	200	.040	↓		1.33	3.34	4.67	5.80

31 05 16 – Aggregates for Earthwork

31 05 16.10 Borrow		Crew	Daily Output	Labor-Hours	Unit	Material	Labor	Equipment	Total	Total Incl O&P
0010	**BORROW**									
0020	Spread, with 200 H.P. dozer, no compaction, 2 mi. RT haul									
0100	Bank run gravel	B-15	600	.047	C.Y.	22	1.68	4.21	27.89	31.50
0300	Crushed stone (1.40 tons per CY), 1-1/2"		600	.047		35	1.68	4.21	40.89	46
0320	3/4"		600	.047		35	1.68	4.21	40.89	46
0340	1/2"		600	.047		30	1.68	4.21	35.89	40.50
0360	3/8"		600	.047		29.50	1.68	4.21	35.39	40
0400	Sand, washed, concrete		600	.047		33	1.68	4.21	38.89	43.50
0500	Dead or bank sand		600	.047		10.55	1.68	4.21	16.44	18.95
0600	Select structural fill	↓	600	.047		12.65	1.68	4.21	18.54	21
0900	For 5 mile haul, add	B-34B	200	.040	↓		1.33	3.34	4.67	5.80

31 06 Schedules for Earthwork

31 06 60 – Schedules for Special Foundations and Load Bearing Elements

31 06 60.14 Piling Special Costs		Crew	Daily Output	Labor-Hours	Unit	Material	Labor	Equipment	Total	Total Incl O&P
0010	**PILING SPECIAL COSTS**									
0011	Piling special costs, pile caps, see Div. 03 30 53.40									
0500	Cutoffs, concrete piles, plain	1 Pile	5.50	1.455	Ea.		58.50		58.50	97
0600	With steel thin shell, add		38	.211			8.50		8.50	14.05
0700	Steel pile or "H" piles		19	.421			16.95		16.95	28
0800	Wood piles	↓	38	.211	↓		8.50		8.50	14.05
1000	Testing, any type piles, test load is twice the design load									
1050	50 ton design load, 100 ton test				Ea.				16,000	17,600
1100	100 ton design load, 200 ton test								21,000	23,100
1200	200 ton design load, 400 ton test				↓				28,500	31,300

31 06 60.15 Mobilization		Crew	Daily Output	Labor-Hours	Unit	Material	Labor	Equipment	Total	Total Incl O&P
0010	**MOBILIZATION**									
0020	Set up & remove, air compressor, 600 C.F.M.	A-5	3.30	5.455	Ea.		180	18.15	198.15	315
0100	1200 C.F.M.	"	2.20	8.182			270	27.50	297.50	475
0200	Crane, with pile leads and pile hammer, 75 ton	B-19	.60	106			4,400	3,225	7,625	10,700
0300	150 ton	"	.36	177	↓		7,350	5,375	12,725	17,800

31 23 Excavation and Fill

31 23 16 – Excavation

31 23 16.13 Excavating, Trench

		Crew	Daily Output	Labor-Hours	Unit	Material	2010 Bare Costs Labor	2010 Bare Costs Equipment	Total	Total Incl O&P
0010	**EXCAVATING, TRENCH**									
0011	Or continuous footing									
0020	Common earth with no sheeting or dewatering included									
1400	By hand with pick and shovel 2' to 6' deep, light soil	1 Clab	8	1	B.C.Y.		33		33	54.50
1500	Heavy soil	"	4	2	"		66		66	109
1700	For tamping backfilled trenches, air tamp, add	A-1G	100	.080	E.C.Y.		2.65	.52	3.17	4.92
1900	Vibrating plate, add	B-18	180	.133	"		4.50	.24	4.74	7.65
2100	Trim sides and bottom for concrete pours, common earth		1500	.016	S.F.		.54	.03	.57	.92
2300	Hardpan	↓	600	.040	"		1.35	.07	1.42	2.30
9000	Minimum labor/equipment charge	1 Clab	4	2	Job		66		66	109

31 23 16.14 Excavating, Utility Trench

		Crew	Daily Output	Labor-Hours	Unit	Material	Labor	Equipment	Total	Total Incl O&P
0010	**EXCAVATING, UTILITY TRENCH**									
0011	Common earth									
0050	Trenching with chain trencher, 12 H.P., operator walking									
0100	4" wide trench, 12" deep	B-53	800	.010	L.F.		.41	.08	.49	.74
0150	18" deep		750	.011			.44	.09	.53	.79
0200	24" deep		700	.011			.47	.09	.56	.84
0300	6" wide trench, 12" deep		650	.012			.51	.10	.61	.91
0350	18" deep		600	.013			.55	.11	.66	.98
0400	24" deep		550	.015			.60	.12	.72	1.07
0450	36" deep		450	.018			.73	.15	.88	1.31
0600	8" wide trench, 12" deep		475	.017			.70	.14	.84	1.24
0650	18" deep		400	.020			.83	.17	1	1.47
0700	24" deep		350	.023			.94	.19	1.13	1.69
0750	36" deep		300	.027	↓		1.10	.22	1.32	1.97
0900	Minimum labor/equipment charge	↓	2	4	Job		165	33	198	296
1000	Backfill by hand including compaction, add									
1050	4" wide trench, 12" deep	A-1G	800	.010	L.F.		.33	.06	.39	.61
1100	18" deep		530	.015			.50	.10	.60	.93
1150	24" deep		400	.020			.66	.13	.79	1.23
1300	6" wide trench, 12" deep		540	.015			.49	.10	.59	.91
1350	18" deep		405	.020			.65	.13	.78	1.21
1400	24" deep		270	.030			.98	.19	1.17	1.82
1450	36" deep		180	.044			1.47	.29	1.76	2.74
1600	8" wide trench, 12" deep		400	.020			.66	.13	.79	1.23
1650	18" deep		265	.030			1	.20	1.20	1.86
1700	24" deep		200	.040			1.32	.26	1.58	2.45
1750	36" deep	↓	135	.059	↓		1.96	.38	2.34	3.64
2000	Chain trencher, 40 H.P. operator riding									
2050	6" wide trench and backfill, 12" deep	B-54	1200	.007	L.F.		.28	.28	.56	.74
2100	18" deep		1000	.008			.33	.33	.66	.89
2150	24" deep		975	.008			.34	.34	.68	.91
2200	36" deep		900	.009			.37	.37	.74	.99
2250	48" deep		750	.011			.44	.45	.89	1.18
2300	60" deep		650	.012			.51	.51	1.02	1.37
2400	8" wide trench and backfill, 12" deep		1000	.008			.33	.33	.66	.89
2450	18" deep		950	.008			.35	.35	.70	.93
2500	24" deep		900	.009			.37	.37	.74	.99
2550	36" deep		800	.010			.41	.42	.83	1.11
2600	48" deep		650	.012			.51	.51	1.02	1.37
2700	12" wide trench and backfill, 12" deep		975	.008			.34	.34	.68	.91
2750	18" deep		860	.009	↓		.38	.39	.77	1.03

31 23 Excavation and Fill

31 23 16 – Excavation

31 23 16.14 Excavating, Utility Trench

		Crew	Daily Output	Labor-Hours	Unit	Material	Labor	Equipment	Total	Total Incl O&P
2800	24" deep	B-54	800	.010	L.F.		.41	.42	.83	1.11
2850	36" deep		725	.011			.46	.46	.92	1.22
3000	16" wide trench and backfill, 12" deep		835	.010			.40	.40	.80	1.06
3050	18" deep		750	.011			.44	.45	.89	1.18
3100	24" deep		700	.011			.47	.48	.95	1.27
3200	Compaction with vibratory plate, add								35%	35%
5100	Hand excavate and trim for pipe bells after trench excavation									
5200	8" pipe	1 Clab	155	.052	L.F.		1.71		1.71	2.81
5300	18" pipe	"	130	.062	"		2.04		2.04	3.34
9000	Minimum labor/equipment charge	A-1G	4	2	Job		66	12.95	78.95	123

31 23 16.16 Structural Excavation for Minor Structures

		Crew	Daily Output	Labor-Hours	Unit	Material	Labor	Equipment	Total	Total Incl O&P
0010	**STRUCTURAL EXCAVATION FOR MINOR STRUCTURES**									
0015	Hand, pits to 6' deep, sandy soil	1 Clab	8	1	B.C.Y.		33		33	54.50
0100	Heavy soil or clay		4	2			66		66	109
0300	Pits 6' to 12' deep, sandy soil		5	1.600			53		53	87
0500	Heavy soil or clay		3	2.667			88.50		88.50	145
0700	Pits 12' to 18' deep, sandy soil		4	2			66		66	109
0900	Heavy soil or clay		2	4			132		132	217
1100	Hand loading trucks from stock pile, sandy soil		12	.667			22		22	36
1300	Heavy soil or clay		8	1			33		33	54.50
1500	For wet or muck hand excavation, add to above				%				50%	50%
6000	Machine excavation, for spread and mat footings, elevator pits,									
6001	and small building foundations									
6030	Common earth, hydraulic backhoe, 1/2 C.Y. bucket	B-12E	55	.291	B.C.Y.		11.25	7.05	18.30	26
6035	3/4 C.Y. bucket	B-12F	90	.178			6.90	7.55	14.45	19.30
6040	1 C.Y. bucket	B-12A	108	.148			5.75	7.40	13.15	17.35
6050	1-1/2 C.Y. bucket	B-12B	144	.111			4.31	7.05	11.36	14.65
6060	2 C.Y. bucket	B-12C	200	.080			3.10	6.60	9.70	12.20
6070	Sand and gravel, 3/4 C.Y. bucket	B-12F	100	.160			6.20	6.80	13	17.35
6080	1 C.Y. bucket	B-12A	120	.133			5.15	6.70	11.85	15.60
6090	1-1/2 C.Y. bucket	B-12B	160	.100			3.88	6.35	10.23	13.20
6100	2 C.Y. bucket	B-12C	220	.073			2.82	6	8.82	11.10
6110	Clay, till, or blasted rock, 3/4 C.Y. bucket	B-12F	80	.200			7.75	8.45	16.20	21.50
6120	1 C.Y. bucket	B-12A	95	.168			6.55	8.45	15	19.75
6130	1-1/2 C.Y. bucket	B-12B	130	.123			4.77	7.80	12.57	16.20
6140	2 C.Y. bucket	B-12C	175	.091			3.54	7.55	11.09	13.95
6230	Sandy clay & loam, hydraulic backhoe, 1/2 C.Y. bucket	B-12E	60	.267			10.35	6.45	16.80	23.50
6235	3/4 C.Y. bucket	B-12F	98	.163			6.35	6.90	13.25	17.70
6240	1 C.Y. bucket	B-12A	116	.138			5.35	6.90	12.25	16.15
6250	1-1/2 C.Y. bucket	B-12B	156	.103			3.97	6.50	10.47	13.50
9000	Minimum labor/equipment charge	1 Clab	4	2	Job		66		66	109

31 23 16.42 Excavating, Bulk Bank Measure

			Crew	Daily Output	Labor-Hours	Unit	Material	Labor	Equipment	Total	Total Incl O&P
0010	**EXCAVATING, BULK BANK MEASURE**	R312316-40									
0011	Common earth piled										
0020	For loading onto trucks, add									15%	15%
0200	Excavator, hydraulic, crawler mtd., 1 C.Y. cap. = 100 C.Y./hr.	R312316-45	B-12A	800	.020	B.C.Y.		.78	1	1.78	2.34
1200	Front end loader, track mtd., 1-1/2 C.Y. cap. = 70 C.Y./hr.		B-10N	560	.021			.85	.77	1.62	2.20
1500	Wheel mounted, 3/4 C.Y. cap. = 45 C.Y./hr.		B-10R	360	.033			1.32	.75	2.07	2.92
5000	Excavating, bulk bank measure, sandy clay & loam piled										
5020	For loading onto trucks, add									15%	15%
5100	Excavator, hydraulic, crawler mtd., 1 C.Y. cap. = 120 C.Y./hr.		B-12A	960	.017	B.C.Y.		.65	.84	1.49	1.95
9000	Minimum labor/equipment charge		B-10L	2	6	Job		238	218	456	620

31 23 Excavation and Fill

31 23 16 – Excavation

31 23 16.46 Excavating, Bulk, Dozer

		Crew	Daily Output	Labor-Hours	Unit	Material	2010 Bare Costs Labor	Equipment	Total	Total Incl O&P
0010	**EXCAVATING, BULK, DOZER**									
0011	Open site									
2000	80 H.P., 50' haul, sand & gravel	B-10L	460	.026	B.C.Y.		1.04	.95	1.99	2.68
2200	150' haul, sand & gravel		230	.052			2.07	1.90	3.97	5.40
2400	300' haul, sand & gravel	↓	120	.100			3.97	3.64	7.61	10.30
3000	105 H.P., 50' haul, sand & gravel	B-10W	700	.017			.68	.94	1.62	2.12
3200	150' haul, sand & gravel		310	.039			1.54	2.13	3.67	4.78
3300	300' haul, sand & gravel	↓	140	.086			3.40	4.72	8.12	10.60
4000	200 H.P., 50' haul, sand & gravel	B-10B	1400	.009			.34	.85	1.19	1.48
4200	150' haul, sand & gravel		595	.020			.80	2	2.80	3.47
4400	300' haul, sand & gravel	↓	310	.039			1.54	3.85	5.39	6.65
5040	Clay	B-10M	1025	.012			.46	1.55	2.01	2.45
5400	300' haul, sand & gravel	"	470	.026	↓		1.01	3.39	4.40	5.35

31 23 23 – Fill

31 23 23.13 Backfill

		Crew	Daily Output	Labor-Hours	Unit	Material	Labor	Equipment	Total	Total Incl O&P
0010	**BACKFILL** R312323-30									
0015	By hand, no compaction, light soil	1 Clab	14	.571	L.C.Y.		18.90		18.90	31
0100	Heavy soil		11	.727	"		24		24	39.50
0300	Compaction in 6" layers, hand tamp, add to above	↓	20.60	.388	E.C.Y.		12.85		12.85	21
0400	Roller compaction operator walking, add	B-10A	100	.120			4.76	1.50	6.26	9.20
0500	Air tamp, add	B-9D	190	.211			7.05	1.35	8.40	13.10
0600	Vibrating plate, add	A-1D	60	.133			4.41	.57	4.98	7.90
0800	Compaction in 12" layers, hand tamp, add to above	1 Clab	34	.235			7.80		7.80	12.80
1000	Air tamp, add	B-9	285	.140			4.70	.79	5.49	8.55
1100	Vibrating plate, add	A-1E	90	.089	↓		2.94	.48	3.42	5.35
1200	Trench, dozer, no compaction, 60 HP	B-10L	425	.028	L.C.Y.		1.12	1.03	2.15	2.91
1300	Dozer backfilling, bulk, up to 300' haul, no compaction	B-10B	1200	.010	"		.40	.99	1.39	1.72
1400	Air tamped, add	B-11B	80	.200	E.C.Y.		7.45	3.67	11.12	15.95
1900	Dozer backfilling, trench, up to 300' haul, no compaction	B-10B	900	.013	L.C.Y.		.53	1.32	1.85	2.30
2000	Air tamped, add	B-11B	80	.200	E.C.Y.		7.45	3.67	11.12	15.95
2350	Spreading in 8" layers, small dozer	B-10B	1060	.011	L.C.Y.		.45	1.12	1.57	1.95
2450	Compacting with vibrating plate, 8" lifts	A-1D	73	.110	E.C.Y.		3.63	.47	4.10	6.45

31 23 23.14 Backfill, Structural

		Crew	Daily Output	Labor-Hours	Unit	Material	Labor	Equipment	Total	Total Incl O&P
0010	**BACKFILL, STRUCTURAL**									
0011	Dozer or F.E. loader									
0020	From existing stockpile, no compaction									
2000	80 H.P., 50' haul, sand & gravel	B-10L	1100	.011	L.C.Y.		.43	.40	.83	1.13
2010	Sandy clay & loam		1070	.011			.44	.41	.85	1.16
2020	Common earth		975	.012			.49	.45	.94	1.27
2040	Clay		850	.014			.56	.51	1.07	1.46
2400	300' haul, sand & gravel		370	.032			1.29	1.18	2.47	3.34
2410	Sandy clay & loam		360	.033			1.32	1.21	2.53	3.43
2420	Common earth		330	.036			1.44	1.32	2.76	3.75
2440	Clay	↓	290	.041			1.64	1.51	3.15	4.27
3000	105 H.P., 50' haul, sand & gravel	B-10W	1350	.009			.35	.49	.84	1.10
3010	Sandy clay & loam		1325	.009			.36	.50	.86	1.12
3020	Common earth		1225	.010			.39	.54	.93	1.21
3040	Clay		1100	.011			.43	.60	1.03	1.35
3300	300' haul, sand & gravel		465	.026			1.02	1.42	2.44	3.19
3310	Sandy clay & loam		455	.026			1.05	1.45	2.50	3.26
3320	Common earth	↓	415	.029	↓		1.15	1.59	2.74	3.57

31 23 Excavation and Fill

31 23 23 – Fill

31 23 23.14 Backfill, Structural		Crew	Daily Output	Labor-Hours	Unit	Material	2010 Bare Costs Labor	Equipment	Total	Total Incl O&P
3340	Clay	B-10W	370	.032	L.C.Y.		1.29	1.78	3.07	4

31 23 23.16 Fill By Borrow and Utility Bedding

0010	**FILL BY BORROW AND UTILITY BEDDING**									
0049	Utility bedding, for pipe & conduit, not incl. compaction									
0050	Crushed or screened bank run gravel	B-6	150	.160	L.C.Y.	25	5.75	2.25	33	39
0100	Crushed stone 3/4" to 1/2"		150	.160		35	5.75	2.25	43	50
0200	Sand, dead or bank		150	.160		10.55	5.75	2.25	18.55	23.50
0500	Compacting bedding in trench	A-1D	90	.089	E.C.Y.		2.94	.38	3.32	5.25
0600	If material source exceeds 2 miles, add for extra mileage.									

31 23 23.17 General Fill

0010	**GENERAL FILL**									
0011	Spread dumped material, no compaction									
0020	By dozer, no compaction	B-10B	1000	.012	L.C.Y.		.48	1.19	1.67	2.07
0100	By hand	1 Clab	12	.667	"		22		22	36
9000	Minimum labor/equipment charge	"	4	2	Job		66		66	109

31 23 23.20 Hauling

0010	**HAULING**									
0011	Excavated or borrow, loose cubic yards									
0012	no loading equipment, including hauling, waiting, loading/dumping									
0013	time per cycle (wait, load, travel, unload or dump & return)									
0014	8 CY truck, 15 MPH ave, cycle 0.5 miles, 10 min. wait/Ld./Uld.	B-34A	320	.025	L.C.Y.		.83	1.26	2.09	2.73
0016	cycle 1 mile		272	.029			.97	1.48	2.45	3.20
0018	cycle 2 miles		208	.038			1.27	1.93	3.20	4.19
0020	cycle 4 miles		144	.056			1.84	2.79	4.63	6.05
0022	cycle 6 miles		112	.071			2.37	3.59	5.96	7.80
0024	cycle 8 miles		88	.091			3.01	4.57	7.58	9.90
0026	20 MPH ave, cycle 0.5 mile		336	.024			.79	1.20	1.99	2.60
0028	cycle 1 mile		296	.027			.90	1.36	2.26	2.95
0030	cycle 2 miles		240	.033			1.10	1.67	2.77	3.63
0032	cycle 4 miles		176	.045			1.51	2.28	3.79	4.96
0034	cycle 6 miles		136	.059			1.95	2.95	4.90	6.40
0036	cycle 8 miles		112	.071			2.37	3.59	5.96	7.80
0044	25 MPH ave, cycle 4 miles		192	.042			1.38	2.09	3.47	4.54
0046	cycle 6 miles		160	.050			1.66	2.51	4.17	5.45
0048	cycle 8 miles		128	.063			2.07	3.14	5.21	6.80
0050	30 MPH ave, cycle 4 miles		216	.037			1.23	1.86	3.09	4.04
0052	cycle 6 miles		176	.045			1.51	2.28	3.79	4.96
0054	cycle 8 miles		144	.056			1.84	2.79	4.63	6.05
0114	15 MPH ave, cycle 0.5 mile, 15 min. wait/Ld./Uld.		224	.036			1.18	1.79	2.97	3.89
0116	cycle 1 mile		200	.040			1.33	2.01	3.34	4.36
0118	cycle 2 miles		168	.048			1.58	2.39	3.97	5.20
0120	cycle 4 miles		120	.067			2.21	3.35	5.56	7.25
0122	cycle 6 miles		96	.083			2.76	4.19	6.95	9.10
0124	cycle 8 miles		80	.100			3.32	5	8.32	10.95
0126	20 MPH ave, cycle 0.5 mile		232	.034			1.14	1.73	2.87	3.77
0128	cycle 1 mile		208	.038			1.27	1.93	3.20	4.19
0130	cycle 2 miles		184	.043			1.44	2.18	3.62	4.74
0132	cycle 4 miles		144	.056			1.84	2.79	4.63	6.05
0134	cycle 6 miles		112	.071			2.37	3.59	5.96	7.80
0136	cycle 8 miles		96	.083			2.76	4.19	6.95	9.10
0144	25 MPH ave, cycle 4 miles		152	.053			1.74	2.64	4.38	5.75
0146	cycle 6 miles		128	.063			2.07	3.14	5.21	6.80

31 23 Excavation and Fill

31 23 23 – Fill

31 23 23.20 Hauling

		Crew	Daily Output	Labor-Hours	Unit	Material	2010 Bare Costs Labor	Equipment	Total	Total Incl O&P
0148	cycle 8 miles	B-34A	112	.071	L.C.Y.		2.37	3.59	5.96	7.80
0150	30 MPH ave, cycle 4 miles		168	.048			1.58	2.39	3.97	5.20
0152	cycle 6 miles		144	.056			1.84	2.79	4.63	6.05
0154	cycle 8 miles		120	.067			2.21	3.35	5.56	7.25
0214	15 MPH ave, cycle 0.5 mile, 20 min wait/Ld./Uld.		176	.045			1.51	2.28	3.79	4.96
0216	cycle 1 mile		160	.050			1.66	2.51	4.17	5.45
0218	cycle 2 miles		136	.059			1.95	2.95	4.90	6.40
0220	cycle 4 miles		104	.077			2.55	3.86	6.41	8.40
0222	cycle 6 miles		88	.091			3.01	4.57	7.58	9.90
0224	cycle 8 miles		72	.111			3.68	5.60	9.28	12.15
0226	20 MPH ave, cycle 0.5 mile		176	.045			1.51	2.28	3.79	4.96
0228	cycle 1 mile		168	.048			1.58	2.39	3.97	5.20
0230	cycle 2 miles		144	.056			1.84	2.79	4.63	6.05
0232	cycle 4 miles		120	.067			2.21	3.35	5.56	7.25
0234	cycle 6 miles		96	.083			2.76	4.19	6.95	9.10
0236	cycle 8 miles		88	.091			3.01	4.57	7.58	9.90
0244	25 MPH ave, cycle 4 miles		128	.063			2.07	3.14	5.21	6.80
0246	cycle 6 miles		112	.071			2.37	3.59	5.96	7.80
0248	cycle 8 miles		96	.083			2.76	4.19	6.95	9.10
0250	30 MPH ave, cycle 4 miles		136	.059			1.95	2.95	4.90	6.40
0252	cycle 6 miles		120	.067			2.21	3.35	5.56	7.25
0254	cycle 8 miles		104	.077			2.55	3.86	6.41	8.40
0314	15 MPH ave, cycle 0.5 mile, 25 min wait/Ld./Uld.		144	.056			1.84	2.79	4.63	6.05
0316	cycle 1 mile		128	.063			2.07	3.14	5.21	6.80
0318	cycle 2 miles		112	.071			2.37	3.59	5.96	7.80
0320	cycle 4 miles		96	.083			2.76	4.19	6.95	9.10
0322	cycle 6 miles		80	.100			3.32	5	8.32	10.95
0324	cycle 8 miles		64	.125			4.14	6.30	10.44	13.65
0326	20 MPH ave, cycle 0.5 mile		144	.056			1.84	2.79	4.63	6.05
0328	cycle 1 mile		136	.059			1.95	2.95	4.90	6.40
0330	cycle 2 miles		120	.067			2.21	3.35	5.56	7.25
0332	cycle 4 miles		104	.077			2.55	3.86	6.41	8.40
0334	cycle 6 miles		88	.091			3.01	4.57	7.58	9.90
0336	cycle 8 miles		80	.100			3.32	5	8.32	10.95
0344	25 MPH ave, cycle 4 miles		112	.071			2.37	3.59	5.96	7.80
0346	cycle 6 miles		96	.083			2.76	4.19	6.95	9.10
0348	cycle 8 miles		88	.091			3.01	4.57	7.58	9.90
0350	30 MPH ave, cycle 4 miles		112	.071			2.37	3.59	5.96	7.80
0352	cycle 6 miles		104	.077			2.55	3.86	6.41	8.40
0354	cycle 8 miles		96	.083			2.76	4.19	6.95	9.10
0414	15 MPH ave, cycle 0.5 mile, 30 min wait/Ld./Uld.		120	.067			2.21	3.35	5.56	7.25
0416	cycle 1 mile		112	.071			2.37	3.59	5.96	7.80
0418	cycle 2 miles		96	.083			2.76	4.19	6.95	9.10
0420	cycle 4 miles		80	.100			3.32	5	8.32	10.95
0422	cycle 6 miles		72	.111			3.68	5.60	9.28	12.15
0424	cycle 8 miles		64	.125			4.14	6.30	10.44	13.65
0426	20 MPH ave, cycle 0.5 mile		120	.067			2.21	3.35	5.56	7.25
0428	cycle 1 mile		112	.071			2.37	3.59	5.96	7.80
0430	cycle 2 miles		104	.077			2.55	3.86	6.41	8.40
0432	cycle 4 miles		88	.091			3.01	4.57	7.58	9.90
0434	cycle 6 miles		80	.100			3.32	5	8.32	10.95
0436	cycle 8 miles		72	.111			3.68	5.60	9.28	12.15
0444	25 MPH ave, cycle 4 miles		96	.083			2.76	4.19	6.95	9.10

31 23 Excavation and Fill

31 23 23 – Fill

31 23 23.20 Hauling		Crew	Daily Output	Labor-Hours	Unit	Material	2010 Bare Costs Labor	Equipment	Total	Total Incl O&P
0446	cycle 6 miles	B-34A	88	.091	L.C.Y.		3.01	4.57	7.58	9.90
0448	cycle 8 miles		80	.100			3.32	5	8.32	10.95
0450	30 MPH ave, cycle 4 miles		96	.083			2.76	4.19	6.95	9.10
0452	cycle 6 miles		88	.091			3.01	4.57	7.58	9.90
0454	cycle 8 miles		80	.100			3.32	5	8.32	10.95
0514	15 MPH ave, cycle 0.5 mile, 35 min wait/Ld./Uld.		104	.077			2.55	3.86	6.41	8.40
0516	cycle 1 mile		96	.083			2.76	4.19	6.95	9.10
0518	cycle 2 miles		88	.091			3.01	4.57	7.58	9.90
0520	cycle 4 miles		72	.111			3.68	5.60	9.28	12.15
0522	cycle 6 miles		64	.125			4.14	6.30	10.44	13.65
0524	cycle 8 miles		56	.143			4.74	7.20	11.94	15.60
0526	20 MPH ave, cycle 0.5 mile		104	.077			2.55	3.86	6.41	8.40
0528	cycle 1 mile		96	.083			2.76	4.19	6.95	9.10
0530	cycle 2 miles		96	.083			2.76	4.19	6.95	9.10
0532	cycle 4 miles		80	.100			3.32	5	8.32	10.95
0534	cycle 6 miles		72	.111			3.68	5.60	9.28	12.15
0536	cycle 8 miles		64	.125			4.14	6.30	10.44	13.65
0544	25 MPH ave, cycle 4 miles		88	.091			3.01	4.57	7.58	9.90
0546	cycle 6 miles		80	.100			3.32	5	8.32	10.95
0548	cycle 8 miles		72	.111			3.68	5.60	9.28	12.15
0550	30 MPH ave, cycle 4 miles		88	.091			3.01	4.57	7.58	9.90
0552	cycle 6 miles		80	.100			3.32	5	8.32	10.95
0554	cycle 8 miles		72	.111			3.68	5.60	9.28	12.15
1014	12 CY truck, cycle 0.5 mile, 15 MPH ave, 15 min. wait/Ld./Uld.	B-34B	336	.024			.79	1.99	2.78	3.47
1016	cycle 1 mile		300	.027			.88	2.23	3.11	3.89
1018	cycle 2 miles		252	.032			1.05	2.65	3.70	4.63
1020	cycle 4 miles		180	.044			1.47	3.71	5.18	6.45
1022	cycle 6 miles		144	.056			1.84	4.64	6.48	8.10
1024	cycle 8 miles		120	.067			2.21	5.55	7.76	9.70
1025	cycle 10 miles		96	.083			2.76	6.95	9.71	12.15
1026	20 MPH ave, cycle 0.5 mile		348	.023			.76	1.92	2.68	3.35
1028	cycle 1 mile		312	.026			.85	2.14	2.99	3.73
1030	cycle 2 miles		276	.029			.96	2.42	3.38	4.22
1032	cycle 4 miles		216	.037			1.23	3.09	4.32	5.40
1034	cycle 6 miles		168	.048			1.58	3.98	5.56	6.95
1036	cycle 8 miles		144	.056			1.84	4.64	6.48	8.10
1038	cycle 10 miles		120	.067			2.21	5.55	7.76	9.70
1040	25 MPH ave, cycle 4 miles		228	.035			1.16	2.93	4.09	5.10
1042	cycle 6 miles		192	.042			1.38	3.48	4.86	6.05
1044	cycle 8 miles		168	.048			1.58	3.98	5.56	6.95
1046	cycle 10 miles		144	.056			1.84	4.64	6.48	8.10
1050	30 MPH ave, cycle 4 miles		252	.032			1.05	2.65	3.70	4.63
1052	cycle 6 miles		216	.037			1.23	3.09	4.32	5.40
1054	cycle 8 miles		180	.044			1.47	3.71	5.18	6.45
1056	cycle 10 miles		156	.051			1.70	4.28	5.98	7.45
1060	35 MPH ave, cycle 4 miles		264	.030			1	2.53	3.53	4.41
1062	cycle 6 miles		228	.035			1.16	2.93	4.09	5.10
1064	cycle 8 miles		204	.039			1.30	3.27	4.57	5.70
1066	cycle 10 miles		180	.044			1.47	3.71	5.18	6.45
1068	cycle 20 miles		120	.067			2.21	5.55	7.76	9.70
1069	cycle 30 miles		84	.095			3.16	7.95	11.11	13.90
1070	cycle 40 miles		72	.111			3.68	9.30	12.98	16.20
1072	40 MPH ave, cycle 6 miles		240	.033			1.10	2.78	3.88	4.85

31 23 Excavation and Fill

31 23 23 – Fill

31 23 23.20 Hauling

		Crew	Daily Output	Labor-Hours	Unit	Material	2010 Bare Costs Labor	Equipment	Total	Total Incl O&P
1074	cycle 8 miles	B-34B	216	.037	L.C.Y.		1.23	3.09	4.32	5.40
1076	cycle 10 miles		192	.042			1.38	3.48	4.86	6.05
1078	cycle 20 miles		120	.067			2.21	5.55	7.76	9.70
1080	cycle 30 miles		96	.083			2.76	6.95	9.71	12.15
1082	cycle 40 miles		72	.111			3.68	9.30	12.98	16.20
1084	cycle 50 miles		60	.133			4.42	11.15	15.57	19.45
1094	45 MPH ave, cycle 8 miles		216	.037			1.23	3.09	4.32	5.40
1096	cycle 10 miles		204	.039			1.30	3.27	4.57	5.70
1098	cycle 20 miles		132	.061			2.01	5.05	7.06	8.80
1100	cycle 30 miles		108	.074			2.46	6.20	8.66	10.80
1102	cycle 40 miles		84	.095			3.16	7.95	11.11	13.90
1104	cycle 50 miles		72	.111			3.68	9.30	12.98	16.20
1106	50 MPH ave, cycle 10 miles		216	.037			1.23	3.09	4.32	5.40
1108	cycle 20 miles		144	.056			1.84	4.64	6.48	8.10
1110	cycle 30 miles		108	.074			2.46	6.20	8.66	10.80
1112	cycle 40 miles		84	.095			3.16	7.95	11.11	13.90
1114	cycle 50 miles		72	.111			3.68	9.30	12.98	16.20
1214	15 MPH ave, cycle 0.5 mile, 20 min. wait/Ld./Uld.		264	.030			1	2.53	3.53	4.41
1216	cycle 1 mile		240	.033			1.10	2.78	3.88	4.85
1218	cycle 2 miles		204	.039			1.30	3.27	4.57	5.70
1220	cycle 4 miles		156	.051			1.70	4.28	5.98	7.45
1222	cycle 6 miles		132	.061			2.01	5.05	7.06	8.80
1224	cycle 8 miles		108	.074			2.46	6.20	8.66	10.80
1225	cycle 10 miles		96	.083			2.76	6.95	9.71	12.15
1226	20 MPH ave, cycle 0.5 mile		264	.030			1	2.53	3.53	4.41
1228	cycle 1 mile		252	.032			1.05	2.65	3.70	4.63
1230	cycle 2 miles		216	.037			1.23	3.09	4.32	5.40
1232	cycle 4 miles		180	.044			1.47	3.71	5.18	6.45
1234	cycle 6 miles		144	.056			1.84	4.64	6.48	8.10
1236	cycle 8 miles		132	.061			2.01	5.05	7.06	8.80
1238	cycle 10 miles		108	.074			2.46	6.20	8.66	10.80
1240	25 MPH ave, cycle 4 miles		192	.042			1.38	3.48	4.86	6.05
1242	cycle 6 miles		168	.048			1.58	3.98	5.56	6.95
1244	cycle 8 miles		144	.056			1.84	4.64	6.48	8.10
1246	cycle 10 miles		132	.061			2.01	5.05	7.06	8.80
1250	30 MPH ave, cycle 4 miles		204	.039			1.30	3.27	4.57	5.70
1252	cycle 6 miles		180	.044			1.47	3.71	5.18	6.45
1254	cycle 8 miles		156	.051			1.70	4.28	5.98	7.45
1256	cycle 10 miles		144	.056			1.84	4.64	6.48	8.10
1260	35 MPH ave, cycle 4 miles		216	.037			1.23	3.09	4.32	5.40
1262	cycle 6 miles		192	.042			1.38	3.48	4.86	6.05
1264	cycle 8 miles		168	.048			1.58	3.98	5.56	6.95
1266	cycle 10 miles		156	.051			1.70	4.28	5.98	7.45
1268	cycle 20 miles		108	.074			2.46	6.20	8.66	10.80
1269	cycle 30 miles		72	.111			3.68	9.30	12.98	16.20
1270	cycle 40 miles		60	.133			4.42	11.15	15.57	19.45
1272	40 MPH ave, cycle 6 miles		192	.042			1.38	3.48	4.86	6.05
1274	cycle 8 miles		180	.044			1.47	3.71	5.18	6.45
1276	cycle 10 miles		156	.051			1.70	4.28	5.98	7.45
1278	cycle 20 miles		108	.074			2.46	6.20	8.66	10.80
1280	cycle 30 miles		84	.095			3.16	7.95	11.11	13.90
1282	cycle 40 miles		72	.111			3.68	9.30	12.98	16.20
1284	cycle 50 miles		60	.133			4.42	11.15	15.57	19.45

31 23 Excavation and Fill

31 23 23 – Fill

31 23 23.20 Hauling

		Crew	Daily Output	Labor-Hours	Unit	Material	2010 Bare Costs Labor	Equipment	Total	Total Incl O&P
1294	45 MPH ave, cycle 8 miles	B-34B	180	.044	L.C.Y.		1.47	3.71	5.18	6.45
1296	cycle 10 miles		168	.048			1.58	3.98	5.56	6.95
1298	cycle 20 miles		120	.067			2.21	5.55	7.76	9.70
1300	cycle 30 miles		96	.083			2.76	6.95	9.71	12.15
1302	cycle 40 miles		72	.111			3.68	9.30	12.98	16.20
1304	cycle 50 miles		60	.133			4.42	11.15	15.57	19.45
1306	50 MPH ave, cycle 10 miles		180	.044			1.47	3.71	5.18	6.45
1308	cycle 20 miles		132	.061			2.01	5.05	7.06	8.80
1310	cycle 30 miles		96	.083			2.76	6.95	9.71	12.15
1312	cycle 40 miles		84	.095			3.16	7.95	11.11	13.90
1314	cycle 50 miles		72	.111			3.68	9.30	12.98	16.20
1414	15 MPH ave, cycle 0.5 mile, 25 min. wait/Ld./Uld.		204	.039			1.30	3.27	4.57	5.70
1416	cycle 1 mile		192	.042			1.38	3.48	4.86	6.05
1418	cycle 2 miles		168	.048			1.58	3.98	5.56	6.95
1420	cycle 4 miles		132	.061			2.01	5.05	7.06	8.80
1422	cycle 6 miles		120	.067			2.21	5.55	7.76	9.70
1424	cycle 8 miles		96	.083			2.76	6.95	9.71	12.15
1425	cycle 10 miles		84	.095			3.16	7.95	11.11	13.90
1426	20 MPH ave, cycle 0.5 mile		216	.037			1.23	3.09	4.32	5.40
1428	cycle 1 mile		204	.039			1.30	3.27	4.57	5.70
1430	cycle 2 miles		180	.044			1.47	3.71	5.18	6.45
1432	cycle 4 miles		156	.051			1.70	4.28	5.98	7.45
1434	cycle 6 miles		132	.061			2.01	5.05	7.06	8.80
1436	cycle 8 miles		120	.067			2.21	5.55	7.76	9.70
1438	cycle 10 miles		96	.083			2.76	6.95	9.71	12.15
1440	25 MPH ave, cycle 4 miles		168	.048			1.58	3.98	5.56	6.95
1442	cycle 6 miles		144	.056			1.84	4.64	6.48	8.10
1444	cycle 8 miles		132	.061			2.01	5.05	7.06	8.80
1446	cycle 10 miles		108	.074			2.46	6.20	8.66	10.80
1450	30 MPH ave, cycle 4 miles		168	.048			1.58	3.98	5.56	6.95
1452	cycle 6 miles		156	.051			1.70	4.28	5.98	7.45
1454	cycle 8 miles		132	.061			2.01	5.05	7.06	8.80
1456	cycle 10 miles		120	.067			2.21	5.55	7.76	9.70
1460	35 MPH ave, cycle 4 miles		180	.044			1.47	3.71	5.18	6.45
1462	cycle 6 miles		156	.051			1.70	4.28	5.98	7.45
1464	cycle 8 miles		144	.056			1.84	4.64	6.48	8.10
1466	cycle 10 miles		132	.061			2.01	5.05	7.06	8.80
1468	cycle 20 miles		96	.083			2.76	6.95	9.71	12.15
1469	cycle 30 miles		72	.111			3.68	9.30	12.98	16.20
1470	cycle 40 miles		60	.133			4.42	11.15	15.57	19.45
1472	40 MPH ave, cycle 6 miles		168	.048			1.58	3.98	5.56	6.95
1474	cycle 8 miles		156	.051			1.70	4.28	5.98	7.45
1476	cycle 10 miles		144	.056			1.84	4.64	6.48	8.10
1478	cycle 20 miles		96	.083			2.76	6.95	9.71	12.15
1480	cycle 30 miles		84	.095			3.16	7.95	11.11	13.90
1482	cycle 40 miles		60	.133			4.42	11.15	15.57	19.45
1484	cycle 50 miles		60	.133			4.42	11.15	15.57	19.45
1494	45 MPH ave, cycle 8 miles		156	.051			1.70	4.28	5.98	7.45
1496	cycle 10 miles		144	.056			1.84	4.64	6.48	8.10
1498	cycle 20 miles		108	.074			2.46	6.20	8.66	10.80
1500	cycle 30 miles		84	.095			3.16	7.95	11.11	13.90
1502	cycle 40 miles		72	.111			3.68	9.30	12.98	16.20
1504	cycle 50 miles		60	.133			4.42	11.15	15.57	19.45

31 23 Excavation and Fill

31 23 23 – Fill

31 23 23.20 Hauling		Crew	Daily Output	Labor-Hours	Unit	Material	2010 Bare Costs Labor	Equipment	Total	Total Incl O&P
1506	50 MPH ave, cycle 10 miles	B-34B	156	.051	L.C.Y.		1.70	4.28	5.98	7.45
1508	cycle 20 miles		120	.067			2.21	5.55	7.76	9.70
1510	cycle 30 miles		96	.083			2.76	6.95	9.71	12.15
1512	cycle 40 miles		72	.111			3.68	9.30	12.98	16.20
1514	cycle 50 miles		60	.133			4.42	11.15	15.57	19.45
1614	15 MPH, cycle 0.5 mile, 30 min. wait/Ld./Uld.		180	.044			1.47	3.71	5.18	6.45
1616	cycle 1 mile		168	.048			1.58	3.98	5.56	6.95
1618	cycle 2 miles		144	.056			1.84	4.64	6.48	8.10
1620	cycle 4 miles		120	.067			2.21	5.55	7.76	9.70
1622	cycle 6 miles		108	.074			2.46	6.20	8.66	10.80
1624	cycle 8 miles		84	.095			3.16	7.95	11.11	13.90
1625	cycle 10 miles		84	.095			3.16	7.95	11.11	13.90
1626	20 MPH ave, cycle 0.5 mile		180	.044			1.47	3.71	5.18	6.45
1628	cycle 1 mile		168	.048			1.58	3.98	5.56	6.95
1630	cycle 2 miles		156	.051			1.70	4.28	5.98	7.45
1632	cycle 4 miles		132	.061			2.01	5.05	7.06	8.80
1634	cycle 6 miles		120	.067			2.21	5.55	7.76	9.70
1636	cycle 8 miles		108	.074			2.46	6.20	8.66	10.80
1638	cycle 10 miles		96	.083			2.76	6.95	9.71	12.15
1640	25 MPH ave, cycle 4 miles		144	.056			1.84	4.64	6.48	8.10
1642	cycle 6 miles		132	.061			2.01	5.05	7.06	8.80
1644	cycle 8 miles		108	.074			2.46	6.20	8.66	10.80
1646	cycle 10 miles		108	.074			2.46	6.20	8.66	10.80
1650	30 MPH ave, cycle 4 miles		144	.056			1.84	4.64	6.48	8.10
1652	cycle 6 miles		132	.061			2.01	5.05	7.06	8.80
1654	cycle 8 miles		120	.067			2.21	5.55	7.76	9.70
1656	cycle 10 miles		108	.074			2.46	6.20	8.66	10.80
1660	35 MPH ave, cycle 4 miles		156	.051			1.70	4.28	5.98	7.45
1662	cycle 6 miles		144	.056			1.84	4.64	6.48	8.10
1664	cycle 8 miles		132	.061			2.01	5.05	7.06	8.80
1666	cycle 10 miles		120	.067			2.21	5.55	7.76	9.70
1668	cycle 20 miles		84	.095			3.16	7.95	11.11	13.90
1669	cycle 30 miles		72	.111			3.68	9.30	12.98	16.20
1670	cycle 40 miles		60	.133			4.42	11.15	15.57	19.45
1672	40 MPH, cycle 6 miles		144	.056			1.84	4.64	6.48	8.10
1674	cycle 8 miles		132	.061			2.01	5.05	7.06	8.80
1676	cycle 10 miles		120	.067			2.21	5.55	7.76	9.70
1678	cycle 20 miles		96	.083			2.76	6.95	9.71	12.15
1680	cycle 30 miles		72	.111			3.68	9.30	12.98	16.20
1682	cycle 40 miles		60	.133			4.42	11.15	15.57	19.45
1684	cycle 50 miles		48	.167			5.55	13.90	19.45	24.50
1694	45 MPH ave, cycle 8 miles		144	.056			1.84	4.64	6.48	8.10
1696	cycle 10 miles		132	.061			2.01	5.05	7.06	8.80
1698	cycle 20 miles		96	.083			2.76	6.95	9.71	12.15
1700	cycle 30 miles		84	.095			3.16	7.95	11.11	13.90
1702	cycle 40 miles		60	.133			4.42	11.15	15.57	19.45
1704	cycle 50 miles		60	.133			4.42	11.15	15.57	19.45
1706	50 MPH ave, cycle 10 miles		132	.061			2.01	5.05	7.06	8.80
1708	cycle 20 miles		108	.074			2.46	6.20	8.66	10.80
1710	cycle 30 miles		84	.095			3.16	7.95	11.11	13.90
1712	cycle 40 miles		72	.111			3.68	9.30	12.98	16.20
1714	cycle 50 miles		60	.133			4.42	11.15	15.57	19.45
2000	Hauling, 8 CY truck, small project cost per hour	B-34A	8	1	Hr.		33	50	83	110

31 23 Excavation and Fill

31 23 23 – Fill

31 23 23.20 Hauling

		Crew	Daily Output	Labor-Hours	Unit	Material	2010 Bare Costs Labor	2010 Bare Costs Equipment	Total	Total Incl O&P
2100	12 CY Truck	B-34B	8	1	Hr.		33	83.50	116.50	146
2150	16.5 CY Truck	B-34C	8	1			33	89	122	152
2175	18 CY 8 wheel Truck	B-34I	8	1			33	105	138	169
2200	20 CY Truck	B-34D	8	1			33	91	124	154
2300	Grading at dump, or embankment if required, by dozer	B-10B	1000	.012	L.C.Y.		.48	1.19	1.67	2.07
2310	Spotter at fill or cut, if required	1 Clab	8	1	Hr.		33		33	54.50
9014	18 CY truck, 8 wheels,15 min. wait/Ld./Uld.,15 MPH, cycle 0.5 mi.	B-34I	504	.016	L.C.Y.		.53	1.66	2.19	2.68
9016	cycle 1 mile		450	.018			.59	1.86	2.45	3.01
9018	cycle 2 miles		378	.021			.70	2.22	2.92	3.58
9020	cycle 4 miles		270	.030			.98	3.11	4.09	5
9022	cycle 6 miles		216	.037			1.23	3.88	5.11	6.25
9024	cycle 8 miles		180	.044			1.47	4.66	6.13	7.50
9025	cycle 10 miles		144	.056			1.84	5.80	7.64	9.40
9026	20 MPH ave, cycle 0.5 mile		522	.015			.51	1.61	2.12	2.60
9028	cycle 1 mile		468	.017			.57	1.79	2.36	2.89
9030	cycle 2 miles		414	.019			.64	2.02	2.66	3.27
9032	cycle 4 miles		324	.025			.82	2.59	3.41	4.18
9034	cycle 6 miles		252	.032			1.05	3.33	4.38	5.35
9036	cycle 8 miles		216	.037			1.23	3.88	5.11	6.25
9038	cycle 10 miles		180	.044			1.47	4.66	6.13	7.50
9040	25 MPH ave, cycle 4 miles		342	.023			.78	2.45	3.23	3.96
9042	cycle 6 miles		288	.028			.92	2.91	3.83	4.70
9044	cycle 8 miles		252	.032			1.05	3.33	4.38	5.35
9046	cycle 10 miles		216	.037			1.23	3.88	5.11	6.25
9050	30 MPH ave, cycle 4 miles		378	.021			.70	2.22	2.92	3.58
9052	cycle 6 miles		324	.025			.82	2.59	3.41	4.18
9054	cycle 8 miles		270	.030			.98	3.11	4.09	5
9056	cycle 10 miles		234	.034			1.13	3.58	4.71	5.80
9060	35 MPH ave, cycle 4 miles		396	.020			.67	2.12	2.79	3.42
9062	cycle 6 miles		342	.023			.78	2.45	3.23	3.96
9064	cycle 8 miles		288	.028			.92	2.91	3.83	4.70
9066	cycle 10 miles		270	.030			.98	3.11	4.09	5
9068	cycle 20 miles		162	.049			1.64	5.20	6.84	8.35
9070	cycle 30 miles		126	.063			2.10	6.65	8.75	10.70
9072	cycle 40 miles		90	.089			2.95	9.30	12.25	15.05
9074	40 MPH ave, cycle 6 miles		360	.022			.74	2.33	3.07	3.76
9076	cycle 8 miles		324	.025			.82	2.59	3.41	4.18
9078	cycle 10 miles		288	.028			.92	2.91	3.83	4.70
9080	cycle 20 miles		180	.044			1.47	4.66	6.13	7.50
9082	cycle 30 miles		144	.056			1.84	5.80	7.64	9.40
9084	cycle 40 miles		108	.074			2.46	7.75	10.21	12.55
9086	cycle 50 miles		90	.089			2.95	9.30	12.25	15.05
9094	45 MPH ave, cycle 8 miles		324	.025			.82	2.59	3.41	4.18
9096	cycle 10 miles		306	.026			.87	2.74	3.61	4.42
9098	cycle 20 miles		198	.040			1.34	4.23	5.57	6.85
9100	cycle 30 miles		144	.056			1.84	5.80	7.64	9.40
9102	cycle 40 miles		126	.063			2.10	6.65	8.75	10.70
9104	cycle 50 miles		108	.074			2.46	7.75	10.21	12.55
9106	50 MPH ave, cycle 10 miles		324	.025			.82	2.59	3.41	4.18
9108	cycle 20 miles		216	.037			1.23	3.88	5.11	6.25
9110	cycle 30 miles		162	.049			1.64	5.20	6.84	8.35
9112	cycle 40 miles		126	.063			2.10	6.65	8.75	10.70
9114	cycle 50 miles		108	.074			2.46	7.75	10.21	12.55

31 23 Excavation and Fill

31 23 23 – Fill

31 23 23.20 Hauling

		Crew	Daily Output	Labor-Hours	Unit	Material	2010 Bare Costs Labor	Equipment	Total	Total Incl O&P
9214	20 min. wait/Ld./Uld.,15 MPH, cycle 0.5 mi.	B-34I	396	.020	L.C.Y.		.67	2.12	2.79	3.42
9216	cycle 1 mile		360	.022			.74	2.33	3.07	3.76
9218	cycle 2 miles		306	.026			.87	2.74	3.61	4.42
9220	cycle 4 miles		234	.034			1.13	3.58	4.71	5.80
9222	cycle 6 miles		198	.040			1.34	4.23	5.57	6.85
9224	cycle 8 miles		162	.049			1.64	5.20	6.84	8.35
9225	cycle 10 miles		144	.056			1.84	5.80	7.64	9.40
9226	20 MPH ave, cycle 0.5 mile		396	.020			.67	2.12	2.79	3.42
9228	cycle 1 mile		378	.021			.70	2.22	2.92	3.58
9230	cycle 2 miles		324	.025			.82	2.59	3.41	4.18
9232	cycle 4 miles		270	.030			.98	3.11	4.09	5
9234	cycle 6 miles		216	.037			1.23	3.88	5.11	6.25
9236	cycle 8 miles		198	.040			1.34	4.23	5.57	6.85
9238	cycle 10 miles		162	.049			1.64	5.20	6.84	8.35
9240	25 MPH ave, cycle 4 miles		288	.028			.92	2.91	3.83	4.70
9242	cycle 6 miles		252	.032			1.05	3.33	4.38	5.35
9244	cycle 8 miles		216	.037			1.23	3.88	5.11	6.25
9246	cycle 10 miles		198	.040			1.34	4.23	5.57	6.85
9250	30 MPH ave, cycle 4 miles		306	.026			.87	2.74	3.61	4.42
9252	cycle 6 miles		270	.030			.98	3.11	4.09	5
9254	cycle 8 miles		234	.034			1.13	3.58	4.71	5.80
9256	cycle 10 miles		216	.037			1.23	3.88	5.11	6.25
9260	35 MPH ave, cycle 4 miles		324	.025			.82	2.59	3.41	4.18
9262	cycle 6 miles		288	.028			.92	2.91	3.83	4.70
9264	cycle 8 miles		252	.032			1.05	3.33	4.38	5.35
9266	cycle 10 miles		234	.034			1.13	3.58	4.71	5.80
9268	cycle 20 miles		162	.049			1.64	5.20	6.84	8.35
9270	cycle 30 miles		108	.074			2.46	7.75	10.21	12.55
9272	cycle 40 miles		90	.089			2.95	9.30	12.25	15.05
9274	40 MPH ave, cycle 6 miles		288	.028			.92	2.91	3.83	4.70
9276	cycle 8 miles		270	.030			.98	3.11	4.09	5
9278	cycle 10 miles		234	.034			1.13	3.58	4.71	5.80
9280	cycle 20 miles		162	.049			1.64	5.20	6.84	8.35
9282	cycle 30 miles		126	.063			2.10	6.65	8.75	10.70
9284	cycle 40 miles		108	.074			2.46	7.75	10.21	12.55
9286	cycle 50 miles		90	.089			2.95	9.30	12.25	15.05
9294	45 MPH ave, cycle 8 miles		270	.030			.98	3.11	4.09	5
9296	cycle 10 miles		252	.032			1.05	3.33	4.38	5.35
9298	cycle 20 miles		180	.044			1.47	4.66	6.13	7.50
9300	cycle 30 miles		144	.056			1.84	5.80	7.64	9.40
9302	cycle 40 miles		108	.074			2.46	7.75	10.21	12.55
9304	cycle 50 miles		90	.089			2.95	9.30	12.25	15.05
9306	50 MPH ave, cycle 10 miles		270	.030			.98	3.11	4.09	5
9308	cycle 20 miles		198	.040			1.34	4.23	5.57	6.85
9310	cycle 30 miles		144	.056			1.84	5.80	7.64	9.40
9312	cycle 40 miles		126	.063			2.10	6.65	8.75	10.70
9314	cycle 50 miles		108	.074			2.46	7.75	10.21	12.55
9414	25 min. wait/Ld./Uld.,15 MPH, cycle 0.5 mi.		306	.026			.87	2.74	3.61	4.42
9416	cycle 1 mile		288	.028			.92	2.91	3.83	4.70
9418	cycle 2 miles		252	.032			1.05	3.33	4.38	5.35
9420	cycle 4 miles		198	.040			1.34	4.23	5.57	6.85
9422	cycle 6 miles		180	.044			1.47	4.66	6.13	7.50
9424	cycle 8 miles		144	.056			1.84	5.80	7.64	9.40

31 23 Excavation and Fill

31 23 23 – Fill

31 23 23.20 Hauling

		Crew	Daily Output	Labor-Hours	Unit	Material	2010 Bare Costs Labor	Equipment	Total	Total Incl O&P
9425	cycle 10 miles	B-34I	126	.063	L.C.Y.		2.10	6.65	8.75	10.70
9426	20 MPH ave, cycle 0.5 mile		324	.025			.82	2.59	3.41	4.18
9428	cycle 1 mile		306	.026			.87	2.74	3.61	4.42
9430	cycle 2 miles		270	.030			.98	3.11	4.09	5
9432	cycle 4 miles		234	.034			1.13	3.58	4.71	5.80
9434	cycle 6 miles		198	.040			1.34	4.23	5.57	6.85
9436	cycle 8 miles		180	.044			1.47	4.66	6.13	7.50
9438	cycle 10 miles		144	.056			1.84	5.80	7.64	9.40
9440	25 MPH ave, cycle 4 miles		252	.032			1.05	3.33	4.38	5.35
9442	cycle 6 miles		216	.037			1.23	3.88	5.11	6.25
9444	cycle 8 miles		198	.040			1.34	4.23	5.57	6.85
9446	cycle 10 miles		180	.044			1.47	4.66	6.13	7.50
9450	30 MPH ave, cycle 4 miles		252	.032			1.05	3.33	4.38	5.35
9452	cycle 6 miles		234	.034			1.13	3.58	4.71	5.80
9454	cycle 8 miles		198	.040			1.34	4.23	5.57	6.85
9456	cycle 10 miles		180	.044			1.47	4.66	6.13	7.50
9460	35 MPH ave, cycle 4 miles		270	.030			.98	3.11	4.09	5
9462	cycle 6 miles		234	.034			1.13	3.58	4.71	5.80
9464	cycle 8 miles		216	.037			1.23	3.88	5.11	6.25
9466	cycle 10 miles		198	.040			1.34	4.23	5.57	6.85
9468	cycle 20 miles		144	.056			1.84	5.80	7.64	9.40
9470	cycle 30 miles		108	.074			2.46	7.75	10.21	12.55
9472	cycle 40 miles		90	.089			2.95	9.30	12.25	15.05
9474	40 MPH ave, cycle 6 miles		252	.032			1.05	3.33	4.38	5.35
9476	cycle 8 miles		234	.034			1.13	3.58	4.71	5.80
9478	cycle 10 miles		216	.037			1.23	3.88	5.11	6.25
9480	cycle 20 miles		144	.056			1.84	5.80	7.64	9.40
9482	cycle 30 miles		126	.063			2.10	6.65	8.75	10.70
9484	cycle 40 miles		90	.089			2.95	9.30	12.25	15.05
9486	cycle 50 miles		90	.089			2.95	9.30	12.25	15.05
9494	45 MPH ave, cycle 8 miles		234	.034			1.13	3.58	4.71	5.80
9496	cycle 10 miles		216	.037			1.23	3.88	5.11	6.25
9498	cycle 20 miles		162	.049			1.64	5.20	6.84	8.35
9500	cycle 30 miles		126	.063			2.10	6.65	8.75	10.70
9502	cycle 40 miles		108	.074			2.46	7.75	10.21	12.55
9504	cycle 50 miles		90	.089			2.95	9.30	12.25	15.05
9506	50 MPH ave, cycle 10 miles		234	.034			1.13	3.58	4.71	5.80
9508	cycle 20 miles		180	.044			1.47	4.66	6.13	7.50
9510	cycle 30 miles		144	.056			1.84	5.80	7.64	9.40
9512	cycle 40 miles		108	.074			2.46	7.75	10.21	12.55
9514	cycle 50 miles		90	.089			2.95	9.30	12.25	15.05
9614	25 min. wait/Ld./Uld.,15 MPH, cycle 0.5 mi.		270	.030			.98	3.11	4.09	5
9616	cycle 1 mile		252	.032			1.05	3.33	4.38	5.35
9618	cycle 2 miles		216	.037			1.23	3.88	5.11	6.25
9620	cycle 4 miles		180	.044			1.47	4.66	6.13	7.50
9622	cycle 6 miles		162	.049			1.64	5.20	6.84	8.35
9624	cycle 8 miles		126	.063			2.10	6.65	8.75	10.70
9625	cycle 10 miles		126	.063			2.10	6.65	8.75	10.70
9626	20 MPH ave, cycle 0.5 mile		270	.030			.98	3.11	4.09	5
9628	cycle 1 mile		252	.032			1.05	3.33	4.38	5.35
9630	cycle 2 miles		234	.034			1.13	3.58	4.71	5.80
9632	cycle 4 miles		198	.040			1.34	4.23	5.57	6.85
9634	cycle 6 miles		180	.044			1.47	4.66	6.13	7.50

31 23 Excavation and Fill

31 23 23 – Fill

31 23 23.20 Hauling

		Crew	Daily Output	Labor-Hours	Unit	Material	2010 Bare Costs Labor	2010 Bare Costs Equipment	Total	Total Incl O&P
9636	cycle 8 miles	B-34I	162	.049	L.C.Y.		1.64	5.20	6.84	8.35
9638	cycle 10 miles		144	.056			1.84	5.80	7.64	9.40
9640	25 MPH ave, cycle 4 miles		216	.037			1.23	3.88	5.11	6.25
9642	cycle 6 miles		198	.040			1.34	4.23	5.57	6.85
9644	cycle 8 miles		180	.044			1.47	4.66	6.13	7.50
9646	cycle 10 miles		162	.049			1.64	5.20	6.84	8.35
9650	30 MPH ave, cycle 4 miles		216	.037			1.23	3.88	5.11	6.25
9652	cycle 6 miles		198	.040			1.34	4.23	5.57	6.85
9654	cycle 8 miles		180	.044			1.47	4.66	6.13	7.50
9656	cycle 10 miles		162	.049			1.64	5.20	6.84	8.35
9660	35 MPH ave, cycle 4 miles		234	.034			1.13	3.58	4.71	5.80
9662	cycle 6 miles		216	.037			1.23	3.88	5.11	6.25
9664	cycle 8 miles		198	.040			1.34	4.23	5.57	6.85
9666	cycle 10 miles		180	.044			1.47	4.66	6.13	7.50
9668	cycle 20 miles		126	.063			2.10	6.65	8.75	10.70
9670	cycle 30 miles		108	.074			2.46	7.75	10.21	12.55
9672	cycle 40 miles		90	.089			2.95	9.30	12.25	15.05
9674	40 MPH ave, cycle 6 miles		216	.037			1.23	3.88	5.11	6.25
9676	cycle 8 miles		198	.040			1.34	4.23	5.57	6.85
9678	cycle 10 miles		180	.044			1.47	4.66	6.13	7.50
9680	cycle 20 miles		144	.056			1.84	5.80	7.64	9.40
9682	cycle 30 miles		108	.074			2.46	7.75	10.21	12.55
9684	cycle 40 miles		90	.089			2.95	9.30	12.25	15.05
9686	cycle 50 miles		72	.111			3.68	11.65	15.33	18.80
9694	45 MPH ave, cycle 8 miles		216	.037			1.23	3.88	5.11	6.25
9696	cycle 10 miles		198	.040			1.34	4.23	5.57	6.85
9698	cycle 20 miles		144	.056			1.84	5.80	7.64	9.40
9700	cycle 30 miles		126	.063			2.10	6.65	8.75	10.70
9702	cycle 40 miles		108	.074			2.46	7.75	10.21	12.55
9704	cycle 50 miles		90	.089			2.95	9.30	12.25	15.05
9706	50 MPH ave, cycle 10 miles		198	.040			1.34	4.23	5.57	6.85
9708	cycle 20 miles		162	.049			1.64	5.20	6.84	8.35
9710	cycle 30 miles		126	.063			2.10	6.65	8.75	10.70
9712	cycle 40 miles		108	.074			2.46	7.75	10.21	12.55
9714	cycle 50 miles		90	.089			2.95	9.30	12.25	15.05

31 23 23.23 Compaction

		Crew	Daily Output	Labor-Hours	Unit	Material	Labor	Equipment	Total	Total Incl O&P
0010	**COMPACTION**									
5000	Riding, vibrating roller, 6" lifts, 2 passes	B-10Y	3000	.004	E.C.Y.		.16	.17	.33	.44
5020	3 passes		2300	.005			.21	.22	.43	.58
5040	4 passes		1900	.006			.25	.27	.52	.70
5050	8" lifts, 2 passes		4100	.003			.12	.13	.25	.32
5060	12" lifts, 2 passes		5200	.002			.09	.10	.19	.26
5080	3 passes		3500	.003			.14	.15	.29	.38
5100	4 passes		2600	.005			.18	.20	.38	.51
5600	Sheepsfoot or wobbly wheel roller, 6" lifts, 2 passes	B-10G	2400	.005			.20	.52	.72	.88
5620	3 passes		1735	.007			.27	.72	.99	1.23
5640	4 passes		1300	.009			.37	.96	1.33	1.63
5680	12" lifts, 2 passes		5200	.002			.09	.24	.33	.41
5700	3 passes		3500	.003			.14	.36	.50	.61
5720	4 passes		2600	.005			.18	.48	.66	.82
7000	Walk behind, vibrating plate 18" wide, 6" lifts, 2 passes	A-1D	200	.040			1.32	.17	1.49	2.36
7020	3 passes		185	.043			1.43	.19	1.62	2.55

31 23 Excavation and Fill

31 23 23 – Fill

31 23 23.23 Compaction

		Crew	Daily Output	Labor-Hours	Unit	Material	2010 Bare Costs Labor	Equipment	Total	Total Incl O&P
7040	4 passes	A-1D	140	.057	E.C.Y.		1.89	.24	2.13	3.38
7200	12" lifts, 2 passes	A-1E	560	.014			.47	.08	.55	.87
7220	3 passes		375	.021			.71	.12	.83	1.29
7240	4 passes		280	.029			.95	.16	1.11	1.72
7500	Vibrating roller 24" wide, 6" lifts, 2 passes	B-10A	420	.029			1.13	.36	1.49	2.19
7520	3 passes		280	.043			1.70	.53	2.23	3.29
7540	4 passes		210	.057			2.27	.71	2.98	4.38
7600	12" lifts, 2 passes		840	.014			.57	.18	.75	1.10
7620	3 passes		560	.021			.85	.27	1.12	1.64
7640	4 passes		420	.029			1.13	.36	1.49	2.19
8000	Rammer tamper, 6" to 11", 4" lifts, 2 passes	A-1F	130	.062			2.04	.36	2.40	3.74
8050	3 passes		97	.082			2.73	.48	3.21	5
8100	4 passes		65	.123			4.07	.72	4.79	7.50
8200	8" lifts, 2 passes		260	.031			1.02	.18	1.20	1.87
8250	3 passes		195	.041			1.36	.24	1.60	2.49
8300	4 passes		130	.062			2.04	.36	2.40	3.74
8400	13" to 18", 4" lifts, 2 passes	A-1G	390	.021			.68	.13	.81	1.26
8450	3 passes		290	.028			.91	.18	1.09	1.70
8500	4 passes		195	.041			1.36	.27	1.63	2.52
8600	8" lifts, 2 passes		780	.010			.34	.07	.41	.63
8650	3 passes		585	.014			.45	.09	.54	.84
8700	4 passes		390	.021			.68	.13	.81	1.26
9000	Water, 3000 gal. truck, 3 mile haul	B-45	1888	.008		1.15	.32	.48	1.95	2.29
9010	6 mile haul		1444	.011		1.15	.42	.62	2.19	2.62
9020	12 mile haul		1000	.016		1.15	.61	.90	2.66	3.22
9030	6000 gal. wagon, 3 mile haul	B-59	2000	.004		1.15	.13	.25	1.53	1.75
9040	6 mile haul	"	1600	.005		1.15	.17	.31	1.63	1.87

31 23 23.24 Compaction, Structural

		Crew	Daily Output	Labor-Hours	Unit	Material	Labor	Equipment	Total	Total Incl O&P
0010	**COMPACTION, STRUCTURAL** R312323-30									
0020	Steel wheel tandem roller, 5 tons	B-10E	8	1.500	Hr.		59.50	19	78.50	116
0050	Air tamp, 6" to 8" lifts, common fill	B-9	250	.160	E.C.Y.		5.35	.90	6.25	9.80
0060	Select fill	"	300	.133			4.47	.75	5.22	8.15
0600	Vibratory plate, 8" lifts, common fill	A-1D	200	.040			1.32	.17	1.49	2.36
0700	Select fill	"	216	.037			1.23	.16	1.39	2.18
9000	Minimum labor/equipment charge	1 Clab	4	2	Job		66		66	109

31 25 Erosion and Sedimentation Controls

31 25 13 – Erosion Controls

31 25 13.10 Synthetic Erosion Control

			Crew	Daily Output	Labor-Hours	Unit	Material	Labor	Equipment	Total	Total Incl O&P
0010	**SYNTHETIC EROSION CONTROL**										
0020	Jute mesh, 100 SY per roll, 4' wide, stapled	G	B-80A	2400	.010	S.Y.	1.04	.33	.12	1.49	1.82
0100	Plastic netting, stapled, 2" x 1" mesh, 20 mil	G	B-1	2500	.010		.79	.32		1.11	1.40
0200	Polypropylene mesh, stapled, 6.5 oz./S.Y.	G		2500	.010		1.70	.32		2.02	2.40
0300	Tobacco netting, or jute mesh #2, stapled	G		2500	.010		.10	.32		.42	.64
1000	Silt fence, polypropylene, 3' high, ideal conditions	G	2 Clab	1600	.010	L.F.	.40	.33		.73	.98
1100	Adverse conditions	G	"	950	.017	"	.40	.56		.96	1.36
1200	Place and remove hay bales	G	A-2	3	8	Ton	164	263	80	507	700
1250	Hay bales, staked	G	"	2500	.010	L.F.	6.60	.32	.10	7.02	7.90

31 31 Soil Treatment

31 31 16 - Termite Control

31 31 16.13 Chemical Termite Control

		Crew	Daily Output	Labor-Hours	Unit	Material	2010 Bare Costs Labor	2010 Bare Costs Equipment	Total	Total Incl O&P
0010	**CHEMICAL TERMITE CONTROL**									
0020	Slab and walls, residential	1 Skwk	1200	.007	SF Flr.	.31	.28		.59	.80
0100	Commercial, minimum		2496	.003		.33	.14		.47	.58
0200	Maximum		1645	.005	↓	.50	.21		.71	.87
0390	Minimum labor/equipment charge		4	2	Job		85		85	138
0400	Insecticides for termite control, minimum		14.20	.563	Gal.	12.75	24		36.75	53
0500	Maximum	↓	11	.727	"	22	31		53	74

31 41 Shoring

31 41 13 - Timber Shoring

31 41 13.10 Building Shoring

		Crew	Daily Output	Labor-Hours	Unit	Material	Labor	Equipment	Total	Total Incl O&P
0010	**BUILDING SHORING**									
0020	Shoring, existing building, with timber, no salvage allowance	B-51	2.20	21.818	M.B.F.	530	725	109	1,364	1,900
1000	On cribbing with 35 ton screw jacks, per box and jack		3.60	13.333	Jack	65	445	66.50	576.50	875
1090	Minimum labor/equipment charge	↓	2	24	Ea.		800	120	920	1,425
1100	Masonry openings in walls, see Div. 02 41 19.16									

31 41 16 - Sheet Piling

31 41 16.10 Sheet Piling Systems

		Crew	Daily Output	Labor-Hours	Unit	Material	Labor	Equipment	Total	Total Incl O&P
0010	**SHEET PILING SYSTEMS**									
0020	Sheet piling steel, not incl. wales, 22 psf, 15' excav., left in place	B-40	10.81	5.920	Ton	1,250	245	355	1,850	2,150
0100	Drive, extract & salvage R314116-40		6	10.667	"	515	440	640	1,595	1,975
1200	15' deep excavation, 22 psf, left in place		983	.065	S.F.	14.65	2.69	3.90	21.24	25
1300	Drive, extract & salvage	↓	545	.117	"	5.75	4.85	7.05	17.65	22
2100	Rent steel sheet piling and wales, first month				Ton	268			268	295
2200	Per added month				"	27			27	29.50
3900	Wood, solid sheeting, incl. wales, braces and spacers,									
3910	drive, extract & salvage, 8' deep excavation	B-31	330	.121	S.F.	1.16	4.27	.64	6.07	9
4520	Left in place, 8' deep, 55 S.F./hr.		440	.091	"	2.09	3.20	.48	5.77	8.10
4990	Minimum labor/equipment charge	↓	2	20	Job		705	106	811	1,275
5000	For treated lumber add cost of treatment to lumber									

31 43 Concrete Raising

31 43 13 - Pressure Grouting

31 43 13.13 Concrete Pressure Grouting

		Crew	Daily Output	Labor-Hours	Unit	Material	Labor	Equipment	Total	Total Incl O&P
0010	**CONCRETE PRESSURE GROUTING**									
0020	Grouting, pressure, cement & sand, 1:1 mix, minimum	B-61	124	.323	Bag	12.40	11.35	2.80	26.55	35
0100	Maximum		51	.784	"	12.40	27.50	6.80	46.70	66
0200	Cement and sand, 1:1 mix, minimum		250	.160	C.F.	25	5.60	1.39	31.99	38
0300	Maximum		100	.400		37	14.05	3.48	54.53	68
0400	Epoxy cement grout, minimum		137	.292		450	10.25	2.54	462.79	515
0500	Maximum	↓	57	.702	↓	450	24.50	6.10	480.60	540
0700	Alternate pricing method: (Add for materials)									
0710	5 person crew and equipment	B-61	1	40	Day		1,400	350	1,750	2,650

31 46 Needle Beams

31 46 13 – Cantilever Needle Beams

31 46 13.10 Needle Beams		Crew	Daily Output	Labor-Hours	Unit	Material	2010 Bare Costs Labor	Equipment	Total	Total Incl O&P
0010	**NEEDLE BEAMS**									
0011	Incl. wood shoring 10' x 10' opening									
0400	Block, concrete, 8" thick	B-9	7.10	5.634	Ea.	45	189	31.50	265.50	395
0420	12" thick		6.70	5.970		53	200	33.50	286.50	425
0800	Brick, 4" thick with 8" backup block		5.70	7.018		53	235	39.50	327.50	485
1000	Brick, solid, 8" thick		6.20	6.452		45	216	36.50	297.50	445
1040	12" thick		4.90	8.163		53	273	46	372	560
1080	16" thick		4.50	8.889		68	298	50	416	620
2000	Add for additional floors of shoring	B-1	6	4		45	135		180	272
9000	Minimum labor/equipment charge	"	2	12	Job		405		405	665

31 48 Underpinning

31 48 13 – Underpinning Piers

31 48 13.10 Underpinning Foundations		Crew	Daily Output	Labor-Hours	Unit	Material	2010 Bare Costs Labor	Equipment	Total	Total Incl O&P
0010	**UNDERPINNING FOUNDATIONS**									
0011	Including excavation,									
0020	forming, reinforcing, concrete and equipment									
0100	5' to 16' below grade, 100 to 500 C.Y.	B-52	2.30	24.348	C.Y.	273	935	234	1,442	2,075
0200	Over 500 C.Y.		2.50	22.400		245	860	215	1,320	1,900
0400	16' to 25' below grade, 100 to 500 C.Y.		2	28		300	1,075	269	1,644	2,375
0500	Over 500 C.Y.		2.10	26.667		283	1,025	257	1,565	2,275
0700	26' to 40' below grade, 100 to 500 C.Y.		1.60	35		325	1,350	335	2,010	2,925
0800	Over 500 C.Y.		1.80	31.111		300	1,200	299	1,799	2,600
0900	For under 50 C.Y., add					10%	40%			

31 62 Driven Piles

31 62 13 – Concrete Piles

31 62 13.23 Prestressed Concrete Piles

31 62 13.23 Prestressed Concrete Piles		Crew	Daily Output	Labor-Hours	Unit	Material	2010 Bare Costs Labor	Equipment	Total	Total Incl O&P
0010	**PRESTRESSED CONCRETE PILES**, 200 piles									
0020	Unless specified otherwise, not incl. pile caps or mobilization									
3100	Precast, prestressed, 40' long, 10" thick, square	B-19	700	.091	V.L.F.	14.85	3.78	2.76	21.39	25.50
3200	12" thick, square		680	.094		18.70	3.89	2.84	25.43	30
3400	14" thick, square		600	.107		22	4.41	3.22	29.63	35
4000	18" thick, square	B-19A	520	.123		42	5.10	4.54	51.64	59

31 62 16 – Steel Piles

31 62 16.13 Sheet Steel Piles

31 62 16.13 Sheet Steel Piles		Crew	Daily Output	Labor-Hours	Unit	Material	2010 Bare Costs Labor	Equipment	Total	Total Incl O&P
0010	**SHEET STEEL PILES**									
0100	Step tapered, round, concrete filled									
0110	8" tip, 60 ton capacity, 30' depth	B-19	760	.084	V.L.F.	9.95	3.48	2.54	15.97	19.35
0120	60' depth		740	.086		11.20	3.57	2.61	17.38	21
0250	"H" Sections, 50' long, HP8 x 36		640	.100		17.05	4.13	3.02	24.20	29
1300	HP14 X 102	B-19A	510	.125		49	5.20	4.63	58.83	67.50

31 62 19 – Timber Piles

31 62 19.10 Wood Piles

31 62 19.10 Wood Piles		Crew	Daily Output	Labor-Hours	Unit	Material	2010 Bare Costs Labor	Equipment	Total	Total Incl O&P
0010	**WOOD PILES**									
0011	Friction or end bearing, not including									
0050	mobilization or demobilization									
0100	Untreated piles, up to 30' long, 12" butts, 8" points	B-19	625	.102	V.L.F.	7.40	4.23	3.09	14.72	18.35

31 62 Driven Piles

31 62 19 – Timber Piles

31 62 19.10 Wood Piles

		Crew	Daily Output	Labor-Hours	Unit	Material	2010 Bare Costs Labor	Equipment	Total	Total Incl O&P
0200	30' to 39' long, 12" butts, 8" points	B-19	700	.091	V.L.F.	7.40	3.78	2.76	13.94	17.30
0800	Treated piles, 12 lb. per C.F.,									
0810	friction or end bearing, ASTM class B									
1000	Up to 30' long, 12" butts, 8" points	B-19	625	.102	V.L.F.	10.90	4.23	3.09	18.22	22.50
1100	30' to 39' long, 12" butts, 8" points		700	.091		10.65	3.78	2.76	17.19	21
2700	Mobilization for 10,000 L.F. pile job, add		3300	.019			.80	.59	1.39	1.94
2800	25,000 L.F. pile job, add		8500	.008			.31	.23	.54	.76

31 62 23 – Composite Piles

31 62 23.13 Concrete Filled Steel Piles

		Crew	Daily Output	Labor-Hours	Unit	Material	Labor	Equipment	Total	Total Incl O&P
0010	**CONCRETE FILLED STEEL PILES** no mobilization or demobilization									
2600	Pipe piles, 50' lg. 8" diam., 29 lb. per L.F., no concrete	B-19	500	.128	V.L.F.	18.45	5.30	3.87	27.62	33.50
2700	Concrete filled		460	.139		19.40	5.75	4.20	29.35	35.50
3500	14" diameter, 46 lb. per L.F., no concrete		430	.149		30.50	6.15	4.50	41.15	48.50
3600	Concrete filled		355	.180		32	7.45	5.45	44.90	53
4100	18" diameter, 59 lb. per L.F., no concrete		355	.180		44.50	7.45	5.45	57.40	67
4200	Concrete filled		310	.206		50.50	8.55	6.25	65.30	76

31 63 Bored Piles

31 63 26 – Drilled Caissons

31 63 26.13 Fixed End Cassion Piles

			Crew	Daily Output	Labor-Hours	Unit	Material	Labor	Equipment	Total	Total Incl O&P
0010	**FIXED END CASSION PILES**										
0015	Including excavation, concrete, 50 lbs reinforcing										
0020	per C.Y., not incl. mobilization, boulder removal, disposal										
0100	Open style, machine drilled, to 50' deep, in stable ground, no	R316326-60									
0110	casings or ground water, 18" diam., 0.065 C.Y./L.F.		B-43	200	.240	V.L.F.	7.60	8.70	12.95	29.25	36.50
0200	24" diameter, 0.116 C.Y./L.F.			190	.253	"	13.55	9.15	13.65	36.35	44.50
0210	4' bell diameter, add			20	2.400	Ea.	167	87	130	384	465
0500	48" diameter, 0.465 C.Y./L.F.			100	.480	V.L.F.	54.50	17.35	26	97.85	117
0510	9' bell diameter, add			2	24	Ea.	1,350	870	1,300	3,520	4,300
1200	Open style, machine drilled, to 50' deep, in wet ground, pulled										
1220	pulled casing and pumping										
1400	24" diameter, 0.116 C.Y./L.F.		B-48	125	.448	V.L.F.	13.55	16.55	23.50	53.60	67.50
1410	4' bell diameter, add		"	19.80	2.828	Ea.	167	104	149	420	515
1700	48" diameter, 0.465 C.Y./L.F.		B-49	55	1.600	V.L.F.	54.50	61.50	67	183	233
1710	9' bell diameter, add		"	3.30	26.667	Ea.	1,350	1,025	1,125	3,500	4,350
2300	Open style, machine drilled, to 50' deep, in soft rocks and										
2320	medium hard shales										
2500	24" diameter, 0.116 C.Y./L.F.		B-49	30	2.933	V.L.F.	13.55	112	123	248.55	330
2510	4' bell diameter, add			10.90	8.073	Ea.	167	310	340	817	1,050
2800	48" diameter, 0.465 C.Y./L.F.			10	8.800	V.L.F.	54.50	335	370	759.50	1,000
2810	9' bell diameter, add			1.10	80	Ea.	1,000	3,075	3,350	7,425	9,725
3600	For rock excavation, sockets, add, minimum			120	.733	C.F.		28	31	59	79
3650	Average			95	.926			35.50	39	74.50	100
3700	Maximum			48	1.833			70.50	77	147.50	198
3900	For 50' to 100' deep, add					V.L.F.				7%	7%
4000	For 100' to 150' deep, add									25%	25%
4100	For 150' to 200' deep, add									30%	30%
4200	For casings left in place, add					Lb.	.97			.97	1.07
4300	For other than 50 lb. reinf. per C.Y., add or deduct					"	1.03			1.03	1.13
4400	For steel "I" beam cores, add		B-49	8.30	10.602	Ton	1,750	405	445	2,600	3,075

31 63 Bored Piles

31 63 26 – Drilled Caissons

31 63 26.13 Fixed End Cassion Piles		Crew	Daily Output	Labor-Hours	Unit	Material	2010 Bare Costs Labor	Equipment	Total	Total Incl O&P
4500	Load and haul excess excavation, 2 miles	B-34B	178	.045	L.C.Y.		1.49	3.75	5.24	6.55
4600	For mobilization, 50 mile radius, rig to 36"	B-43	2	24	Ea.		870	1,300	2,170	2,825
4650	Rig to 84"	B-48	1.75	32			1,175	1,675	2,850	3,750
4700	For low headroom, add								50%	
4750	For difficult access, add								25%	

31 63 29 – Drilled Concrete Piers and Shafts

31 63 29.13 Uncased Drilled Concrete Piers

		Crew	Daily Output	Labor-Hours	Unit	Material	Labor	Equipment	Total	Total Incl O&P
0010	**UNCASED DRILLED CONCRETE PIERS**									
0020	Unless specified otherwise, not incl. pile caps or mobilization									
0800	Cast in place friction pile, 50' long, fluted,									
0810	tapered steel, 4000 psi concrete, no reinforcing									
0900	12" diameter, 7 ga.	B-19	600	.107	V.L.F.	27.50	4.41	3.22	35.13	40.50
1200	18" diameter, 7 ga.	"	480	.133	"	41	5.50	4.03	50.53	58.50
1300	End bearing, fluted, constant diameter,									
1320	4000 psi concrete, no reinforcing									
1340	12" diameter, 7 ga.	B-19	600	.107	V.L.F.	28.50	4.41	3.22	36.13	42
1400	18" diameter, 7 ga.	"	480	.133	"	45.50	5.50	4.03	55.03	63.50

Estimating Tips

32 01 00 Operations and Maintenance of Exterior Improvements

- Recycling of asphalt pavement is becoming very popular and is an alternative to removal and replacement of asphalt pavement. It can be a good value engineering proposal if removed pavement can be recycled either at the site or another site that is reasonably close to the project site.

32 10 00 Bases, Ballasts, and Paving

- When estimating paving, keep in mind the project schedule. If an asphaltic paving project is in a colder climate and runs through to the spring, consider placing the base course in the autumn and then topping it in the spring, just prior to completion. This could save considerable costs in spring repair. Keep in mind that prices for asphalt and concrete are generally higher in the cold seasons.

32 90 00 Planting

- The timing of planting and guarantee specifications often dictate the costs for establishing tree and shrub growth and a stand of grass or ground cover. Establish the work performance schedule to coincide with the local planting season. Maintenance and growth guarantees can add from 20%–100% to the total landscaping cost. The cost to replace trees and shrubs can be as high as 5% of the total cost, depending on the planting zone, soil conditions, and time of year.

Reference Numbers

Reference numbers are shown in shaded boxes at the beginning of some major classifications. These numbers refer to related items in the Reference Section. The reference information may be an estimating procedure, an alternate pricing method, or technical information.

Note: Not all subdivisions listed here necessarily appear in this publication.

32 01 Operation and Maintenance of Exterior Improvements

32 01 13 – Flexible Paving Surface Treatment

32 01 13.61 Slurry Seal (Latex Modified)

		Crew	Daily Output	Labor-Hours	Unit	Material	2010 Bare Costs Labor	Equipment	Total	Total Incl O&P
0010	**SLURRY SEAL (LATEX MODIFIED)**									
3600	Waterproofing, membrane, tar and fabric, small area	B-63	233	.172	S.Y.	11.65	5.95	.69	18.29	23.50
3640	Large area		1435	.028		10.70	.97	.11	11.78	13.50
3680	Preformed rubberized asphalt, small area		100	.400		15.75	13.90	1.62	31.27	41.50
3720	Large area		367	.109		14.35	3.79	.44	18.58	22.50
3780	Rubberized asphalt (latex) seal	B-45	5000	.003		2.31	.12	.18	2.61	2.93
9000	Minimum labor/equipment charge	1 Clab	2	4	Job		132		132	217

32 01 13.62 Asphalt Surface Treatment

		Crew	Daily Output	Labor-Hours	Unit	Material	Labor	Equipment	Total	Total Incl O&P
0010	**ASPHALT SURFACE TREATMENT**									
3000	Pavement overlay, polypropylene									
3040	6 oz. per S.Y., ideal conditions	B-63	10000	.004	S.Y.	1.85	.14	.02	2.01	2.29
3080	Adverse conditions		1000	.040		2.48	1.39	.16	4.03	5.15
3120	4 oz. per S.Y., ideal conditions		10000	.004		1.30	.14	.02	1.46	1.68
3160	Adverse conditions		1000	.040		1.68	1.39	.16	3.23	4.28
3200	Tack coat, emulsion, .05 gal per S.Y., 1000 S.Y	B-45	2500	.006		.48	.24	.36	1.08	1.32
3240	10,000 S.Y.		10000	.002		.38	.06	.09	.53	.62
3270	.10 gal per S.Y., 1000 S.Y.		2500	.006		.90	.24	.36	1.50	1.78
3275	10,000 S.Y.		10000	.002		.90	.06	.09	1.05	1.19
3280	.15 gal per S.Y., 1000 S.Y.		2500	.006		1.32	.24	.36	1.92	2.24
3320	10,000 S.Y.		10000	.002		1.06	.06	.09	1.21	1.36

32 01 13.64 Sand Seal

		Crew	Daily Output	Labor-Hours	Unit	Material	Labor	Equipment	Total	Total Incl O&P
0010	**SAND SEAL**									
2080	Sand sealing, sharp sand, asphalt emulsion, small area	B-91	10000	.006	S.Y.	1.29	.25	.23	1.77	2.06
2120	Roadway or large area	"	18000	.004	"	1.11	.14	.13	1.38	1.58
3000	Sealing random cracks, min 1/2" wide, to 1-1/2", 1,000 L.F.	B-77	2800	.014	L.F.	1.03	.48	.20	1.71	2.13
3040	10,000 L.F.		4000	.010	"	.79	.33	.14	1.26	1.57
3080	Alternate method, 1,000 L.F.		200	.200	Gal.	16	6.65	2.80	25.45	31.50
3120	10,000 L.F.		325	.123	"	12.85	4.10	1.72	18.67	22.50
3200	Multi-cracks (flooding), 1 coat, small area	B-92	460	.070	S.Y.	4.41	2.34	1.23	7.98	10.05
3240	Large area		2850	.011		3.92	.38	.20	4.50	5.15
3280	2 coat, small area		230	.139		14.55	4.67	2.46	21.68	26.50
3320	Large area		1425	.022		13.20	.75	.40	14.35	16.25
3360	Alternate method, small area		115	.278	Gal.	16	9.35	4.93	30.28	38.50
3400	Large area		715	.045	"	15.05	1.50	.79	17.34	19.90

32 01 13.66 Fog Seal

		Crew	Daily Output	Labor-Hours	Unit	Material	Labor	Equipment	Total	Total Incl O&P
0010	**FOG SEAL**									
0012	Sealcoating, 2 coat coal tar pitch emulsion over 10,000 SY	B-45	5000	.003	S.Y.	1.05	.12	.18	1.35	1.55
0030	1000 to 10,000 S.Y.	"	3000	.005		1.05	.20	.30	1.55	1.81
0100	Under 1000 S.Y.	B-1	1050	.023		1.05	.77		1.82	2.43
0300	Petroleum resistant, over 10,000 S.Y.	B-45	5000	.003		1.30	.12	.18	1.60	1.82
0320	1000 to 10,000 S.Y.	"	3000	.005		1.30	.20	.30	1.80	2.08
0400	Under 1000 S.Y.	B-1	1050	.023		1.30	.77		2.07	2.70
0600	Non-skid pavement renewal, over 10,000 S.Y.	B-45	5000	.003		1.50	.12	.18	1.80	2.04
0620	1000 to 10,000 S.Y.	"	3000	.005		1.50	.20	.30	2	2.30
0700	Under 1000 S.Y.	B-1	1050	.023		1.50	.77		2.27	2.92
0800	Prepare and clean surface for above	A-2	8545	.003			.09	.03	.12	.18
1000	Hand seal asphalt curbing	B-1	4420	.005	L.F.	.80	.18		.98	1.18
1900	Asphalt surface treatment, single course, small area									
1901	0.30 gal/S.Y. asphalt material, 20#/S.Y. aggregate	B-91	5000	.013	S.Y.	1.75	.49	.46	2.70	3.21
1910	Roadway or large area		10000	.006		1.61	.25	.23	2.09	2.41
1950	Asphalt surface treatment, dbl. course for small area		3000	.021		3.20	.82	.76	4.78	5.65
1960	Roadway or large area		6000	.011		2.88	.41	.38	3.67	4.24

32 01 Operation and Maintenance of Exterior Improvements

32 01 13 – Flexible Paving Surface Treatment

32 01 13.66 Fog Seal	Crew	Daily Output	Labor-Hours	Unit	Material	2010 Bare Costs Labor	Equipment	Total	Total Incl O&P
1980 Asphalt surface treatment, single course, for shoulders	B-91	7500	.009	S.Y.	1.86	.33	.31	2.50	2.91

32 01 13.68 Slurry Seal

0010 **SLURRY SEAL**									
0100 Slurry seal, type I, 8 lbs agg./S.Y., 1 coat, small or irregular area	B-90	2800	.023	S.Y.	1.53	.81	.80	3.14	3.87
0150 Roadway or large area		10000	.006		1.53	.23	.22	1.98	2.30
0200 Type II, 12 lbs aggregate/S.Y., 2 coats, small or irregular area		2000	.032		3.06	1.13	1.12	5.31	6.45
0250 Roadway or large area		8000	.008		3.06	.28	.28	3.62	4.14
0300 Type III, 20 lbs aggregate/S.Y., 2 coats, small or irregular area		1800	.036		3.61	1.26	1.24	6.11	7.35
0350 Roadway or large area		6000	.011		3.61	.38	.37	4.36	4.99
0400 Slurry seal, thermoplastic coal-tar, type I, small or irregular area		2400	.027		3.11	.94	.93	4.98	5.95
0450 Roadway or large area		8000	.008		3.11	.28	.28	3.67	4.19
0500 Type II, small or irregular area		2400	.027		3.97	.94	.93	5.84	6.90
0550 Roadway or large area		7800	.008		3.97	.29	.29	4.55	5.15

32 01 16 – Flexible Paving Rehabilitation

32 01 16.71 Cold Milling Asphalt Paving

0010 **COLD MILLING ASPHALT PAVING**									
5200 Cold planing & cleaning, 1" to 3" asphalt pavmt., over 25,000 S.Y.	B-71	6000	.009	S.Y.		.35	1.19	1.54	1.87
5280 5,000 S.Y. to 10,000 S.Y.	"	4000	.014	"		.53	1.78	2.31	2.81
5300 Asphalt pavement removal from conc. base, no haul									
5320 Rip, load & sweep 1" to 3"	B-70	8000	.007	S.Y.		.26	.22	.48	.66
5330 3" to 6" deep	"	5000	.011			.42	.34	.76	1.06
5340 Profile grooving, asphalt pavement load & sweep, 1" deep	B-71	12500	.004			.17	.57	.74	.90
5350 3" deep		9000	.006			.23	.79	1.02	1.25
5360 6" deep		5000	.011			.42	1.42	1.84	2.25

32 01 16.73 In Place Cold Reused Asphalt Paving

0010 **IN PLACE COLD REUSED ASPHALT PAVING**									
5000 Reclamation, pulverizing and blending with existing base									
5040 Aggregate base, 4" thick pavement, over 15,000 S.Y.	B-73	2400	.027	S.Y.		1.05	2.03	3.08	3.92
5080 5,000 S.Y. to 15,000 S.Y.		2200	.029			1.15	2.22	3.37	4.27
5120 8" thick pavement, over 15,000 S.Y.		2200	.029			1.15	2.22	3.37	4.27
5160 5,000 S.Y. to 15,000 S.Y.		2000	.032			1.26	2.44	3.70	4.69

32 01 16.74 In Place Hot Reused Asphalt Paving

0010 **IN PLACE HOT REUSED ASPHALT PAVING**									
5500 Recycle asphalt pavement at site									
5520 Remove, rejuvenate and spread 4" deep [G]	B-72	2500	.026	S.Y.	4.39	.98	4.55	9.92	11.40
5521 6" deep [G]	"	2000	.032	"	6.45	1.23	5.70	13.38	15.30

32 06 Schedules for Exterior Improvements

32 06 10 – Schedules for Bases, Ballasts, and Paving

32 06 10.10 Sidewalks, Driveways and Patios

0010 **SIDEWALKS, DRIVEWAYS AND PATIOS** No base									
0020 Asphaltic concrete, 2" thick	B-37	720	.067	S.Y.	6.90	2.32	.21	9.43	11.55
0100 2-1/2" thick	"	660	.073	"	8.75	2.53	.23	11.51	13.95
0110 Bedding for brick or stone, mortar, 1" thick	D-1	300	.053	S.F.	.79	2.01		2.80	4.10
0120 2" thick	"	200	.080		1.97	3.02		4.99	7
0130 Sand, 2" thick	B-18	8000	.003		.27	.10	.01	.38	.48
0140 4" thick	"	4000	.006		.54	.20	.01	.75	.93
0300 Concrete, 3000 psi, CIP, 6 x 6 - W1.4 x W1.4 mesh,									
0310 broomed finish, no base, 4" thick	B-24	600	.040	S.F.	1.62	1.52		3.14	4.24
0350 5" thick		545	.044		2.16	1.68		3.84	5.10

32 06 Schedules for Exterior Improvements

32 06 10 – Schedules for Bases, Ballasts, and Paving

32 06 10.10 Sidewalks, Driveways and Patios

		Crew	Daily Output	Labor-Hours	Unit	Material	2010 Bare Costs Labor	Equipment	Total	Total Incl O&P
0400	6" thick	B-24	510	.047	S.F.	2.51	1.79		4.30	5.65
0450	For bank run gravel base, 4" thick, add	B-18	2500	.010		.51	.32	.02	.85	1.11
0520	8" thick, add	"	1600	.015		1.03	.51	.03	1.57	2
0550	Exposed aggregate finish, add to above, minimum	B-24	1875	.013		.12	.49		.61	.92
0600	Maximum	"	455	.053		.39	2.01		2.40	3.68
0950	Concrete tree grate, 5' square	B-6	25	.960	Ea.	405	34.50	13.50	453	515
0960	Cast iron tree grate with frame, 2 piece, round, 5' diameter		25	.960		1,150	34.50	13.50	1,198	1,325
0980	Square, 5' side		25	.960		1,150	34.50	13.50	1,198	1,325
1700	Redwood, prefabricated, 4' x 4' sections	2 Carp	316	.051	S.F.	4.74	2.10		6.84	8.65
1750	Redwood planks, 1" thick, on sleepers	"	240	.067	"	4.74	2.77		7.51	9.75
2250	Stone dust, 4" thick	B-62	900	.027	S.Y.	3.27	.96	.18	4.41	5.35
9000	Minimum labor/equipment charge	D-1	2	8	Job		300		300	485

32 06 10.20 Steps

		Crew	Daily Output	Labor-Hours	Unit	Material	Labor	Equipment	Total	Total Incl O&P
0010	**STEPS**									
0011	Incl. excav., borrow & concrete base as required									
0100	Brick steps	B-24	35	.686	LF Riser	12.85	26		38.85	56
0200	Railroad ties	2 Clab	25	.640		3.17	21		24.17	38.50
0300	Bluestone treads, 12" x 2" or 12" x 1-1/2"	B-24	30	.800		28	30.50		58.50	80
0490	Minimum labor/equipment charge	D-1	2	8	Job		300		300	485
0500	Concrete, cast in place, see Div. 03 30 53.40									
0600	Precast concrete, see Div. 03 41 23.50									

32 11 Base Courses

32 11 23 – Aggregate Base Courses

32 11 23.23 Base Course Drainage Layers

		Crew	Daily Output	Labor-Hours	Unit	Material	Labor	Equipment	Total	Total Incl O&P
0010	**BASE COURSE DRAINAGE LAYERS**									
0011	For roadways and large areas									
0050	Crushed 3/4" stone base, compacted, 3" deep	B-36C	5200	.008	S.Y.	3.38	.30	.71	4.39	4.98
0100	6" deep		5000	.008		6.80	.32	.74	7.86	8.75
0200	9" deep		4600	.009		10.15	.34	.81	11.30	12.65
0300	12" deep		4200	.010		13.55	.38	.88	14.81	16.45
0301	Crushed 1-1/2" stone base, compacted to 4" deep	B-36B	6000	.011		4.66	.41	.71	5.78	6.60
0302	6" deep		5400	.012		7	.45	.78	8.23	9.30
0303	8" deep		4500	.014		9.30	.54	.94	10.78	12.15
0304	12" deep		3800	.017		14	.64	1.12	15.76	17.65
0310	Minimum labor/equipment charge	B-36	1	40	Job		1,500	1,475	2,975	4,000
0350	Bank run gravel, spread and compacted									
0370	6" deep	B-32	6000	.005	S.Y.	4.29	.22	.34	4.85	5.45
0390	9" deep		4900	.007		6.45	.26	.42	7.13	7.95
0400	12" deep		4200	.008		8.55	.31	.49	9.35	10.50
6900	For small and irregular areas, add						50%	50%		
6990	Minimum labor/equipment charge	1 Clab	3	2.667	Job		88.50		88.50	145
7000	Prepare and roll sub-base, small areas to 2500 S.Y.	B-32A	1500	.016	S.Y.		.63	.84	1.47	1.93
8000	Large areas over 2500 S.Y.	"	3500	.007			.27	.36	.63	.82
8050	For roadways	B-32	4000	.008			.32	.51	.83	1.07
9000	Minimum labor/equipment charge	1 Clab	4	2	Job		66		66	109

32 12 Flexible Paving

32 12 16 - Asphalt Paving

32 12 16.13 Plant-Mix Asphalt Paving		Crew	Daily Output	Labor-Hours	Unit	Material	2010 Bare Costs Labor	Equipment	Total	Total Incl O&P
0010	**PLANT-MIX ASPHALT PAVING**									
0080	Binder course, 1-1/2" thick	B-25	7725	.011	S.Y.	5.30	.41	.33	6.04	6.90
0120	2" thick		6345	.014		7.05	.50	.40	7.95	9.05
0130	2-1/2" thick		5620	.016		8.85	.56	.46	9.87	11.10
0160	3" thick		4905	.018		10.60	.65	.52	11.77	13.25
0170	3-1/2" thick		4520	.019		12.35	.70	.57	13.62	15.35
0300	Wearing course, 1" thick	B-25B	10575	.009		3.51	.33	.27	4.11	4.69
0340	1-1/2" thick		7725	.012		5.90	.45	.36	6.71	7.65
0380	2" thick		6345	.015		7.90	.55	.44	8.89	10.10
0420	2-1/2" thick		5480	.018		9.75	.64	.51	10.90	12.35
0460	3" thick		4900	.020		11.65	.72	.57	12.94	14.60
0470	3-1/2" thick		4520	.021		13.65	.78	.62	15.05	16.95
0480	4" thick		4140	.023		15.60	.85	.68	17.13	19.25
0500	Open graded friction course	B-25C	5000	.010		3.30	.35	.45	4.10	4.69
0800	Alternate method of figuring paving costs									
0810	Binder course, 1-1/2" thick	B-25	630	.140	Ton	65	5	4.07	74.07	84
0811	2" thick		690	.128		65	4.59	3.72	73.31	83
0812	3" thick		800	.110		65	3.96	3.21	72.17	81.50
0813	4" thick		900	.098		65	3.52	2.85	71.37	80.50
0850	Wearing course, 1" thick	B-25B	575	.167		68	6.10	4.87	78.97	90
0851	1-1/2" thick		630	.152		68	5.55	4.45	78	89
0852	2" thick		690	.139		68	5.10	4.06	77.16	87.50
0853	2-1/2" thick		765	.125		68	4.59	3.66	76.25	86.50
0854	3" thick		800	.120		68	4.39	3.50	75.89	86
1000	Pavement replacement over trench, 2" thick	B-37	90	.533	S.Y.	7.30	18.55	1.69	27.54	40
1050	4" thick		70	.686		14.45	24	2.17	40.62	57.50
1080	6" thick		55	.873		23	30.50	2.77	56.27	78
3000	Prime coat, emulsion, .30 gal per S.Y., 1000 S.Y	B-45	2500	.006		2.88	.24	.36	3.48	3.96
3100	Tack coat, emulsion, .10 gal per S.Y., 1000 S.Y	"	2500	.006		.96	.24	.36	1.56	1.85

32 12 16.14 Asphaltic Concrete Paving		Crew	Daily Output	Labor-Hours	Unit	Material	Labor	Equipment	Total	Total Incl O&P
0011	**ASPHALTIC CONCRETE PAVING**, parking lots & driveways									
0020	6" stone base, 2" binder course, 1" topping	B-25C	9000	.005	S.F.	1.82	.20	.25	2.27	2.60
0025	2" binder course, 2" topping		9000	.005		2.24	.20	.25	2.69	3.05
0030	3" binder course, 2" topping		9000	.005		2.63	.20	.25	3.08	3.48
0035	4" binder course, 2" topping		9000	.005		3.02	.20	.25	3.47	3.91
0040	1.5" binder course, 1" topping		9000	.005		1.63	.20	.25	2.08	2.38
0042	3" binder course, 1" topping		9000	.005		2.22	.20	.25	2.67	3.03
0045	3" binder course, 3" topping		9000	.005		3.05	.20	.25	3.50	3.94
0050	4" binder course, 3" topping		9000	.005		3.44	.20	.25	3.89	4.37
0055	4" binder course, 4" topping		9000	.005		3.84	.20	.25	4.29	4.82
0300	Binder course, 1-1/2" thick		35000	.001		.59	.05	.06	.70	.80
0400	2" thick		25000	.002		.76	.07	.09	.92	1.05
0500	3" thick		15000	.003		1.18	.12	.15	1.45	1.65
0600	4" thick		10800	.004		1.55	.16	.21	1.92	2.19
0800	Sand finish course, 3/4" thick		41000	.001		.32	.04	.05	.41	.49
0900	1" thick		34000	.001		.40	.05	.07	.52	.59
1000	Fill pot holes, hot mix, 2" thick	B-16	4200	.008		.84	.26	.16	1.26	1.52
1100	4" thick		3500	.009		1.23	.31	.19	1.73	2.06
1120	6" thick		3100	.010		1.65	.35	.22	2.22	2.63
1140	Cold patch, 2" thick	B-51	3000	.016		.85	.53	.08	1.46	1.90
1160	4" thick		2700	.018		1.62	.59	.09	2.30	2.85
1180	6" thick		1900	.025		2.52	.84	.13	3.49	4.29

32 12 Flexible Paving

32 12 16 – Asphalt Paving

32 12 16.14 Asphaltic Concrete Paving		Crew	Daily Output	Labor-Hours	Unit	Material	2010 Bare Costs Labor	Equipment	Total	Total Incl O&P
3000	Prime coat, emulsion, .30 gal per S.Y., 1000 S.Y	B-45	2500	.006	S.Y.	2.88	.24	.36	3.48	3.96
3100	Tack coat, emulsion, .10 gal per S.Y., 1000 S.Y	"	2500	.006	"	.96	.24	.36	1.56	1.85

32 13 Rigid Paving

32 13 13 – Concrete Paving

32 13 13.23 Concrete Paving Surface Treatment

		Crew	Daily Output	Labor-Hours	Unit	Material	Labor	Equipment	Total	Total Incl O&P
0010	**CONCRETE PAVING SURFACE TREATMENT**									
0015	Including joints, finishing and curing									
0020	Fixed form, 12' pass, unreinforced, 6" thick	B-26	3000	.029	S.Y.	21	1.08	1.10	23.18	26
0100	8" thick		2750	.032		31.50	1.18	1.20	33.88	38
0400	12" thick		1800	.049		44	1.81	1.83	47.64	53.50
0410	Conc pavement, w/jt,fnsh&curing,fix form,24' pass,unreinforced,6"T		6000	.015		19.45	.54	.55	20.54	23
0430	8" thick		5500	.016		28.50	.59	.60	29.69	33
0460	12" thick		3600	.024		41	.90	.92	42.82	47.50
0505	Minimum labor/equipment charge	1 Cefi	1	8	Job		320		320	495
0520	Welded wire fabric, sheets for rigid paving 2.33 lbs/SY	2 Rodm	389	.041	S.Y.	1.08	1.92		3	4.42
0530	Reinforcing steel for rigid paving 12 lbs/SY		666	.024		5.05	1.12		6.17	7.45
0540	Reinforcing steel for rigid paving 18 lbs/SY		444	.036		7.55	1.69		9.24	11.15
0620	Slip form, 12' pass, unreinforced, 6" thick	B-26A	5600	.016		20.50	.58	.62	21.70	24
0624	8" thick		5300	.017		31	.61	.65	32.26	35.50
0630	12" thick		3470	.025		43	.94	1	44.94	50
0632	15" thick		2890	.030		55.50	1.12	1.20	57.82	64
0640	Slip form, 24' pass, unreinforced, 6" thick		11200	.008		19.70	.29	.31	20.30	22.50
0644	8" thick		10600	.008		28	.31	.33	28.64	32
0650	12" thick		6940	.013		40.50	.47	.50	41.47	46
0652	15" thick		5780	.015		51.50	.56	.60	52.66	58
0700	Finishing, broom finish small areas	2 Cefi	120	.133			5.30		5.30	8.25
1000	Curing, with sprayed membrane by hand	2 Clab	1500	.011		.55	.35		.90	1.19

32 14 Unit Paving

32 14 16 – Brick Unit Paving

32 14 16.10 Brick Paving

		Crew	Daily Output	Labor-Hours	Unit	Material	Labor	Equipment	Total	Total Incl O&P
0010	**BRICK PAVING**									
0012	4" x 8" x 1-1/2", without joints (4.5 brick/S.F.)	D-1	110	.145	S.F.	3.27	5.50		8.77	12.40
0100	Grouted, 3/8" joint (3.9 brick/S.F.)		90	.178		3.38	6.70		10.08	14.50
0200	4" x 8" x 2-1/4", without joints (4.5 bricks/S.F.)		110	.145		3.74	5.50		9.24	12.90
0300	Grouted, 3/8" joint (3.9 brick/S.F.)		90	.178		3.45	6.70		10.15	14.60
0500	Bedding, asphalt, 3/4" thick	B-25	5130	.017		.65	.62	.50	1.77	2.27
0540	Course washed sand bed, 1" thick	B-18	5000	.005		.30	.16	.01	.47	.61
0580	Mortar, 1" thick	D-1	300	.053		.66	2.01		2.67	3.95
0620	2" thick		200	.080		1.31	3.02		4.33	6.30
1500	Brick on 1" thick sand bed laid flat, 4.5 per S.F.		100	.160		3	6.05		9.05	13
2000	Brick pavers, laid on edge, 7.2 per S.F.		70	.229		3.18	8.60		11.78	17.35
2500	For 4" thick concrete bed and joints, add		595	.027		1.12	1.01		2.13	2.87
2800	For steam cleaning, add	A-1H	950	.008		.07	.28	.06	.41	.61
9000	Minimum labor/equipment charge	1 Bric	2	4	Job		167		167	269

32 14 Unit Paving

32 14 23 – Asphalt Unit Paving

32 14 23.10 Asphalt Blocks

		Crew	Daily Output	Labor-Hours	Unit	Material	2010 Bare Costs Labor	Equipment	Total	Total Incl O&P
0010	**ASPHALT BLOCKS**									
0020	Rectangular, 6" x 12" x 1-1/4", w/bed & neopr. adhesive	D-1	135	.119	S.F.	7.80	4.47		12.27	15.80
0100	3" thick		130	.123		10.95	4.64		15.59	19.50
0300	Hexagonal tile, 8" wide, 1-1/4" thick		135	.119		7.80	4.47		12.27	15.80
0400	2" thick		130	.123		10.95	4.64		15.59	19.50
0500	Square, 8" x 8", 1-1/4" thick		135	.119		7.80	4.47		12.27	15.80
0600	2" thick		130	.123		10.95	4.64		15.59	19.50
9000	Minimum labor/equipment charge	1 Bric	2	4	Job		167		167	269

32 14 40 – Stone Paving

32 14 40.10 Stone Pavers

		Crew	Daily Output	Labor-Hours	Unit	Material	Labor	Equipment	Total	Incl O&P
0010	**STONE PAVERS**									
1300	Slate, natural cleft, irregular, 3/4" thick	D-1	92	.174	S.F.	7	6.55		13.55	18.25
1350	Random rectangular, gauged, 1/2" thick		105	.152		15.15	5.75		20.90	26
1400	Random rectangular, butt joint, gauged, 1/4" thick		150	.107		16.30	4.02		20.32	24.50
1450	For sand rubbed finish, add					7.60			7.60	8.35
1550	Granite blocks, 3-1/2" x 3-1/2" x 3-1/2"	D-1	92	.174		10.05	6.55		16.60	21.50
1600	4" to 12" long, 3" to 5" wide, 3" to 5" thick	"	98	.163		8.40	6.15		14.55	19.10

32 16 Curbs and Gutters

32 16 13 – Concrete Curbs and Gutters

32 16 13.13 Cast-in-Place Concrete Curbs and Gutters

		Crew	Daily Output	Labor-Hours	Unit	Material	Labor	Equipment	Total	Incl O&P
0010	**CAST-IN-PLACE CONCRETE CURBS AND GUTTERS**									
0300	Concrete, wood forms, 6" x 18", straight	C-2	500	.096	L.F.	3.63	3.89		7.52	10.40
0400	6" x 18", radius	"	200	.240		3.70	9.70		13.40	20
0404	Concrete, wood forms, 6" x 18", straight & concrete	C-2A	500	.096		6.35	3.86		10.21	13.30
0406	6" x 18", radius	"	200	.240		6.40	9.65		16.05	23
0415	Machine formed, 6" x 18", straight	B-69A	2000	.024		6.20	.87	.41	7.48	8.70
0416	6" x 18", radius	"	900	.053		6.35	1.93	.92	9.20	11.10

32 16 13.26 Precast Concrete Curbs

		Crew	Daily Output	Labor-Hours	Unit	Material	Labor	Equipment	Total	Incl O&P
0010	**PRECAST CONCRETE CURBS**									
0550	Precast, 6" x 18", straight	B-29	700	.080	L.F.	10.85	2.86	1.16	14.87	17.80
0600	6" x 18", radius	"	325	.172	"	12.45	6.15	2.50	21.10	26.50

32 16 19 – Asphalt Curbs

32 16 19.10 Bituminous Concrete Curbs

		Crew	Daily Output	Labor-Hours	Unit	Material	Labor	Equipment	Total	Incl O&P
0010	**BITUMINOUS CONCRETE CURBS**									
0012	Curbs, asphaltic, machine formed, 8" wide, 6" high, 40 L.F./ton	B-27	1000	.032	L.F.	1.70	1.08	.29	3.07	3.95
0100	8" wide, 8" high, 30 L.F. per ton		900	.036		2.27	1.19	.32	3.78	4.81
0150	Asphaltic berm, 12" W, 3"-6" H, 35 L.F./ton, before pavement		700	.046		.04	1.54	.41	1.99	3.01
0200	12" W, 1-1/2" to 4" H, 60 L.F. per ton, laid with pavement	B-2	1050	.038		.02	1.28		1.30	2.13

32 16 40 – Stone Curbs

32 16 40.13 Cut Stone Curbs

		Crew	Daily Output	Labor-Hours	Unit	Material	Labor	Equipment	Total	Incl O&P
0010	**CUT STONE CURBS**									
1000	Granite, split face, straight, 5" x 16"	D-13	500	.096	L.F.	12.05	3.82	1.31	17.18	21
1100	6" x 18"	"	450	.107		15.85	4.24	1.46	21.55	26
1300	Radius curbing, 6" x 18", over 10' radius	B-29	260	.215		19.40	7.70	3.13	30.23	37.50
1400	Corners, 2' radius		80	.700	Ea.	65	25	10.15	100.15	123
1600	Edging, 4-1/2" x 12", straight		300	.187	L.F.	6.05	6.65	2.71	15.41	20.50
1800	Curb inlets, (guttermouth) straight		41	1.366	Ea.	145	49	19.80	213.80	260

32 16 Curbs and Gutters

32 16 40 – Stone Curbs

32 16 40.13 Cut Stone Curbs

		Crew	Daily Output	Labor-Hours	Unit	Material	2010 Bare Costs Labor	Equipment	Total	Total Incl O&P
2000	Indian granite (belgian block)									
2100	Jumbo, 10-1/2" x 7-1/2" x 4", grey	D-1	150	.107	L.F.	2.21	4.02		6.23	8.90
2150	Pink		150	.107		2.88	4.02		6.90	9.60
2200	Regular, 9" x 4-1/2" x 4-1/2", grey		160	.100		2	3.77		5.77	8.25
2250	Pink		160	.100		2.76	3.77		6.53	9.10
2300	Cubes, 4" x 4" x 4", grey		175	.091		1.91	3.45		5.36	7.65
2350	Pink		175	.091		2	3.45		5.45	7.75
2400	6" x 6" x 6", pink		155	.103		4.96	3.89		8.85	11.70
2500	Alternate pricing method for indian granite									
2550	Jumbo, 10-1/2" x 7-1/2" x 4" (30 lb), grey				Ton	126			126	139
2600	Pink					167			167	184
2650	Regular, 9" x 4-1/2" x 4-1/2" (20 lb), grey					141			141	156
2700	Pink					193			193	212
2750	Cubes, 4" x 4" x 4" (5 lb), grey					231			231	255
2800	Pink					257			257	283
2850	6" x 6" x 6" (25 lb), pink					193			193	212
2900	For pallets, add					20			20	22

32 17 Paving Specialties

32 17 13 – Parking Bumpers

32 17 13.13 Metal Parking Bumpers

		Crew	Daily Output	Labor-Hours	Unit	Material	Labor	Equipment	Total	Total Incl O&P
0010	**METAL PARKING BUMPERS**									
0015	Bumper rails for garages, 12 Ga. rail, 6" wide, with steel									
0020	posts 12'-6" O.C., minimum	E-4	190	.168	L.F.	17.35	8	.77	26.12	34.50
0030	Average		165	.194		21.50	9.20	.88	31.58	42
0100	Maximum		140	.229		26	10.85	1.04	37.89	49.50
1300	Pipe bollards, conc filled/paint, 8' L x 4' D hole, 6" diam.	B-6	20	1.200	Ea.	400	43	16.90	459.90	530
1400	8" diam.		15	1.600		610	57.50	22.50	690	790
1500	12" diam.		12	2		790	71.50	28	889.50	1,025
1592	Bollards, steel, 3' H, retractable, incl hydraulic controls, min		4	6		39,800	215	84.50	40,099.50	44,200
1594	Max		2	12		44,400	430	169	44,999	49,700
9000	Minimum labor/equipment charge	E-4	2	16	Job		760	73	833	1,450

32 17 13.16 Plastic Parking Bumpers

			Crew	Daily Output	Labor-Hours	Unit	Material	Labor	Equipment	Total	Total Incl O&P
0010	**PLASTIC PARKING BUMPERS**										
1600	Bollards, recycled plastic, 5" x 5" x 5' L	G	B-6	20	1.200	Ea.	57.50	43	16.90	117.40	151
1610	Wheel stops, recycled plastic, yellow, 4" x 6" x 6' L	G		20	1.200		57.50	43	16.90	117.40	151
1620	Speed bumps, recycled plastic, yellow, 3"H x 10" W x 6' L	G		20	1.200		124	43	16.90	183.90	224
1630	3"H x 10" W x 9' L	G		20	1.200		170	43	16.90	229.90	275

32 17 13.19 Precast Concrete Parking Bumpers

		Crew	Daily Output	Labor-Hours	Unit	Material	Labor	Equipment	Total	Total Incl O&P
0010	**PRECAST CONCRETE PARKING BUMPERS**									
1000	Wheel stops, precast concrete incl. dowels, 6" x 10" x 6'-0"	B-2	120	.333	Ea.	57.50	11.15		68.65	81.50
1100	8" x 13" x 6'-0"	"	120	.333		66.50	11.15		77.65	91.50
1540	Bollards, precast concrete, free standing, round, 15" Dia x 36" H	B-6	14	1.714		175	61.50	24	260.50	320
1550	16" Dia x 30" H		14	1.714		175	61.50	24	260.50	320
1560	18" Dia x 34" H		14	1.714		360	61.50	24	445.50	520
1570	18" Dia x 72" H		14	1.714		435	61.50	24	520.50	605
1572	26" Dia x 18" H		12	2		435	71.50	28	534.50	625
1574	26" Dia x 30" H		12	2		645	71.50	28	744.50	855
1578	26" Dia x 50" H		12	2		875	71.50	28	974.50	1,100
1580	Square, 22" x 32" H		12	2		465	71.50	28	564.50	655

32 17 Paving Specialties

32 17 13 – Parking Bumpers

32 17 13.19 Precast Concrete Parking Bumpers		Crew	Daily Output	Labor-Hours	Unit	Material	2010 Bare Costs Labor	Equipment	Total	Total Incl O&P
1585	22" x 41" H	B-6	12	2	Ea.	820	71.50	28	919.50	1,050
1590	Hexagon, 26" x 41" H	↓	10	2.400	↓	855	86	34	975	1,125

32 17 13.26 Wood Parking Bumpers

		Crew	Daily Output	Labor-Hours	Unit	Material	Labor	Equipment	Total	Total Incl O&P
0010	**WOOD PARKING BUMPERS**									
0020	Parking barriers, timber w/saddles, treated type									
0100	4" x 4" for cars	B-2	520	.077	L.F.	1.93	2.58		4.51	6.35
0200	6" x 6" for trucks	"	520	.077	"	3.96	2.58		6.54	8.60

32 17 23 – Pavement Markings

32 17 23.13 Painted Pavement Markings

		Crew	Daily Output	Labor-Hours	Unit	Material	Labor	Equipment	Total	Total Incl O&P
0010	**PAINTED PAVEMENT MARKINGS**									
0020	Acrylic waterborne, white or yellow, 4" wide	B-78	20000	.002	L.F.	.26	.08	.03	.37	.45
0200	6" wide		11000	.004		.20	.15	.05	.40	.52
0500	8" wide		10000	.005		.34	.16	.06	.56	.69
0600	12" wide		4000	.012	↓	.62	.40	.15	1.17	1.51
0620	Arrows or gore lines		2300	.021	S.F.	.69	.69	.26	1.64	2.18
0640	Temporary paint, white or yellow	↓	15000	.003	L.F.	.24	.11	.04	.39	.47
0660	Removal	1 Clab	300	.027			.88		.88	1.45
0680	Temporary tape	2 Clab	1500	.011		1.78	.35		2.13	2.54
0710	Thermoplastic, white or yellow, 4" wide	B-79	15000	.003		.89	.09	.09	1.07	1.23
0730	6" wide		14000	.003		1.29	.10	.10	1.49	1.69
0740	8" wide		12000	.003		1.74	.11	.11	1.96	2.22
0750	12" wide		6000	.007	↓	2.58	.22	.23	3.03	3.45
0760	Arrows		660	.061	S.F.	1.86	2.02	2.09	5.97	7.65
0770	Gore lines		2500	.016		1.26	.53	.55	2.34	2.87
0780	Letters	↓	660	.061	↓	1.55	2.02	2.09	5.66	7.30
1000	Airport painted markings									
1100	Painting, white or yellow, taxiway markings	B-79	4000	.010	S.F.	.26	.33	.34	.93	1.22
1200	Runway markings		3500	.011		.26	.38	.39	1.03	1.34
1300	Pavement location or direction signs		2500	.016	↓	.26	.53	.55	1.34	1.77
1350	Mobilization airport pavement painting	↓	4	10	Ea.		335	345	680	925
1400	Paint markings or pavement signs removal daytime	B-78B	400	.045	S.F.		1.53	.90	2.43	3.50
1500	Removal nighttime		335	.054	"		1.83	1.08	2.91	4.17
1600	Mobilization pavement paint removal	↓	4	4.500	Ea.		153	90.50	243.50	350

32 17 23.14 Pavement Parking Markings

		Crew	Daily Output	Labor-Hours	Unit	Material	Labor	Equipment	Total	Total Incl O&P
0010	**PAVEMENT PARKING MARKINGS**									
0790	Layout of pavement marking	A-2	25000	.001	L.F.		.03	.01	.04	.06
0800	Lines on pvmt, parking stall, paint, white, 4" wide	B-78	440	.109	Stall	6.05	3.63	1.34	11.02	14.05
1000	Street letters and numbers	"	1600	.030	S.F.	.64	1	.37	2.01	2.74

32 31 Fences and Gates

32 31 13 – Chain Link Fences and Gates

32 31 13.20 Fence, Chain Link Industrial

		Crew	Daily Output	Labor-Hours	Unit	Material	Labor	Equipment	Total	Total Incl O&P
0010	**FENCE, CHAIN LINK INDUSTRIAL**									
0011	Schedule 40, including concrete									
0020	3 strands barb wire, 2" post @ 10' O.C., set in concrete, 6' H									
0200	9 ga. wire, galv. steel, in concrete	B-80C	240	.100	L.F.	14.80	3.28	1.03	19.11	22.50
0248	Fence, add for vinyl coated fabric				S.F.	.16			.16	.18
0300	Aluminized steel	B-80C	240	.100	L.F.	18.95	3.28	1.03	23.26	27.50
0500	6 ga. wire, galv. steel		240	.100		23	3.28	1.03	27.31	32
0600	Aluminized steel	↓	240	.100	↓	26.50	3.28	1.03	30.81	35.50

32 31 Fences and Gates

32 31 13 – Chain Link Fences and Gates

32 31 13.20 Fence, Chain Link Industrial

		Crew	Daily Output	Labor-Hours	Unit	Material	2010 Bare Costs Labor	Equipment	Total	Total Incl O&P
0800	6 ga. wire, 6' high but omit barbed wire, galv. steel	B-80C	250	.096	L.F.	22.50	3.15	.99	26.64	31
0900	Aluminized steel, in concrete		250	.096		31.50	3.15	.99	35.64	41
0920	8' H, 6 ga. wire, 2-1/2" line post, galv. steel, in concrete		180	.133		36	4.38	1.38	41.76	48
0940	Aluminized steel, in concrete		180	.133		44	4.38	1.38	49.76	57
1100	Add for corner posts, 3" diam., galv. steel, in concrete		40	.600	Ea.	105	19.70	6.20	130.90	154
1200	Aluminized steel, in concrete		40	.600		125	19.70	6.20	150.90	177
1300	Add for braces, galv. steel		80	.300		28	9.85	3.10	40.95	50.50
1350	Aluminized steel		80	.300		37.50	9.85	3.10	50.45	61
1400	Gate for 6' high fence, 1-5/8" frame, 3' wide, galv. steel		10	2.400		172	79	25	276	345
1500	Aluminized steel, in concrete		10	2.400		211	79	25	315	390
2000	5'-0" high fence, 9 ga., no barbed wire, 2" line post, in concrete									
2010	10' O.C., 1-5/8" top rail, in concrete									
2100	Galvanized steel, in concrete	B-80C	300	.080	L.F.	12.60	2.63	.83	16.06	19.05
2200	Aluminized steel, in concrete		300	.080	"	15.15	2.63	.83	18.61	22
2400	Gate, 4' wide, 5' high, 2" frame, galv. steel, in concrete		10	2.400	Ea.	185	79	25	289	360
2500	Aluminized steel, in concrete		10	2.400	"	208	79	25	312	385
3100	Overhead slide gate, chain link, 6' high, to 18' wide, in concrete		38	.632	L.F.	162	20.50	6.50	189	219
3110	Cantilever type, in concrete	B-80	48	.667		75	23.50	15.55	114.05	138
3120	8' high, in concrete		24	1.333		108	47.50	31	186.50	230
3130	10' high, in concrete		18	1.778		127	63	41.50	231.50	288
9000	Minimum labor/equipment charge		2	16	Job		565	375	940	1,325

32 31 13.25 Fence, Chain Link Residential

		Crew	Daily Output	Labor-Hours	Unit	Material	Labor	Equipment	Total	Total Incl O&P
0010	**FENCE, CHAIN LINK RESIDENTIAL**									
0011	Schedule 20, 11 gauge wire, 1-5/8" post									
3300	Residential, 11 ga. wire, 1-5/8" line post @ 10' O.C.									
3310	1-3/8" top rail									
3350	4' high, galvanized steel	B-80C	475	.051	L.F.	10.50	1.66	.52	12.68	14.85
3400	Aluminized		475	.051	"	10.60	1.66	.52	12.78	15
3600	Gate, 3' wide, 1-3/8" frame, galv. steel		10	2.400	Ea.	86	79	25	190	251
3700	Aluminized		10	2.400	"	97.50	79	25	201.50	263
3900	3' high, galvanized steel		620	.039	L.F.	6.35	1.27	.40	8.02	9.50
4000	Aluminized		620	.039	"	7.80	1.27	.40	9.47	11.10
4200	Gate, 3' wide, 1-3/8" frame, galv. steel		12	2	Ea.	71	65.50	20.50	157	208
4300	Aluminized		12	2	"	79.50	65.50	20.50	165.50	217
9000	Minimum labor/equipment charge	B-1	2	12	Job		405		405	665

32 31 13.26 Tennis Court Fences and Gates

		Crew	Daily Output	Labor-Hours	Unit	Material	Labor	Equipment	Total	Total Incl O&P
0010	**TENNIS COURT FENCES AND GATES**									
3150	Tennis courts, 11 ga. wire, 1-3/4" mesh, 2-1/2" line posts									
3170	1-5/8" top rail, 3" corner and gate posts									
3190	10' high, galvanized steel	B-80	190	.168	L.F.	19.15	5.95	3.92	29.02	35
3210	Vinyl covered 9 ga. wire		190	.168	"	21	5.95	3.92	30.87	37
3240	Corner posts for above, 3" diameter, 10' high		30	1.067	Ea.	201	38	25	264	310

32 31 13.40 Fence, Fabric and Accessories

		Crew	Daily Output	Labor-Hours	Unit	Material	Labor	Equipment	Total	Total Incl O&P
0010	**FENCE, FABRIC & ACCESSORIES**									
8000	Components only, fabric, galvanized, 4' high	B-80C	585	.041	L.F.	3.86	1.35	.42	5.63	6.90
8040	6' high		430	.056	"	5.05	1.83	.58	7.46	9.25
8200	Posts, line post, 4' high		34	.706	Ea.	14.35	23	7.30	44.65	62
8240	6' high		26	.923	"	14.55	30.50	9.55	54.60	76
8400	Top rails, 1-3/8" O.D.		1000	.024	L.F.	1.84	.79	.25	2.88	3.58
8440	1-5/8" O.D.		1000	.024	"	2.09	.79	.25	3.13	3.86

32 31 Fences and Gates

32 31 19 – Decorative Metal Fences and Gates

32 31 19.10 Decorative Fence

		Crew	Daily Output	Labor-Hours	Unit	Material	2010 Bare Costs Labor	Equipment	Total	Total Incl O&P
0010	**DECORATIVE FENCE**									
5300	Tubular picket, steel, 6' sections, 1-9/16" posts, 4' high	B-80C	300	.080	L.F.	30	2.63	.83	33.46	38.50
5400	2" posts, 5' high		240	.100		42	3.28	1.03	46.31	52.50
5600	2" posts, 6' high		200	.120		47.50	3.94	1.24	52.68	60
5700	Staggered picket 1-9/16" posts, 4' high		300	.080		27.50	2.63	.83	30.96	35
5800	2" posts, 5' high		240	.100		45	3.28	1.03	49.31	56
5900	2" posts, 6' high	▼	200	.120	▼	46.50	3.94	1.24	51.68	59
6200	Gates, 4' high, 3' wide	B-1	10	2.400	Ea.	263	81		344	420
6300	5' high, 3' wide		10	2.400		340	81		421	510
6400	6' high, 3' wide		10	2.400		350	81		431	520
6500	4' wide	▼	10	2.400	▼	410	81		491	585

32 31 23 – Plastic Fences and Gates

32 31 23.20 Fence, Recycled Plastic

			Crew	Daily Output	Labor-Hours	Unit	Material	Labor	Equipment	Total	Total Incl O&P
0010	**FENCE, RECYCLED PLASTIC**										
9015	Fence rail, made from recycled plastic, various colors, 2 rail	G	B-1	150	.160	L.F.	4.61	5.40		10.01	13.90
9018	3 rail	G		150	.160		6.15	5.40		11.55	15.65
9020	4 rail	G		150	.160	▼	7.45	5.40		12.85	17.05
9030	Fence pole, made from recycled plastic, various colors, 7'	G	▼	96	.250	Ea.	17.75	8.45		26.20	33.50
9040	Stockade fence, made from recycled plastic, various colors, 4' high	G	B-80C	160	.150	L.F.	30.50	4.92	1.55	36.97	44
9050	6' high	G		160	.150	"	38	4.92	1.55	44.47	51.50
9060	6' pole	G		96	.250	Ea.	18.90	8.20	2.58	29.68	37.50
9070	9' pole	G		96	.250	"	28.50	8.20	2.58	39.28	47.50
9080	Picket fence, made from recycled plastic, various colors, 3' high	G		160	.150	L.F.	30.50	4.92	1.55	36.97	44
9090	4' high	G		160	.150	"	38	4.92	1.55	44.47	51.50
9100	3' high gate	G		8	3	Ea.	59	98.50	31	188.50	260
9110	4' high gate	G		8	3		68.50	98.50	31	198	271
9120	5' high pole	G		96	.250		10.65	8.20	2.58	21.43	28
9130	6' high pole	G	▼	96	.250		13	8.20	2.58	23.78	30.50
9140	Pole cap only	G					3.58			3.58	3.94
9150	Keeper pins only	G				▼	.26			.26	.29

32 31 26 – Wire Fences and Gates

32 31 26.10 Fences, Misc. Metal

		Crew	Daily Output	Labor-Hours	Unit	Material	Labor	Equipment	Total	Total Incl O&P
0010	**FENCES, MISC. METAL**									
0012	Chicken wire, posts @ 4', 1" mesh, 4' high	B-80C	410	.059	L.F.	2.10	1.92	.60	4.62	6.10
0100	2" mesh, 6' high		350	.069		1.89	2.25	.71	4.85	6.55
0200	Galv. steel, 12 ga., 2" x 4" mesh, posts 5' O.C., 3' high		300	.080		3.03	2.63	.83	6.49	8.55
0300	5' high		300	.080		4.04	2.63	.83	7.50	9.65
0400	14 ga., 1" x 2" mesh, 3' high		300	.080		3.21	2.63	.83	6.67	8.75
0500	5' high	▼	300	.080	▼	4.44	2.63	.83	7.90	10.10
1000	Kennel fencing, 1-1/2" mesh, 6' long, 3'-6" wide, 6'-2" high	2 Clab	4	4	Ea.	505	132		637	770
1050	12' long		4	4		605	132		737	880
1200	Top covers, 1-1/2" mesh, 6' long		15	1.067		102	35.50		137.50	171
1250	12' long	▼	12	1.333	▼	164	44		208	253
4500	Security fence, prison grade, set in concrete, 12' high	B-80	25	1.280	L.F.	44.50	45.50	30	120	156
4600	16' high	"	20	1.600	"	53.50	56.50	37.50	147.50	191

32 31 29 – Wood Fences and Gates

32 31 29.20 Fence, Wood Rail

		Crew	Daily Output	Labor-Hours	Unit	Material	Labor	Equipment	Total	Total Incl O&P
0010	**FENCE, WOOD RAIL**									
0012	Picket, No. 2 cedar, Gothic, 2 rail, 3' high	B-1	160	.150	L.F.	6.05	5.05		11.10	14.95
0050	Gate, 3'-6" wide	B-80C	9	2.667	Ea.	52	87.50	27.50	167	231

32 31 Fences and Gates

32 31 29 – Wood Fences and Gates

32 31 29.20 Fence, Wood Rail		Crew	Daily Output	Labor-Hours	Unit	Material	2010 Bare Costs Labor	Equipment	Total	Total Incl O&P
0600	Open rail, rustic, No. 1 cedar, 2 rail, 3' high	B-80C	160	.150	L.F.	5.40	4.92	1.55	11.87	15.70
0650	Gate, 3' wide		9	2.667	Ea.	60	87.50	27.50	175	240
1200	Stockade, No. 2 cedar, treated wood rails, 6' high		160	.150	L.F.	7.50	4.92	1.55	13.97	18
1250	Gate, 3' wide		9	2.667	Ea.	61.50	87.50	27.50	176.50	242
3300	Board, shadow box, 1" x 6", treated pine, 6' high		160	.150	L.F.	10.20	4.92	1.55	16.67	21
3400	No. 1 cedar, 6' high		150	.160		20	5.25	1.65	26.90	32.50
3900	Basket weave, No. 1 cedar, 6' high		160	.150		19.95	4.92	1.55	26.42	32
3950	Gate, 3'-6" wide	B-1	8	3	Ea.	136	101		237	315
4000	Treated pine, 6' high		150	.160	L.F.	13.05	5.40		18.45	23.50
4200	Gate, 3'-6" wide		9	2.667	Ea.	63.50	90		153.50	218
8000	Posts only, 4' high		40	.600		14.75	20.50		35.25	49.50
8040	6' high		30	.800		20.50	27		47.50	67
9000	Minimum labor/equipment charge	1 Clab	2	4	Job		132		132	217

32 32 Retaining Walls

32 32 60 – Stone Retaining Walls

32 32 60.10 Retaining Walls, Stone

		Crew	Daily Output	Labor-Hours	Unit	Material	Labor	Equipment	Total	Total Incl O&P
0010	**RETAINING WALLS, STONE**									
0015	Including excavation, concrete footing and									
0020	stone 3' below grade. Price is exposed face area.									
0200	Decorative random stone, to 6' high, 1'-6" thick, dry set	D-1	35	.457	S.F.	46	17.25		63.25	78.50
0300	Mortar set		40	.400		46	15.10		61.10	75.50
0500	Cut stone, to 6' high, 1'-6" thick, dry set		35	.457		46	17.25		63.25	78.50
0600	Mortar set		40	.400		46	15.10		61.10	75.50
0800	Random stone, 6' to 10' high, 2' thick, dry set		45	.356		46	13.40		59.40	72.50
0900	Mortar set		50	.320		46	12.05		58.05	70.50
1100	Cut stone, 6' to 10' high, 2' thick, dry set		45	.356		46	13.40		59.40	72.50
1200	Mortar set		50	.320		46	12.05		58.05	70.50
9000	Minimum labor/equipment charge		2	8	Job		300		300	485

32 84 Planting Irrigation

32 84 23 – Underground Sprinklers

32 84 23.10 Sprinkler Irrigation System

		Crew	Daily Output	Labor-Hours	Unit	Material	Labor	Equipment	Total	Total Incl O&P
0010	**SPRINKLER IRRIGATION SYSTEM**									
0011	For lawns									
0100	Golf course with fully automatic system	C-17	.05	1600	9 holes	99,000	69,000		168,000	220,500
0200	24' diam. head at 15' O.C incl. piping, auto oper., minimum	B-20	70	.343	Head	20	12.65		32.65	42.50
0300	Maximum		40	.600		50	22		72	91
0500	60' diam. head at 40' O.C. incl. piping, auto oper., minimum		28	.857		65	31.50		96.50	123
0600	Maximum		23	1.043		180	38.50		218.50	261
0800	Residential system, custom, 1" supply		2000	.012	S.F.	.30	.44		.74	1.05
0900	1-1/2" supply		1800	.013	"	.35	.49		.84	1.19

32 91 Planting Preparation

32 91 13 – Soil Preparation

32 91 13.23 Structural Soil Mixing		Crew	Daily Output	Labor-Hours	Unit	Material	2010 Bare Costs Labor	Equipment	Total	Total Incl O&P
0010	**STRUCTURAL SOIL MIXING**									
0100	Rake topsoil, site material, harley rock rake, ideal	B-6	33	.727	M.S.F.		26	10.25	36.25	53.50
0200	Adverse	"	7	3.429			123	48.50	171.50	251
0300	Screened loam, york rake and finish, ideal	B-62	24	1			36	6.75	42.75	65.50
0400	Adverse	"	20	1.200			43	8.10	51.10	78.50
1000	Remove topsoil & stock pile on site, 75 HP dozer, 6" deep, 50' haul	B-10L	30	.400			15.85	14.55	30.40	41
1050	300' haul		6.10	1.967			78	71.50	149.50	203
1100	12" deep, 50' haul		15.50	.774			30.50	28	58.50	80
1150	300' haul		3.10	3.871			154	141	295	400
1200	200 HP dozer, 6" deep, 50' haul	B-10B	125	.096			3.81	9.55	13.36	16.55
1250	300' haul		30.70	.391			15.50	39	54.50	67
1300	12" deep, 50' haul		62	.194			7.70	19.25	26.95	33
1350	300' haul		15.40	.779			31	77.50	108.50	134
1400	Alternate method, 75 HP dozer, 50' haul	B-10L	860	.014	C.Y.		.55	.51	1.06	1.44
1450	300' haul	"	114	.105			4.18	3.83	8.01	10.85
1500	200 HP dozer, 50' haul	B-10B	2660	.005			.18	.45	.63	.77
1600	300' haul	"	570	.021			.84	2.09	2.93	3.63
1800	Rolling topsoil, hand push roller	1 Clab	3200	.003	S.F.		.08		.08	.14
1850	Tractor drawn roller	B-66	10666	.001	"		.03	.02	.05	.07
2000	Root raking and loading, residential, no boulders	B-6	53.30	.450	M.S.F.		16.15	6.35	22.50	33
2100	With boulders		32	.750			27	10.55	37.55	55
2200	Municipal, no boulders		200	.120			4.30	1.69	5.99	8.80
2300	With boulders		120	.200			7.15	2.82	9.97	14.65
2400	Large commercial, no boulders	B-10B	400	.030			1.19	2.98	4.17	5.15
2500	With boulders	"	240	.050			1.98	4.97	6.95	8.60
3000	Scarify subsoil, residential, skid steer loader w/scarifiers, 50 HP	B-66	32	.250			10.35	6.85	17.20	23.50
3050	Municipal, skid steer loader w/scarifiers, 50 HP	"	120	.067			2.75	1.82	4.57	6.30
3100	Large commercial, 75 HP, dozer w/scarifier	B-10L	240	.050			1.98	1.82	3.80	5.15
3500	Screen topsoil from stockpile, vibrating screen, wet material (organic)	B-10P	200	.060	C.Y.		2.38	5.40	7.78	9.70
3550	Dry material	"	300	.040			1.59	3.59	5.18	6.45
3600	Mixing with conditioners, manure and peat	B-10R	550	.022			.87	.49	1.36	1.91
3650	Mobilization add for 2 days or less operation	B-34K	3	2.667	Job		88.50	305	393.50	485
3800	Spread conditioned topsoil, 6" deep, by hand	B-1	360	.067	S.Y.	5.10	2.25		7.35	9.30
3850	300 HP dozer	B-10M	27	.444	M.S.F.	550	17.65	59	626.65	700
4000	Spread soil conditioners, alum. sulfate, 1#/S.Y., hand push spreader	1 Clab	17500	.001	S.Y.	16.75	.02		16.77	18.45
4050	Tractor spreader	B-66	700	.011	M.S.F.	1,850	.47	.31	1,850.78	2,050
4100	Fertilizer, 0.2#/S.Y., push spreader	1 Clab	17500	.001	S.Y.	.09	.02		.11	.13
4150	Tractor spreader	B-66	700	.011	M.S.F.	10	.47	.31	10.78	12.10
4200	Ground limestone, 1#/S.Y., push spreader	1 Clab	17500	.001	S.Y.	.10	.02		.12	.14
4250	Tractor spreader	B-66	700	.011	M.S.F.	11.10	.47	.31	11.88	13.30
4300	Lusoil, 3#/S.Y., push spreader	1 Clab	17500	.001	S.Y.	.53	.02		.55	.61
4350	Tractor spreader	B-66	700	.011	M.S.F.	59	.47	.31	59.78	66
4400	Manure, 18#/S.Y., push spreader	1 Clab	2500	.003	S.Y.	3.06	.11		3.17	3.54
4450	Tractor spreader	B-66	280	.029	M.S.F.	340	1.18	.78	341.96	380
4500	Perlite, 1" deep, push spreader	1 Clab	17500	.001	S.Y.	9.40	.02		9.42	10.40
4550	Tractor spreader	B-66	700	.011	M.S.F.	1,050	.47	.31	1,050.78	1,150
4600	Vermiculite, push spreader	1 Clab	17500	.001	S.Y.	2.95	.02		2.97	3.28
4650	Tractor spreader	B-66	700	.011	M.S.F.	330	.47	.31	330.78	360
5000	Spread topsoil, skid steer loader and hand dress	B-62	270	.089	C.Y.	23	3.18	.60	26.78	31.50
5100	Articulated loader and hand dress	B-100	320	.038		23	1.49	2.62	27.11	31
5200	Articulated loader and 75HP dozer	B-10M	500	.024		23	.95	3.18	27.13	30.50
5300	Road grader and hand dress	B-11L	1000	.016		23	.61	.62	24.23	27
6000	Tilling topsoil, 20 HP tractor, disk harrow, 2" deep	B-66	450	.018	M.S.F.		.73	.49	1.22	1.68

32 91 Planting Preparation

32 91 13 – Soil Preparation

32 91 13.23 Structural Soil Mixing		Crew	Daily Output	Labor-Hours	Unit	Material	2010 Bare Costs Labor	Equipment	Total	Total Incl O&P
6050	4" deep	B-66	360	.022	M.S.F.		.92	.61	1.53	2.11
6100	6" deep	↓	270	.030	↓		1.22	.81	2.03	2.81
6150	26" rototiller, 2" deep	A-1J	1250	.006	S.Y.		.21	.04	.25	.40
6200	4" deep		1000	.008			.26	.06	.32	.49
6250	6" deep	↓	750	.011	↓		.35	.07	.42	.66

32 91 13.26 Planting Beds

0010	**PLANTING BEDS**									
0100	Backfill planting pit, by hand, on site topsoil	2 Clab	18	.889	C.Y.		29.50		29.50	48.50
0200	Prepared planting mix, by hand	"	24	.667			22		22	36
0300	Skid steer loader, on site topsoil	B-62	340	.071			2.53	.48	3.01	4.60
0400	Prepared planting mix	"	410	.059			2.10	.39	2.49	3.81
1000	Excavate planting pit, by hand, sandy soil	2 Clab	16	1			33		33	54.50
1100	Heavy soil or clay	"	8	2			66		66	109
1200	1/2 C.Y. backhoe, sandy soil	B-11C	150	.107			4.06	2.25	6.31	9
1300	Heavy soil or clay	"	115	.139			5.30	2.94	8.24	11.70
2000	Mix planting soil, incl. loam, manure, peat, by hand	2 Clab	60	.267		40.50	8.85		49.35	59
2100	Skid steer loader	B-62	150	.160	↓	40.50	5.75	1.08	47.33	55
3000	Pile sod, skid steer loader	"	2800	.009	S.Y.		.31	.06	.37	.56
3100	By hand	2 Clab	400	.040			1.32		1.32	2.17
4000	Remove sod, F.E. loader	B-10S	2000	.006			.24	.18	.42	.58
4100	Sod cutter	B-12K	3200	.005			.19	.30	.49	.64
4200	By hand	2 Clab	240	.067	↓		2.21		2.21	3.62

32 91 19 – Landscape Grading

32 91 19.13 Topsoil Placement and Grading

0010	**TOPSOIL PLACEMENT AND GRADING**									
0700	Furnish and place, truck dumped, screened, 4" deep	B-10S	1300	.009	S.Y.	3.75	.37	.28	4.40	5
0800	6" deep		820	.015	"	4.80	.58	.45	5.83	6.70
0810	Minimum labor/equipment charge	↓	2	6	Ea.		238	184	422	580
0900	Fine grading and seeding, incl. lime, fertilizer & seed,									
1000	With equipment	B-14	1000	.048	S.Y.	.50	1.67	.34	2.51	3.64
2000	Minimum labor/equipment charge	1 Clab	4	2	Job		66		66	109

32 92 Turf and Grasses

32 92 19 – Seeding

32 92 19.13 Mechanical Seeding

0010	**MECHANICAL SEEDING**	R329219-50									
0020	Mechanical seeding, 215 lb./acre		B-66	1.50	5.333	Acre	565	220	146	931	1,125
0100	44 lb./M.S.Y.		"	2500	.003	S.Y.	.18	.13	.09	.40	.51
0600	Limestone hand push spreader, 50 lbs. per M.S.F.		1 Clab	180	.044	M.S.F.	4.02	1.47		5.49	6.85
9000	Minimum labor/equipment charge		"	4	2	Job		66		66	109

32 92 23 – Sodding

32 92 23.10 Sodding Systems

0010	**SODDING SYSTEMS**									
0020	Sodding, 1" deep, bluegrass sod, on level ground, over 8 MSF	B-63	22	1.818	M.S.F.	246	63	7.35	316.35	380
0200	4 M.S.F.		17	2.353		273	81.50	9.50	364	445
0300	1000 S.F.		13.50	2.963		297	103	11.95	411.95	505
0500	Sloped ground, over 8 M.S.F.		6	6.667		246	232	27	505	675
0600	4 M.S.F.		5	8		273	278	32.50	583.50	785
0700	1000 S.F.		4	10		297	345	40.50	682.50	935
1000	Bent grass sod, on level ground, over 6 M.S.F.	↓	20	2	↓	525	69.50	8.10	602.60	700

32 92 Turf and Grasses

32 92 23 – Sodding

32 92 23.10 Sodding Systems		Crew	Daily Output	Labor-Hours	Unit	Material	2010 Bare Costs Labor	Equipment	Total	Total Incl O&P
1100	3 M.S.F.	B-63	18	2.222	M.S.F.	605	77	9	691	800
1200	Sodding 1000 S.F. or less		14	2.857		695	99.50	11.55	806.05	940
1500	Sloped ground, over 6 M.S.F.		15	2.667		525	92.50	10.75	628.25	740
1600	3 M.S.F.		13.50	2.963		605	103	11.95	719.95	845
1700	1000 S.F.	↓	12	3.333	↓	695	116	13.45	824.45	970

32 93 Plants

32 93 13 – Ground Covers

32 93 13.10 Ground Cover Plants

		Crew	Daily Output	Labor-Hours	Unit	Material	Labor	Equipment	Total	Total Incl O&P
0010	**GROUND COVER PLANTS**									
0012	Plants, pachysandra, in prepared beds	B-1	15	1.600	C	27	54		81	119
0200	Vinca minor, 1 yr, bare root, in prepared beds		12	2	"	40	67.50		107.50	155
0600	Stone chips, in 50 lb. bags, Georgia marble		520	.046	Bag	5.25	1.56		6.81	8.35
0700	Onyx gemstone		260	.092		22.50	3.12		25.62	30
0800	Quartz		260	.092	↓	8.50	3.12		11.62	14.45
0900	Pea gravel, truckload lots	↓	28	.857	Ton	27.50	29		56.50	77.50

32 93 33 – Shrubs

32 93 33.10 Shrubs and Trees

		Crew	Daily Output	Labor-Hours	Unit	Material	Labor	Equipment	Total	Total Incl O&P
0010	**SHRUBS AND TREES**									
0011	Evergreen, in prepared beds, B & B									
0100	Arborvitae pyramidal, 4'-5'	B-17	30	1.067	Ea.	48.50	37.50	24.50	110.50	141
0150	Globe, 12"-15"	B-1	96	.250		21	8.45		29.45	37
0300	Cedar, blue, 8'-10'	B-17	18	1.778		219	62.50	41	322.50	385
0500	Hemlock, canadian, 2-1/2'-3'	B-1	36	.667		33.50	22.50		56	74
0550	Holly, Savannah, 8' - 10' H		9.68	2.479		485	83.50		568.50	670
0600	Juniper, andorra, 18"-24"		80	.300		16.80	10.15		26.95	35
0620	Wiltoni, 15"-18"	↓	80	.300		21	10.15		31.15	39.50
0640	Skyrocket, 4-1/2'-5'	B-17	55	.582		56	20.50	13.45	89.95	109
0660	Blue pfitzer, 2'-2-1/2'	B-1	44	.545		28	18.40		46.40	61.50
0680	Ketleerie, 2-1/2'-3'		50	.480		34	16.20		50.20	63.50
0700	Pine, black, 2-1/2'-3'		50	.480		45.50	16.20		61.70	76.50
0720	Mugo, 18"-24"	↓	60	.400		36	13.50		49.50	61.50
0740	White, 4'-5'	B-17	75	.427		58.50	15	9.85	83.35	99
0800	Spruce, blue, 18"-24"	B-1	60	.400		42.50	13.50		56	69
0840	Norway, 4'-5'	B-17	75	.427		67	15	9.85	91.85	108
0900	Yew, denisforma, 12"-15"	B-1	60	.400		26	13.50		39.50	50.50
1000	Capitata, 18"-24"		30	.800		24	27		51	70.50
1100	Hicksi, 2'-2-1/2'	↓	30	.800	↓	31.50	27		58.50	79.50

32 93 33.20 Shrubs

		Crew	Daily Output	Labor-Hours	Unit	Material	Labor	Equipment	Total	Total Incl O&P
0010	**SHRUBS**									
0011	Broadleaf Evergreen, planted in prepared beds									
0100	Andromeda, 15"-18", container	B-1	96	.250	Ea.	26.50	8.45		34.95	43
0200	Azalea, 15" - 18", container		96	.250		33	8.45		41.45	50
0300	Barberry, 9"-12", container		130	.185		11.75	6.25		18	23
0400	Boxwood, 15"-18", B&B		96	.250		30	8.45		38.45	46.50
0500	Euonymus, emerald gaiety, 12" to 15", container		115	.209		20.50	7.05		27.55	34
0600	Holly, 15"-18", B & B		96	.250		18.65	8.45		27.10	34.50
0900	Mount laurel, 18" - 24", B & B		80	.300		56.50	10.15		66.65	78.50
1000	Paxistema, 9 – 12" high		130	.185		18.30	6.25		24.55	30.50
1100	Rhododendron, 18"-24", container	↓	48	.500		31.50	16.90		48.40	62

32 93 Plants

32 93 33 − Shrubs

32 93 33.20 Shrubs

		Crew	Daily Output	Labor-Hours	Unit	Material	2010 Bare Costs Labor	Equipment	Total	Total Incl O&P
1200	Rosemary, 1 gal container	B-1	600	.040	Ea.	62.50	1.35		63.85	71
2000	Deciduous, planted in prepared beds, amelanchier, 2'-3', B & B		57	.421		90.50	14.20		104.70	123
2100	Azalea, 15"-18", B & B		96	.250		24.50	8.45		32.95	41
2300	Bayberry, 2'-3', B & B		57	.421		28	14.20		42.20	54
2600	Cotoneaster, 15"-18", B & B		80	.300		16.75	10.15		26.90	35
2800	Dogwood, 3'-4', B & B	B-17	40	.800		26.50	28	18.50	73	95
2900	Euonymus, alatus compacta, 15" to 18", container	B-1	80	.300		22.50	10.15		32.65	41
3200	Forsythia, 2'-3', container	"	60	.400		20	13.50		33.50	44
3300	Hibiscus, 3'-4', B & B	B-17	75	.427		15.05	15	9.85	39.90	51.50
3400	Honeysuckle, 3'-4', B & B	B-1	60	.400		22	13.50		35.50	46.50
3500	Hydrangea, 2'-3', B & B	"	57	.421		26	14.20		40.20	52.50
3600	Lilac, 3'-4', B & B	B-17	40	.800		23	28	18.50	69.50	91
3900	Privet, bare root, 18"-24"	B-1	80	.300		12.55	10.15		22.70	30.50
4100	Quince, 2'-3', B & B	"	57	.421		24	14.20		38.20	50
4200	Russian olive, 3'-4', B & B	B-17	75	.427		24	15	9.85	48.85	61
4400	Spirea, 3'-4', B & B	B-1	70	.343		16	11.60		27.60	36.50
4500	Viburnum, 3'-4', B & B	B-17	40	.800		29.50	28	18.50	76	98.50

32 93 43 − Trees

32 93 43.20 Trees

			Crew	Daily Output	Labor-Hours	Unit	Material	Labor	Equipment	Total	Total Incl O&P
0010	**TREES**										
0011	Deciduous, in prep. beds, balled & burlapped (B&B)										
0100	Ash, 2" caliper	G	B-17	8	4	Ea.	115	141	92.50	348.50	455
0200	Beech, 5'-6'	G		50	.640		223	22.50	14.80	260.30	299
0300	Birch, 6'-8', 3 stems	G		20	1.600		114	56.50	37	207.50	257
0500	Crabapple, 6'-8'	G		20	1.600		160	56.50	37	253.50	310
0600	Dogwood, 4'-5'	G		40	.800		63.50	28	18.50	110	136
0700	Eastern redbud 4'-5'	G		40	.800		157	28	18.50	203.50	239
0800	Elm, 8'-10'	G		20	1.600		119	56.50	37	212.50	263
0900	Ginkgo, 6'-7'	G		24	1.333		163	47	31	241	290
1000	Hawthorn, 8'-10', 1" caliper	G		20	1.600		135	56.50	37	228.50	280
1100	Honeylocust, 10'-12', 1-1/2" caliper	G		10	3.200		145	113	74	332	425
1300	Larch, 8'	G		32	1		105	35	23	163	199
1400	Linden, 8'-10', 1" caliper	G		20	1.600		113	56.50	37	206.50	256
1500	Magnolia, 4'-5'	G		20	1.600		71.50	56.50	37	165	210
1600	Maple, red, 8'-10', 1-1/2" caliper	G		10	3.200		241	113	74	428	530
1700	Mountain ash, 8'-10', 1" caliper	G		16	2		182	70.50	46	298.50	365
1800	Oak, 2-1/2"-3" caliper	G		6	5.333		269	188	123	580	735
2100	Planetree, 9'-11', 1-1/4" caliper	G		10	3.200		105	113	74	292	380
2200	Plum, 6'-8', 1" caliper	G		20	1.600		95.50	56.50	37	189	237
2300	Poplar, 9'-11', 1-1/4" caliper	G		10	3.200		90	113	74	277	365
2500	Sumac, 2'-3'	G		75	.427		23	15	9.85	47.85	60
2700	Tulip, 5'-6'	G		40	.800		53	28	18.50	99.50	124
2800	Willow, 6'-8', 1" caliper	G		20	1.600		68.50	56.50	37	162	207
9000	Minimum labor/equipment charge		1 Clab	4	2	Job		66		66	109

32 94 Planting Accessories

32 94 13 – Landscape Edging

32 94 13.20 Edging		Crew	Daily Output	Labor-Hours	Unit	Material	2010 Bare Costs Labor	Equipment	Total	Total Incl O&P
0010	**EDGING**									
0050	Aluminum alloy, including stakes, 1/8" x 4", mill finish	B-1	390	.062	L.F.	3.15	2.08		5.23	6.90
0051	Black paint		390	.062		3.65	2.08		5.73	7.45
0052	Black anodized	↓	390	.062		4.22	2.08		6.30	8.05
0100	Brick, set horizontally, 1-1/2 bricks per L.F.	D-1	370	.043		1.40	1.63		3.03	4.16
0150	Set vertically, 3 bricks per L.F.	"	135	.119		4.24	4.47		8.71	11.85
0200	Corrugated aluminum, roll, 4" wide	1 Carp	650	.012		.72	.51		1.23	1.63
0250	6" wide	"	550	.015		.90	.60		1.50	1.98
0600	Railroad ties, 6" x 8"	2 Carp	170	.094		3.15	3.91		7.06	9.85
0650	7" x 9"	↓	136	.118		3.50	4.89		8.39	11.90
0750	Redwood 2" x 4"	↓	330	.048		2.26	2.01		4.27	5.80
0800	Steel edge strips, incl. stakes, 1/4" x 5"	B-1	390	.062		4.20	2.08		6.28	8.05
0850	3/16" x 4"	"	390	.062	↓	3.32	2.08		5.40	7.05
9000	Minimum labor/equipment charge	1 Carp	4	2	Job		83		83	137

32 94 50 – Tree Guying

32 94 50.10 Tree Guying Systems

		Crew	Daily Output	Labor-Hours	Unit	Material	Labor	Equipment	Total	Total Incl O&P
0010	**TREE GUYING SYSTEMS**									
0015	Tree guying Including stakes, guy wire and wrap									
0100	Less than 3" caliper, 2 stakes	2 Clab	35	.457	Ea.	17.55	15.15		32.70	44.50
0200	3" to 4" caliper, 3 stakes	"	21	.762	"	20	25		45	63.50
1000	Including arrowhead anchor, cable, turnbuckles and wrap									
1100	Less than 3" caliper, 3" anchors	2 Clab	20	.800	Ea.	51	26.50		77.50	99.50
1200	3" to 6" caliper, 4" anchors		15	1.067		51	35.50		86.50	114
1300	6" caliper, 6" anchors		12	1.333		95.50	44		139.50	178
1400	8" caliper, 8" anchors	↓	9	1.778	↓	107	59		166	215

32 96 Transplanting

32 96 23 – Plant and Bulb Transplanting

32 96 23.23 Planting

		Crew	Daily Output	Labor-Hours	Unit	Material	Labor	Equipment	Total	Total Incl O&P
0010	**PLANTING**									
0012	Moving shrubs on site, 12" ball	B-62	28	.857	Ea.		30.50	5.75	36.25	56
0100	24" ball	"	22	1.091	"		39	7.35	46.35	71

32 96 23.43 Moving Trees

0290	**MOVING TREES**, On site									
0300	Moving trees on site, 36" ball	B-6	3.75	6.400	Ea.		229	90	319	470
0400	60" ball	"	1	24	"		860	340	1,200	1,750

32 96 43 – Tree Transplanting

32 96 43.20 Tree Removal

		Crew	Daily Output	Labor-Hours	Unit	Material	Labor	Equipment	Total	Total Incl O&P
0010	**TREE REMOVAL**									
0100	Dig & lace, shrubs, broadleaf evergreen, 18"-24" high	B-1	55	.436	Ea.		14.75		14.75	24
0200	2'-3'	"	35	.686			23		23	38
0300	3'-4'	B-6	30	.800			28.50	11.25	39.75	58.50
0400	4'-5'	"	20	1.200			43	16.90	59.90	88
1000	Deciduous, 12"-15"	B-1	110	.218			7.35		7.35	12.10
1100	18"-24"		65	.369			12.45		12.45	20.50
1200	2'-3'	↓	55	.436			14.75		14.75	24
1300	3'-4'	B-6	50	.480			17.20	6.75	23.95	35
2000	Evergreen, 18"-24"	B-1	55	.436			14.75		14.75	24
2100	2'-0" to 2'-6"		50	.480			16.20		16.20	26.50
2200	2'-6" to 3'-0"	↓	35	.686	↓		23		23	38

32 96 Transplanting

32 96 43 – Tree Transplanting

32 96 43.20 Tree Removal		Crew	Daily Output	Labor-Hours	Unit	Material	2010 Bare Costs Labor	Equipment	Total	Total Incl O&P
2300	3'-0" to 3'-6"	B-1	20	1.200	Ea.		40.50		40.50	66.50
3000	Trees, deciduous, small, 2'-3'	▼	55	.436			14.75		14.75	24
3100	3'-4'	B-6	50	.480			17.20	6.75	23.95	35
3200	4'-5'		35	.686			24.50	9.65	34.15	50
3300	5'-6'		30	.800			28.50	11.25	39.75	58.50
4000	Shade, 5'-6'		50	.480			17.20	6.75	23.95	35
4100	6'-8'		35	.686			24.50	9.65	34.15	50
4200	8'-10'		25	.960			34.50	13.50	48	70.50
4300	2" caliper		12	2			71.50	28	99.50	147
5000	Evergreen, 4'-5'		35	.686			24.50	9.65	34.15	50
5100	5'-6'		25	.960			34.50	13.50	48	70.50
5200	6'-7'		19	1.263			45.50	17.80	63.30	92.50
5300	7'-8'		15	1.600			57.50	22.50	80	118
5400	8'-10'	▼	11	2.182	▼		78	30.50	108.50	160

Estimating Tips

33 10 00 Water Utilities
33 30 00 Sanitary Sewerage Utilities
33 40 00 Storm Drainage Utilities

- Never assume that the water, sewer, and drainage lines will go in at the early stages of the project. Consider the site access needs before dividing the site in half with open trenches, loose pipe, and machinery obstructions. Always inspect the site to establish that the site drawings are complete. Check off all existing utilities on your drawings as you locate them. Be especially careful with underground because appurtenances are sometimes buried during regrading or repaving operations. If you find any discrepancies, mark up the site plan for further research. Differing site conditions can be very costly if discovered later in the project.
- See also Section 33 01 00 for restoration of pipe where removal/replacement may be undesirable. Use of new types of piping materials can reduce the overall project cost. Owners/design engineers should consider the installing contractor as a valuable source of current information on utility products and local conditions that could lead to significant cost savings.

Reference Numbers

Reference numbers are shown in shaded boxes at the beginning of some major classifications. These numbers refer to related items in the Reference Section. The reference information may be an estimating procedure, an alternate pricing method, or technical information.

Note: Not all subdivisions listed here necessarily appear in this publication.

Note: **Trade Service,** *in part, has been used as a reference source for some of the material prices used in Division 33.*

33 01 Operation and Maintenance of Utilities

33 01 30 – Operation and Maintenance of Sewer Utilities

33 01 30.72 Relining Sewers		Crew	Daily Output	Labor-Hours	Unit	Material	2010 Bare Costs Labor	Equipment	Total	Total Incl O&P
0010	**RELINING SEWERS**									
0011	With cement incl. bypass & cleaning									
0020	Less than 10,000 L.F., urban, 6" to 10"	C-17E	130	.615	L.F.	8.75	26.50	.67	35.92	53.50
0050	10" to 12"		125	.640		10.75	27.50	.70	38.95	57
0070	12" to 16"		115	.696		11.05	30	.76	41.81	61.50
0100	16" to 20"		95	.842		12.95	36	.92	49.87	74
0200	24" to 36"		90	.889		13.95	38	.97	52.92	78.50
0300	48" to 72"		80	1		22	43	1.09	66.09	95
0500	Rural, 6" to 10"		180	.444		8.75	19.10	.48	28.33	41
0550	10" to 12"		175	.457		10.75	19.65	.50	30.90	44.50
0570	12" to 16"		160	.500		11.20	21.50	.55	33.25	48
0600	16" to 20"		135	.593		11.85	25.50	.65	38	55
0700	24" to 36"		125	.640		14.15	27.50	.70	42.35	61
0800	48" to 72"		100	.800		22	34.50	.87	57.37	81
1000	Greater than 10,000 L.F., urban, 6" to 10"		160	.500		8.75	21.50	.55	30.80	45
1050	10" to 12"		155	.516		10.60	22	.56	33.16	48.50
1070	12" to 16"		140	.571		11.05	24.50	.62	36.17	52.50
1100	16" to 20"		120	.667		11.85	28.50	.73	41.08	60.50
1200	24" to 36"		115	.696		14.15	30	.76	44.91	65
1300	48" to 72"		95	.842		22	36	.92	58.92	84
1500	Rural, 6" to 10"		215	.372		8.75	16	.41	25.16	36
1550	10" to 12"		210	.381		10.75	16.40	.42	27.57	39
1570	12" to 16"		185	.432		11.05	18.60	.47	30.12	42.50
1600	16" to 20"		150	.533		11.85	23	.58	35.43	50.50
1700	24" to 36"		140	.571		14.15	24.50	.62	39.27	55.50
1800	48" to 72"		120	.667		22.50	28.50	.73	51.73	72

33 05 Common Work Results for Utilities

33 05 16 – Utility Structures

33 05 16.13 Utility Boxes

		Crew	Daily Output	Labor-Hours	Unit	Material	Labor	Equipment	Total	Total Incl O&P
0010	**UTILITY BOXES** Precast concrete, 6" thick									
0050	5' x 10' x 6' high, I.D.	B-13	2	28	Ea.	2,300	1,000	375	3,675	4,550
0350	Hand hole, precast concrete, 1-1/2" thick									
0400	1'-0" x 2'-0" x 1'-9", I.D., light duty	B-1	4	6	Ea.	390	203		593	765
0450	4'-6" x 3'-2" x 2'-0", O.D., heavy duty	B-6	3	8	"	1,200	287	113	1,600	1,900

33 05 23 – Trenchless Utility Installation

33 05 23.19 Microtunneling

		Crew	Daily Output	Labor-Hours	Unit	Material	Labor	Equipment	Total	Total Incl O&P
0010	**MICROTUNNELING**									
0011	Not including excavation, backfill, shoring,									
0020	or dewatering, average 50'/day, slurry method									
0100	24" to 48" outside diameter, minimum				L.F.				800	880
0110	Adverse conditions, add				%				50%	50%
1000	Rent microtunneling machine, average monthly lease				Month				90,000	99,000
1010	Operating technician				Day				560	620
1100	Mobilization and demobilization, minimum				Job				40,000	44,000
1110	Maximum				"				410,000	451,000

33 05 23.20 Horizontal Boring

0010	**HORIZONTAL BORING**									
0011	Casing only, 100' minimum,									
0020	not incl. jacking pits or dewatering									

33 05 Common Work Results for Utilities

33 05 23 – Trenchless Utility Installation

33 05 23.20 Horizontal Boring

		Crew	Daily Output	Labor-Hours	Unit	Material	2010 Bare Costs Labor	Equipment	Total	Total Incl O&P
0100	Roadwork, 1/2" thick wall, 24" diameter casing	B-42	20	3.200	L.F.	117	119	68.50	304.50	400
0200	36" diameter		16	4		186	149	85.50	420.50	545
0300	48" diameter		15	4.267		273	158	91	522	660
0500	Railroad work, 24" diameter		15	4.267		117	158	91	366	490
0600	36" diameter		14	4.571		186	170	97.50	453.50	590
0700	48" diameter		12	5.333		273	198	114	585	750
0900	For ledge, add									20%
1000	Small diameter boring, 3", sandy soil	B-82	900	.018		28	.66	.09	28.75	32
1040	Rocky soil	"	500	.032		28	1.19	.16	29.35	33

33 05 26 – Utility Line Signs, Markers, and Flags

33 05 26.10 Utility Accessories

			Crew	Daily Output	Labor-Hours	Unit	Material	Labor	Equipment	Total	Total Incl O&P
0010	**UTILITY ACCESSORIES**	R312316-40									
0400	Underground tape, detectable, reinforced, alum. foil core, 2"		1 Clab	150	.053	C.L.F.	2.36	1.77		4.13	5.50
0500	6"			140	.057	"	7.20	1.89		9.09	11
9000	Minimum labor/equipment charge			4	2	Job		66		66	109

33 11 Water Utility Distribution Piping

33 11 13 – Public Water Utility Distribution Piping

33 11 13.15 Water Supply, Ductile Iron Pipe

			Crew	Daily Output	Labor-Hours	Unit	Material	Labor	Equipment	Total	Total Incl O&P
0010	**WATER SUPPLY, DUCTILE IRON PIPE**	R331113-80									
0020	Not including excavation or backfill										
2000	Pipe, class 50 water piping, 18' lengths										
2020	Mechanical joint, 4" diameter		B-21A	200	.200	L.F.	16.55	8.25	3.28	28.08	35
3000	Tyton, push-on joint, 4" diameter			400	.100		12.40	4.13	1.64	18.17	22
3020	6" diameter			333.33	.120		15.85	4.95	1.97	22.77	27.50
8000	Fittings, mechanical joint										
8006	90° bend, 4" diameter		B-20A	16	2	Ea.	230	81		311	380
8020	6" diameter			12.80	2.500		315	101		416	505
8200	Wye or tee, 4" diameter			10.67	2.999		350	121		471	580
8220	6" diameter			8.53	3.751		465	152		617	755
8398	45° bends, 4" diameter			16	2		180	81		261	325
8450	Decreaser, 6" x 4" diameter			14.22	2.250		193	91		284	360
8460	8" x 6" diameter			11.64	2.749		247	111		358	450
8550	Piping, butterfly valves, cast iron										
8560	4" diameter		B-20	6	4	Ea.	900	148		1,048	1,225
9600	Steel sleeve with tap, 4" diameter			3	8		415	295		710	935
9620	6" diameter			2	12		450	445		895	1,225

33 12 Water Utility Distribution Equipment

33 12 13 – Water Service Connections

33 12 13.15 Tapping, Crosses and Sleeves

		Crew	Daily Output	Labor-Hours	Unit	Material	Labor	Equipment	Total	Total Incl O&P
0010	**TAPPING, CROSSES AND SLEEVES**									
4000	Drill and tap pressurized main (labor only)									
4100	6" main, 1" to 2" service	Q-1	3	5.333	Ea.		250		250	385
4150	8" main, 1" to 2" service	"	2.75	5.818	"		273		273	420
4500	Tap and insert gate valve									
4600	8" main, 4" branch	B-21	3.20	8.750	Ea.	335	43		378	590
4650	6" branch		2.70	10.370		395	51		446	695
4800	12" main, 6" branch		2.35	11.915		455	58.50		513.50	800

33 21 Water Supply Wells

33 21 13 – Public Water Supply Wells

33 21 13.10 Wells and Accessories		Crew	Daily Output	Labor-Hours	Unit	Material	2010 Bare Costs Labor	Equipment	Total	Total Incl O&P
0010	**WELLS & ACCESSORIES** R221113-50									
0011	Domestic									
0100	Drilled, 4" to 6" diameter	B-23	120	.333	L.F.		11.15	24	35.15	45
0200	8" diameter	"	95.20	.420	"		14.10	30.50	44.60	56.50
1400	Remove & reset pump, minimum	B-21	4	7	Ea.		266	34.50	300.50	470
1420	Maximum	"	2	14	"		530	69	599	940
1500	Pumps, installed in wells to 100' deep, 4" submersible									
1510	1/2 H.P.	Q-1	3.22	4.969	Ea.	455	233		688	860
1520	3/4 H.P.		2.66	6.015		620	282		902	1,125
1600	1 H.P.		2.29	6.987		840	325		1,165	1,425
1700	1-1/2 H.P.	Q-22	1.60	10		1,050	470	410	1,930	2,350
1800	2 H.P.		1.33	12.030		1,275	565	495	2,335	2,825
1900	3 H.P.		1.14	14.035		1,725	660	575	2,960	3,550
2000	5 H.P.		1.14	14.035		2,350	660	575	3,585	4,250
2050	Remove and install motor only, 4 H.P.		1.14	14.035		2,450	660	575	3,685	4,350
5000	Wells to 180 ft. deep, 4" submersible, 1 HP	B-21	1.10	25.455		2,025	965	125	3,115	3,950
5500	2 HP		1.10	25.455		2,900	965	125	3,990	4,925
6000	3 HP		1	28		3,900	1,075	138	5,113	6,175
7000	5 HP		.90	31.111		5,400	1,175	153	6,728	8,050
9000	Minimum labor/equipment charge		1.80	15.556	Job		590	76.50	666.50	1,050

33 31 Sanitary Utility Sewerage Piping

33 31 13 – Public Sanitary Utility Sewerage Piping

33 31 13.15 Sewage Collection, Concrete Pipe

		Crew	Daily Output	Labor-Hours	Unit	Material	Labor	Equipment	Total	Total Incl O&P
0010	**SEWAGE COLLECTION, CONCRETE PIPE**									
0020	See Div. 33 41 13.60 for sewage/drainage collection, concrete pipe									

33 31 13.25 Sewage Collection, Polyvinyl Chloride Pipe

		Crew	Daily Output	Labor-Hours	Unit	Material	Labor	Equipment	Total	Total Incl O&P
0010	**SEWAGE COLLECTION, POLYVINYL CHLORIDE PIPE**									
0020	Not including excavation or backfill									
2000	20' lengths, S.D.R. 35, B&S, 4" diameter	B-20	375	.064	L.F.	1.68	2.36		4.04	5.70
2040	6" diameter		350	.069		3.64	2.53		6.17	8.15
2080	13' lengths, S.D.R. 35, B&S, 8" diameter		335	.072		7.85	2.65		10.50	12.95
2120	10" diameter	B-21	330	.085		11.90	3.22	.42	15.54	18.80
4000	Piping, DWV PVC, no exc/bkfill, 10' L, Sch 40, 4" diameter	B-20	375	.064		3.71	2.36		6.07	7.95
4010	6" diameter		350	.069		7.25	2.53		9.78	12.10
4020	8" diameter		335	.072		12.35	2.65		15	17.85

33 36 Utility Septic Tanks

33 36 13 – Utility Septic Tank and Effluent Wet Wells

33 36 13.10 Septic Tanks

		Crew	Daily Output	Labor-Hours	Unit	Material	Labor	Equipment	Total	Total Incl O&P
0010	**SEPTIC TANKS**									
0011	Not including excavation or piping									
0015	Septic tanks, precast, 1,000 gal	B-21	8	3.500	Ea.	750	133	17.25	900.25	1,050
0020	1,250 gallon		8	3.500		980	133	17.25	1,130.25	1,300
0060	1,500 gallon		7	4		1,175	152	19.70	1,346.70	1,550
0100	2,000 gallon		5	5.600		1,675	213	27.50	1,915.50	2,225
0140	2,500 gallon		5	5.600		2,375	213	27.50	2,615.50	2,975
0180	4,000 gallon		4	7		5,850	266	34.50	6,150.50	6,900
0220	5,000 gal., 4 piece	B-13	3	18.667		9,350	665	249	10,264	11,600

33 36 Utility Septic Tanks

33 36 13 – Utility Septic Tank and Effluent Wet Wells

33 36 13.10 Septic Tanks

		Crew	Daily Output	Labor-Hours	Unit	Material	2010 Bare Costs Labor	Equipment	Total	Total Incl O&P
0300	15,000 gallon, 4 piece	B-13B	1.70	32.941	Ea.	21,100	1,175	710	22,985	25,900
0400	25,000 gallon, 4 piece		1.10	50.909		44,700	1,825	1,100	47,625	53,500
0500	40,000 gallon, 4 piece		.80	70		52,500	2,500	1,500	56,500	63,500
0520	50,000 gallon, 5 piece	B-13C	.60	93.333		60,500	3,325	2,875	66,700	75,000
0540	75,000 gallon, cast in place	C-14C	.25	448		73,500	17,900	121	91,521	110,500
0560	100,000 gallon	"	.15	746		91,000	29,800	202	121,002	149,500
0600	High density polyethylene, 1,000 gallon	B-21	6	4.667		985	177	23	1,185	1,400
0700	1,500 gallon		4	7		1,375	266	34.50	1,675.50	2,000
0900	Galley, 4' x 4' x 4'		16	1.750		275	66.50	8.65	350.15	425
1000	Distribution boxes, concrete, 7 outlets	2 Clab	16	1		80	33		113	143
1100	9 outlets	"	8	2		445	66		511	600
1150	Leaching field chambers, 13' x 3'-7" x 1'-4", standard	B-13	16	3.500		525	125	46.50	696.50	835
1200	Heavy duty, 8' x 4' x 1'-6"		14	4		275	143	53.50	471.50	595
1300	13' x 3'-9" x 1'-6"		12	4.667		1,025	167	62.50	1,254.50	1,475
1350	20' x 4' x 1'-6"		5	11.200		1,075	400	150	1,625	2,000
1420	Leaching pit, 6', dia, 3' deep complete					1,050			1,050	1,150
1600	Leaching pit, 6'-6" diameter, 6' deep	B-21	5	5.600		1,350	213	27.50	1,590.50	1,850
1620	8' deep		4	7		1,575	266	34.50	1,875.50	2,200
1700	8' diameter, H-20 load, 6' deep		4	7		2,225	266	34.50	2,525.50	2,925
1720	8' deep		3	9.333		2,625	355	46	3,026	3,525
2000	Velocity reducing pit, precast conc., 6' diameter, 3' deep		4.70	5.957		1,500	226	29.50	1,755.50	2,050
2200	Excavation for septic tank, 3/4 C.Y. backhoe	B-12F	145	.110	C.Y.		4.28	4.67	8.95	12
2400	4' trench for disposal field, 3/4 C.Y. backhoe	"	335	.048	L.F.		1.85	2.02	3.87	5.20
2600	Gravel fill, run of bank	B-6	150	.160	C.Y.	20	5.75	2.25	28	33.50
2800	Crushed stone, 3/4"	"	150	.160	"	31	5.75	2.25	39	45.50
3000	Effluent filter, 4" dia.	B-21	4.70	5.957	Ea.	44.50	226	29.50	300	445

33 41 Storm Utility Drainage Piping

33 41 13 – Public Storm Utility Drainage Piping

33 41 13.40 Piping, Storm Drainage, Corrugated Metal

			Crew	Daily Output	Labor-Hours	Unit	Material	Labor	Equipment	Total	Total Incl O&P
0010	**PIPING, STORM DRAINAGE, CORRUGATED METAL**	R221113-50									
0020	Not including excavation or backfill										
2000	Corrugated metal pipe, galvanized										
2020	Bituminous coated with paved invert, 20' lengths										
2040	8" diameter, 16 ga.		B-14	330	.145	L.F.	12.30	5.05	1.02	18.37	23
2080	12" diameter, 16 ga.		"	210	.229	"	17.65	7.95	1.61	27.21	34

33 41 13.60 Sewage/Drainage Collection, Concrete Pipe

		Crew	Daily Output	Labor-Hours	Unit	Material	Labor	Equipment	Total	Total Incl O&P
0010	**SEWAGE/DRAINAGE COLLECTION, CONCRETE PIPE**									
0020	Not including excavation or backfill									
1000	Non-reinforced pipe, extra strength, B&S or T&G joints									
1010	6" diameter	B-14	265.04	.181	L.F.	6.25	6.30	1.27	13.82	18.55
1020	8" diameter		224	.214		6.90	7.45	1.51	15.86	21.50
1040	12" diameter		200	.240		9.35	8.35	1.69	19.39	26
2000	Reinforced culvert, class 3, no gaskets									
2010	12" diameter	B-14	150	.320	L.F.	10.65	11.15	2.25	24.05	32.50
2020	15" diameter	"	150	.320	"	13.95	11.15	2.25	27.35	36

33 44 Storm Utility Water Drains

33 44 13 – Utility Area Drains

33 44 13.13 Catchbasins

		Crew	Daily Output	Labor-Hours	Unit	Material	2010 Bare Costs Labor	Equipment	Total	Total Incl O&P
0010	**CATCHBASINS**									
0011	Not including footing & excavation									
1600	Frames & covers, C.I., 24" square, 500 lb.	B-6	7.80	3.077	Ea.	300	110	43.50	453.50	555
1700	26" D shape, 600 lb.	"	7	3.429	"	515	123	48.50	686.50	820
3320	Frames and covers, existing, raised for paving, 2", including									
3340	row of brick, concrete collar, up to 12" wide frame	B-6	18	1.333	Ea.	44	48	18.75	110.75	146

33 44 13.15 Remove and Replace Catch Basin Cover

		Crew	Daily Output	Labor-Hours	Unit	Material	Labor	Equipment	Total	Total Incl O&P
0010	**REMOVE AND REPLACE CATCH BASIN COVER**									
0023	Remove catch basin cover	1 Clab	80	.100	Ea.		3.31		3.31	5.45
0033	Replace catch basin cover	"	80	.100	"		3.31		3.31	5.45

33 49 Storm Drainage Structures

33 49 13 – Storm Drainage Manholes, Frames, and Covers

33 49 13.10 Storm Drainage Manholes, Frames and Covers

		Crew	Daily Output	Labor-Hours	Unit	Material	Labor	Equipment	Total	Total Incl O&P
0010	**STORM DRAINAGE MANHOLES, FRAMES & COVERS**									
0020	Excludes footing, excavation, backfill (See line items for frame & cover)									
0050	Brick, 4' inside diameter, 4' deep	D-1	1	16	Ea.	440	605		1,045	1,450
0100	6' deep		.70	22.857		625	860		1,485	2,050
0150	8' deep		.50	32		800	1,200		2,000	2,825
0200	For depths over 8', add		4	4	V.L.F.	167	151		318	425
1110	Precast, 4' I.D., 4' deep	B-22	4.10	7.317	Ea.	450	281	50.50	781.50	1,000
1120	6' deep		3	10		800	385	69	1,254	1,575
1130	8' deep		2	15		925	575	104	1,604	2,075
1140	For depths over 8', add		16	1.875	V.L.F.	75	72	12.95	159.95	214

33 51 Natural-Gas Distribution

33 51 13 – Natural-Gas Piping

33 51 13.10 Piping, Gas Service and Distribution, P.E.

		Crew	Daily Output	Labor-Hours	Unit	Material	Labor	Equipment	Total	Total Incl O&P
0010	**PIPING, GAS SERVICE AND DISTRIBUTION, POLYETHYLENE**									
0020	Not including excavation or backfill									
1000	60 psi coils, compression coupling @ 100', 1/2" diameter, SDR 11	B-20A	608	.053	L.F.	1.66	2.13		3.79	5.20
1010	1" diameter, SDR 11		544	.059		1.83	2.38		4.21	5.80
1040	1-1/4" diameter, SDR 11		544	.059		2.57	2.38		4.95	6.60
1100	2" diameter, SDR 11		488	.066		3.18	2.65		5.83	7.70
1160	3" diameter, SDR 11		408	.078		6.65	3.17		9.82	12.35
1500	60 PSI 40' joints with coupling, 3" diameter, SDR 11	B-21A	408	.098		6.65	4.05	1.61	12.31	15.45
1540	4" diameter, SDR 11		352	.114		15.20	4.69	1.86	21.75	26
1600	6" diameter, SDR 11		328	.122		47.50	5.05	2	54.55	62
1640	8" diameter, SDR 11		272	.147		64.50	6.05	2.41	72.96	83.50
9000	Minimum labor/equipment charge	B-20	2	12	Job		445		445	725

33 52 Liquid Fuel Distribution
33 52 16 – Gasoline Distribution

33 52 16.13 Gasoline Piping		Crew	Daily Output	Labor-Hours	Unit	Material	2010 Bare Costs Labor	Equipment	Total	Total Incl O&P
0010	**GASOLINE PIPING**									
0020	Primary containment pipe, fiberglass-reinforced									
0030	Plastic pipe 15' & 30' lengths									
0040	2" diameter	Q-6	425	.056	L.F.	5.10	2.73		7.83	9.85
0050	3" diameter		400	.060		6.65	2.91		9.56	11.85
0060	4" diameter	↓	375	.064	↓	8.60	3.10		11.70	14.25
0100	Fittings									
0110	Elbows, 90° & 45°, bell-ends, 2"	Q-6	24	1	Ea.	51.50	48.50		100	132
0120	3" diameter		22	1.091		54	53		107	142
0130	4" diameter		20	1.200		71.50	58		129.50	169
0200	Tees, bell ends, 2"		21	1.143		62.50	55.50		118	155
0210	3" diameter		18	1.333		63	64.50		127.50	169
0220	4" diameter		15	1.600		86	77.50		163.50	215
0230	Flanges bell ends, 2"		24	1		20.50	48.50		69	97.50
0240	3" diameter		22	1.091		26	53		79	111
0250	4" diameter		20	1.200		35.50	58		93.50	129
0260	Sleeve couplings, 2"		21	1.143		13.15	55.50		68.65	100
0270	3" diameter		18	1.333		18.60	64.50		83.10	121
0280	4" diameter		15	1.600		25.50	77.50		103	149
0290	Threaded adapters 2"		21	1.143		17.25	55.50		72.75	105
0300	3" diameter		18	1.333		30.50	64.50		95	134
0310	4" diameter		15	1.600		41	77.50		118.50	165
0320	Reducers, 2"		27	.889		23	43		66	91.50
0330	3" diameter		22	1.091		26.50	53		79.50	111
0340	4" diameter	↓	20	1.200	↓	34.50	58		92.50	128
1010	Gas station product line for secondary containment (double wall)									
1100	Fiberglass reinforced plastic pipe 25' lengths									
1120	Pipe, plain end, 3" diameter	Q-6	375	.064	L.F.	8.75	3.10		11.85	14.40
1130	4" diameter		350	.069		14.35	3.32		17.67	21
1140	5" diameter		325	.074		18.60	3.58		22.18	26
1150	6" diameter	↓	300	.080	↓	19.15	3.87		23.02	27
1200	Fittings									
1230	Elbows, 90° & 45°, 3" diameter	Q-6	18	1.333	Ea.	63	64.50		127.50	169
1240	4" diameter		16	1.500		110	72.50		182.50	234
1250	5" diameter		14	1.714		254	83		337	410
1260	6" diameter		12	2		257	97		354	435
1270	Tees, 3" diameter		15	1.600		93	77.50		170.50	222
1280	4" diameter		12	2		137	97		234	300
1290	5" diameter		9	2.667		277	129		406	505
1300	6" diameter		6	4		290	194		484	620
1310	Couplings, 3" diameter		18	1.333		44	64.50		108.50	149
1320	4" diameter		16	1.500		114	72.50		186.50	238
1330	5" diameter		14	1.714		237	83		320	390
1340	6" diameter		12	2		245	97		342	420
1350	Cross-over nipples, 3" diameter		18	1.333		10.10	64.50		74.60	111
1360	4" diameter		16	1.500		11.85	72.50		84.35	126
1370	5" diameter		14	1.714		17.55	83		100.55	148
1380	6" diameter		12	2		18.40	97		115.40	171
1400	Telescoping, reducers, concentric 4" x 3"		18	1.333		33.50	64.50		98	137
1410	5" x 4"		17	1.412		87.50	68.50		156	203
1420	6" x 5"	↓	16	1.500		210	72.50		282.50	345

33 71 Electrical Utility Transmission and Distribution

33 71 19 – Electrical Underground Ducts and Manholes

33 71 19.17 Electric and Telephone Underground		Crew	Daily Output	Labor-Hours	Unit	Material	2010 Bare Costs Labor	Equipment	Total	Total Incl O&P
0010	**ELECTRIC AND TELEPHONE UNDERGROUND**									
0011	Not including excavation									
0200	backfill and cast in place concrete									
4200	Underground duct, banks ready for concrete fill, min. of 7.5"									
4400	between conduits, center to center									
4600	2 @ 2" diameter	2 Elec	240	.067	L.F.	1.62	3.27		4.89	6.80
4800	4 @ 2" diameter		120	.133		3.24	6.55		9.79	13.60
5600	4 @ 4" diameter		80	.200		6.50	9.80		16.30	22.50
6200	Rigid galvanized steel, 2 @ 2" diameter		180	.089		14.30	4.36		18.66	22.50
6400	4 @ 2" diameter		90	.178		28.50	8.70		37.20	45
7400	4 @ 4" diameter		34	.471		99	23		122	145
9990	Minimum labor/equipment charge	1 Elec	3.50	2.286	Job		112		112	172

Estimating Tips

34 11 00 Rail Tracks
This subdivision includes items that may involve either repair of existing, or construction of new, railroad tracks. Additional preparation work, such as the roadbed earthwork, would be found in Division 31. Additional new construction siding and turnouts are found in Subdivision 34 72. Maintenance of railroads is found under 34 01 23 Operation and Maintenance of Railways.

34 41 13 Traffic Signals
This subdivision includes traffic signal systems. Other traffic control devices such as traffic signs are found in Subdivision 10 14 53 Traffic Signage.

34 71 13 Vehicle Barriers
This subdivision includes security vehicle barriers, guide and guard rails, crash barriers, and delineators. The actual maintenance and construction of concrete and asphalt pavement is found in Division 32.

Reference Numbers
Reference numbers are shown in shaded boxes at the beginning of some major classifications. These numbers refer to related items in the Reference Section. The reference information may be an estimating procedure, an alternate pricing method, or technical information.

Note: Not all subdivisions listed here necessarily appear in this publication.

No part of this publication may be reproduced, stored in a retrieval system, or transmitted in any form or by any means without prior written permission of R.S. Means Company, Inc.

34 71 Roadway Construction

34 71 13 – Vehicle Barriers

34 71 13.26 Vehicle Guide Rails	Crew	Daily Output	Labor-Hours	Unit	Material	2010 Bare Costs Labor	Equipment	Total	Total Incl O&P
0010 **VEHICLE GUIDE RAILS**									
0012 Corrugated stl, galv. stl posts, 6'-3" O.C.	B-80	850	.038	L.F.	21.50	1.33	.88	23.71	27
0200 End sections, galvanized, flared		50	.640	Ea.	90	22.50	14.90	127.40	152
0300 Wrap around end		50	.640	"	135	22.50	14.90	172.40	202
0400 Timber guide rail, 4" x 8" with 6" x 8" wood posts, treated	↓	960	.033	L.F.	25	1.18	.78	26.96	30.50

Estimating Tips

35 20 16 Hydraulic Gates
This subdivision includes various types of gates that are commonly used in waterway and canal construction. Various earthwork items and structural support is found in Division 31, and concrete work in Division 3.

35 20 23 Dredging
This subdivision includes barge and shore dredging systems for rivers, canals, and channels.

35 31 00 Shoreline Protection
This subdivision includes breakwaters, bulkheads, and revetments for ocean and river inlets. Additional earthwork may be required from Division 31, and concrete work from Division 3.

35 41 00 Levees
Information on levee construction, including estimated cost of clay cone material.

35 49 00 Waterway Structures
This subdivision includes breakwaters and bulkheads for canals.

35 51 00 Floating Construction
This section includes floating piers, docks, and dock accessories. Fixed Pier Timber Construction is found in 06 13 33. Driven piles are found in Division 31, as well as sheet piling, cofferdams, and riprap.

Reference Numbers
Reference numbers are shown in shaded boxes at the beginning of some major classifications. These numbers refer to related items in the Reference Section. The reference information may be an estimating procedure, an alternate pricing method, or technical information.

Note: Not all subdivisions listed here necessarily appear in this publication.

No part of this publication may be reproduced, stored in a retrieval system, or transmitted in any form or by any means without prior written permission of R.S. Means Company, Inc.

41 22 Cranes and Hoists

41 22 13 – Cranes

41 22 13.10 Crane Rail		Crew	Daily Output	Labor-Hours	Unit	Material	2010 Bare Costs Labor	Equipment	Total	Total Incl O&P
0010	**CRANE RAIL**									
0020	Box beam bridge, no equipment included	E-4	3400	.009	Lb.	1.19	.45	.04	1.68	2.18
0210	Running track only, 104 lb per yard, 20' piece	"	160	.200	L.F.	21.50	9.50	.91	31.91	42.50
41 22 13.13 Bridge Cranes										
0010	**BRIDGE CRANES**									
0100	1 girder, 20' span, 3 ton	M-3	1	34	Ea.	23,800	1,650	156	25,606	28,900
0125	5 ton		1	34		26,000	1,650	156	27,806	31,300
0150	7.5 ton		1	34		31,000	1,650	156	32,806	36,800
0175	10 ton		.80	42.500		41,000	2,050	196	43,246	48,400
0200	15 ton		.80	42.500		52,500	2,050	196	54,746	61,500
0225	30' span, 3 ton		1	34		24,700	1,650	156	26,506	29,900
0250	5 ton		1	34		27,200	1,650	156	29,006	32,600
0275	7.5 ton		1	34		32,600	1,650	156	34,406	38,500
0300	10 ton		.80	42.500		42,400	2,050	196	44,646	50,000
0325	15 ton		.80	42.500		55,000	2,050	196	57,246	64,000
0350	2 girder, 40' span, 3 ton	M-4	.50	72		40,900	3,450	505	44,855	51,000
0375	5 ton		.50	72		42,800	3,450	505	46,755	53,000
0400	7.5 ton		.50	72		46,900	3,450	505	50,855	57,500
0425	10 ton		.40	90		55,000	4,325	635	59,960	68,000
0450	15 ton		.40	90		74,500	4,325	635	79,460	89,500
0475	25 ton		.30	120		88,000	5,750	845	94,595	107,000
0500	50' span, 3 ton		.50	72		46,500	3,450	505	50,455	57,000
0525	5 ton		.50	72		48,400	3,450	505	52,355	59,000
0550	7.5 ton		.50	72		52,000	3,450	505	55,955	63,000
0575	10 ton		.40	90		59,500	4,325	635	64,460	73,000
0600	15 ton		.40	90		78,000	4,325	635	82,960	93,500
0625	25 ton		.30	120		92,500	5,750	845	99,095	111,500

Assemblies Section

Table of Contents

Table No.	Page
A SUBSTRUCTURE	**493**
A1010 Standard Foundations	
A1010 100 Wall Foundations	494
A1010 200 Column Found. & Pile Caps	495
A1010 300 Perimeter Drainage	496
A1030 Slab on Grade	
A1030 100 Standard Slab on Grade	497
A2010 Basement Excavation	
A2010 100 Excavation for Basements	498
A2020 Basement Walls	
A2020 100 Basement Wall Construction	500
B SHELL	**503**
B1010 Floor Construction	
B1010 200 Upper Floor Construction	504
B1020 Roof Construction	
B1020 100 Flat Roof Construction	511
B1020 200 Pitched Roof Construction	512
B2010 Exterior Walls	
B2010 100 Exterior Wall Construction	515
B2020 Exterior Windows	
B2020 100 Windows	522
B2030 Exterior Doors	
B2030 100 Glazed Doors & Entrances	526
B2030 200 Solid Exterior Doors	528
B2030 400 Overhead Doors	529
B2030 800 Door Hardware	530
B3010 Roof Coverings	
B3010 100 Roof Finishes	532
B3020 Roof Openings	
B3020 200 Roof Hatches	534
C INTERIORS	**537**
C1010 Partitions	
C1010 100 Fixed Partitions	538
C1010 200 Demountable Partitions	547
C1020 Interior Doors	
C1020 100 Interior Doors	548
C2010 Stair Construction	
C2010 100 Regular Stairs	552

Table No.	Page
C3010 Wall Finishes	
C3010 200 Wall Finishes to Int. Walls	553
C3020 Floor Finishes	
C3020 400 Flooring	555
C3030 Ceiling Finishes	
C3030 200 Suspended Ceilings	557
D SERVICES	**563**
D1010 Elevators and Lifts	
D1010 100 Passenger Elevators	564
D2010 Plumbing Fixtures	
D2010 900 Bidets & Other Plumbing Fixt.	565
D3010 Energy Supply	
D3010 500 Hot Water Supply System	576
D3010 900 Other Energy Supply Systems	578
D3020 Heat Generating Systems	
D3020 100 Boilers	580
D3020 300 Auxiliary Equipment	584
D3040 Distribution Systems	
D3040 100 Air Distribution Systems	586
D3040 600 Changeover Dist. Systems	588
D3050 Terminal & Package Units	
D3050 100 Terminal Self-Contained Units	590
D3050 200 Package Units	594
D4010 Sprinklers	
D4010 300 Dry Sprinkler Systems	597
D4010 400 Wet Sprinkler Systems	599
D5010 Electrical Service/Dist.	
D5010 200 Low Tension Service & Dist.	600
D5020 Lighting & Branch Wiring	
D5020 100 Branch Wiring Devices	602
D5020 200 Lighting Equipment	604
D5090 Other Electrical Systems	
D5090 500 Gen. Const. Items (Electrical)	607
E EQUIPMENT & FURNISHINGS	**609**
E1090 Other Equipment	
E1090 300 Food Service Equipment	610

Table No.	Page
G BUILDING SITEWORK	**613**
G1030 Site Earthwork	
G1030 800 Utilities Trenching	614
G2010 Roadways	
G2010 200 Paving & Surfacing	615
G2020 Parking Lots	
G2020 200 Paving & Surfacing	616
G2030 Pedestrian Paving	
G2030 100 Paving & Surfacing	618
G2050 Landscaping	
G2050 400 Seeding & Sodding	619
G2050 500 Planting	620
G3020 Sanitary Sewer	
G3020 600 Septic Tanks	622
G3060 Fuel Distribution	
G3060 300 Fuel Storage Tanks	623
G4020 Site Lighting	
G4020 200 Lighting Poles	627

How to Use the Assemblies Cost Tables

The following is a detailed explanation of a sample Assemblies Cost Table. Most Assembly Tables are separated into three parts: 1) an illustration of the system to be estimated; 2) the components and related costs of a typical system; and 3) the costs for similar systems with dimensional and/or size variations. For costs of the components that comprise these systems or "assemblies," refer to the Unit Price Section. Next to each bold number below is the item being described with the appropriate component of the sample entry following in parenthesis. In most cases, if the work is to be subcontracted, the general contractor will need to add an additional markup (RSMeans suggests using 10%) to the "Total" figures.

System/Line Numbers (B1010 263 1700)

Each Assemblies Cost Line has been assigned a unique identification number based on the UNIFORMAT II classification system.

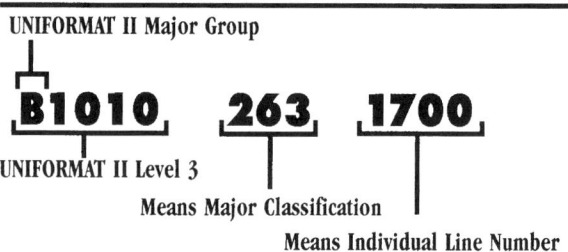

B10 Superstructure

B1010 Floor Construction **1**

This page illustrates and describes a wood joist floor system including wood joist, oak floor, sub-floor, bridging, sand and finish floor, furring, plasterboard, taped, finished and painted ceiling. Lines within System Components give the unit price and total price per square foot for this system. Prices for alternate wood joist floor systems are on Line Items B1010 263 1700 thru 2300. Both material quantities and labor costs have been adjusted for the system listed.

Factors: To adjust for job conditions other than normal working situations use Lines B1010 263 2700 thru 4000.

Example: You are to install the system during off peak hours, 6 P.M. to 2 A.M. Go to Line B1010 263 3800 and apply this percentage to the appropriate INST. costs.

2

6

System Components			COST PER S.F.		
	QUANTITY	UNIT	MAT.	INST.	TOTAL
Wood joists, 2" x 8", 16" O.C., oak floor (sanded & finished), 1/2" sub Floor, 1"x 3" bridging, furring, 5/8" drywall, taped, finished and painted. **3**					
Wood joists, 2" x 8", 16" O.C.	1.000	L.F.	.61	.99	1.60
Subfloor plywood CDX 1/2"	1.000	SF Flr.	.46	.73	1.19
Bridging 1" x 3"	.150	Pr.	.08	.63	.71
Oak flooring, No. 1 common	1.000	S.F.	4.21	3.21	7.42
Sand and finish floor	1.000	S.F.	.86	1.47	2.33
Wood furring, 1" x 3", 16" O.C.	1.000	S.F.	.38	1.56	1.94
Gypsum drywall, 5/8" thick	1.000	S.F.	.31	.61	.92
Tape and finishing	1.000	S.F.	.05	.55	.60
Paint ceiling	1.000	S.F.	.17	.61	.78
		4 **5**	**7**	**8** **9**	
		TOTAL	7.13	10.36	17.49

B1010 263	Floor-Ceiling, Wood Joists	COST PER S.F.		
		MAT.	INST.	TOTAL
1600	For alternate wood joist systems:			
1700	16" on center, 2" x 6" joists	6.95	10.25	17.20
1800	2" x 10"	7.30	10.60	17.90
1900	2" x 12"	7.70	10.60	18.30
2000	2" x 14"	8.70	10.80	19.50
2100	12" on center, 2" x 10" joists	7.50	10.90	18.40
2200	2" x 12"	8	10.95	18.95
2300	2" x 14"	9.25	11.15	20.40

Illustration
At the top of most assembly tables are an illustration, a brief description, and the design criteria used to develop the cost.

System Components
The components of a typical system are listed separately to show what has been included in the development of the total system price. The table below contains prices for other similar systems with dimensional and/or size variations.

Quantity
This is the number of line item units required for one system unit. For example, we assume that it will take .15 pair of 1" × 3" bridging on a square foot basis.

Unit of Measure for Each Item
The abbreviated designation indicates the unit of measure, as defined by industry standards, upon which the price of the component is based. For example, wood joists are priced by L.F. (linear foot) while plywood is priced by S.F. (square foot).

Unit of Measure for Each System (Cost per S.F.)
Costs shown in the three right-hand columns have been adjusted by the component quantity and unit of measure for the entire system. In this example, "Cost per S.F." is the unit of measure for this system, or assembly.

Materials (6.95)
This column contains the Materials Cost of each component. These cost figures are bare costs plus 10% for profit.

Installation (10.25)
Installation includes labor and equipment plus the installing contractor's overhead and profit. Equipment costs are the bare rental costs plus 10% for profit. The labor overhead and profit are defined on the inside back cover of this book.

Total (17.20)
The figure in this column is the sum of the material and installation costs.

Material Cost	+	Installation Cost	=	Total
$6.95	+	$10.25	=	$17.20

A Substructure

No part of this publication may be reproduced, stored in a retrieval system, or transmitted in any form or by any means without prior written permission of R.S. Means Company, Inc.

A10 Foundations

A1010 Standard Foundations

This page illustrates and describes a strip footing system including concrete, forms, reinforcing, keyway and dowels. Lines within System Components give the unit price and total price per linear foot for this system. Prices for alternate strip footing systems are on Line Items A1010 120 1500 thru 2500. Both material quantities and labor costs have been adjusted for the system listed.

Factors: To adjust for job conditions other than normal working situations use Lines A1010 120 2900 thru 4000.

Example: You are to install this footing, and due to a lack of accessibility, only hand tools can be used. Material handling is also a problem. Go to Lines A1010 120 3400 and 3600 and apply these percentages to the appropriate MAT. and INST. costs.

System Components	QUANTITY	UNIT	COST PER L.F.		
			MAT.	INST.	TOTAL
Strip footing, 2'-0" wide x 1'-0" thick, 2000 psi concrete including forms,					
Reinforcing, keyway, and dowels.					
Concrete, 2000 psi	.074	C.Y.	7.47		7.47
Placing concrete	.074	C.Y.		1.71	1.71
Forms, footing, 4 uses	2.000	S.F.	3.76	8.54	12.30
Reinforcing	3.170	Lb.	1.46	1.90	3.36
Keyway, 2" x 4", 4 uses	1.000	L.F.	.15	1.03	1.18
Dowels, #4 bars, 2' long, 24" O.C.	.500	Ea.	.33	1.31	1.64
TOTAL			13.17	14.49	27.66

A1010 120	Strip Footing	COST PER L.F.		
		MAT.	INST.	TOTAL
1400	Above system with the following:			
1500	2'-0" wide x 1' thick, 3000 psi concrete	13.60	14.50	28.10
1600	4000 psi concrete	14.05	14.50	28.55
1800	For alternate footing systems:			
1900	2'-6" wide x 1' thick, 2000 psi concrete	15.45	15.40	30.85
2000	3000 psi concrete	16.05	15.40	31.45
2100	4000 psi concrete	16.60	15.40	32
2300	3'-0" wide x 1' thick, 2000 psi concrete	17.55	16.30	33.85
2400	3000 psi concrete	18.20	16.30	34.50
2500	4000 psi concrete	18.90	16.30	35.20
2700				
2800				
2900	Cut & patch to match existing construction, add, minimum	2%	3%	
3000	Maximum	5%	9%	
3100	Dust protection, add, minimum	1%	2%	
3200	Maximum	4%	11%	
3300	Equipment usage curtailment, add, minimum	1%	1%	
3400	Maximum	3%	10%	
3500	Material handling & storage limitation, add, minimum	1%	1%	
3600	Maximum	6%	7%	
3700	Protection of existing work, add, minimum	2%	2%	
3800	Maximum	5%	7%	
3900	Shift work requirements, add, minimum		5%	
4000	Maximum		30%	

A10 Foundations

A1010 Standard Foundations

This page illustrates and describes a spread footing system including concrete, forms, reinforcing and anchor bolts. Lines within System Components give the unit price and total price on a cost each basis for this system. Prices for alternate spread footing systems are on Line Items A1010 230 1300 thru 2300. Both material quantities and labor costs have been adjusted for the system listed.

Factors: To adjust for job conditions other than normal working situations use Lines A1010 230 2900 thru 4000.

Example: You are to install the system in an existing occupied building. Access to the site and protection to the building are mandatory. Go to Lines A1010 230 3600 and 3800 and apply these percentages to the appropriate MAT. and INST. costs.

System Components	QUANTITY	UNIT	COST EACH		
			MAT.	INST.	TOTAL
Interior column footing, 3' square 1' thick, 2000 psi concrete,					
Including forms, reinforcing, and anchor bolts.					
Concrete, 2000 psi	.330	C.Y.	33.33		33.33
Placing concrete	.330	C.Y.		16.57	16.57
Forms, footing, 4 uses	12.000	SFCA	6.60	60	66.60
Reinforcing	11.000	Lb.	5.06	6.60	11.66
Anchor bolts, 3/4" diameter	2.000	Ea.	1.76	5.38	7.14
TOTAL			46.75	88.55	135.30

A1010 230	Spread Footing	COST EACH		
		MAT.	INST.	TOTAL
1200	Above system with the following:			
1300	3' square x 1' thick, 3000 psi concrete	48.50	88.50	137
1400	4000 psi concrete	50.50	88.50	139
1600	For alternate footing systems:			
1700	4' square x 1' thick, 2000 psi concrete	79	127	206
1800	3000 psi concrete	82.50	127	209.50
1900	4000 psi concrete	86.50	127	213.50
2100	5' square x 1'-3" thick, 2000 psi concrete	150	211	361
2200	3000 psi concrete	157	211	368
2300	4000 psi concrete	164	211	375
2600				
2700				
2900	Cut & patch to match existing construction, add, minimum	2%	3%	
3000	Maximum	5%	9%	
3100	Dust protection, add, minimum	1%	2%	
3200	Maximum	4%	11%	
3300	Equipment usage curtailment, add, minimum	1%	1%	
3400	Maximum	3%	10%	
3500	Material handling & storage limitation, add, minimum	1%	1%	
3600	Maximum	6%	7%	
3700	Protection of existing work, add, minimum	2%	2%	
3800	Maximum	5%	7%	
3900	Shift work requirements, add, minimum		5%	
4000	Maximum		30%	

A10 Foundations

A1010 Standard Foundations

General: Footing drains can be placed either inside or outside of foundation walls depending upon the source of water to be intercepted. If the source of subsurface water is principally from grade or a subsurface stream above the bottom of the footing, outside drains should be used. For high water tables, use inside drains or both inside and outside.

The effectiveness of underdrains depends on good waterproofing. This must be carefully installed and protected during construction.

Costs below include the labor and materials for the pipe and 6" only of gravel or crushed stone around pipe. Excavation and backfill are not included.

System Components	QUANTITY	UNIT	COST PER L.F.		
			MAT.	INST.	TOTAL
Foundation underdrain, outside, PVC 4" diameter.					
PVC pipe 4" diam. S.D.R. 35	1.000	L.F.	1.85	3.86	5.71
Pipe bedding, graded gravel 3/4" to 1/2"	.070	C.Y.	2.70	.82	3.52
TOTAL			4.55	4.68	9.23

A1010 310	Foundation Underdrain	COST PER L.F.		
		MAT.	INST.	TOTAL
0900	For alternate drain systems:			
1000	Foundation underdrain, outside only, PVC, 4" diameter	4.55	4.68	9.23
1100	6" diameter	7.45	5.20	12.65
1400	Porous concrete, 6" diameter	8.35	5.60	13.95
1450	8" diameter	10.25	7.30	17.55
1500	12" diameter	18.85	8.50	27.35
1600	Corrugated metal, 16 ga. asphalt coated, 6" diameter	11.45	9.20	20.65
1650	8" diameter	15.80	9.60	25.40
1700	10" diameter	19.45	10.10	29.55
3000	Outside and inside, PVC, 4" diameter	9.10	9.35	18.45
3100	6" diameter	14.95	10.40	25.35
3400	Porous concrete, 6" diameter	16.65	11.30	27.95
3450	8" diameter	20.50	14.70	35.20
3500	12" diameter	37.50	16.90	54.40
3600	Corrugated metal, 16 ga., asphalt coated, 6" diameter	23	18.35	41.35
3650	8" diameter	31.50	19.30	50.80
3700	10" diameter	39	20	59
4700	Cut & patch to match existing construction, add, minimum	2%	3%	
4800	Maximum	5%	9%	
4900	Dust protection, add, minimum	1%	2%	
5000	Maximum	4%	11%	
5100	Equipment usage curtailment, add, minimum	1%	1%	
5200	Maximum	3%	10%	
5300	Material handling & storage limitation, add, minimum	1%	1%	
5400	Maximum	6%	7%	
5500	Protection of existing work, add, minimum	2%	2%	
5600	Maximum	5%	7%	
5700	Shift work requirements, add, minimum		5%	
5800	Maximum		30%	
5900	Temporary shoring and bracing, add, minimum	2%	5%	
6000	Maximum	5%	12%	

A10 Foundations

A1030 Slab on Grade

This page illustrates and describes a slab on grade system including slab, bank run gravel, bulkhead forms, placing concrete, welded wire fabric, vapor barrier, steel trowel finish and curing paper. Lines within System Components give the unit price and total price per square foot for this system. Prices for alternate slab on grade systems are on Line Items A1030 110 1500 thru 2600. Both material quantities and labor costs have been adjusted for the system listed.

Factors: To adjust for job conditions other than normal working situations use Lines A1030 110 2900 thru 4000.

Example: You are to install the system at a site where protection of the existing building is required. Go to Line A1030 110 3800 and apply these percentages to the appropriate MAT. and INST. costs.

System Components	QUANTITY	UNIT	COST PER S.F.		
			MAT.	INST.	TOTAL
Ground slab, 4" thick, 3000 psi concrete, 4" granular base, vapor barrier Welded wire fabric, screed and steel trowel finish.					
Concrete, 4" thick, 3000 psi concrete	.012	C.Y.	1.28		1.28
Bank run gravel, 4" deep	.074	C.Y.	.29	.09	.38
Polyethylene vapor barrier, 10 mil.	.011	C.S.F.	.08	.16	.24
Bulkhead forms, expansion material	.100	L.F.	.03	.35	.38
Welded wire fabric, 6 x 6 - #10/10	.011	C.S.F.	.15	.40	.55
Place concrete	.012	C.Y.		.30	.30
Screed & steel trowel finish	1.000	S.F.		.86	.86
TOTAL			1.83	2.16	3.99

A1030 110	Interior Slab on Grade	COST PER S.F.		
		MAT.	INST.	TOTAL
1400	Above system with the following:			
1500	4" thick slab, 3000 psi concrete, 6" deep bank run gravel	2.06	2.15	4.21
1600	12" deep bank run gravel	2.58	2.18	4.76
1700				
1800				
1900				
2000	For alternate slab systems:			
2100	5" thick slab, 3000 psi concrete, 6" deep bank run gravel	2.39	2.23	4.62
2200	12" deep bank run gravel	2.91	2.26	5.17
2300				
2400				
2500	6" thick slab, 3000 psi concrete, 6" deep bank run gravel	2.81	2.33	5.14
2600	12" deep bank run gravel	3.33	2.36	5.69
2700				
2900	Cut & patch to match existing construction, add, minimum	2%	3%	
3000	Maximum	5%	9%	
3100	Dust protection, add, minimum	1%	2%	
3200	Maximum	4%	11%	
3300	Equipment usage curtailment, add, minimum	1%	1%	
3400	Maximum	3%	10%	
3500	Material handling & storage limitation, add, minimum	1%	1%	
3600	Maximum	6%	7%	
3700	Protection of existing work, add, minimum	2%	2%	
3800	Maximum	5%	7%	
3900	Shift work requirements, add, minimum		5%	
4000	Maximum		30%	

A20 Basement Construction

A2010 Basement Excavation

This page illustrates and describes continuous footing and trench excavation systems including a wheel mounted backhoe, operator, equipment rental, fuel, oil, mobilization, no hauling or backfill. Lines within System Components give the unit price and total price per linear foot for this system. Prices for alternate continuous footing and trench excavation systems are on Line Items A2010 130 1300 thru 2700. Both material quantities and labor costs have been adjusted for the system listed.

Factors: To adjust for job conditions other than normal working situations use Lines A2010 130 2900 thru 4000.

Example: You are to install the system and supply dust protection. Go to Line A2010 130 3000 and apply this percentage to the appropriate TOTAL costs.

System Components	QUANTITY	UNIT	COST PER L.F.		
			EQUIP.	LABOR	TOTAL
Continuous footing or trench excav. w/ 3/4 C.Y. wheel mntd backhoe Including operator, equipment rental, fuel, oil and mobilization. Trench Is 4' wide at bottom, 3' deep w/sides sloped 1 to 2 and 200' long. Costs Are based on a production rate of 240 C.Y./day. No hauling/backfill incl.					
Equipment operator	4.000	Hr.		270	270
3/4 C.Y. backhoe, wheel mounted	.500	Day	96.25		96.25
Operating expense (fuel, oil)	4.000	Hr.	89.56		89.56
Mobilization	1.000	Ea.	282	120	402
TOTAL			467.81	390	857.81
COST PER L.F.		L.F.	2.34	1.95	4.29

A2010 130	Excavation, Footings or Trench	COST PER L.F.		
		EQUIP.	LABOR	TOTAL
1200	For alternate trench sizes, 4' bottom with sloped sides:			
1300	2' deep, 50' long	7.90	7.80	15.70
1400	100' long	4.12	3.90	8.02
1500	300' long	1.54	1.30	2.84
1600	4' deep, 50' long	8.25	7.80	16.05
1700	100' long	4.45	3.90	8.35
1800	300' long	2.23	2.31	4.54
1900	6' deep, 50' long	8.75	7.80	16.55
2000	100' long	4.66	4.92	9.58
2100	300' long	3.41	4	7.41
2200	8' deep, 50' long	9.35	7.80	17.15
2300	100' long	6.55	6.60	13.15
2400	300' long	4.66	4	8.66
2500	10' deep, 50' long	11.20	10.50	21.70
2600	100' long	8.15	8.95	17.10
2700	300' long	6.15	7.95	14.10
2800				
2900	Dust protection, add, minimum		2%	2%
3000	Maximum		11%	11%
3100	Equipment usage curtailment, add, minimum		1%	1%
3200	Maximum		10%	10%
3300	Material handling & storage limitation, add, minimum		1%	1%
3400	Maximum		7%	7%
3500	Protection of existing work, add, minimum		2%	2%
3600	Maximum		7%	7%
3700	Shift work requirements, add, minimum		5%	5%
3800	Maximum		30%	30%
3900	Temporary shoring and bracing, add, minimum		5%	5%
4000	Maximum		12%	12%

A20 Basement Construction

A2010 Basement Excavation

This page illustrates and describes foundation excavation systems including a backhoe-loader, operator, equipment rental, fuel, oil, mobilization, hauling material, no backfilling. Lines within System Components give the unit price and total price per cubic yard for this system. Prices for alternate foundation excavation systems are on Line Items A2010 140 1600 thru 2200. Both material quantities and labor costs have been adjusted for the system listed.

Factors: To adjust for job conditions other than normal working situations use Lines A2010 140 2900 thru 4000.

Example: You are to install the system with the use of temporary shoring and bracing. Go to Line A2010 140 3900 and apply these percentages to the appropriate TOTAL costs.

System Components	QUANTITY	UNIT	COST PER C.Y.		
			EQUIP.	LABOR	TOTAL
Foundation excav w/ 3/4 C.Y. backhoe-loader, incl. operator, equip Rental, fuel, oil, and mobilization. Hauling of excavated material is Included. Prices based on one day production of 360 C.Y. in medium soil Without backfilling.					
Equipment operator	8.000	Hr.		540	540
Backhoe-loader, 3/4 C.Y.	1.000	Day	192.50		192.50
Operating expense (fuel, oil)	8.000	Hr.	179.12		179.12
Hauling, 12 C.Y. trucks, 1 mile round trip	360.000	L.C.Y.	1,051.20	615.60	1,666.80
Mobilization	1.000	Ea.	282	120	402
TOTAL			1,704.82	1,275.60	2,980.42
COST PER C.Y.		C.Y.	4.72	3.53	8.28

A2010 140		Excavation, Foundation	COST PER C.Y.		
			EQUIP.	LABOR	TOTAL
1400	For alternate size excavations:				
1500					
1600		100 C.Y.	6.25	5.95	12.20
1700		200 C.Y.	5.35	3.83	9.18
1800		300 C.Y.	4.92	3.62	8.54
1900		400 C.Y.	4.65	3.50	8.15
2000		500 C.Y.	4.52	3.46	7.98
2100		600 C.Y.	4.40	3.41	7.81
2200		700 C.Y.	4.34	3.35	7.69
2300					
2400					
2500					
2900	Dust protection, add, minimum			2%	2%
3000	Maximum			11%	11%
3100	Equipment usage curtailment, add, minimum			1%	1%
3200	Maximum			10%	10%
3300	Material handling & storage limitation, add, minimum			1%	1%
3400	Maximum			7%	7%
3500	Protection of existing work, add, minimum			2%	2%
3600	Maximum			7%	7%
3700	Shift work requirements, add, minimum			5%	5%
3800	Maximum			30%	30%
3900	Temporary shoring and bracing, add, minimum			5%	5%
4000	Maximum			12%	12%

A20 Basement Construction

A2020 Basement Walls

This page illustrates and describes a concrete wall system including concrete, placing concrete, forms, reinforcing, insulation, waterproofing and anchor bolts. Lines within System Components give the unit price and total price per linear foot for this system. Prices for alternate concrete wall systems are on Line Items A2020 120 1500 thru 2600. Both material quantities and labor costs have been adjusted for the system listed.

Factors: To adjust for job conditions other than normal working situations use Lines A2020 120 2900 thru 4000.

Example: You are to install this wall system where delivery of material is difficult. Go to Line A2020 120 3600 and apply these percentages to the appropriate MAT. and INST. costs.

System Components			COST PER L.F.		
	QUANTITY	UNIT	MAT.	INST.	TOTAL
Cast in place concrete foundation wall, 8" thick, 3' high, 2500 psi					
Concrete including forms, reinforcing, waterproofing, and anchor bolts.					
Concrete, 2500 psi, 8" thick, 3' high	.070	C.Y.	7.21		7.21
Forms, wall, 4 uses	6.000	S.F.	5.40	37.80	43.20
Reinforcing	6.000	Lb.	2.76	2.52	5.28
Placing concrete	.070	C.Y.		3.18	3.18
Waterproofing	3.000	S.F.	1.32	3	4.32
Rigid insulaton, 1" polystyrene	3.000	S.F.	1.20	2.40	3.60
Anchor bolts, 1/2" diameter, 4' O.C.	.250	Ea.	.42	.71	1.13
TOTAL			18.31	49.61	67.92

A2020 120	Concrete Wall	COST PER L.F.		
		MAT.	INST.	TOTAL
1400	For alternate wall systems:			
1500	8" thick, 2500 psi concrete, 4' high	25	66.50	91.50
1600	6' high	37	99	136
1700	8' high	49.50	132	181.50
1800	3500 psi concrete, 4' high	25.50	66.50	92
1900	6' high	38.50	99	137.50
2000	8' high	51	132	183
2100	12" thick, 2500 psi concrete, 4' high	32	70.50	102.50
2200	6' high	46.50	104	150.50
2300	8' high	61.50	138	199.50
2400	3500 psi concrete, 4' high	33	70.50	103.50
2500	8' high	63.50	138	201.50
2600	10' high	79	172	251
2700				
2900	Cut & patch to match existing construction, add, minimum	2%	3%	
3000	Maximum	5%	9%	
3100	Dust protection, add, minimum	1%	2%	
3200	Maximum	4%	11%	
3300	Equipment usage curtailment, add, minimum	1%	1%	
3400	Maximum	3%	10%	
3500	Material handling & storage limitation, add, minimum	1%	1%	
3600	Maximum	6%	7%	
3700	Protection of existing work, add, minimum	2%	2%	
3800	Maximum	5%	7%	
3900	Shift work requirements, add, minimum		5%	
4000	Maximum		30%	
4100				

A20 Basement Construction

A2020 Basement Walls

This page illustrates and describes a concrete block wall system including concrete block, masonry reinforcing, parging, waterproofing insulation and anchor bolts. Lines within System Components give the unit price and total price per linear foot for this system. Prices for alternate concrete block wall systems are on Line Items A2020 140 1300 thru 2600. Both material quantities and labor costs have been adjusted for the system listed.

Factors: To adjust for job conditions other than normal working situations use Lines A2020 140 2900 thru 4000.

Example: You are to install the system to match an existing foundation wall. Go to Line A2020 140 3000 and apply these percentages to the appropriate MAT. and INST. costs.

System Components			COST PER L.F.		
	QUANTITY	UNIT	MAT.	INST.	TOTAL
Concrete block, 8" thick, masonry reinforcing, parged and Waterproofed, insulation and anchor bolts, wall 2'-8" high.					
Concrete block, 8" x 8" x 16"	2.670	S.F.	8.52	15.62	24.14
Masonry reinforcing	2.000	L.F.	.36	.36	.72
Parging	.193	S.Y.	.75	2.35	3.10
Waterproofing	2.670	S.F.	1.17	2.67	3.84
Insulation, 1" rigid polystyrene	2.670	S.F.	1.07	2.14	3.21
Anchor bolts, 1/2" diameter, 4' O.C.	.250	Ea.	.42	.71	1.13
TOTAL			12.29	23.85	36.14

A2020 140	Concrete Block Wall	COST PER L.F.		
		MAT.	INST.	TOTAL
1200	For alternate wall systems:			
1300	8" thick block, 4' high	18.20	35	53.20
1400	6' high	27	53	80
1500	8' high	36	70	106
1600	Grouted solid, 4' high	23	47.50	70.50
1700	6' high	34.50	71	105.50
1800	8' high	45.50	94.50	140
2100	12" thick block, 4' high	23	51	74
2200	6' high	34	76.50	110.50
2300	8' high	45.50	101	146.50
2400	Grouted solid, 4' high	31	64	95
2500	6' high	46	95.50	141.50
2600	8' high	61.50	127	188.50
2700				
2900	Cut & patch to match existing construction, add, minimum	2%	3%	
3000	Maximum	5%	9%	
3100	Dust protection, add, minimum	1%	2%	
3200	Maximum	4%	11%	
3300	Equipment usage curtailment, add, minimum	1%	1%	
3400	Maximum	3%	10%	
3500	Material handling & storage limitation, add, minimum	1%	1%	
3600	Maximum	6%	7%	
3700	Protection of existing work, add, minimum	2%	2%	
3800	Maximum	5%	7%	
3900	Shift work requirements, add, minimum		5%	
4000	Maximum		30%	

No part of this publication may be reproduced, stored in a retrieval system, or transmitted in any form or by any means without prior written permission of R.S. Means Company, Inc.

B Shell

B10 Superstructure

B1010 Floor Construction

This page illustrates and describes a reinforced concrete slab system including concrete, placing concrete, formwork, reinforcing, steel trowel finish, V.A. floor tile and acoustical spray ceiling finish. Lines within System Components give the unit price per square foot for this system. Prices for alternate reinforced concrete slab systems are on Line Items B1010 225 1500 thru 1800. Both material quantities and labor costs have been adjusted for the system listed.

Factors: To adjust for job conditions other than normal working situations use Lines B1010 225 2700 thru 4000.

Example: You are to install the system to match an existing floor system. Go to Line B1010 225 2800 and apply these percentages to the appropriate MAT. and INST. costs.

System Components	QUANTITY	UNIT	COST PER S.F.		
			MAT.	INST.	TOTAL
Flat slab system, reinforced concrete with vinyl tile floor, sprayed Acoustical ceiling finish, not including columns.					
Concrete, 4000 psi, 6" thick	.020	C.Y.	2.26		2.26
Placing concrete	.020	C.Y.		.65	.65
Formwork, 4 use	1.000	S.F.	1.45	5.70	7.15
Edge form	.120	S.F.	.04	1.10	1.14
Reinforcing steel	3.000	Lb.	1.38	1.80	3.18
Steel trowel finish	1.000	S.F.		1.13	1.13
Vinyl asbestos floor tile	1.000	S.F.	4.79	.98	5.77
Acoustical spray ceiling finish	1.000	S.F.	.26	.16	.42
TOTAL			10.18	11.52	21.70

B1010 225	Floor - Ceiling, Concrete Slab	COST PER S.F.		
		MAT.	INST.	TOTAL
1400	For alternate slab systems:			
1500	Concrete, 4000 psi, 7" thick	10.60	12	22.60
1600	8" thick	11.25	12.65	23.90
1700	9" thick	11.75	13.20	24.95
1800	10" thick	12.35	13.80	26.15
1900				
2000				
2100				
2200				
2300				
2400				
2500				
2700	Cut & patch to match existing construction, add, minimum	2%	3%	
2800	Maximum	5%	9%	
2900	Dust protection, add, minimum	1%	2%	
3000	Maximum	4%	11%	
3100	Equipment usage curtailment, add, minimum	1%	1%	
3200	Maximum	3%	10%	
3300	Material handling & storage limitation, add, minimum	1%	1%	
3400	Maximum	6%	7%	
3500	Protection of existing work, add, minimum	2%	2%	
3600	Maximum	5%	7%	
3700	Shift work requirements, add, minimum		5%	
3800	Maximum		30%	
3900	Temporary shoring and bracing, add, minimum	2%	5%	
4000	Maximum	5%	12%	

B10 Superstructure

B1010 Floor Construction

This page illustrates and describes a hollow core prestressed concrete panel system including hollow core slab, grout, carpet, carpet padding, and sprayed ceiling. Lines within System Components give the unit price and total price per square foot for this system. Prices for alternate hollow core prestressed concrete panel systems are on Line Items B1010 232 1300 thru 1600. Both material and labor costs have been adjusted for the system listed.

Factors: To adjust for job conditions other than normal working situations use Lines B1010 232 2700 thru 3800.

Example: You are to install the system where dust control is a major concern. Go to Line B1010 232 2800 and apply these percentages to the appropriate MAT. and INST. costs.

System Components	QUANTITY	UNIT	COST PER S.F.		
			MAT.	INST.	TOTAL
Precast hollow core plank with carpeted floors, padding And sprayed textured ceiling.					
Hollow core plank, 4" thick with grout topping	1.000	S.F.	6.10	3.36	9.46
Nylon carpet, 26 oz. medium traffic	.110	S.Y.	4.02	.72	4.74
Carpet padding, minimum quality	.110	S.Y.	.48	.36	.84
Sprayed texture ceiling	1.000	S.F.	.06	.58	.64
TOTAL			10.66	5.02	15.68

B1010 232	Floor - Ceiling, Conc. Panel	COST PER S.F.		
		MAT.	INST.	TOTAL
1200	For alternate floor systems:			
1300	Hollow core concrete plank, with grout, 6" thick	11.70	4.55	16.25
1400	8" thick	12.45	4.18	16.63
1500	10" thick	12.80	3.90	16.70
1600	12" thick	13.35	3.68	17.03
1700				
1800				
1900				
2000				
2100				
2200				
2300				
2400				
2500				
2700	Dust protection, add, minimum	1%	2%	
2800	Maximum	4%	11%	
2900	Equipment usage curtailment, add, minimum	1%	1%	
3000	Maximum	3%	10%	
3100	Material handling & storage limitation, add, minimum	1%	1%	
3200	Maximum	6%	7%	
3300	Protection of existing work, add, minimum	2%	2%	
3400	Maximum	5%	7%	
3500	Shift work requirements, add, minimum		5%	
3600	Maximum		30%	
3700	Temporary shoring and bracing, add, minimum	2%	5%	
3800	Maximum	5%	12%	

B10 Superstructure

B1010 Floor Construction

This page illustrates and describes a hollow core plank system including precast hollow core plank, concrete topping, V.A. tile, concealed suspension system and ceiling tile. Lines within System Components give the unit price and total price per square foot for this system. Prices for alternate hollow core plank systems are on Line Items B1010 233 1500 thru 1800. Both material quantities and labor costs have been adjusted for the system listed.

Factors: To adjust for job conditions other than normal working situations use Lines B1010 233 2900 thru 4000.

Example: You are to install the system where equipment usage is a problem. Go to Line B1010 233 3200 and apply these percentages to the appropriate MAT. and INST. costs.

System Components	QUANTITY	UNIT	COST PER S.F.		
			MAT.	INST.	TOTAL
Precast hollow core plank with 2" topping, vinyl composition floor tile And suspended acoustical tile ceiling.					
Hollow core concrete plank, 4" thick	1.000	S.F.	6.10	3.36	9.46
Concrete topping, 2" thick	1.000	S.F.	1.33	5.18	6.51
Vinyl composition tile, .08" thick	1.000	S.F.	1.52	.98	2.50
Concealed Z bar suspension system, 12" module	1.000	S.F.	.76	1.05	1.81
Ceiling tile, mineral fiber, 3/4" thick	1.000	S.F.	2.10	1.82	3.92
TOTAL			11.81	12.39	24.20

B1010 233	Floor - Ceiling, Conc. Plank	COST PER S.F.		
		MAT.	INST.	TOTAL
1400	For alternate floor systems:			
1500	Hollow core concrete plank, 6" thick	12.85	11.95	24.80
1600	8" thick	13.60	11.55	25.15
1700	10" thick	13.95	11.25	25.20
1800	12" thick	14.50	11.05	25.55
1900				
2000				
2100				
2200				
2300				
2400				
2500				
2600				
2700				
2900	Dust protection, add, minimum	1%	2%	
3000	Maximum	4%	11%	
3100	Equipment usage curtailment, add, minimum	1%	1%	
3200	Maximum	3%	10%	
3300	Material handling & storage limitation, add, minimum	1%	1%	
3400	Maximum	6%	7%	
3500	Protection of existing work, add, minimum	2%	2%	
3600	Maximum	5%	7%	
3700	Shift work requirements, add, minimum		5%	
3800	Maximum		30%	
3900	Temporary shoring and bracing, add, minimum	2%	5%	
4000	Maximum	5%	12%	

B10 Superstructure

B1010 Floor Construction

This page illustrates and describes a structural steel w/metal decking and concrete system including steel beams, steel decking, shear studs, concrete, placing concrete, edge form, steel trowel finish, curing, wire fabric, fireproofing, beams and decking, tile floor, and suspended ceiling. Lines within System Components give the unit price and total price per square foot for this system. Prices for alternate structural steel w/metal decking and concrete systems are on Line Items B1010 243 1900 thru 2400. Both material quantities and labor costs have been adjusted for the system listed.

Factors: To adjust for job conditions other than normal working situations use Lines B1010 243 2900 thru 4000.

Example: You are to install the system where material handling and storage are a problem. Go to Line B1010 243 3400 and apply these percentages to the appropriate MAT. and INST. costs.

System Components	QUANTITY	UNIT	COST PER S.F. MAT.	INST.	TOTAL
Composite structural beams, 20 ga. 3" deep steel decking, shear studs, 3000 psi, concrete, placing concrete, edge forms, steel trowel finish, Welded wire fabric fireproofing, tile floor and suspended ceiling.					
Steel deck, 20 gage, 3" deep, galvanized	1.000	S.F.	1.65	.97	2.62
Structural steel framing	.004	Ton	9.70	2.64	12.34
Shear studs, 3/4"	.250	Ea.	.13	.46	.59
Concrete, 3000 psi, 5" thick	.015	C.Y.	1.61		1.61
Place concrete	.015	C.Y.		.48	.48
Edge form	.140	L.F.	.02	.58	.60
Steel trowel finish	1.000	S.F.		1.13	1.13
Welded wire fabric 6 x 6 - #10/10	.011	C.S.F.	.15	.40	.55
Fireproofing, sprayed	1.880	S.F.	1.09	1.86	2.95
Vinyl composition floor tile	1.000	S.F.	1.01	.98	1.99
Suspended acoustical ceiling	1.000	S.F.	2.52	1.44	3.96
TOTAL			17.88	10.94	28.82

B1010 243	Floor - Ceiling, Struc. Steel	COST PER S.F. MAT.	INST.	TOTAL
1800	For alternate floor systems:			
1900	Composite deck, galvanized, 3" deep, 22 ga.	17.70	10.90	28.60
2000	18 ga.	18.20	11	29.20
2100	16 ga.	18.95	11.05	30
2200	Composite deck, galvanized, 2" deep, 22 ga.	17.55	10.75	28.30
2300	20 ga.	17.75	10.80	28.55
2400	16 ga.	18.60	10.90	29.50
2500				
2900	Dust protection, add, minimum	1%	2%	
3000	Maximum	4%	11%	
3100	Equipment usage curtailment, add, minimum	1%	1%	
3200	Maximum	3%	10%	
3300	Material handling & storage limitation, add, minimum	1%	1%	
3400	Maximum	6%	7%	
3500	Protection of existing work, add, minimum	2%	2%	
3600	Maximum	5%	7%	
3700	Shift work requirements, add, minimum		5%	
3800	Maximum		30%	
3900	Temporary shoring and bracing, add, minimum	2%	5%	
4000	Maximum	5%	12%	

B10 Superstructure

B1010 Floor Construction

This page illustrates and describes an open web joist and steel slab-form system including open web steel joist, slab form, concrete, placing concrete, wire fabric, steel trowel finish, tile floor and plasterboard. Lines within System Components give the unit price and total price per square foot for this system. Prices for alternate open web joists and steel slab-form systems are on Line Items B1010 245 1900 thru 2200. Both material quantities and labor costs have been adjusted for the systems listed.

Factors: To adjust for job conditions other than normal working situations use Lines B1010 245 2900 thru 4000.

Example: You are to install the system in a congested commercial area and most work will be done at night. Go to Line B1010 245 3800 and apply this percentage to the appropriate INST. cost.

System Components	QUANTITY	UNIT	COST PER S.F.		
			MAT.	INST.	TOTAL
Open web steel joists, 24" O.C., slab form, 3000 psi concrete, welded					
Wire fabric, steel trowel finish, floor tile, 5/8" drywall ceiling.					
Open web steel joists, 12" deep, 5.2#/L.F., 24" O.C.	2.630	Lb.	1.60	1.13	2.73
Slab form, 28 gage, 9/16" deep, galvanized	1.000	S.F.	.96	.73	1.69
Concrete, 3000 psi, 2-1/2" thick	.008	C.Y.	.86		.86
Placing concrete	.008	C.Y.		.26	.26
Welded wire fabric 6 x 6 - #10/10	.011	C.S.F.	.15	.40	.55
Steel trowel finish	1.000	S.F.		1.13	1.13
Vinyl composition floor tile	1.000	S.F.	1.01	.98	1.99
Ceiling furring, 3/4" channels, 24" O.C.	1.000	S.F.	.20	1.11	1.31
Gypsum drywall, 5/8" thick, finished	1.000	S.F.	.35	1.43	1.78
Paint ceiling	1.000	S.F.	.17	.61	.78
TOTAL			5.30	7.78	13.08

B1010 245	Floor - Ceiling, Steel Joists	COST PER S.F.		
		MAT.	INST.	TOTAL
1800	For alternate floor systems:			
1900	Open web joists, 16" deep, 6.6#/L.F.	5.70	8.05	13.75
2000	20" deep, 8.4# L.F.	6.25	8.45	14.70
2100	24" deep, 11.5# L.F.	7.20	9.10	16.30
2200	26" deep, 12.8# L.F.	7.60	9.40	17
2300				
2400				
2500				
2600				
2700				
2900	Dust protection, add, minimum	1%	2%	
3000	Maximum	4%	11%	
3100	Equipment usage curtailment, add, minimum	1%	1%	
3200	Maximum	3%	10%	
3300	Material handling & storage limitation, add, minimum	1%	1%	
3400	Maximum	6%	7%	
3500	Protection of existing work, add, minimum	2%	2%	
3600	Maximum	5%	7%	
3700	Shift work requirements, add, minimum		5%	
3800	Maximum		30%	
3900	Temporary shoring and bracing, add, minimum	2%	5%	
4000	Maximum	5%	12%	

B10 Superstructure

B1010 Floor Construction

Open Web Wood Joists

This page illustrates and describes floor framing systems including sills, joist hangers, joists, bridging, subfloor and insulation. Lines within systems components give the unit price and total price per square foot for this system. Prices for alternate joist types are on line items B1010 262 0100 through 0700. Both material quantities and labor costs have been adjusted for the system listed.

Factors: To adjust for job conditions other than normal working situations use lines B1010 262 2700 through 4000.

Example: You are to install the system where delivery of material is difficult. Go to line B1010 262 3400 and apply these percentages to the appropriate MAT. and INST. costs.

System Components	QUANTITY	UNIT	COST PER S.F.		
			MAT.	INST.	TOTAL
Open web wood joists, sill, butt to girder, 5/8" subfloor, insul.					
Anchor bolt, hook type, with nut and washer, 1/2" x 8" long	.250	Ea.	.02	.07	.09
Joist & beam hanger, 18 ga. galvanized	.750	Ea.	.10	.25	.35
Bridging, steel, compression, for 2" x 12" joists	.001	C.Pr.	.18	.27	.45
Framing, open web wood joists, 12" deep	.001	M.L.F.	3.05	1.25	4.30
Framing, sills, 2" x 8"	.001	M.B.F.	.46	1.63	2.09
Sub-floor, plywood, CDX, 3/4" thick	1.000	S.F.	.56	.81	1.37
Insulation, fiberglass, kraft face, 9" thick x 23" wide	1.000	S.F.	.77	.40	1.17
TOTAL			5.14	4.68	9.82

B1010 262	Open Web Wood Joists	COST PER S.F.		
		MAT.	INST.	TOTAL
0080	For alternate floor systems:			
0100	12" open web joists, sill, butt to girder, 5/8" sub-floor	5.15	4.68	9.83
0200	14" joists	5.35	4.76	10.11
0300	16" joists	5.35	4.83	10.18
0400	18" joists	5.60	4.91	10.51
0450	Wood struc."I" joists, 24" O.C., to 24' span, 50 PSF LL, butt to girder	4.05	4.35	8.40
0500	55 PSF LL	4.21	4.41	8.62
0600	to 30' span, 45 PSF LL	5	4.28	9.28
0700	55 PSF LL	5.95	4.35	10.30
2700	Cut & patch to match existing construction, add, minimum	2%	3%	
2800	Maximum	5%	9%	
2900	Dust protection, add, minimum	1%	2%	
3000	Maximum	4%	11%	
3100	Equipment usage curtailment, add, minimum	1%	1%	
3200	Maximum	3%	10%	
3300	Material handling & storage limitation, add, minimum	1%	1%	
3400	Maximum	6%	7%	
3500	Protection of existing work, add, minimum	2%	2%	
3600	Maximum	5%	7%	
3700	Shift work requirements, add, minimum		5%	
3800	Maximum		30%	
3900	Temporary shoring and bracing, add, minimum	2%	5%	
4000	Maximum	5%	12%	

B10 Superstructure

B1010 Floor Construction

This page illustrates and describes a wood joist floor system including wood joist, oak floor, sub-floor, bridging, sand and finish floor, furring, plasterboard, taped, finished and painted ceiling. Lines within System Components give the unit price and total price per square foot for this system. Prices for alternate wood joist floor systems are on Line Items B1010 263 1700 thru 2300. Both material quantities and labor costs have been adjusted for the system listed.

Factors: To adjust for job conditions other than normal working situations use Lines B1010 263 2700 thru 4000.

Example: You are to install the system during off peak hours, 6 P.M. to 2 A.M. Go to Line B1010 263 3800 and apply this percentage to the appropriate INST. costs.

System Components			COST PER S.F.		
	QUANTITY	UNIT	MAT.	INST.	TOTAL
Wood joists, 2" x 8", 16" O.C., oak floor (sanded & finished), 1/2" sub Floor, 1" x 3" bridging, furring, 5/8" drywall, taped, finished and painted.					
Wood joists, 2" x 8", 16" O.C.	1.000	L.F.	.61	.99	1.60
Subfloor plywood CDX 1/2"	1.000	SF Flr.	.46	.73	1.19
Bridging 1" x 3"	.150	Pr.	.08	.63	.71
Oak flooring, No. 1 common	1.000	S.F.	4.21	3.21	7.42
Sand and finish floor	1.000	S.F.	.86	1.47	2.33
Wood furring, 1" x 3", 16" O.C.	1.000	L.F.	.38	1.56	1.94
Gypsum drywall, 5/8" thick	1.000	S.F.	.31	.61	.92
Tape and finishing	1.000	S.F.	.05	.55	.60
Paint ceiling	1.000	S.F.	.17	.61	.78
TOTAL			7.13	10.36	17.49

B1010 263	Floor-Ceiling, Wood Joists	COST PER S.F.		
		MAT.	INST.	TOTAL
1600	For alternate wood joist systems:			
1700	16" on center, 2" x 6" joists	6.95	10.25	17.20
1800	2" x 10"	7.30	10.60	17.90
1900	2" x 12"	7.70	10.60	18.30
2000	2" x 14"	8.70	10.80	19.50
2100	12" on center, 2" x 10" joists	7.50	10.90	18.40
2200	2" x 12"	8	10.95	18.95
2300	2" x 14"	9.25	11.15	20.40
2400				
2500				
2700	Cut & patch to match existing construction, add, minimum	2%	3%	
2800	Maximum	5%	9%	
2900	Dust protection, add, minimum	1%	2%	
3000	Maximum	4%	11%	
3100	Equipment usage curtailment, add, minimum	1%	1%	
3200	Maximum	3%	10%	
3300	Material handling & storage limitation, add, minimum	1%	1%	
3400	Maximum	6%	7%	
3500	Protection of existing work, add, minimum	2%	2%	
3600	Maximum	5%	7%	
3700	Shift work requirements, add, minimum		5%	
3800	Maximum		30%	
3900	Temporary shoring and bracing, add, minimum	2%	5%	
4000	Maximum	5%	12%	

B10 Superstructure

B1020 Roof Coverings

This page illustrates and describes a flat roof bar joist system including asphalt and gravel roof, lightweight concrete, metal decking, open web joists, and suspended acoustic ceiling. Lines within System Components give the unit price and total price per square foot for this system. Prices for alternate flat roof bar joist systems are on Line Items B1020 106 1300 thru 1600. Both material quantities and labor costs have been adjusted for the system listed.

Factors: To adjust for job conditions other than normal working situations use Lines B1020 106 2700 thru 4000.

Example: You are to install the system where material handling and storage present some problem. Go to Line B1020 106 3300 and apply these percentages to the appropriate MAT. and INST. costs.

System Components	QUANTITY	UNIT	COST PER S.F. MAT.	INST.	TOTAL
Three ply asphalt and gravel roof on lightweight concrete, metal decking on Web joist 4' O.C., with suspended acoustic ceiling.					
Open web joist, 10" deep, 4' O.C.	1.250	Lb.	.76	.54	1.30
Metal decking, 1-1/2" deep, 22 Ga., galvanized	1.000	S.F.	1.42	.66	2.08
Gravel roofing, 3 ply asphalt	.010	C.S.F.	1.33	1.72	3.05
Perlite roof fill, 3" thick	1.000	S.F.	2.01	.54	2.55
Suspended ceiling, 3/4" mineral fiber on "Z" bar suspension	1.000	S.F.	2.54	3.64	6.18
TOTAL			8.06	7.10	15.16

B1020 106	Steel Joist Roof & Ceiling	COST PER S.F. MAT.	INST.	TOTAL
1200	For alternate roof systems:			
1300	Open web bar joist, 16" deep, 5#/L.F.	8.30	7.25	15.55
1400	18" deep, 6.6#/L.F.	8.50	7.45	15.95
1500	20" deep, 9.6#/L.F.	8.75	7.60	16.35
1600	24" deep, 11.52#/L.F.	9.05	7.80	16.85
1700				
1800				
1900				
2000				
2100				
2200				
2300				
2400				
2500				
2700	Cut & patch to match existing construction, add, minimum	2%	3%	
2800	Maximum	5%	9%	
2900	Dust protection, add, minimum	1%	2%	
3000	Maximum	4%	11%	
3100	Equipment usage curtailment, add, minimum	1%	1%	
3200	Maximum	3%	10%	
3300	Material handling & storage limitation, add, minimum	1%	1%	
3400	Maximum	6%	7%	
3500	Protection of existing work, add, minimum	2%	2%	
3600	Maximum	5%	7%	
3700	Shift work requirements, add, minimum		5%	
3800	Maximum		30%	
3900	Temporary shoring and bracing, add, minimum	2%	5%	
4000	Maximum	5%	12%	

B10 Superstructure

B1020 Roof Coverings

This page illustrates and describes a wood frame roof system including rafters, ceiling joists, sheathing, building paper, asphalt shingles, roof trim, furring, insulation, plaster and paint. Lines within System Components give the unit price and total price per square foot for this system. Prices for alternate wood frame roof systems are on Line Items B1020 202 1900 thru 2700. Both material quantities and labor costs have been adjusted for the system listed.

Factors: To adjust for job conditions other than normal working situations use Lines B1020 202 3300 thru 4000.

Example: You are to install the system while protecting existing work. Go to Line B1020 202 3800 and apply these percentages to the appropriate MAT. and INST. costs.

System Components	QUANTITY	UNIT	COST PER S.F. MAT.	INST.	TOTAL
Wood frame roof system, 4 in 12 pitch, including rafters, sheathing, Shingles, insulation, drywall, thin coat plaster, and painting.					
Rafters, 2" x 6", 16" O.C., 4 in 12 pitch	1.080	L.F.	.44	1.18	1.62
Ceiling joists, 2" x 6", 16 O.C.	1.000	L.F.	.41	.87	1.28
Sheathing, 1/2" CDX	1.080	S.F.	.50	.84	1.34
Building paper, 15# felt	.011	C.S.F.	.07	.16	.23
Asphalt shingles, 240#	.011	C.S.F.	.77	1.10	1.87
Roof trim	.100	L.F.	.16	.24	.40
Furring, 1" x 3", 16" O.C.	1.000	L.F.	.38	1.56	1.94
Fiberglass insulation, 6" batts	1.000	S.F.	.85	.40	1.25
Gypsum board, 1/2" thick	1.000	S.F.	.26	.61	.87
Thin coat plaster	1.000	S.F.	.09	.68	.77
Paint, roller, 2 coats	1.000	S.F.	.17	.61	.78
TOTAL			4.10	8.25	12.35

B1020 202	Wood Frame Roof & Ceiling	MAT.	INST.	TOTAL
1800	For alternate roof systems:			
1900	Rafters 16" O.C., 2" x 8"	4.32	8.30	12.62
2000	2" x 10"	4.52	8.95	13.47
2100	2" x 12"	4.92	9.15	14.07
2200	Rafters 24" O.C., 2" x 6"	3.89	7.75	11.64
2300	2" x 8"	4.05	7.80	11.85
2400	2" x 10"	4.20	8.25	12.45
2500	2" x 12"	4.50	8.40	12.90
2600	Roof pitch, 6 in 12, add	3%	10%	
2700	8 in 12, add	5%	12%	
2800				
2900				
3000				
3100				
3300	Cut & patch to match existing construction, add, minimum	2%	3%	
3400	Maximum	5%	9%	
3500	Material handling & storage limitation, add, minimum	1%	1%	
3600	Maximum	6%	7%	
3700	Protection of existing work, add, minimum	2%	2%	
3800	Maximum	5%	7%	
3900	Shift work requirements, add, minimum		5%	
4000	Maximum		30%	

B10 Superstructure

B1020 Roof Construction

Wood Fabricated Roof Truss System

This page illustrates and describes a sloped roof truss framing system including trusses, truss anchors, sheathing, fascia and insulation. Lines B1020 210 1500 through 1800 are for a flat roof framing system including flat roof trusses, truss anchors, sheathing, fascia and insulation. Lines within system components give the unit price and total price per square foot for this system. Prices for alternate sizes and systems are given in line items B1020 210 0100 through 1800. Both material quantities and labor costs have been adjusted for the system listed.

Factors: To adjust for job conditions other than normal working situations use lines B1020 210 2700 through 4000.

Example: You are to install the system where crane placement and movement will be difficult. Go to line B1020 210 3200 and apply these percentages to the appropriate MAT. and INST. costs.

System Components	QUANTITY	UNIT	COST PER S.F. MAT.	INST.	TOTAL
Roof truss, 2' O.C., 4/12 slope, 12' span, 5/8 sheath., fascia, insul.					
Timber connectors, rafter anchors, galv., 1 1/2" x 5 1/4"	.084	Ea.	.04	.32	.36
Plywood sheathing, CDX, 5/8" thick	1.054	S.F.	.59	.89	1.48
Truss, 2' O.C., metal plate connectors, 12' span	.042	Ea.	2.14	1.68	3.82
Molding, fascia trim, 1" x 8"	.167	L.F.	.26	.41	.67
Insulation, fiberglass, kraft face, 9" thick x 23" wide	1.000	S.F.	.77	.40	1.17
TOTAL			3.80	3.70	7.50

B1020 210	Roof Truss, Wood	MAT.	INST.	TOTAL
0090	For alternate roof systems:			
0100	Roof truss, 2'O.C., 4/12p, 1'ovhg, 12'span, 5/8"sheath, fascia, insul.	3.80	3.70	7.50
0200	20' span	3.49	3.15	6.64
0300	24' span	3.39	2.94	6.33
0400	26' span	4.86	3.95	8.81
0500	28' span	3.25	2.88	6.13
0600	30' span	3.16	2.85	6.01
0700	32' span	3.42	2.80	6.22
0800	34' span	3.41	2.77	6.18
0890	60' span	5.05	3.14	8.19
0900	8/12 pitch, 20' span	3.74	3.39	7.13
1000	24' span	3.86	3.18	7.04
1100	26' span	3.71	3.10	6.81
1200	28' span	3.80	3.09	6.89
1300	32' span	3.93	3.04	6.97
1400	36' span	4.13	2.98	7.11
1500	Flat roof frame, fab.strl.joists, 2' O.C., 15'-24' span, 50 PSF LL	3.47	2.50	5.97
1600	55 PSF LL	3.63	2.56	6.19
1700	24'-30'span, 45 PSF LL	4.41	2.39	6.80
1800	55 PSF LL	5.35	2.46	7.81
2700	Cut & patch to match existing construction, add, minimum	2%	3%	
2800	Maximum	5%	9%	
2900	Dust protection, add, minimum	1%	2%	
3000	Maximum	4%	11%	
3100	Equipment usage curtailment, add, minimum	1%	1%	
3200	Maximum	3%	10%	

B10 Superstructure

B1020 Roof Construction

B1020 210	Roof Truss, Wood	COST PER S.F.		
		MAT.	INST.	TOTAL
3300	Material handling & storage limitation, add, minimum	1%	1%	
3400	Maximum	6%	7%	
3500	Protection of existing work, add, minimum	2%	2%	
3600	Maximum	5%	7%	
3700	Shift work requirements, add, minimum		5%	
3800	Maximum		30%	
3900	Temporary shoring and bracing, add, minimum	2%	5%	
4000	Maximum	5%	12%	

B20 Exterior Enclosure

B2010 Exterior Walls

This page illustrates and describes a masonry concrete block system including concrete block wall, pointed, reinforcing, waterproofing, gypsum plaster on gypsum lath, and furring. Lines within System Components give the unit price and total price per square foot for this system. Prices for alternate masonry concrete block wall systems are on Line Items B2010 108 1500 thru 2200. Both material quantities and labor costs have been adjusted for the system listed.

Factors: To adjust for job conditions other than normal working situations use Lines B2010 108 2700 thru 4000.

Example: You are to install the system and match existing construction at several locations. Go to Line B2010 108 2800 and apply these percentages to the appropriate MAT. and INST. costs.

System Components	QUANTITY	UNIT	COST PER S.F.		
			MAT.	INST.	TOTAL
Concrete block, 8" thick, reinforced every 2 courses, waterproofing gypsum Plaster over gypsum lath on 1" x 3" furring, interior painting & baseboard.					
Concrete block, 8" x 8" x 16", reinforced	1.000	S.F.	4.49	5.95	10.44
Silicone waterproofing, 2 coats	1.000	S.F.	.80	.17	.97
Bituminous coating, 1/16" thick	1.000	S.F.	.47	1	1.47
Furring, 1" x 3", 16" O.C.	1.000	L.F.	.38	2.10	2.48
Gypsum lath, 3/8" thick	.110	S.Y.	.61	.60	1.21
Gypsum plaster, 2 coats	.110	S.Y.	.45	2.58	3.03
Painting, 2 coats	1.000	S.F.	.11	.47	.58
Baseboard wood, 9/16" x 2-5/8"	.100	L.F.	.24	.23	.47
Paint baseboard, primer + 1 coat enamel	.100	L.F.	.01	.06	.07
TOTAL			7.56	13.16	20.72

B2010 108	Masonry Wall, Concrete Block	COST PER S.F.		
		MAT.	INST.	TOTAL
1400	For alternate exterior wall systems:			
1500	8" thick block, fluted 2 sides	10.30	15.45	25.75
1600	Deep grooved	8.20	15.45	23.65
1700	Slump block	9.25	13.45	22.70
1800	Split rib	7.65	15.45	23.10
1900				
2000	12" thick block, regular	6.85	15.75	22.60
2100	Slump block	14.55	14.65	29.20
2200	Split rib	8.50	17.80	26.30
2300				
2700	Cut & patch to match existing construction, add, minimum	2%	3%	
2800	Maximum	5%	9%	
2900	Dust protection, add, minimum	1%	2%	
3000	Maximum	4%	11%	
3100	Equipment usage curtailment, add, minimum	1%	1%	
3200	Maximum	3%	10%	
3300	Material handling & storage limitation, add, minimum	1%	1%	
3400	Maximum	6%	7%	
3500	Protection of existing work, add, minimum	2%	2%	
3600	Maximum	5%	7%	
3700	Shift work requirements, add, minimum		5%	
3800	Maximum		30%	
3900	Temporary shoring and bracing, add, minimum	2%	5%	
4000	Maximum	5%	12%	

B20 Exterior Enclosure

B2010 Exterior Walls

This page illustrates and describes a masonry wall, brick-stone system including brick, concrete block, durawall, insulation, plasterboard, taped and finished, furring, baseboard and painting interior. Lines within System Components give the unit price and total price per square foot for this system. Prices for alternate masonry wall, brick-stone systems are on Line Item B2010 129 1500 thru 2500. Both material quantities and labor costs have been adjusted for the system listed.

Factors: To adjust for job conditions other than normal working situations use Lines B2010 129 3100 thru 4200.

Example: You are to install the system without damaging the existing work. Go to Line B2010 129 3900 and apply these percentages to the appropriate MAT. and INST. costs.

System Components	QUANTITY	UNIT	COST PER S.F. MAT.	INST.	TOTAL
Face brick, 4" thick, concrete block back-up, reinforce every second course, 3/4" insulation, furring, 1/2" drywall, taped, finish, and painted, baseboard					
Face brick, 4" brick	1.000	S.F.	5.15	11.25	16.40
Concrete back-up block, reinforced 8" thick	1.000	S.F.	2.84	6.25	9.09
3/4" rigid polystyrene insulation	1.000	S.F.	.56	.68	1.24
Furring, 1" x 3", wood, 16" O.C.	1.000	L.F.	.38	1.10	1.48
Drywall, 1/2" thick	1.000	S.F.	.26	.55	.81
Taping & finishing	1.000	S.F.	.05	.55	.60
Painting, 2 coats	1.000	S.F.	.17	.61	.78
Baseboard, wood, 9/16" x 2-5/8"	.100	L.F.	.24	.23	.47
Paint baseboard, primer + 1 coat enamel	.100	L.F.	.01	.06	.07
TOTAL			9.66	21.28	30.94

B2010 131		Masonry Wall, Brick - Stone	COST PER S.F. MAT.	INST.	TOTAL
1400	For alternate exterior wall systems:				
1500		Face brick, Norman, 4" x 2-2/3" x 12" (4.5 per S.F.)	10.75	17.80	28.55
1600		Roman, 4" x 2" x 12" (6.0 per S.F.)	10.85	19.95	30.80
1700		Engineer, 4" x 3-1/5" x 8" (5.63 per S.F.)	8.30	19.60	27.90
1800		S.C.R., 6" x 2-2/3" x 12" (4.5 per S.F.)	10.25	18.05	28.30
1900		Jumbo, 6" x 4" x 12" (3.0 per S.F.)	9.85	15.75	25.60
2000		Norwegian, 6" x 3-1/5" x 12" (3.75 per S.F.)	9.05	16.65	25.70
2100					
2200					
2300		Stone, veneer, fieldstone, 6" thick	9.35	18.10	27.45
2400		Marble, 2" thick	50	32	82
2500		Limestone, 2" thick	43	20.50	63.50
3100	Cut & patch to match existing construction, add, minimum		2%	3%	
3200		Maximum	5%	9%	
3300	Dust protection, add, minimum		1%	2%	
3400		Maximum	4%	11%	
3500	Equipment usage curtailment, add, minimum		1%	1%	
3600		Maximum	3%	10%	
3700	Material handling & storage limitation, add, minimum		1%	1%	
3800		Maximum	6%	7%	
3900	Protection of existing work, add, minimum		2%	2%	
4000		Maximum	5%	7%	
4100	Shift work requirements, add, minimum			5%	
4200		Maximum		30%	

B20 Exterior Enclosure

B2010 Exterior Walls

This page illustrates and describes a parapet wall system including wall, coping, flashing and cant strip. Lines within System Components give the unit price and cost per lineal foot for this system. Prices for alternate parapet wall systems are on Lines B2010 142 1300 thru 2400. Both material quantities and labor costs have been adjusted for the system listed.

Factors: To adjust for job conditions other than normal working situations, use Lines B2010 142 2900 thru 4000.

Example: You are to install the system without damaging the adjacent property. Go to Line B2010 142 3600 and apply these percentages to the appropriate MAT. and INST. costs.

System Components	QUANTITY	UNIT	COST PER L.F.		
			MAT.	INST.	TOTAL
Concrete block parapet, incl. reinf., coping, flashing & cant strip, 2' high					
8" concrete block	2.000	S.F.	8.98	11.90	20.88
Masonry reinforcing	1.000	L.F.	.18	.18	.36
Coping, precast	1.000	L.F.	19.30	13.85	33.15
Roof cant	1.000	L.F.	2.45	1.54	3.99
Through wall flashing	1.000	L.F.	2.88	2.43	5.31
Cap flashing	1.000	L.F.	8.30	4.34	12.64
TOTAL			42.09	34.24	76.33

B2010 142	Parapet Wall	COST PER L.F.		
		MAT.	INST.	TOTAL
1200	For alternate systems:			
1300	Concrete block			
1400	12" thick	45.50	45	90.50
1500	Split rib block, 8" thick	42.50	39	81.50
1600	12" thick	43.50	46	89.50
1700	Brick, common brick, 8" wall	48	59	107
1800	12" wall	56	74	130
1900	4" brick, 4" backup block	47	56.50	103.50
2000	8" backup block	49	58	107
2100	Stucco on masonry	42.50	38.50	81
2200	On wood frame	17.85	30.50	48.35
2300	Wood, T 1-11 siding	26.50	37.50	64
2400	Boards, 1" x 6" cedar	28	41.50	69.50
2500				
2600				
2700				
2900	Dust protection, add, minimum	1%	2%	
3000	Maximum	4%	11%	
3100	Equipment usage curtailment, add, minimum	1%	1%	
3200	Maximum	3%	10%	
3300	Material handling & storage limitation, add, minimum	1%	1%	
3400	Maximum	6%	7%	
3500	Protection of existing work, add, minimum	2%	2%	
3600	Maximum	5%	7%	
3700	Shift work requirements, add, minimum		5%	
3800	Maximum		30%	
3900	Temporary shoring and bracing, add, minimum	2%	5%	
4000	Maximum	5%	12%	

B20 Exterior Enclosure

B2010 Exterior Walls

This page illustrates and describes a masonry cleaning and restoration system, including staging, cleaning, repointing. Lines within System Components give the unit price and cost per square foot for this system. Prices for alternate systems are on Lines B2010 143 1300 thru 2500. Both material quantities and labor costs have been adjusted for the system listed.

Factors: To adjust for conditions other than normal working situations, use Lines B2010 143 2900 thru 4200.

Example: You are to clean a wall and be concerned about dust control. Go to line B2010 143 3100 and apply these percentages to the appropriate MAT. and INST. costs.

System Components	QUANTITY	UNIT	COST PER S.F. MAT.	INST.	TOTAL
Repoint existing building, brick, common bond, high pressure cleaning, Water only, soft old mortar.					
Scaffolding, steel tubular, bldg ext wall face, labor only, 1 to 5 stories	.010	S.F.		2.05	2.05
Cleaning, high pressure water only	1.100	S.F.		1.80	1.80
Repoint, brick common bond	1.000	S.F.	.58	5.60	6.18
TOTAL			.58	9.45	10.03

B2010 143	Masonry Restoration - Cleaning	MAT.	INST.	TOTAL
1200	For alternate masonry surfaces:			
1300	Brick, common bond	.58	9.45	10.03
1400	Flemish bond	.61	9.80	10.41
1500	English bond	.61	10.40	11.01
1700	Stone, 2' x 2' blocks	.77	7.25	8.02
1800	2' x 4' blocks	.58	6.40	6.98
2000	Add to above prices for alternate cleaning systems:			
2100	Chemical brush and wash	.04	.61	.65
2200	High pressure chemical and water	.05	.61	.66
2300	Sandblasting, wet system	.63	1.67	2.30
2400	Dry system	.59	.98	1.57
2500	Steam cleaning		.49	.49
2600				
2900	Cut & patch to match existing construction, add, minimum	2%	3%	
3000	Maximum	5%	9%	
3100	Dust protection, add, minimum	1%	2%	
3200	Maximum	4%	11%	
3300	Equipment usage curtailment, add, minimum	1%	1%	
3400	Maximum	3%	10%	
3500	Material handling & storage limitation, add, minimum	1%	1%	
3600	Maximum	6%	7%	
3800				
3900	Protection of existing work, add, minimum	2%	2%	
4000	Maximum	5%	7%	
4100	Shift work requirements, add, minimum		5%	
4200	Maximum		30%	

B20 Exterior Enclosure

B2010 Exterior Walls

This page illustrates and describes a wood frame exterior wall system including wood studs, sheathing, felt, insulation, plasterboard, taped and finished, baseboard and painted interior. Lines within System Components give the unit price and total price per square foot for this system. Prices for alternate wood frame exterior wall systems are on Line Items B2010 150 1700 thru 2700. Both material quantities and labor costs have been adjusted for the system listed.

Factors: To adjust for job conditions other than normal working situations use Lines B2010 150 3100 thru 4200.

Example: You are to install the system with need for complete temporary bracing. Go to Line B2010 150 4200 and apply these percentages to the appropriate MAT. and INST. costs.

System Components	QUANTITY	UNIT	COST PER S.F. MAT.	INST.	TOTAL
Wood stud wall, cedar shingle siding, building paper, plywood sheathing, Insulation, 5/8" drywall, taped, finished and painted, baseboard.					
2" x 4" wood studs, 16" O.C.	.100	L.F.	.32	1.09	1.41
1/2" CDX sheathing	1.000	S.F.	.46	.78	1.24
18" No. 1 red cedar shingles, 7-1/2" exposure	.008	C.S.F.	1.71	1.94	3.65
15# felt paper	.010	C.S.F.	.06	.15	.21
3-1/2" fiberglass insulation	1.000	S.F.	.59	.40	.99
5/8" drywall	1.000	S.F.	.31	.55	.86
Drywall finishing adder	1.000	S.F.	.05	.55	.60
Baseboard trim, stock pine, 9/16" x 3-1/2", painted	.100	L.F.	.24	.23	.47
Paint baseboard, primer + 1 coat enamel	.100	L.F.	.01	.06	.07
Paint, 2 coats, interior	1.000	S.F.	.17	.61	.78
TOTAL			3.92	6.36	10.28

B2010 150	Wood Frame Exterior Wall	COST PER S.F. MAT.	INST.	TOTAL
1600	For alternate exterior wall systems:			
1700	Aluminum siding, horizontal clapboard	4.11	6.55	10.66
1800	Cedar bevel siding, 1/2" x 6", vertical, painted	7.40	6.25	13.65
1900	Redwood siding 1" x 4" to 1" x 6" vertical, T & G	5.60	6.90	12.50
2000	Board and batten	6.55	5.70	12.25
2100	Ship lap siding	4.78	6.05	10.83
2200	Plywood, grooved (T1-11) fir	3.72	6.05	9.77
2300	Redwood	4.88	6.05	10.93
2400	Southern yellow pine	3.52	6.05	9.57
2500	Masonry on stud wall, stucco, wire and plaster	2.89	9.45	12.34
2600	Stone veneer	7.05	12.50	19.55
2700	Brick veneer, brick $275 per M	7.35	15.65	23
3100	Cut & patch to match existing construction, add, minimum	2%	3%	
3200	Maximum	5%	9%	
3300	Dust protection, add, minimum	1%	2%	
3400	Maximum	4%	11%	
3500	Material handling & storage limitation, add, minimum	1%	1%	
3600	Maximum	6%	7%	
3700	Protection of existing work, add, minimum	2%	2%	
3800	Maximum	5%	7%	
3900	Shift work requirements, add, minimum		5%	
4000	Maximum		30%	
4100	Temporary shoring and bracing, add, minimum	2%	5%	
4200	Maximum	5%	12%	

B20 Exterior Enclosure

B2010 Exterior Walls

B2010 190	Selective Price Sheet	COST PER S.F.		
		MAT.	INST.	TOTAL
0100	Exterior surface, masonry, concrete block, standard 4" thick	1.61	5.75	7.36
0200	6" thick	2.29	6.20	8.49
0300	8" thick	2.60	6.60	9.20
0400	12" thick	3.76	8.55	12.31
0500	Split rib, 4" thick	3.48	7.20	10.68
0600	8" thick	4.58	8.25	12.83
0700	Brick running bond, standard size, 6.75/S.F.	5.15	11.25	16.40
0800	Buff, 6.75/S.F.	5.45	11.25	16.70
0900	Stucco, on frame	.68	5.10	5.78
1000	On masonry	.28	4.02	4.30
1100	Metal, aluminum, horizontal, plain	1.90	2.12	4.02
1200	Insulated	2.05	2.12	4.17
1300	Vertical, plain	1.68	2.12	3.80
1400	Insulated	1.75	2.21	3.96
1500	Wood, beveled siding, "A" grade cedar, 1/2" x 6"	5.20	1.85	7.05
1600	1/2" x 8"	6.10	1.65	7.75
1700	Shingles, 16" #1 red, 7-1/2" exposure	2.27	2.66	4.93
1800	18" perfections, 7-1/2" exposure	1.90	2.23	4.13
1900	Handsplit, 10" exposure	1.82	2.18	4
2000	White cedar, 7-1/2" exposure	1.37	2.73	4.10
2100	Vertical, board & batten, redwood	3.41	2.48	5.89
2200	White pine	2.42	1.65	4.07
2300	T. & G. boards, redwood, 1" x 4"	3.41	2.48	5.89
2400	1' x 8"	3.56	1.65	5.21
2500				
2600	Interior surface, drywall, taped & finished, standard, 1/2"	.31	1.13	1.44
2700	5/8" thick	.35	1.13	1.48
2800	Fire resistant, 1/2" thick	.35	1.13	1.48
2900	5/8" thick	.34	1.13	1.47
3000	Moisture resistant, 1/2" thick	.43	1.13	1.56
3100	5/8" thick	.52	1.13	1.65
3200	Core board, 1" thick	.61	2.27	2.88
3300	Plaster, gypsum, 2 coats	.46	2.60	3.06
3400	3 coats	.65	3.13	3.78
3500	Perlite or vermiculite, 2 coats	.46	2.96	3.42
3600	3 coats	.77	3.67	4.44
3700	Gypsum lath, standard, 3/8" thick	.62	.60	1.22
3800	1/2" thick	.60	.64	1.24
3900	Fire resistant, 3/8" thick	.30	.74	1.04
4000	1/2" thick	.30	.79	1.09
4100	Metal lath, diamond, 2.5 lb.	.55	.60	1.15
4200	Rib, 3.4 lb.	.69	.74	1.43
4300	Framing metal studs including top and bottom			
4400	Runners, walls 10' high			
4500	24" O.C., non load bearing 20 gauge, 2-1/2" wide	.30	.91	1.21
4600	3-5/8" wide	.33	.93	1.26
4700	4" wide	.39	.93	1.32
4800	6" wide	.46	.94	1.40
4900	Load bearing 18 gauge, 2-1/2" wide	.62	1.02	1.64
5000	3-5/8" wide	.74	1.04	1.78
5100	4" wide	.83	1.33	2.16
5200	6" wide	.98	1.08	2.06
5300	16" O.C., non load bearing 20 gauge, 2-1/2" wide	.41	1.40	1.81
5400	3-5/8" wide	.44	1.42	1.86
5500	4" wide	.53	1.44	1.97
5600	6" wide	.62	1.46	2.08
5700	Load bearing 18 gauge, 2-1/2" wide	.84	1.42	2.26
5800	3-5/8" wide	1	1.44	2.44

B20 Exterior Enclosure

B2010 Exterior Walls

B2010 190	Selective Price Sheet	COST PER S.F.		
		MAT.	INST.	TOTAL
5900	4" wide	1.05	1.48	2.53
6000	6" wide	1.33	1.50	2.83
6100	Framing wood studs incl. double top plate and			
6200	Single bottom plate, walls 10' high			
6300	24" O.C., 2" x 4"	.24	.88	1.12
6400	2" x 6"	.37	.95	1.32
6500	16" O.C., 2" x 4"	.32	1.09	1.41
6600	2" x 6"	.48	1.22	1.70
6700	Sheathing, boards, 1" x 6"	1.53	1.68	3.21
6800	1" x 8"	1.47	1.43	2.90
6900	Plywood, 3/8" thick	.40	.91	1.31
7000	1/2" thick	.46	.97	1.43
7100	5/8" thick	.56	1.04	1.60
7200	3/4" thick	.64	1.12	1.76
7300	Wood fiber, 5/8" thick	.78	.91	1.69
7400	Gypsum weatherproof, 1/2" thick	.42	1.04	1.46
7500	Insulation, fiberglass batts, 3-1/2" thick, R13	.42	.78	1.20
7600	6" thick, R19	.53	.91	1.44
7700	Poured 4" thick, fiberglass wool, R4/inch	.51	2.73	3.24
7800	Mineral wool, R3/inch	.43	2.73	3.16
7900	Polystyrene, R4/inch	3.40	2.73	6.13
8000	Perlite or vermiculite, R2.7/inch	1.93	2.73	4.66
8100	Rigid, fiberglass, R4.3/inch, 1" thick	.54	.55	1.09
8200	R8.7/inch, 2" thick	1.27	.61	1.88

B20 Exterior Enclosure

B2020 Exterior Windows

This page illustrates and describes an aluminum window system including double hung aluminum window, exterior and interior trim, hardware and insulating glass. Lines within System Components give the unit price and total price on a cost each basis for this system. Prices for alternate aluminum window systems are on Line Items B2020 110 1100 thru 2400. Both material quantities and labor costs have been adjusted for the system listed.

Factors: To adjust for job conditions other than normal working situations use Lines B2020 110 3100 thru 4000.

Example: You are to install the system and cut and patch to match existing construction. Go to Line B2020 110 3200 and apply these percentages to the appropriate MAT. and INST. costs.

System Components	QUANTITY	UNIT	COST EACH		
			MAT.	INST.	TOTAL
Double hung aluminum window, 2'-4" x 2'-6" exterior and interior trim,					
Hardware glazed with insulating glass.					
Double hung aluminum window, 2'-4" x 2' 6"	1.000	Ea.	233.20	39.94	273.14
Exterior and interior trim	1.000	Set	35.50	42	77.50
Hardware	1.000	Set	4.38	23	27.38
Insulating glass	5.830	S.F.	73.46	30.02	103.48
TOTAL			346.54	134.96	481.50

B2020 110		Windows - Aluminum	COST EACH		
			MAT.	INST.	TOTAL
1000	For alternate window systems:				
1100		Aluminum, double hung, 2'-8" x 4'-6"	680	222	902
1200		3'-4" x 5'-6"	1,025	335	1,360
1300		Casement, 3'-6" x 2'-4"	465	163	628
1400		4'-6" x 2'-4"	595	203	798
1500		5'-6" x 2'-4"	740	268	1,008
1600		Projected window, 2'-1" x 3'-0"	345	140	485
1700		3'-5" x 3'-0"	545	201	746
1800		4'-0" x 4'-0"	850	305	1,155
1900		Horizontal sliding 3'-0" x 3'-0"	355	173	528
2000		3'-6" x 4'-0"	540	246	786
2100		5'-0" x 6'-0"	1,125	475	1,600
2200		Picture window, 3'-8" x 3'-1"	390	200	590
2300		4'-4" x 4'-5"	635	305	940
2400		5'-8" x 4'-9"	905	435	1,340
2500					
2600					
2700					
2800					
2900					
3100	Cut & patch to match existing construction, add, minimum		2%	3%	
3200		Maximum	5%	9%	
3300	Dust protection, add, minimum		1%	2%	
3400		Maximum	4%	11%	
3500	Material handling & storage limitation, add, minimum		1%	1%	
3600		Maximum	6%	7%	
3700	Protection of existing work, add, minimum		2%	2%	
3800		Maximum	5%	7%	
3900	Shift work requirements, add, minimum			5%	
4000		Maximum		30%	

B20 Exterior Enclosure

B2020 Exterior Windows

This page illustrates and describes a wood window system including wood picture window, exterior and interior trim, hardware and insulating glass. Lines within System Components give the unit price and total price on a cost each basis for this system. Prices for alternate wood window systems are on Line Items B2020 112 1300 thru 1800. Both material quantities and labor costs have been adjusted for the system listed.

Factors: To adjust for job conditions other than normal working situations use Lines B2020 112 3100 thru 4000.

Example: You are to install the above system where dust control is vital. Go to Line B2020 112 3400 and apply these percentages to the appropriate MAT. and INST. costs.

System Components	QUANTITY	UNIT	COST EACH		
			MAT.	INST.	TOTAL
Wood picture window 4'-0" x 4'-6", exterior and interior trim, hardware, Glazed with insulating glass.					
4'-0" x 4'-6" wood picture window with insulating glass	1.000	Ea.	570	99.50	669.50
Exterior and interior trim	1.000	Set	68.50	91	159.50
Hardware	1.000	Set	4.38	23	27.38
TOTAL			642.88	213.50	856.38

B2020 112	Windows - Wood	COST EACH		
		MAT.	INST.	TOTAL
1200	For alternate window systems:			
1300	Picture window 5'-0" x 4'-0"	725	214	939
1400	6'-0" x 4'-6"	760	223	983
1500	Bow, bay window, 8'-0" x 5'-0", standard	2,525	223	2,748
1600	Deluxe	1,850	223	2,073
1700	Bow, bay window, 12'-0" x 6'-0" standard	3,225	296	3,521
1800	Deluxe	2,600	296	2,896
1900	Double hung, 3'-0" x 4'-0"	350	138	488
2000	4'-0" x 4'-6"	450	183	633
2100	Casement 2'-0" x 3'-0"	288	120	408
2200	2 leaf, 4'-0" x 4'-0"	625	199	824
2300	3 leaf, 6'-0" x 6'-0"	1,625	320	1,945
2400	Awning, 2'-10" x 1'-10"	320	120	440
2500	3'-6" x 2'-4"	410	138	548
2600	4'-0" x 3'-0"	595	183	778
2700	Horizontal sliding 3'-0" x 2'-0"	360	120	480
2800	4'-0" x 3'-6"	430	138	568
2900	6'-0" x 5'-0"	625	183	808
3100	Cut & patch to match existing construction, add, minimum	2%	3%	
3200	Maximum	5%	9%	
3300	Dust protection, add, minimum	1%	2%	
3400	Maximum	4%	11%	
3500	Material handling & storage limitation, add, minimum	1%	1%	
3600	Maximum	6%	7%	
3700	Protection of existing work, add, minimum	2%	2%	
3800	Maximum	5%	7%	
3900	Shift work requirements, add, minimum		5%	
4000	Maximum		30%	

B20 Exterior Enclosure

B2020 Exterior Windows

This page illustrates and describes storm window and door systems based on a cost each price. Prices for alternate storm window and door systems are on Line Items B2020 116 0400 thru 2400. Both material quantities and labor costs have been adjusted for the system listed.

Factors: To adjust for job conditions other than normal working situations use Lines B2020 116 3100 thru 4000 and apply these percentages to the appropriate MAT and INST. costs.

Example: You are to install the system and protect all existing construction. Go to Line B2020 116 3800 and apply these percentages to the appropriate MAT. and INST. costs.

B2020 116	Storm Windows & Doors	COST EACH		
		MAT.	INST.	TOTAL
0350	Storm Window and door systems, single glazed			
0400	Window, custom aluminum anodized, 2'-0" x 3'-5"	99	36.50	135.50
0500	2'-6" x 5'-0"	121	39	160
0600	4'-0" x 6'-0"	249	43.50	292.50
0700	White painted aluminum, 2'-0" x 3'-5"	110	36.50	146.50
0800	2'-6" x 5'-0"	176	39	215
0900	4'-0" x 6'-0"	300	43.50	343.50
1000	Average quality aluminum, anodized, 2'-0" x 3'-5"	93.50	36.50	130
1100	2'-6" x 5'-0"	117	39	156
1200	4'-0" x 6'-0"	139	43.50	182.50
1300	White painted aluminum, 2'-0" x 3'-5"	93.50	36.50	130
1400	2'-6" x 5'-0"	99	39	138
1500	4'-0" x 6'-0"	108	43.50	151.50
1600	Mill finish, 2'-0" x 3'-5"	82.50	36.50	119
1700	2'-6" x 5'-0"	93.50	39	132.50
1800	4'-0" x 6'-0"	109	43.50	152.50
1900				
2000	Door, aluminum anodized 3'-0" x 6'-8"	171	78	249
2100	White painted aluminum, 3'-0" x 6'-8"	320	78	398
2200	Mill finish, 3'-0" x 6'-8"	271	78	349
2300				
2400	Wood, storm and screen, painted	325	121	446
2500				
2600				
2700				
2800				
3100	Cut & patch to match existing construction, add, minimum	2%	3%	
3200	Maximum	5%	9%	
3300	Dust protection, add, minimum	1%	2%	
3400	Maximum	4%	11%	
3500	Material handling & storage limitation, add, minimum	1%	1%	
3600	Maximum	6%	7%	
3700	Protection of existing work, add, minimum	2%	2%	
3800	Maximum	5%	7%	
3900	Shift work requirements, add, minimum		5%	
4000	Maximum		30%	

B20 Exterior Enclosure

B2020 Exterior Windows

This page illustrates and describes a window wall system including aluminum tube framing, caulking, and glass. Lines within System Components give the unit price and total price per square foot for this system. Prices for alternate window wall systems are on Line Items B2020 124 1400 thru 2600. Both material quantities and labor costs have been adjusted for the system listed.

Factors: To adjust for job conditions other than normal working situations use Lines B2020 124 3100 thru 4000.

Example: You are to install the system with need for complete dust protection. Go to Line B2020 124 3400 and apply these percentages to the appropriate MAT. and INST. costs.

System Components	QUANTITY	UNIT	COST PER S.F.		
			MAT.	INST.	TOTAL
Window wall, including aluminum header, sill, mullions, caulking And glass					
Header, mill finish, 1-3/4" x 4-1/2" deep	.167	L.F.	3.11	2.08	5.19
Sill, mill finish, 1-3/4" x 4-1/2" deep	.167	L.F.	3	2.03	5.03
Vertical mullion, 1-3/4" x 4-1/2" deep, 6' O.C.	.191	L.F.	7.52	5.07	12.59
Caulking	.381	L.F.	.06	.71	.77
Glass, 1/4" plate	1.000	S.F.	5.90	8.60	14.50
TOTAL			19.59	18.49	38.08

B2020 124	Aluminum Frame, Window Wall	COST PER S.F.		
		MAT.	INST.	TOTAL
1200	For alternate systems:			
1300				
1400	Mill finish, 2" x 4-1/2" deep, insulating glass	22	21.50	43.50
1500	Thermo-break, 2-1/4" x 4-1/2" deep, insulating glass	23	21.50	44.50
1600	Bronze finish, 1-3/4" x 4-1/2" deep, 1/4" plate glass	22	20	42
1700	2" x 4-1/2" deep, insulating glass	24	23	47
1800	Thermo-break, 2-1/4" x 4-1/2" deep, insulating glass	25.50	23.50	49
1900				
2000	Black finish, 1-3/4" x 4-1/2" deep, 1/4" plate glass	23	21	44
2100	2" x 4-1/2" deep, insulating glass	25	24	49
2200	Thermo-break, 2-1/4" x 4-1/2" deep, insulating glass	26.50	24.50	51
2300				
2400	Stainless steel, 1-3/4" x 4-1/2" deep, 1/4" plate glass	30	25.50	55.50
2500	2" x 4-1/2" deep, insulating glass	30.50	28.50	59
2600	Thermo-break, 2-1/4" x 4-1/2" deep, insulating glass	32.50	29.50	62
2700				
3100	Cut & patch to match existing construction, add, minimum	2%	3%	
3200	Maximum	5%	9%	
3300	Dust protection, add, minimum	1%	2%	
3400	Maximum	4%	11%	
3500	Material handling, & storage limitation, add, minimum	1%	1%	
3600	Maximum	6%	7%	
3700	Protection of existing work, add, minimum	2%	2%	
3800	Maximum	5%	7%	
3900	Shift work requirements, add, minimum		5%	
4000	Maximum		30%	

B20 Exterior Enclosure

B2030 Exterior Doors

This page illustrates and describes a commercial metal door system, including a single aluminum and glass door, narrow stiles, jamb, hardware weatherstripping, panic hardware and closer. Lines within System Components give the unit price and total price on a cost each basis for this system. Prices for alternate commercial metal door systems are on Line Items B2030 125 1300 thru 2500. Both material quantities and labor costs have been adjusted for the system listed.

Factors: To adjust for job conditions other than normal working situations, use Lines B2030 125 3100 thru 4000.

Example: You are to install the system and cut and patch to match existing construction. Go to Line B2030 125 3200 and apply these percentages to the appropriate MAT. and INST. costs.

System Components	QUANTITY	UNIT	COST EACH		
			MAT.	INST.	TOTAL
Single aluminum and glass door, 3'-0"x7'-0", with narrow stiles, ext. jamb. Weatherstripping, 1/2" tempered insul. glass, panic hardware, and closer.					
Aluminum door, 3'-0" x 7'-0" x 1-3/4", narrow stiles	1.000	Ea.	820	685	1,505
Tempered insulating glass, 1/2" thick	20.000	S.F.	500	376	876
Panic hardware	1.000	Set	545	91	636
Automatic closer	1.000	Ea.	122	84	206
TOTAL			1,987	1,236	3,223

B2030 125	Doors, Metal - Commercial	COST EACH		
		MAT.	INST.	TOTAL
1200	For alternate door systems:			
1300	Single aluminum and glass with transom, 3'-0" x 10'-0"	2,675	1,450	4,125
1400	Anodized aluminum and glass, 3'-0" x 7'-0"	2,275	1,475	3,750
1500	With transom, 3'-0" x 10'-0"	3,125	1,725	4,850
1600	Steel, deluxe, hollow metal 3'-0" x 7'-0"	1,500	470	1,970
1700	With transom 3'-0" x 10'-0"	2,000	535	2,535
1800	Fire door, "A" label, 3'-0" x 7'-0"	1,675	470	2,145
1900	Double, aluminum and glass, 6'-0" x 7'-0"	3,000	2,100	5,100
2000	With transom, 6'-0" x 10'-0"	3,900	2,600	6,500
2100	Anodized aluminum and glass 6'-0" x 7'-0"	3,400	2,475	5,875
2200	With transom, 6'-0" x 10'-0"	4,475	3,050	7,525
2300	Steel, deluxe, hollow metal, 6'-0" x 7'-0"	2,350	820	3,170
2400	With transom, 6'-0" x 10'-0"	3,375	950	4,325
2500	Fire door, "A" label, 6'-0" x 7'-0"	2,600	815	3,415
2800				
2900				
3100	Cut & patch to match existing construction, add, minimum	2%	3%	
3200	Maximum	5%	9%	
3300	Dust protection, add, minimum	1%	2%	
3400	Maximum	4%	11%	
3500	Material handling & storage limitation, add, minimum	1%	1%	
3600	Maximum	6%	7%	
3700	Protection of existing work, add, minimum	2%	2%	
3800	Maximum	5%	7%	
3900	Shift work requirements, add, minimum		5%	
4000	Maximum		30%	

B20 Exterior Enclosure

B2030 Exterior Doors

This page illustrates and describes sliding door systems including a sliding door, frame, interior and exterior trim with exterior staining. Lines within System Components give the unit price and total price on a cost each basis for this system. Prices for alternate sliding door systems are on Line Items B2030 150 1100 thru 2400. Both material quantities and labor costs have been adjusted for the system listed.

Factors: To adjust for job conditions other than normal working situations use Lines B2030 150 2700 thru 4000.

Example: You are to install the system with temporary shoring and bracing. Go to Line B2030 150 3900 and apply these percentages to the appropriate MAT. and INST. costs.

System Components	QUANTITY	UNIT	COST EACH		
			MAT.	INST.	TOTAL
Sliding wood door, 6'-0" x 6'-8", with wood frame, interior and exterior Trim and exterior staining.					
Sliding wood door, standard, 6'-0" x 6'-8", insulated glass	1.000	Ea.	1,225	273	1,498
Interior & exterior trim	1.000	Set	29.20	45.40	74.60
Stain door	1.000	Ea.	1.88	27	28.88
Stain trim	1.000	Ea.	2.20	11.60	13.80
TOTAL			1,258.28	357	1,615.28

B2030 150	Doors, Sliding - Patio	COST EACH		
		MAT.	INST.	TOTAL
1000	For alternate sliding door systems:			
1100	Wood, standard, 8'-0" x 6'-8", insulated glass	1,650	455	2,105
1200	12'-0" x 6'-8"	2,700	550	3,250
1300	Vinyl coated, 6'-0" x 6'-8"	1,550	355	1,905
1400	8'-0" x 6'-8"	1,975	455	2,430
1500	12'-0" x 6'-8"	3,275	550	3,825
1700	Aluminum, standard, 6'-0" x 6'-8", insulated glass	1,025	385	1,410
1800	8'-0" x 6'-8"	1,575	490	2,065
1900	12'-0" x 6'-8"	1,725	580	2,305
2000	Anodized, 6'-0" x 6'-8"	1,775	385	2,160
2100	8'-0" x 6'-8"	1,825	490	2,315
2200	12'-0" x 6'-8"	3,325	580	3,905
2300				
2400	Deduct for single glazing	87.50		87.50
2500				
2700	Cut & patch to match existing construction, add, minimum	2%	3%	
2800	Maximum	5%	9%	
2900	Dust protection, add, minimum	1%	2%	
3000	Maximum	4%	11%	
3100	Equipment usage curtailment, add, minimum	1%	1%	
3200	Maximum	3%	10%	
3300	Material handling & storage limitation, add, minimum	1%	1%	
3400	Maximum	6%	7%	
3500	Protection of existing, work, add, minimum	2%	2%	
3600	Maximum	5%	7%	
3700	Shift work requirements, add, minimum		5%	
3800	Maximum		30%	
3900	Temporary shoring and bracing, add, minimum	2%	5%	
4000	Maximum	5%	12%	

B20 Exterior Enclosure

B2030 Exterior Doors

This page illustrates and describes residential door systems including a door, frame, trim, hardware, weatherstripping, stained and finished. Lines within System Components give the unit price and total price on a cost each basis for this system. Prices for alternate residential door systems are on Line Items B2030 240 1500 thru 2200. Both material quantities and labor costs have been adjusted for the system listed.

Factors: To adjust for job conditions other than normal working situations use Lines B2030 240 3100 thru 4000.

Example: You are to install the system with a material handling and storage limitation. Go to Line B2030 240 3600 and apply these percentages to the appropriate MAT. and INST. costs.

System Components	QUANTITY	UNIT	COST EACH		
			MAT.	INST.	TOTAL
Single solid wood colonial door, 3' x 6'-8" with wood frame and trim,					
Stained and finished, including hardware and weatherstripping.					
Solid wood colonial door, fir 3' x 6'-8" x 1-3/4", hinges	1.000	Ea.	475	73	548
Hinges, ball bearing	1.000	Ea.	35		35
Exterior frame, with trim	1.000	Set	158.95	49.47	208.42
Interior trim	1.000	Set	25	45.50	70.50
Lockset	1.000	Set	44	39	83
Sill, oak 8" deep	3.500	Ea.	65.80	38.15	103.95
Weatherstripping	1.000	Set	24	72	96
Stained and finished	1.000	Ea.	4.02	57.50	61.52
TOTAL			831.77	374.62	1,206.39

B2030 240	Doors, Residential - Exterior	COST EACH		
		MAT.	INST.	TOTAL
1400	For alternate exterior door system:			
1500	Single doors			
1600	Hollow metal exterior door, plain	725	350	1,075
1700	Solid core wood door, plain	480	360	840
1800				
1900	Double doors, 6' x 6'-8"			
2000	Solid colonial double doors, fir	1,400	500	1,900
2100	Hollow metal exterior doors, plain	1,175	465	1,640
2200	Hollow core wood doors, plain	695	485	1,180
2300				
2800				
2900				
3100	Cut & patch to match existing construction, add, minimum	2%	3%	
3200	Maximum	5%	9%	
3300	Dust protection, add, minimum	1%	2%	
3400	Maximum	4%	11%	
3500	Material handling & storage limitation, add, minimum	1%	1%	
3600	Maximum	6%	7%	
3700	Protection of existing work, add, minimum	2%	2%	
3800	Maximum	5%	7%	
3900	Shift work requirements, add, minimum		5%	
4000	Maximum		30%	

B20 Exterior Enclosure

B2030 Exterior Doors

This page illustrates and describes overhead door systems including an overhead door, track, hardware, trim, and electric door opener. Lines within System Components give the unit price and total price on a cost each basis for this system. Prices for alternate overhead door systems are on Line Items B2030 410 1300 thru 2300. Both material quantities and labor costs have been adjusted for the system listed.

Factors: To adjust for job conditions other than normal working situations use Lines B2030 410 3100 thru 4000.

Example: You are to install the system and match existing construction. Go to Line B2030 410 3200 and apply these percentages to the appropriate MAT. and INST. costs.

System Components	QUANTITY	UNIT	COST EACH		
			MAT.	INST.	TOTAL
Wood overhead door, commercial sectional, including track, door, hardware And trim, electrically operated.					
Commercial, heavy duty wood door, 8' x 8' x 1-3/4" thick	1.000	Ea.	850	545	1,395
Frame, 2 x 8, pressure treated	1.000	Set	27.36	68.88	96.24
Wood trim	1.000	Set	58.56	54.48	113.04
Painting, two coats	1.000	Ea.	18.72	184.32	203.04
Electric trolley operator	1.000	Ea.	985	273	1,258
TOTAL			1,939.64	1,125.68	3,065.32

B2030 410	Doors, Overhead	COST EACH		
		MAT.	INST.	TOTAL
1200	For alternate sectional overhead door systems:			
1300	Commercial, wood, 1-3/4" thick, 12' x 12'	2,975	1,575	4,550
1400	14' x 14'	4,000	1,875	5,875
1500	Fiberglass & aluminum 12' x 12'	4,000	1,625	5,625
1600	20' x 20'	8,575	4,100	12,675
1700	Residential, wood, 9' x 7'	880	430	1,310
1800	16' x 7'	1,700	640	2,340
1900	Hardboard faced, 9' x 7'	695	430	1,125
2000	16' x 7'	1,375	640	2,015
2100	Fiberglass & aluminum, 9' x 7'	895	500	1,395
2200	16' x 7'	1,825	640	2,465
2300	For residential electric opener, add	660	68.50	728.50
2400				
2500				
2600				
2700				
2800				
2900				
3100	Cut & patch to match existing construction, add, minumum	2%	3%	
3200	Maximum	5%	9%	
3300	Dust protection, add, minimum	1%	2%	
3400	Maximum	4%	11%	
3500	Material handling & storage limitation, add, minimum	1%	1%	
3600	Maximum	6%	7%	
3700	Protection of existing work, add, minimum	2%	2%	
3800	Maximum	5%	7%	
3900	Shift work requirements, add, minimum		5%	
4000	Maximum		30%	

B20 Exterior Enclosure

B2030 Exterior Doors

B2030 810	Selective Price Sheet	COST EACH		
		MAT.	INST.	TOTAL
0100	Door closer, rack and pinion	171	84	255
0200	Backcheck and adjustable power	182	91	273
0300	Regular, hinge face mount, all sizes, regular arm	220	84	304
0400	Hold open arm	264	84	348
0500	Top jamb mount, all sizes, regular arm	215	91	306
0600	Hold open arm	226	91	317
0700	Stop face mount, all sizes, regular arm	209	84	293
0800	Hold open arm	229	84	313
0900	Fusible link, hinge face mount, all sizes, regular arm	248	84	332
1000	Hold open arm	292	84	376
1100	Top jamb mount, all sizes, regular arm	243	91	334
1200	Hold open arm	254	91	345
1300	Stop face mount, all sizes, regular arm	237	84	321
1400	Hold open arm	257	84	341
1500				
1600				
1700	Door stops			
1800				
1900	Holder & bumper, floor or wall	38	17.05	55.05
2000	Wall bumper	12.05	17.05	29.10
2100	Floor bumper	7.70	17.05	24.75
2200	Plunger type, door mounted	30.50	17.05	47.55
2300	Hinges, full mortise, material only, per pair			
2400	Low frequency, 4-1/2" x 4-1/2", steel base, USP	13.85		13.85
2500	Brass base, US10	48		48
2600	Stainless steel base, US32	82.50		82.50
2700	Average frequency, 4-1/2" x 4-1/2", steel base, USP	28		28
2800	Brass base, US10	57.50		57.50
2900	Stainless steel base, US32	85		85
3000	High frequency, 4-1/2" x 4-1/2", steel base, USP	67.50		67.50
3100	Brass base, US10	90.50		90.50
3200	Stainless steel base, US32	140		140
3300	Kick plate			
3400				
3500	6" high, for 3'-0" door, aluminum	28.50	36.50	65
3600	Bronze	55	36.50	91.50
3700	Panic device			
3800				
3900	For rim locks, single door, exit	545	91	636
4000	Outside key and pull	590	109	699
4100	Bar and vertical rod, exit only	785	109	894
4200	Outside key and pull	870	137	1,007
4300	Lockset			
4400				
4500	Heavy duty, cylindrical, passage doors	154	45.50	199.50
4600	Classroom	350	68.50	418.50
4700	Bedroom, bathroom, and inner office doors	198	45.50	243.50
4800	Apartment, office, and corridor doors	264	54.50	318.50
4900	Standard duty, cylindrical, exit doors	94.50	54.50	149
5100	Passage doors	68	45.50	113.50
5200	Public restroom, classroom, & office doors	143	68.50	211.50
5300	Deadlock, mortise, heavy duty	286	60.50	346.50
5400	Double cylinder	292	60.50	352.50
5500	Entrance lock, cylinder, deadlocking latch	145	60.50	205.50
5600	Deadbolt	176	68.50	244.50
5700	Commercial, mortise, wrought knob, keyed, minimum	213	68.50	281.50
5800	Maximum	385	78	463
5900	Cast knob, keyed, minimum	281	60.50	341.50

B20 Exterior Enclosure

B2030 Exterior Doors

B2030 810	Selective Price Sheet	COST EACH		
		MAT.	INST.	TOTAL
6000	Maximum	540	60.50	600.50
6100	Push-pull			
6200				
6300	Aluminum	10.10	45.50	55.60
6400	Bronze	23.50	45.50	69
6500	Door pull, designer style, minimum	78	45.50	123.50
6600	Maximum	405	68.50	473.50
6700	Threshold			
6800				
6900	3'-0" long door saddles, aluminum, minimum	4.58	11.40	15.98
7000	Maximum	43	45.50	88.50
7100	Bronze, minimum	44	9.10	53.10
7200	Maximum	72.50	45.50	118
7300	Rubber, 1/2" thick, 5-1/2" wide	42	27.50	69.50
7400	2-3/4" wide	17.60	27.50	45.10
7500	Weatherstripping, per set			
7600				
7700	Doors, wood frame, interlocking for 3' x 7' door, zinc	17.25	182	199.25
7800	Bronze	27.50	182	209.50
7900	Wood frame, spring type for 3' x 7' door, bronze	24	72	96
8000	Metal frame, spring type for 3' x 7' door, bronze	40.50	182	222.50
8100	For stainless steel, spring type, add	133%		
8200				
8300	Metal frame, extruded sections, 3' x 7' door, aluminum	49.50	182	231.50
8400	Bronze	125	182	307

B30 Roofing

B3010 Roof Coverings

B3010 160	Selective Price Sheet	COST PER S.F.		
		MAT.	INST.	TOTAL
0100	Roofing, built-up, asphalt roll roof, 3 ply organic/mineral surface	.70	1.40	2.10
0200	3 plies glass fiber felt type iv, 1 ply mineral surface	1.14	1.51	2.65
0300	Cold applied, 3 ply		.50	.50
0400	Coal tar pitch, 4 ply tarred felt	2.01	1.80	3.81
0500	Mopped, 3 ply glass fiber	1.65	2	3.65
0600	4 ply organic felt	2.01	1.80	3.81
0700	Elastomeric, hypalon, neoprene unreinforced	2.83	3.33	6.16
0800	Polyester reinforced	2.87	3.95	6.82
0900	Neoprene, 5 coats 60 mils	6.30	11.65	17.95
1000	Over 10,000 S.F.	5.90	6.05	11.95
1100	PVC, traffic deck sprayed	1.85	6.05	7.90
1200	With neoprene	1.95	2.44	4.39
1300	Shingles, fiber cement, strip, 14" x 30", 325#/sq.	3.70	2.27	5.97
1500	Shake, 9.35" x 16" 500#/sq.	3.35	2.27	5.62
1600				
1700	Asphalt, strip, 210-235#/sq.	.81	.91	1.72
1800	235-240#/sq.	.70	1	1.70
1900	Class A laminated	.96	1.11	2.07
2000	Class C laminated	.93	1.25	2.18
2100	Slate, buckingham, 3/16" thick	5.25	2.87	8.12
2200	Black, 1/4" thick	5.25	2.87	8.12
2300	Wood, shingles, 16" no. 1, 5" exp.	3.40	2.18	5.58
2400	Red cedar, 18" perfections	2.92	1.99	4.91
2500	Shakes, 24", 10" exposure	1.82	2.18	4
2600	18", 8-1/2" exposure	1.61	2.73	4.34
2700	Insulation, ceiling batts, fiberglass, 3-1/2" thick, R13	.36	.40	.76
2800	6" thick, R19	.47	.40	.87
2900	9" thick, R30	.77	.48	1.25
3000	12" thick, R38	1.10	.48	1.58
3100	Mineral, 3-1/2" thick, R13	.42	.34	.76
3200	Fiber, 6" thick, R19	.53	.34	.87
3300	Roof deck, fiberboard, 1" thick, R2.78	.46	.62	1.08
3400	Mineral, 2" thick, R5.26	.94	.62	1.56
3500	Perlite boards, 3/4" thick, R2.08	.41	.62	1.03
3600	2" thick, R5.26	.88	.71	1.59
3700	Polystyrene extruded, R5.26, 1" thick,	.54	.33	.87
3800	2" thick R10	.68	.40	1.08
3900	40 PSI compressive strength, 1" thick R5	.56	.33	.89
4000	Tapered for drainage	.78	.36	1.14
4100	Foamglass, 1 1/2" thick R4.55	1.73	.62	2.35
4200	3" thick R9.00	3.18	.71	3.89
4300	Ceiling, plaster, gypsum, 2 coats	.46	2.96	3.42
4400	3 coats	.65	3.48	4.13
4500	Perlite or vermiculite, 2 coats	.46	3.43	3.89
4600	3 coats	.77	4.32	5.09
4700	Gypsum lath, plain 3/8" thick	.62	.60	1.22
4800	1/2" thick	.60	.64	1.24
4900	Firestop, 3/8" thick	.30	.74	1.04
5000	1/2" thick	.30	.79	1.09
5100	Metal lath, rib, 2.75 lb.	.45	.69	1.14
5200	3.40 lb.	.69	.74	1.43
5300	Diamond, 2.50 lb.	.55	.69	1.24
5400	3.40 lb.	.58	.86	1.44
5500	Drywall, taped and finished standard, 1/2" thick	.31	1.43	1.74
5600	5/8" thick	.35	1.43	1.78
5700	Fire resistant, 1/2" thick	.35	1.43	1.78
5800	5/8" thick	.34	1.43	1.77
5900	Water resist., 1/2" thick	.43	1.43	1.86

B30 Roofing

B3010 Roof Coverings

B3010 160	Selective Price Sheet	COST PER S.F.		
		MAT.	INST.	TOTAL
6000	5/8" thick	.52	1.43	1.95
6100	Finish, instead of taping			
6200	For thin coat plaster, add	.09	.68	.77
6300	Finish textured spray, add	.06	.58	.64
6400				
6500	Tile, stapled glued, mineral fiber plastic coated, 5/8" thick	1.97	1.82	3.79
6600	3/4" thick	2.10	1.82	3.92
6700	Wood fiber, 1/2" thick	1.01	1.37	2.38
6800	3/4" thick	1.29	1.37	2.66
6900	Suspended, fiberglass film faced, 5/8" thick	.77	.87	1.64
7000	3" thick	1.68	1.21	2.89
7100	Mineral fiber 5/8" thick, standard face	.86	.81	1.67
7200	Aluminum faced	2	.91	2.91
7300	Wood fiber reveal edge, 1" thick			
7400	3" thick			
7500	Ceiling suspension systems, for tile, "T" bar, class "A", 2' x 4' grid	.75	.68	1.43
7600	2' x 2' grid	.93	.84	1.77
7700	Concealed "Z" bar, 12" module	.76	1.05	1.81
7800				
7900	Plaster/drywall, 3/4" channels, steel furring, 16" O.C.	.31	1.60	1.91
8000	24" O.C.	.20	1.11	1.31
8100	1-1/2" channels, 16" O.C.	.41	1.79	2.20
8200	24" O.C.	.27	1.19	1.46
8300	Ceiling framing, 2" x 4" studs, 16" O.C.	.20	.91	1.11
8400	24" O.C.	.14	.61	.75

B30 Roofing

B3020 Roof Openings

This page illustrates and describes a roof hatch system. Lines within System Components give the unit price and total cost each for this system. Prices for alternate systems are on Lines B3020 230 1500 thru 2400. Both material quantities and labor costs have been adjusted for the system listed.

Factors: To adjust for job conditions other than normal working situations use Lines B3020 230 2900 thru 4000.

Example: You are to install the system and cut and patch to match existing construction. Use line B3020 230 3000 and apply these percentages to the appropriate MAT. and INST. costs.

System Components	QUANTITY	UNIT	COST EACH		
			MAT.	INST.	TOTAL
Roof hatch, 2'-6" x 3'-0", aluminum, curb and cover included.					
Through steel construction.					
Roof hatch, 2'-6" x 3'-0", aluminum	1.000	Ea.	1,075	210	1,285
Cutout decking	11.000	L.F.		28.38	28.38
Frame opening	44.000	L.F.	32.12	374.44	406.56
Flashing, 16 oz. copper	16.000	S.F.	132.80	69.44	202.24
Cant strip, 4 x 4, foamglass	12.000	L.F.	29.40	18.48	47.88
TOTAL			1,269.32	700.74	1,970.06

B3020 230	Roof Hatches, Skylights	COST EACH		
		MAT.	INST.	TOTAL
1400	For alternate systems:			
1500	Roof hatch, aluminum, curb and cover, 2'-6" x 4'-6"	1,175	855	2,030
1600	2'-6" x 8'-0"	2,175	1,225	3,400
1700	Skylight, plexiglass dome, with curb mounting, 30" x 32"	700	665	1,365
1800	30" x 45"	760	835	1,595
1900	40" x 45"	910	1,125	2,035
2000	Smoke hatch, unlabeled, 2'-6" x 3'-0"	1,550	755	2,305
2100	2'-6" x 4'-6"	1,400	915	2,315
2200	2'-6" x 8'-0"	2,350	1,275	3,625
2300				
2400	For steel ladder, add per vertical linear foot	42	34.50	76.50
2500				
2600				
2700				
2900	Cut & patch to match existing construction, add, minimum	2%	3%	
3000	Maximum	5%	9%	
3100	Dust protection, add, minimum	1%	2%	
3200	Maximum	4%	11%	
3300	Equipment usage curtailment, add, minimum	1%	1%	
3400	Maximum	3%	10%	
3500	Material handling & storage limitation, add, minimum	1%	1%	
3600	Maximum	6%	7%	
3700	Protection of existing work, add, minimum	2%	2%	
3800	Maximum	5%	7%	
3900	Shift work requirements, add, minimum		5%	
4000	Maximum		30%	

B30 Roofing

B3020 Roof Openings

B3020 240	Selective Price Sheet	COST PER L.F.		
		MAT.	INST.	TOTAL
0100	Downspouts per L.F., aluminum, enameled .024" thick, 2" x 3"	2.02	3.41	5.43
0200	3" x 4"	2.38	4.39	6.77
0300	Round .025" thick, 3" diam.	1.56	3.23	4.79
0400	4" diam.	2.63	4.39	7.02
0500	Copper, round 16 oz. stock, 2" diam.	7.35	3.23	10.58
0600	3" diam.	6.85	3.23	10.08
0700	4" diam.	9.90	4.24	14.14
0800	5" diam.	15.60	4.73	20.33
0900	Rectangular, 2" x 3"	7.70	3.23	10.93
1000	3" x 4"	10	4.24	14.24
1100	Lead coated copper, round, 2" diam.	24.50	3.23	27.73
1200	3" diam.	24	3.23	27.23
1300	4" diam.	25.50	4.24	29.74
1400	5" diam.	27	4.73	31.73
1500	Rectangular, 2" x 3"	12.10	3.23	15.33
1600	3" x 4"	13.20	4.24	17.44
1700	Steel galvanized, round 28 gauge, 3" diam.	2.46	3.23	5.69
1800	4" diam.	3	4.24	7.24
1900	5" diam.	3.88	4.73	8.61
2000	6" diam.	4.40	5.85	10.25
2100	Rectangular, 2" x 3"	1.41	3.23	4.64
2200	3" x 4"	2.90	4.24	7.14
2300	Elbows, aluminum, round, 3" diam.	2.50	6.15	8.65
2400	4" diam.	7.15	6.15	13.30
2500	Rectangular, 2" x 3"	1.44	6.15	7.59
2600	3" x 4"	4.39	6.15	10.54
2700	Copper, round 16 oz., 2" diam.	14.75	6.15	20.90
2800	3" diam.	12.75	6.15	18.90
2900	4" diam.	18.60	6.15	24.75
3100	Rectangular, 2" x 3"	7.15	6.15	13.30
3200	3" x 4"	14.45	6.15	20.60
3300	Drip edge per L.F., aluminum, 5" wide	.42	1.37	1.79
3400	8" wide	.51	1.37	1.88
3500	28" wide	4.24	5.45	9.69
3600				
3700	Steel galvanized, 5" wide	.53	1.37	1.90
3800	8" wide	.76	1.37	2.13
3900				
4000				
4100				
4200				
4300	Flashing 12" wide per S.F., aluminum, mill finish, .013" thick	.74	3.44	4.18
4400	.019" thick	1.12	3.44	4.56
4500	.040" thick	2.43	3.44	5.87
4600	.050" thick	2.77	3.44	6.21
4700	Copper, mill finish, 16 oz.	8.30	4.34	12.64
4800	20 oz.	9.95	4.53	14.48
4900	24 oz.	12.45	4.75	17.20
5000	32 oz.	15.50	4.99	20.49
5100	Lead, 2.5 lb./S.F., 12" wide	5	3.69	8.69
5200	Over 12" wide	5	3.69	8.69
5300	Lead-coated copper, fabric backed, 2 oz.	2.66	1.51	4.17
5400	5 oz.	3.06	1.51	4.57
5500	Mastic backed, 2 oz.	2.08	1.51	3.59
5600	5 oz.	2.57	1.51	4.08
5700	Paper backed, 2 oz.	1.79	1.51	3.30
5800	3 oz.	2.11	1.51	3.62
5900	Polyvinyl chloride, black, .010" thick	.25	1.75	2

B30 Roofing

B3020 Roof Openings

B3020 240	Selective Price Sheet	COST PER L.F.		
		MAT.	INST.	TOTAL
6000	.020" thick	.36	1.75	2.11
6100	.030" thick	.46	1.75	2.21
6200	.056" thick	1.13	1.75	2.88
6300	Steel, galvanized, 20 gauge	1.31	3.84	5.15
6400	30 gauge	.55	3.12	3.67
6500	Stainless, 32 gauge, .010" thick	4.17	3.22	7.39
6600	28 gauge, .015" thick	5.15	3.22	8.37
6700	26 gauge, .018" thick	6.25	3.22	9.47
6800	24 gauge, .025" thick	8.15	3.22	11.37
6900	Gutters per L.F., aluminum, 5" box, .027" thick	3.03	5.10	8.13
7000	.032" thick	3.01	5.10	8.11
7100	Copper, half round, 4" wide	7.65	5.10	12.75
7200	6" wide	10.95	5.35	16.30
7300	Steel, 26 gauge galvanized, 5" wide	1.75	5.10	6.85
7400	6" wide	3.29	5.10	8.39
7500	Wood, treated hem-fir, 3" x 4"	8.75	5.45	14.20
7600	4" x 5"	10.40	5.45	15.85
7700	Reglet per L.F., aluminum, .025" thick	1.53	2.43	3.96
7800	Copper, 10 oz.	2.88	2.43	5.31
7900	Steel, galvanized, 24 gauge	1.03	2.43	3.46
8000	Stainless, .020" thick	3.47	2.43	5.90
8100	Counter flashing 12" wide per L.F., aluminum, .032" thick	1.79	4.10	5.89
8200	Copper, 10 oz.	5.45	4.10	9.55
8300	Steel, galvanized, 24 gauge	1.29	4.10	5.39
8400	Stainless, .020" thick	5.60	4.10	9.70

C Interiors

C10 Interior Construction

C1010 Partitions

This page illustrates and describes a concrete block wall system including concrete block, horizontal reinforcing alternate courses, mortar, and tooled joints both sides. Lines within System Components give the unit price and total price per square foot for this system. Prices for alternate concrete block wall systems are on Line Items C1010 106 0900 thru 2100. Both material quantities and labor costs have been adjusted for the system listed.

Factors: To adjust for job conditions other than normal working situations use Lines C1010 106 2900 thru 4000.

Example: You are to install the system and protect all existing work. Go to Line C1010 106 3600 and apply these percentages to the appropriate MAT. and INST. costs.

System Components	QUANTITY	UNIT	COST PER S.F.		
			MAT.	INST.	TOTAL
Concrete block partition, including horizontal reinforcing every second Course, mortar, tooled joints both sides.					
8" x 16" concrete block, normal weight, 4" thick	1.000	S.F.	1.61	5.75	7.36
Horizontal reinforcing every second course	.750	L.F.	.14	.14	.28
TOTAL			1.75	5.89	7.64

C1010 106	Partitions, Concrete Block	COST PER S.F.		
		MAT.	INST.	TOTAL
0800	For alternate block partition systems:			
0900	8" x 16" concrete block, normal weight, 6" thick	2.43	6.35	8.78
1000	8" thick	2.74	6.75	9.49
1100	10" thick	3.44	7.05	10.49
1200	12" thick	3.90	8.70	12.60
1300	8" x 16" concrete block, lightweight, 4" thick	1.95	5.80	7.75
1400	6" thick	2.75	6.20	8.95
1500	8" thick	3.31	6.60	9.91
1600	10" thick	3.97	6.85	10.82
1700	12" thick	4.91	8.45	13.36
1800	8" x 16" glazed concrete block, 4" thick	9.50	7.35	16.85
1900	8" thick	10.30	8.15	18.45
2000	Structural facing tile, 6T series, glazed 2 sides, 4" thick	15.40	12.85	28.25
2100	6" thick	20.50	13.55	34.05
2200				
2300				
2400				
2500				
2600				
2900	Cut & patch to match existing construction, add, minimum	2%	3%	
3000	Maximum	5%	9%	
3100	Dust protection, add, minimum	1%	2%	
3200	Maximum	4%	11%	
3300	Material handling & storage limitation, add, minimum	1%	1%	
3400	Maximum	6%	7%	
3500	Protection of existing work, add, minimum	2%	2%	
3600	Maximum	5%	7%	
3700	Shift work requirements, add, minimum		5%	
3800	Maximum		30%	
3900	Temporary shoring and bracing, add, minimum	2%	5%	
4000	Maximum	5%	12%	

C10 Interior Construction

C1010 Partitions

This page illustrates and describes a wood stud partition system including wood studs with plates, gypsum plasterboard – taped and finished, insulation, baseboard and painting. Lines within System Components give the unit price and total price per square foot for this system. Prices for alternate wood stud partition systems are on Line Items C1010 132 1300 thru 2700. Both material quantities and labor costs have been adjusted for the system listed.

Factors: To adjust for job conditions other than normal working situations use Lines C1010 132 2900 thru 4000.

Example: You are to install the system where material handling and storage present a serious problem. Go to Line C1010 132 3400 and apply these percentages to the appropriate MAT. and INST. costs.

System Components	QUANTITY	UNIT	COST PER S.F.		
			MAT.	INST.	TOTAL
Wood stud wall, 2"x4", 16" O.C., dbl. top plate, sngl bot. plate, 5/8" dwl.					
Taped, finished and painted on 2 faces, insulation, baseboard, wall 8' high					
Wood studs, 2" x 4", 16" O.C., 8' high	1.000	S.F.	.34	1.36	1.70
Gypsum drywall, 5/8" thick	2.000	S.F.	.62	1.10	1.72
Taping and finishing	2.000	S.F.	.10	1.10	1.20
Insulation, 3-1/2" fiberglass batts	1.000	S.F.	.59	.40	.99
Baseboard	.200	L.F.	.61	.57	1.18
Paint baseboard, primer + 2 coats	.200	L.F.	.04	.34	.38
Painting, roller, 2 coats	2.000	S.F.	.34	1.22	1.56
TOTAL			2.64	6.09	8.73

C1010 132	Partitions, Wood Stud	COST PER S.F.		
		MAT.	INST.	TOTAL
1200	For alternate wood stud systems:			
1300	2" x 3" studs, 8' high, 16" O.C.	2.62	6	8.62
1400	24" O.C.	2.55	5.75	8.30
1500	10' high, 16" O.C.	2.60	5.75	8.35
1600	24" O.C.	2.53	5.55	8.08
1700	2" x 4" studs, 8' high, 24" O.C.	2.56	5.80	8.36
1800	10' high, 16" O.C.	2.62	5.80	8.42
1900	24" O.C.	2.54	5.60	8.14
2000	12' high, 16" O.C.	2.59	5.80	8.39
2100	24" O.C.	2.50	5.60	8.10
2200	2" x 6" studs, 8' high, 16" O.C.	2.81	6.25	9.06
2300	24" O.C.	2.70	5.90	8.60
2400	10' high, 16" O.C.	2.78	5.95	8.73
2500	24" O.C.	2.67	5.70	8.37
2600	12' high, 16" O.C.	2.74	6	8.74
2700	24" O.C.	2.63	5.70	8.33
2900	Cut & patch to match existing construction, add, minimum	2%	3%	
3000	Maximum	5%	9%	
3100	Dust protection, add, minimum	1%	2%	
3200	Maximum	4%	11%	
3300	Material handling & storage limitation, add, minimum	1%	1%	
3400	Maximum	6%	7%	
3500	Protection of existing work, add, minimum	2%	2%	
3600	Maximum	5%	7%	
3700	Shift work requirements, add, minimum		5%	
3800	Maximum		30%	
3900	Temporary shoring and bracing, add, minimum	2%	5%	
4000	Maximum	5%	12%	

C10 Interior Construction

C1010 Partitions

This page illustrates and describes a non-load bearing metal stud partition system including metal studs with runners, gypsum plasterboard, taped and finished, insulation, baseboard and painting. Lines within System Components give the unit price and total price per square foot for this system. Prices for alternate non-load bearing metal stud partition systems are on Line Items C1010 134 1300 thru 2300. Both material quantities and labor costs have been adjusted for the system listed.

Factors: To adjust for job conditions other than normal working situations use Lines C1010 134 2900 thru 4000.

Example: You are to install the system and cut and patch to match existing construction. Go to Line C1010 134 3000 and apply these percentages to the appropriate MAT. and INST. costs.

System Components	QUANTITY	UNIT	COST PER S.F.		
			MAT.	INST.	TOTAL
Non-load bearing metal studs, including top & bottom runners, 5/8" drywall,					
Taped, finished and painted 2 faces, insulation, painted baseboard.					
Metal studs, 25 ga., 3-5/8" wide, 24" O.C.	1.000	S.F.	.26	.74	1
Gypsum drywall, 5/8" thick	2.000	S.F.	.62	1.10	1.72
Taping & finishing	2.000	S.F.	.10	1.10	1.20
Insulation, 3-1/2" fiberglass batts	1.000	S.F.	.59	.40	.99
Baseboard	.200	L.F.	.49	.45	.94
Paint baseboard, primer + 2 coats	.200	L.F.	.03	.27	.30
Painting, roller work, 2 coats	2.000	S.F.	.34	1.22	1.56
TOTAL			2.43	5.28	7.71

C1010 134	Partitions, Metal Stud, NLB	COST PER S.F.		
		MAT.	INST.	TOTAL
1200	For alternate metal stud systems:			
1300	Non-load bearing, 25 ga., 24" O.C., 2-1/2" wide	2.40	5.25	7.65
1400	6" wide	2.57	5.30	7.87
1500	16" O.C., 2-1/2" wide	2.46	5.45	7.91
1600	3-5/8" wide	2.50	5.45	7.95
1700	6" wide	2.67	5.50	8.17
1800	20 ga., 24" O.C., 2-1/2" wide	2.47	5.45	7.92
1900	3-5/8" wide	2.56	5.45	8.01
2000	6" wide	2.63	5.50	8.13
2100	16" O.C., 2-1/2" wide	2.55	5.70	8.25
2200	3-5/8" wide	2.66	5.70	8.36
2300	6" wide	2.75	5.70	8.45
2400				
2500				
2600				
2700				
2900	Cut & patch to match existing construction, add, minimum	2%	3%	
3000	Maximum	5%	9%	
3100	Dust protection, add, minimum	1%	2%	
3200	Maximum	4%	11%	
3300	Material handling & storage limitation, add, minimum	1%	1%	
3400	Maximum	6%	7%	
3500	Protection of existing work, add, minimum	2%	2%	
3600	Maximum	5%	7%	
3700	Shift work requirements, add, minimum		5%	
3800	Maximum		30%	
3900	Temporary shoring and bracing, add, minimum	2%	5%	
4000	Maximum	5%	12%	

C10 Interior Construction

C1010 Partitions

This page illustrates and describes a drywall system including gypsum plasterboard, taped and finished, metal studs with runners, insulation, baseboard and painting. Lines within System Components give the unit price and total price per square foot for this system. Prices for alternate drywall systems are on Line Items C1010 136 1300 thru 1900. Both material quantities and labor costs have been adjusted for the system listed.

Factors: To adjust for job conditions other than normal working situations use Lines C1010 136 2900 thru 4000.

Example: You are to install the system and control dust in the work area. Go to Line C1010 136 3100 and apply these percentages to the appropriate MAT. and INST. costs.

System Components	QUANTITY	UNIT	COST PER S.F.		
			MAT.	INST.	TOTAL
Gypsum drywall, taped, finished and painted 2 faces, galvanized metal studs					
Including top & bottom runners, insulation, painted baseboard, wall 10' high					
Gypsum drywall, 5/8" thick, standard	2.000	S.F.	.62	1.10	1.72
Taping and finishing	2.000	S.F.	.10	1.10	1.20
Metal studs, 20 ga., 3-5/8" wide, 24" O.C.	1.000	S.F.	.33	.93	1.26
Insulation, 3-1/2" fiberglass batts	1.000	S.F.	.59	.40	.99
Baseboard	.200	L.F.	.49	.45	.94
Paint baseboard, primer + 2 coats	.200	L.F.	.02	.12	.14
Painting, roller 2 coats	2.000	S.F.	.42	1.16	1.58
TOTAL			2.57	5.26	7.83

C1010 136	Partitions, Drywall	COST PER S.F.		
		MAT.	INST.	TOTAL
1200	For alternate drywall systems:			
1300	Gypsum drywall, 5/8" thick, fire resistant	2.48	5.05	7.53
1400	Water resistant	2.82	5.05	7.87
1500	1/2" thick, standard	2.40	5.05	7.45
1600	Fire resistant	2.50	5.05	7.55
1700	Water resistant	2.66	5.05	7.71
1800	3/8" thick, vinyl faced, standard	3.20	6.40	9.60
1900	5/8" thick, vinyl faced, fire resistant	3.44	6.40	9.84
2000				
2100				
2200				
2300				
2400				
2500				
2600				
2700				
2900	Cut & patch to match existing construction, add, minimum	2%	3%	
3000	Maximum	5%	9%	
3100	Dust protection, add, minimum	1%	2%	
3200	Maximum	4%	11%	
3300	Material handling & storage limitation, add, minimum	1%	1%	
3400	Maximum	6%	7%	
3500	Protection of existing work, add, minimum	2%	2%	
3600	Maximum	5%	7%	
3700	Shift work requirements, add, minimum		5%	
3800	Maximum		30%	
3900	Temporary shoring and bracing, add, minimum	2%	5%	
4000	Maximum	5%	12%	

C10 Interior Construction

C1010 Partitions

This page illustrates and describes a load bearing metal stud wall system including metal studs, sheetrock–taped and finished, insulation, baseboard and painting. Lines within System Components give the unit price and total price per square foot for this system. Prices for alternate load bearing metal stud wall systems are on Line Items C1010 138 1500 thru 2500. Both material quantities and labor costs have been adjusted for the system listed.

Factors: To adjust for job conditions other than normal working situations use Lines C1010 138 2900 thru 4000.

Example: You are to install the system using temporary shoring and bracing. Go to Line C1010 138 3900 and apply these percentages to the appropriate MAT. and INST. costs.

System Components	QUANTITY	UNIT	COST PER S.F.		
			MAT.	INST.	TOTAL
Load bearing, 18 ga., 3-5/8", galvanized metal studs, 24" O.C., including Top and bottom runners, 1/2" drywall, taped, finished and painted 2 Faces, 3" insulation, and painted baseboard, wall 10' high.					
Metal studs, 24" O.C., 18 ga., 3-5/8" wide, galvanized	1.000	S.F.	.74	1.04	1.78
Gypsum drywall 1/2" thick	2.000	S.F.	.52	1.10	1.62
Taping and finishing	2.000	S.F.	.10	1.10	1.20
Insulation, 3-1/2" fiberglass batts	1.000	S.F.	.59	.40	.99
Baseboard	.200	L.F.	.49	.45	.94
Paint baseboard, primer + 2 coats	.200	L.F.	.03	.27	.30
Painting, roller 2 coats	2.000	S.F.	.34	1.22	1.56
TOTAL			2.81	5.58	8.39

C1010 138	Partitions, Metal Stud, LB	COST PER S.F.		
		MAT.	INST.	TOTAL
1400	For alternate metal stud systems:			
1500	Load bearing, 18 ga., 24" O.C., 2-1/2" wide	2.69	5.55	8.24
1600	6" wide	3.05	5.60	8.65
1700	16" O.C. 2-1/2" wide	2.85	5.80	8.65
1800	3-5/8" wide	3	5.85	8.85
1900	6" wide	3.30	5.90	9.20
2000	16 ga., 24" O.C., 2-1/2" wide	2.79	5.70	8.49
2100	3-5/8" wide	2.93	5.75	8.68
2200	6" wide	3.21	5.80	9.01
2300	16" O.C., 2-1/2" wide	2.96	6	8.96
2400	3-5/8" wide	3.15	6	9.15
2500	6" wide	3.50	6.10	9.60
2600				
2700				
2900	Cut & patch to match existing construction, add, minimum	2%	3%	
3000	Maximum	5%	9%	
3100	Dust protection, add, minumum	1%	2%	
3200	Maximum	4%	11%	
3300	Material handling & storage limitation, add, minimum	1%	1%	
3400	Maximum	6%	7%	
3500	Protection of existing work, add, minimum	2%	2%	
3600	Maximum	5%	7%	
3700	Shift work requirements, add, minimum		5%	
3800	Maximum		30%	
3900	Temporary shoring and bracing, add, minimum	2%	5%	
4000	Maximum	5%	12%	

C10 Interior Construction

C1010 Partitions

This page illustrates and describes a plaster and lath system including gypsum plaster, gypsum lath, wood studs with plates, insulation, baseboard and painting. Lines within System Components give the unit price and price per square foot for this system. Prices for alternate plaster and lath systems are on Line Items C1010 148 1500 thru 2500. Both material quantities and labor costs have been adjusted for the system listed.

Factors: To adjust for job conditions other than normal working situations use Lines C1010 148 2900 thru 4000.

Example: You are to install the system during evening hours only. Go to Line C1010 148 3800 and apply this percentage to the appropriate INST. costs.

System Components	QUANTITY	UNIT	COST PER S.F.		
			MAT.	INST.	TOTAL
Gypsum plaster, 2 coats, over 3/8" lath, 2 faces, 2" x 4" wood Stud partition, 24"O.C. incl. double top plate, single bottom plate, 3-1/2" Insulation, baseboard and painting.					
Gypsum plaster, 2 coats	.220	S.Y.	.90	5.15	6.05
Lath, gypsum, 3/8" thick	.220	S.Y.	1.22	1.20	2.42
Wood studs, 2" x 4", 24" O.C.	1.000	S.F.	.24	.88	1.12
Insulation, 3-1/2" fiberglass batts	1.000	S.F.	.59	.40	.99
Baseboard, 9/16" x 3-1/2"	.200	L.F.	.49	.45	.94
Paint baseboard, primer + 2 coats	.200	L.F.	.02	.12	.14
Paint, 2 coats	2.000	S.F.	.34	1.22	1.56
TOTAL			3.80	9.42	13.22

C1010 148	Partitions, Plaster & Lath	COST PER S.F.		
		MAT.	INST.	TOTAL
1400	For alternate plaster systems:			
1500	Gypsum plaster, 3 coats	4.20	10.50	14.70
1600	Perlite plaster, 2 coats	3.82	10.10	13.92
1700	3 coats	4.43	11.55	15.98
1800				
1900				
2000	For alternate lath systems:			
2100	Gypsum lath, 1/2" thick	3.80	10.20	14
2200	Foil back, 3/8" thick	3.32	10.30	13.62
2300	1/2" thick	3.34	10.40	13.74
2400	Metal lath, 2.5 Lb. diamond	3.69	10.10	13.79
2500	3.4 Lb. diamond	3.74	10.30	14.04
2600				
2700				
2900	Cut & patch to match existing construction, add, minimum	2%	3%	
3000	Maximum	5%	9%	
3100	Dust protection, add, minimum	1%	2%	
3200	Maximum	4%	11%	
3300	Material handling & storage limitation, add, minimum	1%	1%	
3400	Maximum	6%	7%	
3500	Protection of existing work, add, minimum	2%	2%	
3600	Maximum	5%	7%	
3700	Shift work requirements, add, minimum		5%	
3800	Maximum		30%	
3900	Temporary shoring and bracing, add, minimum	2%	5%	
4000	Maximum	5%	12%	

C10 Interior Construction

C1010 Partitions

C1010 170	Selective Price Sheet	COST PER S.F.		
		MAT.	INST.	TOTAL
0100	Studs			
0200				
0300	24" O.C. metal, 10' high wall, including			
0400	Top and bottom runners			
0500	Non load bearing, galvanized 25 Ga., 1-5/8" wide	.19	.72	.91
0600	2-1/2" wide	.23	.73	.96
0700	3-5/8" wide	.26	.74	1
0800	4" wide	.27	.74	1.01
0900	6" wide	.40	.75	1.15
1000				
1100	Galvanized 20 Ga., 2-1/2" wide	.30	.91	1.21
1200	3-5/8" wide	.33	.93	1.26
1300	4" wide	.39	.93	1.32
1400	6" wide	.46	.94	1.40
1500	Load bearing, painted 18 Ga., 2-1/2" wide	.62	1.02	1.64
1600	3-5/8" wide	.74	1.04	1.78
1700	4" wide	.67	1.06	1.73
1800	6" wide	.98	1.08	2.06
1900	Galvanized 18 Ga., 2-1/2" wide	.62	1.02	1.64
2000	3-5/8" wide	.74	1.04	1.78
2100	4" wide	.67	1.06	1.73
2200	6" wide	.98	1.08	2.06
2300	Galvanized 16 Ga., 2-1/2" wide	.72	1.16	1.88
2400	3-5/8" wide	.86	1.19	2.05
2500	4" wide	.91	1.22	2.13
2600	6" wide	1.14	1.24	2.38
2700				
2800				
2900	24" O.C. wood, 10' high wall, including			
3000	Double top plate and shoe			
3100	2" x 4"	.28	1.03	1.31
3200	2" x 6"	.42	1.10	1.52
3300				
3400				
3500	Furring, 24" O.C. 10' high wall, metal, 3/4" channels	.20	1.11	1.31
3600	1-1/2" channels	.27	1.19	1.46
3700	Wood, on wood, 1" x 2" strips	.14	.50	.64
3800	1" x 3" strips	.19	.50	.69
3900	On masonry, 1" x 2" strips	.14	.55	.69
4000	1" x 3" strips	.19	.55	.74
4100	On concrete, 1" x 2" strips	.14	1.05	1.19
4200	1" x 3" strips	.19	1.05	1.24
4300	Studs			
4400				
4500	16" O.C., metal, 10' high wall, including			
4600	Top and bottom runners			
4700	Non load bearing, galvanized 25 Ga., 1-5/8" wide	.26	1.10	1.36
4800	2-1/2" wide	.31	1.11	1.42
4900	3-5/8" wide	.36	1.14	1.50
5000	4" wide	.36	1.15	1.51
5100	6" wide	.54	1.16	1.70
5200				
5300	Galvanized 20 Ga., 2-1/2" wide	.41	1.40	1.81
5400	3-5/8" wide	.44	1.42	1.86
5500	4" wide	.53	1.44	1.97
5600	6" wide	.62	1.46	2.08

C10 Interior Construction

C1010 Partitions

C1010 170	Selective Price Sheet	COST PER S.F.		
		MAT.	INST.	TOTAL
5700	Load bearing, painted 18 Ga., 2-1/2" wide	.84	1.42	2.26
5800	3-5/8" wide	1	1.44	2.44
5900	4" wide	1.05	1.48	2.53
6000	6" wide	1.33	1.50	2.83
6100	Galvanized 18 Ga., 2-1/2" wide	.84	1.42	2.26
6200	3-5/8" wide	1	1.44	2.44
6300	4" wide	1.05	1.48	2.53
6400	6" wide	1.33	1.50	2.83
6500	Galvanized 16 Ga., 2-1/2" wide	.99	1.61	2.60
6600	3-5/8" wide	1.19	1.66	2.85
6700	4" wide	1.25	1.68	2.93
6800	6" wide	1.57	1.71	3.28
6900				
7000				
7100	16" O.C., wood, 10' high wall, including			
7200	Double top plate and shoe			
7300	2" x 4"	.34	1.19	1.53
7400	2" x 6"	.42	1.07	1.49
7500				
7600				
7700	Furring, 16" O.C. 10' high wall, metal, 3/4" channels	.31	1.60	1.91
7800	1-1/2" channels	.41	1.79	2.20
7900	Wood, on wood, 1" x 2" strips	.20	.74	.94
8000	1" x 3" strips	.29	.74	1.03
8100	On masonry, 1" x 2" strips	.20	.83	1.03
8200	1" x 3" strips	.29	.83	1.12
8300	On concrete, 1" x 2" strips	.20	1.58	1.78
8400	1" x 3" strips	.29	1.58	1.87

C10 Interior Construction

C1010 Partitions

C1010 180	Selective Price Sheet, Drywall & Plaster	COST PER S.F.		
		MAT.	INST.	TOTAL
0100	Lath, gypsum perforated			
0200				
0300	Lath, gypsum, perforated, regular, 3/8" thick	.62	.60	1.22
0400	1/2" thick	.60	.74	1.34
0500	Fire resistant, 3/8" thick	.30	.74	1.04
0600	1/2" thick	.30	.79	1.09
0700	Foil back, 3/8" thick	.36	.69	1.05
0800	1/2" thick	.37	.74	1.11
0900				
1000	Metal lath			
1100	Metal lath, diamond, painted, 2.5 lb.	.55	.60	1.15
1200	3.4 lb.	.58	.69	1.27
1300	Rib painted, 2.75 lb	.45	.69	1.14
1400	3.40 lb	.69	.74	1.43
1500				
1600				
1700	Plaster, gypsum, 2 coats	.46	2.60	3.06
1800	3 coats	.65	3.13	3.78
1900	Perlite/vermiculite, 2 coats	.46	2.96	3.42
2000	3 coats	.77	3.67	4.44
2100	Bondcrete, 1 coat	.43	1.35	1.78
2200				
2500				
2600				
2700	Drywall, standard, 3/8" thick, no finish included	.26	.55	.81
2800	1/2" thick, no finish included	.26	.55	.81
2900	Taped and finished	.31	1.13	1.44
3000	5/8" thick, no finish included	.31	.55	.86
3100	Taped and finished	.35	1.13	1.48
3200	Fire resistant, 1/2" thick, no finish included	.31	.55	.86
3300	Taped and finished	.35	1.13	1.48
3400	5/8" thick, no finish included	.30	.55	.85
3500	Taped and finished	.34	1.13	1.47
3600	Water resistant, 1/2" thick, no finish included	.39	.55	.94
3700	Taped and finished	.43	1.13	1.56
3800	5/8" thick, no finish included	.47	.55	1.02
3900	Taped and finished	.52	1.13	1.65
4000	Finish, instead of taping			
4100	For thin coat plaster, add	.09	.68	.77
4200	Finish, textured spray, add	.06	.58	.64

C10 Interior Construction

C1010 Partitions

This page illustrates and describes a movable office partition system including demountable office partitions of various styles and sizes based on a cost per square foot basis. Prices for alternate movable office partition systems are on Line Items C1010 210 0300 thru 3600.

C1010 210	Partitions, Movable Office	COST PER S.F.		
		MAT.	INST.	TOTAL
0250	Office partition, demountable, no deduction for door opening, add for doors			
0300	Air wall, cork finish, semi acoustic, 1-5/8" thick, minimum	42.50	3.36	45.86
0400	Maximum	40	5.75	45.75
0500	Acoustic, 2" thick, minimum	34	3.58	37.58
0600	Maximum	59	4.85	63.85
0700				
0800				
0900	In-plant modular office system, w/prehung steel door			
1000	3" thick honeycomb core panels			
1100	12' x 12', 2 wall	8.05	.46	8.51
1200	4 wall	15.30	1.38	16.68
1300				
1400				
1500	Gypsum, demountable, 3" to 3-3/4" thick x 9' high, vinyl clad	7.05	2.55	9.60
1600	Fabric clad	17.55	2.78	20.33
1700	1.75 system, vinyl clad hardboard, paper honeycomb core panel			
1800	1-3/4" to 2-1/2" thick x 9' high	11.90	2.55	14.45
1900	Unitized gypsum panel, 2" to 2-1/2" thick x 9' high, vinyl clad	15.20	2.55	17.75
2000	Fabric clad	25	2.78	27.78
2100				
2200				
2300	Unitized mineral fiber panel system, 2-1/4" thick x 9' high			
2400	Vinyl clad mineral fiber	15.20	2.55	17.75
2500	Fabric clad mineral fiber	22.50	2.78	25.28
2600				
2700	Movable steel walls, modular system			
2800	Unitized panels, 48" wide x 9' high			
2900	Baked enamel, pre-finished	17.20	2.02	19.22
3000	Fabric clad	25	2.16	27.16
3100	For acoustical partitions, add, minimum	2.44		2.44
3200	Maximum	11.40		11.40
3300				
3400				
3500	Note: For door prices, see Div. 08			
3600	For door hardware prices, see Div. 08			
3700				
3800				
3900				
4000				

C10 Interior Construction

C1020 Interior Doors

This page illustrates and describes flush interior door systems including hollow core door, jamb, header and trim with hardware. Lines within System Components give the unit price and total price on a cost each basis for this system. Prices for alternate flush interior door systems are on Line items C1020 106 1100 thru 2400. Both material quantities and labor costs have been adjusted for the system listed.

Factors: To adjust for job conditions other than normal working situations use Lines C1020 106 2700 thru 3800.

Example: You are to install the system in an area where dust protection is vital. Go to Line C1020 106 3000 and apply these percentages to the appropriate MAT. and INST. costs.

System Components	QUANTITY	UNIT	COST EACH		
			MAT.	INST.	TOTAL
Single hollow core door, include jamb, header, trim and hardware, painted.					
Hollow core Lauan, 1-3/8" thick, 2'-0" x 6'-8", painted	1.000	Ea.	34	60.50	94.50
Paint door and frame, 1 coat	1.000	Ea.	1.88	27	28.88
Wood jamb, 4-9/16" deep	1.000	Set	114.40	46.56	160.96
Trim, casing	1.000	Set	46.72	72.64	119.36
Hardware, hinges	1.000	Set	13.85		13.85
Hardware, lockset	1.000	Set	24	34	58
TOTAL			234.85	240.70	475.55

C1020 106	Doors, Interior Flush, Wood	COST EACH		
		MAT.	INST.	TOTAL
1000	For alternate door systems:			
1100	Lauan (Mahogany) hollow core, 1-3/8" x 2'-6" x 6'-8"	244	244	488
1200	2'-8" x 6'-8"	249	247	496
1300	3'-0" x 6'-8"	254	256	510
1500	Birch, hollow core, 1-3/8" x 2'-0" x 6'-8"	246	241	487
1600	2'-6" x 6'-8"	259	244	503
1700	2'-8" x 6'-8"	264	247	511
1800	3'-0" x 6'-8"	270	256	526
1900	Solid core, pre-hung, 1-3/8" x 2'-6" x 6'-8"	410	248	658
2000	2'-8" x 6'-8"	425	251	676
2100	3'-0" x 6'-8"	455	261	716
2200				
2400	For metal frame instead of wood, add	50%	20%	
2600				
2700	Cut & patch to match existing construction, add, minimum	2%	3%	
2800	Maximum	5%	9%	
2900	Dust protection, add, minimum	1%	2%	
3000	Maximum	4%	11%	
3100	Equipment usage curtailment, add, minimum	1%	1%	
3200	Maximum	3%	10%	
3300	Material handling & storage limitation, add, minimum	1%	1%	
3400	Maximum	6%	7%	
3500	Protection of existing work, add, minimum	2%	2%	
3600	Maximum	5%	7%	
3700	Shift work requirements, add, minimum		5%	
3800	Maximum		30%	
3900				

C10 Interior Construction

C1020 Interior Doors

This page illustrates and describes interior, solid and louvered door systems including a pine panel door, wood jambs, header, and trim with hardware. Lines within System Components give the unit price and total price on a cost each basis for this system. Prices for alternate interior, solid and louvered systems are on Line Items C1020 108 1200 thru 2400. Both material quantities and labor costs have been adjusted for the system listed.

Factors: To adjust for job conditions other than normal working situations use Lines C1020 108 2900 thru 4000.

Example: You are to install the system during night hours only. Go to Line C1020 108 4000 and apply these percentages to the appropriate INST. costs.

System Components	QUANTITY	UNIT	COST EACH		
			MAT.	INST.	TOTAL
Single interior door, including jamb, header, trim and hardware, painted.					
Solid pine panel door, 1-3/8" thick, 2'-0" x 6'-8"	1.000	Ea.	143	60.50	203.50
Paint door and frame, 1 coat	1.000	Ea.	1.88	27	28.88
Wooden jamb, 4-5/8" deep	1.000	Set	114.40	46.56	160.96
Trim, casing	1.000	Set	46.72	72.64	119.36
Hardware, hinges	1.000	Set	13.85		13.85
Hardware, lockset	1.000	Set	24	34	58
TOTAL			343.85	240.70	584.55

C1020 108	Doors, Interior Solid & Louvered	COST EACH		
		MAT.	INST.	TOTAL
1000				
1100	For alternate door systems:			
1200	Solid pine, painted raised panel, 1-3/8" x 2'-6" x 6'-8"	380	244	624
1300	2'-8" x 6'-8"	385	247	632
1400	3'-0" x 6'-8"	395	256	651
1500				
1600	Louvered pine, painted 1'-6" x 6'-8"	310	238	548
1700	2'-0" x 6'-8"	345	244	589
1800	2'-6" x 6'-8"	360	247	607
1900	3'-0" x 6'-8"	395	256	651
2200	For prehung door, deduct	5%	30%	
2300				
2400	For metal frame instead of wood, add	50%	20%	
2500				
2900	Cut & patch to match existing construction, add, minimum	2%	3%	
3000	Maximum	5%	9%	
3100	Dust protection, add, minimum	1%	2%	
3200	Maximum	4%	11%	
3300	Equipment usage curtailment, add, minimum	1%	1%	
3400	Maximum	3%	10%	
3500	Material handling & storage limitation, add, minimum	1%	1%	
3600	Maximum	6%	7%	
3700	Protection of existing work, add, minimum	2%	2%	
3800	Maximum	5%	7%	
3900	Shift work requirements, add, minimum		5%	
4000	Maximum		30%	

C10 Interior Construction

C1020 Interior Doors

This page illustrates and describes interior metal door systems including a metal door, metal frame and hardware. Lines within System Components give the unit price and total price on a cost each for this system. Prices for alternate interior metal door systems are on Line Items C1020 110 1100 thru 2000. Both material quantities and labor costs have been adjusted for the system listed.

Factors: To adjust for job conditions other than normal working situations use Lines C1020 110 2900 thru 4000.

Example: You are to install the system while protecting existing construction. Go to Line C1020 110 3700 and apply these percentages to the appropriate MAT. and INST. costs.

System Components	QUANTITY	UNIT	COST EACH		
			MAT.	INST.	TOTAL
Single metal door, including frame and hardware.					
Hollow metal door, 1-3/8" thick, 2'-6" x 6'-8", painted	1.000	Ea.	450	73	523
Metal frame, 6-1/4" deep	1.000	Set	174	68.50	242.50
Paint door and frame, 1 coat	1.000	Ea.	1.88	27	28.88
Hardware, hinges	1.000	Set	84		84
Hardware, passage lockset	1.000	Set	24	34	58
TOTAL			733.88	202.50	936.38

C1020 110	Doors, Interior Flush, Metal	COST EACH		
		MAT.	INST.	TOTAL
1000	For alternate systems:			
1100	Hollow metal doors, 1-3/8" thick, 2'-8" x 6'-8"	740	203	943
1200	3'-0" x 7'-0"	725	212	937
1300				
1400	Interior fire door, 1-3/8" thick, 2'-6" x 6'-8"	925	203	1,128
1500	2'-8" x 6'-8"	925	203	1,128
1600	3'-0" x 7'-0"	925	212	1,137
1700				
1800	Add to fire doors:			
1900	Baked enamel finish	30%	15%	
2000	Galvanizing	15%		
2200				
2300				
2400				
2900	Cut & patch to match existing construction, add, minimum	2%	3%	
3000	Maximum	5%	9%	
3100	Dust protection, add, minimum	1%	2%	
3200	Maximum	4%	11%	
3300	Equipment usage curtailment, add, minimum	1%	1%	
3400	Maximum	3%	10%	
3500	Material handling & storage limitation, add, minimum	1%	1%	
3600	Maximum	6%	7%	
3700	Protection of existing work, add, minimum	2%	2%	
3800	Maximum	5%	7%	
3900	Shift work requirements, add, minimum		5%	
4000	Maximum		30%	

C10 Interior Construction

C1020 Interior Doors

This page illustrates and describes an interior closet door system including an interior closet door, painted, with trim and hardware. Prices for alternate interior closet door systems are on Line Items C1020 112 0500 thru 2200. Both material quantities and labor costs have been adjusted for the system listed.

Factors: To adjust for job conditions other than normal working situations use Lines C1020 112 2900 thru 4000.

Example: You are to install the system and match the existing construction. Go to Line C1020 112 2900 and apply these percentages to the appropriate MAT. and INST. costs.

C1020 112	Doors, Closet	MAT.	INST.	TOTAL
0350	Interior closet door painted, including frame, trim and hardware, prehung.			
0400	Bi-fold doors			
0500	Pine paneled, 3'-0" x 6'-8"	430	199	629
0600	6'-0" x 6'-8"	595	267	862
0700	Birch, hollow core, 3'-0" x 6'-8"	202	216	418
0800	6'-0" x 6'-8"	320	288	608
0900	Lauan, hollow core, 3'-0" x 6'-8"	190	199	389
1000	6'-0" x 6'-8"	294	267	561
1100	Louvered pine, 3'-0" x 6'-8"	345	199	544
1200	6'-0" x 6'-8"	460	267	727
1300				
1400	Sliding, bi-passing closet doors			
1500	Pine paneled, 4'-0" x 6'-8"	690	211	901
1600	6'-0" x 6'-8"	810	267	1,077
1700	Birch, hollow core, 4'-0" x 6'-8"	375	211	586
1800	6'-0" x 6'-8"	435	267	702
1900	Lauan, hollow core, 4'-0" x 6'-8"	325	211	536
2000	6'-0" x 6'-8"	375	267	642
2100	Louvered pine, 4'-0" x 6'-8"	580	211	791
2200	6'-0" x 6'-8"	725	267	992
2300				
2400				
2500				
2600				
2700				
2800				
2900	Cut & patch to match existing construction, add, minimum	2%	3%	
3000	Maximum	5%	9%	
3100	Dust protection, add, minimum	1%	2%	
3200	Maximum	4%	11%	
3300	Equipment usage curtailment, add, minimum	1%	1%	
3400	Maximum	3%	10%	
3500	Material handling & storage limitation, add, minimum	1%	1%	
3600	Maximum	6%	7%	
3700	Protection of existing work, add, minimum	2%	2%	
3800	Maximum	5%	7%	
3900	Shift work requirements, add, minimum		5%	
4000	Maximum		30%	

C20 Stairs

C2010 Stair Construction

This page illustrates and describes a stair system based on a cost per flight price. Prices for various stair systems are on Line Items C2010 130 0700 thru 3200. Both material quantities and labor costs have been adjusted for the system listed.

Factors: To adjust for job conditions other than normal working situations use Lines C2010 130 3500 thru 4200.

Example: You are to install the system during evenings only. Go to Line C2010 130 4200 and apply this percentage to the appropriate MAT. and INST. costs.

System Components	QUANTITY	UNIT	COST PER FLIGHT		
			MAT.	INST.	TOTAL
Below are various stair systems based on cost per flight of stairs, no side Walls. Stairs are 4'-0" wide, railings are included unless otherwise noted.					

C2010 130	Stairs	COST PER FLIGHT		
		MAT.	INST.	TOTAL
0700	Concrete, cast in place, no nosings, no railings, 12 risers	269	1,900	2,169
0800	24 risers	540	3,800	4,340
0900	Add for 1 intermediate landing	148	520	668
1000	Concrete, cast in place, with nosings, no railings, 12 risers	980	2,225	3,205
1100	24 risers	1,950	4,450	6,400
1200	Add for 1 intermediate landing	207	550	757
1300	Steel, grating tread, safety nosing, 12 risers	4,025	1,100	5,125
1400	24 risers	8,050	2,200	10,250
1500	Add for intermediate landing	970	294	1,264
1600	Steel, cement fill pan tread, 12 risers	6,000	1,100	7,100
1700	24 risers	12,000	2,200	14,200
1800	Add for intermediate landing	970	294	1,264
1900	Spiral, industrial, 4' - 6" diameter, 12 risers	5,875	785	6,660
2000	24 risers	11,800	1,550	13,350
2100	Wood, box stairs, oak treads, 12 risers	2,225	620	2,845
2200	24 risers	4,450	1,250	5,700
2300	Add for 1 intermediate landing	145	140	285
2400	Wood, basement stairs, no risers, 12 steps	725	252	977
2500	24 steps	1,450	505	1,955
2600	Add for 1 intermediate landing	31	26.50	57.50
2700	Wood, open, rough sawn cedar, 12 steps	2,525	500	3,025
2800	24 steps	5,025	1,000	6,025
2900	Add for 1 intermediate landing	23.50	27.50	51
3000	Wood, residential, oak treads, 12 risers	2,125	2,725	4,850
3100	24 risers	4,225	5,450	9,675
3200	Add for 1 intermediate landing	129	126	255
3500	Dust protection, add, minimum	1%	2%	
3600	Maximum	4%	11%	
3700	Material handling & storage limitation, add, minimum	1%	1%	
3800	Maximum	6%	7%	
3900	Protection of existing work, add, minimum	2%	2%	
4000	Maximum	5%	7%	
4100	Shift work requirements, add, minimum		5%	
4200	Maximum		30%	

C30 Interior Finishes

C3010 Wall Finishes

C3010 210	Selective Price Sheet	COST PER S.F.		
		MAT.	INST.	TOTAL
0100	Painting, on plaster or drywall, brushwork, primer and 1 ct.	.20	.68	.88
0200	Primer and 2 ct.	.28	1.08	1.36
0300	Rollerwork, primer and 1 ct.	.21	.58	.79
0400	Primer and 2 ct.	.28	.92	1.20
0500	Woodwork incl. puttying, brushwork, primer and 1 ct.	.11	1.03	1.14
0600	Primer and 2 ct.	.17	1.36	1.53
0700	Wood trim to 6" wide, enamel, primer and 1 ct.	.11	.58	.69
0800	Primer and 2 ct.	.17	.73	.90
0900	Cabinets and casework, enamel, primer and 1 ct.	.11	1.15	1.26
1000	Primer and 2 ct.	.17	1.42	1.59
1100	On masonry or concrete, latex, brushwork, primer and 1 ct.	.22	.96	1.18
1200	Primer and 2 ct.	.30	1.38	1.68
1300	For block filler, add	.19	1.18	1.37
1400				
1500	Varnish, wood trim, sealer 1 ct., sanding, puttying, quality work	.14	2.07	2.21
1600	Medium work	.11	1.61	1.72
1700	Without sanding	.17	.24	.41
1800				
1900	Wall coverings, wall paper, at low price per double roll, avg workmanship	.41	.73	1.14
2000	At medium price per double roll, average workmanship	.78	.87	1.65
2100	At high price per double roll, quality workmanship	2.72	1.07	3.79
2200				
2300	Grass cloths with lining paper, minimum	.86	1.16	2.02
2400	Maximum	2.74	1.33	4.07
2500	Vinyl, fabric backed, light weight	.76	.73	1.49
2600	Medium weight	.90	.97	1.87
2700	Heavy weight	1.80	1.07	2.87
2800				
2900	Cork tiles, 12" x 12", 3/16" thick	4.92	1.93	6.85
3000	5/16" thick	4.18	1.98	6.16
3100	Granular surface, 12" x 36", 1/2" thick	1.39	1.21	2.60
3200	1" thick	1.79	1.26	3.05
3300	Aluminum foil	1.12	1.69	2.81
3400				
3500	Tile, ceramic, adhesive set, 4-1/4" x 4-1/4"	2.58	4.64	7.22
3600	6" x 6"	3.35	4.41	7.76
3700	Decorated, 4-1/4" x 4-1/4", minimum	3.86	3.26	7.12
3800	Maximum	46	4.90	50.90
3900	For epoxy grout, add	.43	1.10	1.53
4000	Pregrouted sheets	5.20	3.67	8.87
4100	Glass mosaics, 3/4" tile on 12" sheets, minimum	19.05	12.05	31.10
4200	Color group 8	62.50	13.75	76.25
4300	Metal, tile pattern, 4' x 4' sheet, 24 ga., nailed			
4400	Stainless steel	27	2.13	29.13
4500	Aluminized steel	14.45	2.13	16.58
4600				
4700	Brick, interior veneer, 4" face brick, running bond, minimum	5.05	11.50	16.55
4800	Maximum	5.05	11.50	16.55
4900	Simulated, urethane pieces, set in mastic	7.35	3.64	10.99
5000	Fiberglass panels	4.26	2.73	6.99
5100	Wall coating, on drywall, thin coat, plain	.09	.68	.77
5200	Stipple	.09	.68	.77
5300	Textured spray	.06	.58	.64
5400				
5500	Paneling not incl. furring or trim, hardboard, tempered, 1/8" thick	.43	2.18	2.61
5600	1/4" thick	.58	2.18	2.76
5700	Plastic faced, 1/8" thick	.74	2.18	2.92
5800	1/4" thick	.94	2.18	3.12

C30 Interior Finishes

C3010 Wall Finishes

C3010 210	Selective Price Sheet	COST PER S.F.		
		MAT.	INST.	TOTAL
5900	Woodgrained, 1/4" thick, minimum	.63	2.18	2.81
6000	Maximum	1.31	2.57	3.88
6100	Plywood, 4' x 8' shts. 1/4" thick, prefin., birch faced, min.	.98	2.18	3.16
6200	Maximum	2.17	3.12	5.29
6300	Walnut, minimum	3.03	2.18	5.21
6400	Maximum	5.75	2.73	8.48
6500	Mahogany, African	2.76	2.73	5.49
6600	Philippine	1.19	2.18	3.37
6700	Chestnut	5.25	2.91	8.16
6800	Pecan	2.27	2.73	5
6900	Rosewood	5.05	3.41	8.46
7000	Teak	3.54	2.73	6.27
7100	Aromatic cedar, plywood	2.29	2.73	5.02
7200	Particle board	1.11	2.73	3.84
7300	Wood board, 3/4" thick, knotty pine	1.58	3.64	5.22
7400	Rough sawn cedar	2.04	3.64	5.68
7500	Redwood, clear	5.45	3.64	9.09
7600	Aromatic cedar	3.65	3.97	7.62

C30 Interior Finishes

C3020 Floor Finishes

C3020 430	Selective Price Sheet	COST PER S.F.		
		MAT.	INST.	TOTAL
0100	Flooring, carpet, acrylic, 26 oz. light traffic	3.27	.73	4
0200	35 oz. heavy traffic	6.05	.73	6.78
0300	Nylon anti-static, 15 oz. light traffic	2.11	.95	3.06
0400	22 oz. medium traffic	4.16	.73	4.89
0500	26 oz. heavy traffic	5.45	.78	6.23
0600	28 oz. heavy traffic	5.95	.78	6.73
0700	Tile, foamed back, needle punch	4.32	.86	5.18
0800	Tufted loop	2.86	.86	3.72
0900	Wool, 36 oz. medium traffic	14.10	.78	14.88
1000	42 oz. heavy traffic	14.45	.78	15.23
1100	Composition, epoxy, with colored chips, minimum	2.74	4.09	6.83
1200	Maximum	3.32	5.65	8.97
1300	Trowelled, minimum	3.53	4.93	8.46
1400	Maximum	5.15	5.75	10.90
1500	Terrazzo, 1/4" thick, chemical resistant, minimum	5.80	5.90	11.70
1600	Maximum	8.50	7.85	16.35
1700	Resilient, asphalt tile, 1/8" thick	1.31	1.23	2.54
1800	Conductive flooring, rubber, 1/8" thick	5.60	1.56	7.16
1900	Cork tile 1/8" thick, standard finish	6.60	1.56	8.16
2000	Urethane finish	7.85	1.56	9.41
2100	PVC sheet goods for gyms, 1/4" thick	8.55	6.15	14.70
2200	3/8" thick	9.65	8.20	17.85
2300	Vinyl composition 12" x 12" tile, plain, 1/16" thick	1.01	.98	1.99
2400	1/8" thick	1.98	.98	2.96
2500	Vinyl tile, 12" x 12" x 1/8" thick, minimum	4.79	.98	5.77
2600	Maximum	13.95	.98	14.93
2700	Vinyl sheet goods, backed, .093" thick	3.96	2.14	6.10
2900	Slate, random rectangular, 1/4" thick	17.90	6.45	24.35
3000	1/2" thick	16.65	9.25	25.90
3100	Natural cleft, irregular, 3/4" thick	7.70	10.55	18.25
3200	For sand rubbed finish, add	8.35		8.35
3300	Terrazzo, cast in place, bonded 1-3/4" thick, gray cement	3.16	15.70	18.86
3400	White cement	3.60	15.70	19.30
3500	Not bonded 3" thick, gray cement	3.95	16.85	20.80
3600	White cement	4.31	16.85	21.16
3700	Precast, 12" x 12" x 1" thick	20	16.15	36.15
3800	1-1/4" thick	21.50	16.15	37.65
3900	16" x 16" x 1-1/4" thick	23.50	19.40	42.90
4000	1-1/2" thick	21.50	21.50	43
4100	Marble travertine, standard, 12" x 12" x 3/4" thick	13.40	14.70	28.10
4200				
4300	Tile, ceramic, natural clay, thin set	4.61	4.82	9.43
4400	Porcelain, thin set	5.50	4.64	10.14
4500	Specialty, decorator finish	11	4.82	15.82
4600				
4700	Quarry, red, mud set, 4" x 4" x 1/2" thick	8.25	7.35	15.60
4800	6" x 6" x 1/2" thick	7.50	6.30	13.80
4900	Brown, imported, 6" x 6" x 3/4" thick	9.25	7.35	16.60
5000	8" x 8" x 1" thick	9.95	8	17.95
5100	Slate, Vermont, thin set, 6" x 6" x 1/4" thick	7.45	4.90	12.35
5200				
5300	Wood, maple strip, 25/32" x 2-1/4", finished, select	5.95	3.45	9.40
5400	2nd and better	3.82	3.45	7.27
5500	Oak, 25/32" x 2-1/4" finished, clear	2.81	3.45	6.26
5600	No. 1 common	4.38	3.45	7.83
5700	Parquet, standard, 5/16", finished, minimum	5	3.65	8.65
5800	Maximum	5.60	5.70	11.30
5900	Custom, finished, minimum	17	5.45	22.45

C30 Interior Finishes

C3020 Floor Finishes

C3020 430	Selective Price Sheet	COST PER S.F.		
		MAT.	INST.	TOTAL
6000	Maximum	22.50	10.90	33.40
6100	Prefinished, oak, 2-1/4" wide	5.25	3.21	8.46
6200	Ranch plank	7.40	3.77	11.17
6300	Sleepers on concrete, treated, 24" O.C., 1" x 2"	.28	.36	.64
6400	1" x 3"	.49	.47	.96
6500	2" x 4"	.34	.64	.98
6600	2" x 6"	.54	.74	1.28
6700	Refinish old floors, minimum	.86	1.09	1.95
6800	Maximum	1.28	3.34	4.62
6900	Subflooring, plywood, CDX, 1/2" thick	.46	.73	1.19
7000	5/8" thick	.56	.81	1.37
7100	3/4" thick	.64	.87	1.51
7200				
7300	1" x 10" boards, S4S, laid regular	1.62	.99	2.61
7400	Laid diagonal	1.62	1.21	2.83
7500	1" x 8" boards, S4S, laid regular	1.47	1.09	2.56
7600	Laid diagonal	1.47	1.28	2.75
7700	Underlayment, plywood, underlayment grade, 3/8" thick	.85	.73	1.58
7800	1/2" thick	.88	.75	1.63
7900	5/8" thick	1.29	.78	2.07
8000	3/4" thick	1.35	.84	2.19
8100	Particle board, 3/8" thick	.40	.73	1.13
8200	1/2" thick	.43	.75	1.18
8300	5/8" thick	.55	.78	1.33
8400	3/4" thick	.72	.84	1.56

C30 Interior Finishes

C3030 Ceiling Finishes

This page illustrates suspended acoustical board systems including acoustic ceiling board, hangers, and T bar suspension. Lines within System Components give the unit price and total price per square foot for this system. Prices for alternate suspended acoustical board systems are on Line Items C3030 210 1100 thru 2400. Both material quantities and labor costs have been adjusted for the system listed.

Factors: To adjust for job conditions other than normal working situations use Lines C3030 210 2900 thru 4000.

Example: You are to install the system and protect existing construction. Go to Line C3030 210 3800 and apply these percentages to the appropriate MAT. and INST. costs.

System Components	QUANTITY	UNIT	COST PER S.F. MAT.	INST.	TOTAL
Suspended acoustical ceiling board installed on exposed grid system.					
Fiberglass boards, film faced, 2' x 4', 5/8" thick	1.000	S.F.	.77	.87	1.64
Hangers, #12 wire	1.000	S.F.	.56	.53	1.09
T bar suspension system, 2' x 4' grid	1.000	S.F.	.75	.68	1.43
TOTAL			2.08	2.08	4.16

C3030 210	Ceiling, Suspended Acoustical	MAT.	INST.	TOTAL
1000	For alternate suspended ceiling systems:			
1100	2' x 4' grid, mineral fiber board, aluminum faced, 5/8" thick	2.93	2.05	4.98
1200	Standard faced	2.17	2.02	4.19
1300	Plastic faced	2.66	2.58	5.24
1400	Fiberglass, film faced, 3" thick, R11	2.99	2.42	5.41
1500	Grass cloth faced, 3/4" thick	4.47	2.30	6.77
1600	1" thick	3.71	2.34	6.05
1700	1-1/2" thick, nubby face	4.24	2.36	6.60
2200				
2300				
2400	Add for 2' x 2' grid system	.19	.17	.36
2500				
2600				
2700				
2900	Cut & patch to match existing construction, add, minimum	2%	3%	
3000	Maximum	5%	9%	
3100	Dust protection, add, minimum	1%	2%	
3200	Maximum	4%	11%	
3300	Equipment usage curtailment, add, minimum	1%	1%	
3400	Maximum	3%	10%	
3500	Material handling & storage limitation, add, minimum	1%	1%	
3600	Maximum	6%	7%	
3700	Protection of existing work, add, minimum	2%	2%	
3800	Maximum	5%	7%	
3900	Shift work requirements, add, minimum		5%	
4000	Maximum		30%	

C30 Interior Finishes

C3030 Ceiling Finishes

This page illustrates and describes suspended gypsum board systems including gypsum board, metal furring, taping, finished and painted. Lines within System Components give the unit price and total price per square foot for this system. Prices for alternate suspended gypsum board systems are on Line Items C3030 220 1500 thru 1700. Both material quantities and labor costs have been adjusted for the system listed.

Factors: To adjust for job conditions other than normal working situations use Lines C3030 220 2900 thru 4000.

Example: You are to install the system and control dust in the working area. Go to Line C3030 220 3200 and apply these percentages to the appropriate MAT. and INST. costs.

System Components	QUANTITY	UNIT	COST PER S.F.		
			MAT.	INST.	TOTAL
Suspended ceiling gypsum board, 4' x 8' x 5/8" thick,					
On metal furring, taped, finished and painted.					
Gypsum drywall, 4' x 8', 5/8" thick, screwed	1.000	S.F.	.31	.61	.92
Main runners, 1-1/2" C.R.C., 4' O.C.	.500	S.F.	.14	.60	.74
25 ga., channels, 2' O.C.	1.000	S.F.	.20	1.11	1.31
Taped and finished	1.000	S.F.	.05	.55	.60
Paint, 2 coats, roller work	1.000	S.F.	.17	.61	.78
TOTAL			.87	3.48	4.35

C3030 220	Ceilings, Suspended Gypsum Board	COST PER S.F.		
		MAT.	INST.	TOTAL
1400	For alternate drywall ceiling systems:			
1500	Thin coat plaster, 2 coats paint	.74	3	3.74
1600	Spray-on sand finish, no paint	.71	2.90	3.61
1700	12" x 12" x 3/4" acoustical wood fiber tile	1.94	3.69	5.63
1800				
1900				
2000				
2100				
2200				
2300				
2400				
2500				
2600				
2700				
2900	Cut & patch to match existing construction, add, minimum	2%	3%	
3000	Maximum	5%	9%	
3100	Dust protection, add, minimum	1%	2%	
3200	Maximum	4%	11%	
3300	Equipment usage curtailment, add, minimum	1%	1%	
3400	Maximum	3%	10%	
3500	Material handling & storage limitation, add, minimum	1%	1%	
3600	Maximum	6%	7%	
3700	Protection of existing work, add, minimum	2%	2%	
3800	Maximum	5%	7%	
3900	Shift work requirements, add, minimum		5%	
4000	Maximum		30%	

C30 Interior Finishes

C3030 Ceiling Finishes

This page illustrates and describes suspended plaster and lath systems including gypsum plaster, lath, furring and runners with ceiling painted. Lines within System Components give the unit price and total price per square foot for this system. Prices for alternate suspended plaster and lath systems are on Line Items C3030 230 1300 thru 2300. Both material quantities and labor costs have been adjusted for the system listed.

Factors: To adjust for job conditions other than normal working situations use Lines C3030 230 2900 thru 4000.

Example: You are to install the system to match existing construction. Go to Line C3030 230 2900 and apply these percentages to the appropriate MAT. and INST. costs.

System Components	QUANTITY	UNIT	COST PER S.F.		
			MAT.	INST.	TOTAL
Gypsum plaster, 3 coats, on 3.4# rib lath, on 3/4" C.R.C. furring on 1-1/2" main runners, ceiling painted.					
Gypsum plaster, 3 coats	.110	S.Y.	.65	3.46	4.11
3.4# rib lath	.110	S.Y.	.69	.73	1.42
Main runners, 1-1/2" C.R.C. 24" O.C.	.333	S.F.	.27	1.19	1.46
Furring 3/4" C.R.C. 16" O.C.	1.000	S.F.	.31	1.60	1.91
Painting, 2 coats, roller work	1.000	S.F.	.17	.61	.78
TOTAL			2.09	7.59	9.68

C3030 230	Ceiling, Suspended Plaster	COST PER S.F.		
		MAT.	INST.	TOTAL
1200	For alternate plaster ceiling systems:			
1300	Gypsum plaster, 3 coats, on 2.5# diamond lath	1.95	7.55	9.50
1400	On 3/8" gypsum lath	2.01	7.55	9.56
1500	2 coats, on 3.4# rib lath	2	7.60	9.60
1600	On 2.5# diamond lath	1.75	7.05	8.80
1700	On 3/8" gypsum lath	1.81	7.05	8.86
1800	Perlite plaster, 3 coats, on 3.4# rib lath	2.20	8.40	10.60
1900	On 2.5# diamond lath	2.06	8.35	10.41
2000	On 3/8" gypsum lath	2.12	8.35	10.47
2100	2 coats, on 3.4# rib lath	1.90	7.50	9.40
2200	On 2.5# diamond lath	1.76	7.45	9.21
2300	On 3/8" gypsum lath	1.82	7.45	9.27
2400				
2500				
2600				
2700				
2900	Cut & patch to match existing construction, add, minimum	2%	3%	
3000	Maximum	5%	9%	
3100	Dust protection, add, minimum	1%	2%	
3200	Maximum	4%	11%	
3300	Equipment usage curtailment, add, minimum	1%	1%	
3400	Maximum	3%	10%	
3500	Material handling & storage limitation, add, minimum	1%	1%	
3600	Maximum	6%	7%	
3700	Protection of existing work, add, minimum	2%	2%	
3800	Maximum	5%	7%	
3900	Shift work requirements, add, minimum		5%	
4000	Maximum		30%	

C30 Interior Finishes

C3030 Ceiling Finishes

C3030 250	Selective Price Sheet	COST PER S.F.		
		MAT.	INST.	TOTAL
0100	Ceiling, plaster, gypsum, 2 coats	.46	2.96	3.42
0200	3 coats	.65	3.48	4.13
0300	Perlite or vermiculite, 2 coats	.46	3.43	3.89
0400	3 coats	.77	4.32	5.09
0500	Gypsum lath, plain, 3/8" thick	.62	.60	1.22
0600	1/2" thick	.60	.64	1.24
0700	Firestop, 3/8" thick	.30	.74	1.04
0800	1/2" thick	.30	.79	1.09
0900	Metal lath, rib, 2.75 lb.	.45	.69	1.14
1000	3.40 lb.	.69	.74	1.43
1100	Diamond, 2.50 lb.	.55	.69	1.24
1200	3.40 lb.	.58	.86	1.44
1500	Drywall, standard, 1/2" thick, no finish included	.26	.61	.87
1600	Taped and finished	.31	1.43	1.74
1700	5/8" thick, no finish included	.31	.61	.92
1800	Taped and finished	.35	1.43	1.78
1900	Fire resistant, 1/2" thick, no finish included	.31	.61	.92
2000	Taped and finished	.35	1.43	1.78
2100	5/8" thick, no finish included	.30	.61	.91
2200	Taped and finished	.34	1.43	1.77
2300	Water resistant, 1/2" thick, no finish included	.39	.61	1
2400	Taped and finished	.43	1.43	1.86
2500	5/8" thick, no finish included	.47	.61	1.08
2600	Taped and finished	.52	1.43	1.95
2700	Finish, instead of taping			
2800	For thin coat plaster, add	.09	.68	.77
2900	Finish, textured spray, add	.06	.58	.64
3000				
3100				
3200				
3300	Tile, stapled or glued, mineral fiber plastic coated, 5/8" thick	1.97	1.82	3.79
3400	3/4" thick	2.10	1.82	3.92
3500	Wood fiber, 1/2" thick	1.01	1.37	2.38
3600	3/4" thick	1.29	1.37	2.66
3700	Suspended, fiberglass, film faced, 5/8" thick	.77	.87	1.64
3800	3" thick	1.68	1.21	2.89
3900	Mineral fiber, 5/8" thick, standard	.86	.81	1.67
4000	Aluminum	2	.91	2.91
4100	Wood fiber, reveal edge, 1" thick			
4200	3" thick			
4300	Framing, metal furring, 3/4" channels, 12" O.C.	.34	2.21	2.55
4400	16" O.C.	.31	1.60	1.91
4500	24" O.C.	.20	1.11	1.31
4600				
4700	1-1/2" channels, 12" O.C.	.46	2.44	2.90
4800	16" O.C.	.41	1.79	2.20
4900	24" O.C.	.27	1.19	1.46
5000				
5100				
5200				
5300	Ceiling suspension systems, for tile,			
5400	Concealed "Z" bar, 12" module	.76	1.05	1.81
5500	Class A, "T" bar 2'-0" x 4'-0" grid	.75	.68	1.43
5600	2'-0" x 2'-0" grid	.93	.84	1.77
5700	Carrier channels for lighting fixtures, add	.21	1.19	1.40
5800				
5900				
6000				

C30 Interior Finishes

C3030 Ceiling Finishes

C3030 250	Selective Price Sheet	COST PER S.F.		
		MAT.	INST.	TOTAL
6100	Wood, furring 1" x 3", on wood, 12" O.C.	.38	1.56	1.94
6200	16" O.C..	.29	1.17	1.46
6300	24" O.C.	.19	.78	.97
6400				
6500	On concrete, 12" O.C.	.38	2.60	2.98
6600	16" O.C.	.29	1.95	2.24
6700	24" O.C.	.19	1.30	1.49
6800				
6900	Joists, 2" x 4", 12" O.C.	.35	1.12	1.47
7000	16" O.C.	.29	.94	1.23
7100	24" O.C.	.24	.76	1
7200	32" O.C.	.18	.58	.76
7300	2" x 6", 12" O.C.	.54	1.14	1.68
7400	16" O.C.	.44	.94	1.38
7500	24" O.C.	.35	.75	1.10
7600	32" O.C.	.26	.56	.82

D Services

No part of this publication may be reproduced, stored in a retrieval system, or transmitted in any form or by any means without prior written permission of R.S. Means Company, Inc.

D10 Conveying

D1010 Elevators and Lifts

This page illustrates and describes oil hydraulic elevator systems. Prices for various oil hydraulic elevator systems are on Line Items D1010 130 0400 thru 2000. Both material quantities and labor costs have been adjusted for the system listed.

Factors: To adjust for job conditions other than normal working situations use Lines D1010 130 3100 thru 4000.

Example: You are to install the system with limited equipment usage. Go to Line D1010 130 3400 and apply this percentage to the appropriate TOTAL costs.

D1010 130	Oil Hydraulic Elevators	COST EACH		
		MAT.	INST.	TOTAL
0310	Oil-hydraulic elevator systems			
0320	Including piston and piston shaft			
0330				
0400	1500 Lb. passenger, 2 floors	40,600	15,200	55,800
0500	3 floors	55,000	23,500	78,500
0600	4 floors	93,500	33,000	126,500
0700	5 floors	107,500	41,100	148,600
0800	6 floors	120,500	49,000	169,500
0900				
1000	2500 Lb. passenger, 2 floors	44,000	15,200	59,200
1100	3 floors	58,500	23,500	82,000
1200	4 floors	93,000	33,000	126,000
1300	5 floors	107,000	41,100	148,100
1400	6 floors	124,000	49,000	173,000
1500				
1600	4000 Lb. passenger, 2 floors	48,700	15,200	63,900
1700	3 floors	66,500	23,500	90,000
1800	4 floors	101,500	33,000	134,500
1900	5 floors	115,500	41,100	156,600
2000	6 floors	128,500	49,000	177,500
2100				
2200				
2300				
2400				
2500				
2600				
2700				
2800				
2900				
3000				
3100	Dust protection, add, minimum	1%	2%	
3200	Maximum	4%	11%	
3300	Equipment usage curtailment, add, minimum	1%	1%	
3400	Maximum	3%	10%	
3500	Material handling & storage limitation, add, minimum	1%	1%	
3600	Maximum	6%	7%	
3700	Protection of existing work, add, minimum	2%	2%	
3800	Maximum	5%	7%	
3900	Shift work requirements, add, minimum		5%	
4000	Maximum		30%	

D20 Plumbing

D2010 Plumbing Fixtures

This page illustrates and describes a women's public restroom system including a water closet, lavatory, accessories, and service piping. Lines within System Components give the unit price and total price on a cost each basis for this system. Prices for alternate women's public restroom systems are on Line Items D2010 956 1400 thru 1700. Both material quantities and labor costs have been adjusted for the system listed.

Factors: To adjust for job conditions other than normal working situations use Lines D2010 956 2900 thru 4000.

Example: You are to install the system and protect surrounding area from dust. Go to Line D2010 956 3100 and apply these percentages to the MAT. and INST. costs.

System Components	QUANTITY	UNIT	COST EACH		
			MAT.	INST.	TOTAL
Public women's restroom incl. water closet, lavatory, accessories and Necessary service piping to install this system in one wall.					
Water closet, wall mounted, one piece	1.000	Ea.	335	200	535
Rough-in waste and vent for water closet	1.000	Set	1,050	455	1,505
Lavatory, 20" x 18" P.E. cast iron with accessories	1.000	Ea.	390	145	535
Rough-in waste and vent for lavatory	1.000	Set	425	700	1,125
Toilet partition, painted metal between walls, floor mounted	1.000	Ea.	515	182	697
For handicap unit, add	1.000	Ea.	365		365
Grab bar, 36" long	1.000	Ea.	76	55	131
Mirror, 18" x 24", with stainless steel shelf	1.000	Ea.	208	27.50	235.50
Napkin/tampon dispenser, recessed	1.000	Ea.	595	36.50	631.50
Soap Dispenser, chrome, surface mounted, liquid	1.000	Ea.	47	27.50	74.50
Toilet tissue dispenser, surface mounted, stainless steel	1.000	Ea.	19.15	18.20	37.35
Towel dispenser, surface mounted, stainless steel	1.000	Ea.	38.50	34	72.50
TOTAL			4,063.65	1,880.70	5,944.35

D2010 956	Plumbing - Public Restroom	COST EACH		
		MAT.	INST.	TOTAL
1200				
1300	For alternate size restrooms:			
1400	Two water closets, two lavatories	7,200	3,575	10,775
1500				
1600	For each additional water closet over 2, add	2,000	830	2,830
1700	For each additional lavatory over 2, add	1,075	900	1,975
1800				
1900				
2400	NOTE: PLUMBING APPROXIMATIONS			
2500	WATER CONTROL: water meter, backflow preventer,			
2600	Shock absorbers, vacuum breakers, mixer....10 to 15% of fixtures			
2700	PIPE AND FITTINGS: .... 30 to 60% of fixtures			
2800				
2900	Cut & patch to match existing construction, add, minimum	2%	3%	
3000	Maximum	5%	9%	
3100	Dust protection, add, minimum	1%	2%	
3200	Maximum	4%	11%	
3300	Equipment usage curtailment, add, minimum	1%	1%	
3400	Maximum	3%	10%	
3500	Material handling & storage limitation, add, minimum	1%	1%	

D20 Plumbing

D2010 Plumbing Fixtures

D2010 956	Plumbing - Public Restroom	COST EACH		
		MAT.	INST.	TOTAL
3600	Maximum	6%	7%	
3700	Protection of existing work, add, minimum	2%	2%	
3800	Maximum	5%	7%	
3900	Shift work requirements, add, minimum		5%	
4000	Maximum		30%	

D20 Plumbing

D2010 Plumbing Fixtures

This page illustrates and describes a men's public restroom system including a water closet, urinal, lavatory, accessories and service piping. Lines within System Components give the unit price and total price on a cost each basis for this system. Prices for alternate men's public restroom systems are on Line Items D2010 957 1800 thru 2200. Both material quantities and labor costs have been adjusted for the system listed.

Factors: To adjust for job conditions other than normal working situations use Lines D2010 957 2900 thru 4000.

Example: You are to install the system and match existing construction. Go to Line D2010 957 3000 and apply these percentages to the appropriate MAT. and INST. costs.

System Components	QUANTITY	UNIT	COST EACH		
			MAT.	INST.	TOTAL
Public men's restroom incl. water closet, urinal, lavatory, accessories,					
And necessary service piping to install this system in one wall.					
Water closet, wall mounted, one piece	1.000	Ea.	335	200	535
Rough-in waste & vent for water closet	1.000	Set	1,050	455	1,505
Urinal, wall hung	1.000	Ea.	325	385	710
Rough-in waste & vent for urinal	1.000	Set	264	410	674
Lavatory, 20" x 18", P.E. cast iron with accessories	1.000	Ea.	390	145	535
Rough-in waste & vent for lavatory	1.000	Set	425	700	1,125
Partition, painted mtl., between walls, floor mntd.	1.000	Ea.	515	182	697
For handicap unit, add	1.000	Ea.	365		365
Grab bars	1.000	Ea.	76	55	131
Urinal screen, powder coated steel, wall mounted	1.000	Ea.	335	109	444
Mirror, 18" x 24" with stainless steel shelf	1.000	Ea.	208	27.50	235.50
Soap dispenser, surface mounted, liquid	1.000	Ea.	47	27.50	74.50
Toilet tissue dispenser, surface mounted, stainless steel	1.000	Ea.	19.15	18.20	37.35
Towel dispenser, surface mounted, stainless steel	1.000	Ea.	38.50	34	72.50
TOTAL			4,392.65	2,748.20	7,140.85

D2010 957	Plumbing - Public Restroom	COST EACH		
		MAT.	INST.	TOTAL
1600				
1700	For alternate size restrooms:			
1800	Two water closets, two urinals, two lavatories	8,450	5,350	13,800
1900				
2000	For each additional water closet over 2, add	2,000	830	2,830
2100	For each additional urinal over 2, add	925	905	1,830
2200	For each additional lavatory over 2, add	1,075	900	1,975
2300				
2400	NOTE: PLUMBING APPROXIMATIONS			
2500	WATER CONTROL: water meter, backflow preventer,			
2600	Shock absorbers, vacuum breakers, mixer....10 to 15% of fixtures			
2700	PIPE AND FITTINGS: .... 30 to 60% of fixtures			
2800				
2900	Cut & patch to match existing construction, add, minimum	2%	3%	
3000	Maximum	5%	9%	
3100	Dust protection, add, minimum	1%	2%	
3200	Maximum	4%	11%	
3300	Equipment usage curtailment, add, minimum	1%	1%	

D20 Plumbing

D2010 Plumbing Fixtures

D2010 957	Plumbing - Public Restroom	COST EACH		
		MAT.	INST.	TOTAL
3400	Maximum	3%	10%	
3500	Material handling, & storage limitation, add, minimum	1%	1%	
3600	Maximum	6%	7%	
3700	Protection of existing work, add, minimum	2%	2%	
3800	Maximum	5%	7%	
3900	Shift work requirements, add, minimum		5%	
4000	Maximum		30%	

D20 Plumbing

D2010 Plumbing Fixtures

This page illustrates and describes a two fixture lavatory system including a water closet, lavatory, accessories and all service piping. Lines within System Components give the unit price and total price on a cost each basis for this system. Prices for an alternate two fixture lavatory system are on Line Item D2010 958 1900. Both material quantities and labor costs have been adjusted for the system listed.

Factors: To adjust for job conditions other than normal working situations use Lines D2010 958 2900 thru 4000.

Example: You are to install the system while controlling dust in the work area. Go to Line D2010 958 3200 and apply these percentages to the appropriate MAT. and INST. costs.

System Components	QUANTITY	UNIT	COST EACH MAT.	COST EACH INST.	COST EACH TOTAL
Two fixture bathroom incl. water closet, lavatory, accessories and Necessary service piping to install this system in 2 walls.					
Water closet, floor mounted, 2 piece, close coupled	1.000	Ea.	284	219	503
Rough in waste & vent for water closet	1.000	Set	310	380	690
Lavatory, 20" x 18", P.E. cast iron with accessories	1.000	Ea.	390	145	535
Rough in waste & vent for lavatory	1.000	Set	425	700	1,125
Additional service piping					
1/2" copper pipe with sweat solder joints	10.000	L.F.	27.40	79.50	106.90
2" black schedule 40 steel pipe with threaded couplings	12.000	L.F.	166.20	217.80	384
4" cast iron soil pipe with lead and oakum joints	7.000	L.F.	97.65	147	244.65
Accessories					
Toilet tissue dispenser, chrome, single roll	1.000	Ea.	19.15	18.20	37.35
18" long stainless steel towel bar	1.000	Ea.	43.50	23.50	67
Medicine cabinet with mirror, 20" x 16", unlighted	1.000	Ea.	73	39	112
TOTAL			1,835.90	1,969	3,804.90

D2010 958	Plumbing - Two Fixture Bathroom	COST EACH MAT.	COST EACH INST.	COST EACH TOTAL
1800				
1900	Above system installed in 1 wall with all necessary service piping	1,675	1,700	3,375
2000				
2400	NOTE: PLUMBING APPROXIMATIONS			
2500	WATER CONTROL: water meter, backflow preventer,			
2600	Shock absorbers, vacuum breakers, mixer....10 to 15% of fixtures			
2700	PIPE AND FITTINGS: .... 30 to 60% of fixtures			
2800				
2900	Cut & patch to match existing construction, add, minimum	2%	3%	
3000	Maximum	5%	9%	
3100	Dust protection, add, minimum	1%	2%	
3200	Maximum	4%	11%	
3300	Equipment usage curtailment, add, minimum	1%	1%	
3400	Maximum	3%	10%	
3500	Material handling & storage limitation, add, minimum	1%	1%	
3600	Maximum	6%	7%	
3700	Protection of existing work, add, minimum	2%	2%	
3800	Maximum	5%	7%	
3900	Shift work requirements, add, minimum		5%	
4000	Maximum		30%	

D20 Plumbing

D2010 Plumbing Fixtures

This page illustrates and describes a three fixture bathroom system including a water closet, tub, lavatory, accessories and service piping. Lines within System Components give the unit price and total price on a cost each basis for this system. Prices for an alternate three fixture bathroom system are on Line Item D2010 959 1700. Both material quantities and labor costs have been adjusted for the system listed.

Factors: To adjust for job conditions other than normal working situations use Lines D2010 959 2900 thru 4000.

Example: You are to install the system and protect all existing work. Go to Line D2010 959 3800 and apply these percentages to the appropriate MAT. and INST. costs.

System Components	QUANTITY	UNIT	COST EACH		
			MAT.	INST.	TOTAL
Three fixture bathroom incl. water closet, bathtub, lavatory, accessories,					
And necessary service piping to install this system in 1 wall.					
Water closet, floor mounted, 2 piece, close coupled	1.000	Ea.	284	219	503
Rough-in waste & vent for water closet	1.000	Set	288.30	353.40	641.70
Bathtub, P.E. cast iron 5' long with accessories	1.000	Ea.	1,050	264	1,314
Rough-in waste & vent for bathtub	1.000	Set	327.75	532	859.75
Lavatory, 20" x 18" P.E. cast iron with accessories	1.000	Ea.	390	145	535
Rough-in waste & vent for lavatory	1.000	Set	425	700	1,125
Accessories					
Toilet tissue dispenser, chrome, single roll	1.000	Ea.	19.15	18.20	37.35
18" long stainless steel towel bar	2.000	Ea.	87	47	134
Medicine cabinet with mirror, 20" x 16", unlighted	1.000	Ea.	73	39	112
TOTAL			2,944.20	2,317.60	5,261.80

D2010 959	Plumbing - Three Fixture Bathroom	COST EACH		
		MAT.	INST.	TOTAL
1600				
1700	Above system installed in one wall with all necessary service piping	2,900	2,250	5,150
2400	NOTE: PLUMBING APPROXIMATIONS			
2500	WATER CONTROL: water meter, backflow preventer,			
2600	Shock absorbers, vacuum breakers, mixer....10 to 15% of fixtures			
2700	PIPE AND FITTINGS: .... 30 to 60% of fixtures			
2800				
2900	Cut & patch to match existing construction, add, minimum	2%	3%	
3000	Maximum	5%	9%	
3100	Dust protection, add, minimum	1%	2%	
3200	Maximum	4%	11%	
3300	Equipment usage curtailment, add, minimum	1%	1%	
3400	Maximum	3%	10%	
3500	Material handling & storage limitation, add, minimum	1%	1%	
3600	Maximum	6%	7%	
3700	Protection of existing work, add, minimum	2%	2%	
3800	Maximum	5%	7%	
3900	Shift work requirements, add, minimum		5%	
4000	Maximum		30%	

D20 Plumbing

D2010 Plumbing Fixtures

This page illustrates and describes a three fixture bathroom system including a water closet, tub, lavatory, accessories, and service piping. Lines within System Components give the unit price and total price on a cost each basis for this system. Prices for an alternate three fixture bathroom system are on Line Item D2010 960 2000. Both material quantities and labor costs have been adjusted for the system listed.

Factors: To adjust for job conditions other than normal working situations use Lines D2010 960 2900 thru 4000.

Example: You are to install the system and protect the surrounding area from dust. Go to Line D2010 960 3100 and apply these percentages to the appropriate MAT. and INST. costs.

System Components	QUANTITY	UNIT	COST EACH		
			MAT.	INST.	TOTAL
Three fixture bathroom incl. water closet, bathtub, lavatory, accessories,					
And necessary service piping to install this system in 2 walls.					
Water closet, floor mounted, 2 piece, close coupled	1.000	Ea.	284	219	503
Rough-in waste & vent for water closet	1.000	Set	310	380	690
Bathtub, P.E. cast iron, 5' long with accessories	1.000	Ea.	1,050	264	1,314
Rough-in waste & vent for bathtub	1.000	Set	345	560	905
Lavatory, 20" x 18" P.E. cast iron with accessories	1.000	Ea.	390	145	535
Rough-in waste & vent for lavatory	1.000	Set	425	700	1,125
Additional service piping					
1-1/4" copper DWV type tubing, sweat solder joints	6.000	L.F.	64.80	64.50	129.30
2" black schedule 40 steel pipe with threaded couplings	12.000	L.F.	166.20	217.80	384
Accessories					
Toilet tissue dispenser, chrome, single roll	1.000	Ea.	19.15	18.20	37.35
18" long stainless steel towel bar	2.000	Ea.	87	47	134
Medicine cabinet with mirror, 20" x 16", unlighted	1.000	Ea.	73	39	112
TOTAL			3,214.15	2,654.50	5,868.65

D2010 960	Plumbing - Three Fixture Bathroom	COST EACH		
		MAT.	INST.	TOTAL
2000	Above system with corner contour tub, P.E. cast iron	4,550	2,650	7,200
2300				
2400	NOTE: PLUMBING APPROXIMATIONS			
2500	WATER CONTROL: water meter, backflow preventer,			
2600	Shock absorbers, vacuum breakers, mixer....10 to 15% of fixtures			
2700	PIPE AND FITTINGS: .... 30 to 60% of fixtures			
2800				
2900	Cut & patch to match existing construction, add, minimum	2%	3%	
3000	Maximum	5%	9%	
3100	Dust protection, add, minimum	1%	2%	
3200	Maximum	4%	11%	
3300	Equipment usage curtailment, add, minimum	1%	1%	
3400	Maximum	3%	10%	
3500	Material handling & storage limitation, add, minimum	1%	1%	
3600	Maximum	6%	7%	
3700	Protection of existing work, add, minimum	2%	2%	
3800	Maximum	5%	7%	
3900	Shift work requirements, add, minimum		5%	
4000	Maximum		30%	

D20 Plumbing

D2010 Plumbing Fixtures

This page illustrates and describes a three fixture bathroom system including a water closet, shower, lavatory, accessories, and service piping, Lines within System Components give the unit price and total price on a cost each basis for this system. Prices for an alternate three fixture bathroom system are on Line Item D2010 961 2200. Both material quantities and labor costs have been adjusted for the system listed.

Factors: To adjust for job conditions other than normal working situations use Lines D2010 961 2900 thru 4000.

Example: You are to install the system and protect existing construction. Go to Line D2010 961 3700 and apply these percentages to the appropriate MAT. and INST. costs.

System Components	QUANTITY	UNIT	COST EACH		
			MAT.	INST.	TOTAL
Three fixture bathroom incl. water closet, shower, lavatory, accessories,					
And necessary service piping to install this system in 2 walls.					
Water closet, floor mounted, 2 piece, close coupled	1.000	Ea.	284	219	503
Rough-in waste & vent for water closet	1.000	Set	310	380	690
32" shower, enameled stall, molded stone receptor	1.000	Ea.	975	232	1,207
Rough-in waste & vent for shower	1.000	Set	430	565	995
Lavatory, 20" x 18" P.E. cast iron with accessories	1.000	Ea.	390	145	535
Rough-in waste & vent for lavatory	1.000	Set	425	700	1,125
Additional service piping					
1-1/4" Copper DWV type tubing, sweat solder joints	4.000	L.F.	43.20	43	86.20
2" black schedule 40 steel pipe with threaded couplings	6.000	L.F.	83.10	108.90	192
4" cast iron soil with lead and oakum joints	7.000	L.F.	97.65	147	244.65
Accessories					
Toilet tissue dispenser, chrome, single roll	1.000	Ea.	19.15	18.20	37.35
18" long stainless steel towel bar	2.000	Ea.	87	47	134
Medicine cabinet with mirror, 20" x 16", unlighted	1.000	Ea.	73	39	112
TOTAL			3,217.10	2,644.10	5,861.20

D2010 961	Plumbing - Three Fixture Bathroom	COST EACH		
		MAT.	INST.	TOTAL
2100	Above system installed with 36" corner angle shower, enameled steel,			
2200	Molded stone receptor	4,075	2,650	6,725
2300				
2400	NOTE: PLUMBING APPROXIMATIONS			
2500	WATER CONTROL: water meter, backflow preventer,			
2600	Shock absorbers, vacuum breakers, mixer....10 to 15% of fixtures			
2700	PIPE AND FITTINGS: .... 30 to 60% of fixtures			
2800				
2900	Cut & patch to match existing construction, add, minimum	2%	3%	
3000	Maximum	5%	9%	
3100	Dust protection, add, minimum	1%	2%	
3200	Maximum	4%	11%	
3300	Equipment usage curtailment, add, minimum	1%	1%	
3400	Maximum	3%	10%	
3500	Material handling & storage limitation, add, minimum	1%	1%	
3600	Maximum	6%	7%	
3700	Protection of existing work, add, minimum	2%	2%	
3800	Maximum	5%	7%	
3900	Shift work requirements, add, minimum		5%	
4000	Maximum		30%	

D20 Plumbing

D2010 Plumbing Fixtures

This page illustrates and describes a four fixture bathroom system including a water closet, shower, bathtub, lavatory, accessories, and service piping. Lines within System Components give the unit price and total price on a cost each basis for this system. Prices for an alternate four fixture bathroom system are on Line Item D2010 962 1900. Both material quantities and labor costs have been adjusted for the system listed.

Factors: To adjust for job conditions other than normal working situations use Lines D2010 962 2900 thru 4000.

Example: You are to install the system during weekends and evenings. Go to Line D2010 962 4000 and apply these percentages to the appropriate MAT. and INST. costs.

System Components	QUANTITY	UNIT	COST EACH		
			MAT.	INST.	TOTAL
Four fixture bathroom incl. water closet, shower, bathtub, lavatory					
Accessories and necessary service piping to install this system in 2 walls.					
Water closet, floor mounted, 2 piece, close coupled	1.000	Ea.	284	219	503
Rough-in waste & vent for water closet	1.000	Set	310	380	690
32" shower, enameled steel stall, molded stone receptor	1.000	Ea.	975	232	1,207
Rough-in waste & vent for shower	1.000	Set	430	565	995
Bathtub, P.E. cast iron, 5' long with accessories	1.000	Ea.	1,050	264	1,314
Rough-in waste & vent for bathtub	1.000	Set	345	560	905
Lavatory, 20" x 18" P.E. cast iron with accessories	1.000	Ea.	390	145	535
Rough-in waste & vent for lavatory	1.000	Set	425	700	1,125
Accessories					
Toilet tissue dispenser, chrome, single roll	1.000	Ea.	19.15	18.20	37.35
18" long stainless steel towel bar	2.000	Ea.	87	47	134
Medicine cabinet with mirror, 20" x 16" unlighted	1.000	Ea.	73	39	112
TOTAL			4,388.15	3,169.20	7,557.35

D2010 962	Plumbing - Four Fixture Bathroom	COST EACH		
		MAT.	INST.	TOTAL
1900	Above system with 36" corner angle shower, plumbing in 3 walls	5,425	3,475	8,900
2000				
2300				
2400	NOTE: PLUMBING APPROXIMATIONS			
2500	WATER CONTROL: water meter, backflow preventer,			
2600	Shock absorbers, vacuum breakers, mixer....10 to 15% of fixtures			
2700	PIPE AND FITTINGS: .... 30 to 60% of fixtures			
2800				
2900	Cut & patch to match existing construction, add, minimum	2%	3%	
3000	Maximum	5%	9%	
3100	Dust protection, add, minimum	1%	2%	
3200	Maximum	4%	11%	
3300	Equipment usage curtailment, add, minimum	1%	1%	
3400	Maximum	3%	10%	
3500	Material handling & storage limitation, add, minimum	1%	1%	
3600	Maximum	6%	7%	
3700	Protection of existing work, add, minimum	2%	2%	
3800	Maximum	5%	7%	
3900	Shift work requirements, add, minimum		5%	
4000	Maximum		30%	

D20 Plumbing

D2010 Plumbing Fixtures

This page illustrates and describes a five fixture bathroom system including a water closet, shower, bathtub, lavatories, accessories, and service piping. Lines within System Components give the unit price and total price on a cost each basis for this system. Prices for an alternate five fixture bathroom system are on Line Item D2010 963 1900. Both material quantities and labor costs have been adjusted for the system listed.

Factors: To adjust for job conditions other than normal working situations use Lines D2010 963 2900 thru 4000.

Example: You are to install and match any existing construction. Go to Line D2010 963 2900 and apply these percentages to the appropriate MAT. and INST. costs.

System Components	QUANTITY	UNIT	COST EACH		
			MAT.	INST.	TOTAL
Five fixture bathroom incl. water closet, shower, bathtub, 2 lavatories,					
Accessories and necessary service piping to install this system in 1 wall.					
Water closet, floor mounted, 2 piece, close coupled	1.000	Ea.	284	219	503
Rough-in waste & vent for water closet	1.000	Set	310	380	690
36" corner angle shower, enameled steel stall, molded stone receptor	1.000	Ea.	1,825	242	2,067
Rough-in waste & vent for shower	1.000	Set	430	565	995
Bathtub, P.E. cast iron, 5' long with accessories	1.000	Ea.	1,050	264	1,314
Rough-in waste & vent for bathtub	1.000	Set	345	560	905
Countertop, plastic laminate, with backsplash	2.000	Ea.	146	72.80	218.80
Vanity base, 2 doors, 24" wide	2.000	Ea.	253	99.50	352.50
Lavatories, 20" x 18" cabinet mntd, PECI	2.000	Ea.	572	362	934
Rough-in waste & vent for lavatories	1.600	Set	443.20	808	1,251.20
Accessories					
Toilet tissue dispenser, chrome, single roll	1.000	Ea.	19.15	18.20	37.35
18" long stainless steel towel bars	2.000	Ea.	87	47	134
Medicine cabinet with mirror, 20" x 16", unlighted	2.000	Ea.	146	78	224
TOTAL			5,910.35	3,715.50	9,625.85

D2010 963	Plumbing - Five Fixture Bathroom	COST EACH		
		MAT.	INST.	TOTAL
1900	Above system installed in 2 walls with all necessary service piping	6,000	3,850	9,850
2000				
2100				
2200				
2300				
2400	NOTE: PLUMBING APPROXIMATIONS			
2500	WATER CONTROL: water meter, backflow preventer,			
2600	Shock absorbers, vacuum breakers, mixer....10 to 15% of fixtures			
2700	PIPE AND FITTINGS: .... 30 to 60% of fixtures			
2800				
2900	Cut & patch to match existing construction, add, minimum	2%	3%	
3000	Maximum	5%	9%	
3100	Dust protection, add, minimum	1%	2%	
3200	Maximum	4%	11%	
3300	Equipment usage curtailment, add, minimum	1%	1%	
3400	Maximum	3%	10%	
3500	Material handling & storage limitation, add, minimum	1%	1%	
3600	Maximum	6%	7%	

D20 Plumbing

D2010 Plumbing Fixtures

D2010 963 Plumbing - Five Fixture Bathroom

		COST EACH		
		MAT.	INST.	TOTAL
3700	Protection of existing work, add, minimum	2%	2%	
3800	Maximum	5%	7%	
3900	Shift work requirements, add, minimum		5%	
4000	Maximum		30%	

D30 HVAC

D3010 Energy Supply

This page illustrates and describes an oil fired hot water baseboard system including an oil fired boiler, fin tube radiation and all fittings and piping. Lines within System Components give the unit price and total price per square foot for this system. Prices for alternate oil fired hot water baseboard systems are on Line Items D3010 540 1700 thru 2100. Both material quantities and labor costs have been adjusted for the system listed.

Factors: To adjust for job conditions other than normal working situations use Lines D3010 540 2900 thru 4000.

Example: You are to install the system while protecting all existing work. Go to Line D3010 540 3700 and apply these percentages to the appropriate MAT. and INST. costs.

System Components	QUANTITY	UNIT	COST PER S.F. MAT.	INST.	TOTAL
Oil fired hot water baseboard system including boiler, fin tube radiation					
And all necessary fittings and piping.					
Area to 800 S.F.					
Boiler, cast iron, w/oil piping, 97 MBH	1.000	Ea.	2,375	1,406.25	3,781.25
Copper piping	130.000	L.F.	1,196	1,443	2,639
Fin tube radiation	68.000	L.F.	598.40	1,356.60	1,955
Circulator	1.000	Ea.	545	194	739
Oil tank	1.000	Ea.	410	232	642
Expansion tank, ASME	1.000	Ea.	610	82.50	692.50
TOTAL			5,734.40	4,714.35	10,448.75
COST PER S.F.		S.F.	7.17	5.89	13.06

D3010 540	Heating - Oil Fired Hot Water	COST PER S.F. MAT.	INST.	TOTAL
1600	For alternate hot water systems:			
1700	Cast iron boiler, area to 1000 S.F.	6.25	5.10	11.35
1800	To 1200 S.F.	5.40	4.60	10
1900	To 1600 S.F.	6.35	4.59	10.94
2000	To 2000 S.F.	5.35	4.08	9.43
2100	To 3000 S.F.	3.95	3.45	7.40
2200				
2300				
2700				
2800				
2900	Cut & patch to match existing construction, add, minimum	2%	3%	
3000	Maximum	5%	9%	
3100	Dust protection, add, minimum	1%	2%	
3200	Maximum	4%	11%	
3300	Equipment usage curtailment, add, minimum	1%	1%	
3400	Maximum	3%	10%	
3500	Material handling & storage limitation, add, minimum	1%	1%	
3600	Maximum	6%	7%	
3700	Protection of existing work, add, minimum	2%	2%	
3800	Maximum	5%	7%	
3900	Shift work requirements, add, minimum		5%	
4000	Maximum		30%	

D30 HVAC

D3010 Energy Supply

This page illustrates and describes a gas fired hot water baseboard system including a gas fired boiler, fin tube radiation, fittings and piping. Lines within System Components give the unit price and total price per square foot for this system. Prices for alternate gas fired hot water baseboard systems are on Line Items D3010 550 1500 thru 1900. Both material quantities and labor costs have been adjusted for the system listed.

Factors: To adjust for job conditions other than normal working situations use Lines D3010 550 2900 thru 4000.

Example: You are to install the system with minimal equipment usage. Go to Line D3010 550 3400 and apply these percentages to the appropriate MAT. and INST. costs.

System Components	QUANTITY	UNIT	COST PER S.F. MAT.	INST.	TOTAL
Gas fired hot water baseboard system including boiler, fin tube radiation,					
All necessary fittings and piping.					
Area to 800 S.F.					
Cast iron boiler, insulating jacket, gas piping, 80 MBH	1.000	Ea.	2,468.75	2,093.75	4,562.50
Copper piping	130.000	L.F.	1,196	1,443	2,639
Fin tube radiation	68.000	L.F.	598.40	1,356.60	1,955
Circulator, flange connection	1.000	Ea.	545	194	739
Expansion tank, ASME	1.000	Ea.	610	82.50	692.50
TOTAL			5,418.15	5,169.85	10,588
COST PER S.F.		S.F.	6.77	6.46	13.23

D3010 550	Heating - Gas Fired Hot Water	MAT.	INST.	TOTAL
1400	For alternate hot water systems:			
1500	Cast iron boiler, area to 1000 S.F.	5.70	5.65	11.35
1600	To 1200 S.F.	4.96	5.10	10.06
1700	To 1600 S.F.	6.05	4.97	11.02
1800	To 2000 S.F.	5.35	4.46	9.81
1900	To 3000 S.F.	4.15	3.91	8.06
2000				
2100				
2900	Cut & patch to match existing construction, add, minimum	2%	3%	
3000	Maximum	5%	9%	
3100	Dust protection, add, minimum	1%	2%	
3200	Maximum	4%	11%	
3300	Equipment usage curtailment, add, minimum	1%	1%	
3400	Maximum	3%	10%	
3500	Material handling & storage limitation, add, minimum	1%	1%	
3600	Maximum	6%	7%	
3700	Protection of existing work, add, minimum	2%	2%	
3800	Maximum	5%	7%	
3900	Shift work requirements, add, minimum		5%	
4000	Maximum		30%	

D30 HVAC

D3010 Energy Supply

This page illustrates and describes a wood burning stove system including a free standing stove, preformed hearth, masonry chimney, and necessary piping and fittings. Lines within System Components give the unit price and total price on a cost each basis for this system. Prices for alternate wood burning stove systems are on Line Items D3010 991 1600 thru 2000. Both material quantities and labor costs have been adjusted for the system listed.

Factors: To adjust for job conditions other than normal working situations use Lines D3010 991 2900 thru 4000.

Example: You are to install the system and protect existing construction. Go to Line D3010 991 3500 and apply these percentages to the appropriate MAT., and INST. costs.

System Components	QUANTITY	UNIT	COST EACH		
			MAT.	INST.	TOTAL
Cast iron, free standing, wood burning stove with preformed hearth, masonry Chimney, and all necessary piping and fittings to install in chimney.					
Cast iron wood burning stove, stove pipe	1.000	Ea.	1,300	840	2,140
Wall panel or hearth, non-combustible	1.000	Ea.	570	153.90	723.90
16" x 16" brick chimney, 8" x 8" flue	20.000	V.L.F.	587.50	1,337.50	1,925
Foundation	.500	C.Y.	77	114.52	191.52
TOTAL			2,534.50	2,445.92	4,980.42

D3010 991	Wood Burning Stoves	COST EACH		
		MAT.	INST.	TOTAL
1400	For alternate wood burning systems:			
1500				
1600	Installed in existing fireplace	1,600	920	2,520
1700				
1800	Installed with insulated metal chimney			
1900	System including ceiling package,			
2000	Metal chimney to 10 L.F.	2,750	1,300	4,050
2100				
2200				
2300				
2400				
2500				
2600				
2700				
2900	Dust protection, add, minimum	1%	2%	
3000	Maximum	4%	11%	
3100	Equipment usage curtailment, add, minimum	1%	1%	
3200	Maximum	3%	10%	
3300	Material handling & storage limitation, add, minimum	1%	1%	
3400	Maximum	6%	7%	
3500	Protection of existing work, add, minimum	2%	2%	
3600	Maximum	5%	7%	
3700	Shift work requirements, add, minimum		5%	
3800	Maximum		30%	
3900	Temporary shoring and bracing, add, minimum	2%	5%	
4000	Maximum	5%	12%	

D30 HVAC

D3010 Energy Supply

This page illustrates and describes masonry fireplace systems including a brick fireplace, footing, foundation, hearth, firebox, chimney and flue. Lines within System Components give the unit price and total price on a cost each basis for this system. Prices for alternate masonry fireplace systems are on Line Items D3010 992 1300 thru 1500. Both material quantities and labor costs have been adjusted for the system listed.

Factors: To adjust for job conditions other than normal working situations use Lines D3010 992 2700 thru 4000.

Example: You are to install the system with some temporary shoring and bracing. Go to Line D3010 992 3900 and apply these percentages to the appropriate MAT. and INST. costs.

System Components	QUANTITY	UNIT	COST EACH		
			MAT.	INST.	TOTAL
Brick masonry fireplace, including footing, foundation, hearth, firebox, Chimney and flue, chimney 12' above firebox.					
Footing, 4' x 7' x 12" thick, 3000 psi concrete	1.040	C.Y.	171.60	273.73	445.33
Foundation, 12" concrete block	180.000	S.F.	788.40	1,746	2,534.40
Fireplace, brick faced, 6'-0" wide x 5'-0" high	1.000	Ea.	590	2,425	3,015
Hearth	1.000	Ea.	217	485	702
Chimney, 20" x 20", one 12" x 12" flue, 12 V.L.F.	12.000	V.L.F.	510	852	1,362
Mantle, wood	7.000	L.F.	53.20	106.05	159.25
TOTAL			2,330.20	5,887.78	8,217.98

D3010 992	Masonry Fireplace	COST EACH		
		MAT.	INST.	TOTAL
1200	Above system with the following:			
1300	Chimney, 20" x 20", one 12" x 12" flue, 18 V.L.F.	2,575	6,325	8,900
1400	24 V.L.F.	2,850	6,750	9,600
1500	Fieldstone face instead of brick	2,800	5,875	8,675
1600				
1700				
1800				
1900				
2000				
2100				
2200				
2300				
2700	Cut & patch to match existing construction, add, minimum	2%	3%	
2800	Maximum	5%	9%	
2900	Dust protection, add, minimum	1%	2%	
3000	Maximum	4%	11%	
3100	Equipment usage curtailment, add, minimum	1%	1%	
3200	Maximum	3%	10%	
3300	Material handling & storage limitation, add, minimum	1%	1%	
3400	Maximum	6%	7%	
3500	Protection of existing work, add, minimum	2%	2%	
3600	Maximum	5%	7%	
3700	Shift work requirements, add, minimum		5%	
3800	Maximum		30%	
3900	Temporary shoring and bracing, add, minimum	2%	5%	
4000	Maximum	5%	12%	

D30 HVAC

D3020 Heat Generating Systems

This page illustrates and describes an oil fired forced air system including an oil fired furnace, ductwork, registers and hookups. Lines within System Components give the unit price and total price per square foot for this system. Prices for alternate oil fired forced air systems are on Line Items D3020 122 1700 thru 2800. Both material quantities and labor costs have been adjusted for the system listed.

Factors: To adjust for job conditions other than normal working situations use Lines D3020 122 3100 thru 4200.

Example: You are to install the system during evenings and weekends. Go to Line D3020 122 4200 and apply this percentage to the appropriate INST. cost.

System Components	QUANTITY	UNIT	COST PER S.F.		
			MAT.	INST.	TOTAL
Oil fired hot air heating system including furnace, ductwork, registers And all necessary hookups.					
Area to 800 S.F., heat only					
Furnace, oil, atomizing gun type burner, w/oil piping	1.000	Ea.	2,531.25	381.25	2,912.50
Oil tank, steel, 275 gallon	1.000	Ea.	410	232	642
Duct, galvanized, steel	312.000	Lb.	162.24	2,277.60	2,439.84
Insulation, blanket type, duct work	270.000	S.F.	202.50	812.70	1,015.20
Flexible duct, 6" diameter, insulated	100.000	L.F.	262	425	687
Registers, baseboard, gravity, 12" x 6"	8.000	Ea.	144.40	212	356.40
Return, damper, 36" x 18"	1.000	Ea.	143	61.50	204.50
TOTAL			3,855.39	4,402.05	8,257.44
COST PER S.F.		S.F.	4.82	5.50	10.32

D3020 122	Heating-Cooling, Oil, Forced Air	COST PER S.F.		
		MAT.	INST.	TOTAL
1600	For alternate heating systems:			
1700	Oil fired, area to 1000 S.F.	3.88	4.46	8.34
1800	To 1200 S.F.	3.39	3.94	7.33
1900	To 1600 S.F.	2.75	3.29	6.04
2000	To 2000 S.F.	2.64	4.22	6.86
2100	To 3000 S.F.	2.08	3.35	5.43
2200	For combined heating and cooling systems:			
2300	Oil fired, heating and cooling, area to 800 S.F.	7.20	6.55	13.75
2400	To 1000 S.F.	5.90	5.35	11.25
2500	To 1200 S.F.	5.10	4.67	9.77
2600	To 1600 S.F.	4.17	3.86	8.03
2700	To 2000 S.F.	3.79	4.74	8.53
2800	To 3000 S.F.	2.85	3.72	6.57
3100	Cut & patch to match existing construction, add, minimum	2%	3%	
3200	Maximum	5%	9%	
3300	Dust protection, add, minimum	1%	2%	
3400	Maximum	4%	11%	
3500	Equipment usage curtailment, add, minimum	1%	1%	
3600	Maximum	3%	10%	
3700	Material handling & storage limitation, add, minimum	1%	1%	
3800	Maximum	6%	7%	
3900	Protection of existing work, add, minimum	2%	2%	
4000	Maximum	5%	7%	
4100	Shift work requirements, add, minimum		5%	
4200	Maximum		30%	

D30 HVAC

D3020 Heat Generating Systems

This page illustrates and describes a gas fired forced air system including a gas fired furnace, ductwork, registers and hookups. Lines within System Components give the unit price and total price per square foot for this system. Prices for alternate gas fired forced air systems are on Line Items D3020 124 1500 thru 2600. Both material quantities and labor costs have been adjusted for the system listed.

Factors: To adjust for job conditions other than normal working situations use Lines D3020 124 2900 thru 4000.

Example: You are to install the system with material handling and storage limitations. Go to Line D3020 124 3500 and apply these percentages to the appropriate MAT. and INST. costs.

System Components	QUANTITY	UNIT	COST PER S.F. MAT.	COST PER S.F. INST.	COST PER S.F. TOTAL
Gas fired hot air heating system including furnace, ductwork, registers And all necessary hookups.					
Area to 800 S.F., heat only					
Furnace, gas, AGA certified, direct drive, w/gas piping, 44 MBH	1.000	Ea.	793.75	346.25	1,140
Duct, galvanized steel	312.000	Lb.	162.24	2,277.60	2,439.84
Insulation, blanket type, ductwork	270.000	S.F.	202.50	812.70	1,015.20
Flexible duct, 6" diameter, insulated	100.000	L.F.	262	425	687
Registers, baseboard, gravity, 12" x 6"	8.000	Ea.	144.40	212	356.40
Return, damper, 36" x 18"	1.000	Ea.	143	61.50	204.50
TOTAL			1,707.89	4,135.05	5,842.94
COST PER S.F.		S.F.	2.13	5.17	7.30

D3020 124	Heating-Cooling, Gas, Forced Air	MAT.	INST.	TOTAL
1400	For alternate heating systems:			
1500	Gas fired, area to 1000 S.F.	1.78	4.21	5.99
1600	To 1200 S.F.	1.63	3.73	5.36
1700	To 1600 S.F.	1.61	3.49	5.10
1800	To 2000 S.F.	1.60	4.10	5.70
1900	To 3000 S.F.	1.25	3.26	4.51
2000	For combined heating and cooling systems:			
2100	Gas fired, heating and cooling, area to 800 S.F.	4.51	6.20	10.71
2200	To 1000 S.F.	3.80	5.10	8.90
2300	To 1200 S.F.	3.32	4.46	7.78
2400	To 1600 S.F.	3.03	4.06	7.09
2500	To 2000 S.F.	2.75	4.62	7.37
2600	To 3000 S.F.	2.02	3.63	5.65
2900	Cut & patch to match existing construction, add, minimum	2%	3%	
3000	Maximum	5%	9%	
3100	Dust protection, add, minimum	1%	2%	
3200	Maximum	4%	11%	
3300	Equipment usage curtailment, add, minimum	1%	1%	
3400	Maximum	3%	10%	
3500	Material handling & storage limitation, add, minimum	1%	1%	
3600	Maximum	6%	7%	
3700	Protection of existing work, add, minimum	2%	2%	
3800	Maximum	5%	7%	
3900	Shift work requirements, add, minimum		5%	
4000	Maximum		30%	

D30 HVAC

D3020 Heat Generating Systems

Cast Iron Boiler, Hot Water, Gas Fired

This page illustrates and describes boilers including expansion tank, circulating pump and all service piping. Lines within Systems Components give the material and installation price on a cost each basis for the components. Prices for alternate boiler systems are on Line Items D3020 130 1010 thru D3020 138 1040. Material quantities and labor costs have been adjusted for the system listed.

Factors: To adjust for job conditions other than normal working situations use Line D3020 138 2700 thru 4000.

System Components	QUANTITY	UNIT	COST EACH MAT.	INST.	TOTAL
Cast iron boiler, including piping, chimney and accessories					
Boilers, gas fired, std controls, CI, insulated, HW, gross output 100 MBH	1.000	Ea.	2,525	1,825	4,350
Pipe, black steel, Sch 40, threaded, W/coupling & hangers, 10' OC, 3/4" dia	20.000	L.F.	105.98	238.50	344.48
Pipe, black steel, Sch 40, threaded, W/coupling & hangers, 10' OC, 1" dia.	20.000	L.F.	148.35	261.23	409.58
Elbow, 90°, black, straight, 3/4" dia.	9.000	Ea.	28.44	414	442.44
Elbow, 90°, black, straight, 1" dia.	6.000	Ea.	33	297	330
Tee, black, straight, 3/4" dia.	2.000	Ea.	10	143	153
Tee, black, straight, 1" dia.	2.000	Ea.	17.10	161	178.10
Tee, black, reducing, 1" dia.	2.000	Ea.	27.50	161	188.50
Pipe cap, black, 3/4" dia.	1.000	Ea.	3.58	20	23.58
Union, black with brass seat, 3/4" dia.	2.000	Ea.	26.50	99	125.50
Union, black with brass seat, 1" dia.	2.000	Ea.	34.60	108	142.60
Pipe nipples, black, 3/4" dia.	5.000	Ea.	11.78	26.50	38.28
Pipe nipples, black, 1" dia.	3.000	Ea.	10.35	18.23	28.58
Valves, bronze, gate, N.R.S., threaded, class 150, 3/4" size	2.000	Ea.	127	65	192
Valves, bronze, gate, N.R.S., threaded, class 150, 1" size	2.000	Ea.	162	68	230
Gas cock, brass, 3/4" size	1.000	Ea.	12.30	29.50	41.80
Thermometer, stem type, 9" case, 8" stem, 3/4" NPT	2.000	Ea.	372	46	418
Tank, steel, liquid expansion, ASME, painted, 15 gallon capacity	1.000	Ea.	575	68	643
Pump, circulating, bronze, flange connection, 3/4" to 1-1/2" size, 1/8 HP	1.000	Ea.	940	194	1,134
Vent chimney, all fuel, pressure tight, double wall, SS, 6" dia.	20.000	L.F.	1,040	369	1,409
Vent chimney, elbow, 90° fixed, 6" dia.	2.000	Ea.	586	74	660
Vent chimney, Tee, 6" dia.	2.000	Ea.	386	92	478
Vent chimney, ventilated roof thimble, 6" dia.	1.000	Ea.	247	42.50	289.50
Vent chimney, adjustable roof flashing, 6" dia.	1.000	Ea.	69	37	106
Vent chimney, stack cap, 6" diameter	1.000	Ea.	218	24	242
Insulation, fiberglass pipe covering, 1" wall, 1" IPS	20.000	L.F.	24.60	95.60	120.20
TOTAL			7,741.08	4,977.06	12,718.14

D3020 130	Boiler, Cast Iron, Hot Water, Gas	MAT.	INST.	TOTAL
1000	For alternate boilers:			
1010	Boiler, cast iron, gas, hot water, 100 MBH	7,750	4,975	12,725
1020	200 MBH	8,650	5,675	14,325
1030	320 MBH	9,300	6,175	15,475
1040	440 MBH	12,300	7,975	20,275
1050	544 MBH	20,600	12,500	33,100
1060	765 MBH	25,400	13,400	38,800
1070	1088 MBH	28,200	14,300	42,500

D30 HVAC

D3020 Heat Generating Systems

D3020 130	Boiler, Cast Iron, Hot Water, Gas	COST EACH		
		MAT.	INST.	TOTAL
1080	1530 MBH	32,800	19,000	51,800
1090	2312 MBH	38,500	22,300	60,800

D3020 134	Boiler, Cast Iron, Steam, Gas	COST EACH		
		MAT.	INST.	TOTAL
1010	Boiler, cast iron, gas, steam, 100 MBH	6,250	4,800	11,050
1020	200 MBH	12,900	9,075	21,975
1030	320 MBH	14,600	10,200	24,800
1040	544 MBH	22,100	15,900	38,000
1050	765 MBH	25,100	16,600	41,700
1060	1275 MBH	29,900	18,200	48,100
1070	2675 MBH	41,100	26,600	67,700

D3020 136	Boiler, Cast Iron, Hot Water, Gas/Oil	COST EACH		
		MAT.	INST.	TOTAL
1010	Boiler, cast iron, gas & oil, hot water, 584 MBH	20,700	13,900	34,600
1020	876 MBH	30,500	14,900	45,400
1030	1168 MBH	37,100	16,800	53,900
1040	1460 MBH	41,300	20,300	61,600
1050	2044 MBH	47,200	21,000	68,200
1060	2628 MBH	54,500	25,300	79,800

D3020 138	Boiler, Cast Iron, Steam, Gas/Oil	COST EACH		
		MAT.	INST.	TOTAL
1010	Boiler, cast iron, gas & oil, steam, 810 MBH	29,400	18,000	47,400
1020	1360 MBH	32,900	19,200	52,100
1030	2040 MBH	42,900	24,700	67,600
1040	2700 MBH	46,900	27,500	74,400
2700	Cut & patch to match existing construction, add, minimum	2%	3%	
2800	Maximum	5%	9%	
2900	Dust protection, add, minimum	1%	2%	
3000	Maximum	4%	11%	
3100	Equipment usage curtailment, add, minimum	1%	1%	
3200	Maximum	3%	10%	
3300	Material handling & storage limitation, add, minimum	1%	1%	
3400	Maximum	6%	7%	
3500	Protection of existing work, add, minimum	2%	2%	
3600	Maximum	5%	7%	
3700	Shift work requirements, add, minimum		5%	
3800	Maximum		30%	
3900	Temporary shoring and bracing, add, minimum	2%	5%	
4000	Maximum	5%	12%	

D30 HVAC

D3020 Heat Generating Systems

Base Mounted End-Suction Pump

This page illustrates and describes end suction base mounted circulating pumps, including strainer and piping. Lines within system components give the material and installation price on a cost each basis for the components. Prices for alternate circulating pumps are on Line items D3020 330 1010 thru 1050. Material quantities and labor costs have been adjusting for the system listed.

Factors: To adjust for job conditions other than normal working situations uses Lines D3020 330 2700 thru 4000.

System Components	QUANTITY	UNIT	COST EACH		
			MAT.	INST.	TOTAL
Circulator pump, end suction, including pipe, Valves, fittings and guages.					
Pump, circulating, CI, base mounted, 2-1/2" size, 3 HP, to 150 GPM	1.000	Ea.	6,775	645	7,420
Pipe, black steel, Sch. 40, on yoke & roll hangers, 10' O.C., 2-1/2" dia	12.000	L.F.	123.60	310.44	434.04
Elbow, 90°, weld joint, steel, 2-1/2" pipe size	1.000	Ea.	30.50	153.05	183.55
Flange, weld neck, 150 LB, 2-1/2" pipe size	8.000	Ea.	280	612.24	892.24
Valve, iron body, gate, 125 lb., N.R.S., flanged, 2-1/2" size	1.000	Ea.	595	232	827
Strainer, Y type, iron body, flanged, 125 lb., 2-1/2" pipe size	1.000	Ea.	113	232	345
Multipurpose valve, CI body, 2" size	1.000	Ea.	445	80.50	525.50
Expansion joint, flanged spool, 6" F to F, 2-1/2" dia	2.000	Ea.	572	187	759
T-O-L, weld joint, socket, 1/4" pipe size, nozzle	2.000	Ea.	14.70	106.60	121.30
T-O-L, weld joint, socket, 1/2" pipe size, nozzle	2.000	Ea.	14.70	111.86	126.56
Control gauges, pressure or vacuum, 3-1/2" diameter dial	2.000	Ea.	79	40	119
Pressure/temperature relief plug, 316 SS, 3/4" OD, 7-1/2" insertion	2.000	Ea.	188	40	228
Insulation, fiberglass pipe covering, 1-1/2" wall, 2-1/2" IPS	12.000	L.F.	34.56	70.20	104.76
Pump control system	1.000	Ea.	1,200	585	1,785
Pump balancing	1.000	Ea.		283	283
TOTAL			10,465.06	3,688.89	14,153.95

D3020 330		Circulating Pump Systems, End Suction	COST EACH		
			MAT.	INST.	TOTAL
1000	For alternate pump sizes:				
1010	Pump, base mtd with motor, end-suction, 2-1/2" size, 3 HP, to 150 GPM		10,500	3,700	14,200
1020	3" size, 5 HP, to 225 GPM		11,700	4,225	15,925
1030	4" size, 7-1/2 HP, to 350 GPM		12,900	5,100	18,000
1040	5" size, 15 HP, to 1000 GPM		17,800	7,725	25,525
1050	6" size, 25 HP, to 1550 GPM		24,700	9,400	34,100
2700	Cut & patch to match existing construction, add, minimum		2%	3%	
2800	Maximum		5%	9%	
2900	Dust protection, add, minimum		1%	2%	
3000	Maximum		4%	11%	
3100	Equipment usage curtailment, add, minimum		1%	1%	
3200	Maximum		3%	10%	
3300	Material handling & storage limitation, add, minimum		1%	1%	
3400	Maximum		6%	7%	
3500	Protection of existing work, add, minimum		2%	2%	
3600	Maximum		5%	7%	

D30 HVAC

D3020 Heat Generating Systems

D3020 330 Circulating Pump Systems, End Suction

		COST EACH		
		MAT.	INST.	TOTAL
3700	Shift work requirements, add, minimum		5%	
3800	Maximum		30%	
3900	Temporary shoring and bracing, add, minimum	2%	5%	
4000	Maximum	5%	12%	

D30 HVAC

D3040 Distribution Systems

This page illustrates and describes fan coil air conditioners including duct work, duct installation, piping and diffusers. Lines within Systems Components give the material and installation price on a cost each basis for the components. Prices for alternate fan coil A/C unit systems are on Line Items D3040 118 1010 thru D3040 126 1120. Material quantities and labor costs have been adjusted for the system listed.

Factors: To adjust for job conditions other than normal working situations use Lines D3040 126 2700 thru 4000.

Horizontal Fan Coil Air Conditioning System

System Components	QUANTITY	UNIT	MAT.	INST.	TOTAL
Fan coil A/C system, horizontal, with housing, controls, 2 pipe, 1/2 ton					
Fan coil A/C, horizontal housing, filters, chilled water, 1/2 ton cooling	1.000	Ea.	1,200	145	1,345
Pipe, black steel, Sch 40, threaded, W/cplgs & hangers, 10' OC, 3/4" dia.	20.000	L.F.	101.27	227.90	329.17
Elbow, 90°, black, straight, 3/4" dia.	6.000	Ea.	18.96	276	294.96
Tee, black, straight, 3/4" dia.	2.000	Ea.	10	143	153
Union, black with brass seat, 3/4" dia.	2.000	Ea.	26.50	99	125.50
Pipe nipples, 3/4" diam	3.000	Ea.	7.07	15.90	22.97
Valves, bronze, gate, N.R.S., threaded, class 150, 3/4" size	2.000	Ea.	127	65	192
Circuit setter, balance valve, bronze body, threaded, 3/4" pipe size	1.000	Ea.	78.50	32	110.50
Insulation, fiberglass pipe covering, 1" wall, 3/4" IPS	20.000	L.F.	22.80	91.40	114.20
Ductwork, 12" x 8" fabricated, galvanized steel, 12 LF	55.000	Lb.	28.60	401.50	430.10
Insulation, ductwork, blanket type, fiberglass, 1" thk, 1-1/2 LB density	40.000	S.F.	30	120.40	150.40
Diffusers, aluminum, OB damper, ceiling, perf, 24"x24" panel size, 6"x6"	2.000	Ea.	127	77	204
Ductwork, flexible, fiberglass fabric, insulated, 1"thk, PE jacket, 6" dia	16.000	L.F.	41.92	68	109.92
Round volume control damper 6" dia.	2.000	Ea.	51	56	107
Fan coil unit balancing	1.000	Ea.		80	80
Re-heat coil balancing	1.000	Ea.		119	119
Diffuser/register, high, balancing	2.000	Ea.		222	222
TOTAL			1,870.62	2,239.10	4,109.72

D3040 118	Fan Coil A/C Unit, Two Pipe	MAT.	INST.	TOTAL
1000	For alternate A/C units:			
1010	Fan coil A/C system, cabinet mounted, controls, 2 pipe, 1/2 Ton	1,175	1,200	2,375
1020	1 Ton	1,425	1,350	2,775
1030	1-1/2 Ton	1,525	1,375	2,900
1040	2 Ton	2,425	1,675	4,100
1050	3 Ton	3,050	1,750	4,800

D3040 120	Fan Coil A/C Unit, Two Pipe, Electric Heat	MAT.	INST.	TOTAL
1010	Fan coil A/C system, cabinet mntd, elect. ht, controls, 2 pipe, 1/2 Ton	1,675	1,200	2,875
1020	1 Ton	2,025	1,350	3,375
1030	1-1/2 Ton	2,375	1,375	3,750
1040	2 Ton	3,600	1,700	5,300
1050	3 Ton	6,075	1,750	7,825

D30 HVAC

D3040 Distribution Systems

D3040 122 — Fan Coil A/C Unit, Four Pipe

		COST EACH		
		MAT.	INST.	TOTAL
1010	Fan coil A/C system, cabinet mounted, controls, 4 pipe, 1/2 Ton	1,900	2,350	4,250
1020	1 Ton	2,175	2,500	4,675
1030	1-1/2 Ton	2,325	2,525	4,850
1040	2 Ton	2,900	2,625	5,525
1050	3 Ton	3,925	2,800	6,725

D3040 124 — Fan Coil A/C, Horizontal, Duct Mount, 2 Pipe

		COST EACH		
		MAT.	INST.	TOTAL
1010	Fan coil A/C system, horizontal w/housing, controls, 2 pipe, 1/2 Ton	1,875	2,250	4,125
1020	1 Ton	2,425	3,250	5,675
1030	1-1/2 Ton	2,925	4,200	7,125
1040	2 Ton	3,975	5,050	9,025
1050	3 Ton	4,675	6,675	11,350
1060	3-1/2 Ton	4,900	6,925	11,825
1070	4 Ton	5,025	7,850	12,875
1080	5 Ton	5,875	9,950	15,825
1090	6 Ton	5,925	10,700	16,625
1100	7 Ton	6,000	11,100	17,100
1110	8 Ton	6,000	11,300	17,300
1120	10 Ton	6,750	11,800	18,550

D3040 126 — Fan Coil A/C, Horiz., Duct Mount, 2 Pipe, Elec. Ht.

		COST EACH		
		MAT.	INST.	TOTAL
1010	Fan coil A/C system, horiz. hsng, elect. ht, ctrls, 2 pipe, 1/2 Ton	2,125	2,250	4,375
1020	1 Ton	2,750	3,250	6,000
1030	1-1/2 Ton	3,150	4,200	7,350
1040	2 Ton	4,375	5,050	9,425
1050	3 Ton	6,975	6,900	13,875
1060	3-1/2 Ton	7,900	6,925	14,825
1070	4 Ton	8,900	7,850	16,750
1080	5 Ton	10,200	9,950	20,150
1090	6 Ton	10,300	10,700	21,000
1100	7 Ton	10,300	10,600	20,900
1110	8 Ton	11,400	10,800	22,200
1120	10 Ton	12,200	11,100	23,300
2700	Cut & patch to match existing construction, add, minimum	2%	3%	
2800	Maximum	5%	9%	
2900	Dust protection, add, minimum	1%	2%	
3000	Maximum	4%	11%	
3100	Equipment usage curtailment, add, minimum	1%	1%	
3200	Maximum	3%	10%	
3300	Material handling & storage limitation, add, minimum	1%	1%	
3400	Maximum	6%	7%	
3500	Protection of existing work, add, minimum	2%	2%	
3600	Maximum	5%	7%	
3700	Shift work requirements, add, minimum		5%	
3800	Maximum		30%	
3900	Temporary shoring and bracing, add, minimum	2%	5%	
4000	Maximum	5%	12%	

D30 HVAC

D3040 Distribution Systems

Plate Heat Exchanger   Shell and Tube Heat Exchanger

This page illustrates and describes plate and shell and tube type heat exchangers including pressure gauge and exchanger control system. Lines within Systems Components give the material and installation price on a cost each basis for the components. Prices for alternate heat exchanger systems are on Line Items D3040 610 1010 thru D3040 620 1040. Material quantities and labor costs have been adjusted for the system listed.

Factors To adjust for job conditions other then normal working situations use Lines D3040 620 2700 thru 4000.

System Components	QUANTITY	UNIT	COST EACH MAT.	COST EACH INST.	COST EACH TOTAL
Shell and tube type heat exchanger with related valves and piping.					
Heat exchanger, 4 pass, 3/4" O.D. copper tubes, by steam at 10 psi, 40 GPM	1.000	Ea.	4,450	289	4,739
Pipe, black steel, Sch 40, threaded, W/couplings & hangers, 10' OC, 2" dia.	40.000	L.F.	699.43	916.58	1,616.01
Elbow, 90°, straight, 2" dia.	16.000	Ea.	328	1,032	1,360
Reducer, black steel, concentric, 2" dia.	2.000	Ea.	46	111	157
Valves, bronze, gate, rising stem, threaded, class 150, 2" size	6.000	Ea.	1,074	351	1,425
Valves, bronze, globe, class 150, rising stem, threaded, 2" size	2.000	Ea.	930	117	1,047
Strainers, Y type, bronze body, screwed, 150 lb., 2" pipe size	2.000	Ea.	194	99	293
Valve, electric motor actuated, brass, 2 way, screwed, 1-1/2" pipe size	1.000	Ea.	450	50	500
Union, black with brass seat, 2" dia.	8.000	Ea.	288	548	836
Tee, black, straight, 2" dia.	9.000	Ea.	234	954	1,188
Tee, black, reducing run and outlet, 2" dia.	4.000	Ea.	162	424	586
Pipe nipple, black, 2" dia.	21.000	Ea.	6.93	9.08	16.01
Thermometers, stem type, 9" case, 8" stem, 3/4" NPT	2.000	Ea.	372	46	418
Gauges, pressure or vacuum, 3-1/2" diameter dial	2.000	Ea.	79	40	119
Insulation, fiberglass pipe covering, 1" wall, 2" IPS	40.000	L.F.	62	210	272
Heat exchanger control system	1.000	Ea.	2,500	2,050	4,550
Coil balancing	1.000	Ea.		119	119
TOTAL			11,875.36	7,365.66	19,241.02

D3040 610	Heat Exchanger, Plate Type	MAT.	INST.	TOTAL
1000	For alternate heat exchangers:			
1010	Plate heat exchanger, 400 GPM	47,600	14,900	62,500
1020	800 GPM	76,500	19,000	95,500
1030	1200 GPM	115,000	25,600	140,600
1040	1800 GPM	156,000	31,900	187,900

D3040 620	Heat Exchanger, Shell & Tube	MAT.	INST.	TOTAL
1010	Shell & tube heat exchanger, 40 GPM	11,900	7,375	19,275
1020	96 GPM	21,400	13,200	34,600
1030	240 GPM	38,900	17,800	56,700
1040	600 GPM	75,000	25,000	100,000
2700	Cut & patch to match existing construction, add, minimum	2%	3%	
2800	Maximum	5%	9%	
2900	Dust protection, add, minimum	1%	2%	
3000	Maximum	4%	11%	

D30 HVAC

D3040 Distribution Systems

D3040 620	Heat Exchanger, Shell & Tube	COST EACH		
		MAT.	INST.	TOTAL
3100	Equipment usage curtailment, add, minimum	1%	1%	
3200	Maximum	3%	10%	
3300	Material handling & storage limitation, add, minimum	1%	1%	
3400	Maximum	6%	7%	
3500	Protection of existing work, add, minimum	2%	2%	
3600	Maximum	5%	7%	
3700	Shift work requirements, add, minimum		5%	
3800	Maximum		30%	
3900	Temporary shoring and bracing, add, minimum	2%	5%	
4000	Maximum	5%	12%	

D30 HVAC

D3050 Terminal & Package Units

Gas Unit Heater

Hydronic Unit Heater

This page illustrates and describes unit heaters including piping and vents. Lines within Systems Components give the material and installation price on a cost each basis for this system. Prices for alternate unit heater systems are on Line items D3050 120 1010 thru D3050 140 1040. Material quantities and labor costs have been adjusted for the system listed.

Factors: To adjust for job conditions other than normal working situations use Lines D3050 140 2700 thru 4000.

System Components	QUANTITY	UNIT	COST EACH		
			MAT.	INST.	TOTAL
Unit heaters complete with piping, thermostats and chimney as required					
Space heater, propeller fan, 20 MBH output	1.000	Ea.	1,075	136	1,211
Pipe, black steel, Sch 40, threaded, W/coupling & hangers, 10' OC, 3/4" dia	20.000	L.F.	98.91	222.60	321.51
Elbow, 90°, black steel, straight, 1/2" dia.	3.000	Ea.	7.83	129	136.83
Elbow, 90°, black steel, straight, 3/4" dia.	3.000	Ea.	9.48	138	147.48
Tee, black steel, reducing, 3/4" dia.	1.000	Ea.	7.40	71.50	78.90
Union, black with brass seat, 3/4" dia.	1.000	Ea.	13.25	49.50	62.75
Pipe nipples, black, 1/2" dia	2.000	Ea.	7.94	20.50	28.44
Pipe nipples, black, 3/4" dia	2.000	Ea.	4.71	10.60	15.31
Pipe cap, black, 3/4" dia.	1.000	Ea.	3.58	20	23.58
Gas cock, brass, 3/4" size	1.000	Ea.	12.30	29.50	41.80
Thermostat, 1 set back, electric, timed	1.000	Ea.	36.50	77	113.50
Wiring, thermostat hook-up, 25' of #18-3	1.000	Ea.	10.80	25	35.80
Vent chimney, prefab metal, U.L. listed, gas, double wall, galv. st, 4" dia	12.000	L.F.	88.20	195	283.20
Vent chimney, gas, double wall, galv. steel, elbow 90°, 4" dia.	2.000	Ea.	59	65	124
Vent chimney, gas, double wall, galv. steel, Tee, 4" dia.	1.000	Ea.	37.50	42.50	80
Vent chimney, gas, double wall, galv. steel, T cap, 4" dia.	1.000	Ea.	2.89	26.50	29.39
Vent chimney, gas, double wall, galv. steel, roof flashing, 4" dia.	1.000	Ea.	31	32.50	63.50
Vent chimney, gas, double wall, galv. steel, top, 4" dia.	1.000	Ea.	17.80	25	42.80
TOTAL			1,524.09	1,315.70	2,839.79

D3050 120		Unit Heaters, Gas	COST EACH		
			MAT.	INST.	TOTAL
1000	For alternate unit heaters:				
1010	Space heater, suspended, gas fired, propeller fan, 20 MBH		1,525	1,325	2,850
1020		60 MBH	1,800	1,375	3,175
1030		100 MBH	2,025	1,450	3,475
1040		160 MBH	2,350	1,600	3,950
1050		200 MBH	2,800	1,775	4,575
1060		280 MBH	3,500	1,875	5,375
1070		320 MBH	4,600	2,125	6,725

D30 HVAC

D3050 Terminal & Package Units

D3050 130	Unit Heaters, Hydronic	COST EACH		
		MAT.	INST.	TOTAL
1010	Space heater, suspended, horiz. mount, HW, prop. fan, 20 MBH	1,800	1,425	3,225
1020	60 MBH	2,350	1,600	3,950
1030	100 MBH	2,800	1,750	4,550
1040	150 MBH	3,425	1,925	5,350
1050	200 MBH	3,975	2,150	6,125
1060	300 MBH	4,450	2,525	6,975

D3050 140	Cabinet Unit Heaters, Hydronic	COST EACH		
		MAT.	INST.	TOTAL
1010	Unit heater, cabinet type, horizontal blower, hot water, 20 MBH	2,125	1,375	3,500
1020	60 MBH	2,975	1,525	4,500
1030	100 MBH	3,325	1,675	5,000
1040	120 MBH	3,375	1,750	5,125
2700	Cut & patch to match existing construction, add, minimum	2%	3%	
2800	Maximum	5%	9%	
2900	Dust protection, add, minimum	1%	2%	
3000	Maximum	4%	11%	
3100	Equipment usage curtailment, add, minimum	1%	1%	
3200	Maximum	3%	10%	
3300	Material handling & storage limitation, add, minimum	1%	1%	
3400	Maximum	6%	7%	
3500	Protection of existing work, add, minimum	2%	2%	
3600	Maximum	5%	7%	
3700	Shift work requirements, add, minimum		5%	
3800	Maximum		30%	
3900	Temporary shoring and bracing, add, minimum	2%	5%	
4000	Maximum	5%	12%	

D30 HVAC

D3050 Terminal & Package Units

Rooftop Air Conditioner Unit

This page illustrates and describes packaged air conditioners including rooftop DX unit, pipes and fittings. Lines within Systems Components give the material and installation price on a cost each basis for the components. Prices for alternate A/C systems are on Line Items D3050 175 1010 thru 1060. Material quantities and labor costs have been adjusted for the system listed.

Factors: To adjust for job conditions other than normal working situations use Lines D3050 175 2700 thru 4000.

System Components	QUANTITY	UNIT	COST EACH		
			MAT.	INST.	TOTAL
Packaged rooftop DX unit, gas, with related piping and valves.					
Roof top A/C, curb, economizer, sgl zone, elec cool, gas ht, 5 ton, 112 MBH	1.000	Ea.	5,350	2,075	7,425
Pipe, black steel, Sch 40, threaded, W/cplgs & hangers, 10' OC, 1" dia	20.000	L.F.	144.90	255.15	400.05
Elbow, 90°, black, straight, 3/4" dia.	3.000	Ea.	9.48	138	147.48
Elbow, 90°, black, straight, 1" dia.	3.000	Ea.	16.50	148.50	165
Tee, black, reducing, 1" dia.	1.000	Ea.	13.75	80.50	94.25
Union, black with brass seat, 1" dia.	1.000	Ea.	17.30	54	71.30
Pipe nipple, black, 3/4" dia	2.000	Ea.	9.42	21.20	30.62
Pipe nipple, black, 1" dia	2.000	Ea.	6.90	12.15	19.05
Cap, black, 1" dia.	1.000	Ea.	4.32	21.50	25.82
Gas cock, brass, 1" size	1.000	Ea.	19.45	34	53.45
Control system, pneumatic, Rooftop A/C unit	1.000	Ea.	2,575	2,275	4,850
Rooftop unit heat/cool balancing	1.000	Ea.		430	430
TOTAL			8,167.02	5,545	13,712.02

D3050 175	Rooftop Air Conditioner, Const. Volume	COST EACH		
		MAT.	INST.	TOTAL
1000	For alternate A/C systems:			
1010	A/C, Rooftop, DX cool, gas heat, curb, economizer, fltrs, 5 Ton	8,175	5,550	13,725
1020	7-1/2 Ton	11,200	5,750	16,950
1030	12-1/2 Ton	16,200	6,325	22,525
1040	18 Ton	22,600	6,925	29,525
1050	25 Ton	30,400	7,875	38,275
1060	40 Ton	39,700	10,500	50,200
1010	A/C, Rooftop, DX cool, gas heat, curb, ecmizr, fltrs, VAV, 12-1/2 Ton	18,000	6,750	24,750
1020	18 Ton	24,700	7,425	32,125
1030	25 Ton	33,000	8,525	41,525
1040	40 Ton	44,300	11,600	55,900
1050	60 Ton	63,500	15,800	79,300
1060	80 Ton	99,000	19,700	118,700
2700	Cut & patch to match existing construction, add, minimum	2%	3%	
2800	Maximum	5%	9%	
2900	Dust protection, add, minimum	1%	2%	
3000	Maximum	4%	11%	
3100	Equipment usage curtailment, add, minimum	1%	1%	
3200	Maximum	3%	10%	
3300	Material handling & storage limitation, add, minimum	1%	1%	
3400	Maximum	6%	7%	
3500	Protection of existing work, add, minimum	2%	2%	

D30 HVAC

D3050 Terminal & Package Units

D3050 175	Rooftop Air Conditioner, Const. Volume	COST EACH		
		MAT.	INST.	TOTAL
3600	Maximum	5%	7%	
3700	Shift work requirements, add, minimum		5%	
3800	Maximum		30%	
3900	Temporary shoring and bracing, add, minimum	2%	5%	
4000	Maximum	5%	12%	

D30 HVAC

D3050 Terminal & Package Units

Condenser Supply Water Piping

Condenser Return Water Piping

Self-Contained Air Conditioner Unit, Water Cooled

This page illustrates and describes packaged electrical air conditioning units including multizone control and related piping. Lines within Systems Components give the material and installation price on a cost each basis for the components. Prices for alternate packaged electrical A/C unit systems are on Line Items D3050 210 1010 thru D3050 225 1060. Material quantities and labor costs have been adjusted for the system listed.

Factors: To adjust for job conditions other than normal working situations use Lines D3050 225 2700 thru 4000.

System Components	QUANTITY	UNIT	COST EACH		
			MAT.	INST.	TOTAL
Packaged water cooled electric air conditioning unit					
with related piping and valves.					
Self-contained, water cooled, elect. heat, not inc tower, 5 ton, const vol	1.000	Ea.	8,225	1,500	9,725
Pipe, black steel, Sch 40, threaded, W/cplg & hangers, 10' OC, 1-1/4" dia.	20.000	L.F.	188.13	280.58	468.71
Elbow, 90°, black, straight, 1-1/4" dia.	6.000	Ea.	54	318	372
Tee, black, straight, 1-1/4" dia.	2.000	Ea.	27.80	166	193.80
Tee, black, reducing, 1-1/4" dia.	2.000	Ea.	47	166	213
Thermometers, stem type, 9" case, 8" stem, 3/4" NPT	2.000	Ea.	372	46	418
Union, black with brass seat, 1-1/4" dia.	2.000	Ea.	50	111	161
Pipe nipple, black, 1-1/4" dia	3.000	Ea.	13.13	19.58	32.71
Valves, bronze, gate, N.R.S., threaded, class 150, 1-1/4" size	2.000	Ea.	216	86	302
Circuit setter, bal valve, bronze body, threaded, 1-1/4" pipe size	1.000	Ea.	149	43	192
Insulation, fiberglass pipe covering, 1" wall, 1-1/4" IPS	20.000	L.F.	26.60	100	126.60
Control system, pneumatic, A/C Unit with heat	1.000	Ea.	2,575	2,275	4,850
Re-heat coil balancing	1.000	Ea.		119	119
Rooftop unit heat/cool balancing	1.000	Ea.		430	430
TOTAL			11,943.66	5,660.16	17,603.82

D3050 210	AC Unit, Package, Elec. Ht., Water Cooled	COST EACH		
		MAT.	INST.	TOTAL
1000	For alternate A/C systems:			
1010	A/C, Self contained, single pkg., water cooled, elect. heat, 5 Ton	11,900	5,650	17,550
1020	10 Ton	18,800	6,975	25,775
1030	20 Ton	25,600	8,900	34,500
1040	30 Ton	41,500	9,650	51,150
1050	40 Ton	55,000	10,700	65,700
1060	50 Ton	70,500	12,800	83,300

D3050 215	AC Unit, Package, Elec. Ht., Water Cooled, VAV	COST EACH		
		MAT.	INST.	TOTAL
1010	A/C, Self contained, single pkg., water cooled, elect. ht, VAV, 10 Ton	20,300	6,975	27,275
1020	20 Ton	28,100	8,900	37,000
1030	30 Ton	45,400	9,650	55,050
1040	40 Ton	59,500	10,700	70,200
1050	50 Ton	76,000	12,800	88,800
1060	60 Ton	87,000	15,500	102,500

D30 HVAC

D3050 Terminal & Package Units

D3050 220	AC Unit, Package, Hot Water Coil, Water Cooled	COST EACH		
		MAT.	INST.	TOTAL
1010	A/C, Self contn'd, single pkg., water cool, H/W ht, const. vol, 5 Ton	10,500	7,850	18,350
1020	10 Ton	15,700	9,375	25,075
1030	20 Ton	38,700	11,700	50,400
1040	30 Ton	50,500	14,900	65,400
1050	40 Ton	62,000	18,100	80,100
1060	50 Ton	76,500	22,200	98,700

D3050 225	AC Unit, Package, HW Coil, Water Cooled, VAV	COST EACH		
		MAT.	INST.	TOTAL
1010	A/C, Self contn'd, single pkg., water cool, H/W ht, VAV, 10 Ton	17,100	8,825	25,925
1020	20 Ton	24,700	11,400	36,100
1030	30 Ton	39,100	14,300	53,400
1040	40 Ton	54,000	15,600	69,600
1050	50 Ton	67,000	18,300	85,300
1060	60 Ton	80,000	19,900	99,900
2700	Cut & patch to match existing construction, add, minimum	2%	3%	
2800	Maximum	5%	9%	
2900	Dust protection, add, minimum	1%	2%	
3000	Maximum	4%	11%	
3100	Equipment usage curtailment, add, minimum	1%	1%	
3200	Maximum	3%	10%	
3300	Material handling & storage limitation, add, minimum	1%	1%	
3400	Maximum	6%	7%	
3500	Protection of existing work, add, minimum	2%	2%	
3600	Maximum	5%	7%	
3700	Shift work requirements, add, minimum		5%	
3800	Maximum		30%	
3900	Temporary shoring and bracing, add, minimum	2%	5%	
4000	Maximum	5%	12%	

D30 HVAC

D3050 Terminal & Package Units

Thru-Wall Heat Pump   Thru-Wall Air Conditioner

This page illustrates and describes thru-wall air conditioning and heat pump units. Lines within Systems Components give the material and installation price on a cost each basis for the components. Prices for alternate thru-wall unit systems are on Line Items D3050 255 1010 thru 1050. Material quantities and labor costs have been adjusted for the system listed.

Factors: To adjust for job conditions other than normal working conditions use Lines D3050 255 2700 thru 4000.

System Components	QUANTITY	UNIT	COST EACH		
			MAT.	INST.	TOTAL
Electric thru-wall air conditioning unit.					
A/C unit, thru-wall, electric heat., cabinet, louver, 1/2 ton	1.000	Ea.	750	193	943
TOTAL			750	193	943

D3050 255	Thru-Wall A/C Unit	COST EACH		
		MAT.	INST.	TOTAL
1000	For alternate A/C units:			
1010	A/C Unit, thru-the-wall, supplemental elect. heat, cabinet, louver, 1/2 Ton	750	193	943
1020	3/4 Ton	865	232	1,097
1030	1 Ton	935	289	1,224
1040	1-1/2 Ton	1,825	480	2,305
1050	2 Ton	1,850	610	2,460
1010	Heat pump, thru-the-wall, cabinet, louver, 1/2 Ton	2,325	145	2,470
1020	3/4 Ton	2,525	193	2,718
1030	1 Ton	2,775	289	3,064
1040	Supplemental. elect. heat, 1-1/2 Ton	2,800	745	3,545
1050	2 Ton	3,025	770	3,795
2700	Cut & patch to match existing construction, add, minimum	2%	3%	
2800	Maximum	5%	9%	
2900	Dust protection, add, minimum	1%	2%	
3000	Maximum	4%	11%	
3100	Equipment usage curtailment, add, minimum	1%	1%	
3200	Maximum	3%	10%	
3300	Material handling & storage limitation, add, minimum	1%	1%	
3400	Maximum	6%	7%	
3500	Protection of existing work, add, minimum	2%	2%	
3600	Maximum	5%	7%	
3700	Shift work requirements, add, minimum		5%	
3800	Maximum		30%	
3900	Temporary shoring and bracing, add, minimum	2%	5%	
4000	Maximum	5%	12%	

D40 Fire Protection

D4010 Sprinklers

This page illustrates and describes a dry type fire sprinkler system. Lines within System Components give the unit cost of a system for a 2,000 square foot building. Lines D4010 320 1700 thru 2400 give the square foot costs for alternate systems. Both material quantities and labor costs have been adjusted for the system listed.

Factors: To adjust for conditions other than normal working conditions, use Lines D4010 320 2900 thru 4000.

System Components	QUANTITY	UNIT	COST EACH MAT.	COST EACH INST.	COST EACH TOTAL
Pre-action fire sprinkler system, ordinary hazard, open area to 2000 S.F. On one floor.					
Air compressor	1.000	Ea.	850	480	1,330
4" OS & Y valve	1.000	Ea.	575	385	960
Pipe riser 4" diameter	10.000	L.F.	175.50	332.40	507.90
Sprinkler head supply piping, 1"	163.000	L.F.	517.50	911.25	1,428.75
Sprinkler head supply piping, 1-1/4"	163.000	L.F.	481.25	717.75	1,199
Sprinkler head supply piping, 2-1/2"	163.000	L.F.	817	874	1,691
Pipe fittings, 1"	8.000	Ea.	68.40	644	712.40
Pipe fittings, 1-1/4"	8.000	Ea.	111.20	664	775.20
Pipe fittings, 2-1/2"	4.000	Ea.	254	516	770
Pipe fittings, 4"	2.000	Ea.	452	580	1,032
Detectors	2.000	Ea.	1,150	78	1,228
Sprinkler heads	16.000	Ea.	129.60	624	753.60
Fire department connection	1.000	Ea.	470	224	694
TOTAL			6,051.45	7,030.40	13,081.85

D4010 320	Fire Sprinkler Systems, Dry	COST PER S.F. MAT.	COST PER S.F. INST.	COST PER S.F. TOTAL
1700	Ordinary hazard, one floor, area to 2000 S.F./floor	3.05	3.52	6.57
1800	For each additional floor, add per floor	1.51	2.94	4.45
1900	Area to 3200 S.F./ floor	2.57	3.48	6.05
2000	For each additional floor, add per floor	1.61	3.12	4.73
2100	Area to 5000 S.F./ floor	2.91	3.56	6.47
2200	For each additional floor, add per floor	2.07	3.31	5.38
2300	Area to 8000 S.F./ floor	2.29	2.99	5.28
2400	For each additional floor, add per floor	1.69	2.83	4.52
2500				
2600				
2700				
2800				
2900	Cut & patch to match existing construction, add, minimum	2%	3%	
3000	Maximum	5%	9%	
3100	Dust protection, add, minimum	1%	2%	
3200	Maximum	4%	11%	
3300	Equipment usage curtailment, add, minimum	1%	1%	
3400	Maximum	3%	10%	

D40 Fire Protection

D4010 Sprinklers

D4010 320	Fire Sprinkler Systems, Dry	COST PER S.F.		
		MAT.	INST.	TOTAL
3500	Material handling & storage limitation, add, minimum	1%	1%	
3600	Maximum	6%	7%	
3700	Protection of existing work, add, minimum	2%	2%	
3800	Maximum	5%	7%	
3900	Shift work requirements, add, minimum		5%	
4000	Maximum		30%	

D40 Fire Protection

D4010 Sprinklers

This page illustrates and describes a wet type fire sprinkler system. Lines within System Components give the unit cost of a system for a 2,000 square foot building. Lines D4010 420 1900 thru 2600 give the square foot costs for alternate systems. Both material quantities and labor costs have been adjusted for the system listed.

Factors: To adjust for conditions other than normal working conditions, use Lines D4010 420 2900 thru 3600.

System Components	QUANTITY	UNIT	COST EACH MAT.	COST EACH INST.	COST EACH TOTAL
Wet pipe fire sprinkler system, ordinary hazard, open area to 2000 S.F. On one floor.					
4" OS & Y valve	1.000	Ea.	575	385	960
Wet pipe alarm valve	1.000	Ea.	1,400	560	1,960
Water motor alarm	1.000	Ea.	335	156	491
3" check valve	1.000	Ea.	280	375	655
Pipe riser, 4" diameter	10.000	L.F.	175.50	332.40	507.90
Water gauges and trim	1.000	Set	2,300	1,125	3,425
Electric fire horn	1.000	Ea.	64.50	90	154.50
Sprinkler head supply piping, 1" diameter	168.000	L.F.	517.50	911.25	1,428.75
Sprinkler head supply piping, 1-1/2" diameter	168.000	L.F.	572	797.50	1,369.50
Sprinkler head supply piping, 2-1/2" diameter	168.000	L.F.	817	874	1,691
Pipe fittings, 1"	8.000	Ea.	68.40	644	712.40
Pipe fittings, 1-1/2"	8.000	Ea.	138	716	854
Pipe fittings, 2-1/2"	4.000	Ea.	254	516	770
Pipe fittings, 4"	2.000	Ea.	452	580	1,032
Sprinkler heads	16.000	Ea.	129.60	624	753.60
Fire department connection	1.000	Ea.	470	224	694
TOTAL			8,548.50	8,910.15	17,458.65

D4010 420	Fire Sprinkler Systems, Wet	COST PER S.F. MAT.	COST PER S.F. INST.	COST PER S.F. TOTAL
1900	Ordinary hazard, one floor, area to 2000 S.F./floor	4.29	4.47	8.76
2000	For each additional floor, add per floor	1.57	3.01	4.58
2100	Area to 3200 S.F./floor	3.31	4.04	7.35
2200	For each additional floor, add per floor	1.61	3.12	4.73
2300	Area to 5000 S.F./ floor	3.16	3.90	7.06
2400	For each additional floor, add per floor	2.07	3.31	5.38
2500	Area to 8000 S.F./ floor	2.38	3.20	5.58
2600	For each additional floor, add per floor	1.82	3	4.82
2700				
2900	Cut & patch to match existing construction, add minimum	2%	3%	
3000	Maximum	5%	9%	
3100	Dust protection, add, minimum	1%	2%	
3200	Maximum	4%	11%	
3300	Equipment usage curtailment, add, minimum	1%	1%	
3400	Maximum	3%	10%	
3500	Material handling & storage limitation, add, minimum	1%	1%	
3600	Maximum	6%	7%	

D50 Electrical

D5010 Electrical Service/Distribution

This page illustrates and describes commercial service systems including a meter socket, service head and cable, entrance switch, steel conduit, copper wire, panel board, ground rod, wire, and conduit. Lines within System Components give the unit price and total price on a cost each basis for this system. Prices for alternate commercial service systems are on Line Items D5010 210 1500 thru 1700. Both material quantities and labor costs have been adjusted for the system listed.

Factors: To adjust for job conditions other than normal working situations use Lines D5010 210 2900 thru 4000.

Example: You are to install the system with maximum equipment usage curtailment. Go to Line D5010 210 3400 and apply these percentages to the appropriate MAT. and INST. costs.

System Components	QUANTITY	UNIT	COST EACH MAT.	COST EACH INST.	COST EACH TOTAL
Commercial electric service including service breakers, metering 120/208 Volt, 3 phase, 4 wire, feeder, and panel board.					
100 Amp Service					
Meter socket	1.000	Ea.	48.50	188	236.50
Service head	.200	C.L.F.	49.20	120	169.20
Service entrance switch	1.000	Ea.	360	315	675
Rigid steel conduit	20.000	L.F.	90.80	185	275.80
600 volt copper wire #3	1.000	C.L.F.	117	120	237
Panel board, 24 circuits, 20 Amp breakers	1.000	Ea.	920	1,000	1,920
Ground rod	1.000	Ea.	18.40	110	128.40
TOTAL			1,603.90	2,038	3,641.90

D5010 210	Commercial Service - 3 Phase	COST EACH MAT.	COST EACH INST.	COST EACH TOTAL
1400	For alternate size services:			
1500	120/208 Volt, 3 phase, 4 wire service, 60 Amp	1,075	1,575	2,650
1600	200 Amp	3,250	3,600	6,850
1700	400 Amp	7,325	4,500	11,825
1800				
1900				
2000				
2100				
2200				
2300				
2400				
2500				
2900	Cut & patch to match existing construction, add, minimum	2%	3%	
3000	Maximum	5%	9%	
3100	Dust protection, add, minimum	1%	2%	
3200	Maximum	4%	11%	
3300	Equipment usage curtailment, add, minimum	1%	1%	
3400	Maximum	3%	10%	
3500	Material handling & storage limitation, add, minimum	1%	1%	
3600	Maximum	6%	7%	
3700	Protection of existing work, add, minimum	2%	2%	
3800	Maximum	5%	7%	
3900	Shift work requirements, add, minimum		5%	
4000	Maximum		30%	

D50 Electrical

D5010 Electrical Service/Distribution

This page illustrates and describes a residential, single phase system including a weather cap, service entrance cable, meter socket, entrance switch, ground rod, ground cable, EMT, and panelboard. Lines with System Components give the unit price and total price on a cost each basis for this system. Prices for an alternate residential, single phase system are also given. Both material quantities and labor costs have been adjusted for the system listed.

Factors: To adjust for job conditions other than normal working situations use Lines D5010 220 2900 thru 4000.

Example: You are to install the system with a minimum equipment usage curtailment. Go to Line D5010 220 3300 and apply these percentages to the appropriate MAT. and INST. costs.

System Components	QUANTITY	UNIT	COST EACH		
			MAT.	INST.	TOTAL
100 Amp Service, single phase					
Weathercap	1.000	Ea.	14.30	50	64.30
Service entrance cable	.200	C.L.F.	49.20	120	169.20
Meter socket	1.000	Ea.	48.50	188	236.50
Entrance disconnect switch	1.000	Ea.	360	315	675
Ground rod, with clamp	1.000	Ea.	21.50	126	147.50
Ground cable	.100	C.L.F.	21.30	37.50	58.80
Panelboard, 12 circuit	1.000	Ea.	277	500	777
TOTAL			791.80	1,336.50	2,128.30
200 Amp Service, single phase					
Weathercap	1.000	Ea.	39.50	75.50	115
Service entrance cable	.200	C.L.F.	125	172	297
Meter socket	1.000	Ea.	81.50	315	396.50
Entrance disconnect switch	1.000	Ea.	780	465	1,245
Ground rod, with clamp	1.000	Ea.	40	137	177
Ground cable	.100	C.L.F.	29.20	20.80	50
3/4" EMT	10.000	L.F.	10.60	46.30	56.90
Panelboard, 24 circuit	1.000	Ea.	630	780	1,410
TOTAL			1,735.80	2,011.60	3,747.40

D5010 220	Residential Service - Single Phase	COST EACH		
		MAT.	INST.	TOTAL
2800				
2810	100 Amp Service, single phase	790	1,325	2,115
2820	200 Amp Service, single phase	1,725	2,000	3,725
2900	Cut & patch to match existing construction, add, minimum	2%	3%	
3000	Maximum	5%	9%	
3100	Dust protection, add, minimum	1%	2%	
3200	Maximum	4%	11%	
3300	Equipment usage curtailment, add, minimum	1%	1%	
3400	Maximum	3%	10%	
3500	Material handling & storage limitation, add, minimum	1%	1%	
3600	Maximum	6%	7%	
3700	Protection of existing work, add, minimum	2%	2%	
3800	Maximum	5%	7%	
3900	Shift work requirements, add, minimum		5%	
4000	Maximum		30%	

D50 Electrical

D5020 Lighting and Branch Wiring

D5020 180	Selective Price Sheet	COST EACH		
		MAT.	INST.	TOTAL
0100	Using non-metallic sheathed, cable, air conditioning receptacle	22	60	82
0200	Disposal wiring	16.20	67	83.20
0300	Dryer circuit	42	110	152
0400	Duplex receptacle	22	46.50	68.50
0500	Fire alarm or smoke detector	97	60	157
0600	Furnace circuit & switch	25.50	100	125.50
0700	Ground fault receptacle	55.50	75.50	131
0800	Heater circuit	28	75.50	103.50
0900	Lighting wiring	29	37.50	66.50
1000	Range circuit	89	151	240
1100	Switches single pole	18.50	37.50	56
1200	3-way	23.50	50	73.50
1300	Water heater circuit	30.50	120	150.50
1400	Weatherproof receptacle	156	100	256
1500	Using BX cable, air conditioning receptacle	39	72.50	111.50
1600	Disposal wiring	32.50	80.50	113
1700	Dryer circuit	57.50	131	188.50
1800	Duplex receptacle	39	56	95
1900	Fire alarm or smoke detector	97	60	157
2000	Furnace circuit & switch	45	120	165
2100	Ground fault receptacle	71	91.50	162.50
2200	Heater circuit	37.50	91.50	129
2300	Lighting wiring	38.50	45.50	84
2400	Range circuit	143	183	326
2500	Switches, single pole	35.50	45.50	81
2600	3-way	37.50	60	97.50
2700	Water heater circuit	56	143	199
2800	Weatherproof receptacle	166	120	286
2900	Using EMT conduit, air conditioning receptacle	48	90	138
3000	Disposal wiring	43.50	100	143.50
3100	Dryer circuit	58	163	221
3200	Duplex receptacle	48	69	117
3300	Fire alarm or smoke detector	116	90	206
3400	Furnace circuit & switch	51	151	202
3500	Ground fault receptacle	90	112	202
3600	Heater circuit	46.50	112	158.50
3700	Lighting wiring	44	56.50	100.50
3800	Range circuit	103	223	326
3900	Switches, single pole	45	56.50	101.50
4000	3-way	39	75.50	114.50
4100	Water heater circuit	52.50	177	229.50
4200	Weatherproof receptacle	175	151	326
4300	Using aluminum conduit, air conditioning receptacle	60	120	180
4400	Disposal wiring	55.50	134	189.50
4500	Dryer circuit	86	215	301
4600	Duplex receptacle	57.50	92.50	150
4700	Fire alarm or smoke detector	137	120	257
4800	Furnace circuit & switch	70	201	271
4900	Ground fault receptacle	104	151	255
5000	Heater circuit	60.50	151	211.50
5100	Lighting wiring	68.50	75.50	144
5200	Range circuit	136	300	436
5300	Switches, single pole	70	75.50	145.50
5400	3-way	64.50	100	164.50
5500	Water heater circuit	75.50	241	316.50
5600	Weatherproof receptacle	208	201	409
5700	Using galvanized steel conduit	56	128	184
5800	Disposal wiring	44	143	187

D50 Electrical

D5020 Lighting and Branch Wiring

D5020 180	Selective Price Sheet	COST EACH		
		MAT.	INST.	TOTAL
5900	Dryer circuit	79	232	311
6000	Duplex receptacle	56	99	155
6100	Fire alarm or smoke detector	125	128	253
6200	Furnace circuit & switch	58.50	215	273.50
6300	Ground fault receptacle	92.50	159	251.50
6400	Heater circuit	53.50	159	212.50
6500	Lighting wiring	62.50	80.50	143
6600	Range circuit	124	315	439
6700	Switches, single pole	58.50	80.50	139
6800	3-way	53	104	157
6900	Water heater circuit	63	251	314
7000	Weatherproof receptacle	190	215	405
7100				
7200				
7300				
7400				
7500				
7600				
7700				
7800				
7900				
8000				
8100				
8200				
8300				
8400				

D50 Electrical

D5020 Lighting and Branch Wiring

This page illustrates and describes fluorescent lighting systems including a fixture, lamp, outlet box and wiring. Lines within System Components give the unit price and total price on a cost each basis for this system. Prices for alternate fluorescent lighting systems are on Line Items D5020 248 1300 thru 1500. Both material quantities and labor costs have been adjusted for the system listed.

Factors: To adjust for job conditions other than normal working situations use Lines D5020 248 2900 thru 4000.

Example: You are to install the system during evening hours. Go to Line D5020 248 3900 and apply this percentage to the appropriate INST. cost.

System Components			COST EACH		
	QUANTITY	UNIT	MAT.	INST.	TOTAL
Fluorescent lighting, including fixture, lamp, outlet box and wiring.					
Recessed lighting fixture, on suspended system	1.000	Ea.	72	128	200
Outlet box	1.000	Ea.	2.62	33.50	36.12
#12 wire	.660	C.L.F.	8.45	36.30	44.75
Conduit, EMT, 1/2" conduit	20.000	L.F.	13.60	70.80	84.40
TOTAL			96.67	268.60	365.27

D5020 248	Lighting, Fluorescent	COST EACH		
		MAT.	INST.	TOTAL
1200	For alternate lighting fixtures:			
1300	Surface mounted, 2' x 4', acrylic prismatic diffuser	129	283	412
1400	Strip fixture, 8' long, two 8' lamps	88	266	354
1500	Pendant mounted, industrial, 8' long, with reflectors	131	305	436
1600				
1700				
1800				
1900				
2000				
2100				
2200				
2300				
2900	Cut & patch to match existing construction, add, minimum	2%	3%	
3000	Maximum	5%	9%	
3100	Dust protection, add, minimum	1%	2%	
3200	Maximum	4%	11%	
3300	Equipment usage curtailment, add, minimum	1%	1%	
3400	Maximum	3%	10%	
3500	Material handling & storage limitation, add, minimum	1%	1%	
3600	Maximum	6%	7%	
3700	Protection of existing work, add, minimum	2%	2%	
3800	Maximum	5%	7%	
3900	Shift work requirements, add, minimum		5%	
4000	Maximum		30%	

D50 Electrical

D5020 Lighting and Branch Wiring

This page illustrates and describes incandescent lighting systems including a fixture, lamp, outlet box, conduit and wiring. Lines within System Components give the unit price and total price on a cost each basis for this system. Prices for an alternate incandescent lighting system are also given. Both material quantities and labor costs have been adjusted for the system listed.

Factors: To adjust for conditions other than normal working situations use Lines D5020 250 2900 thru 4000.

Example: You are to install the system and cut and match existing construction. Go to Line D5020 250 3000 and apply this percentage to the appropriate INST. costs.

System Components	QUANTITY	UNIT	COST EACH		
			MAT.	INST.	TOTAL
Incandescent light fixture, including lamp, outlet box, conduit and wiring.					
Recessed wide reflector with flat glass lens	1.000	Ea.	105	90	195
Outlet box	1.000	Ea.	2.62	33.50	36.12
Armored cable, 3 wire	.200	C.L.F.	24.40	54.80	79.20
TOTAL			132.02	178.30	310.32
Recessed flood light fixture, including lamp, outlet box, conduit & wire.					
Recessed, R-40 flood lamp with refractor	1.000	Ea.	112	75.50	187.50
150 watt R-40 flood lamp	.010	Ea.	6.65	4.65	11.30
Outlet box	1.000	Ea.	2.62	33.50	36.12
Outlet box cover	1.000	Ea.	1.68	9.40	11.08
Romex, 12-2 with ground	.200	C.L.F.	7.50	54.80	62.30
Conduit, 1/2" EMT	20.000	L.F.	13.60	70.80	84.40
TOTAL			144.05	248.65	392.70

D5020 250	Lighting, Incandescent	COST EACH		
		MAT.	INST.	TOTAL
2810	Incandescent light fixture, including lamp, outlet box, conduit and wiring.	132	178	310
2820	Recessed flood light fixture, including lamp, outlet box, conduit & wire	144	249	393
2900	Cut & patch to match existing construction, add, minimum	2%	3%	
3000	Maximum	5%	9%	
3100	Dust protection, add, minimum	1%	2%	
3200	Maximum	4%	11%	
3300	Equipment usage curtailment, add, minimum	1%	1%	
3400	Maximum	3%	10%	
3500	Material handling & storage limitation, add, minimum	1%	1%	
3600	Maximum	6%	7%	
3700	Protection of existing work, add, minimum	2%	2%	
3800	Maximum	5%	7%	
3900	Shift work requirements, add, minimum		5%	
4000	Maximum		30%	

D50 Electrical

D5020 Lighting and Branch Wiring

This page illustrates and describes high intensity lighting systems including a lamp, EMT conduit, EMT "T" fitting with cover, and wire. Lines within System Components give the unit price and total price on a cost each basis for this system. Prices for alternate high intensity lighting systems are on Line Items D5020 252 1200 thru 1800. Both material quantities and labor costs have been adjusted for the system listed.

Factors: To adjust for job conditions other than normal working situations use Lines D5020 252 2900 thru 4000.

Example: You are to install the system and protect existing construction. Go to Line D5020 252 3700 and apply these percentages to the appropriate MAT. and INST. costs.

System Components	QUANTITY	UNIT	COST EACH		
			MAT.	INST.	TOTAL
High intensity lighting system consisting of 400 watt metal halide fixture And lamp with 1/2" EMT conduit and fittings using #12 wire, high bay.					
Single unit, 400 watt	1.000	Ea.	445	262	707
Electric metallic tubing (EMT), 1/2" diameter	30.000	Ea.	20.40	106.20	126.60
1/2" EMT "T" fitting with cover	1.000	Ea.	16.40	37.50	53.90
Wire, 600 volt, type THWN-THHN, copper, solid, #12	.600	Ea.	7.68	33	40.68
TOTAL			489.48	438.70	928.18

D5020 252	Lighting, High Intensity	COST EACH		
		MAT.	INST.	TOTAL
1000	For alternate high intensity systems:			
1200	High bay: 400 watt, high pressure sodium and lamp	455	440	895
1400	1000 watt, metal halide	685	475	1,160
1500	High pressure sodium	635	475	1,110
1700	Low bay: 250 watt, metal halide	435	365	800
1800	150 watt, high pressure sodium	400	365	765
1900				
2000				
2100				
2200				
2300				
2400				
2500				
2600				
2700				
2900	Cut & patch to match existing construction, add, minimum	2%	3%	
3000	Maximum	5%	9%	
3100	Dust protection, add, minimum	1%	2%	
3200	Maximum	4%	11%	
3300	Equipment usage curtailment, add, minimum	1%	1%	
3400	Maximum	3%	10%	
3500	Material handling & storage limitation, add, minimum	1%	1%	
3600	Maximum	6%	7%	
3700	Protection of existing work, add, minimum	2%	2%	
3800	Maximum	5%	7%	
3900	Shift work requirements, add, minimum		5%	
4000	Maximum		30%	

D50 Electrical

D5090 Other Electrical Systems

This page illustrates and describes baseboard heat systems including a thermostat, outlet box, breaker, and feed. Lines within System Components give the unit price and total price on a cost each basis for this system. Prices for alternate baseboard heat systems are on Line Items D5090 430 1500 thru 1800. Both material quantities and labor costs have been adjusted for the system listed.

Factors: To adjust for job conditions other than normal working situations use Lines D5090 430 2900 thru 4000.

Example: You are to install the system during evenings and weekends only. Go to Line D5090 430 4000 and apply this percentage to the appropriate INST. cost.

System Components	QUANTITY	UNIT	COST EACH		
			MAT.	INST.	TOTAL
Baseboard heat including thermostat, outlet box, breaker and feed.					
Electric baseboard heater, 4' long	1.000	Ea.	59.50	90	149.50
Thermostat, integral	1.000	Ea.	33	37.50	70.50
Romex, 12-3 with ground	.400	C.L.F.	21.40	120	141.40
Panel board breaker, 20 Amp	1.000	Ea.	38.50	75.50	114
Outlet box	1.000	Ea.	2.62	33.50	36.12
TOTAL			155.02	356.50	511.52

D5090 530	Heat, Baseboard	COST EACH		
		MAT.	INST.	TOTAL
1400	For alternate baseboard heating systems:			
1500	Electric baseboard, 2' long	140	340	480
1600	6' long	177	385	562
1700	8' long	197	420	617
1800	10' long	310	450	760
1900				
2000				
2100				
2200				
2300				
2400				
2500				
2600				
2700				
2900	Cut & patch to match existing construction, add, minimum	2%	3%	
3000	Maximum	5%	9%	
3100	Dust protection, add, minimum	1%	2%	
3200	Maximum	4%	11%	
3300	Equipment usage curtailment, add, minimum	1%	1%	
3400	Maximum	3%	10%	
3500	Material handling & storage limitation, add, minimum	1%	1%	
3600	Maximum	6%	7%	
3700	Protection of existing work, add, minimum	2%	2%	
3800	Maximum	5%	7%	
3900	Shift work requirements, add, minimum		5%	
4000	Maximum		30%	

E Equipment & Furnishings

No part of this publication may be reproduced, stored in a retrieval system, or transmitted in any form or by any means without prior written permission of R.S. Means Company, Inc.

E10 Equipment

E1090 Other Equipment

This page illustrates and describes kitchen systems including top and bottom cabinets, custom laminated plastic top, single bowl sink, and appliances. Lines within System Components give the unit price and total price on a cost each basis for this system. Prices for alternate kitchen systems are on Line Items E1090 310 1500 and 1600. Both material quantities and labor costs have been adjusted for the system listed.

Factors: To adjust for job conditions other than normal working situations use Lines E1090 310 2900 thru 4000.

Example: You are to install the system and protect the work area from dust. Go to Line E1090 310 3200 and apply these percentages to the appropriate MAT. and INST. costs.

System Components	QUANTITY	UNIT	COST EACH		
			MAT.	INST.	TOTAL
Kitchen cabinets including wall and base cabinets, custom laminated Plastic top, sink & appliances, no plumbing or electrical rough-in included.					
Prefinished wood cabinets, average quality, wall and base	20.000	L.F.	3,220	730	3,950
Custom laminated plastic counter top	20.000	L.F.	200	364	564
Stainless steel sink, 22" x 25"	1.000	Ea.	680	207	887
Faucet, top mount	1.000	Ea.	67.50	64.50	132
Dishwasher, built-in	1.000	Ea.	234	310	544
Compactor, built-in	1.000	Ea.	585	109	694
Range hood, vented, 30" wide	1.000	Ea.	59	231	290
TOTAL			5,045.50	2,015.50	7,061

E1090 310	Kitchens	COST EACH		
		MAT.	INST.	TOTAL
1400	For alternate kitchen systems:			
1500	Prefinished wood cabinets, high quality	9,025	2,300	11,325
1600	Custom cabinets, built in place, high quality	11,700	2,650	14,350
1700				
1800				
1900				
2000				
2100				
2200	NOTE: No plumbing or electric rough-ins are included in the above			
2300	Prices, for plumbing see Division 22, for electric see Division 26.			
2400				
2500				
2600				
2700				
2900	Cut & patch to match existing construction, add, minimum	2%	3%	
3000	Maximum	5%	9%	
3100	Dust protection, add, minimum	1%	2%	
3200	Maximum	4%	11%	
3300	Equipment usage curtailment, add, minimum	1%	1%	
3400	Maximum	3%	10%	
3500	Material handling & storage limitation, add, minimum	1%	1%	
3600	Maximum	6%	7%	
3700	Protection of existing work, add, minimum	2%	2%	
3800	Maximum	5%	7%	
3900	Shift work requirements, add, minimum		5%	
4000	Maximum		30%	

E10 Equipment

E1090 Other Equipment

E1090 315	Selective Price Sheet	COST EACH		
		MAT.	INST.	TOTAL
0100	Cabinets standard wood, base, one drawer one door, 12" wide	252	44	296
0200	15" wide	262	45.50	307.50
0300	18" wide	284	47	331
0400	21" wide	299	48	347
0500	24" wide	345	49	394
0600				
0700	Two drawers two doors, 27" wide	365	49.50	414.50
0800	30" wide	400	51	451
0900	33" wide	420	52.50	472.50
1000	36" wide	440	54	494
1100	42" wide	470	55	525
1200	48" wide	505	58	563
1300	Drawer base (4 drawers), 12" wide	264	44	308
1400	15" wide	267	45.50	312.50
1500	18" wide	295	47	342
1600	24" wide	335	49	384
1700	Sink or range base, 30" wide	335	51	386
1800	33" wide	360	52.50	412.50
1900	36" wide	380	54	434
2000	42" wide	405	55	460
2100	Corner base, 36" wide	595	60.50	655.50
2200	Lazy susan with revolving door	795	66	861
2300	Cabinets standard wood, wall two doors, 12" high, 30" wide	225	44	269
2400	36" wide	264	45.50	309.50
2500	15" high, 30" high	181	45.50	226.50
2600	36" wide	264	48	312
2700	24" high, 30" wide	305	47	352
2800	36" wide	335	48	383
2900	30" high, 30" wide	335	56.50	391.50
3000	36" wide	385	58	443
3100	42" wide	420	59	479
3200	48" wide	475	59.50	534.50
3300	One door, 30" high, 12" wide	204	49.50	253.50
3400	15" wide	228	51	279
3500	18" wide	252	52.50	304.50
3600	24" wide	293	54	347
3700	Corner, 30" high, 24" wide	256	60.50	316.50
3800	36" wide	325	66	391
3900	Broom, 84" high, 24" deep, 18" wide	635	109	744
4000	Oven, 84" high, 24" deep, 27" wide	945	137	1,082
4100	Valance board, 4' long	54.50	11.05	65.55
4200	6" long	82	16.55	98.55
4300	Counter tops, laminated plastic, stock 25" wide w/backsplash, min.	10	18.20	28.20
4400	Maximum	21.50	22	43.50
4500	Custom, 7/8" thick, no splash	23	18.20	41.20
4600	Cove splash	30	18.20	48.20
4700	1-1/4" thick, no splash	34	19.50	53.50
4800	Square splash	33	19.50	52.50
4900	Post formed	11.80	18.20	30
5000				
5100	Maple laminated 1-1/2" thick, no splash	68.50	19.50	88
5200	Square splash	81.50	19.50	101
5300				
5400				
5500	Appliances, range, free standing, minimum	380	87	467
5600	Maximum	2,100	217	2,317
5700	Built-in, minimum	750	100	850
5800	Maximum	1,650	545	2,195

E10 Equipment

E1090 Other Equipment

E1090 315	Selective Price Sheet	COST EACH		
		MAT.	INST.	TOTAL
5900	Counter top range 4 burner, maximum	330	100	430
6000	Maximum	660	201	861
6100	Compactor, built-in, minimum	585	109	694
6200	Maximum	910	182	1,092
6300	Dishwasher, built-in, minimum	234	310	544
6400	Maximum	350	625	975
6500	Garbage disposer, sink-pipe, minimum	79.50	125	204.50
6600	Maximum	210	125	335
6700	Range hood, 30" wide, 2 speed, minimum	59	231	290
6800	Maximum	900	385	1,285
6900	Refrigerator, no frost, 12 cu. ft.	490	87	577
7000	20 cu. ft.	495	145	640
7100	Plumb. not incl. rough-ins, sinks porc. C.I., single bowl, 21" x 24"	282	207	489
7200	21" x 30"	330	207	537
7300	Double bowl, 20" x 32"	425	242	667
7400				
7500	Stainless steel, single bowl, 19" x 18"	610	207	817
7600	22" x 25"	680	207	887
7700				
7800				
7900				
8000				
8100				
8200				
8300				
8400				

No part of this publication may be reproduced, stored in a retrieval system, or transmitted in any form or by any means without prior written permission of R.S. Means Company, Inc.

G10 Site Preparation

G1030 Site Clearing

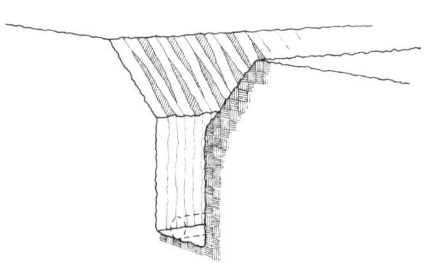

This page illustrates and describes utility trench excavation with backfill systems including a trench, backfill, excavated material, utility pipe or ductwork not included. Lines within System Components give the unit price and total price on a linear foot basis for this system. Prices for alternate utility trench excavation with backfill systems are on Line Items G1030 810 1100 thru 2500. Both material quantities and labor costs have been adjusted for the system listed.

Factors: To adjust for job conditions other than normal working situations use Lines G1030 810 3100 thru 4000.

Example: You are to install the system and protect existing construction. Go to Line G1030 810 3600 and apply these percentages to the appropriate TOTAL cost.

System Components	QUANTITY	UNIT	COST PER L.F.		
			EQUIP.	LABOR	TOTAL
Cont. utility trench excav. with 3/8 C.Y. wheel mtd. backhoe, Backfilling with excav. mat. and comp. in 12" lifts. Trench is 2' wide by 4' dp. Cost of utility piping or ductwork is not incl. Cost based on an Excavation production rate of 150 C.Y. daily no hauling included.					
Machine excavate trench, 2' wide by 4' deep	.296	C.Y.	.73	1.92	2.65
Dozer backfill with excavated material	.296	C.Y.	.33	.53	.86
Compact in 12" lifts, vibrating plate	.296	C.Y.	.08	2.19	2.27
TOTAL			1.14	4.64	5.78

G1030 810	Excavation, Utility Trench	COST PER L.F.		
		EQUIP.	LABOR	TOTAL
1000	Alternate size trenches:			
1100	2' wide x 2' deep	.58	2.32	2.90
1200	3' deep	.86	3.48	4.34
1300	With sloping sides, 2' wide x 5' deep	3.23	13.05	16.30
1400	6' deep	4.32	17.40	21.50
1500	7' deep	5.55	22.50	28
1600	8' deep	6.90	28	35
1700	9' deep	8.40	34	42.50
1800	10' deep	10.05	40.50	50.50
2000	For hauling excavated material up to 2 miles & backfilling w/gravel			
2100	Gravel, 2' wide by 2' deep, add	4.07	3.36	7.43
2200	4' deep, add	8.15	6.70	14.85
2300	6' deep, add	30.50	25.50	56
2400	8' deep, add	49	40.50	89.50
2500	10' deep, add	71.50	59	130.50
2600				
2700				
2800	For shallow, hand excavated trenches, no backfill, to 6' dp, light soil		54.50	54.50
2900	Heavy soil		109	109
3000				
3100	Equipment usage curtailment, add, minimum	1%	1%	
3200	Maximum	3%	10%	
3300	Material handling & storage limitation, add, minimum	1%	1%	
3400	Maximum	6%	7%	
3500	Protection of existing work, add, minimum	2%	2%	
3600	Maximum	5%	7%	
3700	Shift work requirements, add, minimum		5%	
3800	Maximum		30%	
3900	Temporary shoring and bracing, add, minimum	2%	5%	
4000	Maximum	5%	12%	

G20 Site Improvements

G2010 Roadways

This page illustrates and describes driveway systems including concrete slab, gravel base, broom finish, compaction, and joists. Lines within System Components give the unit price and total price on a cost each basis for this system. Prices for alternate driveway systems are on Line Items G2010 240 1300 thru 2700. Both material quantities and labor costs have been adjusted for the system listed.

Factors: To adjust for job conditions other than normal working situations use Lines G2010 240 2900 thru 4000.

Example: You are to install the system and match existing construction. Go to Line G2010 240 3000 and apply these percentages to the appropriate MAT. and INST. costs.

System Components	QUANTITY	UNIT	COST EACH		
			MAT.	INST.	TOTAL
Complete driveway, 10' x 30', 4" concrete slab, 3000 psi, on 6" compacted					
Gravel base, concrete broom finished and cured with 10' x 10' joints.					
Grade and compact subgrade	33.000	S.Y.		63.69	63.69
Place and compact 6" crushed stone base	33.000	S.Y.	245.85	43.23	289.08
Place and remove edgeforms (4 use)	80.000	L.F.	24	276	300
Concrete for slab, 3000 psi	4.000	C.Y.	428		428
Place concrete, 4" thick slab, direct chute	4.000	C.Y.		100.44	100.44
Screed, float and broom finish slab	4.000	C.Y.		231	231
Spray on membrane curing compound	3.000	S.F.	20.25	27.45	47.70
Saw cut 3" deep joints	60.000	L.F.	33	89.40	122.40
TOTAL			751.10	831.21	1,582.31

G2010 240	Driveways	COST EACH		
		MAT.	INST.	TOTAL
1200	For alternate driveway systems:			
1300	Concrete: 10' x 30' with 6" concrete on 6" crushed stone	965	885	1,850
1400	20' x 30' with 4" concrete	1,550	1,575	3,125
1500	With 6" concrete	1,975	1,675	3,650
1600	10' wide, for each additional 10' length over 30', 4" thick, add	246	267	513
1700	6" thick, add	325	287	612
1800	20' wide, for each additional 10' length over 30', 4" thick, add	500	510	1,010
1900	6" thick, add	635	545	1,180
2000	Asphalt: 10' x 30' with 2" binder, 1" topping on 6" cr. st., sealed	660	630	1,290
2100	3" binder, 1" topping	785	645	1,430
2200	20' x 30' with 2" binder, 1" topping	1,325	855	2,180
2300	3" binding, 1" topping	1,575	880	2,455
2400	10' wide for each add'l. 10' length over 30', 2" b + 1" t, add	219	75.50	294.50
2500	3" binder, 1" topping	262	79.50	341.50
2600	20' wide for each add'l. 10' length over 30', 2" b + 1" t, add	440	152	592
2700	3" binder, 1" topping	525	160	685
2800				
2900	Cut & patch to match existing construction, add, minimum	2%	3%	
3000	Maximum	5%	9%	
3100	Dust protection, add, minimum	1%	2%	
3200	Maximum	4%	11%	
3300	Equipment usage curtailment, add, minimum	1%	1%	
3400	Maximum	3%	10%	
3500	Material handling & storage limitation, add, minimum	1%	1%	
3600	Maximum	6%	7%	
3700	Protection of existing work, add, minimum	2%	2%	
3800	Maximum	5%	7%	
3900	Shift work requirements, add, minimum		5%	
4000	Maximum		30%	

G20 Site Improvements

G2020 Parking Lots

This page illustrates and describes asphalt parking lot systems including asphalt binder, topping, crushed stone base, painted parking stripes and concrete parking blocks. Lines within System Components give the unit price and total price per square yard for this system. Prices for alternate asphalt parking lot systems are on Line Items G2020 230 1500 thru 2500. Both material quantities and labor costs have been adjusted for the system listed.

Factors: To adjust for job conditions other than normal working situations use Lines G2020 230 2900 thru 4000.

Example: You are to install the system and match existing construction. Go to Line G2020 230 2900 and apply these percentages to the appropriate MAT. and INST. costs.

System Components	QUANTITY	UNIT	COST PER S.Y.		
			MAT.	INST.	TOTAL
Parking lot consisting of 2" asphalt binder and 1" topping on 6" Crushed stone base with painted parking stripes and concrete parking blocks					
Fine grade and compact subgrade	1.000	S.Y.		1.93	1.93
6" crushed stone base, stone	.320	Ton	7.45	1.31	8.76
2" asphalt binder	1.000	S.Y.	7.80	1.25	9.05
1" asphalt topping	1.000	S.Y.	3.86	.83	4.69
Paint parking stripes	.500	L.F.	.15	.09	.24
6" x 10" x 6' precast concrete parking blocks	.020	Ea.	1.26	.37	1.63
Mobilization of equipment	.005	Ea.		2.01	2.01
TOTAL			20.52	7.79	28.31

G2020 230	Parking Lots, Asphalt	COST PER S.Y.		
		MAT.	INST.	TOTAL
1400	For alternate parking lot systems:			
1500	Above system on 9" crushed stone	24.50	7.90	32.40
1600	12" crushed stone	28	8.05	36.05
1700	On bank run gravel, 6" deep	17.80	7.20	25
1800	9" deep	20	7.35	27.35
1900	12" deep	22.50	7.50	30
2000	3" binder plus 1" topping on 6" crushed stone	24.50	8.15	32.65
2100	9" deep crushed stone	28	8.30	36.30
2200	12" deep crushed stone	32	8.40	40.40
2300	On bank run gravel, 6" deep	21.50	7.55	29.05
2400	9" deep	24	7.75	31.75
2500	12" deep	26.50	7.90	34.40
2600				
2700				
2900	Cut & patch to match existing construction, add, minimum	2%	3%	
3000	Maximum	5%	9%	
3100	Dust protection, add, minimum	1%	2%	
3200	Maximum	4%	11%	
3300	Equipment usage curtailment, add, minimum	1%	1%	
3400	Maximum	3%	10%	
3500	Material handling & storage limitation, add, minimum	1%	1%	
3600	Maximum	6%	7%	
3700	Protection of existing work, add, minimum	2%	2%	
3800	Maximum	5%	7%	
3900	Shift work requirements, add, minimum		5%	
4000	Maximum		30%	

G20 Site Improvements

G2020 Parking Lots

This page illustrates and describes concrete parking lot systems including a concrete slab, unreinforced, compacted gravel base, broom finished, cured, joints, painted parking stripes, and parking blocks. Lines within System Components give the unit price and total price per square yard for this system. Prices for alternate concrete parking lot systems are on Line Items G2020 240 1500 thru 2500. All materials have been adjusted according to the system listed.

Factors: To adjust for job conditions other than normal working situations use Lines G2020 240 2900 thru 4000.

Example: You are to install the system and protect existing construction. Go to Line G2020 240 3800 and apply these percentages to the appropriate MAT. and INST. costs.

System Components	QUANTITY	UNIT	COST PER S.Y.		
			MAT.	INST.	TOTAL
Parking lot, 4" slab, 3000 psi concrete on 6" compacted gravel base					
With 12' x 12' joints, painted parking strips, conc. parking blocks.					
Grade and compact subgrade	1.000	S.Y.		1.93	1.93
6" crushed stone base, stone	.320	Ton	7.45	1.31	8.76
Place and remove edge forms, 4 use	.250	L.F.	.08	.86	.94
Concrete for slab, 3000 psi	1.000	S.F.	11.77		11.77
Place concrete, 4" thick slab, direct chute	1.000	S.F.		2.77	2.77
Spray on membrane curing compound	1.000	S.F.	.61	.58	1.19
Saw cut 3" deep joints	2.250	L.F.	1.24	3.35	4.59
Paint parking stripes	.500	L.F.	.15	.09	.24
6" x 10" x 6' precast concrete parking blocks	.020	Ea.	1.26	.37	1.63
TOTAL			22.56	11.26	33.82

G2020 240	Parking Lots, Concrete	COST PER S.Y.		
		MAT.	INST.	TOTAL
1400	For alternate parking lot systems:			
1500	Above system on 3" crushed stone	22.50	11.25	33.75
1600	9" crushed stone	26.50	11.40	37.90
1700	On bank run gravel, 6" deep	19.80	10.70	30.50
1800	9" deep	22	10.80	32.80
1900	On compacted subgrade only	15.10	9.95	25.05
2000	6" concrete slab, 3000 psi on 3" crushed stone	28	12.60	40.60
2100	6" crushed stone	28	12.60	40.60
2200	9" crushed stone	32	12.70	44.70
2300	On bank run gravel, 6" deep	25.50	12	37.50
2400	9" deep	28	12.15	40.15
2500	On compacted subgrade only	28	12.60	40.60
2600				
2700				
2900	Cut & patch to match existing construction, add, minimum	2%	3%	
3000	Maximum	5%	9%	
3100	Dust protection, add, minimum	1%	2%	
3200	Maximum	4%	11%	
3300	Equipment usage curtailment, add, minimum	1%	1%	
3400	Maximum	3%	10%	
3500	Material handling & storage limitation, add, minimum	1%	1%	
3600	Maximum	6%	7%	
3700	Protection of existing work, add, minimum	2%	2%	
3800	Maximum	5%	7%	
3900	Shift work requirements, add, minimum		5%	
4000	Maximum		30%	

G20 Site Improvements

G2030 Pedestrian Paving

This page illustrates and describes sidewalk systems including concrete, welded wire and broom finish. Lines within System Components give unit price and total price per square foot for this system. Prices for alternate sidewalk systems are on Line Items G2030 105 1700 thru 1900. Both material quantities and labor costs have been adjusted for the system listed.

Factors: To adjust for job conditions other than normal working situations use Lines G2030 105 2900 thru 4000.

Example: You are to install the system and match existing construction. Go to Line G2030 105 2900 and apply these percentages to the appropriate MAT. and INST. costs.

System Components	QUANTITY	UNIT	COST PER S.F.		
			MAT.	INST.	TOTAL
4" thick concrete sidewalk with welded wire fabric					
3000 psi air entrained concrete, broom finish.					
Gravel fill, 4" deep	.012	C.Y.	.29	.09	.38
Compact fill	.012	C.Y.		.02	.02
Hand grade	1.000	S.F.		1.97	1.97
Edge form	.250	L.F.	.08	.86	.94
Welded wire fabric	.011	S.F.	.15	.40	.55
Concrete, 3000 psi air entrained	.012	C.Y.	1.28		1.28
Place concrete	.012	C.Y.		.30	.30
Broom finish	1.000	S.F.		.77	.77
TOTAL			1.80	4.41	6.21

G2030 105	Sidewalks	COST PER S.F.		
		MAT.	INST.	TOTAL
1600	For alternate sidewalk systems:			
1700	Asphalt (bituminous), 2" thick	1.12	2.53	3.65
1800	Brick, on sand, bed, 4.5 brick per S.F.	3.30	9.70	13
1900	Flagstone, slate, 1" thick, rectangular	9.95	10.55	20.50
2000				
2100				
2200				
2300				
2400				
2500				
2600				
2700				
2900	Cut & patch to match existing construction, add, minimum	2%	3%	
3000	Maximum	5%	9%	
3100	Dust protection, add, minimum	1%	2%	
3200	Maximum	4%	11%	
3300	Equipment usage curtailment, add, minimum	1%	1%	
3400	Maximum	3%	10%	
3500	Material handling & storage limitation, add, minimum	1%	1%	
3600	Maximum	6%	7%	
3700	Protection of existing work, add, minimum	2%	2%	
3800	Maximum	5%	7%	
3900	Shift work requirements, add, minimum		5%	
4000	Maximum		30%	

G20 Site Improvements

G2050 Landscaping

This page describes landscaping—lawn establishment systems including loam, lime, fertilizer, side and top mulching. Lines within system components give the unit price and total price per square yard for this system. Prices for alternate landscaping—lawn establishment systems are on Line Items G2050 420 1100 and 1200. Both material quantities and labor costs have been adjusted for the system listed.

Factors: To adjust for job conditions other than normal working situations use Lines G2050 420 2900 thru 4000.

Example: You are to install the system and provide dust protection. Go to Line G2050 420 3100 and apply these percentages to the appropriate MAT. and INST. costs.

System Components	QUANTITY	UNIT	COST PER S.Y. MAT.	INST.	TOTAL
Establishing lawns with loam, lime, fertilizer, seed and top mulching On rough graded areas.					
Furnish and place loam 4" deep	.110	C.Y.	4.13	.89	5.02
Fine grade, lime, fertilize and seed	1.000	S.Y.	.55	3.09	3.64
Hay mulch, 1 bale/M.S.F.	1.000	S.Y.	.11	.53	.64
Rolling with hand roller	1.000	S.Y.		.93	.93
TOTAL			4.79	5.44	10.23

G2050 420	Landscaping - Lawn Establishment	COST PER S.Y. MAT.	INST.	TOTAL
1000	For alternate lawn systems:			
1100	Above system with jute mesh in place of hay mulch	5.80	5.60	11.40
1200	Above system with sod in place of seed	6.65	1.93	8.58
1300				
1400				
1500				
1600				
1700				
1800				
1900				
2000				
2100				
2200				
2300				
2400				
2500				
2600				
2700				
2900	Cut & patch to match existing construction, add, minimum	2%	3%	
3000	Maximum	5%	9%	
3100	Dust protection, add, minimum	1%	2%	
3300	Equipment usage curtailment, add, minimum	1%	1%	
3400	Maximum	3%	10%	
3500	Material handling & storage limitation, add, minimum	1%	1%	
3600	Maximum	6%	7%	
3700	Protection of existing work, add, minimum	2%	2%	
3800	Maximum	5%	7%	
3900	Shift work requirements, add, minimum		5%	
4000	Maximum		30%	

G20 Site Improvements

G2050 Landscaping

G2050 510	Selective Price Sheet	COST PER UNIT		
		MAT.	INST.	TOTAL
0100	Shrubs and trees, Evergreen, in prepared beds			
0200				
0300	Arborvitae pyramidal, 4'-5'	53	87.50	140.50
0400	Globe, 12"-15"	23	13.85	36.85
0500	Cedar, blue, 8'-10'	241	146	387
0600	Hemlock, Canadian, 2-1/2'-3'	37	37	74
0700	Juniper, andora, 18"-24"	18.50	16.65	35.15
0800	Wiltoni, 15"-18"	23	16.65	39.65
0900	Skyrocket, 4-1/2'-5'	61.50	48	109.50
1000	Blue pfitzer, 2'-2-1/2'	31	30.50	61.50
1100	Ketleerie, 2-1/2'-3'	37	26.50	63.50
1200	Pine, black, 2-1/2'-3'	50	26.50	76.50
1300	Mugo, 18"-24"	39.50	22	61.50
1400	White, 4'-5'	64	35	99
1500	Spruce, blue, 18"-24"	47	22	69
1600	Norway, 4'-5'	73.50	35	108.50
1700	Yew, denisforma, 12"-15"	28.50	22	50.50
1800	Capitata, 18"-24"	26	44.50	70.50
1900	Hicksi, 2'-2-1/2'	35	44.50	79.50
2000				
2100	Trees, Deciduous, in prepared beds			
2200				
2300	Beech, 5'-6'	246	53	299
2400	Dogwood, 4'-5'	70	66	136
2500	Elm, 8'-10'	131	132	263
2600	Magnolia, 4'-5'	78.50	132	210.50
2700	Maple, red, 8'-10', 1-1/2" caliper	265	264	529
2800	Oak, 2-1/2"-3" caliper	296	440	736
2900	Willow, 6'-8', 1" caliper	75.50	132	207.50
3000				
3100	Shrubs, Broadleaf Evergreen, in prepared beds			
3200				
3300	Andromeda, 15"-18", container	29	13.85	42.85
3400	Azalea, 15"-18", container	36	13.85	49.85
3500	Barberry, 9"-12", container	12.95	10.25	23.20
3600	Boxwood, 15"-18", B & B	32.50	13.85	46.35
3700	Euonymus, emerald gaiety, 12"-15", container	22.50	11.55	34.05
3800	Holly, 15"-18", B & B	20.50	13.85	34.35
3900	Mount laurel, 18"-24", B & B	62	16.65	78.65
4000	Privet, 18"-24", B & B	20	10.25	30.25
4100	Rhodendron, 18"-24", container	34.50	27.50	62
4200	Rosemary, 1 gal. container	69	2.22	71.22
4300	Deciduous, amalanchier, 2'-3', B & B	99.50	23.50	123
4400	Azalea, 15"-18", B & B	27	13.85	40.85
4500	Bayberry, 2'-3', B & B	30.50	23.50	54
4600	Cotoneaster, 15"-18", B & B	18.45	16.65	35.10
4700	Dogwood, 3'-4', B & B	29	66	95
4800	Euonymus, alatus compacta, 15"-18", container	24.50	16.65	41.15
4900	Forsythia, 2'-3', container	22	22	44
5000	Honeysuckle, 3'-4', B & B	24.50	22	46.50
5100	Hydrangea, 2'-3', B & B	29	23.50	52.50
5200	Lilac, 3'-4', B & B	25	66	91
5300	Quince, 2'-3', B & B	26.50	23.50	50
5400				
5500	Ground Cover			
5600				
5700	Plants, Pachysandra, prepared beds, per hundred	30	88.50	118.50
5800	Vinca minor or English ivy, per hundred	44	111	155

G20 Site Improvements

G2050 Landscaping

G2050 510	Selective Price Sheet	COST PER UNIT		
		MAT.	INST.	TOTAL
5900	Plant bed prep. 18" dp., mach., per square foot	2.47	.39	2.86
6000	By hand, per square foot	2.47	2.81	5.28
6100	Stone chips, Georgia marble, 50# bags, per bag	5.80	2.56	8.36
6200	Onyx gemstone, per bag	25	5.10	30.10
6300	Quartz, per bag	9.35	5.10	14.45
6400	Pea gravel, truck load lots, per cubic yard	30	47.50	77.50
6500	Mulch, polyethylene mulch, per square yard	.24	.43	.67
6600	Wood chips, 2" deep, per square yard	2.12	1.98	4.10
6700	Peat moss, 1" deep, per square yard	2.09	.48	2.57
6800	Erosion control per square yard, Jute mesh stapled	1.14	.68	1.82
6900	Plastic netting, stapled	.87	.53	1.40
7000	Polypropylene mesh, stapled	1.87	.53	2.40
7100	Tobacco netting, stapled	.11	.53	.64
7200	Lawns per square yard, seeding incl. fine grade, limestone, fert. and seed	.55	3.09	3.64
7300	Sodding, incl. fine grade, level ground	2.44	1	3.44
7500	Edging, per linear foot, Redwood, untreated, 1" x 4"	1.28	2.18	3.46
7600	2" x 4"	2.49	3.31	5.80
7700	Stl. edge strips, 1/4" x 5" inc. stakes	5.05	3.31	8.36
7800	3/16" x 4"	4.18	3.31	7.49
7900	Brick edging, set on edge	3.08	7.20	10.28
8000	Set flat	1.54	2.62	4.16
8100				
8200				
8300				
8400				

G30 Site Mechanical Utilities

G3020 Sanitary Sewer

This page illustrates and describes drainage and utilities – septic system including a tank, distribution box, excavation, piping, crushed stone and backfill. Lines within System Components give the unit price and total price on a cost each basis for this system. Prices for alternate drainage and utilities – septic systems are on Line Items G3020 610 1700 thru 2000. Both material quantities and labor costs have been adjusted for the system listed.

Factors: To adjust for job conditions other than normal working situations use Lines G3020 610 3100 thru 4000.

Example: You are to install the system with a material handling limitation. Go to Line G3020 610 3500 and apply these percentages to the appropriate MAT. and INST. costs.

System Components	QUANTITY	UNIT	COST EACH		
			MAT.	INST.	TOTAL
Septic system including tank, distribution box, excavation, piping, crushed Stone and backfill for a 1000 S.F. leaching field.					
Precast concrete septic tank, 1000 gal. capacity	1.000	Ea.	825	235	1,060
Concrete distribution box	1.000	Ea.	88	54.50	142.50
Sch. 40 sewer pipe, PVC, 4" diameter	25.000	L.F.	314.50	656.20	970.70
Sch. 40 sewer pipe fittings, PVC, 4" diameter	8.000	Ea.	99.20	736	835.20
Excavation for septic tank	120.000	C.Y.		216	216
Excavation for disposal field	120.000	C.Y.		524.19	524.19
Crushed stone backfill	76.000	C.Y.	1,672	891.48	2,563.48
Backfill with excavated material	26.000	C.Y.		44.72	44.72
Asphalt felt paper, 15#	6.000	Sq.	38.70	88.50	127.20
TOTAL			3,037.40	3,446.59	6,483.99

G3020 610	Septic Systems	COST EACH		
		MAT.	INST.	TOTAL
1600	For alternate septic systems:			
1700	1000 gal. tank with 2000 S.F. field	4,450	5,025	9,475
1800	With leaching pits	3,350	1,125	4,475
1900	2000 gal. tank with 2000 S.F. field	5,475	5,150	10,625
2000	With leaching pits	4,475	1,650	6,125
2100				
2200				
2300				
2400				
2500				
2600				
2700				
2800				
2900				
3100	Dust protection, add, minimum	1%	2%	
3200	Maximum	4%	11%	
3300	Equipment usage curtailment, add, minimum	1%	1%	
3400	Maximum	3%	10%	
3500	Material handling & storage limitation, add, minimum	1%	1%	
3600	Maximum	6%	7%	
3700	Protection of existing work, add, minimum	2%	2%	
3800	Maximum	5%	7%	
3900	Shift work requirements, add, minimum		5%	
4000	Maximum		30%	

G30 Site Mechanical Utilities

G3060 Fuel Distribution

Fiberglass Underground Fuel Storage Tank, Single Wall

This page illustrates and describes fiberglass tanks including hold down slab, reinforcing steel and peastone gravel. Lines within Systems Components give the material and installation price on a cost each basis for the components. Prices for alternate fiberglass tank systems are on Line Items G3060 310 1010 thru G3060 310 1100. Material quantities and labor costs have been adjusted for the system listed.

Factors: To adjust for job conditions other than normal working situations use Lines G3060 310 2700 thru 4000.

System Components	QUANTITY	UNIT	COST EACH MAT.	COST EACH INST.	COST EACH TOTAL
Double wall underground fiberglass storage tanks including, excavation, hold down slab, backfill, manway extension and related piping.					
Excavating trench, no sheeting or dewatering, 6'-10 ' D, 3/4 CY hyd backhoe	18.000	Ea.		138.96	138.96
Corrosion resistance wrap & coat, small diam pipe, 1" diam, add	120.000	Ea.	200.40		200.40
Fiberglass-reinforced elbows 90 & 45 degree, 2 inch	7.000	Ea.	395.50	525	920.50
Fiberglass-reinforced threaded adapters, 2 inch	2.000	Ea.	37.90	172	209.90
Remote tank gauging system, 30', 5" pointer travel	1.000	Ea.	5,050	320	5,370
Corrosion resistance wrap & coat, small diam pipe. 4 " diam, add	30.000	Ea.	66.60		66.60
Pea gravel	9.000	Ea.	679.50	427.50	1,107
Reinforcing in hold down pad, #3 to #7	120.000	Ea.	52.80	72	124.80
Tank leak detector system, 8 channel, external monitoring	1.000	Ea.	885		885
Tubbing copper, type L, 3/4" diam	120.000	Ea.	505.20	1,020	1,525.20
Elbow 90 degree ,wrought cu x cu 3/4 " diam	4.000	Ea.	14.40	136	150.40
Pipe, black steel welded, sch 40, 3" diam	30.000	Ea.	376.50	855	1,231.50
Elbow 90 degree, steel welded butt joint, 3 " diam	3.000	Ea.	111	525.60	636.60
Fiberglass-reinforced plastic pipe, 2 inch	156.000	Ea.	873.60	661.44	1,535.04
Fiberglass-reinforced plastic pipe, 3 inch	36.000	Ea.	345.60	172.80	518.40
Fiberglass-reinforced elbows 90 & 45 degree, 3 inch	3.000	Ea.	207	300	507
Tank leak detection system, probes, well monitoring, liquid phase detection	2.000	Ea.	940		940
Double wall fiberglass annular space	1.000	Ea.	320		320
Flange, weld neck, 150 lb, 3 " pipe size	1.000	Ea.	38.50	87.61	126.11
Vent protector / breather, 2" dia.	1.000	Ea.	23	20	43
Corrosion resistance wrap & coat, small diam pipe, 2" diam, add	36.000	Ea.	66.24		66.24
Concrete hold down pad, 6' x 5' x 8" thick	30.000	Ea.	84.90	43.20	128.10
600 gallon capacity	1.000	Ea.	7,375	480	7,855
Tank, for hold-downs 500-4000 gal., add	1.000	Ea.	470	153	623
Foot valve, single poppet, 3/4" dia	1.000	Ea.	62.50	35.50	98
Fuel fill box, locking inner cover, 3" dia.	1.000	Ea.	130	129	259
Fuel oil specialties, valve,ball chk, globe type fusible, 3/4" dia	2.000	Ea.	147	64	211
TOTAL			19,458.14	6,338.61	25,796.75

G3060 310	Fiberglass Fuel Tank, Double Wall	MAT.	INST.	TOTAL
1000	For alternate storage tanks:			
1010	Storage tank, fuel, underground, double wall fiberglass, 550 Gal.	19,500	6,325	25,825
1020	2500 Gal.	29,600	8,800	38,400
1030	4000 Gal.	36,100	10,900	47,000
1040	6000 Gal.	37,100	12,300	49,400
1050	8000 Gal.	45,600	14,900	60,500

G30 Site Mechanical Utilities

G3060 Fuel Distribution

G3060 310	Fiberglass Fuel Tank, Double Wall	COST EACH		
		MAT.	INST.	TOTAL
1060	10,000 Gal.	51,500	17,200	68,700
1070	15,000 Gal.	69,000	19,700	88,700
1080	20,000 Gal.	86,000	24,000	110,000
1090	25,000 Gal.	107,500	27,900	135,400
1100	30,000 Gal.	122,500	29,600	152,100
2700	Cut & patch to match existing construction, add, minimum	2%	3%	
2800	Maximum	5%	9%	
2900	Dust protection, add, minimum	1%	2%	
3000	Maximum	4%	11%	
3100	Equipment usage curtailment, add, minimum	1%	1%	
3200	Maximum	3%	10%	
3300	Material handling & storage limitation, add, minimum	1%	1%	
3400	Maximum	6%	7%	
3500	Protection of existing work, add, minimum	2%	2%	
3600	Maximum	5%	7%	
3700	Shift work requirements, add, minimum		5%	
3800	Maximum		30%	
3900	Temporary shoring and bracing, add, minimum	2%	5%	
4000	Maximum	5%	12%	

G30 Site Mechanical Utilities

G3060 Fuel Distribution

Steel Underground Fuel Storage Tank, Single Wall

System Components	QUANTITY	UNIT	COST EACH		
			MAT.	INST.	TOTAL
Double wall underground steel storage tank including, excavation, hold down slab, backfill, manway extension and related piping.					
Fiberglass-reinforced plastic pipe, 2 inch	156.000	Ea.	873.60	661.44	1,535.04
Fiberglass-reinforced plastic pipe, 3 inch	36.000	Ea.	345.60	172.80	518.40
Fiberglass-reinforced elbows 90 & 45 degree, 3 inch	3.000	Ea.	207	300	507
Tank leak detection system, probes, well monitoring, liquid phase detection	2.000	Ea.	940		940
Flange, weld neck, 150 lb, 3" pipe size	1.000	Ea.	38.50	87.61	126.11
Vent protector/breather, 2" dia.	1.000	Ea.	23	20	43
Concrete hold down pad, 6' x 5' x 8" thick	30.000	Ea.	84.90	43.20	128.10
Tanks, for hold-downs, 500-2000 gal, add	1.000	Ea.	166	153	319
Double wall steel tank annular space	1.000	Ea.	330		330
Foot valve, single poppet, 3/4" dia	1.000	Ea.	62.50	35.50	98
Fuel fill box, locking inner cover, 3" dia.	1.000	Ea.	130	129	259
Fuel oil specialties, valve, ball chk, globe type fusible, 3/4" dia	2.000	Ea.	147	64	211
Pea gravel	9.000	Ea.	679.50	427.50	1,107
Reinforcing in hold down pad, #3 to #7	120.000	Ea.	52.80	72	124.80
Tank, for manways, add	1.000	Ea.	1,650		1,650
Tank leak detector system, 8 channel, external monitoring	1.000	Ea.	885		885
Tubbing copper, type L, 3/4" diam	120.000	Ea.	505.20	1,020	1,525.20
Elbow 90 degree, wrought cu x cu 3/4" diam	8.000	Ea.	28.80	272	300.80
Pipe, black steel welded, sch 40, 3" diam	30.000	Ea.	376.50	855	1,231.50
Elbow 90 degree, steel welded butt joint, 3" diam	3.000	Ea.	111	525.60	636.60
Excavating trench, no sheeting or dewatering, 6'-10' D, 3/4 CY hyd backhoe	18.000	Ea.		138.96	138.96
Fiberglass-reinforced elbows 90 & 45 degree, 2 inch	7.000	Ea.	395.50	525	920.50
Fiberglass-reinforced threaded adapters, 2 inch	2.000	Ea.	37.90	172	209.90
500 gallon capacity	1.000	Ea.	4,300	480	4,780
Remote tank gauging system, 30', 5" pointer travel	1.000	Ea.	5,050	320	5,370
TOTAL			17,420.30	6,474.61	23,894.91

G3060 320	Steel Fuel Tank, Double Wall	COST EACH		
		MAT.	INST.	TOTAL
1000	For alternate storage tanks:			
1010	Storage tank, fuel, underground, double wall steel, 500 Gal.	17,400	6,475	23,875
1020	2000 Gal.	24,100	8,500	32,600
1030	4000 Gal.	33,800	11,000	44,800
1040	6000 Gal.	38,500	12,200	50,700
1050	8000 Gal.	44,500	15,100	59,600
1060	10,000 Gal.	51,000	17,400	68,400
1070	15,000 Gal.	61,500	20,300	81,800

G30 Site Mechanical Utilities

G3060 Fuel Distribution

G3060 320	Steel Fuel Tank, Double Wall	COST EACH		
		MAT.	INST.	TOTAL
1080	20,000 Gal.	80,500	24,300	104,800
1090	25,000 Gal.	107,000	27,800	134,800
1100	30,000 Gal.	122,500	29,700	152,200
1110	40,000 Gal.	153,500	37,400	190,900
1120	50,000 Gal.	185,000	44,800	229,800

G3060 325	Steel Tank, Above Ground, Single Wall	COST EACH		
		MAT.	INST.	TOTAL
1010	Storage tank, fuel, above ground, single wall steel, 550 Gal.	11,600	4,550	16,150
1020	2000 Gal.	14,500	4,650	19,150
1030	5000 Gal.	17,400	7,125	24,525
1040	10,000 Gal.	30,700	7,575	38,275
1050	15,000 Gal.	33,500	7,800	41,300
1060	20,000 Gal.	39,300	8,050	47,350
1070	25,000 Gal.	45,800	8,225	54,025
1080	30,000 Gal.	50,000	8,575	58,575

G3060 330	Steel Tank, Above Ground, Double Wall	COST EACH		
		MAT.	INST.	TOTAL
1010	Storage tank, fuel, above ground, double wall steel, 500 Gal.	13,000	4,600	17,600
1020	2000 Gal.	19,500	4,700	24,200
1030	4000 Gal.	26,900	4,800	31,700
1040	6000 Gal.	31,300	7,375	38,675
1050	8000 Gal.	37,100	7,575	44,675
1060	10,000 Gal.	40,100	7,700	47,800
1070	15,000 Gal.	54,500	7,975	62,475
1080	20,000 Gal.	60,000	8,225	68,225
1090	25,000 Gal.	70,000	8,475	78,475
1100	30,000 Gal.	76,000	8,800	84,800
2700	Cut & patch to match existing construction, add, minimum	2%	3%	
2800	Maximum	5%	9%	
2900	Dust protection, add, minimum	1%	2%	
3000	Maximum	4%	11%	
3100	Equipment usage curtailment, add, minimum	1%	1%	
3200	Maximum	3%	10%	
3300	Material handling & storage limitation, add, minimum	1%	1%	
3400	Maximum	6%	7%	
3500	Protection of existing work, add, minimum	2%	2%	
3600	Maximum	5%	7%	
3700	Shift work requirements, add, minimum		5%	
3800	Maximum		30%	
3900	Temporary shoring and bracing, add, minimum	2%	5%	
4000	Maximum	5%	12%	

G40 Site Electrical Utilities

G4020 Site Lighting

This page illustrates and describes light poles for parking or walkway area lighting. Included are aluminum or steel light poles, single or multiple fixture bracket arms, excavation, concrete footing, backfill and compaction. Lines within system components give the unit price and total price on a cost per each basis. Prices for alternate systems are shown on lines G4020 210 0200 through 1440. Both material quantities and labor costs have been adjusted for the system listed.

Factors: To adjust for job conditions other than normal working situations use lines G4020 210 2700 through 3600.

Example: You are to install the system where you must be careful not to damage existing walks or parking areas. Go to line G4020 210 3300 and apply these percentages to the appropriate MAT. and INST. costs.

System Components	QUANTITY	UNIT	COST EACH		
			MAT.	INST.	TOTAL
Light poles, aluminum, 20' high, 1 arm bracket					
Aluminum light pole, 20', no concrete base	1.000	Ea.	965	567.50	1,532.50
Bracket arm for Aluminum light pole	1.000	Ea.	127	75.50	202.50
Excavation by hand, pits to 6' deep, heavy soil or clay	2.368	C.Y.		258.11	258.11
Footing, concrete incl forms, reinforcing, spread, under 1 C.Y.	.465	C.Y.	76.73	122.39	199.12
Backfill by hand	1.903	C.Y.		75.17	75.17
Compaction vibrating plate	1.903	C.Y.		10.20	10.20
TOTAL			1,168.73	1,108.87	2,277.60

G4020 210	Light Pole (Installed)	COST EACH		
		MAT.	INST.	TOTAL
0195	For alternate light pole systems:			
0200	Light pole, aluminum, 20' high, 1 arm bracket	1,175	1,100	2,275
0240	2 arm brackets	1,300	1,100	2,400
0280	3 arm brackets	1,425	1,150	2,575
0320	4 arm brackets	1,550	1,150	2,700
0360	30' high, 1 arm bracket	2,100	1,400	3,500
0400	2 arm brackets	2,225	1,400	3,625
0440	3 arm brackets	2,350	1,425	3,775
0480	4 arm brackets	2,475	1,425	3,900
0680	40' high, 1 arm bracket	2,575	1,875	4,450
0720	2 arm brackets	2,700	1,875	4,575
0760	3 arm brackets	2,825	1,900	4,725
0800	4 arm brackets	2,950	1,900	4,850
0840	Steel, 20' high, 1 arm bracket	1,425	1,175	2,600
0880	2 arm brackets	1,525	1,175	2,700
0920	3 arm brackets	1,550	1,200	2,750
0960	4 arm brackets	1,675	1,200	2,875
1000	30' high, 1 arm bracket	1,625	1,500	3,125
1040	2 arm brackets	1,725	1,500	3,225
1080	3 arm brackets	1,750	1,525	3,275
1120	4 arm brackets	1,875	1,525	3,400
1320	40' high, 1 arm bracket	2,125	2,025	4,150
1360	2 arm brackets	2,225	2,025	4,250
1400	3 arm brackets	2,250	2,050	4,300
1440	4 arm brackets	2,375	2,050	4,425
2700	Cut & patch to match existing construction, add, minimum	2%	3%	

G40 Site Electrical Utilities

G4020 Site Lighting

G4020 210	Light Pole (Installed)	COST EACH		
		MAT.	INST.	TOTAL
2800	Maximum	5%	9%	
2900	Equipment usage curtailment, add, minimum	1%	1%	
3000	Maximum	3%	10%	
3100	Material handling & storage limitation, add, minimum	1%	1%	
3200	Maximum	6%	7%	
3300	Protection of existing work, add, minimum	2%	2%	
3400	Maximum	5%	7%	
3500	Shift work requirements, add, minimum		5%	
3600	Maximum		30%	

Reference Section

All the reference information is in one section, making it easy to find what you need to know ... and easy to use the book on a daily basis. This section is visually identified by a vertical gray bar on the page edges.

In this Reference Section, we've included Equipment Rental Costs, a listing of rental and operating costs; Crew Listings, a full listing of all crews and equipment, and their costs; Historical Cost Indexes for cost comparisons over time; City Cost Indexes and Location Factors for adjusting costs to the region you are in; Reference Tables, where you will find explanations, estimating information and procedures, or technical data; Change Orders, information on pricing changes to contract documents; Square Foot Costs that allow you to make a rough estimate for the overall cost of a project; and an explanation of all the Abbreviations in the book.

Table of Contents

Construction Equipment Rental Costs	631
Crew Listings	643
Historical Cost Indexes	673
City Cost Indexes	674
Location Factors	694
Reference Tables	700
R01 General Requirements	700
R02 Existing Conditions	709
R03 Concrete	711
R04 Masonry	719
R05 Metals	720
R06 Wood, Plastics & Composites	722
R07 Thermal & Moisture Protection	722
R08 Openings	724
R09 Finishes	726
R14 Conveying Equipment	728

Reference Tables (cont.)

R22 Plumbing	729
R23 Heating, Ventilating & Air Conditioning	730
R26 Electrical	731
R31 Earthwork	732
R32 Exterior Improvements	734
R33 Utilities	735
Change Orders	736
Square Foot Costs	741
Abbreviations	751

Equipment Rental Costs

Estimating Tips

- This section contains the average costs to rent and operate hundreds of pieces of construction equipment. This is useful information when estimating the time and material requirements of any particular operation in order to establish a unit or total cost. Equipment costs include not only rental, but also operating costs for equipment under normal use.

Rental Costs

- Equipment rental rates are obtained from industry sources throughout North America-contractors, suppliers, dealers, manufacturers, and distributors.
- Rental rates vary throughout the country, with larger cities generally having lower rates. Lease plans for new equipment are available for periods in excess of six months, with a percentage of payments applying toward purchase.
- Monthly rental rates vary from 2% to 5% of the purchase price of the equipment depending on the anticipated life of the equipment and its wearing parts.
- Weekly rental rates are about 1/3 the monthly rates, and daily rental rates are about 1/3 the weekly rate.

Operating Costs

- The operating costs include parts and labor for routine servicing, such as repair and replacement of pumps, filters and worn lines. Normal operating expendables, such as fuel, lubricants, tires and electricity (where applicable), are also included.
- Extraordinary operating expendables with highly variable wear patterns, such as diamond bits and blades, are excluded. These costs can be found as material costs in the Unit Price section.
- The hourly operating costs listed do not include the operator's wages.

Equipment Cost/Day

- Any power equipment required by a crew is shown in the Crew Listings with a daily cost.
- The daily cost of equipment needed by a crew is based on dividing the weekly rental rate by 5 (number of working days in the week), and then adding the hourly operating cost times 8 (the number of hours in a day). This "Equipment Cost/Day" is shown in the far right column of the Equipment Rental pages.
- If equipment is needed for only one or two days, it is best to develop your own cost by including components for daily rent and hourly operating cost. This is important when the listed Crew for a task does not contain the equipment needed, such as a crane for lifting mechanical heating/cooling equipment up onto a roof.
- If the quantity of work is less than the crew's Daily Output shown for a Unit Price line item that includes a bare unit equipment cost, it is recommended to estimate one day's rental cost and operating cost for equipment shown in the Crew Listing for that line item.

Mobilization/Demobilization

- The cost to move construction equipment from an equipment yard or rental company to the jobsite and back again is not included in equipment rental costs listed in the Reference section, nor in the bare equipment cost of any Unit Price line item, nor in any equipment costs shown in the Crew listings.
- Mobilization (to the site) and demobilization (from the site) costs can be found in the Unit Price section.
- If a piece of equipment is already at the jobsite, it is not appropriate to utilize mobil./demob. costs again in an estimate.

No part of this publication may be reproduced, stored in a retrieval system, or transmitted in any form or by any means without prior written permission of R.S. Means Company, Inc.

01 54 | Construction Aids

01 54 33 | Equipment Rental

		UNIT	HOURLY OPER. COST	RENT PER DAY	RENT PER WEEK	RENT PER MONTH	EQUIPMENT COST/DAY	
10	0010 **CONCRETE EQUIPMENT RENTAL** without operators R015433-10							10
	0200 Bucket, concrete lightweight, 1/2 C.Y.	Ea.	.70	20	60	180	17.60	
	0300 1 C.Y.		.75	24.50	73	219	20.60	
	0400 1-1/2 C.Y.		.95	33	99	297	27.40	
	0500 2 C.Y.		1.05	40	120	360	32.40	
	0580 8 C.Y.		5.65	260	780	2,350	201.20	
	0600 Cart, concrete, self-propelled, operator walking, 10 C.F.		3.05	56.50	170	510	58.40	
	0700 Operator riding, 18 C.F.		5.00	90	270	810	94	
	0800 Conveyer for concrete, portable, gas, 16" wide, 26' long		11.95	123	370	1,100	169.60	
	0900 46' long		12.30	150	450	1,350	188.40	
	1000 56' long		12.45	158	475	1,425	194.60	
	1100 Core drill, electric, 2-1/2 H.P., 1" to 8" bit diameter		1.75	65.50	197	590	53.40	
	1150 11 H.P., 8" to 18" cores		5.60	115	345	1,025	113.80	
	1200 Finisher, concrete floor, gas, riding trowel, 96" wide		12.35	145	435	1,300	185.80	
	1300 Gas, walk-behind, 3 blade, 36" trowel		2.05	19.65	59	177	28.20	
	1400 4 blade, 48" trowel		4.10	27	81	243	49	
	1500 Float, hand-operated (Bull float) 48" wide		.08	13.65	41	123	8.85	
	1570 Curb builder, 14 H.P., gas, single screw		13.55	247	740	2,225	256.40	
	1590 Double screw		14.20	287	860	2,575	285.60	
	1600 Grinder, concrete and terrazzo, electric, floor		2.48	104	312	935	82.25	
	1700 Wall grinder		1.24	52	156	470	41.10	
	1800 Mixer, powered, mortar and concrete, gas, 6 C.F., 18 H.P.		8.20	118	355	1,075	136.60	
	1900 10 C.F., 25 H.P.		10.30	143	430	1,300	168.40	
	2000 16 C.F.		10.60	167	500	1,500	184.80	
	2100 Concrete, stationary, tilt drum, 2 C.Y.		6.30	232	695	2,075	189.40	
	2120 Pump, concrete, truck mounted 4" line 80' boom		23.35	920	2,760	8,275	738.80	
	2140 5" line, 110' boom		30.30	1,225	3,670	11,000	976.40	
	2160 Mud jack, 50 C.F. per hr		6.95	128	385	1,150	132.60	
	2180 225 C.F. per hr		9.00	147	440	1,325	160	
	2190 Shotcrete pump rig, 12 C.Y./hr		14.15	227	680	2,050	249.20	
	2200 35 C.Y./hr		16.75	243	730	2,200	280	
	2600 Saw, concrete, manual, gas, 18 H.P.		6.35	38.50	115	345	73.80	
	2650 Self-propelled, gas, 30 H.P.		12.80	98.50	295	885	161.40	
	2700 Vibrators, concrete, electric, 60 cycle, 2 H.P.		.45	7.35	22	66	8	
	2800 3 H.P.		.67	10.35	31	93	11.55	
	2900 Gas engine, 5 H.P.		1.95	14.35	43	129	24.20	
	3000 8 H.P.		2.65	15.35	46	138	30.40	
	3050 Vibrating screed, gas engine, 8 H.P.		2.72	72.50	217	650	65.15	
	3120 Concrete transit mixer, 6 x 4, 250 H.P., 8 C.Y., rear discharge		59.20	575	1,725	5,175	818.60	
	3200 Front discharge		69.95	710	2,125	6,375	984.60	
	3300 6 x 6, 285 H.P., 12 C.Y., rear discharge		69.00	665	2,000	6,000	952	
	3400 Front discharge		72.20	715	2,150	6,450	1,008	
20	0010 **EARTHWORK EQUIPMENT RENTAL** without operators R015433-10	Ea.						20
	0040 Aggregate spreader, push type 8' to 12' wide		2.90	25.50	77	231	38.60	
	0045 Tailgate type, 8' wide		2.75	33.50	100	300	42	
	0055 Earth auger, truck-mounted, for fence & sign posts, utility poles		18.90	495	1,490	4,475	449.20	
	0060 For borings and monitoring wells		43.85	665	1,995	5,975	749.80	
	0070 Portable, trailer mounted		3.05	25.50	77	231	39.80	
	0075 Truck-mounted, for caissons, water wells		96.05	3,050	9,115	27,300	2,591	
	0080 Horizontal boring machine, 12" to 36" diameter, 45 H.P.		22.80	200	600	1,800	302.40	
	0090 12" to 48" diameter, 65 H.P.		32.65	355	1,060	3,175	473.20	
	0095 Auger, for fence posts, gas engine, hand held		.55	5.65	17	51	7.80	
	0100 Excavator, diesel hydraulic, crawler mounted, 1/2 C.Y. cap.		20.20	375	1,125	3,375	386.60	
	0120 5/8 C.Y. capacity		30.30	535	1,610	4,825	564.40	
	0140 3/4 C.Y. capacity		37.85	625	1,875	5,625	677.80	
	0150 1 C.Y. capacity		47.70	700	2,100	6,300	801.60	
	0200 1-1/2 C.Y. capacity		57.00	935	2,805	8,425	1,017	
	0300 2 C.Y. capacity		76.90	1,175	3,530	10,600	1,321	

01 54 | Construction Aids

01 54 33 | Equipment Rental

		UNIT	HOURLY OPER. COST	RENT PER DAY	RENT PER WEEK	RENT PER MONTH	EQUIPMENT COST/DAY	
0320	2-1/2 C.Y. capacity	Ea.	104.45	1,575	4,740	14,200	1,784	
0325	3-1/2 C.Y. capacity		138.85	2,075	6,240	18,700	2,359	
0330	4-1/2 C.Y. capacity		166.15	2,575	7,705	23,100	2,870	
0335	6 C.Y. capacity		211.30	2,825	8,500	25,500	3,390	
0340	7 C.Y. capacity		216.05	3,050	9,125	27,400	3,553	
0342	Excavator attachments, bucket thumbs		2.95	238	715	2,150	166.60	
0345	Grapples		2.65	207	620	1,850	145.20	
0347	Hydraulic hammer for boom mounting, 5000 ft lbs		12.45	410	1,225	3,675	344.60	
0349	11,000 ft lbs		21.05	735	2,200	6,600	608.40	
0350	Gradall type, truck mounted, 3 ton @ 15' radius, 5/8 C.Y.		38.60	840	2,520	7,550	812.80	
0370	1 C.Y. capacity		43.55	1,025	3,050	9,150	958.40	
0400	Backhoe-loader, 40 to 45 H.P., 5/8 C.Y. capacity		12.80	193	580	1,750	218.40	
0450	45 H.P. to 60 H.P., 3/4 C.Y. capacity		20.35	292	875	2,625	337.80	
0460	80 H.P., 1-1/4 C.Y. capacity		24.45	320	960	2,875	387.60	
0470	112 H.P., 1-1/2 C.Y. capacity		38.55	525	1,570	4,700	622.40	
0482	Backhoe-loader attachment, compactor, 20,000 lb.		5.25	142	425	1,275	127	
0485	Hydraulic hammer, 750 ft lbs		2.95	91.50	275	825	78.60	
0486	Hydraulic hammer, 1200 ft lbs		5.40	185	555	1,675	154.20	
0500	Brush chipper, gas engine, 6" cutter head, 35 H.P.		10.65	108	325	975	150.20	
0550	12" cutter head, 130 H.P.		18.15	180	540	1,625	253.20	
0600	15" cutter head, 165 H.P.		27.75	217	650	1,950	352	
0750	Bucket, clamshell, general purpose, 3/8 C.Y.		1.15	36.50	110	330	31.20	
0800	1/2 C.Y.		1.25	45	135	405	37	
0850	3/4 C.Y.		1.40	55	165	495	44.20	
0900	1 C.Y.		1.45	58.50	175	525	46.60	
0950	1-1/2 C.Y.		2.35	80	240	720	66.80	
1000	2 C.Y.		2.45	90	270	810	73.60	
1010	Bucket, dragline, medium duty, 1/2 C.Y.		.65	23.50	71	213	19.40	
1020	3/4 C.Y.		.70	24.50	74	222	20.40	
1030	1 C.Y.		.70	26.50	80	240	21.60	
1040	1-1/2 C.Y.		1.10	40	120	360	32.80	
1050	2 C.Y.		1.20	45	135	405	36.60	
1070	3 C.Y.		1.80	63.50	190	570	52.40	
1200	Compactor, manually guided 2-drum vibratory smooth roller, 7.5 H.P.		6.20	167	500	1,500	149.60	
1250	Rammer/tamper, gas, 8"		2.60	43.50	130	390	46.80	
1260	15"		2.85	48.50	145	435	51.80	
1300	Vibratory plate, gas, 18" plate, 3000 lb. blow		2.55	23	69	207	34.20	
1350	21" plate, 5000 lb. blow		3.20	30	90	270	43.60	
1370	Curb builder/extruder, 14 H.P., gas, single screw		13.55	247	740	2,225	256.40	
1390	Double screw		14.20	287	860	2,575	285.60	
1500	Disc harrow attachment, for tractor		.40	67	201	605	43.40	
1810	Feller buncher, shearing & accumulating trees, 100 H.P.		34.20	560	1,680	5,050	609.60	
1860	Grader, self-propelled, 25,000 lb.		31.75	475	1,420	4,250	538	
1910	30,000 lb.		36.65	550	1,645	4,925	622.20	
1920	40,000 lb.		55.90	1,050	3,120	9,350	1,071	
1930	55,000 lb.		72.55	1,375	4,155	12,500	1,411	
1950	Hammer, pavement breaker, self-propelled, diesel, 1000 to 1250 lb.		28.25	345	1,040	3,125	434	
2000	1300 to 1500 lb.		42.38	695	2,080	6,250	755.05	
2050	Pile driving hammer, steam or air, 4150 ft lbs @ 225 bpm		9.65	470	1,405	4,225	358.20	
2100	8750 ft lbs @ 145 bpm		11.65	650	1,945	5,825	482.20	
2150	15,000 ft lbs @ 60 bpm		13.20	785	2,360	7,075	577.60	
2200	24,450 ft lbs @ 111 bpm		14.25	880	2,635	7,900	641	
2250	Leads, 60' high for pile driving hammers up to 20,000 ft lbs		3.00	78	234	700	70.80	
2300	90' high for hammers over 20,000 ft lbs		4.50	141	422	1,275	120.40	
2350	Diesel type hammer, 22,400 ft lbs		29.80	635	1,905	5,725	619.40	
2400	41,300 ft lbs		38.60	690	2,070	6,200	722.80	
2450	141,000 ft lbs		56.40	1,175	3,540	10,600	1,159	
2500	Vib. elec. hammer/extractor, 200 kW diesel generator, 34 H.P.		53.70	660	1,980	5,950	825.60	

01 54 | Construction Aids

01 54 33 | Equipment Rental

		UNIT	HOURLY OPER. COST	RENT PER DAY	RENT PER WEEK	RENT PER MONTH	EQUIPMENT COST/DAY	
2550	80 H.P.	Ea.	99.00	970	2,910	8,725	1,374	20
2600	150 H.P.		188.90	1,875	5,650	17,000	2,641	
2700	Extractor, steam or air, 700 ft lbs		17.50	510	1,525	4,575	445	
2750	1000 ft lbs		19.95	625	1,870	5,600	533.60	
2800	Log chipper, up to 22" diameter, 600 H.P.		52.70	445	1,340	4,025	689.60	
2850	Logger, for skidding & stacking logs, 150 H.P.		50.90	835	2,500	7,500	907.20	
2860	Mulcher, diesel powered, trailer mounted		23.50	218	655	1,975	319	
2900	Rake, spring tooth, with tractor		14.34	285	855	2,575	285.70	
3000	Roller, vibratory, tandem, smooth drum, 20 H.P.		8.25	143	430	1,300	152	
3050	35 H.P.		10.40	253	760	2,275	235.20	
3100	Towed type vibratory compactor, smooth drum, 50 H.P.		25.90	380	1,135	3,400	434.20	
3150	Sheepsfoot, 50 H.P.		28.15	440	1,320	3,950	489.20	
3170	Landfill compactor, 220 H.P.		82.85	1,475	4,390	13,200	1,541	
3200	Pneumatic tire roller, 80 H.P.		16.15	345	1,040	3,125	337.20	
3250	120 H.P.		24.20	610	1,835	5,500	560.60	
3300	Sheepsfoot vibratory roller, 240 H.P.		69.85	1,150	3,420	10,300	1,243	
3320	340 H.P.		94.00	1,400	4,220	12,700	1,596	
3350	Smooth drum vibratory roller, 75 H.P.		23.20	550	1,655	4,975	516.60	
3400	125 H.P.		29.90	655	1,970	5,900	633.20	
3410	Rotary mower, brush, 60", with tractor		18.25	260	780	2,350	302	
3420	Rototiller, walk-behind, gas, 5 H.P.		2.12	65.50	196	590	56.15	
3422	8 H.P.		3.30	100	300	900	86.40	
3440	Scrapers, towed type, 7 C.Y. capacity		5.15	107	320	960	105.20	
3450	10 C.Y. capacity		6.45	175	525	1,575	156.60	
3500	15 C.Y. capacity		7.15	192	575	1,725	172.20	
3525	Self-propelled, single engine, 14 C.Y. capacity		103.95	1,475	4,425	13,300	1,717	
3550	Dual engine, 21 C.Y. capacity		153.65	1,700	5,065	15,200	2,242	
3600	31 C.Y. capacity		208.65	2,575	7,750	23,300	3,219	
3640	44 C.Y. capacity		264.30	3,625	10,910	32,700	4,296	
3650	Elevating type, single engine, 11 C.Y. capacity		66.70	1,075	3,255	9,775	1,185	
3700	22 C.Y. capacity		126.05	2,125	6,375	19,100	2,283	
3710	Screening plant 110 H.P. w/5' x 10' screen		36.90	415	1,240	3,725	543.20	
3720	5' x 16' screen		39.05	515	1,550	4,650	622.40	
3850	Shovel, crawler-mounted, front-loading, 7 C.Y. capacity		228.10	2,850	8,555	25,700	3,536	
3855	12 C.Y. capacity		305.30	3,625	10,860	32,600	4,614	
3860	Shovel/backhoe bucket, 1/2 C.Y.		2.20	63.50	190	570	55.60	
3870	3/4 C.Y.		2.25	70	210	630	60	
3880	1 C.Y.		2.35	78.50	235	705	65.80	
3890	1-1/2 C.Y.		2.50	93.50	280	840	76	
3910	3 C.Y.		2.80	127	380	1,150	98.40	
3950	Stump chipper, 18" deep, 30 H.P.		7.56	217	652	1,950	190.90	
4110	Tractor, crawler, with bulldozer, torque converter, diesel 80 H.P.		26.45	375	1,125	3,375	436.60	
4150	105 H.P.		39.30	575	1,730	5,200	660.40	
4200	140 H.P.		42.45	715	2,150	6,450	769.60	
4260	200 H.P.		68.60	1,075	3,215	9,650	1,192	
4310	300 H.P.		82.50	1,550	4,660	14,000	1,592	
4360	410 H.P.		109.70	2,000	6,005	18,000	2,079	
4370	500 H.P.		147.85	2,700	8,135	24,400	2,810	
4380	700 H.P.		245.45	4,400	13,170	39,500	4,598	
4400	Loader, crawler, torque conv., diesel, 1-1/2 C.Y., 80 H.P.		24.10	400	1,205	3,625	433.80	
4450	1-1/2 to 1-3/4 C.Y., 95 H.P.		28.20	515	1,545	4,625	534.60	
4510	1-3/4 to 2-1/4 C.Y., 130 H.P.		42.30	835	2,505	7,525	839.40	
4530	2-1/2 to 3-1/4 C.Y., 190 H.P.		56.50	1,050	3,125	9,375	1,077	
4560	3-1/2 to 5 C.Y., 275 H.P.		81.25	1,500	4,490	13,500	1,548	
4610	Front end loader, 4WD, articulated frame, 1 to 1-1/4 C.Y., 70 H.P.		18.05	208	625	1,875	269.40	
4620	1-1/2 to 1-3/4 C.Y., 95 H.P.		23.95	293	880	2,650	367.60	
4650	1-3/4 to 2 C.Y., 130 H.P.		27.65	345	1,035	3,100	428.20	
4710	2-1/2 to 3-1/2 C.Y., 145 H.P.		26.35	375	1,125	3,375	435.80	

01 54 | Construction Aids

01 54 33 | Equipment Rental

		UNIT	HOURLY OPER. COST	RENT PER DAY	RENT PER WEEK	RENT PER MONTH	EQUIPMENT COST/DAY	
20 4730	3 to 4-1/2 C.Y., 185 H.P.	Ea.	35.85	570	1,705	5,125	627.80	20
4760	5-1/4 to 5-3/4 C.Y., 270 H.P.		56.30	905	2,720	8,150	994.40	
4810	7 to 9 C.Y., 475 H.P.		97.15	1,625	4,900	14,700	1,757	
4870	9 - 11 C.Y., 620 H.P.		129.55	2,600	7,825	23,500	2,601	
4880	Skid steer loader, wheeled, 10 C.F., 30 H.P. gas		9.30	145	435	1,300	161.40	
4890	1 C.Y., 78 H.P., diesel		19.00	232	695	2,075	291	
4892	Skid-steer attachment, auger		.55	91.50	275	825	59.40	
4893	Backhoe		.67	112	336	1,000	72.55	
4894	Broom		.73	121	364	1,100	78.65	
4895	Forks		.24	40	120	360	25.90	
4896	Grapple		.57	95.50	286	860	61.75	
4897	Concrete hammer		1.06	177	532	1,600	114.90	
4898	Tree spade		.82	136	408	1,225	88.15	
4899	Trencher		.79	131	393	1,175	84.90	
4900	Trencher, chain, boom type, gas, operator walking, 12 H.P.		4.75	46.50	140	420	66	
4910	Operator riding, 40 H.P.		19.55	297	890	2,675	334.40	
5000	Wheel type, diesel, 4' deep, 12" wide		82.05	825	2,480	7,450	1,152	
5100	6' deep, 20" wide		86.95	1,900	5,735	17,200	1,843	
5150	Chain type, diesel, 5' deep, 8" wide		36.25	580	1,745	5,225	639	
5200	Diesel, 8' deep, 16" wide		88.95	2,300	6,890	20,700	2,090	
5202	Rock trencher, wheel type, 6" wide x 18" deep		19.55	315	940	2,825	344.40	
5206	Chain type, 18" wide x 7' deep		106.25	2,875	8,605	25,800	2,571	
5210	Tree spade, self-propelled		13.10	267	800	2,400	264.80	
5250	Truck, dump, 2-axle, 12 ton, 8 C.Y. payload, 220 H.P.		33.35	225	675	2,025	401.80	
5300	Three axle dump, 16 ton, 12 C.Y. payload, 400 H.P.		59.35	320	965	2,900	667.80	
5310	Four axle dump, 25 ton, 18 C.Y. payload, 450 H.P.		69.55	470	1,410	4,225	838.40	
5350	Dump trailer only, rear dump, 16-1/2 C.Y.		4.90	135	405	1,225	120.20	
5400	20 C.Y.		5.35	153	460	1,375	134.80	
5450	Flatbed, single axle, 1-1/2 ton rating		24.85	68.50	205	615	239.80	
5500	3 ton rating		29.75	96.50	290	870	296	
5550	Off highway rear dump, 25 ton capacity		74.80	1,325	4,005	12,000	1,399	
5600	35 ton capacity		70.50	1,175	3,515	10,500	1,267	
5610	50 ton capacity		95.65	1,625	4,895	14,700	1,744	
5620	65 ton capacity		99.70	1,625	4,860	14,600	1,770	
5630	100 ton capacity		128.25	2,125	6,350	19,100	2,296	
6000	Vibratory plow, 25 H.P., walking		8.15	61.50	185	555	102.20	
40 0010	**GENERAL EQUIPMENT RENTAL** without operators							40
0150	Aerial lift, scissor type, to 15' high, 1000 lb. cap., electric	Ea.	2.80	55	165	495	55.40	
0160	To 25' high, 2000 lb. capacity		3.20	70	210	630	67.60	
0170	Telescoping boom to 40' high, 500 lb. capacity, gas		19.30	320	965	2,900	347.40	
0180	To 45' high, 500 lb. capacity		20.35	370	1,115	3,350	385.80	
0190	To 60' high, 600 lb. capacity		22.55	490	1,465	4,400	473.40	
0195	Air compressor, portable, 6.5 cfm, electric		.50	13	39	117	11.80	
0196	Gasoline		.75	19.65	59	177	17.80	
0200	Towed type, gas engine, 60 cfm		13.70	48.50	145	435	138.60	
0300	160 cfm		16.00	50	150	450	158	
0400	Diesel engine, rotary screw, 250 cfm		16.30	105	315	945	193.40	
0500	365 cfm		22.20	130	390	1,175	255.60	
0550	450 cfm		28.25	163	490	1,475	324	
0600	600 cfm		49.20	228	685	2,050	530.60	
0700	750 cfm		50.10	237	710	2,125	542.80	
0800	For silenced models, small sizes, add to rent		3%	5%	5%	5%		
0900	Large sizes, add to rent		5%	7%	7%	7%		
0930	Air tools, breaker, pavement, 60 lb.	Ea.	.45	9.65	29	87	9.40	
0940	80 lb.		.45	10.35	31	93	9.80	
0950	Drills, hand (jackhammer) 65 lb.		.55	17.35	52	156	14.80	
0960	Track or wagon, swing boom, 4" drifter		57.35	765	2,295	6,875	917.80	
0970	5" drifter		71.65	825	2,480	7,450	1,069	

R015433-10

01 54 | Construction Aids

01 54 33 | Equipment Rental

		UNIT	HOURLY OPER. COST	RENT PER DAY	RENT PER WEEK	RENT PER MONTH	EQUIPMENT COST/DAY	
0975	Track mounted quarry drill, 6" diameter drill	Ea.	96.10	1,275	3,790	11,400	1,527	40
0980	Dust control per drill		1.04	19	57	171	19.70	
0990	Hammer, chipping, 12 lb.		.50	23.50	70	210	18	
1000	Hose, air with couplings, 50' long, 3/4" diameter		.03	4.33	13	39	2.85	
1100	1" diameter		.04	6.35	19	57	4.10	
1200	1-1/2" diameter		.06	9.65	29	87	6.30	
1300	2" diameter		.11	19	57	171	12.30	
1400	2-1/2" diameter		.12	20	60	180	12.95	
1410	3" diameter		.15	25	75	225	16.20	
1450	Drill, steel, 7/8" x 2'		.05	8	24	72	5.20	
1460	7/8" x 6'		.05	9	27	81	5.80	
1520	Moil points		.02	4	12	36	2.55	
1525	Pneumatic nailer w/accessories		.46	30.50	91	273	21.90	
1530	Sheeting driver for 60 lb. breaker		.04	6	18	54	3.90	
1540	For 90 lb. breaker		.12	8	24	72	5.75	
1550	Spade, 25 lb.		.40	6	18	54	6.80	
1560	Tamper, single, 35 lb.		.53	35	105	315	25.25	
1570	Triple, 140 lb.		.79	52.50	158	475	37.90	
1580	Wrenches, impact, air powered, up to 3/4" bolt		.40	9	27	81	8.60	
1590	Up to 1-1/4" bolt		.50	23	69	207	17.80	
1600	Barricades, barrels, reflectorized, 1 to 50 barrels		.03	4.60	13.80	41.50	3	
1610	100 to 200 barrels		.02	3.53	10.60	32	2.30	
1620	Barrels with flashers, 1 to 50 barrels		.03	5.25	15.80	47.50	3.40	
1630	100 to 200 barrels		.03	4.20	12.60	38	2.75	
1640	Barrels with steady burn type C lights		.04	7	21	63	4.50	
1650	Illuminated board, trailer mounted, with generator		.75	118	355	1,075	77	
1670	Portable barricade, stock, with flashers, 1 to 6 units		.03	5.25	15.80	47.50	3.40	
1680	25 to 50 units		.03	4.90	14.70	44	3.20	
1690	Butt fusion machine, electric		12.80	53.50	160	480	134.40	
1695	Electro fusion machine		9.45	38.50	115	345	98.60	
1700	Carts, brick, hand powered, 1000 lb. capacity		.36	60.50	182	545	39.30	
1800	Gas engine, 1500 lb., 7-1/2' lift		4.01	120	359	1,075	103.90	
1822	Dehumidifier, medium, 6 lb/hr, 150 cfm		.98	60	180	540	43.85	
1824	Large, 18 lb/hr, 600 cfm		1.98	122	366	1,100	89.05	
1830	Distributor, asphalt, trailer mounted, 2000 gal., 38 H.P. diesel		9.40	340	1,015	3,050	278.20	
1840	3000 gal., 38 H.P. diesel		10.75	370	1,105	3,325	307	
1850	Drill, rotary hammer, electric		.88	25	75	225	22.05	
1860	Carbide bit, 1-1/2" diameter, add to electric rotary hammer		.04	7.60	22.75	68.50	4.85	
1865	Rotary, crawler, 250 H.P.		141.75	2,075	6,245	18,700	2,383	
1870	Emulsion sprayer, 65 gal., 5 H.P. gas engine		2.83	99	297	890	82.05	
1880	200 gal., 5 H.P. engine		7.45	165	495	1,475	158.60	
1930	Floodlight, mercury vapor, or quartz, on tripod, 1000 watt		.36	14	42	126	11.30	
1940	2000 watt		.63	26	78	234	20.65	
1950	Floodlights, trailer mounted with generator, 1 - 300 watt light		3.35	71.50	215	645	69.80	
1960	2 - 1000 watt lights		4.45	90	270	810	89.60	
2000	4 - 300 watt lights		4.20	88.50	265	795	86.60	
2005	Foam spray rig, incl. box trailer, compressor, generator, proportioner		32.33	490	1,465	4,400	551.65	
2020	Forklift, straight mast, 12' lift, 5000 lb., 2 wheel drive, gas		25.20	200	600	1,800	321.60	
2040	21' lift, 5000 lb., 4 wheel drive, diesel		20.20	247	740	2,225	309.60	
2050	For rough terrain, 42' lift, 35' reach, 9000 lb., 110 H.P.		26.40	445	1,335	4,000	478.20	
2060	For plant, 4 ton capacity, 80 H.P., 2 wheel drive, gas		15.50	102	305	915	185	
2080	10 ton capacity, 120 H.P., 2 wheel drive, diesel		23.40	178	535	1,600	294.20	
2100	Generator, electric, gas engine, 1.5 kW to 3 kW		3.55	11.35	34	102	35.20	
2200	5 kW		4.60	15.35	46	138	46	
2300	10 kW		8.70	36.50	110	330	91.60	
2400	25 kW		10.20	75	225	675	126.60	
2500	Diesel engine, 20 kW		11.50	75	225	675	137	
2600	50 kW		21.55	103	310	930	234.40	

01 54 | Construction Aids

01 54 33 | Equipment Rental

			UNIT	HOURLY OPER. COST	RENT PER DAY	RENT PER WEEK	RENT PER MONTH	EQUIPMENT COST/DAY	
40	2700	100 kW	Ea.	42.45	132	395	1,175	418.60	40
	2800	250 kW		82.90	233	700	2,100	803.20	
	2850	Hammer, hydraulic, for mounting on boom, to 500 ft lbs		2.25	68.50	205	615	59	
	2860	1000 ft lbs		3.80	112	335	1,000	97.40	
	2900	Heaters, space, oil or electric, 50 MBH		1.69	7	21	63	17.70	
	3000	100 MBH		3.06	9.35	28	84	30.10	
	3100	300 MBH		9.79	31.50	94	282	97.10	
	3150	500 MBH		19.75	41.50	125	375	183	
	3200	Hose, water, suction with coupling, 20' long, 2" diameter		.02	3	9	27	1.95	
	3210	3" diameter		.03	4.67	14	42	3.05	
	3220	4" diameter		.03	5.35	16	48	3.45	
	3230	6" diameter		.11	18.35	55	165	11.90	
	3240	8" diameter		.20	33.50	100	300	21.60	
	3250	Discharge hose with coupling, 50' long, 2" diameter		.01	1.33	4	12	.90	
	3260	3" diameter		.02	2.67	8	24	1.75	
	3270	4" diameter		.02	3.67	11	33	2.35	
	3280	6" diameter		.06	9.65	29	87	6.30	
	3290	8" diameter		.20	33.50	100	300	21.60	
	3295	Insulation blower		.22	6	18	54	5.35	
	3300	Ladders, extension type, 16' to 36' long		.14	23	69	207	14.90	
	3400	40' to 60' long		.20	33.50	100	300	21.60	
	3405	Lance for cutting concrete		3.07	104	313	940	87.15	
	3407	Lawn mower, rotary, 22", 5 H.P.		2.12	65	195	585	55.95	
	3408	48" self propelled		3.28	114	341	1,025	94.45	
	3410	Level, laser type, for pipe and sewer leveling		1.46	97	291	875	69.90	
	3430	Electronic		.75	50	150	450	36	
	3440	Laser type, rotating beam for grade control		1.17	78	234	700	56.15	
	3460	Builders level with tripod and rod		.09	14.65	44	132	9.50	
	3500	Light towers, towable, with diesel generator, 2000 watt		4.20	88.50	265	795	86.60	
	3600	4000 watt		4.45	98.50	295	885	94.60	
	3700	Mixer, powered, plaster and mortar, 6 C.F., 7 H.P.		2.55	18.65	56	168	31.60	
	3800	10 C.F., 9 H.P.		2.70	30	90	270	39.60	
	3850	Nailer, pneumatic		.46	30.50	91	273	21.90	
	3900	Paint sprayers complete, 8 cfm		.69	45.50	137	410	32.90	
	4000	17 cfm		1.28	85	255	765	61.25	
	4020	Pavers, bituminous, rubber tires, 8' wide, 50 H.P., diesel		39.45	1,050	3,165	9,500	948.60	
	4030	10' wide, 150 H.P.		85.85	1,700	5,070	15,200	1,701	
	4050	Crawler, 8' wide, 100 H.P., diesel		83.90	1,800	5,420	16,300	1,755	
	4060	10' wide, 150 H.P.		96.80	2,125	6,365	19,100	2,047	
	4070	Concrete paver, 12' to 24' wide, 250 H.P.		96.25	1,575	4,750	14,300	1,720	
	4080	Placer-spreader-trimmer, 24' wide, 300 H.P.		140.15	2,600	7,785	23,400	2,678	
	4100	Pump, centrifugal gas pump, 1-1/2" diam., 65 GPM		3.60	45	135	405	55.80	
	4200	2" diameter, 130 GPM		5.00	51.50	155	465	71	
	4300	3" diameter, 250 GPM		5.30	53.50	160	480	74.40	
	4400	6" diameter, 1500 GPM		29.25	177	530	1,600	340	
	4500	Submersible electric pump, 1-1/4" diameter, 55 GPM		.41	16	48	144	12.90	
	4600	1-1/2" diameter, 83 GPM		.44	18.35	55	165	14.50	
	4700	2" diameter, 120 GPM		1.25	23	69	207	23.80	
	4800	3" diameter, 300 GPM		2.15	41.50	125	375	42.20	
	4900	4" diameter, 560 GPM		9.55	163	490	1,475	174.40	
	5000	6" diameter, 1590 GPM		14.05	220	660	1,975	244.40	
	5100	Diaphragm pump, gas, single, 1-1/2" diameter		1.10	48	144	430	37.60	
	5200	2" diameter		4.00	60	180	540	68	
	5300	3" diameter		4.00	60	180	540	68	
	5400	Double, 4" diameter		5.35	80	240	720	90.80	
	5450	Pressure Washer 5 GPM, 3000 psi		4.60	40	120	360	60.80	
	5500	Trash pump, self-priming, gas, 2" diameter		4.35	20.50	62	186	47.20	
	5600	Diesel, 4" diameter		11.90	58.50	175	525	130.20	

01 54 | Construction Aids

01 54 33 | Equipment Rental

			UNIT	HOURLY OPER. COST	RENT PER DAY	RENT PER WEEK	RENT PER MONTH	EQUIPMENT COST/DAY	
40	5650	Diesel, 6" diameter	Ea.	40.15	130	390	1,175	399.20	40
	5655	Grout Pump		18.10	95	285	855	201.80	
	5700	Salamanders, L.P. gas fired, 100,000 Btu		3.10	16	48	144	34.40	
	5705	50,000 Btu		2.31	9.35	28	84	24.10	
	5720	Sandblaster, portable, open top, 3 C.F. capacity		.55	27	81	243	20.60	
	5730	6 C.F. capacity		.85	40	120	360	30.80	
	5740	Accessories for above		.12	20	60	180	12.95	
	5750	Sander, floor		.85	19	57	171	18.20	
	5760	Edger		.81	28	84	252	23.30	
	5800	Saw, chain, gas engine, 18" long		2.15	18.35	55	165	28.20	
	5900	Hydraulic powered, 36" long		.65	56.50	170	510	39.20	
	5950	60" long		.65	56.50	170	510	39.20	
	6000	Masonry, table mounted, 14" diameter, 5 H.P.		1.34	56.50	170	510	44.70	
	6050	Portable cut-off, 8 H.P.		2.40	32.50	98	294	38.80	
	6100	Circular, hand held, electric, 7-1/4" diameter		.21	4.67	14	42	4.50	
	6200	12" diameter		.27	7.65	23	69	6.75	
	6250	Wall saw, w/hydraulic power, 10 H.P		9.30	60	180	540	110.40	
	6275	Shot blaster, walk-behind, 20" wide		4.60	298	895	2,675	215.80	
	6280	Sidewalk broom, walk-behind		2.05	61.50	185	555	53.40	
	6300	Steam cleaner, 100 gallons per hour		3.50	53.50	161	485	60.20	
	6310	200 gallons per hour		5.05	63.50	190	570	78.40	
	6340	Tar Kettle/Pot, 400 gallons		4.68	80	240	720	85.45	
	6350	Torch, cutting, acetylene-oxygen, 150' hose		.50	25	75	225	19	
	6360	Hourly operating cost includes tips and gas		9.90				79.20	
	6410	Toilet, portable chemical		.12	19.35	58	174	12.55	
	6420	Recycle flush type		.14	23.50	70	210	15.10	
	6430	Toilet, fresh water flush, garden hose,		.16	27	81	243	17.50	
	6440	Hoisted, non-flush, for high rise		.14	23	69	207	14.90	
	6450	Toilet, trailers, minimum		.24	40	120	360	25.90	
	6460	Maximum		.73	122	365	1,100	78.85	
	6465	Tractor, farm with attachment		17.10	267	800	2,400	296.80	
	6500	Trailers, platform, flush deck, 2 axle, 25 ton capacity		4.95	110	330	990	105.60	
	6600	40 ton capacity		6.40	153	460	1,375	143.20	
	6700	3 axle, 50 ton capacity		6.95	170	510	1,525	157.60	
	6800	75 ton capacity		8.65	223	670	2,000	203.20	
	6810	Trailer mounted cable reel for high voltage line work		4.94	235	706	2,125	180.70	
	6820	Trailer mounted cable tensioning rig		9.80	465	1,400	4,200	358.40	
	6830	Cable pulling rig		65.52	2,625	7,860	23,600	2,096	
	6900	Water tank trailer, engine driven discharge, 5000 gallons		6.45	147	440	1,325	139.60	
	6925	10,000 gallons		8.80	203	610	1,825	192.40	
	6950	Water truck, off highway, 6000 gallons		84.65	805	2,420	7,250	1,161	
	7010	Tram car for high voltage line work, powered, 2 conductor		6.81	128	383	1,150	131.10	
	7020	Transit (builder's level) with tripod		.09	14.65	44	132	9.50	
	7030	Trench box, 3000 lbs. 6' x 8'		.56	93	279	835	60.30	
	7040	7200 lbs. 6' x 20'		1.05	175	525	1,575	113.40	
	7050	8000 lbs., 8' x 16'		.95	158	475	1,425	102.60	
	7060	9500 lbs., 8' x 20'		1.19	199	596	1,800	128.70	
	7065	11,000 lbs., 8' x 24'		1.27	212	637	1,900	137.55	
	7070	12,000 lbs., 10' x 20'		1.71	285	855	2,575	184.70	
	7100	Truck, pickup, 3/4 ton, 2 wheel drive		13.25	56.50	170	510	140	
	7200	4 wheel drive		13.50	71.50	215	645	151	
	7250	Crew carrier, 9 passenger		18.80	86.50	260	780	202.40	
	7290	Flat bed truck, 20,000 lbs. GVW		21.20	127	380	1,150	245.60	
	7300	Tractor, 4 x 2, 220 H.P.		29.50	197	590	1,775	354	
	7410	330 H.P.		43.80	270	810	2,425	512.40	
	7500	6 x 4, 380 H.P.		50.30	315	950	2,850	592.40	
	7600	450 H.P.		60.85	380	1,145	3,425	715.80	
	7620	Vacuum truck, hazardous material, 2500 gallons		11.15	305	920	2,750	273.20	

01 54 | Construction Aids

01 54 33 | Equipment Rental

			UNIT	HOURLY OPER. COST	RENT PER DAY	RENT PER WEEK	RENT PER MONTH	EQUIPMENT COST/DAY	
40	7625	5,000 gallons	Ea.	13.23	430	1,290	3,875	363.85	40
	7640	Tractor, with A frame, boom and winch, 225 H.P.		32.15	273	820	2,450	421.20	
	7650	Vacuum, HEPA, 16 gallon, wet/dry		.84	18	54	162	17.50	
	7655	55 gallon, wet/dry		.86	27	81	243	23.10	
	7660	Water tank, portable		.17	28.50	85.50	257	18.45	
	7690	Sewer/catch basin vacuum, 14 C.Y., 1500 gallons		17.78	645	1,940	5,825	530.25	
	7700	Welder, electric, 200 amp		3.86	17.35	52	156	41.30	
	7800	300 amp		5.72	21.50	64	192	58.55	
	7900	Gas engine, 200 amp		14.15	25.50	76	228	128.40	
	8000	300 amp		16.20	27	81	243	145.80	
	8100	Wheelbarrow, any size		.07	11.65	35	105	7.55	
	8200	Wrecking ball, 4000 lbs.		2.15	71.50	215	645	60.20	
50	0010	**HIGHWAY EQUIPMENT RENTAL** without operators	R015433 -10						50
	0050	Asphalt batch plant, portable drum mixer, 100 ton/hr	Ea.	66.15	1,450	4,355	13,100	1,400	
	0060	200 ton/hr		73.90	1,525	4,600	13,800	1,511	
	0070	300 ton/hr		85.75	1,800	5,425	16,300	1,771	
	0100	Backhoe attachment, long stick, up to 185 H.P., 10.5' long		.33	21.50	65	195	15.65	
	0140	Up to 250 H.P., 12' long		.35	23.50	70	210	16.80	
	0180	Over 250 H.P., 15' long		.48	31.50	95	285	22.85	
	0200	Special dipper arm, up to 100 H.P., 32' long		.98	65	195	585	46.85	
	0240	Over 100 H.P., 33' long		1.22	81	243	730	58.35	
	0280	Catch basin/sewer cleaning truck, 3 ton, 9 C.Y., 1000 gal		40.90	405	1,210	3,625	569.20	
	0300	Concrete batch plant, portable, electric, 200 C.Y./hr		17.80	515	1,540	4,625	450.40	
	0520	Grader/dozer attachment, ripper/scarifier, rear mounted, up to 135 H.P.		3.20	65	195	585	64.60	
	0540	Up to 180 H.P.		3.75	83.50	250	750	80	
	0580	Up to 250 H.P.		4.15	95	285	855	90.20	
	0700	Pvmt. removal bucket, for hyd. excavator, up to 90 H.P.		1.70	50	150	450	43.60	
	0740	Up to 200 H.P.		1.90	71.50	215	645	58.20	
	0780	Over 200 H.P.		2.05	83.50	250	750	66.40	
	0900	Aggregate spreader, self-propelled, 187 H.P.		50.35	695	2,090	6,275	820.80	
	1000	Chemical spreader, 3 C.Y.		3.05	43.50	130	390	50.40	
	1900	Hammermill, traveling, 250 H.P.		63.25	1,875	5,650	17,000	1,636	
	2000	Horizontal borer, 3" diameter, 13 H.P. gas driven		6.00	56.50	170	510	82	
	2150	Horizontal directional drill, 20,000 lb. thrust, 78 H.P. diesel		29.70	690	2,070	6,200	651.60	
	2160	30,000 lb. thrust, 115 H.P.		37.15	1,050	3,155	9,475	928.20	
	2170	50,000 lb. thrust, 170 H.P.		53.00	1,350	4,040	12,100	1,232	
	2190	Mud trailer for HDD, 1500 gallons, 175 H.P., gas		30.75	153	460	1,375	338	
	2200	Hydromulchers, gas power, 3000 gallons, for truck mounting		22.90	258	775	2,325	338.20	
	2400	Joint & crack cleaner, walk behind, 25 H.P.		3.80	50	150	450	60.40	
	2500	Filler, trailer mounted, 400 gallons, 20 H.P.		9.15	217	650	1,950	203.20	
	3000	Paint striper, self-propelled, double line, 30 H.P.		6.95	162	485	1,450	152.60	
	3200	Post drivers, 6" I-Beam frame, for truck mounting		18.75	425	1,280	3,850	406	
	3400	Road sweeper, self-propelled, 8' wide, 90 H.P.		37.80	580	1,740	5,225	650.40	
	3450	Road sweeper, vacuum assisted, 4 C.Y., 220 gallons		73.55	635	1,905	5,725	969.40	
	4000	Road mixer, self-propelled, 130 H.P.		44.05	775	2,325	6,975	817.40	
	4100	310 H.P.		79.75	2,200	6,585	19,800	1,955	
	4220	Cold mix paver, incl. pug mill and bitumen tank, 165 H.P.		96.80	2,150	6,440	19,300	2,062	
	4250	Paver, asphalt, wheel or crawler, 130 H.P., diesel		95.35	2,050	6,155	18,500	1,994	
	4300	Paver, road widener, gas 1' to 6', 67 H.P.		44.65	835	2,500	7,500	857.20	
	4400	Diesel, 2' to 14', 88 H.P.		59.80	1,075	3,190	9,575	1,116	
	4600	Slipform pavers, curb and gutter, 2 track, 75 H.P.		44.25	785	2,355	7,075	825	
	4700	4 track, 165 H.P.		54.95	825	2,475	7,425	934.60	
	4800	Median barrier, 215 H.P.		55.65	855	2,570	7,700	959.20	
	4901	Trailer, low bed, 75 ton capacity		9.35	218	655	1,975	205.80	
	5000	Road planer, walk behind, 10" cutting width, 10 H.P.		3.45	30.50	91	273	45.80	
	5100	Self-propelled, 12" cutting width, 64 H.P.		10.20	127	380	1,150	157.60	
	5120	Traffic line remover, metal ball blaster, truck mounted, 115 H.P.		52.80	770	2,305	6,925	883.40	
	5140	Grinder, truck mounted, 115 H.P.		56.60	835	2,505	7,525	953.80	

01 54 | Construction Aids

01 54 33 | Equipment Rental

			UNIT	HOURLY OPER. COST	RENT PER DAY	RENT PER WEEK	RENT PER MONTH	EQUIPMENT COST/DAY	
50	5160	Walk-behind, 11 H.P.	Ea.	3.90	46.50	140	420	59.20	50
	5200	Pavement profiler, 4' to 6' wide, 450 H.P.		244.25	3,400	10,220	30,700	3,998	
	5300	8' to 10' wide, 750 H.P.		391.35	4,950	14,865	44,600	6,104	
	5400	Roadway plate, steel, 1" x 8' x 20'		.07	11.65	35	105	7.55	
	5600	Stabilizer, self-propelled, 150 H.P.		47.20	615	1,850	5,550	747.60	
	5700	310 H.P.		81.20	1,350	4,070	12,200	1,464	
	5800	Striper, thermal, truck mounted 120 gallon paint, 150 H.P.		62.30	505	1,520	4,550	802.40	
	6000	Tar kettle, 330 gallon, trailer mounted		4.30	61.50	185	555	71.40	
	7000	Tunnel locomotive, diesel, 8 to 12 ton		30.00	580	1,745	5,225	589	
	7005	Electric, 10 ton		24.00	665	1,995	5,975	591	
	7010	Muck cars, 1/2 C.Y. capacity		1.85	23.50	70	210	28.80	
	7020	1 C.Y. capacity		2.05	31.50	95	285	35.40	
	7030	2 C.Y. capacity		2.15	36.50	110	330	39.20	
	7040	Side dump, 2 C.Y. capacity		2.35	45	135	405	45.80	
	7050	3 C.Y. capacity		3.20	51.50	155	465	56.60	
	7060	5 C.Y. capacity		4.55	65	195	585	75.40	
	7100	Ventilating blower for tunnel, 7-1/2 H.P.		1.39	51.50	155	465	42.10	
	7110	10 H.P.		.83	.52	1.55	4.65	6.95	
	7120	20 H.P.		1.66	.67	2	6	13.70	
	7140	40 H.P.		3.31	.97	2.90	8.70	27.05	
	7160	60 H.P.		7.23	152	455	1,375	148.85	
	7175	75 H.P.		9.22	202	607	1,825	195.15	
	7180	200 H.P.		21.05	305	910	2,725	350.40	
	7800	Windrow loader, elevating		52.00	1,350	4,060	12,200	1,228	
60	0010	**LIFTING AND HOISTING EQUIPMENT RENTAL** without operators	R015433-10						60
	0120	Aerial lift truck, 2 person, to 80'	Ea.	28.05	735	2,205	6,625	665.40	
	0140	Boom work platform, 40' snorkel		17.60	265	795	2,375	299.80	
	0150	Crane, flatbed mounted, 3 ton capacity	R015433-15	15.70	195	585	1,750	242.60	
	0200	Crane, climbing, 106' jib, 6000 lb. capacity, 410 fpm		38.95	1,500	4,490	13,500	1,210	
	0300	101' jib, 10,250 lb. capacity, 270 fpm	R312316-45	44.95	1,900	5,690	17,100	1,498	
	0500	Tower, static, 130' high, 106' jib, 6200 lb. capacity at 400 fpm		42.45	1,725	5,190	15,600	1,378	
	0600	Crawler mounted, lattice boom, 1/2 C.Y., 15 tons at 12' radius		35.47	685	2,050	6,150	693.75	
	0700	3/4 C.Y., 20 tons at 12' radius		47.29	855	2,570	7,700	892.30	
	0800	1 C.Y., 25 tons at 12' radius		63.05	1,150	3,420	10,300	1,188	
	0900	1-1/2 C.Y., 40 tons at 12' radius		63.05	1,150	3,445	10,300	1,193	
	1000	2 C.Y., 50 tons at 12' radius		66.90	1,350	4,055	12,200	1,346	
	1100	3 C.Y., 75 tons at 12' radius		71.80	1,575	4,710	14,100	1,516	
	1200	100 ton capacity, 60' boom		81.40	1,800	5,405	16,200	1,732	
	1300	165 ton capacity, 60' boom		103.75	2,100	6,320	19,000	2,094	
	1400	200 ton capacity, 70' boom		125.80	2,675	7,990	24,000	2,604	
	1500	350 ton capacity, 80' boom		177.05	4,075	12,195	36,600	3,855	
	1600	Truck mounted, lattice boom, 6 x 4, 20 tons at 10' radius		36.33	1,225	3,690	11,100	1,029	
	1700	25 tons at 10' radius		39.34	1,350	4,020	12,100	1,119	
	1800	8 x 4, 30 tons at 10' radius		42.56	1,425	4,280	12,800	1,196	
	1900	40 tons at 12' radius		45.29	1,500	4,470	13,400	1,256	
	2000	60 tons at 15' radius		50.61	1,575	4,730	14,200	1,351	
	2050	82 tons at 15' radius		56.35	1,675	5,050	15,200	1,461	
	2100	90 tons at 15' radius		63.07	1,825	5,510	16,500	1,607	
	2200	115 tons at 15' radius		71.12	2,050	6,160	18,500	1,801	
	2300	150 tons at 18' radius		88.35	2,150	6,480	19,400	2,003	
	2350	165 tons at 18' radius		83.09	2,300	6,870	20,600	2,039	
	2400	Truck mounted, hydraulic, 12 ton capacity		41.45	540	1,620	4,850	655.60	
	2500	25 ton capacity		43.70	665	1,990	5,975	747.60	
	2550	33 ton capacity		44.25	680	2,045	6,125	763	
	2560	40 ton capacity		57.20	800	2,405	7,225	938.60	
	2600	55 ton capacity		75.80	1,000	3,015	9,050	1,209	
	2700	80 ton capacity		97.75	1,275	3,815	11,400	1,545	
	2720	100 ton capacity		91.65	1,525	4,600	13,800	1,653	

01 54 | Construction Aids

01 54 33 | Equipment Rental

		UNIT	HOURLY OPER. COST	RENT PER DAY	RENT PER WEEK	RENT PER MONTH	EQUIPMENT COST/DAY	
60	2740	120 ton capacity	Ea.	106.15	1,650	4,920	14,800	1,833	60
	2760	150 ton capacity		124.30	2,250	6,735	20,200	2,341	
	2800	Self-propelled, 4 x 4, with telescoping boom, 5 ton		16.90	235	705	2,125	276.20	
	2900	12-1/2 ton capacity		41.45	540	1,620	4,850	655.60	
	3000	15 ton capacity		39.70	585	1,755	5,275	668.60	
	3050	20 ton capacity		35.20	575	1,720	5,150	625.60	
	3100	25 ton capacity		36.45	605	1,820	5,450	655.60	
	3150	40 ton capacity		57.95	915	2,750	8,250	1,014	
	3200	Derricks, guy, 20 ton capacity, 60' boom, 75' mast		21.70	365	1,100	3,300	393.60	
	3300	100' boom, 115' mast		34.23	630	1,890	5,675	651.85	
	3400	Stiffleg, 20 ton capacity, 70' boom, 37' mast		24.01	475	1,430	4,300	478.10	
	3500	100' boom, 47' mast		36.96	760	2,280	6,850	751.70	
	3550	Helicopter, small, lift to 1250 lbs. maximum, w/pilot		92.16	2,950	8,880	26,600	2,513	
	3600	Hoists, chain type, overhead, manual, 3/4 ton		.10	1	3	9	1.40	
	3900	10 ton		.75	9.35	28	84	11.60	
	4000	Hoist and tower, 5000 lb. cap., portable electric, 40' high		4.82	211	633	1,900	165.15	
	4100	For each added 10' section, add		.10	16.65	50	150	10.80	
	4200	Hoist and single tubular tower, 5000 lb. electric, 100' high		6.48	295	884	2,650	228.65	
	4300	For each added 6'-6" section, add		.17	28.50	85	255	18.35	
	4400	Hoist and double tubular tower, 5000 lb., 100' high		6.93	325	974	2,925	250.25	
	4500	For each added 6'-6" section, add		.19	31.50	95	285	20.50	
	4550	Hoist and tower, mast type, 6000 lb., 100' high		7.53	335	1,010	3,025	262.25	
	4570	For each added 10' section, add		.12	20	60	180	12.95	
	4600	Hoist and tower, personnel, electric, 2000 lb., 100' @ 125 fpm		15.10	895	2,690	8,075	658.80	
	4700	3000 lb., 100' @ 200 fpm		17.31	1,025	3,050	9,150	748.50	
	4800	3000 lb., 150' @ 300 fpm		19.11	1,125	3,410	10,200	834.90	
	4900	4000 lb., 100' @ 300 fpm		19.88	1,150	3,480	10,400	855.05	
	5000	6000 lb., 100' @ 275 fpm		21.55	1,225	3,650	11,000	902.40	
	5100	For added heights up to 500', add	L.F.	.01	1.67	5	15	1.10	
	5200	Jacks, hydraulic, 20 ton	Ea.	.05	2	6	18	1.60	
	5500	100 ton		.35	11.35	34	102	9.60	
	6100	Jacks, hydraulic, climbing w/ 50' jackrods, control console, 30 ton cap.		1.82	121	364	1,100	87.35	
	6150	For each added 10' jackrod section, add		.05	3.33	10	30	2.40	
	6300	50 ton capacity		2.93	195	585	1,750	140.45	
	6350	For each added 10' jackrod section, add		.06	4	12	36	2.90	
	6500	125 ton capacity		7.70	515	1,540	4,625	369.60	
	6550	For each added 10' jackrod section, add		.53	35	105	315	25.25	
	6600	Cable jack, 10 ton capacity with 200' cable		1.53	102	305	915	73.25	
	6650	For each added 50' of cable, add		.18	11.65	35	105	8.45	
70	0010	**WELLPOINT EQUIPMENT RENTAL** without operators	R015433-10						70
	0020	Based on 2 months rental							
	0100	Combination jetting & wellpoint pump, 60 H.P. diesel	Ea.	14.71	300	902	2,700	298.10	
	0200	High pressure gas jet pump, 200 H.P., 300 psi	"	33.40	257	771	2,325	421.40	
	0300	Discharge pipe, 8" diameter	L.F.	.01	.49	1.47	4.41	.35	
	0350	12" diameter		.01	.72	2.16	6.50	.50	
	0400	Header pipe, flows up to 150 GPM, 4" diameter		.01	.44	1.32	3.96	.35	
	0500	400 GPM, 6" diameter		.01	.52	1.57	4.71	.40	
	0600	800 GPM, 8" diameter		.01	.72	2.16	6.50	.50	
	0700	1500 GPM, 10" diameter		.01	.75	2.26	6.80	.55	
	0800	2500 GPM, 12" diameter		.02	1.43	4.29	12.85	1	
	0900	4500 GPM, 16" diameter		.03	1.83	5.50	16.50	1.35	
	0950	For quick coupling aluminum and plastic pipe, add		.03	1.90	5.69	17.05	1.40	
	1100	Wellpoint, 25' long, with fittings & riser pipe, 1-1/2" or 2" diameter	Ea.	.06	3.78	11.35	34	2.75	
	1200	Wellpoint pump, diesel powered, 4" suction, 20 H.P.		6.44	173	520	1,550	155.50	
	1300	6" suction, 30 H.P.		8.72	215	645	1,925	198.75	
	1400	8" suction, 40 H.P.		11.79	295	884	2,650	271.10	
	1500	10" suction, 75 H.P.		17.73	345	1,033	3,100	348.45	

641

01 54 | Construction Aids

01 54 33 | Equipment Rental

			UNIT	HOURLY OPER. COST	RENT PER DAY	RENT PER WEEK	RENT PER MONTH	EQUIPMENT COST/DAY	
70	1600	12" suction, 100 H.P.	Ea.	25.55	550	1,650	4,950	534.40	70
	1700	12" suction, 175 H.P.	"	37.17	605	1,810	5,425	659.35	
80	0010	**MARINE EQUIPMENT RENTAL** without operators							80
	0200	Barge, 400 Ton, 30' wide x 90' long	Ea.	15.50	1,075	3,220	9,650	768	
	0240	800 Ton, 45' wide x 90' long		18.80	1,300	3,900	11,700	930.40	
	2000	Tugboat, diesel, 100 H.P.		36.45	203	610	1,825	413.60	
	2040	250 H.P.		77.10	375	1,120	3,350	840.80	
	2080	380 H.P.		149.05	1,100	3,325	9,975	1,857	

R015433-10

Crews

Crew No.	Bare Costs		Incl. Subs O&P		Cost Per Labor-Hour	
Crew A-1	**Hr.**	**Daily**	**Hr.**	**Daily**	**Bare Costs**	**Incl. O&P**
1 Building Laborer	$33.10	$264.80	$54.35	$434.80	$33.10	$54.35
1 Concrete saw, gas manual		73.80		81.18	9.22	10.15
8 L.H., Daily Totals		$338.60		$515.98	$42.33	$64.50
Crew A-1A	**Hr.**	**Daily**	**Hr.**	**Daily**	**Bare Costs**	**Incl. O&P**
1 Skilled Worker	$42.60	$340.80	$68.90	$551.20	$42.60	$68.90
1 Shot Blaster, 20"		215.80		237.38	26.98	29.67
8 L.H., Daily Totals		$556.60		$788.58	$69.58	$98.57
Crew A-1B	**Hr.**	**Daily**	**Hr.**	**Daily**	**Bare Costs**	**Incl. O&P**
1 Building Laborer	$33.10	$264.80	$54.35	$434.80	$33.10	$54.35
1 Concrete Saw		161.40		177.54	20.18	22.19
8 L.H., Daily Totals		$426.20		$612.34	$53.27	$76.54
Crew A-1C	**Hr.**	**Daily**	**Hr.**	**Daily**	**Bare Costs**	**Incl. O&P**
1 Building Laborer	$33.10	$264.80	$54.35	$434.80	$33.10	$54.35
1 Chain Saw, gas, 18"		28.20		31.02	3.52	3.88
8 L.H., Daily Totals		$293.00		$465.82	$36.63	$58.23
Crew A-1D	**Hr.**	**Daily**	**Hr.**	**Daily**	**Bare Costs**	**Incl. O&P**
1 Building Laborer	$33.10	$264.80	$54.35	$434.80	$33.10	$54.35
1 Vibrating plate, gas, 18"		34.20		37.62	4.28	4.70
8 L.H., Daily Totals		$299.00		$472.42	$37.38	$59.05
Crew A-1E	**Hr.**	**Daily**	**Hr.**	**Daily**	**Bare Costs**	**Incl. O&P**
1 Building Laborer	$33.10	$264.80	$54.35	$434.80	$33.10	$54.35
1 Vibratory Plate, gas, 21"		43.60		47.96	5.45	6.00
8 L.H., Daily Totals		$308.40		$482.76	$38.55	$60.34
Crew A-1F	**Hr.**	**Daily**	**Hr.**	**Daily**	**Bare Costs**	**Incl. O&P**
1 Building Laborer	$33.10	$264.80	$54.35	$434.80	$33.10	$54.35
1 Rammer/tamper, gas, 8"		46.80		51.48	5.85	6.43
8 L.H., Daily Totals		$311.60		$486.28	$38.95	$60.78
Crew A-1G	**Hr.**	**Daily**	**Hr.**	**Daily**	**Bare Costs**	**Incl. O&P**
1 Building Laborer	$33.10	$264.80	$54.35	$434.80	$33.10	$54.35
1 Rammer/tamper, gas, 15"		51.80		56.98	6.47	7.12
8 L.H., Daily Totals		$316.60		$491.78	$39.58	$61.47
Crew A-1H	**Hr.**	**Daily**	**Hr.**	**Daily**	**Bare Costs**	**Incl. O&P**
1 Building Laborer	$33.10	$264.80	$54.35	$434.80	$33.10	$54.35
1 Exterior Steam Cleaner		60.20		66.22	7.53	8.28
8 L.H., Daily Totals		$325.00		$501.02	$40.63	$62.63
Crew A-1J	**Hr.**	**Daily**	**Hr.**	**Daily**	**Bare Costs**	**Incl. O&P**
1 Building Laborer	$33.10	$264.80	$54.35	$434.80	$33.10	$54.35
1 Cultivator, Walk-Behind, 5 H.P.		56.15		61.77	7.02	7.72
8 L.H., Daily Totals		$320.95		$496.57	$40.12	$62.07
Crew A-1K	**Hr.**	**Daily**	**Hr.**	**Daily**	**Bare Costs**	**Incl. O&P**
1 Building Laborer	$33.10	$264.80	$54.35	$434.80	$33.10	$54.35
1 Cultivator, Walk-Behind, 8 H.P.		86.40		95.04	10.80	11.88
8 L.H., Daily Totals		$351.20		$529.84	$43.90	$66.23
Crew A-1M	**Hr.**	**Daily**	**Hr.**	**Daily**	**Bare Costs**	**Incl. O&P**
1 Building Laborer	$33.10	$264.80	$54.35	$434.80	$33.10	$54.35
1 Snow Blower, Walk-Behind		53.40		58.74	6.67	7.34
8 L.H., Daily Totals		$318.20		$493.54	$39.77	$61.69

Crew No.	Bare Costs		Incl. Subs O&P		Cost Per Labor-Hour	
Crew A-2	**Hr.**	**Daily**	**Hr.**	**Daily**	**Bare Costs**	**Incl. O&P**
2 Laborers	$33.10	$529.60	$54.35	$869.60	$32.82	$53.70
1 Truck Driver (light)	32.25	258.00	52.40	419.20		
1 Flatbed Truck, Gas, 1.5 Ton		239.80		263.78	9.99	10.99
24 L.H., Daily Totals		$1027.40		$1552.58	$42.81	$64.69
Crew A-2A	**Hr.**	**Daily**	**Hr.**	**Daily**	**Bare Costs**	**Incl. O&P**
2 Laborers	$33.10	$529.60	$54.35	$869.60	$32.82	$53.70
1 Truck Driver (light)	32.25	258.00	52.40	419.20		
1 Flatbed Truck, Gas, 1.5 Ton		239.80		263.78		
1 Concrete Saw		161.40		177.54	16.72	18.39
24 L.H., Daily Totals		$1188.80		$1730.12	$49.53	$72.09
Crew A-2B	**Hr.**	**Daily**	**Hr.**	**Daily**	**Bare Costs**	**Incl. O&P**
1 Truck Driver (light)	$32.25	$258.00	$52.40	$419.20	$32.25	$52.40
1 Flatbed Truck, Gas, 1.5 Ton		239.80		263.78	29.98	32.97
8 L.H., Daily Totals		$497.80		$682.98	$62.23	$85.37
Crew A-3A	**Hr.**	**Daily**	**Hr.**	**Daily**	**Bare Costs**	**Incl. O&P**
1 Truck Driver (light)	$32.25	$258.00	$52.40	$419.20	$32.25	$52.40
1 Pickup truck, 4 x 4, 3/4 ton		151.00		166.10	18.88	20.76
8 L.H., Daily Totals		$409.00		$585.30	$51.13	$73.16
Crew A-3B	**Hr.**	**Daily**	**Hr.**	**Daily**	**Bare Costs**	**Incl. O&P**
1 Equip. Oper. (med.)	$42.95	$343.60	$67.30	$538.40	$38.05	$60.58
1 Truck Driver (heav.)	33.15	265.20	53.85	430.80		
1 Dump Truck, 12 C.Y., 400 H.P.		667.80		734.58		
1 F.E. Loader, W.M.,2.5 C.Y.		435.80		479.38	68.97	75.87
16 L.H., Daily Totals		$1712.40		$2183.16	$107.03	$136.45
Crew A-3C	**Hr.**	**Daily**	**Hr.**	**Daily**	**Bare Costs**	**Incl. O&P**
1 Equip. Oper. (light)	$41.30	$330.40	$64.70	$517.60	$41.30	$64.70
1 Loader, Skid Steer, 78 H.P.		291.00		320.10	36.38	40.01
8 L.H., Daily Totals		$621.40		$837.70	$77.67	$104.71
Crew A-3D	**Hr.**	**Daily**	**Hr.**	**Daily**	**Bare Costs**	**Incl. O&P**
1 Truck Driver, Light	$32.25	$258.00	$52.40	$419.20	$32.25	$52.40
1 Pickup truck, 4 x 4, 3/4 ton		151.00		166.10		
1 Flatbed Trailer, 25 Ton		105.60		116.16	32.08	35.28
8 L.H., Daily Totals		$514.60		$701.46	$64.33	$87.68
Crew A-3E	**Hr.**	**Daily**	**Hr.**	**Daily**	**Bare Costs**	**Incl. O&P**
1 Equip. Oper. (crane)	$44.40	$355.20	$69.55	$556.40	$38.77	$61.70
1 Truck Driver (heavy)	33.15	265.20	53.85	430.80		
1 Pickup truck, 4 x 4, 3/4 ton		151.00		166.10	9.44	10.38
16 L.H., Daily Totals		$771.40		$1153.30	$48.21	$72.08
Crew A-3F	**Hr.**	**Daily**	**Hr.**	**Daily**	**Bare Costs**	**Incl. O&P**
1 Equip. Oper. (crane)	$44.40	$355.20	$69.55	$556.40	$38.77	$61.70
1 Truck Driver (heavy)	33.15	265.20	53.85	430.80		
1 Pickup truck, 4 x 4, 3/4 ton		151.00		166.10		
1 Truck Tractor, 6x4, 380 H.P.		592.40		651.64		
1 Lowbed Trailer, 75 Ton		205.80		226.38	59.33	65.26
16 L.H., Daily Totals		$1569.60		$2031.32	$98.10	$126.96

Crews

Crew No.	Bare Costs		Incl. Subs O&P		Cost Per Labor-Hour	
Crew A-3G	Hr.	Daily	Hr.	Daily	Bare Costs	Incl. O&P
1 Equip. Oper. (crane)	$44.40	$355.20	$69.55	$556.40	$38.77	$61.70
1 Truck Driver (heavy)	33.15	265.20	53.85	430.80		
1 Pickup truck, 4 x 4, 3/4 ton		151.00		166.10		
1 Truck Tractor, 6x4, 450 H.P.		715.80		787.38		
1 Lowbed Trailer, 75 Ton		205.80		226.38	67.04	73.74
16 L.H., Daily Totals		$1693.00		$2167.06	$105.81	$135.44
Crew A-3H	Hr.	Daily	Hr.	Daily	Bare Costs	Incl. O&P
1 Equip. Oper. (crane)	$44.40	$355.20	$69.55	$556.40	$44.40	$69.55
1 Hyd. crane, 12 Ton (daily)		871.60		958.76	108.95	119.85
8 L.H., Daily Totals		$1226.80		$1515.16	$153.35	$189.40
Crew A-3I	Hr.	Daily	Hr.	Daily	Bare Costs	Incl. O&P
1 Equip. Oper. (crane)	$44.40	$355.20	$69.55	$556.40	$44.40	$69.55
1 Hyd. crane, 25 Ton (daily)		1015.00		1116.50	126.88	139.56
8 L.H., Daily Totals		$1370.20		$1672.90	$171.28	$209.11
Crew A-3J	Hr.	Daily	Hr.	Daily	Bare Costs	Incl. O&P
1 Equip. Oper. (crane)	$44.40	$355.20	$69.55	$556.40	$44.40	$69.55
1 Hyd. crane, 40 Ton (daily)		1258.00		1383.80	157.25	172.97
8 L.H., Daily Totals		$1613.20		$1940.20	$201.65	$242.53
Crew A-3K	Hr.	Daily	Hr.	Daily	Bare Costs	Incl. O&P
1 Equip. Oper. (crane)	$44.40	$355.20	$69.55	$556.40	$41.35	$64.78
1 Equip. Oper. Oiler	38.30	306.40	60.00	480.00		
1 Hyd. crane, 55 Ton (daily)		1611.00		1772.10		
1 P/U Truck, 3/4 Ton (daily)		161.00		177.10	110.75	121.83
16 L.H., Daily Totals		$2433.60		$2985.60	$152.10	$186.60
Crew A-3L	Hr.	Daily	Hr.	Daily	Bare Costs	Incl. O&P
1 Equip. Oper. (crane)	$44.40	$355.20	$69.55	$556.40	$41.35	$64.78
1 Equip. Oper. Oiler	38.30	306.40	60.00	480.00		
1 Hyd. crane, 80 Ton (daily)		2052.00		2257.20		
1 P/U Truck, 3/4 Ton (daily)		161.00		177.10	138.31	152.14
16 L.H., Daily Totals		$2874.60		$3470.70	$179.66	$216.92
Crew A-3M	Hr.	Daily	Hr.	Daily	Bare Costs	Incl. O&P
1 Equip. Oper. (crane)	$44.40	$355.20	$69.55	$556.40	$41.35	$64.78
1 Equip. Oper. Oiler	38.30	306.40	60.00	480.00		
1 Hyd. crane, 100 Ton (daily)		2268.00		2494.80		
1 P/U Truck, 3/4 Ton (daily)		161.00		177.10	151.81	166.99
16 L.H., Daily Totals		$3090.60		$3708.30	$193.16	$231.77
Crew A-3N	Hr.	Daily	Hr.	Daily	Bare Costs	Incl. O&P
1 Equip. Oper. (crane)	$44.40	$355.20	$69.55	$556.40	$44.40	$69.55
1 Tower crane (monthly)		1049.00		1153.90	131.13	144.24
8 L.H., Daily Totals		$1404.20		$1710.30	$175.53	$213.79
Crew A-3P	Hr.	Daily	Hr.	Daily	Bare Costs	Incl. O&P
1 Equip. Oper., Light	$41.30	$330.40	$64.70	$517.60	$41.30	$64.70
1 A.T. Forklift, 42' lift		478.20		526.02	59.77	65.75
8 L.H., Daily Totals		$808.60		$1043.62	$101.08	$130.45
Crew A-4	Hr.	Daily	Hr.	Daily	Bare Costs	Incl. O&P
2 Carpenters	$41.55	$664.80	$68.25	$1092.00	$39.82	$64.73
1 Painter, Ordinary	36.35	290.80	57.70	461.60		
24 L.H., Daily Totals		$955.60		$1553.60	$39.82	$64.73

Crew No.	Bare Costs		Incl. Subs O&P		Cost Per Labor-Hour	
Crew A-5	Hr.	Daily	Hr.	Daily	Bare Costs	Incl. O&P
2 Laborers	$33.10	$529.60	$54.35	$869.60	$33.01	$54.13
.25 Truck Driver (light)	32.25	64.50	52.40	104.80		
.25 Flatbed Truck, Gas, 1.5 Ton		59.95		65.94	3.33	3.66
18 L.H., Daily Totals		$654.05		$1040.35	$36.34	$57.80
Crew A-6	Hr.	Daily	Hr.	Daily	Bare Costs	Incl. O&P
1 Instrument Man	$42.60	$340.80	$68.90	$551.20	$41.40	$66.72
1 Rodman/Chainman	40.20	321.60	64.55	516.40		
1 Laser Transit/Level		69.90		76.89	4.37	4.81
16 L.H., Daily Totals		$732.30		$1144.49	$45.77	$71.53
Crew A-7	Hr.	Daily	Hr.	Daily	Bare Costs	Incl. O&P
1 Chief Of Party	$52.25	$418.00	$82.50	$660.00	$45.02	$71.98
1 Instrument Man	42.60	340.80	68.90	551.20		
1 Rodman/Chainman	40.20	321.60	64.55	516.40		
1 Laser Transit/Level		69.90		76.89	2.91	3.20
24 L.H., Daily Totals		$1150.30		$1804.49	$47.93	$75.19
Crew A-8	Hr.	Daily	Hr.	Daily	Bare Costs	Incl. O&P
1 Chief of Party	$52.25	$418.00	$82.50	$660.00	$43.81	$70.13
1 Instrument Man	42.60	340.80	68.90	551.20		
2 Rodmen/Chainmen	40.20	643.20	64.55	1032.80		
1 Laser Transit/Level		69.90		76.89	2.18	2.40
32 L.H., Daily Totals		$1471.90		$2320.89	$46.00	$72.53
Crew A-9	Hr.	Daily	Hr.	Daily	Bare Costs	Incl. O&P
1 Asbestos Foreman	$46.05	$368.40	$73.85	$590.80	$45.61	$73.15
7 Asbestos Workers	45.55	2550.80	73.05	4090.80		
64 L.H., Daily Totals		$2919.20		$4681.60	$45.61	$73.15
Crew A-10A	Hr.	Daily	Hr.	Daily	Bare Costs	Incl. O&P
1 Asbestos Foreman	$46.05	$368.40	$73.85	$590.80	$45.72	$73.32
2 Asbestos Workers	45.55	728.80	73.05	1168.80		
24 L.H., Daily Totals		$1097.20		$1759.60	$45.72	$73.32
Crew A-10B	Hr.	Daily	Hr.	Daily	Bare Costs	Incl. O&P
1 Asbestos Foreman	$46.05	$368.40	$73.85	$590.80	$45.67	$73.25
3 Asbestos Workers	45.55	1093.20	73.05	1753.20		
32 L.H., Daily Totals		$1461.60		$2344.00	$45.67	$73.25
Crew A-10C	Hr.	Daily	Hr.	Daily	Bare Costs	Incl. O&P
3 Asbestos Workers	$45.55	$1093.20	$73.05	$1753.20	$45.55	$73.05
1 Flatbed Truck, Gas, 1.5 Ton		239.80		263.78	9.99	10.99
24 L.H., Daily Totals		$1333.00		$2016.98	$55.54	$84.04
Crew A-10D	Hr.	Daily	Hr.	Daily	Bare Costs	Incl. O&P
2 Asbestos Workers	$45.55	$728.80	$73.05	$1168.80	$43.45	$68.91
1 Equip. Oper. (crane)	44.40	355.20	69.55	556.40		
1 Equip. Oper. Oiler	38.30	306.40	60.00	480.00		
1 Hydraulic Crane, 33 Ton		763.00		839.30	23.84	26.23
32 L.H., Daily Totals		$2153.40		$3044.50	$67.29	$95.14
Crew A-11	Hr.	Daily	Hr.	Daily	Bare Costs	Incl. O&P
1 Asbestos Foreman	$46.05	$368.40	$73.85	$590.80	$45.61	$73.15
7 Asbestos Workers	45.55	2550.80	73.05	4090.80		
2 Chip. Hammers, 12 Lb., Elec.		36.00		39.60	.56	.62
64 L.H., Daily Totals		$2955.20		$4721.20	$46.17	$73.77

Crews

Crew No.	Bare Costs		Incl. Subs O&P		Cost Per Labor-Hour	
Crew A-12	Hr.	Daily	Hr.	Daily	Bare Costs	Incl. O&P
1 Asbestos Foreman	$46.05	$368.40	$73.85	$590.80	$45.61	$73.15
7 Asbestos Workers	45.55	2550.80	73.05	4090.80		
1 Trk-mtd vac, 14 CY, 1500 Gal.		530.25		583.27		
1 Flatbed Truck, 20,000 GVW		245.60		270.16	12.12	13.33
64 L.H., Daily Totals		$3695.05		$5535.03	$57.74	$86.48

Crew A-13	Hr.	Daily	Hr.	Daily	Bare Costs	Incl. O&P
1 Equip. Oper. (light)	$41.30	$330.40	$64.70	$517.60	$41.30	$64.70
1 Trk-mtd vac, 14 CY, 1500 Gal.		530.25		583.27		
1 Flatbed Truck, 20,000 GVW		245.60		270.16	96.98	106.68
8 L.H., Daily Totals		$1106.25		$1371.04	$138.28	$171.38

Crew B-1	Hr.	Daily	Hr.	Daily	Bare Costs	Incl. O&P
1 Labor Foreman (outside)	$35.10	$280.80	$57.65	$461.20	$33.77	$55.45
2 Laborers	33.10	529.60	54.35	869.60		
24 L.H., Daily Totals		$810.40		$1330.80	$33.77	$55.45

Crew B-1A	Hr.	Daily	Hr.	Daily	Bare Costs	Incl. O&P
1 Laborer Foreman	$35.10	$280.80	$57.65	$461.20	$33.77	$55.45
2 Laborers	33.10	529.60	54.35	869.60		
2 Cutting Torches		38.00		41.80		
2 Set of Gases		158.40		174.24	8.18	9.00
24 L.H., Daily Totals		$1006.80		$1546.84	$41.95	$64.45

Crew B-1B	Hr.	Daily	Hr.	Daily	Bare Costs	Incl. O&P
1 Laborer Foreman	$35.10	$280.80	$57.65	$461.20	$36.42	$58.98
2 Laborers	33.10	529.60	54.35	869.60		
1 Equip. Oper. (crane)	44.40	355.20	69.55	556.40		
2 Cutting Torches		38.00		41.80		
2 Set of Gases		158.40		174.24		
1 Hyd. Crane, 12 Ton		655.60		721.16	26.63	29.29
32 L.H., Daily Totals		$2017.60		$2824.40	$63.05	$88.26

Crew B-2	Hr.	Daily	Hr.	Daily	Bare Costs	Incl. O&P
1 Labor Foreman (outside)	$35.10	$280.80	$57.65	$461.20	$33.50	$55.01
4 Laborers	33.10	1059.20	54.35	1739.20		
40 L.H., Daily Totals		$1340.00		$2200.40	$33.50	$55.01

Crew B-3	Hr.	Daily	Hr.	Daily	Bare Costs	Incl. O&P
1 Labor Foreman (outside)	$35.10	$280.80	$57.65	$461.20	$35.09	$56.89
2 Laborers	33.10	529.60	54.35	869.60		
1 Equip. Oper. (med.)	42.95	343.60	67.30	538.40		
2 Truck Drivers (heavy)	33.15	530.40	53.85	861.60		
1 Crawler Loader, 3 C.Y.		1077.00		1184.70		
2 Dump Trucks 12 C.Y., 400 H.P.		1335.60		1469.16	50.26	55.29
48 L.H., Daily Totals		$4097.00		$5384.66	$85.35	$112.18

Crew B-3A	Hr.	Daily	Hr.	Daily	Bare Costs	Incl. O&P
4 Laborers	$33.10	$1059.20	$54.35	$1739.20	$35.07	$56.94
1 Equip. Oper. (med.)	42.95	343.60	67.30	538.40		
1 Hyd. Excavator, 1.5 C.Y.		1017.00		1118.70	25.43	27.97
40 L.H., Daily Totals		$2419.80		$3396.30	$60.49	$84.91

Crew B-3B	Hr.	Daily	Hr.	Daily	Bare Costs	Incl. O&P
2 Laborers	$33.10	$529.60	$54.35	$869.60	$35.58	$57.46
1 Equip. Oper. (med.)	42.95	343.60	67.30	538.40		
1 Truck Driver (heavy)	33.15	265.20	53.85	430.80		
1 Backhoe Loader, 80 H.P.		387.60		426.36		
1 Dump Truck, 12 C.Y., 400 H.P.		667.80		734.58	32.98	36.28
32 L.H., Daily Totals		$2193.80		$2999.74	$68.56	$93.74

Crew B-3C	Hr.	Daily	Hr.	Daily	Bare Costs	Incl. O&P
3 Laborers	$33.10	$794.40	$54.35	$1304.40	$35.56	$57.59
1 Equip. Oper. (med.)	42.95	343.60	67.30	538.40		
1 Crawler Loader, 4 C.Y.		1548.00		1702.80	48.38	53.21
32 L.H., Daily Totals		$2686.00		$3545.60	$83.94	$110.80

Crew B-4	Hr.	Daily	Hr.	Daily	Bare Costs	Incl. O&P
1 Labor Foreman (outside)	$35.10	$280.80	$57.65	$461.20	$33.44	$54.82
4 Laborers	33.10	1059.20	54.35	1739.20		
1 Truck Driver (heavy)	33.15	265.20	53.85	430.80		
1 Truck Tractor, 220 H.P.		354.00		389.40		
1 Flatbed Trailer, 40 Ton		143.20		157.52	10.36	11.39
48 L.H., Daily Totals		$2102.40		$3178.12	$43.80	$66.21

Crew B-5	Hr.	Daily	Hr.	Daily	Bare Costs	Incl. O&P
1 Labor Foreman (outside)	$35.10	$280.80	$57.65	$461.20	$36.20	$58.52
4 Laborers	33.10	1059.20	54.35	1739.20		
2 Equip. Oper. (med.)	42.95	687.20	67.30	1076.80		
1 Air Compressor, 250 cfm		193.40		212.74		
2 Breakers, Pavement, 60 lb.		18.80		20.68		
2 -50' Air Hoses, 1.5"		12.60		13.86		
1 Crawler Loader, 3 C.Y.		1077.00		1184.70	23.25	25.57
56 L.H., Daily Totals		$3329.00		$4709.18	$59.45	$84.09

Crew B-6	Hr.	Daily	Hr.	Daily	Bare Costs	Incl. O&P
2 Laborers	$33.10	$529.60	$54.35	$869.60	$35.83	$57.80
1 Equip. Oper. (light)	41.30	330.40	64.70	517.60		
1 Backhoe Loader, 48 H.P.		337.80		371.58	14.07	15.48
24 L.H., Daily Totals		$1197.80		$1758.78	$49.91	$73.28

Crew B-6B	Hr.	Daily	Hr.	Daily	Bare Costs	Incl. O&P
2 Labor Foremen (out)	$35.10	$561.60	$57.65	$922.40	$33.77	$55.45
4 Laborers	33.10	1059.20	54.35	1739.20		
1 S.P. Crane, 4x4, 5 Ton		276.20		303.82		
1 Flatbed Truck, Gas, 1.5 Ton		239.80		263.78		
1 Butt Fusion Machine		134.40		147.84	13.55	14.90
48 L.H., Daily Totals		$2271.20		$3377.04	$47.32	$70.36

Crew B-7	Hr.	Daily	Hr.	Daily	Bare Costs	Incl. O&P
1 Labor Foreman (outside)	$35.10	$280.80	$57.65	$461.20	$35.08	$57.06
4 Laborers	33.10	1059.20	54.35	1739.20		
1 Equip. Oper. (med.)	42.95	343.60	67.30	538.40		
1 Brush Chipper, 12", 130 H.P.		253.20		278.52		
1 Crawler Loader, 3 C.Y.		1077.00		1184.70		
2 Chain Saws, gas, 36" Long		78.40		86.24	29.35	32.28
48 L.H., Daily Totals		$3092.20		$4288.26	$64.42	$89.34

Crew B-7A	Hr.	Daily	Hr.	Daily	Bare Costs	Incl. O&P
2 Laborers	$33.10	$529.60	$54.35	$869.60	$35.83	$57.80
1 Equip. Oper. (light)	41.30	330.40	64.70	517.60		
1 Rake w/Tractor		285.70		314.27		
2 Chain Saws, gas, 18"		56.40		62.04	14.25	15.68
24 L.H., Daily Totals		$1202.10		$1763.51	$50.09	$73.48

Crews

Crew No.	Bare Costs		Incl. Subs O&P		Cost Per Labor-Hour	
Crew B-8	Hr.	Daily	Hr.	Daily	Bare Costs	Incl. O&P
1 Labor Foreman (outside)	$35.10	$280.80	$57.65	$461.20	$36.48	$58.58
2 Laborers	33.10	529.60	54.35	869.60		
2 Equip. Oper. (med.)	42.95	687.20	67.30	1076.80		
1 Equip. Oper. Oiler	38.30	306.40	60.00	480.00		
2 Truck Drivers (heavy)	33.15	530.40	53.85	861.60		
1 Hyd. Crane, 25 Ton		747.60		822.36		
1 Crawler Loader, 3 C.Y.		1077.00		1184.70		
2 Dump Trucks, 12 C.Y., 400 H.P.		1335.60		1469.16	49.38	54.32
64 L.H., Daily Totals		$5494.60		$7225.42	$85.85	$112.90
Crew B-9	Hr.	Daily	Hr.	Daily	Bare Costs	Incl. O&P
1 Labor Foreman (outside)	$35.10	$280.80	$57.65	$461.20	$33.50	$55.01
4 Laborers	33.10	1059.20	54.35	1739.20		
1 Air Compressor, 250 cfm		193.40		212.74		
2 Breakers, Pavement, 60 lb.		18.80		20.68		
2 -50' Air Hoses, 1.5"		12.60		13.86	5.62	6.18
40 L.H., Daily Totals		$1564.80		$2447.68	$39.12	$61.19
Crew B-9A	Hr.	Daily	Hr.	Daily	Bare Costs	Incl. O&P
2 Laborers	$33.10	$529.60	$54.35	$869.60	$33.12	$54.18
1 Truck Driver (heavy)	33.15	265.20	53.85	430.80		
1 Water Tank Trailer, 5000 Gal.		139.60		153.56		
1 Truck Tractor, 220 H.P.		354.00		389.40		
2 -50' Discharge Hoses, 3"		3.50		3.85	20.71	22.78
24 L.H., Daily Totals		$1291.90		$1847.21	$53.83	$76.97
Crew B-9B	Hr.	Daily	Hr.	Daily	Bare Costs	Incl. O&P
2 Laborers	$33.10	$529.60	$54.35	$869.60	$33.12	$54.18
1 Truck Driver (heavy)	33.15	265.20	53.85	430.80		
2 -50' Discharge Hoses, 3"		3.50		3.85		
1 Water Tank Trailer, 5000 Gal.		139.60		153.56		
1 Truck Tractor, 220 H.P.		354.00		389.40		
1 Pressure Washer		60.80		66.88	23.25	25.57
24 L.H., Daily Totals		$1352.70		$1914.09	$56.36	$79.75
Crew B-9D	Hr.	Daily	Hr.	Daily	Bare Costs	Incl. O&P
1 Labor Foreman (Outside)	$35.10	$280.80	$57.65	$461.20	$33.50	$55.01
4 Common Laborers	33.10	1059.20	54.35	1739.20		
1 Air Compressor, 250 cfm		193.40		212.74		
2 -50' Air Hoses, 1.5"		12.60		13.86		
2 Air Powered Tampers		50.50		55.55	6.41	7.05
40 L.H., Daily Totals		$1596.50		$2482.55	$39.91	$62.06
Crew B-10	Hr.	Daily	Hr.	Daily	Bare Costs	Incl. O&P
1 Equip. Oper. (med.)	$42.95	$343.60	$67.30	$538.40	$39.67	$62.98
.5 Laborer	33.10	132.40	54.35	217.40		
12 L.H., Daily Totals		$476.00		$755.80	$39.67	$62.98
Crew B-10A	Hr.	Daily	Hr.	Daily	Bare Costs	Incl. O&P
1 Equip. Oper. (med.)	$42.95	$343.60	$67.30	$538.40	$39.67	$62.98
.5 Laborer	33.10	132.40	54.35	217.40		
1 Roller, 2-Drum, W.B., 7.5 H.P.		149.60		164.56	12.47	13.71
12 L.H., Daily Totals		$625.60		$920.36	$52.13	$76.70
Crew B-10B	Hr.	Daily	Hr.	Daily	Bare Costs	Incl. O&P
1 Equip. Oper. (med.)	$42.95	$343.60	$67.30	$538.40	$39.67	$62.98
.5 Laborer	33.10	132.40	54.35	217.40		
1 Dozer, 200 H.P.		1192.00		1311.20	99.33	109.27
12 L.H., Daily Totals		$1668.00		$2067.00	$139.00	$172.25

Crew No.	Bare Costs		Incl. Subs O&P		Cost Per Labor-Hour	
Crew B-10C	Hr.	Daily	Hr.	Daily	Bare Costs	Incl. O&P
1 Equip. Oper. (med.)	$42.95	$343.60	$67.30	$538.40	$39.67	$62.98
.5 Laborer	33.10	132.40	54.35	217.40		
1 Dozer, 200 H.P.		1192.00		1311.20		
1 Vibratory Roller, Towed, 23 Ton		434.20		477.62	135.52	149.07
12 L.H., Daily Totals		$2102.20		$2544.62	$175.18	$212.05
Crew B-10D	Hr.	Daily	Hr.	Daily	Bare Costs	Incl. O&P
1 Equip. Oper. (med.)	$42.95	$343.60	$67.30	$538.40	$39.67	$62.98
.5 Laborer	33.10	132.40	54.35	217.40		
1 Dozer, 200 H.P.		1192.00		1311.20		
1 Sheepsft. Roller, Towed		489.20		538.12	140.10	154.11
12 L.H., Daily Totals		$2157.20		$2605.12	$179.77	$217.09
Crew B-10E	Hr.	Daily	Hr.	Daily	Bare Costs	Incl. O&P
1 Equip. Oper. (med.)	$42.95	$343.60	$67.30	$538.40	$39.67	$62.98
.5 Laborer	33.10	132.40	54.35	217.40		
1 Tandem Roller, 5 Ton		152.00		167.20	12.67	13.93
12 L.H., Daily Totals		$628.00		$923.00	$52.33	$76.92
Crew B-10F	Hr.	Daily	Hr.	Daily	Bare Costs	Incl. O&P
1 Equip. Oper. (med.)	$42.95	$343.60	$67.30	$538.40	$39.67	$62.98
.5 Laborer	33.10	132.40	54.35	217.40		
1 Tandem Roller, 10 Ton		235.20		258.72	19.60	21.56
12 L.H., Daily Totals		$711.20		$1014.52	$59.27	$84.54
Crew B-10G	Hr.	Daily	Hr.	Daily	Bare Costs	Incl. O&P
1 Equip. Oper. (med.)	$42.95	$343.60	$67.30	$538.40	$39.67	$62.98
.5 Laborer	33.10	132.40	54.35	217.40		
1 Sheepsft. Roll., 240 H.P.		1243.00		1367.30	103.58	113.94
12 L.H., Daily Totals		$1719.00		$2123.10	$143.25	$176.93
Crew B-10H	Hr.	Daily	Hr.	Daily	Bare Costs	Incl. O&P
1 Equip. Oper. (med.)	$42.95	$343.60	$67.30	$538.40	$39.67	$62.98
.5 Laborer	33.10	132.40	54.35	217.40		
1 Diaphragm Water Pump, 2"		68.00		74.80		
1 -20' Suction Hose, 2"		1.95		2.15		
2 -50' Discharge Hoses, 2"		1.80		1.98	5.98	6.58
12 L.H., Daily Totals		$547.75		$834.73	$45.65	$69.56
Crew B-10I	Hr.	Daily	Hr.	Daily	Bare Costs	Incl. O&P
1 Equip. Oper. (med.)	$42.95	$343.60	$67.30	$538.40	$39.67	$62.98
.5 Laborer	33.10	132.40	54.35	217.40		
1 Diaphragm Water Pump, 4"		90.80		99.88		
1 -20' Suction Hose, 4"		3.45		3.79		
2 -50' Discharge Hoses, 4"		4.70		5.17	8.25	9.07
12 L.H., Daily Totals		$574.95		$864.64	$47.91	$72.05
Crew B-10J	Hr.	Daily	Hr.	Daily	Bare Costs	Incl. O&P
1 Equip. Oper. (med.)	$42.95	$343.60	$67.30	$538.40	$39.67	$62.98
.5 Laborer	33.10	132.40	54.35	217.40		
1 Centrifugal Water Pump, 3"		74.40		81.84		
1 -20' Suction Hose, 3"		3.05		3.36		
2 -50' Discharge Hoses, 3"		3.50		3.85	6.75	7.42
12 L.H., Daily Totals		$556.95		$844.85	$46.41	$70.40

Crews

Crew No.		Bare Costs		Incl. Subs O&P		Cost Per Labor-Hour	
Crew B-10K	Hr.	Daily	Hr.	Daily	Bare Costs	Incl. O&P
1 Equip. Oper. (med.)	$42.95	$343.60	$67.30	$538.40	$39.67	$62.98
.5 Laborer	33.10	132.40	54.35	217.40		
1 Centr. Water Pump, 6"		340.00		374.00		
1 -20' Suction Hose, 6"		11.90		13.09		
2 -50' Discharge Hoses, 6"		12.60		13.86	30.38	33.41
12 L.H., Daily Totals		$840.50		$1156.75	$70.04	$96.40
Crew B-10L	Hr.	Daily	Hr.	Daily	Bare Costs	Incl. O&P
1 Equip. Oper. (med.)	$42.95	$343.60	$67.30	$538.40	$39.67	$62.98
.5 Laborer	33.10	132.40	54.35	217.40		
1 Dozer, 80 H.P.		436.60		480.26	36.38	40.02
12 L.H., Daily Totals		$912.60		$1236.06	$76.05	$103.01
Crew B-10M	Hr.	Daily	Hr.	Daily	Bare Costs	Incl. O&P
1 Equip. Oper. (med.)	$42.95	$343.60	$67.30	$538.40	$39.67	$62.98
.5 Laborer	33.10	132.40	54.35	217.40		
1 Dozer, 300 H.P.		1592.00		1751.20	132.67	145.93
12 L.H., Daily Totals		$2068.00		$2507.00	$172.33	$208.92
Crew B-10N	Hr.	Daily	Hr.	Daily	Bare Costs	Incl. O&P
1 Equip. Oper. (med.)	$42.95	$343.60	$67.30	$538.40	$39.67	$62.98
.5 Laborer	33.10	132.40	54.35	217.40		
1 F.E. Loader, T.M., 1.5 C.Y.		433.80		477.18	36.15	39.77
12 L.H., Daily Totals		$909.80		$1232.98	$75.82	$102.75
Crew B-10O	Hr.	Daily	Hr.	Daily	Bare Costs	Incl. O&P
1 Equip. Oper. (med.)	$42.95	$343.60	$67.30	$538.40	$39.67	$62.98
.5 Laborer	33.10	132.40	54.35	217.40		
1 F.E. Loader, T.M., 2.25 C.Y.		839.40		923.34	69.95	76.94
12 L.H., Daily Totals		$1315.40		$1679.14	$109.62	$139.93
Crew B-10P	Hr.	Daily	Hr.	Daily	Bare Costs	Incl. O&P
1 Equip. Oper. (med.)	$42.95	$343.60	$67.30	$538.40	$39.67	$62.98
.5 Laborer	33.10	132.40	54.35	217.40		
1 Crawler Loader, 3 C.Y.		1077.00		1184.70	89.75	98.72
12 L.H., Daily Totals		$1553.00		$1940.50	$129.42	$161.71
Crew B-10Q	Hr.	Daily	Hr.	Daily	Bare Costs	Incl. O&P
1 Equip. Oper. (med.)	$42.95	$343.60	$67.30	$538.40	$39.67	$62.98
.5 Laborer	33.10	132.40	54.35	217.40		
1 Crawler Loader, 4 C.Y.		1548.00		1702.80	129.00	141.90
12 L.H., Daily Totals		$2024.00		$2458.60	$168.67	$204.88
Crew B-10R	Hr.	Daily	Hr.	Daily	Bare Costs	Incl. O&P
1 Equip. Oper. (med.)	$42.95	$343.60	$67.30	$538.40	$39.67	$62.98
.5 Laborer	33.10	132.40	54.35	217.40		
1 F.E. Loader, W.M., 1 C.Y.		269.40		296.34	22.45	24.70
12 L.H., Daily Totals		$745.40		$1052.14	$62.12	$87.68
Crew B-10S	Hr.	Daily	Hr.	Daily	Bare Costs	Incl. O&P
1 Equip. Oper. (med.)	$42.95	$343.60	$67.30	$538.40	$39.67	$62.98
.5 Laborer	33.10	132.40	54.35	217.40		
1 F.E. Loader, W.M., 1.5 C.Y.		367.60		404.36	30.63	33.70
12 L.H., Daily Totals		$843.60		$1160.16	$70.30	$96.68
Crew B-10T	Hr.	Daily	Hr.	Daily	Bare Costs	Incl. O&P
1 Equip. Oper. (med.)	$42.95	$343.60	$67.30	$538.40	$39.67	$62.98
.5 Laborer	33.10	132.40	54.35	217.40		
1 F.E. Loader, W.M., 2.5 C.Y.		435.80		479.38	36.32	39.95
12 L.H., Daily Totals		$911.80		$1235.18	$75.98	$102.93

Crew No.		Bare Costs		Incl. Subs O&P		Cost Per Labor-Hour	
Crew B-10U	Hr.	Daily	Hr.	Daily	Bare Costs	Incl. O&P
1 Equip. Oper. (med.)	$42.95	$343.60	$67.30	$538.40	$39.67	$62.98
.5 Laborer	33.10	132.40	54.35	217.40		
1 F.E. Loader, W.M., 5.5 C.Y.		994.40		1093.84	82.87	91.15
12 L.H., Daily Totals		$1470.40		$1849.64	$122.53	$154.14
Crew B-10V	Hr.	Daily	Hr.	Daily	Bare Costs	Incl. O&P
1 Equip. Oper. (med.)	$42.95	$343.60	$67.30	$538.40	$39.67	$62.98
.5 Laborer	33.10	132.40	54.35	217.40		
1 Dozer, 700 H.P.		4598.00		5057.80	383.17	421.48
12 L.H., Daily Totals		$5074.00		$5813.60	$422.83	$484.47
Crew B-10W	Hr.	Daily	Hr.	Daily	Bare Costs	Incl. O&P
1 Equip. Oper. (med.)	$42.95	$343.60	$67.30	$538.40	$39.67	$62.98
.5 Laborer	33.10	132.40	54.35	217.40		
1 Dozer, 105 H.P.		660.40		726.44	55.03	60.54
12 L.H., Daily Totals		$1136.40		$1482.24	$94.70	$123.52
Crew B-10X	Hr.	Daily	Hr.	Daily	Bare Costs	Incl. O&P
1 Equip. Oper. (med.)	$42.95	$343.60	$67.30	$538.40	$39.67	$62.98
.5 Laborer	33.10	132.40	54.35	217.40		
1 Dozer, 410 H.P.		2079.00		2286.90	173.25	190.57
12 L.H., Daily Totals		$2555.00		$3042.70	$212.92	$253.56
Crew B-10Y	Hr.	Daily	Hr.	Daily	Bare Costs	Incl. O&P
1 Equip. Oper. (med.)	$42.95	$343.60	$67.30	$538.40	$39.67	$62.98
.5 Laborer	33.10	132.40	54.35	217.40		
1 Vibr. Roller, Towed, 12 Ton		516.60		568.26	43.05	47.35
12 L.H., Daily Totals		$992.60		$1324.06	$82.72	$110.34
Crew B-11A	Hr.	Daily	Hr.	Daily	Bare Costs	Incl. O&P
1 Equipment Oper. (med.)	$42.95	$343.60	$67.30	$538.40	$38.02	$60.83
1 Laborer	33.10	264.80	54.35	434.80		
1 Dozer, 200 H.P.		1192.00		1311.20	74.50	81.95
16 L.H., Daily Totals		$1800.40		$2284.40	$112.53	$142.78
Crew B-11B	Hr.	Daily	Hr.	Daily	Bare Costs	Incl. O&P
1 Equipment Oper. (light)	$41.30	$330.40	$64.70	$517.60	$37.20	$59.52
1 Laborer	33.10	264.80	54.35	434.80		
1 Air Powered Tamper		25.25		27.77		
1 Air Compressor, 365 cfm		255.60		281.16		
2 -50' Air Hoses, 1.5"		12.60		13.86	18.34	20.17
16 L.H., Daily Totals		$888.65		$1275.19	$55.54	$79.70
Crew B-11C	Hr.	Daily	Hr.	Daily	Bare Costs	Incl. O&P
1 Equipment Oper. (med.)	$42.95	$343.60	$67.30	$538.40	$38.02	$60.83
1 Laborer	33.10	264.80	54.35	434.80		
1 Backhoe Loader, 48 H.P.		337.80		371.58	21.11	23.22
16 L.H., Daily Totals		$946.20		$1344.78	$59.14	$84.05
Crew B-11K	Hr.	Daily	Hr.	Daily	Bare Costs	Incl. O&P
1 Equipment Oper. (med.)	$42.95	$343.60	$67.30	$538.40	$38.02	$60.83
1 Laborer	33.10	264.80	54.35	434.80		
1 Trencher, Chain Type, 8' D		2090.00		2299.00	130.63	143.69
16 L.H., Daily Totals		$2698.40		$3272.20	$168.65	$204.51
Crew B-11L	Hr.	Daily	Hr.	Daily	Bare Costs	Incl. O&P
1 Equipment Oper. (med.)	$42.95	$343.60	$67.30	$538.40	$38.02	$60.83
1 Laborer	33.10	264.80	54.35	434.80		
1 Grader, 30,000 Lbs.		622.20		684.42	38.89	42.78
16 L.H., Daily Totals		$1230.60		$1657.62	$76.91	$103.60

Crews

Crew No.	Bare Costs		Incl. Subs O&P		Cost Per Labor-Hour	
Crew B-11M	Hr.	Daily	Hr.	Daily	Bare Costs	Incl. O&P
1 Equipment Oper. (med.)	$42.95	$343.60	$67.30	$538.40	$38.02	$60.83
1 Laborer	33.10	264.80	54.35	434.80		
1 Backhoe Loader, 80 H.P.		387.60		426.36	24.23	26.65
16 L.H., Daily Totals		$996.00		$1399.56	$62.25	$87.47
Crew B-11S	Hr.	Daily	Hr.	Daily	Bare Costs	Incl. O&P
1 Equipment Operator (med.)	$42.95	$343.60	$67.30	$538.40	$39.67	$62.98
.5 Laborer	33.10	132.40	54.35	217.40		
1 Dozer, 300 H.P.		1592.00		1751.20		
1 Ripper, beam & 1 shank		80.00		88.00	139.33	153.27
12 L.H., Daily Totals		$2148.00		$2595.00	$179.00	$216.25
Crew B-11T	Hr.	Daily	Hr.	Daily	Bare Costs	Incl. O&P
1 Equipment Operator (med.)	$42.95	$343.60	$67.30	$538.40	$39.67	$62.98
.5 Laborer	33.10	132.40	54.35	217.40		
1 Dozer, 410 H.P.		2079.00		2286.90		
1 Ripper, beam & 2 shanks		90.20		99.22	180.77	198.84
12 L.H., Daily Totals		$2645.20		$3141.92	$220.43	$261.83
Crew B-11W	Hr.	Daily	Hr.	Daily	Bare Costs	Incl. O&P
1 Equipment Operator (med.)	$42.95	$343.60	$67.30	$538.40	$33.96	$55.01
1 Common Laborer	33.10	264.80	54.35	434.80		
10 Truck Drivers (hvy.)	33.15	2652.00	53.85	4308.00		
1 Dozer, 200 H.P.		1192.00		1311.20		
1 Vibratory Roller, Towed, 23 Ton		434.20		477.62		
10 Dump Trucks, 8 C.Y., 220 H.P.		4018.00		4419.80	58.79	64.67
96 L.H., Daily Totals		$8904.60		$11489.82	$92.76	$119.69
Crew B-11Y	Hr.	Daily	Hr.	Daily	Bare Costs	Incl. O&P
1 Labor Foreman (Outside)	$35.10	$280.80	$57.65	$461.20	$36.61	$59.03
5 Common Laborers	33.10	1324.00	54.35	2174.00		
3 Equipment Operators (med.)	42.95	1030.80	67.30	1615.20		
1 Dozer, 80 H.P.		436.60		480.26		
2 Rollers, 2-Drum, W.B., 7.5 H.P.		299.20		329.12		
4 Vibratory Plates, gas, 21"		174.40		191.84	12.64	13.91
72 L.H., Daily Totals		$3545.80		$5251.62	$49.25	$72.94
Crew B-12A	Hr.	Daily	Hr.	Daily	Bare Costs	Incl. O&P
1 Equip. Oper. (crane)	$44.40	$355.20	$69.55	$556.40	$38.75	$61.95
1 Laborer	33.10	264.80	54.35	434.80		
1 Hyd. Excavator, 1 C.Y.		801.60		881.76	50.10	55.11
16 L.H., Daily Totals		$1421.60		$1872.96	$88.85	$117.06
Crew B-12B	Hr.	Daily	Hr.	Daily	Bare Costs	Incl. O&P
1 Equip. Oper. (crane)	$44.40	$355.20	$69.55	$556.40	$38.75	$61.95
1 Laborer	33.10	264.80	54.35	434.80		
1 Hyd. Excavator, 1.5 C.Y.		1017.00		1118.70	63.56	69.92
16 L.H., Daily Totals		$1637.00		$2109.90	$102.31	$131.87
Crew B-12C	Hr.	Daily	Hr.	Daily	Bare Costs	Incl. O&P
1 Equip. Oper. (crane)	$44.40	$355.20	$69.55	$556.40	$38.75	$61.95
1 Laborer	33.10	264.80	54.35	434.80		
1 Hyd. Excavator, 2 C.Y.		1321.00		1453.10	82.56	90.82
16 L.H., Daily Totals		$1941.00		$2444.30	$121.31	$152.77
Crew B-12D	Hr.	Daily	Hr.	Daily	Bare Costs	Incl. O&P
1 Equip. Oper. (crane)	$44.40	$355.20	$69.55	$556.40	$38.75	$61.95
1 Laborer	33.10	264.80	54.35	434.80		
1 Hyd. Excavator, 3.5 C.Y.		2359.00		2594.90	147.44	162.18
16 L.H., Daily Totals		$2979.00		$3586.10	$186.19	$224.13

Crew No.	Bare Costs		Incl. Subs O&P		Cost Per Labor-Hour	
Crew B-12E	Hr.	Daily	Hr.	Daily	Bare Costs	Incl. O&P
1 Equip. Oper. (crane)	$44.40	$355.20	$69.55	$556.40	$38.75	$61.95
1 Laborer	33.10	264.80	54.35	434.80		
1 Hyd. Excavator, .5 C.Y.		386.60		425.26	24.16	26.58
16 L.H., Daily Totals		$1006.60		$1416.46	$62.91	$88.53
Crew B-12F	Hr.	Daily	Hr.	Daily	Bare Costs	Incl. O&P
1 Equip. Oper. (crane)	$44.40	$355.20	$69.55	$556.40	$38.75	$61.95
1 Laborer	33.10	264.80	54.35	434.80		
1 Hyd. Excavator, .75 C.Y.		677.80		745.58	42.36	46.60
16 L.H., Daily Totals		$1297.80		$1736.78	$81.11	$108.55
Crew B-12G	Hr.	Daily	Hr.	Daily	Bare Costs	Incl. O&P
1 Equip. Oper. (crane)	$44.40	$355.20	$69.55	$556.40	$38.75	$61.95
1 Laborer	33.10	264.80	54.35	434.80		
1 Crawler Crane, 15 Ton		693.75		763.13		
1 Clamshell Bucket, .5 C.Y.		37.00		40.70	45.67	50.24
16 L.H., Daily Totals		$1350.75		$1795.03	$84.42	$112.19
Crew B-12H	Hr.	Daily	Hr.	Daily	Bare Costs	Incl. O&P
1 Equip. Oper. (crane)	$44.40	$355.20	$69.55	$556.40	$38.75	$61.95
1 Laborer	33.10	264.80	54.35	434.80		
1 Crawler Crane, 25 Ton		1188.00		1306.80		
1 Clamshell Bucket, 1 C.Y.		46.60		51.26	77.16	84.88
16 L.H., Daily Totals		$1854.60		$2349.26	$115.91	$146.83
Crew B-12I	Hr.	Daily	Hr.	Daily	Bare Costs	Incl. O&P
1 Equip. Oper. (crane)	$44.40	$355.20	$69.55	$556.40	$38.75	$61.95
1 Laborer	33.10	264.80	54.35	434.80		
1 Crawler Crane, 20 Ton		892.30		981.53		
1 Dragline Bucket, .75 C.Y.		20.40		22.44	57.04	62.75
16 L.H., Daily Totals		$1532.70		$1995.17	$95.79	$124.70
Crew B-12J	Hr.	Daily	Hr.	Daily	Bare Costs	Incl. O&P
1 Equip. Oper. (crane)	$44.40	$355.20	$69.55	$556.40	$38.75	$61.95
1 Laborer	33.10	264.80	54.35	434.80		
1 Gradall, 5/8 C.Y.		812.80		894.08	50.80	55.88
16 L.H., Daily Totals		$1432.80		$1885.28	$89.55	$117.83
Crew B-12K	Hr.	Daily	Hr.	Daily	Bare Costs	Incl. O&P
1 Equip. Oper. (crane)	$44.40	$355.20	$69.55	$556.40	$38.75	$61.95
1 Laborer	33.10	264.80	54.35	434.80		
1 Gradall, 3 Ton, 1 C.Y.		958.40		1054.24	59.90	65.89
16 L.H., Daily Totals		$1578.40		$2045.44	$98.65	$127.84
Crew B-12L	Hr.	Daily	Hr.	Daily	Bare Costs	Incl. O&P
1 Equip. Oper. (crane)	$44.40	$355.20	$69.55	$556.40	$38.75	$61.95
1 Laborer	33.10	264.80	54.35	434.80		
1 Crawler Crane, 15 Ton		693.75		763.13		
1 F.E. Attachment, .5 C.Y.		55.60		61.16	46.83	51.52
16 L.H., Daily Totals		$1369.35		$1815.48	$85.58	$113.47
Crew B-12M	Hr.	Daily	Hr.	Daily	Bare Costs	Incl. O&P
1 Equip. Oper. (crane)	$44.40	$355.20	$69.55	$556.40	$38.75	$61.95
1 Laborer	33.10	264.80	54.35	434.80		
1 Crawler Crane, 20 Ton		892.30		981.53		
1 F.E. Attachment, .75 C.Y.		60.00		66.00	59.52	65.47
16 L.H., Daily Totals		$1572.30		$2038.73	$98.27	$127.42

Crews

Crew No.		Bare Costs		Incl. Subs O&P	Cost Per Labor-Hour	
Crew B-12N	Hr.	Daily	Hr.	Daily	Bare Costs	Incl. O&P
1 Equip. Oper. (crane)	$44.40	$355.20	$69.55	$556.40	$38.75	$61.95
1 Laborer	33.10	264.80	54.35	434.80		
1 Crawler Crane, 25 Ton		1188.00		1306.80		
1 F.E. Attachment, 1 C.Y.		65.80		72.38	78.36	86.20
16 L.H., Daily Totals		$1873.80		$2370.38	$117.11	$148.15
Crew B-12O	Hr.	Daily	Hr.	Daily	Bare Costs	Incl. O&P
1 Equip. Oper. (crane)	$44.40	$355.20	$69.55	$556.40	$38.75	$61.95
1 Laborer	33.10	264.80	54.35	434.80		
1 Crawler Crane, 40 Ton		1193.00		1312.30		
1 F.E. Attachment, 1.5 C.Y.		76.00		83.60	79.31	87.24
16 L.H., Daily Totals		$1889.00		$2387.10	$118.06	$149.19
Crew B-12P	Hr.	Daily	Hr.	Daily	Bare Costs	Incl. O&P
1 Equip. Oper. (crane)	$44.40	$355.20	$69.55	$556.40	$38.75	$61.95
1 Laborer	33.10	264.80	54.35	434.80		
1 Crawler Crane, 40 Ton		1193.00		1312.30		
1 Dragline Bucket, 1.5 C.Y.		32.80		36.08	76.61	84.27
16 L.H., Daily Totals		$1845.80		$2339.58	$115.36	$146.22
Crew B-12Q	Hr.	Daily	Hr.	Daily	Bare Costs	Incl. O&P
1 Equip. Oper. (crane)	$44.40	$355.20	$69.55	$556.40	$38.75	$61.95
1 Laborer	33.10	264.80	54.35	434.80		
1 Hyd. Excavator, 5/8 C.Y.		564.40		620.84	35.27	38.80
16 L.H., Daily Totals		$1184.40		$1612.04	$74.03	$100.75
Crew B-12S	Hr.	Daily	Hr.	Daily	Bare Costs	Incl. O&P
1 Equip. Oper. (crane)	$44.40	$355.20	$69.55	$556.40	$38.75	$61.95
1 Laborer	33.10	264.80	54.35	434.80		
1 Hyd. Excavator, 2.5 C.Y.		1784.00		1962.40	111.50	122.65
16 L.H., Daily Totals		$2404.00		$2953.60	$150.25	$184.60
Crew B-12T	Hr.	Daily	Hr.	Daily	Bare Costs	Incl. O&P
1 Equip. Oper. (crane)	$44.40	$355.20	$69.55	$556.40	$38.75	$61.95
1 Laborer	33.10	264.80	54.35	434.80		
1 Crawler Crane, 75 Ton		1516.00		1667.60		
1 F.E. Attachment, 3 C.Y.		98.40		108.24	100.90	110.99
16 L.H., Daily Totals		$2234.40		$2767.04	$139.65	$172.94
Crew B-12V	Hr.	Daily	Hr.	Daily	Bare Costs	Incl. O&P
1 Equip. Oper. (crane)	$44.40	$355.20	$69.55	$556.40	$38.75	$61.95
1 Laborer	33.10	264.80	54.35	434.80		
1 Crawler Crane, 75 Ton		1516.00		1667.60		
1 Dragline Bucket, 3 C.Y.		52.40		57.64	98.03	107.83
16 L.H., Daily Totals		$2188.40		$2716.44	$136.78	$169.78
Crew B-13	Hr.	Daily	Hr.	Daily	Bare Costs	Incl. O&P
1 Labor Foreman (outside)	$35.10	$280.80	$57.65	$461.20	$35.74	$57.80
4 Laborers	33.10	1059.20	54.35	1739.20		
1 Equip. Oper. (crane)	44.40	355.20	69.55	556.40		
1 Equip. Oper. Oiler	38.30	306.40	60.00	480.00		
1 Hyd. Crane, 25 Ton		747.60		822.36	13.35	14.69
56 L.H., Daily Totals		$2749.20		$4059.16	$49.09	$72.48

Crew No.		Bare Costs		Incl. Subs O&P	Cost Per Labor-Hour	
Crew B-13A	Hr.	Daily	Hr.	Daily	Bare Costs	Incl. O&P
1 Foreman	$35.10	$280.80	$57.65	$461.20	$36.21	$58.38
2 Laborers	33.10	529.60	54.35	869.60		
2 Equipment Operators (med.)	42.95	687.20	67.30	1076.80		
2 Truck Drivers (heavy)	33.15	530.40	53.85	861.60		
1 Crawler Crane, 75 Ton		1516.00		1667.60		
1 Crawler Loader, 4 C.Y.		1548.00		1702.80		
2 Dump Trucks, 8 C.Y., 220 H.P.		803.60		883.96	69.06	75.97
56 L.H., Daily Totals		$5895.60		$7523.56	$105.28	$134.35
Crew B-13B	Hr.	Daily	Hr.	Daily	Bare Costs	Incl. O&P
1 Labor Foreman (outside)	$35.10	$280.80	$57.65	$461.20	$35.74	$57.80
4 Laborers	33.10	1059.20	54.35	1739.20		
1 Equip. Oper. (crane)	44.40	355.20	69.55	556.40		
1 Equip. Oper. Oiler	38.30	306.40	60.00	480.00		
1 Hyd. Crane, 55 Ton		1209.00		1329.90	21.59	23.75
56 L.H., Daily Totals		$3210.60		$4566.70	$57.33	$81.55
Crew B-13C	Hr.	Daily	Hr.	Daily	Bare Costs	Incl. O&P
1 Labor Foreman (outside)	$35.10	$280.80	$57.65	$461.20	$35.74	$57.80
4 Laborers	33.10	1059.20	54.35	1739.20		
1 Equip. Oper. (crane)	44.40	355.20	69.55	556.40		
1 Equip. Oper. Oiler	38.30	306.40	60.00	480.00		
1 Crawler Crane, 100 Ton		1732.00		1905.20	30.93	34.02
56 L.H., Daily Totals		$3733.60		$5142.00	$66.67	$91.82
Crew B-13D	Hr.	Daily	Hr.	Daily	Bare Costs	Incl. O&P
1 Laborer	$33.10	$264.80	$54.35	$434.80	$38.75	$61.95
1 Equip. Oper. (crane)	44.40	355.20	69.55	556.40		
1 Hyd. Excavator, 1 C.Y.		801.60		881.76		
1 Trench Box		113.40		124.74	57.19	62.91
16 L.H., Daily Totals		$1535.00		$1997.70	$95.94	$124.86
Crew B-13E	Hr.	Daily	Hr.	Daily	Bare Costs	Incl. O&P
1 Laborer	$33.10	$264.80	$54.35	$434.80	$38.75	$61.95
1 Equip. Oper. (crane)	44.40	355.20	69.55	556.40		
1 Hyd. Excavator, 1.5 C.Y.		1017.00		1118.70		
1 Trench Box		113.40		124.74	70.65	77.72
16 L.H., Daily Totals		$1750.40		$2234.64	$109.40	$139.66
Crew B-13F	Hr.	Daily	Hr.	Daily	Bare Costs	Incl. O&P
1 Laborer	$33.10	$264.80	$54.35	$434.80	$38.75	$61.95
1 Equip. Oper. (crane)	44.40	355.20	69.55	556.40		
1 Hyd. Excavator, 3.5 C.Y.		2359.00		2594.90		
1 Trench Box		113.40		124.74	154.53	169.98
16 L.H., Daily Totals		$3092.40		$3710.84	$193.28	$231.93
Crew B-13G	Hr.	Daily	Hr.	Daily	Bare Costs	Incl. O&P
1 Laborer	$33.10	$264.80	$54.35	$434.80	$38.75	$61.95
1 Equip. Oper. (crane)	44.40	355.20	69.55	556.40		
1 Hyd. Excavator, .75 C.Y.		677.80		745.58		
1 Trench Box		113.40		124.74	49.45	54.40
16 L.H., Daily Totals		$1411.20		$1861.52	$88.20	$116.35
Crew B-13H	Hr.	Daily	Hr.	Daily	Bare Costs	Incl. O&P
1 Laborer	$33.10	$264.80	$54.35	$434.80	$38.75	$61.95
1 Equip. Oper. (crane)	44.40	355.20	69.55	556.40		
1 Gradall, 5/8 C.Y.		812.80		894.08		
1 Trench Box		113.40		124.74	57.89	63.68
16 L.H., Daily Totals		$1546.20		$2010.02	$96.64	$125.63

Crews

Crew No.	Bare Costs		Incl. Subs O&P		Cost Per Labor-Hour	
Crew B-13I	Hr.	Daily	Hr.	Daily	Bare Costs	Incl. O&P
1 Laborer	$33.10	$264.80	$54.35	$434.80	$38.75	$61.95
1 Equip. Oper. (crane)	44.40	355.20	69.55	556.40		
1 Gradall, 3 Ton, 1 C.Y.		958.40		1054.24		
1 Trench Box		113.40		124.74	66.99	73.69
16 L.H., Daily Totals		$1691.80		$2170.18	$105.74	$135.64
Crew B-13J	Hr.	Daily	Hr.	Daily	Bare Costs	Incl. O&P
1 Laborer	$33.10	$264.80	$54.35	$434.80	$38.75	$61.95
1 Equip. Oper. (crane)	44.40	355.20	69.55	556.40		
1 Hyd. Excavator, 2.5 C.Y.		1784.00		1962.40		
1 Trench Box		113.40		124.74	118.59	130.45
16 L.H., Daily Totals		$2517.40		$3078.34	$157.34	$192.40
Crew B-14	Hr.	Daily	Hr.	Daily	Bare Costs	Incl. O&P
1 Labor Foreman (outside)	$35.10	$280.80	$57.65	$461.20	$34.80	$56.63
4 Laborers	33.10	1059.20	54.35	1739.20		
1 Equip. Oper. (light)	41.30	330.40	64.70	517.60		
1 Backhoe Loader, 48 H.P.		337.80		371.58	7.04	7.74
48 L.H., Daily Totals		$2008.20		$3089.58	$41.84	$64.37
Crew B-14A	Hr.	Daily	Hr.	Daily	Bare Costs	Incl. O&P
1 Equip. Oper. (crane)	$44.40	$355.20	$69.55	$556.40	$40.63	$64.48
.5 Laborer	33.10	132.40	54.35	217.40		
1 Hyd. Excavator, 4.5 C.Y.		2870.00		3157.00	239.17	263.08
12 L.H., Daily Totals		$3357.60		$3930.80	$279.80	$327.57
Crew B-14B	Hr.	Daily	Hr.	Daily	Bare Costs	Incl. O&P
1 Equip. Oper. (crane)	$44.40	$355.20	$69.55	$556.40	$40.63	$64.48
.5 Laborer	33.10	132.40	54.35	217.40		
1 Hyd. Excavator, 6 C.Y.		3390.00		3729.00	282.50	310.75
12 L.H., Daily Totals		$3877.60		$4502.80	$323.13	$375.23
Crew B-14C	Hr.	Daily	Hr.	Daily	Bare Costs	Incl. O&P
1 Equip. Oper. (crane)	$44.40	$355.20	$69.55	$556.40	$40.63	$64.48
.5 Laborer	33.10	132.40	54.35	217.40		
1 Hyd. Excavator, 7 C.Y.		3553.00		3908.30	296.08	325.69
12 L.H., Daily Totals		$4040.60		$4682.10	$336.72	$390.18
Crew B-14F	Hr.	Daily	Hr.	Daily	Bare Costs	Incl. O&P
1 Equip. Oper. (crane)	$44.40	$355.20	$69.55	$556.40	$40.63	$64.48
.5 Laborer	33.10	132.40	54.35	217.40		
1 Hyd. Shovel, 7 C.Y.		3536.00		3889.60	294.67	324.13
12 L.H., Daily Totals		$4023.60		$4663.40	$335.30	$388.62
Crew B-14G	Hr.	Daily	Hr.	Daily	Bare Costs	Incl. O&P
1 Equip. Oper. (crane)	$44.40	$355.20	$69.55	$556.40	$40.63	$64.48
.5 Laborer	33.10	132.40	54.35	217.40		
1 Hyd. Shovel, 12 C.Y.		4614.00		5075.40	384.50	422.95
12 L.H., Daily Totals		$5101.60		$5849.20	$425.13	$487.43
Crew B-14J	Hr.	Daily	Hr.	Daily	Bare Costs	Incl. O&P
1 Equip. Oper. (med.)	$42.95	$343.60	$67.30	$538.40	$39.67	$62.98
.5 Laborer	33.10	132.40	54.35	217.40		
1 F.E. Loader, 8 C.Y.		1757.00		1932.70	146.42	161.06
12 L.H., Daily Totals		$2233.00		$2688.50	$186.08	$224.04

Crew No.	Bare Costs		Incl. Subs O&P		Cost Per Labor-Hour	
Crew B-14K	Hr.	Daily	Hr.	Daily	Bare Costs	Incl. O&P
1 Equip. Oper. (med.)	$42.95	$343.60	$67.30	$538.40	$39.67	$62.98
.5 Laborer	33.10	132.40	54.35	217.40		
1 F.E. Loader, 10 C.Y.		2601.00		2861.10	216.75	238.43
12 L.H., Daily Totals		$3077.00		$3616.90	$256.42	$301.41
Crew B-15	Hr.	Daily	Hr.	Daily	Bare Costs	Incl. O&P
1 Equipment Oper. (med.)	$42.95	$343.60	$67.30	$538.40	$35.94	$57.76
.5 Laborer	33.10	132.40	54.35	217.40		
2 Truck Drivers (heavy)	33.15	530.40	53.85	861.60		
2 Dump Trucks, 12 C.Y., 400 H.P.		1335.60		1469.16		
1 Dozer, 200 H.P.		1192.00		1311.20	90.27	99.30
28 L.H., Daily Totals		$3534.00		$4397.76	$126.21	$157.06
Crew B-16	Hr.	Daily	Hr.	Daily	Bare Costs	Incl. O&P
1 Labor Foreman (outside)	$35.10	$280.80	$57.65	$461.20	$33.61	$55.05
2 Laborers	33.10	529.60	54.35	869.60		
1 Truck Driver (heavy)	33.15	265.20	53.85	430.80		
1 Dump Truck, 12 C.Y., 400 H.P.		667.80		734.58	20.87	22.96
32 L.H., Daily Totals		$1743.40		$2496.18	$54.48	$78.01
Crew B-17	Hr.	Daily	Hr.	Daily	Bare Costs	Incl. O&P
2 Laborers	$33.10	$529.60	$54.35	$869.60	$35.16	$56.81
1 Equip. Oper. (light)	41.30	330.40	64.70	517.60		
1 Truck Driver (heavy)	33.15	265.20	53.85	430.80		
1 Backhoe Loader, 48 H.P.		337.80		371.58		
1 Dump Truck, 8 C.Y., 220 H.P.		401.80		441.98	23.11	25.42
32 L.H., Daily Totals		$1864.80		$2631.56	$58.27	$82.24
Crew B-17A	Hr.	Daily	Hr.	Daily	Bare Costs	Incl. O&P
2 Laborer Foremen	$35.10	$561.60	$57.65	$922.40	$35.60	$58.24
6 Laborers	33.10	1588.80	54.35	2608.80		
1 Skilled Worker Foreman	44.60	356.80	72.10	576.80		
1 Skilled Worker	42.60	340.80	68.90	551.20		
80 L.H., Daily Totals		$2848.00		$4659.20	$35.60	$58.24
Crew B-18	Hr.	Daily	Hr.	Daily	Bare Costs	Incl. O&P
1 Labor Foreman (outside)	$35.10	$280.80	$57.65	$461.20	$33.77	$55.45
2 Laborers	33.10	529.60	54.35	869.60		
1 Vibratory Plate, gas, 21"		43.60		47.96	1.82	2.00
24 L.H., Daily Totals		$854.00		$1378.76	$35.58	$57.45
Crew B-19	Hr.	Daily	Hr.	Daily	Bare Costs	Incl. O&P
1 Pile Driver Foreman	$42.30	$338.40	$70.15	$561.20	$41.33	$67.06
4 Pile Drivers	40.30	1289.60	66.80	2137.60		
2 Equip. Oper. (crane)	44.40	710.40	69.55	1112.80		
1 Equip. Oper. Oiler	38.30	306.40	60.00	480.00		
1 Crawler Crane, 40 Ton		1193.00		1312.30		
1 Lead, 90' high		120.40		132.44		
1 Hammer, Diesel, 22k ft-lb		619.40		681.34	30.20	33.22
64 L.H., Daily Totals		$4577.60		$6417.68	$71.53	$100.28
Crew B-19A	Hr.	Daily	Hr.	Daily	Bare Costs	Incl. O&P
1 Pile Driver Foreman	$42.30	$338.40	$70.15	$561.20	$41.33	$67.06
4 Pile Drivers	40.30	1289.60	66.80	2137.60		
2 Equip. Oper. (crane)	44.40	710.40	69.55	1112.80		
1 Equip. Oper. Oiler	38.30	306.40	60.00	480.00		
1 Crawler Crane, 75 Ton		1516.00		1667.60		
1 Lead, 90' high		120.40		132.44		
1 Hammer, Diesel, 41k ft-lb		722.80		795.08	36.86	40.55
64 L.H., Daily Totals		$5004.00		$6886.72	$78.19	$107.61

Crews

Crew No.	Bare Costs		Incl. Subs O&P		Cost Per Labor-Hour	
Crew B-20	Hr.	Daily	Hr.	Daily	Bare Costs	Incl. O&P
1 Labor Foreman (out)	$35.10	$280.80	$57.65	$461.20	$36.93	$60.30
1 Skilled Worker	42.60	340.80	68.90	551.20		
1 Laborer	33.10	264.80	54.35	434.80		
24 L.H., Daily Totals		$886.40		$1447.20	$36.93	$60.30

Crew No.	Bare Costs		Incl. Subs O&P		Cost Per Labor-Hour	
Crew B-20A	Hr.	Daily	Hr.	Daily	Bare Costs	Incl. O&P
1 Labor Foreman	$35.10	$280.80	$57.65	$461.20	$40.48	$64.29
1 Laborer	33.10	264.80	54.35	434.80		
1 Plumber	52.05	416.40	80.65	645.20		
1 Plumber Apprentice	41.65	333.20	64.50	516.00		
32 L.H., Daily Totals		$1295.20		$2057.20	$40.48	$64.29

Crew B-21	Hr.	Daily	Hr.	Daily	Bare Costs	Incl. O&P
1 Labor Foreman (out)	$35.10	$280.80	$57.65	$461.20	$38.00	$61.62
1 Skilled Worker	42.60	340.80	68.90	551.20		
1 Laborer	33.10	264.80	54.35	434.80		
.5 Equip. Oper. (crane)	44.40	177.60	69.55	278.20		
.5 S.P. Crane, 4x4, 5 Ton		138.10		151.91	4.93	5.43
28 L.H., Daily Totals		$1202.10		$1877.31	$42.93	$67.05

Crew B-21A	Hr.	Daily	Hr.	Daily	Bare Costs	Incl. O&P
1 Labor Foreman	$35.10	$280.80	$57.65	$461.20	$41.26	$65.34
1 Laborer	33.10	264.80	54.35	434.80		
1 Plumber	52.05	416.40	80.65	645.20		
1 Plumber Apprentice	41.65	333.20	64.50	516.00		
1 Equip. Oper. (crane)	44.40	355.20	69.55	556.40		
1 S.P. Crane, 4x4, 12 Ton		655.60		721.16	16.39	18.03
40 L.H., Daily Totals		$2306.00		$3334.76	$57.65	$83.37

Crew B-21B	Hr.	Daily	Hr.	Daily	Bare Costs	Incl. O&P
1 Laborer Foreman	$35.10	$280.80	$57.65	$461.20	$35.76	$58.05
3 Laborers	33.10	794.40	54.35	1304.40		
1 Equip. Oper. (crane)	44.40	355.20	69.55	556.40		
1 Hyd. Crane, 12 Ton		655.60		721.16	16.39	18.03
40 L.H., Daily Totals		$2086.00		$3043.16	$52.15	$76.08

Crew B-21C	Hr.	Daily	Hr.	Daily	Bare Costs	Incl. O&P
1 Laborer Foreman	$35.10	$280.80	$57.65	$461.20	$35.74	$57.80
4 Laborers	33.10	1059.20	54.35	1739.20		
1 Equip. Oper. (crane)	44.40	355.20	69.55	556.40		
1 Equip. Oper. Oiler	38.30	306.40	60.00	480.00		
2 Cutting Torches		38.00		41.80		
2 Set of Gases		158.40		174.24		
1 Lattice Boom Crane, 90 Ton		1607.00		1767.70	32.20	35.42
56 L.H., Daily Totals		$3805.00		$5220.54	$67.95	$93.22

Crew B-22	Hr.	Daily	Hr.	Daily	Bare Costs	Incl. O&P
1 Labor Foreman (out)	$35.10	$280.80	$57.65	$461.20	$38.43	$62.15
1 Skilled Worker	42.60	340.80	68.90	551.20		
1 Laborer	33.10	264.80	54.35	434.80		
.75 Equip. Oper. (crane)	44.40	266.40	69.55	417.30		
.75 S.P. Crane, 4x4, 5 Ton		207.15		227.87	6.91	7.60
30 L.H., Daily Totals		$1359.95		$2092.36	$45.33	$69.75

Crew B-22A	Hr.	Daily	Hr.	Daily	Bare Costs	Incl. O&P
1 Labor Foreman (out)	$35.10	$280.80	$57.65	$461.20	$37.31	$60.51
1 Skilled Worker	42.60	340.80	68.90	551.20		
2 Laborers	33.10	529.60	54.35	869.60		
.75 Equipment Oper. (crane)	44.40	266.40	69.55	417.30		
.75 S.P. Crane, 4x4, 5 Ton		207.15		227.87		
1 Generator, 5 kW		46.00		50.60		
1 Butt Fusion Machine		134.40		147.84	10.20	11.22
38 L.H., Daily Totals		$1805.15		$2725.61	$47.50	$71.73

Crew B-22B	Hr.	Daily	Hr.	Daily	Bare Costs	Incl. O&P
1 Skilled Worker	$42.60	$340.80	$68.90	$551.20	$37.85	$61.63
1 Laborer	33.10	264.80	54.35	434.80		
1 Electro Fusion Machine		98.60		108.46	6.16	6.78
16 L.H., Daily Totals		$704.20		$1094.46	$44.01	$68.40

Crew B-23	Hr.	Daily	Hr.	Daily	Bare Costs	Incl. O&P
1 Labor Foreman (outside)	$35.10	$280.80	$57.65	$461.20	$33.50	$55.01
4 Laborers	33.10	1059.20	54.35	1739.20		
1 Drill Rig, Truck-Mounted		2591.00		2850.10		
1 Flatbed Truck, Gas, 3 Ton		296.00		325.60	72.17	79.39
40 L.H., Daily Totals		$4227.00		$5376.10	$105.68	$134.40

Crew B-23A	Hr.	Daily	Hr.	Daily	Bare Costs	Incl. O&P
1 Labor Foreman (outside)	$35.10	$280.80	$57.65	$461.20	$37.05	$59.77
1 Laborer	33.10	264.80	54.35	434.80		
1 Equip. Operator (med.)	42.95	343.60	67.30	538.40		
1 Drill Rig, Truck-Mounted		2591.00		2850.10		
1 Pickup Truck, 3/4 Ton		140.00		154.00	113.79	125.17
24 L.H., Daily Totals		$3620.20		$4438.50	$150.84	$184.94

Crew B-23B	Hr.	Daily	Hr.	Daily	Bare Costs	Incl. O&P
1 Labor Foreman (outside)	$35.10	$280.80	$57.65	$461.20	$37.05	$59.77
1 Laborer	33.10	264.80	54.35	434.80		
1 Equip. Operator (med.)	42.95	343.60	67.30	538.40		
1 Drill Rig, Truck-Mounted		2591.00		2850.10		
1 Pickup Truck, 3/4 Ton		140.00		154.00		
1 Centr. Water Pump, 6"		340.00		374.00	127.96	140.75
24 L.H., Daily Totals		$3960.20		$4812.50	$165.01	$200.52

Crew B-24	Hr.	Daily	Hr.	Daily	Bare Costs	Incl. O&P
1 Cement Finisher	$39.70	$317.60	$61.95	$495.60	$38.12	$61.52
1 Laborer	33.10	264.80	54.35	434.80		
1 Carpenter	41.55	332.40	68.25	546.00		
24 L.H., Daily Totals		$914.80		$1476.40	$38.12	$61.52

Crew B-25	Hr.	Daily	Hr.	Daily	Bare Costs	Incl. O&P
1 Labor Foreman	$35.10	$280.80	$57.65	$461.20	$35.97	$58.18
7 Laborers	33.10	1853.60	54.35	3043.60		
3 Equip. Oper. (med.)	42.95	1030.80	67.30	1615.20		
1 Asphalt Paver, 130 H.P.		1994.00		2193.40		
1 Tandem Roller, 10 Ton		235.20		258.72		
1 Roller, Pneum. Whl, 12 Ton		337.20		370.92	29.16	32.08
88 L.H., Daily Totals		$5731.60		$7943.04	$65.13	$90.26

Crews

Crew No.	Bare Costs		Incl. Subs O&P		Cost Per Labor-Hour	
Crew B-25B	Hr.	Daily	Hr.	Daily	Bare Costs	Incl. O&P
1 Labor Foreman	$35.10	$280.80	$57.65	$461.20	$36.55	$58.94
7 Laborers	33.10	1853.60	54.35	3043.60		
4 Equip. Oper. (med.)	42.95	1374.40	67.30	2153.60		
1 Asphalt Paver, 130 H.P.		1994.00		2193.40		
2 Tandem Rollers, 10 Ton		470.40		517.44		
1 Roller, Pneum. Whl, 12 Ton		337.20		370.92	29.18	32.10
96 L.H., Daily Totals		$6310.40		$8740.16	$65.73	$91.04
Crew B-25C	Hr.	Daily	Hr.	Daily	Bare Costs	Incl. O&P
1 Labor Foreman	$35.10	$280.80	$57.65	$461.20	$36.72	$59.22
3 Laborers	33.10	794.40	54.35	1304.40		
2 Equip. Oper. (med.)	42.95	687.20	67.30	1076.80		
1 Asphalt Paver, 130 H.P.		1994.00		2193.40		
1 Tandem Roller, 10 Ton		235.20		258.72	46.44	51.09
48 L.H., Daily Totals		$3991.60		$5294.52	$83.16	$110.30
Crew B-26	Hr.	Daily	Hr.	Daily	Bare Costs	Incl. O&P
1 Labor Foreman (outside)	$35.10	$280.80	$57.65	$461.20	$36.92	$59.90
6 Laborers	33.10	1588.80	54.35	2608.80		
2 Equip. Oper. (med.)	42.95	687.20	67.30	1076.80		
1 Rodman (reinf.)	46.80	374.40	78.65	629.20		
1 Cement Finisher	39.70	317.60	61.95	495.60		
1 Grader, 30,000 Lbs.		622.20		684.42		
1 Paving Mach. & Equip.		2678.00		2945.80	37.50	41.25
88 L.H., Daily Totals		$6549.00		$8901.82	$74.42	$101.16
Crew B-26A	Hr.	Daily	Hr.	Daily	Bare Costs	Incl. O&P
1 Labor Foreman (outside)	$35.10	$280.80	$57.65	$461.20	$36.92	$59.90
6 Laborers	33.10	1588.80	54.35	2608.80		
2 Equip. Oper. (med.)	42.95	687.20	67.30	1076.80		
1 Rodman (reinf.)	46.80	374.40	78.65	629.20		
1 Cement Finisher	39.70	317.60	61.95	495.60		
1 Grader, 30,000 Lbs.		622.20		684.42		
1 Paving Mach. & Equip.		2678.00		2945.80		
1 Concrete Saw		161.40		177.54	39.34	43.27
88 L.H., Daily Totals		$6710.40		$9079.36	$76.25	$103.17
Crew B-26B	Hr.	Daily	Hr.	Daily	Bare Costs	Incl. O&P
1 Labor Foreman (outside)	$35.10	$280.80	$57.65	$461.20	$37.42	$60.52
6 Laborers	33.10	1588.80	54.35	2608.80		
3 Equip. Oper. (med.)	42.95	1030.80	67.30	1615.20		
1 Rodman (reinf.)	46.80	374.40	78.65	629.20		
1 Cement Finisher	39.70	317.60	61.95	495.60		
1 Grader, 30,000 Lbs.		622.20		684.42		
1 Paving Mach. & Equip.		2678.00		2945.80		
1 Concrete Pump, 110' Boom		976.40		1074.04	44.55	49.00
96 L.H., Daily Totals		$7869.00		$10514.26	$81.97	$109.52
Crew B-27	Hr.	Daily	Hr.	Daily	Bare Costs	Incl. O&P
1 Labor Foreman (outside)	$35.10	$280.80	$57.65	$461.20	$33.60	$55.17
3 Laborers	33.10	794.40	54.35	1304.40		
1 Berm Machine		285.60		314.16	8.93	9.82
32 L.H., Daily Totals		$1360.80		$2079.76	$42.52	$64.99
Crew B-28	Hr.	Daily	Hr.	Daily	Bare Costs	Incl. O&P
2 Carpenters	$41.55	$664.80	$68.25	$1092.00	$38.73	$63.62
1 Laborer	33.10	264.80	54.35	434.80		
24 L.H., Daily Totals		$929.60		$1526.80	$38.73	$63.62

Crew No.	Bare Costs		Incl. Subs O&P		Cost Per Labor-Hour	
Crew B-29	Hr.	Daily	Hr.	Daily	Bare Costs	Incl. O&P
1 Labor Foreman (outside)	$35.10	$280.80	$57.65	$461.20	$35.74	$57.80
4 Laborers	33.10	1059.20	54.35	1739.20		
1 Equip. Oper. (crane)	44.40	355.20	69.55	556.40		
1 Equip. Oper. Oiler	38.30	306.40	60.00	480.00		
1 Gradall, 5/8 C.Y.		812.80		894.08	14.51	15.97
56 L.H., Daily Totals		$2814.40		$4130.88	$50.26	$73.77
Crew B-30	Hr.	Daily	Hr.	Daily	Bare Costs	Incl. O&P
1 Equip. Oper. (med.)	$42.95	$343.60	$67.30	$538.40	$36.42	$58.33
2 Truck Drivers (heavy)	33.15	530.40	53.85	861.60		
1 Hyd. Excavator, 1.5 C.Y.		1017.00		1118.70		
2 Dump Trucks, 12 C.Y., 400 H.P.		1335.60		1469.16	98.03	107.83
24 L.H., Daily Totals		$3226.60		$3987.86	$134.44	$166.16
Crew B-31	Hr.	Daily	Hr.	Daily	Bare Costs	Incl. O&P
1 Labor Foreman (outside)	$35.10	$280.80	$57.65	$461.20	$35.19	$57.79
3 Laborers	33.10	794.40	54.35	1304.40		
1 Carpenter	41.55	332.40	68.25	546.00		
1 Air Compressor, 250 cfm		193.40		212.74		
1 Sheeting Driver		5.75		6.33		
2 -50' Air Hoses, 1.5"		12.60		13.86	5.29	5.82
40 L.H., Daily Totals		$1619.35		$2544.53	$40.48	$63.61
Crew B-32	Hr.	Daily	Hr.	Daily	Bare Costs	Incl. O&P
1 Laborer	$33.10	$264.80	$54.35	$434.80	$40.49	$64.06
3 Equip. Oper. (med.)	42.95	1030.80	67.30	1615.20		
1 Grader, 30,000 Lbs.		622.20		684.42		
1 Tandem Roller, 10 Ton		235.20		258.72		
1 Dozer, 200 H.P.		1192.00		1311.20	64.04	70.45
32 L.H., Daily Totals		$3345.00		$4304.34	$104.53	$134.51
Crew B-32A	Hr.	Daily	Hr.	Daily	Bare Costs	Incl. O&P
1 Laborer	$33.10	$264.80	$54.35	$434.80	$39.67	$62.98
2 Equip. Oper. (med.)	42.95	687.20	67.30	1076.80		
1 Grader, 30,000 Lbs.		622.20		684.42		
1 Roller, Vibratory, 25 Ton		633.20		696.52	52.31	57.54
24 L.H., Daily Totals		$2207.40		$2892.54	$91.97	$120.52
Crew B-32B	Hr.	Daily	Hr.	Daily	Bare Costs	Incl. O&P
1 Laborer	$33.10	$264.80	$54.35	$434.80	$39.67	$62.98
2 Equip. Oper. (med.)	42.95	687.20	67.30	1076.80		
1 Dozer, 200 H.P.		1192.00		1311.20		
1 Roller, Vibratory, 25 Ton		633.20		696.52	76.05	83.66
24 L.H., Daily Totals		$2777.20		$3519.32	$115.72	$146.64
Crew B-32C	Hr.	Daily	Hr.	Daily	Bare Costs	Incl. O&P
1 Labor Foreman	$35.10	$280.80	$57.65	$461.20	$38.36	$61.38
2 Laborers	33.10	529.60	54.35	869.60		
3 Equip. Oper. (med.)	42.95	1030.80	67.30	1615.20		
1 Grader, 30,000 Lbs.		622.20		684.42		
1 Tandem Roller, 10 Ton		235.20		258.72		
1 Dozer, 200 H.P.		1192.00		1311.20	42.70	46.97
48 L.H., Daily Totals		$3890.60		$5200.34	$81.05	$108.34
Crew B-33A	Hr.	Daily	Hr.	Daily	Bare Costs	Incl. O&P
1 Equip. Oper. (med.)	$42.95	$343.60	$67.30	$538.40	$40.14	$63.60
.5 Laborer	33.10	132.40	54.35	217.40		
.25 Equip. Oper. (med.)	42.95	85.90	67.30	134.60		
1 Scraper, Towed, 7 C.Y.		105.20		115.72		
1.25 Dozers, 300 H.P.		1990.00		2189.00	149.66	164.62
14 L.H., Daily Totals		$2657.10		$3195.12	$189.79	$228.22

Crews

Crew No.	Bare Costs		Incl. Subs O&P		Cost Per Labor-Hour	
Crew B-33B	Hr.	Daily	Hr.	Daily	Bare Costs	Incl. O&P
1 Equip. Oper. (med.)	$42.95	$343.60	$67.30	$538.40	$40.14	$63.60
.5 Laborer	33.10	132.40	54.35	217.40		
.25 Equip. Oper. (med.)	42.95	85.90	67.30	134.60		
1 Scraper, Towed, 10 C.Y.		156.60		172.26		
1.25 Dozers, 300 H.P.		1990.00		2189.00	153.33	168.66
14 L.H., Daily Totals		$2708.50		$3251.66	$193.46	$232.26
Crew B-33C	Hr.	Daily	Hr.	Daily	Bare Costs	Incl. O&P
1 Equip. Oper. (med.)	$42.95	$343.60	$67.30	$538.40	$40.14	$63.60
.5 Laborer	33.10	132.40	54.35	217.40		
.25 Equip. Oper. (med.)	42.95	85.90	67.30	134.60		
1 Scraper, Towed, 15 C.Y.		172.20		189.42		
1.25 Dozers, 300 H.P.		1990.00		2189.00	154.44	169.89
14 L.H., Daily Totals		$2724.10		$3268.82	$194.58	$233.49
Crew B-33D	Hr.	Daily	Hr.	Daily	Bare Costs	Incl. O&P
1 Equip. Oper. (med.)	$42.95	$343.60	$67.30	$538.40	$40.14	$63.60
.5 Laborer	33.10	132.40	54.35	217.40		
.25 Equip. Oper. (med.)	42.95	85.90	67.30	134.60		
1 S.P. Scraper, 14 C.Y.		1717.00		1888.70		
.25 Dozer, 300 H.P.		398.00		437.80	151.07	166.18
14 L.H., Daily Totals		$2676.90		$3216.90	$191.21	$229.78
Crew B-33E	Hr.	Daily	Hr.	Daily	Bare Costs	Incl. O&P
1 Equip. Oper. (med.)	$42.95	$343.60	$67.30	$538.40	$40.14	$63.60
.5 Laborer	33.10	132.40	54.35	217.40		
.25 Equip. Oper. (med.)	42.95	85.90	67.30	134.60		
1 S.P. Scraper, 21 C.Y.		2242.00		2466.20		
.25 Dozer, 300 H.P.		398.00		437.80	188.57	207.43
14 L.H., Daily Totals		$3201.90		$3794.40	$228.71	$271.03
Crew B-33F	Hr.	Daily	Hr.	Daily	Bare Costs	Incl. O&P
1 Equip. Oper. (med.)	$42.95	$343.60	$67.30	$538.40	$40.14	$63.60
.5 Laborer	33.10	132.40	54.35	217.40		
.25 Equip. Oper. (med.)	42.95	85.90	67.30	134.60		
1 Elev. Scraper, 11 C.Y.		1185.00		1303.50		
.25 Dozer, 300 H.P.		398.00		437.80	113.07	124.38
14 L.H., Daily Totals		$2144.90		$2631.70	$153.21	$187.98
Crew B-33G	Hr.	Daily	Hr.	Daily	Bare Costs	Incl. O&P
1 Equip. Oper. (med.)	$42.95	$343.60	$67.30	$538.40	$40.14	$63.60
.5 Laborer	33.10	132.40	54.35	217.40		
.25 Equip. Oper. (med.)	42.95	85.90	67.30	134.60		
1 Elev. Scraper, 22 C.Y.		2283.00		2511.30		
.25 Dozer, 300 H.P.		398.00		437.80	191.50	210.65
14 L.H., Daily Totals		$3242.90		$3839.50	$231.64	$274.25
Crew B-33K	Hr.	Daily	Hr.	Daily	Bare Costs	Incl. O&P
1 Equipment Operator (med.)	$42.95	$343.60	$67.30	$538.40	$40.14	$63.60
.25 Equipment Operator (med.)	42.95	85.90	67.30	134.60		
.5 Laborer	33.10	132.40	54.35	217.40		
1 S.P. Scraper, 31 C.Y.		3219.00		3540.90		
.25 Dozer, 410 H.P.		519.75		571.73	267.05	293.76
14 L.H., Daily Totals		$4300.65		$5003.02	$307.19	$357.36
Crew B-34A	Hr.	Daily	Hr.	Daily	Bare Costs	Incl. O&P
1 Truck Driver (heavy)	$33.15	$265.20	$53.85	$430.80	$33.15	$53.85
1 Dump Truck, 8 C.Y., 220 H.P.		401.80		441.98	50.23	55.25
8 L.H., Daily Totals		$667.00		$872.78	$83.38	$109.10
Crew B-34B	Hr.	Daily	Hr.	Daily	Bare Costs	Incl. O&P
1 Truck Driver (heavy)	$33.15	$265.20	$53.85	$430.80	$33.15	$53.85
1 Dump Truck, 12 C.Y., 400 H.P.		667.80		734.58	83.47	91.82
8 L.H., Daily Totals		$933.00		$1165.38	$116.63	$145.67
Crew B-34C	Hr.	Daily	Hr.	Daily	Bare Costs	Incl. O&P
1 Truck Driver (heavy)	$33.15	$265.20	$53.85	$430.80	$33.15	$53.85
1 Truck Tractor, 6x4, 380 H.P.		592.40		651.64		
1 Dump Trailer, 16.5 C.Y.		120.20		132.22	89.08	97.98
8 L.H., Daily Totals		$977.80		$1214.66	$122.22	$151.83
Crew B-34D	Hr.	Daily	Hr.	Daily	Bare Costs	Incl. O&P
1 Truck Driver (heavy)	$33.15	$265.20	$53.85	$430.80	$33.15	$53.85
1 Truck Tractor, 6x4, 380 H.P.		592.40		651.64		
1 Dump Trailer, 20 C.Y.		134.80		148.28	90.90	99.99
8 L.H., Daily Totals		$992.40		$1230.72	$124.05	$153.84
Crew B-34E	Hr.	Daily	Hr.	Daily	Bare Costs	Incl. O&P
1 Truck Driver (heavy)	$33.15	$265.20	$53.85	$430.80	$33.15	$53.85
1 Dump Truck, Off Hwy., 25 Ton		1399.00		1538.90	174.88	192.36
8 L.H., Daily Totals		$1664.20		$1969.70	$208.03	$246.21
Crew B-34F	Hr.	Daily	Hr.	Daily	Bare Costs	Incl. O&P
1 Truck Driver (heavy)	$33.15	$265.20	$53.85	$430.80	$33.15	$53.85
1 Dump Truck, Off Hwy., 35 Ton		1267.00		1393.70	158.38	174.21
8 L.H., Daily Totals		$1532.20		$1824.50	$191.53	$228.06
Crew B-34G	Hr.	Daily	Hr.	Daily	Bare Costs	Incl. O&P
1 Truck Driver (heavy)	$33.15	$265.20	$53.85	$430.80	$33.15	$53.85
1 Dump Truck, Off Hwy., 50 Ton		1744.00		1918.40	218.00	239.80
8 L.H., Daily Totals		$2009.20		$2349.20	$251.15	$293.65
Crew B-34H	Hr.	Daily	Hr.	Daily	Bare Costs	Incl. O&P
1 Truck Driver (heavy)	$33.15	$265.20	$53.85	$430.80	$33.15	$53.85
1 Dump Truck, Off Hwy., 65 Ton		1770.00		1947.00	221.25	243.38
8 L.H., Daily Totals		$2035.20		$2377.80	$254.40	$297.23
Crew B-34I	Hr.	Daily	Hr.	Daily	Bare Costs	Incl. O&P
1 Truck Driver (heavy)	$33.15	$265.20	$53.85	$430.80	$33.15	$53.85
1 Dump Truck, 18 C.Y., 450 H.P.		838.40		922.24	104.80	115.28
8 L.H., Daily Totals		$1103.60		$1353.04	$137.95	$169.13
Crew B-34J	Hr.	Daily	Hr.	Daily	Bare Costs	Incl. O&P
1 Truck Driver (heavy)	$33.15	$265.20	$53.85	$430.80	$33.15	$53.85
1 Dump Truck, Off Hwy., 100 Ton		2296.00		2525.60	287.00	315.70
8 L.H., Daily Totals		$2561.20		$2956.40	$320.15	$369.55
Crew B-34K	Hr.	Daily	Hr.	Daily	Bare Costs	Incl. O&P
1 Truck Driver (heavy)	$33.15	$265.20	$53.85	$430.80	$33.15	$53.85
1 Truck Tractor, 6x4, 450 H.P.		715.80		787.38		
1 Lowbed Trailer, 75 Ton		205.80		226.38	115.20	126.72
8 L.H., Daily Totals		$1186.80		$1444.56	$148.35	$180.57
Crew B-34L	Hr.	Daily	Hr.	Daily	Bare Costs	Incl. O&P
1 Equip. Oper. (light)	$41.30	$330.40	$64.70	$517.60	$41.30	$64.70
1 Flatbed Truck, Gas, 1.5 Ton		239.80		263.78	29.98	32.97
8 L.H., Daily Totals		$570.20		$781.38	$71.28	$97.67

Crews

Crew No.	Bare Costs		Incl. Subs O&P		Cost Per Labor-Hour	
Crew B-34N	Hr.	Daily	Hr.	Daily	Bare Costs	Incl. O&P
1 Truck Driver (heavy)	$33.15	$265.20	$53.85	$430.80	$33.15	$53.85
1 Dump Truck, 8 C.Y., 220 H.P.		401.80		441.98		
1 Flatbed Trailer, 40 Ton		143.20		157.52	68.13	74.94
8 L.H., Daily Totals		$810.20		$1030.30	$101.28	$128.79
Crew B-34P	Hr.	Daily	Hr.	Daily	Bare Costs	Incl. O&P
1 Pipe Fitter	$51.90	$415.20	$80.40	$643.20	$42.37	$66.70
1 Truck Driver (light)	32.25	258.00	52.40	419.20		
1 Equip. Oper. (med.)	42.95	343.60	67.30	538.40		
1 Flatbed Truck, Gas, 3 Ton		296.00		325.60		
1 Backhoe Loader, 48 H.P.		337.80		371.58	26.41	29.05
24 L.H., Daily Totals		$1650.60		$2297.98	$68.78	$95.75
Crew B-34Q	Hr.	Daily	Hr.	Daily	Bare Costs	Incl. O&P
1 Pipe Fitter	$51.90	$415.20	$80.40	$643.20	$42.85	$67.45
1 Truck Driver (light)	32.25	258.00	52.40	419.20		
1 Eqip. Oper. (crane)	44.40	355.20	69.55	556.40		
1 Flatbed Trailer, 25 Ton		105.60		116.16		
1 Dump Truck, 8 C.Y., 220 H.P.		401.80		441.98		
1 Hyd. Crane, 25 Ton		747.60		822.36	52.29	57.52
24 L.H., Daily Totals		$2283.40		$2999.30	$95.14	$124.97
Crew B-34R	Hr.	Daily	Hr.	Daily	Bare Costs	Incl. O&P
1 Pipe Fitter	$51.90	$415.20	$80.40	$643.20	$42.85	$67.45
1 Truck Driver (light)	32.25	258.00	52.40	419.20		
1 Eqip. Oper. (crane)	44.40	355.20	69.55	556.40		
1 Flatbed Trailer, 25 Ton		105.60		116.16		
1 Dump Truck, 8 C.Y., 220 H.P.		401.80		441.98		
1 Hyd. Crane, 25 Ton		747.60		822.36		
1 Hyd. Excavator, 1 C.Y.		801.60		881.76	85.69	94.26
24 L.H., Daily Totals		$3085.00		$3881.06	$128.54	$161.71
Crew B-34S	Hr.	Daily	Hr.	Daily	Bare Costs	Incl. O&P
2 Pipe Fitters	$51.90	$830.40	$80.40	$1286.40	$45.34	$71.05
1 Truck Driver (heavy)	33.15	265.20	53.85	430.80		
1 Equip. Oper. (crane)	44.40	355.20	69.55	556.40		
1 Flatbed Trailer, 40 Ton		143.20		157.52		
1 Truck Tractor, 6x4, 380 H.P.		592.40		651.64		
1 Hyd. Crane, 80 Ton		1545.00		1699.50		
1 Hyd. Excavator, 2 C.Y.		1321.00		1453.10	112.55	123.81
32 L.H., Daily Totals		$5052.40		$6235.36	$157.89	$194.85
Crew B-34T	Hr.	Daily	Hr.	Daily	Bare Costs	Incl. O&P
2 Pipe Fitters	$51.90	$830.40	$80.40	$1286.40	$45.34	$71.05
1 Truck Driver (heavy)	33.15	265.20	53.85	430.80		
1 Eqip. Oper. (crane)	44.40	355.20	69.55	556.40		
1 Flatbed Trailer, 40 Ton		143.20		157.52		
1 Truck Tractor, 6x4, 380 H.P.		592.40		651.64		
1 Hyd. Crane, 80 Ton		1545.00		1699.50	71.27	78.40
32 L.H., Daily Totals		$3731.40		$4782.26	$116.61	$149.45
Crew B-35	Hr.	Daily	Hr.	Daily	Bare Costs	Incl. O&P
1 Laborer Foreman (out)	$35.10	$280.80	$57.65	$461.20	$40.92	$65.18
1 Skilled Worker	42.60	340.80	68.90	551.20		
1 Welder (plumber)	52.05	416.40	80.65	645.20		
1 Laborer	33.10	264.80	54.35	434.80		
1 Equip. Oper. (crane)	44.40	355.20	69.55	556.40		
1 Equip. Oper. Oiler	38.30	306.40	60.00	480.00		
1 Welder, electric, 300 amp		58.55		64.41		
1 Hyd. Excavator, .75 C.Y.		677.80		745.58	15.34	16.87
48 L.H., Daily Totals		$2700.75		$3938.78	$56.27	$82.06

Crew No.	Bare Costs		Incl. Subs O&P		Cost Per Labor-Hour	
Crew B-35A	Hr.	Daily	Hr.	Daily	Bare Costs	Incl. O&P
1 Laborer Foreman (out)	$35.10	$280.80	$57.65	$461.20	$39.81	$63.64
2 Laborers	33.10	529.60	54.35	869.60		
1 Skilled Worker	42.60	340.80	68.90	551.20		
1 Welder (plumber)	52.05	416.40	80.65	645.20		
1 Equip. Oper. (crane)	44.40	355.20	69.55	556.40		
1 Equip. Oper. Oiler	38.30	306.40	60.00	480.00		
1 Welder, gas engine, 300 amp		145.80		160.38		
1 Crawler Crane, 75 Ton		1516.00		1667.60	29.68	32.64
56 L.H., Daily Totals		$3891.00		$5391.58	$69.48	$96.28
Crew B-36	Hr.	Daily	Hr.	Daily	Bare Costs	Incl. O&P
1 Labor Foreman (outside)	$35.10	$280.80	$57.65	$461.20	$37.44	$60.19
2 Laborers	33.10	529.60	54.35	869.60		
2 Equip. Oper. (med.)	42.95	687.20	67.30	1076.80		
1 Dozer, 200 H.P.		1192.00		1311.20		
1 Aggregate Spreader		38.60		42.46		
1 Tandem Roller, 10 Ton		235.20		258.72	36.65	40.31
40 L.H., Daily Totals		$2963.40		$4019.98	$74.08	$100.50
Crew B-36A	Hr.	Daily	Hr.	Daily	Bare Costs	Incl. O&P
1 Labor Foreman (outside)	$35.10	$280.80	$57.65	$461.20	$39.01	$62.22
2 Laborers	33.10	529.60	54.35	869.60		
4 Equip. Oper. (med.)	42.95	1374.40	67.30	2153.60		
1 Dozer, 200 H.P.		1192.00		1311.20		
1 Aggregate Spreader		38.60		42.46		
1 Tandem Roller, 10 Ton		235.20		258.72		
1 Roller, Pneum. Whl, 12 Ton		337.20		370.92	32.20	35.42
56 L.H., Daily Totals		$3987.80		$5467.70	$71.21	$97.64
Crew B-36B	Hr.	Daily	Hr.	Daily	Bare Costs	Incl. O&P
1 Labor Foreman (outside)	$35.10	$280.80	$57.65	$461.20	$38.28	$61.17
2 Laborers	33.10	529.60	54.35	869.60		
4 Equip. Oper. (med.)	42.95	1374.40	67.30	2153.60		
1 Truck Driver, Heavy	33.15	265.20	53.85	430.80		
1 Grader, 30,000 Lbs.		622.20		684.42		
1 F.E. Loader, crl 1.5 C.Y.		534.60		588.06		
1 Dozer, 300 H.P.		1592.00		1751.20		
1 Roller, Vibratory, 25 Ton		633.20		696.52		
1 Truck Tractor, 6x4, 450 H.P.		715.80		787.38		
1 Water Tank Trailer, 5000 Gal.		139.60		153.56	66.21	72.83
64 L.H., Daily Totals		$6687.40		$8576.34	$104.49	$134.01
Crew B-36C	Hr.	Daily	Hr.	Daily	Bare Costs	Incl. O&P
1 Labor Foreman (outside)	$35.10	$280.80	$57.65	$461.20	$39.42	$62.68
3 Equip. Oper. (med.)	42.95	1030.80	67.30	1615.20		
1 Truck Driver, Heavy	33.15	265.20	53.85	430.80		
1 Grader, 30,000 Lbs.		622.20		684.42		
1 Dozer, 300 H.P.		1592.00		1751.20		
1 Roller, Vibratory, 25 Ton		633.20		696.52		
1 Truck Tractor, 6x4, 450 H.P.		715.80		787.38		
1 Water Tank Trailer, 5000 Gal.		139.60		153.56	92.57	101.83
40 L.H., Daily Totals		$5279.60		$6580.28	$131.99	$164.51
Crew B-37	Hr.	Daily	Hr.	Daily	Bare Costs	Incl. O&P
1 Labor Foreman (outside)	$35.10	$280.80	$57.65	$461.20	$34.80	$56.63
4 Laborers	33.10	1059.20	54.35	1739.20		
1 Equip. Oper. (light)	41.30	330.40	64.70	517.60		
1 Tandem Roller, 5 Ton		152.00		167.20	3.17	3.48
48 L.H., Daily Totals		$1822.40		$2885.20	$37.97	$60.11

Crews

Crew No.	Bare Costs		Incl. Subs O&P		Cost Per Labor-Hour	
Crew B-38	Hr.	Daily	Hr.	Daily	Bare Costs	Incl. O&P
1 Labor Foreman (outside)	$35.10	$280.80	$57.65	$461.20	$37.11	$59.67
2 Laborers	33.10	529.60	54.35	869.60		
1 Equip. Oper. (light)	41.30	330.40	64.70	517.60		
1 Equip. Oper. (med.)	42.95	343.60	67.30	538.40		
1 Backhoe Loader, 48 H.P.		337.80		371.58		
1 Hyd.Hammer, (1200 lb.)		154.20		169.62		
1 F.E. Loader, W.M., 4 C.Y.		627.80		690.58		
1 Pvmt. Rem. Bucket		58.20		64.02	29.45	32.40
40 L.H., Daily Totals		$2662.40		$3682.60	$66.56	$92.06
Crew B-39	Hr.	Daily	Hr.	Daily	Bare Costs	Incl. O&P
1 Labor Foreman (outside)	$35.10	$280.80	$57.65	$461.20	$34.80	$56.63
4 Laborers	33.10	1059.20	54.35	1739.20		
1 Equip. Oper. (light)	41.30	330.40	64.70	517.60		
1 Air Compressor, 250 cfm		193.40		212.74		
2 Breakers, Pavement, 60 lb.		18.80		20.68		
2 -50' Air Hoses, 1.5"		12.60		13.86	4.68	5.15
48 L.H., Daily Totals		$1895.20		$2965.28	$39.48	$61.78
Crew B-40	Hr.	Daily	Hr.	Daily	Bare Costs	Incl. O&P
1 Pile Driver Foreman (out)	$42.30	$338.40	$70.15	$561.20	$41.33	$67.06
4 Pile Drivers	40.30	1289.60	66.80	2137.60		
2 Equip. Oper. (crane)	44.40	710.40	69.55	1112.80		
1 Equip. Oper. Oiler	38.30	306.40	60.00	480.00		
1 Crawler Crane, 40 Ton		1193.00		1312.30		
1 Vibratory Hammer & Gen.		2641.00		2905.10	59.91	65.90
64 L.H., Daily Totals		$6478.80		$8509.00	$101.23	$132.95
Crew B-40B	Hr.	Daily	Hr.	Daily	Bare Costs	Incl. O&P
1 Laborer Foreman	$35.10	$280.80	$57.65	$461.20	$36.18	$58.38
3 Laborers	33.10	794.40	54.35	1304.40		
1 Equip. Oper. (crane)	44.40	355.20	69.55	556.40		
1 Equip. Oper. Oiler	38.30	306.40	60.00	480.00		
1 Lattice Boom Crane, 40 Ton		1256.00		1381.60	26.17	28.78
48 L.H., Daily Totals		$2992.80		$4183.60	$62.35	$87.16
Crew B-41	Hr.	Daily	Hr.	Daily	Bare Costs	Incl. O&P
1 Labor Foreman (outside)	$35.10	$280.80	$57.65	$461.20	$34.21	$55.90
4 Laborers	33.10	1059.20	54.35	1739.20		
.25 Equip. Oper. (crane)	44.40	88.80	69.55	139.10		
.25 Equip. Oper. Oiler	38.30	76.60	60.00	120.00		
.25 Crawler Crane, 40 Ton		298.25		328.07	6.78	7.46
44 L.H., Daily Totals		$1803.65		$2787.57	$40.99	$63.35
Crew B-42	Hr.	Daily	Hr.	Daily	Bare Costs	Incl. O&P
1 Labor Foreman (outside)	$35.10	$280.80	$57.65	$461.20	$37.14	$61.29
4 Laborers	33.10	1059.20	54.35	1739.20		
1 Equip. Oper. (crane)	44.40	355.20	69.55	556.40		
1 Equip. Oper. Oiler	38.30	306.40	60.00	480.00		
1 Welder	46.90	375.20	85.75	686.00		
1 Hyd. Crane, 25 Ton		747.60		822.36		
1 Welder, gas engine, 300 amp		145.80		160.38		
1 Horz. Boring Csg. Mch.		473.20		520.52	21.35	23.49
64 L.H., Daily Totals		$3743.40		$5426.06	$58.49	$84.78

Crew No.	Bare Costs		Incl. Subs O&P		Cost Per Labor-Hour	
Crew B-43	Hr.	Daily	Hr.	Daily	Bare Costs	Incl. O&P
1 Labor Foreman (outside)	$35.10	$280.80	$57.65	$461.20	$36.18	$58.38
3 Laborers	33.10	794.40	54.35	1304.40		
1 Equip. Oper. (crane)	44.40	355.20	69.55	556.40		
1 Equip. Oper. Oiler	38.30	306.40	60.00	480.00		
1 Drill Rig, Truck-Mounted		2591.00		2850.10	53.98	59.38
48 L.H., Daily Totals		$4327.80		$5652.10	$90.16	$117.75
Crew B-44	Hr.	Daily	Hr.	Daily	Bare Costs	Incl. O&P
1 Pile Driver Foreman	$42.30	$338.40	$70.15	$561.20	$40.67	$66.35
4 Pile Drivers	40.30	1289.60	66.80	2137.60		
2 Equip. Oper. (crane)	44.40	710.40	69.55	1112.80		
1 Laborer	33.10	264.80	54.35	434.80		
1 Crawler Crane, 40 Ton		1193.00		1312.30		
1 Lead, 60' high		70.80		77.88		
1 Hammer, diesel, 15K ft.-lbs.		577.60		635.36	28.77	31.65
64 L.H., Daily Totals		$4444.60		$6271.94	$69.45	$98.00
Crew B-45	Hr.	Daily	Hr.	Daily	Bare Costs	Incl. O&P
1 Equip. Oper. (med.)	$42.95	$343.60	$67.30	$538.40	$38.05	$60.58
1 Truck Driver (heavy)	33.15	265.20	53.85	430.80		
1 Dist. Tanker, 3000 Gallon		307.00		337.70		
1 Truck Tractor, 6x4, 380 H.P.		592.40		651.64	56.21	61.83
16 L.H., Daily Totals		$1508.20		$1958.54	$94.26	$122.41
Crew B-46	Hr.	Daily	Hr.	Daily	Bare Costs	Incl. O&P
1 Pile Driver Foreman	$42.30	$338.40	$70.15	$561.20	$37.03	$61.13
2 Pile Drivers	40.30	644.80	66.80	1068.80		
3 Laborers	33.10	794.40	54.35	1304.40		
1 Chain Saw, gas, 36" Long		39.20		43.12	.82	.90
48 L.H., Daily Totals		$1816.80		$2977.52	$37.85	$62.03
Crew B-47	Hr.	Daily	Hr.	Daily	Bare Costs	Incl. O&P
1 Blast Foreman	$35.10	$280.80	$57.65	$461.20	$36.50	$58.90
1 Driller	33.10	264.80	54.35	434.80		
1 Equip. Oper. (light)	41.30	330.40	64.70	517.60		
1 Air Track Drill, 4"		917.80		1009.58		
1 Air Compressor, 600 cfm		530.60		583.66		
2 -50' Air Hoses, 3"		32.40		35.64	61.70	67.87
24 L.H., Daily Totals		$2356.80		$3042.48	$98.20	$126.77
Crew B-47A	Hr.	Daily	Hr.	Daily	Bare Costs	Incl. O&P
1 Drilling Foreman	$35.10	$280.80	$57.65	$461.20	$39.27	$62.40
1 Equip. Oper. (heavy)	44.40	355.20	69.55	556.40		
1 Oiler	38.30	306.40	60.00	480.00		
1 Air Track Drill, 5"		1069.00		1175.90	44.54	49.00
24 L.H., Daily Totals		$2011.40		$2673.50	$83.81	$111.40
Crew B-47C	Hr.	Daily	Hr.	Daily	Bare Costs	Incl. O&P
1 Laborer	$33.10	$264.80	$54.35	$434.80	$37.20	$59.52
1 Equip. Oper. (light)	41.30	330.40	64.70	517.60		
1 Air Compressor, 750 cfm		542.80		597.08		
2 -50' Air Hoses, 3"		32.40		35.64		
1 Air Track Drill, 4"		917.80		1009.58	93.31	102.64
16 L.H., Daily Totals		$2088.20		$2594.70	$130.51	$162.17
Crew B-47E	Hr.	Daily	Hr.	Daily	Bare Costs	Incl. O&P
1 Laborer Foreman	$35.10	$280.80	$57.65	$461.20	$33.60	$55.17
3 Laborers	33.10	794.40	54.35	1304.40		
1 Flatbed Truck, Gas, 3 Ton		296.00		325.60	9.25	10.18
32 L.H., Daily Totals		$1371.20		$2091.20	$42.85	$65.35

Crews

Crew No.	Bare Costs		Incl. Subs O&P		Cost Per Labor-Hour	
Crew B-47G	Hr.	Daily	Hr.	Daily	Bare Costs	Incl. O&P
1 Laborer Foreman	$35.10	$280.80	$57.65	$461.20	$35.65	$57.76
2 Laborers	33.10	529.60	54.35	869.60		
1 Equip. Oper. (light)	41.30	330.40	64.70	517.60		
1 Air Track Drill, 4"		917.80		1009.58		
1 Air Compressor, 600 cfm		530.60		583.66		
2 -50' Air Hoses, 3"		32.40		35.64		
1 Grout Pump		201.80		221.98	52.58	57.84
32 L.H., Daily Totals		$2823.40		$3699.26	$88.23	$115.60
Crew B-47H	Hr.	Daily	Hr.	Daily	Bare Costs	Incl. O&P
1 Skilled Worker Foreman	$44.60	$356.80	$72.10	$576.80	$43.10	$69.70
3 Skilled Workers	42.60	1022.40	68.90	1653.60		
1 Flatbed Truck, Gas, 3 Ton		296.00		325.60	9.25	10.18
32 L.H., Daily Totals		$1675.20		$2556.00	$52.35	$79.88
Crew B-48	Hr.	Daily	Hr.	Daily	Bare Costs	Incl. O&P
1 Labor Foreman (outside)	$35.10	$280.80	$57.65	$461.20	$36.91	$59.28
3 Laborers	33.10	794.40	54.35	1304.40		
1 Equip. Oper. (crane)	44.40	355.20	69.55	556.40		
1 Equip. Oper. Oiler	38.30	306.40	60.00	480.00		
1 Equip. Oper. (light)	41.30	330.40	64.70	517.60		
1 Centr. Water Pump, 6"		340.00		374.00		
1 -20' Suction Hose, 6"		11.90		13.09		
1 -50' Discharge Hose, 6"		6.30		6.93		
1 Drill Rig, Truck-Mounted		2591.00		2850.10	52.66	57.93
56 L.H., Daily Totals		$5016.40		$6563.72	$89.58	$117.21
Crew B-49	Hr.	Daily	Hr.	Daily	Bare Costs	Incl. O&P
1 Labor Foreman (outside)	$35.10	$280.80	$57.65	$461.20	$38.34	$61.65
3 Laborers	33.10	794.40	54.35	1304.40		
2 Equip. Oper. (crane)	44.40	710.40	69.55	1112.80		
2 Equip. Oper. Oilers	38.30	612.80	60.00	960.00		
1 Equip. Oper. (light)	41.30	330.40	64.70	517.60		
2 Pile Drivers	40.30	644.80	66.80	1068.80		
1 Hyd. Crane, 25 Ton		747.60		822.36		
1 Centr. Water Pump, 6"		340.00		374.00		
1 -20' Suction Hose, 6"		11.90		13.09		
1 -50' Discharge Hose, 6"		6.30		6.93		
1 Drill Rig, Truck-Mounted		2591.00		2850.10	42.01	46.21
88 L.H., Daily Totals		$7070.40		$9491.28	$80.35	$107.86
Crew B-50	Hr.	Daily	Hr.	Daily	Bare Costs	Incl. O&P
2 Pile Driver Foremen	$42.30	$676.80	$70.15	$1122.40	$39.49	$64.52
6 Pile Drivers	40.30	1934.40	66.80	3206.40		
2 Equip. Oper. (crane)	44.40	710.40	69.55	1112.80		
1 Equip. Oper. Oiler	38.30	306.40	60.00	480.00		
3 Laborers	33.10	794.40	54.35	1304.40		
1 Crawler Crane, 40 Ton		1193.00		1312.30		
1 Lead, 60' high		70.80		77.88		
1 Hammer, diesel, 15K ft.-lbs.		577.60		635.36		
1 Air Compressor, 600 cfm		530.60		583.66		
2 -50' Air Hoses, 3"		32.40		35.64		
1 Chain Saw, gas, 36" Long		39.20		43.12	21.82	24.00
112 L.H., Daily Totals		$6866.00		$9913.96	$61.30	$88.52
Crew B-51	Hr.	Daily	Hr.	Daily	Bare Costs	Incl. O&P
1 Labor Foreman (outside)	$35.10	$280.80	$57.65	$461.20	$33.29	$54.58
4 Laborers	33.10	1059.20	54.35	1739.20		
1 Truck Driver (light)	32.25	258.00	52.40	419.20		
1 Flatbed Truck, Gas, 1.5 Ton		239.80		263.78	5.00	5.50
48 L.H., Daily Totals		$1837.80		$2883.38	$38.29	$60.07

Crew No.	Bare Costs		Incl. Subs O&P		Cost Per Labor-Hour	
Crew B-52	Hr.	Daily	Hr.	Daily	Bare Costs	Incl. O&P
1 Carpenter Foreman	$43.55	$348.40	$71.50	$572.00	$38.42	$62.53
1 Carpenter	41.55	332.40	68.25	546.00		
3 Laborers	33.10	794.40	54.35	1304.40		
1 Cement Finisher	39.70	317.60	61.95	495.60		
.5 Rodman (reinf.)	46.80	187.20	78.65	314.60		
.5 Equip. Oper. (med.)	42.95	171.80	67.30	269.20		
.5 Crawler Loader, 3 C.Y.		538.50		592.35	9.62	10.58
56 L.H., Daily Totals		$2690.30		$4094.15	$48.04	$73.11
Crew B-53	Hr.	Daily	Hr.	Daily	Bare Costs	Incl. O&P
1 Equip. Oper. (light)	$41.30	$330.40	$64.70	$517.60	$41.30	$64.70
1 Trencher, Chain, 12 H.P.		66.00		72.60	8.25	9.07
8 L.H., Daily Totals		$396.40		$590.20	$49.55	$73.78
Crew B-54	Hr.	Daily	Hr.	Daily	Bare Costs	Incl. O&P
1 Equip. Oper. (light)	$41.30	$330.40	$64.70	$517.60	$41.30	$64.70
1 Trencher, Chain, 40 H.P.		334.40		367.84	41.80	45.98
8 L.H., Daily Totals		$664.80		$885.44	$83.10	$110.68
Crew B-54A	Hr.	Daily	Hr.	Daily	Bare Costs	Incl. O&P
.17 Labor Foreman (outside)	$35.10	$47.74	$57.65	$78.40	$41.81	$65.90
1 Equipment Operator (med.)	42.95	343.60	67.30	538.40		
1 Wheel Trencher, 67 H.P.		1152.00		1267.20	123.08	135.38
9.36 L.H., Daily Totals		$1543.34		$1884.00	$164.89	$201.28
Crew B-54B	Hr.	Daily	Hr.	Daily	Bare Costs	Incl. O&P
.25 Labor Foreman (outside)	$35.10	$70.20	$57.65	$115.30	$41.38	$65.37
1 Equipment Operator (med.)	42.95	343.60	67.30	538.40		
1 Wheel Trencher, 150 H.P.		1843.00		2027.30	184.30	202.73
10 L.H., Daily Totals		$2256.80		$2681.00	$225.68	$268.10
Crew B-54D	Hr.	Daily	Hr.	Daily	Bare Costs	Incl. O&P
1 Laborer	$33.10	$264.80	$54.35	$434.80	$38.02	$60.83
1 Equipment Operator (med.)	42.95	343.60	67.30	538.40		
1 Rock trencher, 6" width		344.40		378.84	21.52	23.68
16 L.H., Daily Totals		$952.80		$1352.04	$59.55	$84.50
Crew B-54E	Hr.	Daily	Hr.	Daily	Bare Costs	Incl. O&P
1 Laborer	$33.10	$264.80	$54.35	$434.80	$38.02	$60.83
1 Equipment Operator (med.)	42.95	343.60	67.30	538.40		
1 Rock trencher, 18" width		2571.00		2828.10	160.69	176.76
16 L.H., Daily Totals		$3179.40		$3801.30	$198.71	$237.58
Crew B-55	Hr.	Daily	Hr.	Daily	Bare Costs	Incl. O&P
2 Laborers	$33.10	$529.60	$54.35	$869.60	$32.82	$53.70
1 Truck Driver (light)	32.25	258.00	52.40	419.20		
1 Truck-mounted earth auger		749.80		824.78		
1 Flatbed Truck, Gas, 3 Ton		296.00		325.60	43.58	47.93
24 L.H., Daily Totals		$1833.40		$2439.18	$76.39	$101.63
Crew B-56	Hr.	Daily	Hr.	Daily	Bare Costs	Incl. O&P
1 Laborer	$33.10	$264.80	$54.35	$434.80	$37.20	$59.52
1 Equip. Oper. (light)	41.30	330.40	64.70	517.60		
1 Air Track Drill, 4"		917.80		1009.58		
1 Air Compressor, 600 cfm		530.60		583.66		
1 -50' Air Hose, 3"		16.20		17.82	91.54	100.69
16 L.H., Daily Totals		$2059.80		$2563.46	$128.74	$160.22

Crews

Crew No.	Bare Costs		Incl. Subs O&P		Cost Per Labor-Hour	
Crew B-57	Hr.	Daily	Hr.	Daily	Bare Costs	Incl. O&P
1 Labor Foreman (outside)	$35.10	$280.80	$57.65	$461.20	$37.55	$60.10
2 Laborers	33.10	529.60	54.35	869.60		
1 Equip. Oper. (crane)	44.40	355.20	69.55	556.40		
1 Equip. Oper. (light)	41.30	330.40	64.70	517.60		
1 Equip. Oper. Oiler	38.30	306.40	60.00	480.00		
1 Crawler Crane, 25 Ton		1188.00		1306.80		
1 Clamshell Bucket, 1 C.Y.		46.60		51.26		
1 Centr. Water Pump, 6"		340.00		374.00		
1 -20' Suction Hose, 6"		11.90		13.09		
20 -50' Discharge Hoses, 6"		126.00		138.60	35.68	39.24
48 L.H., Daily Totals		$3514.90		$4768.55	$73.23	$99.34
Crew B-58	Hr.	Daily	Hr.	Daily	Bare Costs	Incl. O&P
2 Laborers	$33.10	$529.60	$54.35	$869.60	$35.83	$57.80
1 Equip. Oper. (light)	41.30	330.40	64.70	517.60		
1 Backhoe Loader, 48 H.P.		337.80		371.58		
1 Small Helicopter, w/pilot		2513.00		2764.30	118.78	130.66
24 L.H., Daily Totals		$3710.80		$4523.08	$154.62	$188.46
Crew B-59	Hr.	Daily	Hr.	Daily	Bare Costs	Incl. O&P
1 Truck Driver (heavy)	$33.15	$265.20	$53.85	$430.80	$33.15	$53.85
1 Truck Tractor, 220 H.P.		354.00		389.40		
1 Water Tank Trailer, 5000 Gal.		139.60		153.56	61.70	67.87
8 L.H., Daily Totals		$758.80		$973.76	$94.85	$121.72
Crew B-59A	Hr.	Daily	Hr.	Daily	Bare Costs	Incl. O&P
2 Laborers	$33.10	$529.60	$54.35	$869.60	$33.12	$54.18
1 Truck Driver (heavy)	33.15	265.20	53.85	430.80		
1 Water Tank Trailer, 5000 Gal.		139.60		153.56		
1 Truck Tractor, 220 H.P.		354.00		389.40	20.57	22.62
24 L.H., Daily Totals		$1288.40		$1843.36	$53.68	$76.81
Crew B-60	Hr.	Daily	Hr.	Daily	Bare Costs	Incl. O&P
1 Labor Foreman (outside)	$35.10	$280.80	$57.65	$461.20	$38.09	$60.76
2 Laborers	33.10	529.60	54.35	869.60		
1 Equip. Oper. (crane)	44.40	355.20	69.55	556.40		
2 Equip. Oper. (light)	41.30	660.80	64.70	1035.20		
1 Equip. Oper. Oiler	38.30	306.40	60.00	480.00		
1 Crawler Crane, 40 Ton		1193.00		1312.30		
1 Lead, 60' high		70.80		77.88		
1 Hammer, diesel, 15K ft.-lbs.		577.60		635.36		
1 Backhoe Loader, 48 H.P.		337.80		371.58	38.91	42.81
56 L.H., Daily Totals		$4312.00		$5799.52	$77.00	$103.56
Crew B-61	Hr.	Daily	Hr.	Daily	Bare Costs	Incl. O&P
1 Labor Foreman (outside)	$35.10	$280.80	$57.65	$461.20	$35.14	$57.08
3 Laborers	33.10	794.40	54.35	1304.40		
1 Equip. Oper. (light)	41.30	330.40	64.70	517.60		
1 Cement Mixer, 2 C.Y.		189.40		208.34		
1 Air Compressor, 160 cfm		158.00		173.80	8.69	9.55
40 L.H., Daily Totals		$1753.00		$2665.34	$43.83	$66.63
Crew B-62	Hr.	Daily	Hr.	Daily	Bare Costs	Incl. O&P
2 Laborers	$33.10	$529.60	$54.35	$869.60	$35.83	$57.80
1 Equip. Oper. (light)	41.30	330.40	64.70	517.60		
1 Loader, Skid Steer, 30 H.P., gas		161.40		177.54	6.72	7.40
24 L.H., Daily Totals		$1021.40		$1564.74	$42.56	$65.20

Crew No.	Bare Costs		Incl. Subs O&P		Cost Per Labor-Hour	
Crew B-63	Hr.	Daily	Hr.	Daily	Bare Costs	Incl. O&P
4 Laborers	$33.10	$1059.20	$54.35	$1739.20	$34.74	$56.42
1 Equip. Oper. (light)	41.30	330.40	64.70	517.60		
1 Loader, Skid Steer, 30 H.P., gas		161.40		177.54	4.04	4.44
40 L.H., Daily Totals		$1551.00		$2434.34	$38.77	$60.86
Crew B-64	Hr.	Daily	Hr.	Daily	Bare Costs	Incl. O&P
1 Laborer	$33.10	$264.80	$54.35	$434.80	$32.67	$53.38
1 Truck Driver (light)	32.25	258.00	52.40	419.20		
1 Power Mulcher (small)		150.20		165.22		
1 Flatbed Truck, Gas, 1.5 Ton		239.80		263.78	24.38	26.81
16 L.H., Daily Totals		$912.80		$1283.00	$57.05	$80.19
Crew B-65	Hr.	Daily	Hr.	Daily	Bare Costs	Incl. O&P
1 Laborer	$33.10	$264.80	$54.35	$434.80	$32.67	$53.38
1 Truck Driver (light)	32.25	258.00	52.40	419.20		
1 Power Mulcher (large)		319.00		350.90		
1 Flatbed Truck, Gas, 1.5 Ton		239.80		263.78	34.92	38.42
16 L.H., Daily Totals		$1081.60		$1468.68	$67.60	$91.79
Crew B-66	Hr.	Daily	Hr.	Daily	Bare Costs	Incl. O&P
1 Equip. Oper. (light)	$41.30	$330.40	$64.70	$517.60	$41.30	$64.70
1 Loader-Backhoe		218.40		240.24	27.30	30.03
8 L.H., Daily Totals		$548.80		$757.84	$68.60	$94.73
Crew B-67	Hr.	Daily	Hr.	Daily	Bare Costs	Incl. O&P
1 Millwright	$42.95	$343.60	$67.00	$536.00	$42.13	$65.85
1 Equip. Oper. (light)	41.30	330.40	64.70	517.60		
1 Forklift, R/T, 4,000 Lb.		309.60		340.56	19.35	21.29
16 L.H., Daily Totals		$983.60		$1394.16	$61.48	$87.14
Crew B-68	Hr.	Daily	Hr.	Daily	Bare Costs	Incl. O&P
2 Millwrights	$42.95	$687.20	$67.00	$1072.00	$42.40	$66.23
1 Equip. Oper. (light)	41.30	330.40	64.70	517.60		
1 Forklift, R/T, 4,000 Lb.		309.60		340.56	12.90	14.19
24 L.H., Daily Totals		$1327.20		$1930.16	$55.30	$80.42
Crew B-69	Hr.	Daily	Hr.	Daily	Bare Costs	Incl. O&P
1 Labor Foreman (outside)	$35.10	$280.80	$57.65	$461.20	$36.18	$58.38
3 Laborers	33.10	794.40	54.35	1304.40		
1 Equip Oper. (crane)	44.40	355.20	69.55	556.40		
1 Equip Oper. Oiler	38.30	306.40	60.00	480.00		
1 Hyd. Crane, 80 Ton		1545.00		1699.50	32.19	35.41
48 L.H., Daily Totals		$3281.80		$4501.50	$68.37	$93.78
Crew B-69A	Hr.	Daily	Hr.	Daily	Bare Costs	Incl. O&P
1 Labor Foreman	$35.10	$280.80	$57.65	$461.20	$36.17	$58.33
3 Laborers	33.10	794.40	54.35	1304.40		
1 Equip. Oper. (med.)	42.95	343.60	67.30	538.40		
1 Concrete Finisher	39.70	317.60	61.95	495.60		
1 Curb/Gutter Paver, 2-Track		825.00		907.50	17.19	18.91
48 L.H., Daily Totals		$2561.40		$3707.10	$53.36	$77.23
Crew B-69B	Hr.	Daily	Hr.	Daily	Bare Costs	Incl. O&P
1 Labor Foreman	$35.10	$280.80	$57.65	$461.20	$36.17	$58.33
3 Laborers	33.10	794.40	54.35	1304.40		
1 Equip. Oper. (med.)	42.95	343.60	67.30	538.40		
1 Cement Finisher	39.70	317.60	61.95	495.60		
1 Curb/Gutter Paver, 4-Track		934.60		1028.06	19.47	21.42
48 L.H., Daily Totals		$2671.00		$3827.66	$55.65	$79.74

Crews

Crew No.	Bare Costs		Incl. Subs O&P		Cost Per Labor-Hour	

Crew B-70	Hr.	Daily	Hr.	Daily	Bare Costs	Incl. O&P
1 Labor Foreman (outside)	$35.10	$280.80	$57.65	$461.20	$37.61	$60.37
3 Laborers	33.10	794.40	54.35	1304.40		
3 Equip. Oper. (med.)	42.95	1030.80	67.30	1615.20		
1 Grader, 30,000 Lbs.		622.20		684.42		
1 Ripper, beam & 1 shank		80.00		88.00		
1 Road Sweeper, S.P., 8' wide		650.40		715.44		
1 F.E. Loader, W.M., 1.5 C.Y.		367.60		404.36	30.72	33.79
56 L.H., Daily Totals		$3826.20		$5273.02	$68.33	$94.16

Crew B-71	Hr.	Daily	Hr.	Daily	Bare Costs	Incl. O&P
1 Labor Foreman (outside)	$35.10	$280.80	$57.65	$461.20	$37.61	$60.37
3 Laborers	33.10	794.40	54.35	1304.40		
3 Equip. Oper. (med.)	42.95	1030.80	67.30	1615.20		
1 Pvmt. Profiler, 750 H.P.		6104.00		6714.40		
1 Road Sweeper, S.P., 8' wide		650.40		715.44		
1 F.E. Loader, W.M., 1.5 C.Y.		367.60		404.36	127.18	139.90
56 L.H., Daily Totals		$9228.00		$11215.00	$164.79	$200.27

Crew B-72	Hr.	Daily	Hr.	Daily	Bare Costs	Incl. O&P
1 Labor Foreman (outside)	$35.10	$280.80	$57.65	$461.20	$38.27	$61.24
3 Laborers	33.10	794.40	54.35	1304.40		
4 Equip. Oper. (med.)	42.95	1374.40	67.30	2153.60		
1 Pvmt. Profiler, 750 H.P.		6104.00		6714.40		
1 Hammermill, 250 H.P.		1636.00		1799.60		
1 Windrow Loader		1228.00		1350.80		
1 Mix Paver 165 H.P.		2062.00		2268.20		
1 Roller, Pneum. Whl, 12 Ton		337.20		370.92	177.61	195.37
64 L.H., Daily Totals		$13816.80		$16423.12	$215.89	$256.61

Crew B-73	Hr.	Daily	Hr.	Daily	Bare Costs	Incl. O&P
1 Labor Foreman (outside)	$35.10	$280.80	$57.65	$461.20	$39.51	$62.86
2 Laborers	33.10	529.60	54.35	869.60		
5 Equip. Oper. (med.)	42.95	1718.00	67.30	2692.00		
1 Road Mixer, 310 H.P.		1955.00		2150.50		
1 Tandem Roller, 10 Ton		235.20		258.72		
1 Hammermill, 250 H.P.		1636.00		1799.60		
1 Grader, 30,000 Lbs.		622.20		684.42		
.5 F.E. Loader, W.M., 1.5 C.Y.		183.80		202.18		
.5 Truck Tractor, 220 H.P.		177.00		194.70		
.5 Water Tank Trailer, 5000 Gal.		69.80		76.78	76.23	83.86
64 L.H., Daily Totals		$7407.40		$9389.70	$115.74	$146.71

Crew B-74	Hr.	Daily	Hr.	Daily	Bare Costs	Incl. O&P
1 Labor Foreman (outside)	$35.10	$280.80	$57.65	$461.20	$38.29	$61.11
1 Laborer	33.10	264.80	54.35	434.80		
4 Equip. Oper. (med.)	42.95	1374.40	67.30	2153.60		
2 Truck Drivers (heavy)	33.15	530.40	53.85	861.60		
1 Grader, 30,000 Lbs.		622.20		684.42		
1 Ripper, beam & 1 shank		80.00		88.00		
2 Stabilizers, 310 H.P.		2928.00		3220.80		
1 Flatbed Truck, Gas, 3 Ton		296.00		325.60		
1 Chem. Spreader, Towed		50.40		55.44		
1 Roller, Vibratory, 25 Ton		633.20		696.52		
1 Water Tank Trailer, 5000 Gal.		139.60		153.56		
1 Truck Tractor, 220 H.P.		354.00		389.40	79.74	87.71
64 L.H., Daily Totals		$7553.80		$9524.94	$118.03	$148.83

Crew B-75	Hr.	Daily	Hr.	Daily	Bare Costs	Incl. O&P
1 Labor Foreman (outside)	$35.10	$280.80	$57.65	$461.20	$39.02	$62.15
1 Laborer	33.10	264.80	54.35	434.80		
4 Equip. Oper. (med.)	42.95	1374.40	67.30	2153.60		
1 Truck Driver (heavy)	33.15	265.20	53.85	430.80		
1 Grader, 30,000 Lbs.		622.20		684.42		
1 Ripper, beam & 1 shank		80.00		88.00		
2 Stabilizers, 310 H.P.		2928.00		3220.80		
1 Dist. Tanker, 3000 Gallon		307.00		337.70		
1 Truck Tractor, 6x4, 380 H.P.		592.40		651.64		
1 Roller, Vibratory, 25 Ton		633.20		696.52	92.19	101.41
56 L.H., Daily Totals		$7348.00		$9159.48	$131.21	$163.56

Crew B-76	Hr.	Daily	Hr.	Daily	Bare Costs	Incl. O&P
1 Dock Builder Foreman	$42.30	$338.40	$70.15	$561.20	$41.21	$67.03
5 Dock Builders	40.30	1612.00	66.80	2672.00		
2 Equip. Oper. (crane)	44.40	710.40	69.55	1112.80		
1 Equip. Oper. Oiler	38.30	306.40	60.00	480.00		
1 Crawler Crane, 50 Ton		1346.00		1480.60		
1 Barge, 400 Ton		768.00		844.80		
1 Hammer, diesel, 15K ft.-lbs.		577.60		635.36		
1 Lead, 60' high		70.80		77.88		
1 Air Compressor, 600 cfm		530.60		583.66		
2 -50' Air Hoses, 3"		32.40		35.64	46.19	50.80
72 L.H., Daily Totals		$6292.60		$8483.94	$87.40	$117.83

Crew B-76A	Hr.	Daily	Hr.	Daily	Bare Costs	Incl. O&P
1 Laborer Foreman	$35.10	$280.80	$57.65	$461.20	$35.41	$57.37
5 Laborers	33.10	1324.00	54.35	2174.00		
1 Equip. Oper. (crane)	44.40	355.20	69.55	556.40		
1 Equip. Oper. Oiler	38.30	306.40	60.00	480.00		
1 Crawler Crane, 50 Ton		1346.00		1480.60		
1 Barge, 400 Ton		768.00		844.80	33.03	36.33
64 L.H., Daily Totals		$4380.40		$5997.00	$68.44	$93.70

Crew B-77	Hr.	Daily	Hr.	Daily	Bare Costs	Incl. O&P
1 Labor Foreman	$35.10	$280.80	$57.65	$461.20	$33.33	$54.62
3 Laborers	33.10	794.40	54.35	1304.40		
1 Truck Driver (light)	32.25	258.00	52.40	419.20		
1 Crack Cleaner, 25 H.P.		60.40		66.44		
1 Crack Filler, Trailer Mtd.		203.20		223.52		
1 Flatbed Truck, Gas, 3 Ton		296.00		325.60	13.99	15.39
40 L.H., Daily Totals		$1892.80		$2800.36	$47.32	$70.01

Crew B-78	Hr.	Daily	Hr.	Daily	Bare Costs	Incl. O&P
1 Labor Foreman	$35.10	$280.80	$57.65	$461.20	$33.29	$54.58
4 Laborers	33.10	1059.20	54.35	1739.20		
1 Truck Driver (light)	32.25	258.00	52.40	419.20		
1 Paint Striper, S.P.		152.60		167.86		
1 Flatbed Truck, Gas, 3 Ton		296.00		325.60		
1 Pickup Truck, 3/4 Ton		140.00		154.00	12.26	13.49
48 L.H., Daily Totals		$2186.60		$3267.06	$45.55	$68.06

Crew B-78A	Hr.	Daily	Hr.	Daily	Bare Costs	Incl. O&P
1 Equip. Oper. (light)	$41.30	$330.40	$64.70	$517.60	$41.30	$64.70
1 Line Rem. (metal balls) 115 H.P.		883.40		971.74	110.43	121.47
8 L.H., Daily Totals		$1213.80		$1489.34	$151.72	$186.17

Crews

Crew No.	Bare Costs		Incl. Subs O&P		Cost Per Labor-Hour	
Crew B-78B	Hr.	Daily	Hr.	Daily	Bare Costs	Incl. O&P
2 Laborers	$33.10	$529.60	$54.35	$869.60	$34.01	$55.50
.25 Equip. Oper. (light)	41.30	82.60	64.70	129.40		
1 Pickup Truck, 3/4 Ton		140.00		154.00		
1 Line Rem., 11 H.P., walk behind		59.20		65.12		
.25 Road Sweeper, S.P., 8' wide		162.60		178.86	20.10	22.11
18 L.H., Daily Totals		$974.00		$1396.98	$54.11	$77.61

Crew No.	Bare Costs		Incl. Subs O&P		Cost Per Labor-Hour	
Crew B-79	Hr.	Daily	Hr.	Daily	Bare Costs	Incl. O&P
1 Labor Foreman	$35.10	$280.80	$57.65	$461.20	$33.33	$54.62
3 Laborers	33.10	794.40	54.35	1304.40		
1 Truck Driver (light)	32.25	258.00	52.40	419.20		
1 Thermo. Striper, T.M.		802.40		882.64		
1 Flatbed Truck, Gas, 3 Ton		296.00		325.60		
2 Pickup Trucks, 3/4 Ton		280.00		308.00	34.46	37.91
40 L.H., Daily Totals		$2711.60		$3701.04	$67.79	$92.53

Crew B-79A	Hr.	Daily	Hr.	Daily	Bare Costs	Incl. O&P
1.5 Equip. Oper. (light)	$41.30	$495.60	$64.70	$776.40	$41.30	$64.70
.5 Line Remov. (grinder) 115 H.P.		476.90		524.59		
1 Line Rem. (metal balls) 115 H.P.		883.40		971.74	113.36	124.69
12 L.H., Daily Totals		$1855.90		$2272.73	$154.66	$189.39

Crew B-80	Hr.	Daily	Hr.	Daily	Bare Costs	Incl. O&P
1 Labor Foreman	$35.10	$280.80	$57.65	$461.20	$35.44	$57.27
1 Laborer	33.10	264.80	54.35	434.80		
1 Truck Driver (light)	32.25	258.00	52.40	419.20		
1 Equip. Oper. (light)	41.30	330.40	64.70	517.60		
1 Flatbed Truck, Gas, 3 Ton		296.00		325.60		
1 Earth Auger, Truck-Mtd.		449.20		494.12	23.29	25.62
32 L.H., Daily Totals		$1879.20		$2652.52	$58.73	$82.89

Crew B-80A	Hr.	Daily	Hr.	Daily	Bare Costs	Incl. O&P
3 Laborers	$33.10	$794.40	$54.35	$1304.40	$33.10	$54.35
1 Flatbed Truck, Gas, 3 Ton		296.00		325.60	12.33	13.57
24 L.H., Daily Totals		$1090.40		$1630.00	$45.43	$67.92

Crew B-80B	Hr.	Daily	Hr.	Daily	Bare Costs	Incl. O&P
3 Laborers	$33.10	$794.40	$54.35	$1304.40	$35.15	$56.94
1 Equip. Oper. (light)	41.30	330.40	64.70	517.60		
1 Crane, Flatbed Mounted, 3 Ton		242.60		266.86	7.58	8.34
32 L.H., Daily Totals		$1367.40		$2088.86	$42.73	$65.28

Crew B-80C	Hr.	Daily	Hr.	Daily	Bare Costs	Incl. O&P
2 Laborers	$33.10	$529.60	$54.35	$869.60	$32.82	$53.70
1 Truck Driver (light)	32.25	258.00	52.40	419.20		
1 Flatbed Truck, Gas, 1.5 Ton		239.80		263.78		
1 Manual fence post auger, gas		7.80		8.58	10.32	11.35
24 L.H., Daily Totals		$1035.20		$1561.16	$43.13	$65.05

Crew B-81	Hr.	Daily	Hr.	Daily	Bare Costs	Incl. O&P
1 Laborer	$33.10	$264.80	$54.35	$434.80	$36.40	$58.50
1 Equip. Oper. (med.)	42.95	343.60	67.30	538.40		
1 Truck Driver (heavy)	33.15	265.20	53.85	430.80		
1 Hydromulcher, T.M.		338.20		372.02		
1 Truck Tractor, 220 H.P.		354.00		389.40	28.84	31.73
24 L.H., Daily Totals		$1565.80		$2165.42	$65.24	$90.23

Crew No.	Bare Costs		Incl. Subs O&P		Cost Per Labor-Hour	
Crew B-82	Hr.	Daily	Hr.	Daily	Bare Costs	Incl. O&P
1 Laborer	$33.10	$264.80	$54.35	$434.80	$37.20	$59.52
1 Equip. Oper. (light)	41.30	330.40	64.70	517.60		
1 Horiz. Borer, 6 H.P.		82.00		90.20	5.13	5.64
16 L.H., Daily Totals		$677.20		$1042.60	$42.33	$65.16

Crew B-82A	Hr.	Daily	Hr.	Daily	Bare Costs	Incl. O&P
1 Laborer	$33.10	$264.80	$54.35	$434.80	$37.20	$59.52
1 Equip. Oper. (light)	41.30	330.40	64.70	517.60		
1 Flatbed Truck, Gas, 3 Ton		296.00		325.60		
1 Flatbed Trailer, 25 Ton		105.60		116.16		
1 Horiz. Dir. Drill, 20k lb. thrust		651.60		716.76	65.83	72.41
16 L.H., Daily Totals		$1648.40		$2110.92	$103.03	$131.93

Crew B-82B	Hr.	Daily	Hr.	Daily	Bare Costs	Incl. O&P
2 Laborers	$33.10	$529.60	$54.35	$869.60	$35.83	$57.80
1 Equip. Oper. (light)	41.30	330.40	64.70	517.60		
1 Flatbed Truck, Gas, 3 Ton		296.00		325.60		
1 Flatbed Trailer, 25 Ton		105.60		116.16		
1 Horiz. Dir. Drill, 30k lb. thrust		928.20		1021.02	55.41	60.95
24 L.H., Daily Totals		$2189.80		$2849.98	$91.24	$118.75

Crew B-82C	Hr.	Daily	Hr.	Daily	Bare Costs	Incl. O&P
2 Laborers	$33.10	$529.60	$54.35	$869.60	$35.83	$57.80
1 Equip. Oper. (light)	41.30	330.40	64.70	517.60		
1 Flatbed Truck, Gas, 3 Ton		296.00		325.60		
1 Flatbed Trailer, 25 Ton		105.60		116.16		
1 Horiz. Dir. Drill, 50k lb. thrust		1232.00		1355.20	68.07	74.87
24 L.H., Daily Totals		$2493.60		$3184.16	$103.90	$132.67

Crew B-82D	Hr.	Daily	Hr.	Daily	Bare Costs	Incl. O&P
1 Equip. Oper. (light)	$41.30	$330.40	$64.70	$517.60	$41.30	$64.70
1 Mud Trailer for HDD, 1500 gallon		338.00		371.80	42.25	46.48
8 L.H., Daily Totals		$668.40		$889.40	$83.55	$111.18

Crew B-83	Hr.	Daily	Hr.	Daily	Bare Costs	Incl. O&P
1 Tugboat Captain	$42.95	$343.60	$67.30	$538.40	$38.02	$60.83
1 Tugboat Hand	33.10	264.80	54.35	434.80		
1 Tugboat, 250 H.P.		840.80		924.88	52.55	57.81
16 L.H., Daily Totals		$1449.20		$1898.08	$90.58	$118.63

Crew B-84	Hr.	Daily	Hr.	Daily	Bare Costs	Incl. O&P
1 Equip. Oper. (med.)	$42.95	$343.60	$67.30	$538.40	$42.95	$67.30
1 Rotary Mower/Tractor		302.00		332.20	37.75	41.52
8 L.H., Daily Totals		$645.60		$870.60	$80.70	$108.83

Crew B-85	Hr.	Daily	Hr.	Daily	Bare Costs	Incl. O&P
3 Laborers	$33.10	$794.40	$54.35	$1304.40	$35.08	$56.84
1 Equip. Oper. (med.)	42.95	343.60	67.30	538.40		
1 Truck Driver (heavy)	33.15	265.20	53.85	430.80		
1 Aerial Lift Truck, 80'		665.40		731.94		
1 Brush Chipper, 12", 130 H.P.		253.20		278.52		
1 Pruning Saw, Rotary		6.75		7.42	23.13	25.45
40 L.H., Daily Totals		$2328.55		$3291.49	$58.21	$82.29

Crew B-86	Hr.	Daily	Hr.	Daily	Bare Costs	Incl. O&P
1 Equip. Oper. (med.)	$42.95	$343.60	$67.30	$538.40	$42.95	$67.30
1 Stump Chipper, S.P.		190.90		209.99	23.86	26.25
8 L.H., Daily Totals		$534.50		$748.39	$66.81	$93.55

Crews

Crew No.	Bare Costs		Incl. Subs O&P		Cost Per Labor-Hour	
Crew B-86A	Hr.	Daily	Hr.	Daily	Bare Costs	Incl. O&P
1 Equip. Oper. (med.)	$42.95	$343.60	$67.30	$538.40	$42.95	$67.30
1 Grader, 30,000 Lbs.		622.20		684.42	77.78	85.55
8 L.H., Daily Totals		$965.80		$1222.82	$120.72	$152.85
Crew B-86B	Hr.	Daily	Hr.	Daily	Bare Costs	Incl. O&P
1 Equip. Oper. (med.)	$42.95	$343.60	$67.30	$538.40	$42.95	$67.30
1 Dozer, 200 H.P.		1192.00		1311.20	149.00	163.90
8 L.H., Daily Totals		$1535.60		$1849.60	$191.95	$231.20
Crew B-87	Hr.	Daily	Hr.	Daily	Bare Costs	Incl. O&P
1 Laborer	$33.10	$264.80	$54.35	$434.80	$40.98	$64.71
4 Equip. Oper. (med.)	42.95	1374.40	67.30	2153.60		
2 Feller Bunchers, 100 H.P.		1219.20		1341.12		
1 Log Chipper, 22" Tree		689.60		758.56		
1 Dozer, 105 H.P.		660.40		726.44		
1 Chain Saw, gas, 36" Long		39.20		43.12	65.21	71.73
40 L.H., Daily Totals		$4247.60		$5457.64	$106.19	$136.44
Crew B-88	Hr.	Daily	Hr.	Daily	Bare Costs	Incl. O&P
1 Laborer	$33.10	$264.80	$54.35	$434.80	$41.54	$65.45
6 Equip. Oper. (med.)	42.95	2061.60	67.30	3230.40		
2 Feller Bunchers, 100 H.P.		1219.20		1341.12		
1 Log Chipper, 22" Tree		689.60		758.56		
2 Log Skidders, 50 H.P.		1814.40		1995.84		
1 Dozer, 105 H.P.		660.40		726.44		
1 Chain Saw, gas, 36" Long		39.20		43.12	78.98	86.88
56 L.H., Daily Totals		$6749.20		$8530.28	$120.52	$152.33
Crew B-89	Hr.	Daily	Hr.	Daily	Bare Costs	Incl. O&P
1 Equip. Oper. (light)	$41.30	$330.40	$64.70	$517.60	$36.77	$58.55
1 Truck Driver (light)	32.25	258.00	52.40	419.20		
1 Flatbed Truck, Gas, 3 Ton		296.00		325.60		
1 Concrete Saw		161.40		177.54		
1 Water Tank, 65 Gal.		18.45		20.30	29.74	32.71
16 L.H., Daily Totals		$1064.25		$1460.23	$66.52	$91.26
Crew B-89A	Hr.	Daily	Hr.	Daily	Bare Costs	Incl. O&P
1 Skilled Worker	$42.60	$340.80	$68.90	$551.20	$37.85	$61.63
1 Laborer	33.10	264.80	54.35	434.80		
1 Core Drill (large)		113.80		125.18	7.11	7.82
16 L.H., Daily Totals		$719.40		$1111.18	$44.96	$69.45
Crew B-89B	Hr.	Daily	Hr.	Daily	Bare Costs	Incl. O&P
1 Equip. Oper. (light)	$41.30	$330.40	$64.70	$517.60	$36.77	$58.55
1 Truck Driver, Light	32.25	258.00	52.40	419.20		
1 Wall Saw, Hydraulic, 10 H.P.		110.40		121.44		
1 Generator, Diesel, 100 kW		418.60		460.46		
1 Water Tank, 65 Gal.		18.45		20.30		
1 Flatbed Truck, Gas, 3 Ton		296.00		325.60	52.72	57.99
16 L.H., Daily Totals		$1431.85		$1864.60	$89.49	$116.54
Crew B-90	Hr.	Daily	Hr.	Daily	Bare Costs	Incl. O&P
1 Labor Foreman (outside)	$35.10	$280.80	$57.65	$461.20	$35.41	$57.23
3 Laborers	33.10	794.40	54.35	1304.40		
2 Equip. Oper. (light)	41.30	660.80	64.70	1035.20		
2 Truck Drivers (heavy)	33.15	530.40	53.85	861.60		
1 Road Mixer, 310 H.P.		1955.00		2150.50		
1 Dist. Truck, 2000 Gal.		278.20		306.02	34.89	38.38
64 L.H., Daily Totals		$4499.60		$6118.92	$70.31	$95.61

Crew No.	Bare Costs		Incl. Subs O&P		Cost Per Labor-Hour	
Crew B-90A	Hr.	Daily	Hr.	Daily	Bare Costs	Incl. O&P
1 Labor Foreman	$35.10	$280.80	$57.65	$461.20	$39.01	$62.22
2 Laborers	33.10	529.60	54.35	869.60		
4 Equip. Oper. (med.)	42.95	1374.40	67.30	2153.60		
2 Graders, 30,000 Lbs.		1244.40		1368.84		
1 Tandem Roller, 10 Ton		235.20		258.72		
1 Roller, Pneum. Whl, 12 Ton		337.20		370.92	32.44	35.69
56 L.H., Daily Totals		$4001.60		$5482.88	$71.46	$97.91
Crew B-90B	Hr.	Daily	Hr.	Daily	Bare Costs	Incl. O&P
1 Labor Foreman	$35.10	$280.80	$57.65	$461.20	$38.36	$61.38
2 Laborers	33.10	529.60	54.35	869.60		
3 Equip. Oper. (med.)	42.95	1030.80	67.30	1615.20		
1 Tandem Roller, 10 Ton		235.20		258.72		
1 Roller, Pneum. Whl, 12 Ton		337.20		370.92		
1 Road Mixer, 310 H.P.		1955.00		2150.50	52.65	57.92
48 L.H., Daily Totals		$4368.60		$5726.14	$91.01	$119.29
Crew B-91	Hr.	Daily	Hr.	Daily	Bare Costs	Incl. O&P
1 Labor Foreman (outside)	$35.10	$280.80	$57.65	$461.20	$38.28	$61.17
2 Laborers	33.10	529.60	54.35	869.60		
4 Equip. Oper. (med.)	42.95	1374.40	67.30	2153.60		
1 Truck Driver (heavy)	33.15	265.20	53.85	430.80		
1 Dist. Tanker, 3000 Gallon		307.00		337.70		
1 Truck Tractor, 6x4, 380 H.P.		592.40		651.64		
1 Aggreg. Spreader, S.P.		820.80		902.88		
1 Roller, Pneum. Whl, 12 Ton		337.20		370.92		
1 Tandem Roller, 10 Ton		235.20		258.72	35.82	39.40
64 L.H., Daily Totals		$4742.60		$6437.06	$74.10	$100.58
Crew B-92	Hr.	Daily	Hr.	Daily	Bare Costs	Incl. O&P
1 Labor Foreman (outside)	$35.10	$280.80	$57.65	$461.20	$33.60	$55.17
3 Laborers	33.10	794.40	54.35	1304.40		
1 Crack Cleaner, 25 H.P.		60.40		66.44		
1 Air Compressor, 60 cfm		138.60		152.46		
1 Tar Kettle, T.M.		71.40		78.54		
1 Flatbed Truck, Gas, 3 Ton		296.00		325.60	17.70	19.47
32 L.H., Daily Totals		$1641.60		$2388.64	$51.30	$74.64
Crew B-93	Hr.	Daily	Hr.	Daily	Bare Costs	Incl. O&P
1 Equip. Oper. (med.)	$42.95	$343.60	$67.30	$538.40	$42.95	$67.30
1 Feller Buncher, 100 H.P.		609.60		670.56	76.20	83.82
8 L.H., Daily Totals		$953.20		$1208.96	$119.15	$151.12
Crew B-94A	Hr.	Daily	Hr.	Daily	Bare Costs	Incl. O&P
1 Laborer	$33.10	$264.80	$54.35	$434.80	$33.10	$54.35
1 Diaphragm Water Pump, 2"		68.00		74.80		
1 -20' Suction Hose, 2"		1.95		2.15		
2 -50' Discharge Hoses, 2"		1.80		1.98	8.97	9.87
8 L.H., Daily Totals		$336.55		$513.73	$42.07	$64.22
Crew B-94B	Hr.	Daily	Hr.	Daily	Bare Costs	Incl. O&P
1 Laborer	$33.10	$264.80	$54.35	$434.80	$33.10	$54.35
1 Diaphragm Water Pump, 4"		90.80		99.88		
1 -20' Suction Hose, 4"		3.45		3.79		
2 -50' Discharge Hoses, 4"		4.70		5.17	12.37	13.61
8 L.H., Daily Totals		$363.75		$543.64	$45.47	$67.96

Crews

Crew No.	Bare Costs		Incl. Subs O&P		Cost Per Labor-Hour	
Crew B-94C	Hr.	Daily	Hr.	Daily	Bare Costs	Incl. O&P
1 Laborer	$33.10	$264.80	$54.35	$434.80	$33.10	$54.35
1 Centrifugal Water Pump, 3"		74.40		81.84		
1 -20' Suction Hose, 3"		3.05		3.36		
2 -50' Discharge Hoses, 3"		3.50		3.85	10.12	11.13
8 L.H., Daily Totals		$345.75		$523.85	$43.22	$65.48
Crew B-94D	Hr.	Daily	Hr.	Daily	Bare Costs	Incl. O&P
1 Laborer	$33.10	$264.80	$54.35	$434.80	$33.10	$54.35
1 Centr. Water Pump, 6"		340.00		374.00		
1 -20' Suction Hose, 6"		11.90		13.09		
2 -50' Discharge Hoses, 6"		12.60		13.86	45.56	50.12
8 L.H., Daily Totals		$629.30		$835.75	$78.66	$104.47
Crew C-1	Hr.	Daily	Hr.	Daily	Bare Costs	Incl. O&P
3 Carpenters	$41.55	$997.20	$68.25	$1638.00	$39.44	$64.78
1 Laborer	33.10	264.80	54.35	434.80		
32 L.H., Daily Totals		$1262.00		$2072.80	$39.44	$64.78
Crew C-2	Hr.	Daily	Hr.	Daily	Bare Costs	Incl. O&P
1 Carpenter Foreman (out)	$43.55	$348.40	$71.50	$572.00	$40.48	$66.47
4 Carpenters	41.55	1329.60	68.25	2184.00		
1 Laborer	33.10	264.80	54.35	434.80		
48 L.H., Daily Totals		$1942.80		$3190.80	$40.48	$66.47
Crew C-2A	Hr.	Daily	Hr.	Daily	Bare Costs	Incl. O&P
1 Carpenter Foreman (out)	$43.55	$348.40	$71.50	$572.00	$40.17	$65.42
3 Carpenters	41.55	997.20	68.25	1638.00		
1 Cement Finisher	39.70	317.60	61.95	495.60		
1 Laborer	33.10	264.80	54.35	434.80		
48 L.H., Daily Totals		$1928.00		$3140.40	$40.17	$65.42
Crew C-3	Hr.	Daily	Hr.	Daily	Bare Costs	Incl. O&P
1 Rodman Foreman	$48.80	$390.40	$82.05	$656.40	$42.94	$71.26
4 Rodmen (reinf.)	46.80	1497.60	78.65	2516.80		
1 Equip. Oper. (light)	41.30	330.40	64.70	517.60		
2 Laborers	33.10	529.60	54.35	869.60		
3 Stressing Equipment		28.80		31.68		
.5 Grouting Equipment		80.00		88.00	1.70	1.87
64 L.H., Daily Totals		$2856.80		$4680.08	$44.64	$73.13
Crew C-4	Hr.	Daily	Hr.	Daily	Bare Costs	Incl. O&P
1 Rodman Foreman	$48.80	$390.40	$82.05	$656.40	$47.30	$79.50
3 Rodmen (reinf.)	46.80	1123.20	78.65	1887.60		
3 Stressing Equipment		28.80		31.68	.90	.99
32 L.H., Daily Totals		$1542.40		$2575.68	$48.20	$80.49
Crew C-4A	Hr.	Daily	Hr.	Daily	Bare Costs	Incl. O&P
2 Rodmen (reinf.)	$46.80	$748.80	$78.65	$1258.40	$46.80	$78.65
4 Stressing Equipment		38.40		42.24	2.40	2.64
16 L.H., Daily Totals		$787.20		$1300.64	$49.20	$81.29
Crew C-5	Hr.	Daily	Hr.	Daily	Bare Costs	Incl. O&P
1 Rodman Foreman	$48.80	$390.40	$82.05	$656.40	$45.53	$75.17
4 Rodmen (reinf.)	46.80	1497.60	78.65	2516.80		
1 Equip. Oper. (crane)	44.40	355.20	69.55	556.40		
1 Equip. Oper. Oiler	38.30	306.40	60.00	480.00		
1 Hyd. Crane, 25 Ton		747.60		822.36	13.35	14.69
56 L.H., Daily Totals		$3297.20		$5031.96	$58.88	$89.86

Crew No.	Bare Costs		Incl. Subs O&P		Cost Per Labor-Hour	
Crew C-6	Hr.	Daily	Hr.	Daily	Bare Costs	Incl. O&P
1 Labor Foreman (outside)	$35.10	$280.80	$57.65	$461.20	$34.53	$56.17
4 Laborers	33.10	1059.20	54.35	1739.20		
1 Cement Finisher	39.70	317.60	61.95	495.60		
2 Gas Engine Vibrators		60.80		66.88	1.27	1.39
48 L.H., Daily Totals		$1718.40		$2762.88	$35.80	$57.56
Crew C-7	Hr.	Daily	Hr.	Daily	Bare Costs	Incl. O&P
1 Labor Foreman (outside)	$35.10	$280.80	$57.65	$461.20	$35.73	$57.63
5 Laborers	33.10	1324.00	54.35	2174.00		
1 Cement Finisher	39.70	317.60	61.95	495.60		
1 Equip. Oper. (med.)	42.95	343.60	67.30	538.40		
1 Equip. Oper. (oiler)	38.30	306.40	60.00	480.00		
2 Gas Engine Vibrators		60.80		66.88		
1 Concrete Bucket, 1 C.Y.		20.60		22.66		
1 Hyd. Crane, 55 Ton		1209.00		1329.90	17.92	19.71
72 L.H., Daily Totals		$3862.80		$5568.64	$53.65	$77.34
Crew C-8	Hr.	Daily	Hr.	Daily	Bare Costs	Incl. O&P
1 Labor Foreman (outside)	$35.10	$280.80	$57.65	$461.20	$36.68	$58.84
3 Laborers	33.10	794.40	54.35	1304.40		
2 Cement Finishers	39.70	635.20	61.95	991.20		
1 Equip. Oper. (med.)	42.95	343.60	67.30	538.40		
1 Concrete Pump (small)		738.80		812.68	13.19	14.51
56 L.H., Daily Totals		$2792.80		$4107.88	$49.87	$73.36
Crew C-8A	Hr.	Daily	Hr.	Daily	Bare Costs	Incl. O&P
1 Labor Foreman (outside)	$35.10	$280.80	$57.65	$461.20	$35.63	$57.43
3 Laborers	33.10	794.40	54.35	1304.40		
2 Cement Finishers	39.70	635.20	61.95	991.20		
48 L.H., Daily Totals		$1710.40		$2756.80	$35.63	$57.43
Crew C-8B	Hr.	Daily	Hr.	Daily	Bare Costs	Incl. O&P
1 Labor Foreman (outside)	$35.10	$280.80	$57.65	$461.20	$35.47	$57.60
3 Laborers	33.10	794.40	54.35	1304.40		
1 Equip. Oper. (med.)	42.95	343.60	67.30	538.40		
1 Vibrating Power Screed		65.15		71.67		
1 Roller, Vibratory, 25 Ton		633.20		696.52		
1 Dozer, 200 H.P.		1192.00		1311.20	47.26	51.98
40 L.H., Daily Totals		$3309.15		$4383.39	$82.73	$109.58
Crew C-8C	Hr.	Daily	Hr.	Daily	Bare Costs	Incl. O&P
1 Labor Foreman (outside)	$35.10	$280.80	$57.65	$461.20	$36.17	$58.33
3 Laborers	33.10	794.40	54.35	1304.40		
1 Cement Finisher	39.70	317.60	61.95	495.60		
1 Equip. Oper. (med.)	42.95	343.60	67.30	538.40		
1 Shotcrete Rig, 12 C.Y./hr		249.20		274.12	5.19	5.71
48 L.H., Daily Totals		$1985.60		$3073.72	$41.37	$64.04
Crew C-8D	Hr.	Daily	Hr.	Daily	Bare Costs	Incl. O&P
1 Labor Foreman (outside)	$35.10	$280.80	$57.65	$461.20	$37.30	$59.66
1 Laborer	33.10	264.80	54.35	434.80		
1 Cement Finisher	39.70	317.60	61.95	495.60		
1 Equipment Oper. (light)	41.30	330.40	64.70	517.60		
1 Air Compressor, 250 cfm		193.40		212.74		
2 -50' Air Hoses, 1"		8.20		9.02	6.30	6.93
32 L.H., Daily Totals		$1395.20		$2130.96	$43.60	$66.59

Crews

Crew No.	Bare Costs		Incl. Subs O&P		Cost Per Labor-Hour	
Crew C-8E	Hr.	Daily	Hr.	Daily	Bare Costs	Incl. O&P
1 Labor Foreman (outside)	$35.10	$280.80	$57.65	$461.20	$37.30	$59.66
1 Laborer	33.10	264.80	54.35	434.80		
1 Cement Finisher	39.70	317.60	61.95	495.60		
1 Equipment Oper. (light)	41.30	330.40	64.70	517.60		
1 Air Compressor, 250 cfm		193.40		212.74		
2 -50' Air Hoses, 1"		8.20		9.02		
1 Concrete Pump (small)		738.80		812.68	29.39	32.33
32 L.H., Daily Totals		$2134.00		$2943.64	$66.69	$91.99

Crew C-10	Hr.	Daily	Hr.	Daily	Bare Costs	Incl. O&P
1 Laborer	$33.10	$264.80	$54.35	$434.80	$37.50	$59.42
2 Cement Finishers	39.70	635.20	61.95	991.20		
24 L.H., Daily Totals		$900.00		$1426.00	$37.50	$59.42

Crew C-10B	Hr.	Daily	Hr.	Daily	Bare Costs	Incl. O&P
3 Laborers	$33.10	$794.40	$54.35	$1304.40	$35.74	$57.39
2 Cement Finishers	39.70	635.20	61.95	991.20		
1 Concrete Mixer, 10 C.F.		168.40		185.24		
2 Trowels, 48" Walk-Behind		98.00		107.80	6.66	7.33
40 L.H., Daily Totals		$1696.00		$2588.64	$42.40	$64.72

Crew C-10C	Hr.	Daily	Hr.	Daily	Bare Costs	Incl. O&P
1 Laborer	$33.10	$264.80	$54.35	$434.80	$37.50	$59.42
2 Cement Finishers	39.70	635.20	61.95	991.20		
1 Trowel, 48" Walk-Behind		49.00		53.90	2.04	2.25
24 L.H., Daily Totals		$949.00		$1479.90	$39.54	$61.66

Crew C-10D	Hr.	Daily	Hr.	Daily	Bare Costs	Incl. O&P
1 Laborer	$33.10	$264.80	$54.35	$434.80	$37.50	$59.42
2 Cement Finishers	39.70	635.20	61.95	991.20		
1 Vibrating Power Screed		65.15		71.67		
1 Trowel, 48" Walk-Behind		49.00		53.90	4.76	5.23
24 L.H., Daily Totals		$1014.15		$1551.57	$42.26	$64.65

Crew C-10E	Hr.	Daily	Hr.	Daily	Bare Costs	Incl. O&P
1 Laborer	$33.10	$264.80	$54.35	$434.80	$37.50	$59.42
2 Cement Finishers	39.70	635.20	61.95	991.20		
1 Vibrating Power Screed		65.15		71.67		
1 Cement Trowel, 96" Ride-On		185.80		204.38	10.46	11.50
24 L.H., Daily Totals		$1150.95		$1702.05	$47.96	$70.92

Crew C-11	Hr.	Daily	Hr.	Daily	Bare Costs	Incl. O&P
1 Struc. Steel Foreman	$48.90	$391.20	$89.40	$715.20	$45.89	$81.49
6 Struc. Steel Workers	46.90	2251.20	85.75	4116.00		
1 Equip. Oper. (crane)	44.40	355.20	69.55	556.40		
1 Equip. Oper. Oiler	38.30	306.40	60.00	480.00		
1 Lattice Boom Crane, 150 Ton		2003.00		2203.30	27.82	30.60
72 L.H., Daily Totals		$5307.00		$8070.90	$73.71	$112.10

Crew C-12	Hr.	Daily	Hr.	Daily	Bare Costs	Incl. O&P
1 Carpenter Foreman (out)	$43.55	$348.40	$71.50	$572.00	$40.95	$66.69
3 Carpenters	41.55	997.20	68.25	1638.00		
1 Laborer	33.10	264.80	54.35	434.80		
1 Equip. Oper. (crane)	44.40	355.20	69.55	556.40		
1 Hyd. Crane, 12 Ton		655.60		721.16	13.66	15.02
48 L.H., Daily Totals		$2621.20		$3922.36	$54.61	$81.72

Crew No.	Bare Costs		Incl. Subs O&P		Cost Per Labor-Hour	
Crew C-13	Hr.	Daily	Hr.	Daily	Bare Costs	Incl. O&P
1 Struc. Steel Worker	$46.90	$375.20	$85.75	$686.00	$45.12	$79.92
1 Welder	46.90	375.20	85.75	686.00		
1 Carpenter	41.55	332.40	68.25	546.00		
1 Welder, gas engine, 300 amp		145.80		160.38	6.08	6.68
24 L.H., Daily Totals		$1228.60		$2078.38	$51.19	$86.60

Crew C-14	Hr.	Daily	Hr.	Daily	Bare Costs	Incl. O&P
1 Carpenter Foreman (out)	$43.55	$348.40	$71.50	$572.00	$40.72	$66.57
5 Carpenters	41.55	1662.00	68.25	2730.00		
4 Laborers	33.10	1059.20	54.35	1739.20		
4 Rodmen (reinf.)	46.80	1497.60	78.65	2516.80		
2 Cement Finishers	39.70	635.20	61.95	991.20		
1 Equip. Oper. (crane)	44.40	355.20	69.55	556.40		
1 Equip. Oper. Oiler	38.30	306.40	60.00	480.00		
1 Hyd. Crane, 80 Ton		1545.00		1699.50	10.73	11.80
144 L.H., Daily Totals		$7409.00		$11285.10	$51.45	$78.37

Crew C-14A	Hr.	Daily	Hr.	Daily	Bare Costs	Incl. O&P
1 Carpenter Foreman (out)	$43.55	$348.40	$71.50	$572.00	$41.78	$68.64
16 Carpenters	41.55	5318.40	68.25	8736.00		
4 Rodmen (reinf.)	46.80	1497.60	78.65	2516.80		
2 Laborers	33.10	529.60	54.35	869.60		
1 Cement Finisher	39.70	317.60	61.95	495.60		
1 Equip. Oper. (med.)	42.95	343.60	67.30	538.40		
1 Gas Engine Vibrator		30.40		33.44		
1 Concrete Pump (small)		738.80		812.68	3.85	4.23
200 L.H., Daily Totals		$9124.40		$14574.52	$45.62	$72.87

Crew C-14B	Hr.	Daily	Hr.	Daily	Bare Costs	Incl. O&P
1 Carpenter Foreman (out)	$43.55	$348.40	$71.50	$572.00	$41.70	$68.38
16 Carpenters	41.55	5318.40	68.25	8736.00		
4 Rodmen (reinf.)	46.80	1497.60	78.65	2516.80		
2 Laborers	33.10	529.60	54.35	869.60		
2 Cement Finishers	39.70	635.20	61.95	991.20		
1 Equip. Oper. (med.)	42.95	343.60	67.30	538.40		
1 Gas Engine Vibrator		30.40		33.44		
1 Concrete Pump (small)		738.80		812.68	3.70	4.07
208 L.H., Daily Totals		$9442.00		$15070.12	$45.39	$72.45

Crew C-14C	Hr.	Daily	Hr.	Daily	Bare Costs	Incl. O&P
1 Carpenter Foreman (out)	$43.55	$348.40	$71.50	$572.00	$39.90	$65.55
6 Carpenters	41.55	1994.40	68.25	3276.00		
2 Rodmen (reinf.)	46.80	748.80	78.65	1258.40		
4 Laborers	33.10	1059.20	54.35	1739.20		
1 Cement Finisher	39.70	317.60	61.95	495.60		
1 Gas Engine Vibrator		30.40		33.44	.27	.30
112 L.H., Daily Totals		$4498.80		$7374.64	$40.17	$65.84

Crew C-14D	Hr.	Daily	Hr.	Daily	Bare Costs	Incl. O&P
1 Carpenter Foreman (out)	$43.55	$348.40	$71.50	$572.00	$41.36	$67.81
18 Carpenters	41.55	5983.20	68.25	9828.00		
2 Rodmen (reinf.)	46.80	748.80	78.65	1258.40		
2 Laborers	33.10	529.60	54.35	869.60		
1 Cement Finisher	39.70	317.60	61.95	495.60		
1 Equip. Oper. (med.)	42.95	343.60	67.30	538.40		
1 Gas Engine Vibrator		30.40		33.44		
1 Concrete Pump (small)		738.80		812.68	3.85	4.23
200 L.H., Daily Totals		$9040.40		$14408.12	$45.20	$72.04

Crews

Crew No.	Bare Costs		Incl. Subs O&P		Cost Per Labor-Hour	
Crew C-14E	Hr.	Daily	Hr.	Daily	Bare Costs	Incl. O&P
1 Carpenter Foreman (out)	$43.55	$348.40	$71.50	$572.00	$41.17	$67.96
2 Carpenters	41.55	664.80	68.25	1092.00		
4 Rodmen (reinf.)	46.80	1497.60	78.65	2516.80		
3 Laborers	33.10	794.40	54.35	1304.40		
1 Cement Finisher	39.70	317.60	61.95	495.60		
1 Gas Engine Vibrator		30.40		33.44	.35	.38
88 L.H., Daily Totals		$3653.20		$6014.24	$41.51	$68.34
Crew C-14F	Hr.	Daily	Hr.	Daily	Bare Costs	Incl. O&P
1 Laborer Foreman (out)	$35.10	$280.80	$57.65	$461.20	$37.72	$59.78
2 Laborers	33.10	529.60	54.35	869.60		
6 Cement Finishers	39.70	1905.60	61.95	2973.60		
1 Gas Engine Vibrator		30.40		33.44	.42	.46
72 L.H., Daily Totals		$2746.40		$4337.84	$38.14	$60.25
Crew C-14G	Hr.	Daily	Hr.	Daily	Bare Costs	Incl. O&P
1 Laborer Foreman (out)	$35.10	$280.80	$57.65	$461.20	$37.16	$59.16
2 Laborers	33.10	529.60	54.35	869.60		
4 Cement Finishers	39.70	1270.40	61.95	1982.40		
1 Gas Engine Vibrator		30.40		33.44	.54	.60
56 L.H., Daily Totals		$2111.20		$3346.64	$37.70	$59.76
Crew C-14H	Hr.	Daily	Hr.	Daily	Bare Costs	Incl. O&P
1 Carpenter Foreman (out)	$43.55	$348.40	$71.50	$572.00	$41.04	$67.16
2 Carpenters	41.55	664.80	68.25	1092.00		
1 Rodman (reinf.)	46.80	374.40	78.65	629.20		
1 Laborer	33.10	264.80	54.35	434.80		
1 Cement Finisher	39.70	317.60	61.95	495.60		
1 Gas Engine Vibrator		30.40		33.44	.63	.70
48 L.H., Daily Totals		$2000.40		$3257.04	$41.67	$67.86
Crew C-14L	Hr.	Daily	Hr.	Daily	Bare Costs	Incl. O&P
1 Carpenter Foreman (out)	$43.55	$348.40	$71.50	$572.00	$38.75	$63.36
6 Carpenters	41.55	1994.40	68.25	3276.00		
4 Laborers	33.10	1059.20	54.35	1739.20		
1 Cement Finisher	39.70	317.60	61.95	495.60		
1 Gas Engine Vibrator		30.40		33.44	.32	.35
96 L.H., Daily Totals		$3750.00		$6116.24	$39.06	$63.71
Crew C-15	Hr.	Daily	Hr.	Daily	Bare Costs	Incl. O&P
1 Carpenter Foreman (out)	$43.55	$348.40	$71.50	$572.00	$39.13	$63.73
2 Carpenters	41.55	664.80	68.25	1092.00		
3 Laborers	33.10	794.40	54.35	1304.40		
2 Cement Finishers	39.70	635.20	61.95	991.20		
1 Rodman (reinf.)	46.80	374.40	78.65	629.20		
72 L.H., Daily Totals		$2817.20		$4588.80	$39.13	$63.73
Crew C-16	Hr.	Daily	Hr.	Daily	Bare Costs	Incl. O&P
1 Labor Foreman (outside)	$35.10	$280.80	$57.65	$461.20	$38.93	$63.24
3 Laborers	33.10	794.40	54.35	1304.40		
2 Cement Finishers	39.70	635.20	61.95	991.20		
1 Equip. Oper. (med.)	42.95	343.60	67.30	538.40		
2 Rodmen (reinf.)	46.80	748.80	78.65	1258.40		
1 Concrete Pump (small)		738.80		812.68	10.26	11.29
72 L.H., Daily Totals		$3541.60		$5366.28	$49.19	$74.53
Crew C-17	Hr.	Daily	Hr.	Daily	Bare Costs	Incl. O&P
2 Skilled Worker Foremen	$44.60	$713.60	$72.10	$1153.60	$43.00	$69.54
8 Skilled Workers	42.60	2726.40	68.90	4409.60		
80 L.H., Daily Totals		$3440.00		$5563.20	$43.00	$69.54

Crew No.	Bare Costs		Incl. Subs O&P		Cost Per Labor-Hour	
Crew C-17A	Hr.	Daily	Hr.	Daily	Bare Costs	Incl. O&P
2 Skilled Worker Foremen	$44.60	$713.60	$72.10	$1153.60	$43.02	$69.54
8 Skilled Workers	42.60	2726.40	68.90	4409.60		
.125 Equip. Oper. (crane)	44.40	44.40	69.55	69.55		
.125 Hyd. Crane, 80 Ton		193.13		212.44	2.38	2.62
81 L.H., Daily Totals		$3677.53		$5845.19	$45.40	$72.16
Crew C-17B	Hr.	Daily	Hr.	Daily	Bare Costs	Incl. O&P
2 Skilled Worker Foremen	$44.60	$713.60	$72.10	$1153.60	$43.03	$69.54
8 Skilled Workers	42.60	2726.40	68.90	4409.60		
.25 Equip. Oper. (crane)	44.40	88.80	69.55	139.10		
.25 Hyd. Crane, 80 Ton		386.25		424.88		
.25 Trowel, 48" Walk-Behind		12.25		13.48	4.86	5.35
82 L.H., Daily Totals		$3927.30		$6140.65	$47.89	$74.89
Crew C-17C	Hr.	Daily	Hr.	Daily	Bare Costs	Incl. O&P
2 Skilled Worker Foremen	$44.60	$713.60	$72.10	$1153.60	$43.05	$69.54
8 Skilled Workers	42.60	2726.40	68.90	4409.60		
.375 Equip. Oper. (crane)	44.40	133.20	69.55	208.65		
.375 Hyd. Crane, 80 Ton		579.38		637.31	6.98	7.68
83 L.H., Daily Totals		$4152.57		$6409.16	$50.03	$77.22
Crew C-17D	Hr.	Daily	Hr.	Daily	Bare Costs	Incl. O&P
2 Skilled Worker Foremen	$44.60	$713.60	$72.10	$1153.60	$43.07	$69.54
8 Skilled Workers	42.60	2726.40	68.90	4409.60		
.5 Equip. Oper. (crane)	44.40	177.60	69.55	278.20		
.5 Hyd. Crane, 80 Ton		772.50		849.75	9.20	10.12
84 L.H., Daily Totals		$4390.10		$6691.15	$52.26	$79.66
Crew C-17E	Hr.	Daily	Hr.	Daily	Bare Costs	Incl. O&P
2 Skilled Worker Foremen	$44.60	$713.60	$72.10	$1153.60	$43.00	$69.54
8 Skilled Workers	42.60	2726.40	68.90	4409.60		
1 Hyd. Jack with Rods		87.35		96.08	1.09	1.20
80 L.H., Daily Totals		$3527.35		$5659.28	$44.09	$70.74
Crew C-18	Hr.	Daily	Hr.	Daily	Bare Costs	Incl. O&P
.125 Labor Foreman (out)	$35.10	$35.10	$57.65	$57.65	$33.32	$54.72
1 Laborer	33.10	264.80	54.35	434.80		
1 Concrete Cart, 10 C.F.		58.40		64.24	6.49	7.14
9 L.H., Daily Totals		$358.30		$556.69	$39.81	$61.85
Crew C-19	Hr.	Daily	Hr.	Daily	Bare Costs	Incl. O&P
.125 Labor Foreman (out)	$35.10	$35.10	$57.65	$57.65	$33.32	$54.72
1 Laborer	33.10	264.80	54.35	434.80		
1 Concrete Cart, 18 C.F.		94.00		103.40	10.44	11.49
9 L.H., Daily Totals		$393.90		$595.85	$43.77	$66.21
Crew C-20	Hr.	Daily	Hr.	Daily	Bare Costs	Incl. O&P
1 Labor Foreman (outside)	$35.10	$280.80	$57.65	$461.20	$35.41	$57.33
5 Laborers	33.10	1324.00	54.35	2174.00		
1 Cement Finisher	39.70	317.60	61.95	495.60		
1 Equip. Oper. (med.)	42.95	343.60	67.30	538.40		
2 Gas Engine Vibrators		60.80		66.88		
1 Concrete Pump (small)		738.80		812.68	12.49	13.74
64 L.H., Daily Totals		$3065.60		$4548.76	$47.90	$71.07

Crews

Crew No.	Bare Costs		Incl. Subs O&P		Cost Per Labor-Hour	
Crew C-21	Hr.	Daily	Hr.	Daily	Bare Costs	Incl. O&P
1 Labor Foreman (outside)	$35.10	$280.80	$57.65	$461.20	$35.41	$57.33
5 Laborers	33.10	1324.00	54.35	2174.00		
1 Cement Finisher	39.70	317.60	61.95	495.60		
1 Equip. Oper. (med.)	42.95	343.60	67.30	538.40		
2 Gas Engine Vibrators		60.80		66.88		
1 Concrete Conveyer		194.60		214.06	3.99	4.39
64 L.H., Daily Totals		$2521.40		$3950.14	$39.40	$61.72
Crew C-22	Hr.	Daily	Hr.	Daily	Bare Costs	Incl. O&P
1 Rodman Foreman	$48.80	$390.40	$82.05	$656.40	$46.92	$78.64
4 Rodmen (reinf.)	46.80	1497.60	78.65	2516.80		
.125 Equip. Oper. (crane)	44.40	44.40	69.55	69.55		
.125 Equip. Oper. Oiler	38.30	38.30	60.00	60.00		
.125 Hyd. Crane, 25 Ton		93.45		102.80	2.23	2.45
42 L.H., Daily Totals		$2064.15		$3405.55	$49.15	$81.08
Crew C-23	Hr.	Daily	Hr.	Daily	Bare Costs	Incl. O&P
2 Skilled Worker Foremen	$44.60	$713.60	$72.10	$1153.60	$42.75	$68.72
6 Skilled Workers	42.60	2044.80	68.90	3307.20		
1 Equip. Oper. (crane)	44.40	355.20	69.55	556.40		
1 Equip. Oper. Oiler	38.30	306.40	60.00	480.00		
1 Lattice Boom Crane, 90 Ton		1607.00		1767.70	20.09	22.10
80 L.H., Daily Totals		$5027.00		$7264.90	$62.84	$90.81
Crew C-24	Hr.	Daily	Hr.	Daily	Bare Costs	Incl. O&P
2 Skilled Worker Foremen	$44.60	$713.60	$72.10	$1153.60	$42.75	$68.72
6 Skilled Workers	42.60	2044.80	68.90	3307.20		
1 Equip. Oper. (crane)	44.40	355.20	69.55	556.40		
1 Equip. Oper. Oiler	38.30	306.40	60.00	480.00		
1 Lattice Boom Crane, 150 Ton		2003.00		2203.30	25.04	27.54
80 L.H., Daily Totals		$5423.00		$7700.50	$67.79	$96.26
Crew C-25	Hr.	Daily	Hr.	Daily	Bare Costs	Incl. O&P
2 Rodmen (reinf.)	$46.80	$748.80	$78.65	$1258.40	$36.50	$62.40
2 Rodmen Helpers	26.20	419.20	46.15	738.40		
32 L.H., Daily Totals		$1168.00		$1996.80	$36.50	$62.40
Crew C-27	Hr.	Daily	Hr.	Daily	Bare Costs	Incl. O&P
2 Cement Finishers	$39.70	$635.20	$61.95	$991.20	$39.70	$61.95
1 Concrete Saw		161.40		177.54	10.09	11.10
16 L.H., Daily Totals		$796.60		$1168.74	$49.79	$73.05
Crew C-28	Hr.	Daily	Hr.	Daily	Bare Costs	Incl. O&P
1 Cement Finisher	$39.70	$317.60	$61.95	$495.60	$39.70	$61.95
1 Portable Air Compressor, Gas		17.80		19.58	2.23	2.45
8 L.H., Daily Totals		$335.40		$515.18	$41.92	$64.40
Crew D-1	Hr.	Daily	Hr.	Daily	Bare Costs	Incl. O&P
1 Bricklayer	$41.75	$334.00	$67.15	$537.20	$37.70	$60.63
1 Bricklayer Helper	33.65	269.20	54.10	432.80		
16 L.H., Daily Totals		$603.20		$970.00	$37.70	$60.63
Crew D-2	Hr.	Daily	Hr.	Daily	Bare Costs	Incl. O&P
3 Bricklayers	$41.75	$1002.00	$67.15	$1611.60	$38.79	$62.50
2 Bricklayer Helpers	33.65	538.40	54.10	865.60		
.5 Carpenter	41.55	166.20	68.25	273.00		
44 L.H., Daily Totals		$1706.60		$2750.20	$38.79	$62.50

Crew No.	Bare Costs		Incl. Subs O&P		Cost Per Labor-Hour	
Crew D-3	Hr.	Daily	Hr.	Daily	Bare Costs	Incl. O&P
3 Bricklayers	$41.75	$1002.00	$67.15	$1611.60	$38.65	$62.23
2 Bricklayer Helpers	33.65	538.40	54.10	865.60		
.25 Carpenter	41.55	83.10	68.25	136.50		
42 L.H., Daily Totals		$1623.50		$2613.70	$38.65	$62.23
Crew D-4	Hr.	Daily	Hr.	Daily	Bare Costs	Incl. O&P
1 Bricklayer	$41.75	$334.00	$67.15	$537.20	$37.59	$60.01
2 Bricklayer Helpers	33.65	538.40	54.10	865.60		
1 Equip. Oper. (light)	41.30	330.40	64.70	517.60		
1 Grout Pump, 50 C.F./hr.		132.60		145.86	4.14	4.56
32 L.H., Daily Totals		$1335.40		$2066.26	$41.73	$64.57
Crew D-5	Hr.	Daily	Hr.	Daily	Bare Costs	Incl. O&P
1 Bricklayer	$41.75	$334.00	$67.15	$537.20	$41.75	$67.15
8 L.H., Daily Totals		$334.00		$537.20	$41.75	$67.15
Crew D-6	Hr.	Daily	Hr.	Daily	Bare Costs	Incl. O&P
3 Bricklayers	$41.75	$1002.00	$67.15	$1611.60	$37.85	$60.93
3 Bricklayer Helpers	33.65	807.60	54.10	1298.40		
.25 Carpenter	41.55	83.10	68.25	136.50		
50 L.H., Daily Totals		$1892.70		$3046.50	$37.85	$60.93
Crew D-7	Hr.	Daily	Hr.	Daily	Bare Costs	Incl. O&P
1 Tile Layer	$39.35	$314.80	$61.45	$491.60	$35.27	$55.08
1 Tile Layer Helper	31.20	249.60	48.70	389.60		
16 L.H., Daily Totals		$564.40		$881.20	$35.27	$55.08
Crew D-8	Hr.	Daily	Hr.	Daily	Bare Costs	Incl. O&P
3 Bricklayers	$41.75	$1002.00	$67.15	$1611.60	$38.51	$61.93
2 Bricklayer Helpers	33.65	538.40	54.10	865.60		
40 L.H., Daily Totals		$1540.40		$2477.20	$38.51	$61.93
Crew D-9	Hr.	Daily	Hr.	Daily	Bare Costs	Incl. O&P
3 Bricklayers	$41.75	$1002.00	$67.15	$1611.60	$37.70	$60.63
3 Bricklayer Helpers	33.65	807.60	54.10	1298.40		
48 L.H., Daily Totals		$1809.60		$2910.00	$37.70	$60.63
Crew D-10	Hr.	Daily	Hr.	Daily	Bare Costs	Incl. O&P
1 Bricklayer Foreman	$43.75	$350.00	$70.35	$562.80	$40.89	$65.29
1 Bricklayer	41.75	334.00	67.15	537.20		
1 Bricklayer Helper	33.65	269.20	54.10	432.80		
1 Equip. Oper. (crane)	44.40	355.20	69.55	556.40		
1 S.P. Crane, 4x4, 12 Ton		655.60		721.16	20.49	22.54
32 L.H., Daily Totals		$1964.00		$2810.36	$61.38	$87.82
Crew D-11	Hr.	Daily	Hr.	Daily	Bare Costs	Incl. O&P
1 Bricklayer Foreman	$43.75	$350.00	$70.35	$562.80	$39.72	$63.87
1 Bricklayer	41.75	334.00	67.15	537.20		
1 Bricklayer Helper	33.65	269.20	54.10	432.80		
24 L.H., Daily Totals		$953.20		$1532.80	$39.72	$63.87
Crew D-12	Hr.	Daily	Hr.	Daily	Bare Costs	Incl. O&P
1 Bricklayer Foreman	$43.75	$350.00	$70.35	$562.80	$38.20	$61.42
1 Bricklayer	41.75	334.00	67.15	537.20		
2 Bricklayer Helpers	33.65	538.40	54.10	865.60		
32 L.H., Daily Totals		$1222.40		$1965.60	$38.20	$61.42

Crews

Crew No.	Bare Costs		Incl. Subs O&P		Cost Per Labor-Hour	
Crew D-13	Hr.	Daily	Hr.	Daily	Bare Costs	Incl. O&P
1 Bricklayer Foreman	$43.75	$350.00	$70.35	$562.80	$39.79	$63.92
1 Bricklayer	41.75	334.00	67.15	537.20		
2 Bricklayer Helpers	33.65	538.40	54.10	865.60		
1 Carpenter	41.55	332.40	68.25	546.00		
1 Equip. Oper. (crane)	44.40	355.20	69.55	556.40		
1 S.P. Crane, 4x4, 12 Ton		655.60		721.16	13.66	15.02
48 L.H., Daily Totals		$2565.60		$3789.16	$53.45	$78.94
Crew E-1	Hr.	Daily	Hr.	Daily	Bare Costs	Incl. O&P
1 Welder Foreman	$48.90	$391.20	$89.40	$715.20	$45.70	$79.95
1 Welder	46.90	375.20	85.75	686.00		
1 Equip. Oper. (light)	41.30	330.40	64.70	517.60		
1 Welder, gas engine, 300 amp		145.80		160.38	6.08	6.68
24 L.H., Daily Totals		$1242.60		$2079.18	$51.77	$86.63
Crew E-2	Hr.	Daily	Hr.	Daily	Bare Costs	Incl. O&P
1 Struc. Steel Foreman	$48.90	$391.20	$89.40	$715.20	$45.60	$80.28
4 Struc. Steel Workers	46.90	1500.80	85.75	2744.00		
1 Equip. Oper. (crane)	44.40	355.20	69.55	556.40		
1 Equip. Oper. Oiler	38.30	306.40	60.00	480.00		
1 Lattice Boom Crane, 90 Ton		1607.00		1767.70	28.70	31.57
56 L.H., Daily Totals		$4160.60		$6263.30	$74.30	$111.84
Crew E-3	Hr.	Daily	Hr.	Daily	Bare Costs	Incl. O&P
1 Struc. Steel Foreman	$48.90	$391.20	$89.40	$715.20	$47.57	$86.97
1 Struc. Steel Worker	46.90	375.20	85.75	686.00		
1 Welder	46.90	375.20	85.75	686.00		
1 Welder, gas engine, 300 amp		145.80		160.38	6.08	6.68
24 L.H., Daily Totals		$1287.40		$2247.58	$53.64	$93.65
Crew E-4	Hr.	Daily	Hr.	Daily	Bare Costs	Incl. O&P
1 Struc. Steel Foreman	$48.90	$391.20	$89.40	$715.20	$47.40	$86.66
3 Struc. Steel Workers	46.90	1125.60	85.75	2058.00		
1 Welder, gas engine, 300 amp		145.80		160.38	4.56	5.01
32 L.H., Daily Totals		$1662.60		$2933.58	$51.96	$91.67
Crew E-5	Hr.	Daily	Hr.	Daily	Bare Costs	Incl. O&P
2 Struc. Steel Foremen	$48.90	$782.40	$89.40	$1430.40	$46.19	$82.28
5 Struc. Steel Workers	46.90	1876.00	85.75	3430.00		
1 Equip. Oper. (crane)	44.40	355.20	69.55	556.40		
1 Welder	46.90	375.20	85.75	686.00		
1 Equip. Oper. Oiler	38.30	306.40	60.00	480.00		
1 Lattice Boom Crane, 90 Ton		1607.00		1767.70		
1 Welder, gas engine, 300 amp		145.80		160.38	21.91	24.10
80 L.H., Daily Totals		$5448.00		$8510.88	$68.10	$106.39
Crew E-6	Hr.	Daily	Hr.	Daily	Bare Costs	Incl. O&P
3 Struc. Steel Foremen	$48.90	$1173.60	$89.40	$2145.60	$46.23	$82.50
9 Struc. Steel Workers	46.90	3376.80	85.75	6174.00		
1 Equip. Oper. (crane)	44.40	355.20	69.55	556.40		
1 Welder	46.90	375.20	85.75	686.00		
1 Equip. Oper. Oiler	38.30	306.40	60.00	480.00		
1 Equip. Oper. (light)	41.30	330.40	64.70	517.60		
1 Lattice Boom Crane, 90 Ton		1607.00		1767.70		
1 Welder, gas engine, 300 amp		145.80		160.38		
1 Air Compressor, 160 cfm		158.00		173.80		
2 Impact Wrenches		35.60		39.16	15.21	16.73
128 L.H., Daily Totals		$7864.00		$12700.64	$61.44	$99.22

Crew No.	Bare Costs		Incl. Subs O&P		Cost Per Labor-Hour	
Crew E-7	Hr.	Daily	Hr.	Daily	Bare Costs	Incl. O&P
1 Struc. Steel Foreman	$48.90	$391.20	$89.40	$715.20	$46.19	$82.28
4 Struc. Steel Workers	46.90	1500.80	85.75	2744.00		
1 Equip. Oper. (crane)	44.40	355.20	69.55	556.40		
1 Equip. Oper. Oiler	38.30	306.40	60.00	480.00		
1 Welder Foreman	48.90	391.20	89.40	715.20		
2 Welders	46.90	750.40	85.75	1372.00		
1 Lattice Boom Crane, 90 Ton		1607.00		1767.70		
2 Welders, gas engine, 300 amp		291.60		320.76	23.73	26.11
80 L.H., Daily Totals		$5593.80		$8671.26	$69.92	$108.39
Crew E-8	Hr.	Daily	Hr.	Daily	Bare Costs	Incl. O&P
1 Struc. Steel Foreman	$48.90	$391.20	$89.40	$715.20	$45.92	$81.47
4 Struc. Steel Workers	46.90	1500.80	85.75	2744.00		
1 Welder Foreman	48.90	391.20	89.40	715.20		
4 Welders	46.90	1500.80	85.75	2744.00		
1 Equip. Oper. (crane)	44.40	355.20	69.55	556.40		
1 Equip. Oper. Oiler	38.30	306.40	60.00	480.00		
1 Equip. Oper. (light)	41.30	330.40	64.70	517.60		
1 Lattice Boom Crane, 90 Ton		1607.00		1767.70		
4 Welders, gas engine, 300 amp		583.20		641.52	21.06	23.17
104 L.H., Daily Totals		$6966.20		$10881.62	$66.98	$104.63
Crew E-9	Hr.	Daily	Hr.	Daily	Bare Costs	Incl. O&P
2 Struc. Steel Foremen	$48.90	$782.40	$89.40	$1430.40	$46.23	$82.50
5 Struc. Steel Workers	46.90	1876.00	85.75	3430.00		
1 Welder Foreman	48.90	391.20	89.40	715.20		
5 Welders	46.90	1876.00	85.75	3430.00		
1 Equip. Oper. (crane)	44.40	355.20	69.55	556.40		
1 Equip. Oper. Oiler	38.30	306.40	60.00	480.00		
1 Equip. Oper. (light)	41.30	330.40	64.70	517.60		
1 Lattice Boom Crane, 90 Ton		1607.00		1767.70		
5 Welders, gas engine, 300 amp		729.00		801.90	18.25	20.07
128 L.H., Daily Totals		$8253.60		$13129.20	$64.48	$102.57
Crew E-10	Hr.	Daily	Hr.	Daily	Bare Costs	Incl. O&P
1 Welder Foreman	$48.90	$391.20	$89.40	$715.20	$47.90	$87.58
1 Welder	46.90	375.20	85.75	686.00		
1 Welder, gas engine, 300 amp		145.80		160.38		
1 Flatbed Truck, Gas, 3 Ton		296.00		325.60	27.61	30.37
16 L.H., Daily Totals		$1208.20		$1887.18	$75.51	$117.95
Crew E-11	Hr.	Daily	Hr.	Daily	Bare Costs	Incl. O&P
2 Painters, Struc. Steel	$37.40	$598.40	$70.65	$1130.40	$37.30	$65.09
1 Building Laborer	33.10	264.80	54.35	434.80		
1 Equip. Oper. (light)	41.30	330.40	64.70	517.60		
1 Air Compressor, 250 cfm		193.40		212.74		
1 Sandblaster, portable, 3 C.F.		20.60		22.66		
1 Set Sand Blasting Accessories		12.95		14.24	7.09	7.80
32 L.H., Daily Totals		$1420.55		$2332.45	$44.39	$72.89
Crew E-12	Hr.	Daily	Hr.	Daily	Bare Costs	Incl. O&P
1 Welder Foreman	$48.90	$391.20	$89.40	$715.20	$45.10	$77.05
1 Equip. Oper. (light)	41.30	330.40	64.70	517.60		
1 Welder, gas engine, 300 amp		145.80		160.38	9.11	10.02
16 L.H., Daily Totals		$867.40		$1393.18	$54.21	$87.07
Crew E-13	Hr.	Daily	Hr.	Daily	Bare Costs	Incl. O&P
1 Welder Foreman	$48.90	$391.20	$89.40	$715.20	$46.37	$81.17
.5 Equip. Oper. (light)	41.30	165.20	64.70	258.80		
1 Welder, gas engine, 300 amp		145.80		160.38	12.15	13.37
12 L.H., Daily Totals		$702.20		$1134.38	$58.52	$94.53

Crews

Crew No.	Bare Costs		Incl. Subs O&P		Cost Per Labor-Hour	
Crew E-14	Hr.	Daily	Hr.	Daily	Bare Costs	Incl. O&P
1 Welder Foreman	$48.90	$391.20	$89.40	$715.20	$48.90	$89.40
1 Welder, gas engine, 300 amp		145.80		160.38	18.23	20.05
8 L.H., Daily Totals		$537.00		$875.58	$67.13	$109.45
Crew E-16	Hr.	Daily	Hr.	Daily	Bare Costs	Incl. O&P
1 Welder Foreman	$48.90	$391.20	$89.40	$715.20	$47.90	$87.58
1 Welder	46.90	375.20	85.75	686.00		
1 Welder, gas engine, 300 amp		145.80		160.38	9.11	10.02
16 L.H., Daily Totals		$912.20		$1561.58	$57.01	$97.60
Crew E-17	Hr.	Daily	Hr.	Daily	Bare Costs	Incl. O&P
1 Structural Steel Foreman	$48.90	$391.20	$89.40	$715.20	$47.90	$87.58
1 Structural Steel Worker	46.90	375.20	85.75	686.00		
16 L.H., Daily Totals		$766.40		$1401.20	$47.90	$87.58
Crew E-18	Hr.	Daily	Hr.	Daily	Bare Costs	Incl. O&P
1 Structural Steel Foreman	$48.90	$391.20	$89.40	$715.20	$46.51	$82.79
3 Structural Steel Workers	46.90	1125.60	85.75	2058.00		
1 Equipment Operator (med.)	42.95	343.60	67.30	538.40		
1 Lattice Boom Crane, 20 Ton		1029.00		1131.90	25.73	28.30
40 L.H., Daily Totals		$2889.40		$4443.50	$72.23	$111.09
Crew E-19	Hr.	Daily	Hr.	Daily	Bare Costs	Incl. O&P
1 Structural Steel Worker	$46.90	$375.20	$85.75	$686.00	$45.70	$79.95
1 Structural Steel Foreman	48.90	391.20	89.40	715.20		
1 Equip. Oper. (light)	41.30	330.40	64.70	517.60		
1 Lattice Boom Crane, 20 Ton		1029.00		1131.90	42.88	47.16
24 L.H., Daily Totals		$2125.80		$3050.70	$88.58	$127.11
Crew E-20	Hr.	Daily	Hr.	Daily	Bare Costs	Incl. O&P
1 Structural Steel Foreman	$48.90	$391.20	$89.40	$715.20	$45.76	$80.96
5 Structural Steel Workers	46.90	1876.00	85.75	3430.00		
1 Equip. Oper. (crane)	44.40	355.20	69.55	556.40		
1 Oiler	38.30	306.40	60.00	480.00		
1 Lattice Boom Crane, 40 Ton		1256.00		1381.60	19.63	21.59
64 L.H., Daily Totals		$4184.80		$6563.20	$65.39	$102.55
Crew E-22	Hr.	Daily	Hr.	Daily	Bare Costs	Incl. O&P
1 Skilled Worker Foreman	$44.60	$356.80	$72.10	$576.80	$43.27	$69.97
2 Skilled Workers	42.60	681.60	68.90	1102.40		
24 L.H., Daily Totals		$1038.40		$1679.20	$43.27	$69.97
Crew E-24	Hr.	Daily	Hr.	Daily	Bare Costs	Incl. O&P
3 Structural Steel Workers	$46.90	$1125.60	$85.75	$2058.00	$45.91	$81.14
1 Equipment Operator (med.)	42.95	343.60	67.30	538.40		
1 Hyd. Crane, 25 Ton		747.60		822.36	23.36	25.70
32 L.H., Daily Totals		$2216.80		$3418.76	$69.28	$106.84
Crew E-25	Hr.	Daily	Hr.	Daily	Bare Costs	Incl. O&P
1 Welder Foreman	$48.90	$391.20	$89.40	$715.20	$48.90	$89.40
1 Cutting Torch		19.00		20.90		
1 Set of Gases		79.20		87.12	12.28	13.50
8 L.H., Daily Totals		$489.40		$823.22	$61.17	$102.90
Crew F-3	Hr.	Daily	Hr.	Daily	Bare Costs	Incl. O&P
4 Carpenters	$41.55	$1329.60	$68.25	$2184.00	$42.12	$68.51
1 Equip. Oper. (crane)	44.40	355.20	69.55	556.40		
1 Hyd. Crane, 12 Ton		655.60		721.16	16.39	18.03
40 L.H., Daily Totals		$2340.40		$3461.56	$58.51	$86.54

Crew No.	Bare Costs		Incl. Subs O&P		Cost Per Labor-Hour	
Crew F-4	Hr.	Daily	Hr.	Daily	Bare Costs	Incl. O&P
4 Carpenters	$41.55	$1329.60	$68.25	$2184.00	$41.48	$67.09
1 Equip. Oper. (crane)	44.40	355.20	69.55	556.40		
1 Equip. Oper. Oiler	38.30	306.40	60.00	480.00		
1 Hyd. Crane, 55 Ton		1209.00		1329.90	25.19	27.71
48 L.H., Daily Totals		$3200.20		$4550.30	$66.67	$94.80
Crew F-5	Hr.	Daily	Hr.	Daily	Bare Costs	Incl. O&P
1 Carpenter Foreman	$43.55	$348.40	$71.50	$572.00	$42.05	$69.06
3 Carpenters	41.55	997.20	68.25	1638.00		
32 L.H., Daily Totals		$1345.60		$2210.00	$42.05	$69.06
Crew F-6	Hr.	Daily	Hr.	Daily	Bare Costs	Incl. O&P
2 Carpenters	$41.55	$664.80	$68.25	$1092.00	$38.74	$62.95
2 Building Laborers	33.10	529.60	54.35	869.60		
1 Equip. Oper. (crane)	44.40	355.20	69.55	556.40		
1 Hyd. Crane, 12 Ton		655.60		721.16	16.39	18.03
40 L.H., Daily Totals		$2205.20		$3239.16	$55.13	$80.98
Crew F-7	Hr.	Daily	Hr.	Daily	Bare Costs	Incl. O&P
2 Carpenters	$41.55	$664.80	$68.25	$1092.00	$37.33	$61.30
2 Building Laborers	33.10	529.60	54.35	869.60		
32 L.H., Daily Totals		$1194.40		$1961.60	$37.33	$61.30
Crew G-1	Hr.	Daily	Hr.	Daily	Bare Costs	Incl. O&P
1 Roofer Foreman	$37.40	$299.20	$65.90	$527.20	$33.06	$58.23
4 Roofers, Composition	35.40	1132.80	62.35	1995.20		
2 Roofer Helpers	26.20	419.20	46.15	738.40		
1 Application Equipment		188.40		207.24		
1 Tar Kettle/Pot		85.45		94.00		
1 Crew Truck		202.40		222.64	8.50	9.35
56 L.H., Daily Totals		$2327.45		$3784.68	$41.56	$67.58
Crew G-2	Hr.	Daily	Hr.	Daily	Bare Costs	Incl. O&P
1 Plasterer	$37.30	$298.40	$59.40	$475.20	$34.77	$55.90
1 Plasterer Helper	33.90	271.20	53.95	431.60		
1 Building Laborer	33.10	264.80	54.35	434.80		
1 Grout Pump, 50 C.F./hr		132.60		145.86	5.53	6.08
24 L.H., Daily Totals		$967.00		$1487.46	$40.29	$61.98
Crew G-2A	Hr.	Daily	Hr.	Daily	Bare Costs	Incl. O&P
1 Roofer, composition	$35.40	$283.20	$62.35	$498.80	$31.57	$54.28
1 Roofer Helper	26.20	209.60	46.15	369.20		
1 Building Laborer	33.10	264.80	54.35	434.80		
1 Foam spray rig, trailer-mtd.		551.65		606.82		
1 Pickup Truck, 3/4 Ton		140.00		154.00	28.82	31.70
24 L.H., Daily Totals		$1449.25		$2063.61	$60.39	$85.98
Crew G-3	Hr.	Daily	Hr.	Daily	Bare Costs	Incl. O&P
2 Sheet Metal Workers	$49.10	$785.60	$76.80	$1228.80	$41.10	$65.58
2 Building Laborers	33.10	529.60	54.35	869.60		
32 L.H., Daily Totals		$1315.20		$2098.40	$41.10	$65.58
Crew G-4	Hr.	Daily	Hr.	Daily	Bare Costs	Incl. O&P
1 Labor Foreman (outside)	$35.10	$280.80	$57.65	$461.20	$33.77	$55.45
2 Building Laborers	33.10	529.60	54.35	869.60		
1 Flatbed Truck, Gas, 1.5 Ton		239.80		263.78		
1 Air Compressor, 160 cfm		158.00		173.80	16.57	18.23
24 L.H., Daily Totals		$1208.20		$1768.38	$50.34	$73.68

Crews

Crew No.	Bare Costs		Incl. Subs O&P		Cost Per Labor-Hour	
	Hr.	Daily	Hr.	Daily	Bare Costs	Incl. O&P
Crew G-5						
1 Roofer Foreman	$37.40	$299.20	$65.90	$527.20	$32.12	$56.58
2 Roofers, Composition	35.40	566.40	62.35	997.60		
2 Roofer Helpers	26.20	419.20	46.15	738.40		
1 Application Equipment		188.40		207.24	4.71	5.18
40 L.H., Daily Totals		$1473.20		$2470.44	$36.83	$61.76
Crew G-6A	Hr.	Daily	Hr.	Daily	Bare Costs	Incl. O&P
2 Roofers Composition	$35.40	$566.40	$62.35	$997.60	$35.40	$62.35
1 Small Compressor, Electric		11.80		12.98		
2 Pneumatic Nailers		43.80		48.18	3.48	3.82
16 L.H., Daily Totals		$622.00		$1058.76	$38.88	$66.17
Crew G-7	Hr.	Daily	Hr.	Daily	Bare Costs	Incl. O&P
1 Carpenter	$41.55	$332.40	$68.25	$546.00	$41.55	$68.25
1 Small Compressor, Electric		11.80		12.98		
1 Pneumatic Nailer		21.90		24.09	4.21	4.63
8 L.H., Daily Totals		$366.10		$583.07	$45.76	$72.88
Crew H-1	Hr.	Daily	Hr.	Daily	Bare Costs	Incl. O&P
2 Glaziers	$40.20	$643.20	$64.55	$1032.80	$43.55	$75.15
2 Struc. Steel Workers	46.90	750.40	85.75	1372.00		
32 L.H., Daily Totals		$1393.60		$2404.80	$43.55	$75.15
Crew H-2	Hr.	Daily	Hr.	Daily	Bare Costs	Incl. O&P
2 Glaziers	$40.20	$643.20	$64.55	$1032.80	$37.83	$61.15
1 Building Laborer	33.10	264.80	54.35	434.80		
24 L.H., Daily Totals		$908.00		$1467.60	$37.83	$61.15
Crew H-3	Hr.	Daily	Hr.	Daily	Bare Costs	Incl. O&P
1 Glazier	$40.20	$321.60	$64.55	$516.40	$35.90	$58.08
1 Helper	31.60	252.80	51.60	412.80		
16 L.H., Daily Totals		$574.40		$929.20	$35.90	$58.08
Crew J-1	Hr.	Daily	Hr.	Daily	Bare Costs	Incl. O&P
3 Plasterers	$37.30	$895.20	$59.40	$1425.60	$35.94	$57.22
2 Plasterer Helpers	33.90	542.40	53.95	863.20		
1 Mixing Machine, 6 C.F.		136.60		150.26	3.42	3.76
40 L.H., Daily Totals		$1574.20		$2439.06	$39.35	$60.98
Crew J-2	Hr.	Daily	Hr.	Daily	Bare Costs	Incl. O&P
3 Plasterers	$37.30	$895.20	$59.40	$1425.60	$36.11	$57.36
2 Plasterer Helpers	33.90	542.40	53.95	863.20		
1 Lather	36.95	295.60	58.05	464.40		
1 Mixing Machine, 6 C.F.		136.60		150.26	2.85	3.13
48 L.H., Daily Totals		$1869.80		$2903.46	$38.95	$60.49
Crew J-3	Hr.	Daily	Hr.	Daily	Bare Costs	Incl. O&P
1 Terrazzo Worker	$39.00	$312.00	$60.90	$487.20	$35.45	$55.35
1 Terrazzo Helper	31.90	255.20	49.80	398.40		
1 Terrazzo Grinder, Electric		82.25		90.47		
1 Terrazzo Mixer		184.80		203.28	16.69	18.36
16 L.H., Daily Totals		$834.25		$1179.36	$52.14	$73.71
Crew K-1	Hr.	Daily	Hr.	Daily	Bare Costs	Incl. O&P
1 Carpenter	$41.55	$332.40	$68.25	$546.00	$36.90	$60.33
1 Truck Driver (light)	32.25	258.00	52.40	419.20		
1 Flatbed Truck, Gas, 3 Ton		296.00		325.60	18.50	20.35
16 L.H., Daily Totals		$886.40		$1290.80	$55.40	$80.67

Crew No.	Bare Costs		Incl. Subs O&P		Cost Per Labor-Hour	
	Hr.	Daily	Hr.	Daily	Bare Costs	Incl. O&P
Crew K-2						
1 Struc. Steel Foreman	$48.90	$391.20	$89.40	$715.20	$42.68	$75.85
1 Struc. Steel Worker	46.90	375.20	85.75	686.00		
1 Truck Driver (light)	32.25	258.00	52.40	419.20		
1 Flatbed Truck, Gas, 3 Ton		296.00		325.60	12.33	13.57
24 L.H., Daily Totals		$1320.40		$2146.00	$55.02	$89.42
Crew L-1	Hr.	Daily	Hr.	Daily	Bare Costs	Incl. O&P
1 Electrician	$49.00	$392.00	$75.30	$602.40	$50.52	$77.97
1 Plumber	52.05	416.40	80.65	645.20		
16 L.H., Daily Totals		$808.40		$1247.60	$50.52	$77.97
Crew L-2	Hr.	Daily	Hr.	Daily	Bare Costs	Incl. O&P
1 Carpenter	$41.55	$332.40	$68.25	$546.00	$36.58	$59.92
1 Carpenter Helper	31.60	252.80	51.60	412.80		
16 L.H., Daily Totals		$585.20		$958.80	$36.58	$59.92
Crew L-3	Hr.	Daily	Hr.	Daily	Bare Costs	Incl. O&P
1 Carpenter	$41.55	$332.40	$68.25	$546.00	$45.30	$72.15
.5 Electrician	49.00	196.00	75.30	301.20		
.5 Sheet Metal Worker	49.10	196.40	76.80	307.20		
16 L.H., Daily Totals		$724.80		$1154.40	$45.30	$72.15
Crew L-3A	Hr.	Daily	Hr.	Daily	Bare Costs	Incl. O&P
1 Carpenter Foreman (outside)	$43.55	$348.40	$71.50	$572.00	$45.40	$73.27
.5 Sheet Metal Worker	49.10	196.40	76.80	307.20		
12 L.H., Daily Totals		$544.80		$879.20	$45.40	$73.27
Crew L-4	Hr.	Daily	Hr.	Daily	Bare Costs	Incl. O&P
2 Skilled Workers	$42.60	$681.60	$68.90	$1102.40	$38.93	$63.13
1 Helper	31.60	252.80	51.60	412.80		
24 L.H., Daily Totals		$934.40		$1515.20	$38.93	$63.13
Crew L-5	Hr.	Daily	Hr.	Daily	Bare Costs	Incl. O&P
1 Struc. Steel Foreman	$48.90	$391.20	$89.40	$715.20	$46.83	$83.96
5 Struc. Steel Workers	46.90	1876.00	85.75	3430.00		
1 Equip. Oper. (crane)	44.40	355.20	69.55	556.40		
1 Hyd. Crane, 25 Ton		747.60		822.36	13.35	14.69
56 L.H., Daily Totals		$3370.00		$5523.96	$60.18	$98.64
Crew L-5A	Hr.	Daily	Hr.	Daily	Bare Costs	Incl. O&P
1 Structural Steel Foreman	$48.90	$391.20	$89.40	$715.20	$46.77	$82.61
2 Structural Steel Workers	46.90	750.40	85.75	1372.00		
1 Equip. Oper. (crane)	44.40	355.20	69.55	556.40		
1 S.P. Crane, 4x4, 25 Ton		655.60		721.16	20.49	22.54
32 L.H., Daily Totals		$2152.40		$3364.76	$67.26	$105.15
Crew L-5B	Hr.	Daily	Hr.	Daily	Bare Costs	Incl. O&P
1 Structural Steel Foreman	$48.90	$391.20	$89.40	$715.20	$47.47	$77.98
2 Structural Steel Workers	46.90	750.40	85.75	1372.00		
2 Electricians	49.00	784.00	75.30	1204.80		
2 Steamfitters/Pipefitters	51.90	830.40	80.40	1286.40		
1 Equip. Oper. (crane)	44.40	355.20	69.55	556.40		
1 Equip. Oper. Oiler	38.30	306.40	60.00	480.00		
1 Hyd. Crane, 80 Ton		1545.00		1699.50	21.46	23.60
72 L.H., Daily Totals		$4962.60		$7314.30	$68.92	$101.59
Crew L-6	Hr.	Daily	Hr.	Daily	Bare Costs	Incl. O&P
1 Plumber	$52.05	$416.40	$80.65	$645.20	$51.03	$78.87
.5 Electrician	49.00	196.00	75.30	301.20		
12 L.H., Daily Totals		$612.40		$946.40	$51.03	$78.87

Crews

Crew No.	Bare Costs		Incl. Subs O&P		Cost Per Labor-Hour	
Crew L-7	Hr.	Daily	Hr.	Daily	Bare Costs	Incl. O&P
2 Carpenters	$41.55	$664.80	$68.25	$1092.00	$40.20	$65.29
1 Building Laborer	33.10	264.80	54.35	434.80		
.5 Electrician	49.00	196.00	75.30	301.20		
28 L.H., Daily Totals		$1125.60		$1828.00	$40.20	$65.29
Crew L-8	Hr.	Daily	Hr.	Daily	Bare Costs	Incl. O&P
2 Carpenters	$41.55	$664.80	$68.25	$1092.00	$43.65	$70.73
.5 Plumber	52.05	208.20	80.65	322.60		
20 L.H., Daily Totals		$873.00		$1414.60	$43.65	$70.73
Crew L-9	Hr.	Daily	Hr.	Daily	Bare Costs	Incl. O&P
1 Labor Foreman (inside)	$33.60	$268.80	$55.15	$441.20	$38.04	$63.83
2 Building Laborers	33.10	529.60	54.35	869.60		
1 Struc. Steel Worker	46.90	375.20	85.75	686.00		
.5 Electrician	49.00	196.00	75.30	301.20		
36 L.H., Daily Totals		$1369.60		$2298.00	$38.04	$63.83
Crew L-10	Hr.	Daily	Hr.	Daily	Bare Costs	Incl. O&P
1 Structural Steel Foreman	$48.90	$391.20	$89.40	$715.20	$46.73	$81.57
1 Structural Steel Worker	46.90	375.20	85.75	686.00		
1 Equip. Oper. (crane)	44.40	355.20	69.55	556.40		
1 Hyd. Crane, 12 Ton		655.60		721.16	27.32	30.05
24 L.H., Daily Totals		$1777.20		$2678.76	$74.05	$111.61
Crew L-11	Hr.	Daily	Hr.	Daily	Bare Costs	Incl. O&P
2 Wreckers	$33.10	$529.60	$58.80	$940.80	$37.98	$62.96
1 Equip. Oper. (crane)	44.40	355.20	69.55	556.40		
1 Equip. Oper. (light)	41.30	330.40	64.70	517.60		
1 Hyd. Excavator, 2.5 C.Y.		1784.00		1962.40		
1 Loader, Skid Steer, 78 H.P.		291.00		320.10	64.84	71.33
32 L.H., Daily Totals		$3290.20		$4297.30	$102.82	$134.29
Crew M-1	Hr.	Daily	Hr.	Daily	Bare Costs	Incl. O&P
3 Elevator Constructors	$61.70	$1480.80	$94.75	$2274.00	$58.61	$90.01
1 Elevator Apprentice	49.35	394.80	75.80	606.40		
5 Hand Tools		58.00		63.80	1.81	1.99
32 L.H., Daily Totals		$1933.60		$2944.20	$60.42	$92.01
Crew M-3	Hr.	Daily	Hr.	Daily	Bare Costs	Incl. O&P
1 Electrician Foreman (out)	$51.00	$408.00	$78.40	$627.20	$48.44	$75.32
1 Common Laborer	33.10	264.80	54.35	434.80		
.25 Equipment Operator, Medium	42.95	85.90	67.30	134.60		
1 Elevator Constructor	61.70	493.60	94.75	758.00		
1 Elevator Apprentice	49.35	394.80	75.80	606.40		
.25 S.P. Crane, 4x4, 20 Ton		156.40		172.04	4.60	5.06
34 L.H., Daily Totals		$1803.50		$2733.04	$53.04	$80.38
Crew M-4	Hr.	Daily	Hr.	Daily	Bare Costs	Incl. O&P
1 Electrician Foreman (out)	$51.00	$408.00	$78.40	$627.20	$47.96	$74.60
1 Common Laborer	33.10	264.80	54.35	434.80		
.25 Equipment Operator, Crane	44.40	88.80	69.55	139.10		
.25 Equipment Operator, Oiler	38.30	76.60	60.00	120.00		
1 Elevator Constructor	61.70	493.60	94.75	758.00		
1 Elevator Apprentice	49.35	394.80	75.80	606.40		
.25 S.P. Crane, 4x4, 40 Ton		253.50		278.85	7.04	7.75
36 L.H., Daily Totals		$1980.10		$2964.35	$55.00	$82.34

Crew No.	Bare Costs		Incl. Subs O&P		Cost Per Labor-Hour	
Crew Q-1	Hr.	Daily	Hr.	Daily	Bare Costs	Incl. O&P
1 Plumber	$52.05	$416.40	$80.65	$645.20	$46.85	$72.58
1 Plumber Apprentice	41.65	333.20	64.50	516.00		
16 L.H., Daily Totals		$749.60		$1161.20	$46.85	$72.58
Crew Q-1A	Hr.	Daily	Hr.	Daily	Bare Costs	Incl. O&P
.25 Plumber Foreman (out)	$54.05	$108.10	$83.70	$167.40	$52.45	$81.26
1 Plumber	52.05	416.40	80.65	645.20		
10 L.H., Daily Totals		$524.50		$812.60	$52.45	$81.26
Crew Q-1C	Hr.	Daily	Hr.	Daily	Bare Costs	Incl. O&P
1 Plumber	$52.05	$416.40	$80.65	$645.20	$45.55	$70.82
1 Plumber Apprentice	41.65	333.20	64.50	516.00		
1 Equip. Oper. (medium)	42.95	343.60	67.30	538.40		
1 Trencher, Chain Type, 8' D		2090.00		2299.00	87.08	95.79
24 L.H., Daily Totals		$3183.20		$3998.60	$132.63	$166.61
Crew Q-2	Hr.	Daily	Hr.	Daily	Bare Costs	Incl. O&P
2 Plumbers	$52.05	$832.80	$80.65	$1290.40	$48.58	$75.27
1 Plumber Apprentice	41.65	333.20	64.50	516.00	48.58	75.27
24 L.H., Daily Totals		$1166.00		$1806.40	$48.58	$75.27
Crew Q-3	Hr.	Daily	Hr.	Daily	Bare Costs	Incl. O&P
1 Plumber Foreman (inside)	$52.55	$420.40	$81.40	$651.20	$49.58	$76.80
2 Plumbers	52.05	832.80	80.65	1290.40		
1 Plumber Apprentice	41.65	333.20	64.50	516.00		
32 L.H., Daily Totals		$1586.40		$2457.60	$49.58	$76.80
Crew Q-4	Hr.	Daily	Hr.	Daily	Bare Costs	Incl. O&P
1 Plumber Foreman (inside)	$52.55	$420.40	$81.40	$651.20	$49.58	$76.80
1 Plumber	52.05	416.40	80.65	645.20		
1 Welder (plumber)	52.05	416.40	80.65	645.20		
1 Plumber Apprentice	41.65	333.20	64.50	516.00		
1 Welder, electric, 300 amp		58.55		64.41	1.83	2.01
32 L.H., Daily Totals		$1644.95		$2522.01	$51.40	$78.81
Crew Q-5	Hr.	Daily	Hr.	Daily	Bare Costs	Incl. O&P
1 Steamfitter	$51.90	$415.20	$80.40	$643.20	$46.70	$72.35
1 Steamfitter Apprentice	41.50	332.00	64.30	514.40		
16 L.H., Daily Totals		$747.20		$1157.60	$46.70	$72.35
Crew Q-6	Hr.	Daily	Hr.	Daily	Bare Costs	Incl. O&P
2 Steamfitters	$51.90	$830.40	$80.40	$1286.40	$48.43	$75.03
1 Steamfitter Apprentice	41.50	332.00	64.30	514.40	48.43	75.03
24 L.H., Daily Totals		$1162.40		$1800.80	$48.43	$75.03
Crew Q-7	Hr.	Daily	Hr.	Daily	Bare Costs	Incl. O&P
1 Steamfitter Foreman (inside)	$52.40	$419.20	$81.15	$649.20	$49.42	$76.56
2 Steamfitters	51.90	830.40	80.40	1286.40		
1 Steamfitter Apprentice	41.50	332.00	64.30	514.40		
32 L.H., Daily Totals		$1581.60		$2450.00	$49.42	$76.56
Crew Q-8	Hr.	Daily	Hr.	Daily	Bare Costs	Incl. O&P
1 Steamfitter Foreman (inside)	$52.40	$419.20	$81.15	$649.20	$49.42	$76.56
1 Steamfitter	51.90	415.20	80.40	643.20		
1 Welder (steamfitter)	51.90	415.20	80.40	643.20		
1 Steamfitter Apprentice	41.50	332.00	64.30	514.40		
1 Welder, electric, 300 amp		58.55		64.41	1.83	2.01
32 L.H., Daily Totals		$1640.15		$2514.41	$51.25	$78.58

Crews

Crew No.	Bare Costs		Incl. Subs O&P		Cost Per Labor-Hour	
Crew Q-9	Hr.	Daily	Hr.	Daily	Bare Costs	Incl. O&P
1 Sheet Metal Worker	$49.10	$392.80	$76.80	$614.40	$44.20	$69.13
1 Sheet Metal Apprentice	39.30	314.40	61.45	491.60		
16 L.H., Daily Totals		$707.20		$1106.00	$44.20	$69.13
Crew Q-10	Hr.	Daily	Hr.	Daily	Bare Costs	Incl. O&P
2 Sheet Metal Workers	$49.10	$785.60	$76.80	$1228.80	$45.83	$71.68
1 Sheet Metal Apprentice	39.30	314.40	61.45	491.60		
24 L.H., Daily Totals		$1100.00		$1720.40	$45.83	$71.68
Crew Q-11	Hr.	Daily	Hr.	Daily	Bare Costs	Incl. O&P
1 Sheet Metal Foreman (inside)	$49.60	$396.80	$77.55	$620.40	$46.77	$73.15
2 Sheet Metal Workers	49.10	785.60	76.80	1228.80		
1 Sheet Metal Apprentice	39.30	314.40	61.45	491.60		
32 L.H., Daily Totals		$1496.80		$2340.80	$46.77	$73.15
Crew Q-12	Hr.	Daily	Hr.	Daily	Bare Costs	Incl. O&P
1 Sprinkler Installer	$50.40	$403.20	$77.95	$623.60	$45.35	$70.15
1 Sprinkler Apprentice	40.30	322.40	62.35	498.80		
16 L.H., Daily Totals		$725.60		$1122.40	$45.35	$70.15
Crew Q-13	Hr.	Daily	Hr.	Daily	Bare Costs	Incl. O&P
1 Sprinkler Foreman (inside)	$50.90	$407.20	$78.75	$630.00	$48.00	$74.25
2 Sprinkler Installers	50.40	806.40	77.95	1247.20		
1 Sprinkler Apprentice	40.30	322.40	62.35	498.80		
32 L.H., Daily Totals		$1536.00		$2376.00	$48.00	$74.25
Crew Q-14	Hr.	Daily	Hr.	Daily	Bare Costs	Incl. O&P
1 Asbestos Worker	$45.55	$364.40	$73.05	$584.40	$41.00	$65.75
1 Asbestos Apprentice	36.45	291.60	58.45	467.60		
16 L.H., Daily Totals		$656.00		$1052.00	$41.00	$65.75
Crew Q-15	Hr.	Daily	Hr.	Daily	Bare Costs	Incl. O&P
1 Plumber	$52.05	$416.40	$80.65	$645.20	$46.85	$72.58
1 Plumber Apprentice	41.65	333.20	64.50	516.00		
1 Welder, electric, 300 amp		58.55		64.41	3.66	4.03
16 L.H., Daily Totals		$808.15		$1225.61	$50.51	$76.60
Crew Q-16	Hr.	Daily	Hr.	Daily	Bare Costs	Incl. O&P
2 Plumbers	$52.05	$832.80	$80.65	$1290.40	$48.58	$75.27
1 Plumber Apprentice	41.65	333.20	64.50	516.00		
1 Welder, electric, 300 amp		58.55		64.41	2.44	2.68
24 L.H., Daily Totals		$1224.55		$1870.81	$51.02	$77.95
Crew Q-17	Hr.	Daily	Hr.	Daily	Bare Costs	Incl. O&P
1 Steamfitter	$51.90	$415.20	$80.40	$643.20	$46.70	$72.35
1 Steamfitter Apprentice	41.50	332.00	64.30	514.40		
1 Welder, electric, 300 amp		58.55		64.41	3.66	4.03
16 L.H., Daily Totals		$805.75		$1222.01	$50.36	$76.38
Crew Q-17A	Hr.	Daily	Hr.	Daily	Bare Costs	Incl. O&P
1 Steamfitter	$51.90	$415.20	$80.40	$643.20	$45.93	$71.42
1 Steamfitter Apprentice	41.50	332.00	64.30	514.40		
1 Equip. Oper. (crane)	44.40	355.20	69.55	556.40		
1 Hyd. Crane, 12 Ton		655.60		721.16		
1 Welder, electric, 300 amp		58.55		64.41	29.76	32.73
24 L.H., Daily Totals		$1816.55		$2499.57	$75.69	$104.15

Crew No.	Bare Costs		Incl. Subs O&P		Cost Per Labor-Hour	
Crew Q-18	Hr.	Daily	Hr.	Daily	Bare Costs	Incl. O&P
2 Steamfitters	$51.90	$830.40	$80.40	$1286.40	$48.43	$75.03
1 Steamfitter Apprentice	41.50	332.00	64.30	514.40		
1 Welder, electric, 300 amp		58.55		64.41	2.44	2.68
24 L.H., Daily Totals		$1220.95		$1865.20	$50.87	$77.72
Crew Q-19	Hr.	Daily	Hr.	Daily	Bare Costs	Incl. O&P
1 Steamfitter	$51.90	$415.20	$80.40	$643.20	$47.47	$73.33
1 Steamfitter Apprentice	41.50	332.00	64.30	514.40		
1 Electrician	49.00	392.00	75.30	602.40		
24 L.H., Daily Totals		$1139.20		$1760.00	$47.47	$73.33
Crew Q-20	Hr.	Daily	Hr.	Daily	Bare Costs	Incl. O&P
1 Sheet Metal Worker	$49.10	$392.80	$76.80	$614.40	$45.16	$70.36
1 Sheet Metal Apprentice	39.30	314.40	61.45	491.60		
.5 Electrician	49.00	196.00	75.30	301.20		
20 L.H., Daily Totals		$903.20		$1407.20	$45.16	$70.36
Crew Q-21	Hr.	Daily	Hr.	Daily	Bare Costs	Incl. O&P
2 Steamfitters	$51.90	$830.40	$80.40	$1286.40	$48.58	$75.10
1 Steamfitter Apprentice	41.50	332.00	64.30	514.40		
1 Electrician	49.00	392.00	75.30	602.40		
32 L.H., Daily Totals		$1554.40		$2403.20	$48.58	$75.10
Crew Q-22	Hr.	Daily	Hr.	Daily	Bare Costs	Incl. O&P
1 Plumber	$52.05	$416.40	$80.65	$645.20	$46.85	$72.58
1 Plumber Apprentice	41.65	333.20	64.50	516.00		
1 Hyd. Crane, 12 Ton		655.60		721.16	40.98	45.07
16 L.H., Daily Totals		$1405.20		$1882.36	$87.83	$117.65
Crew Q-22A	Hr.	Daily	Hr.	Daily	Bare Costs	Incl. O&P
1 Plumber	$52.05	$416.40	$80.65	$645.20	$42.80	$67.26
1 Plumber Apprentice	41.65	333.20	64.50	516.00		
1 Laborer	33.10	264.80	54.35	434.80		
1 Equip. Oper. (crane)	44.40	355.20	69.55	556.40		
1 Hyd. Crane, 12 Ton		655.60		721.16	20.49	22.54
32 L.H., Daily Totals		$2025.20		$2873.56	$63.29	$89.80
Crew Q-23	Hr.	Daily	Hr.	Daily	Bare Costs	Incl. O&P
1 Plumber Foreman	$54.05	$432.40	$83.70	$669.60	$49.68	$77.22
1 Plumber	52.05	416.40	80.65	645.20		
1 Equip. Oper. (med.)	42.95	343.60	67.30	538.40		
1 Lattice Boom Crane, 20 Ton		1029.00		1131.90	42.88	47.16
24 L.H., Daily Totals		$2221.40		$2985.10	$92.56	$124.38
Crew R-1	Hr.	Daily	Hr.	Daily	Bare Costs	Incl. O&P
1 Electrician Foreman	$49.50	$396.00	$76.10	$608.80	$43.28	$67.53
3 Electricians	49.00	1176.00	75.30	1807.20		
2 Helpers	31.60	505.60	51.60	825.60		
48 L.H., Daily Totals		$2077.60		$3241.60	$43.28	$67.53
Crew R-1A	Hr.	Daily	Hr.	Daily	Bare Costs	Incl. O&P
1 Electrician	$49.00	$392.00	$75.30	$602.40	$40.30	$63.45
1 Helper	31.60	252.80	51.60	412.80		
16 L.H., Daily Totals		$644.80		$1015.20	$40.30	$63.45

Crews

Crew No.	Bare Costs		Incl. Subs O&P		Cost Per Labor-Hour	
Crew R-2	Hr.	Daily	Hr.	Daily	Bare Costs	Incl. O&P
1 Electrician Foreman	$49.50	$396.00	$76.10	$608.80	$43.44	$67.82
3 Electricians	49.00	1176.00	75.30	1807.20		
2 Helpers	31.60	505.60	51.60	825.60		
1 Equip. Oper. (crane)	44.40	355.20	69.55	556.40		
1 S.P. Crane, 4x4, 5 Ton		276.20		303.82	4.93	5.43
56 L.H., Daily Totals		$2709.00		$4101.82	$48.38	$73.25
Crew R-3	Hr.	Daily	Hr.	Daily	Bare Costs	Incl. O&P
1 Electrician Foreman	$49.50	$396.00	$76.10	$608.80	$48.28	$74.47
1 Electrician	49.00	392.00	75.30	602.40		
.5 Equip. Oper. (crane)	44.40	177.60	69.55	278.20		
.5 S.P. Crane, 4x4, 5 Ton		138.10		151.91	6.91	7.60
20 L.H., Daily Totals		$1103.70		$1641.31	$55.19	$82.07
Crew R-4	Hr.	Daily	Hr.	Daily	Bare Costs	Incl. O&P
1 Struc. Steel Foreman	$48.90	$391.20	$89.40	$715.20	$47.72	$84.39
3 Struc. Steel Workers	46.90	1125.60	85.75	2058.00		
1 Electrician	49.00	392.00	75.30	602.40		
1 Welder, gas engine, 300 amp		145.80		160.38	3.65	4.01
40 L.H., Daily Totals		$2054.60		$3535.98	$51.37	$88.40
Crew R-5	Hr.	Daily	Hr.	Daily	Bare Costs	Incl. O&P
1 Electrician Foreman	$49.50	$396.00	$76.10	$608.80	$42.72	$66.75
4 Electrician Linemen	49.00	1568.00	75.30	2409.60		
2 Electrician Operators	49.00	784.00	75.30	1204.80		
4 Electrician Groundmen	31.60	1011.20	51.60	1651.20		
1 Crew Truck		202.40		222.64		
1 Flatbed Truck, 20,000 GVW		245.60		270.16		
1 Pickup Truck, 3/4 Ton		140.00		154.00		
.2 Hyd. Crane, 55 Ton		241.80		265.98		
.2 Hyd. Crane, 12 Ton		131.12		144.23		
.2 Earth Auger, Truck-Mtd.		89.84		98.82		
1 Tractor w/Winch		421.20		463.32	16.73	18.40
88 L.H., Daily Totals		$5231.16		$7493.56	$59.45	$85.15
Crew R-6	Hr.	Daily	Hr.	Daily	Bare Costs	Incl. O&P
1 Electrician Foreman	$49.50	$396.00	$76.10	$608.80	$42.72	$66.75
4 Electrician Linemen	49.00	1568.00	75.30	2409.60		
2 Electrician Operators	49.00	784.00	75.30	1204.80		
4 Electrician Groundmen	31.60	1011.20	51.60	1651.20		
1 Crew Truck		202.40		222.64		
1 Flatbed Truck, 20,000 GVW		245.60		270.16		
1 Pickup Truck, 3/4 Ton		140.00		154.00		
.2 Hyd. Crane, 55 Ton		241.80		265.98		
.2 Hyd. Crane, 12 Ton		131.12		144.23		
.2 Earth Auger, Truck-Mtd.		89.84		98.82		
1 Tractor w/Winch		421.20		463.32		
3 Cable Trailers		542.10		596.31		
.5 Tensioning Rig		179.20		197.12		
.5 Cable Pulling Rig		1048.00		1152.80	36.83	40.52
88 L.H., Daily Totals		$7000.46		$9439.79	$79.55	$107.27
Crew R-7	Hr.	Daily	Hr.	Daily	Bare Costs	Incl. O&P
1 Electrician Foreman	$49.50	$396.00	$76.10	$608.80	$34.58	$55.68
5 Electrician Groundmen	31.60	1264.00	51.60	2064.00		
1 Crew Truck		202.40		222.64	4.22	4.64
48 L.H., Daily Totals		$1862.40		$2895.44	$38.80	$60.32

Crew No.	Bare Costs		Incl. Subs O&P		Cost Per Labor-Hour	
Crew R-8	Hr.	Daily	Hr.	Daily	Bare Costs	Incl. O&P
1 Electrician Foreman	$49.50	$396.00	$76.10	$608.80	$43.28	$67.53
3 Electrician Linemen	49.00	1176.00	75.30	1807.20		
2 Electrician Groundmen	31.60	505.60	51.60	825.60		
1 Pickup Truck, 3/4 Ton		140.00		154.00		
1 Crew Truck		202.40		222.64	7.13	7.85
48 L.H., Daily Totals		$2420.00		$3618.24	$50.42	$75.38
Crew R-9	Hr.	Daily	Hr.	Daily	Bare Costs	Incl. O&P
1 Electrician Foreman	$49.50	$396.00	$76.10	$608.80	$40.36	$63.55
1 Electrician Lineman	49.00	392.00	75.30	602.40		
2 Electrician Operators	49.00	784.00	75.30	1204.80		
4 Electrician Groundmen	31.60	1011.20	51.60	1651.20		
1 Pickup Truck, 3/4 Ton		140.00		154.00		
1 Crew Truck		202.40		222.64	5.35	5.88
64 L.H., Daily Totals		$2925.60		$4443.84	$45.71	$69.44
Crew R-10	Hr.	Daily	Hr.	Daily	Bare Costs	Incl. O&P
1 Electrician Foreman	$49.50	$396.00	$76.10	$608.80	$46.18	$71.48
4 Electrician Linemen	49.00	1568.00	75.30	2409.60		
1 Electrician Groundman	31.60	252.80	51.60	412.80		
1 Crew Truck		202.40		222.64		
3 Tram Cars		393.30		432.63	12.41	13.65
48 L.H., Daily Totals		$2812.50		$4086.47	$58.59	$85.13
Crew R-11	Hr.	Daily	Hr.	Daily	Bare Costs	Incl. O&P
1 Electrician Foreman	$49.50	$396.00	$76.10	$608.80	$46.14	$71.60
4 Electricians	49.00	1568.00	75.30	2409.60		
1 Equip. Oper. (crane)	44.40	355.20	69.55	556.40		
1 Common Laborer	33.10	264.80	54.35	434.80		
1 Crew Truck		202.40		222.64		
1 Hyd. Crane, 12 Ton		655.60		721.16	15.32	16.85
56 L.H., Daily Totals		$3442.00		$4953.40	$61.46	$88.45
Crew R-12	Hr.	Daily	Hr.	Daily	Bare Costs	Incl. O&P
1 Carpenter Foreman	$42.05	$336.40	$69.05	$552.40	$39.14	$64.77
4 Carpenters	41.55	1329.60	68.25	2184.00		
4 Common Laborers	33.10	1059.20	54.35	1739.20		
1 Equip. Oper. (med.)	42.95	343.60	67.30	538.40		
1 Steel Worker	46.90	375.20	85.75	686.00		
1 Dozer, 200 H.P.		1192.00		1311.20		
1 Pickup Truck, 3/4 Ton		140.00		154.00	15.14	16.65
88 L.H., Daily Totals		$4776.00		$7165.20	$54.27	$81.42
Crew R-13	Hr.	Daily	Hr.	Daily	Bare Costs	Incl. O&P
1 Electrician Foreman	$49.50	$396.00	$76.10	$608.80	$46.84	$72.26
3 Electricians	49.00	1176.00	75.30	1807.20		
.25 Equip. Oper. (crane)	44.40	88.80	69.55	139.10		
1 Equipment Oiler	38.30	306.40	60.00	480.00		
.25 Hydraulic Crane, 33 Ton		190.75		209.82	4.54	5.00
42 L.H., Daily Totals		$2157.95		$3244.93	$51.38	$77.26
Crew R-15	Hr.	Daily	Hr.	Daily	Bare Costs	Incl. O&P
1 Electrician Foreman	$49.50	$396.00	$76.10	$608.80	$47.80	$73.67
4 Electricians	49.00	1568.00	75.30	2409.60		
1 Equipment Oper. (light)	41.30	330.40	64.70	517.60		
1 Aerial Lift Truck		347.40		382.14	7.24	7.96
48 L.H., Daily Totals		$2641.80		$3918.14	$55.04	$81.63

Crews

Crew No.	Bare Costs		Incl. Subs O & P		Cost Per Labor-Hour	
Crew R-18	Hr.	Daily	Hr.	Daily	Bare Costs	Incl. O&P
.25 Electrician Foreman	$49.50	$99.00	$76.10	$152.20	$38.33	$60.78
1 Electrician	49.00	392.00	75.30	602.40		
2 Helpers	31.60	505.60	51.60	825.60		
26 L.H., Daily Totals		$996.60		$1580.20	$38.33	$60.78
Crew R-19	Hr.	Daily	Hr.	Daily	Bare Costs	Incl. O&P
.5 Electrician Foreman	$49.50	$198.00	$76.10	$304.40	$49.10	$75.46
2 Electricians	49.00	784.00	75.30	1204.80		
20 L.H., Daily Totals		$982.00		$1509.20	$49.10	$75.46
Crew R-21	Hr.	Daily	Hr.	Daily	Bare Costs	Incl. O&P
1 Electrician Foreman	$49.50	$396.00	$76.10	$608.80	$48.97	$75.30
3 Electricians	49.00	1176.00	75.30	1807.20		
.1 Equip. Oper. (med.)	42.95	34.36	67.30	53.84		
.1 S.P. Crane, 4x4, 25 Ton		65.56		72.12	2.00	2.20
32.8 L.H., Daily Totals		$1671.92		$2541.96	$50.97	$77.50
Crew R-22	Hr.	Daily	Hr.	Daily	Bare Costs	Incl. O&P
.66 Electrician Foreman	$49.50	$261.36	$76.10	$401.81	$41.60	$65.24
2 Helpers	31.60	505.60	51.60	825.60		
2 Electricians	49.00	784.00	75.30	1204.80		
37.28 L.H., Daily Totals		$1550.96		$2432.21	$41.60	$65.24
Crew R-30	Hr.	Daily	Hr.	Daily	Bare Costs	Incl. O&P
.25 Electrician Foreman (out)	$51.00	$102.00	$78.40	$156.80	$39.37	$62.65
1 Electrician	49.00	392.00	75.30	602.40		
2 Laborers, (Semi-Skilled)	33.10	529.60	54.35	869.60		
26 L.H., Daily Totals		$1023.60		$1628.80	$39.37	$62.65
Crew R-31	Hr.	Daily	Hr.	Daily	Bare Costs	Incl. O&P
1 Electrician	$49.00	$392.00	$75.30	$602.40	$49.00	$75.30
1 Core Drill, Electric, 2.5 H.P.		53.40		58.74	6.67	7.34
8 L.H., Daily Totals		$445.40		$661.14	$55.67	$82.64

Historical Cost Indexes

The table below lists both the RSMeans Historical Cost Index based on Jan. 1, 1993 = 100 as well as the computed value of an index based on Jan. 1, 2010 costs. Since the Jan. 1, 2010 figure is estimated, space is left to write in the actual index figures as they become available through either the quarterly "RSMeans Construction Cost Indexes" or as printed in the "Engineering News-Record." To compute the actual index based on Jan. 1, 2010 = 100, divide the Historical Cost Index for a particular year by the actual Jan. 1, 2010 Construction Cost Index. Space has been left to advance the index figures as the year progresses.

Year	Historical Cost Index Jan. 1, 1993 = 100		Current Index Based on Jan. 1, 2010 = 100		Year	Historical Cost Index Jan. 1, 1993 = 100	Current Index Based on Jan. 1, 2010 = 100		Year	Historical Cost Index Jan. 1, 1993 = 100	Current Index Based on Jan. 1, 2010 = 100	
	Est.	Actual	Est.	Actual		Actual	Est.	Actual		Actual	Est.	Actual
Oct 2010					July 1995	107.6	58.9		July 1977	49.5	27.1	
July 2010					1994	104.4	57.1		1976	46.9	25.7	
April 2010					1993	101.7	55.6		1975	44.8	24.5	
Jan 2010	182.8		100.0	100.0	1992	99.4	54.4		1974	41.4	22.6	
July 2009		180.1	98.5		1991	96.8	53.0		1973	37.7	20.6	
2008		180.4	98.7		1990	94.3	51.6		1972	34.8	19.0	
2007		169.4	92.7		1989	92.1	50.4		1971	32.1	17.6	
2006		162.0	88.6		1988	89.9	49.2		1970	28.7	15.7	
2005		151.6	82.9		1987	87.7	48.0		1969	26.9	14.7	
2004		143.7	78.6		1986	84.2	46.1		1968	24.9	13.6	
2003		132.0	72.2		1985	82.6	45.2		1967	23.5	12.9	
2002		128.7	70.4		1984	82.0	44.8		1966	22.7	12.4	
2001		125.1	68.4		1983	80.2	43.9		1965	21.7	11.9	
2000		120.9	66.1		1982	76.1	41.7		1964	21.2	11.6	
1999		117.6	64.3		1981	70.0	38.3		1963	20.7	11.3	
1998		115.1	63.0		1980	62.9	34.4		1962	20.2	11.1	
1997		112.8	61.7		1979	57.8	31.6		1961	19.8	10.8	
1996		110.2	60.3		1978	53.5	29.3		1960	19.7	10.8	

Adjustments to Costs

The Historical Cost Index can be used to convert National Average building costs at a particular time to the approximate building costs for some other time.

Example:

Estimate and compare construction costs for different years in the same city.

To estimate the National Average construction cost of a building in 1970, knowing that it cost $900,000 in 2010:

INDEX in 1970 = 28.7
INDEX in 2010 = 182.8

Note: The City Cost Indexes for Canada can be used to convert U.S. National averages to local costs in Canadian dollars.

Time Adjustment using the Historical Cost Indexes:

$$\frac{\text{Index for Year A}}{\text{Index for Year B}} \times \text{Cost in Year B} = \text{Cost in Year A}$$

$$\frac{\text{INDEX 1970}}{\text{INDEX 2010}} \times \text{Cost 2010} = \text{Cost 1970}$$

$$\frac{28.7}{182.8} \times \$900,000 = .157 \times \$900,000 = \$141,300$$

The construction cost of the building in 1970 is $141,300.

City Cost Indexes

How to Use the City Cost Indexes

What you should know before you begin

RSMeans City Cost Indexes (CCI) are an extremely useful tool to use when you want to compare costs from city to city and region to region.

This publication contains average construction cost indexes for 316 U.S. and Canadian cities covering over 930 three-digit zip code locations.

Keep in mind that a City Cost Index number is a percentage ratio of a specific city's cost to the national average cost of the same item at a stated time period.

In other words, these index figures represent relative construction factors (or, if you prefer, multipliers) for Material and Installation costs, as well as the weighted average for Total In Place costs for each CSI MasterFormat division. Installation costs include both labor and equipment rental costs. When estimating equipment rental rates only, for a specific location, use 01543 CONTRACTOR EQUIPMENT index.

The 30 City Average Index is the average of 30 major U.S. cities and serves as a National Average.

Index figures for both material and installation are based on the 30 major city average of 100 and represent the cost relationship as of July 1, 2008. The index for each division is computed from representative material and labor quantities for that division. The weighted average for each city is a weighted total of the components listed above it, but does not include relative productivity between trades or cities.

As changes occur in local material prices, labor rates, and equipment rental rates, (including fuel costs) the impact of these changes should be accurately measured by the change in the City Cost Index for each particular city (as compared to the 30 City Average).

Therefore, if you know (or have estimated) building costs in one city today, you can easily convert those costs to expected building costs in another city.

In addition, by using the Historical Cost Index, you can easily convert National Average building costs at a particular time to the approximate building costs for some other time. The City Cost Indexes can then be applied to calculate the costs for a particular city.

Quick Calculations

Location Adjustment Using the City Cost Indexes:

$$\frac{\text{Index for City A}}{\text{Index for City B}} \times \text{Cost in City B} = \text{Cost in City A}$$

Time Adjustment for the National Average Using the Historical Cost Index:

$$\frac{\text{Index for Year A}}{\text{Index for Year B}} \times \text{Cost in Year B} = \text{Cost in Year A}$$

Adjustment from the National Average:

$$\frac{\text{Index for City A}}{100} \times \text{National Average Cost} = \text{Cost in City A}$$

Since each of the other RSMeans publications contains many different items, any *one* item multiplied by the particular city index may give incorrect results. However, the larger the number of items compiled, the closer the results should be to actual costs for that particular city.

The City Cost Indexes for Canadian cities are calculated using Canadian material and equipment prices and labor rates, in Canadian dollars. Therefore, indexes for Canadian cities can be used to convert U.S. National Average prices to local costs in Canadian dollars.

How to use this section

1. Compare costs from city to city.

In using the RSMeans Indexes, remember that an index number is not a fixed number but a ratio: It's a percentage ratio of a building component's cost at any stated time to the National Average cost of that same component at the same time period. Put in the form of an equation:

$$\frac{\text{Specific City Cost}}{\text{National Average Cost}} \times 100 = \text{City Index Number}$$

Therefore, when making cost comparisons between cities, do not subtract one city's index number from the index number of another city and read the result as a percentage difference. Instead, divide one city's index number by that of the other city. The resulting number may then be used as a multiplier to calculate cost differences from city to city.

The formula used to find cost differences between cities for the purpose of comparison is as follows:

$$\frac{\text{City A Index}}{\text{City B Index}} \times \text{City B Cost (Known)} = \text{City A Cost (Unknown)}$$

In addition, you can use RSMeans CCI to calculate and compare costs division by division between cities using the same basic formula. (Just be sure that you're comparing similar divisions.)

2. Compare a specific city's construction costs with the National Average.

When you're studying construction location feasibility, it's advisable to compare a prospective project's cost index with an index of the National Average cost.

For example, divide the weighted average index of construction costs of a specific city by that of the 30 City Average, which = 100.

$$\frac{\text{City Index}}{100} = \% \text{ of National Average}$$

As a result, you get a ratio that indicates the relative cost of construction in that city in comparison with the National Average.

3. Convert U.S. National Average to actual costs in Canadian City.

$$\frac{\text{Index for Canadian City}}{100} \times \text{National Average Cost} = \text{Cost in Canadian City in \$ CAN}$$

4. **Adjust construction cost data based on a National Average.**
When you use a source of construction cost data which is based on a National Average (such as RSMeans cost data publications), it is necessary to adjust those costs to a specific location.

$$\frac{\text{City Index}}{100} \times \text{"Book" Cost Based on National Average Costs} = \text{City Cost (Unknown)}$$

5. **When applying the City Cost Indexes to demolition projects, use the appropriate division installation index.** For example, for removal of existing doors and windows, use Division 8 (Openings) index.

What you might like to know about how we developed the Indexes

The information presented in the CCI is organized according to the Construction Specifications Institute (CSI) MasterFormat 2004.

To create a reliable index, RSMeans researched the building type most often constructed in the United States and Canada. Because it was concluded that no one type of building completely represented the building construction industry, nine different types of buildings were combined to create a composite model.

The exact material, labor and equipment quantities are based on detailed analysis of these nine building types, then each quantity is weighted in proportion to expected usage. These various material items, labor hours, and equipment rental rates are thus combined to form a composite building representing as closely as possible the actual usage of materials, labor and equipment used in the North American Building Construction Industry.

The following structures were chosen to make up that composite model:

1. Factory, 1 story
2. Office, 2-4 story
3. Store, Retail
4. Town Hall, 2-3 story
5. High School, 2-3 story
6. Hospital, 4-8 story
7. Garage, Parking
8. Apartment, 1-3 story
9. Hotel/Motel, 2-3 story

For the purposes of ensuring the timeliness of the data, the components of the index for the composite model have been streamlined. They currently consist of:

- specific quantities of 66 commonly used construction materials;
- specific labor-hours for 21 building construction trades; and
- specific days of equipment rental for 6 types of construction equipment (normally used to install the 66 material items by the 21 trades.) Fuel costs and routine maintenance costs are included in the equipment costs.

A sophisticated computer program handles the updating of all costs for each city on a quarterly basis. Material and equipment price quotations are gathered quarterly from 316 cities in the United States and Canada. These prices and the latest negotiated labor wage rates for 21 different building trades are used to compile the quarterly update of the City Cost Index.

The 30 major U.S. cities used to calculate the National Average are:

Atlanta, GA
Baltimore, MD
Boston, MA
Buffalo, NY
Chicago, IL
Cincinnati, OH
Cleveland, OH
Columbus, OH
Dallas, TX
Denver, CO
Detroit, MI
Houston, TX
Indianapolis, IN
Kansas City, MO
Los Angeles, CA
Memphis, TN
Milwaukee, WI
Minneapolis, MN
Nashville, TN
New Orleans, LA
New York, NY
Philadelphia, PA
Phoenix, AZ
Pittsburgh, PA
St. Louis, MO
San Antonio, TX
San Diego, CA
San Francisco, CA
Seattle, WA
Washington, DC

What the CCI does not indicate

The weighted average for each city is a total of the divisional components weighted to reflect typical usage, but it does not include the productivity variations between trades or cities.

In addition, the CCI does not take into consideration factors such as the following:

- managerial efficiency
- competitive conditions
- automation
- restrictive union practices
- unique local requirements
- regional variations due to specific building codes

City Cost Indexes

		UNITED STATES			ALABAMA												
	DIVISION	ANYTOWN			BIRMINGHAM			HUNTSVILLE			MOBILE			MONTGOMERY			TUSCALOOSA		
		MAT.	INST.	TOTAL	MAT.	INST.	TOTAL	MAT.	INST.	TOTAL	MAT.	INST.	TOTAL	MAT.	INST.	TOTAL	MAT.	INST.	TOTAL
015433	CONTRACTOR EQUIPMENT		100.0	100.0		101.0	101.0		100.9	100.9		97.8	97.8		97.8	97.8		100.9	100.9
0241, 31 - 34	SITE & INFRASTRUCTURE, DEMOLITION	100.0	100.0	100.0	87.0	93.8	91.8	82.7	92.9	89.8	96.1	88.3	90.7	94.1	87.7	89.6	83.1	92.9	89.9
0310	Concrete Forming & Accessories	100.0	100.0	100.0	90.6	70.2	72.9	94.1	67.4	71.0	94.2	59.6	64.3	92.2	47.7	53.9	94.0	51.7	57.4
0320	Concrete Reinforcing	100.0	100.0	100.0	83.4	87.7	85.6	83.4	77.2	80.2	86.0	85.7	85.8	86.0	86.3	86.1	83.4	86.4	84.9
0330	Cast-in-Place Concrete	100.0	100.0	100.0	100.8	68.8	88.7	91.2	65.4	81.4	96.2	72.0	87.1	98.6	49.7	80.1	94.9	52.2	78.8
03	CONCRETE	100.0	100.0	100.0	92.3	74.1	83.8	87.7	69.7	79.3	90.6	70.2	81.1	91.6	57.5	75.7	89.6	60.0	75.8
04	MASONRY	100.0	100.0	100.0	87.6	83.4	85.0	87.1	63.5	72.9	87.8	58.7	70.3	86.2	43.7	60.7	86.2	51.1	65.1
05	METALS	100.0	100.0	100.0	96.6	94.7	95.9	97.8	89.3	94.9	96.7	93.8	95.7	96.5	91.6	94.9	96.9	92.0	95.2
06	WOOD, PLASTICS & COMPOSITES	100.0	100.0	100.0	92.0	67.1	77.9	94.7	68.7	80.0	94.9	59.1	74.6	91.6	47.3	66.5	94.7	49.6	69.2
07	THERMAL & MOISTURE PROTECTION	100.0	100.0	100.0	94.0	83.9	90.2	91.6	75.7	85.6	91.6	75.5	85.5	90.8	66.5	81.6	91.7	69.4	83.3
08	OPENINGS	100.0	100.0	100.0	97.2	73.0	91.2	97.5	65.3	89.5	97.5	64.0	89.2	97.5	57.2	87.5	97.5	63.6	89.1
0920	Plaster & Gypsum Board	100.0	100.0	100.0	98.5	66.6	76.4	95.3	68.2	76.6	95.3	58.4	69.8	98.5	46.2	62.3	95.3	48.6	63.0
0950, 0980	Ceilings & Acoustic Treatment	100.0	100.0	100.0	94.1	66.6	76.6	98.4	68.2	79.2	98.4	58.4	72.9	94.1	46.2	63.6	98.4	48.6	66.7
0960	Flooring	100.0	100.0	100.0	103.1	64.9	92.6	103.1	58.2	90.7	112.8	60.3	98.3	111.1	31.9	89.2	103.1	40.4	85.8
0970, 0990	Wall Finishes & Painting/Coating	100.0	100.0	100.0	96.1	61.7	75.4	96.1	64.9	77.4	100.1	60.3	76.1	96.1	59.2	73.9	96.1	51.0	69.0
09	FINISHES	100.0	100.0	100.0	96.5	67.6	80.8	96.7	65.2	79.6	101.6	58.8	78.4	99.9	44.8	70.0	96.7	48.5	70.5
COVERS	DIVS. 10 - 14, 25, 28, 41, 43, 44	100.0	100.0	100.0	100.0	85.2	96.9	100.0	82.2	96.2	100.0	82.0	96.2	100.0	76.8	95.1	100.0	79.0	95.5
21, 22, 23	FIRE SUPPRESSION, PLUMBING & HVAC	100.0	100.0	100.0	99.8	66.0	85.9	99.7	58.6	82.8	99.7	60.4	83.5	99.8	38.4	74.5	99.8	38.2	74.4
26, 27, 3370	ELECTRICAL, COMMUNICATIONS & UTIL.	100.0	100.0	100.0	108.0	64.3	85.2	99.6	68.4	83.3	99.6	56.2	76.9	102.7	68.2	84.7	99.0	64.3	80.8
MF2004	WEIGHTED AVERAGE	100.0	100.0	100.0	97.4	75.2	87.6	96.0	70.1	84.6	97.1	67.4	84.0	97.1	58.3	80.0	96.0	60.2	80.2

		ALASKA									ARIZONA								
	DIVISION	ANCHORAGE			FAIRBANKS			JUNEAU			FLAGSTAFF			MESA/TEMPE			PHOENIX		
		MAT.	INST.	TOTAL	MAT.	INST.	TOTAL	MAT.	INST.	TOTAL	MAT.	INST.	TOTAL	MAT.	INST.	TOTAL	MAT.	INST.	TOTAL
015433	CONTRACTOR EQUIPMENT		116.2	116.2		116.2	116.2		116.2	116.2		94.9	94.9		98.0	98.0		98.6	98.6
0241, 31 - 34	SITE & INFRASTRUCTURE, DEMOLITION	137.3	131.5	133.2	123.9	131.5	129.2	136.4	131.5	132.9	80.1	99.5	93.6	83.9	101.9	96.5	84.4	103.3	97.5
0310	Concrete Forming & Accessories	135.6	118.5	120.8	141.4	119.1	122.1	137.8	118.5	121.1	100.0	63.3	68.3	100.5	57.4	63.2	101.4	71.4	75.4
0320	Concrete Reinforcing	146.4	106.5	126.0	150.6	106.5	128.1	112.1	106.5	109.3	96.6	79.7	88.0	98.2	79.6	88.2	95.6	79.9	87.6
0330	Cast-in-Place Concrete	162.5	114.5	144.4	142.6	117.1	132.9	163.4	114.5	144.9	91.9	75.7	85.8	98.1	62.0	84.6	98.1	76.3	89.8
03	CONCRETE	136.5	114.2	126.1	123.1	115.3	119.5	132.2	114.2	123.8	119.0	70.8	96.4	105.3	63.7	85.8	105.0	74.6	90.8
04	MASONRY	202.7	119.0	152.4	211.4	118.8	155.7	194.2	119.0	149.0	95.2	60.8	74.5	110.0	45.8	71.4	96.6	64.6	77.3
05	METALS	120.2	101.2	113.8	120.3	101.4	113.9	120.2	101.1	113.8	97.3	73.6	89.3	93.7	73.8	87.0	95.4	75.4	88.7
06	WOOD, PLASTICS & COMPOSITES	114.1	118.6	116.7	117.8	119.3	118.7	114.1	118.6	116.7	105.0	62.3	80.8	98.8	62.5	78.2	99.7	72.8	84.4
07	THERMAL & MOISTURE PROTECTION	185.0	114.3	158.3	183.1	114.4	157.2	183.7	114.3	157.5	95.6	66.2	84.5	102.4	56.0	84.9	102.3	68.1	89.4
08	OPENINGS	121.0	113.5	119.1	118.1	113.6	117.0	118.1	113.5	117.0	101.1	67.9	92.8	96.8	64.8	88.9	97.8	73.6	91.8
0920	Plaster & Gypsum Board	137.8	119.0	124.8	157.2	119.7	131.3	137.8	119.0	124.8	95.4	61.2	71.7	102.4	61.2	73.9	105.2	71.9	82.1
0950, 0980	Ceilings & Acoustic Treatment	121.3	119.0	119.8	126.5	119.7	122.2	121.3	119.0	119.8	126.2	61.2	84.8	116.2	61.2	81.2	123.9	71.9	90.8
0960	Flooring	159.3	122.5	149.1	159.3	122.5	149.1	159.3	122.5	149.1	97.4	42.8	82.3	102.1	53.3	88.7	102.4	55.7	89.5
0970, 0990	Wall Finishes & Painting/Coating	154.8	112.8	129.5	154.8	125.7	137.3	154.8	112.8	129.5	90.4	55.1	69.2	95.0	55.1	71.0	95.0	60.9	74.5
09	FINISHES	147.2	119.2	132.0	148.7	121.0	133.7	146.0	119.2	131.4	105.1	57.7	79.4	103.9	55.8	77.8	106.1	66.9	84.8
COVERS	DIVS. 10 - 14, 25, 28, 41, 43, 44	100.0	110.3	102.2	100.0	110.3	102.2	100.0	110.3	102.2	100.0	82.4	96.3	100.0	77.3	95.2	100.0	84.0	96.6
21, 22, 23	FIRE SUPPRESSION, PLUMBING & HVAC	100.4	101.7	100.9	100.3	104.5	102.1	100.4	99.0	99.8	100.2	76.8	90.6	100.3	62.7	84.8	100.3	76.9	90.6
26, 27, 3370	ELECTRICAL, COMMUNICATIONS & UTIL.	141.6	110.0	125.1	147.9	110.0	128.1	142.0	110.0	125.3	100.8	58.9	78.9	94.2	58.9	75.7	101.4	67.0	83.4
MF2004	WEIGHTED AVERAGE	127.4	112.3	120.7	126.3	113.3	120.5	126.0	111.7	119.7	101.6	70.4	87.9	99.4	64.4	83.9	99.9	74.6	88.7

		ARIZONA						ARKANSAS											
	DIVISION	PRESCOTT			TUCSON			FORT SMITH			JONESBORO			LITTLE ROCK			PINE BLUFF		
		MAT.	INST.	TOTAL	MAT.	INST.	TOTAL	MAT.	INST.	TOTAL	MAT.	INST.	TOTAL	MAT.	INST.	TOTAL	MAT.	INST.	TOTAL
015433	CONTRACTOR EQUIPMENT		94.9	94.9		98.0	98.0		87.4	87.4		106.6	106.6		87.4	87.4		87.4	87.4
0241, 31 - 34	SITE & INFRASTRUCTURE, DEMOLITION	68.9	97.9	89.1	79.9	103.0	96.0	78.6	84.1	82.4	100.1	98.0	98.6	86.5	84.1	84.8	80.5	84.1	83.0
0310	Concrete Forming & Accessories	96.6	47.3	53.9	100.7	70.8	74.9	97.4	39.1	47.0	87.8	53.2	57.9	89.6	65.5	68.9	81.5	65.4	67.5
0320	Concrete Reinforcing	96.6	70.7	83.4	80.4	79.7	80.0	99.4	71.9	85.4	95.0	68.3	81.4	99.6	66.9	82.9	99.5	66.9	82.8
0330	Cast-in-Place Concrete	91.8	56.1	78.3	100.9	76.1	91.5	95.4	64.9	83.9	90.9	59.2	78.9	95.9	65.1	84.2	87.5	65.0	79.0
03	CONCRETE	103.7	55.1	81.0	104.1	74.2	90.2	91.3	55.1	74.4	88.1	59.4	74.7	91.1	66.0	79.3	89.1	65.8	78.2
04	MASONRY	95.5	48.9	67.5	94.6	60.8	74.3	97.8	54.8	71.9	93.3	46.7	65.3	94.6	54.8	70.7	117.9	54.8	80.0
05	METALS	97.3	65.6	86.6	94.6	74.1	87.6	100.8	70.7	90.6	95.2	80.3	90.2	96.9	69.4	87.6	99.2	69.1	89.0
06	WOOD, PLASTICS & COMPOSITES	100.9	45.6	69.6	99.0	72.8	84.2	105.1	34.6	65.2	92.8	54.0	70.8	99.7	70.0	82.9	85.8	70.0	76.9
07	THERMAL & MOISTURE PROTECTION	94.4	52.8	78.7	103.6	63.6	88.5	98.1	54.7	81.7	101.7	54.0	83.7	97.4	58.4	82.7	97.0	58.4	82.5
08	OPENINGS	101.1	50.6	88.6	93.2	73.6	88.3	97.5	43.8	84.2	99.5	57.6	89.1	97.5	63.1	88.9	92.7	63.1	85.4
0920	Plaster & Gypsum Board	92.3	44.0	58.9	108.8	71.9	83.2	84.1	33.2	48.9	94.6	52.8	65.7	79.9	69.6	72.8	77.0	69.6	71.9
0950, 0980	Ceilings & Acoustic Treatment	124.4	44.0	73.2	117.0	71.9	88.3	93.0	33.2	54.9	92.7	52.8	67.3	91.3	69.6	77.5	89.6	69.6	76.9
0960	Flooring	95.8	42.6	81.1	93.0	42.8	79.2	111.6	65.5	98.8	74.7	57.5	70.0	112.9	65.5	99.8	101.3	65.5	91.4
0970, 0990	Wall Finishes & Painting/Coating	90.4	41.0	60.7	95.2	55.1	71.1	98.6	59.8	75.2	87.6	53.7	67.2	98.6	61.3	76.2	98.6	61.3	76.2
09	FINISHES	102.6	44.9	71.3	101.9	63.9	81.3	94.9	45.0	67.8	87.1	53.8	69.0	94.5	65.9	79.0	90.3	65.9	77.1
COVERS	DIVS. 10 - 14, 25, 28, 41, 43, 44	100.0	78.4	95.4	100.0	83.9	96.6	100.0	71.7	94.0	100.0	52.9	90.0	100.0	75.8	94.9	100.0	75.8	94.9
21, 22, 23	FIRE SUPPRESSION, PLUMBING & HVAC	100.2	67.1	86.6	100.2	65.6	86.0	100.2	45.4	77.6	100.4	45.3	77.7	100.1	72.6	88.7	100.1	47.8	78.6
26, 27, 3370	ELECTRICAL, COMMUNICATIONS & UTIL.	100.5	58.8	78.7	96.5	58.4	76.6	97.0	72.4	84.1	104.0	50.2	75.8	102.4	73.6	87.3	96.3	73.6	84.4
MF2004	WEIGHTED AVERAGE	99.1	61.1	82.4	98.2	69.9	85.7	97.3	57.8	79.9	96.9	57.9	79.7	97.2	67.0	83.9	96.8	63.9	82.3

City Cost Indexes

	DIVISION	ARKANSAS TEXARKANA			CALIFORNIA ANAHEIM			BAKERSFIELD			FRESNO			LOS ANGELES			OAKLAND		
		MAT.	INST.	TOTAL	MAT.	INST.	TOTAL	MAT.	INST.	TOTAL	MAT.	INST.	TOTAL	MAT.	INST.	TOTAL	MAT.	INST.	TOTAL
015433	CONTRACTOR EQUIPMENT		88.0	88.0		101.6	101.6		99.3	99.3		99.3	99.3		100.6	100.6		98.7	98.7
0241, 31 - 34	SITE & INFRASTRUCTURE, DEMOLITION	94.8	84.6	87.7	100.6	109.4	106.7	106.3	106.7	106.6	106.0	106.3	106.2	97.8	109.0	105.6	133.5	103.2	112.4
0310	Concrete Forming & Accessories	87.4	43.9	49.7	105.8	125.9	123.2	104.3	125.2	122.3	101.9	127.5	124.0	105.6	126.2	123.4	111.1	143.2	138.8
0320	Concrete Reinforcing	99.1	66.3	82.4	100.6	118.0	109.5	113.0	118.2	115.6	96.0	118.2	107.4	111.3	118.6	115.0	111.5	118.9	115.3
0330	Cast-in-Place Concrete	95.4	46.0	76.7	97.5	119.8	105.9	104.3	118.6	109.7	102.8	114.2	107.1	103.9	118.5	109.4	126.3	117.7	123.1
03	CONCRETE	88.1	49.7	70.1	105.2	121.3	112.7	107.7	120.5	113.7	106.9	120.0	113.0	110.3	121.1	115.3	122.7	128.1	125.2
04	MASONRY	97.8	31.9	58.2	86.3	111.0	101.2	107.0	116.1	112.5	110.0	115.8	113.5	93.1	120.4	109.5	153.5	124.9	136.3
05	METALS	92.5	66.0	83.5	103.7	102.1	103.2	99.0	101.4	99.8	103.6	100.4	102.5	102.2	103.7	102.7	98.2	100.2	98.9
06	WOOD, PLASTICS & COMPOSITES	94.0	46.9	67.3	93.6	127.0	112.5	87.9	127.1	110.1	99.4	129.1	116.2	88.9	126.8	110.4	106.6	147.9	130.0
07	THERMAL & MOISTURE PROTECTION	97.8	43.9	77.4	103.4	115.2	107.8	102.0	107.2	104.0	96.9	111.1	102.3	102.4	118.9	108.7	113.9	122.8	117.3
08	OPENINGS	98.0	49.6	86.0	104.4	120.8	108.5	101.9	118.6	106.0	103.7	118.2	107.3	98.6	123.9	104.9	103.5	132.9	110.7
0920	Plaster & Gypsum Board	81.6	45.9	56.8	107.3	127.7	121.4	102.1	127.7	119.8	102.3	129.8	121.3	100.9	127.7	119.5	106.0	148.9	135.7
0950, 0980	Ceilings & Acoustic Treatment	94.7	45.9	63.6	112.2	127.7	122.1	113.6	127.7	122.6	115.1	129.8	124.5	114.7	127.7	123.0	114.2	148.9	136.3
0960	Flooring	104.0	54.9	90.4	114.5	114.5	114.5	114.4	84.4	106.1	121.1	128.5	123.1	102.4	114.5	105.8	118.2	111.6	116.4
0970, 0990	Wall Finishes & Painting/Coating	98.6	31.5	58.3	104.6	102.2	103.2	106.3	98.2	101.4	124.0	97.1	107.8	100.0	110.0	106.0	108.3	139.6	127.2
09	FINISHES	93.4	44.5	66.9	108.4	121.9	115.7	110.0	115.8	113.1	113.9	125.7	120.3	105.8	122.6	114.9	113.5	138.8	127.3
COVERS	DIVS. 10 - 14, 25, 28, 41, 43, 44	100.0	38.0	86.8	100.0	112.6	102.7	100.0	109.6	102.0	100.0	124.6	105.2	100.0	112.2	102.6	100.0	128.2	106.0
21, 22, 23	FIRE SUPPRESSION, PLUMBING & HVAC	100.2	33.4	72.7	100.1	112.9	105.4	100.1	94.2	97.7	100.2	107.8	103.4	100.2	112.9	105.4	100.2	136.1	115.0
26, 27, 3370	ELECTRICAL, COMMUNICATIONS & UTIL.	98.4	40.3	68.0	93.5	101.1	97.5	105.5	93.8	99.4	92.8	90.7	91.8	101.9	115.8	109.1	100.9	125.0	113.5
MF2004	WEIGHTED AVERAGE	96.2	46.5	74.2	101.1	112.6	106.2	102.9	106.6	104.6	102.8	110.6	106.3	101.7	116.0	108.0	108.5	126.0	116.2

	DIVISION	CALIFORNIA OXNARD			REDDING			RIVERSIDE			SACRAMENTO			SAN DIEGO			SAN FRANCISCO		
		MAT.	INST.	TOTAL	MAT.	INST.	TOTAL	MAT.	INST.	TOTAL	MAT.	INST.	TOTAL	MAT.	INST.	TOTAL	MAT.	INST.	TOTAL
015433	CONTRACTOR EQUIPMENT		98.2	98.2		99.0	99.0		100.3	100.3		98.3	98.3		98.6	98.6		107.5	107.5
0241, 31 - 34	SITE & INFRASTRUCTURE, DEMOLITION	106.4	104.7	105.2	110.4	105.0	106.6	98.9	107.3	104.7	106.4	110.0	108.9	105.1	102.8	103.5	135.8	110.2	118.0
0310	Concrete Forming & Accessories	107.0	125.9	123.3	106.5	127.0	124.2	105.9	125.7	123.0	109.4	128.8	126.2	108.8	110.4	110.2	110.6	145.2	140.5
0320	Concrete Reinforcing	113.0	117.8	115.4	109.6	117.8	113.8	112.0	117.8	114.9	103.4	118.3	111.0	115.0	118.1	116.6	127.0	119.6	123.2
0330	Cast-in-Place Concrete	103.5	119.0	109.4	117.3	113.9	116.0	101.1	119.7	108.1	106.8	115.2	110.0	100.2	106.7	102.7	129.5	120.0	125.9
03	CONCRETE	107.5	120.9	113.8	117.1	119.6	118.2	106.1	121.1	113.1	111.8	120.6	116.0	108.5	110.0	109.2	126.5	130.4	128.3
04	MASONRY	111.8	105.7	108.2	113.7	108.0	110.3	85.0	106.6	98.0	122.8	115.9	118.6	95.4	109.4	103.8	154.1	133.0	141.4
05	METALS	98.8	101.4	99.7	103.4	99.2	102.0	103.9	101.6	103.1	94.0	96.9	95.0	101.8	102.7	102.1	104.8	110.0	106.6
06	WOOD, PLASTICS & COMPOSITES	94.1	127.1	112.8	94.1	129.1	113.9	93.6	127.0	112.5	100.0	130.4	117.2	94.9	108.1	102.4	106.6	148.1	130.1
07	THERMAL & MOISTURE PROTECTION	106.8	111.2	108.5	107.3	106.6	107.0	103.6	108.8	105.5	120.6	112.0	117.4	108.8	104.0	107.0	116.5	136.4	124.0
08	OPENINGS	100.8	120.8	105.8	103.5	119.0	107.3	102.2	120.8	107.4	116.3	120.9	117.4	100.1	109.0	102.3	107.8	136.5	115.0
0920	Plaster & Gypsum Board	107.3	127.7	121.4	105.3	129.8	122.3	106.8	127.7	121.3	100.3	131.0	121.5	115.8	108.1	110.5	108.8	148.9	136.5
0950, 0980	Ceilings & Acoustic Treatment	117.0	127.7	123.8	122.2	129.8	127.0	115.3	127.7	123.2	112.3	131.0	124.2	108.4	108.1	108.2	121.7	148.9	139.0
0960	Flooring	112.9	114.5	113.3	113.7	105.3	111.3	118.5	114.5	117.4	117.6	109.8	115.4	105.7	105.9	105.7	118.2	125.5	120.2
0970, 0990	Wall Finishes & Painting/Coating	105.6	93.2	98.1	105.6	101.2	102.9	102.7	102.2	102.4	105.6	110.5	108.5	104.3	108.1	106.6	108.3	151.6	134.4
09	FINISHES	110.8	120.9	116.3	112.5	122.0	117.7	109.9	121.9	116.4	110.1	124.7	118.1	107.6	109.1	108.4	115.8	143.3	130.7
COVERS	DIVS. 10 - 14, 25, 28, 41, 43, 44	100.0	112.8	102.7	100.0	124.6	105.2	100.0	112.5	102.7	100.0	125.2	105.4	100.0	109.5	102.0	100.0	128.8	106.1
21, 22, 23	FIRE SUPPRESSION, PLUMBING & HVAC	100.1	112.9	105.4	100.1	98.7	99.5	100.1	112.9	105.3	100.1	110.8	104.5	100.2	111.3	104.8	100.2	165.1	126.9
26, 27, 3370	ELECTRICAL, COMMUNICATIONS & UTIL.	97.6	95.6	96.6	101.0	94.3	97.5	93.6	92.6	93.1	96.3	99.7	98.1	101.7	102.5	102.1	101.0	157.9	130.7
MF2004	WEIGHTED AVERAGE	102.5	110.6	106.1	105.4	107.6	106.3	101.1	110.6	105.3	104.9	112.6	108.3	102.4	107.5	104.6	110.7	140.4	123.8

	DIVISION	CALIFORNIA SAN JOSE			SANTA BARBARA			STOCKTON			VALLEJO			COLORADO COLORADO SPRINGS			DENVER		
		MAT.	INST.	TOTAL	MAT.	INST.	TOTAL	MAT.	INST.	TOTAL	MAT.	INST.	TOTAL	MAT.	INST.	TOTAL	MAT.	INST.	TOTAL
015433	CONTRACTOR EQUIPMENT		99.8	99.8		99.3	99.3		99.0	99.0		98.8	98.8		94.7	94.7		99.2	99.2
0241, 31 - 34	SITE & INFRASTRUCTURE, DEMOLITION	140.9	100.9	113.0	106.3	106.7	106.6	104.1	106.1	105.5	106.0	110.0	108.8	90.2	94.7	93.3	89.4	103.2	99.0
0310	Concrete Forming & Accessories	109.1	144.1	139.4	108.0	125.6	123.2	107.1	127.4	124.6	107.8	142.3	137.7	94.9	77.5	79.8	99.9	81.6	84.1
0320	Concrete Reinforcing	100.0	119.4	109.9	113.0	118.0	115.5	113.5	117.8	115.7	111.1	118.6	114.9	101.0	82.2	91.4	101.0	82.3	91.5
0330	Cast-in-Place Concrete	123.1	118.2	121.2	103.1	118.8	109.1	103.1	114.5	107.4	116.3	116.3	116.3	99.8	80.0	92.3	94.5	82.0	89.8
03	CONCRETE	115.4	129.2	121.9	107.4	120.7	113.6	107.4	120.0	113.3	117.5	127.0	122.0	104.6	79.6	92.9	98.8	82.1	91.0
04	MASONRY	143.9	132.7	137.1	108.3	110.4	109.5	114.1	108.2	109.5	87.3	118.9	106.3	97.4	73.1	82.8	99.3	79.1	87.2
05	METALS	98.6	108.5	102.0	99.2	101.4	99.9	100.4	100.8	100.5	97.1	98.0	97.4	105.4	87.0	99.2	108.0	87.4	101.1
06	WOOD, PLASTICS & COMPOSITES	102.0	147.8	127.9	94.1	127.1	112.8	95.7	129.1	114.6	97.3	147.7	125.8	93.1	80.8	86.2	99.7	83.6	90.6
07	THERMAL & MOISTURE PROTECTION	103.2	135.7	115.5	103.2	106.2	104.3	106.8	107.7	107.2	116.2	119.4	117.4	99.6	77.5	91.3	98.5	76.2	90.1
08	OPENINGS	94.1	137.3	104.8	101.9	120.8	106.6	101.1	120.2	105.9	115.9	136.5	121.0	99.7	83.3	95.6	99.5	85.4	96.0
0920	Plaster & Gypsum Board	102.3	148.9	134.5	107.3	127.7	121.4	107.7	129.8	123.0	105.9	148.9	135.6	90.2	80.3	83.3	104.5	83.4	89.9
0950, 0980	Ceilings & Acoustic Treatment	107.5	148.9	133.8	117.0	127.7	123.8	117.0	129.8	125.2	121.9	148.9	139.1	118.1	80.3	94.0	117.3	83.4	95.7
0960	Flooring	106.4	125.5	111.7	114.4	84.4	106.1	114.4	101.2	110.7	126.5	125.5	126.2	103.5	73.5	95.2	108.6	91.3	103.8
0970, 0990	Wall Finishes & Painting/Coating	106.2	141.3	127.3	105.6	93.2	98.1	105.6	101.5	103.2	105.4	123.9	116.5	104.2	45.0	68.6	104.5	74.0	86.1
09	FINISHES	108.1	141.9	126.4	111.5	115.7	113.8	111.4	121.3	116.8	113.6	139.2	127.5	105.3	73.5	88.0	108.2	82.7	94.4
COVERS	DIVS. 10 - 14, 25, 28, 41, 43, 44	100.0	127.8	105.9	100.0	112.8	102.7	100.0	124.7	105.2	100.0	127.0	105.7	100.0	88.1	97.5	100.0	89.6	97.8
21, 22, 23	FIRE SUPPRESSION, PLUMBING & HVAC	100.1	138.5	115.9	100.1	112.9	105.4	100.1	98.8	99.6	100.1	112.0	105.0	100.1	70.8	88.1	100.0	83.2	93.1
26, 27, 3370	ELECTRICAL, COMMUNICATIONS & UTIL.	104.2	142.1	124.0	90.6	96.4	93.7	100.5	107.4	104.1	93.3	111.9	103.0	102.2	80.0	90.6	103.9	85.4	94.2
MF2004	WEIGHTED AVERAGE	105.8	131.3	117.1	101.7	110.5	105.6	103.0	109.7	106.0	103.9	118.8	110.5	101.5	78.7	91.4	101.6	85.1	94.3

City Cost Indexes

		COLORADO											CONNECTICUT		
DIVISION		FORT COLLINS			GRAND JUNCTION			GREELEY			PUEBLO			BRIDGEPORT			BRISTOL		
		MAT.	INST.	TOTAL	MAT.	INST.	TOTAL	MAT.	INST.	TOTAL	MAT.	INST.	TOTAL	MAT.	INST.	TOTAL	MAT.	INST.	TOTAL
015433	CONTRACTOR EQUIPMENT		96.1	96.1		99.2	99.2		96.1	96.1		96.4	96.4		100.6	100.6		100.6	100.6
0241, 31 - 34	SITE & INFRASTRUCTURE, DEMOLITION	100.2	96.6	97.7	118.1	99.3	105.0	87.6	95.4	93.1	110.0	94.5	99.2	97.7	104.4	102.3	96.9	104.3	102.1
0310	Concrete Forming & Accessories	101.0	71.4	75.4	109.5	71.7	76.8	99.2	45.0	52.3	107.1	77.8	81.7	97.7	121.9	118.6	97.7	121.6	118.3
0320	Concrete Reinforcing	102.1	74.0	87.7	103.5	73.9	88.4	102.0	73.5	87.4	99.7	82.1	90.7	110.6	127.2	119.1	110.6	127.1	119.1
0330	Cast-in-Place Concrete	109.7	74.8	96.5	113.4	75.7	99.1	91.7	53.1	77.1	101.6	81.1	93.9	101.0	117.5	107.3	94.7	117.4	103.3
03	CONCRETE	111.6	73.4	93.8	116.7	73.8	96.6	96.4	54.2	76.7	106.3	80.1	94.0	109.2	121.2	114.8	106.1	121.0	113.0
04	MASONRY	109.6	50.7	74.2	136.9	52.7	86.3	104.1	33.6	61.7	100.8	71.5	83.2	107.1	132.0	122.1	99.6	132.0	119.1
05	METALS	103.5	78.4	95.0	100.0	77.6	92.5	103.4	76.9	94.5	101.7	87.8	97.0	100.8	125.0	109.0	100.8	124.7	108.9
06	WOOD, PLASTICS & COMPOSITES	99.5	75.2	85.8	104.0	75.4	87.8	97.2	44.7	67.4	101.5	81.1	90.0	98.8	119.9	110.8	98.8	119.9	110.8
07	THERMAL & MOISTURE PROTECTION	99.4	61.9	85.2	101.9	58.7	85.6	98.8	51.9	81.1	101.5	76.9	92.2	95.9	127.1	107.6	96.0	123.9	106.5
08	OPENINGS	95.4	78.2	91.2	99.7	78.3	94.4	95.4	61.7	87.1	94.2	83.4	91.5	104.3	129.7	110.6	104.3	126.7	109.9
0920	Plaster & Gypsum Board	102.4	74.7	83.3	123.6	74.7	89.8	101.2	43.3	61.1	89.1	80.3	83.0	99.6	119.8	113.6	99.6	119.8	113.6
0950, 0980	Ceilings & Acoustic Treatment	110.4	74.7	87.7	120.1	74.7	91.2	110.4	43.3	67.6	131.3	80.3	98.8	98.7	119.8	112.1	98.7	119.8	112.1
0960	Flooring	108.5	59.0	94.8	118.9	59.0	102.3	107.2	59.0	93.9	115.3	91.3	108.7	97.6	123.1	104.6	97.6	107.5	100.3
0970, 0990	Wall Finishes & Painting/Coating	104.5	44.7	68.5	108.4	74.0	87.7	104.5	27.8	58.4	108.4	41.7	68.3	94.6	118.2	108.8	94.6	118.2	108.8
09	FINISHES	107.0	66.4	85.0	116.0	70.3	91.2	105.7	44.4	72.4	112.9	76.8	93.3	101.1	121.5	112.2	101.1	118.8	110.8
COVERS	DIVS. 10 - 14, 25, 28, 41, 43, 44	100.0	86.3	97.1	100.0	87.3	97.3	100.0	79.5	95.6	100.0	88.8	97.6	100.0	114.1	103.0	100.0	114.1	103.0
21, 22, 23	FIRE SUPPRESSION, PLUMBING & HVAC	100.0	74.5	89.5	100.0	61.4	84.1	100.0	69.8	87.6	100.0	64.1	85.2	100.1	112.8	105.4	100.1	112.8	105.3
26, 27, 3370	ELECTRICAL, COMMUNICATIONS & UTIL.	98.3	82.0	89.8	95.1	59.3	76.4	98.3	82.0	89.8	96.1	74.6	84.9	97.9	109.4	103.9	97.9	109.1	103.7
MF2004	WEIGHTED AVERAGE	102.4	74.4	90.0	105.4	69.3	89.5	99.7	64.5	84.1	101.5	76.9	90.7	101.7	118.2	109.0	100.9	117.5	108.2

		CONNECTICUT																	
DIVISION		HARTFORD			NEW BRITAIN			NEW HAVEN			NORWALK			STAMFORD			WATERBURY		
		MAT.	INST.	TOTAL	MAT.	INST.	TOTAL	MAT.	INST.	TOTAL	MAT.	INST.	TOTAL	MAT.	INST.	TOTAL	MAT.	INST.	TOTAL
015433	CONTRACTOR EQUIPMENT		100.6	100.6		100.6	100.6		101.0	101.0		100.6	100.6		100.6	100.6		100.6	100.6
0241, 31 - 34	SITE & INFRASTRUCTURE, DEMOLITION	94.4	104.3	101.3	97.1	104.3	102.1	96.9	105.1	102.6	97.5	104.4	102.3	98.1	104.4	102.5	97.3	104.4	102.2
0310	Concrete Forming & Accessories	95.6	121.6	118.1	97.9	121.6	118.4	97.4	121.5	118.3	97.7	122.1	118.8	97.6	122.4	119.1	97.7	121.7	118.5
0320	Concrete Reinforcing	110.6	127.1	119.1	110.6	127.1	119.1	110.6	127.1	119.0	110.6	127.3	119.1	110.6	127.4	119.2	110.6	127.2	119.1
0330	Cast-in-Place Concrete	97.1	117.4	104.8	96.2	117.4	104.2	97.9	108.5	101.9	99.4	128.8	110.5	101.0	129.0	111.6	101.0	117.4	107.2
03	CONCRETE	107.1	121.0	113.6	106.8	121.0	113.4	123.6	117.9	121.0	108.4	125.2	116.2	109.2	125.4	116.8	109.2	121.1	114.8
04	MASONRY	99.8	132.0	119.2	101.6	132.0	119.9	99.8	132.0	119.2	99.3	133.6	119.9	100.1	133.6	120.3	100.1	132.0	119.3
05	METALS	105.5	124.7	112.0	97.1	124.7	106.4	97.3	124.7	106.6	100.8	125.4	109.1	100.8	125.7	109.2	100.8	124.9	108.9
06	WOOD, PLASTICS & COMPOSITES	97.4	119.9	110.1	98.8	119.9	110.8	98.8	119.9	110.8	98.8	119.9	110.8	98.8	119.9	110.8	98.8	119.9	110.8
07	THERMAL & MOISTURE PROTECTION	96.7	123.9	106.9	96.0	123.9	106.5	96.0	122.5	106.0	96.0	129.2	108.6	96.0	129.3	108.5	96.0	123.9	106.5
08	OPENINGS	105.0	126.7	110.4	104.3	126.7	109.9	104.3	129.7	110.6	104.3	129.7	110.6	104.3	129.7	110.6	104.3	129.7	110.6
0920	Plaster & Gypsum Board	101.1	119.8	114.1	99.6	119.8	113.6	99.6	119.8	113.6	99.6	119.8	113.6	99.6	119.8	113.6	99.6	119.8	113.6
0950, 0980	Ceilings & Acoustic Treatment	98.7	119.8	112.1	98.7	119.8	112.1	98.7	119.8	112.1	98.7	119.8	112.1	98.7	119.8	112.1	98.7	119.8	112.1
0960	Flooring	97.6	107.5	100.3	97.6	107.5	100.3	97.6	123.1	104.6	97.6	123.1	104.6	97.6	123.1	104.6	97.6	123.1	104.6
0970, 0990	Wall Finishes & Painting/Coating	94.6	118.2	108.8	94.6	118.2	108.8	94.6	118.2	108.8	94.6	118.2	108.8	94.6	118.2	108.8	94.6	118.2	108.8
09	FINISHES	100.8	118.8	110.6	101.1	118.8	110.8	101.2	121.5	112.2	101.2	121.5	112.2	101.2	121.5	112.2	101.0	121.5	112.2
COVERS	DIVS. 10 - 14, 25, 28, 41, 43, 44	100.0	114.1	103.0	100.0	114.1	103.0	100.0	114.1	103.0	100.0	114.1	103.0	100.0	114.3	103.0	100.0	114.1	103.0
21, 22, 23	FIRE SUPPRESSION, PLUMBING & HVAC	100.1	112.8	105.3	100.1	112.8	105.3	100.1	112.8	105.3	100.1	112.8	105.4	100.1	112.9	105.4	100.1	112.8	105.3
26, 27, 3370	ELECTRICAL, COMMUNICATIONS & UTIL.	98.0	110.0	104.3	98.0	109.1	103.8	97.8	109.1	103.7	97.9	107.8	103.1	97.9	156.0	128.3	97.4	109.4	103.7
MF2004	WEIGHTED AVERAGE	101.7	117.6	108.7	100.6	117.5	108.1	102.6	117.6	109.2	101.2	118.7	109.0	101.4	125.5	112.0	101.3	118.0	108.7

		D.C.			DELAWARE			FLORIDA									
DIVISION		WASHINGTON			WILMINGTON			DAYTONA BEACH			FORT LAUDERDALE			JACKSONVILLE			MELBOURNE		
		MAT.	INST.	TOTAL	MAT.	INST.	TOTAL	MAT.	INST.	TOTAL	MAT.	INST.	TOTAL	MAT.	INST.	TOTAL	MAT.	INST.	TOTAL
015433	CONTRACTOR EQUIPMENT		104.4	104.4		117.8	117.8		97.8	97.8		90.4	90.4		97.8	97.8		97.8	97.8
0241, 31 - 34	SITE & INFRASTRUCTURE, DEMOLITION	112.3	92.8	98.7	90.9	112.3	105.8	116.3	89.2	97.5	101.6	77.3	84.7	116.4	88.5	97.0	124.8	88.5	99.5
0310	Concrete Forming & Accessories	101.3	82.1	84.7	100.6	101.9	101.7	94.8	75.7	78.2	93.5	76.5	78.8	94.6	60.1	64.8	91.5	77.5	79.4
0320	Concrete Reinforcing	103.0	89.0	95.8	93.9	102.4	98.2	86.0	80.7	83.3	86.0	80.4	83.1	86.0	62.8	74.1	86.9	80.7	83.8
0330	Cast-in-Place Concrete	126.8	88.3	112.2	89.5	102.7	94.5	93.4	76.7	87.1	98.0	84.6	92.9	94.3	73.8	86.5	112.6	79.3	100.0
03	CONCRETE	117.1	86.8	102.9	95.5	103.2	99.1	89.3	78.0	84.0	91.5	81.1	86.7	89.8	66.7	79.0	101.1	79.7	91.1
04	MASONRY	104.8	83.0	91.7	108.7	98.7	102.7	86.6	73.1	78.4	86.8	76.2	80.4	86.4	68.1	75.4	84.3	77.6	80.3
05	METALS	94.3	108.4	99.0	104.7	117.1	108.9	98.1	93.8	96.7	98.1	95.7	97.3	96.8	85.2	92.8	106.1	93.9	102.0
06	WOOD, PLASTICS & COMPOSITES	99.9	81.8	89.6	101.1	101.2	101.2	95.6	77.5	85.4	92.0	75.2	82.5	95.6	57.9	74.3	91.6	77.5	83.6
07	THERMAL & MOISTURE PROTECTION	102.4	86.1	96.3	99.4	115.3	105.4	93.9	81.8	89.3	93.9	89.9	92.4	94.1	68.4	84.2	94.2	84.4	90.5
08	OPENINGS	107.8	91.0	103.6	95.9	108.8	99.1	100.1	74.2	93.7	97.5	73.3	91.5	100.1	57.2	89.5	99.3	77.9	94.0
0920	Plaster & Gypsum Board	107.5	81.2	89.3	111.7	101.1	104.4	95.3	77.4	82.9	94.9	75.0	81.1	95.3	57.1	68.9	90.8	77.4	81.5
0950, 0980	Ceilings & Acoustic Treatment	107.9	81.2	90.9	92.2	101.1	97.9	98.4	77.4	85.0	98.4	75.0	83.5	98.4	57.1	72.1	93.2	77.4	83.1
0960	Flooring	127.0	97.0	118.7	83.3	108.2	90.2	116.6	79.3	106.3	116.6	74.7	105.0	116.6	69.5	103.6	113.4	79.3	104.0
0970, 0990	Wall Finishes & Painting/Coating	125.0	84.5	100.6	92.5	102.7	98.7	111.0	82.3	93.7	111.0	58.2	77.3	111.0	51.6	75.3	111.0	102.1	105.6
09	FINISHES	108.9	84.6	95.7	97.3	102.5	100.2	105.7	77.0	90.1	103.1	73.5	87.0	105.7	60.3	81.1	103.6	80.4	91.0
COVERS	DIVS. 10 - 14, 25, 28, 41, 43, 44	100.0	98.3	99.6	100.0	101.2	100.2	100.0	85.7	97.0	100.0	89.1	97.7	100.0	81.6	96.1	100.0	87.4	97.3
21, 22, 23	FIRE SUPPRESSION, PLUMBING & HVAC	100.2	93.1	97.3	100.1	115.7	106.5	99.8	66.8	86.2	99.8	68.7	87.0	99.8	49.2	79.0	99.8	69.1	87.2
26, 27, 3370	ELECTRICAL, COMMUNICATIONS & UTIL.	104.1	99.9	101.9	103.7	109.4	106.7	98.4	61.6	79.1	98.4	76.0	86.7	98.1	69.8	83.3	99.6	71.7	85.0
MF2004	WEIGHTED AVERAGE	104.0	92.2	98.8	100.1	108.7	103.9	98.2	75.3	88.1	97.5	77.4	88.7	98.0	66.3	84.0	100.8	78.5	91.0

City Cost Indexes

		FLORIDA																	
	DIVISION	MIAMI			ORLANDO			PANAMA CITY			PENSACOLA			ST. PETERSBURG			TALLAHASSEE		
		MAT.	INST.	TOTAL	MAT.	INST.	TOTAL	MAT.	INST.	TOTAL	MAT.	INST.	TOTAL	MAT.	INST.	TOTAL	MAT.	INST.	TOTAL
015433	CONTRACTOR EQUIPMENT		90.4	90.4		97.8	97.8		97.8	97.8		97.8	97.8		97.8	97.8		97.8	97.8
0241, 31 - 34	SITE & INFRASTRUCTURE, DEMOLITION	105.1	77.1	85.6	119.0	88.3	97.6	131.2	86.5	100.1	130.8	87.8	100.9	117.2	88.1	96.9	114.5	87.5	95.7
0310	Concrete Forming & Accessories	97.7	77.0	79.8	94.5	80.2	82.1	94.0	47.4	53.7	92.0	57.4	62.0	93.0	52.2	57.7	96.9	48.7	55.2
0320	Concrete Reinforcing	86.0	80.5	83.2	86.0	77.7	81.7	89.7	58.9	74.0	92.1	58.8	75.1	89.1	66.2	77.4	86.0	62.4	73.9
0330	Cast-in-Place Concrete	106.3	85.7	98.5	123.1	76.5	105.5	99.1	52.8	81.6	122.4	71.3	103.1	105.7	69.5	92.0	103.9	60.5	87.5
03	CONCRETE	95.9	81.7	89.2	105.3	79.4	93.2	98.0	52.7	76.8	108.9	63.9	87.9	95.7	62.4	80.2	94.6	57.1	77.1
04	MASONRY	86.1	80.2	82.6	90.6	73.1	80.1	90.8	52.2	67.6	109.1	60.5	79.9	132.7	54.5	85.7	89.4	56.1	69.4
05	METALS	102.8	94.4	99.9	103.7	92.2	99.8	97.9	71.4	89.0	99.1	83.5	93.8	101.0	85.8	95.9	91.9	84.5	89.4
06	WOOD, PLASTICS & COMPOSITES	97.3	75.2	84.8	95.4	84.3	89.1	94.6	46.6	67.5	93.0	57.6	73.0	93.3	50.7	69.2	93.2	46.1	66.5
07	THERMAL & MOISTURE PROTECTION	102.7	83.6	95.5	95.2	82.5	90.4	94.3	55.2	79.5	94.2	67.2	84.0	93.8	62.4	82.0	99.8	78.4	91.7
08	OPENINGS	97.5	73.2	91.5	100.1	74.7	93.8	97.7	41.2	83.7	97.7	57.9	87.8	98.7	54.9	87.9	100.6	50.5	88.2
0920	Plaster & Gypsum Board	92.8	75.0	80.5	105.8	84.3	90.9	93.4	45.5	60.3	94.3	56.9	68.4	92.5	49.8	62.9	111.1	45.0	65.4
0950, 0980	Ceilings & Acoustic Treatment	94.1	75.0	81.9	98.4	84.3	89.4	93.2	45.5	62.8	93.2	56.9	70.1	93.2	49.8	65.5	98.4	45.0	64.4
0960	Flooring	124.0	78.6	111.5	116.8	79.3	106.4	116.0	46.3	96.8	110.4	67.8	98.6	115.1	59.4	99.7	116.6	60.6	101.2
0970, 0990	Wall Finishes & Painting/Coating	106.1	58.2	77.3	111.0	61.1	81.0	111.0	52.3	75.7	111.0	50.9	74.9	111.0	53.7	76.5	111.0	48.1	73.2
09	FINISHES	104.4	74.7	88.3	106.9	78.6	91.6	105.5	47.5	74.0	103.5	58.1	78.9	103.6	52.8	76.0	107.2	49.6	75.9
COVERS	DIVS. 10 - 14, 25, 28, 41, 43, 44	100.0	89.8	97.8	100.0	86.4	97.1	100.0	54.2	90.3	100.0	54.4	90.3	100.0	61.5	91.8	100.0	77.8	95.3
21, 22, 23	FIRE SUPPRESSION, PLUMBING & HVAC	99.8	74.4	89.3	99.7	64.1	85.0	99.8	42.1	76.1	99.8	59.1	83.0	99.8	57.2	82.2	99.8	44.0	76.8
26, 27, 3370	ELECTRICAL, COMMUNICATIONS & UTIL.	106.3	78.8	91.9	106.5	48.4	76.1	97.0	46.4	70.6	101.1	58.6	78.9	98.7	51.5	74.0	106.4	46.0	74.8
MF2004	WEIGHTED AVERAGE	100.1	79.4	90.9	102.2	73.2	89.4	99.5	53.1	79.0	102.2	64.4	85.5	101.6	61.6	83.9	99.2	57.7	80.9

		FLORIDA			GEORGIA													
	DIVISION	TAMPA			ALBANY			ATLANTA			AUGUSTA			COLUMBUS			MACON		
		MAT.	INST.	TOTAL	MAT.	INST.	TOTAL	MAT.	INST.	TOTAL	MAT.	INST.	TOTAL	MAT.	INST.	TOTAL	MAT.	INST.	TOTAL
015433	CONTRACTOR EQUIPMENT		97.8	97.8		91.2	91.2		93.1	93.1		92.6	92.6		91.2	91.2		102.1	102.1
0241, 31 - 34	SITE & INFRASTRUCTURE, DEMOLITION	117.6	88.8	97.5	101.9	78.4	85.6	103.6	94.4	97.2	100.0	92.1	94.5	101.8	78.7	85.7	102.3	94.2	96.6
0310	Concrete Forming & Accessories	95.6	84.3	85.8	94.1	47.7	54.0	96.2	75.0	77.9	94.0	68.3	71.8	94.0	59.7	64.4	93.3	56.9	61.8
0320	Concrete Reinforcing	86.0	104.3	95.3	85.8	87.1	86.5	90.5	88.8	89.6	91.5	75.8	83.5	86.0	88.1	87.0	87.1	87.4	87.2
0330	Cast-in-Place Concrete	103.2	75.8	92.9	99.8	49.0	80.6	110.5	73.0	96.3	104.4	51.2	84.2	99.3	58.1	83.7	98.0	51.6	80.5
03	CONCRETE	94.2	86.0	90.4	92.3	57.2	75.9	99.8	77.0	89.1	95.2	64.1	80.7	92.1	65.8	79.8	91.6	62.2	77.9
04	MASONRY	87.8	84.4	85.8	89.5	43.9	62.1	89.5	71.4	78.6	89.6	43.5	61.9	88.7	61.6	72.4	101.2	41.6	65.4
05	METALS	100.0	103.8	101.3	97.5	88.2	94.3	90.9	81.0	87.6	89.8	72.7	84.0	96.9	90.6	94.7	93.2	89.9	92.1
06	WOOD, PLASTICS & COMPOSITES	96.8	86.8	91.1	94.7	42.4	65.1	94.0	76.7	84.2	91.4	73.6	81.4	94.7	58.3	74.1	104.5	57.2	77.7
07	THERMAL & MOISTURE PROTECTION	94.0	94.1	94.0	91.6	62.7	80.7	91.8	76.4	86.0	91.4	58.5	79.0	91.5	63.8	81.0	90.1	65.3	80.7
08	OPENINGS	100.1	83.8	96.1	97.5	49.0	85.5	98.7	73.8	92.5	93.0	65.6	86.2	97.5	59.6	88.1	95.7	58.1	86.4
0920	Plaster & Gypsum Board	95.3	86.9	89.5	95.1	41.2	57.8	113.6	76.7	87.9	112.3	73.4	85.4	95.1	57.5	69.1	104.4	56.4	71.2
0950, 0980	Ceilings & Acoustic Treatment	98.4	86.9	91.1	97.5	41.2	61.6	98.4	76.4	87.4	107.7	73.4	85.9	97.5	57.5	72.1	92.6	56.4	69.6
0960	Flooring	116.6	60.9	101.2	116.6	33.2	93.6	89.4	70.2	84.1	88.3	42.3	75.6	116.6	54.3	99.4	90.9	39.1	76.5
0970, 0990	Wall Finishes & Painting/Coating	111.0	66.2	84.0	106.1	55.5	75.6	97.5	86.2	90.7	97.5	40.6	63.3	106.1	67.5	82.9	108.2	55.5	76.5
09	FINISHES	105.7	77.9	90.6	102.8	43.9	70.8	97.7	75.2	85.5	97.2	61.1	77.6	102.8	58.7	78.8	91.0	52.3	70.0
COVERS	DIVS. 10 - 14, 25, 28, 41, 43, 44	100.0	86.5	97.1	100.0	77.2	95.2	100.0	81.8	96.1	100.0	58.1	91.1	100.0	79.2	95.6	100.0	77.0	95.1
21, 22, 23	FIRE SUPPRESSION, PLUMBING & HVAC	99.8	91.7	96.4	99.7	69.9	87.4	100.0	76.0	90.1	100.0	69.4	87.4	99.8	52.6	80.4	99.8	67.7	86.6
26, 27, 3370	ELECTRICAL, COMMUNICATIONS & UTIL.	98.4	51.5	73.9	93.8	57.0	74.6	98.4	77.1	87.3	99.7	69.3	83.8	94.0	74.3	83.7	92.8	63.6	77.5
MF2004	WEIGHTED AVERAGE	99.2	83.3	92.2	97.1	61.7	81.5	97.4	77.8	88.8	96.0	66.5	83.0	97.0	66.2	83.4	95.8	65.8	82.6

		GEORGIA			HAWAII						IDAHO					
	DIVISION	SAVANNAH			VALDOSTA			HONOLULU			STATES & POSS., GUAM			BOISE			LEWISTON		
		MAT.	INST.	TOTAL	MAT.	INST.	TOTAL	MAT.	INST.	TOTAL	MAT.	INST.	TOTAL	MAT.	INST.	TOTAL	MAT.	INST.	TOTAL
015433	CONTRACTOR EQUIPMENT		92.2	92.2		91.2	91.2		99.8	99.8		166.6	166.6		100.7	100.7		94.6	94.6
0241, 31 - 34	SITE & INFRASTRUCTURE, DEMOLITION	104.4	79.3	86.9	112.2	78.5	88.7	134.6	106.3	114.9	172.3	105.9	126.0	76.7	99.8	92.8	83.2	95.3	91.6
0310	Concrete Forming & Accessories	94.1	53.0	58.6	84.7	46.8	51.9	116.4	136.8	134.0	111.4	62.8	69.3	99.9	72.2	75.9	122.7	60.8	69.1
0320	Concrete Reinforcing	86.8	76.0	81.3	87.8	73.8	80.7	113.0	107.1	110.0	205.9	33.5	117.9	98.3	69.2	83.5	106.5	91.0	98.6
0330	Cast-in-Place Concrete	115.9	50.5	91.2	97.6	53.1	80.8	168.1	121.2	150.3	171.1	113.1	149.1	92.7	77.9	87.1	104.2	86.6	97.5
03	CONCRETE	100.4	57.9	80.5	96.3	55.7	77.3	139.9	124.6	132.7	154.2	75.1	117.2	102.5	73.8	89.1	113.7	75.8	96.0
04	MASONRY	88.6	54.3	68.0	94.3	49.3	67.3	122.9	124.9	124.1	187.2	48.4	103.8	126.2	62.0	87.6	128.1	82.2	100.5
05	METALS	93.6	84.7	90.6	96.5	83.8	92.2	131.3	102.1	121.5	142.2	80.0	121.2	102.5	72.9	92.5	96.4	84.2	92.3
06	WOOD, PLASTICS & COMPOSITES	110.4	50.0	76.2	83.3	41.3	59.5	108.1	141.5	127.0	108.4	64.2	83.3	93.5	72.9	81.8	104.9	53.2	75.6
07	THERMAL & MOISTURE PROTECTION	93.9	58.2	80.4	91.7	63.5	81.1	116.5	123.2	119.0	122.9	70.2	103.0	93.4	68.3	83.9	144.8	75.1	118.5
08	OPENINGS	98.7	50.3	86.7	93.0	44.7	81.0	106.0	129.9	111.9	105.5	54.3	92.8	93.0	68.3	86.9	112.6	60.0	99.6
0920	Plaster & Gypsum Board	94.1	49.0	62.9	87.3	40.1	54.6	133.6	142.5	139.8	211.9	51.6	100.9	89.8	72.2	77.5	157.4	51.8	84.3
0950, 0980	Ceilings & Acoustic Treatment	98.4	49.0	66.9	93.2	40.1	59.4	116.2	142.5	132.9	233.1	51.6	117.5	125.5	72.0	91.4	161.2	51.8	91.5
0960	Flooring	117.4	49.9	98.8	108.8	41.7	90.3	165.2	128.4	155.0	164.9	50.0	133.2	99.5	68.4	90.9	141.8	86.1	126.4
0970, 0990	Wall Finishes & Painting/Coating	108.0	59.1	78.6	106.1	55.5	75.6	104.1	138.8	124.9	105.6	39.9	66.1	97.9	43.8	65.4	118.7	65.2	86.5
09	FINISHES	103.5	52.3	75.7	99.3	45.1	69.9	132.9	137.1	135.1	197.0	58.7	121.9	104.7	69.2	85.4	174.1	64.6	114.7
COVERS	DIVS. 10 - 14, 25, 28, 41, 43, 44	100.0	57.3	90.9	100.0	56.3	90.7	100.0	115.1	103.2	100.0	88.0	97.0	100.0	67.2	93.0	100.0	70.2	93.7
21, 22, 23	FIRE SUPPRESSION, PLUMBING & HVAC	99.8	70.0	83.8	99.8	69.8	87.4	100.1	107.0	102.9	102.4	42.2	77.7	100.0	65.2	85.7	101.0	78.7	91.8
26, 27, 3370	ELECTRICAL, COMMUNICATIONS & UTIL.	97.2	58.1	76.8	92.0	57.7	74.1	112.8	116.6	114.7	161.4	46.2	101.2	96.1	74.1	84.6	85.9	75.0	80.2
MF2004	WEIGHTED AVERAGE	98.4	61.4	82.0	97.1	61.1	81.2	116.9	117.9	117.3	135.6	61.7	103.0	100.4	71.8	87.7	110.1	76.8	95.4

City Cost Indexes

		IDAHO						ILLINOIS										
	DIVISION	POCATELLO			TWIN FALLS			CHICAGO			DECATUR			EAST ST. LOUIS			JOLIET		
		MAT.	INST.	TOTAL	MAT.	INST.	TOTAL	MAT.	INST.	TOTAL	MAT.	INST.	TOTAL	MAT.	INST.	TOTAL	MAT.	INST.	TOTAL
015433	CONTRACTOR EQUIPMENT		100.7	100.7		100.7	100.7		93.4	93.4		101.9	101.9		107.7	107.7		91.4	91.4
0241, 31 - 34	SITE & INFRASTRUCTURE, DEMOLITION	77.6	99.8	93.0	83.9	98.0	93.7	103.4	93.6	96.6	89.5	97.6	95.1	105.2	98.4	100.5	103.2	93.1	96.2
0310	Concrete Forming & Accessories	100.1	71.9	75.7	100.8	30.6	40.0	101.7	154.8	147.6	100.5	114.4	112.6	97.0	111.0	109.1	103.9	158.6	151.2
0320	Concrete Reinforcing	98.7	69.5	83.8	100.7	67.8	83.9	98.8	151.1	125.5	96.7	104.7	100.8	98.6	105.2	102.0	98.8	148.6	124.2
0330	Cast-in-Place Concrete	95.3	77.8	88.7	97.8	40.6	76.1	109.7	143.5	122.5	100.2	107.4	102.9	93.8	112.2	100.7	109.6	147.2	123.8
03	CONCRETE	101.7	73.6	88.6	109.8	42.2	78.2	103.6	149.0	124.8	97.2	110.2	103.3	88.2	111.0	98.9	103.7	151.3	126.0
04	MASONRY	123.5	67.9	90.1	126.3	32.5	69.9	97.7	153.4	131.2	71.8	109.2	94.3	75.3	116.5	100.1	100.7	149.8	130.2
05	METALS	110.4	73.0	97.8	110.5	69.4	96.6	92.0	130.3	105.0	94.8	108.2	99.3	92.4	119.1	101.4	89.8	127.1	102.4
06	WOOD, PLASTICS & COMPOSITES	93.5	72.9	81.8	94.1	29.6	57.6	106.0	154.3	133.3	101.0	114.6	108.7	99.0	108.5	104.4	107.8	159.9	137.3
07	THERMAL & MOISTURE PROTECTION	93.4	63.7	82.2	94.1	39.9	73.7	102.2	143.2	117.7	100.2	103.8	101.6	94.7	110.2	100.6	102.2	141.7	117.1
08	OPENINGS	93.7	62.7	86.0	97.0	36.9	82.1	101.8	153.2	114.6	96.6	112.0	100.4	86.4	114.1	93.3	99.8	155.7	113.6
0920	Plaster & Gypsum Board	87.0	72.0	76.6	86.9	27.4	45.7	95.0	155.8	137.0	99.6	115.0	110.2	95.7	108.7	104.7	92.0	161.7	140.2
0950, 0980	Ceilings & Acoustic Treatment	131.3	72.0	93.5	125.3	27.4	62.9	108.2	155.8	138.5	98.1	115.0	108.8	92.0	108.7	102.6	108.2	161.7	142.2
0960	Flooring	102.7	68.4	93.2	104.0	46.4	88.1	91.3	152.8	108.3	100.9	106.8	102.6	110.4	118.0	112.5	90.9	142.0	105.0
0970, 0990	Wall Finishes & Painting/Coating	97.8	45.6	66.4	97.9	26.9	55.2	93.0	152.1	128.6	96.2	107.7	103.1	103.9	109.2	107.1	91.2	142.2	121.9
09	FINISHES	106.6	69.4	86.4	106.3	32.2	66.1	97.6	155.2	128.9	99.9	113.3	107.1	98.7	112.2	106.1	97.1	155.5	128.8
COVERS	DIVS. 10 - 14, 25, 28, 41, 43, 44	100.0	67.2	93.0	100.0	38.3	86.9	100.0	126.3	105.6	100.0	98.1	99.6	100.0	93.7	98.7	100.0	126.7	105.7
21, 22, 23	FIRE SUPPRESSION, PLUMBING & HVAC	100.0	65.2	85.6	100.0	54.2	81.1	100.1	134.5	114.2	100.0	100.1	100.0	100.0	95.4	98.1	100.0	129.5	112.2
26, 27, 3370	ELECTRICAL, COMMUNICATIONS & UTIL.	93.1	69.9	81.0	89.1	61.1	74.5	95.8	130.5	113.9	96.7	94.3	95.4	93.6	103.5	98.8	95.1	126.5	111.5
MF2004	WEIGHTED AVERAGE	101.2	71.4	88.1	102.5	51.8	80.1	99.0	137.6	116.0	96.5	104.5	100.0	94.0	106.6	99.6	98.5	135.8	115.0

		ILLINOIS									INDIANA								
	DIVISION	PEORIA			ROCKFORD			SPRINGFIELD			ANDERSON			BLOOMINGTON			EVANSVILLE		
		MAT.	INST.	TOTAL	MAT.	INST.	TOTAL	MAT.	INST.	TOTAL	MAT.	INST.	TOTAL	MAT.	INST.	TOTAL	MAT.	INST.	TOTAL
015433	CONTRACTOR EQUIPMENT		101.2	101.2		101.2	101.2		101.9	101.9		96.3	96.3		86.8	86.8		117.5	117.5
0241, 31 - 34	SITE & INFRASTRUCTURE, DEMOLITION	98.2	96.5	97.0	97.4	97.6	97.5	96.6	97.7	97.3	87.7	94.6	92.5	76.4	94.7	89.2	81.0	125.6	112.0
0310	Concrete Forming & Accessories	98.9	114.5	112.4	102.9	132.7	128.7	101.0	112.4	110.8	94.7	79.5	81.6	98.6	74.9	78.1	93.4	79.9	81.7
0320	Concrete Reinforcing	96.7	103.1	100.0	91.5	138.5	115.5	96.7	104.8	100.8	95.7	79.9	87.6	91.2	79.7	85.4	100.0	78.0	88.7
0330	Cast-in-Place Concrete	97.9	112.0	103.2	100.2	118.0	107.0	89.1	106.0	95.5	96.2	81.8	90.7	95.2	78.2	88.8	91.0	87.7	89.7
03	CONCRETE	96.0	111.6	103.3	96.6	128.5	111.5	91.7	108.8	99.7	91.6	80.5	86.6	99.0	76.9	88.7	99.5	82.4	91.5
04	MASONRY	113.4	108.1	110.2	88.0	132.5	114.7	76.6	110.0	96.7	82.6	83.2	83.0	86.0	80.9	83.0	82.2	84.6	83.7
05	METALS	94.8	109.5	99.7	94.7	127.3	105.7	96.9	108.6	100.9	96.4	88.0	93.5	94.8	77.0	88.8	88.7	84.9	87.4
06	WOOD, PLASTICS & COMPOSITES	101.0	111.5	106.9	101.0	131.5	118.2	102.3	110.9	107.2	103.7	78.8	89.6	117.9	72.2	92.0	96.7	77.9	86.1
07	THERMAL & MOISTURE PROTECTION	100.3	110.2	104.1	102.8	125.8	111.5	100.1	109.2	103.5	96.5	80.3	90.4	93.3	80.6	88.5	97.3	85.4	92.8
08	OPENINGS	96.7	110.2	100.1	96.7	131.7	105.4	97.7	110.0	100.8	95.7	80.6	91.9	103.5	77.0	96.9	95.7	79.6	91.7
0920	Plaster & Gypsum Board	99.6	111.7	108.0	99.6	132.3	122.2	95.4	111.1	106.3	92.7	78.6	82.9	87.6	72.0	76.8	83.1	76.6	78.6
0950, 0980	Ceilings & Acoustic Treatment	98.1	111.7	106.8	98.1	132.3	119.9	96.4	111.1	105.8	87.5	78.6	81.8	76.7	72.0	73.7	81.4	76.6	78.4
0960	Flooring	100.9	112.1	104.0	100.9	115.0	104.8	100.9	106.4	102.4	93.2	83.6	90.5	104.5	80.8	98.0	98.0	86.9	94.9
0970, 0990	Wall Finishes & Painting/Coating	96.2	126.2	114.3	96.2	133.7	118.8	96.2	107.3	102.9	93.8	73.8	81.8	90.4	87.8	88.8	96.4	87.2	90.9
09	FINISHES	100.0	115.5	108.4	100.0	130.2	116.4	98.8	111.2	105.5	92.0	79.6	85.3	93.5	76.7	84.4	92.0	81.8	86.5
COVERS	DIVS. 10 - 14, 25, 28, 41, 43, 44	100.0	105.6	101.2	100.0	101.9	100.4	100.0	97.9	99.6	100.0	91.4	98.2	100.0	90.2	97.9	100.0	81.3	96.0
21, 22, 23	FIRE SUPPRESSION, PLUMBING & HVAC	100.0	107.4	103.0	100.1	117.6	107.3	99.9	104.5	101.8	99.8	77.0	90.4	99.7	82.0	92.4	99.9	82.9	92.9
26, 27, 3370	ELECTRICAL, COMMUNICATIONS & UTIL.	95.8	98.9	97.4	96.1	123.9	110.7	101.6	96.2	98.8	86.1	90.4	88.4	100.1	86.5	93.0	95.8	83.9	89.6
MF2004	WEIGHTED AVERAGE	98.7	107.3	102.5	97.5	122.8	108.7	97.0	105.4	100.7	94.5	83.5	89.7	97.4	81.6	90.4	95.1	86.6	91.4

		INDIANA																	
	DIVISION	FORT WAYNE			GARY			INDIANAPOLIS			MUNCIE			SOUTH BEND			TERRE HAUTE		
		MAT.	INST.	TOTAL	MAT.	INST.	TOTAL	MAT.	INST.	TOTAL	MAT.	INST.	TOTAL	MAT.	INST.	TOTAL	MAT.	INST.	TOTAL
015433	CONTRACTOR EQUIPMENT		96.3	96.3		96.3	96.3		91.7	91.7		95.8	95.8		105.1	105.1		117.5	117.5
0241, 31 - 34	SITE & INFRASTRUCTURE, DEMOLITION	88.7	94.6	92.8	88.2	98.0	95.0	87.4	97.9	94.7	76.4	94.8	89.2	90.5	95.0	93.6	82.9	125.8	112.8
0310	Concrete Forming & Accessories	93.5	75.9	78.3	94.8	113.2	110.7	95.4	83.4	85.0	92.2	79.2	81.0	96.2	79.6	81.8	94.3	81.6	83.3
0320	Concrete Reinforcing	95.7	79.0	87.2	95.7	103.7	99.8	95.1	80.0	87.4	101.0	79.9	90.2	95.7	77.4	86.3	100.0	80.1	89.8
0330	Cast-in-Place Concrete	102.2	77.8	93.0	100.5	111.0	104.5	92.5	83.7	89.2	100.1	81.0	92.9	97.4	81.7	91.5	88.1	90.3	88.9
03	CONCRETE	94.5	77.8	86.7	93.7	110.5	101.6	93.6	82.6	88.4	97.9	80.5	89.8	88.3	81.2	85.0	102.2	84.5	93.9
04	MASONRY	86.2	81.2	83.2	83.8	113.0	101.4	89.9	86.0	87.6	88.0	83.2	85.1	82.9	81.4	82.0	88.9	83.4	85.6
05	METALS	96.4	81.5	91.4	96.4	105.2	99.4	98.6	78.7	91.9	96.7	88.1	93.8	96.4	99.5	97.4	89.3	86.9	88.5
06	WOOD, PLASTICS & COMPOSITES	103.4	74.9	87.3	101.7	112.3	107.7	100.5	82.7	90.4	113.3	78.5	93.6	101.1	78.5	88.4	98.5	80.4	88.3
07	THERMAL & MOISTURE PROTECTION	96.3	83.9	91.6	95.4	109.7	100.8	94.7	83.7	90.6	96.0	80.3	90.0	91.1	85.3	88.9	97.4	84.1	92.4
08	OPENINGS	95.7	74.9	90.5	95.7	114.2	100.3	103.3	82.7	98.2	98.5	80.5	94.0	89.2	79.1	86.7	96.3	81.5	92.6
0920	Plaster & Gypsum Board	91.8	74.6	79.9	87.1	113.1	105.1	90.4	82.4	84.9	83.1	78.6	80.0	84.4	78.8	80.6	83.1	79.2	80.4
0950, 0980	Ceilings & Acoustic Treatment	87.5	74.6	79.3	87.5	113.1	103.8	91.0	82.4	85.5	77.5	78.6	78.2	83.2	78.8	80.4	81.4	79.2	80.0
0960	Flooring	93.2	80.7	89.8	93.2	119.0	100.3	93.0	90.0	92.2	97.5	83.6	93.6	91.6	85.7	90.0	98.0	90.0	95.8
0970, 0990	Wall Finishes & Painting/Coating	93.8	78.4	84.5	93.8	115.4	106.8	93.8	87.8	90.2	90.4	73.8	80.4	93.3	85.1	88.4	96.4	88.4	91.5
09	FINISHES	91.8	77.1	83.8	91.3	114.7	104.0	92.9	85.2	88.7	90.7	79.4	84.6	89.5	81.3	85.0	92.0	83.7	87.5
COVERS	DIVS. 10 - 14, 25, 28, 41, 43, 44	100.0	91.9	98.3	100.0	107.8	101.9	100.0	92.8	98.5	100.0	90.8	98.0	100.0	76.9	95.1	100.0	94.2	98.8
21, 22, 23	FIRE SUPPRESSION, PLUMBING & HVAC	99.8	72.5	88.6	99.8	107.6	103.0	99.8	82.9	92.8	99.7	76.9	90.3	99.7	79.7	91.5	99.9	81.9	92.5
26, 27, 3370	ELECTRICAL, COMMUNICATIONS & UTIL.	86.8	76.9	81.6	99.0	103.9	101.5	101.0	90.4	95.5	91.1	76.6	83.5	106.3	88.8	97.2	93.9	86.6	90.1
MF2004	WEIGHTED AVERAGE	95.2	79.6	88.3	96.1	108.3	101.5	97.7	85.7	92.4	96.0	81.5	89.6	95.1	84.7	90.5	95.9	87.9	92.3

City Cost Indexes

DIVISION		IOWA																	
		CEDAR RAPIDS			COUNCIL BLUFFS			DAVENPORT			DES MOINES			DUBUQUE			SIOUX CITY		
		MAT.	INST.	TOTAL	MAT.	INST.	TOTAL	MAT.	INST.	TOTAL	MAT.	INST.	TOTAL	MAT.	INST.	TOTAL	MAT.	INST.	TOTAL
015433	CONTRACTOR EQUIPMENT		96.5	96.5		96.1	96.1		100.0	100.0		101.7	101.7		95.2	95.2		100.0	100.0
0241, 31 - 34	SITE & INFRASTRUCTURE, DEMOLITION	97.7	95.3	96.1	105.4	92.3	96.3	95.5	99.3	98.1	87.9	99.6	96.0	95.5	92.2	93.2	106.8	95.5	99.0
0310	Concrete Forming & Accessories	102.3	73.9	77.7	86.8	51.9	56.6	102.0	84.3	86.7	103.2	69.3	73.9	88.0	65.7	68.7	102.3	59.3	65.1
0320	Concrete Reinforcing	93.8	76.7	85.1	95.7	71.8	83.5	93.8	88.1	90.9	93.8	73.4	83.4	92.6	76.6	84.4	93.8	57.8	75.5
0330	Cast-in-Place Concrete	107.1	75.8	95.3	111.4	73.8	97.2	103.2	88.7	97.7	98.9	89.1	95.2	104.9	85.7	97.6	106.2	51.1	85.4
03	CONCRETE	101.1	75.8	89.3	103.7	64.5	85.4	99.1	87.2	93.6	95.9	77.8	87.4	98.0	75.6	87.5	100.4	57.4	80.3
04	MASONRY	106.9	72.9	86.5	107.7	73.3	87.0	102.8	81.8	90.1	97.9	66.8	79.2	107.9	64.7	81.9	100.3	51.0	70.6
05	METALS	88.3	88.2	88.3	92.9	85.6	90.4	88.3	98.2	91.7	88.5	87.8	88.2	86.9	87.7	87.2	88.3	77.8	84.8
06	WOOD, PLASTICS & COMPOSITES	106.2	73.1	87.5	87.3	45.6	63.7	106.2	83.0	93.0	106.0	69.0	85.1	88.9	65.5	75.6	106.2	59.6	79.8
07	THERMAL & MOISTURE PROTECTION	102.6	74.7	92.0	102.1	61.9	86.9	102.1	83.5	95.1	103.6	71.6	91.5	102.3	65.8	88.5	102.1	50.1	82.5
08	OPENINGS	99.3	76.2	93.6	98.2	60.2	88.8	99.3	85.5	95.9	99.3	73.6	92.9	98.3	74.0	92.2	99.3	57.2	88.8
0920	Plaster & Gypsum Board	101.4	72.6	81.5	92.8	44.2	59.2	101.4	82.5	88.3	90.4	68.2	75.0	93.3	64.7	73.5	101.4	58.5	71.7
0950, 0980	Ceilings & Acoustic Treatment	111.3	72.6	86.6	107.0	44.2	67.0	111.3	82.5	93.0	107.0	68.2	82.3	107.0	64.7	80.1	111.3	58.5	77.7
0960	Flooring	131.3	48.0	108.3	105.1	39.2	86.9	116.0	89.4	108.7	115.7	35.8	93.6	120.4	33.4	96.4	116.5	50.6	98.3
0970, 0990	Wall Finishes & Painting/Coating	105.6	70.8	84.7	98.6	67.9	80.1	102.1	85.5	92.1	102.1	73.8	85.1	104.6	70.8	84.3	103.1	58.9	76.5
09	FINISHES	113.8	68.0	88.9	103.2	49.2	73.9	109.0	85.1	96.0	105.3	62.8	82.2	108.3	59.3	81.7	110.8	57.6	81.9
COVERS	DIVS. 10 - 14, 25, 28, 41, 43, 44	100.0	79.9	95.7	100.0	78.7	95.5	100.0	83.0	96.4	100.0	78.5	95.4	100.0	76.8	95.1	100.0	75.9	94.9
21, 22, 23	FIRE SUPPRESSION, PLUMBING & HVAC	100.1	76.6	90.4	100.1	73.3	89.1	100.1	82.9	93.0	100.0	66.4	86.2	100.1	68.7	87.2	100.1	68.1	87.0
26, 27, 3370	ELECTRICAL, COMMUNICATIONS & UTIL.	95.3	75.5	84.9	100.9	79.2	89.6	93.3	83.9	88.4	108.1	69.9	88.2	99.4	72.1	85.1	95.3	68.7	81.4
MF2004	WEIGHTED AVERAGE	99.5	77.5	89.8	100.2	71.6	87.5	98.4	86.7	93.2	98.7	73.6	87.7	98.6	72.6	87.2	99.1	66.1	84.5

DIVISION		IOWA			KANSAS														
		WATERLOO			DODGE CITY			KANSAS CITY			SALINA			TOPEKA			WICHITA		
		MAT.	INST.	TOTAL	MAT.	INST.	TOTAL	MAT.	INST.	TOTAL	MAT.	INST.	TOTAL	MAT.	INST.	TOTAL	MAT.	INST.	TOTAL
015433	CONTRACTOR EQUIPMENT		100.0	100.0		102.9	102.9		99.7	99.7		102.9	102.9		101.2	101.2		102.9	102.9
0241, 31 - 34	SITE & INFRASTRUCTURE, DEMOLITION	96.0	94.7	95.1	112.8	92.5	98.7	93.0	92.1	92.4	102.5	92.9	95.8	96.7	90.5	92.4	95.2	92.7	93.5
0310	Concrete Forming & Accessories	102.7	41.6	49.8	98.3	58.3	63.7	102.1	98.8	99.2	95.2	47.1	53.6	102.4	44.0	51.9	101.3	48.6	55.7
0320	Concrete Reinforcing	93.8	76.3	84.9	104.7	52.1	77.9	99.5	92.0	95.6	104.1	67.2	85.3	96.7	94.1	95.4	96.7	77.2	86.7
0330	Cast-in-Place Concrete	107.1	41.0	82.9	116.3	50.9	91.5	91.2	98.0	93.8	100.9	44.0	79.3	92.3	48.3	75.6	88.0	52.0	74.4
03	CONCRETE	101.1	50.3	77.4	114.5	55.9	87.1	95.4	97.5	96.4	101.5	51.4	78.1	93.4	56.6	76.2	91.2	56.6	75.0
04	MASONRY	101.8	53.5	72.8	105.5	49.4	71.7	105.5	97.0	100.4	120.7	43.3	74.2	100.0	56.6	73.9	93.6	56.0	71.0
05	METALS	88.3	86.1	87.6	92.0	76.0	86.6	97.7	98.5	98.0	91.8	82.5	88.6	96.9	94.6	96.1	96.9	85.3	93.0
06	WOOD, PLASTICS & COMPOSITES	106.7	41.3	69.7	94.8	64.8	77.8	100.4	100.2	100.3	91.6	49.6	67.8	96.7	41.6	65.5	98.1	47.5	69.4
07	THERMAL & MOISTURE PROTECTION	101.9	46.4	80.9	100.9	55.2	83.6	99.8	97.5	98.9	100.5	50.9	81.8	101.7	62.1	86.7	100.1	57.0	83.8
08	OPENINGS	94.9	53.6	84.7	96.4	55.3	86.2	95.5	92.5	94.7	96.3	49.5	84.7	96.7	57.8	87.1	96.7	55.3	86.5
0920	Plaster & Gypsum Board	101.4	39.6	58.7	96.6	63.6	73.8	92.2	100.1	97.7	95.7	48.0	62.7	94.1	39.7	56.5	91.0	45.8	59.7
0950, 0980	Ceilings & Acoustic Treatment	111.3	39.6	65.7	92.0	63.6	74.0	92.0	100.1	97.2	92.0	48.0	64.0	92.0	39.7	58.7	91.2	45.8	62.3
0960	Flooring	117.9	50.8	99.3	100.3	42.2	84.2	89.8	97.7	92.0	98.3	30.8	79.7	101.9	39.2	84.5	100.7	67.9	91.7
0970, 0990	Wall Finishes & Painting/Coating	103.1	30.9	59.7	96.2	43.2	64.3	103.9	75.2	86.7	96.2	43.2	64.3	96.2	58.4	73.4	96.2	53.3	70.4
09	FINISHES	109.6	41.3	72.6	99.2	54.5	74.9	96.7	95.3	96.0	97.7	43.1	68.1	98.1	42.8	68.1	97.2	52.3	72.8
COVERS	DIVS. 10 - 14, 25, 28, 41, 43, 44	100.0	70.2	93.7	100.0	42.8	87.9	100.0	71.0	93.8	100.0	52.5	89.9	100.0	55.5	90.6	100.0	54.7	90.4
21, 22, 23	FIRE SUPPRESSION, PLUMBING & HVAC	100.1	37.5	74.4	100.0	55.9	81.9	99.9	96.5	98.5	100.0	55.9	81.8	100.0	60.3	83.7	99.8	54.2	81.0
26, 27, 3370	ELECTRICAL, COMMUNICATIONS & UTIL.	95.3	62.4	78.1	96.0	69.0	81.9	101.8	94.9	98.2	95.7	69.0	81.7	107.7	69.3	87.6	103.3	69.0	85.3
MF2004	WEIGHTED AVERAGE	98.4	56.0	79.7	100.4	61.5	83.3	98.6	95.3	97.1	99.1	59.2	81.5	99.0	63.8	83.4	97.7	62.7	82.2

DIVISION		KENTUCKY											LOUISIANA					
		BOWLING GREEN			LEXINGTON			LOUISVILLE			OWENSBORO			ALEXANDRIA			BATON ROUGE		
		MAT.	INST.	TOTAL	MAT.	INST.	TOTAL	MAT.	INST.	TOTAL	MAT.	INST.	TOTAL	MAT.	INST.	TOTAL	MAT.	INST.	TOTAL
015433	CONTRACTOR EQUIPMENT		95.5	95.5		102.1	102.1		95.5	95.5		117.5	117.5		88.0	88.0		88.0	88.0
0241, 31 - 34	SITE & INFRASTRUCTURE, DEMOLITION	69.1	98.7	89.7	76.8	101.1	93.7	66.9	98.5	88.9	81.1	125.8	112.2	101.9	85.4	90.4	116.4	85.1	94.6
0310	Concrete Forming & Accessories	83.8	82.3	82.5	96.7	74.5	77.5	91.4	80.7	82.2	90.5	71.9	74.4	83.5	37.9	44.0	94.9	60.9	65.5
0320	Concrete Reinforcing	90.6	84.0	87.2	100.0	91.5	95.6	100.0	92.0	95.9	90.7	90.6	90.6	101.0	60.2	80.1	104.9	60.3	82.1
0330	Cast-in-Place Concrete	83.1	103.2	90.7	89.7	81.6	86.6	93.6	75.5	86.8	85.7	99.4	90.9	99.7	40.0	77.1	102.9	59.4	86.4
03	CONCRETE	91.3	89.9	90.6	93.4	80.5	87.3	95.0	81.2	88.5	100.4	85.2	93.3	93.4	44.2	70.4	98.0	60.8	80.6
04	MASONRY	89.2	74.0	80.1	86.2	65.7	73.9	87.4	79.0	82.4	85.9	71.5	77.2	115.3	50.4	76.3	106.8	52.2	74.0
05	METALS	92.8	84.2	89.9	94.3	88.5	92.3	101.0	87.9	96.6	85.4	88.5	86.4	91.1	72.8	84.9	103.1	70.0	91.9
06	WOOD, PLASTICS & COMPOSITES	93.0	81.8	86.7	102.0	74.5	86.4	99.7	81.8	89.5	93.4	68.2	79.1	89.1	36.8	59.5	102.7	66.1	82.0
07	THERMAL & MOISTURE PROTECTION	84.9	84.1	84.6	97.6	85.9	93.2	91.5	78.8	86.7	97.4	75.7	89.2	98.2	49.3	79.7	97.5	57.8	82.8
08	OPENINGS	95.3	77.7	90.9	96.3	78.5	91.9	96.3	84.4	93.3	93.4	78.6	89.7	99.6	43.7	85.8	102.6	59.4	91.9
0920	Plaster & Gypsum Board	78.8	81.6	80.7	85.9	73.1	77.1	82.0	81.6	81.7	78.9	66.6	70.4	78.2	35.5	48.7	94.2	65.5	74.4
0950, 0980	Ceilings & Acoustic Treatment	81.4	81.6	81.5	86.6	73.1	78.0	82.3	81.6	81.8	74.5	66.6	69.5	92.2	35.5	56.1	99.9	65.5	78.0
0960	Flooring	93.9	81.6	90.5	98.8	59.8	88.0	97.3	70.3	89.8	96.4	81.6	92.3	101.6	66.0	91.8	107.0	66.0	95.7
0970, 0990	Wall Finishes & Painting/Coating	96.4	71.5	81.4	96.4	69.7	80.3	96.4	79.0	85.9	96.4	98.7	97.7	98.6	35.4	60.6	103.2	40.6	65.5
09	FINISHES	90.3	81.0	85.3	93.9	70.6	81.2	91.5	78.5	84.4	89.5	75.7	82.0	92.2	42.0	64.9	101.7	60.3	79.3
COVERS	DIVS. 10 - 14, 25, 28, 41, 43, 44	100.0	66.1	92.8	100.0	90.9	98.1	100.0	90.4	98.0	100.0	84.2	96.6	100.0	55.5	90.5	100.0	78.6	95.5
21, 22, 23	FIRE SUPPRESSION, PLUMBING & HVAC	99.9	83.1	93.0	99.9	65.0	85.5	99.9	81.0	92.1	99.9	68.8	87.1	100.2	52.1	80.4	99.9	51.2	79.9
26, 27, 3370	ELECTRICAL, COMMUNICATIONS & UTIL.	94.4	76.3	84.9	95.0	73.6	83.8	106.2	76.3	90.6	93.9	82.3	87.8	94.9	53.4	73.2	116.1	60.4	87.0
MF2004	WEIGHTED AVERAGE	93.9	82.6	88.9	95.5	76.3	87.0	97.1	82.4	90.6	94.3	81.7	88.8	97.5	54.0	78.3	103.0	61.2	84.5

City Cost Indexes

DIVISION		LOUISIANA											MAINE					
		LAKE CHARLES			MONROE			NEW ORLEANS			SHREVEPORT			AUGUSTA			BANGOR		
		MAT.	INST.	TOTAL	MAT.	INST.	TOTAL	MAT.	INST.	TOTAL	MAT.	INST.	TOTAL	MAT.	INST.	TOTAL	MAT.	INST.	TOTAL
015433	CONTRACTOR EQUIPMENT		88.0	88.0		88.0	88.0		89.0	89.0		88.0	88.0		100.6	100.6		100.6	100.6
0241, 31 - 34	SITE & INFRASTRUCTURE, DEMOLITION	118.0	84.9	95.0	101.9	84.7	89.9	121.0	87.7	97.8	103.5	84.6	90.4	81.0	101.1	95.0	80.2	98.8	93.1
0310	Concrete Forming & Accessories	97.8	52.1	58.3	83.0	38.4	44.4	100.1	63.7	68.6	96.9	40.7	48.3	96.5	102.2	101.4	92.3	91.5	91.6
0320	Concrete Reinforcing	104.9	58.0	81.0	99.8	60.1	79.5	104.9	61.8	82.9	99.4	60.5	79.5	90.4	103.1	96.9	90.1	102.8	96.6
0330	Cast-in-Place Concrete	105.0	68.8	91.3	99.7	57.6	83.8	104.0	69.0	90.7	99.1	46.0	79.0	84.3	63.7	76.5	78.3	62.3	72.3
03	CONCRETE	99.2	59.7	80.7	93.2	50.2	73.1	98.8	65.8	83.4	93.1	47.4	71.7	103.0	88.5	96.2	99.3	82.9	91.7
04	MASONRY	104.9	59.8	77.8	110.2	48.3	73.0	106.2	57.5	76.9	101.3	49.9	70.4	98.0	63.2	77.1	112.7	59.0	80.4
05	METALS	97.9	69.6	88.4	91.0	70.6	84.1	111.0	74.0	98.5	87.8	70.7	82.0	88.3	85.1	87.2	87.9	80.1	85.3
06	WOOD, PLASTICS & COMPOSITES	107.1	53.7	76.9	88.4	37.2	59.4	103.7	66.1	82.4	104.4	40.4	68.2	97.5	114.0	106.8	92.7	99.9	96.8
07	THERMAL & MOISTURE PROTECTION	99.6	59.3	84.4	98.2	52.2	80.8	100.4	61.1	85.6	97.5	51.1	79.9	95.8	64.5	83.9	96.1	60.1	82.5
08	OPENINGS	102.6	53.2	90.4	99.6	47.8	86.8	103.4	66.4	94.2	97.5	45.8	84.7	101.1	91.4	98.7	101.0	85.1	97.1
0920	Plaster & Gypsum Board	100.9	52.8	67.6	77.8	35.9	48.8	99.6	65.5	76.0	83.6	39.2	52.9	102.3	113.7	110.2	94.7	99.2	97.8
0950, 0980	Ceilings & Acoustic Treatment	99.1	52.8	69.6	92.2	35.9	56.3	98.1	65.5	77.3	91.3	39.2	58.1	90.1	113.7	105.1	90.9	99.2	96.2
0960	Flooring	107.2	78.5	99.3	101.1	61.5	90.2	107.6	66.0	96.1	111.0	66.0	98.6	97.6	59.5	87.1	95.3	59.2	85.3
0970, 0990	Wall Finishes & Painting/Coating	103.2	37.4	63.6	98.6	42.3	64.7	104.7	64.0	80.2	98.6	35.4	60.6	94.6	38.3	60.7	94.6	35.9	59.3
09	FINISHES	102.3	55.3	76.8	92.0	42.2	65.0	102.6	64.2	81.8	95.8	44.1	67.7	98.1	89.1	93.2	96.7	81.2	88.3
COVERS	DIVS. 10 - 14, 25, 28, 41, 43, 44	100.0	77.2	95.2	100.0	50.8	89.6	100.0	80.5	95.9	100.0	70.9	93.8	100.0	91.1	98.1	100.0	105.3	101.1
21, 22, 23	FIRE SUPPRESSION, PLUMBING & HVAC	100.0	52.7	80.5	100.2	46.2	78.0	100.0	64.1	85.2	100.1	54.8	81.5	100.0	69.0	87.3	100.1	64.7	85.5
26, 27, 3370	ELECTRICAL, COMMUNICATIONS & UTIL.	102.2	58.7	79.5	96.9	54.5	74.8	102.8	71.7	86.6	104.4	68.9	85.8	98.9	81.5	89.8	95.8	80.7	87.9
MF2004	WEIGHTED AVERAGE	101.1	60.7	83.3	97.4	53.4	77.9	103.2	68.2	87.8	97.4	57.7	79.9	97.7	81.5	90.6	97.5	77.5	88.7

DIVISION		MAINE						MARYLAND						MASSACHUSETTS					
		LEWISTON			PORTLAND			BALTIMORE			HAGERSTOWN			BOSTON			BROCKTON		
		MAT.	INST.	TOTAL	MAT.	INST.	TOTAL	MAT.	INST.	TOTAL	MAT.	INST.	TOTAL	MAT.	INST.	TOTAL	MAT.	INST.	TOTAL
015433	CONTRACTOR EQUIPMENT		100.6	100.6		100.6	100.6		102.9	102.9		99.2	99.2		106.5	106.5		102.5	102.5
0241, 31 - 34	SITE & INFRASTRUCTURE, DEMOLITION	77.8	98.8	92.4	79.3	98.8	92.9	95.2	94.4	94.6	85.4	90.9	89.2	85.3	108.1	101.2	82.7	104.3	97.8
0310	Concrete Forming & Accessories	96.7	91.5	92.2	95.5	91.5	92.0	98.3	75.9	78.9	89.1	78.1	79.5	100.0	139.7	134.4	100.5	127.1	123.6
0320	Concrete Reinforcing	110.6	102.8	106.6	110.6	102.8	106.6	109.9	84.8	97.1	97.1	75.7	86.2	109.4	147.7	129.0	110.6	147.5	129.5
0330	Cast-in-Place Concrete	79.9	62.3	73.3	101.6	62.3	86.8	103.9	79.1	94.5	88.4	58.6	77.1	115.1	150.6	129.5	96.2	146.0	115.0
03	CONCRETE	98.7	82.9	91.4	109.4	82.9	97.0	105.1	79.7	93.2	90.0	71.9	81.6	112.4	144.0	127.2	109.7	136.7	122.3
04	MASONRY	96.9	59.0	74.1	97.2	59.0	74.2	101.1	70.5	82.7	105.0	81.3	90.7	115.7	156.6	140.3	111.9	149.4	134.4
05	METALS	91.1	80.1	87.4	92.3	80.1	88.2	93.3	96.4	94.3	90.1	91.7	90.6	103.0	128.5	111.6	99.6	125.8	108.5
06	WOOD, PLASTICS & COMPOSITES	98.1	99.9	99.2	97.3	99.9	98.8	95.9	77.5	85.5	85.5	77.3	80.9	101.4	138.9	122.7	100.2	125.9	114.8
07	THERMAL & MOISTURE PROTECTION	96.0	60.1	82.4	99.1	60.1	84.4	94.4	79.6	88.8	93.7	69.9	84.7	96.1	150.2	116.5	96.2	144.2	114.3
08	OPENINGS	104.3	85.1	99.6	104.3	85.1	99.6	93.1	83.7	90.8	90.4	77.9	87.3	100.5	142.6	111.0	98.6	132.7	107.0
0920	Plaster & Gypsum Board	100.4	99.2	99.6	106.9	99.2	101.6	100.5	77.0	84.2	95.3	77.0	82.7	103.3	139.5	128.3	97.6	126.0	117.2
0950, 0980	Ceilings & Acoustic Treatment	98.7	99.2	99.0	96.1	99.2	98.1	93.8	77.0	83.1	94.7	77.0	83.4	97.7	139.5	124.3	99.7	126.0	116.4
0960	Flooring	98.3	59.2	87.5	97.6	59.2	87.0	89.2	75.1	85.3	84.7	84.5	84.7	98.0	168.9	117.6	98.5	168.9	118.0
0970, 0990	Wall Finishes & Painting/Coating	94.6	35.9	59.3	94.6	35.9	59.3	97.0	81.4	87.6	97.0	34.5	59.4	97.2	159.0	134.4	96.7	143.5	124.9
09	FINISHES	99.7	81.2	89.7	99.6	81.2	89.6	92.7	75.7	83.5	90.2	74.7	81.8	99.7	147.6	125.7	99.6	136.7	119.7
COVERS	DIVS. 10 - 14, 25, 28, 41, 43, 44	100.0	105.3	101.1	100.0	105.3	101.1	100.0	88.1	97.5	100.0	91.1	98.1	100.0	120.6	104.4	100.0	116.8	103.6
21, 22, 23	FIRE SUPPRESSION, PLUMBING & HVAC	100.1	64.7	85.5	100.0	64.7	85.5	100.0	87.7	94.9	100.0	87.8	94.9	100.1	129.1	112.0	100.1	111.5	104.8
26, 27, 3370	ELECTRICAL, COMMUNICATIONS & UTIL.	98.0	81.3	89.2	104.5	81.3	92.4	97.5	95.0	96.2	94.8	79.5	86.8	97.1	136.8	117.8	96.2	100.5	98.4
MF2004	WEIGHTED AVERAGE	97.8	77.6	88.9	100.1	77.6	90.2	97.8	85.2	92.2	94.5	81.9	88.9	102.0	136.5	117.2	100.6	123.2	110.6

DIVISION		MASSACHUSETTS																	
		FALL RIVER			HYANNIS			LAWRENCE			LOWELL			NEW BEDFORD			PITTSFIELD		
		MAT.	INST.	TOTAL	MAT.	INST.	TOTAL	MAT.	INST.	TOTAL	MAT.	INST.	TOTAL	MAT.	INST.	TOTAL	MAT.	INST.	TOTAL
015433	CONTRACTOR EQUIPMENT		103.5	103.5		102.5	102.5		102.5	102.5		100.6	100.6		103.5	103.5		100.6	100.6
0241, 31 - 34	SITE & INFRASTRUCTURE, DEMOLITION	81.8	104.4	97.5	79.9	104.2	96.8	83.5	104.3	98.0	82.3	104.3	97.6	80.4	104.4	97.1	83.3	102.7	96.8
0310	Concrete Forming & Accessories	100.5	126.8	123.3	93.4	126.5	122.0	100.5	127.3	123.6	97.7	127.5	123.5	100.5	126.8	123.3	97.7	108.4	106.9
0320	Concrete Reinforcing	110.6	123.3	117.1	88.7	123.2	106.3	109.7	137.2	123.7	110.6	137.2	124.2	110.6	123.3	117.1	92.2	109.7	101.1
0330	Cast-in-Place Concrete	93.0	146.4	113.2	87.8	145.8	109.7	97.0	146.1	115.6	88.2	146.2	110.1	82.0	146.4	106.3	96.2	122.7	106.2
03	CONCRETE	108.2	132.0	119.3	99.8	131.7	114.7	109.9	134.8	121.6	100.3	134.7	116.4	102.7	132.0	116.4	101.6	112.9	106.9
04	MASONRY	112.0	149.3	134.4	110.8	149.4	134.0	111.4	149.4	134.2	97.5	149.5	128.7	110.8	149.3	133.9	98.2	122.1	112.5
05	METALS	99.6	115.8	105.1	96.0	115.3	102.5	96.9	121.7	105.3	96.9	119.5	104.5	99.6	115.8	105.1	96.7	102.3	98.6
06	WOOD, PLASTICS & COMPOSITES	100.2	126.2	114.9	91.6	125.9	111.0	100.2	125.9	114.8	99.8	125.9	114.6	100.2	126.2	114.9	99.8	108.5	104.7
07	THERMAL & MOISTURE PROTECTION	96.2	138.1	112.0	95.9	138.8	112.1	96.2	144.2	114.3	96.0	144.2	114.2	96.2	138.1	112.0	96.0	117.6	104.2
08	OPENINGS	98.6	122.1	104.4	94.7	121.9	101.4	98.6	129.9	106.3	104.3	129.9	110.6	98.6	122.1	104.4	104.3	109.2	105.5
0920	Plaster & Gypsum Board	97.6	126.0	117.2	90.1	126.0	114.9	100.4	126.0	118.1	100.4	126.0	118.1	97.6	126.0	117.2	100.4	108.0	105.7
0950, 0980	Ceilings & Acoustic Treatment	99.7	126.0	116.4	91.9	126.0	113.6	98.7	126.0	116.1	98.7	126.0	116.1	99.7	126.0	116.4	98.7	108.0	104.6
0960	Flooring	98.4	168.9	117.8	94.7	168.9	115.2	97.6	168.9	117.3	97.6	168.9	117.3	98.4	168.9	117.8	97.8	138.1	108.9
0970, 0990	Wall Finishes & Painting/Coating	96.7	143.5	124.9	96.7	143.5	124.9	94.7	143.5	124.1	94.6	143.5	124.0	96.7	143.5	124.9	94.6	108.0	102.7
09	FINISHES	99.6	136.9	119.8	95.5	136.7	117.9	99.3	136.7	119.6	99.3	136.7	119.6	99.5	136.9	119.8	99.3	114.1	107.3
COVERS	DIVS. 10 - 14, 25, 28, 41, 43, 44	100.0	117.5	103.7	100.0	116.8	103.6	100.0	116.9	103.6	100.0	116.9	103.6	100.0	117.5	103.7	100.0	105.6	101.2
21, 22, 23	FIRE SUPPRESSION, PLUMBING & HVAC	100.1	111.3	104.7	100.1	111.2	104.7	100.1	114.4	106.0	100.1	124.9	110.3	100.1	111.3	104.7	100.1	98.7	99.6
26, 27, 3370	ELECTRICAL, COMMUNICATIONS & UTIL.	96.1	100.4	98.4	93.8	100.4	97.3	97.4	128.5	113.7	98.0	128.5	113.9	97.0	100.4	98.8	97.9	93.4	95.6
MF2004	WEIGHTED AVERAGE	100.4	121.1	109.5	97.7	120.9	107.9	100.3	127.0	112.1	99.0	129.0	112.2	99.7	121.1	109.1	99.2	106.1	102.3

City Cost Indexes

DIVISION		MASSACHUSETTS						MICHIGAN											
		SPRINGFIELD			WORCESTER			ANN ARBOR			DEARBORN			DETROIT			FLINT		
		MAT.	INST.	TOTAL	MAT.	INST.	TOTAL	MAT.	INST.	TOTAL	MAT.	INST.	TOTAL	MAT.	INST.	TOTAL	MAT.	INST.	TOTAL
015433	CONTRACTOR EQUIPMENT		100.6	100.6		100.6	100.6		110.3	110.3		110.3	110.3		97.6	97.6		110.3	110.3
0241, 31 - 34	SITE & INFRASTRUCTURE, DEMOLITION	82.8	102.9	96.8	82.7	104.2	97.7	75.3	95.7	89.5	75.1	95.8	89.5	89.0	97.4	94.9	65.7	95.0	86.1
0310	Concrete Forming & Accessories	97.9	109.4	107.9	98.2	122.6	119.3	95.4	106.7	105.2	95.2	115.6	112.8	96.0	115.7	113.0	97.9	93.4	94.0
0320	Concrete Reinforcing	110.6	115.1	112.9	110.6	147.5	129.4	116.3	126.4	121.5	116.3	126.6	121.6	115.5	126.6	121.2	116.3	125.9	121.2
0330	Cast-in-Place Concrete	91.8	124.2	104.0	91.2	146.0	111.9	80.9	117.1	94.6	79.1	114.8	92.6	86.8	114.8	97.4	81.5	98.5	87.9
03	CONCRETE	102.1	114.9	108.1	101.8	134.5	117.1	91.9	114.4	102.4	91.0	117.6	103.4	94.7	116.4	104.8	92.3	102.0	96.8
04	MASONRY	97.8	124.7	114.0	97.3	149.5	128.7	92.3	113.2	104.9	92.2	117.3	107.3	91.5	117.3	107.0	92.4	100.4	97.2
05	METALS	99.5	104.5	101.2	99.6	123.0	107.5	91.2	122.0	101.6	91.3	122.6	101.9	95.0	104.2	98.1	91.3	119.5	100.8
06	WOOD, PLASTICS & COMPOSITES	99.8	108.5	104.7	100.1	120.3	111.5	98.9	105.0	102.3	98.9	115.7	108.4	98.9	115.7	108.4	101.8	91.3	95.9
07	THERMAL & MOISTURE PROTECTION	96.0	118.7	104.6	96.0	138.5	112.1	102.3	114.5	106.9	100.8	119.5	107.9	99.1	119.5	106.8	100.0	101.0	100.4
08	OPENINGS	104.3	110.6	105.9	104.3	129.6	110.6	94.0	108.4	97.6	94.0	114.2	99.0	95.7	114.2	100.3	94.0	98.7	95.2
0920	Plaster & Gypsum Board	100.4	108.0	105.7	100.4	120.2	114.1	97.2	104.1	101.9	97.2	115.2	109.6	97.2	115.2	109.6	98.9	90.1	92.8
0950, 0980	Ceilings & Acoustic Treatment	98.7	108.0	104.6	98.7	120.2	112.4	91.1	104.1	99.4	91.1	115.2	106.4	92.0	115.2	106.8	91.1	90.1	90.4
0960	Flooring	97.5	138.1	108.7	97.6	168.9	117.3	92.7	120.4	100.4	92.4	121.2	100.3	92.4	121.2	100.4	92.5	84.9	90.4
0970, 0990	Wall Finishes & Painting/Coating	96.1	108.0	103.3	94.6	143.5	124.0	93.5	107.5	101.9	93.5	107.6	102.0	95.0	107.6	102.6	93.5	94.9	94.3
09	FINISHES	99.3	114.7	107.7	99.3	133.4	117.8	92.3	109.1	101.4	92.2	116.0	105.1	93.4	116.0	105.7	91.6	91.1	91.3
COVERS	DIVS. 10 - 14, 25, 28, 41, 43, 44	100.0	106.5	101.4	100.0	110.1	102.2	100.0	107.0	101.5	100.0	108.9	101.9	100.0	108.9	101.9	100.0	100.0	100.0
21, 22, 23	FIRE SUPPRESSION, PLUMBING & HVAC	100.1	101.6	100.7	100.1	111.8	104.9	100.1	100.2	100.1	100.1	110.5	104.3	100.1	112.7	105.2	100.1	98.9	99.6
26, 27, 3370	ELECTRICAL, COMMUNICATIONS & UTIL.	98.0	93.4	95.6	98.0	102.9	100.5	94.6	92.4	93.5	94.6	112.2	103.8	95.7	112.1	104.3	94.6	100.2	97.5
MF2004	WEIGHTED AVERAGE	99.6	107.7	103.2	99.6	122.1	109.5	94.9	106.1	99.9	94.7	113.4	103.0	96.4	112.1	103.3	94.6	100.2	97.1

DIVISION		MICHIGAN															MINNESOTA		
		GRAND RAPIDS			KALAMAZOO			LANSING			MUSKEGON			SAGINAW			DULUTH		
		MAT.	INST.	TOTAL	MAT.	INST.	TOTAL	MAT.	INST.	TOTAL	MAT.	INST.	TOTAL	MAT.	INST.	TOTAL	MAT.	INST.	TOTAL
015433	CONTRACTOR EQUIPMENT		103.5	103.5		103.5	103.5		110.3	110.3		103.5	103.5		110.3	110.3		101.0	101.0
0241, 31 - 34	SITE & INFRASTRUCTURE, DEMOLITION	82.9	87.6	86.2	81.7	87.8	85.9	83.5	94.5	91.1	79.4	87.7	85.2	68.4	94.3	86.5	95.2	102.1	100.0
0310	Concrete Forming & Accessories	97.8	82.8	84.8	97.6	93.7	94.3	98.7	92.0	92.9	98.1	92.1	92.9	95.4	91.5	92.0	100.9	121.7	118.9
0320	Concrete Reinforcing	100.0	85.9	92.8	100.0	88.0	93.8	116.3	125.7	121.1	100.6	87.0	93.7	116.3	125.4	121.0	92.2	114.4	103.5
0330	Cast-in-Place Concrete	90.7	98.6	93.7	91.4	103.1	95.9	83.1	96.3	88.1	89.3	99.8	93.3	79.9	95.7	85.9	108.3	109.5	108.7
03	CONCRETE	93.9	88.3	91.3	96.6	95.1	95.9	93.2	100.6	96.6	91.7	93.1	92.3	91.4	100.1	95.5	101.5	116.6	108.6
04	MASONRY	88.6	56.3	69.2	91.5	90.9	91.1	86.3	98.1	93.4	90.3	74.2	80.6	93.7	91.0	92.1	109.6	122.8	117.5
05	METALS	94.3	81.0	89.8	93.2	85.7	90.7	89.9	119.0	99.7	90.9	84.0	88.6	91.3	118.5	100.5	89.4	128.6	102.7
06	WOOD, PLASTICS & COMPOSITES	95.7	82.7	88.3	98.5	94.5	96.3	101.1	90.3	95.0	94.6	93.3	93.8	94.3	91.0	92.5	110.6	122.8	117.5
07	THERMAL & MOISTURE PROTECTION	95.4	60.7	82.3	93.5	91.2	92.6	101.3	95.9	99.3	92.7	79.7	87.8	100.8	97.5	99.5	103.6	121.5	110.4
08	OPENINGS	94.2	71.2	88.5	90.7	86.5	89.7	94.0	98.1	95.0	89.9	87.6	89.4	91.9	98.5	93.6	95.2	126.7	103.0
0920	Plaster & Gypsum Board	82.8	78.1	79.5	84.6	90.3	88.5	95.4	89.0	91.0	69.4	89.0	82.9	97.2	89.7	92.0	93.2	123.9	114.5
0950, 0980	Ceilings & Acoustic Treatment	84.8	78.1	80.5	86.6	90.3	88.9	89.4	89.0	89.1	87.4	89.0	88.4	91.1	89.7	90.2	92.2	123.9	112.4
0960	Flooring	97.8	36.5	80.9	97.7	78.0	92.2	102.5	84.9	97.6	96.9	67.2	88.7	92.7	80.3	89.3	107.4	131.5	114.0
0970, 0990	Wall Finishes & Painting/Coating	96.4	35.8	60.0	96.4	83.8	88.8	105.5	88.6	95.3	95.7	68.5	79.3	93.5	87.8	90.1	89.6	113.4	103.9
09	FINISHES	92.9	69.0	79.9	93.3	89.9	91.5	96.3	89.7	92.7	90.8	85.1	87.7	92.0	88.4	90.0	98.6	123.4	112.1
COVERS	DIVS. 10 - 14, 25, 28, 41, 43, 44	100.0	107.6	101.6	100.0	109.3	102.0	100.0	103.1	100.7	100.0	108.9	101.9	100.0	98.5	99.7	100.0	106.5	101.4
21, 22, 23	FIRE SUPPRESSION, PLUMBING & HVAC	99.9	83.2	93.0	99.9	81.8	92.4	100.0	96.9	98.7	99.7	85.9	94.0	100.1	88.6	95.3	99.8	100.0	99.9
26, 27, 3370	ELECTRICAL, COMMUNICATIONS & UTIL.	104.6	51.5	76.9	94.3	84.5	89.2	100.6	98.2	99.3	94.8	81.4	87.8	92.5	93.5	93.0	99.7	114.9	107.6
MF2004	WEIGHTED AVERAGE	96.4	74.8	86.8	95.3	88.2	92.2	95.7	98.7	97.0	93.9	85.6	90.2	94.2	95.3	94.7	98.6	114.6	105.6

DIVISION		MINNESOTA												MISSISSIPPI					
		MINNEAPOLIS			ROCHESTER			SAINT PAUL			ST. CLOUD			BILOXI			GREENVILLE		
		MAT.	INST.	TOTAL	MAT.	INST.	TOTAL	MAT.	INST.	TOTAL	MAT.	INST.	TOTAL	MAT.	INST.	TOTAL	MAT.	INST.	TOTAL
015433	CONTRACTOR EQUIPMENT		104.7	104.7		101.0	101.0		101.0	101.0		101.6	101.6		98.3	98.3		98.3	98.3
0241, 31 - 34	SITE & INFRASTRUCTURE, DEMOLITION	93.9	107.7	103.5	92.9	101.9	99.2	97.7	102.7	101.2	86.0	105.8	99.8	104.0	88.7	93.4	107.8	88.8	94.6
0310	Concrete Forming & Accessories	102.9	135.3	130.9	103.7	112.4	111.2	98.1	131.8	127.2	90.8	130.6	125.3	93.0	59.1	63.7	81.9	71.4	72.8
0320	Concrete Reinforcing	92.4	125.6	109.4	92.2	125.2	109.0	89.0	125.6	107.7	94.7	125.4	110.4	86.0	60.2	72.8	93.2	67.1	79.9
0330	Cast-in-Place Concrete	112.4	122.9	116.4	110.3	105.0	108.3	114.4	122.2	117.4	97.2	121.9	106.6	108.5	57.2	89.1	109.0	60.5	90.6
03	CONCRETE	105.5	129.4	116.7	102.7	113.1	107.6	108.2	127.7	117.3	91.6	126.9	108.1	96.6	60.2	79.6	99.8	68.1	85.0
04	MASONRY	108.4	133.6	123.5	107.6	120.4	115.3	118.3	133.5	127.5	109.5	131.1	122.5	88.9	48.2	64.4	129.8	53.9	84.2
05	METALS	92.2	136.2	107.1	89.3	134.1	104.4	88.7	135.8	104.6	90.6	134.0	105.3	94.7	84.7	91.3	92.8	87.9	91.2
06	WOOD, PLASTICS & COMPOSITES	113.7	134.9	125.7	113.7	111.0	112.2	106.9	130.3	120.2	95.0	129.1	114.3	95.1	62.8	76.8	80.6	76.3	78.2
07	THERMAL & MOISTURE PROTECTION	103.8	133.1	114.8	104.0	107.9	105.4	105.1	132.4	115.4	103.0	117.5	108.5	91.6	59.7	79.6	91.8	66.1	82.1
08	OPENINGS	98.4	142.2	109.2	95.2	129.3	103.7	92.8	139.7	104.4	91.7	139.0	103.4	97.5	63.2	89.0	96.9	68.6	89.9
0920	Plaster & Gypsum Board	98.4	136.2	124.6	97.9	111.8	107.5	90.5	131.7	119.0	86.3	130.5	116.9	99.2	62.2	73.6	87.3	76.1	79.6
0950, 0980	Ceilings & Acoustic Treatment	95.6	136.2	121.5	93.9	111.8	105.3	90.4	131.7	116.7	69.0	130.5	108.2	98.4	62.2	75.3	93.2	76.1	82.3
0960	Flooring	105.1	130.1	112.0	107.5	93.6	103.7	101.1	130.1	109.1	104.3	130.1	111.5	116.6	56.9	100.1	107.6	56.9	93.6
0970, 0990	Wall Finishes & Painting/Coating	97.3	126.8	115.0	92.5	105.0	100.1	97.3	132.2	118.3	103.7	126.8	117.6	106.1	48.3	71.3	106.1	73.2	86.3
09	FINISHES	99.6	134.1	118.4	99.6	108.9	104.7	96.7	132.0	115.9	93.2	130.7	113.6	103.5	57.7	78.6	98.5	69.8	82.9
COVERS	DIVS. 10 - 14, 25, 28, 41, 43, 44	100.0	111.7	102.5	100.0	104.9	101.0	100.0	110.8	102.3	100.0	110.5	102.2	100.0	65.4	92.6	100.0	69.0	93.4
21, 22, 23	FIRE SUPPRESSION, PLUMBING & HVAC	99.8	124.1	109.8	99.8	99.3	99.6	99.7	115.9	106.4	99.5	124.0	109.6	99.8	53.0	80.5	99.8	54.0	80.9
26, 27, 3370	ELECTRICAL, COMMUNICATIONS & UTIL.	99.2	115.0	107.5	98.2	84.9	91.3	98.1	115.0	106.9	99.2	115.0	107.4	98.9	60.5	78.8	98.2	70.3	83.6
MF2004	WEIGHTED AVERAGE	99.7	126.3	111.5	98.5	107.8	102.6	99.3	123.4	109.9	96.2	124.2	108.6	97.9	62.2	82.1	99.6	67.9	85.6

City Cost Indexes

DIVISION		MISSISSIPPI						MISSOURI											
		JACKSON			MERIDIAN			CAPE GIRARDEAU			COLUMBIA			JOPLIN			KANSAS CITY		
		MAT.	INST.	TOTAL	MAT.	INST.	TOTAL	MAT.	INST.	TOTAL	MAT.	INST.	TOTAL	MAT.	INST.	TOTAL	MAT.	INST.	TOTAL
015433	CONTRACTOR EQUIPMENT		98.3	98.3		98.3	98.3		106.7	106.7		107.7	107.7		105.3	105.3		103.2	103.2
0241, 31 - 34	SITE & INFRASTRUCTURE, DEMOLITION	99.2	88.8	91.9	99.0	89.1	92.1	95.3	93.2	93.8	103.4	95.8	98.1	98.7	97.8	98.1	92.4	97.9	96.2
0310	Concrete Forming & Accessories	92.3	64.3	68.1	80.9	70.8	72.2	93.2	70.9	73.9	92.5	75.4	77.7	103.9	66.1	71.2	102.8	107.2	106.6
0320	Concrete Reinforcing	86.0	59.2	72.3	92.1	59.2	75.3	104.5	79.1	91.5	100.4	112.4	106.5	102.1	67.5	84.4	96.9	113.6	105.4
0330	Cast-in-Place Concrete	100.8	61.3	85.9	102.8	63.9	88.1	94.5	80.4	89.2	89.7	85.2	88.0	98.5	66.6	86.4	91.6	107.5	97.6
03	CONCRETE	92.7	63.7	79.2	92.4	67.5	80.8	93.7	77.4	86.1	85.4	87.4	86.3	97.6	67.7	83.6	93.8	108.8	100.8
04	MASONRY	90.0	53.9	68.3	88.5	58.3	70.4	116.7	74.3	91.2	132.4	82.3	102.3	93.8	56.6	71.4	98.0	108.1	104.1
05	METALS	94.7	84.5	91.2	92.8	84.7	90.0	94.1	102.3	96.9	90.0	118.4	99.6	93.0	84.2	90.1	100.7	113.9	105.1
06	WOOD, PLASTICS & COMPOSITES	94.0	66.7	78.5	79.6	73.0	75.9	87.3	69.0	76.9	94.3	72.1	81.7	99.3	65.2	80.0	98.6	106.5	103.1
07	THERMAL & MOISTURE PROTECTION	93.6	65.1	82.8	91.4	67.7	82.4	100.9	75.1	91.1	94.8	80.4	89.4	98.8	62.3	85.0	98.7	109.8	102.9
08	OPENINGS	97.9	61.2	88.8	96.8	70.8	90.4	98.4	69.3	91.2	91.2	92.3	91.5	90.6	67.3	84.8	96.5	109.0	99.6
0920	Plaster & Gypsum Board	93.4	66.2	74.6	87.3	72.7	77.2	99.9	68.0	77.8	94.8	71.2	78.4	100.4	64.0	75.2	96.5	106.6	103.5
0950, 0980	Ceilings & Acoustic Treatment	94.1	66.2	76.3	93.2	72.7	80.1	87.6	68.0	75.1	92.0	71.2	78.7	89.4	64.0	73.2	95.4	106.6	102.5
0960	Flooring	117.0	56.9	100.4	106.4	56.9	92.7	89.9	64.5	82.9	107.3	86.1	101.4	122.0	43.4	100.3	98.5	107.9	101.1
0970, 0990	Wall Finishes & Painting/Coating	106.1	73.2	86.3	106.1	84.6	93.1	101.0	75.2	85.4	103.9	72.9	85.3	95.5	39.0	61.5	100.5	117.6	110.8
09	FINISHES	102.0	64.2	81.5	97.4	70.3	82.7	94.2	69.3	80.7	97.5	74.4	85.0	104.7	59.1	80.0	99.9	108.2	104.4
COVERS	DIVS. 10 - 14, 25, 28, 41, 43, 44	100.0	67.9	93.2	100.0	70.2	93.7	100.0	56.7	90.8	100.0	96.7	99.3	100.0	65.3	92.6	100.0	101.4	100.3
21, 22, 23	FIRE SUPPRESSION, PLUMBING & HVAC	99.8	59.7	83.3	99.8	62.1	84.3	99.8	84.7	93.6	100.0	86.4	94.4	100.1	49.5	79.3	100.1	112.9	105.3
26, 27, 3370	ELECTRICAL, COMMUNICATIONS & UTIL.	103.3	70.3	86.0	97.8	61.3	78.7	100.5	104.1	102.4	95.1	83.7	89.1	90.5	66.1	77.8	101.4	104.4	103.0
MF2004	WEIGHTED AVERAGE	97.8	67.0	84.2	96.1	68.6	84.0	98.4	83.8	92.0	96.8	88.2	93.0	96.8	65.2	82.9	98.7	108.2	102.9

DIVISION		MISSOURI						MONTANA									
		SPRINGFIELD			ST. JOSEPH			ST. LOUIS			BILLINGS			BUTTE			GREAT FALLS		
		MAT.	INST.	TOTAL	MAT.	INST.	TOTAL	MAT.	INST.	TOTAL	MAT.	INST.	TOTAL	MAT.	INST.	TOTAL	MAT.	INST.	TOTAL
015433	CONTRACTOR EQUIPMENT		102.1	102.1		101.9	101.9		107.8	107.8		99.0	99.0		98.8	98.8		98.8	98.8
0241, 31 - 34	SITE & INFRASTRUCTURE, DEMOLITION	96.9	93.6	94.6	93.9	93.3	93.5	95.2	97.2	96.6	93.7	96.9	95.9	103.4	96.2	98.4	107.3	96.9	100.0
0310	Concrete Forming & Accessories	103.6	71.2	75.6	102.6	76.0	79.6	104.4	103.9	103.9	99.8	63.2	68.2	89.6	66.8	69.9	99.7	63.9	68.7
0320	Concrete Reinforcing	96.7	112.0	104.5	95.8	99.7	97.8	95.6	108.7	102.3	93.9	67.3	80.3	101.8	67.3	84.2	93.8	67.2	80.2
0330	Cast-in-Place Concrete	97.4	74.2	88.6	91.5	101.4	95.3	94.5	110.2	100.4	110.6	66.9	94.0	121.7	67.8	101.3	128.6	65.4	104.7
03	CONCRETE	96.0	80.9	89.0	93.6	90.2	92.0	93.1	107.8	100.0	103.0	66.3	85.8	106.4	68.2	88.5	111.5	66.0	90.3
04	MASONRY	88.9	81.5	84.5	97.7	86.9	91.2	97.6	110.0	105.0	126.2	69.4	92.0	121.6	71.6	91.6	126.0	70.4	92.6
05	METALS	94.0	104.8	97.6	97.0	104.6	99.6	98.8	119.8	105.9	105.6	84.0	98.3	99.0	84.1	94.0	102.0	83.7	95.8
06	WOOD, PLASTICS & COMPOSITES	101.2	69.5	83.3	99.3	71.4	83.5	100.3	101.8	101.1	101.0	62.2	79.0	90.2	67.0	77.1	102.1	62.2	79.5
07	THERMAL & MOISTURE PROTECTION	99.1	75.1	90.0	99.0	89.1	95.3	100.8	107.3	103.2	102.2	70.2	90.1	102.1	71.0	90.4	102.6	65.6	88.6
08	OPENINGS	96.7	78.5	92.2	95.3	86.5	93.1	97.2	108.8	100.1	97.9	61.0	88.7	95.8	62.9	87.7	99.3	60.2	89.6
0920	Plaster & Gypsum Board	97.8	68.5	77.5	101.9	70.4	80.1	107.9	101.8	103.7	95.8	61.3	71.9	94.0	66.3	74.8	101.4	61.3	73.7
0950, 0980	Ceilings & Acoustic Treatment	92.0	68.5	77.0	94.5	70.4	79.2	92.8	101.8	98.5	104.5	61.3	77.0	109.6	66.3	82.0	111.3	61.3	79.5
0960	Flooring	110.0	43.4	91.6	101.5	93.2	99.2	97.3	98.6	97.6	109.4	75.7	100.1	106.1	79.9	98.9	114.7	79.9	105.1
0970, 0990	Wall Finishes & Painting/Coating	98.2	55.8	72.7	96.0	82.5	87.9	101.0	109.0	105.8	100.7	52.0	71.4	98.6	56.9	73.5	98.6	54.0	71.8
09	FINISHES	101.3	63.7	80.9	101.0	78.4	88.7	98.8	102.7	100.9	104.4	63.8	82.4	104.0	68.1	84.5	108.2	65.3	84.9
COVERS	DIVS. 10 - 14, 25, 28, 41, 43, 44	100.0	92.8	98.5	100.0	94.0	98.7	100.0	104.5	101.0	100.0	81.1	96.0	100.0	81.8	96.1	100.0	81.9	96.2
21, 22, 23	FIRE SUPPRESSION, PLUMBING & HVAC	100.0	67.7	86.7	100.1	85.0	93.9	99.8	105.0	102.0	100.0	72.4	88.6	100.1	66.0	86.1	100.1	70.1	87.8
26, 27, 3370	ELECTRICAL, COMMUNICATIONS & UTIL.	99.3	70.3	84.1	99.6	80.7	89.7	103.3	110.5	107.0	93.9	72.6	82.7	101.8	69.7	85.0	93.9	69.7	81.2
MF2004	WEIGHTED AVERAGE	97.7	77.8	88.9	98.0	87.2	93.3	98.6	107.2	102.4	102.0	72.9	89.2	102.0	72.3	88.9	103.4	72.2	89.6

DIVISION		MONTANA						NEBRASKA									
		HELENA			MISSOULA			GRAND ISLAND			LINCOLN			NORTH PLATTE			OMAHA		
		MAT.	INST.	TOTAL	MAT.	INST.	TOTAL	MAT.	INST.	TOTAL	MAT.	INST.	TOTAL	MAT.	INST.	TOTAL	MAT.	INST.	TOTAL
015433	CONTRACTOR EQUIPMENT		98.8	98.8		98.8	98.8		101.2	101.2		101.2	101.2		101.2	101.2		91.8	91.8
0241, 31 - 34	SITE & INFRASTRUCTURE, DEMOLITION	104.4	96.1	98.6	84.1	96.3	92.6	100.9	91.9	94.7	90.6	91.9	91.5	101.9	91.5	94.7	80.8	90.8	87.7
0310	Concrete Forming & Accessories	99.9	63.1	68.1	92.1	65.4	69.0	100.4	79.1	82.0	105.2	62.5	68.3	100.4	78.0	81.0	95.4	69.2	72.7
0320	Concrete Reinforcing	97.1	67.1	81.8	103.5	76.4	89.7	105.3	70.0	87.3	96.7	69.5	82.8	106.7	70.8	88.4	101.3	69.7	85.1
0330	Cast-in-Place Concrete	108.5	63.9	91.6	89.7	65.0	80.3	117.0	71.4	99.7	93.3	74.0	86.0	117.0	68.0	98.4	100.3	75.9	91.1
03	CONCRETE	102.1	65.2	84.9	84.1	68.4	76.7	108.2	75.4	92.8	94.1	68.8	82.3	108.4	73.8	92.2	96.7	71.9	85.1
04	MASONRY	118.2	68.9	88.5	146.2	70.2	100.5	108.1	81.6	92.2	96.9	71.7	81.8	92.3	93.4	93.0	102.4	77.2	87.2
05	METALS	101.1	83.4	95.1	95.6	88.0	93.0	89.9	84.7	88.2	94.8	83.7	91.0	90.2	85.3	88.5	94.1	75.4	87.8
06	WOOD, PLASTICS & COMPOSITES	102.2	62.2	79.5	93.3	65.7	77.7	97.9	81.0	88.4	100.6	58.8	76.9	97.8	81.0	88.3	91.3	68.2	78.2
07	THERMAL & MOISTURE PROTECTION	102.3	70.9	90.5	101.5	70.5	89.8	100.0	77.4	91.5	99.6	72.5	89.3	100.0	84.6	94.2	95.2	79.8	89.4
08	OPENINGS	98.7	59.0	88.9	95.9	64.5	88.1	90.4	74.5	86.5	95.9	58.2	86.6	89.7	73.8	85.8	98.9	68.0	91.2
0920	Plaster & Gypsum Board	93.2	61.3	71.1	94.5	65.0	74.1	95.7	80.3	85.1	91.4	57.4	67.9	95.7	80.3	85.1	94.1	67.7	75.8
0950, 0980	Ceilings & Acoustic Treatment	107.0	61.3	77.9	109.6	65.0	81.2	92.0	80.3	84.6	93.8	57.4	70.6	92.0	80.3	84.6	107.1	67.7	82.0
0960	Flooring	114.7	63.6	100.6	108.3	79.9	100.5	99.2	115.5	103.7	100.9	90.0	97.9	99.1	108.1	101.6	129.0	52.3	107.8
0970, 0990	Wall Finishes & Painting/Coating	98.6	48.0	68.2	98.6	52.9	71.1	96.2	75.1	83.5	96.2	86.5	90.4	96.2	70.8	80.9	167.0	69.3	108.2
09	FINISHES	106.0	61.5	81.8	103.5	66.6	83.5	98.3	85.0	91.1	98.3	69.6	82.7	98.2	82.5	89.7	119.9	65.4	90.3
COVERS	DIVS. 10 - 14, 25, 28, 41, 43, 44	100.0	81.4	96.1	100.0	81.1	96.0	100.0	83.2	96.4	100.0	80.6	95.9	100.0	77.4	95.2	100.0	80.4	95.8
21, 22, 23	FIRE SUPPRESSION, PLUMBING & HVAC	100.0	66.2	86.1	100.1	64.6	85.5	100.0	75.4	89.9	99.9	75.4	89.8	100.0	73.7	89.2	99.7	75.3	89.7
26, 27, 3370	ELECTRICAL, COMMUNICATIONS & UTIL.	99.7	69.7	84.0	99.5	73.7	86.0	89.7	71.1	79.9	103.1	71.1	86.3	91.8	86.4	89.0	102.9	80.0	90.9
MF2004	WEIGHTED AVERAGE	101.9	70.6	88.1	99.2	72.7	87.5	97.9	79.2	89.6	97.8	74.1	87.3	97.2	81.6	90.4	99.8	75.6	89.2

City Cost Indexes

DIVISION		NEVADA									NEW HAMPSHIRE								
		CARSON CITY			LAS VEGAS			RENO			MANCHESTER			NASHUA			PORTSMOUTH		
		MAT.	INST.	TOTAL	MAT.	INST.	TOTAL	MAT.	INST.	TOTAL	MAT.	INST.	TOTAL	MAT.	INST.	TOTAL	MAT.	INST.	TOTAL
015433	CONTRACTOR EQUIPMENT		100.7	100.7		100.7	100.7		100.7	100.7		100.6	100.6		100.6	100.6		100.6	100.6
0241, 31 - 34	SITE & INFRASTRUCTURE, DEMOLITION	68.0	101.5	91.3	62.6	105.1	92.2	60.8	101.5	89.2	81.9	98.8	93.7	84.4	98.8	94.5	78.8	100.2	93.7
0310	Concrete Forming & Accessories	100.7	88.9	90.5	103.0	116.4	114.6	98.1	89.0	90.2	96.0	83.0	84.7	97.9	83.0	85.0	88.3	87.1	87.2
0320	Concrete Reinforcing	103.5	107.8	105.7	95.5	125.7	110.9	98.2	124.2	111.5	110.6	93.4	101.8	110.6	93.4	101.8	88.5	93.5	91.0
0330	Cast-in-Place Concrete	104.8	87.7	98.4	102.0	116.3	107.4	111.5	87.8	102.5	102.0	113.3	106.3	89.1	113.3	98.3	84.5	119.6	97.8
03	CONCRETE	107.1	92.2	100.1	104.7	117.6	110.8	109.5	95.3	102.9	109.6	95.3	102.9	103.3	95.3	99.6	94.7	99.4	96.9
04	MASONRY	120.3	75.0	93.1	114.6	112.5	113.4	121.1	75.0	93.4	98.3	84.5	90.0	98.1	84.5	89.9	93.7	94.3	94.0
05	METALS	96.9	98.6	97.5	103.2	111.1	105.9	97.3	105.4	100.1	99.6	89.5	96.2	99.5	89.5	96.2	95.9	92.5	94.8
06	WOOD, PLASTICS & COMPOSITES	93.3	89.7	91.3	93.3	115.7	106.0	89.8	89.7	89.8	96.5	83.7	89.3	99.8	83.7	90.7	88.2	83.7	85.7
07	THERMAL & MOISTURE PROTECTION	97.0	84.4	92.2	112.1	104.9	109.4	98.3	84.4	93.0	95.5	90.1	93.5	96.1	90.1	93.9	95.8	114.5	102.9
08	OPENINGS	93.6	103.8	96.1	95.2	122.3	101.9	94.0	103.6	96.4	104.3	78.7	98.0	104.3	78.7	98.0	105.0	73.8	97.3
0920	Plaster & Gypsum Board	86.8	89.3	88.5	93.9	116.0	109.2	87.0	89.3	88.6	104.1	82.5	89.1	100.4	82.5	88.0	92.1	82.5	85.5
0950, 0980	Ceilings & Acoustic Treatment	121.0	89.3	100.8	135.6	116.0	123.1	131.3	89.3	104.5	96.1	82.5	87.4	98.7	82.5	88.4	90.9	82.5	85.6
0960	Flooring	104.0	55.1	90.5	97.9	106.4	100.2	103.6	55.1	90.2	98.6	96.4	98.0	97.6	96.4	97.3	92.4	96.4	93.5
0970, 0990	Wall Finishes & Painting/Coating	98.9	84.8	90.4	100.3	122.4	113.6	97.9	84.8	90.0	94.6	101.8	98.9	94.6	101.8	98.9	94.6	101.8	98.9
09	FINISHES	104.8	81.6	92.2	106.4	115.0	111.1	106.2	81.6	92.8	99.3	87.3	92.8	99.7	87.3	93.0	94.9	89.6	92.0
COVERS	DIVS. 10 - 14, 25, 28, 41, 43, 44	100.0	103.7	100.8	100.0	101.7	100.4	100.0	103.7	100.8	100.0	66.4	92.9	100.0	66.4	92.9	100.0	69.8	93.6
21, 22, 23	FIRE SUPPRESSION, PLUMBING & HVAC	99.9	82.3	92.7	100.1	111.1	104.6	100.0	82.5	92.8	100.0	85.6	94.1	100.1	85.6	94.2	100.1	90.5	96.2
26, 27, 3370	ELECTRICAL, COMMUNICATIONS & UTIL.	98.1	95.5	96.7	98.4	123.0	111.3	94.5	95.5	95.0	97.4	80.3	88.4	97.9	80.3	88.7	95.8	80.3	87.7
MF2004	WEIGHTED AVERAGE	99.9	89.5	95.3	101.0	113.8	106.7	100.0	90.5	95.8	100.4	87.0	94.5	99.9	87.0	94.2	97.3	90.9	94.4

DIVISION		NEW JERSEY																	
		CAMDEN			ELIZABETH			JERSEY CITY			NEWARK			PATERSON			TRENTON		
		MAT.	INST.	TOTAL	MAT.	INST.	TOTAL	MAT.	INST.	TOTAL	MAT.	INST.	TOTAL	MAT.	INST.	TOTAL	MAT.	INST.	TOTAL
015433	CONTRACTOR EQUIPMENT		98.7	98.7		100.6	100.6		98.7	98.7		100.6	100.6		100.6	100.6		98.2	98.2
0241, 31 - 34	SITE & INFRASTRUCTURE, DEMOLITION	87.3	104.4	99.2	99.9	105.0	103.4	87.3	105.4	99.9	102.6	105.0	104.2	98.5	105.5	103.4	87.1	105.3	99.8
0310	Concrete Forming & Accessories	97.9	118.8	115.9	104.9	119.4	117.4	97.9	119.5	116.6	94.6	119.4	116.0	96.7	119.4	116.3	95.1	119.0	115.7
0320	Concrete Reinforcing	110.6	121.9	116.4	85.2	123.8	104.9	110.6	123.8	117.4	110.6	123.8	117.4	110.6	123.8	117.4	115.0	115.0	115.0
0330	Cast-in-Place Concrete	77.5	133.4	98.7	85.3	130.4	102.4	77.5	130.5	97.5	99.1	130.4	110.9	98.8	130.4	110.8	88.2	131.4	104.6
03	CONCRETE	97.6	123.4	109.7	100.2	123.3	111.0	97.6	123.2	109.6	108.1	123.3	115.2	108.0	123.3	115.2	102.7	122.4	111.9
04	MASONRY	91.7	128.4	113.8	111.4	129.0	122.0	89.2	129.0	113.1	100.4	129.0	117.6	95.9	129.0	115.8	87.8	128.4	112.2
05	METALS	99.4	106.6	101.8	95.6	112.1	101.2	99.5	109.7	102.9	99.4	112.1	103.7	94.7	112.1	100.6	94.6	107.1	98.8
06	WOOD, PLASTICS & COMPOSITES	99.8	115.9	108.9	112.6	115.9	114.5	99.8	115.9	108.9	98.5	115.9	108.4	101.5	115.9	109.7	96.9	115.9	107.6
07	THERMAL & MOISTURE PROTECTION	95.7	125.0	106.8	96.3	128.6	108.5	95.7	128.6	108.1	95.8	128.6	108.2	96.3	125.3	107.2	91.9	125.4	104.5
08	OPENINGS	104.3	118.0	107.7	105.1	119.7	108.7	104.3	119.7	108.1	110.1	119.7	112.5	110.1	119.7	112.5	104.9	118.2	108.2
0920	Plaster & Gypsum Board	100.4	115.7	111.0	104.2	115.7	112.1	100.4	115.7	111.0	104.1	115.7	112.1	100.4	115.7	111.0	104.1	115.7	112.1
0950, 0980	Ceilings & Acoustic Treatment	98.7	115.7	109.5	90.9	115.7	106.7	98.7	115.7	109.5	96.1	115.7	108.6	98.7	115.7	109.5	96.1	115.7	108.6
0960	Flooring	97.6	136.9	108.4	102.1	145.8	114.1	97.6	145.8	110.9	97.8	145.8	111.0	97.6	145.8	110.9	98.0	145.8	111.2
0970, 0990	Wall Finishes & Painting/Coating	94.6	127.6	114.5	94.5	129.7	115.7	94.6	129.7	115.7	94.5	129.7	115.7	94.5	129.7	115.7	94.6	127.6	114.5
09	FINISHES	100.1	122.8	112.4	100.9	124.7	113.8	100.1	124.7	113.5	100.2	124.7	113.5	100.3	124.7	113.5	99.9	124.3	113.2
COVERS	DIVS. 10 - 14, 25, 28, 41, 43, 44	100.0	111.6	102.5	100.0	118.4	103.9	100.0	118.4	103.9	100.0	118.4	103.9	100.0	118.4	103.9	100.0	111.5	102.4
21, 22, 23	FIRE SUPPRESSION, PLUMBING & HVAC	100.1	126.3	110.9	100.1	127.1	111.2	100.1	129.6	112.3	100.2	127.1	111.3	100.1	129.6	112.2	100.2	126.1	110.9
26, 27, 3370	ELECTRICAL, COMMUNICATIONS & UTIL.	98.3	139.6	119.9	95.0	136.2	116.5	99.4	140.0	120.6	99.4	140.0	120.7	99.4	136.2	118.6	102.5	142.2	123.3
MF2004	WEIGHTED AVERAGE	99.0	122.9	109.5	100.1	123.8	110.6	99.0	124.7	110.3	101.9	124.3	111.8	100.9	124.3	111.2	99.0	123.4	109.8

DIVISION		NEW MEXICO															NEW YORK		
		ALBUQUERQUE			FARMINGTON			LAS CRUCES			ROSWELL			SANTA FE			ALBANY		
		MAT.	INST.	TOTAL	MAT.	INST.	TOTAL	MAT.	INST.	TOTAL	MAT.	INST.	TOTAL	MAT.	INST.	TOTAL	MAT.	INST.	TOTAL
015433	CONTRACTOR EQUIPMENT		113.4	113.4		113.4	113.4		87.8	87.8		113.4	113.4		113.4	113.4		114.6	114.6
0241, 31 - 34	SITE & INFRASTRUCTURE, DEMOLITION	77.2	108.6	99.1	83.6	108.6	101.0	87.8	86.5	86.9	88.2	108.6	102.4	82.9	108.6	100.8	73.0	107.5	97.0
0310	Concrete Forming & Accessories	98.8	68.8	72.8	98.9	68.8	72.9	95.6	67.1	70.9	98.9	68.5	72.6	98.1	68.4	72.7	98.1	96.2	96.5
0320	Concrete Reinforcing	98.9	71.5	84.9	108.1	71.5	89.4	101.6	48.2	74.3	107.2	48.5	77.2	106.1	71.5	88.4	94.7	92.9	93.7
0330	Cast-in-Place Concrete	104.2	76.1	93.6	105.2	76.1	94.2	91.5	67.9	82.6	96.9	76.0	89.0	120.2	76.1	103.5	80.9	103.4	89.4
03	CONCRETE	106.1	72.8	90.5	109.9	72.8	92.6	86.5	64.5	76.2	110.5	68.2	90.7	114.9	72.8	95.2	98.7	98.8	98.8
04	MASONRY	111.2	63.5	82.6	117.6	63.5	85.1	104.5	61.1	78.4	118.9	63.5	85.6	113.7	63.5	83.6	92.1	102.0	98.0
05	METALS	103.1	88.8	98.3	100.9	88.8	96.8	100.8	70.3	90.5	100.9	77.5	93.0	100.9	88.8	96.8	99.0	105.6	101.2
06	WOOD, PLASTICS & COMPOSITES	93.4	69.7	80.0	93.5	69.7	80.0	83.8	68.5	75.1	93.5	69.7	80.0	92.9	69.7	79.8	97.5	94.7	95.9
07	THERMAL & MOISTURE PROTECTION	97.8	72.8	88.4	98.1	72.8	88.5	84.5	67.4	78.1	98.3	72.8	88.7	97.7	72.8	88.3	90.2	96.0	92.4
08	OPENINGS	93.8	71.6	88.3	96.6	71.6	90.4	86.8	63.7	81.0	92.5	64.4	85.6	92.7	71.6	87.5	96.1	88.3	94.2
0920	Plaster & Gypsum Board	92.2	68.4	75.7	83.9	68.4	73.2	86.5	68.4	73.9	83.9	68.4	73.2	93.7	68.4	76.2	106.0	94.3	97.9
0950, 0980	Ceilings & Acoustic Treatment	125.7	68.4	89.2	121.0	68.4	87.5	113.2	68.4	84.7	121.0	68.4	87.5	121.0	68.4	87.5	93.6	94.3	94.0
0960	Flooring	101.2	66.3	91.6	103.6	66.3	93.3	137.0	62.6	116.4	103.6	66.3	93.3	103.6	66.3	93.3	85.0	102.6	89.8
0970, 0990	Wall Finishes & Painting/Coating	102.4	53.1	72.7	97.9	53.1	70.9	90.4	53.1	67.9	97.9	53.1	70.9	97.9	53.1	70.9	82.9	82.7	82.8
09	FINISHES	105.6	66.7	84.5	104.2	66.7	83.8	118.9	65.0	89.7	104.7	66.7	84.1	105.5	66.7	84.4	94.4	95.9	95.2
COVERS	DIVS. 10 - 14, 25, 28, 41, 43, 44	100.0	71.3	93.9	100.0	71.3	93.9	100.0	68.1	93.2	100.0	71.3	93.9	100.0	71.3	93.9	100.0	97.9	99.6
21, 22, 23	FIRE SUPPRESSION, PLUMBING & HVAC	100.1	70.8	88.0	100.0	70.8	88.0	100.3	70.2	87.9	100.0	70.6	87.9	100.0	70.8	88.0	100.1	93.1	97.2
26, 27, 3370	ELECTRICAL, COMMUNICATIONS & UTIL.	89.5	70.9	79.8	87.6	70.9	78.9	88.4	54.9	70.9	88.4	70.9	79.3	100.9	70.9	85.3	104.7	93.4	98.8
MF2004	WEIGHTED AVERAGE	99.8	74.8	88.8	100.4	74.8	89.1	96.8	66.7	83.5	100.4	72.8	88.2	101.8	74.8	89.9	97.7	97.6	97.7

City Cost Indexes

		NEW YORK																	
	DIVISION	BINGHAMTON			BUFFALO			HICKSVILLE			NEW YORK			RIVERHEAD			ROCHESTER		
		MAT.	INST.	TOTAL	MAT.	INST.	TOTAL	MAT.	INST.	TOTAL	MAT.	INST.	TOTAL	MAT.	INST.	TOTAL	MAT.	INST.	TOTAL
015433	CONTRACTOR EQUIPMENT		114.9	114.9		96.3	96.3		115.8	115.8		115.5	115.5		115.8	115.8		117.1	117.1
0241, 31 - 34	SITE & INFRASTRUCTURE, DEMOLITION	95.1	91.5	92.6	97.2	96.7	96.9	114.2	128.9	124.5	128.2	126.8	127.2	115.4	128.9	124.8	73.0	109.1	98.1
0310	Concrete Forming & Accessories	102.4	82.5	85.2	98.3	112.5	110.6	93.2	146.8	139.6	111.6	186.1	176.0	97.4	146.7	140.1	99.6	96.6	97.0
0320	Concrete Reinforcing	93.8	93.1	93.4	93.7	97.5	95.7	95.4	187.7	142.5	102.0	188.0	145.9	97.2	184.6	141.8	94.5	89.8	92.1
0330	Cast-in-Place Concrete	99.8	93.9	97.6	107.6	116.9	111.2	93.4	156.6	117.3	107.6	171.3	131.7	94.9	156.5	118.2	96.8	99.8	97.9
03	CONCRETE	98.5	90.1	94.6	109.2	110.5	109.8	103.3	156.6	128.2	115.3	179.1	145.1	103.9	155.9	128.2	114.6	97.3	106.5
04	MASONRY	107.7	87.3	95.5	107.0	117.0	113.0	111.4	158.5	139.7	107.4	170.5	145.3	117.1	158.5	142.0	103.1	101.6	102.2
05	METALS	95.7	114.8	102.2	99.4	93.2	97.3	105.0	140.0	116.8	115.5	143.4	125.0	105.4	138.8	116.7	96.0	106.4	99.5
06	WOOD, PLASTICS & COMPOSITES	107.3	82.3	93.1	101.2	112.3	107.5	90.1	145.4	121.4	112.8	190.4	156.8	95.2	145.4	123.6	98.1	96.7	97.3
07	THERMAL & MOISTURE PROTECTION	104.9	84.1	97.1	101.7	108.4	104.2	107.4	150.5	123.7	110.9	166.2	131.8	108.0	150.5	124.0	97.2	96.7	97.0
08	OPENINGS	91.0	80.7	88.4	94.5	100.4	96.0	88.9	153.1	104.8	97.4	179.4	117.7	88.9	150.8	104.2	98.3	89.3	96.0
0920	Plaster & Gypsum Board	119.4	81.3	93.0	99.4	112.4	108.4	106.0	146.7	134.2	128.6	192.7	172.9	108.3	146.7	134.9	98.4	96.6	97.2
0950, 0980	Ceilings & Acoustic Treatment	98.7	81.3	87.6	103.9	112.4	109.3	82.4	146.7	123.3	105.3	192.7	161.0	84.1	146.7	123.9	103.1	96.6	99.0
0960	Flooring	95.7	89.1	93.9	91.4	115.0	97.9	95.4	153.7	111.5	93.6	178.4	117.1	96.6	153.7	112.4	81.6	103.5	87.7
0970, 0990	Wall Finishes & Painting/Coating	86.9	86.2	86.5	91.2	111.8	103.6	112.3	149.3	134.5	91.2	156.8	130.7	112.3	149.3	134.5	90.4	96.7	94.2
09	FINISHES	97.6	83.8	90.1	95.0	113.7	105.1	105.8	147.7	128.5	108.5	183.1	149.0	106.8	147.7	129.0	94.6	98.2	96.5
COVERS	DIVS. 10 - 14, 25, 28, 41, 43, 44	100.0	94.1	98.7	100.0	105.2	101.1	100.0	127.3	105.8	100.0	138.7	108.2	100.0	127.3	105.8	100.0	99.7	99.9
21, 22, 23	FIRE SUPPRESSION, PLUMBING & HVAC	100.4	82.6	93.1	100.0	94.7	97.8	99.8	142.6	117.4	100.3	163.8	126.4	99.9	141.2	116.9	100.0	91.9	96.7
26, 27, 3370	ELECTRICAL, COMMUNICATIONS & UTIL.	100.9	91.6	96.0	98.3	95.5	96.8	99.8	148.9	125.4	107.2	178.5	144.5	101.6	148.8	126.3	99.8	90.9	95.1
MF2004	WEIGHTED AVERAGE	98.8	89.5	94.7	100.3	102.8	101.4	101.6	146.4	121.4	107.0	166.3	133.2	102.4	145.9	121.6	99.8	97.4	98.7

		NEW YORK																	
	DIVISION	SCHENECTADY			SYRACUSE			UTICA			WATERTOWN			WHITE PLAINS			YONKERS		
		MAT.	INST.	TOTAL	MAT.	INST.	TOTAL	MAT.	INST.	TOTAL	MAT.	INST.	TOTAL	MAT.	INST.	TOTAL	MAT.	INST.	TOTAL
015433	CONTRACTOR EQUIPMENT		114.6	114.6		114.6	114.6		114.6	114.6		114.6	114.6		115.2	115.2		115.2	115.2
0241, 31 - 34	SITE & INFRASTRUCTURE, DEMOLITION	72.9	107.5	97.0	94.0	106.5	102.7	71.1	104.6	94.4	79.2	107.0	98.5	118.5	121.0	120.3	127.6	120.8	122.9
0310	Concrete Forming & Accessories	102.1	95.9	96.7	101.1	88.4	90.1	102.8	82.4	85.2	89.4	91.7	91.4	111.0	132.7	129.7	111.3	132.7	129.8
0320	Concrete Reinforcing	93.4	92.9	93.1	94.7	90.5	92.5	94.7	90.8	92.7	95.3	90.1	92.6	94.9	187.1	142.0	98.7	187.2	143.9
0330	Cast-in-Place Concrete	91.6	102.8	95.9	92.6	97.7	94.5	84.7	90.9	87.0	98.6	98.4	98.5	94.8	136.0	110.4	106.1	136.0	117.4
03	CONCRETE	104.1	98.4	101.4	102.6	92.9	98.0	100.9	87.9	94.8	114.7	94.5	105.2	103.4	143.0	121.9	114.2	143.0	127.7
04	MASONRY	91.6	101.0	97.2	98.4	98.9	98.7	90.2	83.5	86.2	91.3	97.5	95.0	101.9	134.2	121.3	107.0	134.2	123.4
05	METALS	99.1	105.6	101.3	99.0	103.0	100.3	96.9	102.7	98.9	97.0	102.6	98.9	101.3	136.9	113.3	111.1	137.0	119.9
06	WOOD, PLASTICS & COMPOSITES	103.4	94.7	98.5	103.4	86.8	94.0	103.4	84.0	92.4	86.9	90.6	89.0	114.0	135.4	126.1	113.8	135.4	126.0
07	THERMAL & MOISTURE PROTECTION	90.6	95.5	92.5	101.2	96.4	99.4	90.4	90.3	90.4	90.6	94.6	92.1	111.3	140.5	122.3	111.5	140.5	122.5
08	OPENINGS	96.1	88.3	94.2	94.1	80.2	90.7	96.1	81.0	92.3	96.1	86.1	93.6	91.9	147.7	105.7	95.5	149.5	108.9
0920	Plaster & Gypsum Board	113.7	94.3	100.2	113.7	86.1	94.6	113.7	83.3	92.6	106.3	90.1	95.1	121.9	136.0	131.7	128.1	136.0	133.6
0950, 0980	Ceilings & Acoustic Treatment	98.7	94.3	95.9	98.7	86.1	90.7	98.7	83.3	88.9	98.7	90.1	93.2	81.1	136.0	116.1	103.6	136.0	124.2
0960	Flooring	84.6	102.6	89.6	86.3	89.1	87.1	84.6	93.0	86.9	77.0	93.0	81.5	90.1	162.9	110.2	89.6	162.9	109.9
0970, 0990	Wall Finishes & Painting/Coating	82.9	82.7	82.8	89.1	93.2	91.5	82.9	82.5	82.7	82.9	89.2	86.7	89.2	149.3	125.3	89.2	149.3	125.3
09	FINISHES	96.2	95.7	95.9	97.8	88.8	92.9	96.2	84.3	89.7	93.7	91.8	92.7	100.1	139.7	121.6	106.6	139.7	124.6
COVERS	DIVS. 10 - 14, 25, 28, 41, 43, 44	100.0	97.6	99.5	100.0	96.5	99.2	100.0	91.7	98.2	100.0	90.5	98.0	100.0	121.3	104.5	100.0	122.1	104.7
21, 22, 23	FIRE SUPPRESSION, PLUMBING & HVAC	100.2	92.6	97.1	100.2	85.9	94.3	100.2	82.8	93.1	100.2	78.9	91.5	100.5	124.5	110.4	100.5	124.5	110.4
26, 27, 3370	ELECTRICAL, COMMUNICATIONS & UTIL.	100.9	93.4	97.0	100.9	91.3	95.9	98.8	90.4	94.4	100.9	90.4	95.4	96.6	152.0	125.5	105.1	152.0	129.6
MF2004	WEIGHTED AVERAGE	98.2	97.3	97.8	99.3	93.0	96.5	97.2	88.9	93.5	99.1	91.9	95.9	100.7	136.0	116.3	105.7	136.1	119.2

		NORTH CAROLINA																	
	DIVISION	ASHEVILLE			CHARLOTTE			DURHAM			FAYETTEVILLE			GREENSBORO			RALEIGH		
		MAT.	INST.	TOTAL	MAT.	INST.	TOTAL	MAT.	INST.	TOTAL	MAT.	INST.	TOTAL	MAT.	INST.	TOTAL	MAT.	INST.	TOTAL
015433	CONTRACTOR EQUIPMENT		95.2	95.2		95.2	95.2		100.8	100.8		100.8	100.8		100.8	100.8		100.8	100.8
0241, 31 - 34	SITE & INFRASTRUCTURE, DEMOLITION	106.6	75.1	84.6	110.9	75.1	86.0	107.0	84.0	91.0	105.7	84.1	90.7	106.9	84.1	91.0	111.1	84.2	92.3
0310	Concrete Forming & Accessories	95.4	39.3	46.9	99.4	41.3	49.1	97.6	39.9	47.7	95.0	40.7	48.0	97.4	40.6	48.3	97.8	41.5	49.1
0320	Concrete Reinforcing	100.5	43.0	71.2	101.0	40.1	69.9	102.2	50.3	75.7	104.4	50.4	76.8	101.0	59.0	79.5	101.0	50.4	75.1
0330	Cast-in-Place Concrete	106.9	46.7	84.1	120.1	48.7	93.1	108.6	46.9	85.3	112.1	50.0	88.6	107.8	48.2	85.3	123.2	53.5	96.8
03	CONCRETE	106.1	44.5	77.3	112.2	45.5	81.0	106.6	46.2	78.4	108.1	47.6	79.8	106.0	48.6	79.2	113.6	49.2	83.5
04	MASONRY	88.3	40.3	59.4	95.1	44.1	64.5	93.9	36.3	59.2	92.0	41.3	61.5	91.9	36.9	58.9	94.2	42.2	62.9
05	METALS	85.0	74.3	81.4	88.9	73.4	83.6	97.7	77.5	90.9	101.3	77.6	93.3	92.1	80.9	88.3	89.9	77.6	85.7
06	WOOD, PLASTICS & COMPOSITES	95.3	38.8	63.3	100.3	41.0	66.7	97.5	40.3	65.1	94.2	40.3	63.7	97.2	40.2	64.9	97.5	40.9	65.5
07	THERMAL & MOISTURE PROTECTION	99.3	43.3	78.1	97.0	44.9	77.3	100.0	45.4	79.3	99.2	43.9	78.3	99.8	42.9	78.3	99.1	44.1	78.3
08	OPENINGS	94.3	38.6	80.5	98.5	40.1	84.0	98.5	42.7	84.6	94.4	42.7	81.6	98.5	45.0	85.2	95.2	42.8	82.2
0920	Plaster & Gypsum Board	103.3	36.8	57.2	105.1	39.0	59.3	106.6	38.3	59.3	107.2	38.3	59.5	108.6	38.2	59.9	105.1	39.0	59.3
0950, 0980	Ceilings & Acoustic Treatment	92.6	36.8	57.0	95.2	39.0	59.4	96.1	38.3	59.3	93.5	38.3	58.3	96.1	38.2	59.2	95.2	39.0	59.4
0960	Flooring	106.8	44.9	89.7	110.4	45.0	92.3	110.9	44.9	92.7	106.9	44.9	89.8	110.9	41.5	91.7	110.9	44.9	92.7
0970, 0990	Wall Finishes & Painting/Coating	116.0	35.6	67.6	116.0	40.6	70.6	116.0	37.3	68.6	116.0	35.6	67.6	116.0	34.6	67.0	116.0	39.2	69.8
09	FINISHES	99.4	39.7	67.0	101.2	41.7	68.9	101.9	40.1	68.4	100.1	40.5	67.8	102.2	39.8	68.3	101.6	41.6	69.0
COVERS	DIVS. 10 - 14, 25, 28, 41, 43, 44	100.0	70.6	93.8	100.0	71.1	93.9	100.0	75.6	94.8	100.0	76.4	95.0	100.0	70.8	93.8	100.0	76.8	95.1
21, 22, 23	FIRE SUPPRESSION, PLUMBING & HVAC	100.2	39.0	75.1	100.2	39.2	75.0	100.3	37.8	74.6	100.1	39.0	75.0	100.2	39.1	75.1	100.1	37.3	74.2
26, 27, 3370	ELECTRICAL, COMMUNICATIONS & UTIL.	98.3	38.6	67.1	100.5	50.3	74.2	99.0	48.9	72.8	97.1	44.4	69.6	98.0	38.8	67.1	104.7	38.5	70.1
MF2004	WEIGHTED AVERAGE	97.4	47.2	75.3	99.8	49.7	77.7	100.3	49.7	78.0	100.0	50.1	78.0	99.2	49.2	77.1	100.3	49.4	77.8

City Cost Indexes

		NORTH CAROLINA						NORTH DAKOTA										
	DIVISION	WILMINGTON			WINSTON-SALEM			BISMARCK			FARGO			GRAND FORKS			MINOT		
		MAT.	INST.	TOTAL	MAT.	INST.	TOTAL	MAT.	INST.	TOTAL	MAT.	INST.	TOTAL	MAT.	INST.	TOTAL	MAT.	INST.	TOTAL
015433	CONTRACTOR EQUIPMENT		95.2	95.2		100.8	100.8		98.8	98.8		98.8	98.8		98.8	98.8		98.8	98.8
0241, 31 - 34	SITE & INFRASTRUCTURE, DEMOLITION	108.0	75.0	85.1	107.2	84.1	91.1	99.3	96.7	97.5	98.8	96.7	97.3	108.1	94.1	98.3	105.3	96.7	99.3
0310	Concrete Forming & Accessories	96.6	39.9	47.5	98.6	39.6	47.6	97.2	43.3	50.6	98.5	43.6	51.0	94.6	37.5	45.2	91.1	57.8	62.3
0320	Concrete Reinforcing	101.3	48.6	74.4	101.0	39.9	69.8	101.8	66.3	83.7	93.8	65.6	79.4	100.3	66.7	83.1	103.7	66.7	84.8
0330	Cast-in-Place Concrete	106.4	47.9	84.3	110.5	45.6	86.0	102.7	50.1	82.8	98.9	51.9	81.1	106.7	47.6	84.3	106.7	49.0	84.9
03	CONCRETE	106.0	46.2	78.0	107.4	43.6	77.6	99.8	51.4	77.1	107.1	52.0	81.3	107.2	47.8	79.4	105.7	57.5	83.2
04	MASONRY	77.6	38.2	53.9	92.1	38.0	59.6	104.8	55.8	75.3	106.4	41.8	67.6	107.4	62.7	80.5	107.3	62.7	80.5
05	METALS	84.9	76.6	82.1	89.9	72.8	84.1	92.5	77.7	87.5	95.0	76.6	88.8	92.5	73.7	86.1	92.8	78.4	88.0
06	WOOD, PLASTICS & COMPOSITES	96.9	39.6	64.4	97.2	39.5	64.6	84.7	38.7	58.7	84.6	39.3	58.9	81.7	35.6	55.6	77.6	58.1	66.6
07	THERMAL & MOISTURE PROTECTION	99.3	43.3	78.1	99.8	42.6	78.1	102.4	48.5	82.0	100.8	46.3	80.2	103.2	50.7	83.4	103.2	53.8	83.8
08	OPENINGS	94.5	41.2	81.3	98.5	39.4	83.8	99.3	44.3	85.7	99.2	44.6	85.7	99.3	39.0	84.4	99.4	54.8	88.4
0920	Plaster & Gypsum Board	104.4	37.6	58.1	108.6	37.5	59.4	99.4	37.2	56.3	99.4	37.8	56.7	105.0	34.0	55.9	103.2	57.1	71.3
0950, 0980	Ceilings & Acoustic Treatment	93.5	37.6	57.9	96.1	37.5	58.8	139.6	37.2	74.4	139.6	37.8	74.7	142.2	34.0	73.3	142.2	57.1	88.0
0960	Flooring	107.7	44.9	90.4	110.9	44.9	92.7	115.2	64.7	101.2	114.8	38.7	93.7	113.4	38.7	92.8	110.8	77.6	101.6
0970, 0990	Wall Finishes & Painting/Coating	116.0	34.5	67.0	116.0	36.3	68.0	98.6	32.7	58.9	98.6	67.3	79.8	98.6	25.0	54.3	98.6	27.4	55.7
09	FINISHES	100.0	39.9	67.4	102.2	40.0	68.4	114.9	44.8	76.8	114.8	43.7	76.2	116.3	35.6	72.5	115.0	58.2	84.2
COVERS	DIVS. 10 - 14, 25, 28, 41, 43, 44	100.0	76.1	94.9	100.0	70.5	93.7	100.0	75.3	94.8	100.0	75.4	94.8	100.0	37.9	86.8	100.0	77.6	95.2
21, 22, 23	FIRE SUPPRESSION, PLUMBING & HVAC	100.2	38.7	74.9	100.2	37.7	74.5	100.1	57.4	82.6	100.1	60.9	84.0	100.2	37.4	74.4	100.2	56.9	82.4
26, 27, 3370	ELECTRICAL, COMMUNICATIONS & UTIL.	98.5	37.3	66.5	98.0	42.3	68.9	101.3	63.7	81.6	102.8	63.7	82.3	97.0	60.0	77.7	100.5	64.1	81.8
MF2004	WEIGHTED AVERAGE	96.9	47.6	75.1	99.1	47.8	76.5	100.4	60.4	82.8	101.8	59.5	83.2	101.5	52.8	80.0	101.5	64.6	85.2

		OHIO																	
	DIVISION	AKRON			CANTON			CINCINNATI			CLEVELAND			COLUMBUS			DAYTON		
		MAT.	INST.	TOTAL	MAT.	INST.	TOTAL	MAT.	INST.	TOTAL	MAT.	INST.	TOTAL	MAT.	INST.	TOTAL	MAT.	INST.	TOTAL
015433	CONTRACTOR EQUIPMENT		96.6	96.6		96.6	96.6		101.1	101.1		96.9	96.9		95.6	95.6		96.1	96.1
0241, 31 - 34	SITE & INFRASTRUCTURE, DEMOLITION	91.4	104.0	100.2	91.5	103.9	100.1	75.7	105.7	96.6	91.3	104.7	100.6	90.8	100.9	97.8	74.9	105.8	96.4
0310	Concrete Forming & Accessories	99.7	93.6	94.5	99.7	82.1	84.5	97.5	81.9	84.0	99.8	99.2	99.3	97.1	87.0	88.3	97.5	77.7	80.3
0320	Concrete Reinforcing	101.4	91.4	96.3	101.4	78.8	89.9	93.2	84.5	88.7	102.0	91.8	96.8	93.2	84.5	88.8	93.2	76.6	84.7
0330	Cast-in-Place Concrete	98.2	100.6	99.1	99.2	96.5	98.2	80.2	84.9	82.0	96.4	107.8	100.7	85.9	94.9	89.3	74.6	89.1	80.1
03	CONCRETE	96.5	94.8	95.7	97.0	85.9	91.8	86.4	83.6	85.1	95.7	99.9	97.7	89.4	88.9	89.1	83.8	81.2	82.5
04	MASONRY	90.1	97.0	94.2	90.8	86.5	88.2	76.5	89.2	84.1	94.8	105.0	100.9	94.2	98.3	96.7	76.0	86.1	82.1
05	METALS	93.6	80.2	89.1	93.6	74.4	87.1	95.0	86.2	92.0	95.2	83.8	91.4	95.2	81.4	90.5	94.2	75.8	88.0
06	WOOD, PLASTICS & COMPOSITES	94.0	92.8	93.3	94.3	81.9	87.3	97.2	79.3	87.1	93.0	96.7	95.1	103.3	83.4	92.0	98.6	74.8	85.1
07	THERMAL & MOISTURE PROTECTION	105.0	96.7	101.9	106.0	91.1	100.4	101.5	91.8	97.9	103.9	108.6	105.7	100.1	94.9	98.2	106.2	86.6	98.8
08	OPENINGS	107.4	92.8	103.8	101.6	73.1	94.5	98.4	80.0	93.8	97.8	94.9	97.1	100.9	82.1	96.2	98.7	74.8	92.8
0920	Plaster & Gypsum Board	87.3	92.2	90.7	88.2	81.0	83.2	88.6	79.0	82.0	86.4	96.3	93.2	86.8	82.9	84.0	88.6	74.3	78.7
0950, 0980	Ceilings & Acoustic Treatment	86.0	92.2	90.0	88.2	81.0	82.8	92.0	79.0	83.7	84.3	96.3	91.9	83.6	82.9	83.1	93.0	74.3	81.1
0960	Flooring	102.1	84.7	97.3	102.3	77.7	95.5	106.4	92.1	102.5	102.0	102.8	102.2	92.7	87.9	91.3	109.1	82.3	101.9
0970, 0990	Wall Finishes & Painting/Coating	104.2	104.4	104.3	104.2	84.2	92.1	105.2	87.8	94.7	104.2	111.8	108.8	94.2	93.8	93.9	105.2	82.6	91.6
09	FINISHES	98.1	93.2	95.5	98.3	81.5	89.2	97.5	83.9	90.1	97.6	101.0	99.5	93.0	87.4	90.0	98.6	78.6	87.7
COVERS	DIVS. 10 - 14, 25, 28, 41, 43, 44	100.0	85.0	96.8	100.0	64.2	92.4	100.0	88.5	97.6	100.0	97.8	99.5	100.0	94.9	98.9	100.0	87.7	97.4
21, 22, 23	FIRE SUPPRESSION, PLUMBING & HVAC	99.9	91.9	96.6	99.9	77.1	90.5	100.0	84.6	93.6	99.9	99.4	99.7	99.8	90.9	96.1	100.8	81.0	92.6
26, 27, 3370	ELECTRICAL, COMMUNICATIONS & UTIL.	101.3	84.3	92.4	100.4	91.7	95.9	95.3	81.2	87.9	100.9	106.5	103.8	99.4	84.5	91.6	93.9	81.4	87.4
MF2004	WEIGHTED AVERAGE	98.7	91.9	95.7	98.2	83.7	91.8	94.9	86.4	91.1	98.0	100.2	99.0	96.9	89.8	93.7	94.7	82.9	89.5

		OHIO												OKLAHOMA					
	DIVISION	LORAIN			SPRINGFIELD			TOLEDO			YOUNGSTOWN			ENID			LAWTON		
		MAT.	INST.	TOTAL	MAT.	INST.	TOTAL	MAT.	INST.	TOTAL	MAT.	INST.	TOTAL	MAT.	INST.	TOTAL	MAT.	INST.	TOTAL
015433	CONTRACTOR EQUIPMENT		96.6	96.6		96.1	96.1		98.3	98.3		96.6	96.6		79.5	79.5		80.4	80.4
0241, 31 - 34	SITE & INFRASTRUCTURE, DEMOLITION	90.8	103.5	99.6	75.1	104.3	95.5	90.0	100.6	97.4	91.2	104.5	100.5	107.9	88.8	94.6	103.2	90.3	94.2
0310	Concrete Forming & Accessories	99.8	91.6	92.7	97.5	83.7	85.6	97.1	99.0	98.7	99.7	86.8	88.5	94.2	35.1	43.1	97.3	47.3	54.1
0320	Concrete Reinforcing	101.4	91.7	96.4	93.2	84.0	88.5	93.2	91.3	92.2	101.4	87.1	94.1	99.2	79.2	89.2	99.4	79.7	89.4
0330	Cast-in-Place Concrete	93.5	104.3	97.6	76.6	89.1	81.4	85.9	103.1	92.4	97.3	100.7	98.6	98.4	46.5	78.8	95.2	46.5	76.8
03	CONCRETE	94.2	95.2	94.7	84.6	85.3	84.9	89.3	98.4	93.6	96.1	91.0	93.7	94.9	48.0	73.0	91.2	53.5	73.5
04	MASONRY	86.6	99.9	94.6	76.2	86.1	82.2	103.1	99.8	101.1	90.4	92.9	91.9	103.7	57.8	76.1	97.8	57.8	73.7
05	METALS	94.2	81.9	90.0	94.2	78.4	88.8	95.0	87.2	92.3	93.6	78.9	88.6	94.0	69.8	85.8	97.6	69.8	88.2
06	WOOD, PLASTICS & COMPOSITES	94.0	89.9	91.7	100.2	83.1	90.5	123.3	99.3	101.0	94.0	84.9	88.9	102.4	31.6	62.3	105.1	48.2	72.9
07	THERMAL & MOISTURE PROTECTION	106.0	103.7	105.1	106.1	87.5	99.1	101.6	105.6	103.1	106.2	93.7	101.5	98.4	59.7	83.7	98.2	61.4	84.3
08	OPENINGS	101.6	91.3	99.1	96.6	81.1	92.7	98.9	94.8	97.9	101.6	86.5	97.9	95.7	46.0	83.4	97.5	54.9	86.9
0920	Plaster & Gypsum Board	87.3	89.3	88.7	88.6	82.9	84.7	86.4	99.3	95.3	87.3	84.2	85.1	81.8	30.3	46.1	84.1	47.3	58.6
0950, 0980	Ceilings & Acoustic Treatment	86.0	89.3	88.1	93.0	82.9	86.6	83.6	99.3	93.6	86.0	84.2	84.8	85.3	30.3	50.2	93.0	47.3	63.9
0960	Flooring	102.3	107.5	103.8	109.1	82.8	101.9	91.9	94.9	92.7	102.3	89.5	98.8	108.8	44.3	91.0	111.6	44.3	93.0
0970, 0990	Wall Finishes & Painting/Coating	104.2	111.8	108.8	105.2	82.6	91.6	94.2	102.8	99.4	104.2	93.8	97.9	98.6	56.9	73.6	98.6	56.9	73.6
09	FINISHES	98.2	96.6	97.3	98.6	83.4	90.4	92.7	98.9	96.1	98.2	87.6	92.4	94.2	36.8	63.0	96.5	46.5	69.4
COVERS	DIVS. 10 - 14, 25, 28, 41, 43, 44	100.0	95.1	98.9	100.0	88.6	97.6	100.0	86.5	97.1	100.0	83.0	96.4	100.0	72.8	94.2	100.0	74.7	94.6
21, 22, 23	FIRE SUPPRESSION, PLUMBING & HVAC	99.9	79.7	91.6	100.8	81.0	92.7	99.8	95.1	97.9	99.9	93.1	97.1	100.2	62.0	84.5	100.2	62.0	84.5
26, 27, 3370	ELECTRICAL, COMMUNICATIONS & UTIL.	100.6	91.0	95.6	93.9	79.0	86.1	99.6	100.9	100.3	100.6	82.6	91.2	96.7	69.9	82.7	98.5	64.0	80.5
MF2004	WEIGHTED AVERAGE	97.7	91.5	95.0	94.6	84.3	90.1	97.1	97.1	97.1	98.1	89.6	94.3	97.7	59.8	80.9	97.9	61.7	81.9

City Cost Indexes

DIVISION		OKLAHOMA									OREGON								
		MUSKOGEE			OKLAHOMA CITY			TULSA			EUGENE			MEDFORD			PORTLAND		
		MAT.	INST.	TOTAL	MAT.	INST.	TOTAL	MAT.	INST.	TOTAL	MAT.	INST.	TOTAL	MAT.	INST.	TOTAL	MAT.	INST.	TOTAL
015433	CONTRACTOR EQUIPMENT		88.0	88.0		80.7	80.7		88.0	88.0		99.3	99.3		99.3	99.3		99.3	99.3
0241, 31 - 34	SITE & INFRASTRUCTURE, DEMOLITION	93.7	84.7	87.4	102.6	90.8	94.4	100.3	86.1	90.4	105.3	104.0	104.4	114.2	104.0	107.0	108.2	104.0	105.3
0310	Concrete Forming & Accessories	98.4	31.0	40.1	96.4	41.1	48.5	98.2	40.8	48.6	106.9	95.7	97.2	103.1	95.7	96.7	108.2	96.1	97.7
0320	Concrete Reinforcing	99.2	32.6	65.2	99.4	79.7	89.4	99.4	79.6	89.3	110.1	90.9	100.3	107.2	90.9	98.9	110.9	91.2	100.8
0330	Cast-in-Place Concrete	88.5	35.9	68.6	98.0	49.3	79.6	96.4	45.3	77.1	101.0	99.0	100.3	104.3	99.0	102.3	103.8	99.2	102.1
03	CONCRETE	86.6	34.3	62.1	92.5	51.6	73.4	91.8	51.0	72.7	105.9	95.7	101.1	112.8	95.7	104.8	107.4	96.0	102.1
04	MASONRY	113.6	47.0	73.6	97.5	59.6	74.7	96.9	59.6	74.5	111.6	94.4	101.3	108.7	94.4	100.1	112.5	99.2	104.5
05	METALS	93.9	56.2	81.1	99.4	69.8	89.4	97.2	81.3	91.8	90.3	91.2	90.6	89.9	91.2	90.3	91.1	91.9	91.4
06	WOOD, PLASTICS & COMPOSITES	106.6	31.3	63.9	104.3	38.7	67.2	105.7	39.9	68.4	94.0	95.8	95.0	88.5	95.8	92.6	95.1	95.8	95.5
07	THERMAL & MOISTURE PROTECTION	98.0	41.8	76.8	95.0	61.2	82.2	98.0	56.3	82.3	106.5	88.0	99.5	107.0	90.0	100.6	106.5	96.5	102.7
08	OPENINGS	95.7	30.4	79.5	97.5	49.8	85.7	97.5	50.0	85.7	99.2	96.8	98.6	101.8	96.8	100.5	96.8	96.8	96.8
0920	Plaster & Gypsum Board	84.5	29.8	46.6	79.9	37.6	50.6	84.1	38.7	52.6	104.0	95.5	98.1	102.0	95.5	97.5	107.5	95.5	99.2
0950, 0980	Ceilings & Acoustic Treatment	93.0	29.8	52.7	91.3	37.6	57.1	93.0	38.7	58.4	101.9	95.5	97.8	115.3	95.5	102.7	103.9	95.5	98.5
0960	Flooring	112.7	35.7	91.4	111.6	44.3	93.0	111.4	46.1	93.4	112.6	95.1	107.8	110.1	95.1	105.9	110.0	95.1	105.9
0970, 0990	Wall Finishes & Painting/Coating	98.6	30.5	57.6	98.6	56.9	73.6	98.6	45.0	66.4	110.1	65.5	83.2	110.1	60.0	80.0	109.6	70.3	86.0
09	FINISHES	96.2	32.1	61.4	95.4	41.4	66.1	95.9	41.1	66.2	107.2	92.5	99.2	110.1	91.9	100.2	107.3	93.1	99.6
COVERS	DIVS. 10 - 14, 25, 28, 41, 43, 44	100.0	60.2	91.5	100.0	74.3	94.5	100.0	62.8	92.1	100.0	95.6	99.1	100.0	95.6	99.1	100.0	95.7	99.1
21, 22, 23	FIRE SUPPRESSION, PLUMBING & HVAC	100.2	24.2	68.9	100.1	62.9	84.8	100.2	50.0	79.5	100.1	93.8	97.5	100.1	93.8	97.5	100.1	98.9	99.4
26, 27, 3370	ELECTRICAL, COMMUNICATIONS & UTIL.	96.3	29.5	61.4	103.7	69.9	86.0	98.5	38.5	67.1	100.8	91.4	95.9	104.7	81.7	92.7	101.0	95.9	98.3
MF2004	WEIGHTED AVERAGE	96.9	39.6	71.6	98.6	61.7	82.3	97.7	54.8	78.8	100.9	94.3	98.0	102.7	92.9	98.4	101.1	96.8	99.2

DIVISION		OREGON			PENNSYLVANIA														
		SALEM			ALLENTOWN			ALTOONA			ERIE			HARRISBURG			PHILADELPHIA		
		MAT.	INST.	TOTAL	MAT.	INST.	TOTAL	MAT.	INST.	TOTAL	MAT.	INST.	TOTAL	MAT.	INST.	TOTAL	MAT.	INST.	TOTAL
015433	CONTRACTOR EQUIPMENT		99.3	99.3		114.6	114.6		114.6	114.6		114.6	114.6		113.9	113.9		96.9	96.9
0241, 31 - 34	SITE & INFRASTRUCTURE, DEMOLITION	101.9	104.0	103.3	92.9	106.4	102.3	97.1	106.2	103.5	93.6	106.8	102.8	83.5	105.3	98.7	101.0	98.1	98.9
0310	Concrete Forming & Accessories	105.8	95.9	97.2	100.5	117.6	115.3	87.5	81.0	81.8	100.4	87.4	89.2	93.4	88.5	89.2	102.1	138.8	133.8
0320	Concrete Reinforcing	111.2	91.2	101.0	94.7	111.1	103.1	91.8	91.1	91.5	93.8	93.1	93.4	94.7	101.2	98.0	95.6	153.4	125.1
0330	Cast-in-Place Concrete	91.3	99.1	94.2	83.9	104.7	91.7	93.5	88.1	91.4	91.9	86.2	89.8	93.2	96.3	94.4	99.2	133.2	112.1
03	CONCRETE	101.2	95.9	98.7	97.1	112.6	104.4	91.7	86.8	89.4	90.3	89.4	89.9	99.3	95.0	97.3	100.6	139.2	118.6
04	MASONRY	116.8	99.2	106.3	95.4	104.3	100.8	97.5	71.8	82.0	87.5	90.8	89.5	95.3	92.5	93.6	97.1	133.8	119.2
05	METALS	90.7	91.7	91.0	99.2	124.6	107.8	93.2	112.0	99.6	93.4	112.7	99.9	101.2	119.2	107.3	107.5	138.1	117.8
06	WOOD, PLASTICS & COMPOSITES	91.5	95.8	93.9	102.8	121.8	113.6	82.6	83.9	83.3	98.7	86.3	91.7	97.3	87.7	91.8	101.3	139.6	123.0
07	THERMAL & MOISTURE PROTECTION	104.1	92.4	99.7	101.2	119.9	108.3	100.6	88.6	96.1	101.0	93.4	98.1	107.2	110.6	108.5	99.3	138.5	114.1
08	OPENINGS	98.7	96.8	98.2	94.1	117.9	100.0	88.6	87.8	88.4	88.7	88.8	88.8	94.1	92.8	93.8	98.3	148.5	110.7
0920	Plaster & Gypsum Board	96.2	95.5	95.7	111.1	122.2	118.8	104.4	83.1	89.7	111.1	85.7	93.5	108.4	87.0	93.6	103.5	140.6	129.2
0950, 0980	Ceilings & Acoustic Treatment	102.4	95.5	98.0	90.1	122.2	110.6	93.6	83.1	86.9	90.1	85.7	87.3	91.8	87.0	88.8	94.0	140.6	123.7
0960	Flooring	112.8	95.1	107.9	86.3	94.9	88.6	80.2	49.1	71.6	87.8	73.1	83.7	86.8	86.4	86.7	85.6	143.3	101.5
0970, 0990	Wall Finishes & Painting/Coating	110.1	65.5	83.2	89.1	78.4	82.6	85.1	105.5	97.4	95.1	92.2	93.4	89.1	87.7	88.3	94.0	146.9	125.8
09	FINISHES	106.1	92.5	98.7	95.5	108.9	102.8	94.0	76.7	84.6	96.9	84.6	90.2	94.4	87.7	90.8	97.2	140.4	120.7
COVERS	DIVS. 10 - 14, 25, 28, 41, 43, 44	100.0	95.6	99.1	100.0	107.5	101.6	100.0	96.1	99.2	100.0	100.7	100.2	100.0	98.0	99.6	100.0	122.1	104.7
21, 22, 23	FIRE SUPPRESSION, PLUMBING & HVAC	100.0	93.8	97.5	100.2	113.3	105.6	99.8	85.5	93.9	99.8	93.5	97.2	100.1	91.1	96.4	100.0	131.7	113.0
26, 27, 3370	ELECTRICAL, COMMUNICATIONS & UTIL.	108.2	91.4	99.4	100.1	98.4	99.2	89.5	102.7	96.4	91.4	84.1	87.6	98.8	85.1	91.6	97.1	144.4	121.8
MF2004	WEIGHTED AVERAGE	101.0	95.0	98.4	98.2	110.5	103.6	94.9	90.3	92.9	94.7	93.1	94.0	98.5	95.1	97.0	100.3	134.2	115.3

DIVISION		PENNSYLVANIA												PUERTO RICO			RHODE ISLAND		
		PITTSBURGH			READING			SCRANTON			YORK			SAN JUAN			PROVIDENCE		
		MAT.	INST.	TOTAL	MAT.	INST.	TOTAL	MAT.	INST.	TOTAL	MAT.	INST.	TOTAL	MAT.	INST.	TOTAL	MAT.	INST.	TOTAL
015433	CONTRACTOR EQUIPMENT		112.9	112.9		117.7	117.7		114.6	114.6		113.9	113.9		89.7	89.7		102.2	102.2
0241, 31 - 34	SITE & INFRASTRUCTURE, DEMOLITION	109.4	107.4	108.0	98.4	111.4	107.5	93.3	106.6	102.6	82.1	105.3	98.3	119.8	90.8	99.6	81.2	104.0	97.1
0310	Concrete Forming & Accessories	98.1	96.6	96.8	101.6	89.5	91.1	100.7	90.5	91.9	83.1	88.8	88.0	92.6	18.1	28.1	99.1	117.1	114.7
0320	Concrete Reinforcing	93.7	103.5	98.7	93.9	101.3	97.7	94.7	107.6	101.3	93.8	101.2	97.5	198.7	12.8	103.8	110.6	114.7	112.7
0330	Cast-in-Place Concrete	97.3	98.6	97.8	72.5	95.4	81.2	87.6	93.1	89.7	82.6	96.7	87.9	101.1	31.1	74.6	95.1	115.4	102.8
03	CONCRETE	97.5	99.7	98.5	87.2	95.2	90.9	99.0	96.0	97.6	97.7	95.2	96.5	112.2	22.5	70.2	109.1	115.6	112.1
04	MASONRY	90.3	99.1	95.6	98.7	95.9	97.0	95.8	99.3	97.9	95.4	93.1	94.0	87.9	17.1	45.4	108.0	126.4	119.1
05	METALS	94.2	118.6	102.5	103.7	119.7	109.1	101.3	122.4	108.4	98.7	119.2	105.6	113.3	33.7	86.4	99.6	110.3	103.2
06	WOOD, PLASTICS & COMPOSITES	96.6	94.1	95.2	101.5	87.3	93.5	102.8	87.7	88.5	89.5	87.7	88.5	93.8	17.8	50.8	99.3	116.0	108.7
07	THERMAL & MOISTURE PROTECTION	103.5	101.5	102.7	98.9	112.4	104.0	101.1	100.8	101.0	100.5	110.9	104.1	125.2	21.7	86.1	94.0	117.6	102.9
08	OPENINGS	92.6	103.0	95.2	95.9	96.3	96.0	94.1	94.8	94.3	90.9	92.8	91.3	149.0	15.2	115.8	99.2	115.5	103.3
0920	Plaster & Gypsum Board	96.1	93.7	94.4	107.9	86.8	93.3	113.7	87.4	95.5	107.5	87.0	93.3	116.3	15.3	46.4	98.1	115.8	110.3
0950, 0980	Ceilings & Acoustic Treatment	97.1	93.7	94.9	84.4	86.8	85.9	98.7	87.4	91.5	91.0	87.0	88.5	229.1	15.3	92.9	96.2	115.8	108.7
0960	Flooring	98.4	99.2	98.6	83.7	90.4	85.5	86.3	109.7	92.7	82.1	90.4	84.4	202.3	14.1	150.3	98.0	131.7	107.3
0970, 0990	Wall Finishes & Painting/Coating	97.5	107.2	103.4	92.5	98.6	96.2	89.1	97.6	94.2	89.1	87.7	88.3	203.7	13.3	89.1	96.7	124.2	113.2
09	FINISHES	100.6	97.2	98.8	95.0	89.4	92.0	97.8	94.1	95.8	92.8	88.6	90.5	205.4	17.0	103.1	98.8	120.8	110.8
COVERS	DIVS. 10 - 14, 25, 28, 41, 43, 44	100.0	103.7	100.8	100.0	100.8	100.2	100.0	101.9	100.4	100.0	98.2	99.6	100.0	17.9	82.6	100.0	108.7	101.8
21, 22, 23	FIRE SUPPRESSION, PLUMBING & HVAC	99.9	98.6	99.4	100.1	109.9	104.1	100.0	101.7	100.7	100.2	91.4	96.6	103.6	14.0	66.8	100.0	111.5	104.7
26, 27, 3370	ELECTRICAL, COMMUNICATIONS & UTIL.	96.3	102.7	99.7	99.2	92.6	95.7	100.2	96.7	98.4	93.3	85.1	89.0	124.0	13.3	66.1	97.0	100.5	98.8
MF2004	WEIGHTED AVERAGE	97.7	102.1	99.6	97.9	101.5	99.5	99.0	100.9	99.8	96.5	95.4	96.0	121.2	24.5	78.5	100.2	112.8	105.8

City Cost Indexes

DIVISION		SOUTH CAROLINA																	SOUTH DAKOTA		
		CHARLESTON			COLUMBIA			FLORENCE			GREENVILLE			SPARTANBURG			ABERDEEN		
		MAT.	INST.	TOTAL	MAT.	INST.	TOTAL	MAT.	INST.	TOTAL	MAT.	INST.	TOTAL	MAT.	INST.	TOTAL	MAT.	INST.	TOTAL
015433	CONTRACTOR EQUIPMENT		100.5	100.5		100.5	100.5		100.5	100.5		100.5	100.5		100.5	100.5		98.8	98.8
0241, 31 - 34	SITE & INFRASTRUCTURE, DEMOLITION	101.1	84.2	89.3	102.4	84.1	89.7	112.0	84.1	92.6	106.5	83.8	90.7	106.2	83.8	90.6	93.7	94.9	94.5
0310	Concrete Forming & Accessories	97.2	43.2	50.5	98.8	42.2	49.9	86.2	42.4	48.3	96.8	42.2	49.6	99.7	42.2	50.0	98.1	40.9	48.6
0320	Concrete Reinforcing	101.0	61.8	81.0	101.0	61.6	80.8	100.6	61.7	80.7	100.5	49.2	74.3	100.5	49.2	74.3	100.4	42.2	70.7
0330	Cast-in-Place Concrete	99.0	51.2	80.9	106.0	51.8	85.5	84.3	51.0	71.7	84.3	50.9	71.7	84.3	50.9	71.7	101.7	47.1	81.0
03	CONCRETE	101.6	51.1	78.0	105.2	50.8	79.8	101.7	50.6	77.8	99.9	48.2	75.8	100.1	48.2	75.9	99.8	44.7	74.0
04	MASONRY	100.2	41.0	64.6	94.6	38.8	61.1	82.3	41.0	57.5	80.2	41.0	56.6	82.3	41.0	57.5	108.8	58.1	78.3
05	METALS	85.8	79.2	83.5	85.8	78.3	83.3	84.7	78.6	82.7	84.7	74.1	81.1	84.7	74.1	81.1	98.1	67.1	87.6
06	WOOD, PLASTICS & COMPOSITES	97.2	42.2	66.0	99.6	41.7	66.8	83.7	41.7	59.9	97.2	41.7	65.8	101.0	41.7	67.4	100.1	40.1	66.2
07	THERMAL & MOISTURE PROTECTION	99.4	44.4	78.6	98.6	43.1	77.6	94.4	44.4	78.8	99.6	44.4	78.7	99.7	44.4	78.8	102.0	50.8	82.7
08	OPENINGS	98.5	46.1	85.5	98.5	45.8	85.4	94.4	45.8	82.4	94.3	42.9	81.6	94.3	42.9	81.6	95.0	41.0	81.6
0920	Plaster & Gypsum Board	111.4	40.3	62.2	105.1	39.8	59.9	99.2	39.8	58.1	105.0	39.8	59.8	107.1	39.8	60.5	103.5	38.6	58.6
0950, 0980	Ceilings & Acoustic Treatment	96.1	40.3	60.5	95.2	39.8	59.9	93.5	39.8	59.3	92.6	39.8	58.9	92.6	39.8	58.9	110.0	38.6	64.6
0960	Flooring	110.9	62.7	97.6	107.4	47.0	90.7	100.9	47.0	86.0	108.1	61.6	95.3	109.8	61.6	96.5	115.4	54.2	98.5
0970, 0990	Wall Finishes & Painting/Coating	116.0	43.5	72.4	113.5	43.5	71.4	116.0	43.5	72.4	116.0	43.5	72.4	116.0	43.5	72.4	98.6	41.2	64.1
09	FINISHES	102.9	46.5	72.2	100.5	43.0	69.3	98.9	43.5	68.8	101.2	46.0	71.2	102.0	46.0	71.6	108.0	43.0	72.7
COVERS	DIVS. 10 - 14, 25, 28, 41, 43, 44	100.0	70.6	93.8	100.0	70.3	93.7	100.0	70.3	93.7	100.0	70.3	93.7	100.0	70.3	93.7	100.0	47.6	88.9
21, 22, 23	FIRE SUPPRESSION, PLUMBING & HVAC	100.2	42.1	76.3	100.0	37.3	74.2	100.2	37.3	74.4	100.2	35.8	73.7	100.2	35.8	73.7	100.0	45.1	77.4
26, 27, 3370	ELECTRICAL, COMMUNICATIONS & UTIL.	97.7	99.6	98.7	101.0	50.9	74.8	95.5	48.4	70.9	97.8	52.4	74.0	97.8	52.4	74.0	97.6	57.4	76.6
MF2004	WEIGHTED AVERAGE	98.1	59.7	81.2	98.3	51.2	77.5	96.3	51.2	76.4	96.3	50.8	76.2	96.5	50.8	76.3	100.0	54.0	79.7

DIVISION		SOUTH DAKOTA									TENNESSEE								
		PIERRE			RAPID CITY			SIOUX FALLS			CHATTANOOGA			JACKSON			JOHNSON CITY		
		MAT.	INST.	TOTAL	MAT.	INST.	TOTAL	MAT.	INST.	TOTAL	MAT.	INST.	TOTAL	MAT.	INST.	TOTAL	MAT.	INST.	TOTAL
015433	CONTRACTOR EQUIPMENT		98.8	98.8		98.8	98.8		99.8	99.8		104.0	104.0		104.1	104.1		97.7	97.7
0241, 31 - 34	SITE & INFRASTRUCTURE, DEMOLITION	92.9	95.0	94.3	91.8	95.0	94.0	93.5	96.7	95.7	102.6	98.7	99.9	100.9	95.7	97.3	111.3	87.1	94.5
0310	Concrete Forming & Accessories	96.9	41.8	49.2	103.8	39.6	48.3	97.6	44.5	51.6	94.0	57.2	62.1	88.4	44.3	50.3	82.9	41.1	46.8
0320	Concrete Reinforcing	100.0	53.2	76.1	93.8	53.0	73.0	93.8	53.2	73.1	79.3	65.2	72.1	81.3	61.1	71.0	79.9	61.5	70.5
0330	Cast-in-Place Concrete	93.5	45.2	75.2	98.0	45.8	78.3	93.5	46.2	75.6	100.7	64.4	87.0	104.4	42.0	80.8	81.0	48.8	68.8
03	CONCRETE	95.0	46.5	72.3	96.7	45.7	72.9	92.9	48.1	72.0	90.9	62.9	77.8	93.7	48.7	72.7	96.3	49.8	74.5
04	MASONRY	104.6	57.8	76.4	105.8	53.7	74.5	101.6	55.9	74.1	96.7	54.2	71.2	110.5	38.2	67.0	108.4	37.2	65.6
05	METALS	98.1	72.4	89.4	100.0	72.4	90.7	100.2	72.9	91.0	100.0	89.4	96.4	95.0	84.1	91.3	97.2	86.4	93.5
06	WOOD, PLASTICS & COMPOSITES	98.0	41.0	65.7	103.4	37.2	65.9	98.8	43.9	67.7	101.7	58.2	77.1	86.8	45.8	63.6	76.1	41.2	56.3
07	THERMAL & MOISTURE PROTECTION	102.0	49.2	82.0	102.4	50.7	82.9	104.0	52.5	84.5	95.0	61.2	82.2	91.4	46.0	74.3	90.4	54.3	76.7
08	OPENINGS	98.2	44.8	85.0	99.3	42.8	85.3	99.5	46.4	86.4	100.4	58.2	89.9	101.0	52.8	89.1	96.1	48.8	84.4
0920	Plaster & Gypsum Board	96.5	39.5	57.1	103.2	35.6	56.4	100.2	42.5	60.2	78.3	57.5	63.9	86.7	44.7	57.6	91.7	40.0	55.9
0950, 0980	Ceilings & Acoustic Treatment	104.8	39.5	63.2	113.5	35.6	63.9	107.4	42.5	66.1	95.7	57.5	71.3	93.7	44.7	62.5	91.9	40.0	58.8
0960	Flooring	114.7	38.6	93.7	114.7	68.2	101.8	114.7	73.7	103.4	98.7	61.0	88.3	91.4	20.6	71.8	93.0	43.8	79.4
0970, 0990	Wall Finishes & Painting/Coating	98.6	49.5	69.1	98.6	49.5	69.1	98.6	49.5	69.1	105.0	58.6	77.1	96.3	55.1	71.5	102.6	46.8	69.0
09	FINISHES	105.6	41.3	70.7	108.4	45.3	74.1	106.8	50.1	76.0	94.0	57.9	74.4	93.6	41.1	65.1	95.7	41.3	66.2
COVERS	DIVS. 10 - 14, 25, 28, 41, 43, 44	100.0	71.4	93.9	100.0	71.2	93.9	100.0	71.8	94.0	100.0	47.7	88.9	100.0	49.2	89.2	100.0	49.4	89.3
21, 22, 23	FIRE SUPPRESSION, PLUMBING & HVAC	100.0	44.7	77.2	100.0	45.2	77.5	100.0	43.5	76.7	99.9	61.9	84.3	99.9	47.1	78.2	99.7	49.2	78.9
26, 27, 3370	ELECTRICAL, COMMUNICATIONS & UTIL.	98.7	49.8	73.1	93.9	49.8	70.8	101.1	66.4	83.0	104.5	70.0	86.4	102.1	65.6	83.0	93.8	52.0	71.9
MF2004	WEIGHTED AVERAGE	99.3	54.2	79.4	99.8	54.2	79.7	99.8	57.8	81.2	98.6	66.9	84.6	98.4	56.0	79.7	97.9	54.2	78.6

DIVISION		TENNESSEE									TEXAS								
		KNOXVILLE			MEMPHIS			NASHVILLE			ABILENE			AMARILLO			AUSTIN		
		MAT.	INST.	TOTAL	MAT.	INST.	TOTAL	MAT.	INST.	TOTAL	MAT.	INST.	TOTAL	MAT.	INST.	TOTAL	MAT.	INST.	TOTAL
015433	CONTRACTOR EQUIPMENT		97.7	97.7		101.6	101.6		105.2	105.2		88.0	88.0		88.0	88.0		87.2	87.2
0241, 31 - 34	SITE & INFRASTRUCTURE, DEMOLITION	88.6	87.8	88.1	91.5	91.9	91.8	97.1	99.7	98.9	100.7	85.6	90.2	101.9	85.6	91.4	91.1	85.8	87.4
0310	Concrete Forming & Accessories	93.0	45.8	52.2	93.4	63.2	67.3	94.0	66.8	70.5	95.8	41.7	49.0	96.4	55.3	60.8	90.4	53.8	58.7
0320	Concrete Reinforcing	79.3	63.1	71.1	84.9	66.3	75.4	84.9	65.5	75.0	96.6	53.9	74.8	96.6	50.3	72.9	93.7	50.2	71.5
0330	Cast-in-Place Concrete	94.6	51.7	78.4	90.7	64.3	80.7	92.6	68.4	83.4	100.0	45.3	79.3	99.1	53.4	81.8	87.9	50.4	73.7
03	CONCRETE	88.1	53.1	71.7	87.0	65.7	77.0	88.4	68.4	79.1	93.1	46.4	71.2	92.7	54.5	74.8	82.1	52.7	68.4
04	MASONRY	74.5	57.0	64.0	87.9	67.0	75.3	81.7	62.1	70.0	100.5	53.7	72.3	100.5	51.1	70.8	92.8	50.2	67.2
05	METALS	100.8	87.9	96.4	101.5	90.0	97.6	102.1	88.3	97.5	100.8	68.7	89.9	100.8	67.5	89.5	96.8	64.9	86.0
06	WOOD, PLASTICS & COMPOSITES	88.4	41.2	61.7	92.8	64.5	76.8	99.3	68.5	81.9	102.1	40.6	67.3	100.6	58.0	76.5	93.8	56.6	72.7
07	THERMAL & MOISTURE PROTECTION	88.6	61.1	78.2	89.2	67.2	80.9	94.3	64.1	82.9	98.1	50.8	80.2	98.4	50.6	80.3	87.6	52.7	74.4
08	OPENINGS	93.0	49.1	82.1	99.2	65.0	90.8	98.4	67.4	90.7	93.7	44.3	81.5	93.7	52.6	83.6	100.5	55.2	89.3
0920	Plaster & Gypsum Board	98.9	40.0	58.1	87.2	63.8	71.0	95.2	68.0	76.3	84.1	39.4	53.2	79.4	57.3	64.1	85.2	55.9	64.9
0950, 0980	Ceilings & Acoustic Treatment	92.8	40.0	59.1	92.9	63.8	74.4	95.4	68.0	77.9	93.0	39.4	58.9	91.3	57.3	69.6	85.7	55.9	66.7
0960	Flooring	98.4	58.2	87.3	93.0	38.9	78.1	101.3	69.2	92.4	111.6	67.5	99.4	111.4	61.7	97.7	94.2	42.6	79.9
0970, 0990	Wall Finishes & Painting/Coating	102.6	46.8	69.0	97.5	58.2	73.9	105.2	72.6	85.6	97.2	54.6	71.6	97.2	43.7	65.0	95.3	44.4	64.7
09	FINISHES	90.2	47.1	66.8	89.6	57.7	72.3	99.9	68.1	82.6	95.9	47.3	69.5	95.2	55.3	73.6	87.6	50.5	67.4
COVERS	DIVS. 10 - 14, 25, 28, 41, 43, 44	100.0	53.6	90.1	100.0	81.0	96.0	100.0	76.6	95.0	100.0	76.2	95.0	100.0	77.7	95.3	100.0	77.9	95.3
21, 22, 23	FIRE SUPPRESSION, PLUMBING & HVAC	99.7	64.1	85.0	99.8	72.5	88.6	99.8	78.4	91.0	100.2	39.6	75.3	100.1	52.4	80.5	100.1	55.6	81.8
26, 27, 3370	ELECTRICAL, COMMUNICATIONS & UTIL.	97.5	57.6	77.8	101.6	68.6	84.4	101.4	64.4	82.0	98.6	52.5	74.5	103.3	70.8	86.3	102.7	65.8	83.4
MF2004	WEIGHTED AVERAGE	94.9	61.8	80.2	96.4	71.6	85.5	97.5	73.9	87.1	98.2	52.8	78.2	98.6	60.5	81.8	95.5	59.4	79.6

689

City Cost Indexes

TEXAS

	DIVISION	BEAUMONT			CORPUS CHRISTI			DALLAS			EL PASO			FORT WORTH			HOUSTON		
		MAT.	INST.	TOTAL	MAT.	INST.	TOTAL	MAT.	INST.	TOTAL	MAT.	INST.	TOTAL	MAT.	INST.	TOTAL	MAT.	INST.	TOTAL
015433	CONTRACTOR EQUIPMENT		89.1	89.1		94.9	94.9		97.9	97.9		88.0	88.0		88.0	88.0		98.4	98.4
0241, 31 - 34	SITE & INFRASTRUCTURE, DEMOLITION	96.9	86.0	89.3	123.9	81.4	94.2	115.9	87.1	95.8	98.9	85.1	89.3	103.3	86.0	91.3	120.3	84.5	95.4
0310	Concrete Forming & Accessories	102.2	52.5	59.2	97.0	39.5	47.3	94.6	58.7	63.6	95.1	47.2	53.6	91.9	58.1	62.7	91.8	64.8	68.4
0320	Concrete Reinforcing	98.3	44.6	70.9	92.9	49.5	70.8	94.6	53.9	73.8	96.6	49.4	72.5	96.6	53.8	74.7	100.1	64.7	82.0
0330	Cast-in-Place Concrete	98.0	54.7	81.6	95.0	47.0	76.8	102.0	57.8	85.3	91.8	42.6	73.2	104.3	53.5	85.0	101.1	69.3	89.0
03	CONCRETE	94.8	52.7	75.1	88.0	45.9	68.3	94.0	59.0	77.6	89.0	46.9	69.3	94.9	56.4	76.9	94.2	67.8	81.8
04	MASONRY	99.6	59.4	75.4	82.9	51.9	64.3	106.1	60.9	79.0	98.1	50.5	69.5	95.2	60.9	74.6	94.4	61.0	74.3
05	METALS	99.5	66.0	88.2	96.4	76.2	89.6	100.4	80.7	93.8	100.5	64.1	88.2	97.8	68.9	88.0	103.7	88.6	98.6
06	WOOD, PLASTICS & COMPOSITES	115.8	52.9	80.2	112.7	39.2	71.1	103.0	59.1	78.1	101.7	49.9	72.4	103.5	58.9	78.2	101.4	65.1	80.9
07	THERMAL & MOISTURE PROTECTION	106.5	58.4	88.3	96.7	47.7	78.2	89.8	63.3	79.7	94.4	54.5	79.3	99.1	55.1	82.5	97.8	66.7	86.1
08	OPENINGS	95.7	49.3	84.2	107.6	40.6	91.0	103.3	54.8	91.3	93.7	45.9	81.9	88.2	54.8	79.9	104.7	64.0	94.6
0920	Plaster & Gypsum Board	98.0	52.1	66.2	89.9	37.8	53.8	89.1	58.3	67.8	82.3	49.0	59.3	83.2	58.3	66.0	96.0	64.6	74.3
0950, 0980	Ceilings & Acoustic Treatment	103.0	52.1	70.6	87.4	37.8	55.8	94.2	58.3	71.3	91.3	49.0	64.4	93.0	58.3	70.9	103.6	64.6	78.7
0960	Flooring	110.7	72.7	100.2	106.6	43.0	89.0	100.1	51.4	86.7	111.6	67.2	99.3	144.9	43.8	117.0	98.2	56.8	86.8
0970, 0990	Wall Finishes & Painting/Coating	94.5	52.9	69.4	108.1	62.8	80.8	104.2	55.5	74.9	97.2	38.0	61.6	98.6	55.5	72.7	102.0	61.5	77.6
09	FINISHES	93.5	56.0	73.2	95.2	41.9	66.3	98.4	56.8	75.8	95.2	50.2	70.7	106.6	55.1	78.7	97.8	62.7	78.8
COVERS	DIVS. 10 - 14, 25, 28, 41, 43, 44	100.0	80.0	95.7	100.0	77.1	95.1	100.0	81.4	96.0	100.0	78.1	95.4	100.0	81.0	96.0	100.0	84.2	96.7
21, 22, 23	FIRE SUPPRESSION, PLUMBING & HVAC	100.1	62.6	84.7	100.1	42.3	76.3	100.0	64.8	85.5	100.1	37.4	74.3	100.1	60.0	83.6	100.0	69.2	87.3
26, 27, 3370	ELECTRICAL, COMMUNICATIONS & UTIL.	95.2	62.5	78.1	96.2	53.3	73.8	97.0	71.4	83.6	99.7	54.6	76.1	99.6	63.5	80.7	100.6	81.7	
MF2004	WEIGHTED AVERAGE	98.1	62.1	82.2	97.8	52.7	77.9	99.6	66.8	85.2	97.4	52.7	77.6	98.2	62.7	82.6	100.0	70.3	86.9

TEXAS

	DIVISION	LAREDO			LUBBOCK			ODESSA			SAN ANTONIO			WACO			WICHITA FALLS		
		MAT.	INST.	TOTAL	MAT.	INST.	TOTAL	MAT.	INST.	TOTAL	MAT.	INST.	TOTAL	MAT.	INST.	TOTAL	MAT.	INST.	TOTAL
015433	CONTRACTOR EQUIPMENT		87.2	87.2		97.0	97.0		88.0	88.0		89.8	89.8		88.0	88.0		88.0	88.0
0241, 31 - 34	SITE & INFRASTRUCTURE, DEMOLITION	90.6	85.5	87.1	127.9	83.9	97.3	100.9	86.0	90.5	90.3	89.9	90.0	99.5	86.0	90.1	100.3	85.6	90.0
0310	Concrete Forming & Accessories	91.5	39.7	46.7	93.7	44.5	51.1	95.7	41.0	48.4	91.5	53.3	58.4	96.2	42.3	49.6	96.2	40.2	47.8
0320	Concrete Reinforcing	93.7	49.4	71.1	97.9	53.7	75.3	96.6	53.5	74.6	100.1	48.9	74.0	96.6	50.2	72.9	96.6	53.4	74.5
0330	Cast-in-Place Concrete	73.1	59.5	68.0	100.2	50.0	81.2	100.0	46.0	79.6	71.7	65.5	69.5	90.2	54.7	76.8	96.4	46.8	77.6
03	CONCRETE	79.5	49.4	65.4	92.0	50.0	72.3	93.1	46.3	71.2	79.7	57.6	69.4	88.3	49.2	70.0	91.3	46.2	70.2
04	MASONRY	90.6	49.7	66.0	99.8	48.5	69.0	100.5	45.9	67.7	90.5	58.3	71.1	97.4	58.8	74.2	97.9	59.7	74.9
05	METALS	97.8	63.9	86.4	104.1	80.9	96.2	100.0	67.7	89.1	98.6	66.8	87.8	100.7	66.8	89.2	100.6	69.5	90.1
06	WOOD, PLASTICS & COMPOSITES	93.1	39.1	62.5	101.1	45.4	69.6	102.1	41.1	67.6	93.1	51.6	69.6	108.8	38.2	68.8	108.8	38.1	68.8
07	THERMAL & MOISTURE PROTECTION	92.4	50.6	76.6	88.6	50.8	74.3	98.1	45.7	78.3	92.4	63.3	81.4	98.6	50.6	80.4	98.6	53.9	81.7
08	OPENINGS	101.2	41.2	86.3	104.0	45.7	89.5	93.7	42.9	81.2	103.1	51.7	90.4	88.2	39.2	76.0	88.2	43.5	77.1
0920	Plaster & Gypsum Board	86.4	37.8	52.8	84.6	44.2	56.7	84.1	39.9	53.5	86.4	50.6	61.6	84.1	36.9	51.5	84.1	36.8	51.4
0950, 0980	Ceilings & Acoustic Treatment	87.4	37.8	55.8	94.7	44.2	62.6	93.0	39.9	59.2	87.4	50.6	64.0	93.0	36.9	57.3	93.0	36.8	57.2
0960	Flooring	91.2	43.0	77.9	103.6	41.6	86.5	111.6	41.1	92.1	91.2	68.0	84.8	144.9	38.2	115.4	146.1	77.9	127.2
0970, 0990	Wall Finishes & Painting/Coating	95.3	52.1	69.3	108.4	35.5	64.5	97.2	35.5	60.1	95.3	52.1	69.3	98.6	36.6	61.3	102.1	49.8	70.6
09	FINISHES	87.3	40.6	62.0	98.1	42.3	67.8	95.9	39.7	65.4	87.3	54.8	69.7	106.4	39.7	70.2	107.0	47.4	74.6
COVERS	DIVS. 10 - 14, 25, 28, 41, 43, 44	100.0	74.8	94.6	100.0	76.5	95.0	100.0	77.5	95.2	100.0	79.4	95.6	100.0	78.6	95.5	100.0	75.0	94.7
21, 22, 23	FIRE SUPPRESSION, PLUMBING & HVAC	100.0	40.9	75.7	99.7	45.7	77.5	100.2	37.3	74.3	100.0	65.4	85.8	100.2	54.0	81.2	100.2	48.1	78.8
26, 27, 3370	ELECTRICAL, COMMUNICATIONS & UTIL.	98.1	62.4	79.4	97.2	46.9	70.9	98.7	44.1	70.2	98.5	62.4	79.6	98.7	71.3	84.4	100.7	61.1	80.0
MF2004	WEIGHTED AVERAGE	94.9	53.0	76.4	99.9	53.8	79.6	98.1	49.3	76.5	95.3	63.7	81.3	97.8	58.1	80.3	98.4	56.5	79.9

UTAH / VERMONT

	DIVISION	LOGAN			OGDEN			PROVO			SALT LAKE CITY			BURLINGTON			RUTLAND		
		MAT.	INST.	TOTAL	MAT.	INST.	TOTAL	MAT.	INST.	TOTAL	MAT.	INST.	TOTAL	MAT.	INST.	TOTAL	MAT.	INST.	TOTAL
015433	CONTRACTOR EQUIPMENT		99.9	99.9		99.9	99.9		98.8	98.8		99.9	99.9		100.6	100.6		100.6	100.6
0241, 31 - 34	SITE & INFRASTRUCTURE, DEMOLITION	87.7	98.4	95.2	76.6	98.4	91.8	84.2	96.7	92.9	76.3	98.4	91.7	76.5	97.3	91.0	76.6	97.3	91.1
0310	Concrete Forming & Accessories	104.8	52.9	59.9	104.9	52.9	59.9	106.4	53.0	60.1	107.4	53.0	60.3	95.2	53.4	59.1	98.2	54.0	59.9
0320	Concrete Reinforcing	98.7	70.4	84.3	98.3	70.4	84.1	106.7	70.4	88.2	100.5	70.4	85.1	110.6	83.9	97.0	110.6	83.9	97.0
0330	Cast-in-Place Concrete	88.4	68.2	80.7	89.7	68.2	81.6	88.5	68.2	80.8	98.1	68.2	86.8	95.1	103.3	98.2	89.0	103.5	94.5
03	CONCRETE	110.3	62.1	87.8	99.2	62.1	81.8	109.8	62.1	87.5	119.8	62.1	92.8	106.1	76.8	92.4	103.3	77.1	91.1
04	MASONRY	110.5	55.7	77.6	104.3	55.7	75.1	115.8	55.7	79.7	118.0	55.7	80.6	107.6	51.7	74.0	86.6	51.7	65.6
05	METALS	102.7	72.2	92.4	103.2	72.2	92.7	100.8	72.3	91.2	108.1	72.3	96.0	101.4	78.9	93.8	99.5	79.5	92.8
06	WOOD, PLASTICS & COMPOSITES	84.4	49.9	64.9	84.4	49.9	64.9	85.7	49.9	65.4	86.0	49.9	65.6	100.1	53.4	71.4	100.1	53.4	73.7
07	THERMAL & MOISTURE PROTECTION	96.9	61.4	83.5	95.8	61.4	82.8	98.5	61.4	84.5	102.3	61.4	86.8	95.5	53.5	79.6	95.5	61.6	82.7
08	OPENINGS	88.6	52.2	79.6	88.6	52.2	79.6	92.9	52.2	82.8	90.5	52.2	81.0	104.3	54.6	92.0	104.3	54.6	92.0
0920	Plaster & Gypsum Board	84.2	48.2	59.3	84.2	48.2	59.3	84.6	48.2	59.4	87.0	48.2	60.2	100.2	51.3	66.3	98.9	51.3	65.9
0950, 0980	Ceilings & Acoustic Treatment	121.8	48.2	75.0	121.8	48.2	75.0	121.8	48.2	75.0	116.1	48.2	72.9	92.7	51.3	66.3	93.5	51.3	66.6
0960	Flooring	103.6	51.4	89.2	101.3	51.4	87.5	104.4	51.4	89.7	104.4	51.4	89.8	97.8	59.4	87.2	97.6	59.4	87.0
0970, 0990	Wall Finishes & Painting/Coating	97.9	41.7	64.1	97.9	41.7	64.1	97.9	53.3	71.1	100.6	53.3	72.1	94.6	34.5	58.4	94.6	34.5	58.4
09	FINISHES	105.4	50.6	75.6	103.5	50.6	74.8	106.1	51.9	76.7	104.2	51.9	75.8	97.2	51.6	72.5	97.1	51.6	72.4
COVERS	DIVS. 10 - 14, 25, 28, 41, 43, 44	100.0	60.0	91.5	100.0	60.0	91.5	100.0	60.0	91.5	100.0	60.0	91.5	100.0	77.1	95.1	100.0	77.1	95.1
21, 22, 23	FIRE SUPPRESSION, PLUMBING & HVAC	100.0	68.3	86.9	100.0	68.3	86.9	100.0	68.3	86.9	100.1	68.3	87.0	100.0	65.5	85.8	100.1	65.6	85.9
26, 27, 3370	ELECTRICAL, COMMUNICATIONS & UTIL.	94.5	64.2	78.7	94.9	64.2	78.9	95.2	67.0	80.5	97.9	67.0	81.8	99.7	66.1	82.1	97.8	66.1	81.2
MF2004	WEIGHTED AVERAGE	100.3	65.0	84.7	98.2	65.0	83.6	100.8	65.4	85.2	103.0	65.5	86.5	100.6	67.3	85.9	98.8	67.7	85.0

City Cost Indexes

VIRGINIA

DIVISION		ALEXANDRIA MAT.	ALEXANDRIA INST.	ALEXANDRIA TOTAL	ARLINGTON MAT.	ARLINGTON INST.	ARLINGTON TOTAL	NEWPORT NEWS MAT.	NEWPORT NEWS INST.	NEWPORT NEWS TOTAL	NORFOLK MAT.	NORFOLK INST.	NORFOLK TOTAL	PORTSMOUTH MAT.	PORTSMOUTH INST.	PORTSMOUTH TOTAL	RICHMOND MAT.	RICHMOND INST.	RICHMOND TOTAL
015433	CONTRACTOR EQUIPMENT		102.2	102.2		100.8	100.8		105.4	105.4		106.0	106.0		105.3	105.3		105.3	105.3
0241, 31 - 34	SITE & INFRASTRUCTURE, DEMOLITION	117.5	88.4	97.2	128.2	85.2	98.3	111.0	87.6	94.7	112.1	88.6	95.7	109.4	86.8	93.6	111.9	88.0	95.2
0310	Concrete Forming & Accessories	92.6	76.2	78.4	92.5	73.8	76.3	97.4	69.5	73.3	98.7	69.6	73.6	88.1	57.0	61.2	96.8	62.8	67.4
0320	Concrete Reinforcing	89.7	81.1	85.3	101.3	75.0	87.9	101.0	73.2	86.8	101.0	73.2	86.8	100.6	70.5	85.2	101.0	73.6	87.0
0330	Cast-in-Place Concrete	110.6	82.3	99.9	107.6	79.0	96.8	108.3	62.7	91.1	119.1	62.8	97.8	107.3	62.0	90.1	110.0	57.4	90.1
03	CONCRETE	108.9	80.3	95.5	113.8	76.5	96.4	106.3	69.2	88.9	111.7	69.2	91.8	105.1	62.7	85.3	107.0	64.4	87.1
04	MASONRY	95.0	72.6	81.5	108.4	68.5	84.4	101.1	57.0	74.6	107.0	57.0	77.0	106.4	56.8	76.6	99.3	58.7	74.9
05	METALS	93.3	95.7	94.1	92.0	84.0	89.3	93.3	91.6	92.7	92.3	91.7	92.1	92.3	88.1	90.9	95.5	92.2	94.4
06	WOOD, PLASTICS & COMPOSITES	96.4	76.1	84.9	93.6	76.1	83.7	97.2	74.1	84.1	99.2	74.1	85.0	86.1	57.4	69.9	96.3	65.5	78.8
07	THERMAL & MOISTURE PROTECTION	98.3	81.7	92.0	100.1	71.2	89.2	99.4	59.1	84.2	97.1	59.1	82.7	99.5	56.7	83.3	98.4	59.4	83.7
08	OPENINGS	98.5	77.4	93.3	96.3	71.8	90.3	98.5	67.7	90.8	98.5	67.7	90.8	98.5	58.0	88.5	98.5	64.0	89.9
0920	Plaster & Gypsum Board	108.6	75.2	85.5	106.1	75.2	84.7	108.6	72.4	83.5	105.3	72.4	82.5	101.7	55.3	69.6	114.9	63.5	79.3
0950, 0980	Ceilings & Acoustic Treatment	96.1	75.2	82.8	93.5	75.2	81.9	96.1	72.4	81.0	96.1	72.4	81.0	96.1	55.3	70.1	99.5	63.5	76.6
0960	Flooring	110.9	86.0	104.0	109.0	59.3	95.2	110.9	54.4	95.3	110.4	54.4	94.9	102.0	68.8	92.8	110.4	76.7	101.1
0970, 0990	Wall Finishes & Painting/Coating	130.0	85.5	103.2	130.0	85.5	103.2	116.0	46.7	74.3	116.0	65.7	85.7	116.0	65.7	85.7	116.0	65.3	85.5
09	FINISHES	103.1	78.7	89.8	102.8	73.0	86.6	102.3	64.8	81.9	101.6	66.9	82.7	98.7	59.6	77.5	103.4	65.8	83.0
COVERS	DIVS. 10 - 14, 25, 28, 41, 43, 44	100.0	88.4	97.5	100.0	76.6	95.0	100.0	82.5	96.3	100.0	82.5	96.3	100.0	80.5	95.9	100.0	80.3	95.8
21, 22, 23	FIRE SUPPRESSION, PLUMBING & HVAC	100.2	87.9	95.2	100.2	85.5	94.2	100.2	65.0	85.7	100.1	64.1	85.3	100.2	64.0	85.3	100.1	58.2	82.9
26, 27, 3370	ELECTRICAL, COMMUNICATIONS & UTIL.	98.1	96.6	97.3	95.4	96.3	95.9	97.7	61.3	78.7	104.8	60.9	81.9	95.8	60.9	77.6	104.2	72.5	87.6
MF2004	WEIGHTED AVERAGE	100.3	85.4	93.7	101.2	81.0	92.3	100.0	69.2	86.4	101.4	69.3	87.2	99.4	66.2	84.8	101.0	68.7	86.7

VIRGINIA / WASHINGTON

DIVISION		ROANOKE MAT.	ROANOKE INST.	ROANOKE TOTAL	EVERETT MAT.	EVERETT INST.	EVERETT TOTAL	RICHLAND MAT.	RICHLAND INST.	RICHLAND TOTAL	SEATTLE MAT.	SEATTLE INST.	SEATTLE TOTAL	SPOKANE MAT.	SPOKANE INST.	SPOKANE TOTAL	TACOMA MAT.	TACOMA INST.	TACOMA TOTAL
015433	CONTRACTOR EQUIPMENT		100.8	100.8		103.9	103.9		91.1	91.1		103.8	103.8		91.1	91.1		103.9	103.9
0241, 31 - 34	SITE & INFRASTRUCTURE, DEMOLITION	108.6	84.9	92.1	100.3	113.4	109.4	102.4	90.4	94.0	104.6	112.3	110.0	101.7	90.4	93.8	103.7	114.3	111.1
0310	Concrete Forming & Accessories	97.1	66.2	70.4	109.4	101.1	102.2	124.1	77.1	83.4	101.8	104.7	104.3	128.3	77.1	84.0	101.8	104.1	103.8
0320	Concrete Reinforcing	101.0	70.5	85.4	107.5	99.3	103.3	107.2	82.0	94.3	106.0	99.7	102.8	107.9	82.1	94.7	106.0	99.4	102.7
0330	Cast-in-Place Concrete	121.8	62.3	99.3	102.6	97.4	100.6	107.1	83.2	98.1	107.7	110.9	108.9	111.3	83.2	100.7	105.5	110.6	107.5
03	CONCRETE	112.9	67.0	91.4	95.9	98.9	97.3	103.1	80.1	92.3	98.9	105.3	101.9	105.5	80.1	93.6	97.8	104.9	101.1
04	MASONRY	101.7	58.9	76.0	133.3	92.7	108.9	112.6	78.2	91.9	128.4	104.2	113.9	113.3	80.5	93.6	128.3	99.5	110.9
05	METALS	93.1	88.5	91.5	106.3	89.6	100.6	90.3	78.8	86.4	108.1	93.8	103.3	92.7	79.0	88.1	108.1	91.5	102.5
06	WOOD, PLASTICS & COMPOSITES	97.2	70.1	81.9	99.5	104.1	102.1	97.5	76.5	85.6	91.5	104.1	98.7	105.9	76.5	89.2	90.3	104.1	98.1
07	THERMAL & MOISTURE PROTECTION	99.4	57.7	83.6	103.6	92.5	99.4	155.8	76.9	126.0	103.6	98.3	101.6	152.2	78.4	124.3	103.3	94.5	100.0
08	OPENINGS	98.5	64.3	90.0	103.3	99.5	102.4	110.5	71.4	100.8	105.5	100.9	104.3	111.1	71.4	101.3	104.0	100.9	103.2
0920	Plaster & Gypsum Board	108.6	69.0	81.2	119.7	104.2	109.0	139.3	75.6	95.2	114.5	104.2	107.3	136.0	75.6	94.2	117.3	104.2	108.2
0950, 0980	Ceilings & Acoustic Treatment	96.1	69.0	78.8	110.4	104.2	106.4	108.4	75.6	87.5	112.8	104.2	107.3	105.5	75.6	86.4	113.0	104.2	107.4
0960	Flooring	110.9	42.2	91.9	121.4	98.6	115.1	107.7	37.3	88.3	113.8	106.6	111.9	108.2	67.5	96.9	114.3	98.8	110.0
0970, 0990	Wall Finishes & Painting/Coating	116.0	46.2	74.0	106.3	78.8	89.8	109.4	65.0	82.7	106.3	88.6	95.7	109.3	65.0	82.7	106.3	88.6	95.7
09	FINISHES	102.1	60.2	79.4	112.2	99.0	105.0	121.1	67.8	92.2	110.0	103.4	106.4	120.2	74.0	95.1	110.5	101.7	105.7
COVERS	DIVS. 10 - 14, 25, 28, 41, 43, 44	100.0	77.2	95.2	100.0	93.5	98.6	100.0	71.5	93.9	100.0	102.9	100.6	100.0	71.4	93.9	100.0	102.8	100.6
21, 22, 23	FIRE SUPPRESSION, PLUMBING & HVAC	100.2	50.1	79.6	100.1	90.9	96.3	100.5	91.0	96.6	100.1	111.7	104.9	100.4	77.4	90.9	100.1	94.5	97.8
26, 27, 3370	ELECTRICAL, COMMUNICATIONS & UTIL.	97.7	42.7	68.9	105.0	86.8	95.5	99.3	87.2	93.0	104.8	100.4	102.5	96.4	73.4	84.4	104.8	95.2	99.8
MF2004	WEIGHTED AVERAGE	100.7	61.9	83.6	104.2	95.1	100.1	104.6	81.6	94.5	104.6	104.7	104.6	104.9	77.9	93.0	104.3	99.4	102.2

WASHINGTON / WEST VIRGINIA

DIVISION		VANCOUVER MAT.	VANCOUVER INST.	VANCOUVER TOTAL	YAKIMA MAT.	YAKIMA INST.	YAKIMA TOTAL	CHARLESTON MAT.	CHARLESTON INST.	CHARLESTON TOTAL	HUNTINGTON MAT.	HUNTINGTON INST.	HUNTINGTON TOTAL	PARKERSBURG MAT.	PARKERSBURG INST.	PARKERSBURG TOTAL	WHEELING MAT.	WHEELING INST.	WHEELING TOTAL
015433	CONTRACTOR EQUIPMENT		98.1	98.1		103.9	103.9		100.8	100.8		100.8	100.8		100.8	100.8		100.8	100.8
0241, 31 - 34	SITE & INFRASTRUCTURE, DEMOLITION	115.8	99.9	104.7	106.7	112.2	110.5	104.3	88.2	93.1	107.7	89.4	95.0	112.3	88.2	95.5	113.1	87.4	95.2
0310	Concrete Forming & Accessories	103.2	87.8	89.9	102.2	95.5	96.4	102.5	82.5	85.2	98.5	94.0	94.6	90.1	81.3	82.3	91.5	81.3	82.7
0320	Concrete Reinforcing	107.1	99.0	102.9	106.6	81.6	93.8	101.0	85.2	92.9	101.0	87.9	94.3	99.5	84.1	91.6	98.9	85.8	92.2
0330	Cast-in-Place Concrete	117.9	94.6	109.0	112.8	79.1	100.0	102.9	104.2	103.4	114.3	105.8	111.1	106.8	95.2	102.4	106.8	102.4	105.2
03	CONCRETE	107.9	92.2	100.6	102.7	86.9	95.3	103.9	91.3	98.0	109.3	97.4	103.7	109.3	87.4	99.1	109.3	90.3	100.4
04	MASONRY	128.7	89.4	105.1	120.7	61.0	84.8	98.0	92.3	94.5	101.1	85.6	91.8	84.7	86.2	85.6	109.5	85.9	95.3
05	METALS	105.3	90.5	100.3	106.2	81.2	97.7	93.3	97.8	94.8	93.4	98.3	95.0	92.0	96.9	93.6	92.1	97.9	94.1
06	WOOD, PLASTICS & COMPOSITES	82.8	87.8	85.6	90.6	104.1	98.3	103.2	79.3	89.7	97.2	93.4	95.1	87.8	77.8	82.2	89.4	79.3	83.7
07	THERMAL & MOISTURE PROTECTION	103.4	83.8	96.0	103.4	74.9	92.6	96.9	88.1	93.6	99.6	88.6	95.5	99.4	86.5	94.6	99.8	87.4	95.1
08	OPENINGS	101.9	89.9	99.0	103.4	84.2	98.6	99.7	77.2	94.1	98.5	85.6	95.3	99.1	76.2	93.4	100.0	79.7	95.0
0920	Plaster & Gypsum Board	115.3	87.6	96.1	116.9	104.2	108.1	104.8	78.5	86.6	107.6	93.0	97.5	101.7	77.0	84.6	102.1	78.5	85.8
0950, 0980	Ceilings & Acoustic Treatment	108.8	87.6	95.3	108.7	104.2	105.8	94.3	78.5	84.2	92.6	93.0	92.9	91.7	77.0	82.3	91.7	78.5	83.3
0960	Flooring	120.1	68.6	105.9	115.5	56.0	99.1	110.9	108.1	110.1	110.7	89.9	105.0	105.2	95.7	102.6	106.3	102.1	105.2
0970, 0990	Wall Finishes & Painting/Coating	112.6	61.9	82.1	106.3	65.0	81.5	116.0	87.6	98.9	116.0	87.6	98.9	116.0	87.7	99.0	116.0	88.2	99.3
09	FINISHES	109.2	81.2	94.0	110.0	86.1	97.0	101.1	87.3	93.6	101.1	92.6	96.5	99.1	83.8	90.8	99.5	85.3	91.8
COVERS	DIVS. 10 - 14, 25, 28, 41, 43, 44	100.0	68.0	93.2	100.0	96.5	99.3	100.0	99.2	99.8	100.0	101.9	100.4	100.0	98.9	99.8	100.0	96.1	99.2
21, 22, 23	FIRE SUPPRESSION, PLUMBING & HVAC	100.2	95.2	98.2	100.1	87.7	95.0	100.1	85.6	94.1	100.2	83.7	93.4	100.1	86.2	94.4	100.2	88.9	95.6
26, 27, 3370	ELECTRICAL, COMMUNICATIONS & UTIL.	110.8	93.8	101.9	108.1	87.2	97.2	102.1	88.8	95.2	97.7	99.1	98.4	98.2	91.7	94.8	95.1	92.7	93.8
MF2004	WEIGHTED AVERAGE	105.8	90.8	99.2	104.6	86.1	96.4	99.8	89.1	95.0	100.2	91.7	96.4	99.1	87.8	94.1	100.3	89.2	95.4

City Cost Indexes

		WISCONSIN																	
		EAU CLAIRE			GREEN BAY			KENOSHA			LA CROSSE			MADISON			MILWAUKEE		
DIVISION		MAT.	INST.	TOTAL	MAT.	INST.	TOTAL	MAT.	INST.	TOTAL	MAT.	INST.	TOTAL	MAT.	INST.	TOTAL	MAT.	INST.	TOTAL
015433	CONTRACTOR EQUIPMENT		100.9	100.9		98.8	98.8		99.5	99.5		100.9	100.9		101.5	101.5		89.2	89.2
0241, 31 - 34	SITE & INFRASTRUCTURE, DEMOLITION	93.7	102.7	100.0	97.3	99.0	98.4	101.0	103.6	102.8	87.1	102.7	98.0	94.5	106.5	102.8	95.5	96.8	96.4
0310	Concrete Forming & Accessories	100.7	93.8	94.8	104.9	93.1	94.6	104.1	100.3	100.8	89.8	93.4	92.9	100.2	94.4	95.2	102.9	115.3	113.6
0320	Concrete Reinforcing	92.8	95.4	94.1	90.8	81.5	86.1	93.5	101.1	97.4	92.5	86.4	89.4	93.7	86.8	90.2	93.7	101.8	97.8
0330	Cast-in-Place Concrete	99.5	97.3	98.6	102.9	94.9	99.9	111.5	97.2	106.1	89.4	91.8	90.3	100.3	97.3	99.1	100.1	109.8	103.8
03	CONCRETE	98.2	95.6	97.0	101.1	91.8	96.8	103.1	99.5	101.4	89.6	91.8	90.6	97.3	94.1	95.8	97.5	110.0	103.4
04	MASONRY	93.0	97.5	95.7	125.2	94.7	106.8	102.3	103.7	103.1	92.2	97.6	95.5	103.0	101.2	101.9	105.0	119.4	113.7
05	METALS	91.8	99.4	94.3	93.9	93.0	93.6	98.6	102.5	99.9	91.7	95.1	92.8	100.7	94.5	98.6	101.5	96.4	99.8
06	WOOD, PLASTICS & COMPOSITES	112.4	93.5	101.7	112.5	93.5	101.7	108.8	99.6	103.6	98.9	93.5	95.8	106.1	93.5	99.0	110.8	114.9	113.1
07	THERMAL & MOISTURE PROTECTION	101.2	84.6	94.9	102.5	82.7	95.0	100.9	93.8	98.2	100.8	83.4	94.2	97.5	92.0	95.4	99.3	113.4	104.6
08	OPENINGS	100.7	90.8	98.2	99.0	89.1	96.6	98.7	103.4	99.9	100.6	81.3	95.8	103.4	93.1	100.8	105.6	111.7	107.1
0920	Plaster & Gypsum Board	98.6	93.6	95.1	95.2	93.6	94.1	84.5	100.0	95.2	93.3	93.6	93.5	92.4	93.6	93.3	97.6	115.6	110.0
0950, 0980	Ceilings & Acoustic Treatment	99.2	93.6	95.6	91.5	93.6	92.8	84.7	100.0	94.4	97.5	93.6	95.0	86.7	93.6	91.1	88.2	115.6	105.6
0960	Flooring	96.4	102.5	98.1	115.9	102.5	112.2	121.4	107.7	117.6	89.0	109.1	94.5	97.6	100.9	98.5	106.1	119.3	109.8
0970, 0990	Wall Finishes & Painting/Coating	89.6	88.7	89.1	101.1	72.0	83.6	105.0	98.4	101.0	89.6	69.1	77.3	96.0	90.3	92.6	98.0	114.6	108.0
09	FINISHES	98.4	95.3	96.7	103.6	93.3	98.0	103.2	102.1	102.6	94.7	94.5	94.6	96.3	95.4	95.8	100.2	116.5	109.1
COVERS	DIVS. 10 - 14, 25, 28, 41, 43, 44	100.0	84.0	96.6	100.0	83.4	96.5	100.0	100.2	100.0	100.0	84.1	96.6	100.0	83.8	96.6	100.0	104.8	101.0
21, 22, 23	FIRE SUPPRESSION, PLUMBING & HVAC	100.0	77.6	90.8	100.2	81.0	92.3	100.0	83.5	93.2	100.0	77.4	90.7	99.6	98.8	99.3	99.8	102.9	101.1
26, 27, 3370	ELECTRICAL, COMMUNICATIONS & UTIL.	101.4	84.2	92.4	95.4	80.1	87.4	97.9	99.9	98.9	101.7	84.2	92.6	100.1	85.5	92.5	98.9	100.2	99.6
MF2004	WEIGHTED AVERAGE	98.3	90.5	94.8	100.5	88.5	95.2	100.4	97.6	99.1	96.6	89.0	93.2	99.7	95.5	97.8	100.5	106.6	103.2

		WISCONSIN			WYOMING						CANADA					
		RACINE			CASPER			CHEYENNE			ROCK SPRINGS			CALGARY, ALBERTA			EDMONTON, ALBERTA		
DIVISION		MAT.	INST.	TOTAL	MAT.	INST.	TOTAL	MAT.	INST.	TOTAL	MAT.	INST.	TOTAL	MAT.	INST.	TOTAL	MAT.	INST.	TOTAL
015433	CONTRACTOR EQUIPMENT		101.5	101.5		100.7	100.7		100.7	100.7		100.7	100.7		106.7	106.7		106.7	106.7
0241, 31 - 34	SITE & INFRASTRUCTURE, DEMOLITION	94.6	107.2	103.4	91.7	98.9	96.7	85.1	98.9	94.7	81.5	97.8	92.8	124.3	105.3	111.0	154.1	105.3	120.1
0310	Concrete Forming & Accessories	101.1	100.5	100.6	102.2	42.8	50.8	103.7	58.5	64.6	100.5	38.3	46.7	126.5	86.3	91.7	129.0	86.2	92.0
0320	Concrete Reinforcing	93.7	101.1	97.5	105.2	43.8	73.8	99.4	44.6	71.4	107.3	44.4	75.2	148.3	60.1	103.3	148.3	60.1	103.3
0330	Cast-in-Place Concrete	100.5	97.0	99.2	98.3	72.9	88.7	93.8	73.1	86.0	94.8	53.8	79.3	186.7	97.2	152.8	208.8	97.2	166.6
03	CONCRETE	97.5	99.5	98.4	104.2	54.2	80.9	101.8	61.4	82.9	102.7	45.8	76.1	159.4	85.5	124.8	170.4	85.5	130.7
04	MASONRY	105.0	103.7	104.2	104.8	37.8	64.5	101.5	52.1	71.8	158.7	46.8	91.4	213.9	87.9	138.2	202.5	87.9	133.6
05	METALS	99.7	102.6	100.7	100.4	61.0	87.1	101.7	62.5	88.5	97.5	60.3	84.9	147.7	84.3	126.3	148.3	84.2	126.6
06	WOOD, PLASTICS & COMPOSITES	109.2	99.6	103.8	97.3	40.4	65.1	98.9	60.9	77.4	96.2	37.2	62.8	113.3	85.7	97.7	110.1	85.7	96.3
07	THERMAL & MOISTURE PROTECTION	101.1	95.0	98.8	95.3	49.8	78.1	96.7	55.8	81.2	98.1	47.8	79.1	123.1	84.6	108.5	124.2	84.6	109.2
08	OPENINGS	103.4	103.4	103.4	93.1	39.9	79.9	95.2	51.1	84.3	99.5	38.2	84.3	92.8	76.3	88.7	92.8	76.3	88.7
0920	Plaster & Gypsum Board	96.6	100.0	98.9	102.2	38.4	58.1	90.9	59.6	69.2	93.2	35.1	53.0	172.8	84.8	111.9	165.3	84.8	109.6
0950, 0980	Ceilings & Acoustic Treatment	84.7	100.0	94.4	126.3	38.4	70.3	119.1	59.6	81.2	121.9	35.1	66.7	142.4	84.8	105.7	157.4	84.8	111.2
0960	Flooring	103.3	107.7	104.5	102.6	35.5	84.1	107.0	57.1	93.2	106.0	50.2	90.6	134.5	85.3	120.9	135.6	85.3	121.7
0970, 0990	Wall Finishes & Painting/Coating	96.0	98.4	97.4	97.8	47.8	67.7	101.4	47.8	69.0	97.1	30.6	57.1	110.0	97.3	102.4	110.1	89.1	97.5
09	FINISHES	98.3	102.1	100.3	109.7	41.0	72.4	108.8	57.4	80.9	106.2	38.6	69.5	132.8	87.3	108.4	137.6	86.9	110.1
COVERS	DIVS. 10 - 14, 25, 28, 41, 43, 44	100.0	100.2	100.0	100.0	74.7	94.6	100.0	77.2	95.2	100.0	61.3	91.8	140.0	96.6	130.8	140.0	96.6	130.8
21, 22, 23	FIRE SUPPRESSION, PLUMBING & HVAC	99.8	88.5	95.2	100.1	57.1	82.4	100.0	56.4	82.1	100.0	54.0	81.0	98.2	82.5	91.7	98.2	82.5	91.8
26, 27, 3370	ELECTRICAL, COMMUNICATIONS & UTIL.	97.2	101.1	99.2	100.3	57.0	77.7	97.4	69.9	83.0	93.5	58.0	74.9	126.9	81.4	103.1	117.8	81.4	98.8
MF2004	WEIGHTED AVERAGE	99.7	99.1	99.5	100.5	56.0	80.9	100.0	63.2	83.7	102.2	54.2	81.0	129.4	86.3	110.4	130.6	86.2	111.0

		CANADA																	
		HALIFAX, NOVA SCOTIA			HAMILTON, ONTARIO			KITCHENER, ONTARIO			LAVAL, QUEBEC			LONDON, ONTARIO			MONTREAL, QUEBEC		
DIVISION		MAT.	INST.	TOTAL	MAT.	INST.	TOTAL	MAT.	INST.	TOTAL	MAT.	INST.	TOTAL	MAT.	INST.	TOTAL	MAT.	INST.	TOTAL
015433	CONTRACTOR EQUIPMENT		100.9	100.9		108.6	108.6		103.8	103.8		101.8	101.8		104.0	104.0		103.4	103.4
0241, 31 - 34	SITE & INFRASTRUCTURE, DEMOLITION	99.2	97.9	98.3	115.0	112.9	113.5	101.5	105.0	103.9	94.6	99.5	98.0	114.6	105.2	108.0	101.2	99.4	99.9
0310	Concrete Forming & Accessories	97.4	72.9	76.2	130.1	91.2	96.5	116.3	83.9	88.3	126.7	82.2	88.2	130.2	85.1	91.2	134.9	89.6	95.7
0320	Concrete Reinforcing	155.9	58.2	106.0	161.3	86.0	122.8	110.8	85.9	98.1	158.1	75.8	116.1	131.1	84.6	107.4	146.6	89.4	117.4
0330	Cast-in-Place Concrete	171.3	71.9	133.7	151.9	97.5	131.3	140.8	78.9	117.4	134.5	91.5	118.2	149.3	95.5	129.0	174.6	97.8	145.6
03	CONCRETE	150.3	70.4	112.9	142.1	92.5	118.9	122.0	82.9	103.7	132.8	84.5	110.2	136.5	88.7	114.2	151.4	92.4	123.8
04	MASONRY	173.1	79.7	116.9	180.9	97.7	130.9	161.5	94.0	120.9	160.2	82.1	113.3	182.2	95.3	130.0	167.0	90.1	120.8
05	METALS	118.9	76.7	104.7	125.8	90.3	113.9	115.3	89.9	106.8	105.1	84.6	98.2	123.4	88.2	111.5	122.1	90.4	111.4
06	WOOD, PLASTICS & COMPOSITES	87.4	72.1	78.7	114.3	90.6	100.9	108.7	82.4	93.8	125.9	82.0	101.0	114.3	83.3	96.7	125.9	89.9	105.5
07	THERMAL & MOISTURE PROTECTION	106.5	73.2	93.9	121.6	92.9	110.7	110.5	89.7	102.7	105.3	84.9	97.6	122.4	89.7	110.1	118.4	93.3	108.9
08	OPENINGS	83.5	65.7	79.1	92.8	87.4	91.4	84.0	81.4	83.3	92.8	71.3	87.4	93.9	81.6	90.8	92.8	78.5	89.2
0920	Plaster & Gypsum Board	159.5	71.2	98.4	192.1	90.3	121.6	154.3	81.9	104.2	148.2	81.3	101.9	195.9	82.7	117.6	165.5	89.2	112.7
0950, 0980	Ceilings & Acoustic Treatment	105.1	71.2	83.5	124.8	90.3	102.8	105.1	81.9	90.3	95.6	81.3	86.5	137.7	82.7	102.7	114.5	89.2	98.4
0960	Flooring	110.0	62.0	96.7	134.5	91.0	122.5	127.8	91.0	117.7	132.1	91.7	120.9	134.7	91.0	122.6	134.0	91.7	122.3
0970, 0990	Wall Finishes & Painting/Coating	110.1	71.9	87.1	110.1	97.1	102.3	110.1	86.7	96.0	110.1	85.6	95.3	110.1	93.7	100.2	110.1	100.7	104.4
09	FINISHES	112.8	71.0	90.1	130.1	91.9	109.3	117.6	85.2	100.0	115.9	84.5	98.9	133.6	86.9	108.2	123.2	91.8	106.1
COVERS	DIVS. 10 - 14, 25, 28, 41, 43, 44	140.0	69.2	125.0	140.0	100.6	131.6	140.0	98.7	131.2	140.0	86.1	128.6	140.0	99.2	131.3	140.0	88.5	129.1
21, 22, 23	FIRE SUPPRESSION, PLUMBING & HVAC	97.3	71.6	86.8	98.3	86.6	93.5	97.3	83.7	91.7	97.3	87.3	93.2	98.3	83.8	92.3	98.3	88.9	94.4
26, 27, 3370	ELECTRICAL, COMMUNICATIONS & UTIL.	124.6	71.6	96.9	127.2	92.3	109.0	121.1	89.8	104.7	120.5	71.5	94.9	119.1	89.8	103.8	127.6	82.8	104.2
MF2004	WEIGHTED AVERAGE	117.4	74.6	98.5	121.9	93.2	109.2	113.3	88.5	102.4	113.5	83.8	100.4	120.6	89.6	106.9	120.8	89.7	107.1

City Cost Indexes

		CANADA																	
	DIVISION	OSHAWA, ONTARIO			OTTAWA, ONTARIO			QUEBEC, QUEBEC			REGINA, SASKATCHEWAN			SASKATOON, SASKATCHEWAN			ST CATHARINES, ONTARIO		
		MAT.	INST.	TOTAL	MAT.	INST.	TOTAL	MAT.	INST.	TOTAL	MAT.	INST.	TOTAL	MAT.	INST.	TOTAL	MAT.	INST.	TOTAL
015433	CONTRACTOR EQUIPMENT		103.8	103.8		103.8	103.8		103.8	103.8		99.9	99.9		99.9	99.9		101.5	101.5
0241, 31 - 34	SITE & INFRASTRUCTURE, DEMOLITION	114.1	104.3	107.3	111.9	104.9	107.0	99.8	99.4	99.6	113.0	96.0	101.2	107.3	96.2	99.5	102.4	101.5	101.7
0310	Concrete Forming & Accessories	121.7	83.8	89.0	131.2	87.7	93.6	136.5	89.8	96.1	106.2	56.7	63.4	106.2	56.6	63.3	114.5	88.6	92.1
0320	Concrete Reinforcing	175.4	81.2	127.3	160.0	84.6	121.5	140.7	89.4	114.5	127.1	59.6	92.6	119.7	59.5	89.0	111.7	85.9	98.5
0330	Cast-in-Place Concrete	162.9	83.3	132.8	153.0	94.6	130.9	148.3	98.2	129.3	156.3	66.9	122.5	142.0	66.8	113.6	134.5	97.3	120.4
03	CONCRETE	149.0	83.5	118.3	142.5	89.7	117.8	137.7	92.6	116.7	131.2	61.5	98.6	123.1	61.5	94.3	118.9	91.2	105.9
04	MASONRY	164.6	90.2	119.9	179.7	93.5	127.9	174.6	90.1	123.8	165.9	61.1	102.9	165.5	61.1	102.8	161.0	97.7	122.9
05	METALS	106.9	88.8	100.8	124.3	89.8	112.6	123.6	90.6	112.5	104.8	72.0	93.7	104.9	71.9	93.7	106.0	89.8	100.5
06	WOOD, PLASTICS & COMPOSITES	115.2	82.2	96.5	114.1	87.1	98.8	126.5	90.0	105.8	96.2	55.1	73.0	94.5	55.1	72.2	106.4	87.6	95.8
07	THERMAL & MOISTURE PROTECTION	111.4	83.1	100.7	124.6	89.2	111.3	120.6	93.4	110.4	105.9	61.0	88.9	105.0	59.9	88.0	110.5	89.9	102.7
08	OPENINGS	91.6	82.0	89.2	92.8	84.5	90.7	92.8	85.6	91.0	87.6	53.0	79.1	86.7	53.0	78.4	83.4	84.5	83.7
0920	Plaster & Gypsum Board	157.5	81.6	105.0	234.2	86.7	132.1	193.0	89.2	121.2	166.7	53.8	88.5	146.9	53.8	82.4	136.8	87.2	102.5
0950, 0980	Ceilings & Acoustic Treatment	99.9	81.6	88.3	135.1	86.7	104.3	120.2	89.2	100.4	118.9	53.8	77.4	118.9	53.8	77.4	99.9	87.2	91.8
0960	Flooring	132.6	93.4	121.4	134.5	89.7	122.1	134.5	91.7	122.7	119.8	58.3	102.8	119.8	58.3	102.8	126.1	91.0	116.4
0970, 0990	Wall Finishes & Painting/Coating	110.1	101.4	104.8	110.1	88.4	97.0	110.5	100.7	104.6	110.1	62.4	81.4	110.1	53.2	75.8	110.1	97.1	102.3
09	FINISHES	118.8	87.3	101.7	137.1	88.1	110.5	128.3	91.8	108.5	121.2	57.2	86.4	118.5	56.1	84.7	113.7	90.1	100.9
COVERS	DIVS. 10 - 14, 25, 28, 41, 43, 44	140.0	98.8	131.2	140.0	97.2	130.9	140.0	88.7	129.1	140.0	65.3	124.1	140.0	65.3	124.1	140.0	76.3	126.5
21, 22, 23	FIRE SUPPRESSION, PLUMBING & HVAC	97.3	97.2	97.3	98.3	84.2	92.5	98.2	88.9	94.4	97.5	73.1	87.4	97.4	73.1	87.4	97.3	85.3	92.4
26, 27, 3370	ELECTRICAL, COMMUNICATIONS & UTIL.	122.4	89.1	105.0	118.9	90.7	104.2	118.7	82.8	99.9	123.6	61.8	91.3	123.8	61.8	91.4	123.0	90.8	106.2
MF2004	WEIGHTED AVERAGE	117.0	90.9	105.5	121.5	90.1	107.6	119.4	90.1	106.4	114.2	67.0	93.4	112.7	66.8	92.4	111.4	90.4	102.1

		CANADA																	
	DIVISION	ST JOHNS, NEWFOUNDLAND			THUNDER BAY, ONTARIO			TORONTO, ONTARIO			VANCOUVER, BRITISH COLUMBIA			WINDSOR, ONTARIO			WINNIPEG, MANITOBA		
		MAT.	INST.	TOTAL	MAT.	INST.	TOTAL	MAT.	INST.	TOTAL	MAT.	INST.	TOTAL	MAT.	INST.	TOTAL	MAT.	INST.	TOTAL
015433	CONTRACTOR EQUIPMENT		102.1	102.1		101.5	101.5		103.9	103.9		112.1	112.1		101.5	101.5		105.9	105.9
0241, 31 - 34	SITE & INFRASTRUCTURE, DEMOLITION	115.4	97.8	103.2	107.8	101.4	103.3	138.6	105.7	115.7	121.7	107.4	111.7	97.2	101.1	99.9	119.3	101.0	106.5
0310	Concrete Forming & Accessories	104.1	61.2	67.0	121.6	88.9	93.3	131.4	96.3	101.1	126.0	74.4	81.4	121.6	86.0	90.8	130.7	62.4	71.7
0320	Concrete Reinforcing	167.7	54.9	110.1	99.9	85.3	92.4	157.1	86.6	121.1	158.3	68.8	112.6	109.5	84.6	96.8	148.3	53.3	99.8
0330	Cast-in-Place Concrete	172.4	73.6	135.0	148.0	95.9	128.3	147.4	105.5	131.5	162.9	88.2	134.7	137.7	96.9	122.3	183.4	68.8	140.1
03	CONCRETE	165.2	65.1	118.4	127.8	90.7	110.5	139.3	97.5	119.8	153.1	78.8	118.3	120.6	89.7	106.2	158.0	63.7	114.0
04	MASONRY	163.3	62.6	102.8	161.8	97.5	123.1	193.3	102.7	138.8	177.0	78.0	117.5	161.2	96.1	122.0	179.8	59.8	107.7
05	METALS	108.1	73.8	96.5	105.8	88.8	100.0	126.0	91.1	114.2	162.4	84.7	136.1	105.8	89.5	100.3	155.8	74.2	128.2
06	WOOD, PLASTICS & COMPOSITES	95.8	61.3	76.3	115.2	87.7	99.7	115.2	95.2	103.9	114.9	71.8	90.5	115.2	84.5	97.8	113.2	63.0	84.8
07	THERMAL & MOISTURE PROTECTION	109.3	60.8	91.0	110.7	91.3	103.4	122.6	97.4	113.1	133.0	78.9	112.5	110.5	90.1	102.8	115.7	64.6	96.4
08	OPENINGS	99.0	57.4	88.7	82.4	84.0	82.8	91.6	91.8	91.6	94.5	72.4	89.0	82.2	82.7	82.3	92.8	58.0	84.1
0920	Plaster & Gypsum Board	165.2	60.1	92.4	170.4	87.4	112.9	179.3	95.1	121.0	155.7	70.4	96.7	159.0	84.0	107.1	163.4	61.4	92.8
0950, 0980	Ceilings & Acoustic Treatment	104.2	60.1	76.1	95.6	87.4	90.4	133.5	95.1	109.0	148.3	70.4	98.7	95.6	84.0	88.2	136.1	61.4	88.5
0960	Flooring	114.7	52.7	97.6	132.1	52.4	110.1	135.6	96.5	124.8	134.9	88.4	122.0	132.1	91.7	121.0	133.4	64.7	114.4
0970, 0990	Wall Finishes & Painting/Coating	110.1	62.4	81.4	110.1	90.1	98.0	112.0	101.4	105.6	110.0	85.7	95.4	110.1	89.7	97.8	110.1	46.4	71.8
09	FINISHES	116.0	59.7	85.4	119.1	83.0	99.5	132.8	97.0	113.4	131.9	77.7	102.5	117.3	87.5	101.1	128.9	61.4	92.2
COVERS	DIVS. 10 - 14, 25, 28, 41, 43, 44	140.0	66.5	124.4	140.0	76.7	126.6	140.0	102.4	132.0	140.0	94.1	130.3	140.0	75.8	126.4	140.0	67.2	124.5
21, 22, 23	FIRE SUPPRESSION, PLUMBING & HVAC	97.6	60.4	82.3	97.3	85.4	92.4	98.2	90.3	95.0	98.2	69.0	86.2	97.3	85.0	92.3	98.3	61.8	83.3
26, 27, 3370	ELECTRICAL, COMMUNICATIONS & UTIL.	119.2	63.2	89.9	121.0	89.1	104.4	121.1	93.0	106.4	121.7	72.1	95.8	126.3	90.8	107.7	125.1	64.4	93.4
MF2004	WEIGHTED AVERAGE	119.2	66.1	95.7	112.9	89.1	102.4	122.5	95.8	110.7	128.7	78.6	106.6	112.0	89.4	102.0	127.7	66.7	100.8

Location Factors

Costs shown in RSMeans cost data publications are based on national averages for materials and installation. To adjust these costs to a specific location, simply multiply the base cost by the factor and divide by 100 for that city. The data is arranged alphabetically by state and postal zip code numbers. For a city not listed, use the factor for a nearby city with similar economic characteristics.

STATE/ZIP	CITY	MAT.	INST.	TOTAL
ALABAMA				
350-352	Birmingham	97.4	75.2	87.6
354	Tuscaloosa	96.0	60.2	80.2
355	Jasper	96.3	58.5	79.6
356	Decatur	96.0	61.8	80.9
357-358	Huntsville	96.0	70.1	84.6
359	Gadsden	95.9	59.2	79.7
360-361	Montgomery	97.1	58.3	80.0
362	Anniston	95.2	67.0	82.8
363	Dothan	95.9	53.7	77.3
364	Evergreen	95.4	55.6	77.8
365-366	Mobile	97.1	67.4	84.0
367	Selma	95.6	53.5	77.0
368	Phenix City	96.4	57.2	79.1
369	Butler	95.8	53.9	77.3
ALASKA				
995-996	Anchorage	127.4	112.3	120.7
997	Fairbanks	126.3	113.3	120.5
998	Juneau	126.0	111.7	119.7
999	Ketchikan	139.3	111.7	127.1
ARIZONA				
850,853	Phoenix	99.9	74.6	88.7
851,852	Mesa/Tempe	99.4	64.4	83.9
855	Globe	99.5	60.5	82.3
856-857	Tucson	98.2	69.9	85.7
859	Show Low	99.6	61.6	82.8
860	Flagstaff	101.6	70.4	87.9
863	Prescott	99.1	61.1	82.4
864	Kingman	97.2	67.9	84.3
865	Chambers	97.3	61.8	81.6
ARKANSAS				
716	Pine Bluff	96.8	63.9	82.3
717	Camden	94.1	42.2	71.2
718	Texarkana	96.2	46.5	74.2
719	Hot Springs	93.3	45.3	72.1
720-722	Little Rock	97.2	67.0	83.9
723	West Memphis	95.8	57.4	78.9
724	Jonesboro	96.9	57.9	79.7
725	Batesville	94.2	51.4	75.3
726	Harrison	95.5	52.2	76.4
727	Fayetteville	92.8	51.8	74.7
728	Russellville	94.1	53.8	76.3
729	Fort Smith	97.3	57.8	79.9
CALIFORNIA				
900-902	Los Angeles	101.7	116.0	108.0
903-905	Inglewood	96.8	104.2	100.1
906-908	Long Beach	98.6	104.3	101.1
910-912	Pasadena	98.0	104.2	100.7
913-916	Van Nuys	101.4	104.2	102.6
917-918	Alhambra	100.2	104.2	102.0
919-921	San Diego	102.4	107.5	104.6
922	Palm Springs	98.3	103.0	100.4
923-924	San Bernardino	95.7	103.6	99.2
925	Riverside	101.1	110.6	105.3
926-927	Santa Ana	97.9	104.0	100.6
928	Anaheim	101.1	112.6	106.2
930	Oxnard	102.5	110.6	106.1
931	Santa Barbara	101.7	110.5	105.6
932-933	Bakersfield	102.9	106.6	104.6
934	San Luis Obispo	102.1	101.4	101.8
935	Mojave	98.9	100.4	99.6
936-938	Fresno	102.8	110.6	106.3
939	Salinas	102.9	114.3	107.9
940-941	San Francisco	110.7	140.4	123.8
942,956-958	Sacramento	104.9	112.6	108.3
943	Palo Alto	102.8	119.2	110.0
944	San Mateo	105.8	125.7	114.6
945	Vallejo	103.9	118.8	110.5
946	Oakland	108.5	126.0	116.2
947	Berkeley	107.5	119.2	112.7
948	Richmond	106.8	120.6	112.9
949	San Rafael	108.5	121.4	114.2
950	Santa Cruz	107.7	114.4	110.7

STATE/ZIP	CITY	MAT.	INST.	TOTAL
CALIFORNIA (CONT'D)				
951	San Jose	105.8	131.3	117.1
952	Stockton	103.0	109.7	106.0
953	Modesto	103.0	109.9	106.1
954	Santa Rosa	102.6	123.5	111.9
955	Eureka	104.1	107.3	105.5
959	Marysville	103.3	108.8	105.7
960	Redding	105.4	107.6	106.3
961	Susanville	104.0	107.7	105.6
COLORADO				
800-802	Denver	101.6	85.1	94.3
803	Boulder	98.2	78.9	89.7
804	Golden	100.5	77.8	90.5
805	Fort Collins	102.4	74.4	90.0
806	Greeley	99.7	64.5	84.1
807	Fort Morgan	98.9	77.5	89.5
808-809	Colorado Springs	101.5	78.7	91.4
810	Pueblo	101.5	76.9	90.7
811	Alamosa	102.6	72.5	89.3
812	Salida	102.4	73.4	89.6
813	Durango	102.9	73.2	89.8
814	Montrose	101.6	71.7	88.4
815	Grand Junction	105.4	69.3	89.5
816	Glenwood Springs	102.7	74.8	90.4
CONNECTICUT				
060	New Britain	100.6	117.5	108.1
061	Hartford	101.7	117.6	108.7
062	Willimantic	101.2	117.1	108.2
063	New London	96.9	116.6	105.6
064	Meriden	99.0	118.1	107.4
065	New Haven	102.6	117.6	109.2
066	Bridgeport	101.7	118.2	109.0
067	Waterbury	101.3	118.0	108.7
068	Norwalk	101.2	118.7	109.0
069	Stamford	101.4	125.5	112.0
D.C.				
200-205	Washington	104.0	92.2	98.8
DELAWARE				
197	Newark	99.8	108.7	103.7
198	Wilmington	100.1	108.7	103.9
199	Dover	100.5	108.7	104.1
FLORIDA				
320,322	Jacksonville	98.0	66.3	84.0
321	Daytona Beach	98.2	75.3	88.1
323	Tallahassee	99.2	57.7	80.9
324	Panama City	99.5	53.1	79.0
325	Pensacola	102.2	64.4	85.5
326,344	Gainesville	99.8	69.1	86.2
327-328,347	Orlando	102.2	73.2	89.4
329	Melbourne	100.8	78.5	91.0
330-332,340	Miami	100.1	79.4	90.9
333	Fort Lauderdale	97.5	77.4	88.7
334,349	West Palm Beach	96.0	74.2	86.4
335-336,346	Tampa	99.2	83.3	92.2
337	St. Petersburg	101.6	61.6	83.9
338	Lakeland	98.1	83.0	91.4
339,341	Fort Myers	97.3	73.9	87.0
342	Sarasota	99.4	76.1	89.1
GEORGIA				
300-303,399	Atlanta	97.4	77.8	88.8
304	Statesboro	96.6	54.7	78.1
305	Gainesville	95.4	63.4	81.3
306	Athens	94.8	65.6	81.9
307	Dalton	96.6	61.8	81.2
308-309	Augusta	96.0	66.5	83.0
310-312	Macon	95.8	65.8	82.6
313-314	Savannah	98.4	61.4	82.0
315	Waycross	96.8	61.4	81.2
316	Valdosta	97.1	61.1	81.2
317,398	Albany	97.1	61.7	81.5
318-319	Columbus	97.0	66.2	83.4

Location Factors

STATE/ZIP	CITY	MAT.	INST.	TOTAL
HAWAII				
967	Hilo	113.5	117.9	115.4
968	Honolulu	116.9	117.9	117.3
STATES & POSS.				
969	Guam	135.6	61.7	103.0
IDAHO				
832	Pocatello	101.2	71.4	88.1
833	Twin Falls	102.5	51.8	80.1
834	Idaho Falls	99.7	57.3	81.0
835	Lewiston	110.1	76.8	95.4
836-837	Boise	100.4	71.8	87.7
838	Coeur d'Alene	109.4	73.9	93.7
ILLINOIS				
600-603	North Suburban	98.5	133.3	113.9
604	Joliet	98.5	135.8	115.0
605	South Suburban	98.5	133.3	113.9
606-608	Chicago	99.0	137.6	116.0
609	Kankakee	94.3	124.3	107.5
610-611	Rockford	97.5	122.8	108.7
612	Rock Island	94.8	99.7	96.9
613	La Salle	96.0	119.4	106.3
614	Galesburg	95.8	104.8	99.8
615-616	Peoria	98.7	107.3	102.5
617	Bloomington	95.2	108.5	101.0
618-619	Champaign	98.8	107.3	102.6
620-622	East St. Louis	94.0	106.6	99.6
623	Quincy	94.9	100.0	97.2
624	Effingham	94.2	104.1	98.6
625	Decatur	96.5	104.5	100.0
626-627	Springfield	97.0	105.4	100.7
628	Centralia	92.2	106.1	98.3
629	Carbondale	91.9	101.4	96.1
INDIANA				
460	Anderson	94.5	83.5	89.7
461-462	Indianapolis	97.7	85.7	92.4
463-464	Gary	96.1	108.3	101.5
465-466	South Bend	95.1	84.7	90.5
467-468	Fort Wayne	95.2	79.6	88.3
469	Kokomo	92.0	82.7	87.9
470	Lawrenceburg	91.3	80.4	86.5
471	New Albany	92.5	76.1	85.3
472	Columbus	94.8	82.4	89.3
473	Muncie	96.0	81.5	89.6
474	Bloomington	97.4	81.6	90.4
475	Washington	93.5	83.3	89.0
476-477	Evansville	95.1	86.6	91.4
478	Terre Haute	95.9	87.9	92.3
479	Lafayette	94.5	83.4	89.6
IOWA				
500-503,509	Des Moines	98.7	73.6	87.7
504	Mason City	96.1	59.0	79.7
505	Fort Dodge	96.4	54.9	78.1
506-507	Waterloo	98.4	56.0	79.7
508	Creston	96.7	58.8	80.0
510-511	Sioux City	99.1	66.1	84.5
512	Sibley	97.5	45.7	74.6
513	Spencer	99.2	44.5	75.0
514	Carroll	96.0	49.3	75.4
515	Council Bluffs	100.2	71.6	87.5
516	Shenandoah	96.4	47.4	74.8
520	Dubuque	98.6	72.6	87.2
521	Decorah	97.0	46.8	74.9
522-524	Cedar Rapids	99.5	77.5	89.8
525	Ottumwa	97.0	66.4	83.5
526	Burlington	96.1	67.9	83.6
527-528	Davenport	98.4	86.7	93.2
KANSAS				
660-662	Kansas City	98.6	95.3	97.1
664-666	Topeka	99.0	63.8	83.4
667	Fort Scott	96.7	68.6	84.3
668	Emporia	96.9	57.3	79.4
669	Belleville	98.7	58.7	81.0
670-672	Wichita	97.7	62.7	82.2
673	Independence	98.3	60.9	81.8
674	Salina	99.1	59.2	81.5
675	Hutchinson	93.6	59.0	78.3
676	Hays	97.9	60.1	81.2
677	Colby	98.6	60.0	81.5

STATE/ZIP	CITY	MAT.	INST.	TOTAL
KANSAS (CONT'D)				
678	Dodge City	100.4	61.5	83.3
679	Liberal	97.7	59.9	81.0
KENTUCKY				
400-402	Louisville	97.1	82.4	90.6
403-405	Lexington	95.5	76.3	87.0
406	Frankfort	96.3	76.1	87.4
407-409	Corbin	92.4	63.3	79.6
410	Covington	93.2	97.4	95.1
411-412	Ashland	92.0	98.2	94.8
413-414	Campton	93.3	63.7	80.2
415-416	Pikeville	94.5	78.0	87.2
417-418	Hazard	92.6	56.5	76.7
420	Paducah	91.6	86.7	89.4
421-422	Bowling Green	93.9	82.6	88.9
423	Owensboro	94.3	81.7	88.8
424	Henderson	91.3	87.3	89.5
425-426	Somerset	90.6	67.7	80.5
427	Elizabethtown	90.0	82.3	86.6
LOUISIANA				
700-701	New Orleans	103.2	68.2	87.8
703	Thibodaux	101.1	61.0	83.4
704	Hammond	98.2	55.8	79.5
705	Lafayette	100.9	56.1	81.1
706	Lake Charles	101.1	60.7	83.3
707-708	Baton Rouge	103.0	61.2	84.5
710-711	Shreveport	97.4	57.7	79.9
712	Monroe	97.4	53.4	77.9
713-714	Alexandria	97.5	54.0	78.3
MAINE				
039	Kittery	94.3	81.9	88.9
040-041	Portland	100.1	77.6	90.2
042	Lewiston	97.8	77.6	88.9
043	Augusta	97.7	81.5	90.6
044	Bangor	97.5	77.5	88.7
045	Bath	95.6	81.3	89.3
046	Machias	95.1	80.5	88.7
047	Houlton	95.2	80.7	88.8
048	Rockland	94.2	81.1	88.4
049	Waterville	95.6	81.6	89.4
MARYLAND				
206	Waldorf	100.0	69.2	86.4
207-208	College Park	100.0	83.7	92.8
209	Silver Spring	99.2	75.2	88.6
210-212	Baltimore	97.8	85.2	92.2
214	Annapolis	98.5	79.3	90.0
215	Cumberland	93.3	81.0	87.9
216	Easton	95.0	41.8	71.5
217	Hagerstown	94.5	81.9	88.9
218	Salisbury	95.4	48.8	74.8
219	Elkton	92.3	66.7	81.0
MASSACHUSETTS				
010-011	Springfield	99.6	107.7	103.2
012	Pittsfield	99.2	106.1	102.3
013	Greenfield	96.8	107.3	101.4
014	Fitchburg	95.5	120.6	106.5
015-016	Worcester	99.6	122.1	109.5
017	Framingham	95.0	129.5	110.2
018	Lowell	99.0	129.0	112.2
019	Lawrence	100.3	127.0	112.1
020-022, 024	Boston	102.0	136.5	117.2
023	Brockton	100.6	123.2	110.6
025	Buzzards Bay	94.3	120.9	106.0
026	Hyannis	97.7	120.9	107.9
027	New Bedford	99.7	121.1	109.1
MICHIGAN				
480,483	Royal Oak	92.1	105.7	98.1
481	Ann Arbor	94.9	106.1	99.9
482	Detroit	96.4	112.1	103.3
484-485	Flint	94.6	100.2	97.1
486	Saginaw	94.2	95.3	94.7
487	Bay City	94.4	94.3	94.3
488-489	Lansing	95.7	98.7	97.0
490	Battle Creek	95.0	91.0	93.2
491	Kalamazoo	95.3	88.2	92.2
492	Jackson	92.7	95.5	94.0
493,495	Grand Rapids	96.4	74.8	86.8
494	Muskegon	93.9	85.6	90.2

Location Factors

STATE/ZIP	CITY	MAT.	INST.	TOTAL	STATE/ZIP	CITY	MAT.	INST.	TOTAL
MICHIGAN (CONT'D)					**NEW HAMPSHIRE (CONT'D)**				
496	Traverse City	92.2	79.8	86.8	032-033	Concord	98.5	83.3	91.8
497	Gaylord	93.4	74.2	84.9	034	Keene	96.3	51.3	76.4
498-499	Iron mountain	95.5	88.8	92.5	035	Littleton	96.1	58.2	79.3
					036	Charleston	95.8	48.8	75.0
MINNESOTA					037	Claremont	94.7	48.8	74.5
550-551	Saint Paul	99.3	123.4	109.9	038	Portsmouth	97.3	90.9	94.4
553-555	Minneapolis	99.7	126.3	111.5					
556-558	Duluth	98.6	114.6	105.6	**NEW JERSEY**				
MINNESOTA (CONT'd)					070-071	Newark	101.9	124.3	111.8
559	Rochester	98.5	107.8	102.6	072	Elizabeth	100.1	123.8	110.6
560	Mankato	96.2	105.5	100.3	073	Jersey City	99.0	124.7	110.3
561	Windom	94.8	77.8	87.3	074-075	Paterson	100.9	124.3	111.2
562	Willmar	94.4	84.4	90.0	076	Hackensack	98.8	124.8	110.3
563	St. Cloud	96.2	124.2	108.6	077	Long Branch	98.5	122.4	109.1
564	Brainerd	96.1	107.4	101.1	078	Dover	99.1	124.3	110.2
565	Detroit Lakes	97.9	97.7	97.8	079	Summit	99.0	123.8	110.0
566	Bemidji	97.2	100.1	98.5	080,083	Vineland	97.1	122.2	108.2
567	Thief River Falls	96.7	96.4	96.6	081	Camden	99.0	122.9	109.5
					082,084	Atlantic City	97.6	122.2	108.5
MISSISSIPPI					085-086	Trenton	99.0	123.4	109.8
386	Clarksdale	95.8	57.8	79.0	087	Point Pleasant	99.2	122.2	109.3
387	Greenville	99.6	67.9	85.6	088-089	New Brunswick	99.7	123.6	110.2
388	Tupelo	97.3	60.4	81.0					
389	Greenwood	97.1	57.3	79.5	**NEW MEXICO**				
390-392	Jackson	97.8	67.0	84.2	870-872	Albuquerque	99.8	74.8	88.8
393	Meridian	96.1	68.6	84.0	873	Gallup	100.0	74.8	88.9
394	Laurel	97.1	60.9	81.2	874	Farmington	100.4	74.8	89.1
395	Biloxi	97.9	62.2	82.1	875	Santa Fe	101.8	74.8	89.9
396	Mccomb	95.5	56.8	78.4	877	Las Vegas	98.2	74.6	87.8
397	Columbus	97.0	59.2	80.4	878	Socorro	97.8	74.8	87.6
					879	Truth/Consequences	97.9	68.9	85.1
MISSOURI					880	Las Cruces	96.8	66.7	83.5
630-631	St. Louis	98.6	107.2	102.4	881	Clovis	98.5	72.7	87.1
633	Bowling Green	97.0	86.6	92.4	882	Roswell	100.4	72.8	88.2
634	Hannibal	95.8	76.8	87.4	883	Carrizozo	100.8	74.8	89.3
635	Kirksville	98.9	70.7	86.5	884	Tucumcari	99.3	72.7	87.6
636	Flat River	98.0	87.4	93.3					
637	Cape Girardeau	98.4	83.8	92.0	**NEW YORK**				
638	Sikeston	96.0	74.2	86.4	100-102	New York	107.0	166.3	133.2
639	Poplar Bluff	95.4	74.7	86.3	103	Staten Island	103.1	162.5	129.3
640-641	Kansas City	98.7	108.2	102.9	104	Bronx	100.8	162.5	128.0
644-645	St. Joseph	98.0	87.2	93.3	105	Mount Vernon	100.6	136.0	116.2
646	Chillicothe	94.7	66.1	82.1	106	White Plains	100.7	136.0	116.3
647	Harrisonville	94.2	94.8	94.5	107	Yonkers	105.7	136.1	119.2
648	Joplin	96.8	65.2	82.9	108	New Rochelle	100.9	136.0	116.4
650-651	Jefferson City	95.6	85.0	90.9	109	Suffern	100.7	126.7	112.2
652	Columbia	96.8	88.2	93.0	110	Queens	101.8	162.4	128.6
653	Sedalia	95.7	78.8	88.2	111	Long Island City	103.5	162.4	129.5
654-655	Rolla	94.0	75.8	86.0	112	Brooklyn	103.9	162.4	129.7
656-658	Springfield	97.7	77.8	88.9	113	Flushing	103.4	162.4	129.4
					114	Jamaica	101.5	162.4	128.4
MONTANA					115,117,118	Hicksville	101.6	146.4	121.4
590-591	Billings	102.0	72.9	89.2	116	Far Rockaway	103.6	162.4	129.5
592	Wolf Point	101.5	68.6	87.0	119	Riverhead	102.4	145.9	121.6
593	Miles City	99.0	70.4	86.4	120-122	Albany	97.7	97.6	97.7
594	Great Falls	103.4	72.2	89.6	123	Schenectady	98.2	97.3	97.8
595	Havre	100.3	71.3	87.5	124	Kingston	101.2	117.9	108.6
596	Helena	101.9	70.6	88.1	125-126	Poughkeepsie	100.3	134.3	115.3
597	Butte	102.0	72.3	88.9	127	Monticello	99.6	122.5	109.7
598	Missoula	99.2	72.7	87.5	128	Glens Falls	92.2	90.5	91.5
599	Kalispell	97.8	70.9	86.0	129	Plattsburgh	97.1	87.3	92.8
					130-132	Syracuse	99.3	93.0	96.5
NEBRASKA					133-135	Utica	97.2	88.9	93.5
680-681	Omaha	99.8	75.6	89.2	136	Watertown	99.1	91.9	95.9
683-685	Lincoln	97.8	74.1	87.3	137-139	Binghamton	98.8	89.5	94.7
686	Columbus	95.8	73.4	85.9	140-142	Buffalo	100.3	102.8	101.4
687	Norfolk	97.5	77.1	88.5	143	Niagara Falls	97.7	100.8	99.1
688	Grand Island	97.9	79.2	89.6	144-146	Rochester	99.8	97.4	98.7
689	Hastings	96.8	81.5	90.0	147	Jamestown	96.6	86.9	92.3
690	Mccook	96.5	72.2	85.8	148-149	Elmira	96.3	87.8	92.6
691	North Platte	97.2	81.6	90.4					
692	Valentine	98.5	70.6	86.2	**NORTH CAROLINA**				
693	Alliance	98.7	68.7	85.5	270,272-274	Greensboro	99.2	49.2	77.1
					271	Winston-Salem	99.1	47.8	76.5
NEVADA					275-276	Raleigh	100.3	49.4	77.8
889-891	Las Vegas	101.0	113.8	106.7	277	Durham	100.3	49.7	78.0
893	Ely	99.6	69.1	86.1	278	Rocky Mount	96.3	40.3	71.6
894-895	Reno	100.0	90.5	95.8	279	Elizabeth City	97.3	41.6	72.7
897	Carson City	99.9	89.5	95.3	280	Gastonia	98.2	47.2	75.7
898	Elko	98.2	74.4	87.7	281-282	Charlotte	99.8	49.7	77.7
					283	Fayetteville	100.0	50.1	78.0
NEW HAMPSHIRE					284	Wilmington	96.9	47.6	75.1
030	Nashua	99.9	87.0	94.2	285	Kinston	94.6	41.6	71.2
031	Manchester	100.4	87.0	94.5					

Location Factors

STATE/ZIP	CITY	MAT.	INST.	TOTAL
NORTH CAROLINA (CONT'D)				
286	Hickory	94.9	43.9	72.4
287-288	Asheville	97.4	47.2	75.3
289	Murphy	95.8	34.0	68.5
NORTH DAKOTA				
580-581	Fargo	101.8	59.5	83.2
582	Grand Forks	101.5	52.8	80.0
583	Devils Lake	100.3	55.0	80.3
584	Jamestown	100.4	47.1	76.9
585	Bismarck	100.4	60.4	82.8
586	Dickinson	101.3	57.7	82.0
587	Minot	101.5	64.6	85.2
588	Williston	99.6	57.7	81.1
OHIO				
430-432	Columbus	96.9	89.8	93.7
433	Marion	92.9	80.6	87.5
434-436	Toledo	97.1	97.1	97.1
437-438	Zanesville	93.3	80.0	87.5
439	Steubenville	94.9	89.6	92.6
440	Lorain	97.7	91.5	95.0
441	Cleveland	98.0	100.2	99.0
442-443	Akron	98.7	91.9	95.7
444-445	Youngstown	98.1	89.6	94.3
446-447	Canton	98.2	83.7	91.8
448-449	Mansfield	94.9	87.1	91.5
450	Hamilton	94.6	83.6	89.7
451-452	Cincinnati	94.9	86.4	91.1
453-454	Dayton	94.7	82.9	89.5
455	Springfield	94.6	84.3	90.1
456	Chillicothe	93.1	89.1	91.3
457	Athens	95.6	78.4	88.0
458	Lima	96.0	83.3	90.4
OKLAHOMA				
730-731	Oklahoma City	98.6	61.7	82.3
734	Ardmore	95.1	61.2	80.2
735	Lawton	97.9	61.7	81.9
736	Clinton	96.6	59.8	80.3
737	Enid	97.7	59.8	80.9
738	Woodward	95.3	59.9	79.6
739	Guymon	96.4	30.8	67.4
740-741	Tulsa	97.7	54.8	78.8
743	Miami	93.8	62.6	80.1
744	Muskogee	96.9	39.6	71.6
745	Mcalester	93.5	51.0	74.7
746	Ponca City	94.2	59.2	78.8
747	Durant	94.2	58.6	78.5
748	Shawnee	95.9	57.4	78.9
749	Poteau	93.2	61.1	79.0
OREGON				
970-972	Portland	101.1	96.8	99.2
973	Salem	101.0	95.0	98.4
974	Eugene	100.9	94.3	98.0
975	Medford	102.7	92.9	98.4
976	Klamath Falls	102.4	92.7	98.1
977	Bend	101.2	94.4	98.2
978	Pendleton	95.5	94.7	95.1
979	Vale	93.1	86.3	90.1
PENNSYLVANIA				
150-152	Pittsburgh	97.7	102.1	99.6
153	Washington	94.3	102.1	97.7
154	Uniontown	94.6	100.4	97.2
155	Bedford	95.6	91.0	93.6
156	Greensburg	95.6	100.6	97.8
157	Indiana	94.4	99.4	96.6
158	Dubois	96.1	94.7	95.5
159	Johnstown	95.6	94.9	95.3
160	Butler	92.2	101.6	96.3
161	New Castle	92.2	98.9	95.2
162	Kittanning	92.7	102.6	97.1
163	Oil City	92.2	95.6	93.7
164-165	Erie	94.7	93.1	94.0
166	Altoona	94.9	90.3	92.9
167	Bradford	95.6	93.8	94.8
168	State College	95.2	90.7	93.2
169	Wellsboro	96.3	92.8	94.8
170-171	Harrisburg	98.5	95.1	97.0
172	Chambersburg	95.8	89.7	93.1
173-174	York	96.5	95.4	96.0
175-176	Lancaster	94.4	89.5	92.2

STATE/ZIP	CITY	MAT.	INST.	TOTAL
PENNSYLVANIA (CONT'D)				
177	Williamsport	92.9	81.1	87.7
178	Sunbury	95.2	93.8	94.6
179	Pottsville	94.3	97.7	95.8
180	Lehigh Valley	95.7	114.3	103.9
181	Allentown	98.2	110.5	103.6
182	Hazleton	95.1	96.8	95.9
183	Stroudsburg	95.0	105.1	99.5
184-185	Scranton	99.0	100.9	99.8
186-187	Wilkes-Barre	94.8	97.4	95.9
188	Montrose	94.5	98.4	96.2
189	Doylestown	94.8	122.6	107.1
190-191	Philadelphia	100.3	134.2	115.3
193	Westchester	96.4	127.0	109.9
194	Norristown	95.4	132.6	111.8
195-196	Reading	97.9	101.5	99.5
PUERTO RICO				
009	San Juan	121.2	24.5	78.5
RHODE ISLAND				
028	Newport	99.0	112.8	105.1
029	Providence	100.2	112.8	105.8
SOUTH CAROLINA				
290-292	Columbia	98.3	51.2	77.5
293	Spartanburg	96.5	50.8	76.3
294	Charleston	98.1	59.7	81.2
295	Florence	96.3	51.2	76.4
296	Greenville	96.3	50.8	76.2
297	Rock Hill	95.6	48.9	74.9
298	Aiken	96.5	72.0	85.7
299	Beaufort	97.3	44.1	73.8
SOUTH DAKOTA				
570-571	Sioux Falls	99.8	57.8	81.2
572	Watertown	98.0	53.1	78.2
573	Mitchell	96.8	53.0	77.4
574	Aberdeen	100.0	54.0	79.7
575	Pierre	99.3	54.2	79.4
576	Mobridge	97.5	53.2	77.9
577	Rapid City	99.8	54.2	79.7
TENNESSEE				
370-372	Nashville	97.5	73.9	87.1
373-374	Chattanooga	98.6	66.9	84.6
375,380-381	Memphis	96.4	71.6	85.5
376	Johnson City	97.9	54.2	78.6
377-379	Knoxville	94.9	61.8	80.2
382	Mckenzie	96.4	55.6	78.4
383	Jackson	98.4	56.0	79.7
384	Columbia	94.9	62.3	80.5
385	Cookeville	96.3	60.5	80.5
TEXAS				
750	Mckinney	98.8	51.2	77.8
751	Waxahackie	98.7	56.4	80.0
752-753	Dallas	99.6	66.8	85.2
754	Greenville	98.8	40.0	72.9
755	Texarkana	98.3	52.6	78.1
756	Longview	98.9	41.0	73.4
757	Tyler	99.3	55.6	80.0
758	Palestine	95.1	41.5	71.4
759	Lufkin	95.8	45.6	73.6
760-761	Fort Worth	98.2	62.7	82.6
762	Denton	97.1	49.2	76.0
763	Wichita Falls	98.4	56.5	79.9
764	Eastland	96.8	40.8	72.1
765	Temple	95.1	51.0	75.6
766-767	Waco	97.8	58.1	80.3
768	Brownwood	97.9	38.6	71.7
769	San Angelo	97.4	49.3	76.2
770-772	Houston	100.0	70.3	86.9
773	Huntsville	98.1	39.3	72.2
774	Wharton	99.6	42.9	74.6
775	Galveston	97.4	70.1	85.3
776-777	Beaumont	98.1	62.1	82.2
778	Bryan	94.4	63.0	80.6
779	Victoria	99.6	45.2	75.6
780	Laredo	94.9	53.0	76.4
781-782	San Antonio	95.3	63.7	81.3
783-784	Corpus Christi	97.8	52.7	77.9
785	Mc Allen	97.2	48.0	75.5
786-787	Austin	95.5	59.4	79.6

Location Factors

STATE/ZIP	CITY	MAT.	INST.	TOTAL
TEXAS (CONT'D)				
788	Del Rio	96.9	33.7	69.0
789	Giddings	93.9	41.6	70.8
790-791	Amarillo	98.6	60.5	81.8
792	Childress	97.4	53.4	78.0
793-794	Lubbock	99.9	53.8	79.6
795-796	Abilene	98.2	52.8	78.2
797	Midland	99.6	50.5	77.9
798-799,885	El Paso	97.4	52.7	77.6
UTAH				
840-841	Salt Lake City	103.0	65.5	86.5
842,844	Ogden	98.2	65.0	83.6
843	Logan	100.3	65.0	84.7
845	Price	100.7	44.4	75.8
846-847	Provo	100.8	65.4	85.2
VERMONT				
050	White River Jct.	97.6	56.1	79.3
051	Bellows Falls	95.9	67.2	83.2
052	Bennington	96.3	67.0	83.4
053	Brattleboro	96.7	71.5	85.6
054	Burlington	100.6	67.3	85.9
056	Montpelier	97.1	67.7	84.1
057	Rutland	98.8	67.7	85.0
058	St. Johnsbury	97.6	56.6	79.5
059	Guildhall	96.2	56.4	78.6
VIRGINIA				
220-221	Fairfax	99.6	82.8	92.2
222	Arlington	101.2	81.0	92.3
223	Alexandria	100.3	85.4	93.7
224-225	Fredericksburg	98.2	73.9	87.5
226	Winchester	98.9	64.2	83.5
227	Culpeper	98.8	69.6	85.9
228	Harrisonburg	99.0	65.7	84.3
229	Charlottesville	99.4	65.2	84.3
230-232	Richmond	101.0	68.7	86.7
233-235	Norfolk	101.4	69.3	87.2
236	Newport News	100.0	69.2	86.4
237	Portsmouth	99.4	66.2	84.8
238	Petersburg	99.3	68.7	85.8
239	Farmville	98.4	54.5	79.0
240-241	Roanoke	100.7	61.9	83.6
242	Bristol	98.0	56.3	79.6
243	Pulaski	97.7	52.3	77.7
244	Staunton	98.6	60.7	81.9
245	Lynchburg	98.8	66.1	84.3
246	Grundy	98.1	51.7	77.6
WASHINGTON				
980-981,987	Seattle	104.6	104.7	104.6
982	Everett	104.2	95.1	100.1
983-984	Tacoma	104.3	99.4	102.2
985	Olympia	102.5	99.4	101.1
986	Vancouver	105.8	90.8	99.2
988	Wenatchee	103.9	78.5	92.7
989	Yakima	104.6	86.1	96.4
990-992	Spokane	104.9	77.9	93.0
993	Richland	104.6	81.6	94.5
994	Clarkston	102.8	78.3	92.0
WEST VIRGINIA				
247-248	Bluefield	96.7	77.1	88.1
249	Lewisburg	98.6	82.8	91.6
250-253	Charleston	99.8	89.1	95.0
254	Martinsburg	98.3	80.3	90.3
255-257	Huntington	100.2	91.7	96.4
258-259	Beckley	96.6	86.7	92.2
260	Wheeling	100.3	89.2	95.4
261	Parkersburg	99.1	87.8	94.1
262	Buckhannon	98.2	89.0	94.1
263-264	Clarksburg	98.8	87.7	93.9
265	Morgantown	98.9	88.3	94.2
266	Gassaway	98.1	89.3	94.2
267	Romney	98.1	82.7	91.3
268	Petersburg	97.9	85.4	92.4
WISCONSIN				
530,532	Milwaukee	100.5	106.6	103.2
531	Kenosha	100.4	97.6	99.1
534	Racine	99.7	99.1	99.5
535	Beloit	99.6	92.1	96.3
537	Madison	99.7	95.5	97.8

STATE/ZIP	CITY	MAT.	INST.	TOTAL
WISCONSIN (CONT'D)				
538	Lancaster	96.9	86.8	92.4
539	Portage	95.4	90.6	93.3
540	New Richmond	95.7	89.9	93.2
541-543	Green Bay	100.5	88.5	95.2
544	Wausau	94.8	89.4	92.4
545	Rhinelander	98.3	85.2	92.5
546	La Crosse	96.6	89.0	93.2
547	Eau Claire	98.3	90.5	94.8
548	Superior	95.5	93.6	94.7
549	Oshkosh	95.5	86.6	91.6
WYOMING				
820	Cheyenne	100.0	63.2	83.7
821	Yellowstone Nat'l Park	97.4	56.2	79.2
822	Wheatland	98.5	56.2	79.8
823	Rawlins	100.3	56.0	80.7
824	Worland	98.0	54.2	78.7
825	Riverton	99.2	54.2	79.3
826	Casper	100.5	56.0	80.9
827	Newcastle	97.8	56.0	79.3
828	Sheridan	100.7	58.0	81.9
829-831	Rock Springs	102.2	54.2	81.0
CANADIAN FACTORS (reflect Canadian currency)				
ALBERTA				
	Calgary	129.4	86.3	110.4
	Edmonton	130.6	86.2	111.0
	Fort McMurray	126.0	88.8	109.6
	Lethbridge	119.9	88.3	105.9
	Lloydminster	115.1	85.2	101.9
	Medicine Hat	115.3	84.5	101.7
	Red Deer	115.9	84.5	102.0
BRITISH COLUMBIA				
	Kamloops	116.4	86.9	103.4
	Prince George	117.6	86.9	104.1
	Vancouver	128.7	78.6	106.6
	Victoria	117.7	75.6	99.1
MANITOBA				
	Brandon	115.0	72.9	96.4
	Portage la Prairie	115.0	71.6	95.9
	Winnipeg	127.7	66.7	100.8
NEW BRUNSWICK				
	Bathurst	113.2	64.8	91.8
	Dalhousie	113.2	64.8	91.8
	Fredericton	115.8	69.2	95.2
	Moncton	113.5	65.6	92.3
	Newcastle	113.2	64.8	91.8
	St. John	115.9	69.2	95.3
NEWFOUNDLAND				
	Corner Brook	118.7	65.0	95.0
	St Johns	119.2	66.1	95.7
NORTHWEST TERRITORIES				
	Yellowknife	120.5	82.0	103.5
NOVA SCOTIA				
	Bridgewater	114.8	71.7	95.8
	Dartmouth	116.2	71.7	96.5
	Halifax	117.4	74.6	98.5
	New Glasgow	114.2	71.7	95.5
	Sydney	111.6	71.7	94.0
	Truro	114.2	71.7	95.5
	Yarmouth	114.1	71.7	95.4
ONTARIO				
	Barrie	118.0	89.2	105.3
	Brantford	117.1	92.9	106.4
	Cornwall	116.9	89.5	104.8
	Hamilton	121.9	93.2	109.2
	Kingston	117.8	89.6	105.4
	Kitchener	113.3	88.5	102.4
	London	120.6	89.6	106.9
	North Bay	117.1	87.5	104.0
	Oshawa	117.0	90.9	105.5
	Ottawa	121.5	90.1	107.6
	Owen Sound	118.1	87.5	104.6
	Peterborough	117.1	89.3	104.8
	Sarnia	116.8	93.5	106.5
	Sault Ste Marie	111.7	88.8	101.6

Location Factors

STATE/ZIP	CITY	MAT.	INST.	TOTAL
	St. Catharines	111.4	90.4	102.1
	Sudbury	111.4	89.1	101.6
	Thunder Bay	112.9	89.1	102.4
	Timmins	117.3	87.5	104.1
	Toronto	122.5	95.8	110.7
	Windsor	112.0	89.4	102.0
PRINCE EDWARD ISLAND				
	Charlottetown	116.6	60.5	91.8
	Summerside	116.1	60.5	91.6
QUEBEC				
	Cap-de-la-Madeleine	113.8	84.0	100.7
	Charlesbourg	113.8	84.0	100.7
	Chicoutimi	112.8	89.4	102.5
	Gatineau	113.4	83.8	100.3
	Granby	113.7	83.8	100.5
	Hull	113.6	83.8	100.4
	Joliette	114.1	84.0	100.8
	Laval	113.5	83.8	100.4
	Montreal	120.8	89.7	107.1
	Quebec	119.4	90.1	106.4
	Rimouski	113.3	89.4	102.8
	Rouyn-Noranda	113.4	83.8	100.3
	Saint Hyacinthe	113.2	83.8	100.2
	Sherbrooke	113.8	83.8	100.5
	Sorel	114.0	84.0	100.8
	St Jerome	113.4	83.8	100.4
	Trois Rivieres	114.0	84.0	100.8
SASKATCHEWAN				
	Moose Jaw	112.2	66.9	92.2
	Prince Albert	111.1	65.2	90.9
	Regina	114.2	67.0	93.4
	Saskatoon	112.7	66.8	92.4
YUKON				
	Whitehorse	112.0	65.9	91.7

General Requirements — R0111 Summary of Work

R011105-05 Tips for Accurate Estimating

1. Use pre-printed or columnar forms for orderly sequence of dimensions and locations and for recording telephone quotations.
2. Use only the front side of each paper or form except for certain pre-printed summary forms.
3. Be consistent in listing dimensions: For example, length x width x height. This helps in rechecking to ensure that, the total length of partitions is appropriate for the building area.
4. Use printed (rather than measured) dimensions where given.
5. Add up multiple printed dimensions for a single entry where possible.
6. Measure all other dimensions carefully.
7. Use each set of dimensions to calculate multiple related quantities.
8. Convert foot and inch measurements to decimal feet when listing. Memorize decimal equivalents to .01 parts of a foot (1/8″ equals approximately .01′).
9. Do not "round off" quantities until the final summary.
10. Mark drawings with different colors as items are taken off.
11. Keep similar items together, different items separate.
12. Identify location and drawing numbers to aid in future checking for completeness.
13. Measure or list everything on the drawings or mentioned in the specifications.
14. It may be necessary to list items not called for to make the job complete.
15. Be alert for: Notes on plans such as N.T.S. (not to scale); changes in scale throughout the drawings; reduced size drawings; discrepancies between the specifications and the drawings.
16. Develop a consistent pattern of performing an estimate. For example:
    a. Start the quantity takeoff at the lower floor and move to the next higher floor.
    b. Proceed from the main section of the building to the wings.
    c. Proceed from south to north or vice versa, clockwise or counterclockwise.
    d. Take off floor plan quantities first, elevations next, then detail drawings.
17. List all gross dimensions that can be either used again for different quantities, or used as a rough check of other quantities for verification (exterior perimeter, gross floor area, individual floor areas, etc.).
18. Utilize design symmetry or repetition (repetitive floors, repetitive wings, symmetrical design around a center line, similar room layouts, etc.). Note: Extreme caution is needed here so as not to omit or duplicate an area.
19. Do not convert units until the final total is obtained. For instance, when estimating concrete work, keep all units to the nearest cubic foot, then summarize and convert to cubic yards.
20. When figuring alternatives, it is best to total all items involved in the basic system, then total all items involved in the alternates. Therefore you work with positive numbers in all cases. When adds and deducts are used, it is often confusing whether to add or subtract a portion of an item; especially on a complicated or involved alternate.

R011110-10 Architectural Fees

Tabulated below are typical percentage fees by project size, for good professional architectural service. Fees may vary from those listed depending upon degree of design difficulty and economic conditions in any particular area.

Rates can be interpolated horizontally and vertically. Various portions of the same project requiring different rates should be adjusted proportionately. For alterations, add 50% to the fee for the first $500,000 of project cost and add 25% to the fee for project cost over $500,000.

Architectural fees tabulated below include Structural, Mechanical and Electrical Engineering Fees. They do not include the fees for special consultants such as kitchen planning, security, acoustical, interior design, etc.

Civil Engineering fees are included in the Architectural fee for project sites requiring minimal design such as city sites. However, separate Civil Engineering fees must be added when utility connections require design, drainage calculations are needed, stepped foundations are required, or provisions are required to protect adjacent wetlands.

Building Types	Total Project Size in Thousands of Dollars						
	100	250	500	1,000	5,000	10,000	50,000
Factories, garages, warehouses, repetitive housing	9.0%	8.0%	7.0%	6.2%	5.3%	4.9%	4.5%
Apartments, banks, schools, libraries, offices, municipal buildings	12.2	12.3	9.2	8.0	7.0	6.6	6.2
Churches, hospitals, homes, laboratories, museums, research	15.0	13.6	12.7	11.9	9.5	8.8	8.0
Memorials, monumental work, decorative furnishings	—	16.0	14.5	13.1	10.0	9.0	8.3

General Requirements — R0111 Summary of Work

R011110-30 Engineering Fees

Typical **Structural Engineering Fees** based on type of construction and total project size. These fees are included in Architectural Fees.

Type of Construction	Total Project Size (in thousands of dollars)			
	$500	$500-$1,000	$1,000-$5,000	Over $5000
Industrial buildings, factories & warehouses	Technical payroll times 2.0 to 2.5	1.60%	1.25%	1.00%
Hotels, apartments, offices, dormitories, hospitals, public buildings, food stores		2.00%	1.70%	1.20%
Museums, banks, churches and cathedrals		2.00%	1.75%	1.25%
Thin shells, prestressed concrete, earthquake resistive		2.00%	1.75%	1.50%
Parking ramps, auditoriums, stadiums, convention halls, hangars & boiler houses		2.50%	2.00%	1.75%
Special buildings, major alterations, underpinning & future expansion	↓	Add to above 0.5%	Add to above 0.5%	Add to above 0.5%

For complex reinforced concrete or unusually complicated structures, add 20% to 50%.

Typical **Mechanical and Electrical Engineering Fees** are based on the size of the subcontract. The fee structure for both are shown below. These fees are included in Architectural Fees.

Type of Construction	Subcontract Size							
	$25,000	$50,000	$100,000	$225,000	$350,000	$500,000	$750,000	$1,000,000
Simple structures	6.4%	5.7%	4.8%	4.5%	4.4%	4.3%	4.2%	4.1%
Intermediate structures	8.0	7.3	6.5	5.6	5.1	5.0	4.9	4.8
Complex structures	10.1	9.0	9.0	8.0	7.5	7.5	7.0	7.0

For renovations, add 15% to 25% to applicable fee.

General Requirements — R0121 Allowances

R012153-10 Repair and Remodeling

Cost figures are based on new construction utilizing the most cost-effective combination of labor, equipment and material with the work scheduled in proper sequence to allow the various trades to accomplish their work in an efficient manner.

The costs for repair and remodeling work must be modified due to the following factors that may be present in any given repair and remodeling project.

1. Equipment usage curtailment due to the physical limitations of the project, with only hand-operated equipment being used.
2. Increased requirement for shoring and bracing to hold up the building while structural changes are being made and to allow for temporary storage of construction materials on above-grade floors.
3. Material handling becomes more costly due to having to move within the confines of an enclosed building. For multi-story construction, low capacity elevators and stairwells may be the only access to the upper floors.
4. Large amount of cutting and patching and attempting to match the existing construction is required. It is often more economical to remove entire walls rather than create many new door and window openings. This sort of trade-off has to be carefully analyzed.
5. Cost of protection of completed work is increased since the usual sequence of construction usually cannot be accomplished.
6. Economies of scale usually associated with new construction may not be present. If small quantities of components must be custom fabricated due to job requirements, unit costs will naturally increase. Also, if only small work areas are available at a given time, job scheduling between trades becomes difficult and subcontractor quotations may reflect the excessive start-up and shut-down phases of the job.
7. Work may have to be done on other than normal shifts and may have to be done around an existing production facility which has to stay in production during the course of the repair and remodeling.
8. Dust and noise protection of adjoining non-construction areas can involve substantial special protection and alter usual construction methods.
9. Job may be delayed due to unexpected conditions discovered during demolition or removal. These delays ultimately increase construction costs.
10. Piping and ductwork runs may not be as simple as for new construction. Wiring may have to be snaked through walls and floors.
11. Matching "existing construction" may be impossible because materials may no longer be manufactured. Substitutions may be expensive.
12. Weather protection of existing structure requires additional temporary structures to protect building at openings.
13. On small projects, because of local conditions, it may be necessary to pay a tradesman for a minimum of four hours for a task that is completed in one hour.

All of the above areas can contribute to increased costs for a repair and remodeling project. Each of the above factors should be considered in the planning, bidding and construction stage in order to minimize the increased costs associated with repair and remodeling jobs.

General Requirements — R0129 Payment Procedures

R012909-80 Sales Tax by State

State sales tax on materials is tabulated below (5 states have no sales tax). Many states allow local jurisdictions, such as a county or city, to levy additional sales tax.

Some projects may be sales tax exempt, particularly those constructed with public funds.

State	Tax (%)	State	Tax (%)	State	Tax (%)	State	Tax (%)
Alabama	4	Illinois	6.25	Montana	0	Rhode Island	7
Alaska	0	Indiana	7	Nebraska	5.5	South Carolina	6
Arizona	5.6	Iowa	6	Nevada	6.5	South Dakota	4
Arkansas	6	Kansas	5.3	New Hampshire	0	Tennessee	7
California	8.25	Kentucky	6	New Jersey	7	Texas	6.25
Colorado	2.9	Louisiana	4	New Mexico	5	Utah	4.75
Connecticut	6	Maine	5	New York	4	Vermont	6
Delaware	0	Maryland	6	North Carolina	4.5	Virginia	4
District of Columbia	5.75	Massachusetts	6.25	North Dakota	5	Washington	6.5
Florida	6	Michigan	6	Ohio	7	West Virginia	6
Georgia	4	Minnesota	6.5	Oklahoma	4.5	Wisconsin	5
Hawaii	4	Mississippi	7	Oregon	0	Wyoming	4
Idaho	6	Missouri	4.225	Pennsylvania	6	Average	5.01%

Sales Tax by Province (Canada)

GST - a value-added tax, which the government imposes on most goods and services provided in or imported into Canada. PST - a retail sales tax, which five of the provinces impose on the price of most goods and some services. QST - a value-added tax, similar to the federal GST, which Quebec imposes. HST - Three provinces have combined their retail sales tax with the federal GST into one harmonized tax.

Province	PST (%)	QST (%)	GST (%)	HST (%)
Alberta	0	0	5	0
British Columbia	7	0	5	0
Manitoba	7	0	5	0
New Brunswick	0	0	0	13
Newfoundland	0	0	0	13
Northwest Territories	0	0	5	0
Nova Scotia	0	0	0	13
Ontario	8	0	5	0
Prince Edward Island	10	0	5	0
Quebec	0	7.5	5	0
Saskatchewan	5	0	5	0
Yukon	0	0	5	0

R012909-85 Unemployment Taxes and Social Security Taxes

State Unemployment Tax rates vary not only from state to state, but also with the experience rating of the contractor. The Federal Unemployment Tax rate is 6.2% of the first $7,000 of wages. This is reduced by a credit of up to 5.4% for timely payment to the state. The minimum Federal Unemployment Tax is 0.8% after all credits.

Social Security (FICA) for 2010 is estimated at time of publication to be 7.65% of wages up to $106,800.

General Requirements — R0131 Project Management & Coordination

R013113-40 Builder's Risk Insurance

Builder's Risk Insurance is insurance on a building during construction. Premiums are paid by the owner or the contractor. Blasting, collapse and underground insurance would raise total insurance costs above those listed. Floater policy for materials delivered to the job runs $.75 to $1.25 per $100 value. Contractor equipment insurance runs $.50 to $1.50 per $100 value. Insurance for miscellaneous tools to $1,500 value runs from $3.00 to $7.50 per $100 value.

Tabulated below are New England Builder's Risk insurance rates in dollars per $100 value for $1,000 deductible. For $25,000 deductible, rates can be reduced 13% to 34%. On contracts over $1,000,000, rates may be lower than those tabulated. Policies are written annually for the total completed value in place. For "all risk" insurance (excluding flood, earthquake and certain other perils) add $.025 to total rates below.

Coverage	Frame Construction (Class 1) Range	Frame Construction (Class 1) Average	Brick Construction (Class 4) Range	Brick Construction (Class 4) Average	Fire Resistive (Class 6) Range	Fire Resistive (Class 6) Average
Fire Insurance	$.350 to $.850	$.600	$.158 to $.189	$.174	$.052 to $.080	$.070
Extended Coverage	.115 to .200	.158	.080 to .105	.101	.081 to .105	.100
Vandalism	.012 to .016	.014	.008 to .011	.011	.008 to .011	.010
Total Annual Rate	$.477 to $1.066	$.772	$.246 to $.305	$.286	$.141 to $.196	$.180

R013113-50 General Contractor's Overhead

There are two distinct types of overhead on a construction project: Project Overhead and Main Office Overhead. Project Overhead includes those costs at a construction site not directly associated with the installation of construction materials. Examples of Project Overhead costs include the following:

1. Superintendent
2. Construction office and storage trailers
3. Temporary sanitary facilities
4. Temporary utilities
5. Security fencing
6. Photographs
7. Clean up
8. Performance and payment bonds

The above Project Overhead items are also referred to as General Requirements and therefore are estimated in Division 1. Division 1 is the first division listed in the CSI MasterFormat but it is usually the last division estimated. The sum of the costs in Divisions 1 through 49 is referred to as the sum of the direct costs.

All construction projects also include indirect costs. The primary components of indirect costs are the contractor's Main Office Overhead and profit. The amount of the Main Office Overhead expense varies depending on the the following:

1. Owner's compensation
2. Project managers and estimator's wages
3. Clerical support wages
4. Office rent and utilities
5. Corporate legal and accounting costs
6. Advertising
7. Automobile expenses
8. Association dues
9. Travel and entertainment expenses

These costs are usually calculated as a percentage of annual sales volume. This percentage can range from 35% for a small contractor doing less than $500,000 to 5% for a large contractor with sales in excess of $100 million.

General Requirements — R0131 Project Management & Coordination

R013113-60 Workers' Compensation Insurance Rates by Trade

The table below tabulates the national averages for Workers' Compensation insurance rates by trade and type of building. The average "Insurance Rate" is multiplied by the "% of Building Cost" for each trade. This produces the "Workers' Compensation Cost" by % of total labor cost, to be added for each trade by building type to determine the weighted average Workers' Compensation rate for the building types analyzed.

Trade	Insurance Rate (% Labor Cost) Range	Insurance Rate Average	% of Building Cost Office Bldgs.	% of Building Cost Schools & Apts.	% of Building Cost Mfg.	Workers' Compensation Office Bldgs.	Workers' Compensation Schools & Apts.	Workers' Compensation Mfg.
Excavation, Grading, etc.	4.2% to 17.5%	9.4%	4.8%	4.9%	4.5%	0.45%	0.46%	0.42%
Piles & Foundations	4.8 to 57.1	18.5	7.1	5.2	8.7	1.31	0.96	1.61
Concrete	4.8 to 29.9	13.7	5.0	14.8	3.7	0.69	2.03	0.51
Masonry	4.6 to 29.8	13.5	6.9	7.5	1.9	0.93	1.01	0.26
Structural Steel	4.8 to 170.8	36.0	10.7	3.9	17.6	3.85	1.40	6.34
Miscellaneous & Ornamental Metals	4 to 20.3	10.1	2.8	4.0	3.6	0.28	0.40	0.36
Carpentry & Millwork	4.8 to 53.2	16.9	3.7	4.0	0.5	0.63	0.68	0.08
Metal or Composition Siding	4.8 to 35.7	15.5	2.3	0.3	4.3	0.36	0.05	0.67
Roofing	4.8 to 77.1	28.9	2.3	2.6	3.1	0.66	0.75	0.90
Doors & Hardware	3.4 to 34.5	11.3	0.9	1.4	0.4	0.10	0.16	0.05
Sash & Glazing	4.8 to 36	13.3	3.5	4.0	1.0	0.47	0.53	0.13
Lath & Plaster	3.1 to 30.3	11.9	3.3	6.9	0.8	0.39	0.82	0.10
Tile, Marble & Floors	3.1 to 16.8	8.8	2.6	3.0	0.5	0.23	0.26	0.04
Acoustical Ceilings	4.4 to 32	9.8	2.4	0.2	0.3	0.24	0.02	0.03
Painting	4.8 to 29.6	11.5	1.5	1.6	1.6	0.17	0.18	0.18
Interior Partitions	4.8 to 53.2	16.9	3.9	4.3	4.4	0.66	0.73	0.74
Miscellaneous Items	2.4 to 134.5	14.9	5.2	3.7	9.7	0.78	0.55	1.45
Elevators	2.6 to 13.5	6.3	2.1	1.1	2.2	0.13	0.07	0.14
Sprinklers	2.5 to 13.6	7.4	0.5	—	2.0	0.04	—	0.15
Plumbing	3.9 to 13.3	7.6	4.9	7.2	5.2	0.37	0.55	0.40
Heat., Vent., Air Conditioning	3.9 to 21.2	9.1	13.5	11.0	12.9	1.23	1.00	1.17
Electrical	2.9 to 15.4	6.4	10.1	8.4	11.1	0.65	0.54	0.71
Total	2.4% to 170.8%	—	100.0%	100.0%	100.0%	14.62%	13.15%	16.44%
		Overall Weighted Average	14.74%					

Workers' Compensation Insurance Rates by States

The table below lists the weighted average Workers' Compensation base rate for each state with a factor comparing this with the national average of 14.4%.

State	Weighted Average	Factor	State	Weighted Average	Factor	State	Weighted Average	Factor
Alabama	19.8%	138	Kentucky	16.3%	113	North Dakota	12.9%	90
Alaska	19.6	136	Louisiana	28.2	196	Ohio	13.0	90
Arizona	9.7	67	Maine	14.2	99	Oklahoma	14.5	101
Arkansas	10.4	72	Maryland	14.5	101	Oregon	12.4	86
California	21.4	149	Massachusetts	12.1	84	Pennsylvania	14.7	102
Colorado	9.0	63	Michigan	19.0	132	Rhode Island	11.6	81
Connecticut	20.0	139	Minnesota	26.0	181	South Carolina	19.8	138
Delaware	11.5	80	Mississippi	15.3	106	South Dakota	17.2	119
District of Columbia	11.3	78	Missouri	17.5	122	Tennessee	14.2	99
Florida	12.6	88	Montana	15.0	104	Texas	11.2	78
Georgia	26.8	186	Nebraska	21.2	147	Utah	9.1	63
Hawaii	16.9	117	Nevada	10.3	72	Vermont	15.0	104
Idaho	9.9	69	New Hampshire	19.2	133	Virginia	10.9	76
Illinois	23.0	160	New Jersey	13.3	92	Washington	9.8	68
Indiana	7.2	50	New Mexico	18.4	128	West Virginia	12.0	83
Iowa	11.4	79	New York	6.9	48	Wisconsin	15.0	104
Kansas	8.6	60	North Carolina	16.9	117	Wyoming	5.3	37
			Weighted Average for U.S. is	14.7% of payroll = 100%				

Rates in the following table are the base or manual costs per $100 of payroll for Workers' Compensation in each state. Rates are usually applied to straight time wages only and not to premium time wages and bonuses.

The weighted average skilled worker rate for 35 trades is 14.4%. For bidding purposes, apply the full value of Workers' Compensation directly to total labor costs, or if labor is 38%, materials 42% and overhead and profit 20% of total cost, carry 38/80 x 14.4% = 6.8% of cost (before overhead and profit) into overhead. Rates vary not only from state to state but also with the experience rating of the contractor.

Rates are the most current available at the time of publication.

General Requirements — R0131 Project Management & Coordination

R013113-60 Workers' Compensation Insurance Rates by Trade and State (cont.)

State	Carpentry — 3 stories or less	Carpentry — interior cab. work	Carpentry — general	Concrete Work — NOC	Concrete Work — flat (flr., sdwk.)	Electrical Wiring — inside	Excavation — earth NOC	Excavation — rock	Glaziers	Insulation Work	Lathing	Masonry	Painting & Decorating	Pile Driving	Plastering	Plumbing	Roofing	Sheet Metal Work (HVAC)	Steel Erection — door & sash	Steel Erection — inter., ornam.	Steel Erection — structure	Steel Erection — NOC	Tile Work — (interior ceramic)	Waterproofing	Wrecking
	5651	5437	5403	5213	5221	5190	6217	6217	5462	5479	5443	5022	5474	6003	5480	5183	5551	5538	5102	5102	5040	5057	5348	9014	5701
AL	28.11	16.91	33.75	15.19	9.51	10.24	10.24	10.24	22.09	16.53	10.78	23.87	21.44	24.16	15.68	10.18	43.60	10.03	8.28	8.28	41.21	18.79	13.84	7.13	41.21
AK	11.41	14.35	15.47	11.20	10.66	8.55	12.54	12.54	35.96	31.60	8.46	29.81	16.71	31.00	30.26	10.01	35.75	8.86	8.04	8.04	29.65	32.73	6.25	6.29	29.65
AZ	11.55	6.95	17.61	9.58	5.16	4.96	5.32	5.32	7.21	14.50	5.56	7.35	6.53	11.22	5.24	5.77	15.08	6.92	9.96	9.96	17.31	14.27	3.56	3.13	17.31
AR	9.83	6.03	13.66	8.16	6.58	4.73	6.47	6.47	7.41	9.31	5.61	7.02	9.22	12.87	8.87	5.43	17.31	5.37	5.41	5.41	25.61	31.50	5.46	3.15	25.61
CA	34.52	34.52	34.52	12.92	12.92	9.65	15.56	15.56	18.87	14.65	12.69	21.99	16.27	16.32	24.72	13.26	47.80	16.92	13.77	13.77	22.36	17.89	8.82	16.27	17.89
CO	10.43	5.56	7.98	7.19	5.24	3.48	7.24	7.24	6.53	11.47	4.36	9.05	5.77	10.08	5.59	4.64	17.78	4.59	6.83	6.83	27.27	10.66	5.43	3.46	10.66
CT	14.94	15.49	24.99	28.69	11.11	8.50	13.15	13.15	20.92	17.38	17.82	23.28	16.39	20.36	16.27	10.08	34.95	12.68	15.41	15.41	41.41	23.21	13.64	5.64	41.41
DE	11.96	11.96	8.80	8.99	9.49	4.26	7.16	7.16	10.17	8.80	10.97	10.39	11.94	14.39	10.17	6.10	21.13	7.75	9.69	9.69	21.36	9.69	7.19	10.39	21.36
DC	10.70	10.28	7.97	9.28	7.58	6.22	9.12	9.12	16.60	6.06	6.77	8.27	6.06	12.61	10.50	8.51	14.67	8.61	11.03	11.03	24.14	13.86	13.64	3.40	24.14
FL	12.62	7.98	13.59	13.18	5.99	15.44	7.01	7.01	8.77	8.12	5.53	9.08	8.76	38.24	19.38	5.47	21.29	7.98	8.12	8.12	22.59	10.37	5.16	4.53	22.59
GA	35.73	21.35	25.92	17.90	15.31	10.45	16.30	16.30	19.14	22.15	14.61	24.16	24.99	26.75	24.86	12.18	63.28	17.45	20.26	20.26	71.30	39.53	13.97	8.92	71.30
HI	17.38	11.62	28.20	13.74	12.27	7.10	7.73	7.73	20.55	21.19	10.94	17.87	11.33	20.15	15.61	6.06	33.24	8.02	11.11	11.11	31.71	21.70	9.78	11.54	31.71
ID	9.41	6.65	10.29	10.24	5.84	3.48	5.84	5.84	8.87	8.06	9.22	7.35	7.88	11.56	5.48	5.01	26.07	6.56	7.00	7.00	24.33	9.22	9.07	4.91	24.33
IL	25.30	15.57	23.56	29.89	12.90	10.43	10.77	10.77	18.01	20.90	17.31	21.46	11.94	31.27	17.49	12.01	33.86	14.59	19.11	19.11	79.28	21.91	16.76	5.22	79.28
IN	11.86	4.44	6.93	5.02	3.12	2.85	4.38	4.38	6.11	7.10	22.54	4.56	4.86	5.84	3.09	12.76	10.06	4.48	3.95	3.95	12.69	6.07	3.12	2.39	12.69
IA	13.14	10.98	12.06	14.63	7.56	4.65	7.06	7.06	9.01	6.62	5.84	9.42	7.53	9.92	7.28	6.74	22.71	6.08	5.56	5.56	30.86	26.76	8.01	3.61	30.86
KS	12.20	7.16	9.07	6.95	5.58	3.39	5.22	5.22	9.12	7.26	4.46	7.09	7.12	9.74	5.27	4.53	12.45	5.13	4.53	4.53	24.78	16.44	6.24	4.25	16.44
KY	14.98	12.29	25.40	10.80	7.22	5.36	9.95	9.95	13.90	14.09	9.23	7.30	11.88	22.46	12.14	5.65	31.57	11.42	9.86	9.86	60.00	18.72	16.40	4.59	60.00
LA	22.02	24.93	53.17	26.43	15.61	9.88	17.49	17.49	20.14	21.43	24.51	28.33	29.58	31.22	22.37	8.64	77.12	21.20	18.98	18.98	51.76	24.21	13.77	13.78	66.41
ME	15.11	10.18	25.32	18.95	9.40	6.61	8.79	8.79	14.50	11.15	7.69	13.82	13.31	15.89	12.02	9.12	18.69	8.40	8.34	8.34	32.51	19.53	6.58	5.84	32.51
MD	14.33	11.61	13.75	15.69	5.67	5.60	7.82	7.82	11.43	13.56	8.01	9.92	7.12	20.15	12.77	5.69	29.53	6.93	8.31	8.31	52.61	26.15	7.99	5.44	26.15
MA	7.50	5.93	11.92	21.45	6.62	3.17	4.17	4.17	8.93	8.94	6.53	11.62	5.01	4.99	4.89	3.88	30.35	5.74	6.76	6.76	44.61	39.97	6.45	2.50	29.46
MI	20.53	10.95	22.88	15.72	9.41	5.27	10.95	10.95	14.98	13.04	13.59	14.45	13.57	57.09	12.36	7.21	32.64	11.89	11.07	11.07	57.09	19.42	10.83	6.26	57.09
MN	22.68	23.13	32.18	10.25	13.85	6.68	11.88	11.88	27.48	16.33	12.68	19.35	14.58	22.85	12.68	11.28	54.88	16.13	11.70	11.70	170.80	13.03	11.23	6.85	13.03
MS	16.49	10.30	17.28	10.71	7.75	6.48	11.17	11.17	10.34	10.46	6.84	13.96	12.78	30.91	19.04	7.94	34.45	7.79	12.45	12.45	33.05	15.81	8.83	3.94	33.05
MO	20.17	12.95	15.43	19.32	10.66	7.57	10.57	10.57	11.63	15.03	9.23	16.87	12.93	18.60	16.85	10.79	34.32	9.57	11.61	11.61	33.95	41.29	11.92	6.56	33.95
MT	13.10	10.20	23.32	12.63	10.84	5.93	14.51	14.51	8.57	24.07	7.98	12.12	8.43	22.50	10.60	9.71	37.56	10.63	9.09	9.09	22.59	11.57	9.08	7.91	11.57
NE	23.73	18.63	20.88	26.18	14.23	9.80	16.00	16.00	16.90	23.33	11.75	21.90	17.52	20.73	17.43	12.53	36.95	12.48	15.00	15.00	39.83	36.08	10.53	6.83	55.60
NV	14.08	7.98	12.08	9.61	8.20	6.12	7.77	7.77	10.06	7.54	4.60	7.35	7.54	11.58	6.73	6.12	12.82	9.04	8.35	8.35	22.13	16.86	5.83	5.36	16.86
NH	26.15	11.92	19.85	24.52	14.29	7.33	13.00	13.00	10.32	5.63	9.25	21.17	21.90	7.88	10.06	10.99	45.41	8.04	15.28	15.28	64.41	23.21	11.64	6.10	64.41
NJ	14.79	8.84	14.79	15.65	9.31	4.38	8.18	8.18	8.97	12.55	10.73	13.44	11.40	16.59	10.73	5.79	37.97	6.78	9.33	9.33	20.26	16.09	7.96	6.20	21.39
NM	20.42	7.36	18.00	14.67	12.11	8.39	9.67	9.67	21.18	13.47	8.06	15.88	11.90	20.37	15.53	8.71	39.03	11.93	13.94	13.94	44.89	47.45	7.36	6.06	44.89
NY	7.96	3.39	7.17	8.25	6.46	3.17	4.63	4.63	5.71	3.73	6.49	8.74	5.50	8.04	4.21	3.98	14.20	5.50	5.44	5.44	13.31	7.90	3.26	2.83	5.51
NC	16.57	12.67	16.19	18.29	8.30	10.73	11.86	11.86	14.23	12.38	11.28	11.33	11.44	17.48	14.68	9.46	26.41	11.87	10.88	10.88	62.71	20.69	9.08	5.75	62.71
ND	10.87	10.87	10.87	6.50	6.50	3.55	5.58	5.58	10.87	10.87	7.25	17.45	6.87	19.11	7.25	5.19	21.35	5.19	19.14	19.14	19.14	19.14	10.87	21.35	10.90
OH	13.61	7.75	11.13	9.75	9.79	5.67	8.20	8.20	8.42	15.84	31.95	11.77	13.74	13.44	4.26	6.07	30.45	6.93	8.35	8.35	20.46	11.61	9.82	6.91	11.61
OK	13.42	9.17	11.76	11.76	6.34	5.73	10.83	10.83	16.07	19.47	8.45	10.02	8.47	20.25	12.19	7.09	21.13	9.05	13.76	13.76	39.72	23.97	6.50	5.71	39.72
OR	16.06	7.50	14.54	11.48	7.70	4.95	9.25	9.25	12.89	11.25	7.90	12.08	9.99	16.97	11.59	5.41	21.99	7.70	6.54	6.54	22.55	21.87	8.85	5.56	22.55
PA	14.68	14.68	12.10	15.30	11.22	6.32	9.13	9.13	12.06	12.10	12.06	13.10	14.00	16.59	12.06	7.77	29.88	8.29	15.11	15.11	22.64	15.11	8.45	13.10	22.64
RI	10.79	8.37	10.96	13.14	10.29	5.15	7.50	7.50	13.33	17.64	6.04	11.62	9.27	27.21	9.50	4.42	14.82	7.75	5.68	5.68	20.92	14.98	5.62	4.72	14.98
SC	21.43	18.13	22.49	16.89	9.59	11.45	11.78	11.78	16.99	18.58	11.06	13.93	15.50	20.61	15.75	12.36	50.08	13.06	13.91	13.91	41.22	32.53	10.50	6.33	41.22
SD	21.47	9.62	21.86	25.23	7.35	5.33	9.54	9.54	15.06	13.87	8.26	11.62	10.45	20.56	11.87	10.92	26.74	12.70	8.86	8.86	59.15	31.01	7.69	5.99	59.15
TN	14.91	11.20	13.03	12.97	6.75	5.79	13.34	13.34	12.18	9.21	8.32	12.35	11.84	28.35	13.71	6.79	24.19	9.81	8.15	8.15	25.81	25.53	8.36	4.08	25.81
TX	9.17	8.39	9.17	9.00	7.06	5.76	7.88	7.88	8.56	11.97	5.74	10.31	8.18	16.38	8.18	5.88	19.42	12.05	7.87	7.87	33.20	11.24	5.53	6.19	8.70
UT	11.93	6.73	9.82	7.43	6.42	3.47	8.78	8.78	11.14	9.47	4.81	8.51	9.45	5.50	4.68	18.93	6.11	5.43	5.43	19.45	11.11	5.16	3.24	13.92	
VT	11.28	10.68	17.87	12.39	7.41	6.20	12.96	12.96	15.29	16.13	7.53	16.28	8.35	14.60	9.04	8.41	26.14	9.59	9.47	9.47	36.84	30.71	9.76	6.97	36.84
VA	10.22	7.61	9.05	11.32	5.11	4.92	7.70	7.70	10.13	7.82	11.06	8.34	9.02	9.49	9.78	5.63	22.11	6.32	6.90	6.90	32.08	20.56	5.08	2.91	32.08
WA	7.94	7.94	7.94	7.29	7.29	3.15	6.88	6.88	12.21	8.54	7.94	9.48	11.84	16.40	10.06	4.30	17.24	3.92	7.61	7.61	7.61	7.61	8.83	17.24	7.61
WV	14.45	9.91	13.88	11.18	5.99	6.38	8.45	8.45	10.70	9.85	8.29	10.83	10.72	16.15	10.77	6.65	24.50	7.80	7.57	7.57	32.81	7.91	7.67	3.60	17.91
WI	10.46	10.33	15.46	11.52	9.18	5.80	8.13	8.13	11.32	12.60	5.84	14.90	14.06	19.42	10.64	6.15	34.47	7.77	9.80	9.80	22.52	48.12	15.46	4.83	22.52
WY	4.77	4.77	4.77	4.77	4.77	4.77	4.77	4.77	4.77	4.77	4.77	4.77	4.77	4.77	4.77	4.77	4.77	4.77	4.77	4.77	4.77	4.77	4.77	4.77	4.77
AVG.	15.47	11.27	16.88	13.72	8.74	6.38	9.42	9.42	13.27	13.11	9.79	13.47	11.46	18.54	11.92	7.62	28.89	9.14	10.07	10.07	35.99	20.79	8.78	6.48	30.31

General Requirements — R0131 Project Management & Coordination

R013113-60 Workers' Compensation (cont.) (Canada in Canadian dollars)

Province		Alberta	British Columbia	Manitoba	Ontario	New Brunswick	Newfndld. & Labrador	Northwest Territories	Nova Scotia	Prince Edward Island	Quebec	Saskatchewan	Yukon
Carpentry—3 stories or less	Rate	5.99	3.64	4.17	4.35	5.05	11.00	3.40	9.23	6.23	13.26	4.37	9.42
	Code	42143	721028	40102	723	238130	4226	4-41	4226	401	80110	1226	202
Carpentry—interior cab. work	Rate	1.86	2.62	4.17	4.35	4.59	4.37	3.40	4.33	3.82	13.26	2.57	9.42
	Code	42133	721021	40102	723	238350	4279	4-41	4274	402	80110	B1127	202
CARPENTRY—general	Rate	5.99	3.64	4.17	4.35	5.05	4.37	3.40	9.23	6.23	13.26	4.37	9.42
	Code	42143	721028	40102	723	238130	4299	4-41	4226	401	80110	12.02	202
CONCRETE WORK—NOC	Rate	4.01	3.87	7.01	16.50	5.05	11.00	3.40	5.19	6.23	15.41	5.44	5.23
	Code	42104	721010	40110	748	238110	4224	4-41	4224	401	80100	B1314	203
CONCRETE WORK—flat (flr. sidewalk)	Rate	4.01	3.87	7.01	16.50	5.05	11.00	3.40	5.19	6.23	15.41	5.44	5.23
	Code	42104	721010	40110	748	238110	4224	4-41	4224	401	80100	B1314	203
ELECTRICAL Wiring—inside	Rate	1.95	1.41	1.90	3.25	2.45	2.91	2.76	2.66	3.82	5.92	2.57	5.23
	Code	42124	721019	40203	704	238210	4261	4-46	4261	402	80170	B1105	206
EXCAVATION—earth NOC	Rate	2.34	3.25	4.02	4.68	5.05	3.84	3.57	2.88	3.13	7.55	2.76	5.23
	Code	40604	721031	40706	711	238190	4214	4-43	4214	404	80030	R1106	207
EXCAVATION—rock	Rate	2.34	3.25	4.02	4.68	5.05	3.84	3.57	2.88	3.13	7.55	2.76	5.23
	Code	40604	721031	40706	711	238190	4214	4-43	4214	404	80030	R1106	207
GLAZIERS	Rate	2.56	3.05	4.17	9.25	5.05	6.11	3.40	5.19	3.82	14.17	5.44	5.23
	Code	42121	715020	40109	751	238150	4233	4-41	4233	402	80150	B1304	212
INSULATION WORK	Rate	2.05	6.91	4.17	9.25	4.59	6.11	3.40	5.19	6.23	13.26	4.37	9.42
	Code	42184	721029	40102	751	238310	4234	4-41	4234	401	80110	B1207	202
LATHING	Rate	4.67	6.27	4.17	4.35	4.59	4.37	3.40	4.33	3.82	13.26	5.44	9.42
	Code	42135	721042	40102	723	238390	4279	4-41	4271	402	80110	B1316	202
MASONRY	Rate	4.01	3.50	4.17	11.15	5.05	6.11	3.40	5.19	6.23	15.41	5.44	9.42
	Code	42102	721037	40102	741	238140	4231	4-41	4231	401	80100	B1318	202
PAINTING & DECORATING	Rate	3.30	3.36	3.28	6.75	4.59	4.37	3.40	4.33	3.82	13.26	4.37	9.42
	Code	42111	721041	40105	719	238320	4275	4-41	4275	402	80110	B1201	202
PILE DRIVING	Rate	3.95	3.82	4.02	6.34	5.05	3.84	3.57	5.19	6.23	7.55	5.44	9.42
	Code	42159	722004	40706	732	238190	4129	4-43	4221	401	80030	B1310	202
PLASTERING	Rate	4.67	6.27	4.59	6.75	4.59	4.37	3.40	4.33	3.82	13.26	4.37	9.42
	Code	42135	721042	40108	719	238390	4271	4-41	4271	402	80110	B1221	202
PLUMBING	Rate	1.95	1.90	3.04	3.98	2.45	3.08	2.76	2.52	3.82	6.19	2.57	5.23
	Code	42122	721043	40204	707	238220	4241	4-46	4241	402	80160	B1101	214
ROOFING	Rate	6.41	6.18	7.37	13.30	5.05	11.00	3.40	9.23	6.23	19.71	5.44	9.42
	Code	42118	721036	40403	728	238160	4236	4-41	4236	401	80130	B1320	202
SHEET METAL WORK (HVAC)	Rate	1.95	1.90	7.37	3.98	2.45	3.08	2.76	2.52	3.82	6.19	2.57	5.23
	Code	42117	721043	40402	707	238220	4244	4-46	4244	402	80160	B1107	208
STEEL ERECTION—door & sash	Rate	2.05	12.66	10.05	16.50	5.05	11.00	3.40	5.19	6.23	21.98	5.44	9.42
	Code	42106	722005	40502	748	238120	4227	4-41	4227	401	80080	B1322	202
STEEL ERECTION—inter., ornam.	Rate	2.05	12.66	10.05	16.50	5.05	11.00	3.40	5.19	6.23	21.98	5.44	9.42
	Code	42106	722005	40502	748	238120	4227	4-41	4227	401	80080	B1322	202
STEEL ERECTION—structure	Rate	2.05	12.66	10.05	16.50	5.05	11.00	3.40	5.19	6.23	21.98	5.44	9.42
	Code	42106	722005	40502	748	238120	4227	4-41	4227	401	80080	B1322	202
STEEL ERECTION—NOC	Rate	2.05	12.66	10.05	16.50	5.05	11.00	3.40	5.19	6.23	21.98	5.44	9.42
	Code	42106	722005	40502	748	238120	4227	4-41	4227	401	80080	B1322	202
TILE WORK—inter. (ceramic)	Rate	2.89	3.62	1.95	6.75	4.59	4.37	3.40	4.33	3.82	13.26	5.44	9.42
	Code	42113	721054	40103	719	238340	4276	4-41	4276	402	80110	B1301	202
WATERPROOFING	Rate	3.30	3.35	4.17	4.35	5.05	4.37	3.40	5.19	3.82	19.71	4.37	9.42
	Code	42139	721016	40102	723	238190	4299	4-41	4239	402	80130	B1217	202
WRECKING	Rate	2.34	4.44	6.37	16.50	5.05	3.84	3.57	2.88	6.23	13.26	2.76	9.42
	Code	40604	721005	40106	748	238190	4211	4-43	4211	401	80110	R1125	202

General Requirements — R0131 Project Management & Coordination

R013113-80   Performance Bond

This table shows the cost of a Performance Bond for a construction job scheduled to be completed in 12 months. Add 1% of the premium cost per month for jobs requiring more than 12 months to complete. The rates are "standard" rates offered to contractors that the bonding company considers financially sound and capable of doing the work. Preferred rates are offered by some bonding companies based upon financial strength of the contractor. Actual rates vary from contractor to contractor and from bonding company to bonding company. Contractors should prequalify through a bonding agency before submitting a bid on a contract that requires a bond.

Contract Amount	Building Construction Class B Projects	Highways & Bridges Class A New Construction	Class A-1 Highway Resurfacing
First $ 100,000 bid	$25.00 per M	$15.00 per M	$9.40 per M
Next 400,000 bid	$ 2,500 plus $15.00 per M	$ 1,500 plus $10.00 per M	$ 940 plus $7.20 per M
Next 2,000,000 bid	8,500 plus 10.00 per M	5,500 plus 7.00 per M	3,820 plus 5.00 per M
Next 2,500,000 bid	28,500 plus 7.50 per M	19,500 plus 5.50 per M	15,820 plus 4.50 per M
Next 2,500,000 bid	47,250 plus 7.00 per M	33,250 plus 5.00 per M	28,320 plus 4.50 per M
Over 7,500,000 bid	64,750 plus 6.00 per M	45,750 plus 4.50 per M	39,570 plus 4.00 per M

General Requirements — R0154 Construction Aids

R015423-10   Steel Tubular Scaffolding

On new construction, tubular scaffolding is efficient up to 60' high or five stories. Above this it is usually better to use a hung scaffolding if construction permits. Swing scaffolding operations may interfere with tenants. In this case, the tubular is more practical at all heights.

In repairing or cleaning the front of an existing building the cost of tubular scaffolding per S.F. of building front increases as the height increases above the first tier. The first tier cost is relatively high due to leveling and alignment.

The minimum efficient crew for erecting and dismantling is three workers. They can set up and remove 18 frame sections per day up to 5 stories high. For 6 to 12 stories high, a crew of four is most efficient. Use two or more on top and two on the bottom for handing up or hoisting. They can also set up and remove 18 frame sections per day. At 7' horizontal spacing, this will run about 800 S.F. per day of erecting and dismantling. Time for placing and removing planks must be added to the above. A crew of three can place and remove 72 planks per day up to 5 stories. For over 5 stories, a crew of four can place and remove 80 planks per day.

The table below shows the number of pieces required to erect tubular steel scaffolding for 1000 S.F. of building frontage. This area is made up of a scaffolding system that is 12 frames (11 bays) long by 2 frames high.

For jobs under twenty-five frames, add 50% to rental cost. Rental rates will be lower for jobs over three months duration. Large quantities for long periods can reduce rental rates by 20%.

Description of Component	Number of Pieces for 1000 S.F. of Building Front	Unit
5' Wide Standard Frame, 6'-4" High	24	Ea.
Leveling Jack & Plate	24	
Cross Brace	44	
Side Arm Bracket, 21"	12	
Guardrail Post	12	
Guardrail, 7' section	22	
Stairway Section	2	
Stairway Starter Bar	1	
Stairway Inside Handrail	2	
Stairway Outside Handrail	2	
Walk-Thru Frame Guardrail	2	

Scaffolding is often used as falsework over 15' high during construction of cast-in-place concrete beams and slabs. Two foot wide scaffolding is generally used for heavy beam construction. The span between frames depends upon the load to be carried with a maximum span of 5'.

Heavy duty shoring frames with a capacity of 10,000#/leg can be spaced up to 10' O.C. depending upon form support design and loading.

Scaffolding used as horizontal shoring requires less than half the material required with conventional shoring.

On new construction, erection is done by carpenters.

Rolling towers supporting horizontal shores can reduce labor and speed the job. For maintenance work, catwalks with spans up to 70' can be supported by the rolling towers.

General Requirements — R0154 Construction Aids

R015423-20 Pump Staging

Pump staging is generally not available for rent. The table below shows the number of pieces required to erect pump staging for 2400 S.F. of building frontage. This area is made up of a pump jack system that is 3 poles (2 bays) wide by 2 poles high.

Item	Number of Pieces for 2400 S.F. of Building Front	Unit
Aluminum pole section, 24' long	6	Ea.
Aluminum splice joint, 6' long	3	
Aluminum foldable brace	3	
Aluminum pump jack	3	
Aluminum support for workbench/back safety rail	3	
Aluminum scaffold plank/workbench, 14" wide x 24' long	4	
Safety net, 22' long	2	
Aluminum plank end safety rail	2	

The cost in place for this 2400 S.F. will depend on how many uses are realized during the life of the equipment.

R015433-10 Contractor Equipment

Rental Rates shown elsewhere in the book pertain to late model high quality machines in excellent working condition, rented from equipment dealers. Rental rates from contractors may be substantially lower than the rental rates from equipment dealers depending upon economic conditions; for older, less productive machines, reduce rates by a maximum of 15%. Any overtime must be added to the base rates. For shift work, rates are lower. Usual rule of thumb is 150% of one shift rate for two shifts; 200% for three shifts.

For periods of less than one week, operated equipment is usually more economical to rent than renting bare equipment and hiring an operator.

Costs to move equipment to a job site (mobilization) or from a job site (demobilization) are not included in rental rates, nor in any Equipment costs on any Unit Price line items or crew listings. These costs can be found elsewhere. If a piece of equipment is already at a job site, it is not appropriate to utilize mob/demob costs in an estimate again.

Rental rates vary throughout the country with larger cities generally having lower rates. Lease plans for new equipment are available for periods in excess of six months with a percentage of payments applying toward purchase.

Monthly rental rates vary from 2% to 5% of the cost of the equipment depending on the anticipated life of the equipment and its wearing parts. Weekly rates are about 1/3 the monthly rates and daily rental rates about 1/3 the weekly rate.

The hourly operating costs for each piece of equipment include costs to the user such as fuel, oil, lubrication, normal expendables for the equipment, and a percentage of mechanic's wages chargeable to maintenance. The hourly operating costs listed do not include the operator's wages.

The daily cost for equipment used in the standard crews is figured by dividing the weekly rate by five, then adding eight times the hourly operating cost to give the total daily equipment cost, not including the operator. This figure is in the right hand column of the Equipment listings under Equipment Cost/Day.

Pile Driving rates shown for pile hammer and extractor do not include leads, crane, boiler or compressor. Vibratory pile driving requires an added field specialist during set-up and pile driving operation for the electric model. The hydraulic model requires a field specialist for set-up only. Up to 125 reuses of sheet piling are possible using vibratory drivers. For normal conditions, crane capacity for hammer type and size are as follows.

Crane Capacity	Hammer Type and Size		
	Air or Steam	Diesel	Vibratory
25 ton	to 8,750 ft.-lb.		70 H.P.
40 ton	15,000 ft.-lb.	to 32,000 ft.-lb.	170 H.P.
60 ton	25,000 ft.-lb.		300 H.P.
100 ton		112,000 ft.-lb.	

Cranes should be specified for the job by size, building and site characteristics, availability, performance characteristics, and duration of time required.

Backhoes & Shovels rent for about the same as equivalent size cranes but maintenance and operating expense is higher. Crane operators rate must be adjusted for high boom heights. Average adjustments: for 150' boom add 2% per hour; over 185', add 4% per hour; over 210', add 6% per hour; over 250', add 8% per hour and over 295', add 12% per hour.

Tower Cranes of the climbing or static type have jibs from 50' to 200' and capacities at maximum reach range from 4,000 to 14,000 pounds. Lifting capacities increase up to maximum load as the hook radius decreases.

Typical rental rates, based on purchase price are about 2% to 3% per month. Erection and dismantling runs between 500 and 2000 labor hours. Climbing operation takes 10 labor hours per 20' climb. Crane dead time is about 5 hours per 40' climb. If crane is bolted to side of the building add cost of ties and extra mast sections. Climbing cranes have from 80' to 180' of mast while static cranes have 80' to 800' of mast.

Truck Cranes can be converted to tower cranes by using tower attachments. Mast heights over 400' have been used.

A single 100' high material **Hoist and Tower** can be erected and dismantled in about 400 labor hours; a double 100' high hoist and tower in about 600 labor hours. Erection times for additional heights are 3 and 4 labor hours per vertical foot respectively up to 150', and 4 to 5 labor hours per vertical foot over 150' high. A 40' high portable Buck hoist takes about 160 labor hours to erect and dismantle. Additional heights take 2 labor hours per vertical foot to 80' and 3 labor hours per vertical foot for the next 100'. Most material hoists do not meet local code requirements for carrying personnel.

A 150' high **Personnel Hoist** requires about 500 to 800 labor hours to erect and dismantle. Budget erection time at 5 labor hours per vertical foot for all trades. Local code requirements or labor scarcity requiring overtime can add up to 50% to any of the above erection costs.

Earthmoving Equipment: The selection of earthmoving equipment depends upon the type and quantity of material, moisture content, haul distance, haul road, time available, and equipment available. Short haul cut and fill operations may require dozers only, while another operation may require excavators, a fleet of trucks, and spreading and compaction equipment. Stockpiled material and granular material are easily excavated with front end loaders. Scrapers are most economically used with hauls between 300' and 1-1/2 miles if adequate haul roads can be maintained. Shovels are often used for blasted rock and any material where a vertical face of 8' or more can be excavated. Special conditions may dictate the use of draglines, clamshells, or backhoes. Spreading and compaction equipment must be matched to the soil characteristics, the compaction required and the rate the fill is being supplied.

Existing Conditions — R0241 Demolition

R024119-10 Demolition Defined

Whole Building Demolition - Demolition of the whole building with no concern for any particular building element, component, or material type being demolished. This type of demolition is accomplished with large pieces of construction equipment that break up the structure, load it into trucks and haul it to a disposal site, but disposal or dump fees are not included. Demolition of below-grade foundation elements, such as footings, foundation walls, grade beams, slabs on grade, etc., is not included. Certain mechanical equipment containing flammable liquids or ozone-depleting refrigerants, electric lighting elements, communication equipment components, and other building elements may contain hazardous waste, and must be removed, either selectively or carefully, as hazardous waste before the building can be demolished.

Foundation Demolition - Demolition of below-grade foundation footings, foundation walls, grade beams, and slabs on grade. This type of demolition is accomplished by hand or pneumatic hand tools, and does not include saw cutting, or handling, loading, hauling, or disposal of the debris.

Gutting - Removal of building interior finishes and electrical/mechanical systems down to the load-bearing and sub-floor elements of the rough building frame, with no concern for any particular building element, component, or material type being demolished. This type of demolition is accomplished by hand or pneumatic hand tools, and includes loading into trucks, but not hauling, disposal or dump fees, scaffolding, or shoring. Certain mechanical equipment containing flammable liquids or ozone-depleting refrigerants, electric lighting elements, communication equipment components, and other building elements may contain hazardous waste, and must be removed, either selectively or carefully, as hazardous waste, before the building is gutted.

Selective Demolition - Demolition of a selected building element, component, or finish, with some concern for surrounding or adjacent elements, components, or finishes (see the first Subdivision (s) at the beginning of appropriate Divisions). This type of demolition is accomplished by hand or pneumatic hand tools, and does not include handling, loading, storing, hauling, or disposal of the debris, scaffolding, or shoring. "Gutting" methods may be used in order to save time, but damage that is caused to surrounding or adjacent elements, components, or finishes may have to be repaired at a later time.

Careful Removal - Removal of a piece of service equipment, building element or component, or material type, with great concern for both the removed item and surrounding or adjacent elements, components or finishes. The purpose of careful removal may be to protect the removed item for later re-use, preserve a higher salvage value of the removed item, or replace an item while taking care to protect surrounding or adjacent elements, components, connections, or finishes from cosmetic and/or structural damage. An approximation of the time required to perform this type of removal is 1/3 to 1/2 the time it would take to install a new item of like kind (see Reference Number R220105-10). This type of removal is accomplished by hand or pneumatic hand tools, and does not include loading, hauling, or storing the removed item, scaffolding, shoring, or lifting equipment.

Cutout Demolition - Demolition of a small quantity of floor, wall, roof, or other assembly, with concern for the appearance and structural integrity of the surrounding materials. This type of demolition is accomplished by hand or pneumatic hand tools, and does not include saw cutting, handling, loading, hauling, or disposal of debris, scaffolding, or shoring.

Rubbish Handling - Work activities that involve handling, loading or hauling of debris. Generally, the cost of rubbish handling must be added to the cost of all types of demolition, with the exception of whole building demolition.

Minor Site Demolition - Demolition of site elements outside the footprint of a building. This type of demolition is accomplished by hand or pneumatic hand tools, or with larger pieces of construction equipment, and may include loading a removed item onto a truck (check the Crew for equipment used). It does not include saw cutting, hauling or disposal of debris, and, sometimes, handling or loading.

R024119-20 Dumpsters

Dumpster rental costs on construction sites are presented in two ways.

The cost per week rental includes the delivery of the dumpster; its pulling or emptying once per week, and its final removal. The assumption is made that the dumpster contractor could choose to empty a dumpster by simply bringing in an empty unit and removing the full one. These costs also include the disposal of the materials in the dumpster.

The Alternate Pricing can be used when actual planned conditions are not approximated by the weekly numbers. For example, these lines can be used when a dumpster is needed for 4 weeks and will need to be emptied 2 or 3 times per week. Conversely the Alternate Pricing lines can be used when a dumpster will be rented for several weeks or months but needs to be emptied only a few times over this period.

Existing Conditions — R0265 Underground Storage Tank Removal

R026510-20 Underground Storage Tank Removal

Underground Storage Tank Removal can be divided into two categories: Non-Leaking and Leaking. Prior to removing an underground storage tank, tests should be made, with the proper authorities present, to determine whether a tank has been leaking or the surrounding soil has been contaminated.

To safely remove Liquid Underground Storage Tanks:
1. Excavate to the top of the tank.
2. Disconnect all piping.
3. Open all tank vents and access ports.
4. Remove all liquids and/or sludge.
5. Purge the tank with an inert gas.
6. Provide access to the inside of the tank and clean out the interior using proper personal protective equipment (PPE).
7. Excavate soil surrounding the tank using proper PPE for on-site personnel.
8. Pull and properly dispose of the tank.
9. Clean up the site of all contaminated material.
10. Install new tanks or close the excavation.

Existing Conditions — R0282 Asbestos Remediation

R028213-20 Asbestos Removal Process

Asbestos removal is accomplished by a specialty contractor who understands the federal and state regulations regarding the handling and disposal of the material. The process of asbestos removal is divided into many individual steps. An accurate estimate can be calculated only after all the steps have been priced.

The steps are generally as follows:

1. Obtain an asbestos abatement plan from an industrial hygienist.
2. Monitor the air quality in and around the removal area and along the path of travel between the removal area and transport area. This establishes the background contamination.
3. Construct a two part decontamination chamber at entrance to removal area.
4. Install a HEPA filter to create a negative pressure in the removal area.
5. Install wall, floor and ceiling protection as required by the plan, usually 2 layers of fireproof 6 mil polyethylene.
6. Industrial hygienist visually inspects work area to verify compliance with plan.
7. Provide temporary supports for conduit and piping affected by the removal process.
8. Proceed with asbestos removal and bagging process. Monitor air quality as described in Step #2. Discontinue operations when contaminate levels exceed applicable standards.
9. Document the legal disposal of materials in accordance with EPA standards.
10. Thoroughly clean removal area including all ledges, crevices and surfaces.
11. Post abatement inspection by industrial hygienist to verify plan compliance.
12. Provide a certificate from a licensed industrial hygienist attesting that contaminate levels are within acceptable standards before returning area to regular use.

Existing Conditions — R0283 Lead Remediation

R028319-60 Lead Paint Remediation Methods

Lead paint remediation can be accomplished by the following methods.
1. Abrasive blast
2. Chemical stripping
3. Power tool cleaning with vacuum collection system
4. Encapsulation
5. Remove and replace
6. Enclosure

Each of these methods has strengths and weakness depending on the specific circumstances of the project. The following is an overview of each method.

1. **Abrasive blasting** is usually accomplished with sand or recyclable metallic blast. Before work can begin, the area must be contained to ensure the blast material with lead does not escape to the atmosphere. The use of vacuum blast greatly reduces the containment requirements. Lead abatement equipment that may be associated with this work includes a negative air machine. In addition, it is necessary to have an industrial hygienist monitor the project on a continual basis. When the work is complete, the spent blast sand with lead must be disposed of as a hazardous material. If metallic shot was used, the lead is separated from the shot and disposed of as hazardous material. Worker protection includes disposable clothing and respiratory protection.

2. **Chemical stripping** requires strong chemicals to be applied to the surface to remove the lead paint. Before the work can begin, the area under/adjacent to the work area must be covered to catch the chemical and removed lead. After the chemical is applied to the painted surface it is usually covered with paper. The chemical is left in place for the specified period, then the paper with lead paint is pulled or scraped off. The process may require several chemical applications. The paper with chemicals and lead paint adhered to it, plus the containment and loose scrapings collected by a HEPA (High Efficiency Particulate Air Filter) vac, must be disposed of as a hazardous material. The chemical stripping process usually requires a neutralizing agent and several wash downs after the paint is removed. Worker protection includes a neoprene or other compatible protective clothing and respiratory protection with face shield. An industrial hygienist is required intermittently during the process.

3. **Power tool cleaning** is accomplished using shrouded needle blasting guns. The shrouding with different end configurations is held up against the surface to be cleaned. The area is blasted with hardened needles and the shroud captures the lead with a HEPA vac and deposits it in a holding tank. An industrial hygienist monitors the project, protective clothing and a respirator is required until air samples prove otherwise. When the work is complete the lead must be disposed of as a hazardous material.

4. **Encapsulation** is a method that leaves the well bonded lead paint in place after the peeling paint has been removed. Before the work can begin, the area under/adjacent to the work must be covered to catch the scrapings. The scraped surface is then washed with a detergent and rinsed. The prepared surface is covered with approximately 10 mils of paint. A reinforcing fabric can also be embedded in the paint covering. The scraped paint and containment must be disposed of as a hazardous material. Workers must wear protective clothing and respirators.

5. **Remove and replace** is an effective way to remove lead paint from windows, gypsum walls and concrete masonry surfaces. The painted materials are removed and new materials are installed. Workers should wear a respirator and tyvek suit. The demolished materials must be disposed of as hazardous waste if it fails the TCLP (Toxicity Characteristic Leachate Process) test.

6. **Enclosure** is the process that permanently seals lead painted materials in place. This process has many applications such as covering lead painted drywall with new drywall, covering exterior construction with tyvek paper then residing, or covering lead painted structural members with aluminum or plastic. The seams on all enclosing materials must be securely sealed. An industrial hygienist monitors the project, and protective clothing and a respirator is required until air samples prove otherwise.

All the processes require clearance monitoring and wipe testing as required by the hygienist.

Concrete — R0311 Concrete Forming

R031113-10 Wall Form Materials

Aluminum Forms
Approximate weight is 3 lbs. per S.F.C.A. Standard widths are available from 4″ to 36″ with 36″ most common. Standard lengths of 2′, 4′, 6′ to 8′ are available. Forms are lightweight and fewer ties are needed with the wider widths. The form face is either smooth or textured.

Metal Framed Plywood Forms
Manufacturers claim over 75 reuses of plywood and over 300 reuses of steel frames. Many specials such as corners, fillers, pilasters, etc. are available. Monthly rental is generally about 15% of purchase price for first month and 9% per month thereafter with 90% of rental applied to purchase for the first month and decreasing percentages thereafter. Aluminum framed forms cost 25% to 30% more than steel framed.

After the first month, extra days may be prorated from the monthly charge. Rental rates do not include ties, accessories, cleaning, loss of hardware or freight in and out. Approximate weight is 5 lbs. per S.F. for steel; 3 lbs. per S.F. for aluminum.

Forms can be rented with option to buy.

Plywood Forms, Job Fabricated
There are two types of plywood used for concrete forms.
1. Exterior plyform which is completely waterproof. This is face oiled to facilitate stripping. Ten reuses can be expected with this type with 25 reuses possible.
2. An overlaid type consists of a resin fiber fused to exterior plyform. No oiling is required except to facilitate cleaning. This is available in both high density (HDO) and medium density overlaid (MDO). Using HDO, 50 reuses can be expected with 200 possible.

Plyform is available in 5/8″ and 3/4″ thickness. High density overlaid is available in 3/8″, 1/2″, 5/8″ and 3/4″ thickness.

5/8″ thick is sufficient for most building forms, while 3/4″ is best on heavy construction.

Plywood Forms, Modular, Prefabricated
There are many plywood forming systems without frames. Most of these are manufactured from 1-1/8″ (HDO) plywood and have some hardware attached. These are used principally for foundation walls 8′ or less high. With care and maintenance, 100 reuses can be attained with decreasing quality of surface finish.

Steel Forms
Approximate weight is 6-1/2 lbs. per S.F.C.A. including accessories. Standard widths are available from 2″ to 24″, with 24″ most common. Standard lengths are from 2′ to 8′, with 4′ the most common. Forms are easily ganged into modular units.

Forms are usually leased for 15% of the purchase price per month prorated daily over 30 days.

Rental may be applied to sale price, and usually rental forms are bought. With careful handling and cleaning 200 to 400 reuses are possible.

Straight wall gang forms up to 12′ x 20′ or 8′ x 30′ can be fabricated. These crane handled forms usually lease for approx. 9% per month.

Individual job analysis is available from the manufacturer at no charge.

R031113-40 Forms for Reinforced Concrete

Design Economy
Avoid many sizes in proportioning beams and columns.

From story to story avoid changing column dimensions. Gain strength by adding steel or using a richer mix. If a change in size of column is necessary, vary one dimension only to minimize form alterations. Keep beams and columns the same width.

From floor to floor in a multi-story building vary beam depth, not width, as that will leave slab panel form unchanged. It is cheaper to vary the strength of a beam from floor to floor by means of steel area than by 2″ changes in either width or depth.

Cost Factors
Material includes the cost of lumber, cost of rent for metal pans or forms if used, nails, form ties, form oil, bolts and accessories.

Labor includes the cost of carpenters to make up, erect, remove and repair, plus common labor to clean and move. Having carpenters remove forms minimizes repairs.

Improper alignment and condition of forms will increase finishing cost. When forms are heavily oiled, concrete surfaces must be neutralized before finishing. Special curing compounds will cause spillages to spall off in first frost. Gang forming methods will reduce costs on large projects.

Materials Used
Boards are seldom used unless their architectural finish is required. Generally, steel, fiberglass and plywood are used for contact surfaces. Labor on plywood is 10% less than with boards. The plywood is backed up with 2 x 4′s at 12″ to 32″ O.C. Walers are generally 2 - 2 x 4′s. Column forms are held together with steel yokes or bands. Shoring is with adjustable shoring or scaffolding for high ceilings.

Reuse
Floor and column forms can be reused four or possibly five times without excessive repair. Remember to allow for 10% waste on each reuse.

When modular sized wall forms are made, up to twenty uses can be expected with exterior plyform.

When forms are reused, the cost to erect, strip, clean and move will not be affected. 10% replacement of lumber should be included and about one hour of carpenter time for repairs on each reuse per 100 S.F.

The reuse cost for certain accessory items normally rented on a monthly basis will be lower than the cost for the first use.

After fifth use, new material required plus time needed for repair prevent form cost from dropping further and it may go up. Much depends on care in stripping, the number of special bays, changes in beam or column sizes and other factors.

Costs for multiple use of formwork may be developed as follows:

2 Uses	3 Uses	4 Uses
$\dfrac{\text{(1st Use + Reuse)}}{2} = \text{avg. cost/2 uses}$	$\dfrac{\text{(1st Use + 2 Reuse)}}{3} = \text{avg. cost/3 uses}$	$\dfrac{\text{(1st use + 3 Reuse)}}{4} = \text{avg. cost/4 uses}$

Concrete

R0311 Concrete Forming

R031113-60 Formwork Labor-Hours

Item	Unit	Hours Required			Total Hours	Multiple Use		
		Fabricate	Erect & Strip	Clean & Move	1 Use	2 Use	3 Use	4 Use
Beam and Girder, interior beams, 12" wide	100 S.F.	6.4	8.3	1.3	16.0	13.3	12.4	12.0
Hung from steel beams		5.8	7.7	1.3	14.8	12.4	11.6	11.2
Beam sides only, 36" high		5.8	7.2	1.3	14.3	11.9	11.1	10.7
Beam bottoms only, 24" wide		6.6	13.0	1.3	20.9	18.1	17.2	16.7
Box out for openings		9.9	10.0	1.1	21.0	16.6	15.1	14.3
Buttress forms, to 8' high		6.0	6.5	1.2	13.7	11.2	10.4	10.0
Centering, steel, 3/4" rib lath			1.0		1.0			
3/8" rib lath or slab form	▼		0.9		0.9			
Chamfer strip or keyway	100 L.F.		1.5		1.5	1.5	1.5	1.5
Columns, fiber tube 8" diameter			20.6		20.6			
12"			21.3		21.3			
16"			22.9		22.9			
20"			23.7		23.7			
24"			24.6		24.6			
30"	▼		25.6		25.6			
Columns, round steel, 12" diameter			22.0		22.0	22.0	22.0	22.0
16"			25.6		25.6	25.6	25.6	25.6
20"			30.5		30.5	30.5	30.5	30.5
24"	▼		37.7		37.7	37.7	37.7	37.7
Columns, plywood 8" x 8"	100 S.F.	7.0	11.0	1.2	19.2	16.2	15.2	14.7
12" x 12"		6.0	10.5	1.2	17.7	15.2	14.4	14.0
16" x 16"		5.9	10.0	1.2	17.1	14.7	13.8	13.4
24" x 24"		5.8	9.8	1.2	16.8	14.4	13.6	13.2
Columns, steel framed plywood 8" x 8"			10.0	1.0	11.0	11.0	11.0	11.0
12" x 12"			9.3	1.0	10.3	10.3	10.3	10.3
16" x 16"			8.5	1.0	9.5	9.5	9.5	9.5
24" x 24"			7.8	1.0	8.8	8.8	8.8	8.8
Drop head forms, plywood		9.0	12.5	1.5	23.0	19.0	17.7	17.0
Coping forms		8.5	15.0	1.5	25.0	21.3	20.0	19.4
Culvert, box			14.5	4.3	18.8	18.8	18.8	18.8
Curb forms, 6" to 12" high, on grade		5.0	8.5	1.2	14.7	12.7	12.1	11.7
On elevated slabs		6.0	10.8	1.2	18.0	15.5	14.7	14.3
Edge forms to 6" high, on grade	100 L.F.	2.0	3.5	0.6	6.1	5.6	5.4	5.3
7" to 12" high	100 S.F.	2.5	5.0	1.0	8.5	7.8	7.5	7.4
Equipment foundations		10.0	18.0	2.0	30.0	25.5	24.0	23.3
Flat slabs, including drops		3.5	6.0	1.2	10.7	9.5	9.0	8.8
Hung from steel		3.0	5.5	1.2	9.7	8.7	8.4	8.2
Closed deck for domes		3.0	5.8	1.2	10.0	9.0	8.7	8.5
Open deck for pans		2.2	5.3	1.0	8.5	7.9	7.7	7.6
Footings, continuous, 12" high		3.5	3.5	1.5	8.5	7.3	6.8	6.6
Spread, 12" high		4.7	4.2	1.6	10.5	8.7	8.0	7.7
Pile caps, square or rectangular		4.5	5.0	1.5	11.0	9.3	8.7	8.4
Grade beams, 24" deep		2.5	5.3	1.2	9.0	8.3	8.0	7.9
Lintel or Sill forms		8.0	17.0	2.0	27.0	23.5	22.3	21.8
Spandrel beams, 12" wide		9.0	11.2	1.3	21.5	17.5	16.2	15.5
Stairs			25.0	4.0	29.0	29.0	29.0	29.0
Trench forms in floor		4.5	14.0	1.5	20.0	18.3	17.7	17.4
Walls, Plywood, at grade, to 8' high		5.0	6.5	1.5	13.0	11.0	9.7	9.5
8' to 16'		7.5	8.0	1.5	17.0	13.8	12.7	12.1
16' to 20'		9.0	10.0	1.5	20.5	16.5	15.2	14.5
Foundation walls, to 8' high		4.5	6.5	1.0	12.0	10.3	9.7	9.4
8' to 16' high		5.5	7.5	1.0	14.0	11.8	11.0	10.6
Retaining wall to 12' high, battered		6.0	8.5	1.5	16.0	13.5	12.7	12.3
Radial walls to 12' high, smooth		8.0	9.5	2.0	19.5	16.0	14.8	14.3
2' chords		7.0	8.0	1.5	16.5	13.5	12.5	12.0
Prefabricated modular, to 8' high		—	4.3	1.0	5.3	5.3	5.3	5.3
Steel, to 8' high		—	6.8	1.2	8.0	8.0	8.0	8.0
8' to 16' high		—	9.1	1.5	10.6	10.3	10.2	10.2
Steel framed plywood to 8' high		—	6.8	1.2	8.0	7.5	7.3	7.2
8' to 16' high	▼	—	9.3	1.2	10.5	9.5	9.2	9.0

Concrete — R0321 Reinforcing Steel

R032110-10 Reinforcing Steel Weights and Measures

Bar Designation No.**	Nominal Weight Lb./Ft.	U.S. Customary Units — Nominal Dimensions*			SI Units — Nominal Dimensions*			
		Diameter in.	Cross Sectional Area, in.^{2}	Perimeter in.	Nominal Weight kg/m	Diameter mm	Cross Sectional Area, cm^2	Perimeter mm
3	.376	.375	.11	1.178	.560	9.52	.71	29.9
4	.668	.500	.20	1.571	.994	12.70	1.29	39.9
5	1.043	.625	.31	1.963	1.552	15.88	2.00	49.9
6	1.502	.750	.44	2.356	2.235	19.05	2.84	59.8
7	2.044	.875	.60	2.749	3.042	22.22	3.87	69.8
8	2.670	1.000	.79	3.142	3.973	25.40	5.10	79.8
9	3.400	1.128	1.00	3.544	5.059	28.65	6.45	90.0
10	4.303	1.270	1.27	3.990	6.403	32.26	8.19	101.4
11	5.313	1.410	1.56	4.430	7.906	35.81	10.06	112.5
14	7.650	1.693	2.25	5.320	11.384	43.00	14.52	135.1
18	13.600	2.257	4.00	7.090	20.238	57.33	25.81	180.1

* The nominal dimensions of a deformed bar are equivalent to those of a plain round bar having the same weight per foot as the deformed bar.
** Bar numbers are based on the number of eighths of an inch included in the nominal diameter of the bars.

R032110-80 Shop-Fabricated Reinforcing Steel

The material prices for reinforcing, shown in the unit cost sections of the book, are for 50 tons or more of shop-fabricated reinforcing steel and include:

1. Mill base price of reinforcing steel
2. Mill grade/size/length extras
3. Mill delivery to the fabrication shop
4. Shop storage and handling
5. Shop drafting/detailing
6. Shop shearing and bending
7. Shop listing
8. Shop delivery to the job site

Both material and installation costs can be considerably higher for small jobs consisting primarily of smaller bars, while material costs may be slightly lower for larger jobs.

Concrete — R0322 Welded Wire Fabric Reinforcing

R032205-30 Common Stock Styles of Welded Wire Fabric

This table provides some of the basic specifications, sizes, and weights of welded wire fabric used for reinforcing concrete.

	New Designation Spacing — Cross Sectional Area (in.) — (Sq. in. 100)	Old Designation Spacing — Wire Gauge (in.) — (AS & W)		Steel Area per Foot — Longitudinal		Steel Area per Foot — Transverse		Approximate Weight per 100 S.F.	
				in.	cm	in.	cm	lbs	kg
Rolls	6 x 6 — W1.4 x W1.4	6 x 6 — 10 x 10		.028	.071	.028	.071	21	9.53
	6 x 6 — W2.0 x W2.0	6 x 6 — 8 x 8	1	.040	.102	.040	.102	29	13.15
	6 x 6 — W2.9 x W2.9	6 x 6 — 6 x 6		.058	.147	.058	.147	42	19.05
	6 x 6 — W4.0 x W4.0	6 x 6 — 4 x 4		.080	.203	.080	.203	58	26.91
	4 x 4 — W1.4 x W1.4	4 x 4 — 10 x 10		.042	.107	.042	.107	31	14.06
	4 x 4 — W2.0 x W2.0	4 x 4 — 8 x 8	1	.060	.152	.060	.152	43	19.50
	4 x 4 — W2.9 x W2.9	4 x 4 — 6 x 6		.087	.227	.087	.227	62	28.12
	4 x 4 — W4.0 x W4.0	4 x 4 — 4 x 4		.120	.305	.120	.305	85	38.56
Sheets	6 x 6 — W2.9 x W2.9	6 x 6 — 6 x 6		.058	.147	.058	.147	42	19.05
	6 x 6 — W4.0 x W4.0	6 x 6 — 4 x 4		.080	.203	.080	.203	58	26.31
	6 x 6 — W5.5 x W5.5	6 x 6 — 2 x 2	2	.110	.279	.110	.279	80	36.29
	4 x 4 — W1.4 x W1.4	4 x 4 — 4 x 4		.120	.305	.120	.305	85	38.56

NOTES: 1. Exact W—number size for 8 gauge is W2.1
2. Exact W—number size for 2 gauge is W5.4

Concrete — R0331 Structural Concrete

R033105-10 Proportionate Quantities

The tables below show both quantities per S.F. of floor areas as well as form and reinforcing quantities per C.Y. Unusual structural requirements would increase the ratios below. High strength reinforcing would reduce the steel weights. Figures are for 3000 psi concrete and 60,000 psi reinforcing unless specified otherwise.

Type of Construction	Live Load	Span	Per S.F. of Floor Area				Per C.Y. of Concrete		
			Concrete	Forms	Reinf.	Pans	Forms	Reinf.	Pans
Flat Plate	50 psf	15 Ft.	.46 C.F.	1.06 S.F.	1.71 lb.		62 S.F.	101 lb.	
		20	.63	1.02	2.40		44	104	
		25	.79	1.02	3.03		35	104	
	100	15	.46	1.04	2.14		61	126	
		20	.71	1.02	2.72		39	104	
		25	.83	1.01	3.47		33	113	
Flat Plate (waffle construction) 20" domes	50	20	.43	1.00	2.10	.84 S.F.	63	135	53 S.F.
		25	.52	1.00	2.90	.89	52	150	46
		30	.64	1.00	3.70	.87	42	155	37
	100	20	.51	1.00	2.30	.84	53	125	45
		25	.64	1.00	3.20	.83	42	135	35
		30	.76	1.00	4.40	.81	36	160	29
Waffle Construction 30" domes	50	25	.69	1.06	1.83	.68	42	72	40
		30	.74	1.06	2.39	.69	39	87	39
		35	.86	1.05	2.71	.69	33	85	39
		40	.78	1.00	4.80	.68	35	165	40
Flat Slab (two way with drop panels)	50	20	.62	1.03	2.34		45	102	
		25	.77	1.03	2.99		36	105	
		30	.95	1.03	4.09		29	116	
	100	20	.64	1.03	2.83		43	119	
		25	.79	1.03	3.88		35	133	
		30	.96	1.03	4.66		29	131	
	200	20	.73	1.03	3.03		38	112	
		25	.86	1.03	4.23		32	133	
		30	1.06	1.03	5.30		26	135	
One Way Joists 20" Pans	50	15	.36	1.04	1.40	.93	78	105	70
		20	.42	1.05	1.80	.94	67	120	60
		25	.47	1.05	2.60	.94	60	150	54
	100	15	.38	1.07	1.90	.93	77	140	66
		20	.44	1.08	2.40	.94	67	150	58
		25	.52	1.07	3.50	.94	55	185	49
One Way Joists 8" x 16" filler blocks	50	15	.34	1.06	1.80	.81 Ea.	84	145	64 Ea.
		20	.40	1.08	2.20	.82	73	145	55
		25	.46	1.07	3.20	.83	63	190	49
	100	15	.39	1.07	1.90	.81	74	130	56
		20	.46	1.09	2.80	.82	64	160	48
		25	.53	1.10	3.60	.83	56	190	42
One Way Beam & Slab	50	15	.42	1.30	1.73		84	111	
		20	.51	1.28	2.61		68	138	
		25	.64	1.25	2.78		53	117	
	100	15	.42	1.30	1.90		84	122	
		20	.54	1.35	2.69		68	154	
		25	.69	1.37	3.93		54	145	
	200	15	.44	1.31	2.24		80	137	
		20	.58	1.40	3.30		65	163	
		25	.69	1.42	4.89		53	183	
Two Way Beam & Slab	100	15	.47	1.20	2.26		69	130	
		20	.63	1.29	3.06		55	131	
		25	.83	1.33	3.79		43	123	
	200	15	.49	1.25	2.70		41	149	
		20	.66	1.32	4.04		54	165	
		25	.88	1.32	6.08		41	187	

Concrete

R0331 Structural Concrete

R033105-10 Proportionate Quantities (cont.)

| 4000 psi Concrete and 60,000 psi Reinforcing—Form and Reinforcing Quantities per C.Y. ||||||
Item	Size	Forms	Reinforcing	Minimum	Maximum
Columns (square tied)	10" x 10"	130 S.F.C.A.	#5 to #11	220 lbs.	875 lbs.
	12" x 12"	108	#6 to #14	200	955
	14" x 14"	92	#7 to #14	190	900
	16" x 16"	81	#6 to #14	187	1082
	18" x 18"	72	#6 to #14	170	906
	20" x 20"	65	#7 to #18	150	1080
	22" x 22"	59	#8 to #18	153	902
	24" x 24"	54	#8 to #18	164	884
	26" x 26"	50	#9 to #18	169	994
	28" x 28"	46	#9 to #18	147	864
	30" x 30"	43	#10 to #18	146	983
	32" x 32"	40	#10 to #18	175	866
	34" x 34"	38	#10 to #18	157	772
	36" x 36"	36	#10 to #18	175	852
	38" x 38"	34	#10 to #18	158	765
	40" x 40"	32	#10 to #18	143	692

Item	Size	Form	Spiral	Reinforcing	Minimum	Maximum
Columns (spirally reinforced)	12" diameter	34.5 L.F.	190 lbs.	#4 to #11	165 lbs.	1505 lb.
		34.5	190	#14 & #18	—	1100
	14"	25	170	#4 to #11	150	970
		25	170	#14 & #18	800	1000
	16"	19	160	#4 to #11	160	950
		19	160	#14 & #18	605	1080
	18"	15	150	#4 to #11	160	915
		15	150	#14 & #18	480	1075
	20"	12	130	#4 to #11	155	865
		12	130	#14 & #18	385	1020
	22"	10	125	#4 to #11	165	775
		10	125	#14 & #18	320	995
	24"	9	120	#4 to #11	195	800
		9	120	#14 & #18	290	1150
	26"	7.3	100	#4 to #11	200	729
		7.3	100	#14 & #18	235	1035
	28"	6.3	95	#4 to #11	175	700
		6.3	95	#14 & #18	200	1075
	30"	5.5	90	#4 to #11	180	670
		5.5	90	#14 & #18	175	1015
	32"	4.8	85	#4 to #11	185	615
		4.8	85	#14 & #18	155	955
	34"	4.3	80	#4 to #11	180	600
		4.3	80	#14 & #18	170	855
	36"	3.8	75	#4 to #11	165	570
		3.8	75	#14 & #18	155	865
	40"	3.0	70	#4 to #11	165	500
		3.0	70	#14 & #18	145	765

Concrete — R0331 Structural Concrete

R033105-10 Proportionate Quantities (cont.)

3000 psi Concrete and 60,000 psi Reinforcing—Form and Reinforcing Quantities per C.Y.						
Item	Type	Loading	Height	C.Y./L.F.	Forms/C.Y.	Reinf./C.Y.
Retaining Walls	Cantilever	Level Backfill	4 Ft.	0.2 C.Y.	49 S.F.	35 lbs.
			8	0.5	42	45
			12	0.8	35	70
			16	1.1	32	85
			20	1.6	28	105
		Highway Surcharge	4	0.3	41	35
			8	0.5	36	55
			12	0.8	33	90
			16	1.2	30	120
			20	1.7	27	155
		Railroad Surcharge	4	0.4	28	45
			8	0.8	25	65
			12	1.3	22	90
			16	1.9	20	100
			20	2.6	18	120
	Gravity, with Vertical Face	Level Backfill	4	0.4	37	None
			7	0.6	27	↓
			10	1.2	20	
		Sloping Backfill	4	0.3	31	
			7	0.8	21	
			10	1.6	15	↓

		Live Load in Kips per Linear Foot						
	Span	Under 1 Kip		2 to 3 Kips		4 to 5 Kips		6 to 7 Kips	
		Forms	Reinf.	Forms	Reinf.	Forms	Reinf.	Forms	Reinf.
Beams	10 Ft.	—	—	90 S.F.	170 #	85 S.F.	175 #	75 S.F.	185 #
	16	130 S.F.	165 #	85	180	75	180	65	225
	20	110	170	75	185	62	200	51	200
	26	90	170	65	215	62	215	—	—
	30	85	175	60	200	—	—	—	—

Item	Size	Type	Forms per C.Y.	Reinforcing per C.Y.
Spread Footings	Under 1 C.Y.	1,000 psf soil	24 S.F.	44 lbs.
		5,000	24	42
		10,000	24	52
	1 C.Y. to 5 C.Y.	1,000	14	49
		5,000	14	50
		10,000	14	50
	Over 5 C.Y.	1,000	9	54
		5,000	9	52
		10,000	9	56
Pile Caps (30 Ton Concrete Piles)	Under 5 C.Y.	shallow caps	20	65
		medium	20	50
		deep	20	40
	5 C.Y. to 10 C.Y.	shallow	14	55
		medium	15	45
		deep	15	40
	10 C.Y. to 20 C.Y.	shallow	11	60
		medium	11	45
		deep	12	35
	Over 20 C.Y.	shallow	9	60
		medium	9	45
		deep	10	40

Concrete — R0331 Structural Concrete

R033105-10 Proportionate Quantities (cont.)

3000 psi Concrete and 60,000 psi Reinforcing — Form and Reinforcing Quantities per C.Y.						
Item	Size	Pile Spacing	50 T Pile	100 T Pile	50 T Pile	100 T Pile
Pile Caps (Steel H Piles)	Under 5 C.Y.	24" O.C.	24 S.F.	24 S.F.	75 lbs.	90 lbs.
		30"	25	25	80	100
		36"	24	24	80	110
	5 C.Y. to 10 C.Y.	24"	15	15	80	110
		30"	15	15	85	110
		36"	15	15	75	90
	Over 10 C.Y.	24"	13	13	85	90
		30"	11	11	85	95
		36"	10	10	85	90

		8" Thick		10" Thick		12" Thick		15" Thick	
	Height	Forms	Reinf.	Forms	Reinf.	Forms	Reinf.	Forms	Reinf.
Basement Walls	7 Ft.	81 S.F.	44 lbs.	65 S.F.	45 lbs.	54 S.F.	44 lbs.	41 S.F.	43 lbs.
	8		44		45		44		43
	9		46		45		44		43
	10		57		45		44		43
	12		83		50		52		43
	14		116		65		64		51
	16				86		90		65
	18						106		70

R033105-20 Materials for One C.Y. of Concrete

This is an approximate method of figuring quantities of cement, sand and coarse aggregate for a field mix with waste allowance included.

With crushed gravel as coarse aggregate, to determine barrels of cement required, divide 10 by total mix; that is, for 1:2:4 mix, 10 divided by 7 = 1-3/7 barrels.

If the coarse aggregate is crushed stone, use 10-1/2 instead of 10 as given for gravel.

To determine tons of sand required, multiply barrels of cement by parts of sand and then by 0.2; that is, for the 1:2:4 mix, as above, 1-3/7 x 2 x .2 = .57 tons.

Tons of crushed gravel are in the same ratio to tons of sand as parts in the mix, or 4/2 x .57 = 1.14 tons.

1 bag cement = 94#
4 bags = 1 barrel
1 C.Y. sand or crushed gravel = 2700#
1 ton sand or crushed gravel = 20 C.F.
1 C.Y. crushed stone = 2575#
1 ton crushed stone = 21 C.F.

Average carload of cement is 692 bags; of sand or gravel is 56 tons.

Do not stack stored cement over 10 bags high.

R033105-70 Placing Ready-Mixed Concrete

For ground pours allow for 5% waste when figuring quantities.

Prices in the front of the book assume normal deliveries. If deliveries are made before 8 A.M. or after 5 P.M. or on Saturday afternoons add 30%. Negotiated discounts for large volumes are not included in prices in front of book.

For the lower floors without truck access, concrete may be wheeled in rubber-tired buggies, conveyer handled, crane handled or pumped. Pumping is economical if there is top steel. Conveyers are more efficient for thick slabs.

At higher floors the rubber-tired buggies may be hoisted by a hoisting tower and wheeled to location. Placement by a conveyer is limited to three floors and is best for high-volume pours. Pumped concrete is best when building has no crane access. Concrete may be pumped directly as high as thirty-six stories using special pumping techniques. Normal maximum height is about fifteen stories.

Best pumping aggregate is screened and graded bank gravel rather than crushed stone.

Pumping downward is more difficult than pumping upward. Horizontal distance from pump to pour may increase preparation time prior to pour. Placing by cranes, either mobile, climbing or tower types, continues as the most efficient method for high-rise concrete buildings.

Concrete — R0331 Structural Concrete

R033105-85   Lift Slabs

The cost advantage of the lift slab method is due to placing all concrete, reinforcing steel, inserts and electrical conduit at ground level and in reduction of formwork. Minimum economical project size is about 30,000 S.F. Slabs may be tilted for parking garage ramps.

It is now used in all types of buildings and has gone up to 22 stories high in apartment buildings. Current trend is to use post-tensioned flat plate slabs with spans from 22' to 35'. Cylindrical void forms are used when deep slabs are required. One pound of prestressing steel is about equal to seven pounds of conventional reinforcing.

To be considered cured for stressing and lifting, a slab must have attained 75% of design strength. Seven days are usually sufficient with four to five days possible if high early strength cement is used. Slabs can be stacked using two coats of a non-bonding agent to insure that slabs do not stick to each other. Lifting is done by companies specializing in this work. Lift rate is 5' to 15' per hour with an average of 10' per hour. Total areas up to 33,000 S.F. have been lifted at one time. 24 to 36 jacking columns are common. Most economical bay sizes are 24' to 28' with four to fourteen stories most efficient. Continuous design reduces reinforcing steel cost. Use of post-tensioned slabs allows larger bay sizes.

Concrete — R0341 Precast Structural Concrete

R034105-30   Prestressed Precast Concrete Structural Units

Type	Location	Depth	Span in Ft.		Live Load Lb. per S.F.
Double Tee	Floor	28" to 34"	60 to 80		50 to 80
	Roof	12" to 24"	30 to 50		40
	Wall	Width 8'	Up to 55' high		Wind
Multiple Tee	Roof	8" to 12"	15 to 40		40
	Floor	8" to 12"	15 to 30		100
Plank	Roof or Floor		Roof	Floor	
		4"	13	12	40 for Roof
		6"	22	18	
		8"	26	25	
		10"	33	29	100 for Floor
		12"	42	32	
Single Tee	Roof	28"	40		
		32"	80		40
		36"	100		
		48"	120		
AASHO Girder	Bridges	Type 4	100		
		5	110		Highway
		6	125		
Box Beam	Bridges	15"	40		
		27"	to		Highway
		33"	100		

The majority of precast projects today utilize double tees rather than single tees because of speed and ease of installation. As a result casting beds at manufacturing plants are normally formed for double tees. Single tee projects will therefore require an initial set up charge to be spread over the individual single tee costs.

For floors, a 2" to 3" topping is field cast over the shapes. For roofs, insulating concrete or rigid insulation is placed over the shapes.

Member lengths up to 40' are standard haul, 40' to 60' require special permits and lengths over 60' must be escorted. Over width and/or over length can add up to 100% on hauling costs.

Large heavy members may require two cranes for lifting which would increase erection costs by about 45%. An eight man crew can install 12 to 20 double tees, or 45 to 70 quad tees or planks per day.

Grouting of connections must also be included.

Several system buildings utilizing precast members are available. Heights can go up to 22 stories for apartment buildings. Optimum design ratio is 3 S.F. of surface to 1 S.F. of floor area.

Concrete

R0352 Lightweight Concrete Roof Insulation

R035216-10 Lightweight Concrete

Lightweight aggregate concrete is usually purchased ready mixed, but it can also be field mixed.

Vermiculite or Perlite comes in bags of 4 C.F. under various trade names. Weight is about 8 lbs. per C.F. For insulating roof fill use 1:6 mix. For structural deck use 1:4 mix over gypsum boards, steeltex, steel centering, etc., supported by closely spaced joists or bulb trees. For structural slabs use 1:3:2 vermiculite sand concrete over steeltex, metal lath, steel centering, etc., on joists spaced 2'-0" O.C. for maximum L.L. of 80 P.S.F. Use same mix for slab base fill over steel flooring or regular reinforced concrete slab when tile, terrazzo or other finish is to be laid over.

For slabs on grade use 1:3:2 mix when tile, etc., finish is to be laid over. If radiant heating units are installed use a 1:6 mix for a base. After coils are in place, cover with a regular granolithic finish (mix 1:3:2) to a minimum depth of 1-1/2" over top of units.

Reinforce all slabs with 6 x 6 or 10 x 10 welded wire mesh.

Masonry

R0421 Clay Unit Masonry

R042110-20 Common and Face Brick

Common building brick manufactured according to ASTM C62 and facing brick manufactured according to ASTM C216 are the two standard bricks available for general building use.

Building brick is made in three grades; SW, where high resistance to damage caused by cyclic freezing is required; MW, where moderate resistance to cyclic freezing is needed; and NW, where little resistance to cyclic freezing is needed. Facing brick is made in only the two grades SW and MW. Additionally, facing brick is available in three types; FBS, for general use; FBX, for general use where a higher degree of precision and lower permissible variation in size than FBS is needed; and FBA, for general use to produce characteristic architectural effects resulting from non-uniformity in size and texture of the units.

In figuring the material cost of brickwork, an allowance of 25% mortar waste and 3% brick breakage was included. If bricks are delivered palletized with 280 to 300 per pallet, or packaged, allow only 1-1/2% for breakage. Packaged or palletized delivery is practical when a job is big enough to have a crane or other equipment available to handle a package of brick. This is so on all industrial work but not always true on small commercial buildings.

The use of buff and gray face is increasing, and there is a continuing trend to the Norman, Roman, Jumbo and SCR brick.

Common red clay brick for backup is not used that often. Concrete block is the most usual backup material with occasional use of sand lime or cement brick. Building brick is commonly used in solid walls for strength and as a fire stop.

Brick panels built on the ground and then crane erected to the upper floors have proven to be economical. This allows the work to be done under cover and without scaffolding.

R042110-50 Brick, Block & Mortar Quantities

Type Brick	Nominal Size (incl. mortar) L H W	Modular Coursing	Number of Brick per S.F.	C.F. of Mortar per M Bricks, Waste Included 3/8" Joint	C.F. of Mortar per M Bricks, Waste Included 1/2" Joint	Bond Type	Description	Factor
Standard	8 x 2-2/3 x 4	3C=8"	6.75	10.3	12.9	Common	full header every fifth course	+20%
Economy	8 x 4 x 4	1C=4"	4.50	11.4	14.6		full header every sixth course	+16.7%
Engineer	8 x 3-1/5 x 4	5C=16"	5.63	10.6	13.6	English	full header every second course	+50%
Fire	9 x 2-1/2 x 4-1/2	2C=5"	6.40	550 # Fireclay	—	Flemish	alternate headers every course	+33.3%
Jumbo	12 x 4 x 6 or 8	1C=4"	3.00	23.8	30.8		every sixth course	+5.6%
Norman	12 x 2-2/3 x 4	3C=8"	4.50	14.0	17.9	Header	= W x H exposed	+100%
Norwegian	12 x 3-1/5 x 4	5C=16"	3.75	14.6	18.6	Rowlock	= H x W exposed	+100%
Roman	12 x 2 x 4	2C=4"	6.00	13.4	17.0	Rowlock stretcher	= L x W exposed	+33.3%
SCR	12 x 2-2/3 x 6	3C=8"	4.50	21.8	28.0	Soldier	= H x L exposed	—
Utility	12 x 4 x 4	1C=4"	3.00	15.4	19.6	Sailor	= W x L exposed	-33.3%

Concrete Blocks Nominal Size	Approximate Weight per S.F. Standard	Approximate Weight per S.F. Lightweight	Blocks per 100 S.F.	Mortar per M block, waste included Partitions	Mortar per M block, waste included Back up
2" x 8" x 16"	20 PSF	15 PSF	113	27 C.F.	36 C.F.
4"	30	20		41	51
6"	42	30		56	66
8"	55	38		72	82
10"	70	47		87	97
12"	85	55		102	112

Metals — R0505 Common Work Results for Metals

R050521-20 Welded Structural Steel

Usual weight reductions with welded design run 10% to 20% compared with bolted or riveted connections. This amounts to about the same total cost compared with bolted structures since field welding is more expensive than bolts. For normal spans of 18' to 24' figure 6 to 7 connections per ton.

Trusses — For welded trusses add 4% to weight of main members for connections. Up to 15% less steel can be expected in a welded truss compared to one that is shop bolted. Cost of erection is the same whether shop bolted or welded.

General — Typical electrodes for structural steel welding are E6010, E6011, E60T and E70T. Typical buildings vary between 2# to 8# of weld rod per ton of steel. Buildings utilizing continuous design require about three times as much welding as conventional welded structures. In estimating field erection by welding, it is best to use the average linear feet of weld per ton to arrive at the welding cost per ton. The type, size and position of the weld will have a direct bearing on the cost per linear foot. A typical field welder will deposit 1.8# to 2# of weld rod per hour manually. Using semiautomatic methods can increase production by as much as 50% to 75%.

Metals — R0512 Structural Steel Framing

R051223-10 Structural Steel

The bare material prices for structural steel, shown in the unit cost sections of the book, are for 100 tons of shop-fabricated structural steel and include:
1. Mill base price of structural steel
2. Mill scrap/grade/size/length extras
3. Mill delivery to a metals service center (warehouse)
4. Service center storage and handling
5. Service center delivery to a fabrication shop
6. Shop storage and handling
7. Shop drafting/detailing
8. Shop fabrication
9. Shop coat of primer paint
10. Shop listing
11. Shop delivery to the job site

In unit cost sections of the book that contain items for field fabrication of steel components, the bare material cost of steel includes:
1. Mill base price of structural steel
2. Mill scrap/grade/size/length extras
3. Mill delivery to a metals service center (warehouse)
4. Service center storage and handling
5. Service center delivery to the job site

R051223-15 Structural Steel Estimating for Repair and Remodeling Projects

The correct approach to estimating structural steel is dependent upon the amount of steel required for the particular project. If the project is a sizable addition to a building, the data can be used directly from the cost data book. This is not the case however if the project requires a small amount of steel, for instance to reinforce existing roof or floor structural systems. To better understand this, please refer to the unit price line for a W16x31 beam with bolted connections. Assume your project requires the reinforcement of the structural members of the roof system of a 3 story building required for the installation a new piece of HVAC equipment. The project will need 4 pieces of W16x31 that are 30' long. After pricing this 120 L.F. job using the unit prices for a W16x31, an analysis will reveal that the price is wholly inadequate.

The first problem is apparent if you examine the amount of steel that can be installed per day. The unit price line indicates 900 linear feet per day can be installed by a 5 man crew with an 90 ton crane. This productivity is correct for new construction but certainly is not for repair and remodeling work. Installation of new structural steel considers that each member is installed from the foundation to the roof of the structure in a planned and systematic manner with a crane having unrestricted access to all parts of the project. Additionally each connection is planned and detailed with full field access for fit-up and final bolting. The erection is planned and progresses such that interferences and conflicts with other structural members are minimized, if not completely eliminated. All of these assumptions are clearly not the case with a repair and remodeling job, and a significant decrease in the stated productivity will be observed.

A crane will certainly be needed to lift the members into the general area of the project but in most cases will not be able to place the beams into their final position. An opening in the existing roof may not be large enough to permit the beams to pass through and it may be necessary to bring them into the building through existing windows or doors. Moving the beams to the actual area where they will be installed may involve hand labor and the use of dollies. Finally, hoists and/or jacks may be needed for final positioning.

The connection of new members to existing can often be accomplished by field bolting with accurate field measurements and good planning but in many cases access to both sides of existing members is not possible and field welding becomes the only alternative. In addition to the cost of the actual welding, protection of existing finishes, systems and structure and fire protection must be considered.

New beams can never be installed tight to the existing decks or floors which they must support and the use of shims and tack welding becomes necessary. Additionally, further planning and cost is involved in assuring that existing loads are minimized during the installation and shimming process.

It is apparent that installation of structural steel as part of a repair and remodeling project involves more than simply installing the members and estimating the cost in the same manner as new construction. The best procedure for estimating the total cost is adequate planning and coordination of each process and activity that will be needed. Unit costs for the materials, labor and equipment can then be attached to each needed activity and a final, complete price can be determined.

Metals — R0531 Steel Decking

R053100-10 Decking Descriptions

General - All Deck Products

Steel deck is made by cold forming structural grade sheet steel into a repeating pattern of parallel ribs. The strength and stiffness of the panels are the result of the ribs and the material properties of the steel. Deck lengths can be varied to suit job conditions, but because of shipping considerations, are usually less than 40 feet. Standard deck width varies with the product used but full sheets are usually 12", 18", 24", 30", or 36". Deck is typically furnished in a standard width with the ends cut square. Any cutting for width, such as at openings or for angular fit, is done at the job site.

Deck is typically attached to the building frame with arc puddle welds, self-drilling screws, or powder or pneumatically driven pins. Sheet to sheet fastening is done with screws, button punching (crimping), or welds.

Composite Floor Deck

After installation and adequate fastening, floor deck serves several purposes. It (a) acts as a working platform, (b) stabilizes the frame, (c) serves as a concrete form for the slab, and (d) reinforces the slab to carry the design loads applied during the life of the building. Composite decks are distinguished by the presence of shear connector devices as part of the deck. These devices are designed to mechanically lock the concrete and deck together so that the concrete and the deck work together to carry subsequent floor loads. These shear connector devices can be rolled-in embossments, lugs, holes, or wires welded to the panels. The deck profile can also be used to interlock concrete and steel.

Composite deck finishes are either galvanized (zinc coated) or phosphatized/painted. Galvanized deck has a zinc coating on both the top and bottom surfaces. The phosphatized/painted deck has a bare (phosphatized) top surface that will come into contact with the concrete. This bare top surface can be expected to develop rust before the concrete is placed. The bottom side of the deck has a primer coat of paint.

Composite floor deck is normally installed so the panel ends do not overlap on the supporting beams. Shear lugs or panel profile shape often prevent a tight metal to metal fit if the panel ends overlap; the air gap caused by overlapping will prevent proper fusion with the structural steel supports when the panel end laps are shear stud welded.

Adequate end bearing of the deck must be obtained as shown on the drawings. If bearing is actually less in the field than shown on the drawings, further investigation is required.

Roof Deck

Roof deck is not designed to act compositely with other materials. Roof deck acts alone in transferring horizontal and vertical loads into the building frame. Roof deck rib openings are usually narrower than floor deck rib openings. This provides adequate support of rigid thermal insulation board.

Roof deck is typically installed to endlap approximately 2" over supports. However, it can be butted (or lapped more than 2") to solve field fit problems. Since designers frequently use the installed deck system as part of the horizontal bracing system (the deck as a diaphragm), any fastening substitution or change should be approved by the designer. Continuous perimeter support of the deck is necessary to limit edge deflection in the finished roof and may be required for diaphragm shear transfer.

Standard roof deck finishes are galvanized or primer painted. The standard factory applied paint for roof deck is a primer paint and is not intended to weather for extended periods of time. Field painting or touching up of abrasions and deterioration of the primer coat or other protective finishes is the responsibility of the contractor.

Cellular Deck

Cellular deck is made by attaching a bottom steel sheet to a roof deck or composite floor deck panel. Cellular deck can be used in the same manner as floor deck. Electrical, telephone, and data wires are easily run through the chase created between the deck panel and the bottom sheet.

When used as part of the electrical distribution system, the cellular deck must be installed so that the ribs line up and create a smooth cell transition at abutting ends. The joint that occurs at butting cell ends must be taped or otherwise sealed to prevent wet concrete from seeping into the cell. Cell interiors must be free of welding burrs, or other sharp intrusions, to prevent damage to wires.

When used as a roof deck, the bottom flat plate is usually left exposed to view. Care must be maintained during erection to keep good alignment and prevent damage.

Cellular deck is sometimes used with the flat plate on the top side to provide a flat working surface. Installation of the deck for this purpose requires special methods for attachment to the frame because the flat plate, now on the top, can prevent direct access to the deck material that is bearing on the structural steel. It may be advisable to treat the flat top surface to prevent slipping.

Cellular deck is always furnished galvanized or painted over galvanized.

Form Deck

Form deck can be any floor or roof deck product used as a concrete form. Connections to the frame are by the same methods used to anchor floor and roof deck. Welding washers are recommended when welding deck that is less than 20 gauge thickness.

Form deck is furnished galvanized, prime painted, or uncoated. Galvanized deck must be used for those roof deck systems where form deck is used to carry a lightweight insulating concrete fill.

Wood, Plastics & Comp. — R0611 Wood Framing

R061110-30 Lumber Product Material Prices

The price of forest products fluctuates widely from location to location and from season to season depending upon economic conditions. The bare material prices in the unit cost sections of the book show the National Average material prices in effect Jan. 1 of this book year. It must be noted that lumber prices in general may change significantly during the year.

Availability of certain items depends upon geographic location and must be checked prior to firm-price bidding.

Wood, Plastics & Comp. — R0616 Sheathing

R061636-20 Plywood

There are two types of plywood used in construction: interior, which is moisture-resistant but not waterproofed, and exterior, which is waterproofed.

The grade of the exterior surface of the plywood sheets is designated by the first letter: A, for smooth surface with patches allowed; B, for solid surface with patches and plugs allowed; C, which may be surface plugged or may have knot holes up to 1" wide; and D, which is used only for interior type plywood and may have knot holes up to 2-1/2" wide. "Structural Grade" is specifically designed for engineered applications such as box beams. All CC & DD grades have roof and floor spans marked on them.

Underlayment-grade plywood runs from 1/4" to 1-1/4" thick. Thicknesses 5/8" and over have optional tongue and groove joints which eliminate the need for blocking the edges. Underlayment 19/32" and over may be referred to as Sturd-i-Floor.

The price of plywood can fluctuate widely due to geographic and economic conditions.

Typical uses for various plywood grades are as follows:

AA-AD Interior — cupboards, shelving, paneling, furniture
BB Plyform — concrete form plywood
CDX — wall and roof sheathing
Structural — box beams, girders, stressed skin panels
AA-AC Exterior — fences, signs, siding, soffits, etc.
Underlayment — base for resilient floor coverings
Overlaid HDO — high density for concrete forms & highway signs
Overlaid MDO — medium density for painting, siding, soffits & signs
303 Siding — exterior siding, textured, striated, embossed, etc.

Thermal & Moist. Protec. — R0751 Built-Up Bituminous Roofing

R075113-20 Built-Up Roofing

Asphalt is available in kegs of 100 lbs. each; coal tar pitch in 560 lb. kegs. Prepared roofing felts are available in a wide range of sizes, weights and characteristics. However, the most commonly used are #15 (432 S.F. per roll, 13 lbs. per square) and #30 (216 S.F. per roll, 27 lbs. per square).

Inter-ply bitumen varies from 24 lbs. per sq. (asphalt) to 30 lbs. per sq. (coal tar) per ply, MF4@ 25%. Flood coat bitumen also varies from 60 lbs. per sq. (asphalt) to 75 lbs. per sq. (coal tar), MF4@ 25%. Expendable equipment (mops, brooms, screeds, etc.) runs about 16% of the bitumen cost. For new, inexperienced crews this factor may be much higher.

Rigid insulation board is typically applied in two layers. The first is mechanically attached to nailable decks or spot or solid mopped to non-nailable decks; the second layer is then spot or solid mopped to the first layer. Membrane application follows the insulation, except in protected membrane roofs, where the membrane goes down first and the insulation on top, followed with ballast (stone or concrete pavers). Insulation and related labor costs are NOT included in prices for built-up roofing.

Thermal & Moist. Protec. — R0752 Modified Bituminous Membrane Roofing

R075213-30 Modified Bitumen Roofing

The cost of modified bitumen roofing is highly dependent on the type of installation that is planned. Installation is based on the type of modifier used in the bitumen. The two most popular modifiers are atactic polypropylene (APP) and styrene butadiene styrene (SBS). The modifiers are added to heated bitumen during the manufacturing process to change its characteristics. A polyethylene, polyester or fiberglass reinforcing sheet is then sandwiched between layers of this bitumen. When completed, the result is a pre-assembled, built-up roof that has increased elasticity and weatherablility. Some manufacturers include a surfacing material such as ceramic or mineral granules, metal particles or sand.

The preferred method of adhering SBS-modified bitumen roofing to the substrate is with hot-mopped asphalt (much the same as built-up roofing). This installation method requires a tar kettle/pot to heat the asphalt, as well as the labor, tools and equipment necessary to distribute and spread the hot asphalt.

The alternative method for applying APP and SBS modified bitumen is as follows. A skilled installer uses a torch to melt a small pool of bitumen off the membrane. This pool must form across the entire roll for proper adhesion. The installer must unroll the roofing at a pace slow enough to melt the bitumen, but fast enough to prevent damage to the rest of the membrane.

Modified bitumen roofing provides the advantages of both built-up and single-ply roofing. Labor costs are reduced over those of built-up roofing because only a single ply is necessary. The elasticity of single-ply roofing is attained with the reinforcing sheet and polymer modifiers. Modifieds have some self-healing characteristics and because of their multi-layer construction, they offer the reliability and safety of built-up roofing.

Thermal & Moist. Protec. — R0784 Firestopping

R078413-30 Firestopping

Firestopping is the sealing of structural, mechanical, electrical and other penetrations through fire-rated assemblies. The basic components of firestop systems are safing insulation and firestop sealant on both sides of wall penetrations and the top side of floor penetrations.

Pipe penetrations are assumed to be through concrete, grout, or joint compound and can be sleeved or unsleeved. Costs for the penetrations and sleeves are not included. An annular space of 1" is assumed. Escutcheons are not included.

Metallic pipe is assumed to be copper, aluminum, cast iron or similar metallic material. Insulated metallic pipe is assumed to be covered with a thermal insulating jacket of varying thickness and materials.

Non-metallic pipe is assumed to be PVC, CPVC, FR Polypropylene or similar plastic piping material. Intumescent firestop sealant or wrap strips are included. Collars on both sides of wall penetrations and a sheet metal plate on the underside of floor penetrations are included.

Ductwork is assumed to be sheet metal, stainless steel or similar metallic material. Duct penetrations are assumed to be through concrete, grout or joint compound. Costs for penetrations and sleeves are not included. An annular space of 1/2" is assumed.

Multi-trade openings include costs for sheet metal forms, firestop mortar, wrap strips, collars and sealants as necessary.

Structural penetrations joints are assumed to be 1/2" or less. CMU walls are assumed to be within 1-1/2" of metal deck. Drywall walls are assumed to be tight to the underside of metal decking.

Metal panel, glass or curtain wall systems include a spandrel area of 5' filled with mineral wool foil-faced insulation. Fasteners and stiffeners are included.

Openings — R0813 Metal Doors

R081313-20 Steel Door Selection Guide

Standard steel doors are classified into four levels, as recommended by the Steel Door Institute in the chart below. Each of the four levels offers a range of construction models and designs, to meet architectural requirements for preference and appearance, including full flush, seamless, and stile & rail. Recommended minimum gauge requirements are also included.

For complete standard steel door construction specifications and available sizes, refer to the Steel Door Institute Technical Data Series, ANSI A250.8-98 (SDI-100), and ANSI A250.4-94 Test Procedure and Acceptance Criteria for Physical Endurance of Steel Door and Hardware Reinforcements.

Level	Model	Construction	For Full Flush or Seamless		
			Min. Gauge	Thickness (in)	Thickness (mm)
I	Standard Duty	1 Full Flush / 2 Seamless	20	0.032	0.8
II	Heavy Duty	1 Full Flush / 2 Seamless	18	0.042	1.0
III	Extra Heavy Duty	1 Full Flush / 2 Seamless / 3 *Stile & Rail	16	0.053	1.3
IV	Maximum Duty	1 Full Flush / 2 Seamless	14	0.067	1.6

*Stiles & rails are 16 gauge; flush panels, when specified, are 18 gauge

Openings — R0851 Metal Windows

R085123-10 Steel Sash

Ironworker crew will erect 25 S.F. or 1.3 sash unit per hour, whichever is less.

Mechanic will point 30 L.F. per hour.

Painter will paint 90 S.F. per coat per hour.

Glazier production depends on light size.

Allow 1 lb. special steel sash putty per 16" x 20" light.

Openings — R0853 Plastic Windows

R085313-20 Replacement Windows

Replacement windows are typically measured per United Inch.

United Inches are calculated by rounding the width and height of the window opening up to the nearest inch, then adding the two figures.

The labor cost for replacement windows includes removal of sash, existing sash balance or weights, parting bead where necessary and installation of new window.

Debris hauling and dump fees are not included.

Openings — R0871 Door Hardware

R087120-10 Hinges

All closer equipped doors should have ball bearing hinges. Lead lined or extremely heavy doors require special strength hinges. Usually 1-1/2 pair of hinges are used per door up to 7'-6" high openings. Table below shows typical hinge requirements.

Use Frequency	Type Hinge Required	Type of Opening	Type of Structure
High	Heavy weight ball bearing	Entrances	Banks, Office buildings, Schools, Stores & Theaters
		Toilet Rooms	Office buildings and Schools
Average	Standard weight ball bearing	Entrances	Dwellings
		Corridors	Office buildings and Schools
		Toilet Rooms	Stores
Low	Plain bearing	Interior	Dwellings

Door Thickness	Weight of Doors in Pounds per Square Foot				
	White Pine	Oak	Hollow Core	Solid Core	Hollow Metal
1-3/8"	3psf	6psf	1-1/2psf	3-1/2 — 4psf	6-1/2psf
1-3/4"	3-1/2	7	2-1/2	4-1/2 — 5-1/4	6-1/2
2-1/4"	4-1/2	9	—	5-1/2 — 6-3/4	6-1/2

Openings — R0881 Glass Glazing

R088110-10 Glazing Productivity

Some glass sizes are estimated by the "united inch" (height + width). The table below shows the number of lights glazed in an eight-hour period by the crew size indicated, for glass up to 1/4" thick. Square or nearly square lights are more economical on a S.F. basis. Long slender lights will have a high S.F. installation cost. For insulated glass reduce production by 33%. For 1/2" float glass reduce production by 50%. Production time for glazing with two glaziers per day averages: 1/4" float glass 120 S.F.; 1/2" float glass 55 S.F.; 1/2" insulated glass 95 S.F.; 3/4" insulated glass 75 S.F.

Glazing Method	United Inches per Light							
	40"	60"	80"	100"	135"	165"	200"	240"
Number of Men in Crew	1	1	1	1	2	3	3	4
Industrial sash, putty	60	45	24	15	18	—	—	—
With stops, putty bed	50	36	21	12	16	8	4	3
Wood stops, rubber	40	27	15	9	11	6	3	2
Metal stops, rubber	30	24	14	9	9	6	3	2
Structural glass	10	7	4	3	—	—	—	—
Corrugated glass	12	9	7	4	4	4	3	—
Storefronts	16	15	13	11	7	6	4	4
Skylights, putty glass	60	36	21	12	16	—	—	—
Thiokol set	15	15	11	9	9	6	3	2
Vinyl set, snap on	18	18	13	12	12	7	5	4
Maximum area per light	2.8 S.F.	6.3 S.F.	11.1 S.F.	17.4 S.F.	31.6 S.F.	47 S.F.	69 S.F.	100 S.F.

Finishes

R0920 Plaster & Gypsum Board

R092000-50 Lath, Plaster and Gypsum Board

Gypsum board lath is available in 3/8" thick x 16" wide x 4' long sheets as a base material for multi-layer plaster applications. It is also available as a base for either multi-layer or veneer plaster applications in 1/2" and 5/8" thick–4' wide x 8', 10' or 12' long sheets. Fasteners are screws or blued ring shank nails for wood framing and screws for metal framing.

Metal lath is available in diamond mesh pattern with flat or self-furring profiles. Paper backing is available for applications where excessive plaster waste needs to be avoided. A slotted mesh ribbed lath should be used in areas where the span between structural supports is greater than normal. Most metal lath comes in 27" x 96" sheets. Diamond mesh weighs 1.75, 2.5 or 3.4 pounds per square yard, slotted mesh lath weighs 2.75 or 3.4 pounds per square yard. Metal lath can be nailed, screwed or tied in place.

Many **accessories** are available. Corner beads, flat reinforcing strips, casing beads, control and expansion joints, furring brackets and channels are some examples. Note that accessories are not included in plaster or stucco line items.

Plaster is defined as a material or combination of materials that when mixed with a suitable amount of water, forms a plastic mass or paste. When applied to a surface, the paste adheres to it and subsequently hardens, preserving in a rigid state the form or texture imposed during the period of elasticity.

Gypsum plaster is made from ground calcined gypsum. It is mixed with aggregates and water for use as a base coat plaster.

Vermiculite plaster is a fire-retardant plaster covering used on steel beams, concrete slabs and other heavy construction materials. Vermiculite is a group name for certain clay minerals, hydrous silicates or aluminum, magnesium and iron that have been expanded by heat.

Perlite plaster is a plaster using perlite as an aggregate instead of sand. Perlite is a volcanic glass that has been expanded by heat.

Gauging plaster is a mix of gypsum plaster and lime putty that when applied produces a quick drying finish coat.

Veneer plaster is a one or two component gypsum plaster used as a thin finish coat over special gypsum board.

Keenes cement is a white cementitious material manufactured from gypsum that has been burned at a high temperature and ground to a fine powder. Alum is added to accelerate the set. The resulting plaster is hard and strong and accepts and maintains a high polish, hence it is used as a finishing plaster.

Stucco is a Portland cement based plaster used primarily as an exterior finish.

Plaster is used on both interior and exterior surfaces. Generally it is applied in multiple-coat systems. A three-coat system uses the terms scratch, brown and finish to identify each coat. A two-coat system uses base and finish to describe each coat. Each type of plaster and application system has attributes that are chosen by the designer to best fit the intended use.

Gypsum Plaster	2 Coat, 5/8" Thick		3 Coat, 3/4" Thick		
Quantities for 100 S.Y.	Base	Finish	Scratch	Brown	Finish
	1:3 Mix	2:1 Mix	1:2 Mix	1:3 Mix	2:1 Mix
Gypsum plaster	1,300 lb.		1,350 lb.	650 lb.	
Sand	1.75 C.Y.		1.85 C.Y.	1.35 C.Y.	
Finish hydrated lime		340 lb.			340 lb.
Gauging plaster		170 lb.			170 lb.

Vermiculite or Perlite Plaster	2 Coat, 5/8" Thick		3 Coat, 3/4" Thick		
Quantities for 100 S.Y.	Base	Finish	Scratch	Brown	Finish
Gypsum plaster	1,250 lb.		1,450 lb.	800 lb.	
Vermiculite or perlite	7.8 bags		8.0 bags	3.3 bags	
Finish hydrated lime		340 lb.			340 lb.
Gauging plaster		170 lb.			170 lb.

Stucco–Three-Coat System Quantities for 100 S.Y.	On Wood Frame	On Masonry
Portland cement	29 bags	21 bags
Sand	2.6 C.Y.	2.0 C.Y.
Hydrated lime	180 lb.	120 lb.

Finishes — R0929 Gypsum Board

R092910-10 Levels of Gypsum Drywall Finish

In the past, contract documents often used phrases such as "industry standard" and "workmanlike finish" to specify the expected quality of gypsum board wall and ceiling installations. The vagueness of these descriptions led to unacceptable work and disputes.

In order to resolve this problem, four major trade associations concerned with the manufacture, erection, finish and decoration of gypsum board wall and ceiling systems have developed an industry-wide *Recommended Levels of Gypsum Board Finish*.

The finish of gypsum board walls and ceilings for specific final decoration is dependent on a number of factors. A primary consideration is the location of the surface and the degree of decorative treatment desired. Painted and unpainted surfaces in warehouses and other areas where appearance is normally not critical may simply require the taping of wallboard joints and 'spotting' of fastener heads. Blemish-free, smooth, monolithic surfaces often intended for painted and decorated walls and ceilings in habitated structures, ranging from single-family dwellings through monumental buildings, require additional finishing prior to the application of the final decoration.

Other factors to be considered in determining the level of finish of the gypsum board surface are (1) the type of angle of surface illumination (both natural and artificial lighting), and (2) the paint and method of application or the type and finish of wallcovering specified as the final decoration. Critical lighting conditions, gloss paints, and thin wall coverings require a higher level of gypsum board finish than do heavily textured surfaces which are subsequently painted or surfaces which are to be decorated with heavy grade wall coverings.

The following descriptions were developed jointly by the Association of the Wall and Ceiling Industries-International (AWCI), Ceiling & Interior Systems Construction Association (CISCA), Gypsum Association (GA), and Painting and Decorating Contractors of America (PDCA) as a guide.

Level 0: Used in temporary construction or wherever the final decoration has not been determined. Unfinished. No taping, finishing or corner beads are required. Also could be used where non-predecorated panels will be used in demountable-type partitions that are to be painted as a final finish. .

Level 1 Frequently used in plenum areas above ceilings, in attics, in areas where the assembly would generally be concealed, or in building service corridors and other areas not normally open to public view. Some degree of sound and smoke control is provided; in some geographic areas, this level is referred to as "fire-taping," although this level of finish does not typically meet fire-resistant assembly requirements. Where a fire resistance rating is required for the gypsum board assembly, details of construction should be in accordance with reports of fire tests of assemblies that have met the requirements of the fire rating acceptable.

All joints and interior angles shall have tape embedded in joint compound. Accessories are optional at specifier discretion in corridors and other areas with pedestrian traffic. Tape and fastener heads need not be covered with joint compound. Surface shall be free of excess joint compound. Tool marks and ridges are acceptable.

Level 2 It may be specified for standard gypsum board surfaces in garages, warehouse storage, or other similar areas where surface appearance is not of primary importance.

All joints and interior angles shall have tape embedded in joint compound and shall be immediately wiped with a joint knife or trowel, leaving a thin coating of joint compound over all joints and interior angles. Fastener heads and accessories shall be covered with a coat of joint compound. Surface shall be free of excess joint compound. Tool marks and ridges are acceptable.

Level 3 Typically used in areas that are to receive heavy texture (spray or hand applied) finishes before final painting, or where commercial-grade (heavy duty) wall coverings are to be applied as the final decoration. This level of finish should not be used where smooth painted surfaces or where lighter weight wall coverings are specified. The prepared surface shall be coated with a drywall primer prior to the application of final finishes.

All joints and interior angles shall have tape embedded in joint compound and shall be immediately wiped with a joint knife or trowel, leaving a thin coating of joint compound over all joints and interior angles. One additional coat of joint compound shall be applied over all joints and interior angles. Fastener heads and accessories shall be covered with two separate coats of joint compound. All joint compounds shall be smooth and free of tool marks and ridges. The prepared surface shall be covered with a drywall primer prior to the application of the final decoration.

Level 4 This level should be used where residential grade (light duty) wall coverings, flat paints, or light textures are to be applied. The prepared surface shall be coated with a drywall primer prior to the application of final finishes. Release agents for wall coverings are specifically formulated to minimize damage if coverings are subsequently removed.

The weight, texture, and sheen level of the wall covering material selected should be taken into consideration when specifying wall coverings over this level of drywall treatment. Joints and fasteners must be sufficiently concealed if the wall covering material is lightweight, contains limited pattern, has a glossy finish, or has any combination of these features. In critical lighting areas, flat paints applied over light textures tend to reduce joint photographing. Gloss, semi-gloss, and enamel paints are not recommended over this level of finish.

All joints and interior angles shall have tape embedded in joint compound and shall be immediately wiped with a joint knife or trowel, leaving a thin coating of joint compound over all joints and interior angles. In addition, two separate coats of joint compound shall be applied over all flat joints and one separate coat of joint compound applied over interior angles. Fastener heads and accessories shall be covered with three separate coats of joint compound. All joint compounds shall be smooth and free of tool marks and ridges. The prepared surface shall be covered with a drywall primer like Sheetrock first coat prior to the application of the final decoration.

Level 5 The highest quality finish is the most effective method to provide a uniform surface and minimize the possibility of joint photographing and of fasteners showing through the final decoration. This level of finish is required where gloss, semi-gloss, or enamel is specified; when flat joints are specified over an untextured surface; or where critical lighting conditions occur. The prepared surface shall be coated with a drywall primer prior to the application of final decoration.

All joints and interior angles shall have tape embedded in joint compound and be immediately wiped with a joint knife or trowel, leaving a thin coating of joint compound over all joints and interior angles. Two separate coats of joint compound shall be applied over all flat joints and one separate coat of joint compound applied over interior angles. Fastener heads and accessories shall be covered with three separate coats of joint compound.

A thin skim coat of joint compound shall be trowel applied to the entire surface. Excess compound is immediately troweled off, leaving a film or skim coating of compound completely covering the paper. As an alternative to a skim coat, a material manufactured especially for this purpose may be applied such as Sheetrock Tuff-Hide primer surfacer. The surface must be smooth and free of tool marks and ridges. The prepared surface shall be covered with a drywall primer prior to the application of the final decoration.

Conveying Equipment   R1420 Elevators

R142000-10   Freight Elevators

Capacities run from 2,000 lbs. to over 100,000 lbs. with 3,000 lbs. to 10,000 lbs. most common. Travel speeds are generally lower and control less intricate than on passenger elevators. Costs in the unit price sections are for hydraulic and geared elevators.

R142000-20   Elevator Selective Costs   See R142000-40 for cost development.

A. Base Unit	Passenger		Freight		Hospital	
	Hydraulic	Electric	Hydraulic	Electric	Hydraulic	Electric
Capacity	1,500 lb.	2,000 lb.	2,000 lb.	4,000 lb.	4,000 lb.	4,000 lb.
Speed	100 F.P.M.	200 F.P.M.	100 F.P.M.	200 F.P.M.	100 F.P.M.	200 F.P.M.
#Stops/Travel Ft.	2/12	4/40	2/20	4/40	2/20	4/40
Push Button Oper.	Yes	Yes	Yes	Yes	Yes	Yes
Telephone Box & Wire	"	"	"	"	"	"
Emergency Lighting	"	"	No	No	"	"
Cab	Plastic Lam. Walls	Plastic Lam. Walls	Painted Steel	Painted Steel	Plastic Lam. Walls	Plastic Lam. Walls
Cove Lighting	Yes	Yes	No	No	Yes	Yes
Floor	V.C.T.	V.C.T.	Wood w/Safety Treads	Wood w/Safety Treads	V.C.T.	V.C.T.
Doors, & Speedside Slide	Yes	Yes	Yes	Yes	Yes	Yes
Gates, Manual	No	No	No	No	No	No
Signals, Lighted Buttons	Car and Hall	Car and Hall	Car and Hall	Car and Hall	Car and Hall	Car and Hall
O.H. Geared Machine	N.A.	Yes	N.A.	Yes	N.A.	Yes
Variable Voltage Contr.	"	"	N.A.	"	"	"
Emergency Alarm	Yes	"	Yes	"	Yes	"
Class "A" Loading	N.A.	N.A.	"	"	N.A.	N.A.

R142000-30   Passenger Elevators

Electric elevators are used generally but hydraulic elevators can be used for lifts up to 70' and where large capacities are required. Hydraulic speeds are limited to 200 F.P.M. but cars are self leveling at the stops. On low rises, hydraulic installation runs about 15% less than standard electric types but on higher rises this installation cost advantage is reduced. Maintenance of hydraulic elevators is about the same as electric type but underground portion is not included in the maintenance contract.

In electric elevators there are several control systems available, the choice of which will be based upon elevator use, size, speed and cost criteria. The two types of drives are geared for low speeds and gearless for 450 F.P.M. and over.

The tables on the preceding pages illustrate typical installed costs of the various types of elevators available.

R142000-40   Elevator Cost Development

To price a new car or truck from the factory, you must start with the manufacturer's basic model, then add or exchange optional equipment and features. The same is true for pricing elevators.

Requirement: One-passenger elevator, five-story hydraulic, 2,500 lb. capacity, 12' floor to floor, speed 150 F.P.M., emergency power switching and maintenance contract.

Example:

Description	Adjustment
A. Base Elevator: Hydraulic Passenger, 1500 lb. Capacity, 100 fpm, 2 Stops, Standard Finish	1 Ea.
B. Capacity Adjustment (2,500 lb.)	1 Ea.
C. Excess Travel Adjustment: 48' Total Travel (4 x 12') minus 12' Base Unit Travel =	36 V.L.F.
D. Stops Adjustment: 5 Total Stops minus 2 Stops (Base Unit) =	3 Stops
E. Speed Adjustment (150 F.P.M.)	1 Ea.
F. Options:	
1. Intercom Service	1 Ea.
2. Emergency Power Switching, Automatic	1 Ea.
3. Stainless Steel Entrance Doors	5 Ea.
4. Maintenance Contract (12 Months)	1 Ea.
5. Position Indicator for main floor level (none indicated in Base Unit)	1 Ea.

Plumbing — R2201 Operation & Maintenance of Plumbing

R220105-10 Demolition (Selective vs. Removal for Replacement)

Demolition can be divided into two basic categories.

One type of demolition involves the removal of material with no concern for its replacement. The labor-hours to estimate this work are found under "Selective Demolition" in the Fire Protection, Plumbing and HVAC Divisions. It is selective in that individual items or all the material installed as a system or trade grouping such as plumbing or heating systems are removed. This may be accomplished by the easiest way possible, such as sawing, torch cutting, or sledge hammer as well as simple unbolting.

The second type of demolition is the removal of some item for repair or replacement. This removal may involve careful draining, opening of unions, disconnecting and tagging of electrical connections, capping of pipes/ducts to prevent entry of debris or leakage of the material contained as well as transport of the item away from its in-place location to a truck/dumpster. An approximation of the time required to accomplish this type of demolition is to use half of the time indicated as necessary to install a new unit. For example; installation of a new pump might be listed as requiring 6 labor-hours so if we had to estimate the removal of the old pump we would allow an additional 3 hours for a total of 9 hours. That is, the complete replacement of a defective pump with a new pump would be estimated to take 9 labor-hours.

Plumbing — R2211 Facility Water Distribution

R221113-50 Pipe Material Considerations

1. Malleable fittings should be used for gas service.
2. Malleable fittings are used where there are stresses/strains due to expansion and vibration.
3. Cast fittings may be broken as an aid to disassembling of heating lines frozen by long use, temperature and minerals.
4. Cast iron pipe is extensively used for underground and submerged service.
5. Type M (light wall) copper tubing is available in hard temper only and is used for nonpressure and less severe applications than K and L.
6. Type L (medium wall) copper tubing, available hard or soft for interior service.
7. Type K (heavy wall) copper tubing, available in hard or soft temper for use where conditions are severe. For underground and interior service.
8. Hard drawn tubing requires fewer hangers or supports but should not be bent. Silver brazed fittings are recommended, however soft solder is normally used.
9. Type DMV (very light wall) copper tubing designed for drainage, waste and vent plus other non-critical pressure services.

Domestic/Imported Pipe and Fittings Cost

The prices shown in this publication for steel/cast iron pipe and steel, cast iron, malleable iron fittings are based on domestic production sold at the normal trade discounts. The above listed items of foreign manufacture may be available at prices of 1/3 to 1/2 those shown. Some imported items after minor machining or finishing operations are being sold as domestic to further complicate the system.

Caution: Most pipe prices in this book also include a coupling and pipe hangers which for the larger sizes can add significantly to the per foot cost and should be taken into account when comparing "book cost" with quoted supplier's cost.

Plumbing — R2240 Plumbing Fixtures

R224000-40 Plumbing Fixture Installation Time

Item	Rough-In	Set	Total Hours	Item	Rough-In	Set	Total Hours
Bathtub	5	5	10	Shower head only	2	1	3
Bathtub and shower, cast iron	6	6	12	Shower drain	3	1	4
Fire hose reel and cabinet	4	2	6	Shower stall, slate		15	15
Floor drain to 4 inch diameter	3	1	4	Slop sink	5	3	8
Grease trap, single, cast iron	5	3	8	Test 6 fixtures			14
Kitchen gas range		4	4	Urinal, wall	6	2	8
Kitchen sink, single	4	4	8	Urinal, pedestal or floor	6	4	10
Kitchen sink, double	6	6	12	Water closet and tank	4	3	7
Laundry tubs	4	2	6	Water closet and tank, wall hung	5	3	8
Lavatory wall hung	5	3	8	Water heater, 45 gals. gas, automatic	5	2	7
Lavatory pedestal	5	3	8	Water heaters, 65 gals. gas, automatic	5	2	7
Shower and stall	6	4	10	Water heaters, electric, plumbing only	4	2	6

Fixture prices in front of book are based on the cost per fixture set in place. The rough-in cost, which must be added for each fixture, includes carrier, if required, some supply, waste and vent pipe connecting fittings and stops. The lengths of rough-in pipe are nominal runs which would connect to the larger runs and stacks. The supply runs and DWV runs and stacks must be accounted for in separate entries. In the eastern half of the United States it is common for the plumber to carry these to a point 5' outside the building.

Heating, Ventilating & A.C. — R2331 HVAC Ducts & Casings

R233100-20 Ductwork

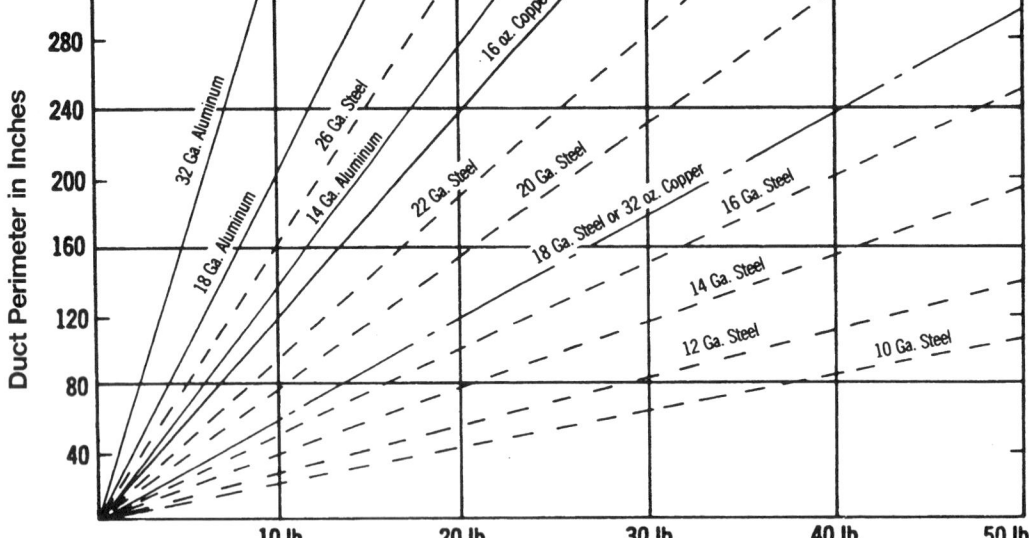

Duct Weight in Pounds per L.F., Straight Runs

Add to the above for fittings; 90° elbow is 3 L.F.; 45° elbow is 2.5 L.F.; offset is 4 L.F.; transition offset is 6 L.F.; square-to-round transition is 4 L.F.; 90° reducing elbow is 5 L.F. For bracing and waste, add 20% to aluminum and copper, 15% to steel.

Electrical — R2601 Operation & Maint. of Elect. Systems

R260105-30 Electrical Demolition (Removal for Replacement)

The purpose of this reference number is to provide a guide to users for electrical "removal for replacement" by applying the rule of thumb: 1/3 of new installation time (typical range from 20% to 50%) for removal. Remember to use reasonable judgment when applying the suggested percentage factor. For example:

Contractors have been requested to remove an existing fluorescent lighting fixture and replace with a new fixture utilizing energy saver lamps and electronic ballast:

In order to fully understand the extent of the project, contractors should visit the job site and estimate the time to perform the renovation work in accordance with applicable national, state and local regulation and codes.

The contractor may need to add extra labor hours to his estimate if he discovers unknown concealed conditions such as: contaminated asbestos ceiling, broken acoustical ceiling tile and need to repair, patch and touch-up paint all the damaged or disturbed areas, tasks normally assigned to general contractors. In addition, the owner could request that the contractors salvage the materials removed and turn over the materials to the owner or dispose of the materials to a reclamation station. The normal removal item is 0.5 labor hour for a lighting fixture and 1.5 labor-hours for new installation time. Revise the estimate times from 2 labor-hours work up to a minimum 4 labor-hours work for just fluorescent lighting fixture.

For removal of large concentrations of lighting fixtures in the same area, apply an "economy of scale" to reduce estimating labor hours.

Electrical — R2605 Common Work Results for Electrical

R260533-22 Conductors in Conduit

Table below lists maximum number of conductors for various sized conduit using THW, TW or THWN insulations.

Copper Wire Size	1/2" TW	1/2" THW	1/2" THWN	3/4" TW	3/4" THW	3/4" THWN	1" TW	1" THW	1" THWN	1-1/4" TW	1-1/4" THW	1-1/4" THWN	1-1/2" TW	1-1/2" THW	1-1/2" THWN	2" TW	2" THW	2" THWN	2-1/2" TW	2-1/2" THW	2-1/2" THWN	3" THW	3" THWN	3-1/2" THW	3-1/2" THWN	4" THW	4" THWN
#14	9	6	13	15	10	24	25	16	39	44	29	69	60	40	94	99	65	154	142	93		143		192			
#12	7	4	10	12	8	18	19	13	29	35	24	51	47	32	70	78	53	114	111	76	164	117		157		163	
#10	5	4	6	9	6	11	15	11	18	26	19	32	36	26	44	60	43	73	85	61	104	95	160	127		163	
#8	2	1	3	4	3	5	7	5	9	12	10	16	17	13	22	28	22	36	40	32	51	49	79	66	106	85	136
#6		1	1	2	4		4	6		7	11		10	15		16	26		23	37		36	57	48	76	62	98
#4		1	1	1	2		3	4		5	7		7	9		12	16		17	22		27	35	36	47	47	60
#3		1	1	1	1		2	3		4	6		6	8		10	13		15	19		23	29	31	39	40	51
#2		1	1	1	1		2	3		4	5		5	7		9	11		13	16		20	25	27	33	34	43
#1				1	1		1	1		3	3		4	5		6	8		9	12		14	18	19	25	25	32
1/0				1	1		1	1		2	3		3	4		5	7		8	10		12	15	16	21	21	27
2/0				1	1		1	1		1	2		3	3		5	6		7	8		10	13	14	17	18	22
3/0				1	1		1	1		1	1		2	3		4	5		6	7		9	11	12	14	15	18
4/0					1		1	1		1	1		1	2		3	4		5	6		7	9	10	12	13	15
250 kcmil							1	1		1	1		1	1		2	3		4	4		6	7	8	10	10	12
300							1	1		1	1		1	1		2	3		3	4		5	6	7	8	9	11
350								1		1	1		1	1		1	2		3	3		4	5	6	7	8	9
400										1	1		1	1		1	1		2	3		4	5	5	6	7	8
500										1	1		1	1		1	1		1	2		3	4	4	5	6	7
600											1		1	1		1	1		1	1		3	3	4	4	5	5
700													1	1		1	1		1	1		2	3	3	4	4	5
750													1	1		1	1		1	1		2	2	3	3	4	4

Earthwork — R3123 Excavation & Fill

R312316-40 Excavating

The selection of equipment used for structural excavation and bulk excavation or for grading is determined by the following factors.
1. Quantity of material
2. Type of material
3. Depth or height of cut
4. Length of haul
5. Condition of haul road
6. Accessibility of site
7. Moisture content and dewatering requirements
8. Availability of excavating and hauling equipment

Some additional costs must be allowed for hand trimming the sides and bottom of concrete pours and other excavation below the general excavation.

Number of B.C.Y. per truck = 1.5 C.Y. bucket x 8 passes = 12 loose C.Y.

$$= 12 \times \frac{100}{118} = 10.2 \text{ B.C.Y. per truck}$$

Truck Haul Cycle:

Load truck, 8 passes	=	4 minutes
Haul distance, 1 mile	=	9 minutes
Dump time	=	2 minutes
Return, 1 mile	=	7 minutes
Spot under machine	=	1 minute
		23 minute cycle

Add the mobilization and demobilization costs to the total excavation costs. When equipment is rented for more than three days, there is often no mobilization charge by the equipment dealer. On larger jobs outside of urban areas, scrapers can move earth economically provided a dump site or fill area and adequate haul roads are available. Excavation within sheeting bracing or cofferdam bracing is usually done with a clamshell and production

When planning excavation and fill, the following should also be considered.
1. Swell factor
2. Compaction factor
3. Moisture content
4. Density requirements

A typical example for scheduling and estimating the cost of excavation of a 15' deep basement on a dry site when the material must be hauled off the site is outlined below.

Assumptions:
1. Swell factor, 18%
2. No mobilization or demobilization
3. Allowance included for idle time and moving on job
4. No dewatering, sheeting, or bracing
5. No truck spotter or hand trimming

Fleet Haul Production per day in B.C.Y.

$$4 \text{ trucks} \times \frac{50 \text{ min. hour}}{23 \text{ min. haul cycle}} \times 8 \text{ hrs.} \times 10.2 \text{ B.C.Y.}$$

$$= 4 \times 2.2 \times 8 \times 10.2 = 718 \text{ B.C.Y./day}$$

is low, since the clamshell may have to be guided by hand between the bracing. When excavating or filling an area enclosed with a wellpoint system, add 10% to 15% to the cost to allow for restricted access. When estimating earth excavation quantities for structures, allow work space outside the building footprint for construction of the foundation and a slope of 1:1 unless sheeting is used.

Earthwork — R3123 Excavation & Fill

R312316-45 Excavating Equipment

The table below lists THEORETICAL hourly production in C.Y./hr. bank measure for some typical excavation equipment. Figures assume 50 minute hours, 83% job efficiency, 100% operator efficiency, 90° swing and properly sized hauling units, which must be modified for adverse digging and loading conditions. Actual production costs in the front of the book average about 50% of the theoretical values listed here.

Equipment	Soil Type	B.C.Y. Weight	% Swell	1 C.Y.	1-1/2 C.Y.	2 C.Y.	2-1/2 C.Y.	3 C.Y.	3-1/2 C.Y.	4 C.Y.
Hydraulic Excavator "Backhoe" 15' Deep Cut	Moist loam, sandy clay	3400 lb.	40%	165	195	200	275	330	385	440
	Sand and gravel	3100	18	140	170	225	240	285	330	380
	Common earth	2800	30	150	180	230	250	300	350	400
	Clay, hard, dense	3000	33	120	140	190	200	240	260	320
Power Shovel Optimum Cut (Ft.)	Moist loam, sandy clay	3400	40	170 (6.0)	245 (7.0)	295 (7.8)	335 (8.4)	385 (8.8)	435 (9.1)	475 (9.4)
	Sand and gravel	3100	18	165 (6.0)	225 (7.0)	275 (7.8)	325 (8.4)	375 (8.8)	420 (9.1)	460 (9.4)
	Common earth	2800	30	145 (7.8)	200 (9.2)	250 (10.2)	295 (11.2)	335 (12.1)	375 (13.0)	425 (13.8)
	Clay, hard, dense	3000	33	120 (9.0)	175 (10.7)	220 (12.2)	255 (13.3)	300 (14.2)	335 (15.1)	375 (16.0)
Drag Line Optimum Cut (Ft.)	Moist loam, sandy clay	3400	40	130 (6.6)	180 (7.4)	220 (8.0)	250 (8.5)	290 (9.0)	325 (9.5)	385 (10.0)
	Sand and gravel	3100	18	130 (6.6)	175 (7.4)	210 (8.0)	245 (8.5)	280 (9.0)	315 (9.5)	375 (10.0)
	Common earth	2800	30	110 (8.0)	160 (9.0)	190 (9.9)	220 (10.5)	250 (11.0)	280 (11.5)	310 (12.0)
	Clay, hard, dense	3000	33	90 (9.3)	130 (10.7)	160 (11.8)	190 (12.3)	225 (12.8)	250 (13.3)	280 (12.0)

Equipment	Soil Type	B.C.Y. Weight	% Swell	Wheel Loaders				Track Loaders		
				3 C.Y.	4 C.Y.	6 C.Y.	8 C.Y.	2-1/4 C.Y.	3 C.Y.	4 C.Y.
Loading Tractors	Moist loam, sandy clay	3400	40	260	340	510	690	135	180	250
	Sand and gravel	3100	18	245	320	480	650	130	170	235
	Common earth	2800	30	230	300	460	620	120	155	220
	Clay, hard, dense	3000	33	200	270	415	560	110	145	200
	Rock, well-blasted	4000	50	180	245	380	520	100	130	180

R312323-30 Compacting Backfill

Compaction of fill in embankments, around structures, in trenches, and under slabs is important to control settlement. Factors affecting compaction are:
1. Soil gradation
2. Moisture content
3. Equipment used
4. Depth of fill per lift
5. Density required

Example:

Compact granular fill around a building foundation using a 21" wide x 24" vibratory plate in 8" lifts. Operator moves at 50 F.P.M. working a 50 minute hour to develop 95% Modified Proctor Density with 4 passes.

Production Rate:

$$\frac{1.75' \text{ plate width} \times 50 \text{ F.P.M.} \times 50 \text{ min./hr.} \times .67' \text{ lift}}{27 \text{ C.F. per C.Y.}} = 108.5 \text{ C.Y./hr.}$$

Production Rate for 4 Passes:

$$\frac{108.5 \text{ C.Y.}}{4 \text{ passes}} = 27.125 \text{ C.Y./hr.} \times 8 \text{ hrs.} = 217 \text{ C.Y./day}$$

Earthwork — R3141 Shoring

R314116-40 Wood Sheet Piling

Wood sheet piling may be used for depths to 20' where there is no ground water. If moderate ground water is encountered Tongue & Groove sheeting will help to keep it out. When considerable ground water is present, steel sheeting must be used.

For estimating purposes on trench excavation, sizes are as follows:

Depth	Sheeting	Wales	Braces	B.F. per S.F.
To 8'	3 x 12's	6 x 8's, 2 line	6 x 8's, @ 10'	4.0 @ 8'
8' x 12'	3 x 12's	10 x 10's, 2 line	10 x 10's, @ 9'	5.0 average
12' to 20'	3 x 12's	12 x 12's, 3 line	12 x 12's, @ 8'	7.0 average

Sheeting to be toed in at least 2' depending upon soil conditions. A five person crew with an air compressor and sheeting driver can drive and brace 440 SF/day at 8' deep, 360 SF/day at 12' deep, and 320 SF/day at 16' deep.

For normal soils, piling can be pulled in 1/3 the time to install. Pulling difficulty increases with the time in the ground. Production can be increased by high pressure jetting.

Earthwork — R3163 Drilled Caissons

R316326-60 Caissons

The three principal types of cassions are:

(1) Belled Caissons, which except for shallow depths and poor soil conditions, are generally recommended. They provide more bearing than shaft area. Because of its conical shape, no horizontal reinforcement of the bell is required.

(2) Straight Shaft Caissons are used where relatively light loads are to be supported by caissons that rest on high value bearing strata. While the shaft is larger in diameter than for belled types this is more than offset by the saving in time and labor.

(3) Keyed Caissons are used when extremely heavy loads are to be carried. A keyed or socketed caisson transfers its load into rock by a combination of end-bearing and shear reinforcing of the shaft. The most economical shaft often consists of a steel casing, a steel wide flange core and concrete. Allowable compressive stresses of .225 f'c for concrete, 16,000 psi for the wide flange core, and 9,000 psi for the steel casing are commonly used. The usual range of shaft diameter is 18" to 84". The number of sizes specified for any one project should be limited due to the problems of casing and auger storage. When hand work is to be performed, shaft diameters should not be less than 32". When inspection of borings is required a minimum shaft diameter of 30" is recommended. Concrete caissons are intended to be poured against earth excavation so permanent forms which add to cost should not be used if the excavation is clean and the earth sufficiently impervious to prevent excessive loss of concrete.

Soil Conditions for Belling		
Good	Requires Handwork	Not Recommended
Clay	Hard Shale	Silt
Sandy Clay	Limestone	Sand
Silty Clay	Sandstone	Gravel
Clayey Silt	Weathered Mica	Igneous Rock
Hard-pan		
Soft Shale		
Decomposed Rock		

Exterior Improvements — R3292 Turf & Grasses

R329219-50 Seeding

The type of grass is determined by light, shade and moisture content of soil plus intended use. Fertilizer should be disked 4" before seeding. For steep slopes disk five tons of mulch and lay two tons of hay or straw on surface per acre after seeding. Surface mulch can be staked, lightly disked or tar emulsion sprayed. Material for mulch can be wood chips, peat moss, partially rotted hay or straw, wood fibers and sprayed emulsions. Hemp seed blankets with fertilizer are also available. For spring seeding, watering is necessary. Late fall seeding may have to be reseeded in the spring. Hydraulic seeding, power mulching, and aerial seeding can be used on large areas.

Utilities — R3311 Water Utility Distribution Piping

R331113-80 Piping Designations

There are several systems currently in use to describe pipe and fittings. The following paragraphs will help to identify and clarify classifications of piping systems used for water distribution.

Piping may be classified by schedule. Piping schedules include 5S, 10S, 10, 20, 30, Standard, 40, 60, Extra Strong, 80, 100, 120, 140, 160 and Double Extra Strong. These schedules are dependent upon the pipe wall thickness. The wall thickness of a particular schedule may vary with pipe size.

Ductile iron pipe for water distribution is classified by Pressure Classes such as Class 150, 200, 250, 300 and 350. These classes are actually the rated water working pressure of the pipe in pounds per square inch (psi). The pipe in these pressure classes is designed to withstand the rated water working pressure plus a surge allowance of 100 psi.

The American Water Works Association (AWWA) provides standards for various types of **plastic pipe.** C-900 is the specification for polyvinyl chloride (PVC) piping used for water distribution in sizes ranging from 4″ through 12″. C-901 is the specification for polyethylene (PE) pressure pipe, tubing and fittings used for water distribution in sizes ranging from 1/2″ through 3″. C-905 is the specification for PVC piping sizes 14″ and greater.

PVC pressure-rated pipe is identified using the standard dimensional ratio (SDR) method. This method is defined by the American Society for Testing and Materials (ASTM) Standard D 2241. This pipe is available in SDR numbers 64, 41, 32.5, 26, 21, 17, and 13.5. Pipe with an SDR of 64 will have the thinnest wall while pipe with an SDR of 13.5 will have the thickest wall. When the pressure rating (PR) of a pipe is given in psi, it is based on a line supplying water at 73 degrees F.

The National Sanitation Foundation (NSF) seal of approval is applied to products that can be used with potable water. These products have been tested to ANSI/NSF Standard 14.

Valves and strainers are classified by American National Standards Institute (ANSI) Classes. These Classes are 125, 150, 200, 250, 300, 400, 600, 900, 1500 and 2500. Within each class there is an operating pressure range dependent upon temperature. Design parameters should be compared to the appropriate material dependent, pressure-temperature rating chart for accurate valve selection.

Change Orders

Change Order Considerations

A Change Order is a written document, usually prepared by the design professional, and signed by the owner, the architect/engineer and the contractor. A change order states the agreement of the parties to: an addition, deletion, or revision in the work; an adjustment in the contract sum, if any; or an adjustment in the contract time, if any. Change orders, or "extras" in the construction process occur after execution of the construction contract and impact architects/engineers, contractors and owners.

Change orders that are properly recognized and managed can ensure orderly, professional and profitable progress for all who are involved in the project. There are many causes for change orders and change order requests. In all cases, change orders or change order requests should be addressed promptly and in a precise and prescribed manner. The following paragraphs include information regarding change order pricing and procedures.

The Causes of Change Orders

Reasons for issuing change orders include:

- Unforeseen field conditions that require a change in the work
- Correction of design discrepancies, errors or omissions in the contract documents
- Owner-requested changes, either by design criteria, scope of work, or project objectives
- Completion date changes for reasons unrelated to the construction process
- Changes in building code interpretations, or other public authority requirements that require a change in the work
- Changes in availability of existing or new materials and products

Procedures

Properly written contract documents must include the correct change order procedures for all parties—owners, design professionals and contractors—to follow in order to avoid costly delays and litigation.

Being "in the right" is not always a sufficient or acceptable defense. The contract provisions requiring notification and documentation must be adhered to within a defined or reasonable time frame.

The appropriate method of handling change orders is by a written proposal and acceptance by all parties involved. Prior to starting work on a project, all parties should identify their authorized agents who may sign and accept change orders, as well as any limits placed on their authority.

Time may be a critical factor when the need for a change arises. For such cases, the contractor might be directed to proceed on a "time and materials" basis, rather than wait for all paperwork to be processed—a delay that could impede progress. In this situation, the contractor must still follow the prescribed change order procedures, including but not limited to, notification and documentation.

All forms used for change orders should be dated and signed by the proper authority. Lack of documentation can be very costly, especially if legal judgments are to be made and if certain field personnel are no longer available. For time and material change orders, the contractor should keep accurate daily records of all labor and material allocated to the change. Forms that can be used to document change order work are available in *Means Forms for Building Construction Professionals.*

Owners or awarding authorities who do considerable and continual building construction (such as the federal government) realize the inevitability of change orders for numerous reasons, both predictable and unpredictable. As a result, the federal government, the American Institute of Architects (AIA), the Engineers Joint Contract Documents Committee (EJCDC) and other contractor, legal and technical organizations have developed standards and procedures to be followed by all parties to achieve contract continuance and timely completion, while being financially fair to all concerned.

In addition to the change order standards put forth by industry associations, there are also many books available on the subject.

Pricing Change Orders

When pricing change orders, regardless of their cause, the most significant factor is when the change occurs. The need for a change may be perceived in the field or requested by the architect/engineer *before* any of the actual installation has begun, or may evolve or appear *during* construction when the item of work in question is partially installed. In the latter cases, the original sequence of construction is disrupted, along with all contiguous and supporting systems. Change orders cause the greatest impact when they occur *after* the installation has been completed and must be uncovered, or even replaced. Post-completion changes may be caused by necessary design changes, product failure, or changes in the owner's requirements that are not discovered until the building or the systems begin to function.

Specified procedures of notification and record keeping must be adhered to and enforced regardless of the stage of construction: *before, during,* or *after* installation. Some bidding documents anticipate change orders by requiring that unit prices including overhead and profit percentages—for additional as well as deductible changes—be listed. Generally these unit prices do not fully take into account the ripple effect, or impact on other trades, and should be used for general guidance only.

When pricing change orders, it is important to classify the time frame in which the change occurs. There are two basic time frames for change orders: *pre-installation change orders*, which occur before the start of construction, and *post-installation change orders*, which involve reworking after the original installation. Change orders that occur between these stages may be priced according to the extent of work completed using a combination of techniques developed for pricing *pre-* and *post-installation* changes.

The following factors are the basis for a check list to use when preparing a change order estimate.

Factors To Consider When Pricing Change Orders

As an estimator begins to prepare a change order, the following questions should be reviewed to determine their impact on the final price.

General

- Is the change order work *pre-installation* or *post-installation*?

  Change order work costs vary according to how much of the installation has been completed. Once workers have the project scoped in their mind, even though they have not started, it can be difficult to refocus.

Consequently they may spend more than the normal amount of time understanding the change. Also, modifications to work in place such as trimming or refitting usually take more time than was initially estimated. The greater the amount of work in place, the more reluctant workers are to change it. Psychologically they may resent the change and as a result the rework takes longer than normal. Post-installation change order estimates must include demolition of existing work as required to accomplish the change. If the work is performed at a later time, additional obstacles such as building finishes may be present which must be protected. Regardless of whether the change occurs pre-installation or post-installation, attempt to isolate the identifiable factors and price them separately. For example, add shipping costs that may be required pre-installation or any demolition required post-installation. Then analyze the potential impact on productivity of psychological and/or learning curve factors and adjust the output rates accordingly. One approach is to break down the typical workday into segments and quantify the impact on each segment. The following chart may be useful as a guide:

Activities (Productivity) Expressed as Percentages of a Workday			
Task	Means Mechanical Cost Data (for New Construction)	Pre-Installation Change Orders	Post-Installation Change Orders
1. Study plans	3%	6%	6%
2. Material procurement	3%	3%	3%
3. Receiving and storing	3%	3%	3%
4. Mobilization	5%	5%	5%
5. Site movement	5%	5%	8%
6. Layout and marking	8%	10%	12%
7. Actual installation	64%	59%	54%
8. Clean-up	3%	3%	3%
9. Breaks—non-productive	6%	6%	6%
Total	100%	100%	100%

Change Order Installation Efficiency

The labor-hours expressed (for new construction) are based on average installation time, using an efficiency level of approximately 60-65%. For change order situations, adjustments to this efficiency level should reflect the daily labor-hour allocation for that particular occurrence.

If any of the specific percentages expressed in the above chart do not apply to a particular project situation, then those percentage points should be reallocated to the appropriate task(s). Example: Using data for new construction, assume there is no new material being utilized. The percentages for Tasks 2 and 3 would therefore be reallocated to other tasks. If the time required for Tasks 2 and 3 can now be applied to installation, we can add the time allocated for *Material Procurement* and *Receiving and Storing* to the *Actual Installation* time for new construction, thereby increasing the Actual Installation percentage.

This chart shows that, due to reduced productivity, labor costs will be higher than those for new construction by 5% to 15% for pre-installation change orders and by 15% to 25% for post-installation change orders. Each job and change order is unique and must be examined individually. Many factors, covered elsewhere in this section, can each have a significant impact on productivity and change order costs. All such factors should be considered in every case.

- Will the change substantially delay the original completion date?

  A significant change in the project may cause the original completion date to be extended. The extended schedule may subject the contractor to new wage rates dictated by relevant labor contracts. Project supervision and other project overhead must also be extended beyond the original completion date. The schedule extension may also put installation into a new weather season. For example, underground piping scheduled for October installation was delayed until January. As a result, frost penetrated the trench area, thereby changing the degree of difficulty of the task. Changes and delays may have a ripple effect throughout the project. This effect must be analyzed and negotiated with the owner.

- What is the net effect of a deduct change order?

  In most cases, change orders resulting in a deduction or credit reflect only bare costs. The contractor may retain the overhead and profit based on the original bid.

Materials

- Will you have to pay more or less for the new material, required by the change order, than you paid for the original purchase?

  The same material prices or discounts will usually apply to materials purchased for change orders as new construction. In some instances, however, the contractor may forfeit the advantages of competitive pricing for change orders. Consider the following example:

  A contractor purchased over $20,000 worth of fan coil units for an installation, and obtained the maximum discount. Some time later it was determined the project required an additional matching unit. The contractor has to purchase this unit from the original supplier to ensure a match. The supplier at this time may not discount the unit because of the small quantity, and the fact that he is no longer in a competitive situation. The impact of quantity on purchase can add between 0% and 25% to material prices and/or subcontractor quotes.

- If materials have been ordered or delivered to the job site, will they be subject to a cancellation charge or restocking fee?

  Check with the supplier to determine if ordered materials are subject to a cancellation charge. Delivered materials not used as result of a change order may be subject to a restocking fee if returned to the supplier. Common restocking charges run between 20% and 40%. Also, delivery charges to return the goods to the supplier must be added.

Labor

- How efficient is the existing crew at the actual installation?

  Is the same crew that performed the initial work going to do the change order? Possibly the change consists of the installation of a unit identical to one already installed; therefore the change should take less time. Be sure to consider this potential productivity increase and modify the productivity rates accordingly.

- If the crew size is increased, what impact will that have on supervision requirements?

  Under most bargaining agreements or management practices, there is a point at which a working foreman is replaced by a nonworking foreman. This replacement increases project overhead by adding a nonproductive worker. If additional workers are added to accelerate the project or to perform changes while maintaining the schedule, be sure to add additional supervision time if warranted. Calculate the hours involved and the additional cost directly if possible.

- What are the other impacts of increased crew size?

  The larger the crew, the greater the potential for productivity to decrease. Some of the factors that cause this productivity loss are: overcrowding (producing restrictive conditions in the working space), and possibly a shortage of any special tools and equipment required. Such factors affect not only the crew working on the elements directly involved in the change order, but other crews whose movement may also be hampered.

  As the crew increases, check its basic composition for changes by the addition or deletion of apprentices or nonworking foreman and quantify the potential effects of equipment shortages or other logistical factors.

- As new crews, unfamiliar with the project, are brought onto the site, how long will it take them to become oriented to the project requirements?

  The orientation time for a new crew to become 100% effective varies with the site and type of project. Orientation is easiest at a new construction site, and most difficult at existing, very restrictive renovation sites. The type of work also affects orientation time. When all elements of the work are exposed, such as concrete or masonry work, orientation is decreased. When the work is concealed or less visible, such as existing electrical systems, orientation takes longer. Usually orientation can be accomplished in one day or less. Costs for added orientation should be itemized and added to the total estimated cost.

- How much actual production can be gained by working overtime?

  Short term overtime can be used effectively to accomplish more work in a day. However, as overtime is scheduled to run beyond several weeks, studies have shown marked decreases in output. The following chart shows the effect of long term overtime on worker efficiency. If the anticipated change requires extended overtime to keep the job on schedule, these factors can be used as a guide to predict the impact on time and cost. Add project overhead, particularly supervision, that may also be incurred.

Days per Week	Hours per Day	Production Efficiency					Payroll Cost Factors	
		1 Week	2 Weeks	3 Weeks	4 Weeks	Average 4 Weeks	@ 1-1/2 Times	@ 2 Times
5	8	100%	100%	100%	100%	100%	100%	100%
	9	100	100	95	90	96.25	105.6	111.1
	10	100	95	90	85	91.25	110.0	120.0
	11	95	90	75	65	81.25	113.6	127.3
	12	90	85	70	60	76.25	116.7	133.3
6	8	100	100	95	90	96.25	108.3	116.7
	9	100	95	90	85	92.50	113.0	125.9
	10	95	90	85	80	87.50	116.7	133.3
	11	95	85	70	65	78.75	119.7	139.4
	12	90	80	65	60	73.75	122.2	144.4
7	8	100	95	85	75	88.75	114.3	128.6
	9	95	90	80	70	83.75	118.3	136.5
	10	90	85	75	65	78.75	121.4	142.9
	11	85	80	65	60	72.50	124.0	148.1
	12	85	75	60	55	68.75	126.2	152.4

Effects of Overtime

Caution: Under many labor agreements, Sundays and holidays are paid at a higher premium than the normal overtime rate.

The use of long-term overtime is counterproductive on almost any construction job; that is, the longer the period of overtime, the lower the actual production rate. Numerous studies have been conducted, and while they have resulted in slightly different numbers, all reach the same conclusion. The figure above tabulates the effects of overtime work on efficiency.

As illustrated, there can be a difference between the *actual* payroll cost per hour and the *effective* cost per hour for overtime work. This is due to the reduced production efficiency with the increase in weekly hours beyond 40. This difference between actual and effective cost results from overtime work over a prolonged period. Short-term overtime work does not result in as great a reduction in efficiency, and in such cases, effective cost may not vary significantly from the actual payroll cost. As the total hours per week are increased on a regular basis, more time is lost because of fatigue, lowered morale, and an increased accident rate.

As an example, assume a project where workers are working 6 days a week, 10 hours per day. From the figure above (based on productivity studies), the average effective productive hours over a four-week period are:

$$0.875 \times 60 = 52.5$$

Depending upon the locale and day of week, overtime hours may be paid at time and a half or double time. For time and a half, the overall (average) *actual* payroll cost (including regular and overtime hours) is determined as follows:

$$\frac{40 \text{ reg. hrs.} + (20 \text{ overtime hrs.} \times 1.5)}{60 \text{ hrs.}} = 1.167$$

Based on 60 hours, the payroll cost per hour will be 116.7% of the normal rate at 40 hours per week. However, because the effective production (efficiency) for 60 hours is reduced to the equivalent of 52.5 hours, the effective cost of overtime is calculated as follows:

For time and a half:

$$\frac{40 \text{ reg. hrs.} + (20 \text{ overtime hrs.} \times 1.5)}{52.5 \text{ hrs.}} = 1.33$$

Installed cost will be 133% of the normal rate (for labor).

Thus, when figuring overtime, the actual cost per unit of work will be higher than the apparent overtime payroll dollar increase, due to the reduced productivity of the longer workweek. These efficiency calculations are true only for those cost factors determined by hours worked. Costs that are applied weekly or monthly, such as equipment rentals, will not be similarly affected.

Equipment

- What equipment is required to complete the change order?

Change orders may require extending the rental period of equipment already on the job site, or the addition of special equipment brought in to accomplish the change work. In either case, the additional rental charges and operator labor charges must be added.

Summary

The preceding considerations and others you deem appropriate should be analyzed and applied to a change order estimate. The impact of each should be quantified and listed on the estimate to form an audit trail.

Change orders that are properly identified, documented, and managed help to ensure the orderly, professional and profitable progress of the work. They also minimize potential claims or disputes at the end of the project.

Square Foot Costs

Estimating Tips

- The cost figures in this Square Foot Cost section were derived from approximately 11,200 projects contained in the RSMeans database of completed construction projects. They include the contractor's overhead and profit, but do not generally include architectural fees or land costs. The figures have been adjusted to January of the current year. New projects are added to our files each year, and outdated projects are discarded. For this reason, certain costs may not show a uniform annual progression. In no case are all subdivisions of a project listed.

- These projects were located throughout the U.S. and reflect a tremendous variation in square foot (S.F.) and cubic foot (C.F.) costs. This is due to differences, not only in labor and material costs, but also in individual owners' requirements. For instance, a bank in a large city would have different features than one in a rural area. This is true of all the different types of buildings analyzed. Therefore, caution should be exercised when using these Square Foot costs. For example, for court houses, costs in the database are local court house costs and will not apply to the larger, more elaborate federal court houses. As a general rule, the projects in the 1/4 column do not include any site work or equipment, while the projects in the 3/4 column may include both equipment and site work. The median figures do not generally include site work.

- None of the figures "go with" any others. All individual cost items were computed and tabulated separately. Thus, the sum of the median figures for Plumbing, HVAC and Electrical will not normally total up to the total Mechanical and Electrical costs arrived at by separate analysis and tabulation of the projects.

- Each building was analyzed as to total and component costs and percentages. The figures were arranged in ascending order with the results tabulated as shown. The 1/4 column shows that 25% of the projects had lower costs and 75% had higher. The 3/4 column shows that 75% of the projects had lower costs and 25% had higher. The median column shows that 50% of the projects had lower costs and 50% had higher.

- There are two times when square foot costs are useful. The first is in the conceptual stage when no details are available. Then square foot costs make a useful starting point. The second is after the bids are in and the costs can be worked back into their appropriate units for information purposes. As soon as details become available in the project design, the square foot approach should be discontinued and the project priced as to its particular components. When more precision is required, or for estimating the replacement cost of specific buildings, the current edition of *RSMeans Square Foot Costs* should be used.

- In using the figures in this section, it is recommended that the median column be used for preliminary figures if no additional information is available. The median figures, when multiplied by the total city construction cost index figures (see City Cost Indexes) and then multiplied by the project size modifier at the end of this section, should present a fairly accurate base figure, which would then have to be adjusted in view of the estimator's experience, local economic conditions, code requirements, and the owner's particular requirements. There is no need to factor the percentage figures, as these should remain constant from city to city. All tabulations mentioning air conditioning had at least partial air conditioning.

- The editors of this book would greatly appreciate receiving cost figures on one or more of your recent projects, which would then be included in the averages for next year. All cost figures received will be kept confidential, except that they will be averaged with other similar projects to arrive at Square Foot cost figures for next year's book. See the last page of the book for details and the discount available for submitting one or more of your projects.

50 17 | Square Foot Costs

		50 17 00	S.F. Costs	UNIT	UNIT COSTS			% OF TOTAL			
					1/4	MEDIAN	3/4	1/4	MEDIAN	3/4	
01	0010	**APARTMENTS Low Rise (1 to 3 story)**		S.F.	66	83.50	111				01
	0020		Total project cost	C.F.	5.95	7.85	9.70				
	0100		Site work	S.F.	5.65	7.70	13.60	6.05%	10.55%	14.05%	
	0500		Masonry		1.30	3.03	5.25	1.54%	3.67%	6.35%	
	1500		Finishes		7	9.65	11.90	9.05%	10.75%	12.85%	
	1800		Equipment		2.16	3.28	4.87	2.73%	4.03%	5.95%	
	2720		Plumbing		5.15	6.60	8.40	6.65%	8.95%	10.05%	
	2770		Heating, ventilating, air conditioning		3.28	4.04	5.95	4.20%	5.60%	7.60%	
	2900		Electrical		3.82	5.10	6.85	5.20%	6.65%	8.40%	
	3100		Total: Mechanical & Electrical		13.25	16.90	21	15.90%	18.05%	23%	
	9000		Per apartment unit, total cost	Apt.	61,500	94,000	138,500				
	9500		Total: Mechanical & Electrical	"	11,600	18,300	23,900				
02	0010	**APARTMENTS Mid Rise (4 to 7 story)**		S.F.	87.50	106	131				02
	0020		Total project costs	C.F.	6.85	9.45	12.90				
	0100		Site work	S.F.	3.51	6.95	12.50	5.25%	6.70%	9.15%	
	0500		Masonry		5.85	8.05	11	5.10%	7.25%	10.50%	
	1500		Finishes		11.05	14.40	18.15	10.55%	13.45%	17.70%	
	1800		Equipment		2.55	3.82	5	2.54%	3.48%	4.31%	
	2500		Conveying equipment		1.89	2.41	2.92	1.94%	2.27%	2.69%	
	2720		Plumbing		5.15	8.25	8.70	5.70%	7.20%	8.95%	
	2900		Electrical		5.80	7.85	9.55	6.65%	7.20%	8.95%	
	3100		Total: Mechanical & Electrical		18.50	23	28	18.50%	21%	23%	
	9000		Per apartment unit, total cost	Apt.	99,000	117,000	193,500				
	9500		Total: Mechanical & Electrical	"	18,700	21,700	27,400				
03	0010	**APARTMENTS High Rise (8 to 24 story)**		S.F.	99.50	115	137				03
	0020		Total project costs	C.F.	9.65	11.25	14.35				
	0100		Site work	S.F.	3.61	5.85	8.15	2.58%	4.84%	6.15%	
	0500		Masonry		5.75	10.45	13	4.74%	9.65%	11.05%	
	1500		Finishes		11.05	13.80	16.30	9.75%	11.80%	13.70%	
	1800		Equipment		3.20	3.93	5.20	2.78%	3.49%	4.35%	
	2500		Conveying equipment		2.26	3.43	4.66	2.23%	2.78%	3.37%	
	2720		Plumbing		7.35	8.65	12.10	6.80%	7.20%	10.45%	
	2900		Electrical		6.85	8.65	11.65	6.45%	7.65%	8.80%	
	3100		Total: Mechanical & Electrical		20.50	26	31.50	17.95%	22.50%	24.50%	
	9000		Per apartment unit, total cost	Apt.	103,500	114,000	158,000				
	9500		Total: Mechanical & Electrical	"	22,400	25,500	27,000				
04	0010	**AUDITORIUMS**		S.F.	103	140	202				04
	0020		Total project costs	C.F.	6.45	9	12.90				
	2720		Plumbing	S.F.	6.25	9.05	11	5.85%	7.20%	8.70%	
	2900		Electrical		8	11.65	18.85	6.80%	9.05%	11.30%	
	3100		Total: Mechanical & Electrical		22.50	45.50	55	24.50%	29%	31.50%	
05	0010	**AUTOMOTIVE SALES**		S.F.	76.50	104	129				05
	0020		Total project costs	C.F.	5.05	6.05	7.85				
	2720		Plumbing	S.F.	3.49	6.05	6.60	2.89%	6.05%	6.50%	
	2770		Heating, ventilating, air conditioning		5.40	8.20	8.90	4.61%	10%	10.35%	
	2900		Electrical		6.15	9.65	13.10	7.25%	9.80%	12.15%	
	3100		Total: Mechanical & Electrical		19.30	27.50	33	19.15%	20.50%	22%	
06	0010	**BANKS**		S.F.	150	187	237				06
	0020		Total project costs	C.F.	10.70	14.55	19.20				
	0100		Site work	S.F.	17.20	26	38	7.85%	12.95%	16.95%	
	0500		Masonry		8.10	15.45	27	3.46%	7.65%	10.10%	
	1500		Finishes		13.90	20.50	25.50	5.85%	8.60%	11.55%	
	1800		Equipment		5.65	12.45	25.50	1%	5.55%	10.50%	
	2720		Plumbing		4.70	6.70	9.80	2.82%	3.90%	4.93%	
	2770		Heating, ventilating, air conditioning		8.95	11.95	15.90	4.86%	7.15%	8.50%	
	2900		Electrical		14.20	18.95	25	8.20%	10.20%	12.20%	
	3100		Total: Mechanical & Electrical		34	46	55	16%	19.40%	23%	
	3500		See also division 11020 & 11030 (MF2004 11 16 00 & 11 17 00)									

50 17 | Square Foot Costs

50 17 00 | S.F. Costs

				UNIT COSTS			% OF TOTAL			
			UNIT	1/4	MEDIAN	3/4	1/4	MEDIAN	3/4	
13	0010	**CHURCHES**	S.F.	101	128	166				13
	0020	Total project costs	C.F.	6.25	7.90	10.40				
	1800	Equipment	S.F.	1.21	2.89	6.15	.95%	2.11%	4.50%	
	2720	Plumbing		3.93	5.50	8.10	3.51%	4.96%	6.25%	
	2770	Heating, ventilating, air conditioning		9.20	11.95	16.95	7.50%	10%	12%	
	2900	Electrical		8.50	11.65	15.65	7.35%	8.75%	10.95%	
	3100	Total: Mechanical & Electrical	↓	26.50	34	45.50	18.30%	22%	24.50%	
	3500	See also division 11040 (MF2004 11 91 00)								
15	0010	**CLUBS, COUNTRY**	S.F.	108	131	164				15
	0020	Total project costs	C.F.	8.75	10.65	14.70				
	2720	Plumbing	S.F.	6.55	9.70	22	5.60%	7.90%	10%	
	2900	Electrical		8.50	11.65	15.20	7%	8.95%	11%	
	3100	Total: Mechanical & Electrical	↓	45.50	56.50	59.50	19%	26.50%	29.50%	
17	0010	**CLUBS, SOCIAL Fraternal**	S.F.	86.50	124	166				17
	0020	Total project costs	C.F.	5.40	8.20	9.75				
	2720	Plumbing	S.F.	5.45	6.75	10.25	5.60%	6.90%	8.55%	
	2770	Heating, ventilating, air conditioning		7.85	9.50	12.20	8.20%	9.25%	14.40%	
	2900	Electrical		6.50	10.70	12.25	6.50%	9.50%	10.55%	
	3100	Total: Mechanical & Electrical	↓	19.20	36.50	46	21%	23%	23.50%	
18	0010	**CLUBS, Y.M.C.A.**	S.F.	109	141	177				18
	0020	Total project costs	C.F.	5	8.35	12.45				
	2720	Plumbing	S.F.	6.85	13.65	15.30	5.65%	7.60%	10.85%	
	2900	Electrical		8.70	10.85	14.95	6.05%	7.60%	9.25%	
	3100	Total: Mechanical & Electrical	↓	33	37	50.50	18.40%	21.50%	28.50%	
19	0010	**COLLEGES Classrooms & Administration**	S.F.	110	150	199				19
	0020	Total project costs	C.F.	8	11.65	17.95				
	0500	Masonry	S.F.	8.05	15.15	18.60	5.65%	8.25%	10.50%	
	2720	Plumbing		5.60	11.40	20.50	5.10%	6.60%	8.95%	
	2900	Electrical		9.15	13.85	18.95	7.70%	9.85%	12%	
	3100	Total: Mechanical & Electrical	↓	36.50	51	60.50	24%	28%	31.50%	
21	0010	**COLLEGES Science, Engineering, Laboratories**	S.F.	204	238	276				21
	0020	Total project costs	C.F.	11.70	17.05	19.35				
	1800	Equipment	S.F.	11.35	25.50	28	2%	6.45%	12.65%	
	2900	Electrical		16.80	23	36.50	7.10%	9.40%	12.10%	
	3100	Total: Mechanical & Electrical	↓	62.50	74	115	28.50%	31.50%	41%	
	3500	See also division 11600 (MF2004 11 53 00)								
23	0010	**COLLEGES Student Unions**	S.F.	130	176	213				23
	0020	Total project costs	C.F.	7.25	9.50	11.75				
	3100	Total: Mechanical & Electrical	S.F.	49	53	62.50	23.50%	26%	29%	
25	0010	**COMMUNITY CENTERS**	S.F.	107	132	178				25
	0020	Total project costs	C.F.	7	10	12.95				
	1800	Equipment	S.F.	2.60	4.39	7	1.48%	3.01%	5.45%	
	2720	Plumbing		5.10	8.90	12.15	4.85%	7%	8.95%	
	2770	Heating, ventilating, air conditioning		8.15	11.90	17	6.80%	10.35%	12.90%	
	2900	Electrical		8.80	11.70	16.80	7.30%	9%	10.45%	
	3100	Total: Mechanical & Electrical	↓	30	37.50	53.50	20%	25%	31%	
28	0010	**COURT HOUSES**	S.F.	155	178	229				28
	0020	Total project costs	C.F.	11.85	14.20	17.90				
	2720	Plumbing	S.F.	7.35	10.30	14.80	5.95%	7.45%	8.20%	
	2900	Electrical		16.45	18.25	27	8.90%	10.45%	11.55%	
	3100	Total: Mechanical & Electrical	↓	41.50	57	63.50	22.50%	27.50%	30.50%	
30	0010	**DEPARTMENT STORES**	S.F.	57	77.50	98				30
	0020	Total project costs	C.F.	3.07	3.98	5.40				
	2720	Plumbing	S.F.	1.78	2.25	3.42	1.82%	4.21%	5.90%	
	2770	Heating, ventilating, air conditioning	↓	5.20	8.05	12.10	8.20%	9.10%	14.80%	

50 17 | Square Foot Costs

50 17 00 | S.F. Costs

			UNIT	UNIT COSTS			% OF TOTAL			
				1/4	MEDIAN	3/4	1/4	MEDIAN	3/4	
30	2900	Electrical	S.F.	6.55	9	10.65	9.05%	12.15%	14.95%	30
	3100	Total: Mechanical & Electrical		11.55	14.80	26	13.20%	21.50%	50%	
31	0010	**DORMITORIES Low Rise (1 to 3 story)**	S.F.	109	141	188				31
	0020	Total project costs	C.F.	6.15	9.90	14.85				
	2720	Plumbing	S.F.	6.55	8.75	11.05	8.05%	9%	9.65%	
	2770	Heating, ventilating, air conditioning		6.90	8.25	11	4.61%	8.05%	10%	
	2900	Electrical		7.20	10.95	14.95	6.40%	8.65%	9.50%	
	3100	Total: Mechanical & Electrical		37.50	40	62.50	22%	25%	27%	
	9000	Per bed, total cost	Bed	46,000	51,000	109,500				
32	0010	**DORMITORIES Mid Rise (4 to 8 story)**	S.F.	133	174	215				32
	0020	Total project costs	C.F.	14.70	16.15	19.30				
	2900	Electrical	S.F.	14.15	16.10	21.50	8.20%	10.20%	11.95%	
	3100	Total: Mechanical & Electrical	"	39.50	41	80.50	25.50%	34.50%	37.50%	
	9000	Per bed, total cost	Bed	19,000	43,300	109,500				
34	0010	**FACTORIES**	S.F.	50.50	75	116				34
	0020	Total project costs	C.F.	3.24	4.83	8				
	0100	Site work	S.F.	5.75	10.50	16.60	6.95%	11.45%	17.95%	
	2720	Plumbing		2.72	5.05	8.35	3.73%	6.05%	8.10%	
	2770	Heating, ventilating, air conditioning		5.30	7.60	10.25	5.25%	8.45%	11.35%	
	2900	Electrical		6.25	9.90	15.15	8.10%	10.50%	14.20%	
	3100	Total: Mechanical & Electrical		17.90	24	36.50	21%	28.50%	35.50%	
36	0010	**FIRE STATIONS**	S.F.	100	138	185				36
	0020	Total project costs	C.F.	5.85	8.05	10.70				
	0500	Masonry	S.F.	14.80	25	34	8.35%	11.55%	16.15%	
	1140	Roofing		3.26	8.85	10.05	1.90%	4.94%	5.05%	
	1580	Painting		2.53	3.78	3.87	1.37%	1.57%	2.07%	
	1800	Equipment		1.24	2.38	4.41	.62%	1.86%	3.54%	
	2720	Plumbing		5.60	9.05	13.05	5.85%	7.35%	9.45%	
	2770	Heating, ventilating, air conditioning		5.55	9	13.90	5.15%	7.40%	9.40%	
	2900	Electrical		7.15	12.55	17	6.80%	8.60%	10.60%	
	3100	Total: Mechanical & Electrical		36.50	47	53	18.40%	23%	26%	
37	0010	**FRATERNITY HOUSES & Sorority Houses**	S.F.	100	129	176				37
	0020	Total project costs	C.F.	9.95	10.40	12.50				
	2720	Plumbing	S.F.	7.55	8.65	15.85	6.80%	8%	10.85%	
	2900	Electrical		6.60	14.25	17.50	6.60%	9.90%	10.65%	
	3100	Total: Mechanical & Electrical		7.80	25	30		15.10%	15.90%	
38	0010	**FUNERAL HOMES**	S.F.	106	144	261				38
	0020	Total project costs	C.F.	10.75	12	23				
	2900	Electrical	S.F.	4.66	8.55	9.35	3.58%	4.44%	5.95%	
	3100	Total: Mechanical & Electrical		16.50	24	33.50	12.90%	12.90%	12.90%	
39	0010	**GARAGES, COMMERCIAL (Service)**	S.F.	60	92.50	128				39
	0020	Total project costs	C.F.	3.93	5.80	8.45				
	1800	Equipment	S.F.	3.37	7.60	11.80	2.21%	4.62%	6.80%	
	2720	Plumbing		4.14	6.40	11.65	5.45%	7.85%	10.65%	
	2730	Heating & ventilating		5.45	7.20	9.75	5.25%	6.85%	8.20%	
	2900	Electrical		5.70	8.65	12.50	7.15%	9.25%	10.85%	
	3100	Total: Mechanical & Electrical		12.55	24	35.50	12.35%	17.40%	26%	
40	0010	**GARAGES, MUNICIPAL (Repair)**	S.F.	88	117	165				40
	0020	Total project costs	C.F.	5.50	6.95	11.95				
	0500	Masonry	S.F.	8.25	16.10	25	5.60%	9.15%	12.50%	
	2720	Plumbing		3.94	7.55	14.25	3.59%	6.70%	7.95%	
	2730	Heating & ventilating		6.75	9.75	18.85	6.15%	7.45%	13.50%	
	2900	Electrical		6.50	10.20	14.70	6.65%	8.15%	11.15%	
	3100	Total: Mechanical & Electrical		30	41.50	60.50	21.50%	25.50%	28.50%	

50 17 | Square Foot Costs

50 17 00 | S.F. Costs

			UNIT	UNIT COSTS 1/4	UNIT COSTS MEDIAN	UNIT COSTS 3/4	% OF TOTAL 1/4	% OF TOTAL MEDIAN	% OF TOTAL 3/4	
41	0010	**GARAGES, PARKING**	S.F.	34	50	85.50				41
	0020	Total project costs	C.F.	3.21	4.36	6.35				
	2720	Plumbing	S.F.	.97	1.50	2.32	1.72%	2.70%	3.85%	
	2900	Electrical		1.87	2.30	3.61	4.33%	5.20%	6.30%	
	3100	Total: Mechanical & Electrical	↓	3.84	5.35	6.65	7%	8.90%	11.05%	
	3200									
	9000	Per car, total cost	Car	14,400	18,100	23,100				
43	0010	**GYMNASIUMS**	S.F.	95.50	127	172				43
	0020	Total project costs	C.F.	4.74	6.45	7.90				
	1800	Equipment	S.F.	2.26	4.24	8.15	1.81%	3.30%	6.70%	
	2720	Plumbing		6	7.15	9.20	4.65%	6.40%	7.75%	
	2770	Heating, ventilating, air conditioning		6.45	9.85	19.80	5.15%	9.05%	11.10%	
	2900	Electrical		7.30	9.95	13.10	6.60%	8.30%	10.30%	
	3100	Total: Mechanical & Electrical	↓	26.50	36	43	19.75%	23.50%	29%	
	3500	See also division 11480 (MF2004 11 67 00)								
46	0010	**HOSPITALS**	S.F.	182	225	310				46
	0020	Total project costs	C.F.	13.85	17.20	24.50				
	1800	Equipment	S.F.	4.63	8.90	15.35	.96%	2.63%	5%	
	2720	Plumbing		15.70	22	28.50	7.60%	9.10%	10.85%	
	2770	Heating, ventilating, air conditioning		23	29	40	7.80%	12.95%	16.65%	
	2900	Electrical		20	27	38	9.90%	11.75%	14%	
	3100	Total: Mechanical & Electrical	↓	56	80.50	122	27%	33.50%	36.50%	
	9000	Per bed or person, total cost	Bed	211,500	291,000	335,500				
	9900	See also division 11700 (MF2004 11 71 00)								
48	0010	**HOUSING For the Elderly**	S.F.	89.50	114	140				48
	0020	Total project costs	C.F.	6.40	8.90	11.35				
	0100	Site work	S.F.	6.25	9.70	14.20	5.05%	7.90%	12.10%	
	0500	Masonry		2.74	10.20	14.95	1.30%	6.05%	11%	
	1800	Equipment		2.17	2.99	4.76	1.88%	3.23%	4.43%	
	2510	Conveying systems		2.19	2.93	3.98	1.78%	2.20%	2.81%	
	2720	Plumbing		6.65	8.50	10.70	8.15%	9.55%	10.50%	
	2730	Heating, ventilating, air conditioning		3.42	4.84	7.25	3.30%	5.60%	7.25%	
	2900	Electrical		6.70	9.10	11.65	7.30%	8.50%	10.25%	
	3100	Total: Mechanical & Electrical	↓	23	27.50	36.50	18.10%	22.50%	29%	
	9000	Per rental unit, total cost	Unit	83,500	97,500	108,500				
	9500	Total: Mechanical & Electrical	"	18,600	21,400	25,000				
50	0010	**HOUSING Public (Low Rise)**	S.F.	75.50	105	136				50
	0020	Total project costs	C.F.	6.70	8.40	10.40				
	0100	Site work	S.F.	9.60	13.85	22.50	8.35%	11.75%	16.50%	
	1800	Equipment		2.05	3.35	5.10	2.26%	3.03%	4.24%	
	2720	Plumbing		5.45	7.20	9.10	7.15%	9.05%	11.60%	
	2730	Heating, ventilating, air conditioning		2.73	5.30	5.80	4.26%	6.05%	6.45%	
	2900	Electrical		4.56	6.80	9.45	5.10%	6.55%	8.25%	
	3100	Total: Mechanical & Electrical	↓	21.50	28	31	14.50%	17.55%	26.50%	
	9000	Per apartment, total cost	Apt.	83,000	94,000	118,500				
	9500	Total: Mechanical & Electrical	"	17,700	21,800	24,100				
51	0010	**ICE SKATING RINKS**	S.F.	64.50	151	166				51
	0020	Total project costs	C.F.	4.74	4.85	5.60				
	2720	Plumbing	S.F.	2.41	4.52	4.62	3.12%	3.23%	5.65%	
	2900	Electrical		6.90	10.60	11.20	6.30%	10.15%	15.05%	
	3100	Total: Mechanical & Electrical	↓	11.45	16.20	20	18.95%	18.95%	18.95%	
52	0010	**JAILS**	S.F.	196	254	325				52
	0020	Total project costs	C.F.	17.70	24.50	29				
	1800	Equipment	S.F.	7.65	22.50	38.50	2.80%	5.55%	10.35%	
	2720	Plumbing		18.90	25.50	33.50	7%	8.90%	13.35%	
	2770	Heating, ventilating, air conditioning		17.70	23.50	45.50	7.50%	9.45%	17.75%	
	2900	Electrical	↓	20.50	27	34.50	8.20%	11.55%	14.95%	

50 17 | Square Foot Costs

		50 17 00	S.F. Costs	UNIT	UNIT COSTS 1/4	MEDIAN	3/4	% OF TOTAL 1/4	MEDIAN	3/4	
52	3100		Total: Mechanical & Electrical	S.F.	53.50	98	116	28%	30%	34%	52
53	0010	**LIBRARIES**		S.F.	124	159	207				53
	0020		Total project costs	C.F.	8.50	10.65	13.60				
	0500		Masonry	S.F.	9.95	17.35	29	5.80%	7.80%	11.80%	
	1800		Equipment		1.70	4.58	6.90	.37%	1.50%	4.07%	
	2720		Plumbing		4.58	6.65	9.05	3.38%	4.60%	5.70%	
	2770		Heating, ventilating, air conditioning		10.15	17.25	22.50	7.80%	10.95%	12.80%	
	2900		Electrical		12.65	16.40	21	8.30%	10.25%	11.95%	
	3100		Total: Mechanical & Electrical		37.50	48	59	20.50%	23%	26.50%	
54	0010	**LIVING, ASSISTED**		S.F.	115	135	159				54
	0020		Total project costs	C.F.	9.65	11.30	12.80				
	0500		Masonry	S.F.	3.38	4.02	4.73	2.37%	3.16%	3.86%	
	1800		Equipment		2.61	3.03	3.89	2.12%	2.45%	2.66%	
	2720		Plumbing		9.60	12.90	13.35	6.05%	8.15%	10.60%	
	2770		Heating, ventilating, air conditioning		11.40	11.95	13.10	7.95%	9.35%	9.70%	
	2900		Electrical		11.25	12.40	14.35	9%	10%	10.70%	
	3100		Total: Mechanical & Electrical		31.50	37	42.50	26%	29%	31.50%	
55	0010	**MEDICAL CLINICS**		S.F.	117	144	183				55
	0020		Total project costs	C.F.	8.55	11.05	14.70				
	1800		Equipment	S.F.	3.16	6.65	10.35	1.05%	2.94%	6.35%	
	2720		Plumbing		7.75	10.95	14.60	6.15%	8.40%	10.10%	
	2770		Heating, ventilating, air conditioning		9.25	12.15	17.85	6.65%	8.85%	11.35%	
	2900		Electrical		10.05	14.25	18.65	8.10%	10%	12.25%	
	3100		Total: Mechanical & Electrical		32	43.50	59.50	22.50%	27%	33.50%	
	3500		See also division 11700 (MF2004 11 71 00)								
57	0010	**MEDICAL OFFICES**		S.F.	110	136	167				57
	0020		Total project costs	C.F.	8.20	11.10	15				
	1800		Equipment	S.F.	3.63	7.15	10.20	.70%	5.10%	7.05%	
	2720		Plumbing		6.05	9.35	12.60	5.60%	6.80%	8.50%	
	2770		Heating, ventilating, air conditioning		7.35	10.60	14	6.10%	8%	9.70%	
	2900		Electrical		8.95	12.80	17.85	7.60%	9.80%	11.70%	
	3100		Total: Mechanical & Electrical		24	34	49.50	19.30%	23%	28.50%	
59	0010	**MOTELS**		S.F.	69.50	100	131				59
	0020		Total project costs	C.F.	6.15	8.25	13.50				
	2720		Plumbing	S.F.	7.05	8.95	10.65	9.45%	10.60%	12.55%	
	2770		Heating, ventilating, air conditioning		4.28	6.40	11.45	5.60%	5.60%	10%	
	2900		Electrical		6.55	8.30	10.30	7.45%	9.05%	10.45%	
	3100		Total: Mechanical & Electrical		21	28	48	18.50%	24%	25.50%	
	5000										
	9000		Per rental unit, total cost	Unit	35,300	67,000	72,500				
	9500		Total: Mechanical & Electrical	"	6,875	10,400	12,100				
60	0010	**NURSING HOMES**		S.F.	109	140	172				60
	0020		Total project costs	C.F.	8.55	10.65	14.60				
	1800		Equipment	S.F.	3.42	4.54	7.55	2.02%	3.62%	4.99%	
	2720		Plumbing		9.30	14.20	17	8.75%	10.10%	12.70%	
	2770		Heating, ventilating, air conditioning		9.80	14.90	19.75	9.70%	11.45%	11.80%	
	2900		Electrical		10.75	13.45	18.30	9.40%	10.55%	12.45%	
	3100		Total: Mechanical & Electrical		25.50	36	60	26%	29.50%	30.50%	
	9000		Per bed or person, total cost	Bed	48,200	60,500	78,000				
61	0010	**OFFICES Low Rise (1 to 4 story)**		S.F.	91.50	119	154				61
	0020		Total project costs	C.F.	6.55	9	11.85				
	0100		Site work	S.F.	7.35	12.70	19.15	6.20%	9.70%	13.60%	
	0500		Masonry		3.54	7.25	13.05	2.62%	5.50%	8.60%	
	1800		Equipment		.97	1.91	5.20	.62%	1.50%	3.51%	
	2720		Plumbing		3.27	5.05	7.40	3.66%	4.50%	6.10%	
	2770		Heating, ventilating, air conditioning		7.25	10.10	14.75	7.20%	10.30%	11.70%	
	2900		Electrical		7.50	10.65	15.10	7.45%	9.60%	11.35%	

50 17 | Square Foot Costs

50 17 00 | S.F. Costs

			UNIT	UNIT COSTS			% OF TOTAL			
				1/4	MEDIAN	3/4	1/4	MEDIAN	3/4	
61	3100	Total: Mechanical & Electrical	S.F.	20.50	28	41	18.15%	22%	27%	61
62	0010	OFFICES Mid Rise (5 to 10 story)	S.F.	97	117	155				62
	0020	Total project costs	C.F.	6.85	8.75	12.45				
	2720	Plumbing	S.F.	2.93	4.54	6.55	2.83%	3.74%	4.50%	
	2770	Heating, ventilating, air conditioning		7.35	10.55	16.80	7.65%	9.40%	11%	
	2900	Electrical		7.20	9.35	12.80	6.35%	7.80%	10%	
	3100	Total: Mechanical & Electrical		18.65	24	45.50	19.15%	21.50%	27.50%	
63	0010	OFFICES High Rise (11 to 20 story)	S.F.	119	150	185				63
	0020	Total project costs	C.F.	8.30	10.40	14.95				
	2900	Electrical	S.F.	7.25	8.85	13.10	5.80%	7.85%	10.50%	
	3100	Total: Mechanical & Electrical		23.50	31.50	52.50	16.90%	23.50%	34%	
64	0010	POLICE STATIONS	S.F.	143	188	237				64
	0020	Total project costs	C.F.	11.35	13.95	19.10				
	0500	Masonry	S.F.	13.40	23.50	29.50	6.70%	9.10%	11.35%	
	1800	Equipment		2.27	9.95	15.80	.98%	3.35%	6.70%	
	2720	Plumbing		8	15.95	19.85	5.65%	6.90%	10.75%	
	2770	Heating, ventilating, air conditioning		12.50	16.60	22.50	5.85%	10.55%	11.70%	
	2900	Electrical		15.60	22.50	29.50	9.80%	11.85%	14.80%	
	3100	Total: Mechanical & Electrical		51	61.50	83.50	25%	31.50%	32.50%	
65	0010	POST OFFICES	S.F.	113	139	177				65
	0020	Total project costs	C.F.	6.80	8.60	9.75				
	2720	Plumbing	S.F.	5.10	6.30	7.95	4.24%	5.30%	5.60%	
	2770	Heating, ventilating, air conditioning		7.95	9.80	10.90	6.65%	7.15%	9.35%	
	2900	Electrical		9.30	13.10	15.55	7.25%	9%	11%	
	3100	Total: Mechanical & Electrical		27	35	40	16.25%	18.80%	22%	
66	0010	POWER PLANTS	S.F.	780	1,000	1,900				66
	0020	Total project costs	C.F.	21.50	47	100				
	2900	Electrical	S.F.	55.50	117	175	9.30%	12.75%	21.50%	
	8100	Total: Mechanical & Electrical		138	450	1,000	32.50%	32.50%	52.50%	
67	0010	RELIGIOUS EDUCATION	S.F.	91	119	146				67
	0020	Total project costs	C.F.	5.05	7.20	9.05				
	2720	Plumbing	S.F.	3.80	5.40	7.65	4.40%	5.30%	7.10%	
	2770	Heating, ventilating, air conditioning		9.60	10.85	15.35	10.05%	11.45%	12.35%	
	2900	Electrical		7.20	10.15	13.45	7.60%	9.05%	10.35%	
	3100	Total: Mechanical & Electrical		29.50	38.50	47	22%	23%	27%	
69	0010	RESEARCH Laboratories & Facilities	S.F.	136	195	284				69
	0020	Total project costs	C.F.	10.25	19.80	24				
	1800	Equipment	S.F.	6.05	11.90	29	.94%	4.58%	9.10%	
	2720	Plumbing		13.65	17.45	28	6.15%	8.30%	10.80%	
	2770	Heating, ventilating, air conditioning		12.25	41	48.50	7.25%	16.50%	17.50%	
	2900	Electrical		15.90	26	43	9.45%	11.15%	15.40%	
	3100	Total: Mechanical & Electrical		52	90.50	131	31.50%	37.50%	42%	
70	0010	RESTAURANTS	S.F.	132	170	221				70
	0020	Total project costs	C.F.	11.10	14.55	19.10				
	1800	Equipment	S.F.	8.50	21	31.50	6.10%	13%	15.65%	
	2720	Plumbing		10.45	12.65	16.65	6.10%	8.15%	9%	
	2770	Heating, ventilating, air conditioning		13.25	18.40	22.50	9.20%	12%	12.40%	
	2900	Electrical		14	17.20	22.50	8.35%	10.55%	11.55%	
	3100	Total: Mechanical & Electrical		43.50	46.50	60	21%	24.50%	29.50%	
	9000	Per seat unit, total cost	Seat	4,850	6,450	7,625				
	9500	Total: Mechanical & Electrical	"	1,225	1,600	1,900				
72	0010	RETAIL STORES	S.F.	61.50	82.50	110				72
	0020	Total project costs	C.F.	4.17	5.95	8.25				
	2720	Plumbing	S.F.	2.23	3.73	6.35	3.26%	4.60%	6.80%	
	2770	Heating, ventilating, air conditioning		4.82	6.60	9.90	6.75%	8.75%	10.15%	
	2900	Electrical		5.55	7.60	10.95	7.25%	9.90%	11.65%	
	3100	Total: Mechanical & Electrical		14.75	18.95	26	17.05%	21%	23.50%	

50 17 | Square Foot Costs

50 17 00 | S.F. Costs

			UNIT	UNIT COSTS 1/4	MEDIAN	3/4	% OF TOTAL 1/4	MEDIAN	3/4	
74	0010	**SCHOOLS Elementary**	S.F.	99.50	123	152				74
	0020	Total project costs	C.F.	6.60	8.40	10.90				
	0500	Masonry	S.F.	9	15.50	23	5.80%	11%	14.95%	
	1800	Equipment		2.75	4.68	8.65	1.89%	3.32%	4.71%	
	2720	Plumbing		5.80	8.15	10.90	5.70%	7.15%	9.35%	
	2730	Heating, ventilating, air conditioning		8.65	13.80	19.25	8.15%	10.80%	14.90%	
	2900	Electrical		9.45	12.50	15.75	8.40%	10.05%	11.70%	
	3100	Total: Mechanical & Electrical		33.50	42.50	52	25%	27.50%	30%	
	9000	Per pupil, total cost	Ea.	11,500	17,100	49,500				
	9500	Total: Mechanical & Electrical	"	3,250	4,125	14,300				
76	0010	**SCHOOLS Junior High & Middle**	S.F.	104	127	155				76
	0020	Total project costs	C.F.	6.60	8.50	9.55				
	0500	Masonry	S.F.	13.25	17.40	20.50	8.60%	11.60%	14.35%	
	1800	Equipment		3.31	5.35	8.05	1.79%	3.09%	4.86%	
	2720	Plumbing		6.05	7.45	9.25	5.30%	6.80%	7.25%	
	2770	Heating, ventilating, air conditioning		12.05	14.65	26	8.90%	11.55%	14.20%	
	2900	Electrical		10.15	12.25	15.80	7.90%	9.35%	10.60%	
	3100	Total: Mechanical & Electrical		33	42.50	52.50	23.50%	27%	29.50%	
	9000	Per pupil, total cost	Ea.	13,100	17,200	23,200				
78	0010	**SCHOOLS Senior High**	S.F.	107	132	165				78
	0020	Total project costs	C.F.	6.50	9.60	15.40				
	1800	Equipment	S.F.	2.83	6.65	9.40	1.86%	2.91%	4.30%	
	2720	Plumbing		6.05	9.05	16.60	5.60%	6.90%	8.30%	
	2770	Heating, ventilating, air conditioning		12.35	14.15	27	8.95%	11.60%	15%	
	2900	Electrical		10.85	14.05	20.50	8.65%	10.15%	12.35%	
	3100	Total: Mechanical & Electrical		36	42	70.50	23.50%	26.50%	29%	
	9000	Per pupil, total cost	Ea.	10,200	20,700	25,900				
80	0010	**SCHOOLS Vocational**	S.F.	87.50	127	157				80
	0020	Total project costs	C.F.	5.40	7.80	10.75				
	0500	Masonry	S.F.	5.10	12.65	19.35	3.53%	6.70%	10.95%	
	1800	Equipment		2.73	6.80	9.45	1.24%	3.10%	4.26%	
	2720	Plumbing		5.60	8.30	12.25	5.40%	6.90%	8.55%	
	2770	Heating, ventilating, air conditioning		7.80	14.55	24.50	8.60%	11.90%	14.65%	
	2900	Electrical		9.10	11.90	16.35	8.45%	10.95%	13.20%	
	3100	Total: Mechanical & Electrical		31.50	35	60	23.50%	27.50%	31%	
	9000	Per pupil, total cost	Ea.	12,200	32,500	48,500				
83	0010	**SPORTS ARENAS**	S.F.	76.50	102	157				83
	0020	Total project costs	C.F.	4.14	7.40	9.55				
	2720	Plumbing	S.F.	4.42	6.70	14.15	4.35%	6.35%	9.40%	
	2770	Heating, ventilating, air conditioning		9.50	11.25	15.65	8.80%	10.20%	13.55%	
	2900	Electrical		7.95	10.80	13.95	8.60%	9.90%	12.25%	
	3100	Total: Mechanical & Electrical		19.75	34.50	46.50	21.50%	25%	27.50%	
85	0010	**SUPERMARKETS**	S.F.	70.50	81.50	95.50				85
	0020	Total project costs	C.F.	3.92	4.74	7.20				
	2720	Plumbing	S.F.	3.93	4.96	5.75	5.40%	6%	7.45%	
	2770	Heating, ventilating, air conditioning		5.80	7.70	9.40	8.60%	8.65%	9.60%	
	2900	Electrical		8.80	10.15	12	10.40%	12.45%	13.60%	
	3100	Total: Mechanical & Electrical		22.50	24.50	32	20.50%	26.50%	31%	
86	0010	**SWIMMING POOLS**	S.F.	114	191	405				86
	0020	Total project costs	C.F.	9.15	11.40	12.45				
	2720	Plumbing	S.F.	10.55	12.05	16.30	4.80%	9.70%	20.50%	
	2900	Electrical		8.55	13.90	20	5.75%	6.95%	7.60%	
	3100	Total: Mechanical & Electrical		21	54	72.50	11.15%	14.10%	23.50%	
87	0010	**TELEPHONE EXCHANGES**	S.F.	151	222	281				87
	0020	Total project costs	C.F.	9.40	15.10	21				
	2720	Plumbing	S.F.	6.40	9.90	14.45	4.52%	5.80%	6.90%	
	2770	Heating, ventilating, air conditioning		14.85	30	37	11.80%	16.05%	18.40%	

50 17 | Square Foot Costs

50 17 00 | S.F. Costs

				UNIT COSTS			% OF TOTAL			
			UNIT	1/4	MEDIAN	3/4	1/4	MEDIAN	3/4	
87	2900	Electrical	S.F.	15.40	24.50	43.50	10.90%	14%	17.85%	87
	3100	Total: Mechanical & Electrical	↓	45.50	86.50	122	29.50%	33.50%	44.50%	
91	0010	**THEATERS**	S.F.	95.50	118	180				91
	0020	Total project costs	C.F.	4.40	6.50	9.55				
	2720	Plumbing	S.F.	3.18	3.44	14.05	2.92%	4.70%	6.80%	
	2770	Heating, ventilating, air conditioning		9.25	11.20	13.85	8%	12.25%	13.40%	
	2900	Electrical		8.35	11.25	23	8.05%	9.35%	12.25%	
	3100	Total: Mechanical & Electrical	↓	21.50	29	34	21.50%	25.50%	27.50%	
94	0010	**TOWN HALLS City Halls & Municipal Buildings**	S.F.	106	135	176				94
	0020	Total project costs	C.F.	9.70	11.65	16.35				
	2720	Plumbing	S.F.	4.44	8.30	15.30	4.31%	5.95%	7.95%	
	2770	Heating, ventilating, air conditioning		8	15.95	23.50	7.05%	9.05%	13.45%	
	2900	Electrical		10.10	14.40	19.65	8.05%	9.45%	11.65%	
	3100	Total: Mechanical & Electrical	↓	35	44.50	68	22%	26.50%	31%	
97	0010	**WAREHOUSES & Storage Buildings**	S.F.	40	59	85				97
	0020	Total project costs	C.F.	2.08	3.25	5.40				
	0100	Site work	S.F.	4.10	8.15	12.30	6.05%	12.95%	19.85%	
	0500	Masonry		2.38	5.65	12.15	3.73%	7.40%	12.30%	
	1800	Equipment		.64	1.37	7.70	.91%	1.82%	5.55%	
	2720	Plumbing		1.32	2.37	4.44	2.90%	4.80%	6.55%	
	2730	Heating, ventilating, air conditioning		1.51	4.26	5.70	2.41%	5%	8.90%	
	2900	Electrical		2.35	4.42	7.30	5.15%	7.20%	10.10%	
	3100	Total: Mechanical & Electrical	↓	6.55	10.05	19.80	12.75%	18.90%	26%	
99	0010	**WAREHOUSE & OFFICES Combination**	S.F.	49	65	89				99
	0020	Total project costs	C.F.	2.50	3.62	5.35				
	1800	Equipment	S.F.	.85	1.64	2.43	.52%	1.20%	2.40%	
	2720	Plumbing		1.89	3.34	4.89	3.74%	4.76%	6.30%	
	2770	Heating, ventilating, air conditioning		2.98	4.65	6.50	5%	5.65%	10.05%	
	2900	Electrical		3.27	4.86	7.65	5.75%	8%	10%	
	3100	Total: Mechanical & Electrical	↓	9.10	14.05	22	14.40%	19.95%	24.50%	

Square Foot Project Size Modifier

One factor that affects the S.F. cost of a particular building is the size. In general, for buildings built to the same specifications in the same locality, the larger building will have the lower S.F. cost. This is due mainly to the decreasing contribution of the exterior walls plus the economy of scale usually achievable in larger buildings. The Area Conversion Scale shown below will give a factor to convert costs for the typical size building to an adjusted cost for the particular project.

The Square Foot Base Size lists the median costs, most typical project size in our accumulated data, and the range in size of the projects.

The Size Factor for your project is determined by dividing your project area in S.F. by the typical project size for the particular Building Type. With this factor, enter the Area Conversion Scale at the appropriate Size Factor and determine the appropriate cost multiplier for your building size.

Example: Determine the cost per S.F. for a 100,000 S.F. Mid-rise apartment building.

$$\frac{\text{Proposed building area} = 100{,}000 \text{ S.F.}}{\text{Typical size from below} = 50{,}000 \text{ S.F.}} = 2.00$$

Enter Area Conversion scale at 2.0, intersect curve, read horizontally the appropriate cost multiplier of .94. Size adjusted cost becomes .94 x $106.00 = $100.00 based on national average costs.

Note: For Size Factors less than .50, the Cost Multiplier is 1.1
For Size Factors greater than 3.5, the Cost Multiplier is .90

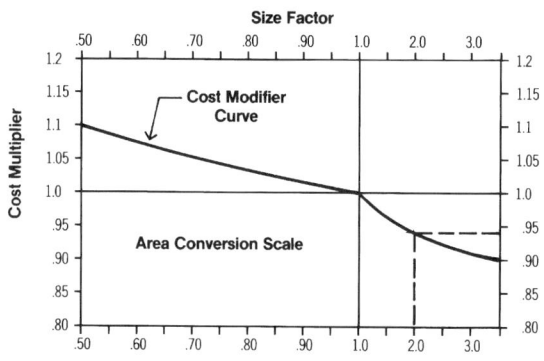

Square Foot Base Size

Building Type	Median Cost per S.F.	Typical Size Gross S.F.	Typical Range Gross S.F.	Building Type	Median Cost per S.F.	Typical Size Gross S.F.	Typical Range Gross S.F.
Apartments, Low Rise	$ 83.50	21,000	9,700 - 37,200	Jails	$ 254.00	40,000	5,500 - 145,000
Apartments, Mid Rise	106.00	50,000	32,000 - 100,000	Libraries	159.00	12,000	7,000 - 31,000
Apartments, High Rise	115.00	145,000	95,000 - 600,000	Living, Assisted	135.00	32,300	23,500 - 50,300
Auditoriums	140.00	25,000	7,600 - 39,000	Medical Clinics	144.00	7,200	4,200 - 15,700
Auto Sales	104.00	20,000	10,800 - 28,600	Medical Offices	136.00	6,000	4,000 - 15,000
Banks	187.00	4,200	2,500 - 7,500	Motels	100.00	40,000	15,800 - 120,000
Churches	128.00	17,000	2,000 - 42,000	Nursing Homes	140.00	23,000	15,000 - 37,000
Clubs, Country	131.00	6,500	4,500 - 15,000	Offices, Low Rise	119.00	20,000	5,000 - 80,000
Clubs, Social	124.00	10,000	6,000 - 13,500	Offices, Mid Rise	117.00	120,000	20,000 - 300,000
Clubs, YMCA	141.00	28,300	12,800 - 39,400	Offices, High Rise	150.00	260,000	120,000 - 800,000
Colleges (Class)	150.00	50,000	15,000 - 150,000	Police Stations	188.00	10,500	4,000 - 19,000
Colleges (Science Lab)	238.00	45,600	16,600 - 80,000	Post Offices	139.00	12,400	6,800 - 30,000
College (Student Union)	176.00	33,400	16,000 - 85,000	Power Plants	1000.00	7,500	1,000 - 20,000
Community Center	132.00	9,400	5,300 - 16,700	Religious Education	119.00	9,000	6,000 - 12,000
Court Houses	178.00	32,400	17,800 - 106,000	Research	195.00	19,000	6,300 - 45,000
Dept. Stores	77.50	90,000	44,000 - 122,000	Restaurants	170.00	4,400	2,800 - 6,000
Dormitories, Low Rise	141.00	25,000	10,000 - 95,000	Retail Stores	82.50	7,200	4,000 - 17,600
Dormitories, Mid Rise	174.00	85,000	20,000 - 200,000	Schools, Elementary	123.00	41,000	24,500 - 55,000
Factories	75.00	26,400	12,900 - 50,000	Schools, Jr. High	127.00	92,000	52,000 - 119,000
Fire Stations	138.00	5,800	4,000 - 8,700	Schools, Sr. High	132.00	101,000	50,500 - 175,000
Fraternity Houses	129.00	12,500	8,200 - 14,800	Schools, Vocational	127.00	37,000	20,500 - 82,000
Funeral Homes	144.00	10,000	4,000 - 20,000	Sports Arenas	102.00	15,000	5,000 - 40,000
Garages, Commercial	92.50	9,300	5,000 - 13,600	Supermarkets	81.50	44,000	12,000 - 60,000
Garages, Municipal	117.00	8,300	4,500 - 12,600	Swimming Pools	191.00	20,000	10,000 - 32,000
Garages, Parking	50.00	163,000	76,400 - 225,300	Telephone Exchange	222.00	4,500	1,200 - 10,600
Gymnasiums	127.00	19,200	11,600 - 41,000	Theaters	118.00	10,500	8,800 - 17,500
Hospitals	225.00	55,000	27,200 - 125,000	Town Halls	135.00	10,800	4,800 - 23,400
House (Elderly)	114.00	37,000	21,000 - 66,000	Warehouses	59.00	25,000	8,000 - 72,000
Housing (Public)	105.00	36,000	14,400 - 74,400	Warehouse & Office	65.00	25,000	8,000 - 72,000
Ice Rinks	151.00	29,000	27,200 - 33,600				

Abbreviations

A	Area Square Feet; Ampere	BTUH	BTU per Hour	Cwt.	100 Pounds
ABS	Acrylonitrile Butadiene Stryrene; Asbestos Bonded Steel	B.U.R.	Built-up Roofing	C.W.X.	Cool White Deluxe
		BX	Interlocked Armored Cable	C.Y.	Cubic Yard (27 cubic feet)
A.C.	Alternating Current; Air-Conditioning; Asbestos Cement; Plywood Grade A & C	°C	degree centegrade	C.Y./Hr.	Cubic Yard per Hour
		c	Conductivity, Copper Sweat	Cyl.	Cylinder
		C	Hundred; Centigrade	d	Penny (nail size)
		C/C	Center to Center, Cedar on Cedar	D	Deep; Depth; Discharge
A.C.I.	American Concrete Institute	C-C	Center to Center	Dis., Disch.	Discharge
AD	Plywood, Grade A & D	Cab.	Cabinet	Db.	Decibel
Addit.	Additional	Cair.	Air Tool Laborer	Dbl.	Double
Adj.	Adjustable	Calc	Calculated	DC	Direct Current
af	Audio-frequency	Cap.	Capacity	DDC	Direct Digital Control
A.G.A.	American Gas Association	Carp.	Carpenter	Demob.	Demobilization
Agg.	Aggregate	C.B.	Circuit Breaker	d.f.u.	Drainage Fixture Units
A.H.	Ampere Hours	C.C.A.	Chromate Copper Arsenate	D.H.	Double Hung
A hr.	Ampere-hour	C.C.F.	Hundred Cubic Feet	DHW	Domestic Hot Water
A.H.U.	Air Handling Unit	cd	Candela	DI	Ductile Iron
A.I.A.	American Institute of Architects	cd/sf	Candela per Square Foot	Diag.	Diagonal
AIC	Ampere Interrupting Capacity	CD	Grade of Plywood Face & Back	Diam., Dia	Diameter
Allow.	Allowance	CDX	Plywood, Grade C & D, exterior glue	Distrib.	Distribution
alt.	Altitude			Div.	Division
Alum.	Aluminum	Cefi.	Cement Finisher	Dk.	Deck
a.m.	Ante Meridiem	Cem.	Cement	D.L.	Dead Load; Diesel
Amp.	Ampere	CF	Hundred Feet	DLH	Deep Long Span Bar Joist
Anod.	Anodized	C.F.	Cubic Feet	Do.	Ditto
Approx.	Approximate	CFM	Cubic Feet per Minute	Dp.	Depth
Apt.	Apartment	c.g.	Center of Gravity	D.P.S.T.	Double Pole, Single Throw
Asb.	Asbestos	CHW	Chilled Water; Commercial Hot Water	Dr.	Drive
A.S.B.C.	American Standard Building Code			Drink.	Drinking
Asbe.	Asbestos Worker	C.I.	Cast Iron	D.S.	Double Strength
ASCE.	American Society of Civil Engineers	C.I.P.	Cast in Place	D.S.A.	Double Strength A Grade
A.S.H.R.A.E.	American Society of Heating, Refrig. & AC Engineers	Circ.	Circuit	D.S.B.	Double Strength B Grade
		C.L.	Carload Lot	Dty.	Duty
A.S.M.E.	American Society of Mechanical Engineers	Clab.	Common Laborer	DWV	Drain Waste Vent
		Clam	Common maintenance laborer	DX	Deluxe White, Direct Expansion
A.S.T.M.	American Society for Testing and Materials	C.L.F.	Hundred Linear Feet	dyn	Dyne
		CLF	Current Limiting Fuse	e	Eccentricity
Attchmt.	Attachment	CLP	Cross Linked Polyethylene	E	Equipment Only; East
Avg., Ave.	Average	cm	Centimeter	Ea.	Each
A.W.G.	American Wire Gauge	CMP	Corr. Metal Pipe	E.B.	Encased Burial
AWWA	American Water Works Assoc.	C.M.U.	Concrete Masonry Unit	Econ.	Economy
Bbl.	Barrel	CN	Change Notice	E.C.Y	Embankment Cubic Yards
B&B	Grade B and Better; Balled & Burlapped	Col.	Column	EDP	Electronic Data Processing
		CO_2	Carbon Dioxide	EIFS	Exterior Insulation Finish System
B.&S.	Bell and Spigot	Comb.	Combination	E.D.R.	Equiv. Direct Radiation
B.&W.	Black and White	Compr.	Compressor	Eq.	Equation
b.c.c.	Body-centered Cubic	Conc.	Concrete	EL	elevation
B.C.Y.	Bank Cubic Yards	Cont.	Continuous; Continued, Container	Elec.	Electrician; Electrical
BE	Bevel End	Corr.	Corrugated	Elev.	Elevator; Elevating
B.F.	Board Feet	Cos	Cosine	EMT	Electrical Metallic Conduit; Thin Wall Conduit
Bg. cem.	Bag of Cement	Cot	Cotangent		
BHP	Boiler Horsepower; Brake Horsepower	Cov.	Cover	Eng.	Engine, Engineered
		C/P	Cedar on Paneling	EPDM	Ethylene Propylene Diene Monomer
B.I.	Black Iron	CPA	Control Point Adjustment		
Bit., Bitum.	Bituminous	Cplg.	Coupling	EPS	Expanded Polystyrene
Bit., Conc.	Bituminous Concrete	C.P.M.	Critical Path Method	Eqhv.	Equip. Oper., Heavy
Bk.	Backed	CPVC	Chlorinated Polyvinyl Chloride	Eqlt.	Equip. Oper., Light
Bkrs.	Breakers	C.Pr.	Hundred Pair	Eqmd.	Equip. Oper., Medium
Bldg.	Building	CRC	Cold Rolled Channel	Eqmm.	Equip. Oper., Master Mechanic
Blk.	Block	Creos.	Creosote	Eqol.	Equip. Oper., Oilers
Bm.	Beam	Crpt.	Carpet & Linoleum Layer	Equip.	Equipment
Boil.	Boilermaker	CRT	Cathode-ray Tube	ERW	Electric Resistance Welded
B.P.M.	Blows per Minute	CS	Carbon Steel, Constant Shear Bar Joist	E.S.	Energy Saver
BR	Bedroom			Est.	Estimated
Brg.	Bearing	Csc	Cosecant	esu	Electrostatic Units
Brhe.	Bricklayer Helper	C.S.F.	Hundred Square Feet	E.W.	Each Way
Bric.	Bricklayer	CSI	Construction Specifications Institute	EWT	Entering Water Temperature
Brk.	Brick			Excav.	Excavation
Brng.	Bearing	C.T.	Current Transformer	Exp.	Expansion, Exposure
Brs.	Brass	CTS	Copper Tube Size	Ext.	Exterior
Brz.	Bronze	Cu	Copper, Cubic	Extru.	Extrusion
Bsn.	Basin	Cu. Ft.	Cubic Foot	f.	Fiber stress
Btr.	Better	cw	Continuous Wave	F	Fahrenheit; Female; Fill
Btu	British Thermal Unit	C.W.	Cool White; Cold Water	Fab.	Fabricated

Abbreviations

FBGS	Fiberglass	H.P.	Horsepower; High Pressure	LE	Lead Equivalent
F.C.	Footcandles	H.P.F.	High Power Factor	LED	Light Emitting Diode
f.c.c.	Face-centered Cubic	Hr.	Hour	L.F.	Linear Foot
f'c.	Compressive Stress in Concrete; Extreme Compressive Stress	Hrs./Day	Hours per Day	L.F. Nose	Linear Foot of Stair Nosing
		HSC	High Short Circuit	L.F. Rsr	Linear Foot of Stair Riser
F.E.	Front End	Ht.	Height	Lg.	Long; Length; Large
FEP	Fluorinated Ethylene Propylene (Teflon)	Htg.	Heating	L & H	Light and Heat
		Htrs.	Heaters	LH	Long Span Bar Joist
F.G.	Flat Grain	HVAC	Heating, Ventilation & Air-Conditioning	L.H.	Labor Hours
F.H.A.	Federal Housing Administration			L.L.	Live Load
Fig.	Figure	Hvy.	Heavy	L.L.D.	Lamp Lumen Depreciation
Fin.	Finished	HW	Hot Water	lm	Lumen
Fixt.	Fixture	Hyd.;Hydr.	Hydraulic	lm/sf	Lumen per Square Foot
Fl. Oz.	Fluid Ounces	Hz.	Hertz (cycles)	lm/W	Lumen per Watt
Flr.	Floor	I.	Moment of Inertia	L.O.A.	Length Over All
F.M.	Frequency Modulation; Factory Mutual	IBC	International Building Code	log	Logarithm
		I.C.	Interrupting Capacity	L-O-L	Lateralolet
Fmg.	Framing	ID	Inside Diameter	long.	longitude
Fdn.	Foundation	I.D.	Inside Dimension; Identification	L.P.	Liquefied Petroleum; Low Pressure
Fori.	Foreman, Inside	I.F.	Inside Frosted	L.P.F.	Low Power Factor
Foro.	Foreman, Outside	I.M.C.	Intermediate Metal Conduit	LR	Long Radius
Fount.	Fountain	In.	Inch	L.S.	Lump Sum
fpm	Feet per Minute	Incan.	Incandescent	Lt.	Light
FPT	Female Pipe Thread	Incl.	Included; Including	Lt. Ga.	Light Gauge
Fr.	Frame	Int.	Interior	L.T.L.	Less than Truckload Lot
F.R.	Fire Rating	Inst.	Installation	Lt. Wt.	Lightweight
FRK	Foil Reinforced Kraft	Insul.	Insulation/Insulated	L.V.	Low Voltage
FRP	Fiberglass Reinforced Plastic	I.P.	Iron Pipe	M	Thousand; Material; Male; Light Wall Copper Tubing
FS	Forged Steel	I.P.S.	Iron Pipe Size		
FSC	Cast Body; Cast Switch Box	I.P.T.	Iron Pipe Threaded	M²CA	Meters Squared Contact Area
Ft.	Foot; Feet	I.W.	Indirect Waste	m/hr.; M.H.	Man-hour
Ftng.	Fitting	J	Joule	mA	Milliampere
Ftg.	Footing	J.I.C.	Joint Industrial Council	Mach.	Machine
Ft lb.	Foot Pound	K	Thousand; Thousand Pounds; Heavy Wall Copper Tubing, Kelvin	Mag. Str.	Magnetic Starter
Furn.	Furniture			Maint.	Maintenance
FVNR	Full Voltage Non-Reversing	K.A.H.	Thousand Amp. Hours	Marb.	Marble Setter
FXM	Female by Male	KCMIL	Thousand Circular Mils	Mat; Mat'l.	Material
Fy.	Minimum Yield Stress of Steel	KD	Knock Down	Max.	Maximum
g	Gram	K.D.A.T.	Kiln Dried After Treatment	MBF	Thousand Board Feet
G	Gauss	kg	Kilogram	MBH	Thousand BTU's per hr.
Ga.	Gauge	kG	Kilogauss	MC	Metal Clad Cable
Gal., gal.	Gallon	kgf	Kilogram Force	M.C.F.	Thousand Cubic Feet
gpm, GPM	Gallon per Minute	kHz	Kilohertz	M.C.F.M.	Thousand Cubic Feet per Minute
Galv.	Galvanized	Kip.	1000 Pounds	M.C.M.	Thousand Circular Mils
Gen.	General	KJ	Kiljoule	M.C.P.	Motor Circuit Protector
G.F.I.	Ground Fault Interrupter	K.L.	Effective Length Factor	MD	Medium Duty
Glaz.	Glazier	K.L.F.	Kips per Linear Foot	M.D.O.	Medium Density Overlaid
GPD	Gallons per Day	Km	Kilometer	Med.	Medium
GPH	Gallons per Hour	K.S.F.	Kips per Square Foot	MF	Thousand Feet
GPM	Gallons per Minute	K.S.I.	Kips per Square Inch	M.F.B.M.	Thousand Feet Board Measure
GR	Grade	kV	Kilovolt	Mfg.	Manufacturing
Gran.	Granular	kVA	Kilovolt Ampere	Mfrs.	Manufacturers
Grnd.	Ground	K.V.A.R.	Kilovar (Reactance)	mg	Milligram
H	High Henry	KW	Kilowatt	MGD	Million Gallons per Day
H.C.	High Capacity	KWh	Kilowatt-hour	MGPH	Thousand Gallons per Hour
H.D.	Heavy Duty; High Density	L	Labor Only; Length; Long; Medium Wall Copper Tubing	MH, M.H.	Manhole; Metal Halide; Man-Hour
H.D.O.	High Density Overlaid			MHz	Megahertz
H.D.P.E.	high density polyethelene	Lab.	Labor	Mi.	Mile
Hdr.	Header	lat	Latitude	MI	Malleable Iron; Mineral Insulated
Hdwe.	Hardware	Lath.	Lather	mm	Millimeter
Help.	Helper Average	Lav.	Lavatory	Mill.	Millwright
HEPA	High Efficiency Particulate Air Filter	lb.; #	Pound	Min., min.	Minimum, minute
		L.B.	Load Bearing; L Conduit Body	Misc.	Miscellaneous
Hg	Mercury	L. & E.	Labor & Equipment	ml	Milliliter, Mainline
HIC	High Interrupting Capacity	lb./hr.	Pounds per Hour	M.L.F.	Thousand Linear Feet
HM	Hollow Metal	lb./L.F.	Pounds per Linear Foot	Mo.	Month
HMWPE	high molecular weight polyethylene	lbf/sq.in.	Pound-force per Square Inch	Mobil.	Mobilization
		L.C.L.	Less than Carload Lot	Mog.	Mogul Base
H.O.	High Output	L.C.Y.	Loose Cubic Yard	MPH	Miles per Hour
Horiz.	Horizontal	Ld.	Load	MPT	Male Pipe Thread

Abbreviations

MRT	Mile Round Trip	Pl.	Plate	S.F.C.A.	Square Foot Contact Area
ms	Millisecond	Plah.	Plasterer Helper	S.F. Flr.	Square Foot of Floor
M.S.F.	Thousand Square Feet	Plas.	Plasterer	S.F.G.	Square Foot of Ground
Mstz.	Mosaic & Terrazzo Worker	Pluh.	Plumbers Helper	S.F. Hor.	Square Foot Horizontal
M.S.Y.	Thousand Square Yards	Plum.	Plumber	S.F.R.	Square Foot of Radiation
Mtd., mtd.	Mounted	Ply.	Plywood	S.F. Shlf.	Square Foot of Shelf
Mthe.	Mosaic & Terrazzo Helper	p.m.	Post Meridiem	S4S	Surface 4 Sides
Mtng.	Mounting	Pntd.	Painted	Shee.	Sheet Metal Worker
Mult.	Multi; Multiply	Pord.	Painter, Ordinary	Sin.	Sine
M.V.A.	Million Volt Amperes	pp	Pages	Skwk.	Skilled Worker
M.V.A.R.	Million Volt Amperes Reactance	PP, PPL	Polypropylene	SL	Saran Lined
MV	Megavolt	P.P.M.	Parts per Million	S.L.	Slimline
MW	Megawatt	Pr.	Pair	Sldr.	Solder
MXM	Male by Male	P.E.S.B.	Pre-engineered Steel Building	SLH	Super Long Span Bar Joist
MYD	Thousand Yards	Prefab.	Prefabricated	S.N.	Solid Neutral
N	Natural; North	Prefin.	Prefinished	S-O-L	Socketolet
nA	Nanoampere	Prop.	Propelled	sp	Standpipe
NA	Not Available; Not Applicable	PSF, psf	Pounds per Square Foot	S.P.	Static Pressure; Single Pole; Self-Propelled
N.B.C.	National Building Code	PSI, psi	Pounds per Square Inch		
NC	Normally Closed	PSIG	Pounds per Square Inch Gauge	Spri.	Sprinkler Installer
N.E.M.A.	National Electrical Manufacturers Assoc.	PSP	Plastic Sewer Pipe	spwg	Static Pressure Water Gauge
		Pspr.	Painter, Spray	S.P.D.T.	Single Pole, Double Throw
NEHB	Bolted Circuit Breaker to 600V.	Psst.	Painter, Structural Steel	SPF	Spruce Pine Fir
N.L.B.	Non-Load-Bearing	P.T.	Potential Transformer	S.P.S.T.	Single Pole, Single Throw
NM	Non-Metallic Cable	P. & T.	Pressure & Temperature	SPT	Standard Pipe Thread
nm	Nanometer	Ptd.	Painted	Sq.	Square; 100 Square Feet
No.	Number	Ptns.	Partitions	Sq. Hd.	Square Head
NO	Normally Open	Pu	Ultimate Load	Sq. In.	Square Inch
N.O.C.	Not Otherwise Classified	PVC	Polyvinyl Chloride	S.S.	Single Strength; Stainless Steel
Nose.	Nosing	Pvmt.	Pavement	S.S.B.	Single Strength B Grade
N.P.T.	National Pipe Thread	Pwr.	Power	sst, ss	Stainless Steel
NQOD	Combination Plug-on/Bolt on Circuit Breaker to 240V.	Q	Quantity Heat Flow	Sswk.	Structural Steel Worker
		Qt.	Quart	Sswl.	Structural Steel Welder
N.R.C.	Noise Reduction Coefficient/ Nuclear Regulator Commission	Quan., Qty.	Quantity	St.; Stl.	Steel
		Q.C.	Quick Coupling	S.T.C.	Sound Transmission Coefficient
N.R.S.	Non Rising Stem	r	Radius of Gyration	Std.	Standard
ns	Nanosecond	R	Resistance	Stg.	Staging
nW	Nanowatt	R.C.P.	Reinforced Concrete Pipe	STK	Select Tight Knot
OB	Opposing Blade	Rect.	Rectangle	STP	Standard Temperature & Pressure
OC	On Center	Reg.	Regular	Stpi.	Steamfitter, Pipefitter
OD	Outside Diameter	Reinf.	Reinforced	Str.	Strength; Starter; Straight
O.D.	Outside Dimension	Req'd.	Required	Strd.	Stranded
ODS	Overhead Distribution System	Res.	Resistant	Struct.	Structural
O.G.	Ogee	Resi.	Residential	Sty.	Story
O.H.	Overhead	Rgh.	Rough	Subj.	Subject
O&P	Overhead and Profit	RGS	Rigid Galvanized Steel	Subs.	Subcontractors
Oper.	Operator	R.H.W.	Rubber, Heat & Water Resistant; Residential Hot Water	Surf.	Surface
Opng.	Opening			Sw.	Switch
Orna.	Ornamental	rms	Root Mean Square	Swbd.	Switchboard
OSB	Oriented Strand Board	Rnd.	Round	S.Y.	Square Yard
O.S.&Y.	Outside Screw and Yoke	Rodm.	Rodman	Syn.	Synthetic
Ovhd.	Overhead	Rofc.	Roofer, Composition	S.Y.P.	Southern Yellow Pine
OWG	Oil, Water or Gas	Rofp.	Roofer, Precast	Sys.	System
Oz.	Ounce	Rohe.	Roofer Helpers (Composition)	t.	Thickness
P.	Pole; Applied Load; Projection	Rots.	Roofer, Tile & Slate	T	Temperature; Ton
p.	Page	R.O.W.	Right of Way	Tan	Tangent
Pape.	Paperhanger	RPM	Revolutions per Minute	T.C.	Terra Cotta
P.A.P.R.	Powered Air Purifying Respirator	R.S.	Rapid Start	T & C	Threaded and Coupled
PAR	Parabolic Reflector	Rsr	Riser	T.D.	Temperature Difference
Pc., Pcs.	Piece, Pieces	RT	Round Trip	Tdd	Telecommunications Device for the Deaf
P.C.	Portland Cement; Power Connector	S.	Suction; Single Entrance; South		
P.C.F.	Pounds per Cubic Foot	SC	Screw Cover	T.E.M.	Transmission Electron Microscopy
P.C.M.	Phase Contrast Microscopy	SCFM	Standard Cubic Feet per Minute	TFE	Tetrafluoroethylene (Teflon)
P.E.	Professional Engineer; Porcelain Enamel; Polyethylene; Plain End	Scaf.	Scaffold	T. & G.	Tongue & Groove; Tar & Gravel
		Sch., Sched.	Schedule		
		S.C.R.	Modular Brick	Th., Thk.	Thick
Perf.	Perforated	S.D.	Sound Deadening	Thn.	Thin
PEX	Cross linked polyethylene	S.D.R.	Standard Dimension Ratio	Thrded	Threaded
Ph.	Phase	S.E.	Surfaced Edge	Tilf.	Tile Layer, Floor
P.I.	Pressure Injected	Sel.	Select	Tilh.	Tile Layer, Helper
Pile.	Pile Driver	S.E.R., S.E.U.	Service Entrance Cable	THHN	Nylon Jacketed Wire
Pkg.	Package	S.F.	Square Foot	THW.	Insulated Strand Wire

Abbreviations

THWN	Nylon Jacketed Wire	USP	United States Primed	Wrck.	Wrecker
T.L.	Truckload	UTP	Unshielded Twisted Pair	W.S.P.	Water, Steam, Petroleum
T.M.	Track Mounted	V	Volt	WT., Wt.	Weight
Tot.	Total	V.A.	Volt Amperes	WWF	Welded Wire Fabric
T-O-L	Threadolet	V.C.T.	Vinyl Composition Tile	XFER	Transfer
T.S.	Trigger Start	VAV	Variable Air Volume	XFMR	Transformer
Tr.	Trade	VC	Veneer Core	XHD	Extra Heavy Duty
Transf.	Transformer	Vent.	Ventilation	XHHW, XLPE	Cross-Linked Polyethylene Wire Insulation
Trhv.	Truck Driver, Heavy	Vert.	Vertical		
Trlr	Trailer	V.F.	Vinyl Faced	XLP	Cross-linked Polyethylene
Trlt.	Truck Driver, Light	V.G.	Vertical Grain	Y	Wye
TTY	Teletypewriter	V.H.F.	Very High Frequency	yd	Yard
TV	Television	VHO	Very High Output	yr	Year
T.W.	Thermoplastic Water Resistant Wire	Vib.	Vibrating	Δ	Delta
		V.L.F.	Vertical Linear Foot	%	Percent
UCI	Uniform Construction Index	Vol.	Volume	~	Approximately
UF	Underground Feeder	VRP	Vinyl Reinforced Polyester	Ø	Phase; diameter
UGND	Underground Feeder	W	Wire; Watt; Wide; West	@	At
U.H.F.	Ultra High Frequency	w/	With	#	Pound; Number
U.I.	United Inch	W.C.	Water Column; Water Closet	<	Less Than
U.L.	Underwriters Laboratory	W.F.	Wide Flange	>	Greater Than
Uld.	unloading	W.G.	Water Gauge	Z	zone
Unfin.	Unfinished	Wldg.	Welding		
URD	Underground Residential Distribution	W. Mile	Wire Mile		
		W-O-L	Weldolet		
US	United States	W.R.	Water Resistant		

Index

A

Abatement asbestos 33
   equipment asbestos 33
ABC extinguisher 292
Abrasive aluminum oxide 42
   floor tile 250
   silicon carbide 42
   tread 250
ABS DWV pipe 359
Absorber shock 363
A/C coils furnace 395
   fan coil system 586, 587
   rooftop DX 592
   rooftop DXVAV 592
   self contained 594, 595
   self contained VAV 594, 595
Access card key 229
   control card type 229
   control facial scanner 229
   control fingerprint scanner .. 229
   control hand scanner 229
   control video camera 437
   door and panels 199
   door basement 60
   door duct 383
   door fire rated 199
   door floor 199
   door metal 199
   door roof 177
   door stainless steel 199
   floor 260
   road and parking area 16
Accessories bathroom 288, 289
   door 230
   drywall 249
   duct 384
   glazing 233
   roof 177
   toilet 288
Accessory drainage 174
   fireplace 290
   masonry 67, 71
Accordion door 196
Acid proof floor 254
Acoustic ceiling board 251
Acoustical ceiling 251, 252
   door 203, 287
   enclosure 332
   folding partition 287
   metal deck 94
   panel 263, 287
   partition 288
   sealant 182, 248
   treatment 263
   underlayment 263
   wallboard 249
   window wall 206
Acrylic ceiling 251
   plexiglass 233
   rubber roofing 168
   sign 281
   wall coating 277
   wallcovering 261
   wood block 255
Adhesive cement 256
   EPDM 148
   neoprene 148
   PVC 148
   roof 164
   wallpaper 261
Adjustable astragal 228
   frequency drives 425
Adjustment factors labor 348
Admixtures concrete 42
Aggregate stone 473

Aids staging 15
Air cleaner electronic 391
   compressor 368
   compressor sprinkler system .. 345
   conditioner direct expansion .. 402
   conditioner fan coil 402
   conditioner gas heat 400
   conditioner receptacle 418
   conditioner removal 380
   conditioner rooftop 400
   conditioner self-contained ... 400
   conditioner window 400
   conditioner wiring 420
   conditioning computer 401
   conditioning fan 386
   conditioning ventilating .. 385-387,
                         398, 399, 401
   cooled condensing unit 398
   entraining agent 42
   extractor 383
   filter 390
   filter mechanical media 390
   filter permanent washable .. 390
   filter roll type 390
   filters washable 390
   filtration 33
   filtration charcoal type ... 390
   handler heater 395
   handler modular 395
   handling fan 386
   handling unit 402
   make-up unit 400
   register 389
   return grille 389
   sampling 33
   supply register 389
   supported structures demo .. 326
   terminal units 388
   tube system 342
   unit make-up 400
   vent roof 177
Air-compressor mobilization .. 440
Airless sprayer 33
Airport painted markings 467
Air-source heat pumps 401
Alarm exit control 436
   fire 436
   residential 420
   sprinkler 437
   valve sprinkler 344
   water motor 345
Alteration fee 8
Alternating tread ladder 106
Aluminum astragal 227
   bench 322
   ceiling tile 252
   column 108, 139
   conduit 602
   coping 82
   cross 312
   diffuser perforated 388
   door 186, 190, 203, 205
   door frame 207
   double acting swing door ... 205
   downspout 173
   drip edge 176
   ductwork 383
   edging 475
   entrance 207
   expansion joint 175
   fence 467
   flashing 170, 366
   foil 152, 155
   gravel stop 172
   grille 389
   gutter 174

   joist shoring 49
   light pole 627
   louver 234
   mansard 161
   nail 116
   oxide abrasive 42
   plank scaffold 15
   reglet 175
   roof 159
   sash 209
   service entrance cable 409
   sheet metal 173
   shingle 156
   siding 161
   siding paint 268
   sliding door 200
   steeple 296
   storefront 208
   storm door 186
   tile 158, 251
   transom 207
   tube frame 205
   weatherstrip 228
   window 209, 522
   window demolition 185
Anchor bolt 69
   brick 70
   buck 69
   channel slot 70
   chemical 61, 87
   epoxy 61, 87
   expansion 87
   framing 117
   hollow wall 87
   machinery 88
   masonry 69
   metal nailing 88
   nailing 88
   nylon nailing 88
   partition 70
   rafter 118
   rigid 70
   roof safety 15
   sill 118
   steel 70
   stone 70
   toggle-bolt 88
   wedge 88
Antenna TV 432
Anti-siphon device 363
Apartment call system 432
Appliance 302
   compactor 612
   dishwasher 612
   garbage disposer 612
   plumbing 369
   range 611
   range hood 612
   refrigerator 612
   residential 302, 420
Apron wood 132
Arc fault circuit-breaker ... 422
Arch laminated 131
   radial 131
Architectural equipment 302
   fee 8, 700
Area clean-up 35
Armored cable 408
Arrow 467
Asbestos abatement 33
   abatement equipment 33
   area decontamination 35
   demolition 35
   disposal 36
   encapsulation 36
   removal 34

   removal process 710
As-built drawings 18
Ash receiver 289
Ashlar stone 80
Asphalt base sheet 164
   block 465
   block floor 465
   coating 147
   curb 465
   cutting 27
   felt 164, 165
   flashing 171
   flood coat 164
   grinding 461
   primer 146, 256
   rubberized 460
   sealcoat rubberized 460
   sheathing 130
   shingle 155
   sidewalk 461
   surface treatment 460
Asphaltic binder 463
   concrete 463
   emulsion 460
   pavement 463
   wearing course 463
Astragal adjustable 228
   aluminum 227
   magnetic 228
   molding 133
   one piece 227
   overlapping 228
   rubber 228
   split 228
   steel 227
Athletic carpet indoor 260
   equipment 311
Atomizer water 33
Attic stair 304
   ventilation fan 387
Auger hole 22
Auto park drain 366
Autoclaved concrete block ... 75
Automatic flush 374
   opener 221
   opener commercial 221
   timed thermostat 381
Automotive equipment 298, 382
   lift 341
   spray painting booth 298
Awning canvas 294, 295
   metal-clad wood window 213
   window 210, 294

B

Backer rod 180
   rod polyethylene 48
Backerboard 148
   cementitious 246
Backfill 443, 444
   compaction 733
   dozer 443
   planting pit 472
   structural 443
   trench 441
Backflow preventer 362
   preventer solar 397
Backhoe 442
Backsplash countertop 318
Backstop basketball 311
   electric 311
Backup block 76
Baffle roof 177
Bag disposable 33

Index

glove	34
Baked enamel door	189
enamel frame	188
Balanced door	207
Balcony fire escape	106
Bale hay	454
Ball check valve	350
valve plastic	350
Ballast fixture	428
Baluster wood stair	136
Balustrade painting	274
Band joist framing	99
molding	132
Bank counter	299
equipment	299
window	299
Bankrun gravel	440
Baptistry	312
Bar bell	311
front	332
grab	288
panic	222
restaurant	320
touch	222
towel	289
tub	288
Zee	249
Barbed wire fence	467
Barber equipment	300
Barometric damper	391
Barricade	17, 486
tape	17
Barrier parking	467
separation	33, 38
X-ray	333
Barriers and enclosures	17
weather	155
Base cabinet	315, 316
carpet	260
column	118
course	462
course drainage layers	462
cove	256
gravel	462
masonry	78
molding	132
quarry tile	250
resilient	256
road	462
sheet	164, 165
sink	315
stone	80, 462
terrazzo	258
vanity	317
wood	132
Baseball scoreboard	312
Baseboard demolition	116
heat	403
heat electric	404
heater electric	404
register	389
Basement bulkhead stair	59
Baseplate scaffold	13
shoring	13
Basic finish materials	238
Basketball backstop	311
scoreboard	312
Bath communal	330
spa	330
steam	330
whirlpool	330
Bathroom	371
accessories	288, 289
exhaust fan	387
faucet	372
fixture	370, 371
heater & fan	387
Bathtub bar	288
removal	348
Bathtubs residential	371
Batt insulation	150
Bead blast demo	239
casing	249
corner	249
parting	133
Beam & girder framing	119
and girder formwork	43
bond	69
bondcrete	245
ceiling	139
concrete	52
concrete placement	55
concrete placement grade	56
decorative	139
drywall	247
fireproofing	178
hanger	117
laminated	131
mantel	138
plaster	245
precast	59
precast concrete	59
precast concrete tee	59
reinforcing	50
wood	119, 127
Beams ld-brng stud wall box	96
Bed molding	132
Bedding brick	464
pipe	444
Bedpan cleanser	375
Belgian block	466
Bell & spigot pipe	364
bar	311
Bench aluminum	322
fiberglass	322
folding	322
greenhouse	331
park	322
players	323
wood	322
work	310
Benches	322
Bend EMT field	411
Berm pavement	465
road	465
Bevel siding	162
Bicycle trainer	311
Bi-fold door	190, 195
Binder asphaltic	463
Bi-passing closet door	195
Birch door	195, 196, 198
molding	133
paneling	134
wood frame	140
Bit drill	63
Bituminous block	465
coating	147
concrete curbs	465
Blanket insulation	381
insulation wall	150
sound attenuation	263
Blast demo bead	239
floor shot	239
Blasting water	66
Bleacher telescoping	322
Blind exterior	141
structural bolt	89
window	314
Block asphalt	465
backup	76
belgian	466
bituminous	465
brick and mortar	719
concrete	75-77
concrete bond beam	76
concrete exterior	77
filler	275
floor	255
floor wood	255
glass	78
glazed concrete	78
grooved	76
insulation	76
lightweight	77
lintel	77
partition	77
profile	76
reflective	79
slump	76
split rib	76
wall removal	24
Blocking carpentry	118
steel	118
wood	118, 122
Blower pneumatic tube	342
Blown-in cellulose	152
fiberglass	152
insulation	152
Blueboard	246
partition	241
Bluegrass sod	472
Bluestone	80
sill	81
step	462
Board & batten siding	162
batten fence	470
ceiling	251
control	310
directory	280, 281
dock	299
gypsum	247
insulation	149
insulation foam	150
paneling	136
partition gypsum	241
partition nlb gypsum	240
sheathing	130
valance	316
verge	133
Body harness	15
Boiler	392
demolition	380
electric	392
electric steam	392
gas fired	393
gas/oil combination	393
hot water	393, 394
oil fired	393
oil-fired	394
removal	380
steam	393
Bollard	90
Bollards pipe	466
Bolt anchor	69
blind structural	89
steel	117
Bolt-on circuit-breaker	421
Bolts remove	86
Bond beam	69
performance	707
Bondcrete	245
Bonding agent	42
Bookcase	293, 316
Bookshelf	321
Booster fan	386
Booth painting	298
restaurant	321
telephone	282
Bored pile	457
Boring and exploratory drilling	22
cased	22
horizontal	478
service	479
Borosilicate pipe	377
Borrow	440
Bosuns chair	15
Bottle storage	305
Boundary and survey marker	22
Bow window	213
Bowl toilet	370
Bowstring truss	131
Box beams ld-brng stud wall	96
distribution	481
electrical pull	412
mail	293
pull	412
stair	136
storage	12
termination	432
vent	72
Boxed beam headers	100
Boxes & wiring device	412
mixing	388
utility	478
VAV	388
Boxing ring	311
Brace cross	90
Bracing	119
let-in	118
load-bearing stud wall	95
metal	118
metal joist	98
roof rafter	102
shoring	13
Bracket scaffold	13
sidewall	15
Brass hinge	225
pipe	352
pipe fittings	352
Brazed connection	410
Break glass station	437
ties concrete	57
Breaker circuit	425
vacuum	363
Brick	75, 553
anchor	70
bedding	464
bedding mortar	461
Brick, block and mortar	719
Brick catch basin	482
chimney simulated	290
common	74
common building	74
concrete	78
demolition	67, 238
economy	74
edging	475
engineer	74
face	74, 719
floor	464
flooring miscellaneous	254
manhole	482
molding	132
oversized	73
paving	464
removal	23
sidewalk	464
sill	81
simulated	82
step	462
structural	74
utility	74
veneer	72
veneer demolition	68

756

Index

veneer thin 73
vent box 72
wall 79, 516
Bridge sidewalk 14
Bridging 119
   load-bearing stud wall 95
   metal joist 99
   roof rafter 102
Broiler 307
Bronze body strainer 382
   plaque 281
   push-pull plate 224
   swing check valve 349
   valve 349
Broom cabinet 316
   finish concrete 57
Brown coat 258
Brownstone 81
Bubbler 376
Buck anchor 69
   rough 121
Buggy concrete 56
Builder's risk insurance 703
Building cmpnts deconst. 29
   deconstruction 25, 28
   demo footing & foundation . 24
   demolition 24
   directory 433
   greenhouse 331
   hardware 221
   insulation 152
   moving 31
   paper 155
   permit 11
   prefabricated 331
   relocation 31
   shoring 455
   sprinkler 345
   temporary 12
Built-up roof 164, 165
   roofing 722
   roofing system 164
Bulb incandescent 430
   tee subpurlin 60
Bulk bank measure excavating . 442
   dozer excavating 443
Bulkhead door 199
Bulkhead/cellar door 199
Bulldozer 443
Bulletproof glass 234
Bumper car 467
   dock 299
   door 224
   parking 467
   plastic parking 466
   wall 224
Burglar alarm indicating panel ... 436
Burlap curing 58
   rubbing concrete 57
Burner gas conversion 394
   gun type 394
   oil 394
   residential 394
Bush hammer 57
   hammer finish concrete 57
Butterfly valve 380, 479
Butyl caulking 181
   expansion joint 175
   flashing 172
   rubber 180
   waterproofing 148
BX cable 602

C

Cabinet base 315, 316
   broom 316
   casework 316
   corner base 315
   corner wall 316
   current transformer 423
   demolition 116
   door 314, 316
   electrical hinged 412
   fire equipment 291
   hardboard 315
   hardware 317
   hinge 317
   hose rack 291
   hotel 289
   key 317
   kitchen 315
   laboratory 317
   medicine 289
   outdoor 317
   oven 316
   shower 285
   stain 271
   standard wood base 611
   storage 317
   strip 433
   transformer 423
   unit heater 591
   varnish 271
   wall 315
Cable armored 408
   electric 408
   electrical 409
   mineral insulated 408
   railing 109
   safety railing 92
   sheathed nonmetallic 409
   sheathed Romex 409
   termination mineral insulated . 408
   UTP 432
Cafe door 197
Caisson concrete 457
   foundation 457
Caissons 734
Calcium chloride 42
Call system apartment 432
Canopy 295
   door 294
   entrance 295
   entry 294
   framing 89, 124
   patio/deck 294
   residential 294
Cant roof 124
Cantilever needle beams 456
Canvas awning 294, 295
Cap concrete placement pile .. 56
   dowel 51
   pile 54
   post 118
   service entrance 409
Car bumper 466, 467
Card key access 229
Carpentry finish 315
Carpet 259, 260
   base 260
   cleaning 260
   computer room 260
   felt pad 259
   floor 260
   nylon 260
   pad 259
   padding 259
   removal 238
   stair 260
   tile 259
   tile removal 238
   urethane pad 259
   wool 260
Carrel 321
Carrier ceiling 243
   channel 252
   fixture 375
Cart concrete 56
   food delivery 306
Carving stone 80
Case good metal 320
   refrigerated 304
   work 316
Cased boring 22
   evaporator coils 395
Casement window 209, 211, 213
   window vinyl 218
Casework cabinet 316
   custom 315
   demolition 116
   ground 126
   metal 317
   painting 271
   varnish 271
Cash register 300
Casing bead 249
   wood 132
Cast in place concrete 52
   in place concrete curbs ... 465
   in place terrazzo 258
   iron bench 323
   iron damper 291
   iron fitting 364
   iron manhole cover 482
   iron pipe 364
   iron pipe fitting 356, 364
   iron radiator 403
   iron trap 365
   trim lock 224
Caster scaffold 13
Castings construction 108
Cast-iron column base 108
Catch basin brick 482
   basin cover remove/replace . 482
   basin precast 482
   basin removal 23
   basins 482
   door 317
Category 3 jack 432
   5 connector 432
   5 jack 432
   5e jack 432
   6 jack 432
Catwalk scaffold 14
Caulking 181
   masonry 181
   polyurethane 181
   sealant 181
Cavity truss reinforcing 71
   wall grout 69
   wall insulation 151, 152
Cedar closet 135
   fence 469
   paneling 136
   post 139
   roof deck 128
   roof plank 128
   shingle 157
   siding 162
Ceiling 251, 560
   acoustical 251, 252
   beam 139
   board 251
   board acoustic 251
   board fiberglass 251
   bondcrete 245
   carrier 243
   demolition 238
   diffuser 388
   drywall 247, 532
   eggcrate 251
   fan 386
   finish price sheet 532, 560
   framing 120
   furring 126, 241, 243
   gypsum board 558
   hatch 199
   insulation 150, 152
   lath 244
   luminous 251, 252, 430
   molding 132
   painting 275, 276
   panels metal 253
   plaster 245, 532
   price sheet 532
   register 389
   stair 304
   support 89
   suspended 243, 251
   suspended acoustical 557
   suspended plaster 559
   suspension system 252
   tile 252, 533
Cell prison 332
Cellar door 199
   wine 305
Cellular concrete 53
   decking 94
Cellulose blown-in 152
   insulation 151
Cement adhesive 256
   flashing 146
   grout 69, 455
   gunite 58
   liner 478
   masonry 68
   masonry unit 76, 77
   mortar 250
   terrazzo portland 258
   underlayment self-leveling . 61
Cementitious backerboard 246
   waterproofing 149
   wood fiber plank 60
Central vacuum 301
Centrifugal fan 386
Ceramic tile 249
   tile countertop 318
   tile demolition 239
   tile floor 249
   tiling 249
Certification welding 11
Chain hoist door 203
   link fence 17
   link fence paint 266
   link industrial fence 467
   trencher 441
Chair barber 300
   bosuns 15
   lifts 340
   molding 133
   movie 309
   rail demolition 116
   restaurant 320
Chalkboard 280
   freestanding 280
   liquid chalk 280
   wall hung 280
Chamber decontamination ... 33, 38
   echo 332
Channel carrier 252

757

Index

frame	188
furring	241, 249
metal frame	188
siding	162
slot	70
slot anchor	70
Charge powder	89
Charges disposal	36
Check valve ball	350
Checkout counter	300
scanner	300
supermarket	300
Chemical anchor	61, 87
cleaning masonry	66
dry extinguisher	292
termite control	455
Chime door	424
Chimney	73
accessories	290
all fuel vent	391
brick	73
demolition	67
fireplace	290
foundation	52
metal	290
screen	290
simulated brick	290
vent	391
vent fitting	391
Chloride accelerator concrete	55
Church equipment	312
pew	322
Chute refuse	341
rubbish	26
Chutes	341
linen	341
Circline fixture	428
Circuit-breaker	425
bolt-on	421
Circulating pump	383
Cladding sheet metal	170
Clamp water pipe ground	410
Clapboard painting	267
Clay fire	82
roofing tile	158
tile	158
tile coping	82
Clean control joint	57
tank	31
Cleaning & disposal equipment	308
masonry	66
rug	260
up	18
Cleanout door	291
floor	350
pipe	350
tee	350
Clean-up area	35
Clip plywood	118
Clock timer	420
Closer concealed	222
electronic	222
floor	222
holder	222
Closers door	222
Closet cedar	135
door	190, 195
pole	133
rod	293
water	370, 373
Closing coordinator door	225
Clothes dryer commercial	301
CMU	75
pointing	66
Coal tar pitch	146, 165
Coat brown	258

glaze	275
hook	288
rack	316
scratch	258
tack	460
Coated roofing foamed	168
Coating bituminous	147
flood	166
roof	146
rubber	149
silicone	149
spray	147
trowel	147
wall	277
water repellent	149
waterproofing	147
Coffee urn	306
Coil cooling	402
Coiling door and grilles	201
Cold milling asphalt paving	461
planing	461
recycling	461
roofing	164
storage door	202
Collection box lobby	293
concrete pipe sewage	480
PVC pipe sewage	480
sewage/drainage	481
Collector solar energy system	396
Colonial door	195, 198
wood frame	140
Coloring concrete	43
Column	139
aluminum	108
base	118
base cast-iron	108
bondcrete	245
brick	73
concrete	52
concrete placement	55
demolition	68
drywall	247
fireproof	178
formwork	44
laminated wood	131
lath	244
plaster	245
plywood formwork	44
precast	59
precast concrete	59
reinforcing	50
removal	68
round fiber tube formwork	44
round fiberglass formwork	44
round steel formwork	44
steel framed formwork	44
tie	70
wood	120, 127, 139
Columns framing	120
structural	90
Combination storm door	195, 198
Command dog	18
Commercial door	189, 194, 199
folding partition	287
gas water heater	369
lavatory	374
water heater	369
water heater electric	369
Commissioning	18
Common brick	74
building brick	74
nail	116
Communal bath	330
Communicating lockset	223
Communication cables & fittings	432
system	432

Compaction	453, 454
backfill	733
soil	443
structural	454
test Proctor	11
Compactor	312
plate	453
residential	303
sheepsfoot	453
tamper	454
Compartments shower	285
toilet	282
Component control	381
Components furnace	395
Composite decking	123
door	189
insulation	154
metal deck	94
rafter	123
Composition floor	259
flooring	259
flooring removal	238
Compound dustproofing	42
Compressed air equip. auto.	298
Compression seals	180
Compressive strength	11
Compressor air	368
reciprocating	368
reciprocating hermetic	368
refrigerant	398
Computer air conditioning	401
floor	260
room carpet	260
Concealed closer	222
Concrete	714-717
admixtures	42
asphaltic	463
beam	52
block	75-77
block autoclaved	75
block back-up	75
block bond beam	76
block decorative	76
block demolition	25
block exterior	77
block foundation	77
block glazed	78
block grout	69
block insulation	151
block interlocking	77
block lintel	77
block partition	77
block wall	501, 515
break ties	57
brick	78
broom finish	57
buggy	56
burlap rubbing	57
bush hammer finish	57
caisson	457
cart	56
cast in place	52, 55
cellular	53
chloride accelerator	55
coloring	43
column	52
coping	82
core drilling	62
crane and bucket	55
curb	465
curing	58, 464
curing compound	58
cutout	25
cylinder	11
demo	23
demolition	25, 42

direct chute	55
drill	413
fiber reinforcing	55
filled steel piles	457
finish	464
float finish	57
floor patching	42
forms	711
formwork shoring	49
furring	126
granolithic finish	57
grout	69
hand trowel finish	57
hole cutting	414
hole drilling	413
hot weather	55
hydrodemolition	22
impact drilling	63
integral colors	42
integral topping	57
integral waterproofing	43
joist	53
lift slab	718
lightweight	53, 719
lightweight insulating	61
machine trowel finish	57
materials	717
membrane curing	58
non-chloride accelerator	55
pan fill stair	50
panel	505
paper curing	59
paving	464
piles prestressed	456
pipe removal	23
placement beam	55
placement column	55
placement footing	55
placement slab	56
placement wall	56
placing	55, 717
plank	506
planter	322
pressure grouting	455
prestressed precast	718
pumped	55
ready-mix	55
removal	23
repair crack	62
retarder	55
sandblast finish	57
scarify	238
septic tank	480
shingle	158
short load	55
sidewalk	461
slab	52, 504
slab conduit	415
slab cutting	62
slabs cutting	62
slabs X-ray	11
stair	54
stair tread insert	49
tile	158
topping	57
truck holding	55
utility vault	478
vertical shoring	49
wall	53, 500
wall patching	42
walls cutting	62
water curing	58
water reducer	55
wheeling	56
winter	55
Condenser pad	404

Index

Condensing unit air cooled 398
Conductive floor 257
   rubber & vinyl flooring 257
Conductor 408
   & grounding 409
Conductors 731
Conduit & fitting flexible 416
   concrete slab 415
   electrical 411
   high installation 408
   in slab PVC 415
   in trench electrical 415
   in trench steel 415
   intermediate steel 411
   rigid in slab 415
Cone traffic 17
Connection brazed 410
   fire-department 344
   standpipe 344
Connections motor 416
Connector flexible 362
   gas 362
   joist 117
   timber 117
Construction castings 108
   management fee 8
   temporary 16
   time 8
Contaminated soil 31
Contingencies 8
Continuous hinge 227
Contractor overhead 18
Control board 310
   component 381
   crack 460
   damper volume 385
   draft 391
   joint 57, 71
   joint clean 57
   joint PVC 72
   joint rubber 71
   joint sawn 57
   joint sealant 57
   package firecycle 345
   system electronic 381
   valve heating 381
Convection oven 307
Conveyor system 341
Cooking equipment 302, 307
   range 302
Cooler 330
   beverage 304
   water 376
Cooling coil 402
   tower 399
   tower forced-draft type 399
   towers fiberglass 399
   towers stainless 399
Coping 80, 82
   aluminum 82
   clay tile 82
   concrete 82
   removal 68
   terra cotta 75, 82
Copper cable 408
   downspout 173
   drum trap 365
   DWV tubing 353
   fitting 353
   flashing 170, 171, 366
   gutter 174
   pipe 352
   reglet 175
   roof 169
   tube fittings 353
   wall covering 261

wire 409
Core drilling concrete 62
   testing 11
Coreboard 248
Cork floor 257
   tile 257
   wall tile 261
Corner base cabinet 315
   bead 249
   guard 288
   wall cabinet 316
Cornerboards PVC 142
Cornice 79
   drain 367
   molding 132
   painting 274
Correspondence lift 341
Corrosion resis. backflow prev. 362
   resistant fan 386
   resistant fitting 377
   resistant pipe 377
   resistant pipe fitting 377
Corrugated metal pipe 481
   roof tile 158
   siding 159-161
Cost mark-up 10
Cot prison 321
Counter bank 299
   checkout 300
   door 201
   flashing 175
   flashing aluminum 536
   flashing copper 536
   flashing steel 536
   top 318
   top demolition 116
Countertop backsplash 318
   laboratory 318
   laminated plastic 611
   sink 370
Course drainage layers base 462
   wearing 463
Court tennis 468
Cove base 256
   base ceramic tile 249
   base terrazzo 258
   molding 132
   molding scotia 132
Cover ground 473
   stair tread 17
CPVC pipe 359
   valve 350
Crack concrete repair 62
   control 460
   filling 460
Crane and bucket concrete 55
   crew daily 12
   daily 12
   mobilization 440
   rail 488
   tower 12
Cranes overhead bridge 488
Crew daily crane 12
   forklift 12
   survey 18
Cross brace 90
   wall 312
Crown molding 132
Cubicle detention 220
   shower 371
   toilet 285
Cultured stone 83
Culvert reinforced 481
Cupola 295
Curb 465
   asphalt 465

concrete 465
   edging 90, 465
   edging angle 90
   granite 465
   inlet 465
   precast 465
   prefabricated 404
   removal 23
   roof 124
   seal 460
   terrazzo 258
Curing compound concrete 58
   concrete 58, 464
   concrete membrane 58
   concrete paper 59
   concrete water 58
   paper 155
Current transformer cabinet 423
Curtain damper fire 385
   divider 312
   gymnasium divider 312
   rod 288
Custom casework 315
Cut stone curbs 465
Cutoff pile 440
Cutout counter 318
   demolition 25
   slab 25
Cutting and drilling 413
   asphalt 27
   block 318
   concrete slabs 62
   concrete walls 62
   masonry 27
   saw 62
   steel 86
   torch 86
   wood 27
Cylinder concrete 11
   lockset 223
   recore 223, 224

D

Daily crane 12
Dairy case 304
Damper 384
   barometric 391
   fire curtain 385
   fireplace 291
   foundation vent 72
   multi-blade 384
   vent 391
Darkroom door 202
   equipment 302
   sink 302
Dead lock 223
Deadbolt 223, 225
Deadlocking latch 223
Deciduous shrub 474
Deck drain 366
   framing 122
   roof 128, 129
   slab form 94
   wood 128
Decking 721
   cellular 94
   composite 123
   floor 94
   form 94
   roof 94
Deconstruction building 25, 28
   building components 29
   material handling 30
Decontamination asbestos area 35

chamber 33, 38
   enclosure 33, 35, 38
   equipment 33
Decorative beam 139
   fence 469
Decorator device 417
   switch 416
Deep therapy room 332
Deep-longspan joist 92
Delicatessen case 304
Delivery charge 16
Demo air supported structures 326
   concrete 23
   footing & foundation building 24
   garden house 326
   geodesic domes 326
   greenhouses 326
   hangars 326
   lightning protection 326
   pre-engineered steel buildings 327
   silos 327
   sound control 327
   special purpose rooms 327
   storage tank 328
   swimming pool 328
   tension structures 328
   thermal & moist. protection 146
   X-ray/radio freq protection 329
Demo. rem. pvemnt and crb 22
Demolition 25, 68, 115, 239, 348, 440, 709
   asbestos 35
   baseboard 116
   boiler 380
   brick 67, 238
   cabinet 116
   casework 116
   ceiling 238
   ceramic tile 239
   chimney 67
   column 68
   concrete 25, 42
   concrete block 25
   disposal 26
   door 185
   drywall 238
   ductwork 380
   electrical 406
   fencing 24
   fireplace 68
   flooring 238
   framing 113
   furnace 380
   glass 185
   granite 68
   gutter 146
   gutting 26
   house 24
   HVAC 380
   joist 113
   lath 238
   masonry 25, 67, 239
   metal 86
   metal stud 239
   millwork 116
   mold contaminated area 38
   paneling 116
   partition 239
   pavement 22
   plaster 238, 239
   plenum 239
   plumbing 348
   plywood 238, 239
   post 114
   rafter 114
   railing 116

Index

roofing 35, 147
saw cutting 27
site 22
steel 86
steel window 185
stucco 239
terrazzo 239
tile 238
torch cutting 28
trim 116
wall 239
walls and partitions 239
window 185
window & door 185
wood 238
wood framing 112
Demountable partition ... 286, 547
Detection intrusion 436
  system 436
  tape 479
Detector explosive 436
  infrared 436
  metal 436
  motion 436
  smoke 437
  temperature rise 437
  ultrasonic motion 436
Detectors fire & heat 437
Detention cubicle 220
  equipment 300
Developing tank 302
Device anti-siphon 363
  decorator 417
  exit 222
  GFI 418
  load management 425
  panic 222
  receptacle 418
  residential 416
  wiring 424
Diamond lath 244
Diaper station 288
Dielectric union 382
Diffuser ceiling 388
  linear 388
  opposed blade damper 388
  perforated aluminum 388
  rectangular 388
  steel 389
  T-bar mount 389
Dimmer switch 416, 424
Direct expansion A/C 402
  expansion water chillers 398
  chute concrete 55
Directional sign 281
Directory 280
  board 280, 281
  building 433
Disappearing stairs 304
Dishwasher 302, 308
Dispenser napkin 289
  soap 289
  ticket 298
  toilet tissue 289
  towel 288
Dispensing equipment food .. 307
Disposable bag 33
Disposal asbestos 36
  charges 36
  field 481
  garbage 303
  or salvage 28
Distribution box 481
  pipe water 479
Divider curtain 312
  strip terrazzo 258

Dock board 299
  bumper 299
  loading 299
  shelter 299
  truck 299
Dog command 18
Dome drain 367
Domed skylights 220
Door accessories 230
  accordion 196
  acoustical 203, 287
  aluminum 186, 190
  and grilles coiling 201
  and panels access 199
  and window matls 185
  automatic entrance 207
  balanced 207
  bell residential 419
  bi-fold 190, 195
  bi-passing closet 195
  birch 196
  blind 141
  bulkhead 199
  bulkhead/cellar 199
  bumper 224
  cabinet 314, 316
  cafe 197
  canopy 294
  casing PVC 142
  catch 317
  chain hoist 203
  chime 424
  cleanout 291
  closer 530
  closers 222
  closet 190, 195, 551
  closing coordinator 225
  cold storage 202
  combination storm 195
  commercial 189, 194
  composite 189
  counter 201
  darkroom 202
  darkroom revolving 203
  demolition 185
  double 190
  dutch 195
  dutch oven 291
  entrance 186, 195, 206
  exterior 197
  fiberglass 197
  fire 193, 201, 437
  fire rated access 199
  flexible 205
  floor 199
  flush 192, 193, 195
  flush wood 191
  folding accordion 196
  frame 187, 188
  frame exterior wood 140
  frame grout 69
  frame interior 140
  french 198
  french exterior 197
  garage 203
  glass 201, 206, 207
  glass sliding 200
  hand carved 191
  handle 317
  hardboard 195
  hardware 221
  hardware accessory 225
  industrial 201
  interior louvered 549
  interior solid 549
  jamb molding 133

kick plate 227
knob 223
knocker 225
labeled 189, 193
metal 186, 550
metal access 199
metal fire 189
mirror 232
molding 133
moulded 194
opener 185, 202, 204, 221
overhead 203
overhead commercial 203
panel 194
paneled 190
partition 287
passage 196, 198
prefinished 192
pre-hung 197, 198
pull 225
refrigerator 202
release 432
removal 185
residential 140, 190, 194
residential garage 203
residential steel 189
revolving 203
rolling 201
roof 177
rough buck 121
sauna 330
sectional 203
sectional overhead 203
shower 286
side light 197
sidelight 190
silencer 225
sill 133, 140, 225
sliding glass 200, 201
sliding panel 208
special 199, 203
stain 271
stainless steel (access) .. 199
steel 188, 189, 724
stop 206, 224, 530
storm 186
swinging glass 207
switch burglar alarm 436
threshold 140
tin clad 194
varnish 271
vertical lift 204
weatherstrip 230
weatherstrip garage 229
wood 191, 194, 197, 548
wood fire 193
wood panel 194
Doorbell system 424
Doors & windows exterior paint . 269
  & windows interior paint . 272, 273
Dormer gable 126
Dormitory furniture 321
Double acting swing door .. 205
  brick 74
  hung window 210, 212, 214
  wall pipe 483
  weight hinge 226
Dowel 70
  cap 51
  reinforcing 50
  sleeve 51
Downspout 173, 174
  aluminum 173, 535
  copper 173, 535
  demolition 146
  elbow 174

  lead-coated copper .. 173, 535
  steel 173
  steel galvanized 535
  strainer 173
Dozer backfill 443
  excavation 443
Draft control 391
  damper vent 391
Drain 367
  deck 366
  dome 367
  floor 366
  main 367
  sanitary 366
Drainage accessory 174
  boot roof 172
  pipe 377, 480, 481
  site 482
  trap 365
Drains roof 367
Drawer kitchen 315
  track 317
  wood 316
Drawings as-built 18
Drill bit 63
  concrete 413
  earth 22, 457
  main 479
  rig 22
  rock 22
  shop 310
  tap main 479
  wood 117
Drilled concrete piers uncased .. 458
Drilling horizontal 479
  plaster 241
  steel 86
Drinking fountain 376
  fountain floor rough-in . 376
  fountain handicap 376
  fountain wall rough-in .. 376
Drip edge 176
  edge aluminum 176, 535
  edge steel galvanized ... 535
Driven pile 456
Driveway 461, 615
  removal 22
Drop pan ceiling 251
Drum trap copper 365
Dry cleaning equipment 300
  fall painting 276
  pipe sprinkler head 345
Drycleaner 300
Dryer commercial clothes .. 301
  darkroom 302
  hand 288
  industrial 300
  receptacle 418
  residential 303
Dry-type transformer 421
Drywall 247, 546
  accessories 249
  column 247
  cutout 26
  demolition 238, 239
  finish 727
  frame 187
  gypsum 247
  interior surface 520
  nail 116
  painting 275, 276
  partition 241
  partition nlb 240
  prefinished 248
  removal 239
Dual flush flushometer ... 373, 374

Index

flush valve toilet 374
Duck tarpaulin 16
Duct access door 383
   accessories 384
   flexible insulated 385
   flexible noninsulated 385
   furnace 396
   heater electric 402
   humidifier 404
   insulation 381
   liner 386
   liner non-fiberous 386
   mechanical 383
   mixing box 388
   silencer 385
   underground 484
   utility 484
Ductile iron fitting 479
   iron fitting mech joint 479
   iron pipe 479
Ductless split system 401
Ducts flexible 385
Ductwork 383, 730
   aluminum 383
   demolition 380
   fabric coated flexible 385
   fiberglass 384
   galvanized 383
   metal 383
   rectangular 383
   removal 380
   rigid 383
Dumbbell 311
Dumbwaiter 336
   electric 336
   manual 336
Dump charges 26
Dumpster 26
Dumpsters 709
Duplex receptacle 424
Dust collector shop 310
Dustproofing compound 42
Dutch door 195
   oven door 291
DWV pipe ABS 359
   pipe PVC 359
   PVC pipe 359
   tubing copper 353

E

Earth drill 22, 457
   vibrator 441, 443, 454
Earthwork equipment 443
Eave vent 176
Ecclesiastical equipment 312
Echo chamber 332
Economizer shwr-head water ... 371
Economy brick 74
Edge drip 176
Edging 475
   aluminum 475
   angle curb 90
   curb 90, 465
Efficiency flush valve high ... 374
Eggcrate ceiling 251
EIFS 154
Ejector pump 367
Elastomeric roof 168
   sheet waterproofing 148
Elbow aluminum 535
   copper 535
   downspout 174
   pipe 358
Electric backstop 311

ballast 428
baseboard heater 404
boiler 392
cabinet 423
cable 408
commercial water heater ... 369
duct heater 402
dumbwaiter 336
fixture 427-430
fixture exterior 429
furnace 394
heating 404
hinge 226
lamp 430
log 290
metallic tubing 411
panelboard 422
relay 425
service 425
stair 304
switch 425
utility 484
water heater residential 368
Elect. & telephone undrgrnd ... 484
   cable 409
   conduit 411
   demolition 406, 731
   field bend 411
   installation drilling 413
   knockout 414
   laboratory 309
   wire 409
Electronic air cleaner 391
   closer 222
   control system 381
Electrostatic painting 275
Elevated conduit 408
   floor 52
   installation add 348
   pipe add 348
   slab 52
   slab formwork 45
   slab reinforcing 50
Elevator 336, 728
   cab 338
   controls 339
   fee 8
   freight 336
   hydraulic freight 337
   hydraulic passenger 338
   options 339
   passenger 336
   residential 337
   shaft wall 240
Embossed print door 194
Emergency equip. lab. 309
   lighting 428, 429
   lighting unit 428
EMT 411
   conduit 602
Emulsion adhesive 257
   asphaltic 460
   pavement 460
Encapsulation asbestos 36
   pipe 36
Encasing steel beam formwork ... 43
Enclosure acoustical 332
   decontamination 33, 35, 38
   swimming pool 331
   telephone 282
Energy circulator air solar ... 396
   system air purger solar 397
   system air vent solar 397
   system balancing valve solar ... 397
   system control valve solar ... 397
   system controller solar 397

system expansion tank solar .. 397
system gauge pressure solar .. 397
system heat exchanger solar .. 397
system storage tank solar 397
system vacuum relief solar ... 397
Engineer brick 74
Engineered stone countertops ... 320
Engineering fee 8
   fees 701
Entrance aluminum 207
   and storefront 207
   canopy 295
   door 186, 195, 198, 206
   door sliding 207
   floor mat 320
   frame 140
   lock 225
   screen 282
Entry canopy 294
EPDM adhesive 148
   roof 167
Epoxy anchor 61, 87
   fiberglass wound pipe 377
   floor 259
   grout 249, 455
   terrazzo 259
   wall coating 277
Equipment 300
   automotive 298
   bank 299
   barber 300
   cleaning & disposal 308
   darkroom 302
   detention 300
   dry cleaning 300
   earthwork 443
   ecclesiastical 312
   fire 344
   food preparation 306
   foundation formwork 45
   gymnasium 311
   health club 311
   insulation 381
   insurance 10
   laboratory 317
   laboratory emergency 309
   laundry 300, 301
   loading dock 299
   mobilization 16
   move 19
   pad 53
   parking 298, 321, 332
   refrigerated storage 304
   rental 708
   security and vault 299
   shop 310
   stage 310
   theater and stage 310
   waste handling 312
Erosion control synthetic 454
Escape fire 106
Estimating 700
Evaporator coils cased 395
Evergreen shrub 473
   tree 473
Excavating bulk bank measure ... 442
   bulk dozer 443
   equipment 733
   trench 441
   utility trench 441
Excavation 441, 442, 732
   dozer 443
   footing 498
   foundation 499
   hand 441-443
   machine 443

planting pit 472
septic tank 481
structural 442
tractor 441
trench 441
utility trench 614
Exercise equipment 311
   ladder 311
   rope 311
   weight 311
Exhaust hood 303
   vent 389
Exit control alarm 436
   device 222
   light 429
   lighting 429
Expansion anchor 87
   joint 48, 175
   joint assembly 182
   joint floor 182
   joint roof 175
   shield 87
   tank 382
   tank steel 382
   water chillers direct 398
Expense office 10
Explosive detection equipment ... 436
   detector 436
Exposed aggregate 462
Exterior blind 141
   concrete block 77
   door 197
   door frame 140
   floodlamp 429
   insulation 154
   insulation finish system 154
   lighting fixture 429
   molding 132
   plaster 245
   pre-hung door 198
   PVC molding 142
   residential door 194
   shutter 141
   siding painting 267
   signage 281
   sprinkler 470
   surface masonry 520
   surface metal 520
   surface wood 520
   wall price sheet 520
   wood frame 140
Extinguisher chemical dry ... 292
   fire 292
   standard 292
Extra work 10
Extractor air 383
Eye wash fountain 376

F

Fabric flashing 171
   overlay 460
   stile 228
   waterproofing 148
   welded wire 51
Fabric-backed flashing 171
Face brick 74, 719
   wash fountain 375
Facing stone 80
   tile structural 75
Factor 8
Fall painting dry 276
Fan 386
   air conditioning 386
   air handling 386

Index

attic ventilation 387
bathroom exhaust 387
booster 386
ceiling 386
centrifugal 386
coil air conditioning 402
corrosion resistant 386
house ventilation 387
in-line 386
kitchen exhaust 387
paddle 420
residential 420
utility set 386
ventilation 420
ventilator 386
wiring 420
Farm type siding 161
Fascia board 123
   board demolition 113
   metal 173
   PVC 142
   wood 132
Fast food equipment 306
Fastener timber 116
   wood 117
Faucet & fitting 372, 374, 375
   bathroom 372
   gooseneck 375
   laundry 372
   lavatory 372
   medical 375
Fee architectural 8
   engineering 8
Felt 165
   asphalt 164, 165
   carpet pad 259
   tarred 166
   waterproofing 148
Fence and gate 467
   board batten 470
   cedar 469
   chain link 17
   chain link industrial 467
   chain link residential 468
   decorative 469
   fabric & accessories 468
   mesh 467
   metal 467
   misc. metal 469
   picket paint 266
   plywood 17
   recycled plastic 469
   security 469
   snow 31
   steel 468, 469
   temporary 17
   tennis 468
   treated pine 470
   tubular 469
   wire 17, 469
   wood rail 469
Fences 266
Fencing demolition 24
Fiber cement shingle 156
   cement siding 163
   reinforcing concrete 55
Fiberboard insulation 153
Fiberglass bench 322
   blown-in 152
   ceiling board 251
   cooling towers 399
   door 197, 203
   ductwork 384
   insulation 149, 150, 152, 381
   panel 17, 160, 263
   planter 322

reinforced plastic panel 262
steeple 296
storage tank 623
waterproofing 148
wool 151
Field bend electrical 411
   bend EMT 411
   disposal 481
   office 12
   personnel 9, 10
Fieldstone 79
Fill 443, 444
   by borrow & utility bedding .. 444
   floor 53
   gravel 440
Filled steel piles concrete 457
Filler block 275
   joint 180, 460
   strip 228
Fillet welding 87
Filling crack 460
Film equipment 302
   security 234
Filter air 390
   mechanical media 390
Filtration air 33
Fin tube radiation 403
Finish carpentry 315
   concrete 464
   floor 255, 277
   hardware 221
   lime 68
   nail 117
Finishes wall 261
Finishing floor concrete 57
   wall concrete 57
Fir column 139
   floor 255
   molding 133
   roof deck 128
   roof plank 128
Fire & heat detectors 437
   alarm 436, 437
   call pullbox 437
   clay 82
   damage repair 266
   damper curtain type 385
   door 193, 201, 437
   door frame 187
   door metal 189
   door wood 193
   equipment 344
   equipment cabinet 291
   escape 106
   escape balcony 106
   escape disappearing 106
   escape stairs 106
   extinguisher 292
   extinguisher portable .. 291
   glass 231
   horn 437
   hose 344
   hose gate valve 344
   hose nozzle 344
   hose rack 344
   hose storage cabinet ... 291
   hose valve 344
   hydrant for building ... 344
   protection 291
   protection kitchen 390
   resistant drywall .. 247, 248
   resistant wall 240
   signal bell 437
   sprinkler head 345
Firebrick 82
Firecycle system control pkg ... 345

Fire-department connection 344
Fireplace accessory 290
   box 82
   built-in 290
   chimney 290
   damper 291
   demolition 68
   form 290
   free standing 290
   mantel 138
   mantel beam 138
   masonry 82
   prefabricated 290
Fireproofing 178
   plaster 178
   sprayed cementitious ... 178
Firestop wood 121
Firestopping 178, 723
Fitting cast iron 364
   cast iron pipe 356, 364
   copper 353
   copper pipe 353
   corrosion resistant 377
   ductile iron 479
   DWV pipe 359
   glass 377
   grooved joint 358
   grooved joint pipe 358
   malleable iron 356
   malleable iron pipe 357
   plastic 359
   PVC 359
   steel 356, 357
   steel carbon pipe 357
   steel pipe 357
   vent chimney 391
Fixed end caisson piles 457
   tackboard 280
Fixture ballast 428
   bathroom 370, 371
   carrier 375
   electric 427
   fluorescent 427, 428
   incandescent 427
   incandescent vaportight .. 428
   interior light 427
   lantern 428
   metal halide 427
   mirror light 428
   plumbing 370, 383
   removal 348
   residential 419, 428
   support handicap 375
   vandalproof 428
Flagging 254
   slate 465
Flange tie 70
Flashing 170-172
   aluminum 170, 366, 535
   asphalt 171
   butyl 172
   cement 146
   copper 170, 171, 366, 535
   counter 175
   fabric 171
   fabric-backed 171
   laminated sheet 171
   lead 535
   lead coated 171
   lead-coated copper 535
   masonry 170
   mastic-backed 171
   membrane 165
   neoprene 172
   paper-backed 171
   plastic sheet 172

polyvinyl chloride 535
PVC 172
sheet metal 170
stainless 170
steel 536
valley 156
vent 366
vent chimney 392
Flat seam sheet metal roofing ... 169
Flexible conduit & fitting 416
   connector 362
   door 205
   ducts 385
   ductwork fabric coated .. 385
   insulated duct 385
   noninsulated duct 385
   sign 281
   sprinkler connector 345
Float finish concrete 57
   glass 230, 232
Floater equipment 10
Floating floor 253
   pin 225
Flood coating 166
Floodlamp exterior 429
Floodlight pole mounted 429
Floor 254
   access 260
   acid proof 254
   asphalt block 465
   brick 254, 464
   carpet 260
   ceramic tile 249
   cleaning 18
   cleanout 350
   closer 222
   composition 259
   concrete finishing 57
   conductive 257
   cork 257
   decking 94
   door 199
   drain 366
   elevated 52
   epoxy 259
   expansion joint 182
   fill 53
   fill lightweight 53
   finish 277
   finish price sheet 555
   framing removal 26
   hardener 42
   hatch 200
   insulation 150
   marble 80
   mat 320
   mat entrance 320
   nail 117
   oak 255
   paint 274
   paint & coating interior .. 273
   parquet 255
   pedestal 260
   plank 127
   plywood 128
   polyethylene 260
   quarry tile 250
   receptor 366
   removal 24
   rubber 256
   sealer 43
   stain 274
   subfloor 128
   terrazzo 258
   tile terrazzo 258
   transition strip 259

Index

underlayment 129
varnish 274
vinyl 257
wood 255
wood composition 255
Flooring 555
composition 259
conductive rubber & vinyl 257
demolition 238
miscellaneous brick 254
Flue chimney metal 391
liner 73
prefab metal 391
screen 290
Fluid heat transfer 397
Fluorescent fixture 427, 428
Fluoroscopy room 332
Flush automatic 374
door 192, 193, 195
flushometer dual 373, 374
het 374
tube framing 205
Flush, urinal het 375
Flush valve 374
wood door 191
Flying truss shoring 49
Foam board insulation 150
core DWV ABS pipe 359
glass insulation 149
insulation 152
roofing 168
Foamed coated roofing 168
Foam-water system components . 345
Fog seal 460
Foil aluminum 152, 155
metallic 152
Folder laundry 300
Folding accordian partition 287
accordion door 196
accordion partition 287
bench 322
gate 286
Food delivery cart 306
dispensing equipment 307
mixer 306
preparation equipment 306
service equipment 307
storage metal shelving 305
warmer 307
Football scoreboard 312
Footing concrete placement 55
formwork continuous 46
keyway 46
keyway formwork 46
reinforcing 50
removal 24
spread 53
Forklift crew 12
Form decking 94
fireplace 290
material 711
Forms 711
concrete 711
Formwork beam and girder 43
column 44
column plywood 44
column round fiber tube 44
column round fiberglass 44
column round steel 44
column steel framed 44
continuous footing 46
elevated slab 45
encasing steel beam 43
equipment foundation 45
footing keyway 46
gas station 46

gas station island 46
girder 43
grade beam 46
interior beam 43
labor hours 712
light base 46
pile cap 46
plywood 44
radial wall 47
retaining wall 47
sign base 46
slab blockout 46
slab box out opening 45
slab bulhead 45, 46
slab curb 45, 46
slab edge 45, 46
slab flat plate 45
slab joist dome 45
slab joist pan 45
slab on grade 46
slab screed 46
slab trench 46
slab void 45, 46
slab with drop panels 45
spandrel beam 43
spread footing 46
wall 47
wall boxout 47
wall brick shelf 47
wall bulkhead 47
wall buttress 47
wall corbel 47
wall liner 47
wall lintel 48
wall pilaster 48
wall plywood 47
wall prefab plywood 48
Fossil fuel boiler 393
Foundation caisson 457
chimney 52
concrete block 77
mat 53
mat concrete placement 56
underdrain 496
vent 72
wall 77
Foundations underpin 456
Fountain drinking 376
eye wash 376
face wash 375
outdoor 329
wash 374
Fountains yard 329
Frame baked enamel 188
door 187, 188
drywall 187
entrance 140
exterior wood door 140
fire door 187
labeled 187
metal 187
metal butt 188
scaffold 13
steel 187
welded 187
window 213
wood 140, 316
Frames shoring 49
Framing anchor 117
band joist 99
beam & girder 119
canopy 89, 124
columns 120
deck 122
demolition 113
heavy 127

joist 120
laminated 131
lightweight 90
lightweight angle 90
lightweight channel 90
load-bearing stud partition 96
load-bearing stud wall 96
metal 520
metal joist 100
miscellaneous wood 121
nlb partition 242
pipe support 91
porch 122
removal 112
roof metal rafter 103
roof metal truss 104
roof rafters 123
roof soffits 103
roof truss 130
sill 124
sleepers 124
slotted channel 90
timber 127
treated lumber 125
tube 205
wall 125
window wall 205
wood 118, 119, 126, 127, 521
Freestanding chalkboard 280
Freezer 304, 330
Freight elevator 336, 728
elevator hydraulic 337
French door 198
exterior door 197
Friction pile 457, 458
Frieze PVC 143
Front end loader 442
FRP panel 262
Fryer 307
Fuel product dispenser 382
Full vision door 186
vision glass 231
Furnace A/C coils 395
components 395
demolition 380
duct 396
electric 394
gas fired 394
hot air 394
oil fired 394
wall 396
Furnishings site 322
Furniture 320
dormitory 321
hospital 321
hotel 320
move 19
restaurant 320
Furring and lathing 241
ceiling 126, 241, 243
channel 241, 249
metal 241
price sheet 544
steel 241
wall 126, 242
Fusible link closer 222

G

Gable dormer 126
Galley septic 481
Galvanized ductwork 383
roof 159
steel conduit 602
steel reglet 175

Galvanizing 91
Garage door 203
door residential 203
door weatherstrip 229
Garbage disposal 303, 308
Garden house demo 326
Gas boiler hw 582
boiler steam 583
connector 362
conversion burner 394
fired boiler 393
fired furnace 394
fired space heater 396
generator 426
generator set 426
log 290
pipe 482
station formwork 46
station island formwork 46
vent 391
water heater 369
water heater commercial . . . 369
water heater residential 369
water heaters instantaneous . . 369
Gas-fired duct heater 396
Gasket glazing 233
neoprene 180
toilet 373
Gaskets joint 180
Gas/oil boiler hw 583
boiler steam 583
combination boiler 393
Gasoline generator 426
piping 483
pump 382
Gate fence 468
folding 286
parking 298
security 286
valve 380
General contractor's overhead . . 703
fill 444
Generator emergency 426
gas 426
gasoline 426
set 426
Geodesic domes demo 326
GFI receptacle 418
Girder formwork 43
reinforcing 50
wood 119, 128
Glass 231-233
bead 205
block 78
break alarm 436
bulletin board 280
bulletproof 234
demolition 185
door 186, 201, 206, 207
door astragal 228
door shower 286
door sliding 201
door swinging 207
fiber rod reinforcing 51
fire 231
fitting 377
float 230
full vision 231
heat reflective 231
insulating 231
laminated 234
lead 332
low emissivity 231
mirror 232, 289
mosaic 553
pipe 377

763

Index

pipe fitting 377
plate 232
reduce heat transfer 231
reflective 232
tempered 230
tile 232
tinted 230
window 232
window wall 206
wire 232
Glaze coat 275
Glazed brick 74
   concrete block 78
   wall coating 277
Glazing 230
   accessories 233
   gasket 233
   plastic 233
   polycarbonate 233
   productivity 725
Glove bag 34
Glued laminated 131
Gooseneck faucet 375
Gore line 467
Grab bar 288
Grade beam concrete placement . . 56
   beam formwork 46
Grading 450
Granite 80
   building 80
   curb 465
   demolition 68
   indian 466
   paving block 465
   sidewalk 465
Granolithic finish concrete 57
Grass cloth wallpaper 261
   lawn 472
Grate precast 462
   tree 462
Grating stair 106
Gravel bankrun 440
   base 462
   fill 440
   stop 172
Grease interceptor 366
   trap 366
Green roof soil mixture 159
   roof system 159
Greenhouse 331
   cooling 331
Greenhouses demo 326
Grid spike 118
Griddle 307
Grille air return 389
   aluminum 389
   decorative wood 138
   painting 274
   roll up 201
   window 215
Grinder shop 310
Grinding asphalt 461
Grooved block 76
   joint fitting 358
   joint pipe 358
Grooving pavement 461
Ground 126
   clamp water pipe 410
   cover 473
   cover edging 621
   fault indicating receptacle . . . 424
   rod 410
   water monitoring 11
   wire 410
Grounding 410
   & conductor 409

wire brazed 410
Group wash fountain 374
Grout 69
   cavity wall 69
   cement 69
   concrete 69
   concrete block 69
   door frame 69
   epoxy 249, 455
   topping 61
   wall 69
Guard corner 288
   gutter 175
   house 331
   service 18
   snow 177
   window 108
Guardrail scaffold 13
   temporary 17
Guards wall & corner 288
Guide rail 486
   rail removal 23
   rail timber 486
   rails vehicle 486
Guide/guard rail 486
Gun type burner 394
Gunite 58
Gutter 174
   aluminum 174, 536
   copper 174, 536
   demolition 146
   guard 175
   lead coated copper 174
   stainless 174
   steel 174, 536
   strainer 175
   vinyl 174
   wood 174
Gutting 26
   demolition 26
Gym mat 311
Gymnasium divider curtain 312
   equipment 311
   floor 260
Gypsum block demolition 25
   board 247
   board accessories 249
   board partition 241
   board partition nlb 240
   board removal 239
   drywall 247
   fabric wallcovering 261
   lath 243
   lath nail 117
   partition 286
   partition nlb 241
   plaster 245
   restoration 238
   shaft wall 240
   sheathing 130
   weatherproof 130

H

H pile 456
Half round molding 133
   round vinyl window 219
Half-round window 214
Hammer bush 57
   pile mobilization 440
Hand carved door 191
   dryer 288
   excavation 441-443
   hole 478
   split shake 157

trowel finish concrete 57
Handicap drinking fountain 376
   fixture support 375
   lever 225
   opener 221
   ramp 53
   tub shower 371
Handle door 317
Handling unit air 402
   waste 312
Handrail wood 133, 138
Handwasher-dryer module 373
Hangars demo 326
Hanger beam 117
   joist 117
Hanging lintel 90
Hardboard cabinet 315
   door 195
   overhead door 203
   paneling 134
   tempered 134
   underlayment 129
Hardener floor 42
Hardware 190, 221
   cabinet 317
   door 221
   finish 221
   panic 222
   price sheet 530, 531
   rough 118
   window 221
Hardwood carrel 321
   floor 255
   grille 138
Harness body 15
Hasp 225
Hat and coat strip 288
Hatch ceiling 199
   floor 200
   roof 177
   smoke 177
Hauling 444
   cycle 444
Hay bale 454
Hazardous waste cleanup 32
   waste disposal 32
   waste handling 312
Head sprinkler 345
Header wood 125
Headers boxed beam 100
   load-bearing stud wall 96
Health club equipment 311
Hearth 82
Heat baseboard 403, 607
   electric baseboard 404
   greenhouse 331
   pump 401
   pump residential 421
   pumps air-source 401
   pumps water-source 402
   recovery air to air 399
   reflective glass 231
   temporary 12, 43
   transfer fluid 397
Heated ready-mix concrete 43
Heater & fan bathroom 387
   air handler 395
   floor mounted space 396
   gas residential water 369
   gas-fired duct 396
   sauna 330
   unit 396
   water residential 369
Heating 351, 392, 396
   component solar 396
   electric 404

gas fired 577
hot air 394
hydronic 393, 394
oil fired 576
Heavy duty shoring 13
   framing 127
   timber 127
Hemlock column 139
Het flush 374
   flush, urinal 375
High efficiency flush valve 374
   installation conduit 408
   intensity discharge lamp 430
   pressure fixture sodium 427
   pressure sodium lighting 427
   rib lath 244
   rise glazing 231
   strength block 76
Highway sign 281
Hinge 225, 530
   brass 225
   cabinet 317
   continuous 227
   electric 226
   hospital 226
   paumelle 226
   prison 226
   residential 226
   school 226
   security 226
   special 226
   stainless steel 225
   steel 225
   wide throw 226
Hinges 725
Hip rafter 124
Hockey scoreboard 312
Hoist 488
   automotive 341
Holder closer 222
Holdown 117
Hole cutting electrical 414
   drilling electrical 413
Hollow core door 191, 196
   metal 187
   metal door 188
   metal stud partition 244
   precast concrete plank 59
   wall anchor 87
Hood and ventilation equipment . 390
   fire protection 390
   range 303
Hook coat 288
   robe 289
Horizontal boring 478
   drilling 479
Horn fire 437
Hose bibb sillcock 373
   fire 344
   rack 344
   rack cabinet 291
Hospital door hardware 221
   furniture 321
   hinge 226
   kitchen equipment 307
   tip pin 226
Hot air furnace 394
   air heating 394
   tub 330
   water boiler 393, 394
   water heating 392, 403
   weather concrete 55
Hotel cabinet 289
   furniture 320
   lockset 223
House demolition 24

Index

guard 331	isocyanurate 149	longspan 93	Lampholder 424
telephone 432	loose fill 151	metal framing 100	Landing metal pan 105
ventilation fan 387	masonry 151	open-web 93	stair 258
Housewrap 155	mineral fiber 151	removal 113	Landscape surface 260
Humidification equipment 331	pipe 350	sister 120	Landscaping 619
Humidifier 404	piping 351	web stiffeners 101	Lantern fixture 428
duct 404	polystyrene 149, 151	wood 120, 130	Latch deadlocking 223
room 404	reflective 152	Jumbo brick 74	set 223, 224
HVAC demolition 380	removal 34, 146	Jute mesh 454	Lateral arm awning retractable .. 294
removal 380	roof 153		Latex caulking 181
Hydrant for building fire 344	roof deck 153	**K**	underlayment 256
for building wall 344	shingle 156		Lath 546
removal 23	spray 152	Kalamein door 194	demolition 238
water 364	vapor barrier 155	Kennel fence 469	gypsum 243
Hydrated lime 68	vermiculite 151	Kettle 307	metal 244
Hydraulic elevator 564	wall 76, 149	Key cabinet 317	Lath, plaster and gypsum board . 726
lift 341	wall blanket 150	Keyless lock 224	Lath rib 244
Hydrodemolition 22	water heater 350	Keyway footing 46	Lathe shop 310
concrete 22	Insurance 10	Kick plate 227, 530	Lattice molding 133
Hydronic heating 393, 394	builder risk 10	plate door 227	Lauan door 196
Hypalon neoprene roofing 168	equipment 10	Kiln vocational 310	Laundry equipment 300, 301
	public liability 10	King brick 74	faucet 372
I	Integral colors concrete 42	Kiosk 332	folder 300
	topping concrete 57	Kitchen cabinet 315	ironer 301
Ice machine 308	waterproofing concrete 43	equipment 307	sink 371
Icemaker 308	Interceptor 366	exhaust fan 387	Lavatory commercial 374
Impact drilling concrete 63	grease 366	selective price sheet 611, 612	faucet 372
Incandescent bulb 430	Interior beam formwork 43	sink 371	pedestal type 370
exterior lamp 430	door frame 140	sink faucet 372	removal 348
fixture 427, 428	light fixture 427	sinks residential 370	residential 370, 374
interior lamp 430	paint doors & windows 271	system 610	support 375
Indian granite 466	paints walls & ceilings 276	unit 301	vanity top 370
Indicating panel burglar alarm ... 436	pre-hung door 198	K-lath 244	wall hung 370
Indoor athletic carpet 260	residential door 195	Knee action mixing valve 375	Lawn bed preparation 471
Industrial address system 432	shutter 314	Knob door 223	grass 472
door 200, 201	wall finish price sheet 553	Knocker door 225	Lazy susan 315
dryer 300	wood door frame 140	Knockout electrical 414	Leaching field chamber 481
lighting 427	Interlocking concrete block 77	Kraft paper 155	pit 481
railing 107	Intermediate metallic conduit ... 411		Lead coated copper downspout . 173
safety fixture 375	Interval timer 416		coated copper gutter 174
window 209	Intrusion detection 436	**L**	coated downspout 173
Inert gas 31	system 436		coated flashing 171
Infrared broiler 307	Inverted bucket steam trap 382	Labeled door 189, 193	flashing 170
detector 436	Iron alloy mechanical joint pipe . 377	frame 187	glass 332
Inlet curb 465	body valve 380	Labor adjustment factors 348	paint encapsulation 36
In-line fan 386	grate 462	formwork 712	paint remediation methods .. 710
Insecticide 455	Ironer laundry 301	modifiers 348	paint removal 37
Inspection technician 11	Ironspot brick 254	Laboratory cabinet 317	plastic 332
Installation add elevated 348	Irrigation system 470	countertop 318	sheets 332
Instantaneous gas water heaters .. 369	system sprinkler 470	equipment 317	Lean-to type greenhouse 331
water heater 368	Isocyanurate insulation 149	safety equipment 309	Let-in bracing 118
Insulated panels 154		sink 309	Letter slot 293
protectors ada 351	**J**	table 309	Leveling jack shoring 13
Insulating glass 231		Ladder alternating tread 106	Lever handicap 225
Insulation 149, 152	Jack pump 13	exercise 311	Lexan 233
batt 150	screw 455	reinforcing 71	Liability insurance 703
batts 521	Jacking 478	ship 106	Library equipment 321
blanket 381	Job condition 9	towel 289	shelf 321
blown-in 152	Joint assembly expansion 182	vertical metal 106	Lift 340, 341
board 149	control 57, 71	Lag screw 88	automotive 341
building 152	expansion 48, 175	Laminated beam 131	correspondence 341
cavity wall 151	filler 180, 460	countertop 318	hydraulic 341
ceiling 150	gaskets 180	countertop plastic 318	slab 54, 718
ceiling batts 532	push-on 479	framing 131	wheelchair 340
cellulose 151	reinforcing 71	glass 234	Lifter platform 299
composite 154	roof 175	roof deck 128	Light base formwork 46
duct 381	sealer 180, 181	sheet flashing 171	exit 429
equipment 381	tyton 479	veneer members 131	fixture interior 427
exterior 154	Jointer shop 310	wood 131	post 428
fiberglass 149, 150, 152, 381	Joist concrete 53	Lamp electric 430	stand 33
finish system exterior 154	connector 117	high intensity discharge ... 430	support 90
floor 150	deep-longspan 92	incandescent exterior 430	temporary 12
foam 152	demolition 113	incandescent interior 430	Lighting 427, 428, 430
foam glass 149	framing 120	metal halide 430	darkroom 302
insert concrete block 76	hanger 117	post 108	emergency 428, 429
			exit 429

765

Index

fixture exterior 429
fluorescent 604
high intensity 606
high pressure sodium 427
incandescent 605
industrial 427
metal halide 427
outlet 419
residential 419
strip 427
track 429
unit emergency 428
Lightning protection demo 326
  suppressor 416
Lightweight angle framing 90
  block 77
  channel framing 90
  concrete 53, 719
  floor fill 53
  framing 90
  insulating concrete 61
Lime finish 68
  hydrated 68
Limestone 80, 472
  coping 82
Line gore 467
Linear diffuser 388
Linen chutes 341
Liner cement 478
  duct 386
  non-fiberous duct 386
  pipe 478
Lint collector 300
Lintel 80
  block 77
  concrete block 77
  hanging 90
  precast concrete 60
  steel 91
Liquid chillers screw 398
Load center residential 416
  management switch 425
Load-brng stud part. framing .. 96
  stud wall bracing 95
  stud wall bridging 95
  stud wall framing 96
  stud wall headers 96
Loader front end 442
  vacuum 33
Loading dock 299
  dock equipment 299
Loam 440, 471
Lobby collection box 293
Lock electric release 224
  entrance 225
  keyless 224
  tile 320
  tubular 223
Locker metal 292
  steel 292
  wire mesh 292
Locking receptacle 419
Lockset 530
  communicating 223
  cylinder 223
  hotel 223
  mortise 223
Log electric 290
  gas 290
Longspan joist 93
Loose fill insulation 151
Louver 234
  aluminum 234
  midget 235
  redwood 235
  ventilation 235

wall 235
wood 138
Louvered door 196, 198
Lubrication equipment 298
Lumber core paneling 135
  product prices 722
  recycled plastic 141
Luminous ceiling ... 251, 252, 430
  panel 251

M

Machine excavation 443
  trowel finish concrete 57
  X-ray 436
Machinery anchor 88
Magnetic astragal 228
  particle test 11
Mahogany door 191
Mail box 293
  box call system 432
  slot 293
Main drain 367
  office expense 10
Maintenance 18
Make-up air unit 400
Mall front 208
Malleable iron fitting 356
  iron pipe fitting 356
Management fee construction 8
Manhole 482
  brick 482
  removal 23
Mansard aluminum 161
Mantel beam 138
  fireplace 138
Manual dumbwaiter 336
Maple countertop 318
Marble 80
  coping 82
  countertop 318
  floor 80
  screen 285
  sill 81
  synthetic 254
  tile 254
Marker boundary and survey 22
Mark-up cost 10
Masking 67
Mason scaffold 13
Masonry accessory 67, 71
  anchor 69
  base 78
  brick 73, 254
  caulking 181
  cement 68
  cleaning 66, 518
  cornice 79
  cutting 27
  demolition 25, 67, 239
  exterior surface 520
  fireplace 82, 579
  flashing 170
  furring 126
  insulation 151
  nail 117
  pointing 66
  reinforcing 71
  removal 23, 67
  restoration 518
  sill 81
  step 462
  toothing 66
  ventilator 72
  wall 76, 470, 515, 516

wall tie 69
Mastic-backed flashing 171
Mat conc. placement found. 56
  floor 320
  foundation 53
  gym 311
  wall 311
Material handling deconstruction . 30
Materials concrete 717
Meat case 304
Mechanical 729
  duct 383
  equipment demolition 380
  fee 8
  media filter 390
  seeding 472
Medical faucet 375
Medicine cabinet 289
Membrane flashing 165
  roofing 164
  waterproofing 148, 460
Mercury vapor lamp 430
Mesh fence 467
  stucco 244
Metal bookshelf 321
  bracing 118
  butt frame 188
  case good 320
  casework 317
  ceiling panels 253
  chimney 290
  deck acoustical 94
  deck composite 94
  demolition 86
  detector 436
  detectors 436
  door 186, 526
  door residential 190
  ductwork 383
  exterior surface 520
  fascia 173
  fence 467
  flue chimney 391
  frame 187
  frame channel 188
  framing parapet 102
  furring 241
  halide fixture 427
  halide lamp 430
  halide lighting 427
  hollow 187
  joist bracing 98
  joist bridging 99
  joist framing 100
  lath 244
  locker 292
  nailing anchor 88
  overhead door 203
  pan ceiling 252
  pan landing 105
  pan stair 105
  parking bumpers 466
  pipe removal 349
  plate stair 105
  rafter framing 103
  roof 159
  roof parapet framing 102
  roof window 220
  sash 210
  screen 210
  sheet 169
  shelf 293
  shelving food storage 305
  shingle 156
  siding 159, 161
  soffit 164

stud demolition 239
stud nlb 242
support assemblies 241
threshold 225
tile 250, 251, 553
toilet partition 282
truss framing 104
window 209, 210
Metallic conduit intermediate 411
  foil 152
Meter socket 423
  water supply 362
  water supply domestic 362
Microtunneling 478
Microwave detector 436
  oven 302
Mill construction 127
Milling pavement 461
Millwork 132, 316
  demolition 116
Mineral fiber ceiling 251, 252
  fiber insulation 151
  insulated cable 408
  roof 168
  wool blown-in 152
Minor site demolition 23
Mirror 232, 288
Mirror ceiling tile 251
Mirror door 232
  glass 232
  light fixture 428
  plexiglass 233
  wall 232, 233
Miscellaneous painting .. 266, 271
Mix planting pit 472
Mixer food 306
Mixing box duct 388
  boxes 388
  boxes VAV 388
  valve 362, 372
Mobilization 440, 458
  air-compressor 440
  or demob. 16
Modification to cost 9
Modified bitumen roofing 723
  bituminous membrane sbs ... 166
Modifiers labor 348
Modular air handler 395
  office 547
  office system 286
Modulating damper motorized .. 384
Module handwasher-dryer 373
  tub-shower 371
Moisture content test 11
Mold abatement 38
  abatement work area 38
  contaminated area demolition .. 38
Molding base 132
  brick 132
  casing 132
  ceiling 132
  chair 133
  cornice 132
  exterior 132
  hardboard 134
  pine 132
  soffit 134
  trim 133
  window and door 133
  wood transition 255
Money safe 299
Monitor support 89
Mop holder strip 289
  roof 166
Mortar 68
Mortar, brick and block 719

Index

Mortar cement 250
    Portland cement 69
    thinset 250
Mortise lockset 223
Mosaic tile 553
Motel swimming pool 329
Motion detector 436
Motor connections 416
    support 89
Motorized modulating damper . . 384
    zone valve 381
Moulded door 194
Mounting board plywood 129
Movable office partition 286
Move equipment 19
    furniture 19
Movie screen 309
Moving building 31
    shrub 475
    structure 31
    tree 475
Mullion vertical 206
Multi-blade damper 384
Multizone air cond. rooftop 400
Municipal swimming pool 329
Muntin window 215
Mylar tarpaulin 17

N

Nail 116
    common 116
Nailer steel 121
    wood 121
Nailing anchor 88
Napkin dispenser 289
Needle beams cantilever 456
Neoprene adhesive 148
    expansion joint 175
    flashing 172
    gasket 180
    roof 168
    waterproofing 148
Newel wood stair 137
No hub pipe 364
    hub pipe fitting 365
Non-chloride accelerator conc. . . . 55
Non-destructive testing 11
Non-metallic sheathed cable 602
Non-removable pin 225
Norwegian brick 74
Nosing rubber 256
    safety 256
    stair 256
Nozzle fire hose 344
Nuts remove 86
Nylon carpet 260
    nailing anchor 88

O

Oak door frame 140
    floor 255
    molding 133
    paneling 135
    threshold 140
Office & storage space 12
    expense 10
    field 12
    floor 260
    overhead 10
    partition 286, 547
    partition movable 286
    safe 299

system modular 287
    trailer 12
Oil burner 394
    fired boiler 393
    fired furnace 394
    water heater 369
    water heater residential 369
Oil-fired boiler 394
Olive knuckle hinge 226
Omitted work 10
One piece astragal 227
One-way vent 177
Opener automatic 221
    commercial automatic 221
    door 185, 202, 204, 221
    handicap 221
Opening roof frame 90
Open-web joist 93
Options elevator 339
Ornamental aluminum rail 109
    glass rail 109
    railing 109
    wrought iron rail 109
OSHA testing 35
Outdoor cabinet 317
    fountain 329
    swimming pool 329
Outlet box plastic 412
    box steel 412
    lighting 419
Oven 307
    cabinet 316
    convection 307
    microwave 302
Overhaul 27
Overhead & profit 10
    bridge cranes 488
    commercial door 203
    contractor 18, 703
    door 203, 529
    office 10
Overlapping astragal 228
Overlay fabric 460
    face door 192, 193
    pavement 460
Oversized brick 73
Overtime 10
Oxygen lance cutting 28

P

P trap 365
    trap running 365
Packaging waste 36
Pad condenser 404
    equipment 53
    prefabricated 404
Padding carpet 259
Paddle fan 420
Paint & coating interior floor 273
    aluminum siding 268
    chain link fence 266
    doors & windows exterior 269
    doors & windows interior . 271-273
    encapsulation lead 36
    fence picket 266
    floor 274
    floor concrete 273, 274
    floor wood 274
    miscellaneous interior 274
    removal 37
    siding 268
    trim exterior 270
    walls masonry exterior 270
Painted markings airport 467

pavement markings 467
Painting 553
    balustrade 274
    booth 298
    casework 271
    ceiling 275, 276
    clapboard 267
    cornice 274
    decking 267
    drywall 275, 276
    electrostatic 275
    exterior miscellaneous 266
    exterior siding 267
    grille 274
    miscellaneous 266, 271
    parking stall 467
    pavement 467
    pipe 274
    plaster 275, 276
    railing 266
    reflective 467
    shutters 267
    siding 267
    stair stringers 267
    steel siding 267
    stucco 267
    temporary road 467
    thermoplastic 467
    trellis/lattice 267
    trim 274
    truss 274
    wall 268, 275, 276
    window 271
Paints & coatings 263
    walls & ceilings interior . . 275, 276
    cabinet & casework 271
Pan shower 171
    slab 53
Panel acoustical 263, 287
    door 194, 195
    fiberglass 17, 160
    fire 437
    FRP 262
    luminous 251
    portable 287
    structural 127
    system 135
    vision 220
    wall 82
Panelboard 422
    electric 422
    w/circuit-breaker 421, 422
Paneled door 190
    pine door 196
Paneling 134, 553
    birch 134
    board 136
    cutout 26
    demolition 116
    hardboard 134
    plywood 134
    wood 134
Panelized shingle 157
Panels insulated 154
    prefabricated 154
Panic bar 222
    device 222, 530
    device door hardware 221
Paper building 155
    sheathing 155
Paper-backed flashing 171
Paperhanging 261
Paperholder 289
Parallel bar 311
Parapet metal framing 102
    wall 517

Park bench 322
Parking barrier 467
    barrier precast 466
    bumper 467
    bumper plastic 466
    equipment 298, 321, 332
    gate 298
    lot asphalt 616
    lot concrete 617
    lots paving 463
    markings pavement 467
    stall painting 467
Parquet floor 255
Particle board underlayment . . . 129
    core door 192
Parting bead 133
Partition 286, 287
    acoustical 288
    anchor 70
    block 77
    blueboard 241
    bulletproof 299
    concrete block 77, 538
    demolition 239
    demountable 286
    door 287
    drywall 241, 541
    folding accordian 287
    folding leaf 288
    framing load bearing stud 96
    framing nlb 242
    gypsum 286
    metal stud 540, 542
    movable office 286, 547
    nlb drywall 240
    nlb gypsum 241
    office 286
    plaster 543
    shower 285
    steel 287
    stud 244
    support 89
    thin plaster 241
    toilet 282-285
    wall nlb 240
    wood frame 121
    wood stud 539
Passage door 196, 198
Passenger elevator 336
    elevator hydraulic 338
    elevators 728
Patch core hole 11
    roof 146
Patching concrete floor 42
    concrete wall 42
Patio 462
    door 201
Patio/deck canopy 294
Paumelle hinge 226
Pavement asphaltic 463
    berm 465
    demolition 22
    emulsion 460
    grooving 461
    markings 466
    markings painted 467
    milling 461
    overlay 460
    painting 467
    parking markings 467
    profiling 461
    pulverization 461
    reclamation 461
    recycling 461
    replacement 463
Paver floor 254

Index

Paving block granite 465
  brick 464
  concrete 464
  parking lots 463
Peastone 473
Pedestal floor 260
  type lavatory 370
Peephole 224
Pegboard 134
Perforated ceiling 251
Performance bond 707
Perlite insulation 149
  plaster 245
  sprayed 277
Permit building 11
Personal respirator 33
Personnel field 9, 10
Pew church 322
  sanctuary 322
Phone booth 282
PIB roof 167
Picket railing 107
Picture and sliding window . 213
  window 212
  window aluminum sash 209
  window steel sash 210
  window vinyl 219
Pier brick 73
Pilaster toilet partition . 285
  wood column 139
Pile bored 457
  cap 54
  cap concrete placement .. 56
  cap formwork 46
  cutoff 440
  driven 456
  driving 708
  friction 457
  H 456
  load test 440
  mobilization hammer 440
  prestressed 458
  sod 472
  steel 456
  steel sheet 455
  step tapered 456
  testing 440
  treated 457
  wood sheet 455
Piles 456
  timber 456
Piling sheet 455
  special costs 440
Pin powder 89
Pine door 195
  door frame 140
  fireplace mantel 138
  floor 255
  molding 132
  roof deck 128
  shelving 293
  siding 162
  stair tread 136
Pipe & fittings ... 349, 352, 353, 355,
              357-359, 362, 367, 372
  add elevated 348
  and fittings 729
  bedding 444
  bedding trench 442
  bollards 466
  brass 352
  cast iron 364
  cleanout 350
  copper 352
  corrosion resistant 377
  corrugated metal 481

covering 351
covering fiberglass 351
CPVC 359
double wall 483
drainage 377, 480, 481
ductile iron 479
DWV PVC 359
elbow 358
encapsulation 36
epoxy fiberglass wound 377
fitting cast iron 356, 364
fitting copper 353
fitting corrosion resistant 377
fitting DWV 359
fitting glass 377
fitting grooved joint 358
fitting malleable iron 357
fitting no hub 364
fitting plastic 359
fitting soil 364
fitting steel 356, 357
fitting steel carbon 357
fitting weld steel 357
fittings brass 352
foam core DWV ABS 359
gas 482
glass 377
grooved joint 358
insulation 350
insulation removal 34
iron alloy mechanical joint 377
liner 478
no hub 364
painting 274
plastic 358, 359
polyethylene 482
PVC 358, 480
rail aluminum 107
rail galvanized 107
rail stainless 107
rail steel 107
rail wall 107
railing 107
reinforced concrete 481
removal 23
removal metal 349
sewage 480, 481
shock absorber 363
single hub 364
soil 364
steel 355
support framing 91
tee 358
weld joint 355
Piping designations 735
Piping gas service polyethylene . 482
  gasoline 483
  insulation 351
  storm drainage 481
Pit excavation 442
  leaching 481
Pitch coal tar 165
  emulsion tar 460
  pocket 177
Pivoted window 209
Placing concrete 55, 717
Plain tube framing 205
Planer shop 310
Planing cold 461
Plank floor 127
  hollow precast concrete . 59
  precast concrete roof ... 59
  precast concrete slab ... 59
  roof 128
  scaffolding 14
  sheathing 130

Plant bed preparation 472
Planter 322
  concrete 322
  fiberglass 322
Planting 475
  price sheet 620
Plant-mix asphalt paving .. 463
Plantsplanting transplanting 475
Plaque 281
Plaster 240, 546
  and drywall price sheet . 546
  beam 245
  ceiling 245
  column 245
  cutout 26
  demolition 238, 239
  drilling 241
  ground 126
  gypsum 245
  painting 275, 276
  partition thin 241
  perlite 245
  thincoat 246
  wall 245
Plasterboard 247
Plastic ball valve 350
  faced hardboard 134
  fence recycled 469
  fitting 359
  glazing 233
  laminate door 192
  laminated countertop 318
  lead 332
  outlet box 412
  pipe 358, 359
  pipe fitting 359, 361
  roof ventilator 386
  sheet flashing 172
  skylight 220
  toilet partition 284
  valve 350
  window 217, 219
Plastic-lam. toilet compart. 283
Plate compactor 453
  glass 232
  heat exchanger 588
  push-pull 225
  shear 118
  steel 91
  stiffener 91
  vibrating 453
  wall switch 424
Platform lifter 299
Plating zinc 117
Players bench 323
Plenum demolition 239
Plexiglass 233
  acrylic 233
  mirror 233
Plug in tandem circuit-breaker .. 422
Plugmold raceway 423
Plumbing 362
  appliance 369
  demolition 348
  fixture 370, 383, 730
  fixtures removal 348
  laboratory 309
  system 565, 567, 569-574
Plywood 722
  clip 118
  demolition 238, 239
  fence 17
  floor 128
  formwork 44
  joist 130
  mounting board 129

paneling 134
sheathing roof & walls 129
shelving 293
sidewalk 17
siding 162
sign 281
soffit 134
subfloor 128
underlayment 129
Pneumatic tube system 342
Pocket door frame 140
pitch 177
Point of use water heater . 368
Pointing CMU 66
  masonry 66
Poisoning soil 455
Pole closet 133
Police connect panel 436
Polycarbonate glazing 233
Polyethylene backer rod ... 48
  floor 260
  pipe 482
  septic tank 481
  tarpaulin 16
  waterproofing 148
Polyolefin roofing thermoplastic . 167
Polystyrene blind 141
  ceiling 251
  ceiling panel 251
  insulation 149-151
Polysulfide caulking 181
Polyurethane caulking 181
  varnish 277
Polyvinyl chloride roof ... 167
  soffit 134, 164
  tarpaulin 16
Porcelain tile 249
Porch framing 122
  molding 133
Portable fire extinguisher 291
  panel 287
  stage 310
Portland cement terrazzo .. 258
Post cap 118
  cedar 139
  demolition 114
  fence 468
  lamp 108
  light 428
  shores 49
  wood 120
Postal specialty 293
Posts sign 281
Potters wheel 310
Poured insulation 151
Powder actuated tool 89
  charge 89
  pin 89
Power temporary 12
  wiring 425
Precast beam 59
  catch basin 482
  column 59
  concrete beam 59
  concrete column 59
  concrete lintel 60
  concrete roof plank 59
  concrete slab plank 59
  concrete stairs 59
  concrete tee beam 59
  concrete wall 60
  concrete window sill 60
  coping 82
  curb 465
  grate 462
  inverted tee beam 59

Index

Term	Page
members	718
parking barrier	466
tee beam	59
terrazzo	258
Pre-engnrd steel bldngs demo	327
Prefab metal flue	391
Prefabricated building	331
fireplace	290
pad	404
panels	154
stair	136
Prefinished door	192
drywall	248
floor	255
shelving	293
Preformed roof panel	159
roofing & siding	159
Pre-hung door	197, 198
Preparation exterior surface	264
interior surface	265
lawn bed	471
plant bed	472
Pressure grouting cement	455
valve relief	349
wash	265
Prestressed concrete piles	456
pile	458
precast concrete	718
Preventer backflow	362
Prices lumber products	722
Primer asphalt	146, 256
Prison cot	321
fence	469
hinge	226
toilet	300
Prisons work in	9
Process air handling fan	386
Proctor compaction test	11
Produce case	304
Product dispenser fuel	382
piping	483
Productivity glazing	725
Profile block	76
Profiling pavement	461
Project sign	18
Projected window	209, 210
Projection screen	309
Propeller unit heaters	403
Protection fire	291
slope	454
stile	228
termite	455
winter	17, 43
worker	33
Protectors ada insulated	351
P&T relief valve	349
Public address system	432
Pull box	412
box electrical	412
door	317
Pulverization pavement	461
Pump	480
circulating	383
end suction	584
gasoline	382
heat	401
in-line centrifugal	383
jack	13
sewage ejector	367
staging	13, 708
submersible	367
sump	303
water	480
Pumped concrete	55
Purlin roof	128
Push button lock	223
Push-on joint	479
Push-pull	531
plate	224
Putlog scaffold	14
Puttying	263
PVC adhesive	148
conduit in slab	415
control joint	72
cornerboards	142
door casing	142
DWV pipe	359
fascia	142
fitting	359
flashing	172
frieze	143
gravel stop	172
molding exterior	142
pipe	358, 480
rake	143
sheet	148, 260
siding	162
soffit	143
waterstop	50

Q

Term	Page
Quarry tile	250
Quarter round molding	133
Quoins	80

R

Term	Page
Raceway	411, 413-415
plugmold	423
surface	423
wiremold	423
Rack coat	316
hose	344
Radial arch	131
wall formwork	47
Radiation fin tube	403
Radiator cast iron	403
Radiography test	11
Rafter	123
anchor	118
composite	123
demolition	114
framing metal	103
metal bracing	102
metal bridging	102
wood	123
Rail aluminum pipe	107
crane	488
dock shelter	299
galvanized pipe	107
guide	486
guide/guard	486
ornamental aluminum	109
ornamental glass	109
ornamental wrought iron	109
stainless pipe	107
steel pipe	107
wall pipe	107
Railing cable	109
demolition	116
industrial	107
ornamental	109
picket	107
pipe	107
wood	133, 136, 138
wood stair	137
Railroad tie	475
tie step	462
track removal	23
Raised floor	260
Rake PVC	143
topsoil	471
Rammer tamper	454
Ramp handicap	53
Ranch plank floor	255
Range cooking	302
hood	303
receptacle	418, 424
restaurant	307
Ready mix concrete	701, 717
Ready-mix concrete	55
concrete heated	43
Receptacle air conditioner	418
device	418
dryer	418
duplex	424
GFI	418
ground fault indicating	424
locking	419
range	418, 424
telephone	419
television	419
trash	322
waste	289
weatherproof	418
Receptor floor	366
Reciprocating compressor	368
hermetic compressor	368
water chiller	398
Reclamation pavement	461
Recore cylinder	223, 224
Rectangular diffuser	388
ductwork	383
Recycled plastic lumber	141
Recycling cold	461
pavement	461
Reduce heat transfer glass	231
Redwood cupola	295
louver	235
paneling	136
siding	162
trim	133
tub	330
wine cellar	305
Refinish floor	255
Reflective block	79
glass	232
insulation	152
painting	467
sign	281
Refrigerant compressor	398
removal	36
Refrigerated case	304
storage equipment	304
wine cellar	305
Refrigeration	330
commercial	304
residential	302
Refrigerator door	202
Refuse chute	341
Register air supply	389
baseboard	389
cash	300
Reglet	175
aluminum	175, 536
copper	536
galvanized steel	175
steel	536
Reinforced concrete pipe	481
culvert	481
plastic panel fiberglass	262
PVC roof	167
Reinforcement welded wire	713
Reinforcing	713
beam	50
column	50
dowel	50
elevated slab	50
footing	50
girder	50
glass fiber rod	51
joint	71
ladder	71
masonry	71
sorting	50
steel	713
steel fiber	52
synthetic fiber	52
truss	71
wall	50
Relay electric	425
Release door	432
Relief valve P&T	349
valve self-closing	349
valve temperature	349
Relining sewer	478
Removal air conditioner	380
asbestos	34
bathtub	348
block wall	24
boiler	380
catch basin	23
concrete	23
concrete pipe	23
curb	23
driveway	22
fixture	348
floor	24
guide rail	23
hydrant	23
insulation	34
lavatory	348
masonry	23
paint	37
pipe	23
pipe insulation	34
plumbing fixtures	348
railroad track	23
refrigerant	36
shingle	146
sidewalk	23
sink	348
sod	472
steel pipe	23
stone	23
tank	31
tree	475
urinal	348
VAT	34
water closet	348
water fountain	349
water heater	349
window	185
Removal/salvage material	28
Remove bolts	86
nuts	86
topsoil	471
Rem./replace catch basin cvr	482
Rendering	8
Repair fire damage	266
slate	156
Repellent water	275
Replacement glass	232
pavement	463
sash	211
windows	184, 724
Resaturant roof	146
Residential alarm	420
appliance	302, 420
applications	416
bathtubs	371

Index

burner 394
canopy 294
closet door 190
device 416
door 140, 190, 194, 528
door bell 419
dryer 303
elevator 337
fan 420
fixture 419, 428
folding partition 287
gas water heater 369
greenhouse 331
gutting 26
heat pump 421
hinge 226
kitchen sinks 370
lavatory 370, 374
lighting 419
load center 416
lock 223
oil water heater 369
overhead door 203
refrigeration 302
roof jack 387
service 416
sinks 370
smoke detector 420
stair 136
storm door 186
swimming pool 329
switch 416
transition 387
wall cap 387
wash bowl 370
washer 303
water heater 420
water heater electric 369
wiring 416, 420
Resilient base 256
  floor 256, 257
Respirator 33
  personal 33
Resquared shingle 157
Restaurant furniture 320
  range 307
Restoration gypsum 238
  window 37
Retaining wall 54, 470
  wall formwork 47
  walls stone 470
Retarder concrete 55
  vapor 155
Retractable lateral arm awning .. 294
Revolving door 203
  door darkroom 203
Rib lath 244
Ribbed waterstop 50
Ridge board 124
  cap 156
  vent 176
Rig drill 22
Rigid anchor 70
  conduit in trench 415
  in slab conduit 415
  insulation 150
  joint sealant 182
Ring boxing 311
  split 118
  toothed 118
Riser rubber 256
  stair 258
  terrazzo 258
  wood stair 137
Road base 462
  berm 465

sign 281
temporary 16
Robe hook 289
Rock drill 22
Rocker switch 424
Rod backer 180
  closet 293
  curtain 288
  ground 410
  tie 90
Roll roof 164
  roofing 168
  type air filter 390
  up grille 201
Roller vibrating 453
Rolling door 201, 208
  service door 202
  topsoil 471
  tower scaffold 14
Roman brick 74
Romex copper 409
Roof accessories 177
  accessory price sheet 535, 536
  adhesive 164
  aluminum 159
  baffle 177
  beam 131
  built-up 164, 165
  cant 124, 165
  clay tile 158
  coating 146
  copper 169
  CSPE 167
  deck 128
  deck insulation 153
  deck laminated 128
  deck wood 128
  decking 94
  drainage boot 172
  drains 367
  elastomeric 168
  EPDM 167
  expansion joint 175
  fiberglass 160
  fill 719
  flashing vent chimney 392
  frame opening 90
  framing removal 26
  hatch 177, 534
  hatch removal 147
  insulation 153
  jack residential 387
  joint 175
  metal 159
  metal rafter framing 103
  metal truss framing 104
  mineral 168
  modified bitumen 166
  mop 166
  nail 117
  panel preformed 159
  patch 146
  PIB 167
  polyvinyl chloride 167
  purlin 128
  rafter 123
  rafter bracing 102
  rafter bridging 102
  rafters framing 123
  reinforced PVC 167
  resaturant 146
  roll 164
  safety anchor 15
  sheathing 129
  sheet metal 169
  shingle 156

single-ply 167
skylight 220
slate 156
soffits framing 103
soil mixture green 159
specialties prefab 173
stainless steel 169
steel 159
system green 159
thermoplastic polyolefin ... 167
tile 158
TPO 167
truss 128, 130, 131
vent 176
ventilator 176, 177
ventilator plastic 386
walkway 166
window metal 220
zinc 169
Roofing & siding preformed ... 159
built-up 532, 722
cold 164
demolition 35, 147
elastomeric 532
finish price sheet 532
flat seam sheet metal 169
membrane 164
modified bitumen 166
roll 168
shingle 532
single ply 167
system built-up 164
Rooftop air conditioner 400
  multizone air cond. 400
Room humidifier 404
Root raking loading 471
Rope exercise 311
  safety line 15
Rosewood door 191
Rough buck 121
  hardware 118
  stone wall 79
Rough-in drinking fountain floor 376
  drinking fountain wall ... 376
  sink countertop 371
  sink raised deck 371
  sink service floor 374
  tub 371
Round rail fence 470
Rubber astragal 228
  base 256
  coating 149
  control joint 71
  floor 256
  floor tile 257
  nosing 256
  riser 256
  sheet 256
  stair 256
  threshold 225
  tile 257
  waterproofing 148
Rubberized asphalt 460
  asphalt sealcoat 460
Rubbish chute 26
  handling 26
Rug cleaning 260

S

Safe office 299
Safety equipment laboratory ... 309
  fixture industrial 375
  line rope 15
  nosing 256

railing cable 92
shower 375
switch 425
Sales tax 9
Salvage or disposal 28
Sampling air 33
Sanctuary pew 322
Sand fill 440
  seal 460
Sandblast finish concrete ... 57
  masonry 66
Sanding 263
  floor 255
Sandstone 81
Sandwich panel skylight ... 221
Sanitary base cove 249
  drain 366
  tee 365
Sash aluminum 209
  metal 210
  replacement 211
  security 210
  steel 209
  wood 212
Sauna 330
  door 330
Saw cutting 62
  shop 310
  table 310
Sawn control joint 57
Sbs modified bit. membrane ... 166
Scaffold aluminum plank ... 15
  baseplate 13
  bracket 13
  caster 13
  catwalk 14
  frame 13
  guardrail 13
  mason 13
  putlog 14
  rolling tower 14
  specialties 14
  stairway 13
  wood plank 13
Scaffolding 707
  plank 14
  tubular 13
Scanner checkout 300
Scarify concrete 238
  subsoil 471
School door hardware ... 221
  equipment 311
  hinge 226
Scissor gate 286
SCR brick 74
Scrape after damage 266
Scraper mobilization 16
Scratch coat 258
Screen chimney 290
  entrance 282
  fence 470
  metal 210
  molding 133
  projection 309
  security 108, 210
  sight 287
  squirrel and bird 290
  urinal 285
  window 210, 215
  wood 215
Screened loam 471
  topsoil 471
Screw jack 455
  lag 88
  liquid chillers 398
Seal curb 460

Index

fog 460
pavement 460
security 228
slurry 461
Sealant 147, 155
  acoustical 182, 248
  caulking 181
  control joint 57
  rigid joint 182
  tape 182
Sealcoat 460
Sealer floor 43
  joint 180, 181
Seals compression 180
Seating movie 309
Sectional door 203, 204
  overhead door 203
Security and vault equipment .. 299
  fence 469
  film 234
  gate 286
  hinge 226
  sash 210
  screen 108, 210
  seal 228
Seeding 472, 734
Selective demolition 25
  demolition fencing 24
  demolition masonry 67
Self-closing relief valve 349
Self-contained air conditioner .. 400
Sentry dog 18
Separation barrier 33, 38
Septic galley 481
  system 480, 622
  tank 480
  tank concrete 480
Service boring 479
  electric 425
  entrance cable aluminum . 409
  entrance cap 409
  residential 416
  sink 374
  sink faucet 375
  station equipment 298
Sewage collection concrete pipe . 480
  collection PVC pipe 480
  ejector pump 367
  pipe 480, 481
Sewage/drainage collection .. 481
Sewer relining 478
Shade 314
Shaft wall 240
Shake wood 157
Shear plate 118
  test 11
  wall 130
Sheathed nonmetallic cable .. 409
  romex cable 409
Sheathing 129, 130
  asphalt 130
  gypsum 130
  paper 155
  roof 129
  roof & walls plywood ... 129
Sheepsfoot compactor 453
Sheet base 165
  metal 169
  metal aluminum 173
  metal cladding 170
  metal flashing 170
  piling 455
  steel piles 456
Sheeting wale 455
  wood 455, 734
Sheets lead 332

Shelf bathroom 288
  library 321
  metal 293
Shellac door 271
Shell/tube heat exchanger .. 588
Shelter dock 299
  rail dock 299
  temporary 43
Shelving 293
  food storage metal 305
  pine 293
  plywood 293
  prefinished 293
  storage 293
  wood 293
Shield expansion 87
Shingle 155, 156
  aluminum 156
  asphalt 155
  cedar 157
  concrete 158
  fiber cement 156
  metal 156
  panelized 157
  removal 146
  roof 156
  steel 156
  strip 155
  wood 157
Ship ladder 106
Shock absorber 363
  absorber pipe 363
  absorbing door ... 185, 205
Shop drill 310
  equipment 310
Shores post 49
Shoring 455
  aluminum joist 49
  baseplate 13
  bracing 13
  concrete formwork 49
  concrete vertical 49
  flying truss 49
  frames 49
  heavy duty 13
  leveling jack 13
  steel beam 49
Short load concrete 55
Shot blast floor 239
Shower by-pass valve 372
  cabinet 285
  compartments 285
  cubicle 371
  door 286
  glass door 286
  pan 171
  partition 285
  safety 375
  spray 372
Shower-head water economizer . 371
Shower/tub control set 372
Shower-tub valve spout set .. 372
Shrub broadleaf evergreen .. 473
  deciduous 474
  evergreen 473
  moving 475
Shrubs and trees 473
Shutter 314
  exterior 141
  interior 314
  wood 141
  wood interior 314
Side light door 197
Sidelight 140
  door 190
Sidewalk 254, 464, 618

asphalt 461
brick 464
bridge 14
concrete 461
removal 23
temporary 17
Sidewalks driveways and patios . 461
Sidewall bracket 15
Siding aluminum 161
  bevel 162
  cedar 162
  fiber cement 163
  fiberglass 160
  metal 159, 161
  nail 117
  paint 268
  painting 267
  plywood 162
  redwood 162
  removal 147
  steel 161
  vinyl 162
  wood 162
Sign 18, 281
  base formwork 46
  directional 281
  flexible 281
  posts 281
  project 18
  reflective 281
  road 281
  traffic 281
Signage exterior 281
Signal bell fire 437
Silencer door 225
  duct 385
Silicon carbide abrasive 42
Silicone 180
  caulking 181
  coating 149
  water repellent 149
Sill 80, 124
  anchor 118
  door 133, 140, 225
  framing 124
  masonry 81
  precast concrete window . 60
  quarry tile 250
  stone 80, 81
  window 213
Sillcock hose bibb 373
Silos demo 327
Silt fence 454
Simulated brick 82
  stone 83
Single hub pipe 364
  hung window 209
  ply roofing 167
  zone rooftop unit 400
Single-ply roof 167
Sink barber 300
  base 315
  countertop 370
  countertop rough-in .. 371
  darkroom 302
  faucet kitchen 372
  kitchen 371
  laboratory 309
  laundry 371
  raised deck rough-in .. 371
  removal 348
  service 374
  service floor rough-in .. 374
  slop 371
Sinks residential 370
Siren 436

Sister joist 120
Site demolition 22
  drainage 482
  furnishings 322
  improvement 322, 470
  preparation 22
Skylight 220, 534
  plastic 220
  removal 146
  roof 220
Skylights domed 220
Skyroof 208
Slab blockout formwork 46
  box out opening formwork . 45
  bulhead formwork .. 45, 46
  concrete 52
  concrete placement 56
  curb formwork 45, 46
  cutout 25
  edge formwork ... 45, 46
  elevated 52
  flat plate formwork 45
  joist dome formwork ... 45
  joist pan formwork 45
  lift 54, 718
  on grade 53, 497
  on grade formwork 46
  pan 53
  screed formwork 46
  stamping 58
  textured 54
  thickened edge 53
  trench formwork 46
  void formwork ... 45, 46
  waffle 52
  with drop panels formwork . 45
Slate 81
  flagging 465
  repair 156
  roof 156
  shingle 156
  sidewalk 465
  sill 81
  stair 81
  tile 254
Slatwall 135, 262
Sleeper 124
Sleepers framing 124
Sleeve and tap 479
  dowel 51
Slide gate 468
Sliding aluminum door 200
  door 527
  door glass 200
  entrance door 207
  glass door 200
  mirror 289
  panel door 208
  window 209, 213
Slop sink 371
Slope protection 454
Slot channel 70
  letter 293
Slotted channel framing 90
Slump block 76
Slurry seal 461
  seal (latex modified) ... 460
Small tools 16
Smoke detector 437
  hatch 177
  vent 177
  vent chimney 391
Snow fence 31
  fencing 31
  guard 177
Soap dispenser 289

Index

holder ... 289
tank ... 289
Socket meter ... 423
Sod ... 472
Sodding ... 472
Sodium high pressure fixture ... 427
   low pressure fixture ... 429
Soffit drywall ... 247
   metal ... 164
   molding ... 134
   plaster ... 245
   plywood ... 134
   polyvinyl ... 134
   PVC ... 143
   stucco ... 245
   vent ... 176, 235
   wood ... 134
Softener water ... 368
Soil compaction ... 443
   decontamination ... 32
   pipe ... 364
   poisoning ... 455
   tamping ... 443
   test ... 11
   treatment ... 455
Solar backflow preventer ... 397
   energy ... 397
   energy circulator air ... 396
   energy system air purger ... 397
   energy system air vent ... 397
   energy sys. balancing valve ... 397
   energy system collector ... 396
   energy system control valve ... 397
   energy system controller ... 397
   energy system expansion tank ... 397
   energy sys. gauge pressure ... 397
   energy sys. heat exchanger ... 397
   energy system storage tank ... 397
   energy system thermometer ... 397
   energy system vacuum relief ... 397
   film glass ... 234
   heating component ... 396
   heating system ... 396
   system solenoid valve ... 397
Solid core door ... 192
   surface countertops ... 318
   wood door ... 194
Sorting reinforcing ... 50
Sound and vid. cable & fttngs ... 432
   attenuation blanket ... 263
   control demo ... 327
   movie ... 309
   system ... 432
Spa bath ... 330
Space heater floor mounted ... 396
   office & storage ... 12
Spandrel beam formwork ... 43
Spanish roof tile ... 158
Special door ... 199, 203
   hinge ... 226
   purpose rooms demo ... 327
Specialties ... 288
   scaffold ... 14
   telephone ... 282
Spike grid ... 118
Spiral stair ... 109, 136
Split astragal ... 228
   rib block ... 76
   ring ... 118
   system ductless ... 401
Spotter ... 450
Spray coating ... 147
   insulation ... 152
   painting booth automotive ... 298
   shower ... 372
   substrate ... 35

Sprayed cemen. fireproofing ... 178
Sprayer airless ... 33
Spread footing ... 53, 495
   footing formwork ... 46
   park soil ... 471
   soil conditioner ... 471
   topsoil ... 471
Spring bolt astragal ... 228
   bronze weatherstrip ... 228
   hinge ... 226
Sprinkler alarm ... 437
   connector flexible ... 345
   head ... 345
   irrigation system ... 470
   system ... 597, 599
   systems ... 345, 470
Stage equipment ... 310
   portable ... 310
Staging aids ... 15
   pump ... 13
   swing ... 15
Stain cabinet ... 271
   door ... 271
   floor ... 274
   lumber ... 131
   siding ... 162
   truss ... 274
Stainless cooling towers ... 399
   duct ... 383
   flashing ... 170
   gutter ... 174
   reglet ... 175
   screen ... 282
   steel cot ... 321
   steel gravel stop ... 172
   steel hinge ... 225
   steel roof ... 169
   steel shelf ... 289
   steel storefront ... 208
Stair ... 552
   basement ... 136
   basement bulkhead ... 59
   carpet ... 260
   ceiling ... 304
   climber ... 340
   concrete ... 54
   concrete pan fill ... 50
   electric ... 304
   grating ... 106
   landing ... 258
   metal pan ... 105
   metal plate ... 105
   nosing ... 256
   parts wood ... 136
   prefabricated ... 136
   railroad tie ... 462
   removal ... 115
   residential ... 136
   riser ... 258
   riser vinyl ... 256
   rubber ... 256
   slate ... 81
   spiral ... 109, 136
   stringer ... 121, 258
   stringer wood ... 121
   temporary protection ... 17
   tread ... 80, 256
   tread insert concrete ... 49
   tread terrazzo ... 258
   tread tile ... 250
   treads and risers ... 256
   wood ... 136
Stairlift wheelchair ... 340
Stairs disappearing ... 304
   fire escape ... 106
   precast concrete ... 59

Stairway door hardware ... 221
   scaffold ... 13
Stairwork and handrails ... 138
Stall toilet ... 282-285
   type urinal ... 373
   urinal ... 373
Stamping slab ... 58
   texture ... 58
Standard extinguisher ... 292
Standing seam ... 169
Standpipe connection ... 344
Starter board & switch ... 422
Station diaper ... 288
Steam bath ... 330
   boiler ... 393
   boiler electric ... 392
   clean masonry ... 67
   jacketed kettle ... 307
   trap ... 382
Steamer ... 307
Steel anchor ... 70
   astragal ... 227
   beam shoring ... 49
   beam w-shape ... 91
   blocking ... 118
   bolt ... 117
   bridging ... 119
   conduit in slab ... 415
   conduit in trench ... 415
   conduit intermediate ... 411
   corner guard ... 288
   cutting ... 86
   demolition ... 86
   diffuser ... 389
   door ... 188, 189, 201, 203, 724
   door residential ... 189
   downspout ... 173
   drilling ... 86
   edging ... 475
   estimating ... 720
   expansion tank ... 382
   fence ... 467-469
   fiber reinforcing ... 52
   fitting ... 356
   flashing ... 170
   frame ... 187
   furring ... 241
   gravel stop ... 172
   gutter ... 174
   hinge ... 225
   joist ... 508
   lath ... 244
   light pole ... 627
   lintel ... 91
   locker ... 292
   members structural ... 91
   nailer ... 121
   outlet box ... 412
   partition ... 287
   pile ... 456
   piles sheet ... 456
   pipe ... 355
   pipe fitting ... 356, 357
   pipe removal ... 23
   plate ... 91
   projects structural ... 92
   reinforcing ... 713
   roof ... 159
   sash ... 209, 724
   sheet pile ... 455
   shingle ... 156
   siding ... 161
   storage tank ... 625, 626
   structural ... 720
   stud nlb ... 242
   window ... 209, 210

window demolition ... 185
Steeple ... 296
   aluminum ... 296
   tip pin ... 226
Step ... 281
   bluestone ... 462
   brick ... 462
   masonry ... 462
   railroad tie ... 462
   stone ... 80
   tapered pile ... 456
Sterilizer barber ... 300
Stiffener plate ... 91
Stiffeners joist web ... 101
Stile fabric ... 228
   protection ... 228
Stone aggregate ... 473
   anchor ... 70
   ashlar ... 80
   base ... 80, 462
   cast ... 82
   countertops engineered ... 320
   cultured ... 83
   curbs cut ... 465
   fill ... 440
   floor ... 80
   ground cover ... 473
   pavers ... 465
   removal ... 23
   retaining walls ... 470
   sill ... 80, 81
   simulated ... 83
   step ... 80
   stool ... 81
   tread ... 80
   wall ... 79, 470, 516
Stool cap ... 133
   stone ... 81
   window ... 81
Stop door ... 133
   gravel ... 172
   valve ... 373
   water supply ... 373
Storage bottle ... 305
   box ... 12
   cabinet ... 317
   metal shelving food ... 305
   shelving ... 293
   tank demo ... 328
Storefront aluminum ... 208
Storm door ... 186
   drainage manholes frames ... 482
   drainage piping ... 481
   window ... 216
   window aluminum residential ... 216
   windows ... 524
Stove ... 291, 307
   wood burning ... 578
   woodburning ... 291
Strainer bronze body ... 382
   downspout ... 173
   gutter ... 175
   roof ... 173
   wire ... 174
   Y type bronze body ... 382
Strap hinge ... 226
   tie ... 118
Straw bale construction ... 332
Strength compressive ... 11
Stringer stair ... 121, 258
   stair terrazzo ... 258
Strip cabinet ... 433
   filler ... 228
   floor ... 255
   footing ... 53, 494
   lighting ... 427

Index

shingle 155
Structural backfill 443
   brick 74
   columns 90
   compaction 454
   excavation 442
   facing tile 75
   fee 8
   insulated panel 127
   panel 127
   soil mixing 471
   steel 507, 720
   steel members 91
   steel projects 92
   tile 75
   welding 87
Structure moving 31
Stucco 245
   demolition 239
   mesh 244
   painting 267
Stud demolition 115
   nlb metal 242
   nlb steel 242
   nlb wall 242
   partition 121, 244
   partition framing load bearing . . 96
   price sheet 544
   wall 121, 241
   wall blocking load-bearing 95
   wall box beams load-bearing . . 96
   wall bracing load-bearing 95
   wall framing load bearing 96
   wall headers load-bearing 96
   wall wood 240
   welded 89
Subcontractor O & P 10
Subfloor 128
   adhesive 128
   plywood 128
Subflooring 556
Submersible pump 367, 480
   sump pump 367
Subpurlin bulb tee 60
Sump pump 303
   pump submersible 367
Supermarket checkout 300
   scanner 300
Supply ductile iron pipe water . . . 479
Support ceiling 89
   framing pipe 91
   lavatory 375
   light 90
   monitor 89
   motor 89
   partition 89
   water closet 375
   X-ray 90
Suppressor lightning 416
Surface countertops solid 318
   landscape 260
   preparation exterior 264
   preparation interior 265
   raceway 423
   treatment 460
Surfacing 460
Surfactant 35
Survey crew 18
   topographic 22
Suspended ceiling 243, 251
   space heater 590, 591
Suspension mounted heater 396
   system ceiling 252
   system ceiling tile 533
Swimming pool demo 328
   pool enclosure 331

pool motel 329
pool municipal 329
pool outdoor 329
pool residential 329
Swing check valve 349
   check valve bronze 349
   clear hinge 226
   staging 15
Swing-up overhead door 204
Switch box 412
   box plastic 412
   decorator 416
   dimmer 416, 424
   electric 425
   general duty 425
   load management 425
   residential 416
   rocker 424
   safety 425
   time 425
   toggle 424
Synthetic erosion control 454
   fiber reinforcing 52
   marble 254
System components foam-water . 345
   irrigation 470
   light pole 627
   septic 480
   solenoid valve solar 397
   sprinkler 345
   tube 342
Systems sprinkler 345, 470

T

Table laboratory 309
   saw 310
Tack coat 460
Tackboard fixed 280
Tamper 441
   compactor 454
   rammer 454
Tamping soil 443
Tank clean 31
   darkroom 302
   developing 302
   disposal 31
   expansion 382
   removal 31
   septic 480
   soap 289
   testing 11
   water storage solar 397
Tap and sleeve 479
Tape barricade 17
   detection 479
   sealant 182
   temporary 467
   tracer 479
   underground 479
Tapping crosses and sleeves 479
   main 479
Tar pitch emulsion 460
   roof 146
Tarpaulin 16, 17
   duck 16
   mylar 17
   polyethylene 16
   polyvinyl 16
Tarred felt 166
Tax 9
   sales 9
   social security 9
   unemployment 9
T-bar mount diffuser 389

Teak floor 255
   molding 133
   paneling 135
Technician inspection 11
Tee cleanout 350
   pipe 358
   precast concrete beam 59
Telephone booth 282
   enclosure 282
   house 432
   receptacle 419
   specialties 282
Telescoping bleacher 322
Television receptacle 419
Temperature relief valve 349
   rise detector 437
Tempered glass 230
   hardboard 134
Tempering valve 350
   valve water 350, 362
Temporary building 12
   construction 12, 16
   facility 17
   fence 17
   guardrail 17
   heat 12, 43
   light 12
   lighting 12
   road 16
   road painting 467
   shelter 43
   tape 467
   utilities 12
   utility 12
Tennis court 468
   court fences and gates 468
Tension structures demo 328
Termination box 432
   mineral insulated cable 408
Termite control chemical 455
   protection 455
Terne coated flashing 171
Terra cotta 75
   cotta coping 82
   cotta demolition 26
Terrazzo base 258
   demolition 239
   epoxy 259
   floor 258
   precast 258
   wainscot 258
Test load pile 440
   moisture content 11
   pile load 440
   soil 11
   ultrasonic 11
Testing 11
   OSHA 35
   pile 440
   tank 11
Texture stamping 58
Textured slab 54
Theater and stage equipment . . . 310
Thermal & moist. protec. demo . 146
Therm. solar energy sys. 397
Thermoplastic 233
   painting 467
   polyolefin roof 167
   polyolefin roofing 167
Thermostat 381
   automatic timed 381
   integral 404
   wire 421
Thin brick veneer 73
   plaster partition 241
Thincoat plaster 246

Thinset ceramic tile 249
   mortar 250
Threshold 225, 531
   door 140
   stone 81
   wood 133
Thru-the-wall A/C 596
   heat pump 596
Ticket dispenser 298
Tie adjustable wall 69
   column 70
   flange 70
   rafter 124
   railroad 475
   rod 90
   strap 118
   wall 70
   wire 70
Tile 249, 256, 553
   aluminum 158
   carpet 259
   ceiling 252
   ceramic 249
   clay 158
   concrete 158
   cork 257
   cork wall 261
   demolition 238
   glass 232
   marble 254
   metal 250, 251
   porcelain 249
   quarry 250
   roof 158
   rubber 257
   slate 254
   stainless steel 251
   stair tread 250
   structural 75
   vinyl 257
   wall 250
   window sill 250
Tiling ceramic 249
Tiling topsoil 471
Timber connector 117
   fastener 116
   framing 127
   guide rail 486
   heavy 127
   laminated 131
   piles 456
Time switch 425
Timer clock 420
   interval 416
   switch ventilator 388
Tin ceiling 253
   clad door 194
Tinted glass 230
Toaster 307
Toggle switch 424
Toggle-bolt anchor 88
Toilet accessories 288
   accessory 288
   bowl 370
   compartment plastic-laminate . 283
   compartments 282
   door hardware 221
   dual flush valve 374
   gasket 373
   partition 282-285
   partition removal 240
   prison 300
   stall 282-285
   tissue dispenser 289
Tool powder actuated 89
Tools small 16

Index

Toothed ring 118
Toothing masonry 66
Top demolition counter 116
   dressing 472
   vanity 319
Topographic survey 22
Topographical surveys 22
Topping concrete 57
   epoxy 259
   grout 61
Topsoil 440
   placement and grading 472
   remove 471
   screened 471
Torch cutting 86
   cutting demolition 28
Touch bar 222
Towel bar 289
   dispenser 288
Tower cooling 399
   crane 12, 708
Tpo roof 167
Tracer tape 479
Track drawer 317
   lighting 429
Tractor 443
Traffic cone 17
   line 467
   sign 281
Trailer office 12
Trainer bicycle 311
Training 18
Transformer cabinet 423
   dry-type 421
Transition molding wood 255
   residential 387
   strip floor 259
Transom aluminum 207
   lite frame 188
   windows 214
Trap cast iron 365
   drainage 365
   grease 366
   inverted bucket steam 382
   steam 382
Trapezoid windows 215
Trash compactor 309
   receptacle 322
Tread abrasive 250
   rubber 256
   stair 80
   stone 81
   vinyl 256
   wood stair 137
Treadmill 311
Treads and risers stair 256
Treated lumber framing 125
   pile 457
   pine fence 470
Treatment acoustical 263
   surface 460
Tree evergreen 473
   grate 462
   guying systems 475
   moving 475
   removal 475
Trees deciduous 474
Trench backfill 441
   excavating 441
   excavation 441
   for disposal field 481
   utility 441
Trencher chain 441
Trim demolition 116
   exterior 132
   painting 274

redwood 133
tile 249
wood 132
Triple brick 75
   weight hinge 226
Trowel coating 147
Truck dock 299
   holding concrete 55
   loading 442
Truss bowstring 131
   framing metal 104
   painting 274
   plate 118
   reinforcing 71
   roof 128, 130, 131
   stain 274
   varnish 274
Tub bar 288
   hot 330
   redwood 330
   rough-in 371
   shower handicap 371
Tube fittings copper 353
   framing 205
   system 342
   system air 342
   system pneumatic 342
Tubing copper 352
   electric metallic 411
Tub-shower module 371
Tubular fence 469
   lock 223
   scaffolding 13
   steel joist 130
Tumbler holder 289
Turned column 139
TV antenna 432
T.V. systems 432
Two piece astragal 228
Tyton joint 479

U

Ultrasonic motion detector 436
   test 11
Uncased drilled concrete piers .. 458
Undereave vent 235
Underground duct 484
   electrical & telephone ... 484
   storage tank removal ... 709
   tape 479
Underlayment 129, 556
   acoustical 263
   latex 256
   self-leveling cement 61
Underpin foundations 456
Unemployment tax 9
Union dielectric 382
Unit heater 396
   heaters propeller 403
   kitchen 301
Urethane foam 263
   wall coating 277
Urinal 373
   removal 348
   screen 285
   stall 373
   stall type 373
   wall hung 373
   watersaving 373
Utilities temporary 12
Utility accessories 479
   boxes 478
   brick 74
   duct 484

electric 484
set fan 386
sitework 484
temporary 12
trench 441
trench excavating 441
UTP cable 432

V

Vacuum breaker 363
   central 301
   cleaning 301
   loader 33
Valance board 316
Valley flashing 156
   rafter 124
Valve 349, 380, 397
   assembly dry pipe sprinkler ... 345
   bronze 349
   butterfly 380
   CPVC 350
   fire hose 344
   flush 374
   gate 380
   heating control 381
   iron body 380
   mixing 362, 372
   motorized zone 381
   plastic 350
   relief pressure 349
   shower by-pass 372
   spout set shower-tub .. 372
   sprinkler alarm 344
   stop 373
   swing check 349
   tempering 350
   water pressure 349
   water supply 373
Vandalproof fixture 428
Vanity base 317
   top 319
   top lavatory 370
Vapor barrier 155
   barrier sheathing ... 130
   retarder 155
Vaportight fixture incandescent .. 428
Variable frequency drives 425
   volume damper 384
Varnish 553
   cabinet 271
   casework 271
   door 271
   floor 274, 277
   polyurethane 277
   truss 274
VAT removal 34
Vaulting side horse 311
VAV boxes 388
   mixing boxes 388
Vct removal 239
Vehicle guide rails 486
Veneer brick 72
   core paneling 135
   granite 80
   members laminated .. 131
   removal 68
Venetian terrazzo 258
Vent box 72
   chimney 391
   chimney all fuel ... 391
   chimney fitting ... 391
   chimney flashing .. 392
   damper 391
   draft damper 391

eave 176
exhaust 389
flashing 366
foundation 72
gas 391
metal chimney 391
one-way 177
ridge 176
ridge strip 235
roof 176
smoke 177
soffit 176, 235
Ventilating air conditioning .. 385-387,
                            398, 399, 401
Ventilation equipment and hood 390
   fan 420
   louver 235
Ventilator fan 386
   masonry 72
   roof 176, 177
   timer switch 388
Verge board 133
Vermiculite insulation 151
Vertical lift door 204
   metal ladder 106
Vibrating plate 453
   roller 453
Vibrator earth 441, 443, 454
   plate 443
Video surveillance 437
Vinyl blind 141
   casement window 218
   composition floor 257
   faced wallboard 247
   floor 257
   gutter 174
   picture window 219
   plastic waterproofing .. 148
   roof 168
   siding 162
   stair riser 256
   stair tread 256
   tile 257
   tread 256
   wall coating 277
   wallpaper 261
   window 217, 219
   window double hung .. 217
   window half round ... 219
   window single hung .. 217
Vision panel 220
Vocational kiln 310
   shop equipment ... 310
Volume control damper .. 385

W

Waffle slab 52
Wainscot molding 133
   quarry tile 250
   terrazzo 258
Wale sheeting 455
Walk 461
Walkway luminaire 429
   roof 166
Wall & corner guards .. 288
   aluminum 161
   blocking load-bearing stud 95
   box beams load-bearing stud .. 96
   boxout formwork .. 47
   bracing load-bearing stud .. 95
   brick 79
   brick shelf formwork ... 47
   bulkhead formwork .. 47
   bumper 224

Index

buttress formwork 47
cabinet 315, 317
canopy 295
cap residential 387
ceramic tile 249
coating 277, 553
concrete 53
concrete block 501, 515
concrete finishing 57
concrete placement 56
corbel formwork 47
cross 312
cutout 25
demolition 239
drywall 247
finishes 261
forms 711
formwork 47
foundation 77
framing 125
framing load bearing stud 96
framing removal 26
furnace 396
furring 126, 242
grout 69
headers load-bearing stud 96
hung lavatory 370
hung urinal 373
hydrant for building 344
insulation 76, 149
lath 243, 244
liner formwork 47
lintel formwork 48
louver 235
masonry 76, 470
mat 311
mirror 232, 233
nlb partition 240
painting 268, 275, 276
panel 82
paneling 135
pilaster formwork 48
plaster 245
plywood formwork 47
precast concrete 60
prefab plywood formwork 48
reinforcing 50
retaining 54, 470
screens 287
shaft 240
shear 130
sheathing 129
stone 79
stucco 245
stud 121, 241
stud nlb 242
switch plate 424
tie 70
tie masonry 69
tile 250
tile cork 261
window 206
wood frame 519
wood stud 240
Wallboard acoustical 249
Wallcovering 261, 553
   acrylic 261
   gypsum fabric 261
Wallpaper 261
   grass cloth 261
   removal 261
   vinyl 261
Walls and partitions demolition .. 239
Walnut floor 255
Wardrobe 293
   wood 316

Wash bowl residential 370
   fountain 374
   fountain group 374
Washable air filter permanent ... 390
   air filters 390
Washer 118
   and extractor 301
   commercial 301
   darkroom 302
   residential 303
Waste cleanup hazardous 32
   disposal hazardous 32
   handling 312
   handling equipment 312
   packaging 36
   receptacle 289
Watchdog 18
Watchman service 18
Water atomizer 33
   blasting 66
   chiller reciprocating ... 398
   closet 370, 373
   closet removal 348
   closet support 375
   cooler 376
   distribution pipe 479
   fountain removal 349
   hammer arrester 363
   heater 420
   heater commercial 369
   heater gas 369
   heater gas residential .. 369
   heater instantaneous 368
   heater insulation 350
   heater oil 369
   heater point of use 368
   heater removal 349
   heater residential electric . 368
   heater under the sink ... 368
   heater wrap kit 350
   heating hot 392, 403
   hydrant 364
   motor alarm 345
   pipe ground clamp 410
   pressure relief valve ... 349
   pressure valve 349
   pump 303, 383, 480
   reducer concrete 55
   repellent 275
   repellent coating 149
   repellent silicone 149
   softener 368
   storage solar tank 397
   supply domestic meter ... 362
   supply ductile iron pipe . 479
   supply meter 362
   supply stop 373
   supply valve 373
   tempering valve 350, 362
   well 480
Waterproofing 149
   butyl 148
   cementitious 149
   coating 147
   demolition 147
   elastomeric sheet 148
   membrane 148, 460
   neoprene 148
   rubber 148
Watersaving urinal 373
Water-source heat pumps 402
Waterstop PVC 50
   ribbed 50
Wearing course 463
Weather barrier or wrap 155
   barriers 155

Weatherproof receptacle 418
Weatherstrip 202, 228
   aluminum 228
   door 230
Weatherstripping 228, 531
Weathervanes 296
Wedge anchor 88
Weight exercise 311
   lifting multi station ... 311
Weld joint pipe 355
Welded frame 187
   structural steel 720
   stud 89
   wire fabric 51, 713
Welder arc 310
Welding certification 11
   fillet 87
   structural 87
Well pump 480
   water 480
Wells & accessories 480
Wheel potters 310
Wheelchair lift 340
   stairlift 340
Whirlpool bath 330
Wide throw hinge 226
Window 209
   & door demolition 185
   air conditioner 400
   aluminum 209
   aluminum residential storm .. 216
   and door molding 133
   awning 210, 294
   awning metal-clad wood .. 213
   bank 299
   blind 314
   casement 209, 211, 213
   demolition 185
   double hung 210, 212, 214
   double hung vinyl 217
   frame 213
   glass 232
   grille 215
   guard 108
   half round vinyl 219
   half-round 214
   hardware 221
   industrial 209
   metal 209
   muntin 215
   painting 271
   picture 212
   picture and sliding 213
   pivoted 209
   plastic 217, 219
   precast concrete sill 60
   projected 209
   removal 185
   restoration 37
   screen 210, 215
   sill 81
   sill tile 250
   single hung vinyl 217
   sliding 213
   steel 209, 210
   steel sash picture 210
   stool 81
   storm 216
   trim set 134
   vinyl 217, 219
   vinyl casement 218
   vinyl clad premium 215
   wall 206, 525
   wall aluminum 206
   wall framing 205
   wood 210-213

wood bow 213
Windows replacement 184
   transom 214
   trapezoid 215
Wine cellar 305
Winter concrete 55
   protection 17, 43
Wire copper 409
   electrical 409
   fabric welded 51
   fence 17, 469
   glass 232
   ground 410
   mesh 245
   mesh locker 292
   reinforcement 713
   strainer 174
   thermostat 421
   THW 409
   tie 70
Wiremold raceway 423
   raceway non-metallic 423
Wireway 412
Wiring air conditioner 420
   device 412, 424
   devices price sheet .. 602, 603
   fan 420
   methods 421
   power 425
   residential 416, 420
Wood base 132
   beam 119, 127
   bench 322
   block floor 255
   block floor demolition .. 239
   blocking 118, 122
   bow window 213
   casing 132
   column 120, 127, 139
   cupola 295
   cutting 27
   deck 128
   demolition 238
   door 191, 194, 197
   door decorator 193
   door frame interior 140
   door residential 194
   double acting swing door . 205
   drawer 316
   exterior surface 520
   fascia 132
   fastener 117
   fiber ceiling 252
   fiber plank cementitious . 60
   fiber sheathing 130
   fiber underlayment 129
   floor 255
   floor demolition 239
   folding partition 287
   frame 140, 316
   frame wall 519
   framing 118, 119, 126
   framing demolition 112
   framing miscellaneous ... 121
   furring 126
   girder 119
   gutter 174
   handrail 133
   interior shutter 314
   joist 120, 130, 510
   laminated 131
   louver 138
   nailer 121
   overhead door 203
   panel door 194
   paneling 134

Index

parquet 255
partition 121
plank scaffold 13
rafter 123
rail fence 469
railing 138
roof deck 128
roof truss 130
sash 212
screen 215
shade 314
shake 157
sheathing 130
sheet piling 734
sheeting 455, 734
shelving 293
shingle 157
shutter 141
sidewalk 462
siding 162
siding demolition 147
sill 124
soffit 134
stair 136
stair baluster 136
stair newel 137
stair parts 136
stair railing 137
stair riser 137
stair stringer 121
stair tread 137
storm door 198
subfloor 128
threshold 133
trim 132
veneer wallpaper 261
wardrobe 316
window 210-213, 523
window demolition 185
Woodburning stove 291
Wool carpet 260
   fiberglass 151
Work extra 10
   in prisons 9
Worker protection 33
Workers' compensation 704-706
Wrestling mat 311

X

X-ray barrier 333
   concrete slabs 11
   equipment 436
   machine 436
   protection 332
   support 90
X-ray/radio freq prot. demo 329

Y

Y type bronze body strainer 382
Yard fountains 329
Yellow pine floor 255

Z

Z bar suspension 252
Zee bar 249
Zinc divider strip 258
   plating 117
   roof 169
   terrazzo strip 258
   weatherstrip 228

Notes

Notes

Notes

Notes

Notes

Notes

Division Notes

Division Notes

	CREW	DAILY OUTPUT	LABOR-HOURS	UNIT	2010 BARE COSTS				TOTAL INCL O&P
					MAT.	LABOR	EQUIP.	TOTAL	

Division Notes

		CREW	DAILY OUTPUT	LABOR-HOURS	UNIT	2010 BARE COSTS				TOTAL INCL O&P
						MAT.	LABOR	EQUIP.	TOTAL	

Division Notes

	CREW	DAILY OUTPUT	LABOR-HOURS	UNIT	MAT.	LABOR	EQUIP.	TOTAL	TOTAL INCL O&P

Division Notes

		CREW	DAILY OUTPUT	LABOR-HOURS	UNIT	2010 BARE COSTS				TOTAL INCL O&P
						MAT.	LABOR	EQUIP.	TOTAL	

Reed Construction Data/RSMeans— a tradition of excellence in construction cost information and services since 1942

Table of Contents
Annual Cost Guides
Reference Books
Seminars
Electronic Data
New Titles
Order Form

For more information visit the RSMeans website at www.rsmeans.com

Book Selection Guide

The following table provides definitive information on the content of each cost data publication. The number of lines of data provided in each unit price or assemblies division, as well as the number of reference tables and crews, is listed for each book. The presence of other elements such as equipment rental costs, historical cost indexes, city cost indexes, square foot models, or cross-referenced indexes is also indicated. You can use the table to help select the RSMeans book that has the quantity and type of information you most need in your work.

Unit Cost Divisions	Building Construction Costs	Mechanical	Electrical	Repair & Remodel	Square Foot	Site Work Landsc.	Assemblies	Interior	Concrete Masonry	Open Shop	Heavy Construction	Light Commercial	Facilities Construction	Plumbing	Residential
1	553	372	385	474		489		301	437	552	483	220	967	386	157
2	737	264	85	676		964		358	215	734	706	434	1154	288	242
3	1506	213	201	850		1357		277	1837	1506	1497	347	1422	182	359
4	897	24	0	697		699		572	1105	873	592	489	1145	0	405
5	1851	201	156	958		795		1021	758	1838	1050	826	1841	276	722
6	1841	67	69	1429		449		1160	299	1820	384	1530	1453	22	2043
7	1438	209	125	1462		524		546	466	1436	0	1143	1500	218	905
8	2048	61	10	2118		294		1737	694	1974	0	1527	2320	0	1397
9	1783	72	26	1598		268		1875	367	1737	15	1511	2027	54	1331
10	986	17	10	597		224		800	157	986	29	481	1072	231	214
11	1020	214	163	487		126		865	30	1005	0	231	1034	173	110
12	547	0	2	303		248		1691	12	536	19	349	1736	23	317
13	733	119	111	247		344		255	69	732	256	80	743	74	79
14	280	36	0	229		28		259	0	279	0	12	299	21	6
21	78	0	16	30		0		250	0	78	0	60	368	373	0
22	1148	7031	155	1106		1466		622	20	1103	1648	823	6940	8763	634
23	1185	7032	609	774		158		679	38	1105	110	660	5089	1869	418
26	1246	493	9875	1003		739		1053	55	1234	523	1080	9811	438	559
27	67	0	210	34		9		59	0	67	35	48	212	0	4
28	86	58	107	61		0		68	0	71	0	40	122	44	25
31	1477	735	608	798		2793		7	1208	1421	2865	600	1538	660	606
32	811	114	0	675		3967		359	274	784	1481	373	1443	226	422
33	519	1025	465	201		1967		0	225	518	1890	117	1553	1196	115
34	109	0	20	4		144		0	31	64	122	0	129	0	0
35	18	0	0	0		166		0	0	18	281	0	83	0	0
41	52	0	0	24		7		22	0	52	30	0	59	14	0
44	98	90	0	0		32		0	0	23	32	0	98	105	0
Totals	23114	18447	13408	16835		18257		14836	8297	22546	14048	12981	46158	15636	11070

Assembly Divisions	Building Construction Costs	Mechanical	Electrical	Repair & Remodel	Square Foot	Site Work Landscape	Assemblies	Interior	Concrete Masonry	Open Shop	Heavy Construction	Light Commercial	Facilities Construction	Plumbing	Asm Div	Residential
A		15	0	182	150	530	598	0	536		570	154	24	0	1	374
B		0	0	809	2471	0	5577	329	1914		368	2031	143	0	2	217
C		0	0	635	865	0	1235	1567	146		0	765	247	0	3	588
D		1066	811	694	1793	0	2422	767	0		0	1271	1025	1078	4	861
E		0	0	85	260	0	297	5	0		0	257	5	0	5	392
F		0	0	0	123	0	124	0	0		0	123	3	0	6	358
G		527	294	311	202	2366	668	0	535		1194	200	293	685	7	300
															8	760
															9	80
															10	0
															11	0
															12	0
Totals		1608	1105	2716	5864	2896	10921	2668	3131		2132	4801	1740	1763		3932

Reference Section	Building Construction Costs	Mechanical	Electrical	Repair & Remodel	Square Foot	Site Work Landscape	Assemblies	Interior	Concrete Masonry	Open Shop	Heavy Construction	Light Commercial	Facilities Construction	Plumbing	Residential
Reference Tables	yes	yes	yes	yes	no	yes	yes	yes	yes	yes	yes	yes	yes	yes	yes
Models					107							46			28
Crews	487	487	487	467		487		487	487	465	487	465	467	487	465
Equipment Rental Costs	yes	yes	yes	yes		yes		yes	yes	yes	yes	yes	yes	yes	yes
Historical Cost Indexes	yes	yes	yes	yes	yes	yes	yes	yes	yes	yes	yes	yes	yes	yes	no
City Cost Indexes	yes	yes	yes	yes	yes	yes	yes	yes	yes	yes	yes	yes	yes	yes	yes

Annual Cost Guides

For more information visit the RSMeans website at www.rsmeans.com

RSMeans Building Construction Cost Data 2010

Offers you unchallenged unit price reliability in an easy-to-use format. Whether used for verifying complete, finished estimates or for periodic checks, it supplies more cost facts better and faster than any comparable source. Over 21,000 unit prices have been updated for 2010. The City Cost Indexes and Location Factors cover over 930 areas, for indexing to any project location in North America. Order and get *RSMeans Quarterly Update Service* FREE. You'll have year-long access to the RSMeans Estimating **HOTLINE** FREE with your subscription. Expert assistance when using RSMeans data is just a phone call away.

$169.95 per copy
Available Sept. 2009
Catalog no. 60010

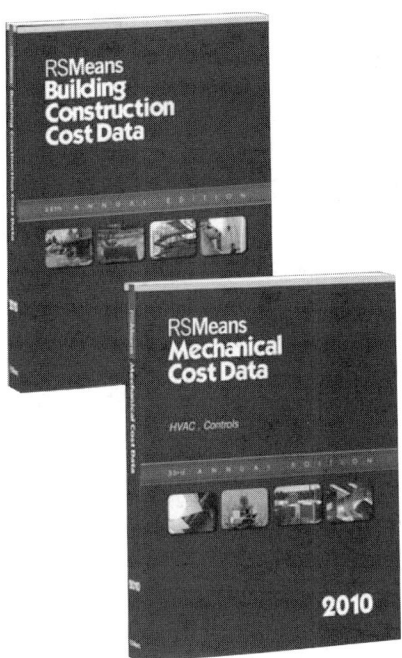

Unit prices organized according to the latest MasterFormat!

RSMeans Mechanical Cost Data 2010
- **HVAC**
- **Controls**

Total unit and systems price guidance for mechanical construction... materials, parts, fittings, and complete labor cost information. Includes prices for piping, heating, air conditioning, ventilation, and all related construction.

Plus new 2010 unit costs for:

- Thousands of installed HVAC/controls assemblies components
- "On-site" Location Factors for over 930 cities and towns in the U.S. and Canada
- Crews, labor, and equipment

$169.95 per copy
Available Oct. 2009
Catalog no. 60020

RSMeans Plumbing Cost Data 2010

Comprehensive unit prices and assemblies for plumbing, irrigation systems, commercial and residential fire protection, point-of-use water heaters, and the latest approved materials. This publication and its companion, *RSMeans Mechanical Cost Data*, provide full-range cost estimating coverage for all the mechanical trades.

Now contains updated costs for potable water and radiant heat systems and high efficiency fixtures.

$169.95 per copy
Available Oct. 2009
Catalog no. 60210

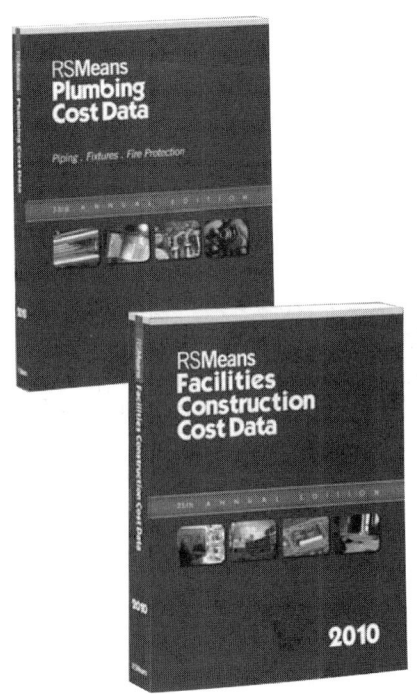

Unit prices organized according to the latest MasterFormat!

RSMeans Facilities Construction Cost Data 2010

For the maintenance and construction of commercial, industrial, municipal, and institutional properties. Costs are shown for new and remodeling construction and are broken down into materials, labor, equipment, and overhead and profit. Special emphasis is given to sections on mechanical, electrical, furnishings, site work, building maintenance, finish work, and demolition.

More than 43,000 unit costs, plus assemblies costs and a comprehensive Reference Section are included.

$419.95 per copy
Available Nov. 2009
Catalog no. 60200

For more information
visit the RSMeans website
at www.rsmeans.com

Annual Cost Guides

RSMeans Square Foot Costs 2010

It's Accurate and Easy To Use!

- **Updated price information** based on nationwide figures from suppliers, estimators, labor experts, and contractors
- "How-to-Use" sections, with **clear examples** of commercial, residential, industrial, and institutional structures
- Realistic graphics, offering true-to-life illustrations of building projects
- Extensive information on using square foot cost data, including sample estimates and alternate pricing methods

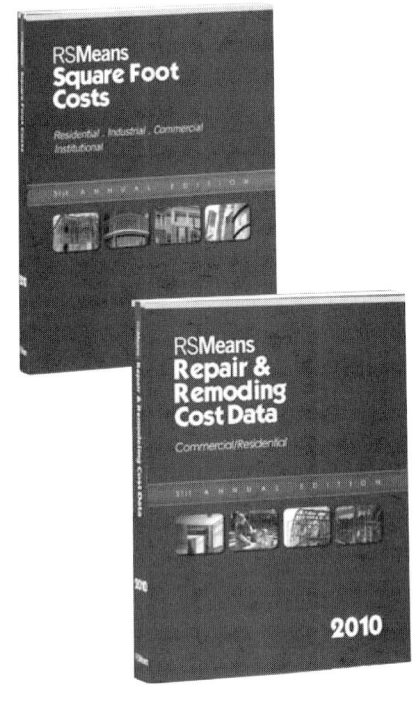

$184.95 per copy
Available Oct. 2009
Catalog no. 60050

RSMeans Repair & Remodeling Cost Data 2010

Commercial/Residential

Use this valuable tool to estimate commercial and residential renovation and remodeling.

Includes: New costs for hundreds of unique methods, materials, and conditions that only come up in repair and remodeling, PLUS:

- Unit costs for over 15,000 construction components
- Installed costs for over 90 assemblies
- Over 930 "on-site" localization factors for the U.S. and Canada.

Unit prices organized according to the latest MasterFormat!

$144.95 per copy
Available Oct. 2009
Catalog no. 60040

RSMeans Electrical Cost Data 2010

Pricing information for every part of electrical cost planning. More than 13,000 unit and systems costs with design tables; clear specifications and drawings; engineering guides; illustrated estimating procedures; complete labor-hour and materials costs for better scheduling and procurement; and the latest electrical products and construction methods.

- A variety of special electrical systems including cathodic protection
- Costs for maintenance, demolition, HVAC/mechanical, specialties, equipment, and more

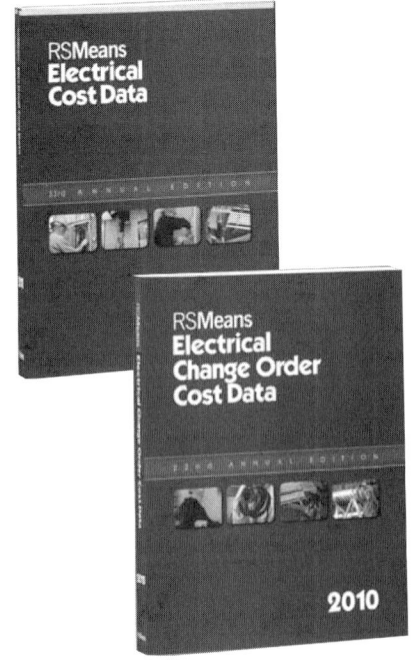

$169.95 per copy
Available Oct. 2009
Catalog no. 60030

Unit prices organized according to the latest MasterFormat!

RSMeans Electrical Change Order Cost Data 2010

RSMeans Electrical Change Order Cost Data provides you with electrical unit prices exclusively for pricing change orders—based on the recent, direct experience of contractors and suppliers. Analyze and check your own change order estimates against the experience others have had doing the same work. It also covers productivity analysis and change order cost justifications. With useful information for calculating the effects of change orders and dealing with their administration.

$169.95 per copy
Available Dec. 2009
Catalog no. 60230

Annual Cost Guides

For more information visit the RSMeans website at www.rsmeans.com

RSMeans Assemblies Cost Data 2010

RSMeans Assemblies Cost Data takes the guesswork out of preliminary or conceptual estimates. Now you don't have to try to calculate the assembled cost by working up individual component costs. We've done all the work for you.

Presents detailed illustrations, descriptions, specifications, and costs for every conceivable building assembly—240 types in all—arranged in the easy-to-use UNIFORMAT II system. Each illustrated "assembled" cost includes a complete grouping of materials and associated installation costs, including the installing contractor's overhead and profit.

$279.95 per copy
Available Sept. 2009
Catalog no. 60060

RSMeans Open Shop Building Construction Cost Data 2010

The latest costs for accurate budgeting and estimating of new commercial and residential construction. . . renovation work. . . change orders. . . cost engineering.

RSMeans Open Shop "BCCD" will assist you to:
- Develop benchmark prices for change orders
- Plug gaps in preliminary estimates and budgets
- Estimate complex projects
- Substantiate invoices on contracts
- Price ADA-related renovations

Unit prices organized according to the latest MasterFormat!

$169.95 per copy
Available Dec. 2009
Catalog no. 60150

RSMeans Residential Cost Data 2010

Contains square foot costs for 30 basic home models with the look of today, plus hundreds of custom additions and modifications you can quote right off the page. With costs for the 100 residential systems you're most likely to use in the year ahead. Complete with blank estimating forms, sample estimates, and step-by-step instructions.

Now contains line items for cultured stone and brick, PVC trim lumber, and TPO roofing.

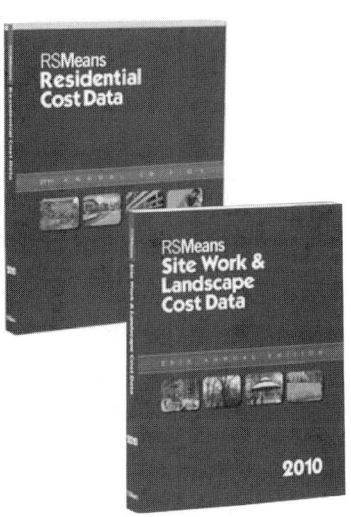

$144.95 per copy
Available Oct. 2009
Catalog no. 60170

Unit prices organized according to the latest MasterFormat!

RSMeans Site Work & Landscape Cost Data 2010

Includes unit and assemblies costs for earthwork, sewerage, piped utilities, site improvements, drainage, paving, trees & shrubs, street openings/repairs, underground tanks, and more. Contains 57 tables of assemblies costs for accurate conceptual estimates.

Includes:
- Estimating for infrastructure improvements
- Environmentally-oriented construction
- ADA-mandated handicapped access
- Hazardous waste line items

$169.95 per copy
Available Nov. 2009
Catalog no. 60280

RSMeans Facilities Maintenance & Repair Cost Data 2010

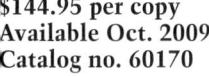

RSMeans Facilities Maintenance & Repair Cost Data gives you a complete system to manage and plan your facility repair and maintenance costs and budget efficiently. Guidelines for auditing a facility and developing an annual maintenance plan. Budgeting is included, along with reference tables on cost and management, and information on frequency and productivity of maintenance operations.

The only nationally recognized source of maintenance and repair costs. Developed in cooperation with the Civil Engineering Research Laboratory (CERL) of the Army Corps of Engineers.

$379.95 per copy
Available Nov. 2009
Catalog no. 60300

RSMeans Light Commercial Cost Data 2010

Specifically addresses the light commercial market, which is a specialized niche in the construction industry. Aids you, the owner/designer/contractor, in preparing all types of estimates—from budgets to detailed bids. Includes new advances in methods and materials.

Assemblies Section allows you to evaluate alternatives in the early stages of design/planning.

Over 11,000 unit costs ensure that you have the prices you need. . . when you need them.

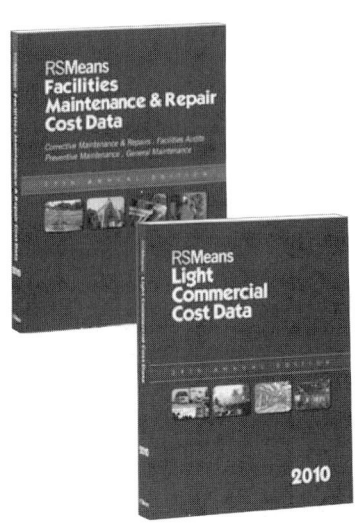

Unit prices organized according to the latest MasterFormat!

$144.95 per copy
Available Nov. 2009
Catalog no. 60180

For more information
visit the RSMeans website
at www.rsmeans.com

Annual Cost Guides

RSMeans Concrete & Masonry Cost Data 2010

Provides you with cost facts for virtually all concrete/masonry estimating needs, from complicated formwork to various sizes and face finishes of brick and block—all in great detail. The comprehensive Unit Price Section contains more than 7,500 selected entries. Also contains an Assemblies [Cost] Section, and a detailed Reference Section that supplements the cost data.

Unit prices organized according to the latest MasterFormat !

$154.95 per copy
Available Dec. 2009
Catalog no. 60110

RSMeans Labor Rates for the Construction Industry 2010

Complete information for estimating labor costs, making comparisons, and negotiating wage rates by trade for over 300 U.S. and Canadian cities. With 46 construction trades listed by local union number in each city, and historical wage rates included for comparison. Each city chart lists the county and is alphabetically arranged with handy visual flip tabs for quick reference.

$379.95 per copy
Available Dec. 2009
Catalog no. 60120

RSMeans Construction Cost Indexes 2010

What materials and labor costs will change unexpectedly this year? By how much?
- Breakdowns for 316 major cities
- National averages for 30 key cities
- Expanded five major city indexes
- Historical construction cost indexes

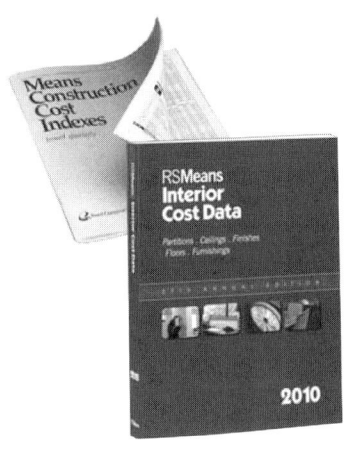

$362.00 per year (subscription)
Catalog no. 50140

$90.50 individual quarters
Catalog no. 60140 A,B,C,D

RSMeans Interior Cost Data 2010

Provides you with prices and guidance needed to make accurate interior work estimates. Contains costs on materials, equipment, hardware, custom installations, furnishings, and labor costs. . . for new and remodel commercial and industrial interior construction, including updated information on office furnishings, and reference information.

Unit prices organized according to the latest MasterFormat !

$169.95 per copy
Available Nov. 2009
Catalog no. 60090

RSMeans Heavy Construction Cost Data 2010

A comprehensive guide to heavy construction costs. Includes costs for highly specialized projects such as tunnels, dams, highways, airports, and waterways. Information on labor rates, equipment, and materials costs is included. Features unit price costs, systems costs, and numerous reference tables for costs and design.

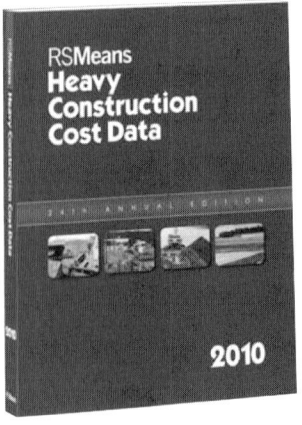

$169.95 per copy
Available Nov. 2009
Catalog no. 60160

Unit prices organized according to the latest MasterFormat!

Reference Books

For more information visit the RSMeans website at www.rsmeans.com

Value Engineering: Practical Applications
For Design, Construction, Maintenance & Operations
by Alphonse Dell'Isola, PE

A tool for immediate application—for engineers, architects, facility managers, owners, and contractors. Includes making the case for VE—the management briefing; integrating VE into planning, budgeting, and design; conducting life cycle costing; using VE methodology in design review and consultant selection; case studies; VE workbook; and a life cycle costing program on disk.

$79.95 per copy
Over 450 pages, illustrated, softcover
Catalog no. 67319A

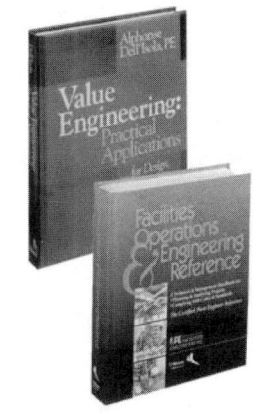

Facilities Operations & Engineering Reference
by the Association for Facilities Engineering and RSMeans

An all-in-one technical reference for planning and managing facility projects and solving day-to-day operations problems. Selected as the official certified plant engineer reference, this handbook covers financial analysis, maintenance, HVAC and energy efficiency, and more.

$54.98 per copy
Over 700 pages, illustrated, hardcover
Catalog no. 67318

The Building Professional's Guide to Contract Documents
3rd Edition
by Waller S. Poage, AIA, CSI, CVS

A comprehensive reference for owners, design professionals, contractors, and students

- Structure your documents for maximum efficiency.
- Effectively communicate construction requirements.
- Understand the roles and responsibilities of construction professionals.
- Improve methods of project delivery.

$64.95 per copy, 400 pages
Diagrams and construction forms, hardcover
Catalog no. 67261A

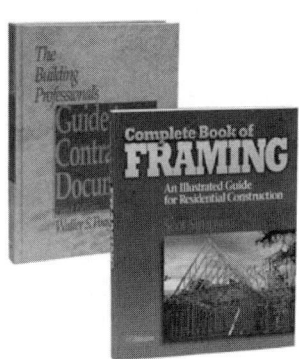

Complete Book of Framing
by Scot Simpson

This straightforward, easy-to-learn method will help framers, carpenters, and handy homeowners build their skills in rough carpentry and framing. Shows how to frame all the parts of a house: floors, walls, roofs, door & window openings, and stairs—with hundreds of color photographs and drawings that show every detail.

The book gives beginners all the basics they need to go from zero framing knowledge to a journeyman level.

$29.95 per copy
352 pages, softcover
Catalog no. 67353

Cost Planning & Estimating for Facilities Maintenance

In this unique book, a team of facilities management authorities shares their expertise on:

- Evaluating and budgeting maintenance operations
- Maintaining and repairing key building components
- Applying *RSMeans Facilities Maintenance & Repair Cost Data* to your estimating

Covers special maintenance requirements of the ten major building types.

$89.95 per copy
Over 475 pages, hardcover
Catalog no. 67314

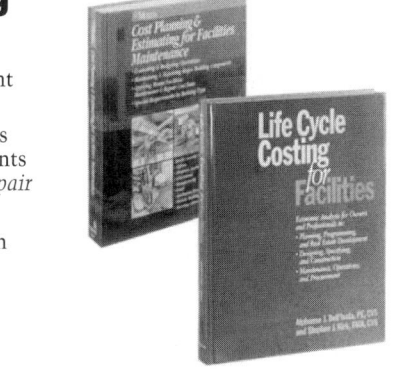

Life Cycle Costing for Facilities
by Alphonse Dell'Isola and Dr. Steven Kirk

Guidance for achieving higher quality design and construction projects at lower costs! Cost-cutting efforts often sacrifice quality to yield the cheapest product. Life cycle costing enables building designers and owners to achieve both. The authors of this book show how LCC can work for a variety of projects — from roads to HVAC upgrades to different types of buildings.

$99.95 per copy
450 pages, hardcover
Catalog no. 67341

Planning & Managing Interior Projects 2nd Edition
by Carol E. Farren, CFM

Expert guidance on managing renovation & relocation projects

This book guides you through every step in relocating to a new space or renovating an old one. From initial meeting through design and construction, to post-project administration, it helps you get the most for your company or client. Includes sample forms, spec lists, agreements, drawings, and much more!

$34.98 per copy
200 pages, softcover
Catalog no. 67245A

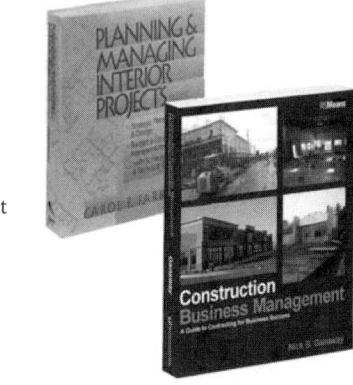

Construction Business Management
by Nick Ganaway

Only 43% of construction firms stay in business after four years. Make sure your company thrives with valuable guidance from a pro with 25 years of success as a commercial contractor. Find out what it takes to build all aspects of a business that is profitable, enjoyable, and enduring. With a bonus chapter on retail construction.

$49.95 per copy
200 pages, softcover
Catalog no. 67352

For more information
visit the RSMeans website
at www.rsmeans.com

Reference Books

Interior Home Improvement Costs 9th Edition

Updated estimates for the most popular remodeling and repair projects—from small, do-it-yourself jobs to major renovations and new construction. Includes: kitchens & baths; new living space from your attic, basement, or garage; new floors, paint, and wallpaper; tearing out or building new walls; closets, stairs, and fireplaces; new energy-saving improvements, home theaters, and more!

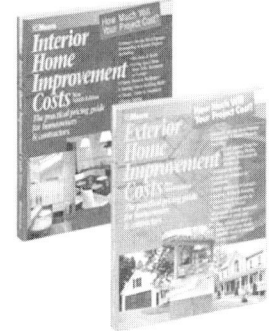

$24.95 per copy
250 pages, illustrated, softcover
Catalog no. 67308E

Exterior Home Improvement Costs 9th Edition

Updated estimates for the most popular remodeling and repair projects—from small, do-it-yourself jobs, to major renovations and new construction. Includes: curb appeal projects—landscaping, patios, porches, driveways, and walkways; new windows and doors; decks, greenhouses, and sunrooms; room additions and garages; roofing, siding, and painting; "green" improvements to save energy & water.

$24.95 per copy
Over 275 pages, illustrated, softcover
Catalog no. 67309E

Builder's Essentials: Plan Reading & Material Takeoff

For Residential and Light Commercial Construction

by Wayne J. DelPico

A valuable tool for understanding plans and specs, and accurately calculating material quantities. Step-by-step instructions and takeoff procedures based on a full set of working drawings.

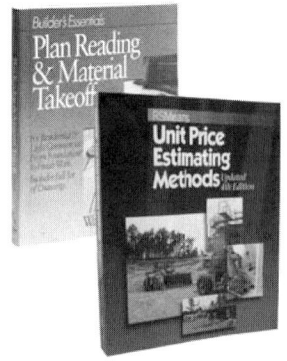

$35.95 per copy
Over 420 pages, softcover
Catalog no. 67307

Means Unit Price Estimating Methods
New 4th Edition

This new edition includes up-to-date cost data and estimating examples, updated to reflect changes to the CSI numbering system and new features of RSMeans cost data. It describes the most productive, universally accepted ways to estimate, and uses checklists and forms to illustrate shortcuts and timesavers. A model estimate demonstrates procedures. A new chapter explores computer estimating alternatives.

$59.95 per copy
Over 350 pages, illustrated, softcover
Catalog no. 67303B

Total Productive Facilities Management

by Richard W. Sievert, Jr.

Today, facilities are viewed as strategic resources... elevating the facility manager to the role of asset manager supporting the organization's overall business goals. Now, Richard Sievert Jr., in this well-articulated guidebook, sets forth a new operational standard for the facility manager's emerging role... a comprehensive program for managing facilities as a true profit center.

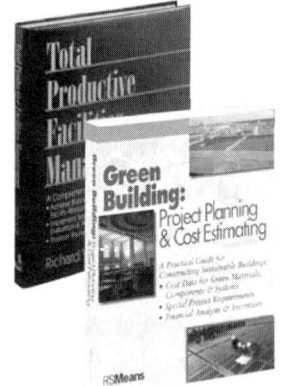

$29.98 per copy
275 pages, softcover
Catalog no. 67321

Green Building: Project Planning & Cost Estimating 2nd Edition

This new edition has been completely updated with the latest in green building technologies, design concepts, standards, and costs. Now includes a 2010 Green Building *CostWorks* CD with more than 300 green building assemblies and over 5,000 unit price line items for sustainable building. The new edition is also full-color with all new case studies—plus a new chapter on deconstruction, a key aspect of green building.

$129.95 per copy
350 pages, softcover
Catalog no. 67338A

Concrete Repair and Maintenance Illustrated

by Peter Emmons

Hundreds of illustrations show users how to analyze, repair, clean, and maintain concrete structures for optimal performance and cost effectiveness. From parking garages to roads and bridges to structural concrete, this comprehensive book describes the causes, effects, and remedies for concrete wear and failure. Invaluable for planning jobs, selecting materials, and training employees, this book is a must-have for concrete specialists, general contractors, facility managers, civil and structural engineers, and architects.

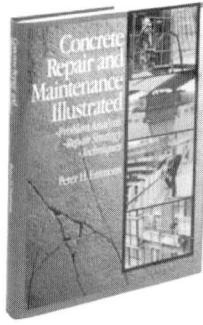

$69.95 per copy
300 pages, illustrated, softcover
Catalog no. 67146

Reference Books

For more information visit the RSMeans website at www.rsmeans.com

Means Illustrated Construction Dictionary Condensed, 2nd Edition

The best portable dictionary for office or field use—an essential tool for contractors, architects, insurance and real estate personnel, facility managers, homeowners, and anyone who needs quick, clear definitions for construction terms. The second edition has been further enhanced with updates and hundreds of new terms and illustrations . . . in keeping with the most recent developments in the construction industry.

Now with a quick-reference Spanish section. Includes tools and equipment, materials, tasks, and more!

$59.95 per copy
Over 500 pages, softcover
Catalog no. 67282A

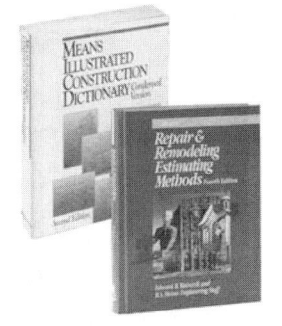

Means Repair & Remodeling Estimating 4th Edition
By Edward B. Wetherill and RSMeans

This important reference focuses on the unique problems of estimating renovations of existing structures, and helps you determine the true costs of remodeling through careful evaluation of architectural details and a site visit.

New section on disaster restoration costs.

$69.95 per copy
Over 450 pages, illustrated, hardcover
Catalog no. 67265B

Facilities Planning & Relocation
by David D. Owen

A complete system for planning space needs and managing relocations. Includes step-by-step manual, over 50 forms, and extensive reference section on materials and furnishings.

New lower price and user-friendly format.

$49.98 per copy
Over 450 pages, softcover
Catalog no. 67301

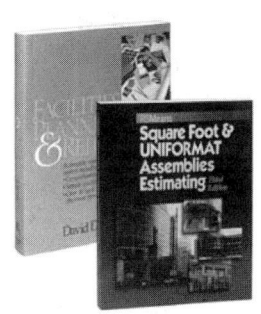

Means Square Foot & UNIFORMAT Assemblies Estimating Methods 3rd Edition

Develop realistic square foot and assemblies costs for budgeting and construction funding. The new edition features updated guidance on square foot and assemblies estimating using UNIFORMAT II. An essential reference for anyone who performs conceptual estimates.

$69.95 per copy
Over 300 pages, illustrated, softcover
Catalog no. 67145B

Means Electrical Estimating Methods 3rd Edition

Expanded edition includes sample estimates and cost information in keeping with the latest version of the CSI MasterFormat and UNIFORMAT II. Complete coverage of fiber optic and uninterruptible power supply electrical systems, broken down by components, and explained in detail. Includes a new chapter on computerized estimating methods. A practical companion to *RSMeans Electrical Cost Data*.

$64.95 per copy
Over 325 pages, hardcover
Catalog no. 67230B

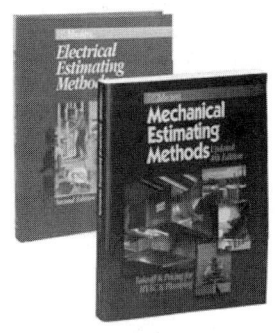

Means Mechanical Estimating Methods 4th Edition

Completely updated, this guide assists you in making a review of plans, specs, and bid packages, with suggestions for takeoff procedures, listings, substitutions, and pre-bid scheduling for all components of HVAC. Includes suggestions for budgeting labor and equipment usage. Compares materials and construction methods to allow you to select the best option.

$64.95 per copy
Over 350 pages, illustrated, softcover
Catalog no. 67294B

Means ADA Compliance Pricing Guide 2nd Edition
by Adaptive Environments and RSMeans

Completely updated and revised to the new 2004 *Americans with Disabilities Act Accessibility Guidelines*, this book features more than 70 of the most commonly needed modifications for ADA compliance. Projects range from installing ramps and walkways, widening doorways and entryways, and installing and refitting elevators, to relocating light switches and signage.

$79.95 per copy
Over 350 pages, illustrated, softcover
Catalog no. 67310A

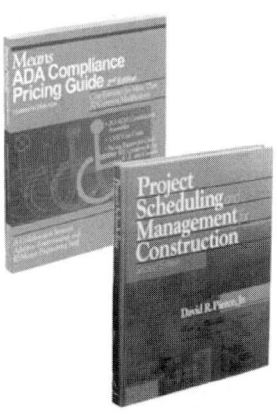

Project Scheduling & Management for Construction 3rd Edition
by David R. Pierce, Jr.

A comprehensive yet easy-to-follow guide to construction project scheduling and control—from vital project management principles through the latest scheduling, tracking, and controlling techniques. The author is a leading authority on scheduling, with years of field and teaching experience at leading academic institutions. Spend a few hours with this book and come away with a solid understanding of this essential management topic.

$64.95 per copy
Over 300 pages, illustrated, hardcover
Catalog no. 67247B

For more information
visit the RSMeans website
at www.rsmeans.com

Reference Books

The Practice of Cost Segregation Analysis
by Bruce A. Desrosiers and Wayne J. DelPico

This expert guide walks you through the practice of cost segregation analysis, which enables property owners to defer taxes and benefit from "accelerated cost recovery" through depreciation deductions on assets that are properly identified and classified.

With a glossary of terms, sample cost segregation estimates for various building types, key information resources, and updates via a dedicated website, this book is a critical resource for anyone involved in cost segregation analysis.

$99.95 per copy
Over 225 pages
Catalog no. 67345

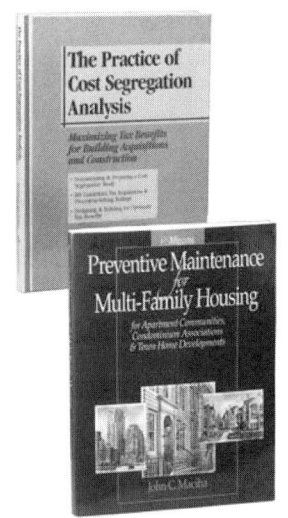

Preventive Maintenance for Multi-Family Housing
by John C. Maciha

Prepared by one of the nation's leading experts on multi-family housing.

This complete PM system for apartment and condominium communities features expert guidance, checklists for buildings and grounds maintenance tasks and their frequencies, a reusable wall chart to track maintenance, and a dedicated website featuring customizable electronic forms. A must-have for anyone involved with multi-family housing maintenance and upkeep.

$89.95 per copy
225 pages
Catalog no. 67346

How to Estimate with Means Data & CostWorks
New 3rd Edition
by RSMeans and Saleh A. Mubarak, Ph.D.

New 3rd Edition—fully updated with new chapters, plus new CD with updated *CostWorks* cost data and MasterFormat organization. Includes all major construction items—with more than 300 exercises and two sets of plans that show how to estimate for a broad range of construction items and systems—including general conditions and equipment costs.

$59.95 per copy
272 pages, softcover
Includes CostWorks CD
Catalog no. 67324B

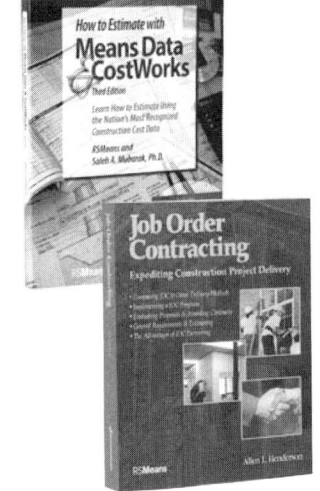

Job Order Contracting
Expediting Construction Project Delivery
by Allen Henderson

Expert guidance to help you implement JOC—fast becoming the preferred project delivery method for repair and renovation, minor new construction, and maintenance projects in the public sector and in many states and municipalities. The author, a leading JOC expert and practitioner, shows how to:

• Establish a JOC program
• Evaluate proposals and award contracts
• Handle general requirements and estimating
• Partner for maximum benefits

$89.95 per copy
192 pages, illustrated, hardcover
Catalog no. 67348

Builder's Essentials: Estimating Building Costs
For the Residential & Light Commercial Contractor
by Wayne J. DelPico

Step-by-step estimating methods for residential and light commercial contractors. Includes a detailed look at every construction specialty—explaining all the components, takeoff units, and labor needed for well-organized, complete estimates. Covers correctly interpreting plans and specifications, and developing accurate and complete labor and material costs.

$29.95 per copy
Over 400 pages, illustrated, softcover
Catalog no. 67343

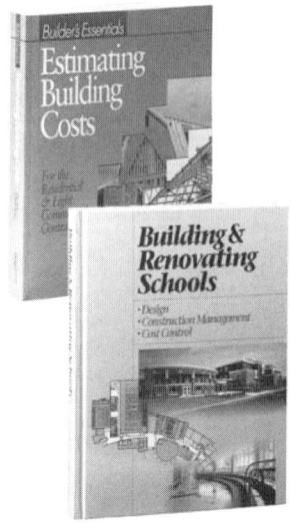

Building & Renovating Schools

This all-inclusive guide covers every step of the school construction process—from initial planning, needs assessment, and design, right through moving into the new facility. A must-have resource for anyone concerned with new school construction or renovation. With square foot cost models for elementary, middle, and high school facilities, and real-life case studies of recently completed school projects.

The contributors to this book—architects, construction project managers, contractors, and estimators who specialize in school construction—provide start-to-finish, expert guidance on the process.

$69.98 per copy
Over 425 pages, hardcover
Catalog no. 67342

Reference Books

For more information visit the RSMeans website at www.rsmeans.com

The Homeowner's Guide to Mold By Michael Pugliese

Expert guidance to protect your health and your home.

Mold, whether caused by leaks, humidity or flooding, is a real health and financial issue—for homeowners and contractors. This full-color book explains:
- Construction and maintenance practices to prevent mold
- How to inspect for and remove mold
- Mold remediation procedures and costs
- What to do after a flood
- How to deal with insurance companies

$21.95 per copy
144 pages, softcover
Catalog no. 67344

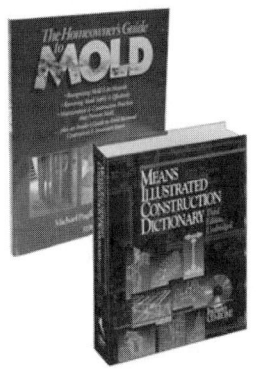

Means Illustrated Construction Dictionary
Unabridged 3rd Edition, with CD-ROM

Long regarded as the industry's finest, *Means Illustrated Construction Dictionary* is now even better. With the addition of over 1,000 new terms and hundreds of new illustrations, it is the clear choice for the most comprehensive and current information. The companion CD-ROM that comes with this new edition adds many extra features: larger graphics, expanded definitions, and links to both CSI MasterFormat numbers and product information.

$99.95 per copy
Over 790 pages, illustrated, hardcover
Catalog no. 67292A

Means Landscape Estimating Methods
New 5th Edition

Answers questions about preparing competitive landscape construction estimates, with up-to-date cost estimates and the new MasterFormat classification system. Expanded and revised to address the latest materials and methods, including new coverage on approaches to green building. Includes:
- Step-by-step explanation of the estimating process
- Sample forms and worksheets that save time and prevent errors

$64.95 per copy
Over 350 pages, softcover
Catalog no. 67295C

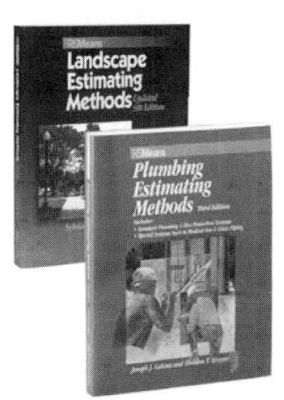

Means Plumbing Estimating Methods 3rd Edition
by Joseph Galeno and Sheldon Greene

Updated and revised! This practical guide walks you through a plumbing estimate, from basic materials and installation methods through change order analysis. *Plumbing Estimating Methods* covers residential, commercial, industrial, and medical systems, and features sample takeoff and estimate forms and detailed illustrations of systems and components.

$29.98 per copy
330+ pages, softcover
Catalog no. 67283B

Understanding & Negotiating Construction Contracts
by Kit Werremeyer

Take advantage of the author's 30 years' experience in small-to-large (including international) construction projects. Learn how to identify, understand, and evaluate high risk terms and conditions typically found in all construction contracts—then negotiate to lower or eliminate the risk, improve terms of payment, and reduce exposure to claims and disputes. The author avoids "legalese" and gives real-life examples from actual projects.

$69.95 per copy
300 pages, softcover
Catalog no. 67350

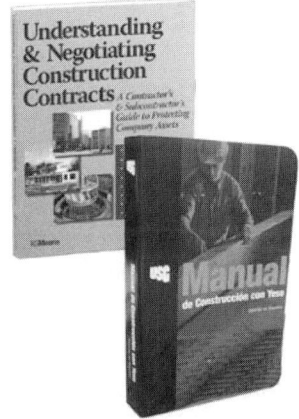

The Gypsum Construction Handbook (Spanish)

The Gypsum Construction Handbook, Spanish edition, is an illustrated, comprehensive, and authoritative guide on all facets of gypsum construction. You'll find comprehensive guidance on available products, installation methods, fire- and sound-rated construction information, illustrated framing-to-finish application instructions, estimating and planning information, and more.

$29.95 per copy
Over 450 pages, softcover
Catalog No. 67357S

Means Spanish/English Construction Dictionary 2nd Edition
by RSMeans and the International Code Council

This expanded edition features thousands of the most common words and useful phrases in the construction industry with easy-to-follow pronunciations in both Spanish and English. Over 800 new terms, phrases, and illustrations have been added. It also features a new stand-alone "Safety & Emergencies" section, with colored pages for quick access. Unique to this dictionary are the systems illustrations showing the relationship of components in the most common building systems for all major trades.

$23.95 per copy
Over 400 pages
Catalog no. 67327A

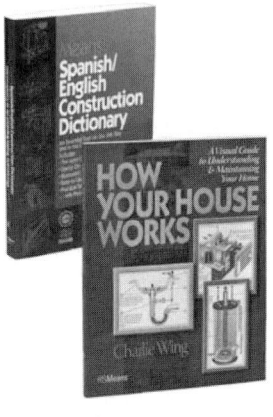

How Your House Works
by Charlie Wing

A must-have reference for every homeowner, handyman, and contractor—for repair, remodeling, and new construction. This book uncovers the mysteries behind just about every major appliance and building element in your house—from electrical, heating and AC, to plumbing, framing, foundations and appliances. Clear, "exploded" drawings show exactly how things should be put together and how they function—what to check if they don't work, and what you can do that might save you having to call in a professional.

$21.95 per copy
160 pages, softcover
Catalog no. 67351

For more information visit the RSMeans website at www.rsmeans.com

Professional Development

Means CostWorks® CD Training

This one-day course helps users become more familiar with the functionality of *Means CostWorks* program. Each menu, icon, screen, and function found in the program is explained in depth. Time is devoted to hands-on estimating exercises.

Some of what you'll learn:
- Searching the database utilizing all navigation methods
- Exporting RSMeans Data to your preferred spreadsheet format
- Viewing crews, assembly components, and much more!
- Automatically regionalizing the database

You are required to bring your own laptop computer for this course.
When you register for this course you will receive an outline for your laptop requirements.

Also offering web training for CostWorks CD!

Unit Price Estimating

This interactive *two-day* seminar teaches attendees how to interpret project information and process it into final, detailed estimates with the greatest accuracy level.

The most important credential an estimator can take to the job is the ability to visualize construction, and estimate accurately.

Some of what you'll learn:
- Interpreting the design in terms of cost
- The most detailed, time-tested methodology for accurate pricing
- Key cost drivers—material, labor, equipment, staging, and subcontracts
- Understanding direct and indirect costs for accurate job cost accounting and change order management

Who should attend: Corporate and government estimators and purchasers, architects, engineers... and others needing to produce accurate project estimates.

Square Foot & Assemblies Estimating

This *two-day* course teaches attendees how to quickly deliver accurate square foot estimates using limited budget and design information.

Some of what you'll learn:
- How square foot costing gets the estimate done faster
- Taking advantage of a "systems" or "assemblies" format
- The RSMeans "building assemblies/square foot cost approach"
- How to create a reliable preliminary systems estimate using bare-bones design information

Who should attend: Facilities managers, facilities engineers, estimators, planners, developers, construction finance professionals... and others needing to make quick, accurate construction cost estimates at commercial, government, educational, and medical facilities.

Repair & Remodeling Estimating

This *two-day* seminar emphasizes all the underlying considerations unique to repair/remodeling estimating and presents the correct methods for generating accurate, reliable R&R project costs using the unit price and assemblies methods.

Some of what you'll learn:
- Estimating considerations—like labor-hours, building code compliance, working within existing structures, purchasing materials in smaller quantities, unforeseen deficiencies
- Identifying problems and providing solutions to estimating building alterations
- Rules for factoring in minimum labor costs, accurate productivity estimates, and allowances for project contingencies
- R&R estimating examples calculated using unit price and assemblies data

Who should attend: Facilities managers, plant engineers, architects, contractors, estimators, builders... and others who are concerned with the proper preparation and/or evaluation of repair and remodeling estimates.

Mechanical & Electrical Estimating

This *two-day* course teaches attendees how to prepare more accurate and complete mechanical/electrical estimates, avoiding the pitfalls of omission and double-counting, while understanding the composition and rationale within the RSMeans Mechanical/Electrical database.

Some of what you'll learn:
- The unique way mechanical and electrical systems are interrelated
- M&E estimates–conceptual, planning, budgeting, and bidding stages
- Order of magnitude, square foot, assemblies, and unit price estimating
- Comparative cost analysis of equipment and design alternatives

Who should attend: Architects, engineers, facilities managers, mechanical and electrical contractors... and others needing a highly reliable method for developing, understanding, and evaluating mechanical and electrical contracts.

Green Building: Planning & Construction

In this *two-day* course, learn about tailoring a building and its placement on the site to the local climate, site conditions, culture, and community. Includes information to reduce resource consumption and augment resource supply.

Some of what you'll learn:
- Green technologies, materials, systems, and standards
- Modeling tools and how to use them
- Cost vs. value of green products over their life cycle
- Low-cost green strategies, and economic incentives and funding
- Health, comfort, and productivity

Who should attend: Contractors, project managers, building owners, building officials, healthcare and insurance professionals.

Facilities Maintenance & Repair Estimating

This *two-day* course teaches attendees how to plan, budget, and estimate the cost of ongoing and preventive maintenance and repair for existing buildings and grounds.

Some of what you'll learn:
- The most financially favorable maintenance, repair, and replacement scheduling and estimating
- Auditing and value engineering facilities
- Preventive planning and facilities upgrading
- Determining both in-house and contract-out service costs
- Annual, asset-protecting M&R plan

Who should attend: Facility managers, maintenance supervisors, buildings and grounds superintendents, plant managers, planners, estimators... and others involved in facilities planning and budgeting.

Practical Project Management for Construction Professionals

In this *two-day* course, acquire the essential knowledge and develop the skills to effectively and efficiently execute the day-to-day responsibilities of the construction project manager.

Covers:
- General conditions of the construction contract
- Contract modifications: change orders and construction change directives
- Negotiations with subcontractors and vendors
- Effective writing: notification and communications
- Dispute resolution: claims and liens

Who should attend: Architects, engineers, owner's representatives, project managers.

Facilities Construction Estimating

In this *two-day* course, professionals working in facilities management can get help with their daily challenges to establish budgets for all phase of a project

Some of what you'll learn:
- Determine the full scope of a project
- Identify the scope of risks & opportunities
- Creative solutions to estimating issues
- Organizing estimates for presentation & discussion
- Special techniques for repair/remodel and maintenance projects
- Negotiating project change orders

Who should attend: Facility managers, engineers, contractors, facility trades-people, planners & project managers

Professional Development

For more information visit the RSMeans website at www.rsmeans.com

Scheduling and Project Management

This *two-day* course teaches attendees the most current and proven scheduling and management techniques needed to bring projects in on time and on budget.

Some of what you'll learn:
- Crucial phases of planning and scheduling
- How to establish project priorities and develop realistic schedules and management techniques
- Critical Path and Precedence Methods
- Special emphasis on cost control

Who should attend: Construction project managers, supervisors, engineers, estimators, contractors... and others who want to improve their project planning, scheduling, and management skills.

Assessing Scope of Work for Facility Construction Estimating

This *two-day* practical training program addresses the vital importance of understanding the SCOPE of projects in order to produce accurate cost estimates for facility repair and remodeling.

Some of what you'll learn:
- Discussions of site visits, plans/specs, record drawings of facilities, and site-specific lists
- Review of CSI divisions, including means, methods, materials, and the challenges of scoping each topic
- Exercises in SCOPE identification and SCOPE writing for accurate estimating of projects
- Hands-on exercises that require SCOPE, take-off, and pricing

Who should attend: Corporate and government estimators, planners, facility managers, and others who need to produce accurate project estimates.

Site Work & Heavy Construction Estimating

This *two-day* course teaches attendees how to estimate earthwork, site utilities, foundations, and site improvements, using the assemblies and the unit price methods.

Some of what you'll learn:
- Basic site work and heavy construction estimating skills
- Estimating foundations, utilities, earthwork, and site improvements
- Correct equipment usage, quality control, and site investigation for estimating purposes

Who should attend: Project managers, design engineers, estimators, and contractors doing site work and heavy construction.

2010 RSMeans Seminar Schedule

Note: call for exact dates and details.

Location	Dates
Las Vegas, NV	March
Washington, DC	April
Denver, CO	May
San Francisco, CA	June
Washington, DC	September
Dallas, TX	September
Las Vegas, NV	October
Atlantic City, NJ	October
Orlando, FL	November
San Diego, CA	December

2010 Regional Seminar Schedule

Location	Dates	Location	Dates
Seattle, WA	January and August	St. Louis, MO	August
Dallas/Ft. Worth, TX	January	Houston, TX	October
Austin, TX	February	Colorado Springs, CO	October
Anchorage, AK	March and September	Baltimore, MD	November
Oklahoma City, OK	March	Sacramento, CA	November
Phoenix, AZ	March	San Antonio, TX	December
Aiea, HI	May	Raleigh, NC	December
Kansas City, MO	May		
Bethesda, MD	June		
Columbus, GA	June		
El Segundo, CA	August		

1-800-334-3509, ext. 5115

For more information visit the RSMeans website at www.rsmeans.com

Professional Development

Registration Information

Register early... Save up to $100! Register 30 days before the start date of a seminar and save $100 off your total fee. *Note: This discount can be applied only once per order. It cannot be applied to team discount registrations or any other special offer.*

How to register Register by phone today! The RSMeans toll-free number for making reservations is **1-800-334-3509, ext. 5115.**

Individual seminar registration fee - $935. *Means CostWorks*® training registration fee - $375. To register by mail, complete the registration form and return, with your full fee, to: RSMeans Seminars, 63 Smiths Lane, Kingston, MA 02364.

Government pricing All federal government employees save off the regular seminar price. Other promotional discounts cannot be combined with the government discount.

Team discount program for two to four seminar registrations. Call for pricing: 1-800-334-3509, ext. 5115

Multiple course discounts When signing up for two or more courses, call for pricing.

Refund policy Cancellations will be accepted up to ten business days prior to the seminar start. There are no refunds for cancellations received later than ten working days prior to the first day of the seminar. A $150 processing fee will be applied for all cancellations. Written notice of cancellation is required. Substitutions can be made at any time before the session starts. **No-shows are subject to the full seminar fee.**

AACE approved courses Many seminars described and offered here have been approved for 14 hours (1.4 recertification credits) of credit by the AACE International Certification Board toward meeting the continuing education requirements for recertification as a Certified Cost Engineer/Certified Cost Consultant.

AIA Continuing Education We are registered with the AIA Continuing Education System (AIA/CES) and are committed to developing quality learning activities in accordance with the CES criteria. Many seminars meet the AIA/CES criteria for Quality Level 2. AIA members may receive (14) learning units (LUs) for each two-day RSMeans course.

Daily course schedule The first day of each seminar session begins at 8:30 a.m. and ends at 4:30 p.m. The second day begins at 8:00 a.m. and ends at 4:00 p.m. Participants are urged to bring a hand-held calculator, since many actual problems will be worked out in each session.

Continental breakfast Your registration includes the cost of a continental breakfast, and a morning and afternoon refreshment break. These informal segments allow you to discuss topics of mutual interest with other seminar attendees. (You are free to make your own lunch and dinner arrangements.)

Hotel/transportation arrangements RSMeans arranges to hold a block of rooms at most host hotels. To take advantage of special group rates when making your reservation, be sure to mention that you are attending the RSMeans seminar. You are, of course, free to stay at the lodging place of your choice. **(Hotel reservations and transportation arrangements should be made directly by seminar attendees.)**

Important Class sizes are limited, so please register as soon as possible.

Note: Pricing subject to change.

Registration Form

ADDS-1000

Call 1-800-334-3509, ext. 5115 to register or FAX this form 1-800-632-6732. Visit our Web site: www.rsmeans.com

Please register the following people for the RSMeans construction seminars as shown here. We understand that we must make our own hotel reservations if overnight stays are necessary.

☐ Full payment of $_____ enclosed.

☐ Bill me

Please print name of registrant(s)

(To appear on certificate of completion)

_____

_____

_____

P.O. #: _____
GOVERNMENT AGENCIES MUST SUPPLY PURCHASE ORDER NUMBER OR TRAINING FORM.

Firm name_____

Address_____

City/State/Zip_____

Telephone no._____ Fax no._____

E-mail address_____

Charge registration(s) to: ☐ MasterCard ☐ VISA ☐ American Express

Account no._____ Exp. date_____

Cardholder's signature_____

Seminar name_____

Seminar City_____

Please mail check to: RSMeans Seminars, 63 Smiths Lane, P.O. Box 800, Kingston, MA 02364 USA

ELECTRONIC PRODUCTS

Harness the full power of RSMeans online with *MeansCostWorks.com*

MeansCostWorks.com provides 24/7 online access to the most used and quoted construction cost data—in tandem with comprehensive estimating and customization functions. Featuring the entire database of RSMeans unit, assembly, and square foot model costs in an online format, **MeansCostWorks.com** allows professionals to create reliable estimates quickly and easily. Estimates can be customized, shared with others, worked on collaboratively, exported to Excel, and saved for future use.

Construction Estimator
Get online access to the entire RSMeans database of unit costs, assemblies, repair and remodeling costs—covering every major category of construction. Create detailed construction cost estimates.

Square Foot Estimator
For easy-to-use conceptual estimating, this tool gives you online access to the extensive RSMeans library of commercial building models and systems. Get a preliminary square foot estimate in just minutes.

MeansCostWorks.com is an easy-to-use, one-stop estimating tool that provides online access to the most used and quoted construction cost data-in tandem with comprehensive estimating and customization functions.

Go to www.MeansCostWorks.com to view the online demo and sign up for your FREE trial subscription.
To speak with a representative call 1.800.334.3509.

MeansData™
Software Integration
CONSTRUCTION COSTS FOR SOFTWARE APPLICATIONS
Your construction estimating software is only as good as your cost data.

Have you ever wondered why so many projects are over budget, over scope, and take longer to design and build than expected?

DProfiler with RSMeans is the solution that reduces your costs, the risks to the owner, and gives you a competitive advantage.

Use DProfiler with RSMeans cost data at the start of a project to establish the project scope and budget.

www.beck-technology.com

CONSTRUCTION COST INFORMATION... PRODUCTS & SERVICES

For more information visit the RSMeans website at www.rsmeans.com

New Titles

The Gypsum Construction Handbook by USG

An invaluable reference of construction procedures for gypsum drywall, cement board, veneer plaster, and conventional plaster. This new edition includes the newest product developments, installation methods, fire- and sound-rated construction information, illustrated framing-to-finish application instructions, estimating and planning information, and more. Great for architects and engineers; contractors, builders, and dealers; apprentices and training programs; building inspectors and code officials; and anyone interested in gypsum construction. Features information on tools and safety practices, a glossary of construction terms, and a list of important agencies and associations. Also available in Spanish.

$29.95
Over 575 pages, illustrated, softcover
Catalog no. 67357A

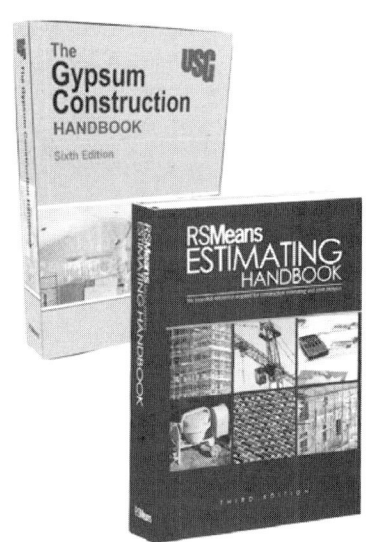

RSMeans Estimating Handbook, 3rd Edition

Widely used in the industry for tasks ranging from routine estimates to special cost analysis projects, this handbook has been completely updated and reorganized with new and expanded technical information.

RSMeans Estimating Handbook will help construction professionals:
- Evaluate architectural plans and specifications
- Prepare accurate quantity takeoffs
- Compare design alternatives and costs
- Perform value engineering
- Double-check estimates and quotes
- Estimate change orders

$99.95 per copy
Over 900 pages, hardcover
Catalog No. 67276B

BIM Handbook
A Guide to Building Information Modeling for Owners, Managers, Designers, Engineers and Contractors
by Chuck Eastman, Paul Teicholz, Rafael Sacks, Kathleen Liston

Building Information Modeling (BIM) is a new approach to design, construction, and facility management where a digital depiction of the building process is used to facilitate the exchange of information in digital format.

The BIM Handbook provides an in-depth understanding of BIM technologies, the business and organizational issues associated with its implementation, and the profound advantages that effective use of BIM can provide to all members of a project team. Also includes a rich set of informative BIM case studies.

$85.00
504 pages
Catalog no. 67356

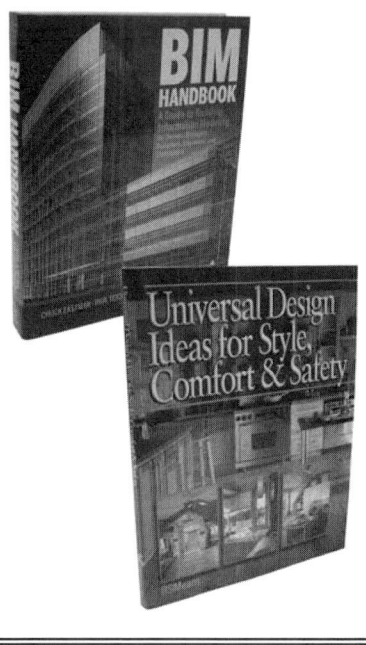

Universal Design Ideas for Style, Comfort & Safety
by RSMeans and Lexicon Consulting, Inc.

Incorporating universal design when building or remodeling helps people of any age and physical ability more fully and safely enjoy their living spaces. This book shows how universal design can be artfully blended into the most attractive homes. It discusses specialized products like adjustable countertops and chair lifts, as well as simple ways to enhance a home's safety and comfort. With color photos and expert guidance, every area of the home is covered. Includes budget estimates that give an idea how much projects will cost.

$21.95
160 pages, illustrated, softcover
Catalog no. 67354

Green Home Improvement
by Daniel D. Chiras, PhD

With energy costs rising and environmental awareness increasing, many people are looking to make their homes greener. This book makes it easy, with 65 projects and actual costs and projected savings that will help homeowners prioritize their green improvements.

Projects range from simple water savers that cost only a few dollars, to bigger-ticket items such as efficient appliances and HVAC systems, green roofing, flooring, landscaping, and much more. With color photos and cost estimates, each project compares options and describes the work involved, the benefits, and the savings.

$24.95
320 pages, illustrated, softcover
Catalog no. 67355

Residential & Light Commercial Construction Standards, 3rd Edition, Updated
by RSMeans and contributing authors

This book provides authoritative requirements and recommendations compiled from leading professional associations, industry publications, and building code organizations. This all-in-one reference helps establish a standard for workmanship, quickly resolve disputes, and avoid defect claims. Updated third edition includes new coverage of green building, seismic, hurricane, and mold-resistant construction.

$59.95
Over 550 pages, illustrated, softcover
Catalog no. 67322B

ORDER TOLL FREE 1-800-334-3509
OR FAX 1-800-632-6732

2010 Order Form

Qty.	Book no.	COST ESTIMATING BOOKS	Unit Price	Total
	60060	Assemblies Cost Data 2010	$279.95	
	60010	Building Construction Cost Data 2010	169.95	
	60110	Concrete & Masonry Cost Data 2010	154.95	
	50140	Construction Cost Indexes 2010 (subscription)	362.00	
	60140A	Construction Cost Index–January 2010	90.50	
	60140B	Construction Cost Index–April 2010	90.50	
	60140C	Construction Cost Index–July 2010	90.50	
	60140D	Construction Cost Index–October 2010	90.50	
	60340	Contr. Pricing Guide: Resid. R & R Costs 2010	39.95	
	60330	Contr. Pricing Guide: Resid. Detailed 2010	39.95	
	60320	Contr. Pricing Guide: Resid. Sq. Ft. 2010	39.95	
	60230	Electrical Change Order Cost Data 2010	169.95	
	60030	Electrical Cost Data 2010	169.95	
	60200	Facilities Construction Cost Data 2010	419.95	
	60300	Facilities Maintenance & Repair Cost Data 2010	379.95	
	60160	Heavy Construction Cost Data 2010	169.95	
	60090	Interior Cost Data 2010	169.95	
	60120	Labor Rates for the Const. Industry 2010	379.95	
	60180	Light Commercial Cost Data 2010	144.95	
	60020	Mechanical Cost Data 2010	169.95	
	60150	Open Shop Building Const. Cost Data 2010	169.95	
	60210	Plumbing Cost Data 2010	169.95	
	60040	Repair & Remodeling Cost Data 2010	144.95	
	60170	Residential Cost Data 2010	144.95	
	60280	Site Work & Landscape Cost Data 2010	169.95	
	60050	Square Foot Costs 2010	184.95	
	62019	Yardsticks for Costing (2009)	154.95	
	62010	Yardsticks for Costing (2010)	169.95	
		REFERENCE BOOKS		
	67310A	ADA Compliance Pricing Guide, 2nd Ed.	79.95	
	67356	BIM Handbook	85.00	
	67330	Bldrs Essentials: Adv. Framing Methods	12.98	
	67329	Bldrs Essentials: Best Bus. Practices for Bldrs	29.95	
	67298A	Bldrs Essentials: Framing/Carpentry 2nd Ed.	15.98	
	67298AS	Bldrs Essentials: Framing/Carpentry Spanish	15.98	
	67307	Bldrs Essentials: Plan Reading & Takeoff	35.95	
	67342	Building & Renovating Schools	69.98	
	67339	Building Security: Strategies & Costs	44.98	
	67353	Complete Book of Framing	29.95	
	67146	Concrete Repair & Maintenance Illustrated	69.95	
	67352	Construction Business Management	49.95	
	67314	Cost Planning & Est. for Facil. Maint.	89.95	
	67328A	Designing & Building with the IBC, 2nd Ed.	99.95	
	67230B	Electrical Estimating Methods, 3rd Ed.	64.95	
	64777A	Environmental Remediation Est. Methods, 2nd Ed.	49.98	
	67343	Estimating Bldg. Costs for Resi. & Lt. Comm.	29.95	
	67276B	Estimating Handbook, 3rd Ed.	$99.95	
	67318	Facilities Operations & Engineering Reference	54.98	
	67301	Facilities Planning & Relocation	49.98	
	67338A	Green Building: Proj. Planning & Cost Est., 2nd Ed.	129.95	
	67355	Green Home Improvement	24.95	

Qty.	Book no.	REFERENCE BOOKS (Cont.)	Unit Price	Total
	67357A	The Gypsum Construction Handbook	29.95	
	67357S	The Gypsum Construction Handbook (Spanish)	29.95	
	67323	Historic Preservation: Proj. Planning & Est.	49.98	
	67308E	Home Improvement Costs–Int. Projects, 9th Ed.	24.95	
	67309E	Home Improvement Costs–Ext. Projects, 9th Ed.	24.95	
	67344	Homeowner's Guide to Mold	21.95	
	67324B	How to Est.w/Means Data & CostWorks, 3rd Ed.	59.95	
	67351	How Your House Works	21.95	
	67282A	Illustrated Const. Dictionary, Condensed, 2nd Ed.	59.95	
	67292A	Illustrated Const. Dictionary, w/CD-ROM, 3rd Ed.	99.95	
	67348	Job Order Contracting	89.95	
	67347	Kitchen & Bath Project Costs	19.98	
	67295C	Landscape Estimating Methods, 5th Ed.	64.95	
	67341	Life Cycle Costing for Facilities	99.95	
	67294B	Mechanical Estimating Methods, 4th Ed.	64.95	
	67245A	Planning & Managing Interior Projects, 2nd Ed.	34.98	
	67283B	Plumbing Estimating Methods, 3rd Ed.	29.98	
	67345	Practice of Cost Segregation Analysis	99.95	
	67337	Preventive Maint. for Higher Education Facilities	149.95	
	67346	Preventive Maint. for Multi-Family Housing	89.95	
	67326	Preventive Maint. Guidelines for School Facil.	149.95	
	67247B	Project Scheduling & Management for Constr. 3rd Ed.	64.95	
	67265B	Repair & Remodeling Estimating Methods, 4th Ed.	69.95	
	67322B	Resi. & Light Commercial Const. Stds., 3rd Ed.	59.95	
	67327A	Spanish/English Construction Dictionary, 2nd Ed.	23.95	
	67145B	Sq. Ft. & Assem. Estimating Methods, 3rd Ed.	69.95	
	67321	Total Productive Facilities Management	29.98	
	67350	Understanding and Negotiating Const. Contracts	69.95	
	67303B	Unit Price Estimating Methods, 4th Ed.	59.95	
	67354	Universal Design	21.95	
	67319A	Value Engineering: Practical Applications	79.95	

MA residents add 6.25% state sales tax		
Shipping & Handling**		
Total (U.S. Funds)*		

Prices are subject to change and are for U.S. delivery only. *Canadian customers may call for current prices. **Shipping & handling charges: Add 7% of total order for check and credit card payments. Add 9% of total order for invoiced orders.

Send order to: ADDV-1000

Name (please print) _____

Company _____

☐ **Company**

☐ **Home   Address** _____

_____

City/State/Zip _____

Phone # _____ P.O. # _____

(Must accompany all orders being billed)

Mail to: RSMeans, P.O. Box 800, Kingston, MA 02364

RSMeans Project Cost Report

By filling out this report, your project data will contribute to the database that supports the RSMeans Project Cost Square Foot Data. When you fill out this form, RSMeans will provide a $30 discount off one of the RSMeans products advertised in the preceding pages. Please complete the form including all items where you have cost data, and all the items marked (✔).

$30.00 Discount per product for each report you submit.

Project Description (No remodeling projects, please.)

✔ Building Use (Office, School...) _____

✔ Address (City, State) _____

_____

✔ Total Building Area (SF) _____

✔ Ground Floor (SF) _____

✔ Frame (Wood, Steel...) _____

✔ Exterior Wall (Brick, Tilt-up...) _____

✔ Basement: (check one) ☐ Full ☐ Partial ☐ None

✔ Number Stories _____

✔ Floor-to-Floor Height _____

✔ Volume (C.F.) _____

% Air Conditioned _____ Tons _____

Total Project Cost $ _____

Owner _____

Architect _____

General Contractor _____

✔ Bid Date _____

✔ Typical Bay Size _____

✔ Capacity _____

✔ Labor Force: _____ % Union _____ % Non-Union

✔ Project Description (Circle one number in each line)

1. Economy 2. Average 3. Custom 4. Luxury
1. Square   2. Rectangular 3. Irregular 4. Very Irregular

Comments _____

_____

A	✔ General Conditions	$	
B	✔ Site Work	$	
C	✔ Concrete	$	
D	✔ Masonry	$	
E	✔ Metals	$	
F	✔ Wood & Plastics	$	
G	✔ Thermal & Moisture Protection	$	
GR	Roofing & Flashing	$	
H	✔ Doors and Windows	$	
J	✔ Finishes	$	
JP	Painting & Wall Covering	$	

K	✔ Specialties	$	
L	✔ Equipment	$	
M	✔ Furnishings	$	
N	✔ Special Construction	$	
P	✔ Conveying Systems	$	
Q	✔ Mechanical	$	
QP	Plumbing	$	
QB	HVAC	$	
R	✔ Electrical	$	
S	✔ Mech./Elec. Combined	$	

Please specify the RSMeans product you wish to receive. Complete the address information.

Product Name _____

Product Number _____

Your Name _____

Title _____

Company _____

☐ Company
☐ Home   Street Address _____

City, State, Zip _____

Return by mail or Fax 888-492-6770.

Method of Payment:

Credit Card # _____

Expiration Date _____

Check _____

Purchase Order _____

RSMeans
a division of Reed Construction Data
Square Foot Costs Department
P.O. Box 800
Kingston, MA 02364-3008

Be sure to keep up-to-date in 2010!

Register now and we'll notify you by email each quarter when the newest cost index is available free online.

Please provide your name, address, and email below and return this card by mail or
Fax: 1-800-632-6732.

To register online: www.rsmeans411.com/register

Name _____

email _____

Company _____

Street _____

City/Town _____ State/Prov. _____ Zip/Postal Code _____

RSMeans

PLACE STAMP HERE

RSMeans
Quarterly Updates
63 Smiths Lane
Kingston, MA 02364-3008
USA